Modern REFRIGERATION and AIR CONDITIONING

by

ANDREW D. ALTHOUSE, B.S. (M.E.), M.A.

Technical-Vocational Education Consultant
Life Member, American Society of
Heating, Refrigerating and Air Conditioning Engineers
Member, American Vocational Association

CARL H. TURNQUIST, B.S., (M.E.), M.A.

Technical-Vocational Education Consultant
Associate Member, American Society of
Heating, Refrigerating and Air Conditioning Engineers
Member, Refrigeration Service Engineers Society
Member, Refrigerating Engineers and Technicians Association
Member, American Vocational Association

ALFRED F. BRACCIANO, B.S., M.Ed., Ed. Sp.

Director of Technical and Vocational Education
Warren, Michigan Consolidated Schools
Member, Refrigeration Service Engineers Society
Member, American Vocational Association

South Holland, Illinois
THE GOODHEART-WILLCOX COMPANY, INC.
Publishers

Library of Congress Cataloging in Publication Data

Althouse, Andrew Daniel.
  Modern refrigeration and air conditioning.

  Includes index.
  1. Refrigeration and refrigerating machinery.
  2. Air conditioning.   I. Turnquist, Carl Harold, joint author.   II. Bracciano, Alfred F., joint author.   III. Title.
TP492.A43 1982   621.5'6   81—20002
ISBN 0—87006—340—5   AACR2

INTRODUCTION

In all types of refrigeration and air conditioning work, a thorough knowledge of the basic principles is important. MODERN REFRIGERATION AND AIR CONDITIONING explains and illustrates these important principles in an easily understood manner.

This text covers the practical application of refrigeration in all its branches: domestic, commercial, air conditioning, heat pumps, automotive air conditioners, thermoelectric, solar energy, special devices and applications. Up-to-date methods of installing, maintaining, diagnosing and repairing are explained.

Written primarily as a textbook, this work is intended for use in refrigeration and air conditioning classes in high schools, technical schools and community colleges. It may also be used in adult education classes and apprenticeship programs. It provides the foundation on which a sound, thorough knowledge of refrigeration and air conditioning may be based.

Beginners and apprentices will find MODERN REFRIGERATION AND AIR CONDITIONING an excellent aid to getting started and pursuing a pleasant and profitable career. Experienced service technicians will find it a valuable guide and reference.

The first chapter of this edition is written using both SI metric units and U.S. conventional units. Metric appears alongside conventional unit throughout the book. Teachers and students familiar with the U.S. conventional system may wish to continue using that system.

Andrew D. Althouse
Carl H. Turnquist
Alfred F. Bracciano

CONTENTS

1 FUNDAMENTALS OF REFRIGERATION 7

2 REFRIGERATION TOOLS AND MATERIALS 37

3 BASIC REFRIGERATION SYSTEMS 73

4 COMPRESSION SYSTEMS AND COMPRESSORS 99

5 REFRIGERANT CONTROLS 141

6 ELECTRICAL — MAGNETIC FUNDAMENTALS 169

7 ELECTRIC MOTORS 209

8 ELECTRIC CIRCUITS AND CONTROLS 243

9 REFRIGERANTS 277

10 DOMESTIC REFRIGERATORS AND FREEZERS 303

11 INSTALLING AND SERVICING
SMALL HERMETIC SYSTEMS 335

12 COMMERCIAL SYSTEMS 393

13 COMMERCIAL SYSTEMS APPLICATIONS 463

14 COMMERCIAL SYSTEMS, INSTALLING AND SERVICING . . 489

15 COMMERCIAL SYSTEMS, HEAT LOADS AND PIPING 555

16 ABSORPTION SYSTEMS,
PRINCIPLES AND APPLICATIONS 603

17 SPECIAL REFRIGERATION SYSTEMS
AND APPLICATIONS 625

18 FUNDAMENTALS OF AIR CONDITIONING 641

19 BASIC AIR CONDITIONING SYSTEMS 673

20 AIR CONDITIONING SYSTEMS,
HEATING AND HUMIDIFYING 699

4

21 AIR CONDITIONING SYSTEMS, COOLING AND DEHUMIDIFYING 747

22 AIR CONDITIONING SYSTEMS, DISTRIBUTING AND CLEANING 765

23 HEAT PUMPS AND COMPLETE AIR CONDITIONING SYSTEMS 800

24 AIR CONDITIONING CONTROLS, CIRCUITS AND INSTRUMENTS 837

25 AIR CONDITIONING SYSTEMS, HEAT LOADS 879

26 AUTOMOBILE AIR CONDITIONING 897

27 SOLAR ENERGY 933

28 TECHNICAL CHARACTERISTICS 947

29 CAREER OPPORTUNITIES IN REFRIGERATION AND AIR CONDITIONING 969

30 DICTIONARY OF TECHNICAL TERMS 971

INDEX 984

ACKNOWLEDGMENTS

The production of a book of this nature would not be possible without the cooperation of the Refrigeration and Air Conditioning Industry. In preparing the manuscript for MODERN REFRIGERATION and AIR CONDITIONING, the industry has been most cooperative. The authors acknowledge the cooperation of these companies with great appreciation:

A C & R Components, Inc.; Abacus International; Abbeon Cal, Inc.; Abrax Instrument Corp.; Acme Electric Corp.; Addison Products Co.; Aeroquip Corp.; Air Balance Inc.; Airco, Inc., Industrial Gases Div.; Airserco Mfg. Co.; Airtemp Applied Machinery Co.; Alco Controls Div., Emerson Electric Co.; Allied Chemical Corp., Industrial Chemicals Div.; Allied Chemical Corp., Specialty Chemicals Div.; Allin Mfg. Co., Inc.; Alnor Instrument Co.; Amana Refrigeration, Inc.; American Automatic Ice Machine Co.; American Motors Corp.; American Panel Corp.; American Society of Heating, Refrigerating and Air Conditioning Engineers, Inc.; American Ultraviolet Co.; Ametek, Inc., U.S. Gauge Div.; Amprobe Instrument Div., SOS Consolidated, Inc.; Anemostat Products Div., Dynamics Corp. of America; Arco Mfg. Corp.; Arkla Air Conditioning Co., Div. of Arkla Industries, Inc.; BDP Co., Div. of Carrier Corp.; Bacharach Instrument Co., Div. of AMBAC Industries, Inc.; Bally Case & Cooler, Inc.; Baltimore Aircoil Co., Inc.; Barber-Colman Co., Environmental Systems Div.; Barnebey-Cheney Co.; R.W. Beckett Corp.; Beckman Instruments, Inc.; Bell & Gosset, ITT; The Bendix Corp.; The Bendix Corp., National Environmental Instruments, Inc., Environmental Science Div.; Bendix-Westinghouse, Refrigeration Products Div.; Bernz-O-Matic Corp.; Bohn Aluminum & Brass Div., Gulf & Western Mfg. Co.; Brasch Mfg. Co., Inc.; Bristol Babcock Inc.; K.G. Brown Mfg. Co., Inc.; Bryant Air Conditioning Co.; Buchbinder; CP Div., St. Regis; Calgon Corp.; The Carlin Co.; Carrier Air Conditioning Group, United Technologies Corp.; Carrier Transicold Co., Div. of Carrier Corp.; Century Electric Co.; Chatleff Controls Div.; Chicago Valve Plate and Seal Co.; Chrysler Corp., Dodge Div.; David Clark Co., Inc.; Cleanweld Products, Inc.; Cleveland Sales Co.; Cleveland Twist Drill Co.; Climatic Air Sales, Inc.; Climatrol Industries, Inc.; The Coleman Co., Inc.; Connor Engineering Corp.; Continental Air Filters Co.; Controls Co. of America, Heating & Air Conditioning Div.; The Cooper Group, Nicholson; Cooper Thermometer Co.; Copeland Corp.; H. Custer Co.; Dairy Equipment Co., Div. of DEC International, Inc.; Danfoss, Inc.; Detroit Edison Co.; Detroit Public Schools; The Dickson Co.; The Henry G. Dietz Co., Inc.; DoALL Co.; Dole Refrigerating Co.; Dongan Electric Mfg. Co.; Dunham-Bush, Inc.; E.I. duPont de Nemours & Co., Inc.; E.I. duPont de Nemours & Co., Inc., Freon Products Div.; Duro Metal Products Co.; Dwyer Instruments, Inc.; Eaton Corp.; Ebco Mfg. Co.; Edison Electric Institute; EG & G Sealol, Inc.; Electro-air Div., Emerson Electric Co.; Electrolux, A.B.; Electronics Corp. of America; Elkay Mfg. Co.; Emerson Electric Co.; The Excelsior Steel Furnace Co.; Farr Co.; Fedders Corp.; Fedders Corp., Norge Div.; FES; Flexonics Div., UOP, Inc.; Ford Div., Ford Motor Co., National Service Office, Ford Motor Co.; Ford Customer Service Div.; Franklin Mfg. Co.; Frick Co.; Frigidaire Co.; Frigiking, Inc.; Fusite Corp.; G & L Adhesives Corp.; The Gates Rubber Co.; General Controls, ITT; General Electric Co.; General Motors Corp., Buick Motor Div.; General Motors Corp., Cadillac Motor Car Div.; General Motors Corp., Delco Air Conditioning Div.; General Motors Corp., Harrison Radiator Div.; General Motors Corp., Pontiac Motor Div.; Gibson Appliance Co.; Hans Goldner & Co.; Gould Inc., Electric Motor Div.; Gould Inc., Fluid Components Div.; H-B Instrument Co.; Halbro Products; Halstead & Mitchell; Handy & Harman; The Hansen Mfg. Co.; Hart & Cooley Mfg. Co., Div. of Allied Thermal Corp.; Harvey-Westbury Corp.; Heico Inc., Field Control Div.; Henry Valve Co.; Hobart Corp.; Honeywell, Inc.; Hotpoint Div., General Electric Co.; Houdaille-Hersey; Howard Engineering Co.; Howe Corp.; Hupp, Inc., Refrigeration Products Div.; Hussmann Refrigeration, Inc.; Imperial-Eastman Corp.; International Harvester Co.; International Metal Products Div., McGraw-Edison Co.; ITT McDonnell & Miller; Jackes-Evans Div., Parker-Hannifin Corp.; Johnson Controls, Inc., Control Products Div.; Johnson Service Co.; George L. Johnston Co.; Kaiser Chemicals, Div. of Kaiser Aluminum & Chemical Corp.; Kalglo Electronics Co., Inc.; Kelvinator, Inc./White Consolidated Industries; Kerotest Mfg. Corp.; Kinney Vacuum Co., General Signal Corp.; Koolatron Industries; La Crosse Cooler Co.; Larchmont Engineering & Irrigation, Inc.; Lennox Industries, Inc.; Lutron Electronics Co., Inc.; Maid-O'-Mist Div.; Marsh Instrument Co.; Marshalltown Instruments; Mast Development Co.; Maurey Mfg. Corp.; Maxitrol Co.; McCall Refrigerator Co.; Mechanical Refrigeration Enterprises; Michigan Farmer; Micro Switch, Div. of Honeywell; Midco International Inc.; Mile High Equipment Co.; Mine Safety Appliances Co.; John E. Mitchell Co.; Monarch Mfg. Works, Inc.; Motor Wheel Corp., Duo-Therm Div.; Mueller Brass Co.; Murray Corp.; Mycom Corp., Mayekawa U.S.A., Inc.; National Cooler Corp.; National Cylinder Gas, Div. of Chemetron Corp.; National Environmental Systems Contractors Assoc. (NESCA); National Gypsum Co.; Nesbitt, ITT, Environmental Products Div.; Owens-Corning Fiberglas Corp.; Packless Industries; Paragon Electric Co., Inc., Sub. of AMF; Parker Seal Co.; Paulin Products Co.; Peerless of America, Inc.; Penn Controls, Inc.; Philco-Ford Corp.; Polymer Corp.; Puffer-Hubbard Refrigerator Div.; Queen Products Div., King-Seeley Thermos Co.; R-Deck, Inc.; Ranco, Inc.; Refrigerating Specialties Div., Flo-Con; Refrigeration Engineering Corp.; Resistoflex Corp.; Revere Solar and Architectural Products, Inc.; The Ridge Tool Co.; Robertshaw Controls Co., Milford Div.; Robinair Mfg. Corp.; Robinair Mfg. Corp., Madden Brass Products Co.; Ross-Temp, Inc.; Rotary Seal Corp.; Royal Industries, Engineered Products Div.; Rubatex Corp.; Sani-Serv; Schaefer Brush Mfg. Co., Inc.; Sealed Unit Parts Co., Inc.; Simicon Co.; Simpson Electric Co.; The Singer Co.; The Singer Co., Climate Control Div.; A.O. Smith Corp.; Snap-on Tools Corp., Electric Motor Div.; Solarex Corp.; Solaron Corp.; Southwest Mfg. Co.; A.W. Sperry Instruments Inc.; Sporlan Valve Co.; Stal Refrigeration AB; Statham Instruments, Inc.; Stevens Appliance Truck Co.; Stewart-Warner Corp.; Sunstrand Hydraulics, Div. of Sunstrand Corp.; Superior Valve Co.; Sweden Freezer Mfg. Co.; Swingline, Inc.; Tappan Air Conditioning Div., Tappan Co.; Taylor Freezer Co.; Taylor Instrument/Consumer Industrial Products, Sybron Corp.; Tecumseh Products Co.; Temprite Products, Eaton Corp., Dispenser Div.; TESA, S.A.; Texas Instruments, Inc., Control Products Div.; Thermal Engineering Co.; Thermal Industries of Florida, Inc.; Thermatron Corp.; Thermo-O-Disc, Inc., Sub. of Emerson Electric Co.; Tjernlund Mfg. Co.; Torin Corp.; Torrington Mfg. Co.; The Trane Co.; Tranter Mfg. Inc., Kold-Hold Div.; TRW Greenfield Tap & Die Div.; Tubing Appliance Co., Inc.; Turbo Refrigerating Co.; Tutco Inc.; Tyler Refrigeration Corp.; Union Carbide Corp., Linde Div.; USON Corp.; Vaco Products Co.; Vickers Div., Sperry-Rand Corp.; Vilter Mfg. Corp.; Virginia Chemicals, Inc.; Volkswagen of America, Inc.; Vortec Corp.; Wabash Corp.; Wagner Electric Corp.; William Wahl Corp.; Walsh-Healy Act of 1969; Warren/Sherer, Div. of Kysor Industrial Corp.; Watsco Inc.; Weather Measure Corp., A Sub. of Systron-Donner Corp.; Webster Electric Co., Inc.; Weil-McLain Co., Inc., Hydronic Div.; Weksler Instruments Corp.; Westclox Div., General Time Corp.; Westinghouse Electric Corp., Air Conditioning Div.; Whirlpool Corp.; White-Rodgers Div., Emerson Electric Co.; The Williamson Co.; The Winterizer Co., Sub. of R.J. Mason, Inc.; York Div., Borg-Warner Corp.

Chapter 1

FUNDAMENTALS OF REFRIGERATION

When studying refrigeration and air conditioning, it is important to first master the basic physical, mechanical, and chemical principles explained in this chapter.

Much of the basic materials will be a review of physics and chemistry for those who have studied these subjects. In addition, however, some engineering principles and formulas have been included. These principles and formulas have been simplified, and are introduced with practical applications.

This chapter uses the SI (Le Systeme International d'Unites) metric units of measure. The early refrigeration industry was developed by engineers in England and the United States. They based their calculations on units of measure and formulas developed for power, mining, automobile and ice-making industries. These units included the horsepower, British thermal unit, foot pound, pounds per square inch, temperatures in Fahrenheit, and the like.

The SI (metric) system is internationally accepted and the United States is moving toward it. The refrigeration industry in this country has recently adopted its usage.

SI units include linear measurement in metres or decimals, mass in kilograms, temperature in degrees Celsius, pressure in pascals, heat in joules, and the like. Complete definitions of the units and U.S. conventional equivalents (equals) are given in Chapter 28.

Users of this text who are unfamiliar with SI metrics have no cause for concern. Conventional measurements are carried alongside the metric. Reference is also made to Chapter 28 whenever problems arise affecting metric usage.

1-1 DEVELOPMENT OF REFRIGERATION

Modern refrigeration has many applications. The first, and probably still the most important, is the preservation of food.

Most foods kept at room temperature spoil rapidly. This is due to the rapid growth of bacteria. At usual refrigeration temperatures of about 4.4°C (40 F.), bacteria grow quite slowly. Food at this temperature will keep much longer. Refrigeration preserves food by keeping it cold.

Other important uses of refrigeration include air conditioning, beverage cooling and humidity control. Many manufacturing processes also use refrigeration.

The refrigeration industry became important commercially during the 18th century. Early refrigeration was obtained by the use of ice. Ice from lakes and ponds was cut and stored in the winter in insulated storerooms for summer use.

The use of natural ice required the building of insulated containers or iceboxes for stores, restaurants, and homes. These units first appeared, on a large scale, during the 19th century.

Ice was first made artificially about 1820 as an experiment. Not until 1834 did artificial ice manufacturing become practical. Jacob Perkins, an American engineer, invented the apparatus which was the forerunner of our modern compression systems. In 1855 a German engineer produced the first absorption type of refrigerating mechanism, although Michael Farady had discovered the principles for it in 1824.

Little artificial ice was produced until shortly after 1890. During 1890 a warm winter resulted in a shortage of natural ice. This helped start the mechanical ice-making industry.

Mechanical domestic refrigeration first appeared about 1910. J. M. Larsen produced a manually operated household machine in 1913. By 1918 Kelvinator produced the first automatic refrigerator for the American market. They sold 67 machines that year and, between 1918 and 1920, 200 more. Now, over 10 million units are sold each year.

The first of the sealed or "hermetic" automatic refrigeration units was introduced by General Electric in 1928. It was named the Monitor Top.

Beginning with 1920, domestic refrigeration became one of our important industries. The Electrolux, which was an automatic domestic absorption unit, appeared in 1927. Automatic refrigeration units, for the comfort cooling part of air conditioning, appeared in 1927.

Fast freezing to preserve food for extended periods was developed about 1923. This marked the beginning of the modern frozen foods industry.

Mechanical refrigeration systems were first connected to heating plants to provide summer cooling in the late 1920's.

By 1940, practically all domestic units were of the hermetic type. Commercial units had also been successfully made and used. These units were capable of refrigerating large commerical food storage systems, comfort cooling of large auditoriums, and producing of low temperatures used in many commercial operations.

From a small, slow start in the late 1930's, air conditioning of automobiles has also grown rapidly.

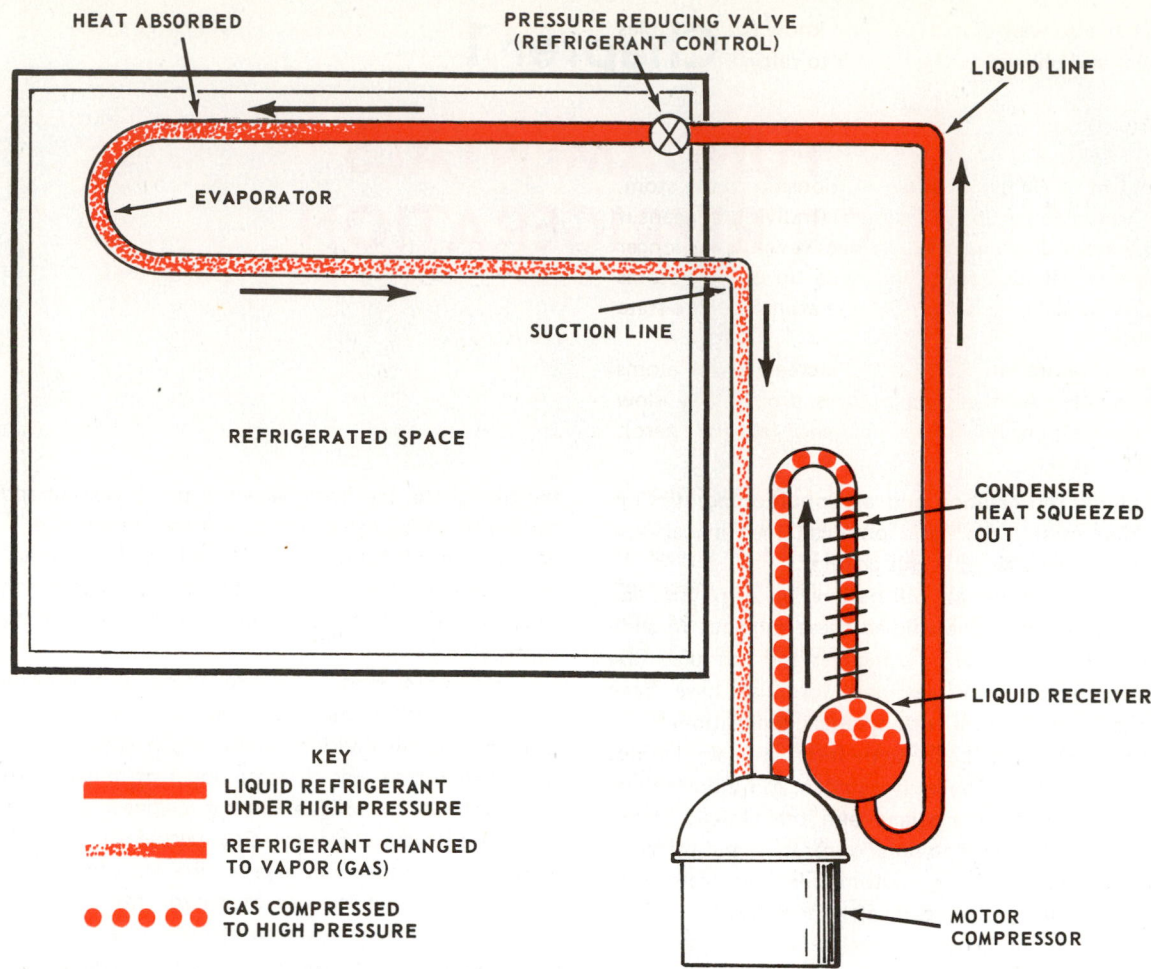

KEY

■■■■■ LIQUID REFRIGERANT
UNDER HIGH PRESSURE

▓▓▓▓▓ REFRIGERANT CHANGED
TO VAPOR (GAS)

●●●●● GAS COMPRESSED
TO HIGH PRESSURE

Fig. 1-1. Elementary mechanical refrigerator. In operation, liquid refrigerant under high pressure (solid red), flows from liquid receiver to pressure reducing valve (refrigerant control) and into evaporator. Here pressure is greatly reduced. Liquid refrigerant boils and absorbs heat from evaporator. Now a vapor, refrigerant (spotted red line) flows back to compressor and is compressed to high pressure (red dots). Its temperature is greatly increased and, in condenser, heat is transferred to surrounding air. Refrigerant cools, becoming liquid again. It flows back into liquid receiver and cooling cycle is repeated.

1-2 HOW A MECHANICAL REFRIGERATOR OPERATES

Removing heat from inside a refrigerator is somewhat like removing water from a leaking canoe. A sponge may be used to soak up the water. The sponge is held over the side, squeezed, and the water is released overboard. The operation may be repeated as often as necessary to transfer the water from the canoe into the lake.

In a refrigerator heat instead of water is transferred. Inside the refrigerating mechanism, heat is absorbed, "soaked up," by evaporating the liquid refrigerant in the evaporator (cooling unit). This occurs as the refrigerant changes from a liquid to a vapor (gas), Fig. 1-1.

After the refrigerant has absorbed heat and has turned into a vapor, it is pumped into the condensing unit located outside the refrigerated space. The condenser works the opposite of the evaporator. In the evaporator, liquid refrigerant enters one end and absorbs heat as it passes through the evaporator. By

the time it reaches the end of the evaporator, it is all a vapor. As this vapor flows through the condenser under a high pressure and high temperature, it gives up its heat to the surrounding air. As it reaches the end of the condenser, the refrigerant, now cooled, has become a liquid again. We say that, in the condenser, the heat is "squeezed out." This cycle repeats until the desired temperature is reached.

Heat enters a refrigerator in many ways. It leaks through the insulated walls or enters when the door is opened. Still more heat is introduced when warm substances are placed in the refrigerator.

Heat is not destroyed to make the refrigerator cold. It is simply removed from the refrigerated space and released outside.

The paragraphs which follow will provide the technical foundation needed to understand the heat removal operation. This background is important for service and repair.

Service managers of refrigerating and air conditioning companies prefer service and installation technicians who are good mechanics. They also want employees who know the principles of mathematics and physics as these apply to refrigeration.

1-3 HEAT

Heat is a form of energy. It has a relationship to the atom, the smallest indivisible part of an element. (Indivisible means if one broke the atom down into more pieces it would no longer be that element.) All substances are made up of tiny atoms which combine to make molecules. All the atoms are in a state of rapid motion.

As the temperature of a substance increases, the atoms move more rapidly. As the temperatures drops, they slow down. If all heat is removed from a substance (absolute zero), all atomic motion stops.

The metric unit of heat is the joule (J). If a substance is warmed, heat is added, if cooled, heat is removed.

The amount of heat in a substance is equal to the mass of the substance multiplied by its temperature. The amount of heat in a substance may greatly affect the nature of the substance. Adding heat causes most substances to expand. Removing heat causes them to contract.

Most substances change their physical state with the addition or removal of heat. For instance, water ice is a solid (under atmospheric pressure at a temperature below $0°C$). By adding heat to the ice, it will melt and become water (a liquid). Further addition of heat will cause the water to turn into a vapor (steam). The basic principle of operation of the compression type refrigeration cycle makes use of this principle in its operation.

The U. S. conventional unit of heat is the British thermal unit (Btu).

1-4 HEAT FLOW

Heat always flows from a warmer to a cooler substance. What happens is that the faster moving atoms give up some of their energy to slower moving atoms. Therefore, the faster atom slows down a little and the slower one moves a little faster.

Heat causes some solids to become liquids or gases, or liquids to become gases. Cooling will reverse the process. This happens because the atoms making up the molecules of these substances act in a different way to temperature. Instead of moving faster or slower, one or more of the atoms in the molecule shift their positions.

1-5 COLD

Cold means low temperature or lack of heat. Cold is the result of removing heat. A refrigerator produces "cold" by drawing heat from the inside of the refrigerator cabinet.

The refrigerator does not destroy the heat, but pumps it from the inside of the cabinet to the outside. Heat cannot travel from a cold body to a hot body, but always travels from a substance at a higher temperature to a substance at a lower temperature (Second Law of Thermodynamics. See Chapter 28.)

1-6 COLD PRESERVES FOOD

Spoiling of food is actually the growth of bacteria in it. As the molecules move slowly, they have an important effect on the bacteria present in most foods. Slowing movement by cooling the molecules makes all organisms more sluggish. Cold, or low temperature, slows up the growth of these bacteria. Foods, thus, do not spoil as fast. If the bacteria can be kept from increasing, the food will be edible longer.

Most foods contain a considerable amount of water. Food, therefore, must be kept slightly above freezing temperatures ($0°C$, 32 F.).

If food is frozen slowly at or near the freezing point of water, the ice crystals formed are large, and their growth breaks down the food tissues. When defrosted, it spoils rapidly; appearance and taste are ruined.

Fast freezing at very low temperatures, -18 to $-26°C$ (0 to 15 F.), forms small crystals which do not injure the food tissues. Food freezers are maintained at or below $-18°C$ (0 F.). Food placed in them will freeze quickly.

Keep in mind the difference between refrigerating and freezing. The correct refrigerating temperature for fresh food is $1.7°C$ (35 F.) to $7.3°C$ (45 F.). To make ice, a temperature lower than $0°C$ (32 F.) is needed.

1-7 TEMPERATURE AND TEMPERATURE MEASUREMENT

Temperature measures the heat intensity or heat level of a substance. Temperature alone does not give the amount of heat in a substance. It indicates the degree of warmth, or how hot or cold the substance or body is.

In the molecular theory of heat, temperature indicates the speed of motion of the molecule. It is important not to use the words "heat" and "temperature" carelessly.

Temperature measures the speed of motion of the atom. Heat is the speed of motion of the atom multiplied by the number of atoms (mass) so affected.

For example, a small copper dish weighing a few grams, heated to $727°C$ (1340 F.) does not contain as much heat as 5 kilograms of copper heated to $140°C$ (284 F.). However, its heat level is higher. Its intensity of heat is greater.

The SI unit of temperature is the kelvin (K). The temperature intervals (space between degrees) on the kelvin scale are the same as Celsius. The U.S. conventional unit of temperature is the degree Fahrenheit.

Temperature is measured with a thermometer usually through uniform expansion of a liquid in a sealed glass tube. There is a bulb at the bottom of the tube and a quantity of liquid (mercury or alcohol) inside.

The glass does not expand or contract as much as the liquid during a temperature change. The liquid will rise and fall in the tube as the temperature changes. The tube is calibrated or marked off in degrees using the desired temperature scale. Fig. 1-2 shows a glass stem thermometer used in refrigeration and air conditioning work.

Some thermometers use metal to measure temperature. Metal will expand and contract as temperature rises and falls.

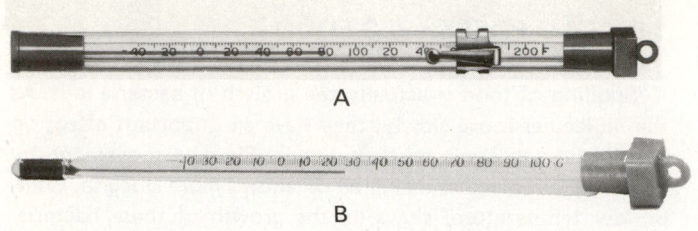

Fig. 1-2. Glass stem thermometers are used in refrigeration work. A—Fahrenheit thermometer has range from −40 to 210 F. B—Celsius thermometer has range from −40 to 100°C. (Marsh Instrument Co.)

This moves an indicator up and down the scale.

Thermometers have also been developed which indicate temperature by measurement of a very small electric voltage generated in a thermocouple. (See Chapters 6 and 18.) These instruments are called potentiometers. They are especially useful in the measurement of high temperatures above the useful range of a glass stem thermometer.

A radiometer is a thermometer which detects infrared rays released by a substance. This thermometer is very easy to use. No contact is needed with the substance whose temperature is to be measured. (See Fig. 28-1.)

Thermistors may also be used to measure temperatures. (See Chapter 6.)

1-8 THERMOMETER SCALES, CELSIUS AND FAHRENHEIT

The two most common thermometer scales are the Celsius, sometimes called the Centigrade scale, and the Fahrenheit. The Celsius scale is named in honor of Anders Celsius, the Swedish astronomer who recommended it.

Two temperatures determine the calibration of a thermometer: 1. The temperature of melting ice. 2. The temperature of boiling water.

On the Celsius thermometer, the temperature of melting ice or the freezing temperature of water is 0°C. The temperature of boiling water is 100°C. There are 100 spaces or degrees on the scale between freezing and boiling.

On the Fahrenheit thermometer, the temperature of melting ice or the freezing temperature of water is 32 F. The temperature of boiling water is 212 F. This provides 180 spaces or degrees between the freezing and boiling temperatures.

For a comparison of the Celsius and Fahrenheit scales, see Fig. 1-3. (Also, see Chapter 28.)

The freezing and boiling points are based on freezing and boiling temperatures of water at standard atmospheric pressure. Effects of pressure on these temperatures is explained in Para. 1-34 and Para. 1-35.

1-9 ABSOLUTE TEMPERATURE SCALES, KELVIN AND RANKINE

Absolute zero is that temperature where molecular motion stops. It is the lowest temperature possible. There is no more

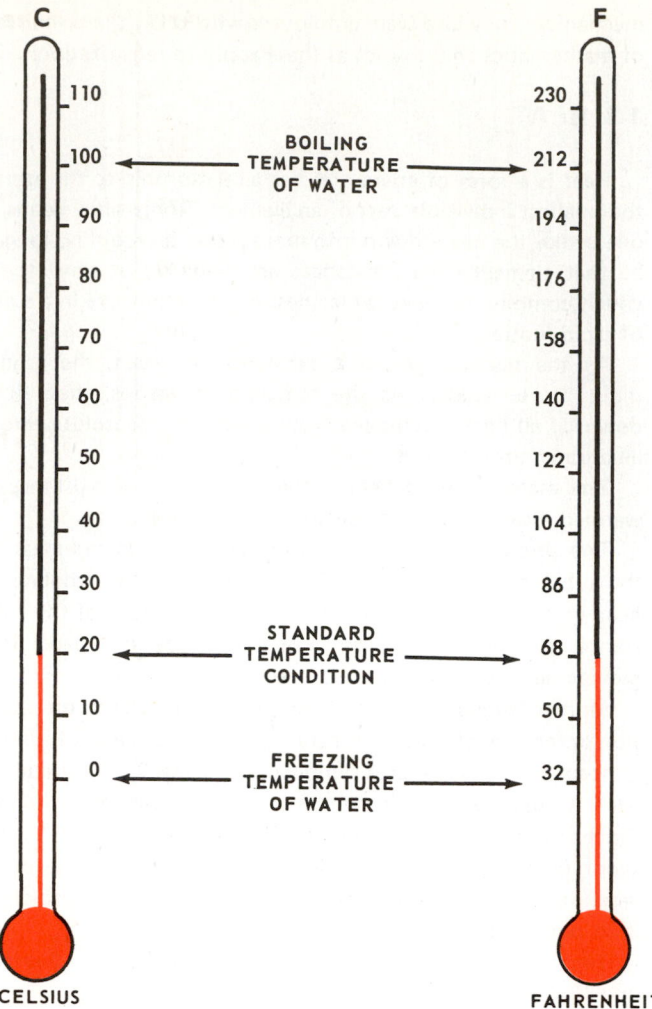

Fig. 1-3. A comparison of Celsius and Fahrenheit thermometer scales.

heat in the substance at this point.

Two absolute temperature scales are used with very low temperature work such as cryogenics. (See Para. 1-56.) These two scales are the Kelvin (Celsius Absolute) scale and the Rankine (Fahrenheit Absolute) scale.

The Kelvin scale uses the same divisions as the Celsius scale. Zero on the Kelvin scale (0 K) is 273 degrees below 0°C.

The Rankine scale uses the same divisions as the Fahrenheit scale. However, zero on this scale (0 R.) is located 460 degrees below 0 F. Fig. 1-4 compares the Celsius, Fahrenheit, Kelvin and Rankine thermometer scales.

Problem: What is the temperature at which water freezes and boils using the Kelvin scale?

Solution, Freezing Point: Water freezes at 0°C. The Kelvin scale zero is 273 degress below 0°C. The freezing temperature of water is, therefore, 273 degrees above zero Kelvin (K), or 273 Kelvin. Freezing temperature is 273 K.

Solution, Boiling Point: Water boils at 100 degrees above 0°C. The boiling point of water on the Kelvin scale will be: 100 + 273 = 373 K. Therefore, the boiling point is 373 K.

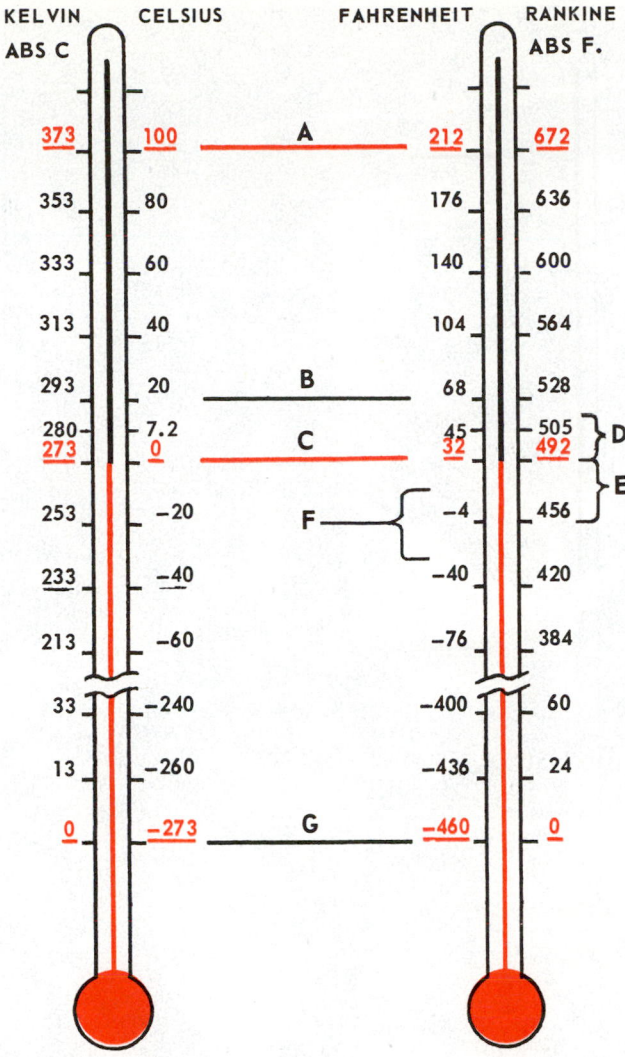

Fig. 1-4. Celsius, Fahrenheit, Kelvin and Rankine thermometer scales are compared. A—Boiling temperature of water. B—Standard conditions temperature. C—Freezing temperature of water. D—Temperature range for fresh food storage. E—Range of evaporator temperatures for food. F—Temperature range for frozen food storage. G—Absolute zero.

1-10 BASIC ARITHMETIC

+ means plus or add.
Example: 4 + 4 = 8

= means equal to or of the same value.

— means minus, subtract or take away.
Example: 4 - 3 = 1

x means multiply by, or times.
Example: 4 x 5 = 20

÷ means divide by.
Example: 12 ÷ 2 = 6

. means multiply by, or times.

( ) means parenthesis; do the arithmetic inside the parenthesis first.
Example: (7 − 3) + 2 = (4) + 2 = 6
Some calculations use parentheses instead of a multiplication sign.
Example: (4) (5) = 20

$( )^2$ means that the number inside the parenthesis is to be multiplied by itself or squared.
Example: $(4)^2 = 4 \times 4 = 16$

$( )^3$ means that the number inside the parenthesis is to be multiplied by itself twice, or cubed.
Example: $(4)^3 = 4 \times 4 \times 4 = 64$

$\dfrac{a}{b}$ means that the top number, "a", is to be divided by the bottom number, "b".
Example:
If "a" = 6, and "b" = 2, $\dfrac{a}{b} = \dfrac{6}{2} = 6 \div 2 = 3$

Δ (delta) means a difference
Example:
$\Delta \underline{T}$ = temperature difference, for instance, 0°C to 40°C.

Most calculations include the use of basic units. Basic units are expressed in digits. In the statement, 7 x 8 = 56, 7 and 8 are digits; 56 is made up of two digits, 5 and 6. In the metric system, multiples of digits are on the basis of 10. For example: The digit 1, if divided by 10, would be 0.1; each subsequent division of 10 would result in 0.01, 0.001, and the like. The prefix (name) for these follow. The digit 1, if multiplied by 10, would be 10; each subsequent multiplication by 10 would result in 100, 1 000, 10 000, 100 000, and the like. *Each level of multiplication or division has a name:*

Symbol	Prefix	Quantity	Pronunciation
M	mega	= 1 000 000	like megaphone
K	kilo	= 1 000	kill'-oh
h	hecto	= 100	heck'-toe
da	deka	= 10	deck'-uh
basic unit		= 1	
d	deci	= 0.1	dess'-ih
c	centi	= 0.01	sen'-tih
m	milli	= 0.001	like military
μ	micro	- 0.000 001	my'-crow

In many calculations, it is difficult to work with numbers using many zeros either ahead of or behind the decimal point. A special number, called a "powers of ten," may be used to express these types of numbers.

"Power of 10" means that the number 10 is multiplied by itself the desired number of times to obtain the required number of zeros. The number of times the ten is to be multiplied by itself is shown by the small number above and to the right of the number 10. This number is also called the "exponent." It works as follows:

For numbers larger than one:
1 000 = 10^3 or (10 x 10 x 10)
100 = 10^2 or (10 x 10)
10 = 10^1 or (10)

For numbers less than one:
0.1 = 10^{-1} or (0.10)
0.01 = 10^{-2} or (0.10 x 0.10)
0.001 = 10^{-3} or (0.10 x 0.10 x 0.10)

ROUNDING NUMBERS

In refrigeration calculations, it is not usually necessary to use fractions or decimals of a unit. When the decimal is less than 5, round to the number, ignoring the decimal. When the decimal is 5 or over, round to the next larger number. For

Mark 1 Space Environmental Chamber located at Arnold Air Force Station, Tennessee. Installation is used for test firing liquid fuel engines (rocket must be tested in vertical position). Massive structure is self-contained, provides living services and is completely air conditioned. Normal refrigeration systems are used, plus cryogenic cooling and pumping systems to lower reference temperature in space chamber to −196°C (−320 F.).

instance, 35.5 becomes 36. If a problem has been carried two or more decimal places and greater accuracy is required, it is acceptable to round such numbers to a single decimal. For instance: 3.52 may be rounded to 4.

1-11 TEMPERATURE CONVERSION FORMULAS

It is sometimes necessary to convert a temperature from one scale to another. Formulas have been developed for this:

°C = temperature in degrees Celsius
F. = temperature in degrees Fahrenheit
K = temperature in degrees Kelvin
R. = temperature in degrees Rankine

1. To convert Celsius degrees into Fahrenheit degrees:
Formula:

$$\text{Temp. F.} = (\frac{180}{100} \times \text{Temp. } °C) + 32 \text{ or}$$

$$\text{F.} = (\frac{9}{5} \times °C) + 32$$

Example: Convert 75 °C into Fahrenheit.
Solution:

$$\text{F.} = (\frac{9}{5} \times 75) + 32 = (135) + 32 = 167 \text{ F.}$$

2. To convert Fahrenheit degrees into Celsius degrees:
Formula:

$$\text{Temp. } °C = \frac{100}{180} \times (\text{Temp. F.} - 32) \text{ or}$$

$$°C = \frac{5}{9} \times (\text{F.} - 32)$$

Example: Convert 212 F. into °C.
Solution:

$$°C = \frac{5}{9} \times (212 - 32) = \frac{5}{9} \times (180) = 100 \text{ °C}$$

3. To convert Fahrenheit degrees into Fahrenheit Absolute (Rankine) degrees:
Formula: Temp. F_A. (R.) = F. + 460
Example: Convert 40 F. to F_A. (R.).
Solution:
   F. = 40 + 460 = 500 degrees F_A. (R.)

4. To convert Rankine degrees to Fahrenheit degrees:
Formula: Temp. F. = R. − 460
Example: Convert 180 R. to F.
Solution: F. = 180 − 460
         F. = −280 F.

5. To convert Celsius degrees to Kelvin:
Formula: Temp. K. = °C + 273
Example: Convert −10 °C to K.
Solution: K. = −10 + 273 = 263 K.

6. To convert Kelvin to Celsius:
Formula: Temp. °C = K. − 273
Example: Convert 400 K. to °C.
Solution: °C = 400 − 273
         °C = 127 °C

7. To convert Rankine to Kelvin or Kelvin to Rankine, use the same ratios as for converting Fahrenheit to Celsius and for converting Celsius to Fahrenheit.

Temperature Difference Calculations:
Calculations which require converting Fahrenheit temperature difference to Celsius temperature difference and Celsius temperature difference to Fahrenheit temperature difference may be computed as follows:

$$°C \text{ temp. diff.} = \frac{5}{9} \text{ F. temp. diff.}$$

$$\text{F. temp. diff.} = \frac{9}{5} °C \text{ temp. diff.}$$

Examples:
Fahrenheit to Celsius:
When the outside temperature is 10 F. and the inside temperature is 75 F., the temperature difference is 65 F. What is the temperature difference in °C?

$$°C \text{ (temp. diff.)} = \frac{5}{9} \times 65 = 36 °C$$

Celsius to Fahrenheit:
When the outside temperature is 10 °C and the inside temperature is 26 °C, the temperature difference is 16 °C. What is the temperature difference in Fahrenheit?

$$\text{F. (temp. diff.)} = \frac{9}{5} \times 16 = 28.8 \text{ or } 29 \text{ F.}$$

Throughout this text temperatures are given in both Celsius and Fahrenheit. Most of the Fahrenheit temperatures are rounded to whole numbers and the equivalent Celsius temperature is shown rounded to the nearest possible whole number. Where the Celsius temperature ends in 0.5 or more, the next higher temperature is used. If the Celsius temperature ends in a 0.4 or less, the next lower Celsius temperature is chosen.
Example: 40 F. is equivalent to 4.4 °C. This is rounded to 4 °C.
Another example is 0 F. Carried to one decimal place, it equals −17.8 °C. This is rounded to −18 °C.

1-12 DIMENSIONS

Dimensions are measurements which are necessary in determing lengths, areas and volumes.

LINEAR MEASUREMENT (Length)
Linear measurement considers only one dimension. Finding the length of a piece of copper tubing is an example.

Metric Units and U. S. Conventional Unit Equivalents:
1 millimetre (mm) = 0.039 in.
10 mm = 1 centimetre (cm) = 0.394 in.
10 cm = 1 decimetre (dm) = 3.937 in.
10 dm = 1 metre (m) = 100 cm = 39.37 in. = 3.28 ft.
1 000 m = 1 kilometre (km) = 3280.8 ft.
2.54 cm = 1 in.

U.S. Conventional Units - Decimals and Fractions of an Inch:

Measurement	How to Express the Measurement
0.001 in.	one thousandth of an inch
0.01 in.	one hundredth of an inch
0.1 in.	one tenth of an inch
1/64 in.	one sixty-fourth of an inch
1/32 in.	one thirty-second of an inch
1/16 in.	one sixteenth of an inch
1/8 in.	one eighth of an inch

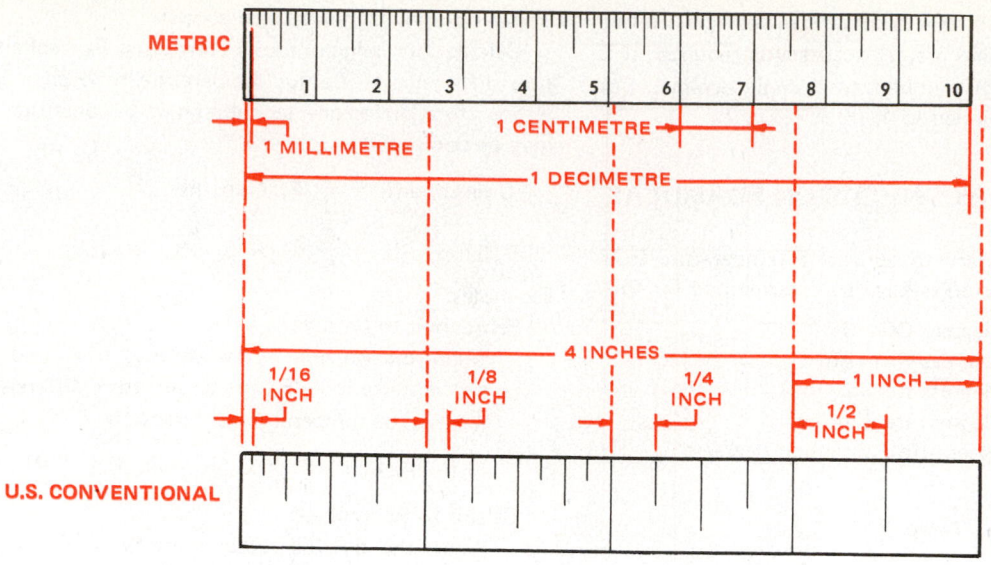

Fig. 1-5. Metric (SI) and U.S. Conventional units of linear measurement are compared.

1/4 in.   one fourth of an inch

1/2 in.   one half of an inch

Sometimes the symbol ('') indicates inches; for example, 6''.
Occasionally the symbol (') indicates feet; for example 6'.

Units of Conventional Linear Measurement:

12 inches = 1 foot

3 feet = 1 yard

5 280 feet = 1 statute mile

6 080 feet = 1 nautical mile

Some linear metric units useful to the service technician are shown in Fig. 1-5. In measuring very tiny particles, the micrometre (μ m) unit has been most used. The micrometre is one-thousandth of a millimetre (mm).

AREA MEASUREMENT

The measurement of area involves the measurement of two-dimensional space.

The area of an object is found by multiplying its length by its width (L x W). For example, the width of a table top is 90 cm and the length of the table is 150 cm. The area of the table top is 90 x 150 = 13 500 cm².

Some special formulas must be used when finding the area of certain objects. For example, the area of a circle is found by using the formula A = πr^2. The symbol, π, is always 3.1416, and r is the radius of a circle. It is equal to 1/2 the diameter. Therefore, this formula may also be expressed as:

$$A = \pi \frac{D^2}{4}$$

Metric Units:

1 square centimetre (cm² or sq. cm) = 0.155 square inch

1 square decimetre (dm² or sq. dm) =

10 cm x 10 cm = 100 cm² = 15.5 sq. in.

1 square metre (m² or sq. m) = 1550 sq. in. =

10 dm x 10 dm = 100 square decimetres (dm²) =

10.76 sq. ft.

These units are shown in view A, Fig. 1-6. The area of a circle is shown in view B., Fig. 1-6.

U. S. Conventional Units:

square inches (sq. in.)

144 sq. in. = 1 square foot

9 sq. ft. = 1 square yard (sq. yd.)

These units are shown in view A of Fig. 1-7. The area of a circle is shown in view B, Fig. 1-7.

VOLUME MEASUREMENT

The measurement of volume involves the measurement of three-dimensional space (cubic).

The volume of an object is determined by multiplying the width by the length by the height. An example is finding the volume of a cube (width x length x height, or W x L x H).

Once again, some special formulas must be used when finding the volume of certain objects. The volume of a cylinder, for example, is determined by multiplying the area of one end (πr^2) by the length (L) of the cylinder.

Metric Units:

1 litre (L) = cubic decimetre (dm³)

= 1.05 quarts (qts.)

= 61 cu. in. = 0.035 cu. ft.

1 000 cubic centimetres (cm³) = 1 cubic decimetre (dm³)

= litre

1 000 cubic decimetres (dm³) = 1 cubic metre (m³)

= 1.3 cu. yd.

These are shown in view A, Fig. 1-8. The volume of a cylinder is shown in view B, Fig. 1-8.

U. S. Conventional Units:

cubic inches (cu. in.)

cubic feet (cu. ft.)

cubic yards (cu. yd.)

1 728 cu. in. = 1 cu. ft.

27 cu. ft. = 1 cu. yd.

1 cu. ft. = 7.48 gal.

These units are shown in view A, Fig. 1-9. The volume of a cylinder is shown in view B, Fig. 1-9.

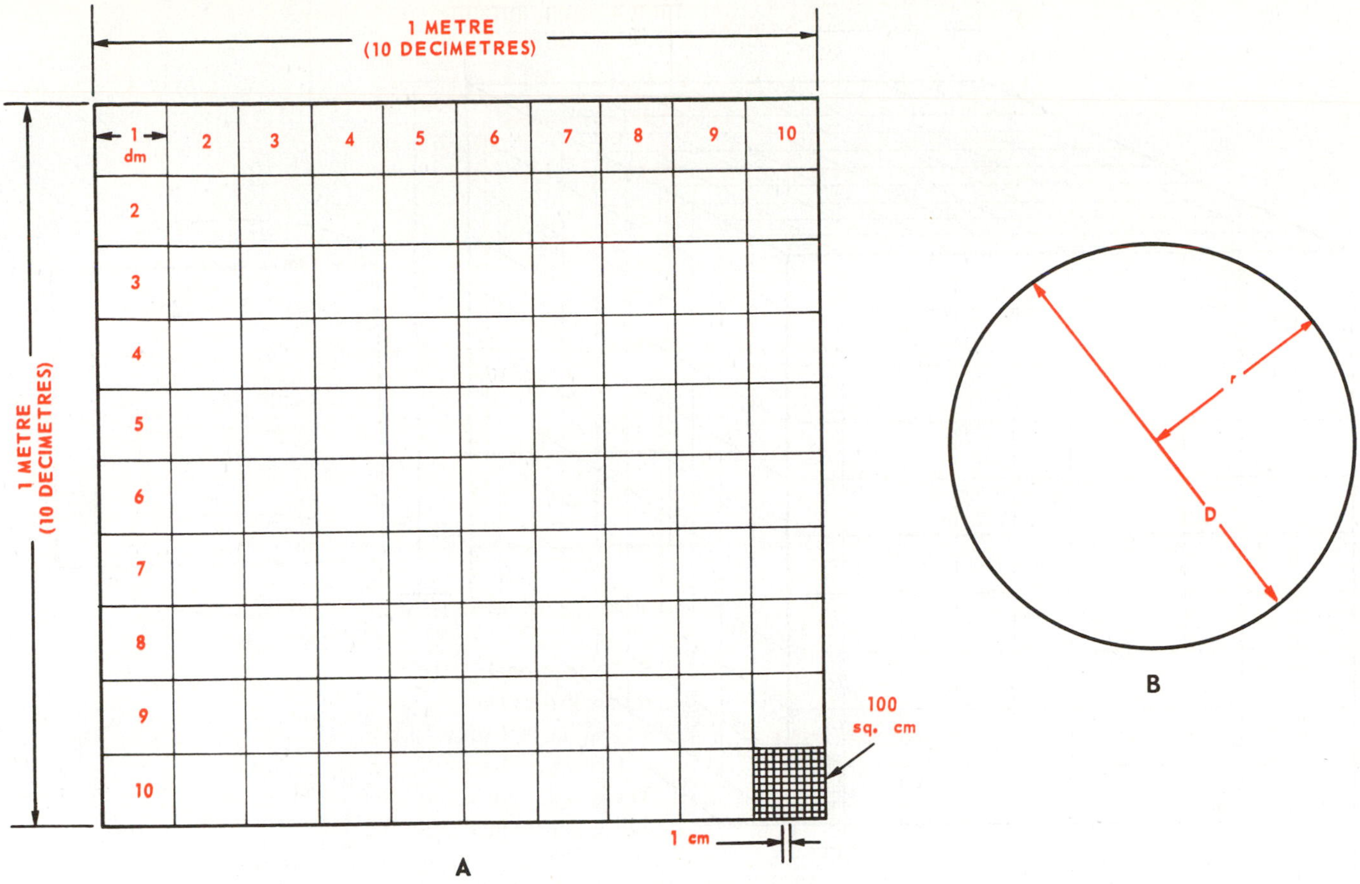

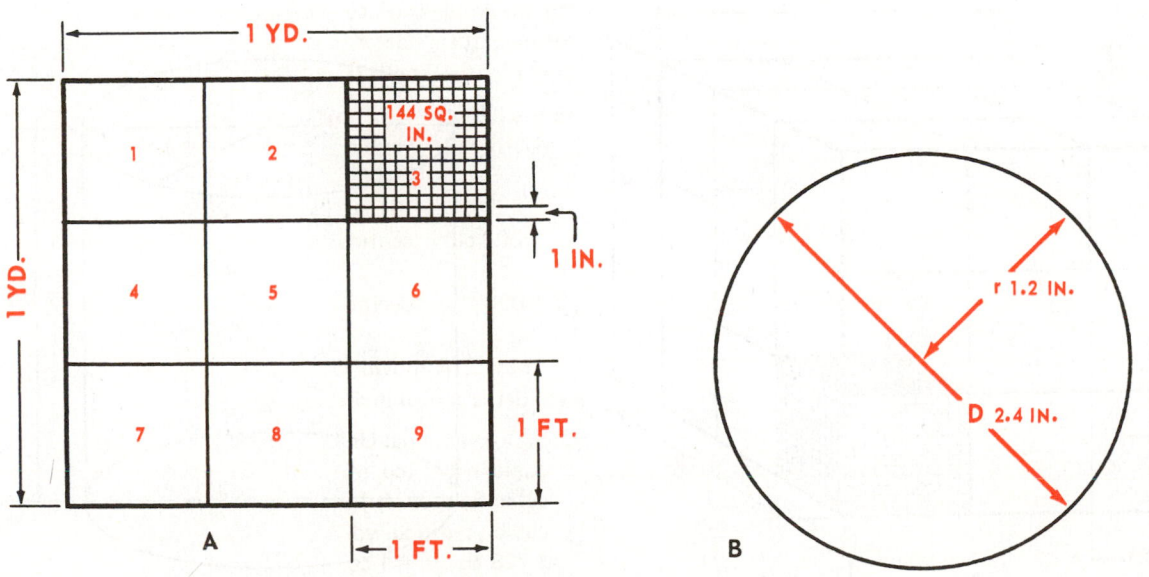

Fig. 1-6. Calculation of standard areas using metric units. A—Area of a rectangle is calculated by multiplying width by length. B—Area of a circle is calculated using the formula πr^2. Value of π is 3.1416. If diameter (D) of circle is 8 cm, radius (r) which is half the diameter, is 4 cm: $r^2 = r \times r = 4 \times 4 = 16$. Area of circle is $3.1416 \times 16 = 50.27$ cm^2.

Fig. 1-7. Calculating standard areas in inches and feet. A—Area of rectangle is calculated by multiplying width by length. Remember, 144 sq. in. equal 1 sq. ft., and 9 sq. ft. equal 1 sq. yd. B—Area of circle is calculated with same formula as in Fig. 1-6, view B.

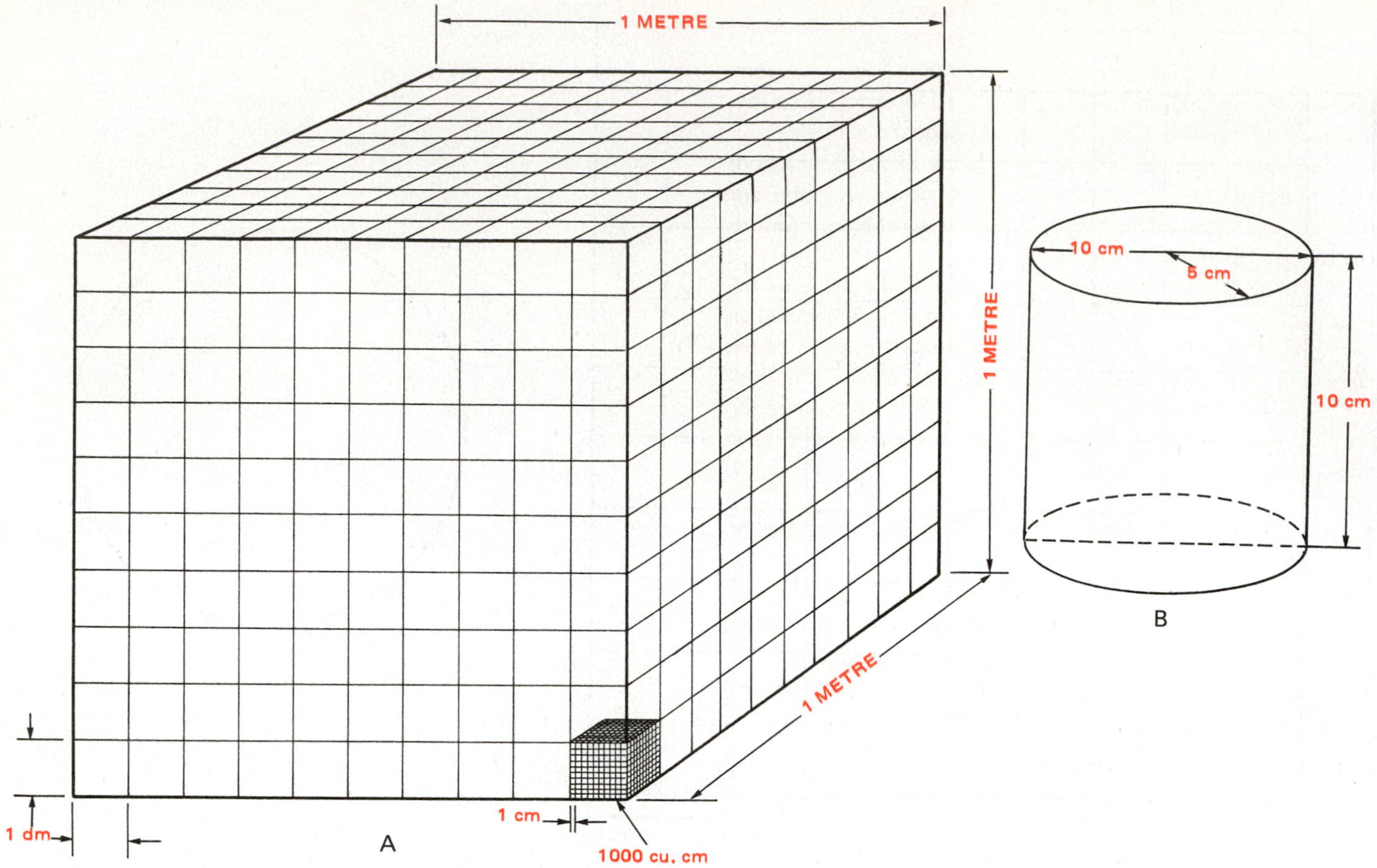

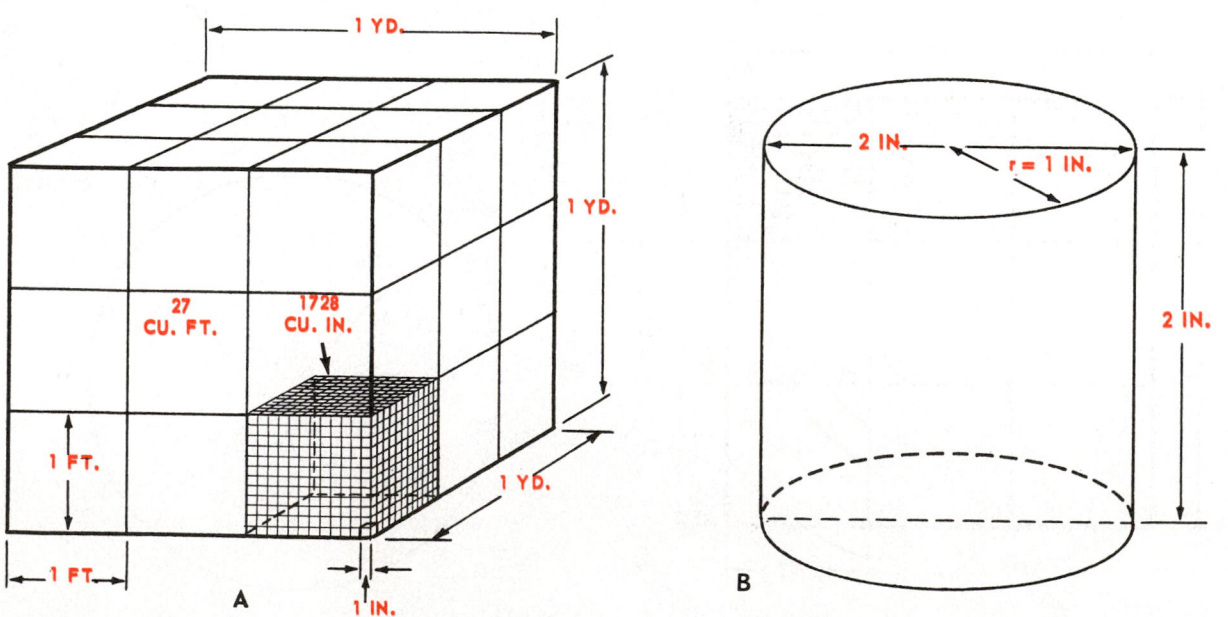

Fig. 1-8. Calculation of standard volumes using metric units. A—Volume is calculated by multiplying width by length by height. B—Volume of cylinder is calculated by multiplying area of end by length.

Fig. 1-9. Calculation of standard volumes in inches and feet. A—Volume is calculated by multiplying width by length by height. B—Volume of cylinder is worked out my multiplying area of end by length. (Cylinder dimensions are usually expressed in inches and decimals of an inch.)

1-13 ANGULAR MEASUREMENT

Circles and arcs of a circle are measured in degrees. A complete circle has 360 degrees. See Fig. 1-10. A degree is further divided into minutes. Sixty (60) minutes equal one degree.

Minutes are further divided into seconds. Sixty (60) seconds equal one minute. These are the same names, minutes and seconds, that are used in measuring time. In angular measurement, they have a different meaning.

The angle does not depend on the size of the circle or the length of the diameter or radius. A part of a circle is called an arc. It is formed by two lines going out from the center of the circle and cutting across the circumference. It can be seen that, for a given central angle, the larger the circle, the longer the arc. The arc of the circle which includes a 90-degree central angle is called a 90-degree arc.

1-14 MASS, WEIGHT AND FORCE OF GRAVITY

In SI metric measurement, *mass* means the same thing that *weight* means in U.S. conventional measure. It is the amount of material in a substance measured in *kilograms* (kg). A cubic centimetre (cm^3) of water at a temperature of greatest density has a mass of 1 gram.

1 gram (g)   = 0.0022 lb. = 15.4 grains (gr)
453.6 g   = 0.4536 kg = 1 lb.
1 kilogram = 1 000 g = 2.2 lb.

Mass is related to another term known as *force of gravity*. This term is described as "the force that, if applied to a body, would give it the acceleration of free fall." In other words, force of gravity is a push in any direction. It can move the

body sideways, lift it, or suspend it. This force is always measured in newtons (N).

Because of gravity, mass itself exerts force of gravity. A mass of 1 kg resting on the earth's surface exerts a force of 9.8 newtons. Thus, to convert the mass of an object to force of gravity, multiply mass by 9.8.

Example: Give the force of gravity of an object which has a mass of 5 kilograms.

Solution: Force of gravity = 5 kg x 9.8 = 49 newtons (N)

All substances have *mass*. But the one condition under which a substance has no force of gravity is when it is falling in a vacuum under the influence of gravity.

U. S. Conventional Units:

The U. S. conventional units for weight are the grain, the ounce, the pound, and the ton.

7 000 grains = 16 ounces (oz.) = 1 pound (lb.)
2 000 pounds (lb.) = 1 ton (ton)

1-15 PRESSURE

Pressure is the force per unit area, and it is expressed in pascals (Pa) or kilopascals (kPa).

The normal pressure of the atmosphere at sea level is 101.3 kPa or 14.7 pounds per square inch (psi). In engineering practice, this is usually rounded to 100 kPa or 15 psi.

A refrigerating system operation depends mainly on pressure differences in the system. Substances always push on the surfaces supporting or containing them. A block of ice (a solid) exerts a pressure on its support. If the support were removed, the block would fall to another supporting level.

A liquid exerts a pressure on the sides and bottom of its container. A gas exerts a pressure on all surfaces of its container. Fig. 1-11 illustrates these types of pressures.

Using the U. S. conventional system, a solid weight of 1 pound with a bottom surface area of 1 inch square would exert a pressure of 1 pound (lb.) per square inch (1 psi) upon a flat surface.

A liquid in a container maintains an increasing pressure on the sides and bottom as the liquid depth increases. The pressure of gas in the container will depend on the quantity of the gas and the temperature.

1-16 PASCAL'S LAW

To honor Pascal, the SI metric system uses the term "pascal" as a unit of pressure. *A pascal is a newton per square metre (N/m^2).*

A newton is the metric unit of force. *One newton is equal to the mass of 1 kilogram being accelerated at a rate of 1 metre per second per second.*

Pascal's Law states that: *Pressure applied upon a confined fluid is transmitted equally in all directions.* It is the basis of operation of most hydraulic and pneumatic systems.

Fig. 1-12 illustrates Pascal's Law. It shows a fluid-filled cylinder. A piston having a cross-sectional area of 645 mm^2 (one square inch) is fitted into a small cylinder connected to the larger cylinder. A force of 690 kPa (100 psi) is applied to the piston in the small cylinder. The pressure gauges show

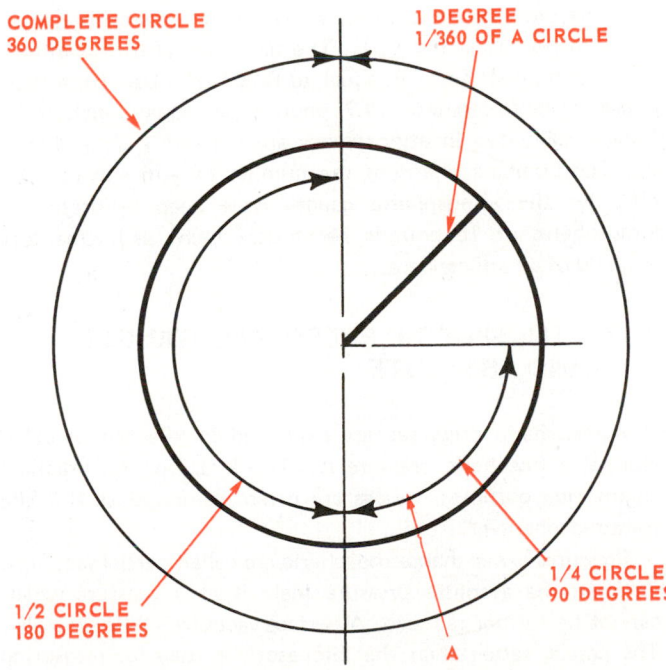

COMPLETE CIRCLE
360 DEGREES

1 DEGREE
1/360 OF A CIRCLE

1/4 CIRCLE
90 DEGREES

1/2 CIRCLE
180 DEGREES

A

Fig. 1-10. Angular measurement. Complete circle consists of 360 degrees. Half circle equals 180 degrees and 1/4 circle equals 90 degrees. "A" indicates 90 degree arc.

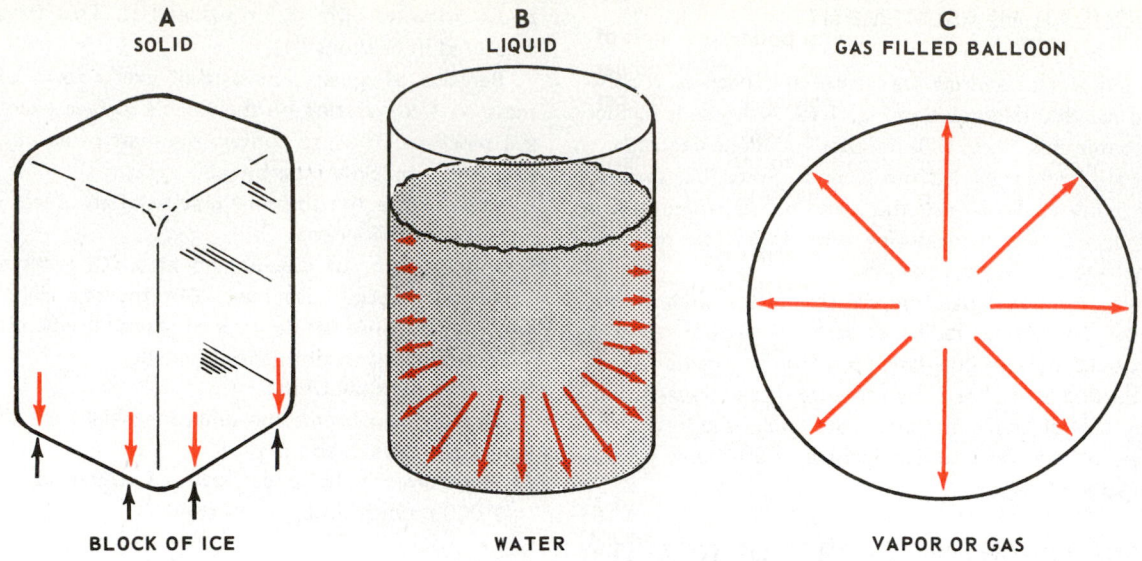

Fig. 1-11. Three states of substance such as water. A—Solid (ice). Mass of ice exerts downward force only. Total force on support equals ice's mass. B—Liquid state. Water exerts pressure on container both vertically and horizontally. C—Vapor or gas in rubber balloon exerts pressure uniformly in all directions.

the pressure being transmitted equally in all directions.

One psi equals 6894.8 pascals (Pa). This can be rounded to 6.9 kPa.

The refrigeration engineer must deal with pressures both above and below atmospheric pressure. Gauges are calibrated so that zero on the gauge means that there is no pressure at all — not even atmospheric pressure. No negative pressures are used. A pressure of 5 kPa, then, is the same as 0.75 pounds per square inch (psi). Expressed in inches of mercury vacuum, it would be about the same as 28.5 inches of Hg vacuum.

In the U. S. conventional units, pressures above atmospheric pressure are measured in pounds per square inch (psi), and pressures below atmospheric are measured in inches of mercury (in.Hg) column. See Para. 1-17.

Some older references may show instruments calibrated in atmospheres, bars, and torrs. The bar is equal to one atmosphere; a millibar (mb) is equal to 0.001 of a bar. An atmosphere is approximately 14.7 pounds per square inch (psi). Gauges calibrated in atmospheres are marked 1, 2, 3, 4, and up. The numbers represent the number of atmospheres. Arbitrarily, the atmospheric gauges have been calibrated in atmospheres of 15 pounds per square inch (psi). One torr is 1/760 of an atmosphere.

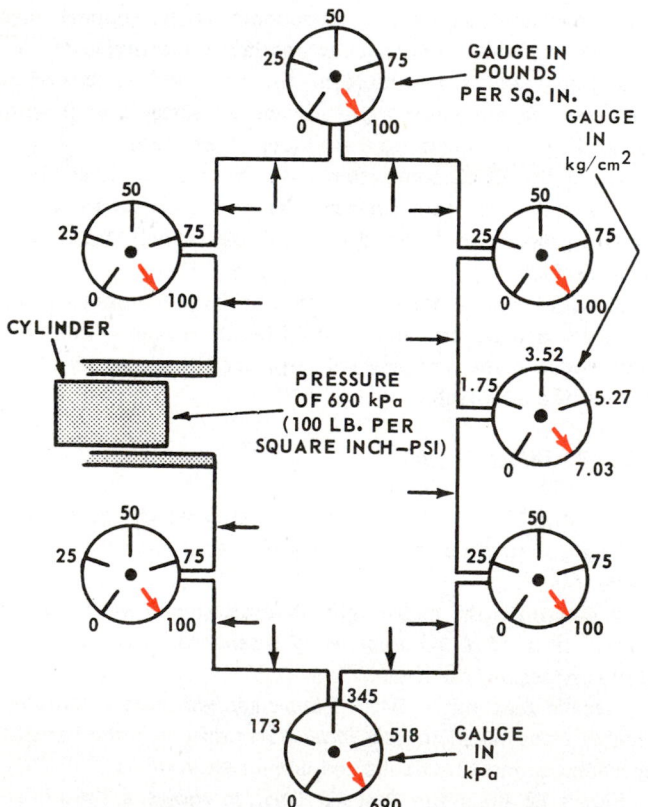

Fig. 1-12. Illustration of Pascal's Law. Pressure of 690 kPa (100 psi) is pressing against piston having head area of 1 sq. in. All gauges have same reading. Center right hand gauge, calibrated in kilograms per square centimetre, reads 7.03 kg/cm² (100 psi). Bottom gauge, calibrated in kilopascals, reads 690 kPa (100 psi).

1-17 ATMOSPHERIC PRESSURES, GAUGES AND ABSOLUTE

Atmospheric pressures are expressed in kPa (kilopascals). Normal atmospheric pressure is 101.3 kPa. But, for practical engineering purposes, gauges are often calibrated at 100 kPa for atmospheric pressure.

Pressures lower than atmospheric are called partial vacuums. Zero on the absolute pressure scale is at a pressure which cannot be further reduced. A perfect vacuum is 0 Pa (pascals). The pascal, rather than the kilopascal, is used for measuring high vacuums (pressure close to an absolute vacuum).

Fig. 1-13, view A, illustrates a pressure gauge used in refrigeration work. It is calibrated in kilopascals rather than in pounds per square inch.

U. S. Conventional Units:

Atmospheric pressures are expressed in pounds per unit of area or in inches of liquid column height. The most popular instruments (gauges) are those that register in pounds per square inch *above* the atmospheric pressure (psig or psi).

A reading of 0 psi on the gauge is equal to the atmospheric pressure, which is about 14.7 psi; (15 pounds per square inch is often used for working out problems).

The absolute pressure scale registers zero at a pressure which cannot be further reduced. A perfect vacuum is 0 pounds per square inch absolute (0 psia).

Pressure may also be indicated in other ways: 1. Inches of mercury (in. Hg). 2. Feet or inches of water column. These gauges may be calibrated either above atmospheric pressure or absolute pressure, depending on the construction. A mercury gauge is often used for measuring below atmospheric pressure.

The barometer in Fig. 1-14 is a mercury gauge. With a vacuum at the closed top of the tube, the atmospheric pressure will support a mercury column of 29.92 in. Hg in height at sea level under standard conditions.

A unit of measure which has been used for reading high vacuums (pressure close to an absolute vacuum) is the *torr.* One torr is equal to a pressure of 1 mm of mercury (mm Hg, 0°C). It is named after the man who invented the mercury barometer. The unit torr may be expressed in fractions of an atmosphere. One torr = 1/760 of an atmosphere. A pressure of one torr is almost a perfect vacuum.

In solving most pressure and volume problems, it is necessary to use absolute pressures (psia). *Absolute pressure is gauge pressure plus atmospheric pressure.*

Example: Calculate absolute pressure when the pressure gauge reading is 21 psi; (psi always indicates gauge pressure, and psia indicates absolute pressure).

Solution: Absolute pressure equals gauge pressure plus atmospheric pressure:

psi + 15 = psia

21 + 15 = psia

21 + 15 = 36 psia

Air pressure or a vacuum can be measured with a column of water. To equal 29.92 in. Hg it would be about 34 feet high. The height is greater because water is so much lighter than mercury.

Since the service engineer must often test both pressures and vacuums in the same system, pressure gauges are made which will measure both. They are called compound gauges.

Compound means they have two or more pressure scales. One measures pressure below atmospheric pressure. The other measures pressures above atmospheric pressure. Fig. 1-13, view B, illustrates such a gauge.

It is also frequently necessary to convert inches of mercury into pounds per square inch absolute or to convert pounds per square inch absolute into inches of mercury. Formulas are available for making an accurate conversion.

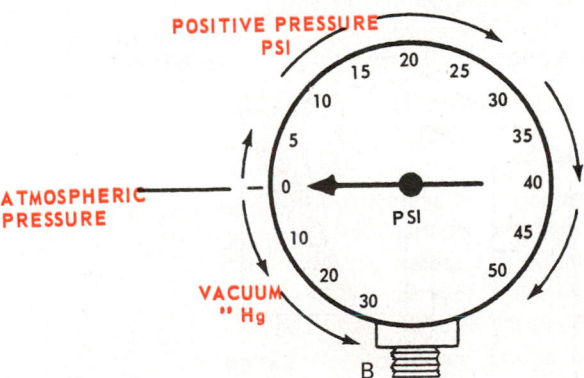

Fig. 1-13. Pressure gauges. A—This one is calibrated in kilopascals. Pressures from 0 to 100 are partial vacuums. Atmospheric pressure is set at 100 kPa. B—Compound gauge measures both pressure above atmospheric in psi and pressure below atmospheric in " Hg. Zero on this scale is atmospheric pressure.

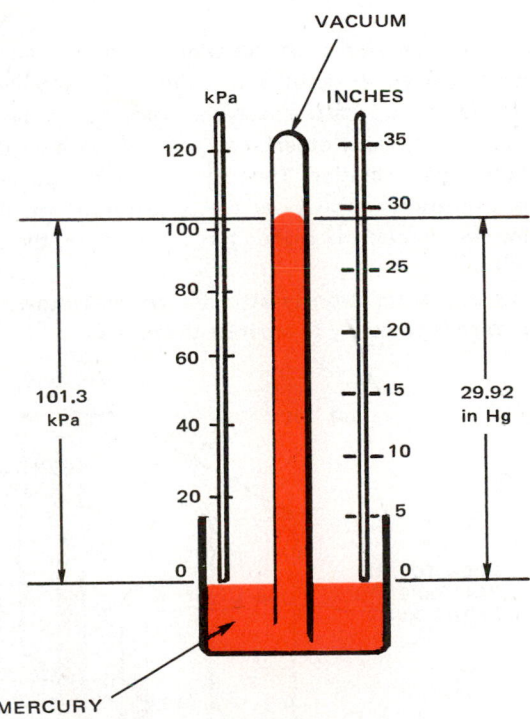

Fig. 1-14. Mercury barometer used for measuring atmospheric pressure. It consists of glass tube sealed at one end and open at other end. Fill tube with mercury, then, with finger sealing open end, invert it in container of mercury. When finger is removed, mercury will drop to level corresponding with atmospheric pressure. Glass tube should be about 86.4 cm (34 in.) long.

Roughly 2 in. Hg equals 1 psi. A chart, Fig. 1-15, makes converting easy.

Water columns are usually designed for measuring small pressures above or below atmospheric pressure. For example, they can be used for pressures in air ducts, gas lines, and the like. A water column 2.3 feet high (or about 28 in.) equals 1 psi.

These pressure measuring devices are called manometers. They are calibrated in inches of water column. See Fig. 1-16.

In some high-pressure refrigerating machines, pressure gauges are calibrated in atmospheres. One atmosphere corresponds to about 15 pounds per square inch (psi), two atmospheres to 30 psi, three atmospheres to 45 psi, and so on.

Fig. 1-17 shows a table comparing various pressure scales.

1-18 THE THREE PHYSICAL STATES

Substances exist in three states, depending on their temperature, pressure and heat content. For example, water at atmospheric pressure is a solid at temperatures below $0^{\circ}C$ (32 F.), and a liquid from $0^{\circ}C$ (32 F.) to $100^{\circ}C$ (212 F.). At $100^{\circ}C$ (212 F.) and above it becomes a vapor (gas).

Water is shown in its three states in Fig. 1-11. In this example, the physical state is controlled both by temperature and pressure. The temperatures just given apply only when atmospheric pressure is at 101.3 kilopascals (kPa) or 14.7 pounds per square inch (psi).

1-19 SOLIDS

A solid is any physical substance which keeps its shape even when not contained. It consists of billions of molecules, all exactly the same size, mass, and shape. These stay in the same relative position to each other. Yet, they are in a condition of rapid motion or vibration. The rate of vibration will depend upon the temperature. The lower the temperature, the slower the molecules vibrate; the higher the temperature, the faster the vibration.

The molecules are strongly attracted to each other. Considerable force is necessary to separate them.

INCHES of Hg	mm of Hg	psia or psi	FT of WATER
30			
(29.92)	760	(14.7)	33.40
29		14.5	
28	711	14	32.2
27		13.5	
26	660	13	29.9
25		12.5	
24	610	12	27.6
23		11.5	
22	559	11	25.3
21		10.5	
20	508	10	23.0
19		9.5	
18	457	9	20.7
17		8.5	
16	408	8	18.4
15		7.5	
14	356	7	16.1
13		6.5	
12	305	6	13.8
11		5.5	
10	254	5	11.5
9		4.5	
8	203	4	9.2
7		3.5	
6	152	3	6.9
5		2.5	
4	102	2	4.6
3		1.5	
2	51	1	2.3
1		0.5	
0	0	0	0

Fig. 1-15. Chart converts inches of mercury (" Hg) into pounds per square inch (psi).

A solid must always be supported by an upward force or it will fall. See view A, Fig. 1-11.

1-20 LIQUIDS

A liquid is any physical substance which will freely take on the shape of its container, view B, Fig. 1-11. Yet, its molecules strongly attract each other.

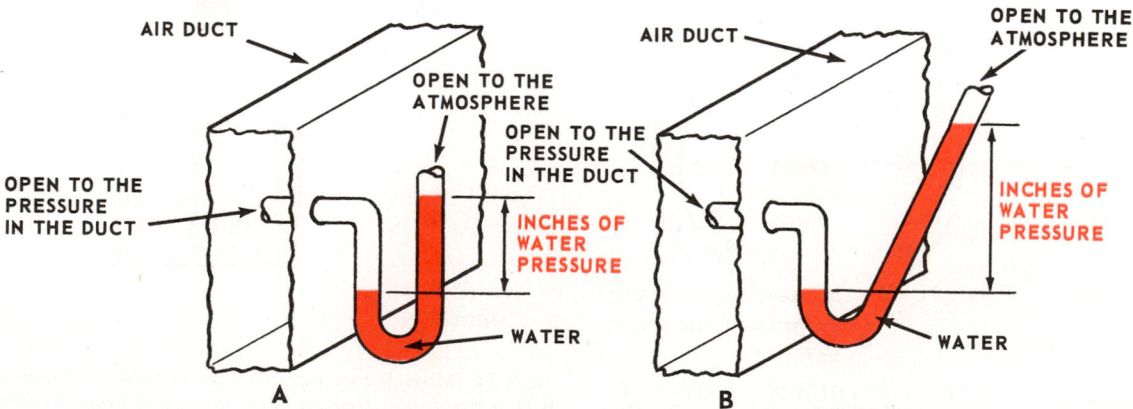

Fig. 1-16. Types of water manometer. This manometer is used to measure low pressure in air ducts and gas lines. Pressure is indicated in inches of water. It is measured by difference in level between surface of water in two branches of tube. B—For easier reading, the end open to atmosphere is often placed at low angle. Red dye in water makes gauges easier to read.

| | POUNDS PER SQUARE INCH | | INCHES MERCURY VACUUM | cm Hg | kPa |
	ABSOLUTE psia	GAUGE psig or psi	"Hg.		
Positive Pressure	105	90			725
	90	75			621
	75	60			518
	60	45			414
	45	30			311
	30	15			207
Atmospheric Pressure	14.7	0	0	0	101.3
Negative Pressure or Vacuum	10	−5	10	25.4	69
	5	−10	20	50.8	35
	0	−15	30	76.2	0

Fig. 1-17. Table compares various pressure scales.

Think of the molecules as swimming among their fellow molecules without ever leaving them. The higher the temperature, the faster the molecules swim. Warmer molecules will move upwards toward the top of the container. This is because they take up more space by their rapid movement. They become lighter than colder molecules.

1-21 GASES

A gas is any physical substance which must be enclosed in a sealed container to prevent its escape into the atmosphere.

The molecules, having little or no attraction for each other, travel (fly) in a straight line. They bounce off each other, off molecules of other substances, or off the container walls. They have little or no attraction for any other substance. The pressures shown in the gas-filled balloon in view C, Fig. 1-11, illustrate how gases behave.

Any substance can be made to exist as a solid, a liquid, or a gas. Any molecule can be made to vibrate, swim, or fly. It depends on two things: temperature and pressure. To understand this change of state, one must study temperature and pressure relationships.

1-22 DENSITY

Some substances are heavier than others. Comparative weights of gases, liquids, and solids may be shown by either density or specific gravity (see Para. 1-24).

Density is a substance's mass per unit of volume.

Density is expressed as kilograms per cubic metre (kg/m^3).

1-23 SPECIFIC VOLUME

When comparing densities of gases, it is common to express the densities in specific volumes. *Specific volume is the volume of one kilogram of a gas at standard conditions.*

Standard conditions are 20°C at 101.3 kPa pressure. The volume of 1 kg of dry, clean air at standard atmospheric conditions is 0.840 m^3. By comparison, 1 kg of hydrogen occupies 11.17 m^3, and 1 kg of the refrigerant ammonia (R-717) occupies 1.311 m^3. Carbon dioxide (R-744) only occupies 0.509 m^3.

If 1 kg of gas occupies a greater space than air, it is called a light gas. If it occupies less space than air, it is classified as a heavy gas. The specific volume is $\frac{1}{density}$.

U. S. Conventional Units:

Standard conditions are 68 F. at 29.92 in. of mercury column pressure. The volume of 1 lb. of dry, clean air at standard atmospheric conditions is 13.454 cu. ft. By comparison, 1 lb. of hydrogen occupies 178.9 cu. ft., and 1 lb. of the refrigerant, ammonia (R-717), occupies 21 cu. ft. Carbon dioxide (R-744) only occupies 8.15 cu. ft.

Equivalents:
1 lb./ft.^{3} = 16 kg/m^3
1 kg/m^3 = 0.0625 lb./ft.^{3}

1-24 SPECIFIC GRAVITY (RELATIVE DENSITY)

Specific gravity (sp. gr.) or relative density is the ratio of the mass of a certain volume of a liquid or a solid as compared to the mass of an equal volume of water.

Water is given a specific gravity of one. Objects which float on water have a specific gravity of less than one. Objects which sink in water have a specific gravity greater than one.

Mixtures of salt and water (brine) have a specific gravity greater than one. A calcium chloride brine adjusted to freeze at −18°C (0 F.) will have a specific gravity of 1.18. See Chapter 28 for a table of brine densities and freezing temperatures.

The relative density of gases is defined as the ratio of the mass of a certain volume of a gas as compared to the mass of an equal volume of hydrogen.

1-25 FORCE

Force applied to a body at rest causes it to move. The unit of force is the newton (N). *The newton is that force which, when applied to a body having a mass of one kilogram, gives it an acceleration of one metre per second per second.* Force may also be called accumulated pressure.

One newton equals one kilogram times one metre divided

by seconds squared: 1 kg x 1 m/s² or 1 kg m/s² = 1 N. (This unit of force is similar to the pound force in the U.S. conventional system.)

Examples: Determine the force on the head of a piston 645 mm² in area and under a pressure of 0.172 MPa. Use the following formula:

F = A x P

in which:

F = force

A = area of the piston head (645 mm²)

P = pressure (0.172 MPa)

Solution:

F = 645 x 0.172

F = 111 newtons (N)

A pascal is the unit of pressure. The newton is the total force, which equals the unit pressure times the area.

U. S. Conventional Units:

To determine the force on the head of a piston 10 square inches (sq. in.) in area and under a pressure of 25 psi use the following formula:

F = force

A = area of the piston head (10 sq. in.)

P = pressure (25 psi)

To solve the problem:

F = 10 x 25

F = 250 pounds (lb.)

Equivalents:

1N = 0.225 lb. force

1 lb. force = 4.45 N

1-26 WORK AND ENERGY

Work (W) is force (F) multiplied by the distance (D) through which it travels.

The unit of work is called the joule (J). The joule is the amount of work done by a force of one newton moving its point of application a distance of one metre.

Example: The propeller on a boat pushes the boat through the water with a force of 200 newtons. If the boat travels 10 km, how much work is done?

Solution:

Work = F x D = 200 x 10 = 2 000 N·km

Change newton per kilometre to newtons per metre:

Work = 2 000 x 1 000 = 2 000 000 N·m = 2MJ

Energy is the capacity or ability to do work.

The electric motor supplies the energy to drive the refrigerator compressor.

There are three kinds of energy.

1. Potential energy is stored energy. Examples are: water behind a dam, electrical energy in a battery, and weight which can fall or drop.
2. Kinetic energy is energy doing work. Examples are: water flowing over a dam, a battery lighting a bulb, and a falling weight.
3. Heat energy (see Para. 1-3).

The work formula is expressed as W = F x D, or newtons times metres (N·m). 1 J = 1 N x 1 m = N·m.

U. S. Conventional Units:

In U. S. conventional measure, the unit of work is called the foot-pound. One foot-pound is the amount of work done in lifting a 1 lb. weight a vertical distance of 1 ft. Work is sometimes expressed in inch-pounds. At such times, the distance through which the force acts is measured in inches.

Example: Calculate the work done in foot-pounds when lifting a weight of 2 000 lb. a vertical distance of 10 ft.

Work = Force x Distance

W = F x D (F = 2 000 lb.; D = 10 ft.)

W = 2 000 x 10 = 20 000 foot-pounds (ft. lb.) or, expressed in inch units,

W = 2 000 x 10 x 12 = 240 000 inch-pounds (in. lb.)

Equivalents:

1 ft. lb. force = 1.356 J = 1.356 N·m

1 J = 1 N·m = 0.737 ft. lb.

1-27 POWER

Power is the time rate of doing work. Work is expressed in joules. A joule is the force of one newton moving through a distance of one metre.

The common unit of mechanical power is the kilowatt (kW). A kilowatt is equal to 1 000 watts. The formula for power is force times distances divided by time. It is expressed in watts (W). 1 W = 1 joule per second = 1 J/s (see Para. 1-26).

Example: What is the power required to lift a mass of 100 kilograms at the rate of 10 metres per second?

Solution:

$$\text{Power} = \frac{\text{Force x Distance}}{\text{Time}} = \frac{\text{newtons x metres}}{\text{seconds}}$$

Force = 100 x 9.8 newtons

Distance = 10 metres

Time = 1 second

$$\text{Power} = \frac{100 \times 9.8 \times 10}{1}$$

= 9 800 W = 9.8 kW

U. S. Conventional Units:

In the U. S. conventional units, the unit of mechanical power is the horsepower. One horsepower (hp) is the equivalent of 33 000 foot-pound (ft. lb.) of work per minute. If a 2 000-lb. weight is lifted 10 ft. in two minutes, the power required would be:

$$\text{Horsepower} = \frac{\text{weight in pounds x distance in feet}}{\text{time in minutes x 33 000}}$$

$$\text{Horsepower} = \frac{2\,000 \times 10}{2 \times 33\,000} = \frac{20\,000}{66\,000} = 0.3 \text{ hp}$$

Equivalents:

1 hp = 746 W

1 W = 0.0013 hp

1-28 FIRST LAW OF THERMODYNAMICS

The First Law of Thermodynamics states that "heat and mechanical energy are mutually convertible."

In Para. 1-3, the joule is defined as the unit of heat. In Para. 1-26, the unit of work is also the joule. This is because work and heat are equal in energy.

Example: 100 joules of work equal 100 joules of heat.

1-29 UNIT OF HEAT

The unit of heat is the joule (J). A joule is a very small unit of heat. For refrigeration work the *kilojoule* (kJ), 1 000 joules, is used. The amount of heat required to raise the temperature of 1 kg of water 1°C is equal to 4.187 kJ. (See Fig. 1-18, view A.) Conversely, the amount of heat removed to lower the temperature of 1 kg of water 1°C is also equal to 4.187 kJ.

The mass in kilograms multiplied by the degrees Celsius temperature difference multiplied by 4.187 kJ equals the amount of heat added or subtracted.

Example: Find the amount of heat required to raise the temperature of 1 kg (approximately 1 litre) of water from 4°C to 27°C:

kJ = 4.187 x mass in kilograms x temperature change in degrees Celsius

= 4.187 x 1 x (27 — 4)

= 4.187 x 1 x 23

= 96.301 kJ

Conversely, if a substance is cooled, heat is removed:

Example: The amount of heat removed to cool 19 kg of water from 27°C to 1°C:

kJ = 4.187 x mass in kilograms x temperature change in degrees Celsius

= 4.187 x 19 x (27 — 1)

= 4.187 x 19 x 26

= 2068.378 kJ

U. S. Conventional Units:

The U. S. conventional unit of heat is the British thermal unit (Btu). The Btu is the amount of heat required to raise the temperature of one pound of water one degree Fahrenheit. (See Fig. 1-18, Part B.)

Whether a substance such as water is cooled or heated, the heat calculation is made in the same way. The temperature difference multiplied by the number of pounds of water gives the number of Btu. Where large heat loads are involved, the unit therm, which equals 100 000 Btu, is often used.

Example: The amount of heat required to raise the temperature of 62.4 lb. (1 cu. ft.) of water from 40 F. to 80 F.:

Btu = wt. in lb. x temperature change in degrees Fahrenheit

= 62.4 lb. x (80 — 40)

= 62.4 x 40

= 2496 Btu

Conversely, if a substance is cooled, heat is removed.

Example: The amount of heat removed to cool 50 lb. of water from 80 F. to 35 F.:

Btu = wt. in lb. x temperature change in degrees Fahrenheit

= 50 lb. x (80 — 35)

= 50 x 45

= 2250 Btu

Another metric unit, the calorie, is the amount of heat required to raise the temperature of one gram of water one degree Celsius. But the calorie is such a small unit that it is no longer used. Most calculations in engineering science are made

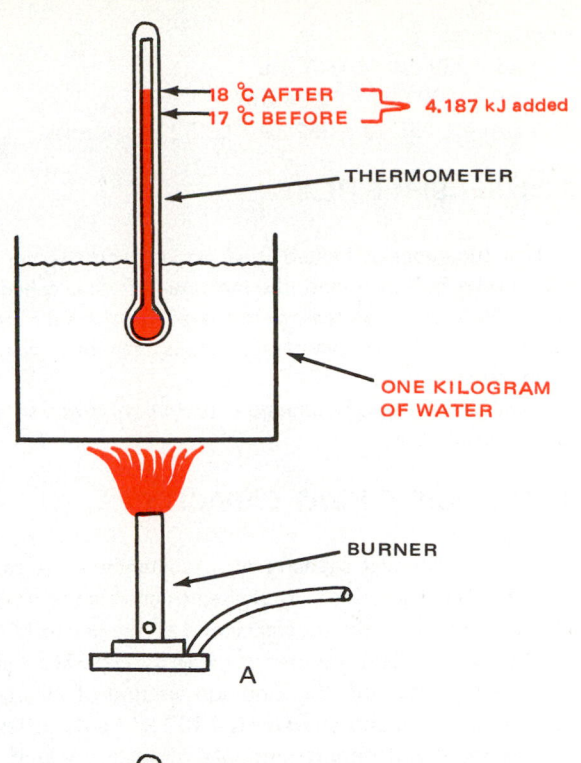

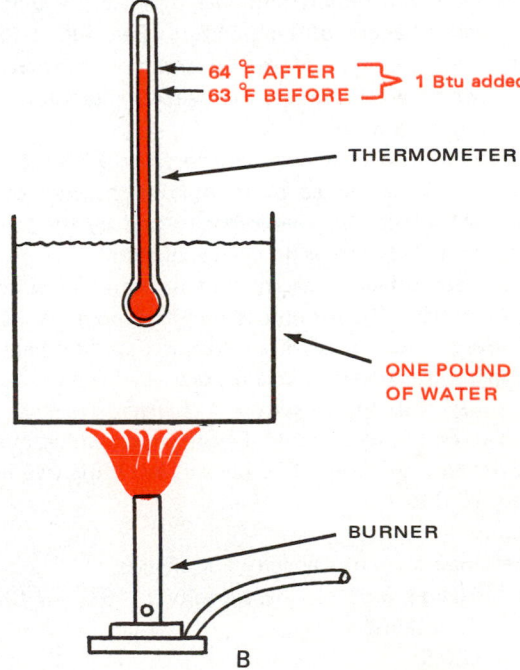

Fig. 1-18. Experiment in heating. A—Raising temperature of 1 kg of water from 17 to 18°C requires 4.187 kJ of heat. B—British thermal unit is equal to heat needed to raise temperature of one pound of water one degree Fahrenheit.

using the kilocalorie which equals 1 000 calories.

Example: Find the amount of heat required to raise the temperature of 150 grams of water from 10°C to 90°C:

Calorie = mass in grams

x temperature change in degrees Celsius

= 150 grams x (90 — 10)

= 150 x 80

= 12 000 calories = 12 kilocalories

Equivalents:

1 kJ = 239 cal = 0.948 Btu

1 cal = 0.004 kJ = 0.004 Btu

1 Btu = 1.055 kJ = 252 cal = 0.252 kilocalories

1-30 SENSIBLE HEAT

If a substance is heated (heat added) and the temperature rises as the heat is added, the increase in heat is called sensible heat. Likewise, heat may be removed from a substance (heat subtracted). If the temperature falls, the heat removed is, again, sensible heat.

Heat which causes a change in temperature in a substance is called sensible heat.

1-31 SPECIFIC HEAT CAPACITY

The specific heat capacity of a substance is the amount of heat that must be added or released to change the temperature of one kilogram of the substance one degree kelvin (K).

The sensible heat required to cause a temperature change in substances varies with the kind and amount of substance. The specific heat capacity of water is 4.187 kJ/kg·K. Different substances require different amounts of heat per unit mass to effect these changes of temperature. See Fig. 1-19 for the specific heat capacity of several common substances. Para. 1-29 shows how to calculate the heat in kilojoules (kJ) added or removed from water.

The amount of heat necessary to cause a desired change of temperature is calculated by multiplying the mass of the substance by the specific heat capacity and by the temperature change (provided there is no change of state).

The specific heat capacity unit is expressed as joules per kilogram kelvin. The formula is J/kg·K. Amount of heat added or removed in kJ = mass of substance x specific heat capacity x temperature change in kelvins, or

kJ = mass x sp. ht. capacity x K change

Example: Find the amount of heat, in kJ, which must be removed to cool 15 kg of 20 percent salt brine (see Fig. 1-19) from 16°C to 7°C.

Solution:

kJ = mass x sp. ht. capacity x K change

kJ = 15 kg x 3.559 sp. ht. capacity x (16°C − 7°C)

kJ = 15 x 3.559 x 9

kJ = 480.5

U. S. Conventional Units:

In the U. S. conventional units, the specific heat capacity of a substance is the amount of heat added or released to change the temperature of one pound of the substance one degree Fahrenheit.

The specific heat capacity of water is 1.0. The specific heat capacity of several common substances is shown in Fig. 1-19.

Para. 1-29 shows how to calculate the heat (Btu) added or removed from water.

To calculate the amount of heat added or removed from substances, use the following formula:

Amount of heat added or removed in Btu = wt. of sub-

MATERIAL	SPECIFIC HEAT CAPACITY	
	kJ/kg·K	Btu/lb./F.
Wood	1.369	0.327
Water	4.187	1.0
Ice	2.110	0.504
Iron	0.540	0.129
Mercury	0.139	0.0333
Alcohol	2.575	0.615
Copper	0.398	0.095
Sulphur	0.741	0.177
Glass	0.783	0.187
Graphite	0.837	0.200
Brick	0.837	0.200
Glycerine	2.412	0.576
R-717 (Liquid ammonia at 40°F)	4.606	1.1
R-744 (Carbon dioxide at 40°F)	2.512	0.6
R-502	1.068	0.255
Salt Brine 20%	3.559	0.85
R-12	0.892	0.213
R-22	1.089	0.26

The above values may be used for computations which involve no "change of state." If a change of state is involved, the specific heat for each state of the substance must be used.

Fig. 1-19. Table of specific heat capacity values for some substances.

stance in lb. x sp. ht. capacity x temperature change in degrees F., or Btu = wt. x sp. ht. capacity x degrees F. change.

Example: The amount of Btu which must be removed to cool 40 lb. of 20 percent salt brine (see Fig. 1-19) from 60 F. to 20 F.:

Solution:

Btu = wt. x sp. ht. capacity x F. change

Btu = 40 lb. x 0.85 sp. ht. capacity x (60 F. − 20 F.)

Btu = 40 x 0.85 x 40

Btu = 1360

Specific Heat Capacity Equivalents:

1 cal/g·°C = 4.187 J/g·K

1 Btu/lb. F. = 4.187 kJ/kg·K (kilojoule per kilogram kelvin)

1 kJ/kg·K = 0.2388 Btu/lb. F.

1-32 LATENT HEAT

All pure substances are able to change their state. Solids become liquids, liquids become gas. These changes of state occur at the same temperature and pressure combinations for any given substance. It takes the addition of heat or the removal of heat to produce these changes.

Heat which brings about a change of state with no change in temperature is called latent (hidden) heat.

In Fig. 1-20, view A, note that considerable heat (335 kJ) was added between points B and C. Even so, the temperature did not change. This heat was required to change the ice to water. This heat is called "latent heat of melting" during a heating operation or "latent heat of fusion" during a cooling operation.

Likewise, between points D and E, 2260 kJ were added and the temperature did not change. This heat was required to change the water to steam. This heat is called "latent heat of vaporization." When cooling the steam to water, the latent heat removed is called the "latent heat of condensation."

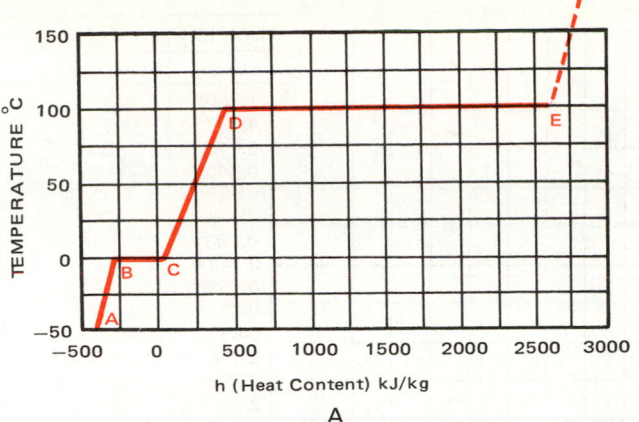

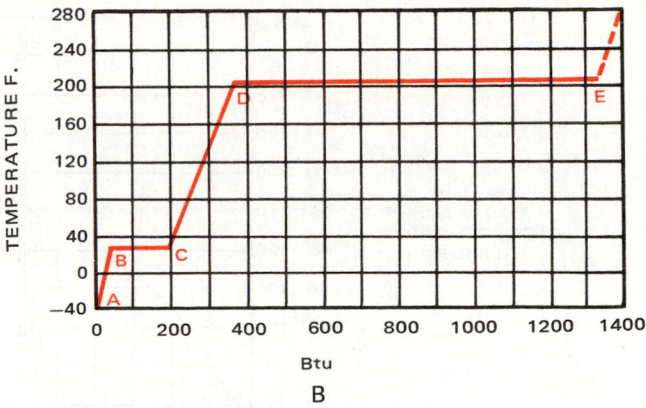

Fig. 1-20. A—Temperature-heat content diagram for one kilogram of water at atmospheric pressure (100 kPa) heated from −50 °C through complete vaporization. From A to B, 100 kJ of heat are added to increase ice temperature from −50 °C to 0 °C. This is 2 kJ/kg °C x 50 °C = 100 kJ/kg. From B to C 335 kJ are added to melt ice without changing its temperature. From C to D, 420 kJ were added to heat water to boiling point (4.2 kJ/kg °C x 100 °C = 420kJ). From D to E, 2260 kJ were added to convert water to steam without changing temperature. More heat increases temperature of steam as shown in dotted line. B—Temperature-heat diagram for one pound of water at atmospheric pressure, heated from −40 F. through complete vaporization. From A to B, 36.3 Btu were added to heat ice from −40 F. to 32 F. (−40 to 32 F. = 72 F. temperature change. Btu = 1 lb. x 0.504 x 72 = 36.288). From B to C, 144 Btu were added to melt ice. Temperature did not change. From C to D, 180 Btu were added to heat water from 32 F. to 212 F. From D to E, 970 Btu were added to vaporize water. Note that temperature did not change.

There are two latent heats for each substance, solid to liquid (melting and freezing) and liquid to gaseous (vaporizing and condensing). Fig. 1-21 shows the latent heat for water and several common refrigerants.

The latent heat of ice is 335 kJ/kg. The latent heat of vaporization of water (at 100 °C) = 2257 kJ/kg.

The change of state is caused by the molecules. As they move under a certain pressure, their motion will increase to a certain speed. Then, if more heat is added, each molecule undergoes a peculiar change. This change takes place within the molecule. It is believed that there is a shift of atoms. It is thought that one or more of the atoms change their position from the outer face of the molecule to the inside, or the reverse.

There are other theories, all based on molecular motion and magnetic attraction. The heat energy needed to make this change is tremendous. It requires as much heat to change 1 kg of ice to 1 kg of water as it does to raise the temperature of that same kg of water from 0 °C to 80 °C.

In U. S. conventional units, it takes as much heat to change 1 lb. of ice to 1 lb. of water as it does to raise the temperature of the same pound of water from 32 F. to 176 F.

All of the basic operations of the compression refrigeration cycle are based upon these two heats, SENSIBLE and

LATENT.
Equivalents:

1 kJ/kg	= 0.4299 Btu/lb.
1 Btu/lb.	= 2.326 kJ/kg
1 kJ/kg	= 0.2388 kcal/kg
1 kcal/kg	= 4.187 kJ/kg
1 kJ/kg	= 0.2388 cal/g
1 cal/g	= 4.187 kJ/kg

1-33 APPLICATION OF LATENT HEAT

In refrigeration work, the physics of latent heat is especially important. Application of this principle give the cold or freezing temperature desired.

As ice melts, its temperature remains constant. Nevertheless, it absorbs a considerable amount of heat in changing from ice to water. To melt one kg of ice, 335 kJ of heat are required. In U. S. conventional units, 144 Btu are required to melt 1 lb. of ice. (288 000 Btu are required to melt 1 ton of ice.)

When a substance passes from a liquid to a vapor, as in a mechanical refrigerator, its ability to absorb heat is very high. This principle is useful in the operation of the refrigerator.

The temperature at which a substance changes its state depends on pressure. The higher the pressure, the higher the

MATERIAL	FREEZING OR MELTING kJ/kg	LATENT HEAT OF VAPORIZATION OR CONDENSATION kJ/kg	FREEZING OR MELTING Btu/lb.	LATENT HEAT OF VAPORIZATION OR CONDENSATION Btu/lb.
Water	335	2257 at 100 °C	144	970.4 at 212 F.
R-717 (Ammonia)	—	1314 at −15 °C	—	565.0 at 5 F.
R-502	—	160 at −15 °C	—	68.96 at 5 F.
R-40 (Methyl chloride)	—	415 at −15 °C	—	178.5 at 5 F.
R-12	—	159 at −15 °C	—	68.2 at 5 F.
R-22	—	217 at −15 °C	—	93.2 at 5 F.

Fig. 1-21. Table of latent heat of vaporization of water and some common refrigerants. Latent heat of fusion is only given for water, as refrigerants do not freeze at temperatures commonly handled by refrigeration service engineer.

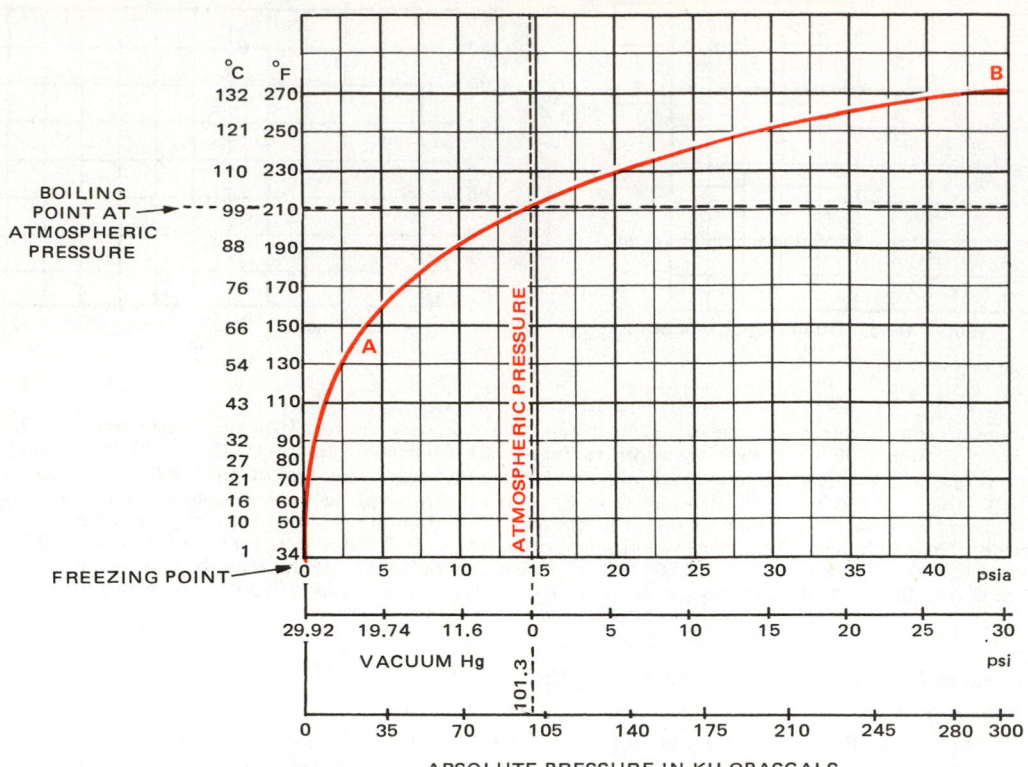

Fig. 1-22. Temperature-pressure curve for water. At atmospheric pressure, water boils at 100°C (212 F.). At point A, with pressure of 14 kPa (24 'Hg), water boils at 62°C (142 F.). Increasing pressure above atmospheric raises boiling temperature. At B, which is at a pressure of 207 kPa (30 psi), boiling temperature is 133°C (271 F.).

temperature needed to bring about the change. Conversely, if the pressure is lowered, the temperature at which the change of state will take place is lowered. This principle is shown in Fig. 1-22.

A liquid under low pressure will boil at a lower temperature. If the vapor resulting from this boiling is then compressed, it will condense back into a liquid at a higher temperature.

Every substance has a different latent heat value because each substance has a different molecular structure. Latent heat temperatures for water and the more common refrigerants are shown in Fig. 1-21. See Chapter 9 for more information concerning refrigerants.

In a modern refrigerator, freezer, or air conditioner, liquid refrigerant is piped under pressure to the evaporator. In the evaporator, the pressure is greatly reduced. The refrigerant boils (vaporizes), absorbing heat from the evaporator. This produces a low temperature and cools the evaporator.

The compressor pumps this vaporized refrigerant out of the evaporator and compresses (squeezes) it into the condenser. Here the heat that was absorbed in the evaporator is "squeezed out," released, to the surrounding atmosphere. Having lost this heat of vaporization, the refrigerant becomes a liquid again. The cycle is then repeated.

1-34 EFFECT OF PRESSURE ON EVAPORATING TEMPERATURES

The evaporating (boiling) temperature for any liquid is

controlled by the pressure placed upon it. Water at atmospheric pressure (100 kPa) or (15 psi) normally boils at 100°C (212 F.). If the pressure is increased to 300 kPa (45 psi), the boiling temperature is raised to 125°C (257 F.). If the pressure is lowered to 50 kPa (7 psi), the boiling temperature will be lowered to 80°C (176 F.), as shown in Fig. 1-22.

Mechanical and absorption refrigerators use the effect of reduced pressure to lower the boiling temperature. Consider the refrigerant, R-12. It boils under atmospheric pressure (100 kPa or 15 psi at −29°C (−20 F.). If the pressure is lowered to 60 kPa (9 psi), the boiling temperature is −41°C (−42 F.). A refrigerator can then cool to −41°C (−42 F.) if the pressure in the evaporator is lowered to 60 kPa (9 psi). But, if the evaporator were to operate at atmospheric pressure, the lowest temperature possible with R-12 would be −29°C (−20 F.). Fig. 1-23 shows the effect of pressure change on the boiling temperature of three substances used in refrigeration work.

1-35 EFFECT OF PRESSURE ON FREEZING TEMPERATURE

The temperature at which water freezes is affected by the pressure on the surface of the water. Increasing the pressures lowers the freezing temperature. Decreasing the pressure raises the freezing temperature. Fig. 1-24 shows this relationship.

1-36 REFRIGERATING EFFECT OF ICE

Ice is still important to the refrigeration industry. As stated

SUBSTANCE	EVAPORATING TEMPERATURE IN °C AT:		
	60 kPa PRESSURE	ATMOSPHERIC PRESSURE 100 kPa	200 kPa PRESSURE
Water	89	100	122
R-12	−41	−29	−10
R-717 (Ammonia)	−40	−38	−18

Fig. 1-23. Effect of pressure is shown on evaporating temperatures of three substances.

before, ice changes to water at 0°C and atmospheric pressure. Heat absorption to produce this change is 335 kJ/kg.

The specific heat capacity of ice = 2.11 kJ/kg·K. Its heat absorption ability, when changing from a temperature below 0°C to 0°C, is 2.11 kJ/kg per degree change. The latent heat of fusion (melting) of ice = 335 kJ/kg. The specific heat of water = 4.19 kJ/kg·K.

Example: How many kJ will be absorbed in changing 93 kg to ice at −20°C to water to 4°C?

Solution (in three steps):

1. To find heat needed to bring ice from −20°C to 0°C:

kJ = mass of ice x sp. ht. capacity of ice x temperature change

   = 93 x 2.11 x 20 = 3 924.5 kJ

2. To find heat needed to melt ice at 0°C:

kJ = mass of ice x latent heat of fusion of ice

   = 93 x 335 = 31 155.0 kJ

3. To find heat needed to raise water temperature from 0°C to 4°C:

kJ = mass of water x sp. ht. capacity of water x temperature change

   = 93 x 4.19 x (4 − 0)

   = 93 x 4.19 x 4 = 1 558.7 kJ

Total heat required to change 93 kg of ice at −20°C to water at 4°C

   = 3 924.6 kJ + 31 155.0 kJ + 1 558.7 kJ = 36 638.3 kJ

U. S. Conventional Units:

In U. S. conventional units, the specific heat equation for changing ice to water is the same:

Heat = wt. of ice x sp. ht. capacity of ice x temperature change

Heat will be in Btu.

Wt. will be in pounds.

Sp. ht. will be in Btu/lb.

The specific heat of water is 1 Btu/lb.

The specific heat of ice is 0.50 Btu/lb.

The latent heat of fusion of ice is 144 Btu/lb.

Example: How many Btu will be absorbed in changing 25 lb. of ice at 5 F. to water at 40 F.?

Solution (in steps):

1. Raise the temperature of ice from 5 F. to 32 F.:

Btu = wt. of ice x sp. ht. capacity of ice x temperature change

   = 25 x 0.50 x (32 − 5)

   = 25 x 0.50 x 27

   = 337.5 Btu

2. To melt the ice at 32 F.

Btu = wt. of ice x latent heat of fusion of ice

   = 25 x 144

   = 3 600 Btu

3. To warm the water from 32 F. to 40 F.:

Btu = wt. of water x sp. ht. capacity of water x temperature change

   = 25 x 1 x (40 − 32)

   = 25 x 1 x 8

   = 200 Btu

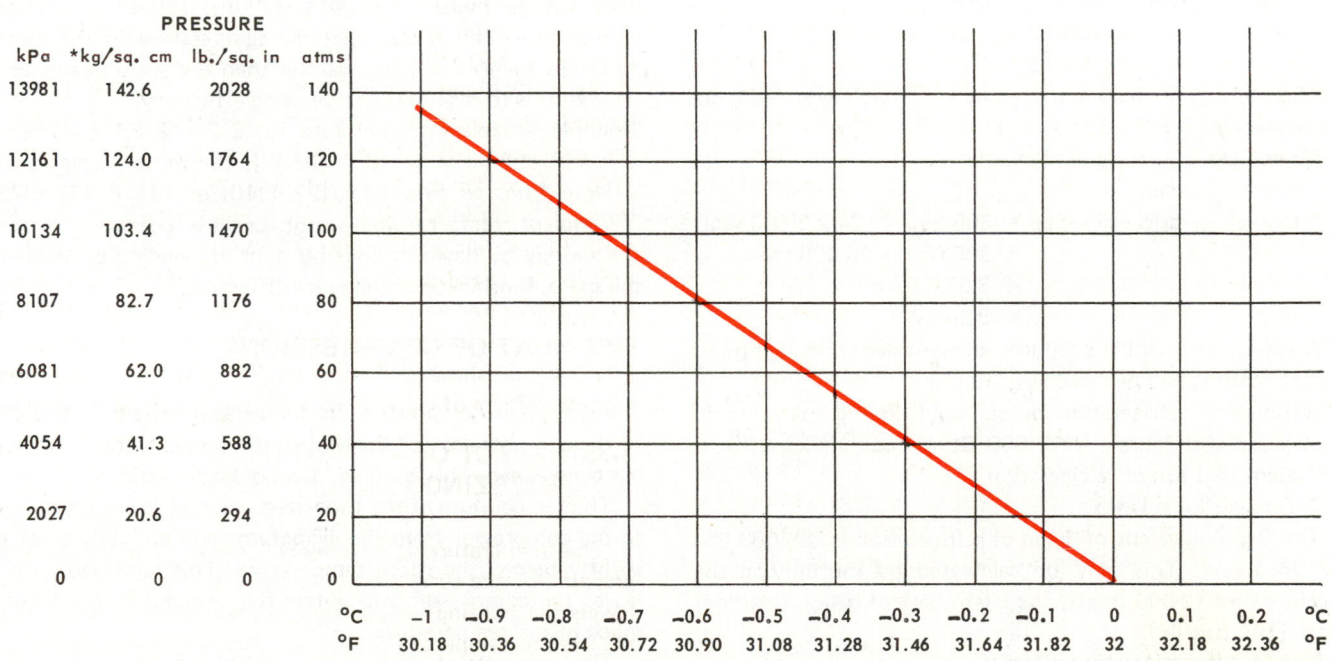

Fig. 1-24. Chart shows effect of pressure on freezing temperature of water.

(*This terminology is not recommended.)

Total heat required to change 25 lb. of ice at 5 F. to water at 40 F.

= 337.5 + 3 600.0 + 200 = 4 137.5 Btu

Equivalents:

1 kJ/kg·K = 0.239 Btu/lb. F.

1 Btu/lb. F. = 4.187 kJ/kg·K

1-37 ICE AND SALT MIXTURES

Refrigerating by ice alone will not provide temperatures below 0°C (32 F.). Therefore, to get the lower temperatures required in some instances, ice and salt mixtures are used. These mixtures, ice and salt (sodium chloride or NaCl) and ice and calcium chloride (CaCl), lower the melting temperature of ice. An ice and salt mixture may be made which will melt at −18°C (0 F.). The reason that the ice and salt mixture produces lower temperatures is that the salt causes the ice to melt faster, and this forced absorption of heat causes the lower temperature which results from the mixture.

1-38 TON OF REFRIGERATION EFFECT

The cooling capacity of older refrigeration units is often indicated in "tons of refrigeration." A ton of refrigeration represents the rate of cooling produced when a ton (2 000 lb.) of ice melts during one 24-hour day. The ice is assumed to be a solid at 0°C (32 F.) initially and becomes water at 0°C (32 F.). The energy absorbed by the ice is the latent heat of ice times the total weight.

The SI metric system has no unit which can be compared with the "ton of refrigeration."

1 ton = approximately 907 kg

latent heat = 337 kJ/kg

energy absorbed = latent heat x weight

= 337 kJ/kg x 907 kg

= 305 659 kJ

The melting of this ice in one day has a cooling or refrigeration capacity of 305 659 kJ

To convert the rating to kilowatts:

1 kW = 1 kJ/sec

1 ton refrigeration capacity = 305 659 ÷ (24 x 3 600 sec)

= 305 659 ÷ 86 400 sec

= 3.54 kJ/sec

= 3.54 kW

A refrigerator which produces an equivalent cooling rate of this ice melting will be rated as a 1-ton unit.

Ratings of refrigeration or air conditioning machines in Btu/hr are also found. A 12 000 Btu/hr cooling capacity is equivalent to 1 ton of refrigeration.

U. S. Conventional Units:

The Btu equivalent of 1 ton of refrigeration is 288 000 Btu per 24 hours. This may be calculated by multiplying the weight of ice (2 000 lb.) by the latent heat of fusion (melting) of ice (144 Btu/lb.).

One ton of refrigeration effect =

2 000 lb./24 hours x 144 =

288 000 Btu/24 hours

A refrigerating or air conditioning mechanism capable of absorbing heat is usually rated in tons per 24 hours by its heat absorbing ability (HA) in Btu divided by 288 000.

T = tons of refrigeration effect

HA = heat absorbing ability

$T = \dfrac{HA}{288\ 000}$ = tons of refrigeration effect

Example: The heat absorbing ability of a refrigerator unit is rated at 1 440 000 Btu per 24 hours. What is its ton rating?

Solution:

$T = \dfrac{1\ 440\ 000}{288\ 000}$ = 5 tons of refrigeration effect

Example: What will be the "ton" rating of a refrigerating mechanism capable of absorbing 1 728 000 Btu in 24 hours?

Solution:

$T = \dfrac{1\ 728\ 000}{288\ 000}$ = 6 tons of refrigeration effect

Example: What is the Btu heat absorbing capacity of a 1/2-ton refrigerating system?

Solution:

1/2 x 288 000 = 144 000 Btu per day, or

6 000 Btu per hour

Most room air conditioners are rated on their heat-absorbing ability in Btu per hour. A 1-ton machine is rated:

$\dfrac{288\ 000}{24}$ = 12 000 Btu per hour

Equivalents:

1 kW = 3415 Btu/hr

1 Btu/hr = 0.3 W

1-39 AMBIENT TEMPERATURE

The term, *ambient temperature,* means the temperature of the air surrounding a motor, a control mechanism, or any other device. A motor, for example, may be guaranteed to deliver its full watt power (horsepower) when its temperature does not go higher than 40°C (72 F.) above the ambient temperature. This means that the temperature of the motor must not be 40°C (72 F.) warmer than the surrounding air if the motor is to maintain its operating efficiency.

Example: If room temperature is 22°C (72 F.) and motor temperature rise is 40°C (72 F.), the motor temperature should not go over 62°C (22 + 40) or 144 F. (72 + 72).

Ambient temperature is not usually constant. It may change day by day and hour by hour, depending on usage of the space, sunshine, and many other factors.

1-40 HEAT OF COMPRESSION

As a gas is compressed, its temperature rises. This is due to the concentration of the heat of the vapor. Little or no heat has been added. This is called "heat of compression."

The temperature of the vaporized (gas) refrigerant returning to the compressor from the evaporator will probably be at, or slightly below, the room temperature. This same vapor, as it leaves the compressor and enters the condenser, will be at a much higher temperature.

The compression of vaporized refrigerant in a refrigerator compressor is considered to be near a state of adiabatic compression (See Chapter 28 for a definition of adiabatic compression.)

Because the compressed vapor in the condenser is now warmer than the temperature of the surrounding air, the heat will be rapidly transferred through the condenser walls to the surrounding air or condenser cooling water. (See Second Law of Thermodynamics, Chapter 28.)

1-41 ENERGY UNITS

In refrigeration work, three common, related forms of energy must be considered: mechanical, electrical, and heat. The study of refrigeration deals mainly with heat energy, but the heat energy is usually produced by a combination of electrical and mechanical energy.

In a compression refrigerating unit, electrical energy flows into an electric motor, and this electrical energy is turned into mechanical energy. This mechanical energy is used to turn a compressor. The compressor, in turn, compresses the vapor to a high pressure and high temperature, transforming mechanical energy into heat energy.

The unit for measuring energy in all three of these forms is the joule. The kilowatt hour is widely used, however, as a measure of electric energy.

U. S. Conventional Units:

In the U. S. conventional units, various units are used for measuring mechanical, heat, and electrical energy. Energy conversion units are expressed as follows:

Mechanical to heat,	1 hp = 2 546 British thermal units (Btu)/hr
	778 ft. lb. = 1 Btu
Mechanical to electrical,	1 hp = 746 watts (W)
Electrical to mechanical,	746 watts = 1 hp
Electrical to heat,	1 watt (1 joule/sec) = 3.412 Btu/hr
	1 kilowatt (kW) = 3 412 Btu/hr
Heat to mechanical,	2 546 Btu/hr = 1 hp
Heat to electrical,	1 Btu/hr = 0.293 watts

These conversion units are used in calculating loads and determining the capacity of equipment required for specific refrigeration applications.

Equivalents:

1 J = 0.7376 ft. lb.
1 ft. lb. = 1.3558 J
1 W = 0.7376 ft. lb./s
1 ft. lb./s = 1.3558 W
1 kW = 1.34 hp = 3 412 Btu/h
1 hp = 0.746 kW

1-42 REFRIGERANT

In refrigerating systems fluids which absorb heat inside the cabinet and release it outside are called refrigerants.

These fluids, in their liquid form, under a reduced pressure, absorb heat in the evaporator and, in absorbing heat, change to a vapor. In their vapor form, the fluids are taken into the compressor where the temperature and pressure are increased. This allows the heat that was absorbed in the evaporator to be released (squeezed out) in the condenser, and the refrigerant is returned to a liquid form.

Chapter 9 describes the refrigerants most used. Their technical characteristics, likewise, are explained.

1-43 ELEMENTARY REFRIGERATOR

In Fig. 1-25, a refrigerant cylinder, A, is shown with the valve closed. The pressure inside is 590 kPa and the temperature is 22°C. All conditions inside the cylinder are balanced. The number of molecules leaving the vapor state, the number diving back into the liquid, and the liquid molecules flying out of the liquid into the vapor state are equal.

In cylinder B, the valve has been opened slightly and some of the vapor is escaping. The results are twofold:

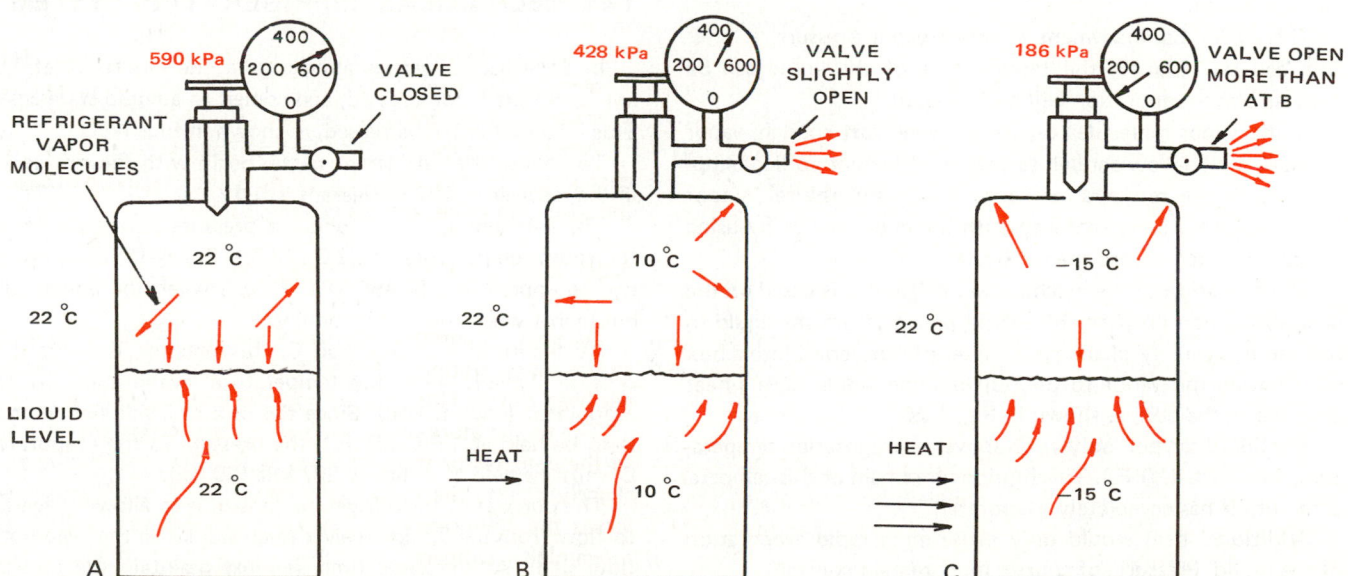

Fig. 1-25. Cooling effect of different pressures operating on surface of liquid refrigerant R-12. A—Cylinder of refrigerant with valve closed. Refrigerant is in state of equilibrium. B—Cylinder of refrigerant with valve slightly opened and refrigerant vapor beginning to leave cylinder. C—Cylinder of refrigerant with valve open wider. Refrigerant vapor is flowing from cylinder.

1. The pressure over the liquid refrigerant in the cylinder is reduced to 428 kPa. This causes change. There is more liquid changing to a vapor than there is vapor changing back into a liquid.
2. With more liquid turning into a vapor than vapor returning to a liquid, heat is absorbed, and the liquid refrigerant and cylinder will be cooled. The temperature of the refrigerant and cylinder is now 10°C (50 F.). Some heat from the surrounding area, which is at 22°C (72 F.), will now flow into the cylinder and the refrigerant.

In cylinder C, the valve has been opened more than in B. Refrigerant vapor now flows out more rapidly. This brings about a still lower pressure on the liquid refrigerant and a more rapid evaporation of the refrigerant.

The increase in the rate of evaporation lowers the temperature of the refrigerant and the cylinder still more. The 22°C (72 F.) air surrouding the cylinder is thus made to give up its heat to the colder cylinder more rapidly.

In cylinder A, there is a state of equilibrium (balance), with all temperatures and pressures in balance.

In cylinder B, there is a slight unbalance due to the vapor escaping through the valve. If this condition were to continue for a considerable time, a condition of balance would again prevail. But, in this new condition, its balance would not be static as in A. Rather, it would be a balance between the rate of heat flow into the cylinder, the evaporation of refrigerant, and the flow of refrigerant vapor out of the cylinder valve. In this condition of balance, the refrigerant is cooling the cylinder and its surroundings.

Cylinder C has a greater balance than cylinder B. The cylinder pressure and temperature, as a result, will be lowered still further.

As long as the valve is open and vapor can escape, the temperature will be lower, because more liquid molecules are becoming vapor molecules than vapor molecules are returning into the liquid.

This vapor bombardment is called vapor pressure. If pressure can be reduced, the temperature of the liquid can be reduced since evaporation will be increased.

If the vapor molecules can be removed fast enough, vapor pressure may be low enough to create refrigerant boiling temperatures in the refrigerating range (low temperature). Vapor molecules are usually removed with a compressor or by using chemicals which absorb the molecules.

The operation of the mechanical refrigerator is based on the heat absorption property of a fluid passing from the liquid to the vapor state. By placing a cylinder of refrigerant into a box and venting the vapor to the outside, one would have a heat absorber in the box, as shown in Fig. 1-26.

The liquid can boil only at or above its evaporation temperature, say −7°C (20 F.). This liquid will remain at this temperature until it has completely evaporated.

Additional heat would only cause more rapid evaporation of the liquid. Pressure, of course, must remain constant.

Since the liquid is at this low temperature, there is a transfer of heat to it from the surrounding objects. This heat increases the evaporation, and the heat itself is carried away as the vapor passes off. Thus, the fluid changing its state to a vapor gets

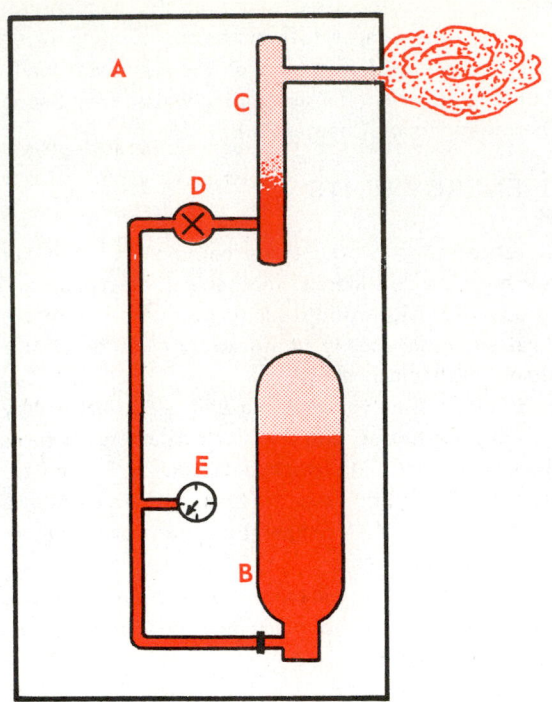

Fig. 1-26. Elementary refrigerator. A—Refrigerated space. B—Cylinder of liquid refrigerant. C—Evaporator. D—Control valve. E—Pressure gauge.

the energy (or heat) for doing this from the objects surrounding it. That same heat is removed with the vapor to the outside of the box.

This type of refrigerating system works nicely. But it is expensive because the refrigerant fluid is lost. Some mobile refrigerating units (trucks) use this method. The refrigerant is usually a liquid nitrogen which is relatively inexpensive. It is called "expendable refrigerant" refrigeration.

1-44 MECHANICAL REFRIGERATING SYSTEM

In a mechanical refrigerating system, the vaporized refrigerant is captured, compressed, and cooled to a liquid state again. This is so that it can be reused, as shown in Fig. 1-27.

To follow the refrigerant cycle, begin with the refrigerant in the receiver B. The refrigerant is R-12.

The refrigerant at B is under a pressure corresponding to the room temperature of 22°C (72 F.). For R-12, this pressure will be approximately 490 kPa (71 psi) when the unit is idle, but higher when the unit is running.

At the refrigerant control, C, this pressure is reduced to provide low-pressure, low-temperature evaporation in the evaporator (cooling coil). Since the box or inside temperature is to be held at 1.5°C (35 F.), the pressure in the evaporator, D, must be held at or below 207 kPa (30 psi).

The purpose of the refrigerant control is to allow refrigerant to flow from the liquid receiver (high side) into the evaporator (low side). At the same time, it must maintain the pressure difference between the high-pressure side (high side) and the low-pressure side (low side).

In the evaporator, D, the liquid refrigerant is now under a much reduced pressure, and it will evaporate or boil very rapidly. This, in turn, cools the evaporator. The compressor, E, creates a low pressure and draws (sucks) the evaporated refrigerant vapor from the evaporator. Then it compresses it back to the high side.

From the compressor, the high-temperature (see heat of compression, Para. 1-40), high-pressure vaporized refrigerant flows into the condenser, F. The temperature of the vapor, as it enters the condenser, will be several degrees warmer than the room temperature. This difference causes a very rapid transfer of the heat from the condenser to the surrouding air.

The vapor, as it flows through the condenser, cools and loses its heat of vaporization. It returns to the liquid state. As a liquid, it flows from the condenser back into the liquid receiver, B.

This refrigeration cycle is repeated over and over until the desired temperature is reached. Then a thermostat opens the electrical circuit to the driving motor. The compressor stops.

The temperature of the evaporator determines the pressure at which the refrigerant is evaporated. The amount of heat removed depends on the amount (mass) of refrigerant vaporized.

1-45 HEAT TRANSFER

Heat may be transferred or moved from one body to another by one of three methods: radiation, conduction, or convection. Some systems of heat transfer use a combination of these three methods.

1-46 RADIATION

The earth receives heat from the sun by radiation. Light rays from the sun turn into heat as they strike opaque or translucent materials which will absorb some or all of the rays. (Opaque means light cannot shine through; translucent, light can pass through but one cannot see through.) (See Chapter 27.)

Air is heated very little as light rays pass through it. Likewise, a glass pane absorbs little heat as rays pass through it.

Sunlight generates more heat when its rays strike dark-colored objects than when they strike light-colored or polished surfaces. This is because light-colored and polished objects reflect the rays. They are not absorbed and changed into heat. Also, rough, dark-colored surfaces will radiate heat better than will light-colored or polished surfaces.

Any heated surface loses heat to cooler surrounding space or surfaces through radiation.

Likewise, a cold surface will absorb radiated heat. Some space heating systems depend on radiant heating sources located in the ceilings, walls, or floors to heat a space or room.

1-47 CONDUCTION

Conduction is the flow of heat between parts of a substance. The flow can also be from one substance to another substance in direct contact.

A piece of iron with one end in a fire will soon become

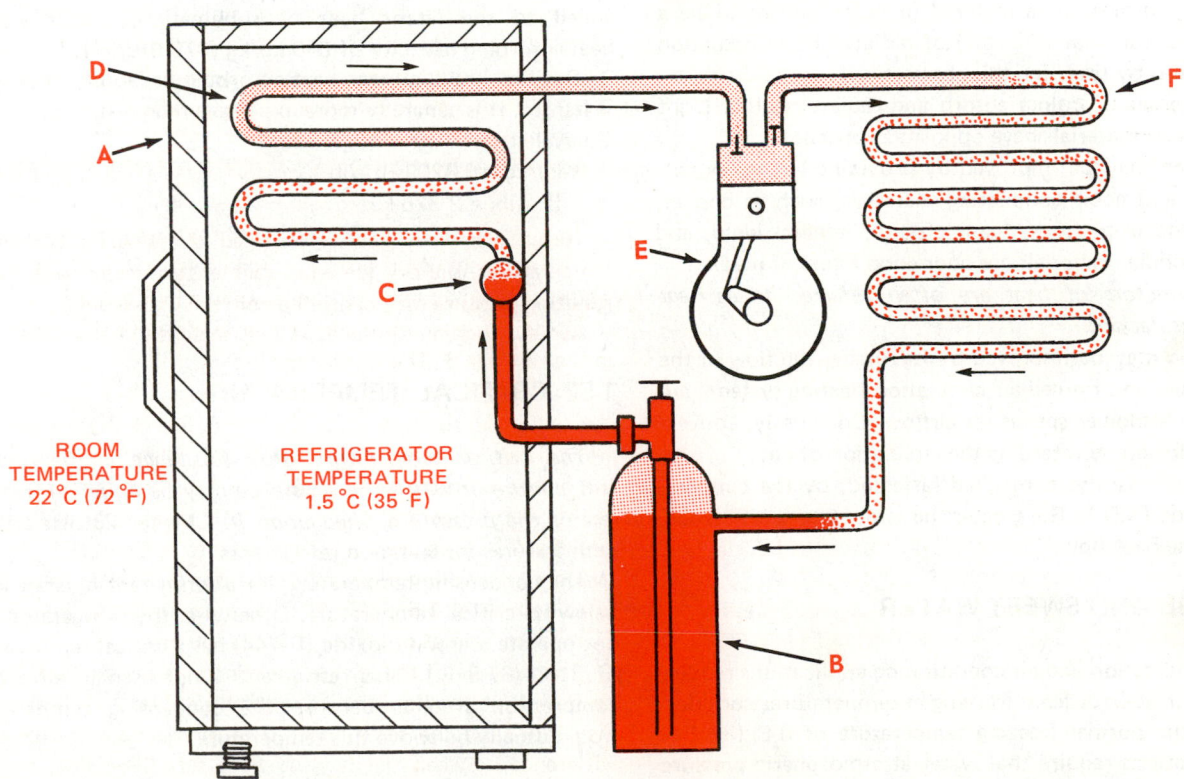

ROOM
TEMPERATURE
22°C (72°F)

REFRIGERATOR
TEMPERATURE
1.5°C (35°F)

Fig. 1-27. Elementary mechanical refrigerator. A—Cabinet. B—Liquid refrigerant receiver. C—Refrigerant control. D—Evaporator. E—Motor-driven compressor. F—Condenser. Refrigerant is recycled, as needed, to get desired temperature.

warm from end to end. This is an example of the transfer of heat by conduction. The heat travels through the iron, using the metal as the conducting medium.

Substances differ in their ability to conduct heat. In general, substances which are good conductors of electricity are also good conductors of heat (Wiedmann-Frank Law).

Substances which conduct heat poorly are called insulators. Such substances are used to insulate refrigerators, homes, and any structure that is to be maintained at a temperature different from its surroundings.

1-48 CONVECTION

Convection is the moving of heat from one place to another by way of fluid or air. For example, heated air moves from a furnace into the rooms of a house. It releases its heat to the rooms. Then cooled air returns through cold air ducts to receive another supply of heat.

The same method may be used to cool a space. Unwanted heat is collected and discharged outside the space.

1-49 CONTROL OF HEAT FLOW

The flow of heat by radiation, conduction, and convection can be controlled. The transfer of heat by each can be increased or cut back according to need.

The transfer of heat by radiation may be improved by making the radiating surfaces out of a material known to be a good radiator of heat. A color known to be a good radiator can also be used. Radiation may also be improved by making the receiving surfaces of a material or color known to be a good absorber (or poor reflector) of radiated heat. Radiation may be reduced by reversing this application.

Dark materials or colors absorb and radiate readily. Light colored or shiny materials have opposite properties.

Conduction may be improved by providing large conducting surfaces and good conducting materials, such as copper, aluminum, and iron. Cork, foam plastics, mineral wool, and many other similar materials are poor conductors of heat.

Poor conductors of heat are often referred to as heat insulators (insulation).

Convection may be improved by increasing the flow of the conveying medium. Forced-air circulation heating systems are an example. A blower speeds up airflow. Conversely, convection can be slowed by retarding the circulation of air.

Heat transfer is also controlled (affected) by the temperature difference (T.D.). The greater the temperature difference, the greater the heat flow.

1-50 BRINE AND SWEET WATER

Some refrigeration and air conditioning applications require that the water be kept from freezing at temperatures considerably below the normal freezing temperature of 0°C (32 F.). Other applications require that water at atmospheric pressure be kept from boiling at temperatures above 100°C (212 F.).

Salt, sodium chloride (NaCl), or calcium chloride (CaCl), added to water, raises the temperature at which the water will boil while lowering the temperature at which it will freeze. See Chapter 28 for tables of brine solutions and their characteristics.

There are some refrigerating and air conditioning installations which use tap water without any salt or other substance added. This is referred to as "sweet water."

1-51 DRY ICE

Solid carbon dioxide (CO_2) is sometimes used for refrigeration. A white crystalline (like crystals) substance, it is formed when liquid carbon dioxide is allowed to escape into a snow chamber (heat-insulated box).

The heat for vaporizing the liquid is drawn from the interior of the chamber so that a very low temperature (-78°C, -108 F.) is formed. As a result, quantities of the carbon dioxide solidify.

This solid is pressed into various shapes and sizes and sold for refrigeration purposes. It is given such names as dry ice, zero ice, and so forth. It remains at a temperature of -78°C (-108 F.) while in a solid state at atmospheric pressure.

Dry ice does not turn into a liquid as it melts. It goes directly from the solid to the vapor state. This is called "sublimation." It has some desirable characteristics. It does not wet the surfaces that it touches, and the vapor given off is a preservative. The low temperature maintained permits handling frozen foods without using a heavily insulated container.

The latent heat of sublimation is 577 kJ/kg (248 Btu/lb.). the heat absorbed by the vapor in passing from -78°C (-108 F.) to 0°C (32 F.) is approximately 63 kJ/kg (27 Btu/lb.). This, added to the latent heat of sublimation, makes a total heat-absorbing capacity of 640 kJ/kg (275 Btu/lb.).

Dry ice has a greater heat-absorbing capability than does water ice. It is generally more expensive than water ice. Equivalents:

1 kJ/kg = 0.4299 Btu/lb.
1 Btu/lb. = 2.326 kJ/kg

Never place dry ice in a sealed container! At ordinary temperatures, the dry ice will sublime (turn into a vapor). The resulting pressure may cause the container to explode.

Avoid touching dry ice. It will instantly freeze the skin.

1-52 CRITICAL TEMPERATURE

The critical temperature of a substance is the highest temperature at which the substance may be liquefied, regardless of the pressure applied upon it. Chapter 28 lists critical temperatures for common refrigerants.

The condensing temperature for a refrigerant must be kept below its critical temperature. Otherwise, the refrigerator will not operate. Carbon dioxide (R-744) has a critical temperature of 31°C (87.8 F.). This refrigerant is not used in air-cooled compression systems because the condensing temperature would usually be above this temperature.

1-53 CRITICAL PRESSURE

The critical pressure of a substance is the pressure at or above which the substance will remain a liquid. Under this pressure it cannot be turned into vapor by adding heat.

1-54 ENTHALPY

Enthalpy is the measure of the energy content of a substance. The amount of enthalpy is determined by both the temperature and the pressure of the substance.

The zero enthalpies for water, refrigerants, and air are arbitrarily taken at some convenient temperature (reference temperature or T_r) and pressure, as shown:

1. For water, zero enthalpy is at $0°C$ and 100 kPa.
2. For refrigerants, $-40°C$ and 100 kPa.
3. For air, $25°C$ and 100 kPa.

The enthalpy is measured in joules (J) or kilojoules (kJ).

Example: What is the total enthalpy of 5 kilograms of water at $80°C$?

Solution: The enthalpy (H) of 5 kilograms of water at $80°C$ is calculated using the following formula:

H = mass (m) x specific heat (C_p) x the temperature change $(T - T_r)$ = kJ

$H = m \times C_p \times (T - T_r)$

m = mass of water = 5 kilograms

C_p = the specific heat of water = 4.19 kJ/kg·K (Para. 1-31)

T = $80°C$ temperature

T_r = $0°C$, the reference temperature for water

$H = 5 \times 4.19 \times 80 = 1676$ kJ

Fig. 1-20, view A, shows the relationship of enthalpy to temperature.

U.S. Conventional Units:

In the U.S. conventional units, enthalpy is all the heat in one pound of a substance calculated from an accepted reference temperature, 32 F. This reference temperature can be used for water and water vapor calculations. For refrigerant calculations, the accepted reference temperature is $-$ 40 F. See Fig. 1-20, view B.

Example: What is the enthalpy of 1 lb. of water at 212 F., assuming 0 enthalpy at 32 F.?

Solution:

Enthalpy at 32 F. = 0

The specific heat of water is C_p = 1 Btu/lb/F.

Heat to raise the temperature of 1 lb. of water from 32 F. to 212 F.:

$212 - 32 = 180$ F.

$H = m \times C_p \times 180 = 1 \times 1 \times 180 = 180$ Btu

Total enthalpy at 212 F. = 180 Btu

1-55 SPECIFIC ENTHALPY

Specific enthalpy is enthalpy per unit mass. It is measured in joules per kilogram (J/kg). Tables of enthalpy of substances and pressure-enthalpy diagrams, such as Fig. 1-20, use specific enthalpy.

Example: If 100 kg of a substance absorb 2 000 kJ of energy when heated from the reference state of 0 kJ, what is the specific enthalpy?

Solution: Specific enthalpy = total enthalpy divided by mass

$$h = \frac{H}{m} = \frac{2\,000}{100} = 20 \text{ kJ/kg}$$

1-56 CRYOGENICS

The frequent use of liquid helium, nitrogen, and liquid hydrogen in refrigeration has increased the common use of the term cryogenics. Cryogenics refers to the use of, or the creating of, temperatures in the range of 116 K down to 0 K ($-157°C$ down to $-273°C$, or -251 F. down to -460 F.). The same term is applied to the low-temperature liquefaction of gases, handling and storage, insulation of containers, instrumentation, and techniques used in such work.

Fig. 1-28 shows the evaporating temperature at atmospheric pressure of some of the common cryogenic fluids. It also indicates the cryogenic range of temperatures. These temperatures are usually in kelvin (K).

BOILING TEMPERATURE AT ATMOSPHERIC PRESSURE

FLUID	CELSIUS °C	KELVIN K (Absolute °C Scale)	FAHREN-HEIT F	RANKINE R (Absolute F Scale)
Water	100	373	212	672
R-12 Refrigerant	-30	243	-22	438
R-22 Refrigerant	-41	230	-41	419
R-744 Refrigerant				
Carbon Dioxide	-78	195	-109	351
R-1150 Refrigerant				
Ethylene	-93	180	-135	325
Beginning of the Cryogenic Range	-157	116	-250	210
Methane	-161	112	-258	202
Oxygen	-183	90	-297	163
Air	-192	81	-313	147
Nitrogen	-196	77	-320	140
Neon	-246	27	-411	49
Hydrogen	-253	20	-423	37
Helium	-270	3	-452	8
Absolute Zero	-273	0	-460	0

Fig. 1-28. Boiling temperatures of some common refrigerants and some other fluids at atmospheric pressure. Note difference between boiling points of some commonly used refrigerants and boiling points of fluids in the cryogenic range.

1-57 PERFECT GAS EQUATION

If a quantity of gas is enclosed in a tight container, the relationship of the pressure to the temperature is: PV = MRT In this equation:

P = the pressure in kilopascals (kPa)

V = the container volume in cubic metres (m^3)

M = the mass of the gas in the container in kilograms (kg)

R = the gas constant which has a value depending on the gas properties. Fig. 1-29 gives values of R for some substances used in refrigeration work.

T = the absolute temperature, kelvin (K) which is $-273 + T°C$

The equation shows that if the container of gas is heated so that the temperature increases, then the pressure will also rise. Cooling a container will reduce both the temperature and the pressure.

MATERIAL	SPECIFIC HEAT kJ/kg·K		GAS CONSTANT (R)	
	c_p	c_v	kJ/kg·K	Btu/lb. °R
Air	1.00	0.71	223.28	53.34
R-717 (Ammonia)	2.13	1.46	515.88	123.24
R-744 (Carbon Dioxide)	0.92	0.71	162.50	38.82
Ether	2.01	1.88	96.74	23.11
Oxygen	0.92	0.67	203.23	48.55
Alcohol	1.88	1.67	173.93	41.55
Water Vapor	2.03	1.55	348.40	83.23

Fig. 1-29. Table lists gas values (constants) for some substances used in refrigeration work.

Example: if 0.2 kg of air at a pressure of 500 kPa is contained in a volume of 50 m³, what is the temperature?

Solution: Solving for T (absolute temperature):

$$T = \frac{PV}{MR}$$

$$P = 500 \text{ kPa} = 500 \text{ kilonewtons(kN)}/m^2$$
$$V = 50 \text{ m}^3$$
$$R = 223.28 \text{ kJ/kg·K (from Fig. 1-29)}$$
$$M = 0.2 \text{ kg}$$

$$T = \frac{500 \times 50}{0.2 \times 223.28} = 559.8 \text{ K}$$

The temperature in Celsius is then:

$$T°C = T \text{ K} - 273 = 559.8 - 273 = 286.8°C$$

Note: The temperature used in the perfect gas equation must always be in kelvin.

Example: If the air in the container is heated until the temperature reaches 500°C, find the new pressure:

Solution: $T = 500 + 273 = 773 \text{ K}$

$$P = \frac{MRT}{V} = \frac{0.2 \times 223.28 \times 773}{50}$$

$$P = 690 \text{ kPa}$$

U. S. Conventional Units:

The relationship between pressure, temperature, and volume may be expressed by the formula:

PV = WRT
P = Pressure in pounds per square foot absolute
V = Volume in cubic feet
W = Weight of gas in pounds
R = Gas constant (R will differ for different gases). (Fig. 1-29 gives the value of R for some common substances.)
T = Absolute temperature R

Example: What will be the volume of 2 lb. of carbon dioxide at 240 F. when the pressure is 185 psi?

Formula: PV = WRT

P = (185 + 15) = 200 psia = 200 × 144 = 28 800 lb. per sq. ft. absolute
W = 2 lb.
R = 38.82
T = (240 + 460) = 700

$$V = \frac{WRT}{P}$$

$$V = \frac{2 \times 38.82 \times 700}{28\ 800}$$

$$V = \frac{54\ 348}{28\ 800} = 1.89$$

$$V = 1.89 \text{ cu. ft.}$$

1-58 DALTON'S LAW

Dalton's Law of partial pressures is the foundation of the principle of operation of one of the absorption type refrigerating system. The law states:

The total pressure of a confined mixture of gases is the sum of the pressures of each of the gases in the mixture.

The total pressure of the air in a compressed air cylinder is the sum of the oxygen gas, the nitrogen gas, the carbon dioxide gas, and the water vapor pressure.

The law further explains that each gas behaves as if it occupies the space alone. To illustrate, the Electrolux refrigerator uses two gases, ammonia and hydrogen. The ammonia, at room temperature, is absorbed by the water in the closed system.

Heating this solution drives out the ammonia. (The hydrogen is not absorbed by the water and remains as a gas.) Due to the pressure it is under, the ammonia condenses into a liquid in the condenser. The pressure is uniform throughout the system. Total pressure in the system is the sum of the vapor pressure of the ammonia plus the hydrogen pressure. When the pressure of the ammonia vapor is below the pressure corresponding to the vapor pressure for ammonia alone, the ammonia continues to evaporate as it tries to reach a vapor pressure corresponding to the temperature in the absorber.

1-59 EVAPORATOR

In this text, the word evaporator is used to indicate that part of the refrigerating mechanism where the liquid refrigerant boils or evaporates and absorbs heat. In some trade literature, the word cooling coil is used to indicate the part in which such cooling takes place. *The correct technical term is evaporator.*

Coils cooled by brine or any fluid which does not evaporate in the coil may properly be called cooling coils.

1-60 VAPOR — GAS

The word *vapor* in this text indicates refrigerant which has become heated, usually in the evaporator, and has changed to a vapor or gaseous state. In some trade and service literature, refrigerant in this state is called a gas. The correct technical term is vapor.

1-61 SATURATED VAPOR

The term "saturated vapor" identifies a condition of balance on an enclosed quantity of a vaporized fluid. The balance is such that some condensate (liquid) will be produced if there is even the slightest lowering of the temperature or increase in pressure.

There is usually some of the substance present in liquid form when the vapor is saturated. In a saturated condition, all of the substance has been vaporized that can be vaporized under the existing conditions of pressure and temperature.

1-62 HUMIDITY — RELATIVE HUMIDITY

The word humidity, as used in connection with refrigeration, air conditioning, and weather information, refers to water vapor or moisture in the air. Air absorbs moisture (water vapor).

The amount depends on the pressure and temperature of the air. The higher the temperature of the air, the more moisture it will absorb. The higher the pressure of the air, the smaller the amount of moisture it will absorb.

The amount of moisture carried in a sample of air, compared to the total amount which it can absorb at the stated pressure and temperature, is called relative humidity. See Chapter 18.

A relative humidity of 50 percent indicates that the air has 50 percent as much moisture as it will hold at that particular temperature and pressure. See Chapter 28 for tables of moisture-holding ability for air at various temperatures and pressures.

1-63 REVIEW OF ABBREVIATIONS AND SYMBOLS

The following is a review of the various abbreviations and symbols studied so far in this chapter. This review includes both the SI and U. S. conventional units.

SI Metric Units:

$^{\circ}$C	=	degrees Celsius
K	=	kelvin
mm	=	millimetre
cm	=	centimetre
cm^2	=	centimetre squared
cm^3	=	centimetre cubed
dm	=	decimetre
dm^2	=	decimetre squared
dm^3	=	decimetre cubed
m	=	metre
m^2	=	metre squared
m^3	=	metre cubed
L	=	litre
g	=	gram
kg	=	kilogram
J	=	joule
kJ	=	kilojoule
N	=	newton
Pa	=	pascal
kPa	=	kilopascal
W	=	watt
kW	=	kilowatt
MW	=	megawatt

U. S. Conventional Units:

Btu	=	British thermal unit
Btu/h	=	British thermal units per hour
F.	=	degrees Fahrenheit
F$_A$	=	degrees Fahrenheit absolute
R	=	degrees Rankine = degrees absolute F.
lb.	=	pound
psi	=	pounds per square inch = lb. per sq. in.
psia	=	pounds per square inch absolute = psi + atmospheric pressure
in.	=	inches = i = "
ft.	=	foot or feet = f = '
sq. in.	=	square inch = in^2
sq. ft.	=	square foot = ft^2
cu. in.	=	cubic inch = in^3
cu. ft.	=	cubic inch = ft^3
ft. lb.	=	foot-pound
ton	=	ton of refrigeration effect
lb./cft.	=	pounds per cubic foot
in. Hg.	=	inches of mercury vacuum = "Hg
hp	=	horsepower
qt.	=	quart
gr	=	grain

Miscellaneous Abbreviations:

P	=	pressure
h	=	hours
sec	=	seconds
T$_r$	=	temperature change
Δ	=	difference
C$_p$	=	specific heat (sp. ht.)
h	=	enthalpy per unit mass
H	=	total enthalpy
D	=	diameter
r	=	radius of circle
A	=	area
π	=	3.1416 (a constant used in determining the area of a circle)
V	=	volume
R	=	gas constant
∞	=	infinity

1-64 REVIEW OF SAFETY

The term "safety," as applied to any refrigeration or air conditioning activity, may have three different applications. It may apply to:

1. *Safety of the operator.* When refrigeration and air conditioning equipment is properly handled, there is relatively little danger to the operator.

   Always pull on a wrench (instead of pushing), to prevent possible slippage of the wrench which could cause rounded corners on nuts and bolts and possible injury to hands. A hoist is recommended for lifting anything weighing over 13 kg (35 lb.).

   Always use leg muscles when lifting objects, never the back muscles. Make certain there is no oil or water on the floor. Always wear safety goggles when working with refrigerants.

   Most refrigerating mechanisms are electrically driven and controlled. When working on electrical circuits, make sure that the circuit is disconnected from the power source. This can usually be accomplished by opening the switch at the power panel. Never work on "hot" electrical circuits.

2. *Safety of the equipment.* Many parts of refrigeration and air conditioning equipment are quite fragile. Parts may be ruined by overtightening nuts and bolts, not tightening them in the correct order, or using the wrong size wrench. Make certain that all connections are tightened before operating a compressor. Before operating open compressors, be sure the flywheel and pulley are in alignment and that guards are in place.

3. *Safety of the contents.* Safety of the contents of the refrigerated space depends entirely on the accuracy and care given the installation and adjustment of the various parts of the system.

Throughout this text, tables are provided which list the proper operating temperatures for various types of refrigerating space. These operating temperatures must be observed if the unit is to provide safe conditions for the refrigerated or air conditioned space.

It is advisable to observe these three points during the service work explained in the chapters following.

Each will have a "Review of Safety" as a reminder of the hazards which may be present when working with the equipment and supplies.

There is no exception to the rule that "The safe way is the right way."

1-65 TEST YOUR KNOWLEDGE

Fundamentals:

1. What is dry ice?
2. Given two objects of the same size, one chrome plated and one painted black, which one will absorb more radiant heat?
3. Name a condition which illustrates the principle of convection.
4. Which material will conduct heat the most rapidly, glass or copper?
5. Express one millimetre in decimals of a centimetre.
6. How many cubic centimetres in a litre?
7. How many centimetres are equal to one inch?
8. What is the piston displacement of:
   a. A compressor having a 30-mm bore and a 45-mm stroke?
   b. A compressor having a 2-in. bore and a 3-in. stroke?
9. What is the advantage of using SI metric units?

Pressure:

10. What determines the temperature at which a refrigerant will vaporize?
11. Should dry ice ever be put in a sealed container? Why?
12. What is the U.S. conventional absolute pressure equivalent in pounds per square inch (psia) of 8 in. of mercury vacuum (8" Hg.)?
13. a. What is the pressure difference between 3 kPa and 35 kPa?
    b. What is the U. S. conventional unit pressure difference in pounds per square inch between 6 in. of mercury vacuum (6" Hg) and 8 pounds per square inch gauge (psig)?
14. If the head pressure is 85 lb. per square inch gauge (psig), what is the total U. S. conventional unit force on one face of a circular disk with a 5 in. diameter?
15. a. What is standard atmospheric pressure in SI units?
    b. Express standard atmospheric pressure in pounds per square feet.

Temperature:

16. Should refrigerants be operated at temperatures above or below their critical temperature?
17. What is the average temperature desired in a domestic cabinet:
    a. in Celsius?
    b. in Fahrenheit?
18. A substance has a temperature of 78 F. What would be the temperature in $^{\circ}$C?
19. A substance has a temperature of 20°C. What would be the temperature in F.?
20. A substance has a temperature of 5 F. (a) What would be the temperature in $^{\circ}$C? (b) What would be the temperature in R?
21. A substance has a temperature of 432 F.
    a. What would be the temperature in $^{\circ}$C?
    b. Give the temperature in K.
22. A substance has a temperature of 14 R. Give its temperature in K.

Heat:

23. What is the unit of heat? What is the unit of heat in U. S. conventional units?
24. a. How much heat will be required to change 3 kg of ice at 0°C into water at 29°C?
    b. How many Btu will be required to change 5 lb. of ice at 32 F. into water at 82 F.?
25. How much heat will 1 kg of ice at 0°C absorb in changing to water at 0°C?
26. How much heat will 1 lb. of ice at 32 F. absorb in changing to water at 32 F.?
27. Calculate the number of joules required to convert 1 kg of ice at 0°C to steam at 100°C.
28. Calculate the number of British thermal units (Btu) that would be required to convert 1 lb. of ice at 0 F. to steam at a temperature of 212 F.
29. What is the enthalpy of 1 kilogram of water if it is at a temperature of 70°C?
30. What is equivalent of 10 ft.-lb. of work in SI metric units?

Chapter 2

REFRIGERATION TOOLS AND MATERIALS

The refrigeration service technician's job consists mainly of performing rather basic mechanical operations using common tools and materials of the field. Four principles should guide the technician:

1. Know what needs to be done.
2. Select the proper tools and materials.
3. Keep all refrigeration mechanisms clean and dry.
4. Follow approved safety procedures.

This chapter has been carefully prepared to give refrigeration and air conditioning service personnel (people) the necessary knowledge of tools and materials used in this industry.

As an example, the job may require brazing of a tube connection. The service person must know how to clean the tubing and the fitting; select the sizes that fit together properly; apply the flux in such a manner that it will not enter the tubing; apply the heat and brazing material in such a way that the brazing material will flow into the joint, making a strong, leakproof connection.

Just as important are the directions for selecting proper tools and materials and the correct and safe handling of tools used in any type of refrigeration work.

2-1 TUBING

Most tubing used in refrigeration and air conditioning is made of copper. However, some aluminum, steel, stainless steel and plastic tubing is being used.

Instructions in this chapter will deal mainly with copper tubing. All tubing used in air conditioning and refrigeration work is carefully processed to be sure that it is clean and dry inside. The service technician must keep the ends sealed to be sure that it remains clean and dry in handling.

Most copper tubing used in air conditioning and refrigeration work is known as Air Conditioning and Refrigeration (ACR) tubing. This designation means that the tubing is intended for air conditioning and refrigeration. Furthermore, it has been processed to give the desired characteristics.

ACR tubing is usually charged with gaseous nitrogen to keep it clean and dry until it is used. Nitrogen should be fed through it during brazing and soldering operations. Take care, it is dangerous to use! See Chapter 11.

This will eliminate the danger of oxidation inside the tube. All tubing should have the ends plugged immediately after cutting a length from the piece.

Copper tubing is available in soft and hard types. Both are available in two wall thicknesses, K and L. Type K is a heavy wall, type L is medium thick. Most ACR tubing used at present is the L thickness. Soft copper tubing is supplied in 25 and 50-foot rolls.

Another type of copper tubing used in heating and plumbing is called "nominal size."

2-2 SOFT COPPER TUBING

Soft copper tubing is used in domestic work and in some commercial refrigeration and air conditioning work. It is annealed (heated and then allowed to cool). This makes it flexible, therefore easy to bend and flare. Being easily bent, this tubing must be supported by clamps or suitable brackets. Soft copper tubing is most often used in connection with flared fittings (Society of Automotive Engineers (SAE) standards) and soft soldered fittings. It is sold in rolls 25+, 50+, and 100 feet long. Sizes most commonly used are 3/16, 1/4, 5/16, 3/8, 7/16, 1/2, 9/16, 5/8, and 3/4 in. outside diameter (OD). Wall thickness is usually specified in thousandths of an inch. Fig. 2-1 is a table of common copper tube diameters and thicknesses.

OUTSIDE DIAMETER	WALL THICKNESS
1/4	.030
3/8	.032
1/2	.032
5/8	.035
3/4	.035
7/8	.045
1 1/8	.050
1 3/8	.055

Fig. 2-1. Copper tube sizes used in refrigeration work. Both soft and hard drawn sizes are the same as the measurements listed in table. OD size for this tubing is the actual outside diameter of tube.

Soft copper tubing may be worked to give it certain properties. It can be hardened by repeated bending or hammering. This is called work hardening. It can also be softened by annealing, as explained earlier.

Tubing must be installed in such a way that there is no strain on it when the job is completed. Horizontal loops may be used to keep vibration from crystallizing the copper, making it crack or break.

2-3 HARD-DRAWN COPPER TUBING

Hard-drawn copper tubing is used in commercial refrigeration and air conditioning applications. Being hard and stiff, it needs few clamps or supports, particularly in larger diameters. *Hard tubing should not be bent. Use straight lengths and fittings to form necessary tubing connections.*

Hard-drawn refrigeration tubing joints should be silver brazed. Soft solder should be used only on water lines. Hard drawn tubing is supplied in 20 ft. lengths. It is available in the same diameters and thicknesses as soft copper tubing.

2-4 STEEL TUBING

Some thin-wall steel tubing is used in refrigeration and air conditioning work, sizes being practically the same as for copper tubing. Connections may be made on steel tubing by using either flared joints or silver brazed joints.

Copper or brass tubing should not be used with refrigerant R-717 (ammonia). Use steel tubing. There is a chance of chemical reaction (corrosion) between ammonia and copper.

Two types of steel tubing are in common use. One type has a double lap brazed construction using SAE 1008 mild steel. The other is butt welded, using the same type steel.

2-5 STAINLESS STEEL TUBING

Stainless steel tubing comes in the usual refrigeration tube sizes. The most common sizes are listed in the table, Fig. 2-2. Stainless steel is strong, very resistant to corrosion and may be easily connected to fittings by either flaring or brazing.

Stainless steel tubing No. 304 is most used. This is a low carbon (C) nickel (Ni) chromium (Cr) stainless steel. It is often required in food processing, ice cream manufacture, milk handling systems and the like.

2-6 METRIC TUBE SIZES

Metric sized tubing is used in some refrigeration and air conditioning systems. The standard sizes are: 6, 8, 10, 12, 14 and 15 millimetres (mm) OD.

2-7 PLASTIC TUBING

Polyethylene is one of the most common substances used in

OUTSIDE DIAMETER	WALL THICKNESS	BURST PRESSURE psi	MINIMUM BEND RADIUS
1/8"	.020	500	1/2"
3/16"	.030	500	1/2"
1/4"	.040	400	1"
5/16"	.062	600	1 1/8"
3/8"	.062	350	1 1/4"
1/2"	.062	250	2 1/2"

Fig. 2-3. Plastic tube specifications. Note there are four different thicknesses used in this size range of plastic tubing. (Imperial-Eastman Corp.)

the manufacture of plastic tubing. Sizes and suggestions for its use are shown in Fig. 2-3.

The usual safe temperature range is from −100 to +175 F. (−73 to 79 C.). Therefore, never use this tubing in installations where fluid temperature goes beyond these limits.

In general, polyethylene tubing is not used in the refrigerating cycle mechanisms but for cold water lines, water-cooled condensers and the like. Being very easy to use, polyethylene tubing may be cut with a knife. It may also be easily bent.

Special fittings are available for connecting polyethylene tubing to refrigeration and air conditioning mechanisms.

2-8 FLEXIBLE TUBING (HOSE)

In many refrigeration and air conditioning applications, the liquid lines and suction lines must be flexible, Fig. 2-4. This is

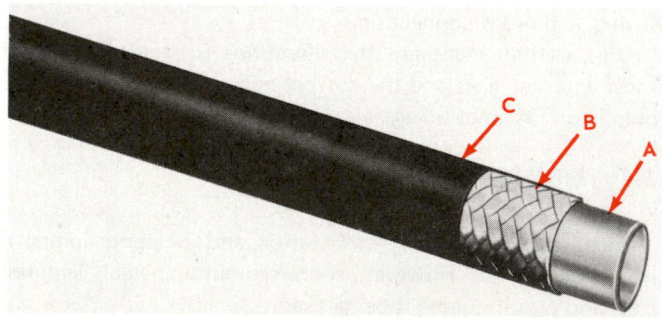

Fig. 2-4. Thermoplastic hose used in refrigeration systems. A—Inner nylon tube. B—Yarn reinforcement. C—Polyethylene cover.

particularly true in many commercial and industrial refrigeration and air conditioning applications.

Air conditioning equipment on motor vehicles require the

	OUTSIDE DIAMETER							WALL THICKNESS
FRACTIONS	1/4	3/8	1/2	5/8	3/4	1	1 1/4	(All of the stainless steel tubing is available in various wall thicknesses (BWG)* — 31 to 20 gage.)
DECIMALS	.250	.375	.500	.625	.750	1.000	1.250	
MILLIMETRES	6.35	9.52	12.7	15.87	19.05	25.40	31.75	

*— Birmingham Wire Gage

Fig. 2-2. Stainless steel tubing sizes are given in fractions, decimals and millimetres.

use of flexible tubing. Hose for this purpose is usually made from a variety of special materials. Such materials do not age, remain flexible, allow very low leakage through the hose wall and they are easy to attach to fittings.

2-9 FLEXIBLE HOSE FITTINGS

Three types of flexible hose fittings are available as described in Fig. 2-5.

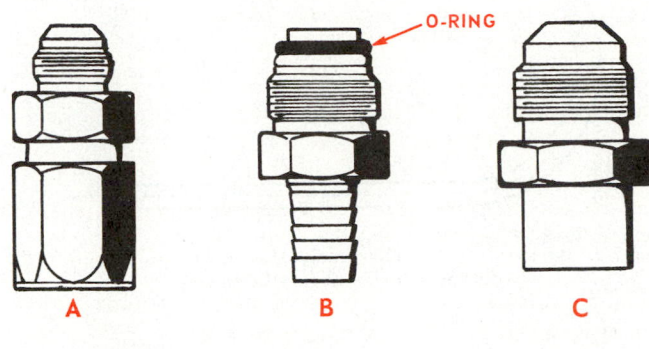

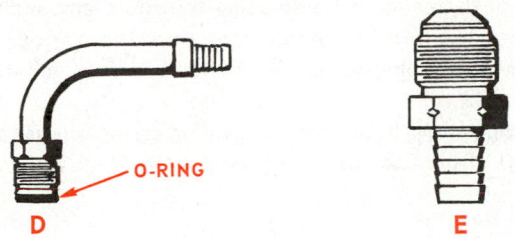

Fig. 2-5. Assorted nylon fittings suitable for use with refrigerant hose: A—Coupling, straight male 45 deg. flare, screw-on reusable. B—Coupling, straight male push-on barb type reusable, with O-ring seal. C—Coupling, straight male 45 deg. flare permanent (crimped-on and not reusable). D—Coupling 90 deg. male push-on barb type reusable, with O-ring seal. E—Coupling, straight male 45 deg. flare push on. (Imperial-Eastman Corp.)

These fittings are usually made of brass; however, nylon fittings are sometimes used. Synthetic rubber O-rings are put on some of these fittings to provide a better seal.

The attachment end of these fittings conforms to the standard SAE fittings specifications.

2-10 NOMINAL SIZE COPPER TUBING

Nominal size copper tubing is a type used on water lines, drains and in other applications but never in connection with refrigerants. It is available in both soft and hard-drawn grades. The table, Fig. 2-6, shows commonly used sizes and wall thicknesses of this tubing.

The classification (type) listed after nominal size is used when referring to copper tubing for heater lines, drains and so forth.

Copper tubing used for such applications is often referred to by its nominal size. If the service technician measures the OD, he will notice that the OD is 1/8 in. larger than that listed under nominal size.

NOMINAL SIZE INCHES	TYPE	OD INCHES		WALL THICKNESS INCHES
1/4	K	0.375	3/8	0.035
	L	0.375	3/8	0.030
3/8	K	0.500	1/2	0.049
	L	0.500	1/2	0.035
1/2	K	0.625	5/8	0.049
	L	0.625	5/8	0.040
5/8	K	0.750	3/4	0.049
	L	0.750	3/4	0.042
3/4	K	0.875	7/8	0.065
	L	0.875	7/8	0.045
1	K	1.125	1 1/8	0.065
	L	1.125	1 1/8	0.050

Fig. 2-6. Nominal size copper tubing. Type K - heavy wall is available in hard and soft temper. Type L - medium wall is available in hard and soft temper. Type K is used where corrosion conditions are severe. Type L is used where conditions may be considered normal. OD sizes indicated by dimension are 1/8 in. (.125) larger than nominal size.

When purchasing fittings for this tubing, it is important that the fitting size be the same as the size tubing purchased. One should order all the tubing, valves, and fittings by nominal size or order all by OD, to avoid problems.

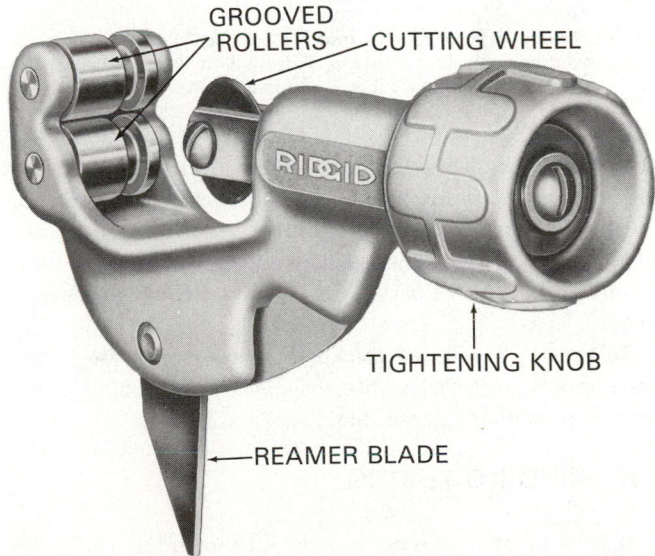

Fig. 2-7. A tube cutter. Note attached reamer which is used to remove burrs from inside of tube after cutting. Grooves in the rollers allow cutter to be used to remove flare from tube with little tubing waste. (The Ridge Tool Co.)

2-11 CUTTING TUBING

To cut tubing, use either a hacksaw or a tube cutter. The tube cutter is usually used on smaller, annealed (soft) copper tubing while the hacksaw is preferred for cutting the larger, hard copper tubing. Fig. 2-7 illustrates a wheel type cutter. The tubing should be straight and cut squarely (90 deg.) to eliminate an off-center flare. After the tubing has been cut, its ends must be scraped or reamed with a pointed tool to remove any sharp burrs on the end of the tubing. Most tube cutters have a reamer.

If a saw is used, a wave set blade of 32 teeth per inch will do the best job. (See Para. 2-65.)

It is important that no filings or chips of any kind enter the tubing.

When cutting tubing with a hacksaw, hold the tubing in such a way that chips will not fall into the section that is to be used. Fig. 2-8 illustrates a sawing fixture.

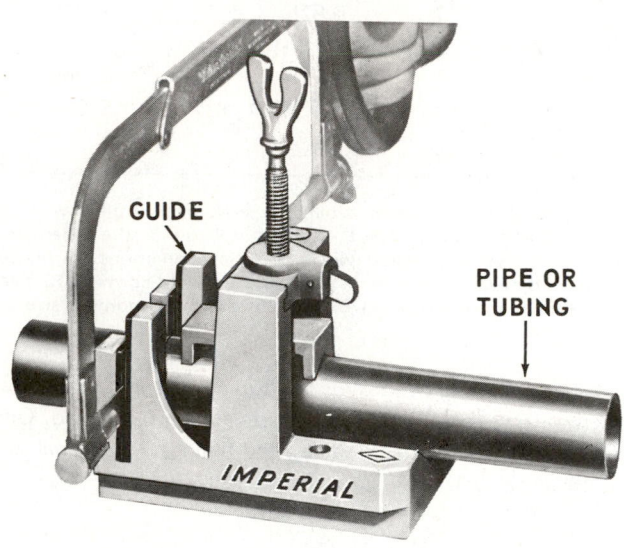

GUIDE

PIPE OR TUBING

IMPERIAL

Fig. 2-8. Sawing fixture used to insure square and accurate cuts when using a hand hacksaw to cut tubing. This method is recommended for cutting hard drawn and steel tubing.

If soft tubing is used, pinching the end of the tube that is not going to be used eliminates the danger of chips entering the tubing. It also seals the tubing against moisture and protects it until used. Again in hard copper tubing installation, the tubing ends of the part not being used should be capped or plugged.

To provide a full-wall thickness at the end of the tubing, many service technicians file the end of the tubing with a smooth or medium cut mill file. (See Para. 2-64.)

2-12 BENDING TUBING

It takes practice to become good at bending tubing. For the smaller sizes used in domestic models, it is not necessary to use special bending tools. However, a much neater job and a much more satisfactory one is possible with such tools.

The tubing should be bent so that it does not place any

strain on the fittings after it is installed. The tubing, at the bend, should not be reduced in cross-sectional area (kinked). Be very careful when bending the tubing to keep it round. Do not allow it to flatten or buckle. The minimum radius for a tubing bend is between five and 10 times the diameter of the tubing as shown in Fig. 2-9.

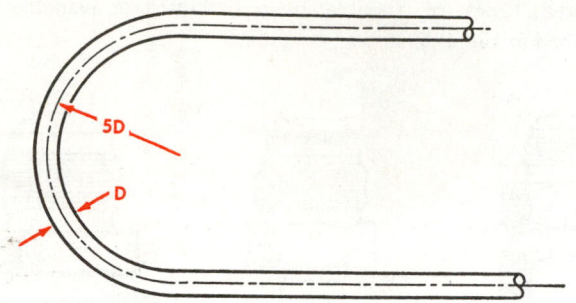

5D

D

Fig. 2-9. Minimum safe bending radius for bending tubing. D is the outside diameter of tube being bent.

Tubes should be bent quite slowly and carefully. It is always wise to use as large a radius as one can. This reduces the amount of flattening. It is also easier to bend a large radius.

Do not try to make the complete bend in one operation, but bend the tubing gradually. There is less danger that the sudden stress will break or buckle the tubing.

An inexpensive tool called a bending spring is illustrated in Fig. 2-10. It may be easily carried in a kit. These are available

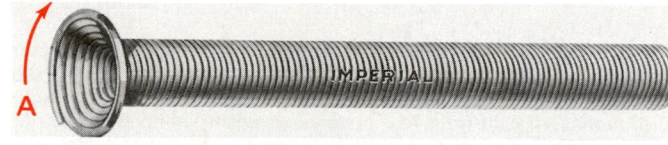

A

IMPERIAL

Fig. 2-10. Tube bending spring may be fitted either outside or inside copper tube while bending tube. Bending spring reduces danger of flattening tube while it is being bent. A—Twist to remove spring.

in a variety of sizes and are made for both external and internal use. The spring bender may be used internally for making bends near the end of the tubing or even after the tubing has been flared.

An external bending spring for 1/4 in. OD tubing may be used as an internal bending spring for 1/2 in. OD tubing. Use the spring bender externally in the middle of long lengths of tubing.

Bending springs tend to bind on the tubing after the bend. It may be easily removed by twisting the spring. This causes the portion on the outside of the bend to expand causing the part of the spring on the inside to contract.

If a bend is to be made near the flare and an external spring is to be used, bend the tubing before making the flare. An internal spring can be used either before or after the flaring operation.

Other tools are available for bending operations. Fig. 2-11 shows such a tool.

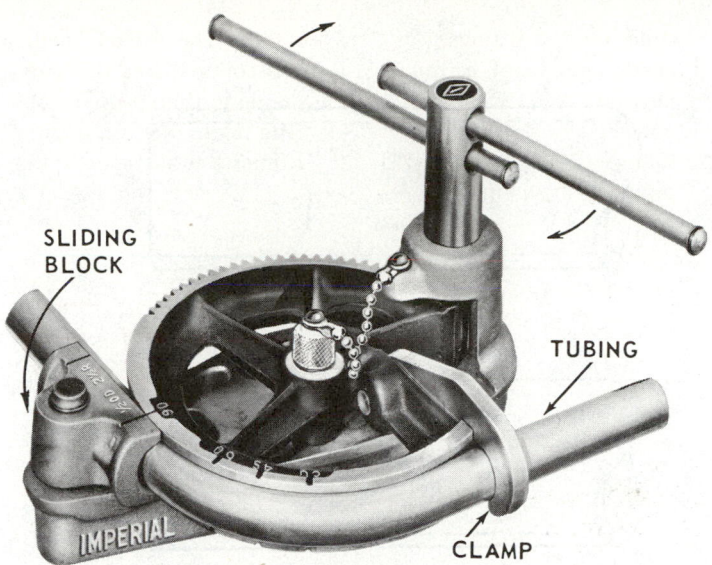

Fig. 2-11. Tube bender which will produce accurate bends and will reduce danger of flattening or buckling tube while it is being bent.

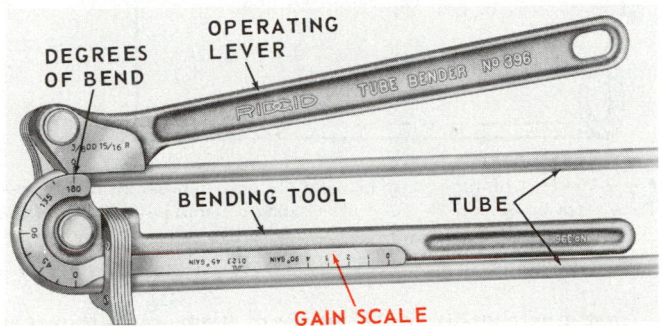

Fig. 2-12. Lever type tube bending tool. As shown, tool is making 180 deg. bend. Tubing is stretched slightly in length during bend. Amount of the stretch (gain) is indicated on the bending tool.
(The Ridge Tool Co.)

A lever-type bender for accurate bending to within 1/32 in. for soft tubing is shown in Fig. 2-12. It can be purchased in six different sizes to match the diameter of the tube to be bent. Always use a bending tool when bending steel tubing. Fig. 2-13 shows some practice bends on tubing.

2-13 CONNECTING TUBING

Since tubing walls are too thin for threading, other methods of joining tubing to tubing and tubing to fittings must be used. The three common methods are:
1. Flared connections.
2. Soldered connections.
3. Silver brazed connections.

2-14 FLARED CONNECTIONS

When connecting tubing to fittings, it is common practice to flare the end of the tube and to use fittings designed to grip the flare for a vapor tight seal. Special tools are used for making flares.

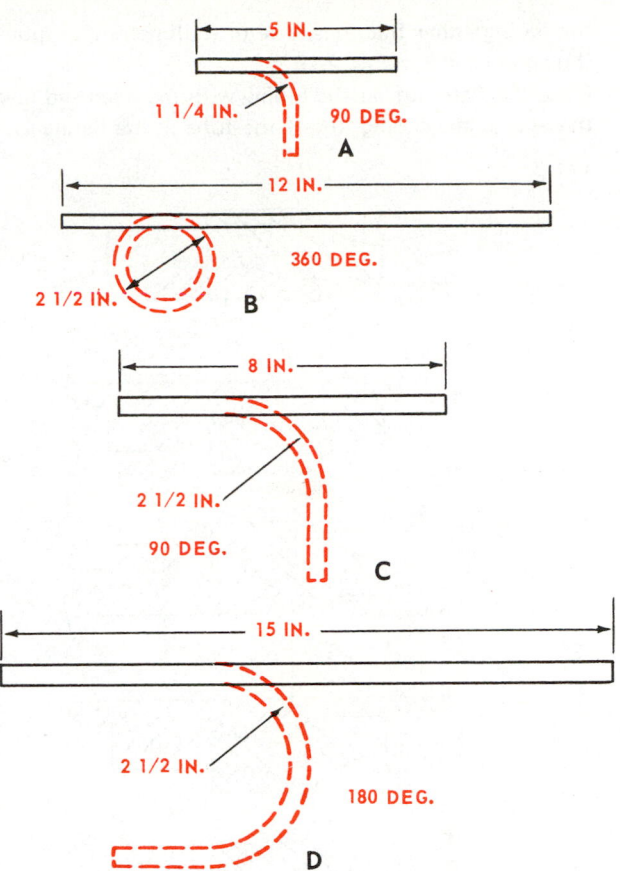

Fig. 2-13. Some practice bends on tubing. A—90 deg. bend on 1/4 in. tubing. B—360 deg. bend on 1/4 in. tubing. C—90 deg. bend on 1/2 in. tubing. D—180 deg. bend on 1/2 in. tubing.

Fig. 2-14 illustrates how a flare is used to form a leakproof joint between a tube and a fitting. It also shows some incorrectly made flares.

Some flares are made which use a single thickness of the tube. Other flares are made with a double thickness of metal in the flare surface. Called "double flares," they are stronger and usually cause less trouble if properly made.

Most flares are made at a 45 deg. angle to the tube. Flares on steel tubing, however, are usually made at a 37 deg. angle. This is because steel tubing is not as easily flared as copper tubing.

2-15 SINGLE THICKNESS FLARE

To make a flare of the correct size using a flaring block, do the following:
1. Carefully prepare the end of the tube for flaring. The end must be straight and square with the tube and the burr from the cutting operation removed by reaming. Fig. 2-15 shows the steps necessary to prepare a tube for flaring.
2. First, use a 10-in. smooth mill file to square the end of the tube. Use great care that no filing enter the tubing. Next, use a burring reamer to remove the slight burr remaining after the cut-off operation.
3. A flaring tool which may be used to make a single thickness flare is shown at view A in Fig. 2-16. A flaring tool suited

for flaring either fractional size or millimetre size tubing is shown at view B in Fig. 2-16.

4. Place the flare nut on the tubing with the open end toward the end of the tubing. Insert the tube in the flaring tool so

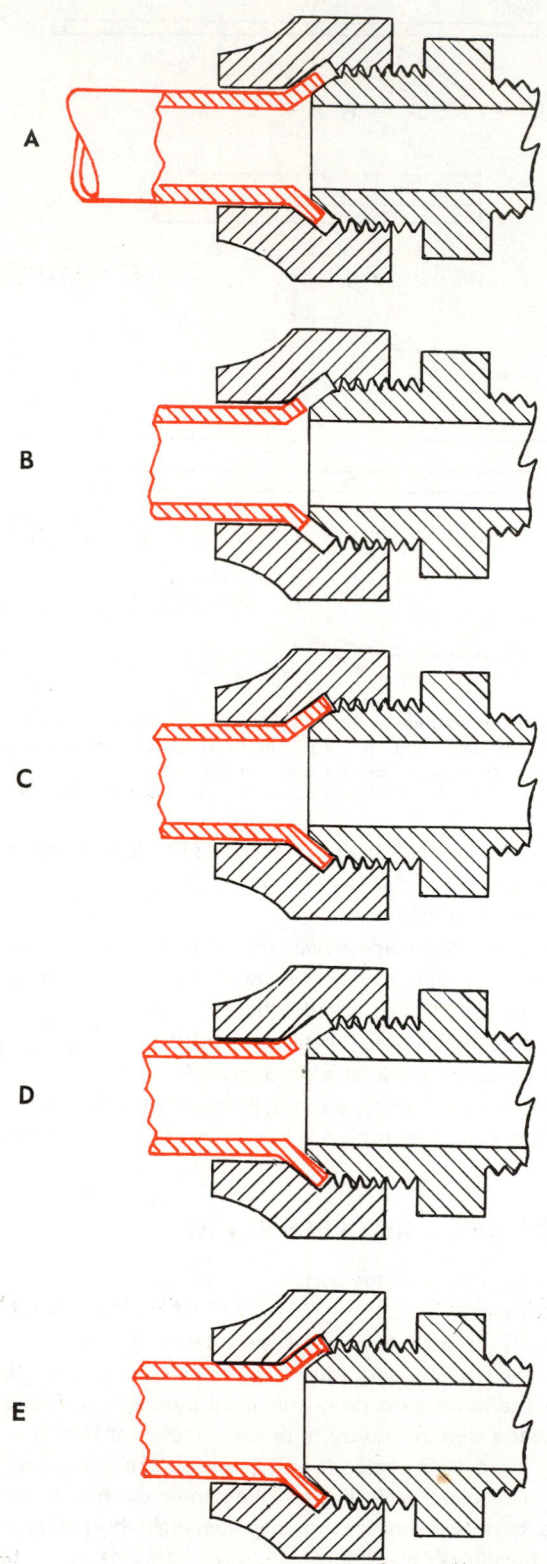

Fig. 2-14. Flared fittings. A—Correctly made flare. B—Flare too small. C—Flare too large. D—Flare is uneven. E—Flare has burrs on edge.

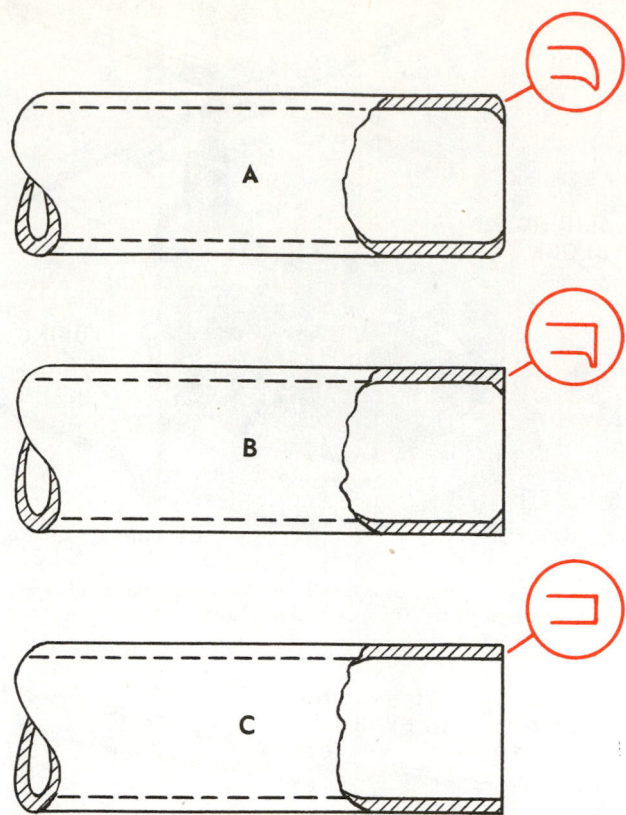

Fig. 2-15. End of tube must be carefully prepared before flaring. A—Tube after being cut. B—Tube after being squared with file. C—Tube filed, reamed and ready for flaring.

that it extends above the surface of the block as shown at view A in Fig. 2-17. This allows enough metal to form a full flare.

5. If the tube extends above the block more than the amount shown, the flare will be too large in diameter and the flare nut will not fit over it. If the tube does not extend above the block, the flare will be too small, and it may be squeezed out of the fitting as the flare nut is tightened. View B in Fig. 2-17 shows appearance of completed flare.

6. To form the flare, first put a drop of refrigerant oil on the flaring tool spinner where it will contact the tubing. Tighten the spinner against the tube end one-half turn and back it off one-quarter turn. Advance it three-quarters of a turn and again back it off one-quarter turn. Repeat the forward movement and backing off until the flare is formed.

Some service technicians make the flare using one continuous motion of the flaring tool; that is, without a back-and-forth motion. It is believed by some that the constant turning of the tool, without back turning, may work harden the tubing and make it more likely to split.

Other technicians like to use a flare which is not completely formed — about seven-eighths complete. They depend on the tightening of the flare nut on the flare to complete it.

Do not tighten up the spinning tool too much because this will thin the wall of the tubing at the flare and weaken it.

Always place the flare nut on the tube in the proper

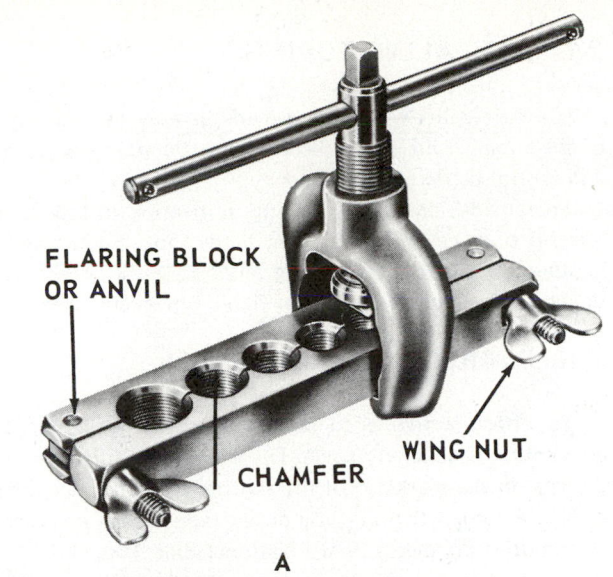

FLARING BLOCK OR ANVIL

CHAMFER

WING NUT

A

FLARE HANDLE

IMPERIAL ♦ EASTMAN

CLAMP TIGHTENING HANDLE

B

Fig. 2-16. Flaring tools. A—Popular style used for making single thickness flares on refrigeration tubing. Flaring block is split, making it easy to insert and clamp tubing in place for flaring. Note 45 deg. chamfer in block which gives the flare its correct shape. (Duro Metal Products Co.) B—Flaring tool having an adjustable tube-holding mechanism which permits flaring tubing 3/16 to 5/8 in. OD and 5 to 16 mm OD. (Imperial-Eastman Corp.)

position before the flare is made. It cannot, in most cases, be installed on the tube after it has been flared.

2-16 DOUBLE THICKNESS FLARE

Double thickness flares are formed with special tools. Fig. 2-18 illustrates a cross-section through a simple block-and-punch type of tool used to make a double flare. The correct shape of the double flare is shown in the final operation in this figure. Some flaring tools are fitted with adapters which makes it possible to form either single or double flares with the same

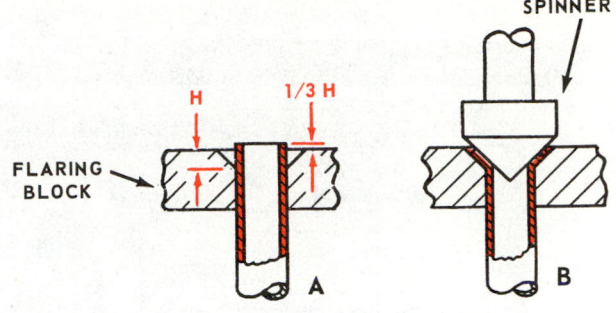

FLARING BLOCK

H   1/3 H

SPINNER

A   B

Fig. 2-17. Tubing to be flared should extend slightly above flaring block to allow sufficient metal to form satisfactory flare. Amount to allow is about a third of the height of flare. A—Proper position of tube in flaring tool before flaring. B—Completed flare.

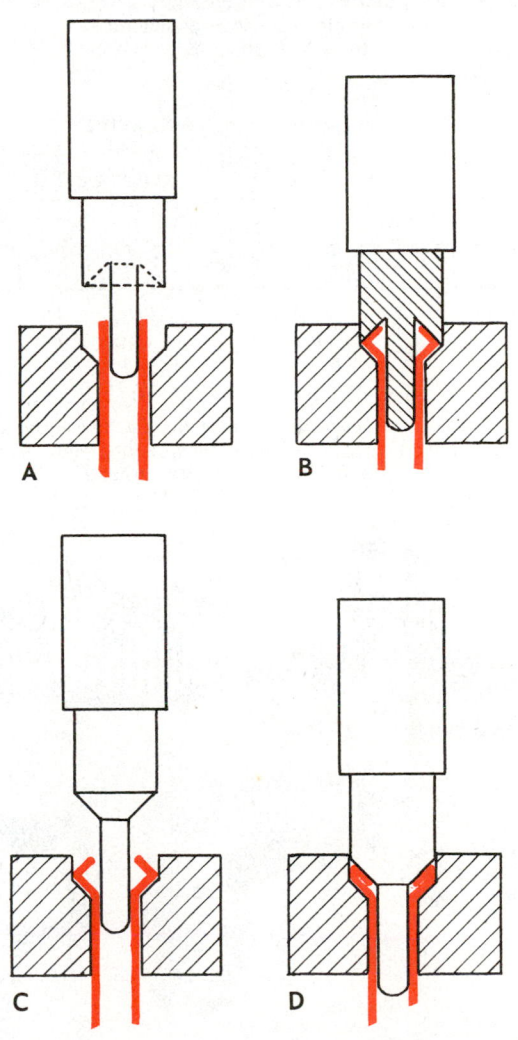

A   B

C   D

Fig. 2-18. Simple block and punch tool for forming double flares on copper tubing. A—Tube is clamped in body of flaring block. B—First female punch bends end of tube inward. C—Shows male punch inserted in partially flared tube. D—In operation, male punch folds end of tube downward to form double thickness and expand flare into final form.

tool, Fig. 2-19.

Fig. 2-20 shows the steps for making a double flare (using the tool shown in Fig. 2-19).

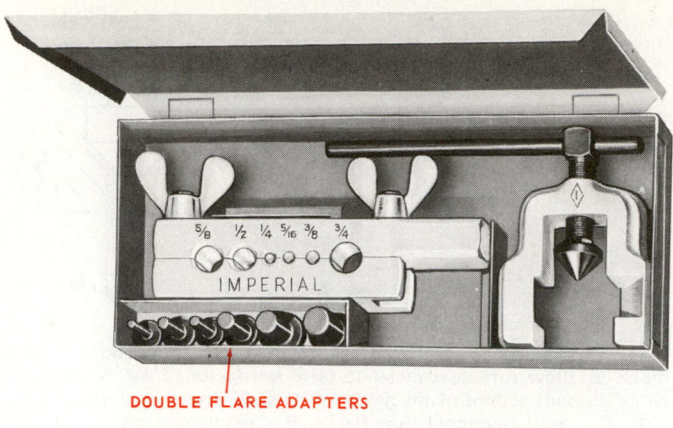

DOUBLE FLARE ADAPTERS

Fig. 2-19. A flaring tool which, with necessary adaptors, is capable of producing either single or double flares. (Imperial-Eastman Corp.)

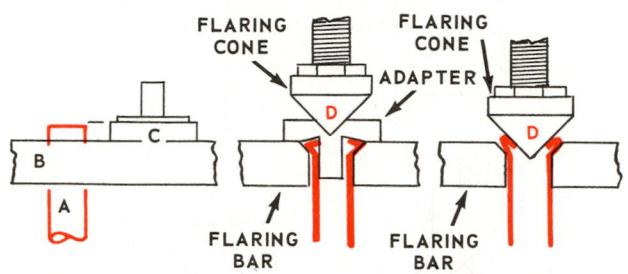

Fig. 2-20. Correct procedure for forming double flare using adaptors with combination single flare-double flare, flaring tool. A—Tubing. B—Block. C—Adaptor. D—Flaring cone.

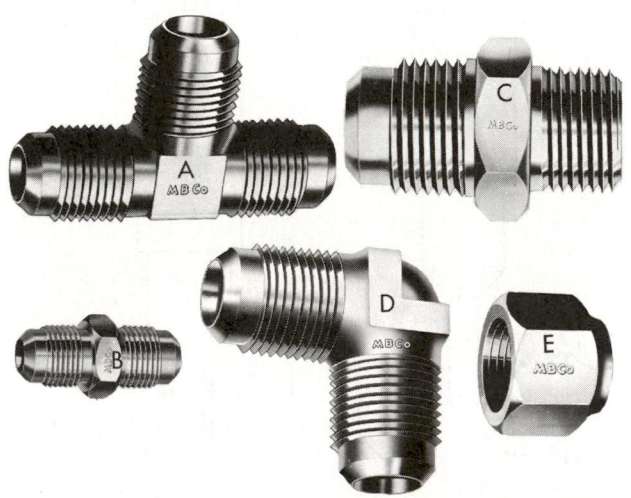

Fig. 2-21. Some of the more common flared type fittings used in refrigeration and air conditioning work. A—Flared Tee fitting, male flare x male flare. B—Flared Union coupling, male flare x male flare. C—Flared Half Union coupling, male flare x male pipe. D—Flared 90 deg. Elbow male flare x male flare. E—Flare Nut.   (Mueller Brass Co.)

Double thickness flares are recommended for only the larger size tubing 5/16 in. and over. Such flares are not easily formed on smaller tubing. The double flare makes a stronger joint than a single flare.

2-17 ANNEALING TUBING

If a flare splits when being made, it may be due to the age of the tubing. Old tubing becomes brittle after a period of use and cannot be flared satisfactorily.

To remedy, anneal the tubing by heating to a dull, cherry red or blue color and allow it to cool. Pounding, rough handling or bending the tubing tends to work harden it. Hard drawn tubing cannot be bent or flared unless annealed.

2-18 FLARED TUBING FITTINGS

To attach a fitting to soft copper tubing, a flared type connection is generally used. There are many different fitting designs on the market, but the accepted standard for refrigeration is a forged fitting using either pipe thread or Society of Automotive Engineers (SAE) National Fine Thread.

The fittings are usually made of drop-forged brass. They are accurately machined to form the National Fine (NF) threads, the National Pipe (NP) threads, the hexagonal shapes for wrench attachment, and the 45 deg. flare for fitting against the tubing flare. These threaded fittings must be carefully handled to prevent damage to them. Fig. 2-21 represents some of the common flared-type fittings used in refrigeration work.

All fitting sizes are based on the tubing size. For example, a 1/4 in. flare nut attaches 1/4 in. tubing to a flared fitting even though it has 7/16 in. NF internal threads and uses a wrench with a 3/4 in. opening to turn it.

Fig. 2-22 is a table of common flared fitting sizes.

Catalog listings of tube fittings usually provide a code number to indicate the size. The code number 3 indicates that the fitting fits 3/16 in. tubing. Code number 4 indicates that it fits 1/4 in. (4/16). Code number 8 fits 1/2 in. tubing (8/16).

Some tubing fittings have pipe threads on one end. Pipe threads taper 1/16 in. in diameter for every inch in length.

With more plastic tubing being used, it has become necessary to provide special fittings. Brass, aluminum and polyethylene materials are commonly used. Fitting connections made on plastic tubing are not flared. Rather, a compression type fitting, as shown in Fig. 2-23, is used.

2-19 METRIC SIZE TUBE FITTINGS

Metric size tubing, as described in Para. 2-6, requires metric size fittings. These are made of the same materials and in the same general styles and shapes as English size fittings and are used in the same way.

Eventually only metric size tubing and fittings will be used. At present, both sizes are in use. The service technician must be careful that English size fittings are not mixed with metric size fittings.

2-20 SOLDERED OR BRAZED TUBING FITTINGS

In modern practice, most tube and fitting connections are made by either soldering or silver brazing. Soldered joints are used for water pipes and drains. Silver brazed joints are used for refrigerant pipes and tubing.

REFRIGERATION FITTINGS (FLARED TYPE) Sizes are based on the Outside Diameter of the Tubing						
Name and Description	1/4	5/16	3/8	7/16	1/2	5/8
Nut Forged	X	X	X	X	X	X
Union (Threads same size)	X	X	X	X	X	X
Half Union (1/8 Pipe)	X	X				
Half Union (1/4 Pipe)	X	X	X	X		
Half Union (3/8 Pipe)					X	
Half Union (1/2 Pipe)						X
Elbow.	X	X	X	X	X	X
Elbow (One 1/8 Pipe).	X	X				
Elbow (One 1/4 Pipe).	X	X	X	X		
Elbow (One 3/8 Pipe).					X	
Elbow (One 1/2 Pipe).						X
Tee (Threads same size)	X	X	X	X	X	X
Tee (One 1/8 Pipe)	X	X				
Tee (One 1/4 Pipe)				X	X	
Tee (One 3/8 Pipe)					X	
Tee (One 1/2 Pipe)						X
Cross	X	X	X	X	X	X
Flared Tube Sealing Plug	X	X	X	X	X	X
Flared Tube Sealing Cap		X	X	X	X	X
Flared Tube Copper Seal Cap . .	X	X	X	X	X	X
Union (Reducing)	5/16-1/4	3/8-1/4	1/2-1/4	1/2-3/8		
Elbow (Reducing)	5/16-1/4	3/8-1/4	1/2-1/4	1/2-3/8	5/8-1/2	
Tee (Reducing)	5/16-1/4	3/8-1/4	1/2-1/4	1/2-3/8	5/8-1/2	

Fig. 2-22. Some of the more popular flared type copper tube fittings commonly used by the refrigeration and air conditioning service engineer. The reducing fittings are used for connecting tubing of different size.

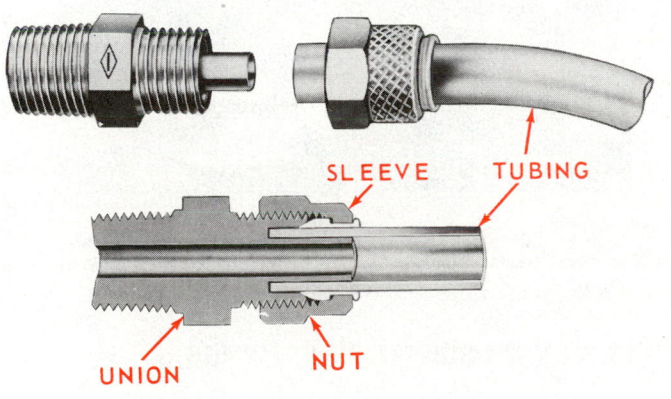

SLEEVE   TUBING

UNION   NUT

Fig. 2-23. A compression type fitting used with polyethylene tubing. Caution: Polyethylene is rather a soft substance and very little tightening is needed. While most fittings are made with flats for wrench tightening, in most polyethylene installations little more than "finger tightness" is necessary.   (Imperial-Eastman Corp.)

Soldered tubing fittings are available in the same sizes and styles as flared fittings. Standard size tubing may be inserted in the fitting. The joint is sealed by heating the fitting and either soldering or silver brazing the tubing to the fitting.

2-21 SOLDERING

Soldering is a process of applying molten (melted) metal to metals that are heated but are not molten. It is an adhesion process. (In adhesion one part is bonded to or is stuck to another by a third material.) The molten solder flows into the pores of the surface of the metals being joined, and as the solder solidifies (hardens) a good bond is obtained.

A fitting soldered to a tube is shown in Fig. 2-24.

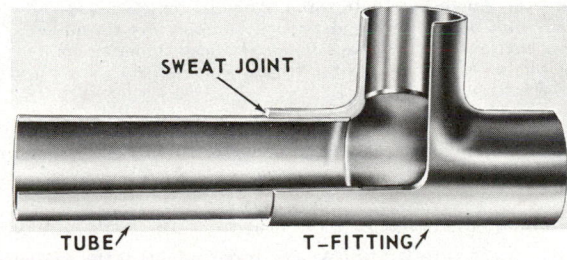

SWEAT JOINT

TUBE   T-FITTING

Fig. 2-24. A cross section of tee fitting soldered to hard ACR tube.

A good sweat joint begins with first cleaning the parts to be joined, then fluxing and assembling them. The assembly is then heated. As soon as the joint reaches the flowing temperature of the solder, solder is added to the joint and flows into it. After the solder cools, it will seal and connect the surfaces. The step-by-step procedure for making a sweat joint is shown in Fig. 2-25.

When assembling swaged (shaped) tube-to-tube joints or tubing to a fitting, it is advisable, after thoroughly cleaning the mating parts, to insert the tube in the fitting 1/16 to 1/8 in.

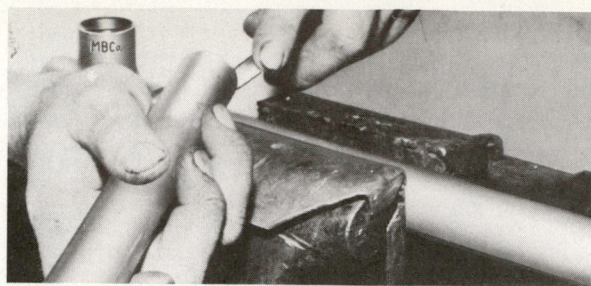

Step 1. Cut tube to length and remove burr with file or scraper.

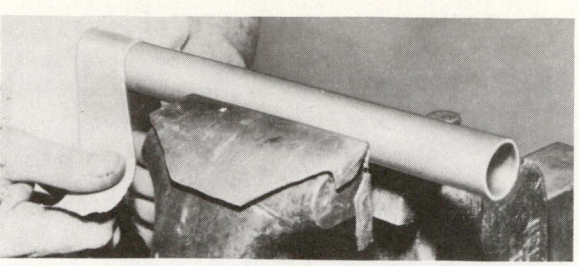

Step 2. Clean outside of tube with clean sandpaper or sand-cloth.

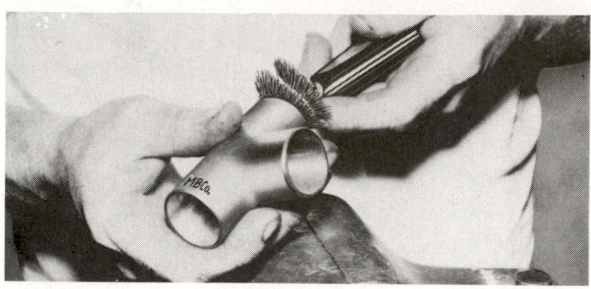

Step 3. Clean inside of fitting with a clean wire brush, sand-cloth or sandpaper. Do not use emery cloth.

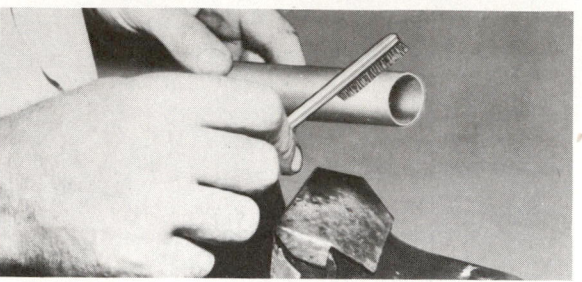

Step 4. Apply flux thoroughly to outside of tube — assemble tube and fitting.

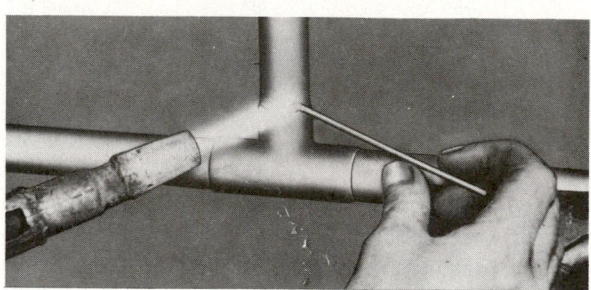

Step 5. Apply heat with torch. When solder melts upon contact with heated fitting, the proper temperature for soldering has been reached. Remove flame and feed solder to the joint at one or two points until a ring of solder appears at the end of the fitting.

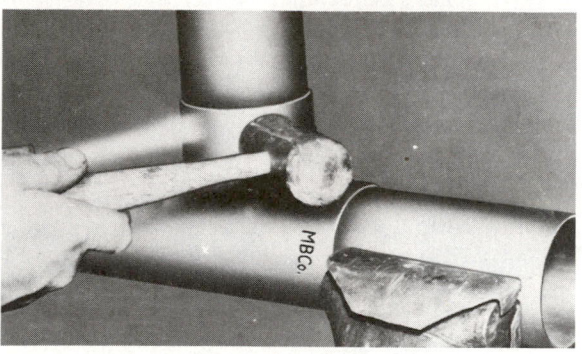

Step 6. Tap larger sized fittings with mallet while soldering, to break surface tension and to distribute solder evenly in joint.

Fig. 2-25. Recommended step-by-step procedures to be followed when soldering tubing. (Mueller Brass Co.)

Then apply the soldering or brazing flux. After the flux is applied, force the mating parts together. Rotate one of the pieces to spread the flux evenly over both the internal and external surface.

Apply the necessary heat for soldering or brazing. Using this procedure will eliminate any possibility of flux entering the system. See Fig. 2-26.

Avoid swaging steel tubing (shaping by hammering). It is harder (less ductile) and may crack or split. Sometimes the process tube of a hermetic motor compressor is made of steel. Many clean the tubing before cutting it to reduce the amount of dirt that may get into the system. One tubing should extend into the other the same distance as the diameter of the larger tubing (i.e., 1/4 in. into 5/16 in. should overlap 5/16 in.).

Fig. 2-27 illustrates some common fittings, which may be

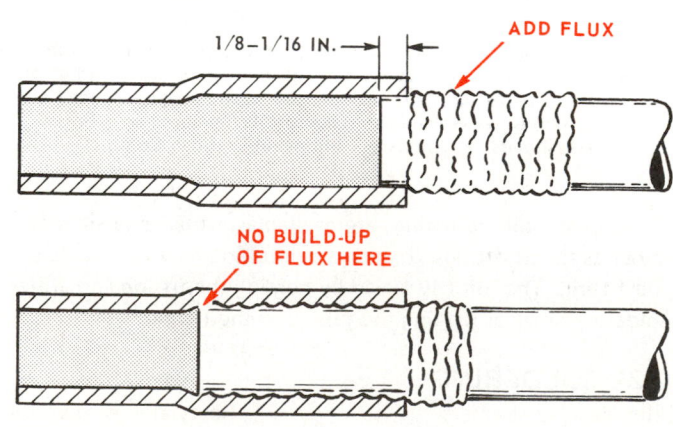

Fig. 2-26. Brazing flux may be a source of corrosion in a system. Apply flux to joints as above so that it will not get into system.

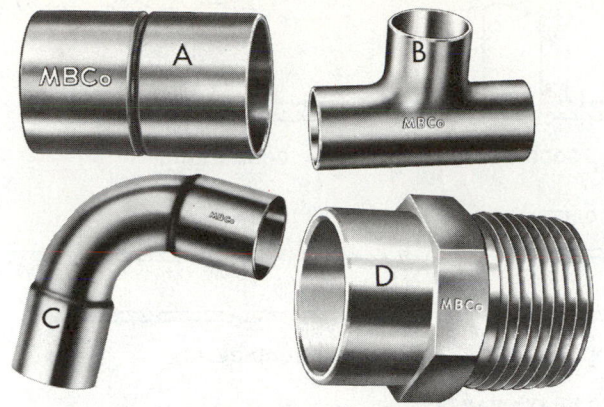

Fig. 2-27. Common fittings which may be either soldered or silver brazed to tubing. A—Coupling with rolled stop, sweat x sweat. B—Tee, sweat x sweat x sweat. C—90 deg. elbow, sweat x sweat. D—Adaptor, sweat x male pipe thread (mpt).

either soldered or silver brazed to tubing. All brass and copper parts may be easily soldered.

To solder:

1. The surfaces to be soldered must be very clean.
2. A good clean flux must be used.
3. A good source of clean heat must be on hand.
4. The parts being soldered must be firmly supported during the soldering operation.

Flux does not clean the metal. It keeps the metal clean once soil has been removed by filing, scraping, sanding, using steel wool or wire brushes. Surfaces being soldered must be free of grease, dirt and oxides.

A 50/50 alloy of tin and lead solder is usually satisfactory for soft soldering. Solders containing as much as 95 percent tin are now being recommended for soldered joints subjected to very low temperatures.

A portable air-acetylene torch, Fig. 2-28, is a practical tool

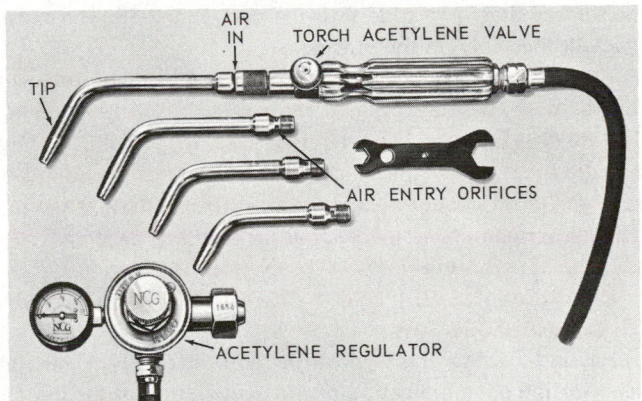

Fig. 2-28. Air-acetylene torch suitable for both soldering and brazing tubing. These torches use the small portable acetylene cylinders. Flame is very clean and hot and torch is easy to use.
(National Cylinder Gas, Div. Chemetron Corp.)

for heating surfaces to be soldered. Acetylene is the fuel recommended for this type of torch. However, liquid propane may be used.

Flux for this type of work should have no corrosive properties. Acid flux should not be used. It tends to corrode fittings making them unsightly and difficult to work on later.

An important fundamental of good soldering is that the metal being joined be hot enough to melt the solder. This is the only way the solder will go into the pores of the metal surface.

Apply the heat to the metal to be soldered; then touch the solder to the metal. Do not overheat. Keep testing the metal with the solder wire. Heat only until solder flows.

If the parts to be soldered are of the correct temperature, have been cleaned and fluxed, the solder will flow over the surface quickly. Do not heat the solder with the torch.

When joining tubing and tubing fittings by soldering, thoroughly clean the surfaces to be soldered. Rolls of clean abrasive paper or abrasive cloth are good cleaning materials. Clean inside and outside brushes may also be used.

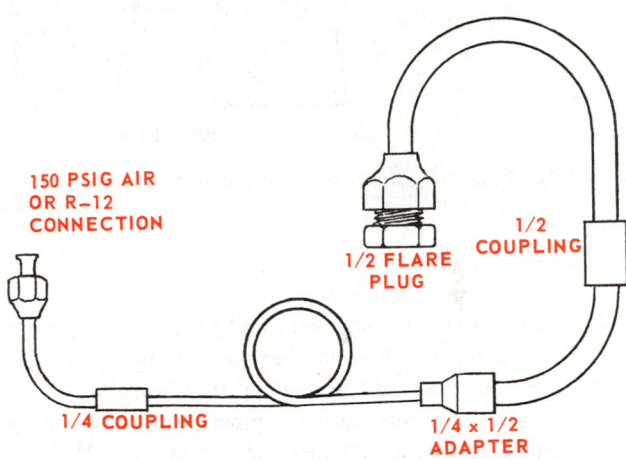

Fig. 2-29. Practice piece used to develop skill in copper tube soldering. Completed assembly should be pressurized as indicated and all of the joints tested for leaks. Use soapsuds.

Fig. 2-29 illustrates a tube soldering practice assembly. This assembly can then be connected to an R-12 cylinder or compressed air line, and the quality of the soldered joint determined by checking for leaks.

Never use oxygen when testing for leaks. Any oil in contact with oxygen under pressure will form an explosive mixture.

Solder in wire form is usually the best because of the difficulty of getting at the soldering surfaces. It is also easy to apply. While soldering, it helps to "wipe" the surfaces after putting some solder on. Use a clean cloth, a brush or the solder wire itself. This action will remove any dirt and will help "tin" the surface.

2-22 SILVER BRAZING

One of the best methods of making leakproof connections while providing maximum strength is to silver braze the joints. These joints are very strong and will stand up under the most extreme temperature conditions. Silver soldering or silver brazing, as it is more correctly called, can be easily done if the

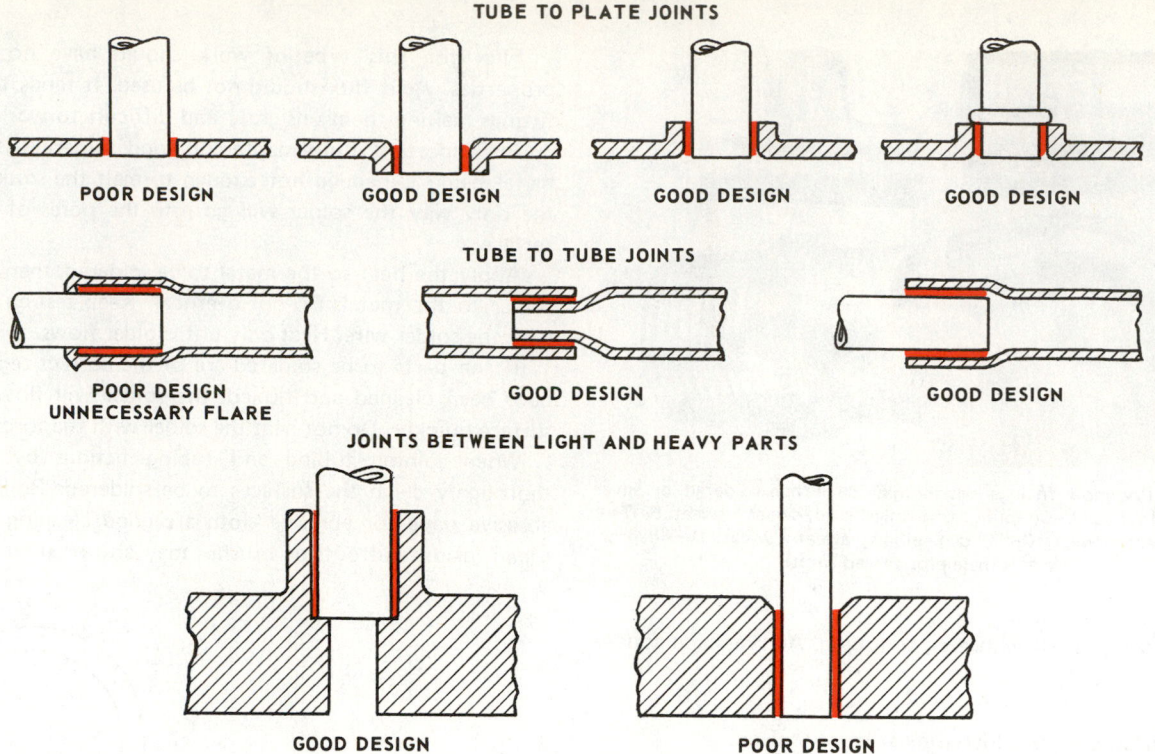

TUBE TO PLATE JOINTS

POOR DESIGN   GOOD DESIGN   GOOD DESIGN   GOOD DESIGN

TUBE TO TUBE JOINTS

POOR DESIGN
UNNECESSARY FLARE   GOOD DESIGN   GOOD DESIGN

JOINTS BETWEEN LIGHT AND HEAVY PARTS

GOOD DESIGN   POOR DESIGN

Fig. 2-30. Suggestions for making joints to be silver brazed. Actual thickness of silver brazing is exaggerated to show its application. (Handy and Harman)

correct procedure is followed:

1. Clean the joints mechanically.
2. Fit the joints closely and support all parts.
3. Apply the clean flux recommended for the silver brazing alloy. Follow the manufacturer's instructions.
4. Heat evenly to recommended temperature. Keep the torch moving constantly in a "figure-eight" motion.
5. Apply silver brazing alloy to the heated parts. Do not heat (melt) the silver brazing alloy with the torch.
6. Cool the joint.
7. Clean the joint thoroughly, using warm water and a brush. Be sure all flux has been removed.

An oxyacetylene torch is an excellent heat source for silver brazing. However, one must have training in its safe use. Be sure to use flashback arrestors at both the acetylene and oxygen regulators.

There are various silver alloys on the market. Most have a 35 to 45 percent silver content. This material usually melts at 1120 F. (604 C.) and flows at 1145 F. (618 C.). Contact the local welding supply house or air conditioning and refrigeration wholesaler for suitable silver brazing supplies.

CAUTION: Carefully check the specifications of the silver brazing alloy used. If it contains any amount of cadmium, be SURE that the work space is well ventilated and that none of the fumes are inhaled or come in contact with the eyes or skin. Cadmium fumes are very poisonous.

The part to be brazed must be fitted accurately and carefully cleaned. Dirt must be removed from any external surface. A fine grade of stainless steel wool is considered suitable for this purpose. Internal circular surfaces can be cleaned with clean stainless steel wire brushes or stainless steel wool rolled on a rod.

The parts must have contacting surfaces of sufficient area, such as a tube sliding into a fitting (not a drive fit), Fig. 2-30. The contacting surfaces need not be very large (three times smallest section).

If the parts are dented or are out of round, these faults must be corrected before the brazing is done. It is important to support all the parts securely during the operation so that the parts will not move.

It is also important to make sure that no flux enters the system during the brazing operation, as it cannot be easily removed. Overfluxing can be avoided by applying the flux to the surface that is to slide into the part, Fig. 2-26. The excess flux will then stay on the outside.

All air must be removed from the tubing being brazed. This can be done best by purging the tubing with either carbon dioxide or nitrogen as shown in Fig. 2-31. If there is any oil inside the tubing or part, the heat of the torch may cause this oil to vaporize. Oil vapor mixed with air will explode if ignited. Using a nonflammable gas such as carbon dioxide or nitrogen will eliminate this hazard. CAUTION — see Chapter 11.

CAUTION: NEVER USE A REFRIGERANT, OXYGEN, OR COMPRESSED AIR.

Heating of the joint must be done carefully. The flux behavior is a good indication of the temperature of the joint as the heating progresses.

1. Keep the joint covered with the flame all during the operation to prevent air getting to it.
2. The flux will dry out; the moisture (water) will boil off at 212 F. (100 C.), then the flux will turn milky in color.
3. Next, it will bubble at about 600 F. (316 C.).
4. At 800 F. (427 C.), the flux lies on the surface and has a milky appearance.

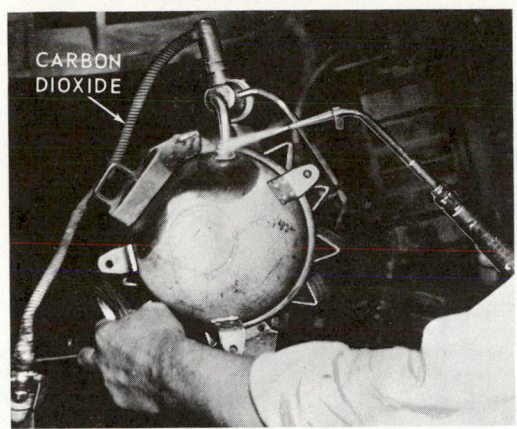

Fig. 2-31. A fitting being silver brazed to compressor dome. Note tube carrying low-pressure carbon dioxide or nitrogen through fitting and into compressor during silver brazing to prevent fire or explosion.

5. Following this, it will turn into a clear liquid at about 1100 F. (593 C.). This point is just short of the brazing temperature.

The silver alloy itself melts at 1120 F. (604 C.) and flows at 1145 F. (618 C.). A torch tip several sizes larger than the one used for soft soldering should be used. Be sure to heat BOTH pieces which are to have the silver alloy adhered to them.

The proper silver brazing temperature will be indicated by the color of the secondary flame. The flame will start to show a green shade as the brazing temperature is reached.

If the tubing is painted, first burn off the paint and then cleanse with clean sand cloth or a clean wire brush.

When heating a copper-to-steel joint, heat the copper first (it takes more heat because it carries it away faster). Put some flux on the brazing rod to help it flow quicker.

When cutting capillary tubing, notch all around with a three-corner file. Break tubing by bending back and forth (small bends). The tubing ID will then remain full size. A tube cutter will reduce inside diameter.

Do not clean the end of the tube or brazing material may run to the end and close or partially close the hole (ID) of the capillary.

When silver brazing, the torch is never held in one spot, but must be moved around the entire area to be brazed. Never hold the heat in one area but keep the torch moving. Many service technicians prefer to move the torch in a "figure-eight" motion. Larger torch tip sizes are recommended to allow a soft, large quantity of heat without excess pressure or "blow." A slight feather on the inner cone is good. See Fig. 2-32.

2-23 CLEANING THE BRAZED JOINT

Thoroughly wash with water and scrub the outside of the completed silver brazed joint. This is always necessary. Flux left on the metals will tend to corrode them or may temporarily stop a leak which will only show up later.

The joint may be cooled quickly or slowly. Cooling with water is allowable. The same water may be used to wash the joint. Visual inspection will quickly reveal any places where

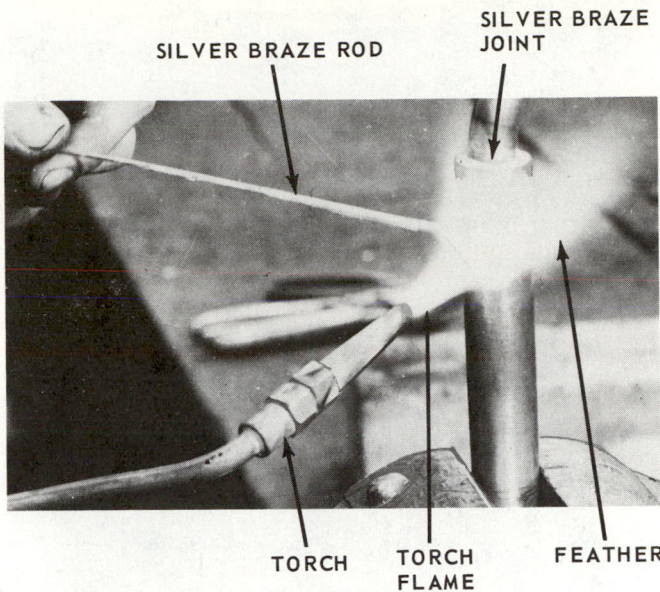

Fig. 2-32. Silver brazing copper tubing connection. Note large soft flame. See text for suggestions on flame movement.

the silver alloy did not adhere. It is advisable to watch for this and make any corrections during the brazing operation.

2-24 SWAGING COPPER TUBING

Swaging, Fig. 2-33, permits two pieces of soft copper tubing of the same diameter to be joined together without the use of fittings.

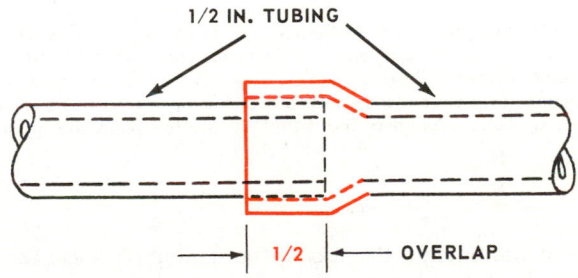

Fig. 2-33. Two pieces of soft copper tubing assembled and ready for soldering or brazing to make connecting joint. Note pieces were of same diameter before one was swaged.

Swaging of copper tubing is often done. It is more convenient to solder one joint than to make two flared connections.

The length of overlap of the two pieces of tubing is important. As a general rule, the length of the overlap should equal the outside diameter (OD) of the tubing.

Two types of swaging tools are commonly used: the punch type and the lever type. In both cases, different tool sizes are available for use with the many sizes of copper tubing.

When using the punch type swaging tool, the copper tubing is inserted into the correct hole size in the anvil block. Then a punch is inserted into the copper tubing and hammered down

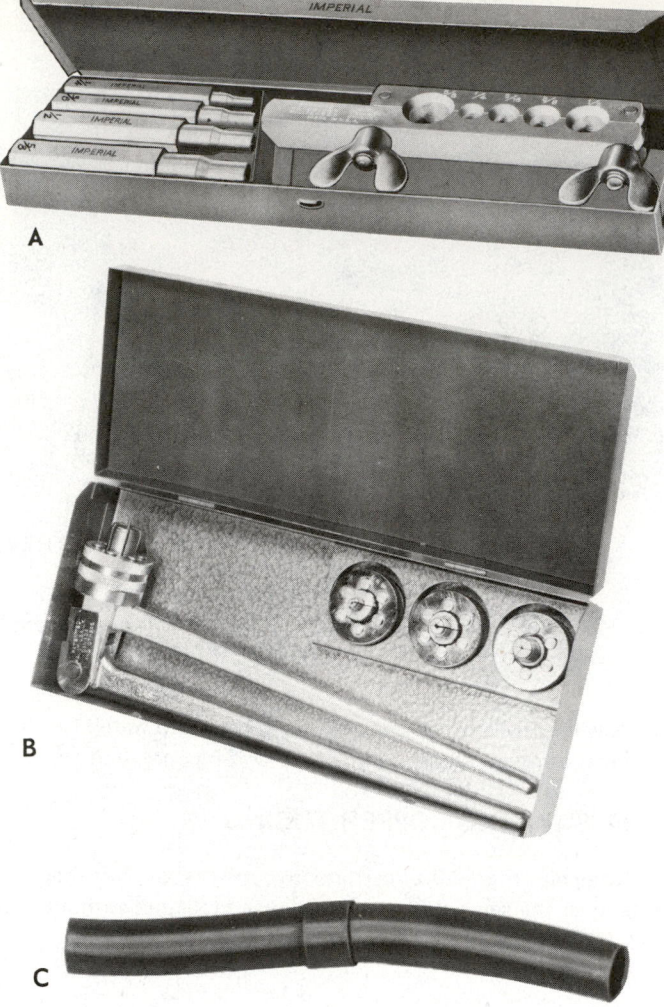

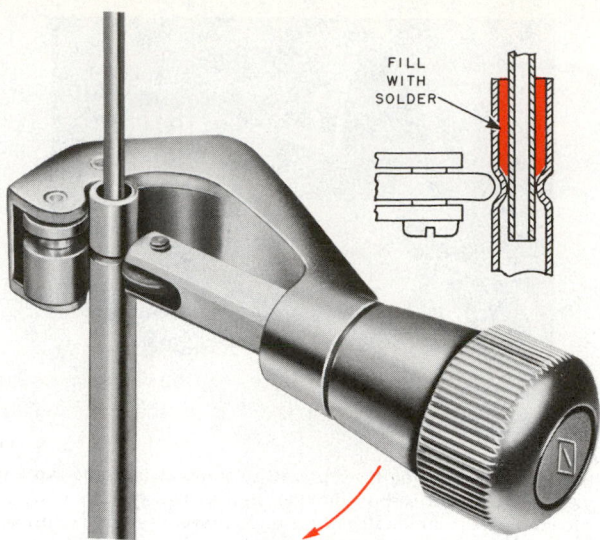

Fig. 2-35. Combination tube cutter and constrictor. Different wheel is used for cutting than for constricting. Insert shows two tubes joined by soldering or brazing. Outer tube has been constricted before soldering.

Fig. 2-34. Swaging tools. A—Punch type swaging tool complete with various size punches and anvil for swaging tubing from 1/4 to 1/2 in. diameter. (Imperial-Eastman Corp.) B—Lever type (swaging tool) tube expander. This expander has various size adaptors to fit various size tubes. C—Tube expanded and fitted to another ready for soldering. (Robinair Mfg. Corp.)

until it has entered the tubing the desired distance. See Fig. 2-34, view A.

When using the lever type tool, the tubing is placed over the expander. Squeezing the lever expands the tube to the proper size, Fig. 2-34, view B. Fig. 2-34, view C, illustrates the end of the tubing expanded and the pieces fitted together, ready for soldering.

2-25 TUBE CONSTRICTOR

In many cases a service technician must make a soldered or brazed joint between two tubes where one tube fits rather loosely inside the other. Good practice demands that the tubes be sized as close as .003 in. to each other. Fig. 2-35 shows a special tool used to constrict the outer tube until it fits the OD of the inner tube. By using this tool the service technician can easily solder or braze the joint without danger of leaks or of flux getting into the system.

2-26 PIPE FITTINGS AND SIZES

Air conditioning and refrigeration installations make wide use of pipe fittings and pipe threads (National Pipe Threads or NP). Taper pipe threads are specially formed V-threads made on a conical spiral. This causes the threads to seal as the fitting is tightened. Pipe threads taper 1/16 in. in diameter for every inch of length. If the threads were not tapered, a gasket or an American Standards Association (ASA) machined shoulder would be necessary to provide a leakproof joint.

Besides being tapered (or in a conial spiral), pipe threads are different from fine-thread series National Fine (NF) and the coarse-thread series National Coarse (NC). NF and NC sizes are based on outside diameter. Pipe thread sizes are based on flow diameter, or roughly the diameter of the hole in the pipe (inside diameter or ID). Fig. 2-36 illustrates a male thread on a 1/2 in. pipe. The external threads are cut with a pipe die. The die is turned by a standard die stock, a ratchet die stock or

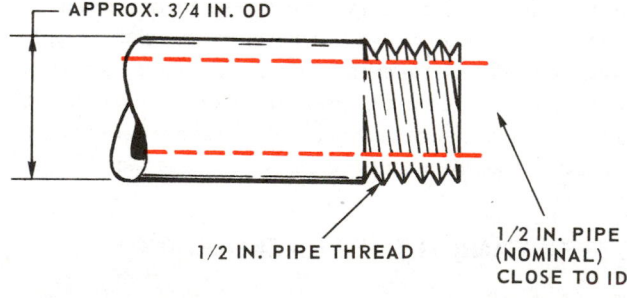

Fig. 2-36. A male thread on a 1/2 in. pipe.

power driven die stock. Fig. 2-37, the pipe thread tap (for cutting the female or internal threads) is turned with a tap wrench. The male threaded pipe should be turned into the female fitting five threads.

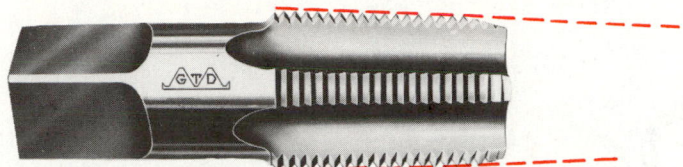

Fig. 2-37. Pipe tap. Note taper of threads.
(TRW Greenfield Tap & Die Div.)

Pipe fittings are supplied with the threads already cut. The most common fittings are the coupling, the reducing coupling, the union, the nipple, the 90 deg. elbow, the reducing elbow, the 45 deg. elbow, and the street ell. A street ell is usually a 90 deg. fitting having a male thread on one end and a female thread on the other.

The threads are made self-sealing by the pressing together of the sharp V-threads as they are assembled. Various commercial compounds are available to help seal these threads. Brushed on pipe threads before assembly, the compound will make a strong, leakproof joint.

Special thread cutting compound should be used when cutting pipe threads. The taps and dies must be kept clean and sharp.

2-27 REPAIRING THREADS

Occasionally the threads of a fitting or fixture — especially pipe threads — may become so worn that they are no longer leakproof. A very convenient and rapid remedy for this condition is to coat the threads with solder and then remove the excess solder by sharply rapping the fitting while it is still hot. The threads will then be coated with a thin film of solder. This will usually remedy the trouble.

2-28 EPOXY RESIN

Epoxy resin may be used to repair cracks and leaks in evaporators and joints. Such resins have good adhesion (sticking) qualities when used with steel, copper, wood and many types of plastics. Epoxy adhesives are available from many manufacturers or refrigeration supply wholesalers.

The most desirable type is the two-part system. This consists of an epoxy resin and a hardener. Combinations of these two will harden at room temperature. The one-part adhesive must be heated in order to cause it to harden.

The two-part system consists of two jars or tubes of a paste-like substance in contrasting colors.

When repairing with epoxy cement, it is necessary to first determine whether a patch is necessary or whether the hole can be repaired by placing the mixed epoxy over the hole or leak. Small leaks or holes up to 1/16 in. diameter can often be successfully sealed by placing the mixed epoxy over the leak and allowing it to cure. The same procedure is recommended for tubing cracks. For larger holes, a patch of the same type of tubing material is recommended.

The refrigeration service technician is cautioned to purchase the epoxy compound from a refrigeration wholesaler because some epoxies available elsewhere are not compatible with refrigerants R-12 and R-22. Furthermore, the shelf life of most epoxy resins is about six months.

The service technician should be careful when using epoxy compounds because they contain chemicals which may irritate the skin. Long contact with the skin should be avoided. (In case of contact, remove the epoxy and clean the skin with rubbing alcohol followed by a thorough washing with soap and water.)

The procedure for using epoxy compound is as follows:

1. Clean the surface or surfaces which are to be bonded by sanding with clean, coarse sandpaper or clean steel wool.
2. Clean surface with recommended solvents such as toluene, acetone or a similar industrial solvent.
3. Mix epoxy on a clean surface such as a piece of cardboard as shown in Fig. 2-38. The two parts of epoxy compounds

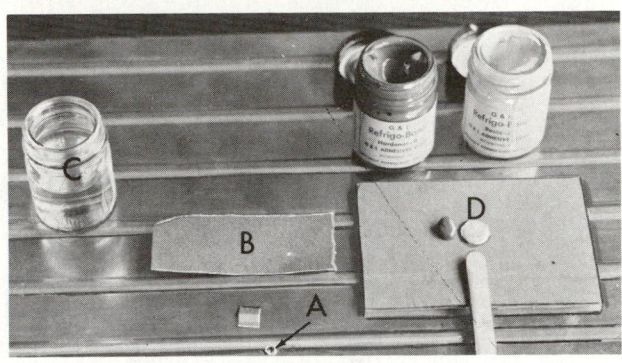

Fig. 2-38. Refrigeration leak undergoing repairs with epoxy compounds. A—The leak. B—Sandpaper. C—Cleaning solvent. D—Epoxy compounds ready for mixing on piece of cardboard.
(G & L Adhesives Corp.)

are mixed together (equal parts of each) until all color streaks have been eliminated and the substance has a uniform color.

4. Apply the epoxy mixture to the surface if it is a small hole or apply to mating surfaces if a patch is to be used as shown in Fig. 2-39. Epoxy compounds should be used immediately after mixing, as chemical hardening, or setting starts immediately.

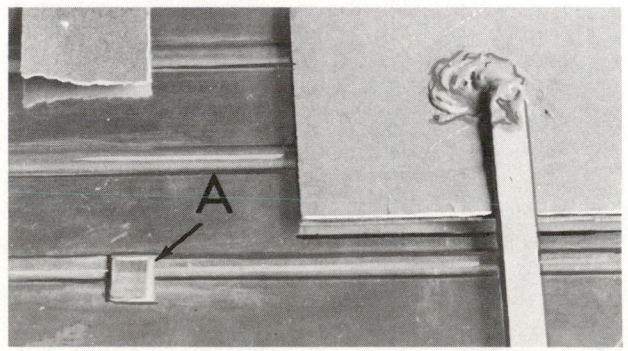

Fig. 2-39. Epoxy compound is used to join the patch material to the injured part. A—Completed patch.

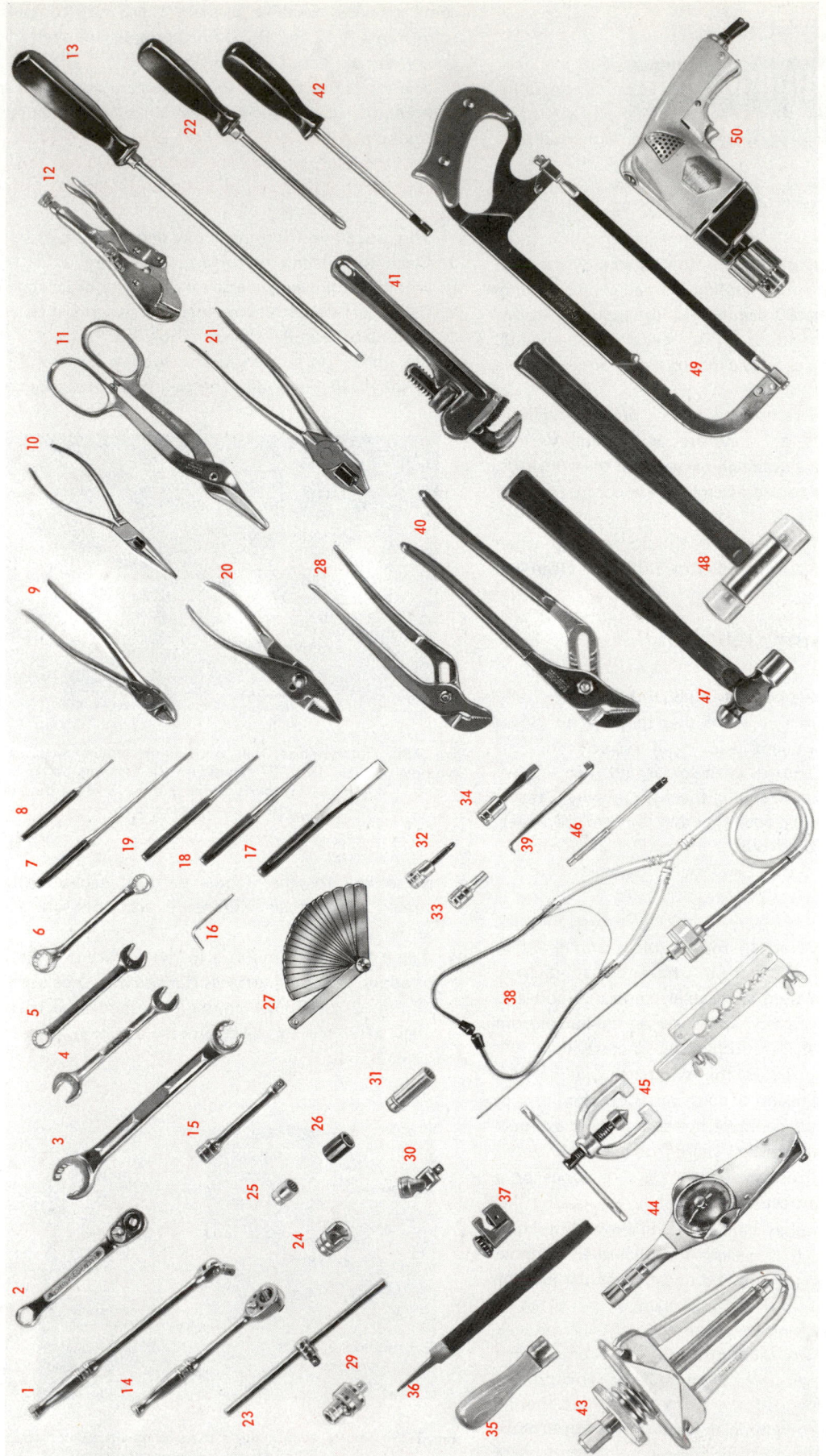

Fig. 2-40. Basic hand tool assortment for refrigeration service. 1—Nut spinner. 2—Wrench, open end. 3—Wrench, refrigeration ratchet. 3—Wrench, flare nut. 4—Wrench, open end. 5—Wrench, double hex, combination. 6—Wrench, box socket – double hex offset. 7—Punch, taper. 8—Punch, pin. 9—Pliers, diagonal cutting. 10—Pliers, needle-nose. 11—Snip, tinner's. 12—Pliers, pinch-off. 13—Screwdriver, standard tip. 14—Handle ratchet. 15—Extension, 4-in. 16—Wrench, hex head (Allen). 17—Chisel, cold. 18—Punch, center. 19—Punch, starter. 20—Pliers, gripping – slip joint. 21—Pliers, lineman. 22—Screwdriver, Phillips. 23—Handle, sliding bar. 24—Socket, Weatherhead. 25—Socket, double hex. 26—Socket, magnetic. 27—Gage, feeler. 28—Pliers, interlocking joint. 29—Adaptor. 30—Universal joint. 31—Socket, double hex deep. 32—Socket, Phillips screwdriver. 33—Socket, clutch screwdriver. 34—Socket, standard screwdriver. 35—File handle. 36—File, half round. 37—Cutter, tube. 38—Stethoscope, mechanic's. 39—Screwdriver, offset. 40—Pliers, large, interlocking joint. 41—Wrench, pipe. 42—Screwdriver, clutch. 43—Puller, two-jaw. 44—Torque wrench, English-metric. 45—Flaring tool. 46—Screw starter. 47—Hammer, ball peen. 48—Hammer, plastic tip. 49—Hacksaw, hand. 50—Drill, electric. (Snap-on Tools Corp.)

2-29 HAND TOOLS

The refrigeration and air conditioning service engineer performs his work chiefly with hand tools. To be successful he must pick good ones, take good care of them and be skilled in their use. Most service failures can be traced to poor hand tool skills.

The refrigerating mechanism, in comparison to an automobile engine, is relatively light. It can easily be damaged by abuse. Great care is necessary to avoid damaging the units.

Fig. 2-40 shows an assortment of basic hand tools needed by the service technician.

The following paragraphs provide useful suggestions for the selection, care and use of hand tools.

2-30 WRENCHES

Most refrigeration and air conditioning installation and service work requires the use of various types of wrenches. Many fasteners and parts are copper or brass and therefore are rather soft. Never use pliers on parts designed to be handled with wrenches.

A service technician needs wrenches of several types and sizes. Wrenches should be made of good alloy steel, properly heat treated, accurately machined and ground to fit the assembly devices. The wrench must fit the nut or bolt head accurately and it must fit as much of the hexagon as possible. For these reasons, the wrench types are listed in the order preferred.

1. Socket wrenches.
2. Box wrenches.
3. Open end wrenches.
4. Adjustable wrenches.

A loose or worn wrench may slip and spoil the corners on nuts. Proper servicing then becomes impossible without replacing the part.

It is important to use wrenches properly so that they fit completely on the nut or bolt. Always pull on a wrench rather than push on it. Otherwise, sudden loosening may cause a serious hand injury.

The socket should be inserted all the way on the nut or bolt head.

Fig. 2-49 shows the proper direction to pull on an adjustable wrench.

Avoid pounding on a wrench to obtain greater turning movement or torque. Avoid using a length of pipe or another wrench for more turning force or torque.

A tight bolt or nut may be loosened safely by soaking the threads with a penetrating oil or by heating the nut or bolt. Some service technicians tap a nut or bolt lightly with a hammer. Any of these methods can be used to loosen corroded threads.

2-31 SOCKET WRENCHES

If the nut or bolt head has enough room around it, the six or twelve-point socket is the best wrench to use. These sockets are usually made of chromium-vanadium steel and are turned

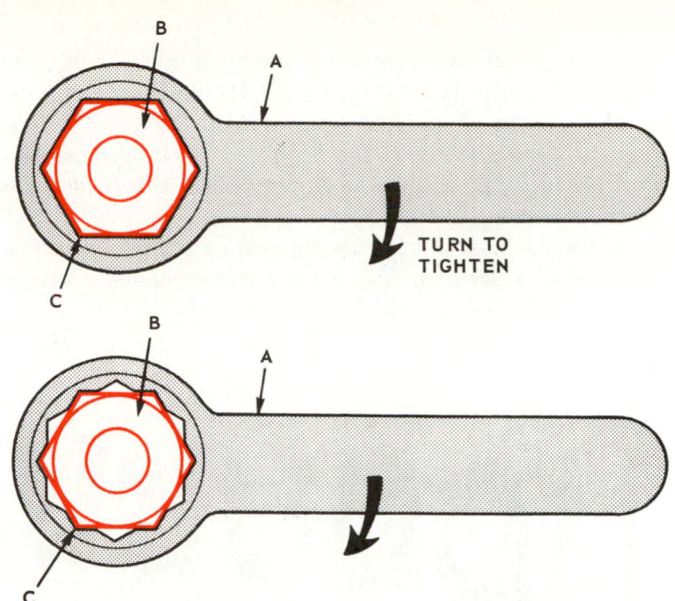

Fig. 2-41. Box end socket wrenches as they appear fitted over hex nuts. Upper wrench is a 6 point box end. Lower is a 12 point box end. A—Wrench handle. B—Nut. C—Contact points (6 contact points).

by handles that have a 1/4, 3/8 or 1/2-in. square drive, Fig. 2-42. The handles come in a variety of designs as shown in the illustration.

Sockets are now available which will hold the nut or cap screw firmly while it is being aligned and threading started. This is especially important where the nut or cap screw will cause problems if it drops out of the socket.

Fig. 2-42. Typical socket wrenches and handles. A—T-handle. B—Ratchet handle. C—Swivel 6-point sockets. D—12-point deep sockets. E—12-point sockets. F—6-point sockets.

Box end and socket wrenches are more usable if they are double broached (12-point). Fig. 2-41 illustrates both the 6-point and the 12-point box end wrench. The 12-point type wrench is easier to use if the handle must be operated in a small or restricted space. The 6-point socket is best for worn hex nuts or bolts.

Metric size nuts and bolts require metric size wrenches. Fig. 2-43 shows a set of metric 6-point sockets commonly used

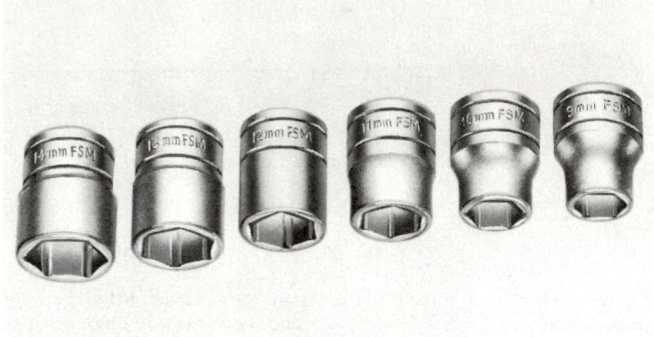

Fig. 2-43. A set of metric size sockets. Note that the size marked on each socket corresponds to the diameter in millimetres (mm) of the bolt or capscrew. Black rings indicate metric sockets. (Snap-on Tools Corp.)

when working with metric size nuts and bolts. The size marked on the socket corresponds to the diameter of the cap screw or bolt and is not the distance across the flats, as is the general practice with common fractional-inch wrenches.

2-32 BOX WRENCHES

When the nuts or bolt heads are in close spaces where one cannot use a socket wrench, the box wrench is satisfactory. Box wrenches are usually 12-point and provide a powerful non-injuring grip on the nut or bolt, Fig. 2-44.

Fig. 2-44. An alloy steel box wrench with 12-point or double hex ends. Ends are offset (double offset) to provide gripping or swinging clearance above mechanism. Next to the socket wrench, box wrenches are safest. They are less likely to slip than open end wrenches.

Box wrenches may be either straight, offset, or double offset. Most box wrenches are double ended. Both ends may be of the same size with one end offset, or they may be of the same pattern and different sizes.

The refrigeration service engineer will find that, for standard bolts and nuts, the table in Fig. 2-45 will give the size of the wrench across the flats for the sizes of bolts and nuts in most common use. Below 1/2 in. bolt size, the wrench size is 3/16 in. larger than the bolt size. A 1/4 in. bolt uses 1/4 +

WRENCH SIZE		
NOMINAL BOLT SIZE	HEAD AND NUT WIDTH ACROSS FLATS	
1/4	7/16	
5/16	1/2	WRENCH IS
3/8	9/16	3/16" LARGER
7/16	5/8	THAN THE BOLT
1/2	3/4	
9/16	13/16	WRENCH IS
5/8	7/8	1/4" LARGER
3/4	1	THAN THE BOLT

Fig. 2-45. Table of wrench openings for standard bolt heads and nuts.

3/16 = 7/16 in. wrench size.

At 1/2 in. bolt size and larger, the wrench size is 1/4 in. larger than the bolt's size. For example, on a 5/8-in. bolt: 5/8 + 1/4 = 5/8 + 2/8 = 7/8 in. wrench size.

The size of the wrench opening across the flats is marked at each wrench opening. Box wrenches having both flat and 15 deg. handles are necessary for a complete tool kit.

2-33 FLARE NUT WRENCHES

Flare nuts used on SAE flared connections require special wrenches. Since the nut is on a fitting connected to tubing, the common box wrench cannot be used. Special flare nut wrenches have been developed for this purpose. An opening permits the box end to slip over the tubing. See Fig. 2-46.

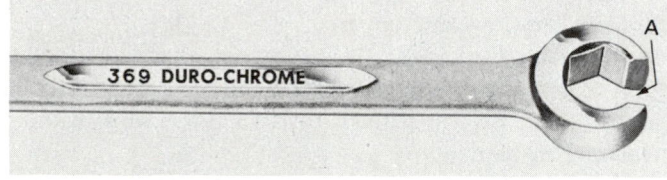

Fig. 2-46. Flare nut wrench used when turning SAE flare nuts. A—Note opening for tubing. (Duro Metal Products Co.)

Since the flare nut wrench is often used in limited space, special wrenches have been devised. Fig. 2-47 illustrates a ratchet type flare nut wrench. A strong, easy-to-operate

Fig. 2-47. Ratchet type flare nut wrench is useful when space for movement is limited. (Tubing Appliance Co., Inc.)

opening type flare nut wrench is shown in Fig. 2-48.

Forged flare nut sizes are an SAE standard used in automotive, marine and refrigeration service. See Para. 27-61 for a table of flare nut wrench sizes.

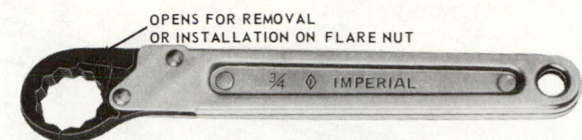

Fig. 2-48. Special type of flare nut wrench opens to pass over the tubing and closes on flare nut to give positive contact. (Imperial-Eastman Corp.)

2-34 OPEN END WRENCHES

Open end wrenches can slide on the nut or bolt head from the side and are used in close spaces on unions and other places where the socket wrench and box wrench cannot be used on the assembly device.

End wrenches should not be used for refrigeration work when the jaws are spread or have burrs. End wrenches used in servicing work should have a thick jaw. Care must be taken when using thin wrench jaws as they have a tendency to bite into soft brass and copper parts.

Popular sizes for open end wrenches are:

1. The 7/16 in. across flats often needed for 1/4-in. screws and bolts.
2. The 1/2 in. across flats for 5/16-in. NC and NF cap screws commonly used on compressors and expansion valves.
3. The 3/4 in. across flats used for 1/4-in. flare nuts.
4. The 1 in. across flats which fit the 1/2-in. flare nuts.

A typical open end wrench, No. 4, is shown in the assortment making up Fig. 2-40.

Another very popular wrench used in refrigeration work is the combination open end and box socket. Both ends are the same size. This wrench is illustrated in Fig. 2-40, No. 5.

2-35 ADJUSTABLE WRENCHES

Wrenches with adjustable jaws, Fig. 2-49, are necessary in the tool kit because of the odd size nuts and bolts often found in this work. Adjustable wrenches must be kept in good repair.

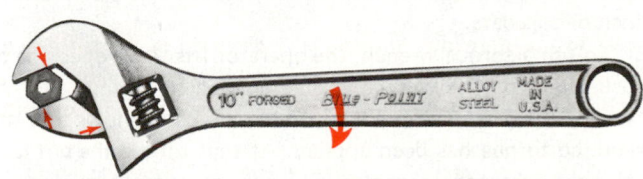

Fig. 2-49. A popular type of adjustable wrench. Handle should be pulled as shown by direction of arrow on handle. Note that wrench is adjusted to fit nut tightly. The red arrows show the pressure of the wrench against the corners of the nut. Turning wrench in the the direction shown tends to press movable jaw against wrench body.

If the wrench does not fit tightly, it may slip and result in a ruined wrench, bruised hand and a ruined nut or bolt head.

The direction of the forces operating on the jaws of the adjustable wrench should be in the direction which will give solid support against both the nut and the body of the wrench, Fig. 2-49.

2-36 PIPE WRENCHES

The pipe wrench is designed to grip pipes, studs and other cylindrical (round) surfaces. The greater the effort applied on the wrench handle, the tighter the wrench will grip the object. Pipe wrenches should not be used on nuts or bolt heads. The typical pipe wrench is pictured in Fig. 2-40, No. 41.

An internal type pipe wrench, Fig. 2-50, may be used for installing pipe, nipples or fittings.

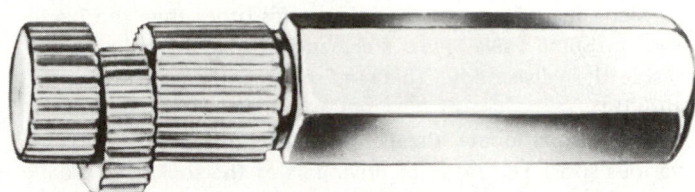

Fig. 2-50. Internal type pipe wrench. It grips the pipe from the inside. (Snap-on Tools Corp.)

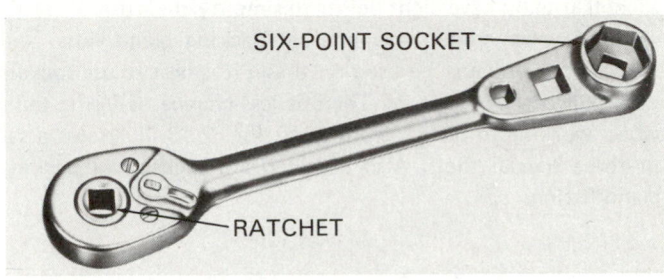

Fig. 2-51. Refrigeration service valve wrench. Fixed end is for "cracking" valves. Ratchet end is for rapid valve stem operation. Left end has 1/4 in. square drive for use with valve stem and packing gland nut sockets. Other openings are 3/16, 1/4 and 5/16 in. square. Six point socket fits 3/8 in. nuts. (Duro Metal Products Co.)

2-37 SERVICE VALVE WRENCHES

Service valve stems are usually constructed with a square end milled on the valve shaft. A special wrench is needed to turn them, Fig. 2-51. This tool, called a service valve wrench, usually has a ratchet and a fixed end for this kind of work.

When "cracking" valves, the fixed end only should be used. By "cracking" is meant, the slight opening required to cause the valve needle or plunger to leave its seat but allow only a very little flow of refrigerant.

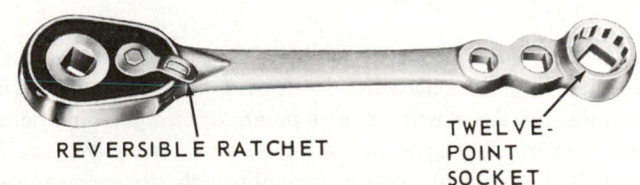

Fig. 2-52. Reversible ratchet wrench. Fixed end has openings for three valve stem sizes, 5/16, 1/4 and 3/16. The 12-point opening is for 5/16-in. nuts and bolts (1/2 in. across the flats). (Duro Metal Products Co.)

Using the fixed end of the wrench gives the service technician quick control of the slight opening and closing of a valve. When rapidly opening and closing valves, the ratchet end may be used.

Some service valve wrenches have a reversible ratchet which enables the operator to reverse the turning without removing the wrench from the stem. Fig. 2-52 is a photograph of such a wrench.

2-38 SERVICE VALVE WRENCH ADAPTORS

Many manufacturers use valve stems other than the 1/4 in. square. Some valve stems are made so that the milled end is inside the valve body. This requires a good socket wrench to turn it.

To accommodate these valves, adaptors are available in various sizes. The male, or drive part of the socket, is usually 1/4 in. square. There are a few which use a larger drive (9/32 in.). The socket or opening which fits the valve stem, comes in five sizes: 3/16 in., 7/32 in., 1/4 in., 5/16 in. and 3/8 in. These sockets usually have eight points to simplify their use.

Most valve stems have internal packing gland nuts, and special sockets must be used on these. It is best to use sockets with ball bearing grippers. There is less chance of losing tools when working in difficult positions. Fig. 2-53 illustrates a set of these special tools. Also included are sockets for packing gland fittings.

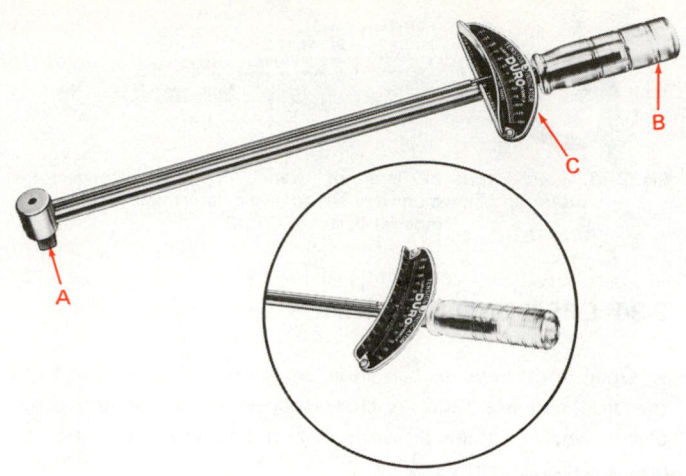

Fig. 2-54. Torque wrench used to measure the amount of tightness of nuts and screws. This wrench is made to be used with standard sockets. A—Socket drive. B—Handle. C—Torque scale.

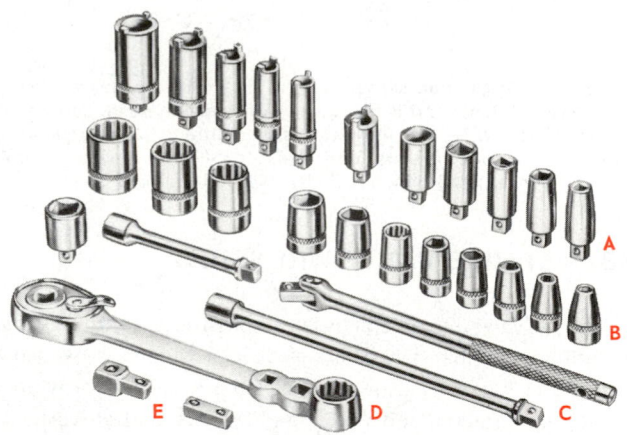

Fig. 2-53. Special service valve set. A—Packing gland sockets and valve stem sockets. B—Variety of 6-point and 12-point sockets. C—Handles. D—Ratchet wrench. E—Adaptors.

2-39 TORQUE WRENCHES

All materials are elastic (will stretch, compress and twist). Even cast iron and hardened steels used in the construction of compressors are elastic up to a point. When tightening bolts, nuts and other attachments on compressor parts and assemblies, it is important that the amount of tightness be measured to avoid warpage or other damage to parts. To measure the amount of tightness, a torque wrench is used, Fig. 2-54. Torque is twisting force.

Torque wrenches are usually wrench handles only and are

made to be used with sockets of different sizes. The handle is equipped with a dial or pointer which measures the foot-pounds or inch-pounds of torque.

One can work out the torque by multiplying the length of the handle in feet by the pull in pounds applied to the handle (foot-pounds). A wrench handle one foot long and pulled by a spring scale reading 50 pounds will produce a torque of 50 foot-pounds.

(Technically, foot-pounds is the wrong term. The correct term should be pounds-feet. The foot-pound is a unit of work. However, popular usage has made the term foot-pound acceptable for the measurement of torque.)

Inch-pounds are calculated by multiplying the length of the handle in inches by the pull on the handle in pounds.

The manufacturers of equipment such as automobiles, airplanes, refrigerating equipment and the like, are able to determine the proper torque (twist) that should be applied to the fasteners on their various mechanisms. The recommended torque for the many parts of refrigerating mechanisms are specified as part of the service data. Factory service manuals include this data.

To use a torque wrench, the operator fits the proper size of socket into it. The wrench is then applied to the nut and the handle of the wrench pulled until the indicator shows that the required torque has been applied. At that torque the nut is at the right tightness recommended by the manufacturer.

2-40 HAMMERS

A hammer is a necessity in the refrigeration shop and the 12 or 16-ounce ball peen hammer is a useful tool. See Fig. 2-40, Part 47. It is important that the hammer head be firmly fastened to the handle and that the handle be in good condition.

To use the hammer, grasp the handle about two-thirds of the way back from the head. For light accurate blows, hold the hammer with the index finger on the top of the handle and use wrist action. For heavy blows, hold the hammer with fingers around the handle and use elbow muscles.

A carpenter's claw hammer may also be needed for mounting pipe supports and fastening sheet metal to wood.

2-41 MALLETS

In refrigeration and air conditioning service work, the mallet is often needed to drive parts into place or to separate them without injury to their surfaces. For such work, a 1 1/2 lb. to 2 lb. mallet is desirable. A rawhide, rubber, wood, plastic or lead mallet should be used. See Fig. 2-40, Part 48.

2-42 PLIERS

Pliers are universal tools. Many different types are available.
1. Gas pliers are slip joint combination pliers, handy for general use. However, they should not be used on nuts, bolts or fittings. They could slip and injure the surface. See Fig. 2-40, Part 20.
2. Cutting pliers are mostly used when working on the wiring of the refrigerator. One type, called the lineman's pliers, is a powerful cutting and gripping tool. Another type called "diagonal" pliers is used to cut in close quarters. See Fig. 2-40, Part 9.
3. Nut pliers are used to good advantage on some jobs. Jaws are parallel and some have an adjustable cam action that locks the jaws on the nut or bolt.

In general, it is not considered good practice to use pliers to hold bolts or nuts. However, on a job such as holding the head of a bolt while turning the nut with a wrench, the use of nut pliers is permissible.
4. Slim nose pliers and duckbill pliers are frequently used in hard-to-reach places. See Fig. 2-40, Part 10.
5. Round nose pliers are used to shape wire into loops and to bend sheet metal edges.

Pliers are made of alloy steel, usually with manganese, although some are chrome-vanadium steel. When of top quality they are usually drop forged. Only those with insulated handles should be used when working on electrical parts.

2-43 SCREWDRIVERS

Screwdrivers are widely used in refrigeration service work both for installation and for shop work. A complete set will be found very necessary. The length of a screwdriver is measured from the blade tip to the handle. Handles are not measured. The recommended average sizes are 2 1/2, 4, 6 and 8 in.

The types of screwdrivers are named for the shape of the end of the blade or bit. See Fig. 2-55. Most popular is the straight blade, slot blade or regular screwdriver. The screwdriver bit should fit the screw slot snugly while the blade should be wide enough to fill the screw slot end-to-end. Also see Fig. 2-40, Part 12.

The Phillips type has a tip which fits a recessed cross in the head of the screw. Phillips screwdrivers are available in four sizes: the 3 in. size for No. 4 and smaller screws; the 4 in. size for No. 5 to No. 9 screws; the 5 in. size for No. 10 to No. 16 screws and the 8 in. size for No. 18 screws and larger.

STANDARD TYPES OF SCREW DRIVER BITS AND SCREW OPENINGS

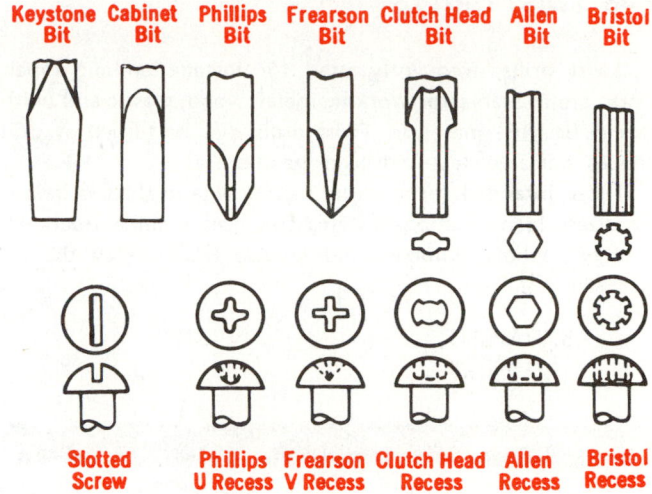

Keystone Bit — Cabinet Bit — Phillips Bit — Frearson Bit — Clutch Head Bit — Allen Bit — Bristol Bit

Slotted Screw — Phillips U Recess — Frearson V Recess — Clutch Head Recess — Allen Recess — Bristol Recess

Fig. 2-55. Several types of screwdrivers. Flat bladed Keystone and Cabinet bits and the Phillips bit are most popular.
(Vaco Products Co.)

Stubby (short) screwdrivers are available for working in small spaces. Some screwdrivers may be equipped with a clip that holds screws while starting them. Better quality screwdrivers have strong handles firmly bonded to the blade. Plastic handles are popular.

An offset screwdriver is necessary in refrigeration work. There are many places where it alone can be used.

Never use a hammer to pound on a screwdriver. If a screwdriver is needed for heavy service, use one with a solid steel handle.

2-44 STAMPS

It is good practice for refrigeration service technicians to stamp their names and the date on units sold or serviced. This often eliminates arguments. Many companies have a code system allowing their own employees to interpret the information stamped on the unit.

This stamping is done with hardened steel stamps which can be purchased in a variety of sizes, letters, figures or symbols. One-eighth in. letters are a popular size. Such stamps are not suitable for use on hardened materials and tools, however. An electric marking tool can be used in these cases.

2-45 VISES

Sturdy machinist's vises are necessary in the shop. They are particularly convenient for holding parts during drilling, filing or assembling. Also useful is a pipe vise.

One vise should be large enough to hold most compressor bodies. Useful for a large shop, is a special pipe vise that has a hacksaw blade slot for accurate cutting of pipe and tubing.

Always use soft jaws made of sheet copper or aluminum

when clamping a part which must not be marred. These are available as inserts which fit over regular vise jaws.

2-46 TWIST DRILLS

Twist drills, frequently used for installation and repair work, are available for working metal, wood, plastic and (with special designs) masonry. Twist drills may be turned by drill presses, portable electric drills or hand braces.

Those intended for working metal come in three different set sizes. Identification systems for sizes include fractional numbers, whole numbers and letters. Sizes go by the bit diameter.

STRAIGHT SHANK   FLUTE

Fig. 2-56. Straight shank twist drill for use on metal. (Cleveland Twist Drill Co.)

Usually, twist drills are of the straight shank type. This means that the section gripped by a three-jaw chuck is straight and perfectly round (cylindrical) in shape. See Fig. 2-56.

The shank carries a stamped identification giving the kind and size of the drill. Depending on quality and use, twist drills may be made from either high carbon steel or alloy steel (HSS) for high speed use.

Fractional sizes come in sets usually beginning with 1/16 in. and going up to 1/2 in. in steps of 1/64 in. Larger fractional sizes are also available.

Numbered sets begin with No. 1 and go up through No. 80 (.228—.0135 in.). However, most commonly used are No. 1 through No. 60. The higher the number, the smaller the drill.

Letter size twist drills are larger than 1/4 in. diameter and vary from .234 for the "A" to .413 for the "Z" drill.

Note that the numbered twist drills cover a range of sizes approximately .013 through 1/4 in. Letter sizes range from approximately 1/4 in. to nearly 1/2 in. These two twist drill sets are often used as tap drills in making holes for inside threads. They provide a greater range of sizes than the fractional twist drills. For a table of various drill diameters, see Chapter 28.

Speed of drilling depends upon the type of material being drilled and the diameter of the hole. In general, the smaller the twist drill the faster it should be turned.

Most twist drills have two cutting edges or "lips." These edges must be sharp, equal in length and must have clearance and a rake angle. See Fig. 2-57.

Twist drills have flutes which remove chips from the hole. Most flutes are spiraled at an angle which automatically provides a rake angle for the cutting edges.

Always be sure the drill is cutting when it is being used. If the cutting edges are just rubbing against the stock, they will quickly heat up, destroying the hardness of the drill.

To insure that the drill forms the correct size hole, both cutting lips must be exactly the same length and angle. See

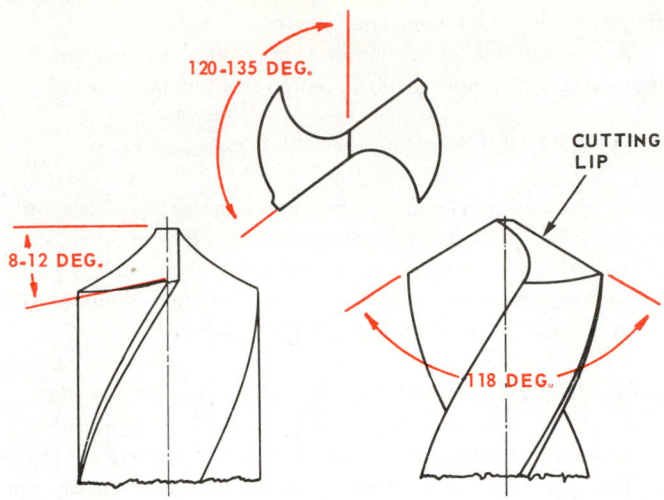

Fig. 2-57. Correctly ground twist drill point for steel. Clearance angle shown - 8-12 deg. - is rake angle for mild steel and cast iron for drills in 1/2 in. range. As diameters are reduced, clearance angles increase. A 1/16 in. diameter twist drill should have a clearance angle of about 20 deg. (Cleveland Twist Drill Co.)

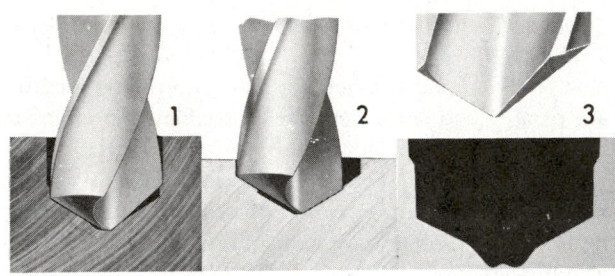

Fig. 2-58. An illustration of what happens when a drill is incorrectly sharpened. 1—Lips are equal in length but at different angles. 2—Lips are at equal angles but are of different lengths. 3—Lips are at different angles and at different lengths.

Fig. 2-58. If one lip is longer, the hole being drilled will be oversize; or, if one lip has a smaller angle, it will do all the cutting and soon grow dull.

Always wear safety glasses to protect eyes from flying chips when using either a drill press or portable drill. See Fig. 26-78.

Electric drills should be grounded for safety. The metal frame of the drill should be electrically connected to a good ground (water pipe or a ground rod). Most electric drills are equipped with a three-prong grounded plug. If the circuit to which the drill is connected is not provided with a three-prong grounded socket, a grounded adaptor should be used.

2-47 TAPS

Many assemblies of metal parts are fastened with machine screws or tap screws threaded into tapped holes.

A tap is used for making threads inside a hole. Taps are of hard alloy steel accurately made with clearance pockets provided for chips. The threads are made with small clearance to provide good cutting.

There are taps for every size or diameter thread and also for each kind of thread — National Fine (NF), National Coarse

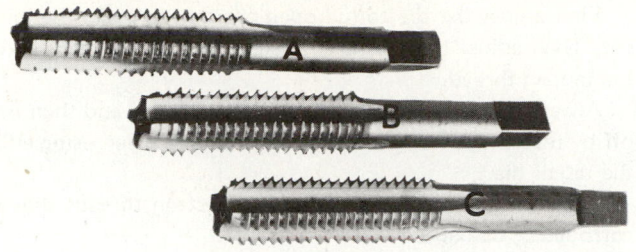

Fig. 2-59. Set of taps for cutting 1/2 in. NC threads. A—Taper tap. B—Plug tap. C—Bottoming tap.

(NC), American Standard Taper Pipe Thread (ASA), or metric. Taps are of three types: taper, plug and bottoming. Most common is the plug type. See Fig. 2-59.

Taper taps are used for starting a cut and for tapping thin pieces in which the tapped hole goes all the way through. Plug types are used to do most of the cutting in blind holes, while bottoming taps are used to cut full threads to the bottom of a blind hole.

The shank of the tap is ground to a square at the end, and a tap wrench is used to turn it. Power tools may also be used for driving. However, a special tap-driving accessory must also be used.

Because tapping is basically a cutting operation, the general rules for cutting metals apply. Most taps have four cutting edges for each thread. These edges must be kept sharp. They

TAP	TAP DRILL	TAP	TAP DRILL
4/36	No. 43	14-20	No. 9
	No. 44		No. 10
	No. 45		No. 11
4-40	3/32	14-24	No. 6
	No. 43		No. 7
	No. 44		No. 8
4-48	No. 41	1/4-20	No. 5
	No. 42		No. 6
5-40	No. 37		13/64
	No. 38		No. 7
	No. 39		No. 8
5-44	No. 36	1/4-28	7/32
	No. 37		No. 3
	No. 38	5/16-18	17/64
6-32	No. 33		G
	No. 34		F
	7/64	5/64-24	J
	No. 36		I
6-40	No. 32	3/8-16	O
	No. 33		5/16
8-32	No. 29	3/8-24	R
8-36	No. 28		Q
	No. 29	7/16-14	3/8
10-24	No. 24		U
	No. 25	7/16-20	25/64
	No. 26		W
10-32	No. 19	1/2-13	27/64
	No. 20	1/2-20	29/64
	No. 21	9/16-12	31/64
	No. 22	9/16-18	33/64
12-24	No. 15	5/8-11	17/32
	No. 16	5/8-18	37/64
	No. 17	3/4-10	21/32
12-28	3/16	3/4-16	11/16
	No. 13	A	B
	No. 14		
	No. 15		

Fig. 2-61. Tap drill sizes recommended for common tapping operations. Note that for certain sizes the tap drill may be a fractional size, a number size or a letter size drill bit. A—Outside diameter. B—Number of threads per inch.

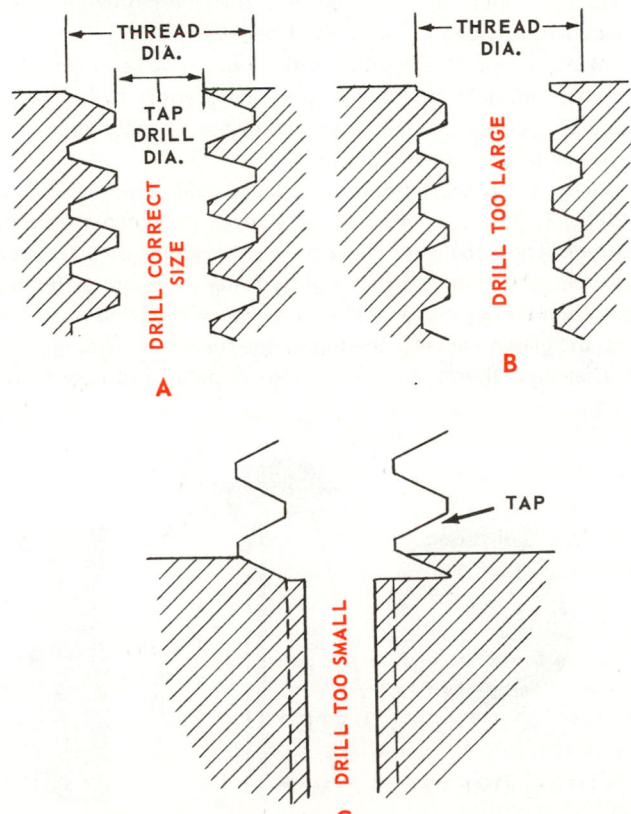

Fig. 2-60. Hole size is important when tapping threads in metal. A—Tap drill correct size, correct thread depth. B—Tap drill too large, threads not full depth. C—Tap drill too small, tap likely to break.

must have a ground cutting face and cutting clearance.

Always use a special thread cutting lubricant when doing any kind of threading. The single exception is gray cast iron, which contains enough graphite to provide necessary lubrication. Thread cutting lubricants, if applied generously, also serve as a coolant.

2-48 TAP-DRILL SIZES

It is very important that the hole to be tapped is first drilled to the correct size. If the hole is oversize, threads will not be full size. If the hole is undersize, the tap must remove too much metal and will probably break. See Fig. 2-60.

The tap drill should be slightly larger in diameter than the root diameter of the threads for which the hole is being drilled. Generally speaking, threads 75 percent of full size are considered satisfactory. Always refer to tap-drill size tables for the correct size drill. For most refrigeration and air conditioning work, the tap-drill table, Fig. 2-61, will be satisfactory.

2-49 DIES

Dies cut external threads on round stock. The threads match those cut by the tap. Because a tap is nonadjustable, dies can usually be adjusted to permit careful matching of the threads. Dies are held and turned by a diestock, Fig. 2-62.

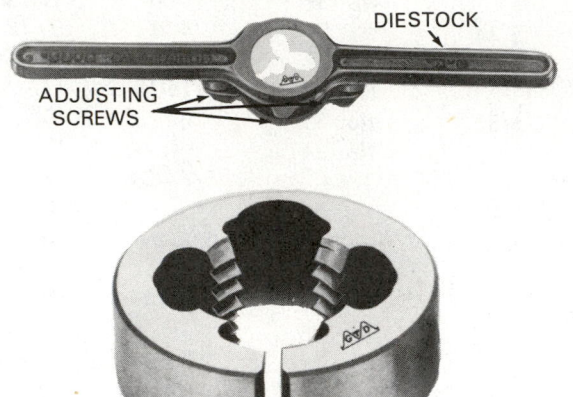

Fig. 2-62. Adjustable diestock has three screws. Two hold die and apply contracting pressure. Third expands die at split.
(TRW Greenfield Tap and Die Div.)

As with taps, there are dies for each type of thread and size. Because they are cutting tools, they, like taps, must be made of tool steel. They, too, must be carefully shaped to cut correctly.

Special precautions should be taken to start straight threading. Guides are available for this purpose. See Fig. 2-63.

Round stock must also be accurately sized. Even if only a few thousandths of an inch oversize, it might break the die.

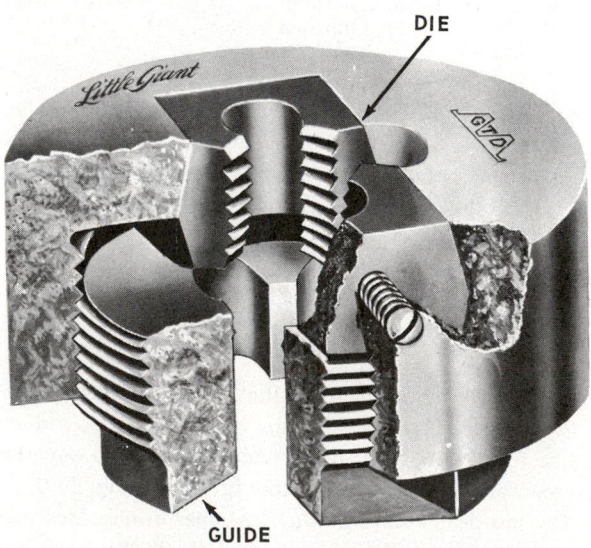

Fig. 2-63. Section through threading die showing guide in place.

First adjust the die to full open position to make the initial cut. Then adjust the die to cut deeper until the threads match the tapped thread.

Always advance the die a quarter to half turn and then back off by reversing the direction of the diestock when using either the tap or die.

Taps and dies may also be used to clean threads that are corroded or damaged.

2-50 INSTRUMENTS AND GAUGES

The refrigeration service technician must use instruments and gauges to determine conditions (pressure and temperature) inside the operating mechanism. The most common instruments are thermometers, pressure gauges and vacuum gauges. Later chapters will treat special instruments such as recording thermometers, hygrometers, ammeters, voltmeters, ohmmeters and others.

Instruments must be kept in good condition and carefully handled if they are to remain accurate. If accuracy is doubtful, the instrument should be sent to repair company for testing and calibration.

2-51 THERMOMETERS

The thermometer records the temperature of the evaporator, refrigerator cabinet, liquid line, suction line, and condensing unit. An ice and water bath may be used to determine its accuracy. When in this solution, the thermometer should check within 1 deg. F. of 32 F. (1 deg. C. of 0 C.).

Many sizes and types of thermometers have been developed for the refrigeration service and installation technician. A popular type has the glass stem mounted in a metal case and is fitted with a pocket clip as in Fig. 1-2.

Glass stem thermometers usually read from −30 F. to 120 F. (−35 C. to 49 C.), in 2 deg. increments (marked spaces). The tube may contain either mercury or a red fluid. The mercury-filled thermometer is faster but more difficult to read. Some thermometers have a special magnifying front built into the glass to enlarge the liquid line for easier reading.

Dial-stem thermometers are also popular and easy to use.

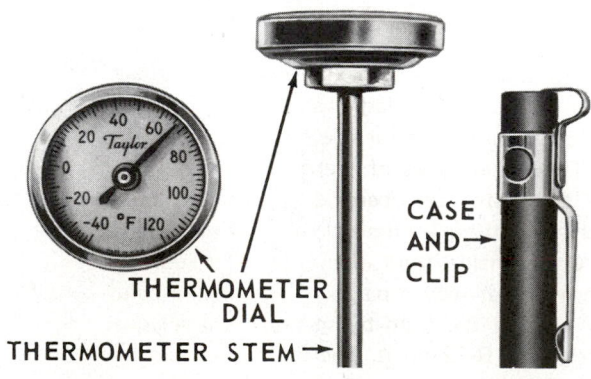

Fig. 2-64. Dial-stem thermometer is calibrated in 2-degree increments from −40 F. to +120 F. This is the temperature range most used by technicians in refrigeration and air conditioning work.
(Taylor Instrument/Consumer Industrial Products, Sybron Corp.)

The one shown in Fig. 2-64 is typical. It may be operated either by a bimetal strip or by a bellows charged with a volatile (vaporizes readily) fluid. Some dial thermometers have a remote sensing bulb connected to the bellows by means of a capillary tube, Fig. 2-65.

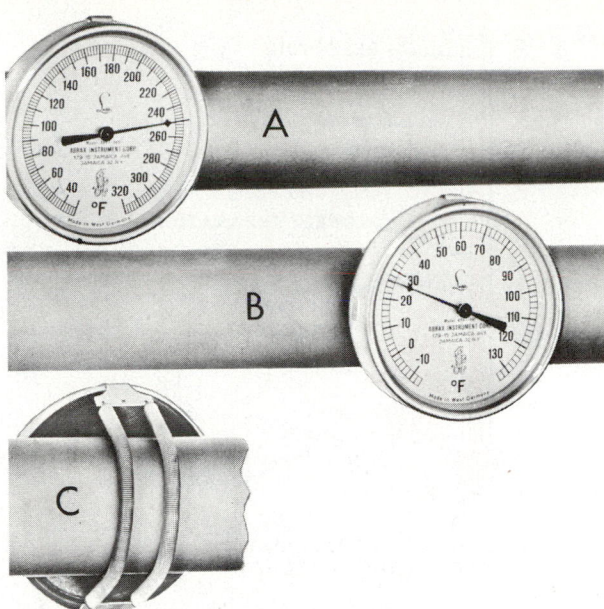

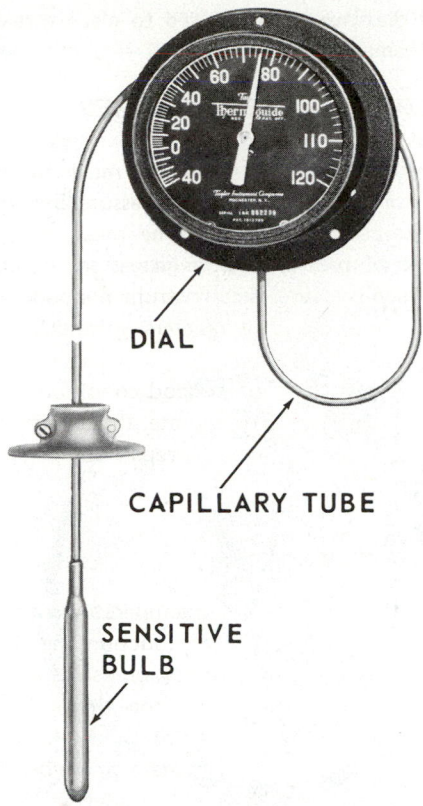

Fig. 2-65. Dial type thermometer with remote temperature sensing bulb. Range is from −40 F. to +120 F.

Fig. 2-66. Dial thermometers are easily clamped to pipes to indicate temperature of the pipe. A—Shows calibration from 40 F. to 320 F. by two-degree increments. Temperature of 250 F. indicates this thermometer is attached to steam or hot water pipe. B—Calibration is from −10 F. to 130 F. by two-degree increments. Temperature (26 F.) indicates thermometer is attached to evaporator suction line. C—Spring arrangement for attaching thermometer to pipe.

The temperature range for this instrument varies, but is usually from −40 to 120 F. (−40 C. to 49 C.) in 2 deg. increments. When using any kind of thermometer, never expose it to temperatures beyond the limits of the scale. To do so may ruin the instrument.

Dial thermometers are very useful for taking pipe temperatures. Fig. 2-66 illustrates two of these thermometers and shows how the instrument may be clamped to pipes or tubes to check temperatures.

Fig. 2-67 shows a maximum and minimum thermometer. This type is particularly useful when attached to a system which is unattended for a few cycles.

Two other types of thermometers, the thermocouple and the thermistor, are operated by electrical current. The fundamentals of these are described in Chapter 6.

Occasionally, fluid in the liquid column of the glass-stem thermometer may separate. To make the column solid again, try cooling the bulb by spraying a small quantity of liquid refrigerant R-12 on it. The column will shrink into the bulb, and when it re-expands, the break should have disappeared.

CAUTION: If the mercury is frozen into a solid, the thermometer will break.

Another way to re-connect the column is to heat the

thermometer as in Fig. 2-68. Heat the thermometer very, very slowly. Do not allow the fluid to reach the top end of the stem. If overheated, the thermometer will burst. Wear goggles.

Fig. 2-67. Maximum and minimum thermometer. A—Hand indicating highest temperature reached. B—Hand indicating present temperature. C—Hand registering lowest temperature reached.
(Weksler Instruments Corp.)

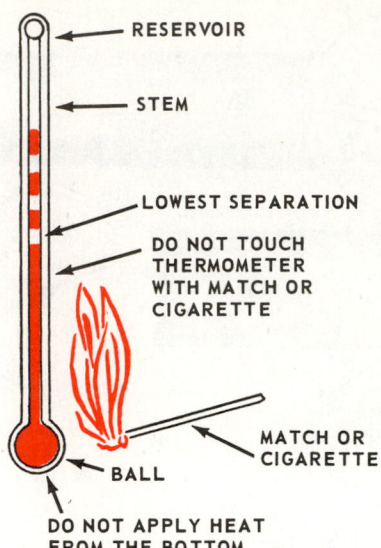

Fig. 2-68. Using heat to connect a break in the liquid column of a glass stem thermometer. Be careful. Wear goggles. Avoid putting match below bulb. Keep flame moving to avoid hot spots. (White-Rodgers Div., Emerson Electric Co.)

2-52 PRESSURE GAUGES

Pressure gauges are used by the technician to help determine what is happening inside the system.

Gauges use a Bourdon tube as the operating element. The Bourdon tube is a flattened metal tube (usually copper alloy) sealed at one end, curved and soldered to the gauge fitting at the other end. This construction is shown in Fig. 2-69.

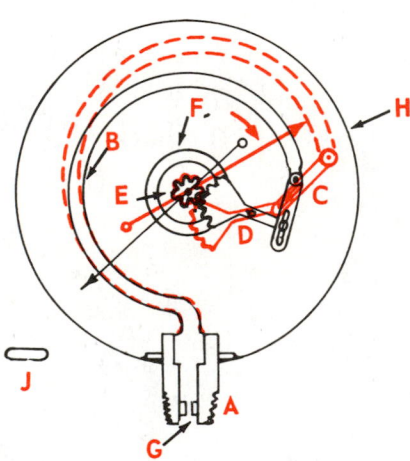

Fig. 2-69. Internal construction of pressure gauge. A—Adaptor fitting, usually 1/8 in. pipe thread. B—Bourdon tube. C—Link. D—Gear sector. E—Pointer shaft gear. F—Calibrating spring. G—Restrictor. H—Case. J—Cross-section of Bourdon tube. Red line indicates how pressure in Bourdon tube causes it to straighten and operate gauge.

A pressure rise in a Bourdon tube makes it tend to straighten. This movement will pull on the link, which will turn the gear sector counterclockwise. The pointer shaft will then turn clockwise to move the needle.

Some gauges have a retarder to permit accurate readings in the usual operating range. The retarder uses an extra spring at pressures above normal. These gauges are easily recognized by the change in graduations at the higher readings of the positive pressure scale.

Most popular gauges have a 2 1/2-in. dial and are connected into the refrigerating system with a 1/8-in. male pipe thread. Some gauges, however, have a 1/8-in. female pipe thread.

It is advisable to use a 1/8-in. pipe nipple on any gauge, as the continual installing and removing quickly wears the threads. Maximum pressure at which a gauge should be continuously used should be no greater than 75 percent of the full-scale range.

Three types of pressure gauges are used in refrigeration service work: high-pressure, vacuum and compound gauge.

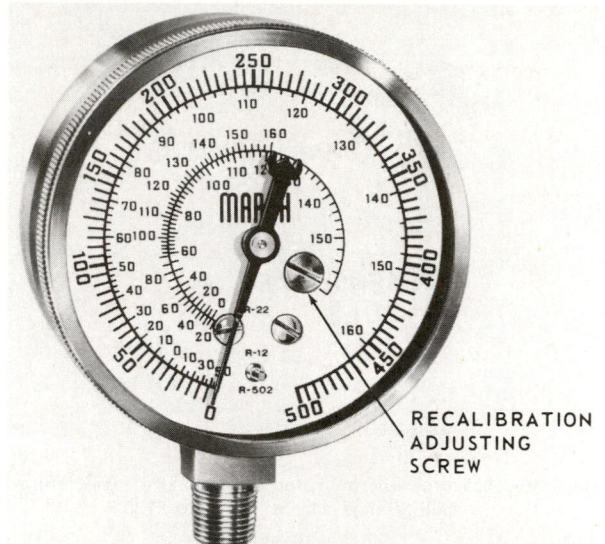

A

B

Fig. 2-70. High-pressure gauges. A—Gauge calibrated in English units. Note pressure-temperature scales for refrigerants R-502, R-12 and R-22. (Marsh Instrument Co.) B—Calibrated in both English and metric units. (AMETEK/U.S. Gauge)

2-53 HIGH-PRESSURE GAUGES

The high-pressure gauge has a single continuous scale, usually calibrated (marked off) to read from 0 to 500 psi. The scale is usually in either 2-lb. or 5-lb. increments and is usually connected into the high-pressure side of the refrigerating mechanism. Fig. 2-70 shows typical gauges.

The one at A is calibrated in English units. The gauge in B is calibrated both in English and metric units.

2-54 VACUUM GAUGES

The vacuum gauge measures lower-than-atmosphere pressure. It will have one of four calibrations: inches of mercury (Hg.); pounds per square inch absolute (psia); millimetres of mercury (mmHg.); or, for very high vacuum, torrs or microns. The micron is explained in Chapter 11.

Generally, the mercury barometer, Fig 1-14, measures vacuum in the normal ranges for refrigeration work. For measurement of very high vacuums, a special instrument, the McLeod gauge, Fig. 2-71, is usually used. Such instruments are calibrated in millimetres of mercury (torr). See Para. 1-17 for

definition of a torr. The vacuum calibration on the compound gauge (inches of Hg.) is most used in refrigeration work for measuring pressures below atmospheric.

For very high vacuums, the thermocouple gauge, Fig. 11-108, should be used. It is accurate between 1 and 1000 microns. A mercury manometer is accurate from 1000 microns and above.

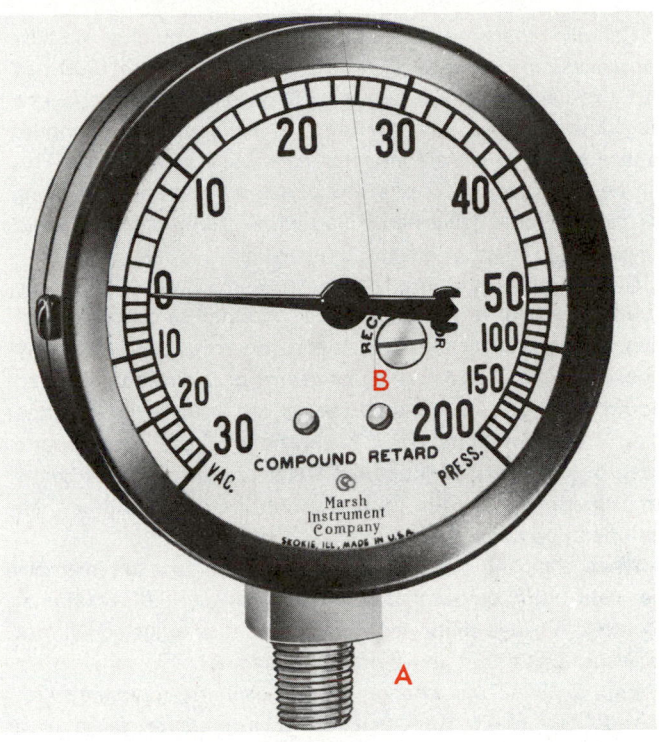

Fig. 2-72. Compound gauge with scale of 30″ Hg. vacuum to 0 psi to 200 psi. Note that scale is greatly shortened (retarded) between 50 psi and 200 psi. This is accomplished by use of retarder spring. A—1/8 in. npt connection. B—Calibration adjustment.   (Marsh Instrument Co.)

2-55 COMPOUND GAUGES

The compound gauge, Fig. 2-72, measures both pressure and vacuum. It is usually calibrated from 0 to 30″ Hg., and from 0 to 200 psi.

Some compound gauges have scales calibrated according to the evaporating temperature of various refrigerants, such as R-12, R-22, R-502 and the like. With this extra scale it is unnecessary to refer to pressure-temperature tables and curves for common refrigerants in order to check for correct pressure-temperature relationships.

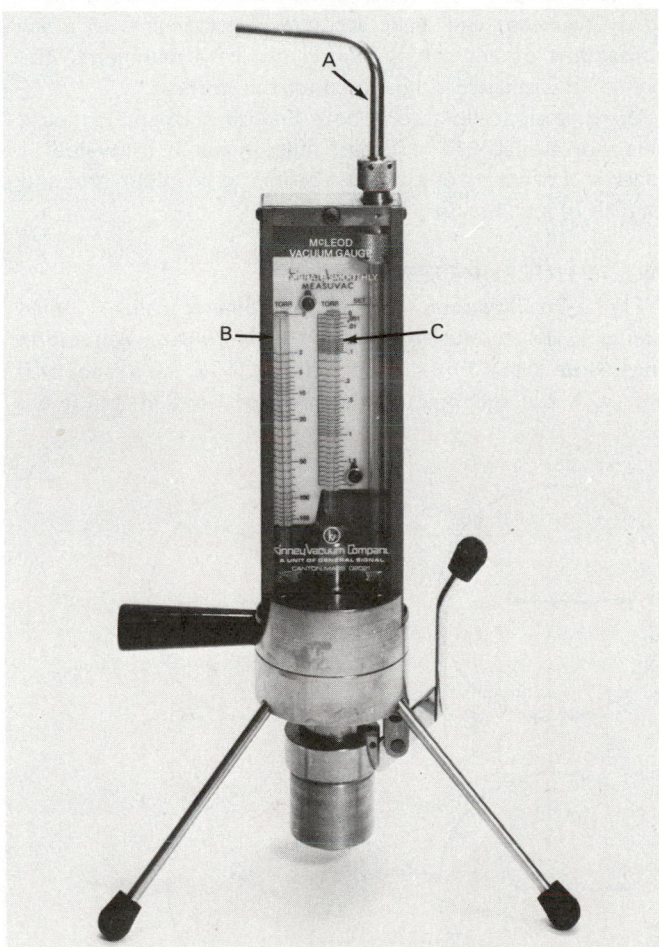

Fig. 2-71. A very high vacuum gauge using the McLeod principle. It can measure from 150 torr (150 mm) to 1 millitorr. A—Connecting tube. B—0-150 mm Hg. scale. C—0-20 mm Hg. scale. (Kinney Vacuum Co.)

ELEVATION	PRESSURE
Sea Level	14.7 psi
2,000 ft.	13.7 psi
4,000 ft.	12.9 psi
5,000 ft.	12.2 psi

Fig. 2-73. Atmospheric pressure change with altitude.

A 30-inch Hg.-0-200 psi gauge should be used when connecting to a refrigerating mechanism in which the high pressure may back up through the compressor or balance through the refrigerant control while the compressor is stopped. Never use compound gauges continuously on the high pressure side of the system.

2-56 CARE AND CALIBRATION OF GAUGES

Rapidly changing fluctuating pressures quickly destroy gauge accuracy. A sudden release of high pressure (300 psi) into a gauge also may injure it. If it is necessary to connect a gauge into a rapidly fluctuating pressure condition, it should be attached through a connector having a very tiny bore. This will help to dampen (choke) the pressure fluctuations entering the gauge. Some gauges are filled with liquid, which tends to prevent rapid fluctuations in the instrument.

Gauges that are used in refrigeration work must be accurate. When it is time for a periodic re-calibration of gauges, instruments are available to do this accurately. Any shop that uses a large number of gauges or is remotely situated should have one. Such instruments are usually constructed using "dead weights" for calibrations above atmospheric pressure and a mercury column indicator for pressures below atmospheric or vacuum. Gauges should be checked over their full operating range or scale.

When checking gauge accuracy it is important to remember that calibrating equipment is made to show a "0" reading at sea level. A gauge calibrated on equipment so adjusted will not be accurate at either above or below sea level.

Fig. 2-73 shows change in atmospheric pressure with altitude. To make sure that either a pressure gauge or a vacuum gauge is calibrated for elevation:

1. For a pressure gauge, set the needle at 0 when the gauge is unconnected.
2. A vacuum gauge, the needle should rest at 0 when the instrument is not connected to any vacuum source.

3. A pressure gauge or vacuum gauge should be adjusted to 0 for the location at which the gauge is being used.

The refrigerating machine pressures recorded by the gauge will then be accurate enough for the service technician to use.

A compound gauge is accurate to about 28 in. vacuum; a mercury manometer to 1 mm of mercury (1000 microns).

There are many dial scales in use; some common ones are shown in Fig. 2-74. One must be careful when reading a pressure gauge to use the correct scale spacings and values.

To speed installation of gauges, one may use a quick coupler system, as in Fig. 2-75.

2-57 MEASURING RULES AND TAPES

A 9 in. or 12 in. steel rule is frequently needed when overhauling or installing units. The rule should be graduated in 1/32 in. and, if possible, should be stainless steel to avoid rusting. Numerals and graduations should be clearly visible. Installation workers will find a 6 ft. flexible steel tape useful when laying out a job.

2-58 MICROMETERS

Often, refrigeration service technicians must inspect, check sizes, dimensions and make accurate measurements to a few thousandths of an inch or hundredths of a millimetre. The common micrometer is most used for this purpose.

Micrometers, calibrated in both English units and in metric units, are available. The English micrometer is calibrated in inches and decimals of an inch; the metric in millimetres and decimals of a millimetre.

ENGLISH MICROMETERS

Fig. 2-76 illustrates a 1 in. micrometer caliper. When reading it, be careful to observe the micrometer calibration range. Note that a 1 in. micrometer, Fig. 2-76, has a range of 0 to 1 in. A 2 in. micrometer has a range of 1 to 2 in. and so on.

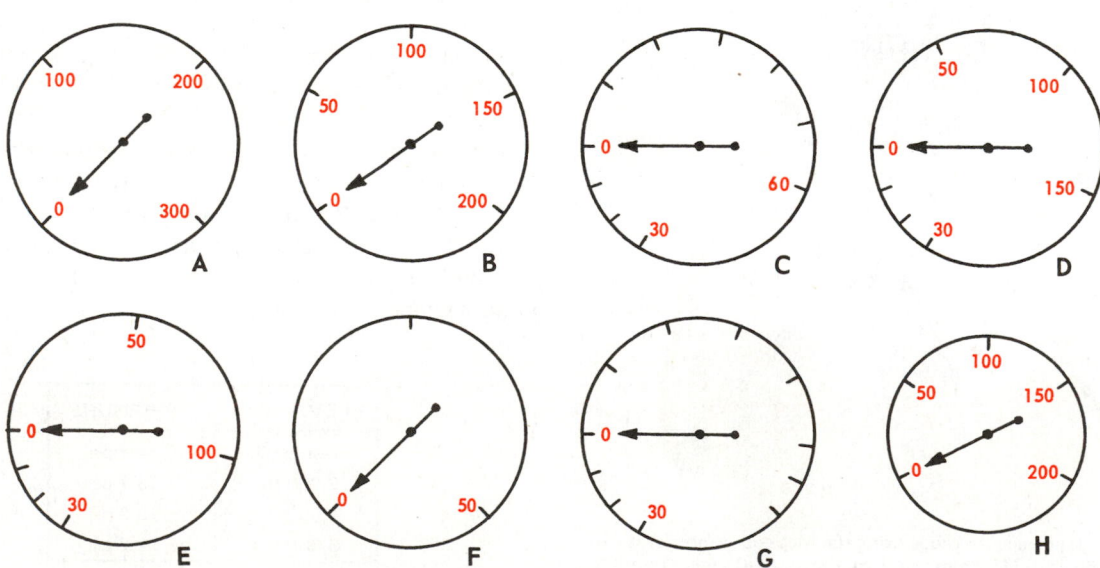

Fig. 2-74. Some common pressure gauge dials. A, B, F and H are pressure gauges. C, D, E and G are compound gauges.

ATMOSPHERIC

120 psi

NIPPLE

QUICK COUPLER

Fig. 2-75. Compound refrigeration gauge shows evaporating temperature for refrigerant R-12 and R-22 corresponding to gauge pressure. Gauge reads from 30 in. vacuum to 0 psi to 120 psi. On these gauges the 30 in. vacuum and the 120 psi reading are on same spot on dial. However, hand must travel a complete turn clockwise from the 30 in. to reach 120 psi reading. (Robinair Mfg. Corp.)

If a 2 in. micrometer is being used, 1 in. must be added to the reading. If it is a 3 in. micrometer, 2 in. must be added to the micrometer reading to compensate for the two-inch gap left uncalibrated on the micrometer.

The following points may help the beginner learn to read the micrometer.

1. The divisions on the sleeve are tenths of an inch.

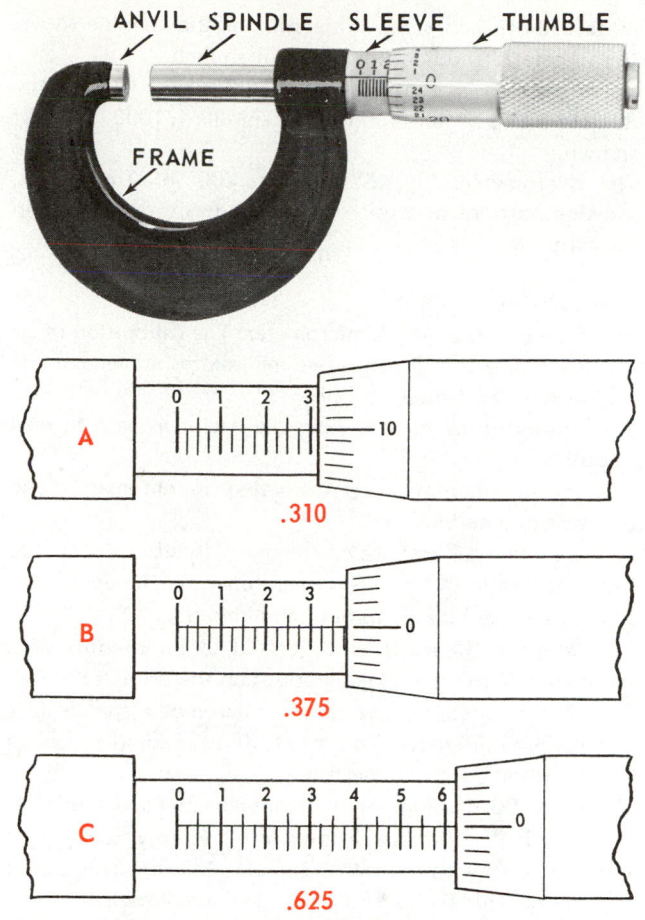

ANVIL SPINDLE SLEEVE THIMBLE

FRAME

A   .310

B   .375

C   .625

Fig. 2-76. Reading a micrometer caliper. Can you tell why the reading in the photograph is .200? Note drawings A, B and C and see how readings are made. Answers (in red) are in thousandths of an inch.

2. The four divisions between the tenths markers are 1/40 in. or .025 (twenty-five thousandths of an inch).

3. The thimble (right hand end of the micrometer) carries the spindle and is threaded into the micrometer body on a

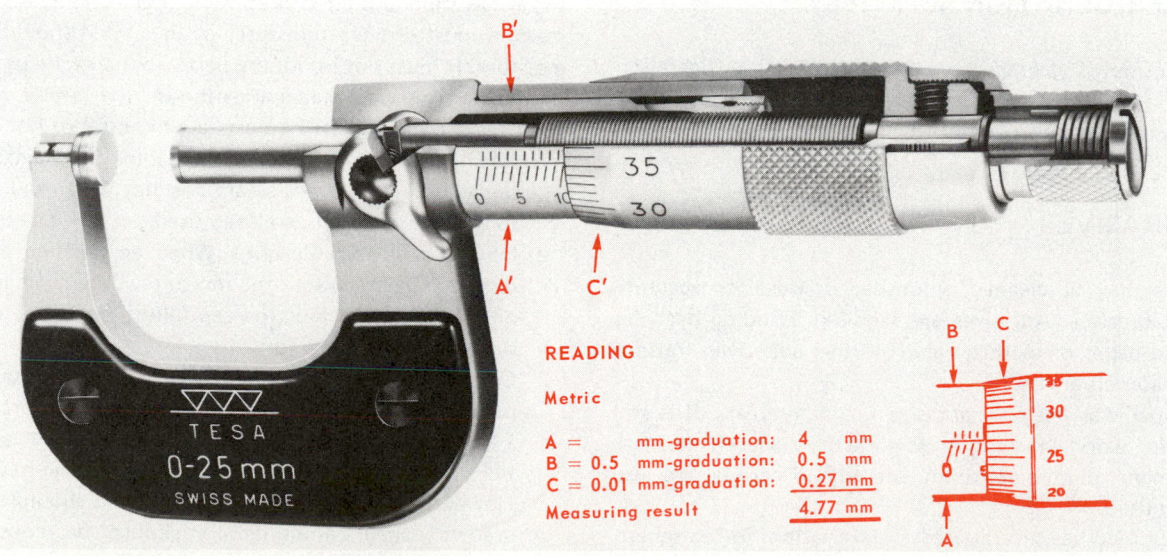

B'

A'   C'

TESA
0-25mm
SWISS MADE

READING

Metric

A =   mm-graduation: 4   mm
B = 0.5   mm-graduation: 0.5   mm
C = 0.01 mm-graduation: 0.27 mm
Measuring result   4.77 mm

B   C

A

Fig. 2-77. A 0-25 mm micrometer. The reading, as shown in the insert, is 4.77 mm. Scales on the photograph A[1], B[1], C[1], show 10.33 mm. (TESA, S.A.)

40-thread-per-inch screw. One turn of the thimble moves the spindle 25 thousandths (.025) of an inch.

4. There are 25 graduations on the thimble. Turning the thimble one graduation moves the spindle 1/1000 (.001) of an inch.

5. The micrometer, Fig. 2-76, reads .200 in. The inserts, showing portions of a micrometer, read as shown in red in the caption.

METRIC MICROMETERS

Fig. 2-77 shows a metric micrometer. The calibration range is 0 to 25 millimetres. Twenty-five millimetres are almost one inch (1 inch = 25.4 mm).

The following points may help the beginner learn to read the metric scale:

1. The metric micrometer is calibrated in millimetres and decimals of a millimetre.

2. There are two calibrations on the sleeve (see Fig. 2-77). The lower calibration (A scale) is in millimetres. The upper calibration (B scale) is half (0.5) of a millimetre.

3. The thimble (C scale) is calibrated in hundredths of a millimetre. There are 50 graduations on the thimble.

4. The thread pitch on the metric micrometer spindle is 2 threads per millimetre. This means that the spindle moves a half (0.5) millimetre for each turn.

5. There are 50 divisions on the thimble. Since the thimble moves a half millimetre in one turn, one division on the thimble (1/50) moves the spindle 1/2 of 1/50 of a millimetre. This equals 1/100 (0.01) of a millimetre.

6. When reading a metric micrometer, it is not necessary to observe the micrometer calibration range. The lower calibration on the sleeve starts with a millimetre reading corresponding to the gap between the anvil and the spindle. As an example, the 0 to 25 mm range starts with 0 on the sleeve; the 25 to 50 mm range starts with 25 on the sleeve; and the 50 to 75 mm range starts with the number 50 on the sleeve.

2-59 REFRIGERATION SUPPLIES

Space prevents giving specifications for all refrigeration supplies. However, a few of the basic items will be named and some information given about their specifications, handling and use.

2-60 ABRASIVES

Surfaces may be cleaned, smoothed or made to accurate size with abrasives. Abrasives are sand-like grinding particles attached to paper or cloth by glue or other adhesives. Various abrasive materials are used.

Sandpaper was the only abrasive for many years. It is still excellent for wood finishing or where a dry surface is wanted. Today, emery, aluminum oxide and silicon carbide are also commonly used.

Each abrasive has several grades or variations in coarseness. Emery cloth: 0000 (finest), 000 (extra fine), 00 (very fine), 0 (fine), 1/2 (medium fine) and 1 (medium).

Silicon carbide: 500 (finest), 360 (very fine), 320 (fine), 220 (medium fine) and 180 (medium).

Aluminum oxides: 320 (extra fine), 240 (fine), 150 (medium fine) and 100 (medium).

These abrasives come in 9 by 11 in. sheets or in rolls usually 1 in. wide.

Sheet abrasives, whether paper or cloth, should be backed by a block of wood, metal, felt or rubber. Special sanding blocks may also be used. Always use clean abrasive paper.

2-61 BRUSHES

A clean steel wire brush is an excellent tool to prepare copper and steel surfaces for welding or brazing. These brushes can be bought in a variety of shapes and sizes. They should have fine steel wire bristles, thickly set and the handle should be comfortable. Special cylindrical brushes, available in all sizes are good for cleaning outside and inside surfaces of tubing and fittings. See Fig. 2-78.

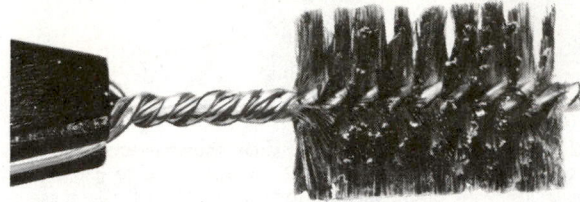

Fig. 2-78. Wire brush used for cleaning inside surfaces of fittings before a soldering or brazing operation. (Schaefer Brush Mfg. Co., Inc.)

2-62 CLEANING SOLVENTS

Any refrigeration mechanism must be thoroughly cleaned before and after repair. Many methods have been used. Some do not do a thorough job; some are dangerous.

Whatever the method, it must remove oil, grease and sludge. In refrigeration and air conditioning applications, the cleaning method must remove moisture, or at least it should not add moisture. It must not injure the parts nor harm the user.

There are several cleaning methods:

1. Steam Cleaning. If the parts are exposed to hot water or steam, the grease will usually become fluid and float off the surface. However, steam and hot water will burn the service technician if carelessly used.

2. Caustic Solution Cleaning. When an alkaline cleaner is dissolved in hot water, the mixture will remove grease and oil. This solution must be carefully used or the user may suffer burns or eye injury.

3. Oleum or Mineral Spirits. This petroleum product is popular for cleaning. It has a flash point of approximately 140 F. (60 C.), kerosene has a flash point of 130 F. (54 C.). It cleans well and leaves a smudge-free surface. However, it presents a fire hazard and should always be used in small amounts. It should be contained in self-closing tanks. The container should be exhaust ventilated (have a hood and an explosion-proof exhaust fan).

4. The refrigerant R-11, is a good cleaning solvent for flushing systems contaminated by hermetic motor burn-out. R-11 has a high boiling point, 74.8 F. (24 C.). It is nontoxic and nonflammable. It leaves no noncondensable residue and has no reaction with the electrical insulation. It can be filtered and reused.

5. Alcohol is also a good cleaning fluid. However, it is both flammable and toxic. Special precautions — such as excellent ventilation, no open flames, use in small amounts — must be taken.

6. Vapor degreasing is a system using a cleaning fluid contained in a tank. The fluid is warmed, filling the upper part of the tank with vapors of the cleaner. Any parts suspended in this cleaning vapor are quickly and thoroughly cleaned. Such a tank must be specially vented.

7. There are several patented cleaning fluids available. When these are used, one should read and carefully follow the manufacturer's instructions.

8. Carbon tetrachloride. Carbon tetrachloride should never be used in cleaning refrigeration or air conditioning mechanisms. It is toxic to the respiratory system and to the skin.

9. Never use gasoline for cleaning. It has a low flash point, its fumes are heavy and may travel far to ignition sources causing an explosion or flash fire.

10. Do not use a propane or LP cylinder to clean parts. The liquid or gas will quickly evaporate and may burn or explode.

Refer to Chapter 28 for more information on cleaning.

2-63 COLD CHISELS AND PUNCHES

A cold chisel is used on various jobs. As an example, one may find the assembly devices on evaporators corroded and a chisel is needed to remove the nut or screw.

A 3/4 in. flat cold chisel is a popular size. Be sure to keep the head (hammering end) of the chisel free from "mushrooming." Flying pieces from a mushroomed head may cause injuries. Always wear goggles or head shield. See Fig. 2-40, Part 17.

Four common types of punches are the prick punch, center punch, pin punch, and drift punch. See Fig. 2-40, Parts 7, 8, 18 and 19.

The prick punch has a long, sharp point and is used only for layout work.

The center punch has a 60 deg. to 90 deg. point and is used for center punching the location of a hole to be drilled. A heavy blow on the punch makes a depression in which the point of the drill may be started. Center punches may also be used to make alignment marks on refrigeration parts before dismantling.

The pin punch is used for driving retainer pins in or out. The blunt end is called the bill. Pin punches are measured in overall length, by diameter of the stock and by diameter of the bill. Bill diameters are available in sizes from 3/32 to 5/16 in.

A drift punch is used to drive out keys and to line up holes in mating surfaces. The punch tapers from its flat point to the stock diameter.

Punches are available in various lengths and are usually made of carefully heat-treated chrome-alloy steel. The cutting edge or point is hard, while the head is tough and shatterproof. Always grind away any mushroom head that forms. A fairly heavy 6 in. punch will be the most satisfactory.

Fig. 2-79 illustrates the shapes of various punches.

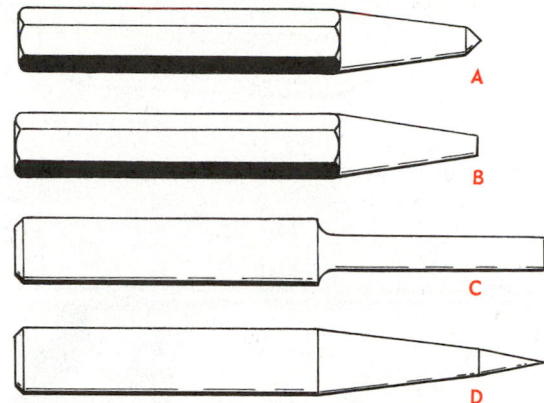

Fig. 2-79. Punches most used in refrigeration and air conditioning. A—Center punch. B—Drift punch. C—Pin punch. D—Prick punch.

2-64 FILES

Various sizes and types of files are needed for cleaning metal surfaces and shaping metal parts. They are classified according to tooth size, shape and the number of directions the teeth are cut on the file.

Files are either single or double cut, Fig. 2-80. The single cut is used for finishing surfaces and double cut for fast metal removal.

Files come in different lengths: 4, 6, 8, 10, 12 in. and so on. The longer, the coarser the teeth. The size of the teeth varies from dead smooth, smooth, second cut, bastard, to coarse. A second cut 6 in. file has finer teeth than one that is second cut 12 in.

File shapes are many: rectangular, half round, round, triangular, square, wedge shape and so on.

One oddity is that there are three types of rectangular cross-section files: mill, hand and flat. The mill file has only single cut teeth; the mill file is uniform in thickness but tapers slightly in width. The hand and flat files have double cut teeth, but the hand file has one edge that has no teeth, called a safe edge. The edges are parallel but the thickness varies slightly. The flat file has teeth on all four surfaces. See Fig. 2-40, Parts 35 and 36.

Use file brushes and file cards to clean file teeth which quickly become filled with metal. If clogging material is not removed, the files become useless. Do not use a file card for any other purpose. The bristles may become clogged with dirt.

2-65 HACKSAWS

A hand hacksaw is used for cutting tubing and other installation and maintenance work requiring metal cutting.

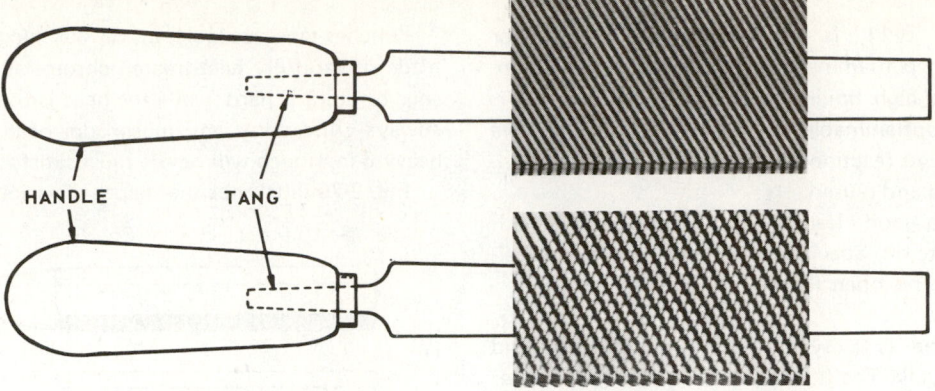

Fig. 2-80. Hand files may be either single cut (upper) or double cut (lower). Files should always be equipped with handles to avoid hand injuries. (The Cooper Group, Nicholson)

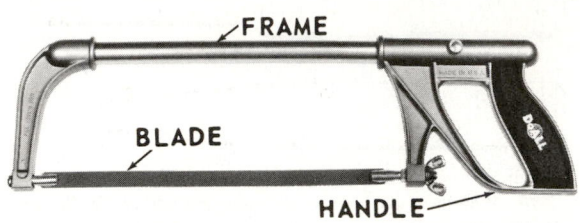

Fig. 2-81. Hand hacksaw must have rigid frame to hold blade in proper tension. (DoALL Co.)

Fig. 2-81 illustrates a popular saw with a rigid frame and a 10 in. blade. Blades are available with different numbers of teeth per inch. Blades with 14 teeth per in. are used for soft metal and wide cuts; 18 teeth per in. for medium soft metals; 24 teeth per in. for general work and 32 teeth per in. for thin metal, tubing or hard metal.

A hacksaw blade should not be stroked over 60 strokes per minute. Most blades are made of high carbon steel and the cutting edges (points) are very sharp and very small. Too rapid use will cause these points to overheat and lose their temper.

Always lift the blade slightly on the back stroke. This helps to keep the cutting edges sharp. If the blade is not lifted, particles of metal may roll between the work and the cutting edge of the blade, dulling the teeth.

With most blades, the teeth are hardened while the back of the blade is soft and flexible. Some higher quality blades may be made of tungsten or molybdenum steel alloy. Fig. 2-82 illustrates a hand hacksaw blade.

Fig. 2-82. Hand hacksaw blade. This is a tungsten alloy blade. The markings (10 x 1/2 x .025-18T) indicate that it is 10 in. long, 1/2 in. wide, .025 in. thick and that it has 18 teeth per in.

Special hacksaw frames are available for working in small holes. There is also a stub hacksaw blade and an adapter drive to fit electric drills.

2-66 FASTENING DEVICES

Many items are made in one piece today that were considered impossible to fabricate a short time ago. However, many mechanisms must still be made of several pieces and then assembled. If there is motion in the mechanism — such as a piston in a cylinder — the apparatus must be made of two or more pieces.

Many clever ways have been developed to fasten pieces together. In metal work, soldering, brazing, welding, crimping, rivets, bolts, machine screws, pins, spring fasteners and force fits have been used with success.

The type of assembly device used depends first on the kind and condition of the metal, and secondly on how frequently the pieces must be dismantled.

If the parts are to be put together permanently, riveting, welding, soldering and brazing are popular assembly methods.

If the parts must be dismantled for frequent repair or service, assembly devices must be used that can be easily removed without injuring the parts. Nuts and bolts, cap screws, machine screws and set screws are then used. Fig. 2-83 shows an assortment of fastening devices. In the metric system, the diameter of fastening sizes is specified in millimetres.

2-67 MACHINE SCREWS, BOLTS AND CAP SCREWS

Many small parts are assembled using specially threaded devices called machine screws. Machine screws may be made of steel, stainless steel, brass, Monel metal or other materials. They are available in a variety of head shapes. Various methods have been used to turn these screws, see Figs. 2-55 and 2-83. Some less commonly used screw heads are shown in Fig. 2-84.

Machine screws come in various diameters. Eight are in the numbered sizes, three in the fractional sizes. Each size may have either fine or coarse threads; the larger the number, the larger the diameter. A table of machine screw sizes and threads is given in Fig. 2-85.

In general, bolts and cap screws are used in sizes 1/4 in. and larger. The length of threads on a bolt are usually two times the bolt diameter. Bolts are usually provided with nuts.

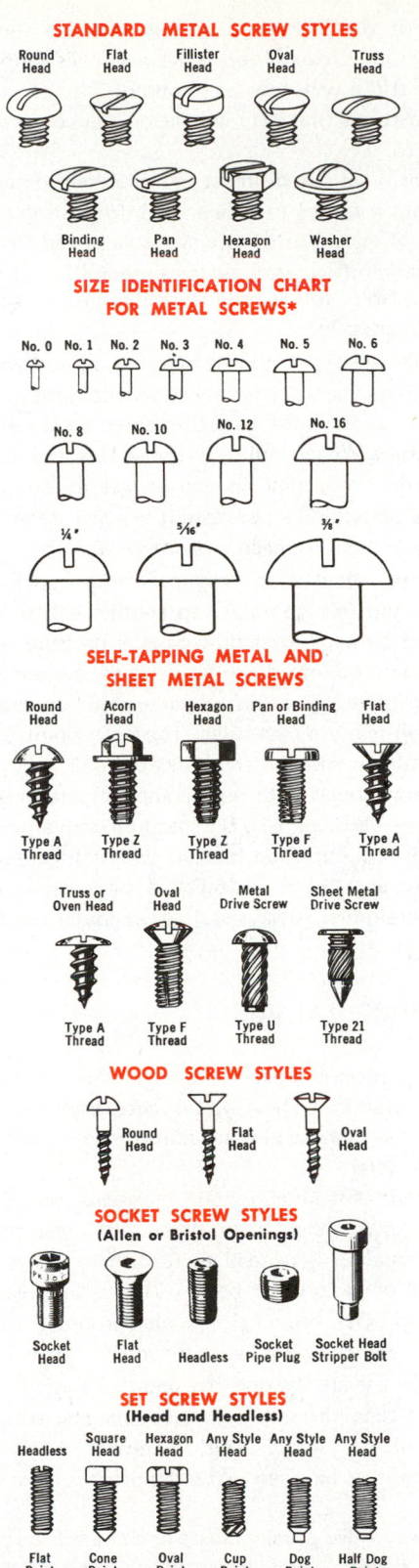

STANDARD METAL SCREW STYLES

Round Head   Flat Head   Fillister Head   Oval Head   Truss Head

Binding Head   Pan Head   Hexagon Head   Washer Head

SIZE IDENTIFICATION CHART FOR METAL SCREWS*

No. 0   No. 1   No. 2   No. 3   No. 4   No. 5   No. 6

No. 8   No. 10   No. 12   No. 16

1/4″   5/16″   3/8″

SELF-TAPPING METAL AND SHEET METAL SCREWS

Round Head   Acorn Head   Hexagon Head   Pan or Binding Head   Flat Head

Type A Thread   Type Z Thread   Type Z Thread   Type F Thread   Type A Thread

Truss or Oven Head   Oval Head   Metal Drive Screw   Sheet Metal Drive Screw

Type A Thread   Type F Thread   Type U Thread   Type 21 Thread

WOOD SCREW STYLES

Round Head   Flat Head   Oval Head

SOCKET SCREW STYLES
(Allen or Bristol Openings)

Socket Head   Flat Head   Headless   Socket Pipe Plug   Socket Head Stripper Bolt

SET SCREW STYLES
(Head and Headless)

Headless   Square Head   Hexagon Head   Any Style Head   Any Style Head   Any Style Head

Flat Point   Cone Point   Oval Point   Cup Point   Dog Point   Half Dog Point

Fig. 2-83. Fasteners must be carefully used and driven with proper tools. (Vaco Products Co.)

The threading on a cap screw is usually longer than the threading on a bolt and sometimes extends up to the head of the cap screw. Cap screws are threaded into a part of the

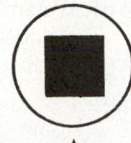

A   B   C

Fig. 2-84. Less commonly used screw heads. A—Square countersunk head (Scrulox). B—Reed Prince head (similar to Phillips). C—Spline.

MACHINE SCREW NUMBER	DIAMETER IN.	THREADS PER INCH	
		COARSE	FINE
2	.086	56	64
3	.099	48	56
4	.112	40	48
5	.125	40	44
6	.138	32	40
8	.164	32	36
10	.190	24	32
12	.216	24	28
1/4	.250	20	28
5/16	.3125	18	24
3/8	.375	16	24

Fig. 2-85. Common machine screw sizes.

mechanism and do not require a nut.

Diameters of metric bolts, nuts and screws, as well as the thread pitches are in millimetres. For example: 10 mm diameter with a 1.50 mm pitch.

The wrench size is the same as the size given for the diameter of the bolt. A 10 mm wrench fits a 10 mm bolt.

The metric system uses:
1. Extra fine thread screws.
2. Fine thread screws.
3. Average thread screws.
4. Coarse thread screws.

The popular diameters and thread pitch sizes are shown in Fig. 2-86.

DIAMETER mm	PITCH mm	WRENCH SIZE mm
3	0.60	3
4	0.70	4
4	0.75	4
5	0.80	5
5	0.90	5
6	1.00	6
7	1.00	7
8	1.00	8
8	1.25	8
9	1.00	9
9	1.25	9
10	1.25	
10	1.50	
11	1.50	
12	1.25	
12	1.50	
12	1.75	

Fig. 2-86. Table of metric screws.

It is believed that the new United States metric system for fasteners will use only one thread pitch for each fastener diameter and that the fastener head will be a 12-spline device and the wrench to turn it will be a 12-point socket.

2-68 GASKETS

Most mating surfaces are somewhat rough. To make a leakproof joint, gaskets are used between surfaces. Gaskets, being soft, seal joints between assembled parts. They keep the refrigerant from leaking out, prevent oil leakage and keep air from leaking into the system. They are used between the valve plate and the compressor body, between the service valve and the compressor body and between the valve plate and the compressor head. Gaskets are also used on the crankcase and at the crankshaft seal on open or external drive units.

Metal is the common gasket material. Lead is popular, being soft and noncorrosive. Aluminum has also been used but composition gaskets made of plastic impregnated paper are also popular.

Gaskets must not restrict the openings. They must not lose their compressibility and replacement gaskets must not be thicker than the original gaskets.

The surfaces of parts that contact the gasket must be free of burrs, bruises and foreign matter.

2-69 REFRIGERANTS

The refrigeration service technician is required to handle and charge refrigerants into refrigerating mechanisms. Several different refrigerants are carefully described in Chapter 9 along with safe handling methods.

Refrigerants must be kept DRY and CLEAN. Remember, all exposed surfaces absorb moisture if left in the open. If a compressor is torn down, overhauled and reassembled, it must first be completely dried before it can be charged with refrigerant. See Chapter 11 for detailed instructions.

Refrigerants are stored and handled by the service technician using portable refrigerant cylinders. Refrigerants are identified by a cylinder color code. See Chapter 9 for a table of color codes for common refrigerants.

Cylinders for different refrigerants must not be interchanged. Refrigerants should always be stored in the cylinder specified (color code).

Never fill refrigerant cylinders over 85 percent of capacity. With a temperature increase, hydrostatic pressure may burst the overfilled cylinder.

2-70 REFRIGERATION OIL

In the mechanical refrigerating system, moving parts must be lubricated with oil for long life and efficient performance.

Refrigerant oil is a specially prepared mineral oil. Refining steps are taken to remove excess wax, moisture, sulphur and other impurities. Most have a foaming inhibitor added.

Use oils that have a low pour point. This will avoid wax separation at the lowest temperature in the system. Wax could clog the refrigerant control orifice.

Because of the low temperatures at which they operate, food freezer and frozen food units need oils with extra low pour points and a very low wax content. Nor can the oil have any hydro-carbons that may collect on the compressor valves or other parts.

The viscosity of the oil must be accurately determined for the temperature ranges to which the refrigerating system may be exposed. Three viscosities are available — 150, 300 and 500. Most automatic refrigerating systems use 300.

When possible follow the manufacturer's recommendations. See Chapter 28.

It is highly important that refrigerant oil be kept in sealed containers, that it be transferred in chemically clean containers and lines and that it not be left exposed to air where it will absorb moisture. When refilling a refrigerator, use new oil.

Refrigeration oils come in one or five gallon cans and in barrels. It is advisable to purchase it in small sealed containers holding only enough for each separate service operation.

Unused oil allowed to remain in the container or oil transferred from one container to another will pick up some moisture and perhaps even dirt. Special pressure pumps may be used to pump oil into the low side of the system.

The price of refrigeration oil varies with the grade. A low pour point oil is more expensive. The pour point of any oil is the temperature at which it starts to flow.

Domestic machines with refrigerant temperatures as low as 0 F. to 5 F. (−18 C. to −15 C.) need oil with a pour point of −20 F. (−29 C.). For food freezers with refrigerant temperatures as low as −50 F. (−46 C.), a pour point of −60 F. (−51 C.) is desirable. Always seal an oil container after having drawn oil from it.

2-71 SERVICE VALVES

Service technicians must be familiar with manual valves in refrigerating systems. These valves enable them to seal off parts of the system while installing gauges, recharging or discharging a system.

Several kinds of manual or hand valves are used. Such valves may have handwheels on their stems, but most are so made that a valve wrench is needed to turn them. Valve stems are made of steel or brass while body of the valve is usually made of drop forged brass. Packing is installed around the valve stem and a packing adjusting nut keeps the joint from leaking.

In common use are the one-way and two-way service valves. "One-way" means there is only one opening which can be either opened or closed. The "two-way" valve has two openings. One may be open while the other is closed. Or both may be open.

The two-way valve usually closes or shuts off the refrigerant flow in the system when the stem is turned all the way in (clockwise). It shuts off the charging, discharging or gauge opening when the valve stem is turned all the way out (counterclockwise). When the valve stem is turned part way, both of the openings allow the fluid (refrigerant) to flow through, Fig. 2-87.

The tubing or pipe is fastened to valves by flare connection or by brazing. The valve may also be attached to the refrigerating mechanism by either pipe threads or by bolted flanges.

It is good practice to open a valve by first "cracking" it (opening it 1/16 or 1/8 turn). This prevents a shock pressure rush which may injure mechanisms, gauges, flush oil in abnormal amounts or injure the service technician.

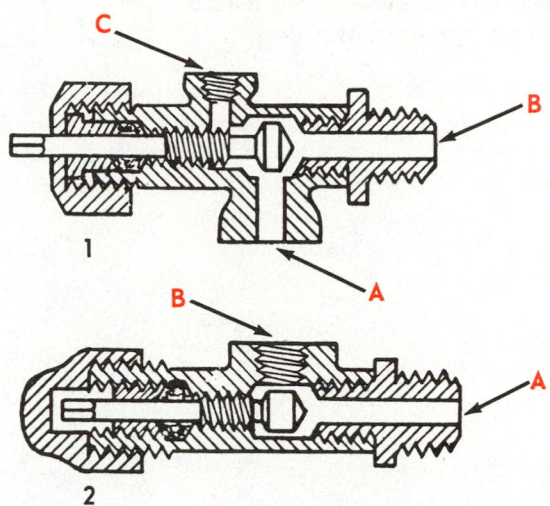

Fig. 2-87. Two typical service valve designs. No. 1 is a two-way valve. A—Opening to compressor. B—Opening to the refrigerant line. C—Service opening. No. 2 is a one-way valve. A—Opening to the liquid line. B—Opening to the liquid receiver. Valve No. 1 has an open valve stem. Valve No. 2 has a cap over the valve stem.

Be sure the valve stem is clean before turning it in (clockwise). A scarred or dirty valve stem will ruin the valve packing. See Chapter 14.

2-72 PURGING

Purging is a term used to describe the process of removing unwanted air, vapors, dirt or moisture from the system. A neutral gas or the recommended refrigerant is allowed to flow through the refrigerator part or tubing, forcing unwanted air and vapors out.

2-73 EVACUATING

A refrigerating system must contain only the refrigerant in either liquid or vapor state along with dry oil. All other vapors, gases and fluids must be removed.

These substances can be removed best by connecting the system to a vacuum pump and allowing the pump to run continuously for some time while a deep vacuum is drawn on the system.

It is sometimes necessary to warm the parts to 120 F. (49 C.) while under a high vacuum, in order to remove all unwanted moisture. Heat the parts using warm air, heat lamps or warm water.

Never use a torch!

2-74 A REVIEW OF SAFETY IN HANDLING REFRIGERATION SUPPLIES AND EQUIPMENT

Before working with tools, always review the safety steps. The following are only a few of the items to remember:

1. Tubing should be bent in as large a radius as possible.
2. Epoxy bonding materials may irritate the skin or many membranes of the user.
3. "Mushroom" heads should be ground from chisels and punches as these particles may fly when struck with the hammer causing serious injury to the operator or a bystander.
4. Files should never be used without handles. The tang may injure the hands.
5. Wear goggles when drilling. Sometimes chips may fly. Eyes should therefore always be protected.
6. Emery cloth should not be used to clean tubing preparatory to soldering. It may leave an oily deposit. Also, the grit is hard and could cause considerable damage if allowed to enter the refrigerating mechanism.
7. When pressure testing for leaks in tubing circuits, carefully and safely use low or medium pressure carbon dioxide or nitrogen. Never use oxygen. CAUTION — see Chapter 11.
8. Carbon tetrachloride should not be used for any purpose as it is toxic and harmful to the skin.
9. Silver brazing materials sometimes contain cadmium. Fumes from heated cadmium are very poisonous. Be sure that the work space is well ventilated. If at all possible, use silver brazing alloys which DO NOT contain cadmium.
10. It is recommended that refrigerant cylinders never be filled above 85 percent of their capacity. If overfilled, hydrostatic pressure may cause them to burst.
11. Wrenches used on refrigeration line fittings should always fit snugly. Poorly fitted wrenches will ruin nuts and bolt heads. Also, the wrench may slip and cause an injury to the technician.
12. Always "crack" service valves and cylinder valves before opening. This gives quick control of the flow of gases if there is any danger.
13. Moisture is always a hazard to refrigerating mechanisms. Keep everything connected with refrigerating mechanism thoroughly dry.

2-75 TEST YOUR KNOWLEDGE

1. List the fittings to be used when connecting a compound gauge with 1/8 in. pipe male thread to a service valve having 1/4 in. pipe female thread.
2. When cutting tubing with a hacksaw, what two precautions must be taken?
3. What are the SAE thread specifications of 1/4, 3/8 and 1/2 in. tubing flare nuts?
4. What is the thickness of refrigeration tubing? What is the inside diameter of 1/4 in. tubing?
5. What may cause copper tubing to harden?
6. Why do some service technicians file the ends of copper tubing?

7. How may an external bending spring be easily removed from the tubing after making a bend?

8. Is it enough to clean metal for soldering with flux only? Why?

9. When soft soldering, what temperature must the soldered metal be?

10. Why must a soldered joint be cleaned after soldering?

11. How may one tell when the correct silver brazing temperature is reached?

12. What type of wrench should be used on a valve stem?

13. Should one push or pull a wrench?

14. Describe an easy way to check the accuracy of a thermometer.

15. What is the purpose of a compound gauge?

16. Why must refrigerant oil be practically wax free?

17. What is a very important precaution one must take when filling refrigerant cylinders?

18. What is a double cut file?

19. Name the tools shown in Fig. 2-40.

20. Why are torque wrenches used?

Chapter 3

BASIC
REFRIGERATION SYSTEMS

This chapter describes and illustrates basic refrigeration systems. Refrigeration systems may be classified in several ways: by type of refrigerant control, type of motor control, compression system, absorption systems and so on. Similar systems will be grouped together.

Each basic refrigeration system explanation:
1. Names the system.
2. Shows a schematic diagram of the system.
3. Describes how the system works.
4. Names some common uses.
5. Names the motor controls used.

It is important that one become familiar with the fundamental operating principles of the common systems described in this chapter.

The illustrations are not intended to show the exact parts and uses of actual units. Instead, they explain the fundamentals of construction and operation. Details and actual uses will be shown and described in later chapters.

3-1 ICE REFRIGERATION

For years, ice (frozen water) was the only refrigerating means available. It is still used in many refrigerating applications. The usual ice refrigerator, Fig. 3-1, is an insulated cabinet equipped with a tray or tank at the top for holding blocks or pieces of ice (see blue).

Shelves for food are located below the ice compartment. Cold air (see yellow) flows downward from the ice compartment and cools the food on the shelves below. The air becomes warmer and rises from the bottom of the cabinet (see red), up the sides and back of the cabinet, flows over the ice, cools, and again flows down over the shelves.

Ice refrigeration has the advantage of maintaining the interior of the cabinet at a fairly high humidity. Food stored in this type refrigerator does not dry out rapidly.

Until the development of the mechanical refrigerator, natural ice refrigeration was quite widely used. Since then, artificial ice has been manufactured for refrigeration. Temperatures inside an ice refrigerator are controlled by the flow of air over the ice and through the cabinet. Temperatures will usually range between 40 and 50 F.

When it is necessary to use ice for cooling temperatures below 32 F. (0 C.), ice and salt mixtures may be used.

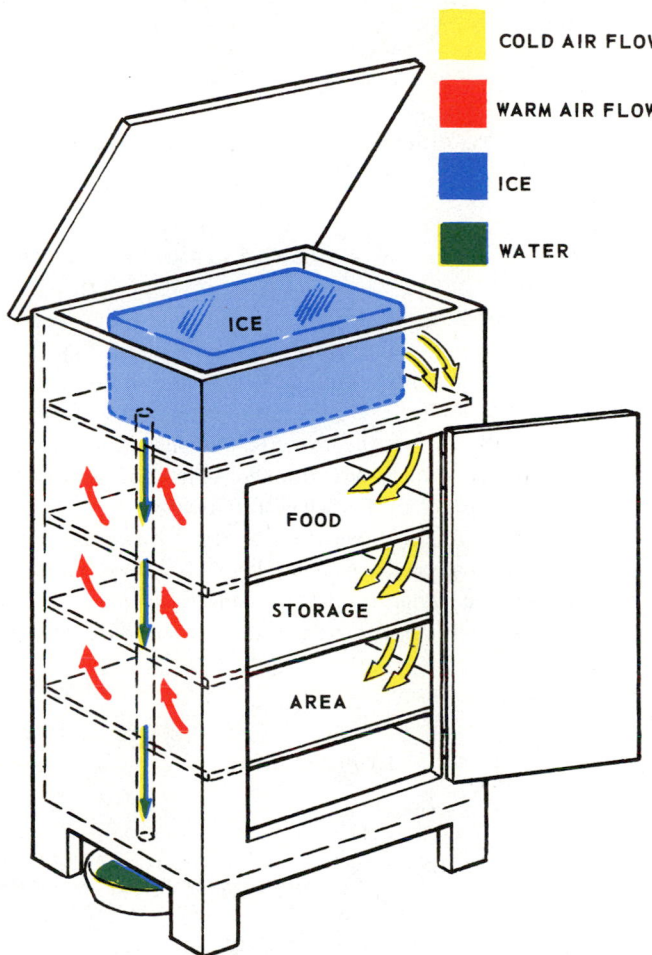

COLD AIR FLOW

WARM AIR FLOW

ICE

WATER

Fig. 3-1. Basic design and operation of ice refrigerator.

Temperatures down to 0 F. (-18 C.) may be easy to get with ice and salt mixtures. See Chapter 28 for a table of ice and salt mixtures.

3-2 EVAPORATIVE REFRIGERATION (DESERT BAG)

When a fluid evaporates heat is absorbed. Evaporation of water is an example. This is why humans and animals perspire.

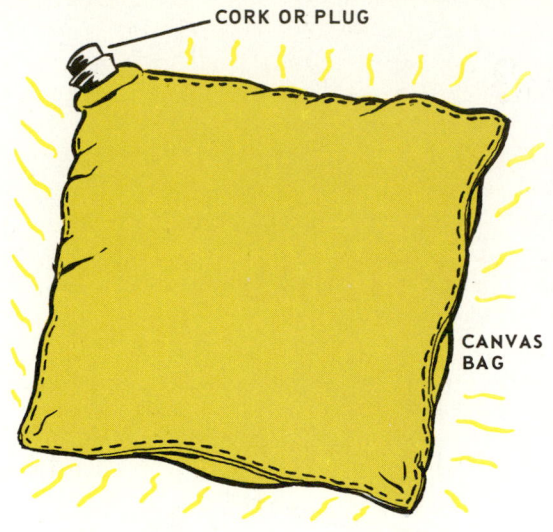

Fig. 3-2. The "desert bag" is an example of cooling by evaporative refrigeration.

3-3 EVAPORATIVE REFRIGERATION (SNOW-MAKING)

Another common application of water evaporation refrigeration is the method of making artificial snow for ski slopes. This device (snow machine) consists of a water nozzle into which a high-pressure jet of air is inserted. Water (see red) flows from the nozzle, as shown in Fig. 3-3, and the air (see blue), under high pressure, causes the water to break up into tiny droplets (almost a fog). If the surrounding air temperature is near freezing or below freezing, the droplets of water tend to evaporate and rapidly cool to the point where tiny drops of ice are formed. Using this method, artifical snow can be made when the temperature of the surrounding air is 32 F. (0 C.) or lower.

If the humidity is low, artificial snow can be made when the temperature is as high as 34 F. (1 C.). This is possible because of the rapid evaporation and evaporative cooling caused by the low humidity.

Evaporative condensers, often used in connection with air conditioners, are another example of evaporative cooling. See Chapter 12. The evaporation of water helps cool the condenser.

3-4 COMPRESSION SYSTEM USING LOW-SIDE FLOAT REFRIGERANT CONTROL

The "low-side float" refrigerant control system was very popular in many of the early refrigerating mechanisms. It is also known as a flooded system.

Fig. 3-4 is a schematic diagram of this system. The liquid refrigerant flows from the liquid receiver through the liquid line, up to the low-side float needle. The evaporator in this system consists of a finned tank (evaporator). It contains a float and needle control, which maintains a constant level of liquid refrigerant under a low-side pressure.

Evaporation of moisture from the skin surface helps to keep one cool.

Another common application of this principle is the "desert bag" used to keep drinking water cool. This bag, Fig. 3-2, made of a tightly woven fabric, is filled with drinking water. Since the bag is not waterproof, some water seeps through. Thus, the outside surface of the bag remains moist. Under desert conditions, which are usually both hot and dry, moisture on the surface of the bag evaporates rapidly.

A large part of the heat necessary to cause this evaporation comes from the bag and the water in it. This heat removal cools the drinking water inside the canvas, keeping it at a temperature several degrees below the temperature of the surrounding air.

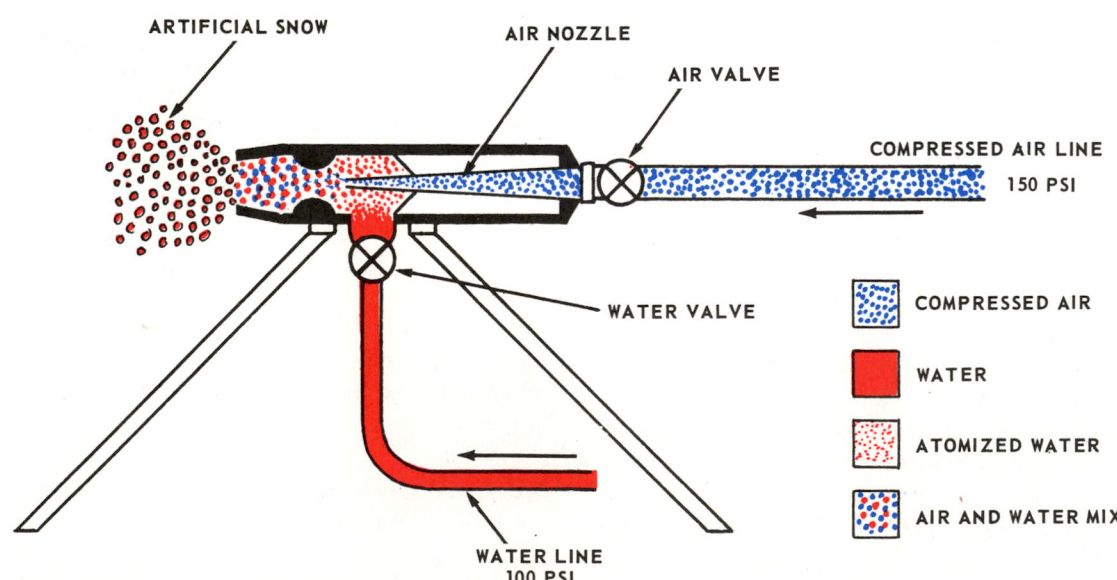

Fig. 3-3. Water-compressed air nozzle is used for making artificial snow.

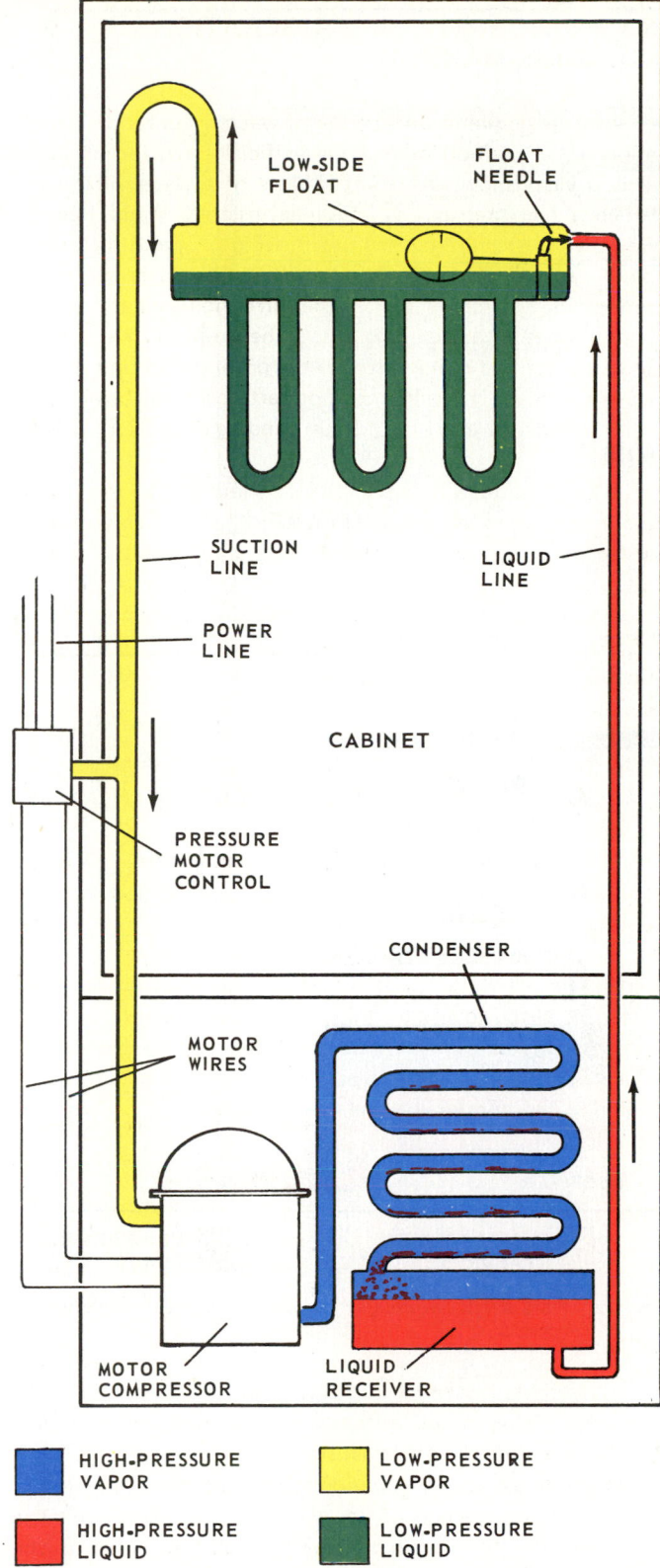

LOW-SIDE FLOAT

FLOAT NEEDLE

SUCTION LINE

LIQUID LINE

POWER LINE

CABINET

PRESSURE MOTOR CONTROL

CONDENSER

MOTOR WIRES

MOTOR COMPRESSOR

LIQUID RECEIVER

■ HIGH-PRESSURE VAPOR

□ LOW-PRESSURE VAPOR

■ HIGH-PRESSURE LIQUID

■ LOW-PRESSURE LIQUID

Fig. 3-4. Compression system using low-side float refrigerant control.

This refrigerant, since it is a liquid on the low side, is at a low temperature. The cold liquid refrigerant will absorb considerable heat on both the on cycle and the off cycle.

Vaporized refrigerant moves through the suction (vapor)

line to the compressor where it is compressed to a high pressure and discharged into the condenser. It is cooled by the condenser, returns to a liquid and flows into the liquid receiver. The operation continues until the desired low temperature is reached.

The pressure on the low side in a flooded system such as this will vary with the temperature. The higher the temperature, the higher the low-side pressure.

The system shown in Fig. 3-4 uses a pressure motor control. A spring-loaded pressure sensitive device on the suction line, or on the evaporator, activates a motor control switch. As the motor drives the compressor, the pressure and temperature in the evaporator will be reduced. At a given pressure setting, the motor compressor will stop.

The cycle will be repeated as soon as the pressure in the evaporator rises to a level corresponding to the refrigerant temperature at which the motor compressor is to start again.

The cabinet temperature may be controlled by a temperature control switch. In this case, the temperature sensitive element may be clamped to the fins on the evaporator.

This refrigerating cycle is often used on drinking fountains and other installations, where a constant temperature is desired.

Since the pressures do not balance on the off cycle, it is necessary to use a motor which will start under a load.

Such a system requires a rather large refrigerant charge, as there is liquid refrigerant in both the liquid receiver and in the evaporator.

All flooded systems are quite efficient since cold liquid refrigerant wets the evaporator surfaces providing excellent heat transfer. These systems are easy to service.

The float needle and seat must be kept in good condition to avoid possible flooding of the low side.

3-5 OPEN – EXTERNAL DRIVE – REFRIGERATING SYSTEM

In this system, the compressor is usually belt driven from an electric motor. The speed of the compressor is usually considerably less than the speed of the motor. This is done by using a small pulley on the motor shaft and a larger pulley (flywheel) on the compressor shaft. Most early refrigerating systems were like this. Fig. 3-5 illustrates an open system.

The liquid refrigerant (see red) flows through the thermostatic expansion valve into the evaporator where it is under low pressure. It boils, vaporizes and absorbs heat in the evaporator.

When the compressor is running, the vaporized refrigerant (see yellow) is drawn through the suction line to the compressor. It is compressed to a high pressure (see blue) before being discharged into the condenser.

In the condenser, the vapor gives up its latent heat of vaporization, is cooled and returns to a liquid (see red). From here the cycle is repeated.

A pressure motor control is shown. The starting mechanism on external drive (open) system motors is usually built into the motor.

This system requires a crankshaft seal on the compressor.

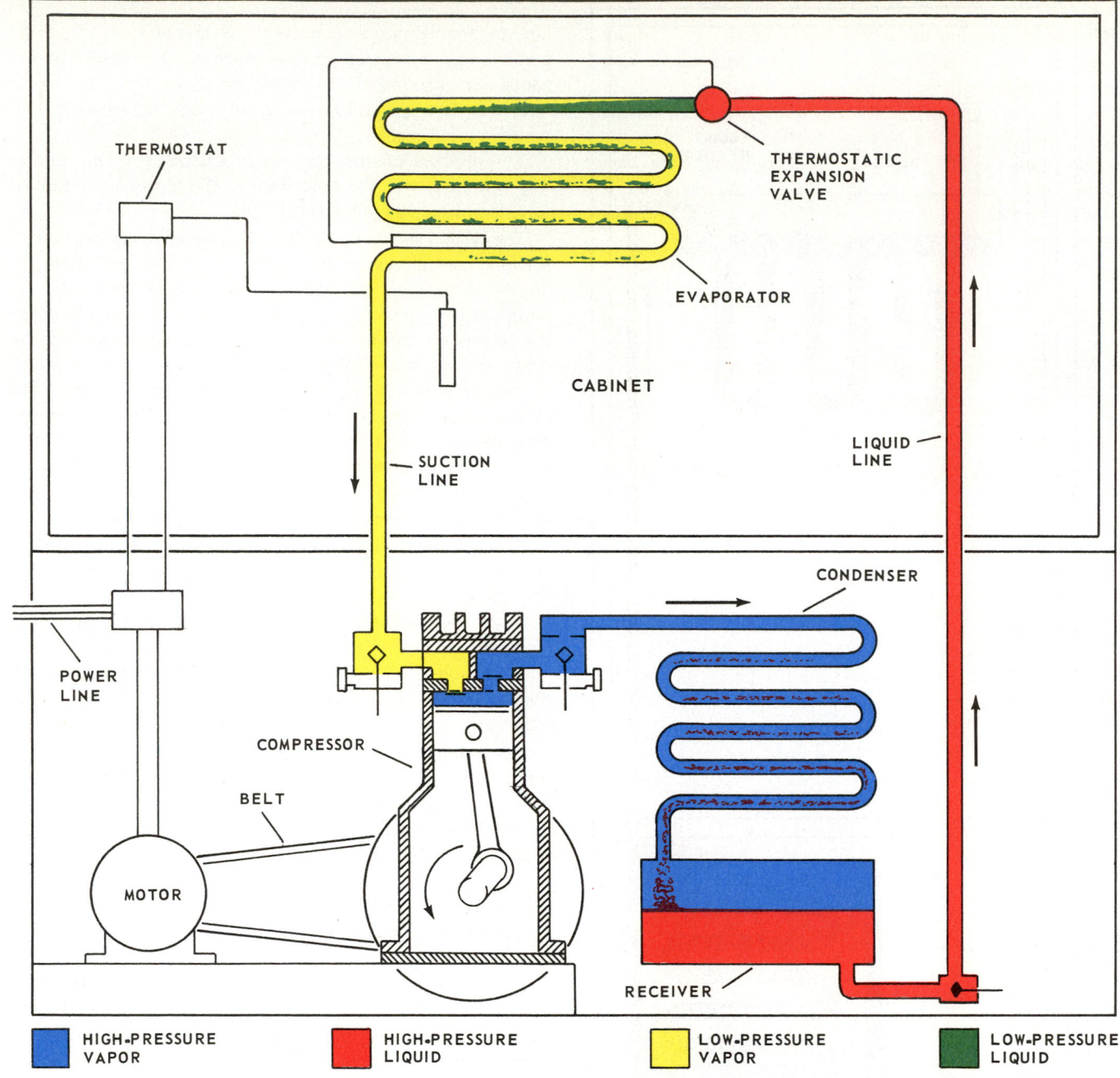

THERMOSTAT

THERMOSTATIC
EXPANSION
VALVE

EVAPORATOR

CABINET

LIQUID
LINE

SUCTION
LINE

POWER
LINE

CONDENSER

COMPRESSOR

BELT

MOTOR

RECEIVER

| HIGH-PRESSURE VAPOR | HIGH-PRESSURE LIQUID | LOW-PRESSURE VAPOR | LOW-PRESSURE LIQUID |

Fig. 3-5. Compression system using external drive (open) type compressor. Crankshaft seal is required at the place where the crankshaft extends through the crankcase of compressor.

The motor and the compressor drive are at atmospheric pressure. The pressure inside the crankcase will vary depending upon the refrigerant used and the temperature. Sometimes it may be considerably above atmospheric pressure; at other times, it may be below. Refrigerant vapor cannot be allowed to flow out or air to flow into the crankcase. Either would quickly ruin the operation of the system.

3-6 COMPRESSION SYSTEM USING HIGH-SIDE FLOAT REFRIGERANT CONTROL

The high-side float system is a flooded system. The evaporator is always filled with liquid refrigerant.

Fig. 3-6 is a schematic diagram of this system. As the compressor runs, liquid refrigerant flows from the liquid line into the high-side float mechanism.

As soon as enough liquid refrigerant has entered the high-side float mechanism, it will raise the float ball. The refrigerant will then begin to flow through the control to the evaporator. Since the evaporator is under low pressure, the tubing connecting the high-side float and the evaporator should be insulated. A capillary tube refrigerant line is frequently used.

If a different size line is used, it should have a weight valve at the evaporator to prevent the refrigerant from evaporating in the connecting line. Fig. 3-6 shows a weight valve in the connecting line.

Refrigerant entering the evaporator is under low pressure (see green). It will rapidly evaporate (boil) and absorb heat from the evaporator.

The vapor (see yellow) then flows through the suction line to the compressor, where it is compressed (squeezed) to the high-side pressure (see blue). In the condenser, the heat absorbed in the evaporator is removed and the refrigerant is returned to the liquid state (see red). It flows into the high-side mechanism where the cycle is repeated.

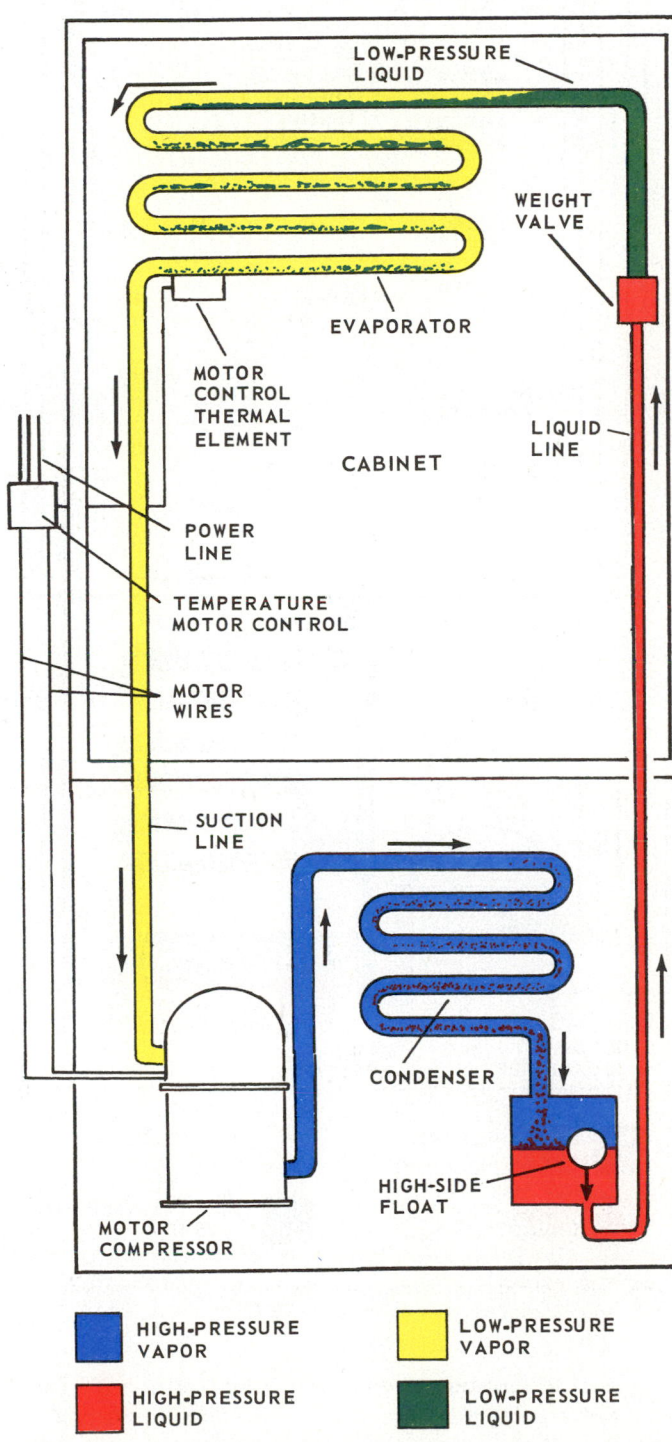

HIGH-PRESSURE VAPOR

LOW-PRESSURE VAPOR

HIGH-PRESSURE LIQUID

LOW-PRESSURE LIQUID

Fig. 3-6. Compression system using high-side float refrigerant control.

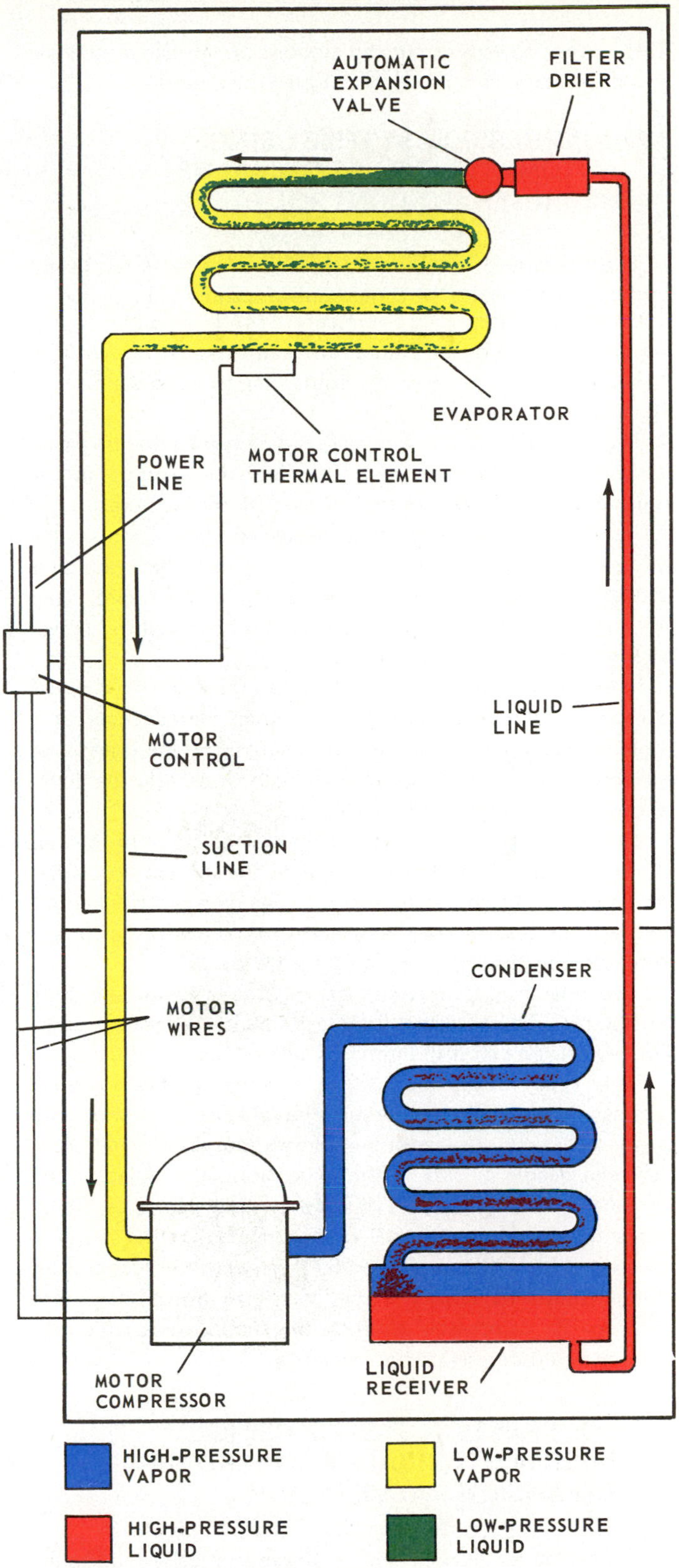

Fig. 3-7. Compression system using automatic expansion valve refrigerant control (AEV).

HIGH-PRESSURE VAPOR

LOW-PRESSURE VAPOR

HIGH-PRESSURE LIQUID

LOW-PRESSURE LIQUID

Either a temperature or a pressure motor control may be used on this refrigeration cycle. Fig. 3-6 shows a temperature motor control located in the refrigerated space.

This system is most used in commercial applications where high operating efficiency is desired.

Basic Refrigeration Systems / 77

It is easy to service but the amount of refrigerant charged into the system must be very accurately measured.

3-7 COMPRESSION SYSTEM USING AUTOMATIC EXPANSION VALVE REFRIGERANT CONTROL (AEV)

The operation of an automatic expansion valve refrigerant control (AEV) refrigerating mechanism is shown in Fig. 3-7.

Compressor, motor and condenser (condensing unit) are in the base of the cabinet. Liquid refrigerant (see red) flows from the liquid receiver through the liquid line, through the filter to the automatic expansion valve.

The automatic expansion valve is designed so that no liquid refrigerant will flow through it unless the pressure in the evaporator is reduced by the running of the compressor. As the compressor runs and liquid refrigerant flows through the automatic expansion valve, it is sprayed into the evaporator (see green). Here, due to low pressure, it boils rapidly and absorbs heat. This vaporized refrigerant (see yellow) moves back to the compressor through the suction line.

In the compressor, it is compressed to the high-side pressure as vapor (see blue). While flowing through the condenser it is cooled, gives up the heat that it absorbed in the evaporator and returns to a liquid (see red). It then flows into the liquid receiver ready to repeat the cycle.

The motor control thermal element is clamped to the end of the evaporator at the beginning of the suction line. After the evaporator is cooled to its proper temperature, the control bulb pressure causes the motor control to turn off the current to the driving motor. The compressor is stopped.

The operating characteristics of this system are quite satisfactory. The refrigerant oil is circulated without trouble. The temperature control limits can also be kept quite close.

This type refrigeration cycle is used widely in small commercial applications. Because the pressures do not balance on the off cycle, the motor compressor must start under load.

If the needle or seat in the expansion valve is faulty and refrigerant leaks through the valve on the off cycle, liquid refrigerant may flow into the suction line. When the compressor starts, this will be indicated by frosting of the suction line. If the trouble is severe, it may result in liquid refrigerant entering the compressor through the suction line. This may cause the compressor to knock severely.

3-8 COMPRESSION SYSTEM USING THERMOSTATICALLY CONTROLLED EXPANSION VALVE (TEV)

A schematic diagram of a thermostatically controlled expansion valve (TEV) refrigeration cycle is shown in Fig. 3-8. The liquid refrigerant (see red) flows from the liquid receiver through the liquid line to the filter-drier and to the thermostatic expansion valve.

The operation of the thermostatic expansion valve is controlled by both the temperature of the TEV control bulb and the pressure in the evaporator. The temperature of the TEV control bulb must be higher than the evaporator

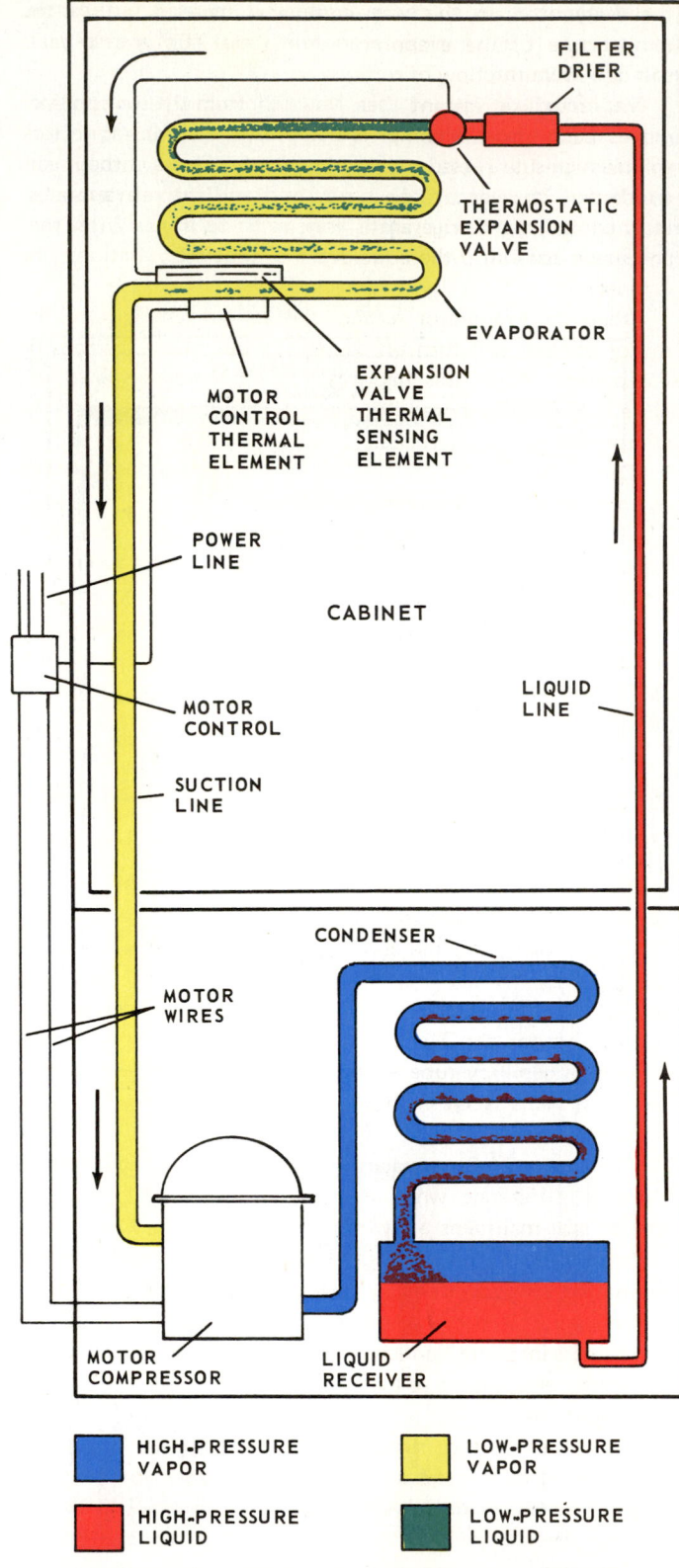

■ HIGH-PRESSURE VAPOR	■ LOW-PRESSURE VAPOR
■ HIGH-PRESSURE LIQUID	■ LOW-PRESSURE LIQUID

Fig. 3-8. Compression system using thermostatically controlled expansion valve (TEV).

refrigerant temperature before the valve will open. The amount of opening will be governed by the temperature of the evaporator. If the evaporator is quite warm, the needle will open quite wide allowing a rapid flow of liquid (see green) into the evaporator. In this way cooling is speeded up. As the temperature of the evaporator drops, the TEV needle valve will cut down the flow of refrigerant.

Vaporized refrigerant (see yellow) from the evaporator moves back into the compressor where it is compressed back to the high-side pressure (see blue). As it flows through the condenser, it gives up the heat absorbed in the evaporator. Now cooled, the refrigerant is condensed to a liquid (see red) and flows back into the liquid receiver. The refrigerating cycle is repeated.

When the evaporator reaches the desired temperature, the motor control will turn off current to the motor to stop the compressor. When this happens, the TEV needle valve will close, allowing no more refrigerant to flow through it until the compressor again lowers the pressure in the evaporator.

This system is used on large commercial refrigerators as well as on many air conditioning applications.

Since pressures do not balance on the off cycle, it is necessary to provide a motor compressor which will start under load.

The TEV control remains closed unless the evaporator is under reduced pressure and the temperature is above normal. A leaking valve will usually be indicated by a frosted or sweating suction line.

3-9 COMPRESSION SYSTEM USING CAPILLARY TUBE REFRIGERANT CONTROL

The "capillary tube" system, Fig. 3-9, is one of the most popular of the compression type systems. Liquid refrigerant (see red) flows from the condenser up through the liquid line to the filter (which may also be a drier). From the filter, refrigerant flows through the capillary tube refrigerant control into the evaporator. The pressure of the liquid refrigerant, as it enters the capillary tube at the filter end, is at a high pressure (see red). This is the high-pressure side. The pressure in the evaporator is low.

The design of the capillary tube is such that it maintains a pressure difference while the compressor is operating. The compressor maintains a low pressure in the evaporator and the refrigerant boils, rapidly absorbing heat. The vaporized refrigerant (see yellow) moves through the suction line back to the compressor. Here it is compressed to a high pressure and discharged into the condenser (see blue). It is cooled in the condenser and returns to a liquid (see red) and again flows into the liquid line.

This operation continues until the thermal element has been cooled to a preset low temperature. When that temperature is reached, the thermal element operates the motor control mechanism and turns off power to the motor. The refrigeration cycle stops. It will remain stopped until the thermal element warms up and the thermal bulb pressure closes the motor control contacts to again operate the compressor.

On the off cycle the capillary tube allows the pressures to balance between the high and low sides. It is not usually necessary, then, to use a motor with a high starting torque.

This system is commonly used in household refrigerators,

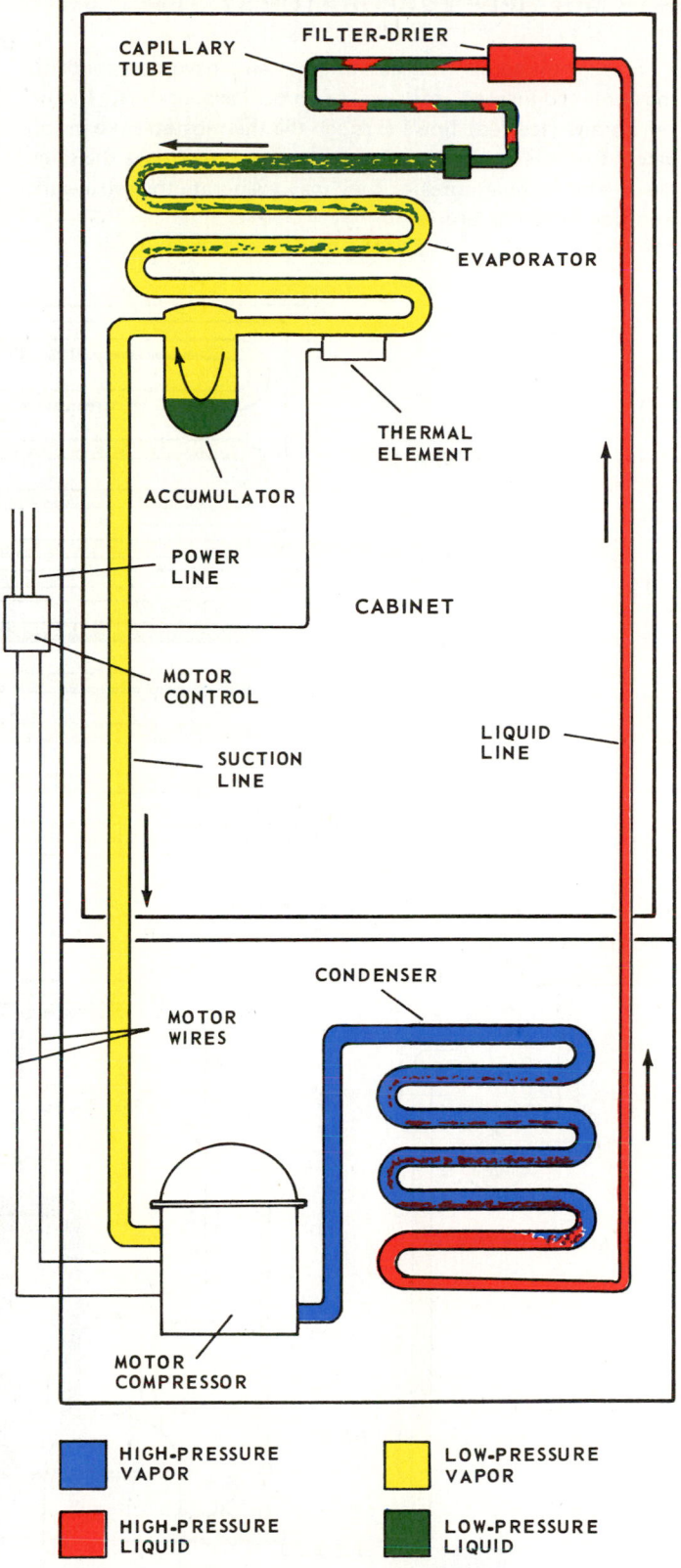

Fig. 3-9. Compression system using capillary tube type refrigerant control.

freezers, air conditioners, dehumidifiers and many small commercial applications. This type of cycle is quite satisfactory for most refrigerating applications.

3-10 MULTIPLE EVAPORATOR SYSTEM

Some commercial refrigerating systems have one condensing unit connected to two or more evaporators. Liquid refrigerant (see red) flows through the thermostatic expansion valves to the evaporators. The evaporators may have the same evaporator temperatures or they may evaporate the refrigerant at different temperatures.

If the evaporator temperatures are the same, the system uses only a low-side float or the thermostatic expansion valve to control the refrigerant.

If two or more evaporating temperatures are desired, a frozen foods temperature and a water cooling temperature (for example), a device must be used to keep one of the evaporators at a higher low-side pressure. Look at schematic, Fig. 3-10. A two-temperature valve in the suction line (upper

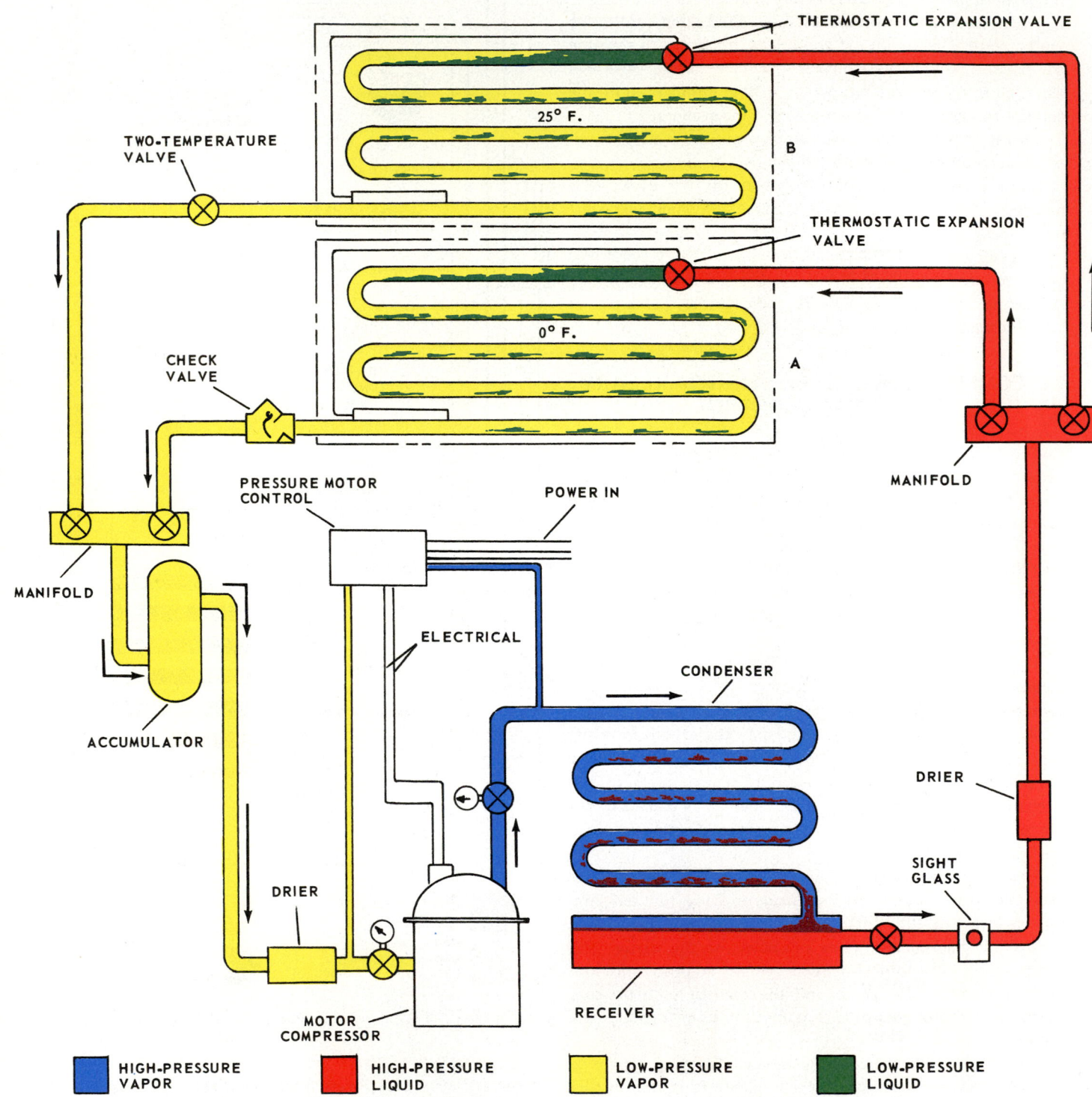

| HIGH-PRESSURE VAPOR | HIGH-PRESSURE LIQUID | LOW-PRESSURE VAPOR | LOW-PRESSURE LIQUID |

Fig. 3-10. A multiple evaporator system. Evaporator A operates at 0 F. (—18 C.). Evaporator B operates at 25 F. (—4 C.).

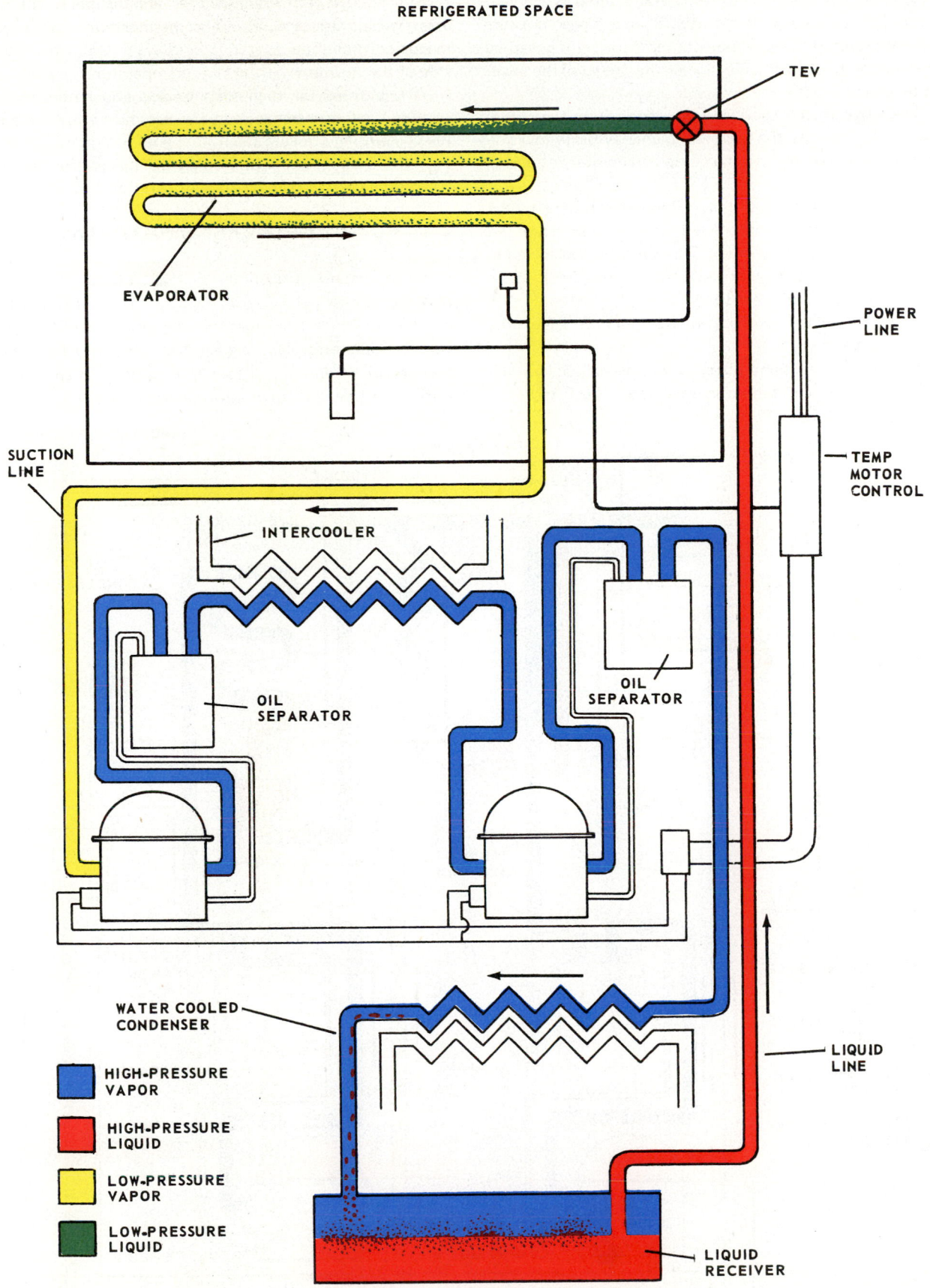

REFRIGERATED SPACE

TEV

EVAPORATOR

POWER LINE

SUCTION LINE

TEMP MOTOR CONTROL

INTERCOOLER

OIL SEPARATOR

OIL SEPARATOR

WATER COOLED CONDENSER

LIQUID LINE

HIGH-PRESSURE VAPOR

HIGH-PRESSURE LIQUID

LOW-PRESSURE VAPOR

LOW-PRESSURE LIQUID

LIQUID RECEIVER

Fig. 3-11. Compound refrigerating system.

left) keeps the low-side pressure refrigerant liquid (see green) and vapor (see yellow) in evaporator B at a higher pressure than at evaporator A. The evaporator temperature is governed by the evaporating pressure. The lower the pressure, the lower the temperature.

A check valve in the suction line coming from the colder evaporator, A, prevents the warmer, higher pressure low-side (see yellow) from moving into the colder evaporator, A, during the off cycle.

The vaporized refrigerant (see yellow) is returned to the motor compressor. It is compressed to a high-pressure and high-temperature vapor (see blue). This vapor is cooled in the condenser becoming a high-pressure liquid (see red) to be stored in the receiver until needed.

Note the filter-drier on the liquid line. It keeps the refrigerant clean and dry.

A liquid indicator (sight glass) is often included in the liquid line. The service technician may then check to see if there is enough refrigerant in the system. Bubbles will indicate a refrigerant shortage. This system, as shown, uses a pressure motor control. The operating pressure is taken from the low side of the system.

A line from the high-pressure side also enters the motor control. This operates a safety device which stops the motor if the condensing pressure (high side) goes too high.

Multiple (two or more) evaporator refrigeration systems are commonly used in commercial refrigeration applications.

3-11 COMPOUND REFRIGERATING SYSTEMS

In compound refrigerating systems, two or more compressors are connected in series, Fig. 3-11. In this illustration, compressor No. 1 discharges into the intake side of compressor No. 2. Compressor No. 2 then discharges into the condenser (see blue). Here the vapor condenses. Then the liquid refrigerant (see red) flows into the liquid receiver.

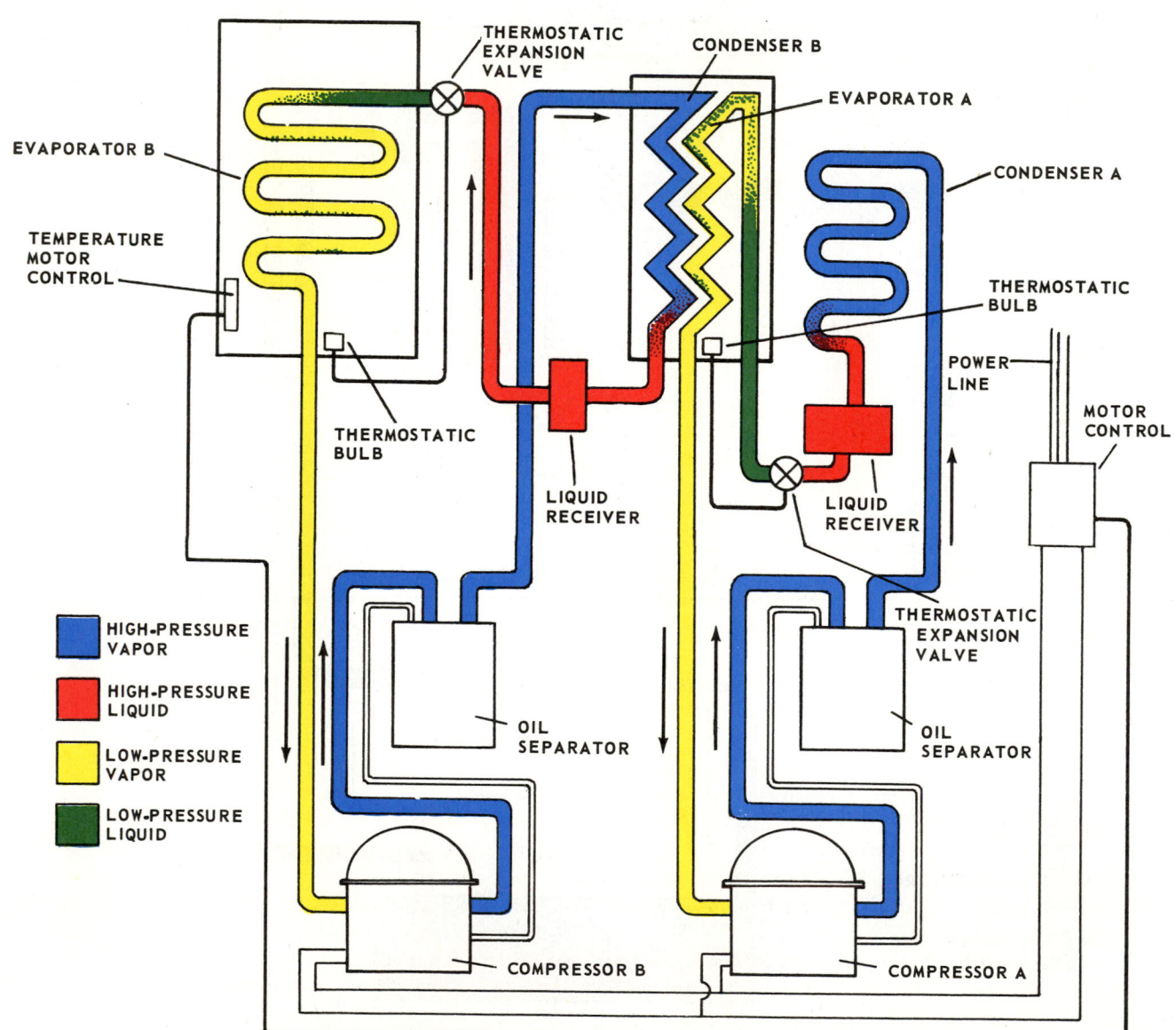

Fig. 3-12. Cascade refrigerating system.

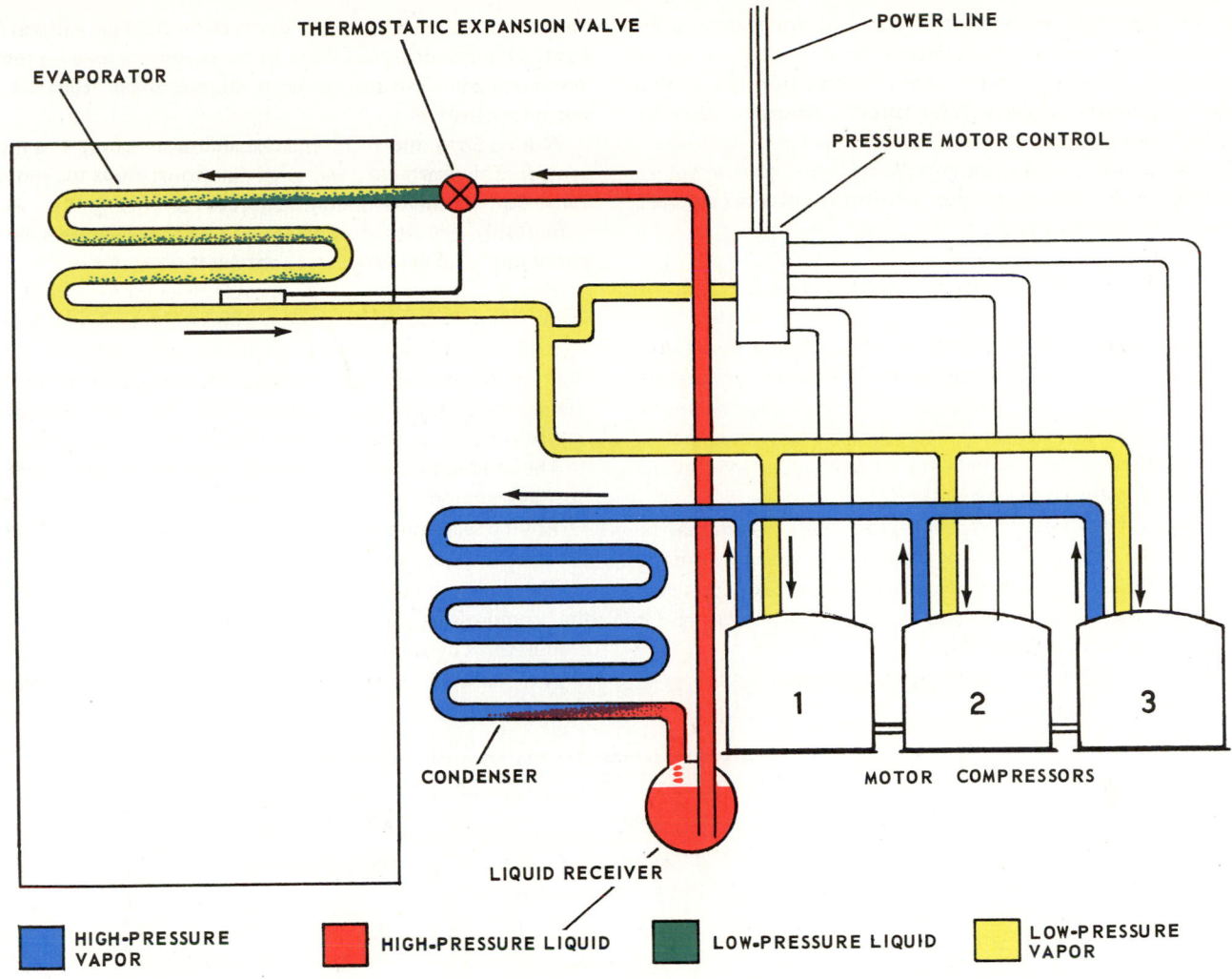

THERMOSTATIC EXPANSION VALVE

EVAPORATOR

POWER LINE

PRESSURE MOTOR CONTROL

CONDENSER

LIQUID RECEIVER

MOTOR   COMPRESSORS

1   2   3

| | HIGH-PRESSURE VAPOR | | HIGH-PRESSURE LIQUID | | LOW-PRESSURE LIQUID | | LOW-PRESSURE VAPOR |

Fig. 3-13. Modulating refrigeration cycle mechanism, which uses three motor compressors. Pressure motor control is arranged to operate one or more compressors as needed.

From the liquid receiver the liquid refrigerant (see red) flows up to the thermostatic expansion valve and into the evaporator. In the evaporator (see green) the refrigerant boils and absorbs heat (see yellow). From the evaporator the vaporized refrigerant flows back to compressor No. 1. From here the cycle is repeated.

Such a compound system increases capacity when pulling down to such a low pressure (low temperature) that one compressor cannot do it well.

Refrigerant vapor is not condensed between compressors. An intercooler lowers the vapor temperature. This type of installation usually requires an oil separator for each compressor.

A single-temperature motor control operates all motors and a thermostatic expansion valve controls the liquid refrigerant flow into the evaporator.

Since the pressures do not balance on the off cycle, motors capable of starting under load are required.

Compound installations usually operate under rather heavy service requirements. Condensers and refrigerant must be kept clean. Compressor valves must be kept in good condition.

3-12 CASCADE REFRIGERATING SYSTEMS

In a cascade refrigerating installation, two or more refrigerating systems are connected as shown in Fig. 3-12. Both systems operate at the same time. System A (on the right) has its evaporator, A, (heat absorbing part) arranged to cool the condenser B for the system B. The evaporator for system B supplies the cooling effect desired. Each system has a thermostatic expansion valve (TEV) for refrigerant control.

The low-pressure liquid (see green) of system A cools the high-pressure vapor (see blue) of system B.

Cascade systems are often used in industrial processes where objects must be cooled to temperatures below −50 F. (−46 C.).

One motor control is used for both motors. It is connected to a temperature sensing bulb on evaporator B.

Motors used on cascade systems must be capable of starting under load. With the use of thermostatic expansion valves, the pressures do not balance on the off cycle.

The condenser-evaporator is usually of the shell-and-tube flooded evaporator type.

Since these systems operate at very low temperatures, the refrigerant must be very dry. Otherwise, any moisture would condense at the needle-seat of the TEV and stop the flow of refrigerant. System A must have special refrigerant oil (wax free, moisture free and flowable at extra low temperatures).

Oil separators should be installed in the compressor-to-condenser lines on both of these condensing units to help keep the oil in the compressors.

3-13 MODULATING REFRIGERATION CYCLE

In most refrigeration installations, the cooling or refrigerating capacity is enough to maintain the desired temperature under the heaviest load. This temperature is maintained by the motor control. It starts the motor compressor when cooling (or heat removal) is required and shuts it off as soon as the desired temperature is reached.

However, if the heat load is light, this single system may be over capacity for the job. The operating expense is greater than it would be if the machine capacity more nearly matched the needed load. The system also tends to cool too fast and it operates on and off too quickly.

A modulating (varying capacity) system has been developed

to fit the machine capacity more closely to the needed heat load. This is sometimes done by using two or more compressors connected in parallel. Each compressor is operated by a motor control.

During operation, if the heat load increases and the temperature starts to rise, one compressor will continue to run. But if the temperature keeps on rising, the second compressor will start to operate. Additional compressors may cut in until enough capacity is obtained.

Fig. 3-13 illustrates a typical cycle diagram for a modulated installation. This installation has three compressors. A pressure control connected to the suction lines operates the motors. The control contains a special switching device which rotates the service of the various compressors. Thus, each compressor will be used about the same amount of time.

The modulating cycle maintains uniform temperatures and operates economically.

Any conventional refrigerant control can be used. However, the thermostatic expansion valve is most common.

The same condenser and liquid receiver may be used by all the compressors, or each may have its own. The same evaporator is connected to all the compressors.

A modulating system may use a multiple cylinder compres-

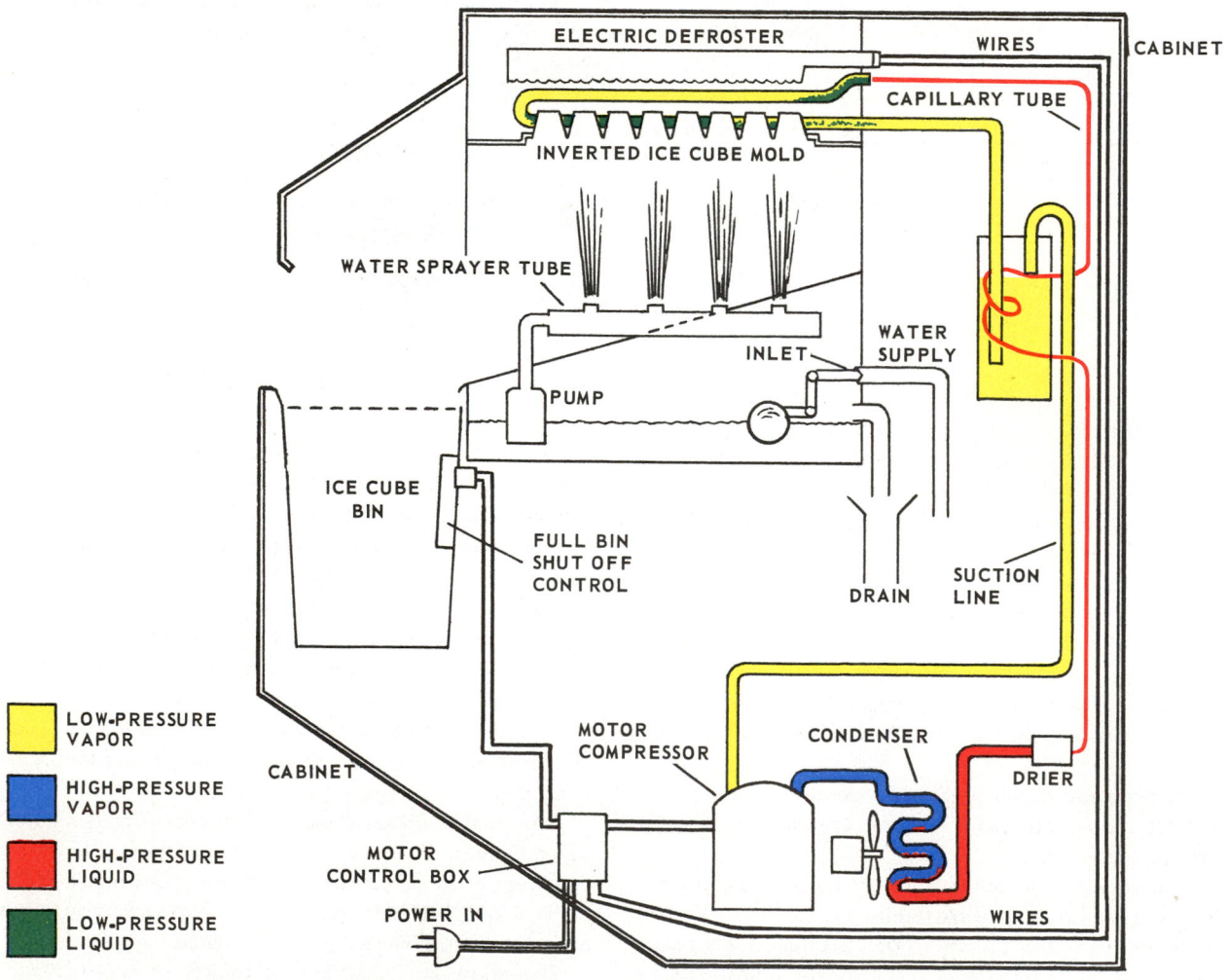

Fig. 3-14. Ice maker. Water is sprayed into ice cube molds to produce clear ice cubes.

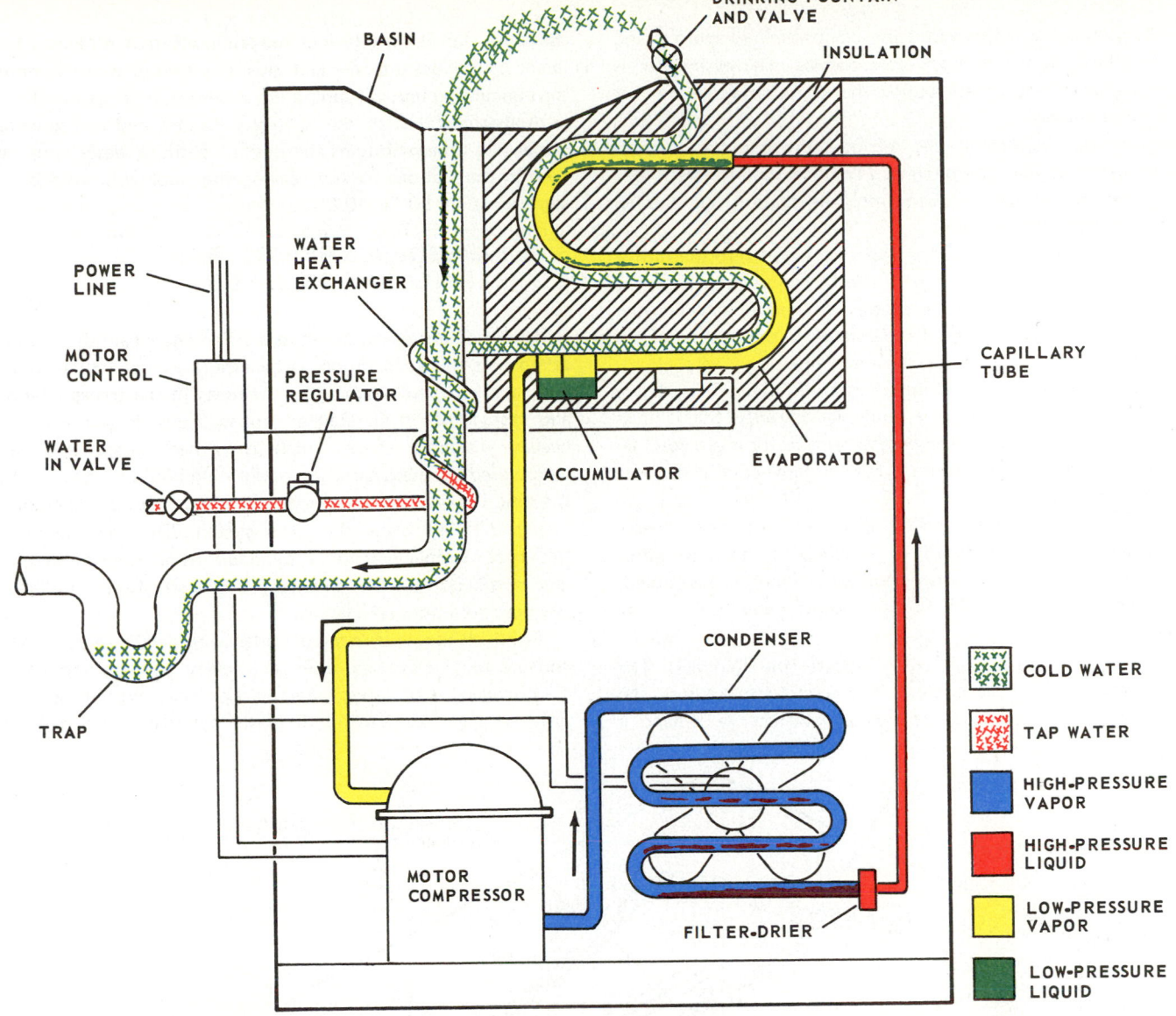

COLD WATER

TAP WATER

HIGH-PRESSURE VAPOR

HIGH-PRESSURE LIQUID

LOW-PRESSURE VAPOR

LOW-PRESSURE LIQUID

Fig. 3-15. A drinking fountain cooled by a compression system refrigerating mechanism.

sor, each cylinder being equipped with an unloader device. Variable speed motors are also used to provide a modulated refrigeration capacity.

3-14 ICE MAKER

Ice makers use various types of refrigerating systems. The simple unit in Fig. 3-14 operates as follows: the motor compressor and condenser are usually located in the bottom of the cabinet. Liquid refrigerant (see red) flows from the bottom of the condenser up through a filter-drier. It enters the evaporator through a capillary tube. The evaporator surrounds inverted (upside-down) ice cube molds.

From the evaporator, the refrigerant vapor (see yellow) flows into an accumulator. This is a type of container which has a coil from the liquid refrigerant line in it or around it. Such an arrangement serves as a heat exchanger. The refrigerant vapor (see yellow) is drawn from the accumulator back to the compressor. Here it is compressed up to the high-side pressure (see blue) and is forced into the

condenser. From here the cycle is repeated.

The mechanism which makes and handles the ice is also shown. Cold water is sprayed into the inverted ice cube molds. The temperature of the molds is very cold. Water striking the molds freezes to the mold surface and gradually builds up until complete ice cubes are formed. Then the refrigerating cycle is stopped. Now an electric heating unit heats the ice cube molds until the cubes fall out and slide down a chute into the ice cube bin. Most surfaces in contact with water and ice are stainless steel for cleanliness.

3-15 DRINKING WATER COOLER

The water cooler is a special use of a refrigerating mechanism. It is used to cool water "on tap" at a drinking fountain. The usual hermetic (airtight) compression refrigerating system is used. The refrigerant control is a capillary tube. The cycle is shown in Fig. 3-15.

Liquid refrigerant flows from the bottom of the condenser through the liquid line, into a filter-drier and into the capillary tube. As it flows into the evaporator, it vaporizes and absorbs heat from the evaporator surface. The evaporator is either adjacent to or surrounds the drinking water coil or water cooling tank.

From the evaporator, the refrigerant vapor goes into an accumulator in the suction line. (The accumulator stops any liquid refrigerant from flowing into the suction line and on into the motor compressor.)

From the accumulator, the vapor is drawn into the motor compressor where it is pumped into the condenser. Here the heat picked up in the evaporator is released. Meanwhile, the refrigerant returns to a liquid and collects in the bottom of the condenser. From here, the cycle is repeated.

Since the demand on a drinking fountain is very irregular, it is necessary that it have some hold-over capacity. Still it must not over-cool the water. The necessary capacity is provided by using either an insulated storage tank or large cooling surfaces in the evaporator.

To increase the mechanism's efficiency, the waste water flows down a tube alongside or attached to the fresh water inlet. In this way, the warmer fresh water (water-in) is cooled, to some extent, by the cooler waste water leaving the fountain.

A water pressure regulator adjusts the bubbler. The condensing unit is air cooled. However, to make sure the fountain can deliver enough cold water under heavy demand, a condenser fan is used to increase the condenser capacity. The fan is connected into the electrical circuit and runs whenever the condensing unit is running.

A thermostat with the control bulb attached to the water dispensing tube maintains the desired drinking water temperature in the bubbler. Water leaving the bubbler should be at approximately 50 F. (10 C.).

3-16 EXPENDABLE REFRIGERANT REFRIGERATION SYSTEM

This simple system, sometimes called chemical refrigeration or open cycle refrigeration, is becoming increasingly popular. It is used on trucks and other vehicles in the transportation and storage of refrigerated or frozen foods. Basically it is a heavily insulated space which is cooled either by being surrounded by tubes carrying evaporating liquid nitrogen or by spraying liquid nitrogen directly into the space to be cooled.

Fig. 3-16 illustrates the spray system. The liquid nitrogen (see red), supplied from a cylinder inside the refrigerated space, is kept under pressure (200 psi). Green indicates low-pressure liquid refrigerant.

Although the pressurized cylinder is insulated, an automatic pressure relief valve will open as a safety measure and allows the nitrogen vapor to escape should pressure exceed the relief valve setting. Heat surrounding the cylinder will sometimes

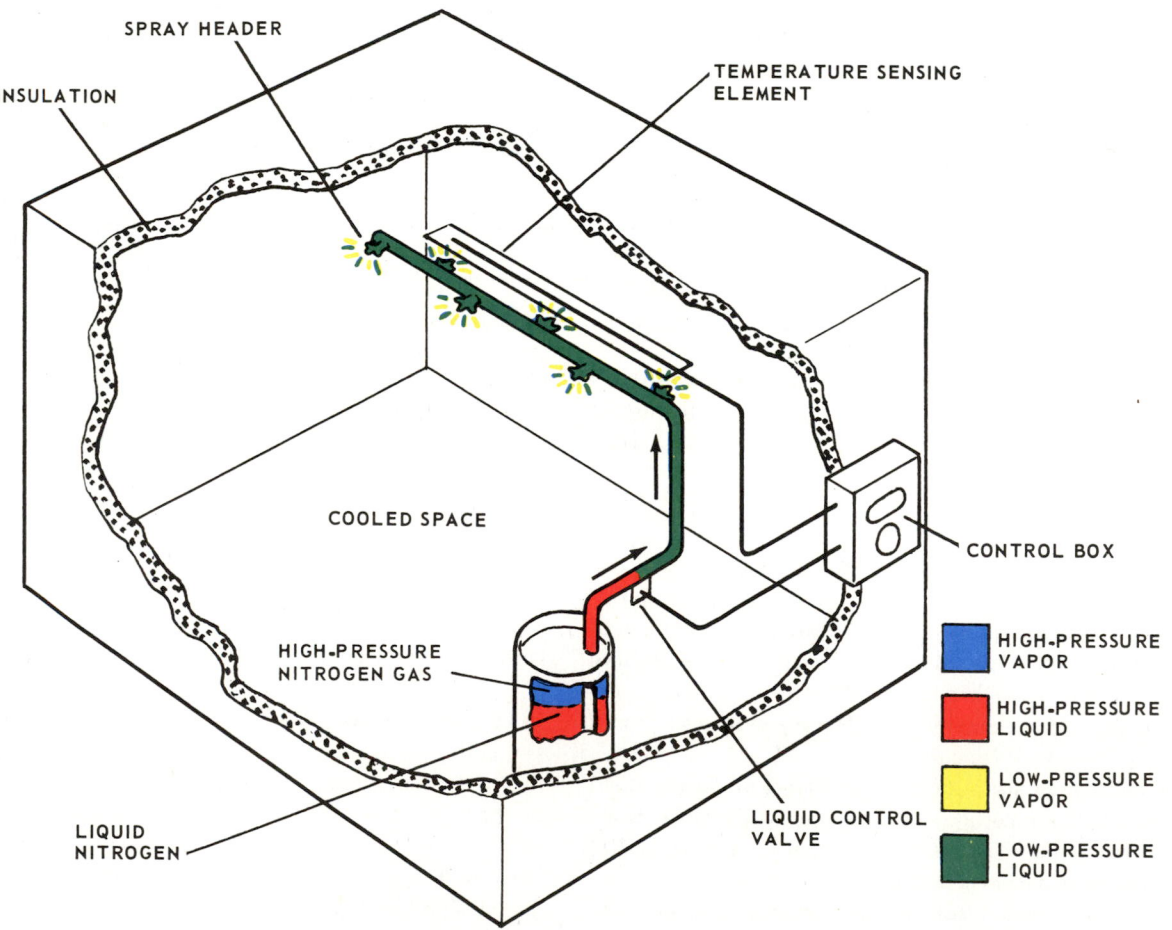

Fig. 3-16. Expendable refrigerating system.

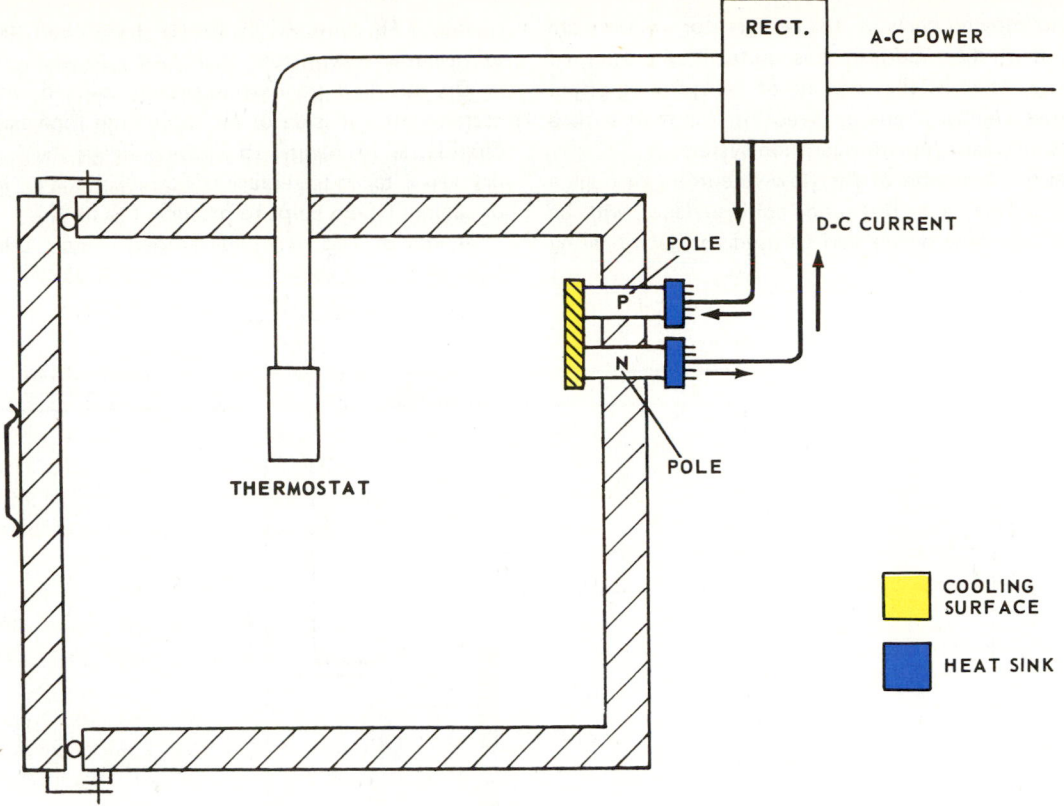

Fig. 3-17A. Diagram of simple thermoelectric couple, used for refrigerating an insulated space. Heat absorbed by thermoelectric couple is released to outside by fins attached to heat radiating surface (heat sink).

make the vapor pressure rise above the automatic pressure release setting. Cold nitrogen vapor, released by the automatic pressure release valve, is discharged into the refrigerated space, or into the refrigerating tubes, depending on the system being used.

A temperature sensing element, control box and liquid control valve, control the flow of liquid nitrogen from the nozzles. They maintain the desired temperatures inside the refrigerated space.

Liquid nitrogen (see red) vaporizes (boils and turns into a gas) at a temperature of −320 F. (−196 C.) at atmospheric pressure (Fig. 1-28). This type system is excellent for shipping frozen foods. Temperatures may be kept as low as desired — usually about −20 F. (−29 C.).

Simple construction such as this demands little attention other than to replace or recharge the nitrogen storage cylinder. Another advantage is its ability to operate without a power source. Safety devices in spaces refrigerated by liquid nitrogen shut off the flow of nitrogen when one opens a door to the space. See Chapter 17.

3-17 THERMOELECTRIC REFRIGERATION

The physical principle (Peltier effect), upon which thermoelectric refrigeration is based, has been known since 1834. This system of transferring heat energy from one place to another uses electrons instead of a refrigerant.

Fig. 3-17A represents a simple thermoelectric couple. The couple moves heat from the inside of an insulated space to a heat exchanger on the outside. Electrons, rather than refriger-

ants, carry away the heat.

Fins on the evaporator (see yellow) increase the heat flow. Fins on the outside of the heat exchanger (see blue) help give off the heat to the surrounding air.

The thermoelectric couple works because of the difference in the energy level of the two semiconductors P and N. The P and N refer to the two different semiconductor materials used (and their characteristics). This should not be confused with positive and negative polarity of an electrical circuit.

Semiconductors are metallic alloys and oxides. These vary greatly in energy levels. The electrical qualities of semiconductors lie between those of insulator and conductors. They have properties of both, thus they do not carry electric current as well as conductors and they do not stop flow of current as well as insulators. The choice of materials for the semiconductors P and N determines the efficiency of the device.

The materials P and N are not good conductors of electricity. They are, therefore, made quite large in cross-section to reduce the electrical resistance. The large size also means less heat is generated by the current flow through them.

Not much heat can be transferred by a single couple. To increase the cooling effect, several couples may be connected in series. This group of couples is called a module. See Fig. 3-17B. Groups of modules may be connected together in parallel, to increase the capacity still further. (See Chapter 6 for series-parallel connections.)

A thermostat inside the refrigerated space controls the current flow through the transformer-rectifier which supplies a controlled d-c current to the modules. In this way, the temperature inside the refrigerator is controlled.

There are no moving parts in this refrigerator. Aside from the construction of the modules, it is quite simple. Thermal efficiency is low. That is, the amount of refrigerating effect obtained for the electrical energy spent is less than with a conventional compressor type refrigeration system.

By reversing the direction of the flow of current through a thermoelectric device, the hot and cold surfaces will be reversed. Thus, the same device can be used for both heating

Fig. 3-18, view A, illustrates a common method of using dry ice as a frozen food refrigerating device.

Dry ice (see yellow) is usually packed with frozen food cartons either beside or on top of the food packages. Carbon dioxide, as it changes to a vapor, keeps the food frozen. The dry vapor tends to replace the atmospheric air in the container or cabinet which helps to preserve the food.

A device has been developed which uses dry ice for

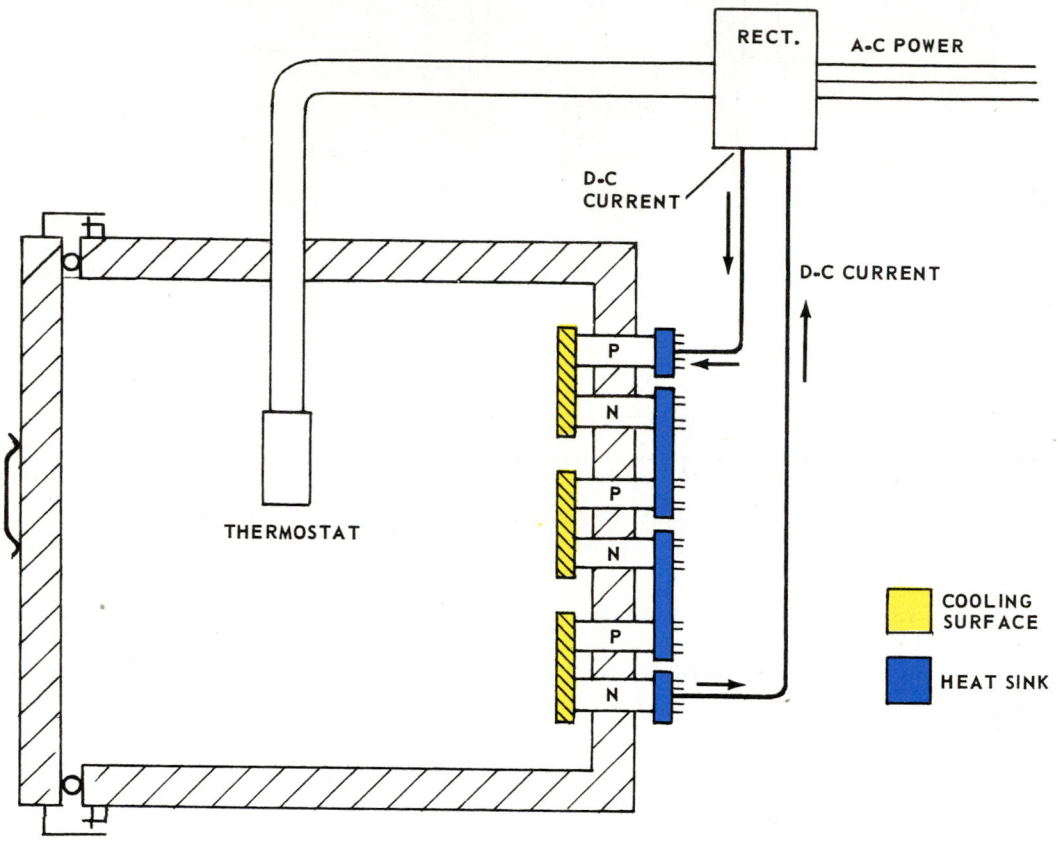

Fig. 3-17B. Thermoelectric module cooling device. Three couples are connected in series to increase heat absorbing effect.

and cooling an insulated space.

One application of this thermoelectric device has been in the air conditioning and heating of nuclear submarines. It is also used extensively to control temperatures in electronic equipment (computers, aerospace devices and so forth).

Refer to Chapter 17 for further technical information concerning thermoelectric refrigeration and air conditioning devices.

3-18 DRY ICE REFRIGERATION

Dry ice is solid carbon dioxide. It may be pressed into various sizes and shapes, blocks or slabs. As it absorbs heat, it changes directly from a solid to a vapor. It does not go through the liquid state. It is, therefore, said to sublime. At atmospheric pressure it vaporizes at —109 F. (—78 C.).

refrigerating materials carried on aircraft. A closed refrigerating circuit, containing a common refrigerant, is connected to an evaporator in the space to be refrigerated and a condenser located in an insulated bin. The bin holds dry ice pellets.

In operation, the very low temperature in the condenser (—109 F. or —78 C.), causes refrigerant vapor entering the condenser to condense quickly to a liquid. This liquid flows by gravity into the evaporator. There, it absorbs heat as it vaporizes and flows upward into the condenser. From here the cycle is repeated.

A thermostatically operated control valve, located in the liquid line, controls the flow of refrigerant into the evaporator. This device is illustrated in view B of Fig. 3-18.

Dry ice is usually stored in heavily insulated cabinets. Never handle it with bare hands. It will cause instant freeze burns. Always wear heavy gloves.

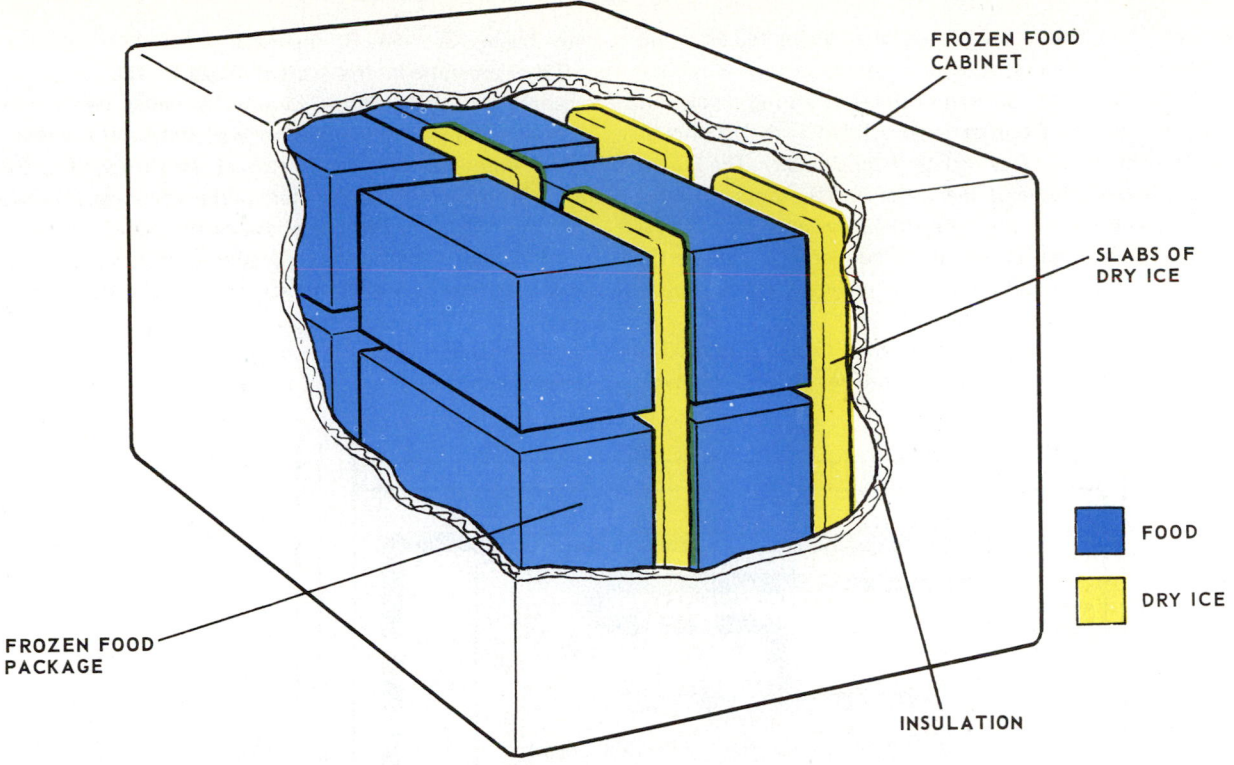

FROZEN FOOD
CABINET

SLABS OF
DRY ICE

FOOD

DRY ICE

INSULATION

FROZEN FOOD
PACKAGE

Fig. 3-18A. Dry ice frozen food container.

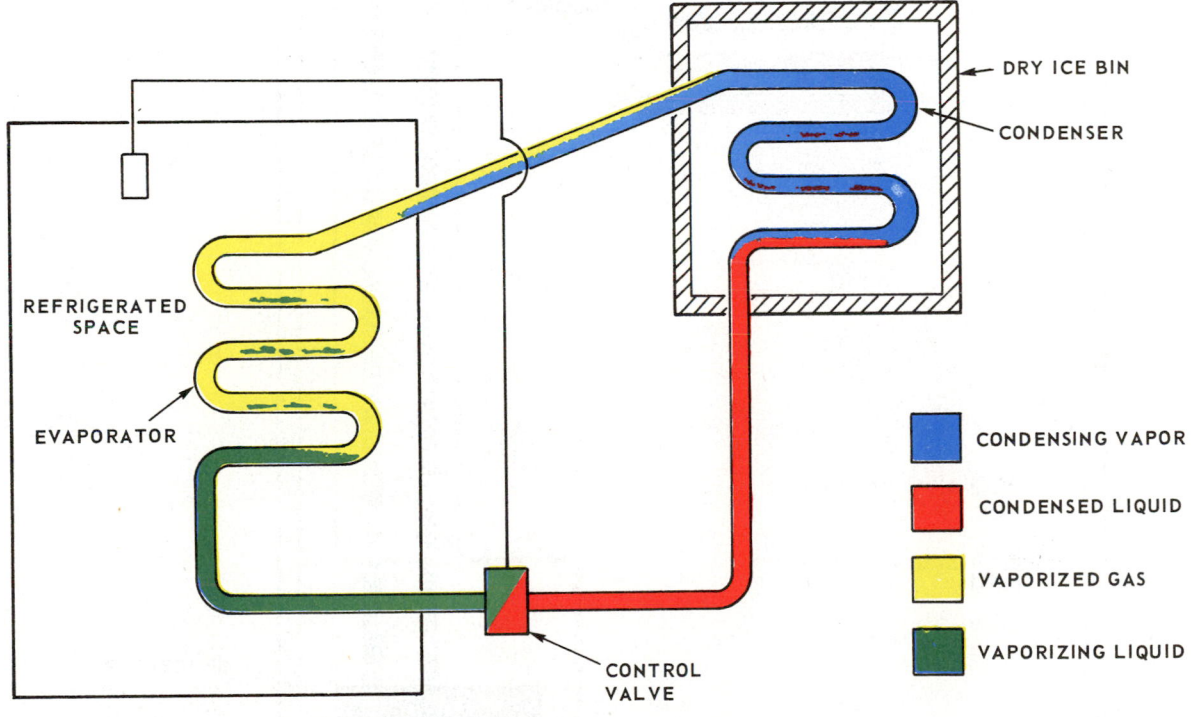

DRY ICE BIN

CONDENSER

REFRIGERATED
SPACE

EVAPORATOR

CONDENSING VAPOR

CONDENSED LIQUID

VAPORIZED GAS

VAPORIZING LIQUID

CONTROL
VALVE

Fig. 3-18B. Dry ice refrigerator does not require a compressor.

3-19 AN INTERMITTENT ABSORPTION SYSTEM

The intermittent absorption system uses a generator charged with water and ammonia. A heat source, usually a kerosene flame, heats this solution in the generator. The ammonia becomes vaporized and is driven off.

A condenser, at the top of the system, condenses the ammonia vapor into a liquid. The liquid flows, by gravity, into the liquid receiver and then into the evaporator. During the generating cycle, as explained above, little or no refrigerating effect is taking place. As the system cools, pressure drops, causing the liquid ammonia in the evaporator to boil and absorb heat. The cycle is completed when vaporized ammonia is reabsorbed in the generator.

Fig. 3-19, view A, diagrams the generating cycle. In operation, the kerosene burner tank is filled with just enough kerosene for one cycle. This usually is once a day. The burner is filled and lighted. It heats the water and ammonia mixture (see brown) in the generator. The ammonia vapor (see blue) is driven off through tube, A, up to the condenser, C, where the ammonia gas is cooled and condensed to liquid ammonia (see red). The liquid flows into the receiver.

When the kerosene has all been burned (usually from 20 to 40 minutes), the generating cycle ends. The refrigeration cycle now begins. See view B, Fig. 3-19.

The pressure in the system drops as the water cools and absorbs ammonia vapor. Liquid ammonia (see green) flows into the evaporator, begins to evaporate and cools it. Evaporated ammonia (see yellow) flows back through tube, B, and is again absorbed by the water in the generator. Refrigeration continues, usually until the next firing of the kerosene burner.

This type of refrigerating system is quite simple. The piping is welded steel since the pressures on the generating cycle are quite high. The refrigerating ability is quite good. Kerosene flame heated absorption refrigerators are popular in areas where electric power is not available.

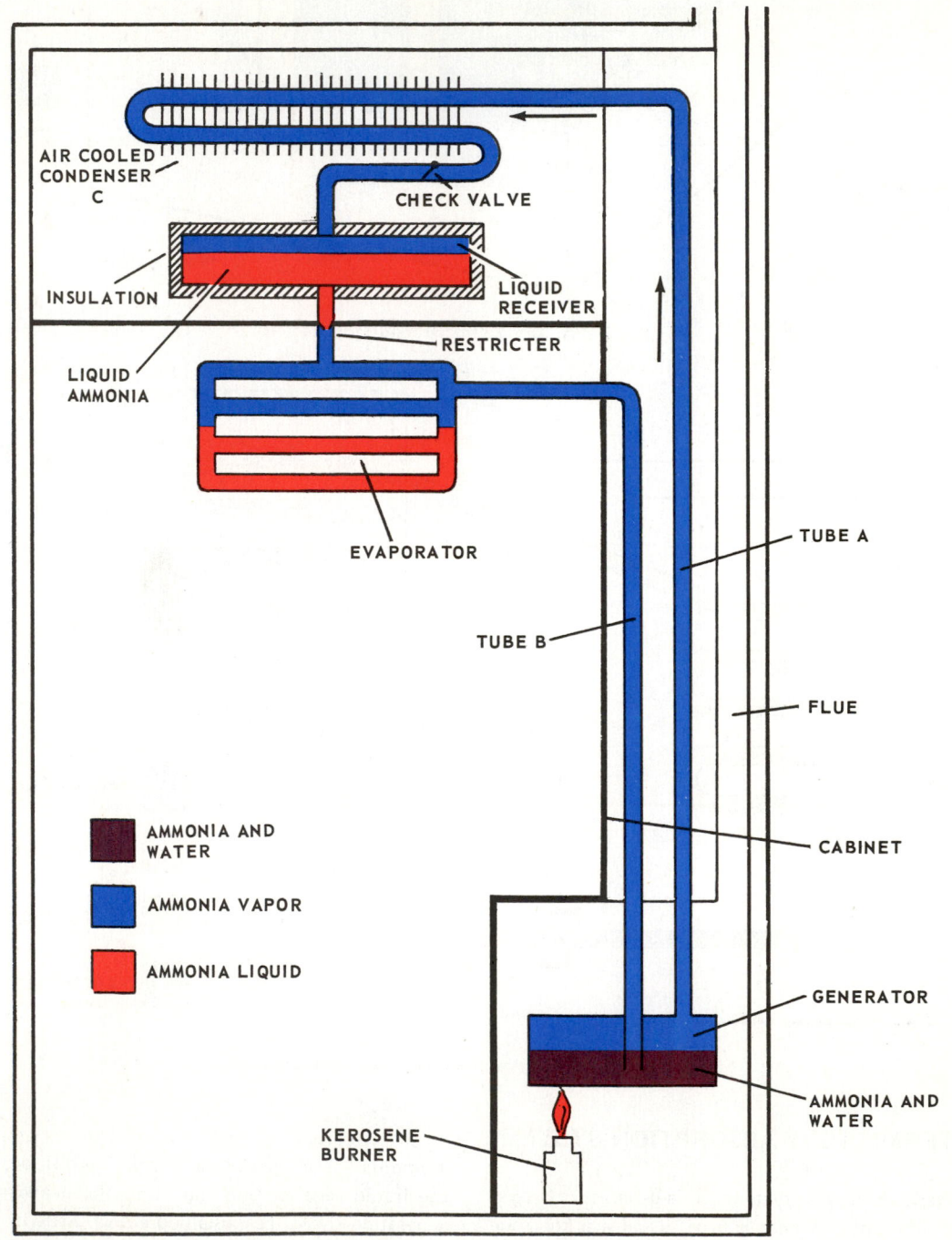

Fig. 3-19A. Intermittent absorption system during generating cycle. System is under high or condensing pressure.

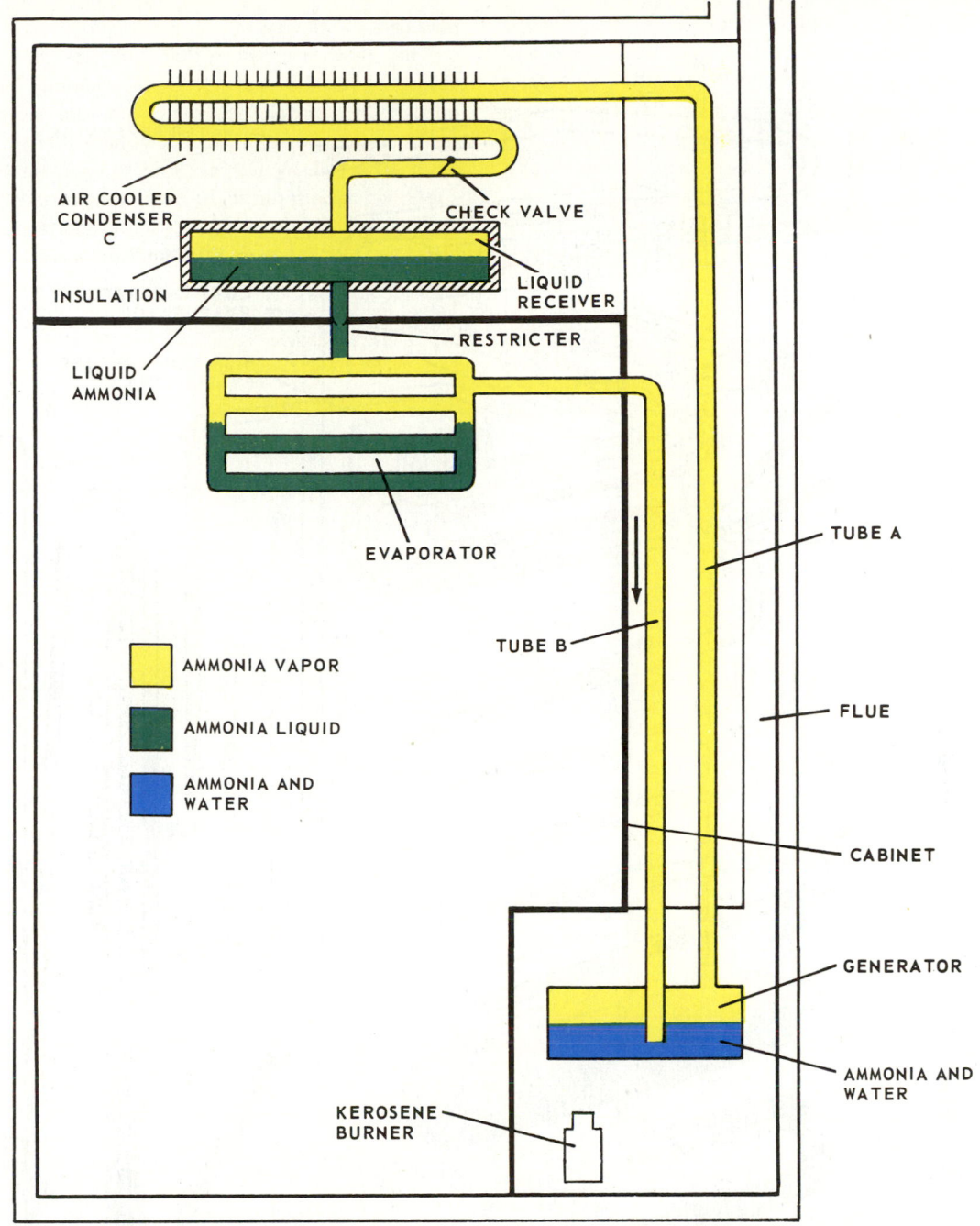

Fig. 3-19B. Intermittent absorption system during refrigerating cycle. System is under low or refrigerating pressure.

3-20 A CONTINUOUS CYCLE ABSORPTION SYSTEM

The continuous cycle absorption system generally uses heat for cycling. The heat source usually is either gas, electricity or kerosene. The operation of this refrigerating mechanism is based on Dalton's Law. See Para. 1-59.

There are four main parts to the unit (Fig. 3-20): the boiler (generator), condenser, evaporator and absorber. When the unit operates on kerosene or gas, a burner supplies heat (at the bottom of the central tube, A). This becomes the boiler.

When the unit operates on electricity, the heating element is placed in pocket, B.

The boiler contains a solution of water and ammonia. When the solution is heated, the ammonia vapor and a weak solution are driven out and up the tube, C. The weak solution passes into the tube, D, while the ammonia vapor passes into the outer tube, E. It moves on to point F. Here it is enriched by bubbling through the liquid before rising into the vapor pipe, G, and flowing on to the water separator.

As it flows into the separator, any water vapor (see blue) present condenses and flows back through tube, G, into the absorber. The ammonia vapor (red dots) flows up through tube, H, to the condenser, where it is cooled and condensed to a liquid (see red). The liquid ammonia flows into the evaporator. The evaporator and part of the absorber contain hydrogen (see yellow).

Pressure throughout the system is about 200 psi. According

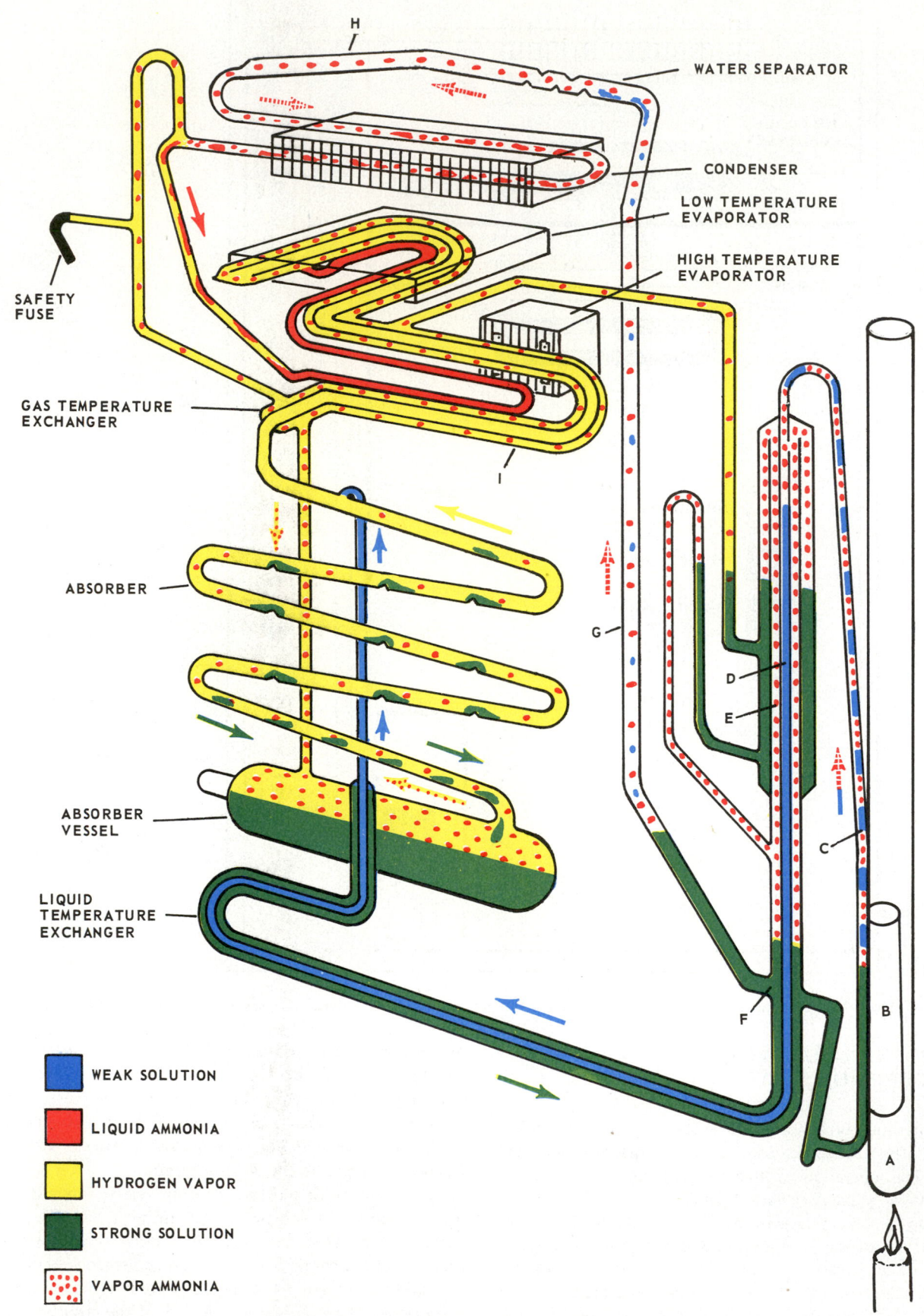

SAFETY
FUSE

WATER SEPARATOR

CONDENSER

LOW TEMPERATURE
EVAPORATOR

HIGH TEMPERATURE
EVAPORATOR

GAS TEMPERATURE
EXCHANGER

ABSORBER

ABSORBER
VESSEL

LIQUID
TEMPERATURE
EXCHANGER

H

I

G

D

E

C

F

B

A

WEAK SOLUTION

LIQUID AMMONIA

HYDROGEN VAPOR

STRONG SOLUTION

VAPOR AMMONIA

Fig. 3-20. Continuous cycle absorption system. (Electrolux AB)

to Dalton's Law, in a mixture of gases, each gas develops its own vapor pressure. So in this refrigerator the ammonia in the spaces which include hydrogen will evaporate at a low pressure and low temperature. Since the hydrogen is a large part of the gas in the space, the ammonia keeps on boiling or vaporizing at its own low pressure, absorbing heat.

The vaporized ammonia and hydrogen mixture from the evaporator are heavier than hydrogen. Gravity flow brings these vapors down through the tube, I, back to the absorber. In the absorber, the water is relatively cool. It absorbs the ammonia vapor. Then the cool water and ammonia solution returns to the generator.

The hydrogen which flows down to the absorber with the ammonia is not absorbed and, being very light, returns to the evaporator through the heat exchanger.

This cycle operates continuously as long as the boiler is heated. A thermostat which controls the heat source regulates temperature of the refrigerated space.

Since the refrigerant is ammonia, it can produce quite low temperatures. Most systems require electrical devices, so both gas and electricity must be supplied. With the exception of the thermostatic controls and (in some cases) fans, there are no moving parts.

This refrigerating device is widely used in domestic refrigerators, recreation vehicles and in year-around air conditioning of both homes and larger buildings.

Service is usually quite simple. The burner and stack must be kept clean. The refrigerator should be carefully leveled before being placed in operation.

Modern absorption systems are illustrated in Chapter 16.

3-21 SOLID ABSORBENT REFRIGERATION

Various kinds of solid absorbent refrigerators have been developed. All have depended on the original Faraday experiment.

In 1824, Michael Faraday tried to liquefy certain "fixed" gases — gases which certain scientists believed could exist only in vapor form. Among them was ammonia, then regarded as a "fixed" gas.

Faraday knew that silver chloride, a white powder, could absorb large amounts of ammonia vapor. He exposed silver chloride to dry ammonia vapor.

When the powder had absorbed all of the vapor it would take, he sealed the ammonia-silver chloride compound in a test tube which was shaped like an inverted "V". (See Fig. 3-21, view A.)

He then heated the end of the tube containing the powder (see red), and at the same time cooled the opposite end with water. The heat released ammonia vapor. Drops of a colorless liquid soon began to appear in the cool end of the tube. It was liquid ammonia.

Faraday continued the heating process until he had enough liquid ammonia for his purpose. Then, he took away the heat, removed the cooling water and watched the newly discovered substance.

Moments later, Faraday saw something unusual. The liquid ammonia, instead of remaining quietly in the sealed test tube, began to bubble and then to boil violently, Fig. 3-21, view B. The liquid was rapidly changing back into a vapor and the vapor was being reabsorbed by the powder.

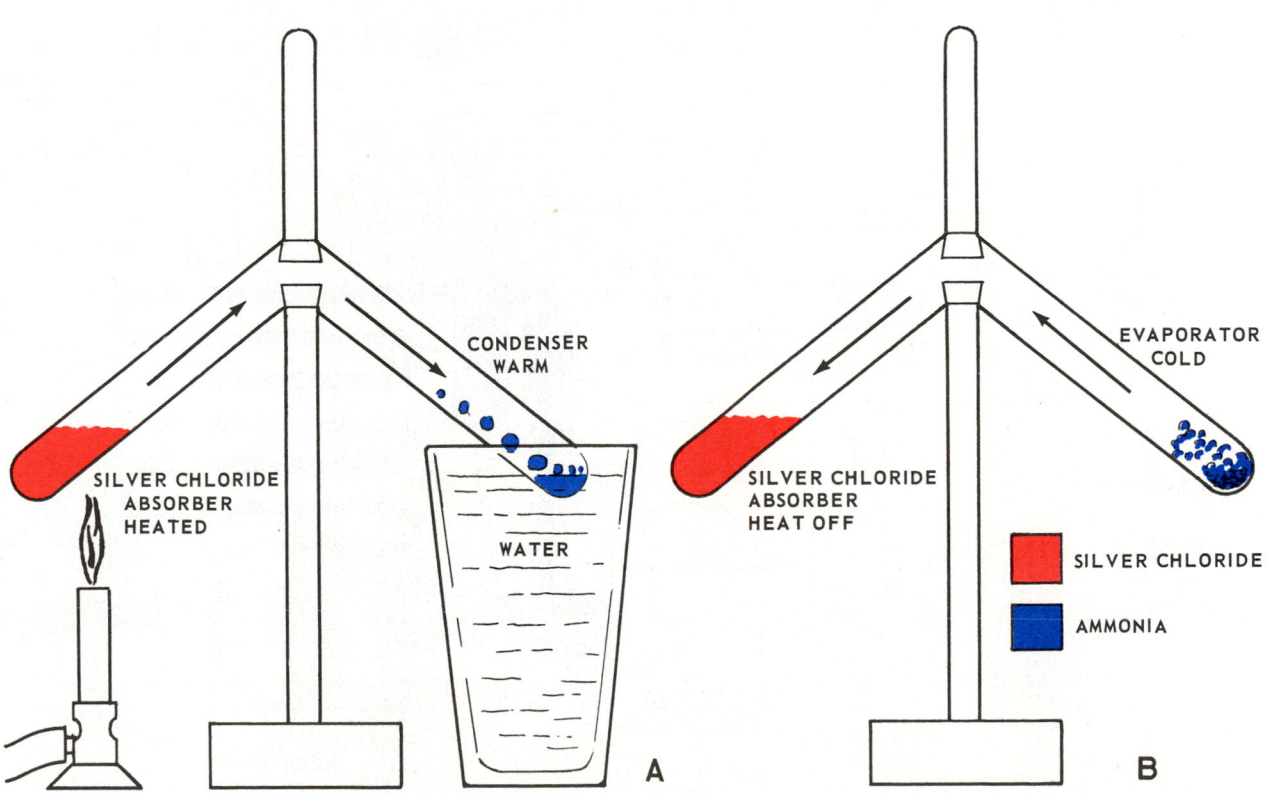

Fig. 3-21. Solid absorbent refrigerator principle as performed by Michael Faraday.

Touching the end of the tube containing the boiling liquid, Faraday found it intensely cold. Ammonia, in changing from liquid to vapor form, had removed heat. It took this heat from the nearest thing at hand — the test tube itself.

At one time many refrigerating cycles used this principle. These are not in common use at present. However, cooling mechanisms have been developed on this principle. They use water as the refrigerant and lithium bromide or lithium chloride as the absorbent. See Chapter 16.

3-22 SOPHISTICATED COMMERCIAL SYSTEMS

Up to now, the compression type refrigerating systems described have been quite simple. Different conditions and refrigeration requirements require accessory (add-on) devices. Pressure regulators, vibration dampeners, crankcase heaters and other mechanisms make the refrigerating systems more efficient and safer. Fig. 3-22 illustrates a small commercial type refrigerating system using a variety of accessory devices.

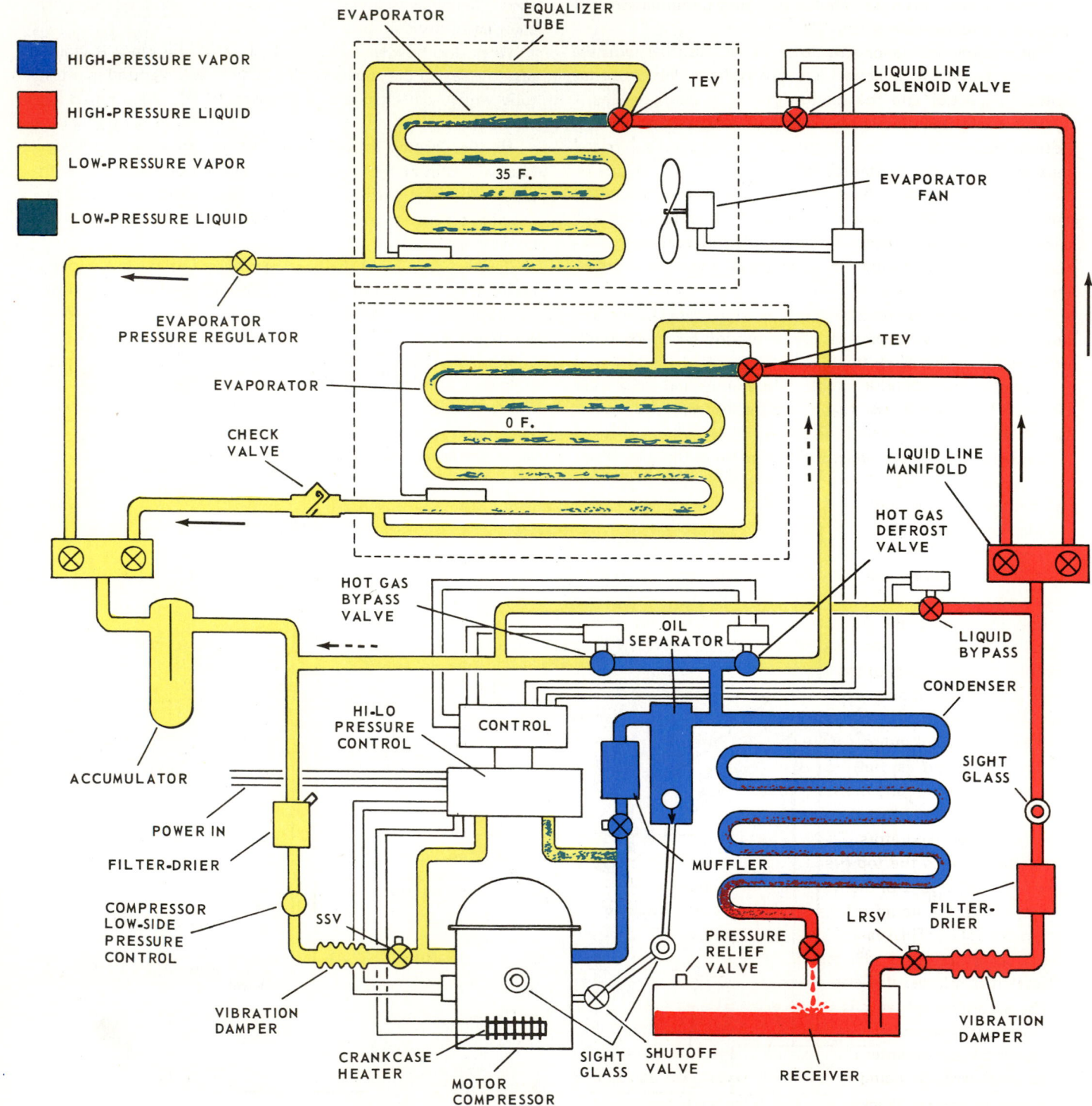

Fig. 3-22. Accessory parts on this commercial system make unit work better. It is also easier to service.

Beginning with the evaporator in the top (warmer) cabinet, refrigerant evaporates (see yellow) and flows back through the suction line toward the motor compressor. The suction line then enters an evaporator pressure regulator. From there it leads to the suction line accumulator. Any liquid refrigerant which may come from the evaporator will stay here and evaporate, preventing it from slugging into or entering the compressor.

The vapor then goes through a suction line filter-drier, which traps any moisture or solid impurities.

A compressor pressure regulator protects the compressor from excessive low-side pressures.

The suction line vapor then enters a vibration dampener. This is a flexible connection on the low side between the motor compressor and the suction line. The sensing element for the motor compressor control is also attached to the suction line.

A service valve is located at the entrance to the low side of the motor compressor. Along with the service valve on the high side, this makes servicing the motor compressor easy.

A crankcase heater keeps refrigerant from liquefying in the motor compressor during the off cycle when the unit is operating in a cold space.

From the high-side service valve, the compressed vapor (see blue) enters an oil separator. Oil removed from the high-pressure refrigerant is returned to the compressor crankcase.

From the oil separator, the high-pressure vapor enters the condenser. The condenser has a head pressure control. When head pressure gets too high, it shuts off the system.

A service valve is placed in the line between the condenser and the liquid receiver. The liquid receiver is a reservoir for liquid refrigerant (see red).

Another service valve is located at the outlet of the liquid receiver. This makes it possible to remove the receiver from the system or store the refrigerant in the receiver during service operations.

The line leaving the liquid receiver is also fitted with a vibration dampener. This flexible tube stops carry over of vibration to other parts of the system.

As the liquid moves on, it passes through a filter-drier which helps keep the refrigerant clean and dry. Then a moisture and liquid indicator allows visual inspection to see if enough refrigerant is flowing. At the same time, its color indicates presence of any moisture in the refrigerant.

A manifold with hand valves allows the liquid to pass into one of the two evaporators. Then the refrigerant flows through a solenoid valve to the top evaporator. This makes it possible to automatically control the flow into the evaporator.

From here, the liquid refrigerant flows into a thermostatic expansion valve. This valve regulates the rate of flow (see green) depending upon the temperature and pressure of the refrigerant as it leaves the evaporator.

The second evaporator is fitted with an electrically operated defrost control device. When opened, it allows hot compressed vapor to enter the evaporator and flow back to the compressor without going through the expansion valve. This heats the evaporator quickly and any frost accumulation on it will be quickly melted.

A hot gas bypass solenoid valve is used to allow hot refrigerant vapor to enter the suction line in case the suction line gets too cold and may allow liquid refrigerant to enter the compressor.

This brief description of some of the accessories and their purpose may help in understanding the commercial systems as described later in the text.

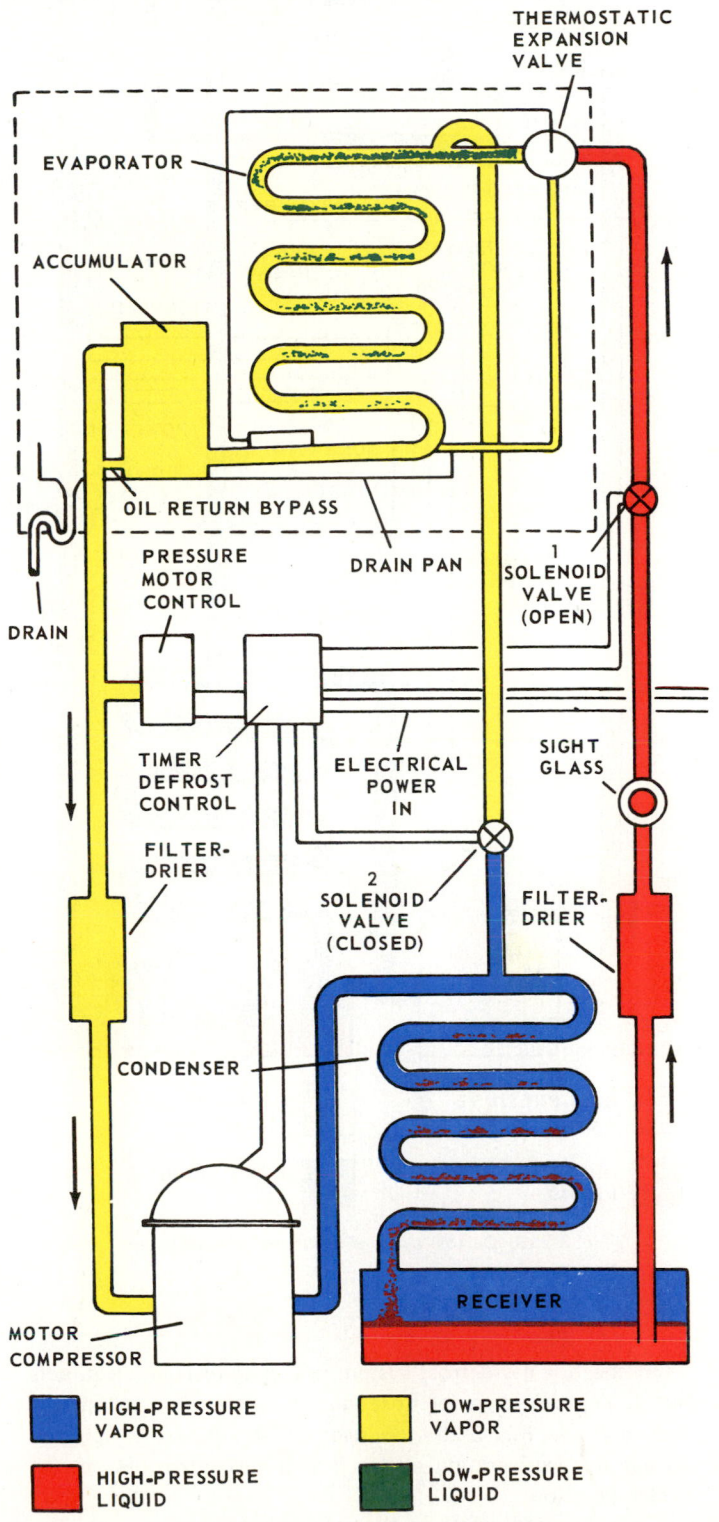

Fig. 3-23A. Normal refrigerating cycle of hot gas defrost system.

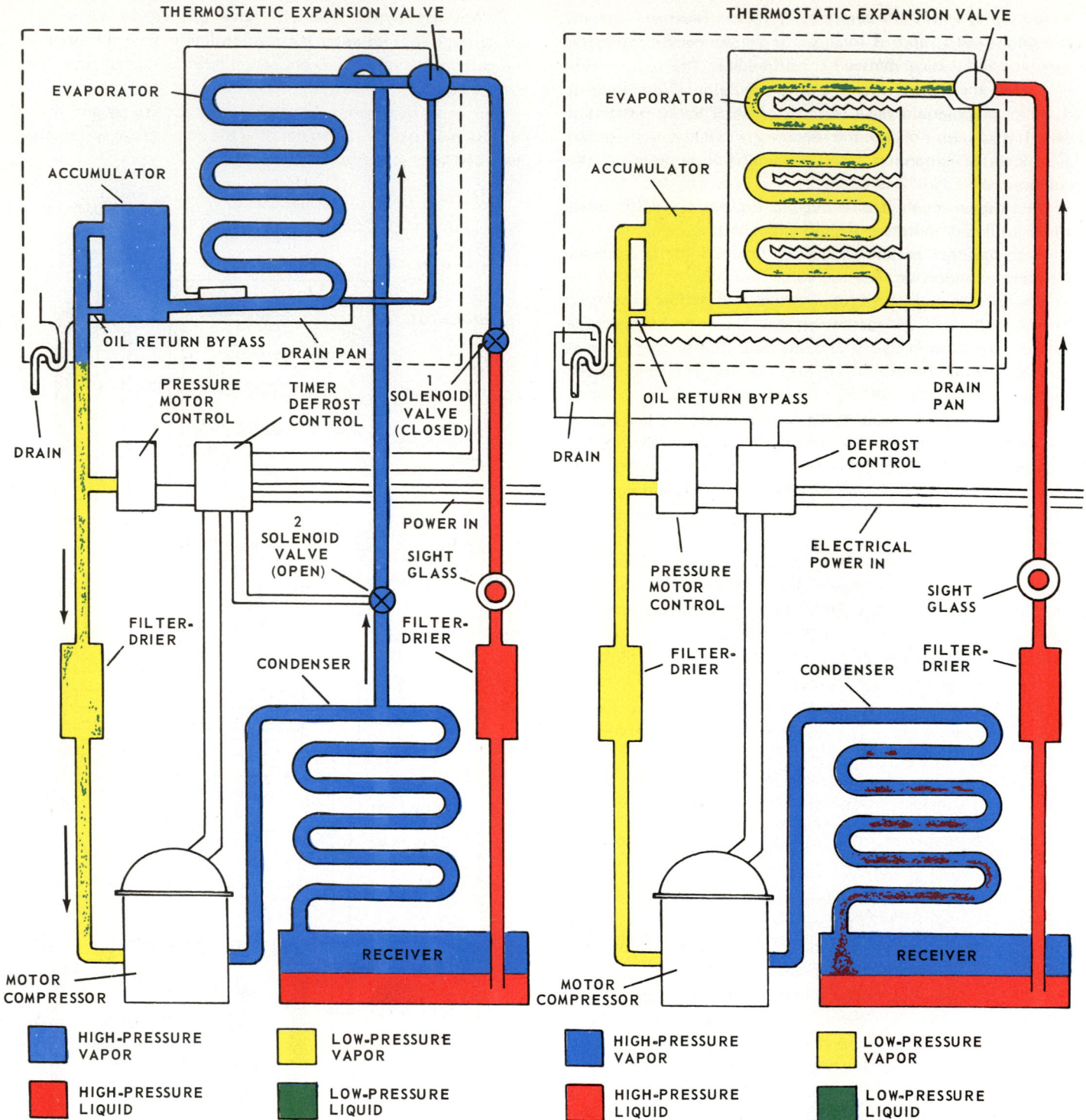

THERMOSTATIC EXPANSION VALVE

EVAPORATOR

ACCUMULATOR

OIL RETURN BYPASS

DRAIN PAN

DRAIN

PRESSURE MOTOR CONTROL

TIMER DEFROST CONTROL

1 SOLENOID VALVE (CLOSED)

2 SOLENOID VALVE (OPEN)

POWER IN

SIGHT GLASS

FILTER-DRIER

CONDENSER

FILTER-DRIER

MOTOR COMPRESSOR

RECEIVER

| HIGH-PRESSURE VAPOR | LOW-PRESSURE VAPOR |
| HIGH-PRESSURE LIQUID | LOW-PRESSURE LIQUID |

Fig. 3-23B. Defrost cycle of a hot gas defrost system.

THERMOSTATIC EXPANSION VALVE

EVAPORATOR

ACCUMULATOR

OIL RETURN BYPASS

DRAIN PAN

DRAIN

DEFROST CONTROL

PRESSURE MOTOR CONTROL

ELECTRICAL POWER IN

FILTER-DRIER

SIGHT GLASS

CONDENSER

FILTER-DRIER

MOTOR COMPRESSOR

RECEIVER

| HIGH-PRESSURE VAPOR | LOW-PRESSURE VAPOR |
| HIGH-PRESSURE LIQUID | LOW-PRESSURE LIQUID |

Fig. 3-24A. An electric defrost system during the refrigerating cycle.

3-23 HOT GAS DEFROST

In the hot gas defrost system, a timing mechanism directs hot high-pressure vapor through the evaporator to remove frost and ice. Fig. 3-23A on page 95 shows how it operates during the refrigerating cycle; Fig. 3-23B illustrates defrost cycle operation.

Two solenoid valves in the refrigerant circuit control the system to provide either the refrigerating cycle or the defrost

cycle. During the refrigerating cycle, view A of Fig. 3-23, solenoid valve No. 1 is open. The refrigerator is operating normally for a refrigerator using a thermostatically controlled expansion valve.

The liquid refrigerant flows from the liquid receiver up through the liquid line, through solenoid valve No. 1, through the thermostatic expansion valve and into the evaporator. It evaporates under low pressure, absorbs heat and returns as a vapor first through the accumulator and then through the suction line back to the compressor. From the compressor, the hot compressed vapor is forced into the condenser. Its heat is removed and it is condensed back into a liquid.

During the defrost cycle, B in Fig. 3-23, solenoid valve No. 1 is closed. Solenoid valve No. 2 is open. Since solenoid valve No. 1 is closed, no liquid refrigerant is flowing through the thermostatic expansion valve into the evaporator. Since solenoid valve No. 2 is open, the hot compressed refrigerant vapor flows through it directly into the evaporator. It passes through the evaporator, through the accumulator and back through the suction line to the suction side of the hermetic (airtight) compressor. The hot vapor, as it passes through the evaporator, melts the ice from the evaporator surface.

The vaporized refrigerant picks up some heat passing through the suction line. The heat of compression, as it passes through the compressor, raises the temperature to such a level that it continues to heat the evaporator and remove the frost. Little or no condensation of vaporized refrigerant takes place during this cycle.

3-24 ELECTRIC DEFROST

Electric heating elements placed alongside the evaporator surfaces heat up to melt the frost and ice buildup from the evaporator. A timer or control mechanism operates the heater during the time that the refrigerating mechanism is on the off cycle. View A in Fig. 3-24 shows the refrigerating cycle; Fig. 3-24, view B, shows the defrost cycle.

In A of Fig. 3-24, the electric heating mechanism is in the refrigerating part of the cycle. Liquid refrigerant is vaporized in the evaporator. It absorbs heat and becomes a vapor. While in the evaporator, it flows through an accumulator, passes on to the suction line back to the compressor. In the compressor, it is compressed to a high pressure, high temperature and flows into the condenser. Here the heat of vaporization is removed and the refrigerant returns to a liquid flowing into the liquid receiver. From here the cycle is repeated.

In B of Fig. 3-24, the same system is in the defrost cycle. The compressor is stopped and the defrost control mechanism lets electric current flow through the resistance heating elements alongside the evaporator surface. Heat warms the evaporator surfaces until the frost and ice are melted and the moisture empties into a drain pan.

The operation of the resistance units is usually timed to control both the frequency and the duration of the electric heating. This timing provides for adequate frost removal and the system is helped to operate efficiently with little or no frost on the evaporator surfaces.

3-25 TEST YOUR KNOWLEDGE

1. List some advantages of the old icebox type of refrigeration.
2. What is the principle upon which the "desert bag" water cooling device operates?
3. What are some of the advantages of the capillary tube type refrigerant control?
4. What causes the needle valve to open in an automatic expansion valve refrigerant control?

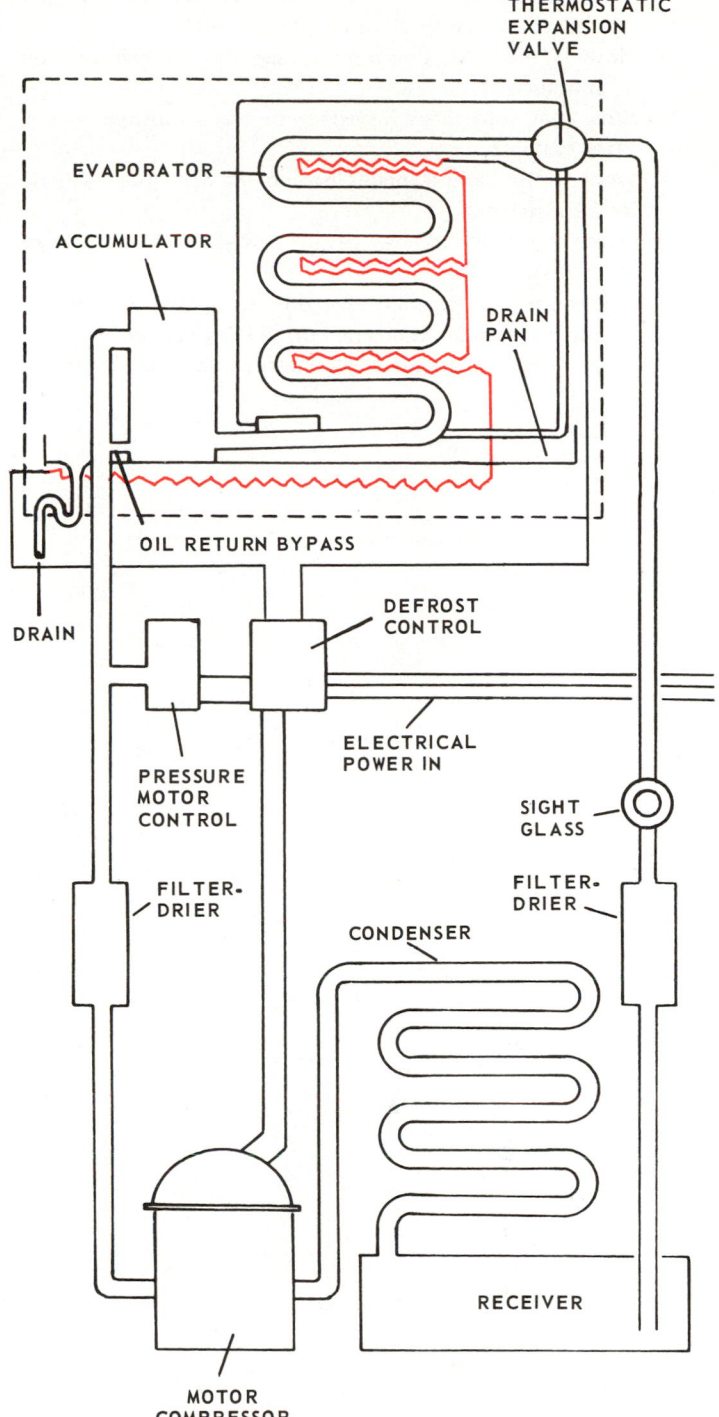

Fig. 3-24B. An electric defrost system during the defrost cycle.

5. What causes the needle valve to open in the thermostatically controlled expansion valve?
6. What controls the cabinet temperature in a system which uses a thermostatically controlled expansion valve?
7. What substances are most used in absorption type refrigerators?
8. What did Faraday discover in his famous experiment?
9. What is the chief advantage of a compound system?
10. What is the chief advantage of a cascade system?

11. In a thermoelectric refrigeration system, how is heat carried out of the space to be refrigerated?

12. How is modulation (varying capacity) in refrigeration accomplished?

13. Why is it sometimes necessary to use a multiple evaporator system?

14. What is the basic principle behind the operation of most ice cube makers?

15. Name some advantages of the low-side float type of refrigerant control.

16. Is amount of refrigerant charged into system very important in a high-side float type refrigerating system?

17. What principle is used in the continuous type absorption refrigerator?

18. Where is "expendable refrigerant" cooling most used?

19. What is the chief disadvantage of thermoelectric refrigeration?

20. What are some of the advantages of dry ice refrigeration?

21. In what kind of refrigerating mechanism are semiconductors used?

22. What is the refrigerant most commonly used in "expendable refrigerant" cooling systems?

23. In "cascade" systems, are the refrigerating temperatures usually above or below 0 F. (—18 C.)?

24. What is another name for an "open" refrigerating system?

25. Why is hydrogen used in some absorption systems?

Chapter 4

COMPRESSION SYSTEMS AND COMPRESSORS

The compression system is the basis of operation of the refrigeration units described in this chapter. One must understand the system to accurately diagnose (identify) mechanical difficulties. Many types of compression mechanisms are explained to help the service technician become familiar with their basic operations.

The arrangement of the units of study in this chapter will be in the same order as the arrangement of the parts in a refrigerating system. Beginning with the evaporator (where the heat is absorbed), the sequence (order of study) is through the suction line, into the compressor; then through the condenser, liquid receiver, filter-drier, liquid line, refrigerant control and back into the evaporator.

Necessary details concerning the construction and operation of the parts — as well as of various types of these mechanisms — are explained fully. Included are necessary temperature and motor controls.

4-1 LAWS OF REFRIGERATION

All refrigerating systems depend on five thermal laws:

1. Fluids absorb heat while changing from a liquid state to a vapor state and give up heat in changing from a vapor to a liquid.
2. The temperature at which a change of state occurs is constant during the change provided the pressure remains constant.
3. Heat flows only from a body which is at a higher temperature to a body which is at a lower temperature (hot to cold).
4. Metallic parts of the evaporating and condensing units use metals which have a high heat conductivity (copper, brass, aluminum).
5. Heat energy and other forms of energy are interchangeable. For example, electricity may be converted to heat; heat to electrical energy and heat to mechanical energy.

4-2 COMPRESSION CYCLE

The compression cycle is so named because it is the compressor which changes the refrigerant vapor from low pressure to high pressure. This pumping causes the transfer of heat energy from the inside of the cabinet to the outside. Since the compression machine transfers heat from one place

to another, it may also be called a heat pump.

A refrigerating system consists principally of a high-pressure side and a low-pressure side, Fig. 4-1.

A refrigeration cycle follows these steps: From the liquid receiver, G, liquid refrigerant, at a high pressure, flows through the refrigerant control, A, (a pressure reducer). It moves into the evaporator, B. The evaporator is under a low pressure. Here the liquid refrigerant vaporizes (boils) and absorbs heat.

The vapor then flows into the compressor through the intake valve, C, back into the compressor cylinder. The piston, D, on the compression stroke, squeezes the vapor into a small

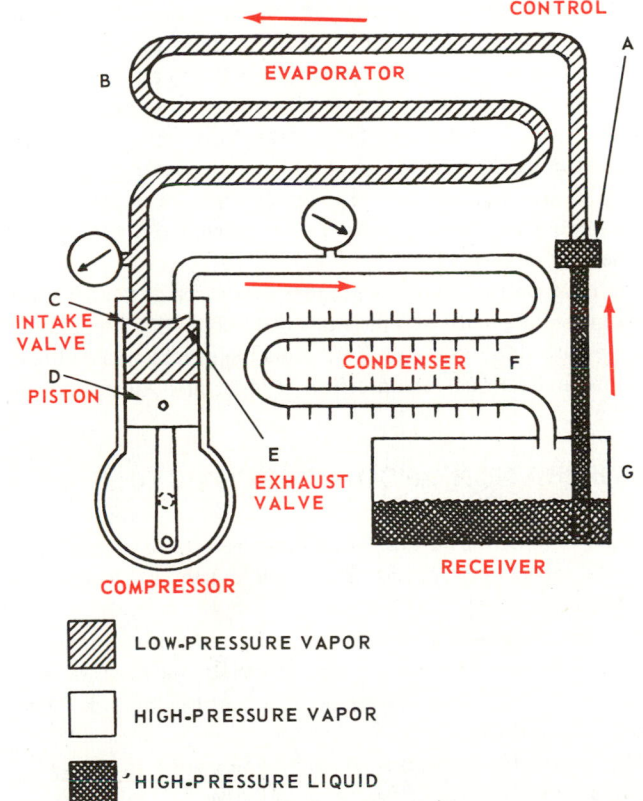

LOW-PRESSURE VAPOR

HIGH-PRESSURE VAPOR

HIGH-PRESSURE LIQUID

Fig. 4-1. Compression cycle showing the two pressure conditions. Low-pressure side extends from refrigerant control, A, through evaporator, B, to the compressor intake valve, C. High-pressure side begins in the cylinder above the piston, D, on the compression stroke. It extends from exhaust valve, E, through condenser, F, liquid receiver, G, and liquid line, to refrigerant control, A.

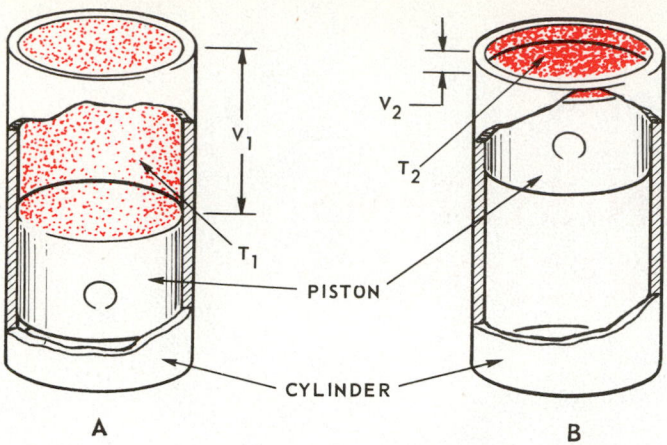

Fig. 4-2. Heat of the vapor compressed into a small space raises vapor temperature greatly. A—V_1 = volume of vapor at the end of the intake stroke = 8 cu. in. (131 cm^3). T_1 = temperature of vapor at end of intake stroke = 50 F. (10 C.). B—V_2 = the volume of the vapor at the end of compression stroke = 1/2 cu. in. (8.2 cm^3). T_2 = the temperature of the vapor at the end of the compression stroke = 250 F. (121 C.).

space with an increase in temperature. Fig. 4-2 illustrates this principle.

In Fig. 4-1, the compressed high-temperature vapor is pushed through the exhaust valve, E, into the condenser, F. In the condenser, heat from the refrigerant is passed on to the surrounding air. In giving up this heat, it returns to a liquid and is stored in the receiver. From here the cycle is repeated.

In operation, the apparatus transfers heat from one place to another place. That is, it takes heat from the inside of a refrigerator to the outside air or from the water of a water cooler to the outside air. This action may be compared to using a sponge to pick up water in one place and releasing it in another by squeezing it.

To have a transfer of heat, there must be a temperature difference. To get the temperature difference, there must be a low-pressure side (heat absorber) and a high-pressure side (heat dissipator). Various compression cycles are illustrated in Chapter 3.

4-3 OPERATION OF COMPRESSION CYCLE

Fig. 4-3 illustrates a typical compression cycle as used in a domestic refrigerator. It has certain necessary parts which will be explained.

In any compression refrigeration system, there are two different pressure conditions. One is called the low side and the other the high side. The evaporator is in the low side. Heat is absorbed in the low side.

The accumulator, suction line and entrance to the compressor suction valve are also on the low side.

The condenser is in the high side, where the heat is released from the refrigerant. The compressor exhaust valve, liquid receiver (if used), liquid line filter-drier, liquid line and the refrigerant control are also on the high side.

A thermostat maintains correct operating temperature by controlling the motor electrical circuit.

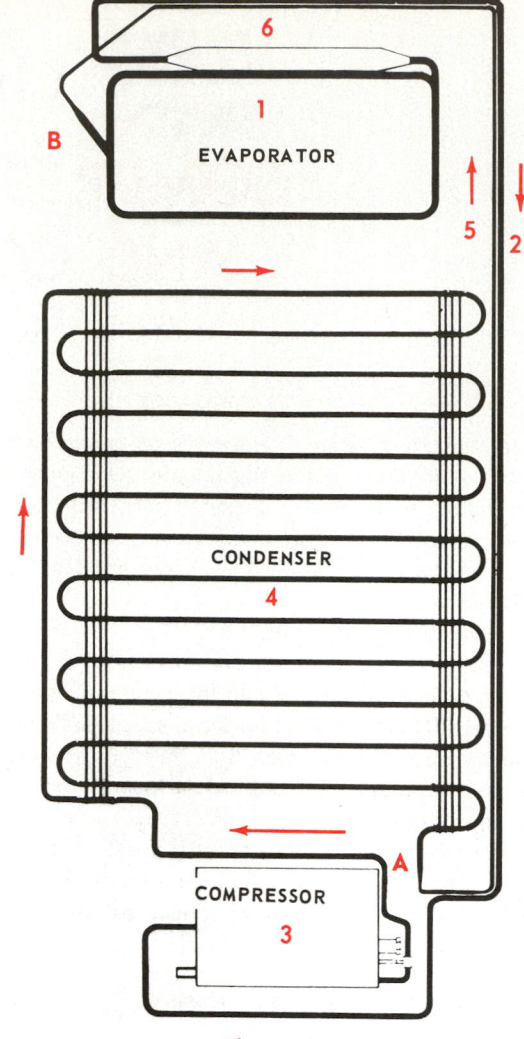

Fig. 4-3. Compression cycle showing the flow of refrigerant. 1—Evaporator. 2—Suction line. 3—Compressor. 4—Condenser. 5—Capillary tube, A to B. 6—Accumulator. (Hotpoint Div., General Electric Co.)

4-4 TEMPERATURE AND PRESSURE CONDITIONS IN THE COMPRESSION CYCLE

When the compressor starts, it moves molecules of refrigerant from the low-pressure side to the high-pressure side without much difficulty. After this change, the molecules are not moving about much faster. See Fig. 4-4A. These molecules of refrigerant enter the condenser from the compressor through opening at 1. The temperatures are the same (70 deg. F.) inside and out.

Pressure is the sum of the bombarding molecules and temperature is the speed of molecular motion. (How fast the molecules move to and fro.)

It is necessary then to speed up the molecules. This will raise their temperature to where they can give up heat to surrounding cooling surfaces (air and water).

The longer the compressor runs, the more vapor molecules it squeezes into the condenser. With each stroke, the pressure

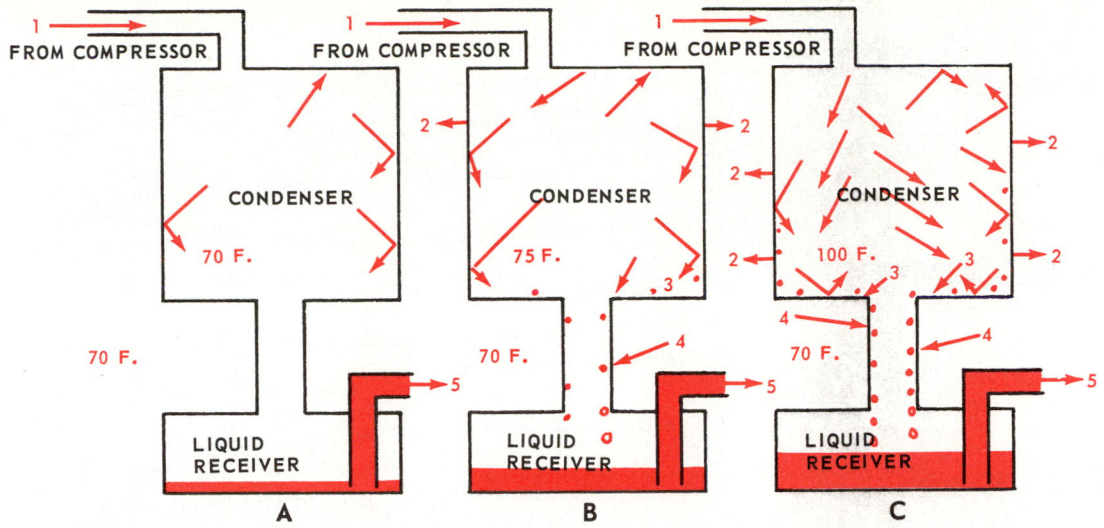

Fig. 4-4. Refrigerant condition as it changes from vapor to liquid in condenser. At A, condensing unit is just starting; B, unit has been operating long enough to condense some refrigerant vapor. At C, unit is in state of equilibrium (balance), heat is being removed and vaporized refrigerant is being condensed at same rate it is being pumped into condenser. A, B, C—1. Vapor enters under pressure. B—2. Heat moving from condenser (small amount). B—3. Vapor losing heat and condensing to liquid (small amount). B—4. Condensed refrigerant enters liquid receiver (small amount). C—2. Heat moving from condenser (large amount). C—3. Condensed refrigerant droplets (large amount). C—4. Condensed refrigerant flowing into liquid receiver (larger amount). Refrigerant liquid line to refrigerant control is shown at 5.

and temperature increase since more molecules are hitting the sides of the container. The compressor piston, pushing the vapor molecules against the higher pressure, hits them harder, speeding them up (increasing temperature).

During compression, the pressure increases (due to Boyle's Law, Chapter 28). At the same time, the temperature increases (explained by Charles's Law, Chapter 28) until the vapor temperature is higher than the temperature of the condenser cooling medium.

The higher temperature, as shown at B, Fig. 4-4, causes a flow of heat to the surrounding metal and air. Heat now moves from the vaporized refrigerant as shown at B2 to the cooling medium.

This cooling continues until enough heat loss makes some vapor molecules become liquid molecules as in B3. As these collect, they flow into the liquid receiver. See B4.

The temperature and pressure will continue to rise until a balance is reached. That is, just as many vapor molecules

condense into a liquid as the compressor pumps into the condenser, Fig. 4-4C.

If anything changes this balance, the condensing pressure and temperature will adjust accordingly. For example, if the room were to get warmer, the pressure and temperature will rise again until just as many vapor molecules are condensing as are being pumped into the condenser.

After condensing (liquefying), the refrigerant is stored in the liquid receiver until needed. When needed it passes through the high-pressure liquid line, 5, to the refrigerant control. Here refrigerant pressure is reduced to allow evaporation of the liquid at a low temperature.

The evaporating liquid absorbs much heat and so furnishes refrigeration. The increase in its volume, as it evaporates, pushes the vaporized refrigerant through the "suction" line. It comes, finally, to the compressor intake, where pressure is greatly reduced. It then passes through the intake valve of the compressor into the cylinder where a new cycle begins.

4-5 EVAPORATOR

The liquid refrigerant entering the evaporator from the refrigerant flow control is suddenly under low pressure. This makes it vaporize (boil) and absorb heat. The vapors move on into the suction line. If all of the liquid refrigerant has not vaporized in the evaporator, there is usually a cylinder (accumulator) to prevent liquid refrigerant from flowing into the suction line.

Evaporators are mainly of two types, a dry system and the flooded system. Refrigerant is fed into the dry system evaporator only as fast as is needed to maintain the temperature wanted. In the flooded system, the evaporator is always filled with liquid refrigerant. The type of refrigerant control used determines the type of evaporator to be used.

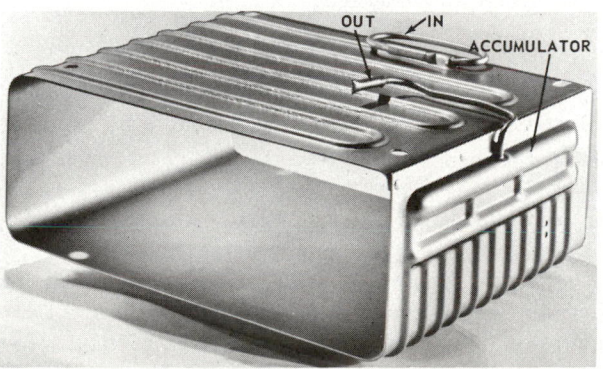

Fig. 4-5. Shell type evaporator used with capillary tube or high-side float type refrigerant control. (Houdaille-Hersey)

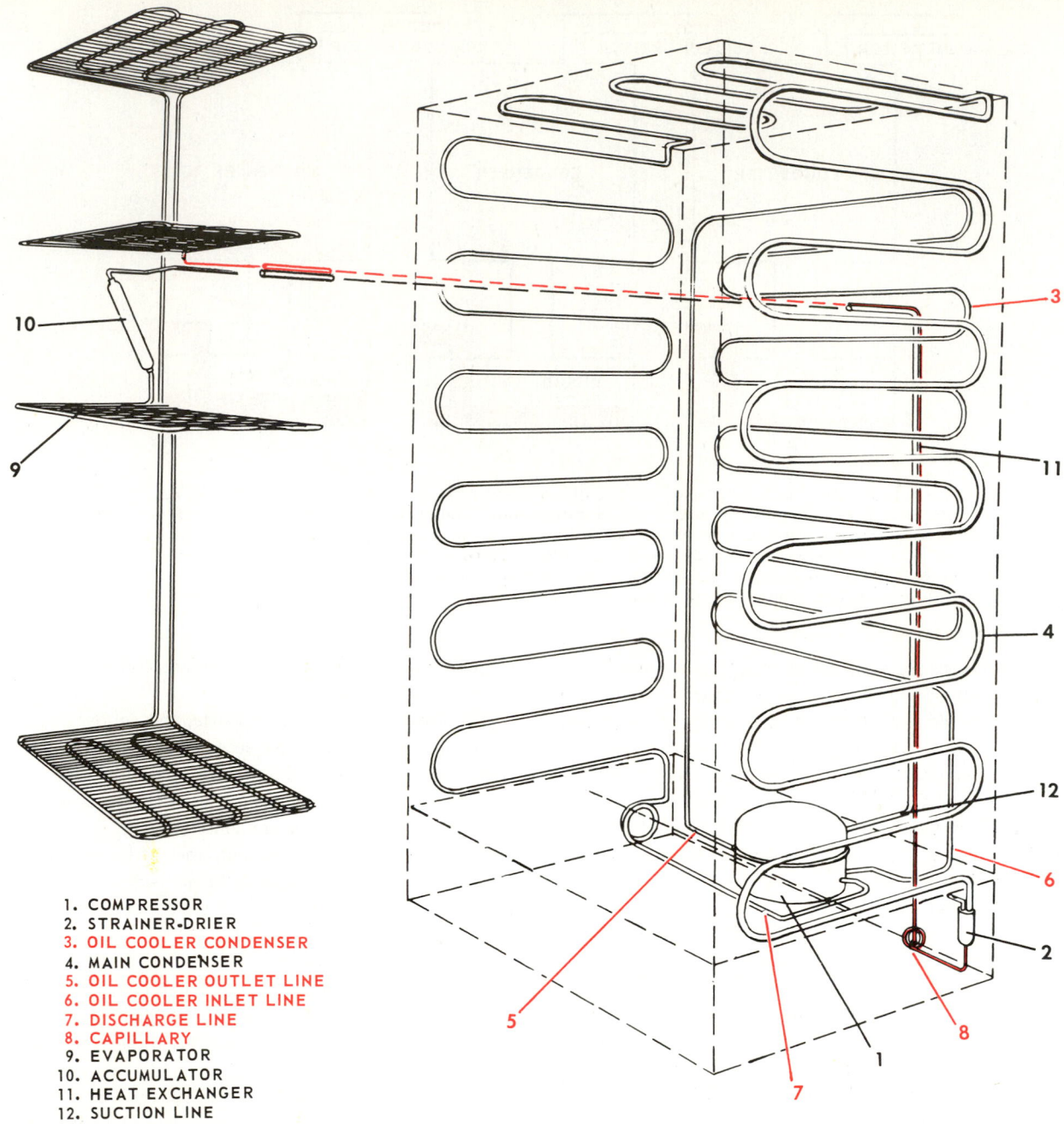

1. COMPRESSOR
2. STRAINER-DRIER
3. OIL COOLER CONDENSER
4. MAIN CONDENSER
5. OIL COOLER OUTLET LINE
6. OIL COOLER INLET LINE
7. DISCHARGE LINE
8. CAPILLARY
9. EVAPORATOR
10. ACCUMULATOR
11. HEAT EXCHANGER
12. SUCTION LINE

Fig. 4-6. Shelf type evaporator 9. This shows evaporator as it forms the shelf in upright freezer. Accumulator 10 is located at outlet of evaporator. This is a small reservoir to catch refrigerant not needed in evaporator.

Evaporators are made in four different styles:

1. Shell type, Fig. 4-5.
2. Shelf type, Fig. 4-6.
3. Wall type, used in chest type freezer, Fig. 4-7.
4. Fin tube type with forced circulation. This type of evaporator is most used with frost-free construction. (See views A and B of Fig. 4-8.)

Frost-free refrigerators usually need a fan or fans. They circulate the air over the evaporator and distribute cold air throughout the cabinet.

There are many types of commercial system evaporators. Air cooling and liquid cooling are two of the basic designs.

They are constructed of plain or finned tubing or of plate.

Details of commercial system evaporators are explained in Chapter 12.

4-6 ACCUMULATOR

The accumulator is a safety device to prevent liquid refrigerant from flowing into the suction line and into the compressor. Liquid refrigerant, if it were to flow into the compressor, could cause considerable knocking and damage to the compressor.

A typical accumulator, Fig. 4-5 and 4-6, has the outlet at

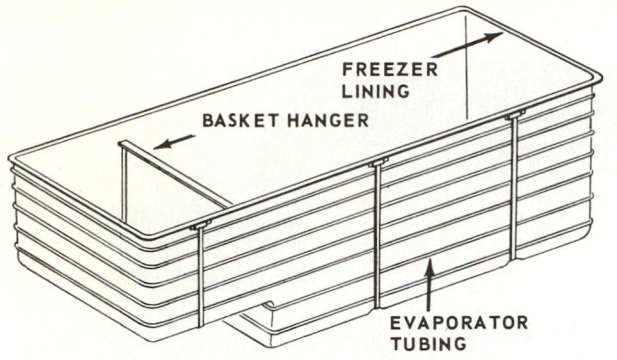

Fig. 4-7. Wall type evaporator. Note evaporator tubing is attached to lining of freezing cabinet. This arrangement provides smooth inside surface with uniform cooling throughout cabinet.

the top. Any liquid refrigerant that flows into the accumulator will be evaporated. Then vapor only will flow into the suction line. Since the accumulator is located inside the cabinet, it also provides some refrigeration.

Not all refrigerating systems have accumulators. Commercial system accumulators are explained in Chapter 12.

4-7 SUCTION LINE

The suction line carries the refrigerant vapor from the evaporator to the compressor. The line must be large enough to carry the vaporized refrigerant with minimal flow resistance. It should slope from the evaporator or accumulator down to the compressor. Otherwise, pockets of oil collect.

In many units, the liquid line is in contact with all or part of the length of the suction line. This cools the liquid refrigerant, helping to reduce flash gas in the evaporator. It also adds some superheat to the refrigerant vapor entering the compressor. See Para. 5-10 for an explanation of superheat.

4-8 LOW SIDE FILTER-DRIER

Some systems include a low-side filter-drier at the compressor end of the suction line. These may be a part of the

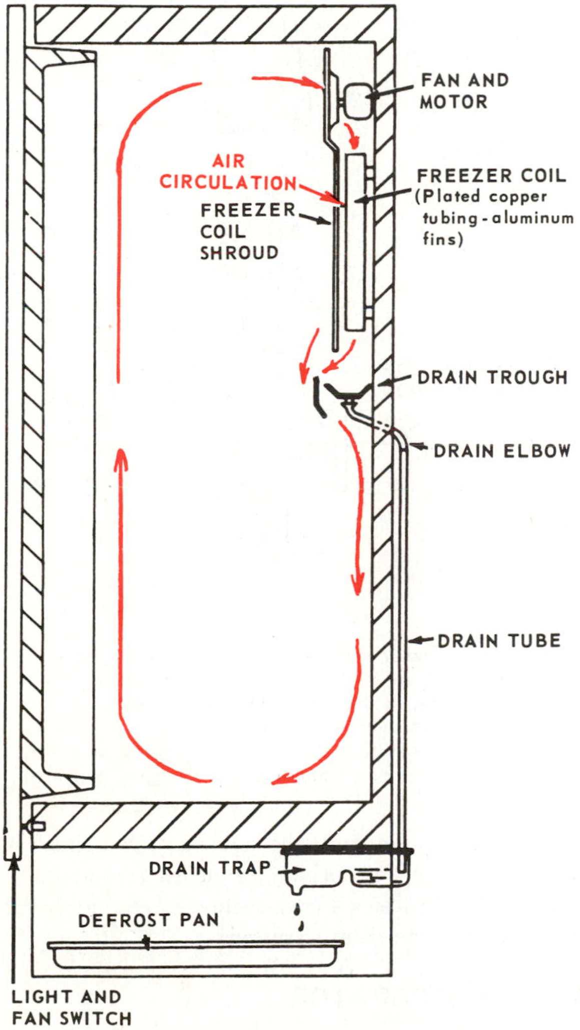

Fig. 4-8. Forced convection evaporator. A—A forced circulation evaporator is used in this upright freezer cabinet. Door switch stops fan when cabinet door is opened. (Gibson Products Corp.) B—Fin type evaporator. Note the trough to collect and carry away the defrost moisture which drains from the evaporator during the defrost part of the cycle. (Kelvinator, Inc.)

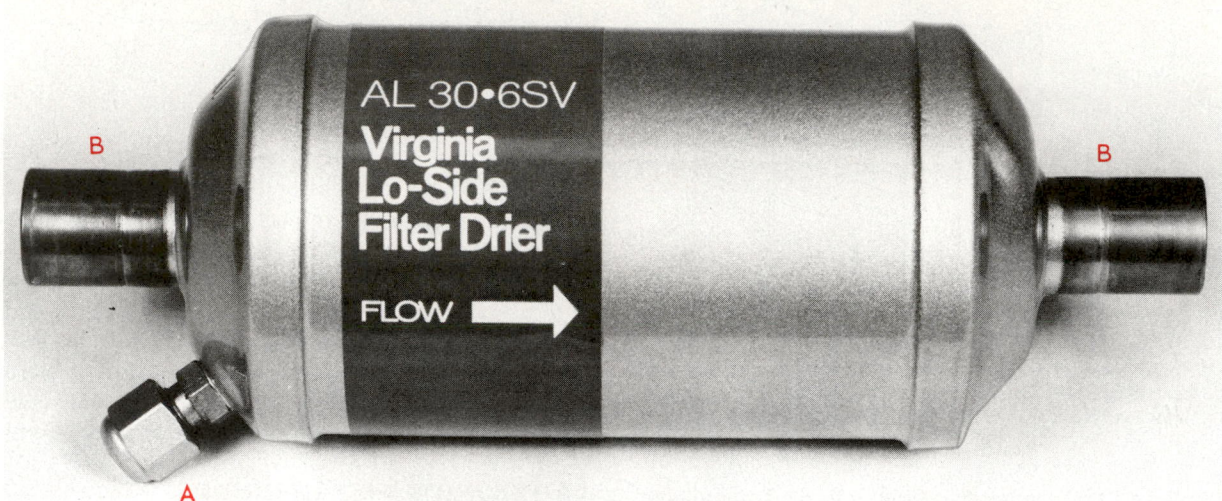

Fig.4-9. Suction line filter-drier. Direction of refrigerant vapor flow is indicated. A—Service connection. B—Tubing connection. (Virginia Chemicals, Inc.)

original system or may be placed in the system, for a short time, to clean it up. Fig. 4-9 is a typical suction line filter-drier. The filter-drier used in the suction line should offer little resistance to the flow of the vaporized refrigerant. This is because the pressure difference between the pressure in the evaporator and the inlet to the compressor should be small.

4-9 COMPRESSOR LOW SIDE OR SUCTION SERVICE VALVE

Many systems have some means that allow the service technician to connect gauges to the system, check pressures and add or take out refrigerant or oil.

A typical compressor suction service valve is pictured in Fig. 4-10. This valve is connected to the compressor at the compressor inlet union. The suction line from the evaporator is attached at the low-side inlet. Sealing caps protect the charging and gauge opening port and the valve stem when the valve is not in use.

More recent domestic models do not have service valves. The service technician must use a saddle valve. See Chapter 11.

4-10 COMPRESSOR

The refrigeration compressor is a motor-driven device which moves the heat-laden vapor refrigerant from the evaporator and compresses (squeezes) it into a small volume and to a high temperature. The various types of pumping mechanisms (compressors) used are explained later in this chapter.

4-11 COMPRESSOR HIGH-SIDE SERVICE VALVE

The compressor high-side service valve provides a shutoff between the compressor and the condenser. It also provides an opening for a high-pressure gauge or a gauge manifold.

With the valve closed (all the way in), it is possible to disconnect the compressor from the condenser without any

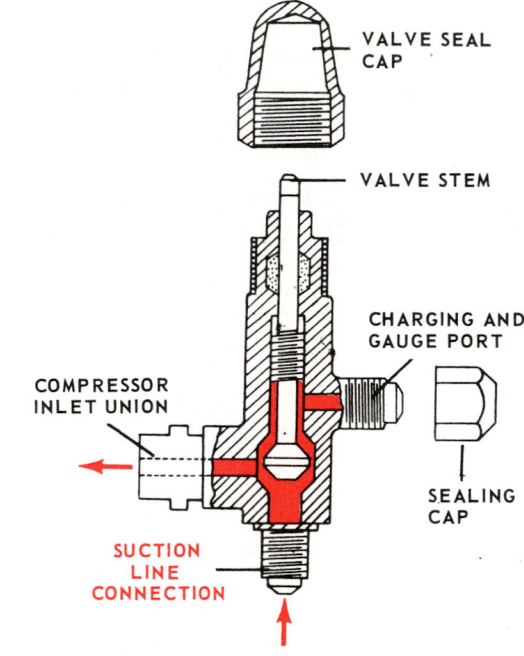

Fig. 4-10. Compressor low-side or suction service valve. If valve stem is turned all the way in, it closes off the connection from the compressor to the suction line. In this position, if valve is removed from compressor, suction line remains sealed. If valve stem is turned out as far as possible, it closes off both the charging and gauge port. This makes it possible to install compound gauge or charging line.

leakage of refrigerant from the condenser. When the valve stem is all the way out, the opening for the gauge is closed.

Fig. 4-11 illustrates a cross-section of the service valve. It is not used on all refrigerating systems.

4-12 OIL SEPARATOR

Refrigeration compressors get their lubrication from a small amount of special lubricating oil placed inside the compressor crankcase or housing. This oil is circulated to various

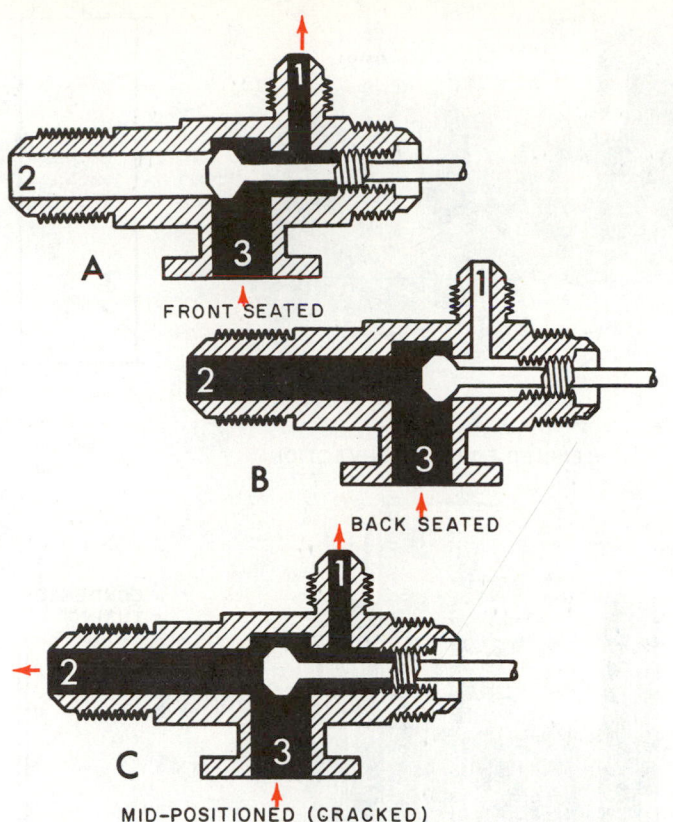

Fig. 4-11. A compressor high-side service valve. If valve is turned all the way in, as in view A, it shuts off connection between compressor, 3, and condenser, 2. If valve is turned all the way out, as in view B, it closes off connection to gauge port, 1. At mid-position, view C, all passages are open. (Murray Corp.)

compressor parts. In a hermetic (airtight) system, this oil also lubricates the motor bearings.

When the compressor operates, small amounts of oil will be pumped out with the hot compressed vapor. A small amount of oil throughout the system does no harm. However, too much oil in such parts as the condenser, refrigerant flow controls, evaporator and filters interferes with their operation.

It is possible to separate the oil from the hot compressed vapor. This involves placing an oil separator between the compressor exhaust and the condenser. The location and operation of such a separator is shown in Fig. 4-12. The separator is enlarged in the illustration to help show details.

The oil separator is a tank or cylinder with a series of baffles or screens which collect the oil. The oil separated from the hot, compressed vapors, drops to the bottom of the separator.

A float arrangement controls a needle valve which opens an oil return line to the compressor crankcase. When the oil level is high enough, the float rises and opens the needle valve. This oil returns quickly to the compressor crankcase, as the pressure in the separator is considerably higher than the pressure in the compressor crankcase.

Oil separators are quite efficient. Very little oil passes on into the system. They are most commonly used in large commercial installations.

4-13 CONDENSER

As mentioned in Para. 4-2, the condenser in the refrigeration cycle removes the condensation heat from the refrigerant

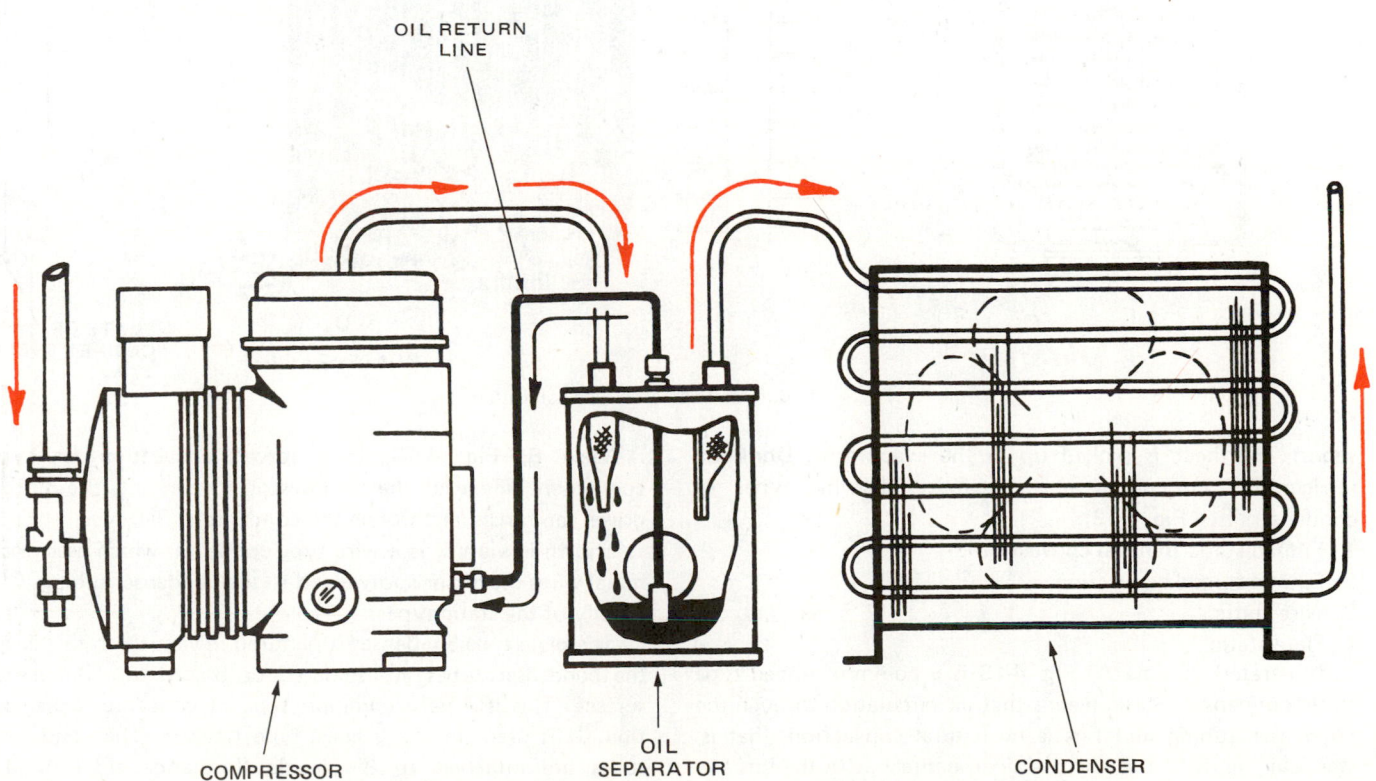

Fig. 4-12. An oil separator located in the discharge line. Note flow of refrigerant and oil. (A C and R Components, Inc.)

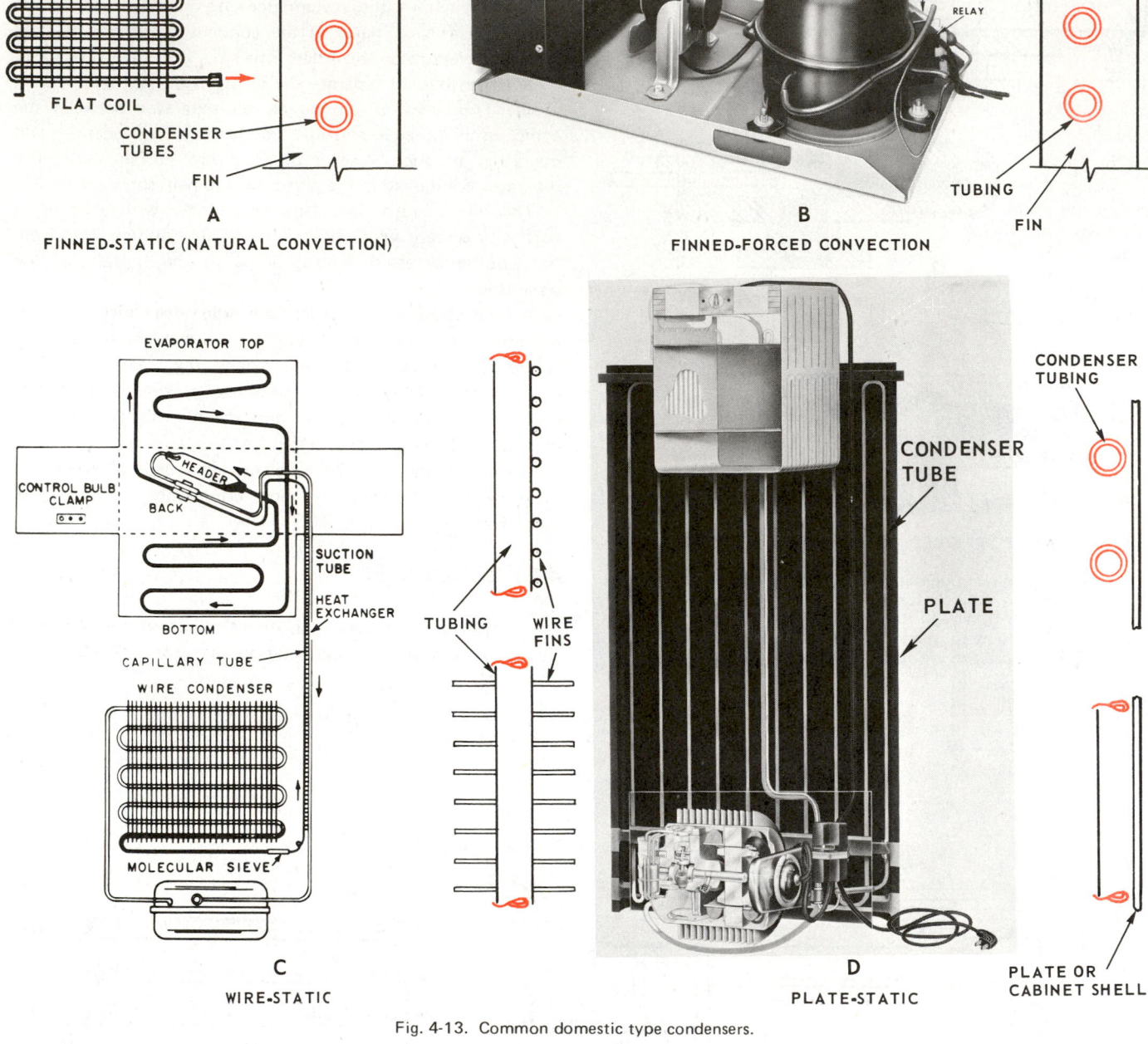

A

FINNED-STATIC (NATURAL CONVECTION)

B

FINNED-FORCED CONVECTION

C

WIRE-STATIC

D

PLATE-STATIC

Fig. 4-13. Common domestic type condensers.

vapor. This heat is picked up in the evaporator. Domestic refrigerators commonly use the four following types of condensers (see Fig. 4-13):

1. Finned-static (natural convection).
2. Finned-forced convection.
3. Wire-static.
4. Plate-static.

Illustrated in view A, Fig. 4-13 is a common finned type static condenser. Static means that air circulation through the condenser tubing and fins is by natural convection; that is, warm air tends to rise. As the air in contact with the fins and tubes becomes heated, it rises and cooler air takes its place. The tubes and fins are usually made of copper or steel.

View B, Fig. 4-13, is a forced convection fin type condenser. Whenever the compressor is operating, the motor-driven fan forces air through the condenser.

Shown in view C is a wire type condenser which uses small metal wires brazed or spotwelded to the condenser tubing. It is usually of the static type.

The plate type condenser is pictured in view D. In this type, the condenser tubes are soldered or brazed to a flat metal surface. This is a very common type of condenser construction. It is used on many chest-type freezers. The condenser tubes are attached to the inside (insulation side) of the freezer's outer shell. This type of condenser is very easy to keep clean. It is only necessary to wipe off the surface of the cabinet shell. To get proper removal of heat from the refrigerant vapor, always keep the condenser clean.

Commercial systems use three types of condensers:

1. Finned-static, air-cooled.
2. Finned-forced convection, air-cooled.
3. Water-cooled, tube-in-a-tube and shell.

The finned-static and the finned-forced convection condensers are built much the same as the domestic condensers, except that they are larger. Forced convection finned condensers are used on many commercial refrigeration installations. Applications of these are shown in Chapter 12.

Water-cooled condensers usually consist of two tubes, one within the other. Water circulates through the inside tube. Hot compressed vapor circulates through the space between the tubes. These condensers are usually called tube-within-a-tube condensers. They are very efficient.

With the rapid development of commercial refrigeration and air conditioning, many communities have difficulty in supplying enough water for water-cooled condensers. Thus, more and more forced convection air-cooled condensers and "cooling towers" are being used in connection with water-cooled condensers.

See Chapter 12 for illustrations of cooling tower applications and water-cooled condensers. Fig. 4-14 shows a typical air-cooled condenser installation and a diagram of a water-cooled condenser.

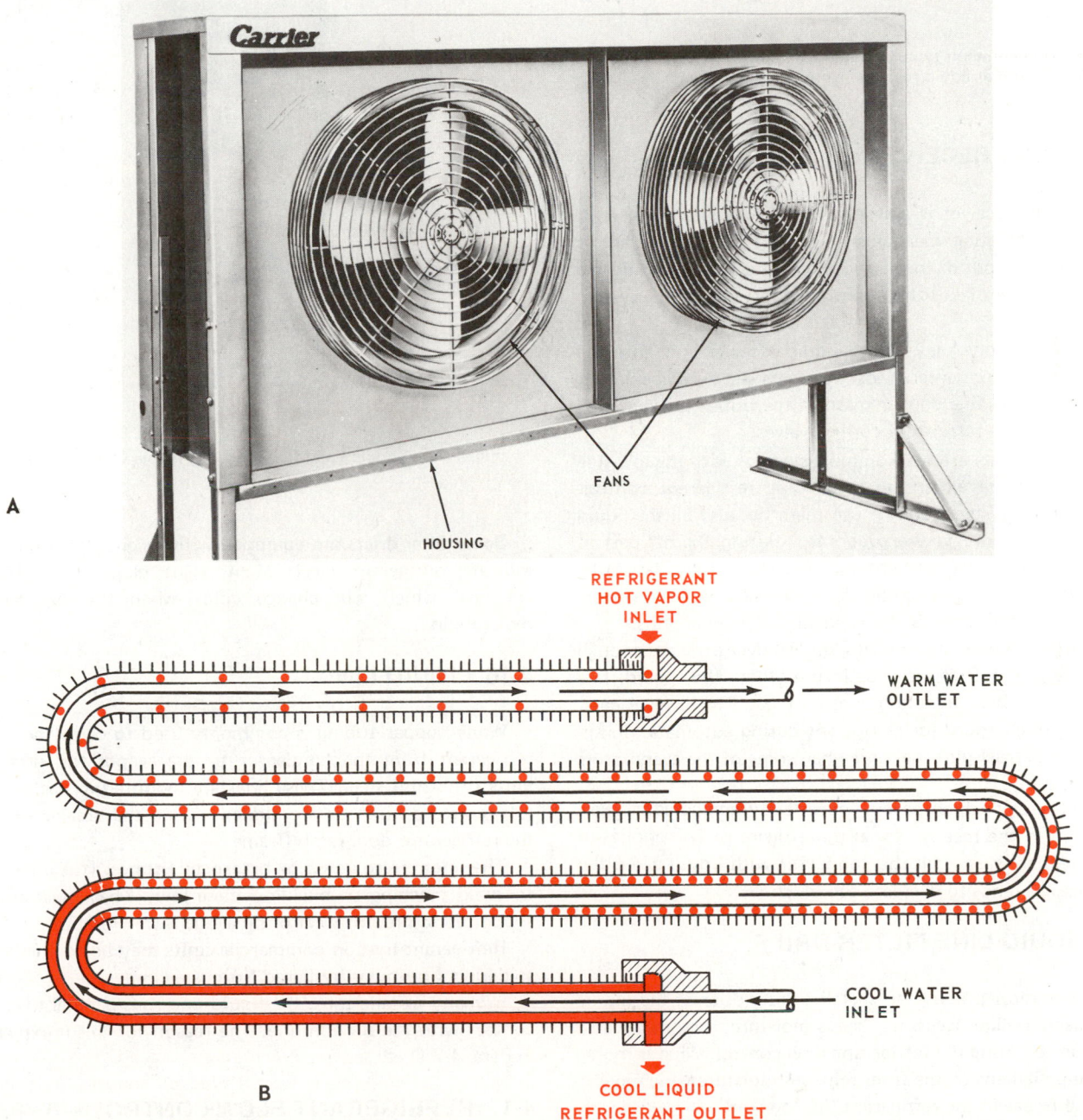

Fig. 4-14. Typical commercial refrigeration condensers. A—Air-cooled condenser which uses motor-driven fans to increase air flow over condenser surfaces. (Carrier Air Conditioning Group, United Technologies Corp.) B—Water-cooled condenser flow diagram. Refrigerant flows through the condenser in opposite direction of water.

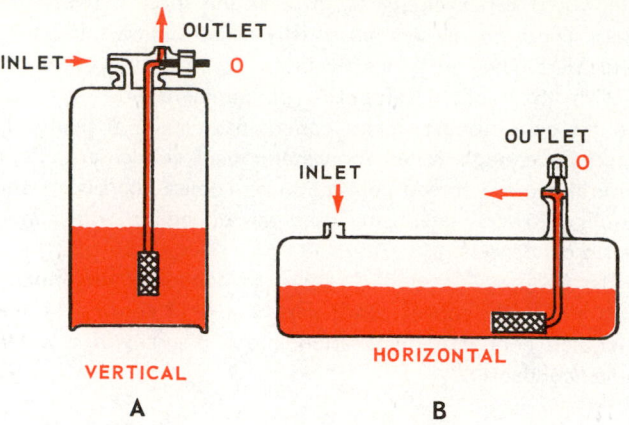

Fig. 4-15. Two common types of liquid receivers. A—Vertical liquid receiver. B—Horizontal liquid receiver. Note liquid line service valve at O.

4-14 LIQUID RECEIVER

The liquid receiver is a storage tank for liquid refrigerant. When a refrigerating mechanism has one, the refrigerant is usually pumped out of the various parts and stored in it during servicing. Its use makes the quantity of refrigerant in a system less critical.

Occasionally one may find a liquid receiver built into the bottom of the condenser. Most receivers have service valves. See Fig. 4-15. A fine copper mesh in the outlet prevents dirt from entering the refrigerant control valves.

Liquid receivers are used on most systems with the low-side float type or the expansion valve type refrigerant control. Capillary tube systems do not use them because all the liquid refrigerant is stored in the evaporator during the off part of the cycle. Greater use of hermetic systems and capillary tube refrigerant controls has removed the need for liquid receivers in domestic systems and in many small commercial units.

On larger commercial systems the receiver provides enough reserve liquid refrigerant to insure that the liquid line refrigerant is subcooled and free of flash gas. The receiver must provide enough room for refrigerant during automatic pump-downs (for defrost purposes and when some of the evaporators are not in use).

Some systems which have an outdoor air-cooled condenser need room in the receiver for extra refrigerant. This applies to systems which partly fill the condenser with liquid when the head pressure tends to decrease too much.

4-15 LIQUID LINE FILTER-DRIER

It is common practice to install a filter-drier in the liquid line. This tank-like accessory keeps moisture, dirt, metal and chips from entering the refrigerant flow control. What is more, the drying element in the filter removes moisture which might otherwise freeze in the refrigerant flow control.

Moisture is also harmful when mixed with oil in a system. It forms sludges and acids. It is especially harmful to hermetic units. A liquid line filter-drier is shown in Fig. 4-16.

Fig. 4-16. A liquid line filter-drier.   (Virginia Chemicals, Inc.)

Some filter-driers are equipped with a sight glass which will indicate refrigerant level. Many sight glasses also have a chemical which will change color when the system has moisture in it.

4-16 LIQUID LINE

While copper tubing is commonly used to carry the liquid refrigerant from the condenser to the evaporator, domestic units often use steel. These lines are mounted in back of the refrigerator cabinet or are hidden behind the breaker strip at the refrigerator door jamb (frame).

The lines are soldered or brazed to fittings. It is important to avoid pinching or buckling these lines. They must also be supported to prevent wear or breakage from vibration.

Refrigerant lines in commercial units may be connected by soldering, brazing or by flared fittings.

In many installations, the liquid line runs parallel to and in contact with the suction line. The reason for this is explained in Para. 4-7.

4-17 REFRIGERANT FLOW CONTROL — TYPES

The refrigerant flow control has two jobs. It allows liquid refrigerant to enter the evaporator and, at the same time,

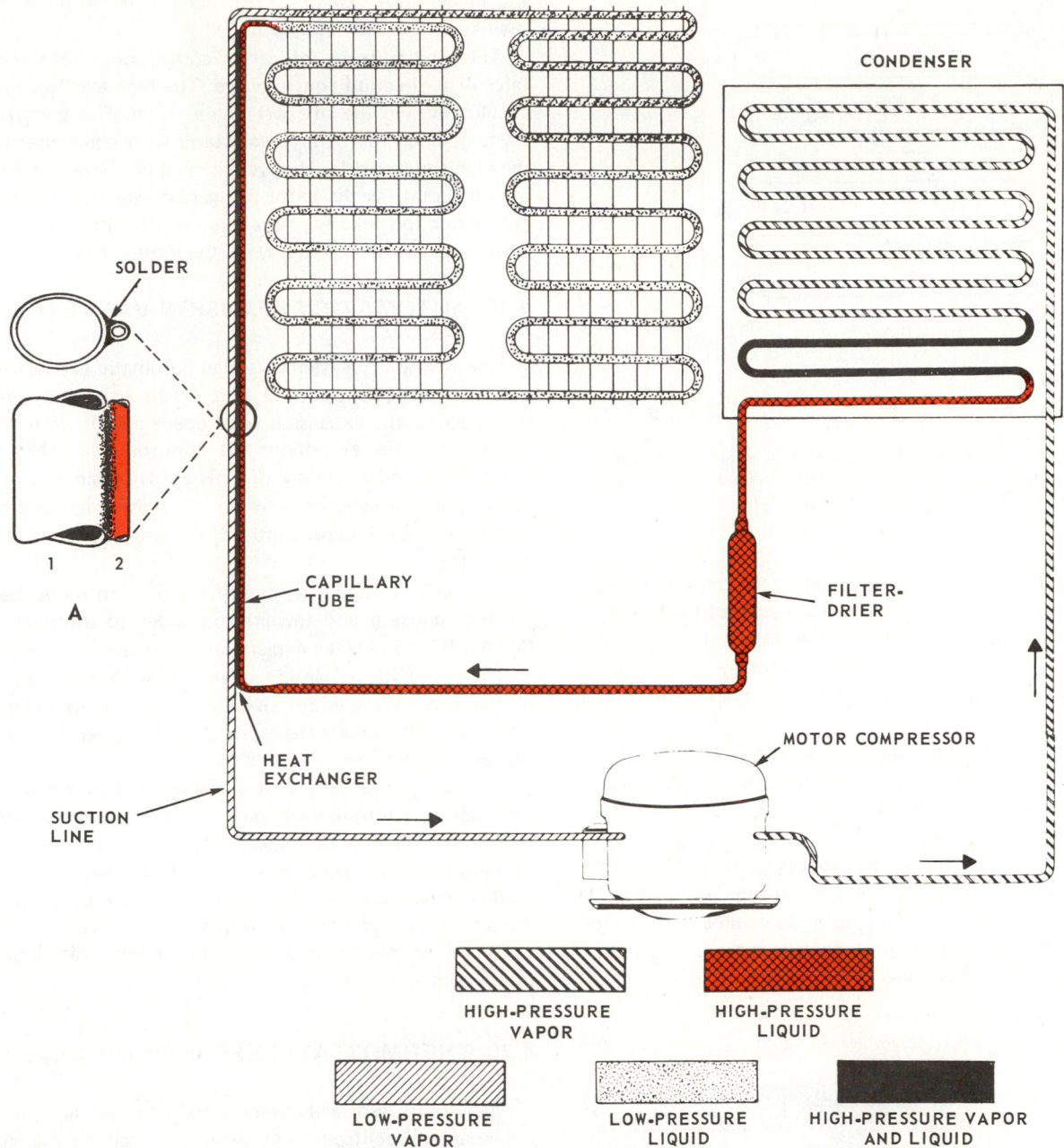

EVAPORATOR

CONDENSER

SOLDER

1 2

A

CAPILLARY
TUBE

FILTER-
DRIER

MOTOR COMPRESSOR

HEAT
EXCHANGER

SUCTION
LINE

HIGH-PRESSURE
VAPOR

HIGH-PRESSURE
LIQUID

LOW-PRESSURE
VAPOR

LOW-PRESSURE
LIQUID

HIGH-PRESSURE VAPOR
AND LIQUID

Fig. 4-17. Refrigerating system using capillary tube refrigerant control. Filter-drier is located in liquid line ahead of connection to capillary tube. Most of the capillary tube is fastened to suction line which provides heat exchange. A—Enlarged cross-section of suction line (1) and capillary tube (2) showing how they are soldered or brazed together. (Kelvinator, Inc.)

maintains the required evaporating pressure in the evaporator.

Several types of refrigerant flow controls are used in modern refrigerating mechanisms. These controls and their characteristics are explained in Chapter 5.

There are five principal types:

1. Capillary Tube (Cap.) — (dry system).
2. Automatic Expansion Valve (AEV) — (dry system).
3. Thermostatic Expansion Valve (TEV) — (dry system).
4. Low-Side Float (LSF) — (flooded system).
5. High-Side Float (HSF) — (flooded system).

4-18 CAPILLARY (CAP.) TUBE

The capillary tube is the most common refrigerant flow control for domestic refrigerators, freezers, room air conditioners and small commercial installations. A typical installation is diagrammed in Fig. 4-17.

The capillary tube is a long length of small diameter tubing. It reduces pressure, by reducing the flow of refrigerant through its length.

Fig. 4-18 is another drawing of a capillary tube refrigerant

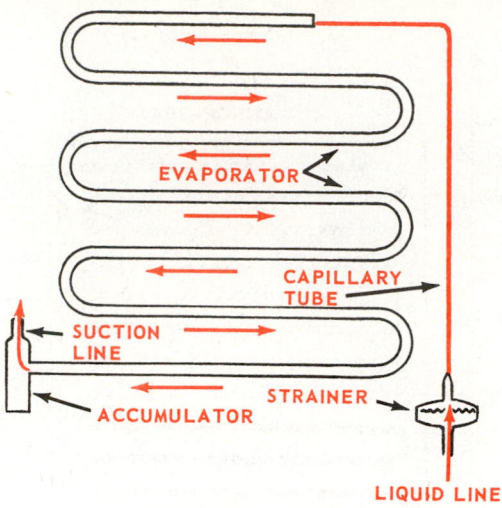

Fig. 4-18. Capillary tube refrigerant control. A strainer is located in liquid line at entrance to capillary tube. Note accumulator. This serves as receiver for any liquid refrigerant overflow.

control. The tube's inside diameter may vary, depending upon the refrigerant, capacity of the unit and length of the line.

The tube is placed between the liquid line and the evaporator. Just enough liquid passes through it to make up for the amount that is vaporized in the evaporator as the compressor operates.

It reduces the liquid refrigerant from its high pressure to its evaporating pressure. There is no change in the liquid except a slight drop in pressure for about the first two-thirds of the length of the capillary tube.

Then some of the liquid starts to change to vapor. By the time the refrigerant reaches the end of the tube, from 10 to 20 percent of it has vaporized. The increased volume of the vapor causes most of the pressure drop to take place in the end of the tube nearest the liquid line.

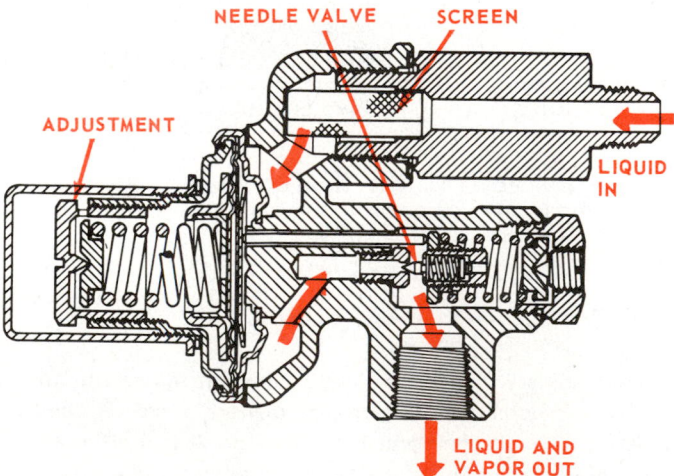

Fig. 4-19. Cross section of an automatic expansion valve shows flow of refrigerant through valve. This valve is designed to control the flow of liquid refrigerant into evaporator. It also maintains constant low pressure in evaporator while compressor is running.

A recent development in capillary tube design uses a larger and longer tube (20 to 30 ft.). Being larger in diameter, it is less likely to become plugged.

The capillary tube refrigerant control does not use a check valve or a direction control valve. The high and low pressures equalize during the off part of the cycle. This permits easier starting, since the compressor starts with equal pressures on the high and low side. The system must not have an overcharge of refrigerant, as the extra refrigerant would tend to fill the evaporator too full. An overcharge will be indicated by severe frosting of the suction line when the motor starts.

4-19 AUTOMATIC EXPANSION VALVE (AEV)

One of the dry systems uses an automatic expansion valve (AEV) as refrigerant control (Fig. 4-19). As pressure drops on the low side, the expansion valve opens and liquid refrigerant flows into the evaporator. It absorbs heat then, while evaporating under low pressure. The valve maintains constant pressure in the evaporator when the system is running. This system operates independently of the amount of refrigerant in the system.

The AEV is, therefore, one of the division points between the high-pressure and low-pressure sides of the system. See Chapter 5 for a detailed explanation of expansion valves.

The automatic expansion valve, Fig. 4-20, may be adjusted to the correct evaporator pressure. Turning the adjustment clockwise will increase the rate of flow and, therefore, increase the low-side pressure.

The rate of refrigerant flow through the automatic expansion valve is controlled by the pressure in the evaporator. No refrigerant will flow through the valve unless the compressor is running and the evaporator is under a low pressure.

Remember, lowering the pressure in the evaporator lowers the temperature at which the refrigerant evaporates.

This valve may be used only with the temperature-operated motor control.

4-20 THERMOSTATIC EXPANSION VALVE (TEV)

Many units, especially commercial ones, are equipped with a temperature controlled expansion valve called a thermostatic expansion valve (TEV). Fig. 4-21 shows a cross-section of a thermostatic expansion valve.

This valve has a temperature sensing bulb mounted on the outlet of the evaporator. The bulb temperature controls the opening of the thermostat valve needle.

Addition of this thermal element to the valve lets the evaporator fill more quickly and permits more efficient cooling. The thermostatic expansion valve keeps the evaporator full of liquid refrigerant when the system is running. See Chapter 5 for a more detailed explanation of its operation.

As the evaporator becomes colder, the TEV reduces the rate of flow of refrigerant into the evaporator. There is no flow at all unless the compressor is running.

Fig. 4-22 demonstrates how a thermostatic expansion valve is connected into a refrigerating system. The valve may be used with either a pressure or temperature operated motor control.

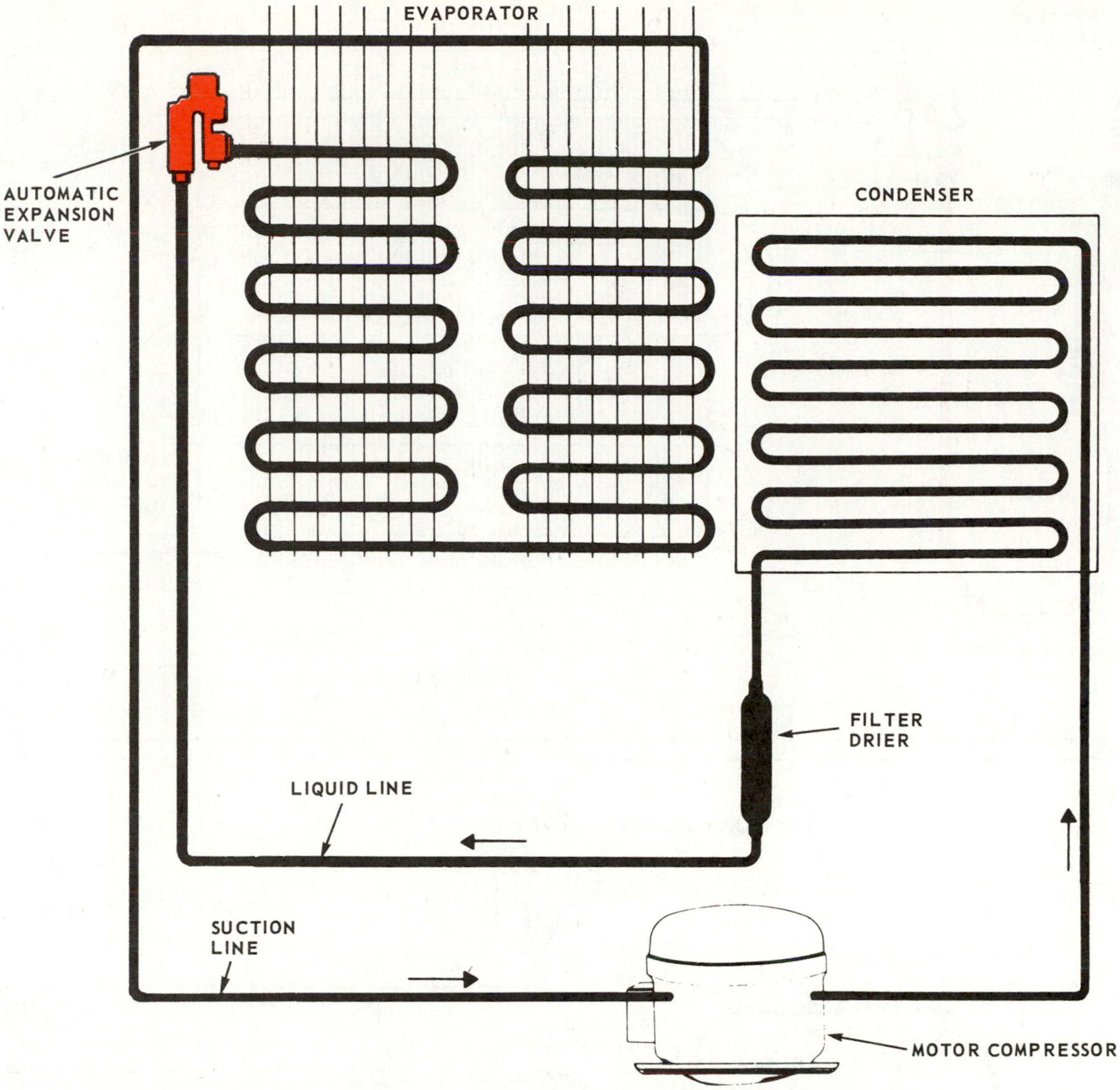

EVAPORATOR

CONDENSER

AUTOMATIC
EXPANSION
VALVE

FILTER
DRIER

LIQUID LINE

SUCTION
LINE

MOTOR COMPRESSOR

Fig. 4-20. A refrigerating system using an automatic expansion valve refrigerant control. A filter-drier is located in liquid line ahead of automatic expansion valve.

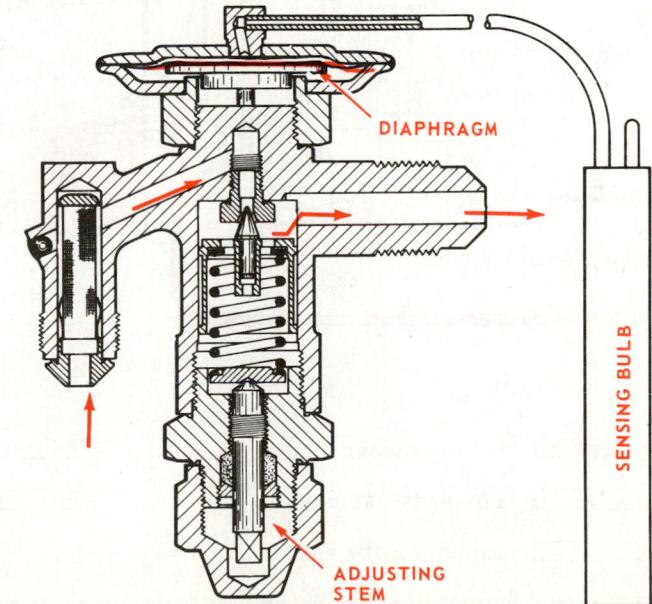

DIAPHRAGM

SENSING BULB

ADJUSTING STEM

Fig. 4-21. Diaphragm type thermostatic expansion valve. Sensing bulb pressure operates on upper surface of diaphragm. As sensing bulb temperature increases, pressure on top of diaphragm increases and tends to open valve allowing liquid refrigerant to enter evaporator. Note screen at liquid line connection and direction of refrigerant flow through valve.
(Sporlan Valve Co.)

Compression Systems and Compressors / 111

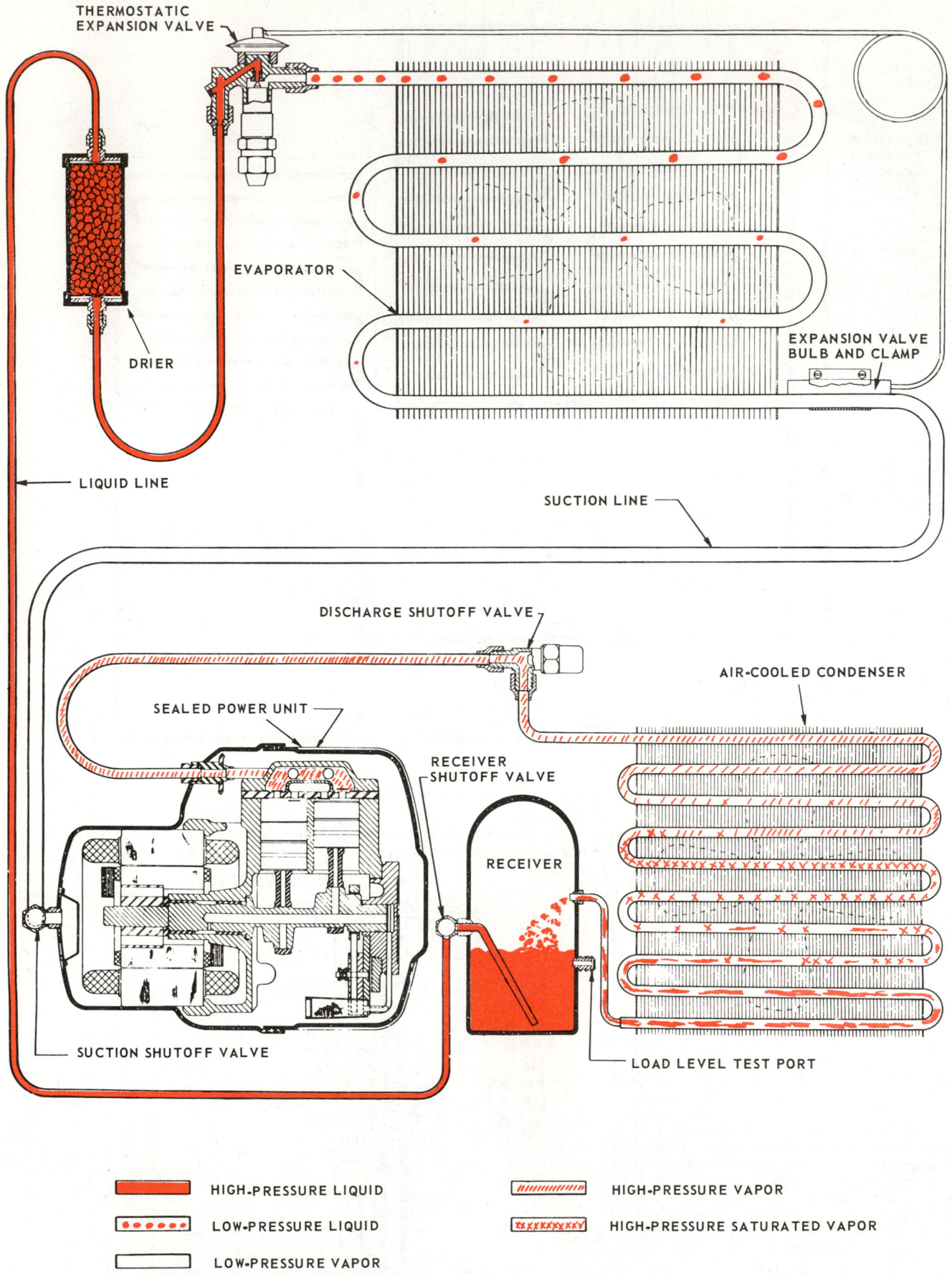

THERMOSTATIC
EXPANSION VALVE

EVAPORATOR

EXPANSION VALVE
BULB AND CLAMP

DRIER

LIQUID LINE

SUCTION LINE

DISCHARGE SHUTOFF VALVE

AIR-COOLED CONDENSER

SEALED POWER UNIT

RECEIVER
SHUTOFF VALVE

RECEIVER

SUCTION SHUTOFF VALVE

LOAD LEVEL TEST PORT

HIGH-PRESSURE LIQUID

HIGH-PRESSURE VAPOR

LOW-PRESSURE LIQUID

HIGH-PRESSURE SATURATED VAPOR

LOW-PRESSURE VAPOR

Fig. 4-22. Typical thermostatic expansion valve system. Note the two-cylinder hermetic compressor. (Bendix-Westinghouse)

It can be used also in a multiple evaporator system.

In a recent development, a thermoelectric element replaces the temperature sensing element in the thermostatic expansion valve. More will be said about this valve in Para. 5-15.

4-21 LOW-SIDE FLOAT (LSF)

This control is used on a flooded system where the evaporator is flooded with refrigerant and the refrigerant level is controlled by a float valve. A float is used for control. As the refrigerant evaporates, the liquid level falls. This lowers a float and, in turn, opens the needle valve connected to it. More liquid enters from the high-pressure liquid line taking the place of the evaporated liquid.

Fig. 4-23 suggests the outside appearance and the inside construction of a flooded evaporator using a low-side float. Either a temperature or pressure motor control switch may be used.

A low-side float system usually has a large liquid receiver. It must be large enough to store all the refrigerant in the system.

Oil picked up by the vapor is normally returned through a small opening at a predetermined level in the suction return tubing. Because of the small diameter of the hole, if the unit is not level, the oil will not return to the compressor and "oil binding" may result.

When this occurs, the oil forms a layer on the surface of the liquid refrigerant. It prevents the refrigerant from evaporating at a rapid rate or at the temperature corresponding to the pressure.

The low-side float refrigerant control can be used in multiple evaporator systems.

4-22 HIGH-SIDE FLOAT (HSF)

A float located in the liquid receiver tank or in a chamber in the high-pressure side make this system operate. When enough liquefied refrigerant has collected in the float chamber, the float will rise enough to open the needle valve. Liquid flows into the low-pressure side or evaporator. The float controls the level of liquid refrigerant on the high-pressure side.

The amount of refrigerant in the system must be carefully measured if the evaporator is to receive the correct amount and if the system is to operate correctly.

Extra refrigerant will overcharge the evaporator and cause frosting of the suction line.

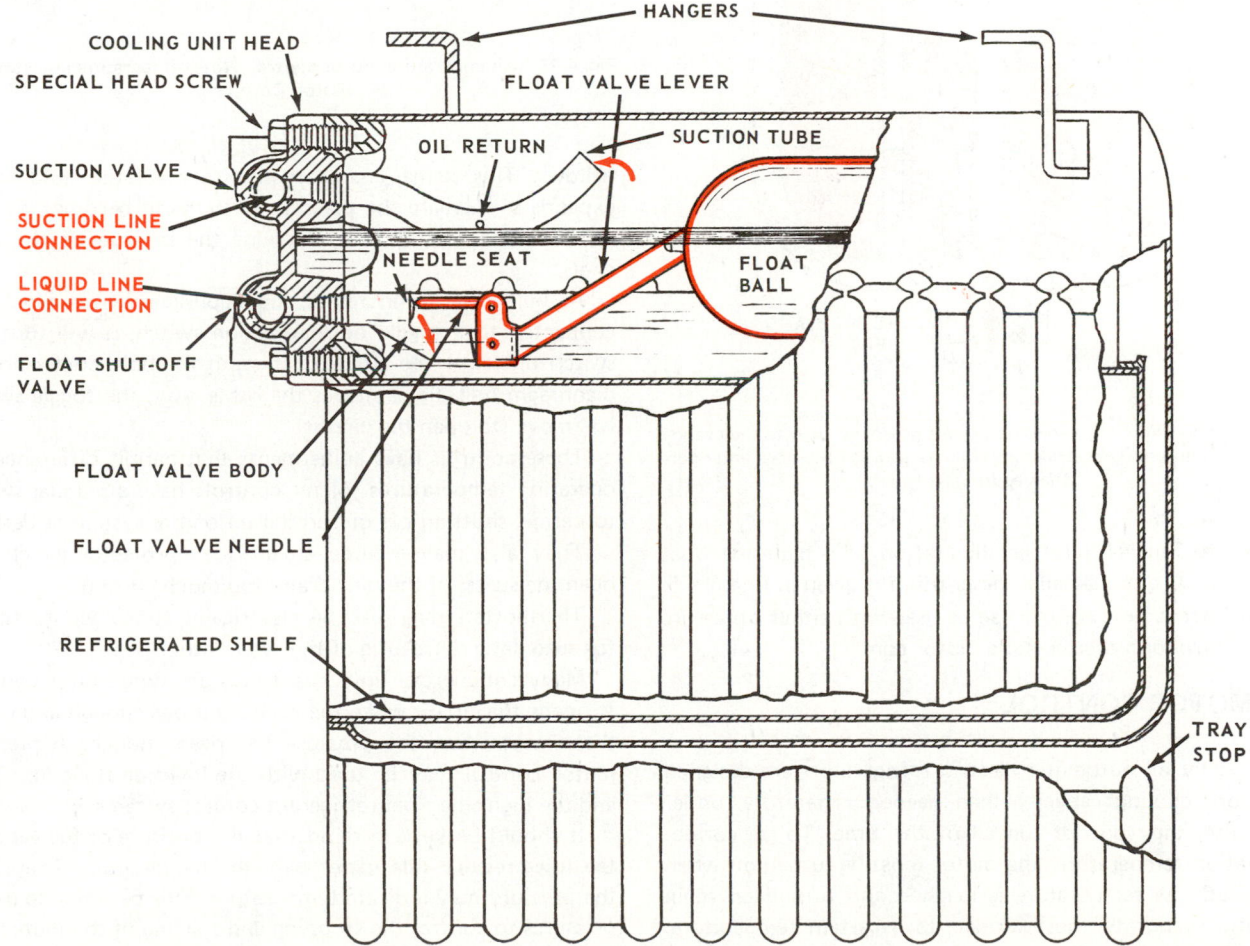

Fig. 4-23. Low-side float refrigerant control. Note suction line and liquid line connections. Float and needle mechanism maintain constant level of liquid refrigerant in evaporator.

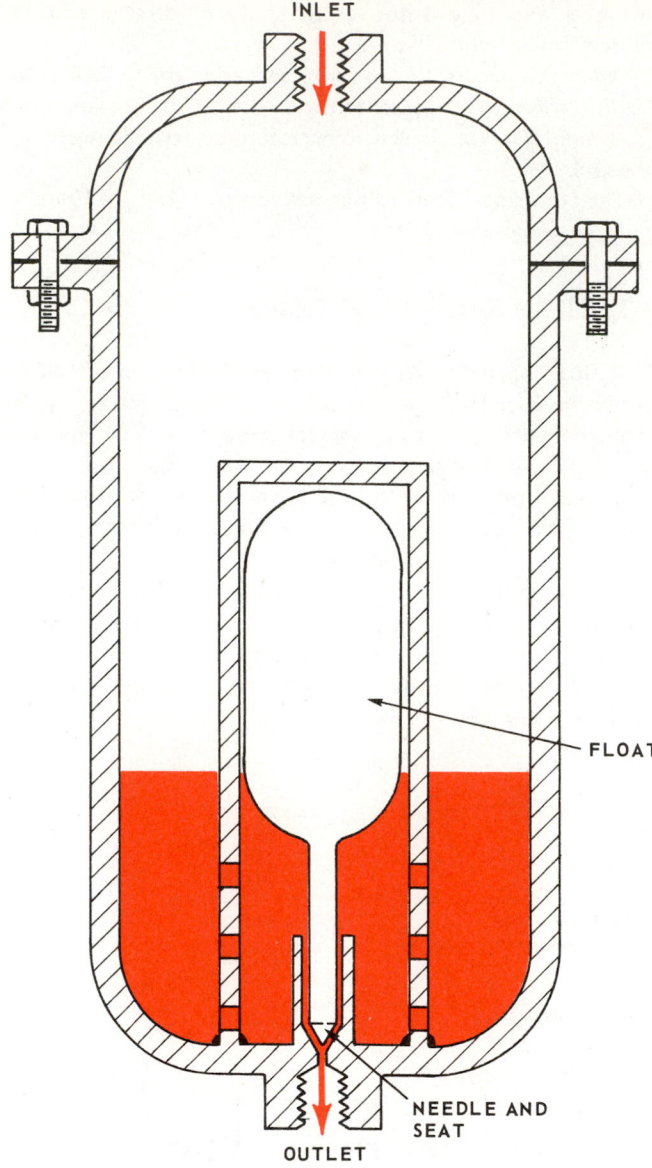

Fig. 4-24. High-side float allows liquid refrigerant to flow to the evaporator only when enough refrigerant collects to raise the float and open the needle valve.

Refer to Fig. 4-24 for an illustration of a high-side float mechanism. A more detailed description is given in Chapter 5. This refrigerant flow control can be used with either a pressure motor control or a thermostatic motor control.

4-23 MOTOR CONTROL

Practically all automatic electric refrigerators are designed with more cooling capacity than needed. Therefore, under normal use, they do not run all of the time. To get correct refrigeration temperature, the motor must be turned off when the desired low temperature is reached and turned on again when the evaporator has warmed to a certain temperature. Two principal types of motor controls are used to turn the motors on and off:

1. Temperature motor control (thermostatic).

2. Pressure motor control (low-side pressure).

The temperature control is the most popular, especially on small installations.

The thermostatic temperature control, Fig. 4-25, has a sensing bulb connected by a capillary tube to a diaphragm or

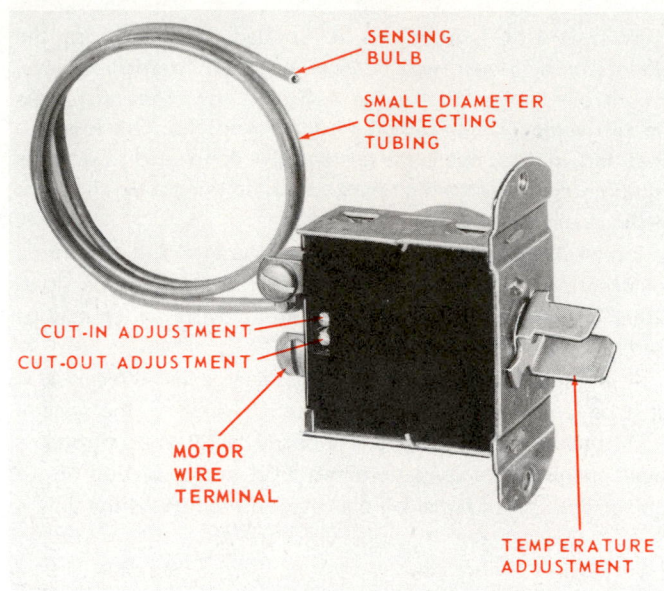

Fig. 4-25. A temperature motor control. Note temperature adjustment. (Eaton Corp.)

bellows. This element is charged with a volatile fluid which expands to increase the pressure as the bulb becomes warmer and will contract again to decrease the pressure as the bulb cools.

As bulb pressure increases, the diaphragm moves. Since it is connected to a toggle or snap action switch, it will turn this switch on (close the circuit). Then, as the bulb cools and the diaphragm or bellows moves the other way, the toggle switch will move (to open the circuit).

These controls have adjustments that permit differences in operating temperatures. Many controls have a manual switch to permit shutting off or turning on of the system as desired.

They also may include an overload protector which will open the switch if the unit draws too much current.

Thermostats may also be electrically connected to timers for automatic defrosting of the evaporator.

Many commercial units use a pressure type motor control. It opens the circuit when the pressure drops enough and closes the circuit when the pressure has risen enough. A pressure motor control may be used with the low-side float, the TEV and the high-side float refrigerant control systems.

It should be kept in mind that the pressure of the vapor in the low-pressure side varies with the temperature. Therefore, the pressure may indicate temperature. This permits the use of pressures to control the stopping and starting of the motor and thus controls the temperature of the cabinet. The details concerning the operation of these controls is explained fully in Chapter 8.

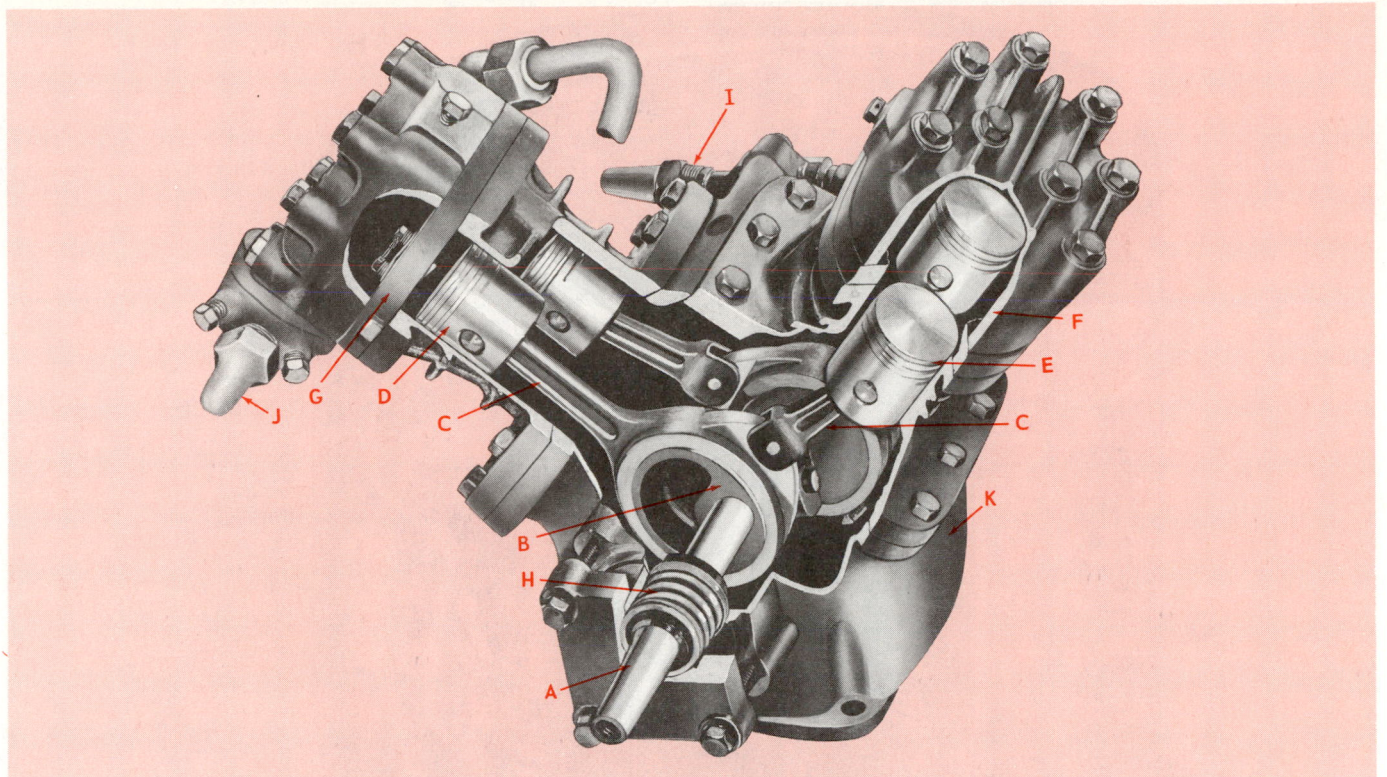

Fig. 4-26. A four-cylinder external drive V type compressor, air cooled. A—Crankshaft. B—Eccentric. C—Connecting rod. D—Piston. E—Piston rings. F—Cylinder. G—Valve plate. H—Crankshaft seal. I—Suction service valve. J—Exhaust service valve. K—Crankcase.   (Frick Co.)

4-24 EXTERNAL DRIVE COMPRESSORS

The purpose of a compressor is described in Para. 4-10.

An external drive (open) compressor is bolted together. Its crankshaft extends through the crankcase. The crankshaft is driven by a flywheel (pulley) and belt or it can be driven directly by an electric motor.

Fig. 4-26 illustrates a cross section through an open compressor. This is a four-cylinder V-type. An eccentric type crankshaft is used. The pistons are fitted with rings.

A master connecting rod is mounted on each eccentric and is connected to a piston in one bank of the "V". The connecting rod which is attached to the piston in the other bank is connected with a pin through a flange on the master connecting rod. These connecting rods are somewhat shorter than the master connecting rod and are called articulated connecting rods.

A crankshaft seal is required where the crankshaft comes through the crankcase.

4-25 HERMETIC COMPRESSORS

The motor in a hermetic compressor is sealed inside a dome or housing with the compressor and is directly connected to the compressor. A crankshaft seal is not needed.

A motor rotor is usually a press fit on the compressor crankshaft. Some motor compressors are made with the motor at the top, while others have the motor at the bottom and the compressor at the top.

The unit is usually spring mounted inside the hermetic dome. This prevents most of the compressor vibration from being felt outside of the dome.

The exhaust and suction lines inside the dome are made flexible. A connection through the dome provides means of fastening the compressor lines to the remainder of the system. The electrical connections to the motor pass through the dome by means of an insulated leakproof seal.

To lubricate the compressor, the return suction gas is fed into a hollow disk mounted on the motor compressor shaft. Centrifugal force throws the oil and a liquid refrigerant to the outer rim of the disk and flows over the motor windings. (Centrifugal force action rotates things to pull spinning particles away from the center of the rotation.) Only the vapor refrigerant remains at the center and is drawn into the cylinders of the compressor.

A hermetic motor compressor usually requires an outside electrical relay starting mechanism. Fig. 4-27 shows a section through such a motor.

Some motor compressors are two-speed. These are popular in large systems and in air conditioning where heat loads change.

4-26 TYPES OF COMPRESSORS

There are four basic types of compressors in use:
1. Reciprocating (piston-cylinder).
2. Rotary.
3. Screw type.
4. Centrifugal.

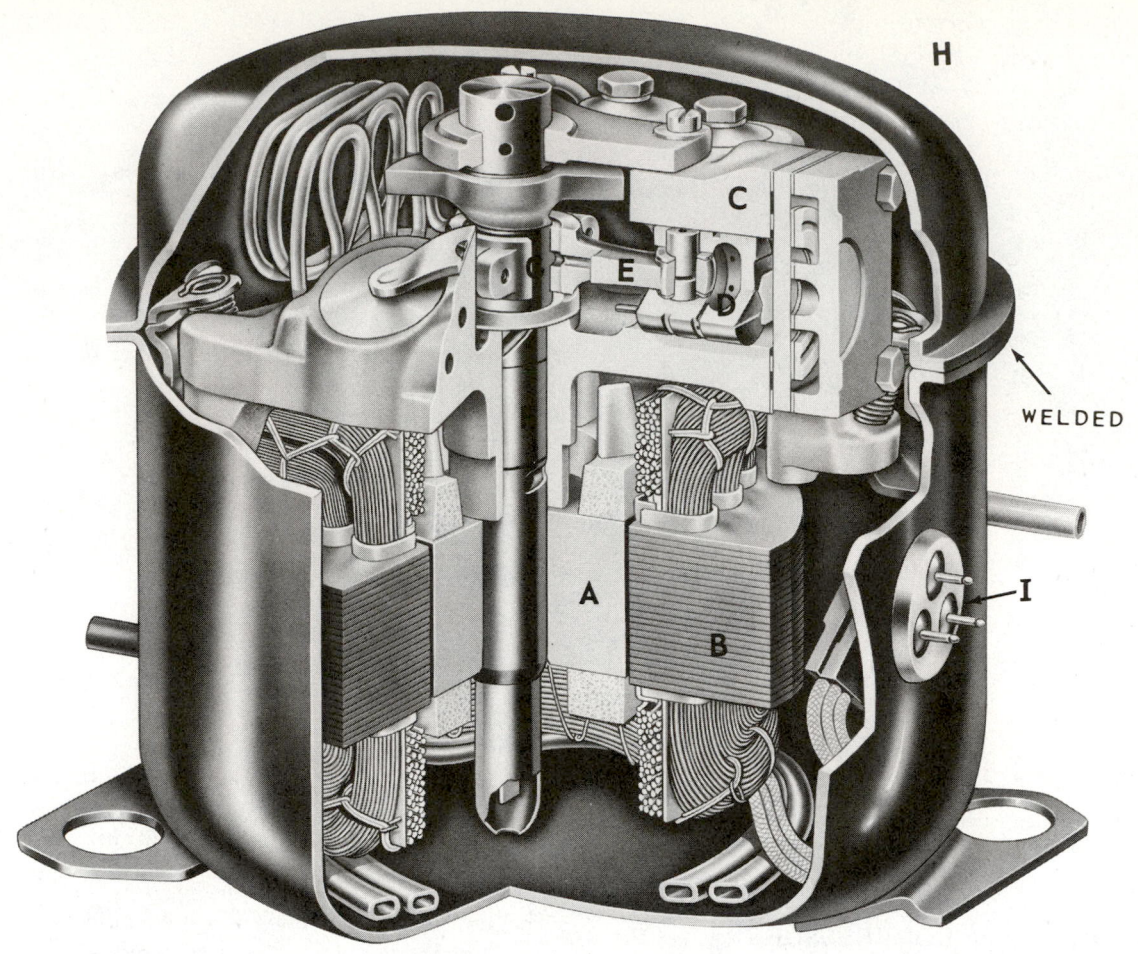

Fig. 4-27. Reciprocating hermetic compressor. Compressor is at top and motor at bottom. Assembly is mounted on springs inside dome. A—Motor rotor. B—Motor stator. C—Compressor cylinder. D—Compressor piston. E—Connecting rod. F—Crankshaft. G—Crank throw. H—Compressor shell. I—Glass sealed electrical connections through compressor dome.   (Tecumseh Products Co.)

Fig. 4-28. Large capacity external drive two-cylinder reciprocating compressor. Note eccentric type crankshaft and pistons fitted with rings.

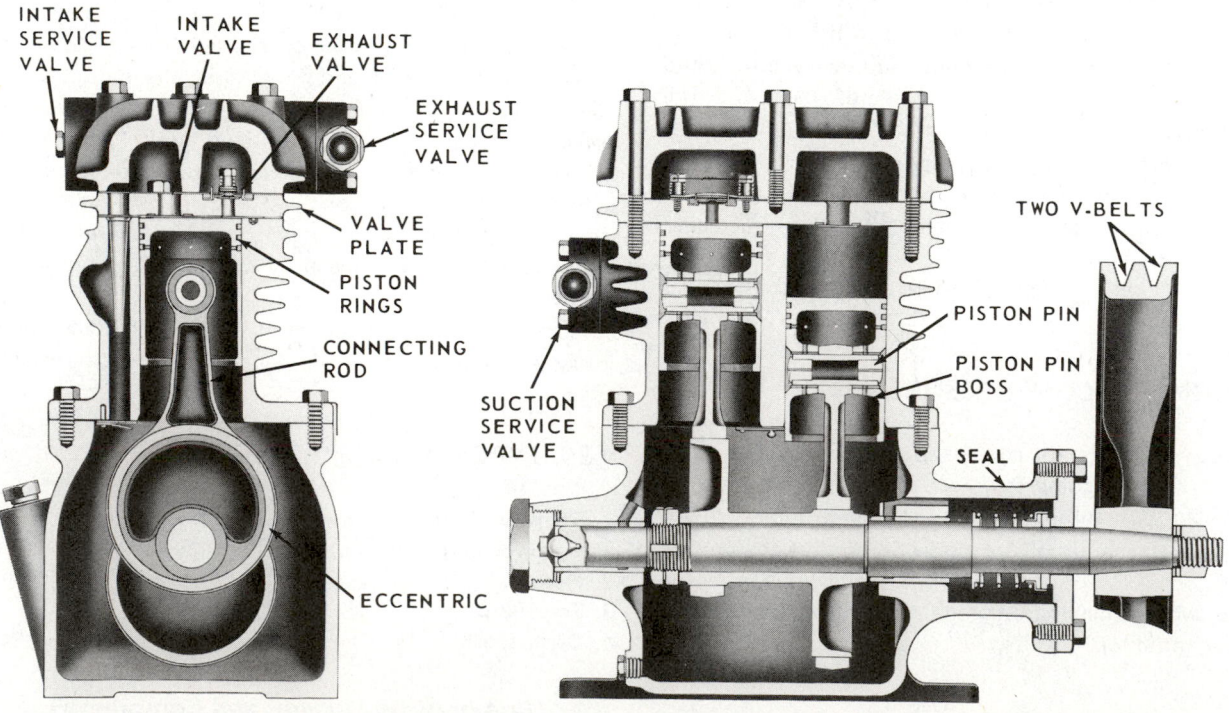

4-27 RECIPROCATING COMPRESSORS

The original energy source is usually an electric motor. Its rotary motion must be changed to reciprocating motion. This change is usually made by a crank and a rod connecting the crank to the piston. The complete mechanism is housed in a leakproof container called a crankcase. It is very efficient. Its construction resembles, in many ways, that of the automobile engine. A typical external drive reciprocating compressor is shown in Fig. 4-28.

Basically, this compressor is a cylinder and a piston. Fig. 4-29 shows the principle of operation of a reciprocating compressor. In illustration No. 1, the piston, B, has moved downward in cylinder, A, and has moved refrigerant vapor from the suction line, C, through the intake valve, E, and into the cylinder space, G. In illustration No. 2, the piston has moved upward and has compressed the vaporized refrigerant into a much smaller space (clearance space) marked H, and has pushed the compressed vapor through valve, F, into the condenser.

4-28 PISTON CYLINDER CRANK ARRANGEMENTS

In compressors having more than one cylinder (multi-cylinder), the crankshaft and cylinders are usually arranged to make the compressor as compact as possible. At the same time, it should provide more pumping capacity for each revolution of the crankshaft. Most two-cylinder compressors use a side-by-side arrangement of the cylinders and a 180 deg. crankshaft. See Fig. 4-28. While one piston is at the top of the stroke, the

other piston is at the bottom. Other two-cylinder compressors have two cylinders at a 90 deg. V. See Fig. 4-26. With this type of cylinder arrangement, a single throw crank is used.

Fig. 4-30 illustrates some common cylinder and crankshaft arrangements.

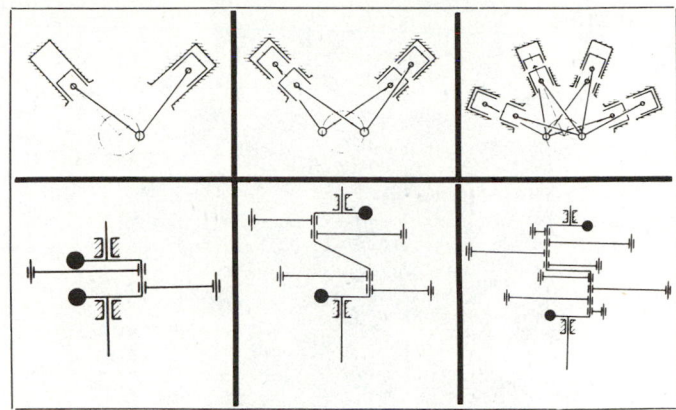

Fig. 4-30. Compressor piston, cylinder and crankshaft arrangements. Piston, cylinder and crankshaft arrangements for two, four and eight-cylinder compressors.   (CP Division, St. Regis)

4-29 CYLINDERS

Compressor cylinders for external drive compressors are usually made of cast iron. The cast iron must be dense enough to prevent the seepage of refrigerant through it. Some nickel is usually added to give the casting this density.

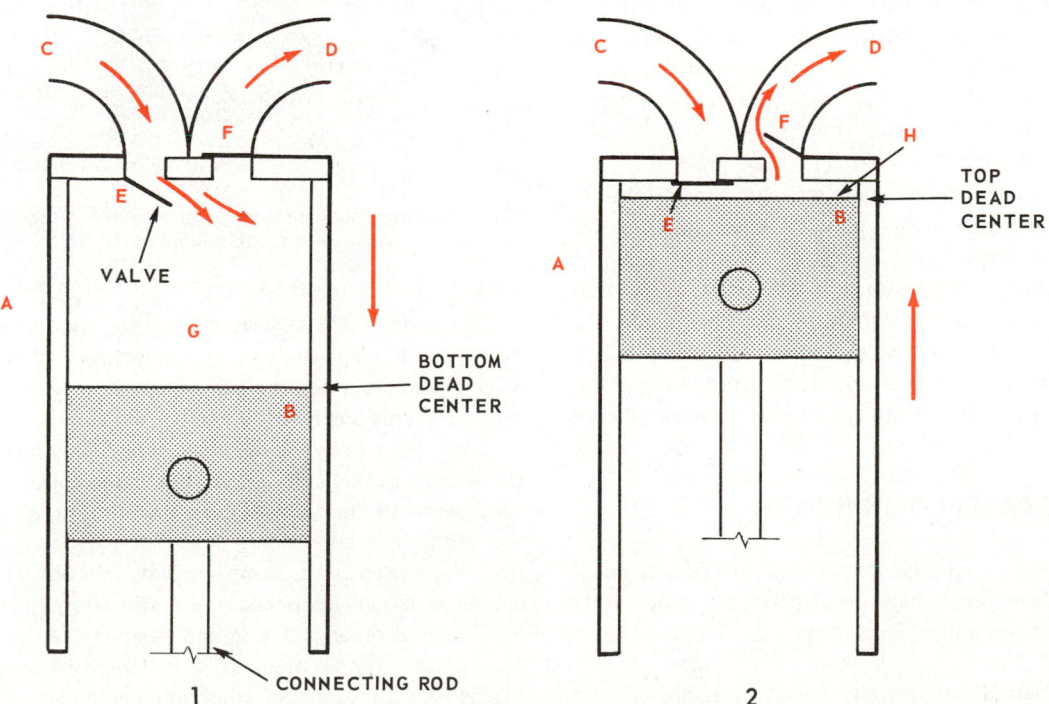

Fig. 4-29. Basic construction of reciprocating compressor. A—Cylinder. B—Piston. C—Intake port from suction line. D—Exhaust port to condenser. E—Intake valve. F—Exhaust valve. G—Piston displacement indicates volume of vapor drawn into cylinder on intake stroke. H—Clearance space at end of compression stroke. Left-hand illustration shows intake stroke; right-hand the exhaust stroke.

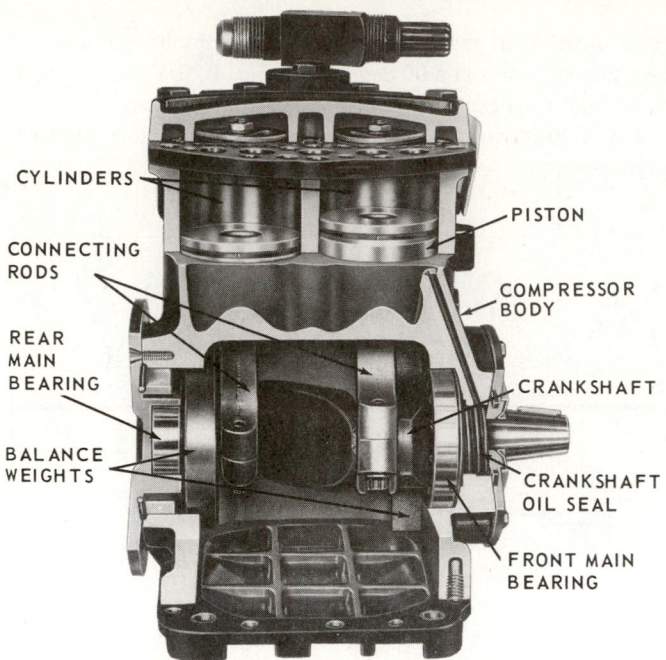

Fig. 4-31. Cutaway view of small, external drive, two-cylinder reciprocating compressor. The body is a casting using light-weight alloy. Cast iron cylinder liners are permanently cast into crankcase body. (York Div., Borg-Warner Corp.)

Small compressors usually have fins cast with the cylinders to provide better air cooling. Larger compressors may have water jackets surrounding the cylinders for cooling. Some compressors are built with cylinder liners or sleeves which may be replaced when worn.

Usually, the crankcase is part of the same casting as the cylinder. This practice cuts down the number of joints that might leak. It also permits close alignment between crankshaft main bearings and cylinder. The main bearings are ball type. Construction is shown in Fig. 4-31. The cylinder arrangement is commonly used on open compressors.

This compressor is designed for use on vehicle air conditioning. However, the same design may be used in other air conditioning applications.

Hermetic (sealed) compressors usually have cast iron cylinders. Some may be of aluminum or other materials. (Typical of hermetic compressor cylinders is C in Fig. 4-27.) Another type of hermetic compressor is pictured in Fig. 4-32. This is a bolted type hermetic and can be dismantled easily for servicing.

4-30 PISTONS AND PISTON RINGS

Pistons used in external drive compressors are usually made of cast iron, while in small high-speed hermetic compressors they are of die-cast aluminum. Smaller sizes do not have piston rings.

Since the temperature of pistons seldom goes higher than 250 F. (121 C.) there is not much expansion of either piston or cylinder. Pistons may be fitted with as little as .0002 in. (.0051 mm) clearance for each inch in diameter.

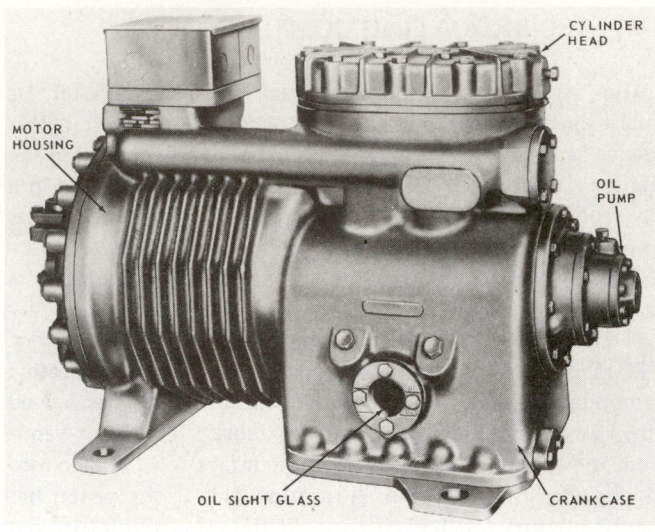

Fig. 4-32. Bolted type hermetic motor compressor assembly. Motor is at left and compressor at right. (Copeland Corp.)

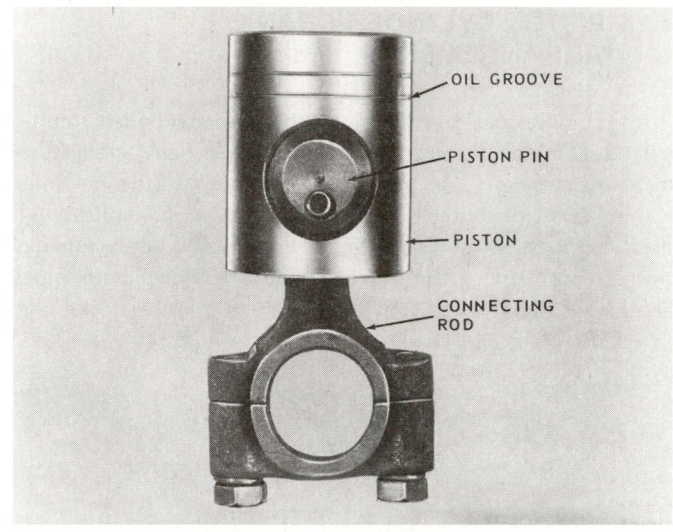

Fig. 4-33. Piston and connecting rod assembly. Note oil grooves cut in piston. (General Electric Co.)

The smaller pistons have oil grooves cut in them. Fig. 4-33 illustrates a common piston connecting rod assembly. Fig. 4-34 illustrates a commercial type piston and connecting rod assembly. This one is fitted with piston rings.

There are two types of piston rings. The upper ring or rings are known as compression rings and the lower is designed to control the oil "flow" past the piston. It is an oil ring.

Piston rings are usually made of cast iron. Some bronze rings have been used, however. Rings should be fitted to the groove as closely as possible and still allow movement. A 45 deg. tapered or angled ring gap permits the ring to push out against the cylinder wall. This gap should be about .001 in. (.0254 mm) for each inch of piston diameter.

Piston pins are made of case hardened high carbon steel accurately ground to size. They are hollow to reduce weight.

Piston pins are usually of the full floating type. This means

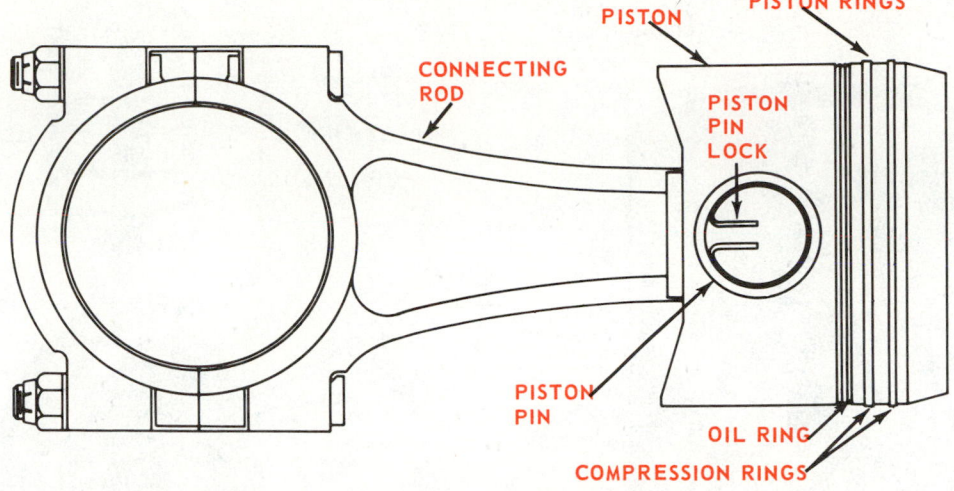

NOTE: PISTON, ROD, AND PIN ARE A MATCHED SET.

Fig. 4-34. Compressor piston and connecting rod assembly. Note how connecting rod's lower (left) end is split and then bolted together to provide bearing to fit crankshaft journal. (Fedders Corp.)

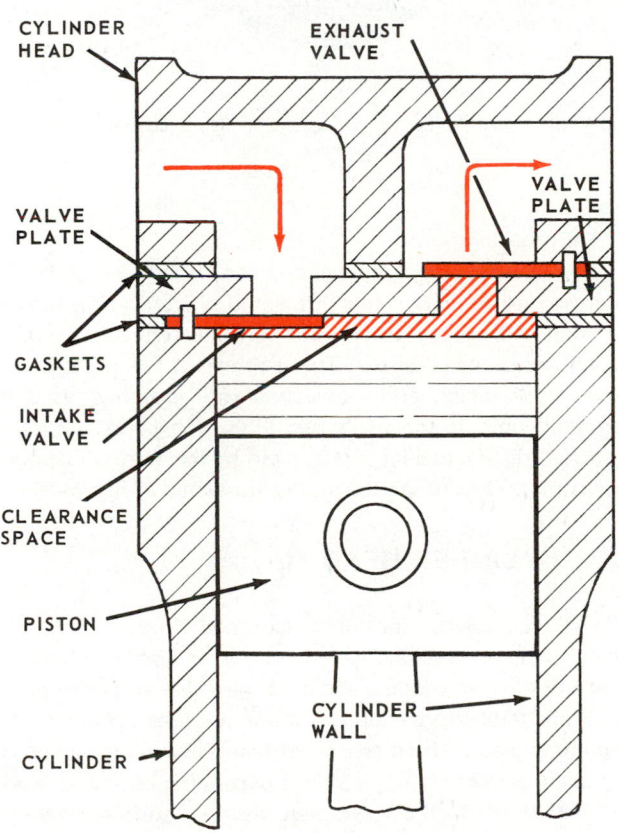

Fig. 4-35. Cross section through compressor cylinder showing cylinder, piston, valve plate, valves, gaskets and cylinder head.

the pin is free to turn in both the connecting rod bushing and the piston bushings.

The piston is designed to come as close as possible to the cylinder head without touching it. This is to press as much of the vapor into the high-pressure side as possible.

When the piston is at upper dead center of its stroke, the clearance between the piston and cylinder head is approximately .010 in. to .020 in. (0.254 mm to 0.508 mm). The volume of space created is called clearance space. (It is illustrated at H in the right half of Fig. 4-29.)

There is a valve plate under the cylinder head with both the intake and exhaust valve located in it, Fig. 4-35.

In hermetic systems, the construction of the pistons and rings, if used, are much the same as those used in external drive compressors. However, since the hermetic compressors usually run at a higher speed than external drive compressors, the pistons are smaller in diameter and are made as light as possible.

Fig. 4-36 illustrates a hermetic compressor. It uses a Scotch yoke piston arrangement. Fig. 4-37 shows still another hermetic motor compressor.

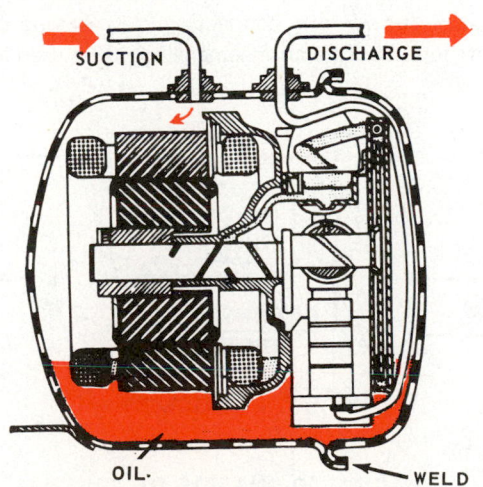

Fig. 4-36. Single cylinder hermetic compressor. Cool refrigerant vapor from suction line flows over motor windings to aid in cooling motor. Compressor uses Scotch yoke piston crank mechanism. Compressor is inverted and a horizontal motor shaft is used.

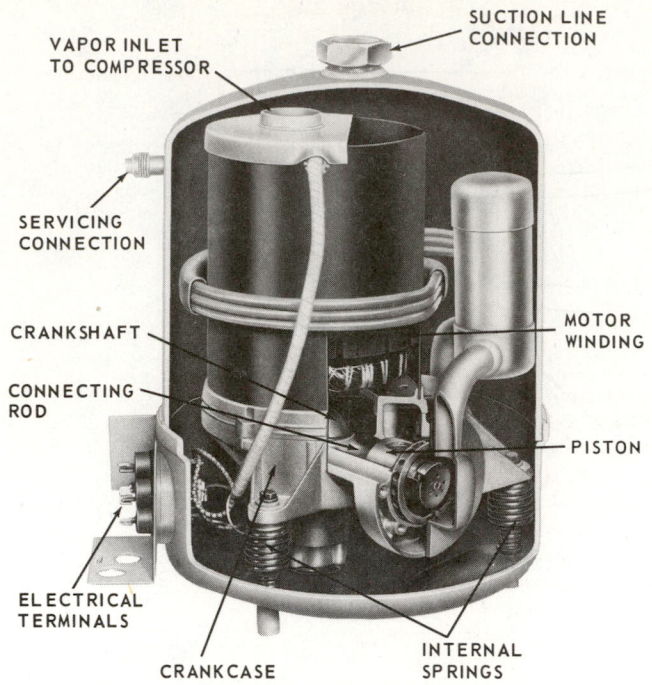

Fig. 4-37. Single cylinder hermetic motor compressor. Note position of motor crankshaft, piston, cylinder and connecting rod. Compressor is spring mounted within dome to minimize vibration. Refrigerant vapor inlet at top protects compressor from possible surge of liquid refrigerant. (General Electric Co.)

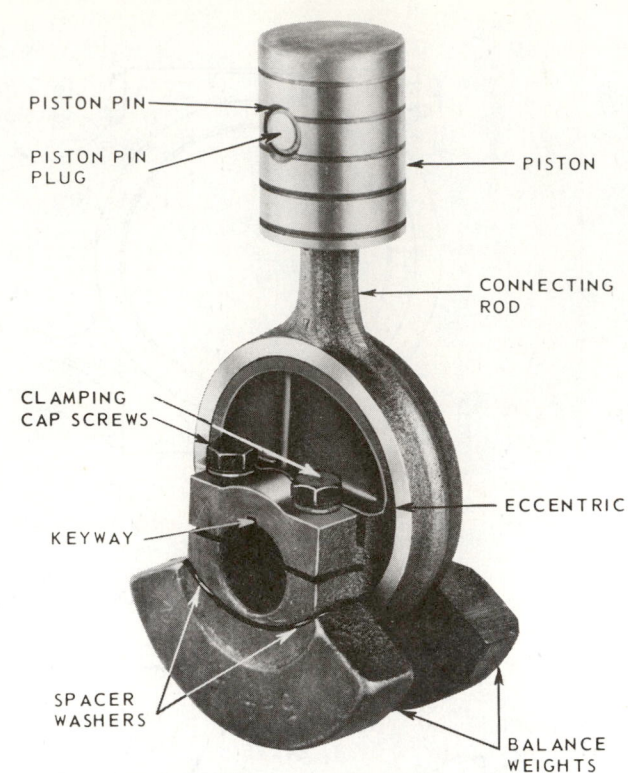

Fig. 4-39. Eccentric crankshaft assembly. Note clamping cap screws and balance weights.

4-31 CONNECTING RODS

The connecting rod attaches the piston to the crankshaft. A conventional connecting rod is shown in Fig. 4-34.

Connecting rods for external drive compressors are usually made of drop forged steel. Crankshafts with a throw use a connecting rod having a split lower end that clamps around the crankshaft journal. The rod bearing must be fitted to a clearance of about .001 in. (.0254 mm). It is, therefore, important that the bolts be carefully tightened (torqued).

Fig. 4-38 shows four connecting rods mounted on one crankshaft journal. This arrangement could be used in a radial

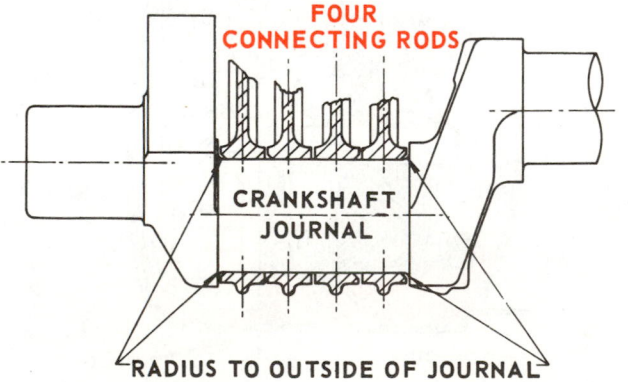

Fig. 4-38. Four connecting rods mounted on one crank journal. This arrangement could be used on radial type compressor. (Airtemp Applied Machinery Co.)

or a V type compressor.

The eccentric type connecting rod usually has a cast iron bearing surface. The crank throw end is a solid ring and must be mounted on the eccentric before the crankshaft proper is assembled to the eccentric. This is shown in Fig. 4-39.

A small piston and connecting rod assembly, used in a hermetic unit, is shown in Fig. 4-33. The connecting rod is attached rigidly to a large piston pin by means of a locking pin and spring. The dismantled unit is illustrated in Fig. 4-40.

4-32 CYLINDER HEAD

Cylinder heads for both external drive and hermetic compressors are usually made of cast iron. The head serves as a pressure plate to support and hold the valves and valve plate in position. It also provides the vapor passages into and out of the compressor. The pressures of compression may amount to as much as 300 psi (21 kg/cm^2) depending upon the kind of refrigerant used. The valve plate must, therefore, have good support so that there will be no leakage at the gaskets on either side of the valve.

In some hermetic systems, the entire compressor housing is inside a dome. The entire space within the dome is open to the suction line. Consequently, the whole dome is under low-side pressure. In such systems, no intake manifold is required — merely an opening into the intake valve or valves.

The cylinder head is usually attached to the cylinder with capscrews. Fig. 4-41 is a cutaway view of a commercial multi-cylinder reciprocating type hermetic compressor.

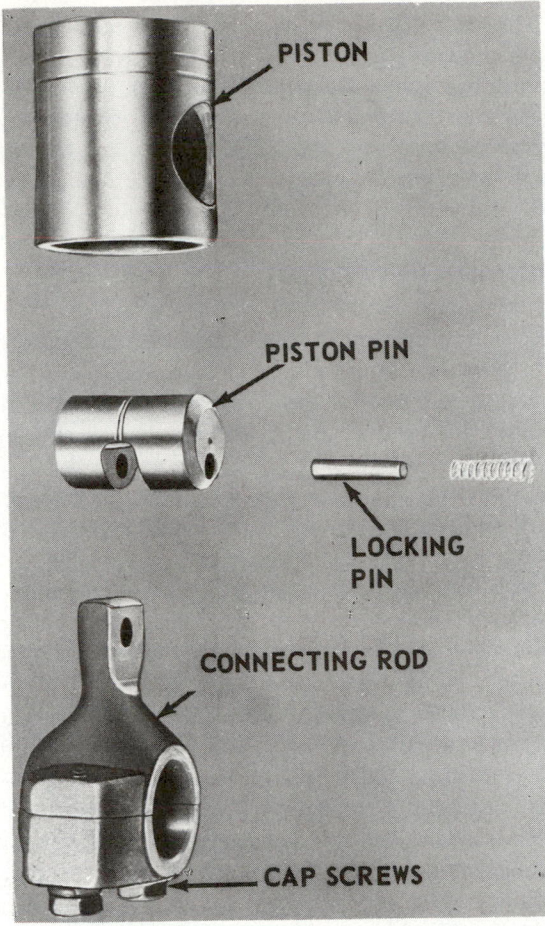

Fig. 4-40. A piston and connecting rod assembly dismantled. This is the same assembly shown in Fig. 4-33. Note small pin which locks connecting rod to large piston pin and the two cap screws used to attach cap to crankshaft end of connecting rod.

The suction line connects to the shutoff valve on the right end. The exhaust, which connects to the condenser, is connected to the discharge shutoff valve at the left end.

Note the location of the crankcase heater (see Para. 4-60). Note, also, the discharge header safety spring. This relieves the pressure in the cylinder should it exceed a safe working limit (due to slugging of liquid refrigerant or refrigerant oil).

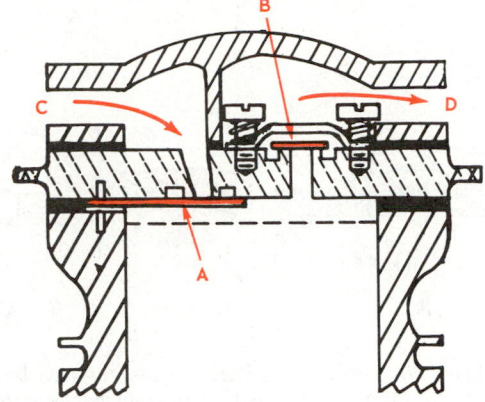

Fig. 4-42. Typical compressor valve plate. A—Intake valve. B—Exhaust valve. Heavy springs on exhaust valve cage permit a greater valve lift to protect compressor in case of severe liquid refrigerant or oil pumping.

4-33 VALVES AND VALVE PLATES

The usual valve assembly consists of a valve plate, an intake valve, an exhaust valve and the valve retainers. See Fig. 4-42.

Valve plates are sometimes made of cast iron, but hardened steel is also used as plates can be thinner with longer wearing valve seats.

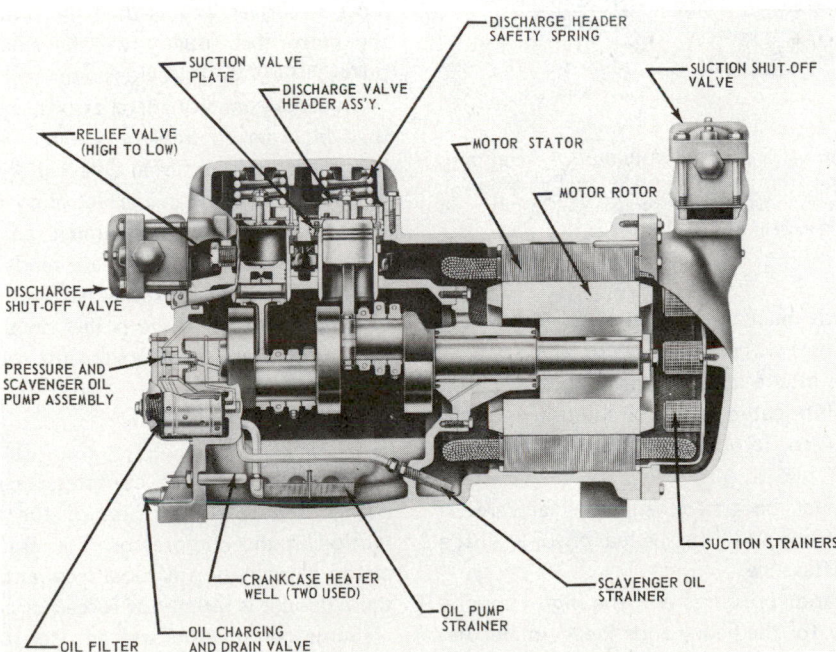

Fig. 4-41. A serviceable hermetic reciprocating type commercial compressor. A bolted type, it has four banks of two cylinders each (four connecting rods on each crankthrow). (Airtemp Applied Machinery Co.)

Compressor valves are usually made of high carbon alloy steel. They are heat treated to give them the properties of spring steel and ground to a perfectly flat surface.

The intake valve is usually kept in place by small pins or the clamping action between the compressor head and valve plate. The exhaust valve may be clamped in the same way.

Some different valve designs are displayed in Fig. 4-43. Fig. 4-44 shows a typical valve plate assembly.

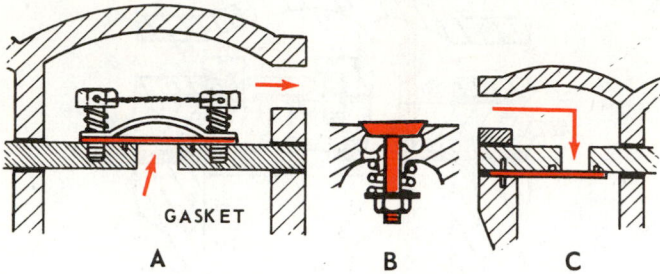

Fig. 4-43. Some typical compressor valve designs. A—Reed valve, spring closed. B—Poppet valve, spring closed. Used on some large compressors. C—Reed valve. Pressure difference keeps valve closed.

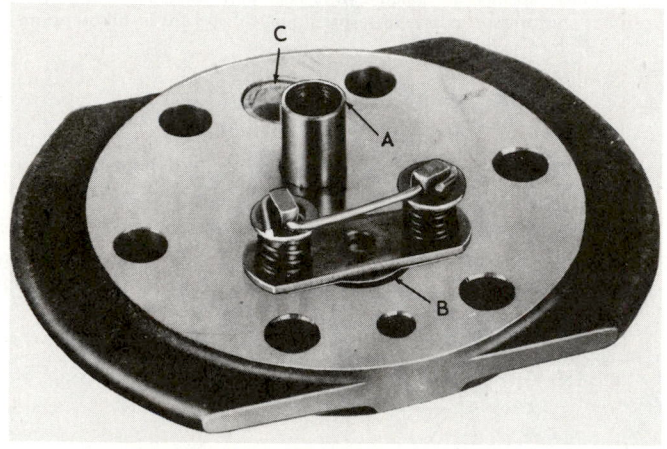

Fig. 4-44. A typical cast iron valve plate construction. A—Inlet port. Note extension to prevent oil from entering intake. B—Exhaust port. Note wire lock used in the two exhaust valve screws. C—Oil return to the crankcase.

The valve disks or reeds must be perfectly flat. A defect of only .0001 in. (0.00254 mm) will cause valves to leak.

Of the two valves, the intake gives the least trouble. This is because it is constantly lubricated by oil circulating with the cool refrigerant vapors. Also, it operates at a relatively cool temperature.

The exhaust valve must be fitted with special care. It operates at high temperatures and must be leakproof against a relatively high pressure difference.

Because of the high vapor pressures and the high temperatures, there is a tendency for the heavy ends (heavy molecules of hydrocarbon oils) to collect on the valve and valve seat as carbon.

The valves open about .010 in. (0.254 mm). If the move-

ment is more, a valve noise develops. If the movement is too little, not enough vapor can move past the valve.

In small high-speed hermetic compressors, the intake valves are made very light and as large as possible. Greater amounts of refrigerant vapor are thus allowed to enter the cylinder during the very small fraction of a second that the intake valve is open.

4-34 CRANKSHAFT SEAL

Refrigerating systems that use an external motor (open type) to drive the compressor need a leakproof joint where the crankshaft comes out of the compressor crankcase. This is absolutely necessary as the pressures vary greatly in the crankcase.

This joint requires seals that are carefully designed and installed, for it is a place where the shaft rotates part of the time and then is at rest part of the time.

All seals use two rubbing surfaces. One surface turns with the crankshaft and is sealed to the shaft with an O-ring of synthetic material. The other surface is stationary and mounted on the housing with leakproof gaskets. The surface materials (accurate to .000001 in. (.0000254 mm) and optically flat) are made of these different combinations: hardened steel and bronze, ceramics and carbon. The two rubbing surfaces must be lubricated or they will wear and start to leak.

Teflon is often used as a gasket material on automobile air conditioning compressors. The crankshaft seal must operate at a high temperature. It is usually made of carbon and ceramic.

Fig. 4-45 illustrates some popular crankshaft seals.

4-35 COMPRESSOR DRIVE (EXTERNAL DRIVE)

External drive compressors are usually driven by a V-belt. Most are driven at less than the motor speed. This means that the motor belt pulley will be smaller than the compressor pulley. The V-belt provides a quiet, efficient drive.

In large capacity installations, more than one belt may be used in order to transmit the required horsepower. Fig. 4-46 illustrates a conventional external drive system.

Pulleys must be in perfect alignment and the pulley shafts (motor and compressor) must be exactly parallel to each other. Most V-belt pulleys are made of cast iron but some are built up from stamped steel parts.

The size of the drive pulley on the motor and the flywheel on the compressor will govern the speed of the compressor.

4-36 CRANKSHAFT

Reciprocating type compressors must use some means to change the rotary motion of the motor into reciprocating motion in the compressor. The crank throw-connecting rod-piston combination is most frequently used. The crankshaft in these designs is usually of forged or cast steel. See Fig. 4-47.

Some compressors use an eccentric fastened to a straight shaft in place of the conventional crankshaft (eccentric shaft). (An eccentric is a section of a shaft which is larger and has a different center than the shaft.) This construction is used to

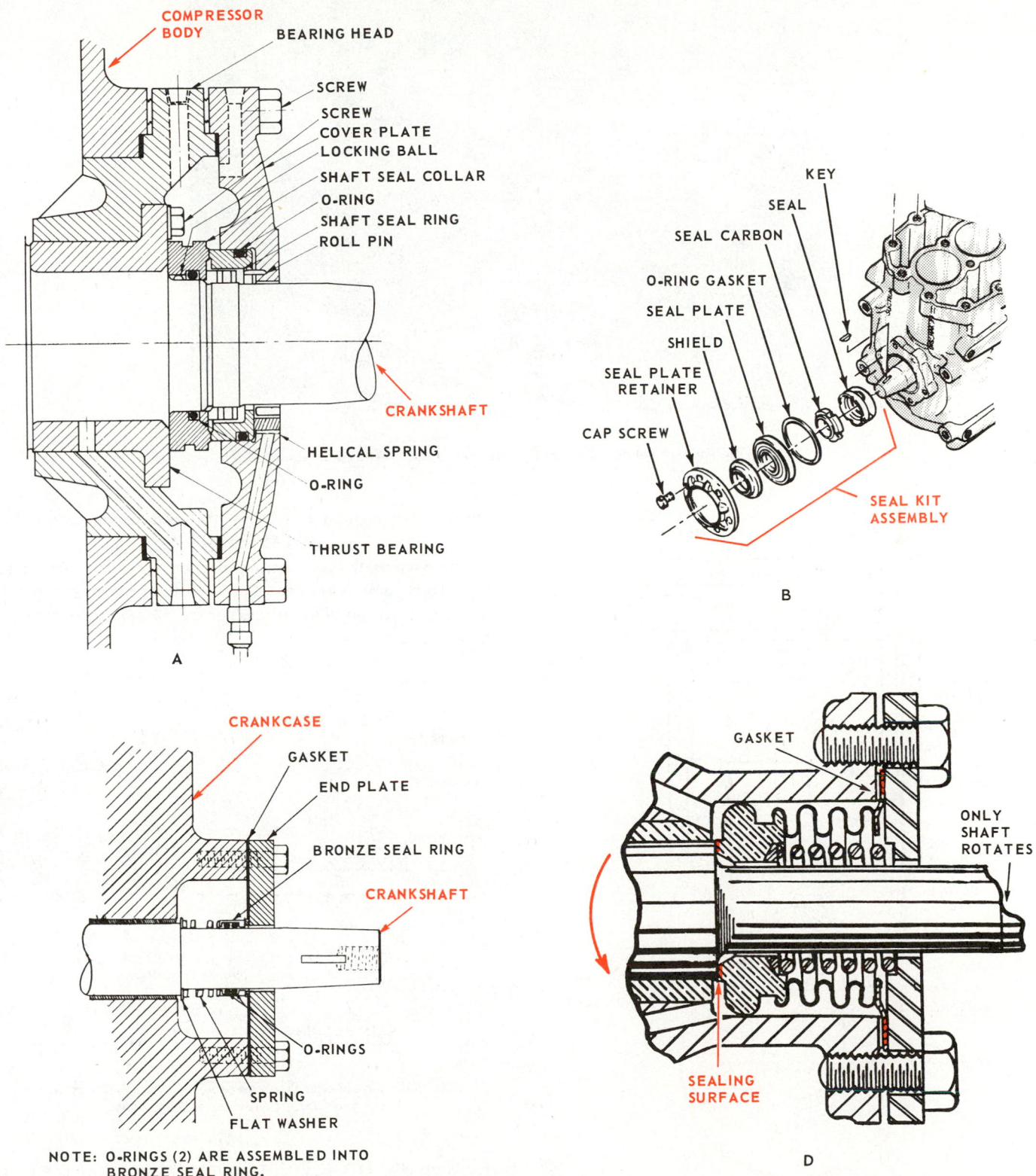

COMPRESSOR BODY
BEARING HEAD
SCREW
SCREW
COVER PLATE
LOCKING BALL
SHAFT SEAL COLLAR
O-RING
SHAFT SEAL RING
ROLL PIN
CRANKSHAFT
HELICAL SPRING
O-RING
THRUST BEARING

A

KEY
SEAL
SEAL CARBON
O-RING GASKET
SEAL PLATE
SHIELD
SEAL PLATE RETAINER
CAP SCREW
SEAL KIT ASSEMBLY

B

CRANKCASE
GASKET
END PLATE
BRONZE SEAL RING
CRANKSHAFT
O-RINGS
SPRING
FLAT WASHER

NOTE: O-RINGS (2) ARE ASSEMBLED INTO BRONZE SEAL RING.

C

GASKET
ONLY SHAFT ROTATES
SEALING SURFACE

D

Fig. 4-45. Crankshaft seal construction for external drive compressors. A—Seal used in commercial compressors. (Mycom-Mayekawa U.S.A., Inc.) B—Seal used with an automobile air conditioning compressor. (Ford Division) C—Replacement seal. (Chicago Valve Plate & Seal Co.) D—Bellows type seal.

Fig. 4-46. An external drive compressor, five-belt drive. (The Gates Rubber Co.)

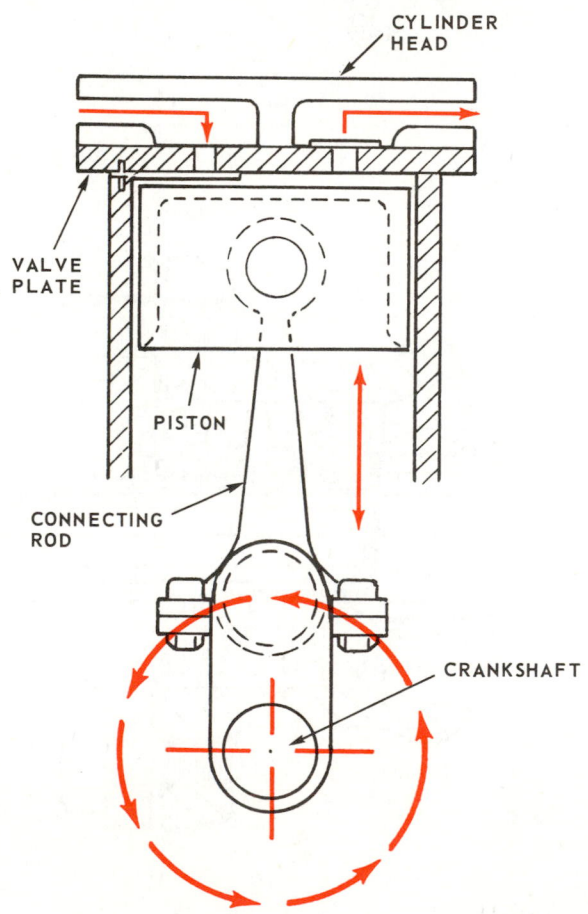

Fig. 4-47. Crank throw type crankshaft. As crankshaft revolves, piston reciprocates (moves up and down). Piston pin oscillates (swings back and forth) as it reciprocates with the piston. Lower end of the connecting rod rotates with crankshaft.

reduce vibration and to remove the need for connecting rod caps and bolts. See Figs. 4-48 and 4-39.

The crankshaft main bearings support the crank. They also must carry any end load. Crankshaft and connecting rod bearings are fitted with great accuracy. Clearance for lubrica-

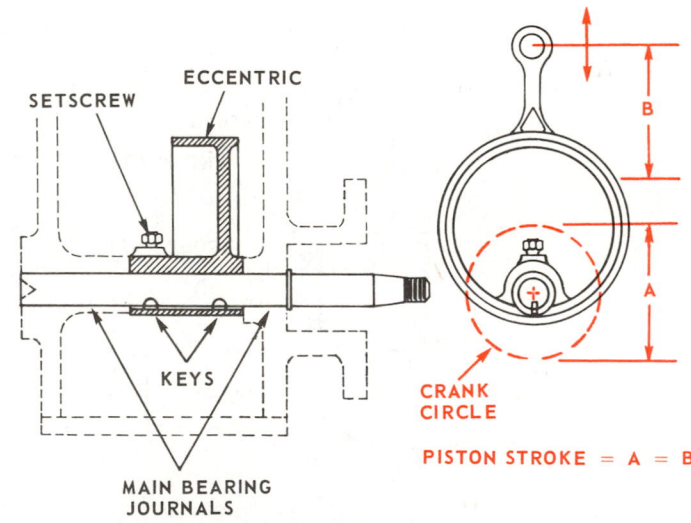

Fig. 4-48. Eccentric type crank mechanism. Note that eccentric is attached to shaft with key and set screw.

tion is usually .001 in. (.0254 mm). In external drive compressors, the drive pulley is usually attached to the crankshaft with a standard taper, a key and a nut-lock washer combination.

Many hermetic systems use the crank throw-crankshaft-connecting rod-piston arrangement, Fig. 4-27.

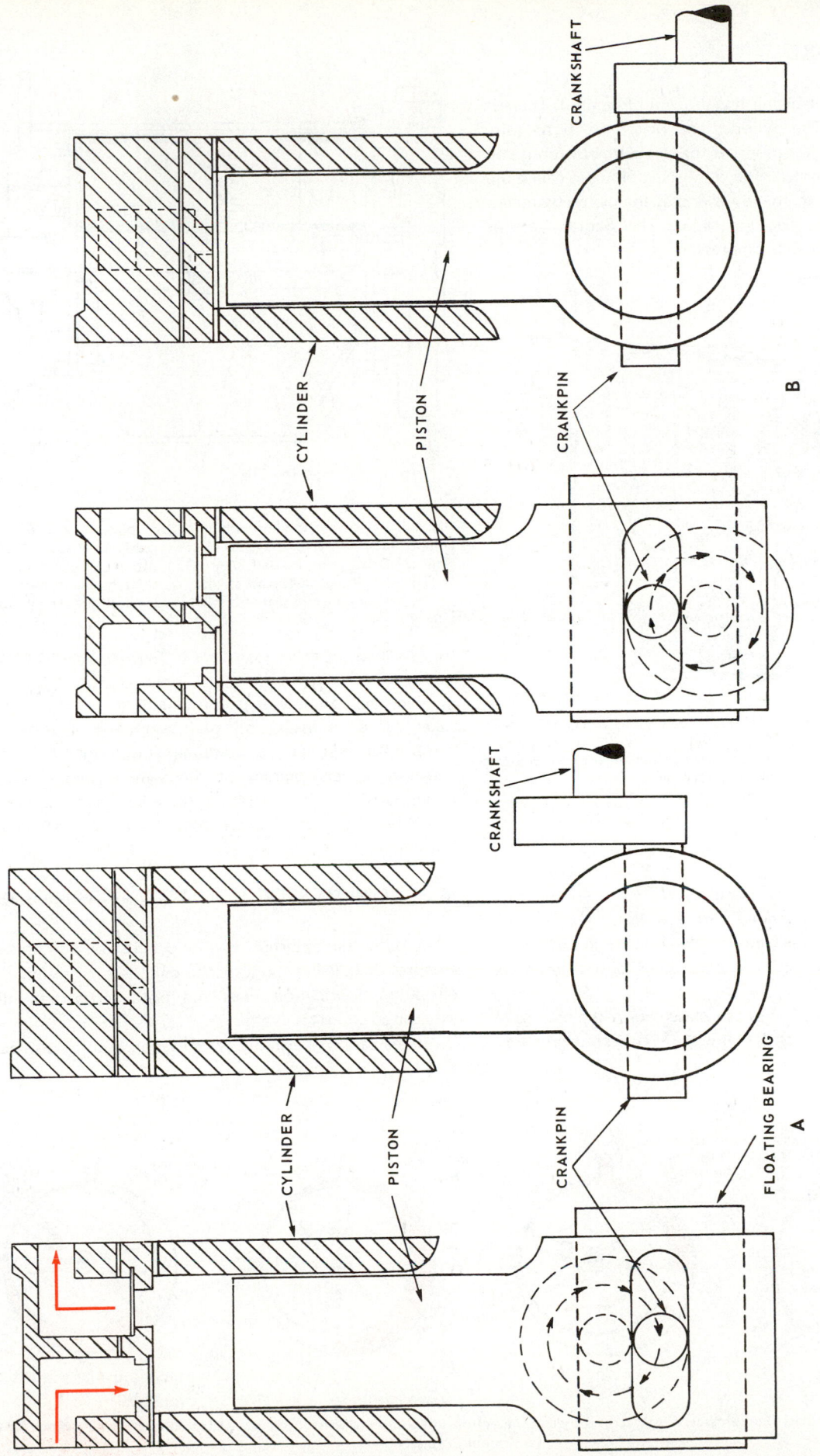

Fig. 4-49. Scotch yoke mechanism used to connect piston to crankshaft. No connecting rod is used. Piston extends to yoke mechanism and compressor cylinder serves as guide. A—Shows piston at bottom of stroke (end of intake stroke). B—Shows piston at top of stroke (end of exhaust stroke).

4-37 SCOTCH YOKE

The Scotch yoke mechanism is shown in Fig. 4-49. There is no connecting rod. The cylinder and piston are both quite long, and even at the lower end of the stroke the piston is still guided by the cylinder wall. The crankshaft pin, also called the crankthrow, connects to the lower end of the piston by means of a floating bearing. See Fig. 4-50. The Scotch yoke is popular in small high speed compressors.

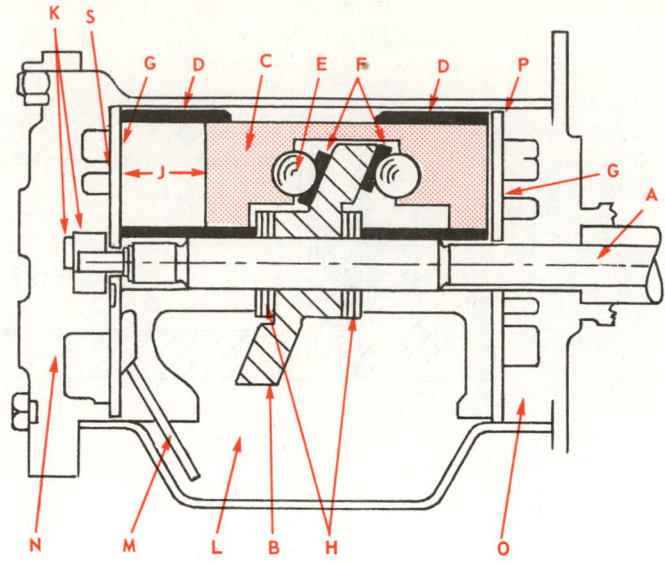

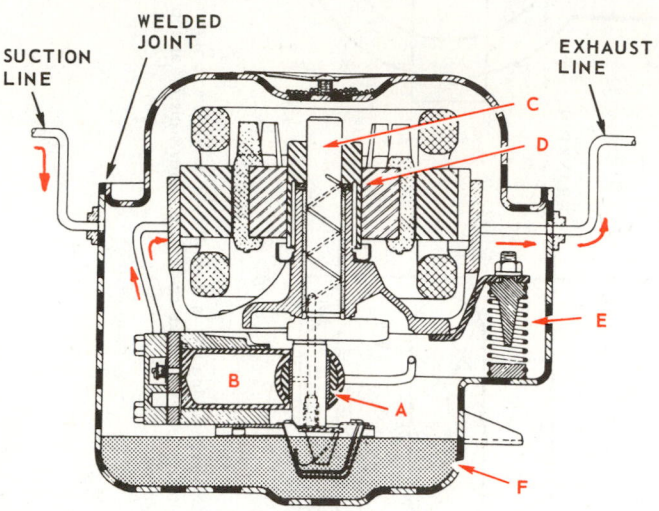

Fig. 4-50. Hermetic compressor using Scotch yoke mechanism. A—Crank throw and Scotch yoke. B—Hollow piston. C—Combined motor shaft and crankshaft. D—Crankshaft thrust bearing. E—Internal mounting spring. F—Oil reservoir.

4-38 SWASH PLATE

A popular type of reciprocating compressor used on many automobile air conditioning systems is known as a "swash" plate or "wobble" plate compressor. No connecting rod is used in this type of compressor. The cylinder and pistons are mounted as in Fig. 4-51.

As the shaft revolves, the swash plate causes the pistons to reciprocate in the cylinders. Usually the swash plate compres-

Fig. 4-51. Cross section through a "swash" plate type of reciprocating compressor. A—Drive shaft. B—Swash plate. C—Piston. D—Cylinder wall. E—Drive ball. F—Ball shoe. G—Valve plate (valve not shown). H—Thrust bearing. J—Piston stroke. As driveshaft and swash plate revolve, double end piston is moved back and forth in cylinder.

sor has three or more cylinders arranged in a circle around the drive shaft.

Since the compressor is double acting, that is, compression takes place at each end of the stroke, a three-cylinder compressor will give a pumping action like a six-cylinder conventional compressor of the same cylinder and stroke dimensions. This is an external drive compressor. It requires a seal where the drive shaft passes through the compressor housing.

4-39 COMPRESSOR HOUSING — CRANKCASE

In both the external drive and hermetic compressor, the compressor housing supports the cylinders, crankshaft, valves, oil pump, lubrication lines and both refrigerant inlet and exhaust openings.

In the case of hermetic systems, the housing also supports

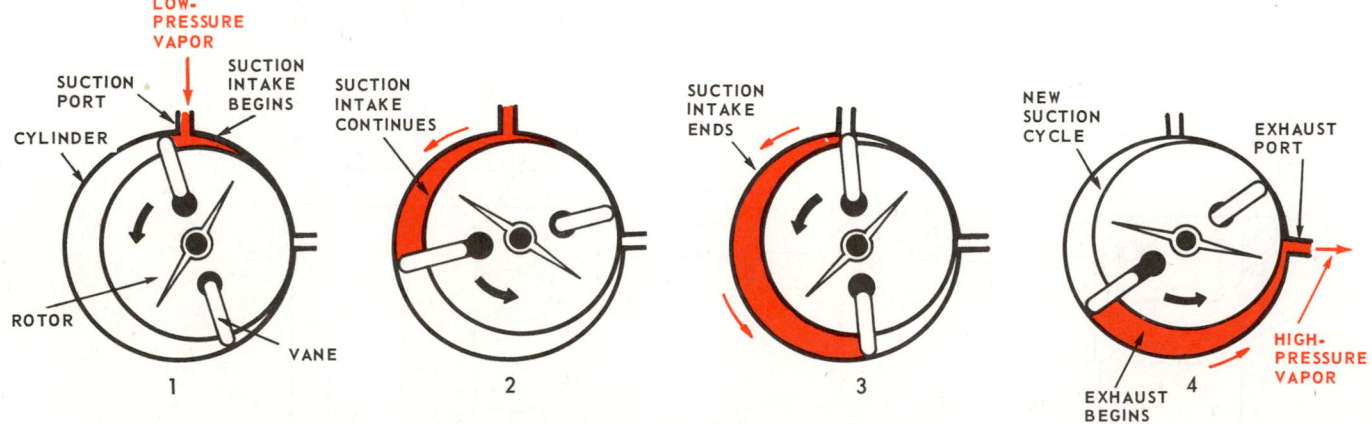

Fig. 4-52. A rotary blade compressor. Black arrows indicate direction of rotation or rotor. Red arrows indicate refrigerant vapor flow. (Whirlpool Corp.)

and aligns the driving motor. Typical compressor housing designs are shown in Figs. 4-32 and 4-50.

Some hermetic designs are bolted together and are provided with service valves. These are called serviceable hermetics, Fig. 4-32. Many hermetic housings, particularly in the smaller sizes, are welded together. See Figs. 4-27 and 4-36.

4-40 INTAKE AND EXHAUST PORTS

Conventional external drive compressors provide inlet and exhaust ports as part of the cylinder head. These ports are usually fitted with service valves. See Fig. 4-28.

Some hermetic compressors. also have service valves. In the small hermetic compressors in which the motor-compressor mechanism is enclosed in a welded dome, the inlet and exhaust lines go directly from the inlet and exhaust port of the compressor through the compressor dome and are not generally supplied with service valves. See Fig. 4-50.

4-41 ROTARY COMPRESSOR

There are two basic types of rotary compressors. One has blades that rotate with the shaft. The other has a stationary blade. The rotary blade compressor using two blades, shown in Fig. 4-52, is typical of this type. The low-pressure vapor from the suction line is drawn into the opening and fills the space behind the blade as it revolves. As the blades revolve, trapped vapor in the space ahead of the blade is compressed until it can be pushed into the exhaust line to the condenser.

A commercial rotary blade compressor, using eight blades, is pictured in Fig. 4-53. The basic operation of the eight-blade compressor is the same as the two-blade.

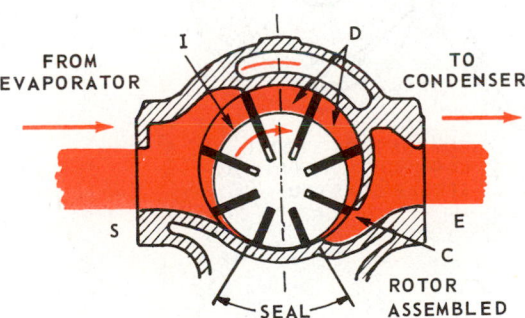

NOTE: SEAL AT BOTTOM OF ROTOR FROM DISCHARGE TO INLET IS A COUSTANT MINIMUM CLEARANCE TO DECREASE LEAKAGE

Fig. 4-53. Eight bladed rotary sliding vane compressor. Black arrows indicate direction of rotation. Red arrows indicate direction of flow of vapor. C—Exhaust port. D—Displacement. E—Exhaust. I—Inlet port. S—Suction. Inlet port is considerably larger than exhaust port. This is because inlet pressure (suction) is considerably less than exhaust pressure. (FES)

Fig. 4-54 illustrates a section through an eight-blade rotary blade compressor. This is an external drive compressor. The shaft seal is shown at the right end.

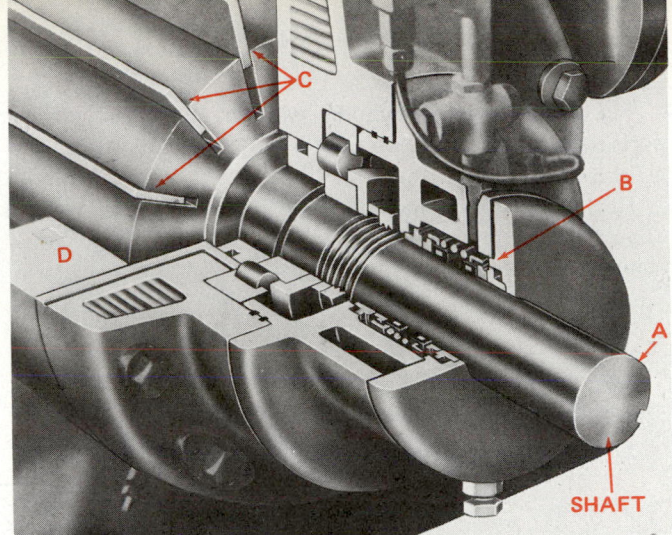

Fig. 4-54. Section through rotary sliding vane compressor. This is an external drive compressor. Note shaft seal at the right end. A—Shaft. B—Shaft seal. C—Blades. D—Housing.   (FES)

Fig. 4-55. Cylinder and rotor from Fig. 4-54 in detail. A—The inside of the cylinder showing port openings. Note intake ports are longer than exhaust ports. B—Relative position of rotor and blades inside cylinder. (FES)

Fig. 4-55 shows the cylinder with the intake and exhaust ports in Part A. The relative positions of the rotor and cylinder are shown in Part B.

Rotating vane compressors are frequently used as the "booster" compressor in cascade systems. This is the name commonly given to the first compressor in a cascade system.

These compressors have three advantages:
1. They provide a large size opening into the suction line.
2. They provide large inlet port openings.
3. They have a very small clearance volume.

Since the low-side pressure may be quite low, this means that the low-side vapor will be drawn into the compressor

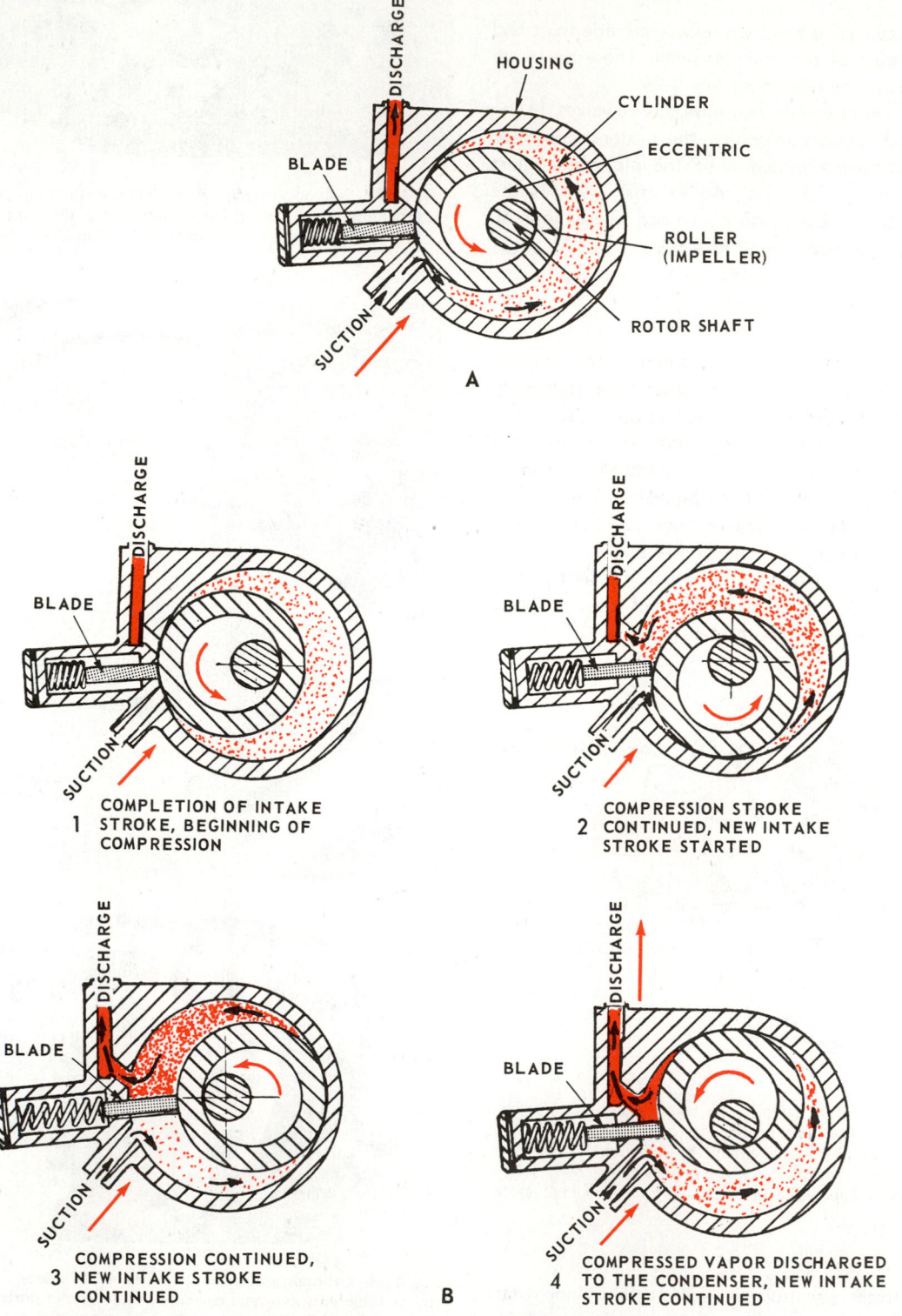

Fig. 4-56. Rotary compressor. Stationary blade or divider block is in contact with a roller (impeller). A—Identification of parts. B—Operation.

under a very small pressure difference. Because these compressors provide a large opening into the compressor from the low side, more vapor will be drawn in on the intake stroke. The small clearance space provided in these compressors means that all the vapor drawn in on the intake stroke will be pushed out on the exhaust stroke. This increases the compressor efficiency. The cascade system is explained in Chapter 17.

Fig. 4-56 represents a stationary blade (often called a divider block) rotary compressor. An eccentric shaft rotates an impeller in a cylinder. This impeller constantly rubs against the outer wall of the cylinder.

As the impeller (or roller) revolves, the blade traps quantities of vapor. The vapor is compressed into a smaller and smaller space, building up the pressure and temperature. Finally the vapor is forced through the exhaust port into the high-pressure side of the system (condenser).

The compression action on one quantity of vapor takes place at the same time another quantity of vapor is filling the cylinder on the intake stroke. All of the parts must be fitted to extremely close tolerances and clearances. The dimensions are so accurate and the surfaces so smooth that no gaskets are needed in the compressor assembly.

Fig. 4-57 shows a hermetic rotary compressor using a stationary blade (dividing block).

In rotary compressors, check valves are usually used in the suction line to prevent the high-pressure vapor and compressor oil from flowing back into the evaporator.

4-42 ROTARY CYLINDER CONSTRUCTION

Rotary cylinders are usually of cast iron. Each is accurately machined, honed, lapped on the inner surface and on the ends.

All cylinders have intake and exhaust ports; some models have oil passages for lubrication. They are usually mounted on an end plate that is part of the main crankcase of the compressor. Refrigerant passages continue into this part.

The exhaust valve reed is mounted on the exhaust port outlet of the compressor as close to the compression chamber as possible. Four or more bolts hold the cylinder to the main part of the compressor.

There are also one or more steel dowel pins to help align the cylinder on the back plate. Another accurately finished plate seals the other end of the cylinder.

4-43 ROTOR — COMPRESSOR CONSTRUCTION

In the rotating blade compressor, the rotor is a fixed part of the shaft. The rotor length must be accurate to .0005 in. (.0127 mm). Usually the slots are on a radius to the center of the shaft. To lower the starting load, one company puts the slots at an angle to prevent the blades from touching the cylinder until the compressor nears its operating speed.

In the stationary blade compressor, the rotor, sometimes called the impeller, accurately fits the eccentric, a fixed part of the shaft.

Fig. 4-54 illustrates a popular type of rotor construction used on external drive commercial compressors. Fig. 4-58

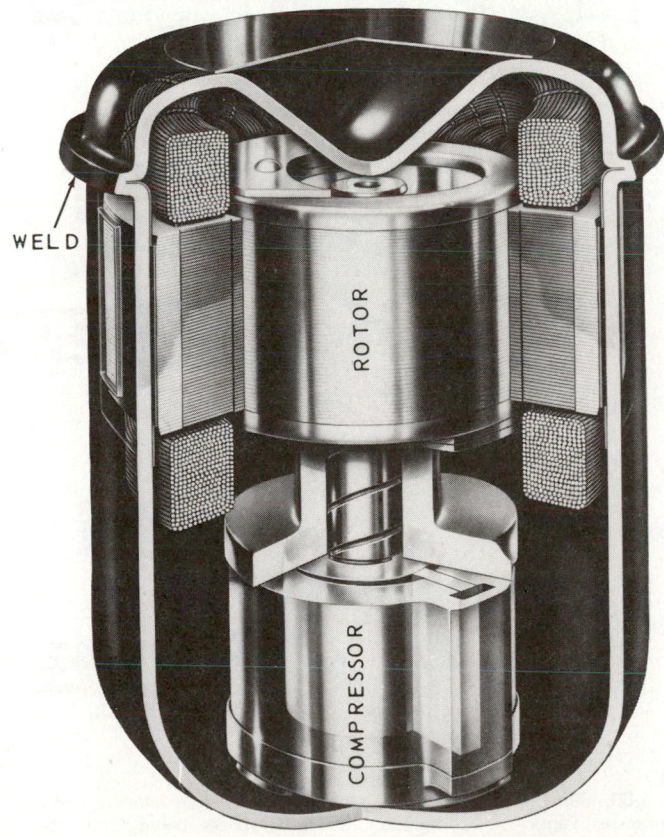

Fig. 4-57. Hermetic rotary compressor. This is a single, stationary blade type compressor. (Frigidaire Co.)

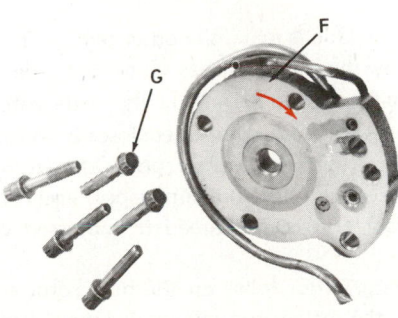

Fig. 4-58. A stationary blade rotary hermetic compressor. A—Cylinder. B—Eccentric. C—Roller. D—Single blade. E—Rotor bearing. F—Cylinder end plate. G—End plate attaching cap. H—Suction line from evaporator. I—Exhaust to the condenser. J—Motor winding.

Compression Systems and Compressors / 129

illustrates a stationary blade rotary compressor. This is a hermetic compressor used both in refrigeration and air conditioning units.

4-44 BLADE (VANE) CONSTRUCTION

Rotating blade compressors use two or more blades. Fig. 4-55, Part B, shows a rotor with eight of them. These blades may be made of either cast iron, steel, aluminum or carbon.

The efficiency of the compressor depends to a great extent on the condition of the edge of the blade where it rubs on the cylinder. The blade must be very accurately ground to fit the slots, the ends of the cylinder and other contact surfaces with the cylinder.

4-45 SCREW TYPE COMPRESSOR

The screw type compressor uses a pair of special helical rotors. They trap and compress air as they revolve in an accurately machined compressor cylinder. These compressors are available in either external drive or hermetic construction. They are used in large systems (20 tons and up).

Fig. 4-59 illustrates a cross section. The two rotors are not

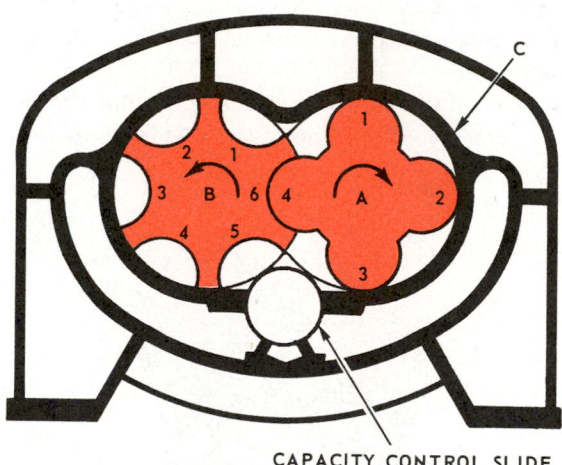

Fig. 4-59. Cross section of screw-type compressor. A—Male rotor. B—Female rotor. C—Cylinder. Vaporized refrigerant enters at one end and exhausts at other end.   (Stal Refrigeration AB)

the same shape. One is male, the other female. The male rotor, A, is driven by the motor. It has four lobes. The female rotor, B, meshes with and is driven by the male rotor. It has six interlobe spaces. The cylinder, C, encloses both rotors.

In operation, the refrigerant vapor is drawn in as shown in Fig. 4-60. The intake (low-pressure vapor) enters at one end of the compressor and is discharged (compressed vapor) at the opposite end.

Since there are four lobes on the male rotor and six on the female rotor, the male rotor will revolve more rapidly than the female rotor. The rotors are helixes so the pumping action will be a continuous action rather than pulsating as with a reciprocating compressor. In the absence of this reciprocating motion, there is very little vibration during operation.

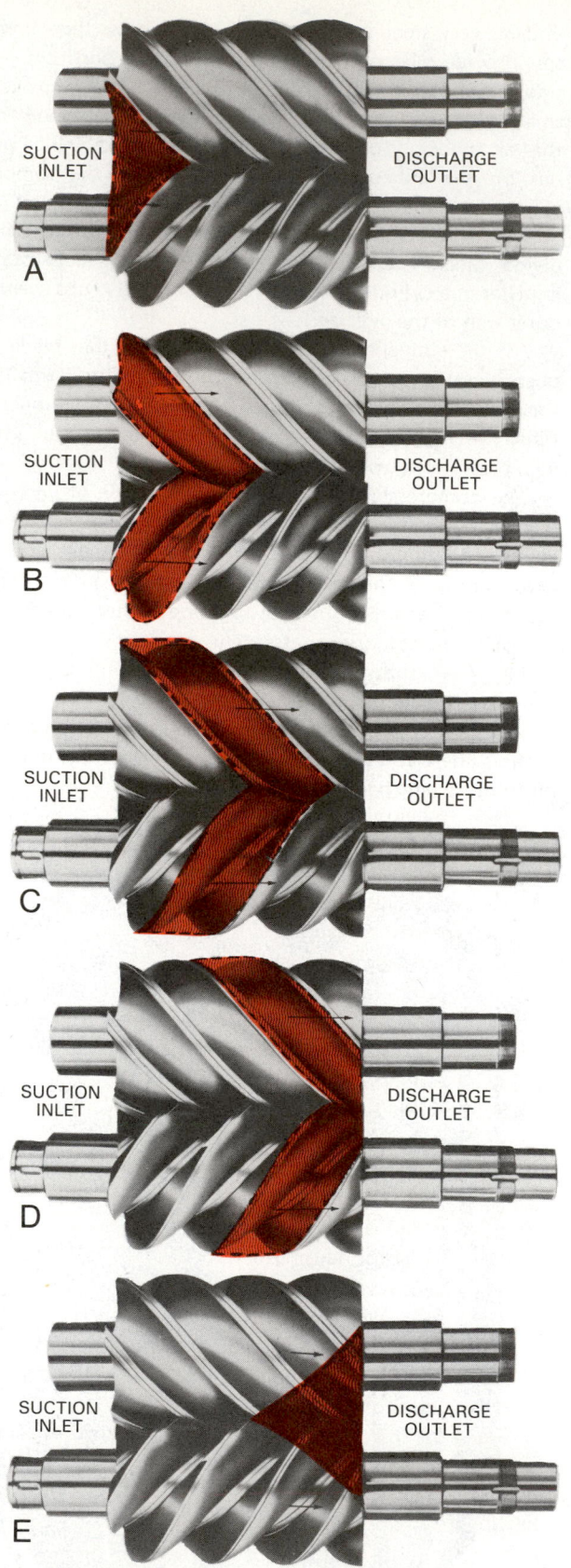

Fig. 4-60. Basic operation of screw-type compressor. Revolving rotor compresses vapor: A—Compressor interlobe spaces being filled. B—Beginning of compression. C—Full compression of trapped vapor. D—Beginning of discharge of compressed vapor. E—Compressed vapor fully discharged from interlobe spaces.   (Dunham-Bush, Inc.)

Fig. 4-61. Screw type compressor which uses matched set of helical rotors. A—Suction inlet. B—Female idler rotor. C—Female section of compressor cylinder. D—Shaft seal. E—Drive shaft. F—Discharge outlet. G—Male section of compressor cylinder. H—Motor driven male rotor. J—Capacity control. (Dunham-Bush, Inc.)

Fig. 4-61 illustrates a cutaway view of an external drive screw type compressor. It is powered by an external electric motor, operating off the shaft marked ''E''. The motor drives the male rotor. Two matched helical rotors, male and female, trap and compress the refrigerant as they turn together. The rotor marked ''C'' in the illustration is mounted on the drive shaft. The other rotor is made to turn by the action of the first rotor. A control device mounted outside the housing regulates the capacity of the unit. The device can be seen at the left.

A hermetic screw type compressor is shown in Fig. 4-62. Its capacity control device is mounted inside the housing. Its inlet port is located at right angles to the rotors. Outlet port is through the motor housing.

Many screw type compressors operate with oil injection. This seals the clearance between the rotors and between the rotors and the cylinder. It also helps cool the compressor. The efficiency of these compressors is quite good. They may be used with most of the common refrigerants.

Fig. 4-63 illustrates a pair of screw type rotors in operating position. Since the screw type compressor operates at fairly high speed, adequate bearings must be provided for good rotor bearing life.

4-46 CENTRIFUGAL COMPRESSORS

Centrifugal compressors are used successfully in large refrigerating systems. In this type compressor, vapor, as it is moved rapidly in a circular path, moves outward. This action is

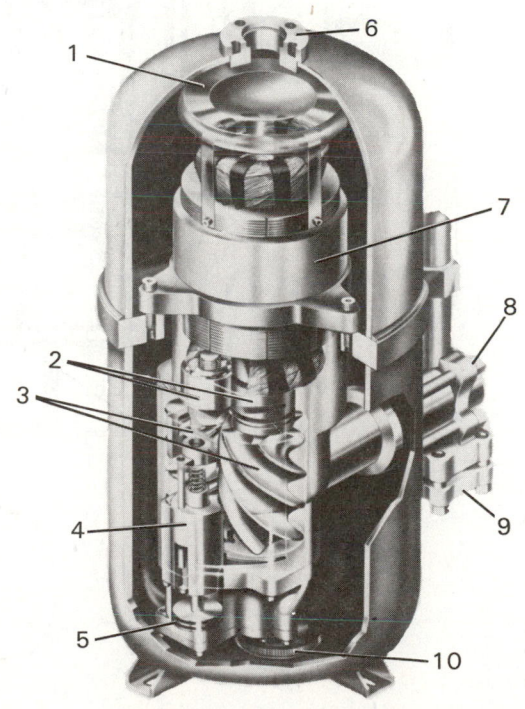

Fig. 4-62. Screw type hermetic motor compressor in cutaway view. Note helical rotors and capacity control slide valve. 1—Oil separation plate. 2—Inlet main bearings. 3—Rotors. 4—Slide valve. 5—Unloader piston. 6—Discharge port. 7—Hermetic motor housing. 8—Suction service valve. 9—Suction port. 10—Oil strainer. (Dunham-Bush, Inc.)

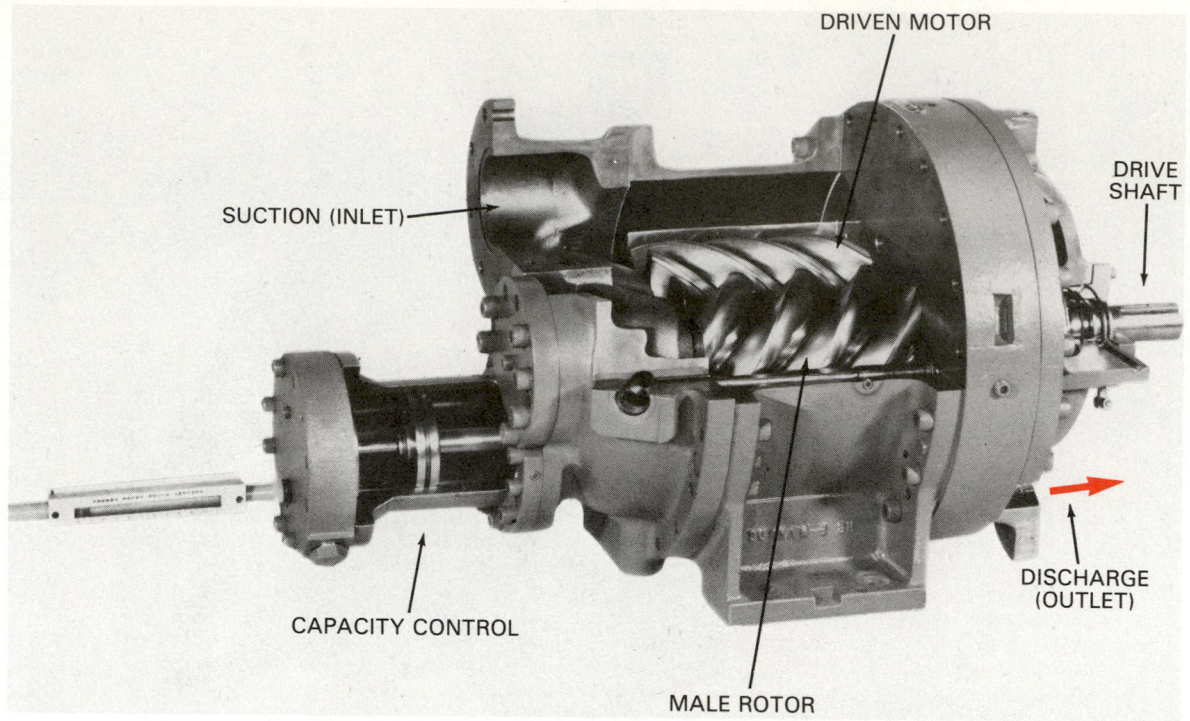

Fig. 4-63. Matched set of helical rotors in screw type compressor. Male rotor drives female rotor. Refrigerant vapor is compressed as it flows from left to right through screw compressor. (Dunham-Bush, Inc.)

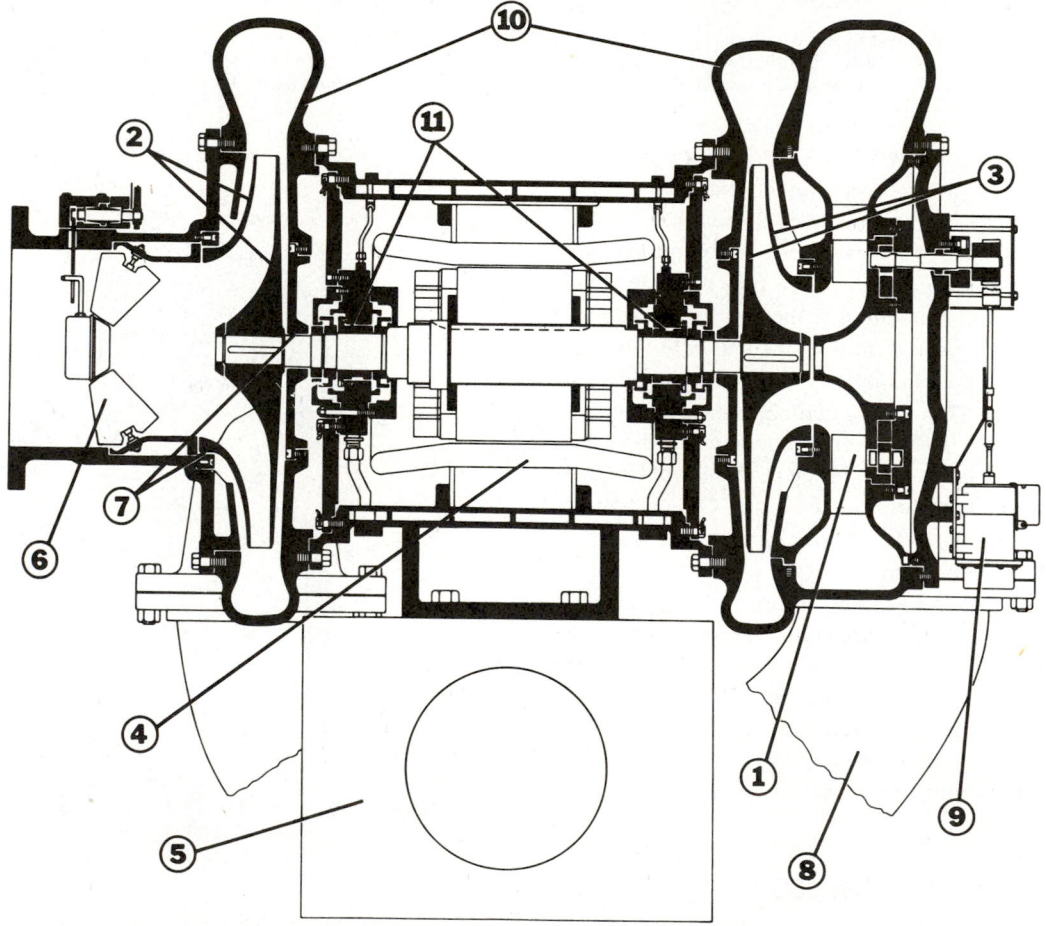

Fig. 4-64. Two-stage centrifugal compressor. 1—Second stage variable inlet guide vane. 2—First stage impeller. 3—Second stage impeller. 4—Water-cooled motor. 5—Base, oil tank and lubricating oil pump assembly. 6—First stage guide vanes and capacity control. 7—Labyrinth seal. 8—Cross-over connection. 9—Guide vane actuator. 10—Volute casing. 11—Pressure lubricated sleeve bearing. Note discharge opening is not shown. (Airtemp Applied Machinery Co.)

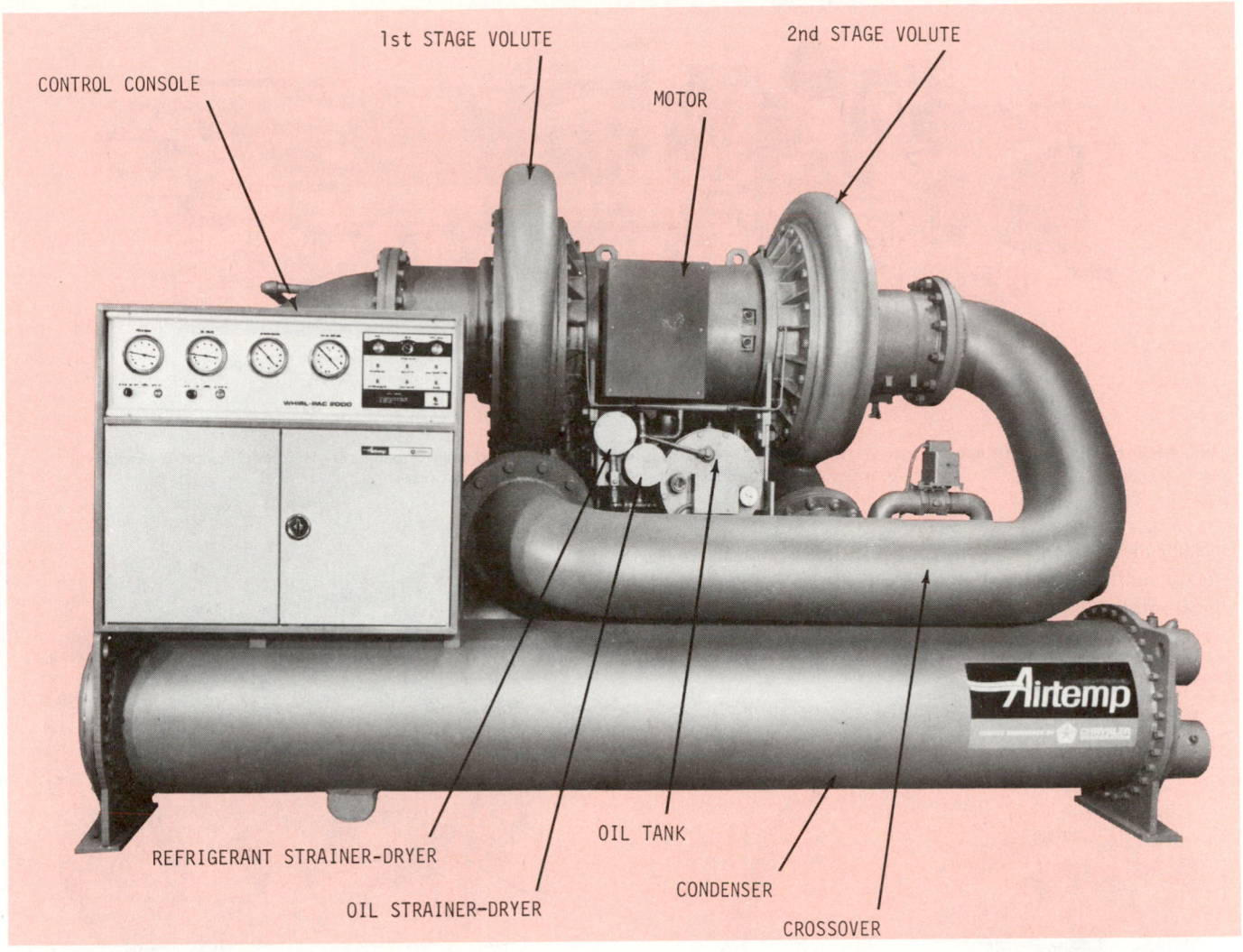

CONTROL CONSOLE

1st STAGE VOLUTE

2nd STAGE VOLUTE

MOTOR

REFRIGERANT STRAINER-DRYER

OIL STRAINER-DRYER

OIL TANK

CONDENSER

CROSSOVER

Fig. 4-65. Two-stage centrifugal compressor. System uses water-cooled condenser. Also note control panel mounted directly on system. (Airtemp Applied Machinery Co.)

called centrifugal force.

The vapor is fed into a housing near the center of the compressor. A disk with radial blades (impellers) spins rapidly in this housing forcing vapor against the outer diameter.

The pressure gained is small so that several of these compressor wheels or impellers are put in series to create greater pressure difference and to pump a sufficient volume of vapor. This type of compressor looks like a steam turbine or an axial flow air compressor for a gas turbine engine.

The centrifugal compressor has the advantage of simplicity. There are no valves or pistons and cylinders. The only wearing parts are the main bearings. Pumping efficiency increases with speed so compressors operate at high speeds.

Fig. 4-64 is a cross-section through a two-stage centrifugal type compressor. The driving motor is mounted between stages. The inlet is at the left on the illustration. The discharge is in the back at the right end of the illustration and is not shown. Fig. 4-65 pictures a centrifugal compressor mounted in the refrigerating system. Fig. 4-66 shows a section through a hermetic centrifugal compressor. These compressors operate at

a high speed and are usually driven by an electric motor or steam turbine.

4-47 STATOR CONSTRUCTION

The stator or casing of a centrifugal compressor is usually made of cast iron. It has a changing radius inside to adapt itself to the vapor pickup by the impellers.

The casing (cylinder) also holds the main bearings, the oil pressure producing pump as well as the refrigerant vapor intake and exhaust ports. It also holds the shaft seal where the shaft extends or sticks out from the casing for the power drive, when an external motor is used. Both the first stage and second stage have adjustable inlet vanes to control the capacity of the pump.

4-48 ROTOR CONSTRUCTION

The rotor or impeller in a centrifugal compressor is keyed to the compressor shaft. It is made of cast iron or steel and is

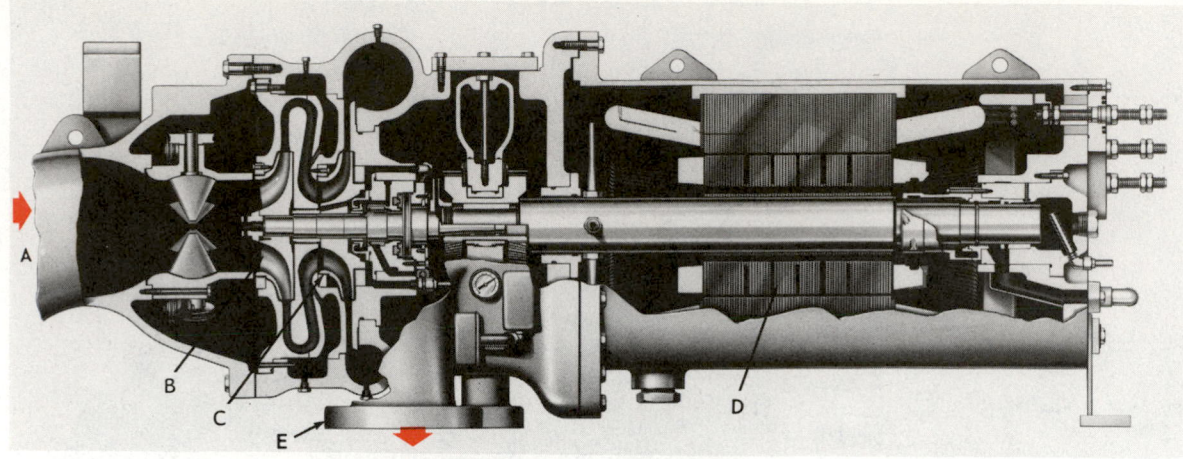

Fig. 4-66. Hermetic centrifugal compressor. A—Intake. B—First stage impeller. C—Second stage impeller. D—Hermetic motor. E—Exhaust. (Carrier Air Conditioning Group, United Technologies Corp.)

specially designed to move the vapors without going above gas velocity limits and without having vapor trapping pockets. A typical rotor is shown in Fig. 4-67.

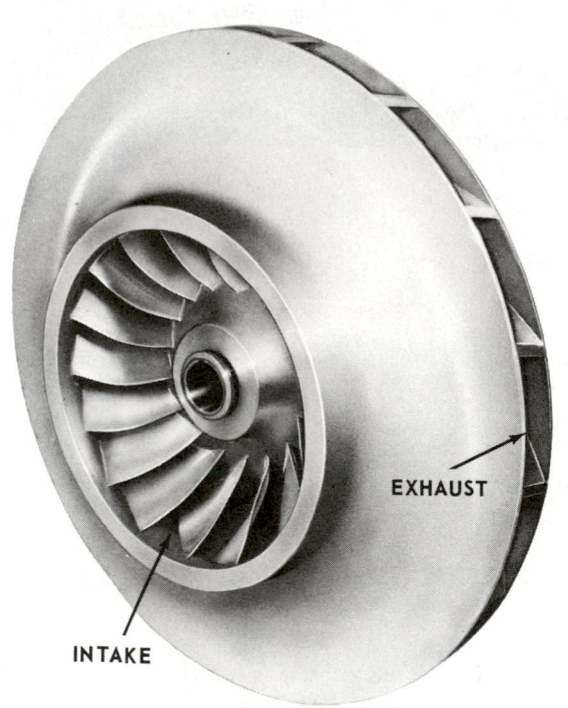

EXHAUST

INTAKE

Fig. 4-67. Impeller from centrifugal compressor. See B of Fig. 4-66. (Carrier Air Conditioning Group, United Technologies Corp.)

4-49 MOTORS

Two different types of motors are used for driving refrigeration compressors.

External drive compressors use conventional motors and usually drive the compressor with one or more V-belts or by direct drive. A sectional view of the kind of motor used on small external drive compressors is shown in Fig. 4-68.

Fig. 4-68. Motor used on external drive compressor. A—Motor shaft. B—Motor rotor. C—Motor stator. D—Starting mechanism. E—Motor bearing. F—Motor frame. (A. O. Smith Corp.)

In hermetic compressors the motor is mounted under the same dome as the compressor. These motors are lubricated by the oil carried in the refrigerant.

Hermetic motors do not use brushes or open points inside the dome. Arcing would cause pollution in both the oil and the refrigerant. This would lead to an electrical burnout.

A special electrical starting device is located outside the dome. Fig. 4-69 shows the three main parts of a motor used in a hermetic refrigeration compressor.

See Chapter 7 for information concerning motors. Chapter 8 has more information about starting devices and motor controls.

4-50 SERVICE VALVES

Service valves are shown in Figs. 4-10 and 4-11.

Many small hermetic systems do not have service valves of

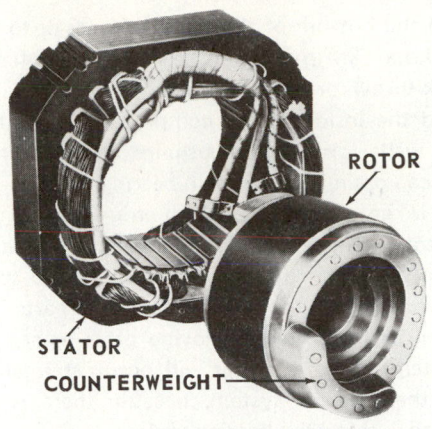

Fig. 4-69. Hermetic unit motor stator and rotor. Rotor is mounted directly on compressor crankshaft. Note counterweight which balances weight of crank, connecting rod and piston.

any type. To attach gauges and service manifolds to these systems, refrigerant lines must be tapped. Special tapping valves are available. These are clamped to a tube so that the valves pierce the tube and, at the same time, provide for necessary gauge and service connections.

4-51 MUFFLERS

Most hermetic units and many external drive systems have noise reducing devices called mufflers on both the intake and the exhaust openings of compressors. Mufflers reduce the sharp gasping sound on the intake stroke and the even sharper puff of the exhaust. These mufflers are brazed cylinders with baffle plates mounted inside. Fig. 4-70 shows a unit equipped with a suction "intake" muffler.

In one design, suction and exhaust mufflers are connected directly to the compressor cylinder head as in Fig. 4-71.

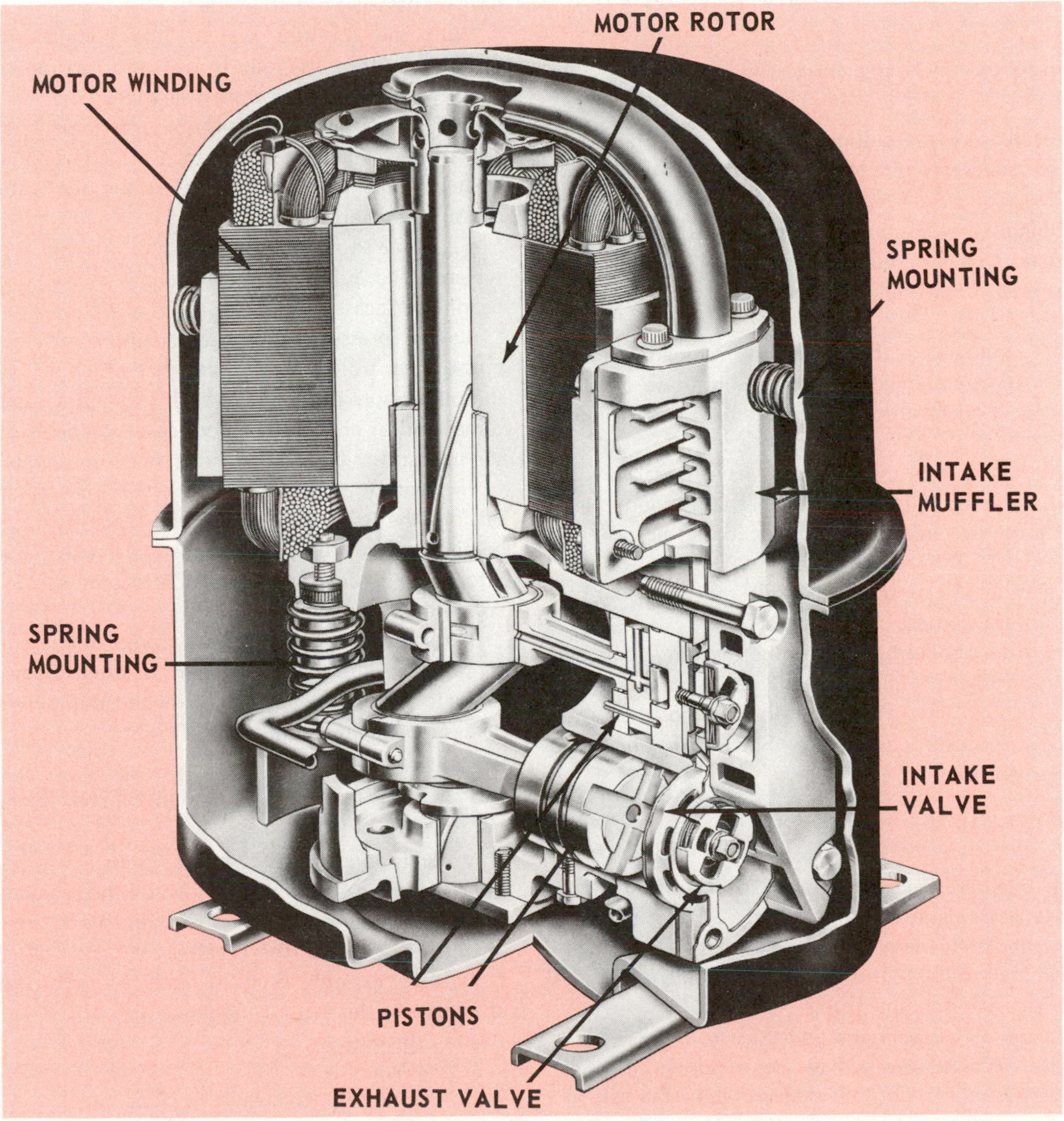

Fig. 4-70. A two-cylinder hermetic compressor suitable for use either in a commercial application or as part of a residence air conditioning system. Note intake muffler. (Tecumseh Products Co.)

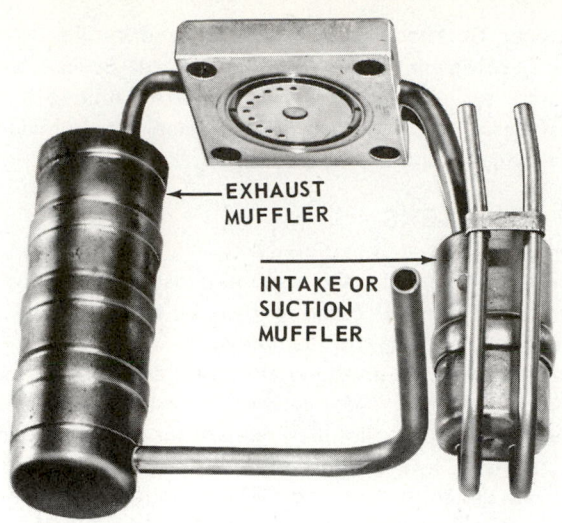

Fig. 4-71. Hermetic compressor cylinder head. Mufflers are attached to suction and exhaust openings.

4-52 COMPRESSOR COOLING

The temperature of the compressor is greatly affected by the heat of compression, which raises the temperature of the vapor as it is "squeezed" and forced into the condenser. Friction (rubbing) between moving parts also adds to compressor temperature. This heat must be removed to prevent loss of efficiency of the pump and to maintain the lubricating qualities of the oil.

The oil that circulates in the compressor removes much of the heat from motor and compressor. As it flows over the heated surfaces, it picks up and carries this heat to cooler surfaces.

Many compressors and many domes have metal fins on their outer surfaces to help carry this heat away. Some units even use a motor driven fan to force cooling air over the compressor.

When a water-cooled condenser is used, this water is often also used to cool the compressor or dome.

Motors are often cooled by passing the suction vapors and return oil over the windings. Some units circulate the crankcase oil through an air-cooled coil. The cooled oil then helps to cool the motor compressor. Some larger hermetic units are water cooled.

4-53 LUBRICATION

Lubricating oils have been developed especially for reciprocating and rotary refrigeration compressors. Usually, these are mineral oils, completely dehydrated, wax-free and nonfoaming. They have a viscosity found to be best for the refrigerant and for the refrigeration temperatures.

Some refrigerating oil contains additives to improve lubricating qualities. The additives may also improve the oil's viscosity properties (ability to flow at given temperatures).

Reciprocating compressors may be lubricated either by a splash or by a pressure (force feed) system.

In the splash system, the crankcase is filled with the correct oil up to the bottom of the main bearings or to the middle of the crankshaft main bearings. At each crankshaft revolution, the crank throw, or the eccentric, dips into the oil and splashes it around the inside of the compressor. Oil is thrown on to cylinder walls, piston pin bushings and into small openings where it can drain into the main bearings.

This is an excellent system for normal use in small compressors. Some compressor connecting rods have little dips or scoops attached to the lower ends. These scoop up small amounts of oil and sling it around to other parts.

Clearances between the moving parts must be less in this type system. Noisy bearings will occur at smaller clearances than in the pressure system, because there is no oil under pressure to cushion the bearing surface.

The force feed or pressure system uses a small oil pump to force oil to the main bearings, lower connecting rod bearings and in some cases, piston pins. It is a more expensive system. It needs a pump and the crankshaft and connecting rod must have oil passages drilled in them.

With the pressure system, the compressor gets better protection from the oil. It will also run more quietly even though it has greater bearing clearances.

The oil pump is usually mounted on one end of the crankshaft. Whenever used, an overload relief valve must be built into the pump to protect it and the system against oil pressures that are too high.

Fig. 4-72 shows the oil path in a pressure type lubrication system. The oil pump delivers oil, under pressure, to all bearing surfaces.

Larger pressure lubricated compressors sometimes use pressure controlled electric switches which will stop the unit if the oil pressure drops too low. It is sometimes necessary to use some kind of an unloading device which enables the compressor to start easily with no vapor pressure load in the cylinder.

In a rotary compressor, it is best to have a constant film of oil on the cylinder, roller and blade surfaces. When the compressor operates, the oil feeds through the main bearings into the cylinder. The cylinder is located so the oil level half covers the main bearings.

In larger units and even in some of the smaller units, a forced feed lubrication system is used. Some units use a separate oil pump, but some use the pumping action of the blades moving in and out of their slots.

4-54 COMPRESSOR VOLUMETRIC EFFICIENCY

Volumetric efficiency of a compressor is the actual volume of refrigerant gas pumped, divided by the calculated volume. A compressor may be designed to pump 10 cu. in. of vapor each revolution or stroke (this is called the piston displacement).

If it pumps only 6 cu. in. each revolution, the volumetric efficiency of the pump is 60 percent (6/10):

$$\frac{6 \text{ cu. in.}}{10 \text{ cu. in.}} = .6:$$

.6 x 100 (to change to percentage) = 60 percent.

Metric: $\frac{98 \text{ cm}^3}{164 \text{ cm}^3} = .6.$ Thus, .6 x 100 = 60 percent.

For efficient operation, the volumetric efficiency must be

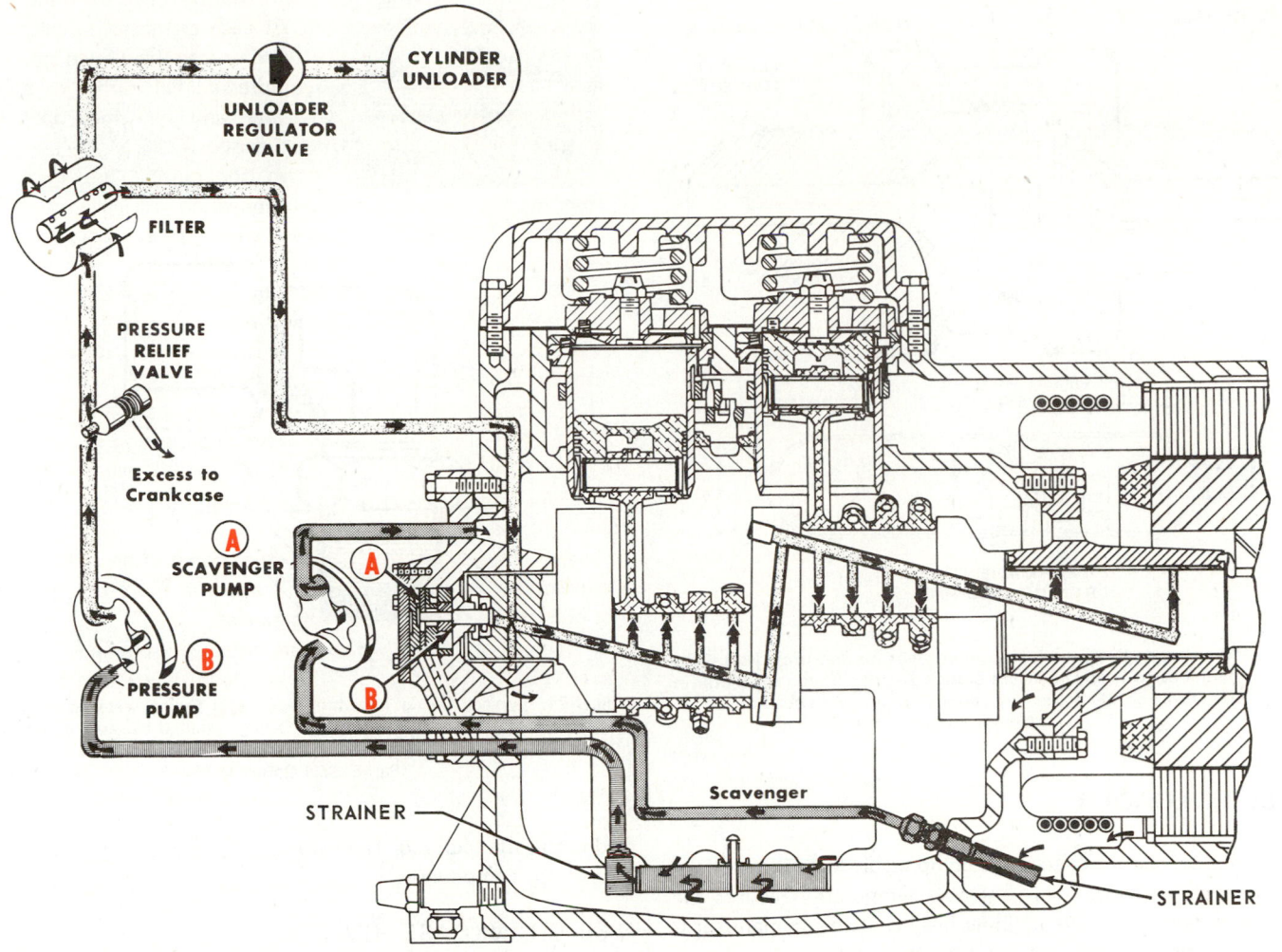

Fig. 4-72. A pressure lubricated hermetic radial multi-cylinder reciprocating refrigerator compressor. Scavenger pump returns oil from motor end of the motor compressor back to the lower portion of the crankcase at the cylinder which serves as an oil sump to store the oil charge. Note pressure relief valve in pressure line. Cylinder unloader is operated by the oil pressure.   (Airtemp Applied Machinery Co.)

as high as possible. Several things affect this efficiency.

First, if the head pressure (the pressure the compressor must pump against) increases, the amount pumped per stroke will decrease. This is because the compressed vapor in the clearance space will expand on the intake stroke and fresh vapor cannot move into the cylinder until the pressure in the cylinder is lower than the pressure in the suction line. The higher the compressed pressure, the greater the compressed vapor in the clearance space will expand.

Second, if the low-side pressure decreases, it is more difficult for the vapors to fill the cylinder and the amount pumped per stroke will decrease.

Third, if the clearance pocket is enlarged, the amount pumped per stroke will decrease. The clearance space is the space left in the cylinder when the roller or piston is at the end of its pumping stroke as shown in Fig. 4-35.

The efficiency of a compressor also depends on the size of the valve openings. If the intake valve reduces the flow of low-side vapor into the cylinder, the cylinder will not be filled and the efficiency of the compressor will be lowered. Also, if the exhaust valves stick or if the line from the compressor to

the condenser is pinched, this extra pressure in the cylinder will cut down the compressor's pumping efficiency.

4-55 COMPRESSION RATIO

Compression ratio in refrigeration is the relationship of the absolute pressure of the low side to the absolute pressure of the high-pressure side to the total volume of the cylinder. Ratios vary up to 10 to 1 for single stage compressors. If the ratio is higher, two-stage compressors must be used. This factor is discussed in Chapter 28.

4-56 CHECK VALVES

Check valves, Fig. 4-73, allow fluid flow in only one direction in a system. Sometimes they are put in refrigerant suction lines. They prevent refrigerant vapor, oil or even liquid refrigerant from backing up into the evaporator or other devices where it might condense or lodge. Check valves are often used in multiple evaporator installations and in heat pump installations.

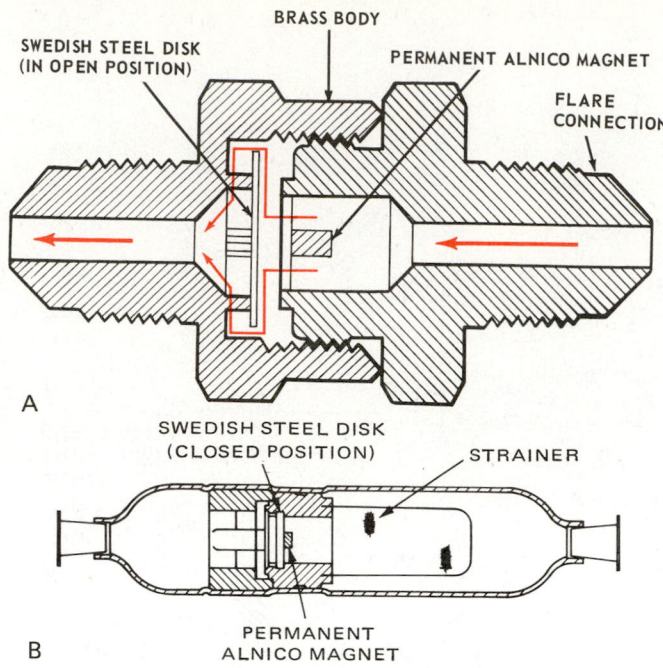

Fig. 4-73. Check valves in which a permanent magnet provides a slight pull. This tends to keep the valves closed. A—Flared type connector in open position. B—Solder type connector in closed position. (Watsco, Inc.)

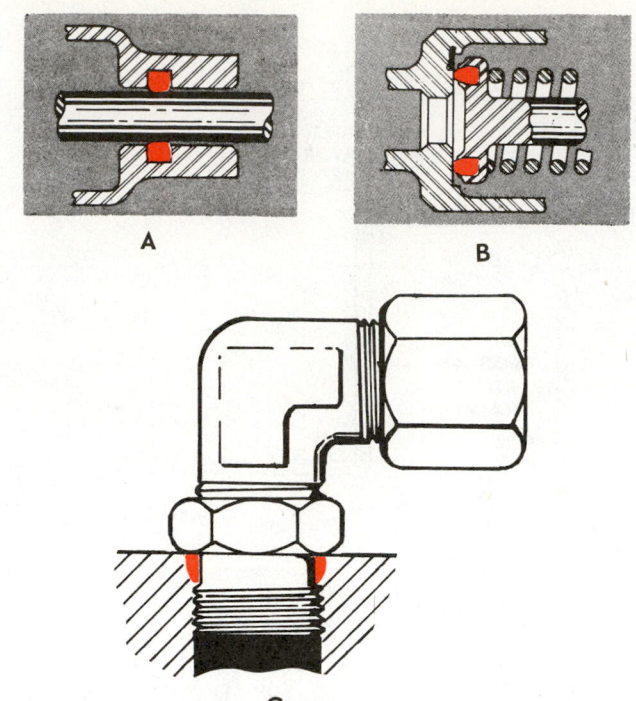

Fig. 4-74. Typical O-ring installations. A—An O-ring installed as seal between a shaft and its housing. B—O-ring installed to serve as a seat seal. C—An O-ring installed as a pipe fitting assembly seal. (Parker Seal Company)

4-57 UNLOADER

To make it easier to start the compressor, some installations provide an unloader. This unloader temporarily reduces the high-side pressure at the cylinder head while the compressor is starting. The unloader may be operated mechanically, electrically, hydraulically or by a solenoid valve. One unloader design is shown in Fig. 4-72.

On small systems which use a capillary tube refrigerant control, the balancing of the low and high pressures serves as an unloader. Unloader mechanisms may also be used to vary the pumping capacity in case of a changing heat load, such as in an air conditioner.

4-58 GASKETS

On external drive compressors, the joining surfaces between bolted parts, such as cylinder heads, valve plates, crankcase openings and the like, are usually sealed with gaskets.

The gaskets may be made of either special paper, synthetic material or lead. Some are made of a plastic substance. Gaskets must be completely free from moisture before use.

4-59 O-RINGS

O-rings are commonly used as a sealing device, particularly in assembling parts where there may be some motion between the assembled parts. Fig. 4-74 illustrates a typical O-ring installation.

The materials used in making O-rings will depend on various factors such as temperature, pressure, fluids to be controlled

and useful life required. They are usually made of oil-resistant, soft plastic compounds.

4-60 CRANKCASE HEATER

In many condensing unit applications, it is necessary to heat the compressor crankcase to evaporate the liquid refrigerant trapped in the oil. Most large compressors used in commercial applications are fitted with a crankcase heater at the time the compressor is manufactured — especially if the compressor may be exposed to cold temperatures (outdoor units). These crankcase heaters may be operated during the off cycle or they may be thermostatically controlled.

For smaller installations which do not usually require a built-in crankcase heater, an accessory heater may be purchased and attached to the crankcase. Crankcase heaters are generally required on remote installations in which the compressor may at times operate at an ambient temperature which is lower than the evaporator temperature.

A compressor with a built-in crankcase heater can be seen in Fig. 4-41. Fig. 4-75 shows a crankcase heater which may be attached to the housing of a hermetic compressor.

4-61 REVIEW OF SAFETY

The service technician should always be alert to avoid service procedures which may:
1. Be a hazard to the technician.
2. Be injurious to the equipment.

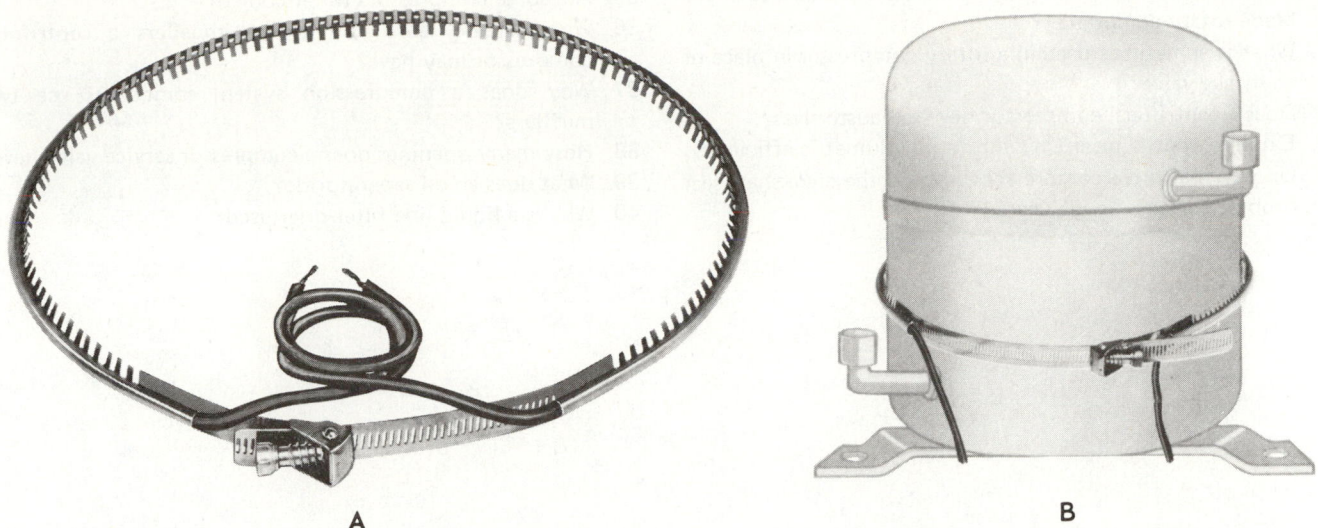

Fig. 4-75. Crankcase heater. A—Accessory type electrically operated crankcase heater. B—Crankcase heater installed on hermetic compressor.
(Tutco, Inc.)

3. Cause refrigeration to fail.

Normal servicing of refrigerator mechanisms is not considered hazardous. There are, however, recommended procedures which should be followed to be sure that the service operations are performed under the safest possible conditions. If the service technician knows and understands the construction and operation of all the parts in a refrigerating system, it will help one service it safely.

The technician should always wear goggles when working on refrigerating systems.

As an installation or servicing technician, one should always handle the parts with care, keeping every dismantled part clean and dry.

It is the little things that count most in servicing refrigerator mechanisms. Care in tightening a tube connection, installing a gasket, replacing an electrical terminal, soldering a fitting, often determines whether the entire job will be safe and satisfactory.

Even a slight amount of moisture allowed to enter a refrigerating mechanism may cause the system to fail either through the formation of ice in the refrigerant control device or the formation of a "sludge" in combination with the refrigerant oil.

One important purpose of refrigeration is to preserve food. If the refrigerator does not maintain the desired temperatures, food may spoil. The service technician must make the necessary settings and adjustments correctly.

4-62 TEST YOUR KNOWLEDGE

1. Name the eight most important parts found in all compression cycle refrigerators.
2. Name the main parts commonly located in the low-pressure side.
3. Name the main parts commonly located in the high-pressure side.
4. Name four types of compressors.
5. What kind of condensing units do not use a crankshaft seal?
6. Why does the temperature rise as a gas is being compressed?
7. Why are check valves used in the suction line on some rotary compressors?
8. Why should condensers be cleaned?
9. What name is applied to the type of evaporator that uses an expansion valve?
10. What control determines the refrigerator cabinet temperature?
11. Name five types of refrigerant controls.
12. How much clearance is allowed between the piston and the cylinder on small compressors?
13. How much lift is allowed the intake and exhaust valves?
14. What does a low-side float control?
15. Why are expansion valves (AEV) usually adjustable?
16. What is a high-side float control?
17. What is the purpose of the crankshaft seal?
18. How does a capillary tube reduce the pressure?
19. When does the condensing pressure stop rising?
20. What basic conditions are necessary to produce refrigeration?
21. When do the molecules changing from liquid to vapor equal the molecules changing from vapor to liquid?
22. What is the purpose of a compressor?
23. What is a full floating piston pin?
24. Why must compressor valves be light weight?
25. How is the crankshaft joint sealed where it leaves the compressor body?
26. Of what materials are compressor pistons made?
27. Why must the clearance pocket volume be kept as small as possible?
28. How are compressor cylinders cooled?
29. How are hermetic motors usually cooled?

30. What is the purpose of the stationary blade in a stationary blade rotary compressor?

31. What is sometimes used in a rotary compressor in place of an intake valve?

32. Does a centrifugal compressor have exhaust valves?

33. Explain what is meant by the term volumetric efficiency.

34. On external drive compressors, should the crankshaft seal rubbing surfaces be lubricated?

35. Are some compressors water-cooled?

36. What is the least number of impellers a centrifugal compressor may have?

37. Why does a compression system sometimes use two mufflers?

38. How many openings does a compressor service valve have?

39. What does an oil separator do?

40. Why is a liquid line filter-drier used?

Chapter 5

REFRIGERANT CONTROLS

Refrigerant in the evaporator must be at a low pressure so it will evaporate at a low temperature. The liquid refrigerant in the condensing unit is at a relatively high pressure.

So that the refrigerating unit may operate automatically, an automatic refrigerant flow control must be placed in the circuit between the liquid line and the evaporator. This control reduces the high pressure in the liquid line to the low pressure in the evaporator.

There are six main types of automatic refrigerant flow controls:

NAME OF CONTROL	ABBREVIATION
A. Automatic Expansion Valve	AEV or AXV
B. Thermostatic Expansion Valve	TEV or TXV
C. Thermal-Electric Expansion Valve	THEXV
D. Low-Pressure Side Float	LSF
E. High-Pressure Side Float	HSF
F. Capillary Tube	Cap. Tube

The basic systems using these controls are described in Chapter 3 and Chapter 4.

In the following paragraphs, in addition to naming the controls, some information and terminology is listed. This will help the student in understanding the operation of each type of refrigerant control. Also, the meaning of the terms "flash gas," "superheat," "heat exchanger," "hunting," and the like, are explained as they apply to refrigeration.

5-1 COMPRESSION SYSTEM REFRIGERANT CONTROLS

Modern refrigeration systems are automatic in operation. Devices have been developed for controlling refrigerant flow into the evaporator and also for controlling the electric motor which drives the mechanism.

The refrigerant controls may be divided into three principal classes:
1. Control based on pressure changes.
2. Control based on temperature changes.
3. Control based on volume or quantity changes.
   A combination of controls may also be used.

Automatic refrigerant and motor controls are needed that will maintain the temperature in the refrigerated space within specific limits. Automatic motor controls are explained in Chapter 8.

Some popular temperature ranges desired are:

1. For fresh food storage, temperatures are usually maintained at 35 to 45 F. (2 to 7 C.).
2. For frozen food storage, usual temperature is between −10 and 0 F. (−23 and −18 C.).
3. For comfort cooling, temperatures are usually maintained not to exceed 10 to 12 F. (5.5 to 6.6 C.) below the ambient temperature.

5-2 AUTOMATIC EXPANSION VALVE PRINCIPLES

An automatic expansion valve (AEV) (AXV) or pressure controlled expansion valve is a refrigerant control operated by the low-side pressure of the system. Its purpose is to throttle the liquid refrigerant in the liquid line down to a constant pressure on the low-pressure side while the compressor is running.

The valve acts like a spray nozzle. While the compressor is running, the liquid refrigerant is sprayed into the evaporator. A system using an automatic expansion valve is sometimes called a "dry" system. This is because the evaporator is never filled with liquid refrigerant, but with a mist or fog.

5-3 AUTOMATIC EXPANSION VALVE DESIGN

The diagram in Fig. 5-1 is of a flexible bellows linked up to a needle valve with evaporator pressure, P_2, on the inside and atmospheric or confined gas pressure, P_1, on the outside. F_1 spring force tends to open the valve, while spring force, F_2, tends to close the valve.

From the illustration it may be seen that, as the pressure in the evaporator decreases, the difference in pressures will force the bellows toward the valve body. Since it is attached to the needle, it will open the needle valve and some liquid refrigerant will spray into the evaporator. As the refrigerant evaporates at a constant low pressure, it keeps the evaporator and cabinet temperature within their design limits.

The expansion valve opens only when the evaporator pressure drops. Pressure drop occurs only when the compressor is running. The expansion valve, however, will not flood the low side when the compressor is running. As soon as the evaporating refrigerant, liquid, spray and vapor reach the end of the evaporator, the motor control (sensing bulb) attached to the suction line will cool. This opens the switch and stops

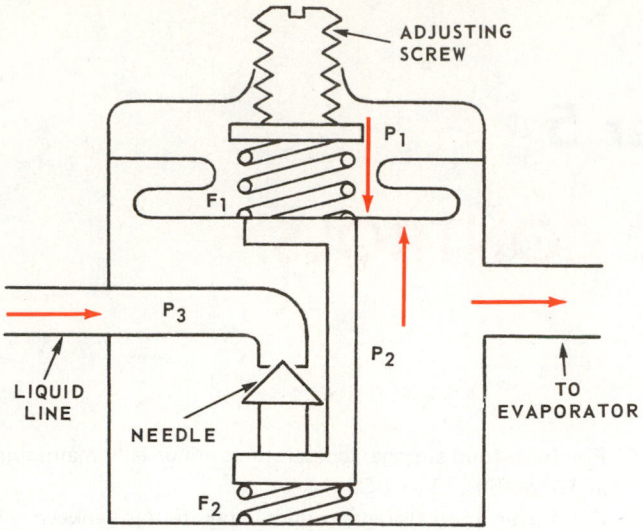

Fig. 5-1. Automatic expansion valve shows various pressures inside the valve which cause it to operate. P_1—Atmospheric pressure. P_2—Suction or evaporator pressure. P_3—Liquid line pressure. F_1—Adjustable spring. F_2—Nonadjustable pressure spring.

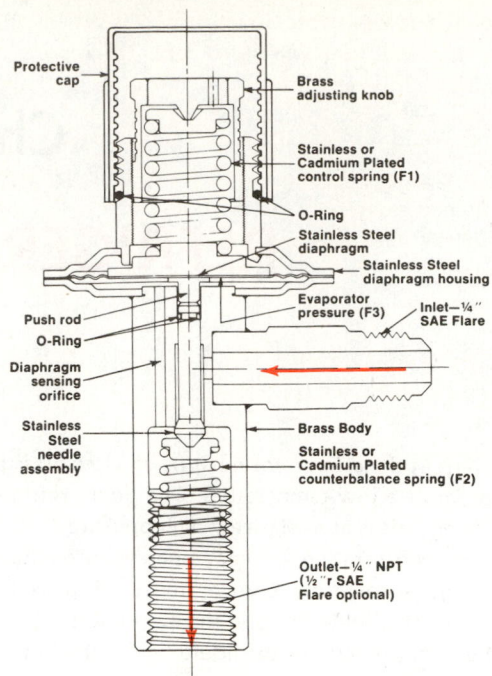

Fig. 5-2. Diaphragm type pressure expansion valve. F1—Control spring of diaphragm. F2—Counterbalance spring. F3—Evaporator pressure. (Jackes-Evans Div., Parker-Hannifin Corp.)

the motor. The low-side pressure will then immediately build up enough to close the expansion valve.

These valves are adjustable to permit opening of the needle valve over a wide range of pressures. Expansion valves must be adjusted for atmospheric pressure, P_1, which affects their operation. High altitudes will cause a decrease in atmospheric pressure. The adjusting screw must be turned in to make up for the lower atmospheric pressure. Various refrigerants have different expansion valve settings. Their evaporating pressures are not the same.

Automatic expansion valves are of many different designs. The flexible part is either a diaphragm or a bellows. Usually it is made of phosphor bronze soldered or brazed to the valve body. Flexible elements must move in and out time after time without losing flexibility. The valve body is usually drop forged brass, but sometimes it is cast. It must be seepage (leak) proof.

The liquid inlet has either a soldered connection, a standard flange, a flared connection, or a pipe thread. It usually has a screen designed for easy removal. The screen is made of brass or stainless steel wire 60 to 80 mesh. (This means 60 to 80 openings per square inch.)

An AEV with a double spring on the needle balances the forces and gives smoother control of the refrigerant flow. Such a unit is shown in Fig. 5-2.

It is important to remember that the liquid refrigerant flowing past the expansion valve needle is the same weight as the vapor (gas) pumped by the compressor. Valve capacity should equal pump capacity.

Use a "one ton valve" with a one ton capacity condensing unit. An under-capacity valve tends to "starve" an evaporator (too little refrigerant gets through). An over-capacity valve will tend to allow too much refrigerant into the evaporator when the valve opens. This may cause sweat backs or frost backs on the suction line.

5-4 BELLOWS TYPE AUTOMATIC EXPANSION VALVE

A bellows type automatic expansion valve, Fig. 5-3, usually has valve seats softer than the needles to eliminate, as much as possible, the wearing of a shoulder on the needle. Usually, needles are Stellite and seats Monel metal. The spring, C, is attached at both ends. It can be adjusted for either pressure or tension. This eliminates the need to have a spring inside the refrigerant space. The needle, L, is mounted in a ball and socket to insure full seating by allowing the needle to align with the seat, M. The bellows, D, made of special brass, is flexible and is soldered to both the body and the disk. This bellows is made of metal .005 to .010 in. thick.

The liquid line is usually 1/4 in. outside diameter (OD) and is fastened to the valve, H, by a special nut permitting easy

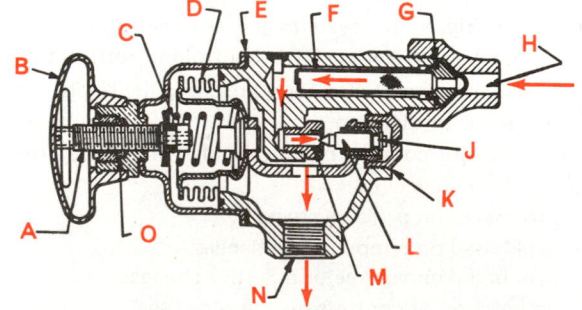

Fig. 5-3. Bellows type automatic expansion valve. A—Adjusting screw. B—Rubber cap. C—Adjustment spring. D—Bellows. E—Gasket. F—Screen. G—Shipping plug. H—Liquid refrigerant inlet. J—Needle shoulder. K—Seating plug. L—Needle. M—Needle valve seat. N—Refrigerant outlet. O—Packing gland. Arrows show direction of refrigerant flow through valve.

removal of the screen. Note how the valve is sealed at O and B to insure that moisture will not enter. If present, it could freeze on the bellows and interfere with the accurate operation of the valve.

The valve may be attached to the evaporator either by means of a threaded fitting or by a two-bolt flange. Flanges sealed with lead gaskets are usually preferred.

This automatic expansion valve is used chiefly on domestic air conditioning units, vending machines and as a replacement for capillary tubes.

5-5 DIAPHRAGM TYPE AUTOMATIC EXPANSION VALVE

A diaphragm type automatic expansion valve is illustrated in Fig. 5-4. The valve has stops to prevent too great a movement of the diaphragm. Note that the diaphragm has concentric corrugations (ripples) to improve its flexibility.

Fig. 5-5 is an external view of an automatic expansion valve. The word IN is marked on the inlet port to insure

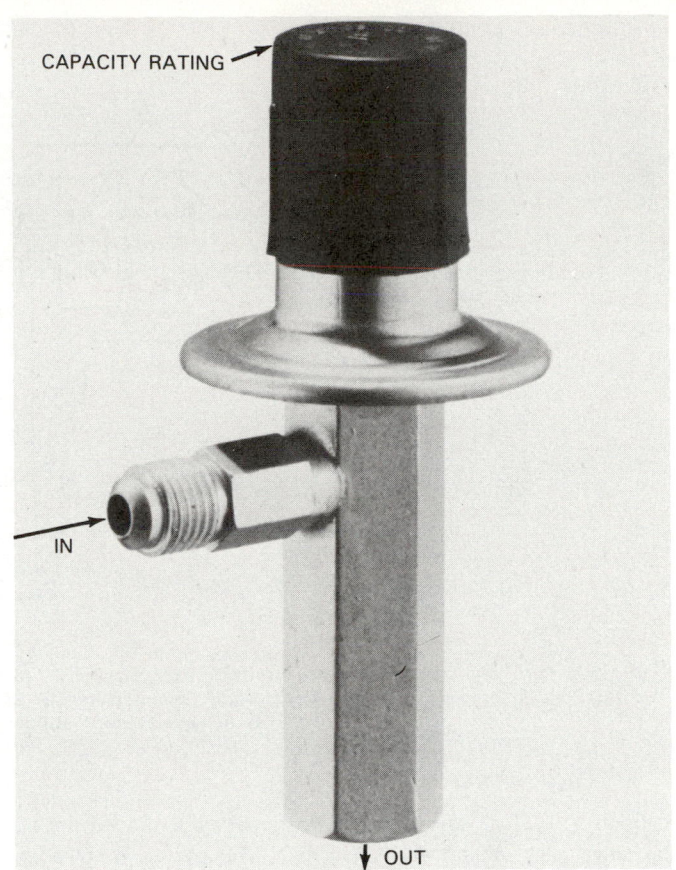

Fig. 5-5. Diaphragm type automatic expansion valve. Note the refrigerant flow direction. (Jackes-Evans Div., Parker-Hannifin Corp.)

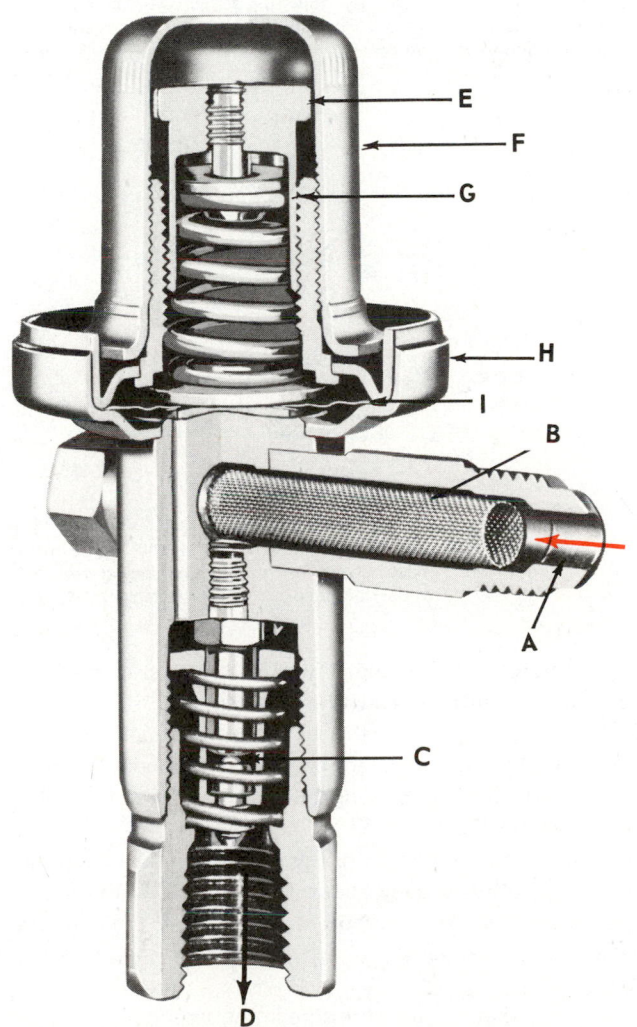

Fig. 5-4. Diaphragm type automatic expansion valve. A—Liquid refrigerant inlet. B—Screen. C—Refrigerant needle and seat. D—Refrigerant outlet leading to the evaporator. E—Valve adjustment. F—Bonnet. G—Adjustment spring. H—Valve housing. I—Control diaphragm.

proper installation (not shown in Fig. 5-5). Its capacity, using a particular refrigerant, is often marked on the cap.

Another diaphragm expansion valve design is shown in Fig. 5-6. The diaphragm movement is limited by the body and the cap. Threads fasten the diaphragm assembly to the body of the valve. The cap or cover plate, protecting the pressure adjustment, is tightly fitted over the entire assembly. It can be removed to adjust the valve.

The diaphragm has a disk on its low-pressure side which presses on a pin that moves the ball valve away from the seat. When the low-side pressure increases, the diaphragm moves against the adjustment spring. This allows the spring at the ball valve to push the ball valve against the seat. The inlet is a 1/4 in. male flare connection. The outlet is a 3/8 in. male flare.

5-6 AUTOMATIC EXPANSION VALVE BYPASS

Many motor-compressor units are designed to start under a low load (torque) condition, such as when the low-side and high-side pressures are equal (balanced). The equal pressures allow the compressor to start without pushing against a high pressure. The motor will therefore require less starting torque. Capillary tube systems have balanced pressures during the off cycle.

Automatic expansion valves seal the refrigerant orifice during the off part of the cycle. To balance pressures an opening is designed into the valve to permit refrigerant to pass

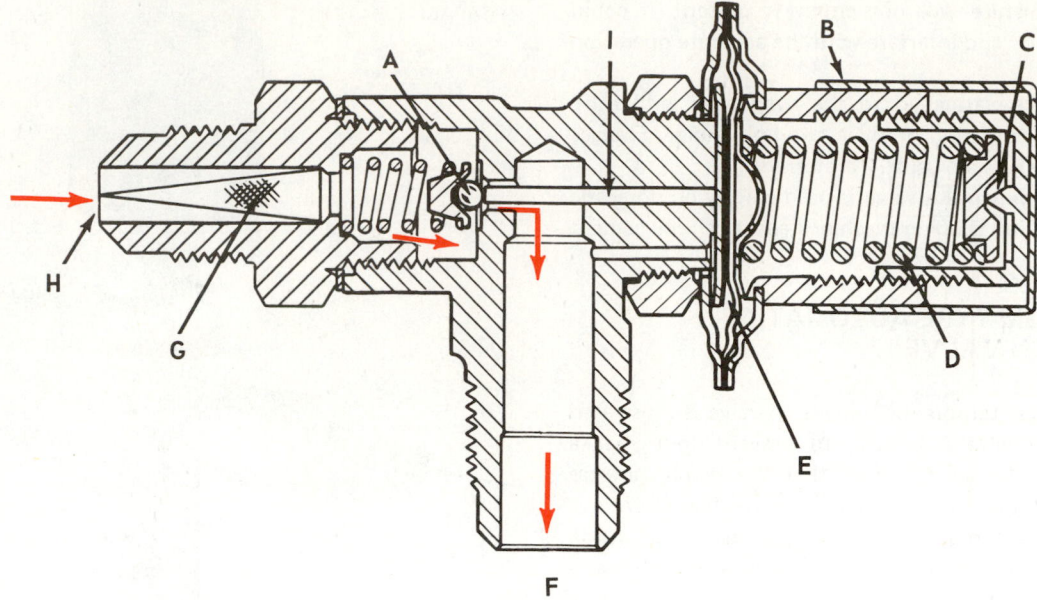

Fig. 5-6. Parts and operation of an automatic expansion valve. A—Valve (ball) and seat. B—Metal cap. C—Adjusting screw. D—Adjusting spring. E—Diaphragm. F—Outlet to evaporator (low-pressure). G—Screen. H—Liquid refrigerant inlet (high-pressure). I—Valve opening pin. Red arrows indicate direction of flow of refrigerant through expansion valve.

through. A typical way to do this is to make a V-shaped slot in the valve seat as shown in Fig. 5-7. The bypass or bleeder openings are so small that they do not interfere with the operation of the valve when the compressor is running.

When using this type of expansion valve, the right amount of refrigerant charge must be used. The evaporator outlet must have an accumulator; otherwise, liquid refrigerant may travel down the suction line and cause sweating or frost on the suction line; also, dangerous liquid refrigerant slugging in the compressor may occur.

5-7 THERMOSTATIC EXPANSION VALVE (TEV) PRINCIPLES

Thermostatic expansion valves are of two basic types:
1. Sensing bulb.
2. Thermal-electric.

The sensing bulb type is further divided into four types:
1. Liquid charged.
2. Gas charged.
3. Liquid cross charged.
4. Gas cross charged.

In the liquid charged and the gas charged elements, the refrigerant is the same as is used in the system. Cross charged means that the fluid in the sensing bulb is different than the refrigerant in the system.

In the automatic expansion valve, the refrigerant flow (through the valve and into the evaporator) is controlled only by the pressure in the evaporator.

In the thermostatic expansion valve, the flow through the valve and into the evaporator is controlled by both the low-side pressure and the temperature of the evaporator outlet. The valve provides a rather high rate of flow

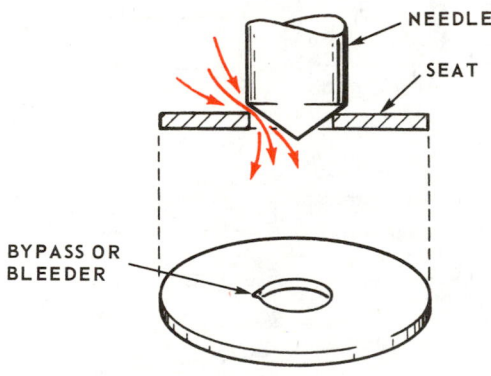

Fig. 5-7. Needle valve and seat used on bypass or "bleeder" type automatic expansion valve. Bypass permits pressures to balance during off part of cycle.

if the evaporator is quite empty (warm). It slows up the flow as the evaporator fills (cools) with refrigerant.

The sensing bulb type thermostatic expansion valve is operated by the accumulated pressure difference or force difference between the sensing bulb bellows and the valve low-side pressure. See Fig. 5-8.

With the unit running, the refrigerant, T_1, in the expansion valve sensing bulb is usually about 10 F. (5.6 C.) warmer than the refrigerant in the evaporator, T_2. This temperature difference produces the different pressures and therefore the different forces.

This means that the unit pressure in the sensing bulb, P_1, is greater than the unit pressure, P_2, in the evaporator. This temperature difference is often described as the superheat of the bulb over the refrigerant temperature inside the evaporator. It should be noted that, as the temperatures increase or

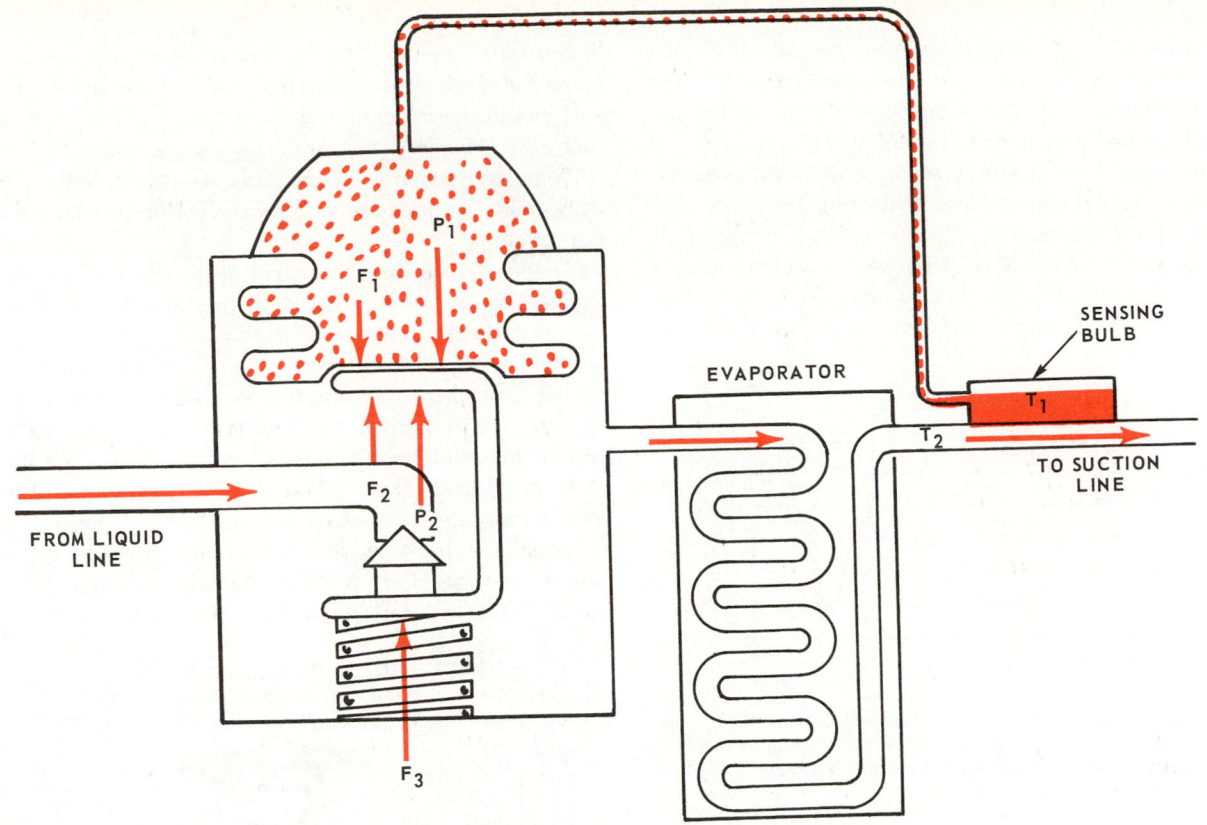

Fig. 5-8. Thermostatic expansion valve showing various pressures and temperatures within the valve which cause it to operate. F_1—Sensing bulb pressure (force) tending to open valve. F_2—Low-side pressure (force) tending to close valve. F_3—Spring force tending to close valve. P_1—Sensing bulb pressure tending to open valve. P_2—Suction pressure (low side) tending to close valve. T_1—Sensing bulb temperature. T_2—Evaporator refrigerant temperature (low side). Valve opens when F_1 is greater than combined force of F_2 and F_3. The valve closes when the combined F_2 and F_3 forces are greater than F_1.

decrease, the pressure will also increase or decrease.

When the compressor stops, the low-side pressure and the sensing bulb pressure tend to equalize. The total expansion valve internal force, F_2, plus F_3 overpowers the sensing bulb force, F_1, and the needle is forced firmly into its seat. Refrigerant flow stops. The needle will stay closed until the

sensing bulb force overcomes the low-side force.

This valve opening action should only happen when the unit is running. If the valve is adjusted correctly, the closing of the valve while the compressor is idle prevents the flooding of the low side with liquid refrigerant.

The thermostatic expansion valve does not regulate the

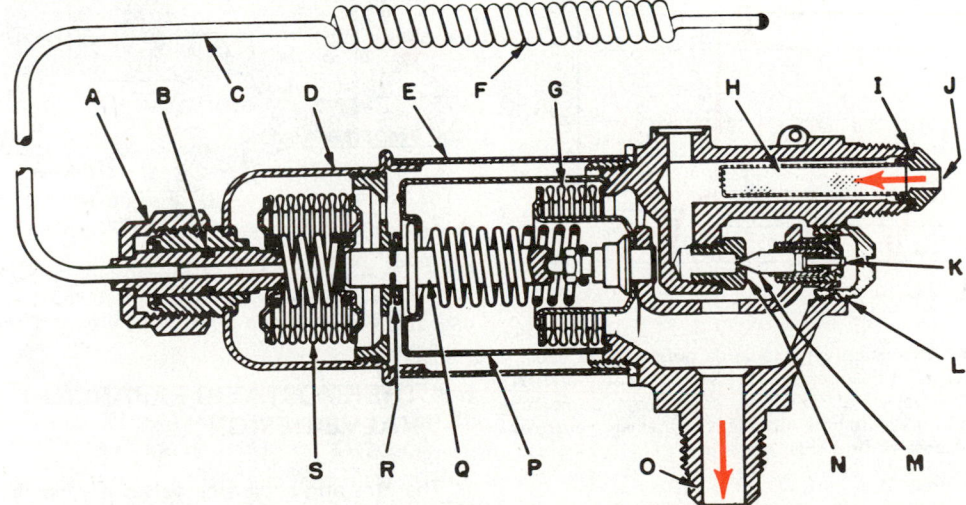

Fig. 5-9. Thermostatic expansion valve. A—Adjusting nut. B—Seal ring. C—Capillary tube. D—Bellows housing. E—Housing spacer. F—Temperature sensing bulb. G—Body bellows. H—Screen. I—Gasket. J—Refrigerant inlet. K—Needle pin. L—Sealed fitting. M—Needle. N—Seat. O—Evaporator connection. P—Inner spacer. Q—Spacer rod. R—Snap ring. S—Thermal bellows.

Refrigerant Controls / 145

low-side pressure, but rather controls the filling of the evaporator with refrigerant. The pumping action of the compressor establishes the low-side pressure. Fig. 5-9 illustrates a bellows type thermostatic expansion valve.

The adjustment enables one to set the valves so the needle can seat itself sooner while the unit is running. The needle will then close even though there is a greater temperature difference (about 15 F. or 8.3 C.) between the refrigerant in the evaporator and that in the sensing bulb, as shown in Fig. 5-10. The evaporator liquid refrigerant will not reach the

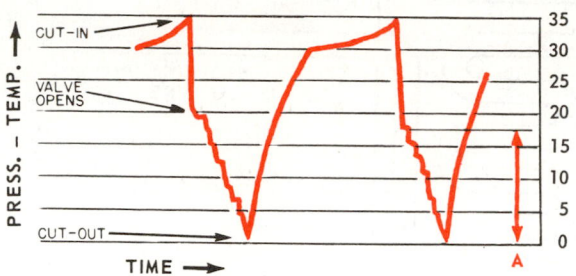

Fig. 5-10. A thermostatic expansion valve low-side pressure-time cycle diagram using high superheat adjustment. A—Pressure drop on the low side between opening of the valve and cut-out point.

sensing bulb location in this case. The temperature of only the low-pressure vapor will be cold enough to reduce the sensing bulb temperature (and therefore the pressure) to the closing point. The needle will close before the evaporator becomes full of liquid refrigerant and the evaporator will be "starved."

If the adjustment is turned in too much the other way (one or two revolutions) to move the needle away from the seat, the temperature of the sensing bulb refrigerant will become closer to the temperature of the evaporator refrigerant 5 to 7 F. (2.8 to 3.9 C.). See Fig. 5-11.

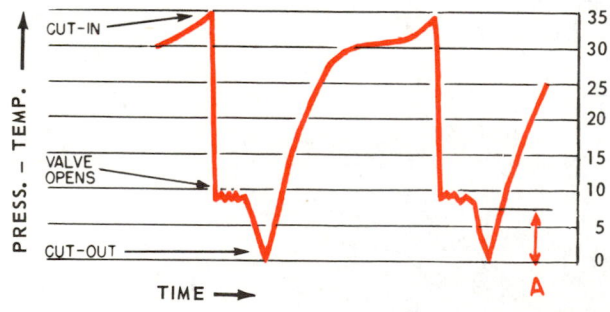

Fig. 5-11. A thermostatic expansion valve low-side pressure-time cycle diagram using a low superheat adjustment. A—Pressure drop on low side between the opening of the valve and the cut-out point. Note: Compare with A, in Fig. 5-10, to understand how superheat adjustment controls the evaporator (low side) pressure.

The evaporator must now become more than full of liquid refrigerant droplets to bring the temperature (pressure) difference down to this value. The evaporator will be completely flooded. Some liquid droplets may even go into the suction

line causing a sweating or frosting of the suction line and may harm the compressor (slugging). This adjustment is sensitive and should never be turned more than one-quarter of a turn each 10 to 15 minutes while the unit is operating.

Some thermostatic expansion valves use diaphragms instead of bellows. In either design, the valve is closed when the unit is not running.

Chapter 3 describes the basic principles of a TEV system.

See Chapter 14 for instructions on installing and servicing thermostatic expansion valves.

In some large refrigeration installations (50 tons and over), a pilot-controlled thermostatic expansion valve may be used. In these installations, a conventional thermostatic expansion valve is mounted on a large auxiliary valve body. The auxiliary valve provides a larger pressure operated needle and orifice. The conventional thermostatic refrigerant control (pilot) regulates the pressure which operates the large refrigerant orifice control. Fig. 5-12 illustrates one type of pilot-controlled thermostatic expansion valve.

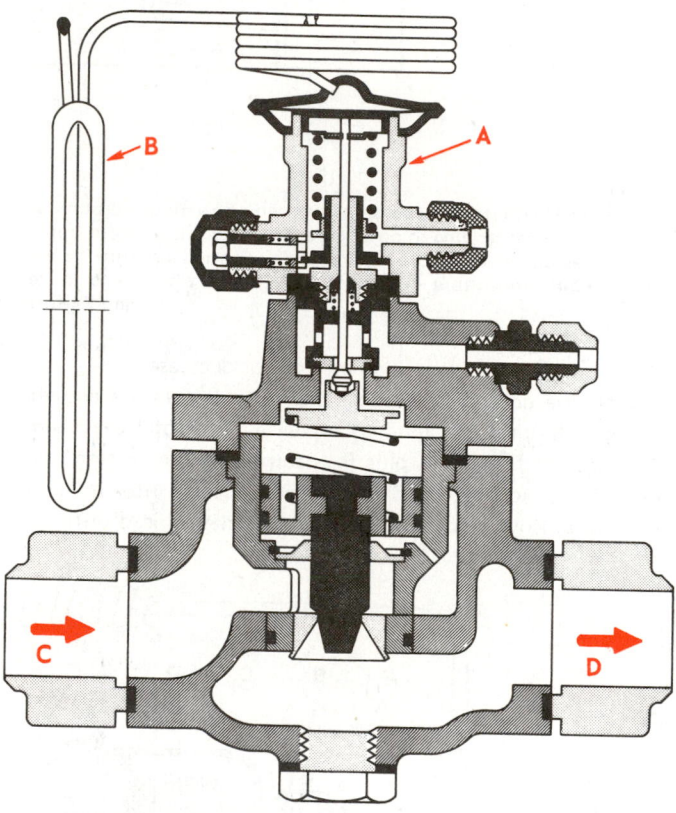

Fig. 5-12. A pilot-controlled thermostatic expansion valve. A—Pilot valve body. B—Temperature sensing bulb. C—Refrigerant line, in. D—Connection to the evaporator. (Danfoss, Inc.)

5-8 THERMOSTATIC EXPANSION VALVE DESIGN

Thermostatic expansion valves are usually used on multiple evaporator systems. However, the low-side float may also be used on multiple systems. It is possible in a multiple system using thermostatic expansion valves to provide a variety of temperatures in the various cabinets. This valve is also commonly used on air conditioning systems.

One must choose the correct sensing element charge and the correct size valve for each installation.

The thermostatic expansion valve consists of a brass body into which the liquid line and evaporator line are connected. The needle and seat are inside the body. The needle is joined to a flexible metal bellows or diaphragm. This bellows, in turn, is made to move by a rod connected at the other end to a sealed bellows or diaphragm (power element) which is joined to the sensing bulb by means of a capillary tube.

Fig. 5-13 shows the refrigerant behavior in the sensing element. Each manufacturer has a code for identifying the fluid that charges the sensing element. Some use letters; others use colors or numbers to identify the charge. Some valves are marked with the refrigerant number. Some valves intended for use with refrigerant R-12 are color coded yellow.

The valve is sealed to prevent moisture seeping into any part. A strainer (screen) is always located between the liquid line connection and the orifice to keep dirt away from the needle and seat. See Fig. 5-14.

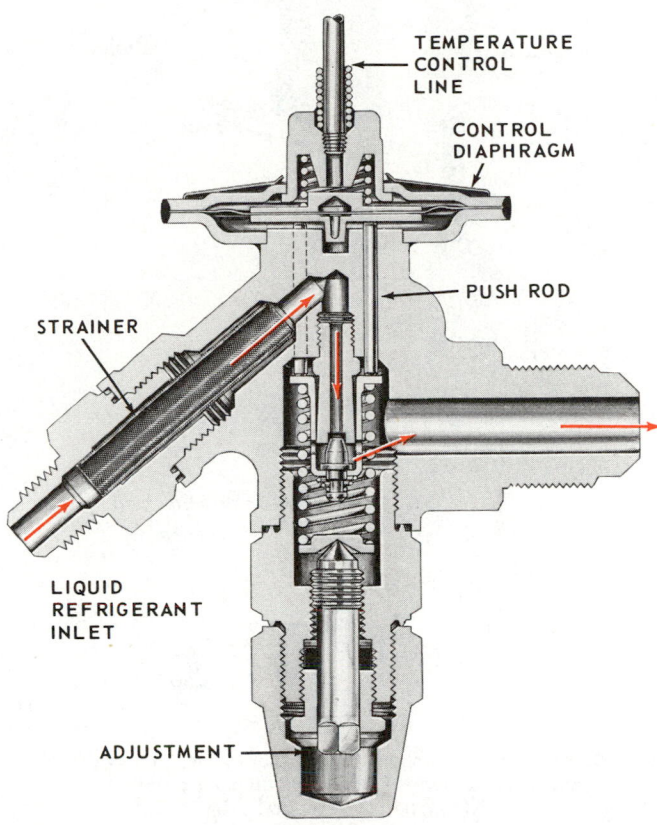

Fig. 5-14. Diaphragm type thermostatic expansion valve using flared inlet and outlet connections. Turn adjustment screw in to "starve" the evaporator. Turn it out to flood evaporator. (Controls Co. of America)

Fig. 5-13. Effect of temperature on sensing bulb of thermostatic expansion valve. A—Sensing bulb is cold, the pressure is low, and a considerable quantity of control fluid is shown as a condensed liquid in the bulb (solid red). B—Sensing bulb is warmer and some of control fluid has evaporated (red dots) and is exerting pressure on expansion valve diaphragm, which will cause it to admit more refrigerant into the evaporator. Note that for accurate control, there is enough control fluid in the sensing bulb to insure control fluid in the sensing bulb at all times. Red cross-hatching indicates suction line pressure.

Some large air conditioning systems may use as many as six thermostatic expansion valves on one evaporator. In this way it is possible:

1. To maintain constant pressures and temperatures.
2. To make sure that all of the evaporator has a full charge of refrigerant.
3. To reduce pressure drop through the evaporator.

Fig. 5-15 is an exploded view of a thermostatic expansion valve. The body is usually made of brass.

Note that the inlet flare surface is mounted on the strainer. The pin or needle is usually made of Stellite, Hastelloy or stainless steel. The needles are usually sharp pointed cones, but spherical valves (balls) and flat orifice closers may also be used. The cone needle is popular for small capacity valves, while the ball type or the flat type is used in larger capacity valves.

Fig. 5-16 illustrates a large capacity flat valve seat. It is

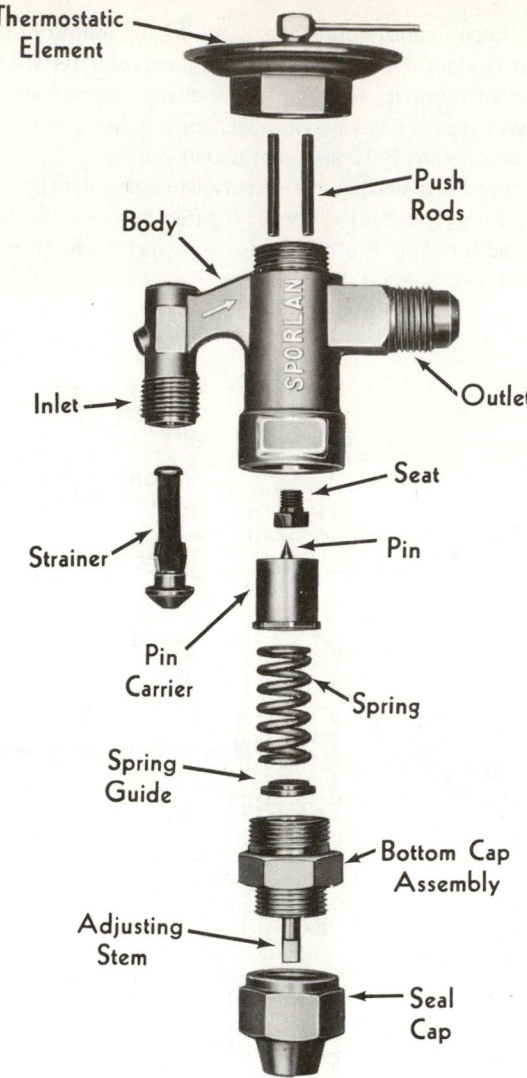

Fig. 5-15. Parts of a thermostatic expansion valve. In this valve the thermostatic element is threaded on body of valve. (Sporlan Valve Co.)

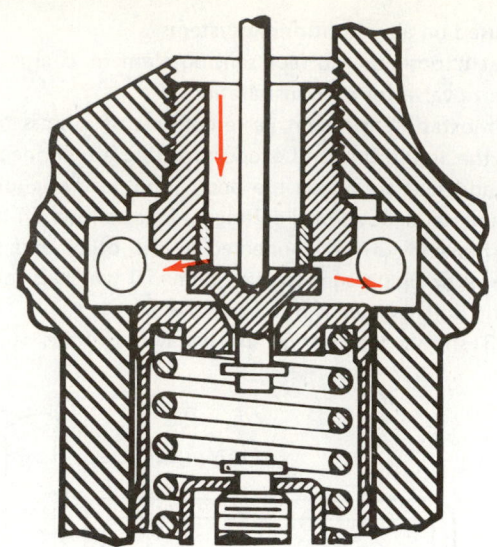

Fig. 5-16. This cutaway shows a thermostatic valve orifice that has been designed for large capacity. Note that both valve and valve seat are formed by flat surfaces. Arrows indicate direction of flow of refrigerant through valve seat mechanism.

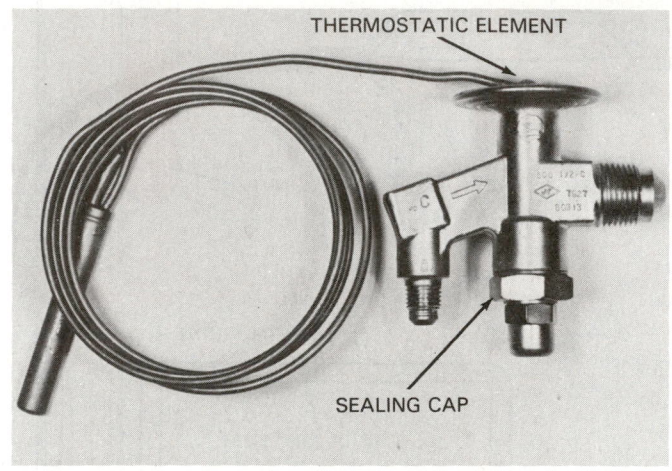

Fig. 5-17. Exterior view of thermostatic expansion valve. Valve may be adjusted after removing sealing cap. (Jackes-Evans Div., Parker-Hannifin Corp.)

always good practice to place a filter-drier in the liquid line immediately ahead of the thermostatic expansion valve.

Fig. 5-17 shows the outside of a diaphragm type TEV with a ball type valve. Fig. 5-18 shows a cutaway of same.

Another type thermostatic expansion valve is shown in Fig. 5-19. It is the single diaphragm type designed particularly for service in air conditioning applications. It is equipped with a bleed valve for rapid pressure balancing (RPB). The rapid pressure bleed mechanism is shown at B in the illustration. The bleed mechanism works only on the off cycle. When the compressor starts up again, the secondary bleed port closes and the valve operates in a normal manner. Rods carry the diaphragm action to the needle. The liquid inlet is on the left; the evaporator connection on the right.

5-9 FLASH GAS

The term "flash gas" is used to indicate that portion of the refrigerant which evaporates instantly (flashes) and turns into a vapor as it passes through the refrigerant control orifice. The instant vaporizing of some of the liquid refrigerant (flash gas) cools the rest of the liquid to the evaporating temperature.

The amount of flash gas depends on temperature of refrigerant in the liquid line and the pressure inside the evaporator. Flash gas reduces the valve capacity. See Chapter 15.

One method used to reduce "flash gas" is to clamp the liquid line to the suction line. This is often called a heat exchanger.

Since the liquid coming from the condenser is often quite warm and the vapor coming from the evaporator is quite cold, clamping the two lines together causes a heat transfer from the liquid line to the suction line.

Cooling the liquid in the liquid line decreases the "flash gas" and increases the heat absorbing capacity of the refriger-

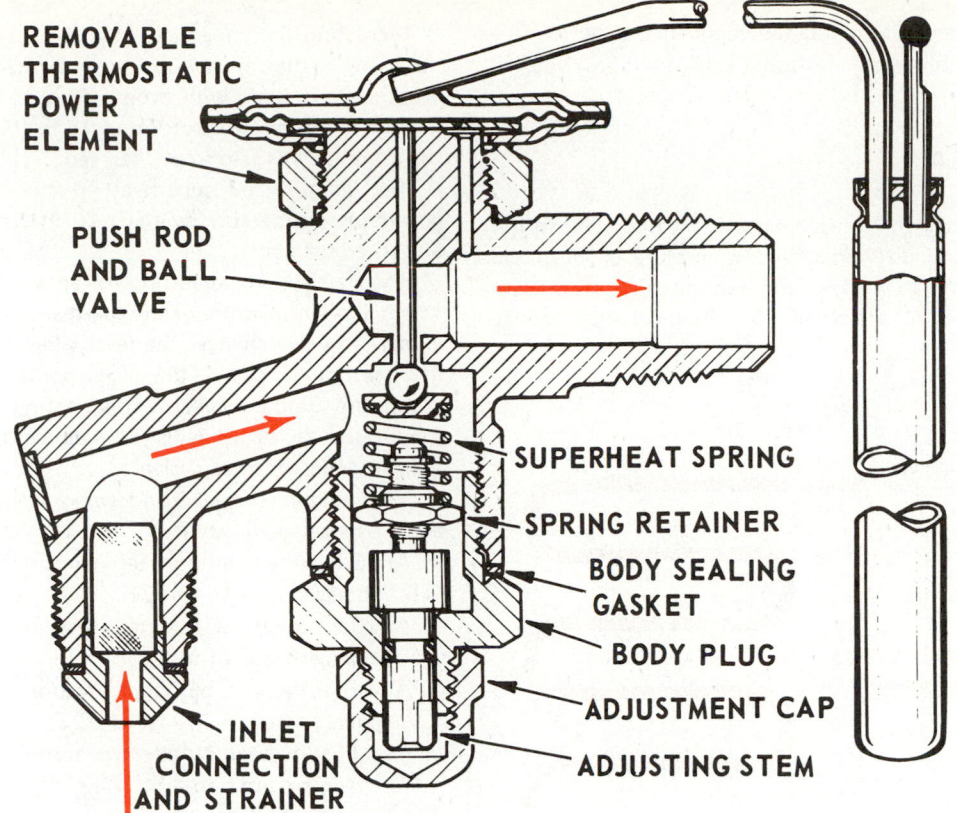

REMOVABLE
THERMOSTATIC
POWER
ELEMENT

PUSH ROD
AND BALL
VALVE

SUPERHEAT SPRING

SPRING RETAINER

BODY SEALING
GASKET

BODY PLUG

ADJUSTMENT CAP

ADJUSTING STEM

INLET
CONNECTION
AND STRAINER

Fig. 5-18. Diaphragm type thermostatic expansion valve. Note a ball is used as the valve, replacing needle. Direction of refrigerant flow through expansion valve is marked by arrows.

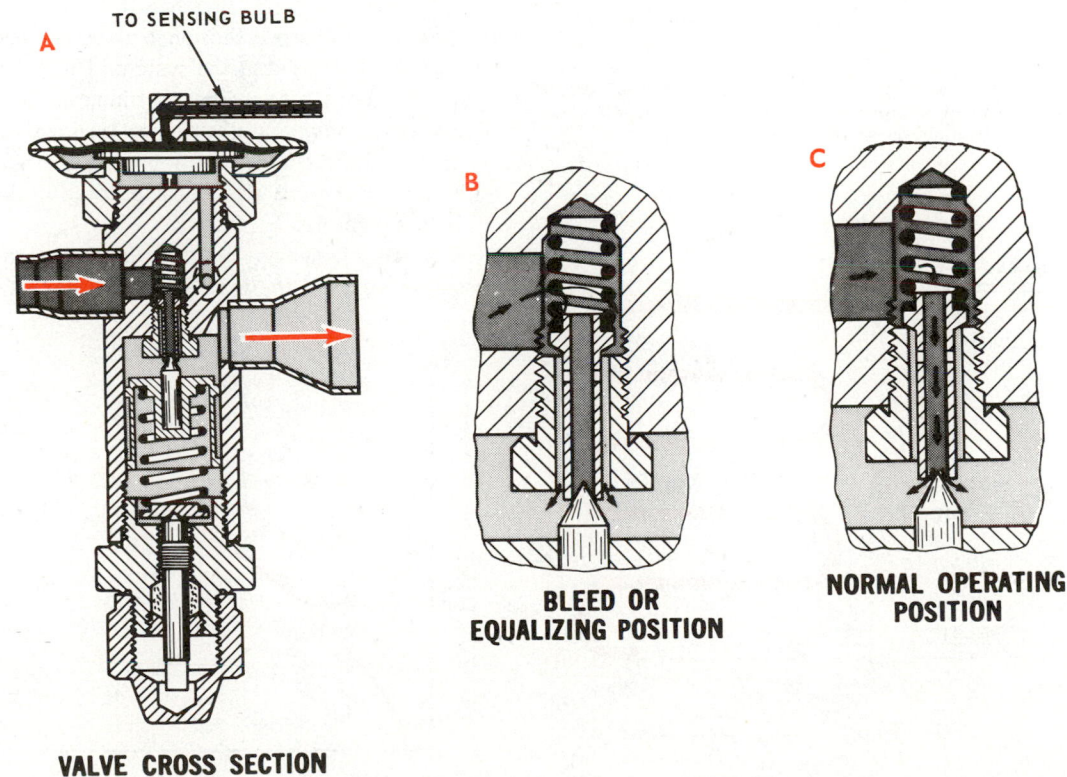

A

TO SENSING BULB

B

C

VALVE CROSS SECTION

BLEED OR
EQUALIZING POSITION

NORMAL OPERATING
POSITION

Fig. 5-19. Diaphragm type thermostatic expansion valve. Sensing bulb pressure operates on the top surface of the diaphragm. As sensing bulb temperature increases, pressure on the top of diaphragm tends to open valve allowing refrigerant to enter evaporator. Note the direction of refrigerant flow through the valve. A—Cross section through the entire valve. B—Bleed valve shown in normal operation. C—Needle valve in normal operating position. (Sporlan Valve Co.)

ant. Raising the temperature of the vapor in the suction line decreases the possibility of any liquid refrigerant entering the compressor.

5-10 SUPERHEAT

The term "superheat," as used with thermostatic expansion valves, refers to the difference in temperature between the vapor in the low side and in the sensing bulb. A system adjusted to operate at a normal 10 F. (5.6 C.) superheat is shown in Fig. 5-20.

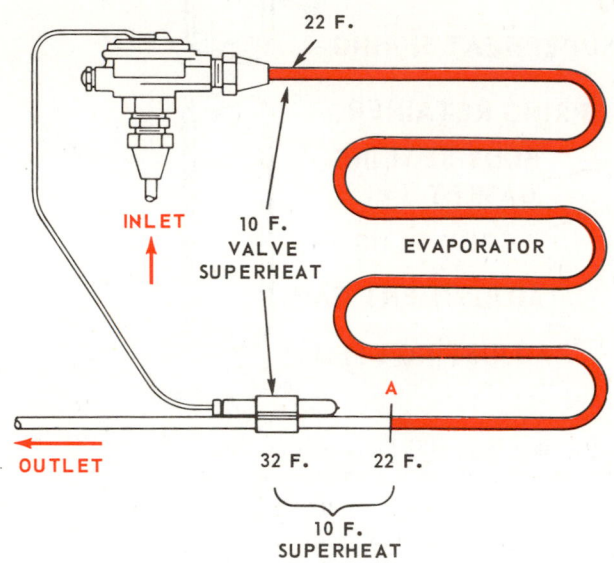

Fig. 5-20. A thermostatic expansion valve adjusted to give a normal 10 F. superheat. Liquid refrigerant will reach Point A before the valve controlling the refrigerant closes.

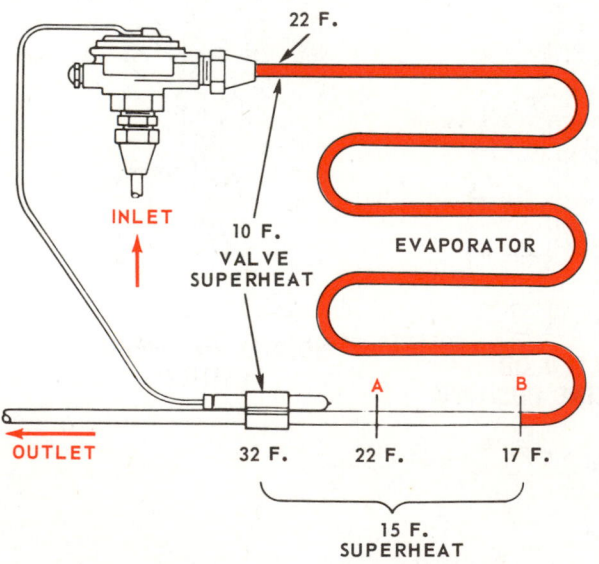

Fig. 5-21. A "starved" evaporator results when there is a pressure drop. Liquid reaches point B at 17 F. and vapor warms to 22 F. Valve closes at this point with too little liquid in the evaporator.

Increasing the superheat tends to "starve" the evaporator. "Starving" the evaporator means that only part of the evaporator is filled with drops of liquid refrigerant. Fig. 5-21 illustrates a superheat setting of 15 F. (8.3 C.). At this setting, the evaporator is said to be "starved."

The amount of superheat in the suction line will be governed considerably by the type of refrigerant control used and its adjustment.

The best superheat setting for an evaporator is the point at which the temperature of the thermal bulb of the thermostatic expansion valve changes the least when the system is running. This setting is called the MSS point or setting. It means Minimum Stable Signal. This setting is a result of the evaporator flow behavior as well as the behavior of the thermostatic expansion valve.

For example, if a valve and evaporator combination behaves as follows when adjusted:

At 12 F. superheat, bulb temperature
changes from 14 to 10 F.
At 10 F. superheat, bulb temperature
changes from 11 to 9 F.
At 8 F. superheat, bulb temperature
changes from 8 1/2 to 7 1/2 F.
At 6 F. superheat, bulb temperature
changes from 8 to 4 F.

The least change (most stable condition) is at 8 deg. superheat setting.

5-11 LIQUID CHARGED SENSING ELEMENT

The liquid charged sensing bulb is charged with the same refrigerant as is used in the system. Thus, the valve is able to maintain a constant superheat setting even though the low-side pressures and temperatures change as shown in Fig. 5-22.

In the liquid charged sensing bulb, the quantity of fluid is sufficient so there is always some liquid in the bulb regardless of its temperature.

The sensing element will always control thermostatic valve operation, even if the temperature of the valve body is lower

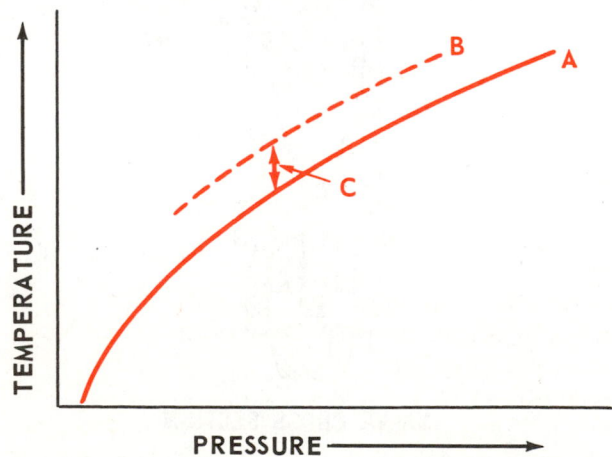

Fig. 5-22. A liquid charged thermostatic expansion valve superheat. A—Vapor pressure curve of refrigerant in system. B—Vapor pressure curve of charge in sensing bulb. C—Superheat (normally 10 F., 5.6 C.).

than the temperature of the sensing element.

As the temperature of the evaporator drops, the amount of superheat increases.

This valve may cause some evaporator flooding when pulling down from normal ambient temperatures.

These elements are designed for a temperature range of from approximately −20 to 40 F. (−28.9 to 4.4 C.).

Air conditioners usually use this type of expansion valve. However, normal refrigerating systems may use them too.

5-12 LIQUID CROSS CHARGED SENSING ELEMENT

The liquid cross charged sensing bulb uses a liquid different from the refrigerant in the system. It may use a mixture of fluids to give the desired operating characteristics.

The amount of charge is such that some liquid will remain in the sensing element under all temperature conditions.

The sensing element will always control the valve operation even if the valve body temperature is lower than the temperature of the sensing element.

The valve closes quickly when the compressor stops because the evaporator pressure increases more rapidly than the sensing bulb pressure as the evaporator warms up.

The load on the compressor is reduced at start-up. As the suction pressure is reduced, the superheat is reduced, thus utilizing maximum evaporator surface.

"Hunting" is reduced because the flatter pressure-temperature curve of the sensing element pressure makes the valve more responsive to changes in suction pressure than to changes in sensing bulb temperature.

The liquid cross charged elements are designed for a temperature range of from −40 to 40 F. (−40 to 4.4 C.). These valves are usually used for either commercial low-temperature applications or with extremely low-temperature systems.

Fig. 5-23 shows the operating characteristics of the liquid cross charged element in graph form.

Fig. 5-24 shows the difference in the superheat curve of the cross charged element as compared to a normal charged element.

5-13 GAS (VAPOR) CHARGED SENSING ELEMENT

The gas charged sensing bulb is charged with the same refrigerant used in the system. The amount of charge is such that, at a predetermined temperature, all the liquid has vaporized.

Increasing the temperature above this point does not cause an increase in element pressure. Consequently the expansion valve does not open more with an increase in the cabinet temperature.

However, if the valve body becomes colder than the sensing bulb, the vaporized control fluid will condense in the body of the valve and control will be lost. The valve will close.

For example, if just enough control fluid is put in the element to produce a maximum pressure of 40 psi (2.8

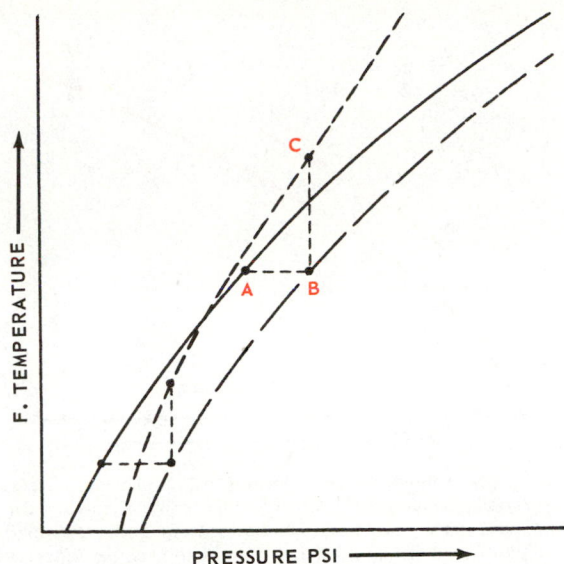

Fig. 5-23. A graph of the action of a liquid cross charged element designed for application within a specific temperature range. It provides a rapid pull-down and is used for normal refrigeration. Superheat setting at the top end is wide to prevent flooding of unit. A—Given evaporator pressure and corresponding saturation temperature. B—Evaporator pressure plus equivalent superheat spring pressure. C—Corresponding sensing bulb and power assembly pressure. B—C—Superheat setting for this evaporator pressure and particular superheat spring setting.

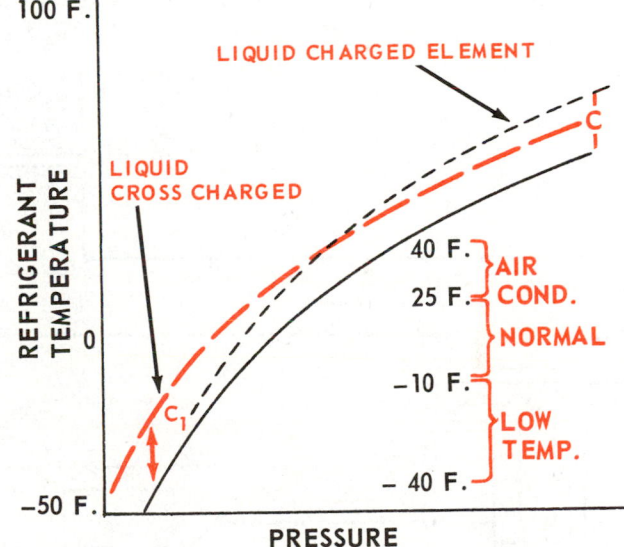

Fig. 5-24. Graph showing constant superheat of liquid cross charged power element designed to be used for all three applications — low temperature, normal and air conditioning — as compared to a liquid charged power element during wide temperature ranges. Red dotted line shows degrees of superheat for a liquid cross charged thermostatic power element. The black dotted line shows degrees of superheat for liquid charge. The C liquid charge changes as temperature drops but the C_1 liquid cross charge remains constant as temperature drops.

kg/cm^2) the element pressure will never exceed this pressure no matter how warm the bulb is. When the low side exceeds this pressure, the valve will not open. See Fig. 5-25. Thus, low-side pressure will not have to operate above 40 psi.

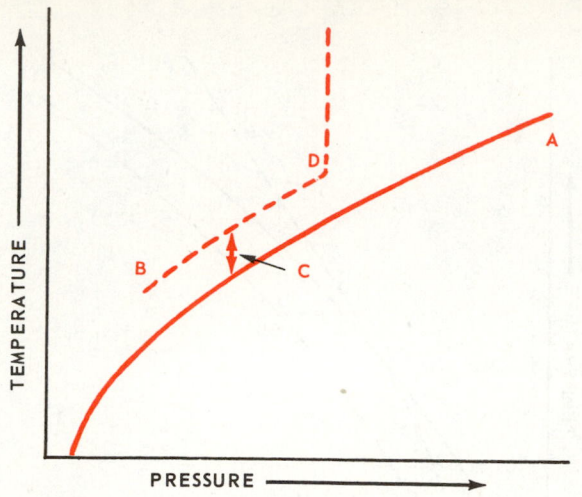

Fig. 5-25. A gas (vapor) charged thermostatic expansion valve superheat. A—Vapor pressure curve of the refrigerant in the system. B—Vapor pressure curve of charge in the sensing bulb. C—Superheat. D—The point at which all fluid in sensing bulb becomes vaporized.

Vapor charged elements are designed for a temperature range of from 30 to 60 F. (−1.1 to 15.6 C.).

5-14 GAS (VAPOR) CROSS CHARGED SENSING ELEMENT — ADSORPTION

The gas cross charged sensing bulb is charged with a liquid different than the refrigerant in the system. The amount of charge is such that, at the desired temperature, all the liquid has vaporized. Increasing the temperature above this point does not cause a usable increase in element pressure. The superheat characteristics are like the liquid cross charged element.

Some types of gas cross charged sensing elements depend upon a different principle than the one explained above. In these thermostatic expansion valves, the sensing element contains two substances. One is a noncondensing gas, such as carbon dioxide, which provides the pressure in the element. The other is a solid such as carbon, silica gel or charcoal.

These substances have the ability to adsorb gas depending upon their temperature. Adsorption means the adhesion of a layer of gas one molecule thick over the surface of a solid substance. There is no chemical combination between the gas and the solid substance (adsorber).

The ability of a substance to adsorb gas depends upon the temperature of the substance. Substances more readily adsorb gas at low temperatures. As the sensing element warms, the pressure in the element will increase due to release of the adsorbed gas. As the sensing element cools, its pressure will decrease due to the adsorption of gas back to the adsorbing substance. This pressure change is used to control the refrigerant needle valve opening in the thermostatic expansion valve.

The appearance and general construction of these thermostatic expansion valves is the same as with the usual gas cross charge. The only difference is in the gas and substances

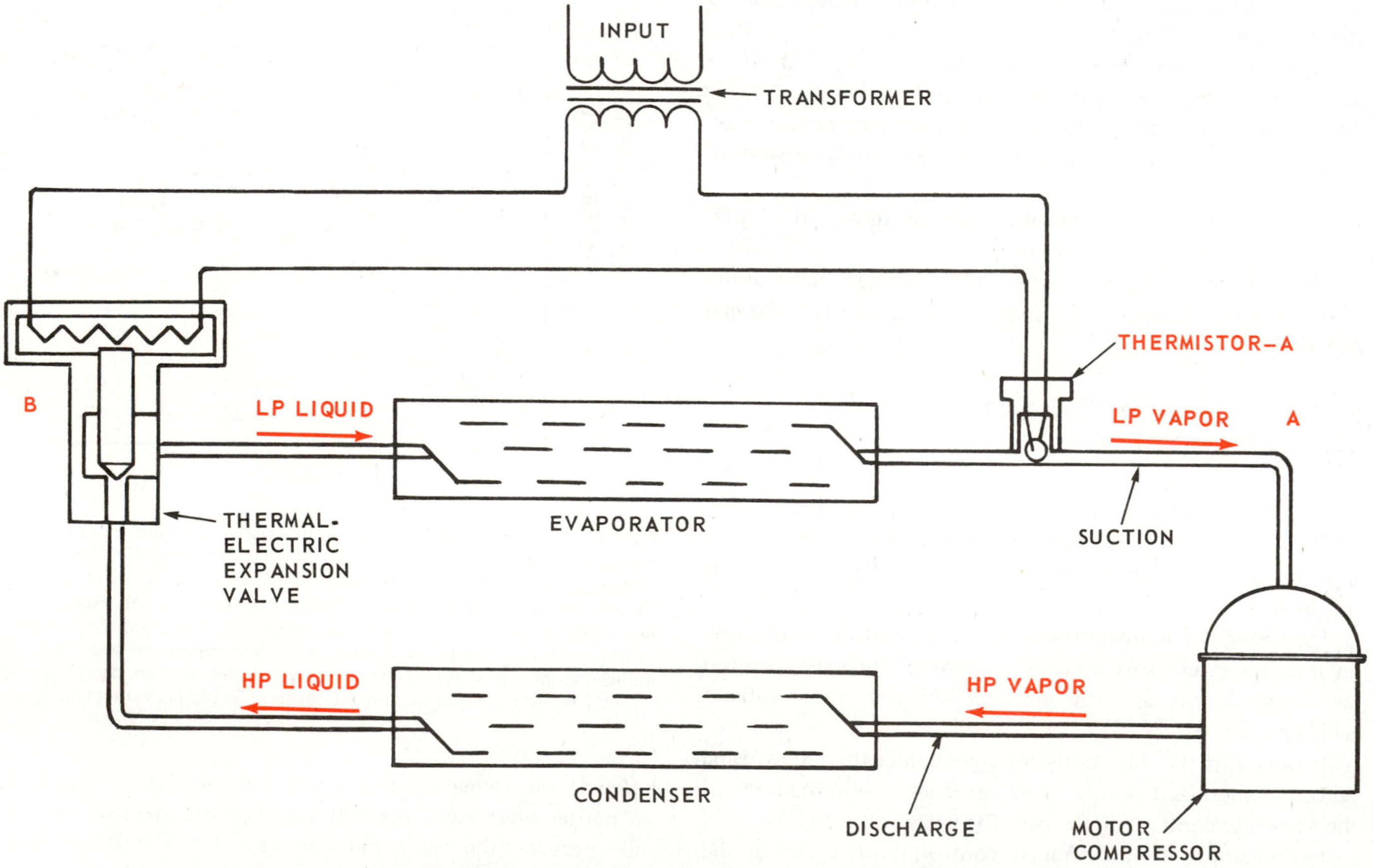

Fig. 5-26. A thermal-electric expansion valve installation. (Controls Co. of America)

contained in the sensing element to control the sensing element pressures.

These thermostatic expansion valves have the advantage of a pressure-temperature lag in their operation. They have very wide temperature applications and can be used on any type refrigerating or air conditioning system. Their range is sufficient to cover most any refrigerating application. The gas in the sensing element is a noncondensing gas and remains a gas throughout the operating range of the valve.

5-15 THERMAL-ELECTRIC (SOLID STATE) EXPANSION VALVE

The thermal-electric controlled expansion valve depends upon the use of thermistors (see Chapter 6), directly exposed to the refrigerant in the suction line, to control the expansion

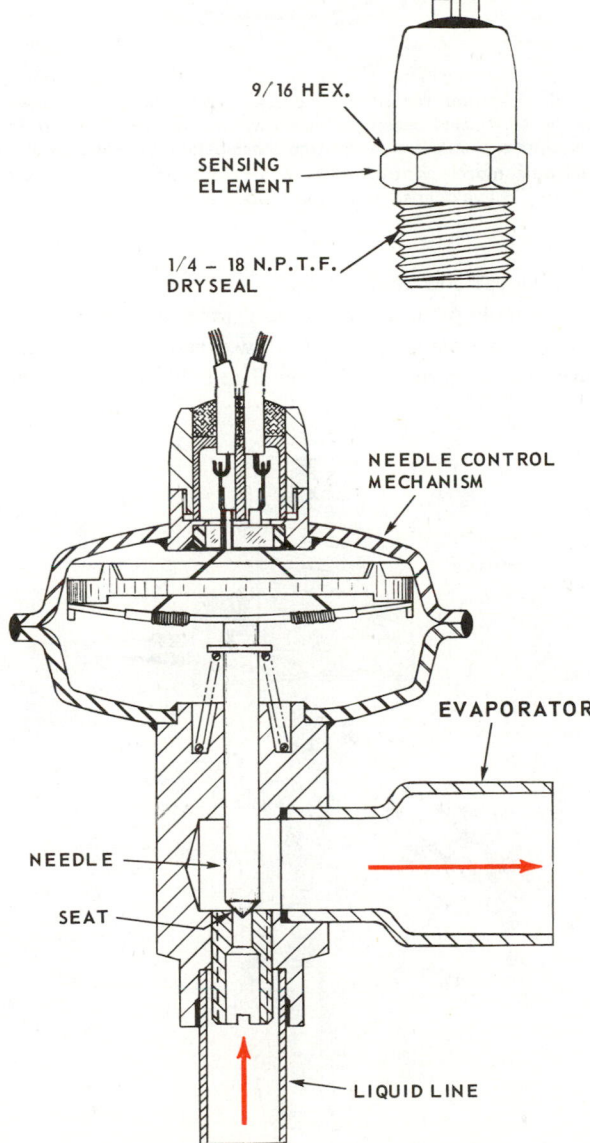

Fig. 5-27. A thermal-electric expansion valve.
(Controls Co. of America)

valve needle opening. It does not use a pressure element as in the thermostatic expansion valve.

The resistance to electrical flow in the thermistor changes with its temperature. Increasing temperature reduces resistance. Therefore, with a given voltage, it increases the current rate of flow. This increase current flow heats the bimetal in the valve body and makes the bimetal bend, opening the valve.

Fig. 5-26 illustrates a typical thermal-electric expansion valve installation. The thermistor—A, is placed in immediate contact with the refrigerant vapor inside the suction line from the evaporator.

A low voltage transformer is the power source. This is connected in series to the expansion valve control mechanism at B in such a way that increasing current flow through the thermistor increases the opening of the expansion valve and, therefore, increases the rate of flow of the refrigerant into the evaporator.

Increasing the current causes the valve needle to open, while a decrease closes the valve. Thereby the refrigerant flow is controlled. The thermal-electric expansion valve is not dependent on the pressure in the evaporator. It restricts the flow of refrigerant and controls the suction line superheat in order to prevent flooding of the compressor.

Fig. 5-27 shows a cross-section of such a valve. This illustrates, in some detail, the electrical connections and the mechanisms which control the operation of this expansion valve. A complete thermal-electric expansion valve and thermistor ready for installation is shown in Fig. 5-28.

Off cycle operation is possible in one of two ways. In one case, the thermal-electric expansion valve may be electrically connected in parallel to the operating system. During the off cycle, the thermistor will become warmed and the thermal-electric expansion valve will remain open. This unloading is similar to the pressure-balancing effect when a capillary tube

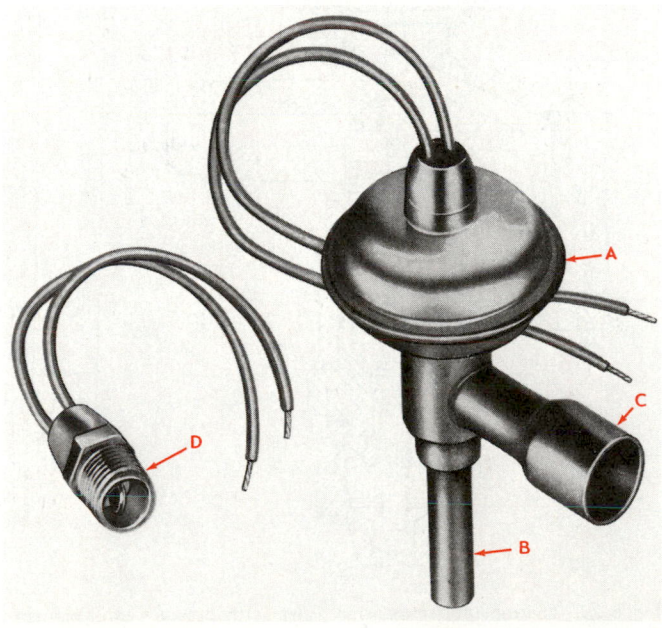

Fig. 5-28. A thermal-electric expansion valve. A—Control mechanism. B—Liquid line. C—Connection to the evaporator. D—Thermistor. (Controls Co. of America)

refrigerant control is used.

In the second case, it may be electrically connected into the motor circuit (interlocked) in such a way that it is only energized when the compressor is running. With this type of connection, the valve will be closed on the off cycle.

5-16 PRESSURE LIMITERS

Sometimes a pressure-limiting expansion valve is used to prevent overloading the condensing unit. It is designed for systems in which the evaporator pressure must not exceed a safe operating limit.

One of these devices is placed between the sensing element mechanism and the needle mechanism. It is a second diaphragm and a spring located beneath the control diaphragm and is designed to collapse at a certain force. Thus, if this element is designed to collapse at 40 psi (2.8 kg/cm^2), the needle will close if the low-side pressure ever exceeds this amount. It does not matter what the evaporator temperature-pressure is. These valves offer rapid pulldown on start-up and maintain the evaporating pressure within the prescribed safe limits. See Fig. 5-29.

A gas charged collapsible element can also be used to provide a limit to the pressure which will open the valve. When the pressure in the low side exceeds a certain set value, the diaphragm will collapse. The gas used is noncondensable and

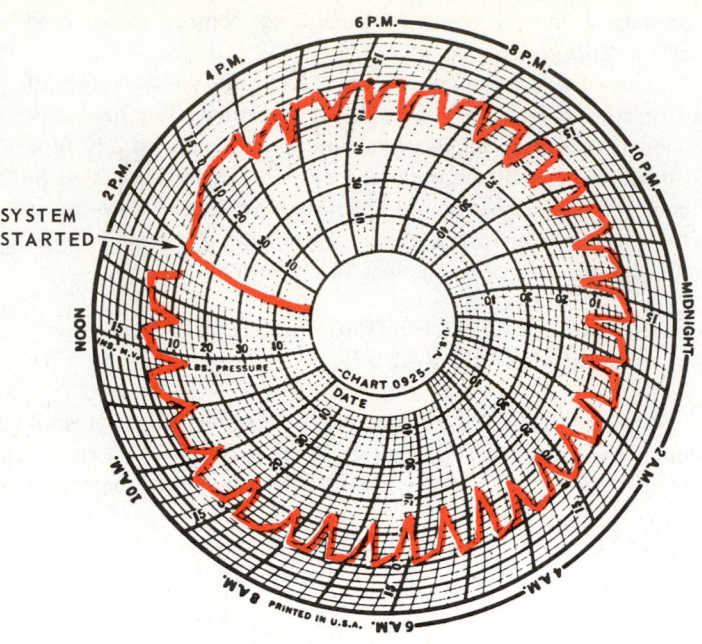

Fig. 5-30. A 24-hour record of the pressure-time relationship for a food freezer as it is cycled beginning with a warm condition at 2 p.m. Freezer is equipped with pressure limiting thermostatic expansion valve and thermostatic motor control. Note quick reduction of pressure until valve opened at 2 p.m. (Sporlan Valve Co.)

obeys Charles's and Boyle's Laws.

An example of how the pressure limiting device prevents a long running time at excessive low-side pressures when the refrigerator is warm is shown in Fig. 5-30. The cycle record

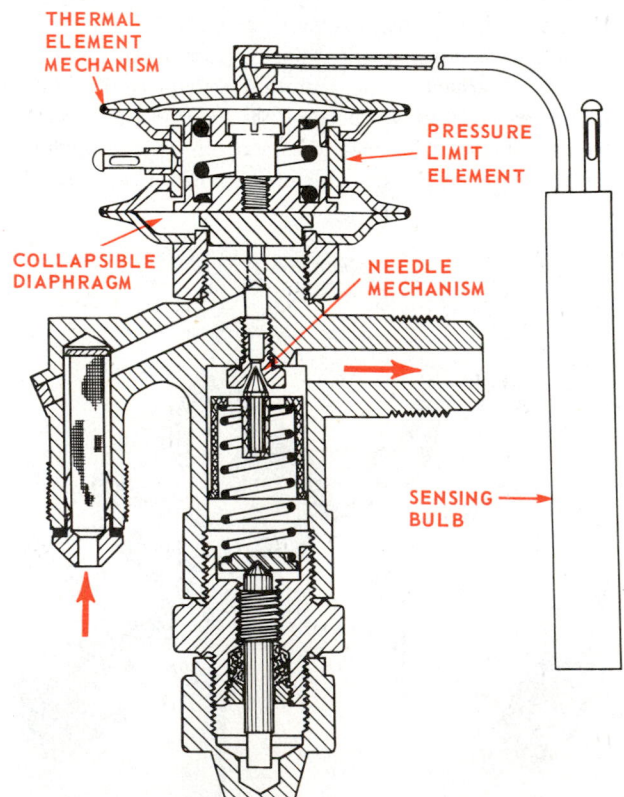

Fig. 5-29. Thermostatic expansion valve with pressure limit element. The element limits pressure by means of two diaphragms and a spring. Whenever suction pressure gets near the motor overload point, the spring between two diaphragms compresses and the valve reduces the flow of refrigerant to the evaporator.

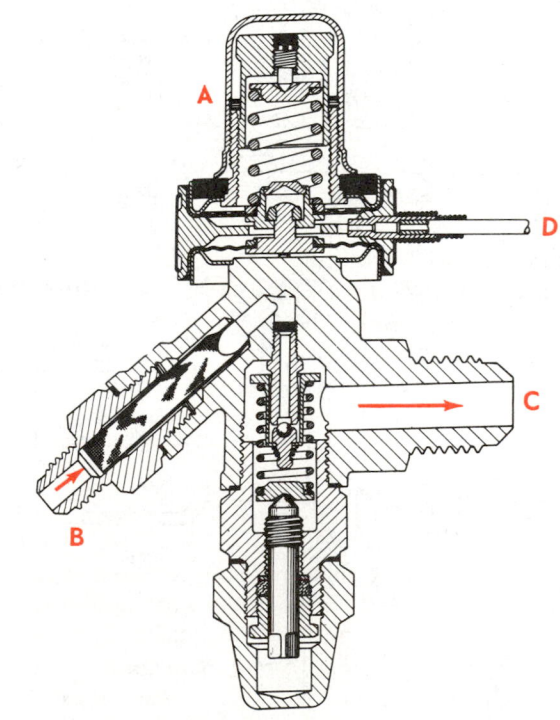

Fig. 5-31. Thermostatic expansion valve with adjustable pressure limiter device. A—Pressure limiter adjustment. B—Liquid refrigerant inlet. C—Evaporator connection. D—Sensing bulb connection.

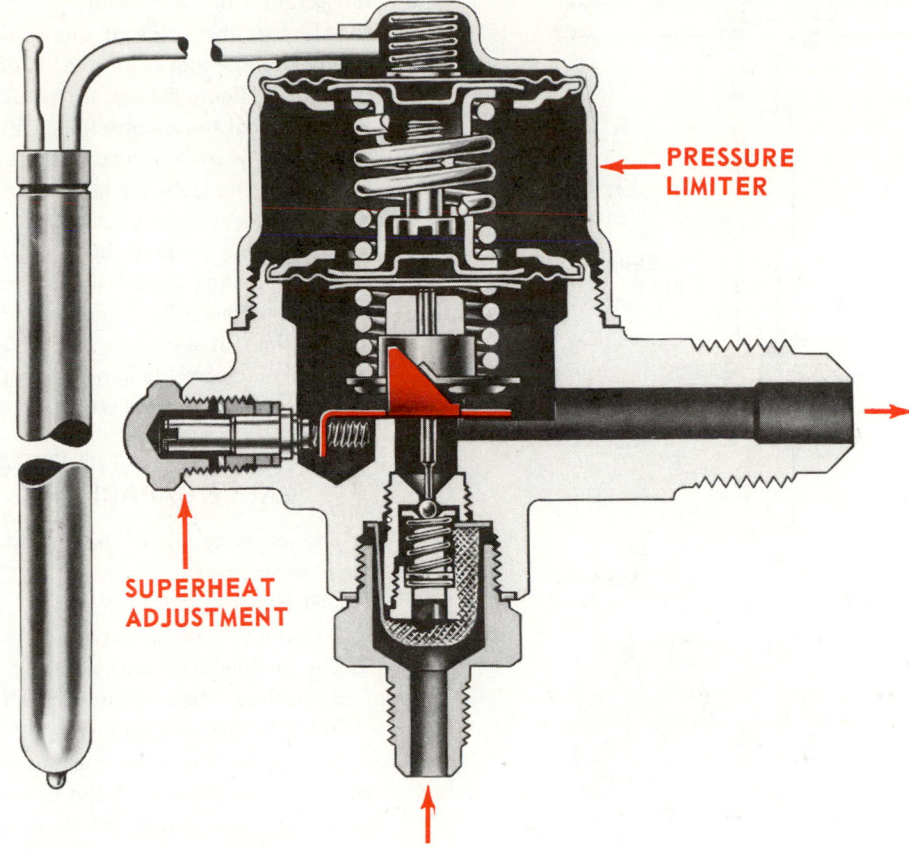

Fig. 5-32. Thermostatic expansion valve with spring loaded pressure limiter. Note that the superheat adjustment operates on an inclined plane which regulates superheat spring tension.

shows a pressure drop from over 50 psi to 10 psi (3.5 to .7 kg/cm²) in just a few minutes. The unit then runs for two hours before it shuts off.

Still another type of pressure limiter thermostatic expansion valve is shown in Fig. 5-31. This valve has an adjustable pressure limiter, A. Above a certain pressure setting at A, the spring above the diaphragm will compress instead of the valve needle being opened. The liquid inlet is at B, and the evaporator connection is at C. The capillary tube and sensing bulb are not shown, but the connection is at D. Another spring loaded pressure limiter is shown in Fig. 5-32. The pressure limiter assembly is shown in Fig. 5-33. It will act as a rigid rod, opening and closing the valve until the pressures in the low-pressure side and in the sensing element are greater than the force of the spring.

5-17 SENSING BULB MOUNTING

The location and the actual mounting of the sensing bulb is very important. It must be in good thermal contact with the evaporator outlet, Fig. 5-34. The bulb should be mounted on the top of the suction line so the liquid in the bulb is close to the suction line as shown at A in the illustration. If it is necessary to mount the bulb on a vertical suction line, the capillary tube of the bulb should always enter from the top of the bulb as shown at B. Never from the bottom!

To keep the bulb from being affected by the air or liquid being cooled, it should be wrapped in insulation, as in C. Special insulation forms are available or plastic tape can be

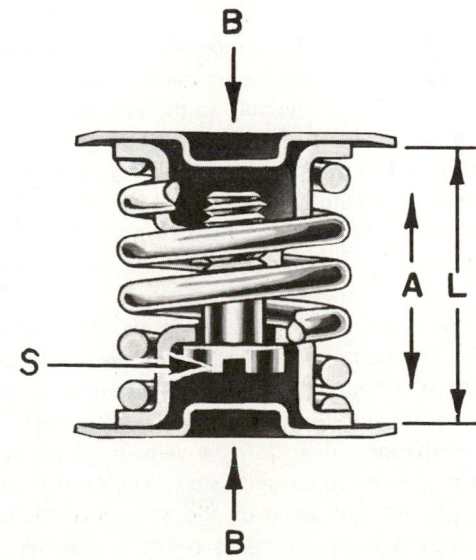

Fig. 5-33. Pressure limiter device used on thermostatic expansion valve shown in cutaway for Fig. 5-32. Adjusting screw, S, is set to proper length, L. When forces (or pressure) at B exceed forces, (or pressure) at A, spring will be compressed and refrigerant orifice in expansion valve will close. This closes the valve.

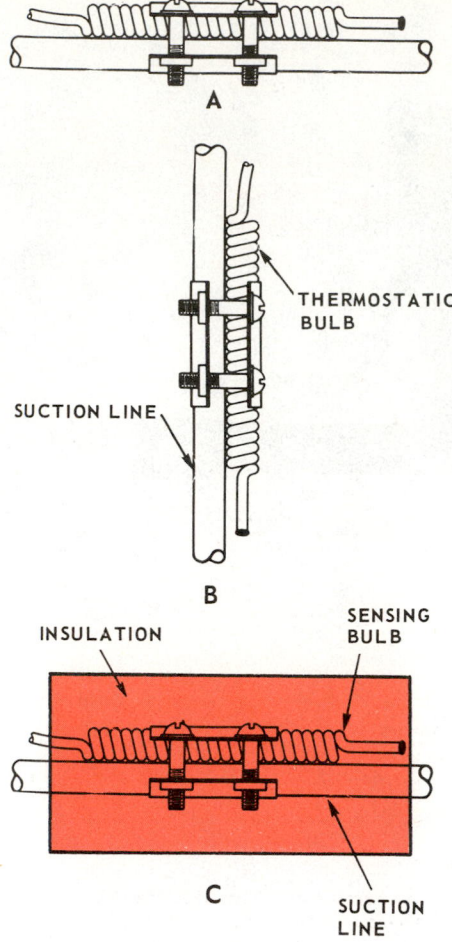

Fig. 5-34. Correct way to attach thermostatic expansion valve sensing bulb to suction line. A—Thermostatic sensing bulb mounted in horizontal position. B—Thermostatic sensing bulb mounted in vertical position. C—Insulated sensing bulb installation. Note that thermal bulb is best mounted in horizontal position and on top of suction line.

used so that only suction line temperature affects the bulb.

Copper straps and nonrusting machine screws and nuts should be used to fasten the bulb to the suction line. The bulb must have excellent thermal contact with the suction line. The connection must be clean and tight. One should clean both the suction line and the bulb with steel wool before assembling.

The suction line carries chiefly vaporized refrigerant. However, there will be some droplets of liquid refrigerant and some oil.

Fig. 5-35 illustrates conditions inside the suction line, particularly on installations which require the line to be a large diameter. At A, refrigerant vapor and some droplets of liquid refrigerant are flowing through a rather large diameter suction line. Due to the large diameter, the velocity of the vaporized refrigerant at times will be quite slow. The droplets of liquid refrigerant and oil will settle on the bottom of the line. The suction area at B is smaller. As a result, the velocity of the vaporized refrigerant will be higher than at A. This means less separating of the oil and liquid refrigerant from the flowing vapor. The inside of the tube will be rather uniformly coated with oil. The suction line at C is shown in vertical position. In

this position, there will be no separation of the droplets of refrigerant from the vapor. However, the oil will rather uniformly coat the inside of the suction line.

The temperature of the vaporized refrigerant and the droplets of liquid refrigerant will be a few degrees colder than the surface of the suction line. This is because of the insulating quality of the oil which coats the inside of the suction line.

The sensing bulb on large suction lines should be located near the under side of the line rather than on top. This is better because droplets of liquid refrigerant tend to separate from the flowing vapors. Also, as mentioned before, oil has an insulating effect. The recommended bulb position (on large suction lines) is shown at D, Fig. 5-35. Some sensing bulbs are crimped or creased lengthwise to provide double contact and to help align the sensing bulb with the suction line.

5-18 THERMOSTATIC EXPANSION VALVE CAPACITIES

The capacity of a thermostatic expansion valve (TEV) varies according to:
1. Orifice size.
2. Pressure difference between the high side and the low side.
3. The temperature and condition of the refrigerant in the liquid line. The amount of flash gas will increase with a rise of liquid line temperature.

The capacity of most thermostatic expansion valves may be selected from the size of the orifice and needle assembly. The same body may be used for many capacities.

The larger the orifice, the more liquid refrigerant that can be fed into the evaporator for each unit of time.

Valves are rated in tons of refrigeration. However, the same orifice usually has three different tonnage capacities. This range of capacity depends upon the difference in pressure between the high side and the low side.

Increasing this pressure difference will increase the rate of refrigerant flow. Therefore, if the valve is used on an R-12 refrigerant system, a valve that has a 1/2 ton (.455 metric ton) rating at 13 psi (.9 kg/cm^2) pressure on the low side will have a 3/4 to 1 ton (.72 to .98 metric ton) capacity at 5 in. (12.7 cm) Hg. vacuum on a frozen food unit. A 1/3 ton (.3 metric ton) will have only a capacity at a low-side pressure of 40 psi (2.8 kg/cm^2) on an air conditioning system.

In the first case, there is a 130 − 13 = 117 psi (8.2 kg) pressure difference, assuming a 130 psi (9.1 kg/cm^2) head pressure. In the second case, it is a matter of 130 plus 2 1/2 psi (5 in. of vacuum = 2 1/2 psi = .18 kg/cm^2) = 132 1/2 psi (9.3 kg/cm^2) pressure difference. In the last case, it is 130 − 40 = 90 psi (6.3 kg/cm^2) pressure difference.

It is important to use a valve of the correct capacity. If the valve orifice is undersize, the evaporator will be starved regardless of the superheat setting. The full capacity of the evaporator cannot be reached.

If the orifice is oversize, the valve will "hunt" or surge. When the valve opens, too much refrigerant will pass into the evaporator and the suction line will sweat or frost before the thermal element can close the valve. If one tries to correct this condition by increasing the superheat setting, the evaporator will be starved much of the time.

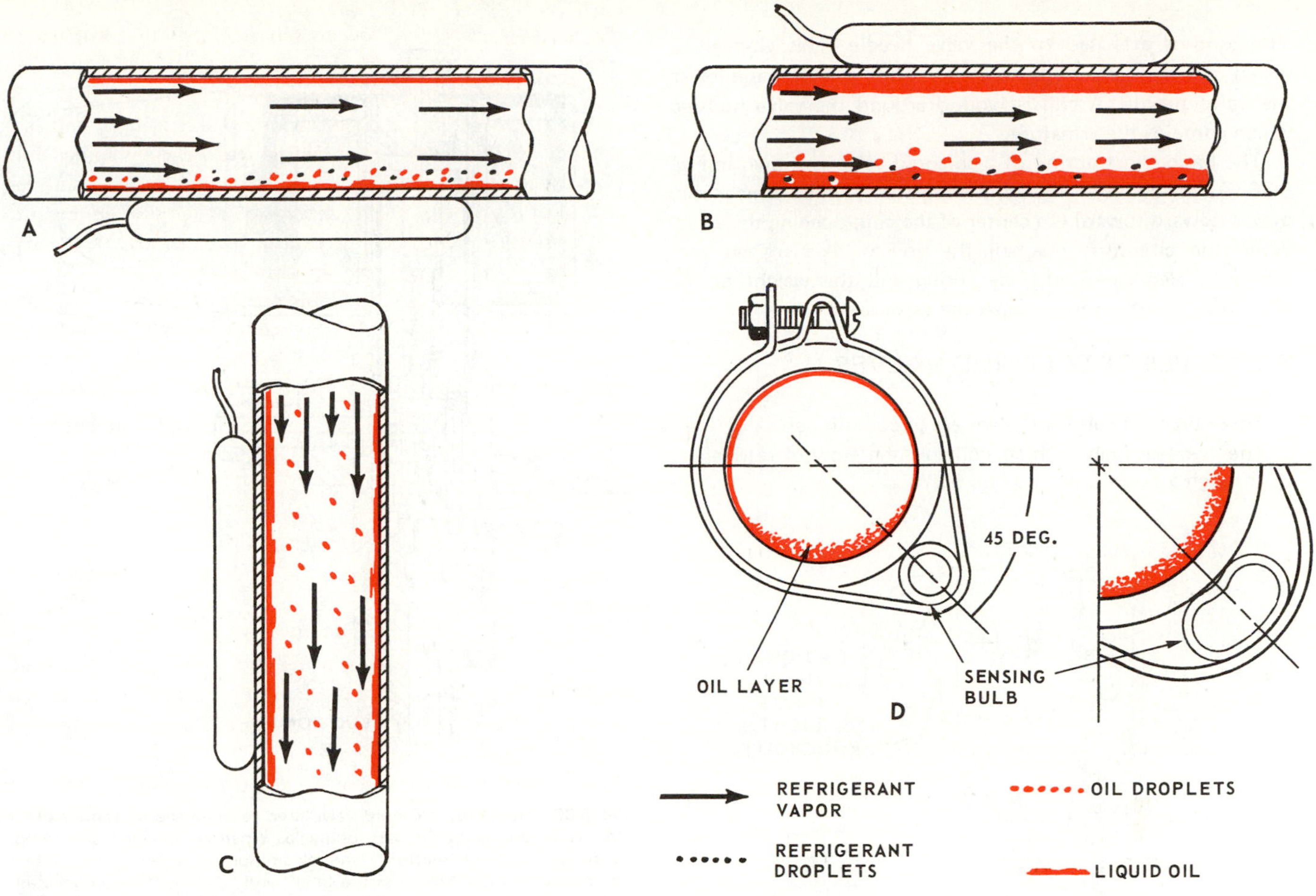

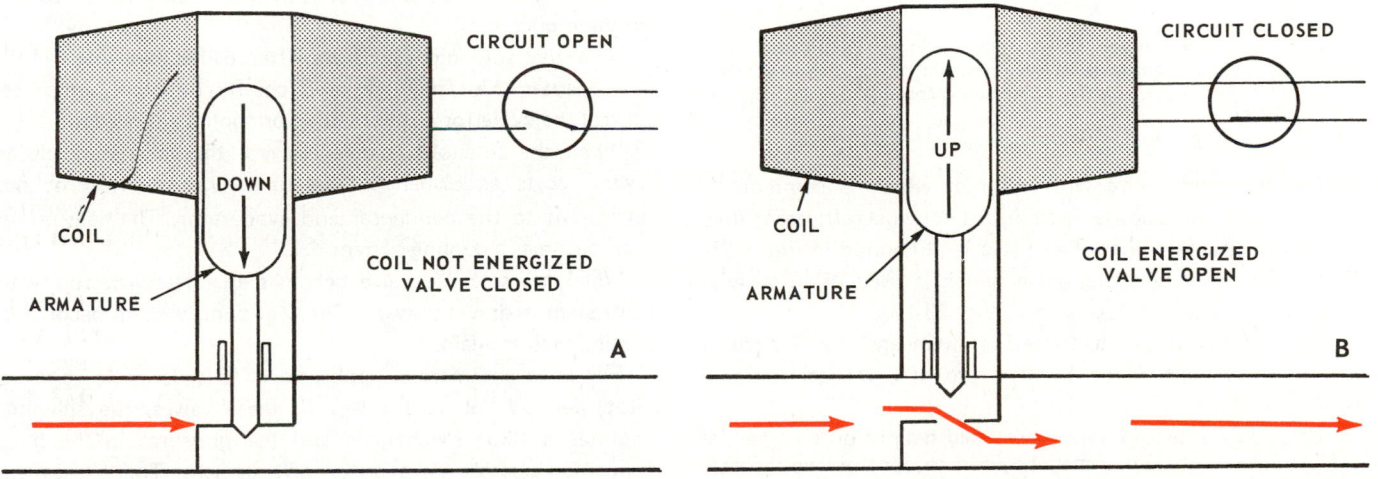

Fig. 5-35. Flow conditions in suction line. A—Horizontal suction line, (large). B—Horizontal suction line, (small). C—Vertical suction line. D—Sensing bulb on large diameter suction line. Oil is shown in red, refrigerant vapor by black arrows. Refrigerant droplets, by black dots.

Fig. 5-36. Solenoid valve. A—Circuit is open, coil is not magnetized and armature, due to its weight, drops and closes the valve. B—Circuit is closed, coil is magnetized and armature, due to magnetic field, is pulled up and opens the valve.

5-19 SOLENOID VALVE PRINCIPLES

In many refrigerating applications it is necessary to automatically close off or open refrigerant circuits to get the desired refrigerating effect. A solenoid valve usually does this.

It is easily installed and uses only simple electrical control circuits.

A solenoid valve is simply an electromagnet with a movable core or center. The basic construction of an electrical solenoid valve includes a movable armature. This is made of an iron alloy and is attached to the valve needle. This element is sealed into the valve body so the armature can raise and lower the valve needle. A coil is wound around the valve housing which contains the armature.

The basic construction of a solenoid valve is shown in Fig. 5-36. As the coil is energized, the armature, which is magnetic, moves upward toward the center of the coil, opening the valve. When the circuit is opened, the coil is de-energized and, therefore, demagnetized. The spring and the weight of the armature force the valve against the valve seat.

5-20 TYPES OF SOLENOID VALVES

Three types of solenoid valves are in common use:
1. The two-way valve which controls the flow of refrigerant through a single line. See Fig. 5-37.

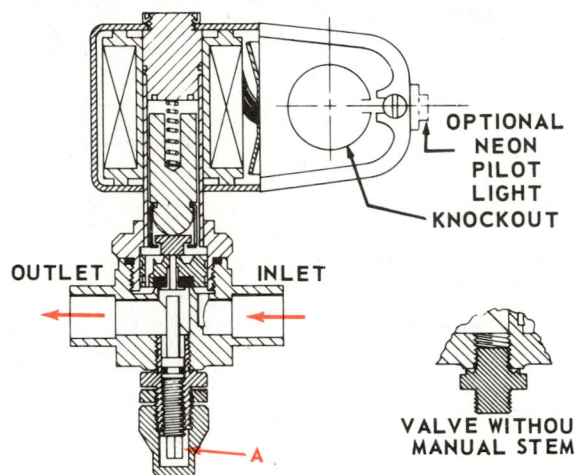

Fig. 5-37. Two-way solenoid valve. Valve is in closed position. A—Manual stem enables one to hold valve open for service or emergencies. (Alco Controls Div., Emerson Electric Co.)

2. The three-way valve with an inlet which is common to either of two opposite openings. It controls refrigerant flow in two different lines. This valve is illustrated in Fig. 5-38.
3. The four-way reversing valve which is used often on heat pumps. These are illustrated in Fig. 23-16.

Solenoid valves may be turned on by means of a thermostat and used to control the temperature of a refrigerator or a room.

Three-way solenoid valves are used mainly on commercial refrigerating units. They may be used to control two separate refrigerant circuits for defrosting, two-temperature evaporators and so on.

In the three-way solenoid valve shown in Fig. 5-38, opening A is the common opening and is never closed. As the electromagnet is de-energized, the weight of the solenoid plunger assembly and the force of the upper spring hold the valve firmly against the lower seat. This action closes port B to the common port and opens port C.

When the electrical circuit is closed, the solenoid becomes energized. The piston with the two valves attached to it will

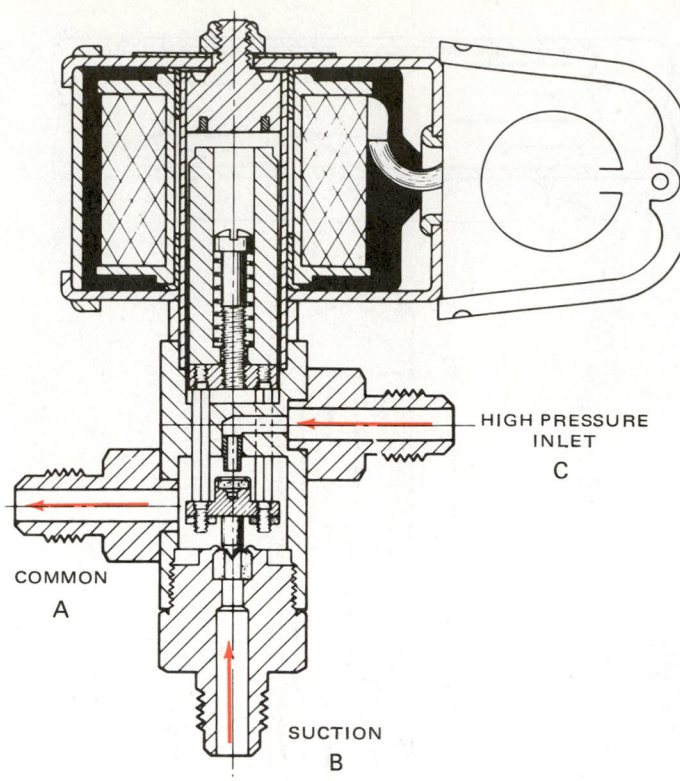

Fig. 5-38. Three-way solenoid valve used to close thermostatic expansion valve during "off" part of cycle. When compressor is running, openings marked "suction" and "common" are connected. When compressor is off, "high pressure inlet" and "common" are connected. (Sporlan Valve Co.)

move in the opposite direction. This action opens port B in Fig. 5-38 to common port A and, thereby, closes port C to the common port.

Four-way solenoid valves are often called reversing valves. See Chapter 23. They are used chiefly on heat pumps to control the cycle for either heating or cooling, as needed.

When the solenoid is de-energized, the valve stem closes several ports and opens others to reverse the flow of the refrigerant to the condenser and evaporator. The heat pump then becomes a cooling system.

When the solenoid valve becomes energized, the four-way valve stem is drawn upward. The heat pump system becomes a heating system again.

For large commercial applications, it is desirable to use a pilot-operated solenoid valve. In these valves, the solenoid operates a pilot mechanism and the pressures in the lines concerned operate on a piston arrangement. This causes the opening and closing of the main or large valve.

Fig. 5-39 is a cutaway of a pilot-operated solenoid valve. When the solenoid A is energized, the plunger B will be pulled from its seat and the pressure in D will leave the cylinder and will cause the piston E to move up. The movement of the piston controls the opening and closing of the main valve F.

When the solenoid valve is de-energized, plunger B returns to its seat. Pressure from G goes through small opening H and builds up pressure in control cylinder D. The spring then closes control valve F.

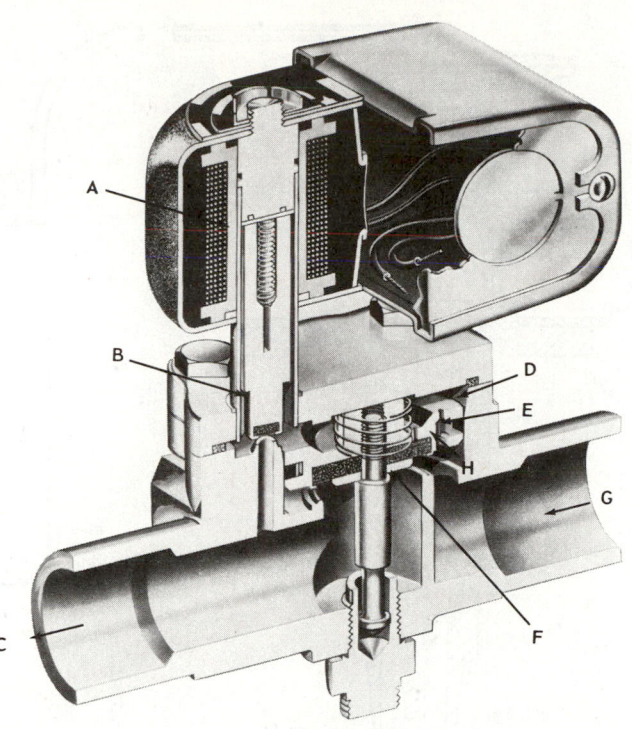

Fig. 5-39. A pilot operated solenoid valve. A—Solenoid. B—Solenoid plunger. C—Pressure at the left end. D—Control cylinder. E—Control piston. F—Control valve. G—Pressure at the right end. H—Opening. (Alco Controls Div., Emerson Electric Co.)

5-21 EQUALIZERS

An equalizer is a small tube — usually 1/4 in. OD — which joins the suction line at the outlet of the evaporator. The other end opens beneath the expansion valve diaphragm. This tube is shown in Fig. 5-40.

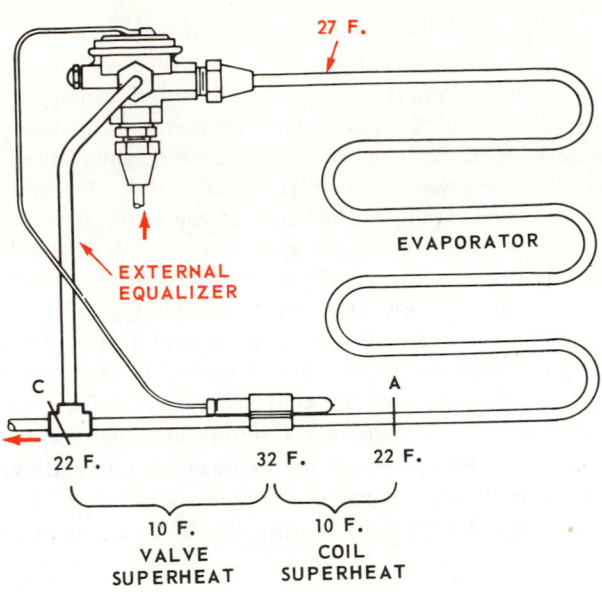

Fig. 5-40. A thermostatic expansion valve fitted with an equalizer tube. Equalizer connects the suction line pressure at the sensing bulb to underside of valve bellows or diaphragm (low pressure side). This enables the low side pressure operating the valve to be the same as the pressure at sensing bulb. This compensates for any pressure drop through evaporator while compressor is running.

There is always some pressure drop through the evaporator. It is recommended that an equalizer device be used if the pressure drop between the inlet of the evaporator and the outlet is more than 4 psi (.3 kg/cm^2). In operating the control valve, the equalizing tube provides the same pressure as is in the suction line at the sensing bulb location. This equalizing of pressure will permit accurate superheating adjustments. Pressure drop in the evaporator always tends to increase

Fig. 5-41. A three-way solenoid valve is used to keep thermostatic expansion valve tightly closed during the "off" cycle. A—When motor compressor is on the "on" part of the cycle, the valve is closed. Pressure in the suction line is transmitted through the equalizer tube to the thermostatic expansion valve. B—With motor compressor on the "off" part of the cycle, valve opens so high pressure can flow up the equalizer tube to close the expansion valve. (Sporlan Valve Co.)

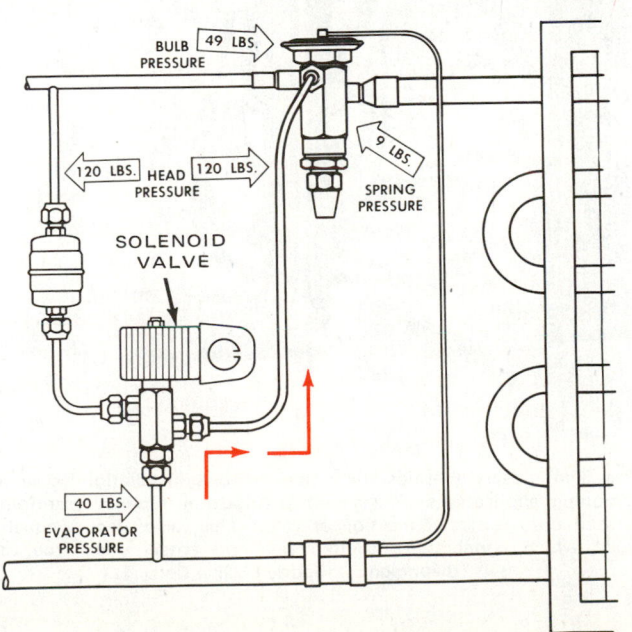

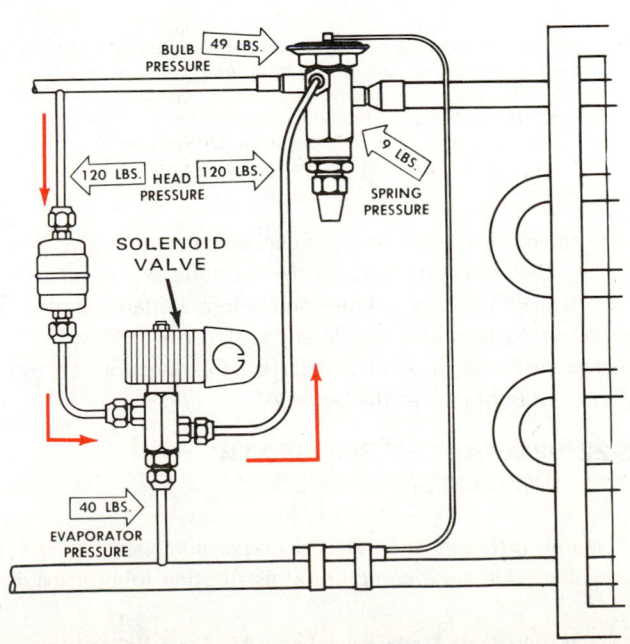

the superheat effect and tends to starve the evaporator.

The thermostatic expansion valve may tend to open intermittently (opening and closing frequently) during the off cycle. This may be caused by temperature fluctuations (changes) which occur during cabinet opening and closing. To prevent this, one may use the high-side pressure on the valve to force it closed during the off part of the cycle. A solenoid valve may be used to control this pressure.

A special solenoid valve connected into the equalizer tube is shown in Fig. 5-41. Note valve construction in Fig. 5-38.

The electrical circuit to the solenoid valve is opened at the time the motor-compressor circuit opens. On opening the circuit, the solenoid core falls and closes the equalizer tube to the suction line. The high-pressure refrigerant enters the top of the solenoid valve and passes up the equalizer tube and forces the thermostatic diaphragm up, closing the valve.

The thermostatic expansion valve, Fig. 5-42, uses an unusual

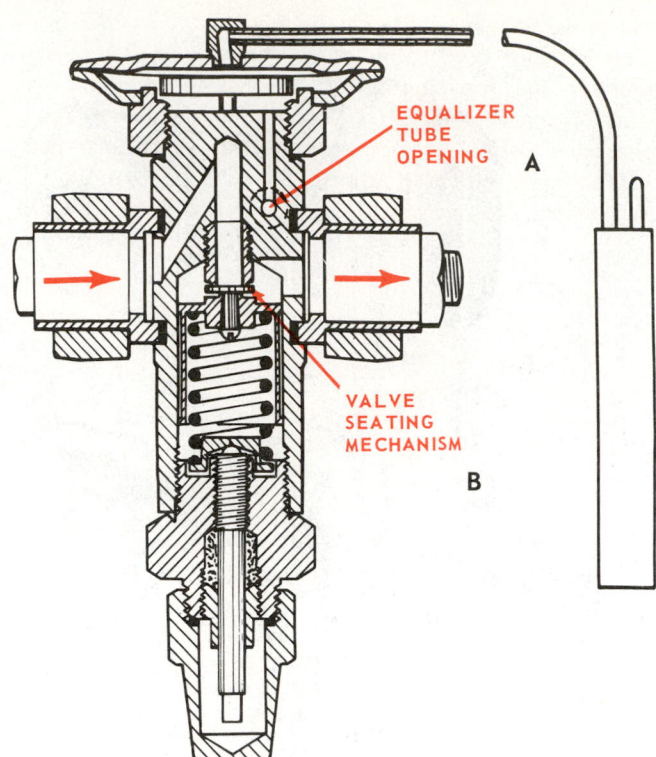

Fig. 5-43. A large capacity thermostatic expansion valve. It is fitted with an equalizer connection. A—Equalizer tube opening. B—Valve seating mechanism. (Sporlan Valve Co.)

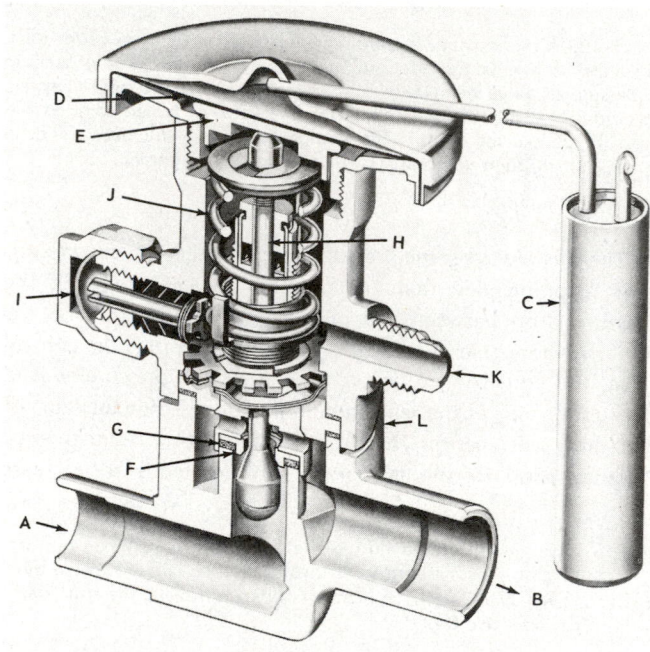

Fig. 5-42. A thermostatic expansion valve which uses an equalizer tube. A—Liquid line connection. B—Suction line connection. C—Sensing bulb. D—Diaphragm. E—Diaphragm contact element. F—Valve. G—Valve seat. H—Push rod. I—Superheat adjustment. J—Superheat spring. K—Equalizer tube connection. L—Valve body.
(Alco Controls Div., Emerson Electric Co.)

adjustment. The equalizer tube connects into the valve at K.

Fig. 5-43 is a large capacity thermostatic expansion valve. It has flanged refrigerant line connections bolted together and gasketed. Instead of a needle and seat, it has a flat (disk) valve surface and seat, B. An equalizer tube connection is shown in the upper right part of the body, A.

5-22 SPECIAL THERMOSTATIC EXPANSION VALVES

Many different thermostatic expansion valve designs are available. One type combines a distributing tube or manifold

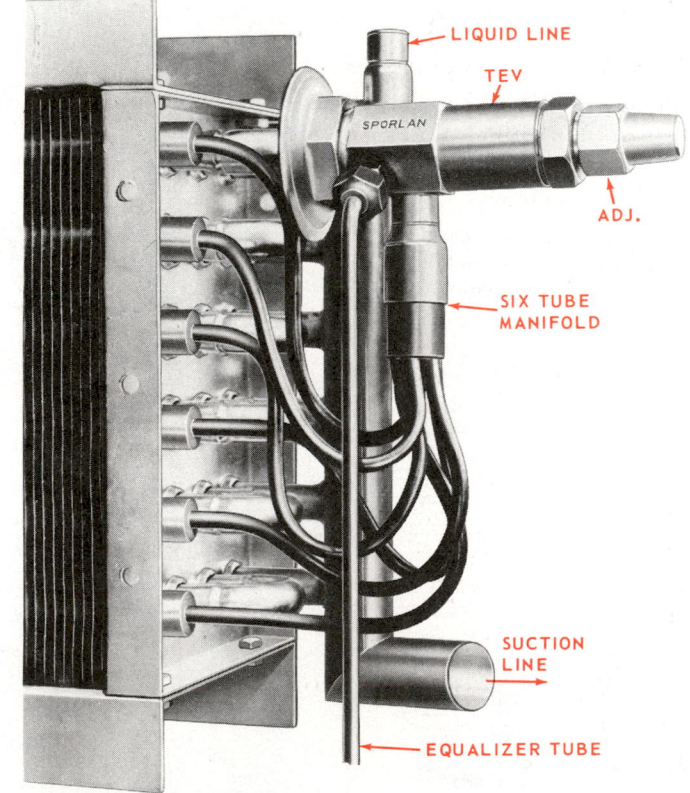

Fig. 5-44. Special thermostatic expansion valve installation for air conditioning applications. Refrigerant distribution tube or manifold is brazed onto outlet connection of valve. This valve uses an equalizer tube which connects to suction line and enters under the valve diaphragm. (Sporlan Valve Co.)

in the body of the valve. See Fig. 5-44.

This design is used to reduce the pressure drop in a large evaporator by providing several parallel refrigerant paths through the evaporator. It is popular for air conditioning applications. Careful engineering is needed as each tube must receive an equal amount of refrigerant.

Fig. 5-45 shows a valve with many outlets, a distributor and

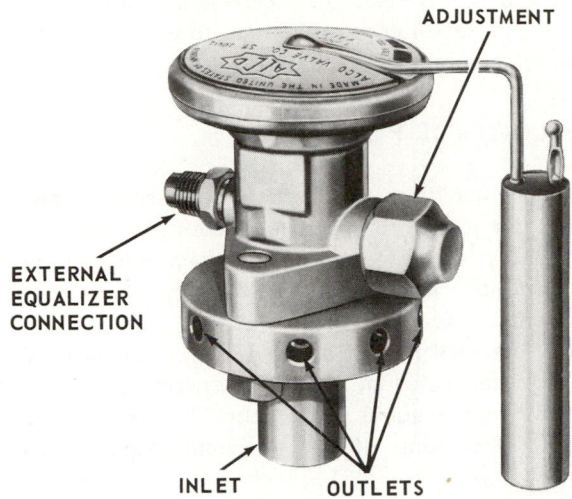

Fig. 5-45. Thermostatic expansion valve having multiple (many) outlets to evaporator. Valve is fitted with connection for an equalizer tube and has superheat adjustment.   (Alco Controls Div., Emerson Electric Co.)

an external equalizer. A diagram of a special thermostatic expansion valve, which provides multiple connections to the evaporator, is shown in Fig. 5-46. Fig. 5-47 shows a distributor used on large capacity evaporators.

5-23 HUNTING

The term "hunting," applied to any type of mechanism, means that the mechanism is first going too far in one direction, then returning too far in the other direction. This is

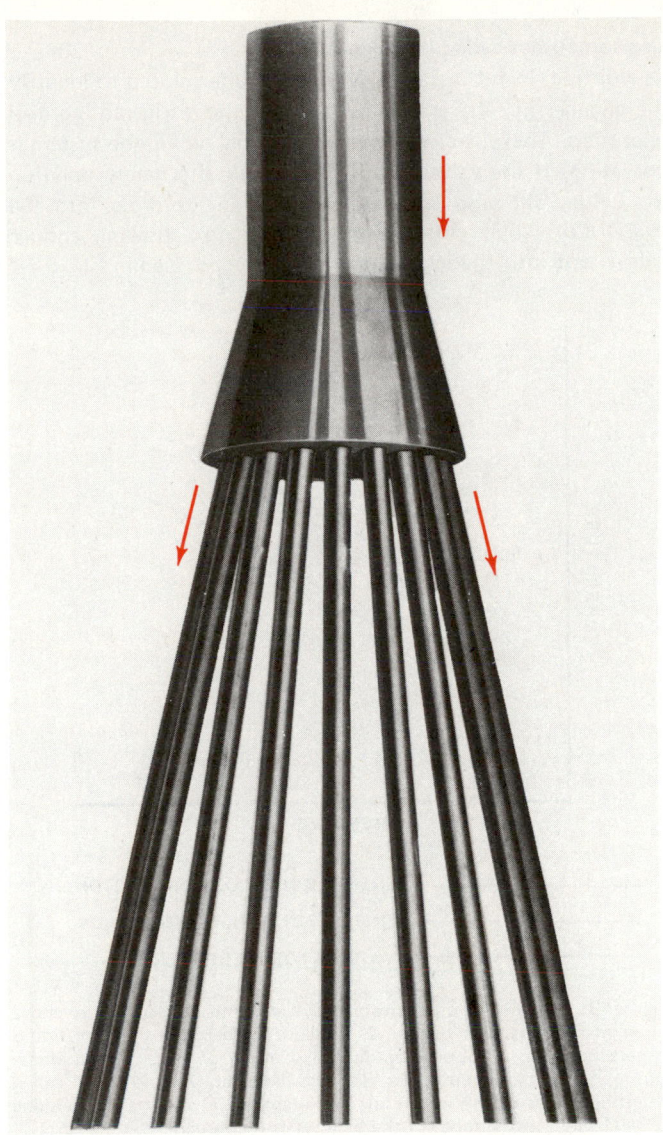

Fig. 5-47. A distributor tube. The thermostatic expansion valve supplies refrigerant at the top. In turn, the distributor tube supplies equal quantities of refrigerant to parallel paths through the evaporator (bottom). (Alco Controls Div., Emerson Electric Co.)

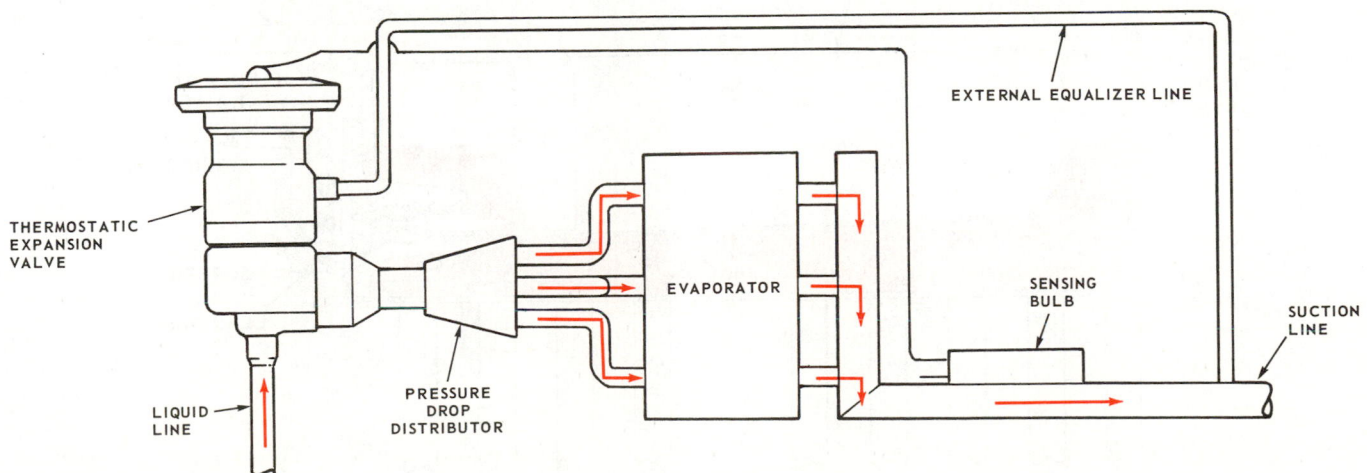

Fig. 5-46. A thermostatic expansion valve which provides multiple connections to the evaporator.   (Alco Controls Div., Emerson Electric Co.)

also sometimes called "surging."

Hunting, in refrigerating systems, is a term used to identify the changes in refrigerant flow through the refrigerant control operation. There is always some "hunting" while the system is operating. If the valves "hunt," they will alternately open up too wide, allowing too much refrigerant to flow into the evaporator, then close down too far, not allowing enough refrigerant into the evaporator. This effect is seen in Fig. 5-48.

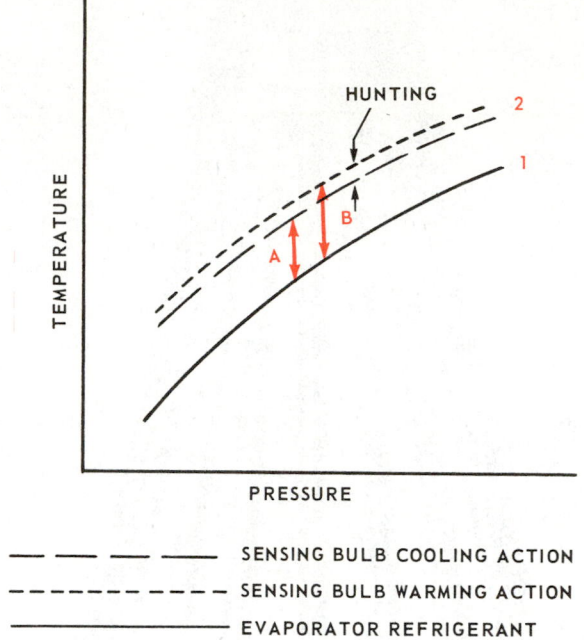

Fig. 5-48. Thermostatic expansion valve superheat. 1—Vapor pressure curve of refrigerant in system. 2—Pressure-temperature curve of liquid in sensing bulb. A—Superheat during warming of bulb. B—Superheat during cooling of bulb. The distance between A and B is called "hunting." It is mainly the result of weight of valve parts and bending resistance of the bellows or diaphragm.

The less "hunting" the more effective the system will be. During the interval that the valve is "hunting" too much, it is not providing a uniform amount of refrigerant to the evaporator. It may even allow liquid refrigerant to reach the compressor and cause compressor damage. This condition may sometimes be caused by a valve that is too large for the system.

Each thermostatic expansion valve and evaporator combination has to be adjusted. The superheat adjustment should be used to reduce surging and hunting to a minimum but still permit full evaporator use.

5-24 LOW-SIDE FLOAT

The low-side float is an efficient, yet simple, refrigerant control. Its job is to maintain a constant level of liquid refrigerant in the evaporator. The system is an efficient heat transfer device. Heat moves easily from the evaporator to the liquid refrigerant.

The float itself may be a sealed ball, a cylinder or an open pan. It is connected by levers to a needle which closes an orifice when the liquid level reaches the correct height and opens it when some of the refrigerant evaporates and the liquid level drops.

The low-side float has the possible disadvantage of "oil binding." Because the mechanism will float in either oil or liquid refrigerant, special provisions must be made to return any excess oil to the compressor. This is sometimes done by the use of a wick or a small bypass at the surface of the liquid refrigerant.

Such a system uses a liquid receiver. Extra refrigerant is stored in the liquid receiver.

The basic principles of the low-side float system are explained in Chapter 3.

The low-side float may be connected either to a needle or a

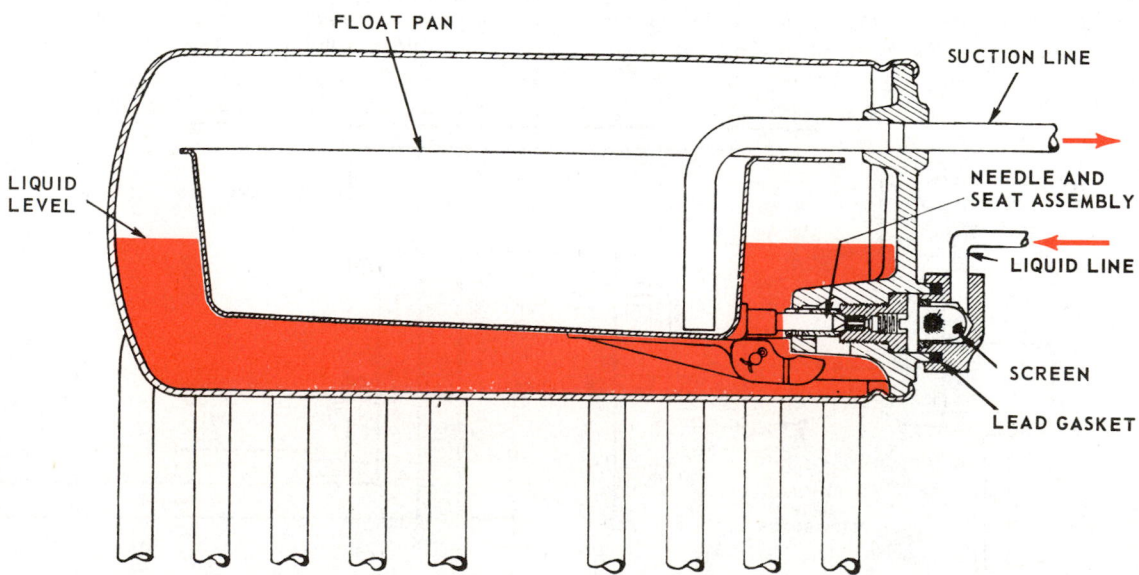

Fig. 5-49. Low-pressure side float refrigerant control. A bucket or pan type float is used in this refrigerant level control. Suction line dips to the bottom of open float in order to remove oil which might otherwise accumulate in open pan.

ball valve. It is calibrated so that the valve will close when the float is at the proper level — that is, when there is a certain level of liquid refrigerant in the evaporator.

The suction tube on these evaporators extends into the float chamber and, with pan-type floats, extends to the bottom of the pan. See Fig. 5-49. This design insures a more positive oil return than the ball type float.

Low-side float systems are used in large industrial systems and in some water cooling systems.

Either a pressure operated motor control or a thermostatic motor control may be used with a low-side float system.

Either the thermostatic expansion valve (TEV) or the low-side float may be used in multiple evaporator systems. This type float control is also used for controlling water level in cooling towers, evaporative condensers and in humidifiers.

5-25 HIGH-SIDE FLOAT

A high-side float refrigerant control is somewhat like the low-side float mechanism. However, it is located in the high-pressure side.

When the compressor is running, the condensed refrigerant from the condenser flows directly to the high-side float chamber. No liquid receiver is used. As the liquid refrigerant level rises, the float inside the chamber opens a valve allowing the liquid refrigerant to flow into the evaporator.

In this type of refrigerant control, the liquid refrigerant is stored in the evaporator, which makes it a flooded evaporator. It is designed accordingly.

As the liquid level falls in the float chamber, the float will move down and close the valve opening into the evaporator. In this way the pressure difference between the high side and the low side is maintained. This float maintains a constant level of liquid refrigerant on the high-pressure side.

Floats are made of either copper or steel. In the hermetic units, steel is usually used. Such units cannot use a liquid receiver unless the float is located within it.

These float controls do not have as much difficulty with oil binding as the low-side float control, because, at the higher pressure, the oil is dissolved more readily in the liquid refrigerant and circulates with it. However, the evaporator used in connection with a high-side float control must be equipped with a special oil return or oil binding will occur.

The float may be connected directly to the needle, or it may operate the needle with a simple lever, as shown in Fig. 5-50. The needles and seats are made of long-wearing alloys such as stainless steel or hard-surfaced alloys.

Some high-side float systems use a weight check valve when the float chamber is located a long way from the evaporator. Chapter 3 explains how this type system operates.

Either a thermostatic or pressure motor control may be used with this refrigerant control.

5-26 CAPILLARY TUBE

The capillary tube is a popular type of refrigerant control. This control is simply a length of tubing with a small inside diameter. It acts as a constant throttle on the refrigerant.

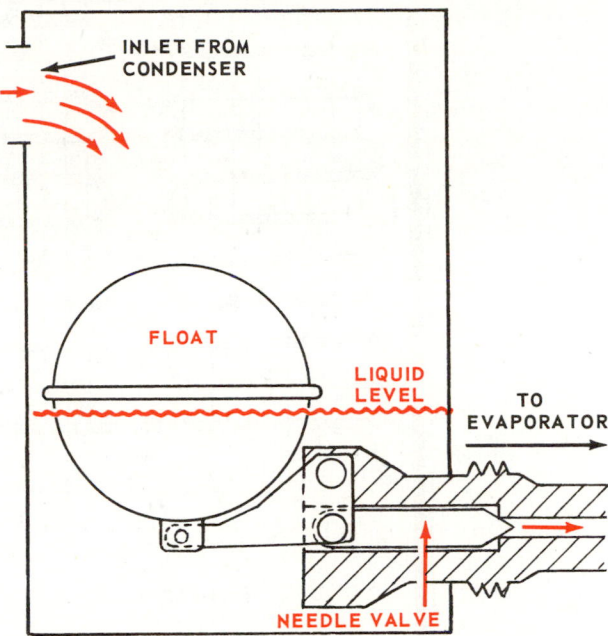

Fig. 5-50. High-pressure side float refrigerant control mechanism. Liquid refrigerant flowing in from condenser will cause float to rise and open needle valve. Then liquid refrigerant flows into evaporator.

It is made of seamless tubing and has a small and accurate inside diameter. It usually is equipped with a fine filter or filter-drier at its inlet. This removes any moisture or dirt from the refrigerant.

The amount of refrigerant in the system must be carefully calibrated, since all of the liquid refrigerant will move into the low-side during the off-cycle as the pressures balance or equalize.

Too much refrigerant will cause the unit to frost back on the low side. This control must be used with a thermostatic motor control.

There are several theories concerning the principle of operation of the capillary tube. The tube resists fluid flow. Pressure goes down as the small liquid flow moves through the tube — until the liquid starts to evaporate in the tube. This vapor formation provides a sudden pressure and temperature drop in approximately the last quarter of the length of the tube. The refrigerant finally is cooled to evaporator temperature and its pressure is reduced to evaporator pressure.

This vapor formation in the capillary tube is called "vapor lock". The design of the capillary tube depends on four variables:

1. Tube length.
2. Inside diameter.
3. Tightness of tube windings.
4. Temperature of tubing.

A typical capillary tube refrigerant control is shown in Fig. 5-51.

The capillary tube refrigerant control has no moving parts. Therefore, it has several advantages. First, there are no parts to wear or stick. Second, the pressures balance in the system when the unit stops. This condition places a minimum starting load on the motor.

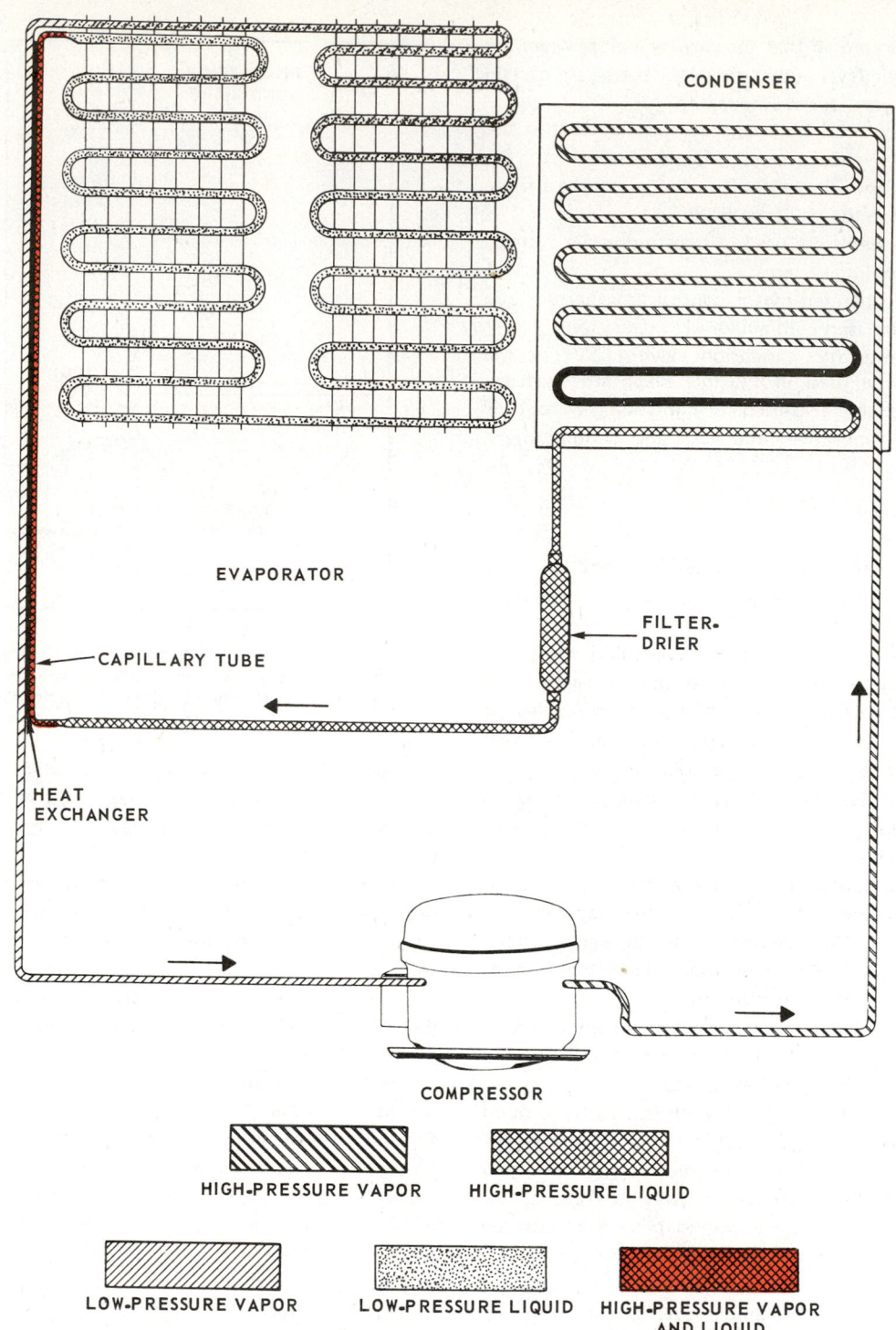

CONDENSER

EVAPORATOR

CAPILLARY TUBE

FILTER-DRIER

HEAT EXCHANGER

COMPRESSOR

HIGH-PRESSURE VAPOR

HIGH-PRESSURE LIQUID

LOW-PRESSURE VAPOR

LOW-PRESSURE LIQUID

HIGH-PRESSURE VAPOR AND LIQUID

Fig. 5-51. A capillary tube refrigerant control. Note heat exchanger which cools capillary tube by transferring heat from capillary tube to suction line. (Kelvinator, Inc.)

The capillary tube may be coiled for part of its length. It is usually attached to the suction line, permitting the suction line vapor to cool the liquid refrigerant in the capillary tube.

A recent development in the design of capillary tubes uses a larger diameter and a longer tube. The larger diameter tube is less likely to become plugged with dirt, ice or wax than the smaller diameter tube. These larger diameter capillary tubes are most used on air conditioning applications.

5-27 CAPILLARY TUBE CAPACITIES

Popular capillary tube sizes are shown in Fig. 5-52. Tubing .114 OD by .049 ID is often used for refrigerant R-12 in domestic units. This is suitable for average temperatures and frozen food temperatures, dependent on the power of the unit and the use. Approximate sizes for capillary tube installations for R-12 are shown in Fig. 5-53.

CAPILLARY TUBE DIAMETERS,

Outside Diameter	Inside Diameter
.083	.031
.094	.036
.109	.042
.114	.049
.120	.055
.130	.065

Fig. 5-52. Some common capillary tube diameters. Other application tables are listed in Chapter 11 and 21.

connections. Detail 1 shows the use of a special nut that squeezes against both the capillary tube and the fitting. Because the nose section is deformed as the nut is tightened, the nut should always be replaced when the capillary tube is serviced.

Detail 2 shows the capillary tube silver brazed to a 1/4 OD soft copper tubing. The larger tube may then be connected to the system by the usual flared fitting.

Detail 3 shows a larger tube silver brazed to the capillary tube. The larger tube connection is then made with a flared fitting. Mechanical connections are often substituted during overhaul using special capillary tube fittings, see Fig. 5-55.

HORSEPOWER	TEMP.	REQUIRED LENGTH OF CAPILLARY TUBING IN FEET						
		.031 ID	.036 ID	.040 ID	.042 ID	.049 ID	.055 ID	.065 ID
1/8	H	1.1	2.2	3.5	4.5	9	15	
	M	4	8	13	16	32	56	
	L	9	18	29	36	72	126	
1/5	H						10	
	M	2.2	4.4	7.0	9	18	31	
	L	5.2	10.5	17	21	42	73	
1/4	H						5	
	M	1.1	2.2	3.5	4.5	9	15	
	L							7.5
1/3	M						9.5	
	L	1.75	3.5	5.6	7	14	2.5	

Fig. 5-53. Capillary tube sizes and applications. Temperatures are shown as high, "H," medium, "M," or low, "L". Required length of tube in feet is indicated for each tube inside diameter (ID). The length in feet is for R-12 refrigerant.

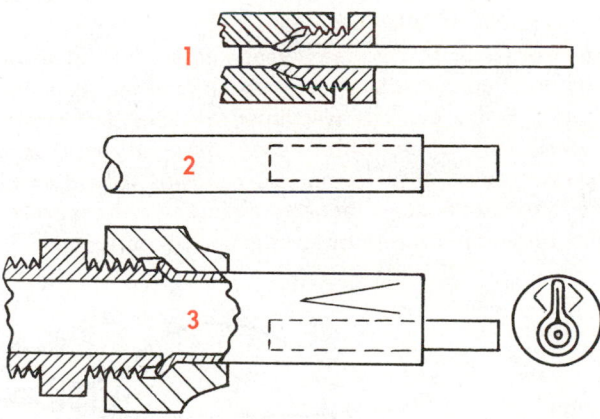

Fig. 5-54. Some typical capillary tube connections.

5-28 CAPILLARY TUBE FITTINGS

The capillary tube connections are most often silver brazed at both the condenser and the evaporator end. The capillary tube may also be attached to the evaporator and condenser or drier with fittings that are leakproof and vibration-proof.

Fig. 5-54 illustrates three popular ways to make these

5-29 COMPARING REFRIGERANT CONTROLS

The pressure-time diagram for one operating cycle for any particular refrigerant control will show its operating characteristics for one running cycle of the refrigerator. The pressure-

Fig. 5-55. Some mechanical capillary tube connections. (Watsco, Inc.)

time diagram will vary with the refrigerant used. Fig. 5-56 shows the pressure-time characteristics for six different refrigerant controls using refrigerant R-12.

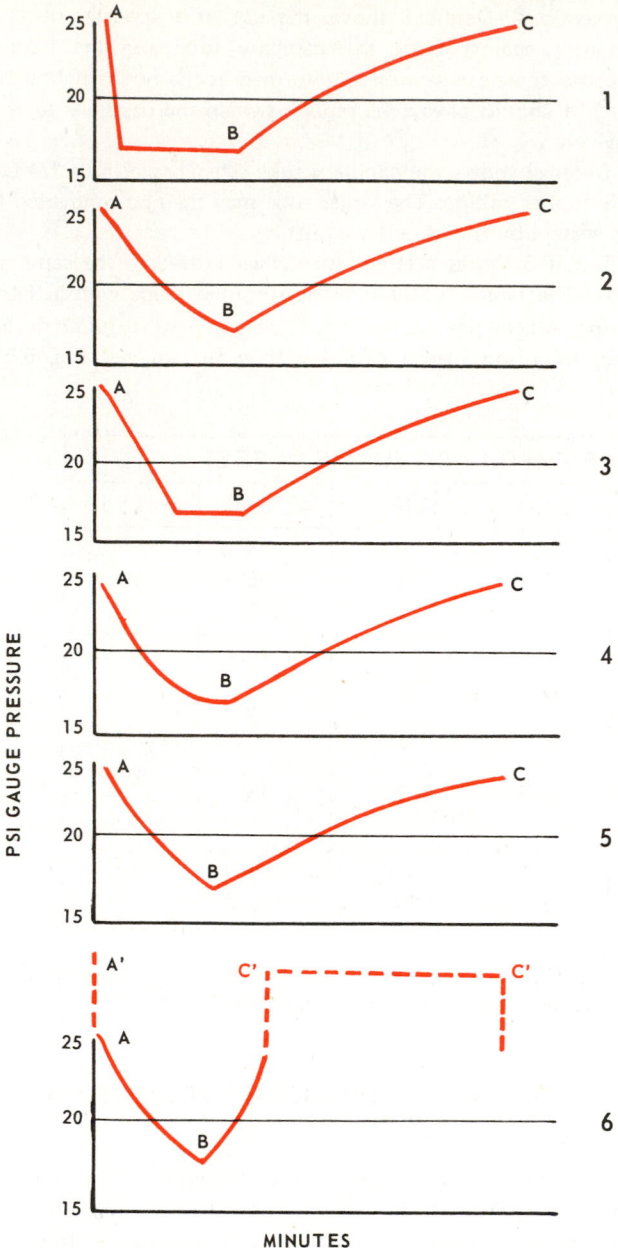

Fig. 5-56. A comparison of the low-side pressure-time cycles for various refrigerant controls using R-12 refrigerant. 1—Automatic expansion valve. 2—Thermostatic expansion valve. 3—Thermal-electric expansion valve. 4—Low-side float. 5—High-side float. 6—Capillary tube. Note that pressure goes up until it balances with high-side pressure. Point A—Cut-in pressure. Point B—Compressor stops. Point C—Cycle repeats. Pressures at Point A' and C' are pressures at cut-in.

5-30 CHECK VALVE

Check valves allow fluid to flow through them in only one direction. Rotary and gear compressors have check valves in the suction line. This is to prevent the high-pressure vapor and the refrigerant oil from backing up into the evaporator during the off part of the cycle.

A typical check valve is shown in Chapter 4, Fig. 4-73. These check valves may use either a disk or a solid ball in their construction.

They may also operate differently. Some use a spring or magnet to keep the valve against the seat. Others are mounted so that the weight of the valve keeps it against its seat.

Multiple systems have evaporators which operate at different temperatures. They use check valves to keep the refrigerant vapors in warmer evaporators from backing up into the colder evaporators. This is explained in Chapter 12.

5-31 SUCTION PRESSURE VALVES

Many systems use bellows or diaphragm-operated pressure regulating valves on the low side of the system. These valves are required on multiple systems in which the evaporators operate at different temperatures.

Some are used to keep the compressor suction pressure at a safe level to avoid overloading the compressor. A modified suction pressure valve is required on most automobile air conditioning systems because the compressor is operating at various speeds. The valve is needed to maintain a fairly constant temperature in the evaporator.

Suction pressure valves may be controlled from the evaporator pressure. These control the evaporator temperature. Some operate from crankcase pressure. These keep the compressor from being overloaded.

As a crankcase pressure regulator, the valve modulates (adjusts) suction pressure to the compressor intake to provide overload protection for the condensing unit motor. Fig. 5-57 illustrates a low-pressure side pressure regulator installation.

Additional information concerning these valves is given in Chapter 12.

5-32 REVIEW OF SAFETY

Perhaps the most common damaging condition of refrigeration operation is allowing liquid refrigerant to enter the compressor. This is called refrigerant slugging. Liquid refrigerant is not compressible. The piston striking the liquid refrigerant in the cylinder will cause knocking and considerable overloading of the piston, connecting rod, bearings and compressor valves. Most compressors provide some kind of a safety overload device in the valve arrangement to attempt to protect the compressor under severe slugging conditions.

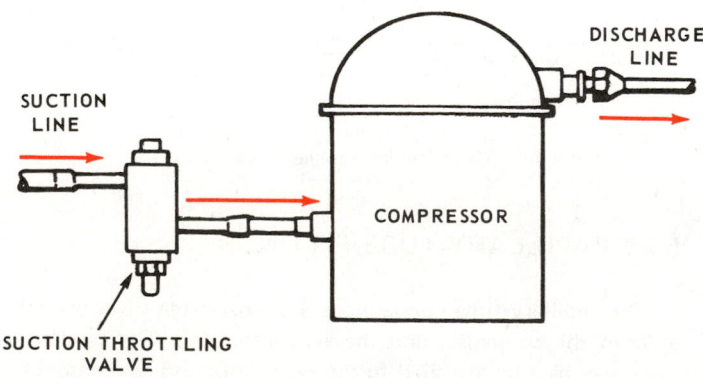

Fig. 5-57. A low-side pressure control valve located in the suction line.

A 24-hour pressure time recorder attached to any new installation is an excellent safety device to make sure that the system is operating within safe pressure limits. A 24-hour temperature recorder may also be used for this purpose.

Keep the floor clear of debris — oil, water or other slippery material.

Wear safety goggles when working on refrigerating systems.

Avoid lifting objects which weigh over 35 pounds (15.9 kg). Use leg muscles when lifting.

Always have good ventilation and good lighting when working on a refrigerating system.

In order to avoid the possibility of electrical shock, make sure that all electrical circuits are well insulated.

All metal parts of refrigerating mechanisms should be grounded.

When removing a valve from a system, always use two wrenches to avoid twisting the valve or the tubing.

5-33 TEST YOUR KNOWLEDGE

1. How many pressures or forces influence the needle movement of an automatic expansion valve needle?
2. Why are expansion valves adjustable?
3. Which refrigerant controls can operate satisfactorily with a varying amount of refrigerant in the system?
4. What is thermostatic expansion valve superheat?
5. What is meant by a gas charged element used in connection with a thermostatic expansion valve?
6. What mesh screens are used in refrigerant controls?
7. What is the most common expansion valve body material?
8. How is the liquid line usually attached to the expansion valve.
9. Why does the suction line extend down into the pan type low-side float?
10. What materials are high-side floats usually made of?
11. What is the purpose of an automatic refrigerant control?
12. What is meant by the term "pilot-operated" solenoid valve?
13. How does a capillary tube operate?
14. Does a capillary tube system usually use a liquid receiver? Why?
15. Does a capillary tube need a filter or a screen at its inlet?
16. What type of system needs a check valve?
17. What energy opens a solenoid valve?
18. Why is a thermostatic expansion valve equalizer tube used?
19. What may be the trouble if a capillary tube unit frosts down the suction line?
20. What is meant by a four-way solenoid valve?
21. How is a capillary tube usually fastened to a 3/8 OD copper tubing?
22. What is the most popular TEV superheat setting?

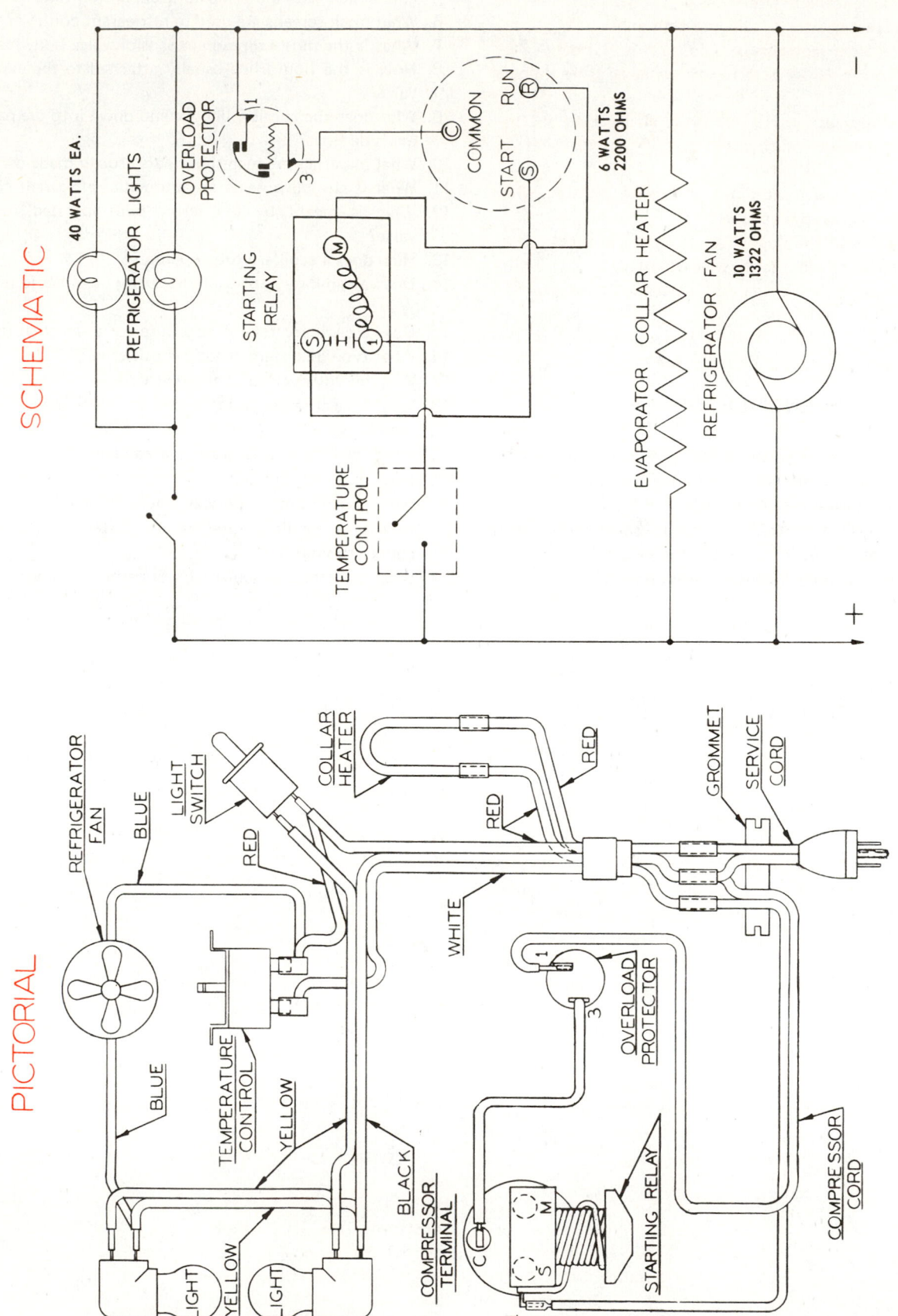

SCHEMATIC

PICTORIAL

Fig. 6-1. Typical wiring diagram of a refrigerating system. Note use of color code for leads (wires). (Franklin Mfg. Co.)

Chapter 6

ELECTRICAL-MAGNETIC FUNDAMENTALS

Electricity is a form of energy. It is used for many purposes. In refrigeration and air conditioning, it drives motors, operates controls and produces heat and light.

There may be some confusion in the understanding of the terms "electricity" and "electronics." Electronics is the branch of physics covering the behavior and effects of electron flow in vacuum tubes, gases and semiconductors. Electricity is the branch of physics concerning the behavior of a natural phenomenon known only by its effects: electric charge, electric current, electric field and electromagnetism.

Since electricity cannot be seen, it can only be studied through what it does. Its effect can be seen and measured by the use of instruments. In all cases, electricity is created by free electrons moving from one substance to another. (See Para. 28-23.) This theory, and other necessary fundamentals of electricity, will be explained in the following paragraphs. A knowledge of the fundamentals will help one better understand the construction and operation of electric motors, controls and circuits used in refrigeration and air conditioning.

6-1 BASIC CURRENT ELECTRICITY

Electricity is usually supplied to homes and industries by electric utility companies. The supply from the utility company to the home ends in a control panel. This panel contains fuses or circuit breakers for each of the separate circuits.

The electricity supplied to most homes is at 120 or 240 volts (V) alternating current. Most lighting and appliances use 120 volts (V). Electric stoves, water heaters, air conditioners and refrigerating systems (the larger sizes) use 240 volts (V).

Service technicians should understand the fundamentals of electricity. This will help them service the complete electrical system of a refrigerator or air conditioner. See Fig. 6-1.

Terms such as "voltage," "direct current," "cycles" and "Hz" are explained in later paragraphs.

6-2 TYPES OF ELECTRICITY

Refrigeration and air conditioning is concerned with two common types of electricity. These are static electricity and current electricity. Static electricity is often defined as electricity at rest. Current electricity is electricity flowing through conductors (wires).

Lightning is created through the discharge of static electricity. Under certain conditions, materials such as sheets of paper and clothing may become charged with static electricity. That is why they sometimes cling together. Static electricity is often produced by the friction between two surfaces rubbing.

Current electricity is commonly used in homes and in industries to drive motors, weld and start engines. It is usually produced by coiled wires moving in an electromagnetic field.

6-3 GENERATING ELECTRICITY

The electricity used in the refrigeration and air conditioning industry is usually generated by electromechanical equipment (generator). This may be either direct or alternating current.

Electricity also may be generated chemically. Dry cells used in flashlights are an example. The voltage of a dry cell is approximately 1.5 volts (V). Chemically generated electricity is always direct current. The chemicals combine or react to generate electricity and, in the process, the chemicals are used up. The cell is discharged.

The automobile storage battery does not generate electricity, but merely stores it. Electricity from the vehicle's generator flows into the storage battery, causing a reversible chemical action between the electrolyte (acid solution) and the battery plates (lead or lead oxide).

Electricity can be caused to flow from other energy forms, such as heat energy, friction, mechanical energy, light, chemistry and magnetism. Any method which produces a movement of free electrons causes an electrical potential (pressure), and free electrons will flow if a conductor (wire) is present.

6-4 ELECTROSTATIC ELECTRICITY

There are two kinds of electrostatic charges — positive and negative. Substances charged with the same kind of electrostatic electricity tend to push apart or repel each other. Two substances, one with a positive charge and the other with a negative charge, tend to be attracted to one another.

A common electronic device used to indicate electrostatic charge is an electrometer. Another is the electroscope. A simple electroscope consists of two small strips of aluminum foil attached to a metal rod. The rod is extended through the top of a stoppered glass bottle. When this rod contacts a body having an electrostatic charge the two strips of aluminum immediately move away from each other. They have both become charged with the same kind of electrical charges. They are either negatively or positively charged.

Placing like poles of two magnets close to each other demonstrates the same reaction. The magnets act as though a strong invisible spring had come between them preventing contact.

A common example of electrostatic generation occurs when one walks across a carpet. This charges the entire body with static electricity. If one then touches a filing cabinet, a faucet or door knob, this charge will be quickly dissipated (spent). There may be a visible and easily heard spark discharge.

In another example, silk rubbed over glass produces a positive charge of static electricity on the surface of the glass. Plastic rubbed with wool or fur will generate a negative charge.

The refrigeration service technician is concerned with static electricity in two applications. It applies to condensers used with capacitor type electric motors. The term "capacitor" means practically the same as "condenser." The unit for measuring static electricity if the farad. See Para. 6-44. Electrostatic electricity is also used in electrostatic air cleaners.

6-5 CURRENT ELECTRICITY

Current electricity is the movement of electrons along an electrical conductor. For example, pushing the button on a doorbell closes (completes) the circuit and causes electrons to flow through the circuit. This electrical flow rings the bell.

There are two common types of electric current:
1. Direct current (d-c).
2. Alternating current (a-c).
Direct current is the continuous flow of electrons in the same direction. It is the type of current used in automobile starting, lighting and ignition, and in most solid states circuits.

Direct current is used on cordless electric appliances such as toothbrushes, electric shavers and drills. It is also used extensively in electronics. Direct current is needed for battery charging since batteries produce only direct current.

Alternating current is the flow of electrons along a conductor first in one direction, then in the other. It is the type of current used for most power and light applications.

6-6 DIRECT CURRENT

Direct current (d-c) is electron flow along a conductor in one direction. It is the type of current produced by batteries. Vehicles — except some larger buses and earth-moving equipment — operate on a 12V direct current circuit supplied by the storage battery.

Some power companies still produce and sell direct current. However, its use is fast disappearing for power and lighting. At present its chief uses are in electronics, elevator service, electric welding and automobiles.

Generally, in both elevator operation and electric welding, the direct current is generated at the site. It is made by using an alternating current rectifier, by driving a d-c generator with an a-c motor, or by driving a d-c generator with a gasoline or diesel engine.

With the development of new lightweight storage cells and

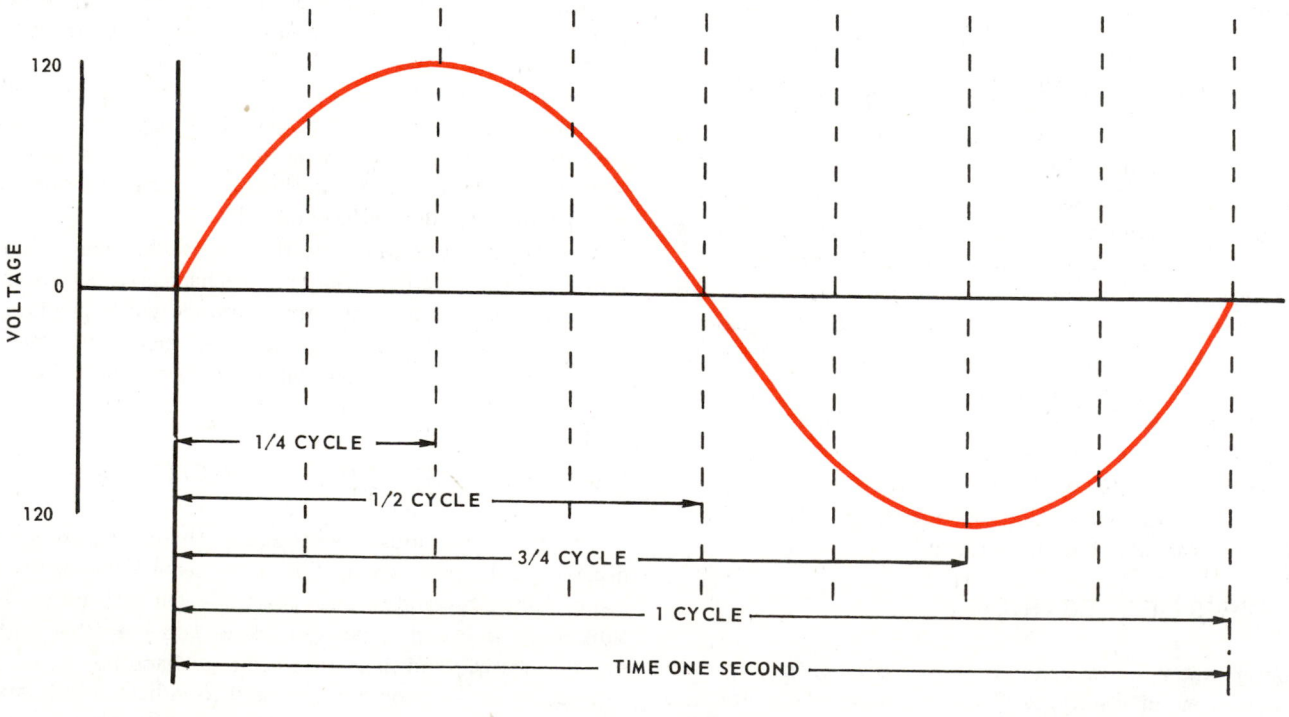

Fig. 6-2. Red line represents one complete cycle of alternating current flow per second in a 120V circuit. Most a-c is at rate of 60 cycles per second. Note that voltage varies from 0 to 120V twice each cycle.

rectifiers, many small appliances are using "cordless" d-c power. Motors in some appliances operate on either a-c or d-c. These units are called universal motors.

6-7 ALTERNATING CURRENT

Alternating current (a-c) is electron flow along a conductor, first in one direction, then in the other. Because the current alternates its direction of flow, it is called alternating current.

A cycle diagram for an alternating current is shown in Fig. 6-2. In this illustration, the voltage starts at zero, reaches its maximum of 120V at a quarter of the cycle and is again zero at a half cycle, reaches it maximum in the opposite direction at three-quarters of the cycle, is again zero at the end.

This diagram pictures the ideal alternating current voltage cycle. Since most alternating current is 60 cycle, it means that this pattern is being repeated in the alternating current circuit at the rate of 60 times per second. Para. 28-67 indicates the root mean squared (rms) values of voltage and current when the current and voltage are not in phase.

Hertz (Hz) is now the accepted word for electrical cycles. Household electricity in most countries is 60 Hz.

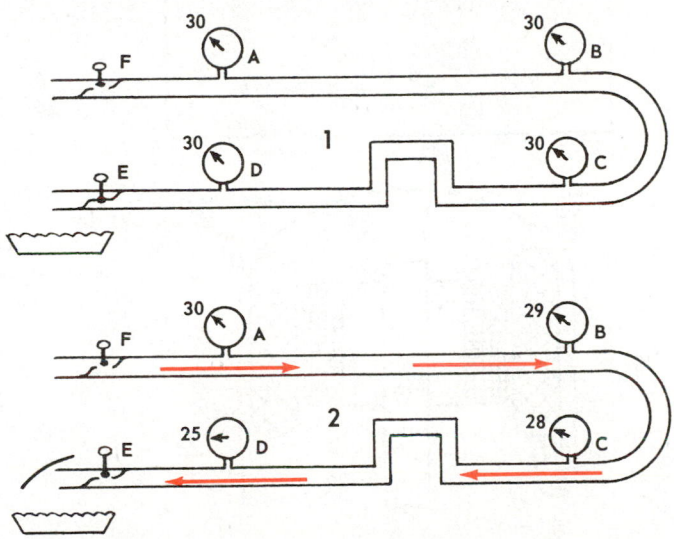

Fig. 6-3. Water flow compared to electrical flow. 1—Valve E closed, valve F open. No water flow, all gauges read the same. 2—Valve E open, valve F open, water flowing. All gauges show a pressure drop. Most of drop is between C and D because of bend in pipe.

6-8 CIRCUIT FUNDAMENTALS

In order to understand an electrical circuit's operation, compare it to a water system. Just like water in pipes, the wires must be large enough to carry the current (amount).

There is always a loss of pressure (volts) in a wire when electricity flows along it. This action is due to resistance in the wire. To relate this to water pressure, view 1 in Fig. 6-3 shows a water pipe with the inlet valve open and the outlet valve closed. Pressure produced by the water pumps is the same throughout the system when the water is not flowing. When the outlet valve is opened, water flows and the pressure drops.

Notice in view 2 in Fig. 6-3 that the pressure drops a little for increased distance from the pump. This pressure drop indicates energy lost pushing the water that distance. Pressure drop between gauges C and D is greatest because additional effort is needed to push water around the four bends.

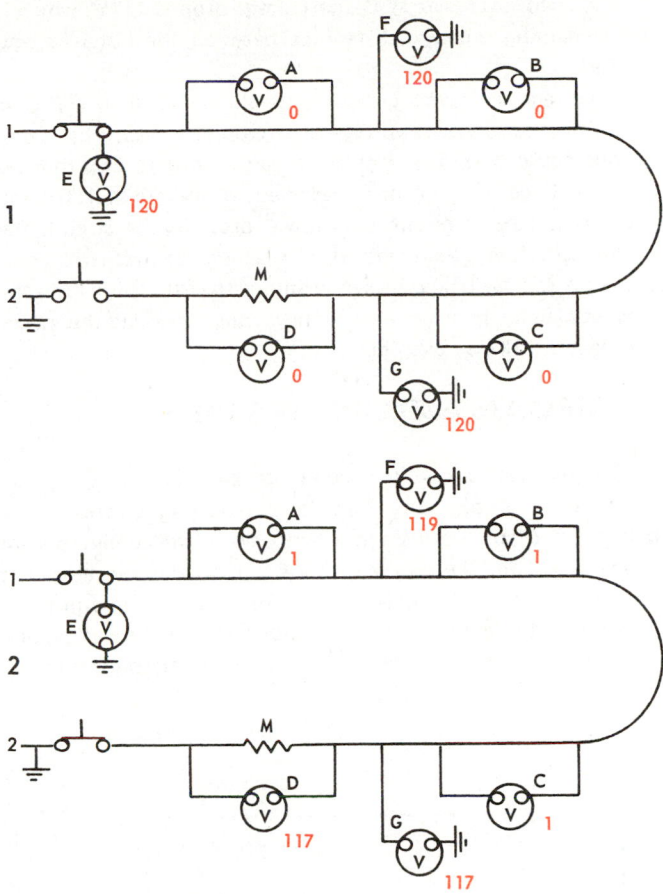

Fig. 6-4. Voltmeter tests of electricity flow show how potential (pressure), resistance and potential drop operate in a system. View 1—Electricity not flowing, with switch No. 1 closed, switch No. 2 open. Pressure to ground is 120V, but other pressure drops are zero. View 2—Electricity is flowing, with switch No. 1 and No. 2 closed. Voltage drop at A, B and C is 1V each and resistance at D has a 117V drop.

In Fig. 6-4 a similar electrical system setup is shown (drawing is in symbols). In both views, a wire extends from switch No. 1 to the resistance, light bulb or motor at M. In view 1, switch No. 1 is closed, but switch No. 2 is open. The potential (voltage) is 120V up to switch No. 2. However, no electricity is flowing (open circuit). Therefore, there is no voltage drop along the line. Voltmeters A, B, C and D show no voltage (no pressure difference) between their lead connections to the main line. However, the voltmeters at E, F and G show the full 120V, because there *is* an electric pressure difference between the line and the ground.

View 2 in Fig. 6-4 shows switch No. 2 closed and current flowing through the circuit (closed circuit). Note that now there is a small voltage drop at A. This drop takes place in any line in which current is flowing. If the line is large enough to carry the current flow, the voltage drop will be very small

(.001 to .0001V). If the line is too small, voltage drop will be greater. Usually, an undersize wire may be detected by the fact that it gets warmer than usual while current is flowing.

Note in Fig. 6-4 that the voltage difference between the line and the ground is less at points F and G in view 2 than in view 1 because of voltage drop in the line up to these measuring points. Also note that at D the voltage drop is 117V, which is the remaining pressure difference between the hot wire and ground.

If voltage drop was greater than shown at A, B and C at view 2 in Fig. 6-4, the voltage at D would be less. This is one serious cause of motor trouble because voltage drop reduces motor voltage. If a motor is designed to operate at 120V and there is a large amount of voltage drop in the circuit, the motor will lose speed and start slipping its magnetic fields (slow up too far below its synchronous speed). This causes the magnetic fields to grow large at the wrong time, and the motor will heat up. It may even burn up.

6-9 CIRCUITS AND CIRCUIT SYMBOLS

An electrical circuit is a complete path (or paths) for the electrons to follow. (See Para. 28-23.) It might consist of a battery, two conductors and a lamp. When the conductors are connected from the two lamp terminals to the two terminals on the battery and the switch is closed, the lamp will light. This shows that current is flowing out of the battery along one conductor, Fig. 6-5, and returns along the other conductor.

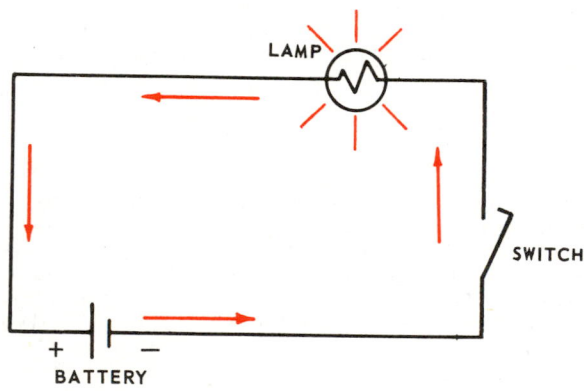

Fig. 6-5. Simple electrical circuit using a one-cell battery, three conductors, a switch and one lamp. Arrows show direction of electron (current travel) from − to +.

There are two terms that should be understood: open circuit and closed circuit. An open circuit means that an electrical switch or other device is open or disconnected and current cannot flow. A closed circuit is when an electrical switch or other device is closed or connected, permitting the electrons to leave from the original source and return to it (a closed path for electron flow). A closed circuit may also be called a continuous circuit.

Fig. 6-6 illustrates three types of circuit troubles:

No. 1 shows an open (disconnected) circuit, meaning it has an open switch or a broken conductor.

No. 2 shows a short circuit. The electrons have taken a short cut back to their source. For example, if a small conductor is placed across the terminals of a lamp, most of the electrons will flow along the path of least resistance (the new conductor) and the light will go out. Because of low resistance,

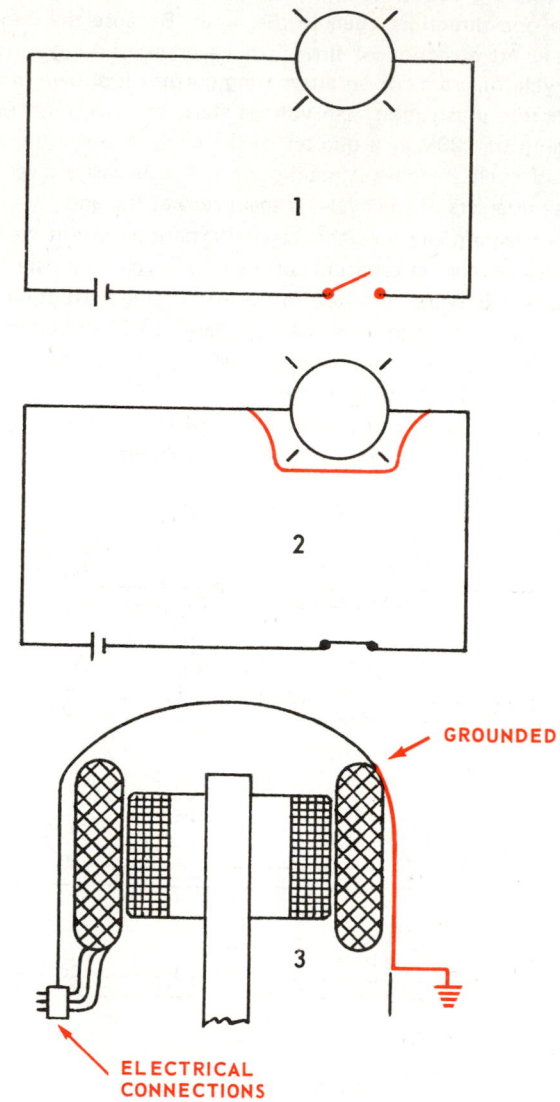

Fig. 6-6. Three common electrical circuit troubles. 1—Open circuit (open switch shown here or a broken conductor). 2—Shorted circuit. 3—Grounded circuit.

too many electrons may flow and the wires may overheat. Amperage flow will greatly increase (because of a decrease in resistance), and trouble usually occurs. Another more common example of a "short" is the touching of two adjacent conductors.

No. 3 shows a ground condition, which occurs when a conductor touches the metal structure of the device. For example, a bare field winding conductor may touch the frame of the motor.

Electrical wiring diagrams use symbols for many of the electrical parts. Fig. 6-7 illustrates standard electrical symbols.

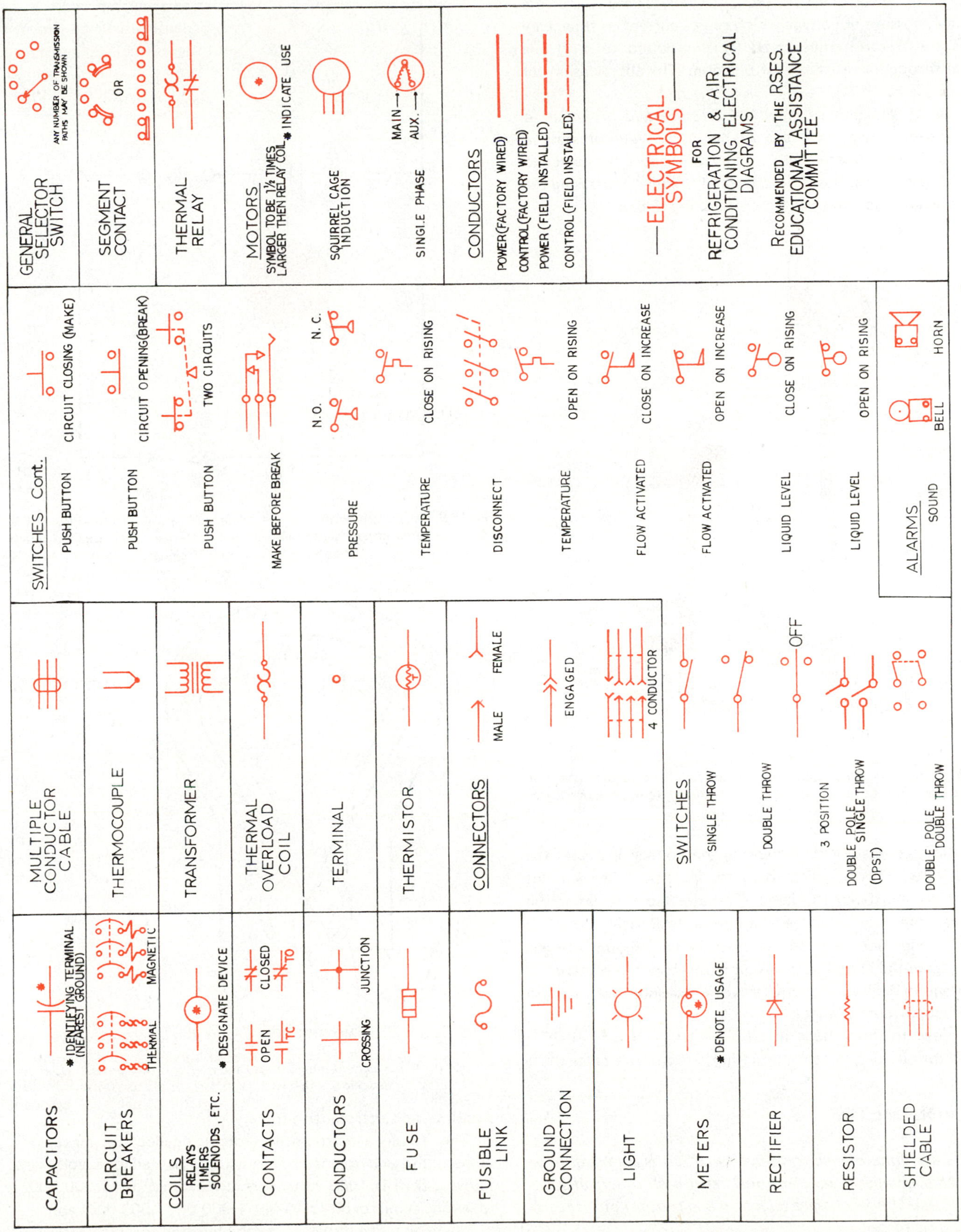

CAPACITORS

CIRCUIT BREAKERS

COILS — RELAYS, TIMERS, SOLENOIDS, ETC.

CONTACTS

CONDUCTORS

FUSE

FUSIBLE LINK

GROUND CONNECTION

LIGHT

METERS

RECTIFIER

RESISTOR

SHIELDED CABLE

MULTIPLE CONDUCTOR CABLE

THERMOCOUPLE

TRANSFORMER

THERMAL OVERLOAD COIL

TERMINAL

THERMISTOR

CONNECTORS — MALE, FEMALE, ENGAGED, 4 CONDUCTOR

SWITCHES — SINGLE THROW, DOUBLE THROW, 3 POSITION, DOUBLE POLE SINGLE THROW (DPST), DOUBLE POLE DOUBLE THROW

SWITCHES Cont. — PUSH BUTTON (CIRCUIT CLOSING (MAKE)), PUSH BUTTON (CIRCUIT OPENING (BREAK)), PUSH BUTTON (TWO CIRCUITS), MAKE BEFORE BREAK, PRESSURE (N.O. / N.C.), TEMPERATURE (CLOSE ON RISING), DISCONNECT, TEMPERATURE (OPEN ON RISING), FLOW ACTIVATED (CLOSE ON INCREASE), FLOW ACTIVATED (OPEN ON INCREASE), LIQUID LEVEL (CLOSE ON RISING), LIQUID LEVEL (OPEN ON RISING)

GENERAL SELECTOR SWITCH — ANY NUMBER OF TRANSMISSION PATHS MAY BE SHOWN

SEGMENT CONTACT

THERMAL RELAY

MOTORS — SYMBOL TO BE 1¼ TIMES LARGER THEN RELAY COIL. * INDICATE USE; SQUIRREL CAGE INDUCTION; SINGLE PHASE (MAIN / AUX.)

CONDUCTORS — POWER (FACTORY WIRED), CONTROL (FACTORY WIRED), POWER (FIELD INSTALLED), CONTROL (FIELD INSTALLED)

ALARMS — SOUND, BELL, HORN

ELECTRICAL SYMBOLS

FOR REFRIGERATION & AIR CONDITIONING ELECTRICAL DIAGRAMS

Recommended by the RSES EDUCATIONAL ASSISTANCE COMMITTEE

Fig. 6-7. Electrical symbols commonly used in wiring diagrams.

6-10 ELECTROMOTIVE FORCE (EMF) — VOLT

The term "electromotive force," or emf, is used to indicate electrical pressure or voltage which causes current to flow. One volt is the electromotive force (emf) required to send one ampere through a resistance of one ohm. The abbreviation for volt (emf) is E.

A volt is the unit of electrical pressure and is similar to pressure used to make gases or liquids flow. Water pressure in psi (pounds per square inch) or kg/cm^2 causes the water to flow through pipes, hoses, nozzles and the like. Increasing the water pressure increases the water flow, Fig. 6-8.

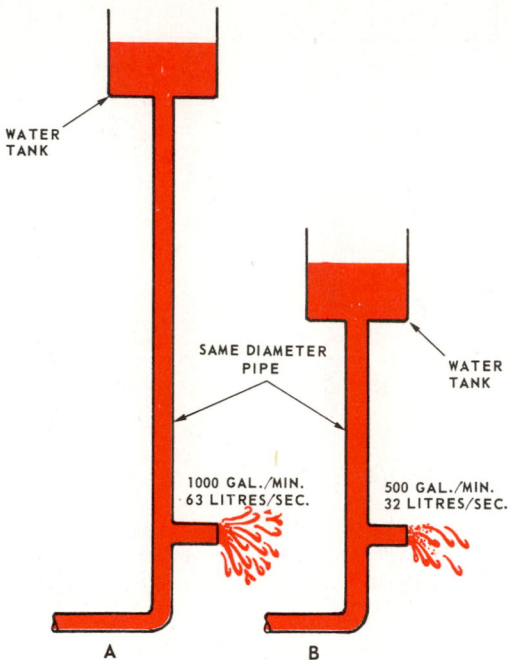

Fig. 6-8. Rate of flow depends upon pressure. A—Higher tank of water provides greater pressure, greater flow. B—Lower tank provides lower pressure, lesser flow.

In the electrical field, increasing the voltage increases the current flow. This is indicated by the increase in the amount of light coming from the lamp. The effect of increasing the electrical pressure — voltage (emf) — on electrical circuit is shown in Fig. 6-9. Note the increase in the amount of light coming from the lamp in the circuit with a two-cell battery.

Voltage in a direct current circuit is measured with the use of a direct current voltmeter.

To measure the voltage in an alternating current, such as home lighting current, an alternating current voltmeter must be used.

6-11 VOLTMETER

Volts are units of electromotive force. Volts are pressure or force. Like charges (two electrons) repel each other; this is a force (volts). Unlike charges (electron and proton) attract each other; this is force.

Instruments called voltmeters have been developed to

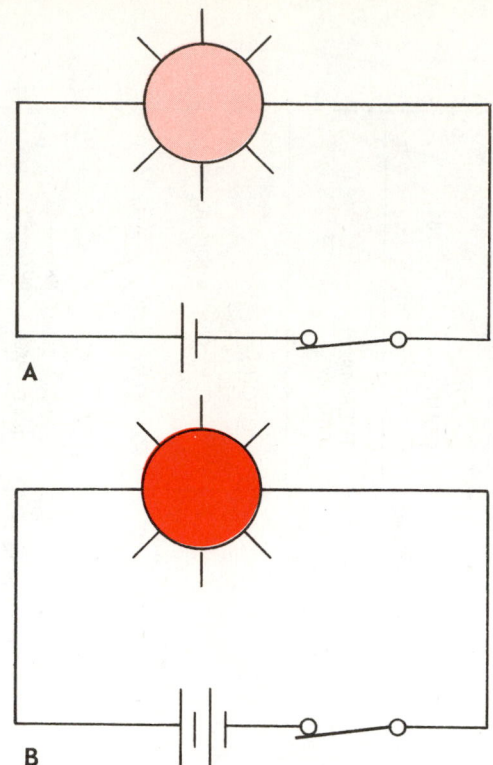

Fig. 6-9. A—Circuit with one-cell battery (1 1/2V). Bulb lights but is dim. B—Same circuit with two one-cell batteries. (3V). Light is much brighter (more current flowing).

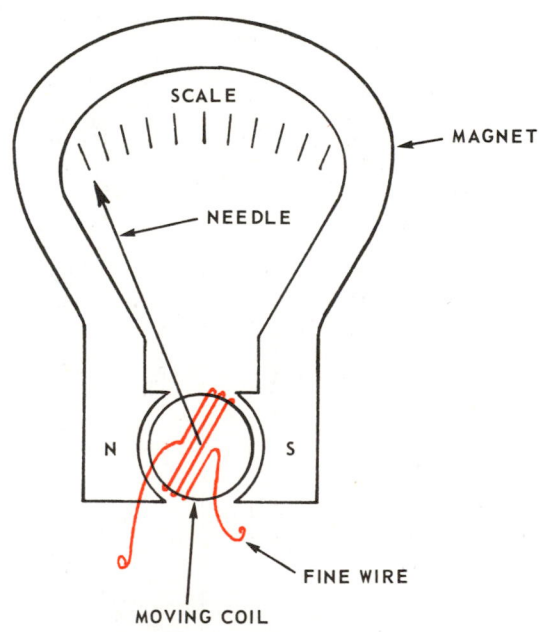

Fig. 6-10. A permanent magnet and moving coil instrument. Used for testing d-c circuits (a-c with rectifier), or a-c to 10 KHz.

measure the electromotive force.

The force (electromotive force) needed to move one ampere through a one-ohm resistance is called a volt. One kilovolt (kV) is 1000 volts. A millivolt (mV) is 1/1000 (.001) of a volt. A microvolt (μV) is 1/1,000,000 (.000 001) volt.

There are five types of voltmeters for measuring electromotive force. The first four use forces made by current flow

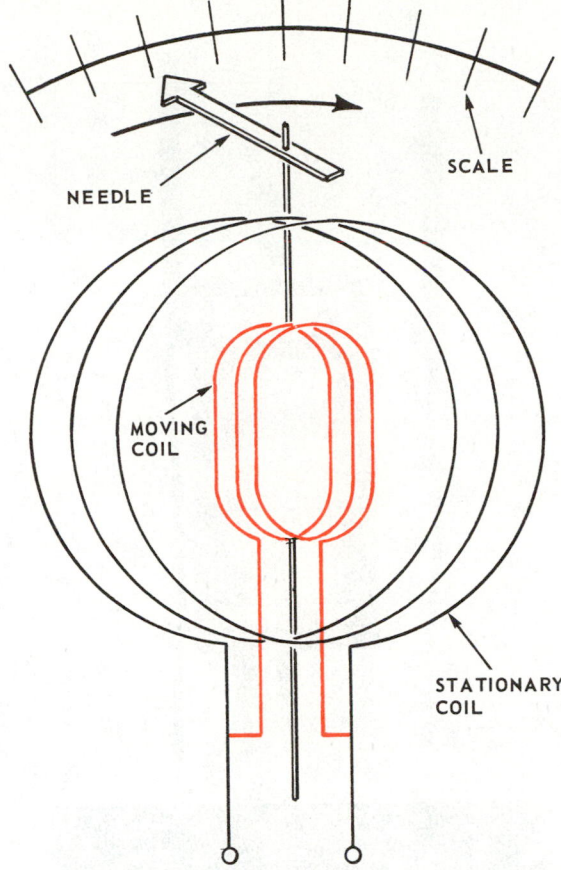

Fig. 6-11. Electrodynamic instrument. Two coils create magnetism and inner coil movement is related to electron flow. Used with a-c or d-c to 200 Hz. Single coil type is used for voltage, current or power. Double coil type is used for power, motor phase angle, frequency and capacity.

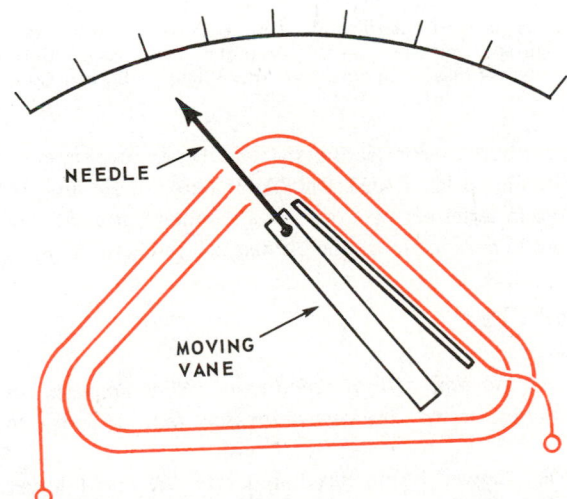

Fig. 6-12. This moving iron or moving vane instrument is used for a-c or d-c to 125 Hz.

and magnetism. The fifth uses electrostatic forces:

1. Permanent magnet-moving coil (D'Arsonval), Fig. 6-10.
2. Electrodynamic (dynamotor movement), Fig. 6-11.
3. Moving vane (iron), Fig. 6-12.
4. Moving magnet (polarized iron or iron vane), Fig. 6-13.

5. The one using the electrostatic type is shown in Fig. 6-14.

Voltmeters are always connected in parallel with the circuit. The method of connecting them is shown in Fig. 6-15.

Digital Voltmeters:

Many new meters use electronic circuitry instead of electromagnetic effects. They have several advantages over the electromagnetic effect devices:

1. No moving mechanical parts.
2. Easy readability.
3. Smaller size.

These devices use solid state semiconductors which withstand shock and vibration better than delicate meter movements. They require no lubrication or compensation for the position of the instrument. The display is a number (digit) instead of a pointer, which allows rapid reading. They are also smaller in size and more portable. Electrical power require-

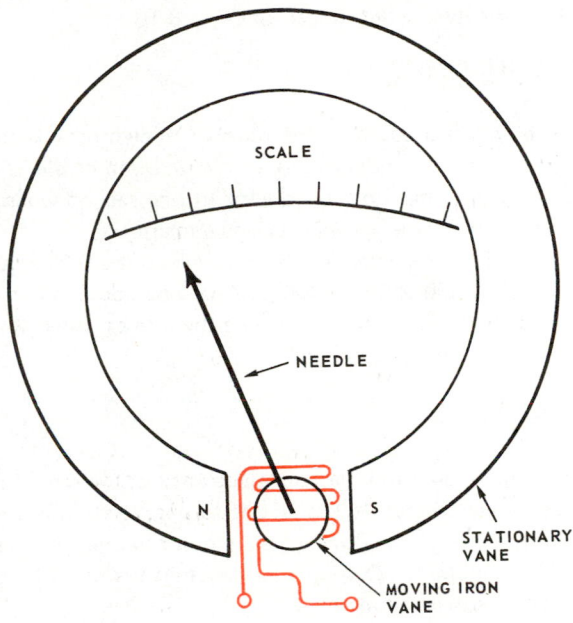

Fig. 6-13. A moving magnet type voltmeter is designed for use with d-c.

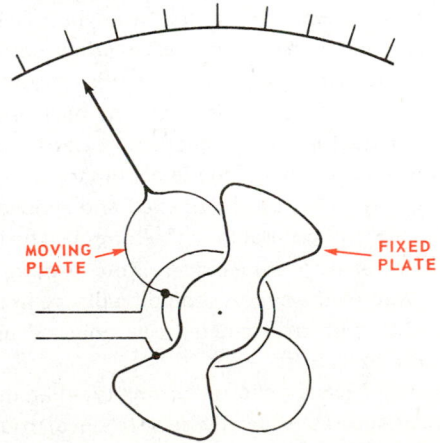

Fig. 6-14. Electrostatic type meter. Used for testing voltage in a-c or d-c circuits over 10V.

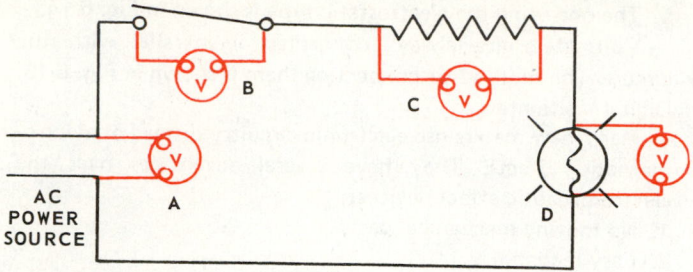

Fig. 6-15. Correct connections for testing with voltmeter. Meter will show: A—115 to 120V if power is on. B—Whether switch is open or closed (no reading if closed, 120V if open). C—Voltage drop across resistor. D—Voltage drop across lamp (no reading if shorted, 120V if broken or open). Voltages of B + C + D should equal A when switch is closed.

ments, when needed, are much smaller. Batteries used in portable instruments last longer. See Fig. 6-16.

6-12 COULOMBS

A coulomb is a count of the number of electrons passing a given point on a conductor. It is the *amount* of electricity passing a given point in a conductor in one second when the rate of current flow is constant at one ampere.

The number of electrons in a coulomb equals 6.24×10^{18}, or 6,240,000,000,000,000,000. A one-coulomb flow per second equals one ampere. This is like the rate of water flow in gallons per minute (gpm).

6-13 AMPERE

The ampere does not measure electrons but the rate of flow of current electricity. It has a one-to-one relationship with coulombs. Ten amperes flowing past a point in one second also equals 10 coulombs. "Current" usually means rate of flow. It is measured with an ammeter.

6-14 AMMETER

An ammeter is an instrument that measures the rate of current flow in amperes. There are two types. One measures direct current flow; the other, alternating current. Most refrigeration and air conditioning work requires an a-c ammeter. Its operation depends upon the magnetic effect of current in a conductor. The magnetism causes a current flow in the tongs (clamp) surrounding the conductor.

The tongs, Fig. 6-17, can be opened and slipped around a conductor. Electricity generated in the tongs by the magnetism moves the indicator to give the ampere flow reading.

Only one wire of the circuit should be placed in the jaws of the meter. This type of ammeter can only be used on an alternating current circuit.

If the reading from a "clip on" (tong-type) ammeter is too low to be read accurately, wrap the wire once around one of the jaws to double the reading (wire goes between jaws twice). Divide this reading by two to get the correct reading. Some of these meters have "multipliers" available as accessories.

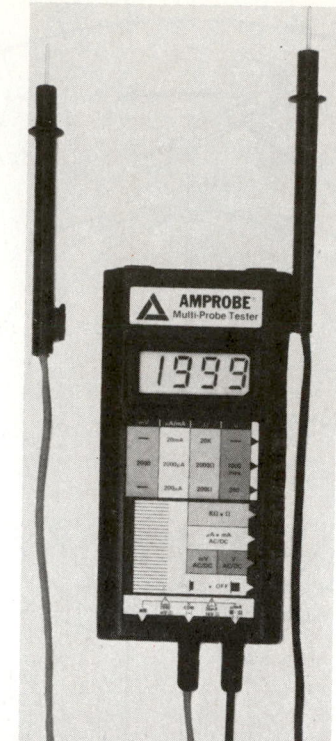

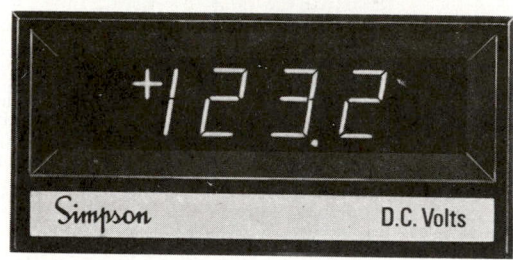

Fig. 6-16. Digital test instruments. Top. Hand-held meter measures voltage, current and resistance. (Amprobe Instruments) Bottom. Voltmeter for measuring direct current. (Simpson Electric Co.)

The method of connecting the two types of ammeters is shown in Fig. 6-18. CAUTION: Ammeters should always be connected in series with a circuit. If an ammeter is accidentally connected in parallel, it will be burned up. See Para. 6-28.

6-15 WATTS

Power is the time rate of doing work. When amperes flow (coulombs per second) at a certain pressure (emf or volts), this is power.

Electrical power is measured in watts (W) and kilowatts (kW). See Para. 6-64. When computing electrical horsepower (hp), 746W or 0.746kW equal one horsepower (hp).

In direct current circuits, the wattage may be obtained by multiplying the current in amperes by the emf (volts), $I \times E = W$.

In alternating current circuits, the wattage may be less than the amount obtained by multiplying the amperes and volts together. This is because the current in an a-c circuit may not be exactly in phase with the voltage.

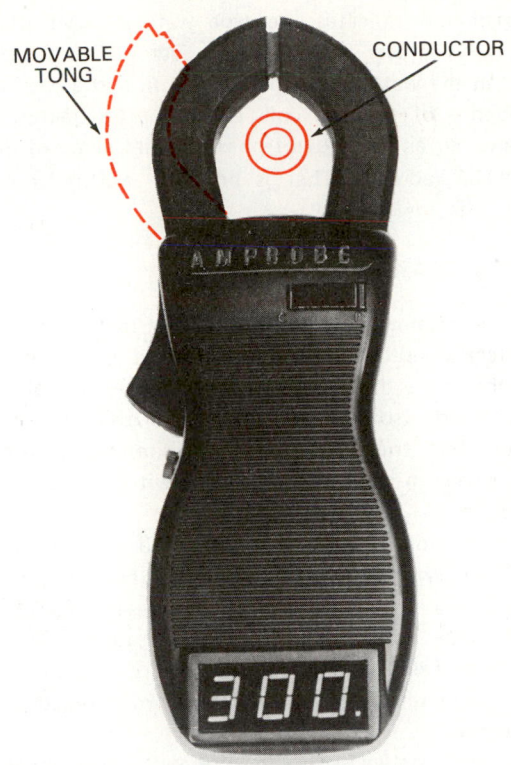

Fig. 6-17. Digital ammeter-voltmeter-ohmmeter, measures current with current carrying conductor inside movable tongs. Red lines show tongs open to receive conductor.   (Amprobe Instrument)

6-16 POWER FACTOR

Current flow usually lags behind the emf, depending on type of circuit. That is, when the emf is at 50 percent of its maximum, the current flow may be 35 or 40 percent of its maximum. Therefore, the wattage or power of an a-c circuit must be obtained by multiplying the voltage times the current times the power factor, (E x I x P.F. = watts).

View A in Fig. 6-19 shows an alternating current cycle in which the voltage and the current are in phase. The current is at its maximum at the same time that the voltage is at its maximum. The voltmeter reads 120V and an ammeter

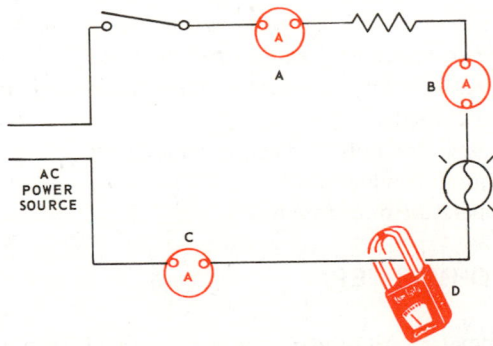

Fig. 6-18. Ammeter connections. Ammeters A, B, C and D should all read the same: Zero when switch is open, 5A when switch is closed. D is special induction type ammeter described in Fig. 6-17.

connected into the circuit reads 10A. The product of the voltage times the current equals 1200W. A wattmeter connected to this circuit will also read 1200W. This indicates a power factor of 100 percent.

View B in Fig. 6-19 shows an alternating current cycle out of phase. The current is lagging behind the voltage one-eighth of a cycle. A voltmeter connected to the circuit reads 120V and an ammeter connected to the circuit reads 10A. The product of the voltmeter reading multiplied by the ammeter reading will be 1200. However, a wattmeter applied to this circuit will read 1000W. The power factor is the ratio of the wattmeter reading over the calculated wattage, which is the current multiplied by the voltage.

$$\text{Power Factor} = PF = \frac{\text{wattmeter reading}}{\text{ammeter reading x voltmeter reading}}$$

Example: wattmeter reading   1000
voltmeter reading   120
ammeter reading   10

$$PF = \frac{1000}{120 \times 10} = \frac{1000}{1200} = .833$$

To change to percentage: .833 x 100 = 83.3 percent

To calculate the wattage, multiply the voltmeter reading by the ammeter reading. The answer must be multiplied by .83 (the power factor), to give the correct wattage:

120 x 10 x .83 = 1000.

Various things in the electrical circuit affect the power factor. If the electrical load is entirely a resistance load, such as electric heating and the like, this is called a noninductive load. It does not seriously affect the power factor.

Most electric motors, transformer welders and the like, do affect the power factor. These are called inductive loads. It

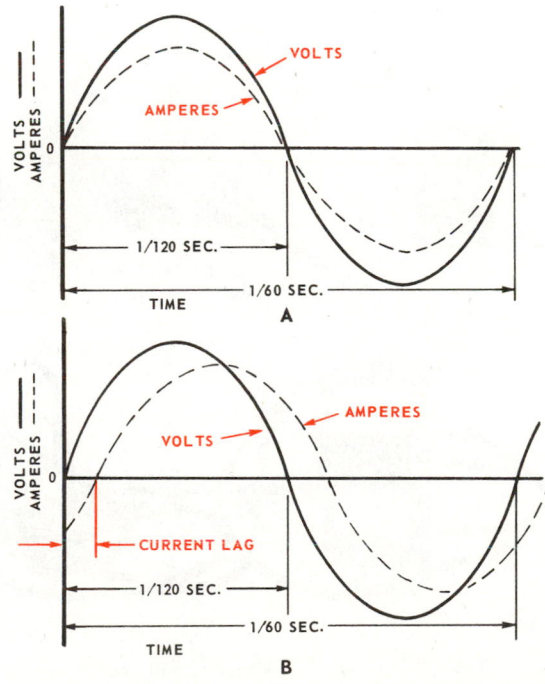

Fig. 6-19. Alternating current cycles. A—Current and voltage in phase. B—Current and voltage not in phase.

takes time for current to build up in any coil. This is due to the counter emf generated in the coil. This lag affects the power factor. See Para. 6-49 which discusses the effects of counter electromotive force.

6-17 WATTMETER

A wattmeter is an instrument used to measure the wattage consumed by an electric motor or an electrical device.

It is connected in series with the circuit being measured. (See Para. 6-28.) Fig. 6-20 illustrates a portable wattmeter and shows how it is connected to measure wattage drawn by an electric drill. This is a combination instrument and may be

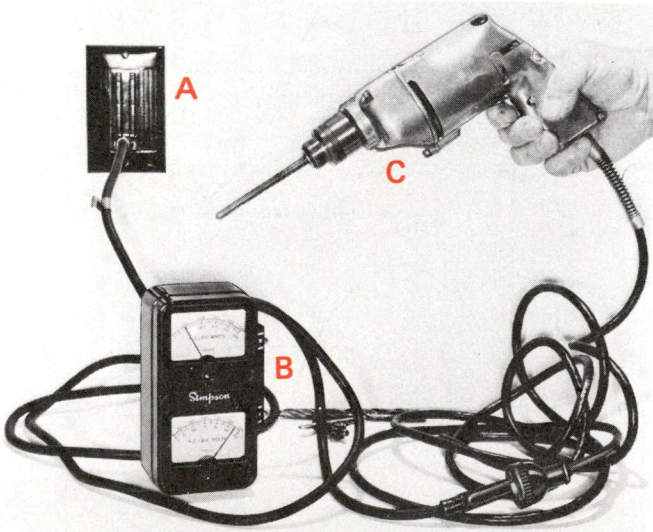

Fig. 6-20. Top. Combination voltmeter and wattmeter. A—Power source connection. B—Electrical load connection. Bottom. Volt-wattmeter being used to test an electric portable drill. A—Power source. B—Volt-wattmeter. C—Electric load. (Simpson Electric Co.)

used for measuring volts or watts.

A wattmeter indicates the true wattage in a circuit. It automatically adjusts for the power factor. This is done by two coils in the wattmeter — a voltage coil and a current coil. The influence of the two coils drives the wattmeter. If the voltage and current are out of phase, the influence of the two coils will be reduced. That is how the wattmeter reading adjusts for the power factor.

6-18 RESISTANCE, IMPEDANCE

Most electrical conductors are made of metal. No material is a perfect conductor of electricity, but some metals are better conductors than others. Silver, copper and aluminum are very good. Iron, steel and carbon will also conduct electricity, but their resistance is quite high. Carbon is sometimes used in electrical circuits, but it is not a very good conductor.

Extremely poor conductors are called resistors or resistances. They have no free electrons or, at best, very few free electrons in the atom. It is difficult for these free electrons to travel through and around the other atoms. The word impedance is often used in place of the word resistance when a part of an a-c circuit or a complete a-c circuit resists the flow of free electrons.

The total resistance is the sum of the resistance of each part of the circuit and the capacitance of an a-c circuit.

The harder it is for the free electrons to move, the greater the heat generated in the conductor. This heating action is shown best in iron, steel and steel alloys used for electric heating purposes. See Chapter 28 for a more complete discussion of the resistance values of conductors, semiconductors and nonconductors.

All materials have some resistance to the flow of electricity. The resistance of electrical conductors usually increases with an increase in temperature or an increase in the length of the conductor. Also, the resistance increases as the diameter (thickness) of the conductor decreases. In the case of some semiconductors, increasing the temperature increases their ability to conduct electricity.

6-19 OHMS

Electrical resistance is measured in ohms. An ohm is the amount of resistance in an electrical circuit which allows an emf of 1 volt to cause 1 ampere to flow through the circuit.

The resistance in a conductor depends upon four things:
1. Material used.
2. Diameter or size (thickness) of conductor.
3. Length of conductor.
4. Temperature of conductor.

6-20 OHMMETER

Ohmmeters are used to check circuits for resistance, open circuits and grounds. **Power must be off when the ohmmeter is being used.** The instrument may be ruined otherwise. The ohmmeter, Fig. 6-21, has a small battery or a magneto as a

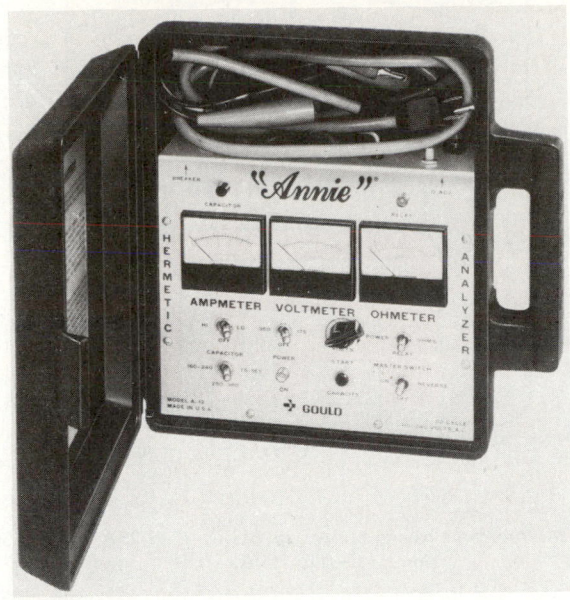

Fig. 6-21. Combination ammeter-voltmeter-ohmmeter. Note test leads and power cord. (Gould Inc., Fluid Components Div.)

power source. Fig. 6-22 shows the correct method of connecting an ohmmeter into a circuit to measure the resistance.

6-21 INSTRUMENT CONNECTING AND HANDLING

Basic electrical instruments may be used to measure either volts, amperes or ohms. The difference is in the external wiring of the instrument. The scale and use of instruments vary depending upon the wiring inside of the instrument. Any

instrument needs only a very few electrons to activate the pointer mechanism.

The voltmeter has a high internal resistance. The ammeter is designed to bypass (shunt around) most of the current outside of the instrument. The ohmmeter allows only a few electrons in the circuit.

The mechanisms for each type of meter vary widely. These instruments must be delicate enough to read accurately. Therefore, they should not be dropped, used above their maximum reading, or carelessly handled or stored.

The voltmeter is connected in parallel with the circuit. A high value resistor in the instrument coil circuit will allow a maximum of 5/1,000,000 of an ampere, or 5 microamperes, to flow into the moving coil through the springs.

Be careful! If a 120V voltmeter is used to measure a 240V circuit, the meter will be ruined. (Double the milliamperes will flow through the meter.)

A low resistance shunt is placed across the terminals of an ammeter. The instrument is placed in series with the circuit when current is being measured.

An ohmmeter is connected in parallel with the circuit while a known voltage is applied (usually a low voltage battery).

6-22 SHUNT

As indicated in Para. 6-21, internal construction of many electrical meters is very much alike. The method of connecting them into the circuit and the accessory devices used enables the same basic instrument to be used as either an ammeter, voltmeter or ohmmeter. See Fig. 6-23.

Voltmeters are connected across or in parallel with the circuit in which the voltage is being measured. Ammeters are connected in series with the circuit. This would indicate that

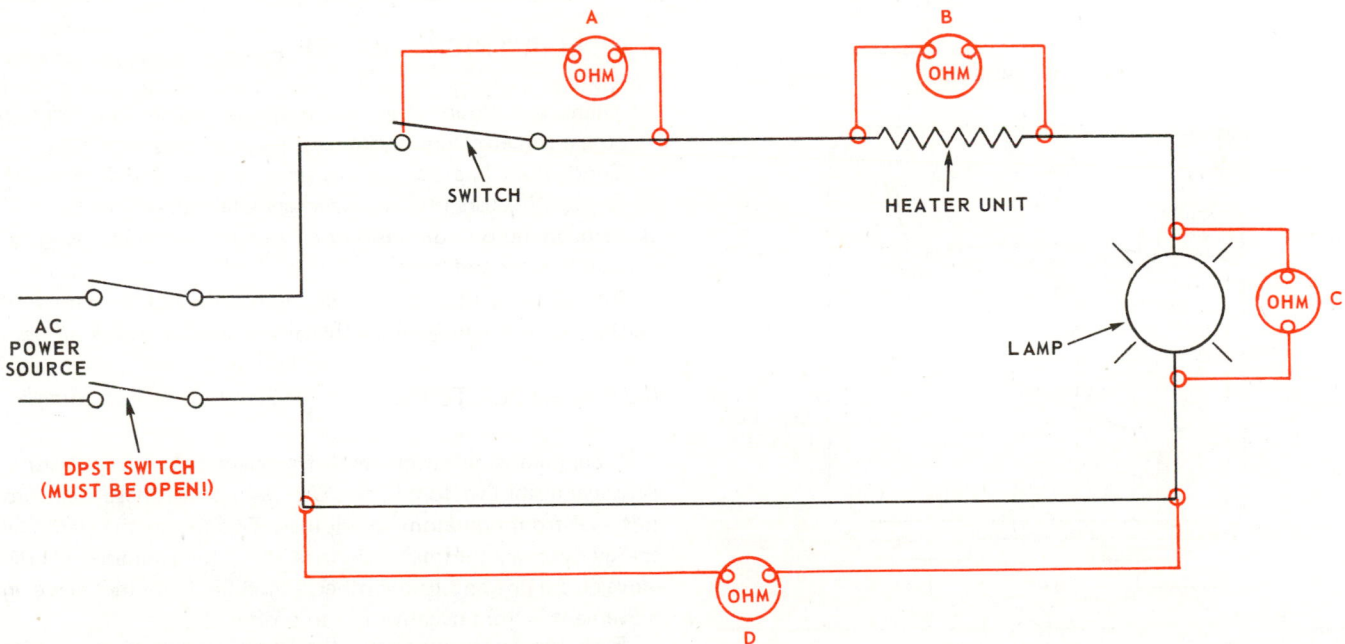

Fig. 6-22. Ohmmeter connections. There must be no power in circuits. A—Testing resistance across a switch in open, then closed position. With switch closed, resistance indicated on meter indicates a poor connection somewhere in switch. B—Testing resistance (ohms) across heater unit. C—Testing resistance across a lamp. D—Measuring continuity of a conductor. If meter reads zero, conductor is not broken and connections are good. If meter reads infinity (maximum resistance), conductor or a connection is broken.

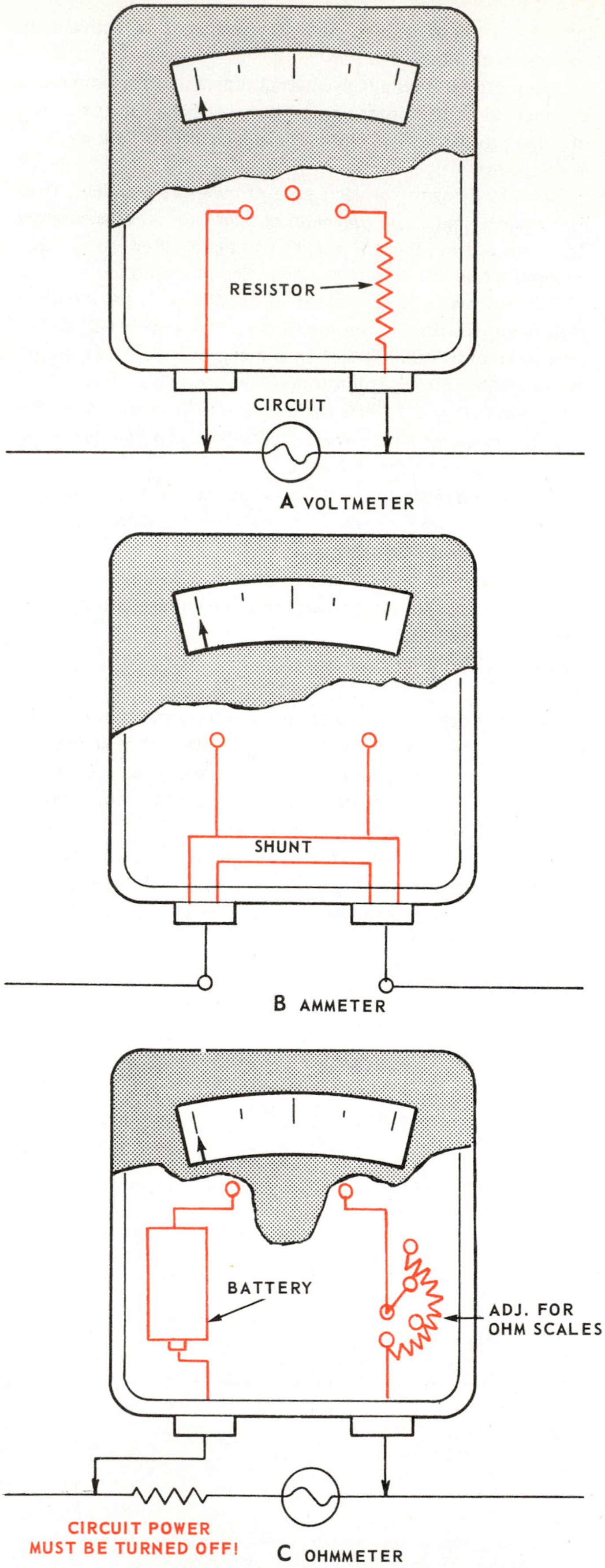

A VOLTMETER

RESISTOR

CIRCUIT

B AMMETER

SHUNT

C OHMMETER

BATTERY

ADJ. FOR
OHM SCALES

**CIRCUIT POWER
MUST BE TURNED OFF!**

Fig. 6-23. How the basic meter movement is usually wired inside the voltmeter, ammeter and ohmmeter.

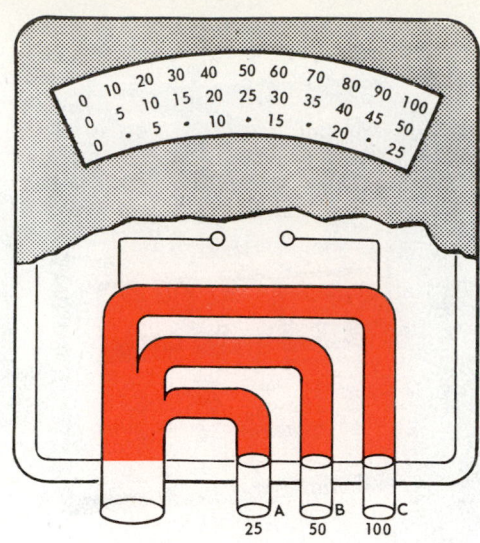

Fig. 6-24. Ammeter having built-in shunts. A—0 to 25A. B—0 to 50A. C—0 to 100A.

all of the current being measured by the ammeter must go into the meter. This is not true. A shunt is placed in parallel with the instrument, and the greater portion of the current goes through the shunt rather than through the meter.

Several different shunts can be used with the same meter. Shunts are calibrated for the current range in which each is to be used. Sometimes shunts are built into the meter. In such cases, the electrical connection to the meter should be made to terminals within the range of the meter.

If in doubt about the voltage or amperage of a circuit, connect first to the upper range of the instrument. If the reading is within the lower range, a more accurate reading may be made by making a connection which uses almost the full scale of the instrument. See Fig. 6-24.

6-23 ELECTRICAL MATERIALS

There are three physical materials which are used in electrical and electronic systems:
1. Conductors (metals, such as silver, copper and aluminum).
2. Semiconductors (metal oxides or metal compounds).
3. Nonconductors or insulators (nonmetals such as glass, wood, paper and mica).

Any of these three may exist in any of the three forms of matter, but most are solids rather than liquids or gases.

6-24 CONDUCTORS

A conductor has atoms with free electrons in its structure. Any electromotive force (pressure) will cause these electrons to travel from one atom to another. This moves the electrical energy through the material. In a wire, for example, energy moves from one end to the other, Fig. 6-25. Note that electron movement is from negative (—) to positive (+).

Each type of atom allows the free electrons to move with varying ease. In this respect, gold and silver atoms are excellent; copper, mercury and aluminum atoms are very good, also. Most good conductors are metals, yet the movement of free electrons in iron is somewhat more difficult.

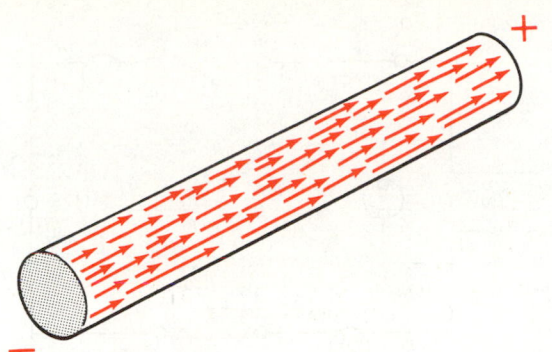

Fig. 6-25. A wire with free electrons traveling from negative (−) to positive (+).

Conductivity of metal is usually expressed in ohms per circular mil foot at standard temperature. Wire with a circular mil cross-sectional area has a diameter of 0.001 in. Standard temperature for measuring conductivity is 68 F. (20 C.).

Wires (solid conductors) are popular for carrying electricity from one electrical device to another. Most electricians call wires "electrical leads" (pronounced like lead the way).

6-25 SEMICONDUCTORS

Conductors such as metals conduct heat and electricity readily. Insulators conduct heat and electricity very poorly. In between these materials are semiconductors. These materials are ordinary insulators but under certain conditions can be made to conduct electricity easily. These materials are the basis of the present electronics industry. The term "solid state electronics" refers to electronic devices which are made up of semiconductor elements.

Common devices which are semiconductors are transistors, diodes, and photocells. Many modern motor controls consist of silicon controlled rectifiers (SCR's), which are semi-conductive switching devices.

The conductivity of a semiconductor can be controlled by an electrical signal, light intensity, pressure, temperature, and other signaling devices. Semiconductors can, therefore, serve as relays and switches.

The a-c current produced by an alternator is converted to d-c current by semiconductor diodes. These are the sensitive devices which can be damaged if one starts a car with the wrong jumper connection between batteries.

Photocells used on automatic door openers are semiconductor switching devices activated by light.

In troubleshooting of motors which fail to operate, the most likely cause of motor control failure is a burned-out semiconductor diode or SCR. See Para. 6-53 for some semiconductor applications.

6-26 NONCONDUCTORS (INSULATORS)

Nonconductors resist electron flow. The atoms have virtually no free electrons. A perfect vacuum is also a nonconductor.

Nonconductors are as useful in electrical systems as conductors or semiconductors. There are many parts of the system in which electrical flow must be stopped (where insulation is needed).

Some common nonconductors are quartz, ceramics, mica, glass and organic substances (rubber, wood, paper and plastics).

The resistance values for nonconductors (insulators) range between 10^9 to 10^{18} ohms.

6-27 OHM'S LAW

The relationship between the volt, the ampere and the ohm is known as Ohm's Law.

If a conductor 1 ft. long will allow 1A to flow with an emf of 1V, this same conductor will allow 2A to flow if the emf is 2V. Also, if the conductor is 2 ft. long and the emf is 1V, only 1/2A will flow. Therefore, it is evident that emf is the product of the current intensity (amperes) multiplied by the resistance (ohms); or:

where E = Electromotive force (emf) in volts
I = Intensity of current in amperes
R = Resistance in ohms.

English System:

$$\frac{Emf}{(Volts)} = \frac{Intensity}{(Amperes)} \times \frac{Resistance}{(Ohms)}$$

$$E = IR$$

and therefore $I = \dfrac{E}{R}$

or $R = \dfrac{E}{I}$

Fig. 6-26 shows the relationship of I, E and R.

SI Metric System:

Voltage = Amperes x Ohms

V = A x Ω (This particular symbol is the Greek letter omega. It is the symbol used for ohms.)

From this basic law, one may learn that, if the resistance

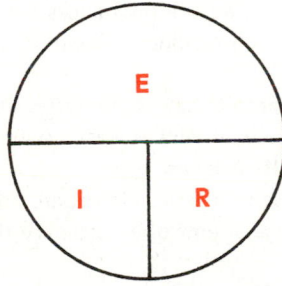

If solving for E, cover E with finger and the circle shows

E = I x R   E =

If solving for R, cover R with finger and the circle shows

$R = \dfrac{E}{I}$ R =

If solving for I, cover I with finger and the circle shows

$I = \dfrac{E}{R}$ I =

Fig. 6-26. Ohm's Law equations shown in graphic form.

stays constant, the current can only be increased by increasing the emf. Also, if the resistance in a circuit becomes low, the amperage will become high.

For example, if a 240W lamp draws 2A at 120V, its resistance is as follows:

$$R = \frac{E}{I}$$

$$R = \frac{120V}{2A}$$

$$R = 60 \text{ ohm } (\Omega)$$

The formula $E = I \times R$ also means that if one doubles the emf in a circuit (the resistance remaining the same), the current flow will double and may cause trouble. An example would be to connect a 120V motor into a 240V power outlet. If the fuse does not "blow," the motor windings will become too hot carrying this excessive current. The insulation on these wires may be destroyed and the motor may be ruined.

If the resistance is increased (putting in a smaller diameter conductor or having a dirty or loose connection) and the emf stays the same, the current will decrease. This causes a loss of power and the wire will heat or the poor connection will become very warm (and may even cause a fire).

6-28 CIRCUITS, SERIES

A series circuit is an arrangement of electrical parts which requires all the electrons to flow through each and every part. See Fig. 6-27.

In Fig. 6-27, each switch, each resistance, each light bulb and each coil has all of the electrons flowing through (or along) it. In a series circuit, if one of the objects — switch, light bulb or whatever — is not allowing the electrons to flow through it, the circuit will not operate.

In a series circuit, all the resistances are added together to determine the total resistance. There is only one intensity (current) value.

Practically all circuits have some series sub-circuits in them. For example, there is usually at least a switch and a resistance, a coil or a light bulb in series.

Applied voltages in a series circuit are added together. For example, the voltage or emf of an ordinary flashlight dry cell is

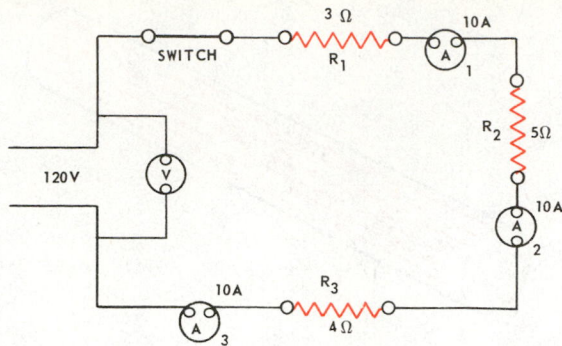

Fig. 6-27. A series circuit with three resistances. R_1 = 3 ohms, R_2 = 5 ohms and R_3 = 4 ohms. Total resistance equals 12 ohms. All three ammeters will read the same, 10A.

1.5V. Three flashlight cells placed end to end (series) produces 4.5V ($1.5 \times 3 = 4.5$).

6-29 CIRCUITS, PARALLEL

A circuit which allows the electrons to flow along either one of two or more conductors or electrical paths at the same time is called a parallel circuit, Fig. 6-28. The sum of the two electron flows equals the total flowing in the main leads ($A_T = A_L + A_M$). The electron flow is based on the resistance of the energy absorbers. If the lamp has one-fourth of the resistance of the motor, three-fourths of the electrons will flow through the lamp and one-fourth through the motor.

The voltage (E) is the same across each of the parallel circuits.

6-30 CIRCUIT, SERIES — PARALLEL

A combination of series and parallel circuits is sometimes used. Some automatic defrost systems do so. Two or more heating elements are connected in parallel. These heating elements in turn are connected into the system when the compressor motor is stopped. A timed switch controls the heating elements. This switch is in series with the heating elements. See Fig. 6-29.

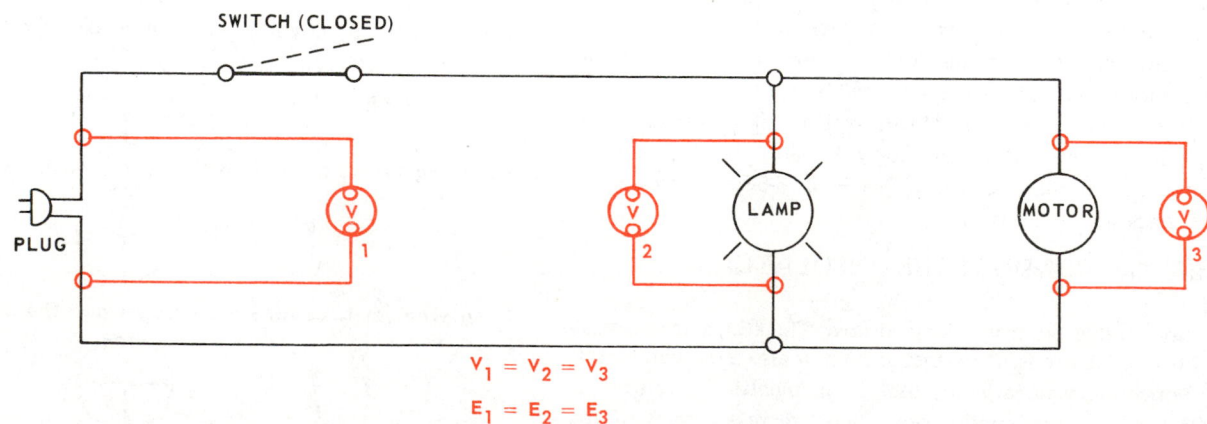

$$V_1 = V_2 = V_3$$
$$E_1 = E_2 = E_3$$

Fig. 6-28. Parallel circuit with current dividing, some going through lamp and some through motor.

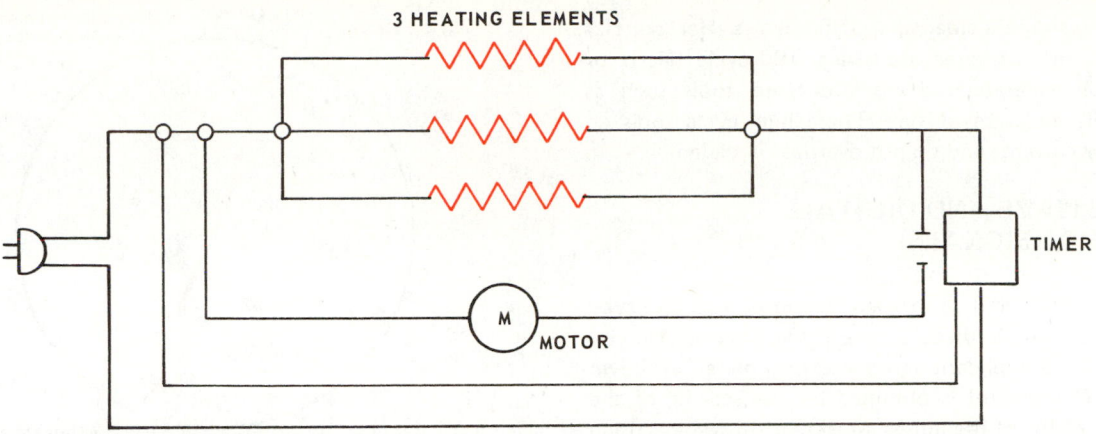

Fig. 6-29. Series parallel circuit with a timer divides three ways, some current going through each of three resistance (heating) elements.

6-31 VOLTAGE DROP (IR)

The sum of the voltage drops in a circuit is always equal to the voltage applied. The voltage drop across any part of a circuit is equal to the current (amperes) multiplied by the resistance (ohms) across that part of the circuit.

In Fig. 6-30, for example, the voltmeter indicates an applied voltage of 120V in a typical refrigeration circuit. The alternating current ammeter indicates a current draw of 5A.

Equivalent resistance of circuit wiring is 0.5 Ω .	5 x .5 =	2.5
Equivalent resistance of thermostat is 0.5 Ω .	5 x .5 =	2.5
Equivalent resistance of starting relay is 1.0 Ω .	5 x 1 =	5.0
Equivalent resistance of motor compressor is 22.0 Ω .	5 x 22 =	110.0
Therefore, total voltage drop is		120.0V

There is always some electrical resistance across any electrical control, relay or circuit wiring. This resistance may be so slight that it will require a very sensitive instrument to measure it.

6-32 POWER LOSS (I²R)

The power loss in a circuit due to resistance is equal to the square of the current multiplied by the resistance. This is usually expressed as I^2R loss = I^2 x R = Power Loss.

Referring to Fig. 6-30, the power applied is at 120V. The total current in the circuit is 5A; the total resistance in the circuit is 24 Ω :

5^2 = 25

25 x 24 = 600W power loss.

Power applied is equal to voltage times amperage:

5A x 120V = 600W.

Power loss results in the generation of heat. One watt converted into heat equals:

3.415 Btu = .861 calories. (1 Btu = .252 calories.)

6-33 ALTERNATING CURRENT CYCLES (HERTZ or Hz)

If electricity flows for 1/120 second in one direction and then for 1/120 second in the other direction, this is called one cycle. Most alternating current is generated in 60 cycles (Hertz). This means that in one second, the electricity flows 60 times one way and 60 times the other, or for 1/120 of a second it flows in one direction and then for 1/120 of a second in the other direction. The current, therefore, makes a complete cycle (Hertz) in 1/60 of a second.

This reversal of the current flow can be shown by a graph, as shown in Fig. 6-31.

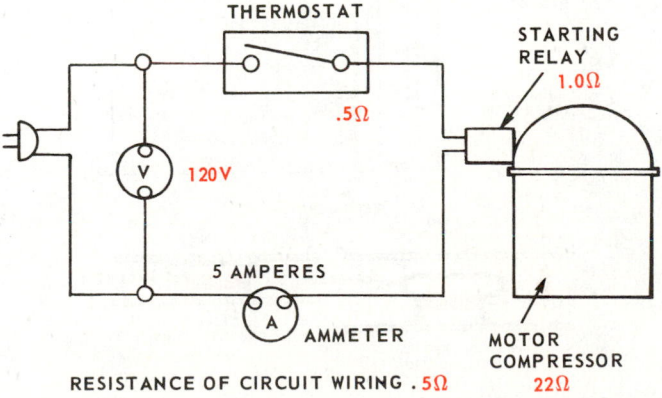

RESISTANCE OF CIRCUIT WIRING .5Ω

Fig. 6-30. Voltage drop in an electrical circuit is equal to voltage applied.

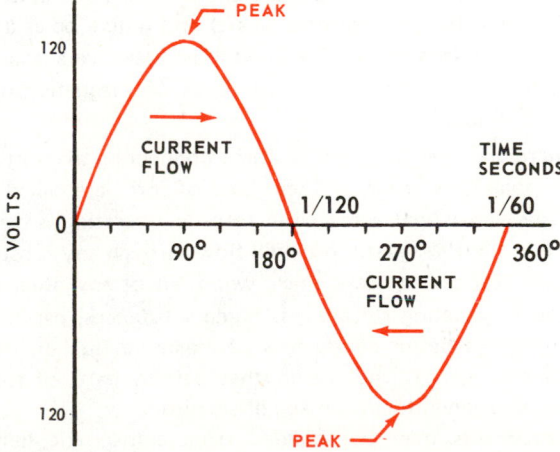

Fig. 6-31. A graph showing flow of 60 Hz alternating current (a-c).

This drawing shows a sine wave of 60 cycles (Hertz or Hz) alternating current. In some industries, 180 cycle (Hertz or Hz) current is generated to operate hand tools such as wrenches, drills and screwdrivers. These high cycle tools do not draw heavy currents and do not overheat if stalled.

6-34 PULSE WAVE AND DIGITAL CONTROL SIGNALS

In computer or digital control applications, a second type of alternating current is used. This is pulse wave electronics. The signals in these applications are electrical pulses, as shown in Fig. 6-32. The control is obtained by the spacing of the pulses and the width of the pulses. Most control systems using computers have 3-volt pulses. If they are used in motor control, the voltage is amplified to the voltage required by the motor.

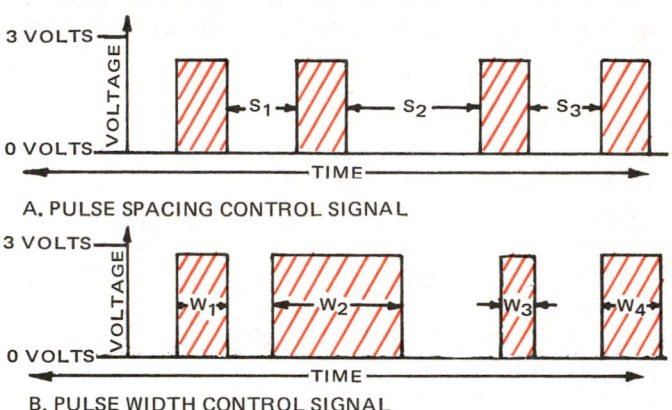

Fig. 6-32. Digital control signals used in pulse wave applications. A—Control using variable spacing between pulses, (S). B—Control using variable pulse width (W).

6-35 MAGNETISM

Operation of the magnetic compass depends on the fact that the earth is a magnet. The north magnetic pole is near the north geographic pole and the compass needle is free to turn and "line up" with the earth's magnetic field. See Fig. 6-33.

All magnets have a north pole and a south pole. Like poles repel each other (try to move apart) and unlike poles attract (pull toward each other). Fig. 6-34 illustrates several shapes of magnets. The attraction and repulsion of magnetic poles is shown in Fig. 6-35.

There are lines of magnetic force connecting the north and south poles of a magnet. These lines of force are called flux. The space in which a magnetic force is operating is called a magnetic field. Magnetic flux will flow through any substance. It is not stopped by glass, mica, wood, air or any other usual electrical insulating substances. Some substances, particularly soft iron, are better conductors of magnetic flux than other substances. This is the reason that certain parts of electric motors and generators are made of soft iron.

Instruments may be shielded from a magnetic field by encasing them in a ring or case of soft iron. The magnetic field passes around the instrument due to the fact that the soft iron

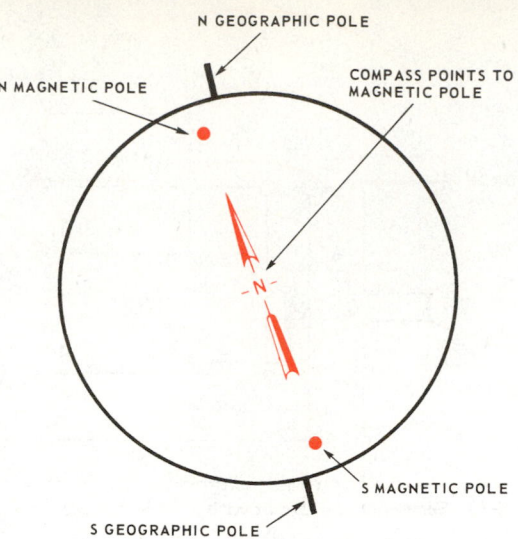

Fig. 6-33. A magnetic compass is a bar of magnetized steel mounted to turn freely on a vertical axis. It points in direction of earth's magnetic lines of force.

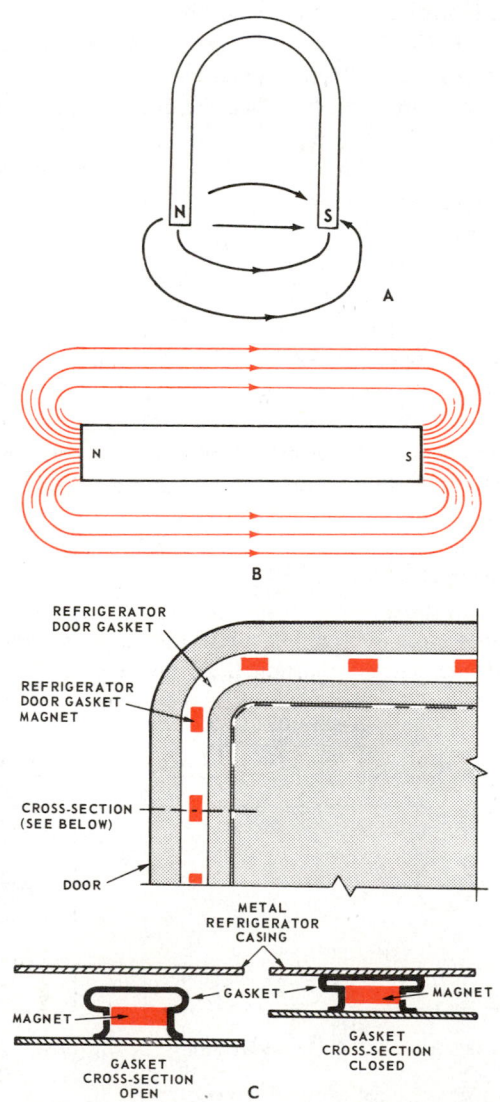

Fig. 6-34. Various shapes of magnets. A—Horseshoe shaped magnet. B—Bar magnet. C—Refrigerator door gasket magnet.

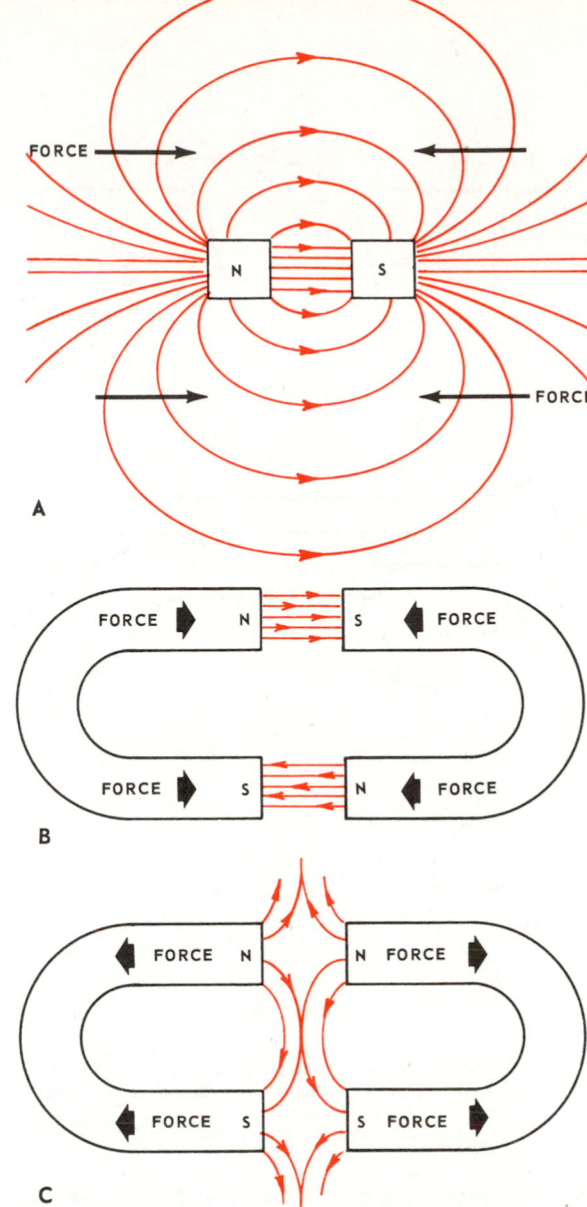

Fig. 6-35. Attraction and repulsion of magnetic poles. A—Magnetic flux around poles of a horseshoe magnet, shown end on. B—Magnetic flux provides a force which tends to pull unlike poles of magnets together (attract). C—Magnetic flux provides a force which tends to push like poles away from each other (repel).

is a good conductor of magnetic flux. Some nonmagnetic watches use this method.

6-36 PERMANENT MAGNETISM

Permanent magnets are usually of hardened steel. Once magnetized, they remain magnetized. There are some patented alloys of iron, aluminum, nickel and cobalt which make strong and super-permanent magnets.

Magnetic lines (flux) are like stretched rubber bands. They tend to become as short as possible. This shortening force has many industrial applications.

Permanent magnets are used in some controls to provide a snap action for electrical contacts. They are also used in small

control motors. In these motors, either the stator or rotor has permanent magnets to provide precision movement of a motor rotor (d-c, servo motors).

6-37 INDUCED MAGNETISM

A property of magnetism is that it may produce magnetism in some other metals near it.

In Fig. 6-36, the permanent magnet is surrounded by its magnetic field, but the soft (low carbon), mild steel piece has no magnetism. If the mild steel piece is placed near the permanent magnet as in view B, or is touching it, as in view C, it too becomes a magnet.

Thus any material capable of being magnetized will become a magnet if it is placed in a magnetic field. This is called induced magnetism. Since the bar is soft iron, it will lose its magnetism as soon as it is removed from the magnetic field.

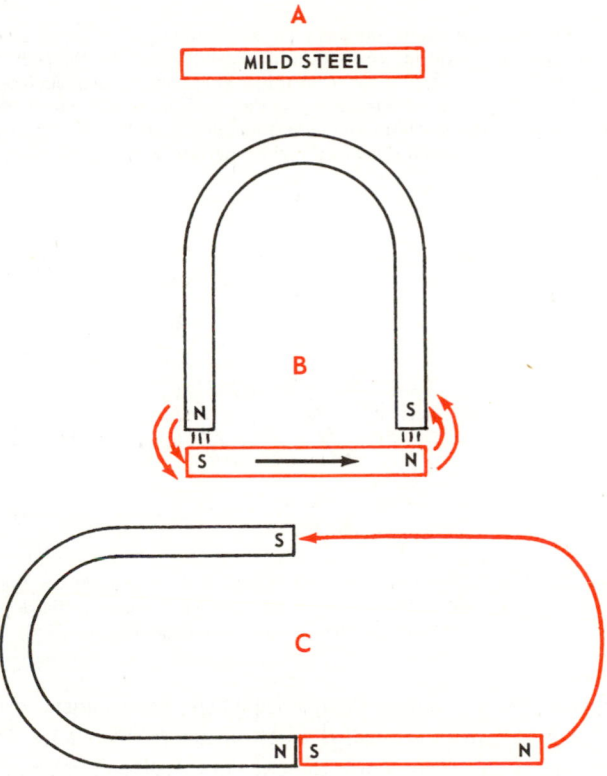

Fig. 6-36. Induced magnetism. A—Nonmagnetized soft iron bar. B—Bar has become a magnet by being placed in a magnetic field. C—Bar is again magnetized by being placed against one pole of permanent magnet.

6-38 ELECTROMAGNETISM

If a current of electricity is passed through a conductor, the conductor becomes surrounded by a magnetic field, Fig. 6-37. If the current is turned off, the magnetic field will cease to be.

If the conductor is wound around a piece of soft iron (a good magnetic conductor) and current is passed through the conductor, the soft iron becomes a magnet. Turning off the current (opening the circuit) stops the magnetic effect. The

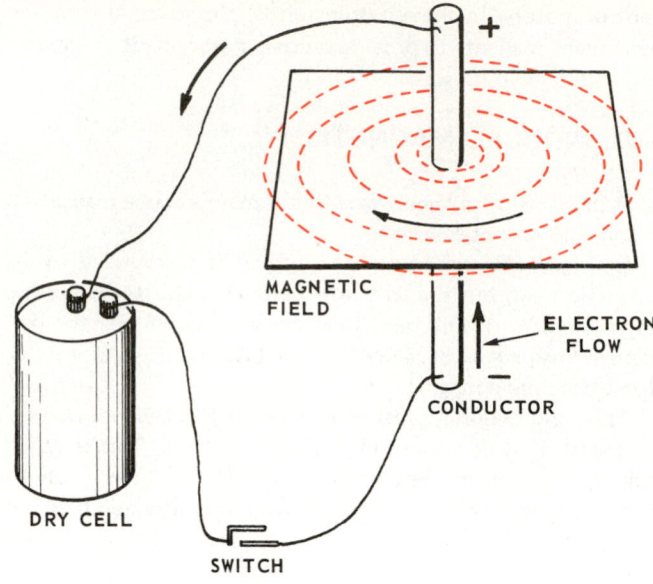

Fig. 6-37. Magnetic field encircles current flowing in a straight conductor. To demonstrate, conductor is passed vertically through the center of a sheet of cardboard. Iron filings are sprinkled over surface of cardboard. When ends of vertical conductor are connected to a dry cell, iron filings form pattern of circles around conductor. Changing placement of cardboard will not change pattern of magnetic field.

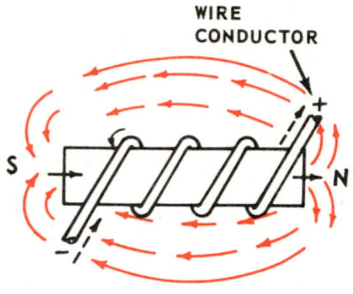

Fig. 6-38. Simple electromagnet has several turns of conductor (wire) placed on soft iron core. Individual circular magnetic fields are combined in core to form one magnet.

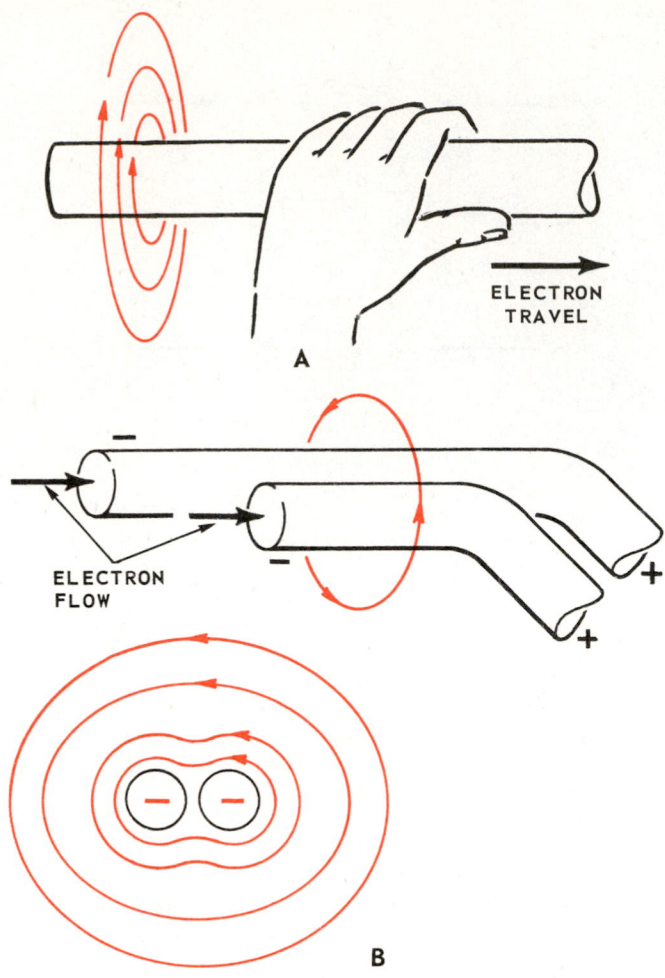

Fig. 6-39. Magnetic fields. A—Around a single wire. B—Around groups of wires. Note that left-hand rule establishes direction of magnetic field travel.

magnetic effect is called electromagnetism, and magnets formed in this manner are called electromagnets, Fig. 6-38. The iron part is called the core; the current-carrying conductor is called the winding.

The characteristics of a magnetic field surrounding a current-carrying conductor (wire) are shown at views A and B in Fig. 6-39. One very important characteristic is that the magnetic field travels around the wire in a specific direction, depending on direction of electron flow in the wire. This, in turn, determines which end of the electromagnet is the north pole. To determine direction of magnetic field travel around a wire, use left hand rule as illustrated in A in Fig. 6-39.

Left hand rule: If one wraps the left hand fingers around a wire with the thumb pointing in the direction of electron flow (— to +), the fingers will point in the direction of the magnetic field travel.

The strength of an electromagnet is determined by the number of turns in the winding around the core and the number of amperes flowing through the winding. This strength is indicated by the term "ampere-turns." This is worked out by multiplying the number of turns in the winding by the amperes flowing through the winding.

Electromagnets are used in motors, relays, solenoids and in many other electromagnetic applications. Fig. 6-40 shows magnetic flow through the field poles and rotor of a four-pole motor. Note that the stator to which the pole shoes are attached carries the same magnetic field as the pole shoes. This is the reason why the iron in the motor stator needs to be so thick and heavy.

6-39 MAGNETIC FIELD STRENGTH

Magnetic field strength depends upon the density of magnetic lines of force. As the lines of force become less dense away from the magnet, the strength of the magnetic field rapidly decreases. To illustrate this effect, magnets placed 1/2 in. from each other may be pulled together with a force of 20 lb. At a distance of 1 in., pull would be only 5 lb. The magnetic field density is measured in units of Gauss.

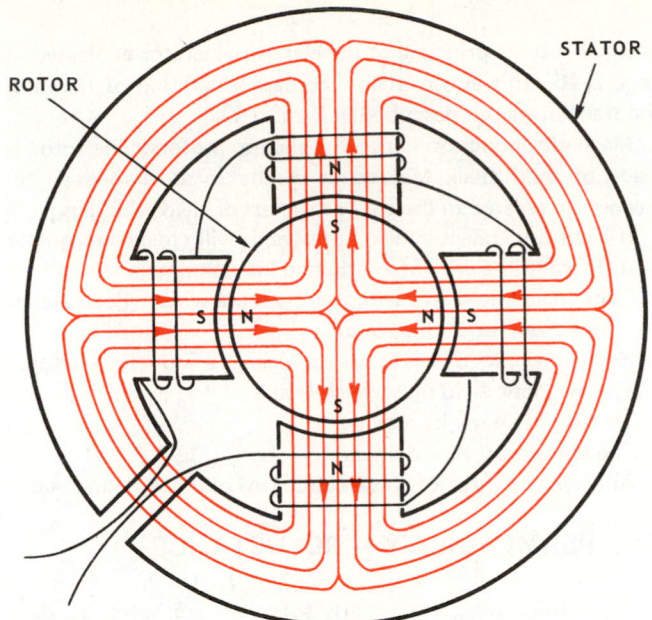

Fig. 6-40. Magnetic lines of force flowing through rotor and stator on a four-pole motor.

6-40 SOLENOID

If a coil is wound on a nonmagnetic substance (paper or plastic) and the coil is made to carry a current of electricity, the coil will become a magnet. See view A in Fig. 6-41. However, the magnetic effect will not be as great as it would be if an iron core were placed inside the coil, as at view B.

If this electromagnet is fitted with a movable iron core and the coil is made to carry a current of electricity, the soft iron core will be quickly drawn into the coil as shown at view C in Fig. 6-41. This is because magnetic lines of force are like stretched rubber bands. They like to be as short as possible.

The coil shown at views A, B and C are commonly known as solenoids. The solenoid principle is used in many places in refrigeration and air conditioning work. They may be used to open and close valves, operate dampers and make other desired movements in a mechanism.

A solenoid may be used on either alternating or direct current, since the lines of magnetic force tend to become as short as possible in either case.

6-41 POLARITY OF ELECTROMAGNETS

The polarity of an electromagnet is determined by the direction of flow of current (or electrons) through the winding on the core of the magnet. If the electromagnet is held in the left hand, Fig. 6-42, with the fingers grasping the magnet in the direction of flow of the electric current, the thumb will point to the north pole of the magnet. This is known as the left hand rule.

6-42 ELECTROMAGNETIC INDUCTION

A magnet may cause another substance to become a magnet. This is shown in Fig. 6-36. Fig. 6-38 illustrates how the flow of electricity may produce magnetism. Electromagnetic induc-

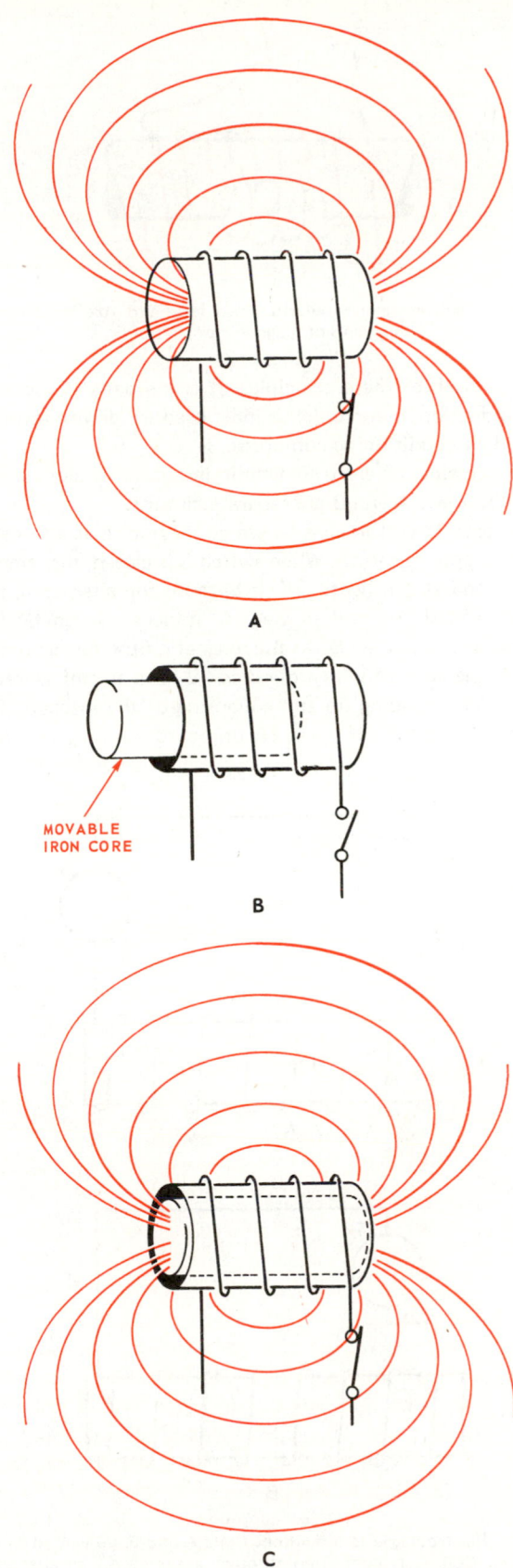

Fig. 6-41. A—Magnetic field is created if a solenoid is magnetic material and current is flowing through windings. B—Solenoid fitted with an iron core with no electric current flowing. C—With current flowing, iron core is drawn into solenoid winding.

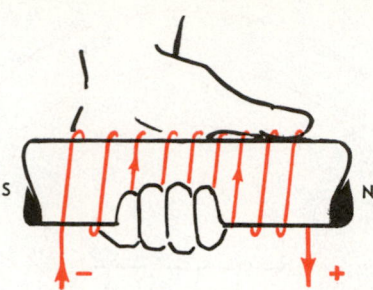

Fig. 6-42. Electromagnetic polarity. Use left-hand rule to determine which end of magnet is north pole.

tion uses both of these principles. It is the basis of operation of the induction motor which is most used for driving refrigeration and air conditioning compressors.

The principle of electromagnetic induction is illustrated in Fig. 6-43. Views A and B are separate circuits.

The leads to coil in view A are connected to an alternating current supply as shown. When switch S is closed, the lamps in views A and B will light. This is because the alternating field created around the coil in view A induces a magnetic field around the coil in view B. As the magnetic flux builds up, then reverses, the coils of wire around coil B have an emf generated in them. The building up and collapsing of the magnetic field produces the effect of wires cutting across a magnetic field.

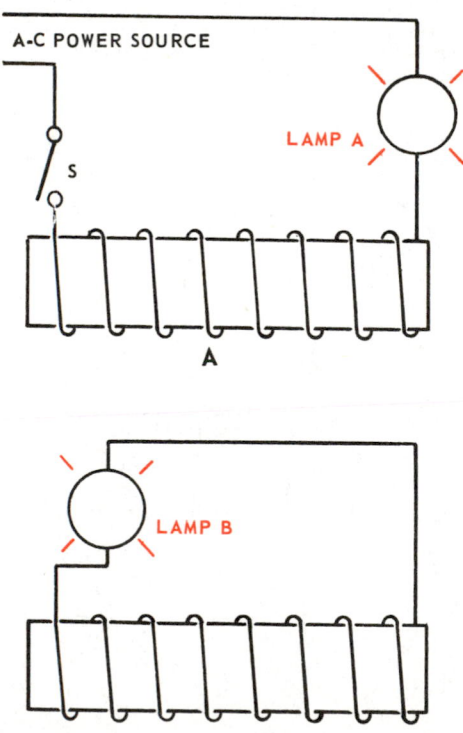

Fig. 6-43. Electromagnetic induction. Coils A and B are wound on soft iron cores. Coil A is connected through a switch to an alternating current supply. Lamp in circuit controls amount of current flowing through coil A. Coil B is wound on a core adjacent to coil A, but they are not connected electrically. A low voltage lamp is connected to terminals of coil B. When switch to coil A is closed, both lights will light, indicating that a current is being generated in coil B by electromagnetic induction.

This is the basic principle of the electric generator explained in Para. 6-46. This also explains the basic operation of the electric transformer, as described in Para. 6-69.

Magnetic induction is used in electric motors. The rotor is made of mild steel. Magnetism in the rotor is induced by magnetism created in the stator magnets or field windings.

If a current travels in a coil of wire, it will create a magnetic field. If there is a wire or another coil close enough to the first coil, the magnetic field cutting across these wires will create an emf in this wire or in this coil.

Remember: In order to create magnetic induction, either:
1. The magnetic field must be changing.
2. The magnet must be moving.
3. The wires must be moving.
4. Any two or three of these conditions must be taking place.

6-43 PERMEABILITY — RELUCTANCE

Some substances, particularly soft iron, are better conductors of magnetic flux than other substances. The property of the material which determines its flux density under a magnetic field is called its magnetic permeability. It indicates the ease with which a material may be magnetized. Air is considered to have a permeability of one. Soft iron has a much greater permeability than air.

Reluctance is the resistance offered to the passage of magnetic lines of force through a substance.

Fig. 6-44 shows some fundamentals of magnetic behavior.

6-44 CAPACITANCE — CAPACITORS

Capacitance may be defined as a system of conductors and insulators which permit the storage of electricity (free electrons). This ability is indicated by the letter C. In referring to the capacities of a series or group of capacitors, they are usually designated as C_1, C_2, C_3 and so on.

The unit of capacity is the farad. The symbol for the farad is the capital letter F. A farad may be defined as a charge of one coulomb on the capacitor surface with a potential difference of one volt between the plates.

A farad is a rather large unit of capacity. Most capacitors (condensers) used in the refrigeration industry are rated in microfarads. A microfarad is one-millionth (.000 001) of a farad, or 1^{-6}. The symbol for the microfarad is (μF). The Greek letter μ (mu) is used instead of the English m.

Any device that has capacitance (stores free electrons) is a capacitor. Large capacitors are large metal surfaces such as aluminum foil separated by insulating material (dielectric). See Fig. 6-45. The capacitor, in series with the load, changes the sine wave and makes the current wave (lead) the voltage curve of an a-c circuit.

Capacitors are used to help start motors, increase their efficiency and improve the power factor.

The capacity value of capacitors in series may be expressed by the formula:

$$\frac{1}{\mu F} = \frac{1}{C_1} + \frac{1}{C_2}$$

μF = field capacitance (effective value)

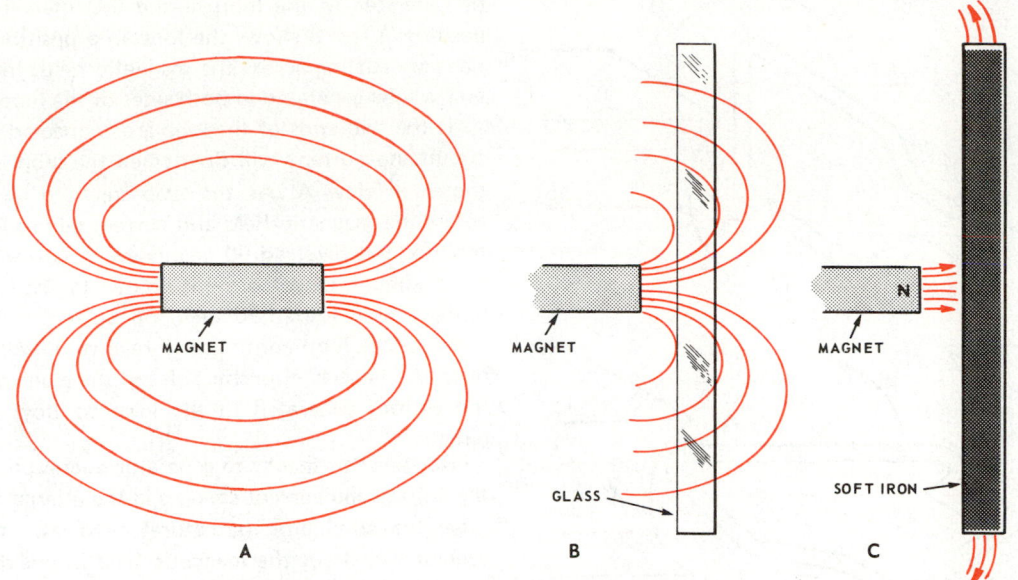

Fig. 6-44. Fundamentals of magnetic behavior. A—Typical magnetic field surrounding a magnet. B—Plate of glass has no effect on shape of magnetic field. C—Effect of a soft iron plate placed in a magnetic field.

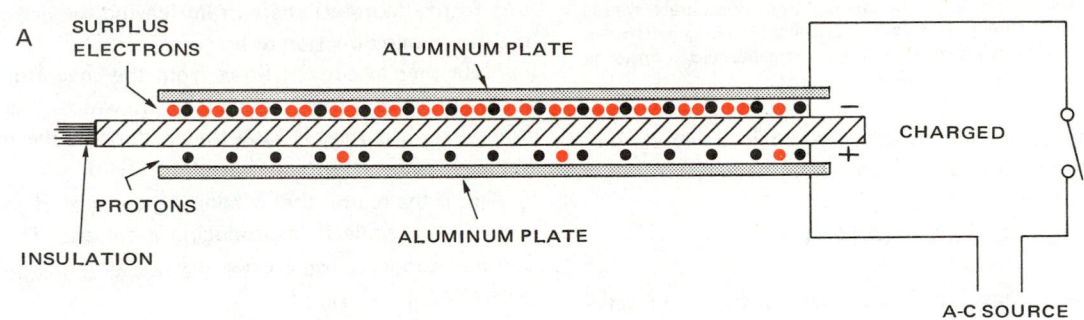

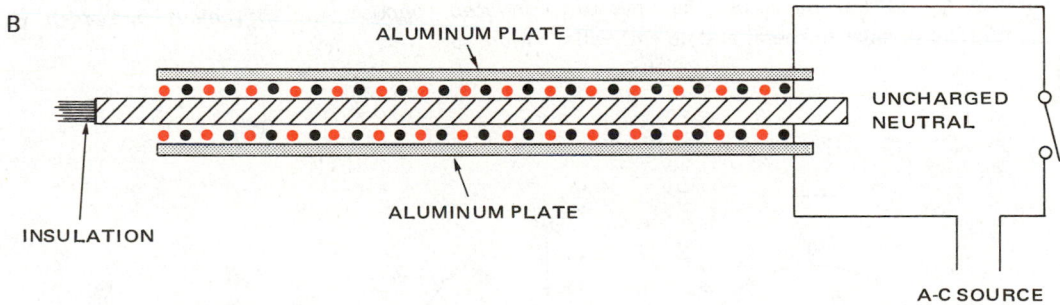

Fig. 6-45. Capacitor construction. A—Charged condition. B—Discharged condition.

C_1 = capacity of capacitor No. 1 in microfarads
C_2 = capacity of capacitor No. 2 in microfarads
The capacity of capacitors in parallel may be expressed by the formula:

$$\mu F = C_1 + C_2$$

or by simply adding together the values of all of the capacitors connected in parallel.

Large capacitors in a machine can store up dangerously high voltages. Before handling or replacing them, drain off the charge. A metal bar with an insulated handle may be used. A high voltage capacitor may store electricity at 600 volts in a three-phase electrical motor circuit. This is dangerous.

6-45 REACTANCE

Reactance is the opposition to an alternating current flow in a circuit.

There are two kinds of reactance: capacitive reactance and inductive reactance. Capacitive reactance is caused by the capacity or condenser effect in the circuit. Inductive reactance

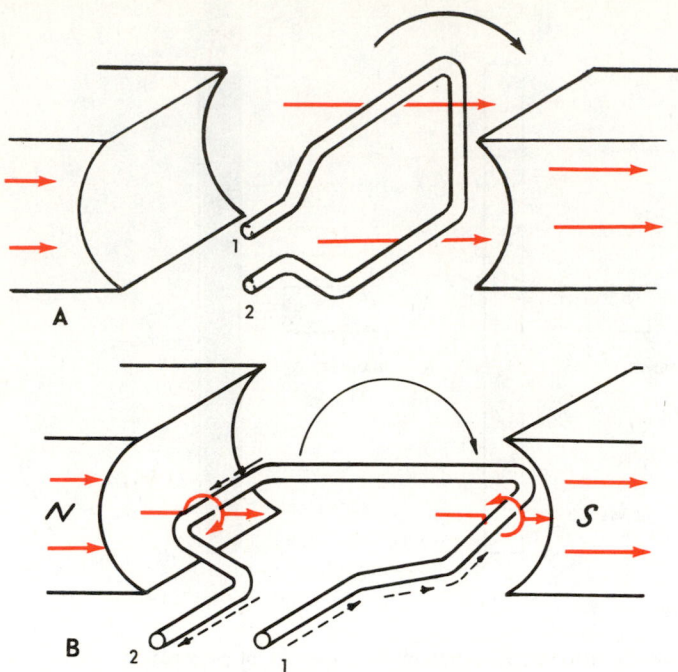

Fig. 6-46. Generation of current. Conductor loop has two sides, 1 and 2, and it is revolving clockwise. A—In this position, wires are traveling more or less parallel to lines of force and no emf is being generated. B—Loop has revolved 90 deg. and current is being generated in opposite direction.

is caused by the generation of counter-emf in a circuit. It is usually produced by a wire coil or an electromagnet.

6-46 ELECTRICAL GENERATOR

If a conductor is moved across a magnetic field, an electrical potential (emf) will be generated in the conductor. This is illustrated at views A and B in Fig. 6-46.

The loop in view A is not cutting across the lines of magnetic force. It is running parallel to them, and no emf will

be generated in the loop during the interval that it is in this position. View B shows the loop in a position so that the two sides are cutting across the magnetic field. In this position, an emf will be generated in both sides of the loop.

If the two ends of the loop are connected into an electrical circuit, no current will flow when the loop is in the position shown in view A. As the loop starts to revolve, it will cut across the magnetic field and current will begin to flow. When the loop has revolved 90 deg., view B, current flow will reach its maximum. Once past this point, current flow will diminish until, after another 90 deg., no current will be generated. Then, as the loop continues to revolve, it will cut through the magnetic field at opposite poles of the magnet. This will cause the current generated in the loop to flow in the opposite direction.

Relating this theory to generator operation, Fig. 6-47 shows the voltage and current changes in the alternating circuit as the generator conductor (armature) revolves. The wires of the generator first cut the magnetic field in one direction, then in the other to generate an alternate (or opposite) flow. This happens each revolution of the armature as the conductors pass the magnetic poles and generate alternating current (a-c). However, commutators and brushes on direct current generators rectify (correct) the current leaving the generator so that it flows in one direction only.

Note that as current flows from the generator, a magnetic field surrounds the conductor in which the current is generated. This field is such that it opposes the movement of the conductor across the generator field.

This is the reason that it takes a great deal of power to drive a generator while it is producing a current. The greater the current produced, the greater the power required to drive the generator.

This principle is stated in Lenz's Law: "The magnetic effect, surrounding the conductor in which a current is induced, opposes the movement by which the current is induced."

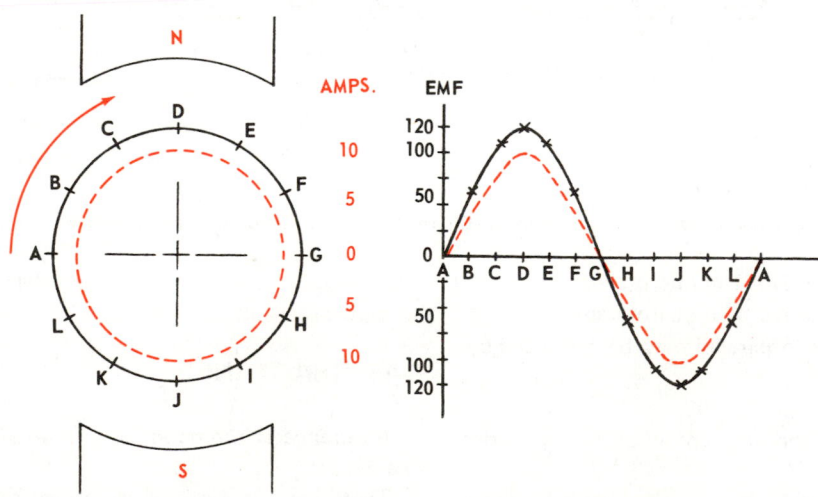

Fig. 6-47. Current (amps.) and voltage (emf) changes in an alternating circuit are charted for one revolution of armature.

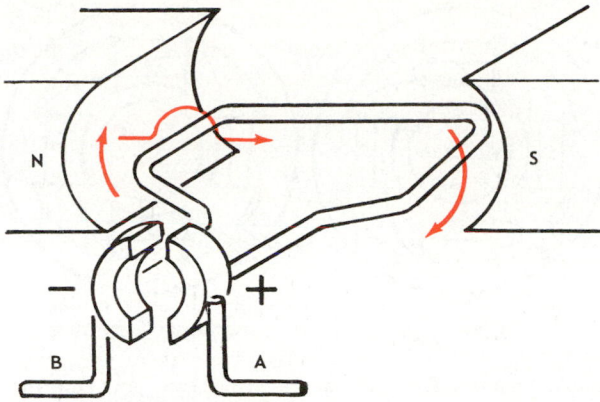

Fig. 6-48. A generator commutator. Brushes A and B contact ends of generating loop. Current from brushes always flows in same direction.

6-47 COMMUTATORS

A conductor moving across a magnetic field will have an emf generated in it. In Fig. 6-47, the generated current in the wires of an armature will flow in one direction, then in the other as the conductors move from one field pole to the other. Alternating current is being produced.

In order to produce direct current, it is necessary to use a commutator and brushes in the generator. Fig. 6-48 illustrates an elementary type of armature fitted with a commutator and brushes.

The commutator and brushes provide a movable electric contact between the rotating armature wires and the stationary electrical device. The movable contact surface on the rotating armature is called the commutator, and the parts that contact the rotating parts of the commutator are called the brushes.

As the armature revolves, the commutator contacts are made in such a way that brush A always carries current into the commutator in a negative to positive direction. Brush B always carries current from the commutator to the charging circuit. Therefore, direct current generators are always fitted with commutators and brushes.

The starting mechanism on some a-c motors with wound motor armatures also are fitted up with commutators and brushes. On some motors, these commutators and brushes are used only during the interval that the motor is starting. Universal motors (both a-c and d-c operation) use the commutators and brushes whenever the motor is in use. Motors of this type are used in electric drills and mixers.

6-48 THE ELEMENTARY ELECTRIC MOTOR

Electrical energy is changed to mechanical energy in an electric motor. First, electrical energy becomes magnetism. Magnetism may then be used to cause motion.

Because like poles repel (N repels N and S repels S), and because unlike poles atttract (N attracts S and S attracts N), motion can be produced by putting one magnet on a shaft and mounting another in a fixed position as shown in Fig. 6-49.

The bar magnet on the shaft will turn until its S pole is near

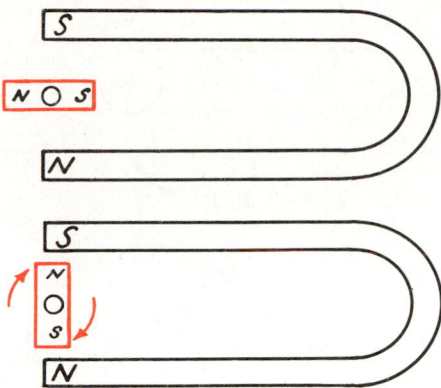

Fig. 6-49. Magnet mounted on an axis will turn when put in another magnetic field. Note direction of movement of pivoted magnet.

the fixed N pole. This movement places the shaft bar magnets N pole near the fixed S pole. (The fixed magnets, which are stationary, are called the stator. The other magnets, which rotate, are called the rotor or armature.)

The rotor will stay in the vertical position until the magnetism of the stator or rotor is reversed. Then the rotor will rotate another half turn.

The magnetism may be reversed by using electromagnets instead of permanent magnets, Fig. 6-50. With this setup, when the rotor reaches the vertical position, the alternating current reverses (due to its cycling) and stator polarity reverses. The rotor S pole is now near the S pole of the stator, and they will repel each other. The rotor will now turn or revolve one half a revolution or 180 deg.

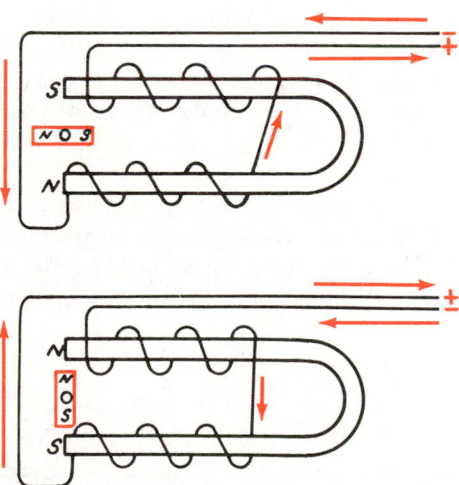

Fig. 6-50. Elementary electric motor. By reversing flow of current through field (stator) winding, its magnetic polarity will be reversed. This will cause rotor magnet to turn on its axis.

The direction of movement of the armature (or rotor) depends on polarity, as shown at A in Fig. 6-51. The direction of movement of a current carrying conductor in a magnetic field may be determined by what is popularly called the "left hand motor rule." To use this rule, place the thumb and

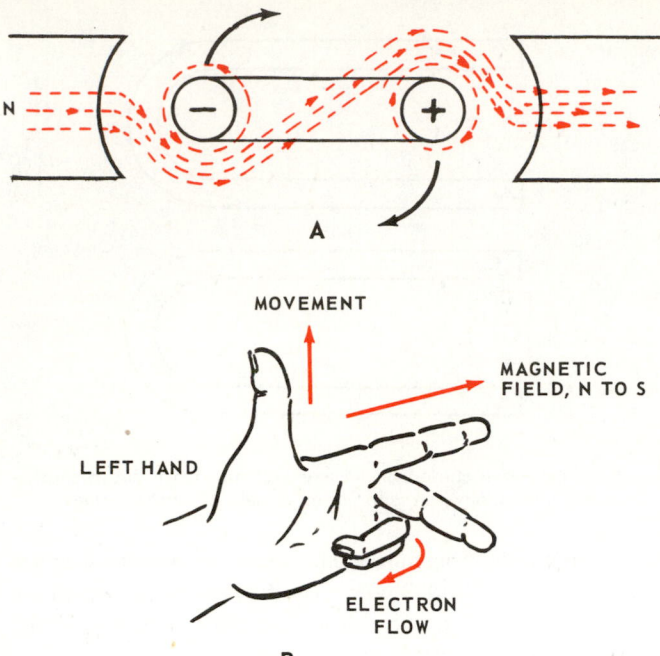

Fig. 6-51. Left hand motor rule. A—Magnetic lines of force from north pole to south pole merge with magnetic field around conductor to cause movement in an electric motor. B—Position fingers of left hand as shown and thumb will point in direction of movement.

first two fingers of the left hand at right angles to each other with the index finger pointed in the direction of the magnetic force, the middle finger pointed in the direction of the current flow in the conductor, and the thumb pointed in the direction of the force or the movement of the conductor. This is shown at view B in Fig. 6-51.

If the rotor turns a half revolution during one half of 60 Hz (cycles), then it turns a half turn during 1/120 of a second. It will then turn 60 revolutions per second or 60 x 60 seconds or 3600 revolutions in one minute. The two-pole motor (3600 rpm) has become a popular hermetic motor for refrigerating and air conditioning units. Its construction is shown in Fig. 6-52.

If four poles are used in the stator, the motor will turn only 1800 rpm. That is, the rotor will only turn one fourth revolution in 1/120 of a second, as shown in Fig. 6-53. Most open motors are of this design and some hermetic motors also use

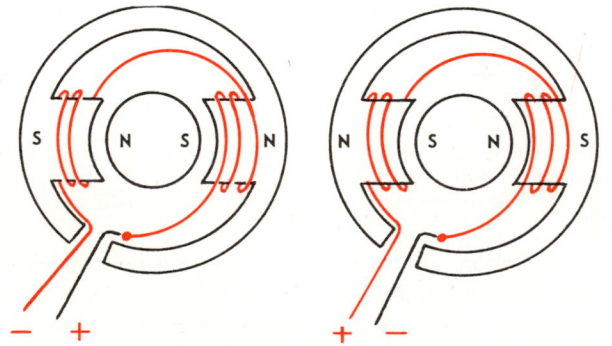

Fig. 6-52. Two-pole stator motor. Rotor will make a half turn with each half of current cycle.

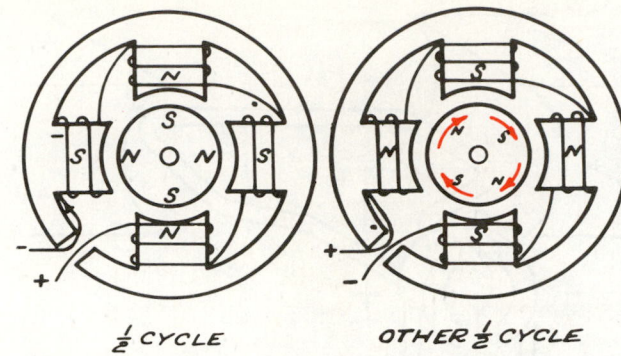

Fig. 6-53. Four-pole stator. It produces one quarter turn of armature as current completes one cycle in field windings.

the four-pole design principle.

Speeds of 3600 rpm and 1800 rpm are called synchronous speeds. Under actual conditions, the 3600 rpm motor usually operates at approximately 3450 rpm, while the 1800 rpm motor operates at approximately 1750 rpm. This reduction in speed is due to a slight magnetic slippage, depending on the load. An overloaded motor will slow down slightly.

The electric motor shown in Fig 6-54 is an open motor of the capacitor-start type. It can be used on external drive refrigeration compressors, pumps and the like. See Chapter 7 for further information on various types of motors used on refrigeration and air conditioning machines.

6-49 COUNTER EMF

A running motor develops an emf in the rotor bars or windings. This is called a counter emf. It opposes the applied emf which is driving the motor.

Counter emf depends upon the speed of the rotor. Under no-load, synchronous speed, the counter emf practically balances the applied emf. As the load increases and the speed decreases, the counter emf drops. As a result, the applied emf sends more current through the windings, tending to maintain the speed constant.

If the motor is slowed down considerably by a heavy load, the current supplied will increase greatly due to the reduced counter emf. The motor then will overheat. Under continuous overload, it is likely to burn out.

If the rotor is locked so it cannot turn and the current is applied, the current will be very high. A motor will quickly burn out under this "locked rotor condition." Counter emf also occurs in all coils and electromagnets.

6-50 INDUCTANCE

A magnet, formed by winding a coil of wire (conductor) on a soft iron core, will become magnetized if an electrical current is passed through the conductor. The entire coil is surrounded and saturated by magnetic lines of force (flux).

As the current is switched on in such a coil, the magnetism is not built up instantly. There is a delay of perhaps a few hundredths of a second during which time the current continues to increase until it reaches its full value. This value

Fig. 6-54. Section through motor suitable for use on external drive refrigeration and air conditioning installations. A—Sleeve bearing. B—Shaft extension end. C—Ball bearing assembly at other end of shaft. (Gould Inc., Electric Motor Div.)

depends on the resistance and emf of the circuit. Likewise, when the switch is opened and the current is turned off, it does not stop flowing instantly. The magnetic lines build up as the switch is turned on and collapse as the switch is turned off.

There is a tendency to generate within the coil an electromotive force which counteracts the change in the current flow. This counteracting force is the counter electromotive force or counter emf (discussed in Para. 6-49).

This principle of inducing a voltage in a coil due to the change in the rate of flow of current in the coil is called inductance. It acts much like a flywheel on a piece of machinery. The flywheel requires power to give it a rotating motion and, likewise, it gives up power if it is forced to stop.

6-51 INDUCTORS

Induced magnetism is useful in electric motors. By putting electrical windings on the field poles, magnetism can be created or induced in the rotor. There is a slight time delay in this induction. Therefore, if the rotor has once started to turn, it will continue to do so because the field pole (stator) always has the opposite polarity.

During the time the field pole changes its polarity, it induces an opposite pole in the stator. This repels the rotor pole toward the next stator pole, which has become an opposite pole in the meantime. The rotor must turn the full distance between the two poles before the induced magnetism can be changed or the motor will not keep running. Consider that it takes only 1/120 of a second to change the magnetism from full strength N to a full strength S pole.

This principle is used in the design of the split-phase motor. There are two windings in this motor. One is a starting winding and the other a running winding. The starting winding is a smaller diameter wire than the running winding. However, it has a greater number of turns. As a result, its magnetic inductance will be greater than for the running winding. This

means that the starting winding is always behind the running winding in both building up and stopping its magnetic field. This type of inductance is generally known as self inductance.

Mutual inductance, on the other hand, is another type of flow of electricity and a magnetic field in another coil that is close to it. The principle of mutual induction is used in all induction type motors.

In induction motors, the magnetic effect of the current flowing in the field windings induces the current in conductors on the rotor. The magnetic effect of this induced rotor current causes the rotor to revolve. This characteristic of the flow of current, in an electromagnet, to resist the flow when the current is turned on and to resist the stopping of the flow when the current is turned off, depends on Lenz's Law. The law states that the polarity of an induced voltage is such that it opposes the motion of the flux inducing it. (See Para. 6-46.)

6-52 ELECTRONICS

Electronics concerns electron flow through gases, vacuums and semiconductors. Electronics was first developed with the discovery of vacuum tubes and gas filled tubes.

It was found that electrons would flow from a heated element in a tube to another element if a potential difference existed between the two elements.

Fig. 6-55 shows a simple circuit of this type. The first radios used the vacuum tube to control and amplify the radio signals. These tubes were fragile and relatively large. The invention of the solid state semiconductor as a replacement for tubes has greatly enlarged the application of electronics.

In the following paragraphs, some electronics will be explained. This includes parts such as:
1. Semiconductors.
2. Diodes.
3. Rectifiers.
4. Transistors.
5. Sensors.
6. Thermistors and Varistors.
7. Amplifiers.
8. Transducers.

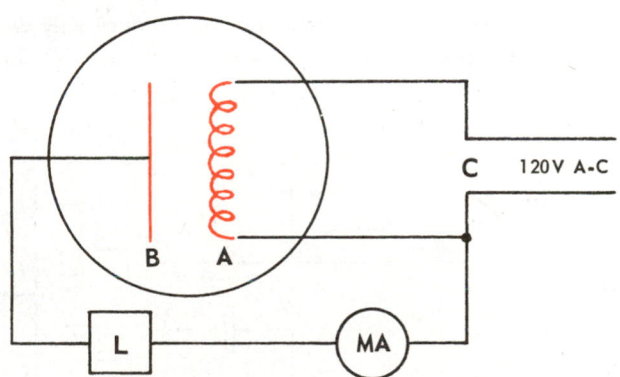

Fig. 6-55. Diagram of vacuum tube rectifier. A—Heating element from which electrons will flow. B—Plate to which electrons flow from A. C—120V a-c power source. L—Direct current load. This might be a storage cell on charge. MA—Milliammeter, which shows rate of d-c flow.

9. Photoelectricity.

With the development of rather complicated automatic control used on refrigerating and air conditioning appliances, more and more electronic devices are being used. It will be necessary to understand these devices in order to understand the circuits in which they are used.

6-53 SEMICONDUCTOR APPLICATIONS

Solid state electronics make up most of the modern electrical control system. They are important in computer control systems and in devices which must withstand hard use.

Semiconductors are of two general types:

1. Intrinsic semiconductors, which are pure substances like silicon and germanium or combined substances like lead sulfide. These are particularly useful as thermometers and as other temperature sensing devices.

2. Extrinsic semiconductors, which are combinations of intrinsic semiconductors with very small "impurities." These semiconductors are very sensitive to electrical forces and are the basic materials for electronic control use. Thermoelectric refrigerators (discussed in Para. 3-17) use them to produce cooling.

6-54 DIODES

A solid state diode is a two-material, solid material wafer or capsule which allows electrons to flow through it in one direction only, Fig. 6-56. The diode acts as an electron flow check valve.

Vacuum tubes can also serve as diodes, Fig. 6-57. Electrons will flow from the pointed filament to the plate on half the cycle. During the other half cycle, the electrons will not flow from the plate to the pointed filament.

6-55 RECTIFIERS

Rectifiers are electronic valves which permit the flow of current through them in one direction only. These devices change alternating current to an output of direct current.

A simple rectifier uses only one half of the sine wave. To use both halves of the sine wave and still produce only direct current, one must use four diodes as shown in Fig. 6-58. Some

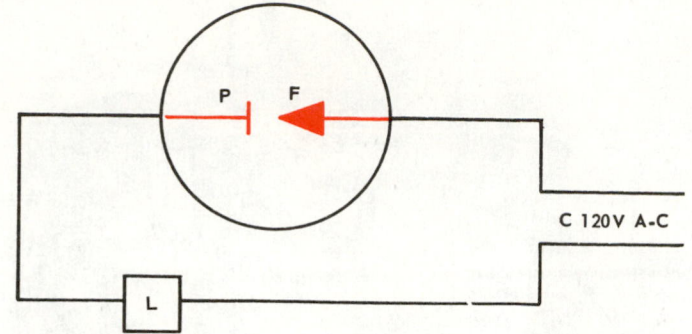

Fig. 6-57. Diagram of a vacuum tube serving as an alternating current rectifier or diode. Note that electrons will flow only from pointed filament F to plate P. L indicates direct current load, which may be a storage battery being charged.

rectifiers use silicon for the rectifying substance. These are called silicon controlled rectifiers (SCR). These are used extensively in electrical motor controls. They are also the basic elements used to convert a-c voltage to d-c voltage in inverter devices. SCR devices are now used in many applications where relays were formerly used for switching.

Four vacuum tubes can be used in the same manner to produce a constant flow of d-c.

6-56 INVERTER

Electrical energy stored in a battery is available as direct current (d-c) energy. The voltage supplied by the battery is a steady voltage, gradually decreasing with time as the charge is drained from the battery. An electric motor powered by a battery must be a d-c motor.

Since d-c motors are heavier and usually more expensive than a-c motors, it is often advantageous to change the voltage from the battery so that an a-c motor can be used. The device to do this is called an inverter. The device does the opposite of the rectifier discussed in Para. 6-55, which converts a-c power to d-c power. Fig. 6-59 indicates the inverter function, converting direct current (d-c) to alternating currect (a-c).

Older electrical systems used a combination of a d-c motor connected to an a-c generator to do this inverting. Newer solid state electronic devices do this without any mechanically mov-

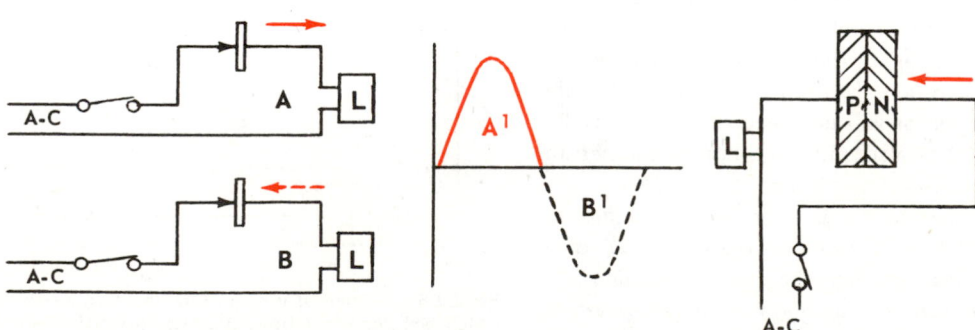

Fig. 6-56. When alternating current is imposed on a diode wafer, it allows only current represented by that part of sine wave shown at A^{1}to pass through. Other half cycle B^1 is stopped from passing through. This provides an intermittent flow of direct current.

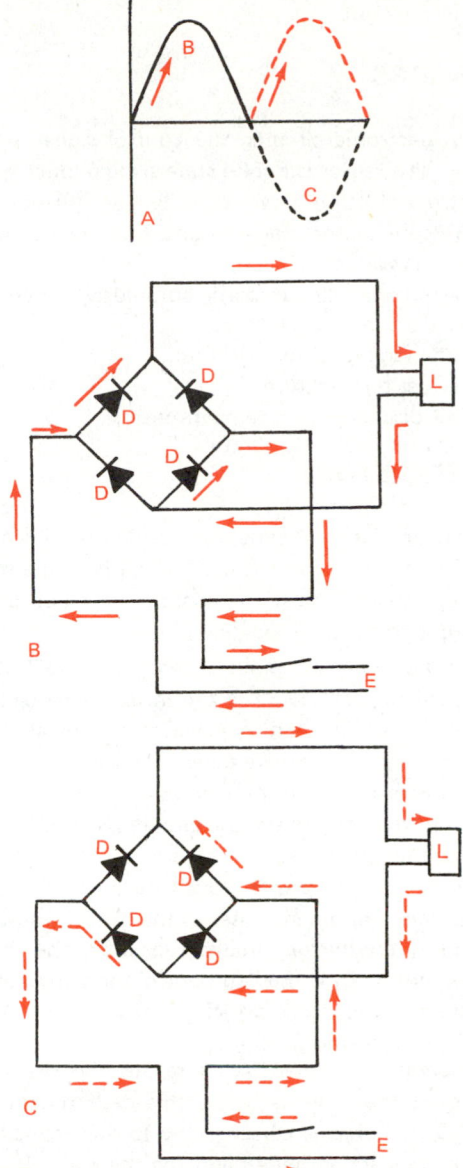

Fig. 6-58. Circuit for a full wave rectifier: A—Note how two halves of an a-c cycle are made to provide d-c flowing in the same direction. B—Solid red arrows show current flow for one-half wave. C—Dotted red arrows show current flow for other half wave. D—Four diodes are needed. E—60 cycle a-c power supply. L—The d-c load.

ing parts. The basic elements used in a solid state inverter are:

1. A crystal which oscillates at the frequency of the a-c power required.
2. A switching circuit which uses this oscillation to switch the d-c power on and off.

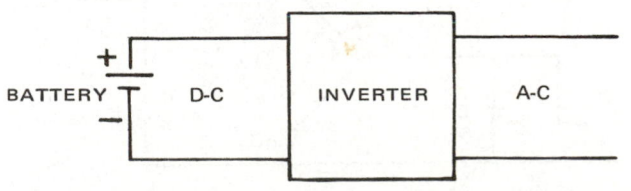

Fig. 6-59. Inverter converts direct current (d-c) supplied by a battery to alternating current (a-c).

3. A set of silicon controlled rectifiers (SCR) diodes to provide shaping of the waveform produced.

A simple inverter, using a set of standard diodes, produces a square wave output. See schematic, Fig. 6-60.

Most a-c motors and controls are designed to operate only with alternating (a-c) power, similar to that provided by the power company. These devices will operate with a square wave but not as efficiently and their lifetimes will usually be reduced.

An inverter is usually required in solar electric energy systems, since the output of solar cells is d-c power.

6-57 TRANSISTORS

A transistor is a three-layer sandwich of two different components which consist chiefly of germanium semiconductor material. Electrically, the three wafers are connected as shown in Fig. 6-61.

The materials are labeled for their properties. P is for positive, meaning a lack of electrons (it has "holes" ready to receive electrons). N is for negative, meaning the material has a surplus of electrons.

Three conductors are connected to the transistor. One attaches to the base or middle wafer, one connects to one of the outer wafers, called the emitter. The third connects to the collector. The outer two wafers are of the same material. The base material (middle wafer) is different.

A small electron flow from the emitter to the base will control a large electron flow from the emitter to the collector. The device, therefore, acts as a valve and as a relay. An electron signal circuit emitter to the base may control electron flow as much as 1000 times larger than emitter to collector.

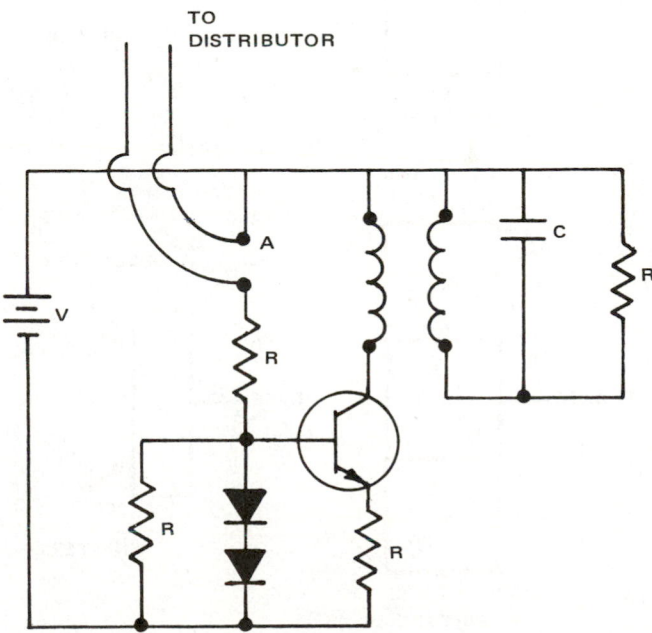

Fig. 6-60. Simple inverter circuit used for converting direct current, shown at V, from battery to alternating current, shown at A. A circuit similar to this is used in automotive ignition system to produce alternating voltage for spark.

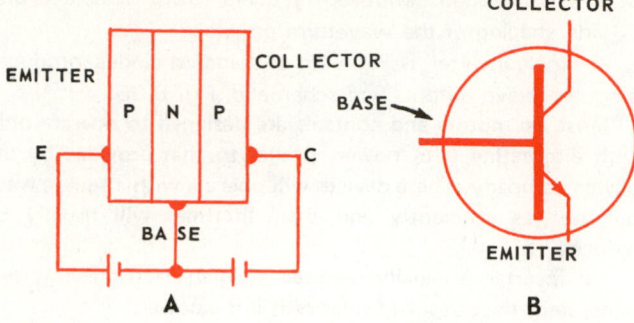

Fig. 6-61. A—Transistor in a circuit. B—Symbol for a transistor.

All three parts are semiconductors having added substances to give the desired characteristics. View A in Fig. 6-61 shows, in a schematic way, the basic construction of a transistor. The two basic types of transistors are shown in Fig. 6-62.

A transistor connected as an amplifier is shown in Fig. 6-63. The low energy signal enters the circuit at the left. The signal is amplified by the action of the transistor and energy supplied by the batteries. The amplified signal (load) is shown leaving

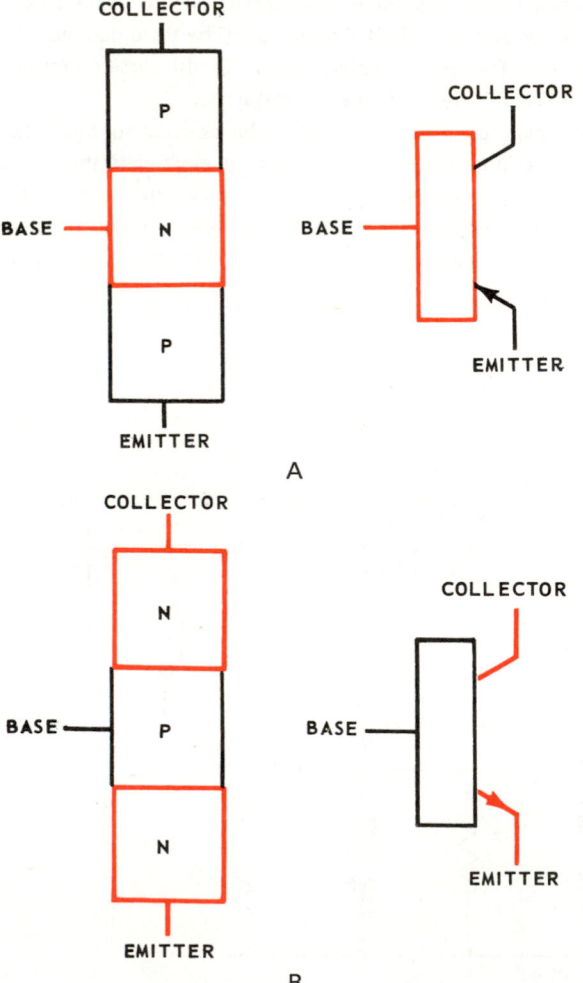

Fig. 6-62. Two basic types of transistors. A—PNP transistor. B—NPN transistor. In A, current flow is controlled by negative charge carried in base. In B, current flow is controlled by positive charge in base.

the circuit at the right.

6-58 SENSORS

In many electronic circuits, the control signal is triggered by a sensor. This sensor is a solid state semiconductor material which controls electron flow as its temperature or pressure changes. Pressure sensors can respond to pressure changes in the refrigerating system.

A typical, completely automatic automotive air conditioner uses three sensors in its circuitry:
1. For outside (ambient) temperature.
2. For in-the-car temperature.
3. For the air discharge duct temperature.

6-59 THERMISTORS

A thermistor is a solid state semiconductor which permits or allows fewer electrons to flow through it as the material's temperature increases. Most thermistors are made of lithium chloride or doped barium titanate.

The resistance changes about 3 percent for each degree F. change (6 percent for 1 C.). The thermistor must be kept free of moisture. It can be used in place of a bimetal strip or in place of a temperature sensitive power element.

The thermistor is used in three ways:
1. A temperature-operated electric circuit control.
2. To measure temperatures.
3. To stop the electric power flow to a motor if the motor windings' temperature increases to the danger point.

In a typical thermistor circuit, Fig. 6-64, the thermistor temperature sensor, C, is used to control the temperature of a room or conditioned space which is heated by a 120V a-c electric resistance heater.

The temperature control, A, is set to the desired room temperature. If the room is below this desired temperature, the sensor, C, will change current flow to transistors D and E. Here the changes are amplified and the thermal relay heat, I, will cause the contact points at J to close. This action brings the electric space heater, K, into operation.

As soon as the desired temperature is reached, the sensor, C, will cause the current to the heater coil to be switched off. Then contact points at J will open and stop the flow of current to the space heater. These devices are very sensitive and will maintain the space temperature within a fraction of a degree.

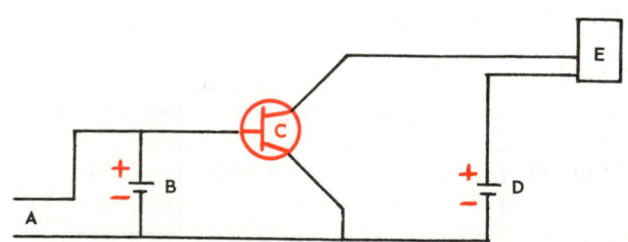

Fig. 6-63. Circuit diagram showing transistor used in amplifier circuit. A—Signal (current) to be amplified enters here. B—Battery. C—Transistor. D—Battery. E—Amplified current load.

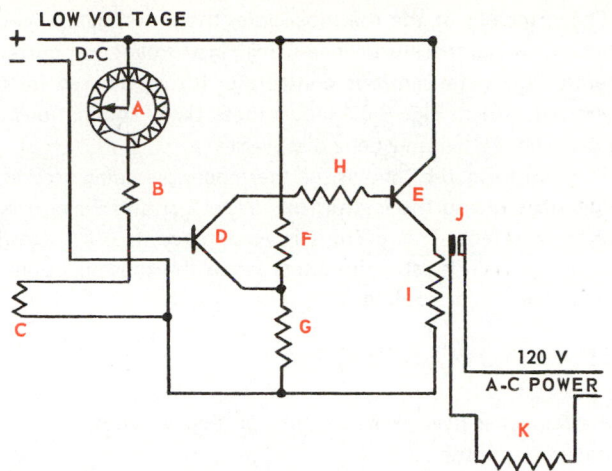

Fig. 6-64. Typical thermistor circuit. A—Temperature control knob, which controls variable resistance. B—Fixed resistance. C—Thermistor temperature sensor. D and E—Transistors. F and G—Bias resistors. H—Current limiting resistor. I—Thermal relay heater. J—Thermal relay. K—120V a-c electric space heater. This sensor will maintain temperature of heated space within close limits.

There is a special thermistor which increases its resistance as the temperature rises. However, it changes from low resistance to high resistance within two deg. It can, therefore, be used as a switch. At present, these operate between 150 F. (65 C.) and 356 F. (180 C.).

It may be used to control crankcase heaters (shuts off current when oil temperature reaches design conditions). It also may be used to sequence heating systems. Number of heaters in operation would increase or decrease as heating need increased or decreased.

It can do the same for a cooling system. It may be used to control a defrosting system on an ice cube maker release circuit. It is small enough to be put in motor windings to protect them from too much heat.

6-60 AMPLIFIERS

An amplifier is an electronic device in which a small current flow is used to control a large current flow. Either a vacuum tube or a semiconductor circuit may be used.

In Fig. 6-65, a small signal received at A is amplified by the battery at E to produce a voltage at F which is high enough to close a relay. The relay switches the "power in" to the "power out." Some systems use a combination of tubes and solid state semiconductors.

6-61 TRANSDUCERS

The term, transducer, is used to identify a great variety of devices which are sensitive to changes in the intensity of some form of energy. The transducer then responds by controlling the intensity of some other form of energy.

Transducers may be operated by pressure, temperature, fluid flow, vibration, electrical potential and others.

A small varying current flow through a transducer may be amplified by an amplifier. The amplified current may then operate a control circuit.

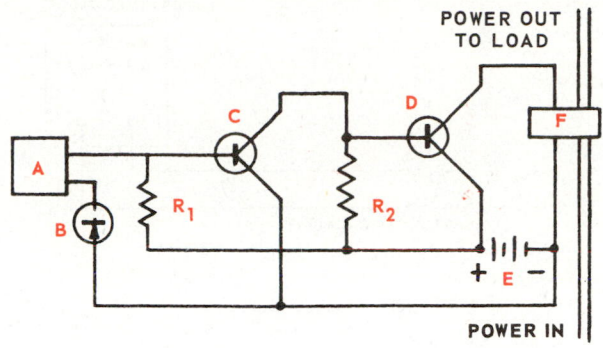

Fig. 6-65. Typical amplifier circuit is shown. A—Sensor. B—Diode. C and D—Transistor. E—Low emf d-c supply. F—Relay R_1 and R_2—emf control resistors.

An application of a transducer is shown in Fig. 6-66. In this application, a pressure transducer is connected to a pipe, A, which is carrying fluid under pressure. The transducer, B, will change pressure variation into electric current variation.

In the amplifier, C, electric current variations are amplified and connected to D. This is a relaying device which may translate what was a weak pipe pressure variation into E. E may be a recorder, pressure gauge, signal light, oscilloscope or some other signal indicating device.

6-62 THERMOCOUPLE AND THERMOELECTRIC

Thermocouples may be used to measure temperatures or to operate controls. Their principle of operation depends on the fact that if two dissimilar (different) metals are connected together and the point of connection heated, an emf (voltage) difference will be developed across the other ends of the two metals. Copper and iron may be used as the two different metals. However, other thermocouples have been developed using tungsten and rhenium as well as other materials.

If a thermocouple is connected to a sensitive voltmeter, the voltage indicated will vary with the temperature of the junction. This principle is the basis of operation of the thermocouple thermometer. The instrument is calibrated to either Fahrenheit or Celsius.

If two or more such junctions are connected in series and connected to a sensitive voltmeter, the voltage indicated will

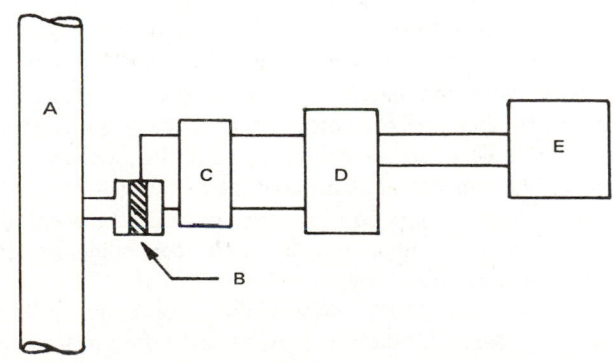

Fig. 6-66. This transducer application indicates pressure in a pipe. A—Water pipe. B—Transducer. C—Amplifier. D—Relay. E—Indicator.

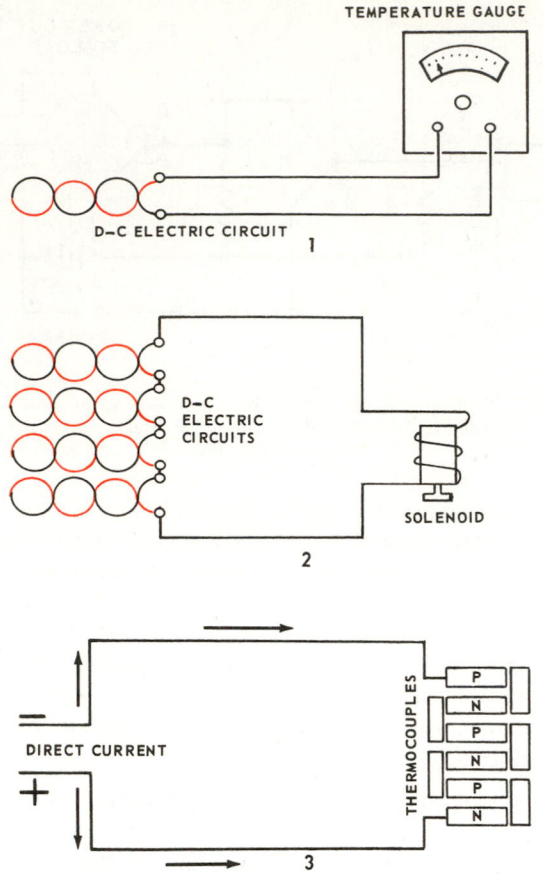

TEMPERATURE GAUGE

D–C ELECTRIC CIRCUIT

1

D–C ELECTRIC CIRCUITS

SOLENOID

2

DIRECT CURRENT

THERMOCOUPLES

P
N
P
N
P
N

3

Fig. 6-67. Three basic uses of thermocouple principle. 1—Temperature measurement (used for −300 to 1700 F. temperatures). 2—Generating d-c electricity to operate solenoid. Solenoid then operates electric circuit or gas valve. 3—Generating heat and cold. When d-c is passed through thermocouple one wire will become hot while other wire becomes cold, depending on direction of current flow.

vary directly with the number of the junctions. A multiple thermocouple installation can generate as much as 500mV. It can, therefore, operate special solenoids which will operate valves on a gas furnace without the need of connecting the system to an electric utility. It may also be used as a safety device to shut off a gas supply if a pilot light is extinguished.

The electrical efficiency of a thermocouple is quite low. It is not an efficient way to generate electricity.

Typical thermocouple applications are shown in Fig. 6-67. In 1820, the German physicist, Thomas J. Seebeck, discovered that when a closed circuit is made through two different metals in contact with each other, an electric current will flow in the circuit when heat is applied to one of the junctions.

In 1834, Jean Peltier discovered that, if direct current is passed through a junction of two dissimilar metals, the junction becomes either hot or cold, depending on the direction of the current flow.

Emil Lenz, in 1837, showed the importance of both Peltier's and Seebeck's discovery. He placed a drop of water on the junction of two dissimilar metals. When current passed in one direction, the drop of water froze. When the current was reversed, the ice melted and the water was warmed.

The principle of the thermocouple, therefore, can be used either as a temperature measuring instrument, a current generator for some sensitive controls, or it may be used for refrigerating effect. Fig. 6-67 shows these three applications of the Seebeck, Peltier and Lenz discoveries.

View 1 of Fig. 6-67 shows the thermocouple being used as a temperature measuring instrument. View 2 shows the current-generating effect. This current is used to control a solenoid valve. View 3 illustrates the effect when the thermocouple is used for heating or cooling.

6-63 PHOTOELECTRICITY

Photoelectric devices are of three particular types:
1. Photoconductor.
2. Photovoltaic.
3. Photoemissive.

Photoconductors are semiconductor devices which increase their conductivity of electricity when they are illuminated. They are used in electric eye devices and in infrared camera devices.

Photovoltaic devices include solar cells. These are semiconductor devices which produce electrical energy when they absorb light. They are used in solar energy conversion and in light meters for photography.

Photoemissive devices give off light when electrical energy is added. The semiconductor in a "light emitting diode" gives off the light which shows the numbers in a calculator or wristwatch. Other light emitting devices include flourescent lights and lasers.

6-64 ELECTRICAL POWER

Electrical power is measured in watts (W), kilowatts (kW), and megawatts (MW). *A watt is the rate at which energy is produced by a current of one ampere flowing under an electrical potential of one volt.* A simple example is the rate at which heat is given off by a wire connected to a 1 volt battery when the current in the wire is 1 ampere.

This can be expressed mathematically:
Power (watts) = current (amps) x electrical potential (volts)
$P = I \times V$

Example: What is the power used by an electric motor that draws a current of 20 amperes (A) from a 120-volt (V) power source?

Solution:
$P = 20 \times 120 = 2400$ W = 2.4 kW

The electrical potential is sometimes called the electromotive force (emf).

In Para. 6-32, the power loss was indicated as I^2R. If the load on a circuit is only a resistance load, then the power is lost as heat. The power loss can then be calculated as either I^2R or $I \times V$, where $V = I \times R$, from Para. 6-31.

A simple electrical circuit which represents a speed control for a d-c fan motor is shown in Fig. 6-68. When the switch is at A, the fan is off, and no current flows. When the switch is at C, the fan is on high speed and the power used is the current I_c times the motor resistance R_m.

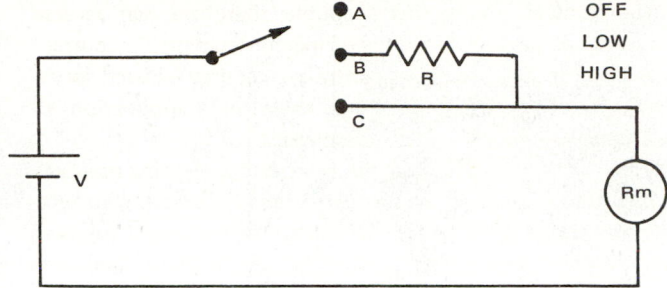

Fig. 6-68. Example of a two-speed fan motor circuit. A—Off. B—Low speed. C—High speed.

$$P_C = I_C^2 R_m$$

The current $I_C = \dfrac{V}{R_m}$

Then $P_C = V I_C = \dfrac{V^2}{R_m}$

If the switch is moved to B, the fan is on low speed. The power is then:

$$P_B = I_B^2 (R + R_m)$$

The current is $I_B = \dfrac{V}{R + R_m}$

Then, $P_B = V I_B$

$$P_B = \dfrac{V^2}{R + R_m}$$

The power (P_B) used with the resistance R in the circuit is less than the power (P_C) used without the resistance. With the switch on low speed, the power used by the fan motor is lower.

6-65 POWER FACTOR

If voltage and current vary within the cycle as shown in Fig. 6-69, the power must be calculated differently. The fact that the current and voltage are not varying together means that the product of the voltage and current varies with time in the cycle. An average voltage times current product must then be used to calculate the power.

An average voltage for a voltage that varies as shown is called the root mean square (rms) voltage (Vrms). It is equal to the maximum voltage (Vmax) times a constant.

Vrms = Vmax x 0.707

Usually, a-c circuits are designated by the rms voltage, such as 120, etc.

The average current is also:

Irms = Imax x 0.707.

If voltage and current vary so that the maximum voltage and current occur at the same time, as shown in Fig. 6-70, the power will be:

P = Vrms x Irms.

If the voltage and current maximum do not occur at the same time, the power is multiplied by the power factor, PF.

P = Vrms x Irms x PF

The voltage and current are then said to be "out of phase." If the electrical load in a circuit contains inductor or capacitor elements, then the voltage and current are out of phase. Inductive loads include electrical motor windings, transformers, solenoids, relays, and electrical coils. Capacitor loads include condensers and crystals. Elements that produce out-of-phase voltage and current, store electrical energy during part of the cycle and release it later. This produces the out-of-phase behavior.

The power factor goes from 1 when the voltage and current are in phase to 0.64 for the situation shown in Fig. 6-69, where the voltage and current are not in phase.

The efficiency of inductive loads, like motors, can be improved by increasing the power factor. This can be done by connecting the proper value capacitor across the motor terminals. The increase in the power factor reduces the current flow (Irms) through the motor resistance, so that the product

P = Vrms x Irms x PF

is smaller.

Electrical utilities usually limit the power factor allowable in industrial and commercial loads. Normally a power factor of at least 0.85 is required.

To increase motor efficiency, the power factor should be brought as close as possible to one. To do this, a wattmeter should be connected to the motor. Capacitors should then be connected across the motor terminals until a minimum wattage reading is obtained on the wattmeter. Power draw by the

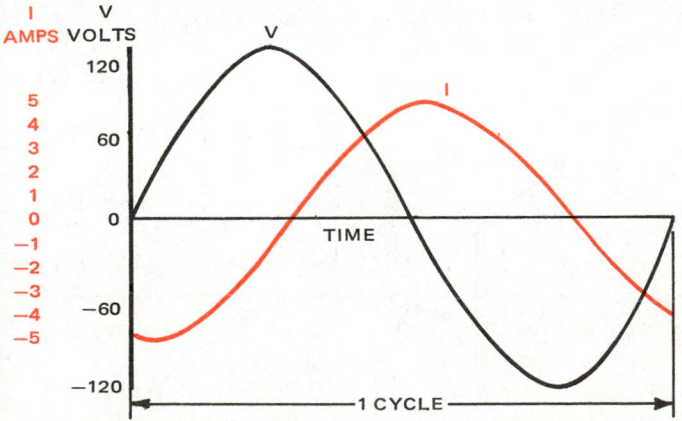

Fig. 6-69. Combined voltage and current flow curves for an inductive load. Current is at maximum when voltage is zero. This is the situation in a low resistance motor or coil. Power factor is a minimum.

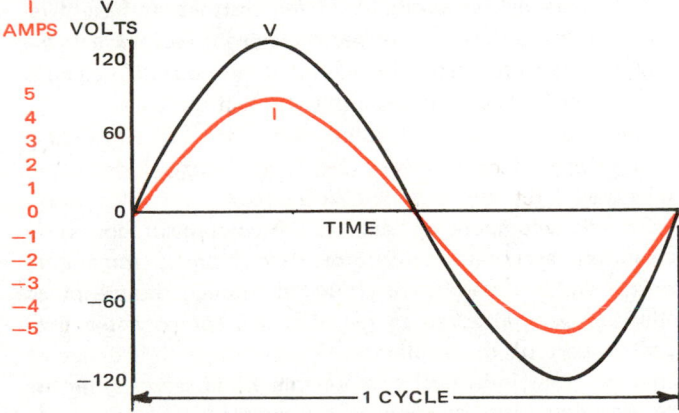

Fig. 6-70. Combined voltage and current flow curves for a resistive load. Current is at maximum when the voltage is at maximum. Power factor is one.

motor will then be at a minimum and the power factor as high as possible.

6-66 GROUNDING

Most soil (ground) is a fairly good conductor of electricity. Moist ground is a better conductor than dry ground. In the development of telephone and early power distribution systems, the ground was frequently used for the return circuit.

A wire for the return circuit was merely extended into the ground, and the electricity flowed through the ground to the end of the circuit. The symbol for the ground became as shown in Fig. 6-71.

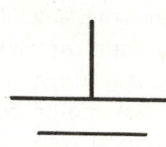

Fig. 6-71. Ground symbol indicates that conductor is attached to frame, shell or other structural part of a mechanism.

The wire ground now, however, refers to an electrical circuit which is attached to the frame, shell or other structural part of mechanism. The symbol for this type of ground may be the same as shown in Fig. 6-71. To ground a receptacle, it is common practice to electrically connect the ground connection to a convenient water pipe. But the best ground is a metal stake driven about 8 ft. into the ground.

Since the 1970 models, all major household appliances are required by Underwriter's Laboratories to be grounded. Therefore, all units produced after September 2, 1969 have a three-wire grounded service cord.

When installing a grounded appliance in a home that does not have a three-wire grounded receptacle, under no conditions is the grounding prong to be cut off or removed. It is the responsibility of the customer to contact a qualified electrician and have a properly grounded three-prong wall receptacle installed in accordance with the appropriate electrical code.

Should a two-prong adaptor plug be required temporarily, it is the personal responsibility of the customer to have it replaced with a properly grounded three-prong receptacle or the two-prong adaptor properly grounded by a qualified electrician in accordance with the appropriate electrical code.

The standard accepted color coding for ground wires is green or green with yellow stripe. These ground leads are not to be used as current carrying conductors.

Electrical components, such as the compressor, condenser fan motor, evaporator fan motor, defrost timer, temperature control and ice maker, are grounded through the use of an individual wire attached to the electrical component and to another part of the appliance. Ground wires should not be removed from individual components while servicing, unless the component is to be removed and replaced.

Should any of the grounded components require servicing which would necessitate removal of the ground wire, it is extremely important that the service technician replace any

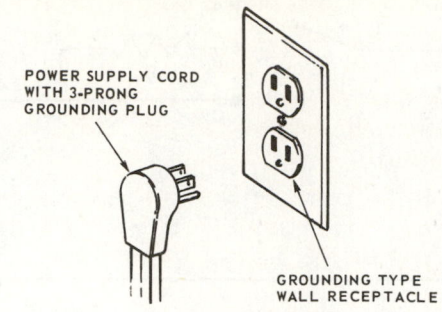

Fig. 6-72. An approved appliance cord receptacle. Slotted openings are for hot and neutral wire. Round opening is ground connection.

and all grounds prior to completion of a service call. Under no conditions should a ground wire be left off. It is a potential hazard to the service technician and the customer.

Fig. 6-72 illustrates a properly grounded receptacle. A grounding adaptor which may be used temporarily is illustrated in Fig. 6-73. A method used to ground old style ungrounded receptacles is shown in Fig. 6-74.

A device called a ground fault protector or interrupter (GFI) is recommended for use on certain circuits. It is a circuit breaker and it will open electrical supply circuits if as little as 5 milliamperes (.005 amperes) is leaking out of the circuit and into the ground.

A GFI is required on outdoor outlets, outdoor lighting, swimming pools and the like. It should also be used by the service technician when using portable tools on extension

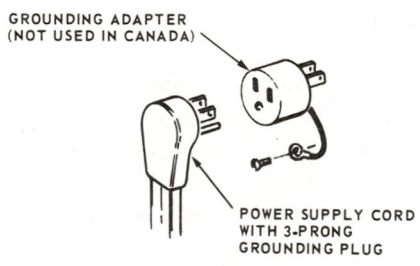

Fig. 6-73. Typical grounding adaptor.

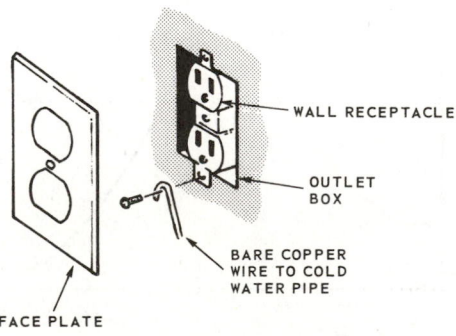

Fig. 6-74. Properly grounded wall receptacle.

cords out-of-doors. These devices are small and they may be plugged into an outlet. This device gives the best protection against shock. There is one exception to this regulation. Certain portable electrical tools, such as electric drills, have the electrical motor insulated from the case. These tools do not require grounding.

6-67 SINGLE-PHASE — THREE-PHASE

A single-phase cycle is explained and illustrated in Para. 6-7. As shown in that paragraph, the voltage and current start at 0, rise to a maximum, and fall to 0 again as the cycle repeats. This means that there is no power produced during the instant that the voltage and current are 0.

Other cycles may be imposed on the above cycle in such a manner that there is, at all times, voltage and current flowing through the circuit. Such an arrangement is called polyphase. The two most common phases in use are single phase and three phase.

The voltage and current characteristics of the three-phase cycle are shown in Fig. 6-75. It should be noted that, in the three-phase system, there is always a considerable voltage applied.

Three-phase motors are generally more efficient than single-phase motors. Three-phase motor sizes begin at about 1/2 hp and extend upward. Small fractional hp motors are usually single-phase motors, and these are not commonly used above 1 hp.

6-68 POWER CIRCUITS

The electric motors used in an air conditioning or refrigeration system must be designed to be used with the electric power furnished by the electric utility.

These motor properties must match the power source in:
1. Emf (volts).
2. Cycle (Hertz).
3. Phase.

The wires must be large enough to carry the full or maximum current that the motor will use. The voltage may be:

1. 110V.
2. 115V.
3. 120V.
4. 208V.
5. 220V.
6. 230V.
7. 240V.
8. 277V.

The cycle (Hertz) may be:
1. 25.
2. 50.
3. 60.

The phase may be:
1. Single phase.
2. Two phase.
3. Three phase.
4. Four phase.

Some popular power electrical sources are:
115V, 60 cycle, single phase.
120V, 60 cycle, single phase.
208V, 60 cycle, single phase.
230V, 60 cycle, single phase.
240V, 60 cycle, single phase.
230V, 60 cycle, three phase.
240V, 60 cycle, three phase.
440V, 60 cycle, three phase.

Carefully check the power source before purchasing or installing equipment. Check with the electrical utility before installing equipment of any sizable horsepower.

One of the concerns of the electrical utility is the flicker in lights, television sets and radios when a motor compressor starts. This flicker is especially critical if the motor compressor starts more than four times in an hour. This flicker is worsened when the electrical flow to the motor is not strong enough (wires too small or too long).

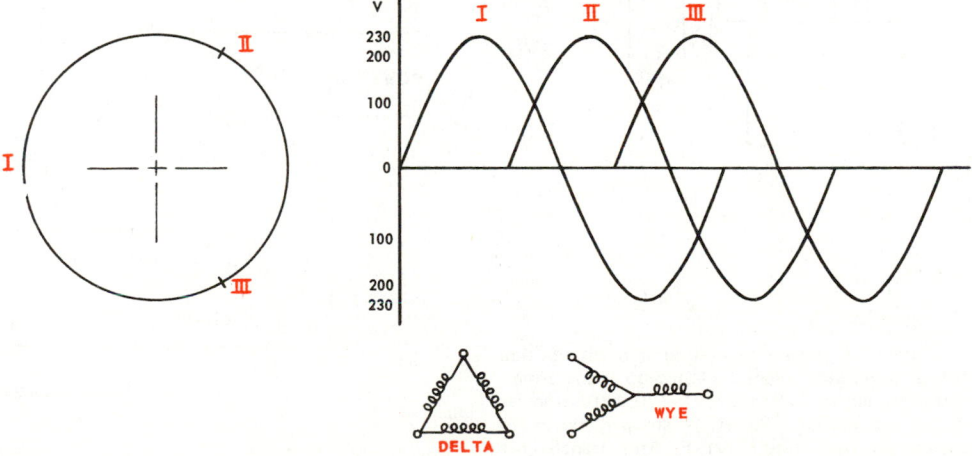

Fig. 6-75. Three sine curves of three phase circuit. System can be wired for either Delta design or Wye (star) design.

6-69 TRANSFORMER PRINCIPLES

Several types of transformers are used by electrical utilities to transform high voltage to electrical power satisfactory to the user. The systems used are dependent on whether the electrical power is mainly for residential, for industrial power or for commercial lighting.

At the generating station, the power is stepped up to a voltage considerably above that used by appliances and motors for either domestic service or by industry. This current flow is usually sent across country over a high voltage transmission line at 120,000V. A schematic diagram, Fig. 6-76, illustrates the fundamentals of a power distributing system.

Step-down transformer stations (region stations) are located along the high voltage transmission line in areas to be served with power. At these region stations, step-down transformers reduce the high voltage transmission line electricity tc 40,000V. This is carried to the communities to be served where it is again stepped down to 13,200V or 4800V. Circuits of this voltage must be handled by certified utility line workers.

Electricity is carried to the customers by the primary distributing lines. At the customer's business or home, the electricity is again stepped down in emf. This is 120V, 240V and 440V, depending on the customer's needs.

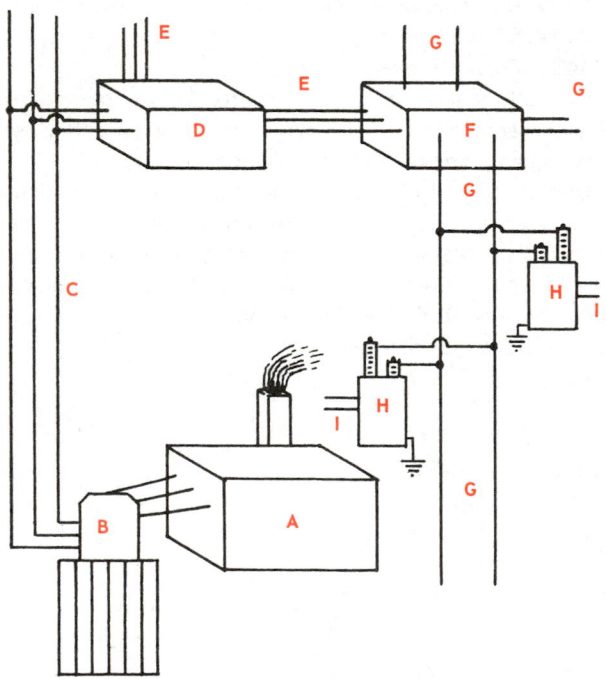

Fig. 6-76. Schematic diagram of power generating and distribution system. A—Steam power generating plant. B—Step-up transformer. Generated power is stepped up to 120,000V. C—120,000V transfer lines. D—Region transformer station. Power is stepped down to 40,000V. E—Sub-transmission line 40,0000V. F—Area transformer station. Voltage is stepped down to 13,200 or 4800V. G—Primary distribution circuits. H—House or neighborhood transformers. I—Secondary circuit to homes, businesses and industries: 120, 240 or 440V.

6-70 TRANSFORMER AND MOTOR CIRCUITS (CHARACTERISTICS)

Transformers are required to step down high voltage to the final voltage to be used by the consumer.

The power coming into the transformer windings is called the primary, the power going out of the transformer windings is called the secondary, regardless of which of the two are of the higher voltage.

Common types of transformers are:
1. Delta type (from the Greek letter *Delta* Δ).
   A. Open Delta.
   B. Closed Delta.
2. Wye (Star) type (Y).

Fig. 6-77 shows a wiring diagram for the Open Delta type transformer. The Closed Delta type will be explained later.

The input voltage from the power station can be one of several — but usually 4200V — and the circuit is three phase (three hot wires). The Open Delta system is different from the Closed Delta system in that two connected transformers are used (1 and 2) instead of three.

The voltage of each secondary outlet is designed to be 240V (A to B or B to D). A ground wire is connected to the middle of the secondary winding of 1. Therefore, the emf between D and C is 120V and B and C is 120V. A special voltage is created between C and A. Because the secondary winding is all of 2 and then angles half way down the secondary winding of 1, the emf between A and C (a geometric change of angle) is 208V.

This 208V is popular where the main electrical load of a building is the lighting load. It is not a good motor voltage. Many motor compressors, even so, are connected to this type circuit. The motor must be designed to operate at this voltage or a correction type line voltage transformer must be used. (See Para. 6-71.)

The longer the wire, the larger its diameter must be to safely carry the current. The voltage should be measured at the motor compressor, not at the power panel.

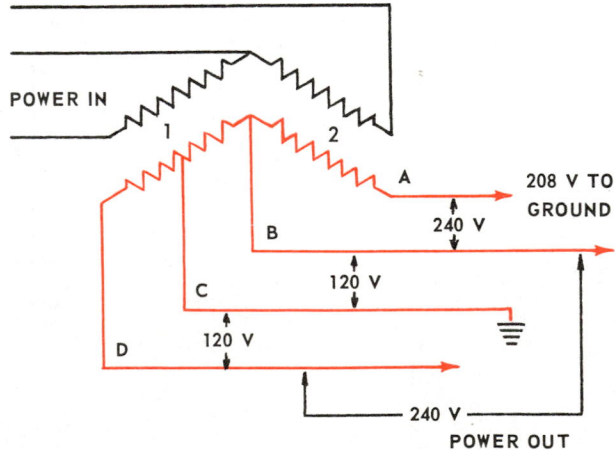

Fig. 6-77. Circuit diagram of Open Delta transformer. Power may be obtained from four taps, A, B, C and D.

INPUT VOLTAGE	CURRENT	TORQUE	TEMPERATURE		CYCLE SLIP	EFFICIENCY
+10%	– 7%	+21%	–3 deg. C. = (5.5 deg. F.)		17% decrease	1% increase
–10%	+11%	–15%	+7 deg. C. = (12.5 deg. F.)		23% increase	2% decrease

Fig. 6-78. Effect of an input voltage on operating characteristics of motor.

A dangerous condition may develop if there is a voltage drop of over 5 percent (less than 95 percent of the desired voltage) at the motor compressor. For example, if a 208V circuit, due to its length, conductor size and ampere flow load, has only (208V x 95 percent = 208V x .95 = 197.6V), it is at the very lowest usable voltage. If the voltage goes below this, the motor may work poorly or burn out.

A serious situation may occur if a 240V motor compressor should happen to be connected to a 208V line (240 x 95 percent = 240 x .95 = 228V.) A voltage of 228V is the lowest voltage on which a 240V motor will work satisfactorily. The 208V circuit is much below the 228V circuit desired. The 240V motor will operate poorly, the overload protection will operate or the motor may fail.

A 230V motor will operate down to 218.5V (230 x 95 percent = 230V x .95 = 218.5V). Therefore, a 208V circuit is dangerous if used with other than 208V motors.

A 220V motor will operate down to 209V (220V x 95 percent = 220V x .95 = 209V) and this motor may operate on 208 volts, but it is a border line case and the efficiency will be low.

However, motors can use a voltage slightly over their rating. In fact, a 208V motor connected to a 220V line will operate very well, will start faster, and will give more power to the compressor. A 10 percent over-normal voltage will let a motor carry a 20 percent overload. However, a 208V motor on a 240V line will be much noisier — a handicap in air conditioning.

A 220V motor operates very well on a 240V line. A 220V motor can operate on a 208V line, but will have lower torque. These are single-phase motors. When only single phase is available and more than 3 horsepower is needed, two or more motor compressors should be used in the system.

Fig. 6-78 is a table of changes in a motor's operating characteristics as the input voltage changes.

A 240V motor compressor works with the same efficiency as a 120V motor compressor. The notion that a 240V motor uses less kilowatt hours to do the same amount of work is false. For example, a 120V motor uses 5A to create 600W (120 x 5) = 600W or .600kW. To provide .6kW, a 240V circuit must carry 2.5A but the electrical cost is the same. The only advantage is that the 240V unit may use smaller size conductors from the meter box to the unit. There will be less voltage drop between the meter box and the motor.

Some motors are labeled 208—220V to indicate that they may be used with either voltage. However, these motors are sensitive to voltages below 208V, and voltages of 195V or lower must not be used.

For example, in a home air conditioner, the power circuit is usually 240V, one phase, three wire. This means there is 240V between the two hot conductors and there is 120V between the hot conductors and the ground conductor, as shown in Fig. 6-79.

These voltages should be checked (verified) with a voltmeter. Do not trust a test light and guess at its bulb brilliance.

When studying a building wiring system to decide on the circuit necessary for a motor compressor, the wire sizes should be checked for capacity. All electrical appliances should be on at the same time. Also, the average load per day should be checked. A recording ammeter may be used for this purpose if the utility does not already have the data. A demand meter will usually be put in by the electrical utility if requested.

6-71 LINE VOLTAGE TRANSFORMER

Improper line voltage may cause refrigeration and air conditioning motors to burn out. Line voltages may vary as much as 5 to 10 percent. Equipment designed to operate on 120V plus or minus 10V may not operate well where the voltage drops to 100V or lower.

When equipment designed for 120V service is used on 240V lines, a line voltage transformer is used to overcome the problem of voltage fluctuation. The transformer is designed to increase and decrease the input voltage to the motor as needed.

Fig. 6-80 shows a basic wiring diagram of a "boost-and-buck" voltage transformer. It is being used to correct the line voltage to 120V. Fig. 6-81 shows a wiring diagram of a single coil transformer designed to change a 208V circuit to 240 volts. This is usually called an autotransformer.

A voltage of 208V is quite common where the main electrical load is lighting. A line transformer will be required if 120V or 240V motors are to be used on the 208V current supply.

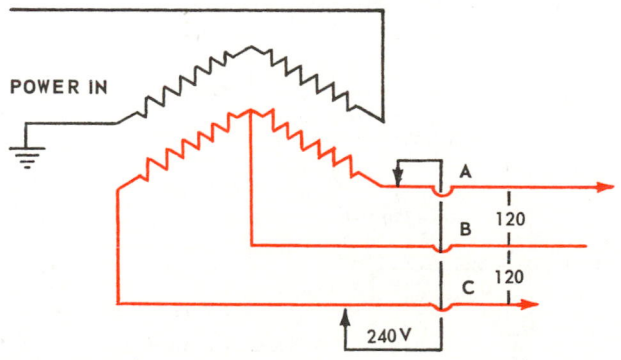

Fig. 6-79. Schematic circuit diagram of transformer used to serve average home. Note that 120V are available between A and B, and between B and C; B is a ground. Also, 240V are available between lines A and C.

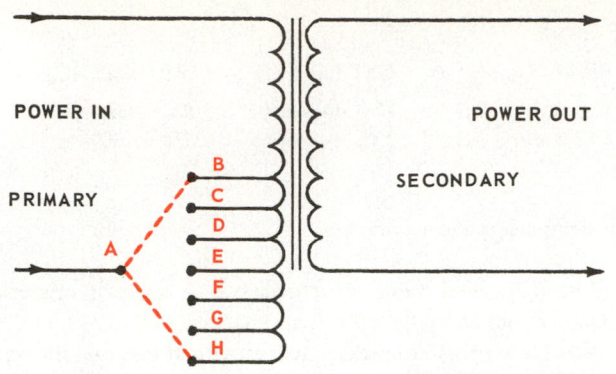

Fig. 6-80. Wiring diagram of transformer designed to step up or step down voltages to provide correct voltage for driving motor. Connecting terminal A to B, C or D will step up power-out voltage. Connecting A to F, G or H will step down power-out voltage. Connecting A to E will not change voltage but will tend to stabilize it.

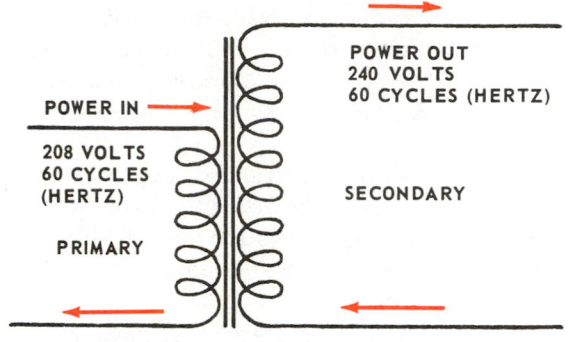

Fig. 6-81. Wiring diagram of voltage transformer used to step up 208V to 240V.

This also is sometimes called an autotransformer. See Fig. 6-82.

A 120V transformer output can be produced through the use of various connections on the input side ranging from 95 to 260V. The service technician must first determine the required voltage from the motor identification plate and

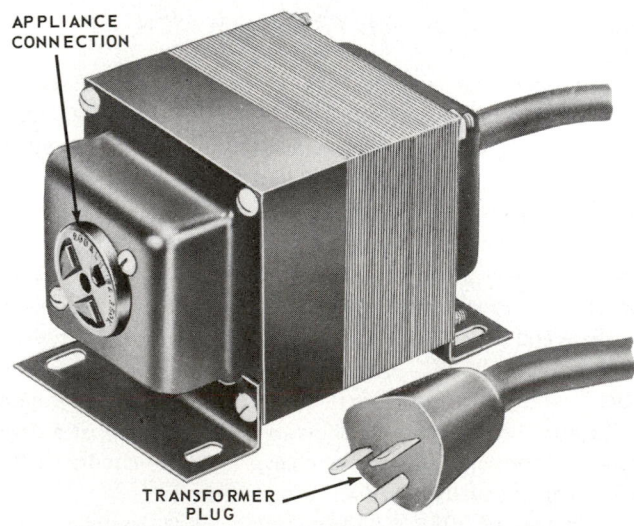

Fig. 6-82. Line transformer used to transform 208V a-c to 240V a-c. (Acme Electric Corp.)

specifications.

A voltmeter is used to measure the voltage and ampere flow at the motor compressor when the unit is running. A transformer is then selected which will carry the current. It is adjusted to raise or lower the voltage as needed.

Transformers are rated in KVA output. KVA means "kilovolt amperes." To determine KVA, one must multiply the output voltage by the amperes.

$$KVA = \frac{Volts \times Amperes}{1000}$$

The total load of the circuits to be fed by the transformer must not be more than the transformer output. The transformer is connected into the circuit between the electrical service and the motor. The service technician should check the completed installation for both ampere flow and voltage. If these values are different than the motor ratings, further adjustments will be necessary.

6-72 THREE-PHASE, FOUR-WIRE TRANSFORMER

Another transformer design used by utilities is the Wye type shown in Fig. 6-83. This system uses one transformer, connected as shown. Note that a 208V circuit is also possible

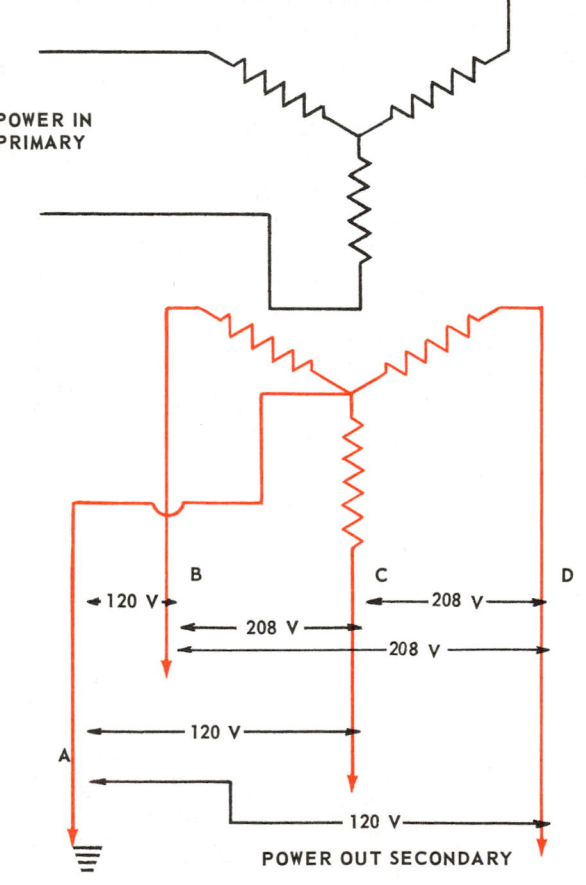

Fig. 6-83. Wye-type transformer. Note that center of Wye is grounded in secondary circuit. Following emf may be obtained from this transformer by connecting terminals as indicated: A and B—120V. A and C—120V. A and D—120V. B and C—208V. C and D—208V.

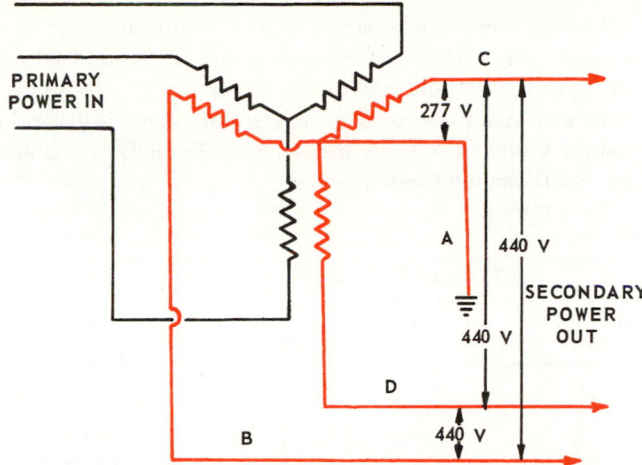

Fig. 6-84. Schematic of Wye-type transformer used in serving industrial and commercial applications. Following voltages are available from this transformer. Between A and C, 277V. Between B and D, 440V. Between C and D, 440V.

from this system. In fact, any two of the three conductors produces 208V. This is made possible by the angle of the two secondary windings.

Some utilities recommend a 277 — 440V system. It is the least critical and therefore is good for commercial-industrial use. This system is shown in Fig. 6-84.

Utilities circuits may show as much as 212V at the meter on a 208V system, or as much as 250V on a 240V system. However, because an increase in voltage is as much a disadvantage to light bulb life as it is an advantage for a motor compressor, the utilities keep their circuit voltages as close to the required value as possible.

The Closed Delta system, Fig. 6-85, is sometimes found in large industries. The transformer is not grounded.

A 277V circuit is used for some refrigeration or air conditioning units of 15,000 to 35,000 Btu/hour capacity. It may be obtained by an angular tap from a Wye type transformer, when one lead is from any of the three legs and the other lead is from the center of the Wye. The secondary voltage between A and B and A and C is 480V.

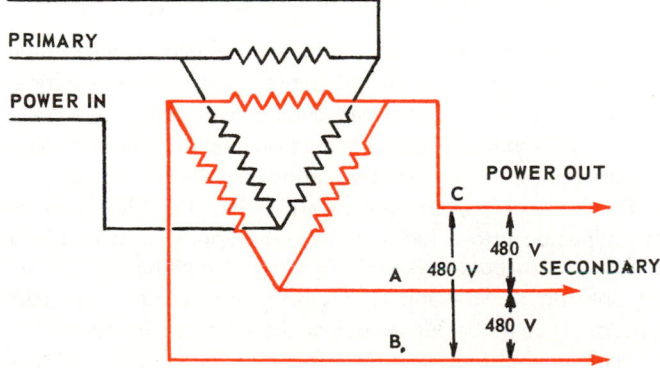

Fig. 6-85. Diagram of Closed Delta type transformer which will deliver 480V between either A and B or A and C.

6-73 RELAYS

Various types of relays are used in electrical circuits on both refrigerating and air conditioning mechanisms. Fan relays, motor starting relays and main power line relays are very popular in air conditioning and refrigerating electrical systems.

These relays are usually an enclosed switch operated by an electromagnet. Relays are designed to protect motors by disconnecting them from the line when they become overheated or overloaded. The amount of current used to operate a relay is very small. The contact points in the relay then operate circuits which can carry quite heavy currents. Fig. 6-86 shows the electrical circuit for a typical relay.

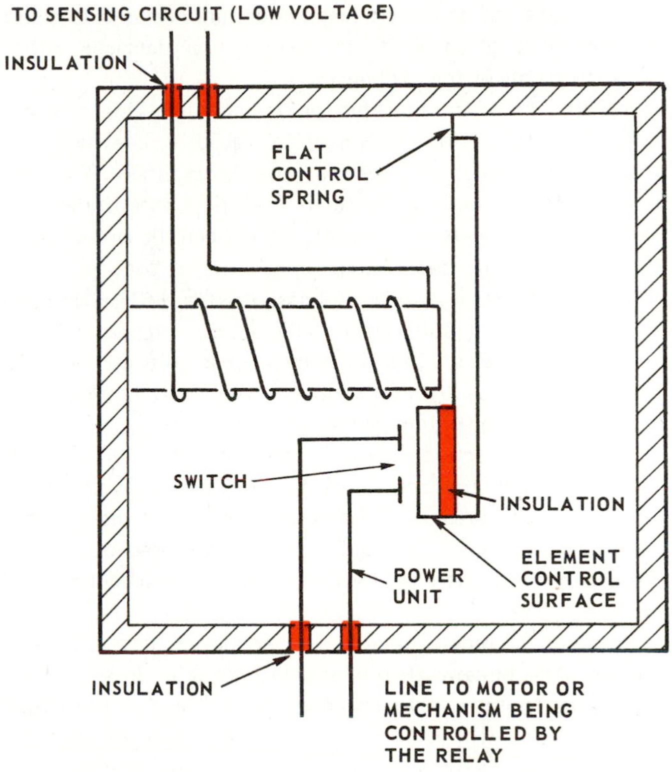

Fig. 6-86. Electrical relay. Low voltage thermostat or sensing instrument controls operation of relay which, in turn, switches power circuit on and off. Power circuit is normally off. It is on only when relay low voltage circuit is energized.

6-74 ELECTRICAL CODES

The National Electrical Code establishes rules and regulations covering materials and methods used in installing electrical systems. In addition, many cities and communities have supplementary local codes. All electrical installations should be made in conformity with national and local codes.

The National Electrical Code provides for a type of wiring called Class 2. These circuits are usually used for controlling relays, bells, signal systems, communications and the like. The current supply is usually from a small transformer having two windings. These transformers have a capacity of about 100 voltamperes.

Refrigeration and air conditioning service technicians are permitted to make Class 2 connections and installations. Under this code, the current from the secondary winding of the transformer is limited according to the voltage: below 15V, up to 5A; 15 to 30V, up to 3A; 30 to 60V, up to 1.5A; above 60V, not over 1A.

Underwriters Laboratory tests and approves electrical components such as switches, extension cords and relays.

6-75 CIRCUIT PROTECTION

Previous paragraphs explain the heating and magnetic effect of current flowing through electrical circuits. Instruments and appliances may be seriously injured or ruined by either overheating or, in the case of instruments, too much magnetism.

Four types of safety devices are commonly used to avoid the possibility of an accidental current surge damaging either the instruments or the appliances.

The most common protection is the fuse. This is a soft metal conductor wire which may be placed in series with the circuit with either a plug or a cartridge arrangement. The size, length and material in the fuse are such that its resistance will cause enough heat to melt the fuse and open the circuit if the current exceeds the fuse rating.

Fuses are rated in amperes. Those used in refrigeration and air conditioning circuits are usually designed to carry 5, 10, 15, 20 or 30A. Fuses used in electronics circuits are available from 1/500A to 2A. Some electronic fuses, however, are available with higher ampere ratings.

Another protective device is the circuit breaker. Current flowing through the circuit being protected passes through a solenoid in the breaker. The magnetic effect of the solenoid current will cause the solenoid to trip a spring-loaded switch in the event the current in the circuit exceeds a predetermined rate (amperage).

Another common circuit breaker device has a bimetal breaker. The bimetal strip is usually connected in series with the breaker point and a resistance heater. The current flowing in the circuit will cause the resistance heater to heat the bimetal strip, causing it to bend. The heating effect will depend upon the current flow — the more current, the more heat. The instrument is calibrated to open the points when the current flow is greater than safely allowed.

The fourth type of circuit protection makes use of the thermistor. Thermistors have been developed for circuit protection which either regulate (modulate) the current flow or which may, in some cases, cause the current flow to be reduced to a safe value.

CAUTION: Never connect alternating current appliances or instruments into direct current circuits. Never connect direct current appliances or instruments into alternating circuits.

6-76 WIRE SIZES

The current carrying capacity of a solid conductor depends on its diameter. Larger wires may carry a heavier current than smaller wires. Information following is based on the use of copper conductors (wires).

The electrical supply wire to outlet receptacles must be of adequate size. Fifteen-ampere outlets should be supplied with No. 12 wire, 30A outlets with No. 10 wire.

Wire sizes are measured in circular mils. A mil is the area of a circle 1/1000 (.001) in. in diameter. Fig. 6-87 is a table of the more common conductor sizes.

WIRE SIZE AWG*	CONDUCTOR DIAMETER IN INCH DECIMALS	CIRCULAR MILS	OHMS PER 1000 FT. AT 68 F.
14	.0641	4,110	2.52
13	.0720	5,180	2.00
12	.0808	6,530	1.59
11	.0907	8,230	1.26
10	.1019	10,380	1.00
9	.1144	13,090	.7925
8	.1285	16,510	.6281
7	.1443	20,820	.4981
6	.1620	26,240	.3952
5	.1819	33,090	.3134
4	.2043	41,740	.2485
3	.2294	52,620	.1971
2	.2576	66,360	.1563
1	.2893	83,690	.1239
0	.3249	105,600	.09825

* American Wire Gauge

Fig. 6-87. Wire size data for rubber or thermoplastic covered conductors.

The cross-sectional area of a No. 10 wire is about 10,000 circular mils. Its resistance is about 1.0 ohm per 1000 ft. A No. 7 wire has approximately double the circular mil area of a No. 10 wire, and half the resistance (approximately 0.5 ohm per 1000 ft.). A No. 4 wire has a circular mil area of approximately 40,000 and its resistance is approximately .25 ohms per 1000 ft. This indicates that the circular mil area doubles every third size as the wire becomes larger and the resistance is cut in half. From this data, the electrician can easily approximately calculate the wire size and resistance for commonly used conductors.

6-77 ATTACHMENT PLUG CONFIGURATIONS (TERMINALS)

It is sometimes necessary to connect electrical devices using flexible leads and attachment plugs. Since most electrical devices are designed for a particular power supply specification, it is important that connections to the power supply conform to the electrical specifications for the equipment.

For instance — if an appliance designed for 120V were to be connected into a 240V circuit, the appliance would very quickly "burn out." Likewise, if an appliance has protection for only up to 15A and is connected into a circuit of 30A capacity, it could be burned out or the safety device ruined.

The National Electrical Manufacturers Association (NEMA) has established attachment plug configurations (terminals). Fig. 6-88 shows some common receptacle and plug configurations for the most commonly used refrigeration and air conditioning equipment.

		15 AMPERE		20 AMPERE	
		RECEPTACLE	PLUG	RECEPTACLE	PLUG
2-POLE	3-WIRE GROUNDING	125V			
		250V			
		277V			
3-POLE 3-WIRE		3-PHASE 250V			
3-POLE 4-WIRE GROUNDING		3-PHASE 250V			
4-POLE 4-WIRE		3-PHASE 120/208V			

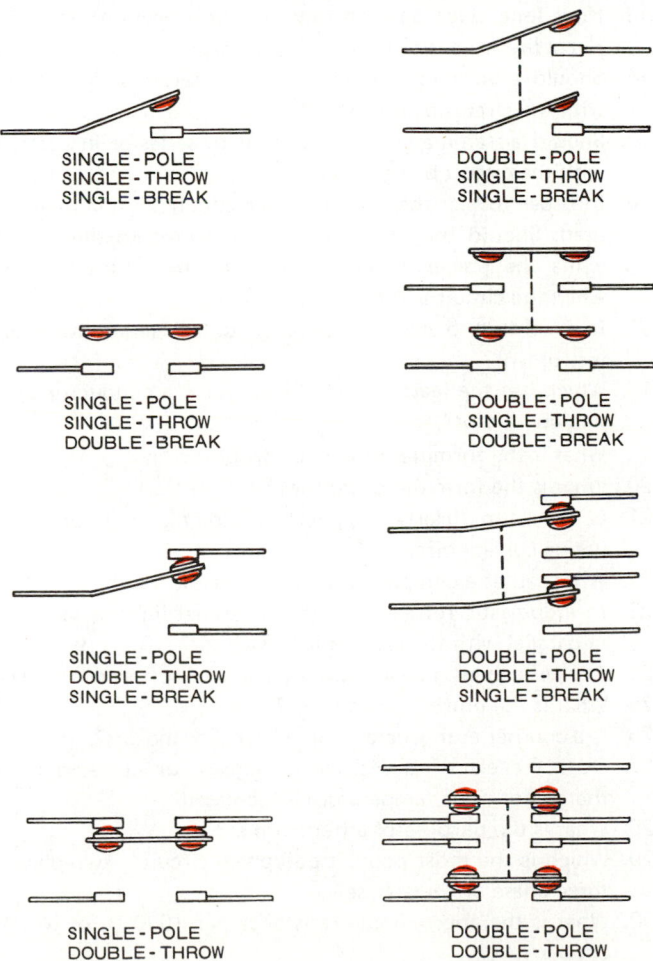

Fig. 6-88. NEMA configurations are illustrated for general purpose nonlocking plugs and receptacles.

SINGLE - POLE
SINGLE - THROW
SINGLE - BREAK

DOUBLE - POLE
SINGLE - THROW
SINGLE - BREAK

SINGLE - POLE
SINGLE - THROW
DOUBLE - BREAK

DOUBLE - POLE
SINGLE - THROW
DOUBLE - BREAK

SINGLE - POLE
DOUBLE - THROW
SINGLE - BREAK

DOUBLE - POLE
DOUBLE - THROW
SINGLE - BREAK

SINGLE - POLE
DOUBLE - THROW
DOUBLE - BREAK

DOUBLE - POLE
DOUBLE - THROW
DOUBLE - BREAK

Fig. 6-89. Types of switches. Red contacts indicate movable breaker points. (Micro Switch, Div. of Honeywell)

6-78 SWITCHES

Any device used to open or close an electrical circuit is called a switch. Switches are made of many materials and there are numerous designs. Some are manually operated, while some are operated automatically.

Fig. 6-89 shows the basic types of switches. All are open contact point. Other types, such as knife blade contact and mercury in a tube contact, use the same basic arrangements.

Normally closed (NC) switches have the contacts touching when there is no power in the circuit.

Normally open (NO) switches have the contacts separated when there is no power in the circuit.

6-79 CIRCUIT TESTING INSTRUMENTS

A test light, Fig. 6-90, is one of the easiest ways to test electrical circuits without causing damage. The test light with the two prongs may be used when the device being tested is connected to electrical power.

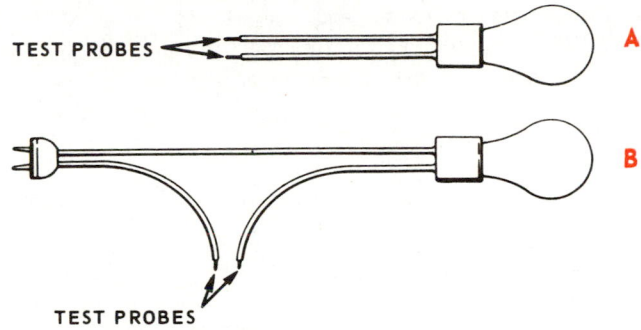

TEST PROBES

A

B

TEST PROBES

Fig. 6-90. Two types of test lights. A—Light bulb with two probes is used to test circuits which have power on. B—Test light connected to electric power source and used to check circuits not connected to power source.

For example, if the refrigerator, air conditioner or heating system is plugged in and will not start, this test light (about 25 to 50W) may be used to determine if power is coming to the wall outlet. The lighting of the light will indicate power is available up to the wire probes. If the wall outlet has power, then the open circuit is in the refrigerator wiring.

The test light at view B in Fig. 6-90 may be used to check electrical devices not connected to power. This test light should be connected to a power source. If the bulb lights when the probes are touched together, the test light is functioning.

To use this light, put one probe on one end of a wire and the other probe on whatever that wire is supposed to connect to. If the bulb lights, the circuit is continuous (there is continuity).

Fig. 6-91 shows a special test light used to test a three-phase circuit. It may be used to determine which leads of a three-phase circuit connect to the three-phase power lines. If the tester glows, the rotation is 1, 2, 3. Reverse any of the two leads and the tester should not glow. If the tester glows, one of

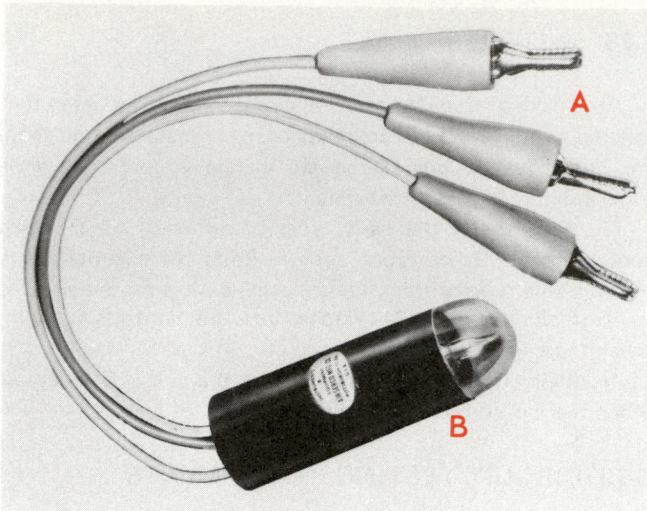

Fig. 6-91. Test light for testing a three-phase circuit. Connect three leads to three-phase terminals. If tester bulb glows, rotation is 1, 2, 3. If it does not glow, reverse pairs of leads until it does glow. If tester glows regardless of connections, one circuit is open. A—Terminals. B—Bulb. (Airserco Mfg. Co.)

the three-phase circuits is open.

These test lights should not be used on solid state circuits. They may damage diodes, transistors or other parts.

6-80 REVIEW OF SAFETY

There are two different factors to be considered in reviewing safety in this chapter.
1. Safety to the operator.
2. Safety to the equipment.

One must have great respect for electricity. It only requires 25mA to seriously injure or kill a person. Voltages as low as 30V can force this much current through vital parts of a person (the heart and brain are the most delicate). Never handle live circuits when in contact with pipes, other wires, damp floors, damp ground or water.

REMEMBER: Electricity cannot be seen. One can only see its effect in light, heat, magnetism and so on. It is too late to learn that a circuit is live after one has touched it. Use instruments to find out if the wires or equipment are safe before starting to work.

Always use insulated tools (or dry rubber gloves) when working on circuits of 30V or higher. Even a large battery can injure one if a tool, a wrist watch or a ring should short circuit the battery. The metal will become hot enough to cause severe burns.

Before working on household, commercial or industrial electrical circuits, disconnect the power. Make sure no one can turn the power on while the circuit is being worked on (use signs, locks or other devices).

Instruments and equipment used in refrigeration and air conditioning work, while durable, are quite sensitive to abuse. When connecting an electrical instrument into a circuit, make sure that the instrument and its setting are within the voltage and current range which may be applied to the instrument. An instrument adjusted to measure a voltage over 150V range will be ruined if connected into a 440V circuit.

In making capacitor tests, the capacitor should not be left in the circuit any length of time. Usually a second or two is sufficient. A longer period may destroy the capacitor.

6-81 TEST YOUR KNOWLEDGE

1. Name the unit of electromotive force.
2. What, in electricity, is similar to gallons per hour of water?
3. Why is the statement "d-c current" incorrect?
4. What is the unit of electrical power?
5. In a-c, does the power equal the emf (volts) times the current (amperes)?
6. In Ohm's Law, the letter I stands for what word?
7. Write the three formulas of Ohm's Law.
8. Can a mechanized motion (rubbing) produce electricity?
9. Describe an electrical circuit.
10. Where is a thermistor used?
11. What is voltage drop?
12. What is electrical continuity?
13. How long does current flow in one direction when 60 Hertz (cycle) current is used?
14. Should a voltmeter be connected in series or in parallel with the circuit being tested?
15. Should an ammeter be connected in series or in parallel with the circuit being tested?
16. In order to test the resistance in a circuit, an ohmmeter is used. Should the ohmmeter be connected to the circuit while the power is being applied to the circuit or only when the circuit is turned off?
17. Is a shunt placed in parallel or in series with an instrument?
18. Which has the least electrical resistance, a conductor or a semiconductor?
19. What is the formula for voltage drop?
20. What is the formula for power loss?
21. Is the term "Hertz" applied to direct current or alternating current circuits?
22. What causes a compass to point north and south?
23. In a domestic refrigerator, is the cabinet light in series or in parallel with the cabinet light switch?
24. Are permanent magnets usually made of soft iron?
25. What is the unit of capacitance?
26. Is a counter emf generated in all running motors?
27. Does the electrical resistance increase or decrease in a thermistor as its temperature is increased?
28. What is the purpose of a transformer?
29. Which is the most popular polyphase circuit — two-phase, three-phase or four-phase?
30. What is the approximate resistance per 1000 ft. of No. 1 copper wire?

Chapter 7

ELECTRIC MOTORS

The compression type refrigerating system must have a power or energy source to turn the compressor. The electric motor is the most popular for the small and medium size units. It is simple, quiet and easily set up for automatic control. The electric motor changes electrical energy into mechanical energy.

Electric motors also drive fans, pumps and other devices in refrigeration and air conditioning systems. It is, therefore, important that the refrigeration service technician thoroughly understand electricity, magnetism, electric motors and electrical circuits.

7-1 ELECTRIC MOTOR APPLICATIONS

In this chapter, power supply voltages are sometimes indicated as 115V and 230V and, at other times, 120V and 240V. The voltages supplied vary in different parts of the country. The service technician should know what voltages are supplied in the community so that equipment matches these voltages.

Refrigerating systems operate either with open or sealed-in (hermetic) motors. The open motor drives the compressor directly off the shaft or by means of a belt. The sealed-in or hermetic type is built inside the compressor dome. It usually drives the compressor directly. Open electric motors are also often used to drive many accessory devices.

Motors may be grouped into four general classifications according to use:
1. To drive compressors.
   A. Open belt drive (external drive).
   B. Hermetic direct drive.
2. To drive fans for:
   A. Condensers.
   B. Evaporators.
   C. Air circulation.
   D. Induced draft.
   E. Forced draft.
3. To drive pumps.
   A. Condensate pumps.
   B. Chilled water pumps.
   C. Condenser water pumps.
   D. Ice making machine water pumps.
   E. Oil pumps.
4. To drive miscellaneous devices.
   A. Vending machines.
   B. Automatic ice cube makers.

7-2 THE MOTOR STRUCTURE

All motors have a similar basic construction.
Each has two main parts:
1. The stator.
2. The rotor.

The stator may also be known as the frame. This frame is usually cylindrical in shape. The field poles with field windings on them are part of the stator. The identification plate is also mounted on the stator.

The rotor is mounted on a shaft which has two journal bearings, one at each end.

The frame has end bells or plates attached to it. These hold the bearings. When the rotor shaft journals are mounted in the bearings, the bells support the rotor.

The bearings are accurately machined to provide the proper amount of end play for the rotor. There is a clearance of .001 to .002 between the motor shaft and the bearing. In hermetic units built into the compressor dome, the compressor bearings may also serve as rotor bearings.

The windings are insulated copper wire. This insulation is usually a polyester material. It is resistant to moisture and has considerable dielectric and mechanical strength.

7-3 TYPES OF ELECTRIC MOTORS

Both alternating current (a-c) and direct current (d-c) may be used to operate electric motors. Alternating current is most commonly used, but, direct current motors are found in areas supplied with direct current only.

Following are the alternating current motor types used to drive compressors. Those listed in red are most common and will be explained in greater detail:
1. Basic types:
   A. Single-phase.
   B. Two, three and four-phase (polyphase).
2. The open motor. These are used on open, belt driven compressors. Types include:
   A. Repulsion-start induction.
   B. Capacitor-start induction.
   C. Capacitor-start, capacitor run.
   D. Capacitor-run.
   E. Permanent split capacitor.
   F. Induction polyphase.
3. Hermetic motors. These types are used on sealed systems in which the motor and compressor are enclosed in a common

housing or dome:

A. Capacitor-start induction.
B. Capacitor-start, capacitor-run.
C. Capacitor-run.
D. Permanent split capacitor.
E. Induction two-phase and polyphase.

4. The condenser and evaporator fan motor types:
A. Split-phase.
B. Shaded-pole.
C. Capacitor.
D. Permanent split capacitor.

Today, the capacitor motor is being used on most applications. It is one of the most popular for single-phase hermetic machines.

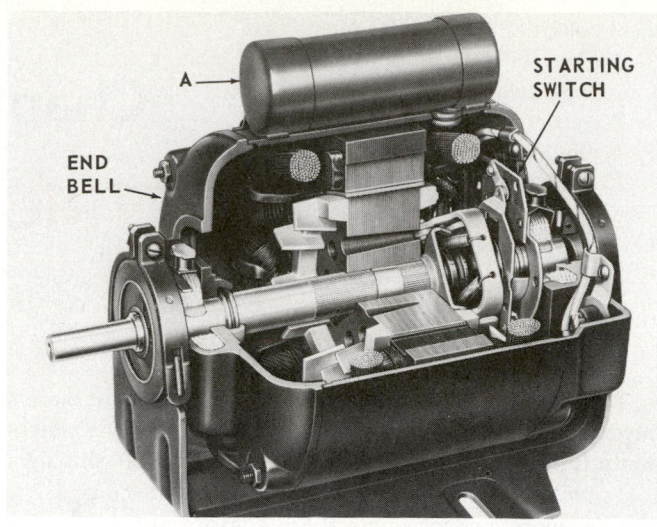

Fig. 7-2. Capacitor-start induction motor. Motor has rubber mounts at each end which provide flexibility. A—Capacitor is mounted in housing on top of motor. Wick oiling is used.

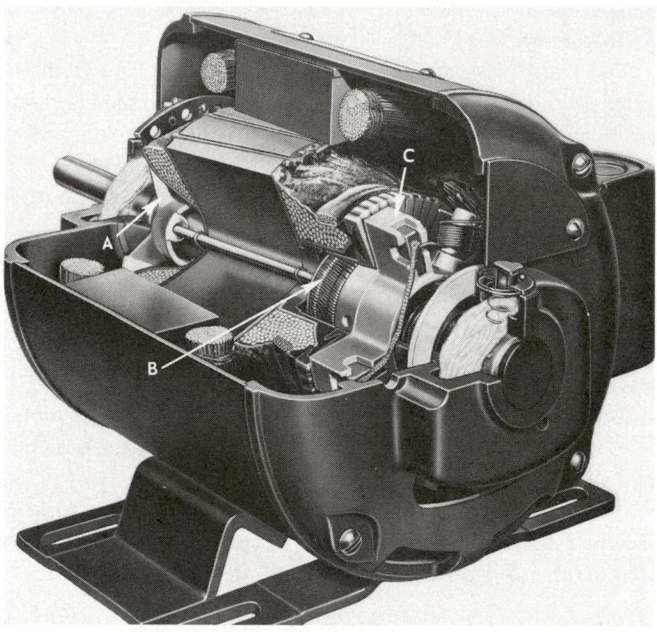

Fig. 7-1. Section view of repulsion-start induction 1/2 hp motor. Note that this motor has wound rotor and centrifugal mechanism which shorts the rotor winding and raises brushes off commutator as soon as motor attains approximately 75 percent of running speed. A—Centrifugal mechanism. B—Commutator shorting segments. C—Brushes and brush holder. Wick oilers are used at bearings.
(Wagner Electric Corp.)

7-4 EXTERNAL DRIVE MOTORS

Three main types of motors are used to drive external drive compressors:

1. Repulsion start-induction run motor, as shown in Fig. 7-1.
2. Capacitor start-induction run motor, Fig. 7-2.
3. Three-phase motors.

V-belts connect the motors to the compressor or a direct connection is made with a coupling. The speed reduction for belt drives is usually about three to one. This means that the compressor flywheel diameter is three times larger than the motor pulley diameter. Para. 7-5 through 7-12 explains features common to most external drive motors. Fig. 7-3 shows the parts of a capacitor type motor.

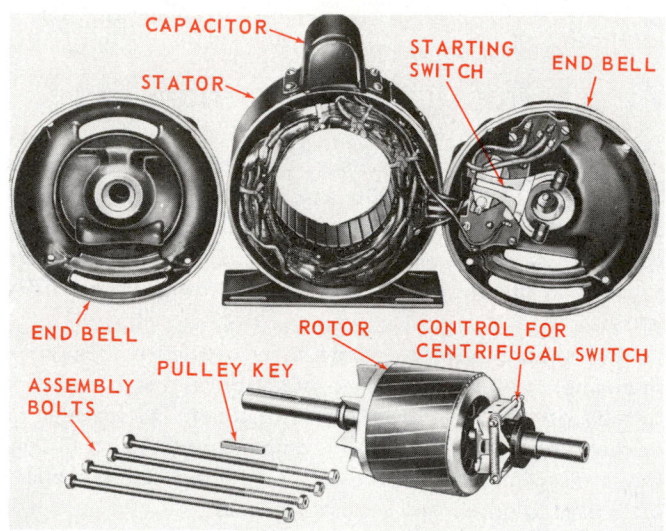

Fig. 7-3. Parts of a capacitor type motor used on open compressor.
(Emerson Electric Co.)

7-5 INDUCTION MOTORS

Induction motors have no windings on the armature (rotor). However, the rotor does have copper bars or other conducting material. These are in its outer surface and lie about parallel to the motor shaft. When current is induced (made to flow) in them, turning power or torque is produced by the magnetism that is created.

One or more field windings are mounted in the stator. Alternating current, passing through these windings, creates a changing magnetic field. This magnetism passes through the rotor, inducing (building up) current in the rotor bars.

The effect of the induced current is to create an opposite magnetic field in the rotor. The opposing magnetism causes the rotor to revolve as it tries to keep up with the changing field polarities in the field windings.

All hermetic machines use induction motors. Most of these motors use two field windings:
1. A starting winding
2. A running winding.

In hermetic motors, it is often necessary to use a starting relay. This is always located outside of the motor compressor dome. This is because electrical contacts would quickly fail from the oil and refrigerant mist inside the dome.

Para. 8-23 describes the various types of starting relays and how they are used.

The starting relay temporarily connects the starting winding of the motor to the power circuit. As soon as the motor reaches about 75 percent of operating speed, the relay opens the circuit and disconnects this winding from the power line.

7-6 SPLIT-PHASE INDUCTION MOTOR

This is the basic motor used for most fractional horsepower appliance operation. It is cheap and simple to operate. Two stator windings are used, one for starting and one for running.

The running windings are made into two or four coils in series depending on the motor speed desired. These windings are energized (have current flowing in them) during the whole time the motor is running.

The starting windings have the same number of coils usually rotated several degrees from the running windings. See Para. 1-13 for information about degrees of arc measurement.

A smaller gage of wire is used in the starting winding. However, it has more turns than the running winding. Counter emf in the greater number of turns causes the current to build up more slowly in this winding. Therefore, the magnetic effect will be several electrical degrees behind the running winding. This will create torque on the rotor to make it start turning in the correct direction. The starting windings are disconnected when the motor reaches approximately 75 percent of its running speed.

There are no windings on the rotor, but magnetism builds up around the rotor bars. Some rotors have heavy copper bars fitted into slots in the laminated iron. The ends of these copper bars are braze welded to heavy copper rings at each end. This completes the induced (by magnetism) electrical circuit. Such an arrangement is often called squirrel-cage winding.

The term "electrical degrees" means that the maximum magnetic effect on the rotor is a few degrees away from (behind) the magnetic effect in the stator.

7-7 REPULSION-START INDUCTION MOTOR

Repulsion-start induction motors were once widely used where motors must start under a heavy load. Torque or twist power was needed at once. Typical applications were external drive refrigerators and air compressors. In many cases, they have been replaced by capacitor type motors.

A special winding in the armature gives it a high starting torque. The motor starts as a repulsion motor, using brushes against a commutator in the armature winding circuit. This

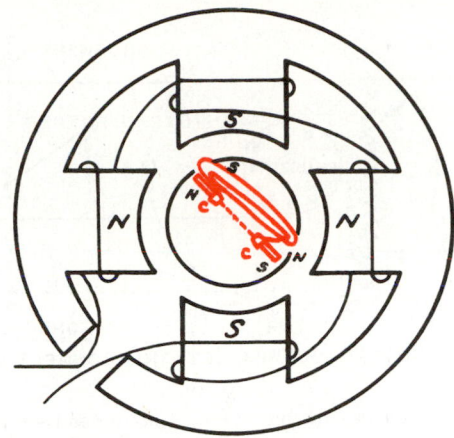

Fig. 7-4. Diagrammatic sketch of repulsion-start, induction-run motor. C—Commutator. Brushes that contact commutator are grounded and complete circuit between two commutator bars.

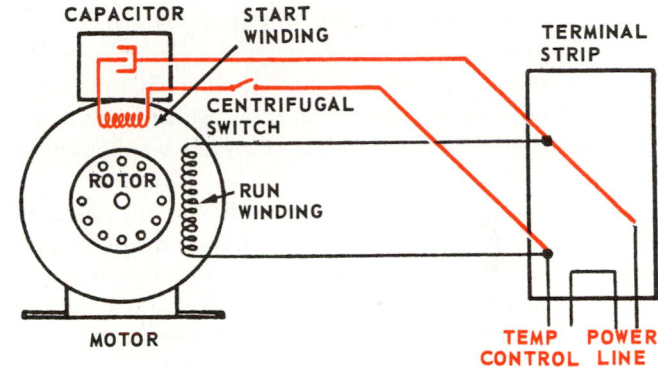

Fig. 7-5. Wiring diagram of capacitor-start induction motor. Note that capacitor is in series with starting field winding as motor starts. Rotor-operated centrifugal switch opens this winding as soon as motor reaches about 75 percent of full speed. (Stewart-Warner Corp.)

increases the induced electrical flow in the armature and produces a more magnetic power. As soon as it reaches a certain speed, the armature windings are shorted. Then the brushes are usually lifted from the commutator and the motor operates as an induction motor, Fig. 7-4. Para. 6-47 explains the operation of commutators and brushes.

7-8 CAPACITOR INDUCTION MOTOR

The open capacitor motor is a popular type. The name comes from the use of a capacitor or condenser in the starting winding. (One may recall that a capacitor is a device for storing electrical energy.) One type (Fig. 7-5) becomes a two-phase motor when starting. More torque is provided that way. Then it becomes a single-phase motor at about 75 percent of its full speed.

During starting, the capacitor changes the phase angle of the current in the starting winding to produce two-phase electrical characteristics.

The capacitor motor is built quite like the repulsion start-induction motor. The outside appearance, the mountings

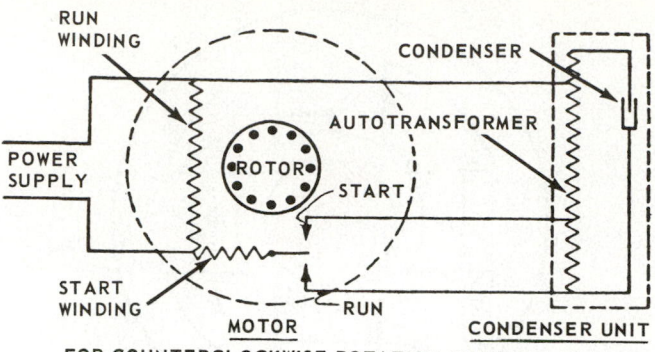

Fig. 7-6. Wiring diagram of combination capacitor and autotransformer motor. Note double-throw single-pole switch for changing transformer output when centrifugal switch moves from start to run. (Norge Div., Fedders Corp.)

and the bearing designs are much the same.

However, the capacitor type uses a capacitor instead of a starting winding in the armature. It is usually placed on top of the motor. The capacitor is connected to a centrifugal switch built into the motor and to a starting winding in the stator. See Fig. 7-6. (A centrifugal switch is one which whirls around with the shaft and weights are moved by the whirling motion to open the contacts.)

Operation is quite simple. When the motor starts, the centrifugal switch (closed when the motor is not running) causes the current to pass through both the starting and the running winding. The starting winding is connected in series with the capacitor. This capacitor puts the electrical surges in the starting winding out of step or out of phase with those of the running winding. The motor then acts as a temporary two-phase motor and has a very high starting torque.

At about 75 percent of the motor's rated speed, the centrifugal switch opens and disconnects the starting winding. The unit, however, continues to run as an induction motor.

These motors are also called capacitor-start induction-run motors. The capacitor is usually mounted on top of the motor in a metal or plastic cylinder. It has two terminals. One connects to the centrifugal switch terminal and the other to the common running and starting winding terminal.

The simplest method for producing greater torque is to change the single-phase motor into a two-phase during starting and/or running. A capacitor is used.

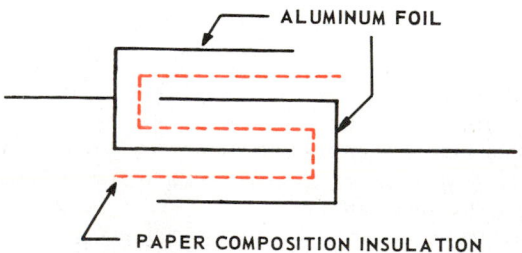

Fig. 7-7. Capacitor construction. Two layers of metal foil are separated by special insulating paper. The two sheets of foil are connected to the two terminals of the condenser. Capacitors used on motors are rolled or folded into compact package and installed in metal housing.

There are two types of capacitors:
1. Starting capacitor.
2. Running capacitor.

The starting capacitor is usually a dry type. It is constructed of two sheets of conductor metal separated by an insulator, as shown in Fig. 7-7. Typical of capacitors is the one shown at the top of Fig. 6-54.

A condenser or capacitor placed in an alternating current line is charged during the buildup of the voltage and current. The surge of power in the line causes this buildup. Then, during the decrease in current flow in the power line, as the a-c reverses, the capacitor discharges. This causes another power surge in the motor starting windings, as shown in Fig. 7-8.

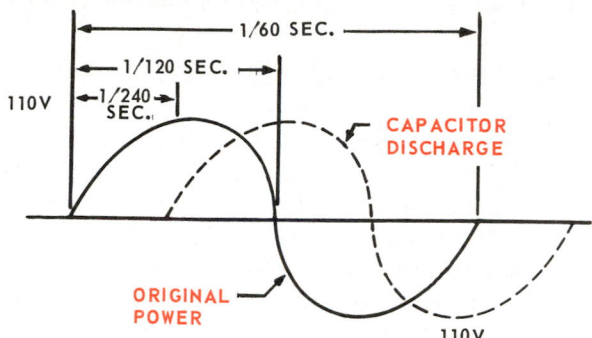

Fig. 7-8. Cycle diagram showing effect of capacitor in series with power flow in single-phase circuit.

The running capacitor operates in the same way, except that it keeps on operating when the motor is running. This capacitor is usually smaller and is made to provide better heat removal. Insulation, plates and terminals are designed to last.

7-9 MOTOR SPEEDS

Induction motors — including split-phase, repulsion start and capacitor motors — are made for use on 25, 50 or 60 cycle (Hz) current. The speed of a motor depends on the cycles (Hz) and number of field poles. Motor speeds are calculated from the synchronous speed.

Synchronous speed has a relationship to cycles per second. Speed is said to be synchronous when the polarity of the field poles changes direction with the same regularity as the alternating current. In other words, it changes the same number of cycles per second (Hz) as the current. The synchronous speed for two and four-pole motors is shown in Fig. 7-9.

Motors do not operate exactly at synchronous speed. They are operating under some load and there is some magnetic slippage. Thus, motors are not rated at the synchronous speed, but rather at a speed which corresponds with a normal load. Fig. 7-9 also shows operating speed for two and four-pole motors operating at 60, 50 and 25 Hz.

There are hermetic motor compressors which may be made to operate as either two-pole or four-pole motors. Such an arrangement makes it possible to operate the motor compres-

MOTOR SPEED — RPM						
NO. OF POLES	60 Hz		50 Hz		25 Hz	
	SYN.	OP.	SYN.	OP.	SYN.	OP.
2	3600	3450	3000	2850	1500	1450
4	1800	1750	1500	1450	750	700

Fig. 7-9. Synchronous and operational speed for two and four-pole motors. Note the operational (op.) speed is approximate.

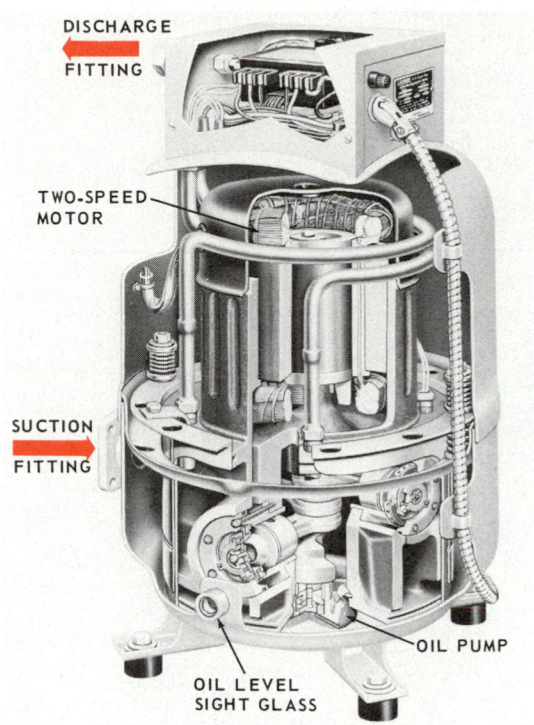

Fig. 7-10. Two-speed hermetic motor compressor. It uses three-phase electrical power. (Lennox Industries, Inc.)

sor either at 3450 rpm or 1750 rpm.

A two-speed three cylinder motor compressor is shown in Fig. 7-10. The motor operates as a two-pole motor at high speed and as a four-pole at low speed. The circuit diagram for a two-speed motor compressor is shown in Fig. 7-11. Solid state circuits control the compressor speeds.

Motors designed and constructed to operate on a certain Hz will not operate at a different Hz. This is because the number of turns of wire on the field poles and the amount of iron in the magnetic circuit is different for each Hz. Some motors have field connections in either series or parallel, allowing either 120V or 240V operation. On 120V, the field winding would be connected in parallel. Operating on 240V, the field winding would be connected in series.

It is possible to build motors having six, eight or more poles. However, such motors are not in common use.

7-10 STARTING AND RUNNING WINDINGS

As mentioned in previous paragraphs, most motors have a running winding and a starting winding. These windings are mounted on the stator.

During starting, current goes through both windings. When the motor reaches 60 to 75 percent of its running speed, the starting winding circuit is opened and the motor operates on the running winding only.

A starting winding is found on split-phase motors and on capacitor motors of all types. It has the same number of coils as the running winding but it has smaller diameter wire and a greater number of turns. Its induction action, therefore, splits the phase creating off-center magnetism.

Electricity passes through the running winding all the time the motor is in operation. Heavier wire is used than in the starting winding. It is installed on the field poles on both the two-pole and the four-pole motors.

Fig. 7-12 shows the electrical circuits of a split-phase electric motor while the motor is starting. When the current goes into the R terminal (during one-half of the cycle), the electrons separate. Most will go through the running winding but some will go through the starting winding. Magnetism builds up faster in the running winding. Thus, when the magnetism is created a few thousandths of a second later in the starting winding, a turning force is exerted upon the rotor.

When current enters terminal R during the other half cycle, most will go through the running winding. Some will go through the starting winding. This action reverses the polarity of the electromagnets and, again, there is a delay of magnetic buildup in the starting winding. The rotor is attracted and repulsed in the same direction of rotation as in the first half of the cycle.

To change the direction of rotation on some of these types of motors, disconnect the starting winding leads from the two terminals and reverse them. That is, put the old S connection (lead) on terminal R and put the old R connection (lead) on terminal S. Reversing the two main leads will not reverse the rotation of the motor.

If the starting winding is left in the circuit, it may overheat. Therefore, a switch is mounted in the starting winding circuit. It remains open as long as the motor is running, as shown in Fig. 7-13.

7-11 STARTING CURRENT

The instant the starting circuit is closed on a motor, there is little or no counter emf. Therefore, the amount of current starting to flow will be determined by the resistance in the circuit. At the instant of starting, the current flow will be quite high (two to four times the running current).

This initial high flow of amperage is called "locked rotor amperage." During this high amperage flow, the voltage drops to its lowest reading.

Frequent starting of a motor causes these high current flows to happen too often. The motor will overheat — especially those of 1/2 hp or more.

Fuses and circuit breakers must be designed with this fact in mind. If the protective devices are set to allow only running current, they will open at the instant of start. The protective device must have delayed action — usually a heating strip and a bimetal switch.

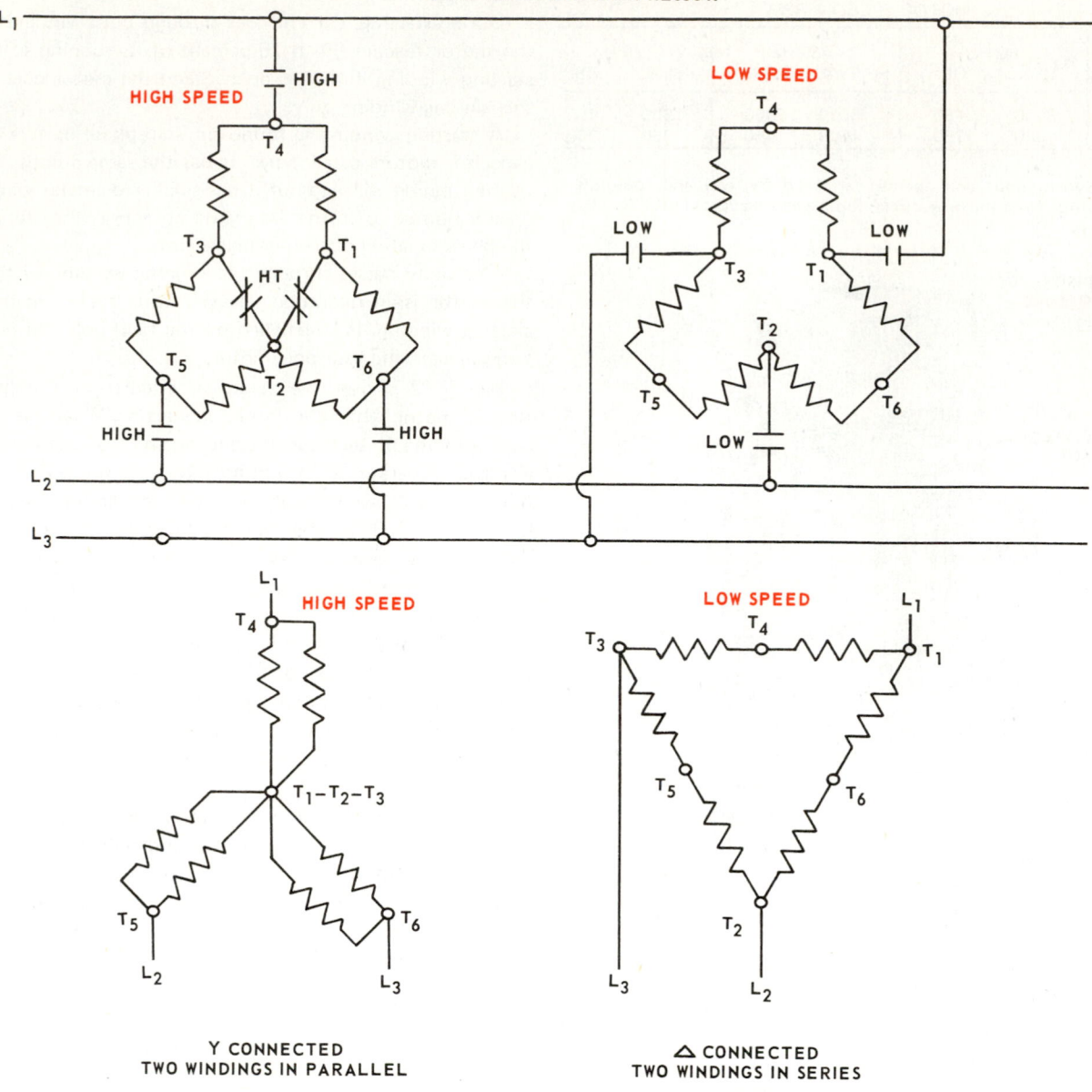

Fig. 7-11. Circuit diagram of a two-speed, three-phase hermetic motor compressor. (Lennox Industries, Inc.)

7-12 MOTOR CONNECTIONS

Except for those that are reversible or of some special design, motors usually come with two or four leads to connect the motor to one of two different voltages.

Provided the correct voltage is maintained at the motor terminals, 120V, 240V (or other corresponding voltages) are equally good. These motors may be used on circuits having voltage of 105 to 125 or 210 to 250 volts.

Most electrical power companies are now providing 120V and 240V electrical power. These voltages are safe to use on motors labeled 110-220V or 115-230V.

It is recommended that, when possible, all motors be connected to 240V service. If the wires are of good size, this will result in a good voltage at the motor terminals, especially if the motor is heavily loaded.

Most repulsion-start induction motors use four motor leads. These leads are connected to field windings, two leads to a winding. The windings are connected in parallel if the voltage to be used is 120V. The leads are connected in series if the voltage used is 240V. Since the power is the product of the emf in volts multiplied by the amperage, the amperage flow for any specified power using 240V will be half the current flow if using 120V.

Fan motors and similar units are usually wired either for 120V or 240V only. Only two leads are connected to the field windings. Air conditioning evaporator fans often have two or three speeds. These motors will have three to four leads.

One must be careful to avoid connecting a 120V motor to a 240V circuit. If this is done, the motor will overheat and "burn out." The maintenance of proper voltage at the motor terminals is absolutely necessary.

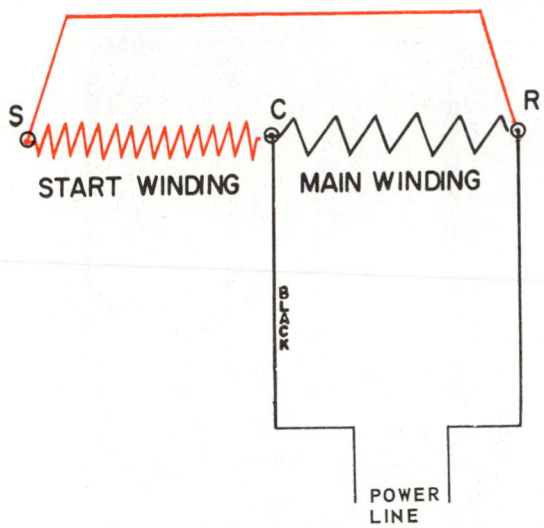

Fig. 7-12. Circuit diagram of a split-phase motor. C—Common terminal for both windings. C to R—Main or running winding. C to S—Starting winding. R—Running winding terminal. S—Starting winding terminal.

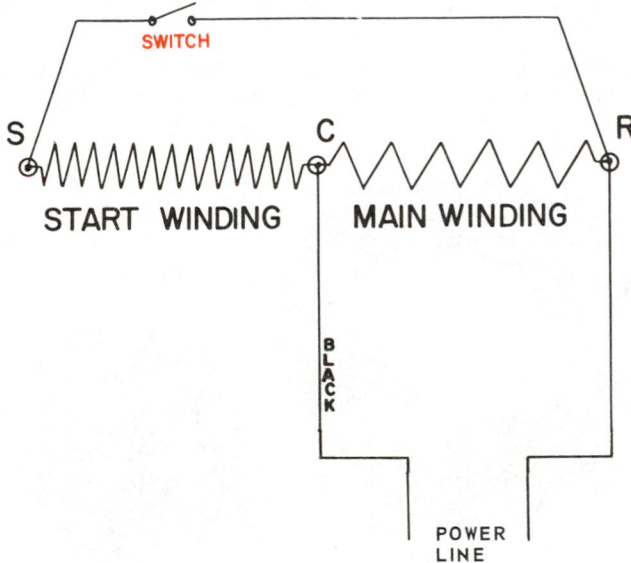

Fig. 7-13. Circuit diagram of split-phase motor using switch in starting winding circuit. Switch is operated by rotor speed or by relay and is closed as motor starts and will open when motor reaches approximately 75 percent of its running speed. (Copeland Corp.)

The power that an alternating current motor will develop varies directly to the square of the voltage at the motor terminals. Low voltage will also result in a rapid drop in the horsepower capacity of any motor.

See Chapter 28 for information on the wire size to be used. Loose connections cause excessive voltage drop and are a fire hazard. All connections should be made with solder or with Underwriters approved pressure connectors.

A 120V motor supplied with current at 100V will develop only about 75 percent as much power as it would if supplied with 120V current. Therefore, to do the job and to develop its

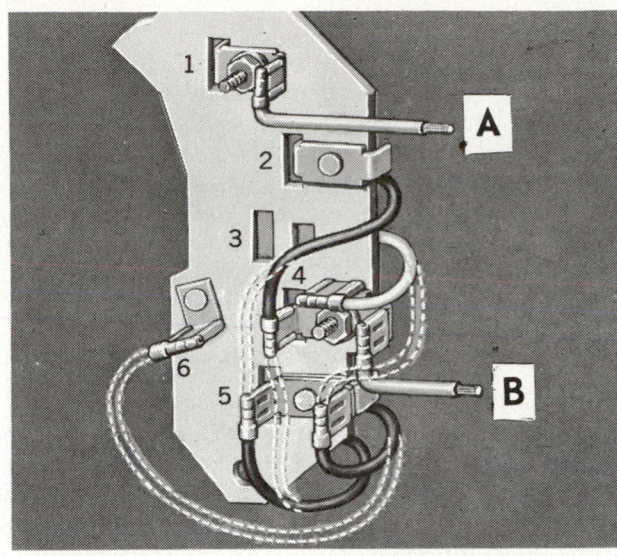

Fig. 7-14. How motor leads may be made to give desired direction of rotation and connect the motor to voltage supplied. Power in leads are shown at A and B. Wiring diagram inside motor cover plate indicates correct connection for each voltage and direction of rotation. Dotted lines show method of reversing. (Century Electric Co.)

full horsepower, it must draw more current. This increase in current may overheat the "power in" leads and/or the motor windings causing insulation failure.

Fig. 7-14 shows a method of making connections for single-phase motors. Note how the direction of rotation may be changed. Special motors requiring other connections are shipped with diagrams and instructions showing proper connections.

A 120-240V motor is connected to a 120V line through an across-the-line switch. This switch is usually operated by the cabinet temperature, although low-side pressure may also be used. The same motor control may be used to operate a 120-240V motor installed on either the 120V or 240V circuit.

The field winding leads come out of the motor frame through an insulated opening. The wire is normally protected with a rubber, fabric or plastic grommet. A small metal box is usually mounted over the leads to protect the electrical connections.

7-13 HERMETIC SYSTEM MOTORS

When the motor and the compressor are placed inside a dome or housing, it is called a hermetic unit. In such systems, the motor must drive the compressor directly. This requires good, leakproof electrical connections. The motor must have good power characteristics and must be of the induction type.

Motors using rotor windings requiring either brushes or slip rings cannot be used. Development of the split-phase motor and the capacitor motor made the hermetic motor possible.

The original hermetic motors were four-pole operating at approximately 1750 rpm.

To calculate the motor speed, the formula is:

$$N = \frac{120f}{P} \text{ where } N = \text{rpm}$$

$$f = \text{frequency (cycle)}$$

P = number of poles

$$N = \frac{120 \times 60}{4} = 30 \times 60 = 1800 \text{ revolutions}$$
$$\text{per minutes (rpm)}$$

The 120 in the formula is the 60 seconds per minute x 2 (the magnetism or polarity changes 2 times per cycle).

Using the above formula, the speed of a two-pole motor will be:

$$N = \frac{120 \times f}{P}$$

$$N = \frac{120 \times 60}{2} = 60 \times 60 = 3600 \text{ rpm}$$

synchronous speed or about 3400 rpm actual speed.

The reason the motor is almost twice as fast is that the motor rotor has to travel one-half revolution during one-half cycle instead of one-fourth revolution as with the four-pole motor. These two-pole motors are about two-thirds the size of the four-pole motors of the same power.

Hermetic motors present some problems not found with open motors:

1. Special cooling provisions must be made.
2. Wiring insulation must be resistant to oil and chemicals in the refrigerant (particularly in the presence of moisture and/or high temperatures).
3. Manufacturing standards must provide exact alignment of the stator, rotor and compressor.
4. Electrical connections through the dome must be electrically perfect and leakproof.

The motors may be cooled by several methods. One successful method is to press the stator into the dome and then put cooling fins on the dome.

Another method uses a water coil to cool the motor windings while the unit is running. Some designs pass partly cooled condenser refrigerant around the motor housing to help cool the electric motor. Most systems pull the cool return suction line vapor and oil over the motor windings.

Most manufacturers have developed special synthetic wire coatings (usually synthetic enamels) which have good insulating qualities while being safe to use with most of the popular refrigerants. The air gap between the rotor and the stator is only a few thousandths of an inch. It must be equal on all sides; otherwise, the motor may hum. Note, in particular, that many motor compressors are mounted on springs inside the dome.

Fig. 7-15 shows a hermetic motor compressor unit. The stator and rotor of a hermetic motor as shown in Fig. 7-16.

7-14 HERMETIC MOTOR ELECTRICAL CHARACTERISTICS

The design characteristics of electric motors used in hermetic machines will depend on whether the unit starts under load, no-load or a balanced pressure condition.

The basic operation of the motor is the same as for open-type motors. All torque must be developed by induction only.

The higher starting torque (turning effort) needed for those units which start under load requires the use of larger conductors in the starting circuit. Usually, manufacturers try

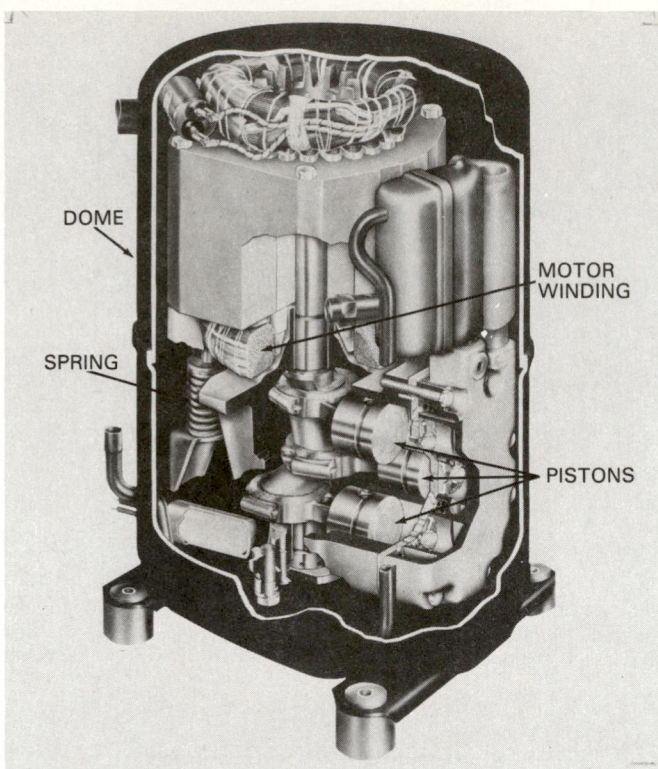

Fig. 7-15. Hermetic motor compressor unit. Note that compressor is mounted on springs inside the dome. (Tecumseh Products Co.)

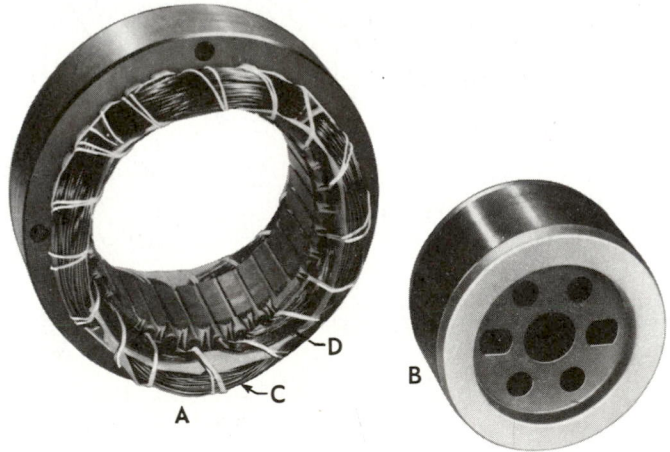

Fig. 7-16. Rotor and stator from a 1/4 hp split-phase hermetic motor: A—Stator. B—Rotor. C—Running winding. D—Starting winding. (Wagner Electric Corp.)

to provide starting power equal to twice the running power. In other words, a 1/6 hp motor is designed to produce 1/3 hp during starting. Various methods are used to shut off the special starting devices after the motor has reached full speed.

The external circuit of a capacitor start hermetic motor including a starting relay is shown in Fig. 7-17.

7-15 HERMETIC MOTOR TYPES

Hermetic motors are either single-phase or polyphase. Four types of single-phase induction motors are used:
1. Split-phase (SP).

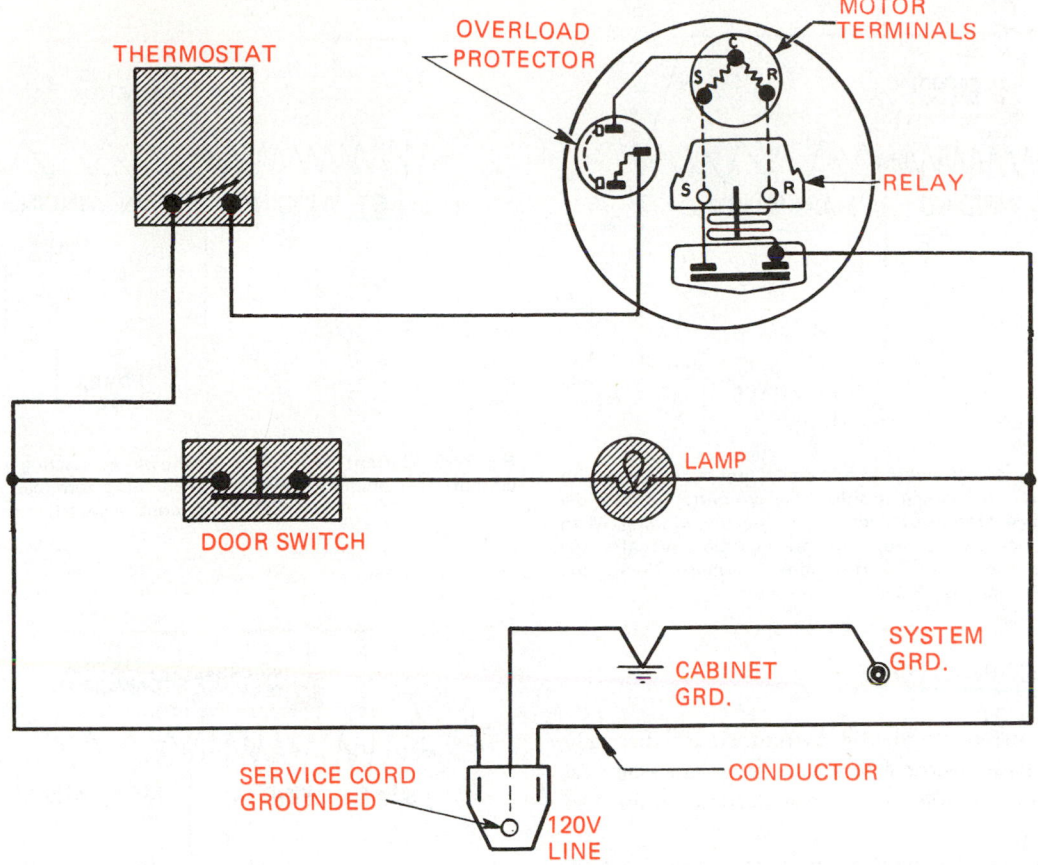

THERMOSTAT

OVERLOAD PROTECTOR

MOTOR TERMINALS

RELAY

LAMP

DOOR SWITCH

SYSTEM GRD.

CABINET GRD.

CONDUCTOR

SERVICE CORD GROUNDED

120V LINE

Fig. 7-17. External circuits used on a hermetic motor compressor. Motor terminals are brought out through an insulated block on side of the motor compressor. Starting relay and overload protector are located on outside of the motor compressor. (Kelvinator, Inc.)

2. Capacitor-start, induction-run (CSIR).
3. Capacitor-start, capacitor-run (CSR).
4. Permanent-split capacitor (PSC).

7-16 HERMETIC SPLIT-PHASE INDUCTION MOTOR

Split-phase induction is the basic type motor for small hermetic condensing units. It is cheaper, since no extra devices are needed.

The principle of operation is simple. There are two windings — one for running and one for starting. Since starting torque is low, such motors must be used on systems with low starting load.

The split-phase motor is very popular on systems which use the capillary tube refrigerant control since pressures balance on the off cycle. Thus, the compressor is not required to start under a load.

These motors may also be used where the system provides an electrical, mechanical or hydraulic pressure unloading device. In these installations, any type refrigerant control may be used.

A split-phase motor used in hermetic motor compressor systems must have some type of outside starting relay. This may be a thermal, current or potential type relay. See Chapter 8 for information of these relays.

A schematic wiring diagram for a hermetic split-phase induction motor with a current type starting relay is shown in Fig. 7-18. Fig. 7-19 shows the same motor using a potential type relay. These motors are most used on small units — 1/10, 1/6, to 1/3 hp.

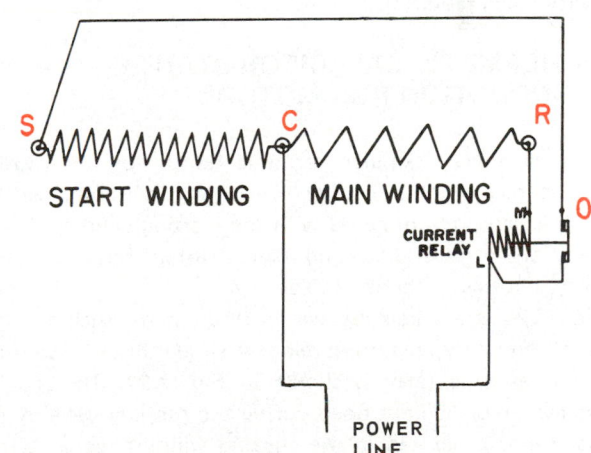

S   C   R

START WINDING   MAIN WINDING   O

CURRENT RELAY

POWER LINE

Fig. 7-18. Wiring diagram of hermetic split-phase induction motor. Current relay control switch O is closed and start winding is cut in. When motor starts, current flowing through relay control winding and main winding creates a magnetic effect in relay. This opens switch O to cut out start winding, and motor operates on main winding. C—Common terminal. R—Running terminal. S—Starting terminal. (Copeland Corp.)

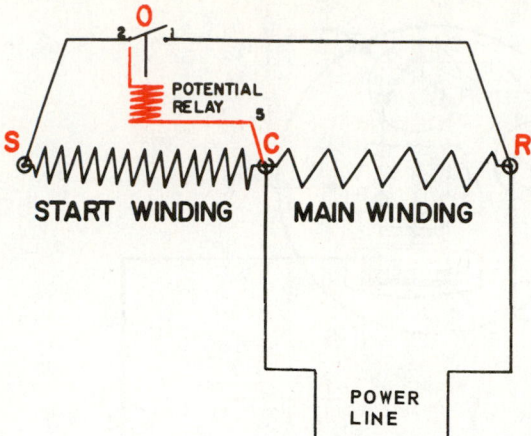

Fig. 7-19. Wiring diagram of hermetic split-phase induction motor with a potential type relay in running position. Relay control winding develops current from generation of electricity by starting winding. When enough magnetic effect is produced, it will open switch O just after the motor starts. C—Common terminal. R—Running terminal. S—Starting terminal. O—Relay control switch.

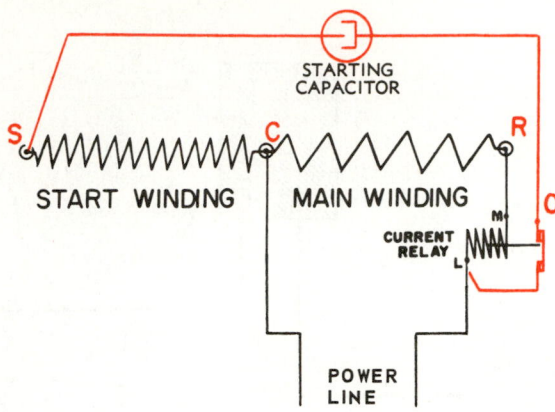

Fig. 7-20. Current type relay is shown in starting position (points closed). C—Common terminal. R—Running terminal. S—Starting terminal. O—Relay control switch.

7-17 MOTOR CAPACITORS

The capacitor in an alternating current circuit basically changes a single-phase electrical flow to a two-phase electrical flow. The capacitor is usable in both the starting winding and the running winding.

If used in the starting winding only, it is called a start capacitor. Being used only for a few seconds at a time, it should have no cooling problems. However, if a motor is started too often or if the starting winding is used longer than it is designed for, the capacitor insulation will overheat and the capacitor may fail.

If the capacitor is used for the running winding, it is carefully designed to discard any heat generated in it during operation. Never use a starting capacitor in the running circuit.

Always use the correct voltage and the correct microfarad capacity when replacing a capacitor.

7-18 HERMETIC CAPACITOR-START, INDUCTION-RUN MOTOR

This is a very popular hermetic motor for refrigerating units. It has a good starting torque which is obtained by placing a capacitor in series with the starting winding. It can use any one of several starting relays: current relay, potential relay or a hot wire (thermal) relay.

Fig. 7-20 is a schematic wiring diagram for such a motor using a current type starting relay. A similar installation using a potential type relay is shown in Fig. 7-21. The starting capacitor circuit is kept open during the running cycle by the induced emf generated in the starting winding (across terminals S and C). Induced voltage pushes enough current through the potential relay coil to produce a magnetic effect. The magnetic force keeps the relay points open.

See Chapter 8 for detailed information concerning construction and operation of the different types of hermetic motor controls.

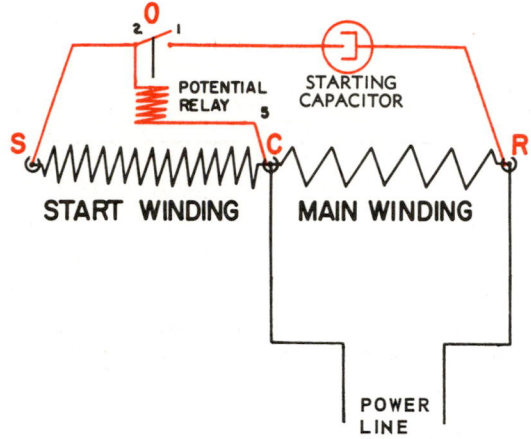

Fig. 7-21. Capacitor-start, induction-run motor with potential type relay in running position (points open). C—Common terminal. R—Main or running terminal. S—Starting terminal. O—Relay control switch. (Copeland Corp.)

7-19 HERMETIC CAPACITOR-START, CAPACITOR-RUN MOTOR

The capacitor-start, capacitor-run motor generally uses two capacitors. Both are in the starting winding circuit but only one is controlled by the relay switch.

When the motor is started, the capacitors turn the motor power surges into two-phase power and produce a high starting torque. After the motor reaches two-thirds or three-fourths of its rated speed, the relay opens the circuit to the starting capacitor. The running capacitor is left in the circuit.

This action produces a two-phase motor that is very efficient. The power factor is much improved. The larger hermetic units in commercial systems use this type of motor.

Fig. 7-22 shows the wiring diagram. Note that the running capacitor is in series with the starting winding. In Fig. 7-23, the motor wiring circuit shows two starting capacitors and two running capacitors. These are connected in series to increase the voltage capacity. (Two 120V capacitors in series can be used in a 240V circuit.)

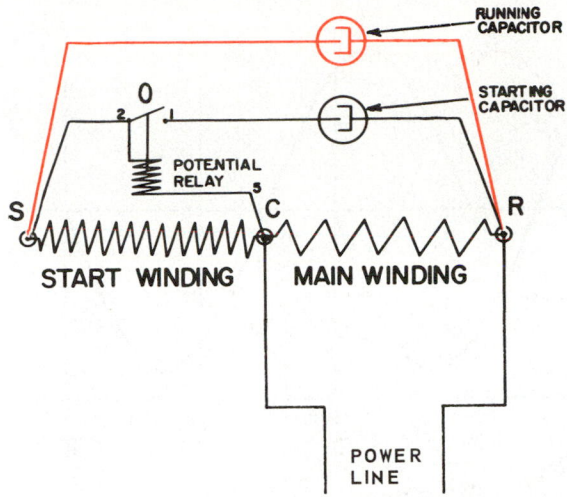

Fig. 7-22. Schematic wiring diagram of capacitor-start, capacitor-run motor using a potential relay. Relay is in running position (points open). C—Common terminal. R—Main or running terminal. S—Starting terminal. O—Relay control switch.

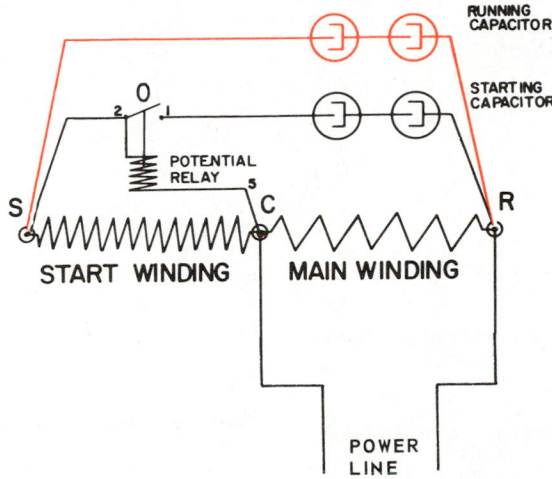

Fig. 7-23. Schematic wiring diagram of hermetic motor. This circuit has a potential relay in running position. C—Common terminal. R—Running terminal. S—Starting terminal. O—Relay control switch.

7-20 PERMANENT SPLIT CAPACITOR MOTOR (PSC)

The permanent split capacitor motor is popular for air conditioning systems. It does not use a relay. Current flows through both the running winding and the starting winding when the power is on. (See diagram, Fig. 7-24.) A running capacitor is connected between the running (R) and starting (S) terminals and is in series with the starting winding.

Such motors are sensitive to line voltage. A 5 to 10 percent drop will cause starting difficulty and overheating. To prevent damage, thermal protectors will open the circuit.

Starting torque is low. Thus, if it tries to start when the system's pressures are not balanced (equal), the motor will overheat and thermal protectors will open the circuit.

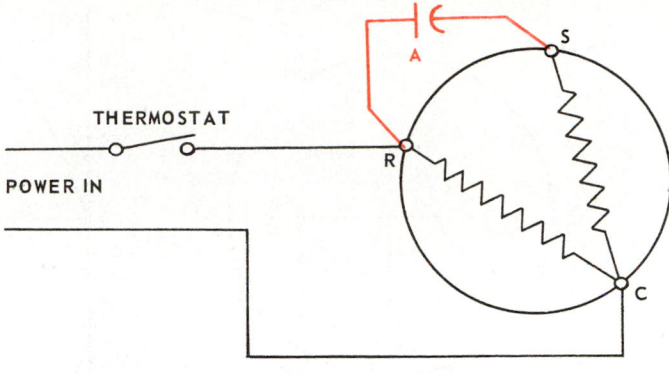

Fig. 7-24. A PSC (permanent split capacitor) motor. A—Running capacitor.

Chapter 21 explains how to change a PSC into a capacitor-start, capacitor-run motor in an air conditioner.

7-21 HERMETIC POLYPHASE MOTOR

Large hermetic compressors usually are driven by three-phase motors. The surges of current travel in these motors are closer together than with single-phase current supply. Therefore, they are more efficient power sources. The sine curves for a two-phase motor are shown in Fig. 7-25, while the sine

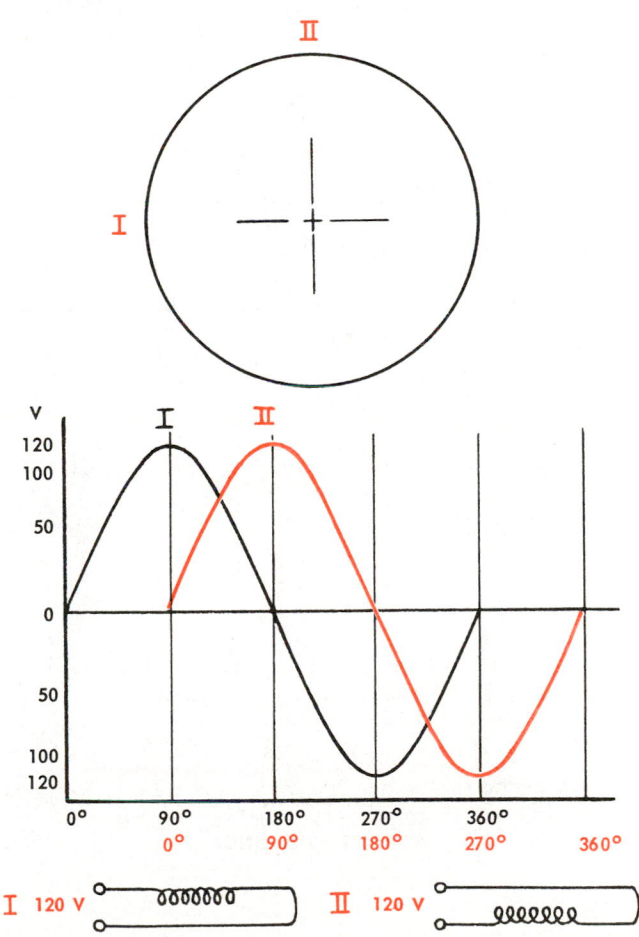

Fig. 7-25. Two power sine curves for two-phase power circuit. Note that there are two separate circuits. They are 90 deg. out of phase.

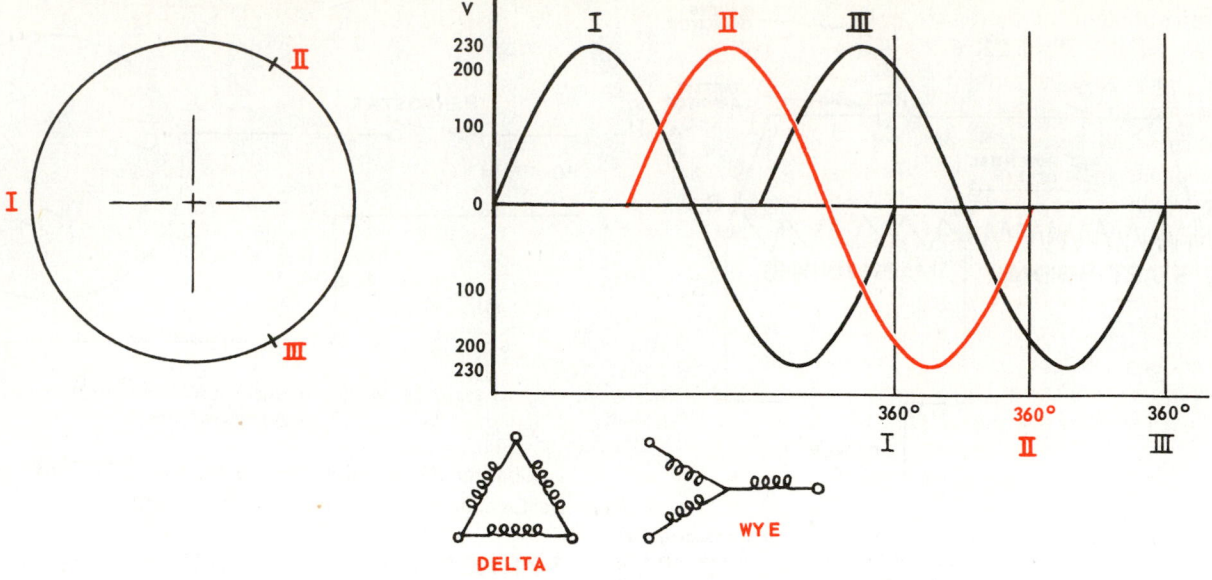

Fig. 7-26. Three power sine curves of three-phase circuit. System can be wired for either Delta or Wye design.

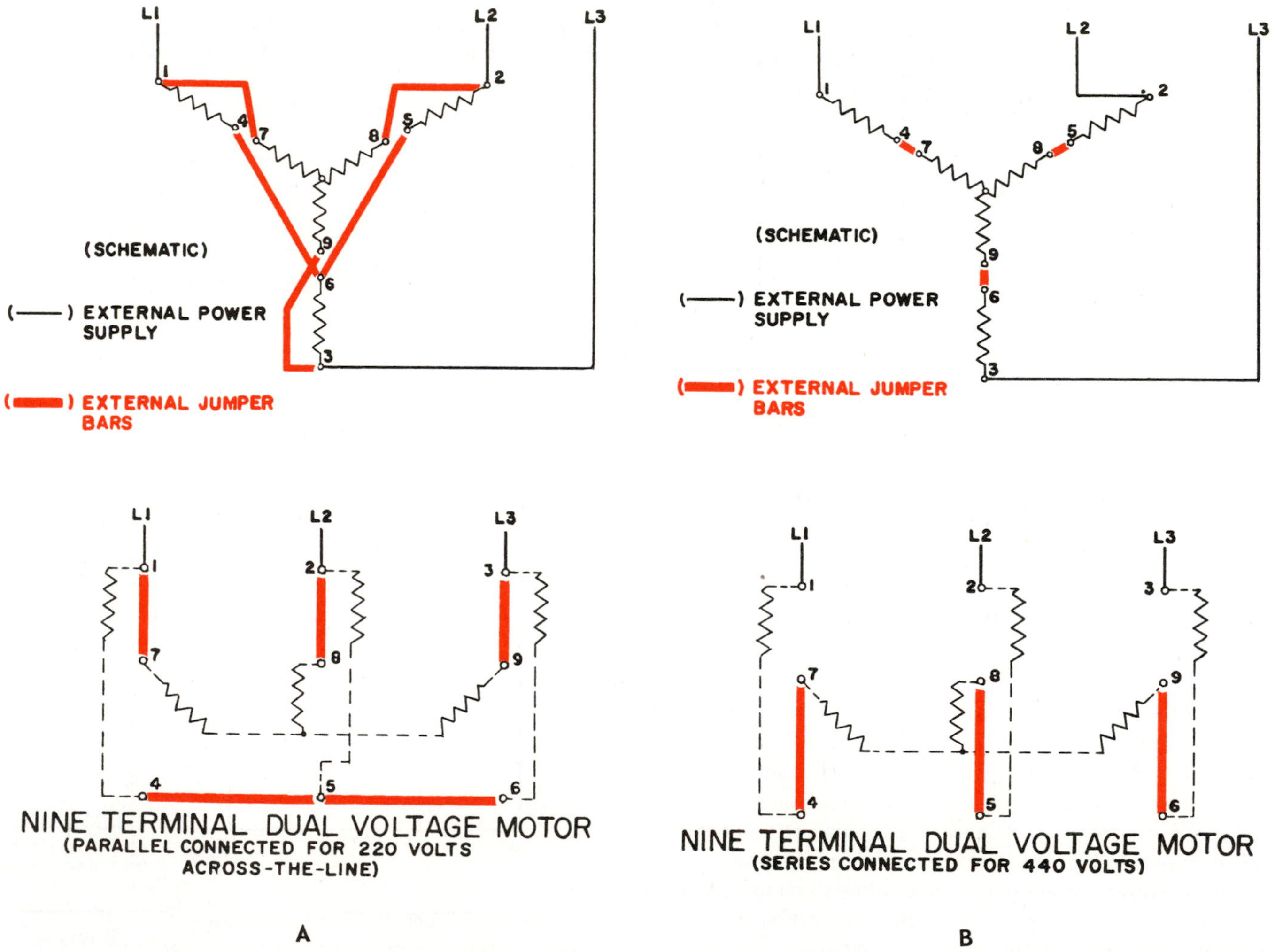

(SCHEMATIC)

(———) EXTERNAL POWER SUPPLY

(▬▬) EXTERNAL JUMPER BARS

(SCHEMATIC)

(———) EXTERNAL POWER SUPPLY

(▬▬) EXTERNAL JUMPER BARS

NINE TERMINAL DUAL VOLTAGE MOTOR
(PARALLEL CONNECTED FOR 220 VOLTS ACROSS-THE-LINE)

A

NINE TERMINAL DUAL VOLTAGE MOTOR
(SERIES CONNECTED FOR 440 VOLTS)

B

Fig. 7-27. Schematic wiring diagram showing circuits and connections for three-phase hermetic motor. A—Circuit as connected for 220V. B—Circuit as connected for 440V. Note that L_1, L_2 and L_3 are the three-phase line connections. Numbers 1-2-3, 4-5-6, 7-8-9 are the connections to the motor windings. Each motor winding coil is designed for 220V. Example, the coil between terminals 1 and 4.

curves for a three-phase motor are shown in Fig. 7-26. These motors are usually of the 220V or 440V type. Since the dome terminal block has nine terminals, the service technician may wire the motor for either 220V or 440V. (See Fig. 7-27.) Some of these motors use a 550V supply.

Three-phase motors are available from 1/2 hp up. The building in which the unit is to be placed must be wired for three-phase service. Very few residences have three-phase, but most industries and some commercial buildings do.

The higher volts are very dangerous. Disconnect the power and lock the switch open (use an actual lock) before starting to service the machines.

Three-phase motors use contactors or motor starters. They do not have the usual starting relays.

Fig. 7-28 shows a three-phase motor wiring circuit with its starting and protector circuit. It is always best to have an electrical journeyman do the electrical work on these units. Since each unit may have certain differences, when servicing the system, it is important to get the wiring diagram for the unit from the manufacturer.

Operating characteristics of two types of polyphase motors are shown in Fig. 7-29.

Occasionally, a three-phase motor may blow a fuse or open a circuit breaker on one phase only. The motor will attempt to operate on the remaining two phases. If there is much load on the motor, it will quickly overheat and may burn out. This is because the two windings must carry all the load so each one will need one and a half times more current.

Each phase of a three-phase motor must be tested individually using a voltmeter. There will be about 50 volts difference between the open line and one of the other lines. The circuit having the "blown" fuse will indicate below-normal voltage. The direction of rotation of a three-phase motor may be reversed by changing any two of the power leads to the motor.

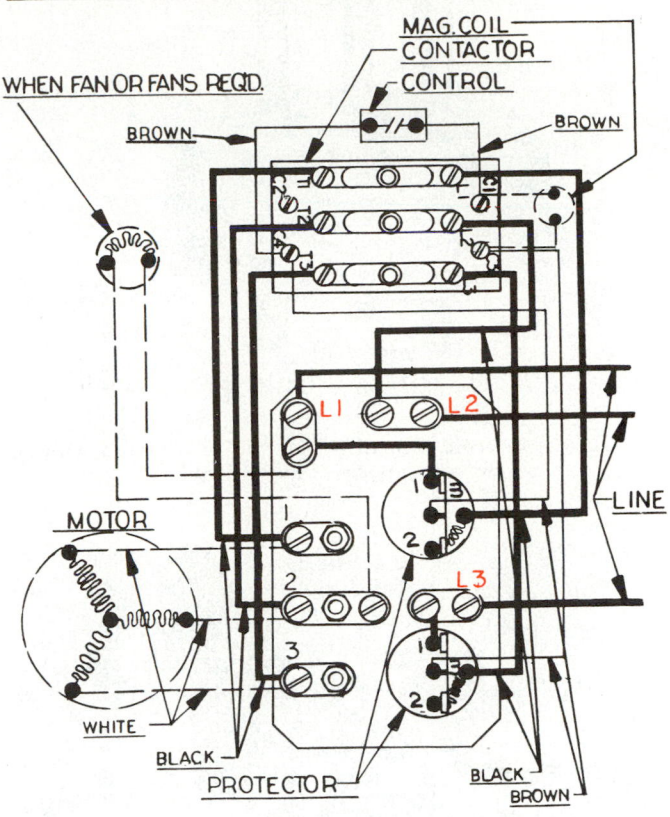

Fig. 7-28. Three-phase hermetic motor circuit wiring diagram. L_1, L_2 and L_3 are the three-phase line connections. Overload protector is shown in L_1 and L_3 circuits and operates magnetic coil of starter switch (top part of drawing). Special magnetic type starter is required for this installation. Automatic temperature control would be connected to control at top of illustration. (Copeland Corp.)

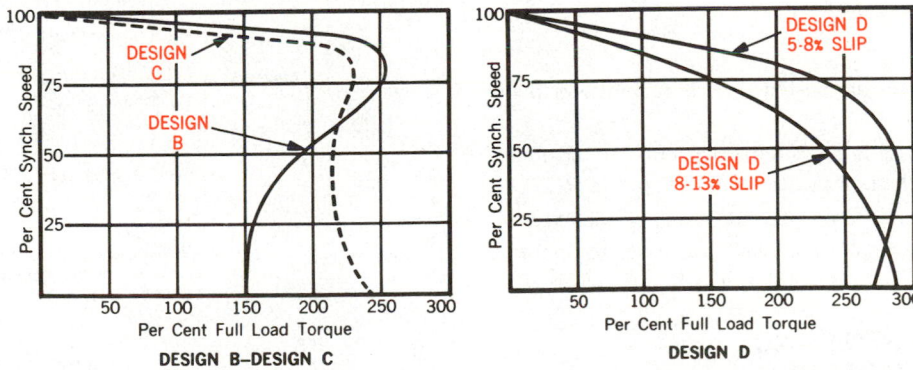

Fig. 7-29. Speed torque curves compared for two polyphase motor designs. Note that Design B and C motors provide starting torque of 150 and 250 percent of full load torque. Design D motors provide starting torque of 260 to 280 percent of full load torque. However, speed of Design D motors will fluctuate much more than Design B and C motors as load changes. (Century Electric Co.)

7-22 HERMETIC MOTOR TERMINALS

The electrical terminals which carry the circuit through the dome must be electrically insulated from the dome or housing. They must also be leakproof.

Most motor terminals are fused to glass while the glass, in turn, is fused to a metal disk. This assembly may be welded to the hermetic dome or housing, as shown in Fig. 7-30. The terminal must be leakproof after thousands of heating and cooling, expansion and contraction cycles. Furthermore, it must have a high insulating value.

A fused glass multiple terminal installation is shown in Fig.

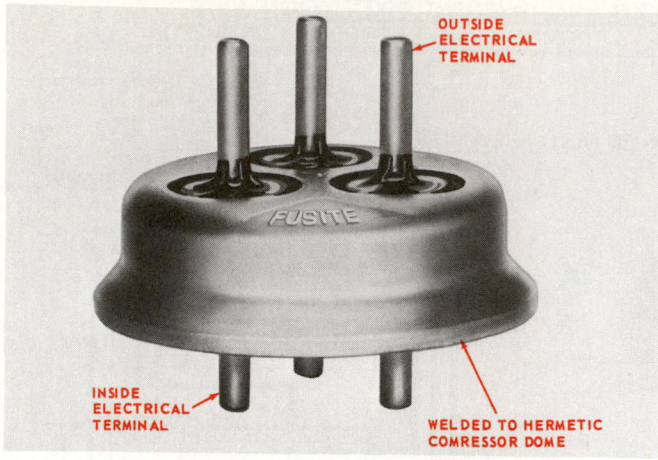

Fig. 7-30. Metal electric terminals which may be welded to hermetic compressor dome. (Fusite Corp.)

Fig. 7-31. A cross-sectional view of a metal-glass fused hermetic electrical terminal installed. Wires to control relay are connected to the terminals by means of spring loaded clips. Note metal structure surrounding terminals to protect them from abuse.

7-31. The wire terminals are spring clips that tightly grip the hermetic terminals.

Some compressors use what is known as built-up terminals. These are attached to the compressor dome, as in Fig. 7-32.

Replacement terminals are used by many service technicians. Gaskets of synthetic material are used to make a leakproof joint.

7-23 DIRECT CURRENT AND UNIVERSAL MOTORS

Areas with d-c power must use direct current motors in refrigerators. These motors are compound wound and have a mechanical likeness to both the capacitor and repulsion-start induction type motors. However, the electrical circuits are quite different. A circuit diagram is shown in Fig. 7-33.

Compound wound motors have two types of field windings; one is in parallel while the other is in series with the armature winding. Because d-c is used, the field poles always have the same magnetic polarity. Also since a d-c current is going

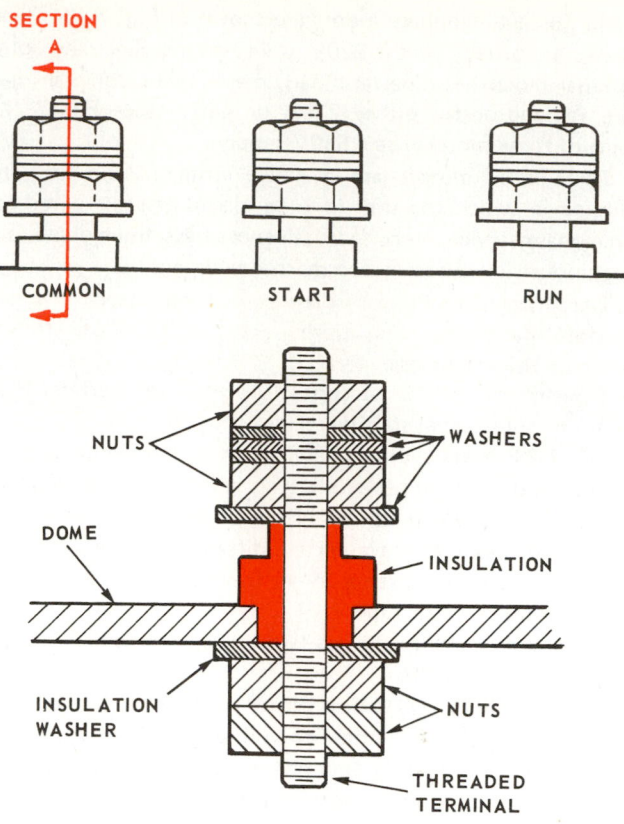

Fig. 7-32. Built-up hermetic motor terminals. A—Cross-section of one of the terminals.

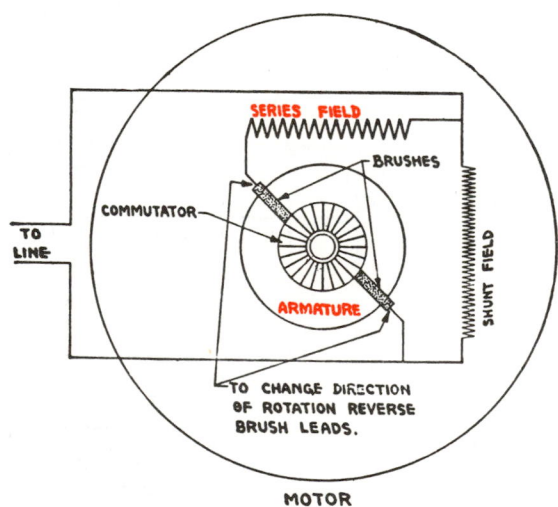

Fig. 7-33. Wiring diagram of compound wound direct current motor. Note that current going through armature must pass through series field. Shunt field winding is in parallel with the armature, series field winding circuit. Series winding gives motor high starting torque. (Norge Div., Fedders Corp.)

through the armature coil, the magnetic polarity of the armature will remain constant. It is positioned to cause a turning effect or torque in the armature. The series field helps to keep the motor speed constant.

The series winding strength builds up to increase the rpm if the motor tends to slow down or weakens to reduce the rpm if the motor tends to speed up. One has only to reverse the brush leads to reverse the direction of rotation of these motors. (It reverses the armature magnetic polarity.)

The possibilities for wear are slightly greater in the direct current motors than in the others because of the armature design. The brushes are in constant contact with the commutator, giving rise to troubles such as dirty commutator, worn brushes, high mica insulation between the bars, shorted armature and squeaky brushes.

Direct current may be used only on open type systems.

7-24 MOTOR HORSEPOWER AND MOTOR CHARACTERISTICS

Energy, work and power are explained in Chapter 1. Motors are rated by horsepower. One horsepower is equal to lifting 33,000 ft. lb. per minute or 550 ft. lb. per second. At 100 percent efficiency, 778 W equals one hp. Refrigeration motors of 1/100 hp to several hundred hp are in use. Motors as small as 1/20 hp have been used to drive compressors.

In refrigeration, a motor's torque is as important as its horsepower. Torque is the force of a twisting or turning action, (such as turning the crankshaft of a compressor).

It is important to know the properties of single-phase motors, the most popular of all motors. Fig. 7-34 shows the

	AMPERAGE			
	120 VOLTS		240 VOLTS	
HP	FULL LOAD	LOCKED ROTOR	FULL LOAD	LOCKED ROTOR
1/6	4.4	26.4	2.2	13.2
1/4	5.8	34.8	2.9	17.4
1/3	7.2	43.2	3.6	21.6
1/2	9.8	58.8	4.9	29.4
3/4	13.8	82.8	6.9	41.4
1	16.0	96.0	8.0	48.0
1 1/2	20.0	120.0	10.0	60.0
2	24.0	144.0	12.0	72.0
3	34.0	204.0	17.0	102.0

Fig. 7-35. Full load and locked rotor amperage for a-c motors.

normal conditions, to allow a motor to start and reach its rated speed fully loaded.

Some motors have to develop maximum rated capacity for short periods while starting quickly under a heavy load. If the motor is being started and operated under the direct supervision of an attendant, a fuse may be used having a short time period rated capacity of 125 percent.

In no case should the fuse be larger than 150 percent of the rated full-load current of the motor. The above applies to plug and cartridge type fuses approved by the National Board of Fire Underwriters. Link and wire fuses often do not provide

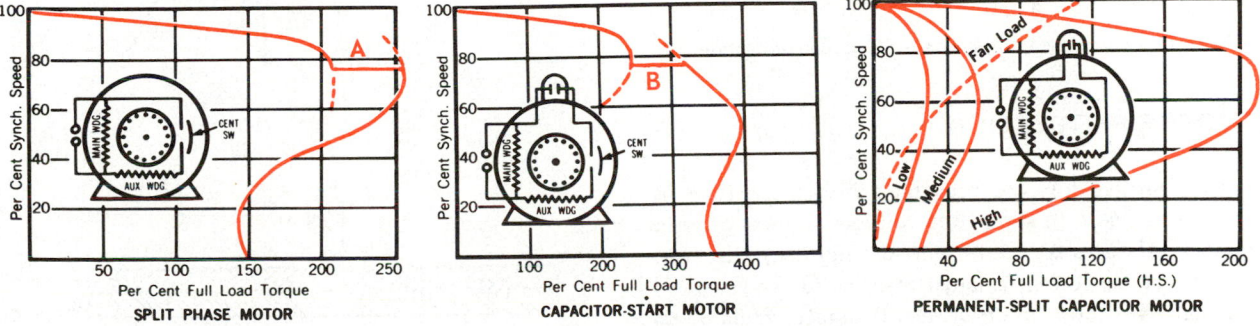

Fig. 7-34. Torque and speed characteristics of three types of single-phase fractional horsepower motors: Split-phase, left. Starting winding is disconnected by rotor-operated centrifugal switch at 75 percent of synchronous speed. See A. This type provides fair starting torque. It is available in horsepower ratings up to 3/4 hp and for either 120V or 240V. Capacitor-start, center. Starting winding is disconnected by rotor operated centrifugal switch at approximately 75 percent of synchronous speed. See B. Higher starting torque is obtained by adding capacitor in starting circuit. Generally available in horsepower rating up to 3 hp and for either 120V or 240V. Permanent-split capacitor, right. Capacitor stays in the running winding and auxiliary winding whenever the motor is running. Its use is limited to easy-to-start loads. These units may be operated as multi-speed motors using simple and inexpensive controls. Generally used in low horsepower applications. Available for either 120V or 240V current. (Century Electric Co.)

operating curves of the various types based on motor speed and percent of full load torque.

The table in Fig. 7-35 gives the average full load amperage and locked rotor amperage for various sizes of a-c motors.

7-25 FUSES AND CIRCUIT BREAKERS

A motor that is started and stopped automatically, and is placed or controlled from a distance, should use a fuse with a rated capacity of not more than 100 to 110 percent of the rated ampere capacity of the motor. This size is ample, under

enough protection for refrigeration unit motors.

A plug fuse will not protect an overloaded motor because the fuse must be of enough capacity to withstand the starting load of the motor. This can be from two to six times more than the running current. If the motor running winding draws this extra load long enough, it may burn out.

Many homes and business establishments now use circuit breakers rather than fuses. It is an automatic switch which will open a circuit if the current draw is too great.

Circuit breakers are usually rated the same as fuses. An "opened" circuit breaker must be manually "reset." Just as with fuses, a circuit which is continually opening the breaker should be carefully examined. If the breaker has sufficient capacity there may be a short or other trouble in the circuit.

7-26 ELECTRIC MOTOR GROUNDING

Electric motors used on refrigeration and air conditioning systems must be grounded. Grounding of electric motors and other refrigerator parts and appliances is explained in some detail in Para. 6-66.

In all refrigeration and electrical circuits, the ground wire is green. This is never used as a current carrying conductor. Its main purpose is to provide protection to the operator in the event of an accidental ground in one of the mechanisms.

HP	A-C Motors Single-Phase Split-Phase or Capacitor		D-C Motors Compound Wound	
	120V	240V	120V	240V
1/6	4.4	2.2	- - -	- - -
1/4	5.8	2.9	2.9	1.5
1/3	7.2	3.6	3.6	1.8
1/2	9.8	4.9	5.2	2.6
3/4	13.8	6.9	7.4	3.7
1	16.0	8.0	9.4	4.7
1 1/2	20.0	10.0	13.2	6.6

Fig. 7-36. Maximum fuse ratings for motor running protection.

7-27 MOTOR PROTECTION

Originally refrigerators had only plug or cartridge fuses to protect them. (Fig. 7-36 shows the full-load current draw in amperes for various sizes of a-c and d-c motors.) However, because the starting current is larger than the running current, the fuse was often too large to protect the motor while it was running. If the running winding of the motor were to draw the full fuse amperage for any length of time, the motor would overheat and be ruined.

To eliminate this danger, time fuses may be installed in the circuit. These devices allow excessive current flow for a very short time (such as in starting). However, they will break the circuit if the high current flow lasts for any length of time.

Three designs are used:

1. One type uses a gear soldered to a shaft. A lever under a spring load is attached to this gear. If the motor takes too much current for too long, the heat from a resistance coil will melt the solder. Then the gear will turn, allowing the spring loaded shaft to open the circuit.

2. Another type uses a heating coil and a bimetal strip. The bimetal will bend if the heating coil becomes warm opening the circuit.

3. Still another type uses an overload cut out device consisting of a heating coil and spring-loaded ratchet throw-out arrangement. Some are built into the thermostat and some

are separate units.

Bimetal overload devices are now the most common safety device and are located at important places in the unit. If these parts overheat for any reason, the circuit will be opened by the bimetal snap switches.

All these devices, however, will only stop the mechanism if the current load is too high. If the motor should overheat from other troubles, the unit may still run and cause damage. Other sources of excessive heat may be: high exhaust temperatures, poor air circulation, poor refrigerant circulation and friction.

Present refrigerating systems have safety devices installed which open the electrical circuit if the motor draws too much current, overheats for any reason or if the compressor becomes too hot.

Fig. 7-37 illustrates the action of a bimetal device. The device opens the circuit if the bimetal disk should reach a temperature that will cause it to snap in the other direction.

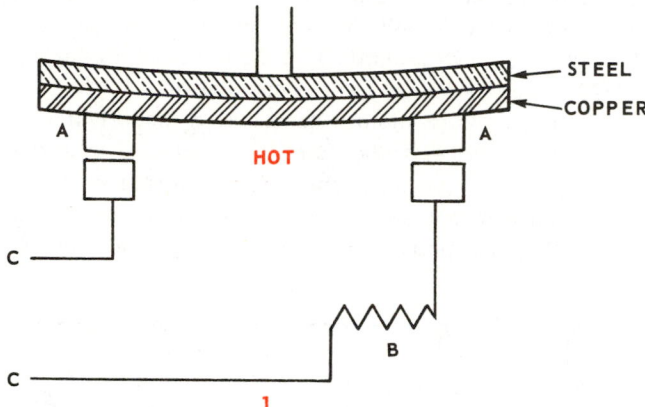

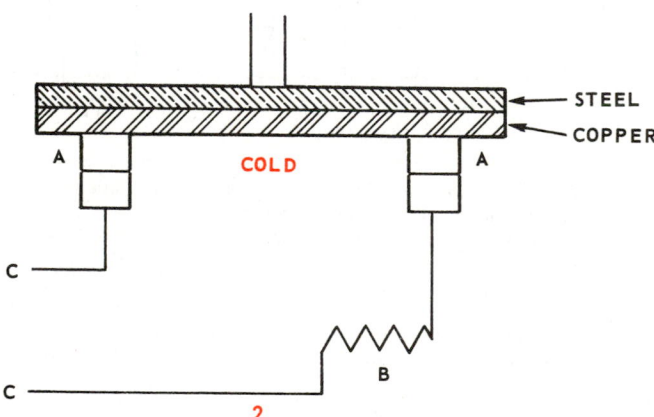

Fig. 7-37. Snap action type bimetal motor overload protector. A—Contact points. B—Heater coil. C—Motor winding connections. 1—Circuit open because of increased temperature or voltage. 2—Points closed in normal operation.

Fig. 7-38 shows the construction of a motor protector.

Considerable damage may result should the compressor or the motor overheat. This overheating could occur without the current draw becoming excessive. It is, therefore, necessary to use heat sensitive overload devices which will open the circuit

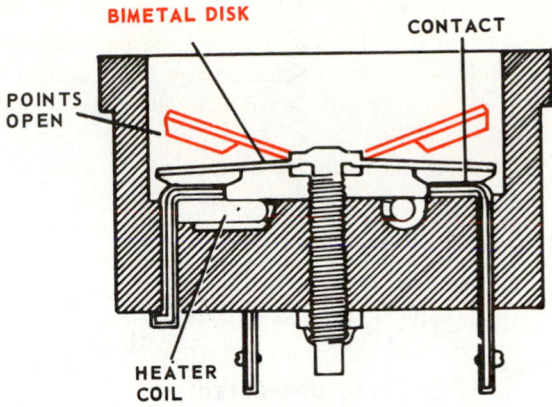

Fig. 7-38. Motor overload protector. Excessive heat will cause bimetal disk to snap (oil-can effect) and open contact points. (Texas Instruments, Inc.)

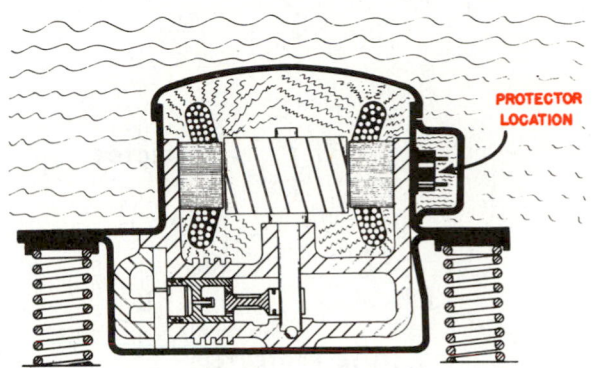

Fig. 7-39. A suitable location for motor compressor overload or excessive temperature protector. (Franklin Manufacturing Co.)

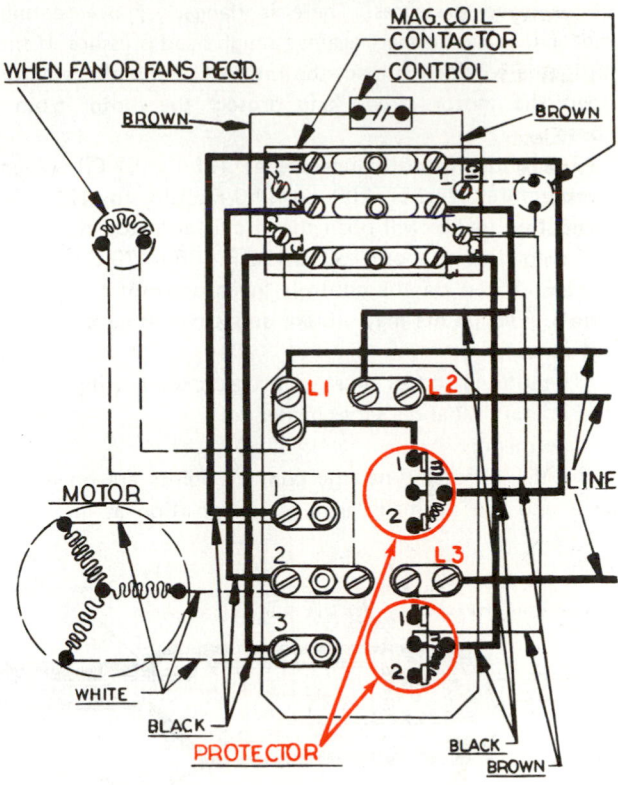

Fig. 7-40. Wiring circuit diagram for three-phase hermetic motor which uses two external motor protectors. (Copeland Corp.)

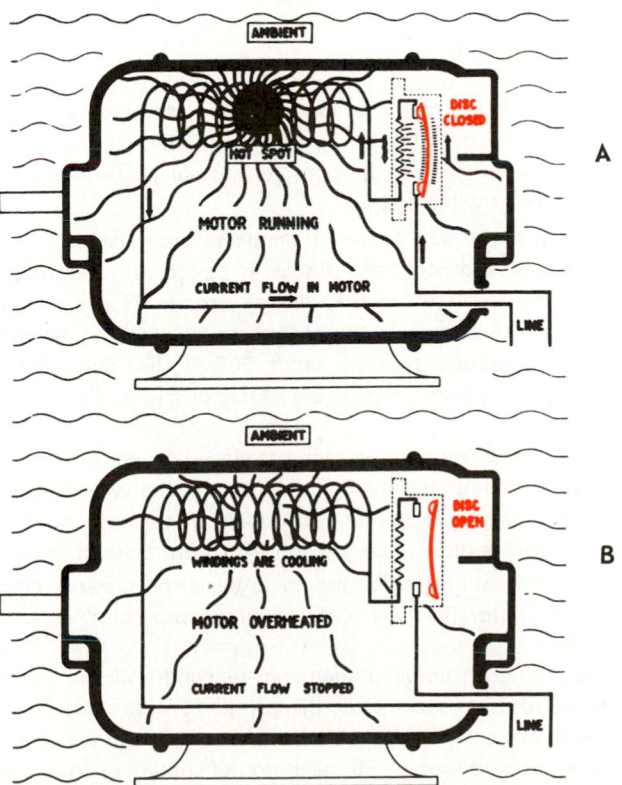

Fig. 7-41. Bimetal overload protector installed inside open type electric motor. A—Current flowing, motor running. B—Current interrupted, motor stopped.

before heat can cause damage. These controls are temperature operated and receive their heat from the motor, compressor and the current draw of the motor. Fig. 7-39 shows an overload protector located on the motor compressor dome or housing. An external overload protector used on polyphase motors is shown in Fig. 7-40.

The action of a bimetal snap action current-actuated and heat-actuated overload protector as used on an open type motor is shown in Fig. 7-41.

Some of the temperature sensitive motor protectors are quite compact and may be easily installed inside the motor winding. Such a protector is described in Para. 7-28.

7-28 MOTOR INTERNAL OVERLOAD AND OVERHEATING PROTECTION

Internal overload motor protectors are mainly used on hermetic motors. They were developed, along with the two-pole motor, for large units.

The motor may overheat if there is too little refrigerant flow. (The refrigerant gas cools the motor.) Or, it may overheat if the unit has to start too soon after shutting off. Too much current draw may be caused by a stuck or locked rotor or compressor.

In most refrigeration and air conditioning equipment, the motor compressor unit is designed to start under a condition of balanced pressures. There is danger of overheating the motor if it has to start against a high head pressure. If there is an increased starting load, the internal overload protector will open the motor circuit and protect the motor from such abuse.

The units normally operate at 125 F. (52 C.). When the temperature reaches 200 to 250 F. (93 to 121 C.), the protection device will open the circuit and stop the motor. It will then close at about 150 to 175 F. (66 to 79 C.).

Do not tap on the controls in an attempt to operate the points. The points may vibrate and arc causing them to burn out quickly.

The internal motor overload protectors are of two types:
1. Bimetal disk and a series heater coil.
2. Thermistor.

In the bimetal type, the contact points are on a bimetal strip and are normally in a closed position as in Fig. 7-42.

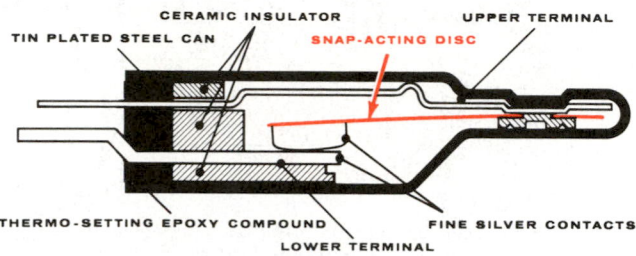

Fig. 7-42. Compact motor protector designed to be fitted into motor windings.

When there is too high a temperature rise, the points will open the circuit. When the temperature in the disk decreases enough, the strip will return to its normal position and the contact points will close.

In the thermistor type, an electronic material, similar to that used in transistors and diodes, is placed in a capsule and the capsule is placed in the control circuit. As the material temperature increases, its current carrying capacity increases. If the temperature rises to about 200 F. (93 C.) — the increased current flow will operate a relay circuit and the circuit will open.

The internal overload protector is placed inside the hermetically sealed compressors, directly on or in the windings. The internal overload protector will open if there is excessive current draw, excessive temperature or both. Loss of refrigerant, a restriction in the system or low suction pressure could lead to a motor burnout if the overload protector were not installed.

Depending on ambient temperature conditions, it may be an hour to two hours after the protector opens the circuit before it will close. Cooling the dome with forced air, dry ice or carbon dioxide spray will speed up the cooling process.

The leads to these protectors must never be shorted, because even a few moments of operating a unit without this protector may burn out the motor. These protectors cannot be

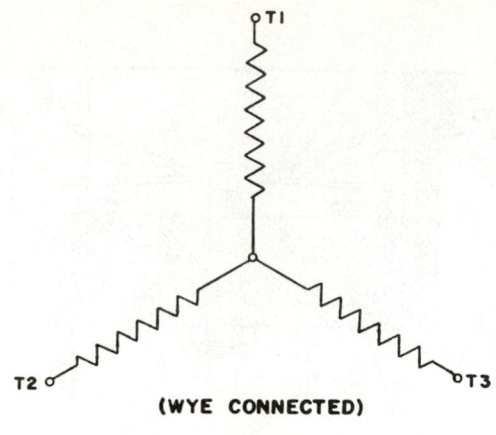

(WYE CONNECTED)

THREE PHASE MOTOR

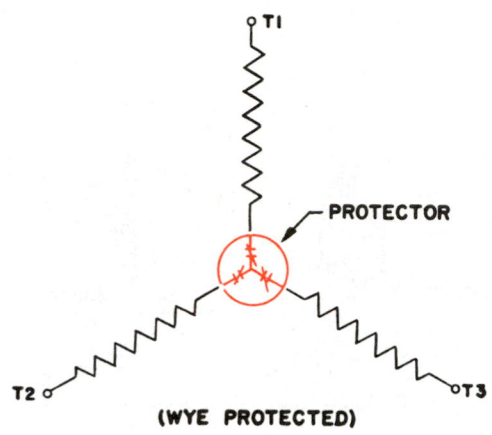

(WYE PROTECTED)

THREE PHASE MOTOR

Fig. 7-43. A method of connecting motor protector in three-phase Y type motor. (Copeland Corp.)

taken out of the circuit. They are the best possible protection for a hermetic compressor. Motors having this protection are usually labeled "Internal Overload Protected."

In Wye type three-phase motors, the internal overload is placed at the common terminal of the three windings and it will open all three circuits when its points open, Fig. 7-43.

These internal protectors are very reliable. Cases of failure are almost unknown. Fig. 7-44 shows an instrument used to check the three circuits of a three-phase system.

7-29 MOTOR TEMPERATURE

The temperature of the hottest part of motor should not be more than 72 F. (40 C.) over the room temperature. This means an average maximum temperature of approximately 150 F. (66 C.).

This temperature is difficult to measure except in a laboratory; so, it is better to check the ambient (surrounding) temperature to be sure it is not too high. Check the motor for cleanliness and airflow. Then check the current draw of the motor. If draw exceeds the rating on the motor identification plate or in the motor manual specifications, the motor will

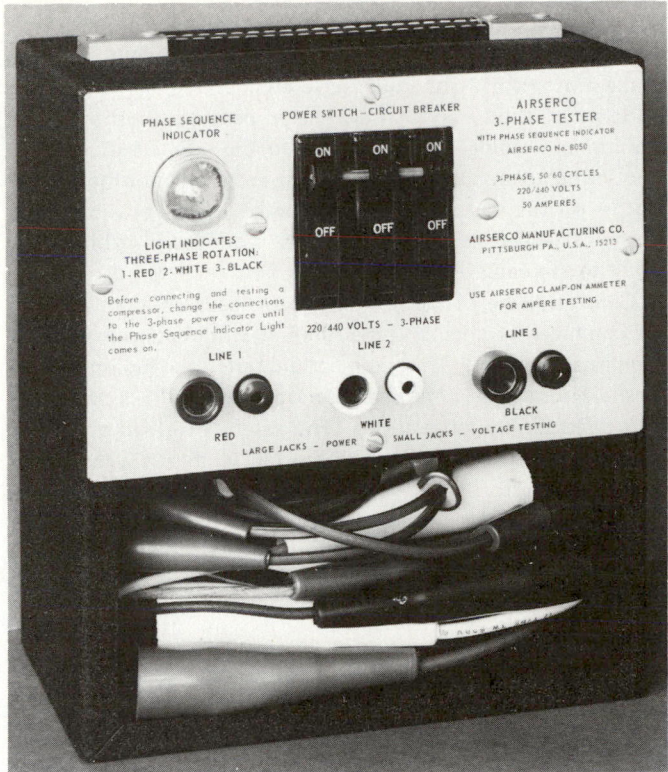

Fig. 7-44. Instrument which may be used to test all three circuits of a three-phase system. (Airserco Mfg. Co.)

overheat. Current overdraw may be due to overloading the motor or to a shorting in the motor windings.

Always measure motor temperature with a thermometer. A motor with too high a frame or stator temperature may have motor winding temperatures so high that the insulation on the wires may fail.

The thermal overload protector for motors usually opens the circuit when the temperature reaches 200 F. (93 C.) and closes the circuit when the temperature drops to about 150 F. (66 C.).

Motors depend on rather cool ambient air for cooling. If this air is too warm or if the airflow is restricted, the motor will overheat.

7-30 STANDARD MOTOR DATA

The data on the motor identification plate usually gives the following information:
1. Required voltage (emf) supply.
2. Hertz (cycle).
3. Running current (amperes).
4. Locked current draw. The locked current draw indicates the condition of the internal circuit when the rotor is locked so it cannot turn. It is also the starting current draw.
5. Temperature rise. The temperature rise is usually specified in degrees Celsius.
6. Energy efficiency ratio (EER - See Chapter 15).

Compressor speed on external drive systems can be controlled by using either two or four-pole motors and by changing pulley sizes. Direct connected compressors must operate at

motor speed. If a four-pole motor is used to replace a two-pole motor, a compressor of greater displacement must be used with the slower rpm motor. Because the unit is running at half its former speed, the compressor displacement per stroke must be doubled.

The size of the conductor used in the refrigeration mechanism is very important. If too small or too long, it will heat up and eventually cause a fire. Long circuits also add an unnecessary resistance to the flow of electricity. This causes excessive voltage drop. If the voltage to the motor is less than 90 percent of the rating of the motor, there is danger of the motor being overheated and ruined. A table of wire sizes recommended for 120V circuits is given in Fig. 7-45.

Wire No.	Diameter of Wire in Inches	Ampere Capacity Plastic Insulation
18	.040	5
16	.051	10
14	.064	20
12	.081	25
10	.102	30
8	.128	50
6	.162	70
4	.204	90
2	.258	125

Fig. 7-45. Recommended wire sizes for various ampere capacity circuits. This table is calculated on the basis that it is used on 120V circuit.

The efficiency of small motors is only 50 to 60 percent because of clearances and efficiencies of the winding. Therefore, they consume nearly twice as much current as they should, compared to larger motors such as 1/2 hp and over. A 1/6 hp motor, which should theoretically use only 124W or 1 1/4A, will be found to need approximately 2 1/2 to 3A (280 to 400W) to develop the 1/6 hp.

When making electrical connections to a domestic refrigerator, the thermostat should be connected into the hot wire of the circuit. This wire has black insulation.

The other wire is called the ground wire of the circuit. It has white insulation. It should be run directly to the motor.

The white wire (ground) carries the same amount of current as is carried in the black wire.

A system of green grounding wires grounds all of the mechanisms in a refrigerator or air conditioner. This ground is *not* a current carrying wire. It is for safety only and is used to avoid the possibility of an electric shock should a short circuit or a ground occur in the electrical system.

7-31 FAN MOTORS

Many hermetic units use motor-driven fans to:
1. Force condenser cooling air through the ducts and over the condenser and condensing units.
2. Circulate air in refrigerated parts of domestic and commercial units. See Fig. 7-46.

To create efficient air movement, the fan and condenser are

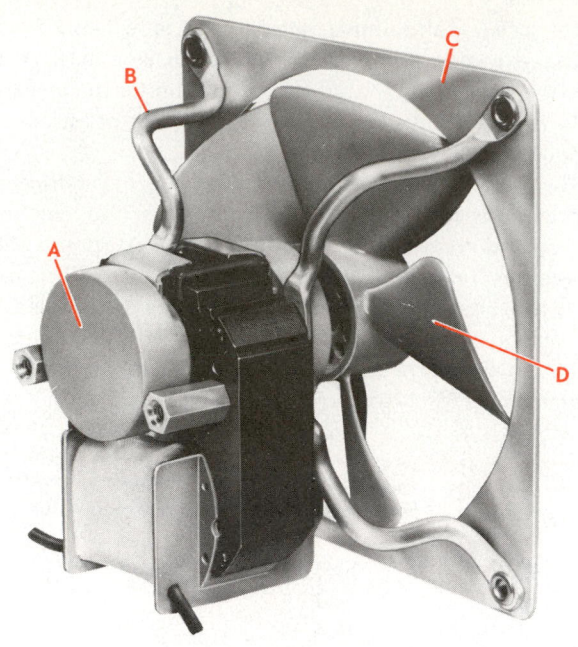

Fig. 7-46. A shaded-pole motor and fan. A—Motor. B—Motor support brackets. C—Mounting plate. D—Molded fan.   (General Electric Co.)

in a housing of sheet metal or plastic. The fans are carefully balanced and run almost noiselessly. They are usually attached to the motor shaft with Allen setscrews.

Some of these motors have sealed bearings (bushings) and require no oiling. Others need oiling (SAE 10 or 20) amounting to one drop per bearing each six months. A few motors on the market have only one bearing. The motors are usually attached to brackets and are mounted in rubber. Fig. 7-47 shows a replacement condenser fan.

Generally, the condenser fan motor leads are connected to the common terminal and the running winding terminal of the compressor motor. This connection puts the fan motor in parallel with the compressor motor and allows it to be controlled by the thermostat. The safety overload cut-out is also put in the circuit ahead of the fan so that it will also cut out the fan motor.

Some fan motors have their own thermal safety controls and many are of the two or three speed type. The variation in speed may be obtained by using extra poles in the stator or by using a solid state control.

The speed of a fan motor is quite sensitive to the applied voltage. As the voltage drops, so will the fan speed. Fig. 7-48 shows the relationship between voltage and fan speed.

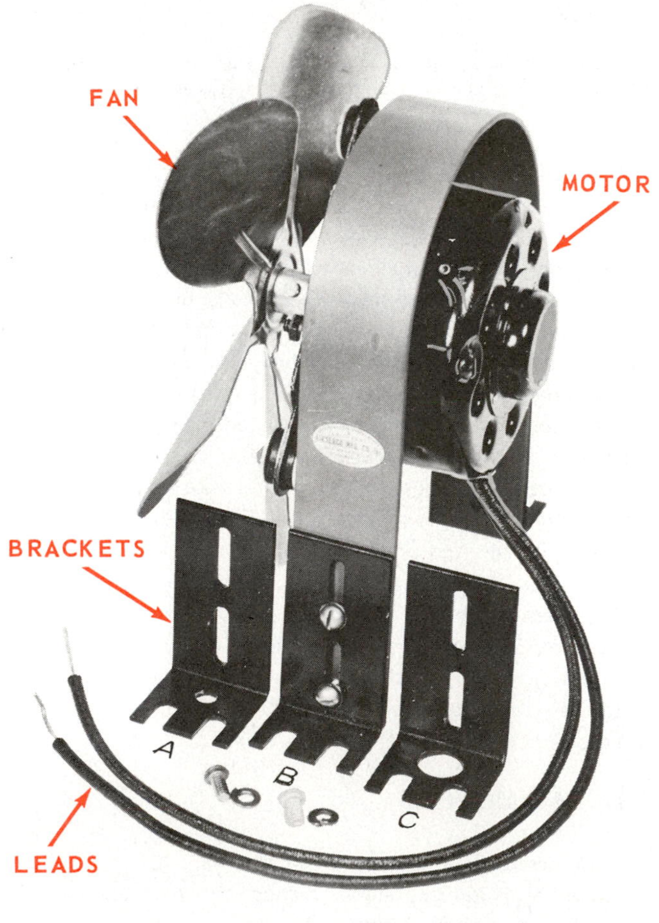

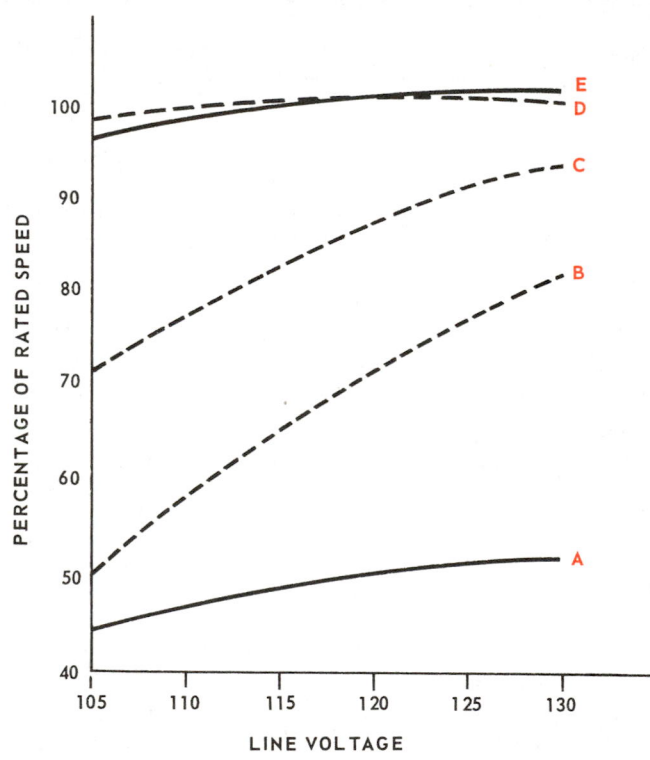

Fig. 7-48. How voltage affects fan speed for six and twelve-pole motors. A—Twelve-pole. B—Low speed. C—Medium speed. D—High speed. E—Six-pole.

Fig. 7-47. Replacement condenser motor and fan with universal mounting brackets (A,B,C) which may be used with a variety of condensing unit designs.

Fig. 7-49 is a schematic of some of the common fan motor circuits. One, two and three speed motor circuits are shown. Refer to Para. 7-46 for information on voltage drop tests, troubleshooting and how to service fan motors.

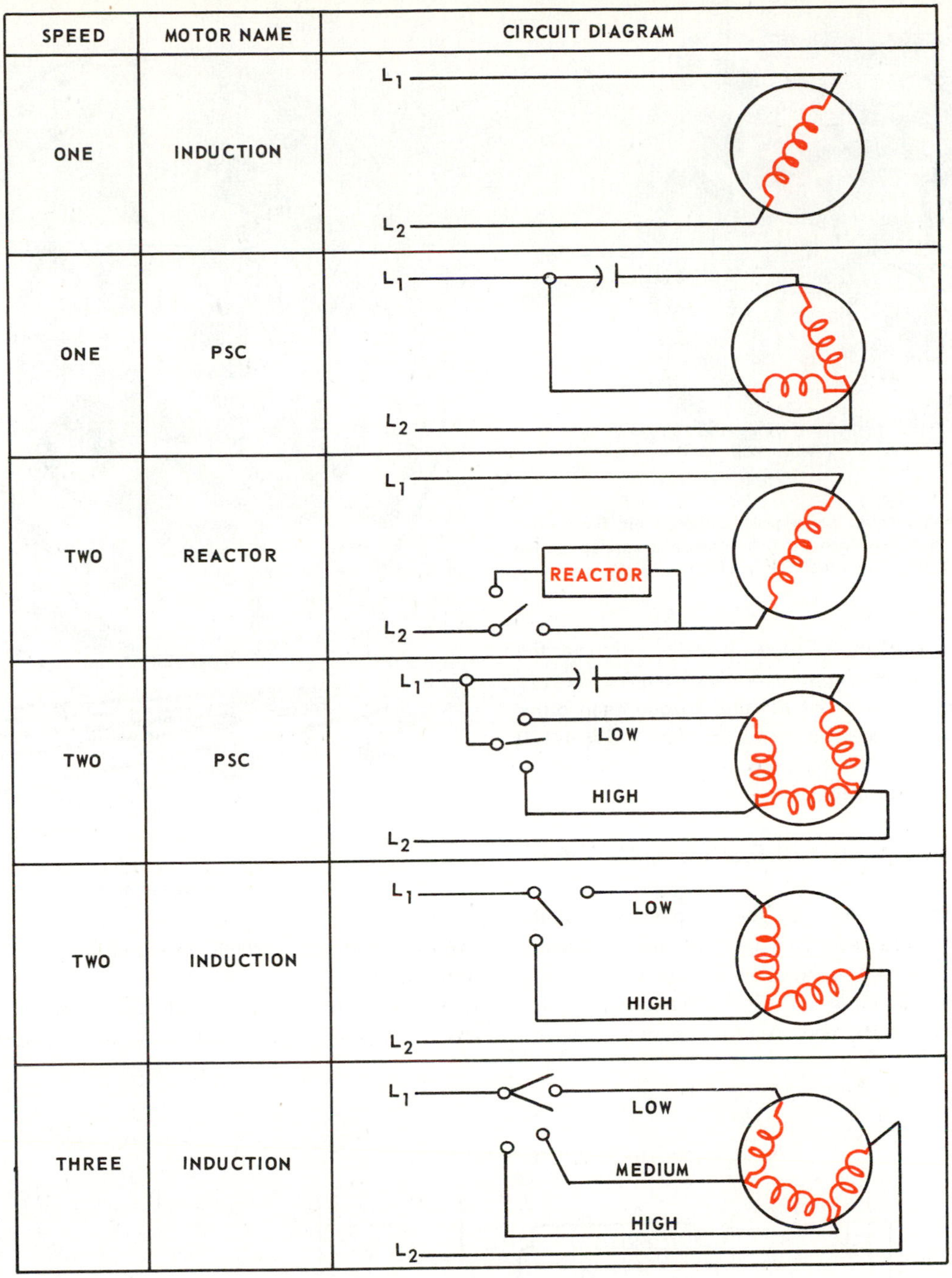

SPEED	MOTOR NAME	CIRCUIT DIAGRAM
ONE	INDUCTION	L_1 ... L_2
ONE	PSC	L_1 ... L_2
TWO	REACTOR	L_1 ... REACTOR ... L_2
TWO	PSC	L_1 ... LOW / HIGH ... L_2
TWO	INDUCTION	L_1 ... LOW / HIGH ... L_2
THREE	INDUCTION	L_1 ... LOW / MEDIUM / HIGH ... L_2

Fig. 7-49. Schematic of some common fan motor circuits.

7-32 SHADED-POLE MOTORS

The construction of a shaded-pole motor is somewhat different than the construction of the motors previously described. The shaded-pole produces a moving magnetic field perpendicular to the field pole and starts the rotor turning as shown in Fig. 7-50.

Approximately half of each pole face has a small copper plate insert, A, with a small winding. This insert slows down the buildup of the magnetic field through the copper plate just enough to cause a magnetic motion toward the copper plate.

This action produces a lag for induced magnetism in the rotor (opposite magnetism). The rotor turns as it is attracted

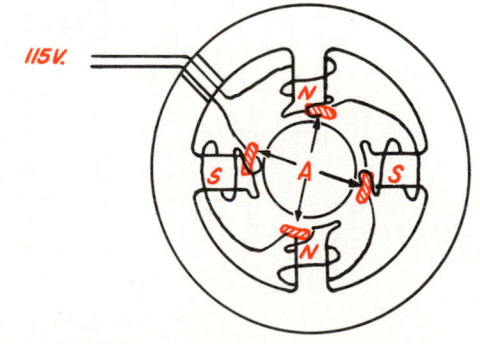

Fig. 7-50. Shaded-pole fan motor. S—South polarity. N—North polarity. A—Shaded pole (copper).

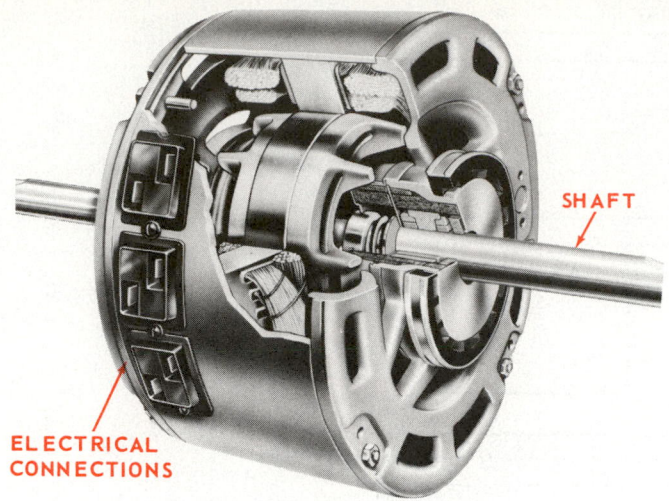

Fig. 7-51. Low starting torque shaded-pole, double-shaft fan motor. Three terminals for electrical connections provide a variety of fan speeds. (General Electric Co.)

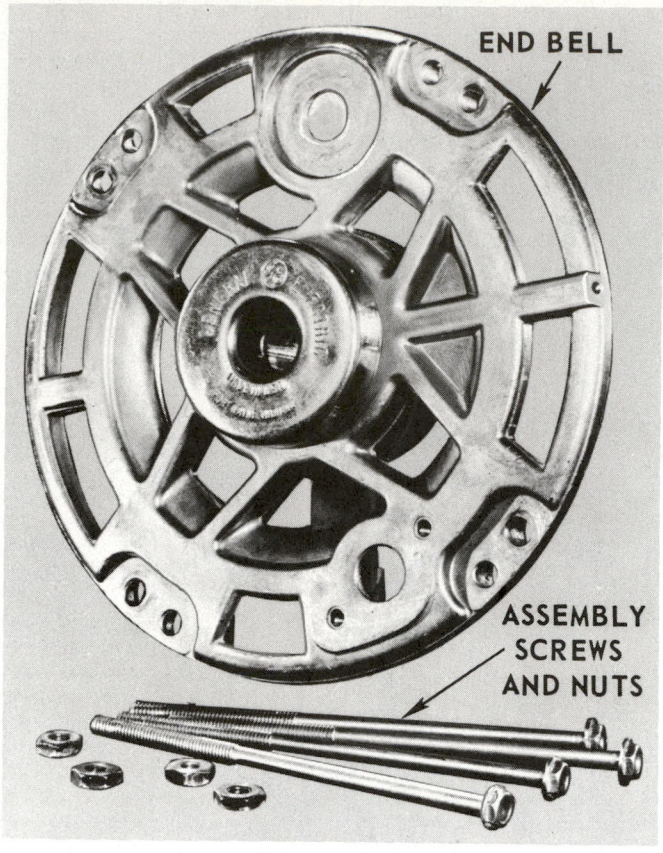

Fig. 7-52. Aluminum end bell for fan motor. Note assembly screws and nuts.

by the magnetism. Movement of the rotor will continue as the alternating current changes the polarity of the poles and rotor.

Although this design has less starting torque than other type motors, it is very successful in small motors 1/6 to 1/100 hp. Fig. 7-51 shows a double-shaft fan motor. The end bell of this motor is shown in Fig. 7-52.

7-33 ELECTRONIC VARIABLE SPEED MOTORS

A method employing transistor switching instead of the brush or commutator has been developed for low horsepower motors. Brushless motors operate with silicon rectifiers, transistors and special circuitry. The advantages of the transistorized motor are its high speed compactness, perfor-

mance, durability, variable speed, elimination of sparking and brush noise.

Speed is changed by adding a small variable resistance (potentiometer) which will vary the resistance within the

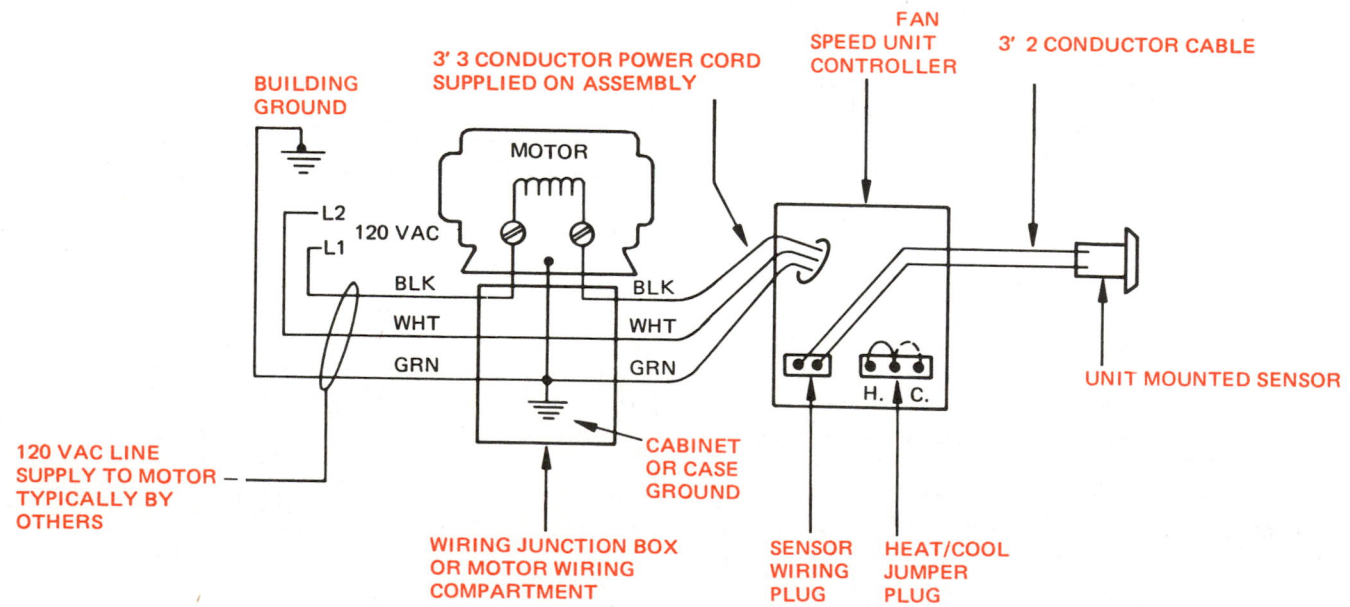

Fig. 7-53. Installation wiring diagram for solid state motor control. (Barber-Colman Co.)

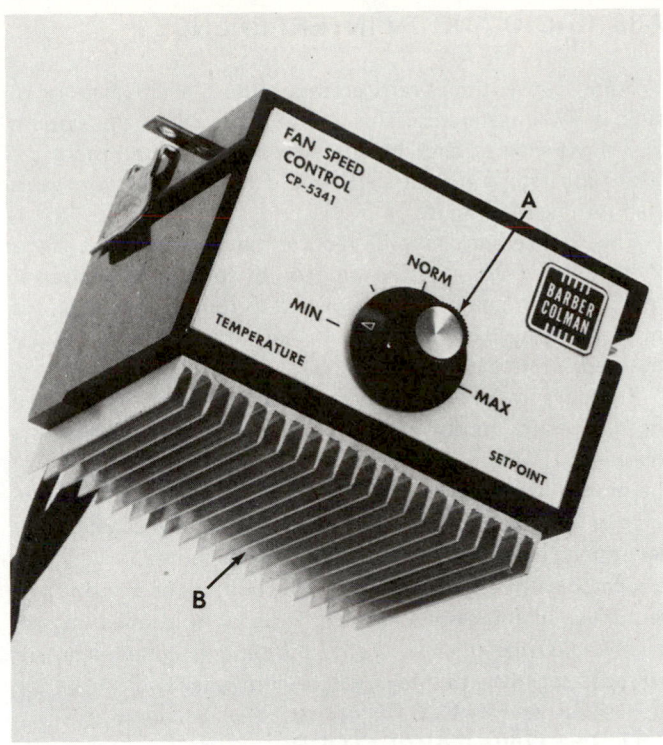

Fig. 7-54. Solid state fan speed control. A—Adjustment. B—Heat sink.

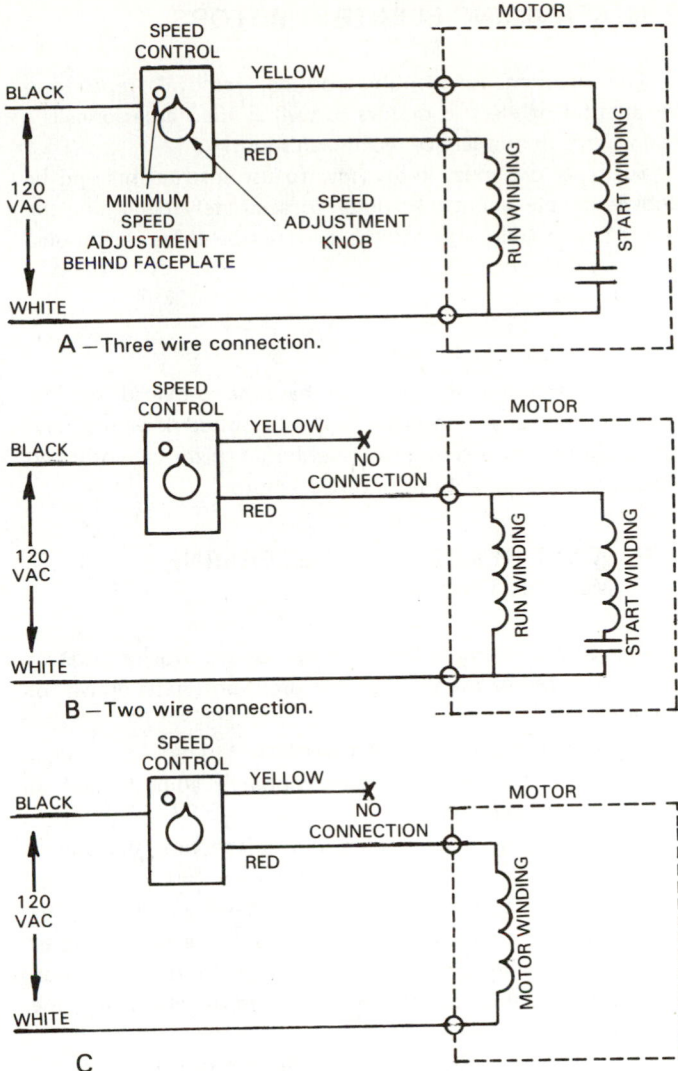

A—Three wire connection.

B—Two wire connection.

C

Fig. 7-55. A and B—Connecting an electronic variable speed control to a permanent split capacitor motor. C—Same control connected to a shaded-pole motor. (Lutron Electronics Co., Inc.)

circuitry of the control. The direction of the motor's rotation can be quickly reversed by manipulating the motor control's switching devices.

Fig. 7-53 is the wiring diagram for a solid state motor speed control. The control reacts to signals from the unit mounted sensor. Fig. 7-54 illustrates the same unit complete with case and controls.

The diagram in Fig. 7-55 is for yet another control which is commonly used to regulate motor speed. Another wiring diagram of a control mechanism is shown in Fig. 7-56.

The circuit operates as follows:

Resistances R_2, R_3, R_4 and capacitor C_2 form an R-C (resistance-capacitance) charging network. The effective resistance of R_2, R_3 and R_4 charges the capacitor, C_2, up to a voltage that causes diac, D_1, to conduct an electric current and fire triac, Q_1.

When Q_1 is fired, it conducts until the current through it drops below the holding current (25 milliamperes) of the triac. At that point, the triac turns off. The control's chargeup, fire, conduct and turn off process repeats every half cycle of alternating current power.

The speed of the motor is controlled by varying the resistance of R_4. The variable resistor is operated by the knob on the front of the control.

The lower the resistance of R_4, the faster the motor will turn. The minimum speed of the motor can be set by adjusting the resistance of trimmer, R_3.

R_1, C_1 and L_1 provide suppression of radio frequency interference caused by the fast switching of Q_1. This circuit can be used with permanent split capacitor, shaded pole and universal motors.

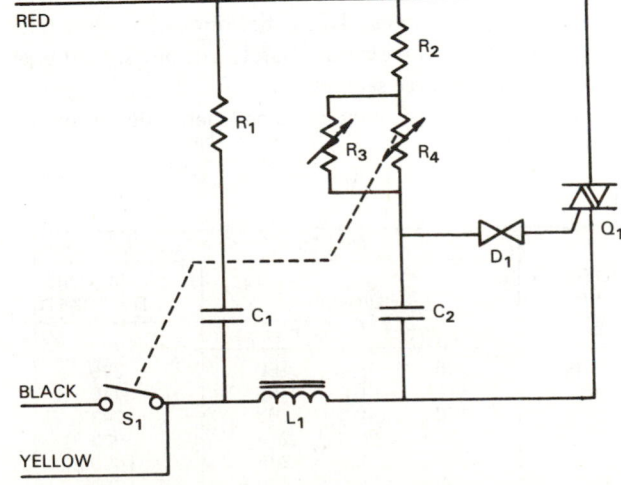

Fig. 7-56. The electronic circuit for a variable speed control used with permanent split capacitor (PSC) motors, shaded-pole motors and universal motors. (Lutron Electronics Co., Inc.)

7-34 SERVICING ELECTRIC MOTORS

The maintenance, troubleshooting, removal, repair and installation of electric motors as well as their accessories is a major portion of a service technician's job.

Such persons must know how to use instruments and be knowledgeable about electricity to accurately determine the trouble. The following paragraphs describe these operations:
1. General service.
2. Open motors.
3. Hermetic motors.
4. Fan motors.

It is important that a solid base be provided for the installation of any motor and that it be bolted down securely. The armature shaft should be level for a horizontal motor and exactly vertical (plumb) for a vertical motor.

7-35 WATT READING TO DETERMINE MOTOR TROUBLES

One way of learning the condition of a motor compressor unit is to observe the wattage consumption (watts drawn) of the unit.

The meter will give two different wattage readings:
1. Combined starting and running winding reading (only 1 to 1 1/2 seconds long).
2. The running winding reading during the time the unit is running.

Note: The wattmeter needle will overswing slightly at the instant that the motor compressor starts. Since the combined starting and running winding watt reading is for only a fraction of a second, the service technician must allow for the overswing.

When the thermostat contacts close, the wattmeter indicator will swing to the right; then it will quickly move back to the combined reading. In a few seconds, the pointer will fall to the running winding reading only.

If the starting winding circuit is open, the wattmeter pointer will swing to the right and then move back to the running winding value only. This action indicates a bad relay or starting winding. The overload safety cut-out should open the circuit in two or three seconds.

Approximate watt readings for small hermetic motors are shown in Fig. 7-57.

MOTOR HP	WATTS AT 120 VOLTS		
	RUNNING		STARTING OR LOCKED
	AT 70 F.	AT 110 F.	
1/16	66	100	375
1/9	117	160	740
1/8	108	163	743
1/7	160	218	970
1/5	242	295	1450
1/4	235	320	1250

Fig. 7-57. Approximate watt readings for small hermetic motor compressors. Indicated temperatures are ambient.

7-36 RADIO AND TV INTERFERENCE

Some conventional refrigerators cause a slight amount of radio or TV interference. This interference will usually amount to a slight snap or click in the radio or TV at the instant the refrigerator stops or starts. It should be no more noticeable than turning off a light.

This interference may be reduced by grounding the frame of the motor to a water pipe or by placing a condenser between the frame of the motor and a ground. In general, the interference is not disagreeable and may usually be disregarded.

Excessive radio interference of a continuing nature when the refrigerator motor operates, or when it starts, indicates a loose electrical connection or some fault in the mechanism of the motor. These include worn brushes, badly pitted commutator or loose connections. The particular trouble can be easily determined by a careful examination of the motor.

Occasionally a static charge will be built up on the belt of a belt-driven compressor. The discharge of this build-up will cause radio interference. If the motor and compressor are grounded together, this noise will be eliminated.

7-37 TESTING CAPACITORS

When a motor or motor compressor does not start or run properly, there is a good possibility that the trouble is in the capacitor.

Most motors have only one — a starting capacitor — but some have two or more. In such cases, one might be a starting capacitor and the other a running capacitor. Some motors have two or more capacitors connected in parallel for additional capacitance or in series for additional voltage.

The starting capacitor is connected in series with the starting winding. It is usually wired into the circuit between the relay and the starting winding terminal of the motor (off-and-on operation).

The running capacitor is also in the starting winding circuit but stays in operation while the unit runs (continuous operation).

There are two types of capacitors:
1. The dry capacitor (for intermittent operation).
2. The electrolytic capacitor (for continuous operation).
Both types may be tested in the same way. (The dry capacitor is described in Para. 7-8.)

The simplest capacitor test is to substitute a good capacitor for the one being tested. If the motor operates, the old capacitor is faulty. The replacement capacitor should be the same capacity as the old. If one must be used of a different capacity, it should be 5 to 10 percent over capacity rather than under.

Fig. 7-58 illustrates a commercial model of a capacitor tester. These are recommended. Testing by shorting is not a good practice.

IMPORTANT: Never place fingers across the terminals of a capacitor. It may be charged and give a shock. Always short it with a coated wire before handling it.

Another way to check a capacitor is shown in Fig. 7-59.

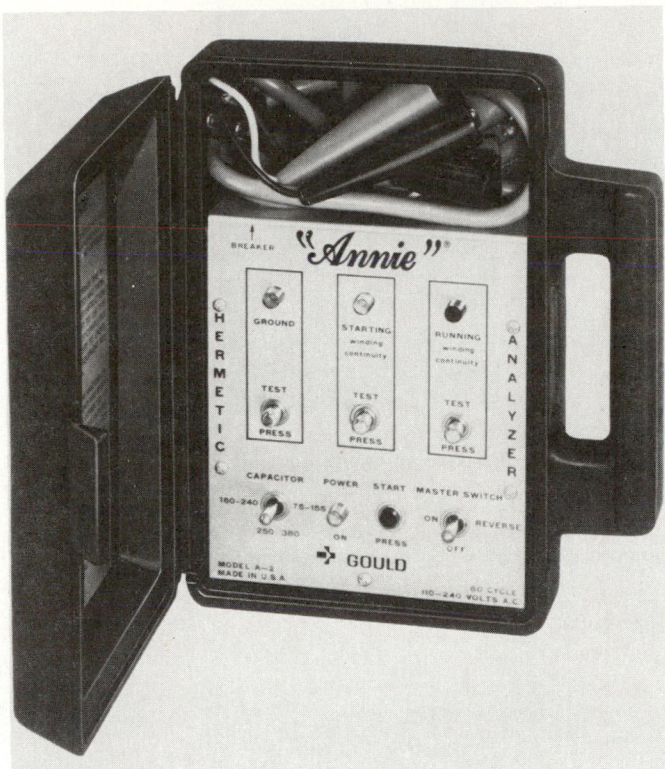

Fig. 7-58. Capacitor tester and analyzer used to detect hermetic motor compressor troubles. It may also be used to reverse the motor. (Gould Inc., Fluid Components Div.)

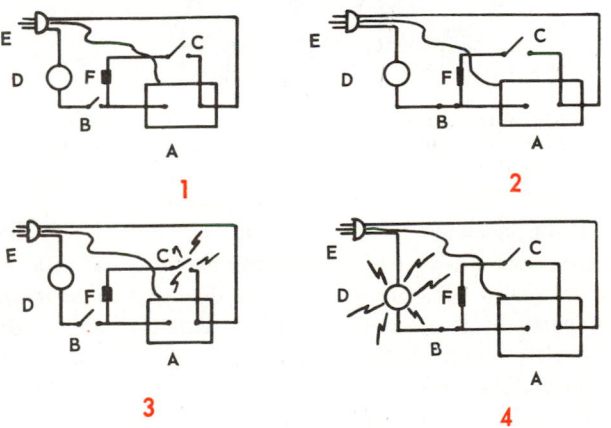

Fig. 7-59. One method of testing a capacitor. A—Capacitor being tested. B—Charging switch. C—Shorting switch. D—Fuse or circuit breaker. E—Attachment plug to 120V circuit. The middle prong of the ground or green wire. F—Resistor. 1—Switches open. 2—Charging switch closed. 3—Good capacitor. 4—Shorted capacitor.

Place the capacitor to be tested as shown at A. Be sure a fuse is in the circuit as shown.

The capacitor should be put in a protective case while testing it, because a shorted capacitor may explode when put in a circuit.

If the capacitor is good, it will spark as in view 3, Fig. 7-59.

If it is shorted or grounded, the fuse will blow as in view 4.

If the condenser does not take a charge, it will not spark. This indicates an open circuit.

In view 1, the tester is connected to the terminals of the capacitor at A. Spring clips should be rubber covered. Use ground leads. Switches at B and C are open. Button switches similar to doorbell switches that are spring loaded to stay open may be used. In view 2, the charging switch, B, is closed momentarily (press it down and remove finger quickly). Note that switch, C, stays open.

If the capacitor is shorted, as at D in view 4, the fuse or circuit breaker will blow. If nothing happens, then proceed as in view 3. With the switch open at B, touch the shorting switch. If the capacitor is good, the switch will spark. If it does not spark the first time, try it three or four times before scrapping the capacitor.

7-38 SERVICING CAPACITORS

Some capacitors have resistors connected across the terminals. This resistor, visible at the capacitor terminal, slowly bleeds the capacitor of its charge to lessen the arcing of the motor control points. Such arcing may occur if the unit cycles frequently.

When testing such a capacitor, remove the resistor from one terminal. To discharge a capacitor, use a 20,000 ohm, 2W resistor in the circuit. Avoid shorting the terminals. The sudden discharge may rupture the thin metal foil in the capacitor.

Capacitor size must be accurately suited to the motor and the motor load. It is general practice to permit up to 25 percent over capacity. For example, a 125 μ f can be used for a 100 μ f capacitor, but an undersized capacitor should never be used. If at all possible, use an exact replacement. The make, model and model number are usually placed on each capacitor.

If this information is unavailable, or if an emergency capacitor must be used temporarily, there are several ways to determine the proper size. One of the quickest ways is to put the temporary capacitor in the circuit. If it brings the unit up to running speed and the relay operates in three seconds, the capacitor is very nearly correct.

A better method is to use a specially designed capacitor selector unit. This selector has a variable capacitance and increasing amounts of microfarads are put in the circuit (in series) until the correct voltage reading is reached. The capacitance registered on the selector indicates the capacitor size that should be put in the circuit.

Another method is as follows: connect the capacitor to a double-fused line and connect the voltage leads of the meter across the capacitor. The ammeter reading is taken by placing the heavy probes of the ammeter around the wire.

Some of the capacitors have mechanical connectors (machine screws) while some have solder-type leads. It is necessary to use a small electric soldering copper or soldering gun to connect the second type.

Some running capacitors use polychlorinated biphenyl dielectric fluid. This fluid is dangerous. Do not open the shell of this capacitor. If the shell is accidentally pierced or broken, be very cautious not to touch the fluid or breathe its fumes. To properly dispose of capacitors containing this dielectric fluid, it is best to bury them.

7-39 SERVICING EXTERNAL DRIVE MOTORS

Troubles found in open type motors are few and they can be classified as:
1. Electrical troubles.
2. Mechanical troubles.

The electrical troubles found in electric motors may be:
1. An open circuit, short circuit or ground may occur in the field windings. In these cases, replacing the motor is recommended.
2. Frequent starting of the motor may result in overheating. Overheating the capacitor may cause the switch and the insulation to fail. If the motor will not start until the pulley is spun but has the characteristic a-c hum, it is a sign that the capacitor or the centrifugal switch points have failed. It is easy to replace the old capacitor with a good one of the same capacity.

If the motor still will not start, the trouble is probably in the centrifugal switch. If the points are dirty, pitted, or dark from overheating, do not try to repair them. Filing or sanding does little good as the contact material is worn away. Repaired points usually last just a few hours and a call back may have to be made. It is best to replace them. Mechanical troubles are almost the same in all open type motors:
1. There is a possibility that the centrifugal switch used for connecting and disconnecting the condenser and/or the starting winding may become worn. Replacement of the switch is necessary, in such cases. Fig. 7-60 shows a centrifugal switch for a capacitor start type electric motor.
2. Other troubles may include bearing wear, end play, excessive vibration, misalignment of the motor with the compressor and improper air gap between the rotor and stator.

Repair and testing of motors is covered in Chapter 14.

7-40 MOTOR BEARINGS AND LUBRICATION

Open motors may be lubricated in various ways. It depends on the type of bearing used and the position of the motor. Open motors using bronze bushings, plain or sleeve, may be lubricated in two different ways:
1. Wick system.
2. Slip ring system.

The wick system uses a well or reservoir in the end bell. A wick (cotton or wool yarn) carries oil from this well to the bushing and shaft. This system allows long intervals between servicing bearings and prevents the bearing getting too much oil. This type of lubrication is shown in Fig. 7-61.

Motors with this lubrication system have the cotton or wool yarn saturated with oil when shipped from the factory. However, before starting the motor, the oil wells should be filled by adding the amount of oil designated or until oil appears in the lower oil level cup.

If the bearing is to be removed from the shaft or if the bushing is to be removed from the end bell, the yarn should be lifted clear of the bearing to prevent the yarn being forced between the shaft and the bearing upon replacement.

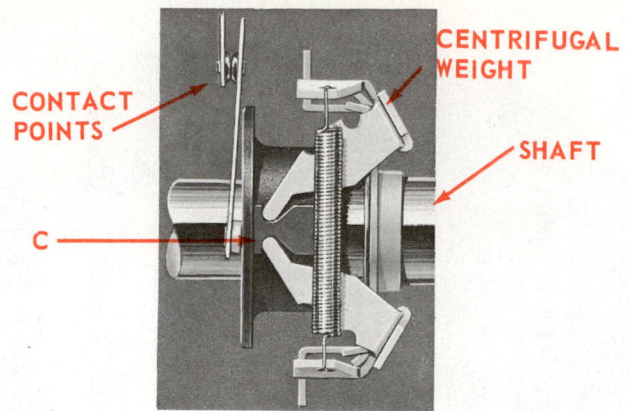

Fig. 7-60. Centrifugal switch mechanism may be used on "open type" unit compressor motor. Centrifugal weights and spool, C, revolve with motor shaft (rotor). At about 75 percent of full operating speed, centrifugal weights cause spool, C, to move to right allowing contact points to open, breaking the starting winding circuit. (Century Electric Co.)

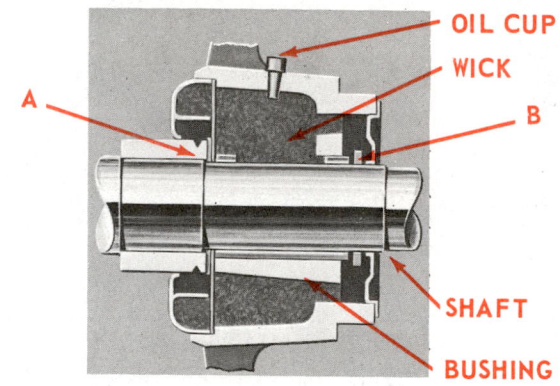

Fig. 7-61. Motor end bearing which uses wick oiler. Oil slingers A and B prevent oil from leaking into motor or out along motor shaft.

When replacing the yarn, pack equal amounts on each side of the bearing, and over the slot of the bearing so the spring on the oil well cover will push the yarn down on the shaft. Wick lubricated bearings should be oiled with one or two drops every six months.

Some larger refrigeration motors use the ring lubricating method. A brass ring rests on the motor shaft through a slot in the top of the bearing. The ring is large enough to dip into the oil pocket below. As the motor shaft turns, the ring turns slowly and the wet portion lubricates the bearing.

Be sure to check the rings when working on these motors. Use a medium viscosity nondetergent oil such as SAE 20 or SAE 30 (200 to 300 viscosity).

Some motors use ball bearings, as in Fig. 7-62. These bearings are grease lubricated. Most are sealed and do not need any lubrication service. Some, however, come with grease cups and can be lubricated with a grease gun.

These motors, when new, are supplied with enough grease in the bearings to lubricate them for a number of months. A small amount of grease should be added every two or three months. Use a high grade "medium" grease on fully enclosed motors. Too much grease may cause the bearings to overheat.

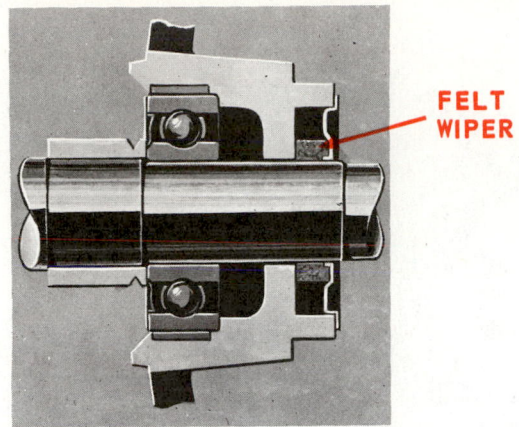

Fig. 7-62. Motor shaft mounted on grease lubricated ball bearings. Felt wiper keeps out dirt and dust.

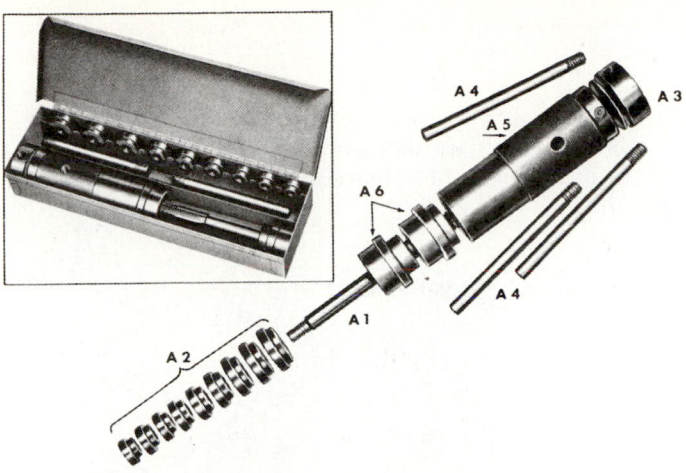

Fig. 7-63. Special motor bearing tool kit contains necessary tools for removing and replacing motor bearings. A1—Main shaft. A2—Bushing puller disks of various sizes. A3—Take-up nut. A4—Nut turning handles. A5—Main spacer. A6—Bushing socket spacers. (Cleveland Sales Co.)

The life of the bearings depends, to a considerable extent, on cleanliness. Use only clean grease and keep all dirt out of the bearing. Clean all fittings before using the grease gun.

Most greases and oil oxidize and will collect dirt while in use. When a motor is reconditioned, the old lubricant must be discarded, the lubricated portions thoroughly cleaned and new lubricant used.

Another method of lubrication presently used is the oilless bushing. In this arrangement, the shaft passes through a sintered (porous bronze) bushing which has been impregnated with oil at the factory. The total tolerance between the shaft and bushing is less than .001 in. and may have a tolerance of as little as .0003 in. As the bushing-shaft tolerance increases, due to wear, the motor will become noisy.

The oilless bushing is considered to be permanently lubricated. It is often used on fan motors and on other low horsepower applications.

Fan motors may become very cold when idle. The bearing oil may become very thick and the motor (usually low torque) will start with difficulty. It may even burn out or activate the overload switch. Be sure to use oil that will remain quite fluid at the temperatures — 0 F. (—18 C.) to —40 F. (—40 C.) — the bearings may reach during idle time.

7-41 SERVICING MOTOR BEARINGS

In any service to motors, bearings should be checked to see if they are worn. If the rotor is hitting the stator, bearings are worn out and must be replaced.

Clearance between rotor and stator varies from .015 in. to .030 in., depending on the size of the motor. This clearance should be the same all the way around the rotor. A heavy rumbling sound at starting usually indicates that the bearing is badly worn, even though the rotor may not be touching the field poles.

Bearings are usually made of phosphor bronze and are pressed into the end brackets or end bells. Sometimes they are locked in by a pin pressed through the bearing housing and into the bearing. The bearing must always be pressed inward to remove it.

Care should be taken not to put an out-of-line force on the end bell when pressing bearings out. This would probably crack the end bell. A special tool, Fig. 7-63, can be used to remove or install bushings or sleeve bearings.

After the new bearing is pressed into the bracket, the bearing must be reamed. It is best to ream the two in-line with adjustable reamers.

The surface of the shaft in contact with the bearing must be perfectly smooth. A scored shaft may be repaired in a lathe using a grinder mounted on the tool post.

If a bearing is overheating, any one of the following may be the cause:

1. Oil too heavy.
2. Oil too thin. Select a good grade of mineral lubrication oil which is not greatly affected by a change in temperature and which does not foam or bubble too freely.
3. Dirt or grit in the oil.
4. Belt too tight.
5. Pulley hub rubbing against the bearing.
6. Motor not properly lined up, causing the armature shaft shoulder to pull on or be pushed against one bearing.

7-42 CLEANING MOTORS

While in service, the motor should be cleaned regularly. Dust and lint in the motor will prevent proper air circulation. Compressed air or a hand bellows should be used frequently to blow dirt out of the motor.

Any oil which may overflow from the bearings should be wiped off. A little attention will result in efficient operation with the motor giving good service for many years.

If the motor must be dismantled, all parts should be carefully cleaned before being worked on or reassembled. Cleaning fluids must be used that are not harmful to the electrical insulation material or to the technician's health. There are many cleaning fluids on the market. Be sure to check the one being used for safety.

Motor interiors are well designed but rough handling may harm them. Fig. 7-64 illustrates typical internal wiring construction.

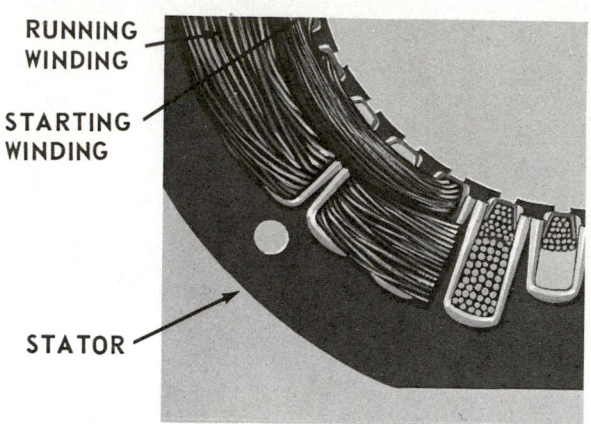

RUNNING WINDING

STARTING WINDING

STATOR

Fig. 7-64. Section through stator windings showing method of installing, insulating and retaining windings. Wire insulation is heavy, modified polyester. Slots are lined and "cuffed" ends prevent grounds where coils cross corners of core. Slots are closed with shaped sticks of heavy electrical paper.

7-43 PULLEYS

Motor shaft pulleys are available in many sizes and types of construction. Some are made of cast iron and some of steel stampings. They come in various shaft size openings and diameters. Pulleys are made with and without fans.

The most popular shaft sizes for fractional horsepower motors are 1/2 in., 5/8 in. and 3/4 in. diameter. Practically all pulleys have a keyway and a setscrew. Setscrews usually have a 5/16 in. NC thread.

Pulley diameters vary from 2 in. to 5 1/4 in. with 1/4 in. between sizes. Larger sizes, up to 15 in. in diameter, are available.

The V-pulleys come in two popular widths. The A width is for belts up to 17/32 in. width while the B width fits belts 1/2 in. to 11/16 in. wide.

Multiple-groove pulleys are available for units that use two or more belts to drive the flywheels. Some air conditioning units use step pulleys for driving the air movement fan. By changing the belt from one groove to another, the speed of the fan can be changed.

Special variable pitch pulleys are also available. These are made with half of the pulley threaded on the hub of the other half, as shown in Fig. 7-65. A setscrew locks the variable half in place when it is properly adjusted.

By turning the variable half, the V-groove can be widened to let the belt ride closer to hub. This reduces speed of a driven flywheel or fan. The speed of the driven unit can be varied by as much as 30 percent using these pulleys. Fig. 7-66 illustrates a double-groove, variable-pitch pulley.

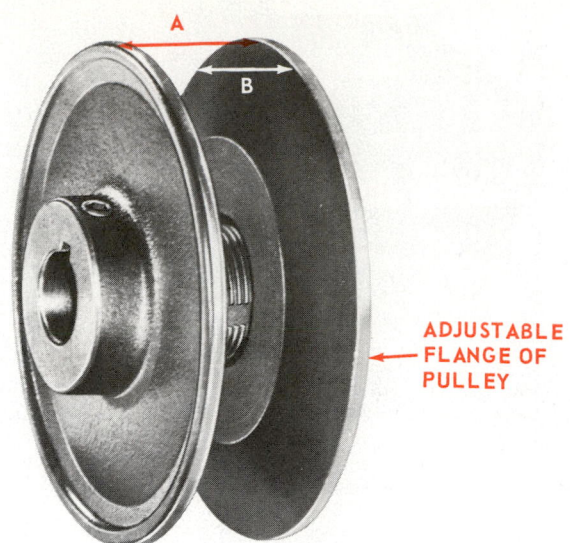

A

B

ADJUSTABLE FLANGE OF PULLEY

Fig. 7-65. Adjustable V-pulley is used to change speed of belt-driven appliances. A—Moving the adjustable flange away from the other flange widens the V and gives the effect of using smaller diameter pulley. B—Narrowing V by moving flange in gives effect of using larger diameter pulley. (Maurey Mfg. Corp.)

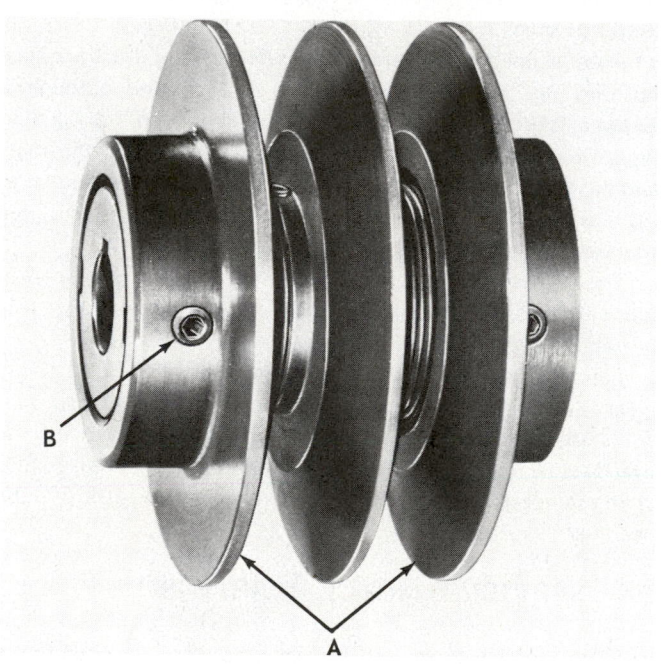

B

A

Fig. 7-66. Adjustable pulley with double V groove. Both grooves are adjustable and must be evenly spaced or one belt will take all the load. A—Adjustable pulley halves. B—Setscrew.

Bushings are available to adapt large bore pulleys to small shafts. For example, a bushing can reduce a 3/4 in. bore to a 1/2 in. bore.

7-44 BELTS

The V-belt is the most popular way to drive the open compressor and large fans. These belts are made in layers of

Fig. 7-67. Cross section through a V-belt. Note cords running length-wise through belt at neutral axis. (The Gates Rubber Co.)

rubber, fabric and cord. Some belts are a mixture of natural and synthetic rubber. Fig. 7-67 is a cross section through a V-belt.

The belts are made in many lengths from 15 in., outside length, to 364 in. outside length. (Outside length is the distance around the outside of the belt.) A steel tape, cloth tape or a special belt measuring fixture can be used to quickly determine this length.

Most belts fall into one of three standard widths. The widths are A, 1/2 in., B, 5/8 in. and C, 15/16 in. Some special belts have been made in widths of 33/64, 9/16, 37/64, 19/32, 41/64, 23/32, 3/4, 7/8, 29/32 and 31/32 in. Measurement is made at the belt's greatest width.

When the motor is belted to the driven machine, both shafts must be parallel to make the belt ride properly on the pulleys.

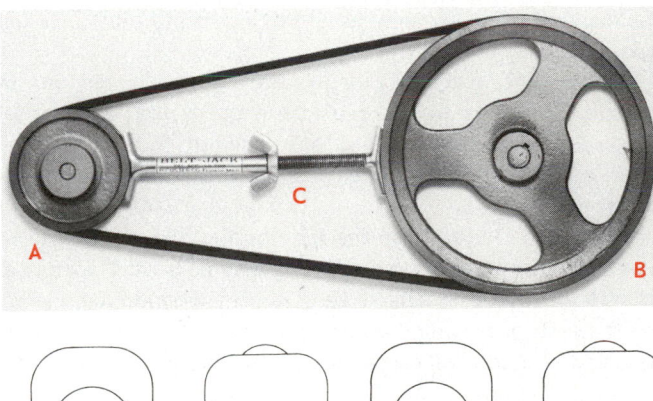

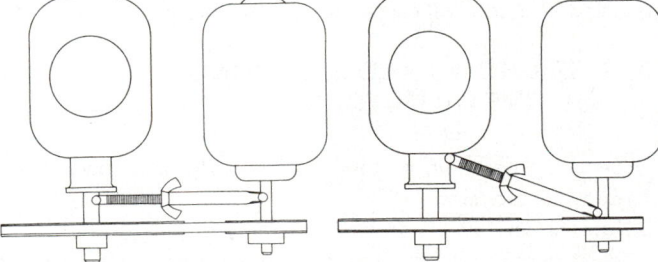

Fig. 7-68. Belt adjusting and aligning tool. Top, tool is applied between pulley to tension belt. A—Pulley. B—Flywheel. C—Tool. Lower panel, tool applies pressure at bases to align and tension belt drives. (H. Custer Co.)

Whenever possible, the installation should be arranged so that the belt pulls on the bottom side of the pulley.

The development of new rubber, new cording design and material permits the use of belts with a small cross-section. When installing belts, be careful to adjust them for proper tension and alignment. They should be snug but not tight. One should be able to depress a properly tensioned belt about 1/2 in. with a 10 lb. force.

The compressor flywheel and the motor pulley must be in line with each other in two different ways to give long life to the belt and to the electric motor. First, the center line of the compressor must be parallel with the center line of the electric motor shaft. Secondly, the pulley grooves must be in line with each other.

A poorly aligned belt will shorten the life of the motor because the motor is not designed to stand an excessive end load. A noisy, poorly operating motor may be the result. Fig. 7-68 pictures a tool which may be used in adjusting and aligning belt drives.

Automobile air conditioner belts are especially designed to transmit power to the compressor. It is very important to follow factory instructions when adjusting these belts. See Chapter 26.

7-45 MOTOR TESTING STAND

Worn electric motors are frequently taken to an electric motor repair shop for overhaul. However, some refrigeration service companies prefer to repair as many of these motors as possible. When rewinding is necessary, this should be done in a shop equipped for such work.

In each case the motor should be carefully tested to determine whether it needs rewinding or simple repair. The technician must decide by studying operating characteristics. To do this, a torque testing stand should be used.

Such a unit consists of a stand and a group of pulleys with equal outside diameters but with shaft diameters to fit various size motor shafts. The faces of these pulleys are smooth and flat. A torque arm lined with automobile brake lining is arranged to fit the pulleys. The torque arm is exactly one foot long between its point of support on the scales and the center of the motor pulley. A spring-loaded adjustment mechanism is arranged on the friction surfaces of the torque arm to enable various frictions to be produced between the torque arm and the motor pulley.

The end of the torque arm rests either on a spring scale or, preferably, on a platform scale. The arm should be balanced to remove any prior loading of the scale.

At the time the readings are taken, the torque arm must be level. The motor torque can then be very accurately checked for stall condition and full-speed load.

The torque obtained in this way may then be read directly from the scales in pound-feet. This data, when checked with the manufacturer's torque ratings, will show the condition of the motor.

Switch and brush mechanisms of these motors may also be checked on this stand. Ammeters, voltmeters and wattmeters will determine electrical characteristics of the motor.

Temperature rise of the bearings should be carefully noted. A thermometer placed in the bearing oil reservoir is recommended for this purpose. Temperature rise should not be more than 72 F. (40 C.) above the room temperature.

Electric dynamometers are available for measuring the power output of electric motors. They can also motorize the motor to determine its friction losses.

Electric meters should be installed in a panel in such a way that push switches can put an ammeter, voltmeter or wattmeter into the circuit.

7-46 SERVICING FAN MOTORS

The most common fan motor troubles are:
A. Loose connections.
B. Dry bearings.
C. Worn bearings.
D. Burned-out motor.
E. Loose fan.
F. Out-of-balance fan.
G. Fan blades touching the housing.

Loose or dirty connections will cause too much voltage drop at the motor and the fan motor will lose speed, hum loudly and overheat. A sensitive voltmeter or an ohmmeter will quickly locate the faulty connection. Do not rely on visual inspection.

A dry bearing will cause the same symptoms but this condition will last only a short time before the bearings will either seize (bind) or become badly worn.

Occasionally, the end play of the rotor becomes excessive (see D in Fig. 7-69) and causes the rotor to shift back and

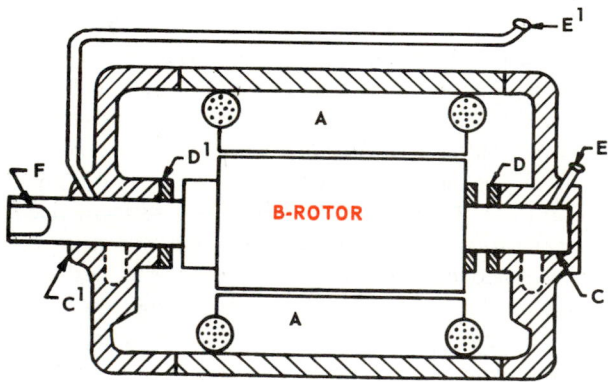

Fig. 7-69. Rotor running in its magnetic center. A—Stator. B—Rotor. C—Bearings. D—End play bearings. E—Oil cups. F—Pulley setscrew contact surface. E and E¹—Oilers for bearings.

forth. As it shifts it produces a distinct knock. Occasionally, the bearing (bushing) inserts in a reconditioned motor are out of position. This may force the rotor out of the magnetic center along its shaft.

When the motor is running, it should float between the extremes of its end play. One may check this by lightly touching the end of the rotor shaft with a wooden stick as the motor is running. It should move back and forth and then

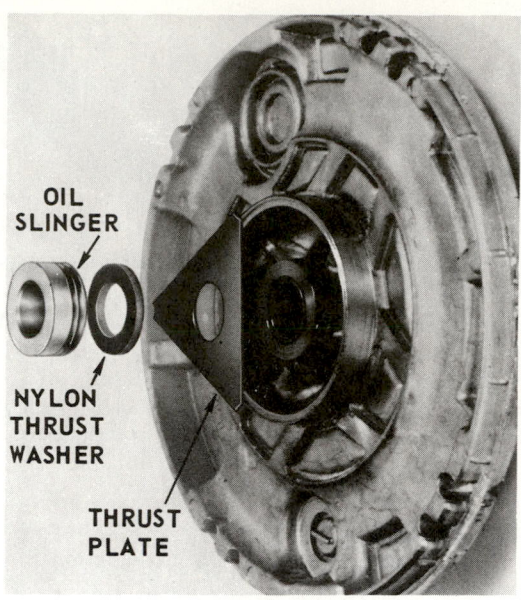

Fig. 7-70. A fan motor bearing and thrust bearing washer. (General Electric Co.)

settle in between the extremes of the end play.

If the rotor cannot assume its magnetic center, it will hum excessively and heat. When running, the heaviest magnetic flow from the stator tries to line up with the heaviest magnetic flow from the rotor, B. This aligning must take place with end play clearance at D and D¹. The total clearance is usually about .030 or 1/32 in. Fig. 7-70 shows the bushing and thrust bearing washer on a fan motor.

A rattle in the fan motor may sometimes be nothing more than a loose fan on the motor shaft. This noise can be remedied by tightening the setscrew that fastens the fan hub to the shaft. Smaller fans have either a round shaft or a flat spot milled on the shaft.

If the fan is abused, the blades may be forced out of position and one or more blades may vibrate. The easiest repair is to replace the fan. Any attempt to rebalance the blades is difficult unless special static and dynamic balancers are available.

If the fan blades touch the fan housing, the motor may be out of line or the shroud or housing may be bent. The contact spot is usually easily detected and remedied by moving the fan on the shaft or moving the shroud or housing. Do not bend the fan blades, as this will cause the unit to vibrate.

7-47 SERVICING AND REPAIRING HERMETIC MOTORS

Servicing hermetic motors involves two major operations:
1. External servicing.
2. Internal servicing.

Most hermetic motor troubles are external — either in the wiring or in the motor control devices. It is important to find out exactly where the electrical troubles are before deciding whether the motor is at fault.

Furthermore, it is essential that any outside trouble be

remedied as soon as possible, because eventually it may cause the motor to fail. The proper steps for internal servicing of motors are explained in Chapter 11.

To help in checking electrical circuits, many companies list the volt ampere (VA) values for their equipment. This data does away with the need for a wattmeter reading. The Power Factor need not be considered or calculated.

For example, a manufacturer may list the motor as a 120V single-phase motor with 2.9 VA locked rotor (starting load) and 1.1 VA full load. The service technician can then check the motor with a combination volt-ampmeter to find out if the motor is operating correctly.

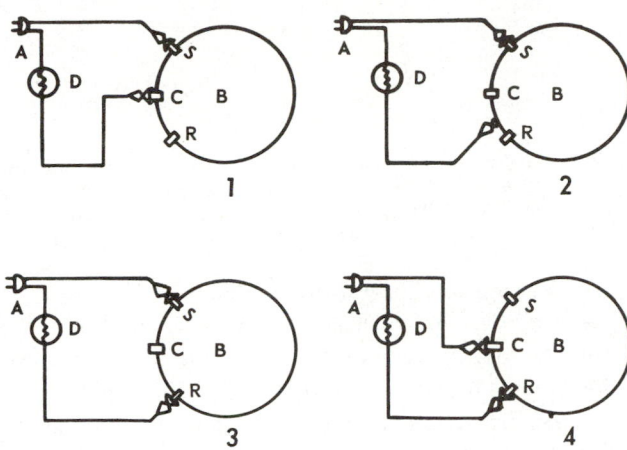

Fig. 7-71. Method of testing motor electrical circuits. A—Power circuit plug 120V a-c. B—Motor being tested. C—Motor common terminal. D—25W electric light bulb. S—Starting winding terminal. R—Running winding terminal.

7-48 EXTERNAL TESTING AND SERVICING HERMETIC MOTORS

The condition of the refrigeration motor can be readily and easily determined with the use of instruments, without the necessity of opening the unit. This may be done by disconnecting the wires to the motor terminals and then testing the motor independent of all its outside electrical connections (Fig. 7-71). Point A is the plug, B is the hermetic motor and D is the light bulb (25W). The starting winding terminal is S, the common terminal is at C and the running or main winding terminal is at R.

In view 1, the starting winding is being checked for continuity. If the bulb, D, lights, it means the current is flowing through the starting winding from S to C. In this case, there is no open spot in the windings.

In view 2 of Fig. 7-71, the windings are being checked for a ground. Grounding means that some part of the internal wiring or the terminals are touching or making electrical contact with the metal parts of the unit.

If the bulb, D, lights when one of the electrical leads is touched to any of the terminals, and when the other lead or clip is touched to the clean or bare metal body of the dome, it means that electricity is flowing along the internal wires and

through a grounded wire into the metalwork. Be sure the terminals are clean and dry during the test. They may be temporarily grounded by dirt.

Insulation on the windings should show no breaks. There should be infinite resistance between the terminal and the casing (dome) at 1000 to 1500V. The instrument in Fig. 7-72

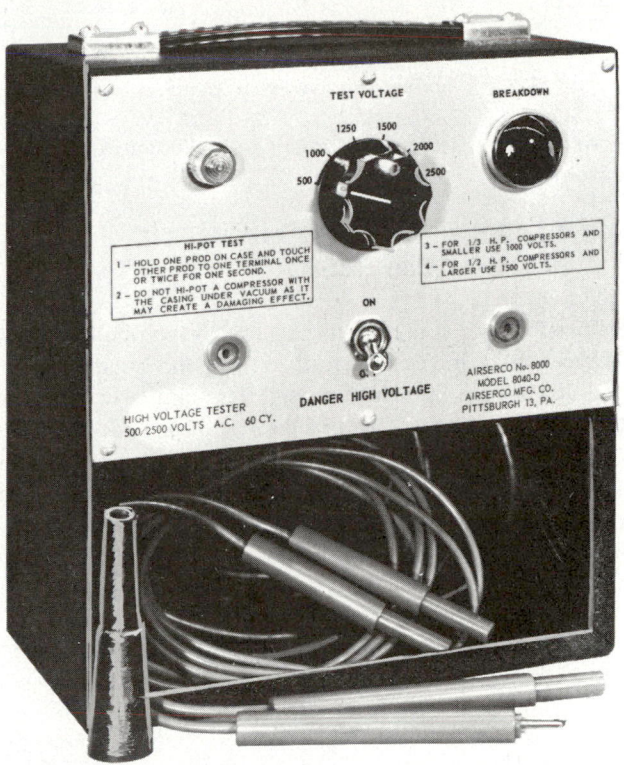

Fig. 7-72. High voltage test instrument for checking motor windings for shorts or grounds. Use 1000V for units up to 1/3 hp and 1500V for units 1/2 hp and higher. Duration of test should not be longer than one second. (Airserco Mfg. Co.)

checks motor windings at high voltages. One must be extremely cautious not to touch the parts when power is on.

In view 3 of Fig. 7-71, the continuity of both the running and starting windings is being checked. In view 4, the running winding only is being checked for continuity.

Sometimes — especially if the unit has been overheated — motor windings short out without a ground being formed. Any shorting of the motor windings will increase the current draw, decrease the power and overheat the unit. A shorted unit can sometimes be detected by an interruption in the steady hum of the motor when it is running. This is a noticeable beat added to the steady hum of the motor. To check for this short, one can roughly determine its existence by the test light, D, as shown in all views of Fig. 7-71. The test light will be brighter than normal if some of the windings are shorted.

A better way to check is to use an ohmmeter and determine the resistance of the coils. As models are checked, record the data. Many service technicians use an ohmmeter to check for continuity, shorts and grounds. An ohmmeter is more accurate than a test light. Repeated tests have shown that the

Ohmmeter Readings

HP	Running Winding	Starting Winding
1/8	4.7	18
1/6	2.7	17
1/5	2.3	14
1/4	1.7	17

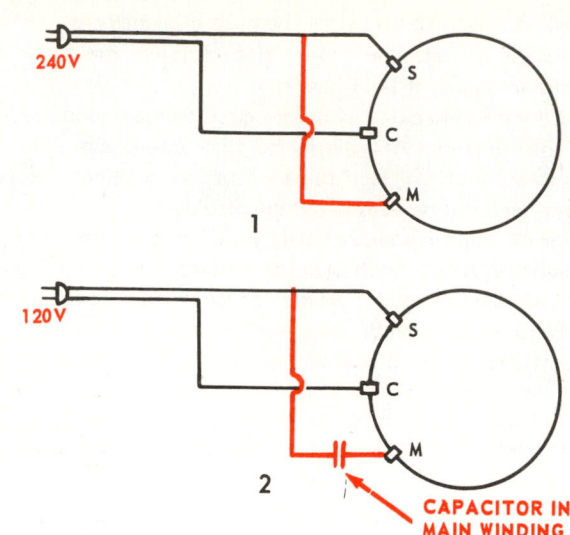

Fig. 7-73. Approximate ohmmeter readings for a typical fractional horsepower, single-phase motor.

approximate resistance of domestic unit windings are as shown in Fig. 7-73.

7-49 STARTING A STUCK HERMETIC MOTOR COMPRESSOR

Occasionally, a unit will not start even though all the electrical tests indicate that it is in good condition. This condition may have several causes. The unit may have been idle for a considerable time or a particle of dirt has gotten into it. On the other hand, some electrolytic plating may have taken place. A more than normal amount of liquid refrigerant in the compressor may also bind the unit. Three methods are recommended to break loose a stuck unit:

1. Connect the power line directly to the motor connections eliminating the starting relay, as in Fig. 7-74.
2. Use above-normal voltage, such as 240V on a 120V circuit, to break it loose, Fig. 7-75, view 1. This method can only

Fig. 7-75. Methods which may be used to start a stuck compressor. 1—Using above normal voltage. 2—Using capacitor in series with running winding. CAUTION: Close the circuit for only one or two seconds at a time.

be used for a very short period of time.

3. Reverse the unit to make it run backward. This reversal rotation may be done by putting a capacitor in series with the running winding, as shown in Fig. 7-75, view 2.

It is important to be continually on the alert when working with high voltage circuits. Short circuit capacitors to discharge them before handling.

ELECTRIC MOTORS

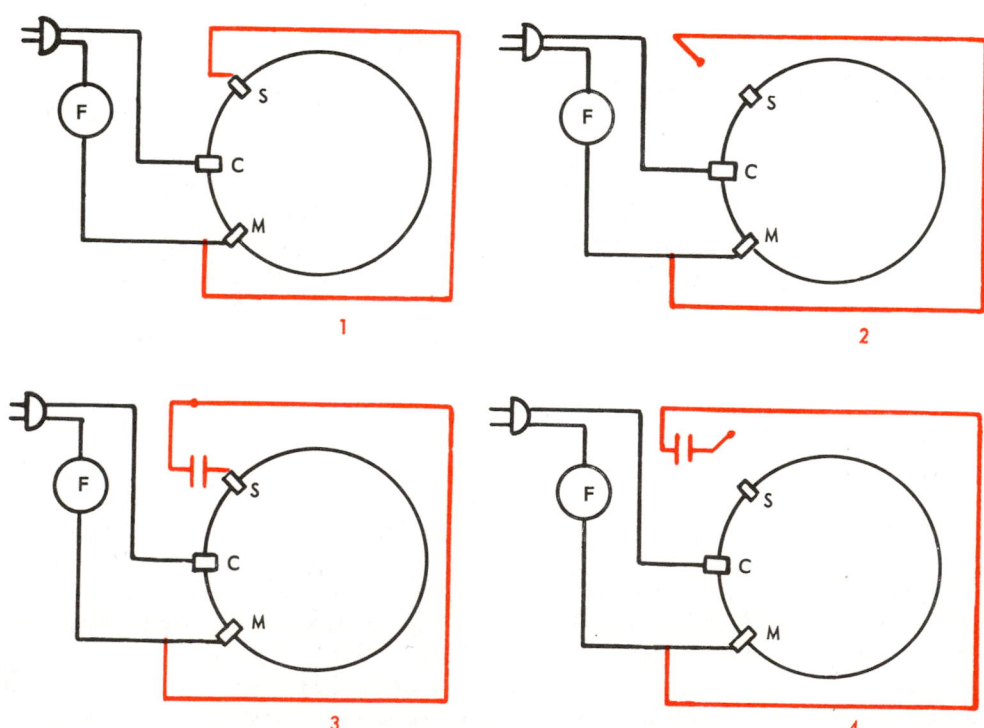

Fig. 7-74. Some methods which may be used to break loose a stuck hermetic compressor. Views 1 and 2—Without a capacitor in the starting circuit, touch S terminal for one or two seconds at the most and then open circuit as shown in view 2. View 3 and 4—With a capacitor in the starting circuit, touch S terminal for one or two seconds at the most, as shown in view 4.

7-50 REVIEW OF SAFETY

Electrical hazards can be considered in two parts:
1. Electricity as a source of ignition to start fires.
2. Electrical shock.

When electricity is passed through a part of an animal or human, it causes muscle spasms. If it passes across the heart or brain, it can be fatal. If enough current (amperes) is present, the electric current can actively overheat the body, cause burns and high temperature which may result in permanent body damage. As little as .25mA can kill a human being!

Disconnect the electrical power source before any repair or service to electrical parts. Lock the switches open to prevent someone from closing them during installation or service operations. The switch should also be tagged to warn other people.

Always short a capacitor before touching its terminals. If it is charged, it may discharge 200 to 500 volts into the body.

Persons subjected to electrical shock should be made to lie down. Keep them warm. If they are unconscious, give artificial respiration. Always call a physician if someone has suffered severe electrical shock.

Avoid touching moving belts. One may be seriously injured if the fingers or hand are caught between the belt and pulley or flywheel. Revolving fans can badly mutilate the fingers and hand.

7-51 TEST YOUR KNOWLEDGE

1. List the most common types of motors used in domestic refrigeration.
2. How many field windings are used in a 120—240V repulsion-start induction motor?
3. Why must an external drive motor be cleaned regularly?
4. Why do some capacitors explode when charged?
5. Why do some capacitors have resistors?
6. What is the running winding wattage of a 1/8 hp hermetic motor?
7. Why is a plug fuse inadequate protection for a motor?
8. What advantage does a capacitor motor have over a repulsion-start induction motor?
9. How often should external drive motor bearings be oiled?
10. What kind of electric motors are usually used in hermetic refrigerators?
11. What are the common voltages used on direct current motors?
12. How can radio interference caused by the refrigerator motor be reduced?
13. How many terminals does a domestic hermetic motor usually have?
14. How many windings does a single-phase hermetic motor stator have?
15. Why are some motors called split-phase motors?
16. What material is used in the construction of replacement terminals?
17. Why are hermetic motor field windings insulated in a special way?
18. What is a capacitor?
19. Is the starting capacitor connected in series with the starting winding?
20. What usually happens in the starting winding circuit when the motor reaches the correct speed?
21. What type motors are used to power the fans used on hermetic systems?
22. How is the fan motor connected electrically to the compressor motor?
23. How may a shorted capacitor be detected?
24. How much may the voltage drop at the motor terminals before it causes trouble?
25. How should the wiring be marked for easy circuit tracing?
26. What is the magnetic center of a motor?
27. Why do some systems use two running capacitors connected in series?
28. What information is usually recorded on the motor identification plate?
29. What is the color of the grounding wire?

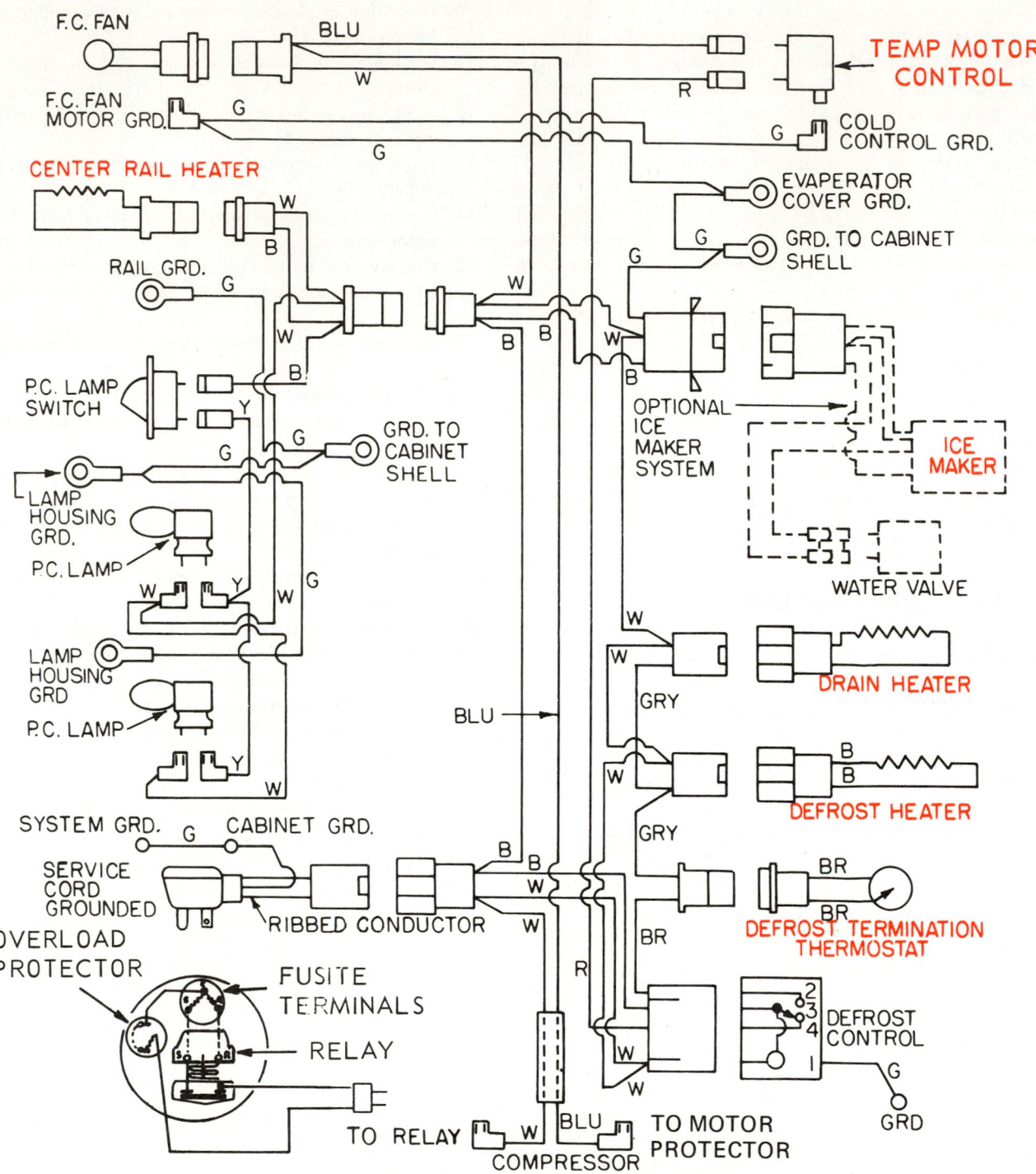

Fig. 8-1. Typical wiring diagram for domestic refrigerator. In addition to temperature control, this circuit includes defrost heater, drain heater, defrost thermostat, ice maker and center rail heater. Service cord attachment plug contains a ground terminal. Grounded parts are connected by green wire.   (Kelvinator, Inc.)

Chapter 8

ELECTRIC CIRCUITS AND CONTROLS

This chapter explains the more common electrical controls as well as the many electrical circuits used in refrigerating and air conditioning systems.

Elementary electricity and the use of electrical test instruments are explained, in some detail, in Chapter 6. One may have difficulty understanding some of the principles concerning electric circuit controls as explained in this chapter. In this case, a review of Chapter 6 may help.

All electrical circuits have a master switch or control. The service technician should know where this is located. Current must be turned off before performing any service operations.

8-1 ELECTRICAL CIRCUITS

A complete electrical circuit diagram for a modern domestic refrigerator is shown in Fig. 8-1. To the beginner, this may appear to be a rather complicated diagram. However, the various circuits and controls may be fairly easily broken down into several individual circuits. The parts should first be identified:
1. Temperature control.
2. Freezer compartment fan.
3. Center rail heater (mullion).
4. Motor compressor.
5. Motor control.
6. Starting relay.
7. Cabinet light.
8. Cabinet light switch.
9. Ice maker.
10. Water valve.
11. Defrost heater.
12. Drain heater.
13. Defrost control.
14. Defrost terminating thermostat.
15. Cabinet ground.
16. Starting relay.
Other devices sometimes found in these circuits include:
1. Condenser fan motor.
2. Evaporator fan.
3. Condenser fan.
4. Ultra-violet ray lamp.
5. Butter warmer.
6. Humidity controls (see Chapter 20).
A similar electrical circuit diagram, showing a condenser fan motor, is shown in Fig. 8-2.

Electrical power is obtained through an insulated extension cord. Usually this is a No. 18 stranded wire for domestic refrigerators. Sizes up to No. 12 are used for the heavier duty air conditioners.

These cords are insulated to withstand at least twice their normal voltage. All power-in cords are of the three-wire (one green ground wire) type.

The cord is usually connected to a junction box mounted in the condensing unit compartment. Older models may only have a two-wire power cable.

In all cases, the green wire is the ground. It is never to be used as a current-carrying conductor. There is no accepted color code for the various electrical circuits of a domestic refrigerator except the green for ground. Standard extension cord plug configurations are explained in Para. 6-77. These various circuits will be identified and described in some detail later in this chapter.

8-2 FUNDAMENTAL ELECTRICAL CIRCUIT CONTROLS

There are several types of electrical circuit controls commonly used on refrigerating and air conditioning mechanisms. These include:
1. Motor cycling controls.
   A. Thermostats.
   B. Pressure motor controls.
2. Motor starting relays.
   A. Current (amperage).
   B. Potential (voltage).
   C. Thermal (hot wire).
3. Defrost timers.
4. Ice-making controls.
   Special air conditioning controls are explained in Chapter 20 (Heating) and Chapter 21 (Cooling).
   Special commercial refrigeration controls are explained in Chapter 12.

8-3 THERMOSTATIC MOTOR CONTROLS

Automatic refrigeration and air conditioning is designed to provide correct temperatures with the least attention. To produce these temperatures under all conditions, a refrigerating unit must have more capacity than is needed under average conditions. This unit would over-refrigerate or over-

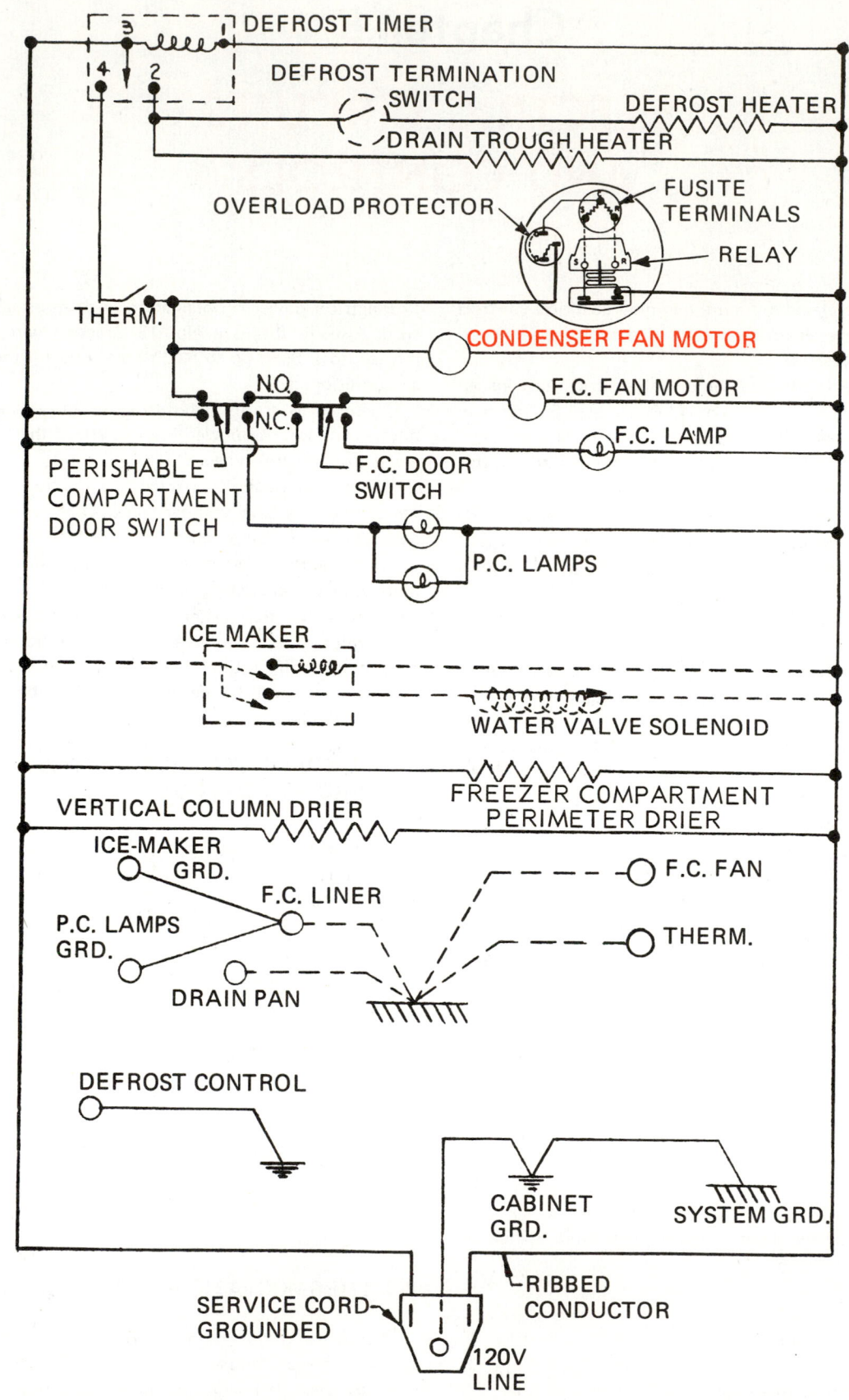

Fig. 8-2. Typical wiring diagram for a domestic refrigerator which uses a condenser fan motor.

heat if it ran all the time.

The most common method of controlling the temperature in the cabinet is a thermostatic (temperature sensitive) motor control. The thermostat starts the motor when the cabinet becomes warm and reaches a predetermined setting. This setting is called "cut-in" temperature. As the motor runs, the refrigerating unit cools the interior of the cabinet. When the cabinet temperature reaches the desired low temperature or "cut-out," the thermostat disconnects the motor from the power source. The refrigerating mechanism stops.

In domestic refrigeration, for the northern temperate latitudes, the unit will run 35 to 40 percent of the time. In semi-tropic latitudes, it will run about 50 percent of the time. Domestic refrigerators usually run 5 to 10 minutes and are idle 10 to 20 minutes.

Most refrigerator manufacturers design their units to operate only 8 to 14 hours out of each 24. This is about 40 percent of the time.

This 14-hour operating time is based on average use of the cabinet. The ambient (surrounding) temperature also affects the running time. A refrigerator in a room at 95 F. (35 C.) will run longer than the same refrigerator operating in a room at 75 F. (24 C.).

Frost free refrigerators and some automatic defrost refrigerators may run a little longer or more often than the conventional older style machine. Defrosting energy adds to the heat load.

Domestic refrigerator cabinets usually have a temperature range between 35 F. (2 C.) to 45 F. (7 C.). The adjustment on the motor control allows the owner to select the desired temperature within this range.

8-4 THERMOSTATIC MOTOR CONTROL PRINCIPLES

Three basic types of mechanisms are used in motor control thermostats:

1. Sensing bulb.
2. Bimetal.
3. Thermistor.

The first type depends upon the vapor pressure of a volatile fluid acting on a flexible bellows or diaphragm to control the mechanism in the thermostat. (A volatile fluid is one that becomes vapor at low temperatures.) This vapor pressure thermostat is most often used on refrigerating mechanisms. Fig. 8-3 shows how it is built.

The volatile liquid is located in the sensing bulb. The bulb is in contact with the refrigerator evaporator. A capillary tube connects the sensing bulb to the operating mechanism in the thermostat. This thermostatic element is pictured in Fig. 8-4.

The second type uses a bimetal strip. Two different metals (commonly copper and iron) are soldered together, Fig. 8-5.

Copper has a greater coefficient of expansion (expands more with an increase in temperature) than does iron. This causes the bimetal strip to bend as the temperature changes. Bending action of the strip opens and closes contact points in the electrical circuit.

This thermostat is most used on house heating, air

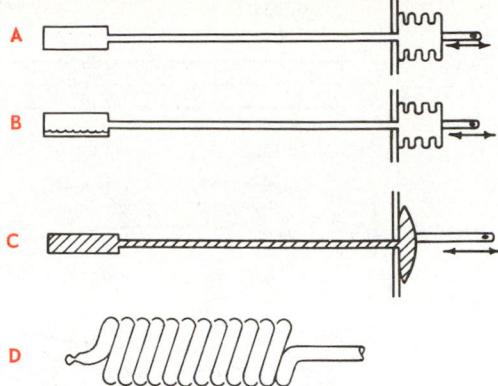

Fig. 8-3. Methods used to obtain motion from temperature changes. A—Gas-charged temperature response bellows. B—Vapor pressure temperature response bellows. C—Liquid charge temperature response diaphragm. D—Capillary tube coil used as a bulb.

conditioning and furnace controls. It is also good for electric motor protection. The bimetal control is placed inside the motor winding.

A thermistor is used as a sensor in a solid state electronic circuit.

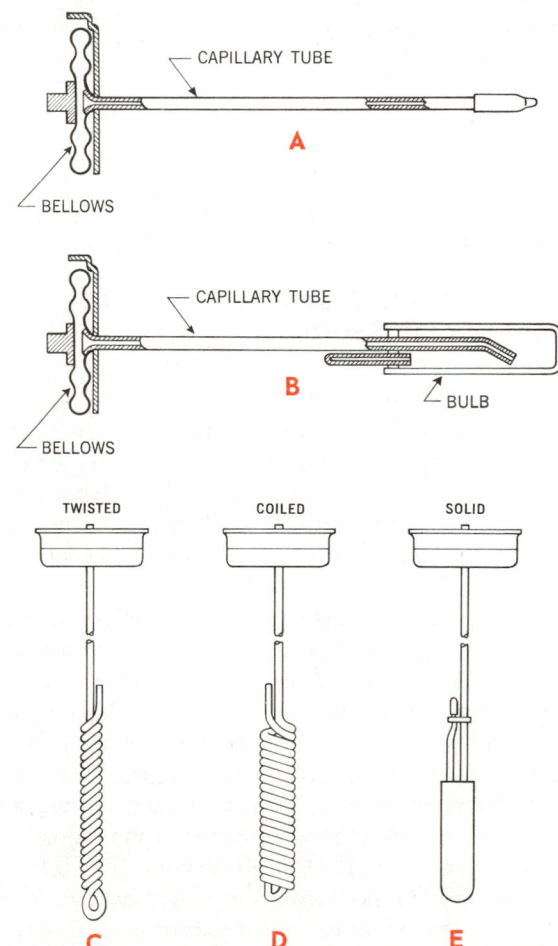

Fig. 8-4. Several thermal element designs. A—Capillary tube sensitive element. B—Bulb sensitive element. C—Twisted sensitive element. D—Coiled sensitive element. E—Solid (bulb) sensitive element. (Ranco, Inc.)

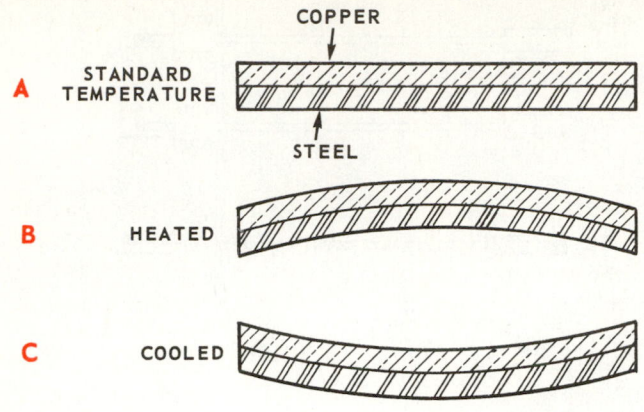

Fig. 8-5. A bimetal strip bends or warps with temperature change. A—Strip at controlled temperature. B—Strip warmed above control temperature. C—Strip cooled below control temperature.

8-5 THERMOSTATIC MOTOR CONTROL APPLICATIONS

Thermostatic motor controls are available for many different applications:

1. Domestic refrigerators.
2. Freezers.
3. Air conditioners.
4. Water coolers.
5. Ice makers.
6. Heat pumps.

In the following paragraphs, the construction and operation of these controls are explained.

8-6 THERMOSTATIC MOTOR CONTROL — VAPOR PRESSURE

Many mechanisms are used in vapor pressure type thermostats. The action of the contact points needs to be quite rapid. If contact points are allowed to open very slowly, there will be some electric arcing as the current tends to jump across the tiny gap. This arcing action would very quickly burn the contact points and ruin their ability to make a good electrical contact.

There are two ways to get this snap action. One is through a snap action toggle mechanism. The other is with a permanent magnet. These two mechanisms are shown in Fig. 8-6.

The toggle mechanism is shown in view 1. In this mechanism the fulcrum points at A and B are under some pressure. This tends to push them together. As the bellows expand, due to warming of the sensing bulb, the toggle point, C, will move down. The instant it passes the center point, it will snap into position, D, closing the contact points P.

As the sensing bulb cools and the bellows contract, (shrinks) the toggle will be moved toward point C. As soon as it passes the center point, it snaps open very quickly. An adjustment at E, controls the pressure tending to move points A and B together. Increasing this pressure will lengthen the running time. This toggle arrangement gives good snap action.

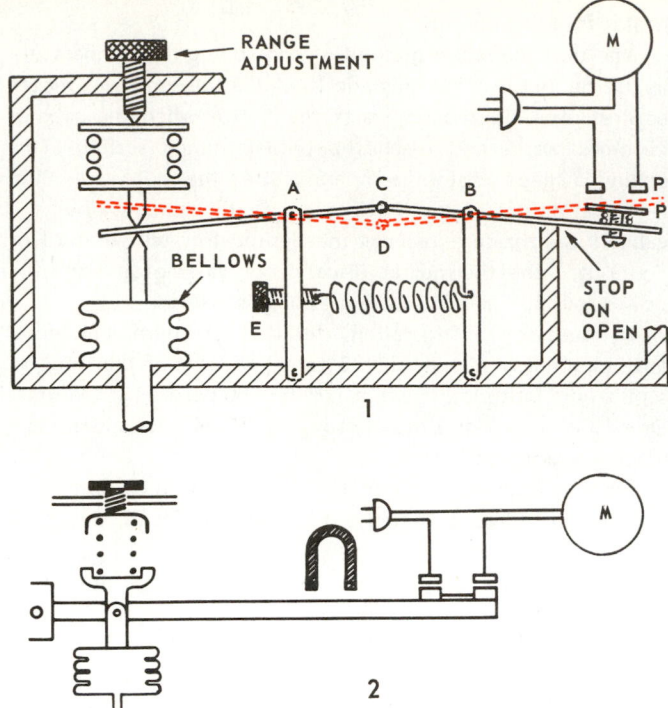

Fig. 8-6. How bellows are connected to snap action motor control switch. 1—Toggle snap action. 2—Magnet snap action.

The system with the magnet snap action is shown at 2 in Fig. 8-6. The bar carrying the contact points is made of magnetic metal (iron).

The permanent magnet tries to draw this material toward it. Magnetic lines of force always try to be as short as possible. As the pressure in the bellows tends to close the points, the magnetic effect increases as the iron bar approaches the magnet. This causes a snap action which closes the points quickly.

Likewise, as the sensing element cools and the bellows contract, it will take force to pull the points apart. However, the magnetic pull decreases rapidly as soon as the points separate, which allows a quick opening of the points.

With this type of snap action, the running interval can be shortened by moving the magnet away from the iron bar. Or, it can be lengthened by bringing the magnet closer to the iron bar. The magnet should never touch the iron bar.

A typical vapor pressure type motor control is shown in Fig. 8-7. The sensing bulb is at the end of the coiled capillary tube. It has a range control which the owner may use to adjust the cabinet temperature.

8-7 RANGE ADJUSTMENT

Range adjustment provides for the correct minimum and maximum temperature or pressure in an automatically operated system. Examples: the range adjustment will keep a refrigerator between certain temperatures, an air conditioned room between certain temperatures and an air compressor between certain pressures.

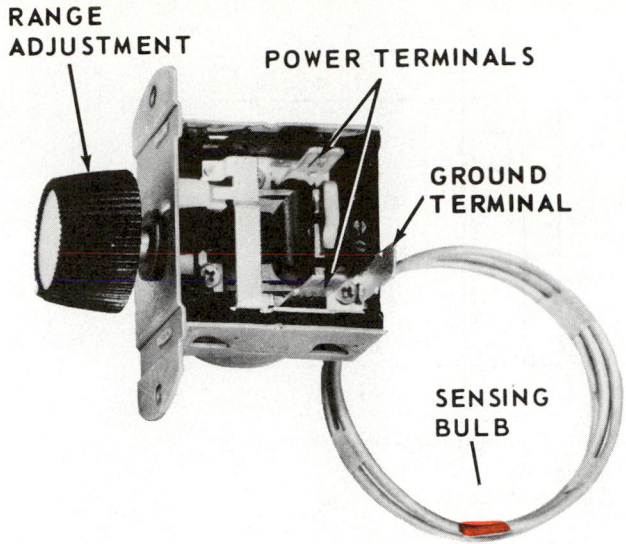

Fig. 8-7. Vapor pressure type thermostatic motor control. (Eaton Corp.)

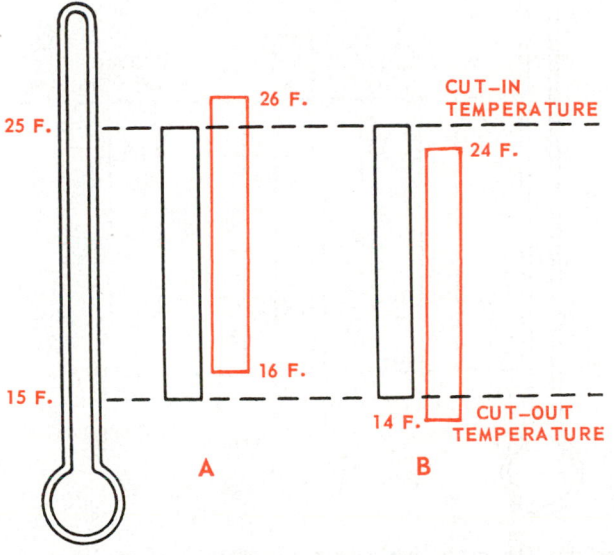

Fig. 8-8. Typical range adjustments. The basic range is shown in black.

It is very difficult to keep any device at one particular temperature and/or pressure, such as 34.4 F. (1.3 C.) or 150.2 psi (10.7 kg/cm²). That is why a range adjustment is used.

A range adjustment is shown in Fig. 8-8. The unit cuts in at 25 F. and cuts out at 15 F. on the evaporator temperature. To make the cabinet operate at a warmer temperature, the range adjustment may be changed so the cut-out becomes 16 F. and the cut-in 26 F., as shown in A of Fig. 8-8.

Note that the temperatures are higher but the temperature distance between the two has not been changed. This distance is still 10 F. The new settings will only slightly affect the running time of the unit. This is because the condition desired is not as cold and, therefore, the condensing unit will not have to do as much work.

The temperature of the refrigerator may be lowered a degree by making the range adjustment cut in at 24 F. and cut out at 14 F., as shown in B of Fig. 8-8.

8-8 RANGE ADJUSTMENT MECHANISMS

The range adjustment is easily recognized. It is an adjustable force pressing directly upon the bellows or diaphragm which operates the switch. This force is always being exerted on the bellows whether the switch is in either the cut-out or the cut-in position.

The adjustable force may be either an adjustable weight that always presses against the bellows or, more commonly, it is a spiral spring with an adjustable screw. Turning the screw changes the pressure or the tension of the spring. The spring may press or pull either directly on the bellows or diaphragm or on a lever attached to the bellows. See Fig. 8-9.

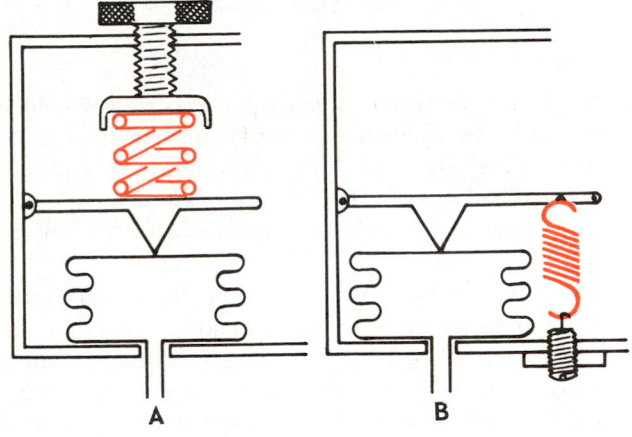

Fig. 8-9. Adjusting range settings of motor control. A—Compression spring range adjustment. B—Tension spring range adjustment.

Most of the range adjusting screws have a calibrated dial or a pointer connected to them. It will indicate the direction the screw should be turned for a warmer or colder setting.

8-9 DIFFERENTIAL ADJUSTMENT

The differential adjustment is built into the temperature control mechanism. It is not adjusted or changed by the operator when making a temperature selection by use of the control knob. The differential adjustment should be made by a service technician who understands the working of the differential adjustment mechanism.

The differential adjustment controls the temperature difference between the cut-out and the cut-in settings.

If the evaporator is set to cut in at 25 F. and cut out at 15 F., the difference, or differential, is 10 F. Whenever the differential (the distance between the settings) is changed, the range is also changed. If the range only is adjusted, the differential will not be affected.

The thermostat differential for capillary tube systems must be large enough to allow the pressures to equalize in the

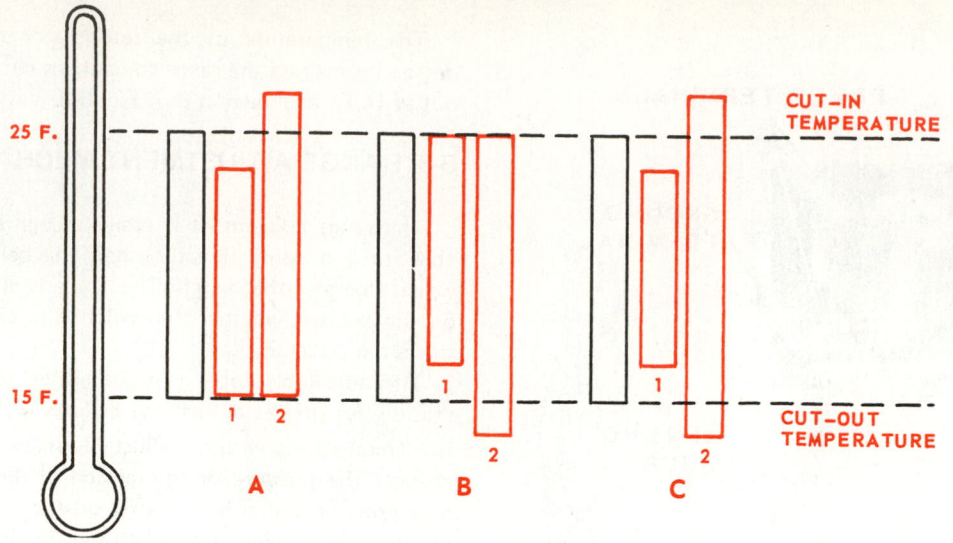

Fig. 8-10. Effect of different types of differential control adjustments. A—Cut-in setting type. B—Cut-out setting type. C—Cut-in and cut-out type. In each illustration, normal setting is shown in black.

system. This increases the off cycle, but the change should not be too much. If it is, the fixture temperature will rise too high. A differential of about 14 F. (8 C.) is common for fresh food refrigerators and about 11 F. (6 C.) for freezers.

Three types of differential adjustments are shown above in Fig. 8-10:

1. Cut-in type. The cut-in point may be moved without changing the cut-out as shown at 1 and 2 in view A.
2. Cut-out type. The cut-out point may be moved without changing the cut-in setting as shown at 1 and 2 in view B.
3. Double type. The cut-in and cut-out settings may be brought closer together or moved farther apart, as shown at 1 and 2 in view C.

One may learn to pick out cut-in and cut-out differential adjustments by the way that they affect the operation of the switch mechanism. In the first case, the effect is seen only when the switch is in the cut-out position. In the second case, only when the switch is in the cut-in position.

The third type is usually an adjustable arrangement which affects the effort of the toggle to snap off and on. This adjustment affects both the cut-in and the cut-out. They may be adjusted farther apart or closer together.

To get a certain cabinet temperature, it is usually necessary to adjust both the range and the differential.

To keep the same average cabinet temperature, the control may be adjusted to 14 and 26 F., shown in A1 of Fig. 8-11. With this setting, the compressor will run longer but will not cycle as often.

If this control range adjustment is set to cut in at 25 F. and cut out at 13 F., as at A2 in Fig. 8-11, the compressor will run longer than normal. The cabinet temperature also will be lower than normal.

Now if this control is set to cut in at 27 F. and cut out at 15 F., as shown at A3 in Fig. 8-11, the compressor will run less and the cabinet temperature will be higher. In each case, the differential has been 12 F., but the range has changed.

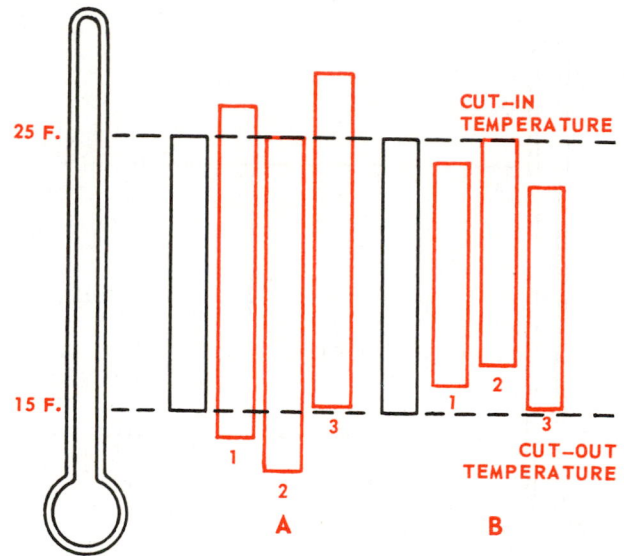

Fig. 8-11. The cut-in and cut-out and differential adjustment. A1—Differential increased. A2—Range lowered. A3—Range raised. B1—Differential decreased. B2—Range raised. B3—Range lowered.

If the control is adjusted to cut in at 24 F. and cut out at 16 F., as shown at B1 in Fig. 8-11, the cabinet temperature will be normal. However, both the on and off cycles will be shorter and the cabinet temperature will vary less than normal. The differential is now 8 F. and the range has been changed.

With the control adjusted to cut in at 25 F. and to cut out at 17 F., as shown in B2 of Fig. 8-11, the cabinet temperature will be warmer than normal. The on and off cycles will be shorter than normal. Cabinet temperature will vary less than normal.

With the control adjusted to cut in at 23 F. and cut out at 15 F., as shown at B3 in Fig. 8-11, cabinet temperature will be below normal. But it will vary less than normal. The on and off cycles will be shorter than normal.

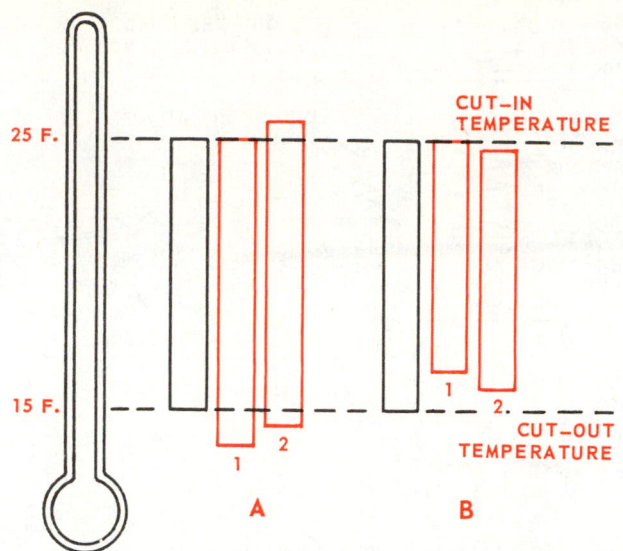

Fig. 8-12. Cut-out differential adjustments. A1—Differential lowered from 15 F. to 14 F. cut-out. A2—Range is then adjusted to 14.5 F. cut-out and 26.5 F. cut-in. B1—Differential is changed from 15 F. to 16 F. cut-out. B2—Range is then adjusted to 15.5 F. cut-out and 24.5 F. cut-in.

For a slightly lower cabinet temperature with little or no increase in running time, the cut-out type differential may be adjusted to cut out at 14 F., as in A1 of Fig. 8-12.

To increase running time, the range adjustment may be set as shown in A2 of Fig. 8-12. A cut-out differential adjustment as shown in B1 of Fig. 8-12, will give a slightly warmer cabinet temperature and a shorter cycle interval. Also, the cabinet temperature will vary less than normal.

A range adjustment as shown at B2 of Fig. 8-12, will provide a normal cabinet temperature, which varies less than normal. The cycling interval will be shorter than normal.

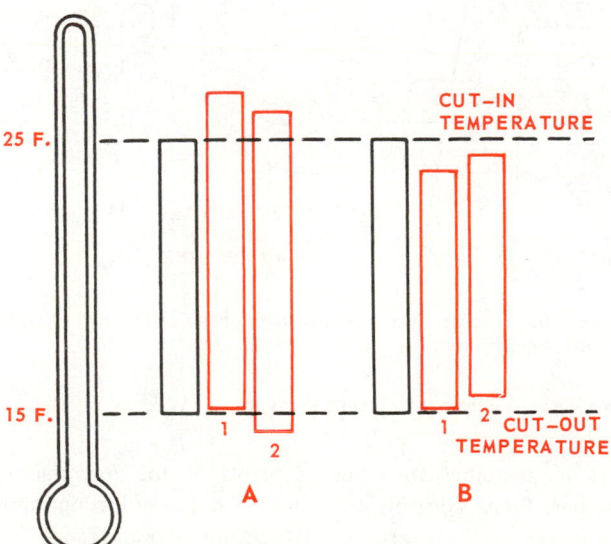

Fig. 8-13. Effect of a cut-in differential adjustment. A1—Cut-in differential raised. A2—Range lowered. B1—Cut-in differential lowered. B2—Range raised.

A typical control adjustment follows:

1. To increase the cycling time when a control with a cut-in differential is used, (a) adjust the differential from a 25 F. cut-in to a 27 F. cut-in, as shown at A1 of Fig. 8-13; (b) adjust the range to move the settings to 26 F. cut-in, and 14 F. cut-out, as shown at A2 of Fig. 8-13.

2. To decrease the cycling time (interval) when a control with a cut-in differential is used, (a) adjust the cut-in differential to cut in at 23 F., as shown at B1 of Fig. 8-13; (b) adjust the range to cut in at 24 F. and cut out at 16 F.

8-10 DIFFERENTIAL ADJUSTMENT MECHANISMS

Two types of differential adjustments have a spring that affects the movement of the bellows, either just before the contact points cut out, or cut in, but not both. This limited action is controlled by a stop on the spring as shown in Fig. 8-14. View A is a cut-in differential; view B is a cut-out differential and view C a cut-in differential.

A third type of differential adjustment affects both the cut-out and cut-in. A spring makes it easier for the points to open and close, or more difficult for the control to open or close electrical contact points. (See view D, Fig. 8-14.)

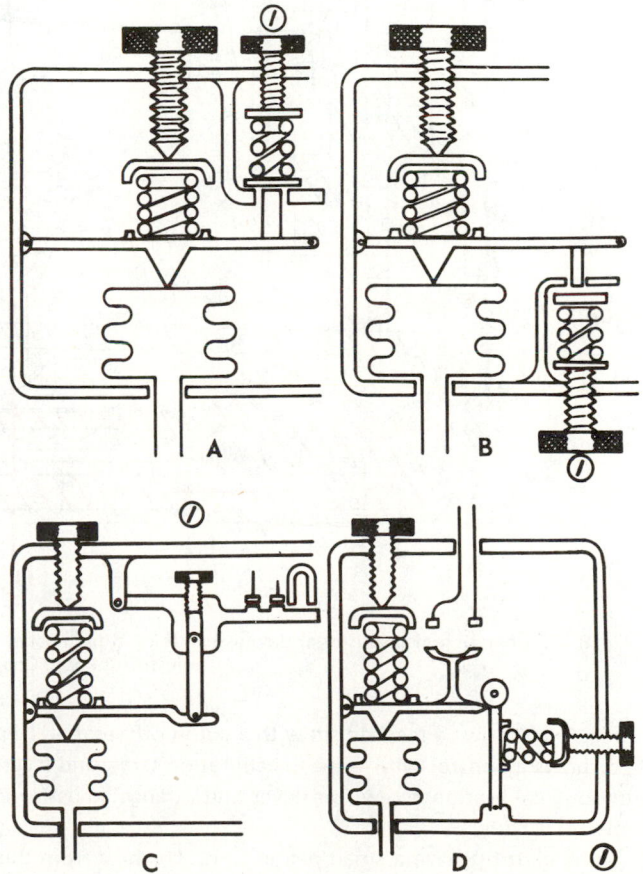

Fig. 8-14. Various types of differential adjustment mechanisms. A—Cut-in type. B—Cut-out type. C—Cut-in type using a slot and magnet. D—Double type. 1—Differential adjustment.

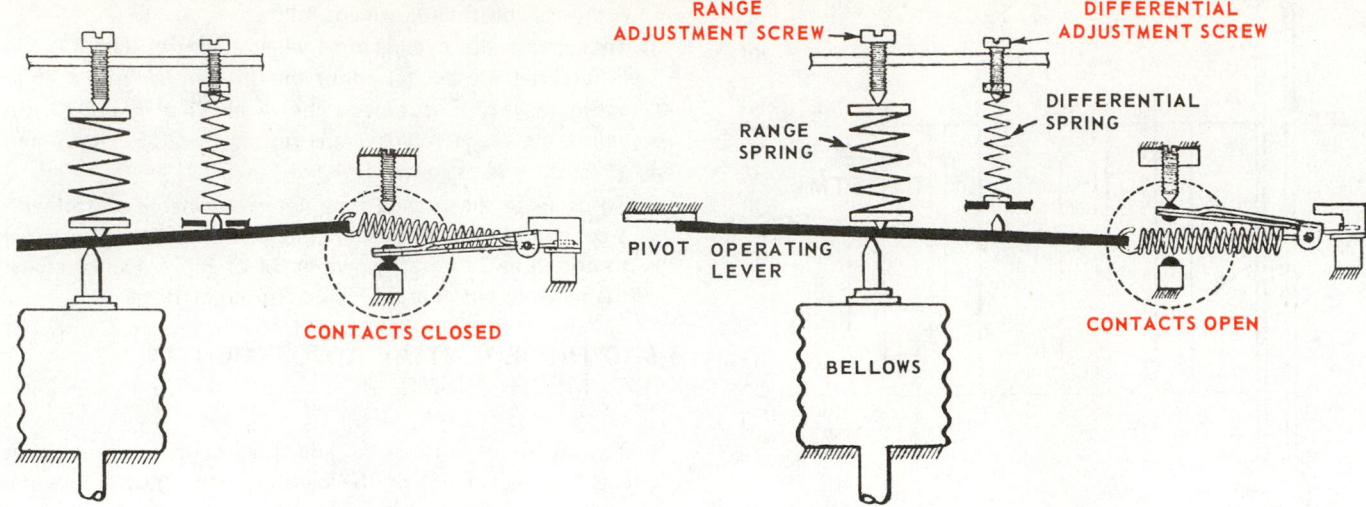

Fig. 8-15. Toggle switch mechanism showing range adjustment and a cut-in type differential adjustment.

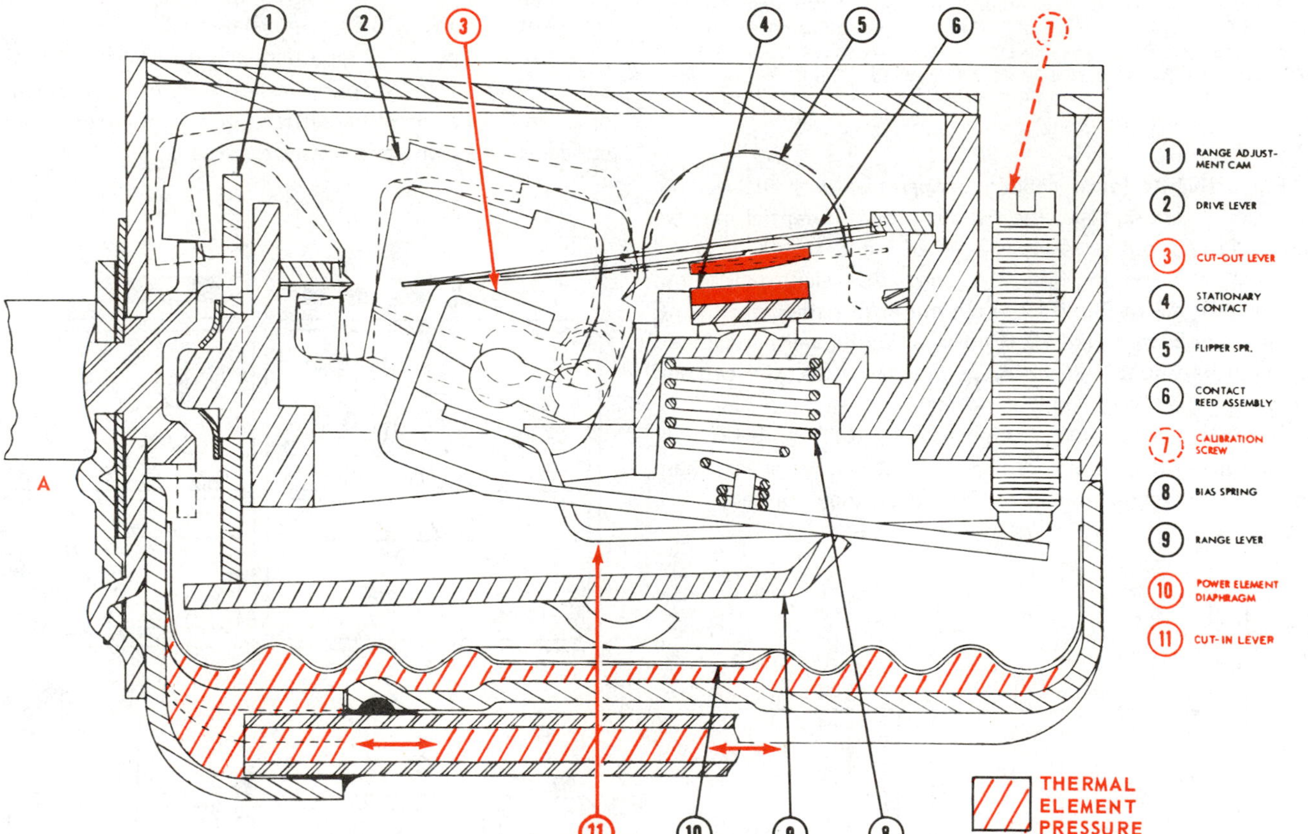

Fig. 8-16. Operating mechanisms and adjustments of a thermostat. 1—Range adjustment (cam). 7—Cut-out differential adjustment. A—Range adjustment knob. Contact points are shown in red.

Fig. 8-15 shows a mechanism with a cut-in differential. Fig. 8-16 shows a control with a cut-in calibration screw and lever, a cut-out calibration screw and lever and a range adjustment screw.

Some controls have a small heater unit. The heat from this unit keeps the bellows and diaphragm from becoming too cold. A too-cold thermostat body will not cut in as soon as it should, and may cause erratic cabinet temperatures.

A complete control is shown in Fig. 8-17.

8-11 ADJUSTING CONTROLS

It is advisable to adjust controls as the manufacturer specifies. Some controls are marked "cut-in adjusting screw" and "cut-out adjusting screw." Others are marked "range."

1. Some controls have a cut-in differential with the control marked "step one, cut-out adjustment" for one adjusting screw, and "step two, cut-in adjustment" for a second adjusting screw. In such cases, the cut-in screw is the

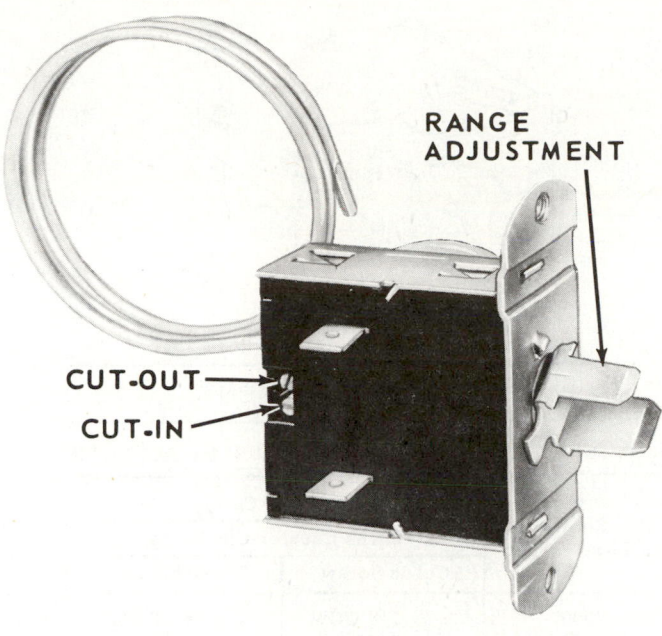

Fig. 8-17. Domestic cabinet thermostat. Range adjustment used by the owner is at right. Note cut-in and cut-out adjustment at left. (Eaton Corp.)

differential adjustment and the cut-out screw is the range.

2. If the control has a cut-out differential, the opposite applies. The cut-in will be the range and the cut-out, the differential.

3. Some controls have a differential adjustment which controls the distance between the cut-out and the cut-in. The

first setting should be made by using the differential adjustment to get the correct distance between the settings. Then the range adjustment is turned to obtain the correct settings.

The adjustment used by the owner is usually a limited range adjustment. However, some models allow the owner to adjust only the cut-out setting. This design insures a safe cut-in temperature at all times.

Contact points may chatter as they open or close. Operate control to check for this condition. If there is any visual indication of pitting or burning of points, replace the control.

8-12 EFFECT OF ALTITUDE ON REFRIGERATOR TEMPERATURES

If a refrigerator is adjusted to operate at an elevation near sea level, it will be too cold at 5000 ft. above sea level.

At high altitudes, the atmospheric pressure drops. Then pressure on the control diaphragm or bellows is lowered enough to affect the settings. The altitude adjustment and range control pressure for the bellows or diaphragm should, therefore, be increased to make up for the lower atmospheric pressure.

Fig. 8-18 shows such a control and its altitude adjustment table. Note that the dial is divided into 60 equal calibrations.

Fig. 8-19 shows two controls and a table of adjustments used to correct each for various altitudes.

8-13 FOOD FREEZER MOTOR CONTROLS

Food freezers are usually designed to operate from about 5 F. (−15 C.) to −30 F. (−34 C.). The motor control must be designed for food freezer use. Most of these controls provide for an adjustment which the owner can make in selecting the temperature range desired.

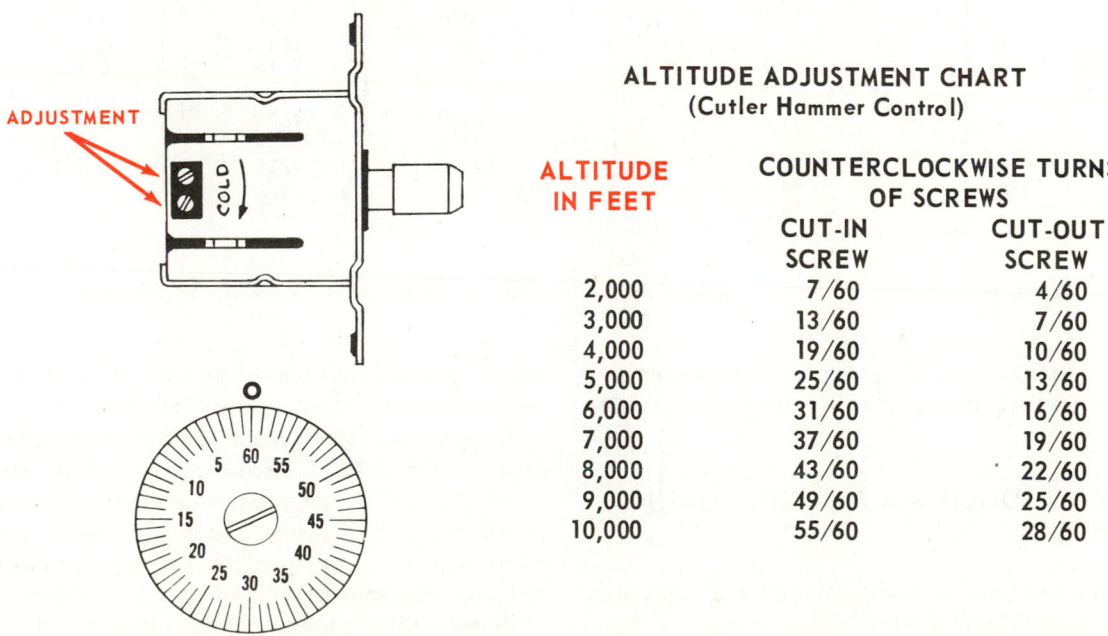

ALTITUDE ADJUSTMENT CHART
(Cutler Hammer Control)

ALTITUDE IN FEET	COUNTERCLOCKWISE TURNS OF SCREWS	
	CUT-IN SCREW	CUT-OUT SCREW
2,000	7/60	4/60
3,000	13/60	7/60
4,000	19/60	10/60
5,000	25/60	13/60
6,000	31/60	16/60
7,000	37/60	19/60
8,000	43/60	22/60
9,000	49/60	25/60
10,000	55/60	28/60

Fig. 8-18. Thermostat equipped with altitude adjustment. Table indicates correct setting.   (Gibson Appliance Co.)

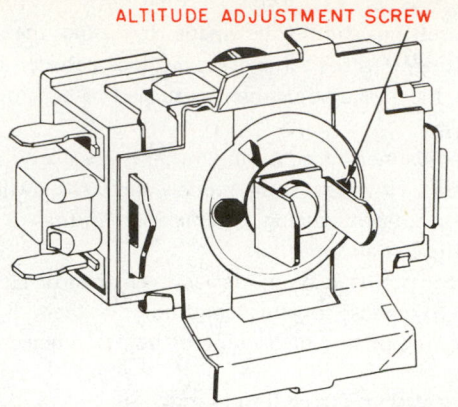

ALTITUDE ADJUSTMENT SCREW

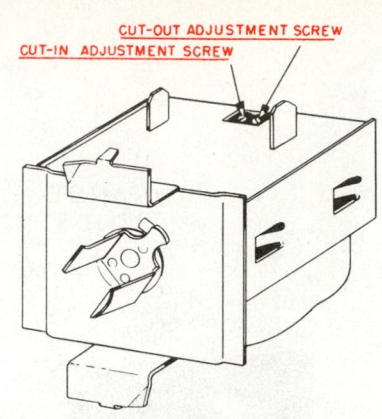

CUT-OUT ADJUSTMENT SCREW
CUT-IN ADJUSTMENT SCREW

Altitude Above Sea Level - Feet	Constant Cut-In
	Altitude Screw Adjustment (Turns Clockwise)
1000	No Change
2000	1/16
3000	1/8
4000	3/16
5000	1/4
6000	5/16
7000	3/8
8000	3/8
9000	-

Altitude Above Sea Level - Feet	Variable Cut-In
	Range Screw Adjustment (Turns Clockwise)
1000	3/32
2000	3/16
3000	7/32
4000	1/4
5000	3/8
6000	7/16
7000	15/32
8000	1/2
9000	9/16

ALTITUDE ADJUSTMENT

Both Cut-In And Cut-Out Screws Must Be Adjusted

Altitude Above Sea Level - Feet	Constant Cut-In	
	Turns Counter-Clockwise	
	Cut-In Screw	Cut-Out Screw
2000	1/8 CCW	1/16 CCW
3000	7/32	1/8
4000	5/16	5/32
5000	13/32	7/32
6000	1/2	1/4
7000	19/32	5/16
8000	11/16	3/8
9000	13/16	13/32
10000	15/16	7/16

Altitude Above Sea Level - Feet	Variable Cut-In	
	Turns Counter-Clockwise	
	Cut-In Screw	Cut-Out Screw
2000	1/16 CCW	1/16 CCW
3000	1/8	1/8
4000	5/32	5/32
5000	3/16	3/16
6000	1/4	1/4
7000	5/16	5/16
8000	3/8	3/8
9000	13/32	13/32
10000	7/16	7/16

Fig. 8-19. Two domestic type thermostats are equipped with altitude adjustments. Left—General Electric. Right—Cutler-Hammer. (Kelvinator, Inc.)

The principle of operation of the motor control used on food freezers is exactly the same as that used on domestic refrigerators.

8-14 COMFORT COOLING AIR CONDITIONING CONTROLS

Comfort cooling air conditioners have two standard controls:
1. Thermostat, to control the temperature.
2. Defrost control, to eliminate icing of the evaporator.

The thermostat sensing element is usually located in the return air duct of the air conditioner. When this incoming air cools enough, the thermostat will stop the unit.

A two stage (two-switch) motor control thermostat is shown in Fig. 8-20. The sensing element reacts to the return air temperature. A rise in temperature will first cause the fan switch to cut in. A further rise in temperature will cause the compressor switch to cut in. As the room temperature drops, this operation will be reversed. Fig. 8-21 shows a schematic which helps explain the operation of this control.

A control knob provides an adjustment which the owner may use in setting the range of the air conditioner. If the room

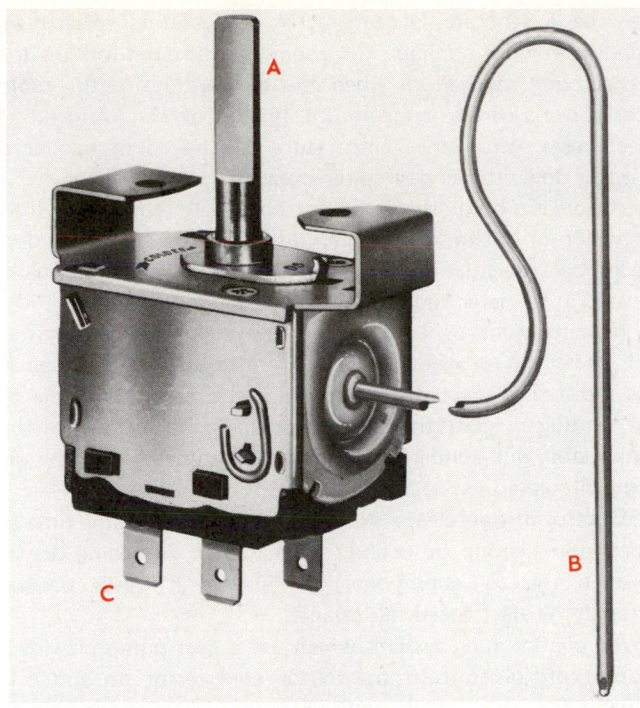

Fig. 8-20. A comfort cooling air conditioner motor control thermostat. It controls operation of fan and compressor. A—Range adjustment. B—Sensing bulb (located in return air duct). C—Electrical terminals. (Ranco, Inc.)

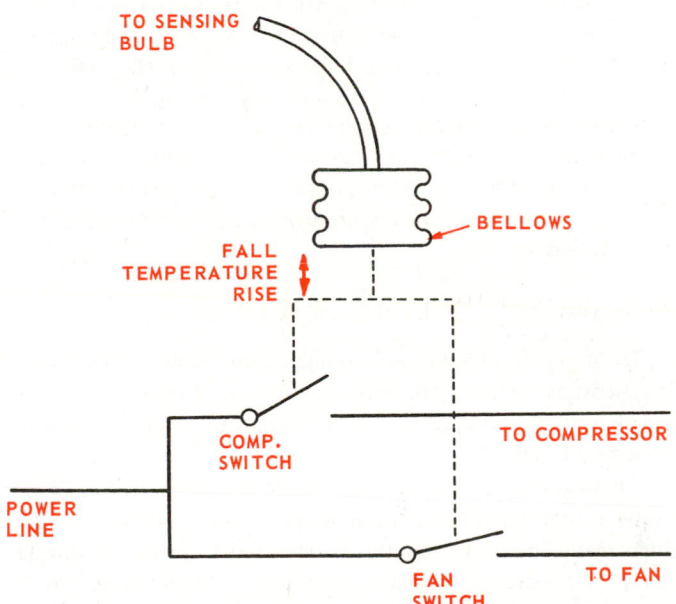

Fig. 8-21. Schematic showing how a room air conditioner thermostat operates. (Ranco, Inc.)

is too cold, adjustment may be made to reduce the running time of the air conditioner.

The defrost control, Fig. 8-22, prevents ice formation on the air conditioner evaporator. This control opens the circuit if the evaporator temperature is close to freezing. It has a factory adjustment only and should not require adjusting on the job.

Fig. 8-22. Air conditioner control unit used to prevent formation of ice on the evaporator.

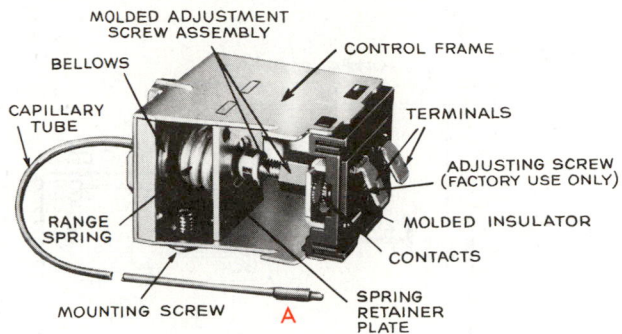

Fig. 8-23. Mechanism of air conditioner defrost control. In operation, control bulb, A, is mounted on evaporator.

This control, Fig. 8-23, is a SPST (single pole, single throw) control if used for defrost only. When defrost heating units are used on the off cycle, it is a SPDT (single pole, double throw) control.

The same controls are used in refrigerators and heat pumps. Fig. 8-24 is a schematic diagram of the control which shows how it is connected into the air conditioning electrical and refrigerating circuits. At A in Fig. 8-24, it is shown without a defrost heater, at view B with a defrost heater.

8-15 CENTRAL AIR CONDITIONING CONTROLS

Central air conditioning implies that the system can provide heating, cooling, humidification, dehumidification and, in many systems, electrostatic air cleaning. Each of these systems has electrical controls.

The heating controls, which usually operate from a room thermostat, govern the running time of the heating device. The thermostat, then, must make suitable electrical contact and operate the heating mechanism until the desired room temperature is reached. Then, the electrical circuit is opened and the heating device stopped. Details concerning installation and operation are given in Chapters 19, 20, 21 and 22.

Cooling is usually controlled by a room thermostat. This thermostat may be built in with the heating thermostat, or it

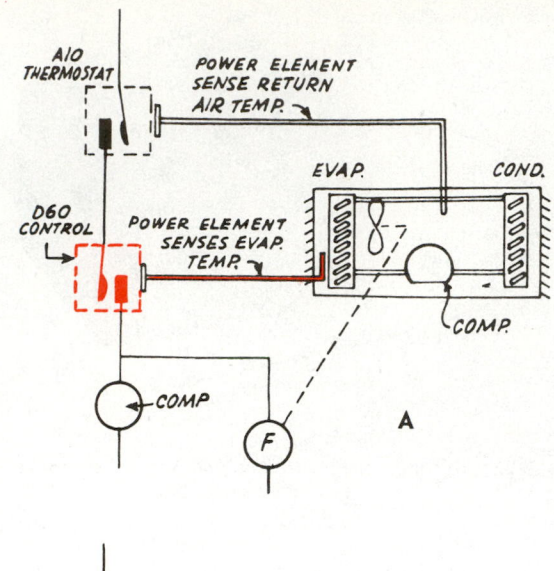

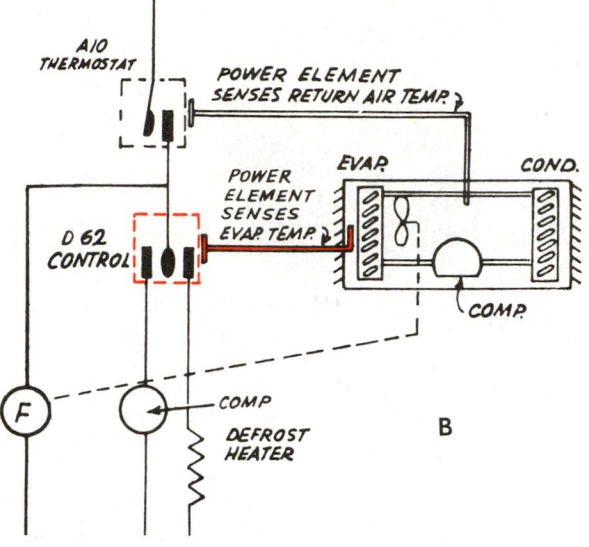

Fig. 8-24. Schematic diagram of air conditioning defrost control. A—Control without defrost heater. B—Control with defrost heater.

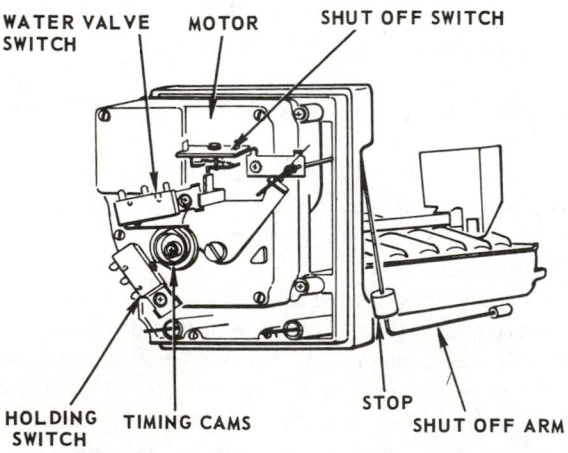

Fig. 8-25. One type of ice cube maker used on domestic refrigerators. Note the shut-off arm which stops the cycling when the cube bin is full. (Kelvinator, Inc.)

may be a separate instrument. A refrigerating mechanism usually provides cooling. The room thermostat turns on the refrigerating mechanism when the temperature of the room goes above a certain temperature. It shuts off the refrigerating mechanism when the temperature of the room has been brought down to the desired temperature.

Automatic humidification controls usually govern the flow of water or steam into a humidifier. If the air is too dry, additional moisture is provided. The humidistat is usually located in the same room as the temperature controls.

In some central air conditioning systems, the humidistat will cause the refrigerating mechanism to operate if the air is too moist. The evaporator is located in the plenum chamber of the central air conditioning system. The cold surfaces of the evaporator will condense out moisture, thus drying the air being circulated.

Electrostatic air cleaners usually operate at the same time as either the heating or cooling mechanisms. Air being drawn through the air conditioner is "scrubbed" while passing through the electrostatic air cleaner.

Air conditioning systems which use a heat pump provide a motor control to turn the motor compressor on and off. Usually, the same motor compressors are used for both heating and cooling. Their operation is controlled by the room thermostat.

During summer cooling, this system works without supplementary controls. During winter heating, however, more controls are required.

In addition to the motor controls used on the heat pump, an electric resistance heater is usually required at the evaporator. This heater will operate if the evaporator surface starts to ice over. (Ice would keep air from flowing through it.)

During certain weather conditions when it is not possible to extract sufficient heat from the outside air, some systems use an electric heating unit. The operation of this auxiliary electric heating unit is automatically controlled by the temperature of the condenser.

8-16 WATER COOLER CONTROLS

Drinking fountains are cooled with small refrigerating mechanisms. The evaporator is so placed that the water flowing to the outlet faucet is cooled to a temperature of about 50 F. (10 C.).

These water coolers usually provide storage cylinders where water is maintained at a lower temperature. In some cases, ice if formed. The motor control is operated from the accumulator which operates to keep a "cold reserve" available to supply needed water to the faucet. Details of some of these water coolers are shown in Chapter 13.

8-17 ICE MAKER CONTROLS

Many domestic refrigerators now have an automatic ice maker. A small fresh water line is connected to the refrigerator. A combination of solenoid valve, timer, evaporator, defroster, and bin level control lets the refrigerator keep a storage bin full of ice cubes at all times. Fig. 8-25 shows the construction of a typical unit.

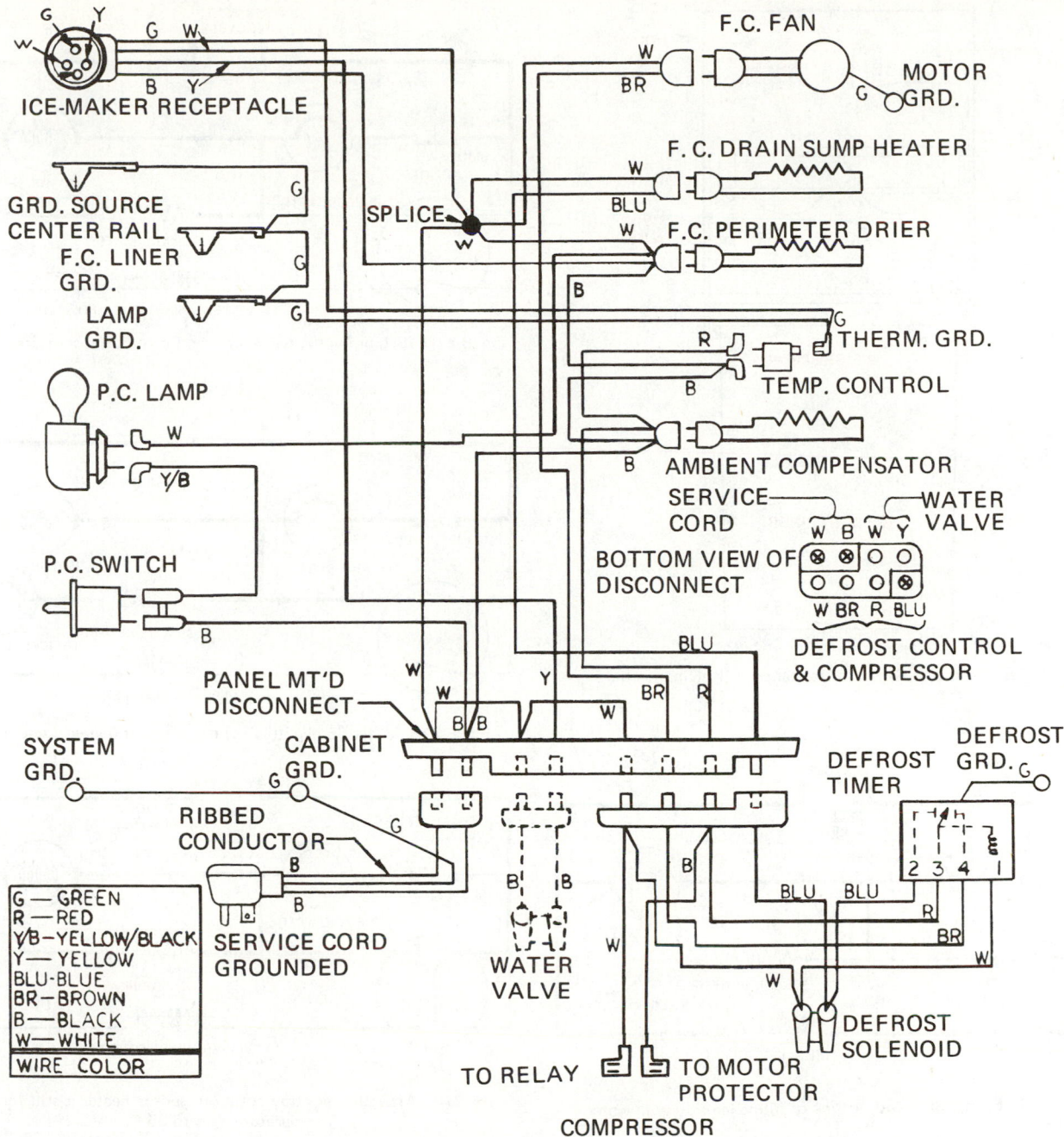

Fig. 8-26. Wiring diagram of domestic compression system refrigerator, including ice cube maker. Note that green is the ground wire.

The electrical wiring diagram, Fig. 8-26, shows an ice maker connected to a refrigerator mechanism. A control circuit works with a temperature sensing device called a sensor. This senses the amount of time required to freeze the ice cubes and start the harvest cycle. A commutator and brush assembly moves the mechanism through the refill, freezing and harvest operation.

Fig. 8-27A illustrates the important parts of a common ice maker. The step-by-step operation of this ice maker is shown in a sequence of drawings, B through Fig. 8-27H.

The filling time in Fig. 8-27B is controlled by the motor-driven commutator. The water valve solenoid is ener-

gized and water flows into the ice tray for 7.5 seconds.

During the freezing cycle, Fig. 8-27C, the temperature in the freezing compartment is at 8 F. or lower.

Start of the harvest cycle is shown at Fig. 8-27D. A "cold" cut-in temperature thermostat (T_1), and high cut-out temperature thermostat (T_2), is part of the temperature sensor and subject to the same temperature and conditions as the ice tray.

The water and temperature sensor are gradually cooled at the same rate to 11 F. (plus or minus 3 deg.). The cold cut-in thermostat closes to start a harvest cycle. The drive motor starts and rotates the ice tray, at a speed of 1/4 rpm. A 22W heater in the temperature sensor is energized and heats the

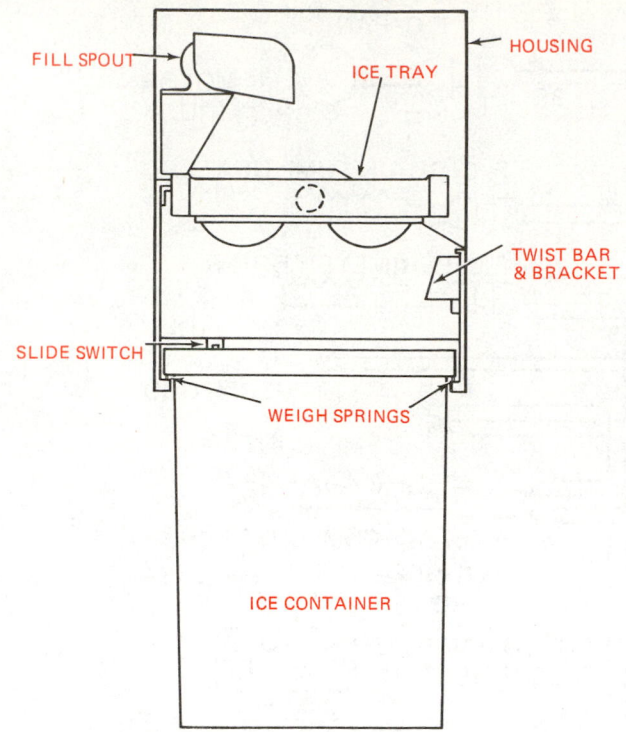

Fig. 8-27A. Parts of a common ice maker. (Kelvinator, Inc.)

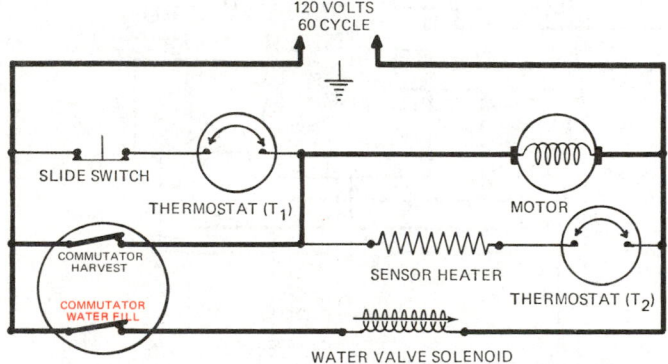

Fig. 8-27B. Electrical circuit at time of filling ice tray with water.

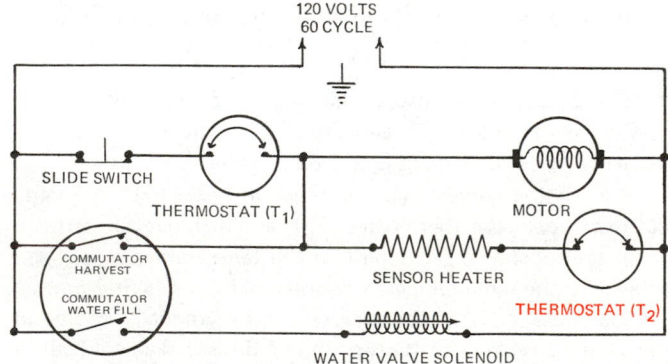

Fig. 8-27C. Electrical circuit during freezing cycle. Note that thermostat (T_2) is closed.

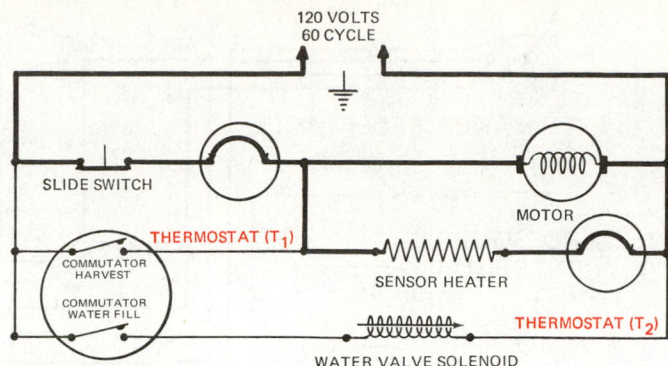

Fig. 8-27D. Beginning the harvest cycle. Both thermostats are closed.

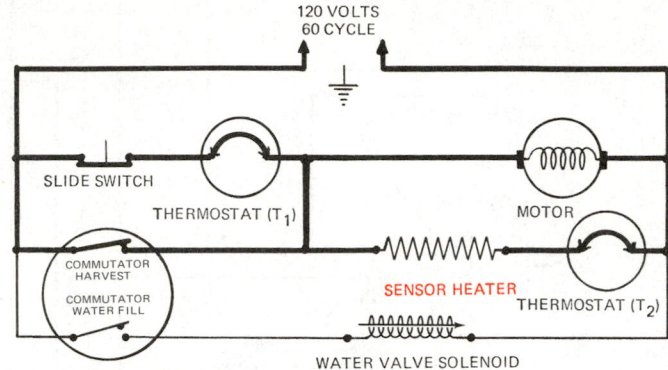

Fig. 8-27E. After 35 deg. tray rotation, sensor heater is energized.

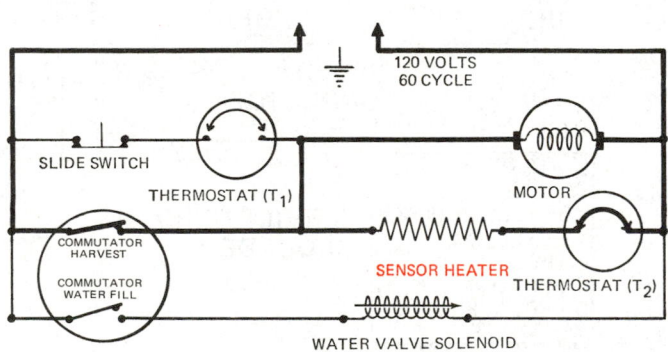

Fig. 8-27F. After 90 deg. tray rotation, sensor heater is still energized. Temperature rises to 35 F.

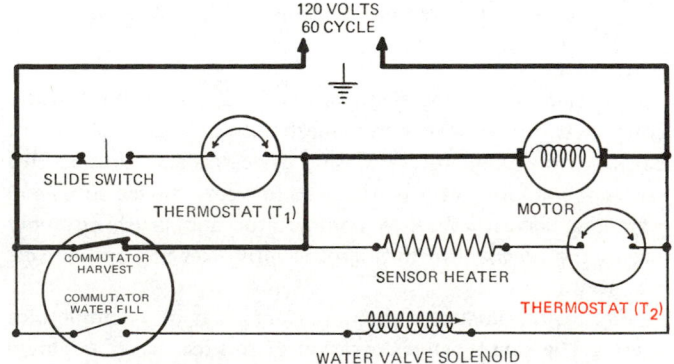

Fig. 8-27G. After 120 deg. tray rotation, thermostat (T_2) opens, breaks circuit to sensor heater. Temperature is now 59 F.

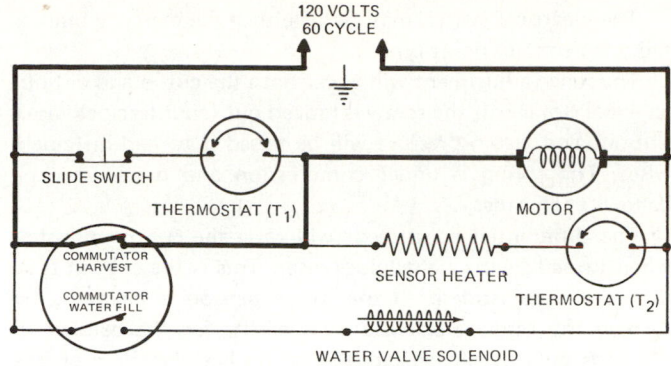

120 VOLTS
60 CYCLE

SLIDE SWITCH

THERMOSTAT (T₁)

COMMUTATOR HARVEST

COMMUTATOR WATER FILL

SENSOR HEATER

MOTOR

THERMOSTAT (T₂)

WATER VALVE SOLENOID

Fig. 8-27H. Completion of harvest cycle. Ice is dumped into container.

temperature sensor.

The harvest cycle continues in Fig. 8-27E. After a 35 deg. tray rotation, a parallel circuit to the motor and temperature sensor is made through the commutator.

In Fig. 8-27F, the harvest cycle is still in operation. After a 90 deg. tray rotation, the "cold" cut-in temperature thermostat opens, the motor and sensor heater remain energized. The sensor temperature rises to 35 F.

At Fig. 8-27G, the tray has rotated 120 deg. The temperature rises to 59 F. and the thermostat (T_2) opens the temperature sensor heater circuit.

The harvest cycle is complete in Fig. 8-27H, after 180 deg. of tray rotation. The rear bracket of the ice tray comes to rest on top of the twist bar. For approximately 18 seconds, the rear portion of the ice tray is stopped, while the front portion continues to rotate. This twisting action loosens the ice cubes. The cam action to the tray front bracket moves the twist bar forward, releasing the rear tray bracket. This dumps the ice cubes into the container.

Many types of commercial ice cube makers are described in Chapters 12 and 13. These systems have normal units except for the evaporator and cycling thermostat. In each case, ice making continues until the ice cube storage bin has enough cubes stored. A lever-operated control stops the process when the bin is full by opening a switch. It closes the switch again when the level drops to about one-fourth to one-third full.

8-18 REMOTE TEMPERATURE SENSING ELEMENT REQUIREMENTS

Remote temperature sensing elements are available in several designs. One major difference in type is based on the pressure in the element. There are two common designs:
1. Above atmospheric pressure in the element (used for controlling refrigeration temperatures).
2. Below atmospheric pressure in the element (used for controlling heating units).

The above-atmospheric-pressure element is used for controls which close the electrical circuit on temperature rise. If the element loses its charge, the unit is unable to start. This is known as a "fail safe" action. This element is used where continuous running would be harmful such as refrigerators and comfort cooling units.

The below-atmospheric-pressure element is used for controls which open the electrical circuit on temperature rise. It is found on electric heating and electric defrost units. If the element loses its charge, the points will open (they are designed to close on temperature drop and open on temperature rise). As in above atmospheric pressure units, this is a "fail safe" device that prevents an electric heating coil from overheating.

The temperature of the mechanism at the point of contact with the sensing element, in most cases, controls the motor operation. This sensing element may be used to stop the motor if the condensing temperature rises too high. Sensing elements may also be used as safety devices to control the motor operation. Commercial installations, in particular, may use them to stop the motor when:
1. The flow of water through the condensing unit stops.
2. The oil pressure in the compressor is too low.
3. The head pressure is too high.
4. The desired low temperature in the cabinet is reached.

Many motor controls may contain several of these requirements. This means that there may be more than one capillary tube entering the motor control.

8-19 THERMOSTATIC MOTOR CONTROLS — BIMETAL

How the bimetal thermostat operates is explained in Para. 8-4. These thermostats are used, to some extent, in refrigeration but not as often as in air conditioning.

Bimetal operating thermostats are shown in Fig. 8-28.

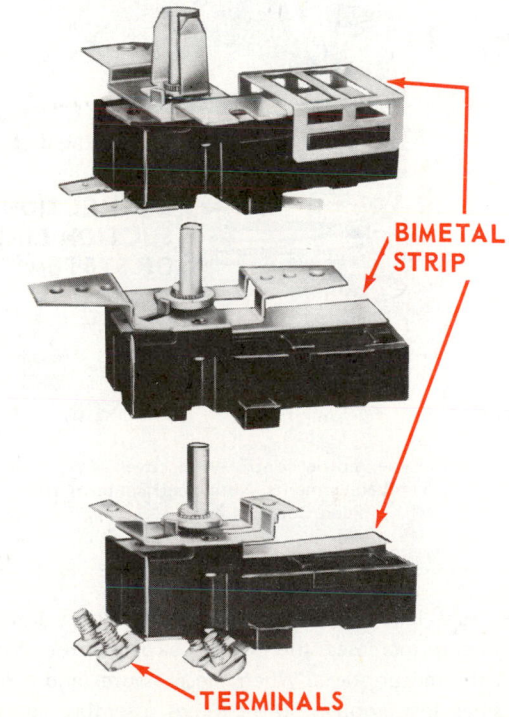

BIMETAL STRIP

TERMINALS

Fig. 8-28. Three types of bimetal operated thermostats. These units are used for control in refrigeration, electric heat and air conditioning. (Therm-O-Disc, Inc.)

These thermostats use bimetal strips for fast snap action. The bimetal control usually consists of a dished disk. Its construction is such that it is dished in one direction when it is cold; and, as it warms, it suddenly snaps into a dish position in the other direction. Snap action disks are also used on pilot lights and gas burner controls.

8-20 PRESSURE MOTOR CONTROLS

A low pressure must be maintained in the evaporator to permit evaporation of the refrigerant at a low temperature. Therefore, automatic control of the motor may be based on pressure differences in the evaporator. This control is used, quite generally, on commercial systems. A bellows-operated low pressure control is shown in Fig. 8-29.

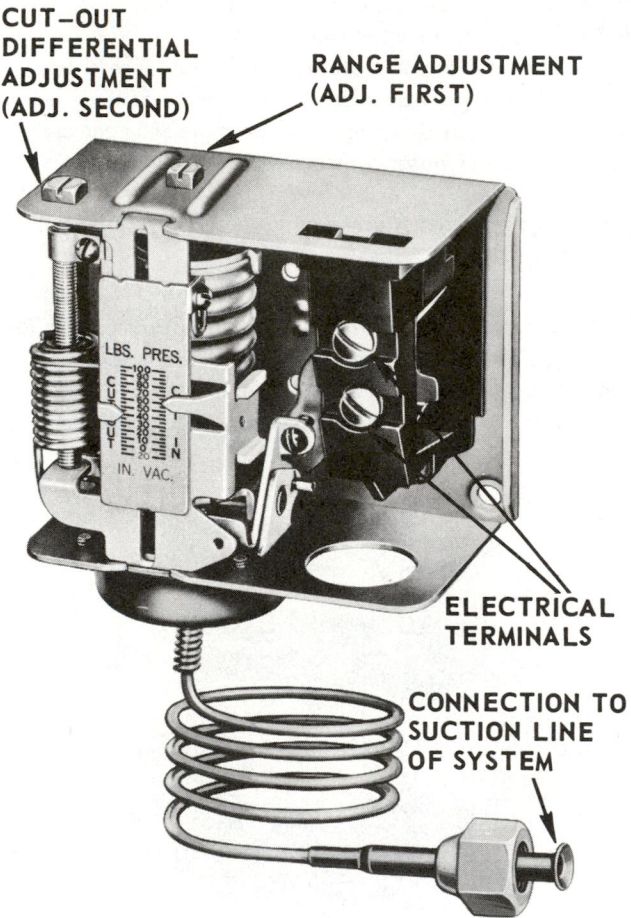

Fig. 8-29. Pressure type motor control with cover removed to show mechanism and adjustments. Note electrical terminals. (Penn Controls, Inc.)

This is how it operates: as the evaporator warms, the low-side pressure increases, the bellows expands, the switch is closed and the motor starts. When the pressure (and temperature) becomes low enough, the bellows assembly contracts, contacts open and the motor automatically shuts off. A pressure motor control with accessible range and differential adjustments is shown in Fig. 8-30.

The electrical switch may be of either the mercury bulb or the open-contact point type.

The range adjustment will lower both the cut-in and cut-out an equal distance if the screw is moved out (counterclockwise). Cut-out and cut-in pressure will be raised if turned in (clockwise). The spring is under compression and presses on the bellows at all times.

The differential adjustment will raise the cut-out pressure when turned to the right (clockwise). This is the cut-out type differential adjustment. If the spring tension is increased by turning the screw clockwise, it is harder for the bellows to reach its cut-out setting but the spring has no effect on the cut-in setting.

Some models of the pressure control are also equipped to act as a safety control. A bellows construction with a pressure tap to the high-pressure side of the compressor is used. If the compression pressure or head pressure should become too high, the bellows will expand. This movement will open the switch and stop the motor.

Such a control is a safety device for the motor. It is especially necessary when a water-cooled unit is used. See Chapter 12 for more information on commercial controls.

The low-pressure type of control is easy to adjust on the job. It can also be easily adjusted with a vacuum pump and a compound gauge after it has been removed from the unit.

The adjustment of these controls is important, because satisfactory operation of the unit depends, to a great extent, on their proper operation.

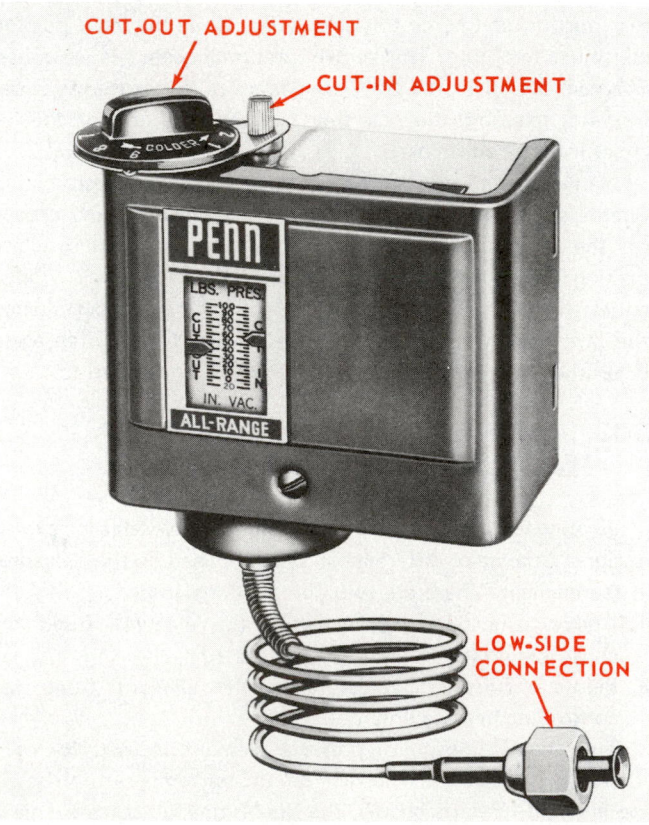

Fig. 8-30. Typical pressure motor control. It is possible to adjust cut-in and cut-out pressure on this pressure type motor control.

8-21 MOTOR SAFETY CONTROLS

Several safety mechanisms are used in the electrical circuit to make sure that the motor compressor is protected. The three most common safety controls are:

1. Head pressure safety cut-out.
2. Low pressure safety cut-out.
3. Oil pressure safety cut-out.

One of the most harmful things that can happen to a hermetic system is to have high head (condensing) pressures. These high pressures raise the temperature of the vapor and oil moving past the compressor exhaust valve to a point which may cause oil and refrigerant breakdown.

This condition is worsened if a little moisture and dirt are present. Carbon, acids and sludges may be formed.

It is important to stop the system before these dangerous temperatures are reached. A high pressure safety cut-out is often used for this purpose. If pressure exceeds a certain amount, the current to the motor will be shut off and the motor will be stopped.

Several conditions might cause this control to operate:

1. Lack of proper air circulation through the condenser.
2. Lack of flow of water through a water-cooled condenser.
3. A greatly increased refrigeration load of some kind. (See Chapter 12.)

A low pressure switch can be used to cycle a refrigerating system using a TEV. It can also be used as a safety device. The cooling of a motor compressor depends on the amount and temperature of the suction vapor. If this vapor pressure is too low (system low on refrigerant or flow restricted), the motor compressor may overheat and burn out. The low pressure safety control will stop the motor compressor before it is damaged.

The schematic wiring diagram for a low-side or suction pressure control is shown in Fig. 8-31.

Some larger units use a safety control device connected to the compressor lubrication system. These shut off the unit if the oil pressure decreases or falls below a predetermined safe pressure above the low-side pressure. The construction is similar to a low-pressure control but usually with a fixed or nonadjustable differential. This control has one connection to the oil pump and one connection to the low-pressure side. See Chapter 12.

8-22 OVERLOAD PROTECTION

All refrigeration and air conditioning units ought to be connected to separate circuits from the control panel. This applies to both domestic units and commercial units. The fuse or circuit breaker in the individual circuit should have enough capacity to provide a continuous flow of current under normal operating conditions. But they should open the circuit in the event of continuous overload of over 25 percent.

At the instant of starting, all motors draw an overload of current. This may amount to 600 percent. However, this is for a very short time and the circuit breaker or fuse should not open the circuit during this short time.

High horsepower motors usually incorporate a starting device in the electrical circuit. This starter does not throw the motor directly on the line. It brings into use a resistance or induction unit which restricts the flow of current at the instant of starting but allows an increase as the motor gains speed.

All starting relays have some type of overload protection. The most popular type is a bimetal control in series with the power supply to the motor. A resistance heating unit, also in series with the power supply, is alongside the bimetal control. This resistance unit will heat up if the motor is overloaded. The bimetal safety device, reacting to the heat of the resistance unit, will bend. The points at the end of the bimetal strip will open and the motor stops. It will not restart until the safety device cools down.

Three-phase circuits and motor compressors are often used in sizes above 1 hp. The three voltages of the three phases should be kept within 10 percent of each other. This is to prevent damage to the motor and to prevent motor reversing.

Controls are used to open the circuit if the voltage in any of the three circuits changes over 10 percent, or if the motor reverses. Some controls will close the circuit again when normal conditions are reached. These controls usually act from about 0.1 of a second up to two seconds (if desired).

These protective devices are basically voltage controlled relays. Transformers are used to step the voltage down for each potential relay. The stepped-down voltage is then rectified and is fed to the solenoid coil. The same voltage is also fed to the control of a transistor. If the voltage varies too much, the transistor will bypass the solenoid, causing circuit to open. Fig. 8-32 shows a three-phase circuit protector.

8-23 MOTOR STARTING RELAYS

Some motor controls for hermetic systems are different from those used on external drive systems.

Starting relays are found on the outside of hermetic systems. These relays are usually one of the following types:

1. Current (magnetic).
2. Potential (magnetic).

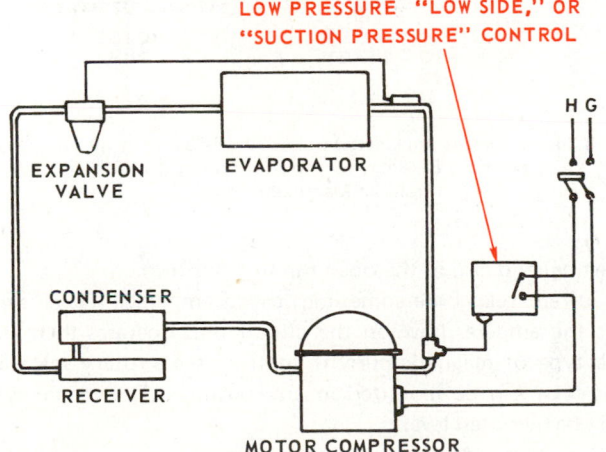

Fig. 8-31. Schematic of a low-pressure motor control installed in a refrigerating system.

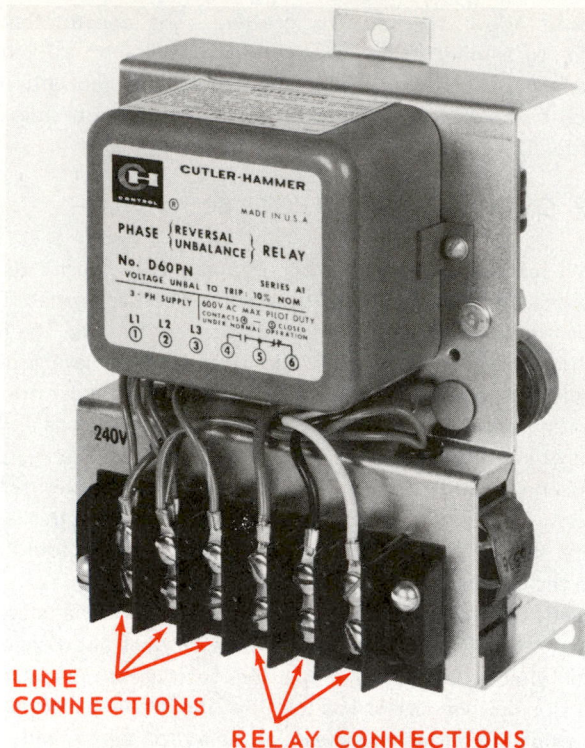

LINE
CONNECTIONS

RELAY CONNECTIONS

Fig. 8-32. A three-phase protection system with cover removed.
(Eaton Corp.)

3. Thermal.

4. Solid state electronic.

The relay permits electricity to flow through the starting winding of the motor until the motor reaches about two-thirds of its rated speed. It then disconnects or opens the starting winding circuit.

The starting winding should be energized only for three or four seconds at a time. If current flows through it for a longer period, the winding may overheat. Many relays have current and/or thermal protection devices to prevent the starting winding from abuse.

To operate correctly, these relays must be the right size for the motor. When replacing one, be sure it has the same electrical specifications as the original.

It is impossible to use open electrical contacts inside a sealed system.

8-24 CURRENT (MAGNETIC) RELAY

Current relays are usually found on low torque, smaller horsepower motors.

The magnetic type relay uses the electrical characteristics of the motor to operate it. The running winding consumes more current when the rotor is not running, or is turning slowly, than it does when it reaches full speed. As the rotor picks up speed, the magnetic fields build up and collapse in the motor, producing a bucking or counter electromotive force (emf) or voltage on the running winding. Current-operated relay switches, used to close and open the starting winding, operate on the change in current flow of the running winding, as it goes from

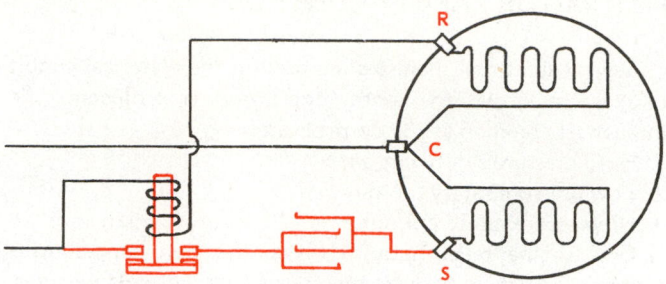

Fig. 8-33. Magnetic type current relay schematic. Relay is shown in open position. R—Running winding terminal. S—Starting winding terminal. C—Common terminal.

a start condition to run.

The magnetic relay is an electromagnet much like a solenoid valve. Either a weight or a spring holds the starting winding contact points open when the system is idle. Fig. 8-33 is a schematic of a weight operated unit.

When the motor control (thermostat or pressurestat) contacts close and the high current flows into the running winding, the magnetic current relay coil is heavily magnetized. It lifts the weight or overcomes the spring pressure and closes the contacts.

This action closes the starting winding circuit and the motor will quickly accelerate (speed up) to two-thirds or three-fourths of the rated speed. As it does so, the amperage draw of the running winding of the motor decreases. This decreases the magnetic strength of the magnetic current relay enough to allow the weight or the spring to open the points. Fig. 8-34 shows a magnetic current relay in the closed position

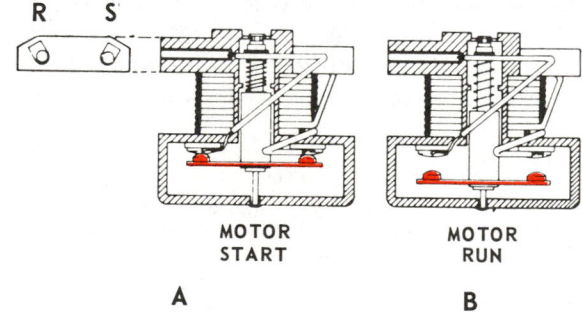

Fig. 8-34. A current (magnetic) relay. A—Relay is in motor starting position. B—Relay is in motor running position.
(Franklin Manufacturing Co.)

(starting) and also in the open running position.

Current relays are sometimes called amperage relays, since it is the ampere draw on the circuit that operates the relay. One type of magnetic current control uses a rotary solenoid. This type can be mounted in any position. The weight type must be mounted level.

One type of weight operated magnetic current relay is shown in Fig. 8-35. A spring operated type is shown in Fig. 8-36. A method of mounting a current starting relay is shown in Fig. 8-37.

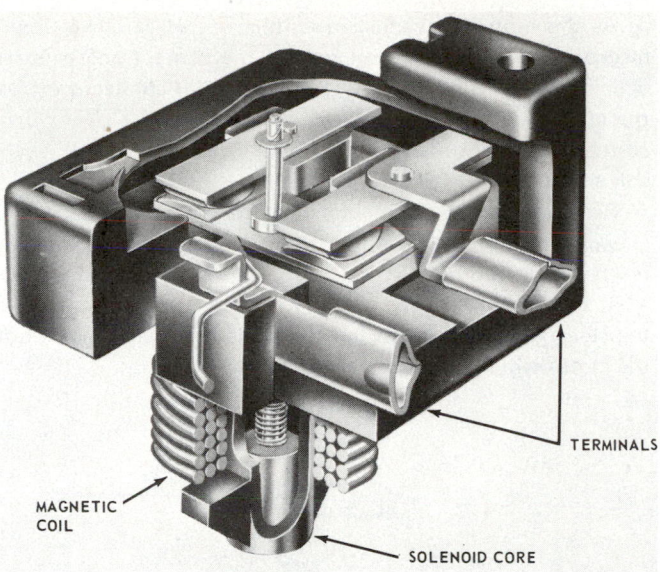

Fig. 8-35. Current type magnetic relay. Plastic housing contains solenoid which has movable center. Heavy current draw on starting raises center plunger and closes starting winding circuit.
(Tecumseh Products Co.)

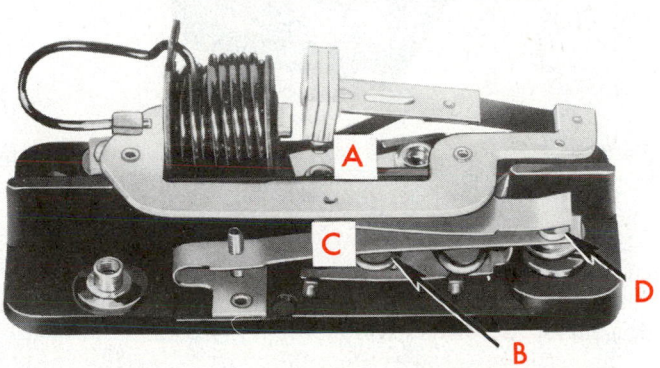

Fig. 8-36. Current type magnetic starting relay with an overload safety switch. Starting winding points are kept open by means of cantilever spring at A. At instant of starting, magnetism pulls spring down and closes starting points at D. If current draw is too great, resistance wire at B will heat and cause bimetal strip at C to bend and open circuit at D.
(Copeland Corp.)

These relays are available in a number of capacities. The difference between closing amperage and opening amperage settings is small. It is this small difference in current flow which will close the starting circuit and then open it when the motor reaches approximately three-fourths speed. Fig. 8-38 is a circuit using a current relay.

The utility companies sometimes reduce line voltage to save current. The current type starting relay gives some protection to the motor under these conditions. However, there is still some danger of a motor "burnout."

8-25 POTENTIAL (MAGNETIC) RELAY

Potential relays, sometimes called voltage relays, are usually used with high torque, capacitor-start motors. They look

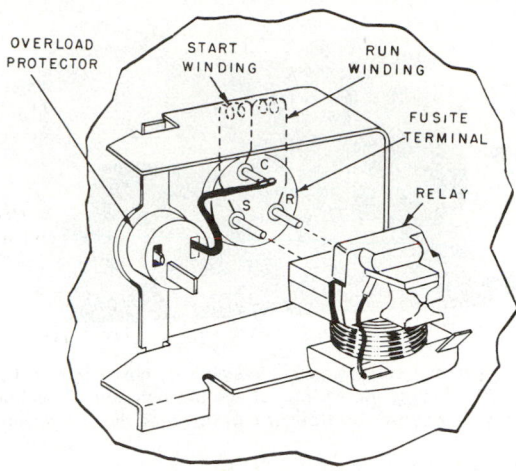

Fig. 8-37. A starting relay mounted on the compressor housing.
(Kelvinator, Inc.)

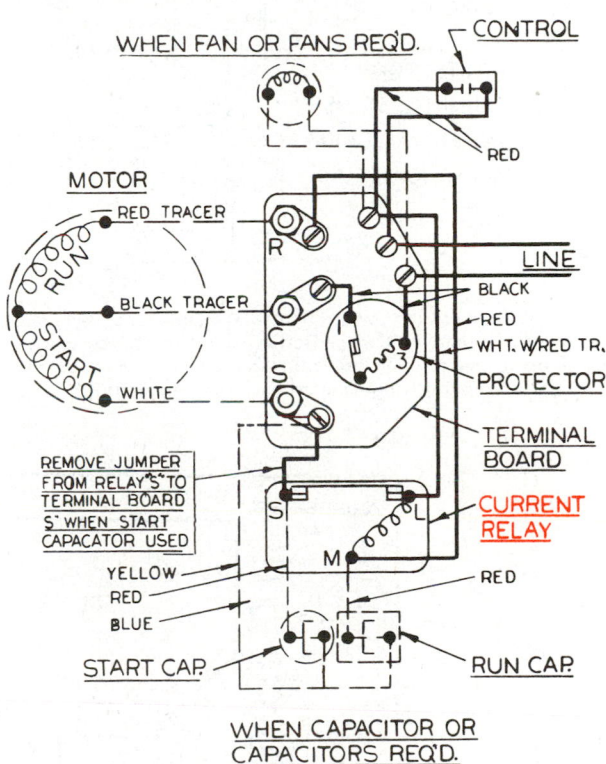

Fig. 8-38. Complete wiring diagram of system using current relay.

somewhat like the amperage relay. However, its operation is based on the increase in the voltage as the unit approaches and reaches its rated speed. Fig. 8-39 is a potential magnetic relay.

The contact points remain closed during the off part of the cycle. This feature is its biggest advantage. If the points are closed as the thermostat closes the power circuit, there will be no arcing of the relay points. This quite often occurs with the current relay.

Fig. 8-39. Potential starting relay. Weight, A, closes points, C, during off cycle. On starting, increasing voltage through coil, B, will pull contact points apart and stop flow of current through starting winding.

As the motor speed increases, higher voltage creates more magnetism in the relay coil pulling the contact points apart, opening the starting circuit. The relay coil is connected across the starting winding. It is made of small wire, so very little current passes through it. This minimizes the heating of the coil and core.

Resistance of the contact points to voltage must be high enough to prevent the points from opening before the motor reaches 80 to 90 percent of its full speed. But such resistance must be low enough to positively open the points and remove the starting winding from the circuit at the right time. If not, the motor will overheat.

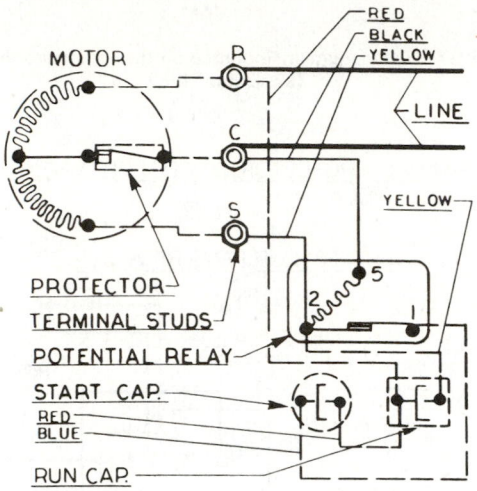

Fig. 8-40. Wiring diagram for a potential (voltage) type magnetic starting relay. Note that starting capacitor circuit is opened when relay contacts open, but running capacitor still is connected across starting and running winding in series.   (Copeland Corp.)

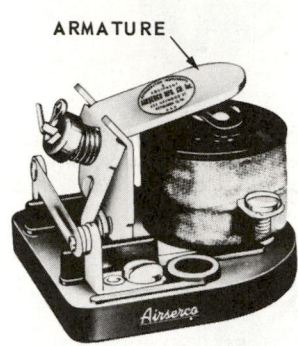

Fig. 8-41. Potential type relay. As armature lever is pulled down by electromagnet, lever at left will open starting circuit points.

Fig. 8-40 is a circuit diagram of a unit with a potential relay. The relay itself is shown in Fig. 8-41. The wiring diagram for its installation is shown in Fig. 8-42.

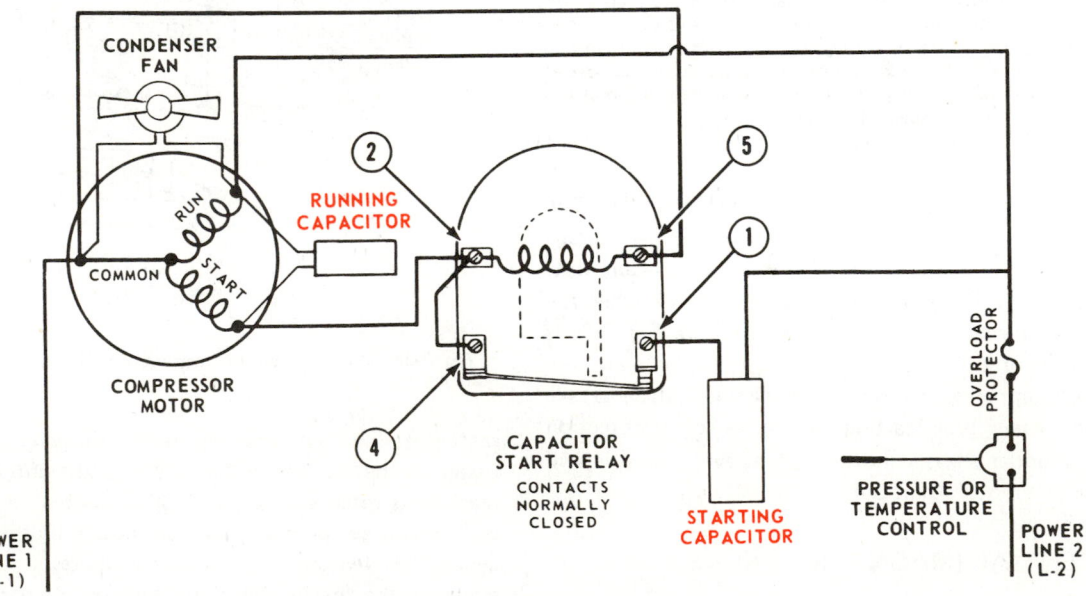

Fig. 8-42. Wiring diagram for potential relay shown in Fig. 8-41. Note that this is a capacitor-start, capacitor-run motor.

8-26 THERMAL RELAY

There are two types of thermal relays. One type uses two bimetal strips to control the contact points. The other controls the contact points through a resistance wire which is under tension.

In the first type, one strip controls the starting winding and the other the running windings. See Fig. 8-43.

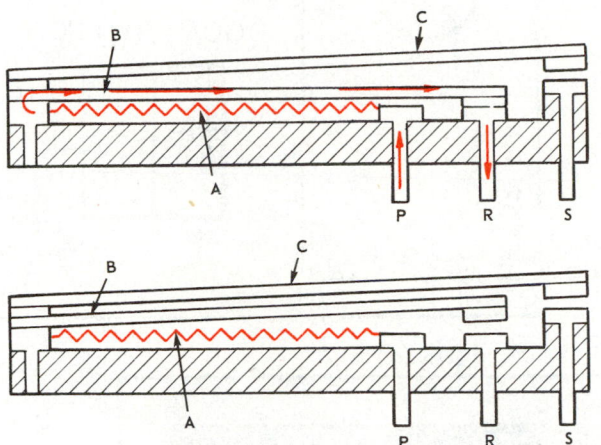

Fig. 8-43. Thermal starting relay using two bimetal strips. A—Heating wire. B—Running winding bimetal. C—Starting winding bimetal. P—Power wire connection. R—Running winding connection. S—Starting winding connection. Above, starting circuit is open, unit is operating on running winding only. Below, both circuits are open and have been drawing too much current.

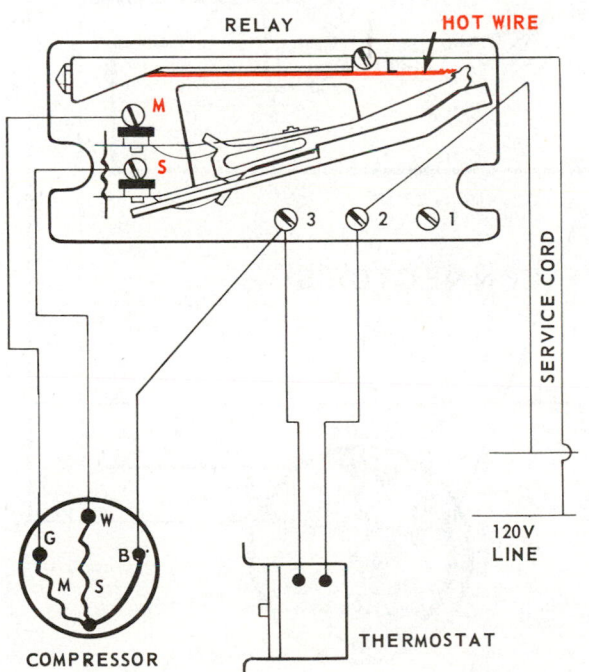

Fig. 8-44. The circuit diagram for a thermal (hot wire) relay. Current draw will cause the "hot wire" to heat and stretch. A slight heating will cause the starting points at S to open. Further heating by a too-heavy current flow will cause the points at M to open and stop the unit. (Kelvinator, Inc.)

When cold, both sets of contact points are closed. A resistance wire is mounted near the bimetal strips. It is in series with both the starting and the running winding. It is the right size and distance from the bimetal strip so that its heat opens the starting winding contact points as soon as the motor reaches its proper operating speed.

This control also serves as a safety cut-out. If the motor should use too much current, the resistance wire will heat the bimetal strip, open the contact points and stop the motor.

In the second type, the resistance wire is attached in series with both the starting and the running windings. The tension of this wire, when cold, keeps both sets of contact points closed. See Fig. 8-44.

While the current passes through, the resistance wire is heated and expands or stretches. At a predetermined setting, the stretched wire opens the starting winding contacts. This control also serves as a safety cut-out. If the motor should use too much current, the wire will stretch enough to open the running winding contact points.

The complete wiring diagram for a refrigerator using a thermal relay is shown in Fig. 8-45.

8-27 SOLID STATE ELECTRONIC RELAYS

Relays using solid state transistors, diodes and triacs are now being used to control starting of hermetic motors. Changes in voltage in the motor, as it starts and then gathers speed, are used to open the starting winding circuit at the correct time. These relays are not as sensitive to the size of the motor as other relays. The same solid state unit can be used for motors varying from 1/12 to 1/3 hp.

8-28 CHECKING AND TESTING RELAYS

In general, relays should be replaced, not repaired. The service technician's job is mainly to determine if the relay is defective. If so, it is replaced with an exact duplicate. Fig. 8-46 shows the wiring connections to three types:
A. Klixon magnetic.
B. General Electric magnetic.
C. Delco hot wire.

The wire size, the contact point area and the spring tension or weight plus the air gaps, must be accurately set for each unit. A slight difference in weight or spring tension, for example, might result in 100 motor revolutions difference.

The most effective way to determine if the relay is causing the trouble is to check the other parts of the circuit first. Check the motor, the capacitor, the overload cut-out and the thermostat. Only if these parts test all right, should the relay be replaced.

A tester in Fig. 8-47 will check out either a current relay or potential relay. It will also check the relay for line voltage. The procedure is simple and quickly done. Connect the numbered leads from the numbered jacks on the tester to the same numbered terminals on the relay. Test on the 120V switch before using the 208—240V switch.

Keep the relay cover in place; never discard it. Dust collecting on the contact points will quickly cause them to

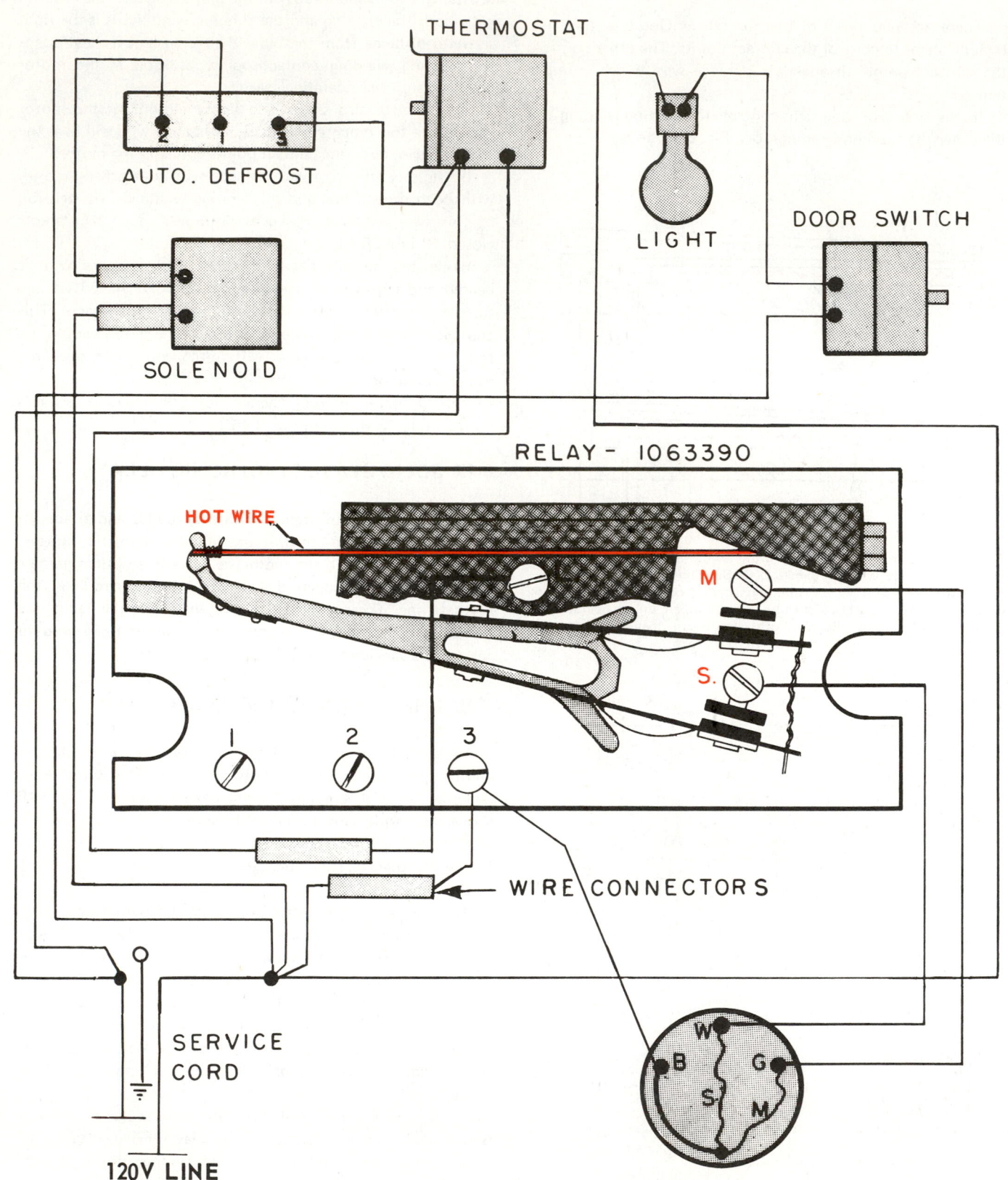

THERMOSTAT

AUTO. DEFROST

LIGHT

DOOR SWITCH

SOLENOID

RELAY - 1063390

HOT WIRE

M

S.

1 2 3

WIRE CONNECTORS

SERVICE CORD

120V LINE

W

B G

S M

Fig. 8-45. Electrical circuit for refrigerator using thermal relay.

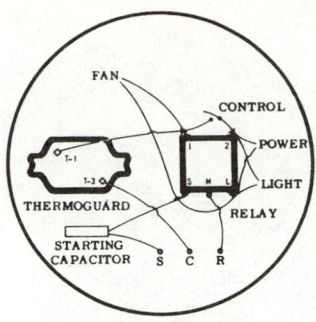
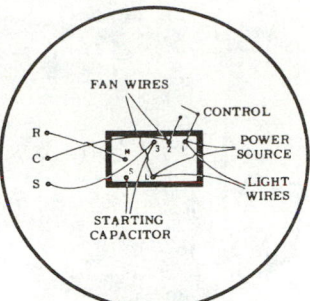

KLIXON MAGNETIC

L and 2	Power Wire
L and 2	Light Wire
1 and 2	Control
L and 1	Fan
T-1 and 1	Then T-3 to common motor terminal

S and Starting Terminal for starting capacitor

If motor is split phase, starting motor terminal is attached direct to S on relay

M to Running Motor Terminal

If control is not used, use jumper between 1 and 2.

Note: Posts 1 and 2 are not on all Klixons, but all are interchangeable

MAGNETIC G-E

C Post and Control Power Wire
Light Directly across Power Wires
Control in Series with Post 4 and Hot Wire

4 and 5	Fan
2 and 3	Running Capacitor if used
1 and 2	Starting Capacitor if used
R and 3	Running Wire
C and 5	Common Wire
S and 2	Starting Wire if motor is capacitor start
S and 1	Starting Wire if motor is split phase

DELCO HOT WIRE

L adn 1	Power Wires
L and 1	Light Wires
1 and 2	Control Wires
L and 2	Fan if used
M and 3	Running Capacitor if used
S and 3	Starting Capacitor if used
R and M	Running Wire
C and 2	Common Wire
S and 3	Starting Wire if motor is capacitor start
S and S	Starting Wire if motor is split phase

A   **B**   **C**

Fig. 8-46. Electrical circuit diagram for three starting relays. A—Current type relay. B—Potential (voltage) type relay. C—Hot wire type relay.

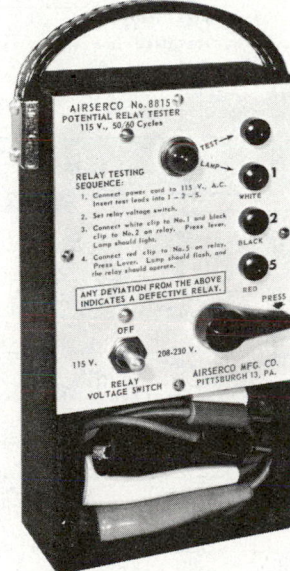

Fig. 8-47. Motor starting relay tester will test both current and potential type relays. May be used on both 120 and 240V circuits. (Airserco Mfg. Co.)

burn. This causes excessive voltage drop across the points and a poorly working control.

The weight-type amperage relay must be mounted in a straight up-and-down position. Otherwise, the plunger will rub and bind against the sides of the relay body. The relay should open in about three seconds if it is working correctly.

When replacing the relay, it is important to disconnect the power supply. Label each wire as it is disconnected and use the correct size screwdriver. The terminals are usually numbered so a tag or clip on each wire with the corresponding number

makes it easier to connect the new relay. Masking tape and a marking pencil are useful for such labeling.

If an exact potential relay replacement is not available, use one of a lower rather than a higher voltage rating (90 percent of rating). Because capacitors can discharge at 300V or higher, potential relay points may be burned (fused) by this discharge if the unit short cycles. To eliminate this trouble, use capacitors equipped with resistors across the capacitor terminals. Or, use a time delay switch to prevent the short cycling.

Frequently a unit will short cycle because the thermostat is exposed to vibration (on a shaky wall or a stairs). Be sure to mount the thermostat to something solid. Avoid tapping a relay to check it. Such tapping may cause the points to touch. This brief contact may ruin the points and damage the motor. The relay must function correctly without being tapped or it should be replaced.

8-29 AUTOMATIC DEFROST CONTROLS

A number of refrigerators have a standard temperature section and a frozen foods section in the cabinet. These dual-purpose cabinets need a special series of motor controls. First, the controls must give correct temperatures in both sections. Second, the controls must provide completely automatic defrost.

One type of control is shown in Fig. 8-48. The wiring diagrams for it are shown in Fig. 8-49.

There are four basic means of controlling the defrosting interval:

1. An electric timer which defrosts the unit at certain time intervals.
2. A device which defrosts the units based on the number of times the refrigerator door is opened.

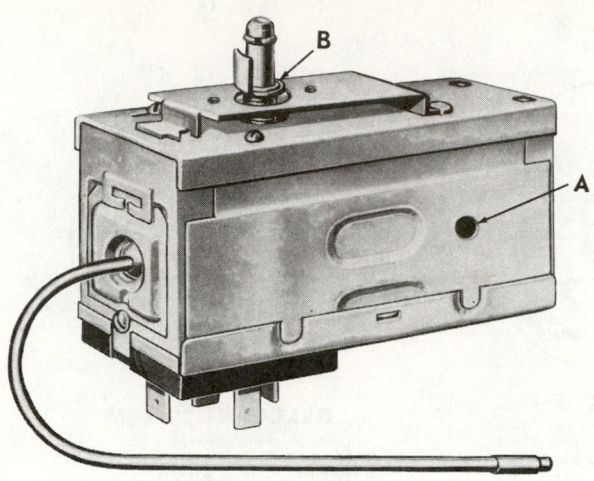

Fig. 8-48. Fully automatic defrost control. A—Electric timer motor. B—Range adjustment.

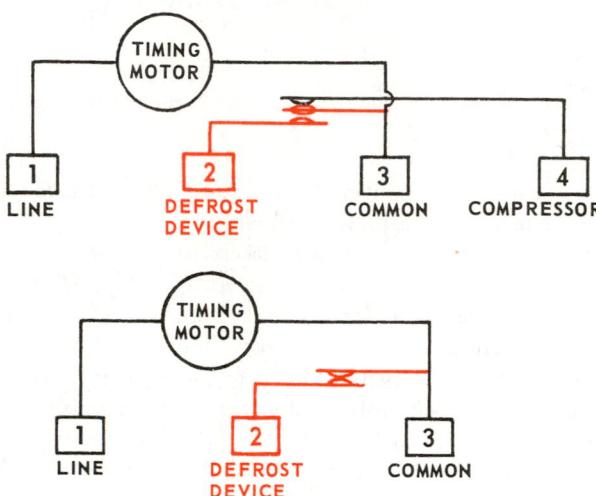

Fig. 8-49. Wiring diagram of an automatic defrost control. Control terminals are labeled. Above, an electric defrost system in which the motor circuit is broken during defrost time. Lower diagram is for hot gas defrost system, where compressor continues to run in defrost cycle. (Ranco, Inc.)

3. A clock which runs only when the unit is running. It defrosts after a predetermined number of hours of running time.

4. A no-frost system which uses forced convection evaporators and defrosts these evaporators during each off-portion of the operating cycle. Either hot gas or electric heating elements are used.

For example, a defrost system which uses a timer to operate the defrost cycle every 12 hours, works this way:

1. It shuts off the compressor and the evaporator fans and starts the electric heaters. The heaters will be on for about 15 minutes.

2. Then it shuts off the electric heaters and starts the compressor.

3. Evaporator fans start after compressor has run about four minutes, and unit returns to normal operations.

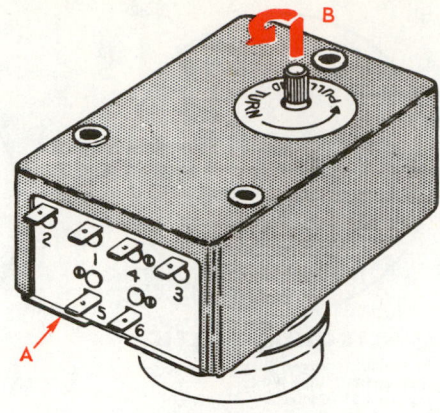

Fig. 8-50. Defrost timer used to operate defrost cycle in three-step nonfrost refrigerator. A—Terminals. B—Adjustment.

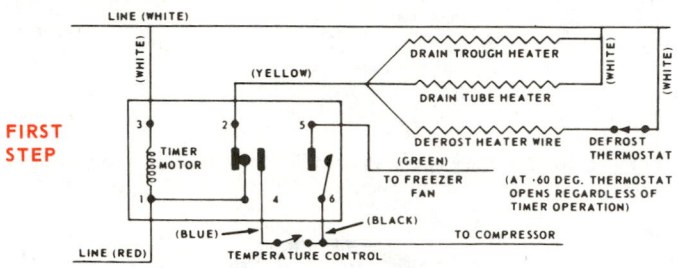

FIRST STEP

FIRST CLICK – DEFROST OPERATION (APPROXIMATELY 16 MIN.)

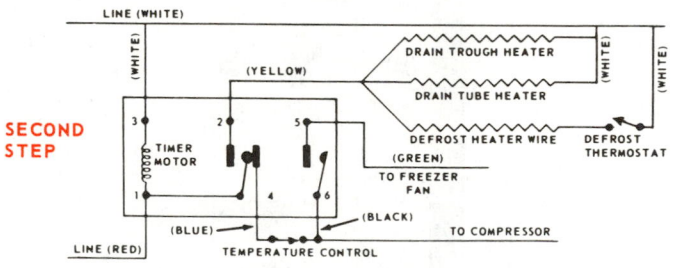

SECOND STEP

SECOND CLICK – FAN DELAY PERIOD (APPROXIMATELY 4 MIN.)

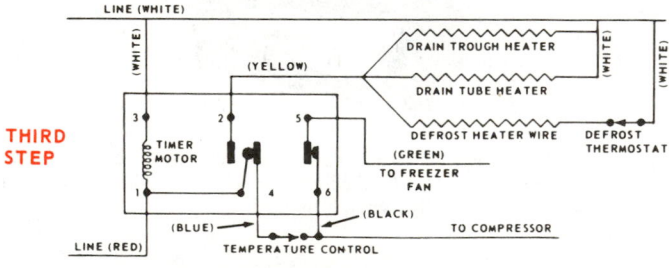

THIRD STEP

THIRD CLICK – NORMAL OPERATION (APPROXIMATELY 11 HRS. 40 MIN.)

Fig. 8-51. Electric circuits used on three-step defrost method. First step—Timer stops compressor and freezer fan, closes circuit to three heaters. Second step—Timer stops heater and starts compressor, but freezer fan does not start. Third step—Freezer fan circuit closed. (Gibson Appliance Co.)

Fig. 8-50 shows such a timer, while Fig. 8-51 shows its electrical circuits. The thermostat controlling the on-and-off circuit for the electric heater elements is shown in Fig. 8-52.

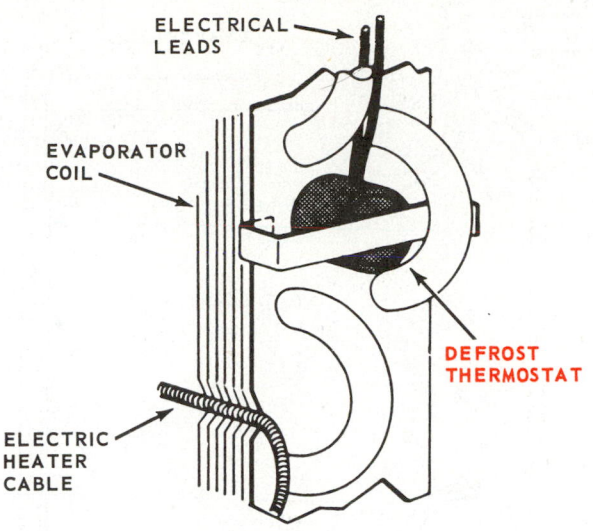

Fig. 8-52. Bimetal defrost thermostat. Control closes at 20 F. and opens at 50 F., during defrost.

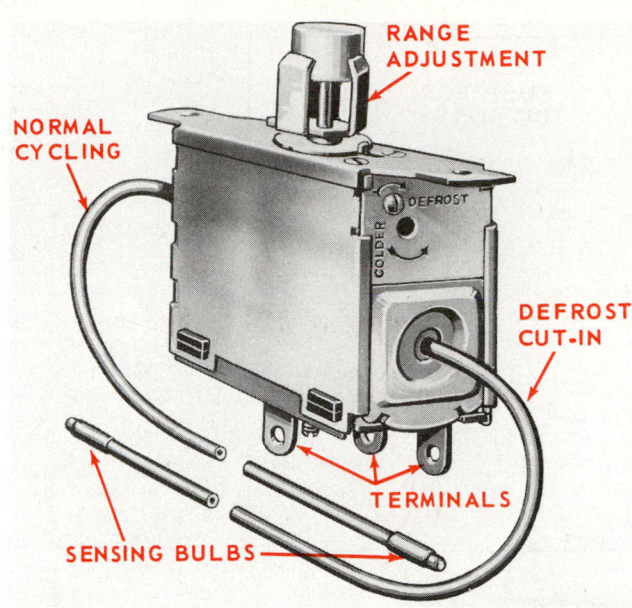

Fig. 8-54. Combination temperature and defrost control.

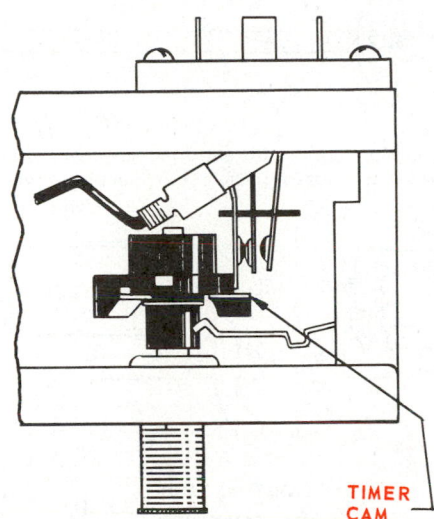

Fig. 8-53. Timer-operated cam which returns unit to normal cycling after about 45 minutes.

One control has a timer operated cam, as shown in Fig. 8-53. This turns the refrigerating circuit to normal operation after a certain time interval even though the thermostat may not call for cooling. This device prevents too long a defrost interval and also serves as a safety device.

8-30 SEMIAUTOMATIC DEFROST CONTROLS

Some domestic refrigerators use semiautomatic defrosting controls. These devices do two things:
1. Defrost the unit when the owner presses the button.
2. Return to regular operation automatically after the unit has defrosted.

A second system raises the range a fixed amount when a button is pressed. The evaporators will run at a temperature warm enough to permit defrosting but will still give satisfac-

tory refrigeration. Pulling on the button returns the system to regular operation.

Refrigerators with manual defrost frequently use a double capillary tube control, Fig. 8-54. This control has a power element for normal cycling and another element as a cut-in control for the defrost. Defrosting starts when the control knob is pushed in. This movement opens the motor circuit and closes the circuit to either a defrost solenoid (hot gas) or to electric heater elements.

When the coils are defrosted, the defrost capillary tube will create enough bellows pressure to open the defrost circuit and return the control connections to the motor circuit.

A wiring diagram of the control connected to a hot gas defrost system is shown in Fig. 8-55. The circuits used for an electric defrost system are shown in Fig. 8-56.

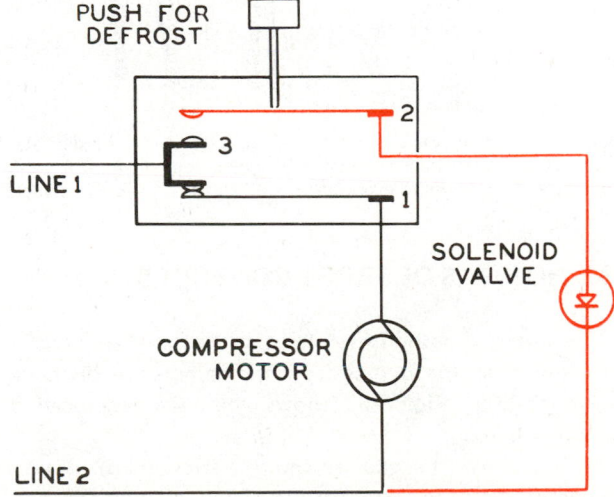

Fig. 8-55. Wiring diagram for "hot gas" defrosting device which uses combination control. Control is shown in refrigerating position. Defrost system is a solenoid valve controlled hot gas defrost.

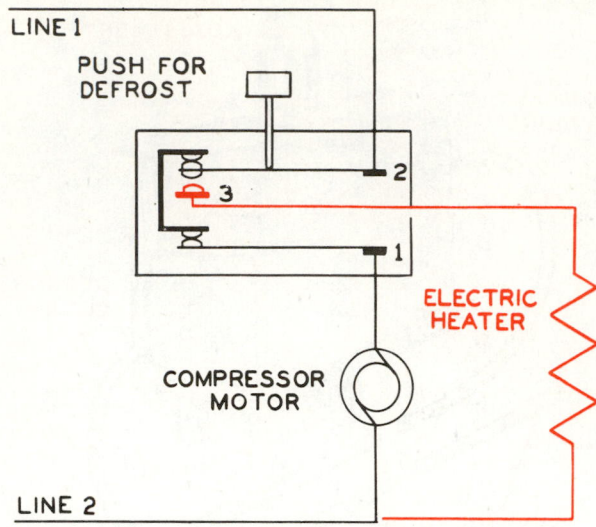

Fig. 8-56. Wiring diagram for electric heater defrost system using combination control. Control is shown in refrigerating position. Heater defrost circuit is shown in red.

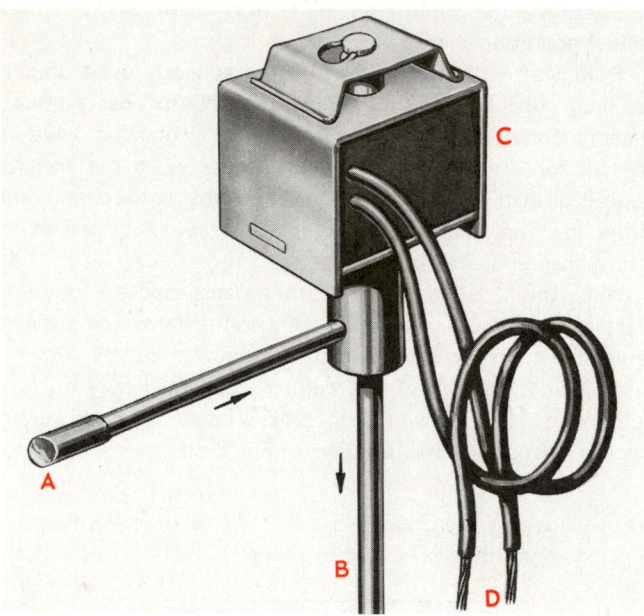

Fig. 8-57. Solenoid valve which may be used with thermostats for either hot gas defrosting or secondary system control. A—Hot gas from condenser. B—Hot gas to evaporator. C—Solenoid. D—Electrical leads.

8-31 HOT GAS DEFROST CONTROLS

The hot gas method of defrosting uses a solenoid valve to open and close the bypass from the compressor discharge to the evaporator. Fig. 8-57 shows the valve equipped with connector lines.

It is similar to the solenoid valves used to control refrigerant flow in secondary systems of two-temperature refrigerators. The valve must be mounted in a vertical position to work correctly.

Fig. 8-58 shows the solenoid valve installed in a hot gas

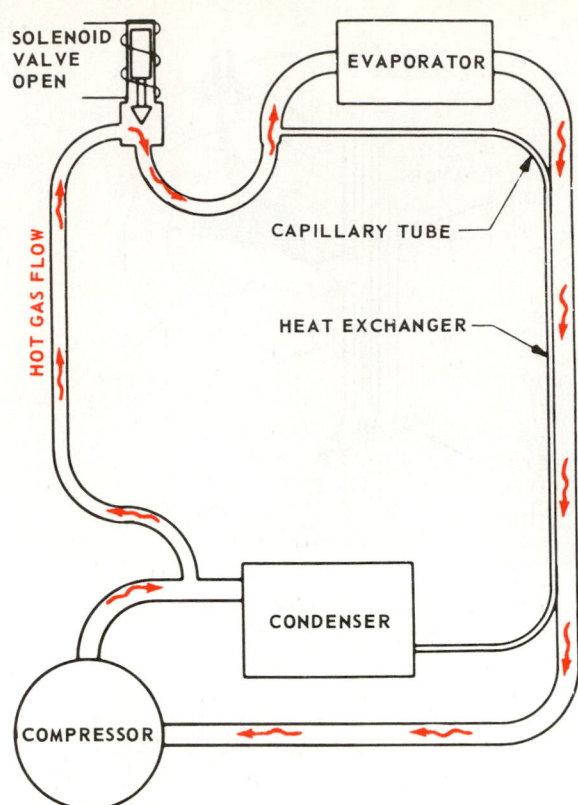

Fig. 8-58. A hot gas defrost system using solenoid valve. Illustration shows valve open and hot gas passing from compressor into evaporator.

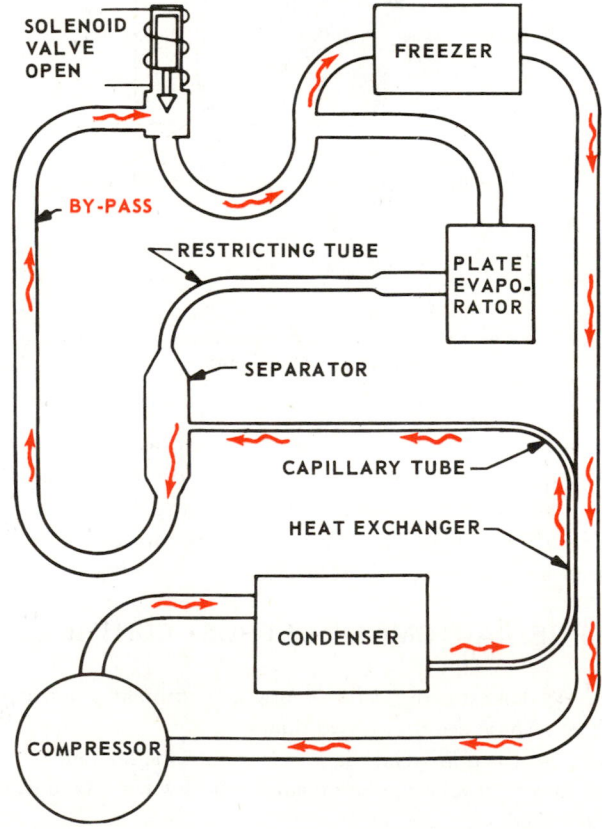

Fig. 8-59. Bypass system using solenoid valve. With valve open as shown, main refrigerating effect is in freezing compartment. With valve closed, plate evaporator and freezer evaporator will be refrigerated.

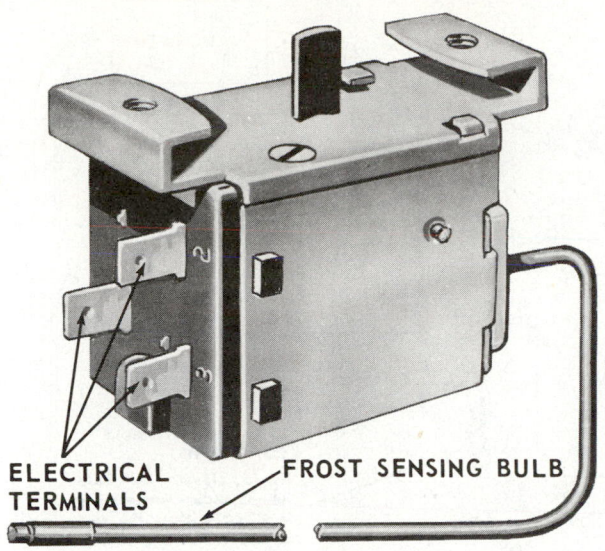

ELECTRICAL
TERMINALS ——— FROST SENSING BULB

Fig. 8-60. Defrost control used in addition to the regular temperature control. (Ranco, Inc.)

defrost system, while Fig. 8-59 is an installation in a bypass refrigeration system.

Some refrigerators use two separate controls. One control is the regular thermostat while the other control is the defrost control. A single control is shown in Fig. 8-60.

It can be used with either the hot gas or electric defrost systems. The control is designed with a vacuum in the sensitive element. If the bellows loses its charge, the pressure will rise and open the circuit (a fail safe control).

8-32 ICE BANK CONTROLS

Milk coolers and vending machines sometimes use a motor control which operates when ice collects on an evaporator. The control makes use of the difference in conductivity of electricity between ice and water. The thicker the ice, the greater the resistance to current flow. It is used on refrigerating machines which use a bank of ice as a refrigeration reserve

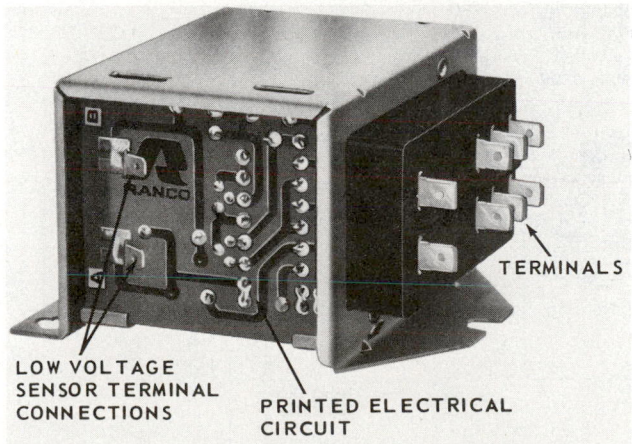

LOW VOLTAGE
SENSOR TERMINAL
CONNECTIONS   PRINTED ELECTRICAL
                      CIRCUIT

Fig. 8-61. Electronic ice bank control. Increased electrical resistance of ice compared to water causes control to start and stop compressor.

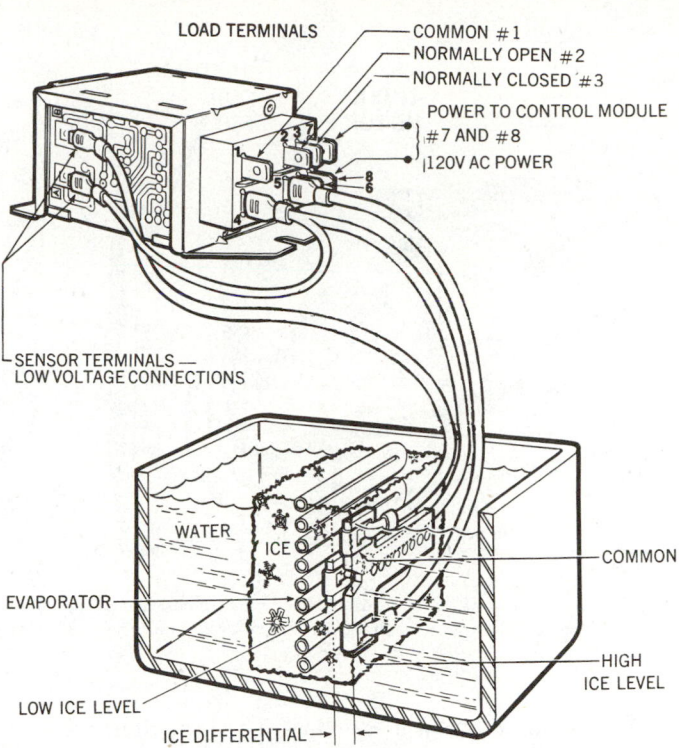

LOAD TERMINALS ——— COMMON #1
              ——— NORMALLY OPEN #2
              ——— NORMALLY CLOSED #3
              POWER TO CONTROL MODULE
              }#7 AND #8
              }120V AC POWER

SENSOR TERMINALS —
LOW VOLTAGE CONNECTIONS

WATER
ICE
EVAPORATOR ——— COMMON

LOW ICE LEVEL ——— HIGH ICE LEVEL

ICE DIFFERENTIAL →

Fig. 8-62. Circuit diagram for ice bank control.

where the service requirement fluctuates greatly (changes often).

Fig. 8-61 shows an ice sensitive element and the electronic control. The printed circuit can be seen on the side of the control. The control is shown connected into the electrical circuit in Fig. 8-62.

The sensor element is mounted on the evaporator. When this part of the evaporator freezes, the water on the sensor element plate (probe) freezes between the two electrodes, as shown in Fig. 8-63. The change (increase) in electrical resistance of the ice reduces the current flow to the control (probe assembly) and the circuit to the motor compressor is opened. When the ice melts on the sensor element, the increased current flow will operate the control, the motor compressor circuit will close and the unit will start.

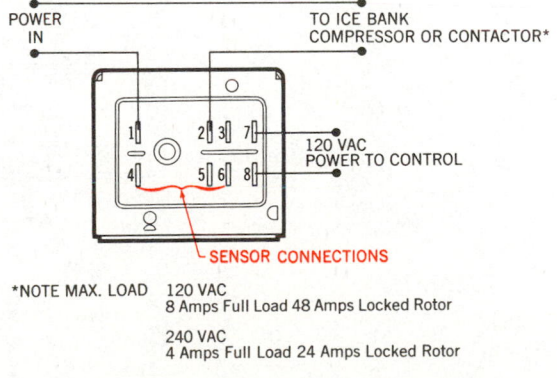

POWER IN   TO ICE BANK
                        COMPRESSOR OR CONTACTOR*

120 VAC
POWER TO CONTROL

SENSOR CONNECTIONS

*NOTE MAX. LOAD   120 VAC
                 8 Amps Full Load 48 Amps Locked Rotor

                 240 VAC
                 4 Amps Full Load 24 Amps Locked Rotor

Fig. 8-63. Ice bank control wiring diagram.

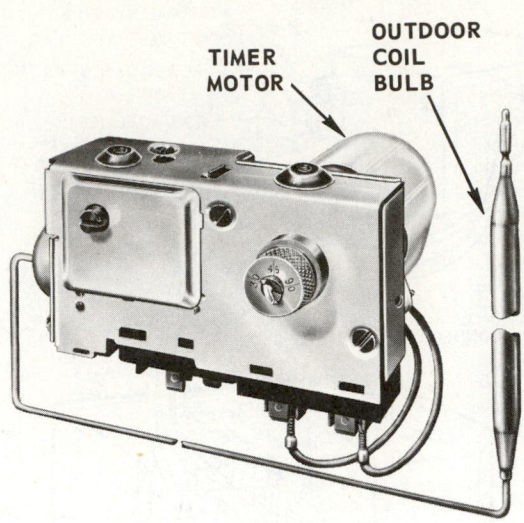

TIMER
MOTOR

OUTDOOR
COIL
BULB

Fig. 8-64. Heat pump de-ice control. Control reverses heat pump cycle and defrosts outdoor coil each 30 to 90 minutes if outdoor coil is at 26 F. or below.

The sensor probe is low-voltage a-c electricity, obtained from a transformer built into the control box.

8-33 DE-ICE CONTROLS

Air-to-air heat pumps which have an outdoor coil sometimes have an icing problem. Ice accumulates on the outdoor

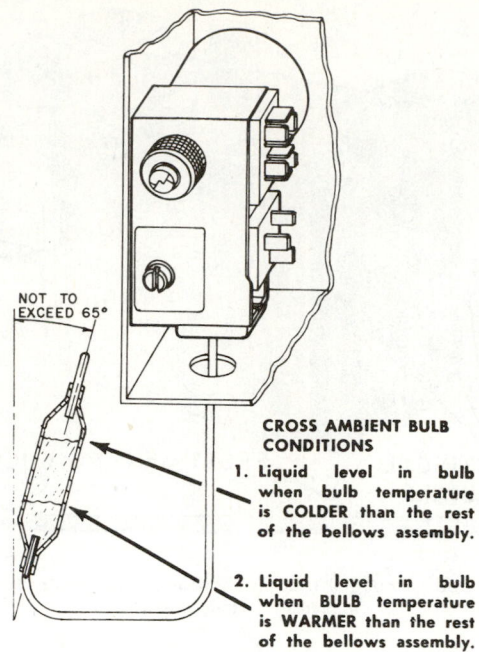

NOT TO
EXCEED 65°

CROSS AMBIENT BULB CONDITIONS

1. **Liquid level in bulb when bulb temperature is COLDER than the rest of the bellows assembly.**

2. **Liquid level in bulb when BULB temperature is WARMER than the rest of the bellows assembly.**

Fig. 8-65. De-icer sensitive bulb operation. Note how the temperature sensitive bulb is positioned.

coil during the cold season when the heat pump is used as a heating unit. This ice reduces the heat flow into the outdoor coil and it also tends to block the airflow through the coil.

A special de-ice control is used to prevent ice accumulation

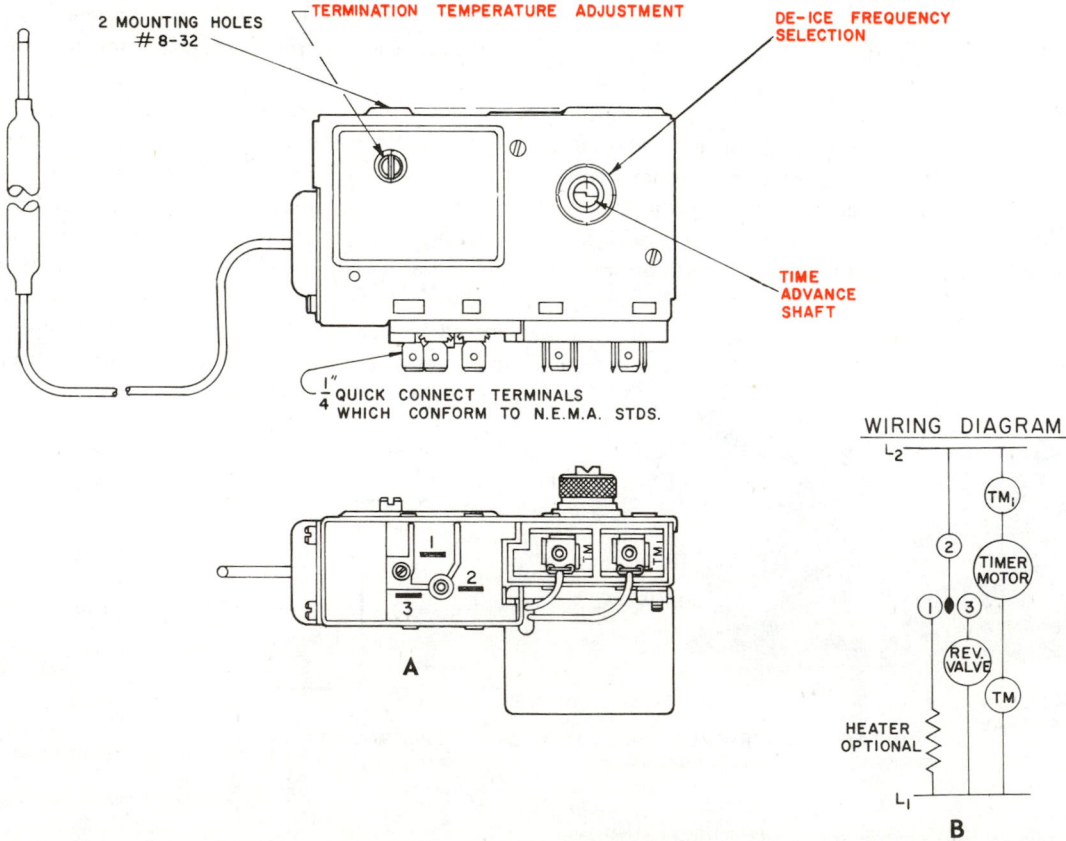

2 MOUNTING HOLES # 8-32

TERMINATION TEMPERATURE ADJUSTMENT

DE-ICE FREQUENCY SELECTION

TIME ADVANCE SHAFT

¼" QUICK CONNECT TERMINALS WHICH CONFORM TO N.E.M.A. STDS.

A

WIRING DIAGRAM

L₂

TM₁

TIMER MOTOR

REV. VALVE

TM

HEATER OPTIONAL

L₁

B

Fig. 8-66. De-icer control. Mechanism of control and its adjustments are shown at left. Wiring diagram is shown at right.

(buildup). The heat pump de-ice control, shown in Fig. 8-64, is a combination timer and thermostat.

The timer is adjustable for a coil defrost cycle of 30, 45 or 90 minutes. The sensitive bulb is cross charged to insure constant bulb control. This control permits the defrost cycle only if the coil is at 26 F. (—3 C.) or below. If above this temperature, the defrost cycle is skipped until the next defrost interval. The sensitive bulb is usually mounted at the place where the ice last melts from the coil.

Fig. 8-65 shows the sensitive bulb details. The temperature cut-in adjustment, the dimensions and the wiring diagram are shown in Fig. 8-66. Note that the defrost cycle reverses the heat pump and the outdoor coil temporarily acts as a condenser during the defrost cycle.

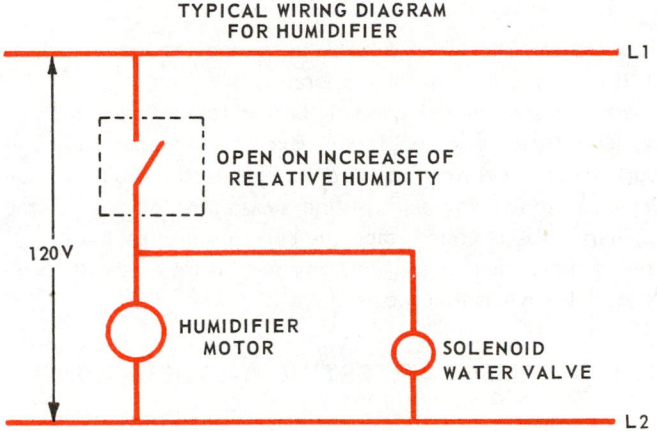

Fig. 8-67. Schematic wiring diagram of humidifier circuit. Control is adjustable to maintain relative humidity between 20 and 80 percent.

8-34 HUMIDITY CONTROLS

Humidity controls are used to control humidifiers and dehumidifiers.

In humidifiers the humidistat (see diagram, Fig. 8-67) closes the circuit when humidity drops or decreases. Chapter 20

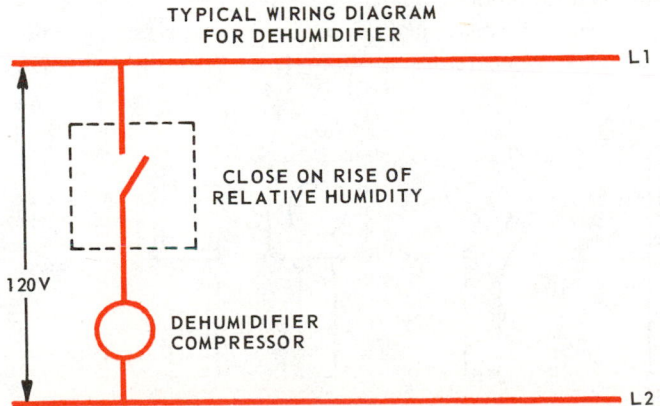

Fig. 8-68. Schematic wiring diagram of dehumidifier circuit. Control is adjustable between 20 and 90 percent relative humidity.

explains the design and operation of humidifiers.

For dehumidifiers, the humidistat (see diagram, Fig. 8-68) closes the circuit when the humidity increases. Chapter 21 explains in more detail.

8-35 DEFROSTING CLOCKS

To ease the burden of the user and to permit more efficient operation of the refrigerating unit, several companies have made defrosting clocks standard equipment on their domestic refrigerators (Para. 8-29).

Defrosting clocks (Fig. 8-69) are available for any make of refrigerator. Also called defrost timers, such clocks may be used on both hot gas defrost systems and with electric heating defrost system.

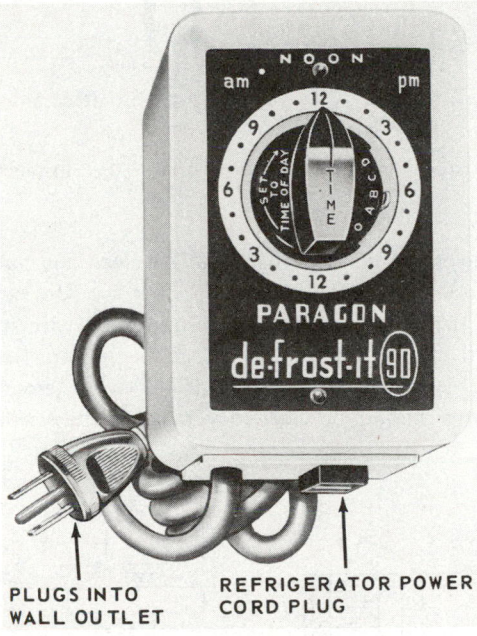

Fig. 8-69. Defrosting clock may be plugged into power circuit. Refrigerator power cord is connected to clock. It may be set to defrost the refrigerator at any selected time. Clock has a 24-hour dial. Time and length of defrost may be selected.

8-36 SPECIAL THERMOSTATS

Fig. 8-70 shows a special thermostat using a mercury-filled thermometer. Two electrodes are fused through the glass and contact the mercury column. When the mercury column extends from one electrode to the other, a very small electric current will flow. This current operates relays or transistors which, in turn, connect power to operate the refrigerating unit or solenoid valves. This control has the special ability to maintain extremely small temperature differences. Differentials as small as .009 F. are possible. Such temperature control is useful in research and laboratory work.

8-37 INSTALLING THERMOSTATS

Thermostats must be correctly installed or the system will not operate accurately or regularly.

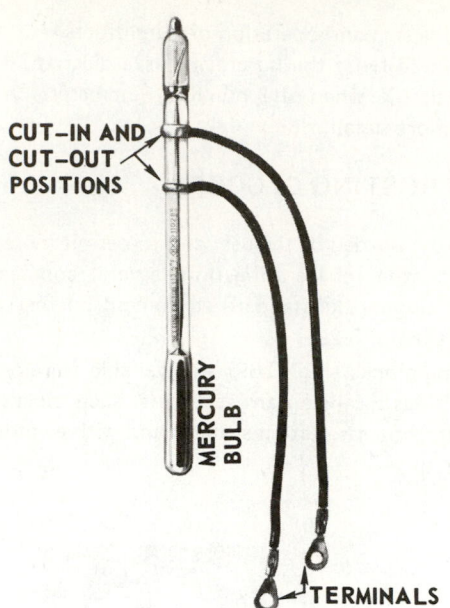

Fig. 8-70. Thermometer type thermostat. (H-B Instrument Co.)

The electrical connections must be clean and tight. Wrapping stranded wires around a terminal screw does not make a good or permanent connection. Strands of wire may work loose and ground the wire or short the terminals. Use only U-shaped (spade) or O-shaped (ring) wire terminals. The connections should be cleaned with clean steel wool before

installing wires.

Many types of terminals have been developed to do this job. The terminal must be large enough to carry the current used in the wire (lead).

Some terminals are connected to posts which either have screws or threaded studs with a nut and washer. See Fig. 8-71. The wire (lead) terminal may be one of several types. Some are designed for easy removal. See Fig. 8-72.

To help service technicians remove and replace wire leads quicker, quick disconnect terminals have been developed and are becoming very popular. These are made in many designs. They must be clean and tight.

The sensing bulb of the thermostat must be very carefully mounted. It should be attached tightly to the evaporator or the tubing. The sensing bulb and the place to which it is clamped should be cleaned with clean steel wool before assembly.

The best place for attaching the thermostat sensing bulb is on the last one-third of the evaporator.

When installing the control, be careful not to bend the capillary tube back and forth. This small copper tube will work harden and may break. Also, be sure the capillary tube does not touch any part of the evaporator. If any of the capillary tube is coiled, tape the coil to prevent vibration. If any of the tubing rubs against any part, it may wear through or work harden at that spot and crack.

8-38 THERMOSTAT TESTING AND SERVICING

A mixture of crushed ice and water may be used to check and adjust a refrigerator thermostat. With these materials, it is easy to determine the operation of the thermostat at 32 F. (0 C.), since this is the temperature of melting ice.

To use this method, place the thermostat control bulb in the ice and water mixture. It is well to use a thermometer when making this test. With the control bulb in the ice and water mixture, set the control dial temperature reading to 32 F. (0 C.). The control contact points should be open. After a few minutes, lift the control bulb from the mixture; and, as the control bulb warms, the points should again close. Use a test cord with a small light bulb to check the opening and closing of the contact points.

To check further, place the control bulb in water and adjust the water temperature to 45 F. (7 C.). Set the thermostat

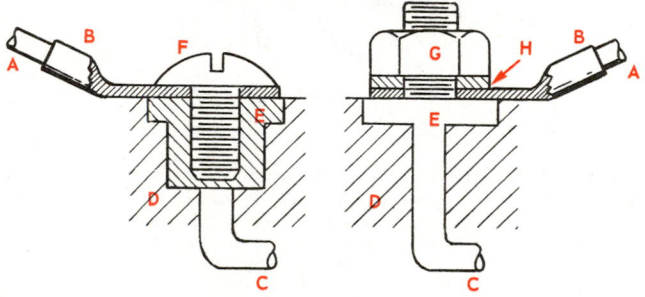

Fig. 8-71. Two types of solderless electrical terminal assemblies. Screw type is at left, nut type at right. A—Wire (lead). B—Wire terminal. C—Wire (lead) to inside of control. D—Insulation. E—Terminal insulated block. F—Machine screw (round head). G—Nut. H—Washer.

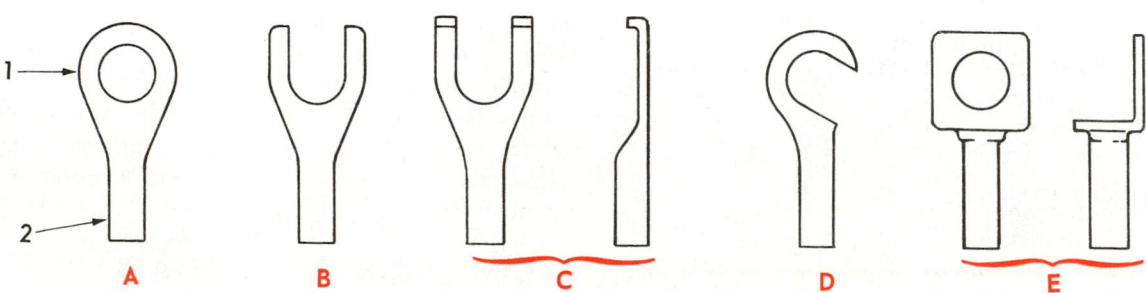

Fig. 8-72. Several different types of terminals which are used to connect leads (wires) to terminal posts. A—Ring. B—Spade. C—Flanged spade. D—Hook. E—Flag. 1—Tongue of terminal. 2—Barrel of terminal.

control to 45 F. At this setting, the points should be open.

Lift the control bulb out of the water, let it warm up for a few minutes and the points should close. If the points do not open and close properly, the thermostat should be replaced.

Freezer cabinet temperatures are usually in the 0 F. to −20 F. range. Ice and water mixtures cannot be used in setting these thermostats. A thermostatic control tester which uses the refrigerant R-12 or R-22 is recommended. Fig. 8-73 illustrates one of these instruments.

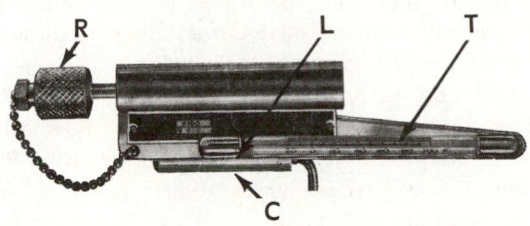

Fig. 8-73. Thermostat testing and adjusting instrument. T—Accurate thermometer. C—Clamp for holding the control bulb. R—Fitting for attaching a refrigerant cylinder. L—Shallow trough for liquid refrigerant.

In operation, the freezer thermostat control bulb is placed in the tester and refrigerant is allowed to flow into the control pressure chamber. By adjusting the refrigerant pressure, it is possible to obtain any desired temperature. Freezer thermostats have a predetermined cut-in and cut-out temperature much the same as refrigerators. However, the freezer temperatures are much lower. Also, the differential is usually a little greater — 10 to 12 degrees.

To test a freezer thermostat, with the freezer plugged in and operating, place the control bulb in the control tester. Adjust the refrigerant flow and pressure until the desired cut-out temperature is indicated on the thermometer. If the freezer control cuts out before the desired cut-out temperature is reached, the control may be adjusted to cut out at a lower temperature.

Leave the freezer connected and open the cabinet door. Adjust the control tester to the desired cut-in temperature. If the thermostat does not cut in at the desired temperature, an adjustment should be made to the thermostat. If it is not possible to get a satisfactory thermostat adjustment, the thermostat should be replaced.

To check completely the operation of the freezer thermostat, it is recommended that a recording thermometer be used to chart the temperature and time for the freezer over a 24-hour period.

Several kinds of trouble may be encountered in thermostats:
1. Corrosion occurring at the contact points causing a poor electrical circuit. Use a test light to check electrical circuits. Once contact points become worn and cause trouble, it is best to replace them. Any repair would only be temporary and a repeat call (call back) would soon be necessary.

Contact points must be clean. Corroded or pitted points (best detected by using an ohmmeter) usually should not be repaired. The control should be replaced. In an emergency, the points can be cleaned with a small, clean file. A single cut or mill file is best, but even a clean nail file may do a temporary job.

Thermostats must have good electrical connections. They must also be adjusted to correct temperatures and the sensitive element must accurately sense the temperature of the evaporator or the cabinet temperature.

Electrical connections must be clean and tight. Only metal terminals should be used. Clean the terminals, the terminal jacks or the screw posts with clean steel wool.
2. The overload protection devices may also have dirty contact points and low current flow may result.
3. The power element and bellows may lose its charge. This may be detected by a simple check. If the charge is lost, the bellows are very easily compressed. If the bellows were charged, the pressure inside would probably be around 75 psi (5.3 kg/cm^2) or more and finger pressure would not affect it. In the event of leakage, replace this part of the control or the complete control.
4. Frequently the control bulb is not attached tightly to the evaporator, requiring a great change of temperature range before the motor will cut in and cut out.

The power element must be firmly clamped to the evaporator. Good thermal contact must be obtained between the thermal bulb and the evaporator. Many evaporators have metal sockets into which the thermal elements are inserted. The contact surfaces must be clean.

To adjust to the correct temperature settings, mount a thermometer at the sensing bulb, then cycle the unit. It is difficult to obtain accurate settings until the unit cycles several times. Time may be saved by using a thermal bath or a thermostat adjusting tool.

The method of checking and adjusting thermostats is explained in Chapter 11.

8-39 TEMPERATURE ALARM SYSTEM

In some installations, such as food freezers, an electrical alarm system will be sounded if the temperature in the cabinet rises above an upper safe limit. These systems sometimes operate from the electrical circuit powering the compressor. Others are provided with a dry cell arrangement. With the dry cell arrangement, if the dry cells are in good condition, the alarm will work even though the electrical power supply may be interrupted or fail.

8-40 LOW-SIDE PRESSURE LIMITER

A popular refrigerator and air conditioning safety device incorporates a low-side pressure limiter. The condensing unit cannot be overloaded if the low-side pressure is maintained at a low enough limit.

Low-side pressure limiters consist of a pressure sensitive element such as a diaphragm or bellows placed in series with a condensing unit circuit. The electrical circuit will be open if the low-side pressure is higher than a desired limit. Some low-side pressure limiters operate through a relay. The low-side pressure operates the relay and the relay, in turn, controls the electrical circuit.

8-41 FAN CONTROLS

In some refrigerating systems, a fan circulates the cold air through the refrigerator. The fan control may be an on-and-off control; or, in some applications, it may be a solid state fan speed control.

8-42 GROUNDING

See Para. 6-66 which explains the requirements for grounding the various parts of refrigerator mechanisms. To avoid any possibility of electrical shock, refrigerators should always be properly grounded. Fig. 8-74 illustrates a typical refrigerator mechanism and the required grounds.

8-43 REVIEW OF SAFETY

Anything in motion, anything which holds back a pressure, anything which can conduct electricity or heat, anything which is rough or sharp and anything which can be dropped is a potential safety hazard.

Most accidents are the result of carelessness. When one is concentrating on a job, or on getting a job done, one tends to, momentarily, neglect safety.

Therefore, service technicians must train themselves to do things safely. They must study the job for its safety problems and their solutions before starting the job.

Think about the safety aspects before each step of a job.

Always disconnect the electrical power and make sure no one can turn it on while working on the electrical parts of a refrigerator or air conditioning system. Replace worn electrical wires or wires which have brittle insulation (brittle insulation cracks when the wire is bent into a loop).

Use only screwdrivers with insulated handles (wood or plastic). Use only wrenches and pliers with insulated handles. This habit is double insurance against shocks.

If one must work in a damp or wet room — stand on a dry, insulated platform.

Always use instruments to check a circuit to see if it is electrically charged before handling wires, terminals or parts.

The electrical resistance of the human body is too low to provide any great protection against the flow of electrical current. In general, low voltages up to 12 or 15 volts are not high enough to allow any appreciable amount of current to flow through the human body. Current flowing through the human body does the damage, not the voltage.

If the skin on the hands is dry, the resistance will be high enough so that relatively little current will flow through. However, if the skin is wet, current flows quite easily. Rubber gloves and rubber-soled shoes or rubber boots or rubbers provide considerable protection.

Most of the resistance to electrical flow is on the skin or surface of the skin. The human body is made up of many fluids which have fairly high electrical conductivity. When a service technician is working in a warm place where his body is wet with perspiration, he is much more likely to receive an electrical shock than if his body were dry. Many technicians carry a 3 ft. by 3 ft. rubber mat to stand on when doing electrical service work.

Always short circuit any capacitors in the circuits before working on them. This will discharge them.

Water pipes, radiators, heating plants and the like are grounded. If the "live" wire (black) comes in contact with one of these grounds, either a circuit breaker or a fuse will be blown. A person touching both a live wire and a ground will receive a very severe shock.

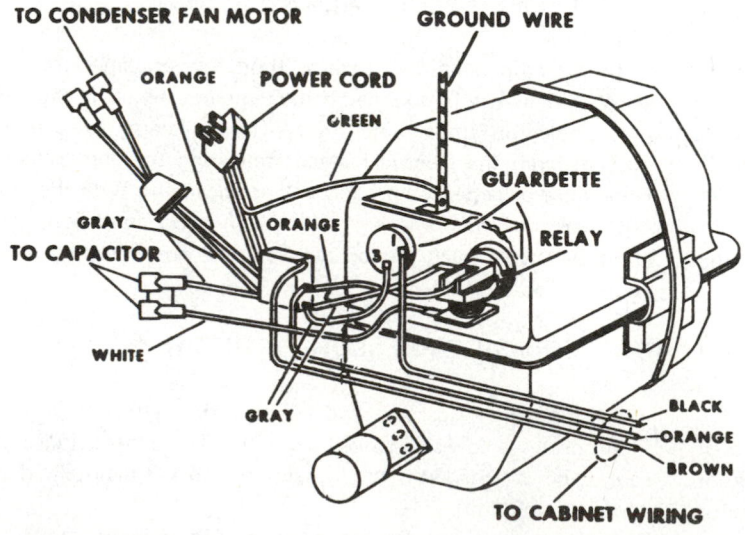

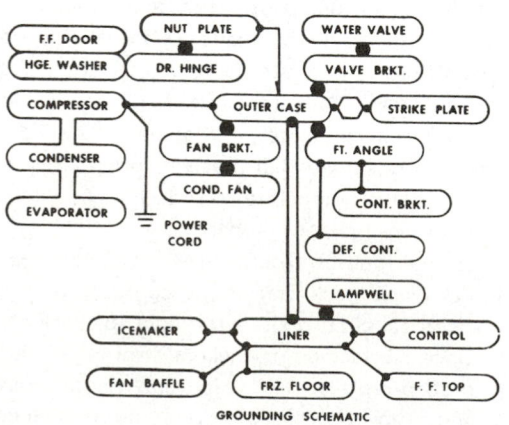

Fig. 8-74. All parts that could cause electrical shock are grounded. Note the green ground wire of the power cord and the ground wire connected to the motor compressor. (General Electric Co.)

8-44 TEST YOUR KNOWLEDGE

1. Can a motor control be operated by low-side pressure? High-pressure? Both?
2. In what ways are pressure and thermostatic types of motor controls essentially similar?
3. What two purposes does a motor control serve?
4. How many types of differential adjustments are there?
5. What are the thermostatic motor control bulbs charged with?
6. Why are starting relays used in connection with most hermetic motor controls?
7. What does the range adjustment control?
8. What does the differential adjustment control?
9. What devices are built into motor controls to protect the motor from using too much current?
10. Will turning the cut-in differential adjustment affect the cut-out temperature?
11. Will turning the double type differential adjustment affect the cut-out temperature?
12. What is the purpose of a timer on a domestic refrigerator?
13. How many types of relays are in use?
14. How may a thermostat be checked quickly?
15. How does a bimetal disk respond to a temperature change?
16. What is voltage drop?
17. Why does a loose or corroded connection in an electrical circuit become warm?
18. Does a hot-wire relay have to be mounted level? Why or why not?
19. Are current relay contact points open or closed during the off part of the cycle?
20. Why do electrical wires have different colored insulations?
21. Is the current flow more when the motor is starting or when it is running?
22. Is the voltage drop more when the motor is starting or when it is running?
23. How does altitude affect a diaphragm or bellows type thermostat?
24. Why are some thermostat bodies heated?
25. Are potential relay contact points open or closed during the off cycle?
26. Are current relay contact points open or closed when the system is running?
27. What size wires are most commonly used on domestic refrigerators?
28. What type system uses a de-ice control?
29. Is there a difference in electrical resistance between water and ice?
30. Can a test light be used to test continuity?

Chapter 9

REFRIGERANTS

A heat carrier must be used to move heat from the interior of a cabinet or room to the outside. In a standard mechanical cooling system, evaporating liquid refrigerant in the evaporator removes heat. The vapor absorbs the heat; then it moves to the condenser which removes this heat from the refrigerant.

Fluids are used which can be changed easily from a liquid to a vapor, and then from a vapor back to a liquid again. Some fluids are better suited to this purpose than others. The most common ones will be discussed in this chapter.

Metric equivalent values are listed for only a few of the refrigerants because the United States has not confirmed which of the SI units will be used.

9-1 REQUIREMENTS FOR REFRIGERANTS

Fluid used as a refrigerant should have certain properties:
1. It should be nonpoisonous.
2. It should be nonexplosive.
3. It should be noncorrosive.
4. It must be nonflammable.
5. Leaks should be easy to detect.
6. Leaks should be easy to locate.
7. It should operate under low pressure (low boiling-point).
8. It should be a stable gas.
9. Parts moving in the fluid should be easy to lubricate.
10. It should be nontoxic (not harmful if inhaled or if spilled on skin).
11. It should have a high liquid volume per pound to provide durable refrigerant controls.
12. It should have a high latent heat per pound to produce good cooling effect per pound of vapor pumped.
13. It should have a low vapor volume per pound. This will reduce compressor displacement needed.
14. The pressure difference between evaporating pressure and condensing pressure should be as little as possible to increase pumping efficiency.

It is desirable to keep normal pressures in the refrigerator as close to atmospheric pressure as possible, because excessive differences may cause leaks, overwork the compressor and decrease the efficiency of the valves.

The standard comparison of refrigerants, as used in the refrigeration industry, is based on an evaporating temperature of 5 F. (−15 C.) and a condensing temperature of 86 F. (30 C.). In this chapter, each refrigerant discussed is compared on this basis.

9-2 IDENTIFYING REFRIGERANTS BY NUMBER

Refrigerants are identified by number. One should, therefore, become familiar with refrigerant numbers, as well as with the names.

The number follows the letter R, which means refrigerant.

REFRIGERANT NO.	NAME AND CHEMICAL FORMULA
R-11	Trichloromonofluoromethane CCl_3F
R-12	Dichlorodifluoromethane CCl_2F_2
R-22	Monochlorodifluoromethane $CHClF_2$
R-500	Azeotropic mixture of 73.8% of (R-12) and 26.2% of (R-152a)
R-502	Azeotropic mixture of 48.8% of (R-22) and 51.2% of (R-115)
R-503	Azeotropic mixture of 40.1% of (R-23) and 59.9% of (R-13)
R-504	Azeotropic mixture of 48.2% of (R-32) and 51.8% of (R-115)
R-717	Ammonia NH_3

Fig. 9-1. The most commonly used refrigerants.

The identifying system of numbering has been standardized by the American Society of Heating, Refrigerating and Air Conditioning Engineers (ASHRAE).

Some refrigerants in common use are shown in Fig. 9-1. See Chapter 28 for a more complete list.

9-2A STANDARD EVAPORATOR AND CONDENSER TEMPERATURES

In domestic refrigerators, the service technician usually adjusts the controls using a temperature of −15°C (5 F.) in the evaporator. Also, the controls are adjusted to give a standard condensing temperature of 30°C (86 F). See Fig. 9-1A.

9-2B USE OF PRESSURE-TEMPERATURE CURVES

The pressure-temperature curves shown in Fig. 9-2 (and in Fig. 9-17) show the refrigerants in a state of equilibrium. (See Fig. 1-25, view A.) The vertical scale is the temperature in degrees Fahrenheit. The horizontal scale is pressure in psi.

To find the pressure of the refrigerant at any particular temperature, read horizontally from the temperature reading until the curve of the particular refrigerant is reached, then move directly down to the pressure reading.

For example, the vapor-pressure of R-12 refrigerant at a temperature of 100 F. is 116.9 psi (38°C is 8.2 kg/cm²). The temperature is always the temperature of the refrigerant. The same curve may be used for determining both the condensing and evaporating temperatures and pressures. The condensing values for both temperature and pressure are higher than the equilibrium state.

When using this chart, several things must be kept in mind:
1. The temperature of the refrigerant in the evaporator is about 8 F. to 12 F. (4°C to 7°C) colder than the evaporator when the compressor is running.
2. The temperature of the refrigerant in the evaporator is the same as the evaporator temperature when the compressor is not running.
3. The temperature of the refrigerant in an air-cooled condenser is approximately 30 F. to 35 F. (17°C to 19°C) warmer than the room temperature.
4. The temperature of the refrigerant in a water-cooled con-

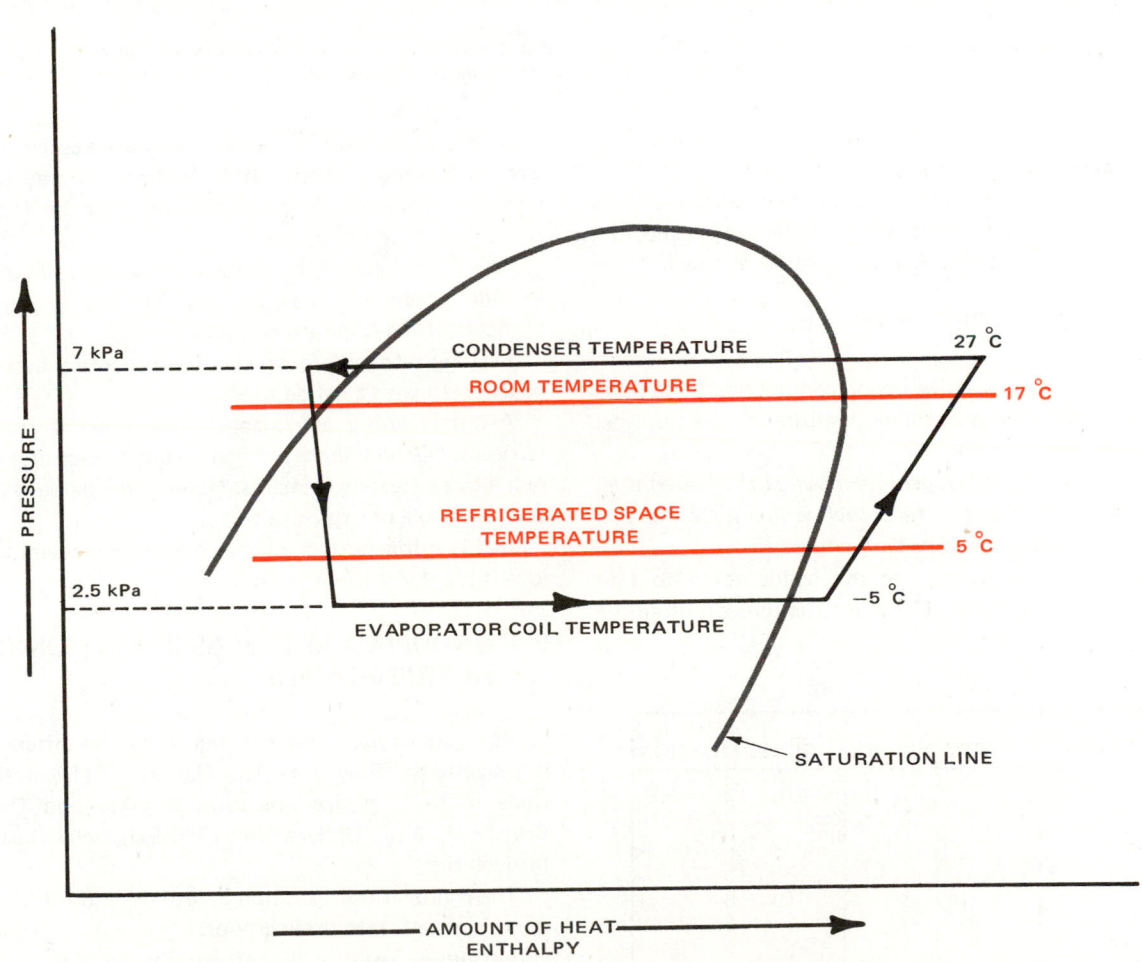

Fig. 9-1A. A typical refrigeration cycle with temperatures of one evaporator and condenser shown in comparison to the room and refrigerated space temperatures.

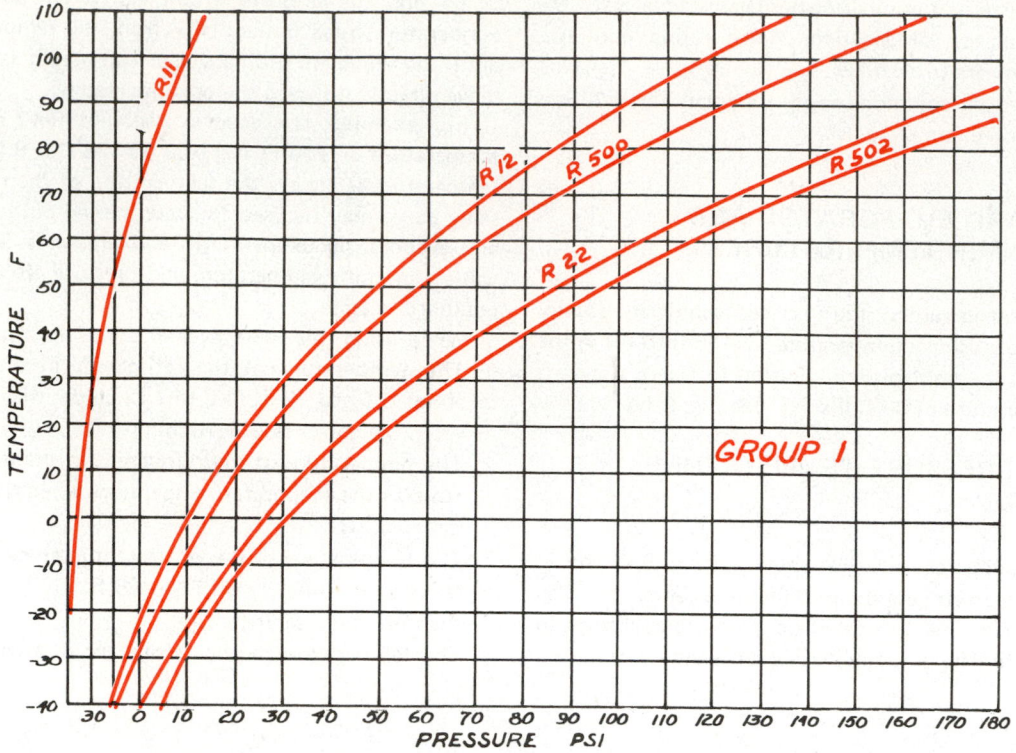

Fig. 9-2. Pressure-temperature curves for some popular Group One refrigerants. R-11 is known as a low-pressure refrigerant and R-502 a medium-pressure refrigerant.

denser is approximately 20 F. (11°C) warmer than the water temperature at the drain outlet.

5. The temperature of the refrigerant in the condenser will be about the same as the temperature of the cooling medium after the unit has been shut off for 15 to 30 minutes.

9-3 HALIDE REFRIGERANTS

Many refrigerants are made up of compounds containing one or more of the halogens. Halogen substances are fluorine, chlorine, iodine and bromine.

It is a property of the halogen compounds that when they are exposed to a hot copper surface, such as in a torch flame, a bright green color is produced in the flame.

This makes it possible to use the halide torch to find refrigerant leak. See Chapter 11 for instructions on using the halide torch.

REFRIGERANT	NRSC GROUP	NBFU CLASSIFICATION
R—11	1	6
R—12	1	6
R—22	1	5
R—500	1	6
R—502	1	6
R—503	1	6
R—504	1	6
R—744	1	5
R—717	2	2

Fig. 9-3. Popular refrigerants listed by the National Refrigeration Safety Code and National Board of Fire Underwriters.

In the presence of a flame or an incandescent (bright red) electric heating element, these compounds may break down into their elements. A sharp, pungent odor develops when the compound decomposes.

This is the basis of the halide leak detector. These elements, in large quantities, may be very harmful to human tissue, particularly the respiratory system.

Leaking refrigerants which do not contain halogen cannot be detected with a halide torch.

Avoid releasing any sizable quantities of these halogen refrigerants where there are flames and/or incandescent (bright red) electric heating elements. None of the halogen refrigerants are flammable or explosive.

Halide refrigerants are heavier than air and will settle to the lowest place in a room or container.

9-4 GROUPING AND CLASSIFICATION OF REFRIGERANTS

Refrigerants have been catalogued by two different national organizations. They are: The National Refrigeration Safety Code (NRSC), groups one through three, and The National Board of Fire Underwriters (NBFU), classifications one through six.

The National Refrigeration Safety Code (NRSC) catalogs all the refrigerants into three groups:

Group One — Safest of the refrigerants.

Group Two — Toxic and somewhat flammable refrigerants.

Group Three — Flammable refrigerants.

The National Board of Fire Underwriters (NBFU) classifies

refrigerants mainly on their degree of toxicity. (Toxic means poisonous or injurious to health.) There are six classifications in this scale. Class One is the most toxic, while Class Six is the least toxic. Fig. 9-3 shows the grouping and classification of some common refrigerants.

9-5 GROUP ONE REFRIGERANTS

The characteristics of the most-used Group One refrigerants are explained in Para. 9-6 through 9-13.

(The pressure-temperature curves for five of the common Group One refrigerants are shown in Fig. 9-2.)

The refrigerants in this group may be used in the greatest quantities in any installation. The allowable quantities are specified by the "American Standard Safety Code for Mechanical Refrigeration." The amounts are:

1. Up to 20 lb. in hospital kitchens.
2. Up to 50 lb. (indirect system) in public assemblies.
3. Up to 50 lb. in residential use if precautions are taken.
4. Up to 20 lb. in residential air conditioning systems.

Some refrigerants in Group One are:

R-11   Trichloromonofluoromethane CCl_3F
R-12   Dichlorodifluoromethane CCl_2F_2
R-22   Monochlorodifluoromethane $CHClF_2$
R-500   73.8 percent R-12 and 26.2 percent R-152a
R-502   48.8 percent R-22 and 51.2 percent R-115
R-503   41.1 percent R-23 and 59.9 percent R-13
R-504   48.2 percent R-32 and 51.8 percent R-115
R-744   Carbon Dioxide CO_2

9-6 R-12 DICHLORODIFLUOROMETHANE (CCl_2F_2)

R-12 is a very popular refrigerant. It is a colorless, almost odorless liquid with a boiling point of −21.7 F. (−29 C.) at atmospheric pressure. It is nontoxic, noncorrosive, non-irritating and nonflammable.

Chemically, it is inert at ordinary temperatures and thermally stable to above 800 F. (427 C.). This temperature is well above the safe operating temperatures of most refrigerating mechanism materials and lubricants. A table of properties of R-12 is shown in Fig. 9-4.

R-12 has a relatively low latent heat value. In the smaller refrigerating machines, this is an advantage. The large amount of refrigerant circulated will permit the use of less sensitive and more positive operating and regulating mechanisms. It is used in reciprocating, rotary and large centrifugal compressors. It operates at a low but positive head and back pressure and with a good volumetric efficiency.

R-12 has a pressure of 26.5 psia or 11.8 psi (.830 kg/cm^2) at 5 F. (−15 C.), and a pressure of 108.0 psia, 93.3 psi (6.56 kg/cm^2) at 86 F. (30 C.). The latent heat of R-12 at 5 F. (−15 C.) is 68.2 Btu/lb. (159 J/g).

An R-12 leak may be detected by several means:

1. A soap solution.
2. A halide lamp.
3. Colored oil added to the system.
4. An electronic leak detector.

See Chapter 11 about the use of leak detectors.

Temp. F.	PRESSURE Psia	PRESSURE Psig	VOLUME VAPOR Cu. Ft./Lb.	DENSITY LIQUID Lb./Cu. Ft.	HEAT CONTENT BTU/LB. Liquid	HEAT CONTENT BTU/LB. Vapor
−150	0.154	29.61*	178.65	104.36	−22.70	60.8
−125	0.516	28.67*	57.28	102.29	−17.59	63.5
−100	1.428	27.01*	22.16	100.15	−12.47	66.2
− 75	3.388	23.02*	9.92	97.93	− 7.31	69.0
− 50	7.117	15.43*	4.97	95.62	− 2.10	71.8
− 25	13.556	2.32*	2.73	93.20	3.17	74.56
− 15	17.141	2.45	2.19	92.20	5.30	75.65
− 10	19.189	4.49	1.97	91.69	6.37	76.2
− 5	21.422	6.73	1.78	91.18	7.44	76.73
0	23.849	9.15	1.61	90.66	8.52	77.27
5	26.483	11.79	1.46	90.14	9.60	77.80
10	29.335	14.64	1.32	89.61	10.68	78.335
25	39.310	24.61	1.00	87.98	13.96	79.9
50	61.394	46.70	0.66	85.14	19.51	82.43
75	91.682	76.99	0.44	82.09	25.20	84.82
86	108.04	93.34	0.38	80.67	27.77	85.82
100	131.86	117.16	0.31	78.79	31.10	87.03
125	183.76	169.06	0.22	75.15	37.28	88.97
150	249.31	234.61	0.16	71.04	43.85	90.53
175	330.64	315.94	0.11	66.20	51.03	91.48
200	430.09	415.39	0.08	60.03	59.20	91.28

*Inches of mercury below one atmosphere.

Fig. 9-4. Properties of liquid and saturated vapor of refrigerant R-12. Note pressures corresponding to standard evaporating temperature of 5 F. (−15 C.) and condensing temperature of 86 F. (30 C.).
(Freon Products Div., E.I. du Pont de Nemours & Co., Inc.)

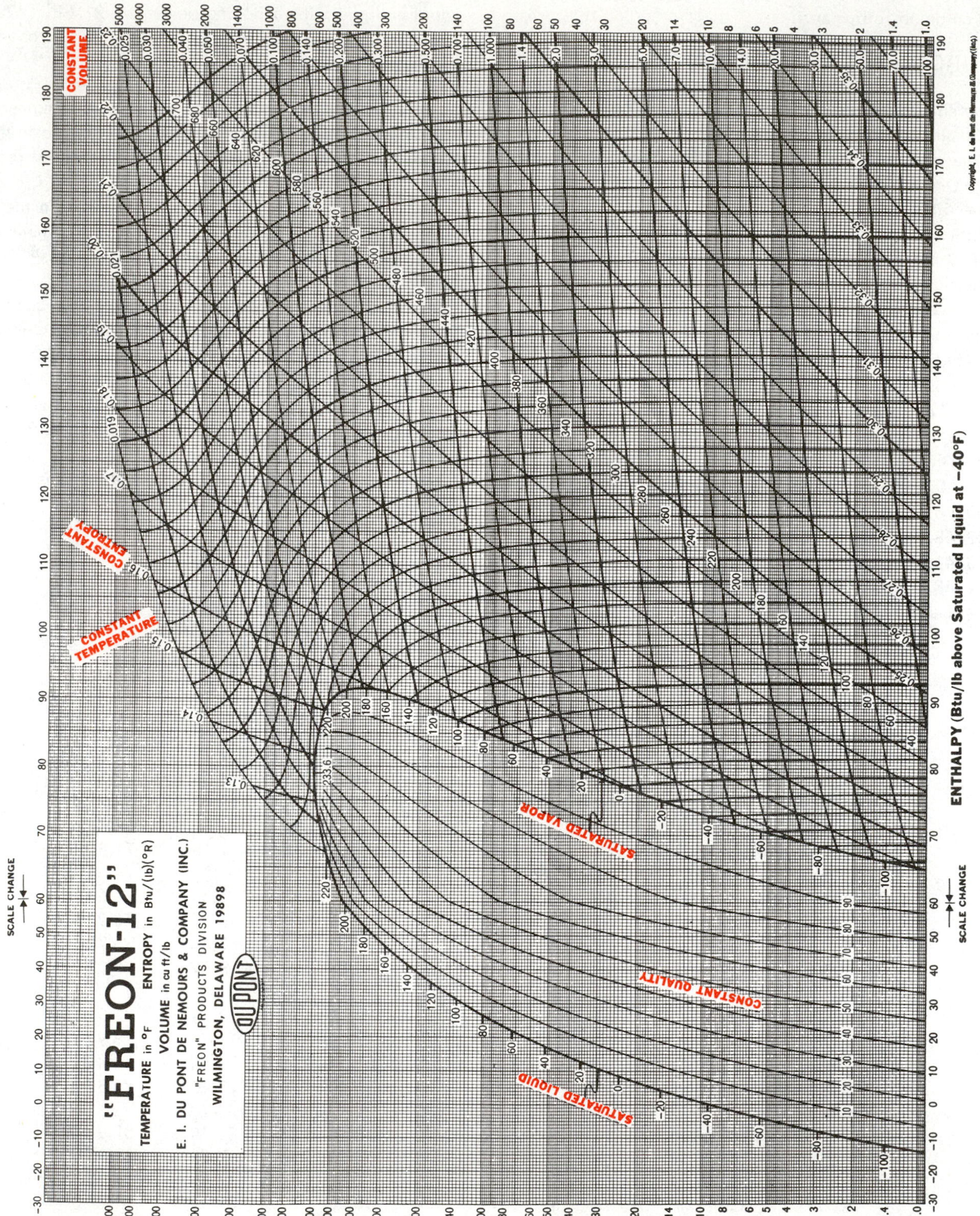

Fig. 9-5. Pressure-enthalpy diagram for refrigerant R-12. Note that 0 of enthalpy scale is taken at −40 F. (−40 C.).

Water is only slightly soluble in R-12. At 0 F. (−18 C.), it will only hold six parts per million by weight. The solution formed is only very slightly corrosive to any of the common metals used in refrigerator construction. The addition of mineral oil to the refrigerant has no effect upon the corrosive action, except to lessen the amount of discoloration caused by the free water.

R-12 is more critical as to its moisture content when compared to R-22 and R-502. R-12 is soluble in oil down to −90 F. (−68 C.). This helps the oil flow in very cold evaporators. The oil will begin to separate at this temperature and, because it is lighter than the refrigerant, will collect on the surface of the liquid refrigerant.

The pressure-heat (enthalpy) diagram for this refrigerant is shown in Fig. 9-5. See Para. 1-54 for an explanation of enthalpy. Metric unit pressure-chart diagrams for R-12 and R-22 are shown in Chapter 28.

It is safe to use 30 lb. of this refrigerant for each 1000 cu. ft. of air conditioned space.

A typical R-12 cycle for a frozen foods unit is shown in Fig. 9-6. The dotted line labeled C'' to D' shows the increased refrigerating effect when the liquid is sub-cooled in the condenser or liquid line. This sub-cooling is done by having a low ambient condition or by using a heat exchanger.

The refrigerant is available in a variety of cylinder sizes and may also be obtained in hermetically sealed throw-away cans. The cylinder code color is white. This is an important fact to remember when purchasing or using refrigerants. See Para. 9-26 for more color code information.

A new, patented refrigerant, Genetron 12/31, is now available. In some installations it may be used as a replacement for R-12. It is a mixture of R-12 and R-31. R-31 is monochloromonofluoromethane. The chemical composition is CCl_2F_2 (78 percent) and CH_2ClF (22 percent). Its latent heat

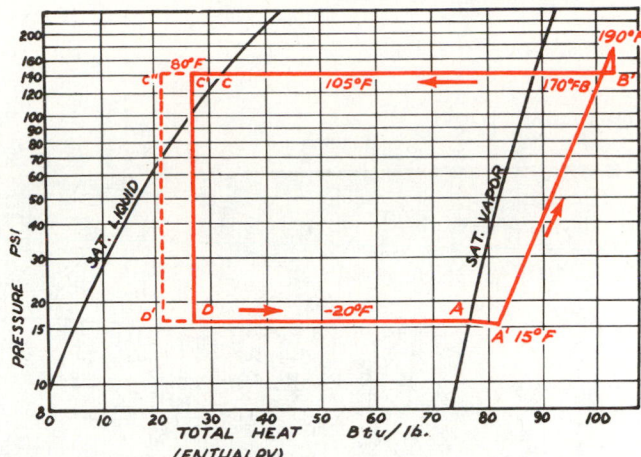

Fig. 9-6. Pressure-heat diagram for freezer application using R-12 as refrigerant. Heat is absorbed in evaporator from D to A. Vapors are compressed by compressor A' to B'. Heat is given off in the condenser B' to C'. Pressure drops from C' to D as refrigerant passes through refrigerant control without change in heat content.

of vaporization is slightly higher than for R-12. Its head pressure is also a little higher than for R-12. Critical temperature is 244 F. (118 C.).

9-7 R-22 MONOCHLORODIFLUOROMETHANE (CHCIF₂)

R-22 is a man-made refrigerant developed for refrigeration installations that need a low evaporating temperature. (See Fig. 9-7.) One application is in fast freezing units which maintain a temperature of −20 F. to −40 F. (−29 to −40 C.). It has also been successfully used in air conditioning units and

Fig. 9-7. Properties of liquid and saturated vapor of refrigerant R-22. Note pressures corresponding to standard evaporating temperature of 5 F. (−15 C.) and condensing temperature of 86 F. (30 C.).

	PRESSURE		VOLUME VAPOR	DENSITY LIQUID	HEAT CONTENT BTU/LB.	
Temp. F.	Psia	Psig	Cu. Ft./Lb.	Lb./Cu. Ft.	Liquid	Vapor
−150	0.272	29.37*	141.23	98.24	−25.97	87.52
−125	0.886	28.12*	46.69	96.04	−20.33	90.43
−100	2.398	25.04*	18.43	93.77	−14.56	93.37
− 75	5.610	18.50*	8.36	91.43	− 8.64	96.29
− 50	11.674	6.15*	4.22	89.00	− 2.51	99.14
− 25	22.086	7.39	2.33	86.48	3.83	101.88
− 15	27.865	13.17	1.87	85.43	6.44	102.94
− 10	31.162	16.47	1.68	84.90	7.75	103.46
− 5	34.754	20.06	1.52	84.37	9.08	103.96
0	38.657	23.96	1.37	83.83	10.41	104.47
5	42.888	28.19	1.24	83.28	11.75	104.96
10	47.464	32.77	1.13	82.72	13.10	105.44
25	63.450	48.75	0.86	81.02	17.22	106.84
50	98.727	84.03	0.56	78.03	24.28	108.95
75	146.91	132.22	0.37	74.80	31.61	110.74
86	172.87	158.17	0.32	73.28	34.93	111.40
100	210.60	195.91	0.26	71.24	39.27	112.11
125	292.62	277.92	0.18	67.20	47.37	112.88
150	396.19	381.50	0.12	62.40	56.14	112.73

*Inches of mercury below one atmosphere.

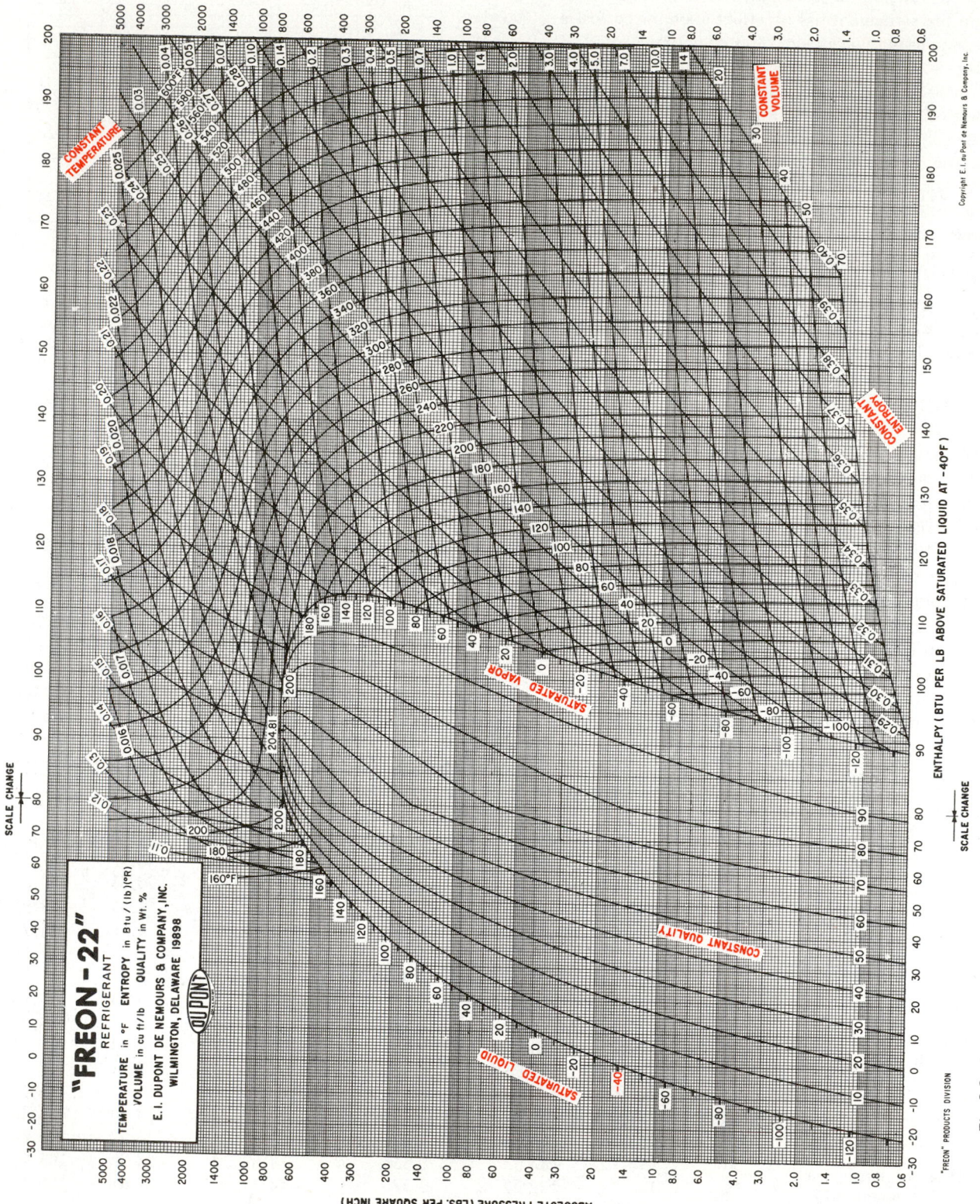

Fig. 9-8. Pressure-enthalpy diagram for refrigerant R-22. Note that 0 of enthalpy scale is taken at −40 F. (−40 C.). Also note that heat (enthalpy) scale changes at 80 Btu per lb. to help one read superheat values more easily. (Freon Products Div., E.I. du Pont de Nemours & Co., Inc.)

in household refrigerators. It is used with both reciprocating and centrifugal compressors. It is not necessary to use R-22 at below-atmosphere pressures in order to obtain these low temperatures.

R-22 has a boiling point of −41 F. (−41 C.) at atmospheric pressure. It has a latent heat of 93.21 Btu/lb. (2165J/g) at 5 F. (−15 C.). The normal head pressure at 86 F. is 172.87 psia, 158 psi (11.1 kg/cm^2), as shown in the table in Fig. 9-7. This refrigerant is stable and is nontoxic, noncorrosive, nonirritating, and nonflammable. The evaporator pressure of R-22 is 43 psia or 28 psi (1.96 kg/cm^2), at 5 F.

Water mixes better with R-22 than R-12 by a ratio of 3 to 1 or 19.5 ppm by weight; (ppm means parts per million). Water must be kept at a minimum, and driers (desiccants) should be used to remove most of the moisture.

Because of the ability of water and R-22 to mix, more desiccant is needed to dry it. R-22 has good solubility in oil down to 16 F. (−9 C.). However, the oil remains fluid enough to flow down the suction line at temperatures down as low as −40 F. (−40 C.). The oil will begin to separate at this point.

Because oil is lighter it will collect on the surface of the liquid refrigerant. Leaks may be detected with a soap solution, a halide torch or with an electronic leak detector. Some of the properties of R-22 are shown in Fig. 9-8.

The cylinder code color is green.

9-8 R-11 TRICHLOROMONOFLUOROMETHANE (CCl$_3$F)

R-11 is a synthetic chemical product which can be used as a refrigerant. It is stable, nonflammable and nontoxic. This means it will not burn and is not a poison. Considered to be a low-pressure refrigerant, it has a low-side pressure of 24 in. vacuum (609.6 mmHg.) at 5 F. (−15 C.), and a high-side pressure of 18.3 psia (1.28 kg/cm^{2}a) at 86 F. (30 C.). The latent heat at 5 F. (−15 C.) is 84.0 Btu/lb. (195 J/g).

This refrigerant is extensively used in large centrifugal compressor systems. As much as 35 lb. of this refrigerant may be used for each 1000 cu. ft. (28.3 m^3) of air conditioned space. (This would be a room about 10 ft. (3 m) by 12.5 ft. (3.8 m) by 8 ft. (2.4 m.) Leaks may be detected by using a soap solution, a halide torch, or by using an electronic detector.

R-11 is often used by service technicians as a flushing agent for cleaning the internal parts of a refrigerator compressor when overhauling systems. It is useful after a system has had a motor burnout or after it has had a great deal of moisture in the system. By flushing moisture from the system with R-11, evacuation time is shortened. R-11 is one of the safest cleaning solvents that can be used for this purpose.

The cylinder code color is orange.

	PRESSURE		VOLUME VAPOR	DENSITY LIQUID	HEAT CONTENT BTU/LB.	
Temp. F.	Psia	Psig	Cu. Ft./Lb.	Lb./Cu. Ft.	Liquid	Vapor
−40	10.95	7.62*	4.0	84.28	0.00	87.74
−30	14.10	1.22*	3.15	83.35	2.38	89.04
−20	17.92	3.23	2.52	82.40	4.79	90.31
−10	22.52	7.82	2.03	81.44	7.22	91.57
0	27.98	13.3	1.66	80.46	9.71	92.81
5	31.07	16.4	1.501	79.96	10.96	93.42
10	34.43	19.7	1.36	79.46	12.23	94.03
20	41.96	27.3	1.13	78.45	14.79	95.22
30	50.70	36.0	0.94	77.41	17.40	96.39
40	60.75	46.1	0.79	76.34	20.05	97.53
50	72.26	57.6	0.67	75.26	22.75	98.64
60	85.33	70.6	0.57	74.14	25.48	99.71
70	100.1	85.4	0.48	72.98	28.28	100.75
80	116.7	102.0	0.42	71.80	31.12	101.75
86	127.6	113.0	0.38	71.06	32.85	102.33
90	135.3	121.0	0.36	70.56	34.01	102.70
100	155.9	141.0	0.31	69.28	36.97	103.60
110	178.8	164.0	0.27	67.95	40.00	104.44
120	204.1	189.0	0.23	66.55	43.10	105.22
130	231.9	217.0	0.20	65.08	46.29	105.91
140	262.4	248.0	0.17	63.51	49.58	106.51

* Inches of mercury vacuum.

Fig. 9-9 Properties of liquid and saturated vapor of refrigerant R-500. Note pressures corresponding to evaporating temperature of 5 F. (−15 C.) and condensing temperature of 86 F. (30 C.). (Allied Chemical Corp.)

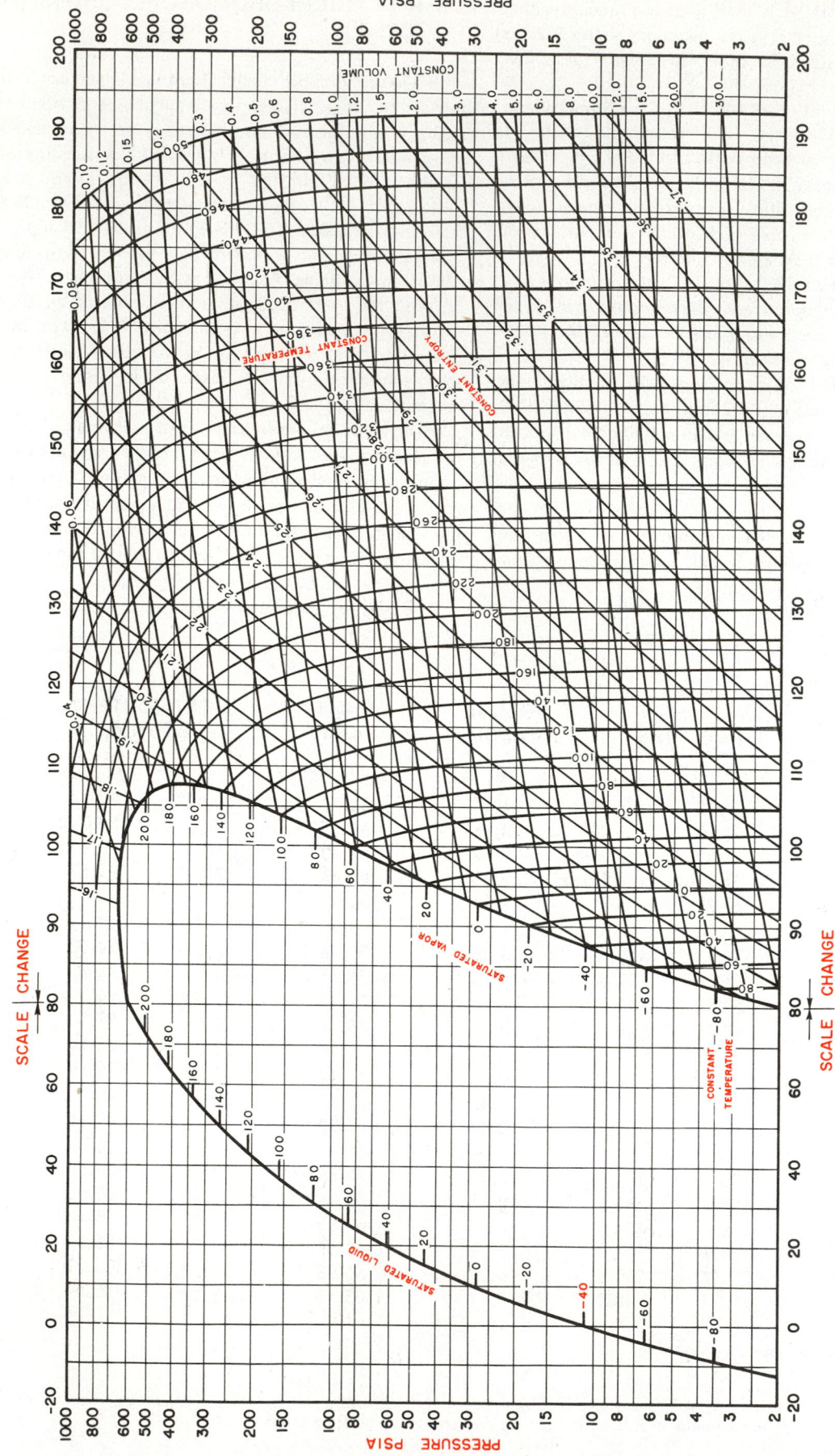

Fig. 9-10. Pressure-enthalpy diagram for refrigerant R-500. Note that 0 of enthalpy scale is taken at −40 F. (−40 C.). Also note that enthalpy scale changes at 80 Btu per lb. to enable one to read the superheat values more easily.

TEMPERATURE IN °F   VOLUME IN CU FT PER LB   ENTROPY IN BTU PER LB PER °F

ENTHALPY (BTU PER LB ABOVE SATURATED LIQUID AT −40 F)

9-9 AZEOTROPIC MIXTURES

Azeotropic refrigerants are liquid mixtures of refrigerants which exhibit a constant maximum and minimum boiling point. However, these mixtures act as a single refrigerant.

Four commonly used azeotropic refrigerants are:

R-500—composed of 73.8 percent R-12 and
26.2 percent R-152a.

R-502—composed of 48.8 percent R-22 and
51.2 percent R-115.

R-503—composed of 41.1 percent R-23 and
59.9 percent R-13.

R-504—composed of 48.2 percent R-32 and
51.8 percent R-115.

These are patented refrigerants and the manufacturing process is rather complicated. The service technician should never attempt to make his own mixtures.

Properties of these refrigerants are explained in the following paragraphs.

Azeotropic mixtures are chiefly used with reciprocating compressors.

9-10 R-500 REFRIGERANT
(R-152a + R-12) (CCl$_2$F$_2$/CH$_3$CHF$_2$)

Refrigerant R-500 is an azeotropic mixture of 26.2 percent R-152a and 73.8 percent R-12. It is used in both industrial and commercial applications but only in systems with reciprocating compressors. It has a fairly constant vapor-pressure temperature curve which is different from the vaporizing curves for either R-152a or R-12.

R-500 offers about 20 percent greater refrigerating capacity than R-12 for the same size motor when used for the same purpose. The evaporator pressure of R-500 is 31.21 psia or 19.7 psi (1.4 kg/cm^2) at 5 F. (−15 C.). It has a boiling point, at atmospheric pressure, of −28 F. (−33 C.). Its condensing pressure is 127.6 psia or 113 psi (7.94 kg/cm^2) at 86 F. (30 C.). Its latent heat at 5 F. is 82.45 Btu/lb. (192 J/g), as shown in table in Fig. 9-9, page 283.

R-500 can be used whenever a higher capacity than that obtained with R-12 is needed. There is little change in condensing temperatures, as shown in Fig. 9-10. R-500 is also recommended where electrical service varies from 60 cycle to 50 cycle (Hz).

The solubility (mixing with or going into solution) of water in R-500 is highly critical. R-500 has fairly high solubility with oil. Use a halide leak detector, an electronic leak detector, soap solution or a colored tracing agent to detect leaks.

Servicing refrigerators using this refrigerant does not present any unusual problem. Water is quite soluble in this refrigerant. It is necessary to keep moisture out of the system by careful dehydration and by using driers.

The cylinder code color is yellow.

9-11 R-502 REFRIGERANT
(R-22 + R-115) (CHClF$_2$/CClF$_2$CF$_3$)

Refrigerant R-502 is an azeotropic mixture of 48.8 percent R-22 and 51.2 percent R-115. It has been used since 1961. It

Fig. 9-11. Properties of liquid and saturated vapor of refrigerant R-502. Note pressures corresponding to standard evaporating temperature of 5 F. (−15 C) and condensing temperature of 86 F. (30 C). (E.I. duPont de Nemours & Co., Inc.)

Temp F.	PRESSURE Psia	PRESSURE Psig	VOLUME VAPOR Cu. Ft./Lb.	DENSITY LIQUID Lb./Cu. Ft.	HEAT CONTENT BTU/LB. Liquid	HEAT CONTENT BTU/LB. Vapor
−100	3.261	23.281*	10.461	97.857	−12.548	65.885
− 75	7.281	15.097*	4.959	95.234	− 7.597	68.919
− 50	14.602	0.190*	2.596	92.513	− 2.251	71.928
− 25	26.817	12.121	1.465	89.673	3.496	74.866
− 20	30.006	15.310	1.317	89.088	4.693	75.442
− 15	33.480	18.784	1.187	88.496	5.905	76.012
− 10	37.256	22.560	1.073	87.898	7.133	76.577
− 5	41.349	26.653	0.973	87.293	8.376	77.137
0	45.775	31.079	0.881	86.681	9.633	77.690
5	50.553	35.857	0.801	86.062	10.906	78.237
10	55.697	41.001	0.731	85.434	12.193	78.777
15	61.225	46.529	0.666	84.797	13.494	79.310
20	67.155	52.459	0.612	84.152	14.809	79.836
25	73.503	58.807	0.557	83.497	16.138	80.353
50	112.12	97.42	0.367	80.058	22.977	82.800
75	163.81	149.11	0.248	76.269	30.122	84.958
86	191.28	176.59	0.210	74.453	33.359	85.789
100	230.89	216.19	0.171	71.967	37.563	86.711
125	316.04	301.35	0.118	66.838	45.361	87.834
150	423.06	408.35	0.079	60.092	53.850	87.757
160	473.38	458.69	0.066	56.429	57.732	87.013

* Inches of mercury below one atmosphere.

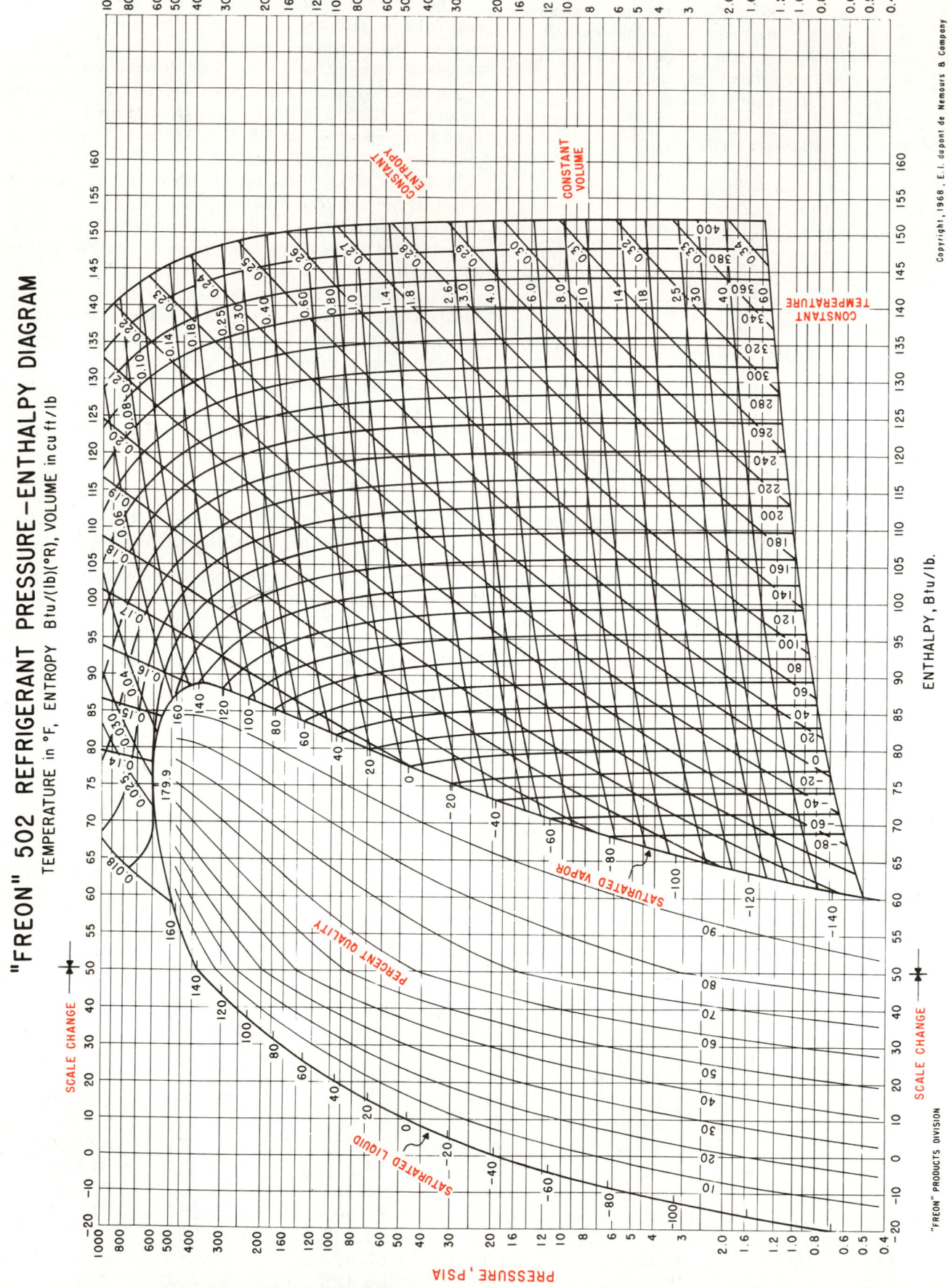

"FREON" 502 REFRIGERANT PRESSURE–ENTHALPY DIAGRAM

TEMPERATURE in °F, ENTROPY Btu/(lb)(°R), VOLUME in cu ft/lb

ENTHALPY, Btu/lb.

PRESSURE, PSIA

Copyright, 1968, E.I. duPont de Nemours & Company

"FREON" PRODUCTS DIVISION

Fig. 9-12. Pressure-enthalpy diagram for refrigerant R-502. Note that 0 of enthalpy scale is at −40 F. (40 C). Also note that scale changes at 50 Btu per lb. above saturated liquid to permit easier reading of values in the superheat area. (E.I. duPont de Nemours & Co., Inc.)

is a nonflammable, noncorrosive, practically nontoxic liquid. A good refrigerant for obtaining medium and low temperatures, it is suitable where temperatures from 0 to −60 F. (−18 to −51 C.) are needed. It is often used in frozen food lockers, frozen food processing plants, frozen food display cases and in storage units for frozen foods and ice cream. It is only used with reciprocating compressors.

Boiling point is −50.1 F. (−46 C.) at atmospheric pressure. Condensing pressure is 198.9 psia or 175.1 psi (12.31 kg/cm^2) at 86 F. Its evaporating pressure at 5 F. is 50.68 psia or 35.99 psi (2.53 kg/cm^2). Its latent heat at −20 F. is 72.5 Btu/lb. (168.6 J/g) as shown in the tables of properties in Fig. 9-11.

Refrigerant R-502 combines many of the good properties of both R-12 and R-22. It gives a machine the approximate capacity of R-22 with just about the condensing temperature of a system using R-12. A pressure-enthalpy diagram of the refrigerant is shown in Fig. 9-12.

This refrigerant's relatively low condensing pressure and temperature increases the life of the compressor valves and other parts. Better lubrication is possible because of the increased viscosity of the oil at the lower condensing temperature.

Because of the lower condensing pressures, it is possible to eliminate liquid injection to cool the compressor. This is often necessary with R-22.

R-502 has all the qualities found in the other halogenated (fluorocarbon) refrigerants. It is nontoxic, nonflammable, nonirritating, stable, and noncorrosive. Leaks are detected with soap solution, halide torch or electronic leak detector.

R-502 will hold 1.5 times more moisture at 0 F. (−18 C.) that R-12 (12.0 ppm by wt.). R-502 has fair solubility in oil above 180 F. (82 C.). Below this temperature, the oil tries to separate and tends to collect on the surface of liquid refrigerant. However, oil is carried back to the compressor at temperatures down to −40 F. (−40 C.). Special devices are sometimes used to return the oil to the compressor.

The cylinder code color is orchid.

9-12 R-503 REFRIGERANT (R-23 + R-13) (CHF$_3$/CClF$_3$)

Refrigerant R-503 is an azeotropic mixture of 40.1 percent R-23 and 59.9 percent R-13.

This is a nonflammable, noncorrosive, practically nontoxic liquid classified under Group 6 in the Underwriter's Laboratories Classification Scale. See Fig. 9-3.

Its boiling temperature at atmospheric pressure is −126 F. (−88 C.). This is lower than either R-23 or R-13.

Its evaporating pressure at 5 F. is 264 psia or 249.9 psi (17.5 kg/cm^2). Its critical temperature is 67 F. (20 C.), and its critical pressure is 607 psia or 592 psi (42 kg/cm^2).

This is a low temperature refrigerant and good for use in the low state of cascade systems which require temperatures in the −100 F. to −125 F. range (−73 to −87 C.). Properties of R-503 are shown in Fig. 9-13. A pressure enthalpy diagram for this refrigerant is shown in Fig. 9-14.

	PRESSURE		VOLUME VAPOR	DENSITY LIQUID	HEAT CONTENT BTU/LB	
Temp. F.	Psia	Psig	Cu. Ft./Lb.	Lb./Cu. Ft.	Liquid	Vapor
−140	9.94	9.69*	3.88	96.0	−26.45	52.88
−130	13.67	2.09*	2.89	94.7	−23.93	53.84
−120	18.45	3.57	2.19	93.4	−21.36	54.77
−110	24.48	9.78	1.69	92.0	−18.77	55.66
−100	31.97	17.3	1.32	90.6	−16.16	56.52
− 90	41.15	26.5	1.04	89.2	−13.53	57.35
− 80	52.27	37.6	0.83	87.7	−10.87	58.13
− 70	65.59	50.9	0.67	86.1	− 8.19	58.86
− 60	81.38	66.7	0.54	84.5	− 5.49	59.54
− 50	99.90	85.2	0.44	82.8	− 2.76	60.16
− 40	121.5	107.0	0.37	81.1	0	60.72
− 30	146.3	132.0	0.30	79.2	2.81	61.20
− 20	174.8	160.0	0.25	77.2	5.66	61.60
− 10	207.1	192.0	0.21	75.2	8.59	61.89
0	243.7	229.0	0.18	72.9	11.60	62.05
10	284.7	270.0	0.15	70.5	14.74	62.04
20	330.5	316.0	0.12	67.8	18.05	61.82
30	381.3	367.0	0.10	64.7	21.60	61.31
40	437.3	423.0	0.08	61.1	26.03	60.45
50	499.0	484.0	0.07	56.6	29.69	58.95
60	566.4	552.0	0.05	49.8	34.32	55.77

*Inches of mercury below one standard atmosphere.

Fig. 9-13. Properties of liquid and saturated vapor of refrigerant R-503. Note that temperature of −120 F. (−84 C.) with evaporator pressure above atmospheric. To operate in above 0 F. (−18 C.) temperature range, evaporator pressures in excess of 230 psi will be required.

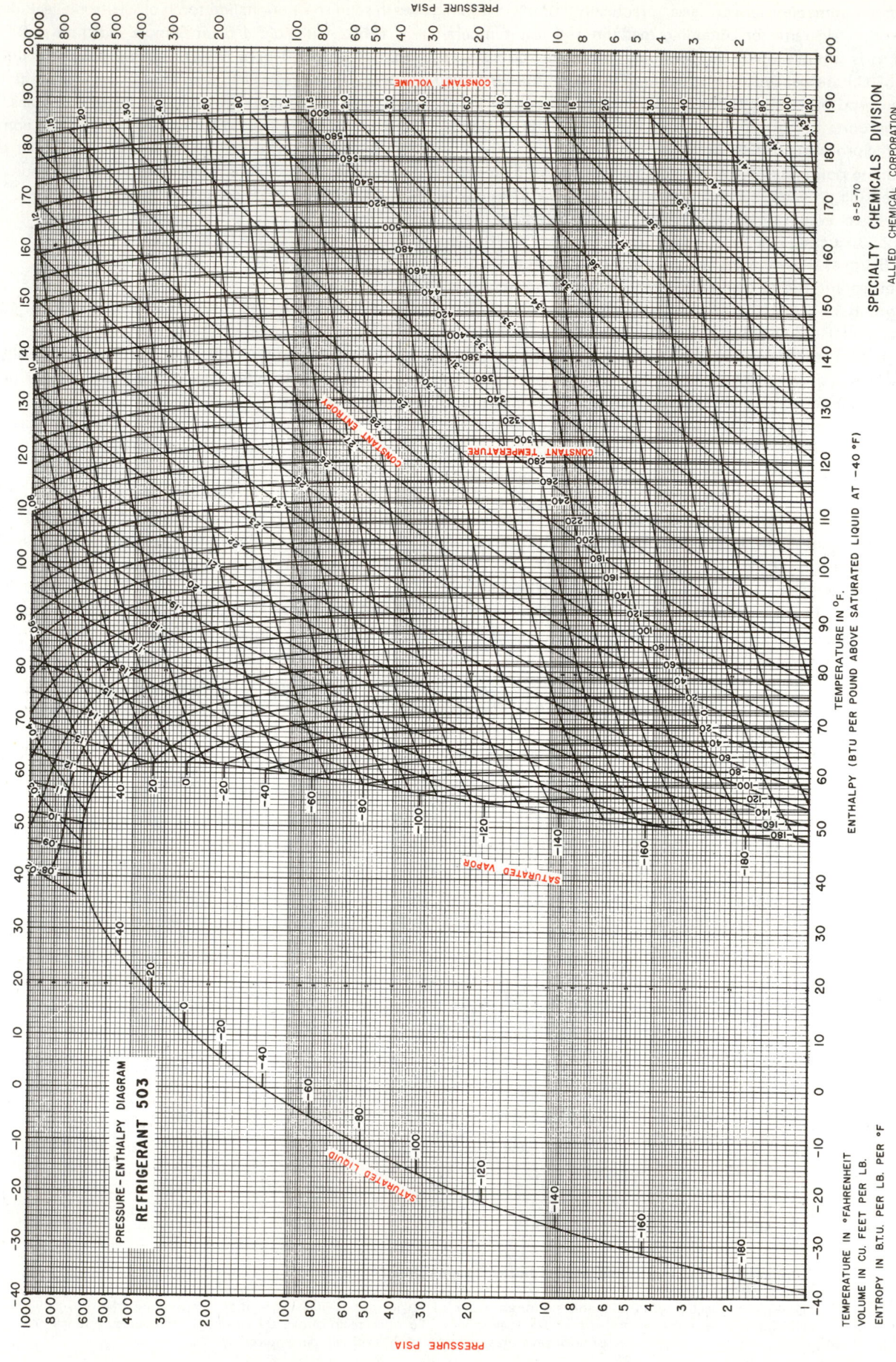

Fig. 9-14. Pressure-enthalpy diagram for refrigerant R-503. (Allied Chemical Corp.)

The latent heat of vaporization at atmospheric pressure (—128 F. or —88 C.) is 74.2 Btu/lb. (173 J/g).

Leaks in R-503 systems may be detected with the use of soap solution, a halide torch, or an electronic leak detector. This refrigerant will hold more moisture than some other low temperature refrigerants. It should be remembered, however, that all low temperature applications must have extreme dryness, since any moisture not in solution with the refrigerant is likely to form ice at the refrigerant control devices.

Oil does not circulate well at low temperatures. Cascade and other low-temperature equipment are usually fitted with oil separators and other devices for returning the oil to the compressor.

The code color for R-503 cylinders is aquamarine.

9-13 R-504 REFRIGERANT
(R-32 + R-115)
(CH_2F_2/CF_3CClF_2)

R-504 is an azeotropic refrigerant made up of 48.3 percent R-32 and 51.7 percent R-115. It is a nonflammable, non-corrosive and nontoxic liquid classified under Group 6 in the Underwriter's Laboratories classification scale. See Fig. 9-3.

The boiling temperature at atmospheric pressure is —70 F. (—57 C.).

Its evaporating pressure at 5 F. is 85.93 psia (6 kg/cm²a) and its critical pressure is 690 psia (49 kg/cm²a). Properties of R-504 refrigerant are shown in Fig. 9-15. A pressure-enthalpy

diagram for this refrigerant is shown in Fig. 9-16.

As with all low-temperature refrigerants, some difficulty may be experienced with the oil circulation. With the addition of 2 percent to 5 percent R-170 (ethane), the oil will be taken into the solution with the refrigerant and will circulate through the system with it.

Leaks in the R-504 system may be easily detected with soap solution, a halide torch or with an electronic leak detector.

This refrigerant is used in industrial processes where a low-temperature range of —40 F. to —80 F. (—40 C. to —62 C.) is desired.

The code color for R-504 cylinders is tan.

9-14 GROUP TWO REFRIGERANTS

The Group Two refrigerants are toxic. They are irritating to breathe and may be slightly flammable. Methyl chloride is quite toxic. Refrigerants in this group include:

R-717	Ammonia	NH_3
R-1130	Dichloroethylene	$C_2H_2Cl_2$
R-160	Ethyl Chloride	C_2H_5Cl
R-40	Methyl Chloride	CH_3Cl
R-611	Methyl Formate	$C_2H_4O_2$
R-764	Sulphur Dioxide	SO_2

Pressure-temperature curves for some common Group 2 refrigerants are shown in Fig. 9-17.

R-717 was one of the first refrigerants used. However, with

	PRESSURE		VOLUME VAPOR	DENSITY LIQUID	HEAT CONTENT BTU/LB.	
Temp. F.	Psia	Psig	Cu. Ft./Lb.	Lb./Cu. Ft.	Liquid	Vapor
—100	5.96	17.8*	8.08	90.49	—16.39	89.31
— 90	8.30	13.0*	5.94	89.22	—13.78	90.59
— 80	11.30	6.9*	4.46	87.93	—11.12	91.84
— 70	15.12	0.43	3.40	86.63	— 8.42	93.06
— 60	19.89	5.20	2.63	85.31	— 5.65	94.25
— 50	25.77	11.1	2.07	83.96	— 2.85	95.39
— 40	32.91	18.2	1.64	82.60	0	96.50
— 30	41.50	26.8	1.32	81.22	2.90	97.56
— 20	51.72	37.0	1.07	79.81	5.85	98.58
— 10	63.75	49.1	0.88	78.37	8.86	99.54
0	77.80	63.1	0.73	76.90	11.91	100.45
10	94.06	79.4	0.61	75.40	15.03	101.30
20	112.8	98.1	0.51	73.85	18.22	102.09
30	134.1	119.0	0.43	72.27	21.47	102.81
40	158.3	144.0	0.36	70.63	24.81	103.44
50	185.6	171.0	0.31	68.94	28.24	103.98
60	216.2	202.0	0.26	67.18	31.78	104.41
70	250.3	236.0	0.22	65.35	35.44	104.70
80	288.3	274.0	0.19	63.41	39.25	65.60
90	330.2	316.0	0.16	61.36	43.23	104.80
100	376.4	362.0	0.14	59.15	47.43	104.49

*Inches of mercury below one standard atmosphere.

Fig. 9-15. Properties of liquid and saturated vapor of refrigerant R-504. Density of liquid decreases very rapidly as temperature increases.

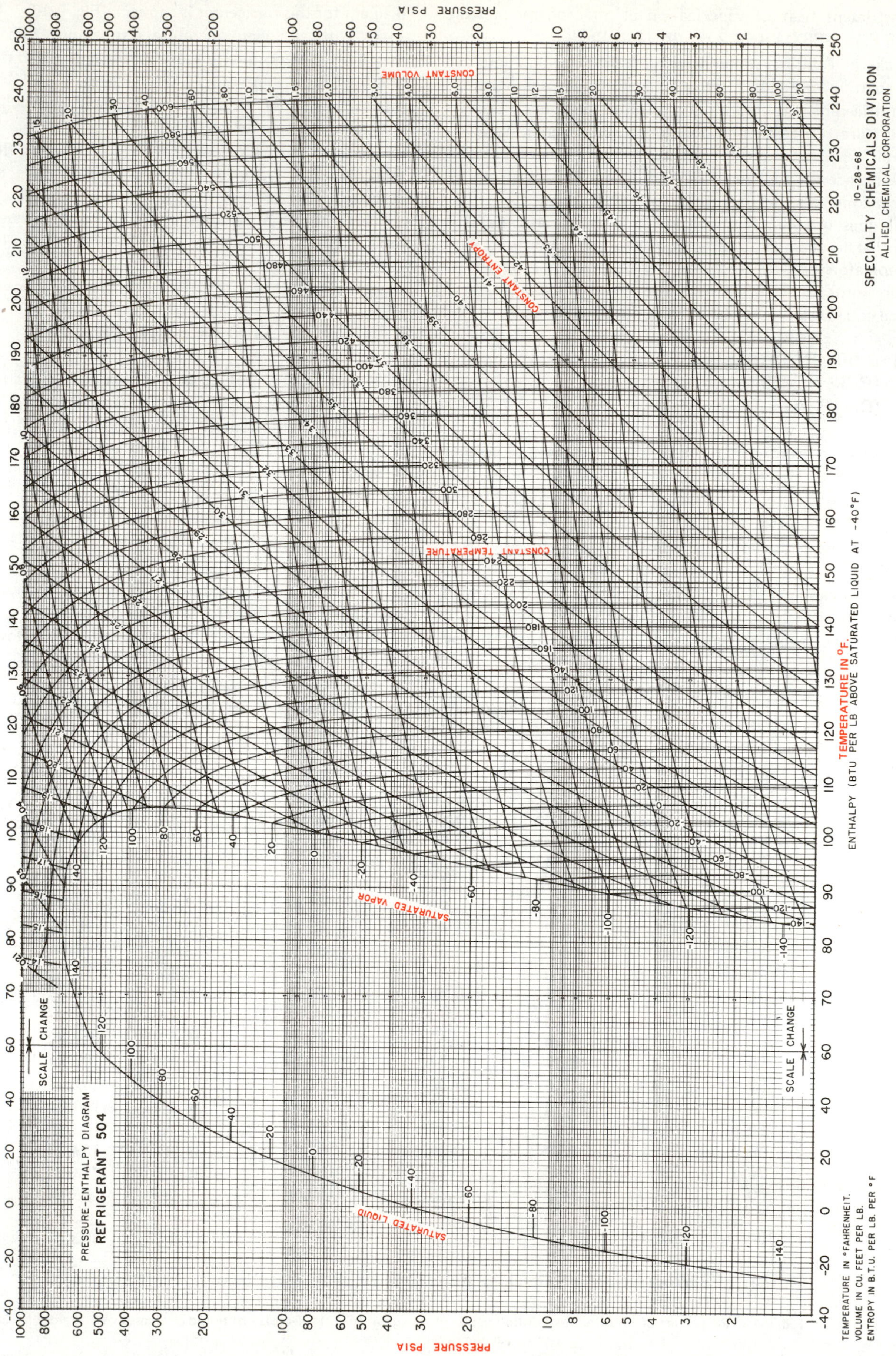

Fig. 9-16. Pressure-enthalpy diagram for refrigerant R-504. The superheat portion of the scale has been expanded.

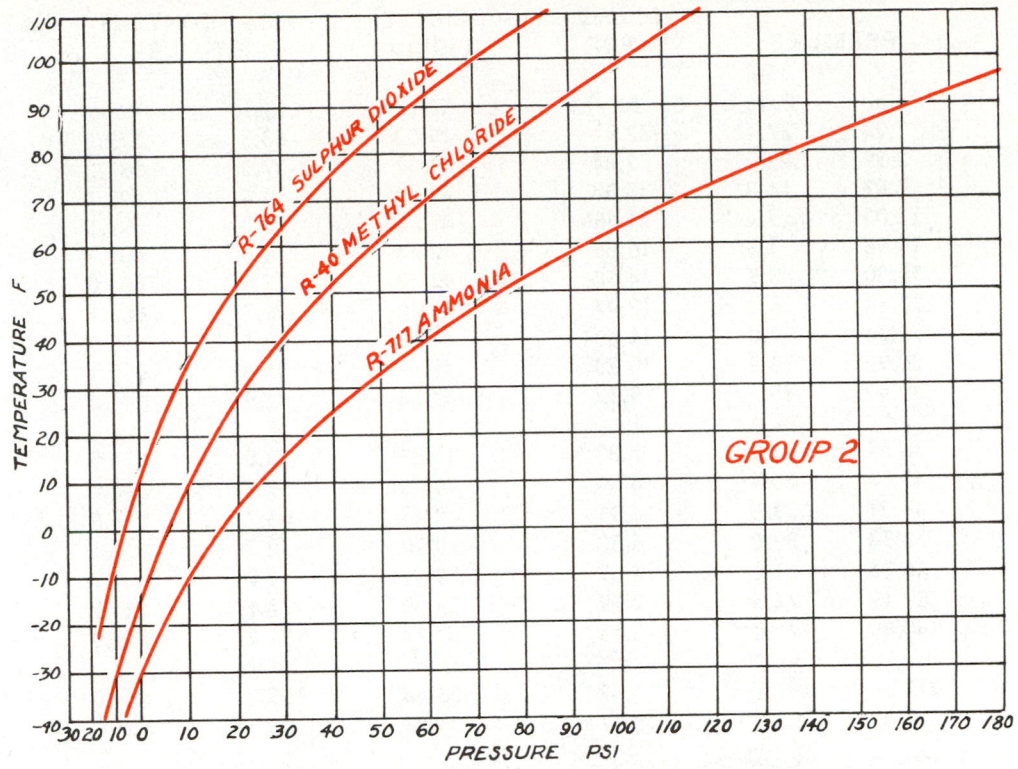

Fig. 9-17. Saturation temperature-pressure curves for some Group 2 refrigerants.

the exception of the absorption refrigerators, it is now used only in large industrial installations.

R-764 and R-40 have practically disappeared although at one time R-764 was the refrigerant most used in domestic refrigerators. However, there are many sulphur dioxide (R-764) and methyl chloride (R-40) charged units still in use.

Additional information about Group Two refrigerants can be found in Chapter 28.

9-15 R-717 AMMONIA (NH₃)

R-717 is commonly used in industrial systems. It is a chemical compound of nitrogen and hydrogen (NH₃) and under ordinary conditions is a colorless gas. Its boiling temperature at atmospheric pressure is —28 F. (—33 C.) and its melting point from the solid is —108 F. (—78 C.).

The low boiling point makes it possible to have refrigeration at temperatures considerably below zero without using pressures below atmospheric in the evaporator. Its latent heat is 565 Btu/lb. at 5 F. (—15 C.). Thus, large refrigerating effects are possible with relatively small-sized machinery. Condensers for R-717 are usually of the water-cooled type, although air-cooled condensers are being developed. The condenser pressure is 154.5 psi at 86 F. (11 kg/cm² at 30 C.) as shown in Fig. 9-18.

R-717 is somewhat flammable and, with the proper proportions of air, will form an explosive mixture. Accidents from this source, however, are rare.

While not classed as poisonous, its effect on the respiratory system is so violent that only very small quantities of it can be breathed safely. About .35 volumes per 100 volumes of air is the strongest concentration one can bear for any length of time. Because of its pronounced and distinguishable odor, it is easily detected in the air.

At 3 to 5 ppm, ammonia is identified by smell. At 15 ppm, the odor is quite irritating. At 30 ppm, the service technician will need a respirator. Exposure of 5 minutes to 50 ppm is the maximum allowed by OSHA. It becomes a hazard to life at 5000 ppm, and is flammable at 150,000 to 270,000 ppm.

Always stand to one side when operating an ammonia valve. A small stem leak may burn, damage the eyes and cause almost instant loss of consciousness. Wear a tight fitting mask. R-717 leaks may be quickly and easily detected by the white smoke-like fumes that it forms in the presence of sulphur candle or sulphur spray vapor.

R-717 attacks copper and bronze in the presence of a little moisture but does not corrode iron or steel. It presents no special problems in connection with lubrication unless extreme temperatures are encountered. R-717 is lighter than oil and no problems are encountered in dealing with the separation of the two. Excess oil in the evaporator may be removed by opening a valve in the bottom of the evaporator. The solubility of oil in liquid R-717 is only 20 ppm at 5 F. (—15 C.) and only 125 ppm at 86 F. (30 C.). R-717 vapor is extremely soluble in water. It is used in large compression machines using reciprocating compressors and in many absorption type systems.

The basic properties of R-717 are shown in Chapter 28.

A recognized safety code for the use and handling of ammonia in refrigerating systems is supplied by The International Institute of Ammonia Refrigeration.

Temp. F.	Pressure Psia	Pressure Psig	Volume Vapor Cu. Ft./Lb.	Density Liquid Lb./Cu. Ft.	Heat Content BTU/LB. Liquid	Heat Content BTU/LB. Vapor	Latent Heat BTU/LB.
−100	1.24	27.4*	182.4	45.52	−63.3	572.5	635.8
−75	3.29	23.2*	72.81	44.52	−37.0	583.3	620.3
−50	7.67	14.3*	33.08	43.49	−10.6	593.7	604.3
−35	12.05	5.4*	21.68	42.86	5.3	599.5	594.2
−25	15.98	1.3	16.66	42.44	16.0	603.2	587.2
−20	18.30	3.6	14.68	42.22	21.4	605.0	583.6
−15	20.88	6.2	12.97	42.00	26.7	606.7	580.0
−10	23.74	9.0	11.50	41.78	32.1	608.5	576.4
−5	26.92	12.2	10.23	41.56	37.5	610.1	572.6
0	30.42	15.7	9.12	41.34	42.9	611.8	568.9
5	34.27	19.6	8.15	41.11	48.3	613.3	565.0
10	38.51	23.8	7.30	40.89	53.8	614.9	561.1
15	43.14	28.4	6.56	40.66	59.2	616.3	557.1
20	48.21	33.5	5.91	40.43	64.7	617.8	553.1
25	53.73	39.0	5.33	40.20	70.2	619.1	548.9
35	66.26	51.6	4.37	39.72	81.2	621.7	540.5
50	89.19	74.5	3.29	39.00	97.9	625.2	527.3
75	140.5	125.8	2.13	37.74	126.2	629.9	503.7
86	169.2	154.5	1.77	37.16	138.9	631.5	492.6
100	211.9	197.2	1.42	36.40	155.2	633.0	477.8
125	307.8	293.1	0.97	34.96	185.1	634.0	448.9

*Inches of mercury below one atmosphere.

Fig. 9-18. Properties of refrigerant R-717 (ammonia). Note high latent heat.

9-16 GROUP THREE REFRIGERANTS

Group Three refrigerants may form a combustible mixture when mixed with air. The more common refrigerants in this group are:

R-600   Butane C_4H_{10}
R-170   Ethane C_2H_6
R-290   Propane C_3H_3

These refrigerants are no longer commonly used. Their characteristics are not covered in detail in this chapter. See Chapter 28 for further information.

9-17 EXPENDABLE REFRIGERANTS

Expendable refrigerants are used to cool a substance or evaporator and the refrigerant is then released to the atmosphere. It is not collected and recondensed, as is the case with the usual compression system. It is used only once. Systems using expendable refrigerants are sometimes referred to as chemical refrigeration or open cycle refrigeration. Refrigerants of this type have a low boiling temperature.

The most common expendable refrigerants are:

1. Liquid nitrogen (R-728); boiling temperature at atmospheric pressure, −320 F. (−196 C.).
2. Liquid helium (R-704); boiling temperature at atmospheric pressure, −452 F. (−269 C.).
3. Carbon dioxide (R-744); boiling temperature at atmospheric pressure, −109 F. (−78 C.) either in the solid or liquid state.

9-18 WATER AS A REFRIGERANT

Water is never used in the compression cycle refrigerating mechanism. However, it is the refrigerant for steam jet refrigeration used in connection with air conditioning systems (see Chapter 19). At atmospheric pressure, water boils at 212 F. (100 C.). One pound of water absorbs 970.3 Btu in changing from a liquid to a vapor (steam) at 212 F. The usual temperature range in using water as a refrigerant is above 45 F. Water, in changing from a liquid to a vapor, absorbs a considerable amount of heat.

The volume of vapor formed is large. At 45 F. (7 C.), one pound of water will turn into 2037 cu. ft. (57.7 m^3) of vapor. Water vaporizing at 29.6202" Hg. vacuum, or at .15 psia (7800 microns), will produce a temperature of 45 F. (7 C.).

9-19 FOOD FREEZANTS

When processing frozen foods, it is best to complete the freezing operation in the shortest possible time. Many commercial food freezing companies submerge the food to be frozen in a liquid refrigerant. The United States Department of Agriculture has approved certain refrigerants for this purpose. They are of a high purity and are designated by the name "Food Freezants."

This method is very rapid, since the heat transfer from the food to be frozen to the liquid is much faster than the heat transfer would be if the food were surrounded by air at the same temperature. The refrigerant used in this process does not in any way affect the wholesomeness of the food.

See Chapter 13 for description of the equipment required for the use of Food Freezants.

9-20 CRYOGENIC FLUIDS

Use of cryogenic fluids is becoming quite general in modern industry. They range in temperature from −250 F. (−157 C.) to absolute 0 (−459.69 F. or −273.15 C.). This is called the cryogenic range. Chapter 28 has a listing of these temperature ranges using various temperature scales.

Such low temperatures may be easily reached by evaporating cryogenic fluids. Common cryogenic fluids are:

R-702   Hydrogen
R-704   Helium
R-720   Neon
R-728   Nitrogen
R-729   Air
R-732   Oxygen
R-740   Argon

Fig. 1-28 lists the boiling temperature of cryogenic fluids.

Containers must be of special materials able to withstand extremely low temperatures without losing their strength. Insulation is very heavy as the temperature of the fluids inside is very low.

Small containers are of a thermos bottle type construction. Pressures are kept at a relatively low level which corresponds to the vapor pressure of the fluid.

No attempt is made to seal the fluid in a pressure-tight container. As a result, some of the fluid is continually boiling. This maintains the rest of the fluid at a very low temperature. The vapor from the boiling fluid is allowed to escape.

In general, these fluids for low temperature application are expendable. This means that they are used but once and the vapor is vented to the atmosphere.

There are certain cautions which must be observed by anyone handling these fluids:

Do not attempt to use any of these fluids in any container or mechanism which was not designed for its use.

An additional caution: Cryogenic fluids must never be allowed to touch the skin. Such a contact would result in immediate freezing of the flesh. Persons handling cryogenic fluid must have their entire body protected by suitable clothing, helmets, gloves and the like.

See Para. 1-56 and Fig. 1-28, also Chapter 28, for further information on cryogenic fluids and temperatures.

9-21 REFRIGERANT CYLINDERS

There are three types of refrigerant cylinders:
1. Storage cylinder.
2. Returnable service cylinder.
3. Disposable (throw-away) cylinder.

Cylinders are made of steel or aluminum. The larger ones usually have a fusible plug safety device threaded into the concave bottom as a protection against overheating or excessive pressures. A valve at the top provides a connection for charging or discharging service cylinders.

Regulations are prescribed by the Department of Transportation (DOT) to insure the safety of those working with cylinders containing refrigerants.

The DOT regulation requires that cylinders which have contained a corrosive refrigerant must be checked every five years. Cylinders containing noncorrosive refrigerants must be checked every 10 years. All cylinders over a 4 1/2 in. diameter and 12 in. long must contain some type of pressure release protective device, a fusible plug or a spring-operated relief valve, for example.

Never recharge a disposable service cylinder. It may explode.

9-22 STORAGE CYLINDERS

It is cheaper to purchase refrigerants in 100 and 150 lb. cylinders. These become storage cylinders which are frequently positioned upside down with the valve at the bottom. This makes charging service cylinders much easier.

The transferring of refrigerant from the large cylinder to the smaller service cylinder should be done carefully and a record kept of the quantity of refrigerant removed. To the total figure, add three percent to account for vapor losses.

The storage cylinders should be dated and stamped with a DOT stamp (Department of Transportation). No cylinder should be used beyond six years from the date on it. Refrigerant manufacturers request that all cylinders be returned to them every six months, or more often, so that the valve fittings and the complete cylinder may be carefully checked. This service helps to assure safe cylinders.

Storage cylinders are fitted with a valve and usually a protective cap, which may be screwed over the valve for shipment. This is shown in Fig. 9-19.

The cylinder valves, usually of the packed one-way type, should receive the same care as the service valves of a refrigeration system. The packing nut should be kept tight unless the valve is being used. The refrigerant opening should be sealed with a plug or cap when not in use.

It is best to use a hoist to lift and move cylinders which weigh over 35 pounds (16 kg).

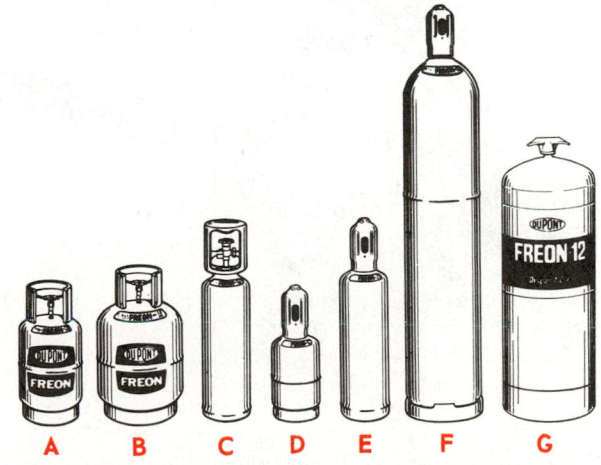

Fig. 9-19. Common refrigerant cylinders. A through E—Service cylinders. F—Supply cylinder. G—"Throw-away" container.

9-23 SERVICE CYLINDERS

The service technician carries small returnable service cylinders, 4 to 25 lb., which he uses when charging refrigerating systems. The cylinder valve is usually fitted with 1/4 in. male flare. The service cylinders are usually filled from the storage cylinders located at the shop.

CAUTION: Never completely fill a refrigerant cylinder with liquid refrigerant. Allow space for expansion. Liquid refrigerant expands with an increase in temperature. A cylinder completely filled with cold or cool refrigerant will burst if allowed to warm up. To check, shake the cylinder endwise. If there is unfilled space, the liquid refrigerant will be heard as it sloshes from end to end. A completely filled cylinder will make no sound. The safe limit is 85 percent full.

Service cylinders should be weighed before and after filling. In this way the amount of refrigerant in the cylinder may be readily determined. During charging, the service cylinder should be placed on an accurate scale. Only the specified weight of refrigerant should be charged into it.

9-24 RETURNABLE SERVICE CYLINDERS

Most refrigeration supply houses provide service cylinders on an exchange basis. Empty cylinders are returned and full ones are provided as a replacement. The supply house then arranges to refill the empty service cylinder.

Some refrigerant service cylinders are fitted with a carrying handle. This may also serve as a stand for direct liquid charging into the high side if the cylinder is turned upside down. These cylinders have a protective ring around the valve, as shown at A, B and C, Fig. 9-19. All of the cylinders shown, except G, have either fusible plugs or pressure-relief safety valves.

A new liquid-vapor valve, Fig. 9-20, is available on standard size cylinders. This valve enables the service technician to charge a system in the usual manner from a standard valve fitting either as a vapor or as a liquid without inverting the cylinder.

To transfer refrigerant either as a liquid or vapor, use one or the other of the two valve stems. The vapor valve stem is located at the top of the valve and is marked "vapor." The liquid valve stem is located on the side of the valve and is marked "liquid." It is attached to a tube which extends inside to the bottom of the cylinder. The valve has only one valve stem wheel to prevent opening both stems at the same time. A standard hood cap is placed over the valve when it is not in use. Fig. 9-21 is a cross-section of the liquid-vapor valve.

The use of returnable service cylinders is being reduced by the increase in the use of disposable service cylinders.

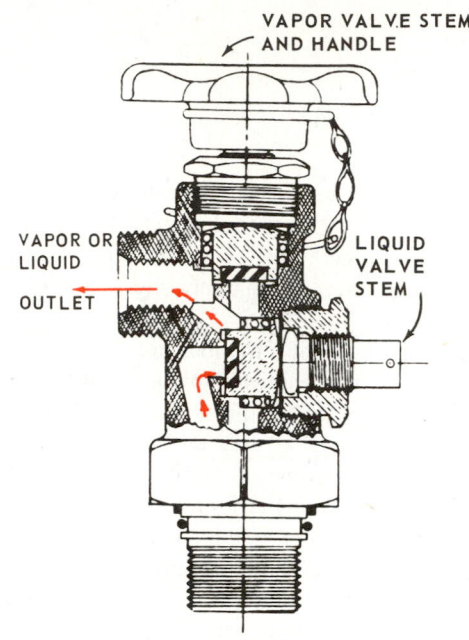

Fig. 9-21. A cross-section of a refrigerant cylinder valve which allows either refrigerant liquid or refrigerant vapor to leave the cylinder. (Kaiser Chemicals, Div. of Kaiser Aluminum & Chemical Corp.)

9-25 DISPOSABLE CYLINDERS

Many popular refrigerants are available in small quantities — from a few ounces up to 30 lb., in "throw-away" (disposable) cylinders. These containers are easy to handle and they eliminate the problem of refilling. Fig. 9-22 illustrates a popular throw-away disposable cylinder.

Most disposable cylinders are fitted with relief valves. Usually these are located in the valve body. Some "throw-away" refrigerant containers are sealed cans. The top is made in such a way that a special service valve can be tightly clamped to the top of the can. This valve, when clamped on the can, can be made to puncture it and provide a means of drawing refrigerant from the can. Fig. 9-23 illustrates such a valve.

Some of these valves provide important safety devices. Fig. 9-24, view A, illustrates such a valve. It has several safety features. These features are shown in B through E, of Fig.

Fig. 9-20. Refrigerant storage cylinder that uses two valve stems. The refrigerant may be drawn from cylinder either as a liquid or as a vapor by using the correct valve stem.

Fig. 9-22. A disposable refrigerant cylinder. A—Handle and safety guard for refrigerant valve. B—Refrigerant valve.

9-24. At B, the valve is shown attached to the cylinder and the cylinder pierced. The service valve is closed, and no refrigerant is escaping from the cylinder. This valve has a safety release mechanism shown at C. Should the pressure in the cylinder exceed a safe working limit, the safety release valve (see arrows) allows refrigerant to escape. This escaping vapor is directed down over the body of the valve, thus cooling the valve. At the same time, refrigerant flow is directed away from the service technician's face or body.

At D, the service valve is opened and the refrigerant is flowing out of the cylinder. At E, the service valve is opened. However, no refrigerant is flowing from the cylinder. This is because the pressure at the outlet of the valve is greater than the cylinder pressure. The ball check valve stops any reverse flow back into the storage cylinder. This could happen when charging a hot automobile air conditioning system.

Disposable cylinders should not be recharged nor used to store refrigerant removed from a system.

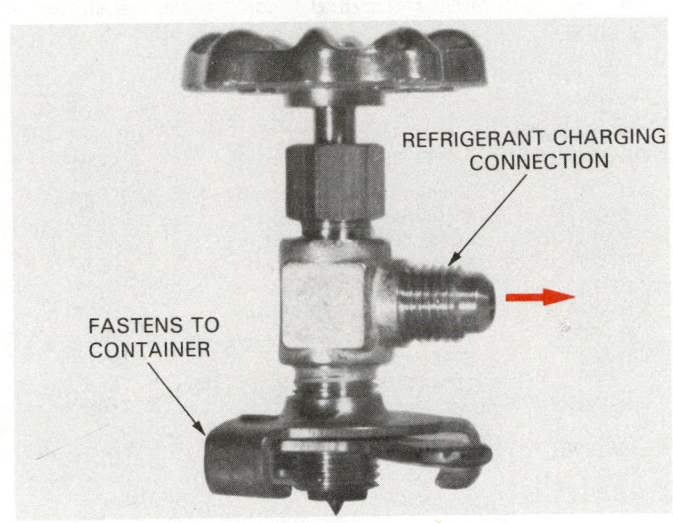

Fig. 9-23. Hand valve which may be attached to throw-away refrigerant containers. Valve clamps to top of refrigerant container. (Robinair Mfg. Corp.)

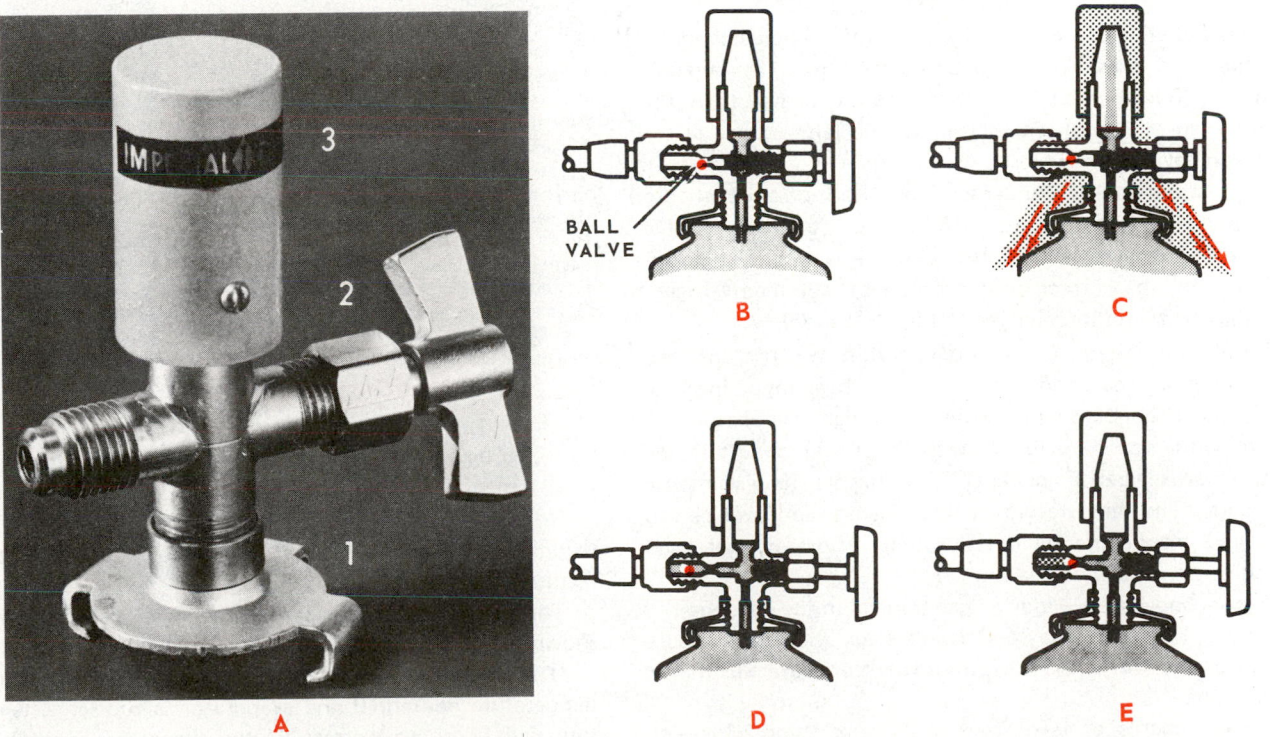

Fig. 9-24. Service valve for "throw-away" refrigerant cylinder. A—Exterior view, (1) clamp, (2) valve, (3) pressure relief valve. Views B through E show cross-sections of valve attached to cylinder. (Imperial-Eastman Corp.)

9-26 COLOR CODE FOR REFRIGERANT CYLINDERS

All cylinders used for transporting refrigerants are color coded to permit easy identification of the refrigerant in the cylinders. This practice helps to prevent accidental mixing of refrigerants within a system.

However, one must always read the label and identify the refrigerant before using the cylinder. The color code shown is not a requirement for all manufacturers. Popular refrigerants, with their R-number and cylinder color code, are shown in Fig. 9-25.

REFRIGERANT NUMBER	REFRIGERANT NAME	CYLINDER COLOR CODE
R—11	Trichloromonofluoromethane	Orange
R—12	Dichlorodifluoromethane	White
R—13	Monochlorotrifluoromethane	Pale Blue
R—22	Monochlorodifluoromethane	Green
R—113	Trichlorotrifluoroethane	Purple
R—114	Dichlorotetrafluoroethane	Dark Blue
R—500	Refrigerant 152A/12	Yellow
R—502	Refrigerants 22/115 (48.8% R—22/51.2% R—115)	Orchid
R—503	Refrigerants 23/13 (40.1% R—23/59.9% R—13)	Aquamarine
R—504	Refrigerants 32/115 (48.3% R—32/51.7% R—115)	Tan
R—717	Ammonia	Silver

Fig. 9-25. Cylinder color code for some common refrigerants.

9-27 HEAD PRESSURES (HIGH SIDE)

Pressures will vary with refrigerants. In air-cooled condensers, the head pressure should correspond to a temperature between 30 F. (17 C.) and 35 F. (19 C.) higher than the ambient (surrounding) temperature or the temperature of the air passing over the condenser.

In a water-cooled condenser, the head pressure should correspond to a temperature 15 F. (8 C.) to 20 F. (11 C.) above the exhaust temperature of the water. Using this information, the correct head pressure for common refrigerants may be found by referring to Figs. 9-2 and 9-17.

In all cases, the condensing temperature will rise until the heat loss from the condenser equals the heat input into the condenser. If condensing pressure is too high, the compressor has to work too hard. Too much vapor will be left in the compressor clearance pocket. This lowers its volumetric efficiency. The temperature of the exhaust vapor will be too high and may cause oil deterioration. Above normal head pressures are usually caused by:

1. A noncondensable vapor or gas trapped in the condenser — air, for example. Due to Dalton's Law, the head pressure will be the sum of the refrigerant vapor pressure plus the air pressure.
2. An overcharge of refrigerant in systems using a low-side float, an expansion valve or a thermostatic expansion valve. Some of the heat radiating space in the condenser will fill with liquid refrigerant and reduce the condenser's heat radiating ability.
3. Either the inside or outside of the condenser is dirty. This dirt will act as an insulator lowering the heat radiating capacity of the condenser. Then, the condenser temperature will rise.
4. If the air movement or the water movement through the condenser is reduced by blocked passages or poor water flow, there will not be enough heat-removing material to remove the heat from the condenser.
5. A restriction in the system, for example, a clogged capillary tube or a stuck refrigerant control, may temporarily cause a high head pressure.
6. If the low-side pressure is above normal, the head pressure will be higher than normal.

9-28 REFRIGERATOR TEMPERATURES

The low-side pressure in a refrigerating system determines the temperature in the evaporator.

One must first determine the temperature that is wanted in the cabinet or fixture, then adjust the motor control until this temperature is maintained. However, there are many cases where both a certain evaporator temperature and a cabinet temperature relationship should exist.

Cabinet temperatures are fairly standard. Fig. 9-26 shows

FIXTURE (Cabinet)	TEMP. F.	TEMP. C.
Back Bar	37—40	3—4
Beverage Cooler	37—40	3—4
Beverage Precooler	35—40	2—4
Candy Case (Display)	60—65	16—18
Candy Case (Storage)	58—65	15—18
Dairy Display Case	36—39	2—3
Double Display Case	36—39	2—3
Delicatessen Case	36—40	2—4
Dough Retarding Refrigerator	34—38	1—3
Florist Display Refrigerator	40—50	4—10
Florist Storage Case	38—45	3—7
Frozen Food Cabinet (Closed)	−10 to −5	−23 to −21
Frozen Food Cabinet (Open)	−7 to −2	−22 to −19
Grocery Refrigerator	35—40	2—4
Retail Market Cooler	34—39	1—3
Pastry Display Case	45—50	7—10
Restaurant Service Refrigerator	36—40	2—4
Restaurant Storage Cooling	35—39	2—3
Top Display Case (Closed)	35—42	2—6
Vegetable Display Refrigerator (Closed)	38—42	3—6
(Open)	38—42	3—6

Fig. 9-26. Recommended fixture (cabinet) temperatures.

recommended fixture (cabinet) temperatures for some common fixtures (cabinets).

The recommended temperature for various applications is shown in Fig. 9-27.

It is necessary to have the correct size evaporator for the temperature desired. If the evaporator is too large, temperature will be above normal. If the evaporator is undersized, temperature will be below normal.

The evaporator will have a lower temperature than the

APPLICATION	TEMP. F.	TEMP. C
Service	34–38	1–3
Meats	30–34	–1 to 1
Bananas	60–65	16–18
Fresh Meats	28–32	–2 to 0
Aging Room	30–34	–1 to 1
Chill Room	35–39	2–3
Curing Room	32–36	0–2
Freezer Room	–15	–26
Poultry	30–34	–1 to 1
Vegetables, Fresh	36–42	2–6
Ice Cream Hardening	–25	–32
Ice Cream Storage	–20 to –10	–29 to –23
Plants and Flowers	38–50	3–10
Fur Storage	33–37	0–3
Locker Room	–5 to 0	–21 to –18

Fig. 9-27. Recommended temperatures for various refrigeration applications.

fixture temperature (a temperature difference is needed for heat flow).

Normally the refrigerant will be 10 F. (5.5 C.) colder than the evaporator temperature when the unit is running. The refrigerant and the evaporator will become the same temperature during the off cycle.

The evaporator surface temperature depends on its size and the rate at which heat is being removed from the fixture.

The temperature of a typical frosting type evaporator (domestic type) will vary from 0 F. to 25 F. (–18 C. to –4 C.), and the refrigerant temperature will be about 10 F.

(5.5 C.) lower than this or in the range of –10 F. to 15 F. (–23 C. to –9.4 C.) while the unit is running.

The table in Fig. 9-28 gives the pressure corresponding to the evaporating temperatures and condensing temperatures for five popular refrigerants. The recommended average low-side pressure is shown in red opposite "Ev." The recommended high-side pressure is shown in red opposite "WC" for water-cooled systems and opposite "AC" for air-cooled systems.

To change psi to the metric kg/cm^2, multiply the psi value by .0703.

Example: R-12 at 5 F. has a pressure of 11.8 psi. The metric pressure will be: 11.8 x .0703 = .83 kg/cm^2.

9-29 USE OF PRESSURE-TEMPERATURE TABLES

The pressure-temperature relationship of the refrigerants under saturated conditions (holding as much vapor as they can) can be shown in tables.

The table can also be used to show the volume of one pound of the vapor at that temperature, as well as the latent heat, the specific heat and the density of the liquid.

Tables of these values for some popular refrigerants are shown in Figs. 9-4, 9-7, 9-9, 9-11, 9-13, 9-15 and 9-18. These values are of great value to a service technician and an engineer.

To use the tables, find in the left-hand, vertical column, the

TEMPERATURE PRESSURE CHART

TEMPERATURE °F.	12	22	500	502	717
–60	19.0	12.0	---	7.0	18.6
–55	17.3	9.2	---	3.6	16.6
–50	15.4	6.2	---	0.0	14.3
–45	13.3	2.7	---	2.1	11.7
–40	11.0	0.5	7.9	4.3	8.7
–35	8.4	2.6	4.8	6.7	5.4
–30	5.5	4.9	1.4	9.4	1.6
–25	2.3	7.4	1.1	12.3	1.3
–20	0.6	10.1	3.1	15.5	3.6
–18	1.3	11.3	4.0	16.9	4.6
–16	2.0	12.5	4.9	18.3	5.6
–14	2.8	13.8	5.8	19.7	6.7
–12	3.6	15.1	6.8	21.2	7.9
–10	4.5	16.5	7.8	22.8	9.0
–8	5.4	17.9	8.9	24.4	10.3
–6	6.3	19.3	9.8	26.0	11.6
–4	7.2	20.8	11.0	27.7	12.9
–2	8.2	22.4	12.1	29.4	14.3
0	9.2	24.0	13.3	31.2	15.7
1	9.7	24.8	13.9	32.2	16.5
2	10.2	25.6	14.5	33.1	17.2
3	10.7	26.4	15.1	34.1	18.0
4	11.2	27.3	15.7	35.0	18.8
Ev 5	11.8	28.2	16.4	36.0	19.6
6	12.3	29.1	17.0	37.0	20.4
7	12.9	30.0	17.7	38.0	21.2
8	13.5	30.9	18.4	39.0	22.1
9	14.0	31.8	19.0	40.0	22.9
10	14.6	32.8	19.8	41.1	23.8
11	15.2	33.7	20.5	42.2	24.7

TEMPERATURE °F.	12	22	500	502	717
12	15.8	34.7	21.2	43.2	25.6
13	16.4	35.7	21.9	44.3	26.5
14	17.1	36.7	22.6	45.4	27.5
15	17.7	37.7	23.4	46.6	28.4
16	18.4	38.7	24.2	47.7	29.4
17	19.0	39.8	24.9	48.9	30.4
18	19.7	40.8	25.7	50.1	31.4
19	20.4	41.9	26.5	51.2	32.4
20	21.0	43.0	27.3	52.4	33.5
21	21.7	44.1	28.2	53.7	34.6
22	22.4	45.3	29.0	54.9	35.7
23	23.2	46.4	29.8	56.2	36.8
24	23.9	47.6	30.7	57.4	37.9
25	24.6	48.8	31.6	58.7	39.0
26	25.4	49.9	32.4	60.0	40.2
27	26.1	51.2	33.3	61.4	41.4
28	26.9	52.4	34.3	62.7	42.6
29	27.7	53.6	35.2	64.1	43.8
30	28.4	54.9	36.1	65.4	45.0
31	29.2	56.2	37.0	66.8	46.3
32	30.1	57.5	38.0	68.2	47.6
33	30.9	58.8	39.0	69.7	48.9
34	31.7	60.1	40.0	71.1	50.2
35	32.6	61.5	41.0	72.6	51.6
36	33.4	62.8	42.0	74.1	52.9
37	34.3	64.2	43.1	75.6	54.3
38	35.2	65.6	44.1	77.1	55.7
39	36.1	67.1	45.2	78.6	57.2
40	37.0	68.5	46.2	80.2	58.6
41	37.9	70.0	47.2	81.8	60.1

TEMPERATURE °F.	12	22	500	502	717
42	38.8	71.4	48.4	83.4	61.6
43	39.8	73.0	49.6	85.0	63.1
44	40.7	74.5	50.7	86.6	64.7
45	41.7	76.0	51.8	88.3	66.3
46	42.6	77.6	53.0	90.0	67.9
47	43.6	79.2	54.2	91.7	69.5
48	44.6	80.8	55.4	93.4	71.1
49	45.7	82.4	56.6	95.2	72.8
50	46.7	84.0	57.8	96.9	74.5
55	52.0	92.6	64.1	106.0	83.4
60	57.7	101.6	71.0	115.6	92.9
65	63.8	111.2	78.1	125.8	103.1
70	70.2	121.4	85.8	136.6	114.1
75	77.0	132.2	93.9	148.0	125.8
80	84.2	143.6	102.5	159.9	138.3
W.C. 85	91.8	155.7	111.5	172.5	151.7
90	99.8	168.4	121.2	185.8	165.9
95	108.2	181.8	131.3	199.8	181.1
100	117.2	195.9	141.9	214.4	197.2
A.C. 105	126.6	210.8	153.1	229.8	214.2
110	136.4	226.4	164.9	245.8	232.3
115	146.8	242.7	177.4	262.7	251.5
120	157.6	259.9	190.3	280.3	271.7
125	169.1	277.9	204.0	298.7	293.1
130	181.0	296.8	218.2	318.0	---
135	193.5	316.6	233.2	338.1	---
140	206.6	337.2	248.8	359.1	---
145	220.3	358.9	265.2	381.1	---
150	234.6	381.5	282.3	403.9	---
155	249.5	405.1	300.2	427.8	---

NOTE: Vacuum — Italic Figures, in Hg.
Gauge Pressure — Bold Figures

Fig. 9-28. A temperature-pressure chart which may be used to determine the operating pressures for various temperatures and for various refrigerants. Ev.—Average low-side pressures. WC—Average water-cooled system head pressures. AC—Average air-cooled systems head pressures. (Sporlan Valve Co.)

temperature being investigated. Then move across to the other columns horizontally to determine the pressure.

9-30 REFRIGERANT APPLICATIONS

Some popular refrigerant applications are shown in Fig. 9-29. One type of refrigerant may be used in a number of applications. Some refrigerant applications recommended for

Domestic refrigerator . . . . R—12, R—22
Domestic food freezers . . . . R—12, R—22, R—502
Automobile air conditioning . . . . R—12
Cryogenic applications . . . . R—13, R—503
Home air conditioning . . . . R—22, R—500
Public building air conditioning:
Low capacity . . . . R—12, R—22
Medium capacity . . . . R—11, R—12, R—22
High capacity . . . . R—11, R—12
Ship-board air conditioning . . . . R—11, R—12, R—22
Frozen food delivery service . . . . R—22, Solid carbon
                                             dioxide
Metal shrinking . . . . Nitrogen
Industrial process . . . . R—11
Food freezants . . . . Carbon dioxide,
                                           nitrogen
Cleaning, absorbing moisture . . . . R—11

Fig. 9-29. Some popular refrigerant applications.

different types of compressors are shown in Fig. 9-30.

The selection of the type of refrigerant to be used in a given system is determined by the manufacturer. Several items are considered in the selection of the refrigerant, such as:
1. The capacity, governed by the refrigerant boiling point.
2. The volume of the vapor pumped to provide the necessary refrigeration.
3. The latent heat of the refrigerant.
4. The operating temperatures required.
5. The size of the equipment.

9-31 REFRIGERATION OIL

Oil circulates through the system with the refrigerant. The oil provides lubrication and cools the compressor's moving parts.

Refrigerant oil must have certain properties, because it is mixed with the refrigerant. The oil comes in direct contact with hot motor windings in hermetic units. Thus, it must be able to withstand extreme temperatures and be harmless to refrigerants and equipment.

Oil in the refrigeration system is cooled to low temperatures. Yet it must be able to withstand high temperatures at the compressor. It must remain fluid in all parts of the system.

The fluidity of an oil-refrigerant mixture is determined by the refrigerant used; the temperature; properties of the oil; its solubility in refrigerant and the solubility of refrigerant in the oil (to keep oil fluid at low temperatures).

The properties of a good refrigerant oil are:
1. Low wax content. Separation of wax from the refrigerant oil mixture may plug refrigerant control orifices (openings).
2. Good thermal stability. It should not form hard carbon deposits at hot spots in the compressor (such as valves or discharge ports).
3. Good chemical stability. There should be little or no chemical reaction with the refrigerant or materials normally found in a system.
4. Low pour point. Ability of the oil to remain in a fluid state at the lowest temperature in the system.
5. Low viscosity. This is the ability of the oil to maintain good oiling properties at high temperatures and good fluidity at low temperatures; to provide a good lubricating film at all times.

Looking to improve the performance of the oil, many manufacturers add chemicals which are designed to inhibit (slow down or stop) the formation of sludge or foaming. (Oil which contains moisture or air will form sludge or varnish and may cause damage to the unit.)

REFRIGERANT	COMPRESSOR TYPE	APPLICATION
R—11	Centrifugal	Large air conditioning systems ranging from 200 to 2000 tons in capacity. Refrigerating systems for industrial process water and brine.
R—12	Reciprocating Centrifugal Rotary	Large air conditioning and refrigeration systems. Small household refrigerators including frozen food and ice cream cabinets, food locker plants, water coolers, room and window air conditioners and others. Principal refrigerant in automobile air conditioning.
R—22	Reciprocating Centrifugal	Residential and commercial air conditioning. Food-freezing plants, frozen-food storage and display cases and many other medium and low-temperature applications.
R—500	Reciprocating	Small home and commercial air conditioning equipment and in household refrigeration — especially in areas where 50 cycle current is common.
R—502	Reciprocating	Frozen food and ice cream display cases, warehouses and food freezing plants. Medium-temperature display cases, truck refrigeration and heat pumps.
R—503	Reciprocating	Low-temperature systems down to about −130 F. (−90 C.).
R—13	Reciprocating	Low-temperature systems down to about −130 F. (−90 C.) in cascade systems.
R—113	Centrifugal	Small to medium air conditioning systems and industrial cooling.

Fig. 9-30. Some refrigerants suitable for different applications of reciprocating, centrifugal and rotary compressors.

Oil removed from a system should be clear. Discoloration means that it is impure. When this has happened, new driers and filters should be placed in the system to keep the new oil clean.

Another indicator of contaminated oil is odor. Dirty oil from a hermetic system may be acidic and will burn the hands. For additional information on oil, see Chapters 2, 11 and 14.

Only oil recommended by manufacturers of the equipment should be used. When the manufacturer's recommendation cannot be found, the viscosity of oil indicated in Fig. 9-31

SERVICE CONDITION	REFRIGERANT	VISCOSITY
Compressor Temperature		
Normal	All	150
		150/additives
High	Halogen	150/additives
	Ammonia	300
		300/additives
Evaporator Temperature		
Above 0 F.	Halogen	150
		150/additives
	Ammonia	300
0 F. to −40 F.	Halogen	150
		150/additives
	Ammonia	150
		150/additives
Below −40 F.	Halogen	150
		150/additives
	Ammonia	150
		150/additives
Automotive Compressors	Halogen	500

Fig. 9-31. Refrigerant oils should be selected according to compressor temperature, evaporator temperature and kind of refrigerant used.

may be used for most applications. Refrigerant oil containers must always be kept tightly sealed. Oil exposed to the atmosphere will absorb moisture.

9-32 MOISTURE IN REFRIGERANT

Moisture (water) in a system may freeze at the refrigerant control and clog or partly clog it. Moisture and some refrigerants in the presence of high compressor temperatures may cause the refrigerant to break down and may form harmful acids. It may cause rusting, corrosion or oil sludging, which could result in a motor burnout in hermetic systems.

Refrigerants must be kept in sealed containers and must be kept completely dry. Methods of drying refrigerants are explained in Chapters 11, 14 and 28.

Moisture indicators placed in the liquid line will help the service technician determine if moisture is present in the refrigerant. The indicator is a chemical which changes color depending upon the amount of moisture present. A sight glass or "window" in the liquid line helps one see the change in color.

It may be impossible to remove all moisture from a refrigerant. However, the amount of moisture must be kept very low. The maximum amount of moisture which may be allowed in a system varies with the kind of refrigerant and the low-side temperature.

Most refrigerant manufacturers supply refrigerants that are dry. The moisture content never exceeds five parts per million (ppm). Liquid refrigerants can hold more moisture (in solution) as the low-side temperature rises.

This enables the refrigerant to circulate without danger of the moisture separating from the refrigerant and freezing or forming harmful compounds. For example, R-12 is safe to use at 20 F. (−7 C.) with 17 ppm, at 0 F. (−18 C.) with 8.3 ppm, at −20 F. (−29 C.) with 3.8 ppm and at −40 F. (−40 C.) with only 1.7 ppm.

A table of safe moisture content for certain refrigerants is shown in Fig. 9-32. Any amount of water at or above the value

REFRIGERANT	DRY COLOR	WET COLOR
12	Below 5	Above 15
22	Below 30	Above 100
502	Below 15	Above 50

Fig. 9-32. Safe and unsafe moisture content for three types of refrigerant. The water concentration is shown as parts per million at 75 F. (24 C.). Dry color column indicates the allowable concentration. Wet color column shows concentrations that will cause trouble. When these quantities are present, the water will start freezing in the low side and cause contamination of the refrigerant.

in the "wet color" will be harmful to the system. The service technician must depend on the moisture indicator to determine the amount of moisture in the system.

If the moisture indicator shows a "wet" color, a new drier should be placed in the line and the system operated until the moisture indicator indicates a "dry" color. It may sometimes be necessary to replace the drier several times to dry up the system.

When servicing, avoid exposing cold internal parts of the system to air. Moisture from the air will condense on the parts and get inside the system. Warm the parts to room temperature with a heat lamp before opening the system.

9-33 REFRIGERANT IDENTIFICATION

A pressure-temperature chart is the basic method used to identify refrigerants. Identifying by smell or color is very difficult. The two exceptions are R-764 (sulphur dioxide) and R-717 (ammonia).

The Airserco Refrig-I-Dent, Fig. 9-33, may be used to help identify refrigerants. With this instrument, the service technician may test and identify an unknown refrigerant either from an operating system or from a refrigerant cylinder. It combines a thermometer and pressure gauge attached to a special manifold. A pressure-temperature chart is also supplied.

The refrigerant sample taken from the system or cylinder is trapped in the manifold of the instrument. The pressure and temperature registered by the instrument is then compared with the pressure-temperature chart. The "R" number of the refrigerant is determined.

Fig. 9-33. This instrument may be used to identify the refrigerant in use in a system.

9-34 CHANGING REFRIGERANTS

It is not good practice to change the kind of refrigerant in a unit. Engineering features designed into the unit to make it efficient may be wasted.

If for some reason the refrigerant must be changed, it is best to use an expansion valve designed for the new refrigerant. If the unit uses a capillary tube, and the density or pressure differences are changed, the tube should be replaced to match the new refrigerant.

Be sure to clean the mechanism thoroughly and replace the oil. A new filter-drier should be installed when a system is changed from one refrigerant to another.

9-35 AMOUNT OF REFRIGERANT REQUIRED IN A SYSTEM

The amount of refrigerant that should be used varies with the type of system. The low-side float, automatic expansion valve and thermostatic expansion valve systems are not sensitive to the amount of refrigerant in the system. High-side float systems and capillary tube systems are very sensitive to the amount of refrigerant charge.

A sure way to determine if the system has sufficient refrigerant is to put a sight glass in the liquid line. The appearance of vapor bubbles in this glass is a sign that the system is short of refrigerant. It should be charged until the vapor bubbles disappear.

Methods of charging refrigerant into different types of systems are explained in Chapters 11 and 14.

9-36 REVIEW OF SAFETY

When a leak is suspected, make certain that the room is thoroughly ventilated before starting to work on the unit. Always check for recommended operating pressures for each refrigerant and install gauges to find the pressures in the system.

Check refrigerant R-number before charging to avoid mixing refrigerants. If the refrigerant is a fluorocarbon, make certain there are no lighted flames near a system that is suspected of having a bad leak. The refrigerant may break down and produce dangerous gases. Always check the ICC cylinder stamp to make sure it is a safe cylinder.

Wear goggles and gloves at all times, especially when charging or discharging. These will protect the eyes, skin and hands in case of a sudden leak.

Always charge refrigerant vapor into the low side of the system. Liquid refrigerant entering a compressor may injure the compressor and may cause the unit to burst.

Liquid refrigerant on the skin may freeze the skin surface and cause "frost bite." If this should happen, quickly wash away the refrigerant with water. Treat the damaged surface for "frost bite."

Accidents involving refrigerants should be immediately referred to a doctor.

Make sure that a refrigerant service cylinder is never completely filled with liquid refrigerant. If a service cylinder is completely filled with liquid refrigerant and then allowed to become warm, the hydrostatic pressure inside the cylinder will cause it to burst.

Refrigerant cylinders should always be stored in a cool, dry place. Carefully assemble fittings and tubing to cylinders. Stripped threads are dangerous and costly. Never use refrigerant cylinders for supports or for rollers.

Refrigerant cylinders should be used only for storing the refrigerant marked on the label. Using the cylinder for compressed air is very dangerous. The cylinder may explode when exposed to these high pressures.

Refrigerants R-717 and R-764 are very irritating to the eyes and lungs. The service technician must always avoid exposure to these refrigerants.

Refrigerant oil contained in a hermetic compressor which has had a burnout may be very acidic. This oil should never be allowed to touch the skin. It may cause an acid burn.

If moisture is allowed to enter a refrigerating system, it is likely to cause considerable damage to the system. All parts of refrigerating mechanisms must be kept dry at all times. Containers of oil must always be kept tightly sealed to eliminate the possibility of absorbing moisture from the air.

Many refrigerants have no disagreeable odor and it might be possible to work in an area in which there is a considerable amount of refrigerant vapor. Also, many refrigerants are heavier than air and will replace the air in a room. This can be very dangerous because if the air one breathes does not contain at least 19.1 percent oxygen, the person breathing the mixture will loose consciousness. Instruments are available which will warn the person if the oxygen content of the air is below a safe limit.

9-37 TEST YOUR KNOWLEDGE

1. What does "halide" mean?
2. How may one test for R-502 leaks?
3. What is a common head pressure for air-cooled R-12 refrigerating systems?
4. What is the method used to locate R-22 leaks?
5. What are the low-side pressures at 5 F. (—15 C.) for R-22 and R-12: In English units? In Metric units?
6. What does toxic mean?
7. What oil properties are required with R-12?
8. What is the pressure of R-12 at 105 F. (41 C.): In English units? In Metric units?
9. What is the evaporating temperature of R-12 under a pressure of 20 pounds per square inch gauge?
10. What is the pressure of R-22 at 0 F. (—18 C.): In English units? In Metric units?
11. Is it advisable to substitute refrigerants in a system?
12. How may one determine the refrigerant temperature in an air-cooled condenser?
13. Name the refrigerants that may be tested for leaks with the halide torch.
14. What will be the high-side pressure (gauge) in an air-cooled condenser using R-502 while the condensing unit is running, if the room temperature is 75 F. (24 C.)?
15. What is the normal head pressure for an air-cooled R-22 system: In English units? In Metric units?
16. Is the refrigerant temperature in the evaporator the same temperature as the evaporator when the unit is running?
17. What effect does air in the system have on the head pressure?
18. Name two refrigerants that may not be tested with the halide torch.
19. How does one determine the correct head pressure of a water-cooled condenser?
20. What is the meaning of Group 1, Group 2 and Group 3 refrigerants?
21. May carbon dioxide be used as a refrigerant?
22. How many divisions are there in the National Board of Fire Underwriters classification of refrigerants?
23. Which number in the above classification is the least toxic?
24. Must an automatic expansion valve system have an exact amount of refrigerant charge?
25. If a capillary tube system has a frosting suction line, what does it indicate?
26. Can an R-22 system safely have more moisture in it than an R-12 system?
27. Why is it dangerous to fill a refrigerant cylinder completely full of liquid refrigerant?
28. How will the head pressure be affected by a covering of lint on the air-cooled condenser?
29. What are soap solutions used for in relation to refrigeration service?
30. How many pounds of R-12 must be evaporated at 5 F. to change 50 lb. of water from 72 F. to ice at 5 F., if the container for the water is made of copper and weighs 3 lb.?

Chapter 10

DOMESTIC REFRIGERATORS AND FREEZERS

A modern domestic fresh food refrigerator or food freezer consists primarily of three parts:
1. The cabinet.
2. The mechanism (condensing unit and evaporator).
3. The electrical circuit.

The cabinet contains and supports the evaporator and condensing unit; it also supplies shelving and storage space for the foods or beverages.

In the evaporator, the liquid refrigerant expands and becomes a vapor. This vapor absorbs heat from the foods or beverages in the cabinet.

The condensing unit removes the heat absorbed in the evaporator. The liquid refrigerant then returns to the evaporator to repeat the refrigerating cycle. It is suggested that one carefully review Chapter 4 before continuing study of this chapter.

The basic fresh food refrigerator described here may have many variations and accessories. (Accessories are parts which are not necessary but which add to the convenience of something else.) The frozen food freezer may be more complicated than the basic freezer mechanism and cabinet. Many companies design mechanisms to eliminate frost accumulations, to provide additional frozen food compartments and to provide automatic ice makers.

In this chapter, each type of refrigerating mechanism is named and illustrated. Under each type of refrigerator, the cabinets, mechanisms and electrical circuits are described and illustrated in some detail.

10-1 PRESERVING FOODS BY REFRIGERATION AND FREEZING

Foods (vegetables and fruits) last longer when kept at temperatures just above freezing. The lower temperatures slow down oxidation of the food, reduce the multiplication of the bacteria in the living cells and fibers and reduce the aspiration (removal of fluid) of the food.

10-2 HOW COLD PRESERVES FOOD

Food is made up of microorganisms, enzymes, colloids and water. If not kept cold, food will spoil.

Enzymes, which cause food spoilage, are controlled by low temperatures. Research has discovered that to preserve some

foods for long periods (a year or more) the temperature must be well below 0 F. (−18 C.). For best results, it should be −20 F. (−29 C.).

Enzymes are tiny particles of matter that exist in food substances. They are not destroyed by fast freezing but their increase is slowed down by the low temperatures. They seem to serve as catalytic agents (stimulants) of organic change. They are destroyed by pasteurization.

Colloids are found in flesh foods. They are tiny cells in meats, fish and poultry. If they are abused in any way, such as cell disruption (breaking), the food quickly becomes rancid. They seem to be cell containers or capsules. If the container is broken, the food rapidly deteriorates.

Meat, poultry and fish have important colloidal (miniature cell) changes that can be slowed by using low temperatures.

Water in food forms ice crystals when frozen. Fast freezing produces small ice crystals; slow freezing allows time for larger crystal growth. The larger the ice crystals, the more the food cell walls are ruptured.

10-3 STORAGE OF FRESH FOODS IN THE REFRIGERATOR

The air in a fresh food refrigerator is always quite dry. What moisture there is in the refrigerator tends to collect and condense on the evaporator surfaces. Therefore, food containers should be covered and as airtight as possible to keep food moist.

The temperature inside the fresh food cabinet should be kept at 35 to 40 F. (2 to 4 C.).

Most fresh foods may be kept from three days to a week at the above temperatures. Unfrozen meat and fish should be stored at as close to 32 F. (0 C.) as possible. The recommended storage temperatures for various foods are listed in Chapter 15.

Fruits and vegetables should be cleaned and prepared for the table before being refrigerated.

10-4 STORAGE OF FROZEN FOOD IN THE FREEZER

The air in a food freezer, as in a refrigerator, is very dry. Any moisture in the air in the freezer quickly condenses on the evaporator surfaces. It is very important, therefore, that all frozen foods be packaged in moistureproof containers.

When packaging food for the freezer, as much air as possible should be removed from the packaging.

Frozen food packages must be tightly sealed. Ordinary paper is too porous for freezer use. If not properly packaged, frozen food will develop "freezer burn."

Freezer burn is indicated by a change in color of the food. Food value is not affected but there is a change in color and outside appearance.

Most frozen foods, if kept at temperatures of 0 F. to −10 F. (−18 C. to −23 C.), may be kept for several weeks. Food to be kept for a year or more should be frozen at −20 F. (−29 C.) or lower. Some frozen foods keep better than others; beef keeps better than pork. Commercial systems for food freezing are explained in Chapter 13.

10-5 REFRIGERATOR AND FREEZER INSULATION

Insulation lines the walls of the refrigerator and freezer cabinet to keep heat from leaking through.

The insulation materials most used in household refrigerators and freezers are urethane foam or fiber glass. Other insulating materials are used in some commercial and industrial systems.

Tables giving the insulating properties of various materials are shown in Chapter 28.

10-6 REFRIGERATOR, FRESH FOOD, MANUAL DEFROST

The simple fresh-food refrigerator consists essentially of an evaporator placed either across the top or in one of the upper corners of the cabinet. Fig. 10-1 is typical. The evaporator is

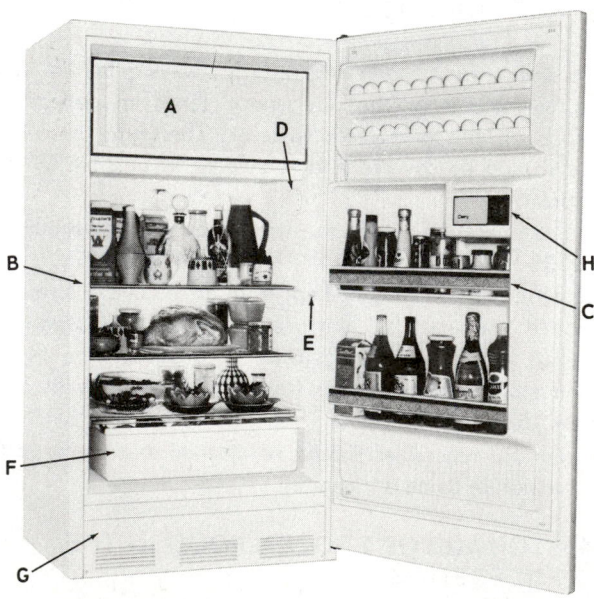

Fig. 10-1. Fresh food refrigerator. A—The evaporator at the top of the cabinet. B—Cabinet shelving. C—Door shelving. D—Thermostat control knob. E—Light switch. F—Crisper. G—Condensing unit. H—Butter conditioner. (Kelvinator, Inc.)

across the top of the cabinet and has a place to store frozen food for a short period of time. The condensing unit is in the bottom of the cabinet.

Some additional food storage is provided in the door. A vegetable crisper drawer is located below the bottom shelf, a butter conditioner in the door. This conditioner has a door, closing it off from the refrigerated compartment, to keep the butter at a slightly higher temperature than the rest of the food in the cabinet.

10-7 CABINETS FOR REFRIGERATED FRESH FOOD

Refrigerator cabinets are made of pressed steel. The seams are welded. The outside shell must be smooth and vapor-proof. The inside shell provides a surface for the interior finish of the cabinet. It also provides brackets for mounting shelves, lights, thermostats, temperature controls and the like.

The insulation is installed between the outer and the inside shell. Urethane foam, when used, is expanded in-place and, therefore, fits with no crevices or open places. The hinge arrangement is usually a part of the outside shell and door assembly

In the simple fresh-food refrigerator, the cabinet provides a space for the evaporator along the top or upper corner. The cold air from the evaporator flows by natural circulation through the refrigerated space. The shelves are usually so constructed that air can circulate freely past the ends and sides. In such an installation, there is no need to use a fan. The crisper for fresh vegetables is usually located in the bottom shelf of the refrigerator. It generally has a cover in order to maintain fairly high humidity around the vegetables.

In most cases, the light switch is located at the hinge side of the door. The light is turned on and off as the door is opened and closed.

To reduce the heat flow from the outer shell to the inner shell at the door opening, a connecting trim of special plastic is usually used. This plastic is sometimes called a "cold ban," being a poor conductor of heat. This trim is usually attached to the refrigerator liner and shell with trim clips. The finish on fresh food refrigerators — both outside and inside — is usually a good grade of baked-on enamel. Porcelain enamel is found on steel cabinet liners.

10-8 MECHANISMS FOR FRESH-FOOD REFRIGERATORS

The typical mechanism in a simple fresh food refrigerator consists of a hermetic compressor placed in the cabinet base. The condenser is either at the bottom or at the back of the cabinet. An evaporator is placed inside the cabinet at the top. A typical mechanism is illustrated in Fig. 10-2.

In this illustration, the parts are shown in diagram fashion. The cycle operation is as follows:

1. The liquid refrigerant (usually R-12) enters the evaporator, A.

2. As the refrigerant boils and absorbs heat in the evaporator, the vapor is drawn through the suction line, B, back to the

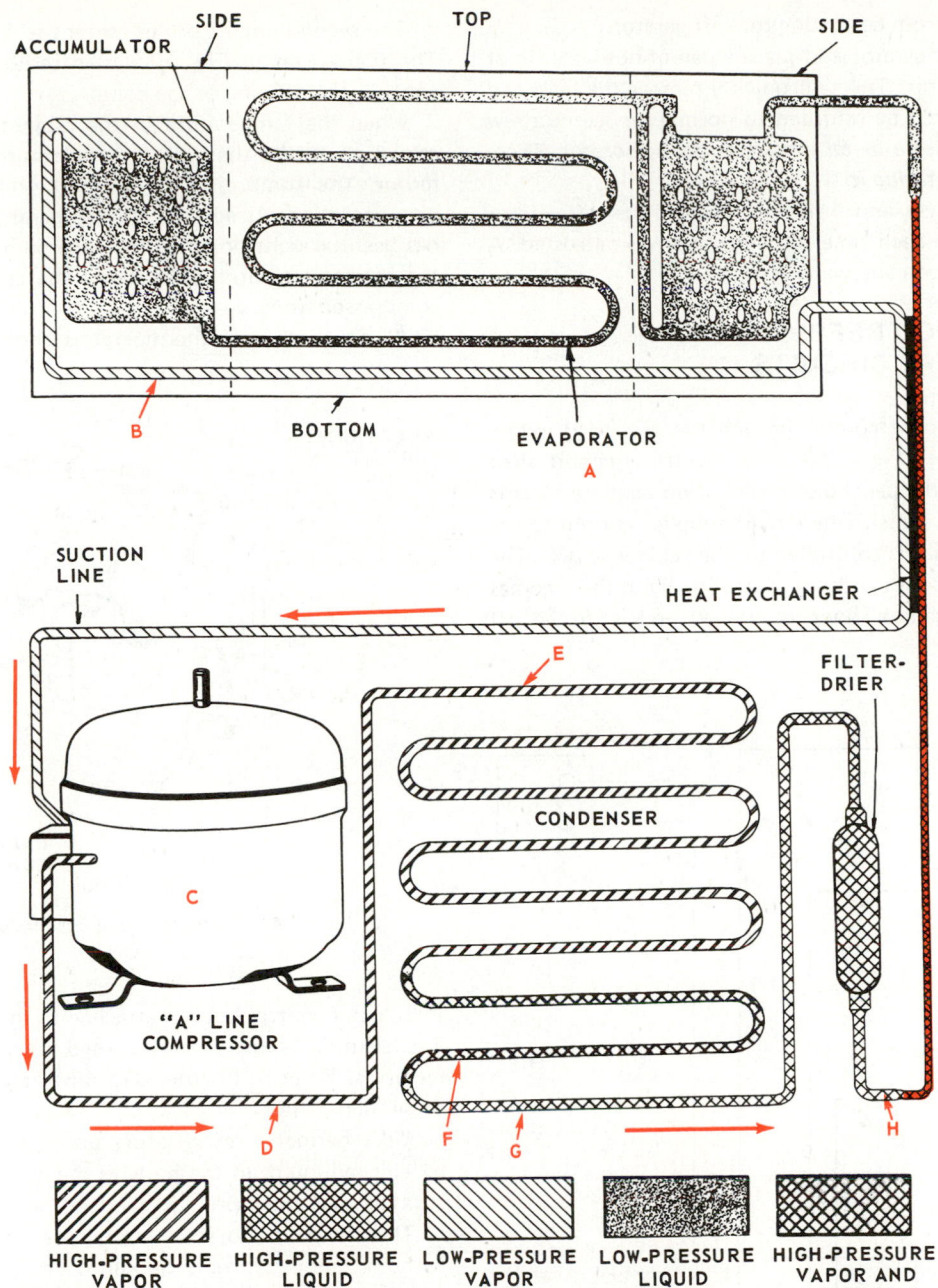

Fig. 10-2. Cycle diagram for a fresh food refrigerator.

compressor, C.

3. In the compressor, it is compressed to a high pressure. Its temperature is increased and the compressed vapor flows through the high pressure vapor line, D, into the condenser, E. The condenser, in this case, is a vertical natural draft, wire-tube type.

4. In the condenser, the high-pressure high-temperature vapor, F, gives up its heat to the surrounding air and the vapor is condensed back to a liquid. The liquid is shown at G in the bottom of the condenser.

5. It then flows through the filter-drier and enters the capillary tube at point, H. The capillary tube refrigerant control, J, is attached to the suction line at the heat exchanger.

6. The warm refrigerant passing through the capillary tube gives up some of its heat to the cold suction line vapor. This increases the heat absorbing ability of the liquid refrigerant slightly and increases the superheat of the vapor entering the compressor.

7. The low-pressure liquid now enters the evaporator, A, and the cycle is repeated.

This is the simplest type of automatic domestic refrigerator. These refrigerators are manually defrosted. As frost builds up on the evaporator it will be necessary, from time to time, to remove it. The ice accumulation on the evaporator greatly reduces the refrigeration effect.

There are two common methods for manually defrosting these refrigerators:

1. The refrigerator is turned off and allowed to remain off overnight. A drip pan must be used to catch the condensa-

tion that comes from defrosting the refrigerator.

2. Turn off the refrigerator and place a pan of hot water in or near the evaporator. This will quickly remove the frost and the refrigerator can be returned to normal service in a few minutes. *Never use a metal scraper on an evaporator. There is danger of puncturing it.*

Evaporator surfaces and inside surfaces of the refrigerator ought to be cleaned each time the mechanism is defrosted. A solution of baking soda and water is best.

10-9 FRESH FOOD REFRIGERATOR — ELECTRICAL CIRCUITS

The electrical supply comes through the grounded extension cord and plug, Fig. 10-3. The electrical circuit then supplies current to the panel disconnect. Two separate circuits lead away from this panel. One circuit supplies current to the cabinet light. The light is controlled by the cabinet switch. The cabinet switch is in series with the light. The light, then, comes on when the refrigerator door is opened and is turned off when the door is closed.

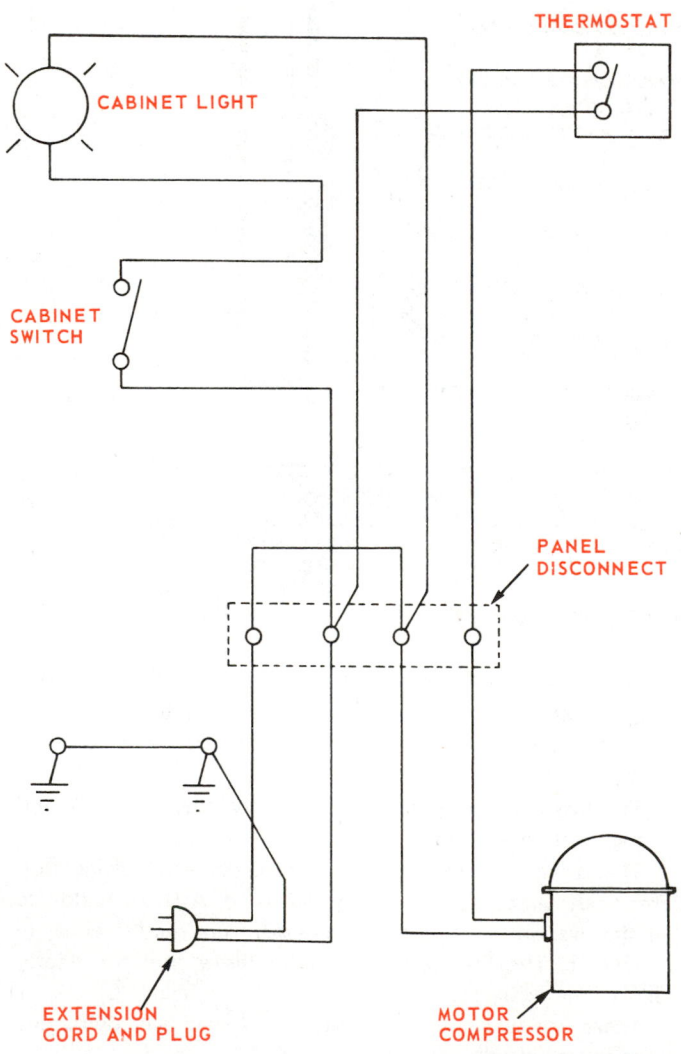

Fig. 10-3. Wiring diagram, fresh food refrigerator.

The second circuit brings current to the motor compressor. The thermostat in Fig. 10-3 is in series with this circuit and controls the running of the compressor.

When the temperature in the cabinet rises above a determined point, the thermostat completes the circuit through the motor. The compressor runs and the refrigeration cycle goes into operation. As soon as the temperature inside the cabinet has been brought down to the minimum desired temperature, the thermostat turns off (opens) the current and the motor compressor stops.

A motor control thermostat is shown in Fig. 10-4. The

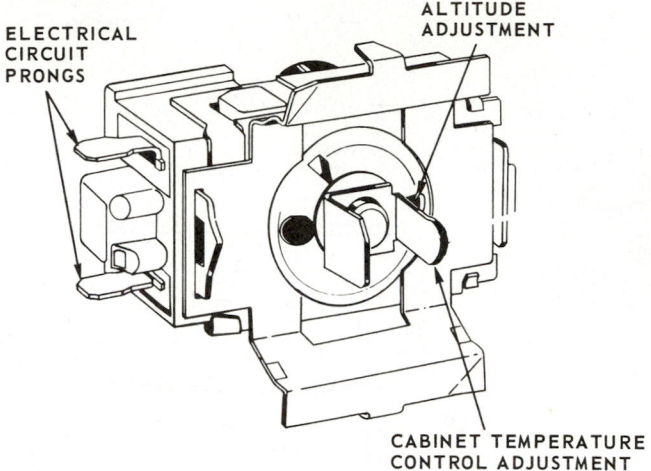

Fig. 10-4. Thermostat for a fresh food refrigerator.

thermostat control knob is attached to the control adjustment. The altitude adjustment is turned clockwise as the altitude increases. Refer to Chapter 8 for instructions on how to adjust these thermostats.

Most hermetic refrigerators use a starting relay which is usually mounted on the body of the motor compressor. These starting relays also provide overload protection for the motor.

The overload protector contains a resistor wired in series with the running current. Should the current draw be too great (overload), the resistor will heat up and cause a bimetal contact to break the circuit.

Fig. 10-5 illustrates the electrical circuits for this refrigerator. The starting relay is shown at A. It connects both the starting winding and the running winding to the power circuit. Then it disconnects the starting winding from the power circuit as soon as the compressor motor has reached about 75 percent of running speed.

10-10 REFRIGERATOR WITH FROZEN FOOD COMPARTMENT AND MANUAL DEFROST

This unit, Fig. 10-6, consists essentially of two refrigerated spaces. It has a frozen food compartment across the top of the cabinet where temperature is kept at approximately 0 F. (−18 C.). A fresh food compartment below maintains a temperature of about 35 to 40 F. (2 to 4 C.).

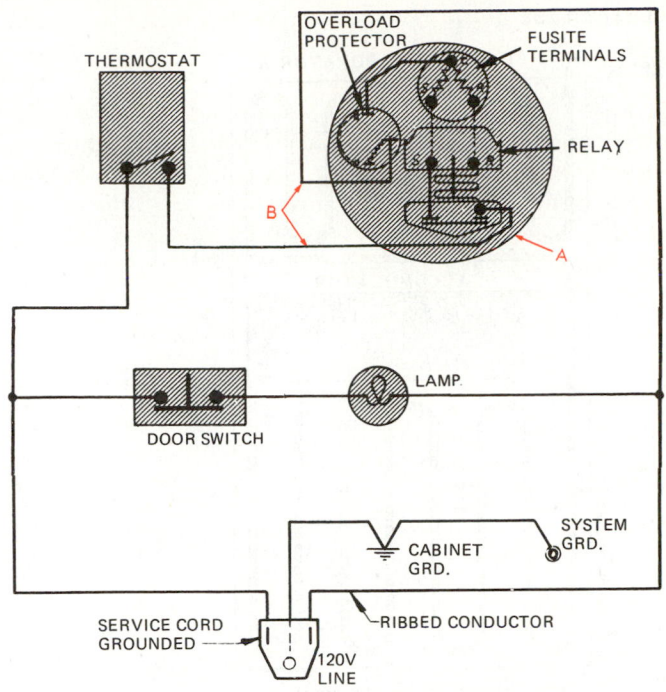

Fig. 10-5. Motor control relay for a fresh food refrigerator. This illustrates the electrical connections to thermostat and motor control relay. This motor compressor relay has an overload protector. A—Starting relay. B—Power supply leads. S—Starting winding terminal. R—Running winding terminal. C—Common terminal.

Fig. 10-6. Two-door model refrigerator-freezer with freezer at top. Note evaporator in upper part of lower compartment. (Kelvinator, Inc.)

Each of these compartments has a separate door. The condensing unit is usually in the bottom of the cabinet with the condenser either in the bottom or at the back.

These refrigerators provide shelves in both compartments. A butter conditioner is usually located in the door of the fresh food compartment. The same door and also the frozen food compartment door is often fitted with narrow shelves for storage of small containers.

10-11 CABINETS FOR REFRIGERATORS WITH FROZEN FOOD COMPARTMENT, MANUAL DEFROST

Cabinet construction for this refrigerator is much the same as for the simple fresh food refrigerator. However, since there is a frozen food compartment, the insulation is usually thicker so that a low temperature may be maintained easily in the frozen food compartment. A separate door is also provided.

The motor control, temperature control, light switch, shelf support and crisper are the same as those on the fresh food refrigerator. The finish is usually a good grade of lacquer, although some cabinets have porcelain finished interiors.

10-12 MECHANISMS FOR REFRIGERATOR WITH FROZEN FOOD COMPARTMENT, MANUAL DEFROST

Refrigerators with a frozen food compartment have a hermetic compressor in the base of the cabinet. The condenser is either at the bottom or the back of the cabinet. The liquid refrigerant flows from the capillary tube into the evaporator in the freezing compartment.

A typical mechanism is illustrated in Fig. 10-7. The refrigerant charge is usually sufficient to keep the freezer evaporator, A, filled. It also has enough for spill-over from the freezer evaporator into the fresh food compartment evaporator, B. Usually located at the back of the fresh food compartment, this evaporator has a rather large accumulator. Any possible spill-over from this enters a third evaporator, C, usually located at the side of the fresh food compartment. This third evaporator is also fitted with an accumulator. This assures that all the refrigerant is evaporated before the vapor is allowed to enter the suction line.

From this third coil accumulator, the vaporized refrigerant is drawn back to the suction line into the compressor. The vapor is compressed and pumped first into a small condensing coil D. From here the high-pressure vapor is pumped through a loop in the base of the compressor which serves as an oil cooler. From here the compressed vapor flows into condenser, E, at the bottom or at the back of the refrigerator.

Here the heat of the vapor is radiated to the surrounding air and the refrigerant is condensed back to a liquid. The liquid flows from the bottom of the condenser through the filter-drier and into the capillary tube attached to the suction line. The capillary tube controls the refrigerant flow into the freezer evaporator A. The cycle is then repeated.

This refrigerator is manually defrosted.

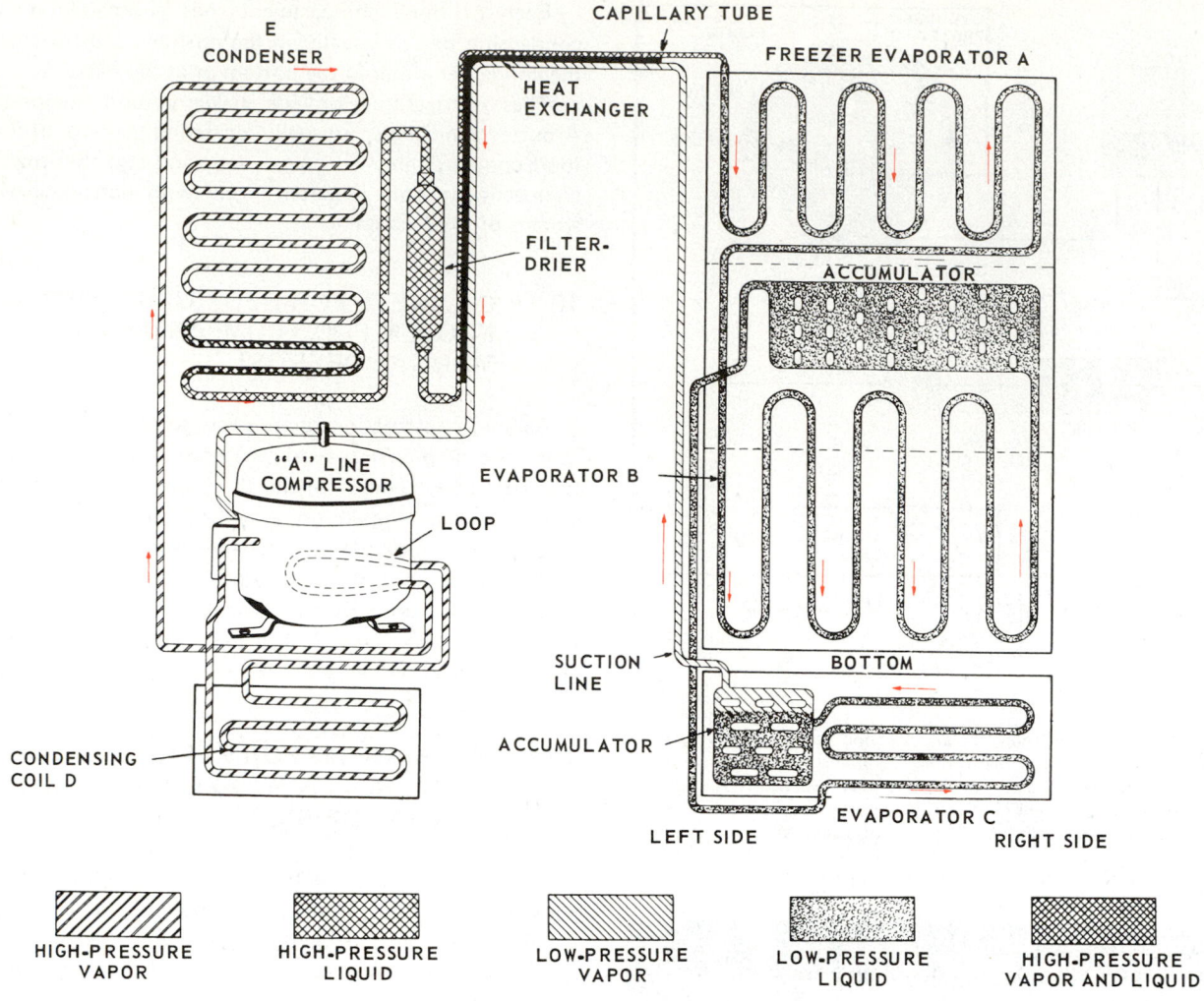

CAPILLARY TUBE

E
CONDENSER

FREEZER EVAPORATOR A

HEAT
EXCHANGER

FILTER-
DRIER

"A" LINE
COMPRESSOR

LOOP

ACCUMULATOR

EVAPORATOR B

SUCTION
LINE

CONDENSING
COIL D

ACCUMULATOR

EVAPORATOR C

LEFT SIDE

BOTTOM

RIGHT SIDE

HIGH-PRESSURE
VAPOR

HIGH-PRESSURE
LIQUID

LOW-PRESSURE
VAPOR

LOW-PRESSURE
LIQUID

HIGH-PRESSURE
VAPOR AND LIQUID

Fig. 10-7. Cycle diagram, refrigerator with frozen food compartment, manual defrost.

10-13 ELECTRICAL CIRCUITS FOR REFRIGERATOR WITH FROZEN FOOD COMPARTMENT, MANUAL DEFROST

The electrical supply is through the grounded extension cord and plug at 1 in Fig. 10-8. The electrical circuit supplies current to the panel disconnect at 2.

From the panel mounted disconnect, two separate circuits are provided. One circuit supplies current to the cabinet light at 3. It is controlled by cabinet switch 4. The cabinet switch is in series with the light. The light comes on or goes off as the refrigerator door is opened and closed.

The second circuit goes to the motor compressor. The thermostat at 5 is in series with the circuit and controls the running of the compressor at 6.

When the temperature in the cabinet climbs to a determined temperature, the thermostat completes the circuit through the motor. The compressor runs and the refrigeration cycle comes into operation. As soon as the temperature inside the cabinet has been brought down to the minimum desired temperature, the thermostat turns off the current and the motor compressor stops.

Two additional electrical devices are usually used on this type of refrigerator. One is an electrical resistance heat wire. Also called a perimeter drier, it is shown at 7. It is placed in the trim of the freezer door.

This heater operates all the time. It provides enough "warming effect" to stop condensation on the exterior of the cabinet and around the freezing compartment door.

The second electrical device is an ambient compensator shown at 8. This is also an electrical resistance heat wire. It draws 15 to 20W, is attached to the insulation side of the fresh food compartment thermostat and is energized only on the "off" cycle (thermostat contacts open).

The purpose of the ambient compensator is to provide a continuous small heat flow into the fresh food compartment. This will cause the refrigerator to cycle in the event the ambient temperature of the room in which the refrigerator is located is lower than the normal thermostat setting.

Fig. 10-9 illustrates the electrical circuits for this refrigerator. The starting relay is shown at A. It connects both the starting winding and the running winding to the power circuit and disconnects the starting winding as soon as the compressor motor has reached approximately 75 percent of running speed.

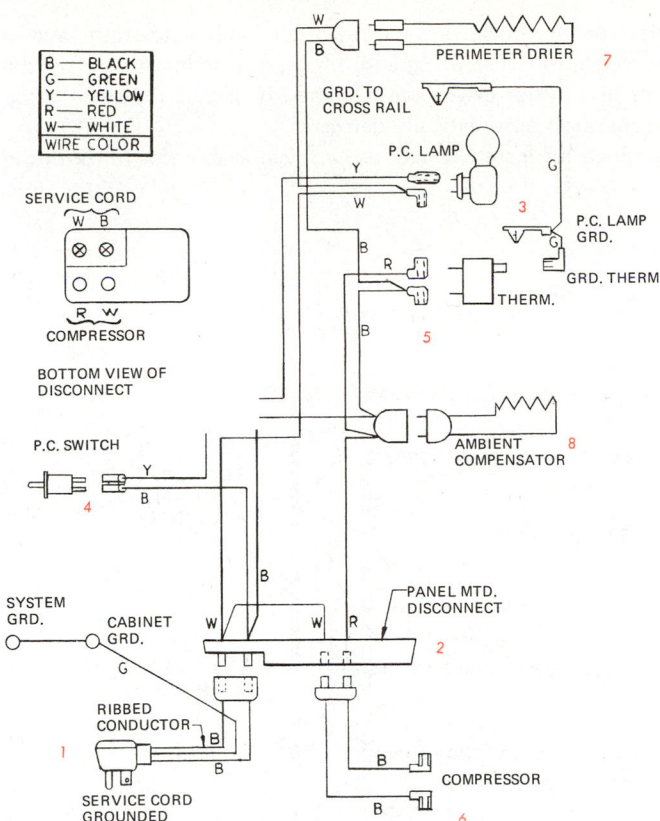

Fig. 10-8. Wiring diagram, refrigerator with frozen food compartment, manual defrost.

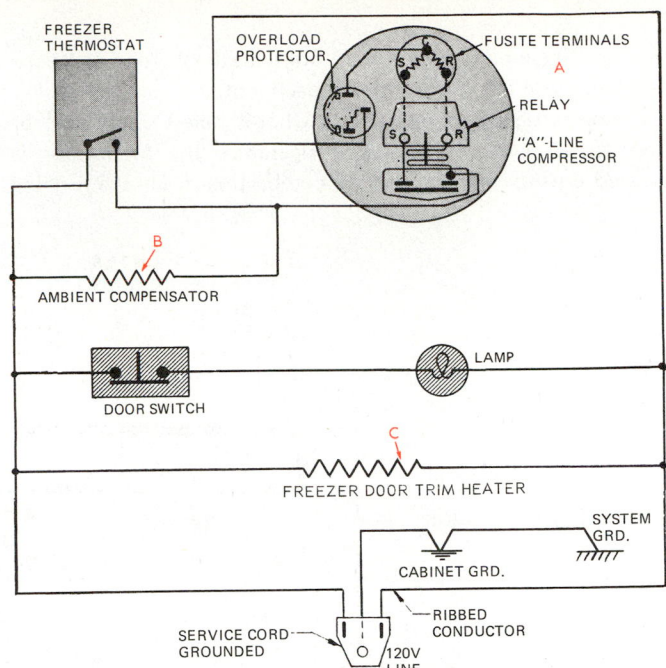

Fig. 10-9. Motor control relay, refrigerator with frozen food compartment, manual defrost. This illustrates the electrical connections to the thermostat and the motor control relay. This motor compressor relay is provided with an overload protector.

This starting relay also provides overload protection for the motor. The overload protector contains a resistor in series with the running current. In the event the current draw is too great (overload), the resistor will heat up and cause a bimetal contactor to break the circuit. The ambient compensator is shown at B. The freezer door trim heater is shown at C.

10-14 REFRIGERATOR WITH FROZEN FOOD COMPARTMENT, AUTOMATIC DEFROST

All air contains some moisture. When air comes in contact with an evaporator surface which is below the freezing temperature, the moisture will condense and form ice on the evaporator.

As mentioned in the earlier paragraphs, it is frequently necessary to defrost the evaporator in order to maintain good refrigerating efficiency. This applies to the evaporator for both the fresh food and the frozen food compartments.

The owner considers it a chore to manually defrost the refrigerator. As a result, many refrigerators provide a system for automatic defrosting. There are two basic systems used in automatic defrosting. The hot gas system, through the use of solenoid valves, uses the heat in the vapor from the compressor discharge line and the condenser to defrost the evaporator.

The other system uses electric heaters to melt the ice on the evaporator surface. A paragraph in this chapter will be devoted to each of these systems.

Fig. 10-10 illustrates a typical refrigerator with a frozen

food compartment. This refrigerator uses an automatic defrost. As is the case with most domestic refrigerators, the condensing unit is mounted in the base of the cabinet, with the condenser either in the bottom or at the back of the

Fig. 10-10. Refrigerator with frozen food compartment, automatic defrost. Evaporator in frozen food compartment serves as fast-freezing shelf. The fresh food cabinet provides butter conditioner, fresh meat storage and vegetable crisper. (Amana Refrigeration, Inc.)

cabinet. Shelving is provided in both the fresh food compartment and in the frozen food compartment.

Some refrigerators operate on what is called "frost free" or "no frost" cycle. In these refrigerators the evaporator is located outside the refrigerated compartment. On the running part of the cycle, air is drawn over this evaporator and is forced into the freezing and refrigerator compartment by the use of a motor driven fan. On the off part of the cycle, these evaporators automatically defrost.

Such refrigerators may use a single evaporator for both the

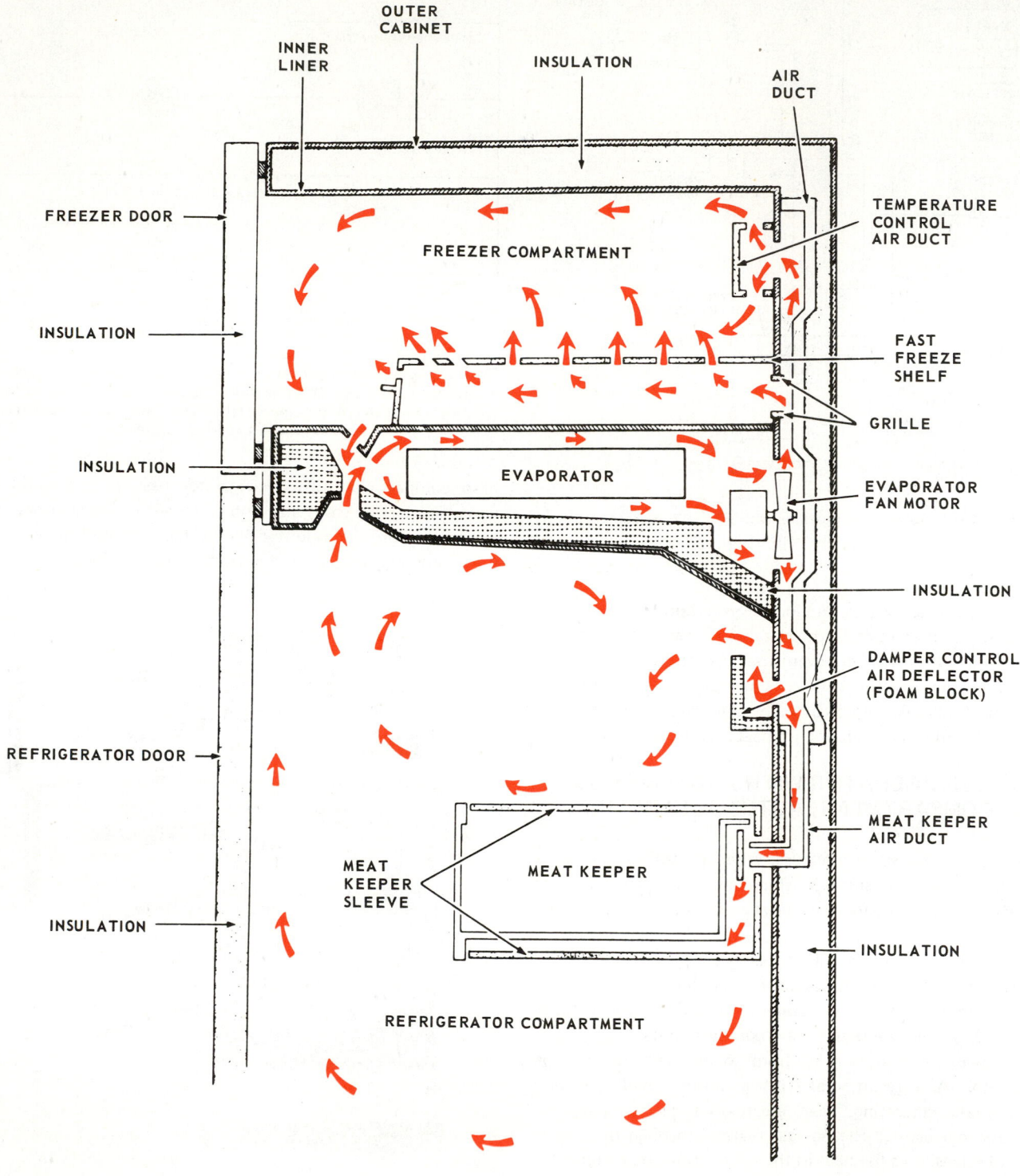

TOP MOUNT AIRFLOW PATTERN

Fig. 10-11. Air circulation in refrigerator with frozen food compartment. Evaporator fan forces circulation of cooled air through various compartments. Dampers control air duct openings. Therefore, they control various cabinet temperatures.

frozen food compartment and the fresh food compartment or a separate evaporator may be used in each compartment. The condensation from the evaporator which melts off during the off cycle is carried to an evaporating pan or collecting surface directly over the compressor and condenser. The heat from the compressor evaporates this moisture and it is returned to the room's atmosphere. There is never any visible frost accumulation in this type of frost control.

10-15 CABINETS FOR REFRIGERATORS WITH FROZEN FOOD COMPARTMENT AND ELECTRIC AUTOMATIC DEFROST

Automatic defrosting requires some important differences in the cabinet construction. Since the condensation which collects on the evaporator must be melted from time to time, different methods are used for disposing of this condensation. Usually, it is collected on a plate or surface just over the motor compressor. This surface, or plate, is heated by the motor compressor and by heat from the condenser. The moisture then reevaporates and goes back into the room. Provisions must be made in the cabinet for tubing to conduct this moisture to the top of the motor compressor.

Some refrigerators use electric defrost. Provisions must be made in the cabinet for housing electric heaters and their controls.

Fig. 10-11 shows a refrigerator with a frozen food compartment that provides other temperatures besides the temperature of the food freezer and the temperature of the fresh food compartment. The evaporator is beneath the fast-freezing shelf at the bottom of the freezing compartment.

All of the refrigerating effect comes from this evaporator. An air circulation system, consisting of a fan, ducts and a damper at the back of the cabinet, provides the necessary airflow to give the temperature desired in each compartment. The condensing unit is located in the bottom of the cabinet.

10-16 MECHANISMS FOR REFRIGERATOR WITH FROZEN FOOD COMPARTMENT, AUTOMATIC ELECTRIC DEFROST

Fig. 10-12 is typical of a mechanism which provides for more than two temperatures in the refrigerator with a frozen food compartment.

EVAPORATOR

This is located at the bottom of the shelf which separates the freezing compartment from the fresh food compartment. Refrigerant is R-12. The evaporation of refrigerant in the evaporator provides all of the heat absorption (cooling) required in the cabinet. A motor-driven fan which forces air over the evaporator surface and through the various ducts provides all the necessary refrigerator temperatures for the compartments.

MOTOR COMPRESSOR

The suction line from the evaporator extends down the wall of the cabinet to the inlet side of the hermetic motor compressor, which is located in the base of the cabinet.

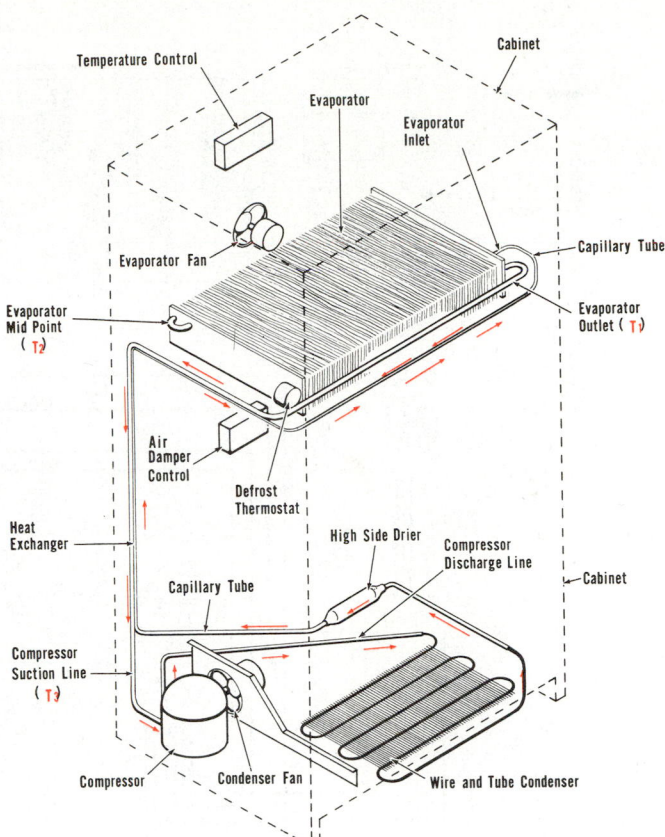

REFRIGERANT SYSTEM SCHEMATIC AND TEMPERATURE TEST POINTS

Fig. 10-12. Cycle diagram for a refrigerator with frozen food compartment and automatic electrical defrost. Proper operation may be checked by determining temperatures at test points, T_1, T_2 and T_3. Recommended operating temperatures are: $T_1 = -17$ to -16 F. (-27 C.). $T_2 = -17$ to -16 F. (-27 C.). $T_3 = 75$ to 95 F. (24 to 35 C.). (Amana Refrigeration, Inc.)

CONDENSER

The condenser is a wire and tube type. Forced air circulation is provided by a motor and fan located in the compartment between the compressor and the condenser.

CAPILLARY TUBE

As the refrigerant is condensed in the condenser, it flows through a high-side filter-drier into a capillary tube attached to a section of the suction line. This provides a heat exchange between the capillary tube and the suction line. The refrigerant from the capillary tube flows into the evaporator and the cooling cycle is completed.

FEATURES

Only the fresh food compartment has a light. It is operated by a switch controlled by the door. The butter conditioner temperature is slightly above the cabinet temperature. There are control dampers for the refrigerator and the freezer compartments. These dampers regulate flow of cold air from the evaporator.

HEATERS

Several heating devices are used as driers. An electrical resistance heater at the top of the cabinet, inside the outer case, keeps the outside of the cabinet warm enough so that it will not collect condensation during damp days.

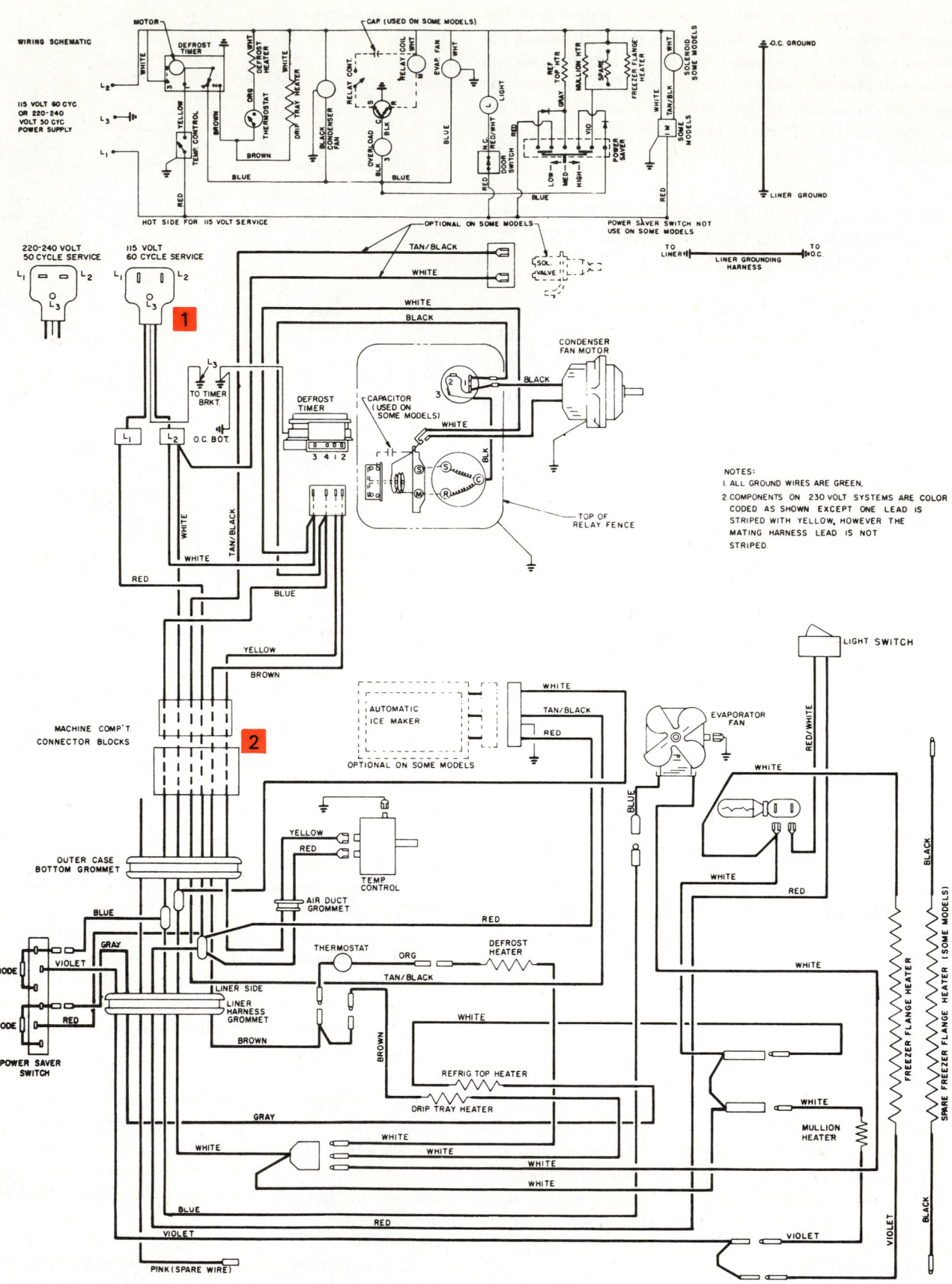

Fig. 10-13. Wiring diagram for refrigerator with frozen food compartment and automatic electric defrost.
(Amana Refrigeration, Inc.)

A second drier wire is placed inside the center mullion to keep its surface dry. A third is around the freezer flange (freezer door opening). A fourth is placed in the drip pan to help evaporate the condensation which flows into the drip pan after automatic defrost.

A "power saver" switch inside the cabinet provides a means of disconnecting these heaters when temperature and humidity conditions allow it. Normally, the heaters are in continuous operation.

AUTOMATIC DEFROST

The evaporator is automatically defrosted by an electric resistance heater every six hours of compressor running time. The heater is located in the fin area on the underside of the evaporator. A timer energizes the switch which turns on the defroster. A thermostat attached to the evaporator opens the defrost heater circuit to end defrosting as soon as the evaporator reaches a temperature of 50 F. (10 C.), plus or minus six degrees F. (three degrees C.). Twenty-eight minutes after the start of the defrost cycle the timer restores the operation of the compressor and the air circulating fan. The defrost terminator contacts close at 20 F. (−7 C.), plus or minus eight degrees F. (four degrees C.). The temperatures of the cabinet are regulated by the temperature control mounted in the rear wall of the freezer compartment. The temperature of the evaporator tubing near the end of a running cycle may vary from −13 F. (−24 C.) to −25 F. (−32 C.). The difference between the temperature of the inlet and the outlet to the evaporator will not vary more than three degrees F. (two degrees C.).

10-17 ELECTRICAL CIRCUITS FOR REFRIGERATORS WITH FROZEN FOOD COMPARTMENT AND AUTOMATIC ELECTRIC DEFROST

The electrical supply is through the grounded extension cord and plug shown at 1 in Fig. 10-13. The compressor, defrost timer and light switch are energized directly from the grounded extension cord.

A machine compartment connector block, shown at 2, is located in the bottom of the cabinet. It provides electrical connections to the evaporator fan, cabinet light, top cabinet heater, defrost heater, drip tray heater, defrost terminator, mullion heater, freezer flange heater and the power saver switch. If the cabinet is equipped with an automatic ice maker, it will be wired into the same connector block. A schematic of the wiring circuit is shown at the top of Fig. 10-13.

10-18 CABINETS FOR REFRIGERATORS WITH FROZEN FOOD COMPARTMENT AND HOT GAS AUTOMATIC DEFROST

These cabinets provide the usual facilities for both fresh and frozen food storage. The thermostat is located in the fresh food compartment.

The defrost timer and the solenoid valve are housed in the cabinet base. In addition, many refrigerators pipe compressed vapor through tubing located on the inside surface of the

outside shelf, particularly around the door openings. The purpose of this is to supply some heat to these surfaces in order to eliminate any possible condensation.

Since their purpose is to keep the surface of the refrigerator cabinet dry, these coils are sometimes referred to as driers. The same term may be applied to the electric heating units installed in the cabinet for the same purpose. *This use of the word drier should not be confused with the term drier as applied to liquid and suction lines.*

Fig. 10-14 illustrates a refrigerator with a frozen food compartment and "hot gas" defrost.

Fig. 10-14. Refrigerator with frozen food compartment and hot gas defrost. The evaporator in frozen food compartment is fast-freezing shelf. In addition, the evaporator extends along the back and ends of frozen food space. Thermostat is in fresh food compartment. Fresh food cabinet provides for butter conditioner and two vegetable crispers. (Kelvinator, Inc.)

10-19 MECHANISMS FOR REFRIGERATORS WITH FROZEN FOOD COMPARTMENT AND HOT GAS AUTOMATIC DEFROST

The mechanism illustrated in Fig. 10-15 is typical of those used in refrigerators with a frozen food compartment and fresh food compartment using "hot gas" defrost.

EVAPORATOR

Two are used with R-12 as the refrigerant. Evaporator A is located in the freezing compartment. It extends along the back and ends of the compartment as well as along the bottom. Refrigerant (both vapor and liquid) leaving this evaporator flows into evaporator, B, located at the back of the fresh food compartment. This evaporator has a rather large accumulator. This is necessary since all refrigerant leaving the fresh food compartment evaporator should be in vapor form.

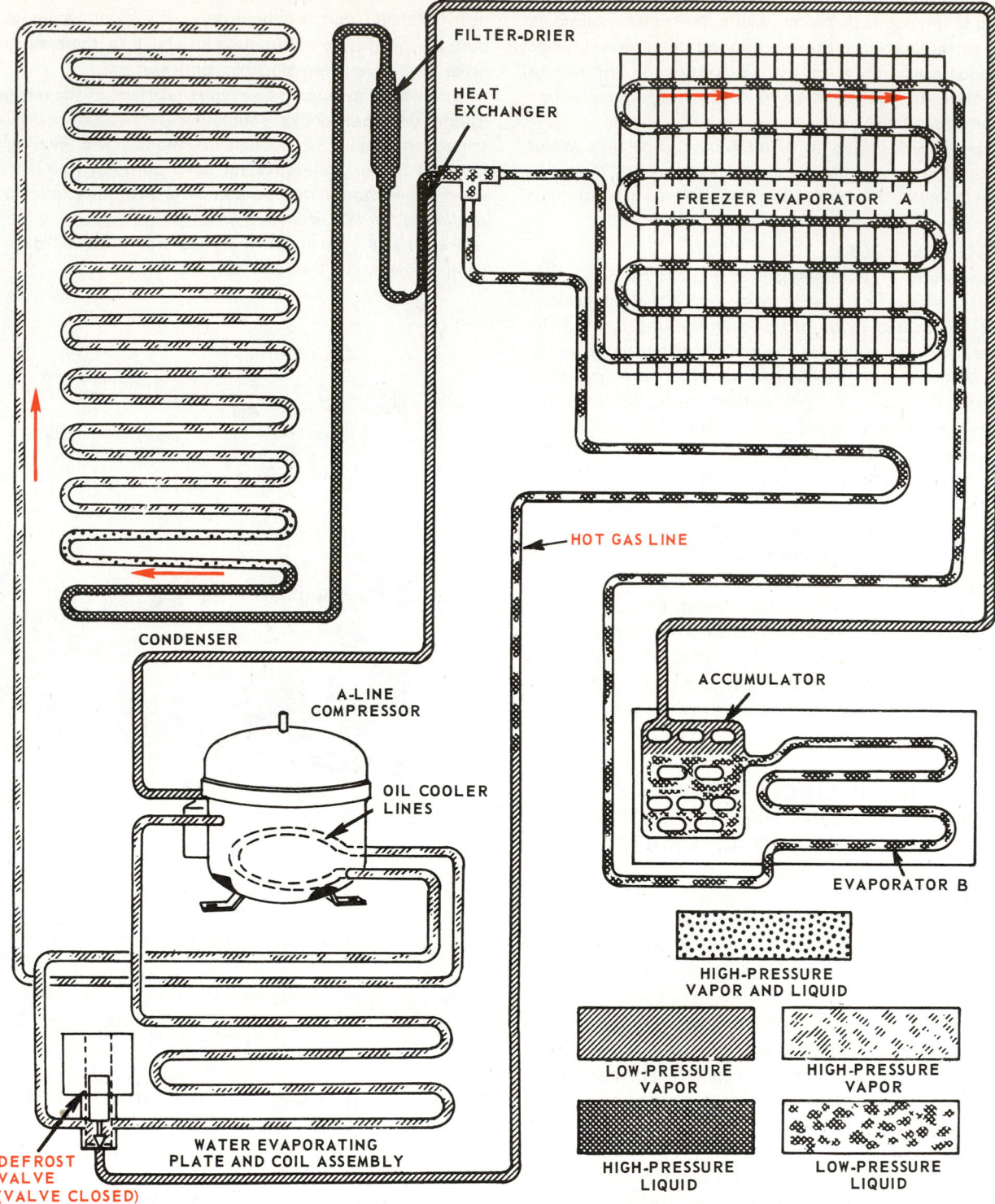

FILTER-DRIER

HEAT EXCHANGER

FREEZER EVAPORATOR A

HOT GAS LINE

CONDENSER

A-LINE COMPRESSOR

OIL COOLER LINES

ACCUMULATOR

EVAPORATOR B

HIGH-PRESSURE VAPOR AND LIQUID

LOW-PRESSURE VAPOR

HIGH-PRESSURE VAPOR

HIGH-PRESSURE LIQUID

LOW-PRESSURE LIQUID

DEFROST VALVE (VALVE CLOSED)

WATER EVAPORATING PLATE AND COIL ASSEMBLY

Fig. 10-15. Refrigerating cycle for a hot gas defrost. Solenoid valve is closed. This closes off bypass line.

MOTOR COMPRESSOR

Vapor is drawn back to the compressor where it is discharged directly into a water evaporating plate and coil assembly. This is located over the compressor and serves to evaporate the moisture drained from the evaporators during the defrost cycle. In normal operation, this compressed vapor next flows through the oil cooler line in the bottom of the compressor.

CONDENSER

From the oil cooler line, vapor flows to the vertical wire tube type condenser. Heat is given off to the surrounding air and the compressed vapor returns to a liquid state. From the bottom of the condenser, the liquid flows through the filter-drier.

CAPILLARY TUBE

From the filter-drier, the liquid then flows into the

capillary tube refrigerant control soldered to the suction line. This serves as a heat exchanger. It reduces the temperature of the liquid refrigerant in the capillary tube and increases the superheat of the vapor in the suction line. From the capillary tube, the refrigerant enters the freezer evaporator at a reduced pressure. It evaporates, absorbs heat from the inside of the cabinet, and the cycle is completed.

HOT GAS DEFROST

Fig. 10-16 shows the solenoid valve open and the hot gas defrost cycle in operation. In this cycle, starting with the compressor, the vapor from the evaporators is drawn into the compressor. It discharges into the water evaporating plate, helping to heat this surface. Then, hot compressed vapor flows back through the drain sump bypass line, into the freezer

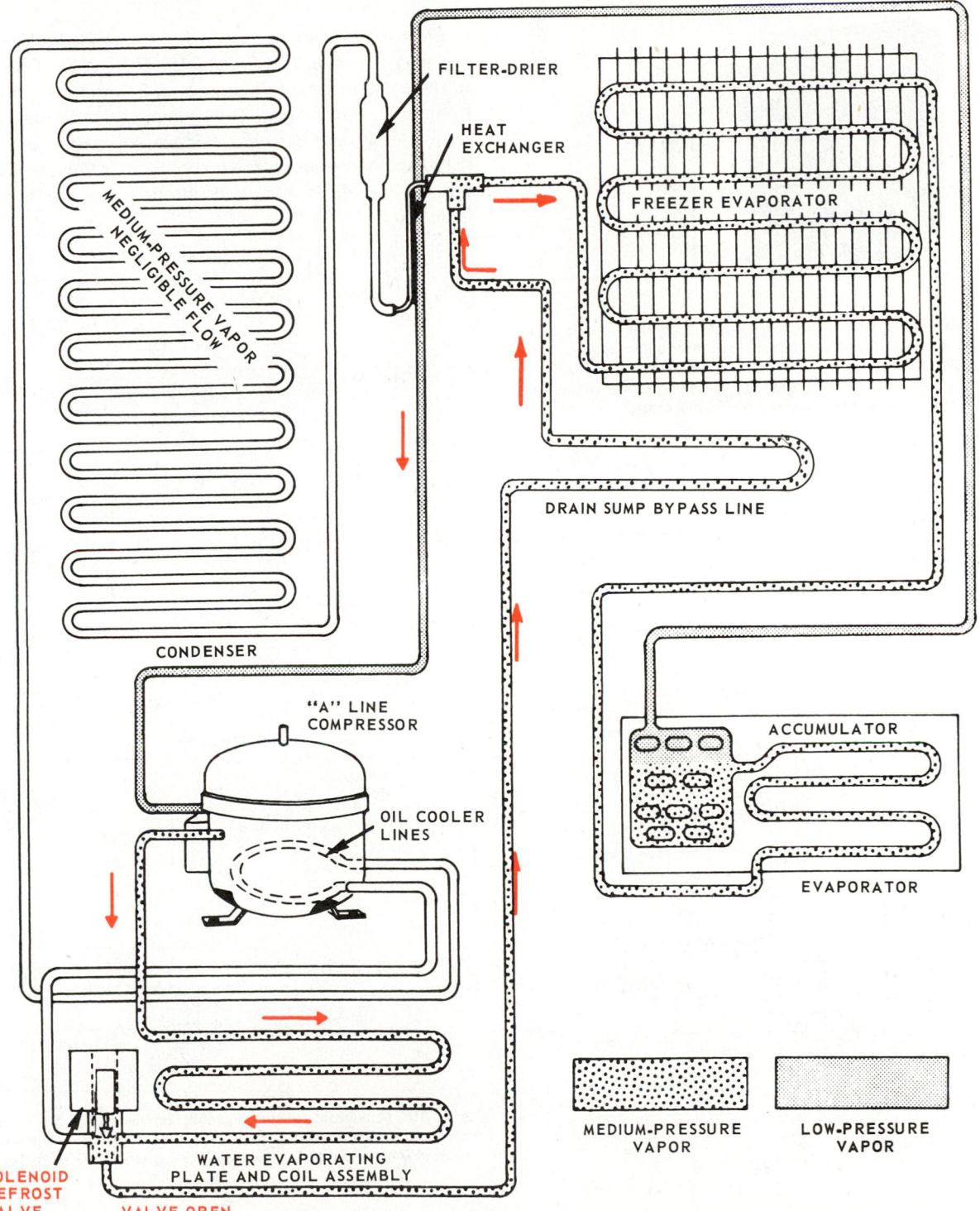

Fig. 10-16. Defrost cycle, hot gas defrost. Solenoid valve is open and hot compressed vapor is traveling through bypass line. The heat from compressed vapor keeps drain sump ice free. It melts ice from both freezer evaporator and fresh food evaporator.

evaporator. From the freezer evaporator, the vapor flows into the fresh food compartment evaporator through the accumulator and back into the suction side of the compressor. Since this compressed vapor is quite warm, it quickly defrosts the evaporators, the heat being supplied directly to the inside surfaces of the evaporators.

A defrost timer, Fig. 10-17, controls the solenoid defrost

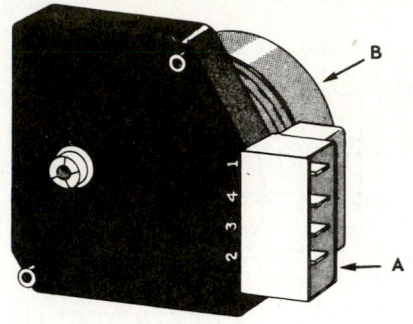

Fig. 10-17. Automatic defrost timer. A—Timer is connected into motor compressor circuit and provides both a clock mechanism and switching mechanism. Clock energizes switch mechanism in such a way that, after eight hours of compressor operation, fan motor is turned off and defrost solenoid is energized. B—Timer motor.

valve. The timer is located at the back lower left corner of the refrigerator cabinet. It is driven by a self-starting electric motor geared to turn the shaft slowly. The shaft completes one revolution every eight hours of compressor operation. When the timer opens the solenoid valve, the defrost cycle

continues for 17 minutes. Then, the solenoid valve closes and the refrigerating cycle operates. The freezer compartment fan is not operating during the defrost cycle.

10-20 ELECTRICAL CIRCUITS FOR REFRIGERATORS WITH A FROZEN FOOD COMPARTMENT AND HOT GAS AUTOMATIC DEFROST

Electricity is supplied through the grounded extension cord and plug shown at 1 in Fig. 10-18. The electrical supply goes to the panel mounted disconnect at 2. Various circuits are fed from this disconnect. The compressor, defrost solenoid, (shown at 3) and defrost timer are energized from the four right-hand connections. If an automatic ice maker is used, the water valve, 4, is energized through the two terminals as indicated.

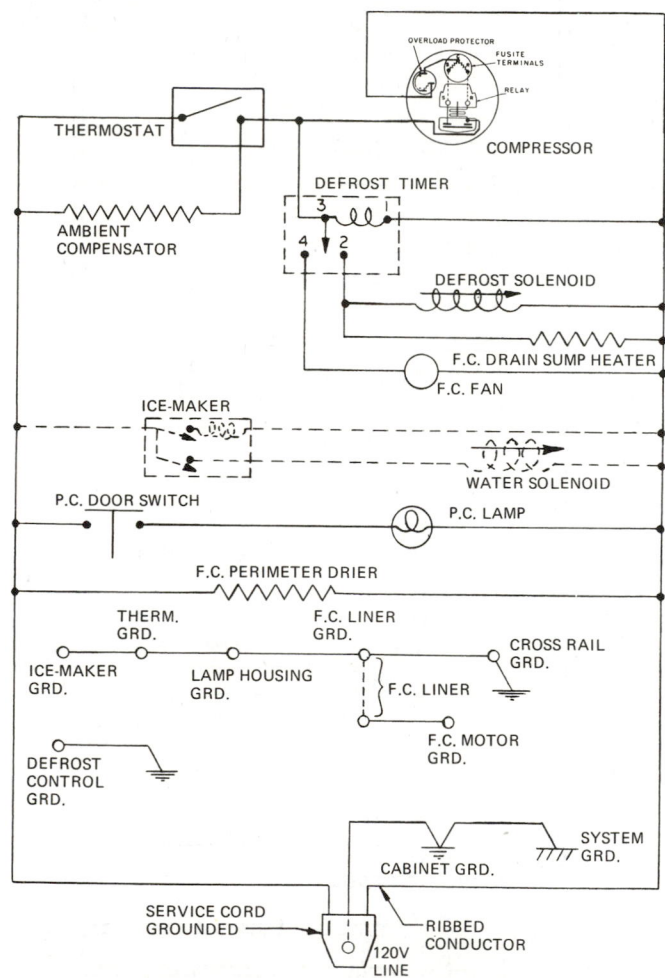

Fig. 10-19. Schematic wiring diagram, refrigerator with frozen food compartment, automatic defrost, hot gas. Note the provisions for the necessary grounding of the various components.   (Kelvinator, Inc.)

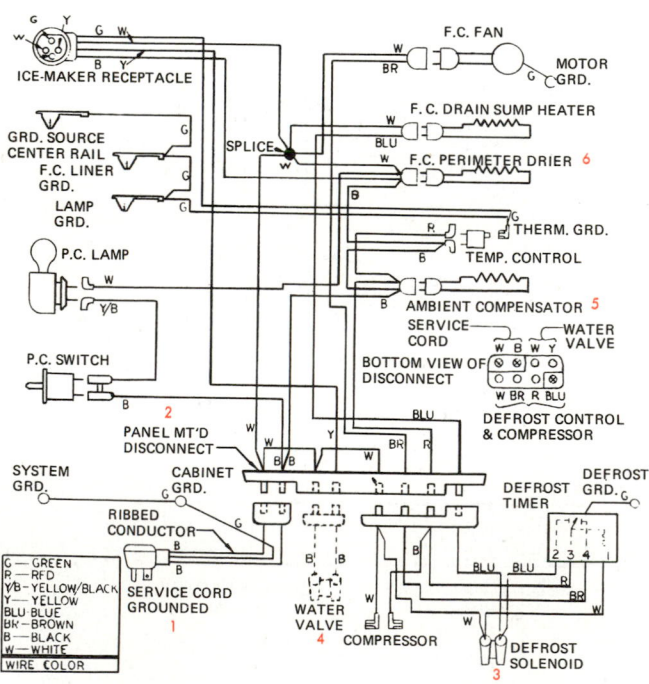

Fig. 10-18. A wiring diagram for a refrigerator with frozen food compartment and hot gas automatic defrost.

In addition to the hot gas defrost, there are three electrical resistance heaters in the system. The ambient compensator at 5 is attached to the insulation side of the fresh food compartment. The perimeter drier, 6, is installed in the trim of the freezer door. The drain sump heater prevents freezing of condensation from the evaporator as it moves down the drain tube to the moisture evaporating pan located over the motor.

This wiring diagram provides a complete grounding system. Each metal part of the mechanism and cabinet has its own ground as specified in Para. 6-66. Fig. 10-19 is a schematic of this wiring diagram.

10-21 FROST-FREE REFRIGERATOR WITH FROZEN FOOD COMPARTMENT

Modern refrigerators are designed to eliminate the task of defrosting the evaporator. Frost-free refrigerators are very popular. Para. 10-14 explains how they work.

10-22 CABINETS FOR FROST-FREE REFRIGERATORS WITH FROZEN FOOD COMPARTMENTS

The refrigerator in Fig. 10-20 stores frozen food at the top and fresh food at the bottom.

The evaporator is in the upper back part of the cabinet. The condenser is along the lower back part. A fan moves cold air from the evaporator in the frozen food compartment into the fresh food compartment. Another fan circulates room air

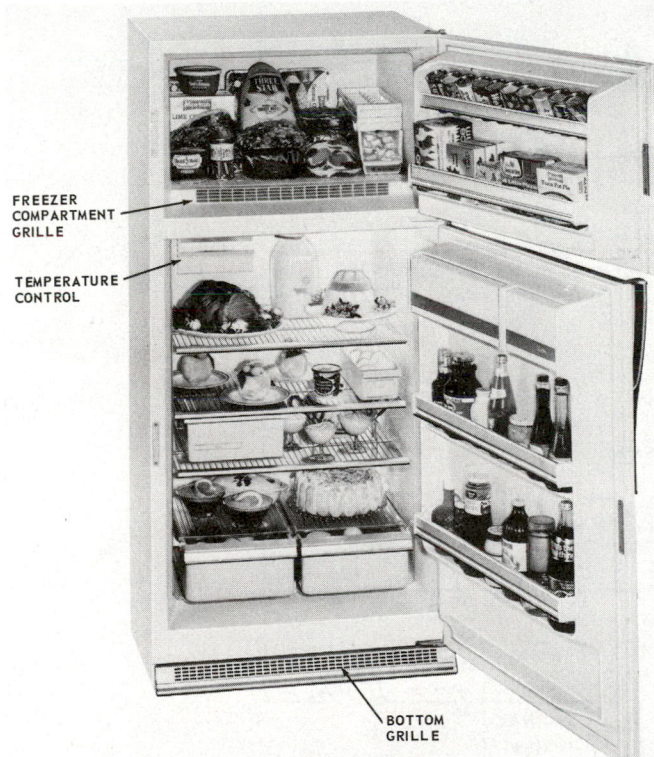

Fig. 10-20. Refrigerator with frozen food compartment, frost-free. The frozen food compartment is cooled by very cold air from evaporator which enters through freezing compartment grille. Condenser is cooled by air which circulates in and out through bottom grille. Evaporator fan is turned off automatically when freezing compartment door is opened. (General Electric Co.)

through the grille at the bottom of the cabinet and over the condenser. *With this type of condenser, it is not necessary to provide any clearance space at the sides or top of the cabinet for air circulation.*

Both doors are held shut by magnets. This maintains a tight seal without the need for a mechanical latch.

The cabinet is mounted on wheels making it easy to roll the refrigerator out for cleaning.

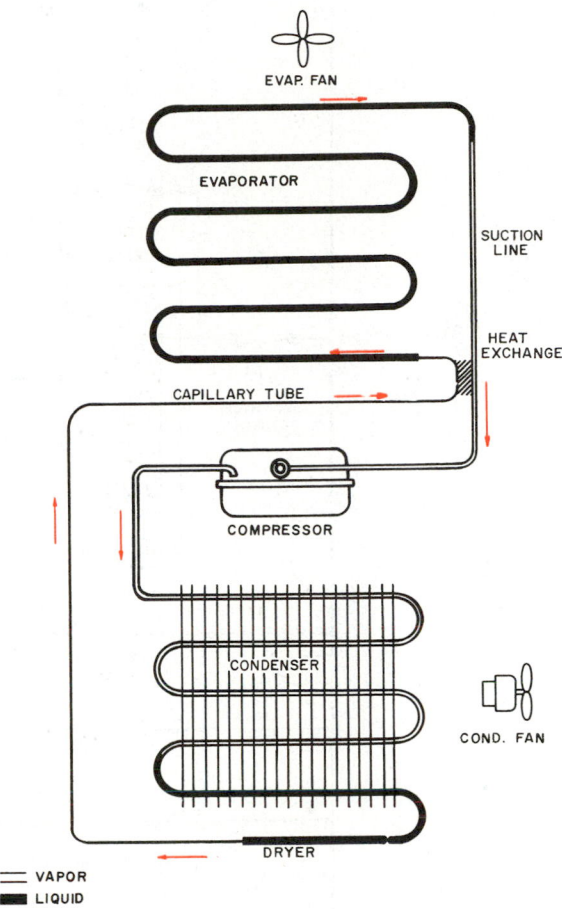

Fig. 10-21. Refrigerator with frozen food compartment, frost-free schematic diagram shows operating cycle.

10-23 FROST-FREE MECHANISMS FOR REFRIGERATORS WITH FROZEN FOOD COMPARTMENTS

Fig. 10-21 is a diagram of the mechanism for this refrigerator. This illustrates the compressor, condenser, capillary tube, heat exchange and evaporator. Note that fans are used on both the condenser and evaporator.

10-24 ELECTRICAL CIRCUITS FOR FROST-FREE REFRIGERATORS WITH FROZEN FOOD COMPARTMENTS

Electricity is supplied through a grounded extension cord and plug shown at 1 in Fig. 10-22. There are several electrical

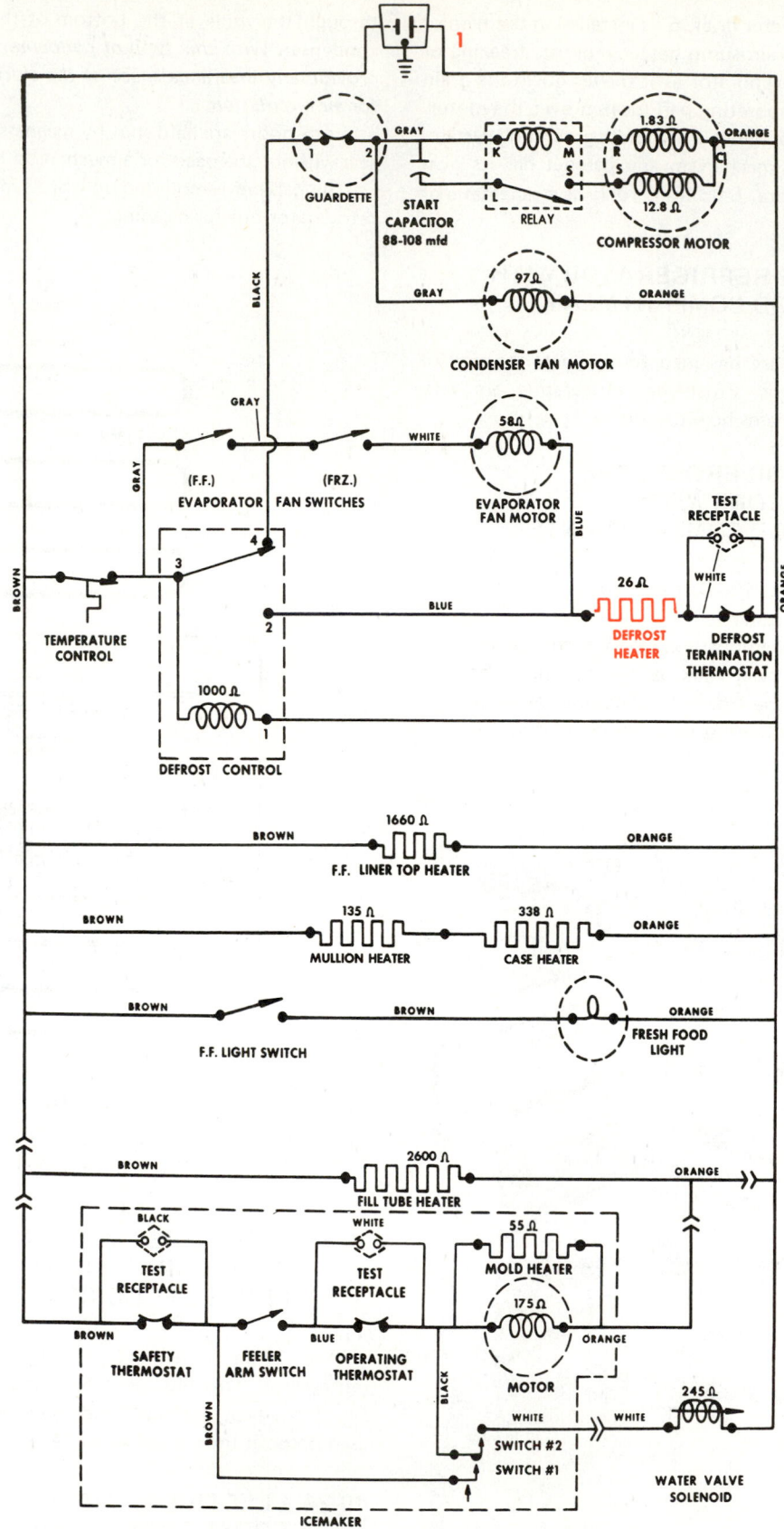

Fig. 10-22. A schematic wiring diagram for a frost-free refrigerator with a frozen food compartment above and a fresh food compartment below.

resistance heating elements (driers) used in this cabinet. Some of these are used to warm certain surfaces of the cabinet to eliminate moisture.

The evaporator defrost and the door mullion are electrically heated. In this illustration, the resistance of the various electrical components is indicated in ohms (Ω).

If an automatic ice maker is used, this electrical circuit provides for easily connecting it into the cabinet.

10-25 FROST-FREE REFRIGERATOR-FREEZERS (SIDE-BY-SIDE) WITH ICE MAKER

The side-by-side refrigerator-freezer arrangement is very popular. The frozen food compartment stores frozen foods at a satisfactory temperature of 0 F. (−18 C.) or below.

10-26 CABINETS FOR FROST-FREE REFRIGERATOR-FREEZERS (SIDE-BY-SIDE) WITH ICE MAKER (GENERAL ELECTRIC)

The side-by-side refrigerator-freezer in Fig. 10-23 provides a fresh food compartment on the right and a frozen food compartment on the left.

The temperature controls for both compartments are located at the top of the fresh food compartment.

Fig. 10-23. A side-by-side frost-free refrigerator-freezer, with automatic ice maker.

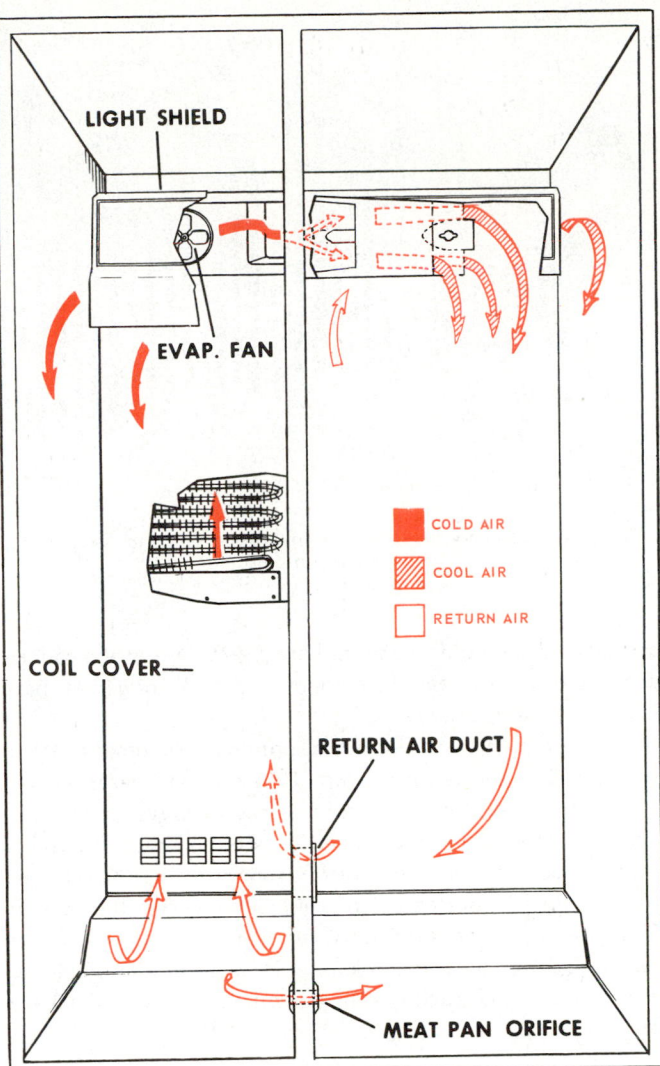

Fig. 10-24. Airflow diagram for a frost-free refrigerator-freezer, side-by-side.

10-27 MECHANISMS FOR FROST-FREE REFRIGERATOR-FREEZERS (SIDE-BY-SIDE) WITH ICE MAKER (GENERAL ELECTRIC)

Evaporator, compressor and condenser are at the back. A fan circulates air over the condenser. This air enters and leaves through the bottom grille.

A fan circulates very cold air from the evaporator into the freezer compartment. Damper arrangements allow some of this cold air to flow from the freezer compartment into the fresh food compartment. Warmer air in the fresh food compartment returns to the evaporator compartment. The airflow is shown in Fig. 10-24.

Refrigerant control is by capillary tube. This capillary tube is attached to the suction line, as shown in Fig. 10-25.

Door switches operate the lights in both freezer and fresh food compartments. A door switch also operates a fan in the freezer. When the door is open, the fan does not operate.

A resistance wire or wires, placed inside the center mullion

Fig. 10-25. Capillary tube and suction line connected to evaporator. (General Electric Co.)

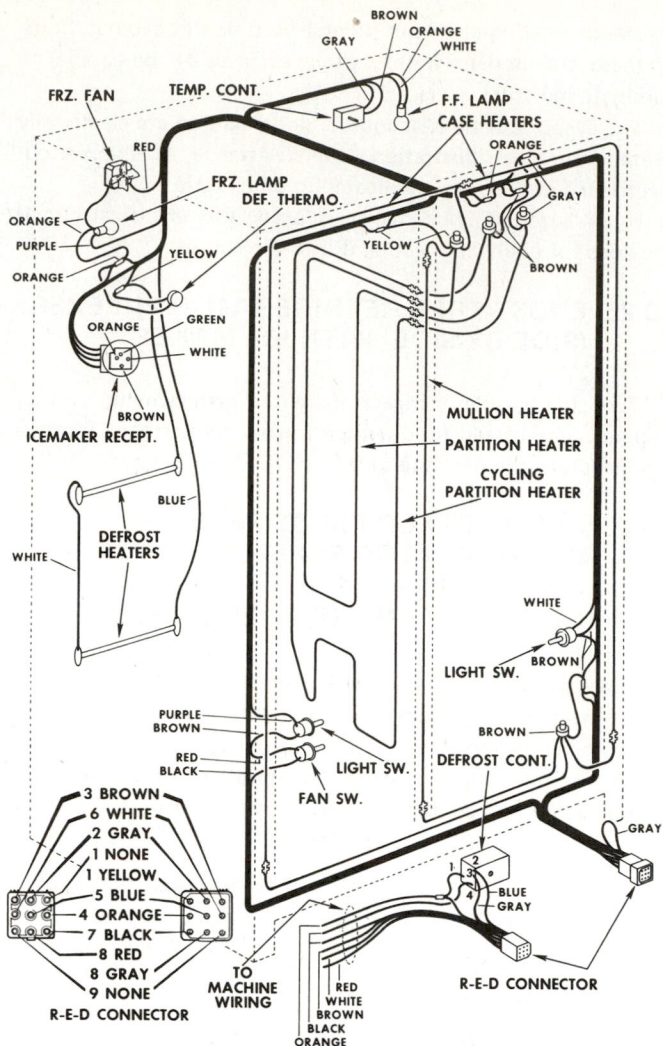

Fig. 10-26. Schematic wiring diagram of side-by-side refrigerator with automatic ice maker.

and around the door openings, heat the surface of the cabinet at these places and eliminate condensation. Door gaskets have magnets on all four sides.

An automatic ice maker is optional equipment. When included, it is located at the top of the cabinet freezer section.

The fresh food compartment defrosts on every off cycle. The freezer evaporator defrosts for about 25 minutes after a six-hour accumulated compressor running time. The freezer defrost timer is located on the cabinet front near the bottom. Moisture from the evaporator surface flows down to pan resting on top of the condenser. Heat from the condenser evaporates it. The cabinet is fitted with rollers, which makes it easy to move.

10-28 ELECTRICAL CIRCUITS FOR FROST-FREE (SIDE-BY-SIDE) REFRIGERATOR-FREEZERS WITH ICE MAKER (GENERAL ELECTRIC)

A grounded extension cord and plug supply electricity. Fig. 10-26 is a schematic illustration of the wiring. The individual circuits can be worked on at any time since wiring is not "foamed" in place. NOTE there are a considerable number of heat elements in the electrical circuit for defrost heaters, mullion heaters, partition heater and a partition cycling heater. Color coding locates the wire going to each of the electrical circuits.

The automatic ice maker controls are plugged into an ice maker receptacle. This eliminates the need to disturb any of the circuits when installing or removing the ice maker.

10-29 CABINET FOR FROST-FREE (SIDE-BY-SIDE) REFRIGERATOR-FREEZERS WITH ICE MAKER (AMANA)

The refrigerator in Fig. 10-27 provides a fresh food compartment on the right and a frozen food compartment on the left. Shelves in the door of the frozen food compartment

make the handling of small items quite convenient.

Airflow through the cabinet is shown in Fig. 10-28. A fan draws air through the single evaporator and distributes it throughout the cabinet. Dampers control temperatures in the various parts of the cabinet.

Door switches control two lights in the refrigerator compartment and one light in the freezer compartment. A resistance wire or wires are placed inside the center mullion and around the door openings to control condensation.

Other features include magnetic door gaskets on four sides. An automatic ice maker is optional and rollers make the unit easy to move.

10-30 MECHANISMS FOR A SIDE-BY-SIDE FROST-FREE REFRIGERATOR-FREEZER WITH ICE MAKER (AMANA)

The evaporator is behind the frozen food compartment, while the compressor and condenser are in the bottom. Air, circulated over the condenser by a fan, enters and leaves

Fig. 10-27. A side-by-side refrigerator-freezer with ice maker.
(Amana Refrigeration, Inc.)

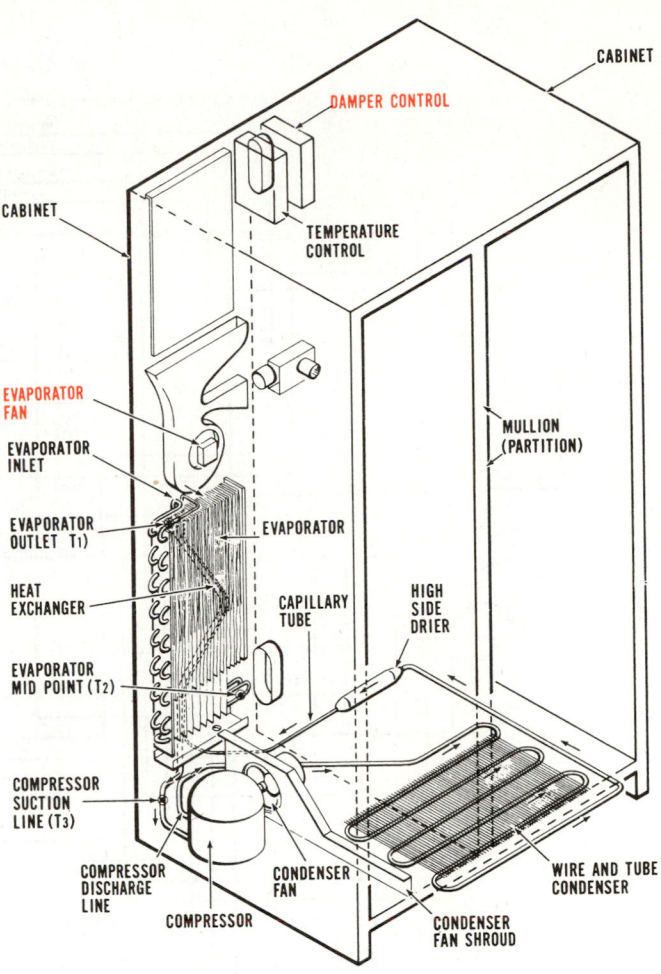

Fig. 10-29. Schematic of side-by-side refrigerator-freezer. Note the condenser lies flat underneath bottom of refrigerator. Compressor is under freezing compartment at back. Evaporator is behind freezer compartment. The various cabinet temperatures are obtained by use of dampers which control flow of cold air into various compartments.
(Amana Refrigeration, Inc.)

through the bottom grille. A fan on the evaporator circulates very cold air in the freezer compartment.

Damper arrangements allow some of this cold air to flow into the fresh food compartment. The fresh food compartment acts as a return air duct from the freezing compartment back into the evaporator compartment.

Refrigerant control is by capillary tube. This tube is attached to the suction line as in Fig. 10-29. It is called the heat exchanger.

The refrigerator automatically defrosts every six hours of compressor running time. The defroster is an electric heater. Attached to the evaporator, it is energized by a switch which is turned on and off by a timer. A defrost terminator (thermostat) is attached to the left end plate of the evaporator and opens the heater circuit at approximately 50 F. (10 C.). Twenty eight minutes after the start of the defrost cycle, the timer restores the operation of the compressor and air circulating fan. The defrost terminator contacts close (reset) at about 20 F. (−7 C.).

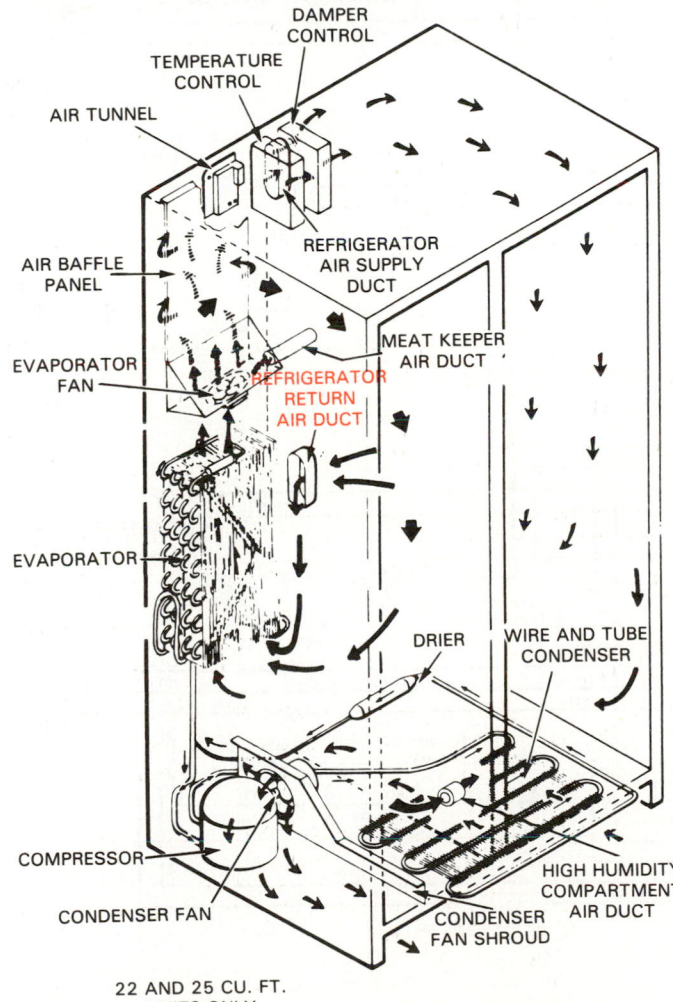

Fig. 10-28. Air movement pattern in side-by-side refrigerator-freezer.
(Amana Refrigeration, Inc.)

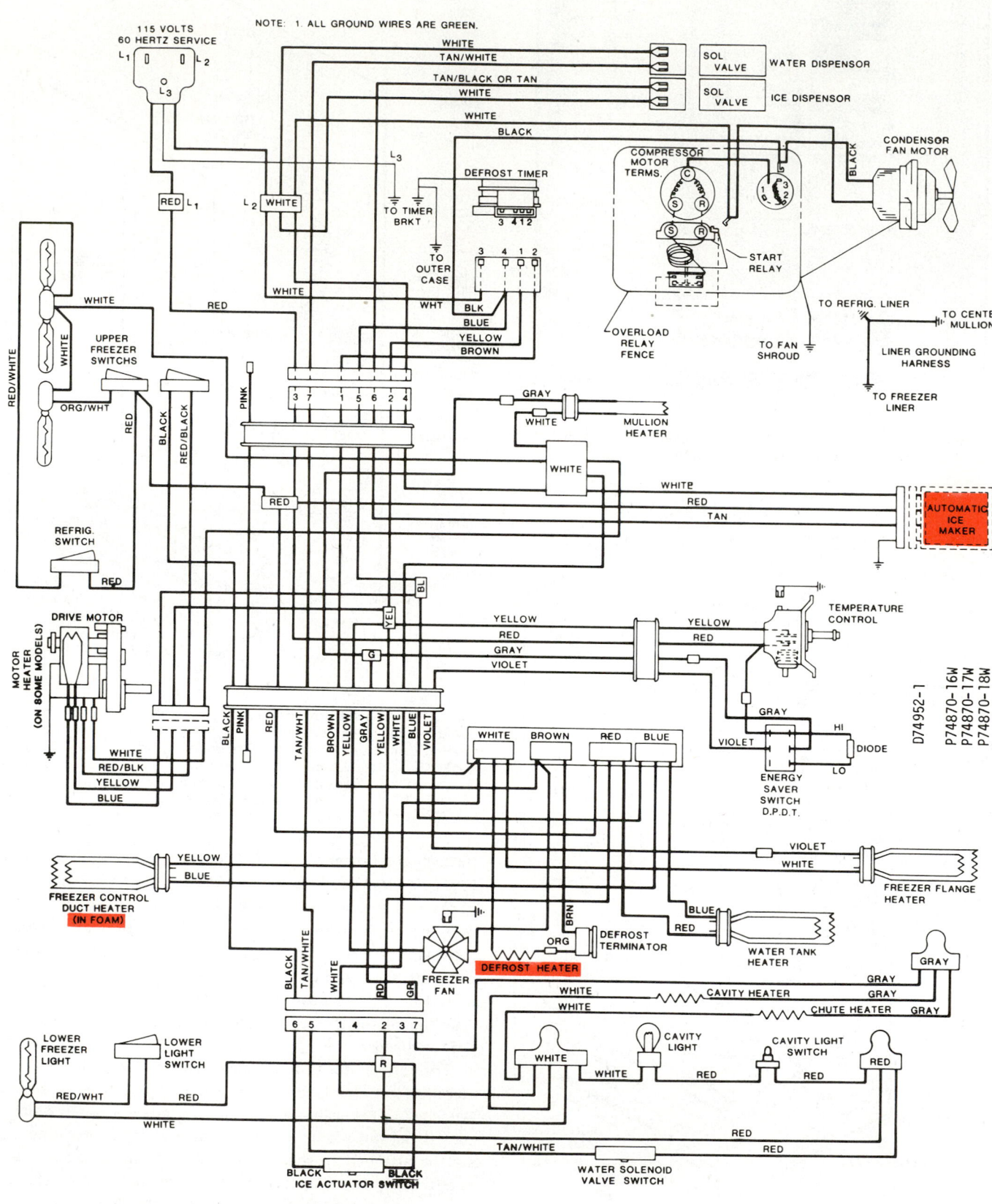

Fig. 10-30. Wiring diagram for side-by-side refrigerator with automatic defrost and automatic icemaker. (Amana Refrigeration, Inc.)

10-31 ELECTRICAL CIRCUITS FOR SIDE-BY-SIDE, FROST-FREE REFRIGERATOR-FREEZERS WITH ICE MAKER (AMANA)

Electricity is supplied through a grounded extension cord and plug. Note electrical circuits for this refrigerator in Fig. 10-30. Wiring is located in the foamed-in-place insulation.

There are a considerable number of heaters in the electrical circuit — defrost heaters, mullion heater, partition heater and partition cycling heater.

An auxiliary heater is foamed in place next to the connected heater. Should the original fail, the auxiliary may be connected into the circuit by disconnecting the existing plastic-covered male and connecting it to the extra plastic-covered male terminal. The electrical connections are located behind the evaporator covers. These heaters are in series with a switch on top of the freezer control. They may be turned off if the refrigerator is being operated in areas of extremely low humidity. This is called a power saver switch.

A color code identifies the wires going to each of the electrical circuits in the cabinet. The automatic ice maker controls are plugged into an ice maker receptacle. This eliminates the need to disturb any of the circuits in order to remove or restore the ice maker.

10-32 CHEST-TYPE FREEZERS

The chest-type freezer has certain advantages. Since cold air is heavier than warm air, the very cold air in a chest-type freezer does not spill out each time the lid is opened. This stops a considerable amount of moisture from entering the cabinet. There is little air change when the cabinet is opened.

To make chest-type freezers more convenient to use, they are usually fitted with baskets that may be lifted out to provide access to frozen food packages near the bottom. Also, the lids usually have a counterbalancing mechanism which makes them easy to open. A light in the lid gives good illumination. The chest-type freezer provides the most economical type of food freezing mechanism.

Most chest-type freezers require a manual defrost. But since so little moisture enters the freezer, defrosting is usually not needed more than once or twice a year.

Defrosting may be accomplished best by unplugging the condensing unit. Then remove the stored food and place either an electric space heater or a pail or two of hot water in the cabinet. With the cabinet closed, the ice will soon drop away from the evaporator surface and will be easy to remove.

Most chest-type freezers have a drain which makes removing the moisture from the cabinet quite easy. Remaining moisture must be wiped out of the cabinet.

Cabinets are available in various capacities. Height and width are quite uniform. However, the length will vary with the capacity of the freezer.

10-33 CABINETS, MECHANISMS AND ELECTRICAL CIRCUITS FOR CHEST-TYPE FOOD FREEZER (KELVINATOR)

The outer and inner shells of the chest-type freezer (Fig. 10-31) are metal. The evaporator surrounds the inner liner and is attached to it. The condenser is attached to the inside of the outer shell and completely surrounds the cabinet.

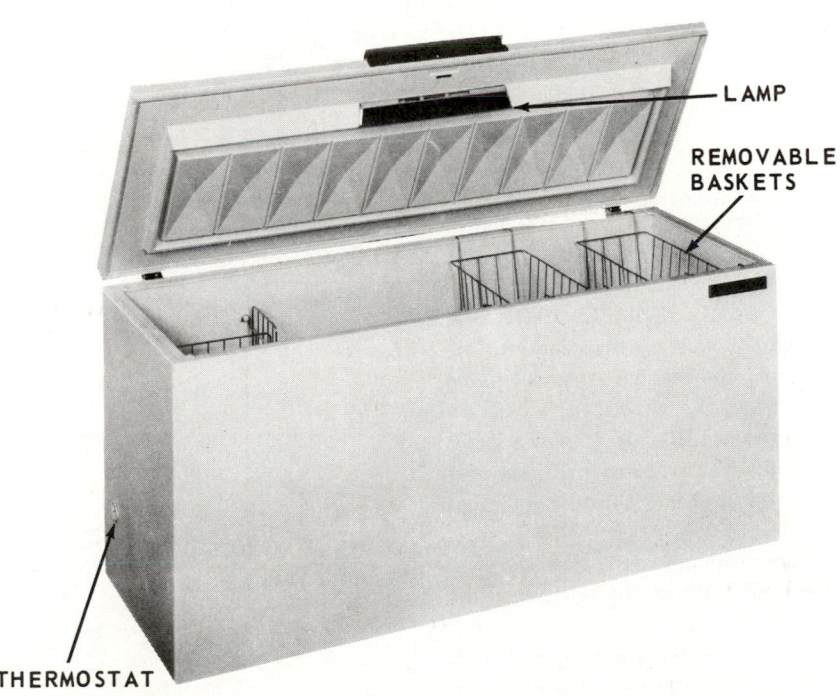

Fig. 10-31. Chest type food freezer. The use of wire baskets makes it easy to reach articles stored in the bottom of the freezer.
(Kelvinator, Inc.)

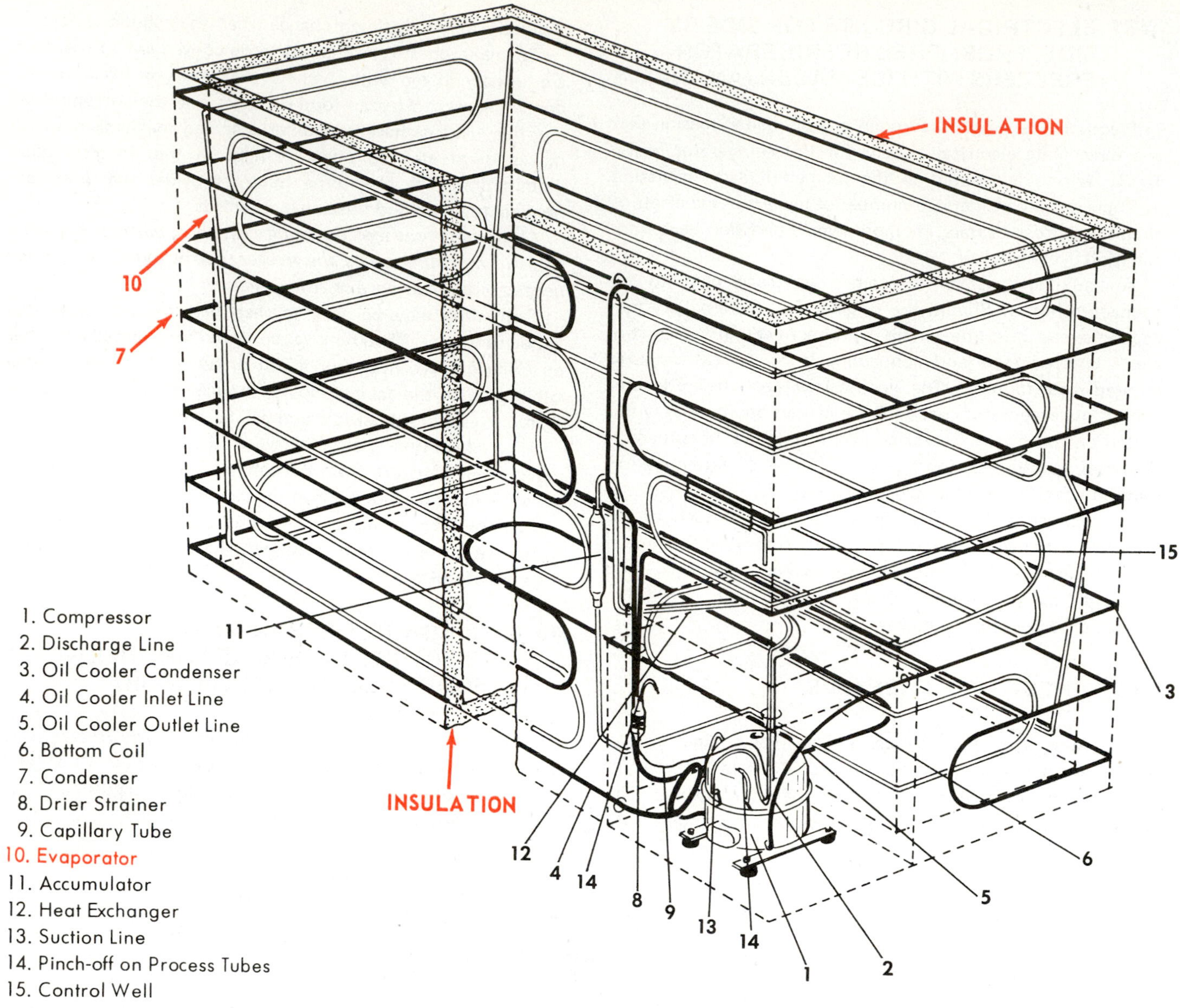

INSULATION

INSULATION

1. Compressor
2. Discharge Line
3. Oil Cooler Condenser
4. Oil Cooler Inlet Line
5. Oil Cooler Outlet Line
6. Bottom Coil
7. Condenser
8. Drier Strainer
9. Capillary Tube
10. Evaporator
11. Accumulator
12. Heat Exchanger
13. Suction Line
14. Pinch-off on Process Tubes
15. Control Well

Fig. 10-32. Chest type freezer evaporator and condensing unit. Note: Special oil cooler condenser at 3 located at bottom of condenser, oil cooler inlet line, shown at 4, and oil cooler outlet line at 5.

Mechanisms consist of common parts. The hermetic compressor is located at the lower right end of Fig. 10-32. The liquid refrigerant flows through the capillary tube and into the evaporator. There evaporation of the refrigerant and cooling takes place. The compressor draws the vaporized refrigerant through the compressor and pumps it into the precooler condenser on the back wall of the freezer. Here it releases part of its latent heat of vaporization and sensible heat of compression.

Since chest-type freezers are manually defrosted, condensate (water) is usually drained out through the bottom of the cabinet, as in Fig. 10-33.

From the precooler condenser, the refrigerant passes back to the machine compartment and through the oil cooling coil in the compressor dome (see Fig. 10-34). Here, additional heat is picked up from the oil. The compressed vapor then flows back to the main condenser where additional heat is released

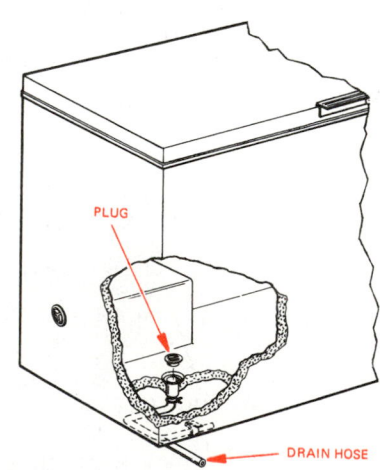

PLUG

DRAIN HOSE

Fig. 10-33. Drain system, chest type freezer. (Kelvinator, Inc.)

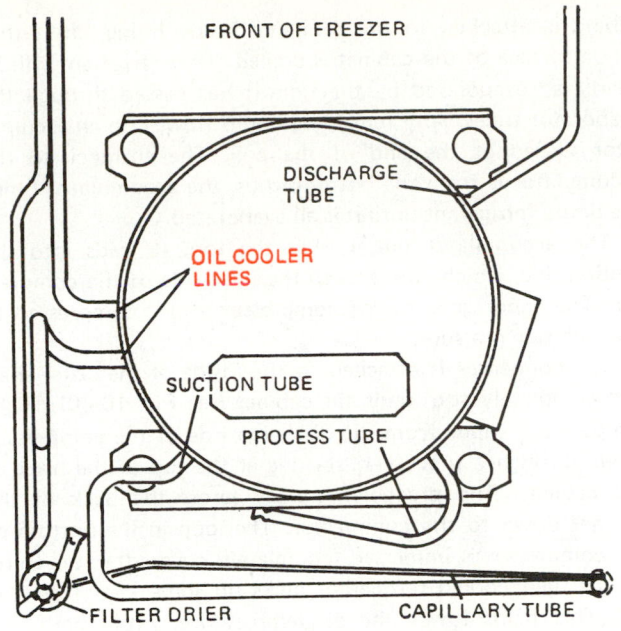

Fig. 10-34. Compressor dome showing oil cooler connections.

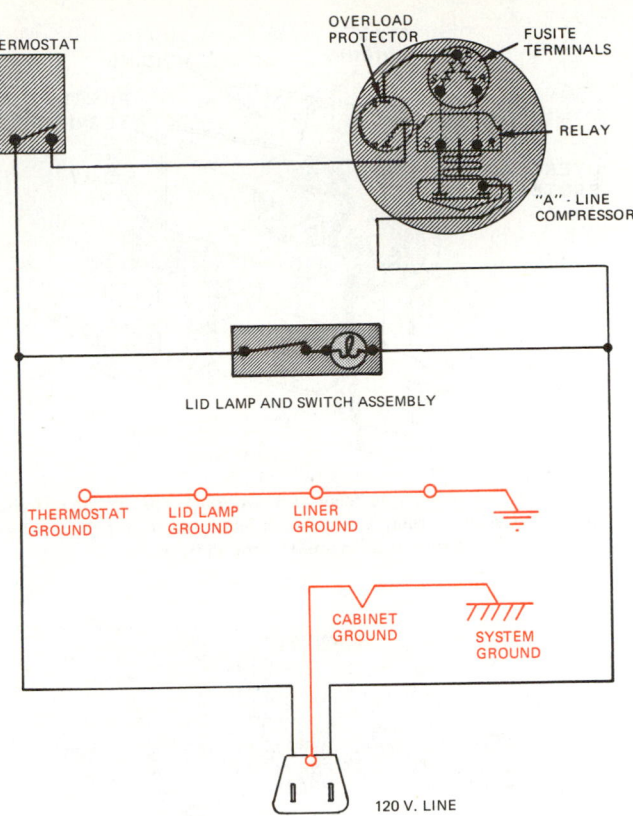

Fig. 10-36. Schematic wiring diagram for chest type freezer.

to the atmosphere. The refrigerant condenses from a high-pressure vapor to a high-pressure liquid.

Since the condenser tubes are in contact with the outer shell of the cabinet, heat from the condenser passes into the outer shell and warms it slightly. This causes a natural flow of warm air upward over the cabinet shell preventing sweating. The liquefied refrigerant collects in the bottom of the condenser tubing, flows into the filter-drier, moves into the capillary tube, on into the evaporator and the cycle repeats. The cycle is shown in Fig. 10-35.

Electrical power is supplied through a grounding type three-prong extension plug and cord. Fig. 10-36 is a schematic

wiring diagram for this freezer. The starting relay and overload protector are attached to the motor by a "push-on" type mount (Fig. 10-37). Note that all parts of the mechanism and the cabinet are grounded. The thermostat is at the end of the cabinet near the top of the compressor compartment. The dial is marked for "off," "normal" and "cold" positions.

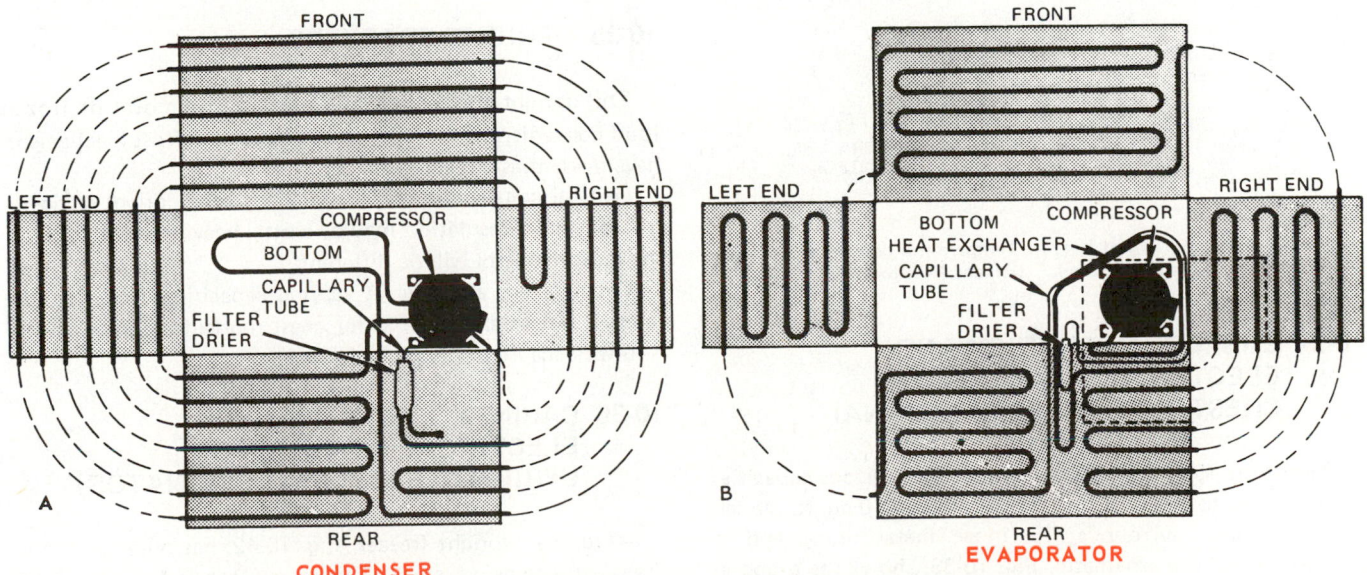

Fig. 10-35. Refrigeration cycle diagram for chest type freezer. A—High side of cycle. Heat absorbed in evaporator is now released by condenser into surrounding atmosphere. B—Low side of cycle. Heat is absorbed by evaporator in cabinet.   (Kelvinator, Inc.)

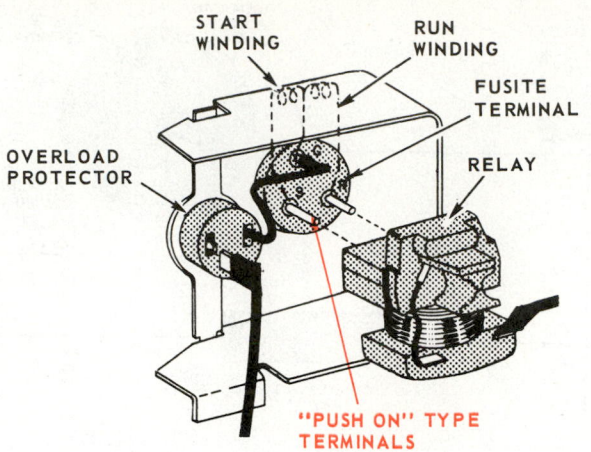

Fig. 10-37. A motor starting relay and overload. The motor terminals are to receive starting relay using "push-on" type terminals. Overload protector is connected in the same way.

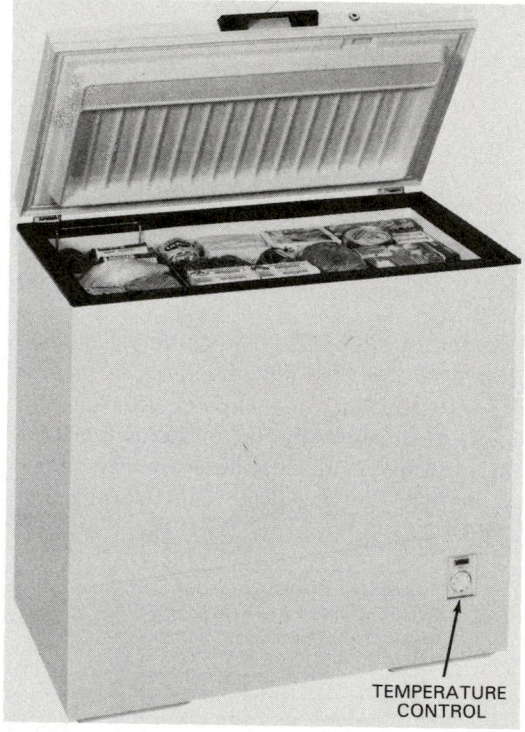

Fig. 10-38. Chest type food freezer. Temperature control is located at the lower right-hand corner. (Amana Refrigeration, Inc.)

10-34 CABINET, MECHANISM AND ELECTRICAL CIRCUITS FOR CHEST-TYPE FREEZER (AMANA)

Chest-type food freezers are available in various capacities from 7 cu. ft. to 28 cu. ft. Fig. 10-38 shows a 10 cu. ft. model. The evaporator surrounds the inner metal lining and is attached to it. The schematic, Fig. 10-39, shows the evaporator and the low-side refrigerant circuit.

When the unit operates, the refrigerant flows through the capillary tube into the evaporator tubes. Since the evaporator

tubing is attached to the cabinet's inside lining, the entire inside surface of the cabinet is cooled. The refrigerant will be nearly all evaporated by the time it has passed through the evaporator tubes. Any remaining liquid flows into an accumulator placed at the end of the coil. The entrance to the accumulator is from the bottom; thus, the accumulator holds the liquid refrigerant until it is all evaporated.

The accumulator outlet is at the top. It leads into the suction line, which connects to the inlet side of the compressor. The vapor goes to the compressor and is compressed to the high-side pressure.

The condenser is attached to the inside of the outer shell and completely surrounds the cabinet (see Fig. 10-40). High-temperature vapor from the discharge side of the compressor flows through a precooler, starting at the top of the back of the cabinet. The precooler zigzags across the back of the cabinet down to the compressor. The loop in the bottom of the compressor is immersed (completely covered) in oil. Here, the partially-cooled refrigerant picks up some heat from the oil. This helps lower the oil temperature. From here, the high-pressure vapor is carried to the top end of the cabinet. It zigzags across the ends and front of the cabinet and returns to the filter-drier, completing the high-side circuit.

Electrical power comes through a grounded type three-prong extension plug and cord. See wiring schematic in Fig. 10-41. Note warning light connected into the electrical circuit. *This warning light only indicates whether or not the electrical circuit is "hot." It does not indicate whether cabinet temperature is satisfactory.*

The temperature control is wired into the motor circuit. It controls the running time of the compressor to maintain desired cabinet temperatures. The normal operating temperature range for this freezer should be between −14 F. (−26 C.) and 6.5 F. (−15 C.). All assemblies are grounded through the green wire. This is connected to the grounding terminal of the attachment plug.

10-35 UPRIGHT FREEZERS

The upright freezer makes storage and removal of frozen food convenient. Frost-free and automatic defrost mechanisms have made these freezers very satisfactory.

General construction is very similar to the upright refrigerator. However, insulation may be a little heavier and, of course, the motor control will be different.

Cabinets are available in varying capacities. However, the range is not as extensive as for chest model freezers. Capacity is lower because of height limitations.

10-36 CABINET, MECHANISM AND ELECTRICAL CIRCUITS FOR UPRIGHT FREEZERS (KELVINATOR)

A popular upright freezer, Fig. 10-42, has outer and inner shells of enameled steel. The evaporator is located at the bottom of the cabinet. The wrap-around condenser is inside the outer shell on the sides, back and top. Door trim is a plastic material which conducts heat poorly.

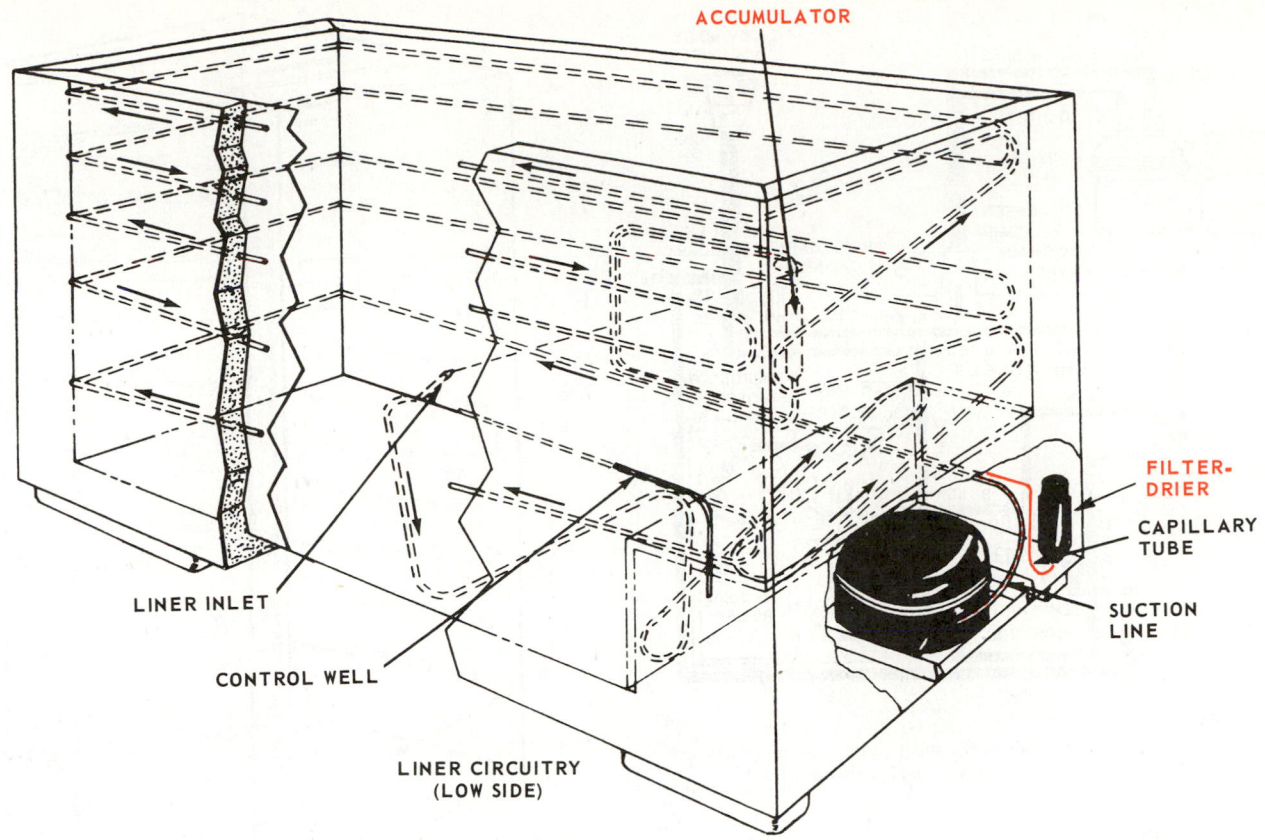

Fig. 10-39. Chest type freezer evaporator coil. Evaporator is attached to cabinet lining. Capillary tube refrigerant control extends from the bottom filter-drier to inlet of evaporator coil.

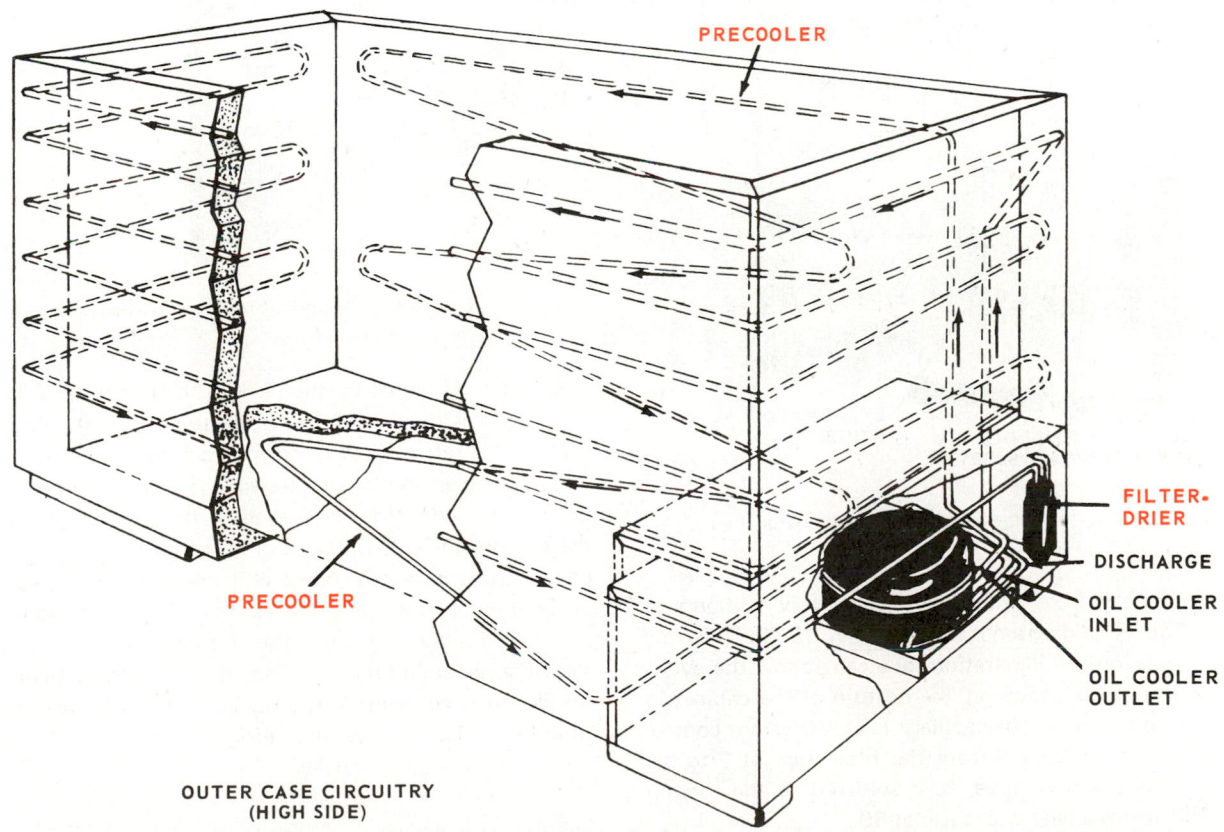

Fig. 10-40. Chest type freezer condenser. Coil is attached to the inside surface of the outer shell. Note precooler coil.

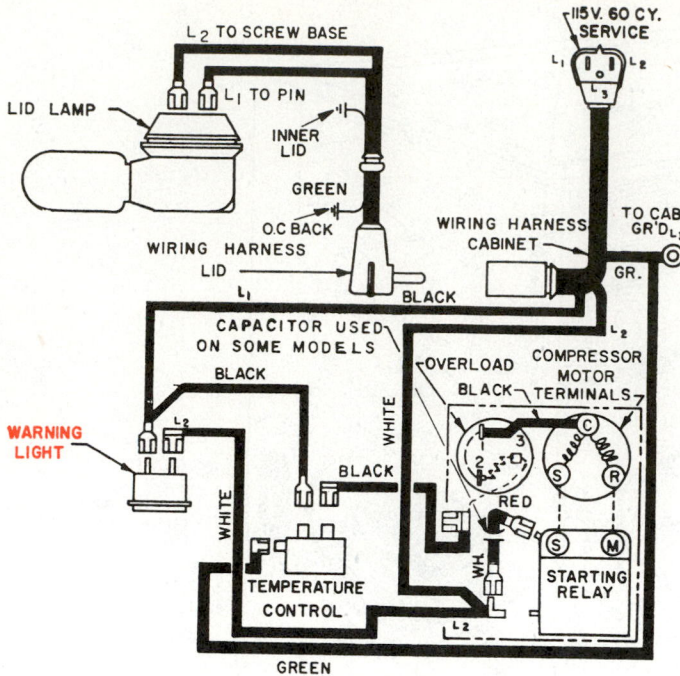

Fig. 10-41. Schematic wiring diagram, chest type freezer. Parts are identified on illustration. (Amana Refrigeration, Inc.)

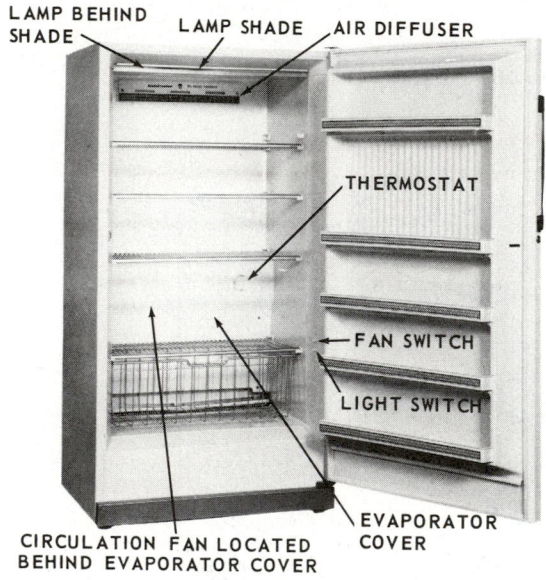

Fig. 10-42. Upright domestic freezer. (Kelvinator, Inc.)

An automatic defrost cycle is activated every 12 hours of operation. The cycle diagram is shown in Fig. 10-43. Referring to the numbers on the illustration, one can follow the cycle. The evaporator, 9, is shown at the bottom of the cabinet in front of the compressor. The capillary tube refrigerant control carries the liquid refrigerant from the filter-drier at 2 to the evaporator. The capillary tube, 8, is soldered to the suction line, 11. This forms a heat exchanger at 10.

Upon entering the evaporator, the liquid refrigerant evaporates and absorbs heat. A fan draws the cold air through the

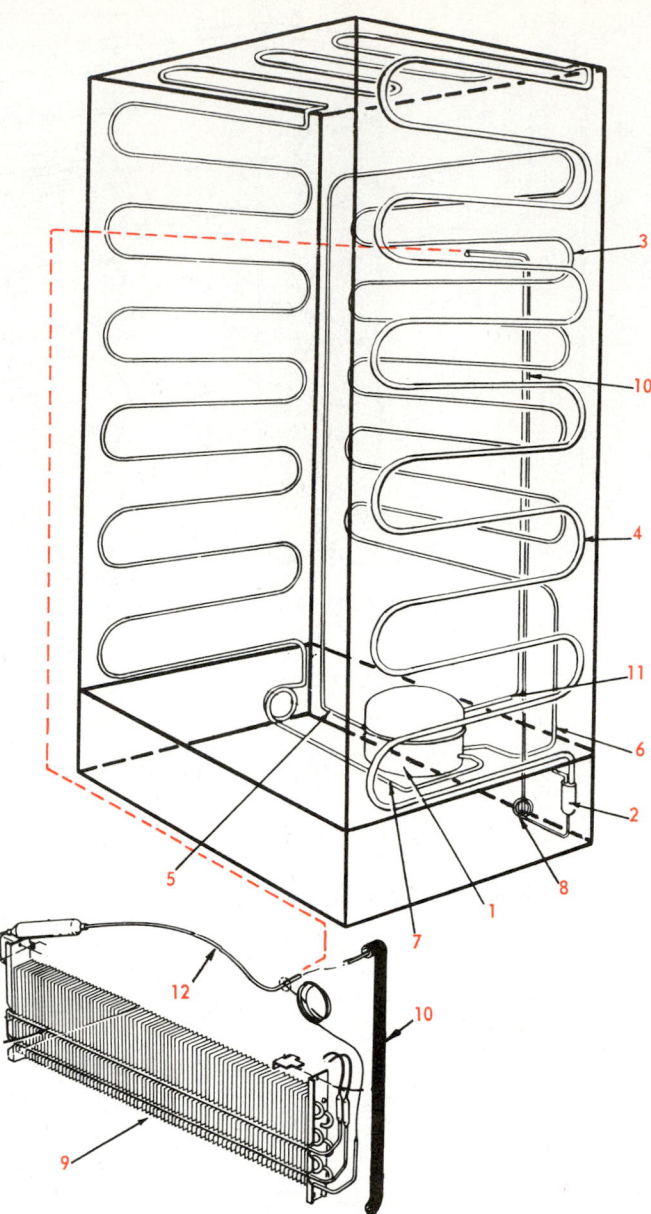

Fig. 10-43. An upright freezer cycle diagram.

evaporator forcing it up the back of the refrigerator. It is then discharged at the top of the refrigerated space. Air flows down through the refrigerated space and back to the evaporator.

A precooler condenser, shown at 3 and 6, extends from the compressor along the back shell of the cabinet. It returns to the compressor at 5 and passes refrigerant on to the oil cooling coil in the compressor dome at 1. Here, the refrigerant picks up additional heat from the oil. The refrigerant vapor is then pumped through the tubing at 7 to the main condenser, 4. Here the remaining heat is released to the atmosphere and the refrigerant is condensed to a liquid. It flows by gravity to the filter-drier. This brings the refrigerant back to the capillary tube where the cycle repeats.

The condenser is attached to the outer shell and warms it slightly. This causes a natural convection of airflow upward over the cabinet shell. The heat from the condenser prevents sweating of the shell. Generally it is desirable to leave some

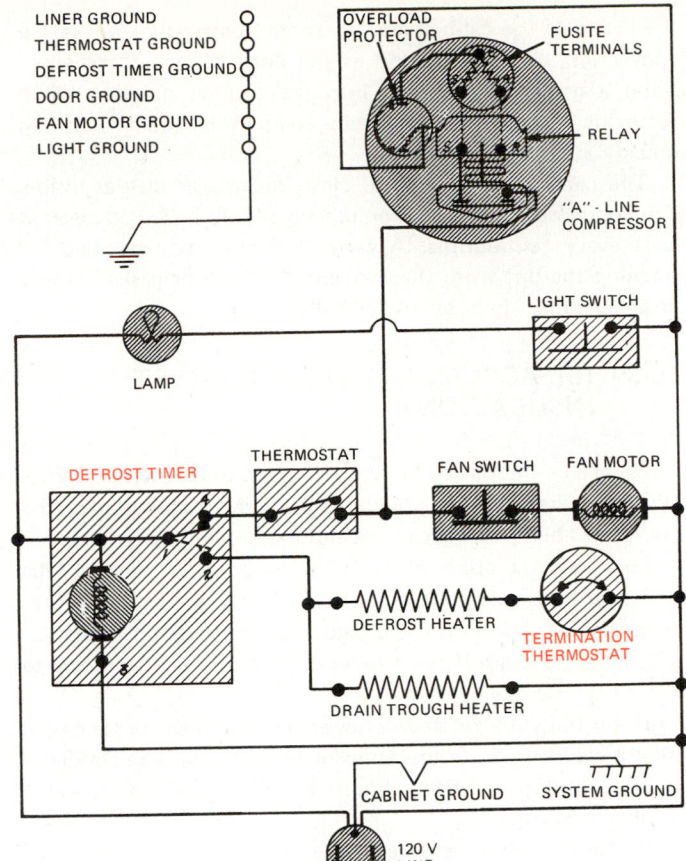

Fig. 10-44. Upright freezer, schematic wiring diagram. Note all parts are grounded. Compressor and fan circuits are open during defrost cycle. A thermostat, as well as a timer, control the defrost interval. (Kelvinator, Inc.)

space around the freezer. Otherwise natural convection (air-flow) will be restricted.

Electrical power is supplied to the freezer through a grounding type three-prong extension plug and cord. See schematic wiring diagram in Fig. 10-44. Note that all parts of the mechanism and the cabinet are grounded (green wire extends from plug to system ground). The cabinet temperature thermostat is located at the back of the cabinet behind the second shelf from the bottom. It provides three settings, off, normal and cold. The defrost timer is located in the compressor compartment, Fig. 10-45.

10-37 UPRIGHT FREEZERS — CABINETS AND ELECTRIC CIRCUITS (WHIRLPOOL)

The upright freezer cabinet arrangement is similar to the one-door refrigerator. See Fig. 10-46. The door often has a lock. The evaporator of the manual defrost unit is usually a part of the cabinet shelves.

No-frost systems use an evaporator located behind a baffle. A fan circulates the air through the evaporator and then through the foods section. The filter and drier are combined. See Fig. 10-46. Signal lights are often used to indicate when power is on and to warn the owner when cabinet temperature is above normal.

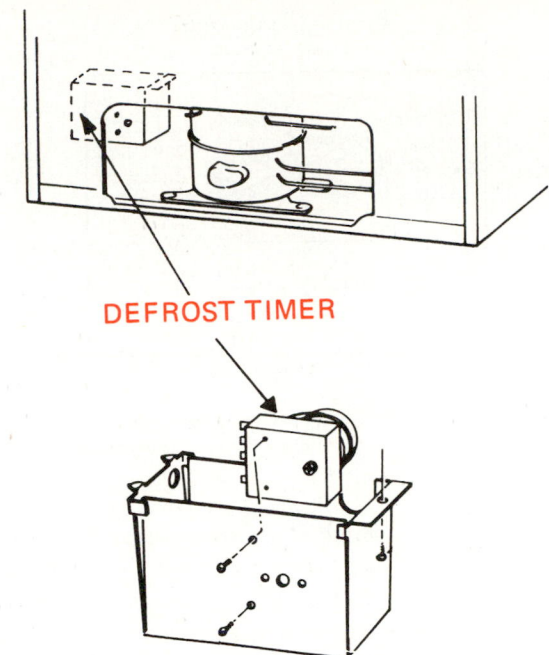

Fig. 10-45. Note defrost timer located in bottom of freezer in compressor compartment.

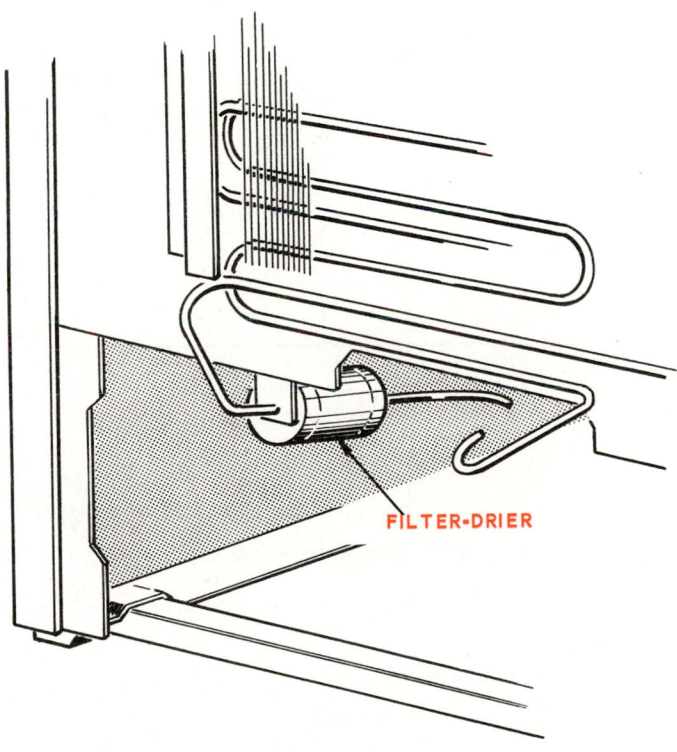

Fig. 10-46. Some domestic upright freezer units use a combination filter-drier. (Whirlpool Corp.)

The mechanism is similar to the regular refrigerating system with a motor compressor designed for low evaporator pressures. A forced convection condenser is used with both air-in and air-out grilles located at the lower front of the cabinet.

The wiring is also similar to that on the refrigerator system.

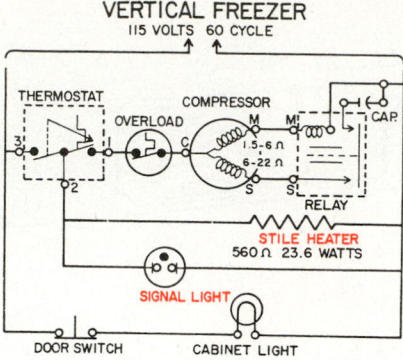

Fig. 10-47. Wiring diagram for a manually defrosted vertical freezer. Note stile heater and signal light which indicates power is on.

Fig. 10-47 shows the wiring of a system designed for manual defrost. Note the cabinet light, switch and signal light which indicates that the system has electrical power. A 23.6W stile heater is used. Fig. 10-48 shows a system which has a polarized circuit (ground wire, etc.). The cabinet and all other metal parts are grounded.

10-38 CARE OF REFRIGERATOR OR FREEZER

The owner of any domestic refrigerator or freezer should be instructed not to allow the door to remain open when the unit is in operation. The great difference in temperature between the inside of the cabinet and the room temperature will set up convection currents as soon as the door is opened. This will bring a great deal of heat into the cabinet quickly. When removing articles from, or placing them in the cabinet, do it as quickly as possible.

The cabinet must be kept clean outside as well as inside. The condenser and motor compressor should be wiped clean at least every six months. A vacuum cleaner may be used for cleaning the lint from the condenser. The door gasket should be checked for tightness periodically.

10-39 ICE ACCUMULATION IN CABINET INSULATION

Ice accumulation is one of the main troubles with improperly or carelessly assembled units. Ice also reduces the insulating ability of the cabinet and the unit will run more.

Such accumulation of ice in a freezer will generally be indicated by a cold spot on the outside surface of the freezer or by condensation on the outside surface. Iced up insulation will melt and drain if the freezer is shut down and allowed to warm up for a few days.

If a refrigerator or freezer has an air leak in the outer casing (shell), moisture from the atmosphere will enter and condense in the insulation. This condition will cause considerable trouble in a freezer.

If the condition is severe, the condensing unit may run continuously. Should cold (sweat) spots appear, remove

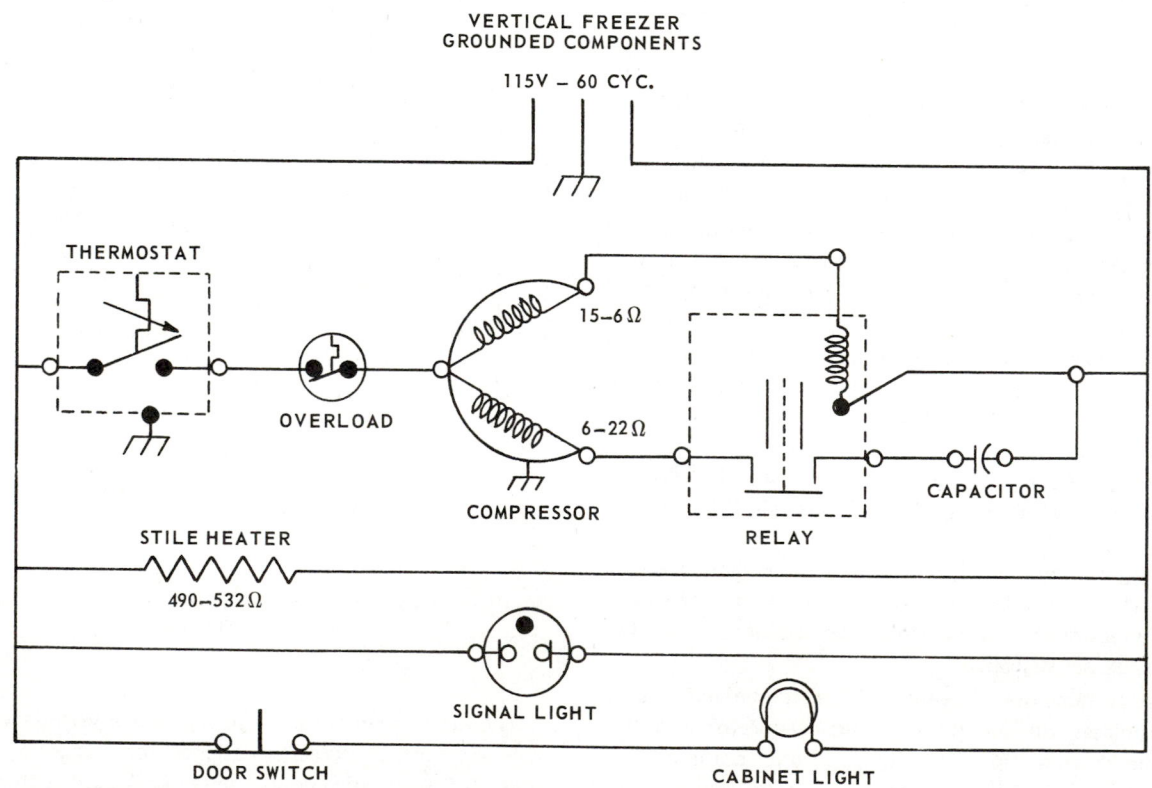

Fig. 10-48. Wiring diagram for a vertical freezer made according to the latest electrical codes. The middle wire is grounded by a wire network to all the metal parts of the cabinet, hardware and refrigerating system.

breaker strips and insert lightly packed fine fiber glass insulation to fill air pockets.

Some freezers provide an opening in the inner lining which will allow any moisture in the insulation to escape into the freezer compartment and be condensed on the cold surface of the evaporator. This tends to keep the insulation dry.

10-40 BUTTER CONDITIONER

Various devices are being used to temperature-condition butter stored in the fresh food cabinet. The most common is to provide a recess in the refrigerator door which will hold a quarter lb. of butter. This recess is separated from the refrigerator compartment by a small door. Being close to the outer shell of the door, the recess provides little insulating effect between butter and room temperature. Some refrigerators provide an electrical resistance unit to aid in warming the butter compartment.

10-41 CABINET HARDWARE

Cabinet hinges usually use ball or nylon bearings. These require little or no lubrication. Usually they are adjustable so that the doors can be made to fit the cabinet properly. Fig. 10-49 illustrates a hinge commonly used in upright refrigerators.

Many of the older refrigerators used complicated door latch mechanisms. They were designed to draw the door firmly into place and securely latch it. Since 1958, a federal law requires that a refrigerator cabinet must be designed to be opened from the inside with no more than a 15 lb. force. Most modern refrigerators, therefore, use magnets to close and hold the door

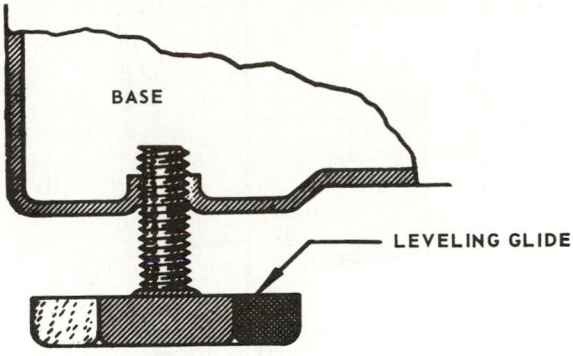

Fig. 10-50. Detail of leveling glide.

shut. No positive latching mechanism is used.

Refrigerator cabinets should be carefully leveled in order that the shelves are not tilted. Most have adjustable feet for this purpose. It is possible to adjust these feet in such a way that the weight of the door will tend to swing it closed from any open position. Most users like this setting. Fig. 10-50 gives detail of a leveling adjustment.

Many modern refrigerator-freezers are mounted on rollers to help move the unit for cleaning and servicing.

The front rollers are usually adjustable. After the refrigerator is in place, these rollers are used to level it. See Fig. 10-51.

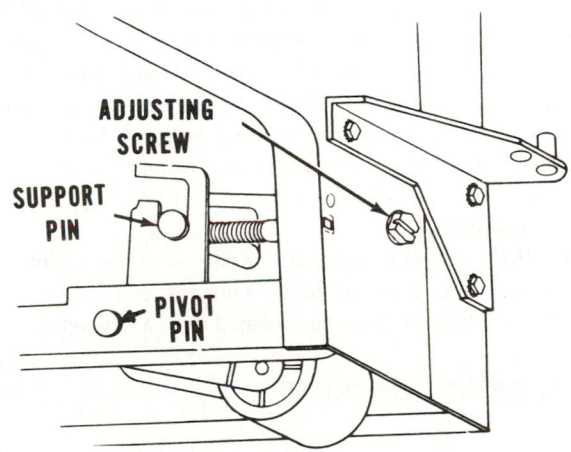

Fig. 10-51. Adjustable front roller assembly for a refrigerator cabinet.

A thumbscrew which is adjusted to touch the floor keeps the refrigerator from moving once it is in place.

Breaker strips, usually made of plastic, connect the metal outer shell (case) to the metal liner at the part of the cabinet that contacts the door. (Some refrigerators have a plastic liner.)

These breaker strips are molded and shaped in such a way that they snap in place. See Fig. 10-52.

A wide bladed putty knife may be inserted at A in Fig. 10-52 to remove the outer edge of the breaker strip. (Wrap the putty knife blade with tape to prevent scratching the parts.) Sometimes pressing with the heel of the hand will separate the

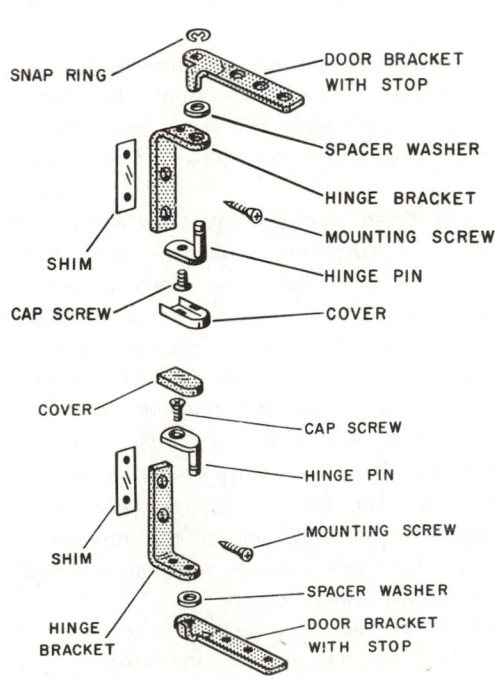

Fig. 10-49. Top and bottom hinges for refrigerator or upright freezer. (General Electric Co.)

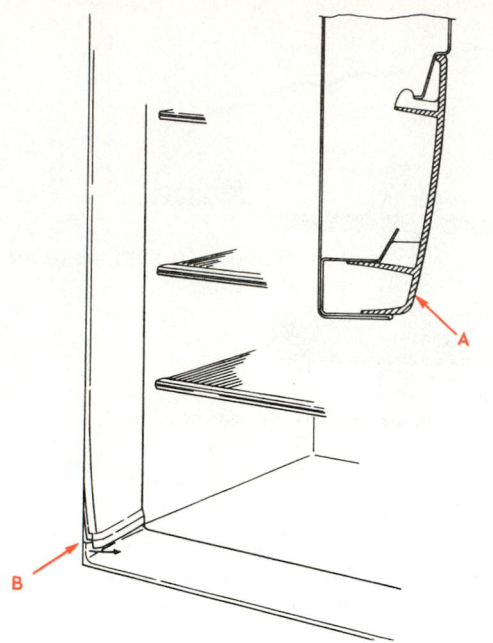

Fig. 10-52. Detail of typical breaker strip. (General Electric Co.)

joint as illustrated at B.

The liner should be room temperature to 90 F. (never warmer) to make it flexible. Use a warm dampened wiping cloth to warm it. Some breaker strips (installed before the foamed-in-place insulation is put in) are fused to the foam insulation. These strips must be broken or cut to be removed.

Chest-type freezers usually have the lid counterbalanced so that it is not necessary to hold the lid open when placing articles in or removing them from the cabinet. However, the adjustment on these counterbalances is usually made so that, as the operator closes the lid, it tends to make a final closing by its own weight.

Very little cabinet hardware repair is done on modern refrigerators. If a latch or hinge fails for any reason, it is recommended that the part be replaced with a new one.

10-42 CABINET GASKETS

Door gaskets are usually made of flexible plastic. They usually have an air cushion design. See Fig. 10-53. Many have magnets built into the plastic to hold the door closed. These are used in place of a latch and striker plate.

Door seals may be checked using .003 in. thick plastic feeler gauge or a thin piece of paper. To check the fit of the gasket, insert the feeler gauge or the strip of paper at several places around the door opening. With the door closed it should require a little pull on the gauge if the gasket is properly fitted. If the paper or feeler gauge falls out of its own weight, the gasket is not fitted properly.

Door gaskets may be quickly and easily checked as follows:
1. Open the refrigerator door (note carefully the pull required to open the door) about half way and allow it to remain open for approximately 10 seconds.
2. Close the refrigerator door and allow it to remain closed for

about 15 seconds.
3. Open the refrigerator door and note carefully the pull required. It should require more effort to open the door now than it did at the time of the first opening.

This is because, during the time that the door was opening, the cold air came out and warm air replaced it. On closing the door, this warmer air is cooled and contracts with the result that the pressure inside the refrigerator cabinet will be slightly less than the room atmospheric pressure. Timing is an important element in this test, as the pressures tend to quickly balance.

With some refrigerators, it is possible to adjust the hinges and the door latch to correct a poorly fitted gasket.

Fig. 10-54 illustrates a door that is either warped or has improperly adjusted hinges. A warped door can usually be straightened by twisting. To check for hinge adjustment, use the test just described for proper gasket seal.

Door gaskets deteriorate with age. If the material has become hard, cracked or broken, it should be replaced.

10-43 REPAIRING FINISHES

As indicated in Para. 10-7, cabinet finishes may be either baked-on enamel or porcelain. To repair enamel finishes, carefully sand the damaged area down to the bare metal. Be sure to remove all wax and rust.

Sand the edges of the damaged area with fine waterproof paper (6/0). The old finish should slope evenly from the surface to the center of the damaged area. This is called "featheredging."

Carefully clean the sanded surface. Apply metal primer to the exposed bare metal using either a brush or a spray gun. After the primer has dried thoroughly, sand again using 6/0 waterproof paper and soapy water as a sanding lubricant. Dry thoroughly.

Spray or brush on the enamel, blending the new coat to the old finish as smoothly as possible. Allow the enamel to dry thoroughly, then sand with 6/0 paper until the edges are invisible. Fig. 10-55 shows the steps in finishing a chipped surface.

For a gloss finish, rub the entire surface including the patched area, with rottenstone, paraffin oil or rubbing oil. When the desired gloss has been reached, wipe off the rottenstone and oil by using a moistened cloth or chamois.

Paint spraying must be done in a fireproof, well ventilated room or booth. The air must be free from dust and the compressed air used in the spraying gun must be free of moisture and oil. Be sure to follow the paint manufacturer's recommendations when refinishing a cabinet surface.

To repair porcelain finish, a special porcelain patching material should be used. Because of the inherent nature of porcelain, its color shade will vary somewhat. A patching kit may be obtained with several colors.

Surface to be patched must be cleaned and warmed. Apply the patching material with a small, fine brush or air brush it. When dry, the finish should be smoothed with waterproof abrasive paper and polished with a soft cloth or with rottenstone and rubbing oil.

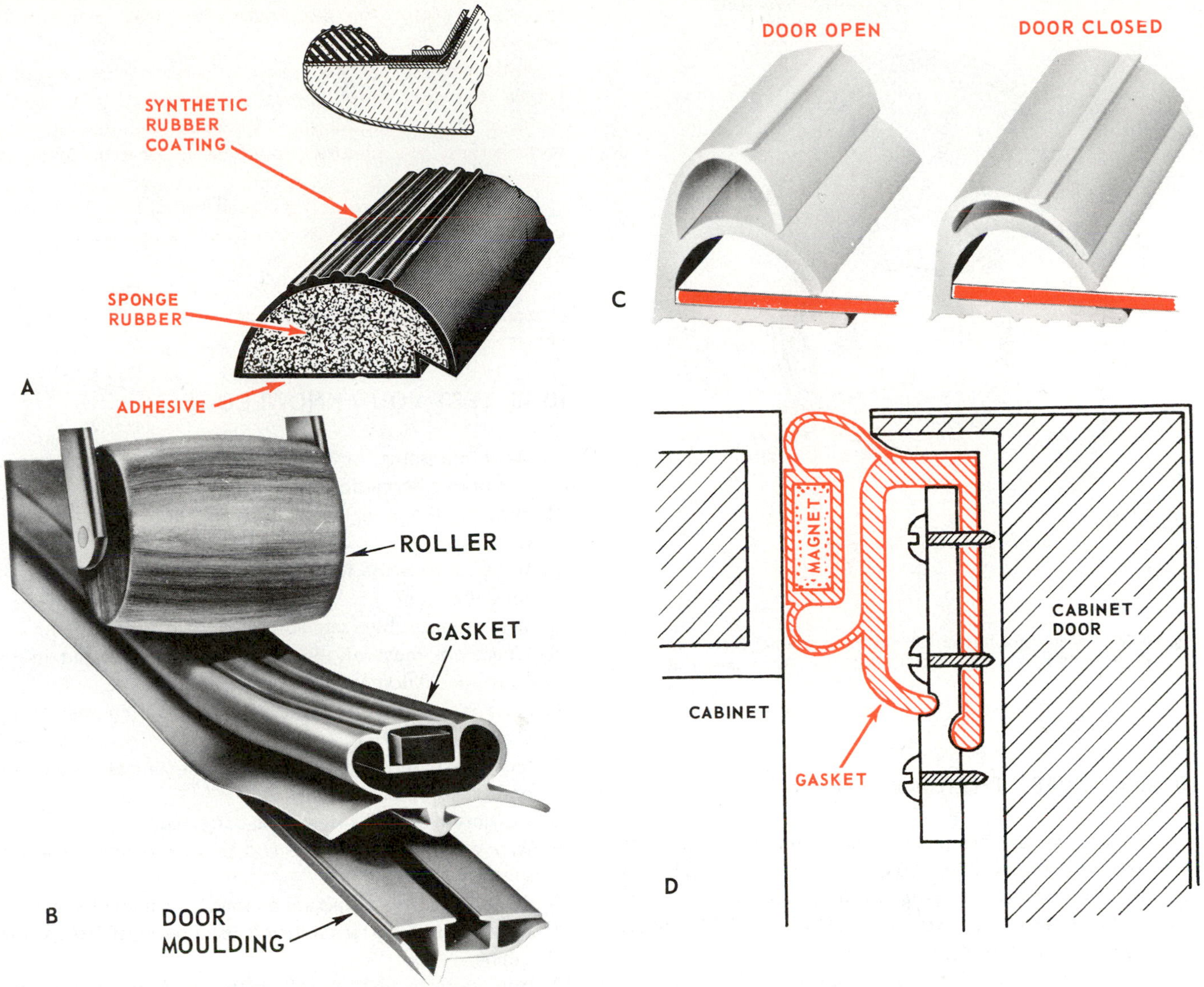

SYNTHETIC RUBBER COATING

DOOR OPEN **DOOR CLOSED**

SPONGE RUBBER

ADHESIVE

A

C

ROLLER

GASKET

MAGNET

CABINET DOOR

CABINET

GASKET

B **DOOR MOULDING**

D

Fig. 10-53. Replacement door gaskets. A—Gasket attached with an adhesive. B—Gaskets used with magnetic door latches (not magnetic gaskets). C—Gasket clipped to door molding. D—Magnetic gasket.

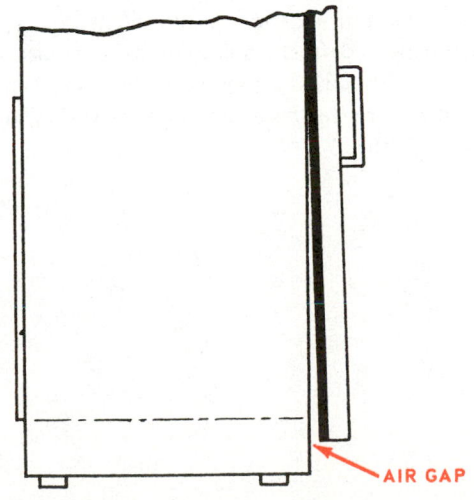

AIR GAP

Fig. 10-54. An improperly hung or warped door.

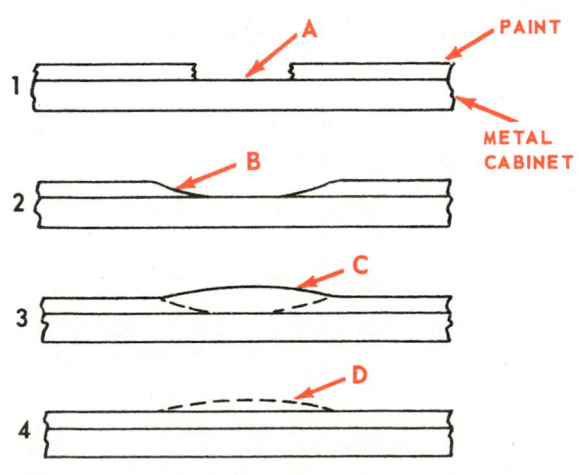

A **PAINT**

1

METAL CABINET

B

2

C

3

D

4

Fig. 10-55. Steps in finishing a chipped surface: A—Paint chipped from cabinet. B—Surface sanded in preparation for repairing. C—New finish applied. D—Excess finish must be sanded down to level.

Domestic Refrigerators and Freezers / 333

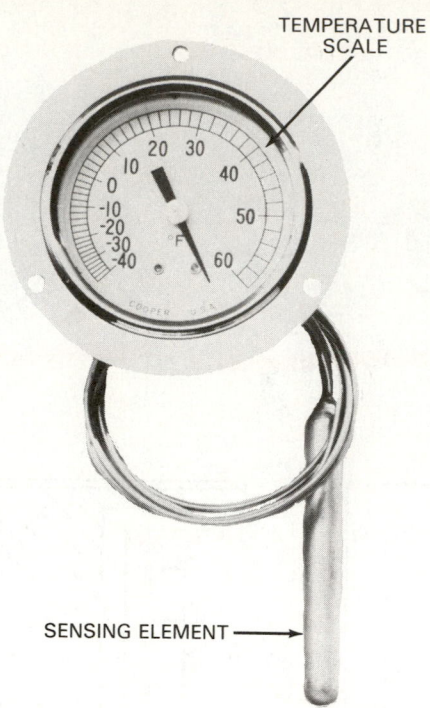

TEMPERATURE SCALE

SENSING ELEMENT

Fig. 10-56. Refrigerator-freezer cabinet thermometer. Scale reads from —40 F. to 60 F. (—40 C to 16 C).   (Cooper Thermometer Co.)

10-44 CABINET THERMOMETERS

The recommended temperature range for fresh food compartment is between 32 F. (0 C.) and 50 F. (10 C.). The recommended temperature range for the frozen food compartment is 0 F. (—18 C) to —10 F. (—23 C). Fig. 10-56 displays a thermometer which indicates these ranges.

10-45 REVIEW OF SAFETY

Avoid putting your hands near revolving fans.

Any spray painting must be done in an approved spray room or booth.

Refrigerators and freezers must be carefully handled in order not to injure those handling them or to damage the cabinets or mechanisms.

Frost and ice should be removed by heating (hot water or electric heat). *Never use a pointed or sharp metal tool to remove ice from the evaporator.* One may puncture the refrigerating unit.

Remember, if an old style refrigerator is taken out of service, the door must be removed immediately. Letting an out-of-service refrigerator stand where children may play in it, is dangerous. The children could be suffocated. Avoid this tragic mistake.

Always disconnect electric power (open the switch or pull the plug) before working on the electrical parts of the system to avoid unpleasant and perhaps fatal shocks.

Always make certain that the electrical system is grounded properly to the receptacle if an approved three-wire grounding plug is not used.

10-46 TEST YOUR KNOWLEDGE

1. What are some of the safety devices in the electrical circuit of a hermetic domestic refrigerator?
2. How is the temperature controlled in a butter compartment?
3. What controls the flow of hot gas through the evaporator for defrosting?
4. What is meant by a semiautomatic defrosting system?
5. Where are most of the refrigerant lines located in the latest model domestic refrigerators?
6. How is the defrost water removed during the defrost cycle?
7. What is the most common refrigerant control system used on frozen food cabinets?
8. How often should a freezer be defrosted?
9. Why are the suction line and the liquid line sometimes soldered together?
10. What is the best temperature range for a home freezer?
11. What kind of motor control is used on most frozen food cabinets?
12. How does a no-frost refrigerator prevent sweating and frosting?
13. What is a drain heater?
14. What are some possible locations of the condensing unit?
15. Why must an evaporator be defrosted?
16. How are cabinet door joints made airtight?
17. List two evaporator locations in the cabinet.
18. What happens to defrost water in modern refrigerators?
19. What type of insulation is used in modern freezers?
20. What is meant by featheredging when repairing an enamel or lacquer finish?

Chapter 11

INSTALLING AND SERVICING SMALL HERMETIC SYSTEMS

This chapter covers installation and service of small hermetic systems. It explains the more common service operations for domestic refrigerators, freezers and small commercial cabinets. Three main areas will be covered:

1. Installation.
2. External mechanism service.
3. Internal mechanism service.

There are six basic qualities in a good refrigerator service technician:

1. Must be skilled in the use of tools and know the supplies used in refrigeration work.
2. Must know basic refrigeration principles and cycles.
3. Must know various refrigerator mechanisms and their operation.
4. Must be able to locate refrigerator troubles quickly and accurately.
5. Must keep all refrigerator parts and mechanisms clean and dry.
6. Must know and practice all the safety regulations.

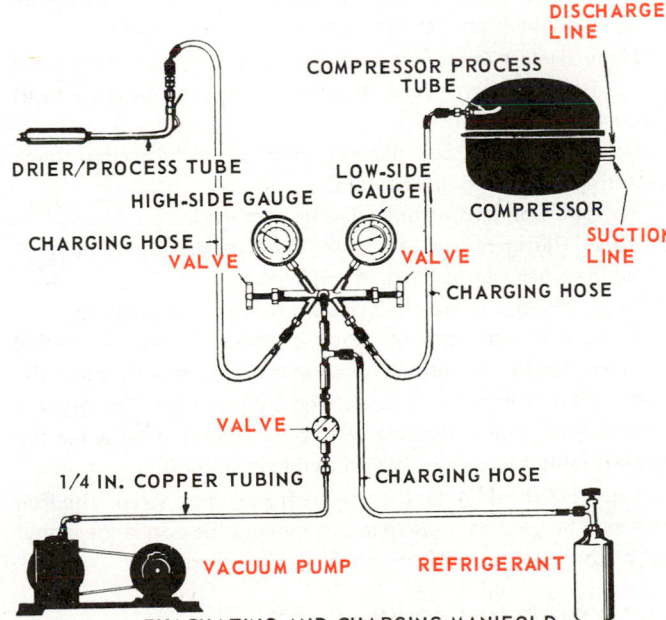

Fig. 11-1. Typical installation of service equipment manifold on a hermetic system. Note refrigerant cylinder, vacuum pump, valve arrangement and the adaptor connection to compressor process tube and to condenser outlet. (Amana Refrigeration, Inc.)

11-1 INSTRUMENTS, TOOLS AND SUPPLIES

In addition to the tools and supplies listed in Chapter 2, special instruments, tools and supplies will be needed for domestic refrigerator service work. These are listed in the following paragraphs under Instruments, Tools and Supplies.

11-2 INSTRUMENTS

Many special testing instruments are needed for refrigeration service work. Their use will be explained as the various service operations are described. Fig. 11-1 shows a small hermetic system with a gauge manifold, vacuum pump and charging cylinder connected to it. Some of the basic instruments required are listed below:

BASIC INSTRUMENTS

Pressure recorder.
Temperature recorder.
Off-on recorder.
Watt recorder.
Electronic sound tracer.
Electronic leak detector.
Compound gauge.
Pressure gauge.
Thermometer, −20 F. to 212 F. (−29 C. to 100 C.).
Voltmeter.
Ammeter.
Ohmmeter.

11-3 TOOLS

REFRIGERANT TOOLS

High-vacuum pump.
Service cylinders for R-12, R-22, R-502.
1 purging line 1/4 in. dia. by 15 ft. equipped with hand shut-off needle valve and check valve.
Capillary tube cleaner.
Capillary tube sizing kit.
Soldering-brazing torch, either LP fuel-air, acetylene-air or oxyacetylene.
Hand vacuum cleaner.
Gauge manifold.
Process tube adaptors.
Bending springs.

WRENCHES

Torque handle 3/8 in. drive.

Speed handle 3/8 in. drive.

Swivel handle 3/8 in. drive.

T-handle 3/8 in. drive for sockets.

8 in. adjustable open end wrench.

Set of Allen setscrew wrenches.

Refrigeration ratchet wrench, 3/16, 7/32 and 1/4 in. with square openings.

1/2 in. 15 deg. open-end wrench.

3/4 in. 15 deg. open-end wrench.

7/8 in. 15 deg. open-end wrench.

1 in. 15 deg. open-end wrench.

1/2 in. box wrench.

1/2 in. T-socket wrench.

Set of sockets, 12 point, 7/16 to 1 in. with 3/8 in. drive.

PLIERS

6 in. combination.

Wire cutters.

Slim nose.

HAMMERS

TUBING TOOLS

Bending springs for 1/4, 3/8 and 1/2 in. OD tubing.

Flaring tool 3/16 to 1/2 in. capacity.

Tube cutter.

Pinch-off tool.

Swaging tool.

SCREWDRIVERS

3 in. regular screwdriver, insulated handle.

6 in. regular screwdriver, insulated handle.

8 in. regular screwdriver, insulated handle.

3 in. Phillips screwdriver, insulated handle.

6 in. Phillips screwdriver, insulated handle.

8 in. Phillips screwdriver, insulated handle.

11-4 SUPPLIES

Some special supplies will be needed by the service technician who makes house calls answering customer complaints. In general, these supplies will include:

60-40 wire solder.

95-5 wire solder.

Solder flux.

Silver brazing wire 45 percent no cadmium.

Phosphorous silver wire.

Silver brazing flux.

Steel wool.

Medium grade sandpaper.

Plastic tape.

Disposable cylinders, R-12.

Disposable cylinders, R-22.

Disposable cylinders, R-502.

Coil of 1/4 in. soft copper tubing.

Coil of 5/16 in. soft copper tubing.

Coil of 3/8 in. soft copper tubing.

Coil of 1/2 in. soft copper tubing.

Copper pipe as needed.

Capillary tubing.

Filter-drier cartridges.

Can refrigerant oil, (spout type).

Can refrigerant oil, 300 viscosity.

Can refrigerant oil, 150 viscosity.

Cleaning cloths.

Relays.

Capacitors.

Motor controls.

Refrigerant controls.

Overload protectors.

Light switches.

Sealing compound.

Driers (flared and soldered fittings).

Drier-filters (flared and soldered fittings).

Sight glasses (flared and soldered fittings).

Flared fittings (SAE) — all sizes and shapes.

Soldered fittings — all sizes and shapes.

Piercing valves and valve adaptors.

Valve cores.

11-5 INSTALLING REFRIGERATORS AND FREEZERS

Proper installation is very important to the operation of a refrigerator or freezer. This includes leveling of the cabinet, correct electrical power and good ventilation. The manufacturer ships the units carefully crated and with full written instructions on how to ship, how to uncrate and how to install.

11-6 UNCRATING A REFRIGERATOR OR FREEZER

The carton usually has printed instructions on safe handling. These instructions should be carefully followed.

Many dealers uncrate the cabinet at the store; others do it just outside the home because most crates are too large to fit through standard doors.

Certain areas of the cabinet may be easily abused when uncrating or moving it. They are:

1. Bottom. Condensing unit may be damaged.
2. Back. Refrigeration condenser may be damaged. (Not all cabinets have condensers in the back.)
3. Door. The doors may be forced out of line or buckled.

There are two types of motor compressor dome shipping methods. One is the use of removable shipping bolts when the dome is mounted on or is suspended from springs. The other is loosening the dome shipping bolts two or three turns when the dome is mounted on synthetic flexible grommets.

Fig. 11-2 shows a hand truck with a hold-on strap. The side rails may be used as skids to aid in moving the appliance in and out of the truck and in and out of the building.

11-7 LOCATING POSITION OF REFRIGERATOR-FREEZER

It is advisable to locate the refrigerator-freezer where it will not be in direct sunlight. Neither should it be near an oven,

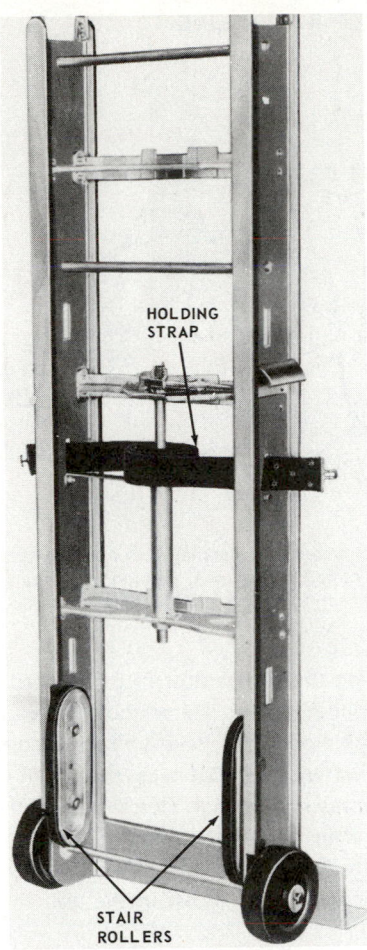

Fig. 11-2. Appliance hand truck. Strap is wound around appliance and is tightened by a wrench powering a ratchet gear. (Stevens Appliance Truck Co.)

at the refrigerator outlet drops more than 10V, the wiring in the circuit is not heavy enough. A flicker in the lights at the instant the refrigerator starts is a sure sign of a poor electrical supply.

It is very important to ground the refrigerator. All removable electrical parts — fans, thermostats, timers, etc. — are already safety grounded. If the wall outlet has a three-wire socket and the unit has a three-wire prong power cord, there is good grounding. Otherwise, a wire must be attached between a metal part of the cabinet and a good ground such as a water pipe. The type of male plug used on the appliance's power cord is an indication of the voltage and grounding. See Fig. 6-83 for these different plug designs.

The service technician should always check to see that there is proper grounding in the outlet box which supplies current to any refrigerating mechanism being serviced. Take a voltmeter reading from a "live" connection in the wall receptacle to the ground connection of the receptacle. A full voltage reading will indicate that the outlet is properly grounded. Fig. 11-3 shows how this test is performed.

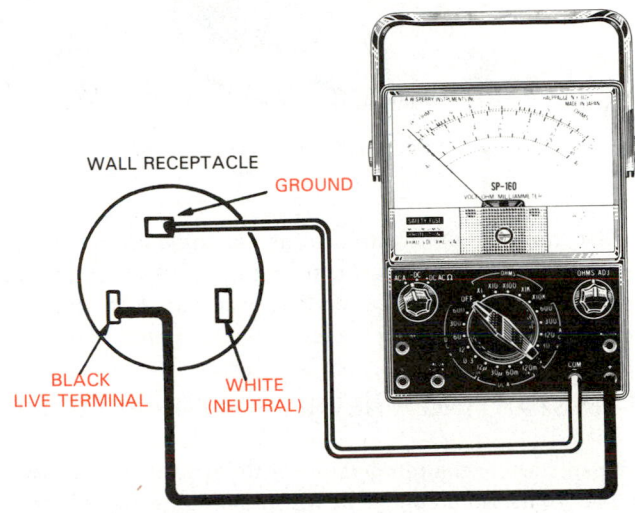

Fig. 11-3. Using a voltmeter to make sure that the ground terminal is actually properly connected to a ground. (A.W. Sperry Instruments, Inc.)

heat radiator or warm air register. Should it be necessary to locate the refrigerator-freezer near an oven or radiator, a strip of aluminum foil large enough to cover the side of the refrigerator-freezer should be hung between the refrigerator-freezer and the heat source.

The room should be large enough to provide enough air to cool the condenser.

11-8 ELECTRICAL SUPPLY

The electrical outlet for the refrigerator-freezer must provide the correct electrical supply. Be sure to read the electrical ratings on the refrigerator and check these against the electrical supply at the wall outlet. The modern household refrigerator-freezer may need more current than the older, simpler refrigerators and freezers.

It is best to have a separate circuit from the fuse or circuit breaker box to the refrigerator-freezer outlet. Avoid using an extension cord between the refrigerator power cord and the wall outlet. The resulting voltage may be too low.

Voltage at the refrigerator outlet can be quite easily checked with a voltmeter. The circuit capacity (wire size, etc.) is checked as follows: If at the instant of starting, the voltage

11-9 INSTALLING A REFRIGERATOR OR FREEZER

Since domestic refrigerators are air-cooled, proper ventilation is very important. Yet, many kitchens are designed without providing any space for air movement around the refrigerator.

In these installations, the refrigerator must have fans to draw the cool air in at the floor level, circulate it over the condenser, exhausting the warm air back into the kitchen at, or near, floor level. It is important that nothing be placed in front of these air openings to block air flow.

Condensers of many domestic refrigerators are mounted on the back. Some are protected by a shroud which provides a chimney effect to increase the rate of air flow over the condenser. With this type, air circulation must be provided at the bottom, back and top.

Many freezers and some refrigerators use the outside shell as the condenser surface. In these cases not less than two inches of space must be provided between the refrigerator cabinet and surrounding surfaces.

The refrigerator must be carefully leveled during installation. A spirit level should be used.

First check the floor where the rear supports or legs of the refrigerator are to rest. If not level, use wood spacers. Usually, the front supports are adjustable and may be used to properly level the cabinet. Fig. 11-4 illustrates the manner of adjusting these levelers.

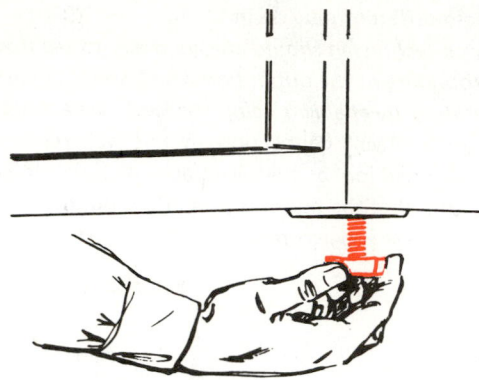

Fig. 11-4. Adjusting a leg leveler.

Test the wall outlet with a voltmeter to be certain there is electrical power. Put the temperature control in the off position. Connect electrical cord to wall outlet and test unit for running before moving the refrigerator into position.

11-10 STARTING A REFRIGERATOR-FREEZER

When starting the refrigerator for the first time, it is best to set the temperature control at the middle of its range. After a few hours of operation, a thermometer in the fresh foods compartment may then be used to adjust the temperature control setting to the customer's requirements.

If the refrigerator does not start, make sure that the electrical circuit is in good condition before checking for mechanical trouble. Open and close the doors to make sure that the interior lights are functioning properly.

CAUTION: If a refrigerator-freezer which uses a capillary tube refrigerant control is stopped and then started immediately, it may fail to operate. This is not necessarily a malfunction. The motor in this type of refrigerator does not provide a large enough starting torque to overcome a high head pressure. Disconnect the refrigerator for a few minutes until the refrigerant pressure has had time to balance between the high side and the low side. Then start the system again.

11-11 INSTALLING ICE CUBE MAKER

Many domestic refrigerators have automatic ice cube makers. See Chapter 10. These units are connected to the nearest cold water line by a 1/4-in. copper tubing. See Fig. 11-5.

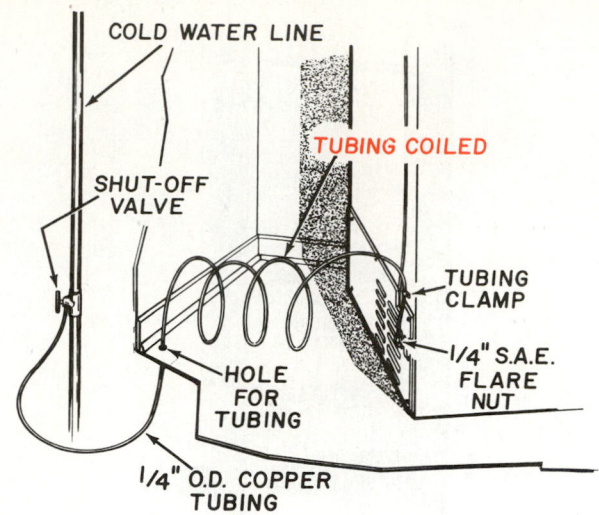

Fig. 11-5. Typical water line installation for automatic ice cube maker. Note tubing is coiled to permit moving refrigerator without disconnecting tubing. (General Electric Co.)

Before putting the refrigerator in its selected place, run the 1/4-in. copper line to the nearest cold water line. (Cabinet partitions or the floor may have to be drilled.) Mount a tap valve on the water line. Connect the tubing to the valve (usually a compression fitting). Connect the other end of the tubing to the refrigerator water line fitting. Allow several large loops of the tubing in back of the refrigerator to permit moving it in or out of its recess in the wall for cleaning and service.

Turn the tap valve stem in slowly and pierce the cold water pipe. Check for water leaks. Gently move the refrigerator into its recess. Be careful to avoid kinking or buckling the 1/4-in. tubing.

11-12 LOCATING AND REMOVING NOISE

Most outside noise in the refrigerator comes from rattles. Loose baffles or ducts, tubing touching something while vibrating, uneven floor which may cause a list (leaning to one side) of the condensing unit, and fan and motor vibration — all are sources of noise. A stethoscopic type listening device or electronic sound tracer, like the one shown in Fig. 11-6, is useful for locating such sounds. This instrument has a built-in sound amplifier. Fig. 11-7 shows a simple, homemade stethoscope.

A loose evaporator unit door, loose articles on shelves and shelves not seated properly on supports, may also cause annoying rattles. Such noise originating in the unit may indicate that it is laboring harder than it should. To determine this, test the electrical load with an ammeter or wattmeter. One may sometimes determine if the unit is overloaded by its starting behavior. Three seconds to operate relay is the average time. A slower start indicates an overload.

Tubing, rattling against refrigerator parts, should be carefully bent away from contact. If the tubing is rigid but has a vibration or hum (harmonic vibration), this noise can be reduced by clamping rubber blocks on the tubing. See Fig. 11-8.

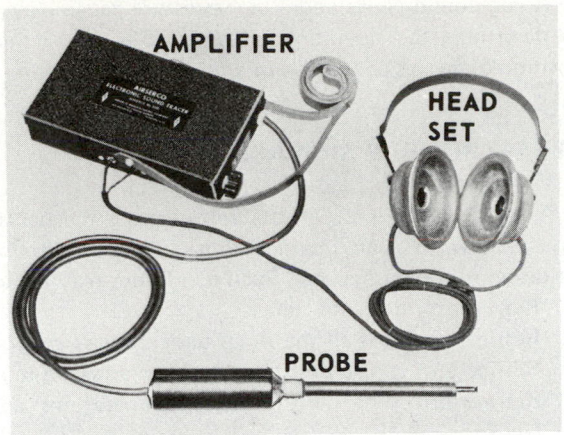

Fig. 11-6. Electronic sound tracer. Note headset, amplifier and probe. To use, touch suspected part with probe. Sound from headset will indicate if probe is at source of sound. (Airserco Mfg. Co.)

Another very helpful device for inspecting a refrigerator is an adjustable mirror on a long flexible extension, Fig. 11-9.

11-13 TEMPERATURE-PRESSURE CONDITIONS

Before servicing a refrigerator, a service technician should know:

1. The normal temperature in the evaporator during the operating cycle.
2. Normal pressure on the low-pressure side during the operating cycle.
3. Normal temperature of the condenser during the operating cycle.
4. Normal pressure on the high-pressure side during the operating cycle.

The above conditions depend upon the refrigerant being used. Temperature-pressure properties of refrigerants vary.

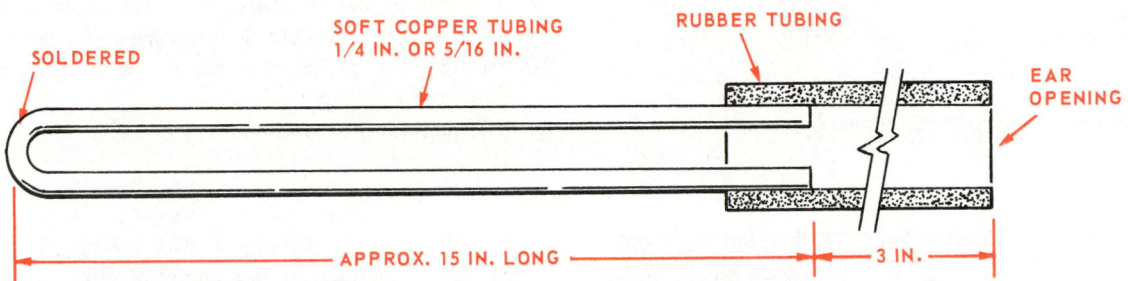

Fig. 11-7. A simple stethoscope made from a section of 1/4 in. copper tubing. It is excellent for locating source of many noises. Place closed end against mechanism and then put rubber tube opening to ear. (Use care around moving parts.)

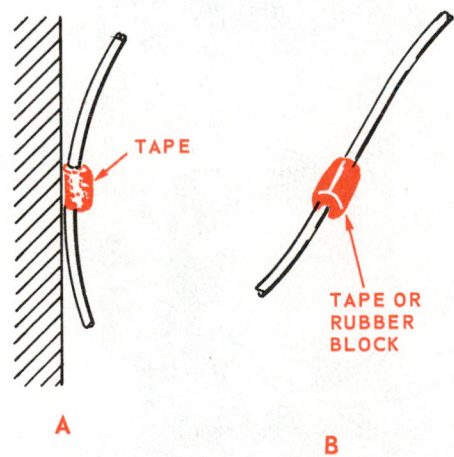

Fig. 11-8. Two ways are shown to reduce noise caused by vibrating tubing. A—Wind tape around tubing where tubing touches cabinet. B—Put tape or rubber block on tubing in center of vibrating section.

Loose baffles and ducts can be secured with self-tapping sheet metal screws.

If the sound tracer indicates that the noise is coming from inside the motor compressor, one should not attempt any service operations until after studying paragraphs 11-48, 11-49 and 11-50.

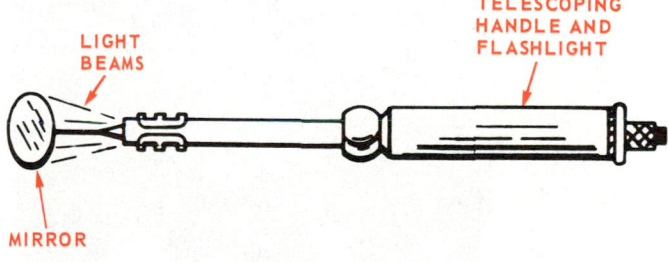

Fig. 11-9. An inspection mirror is often needed to see into hard-to-reach places. It is used to check fan alignment, motor condition, tubing location, cleanliness of condenser, condition of evaporator, etc. Avoid touching moving parts with the mirror.

Fig. 11-10 list the average temperature-pressure conditions for a domestic refrigerator-freezer evaporator and condenser. When using the table, be sure that the correct ambient temperature is used.

The most popular hermetic refrigerant is R-12, except for low-temperature (frozen foods) units and air conditioners where R-22 is quite popular. The properties of R-12 are described in Para. 9-6.

To determine the operating temperatures, a thermocouple thermometer may be used. Fig. 11-11 illustrates such a thermometer in use.

Installing and Servicing Small Hermetic Systems / 339

FRESH FOODS	R-12	
EVAPORATOR TEMPERATURE	AMBIENT TEMP.	
	70 F.	90 F.
START OF CYCLE	15	15
MIDDLE OF CYCLE	5	5
END OF CYCLE	0	0
FRESH FOODS		
EVAPORATOR PRESSURE, psi		
START OF CYCLE	12	12
MIDDLE OF CYCLE	8	8
END OF CYCLE	5	5
CONDENSER TEMPERATURE		
START OF CYCLE	70	90
MIDDLE OF CYCLE	100	120
END OF CYCLE	100	120
CONDENSER PRESSURE, psi		
START OF CYCLE	70	85
MIDDLE OF CYCLE	120	158
END OF CYCLE	120	158

Fig. 11-10. Average temperature and pressure conditions in domestic refrigerator using R-12 refrigerant. Unit has freezer compartment.

A compound gauge and a high-pressure gauge may be used to determine the operating pressures. Fig. 11-12 shows the position of gauges connected to a refrigeration system.

11-14 TROUBLE SIGNALS

A fault in any part of the refrigerating mechanism will usually show up as an unsatisfactory temperature or operating condition of the refrigerator. Such conditions may include:

1. Refrigerator does not run.
2. Refrigerator runs all the time; temperatures are too cold.
3. Refrigerator runs all the time; temperatures are too warm.
4. Refrigerator runs all the time; temperatures are satisfactory.
5. Refrigerator cycles but food compartment is too warm; freezing compartment is satisfactory.
6. Refrigerator cycles but freezing compartment is too cold.
7. Motor control cuts out.
8. Refrigerator cycles satisfactorily, refrigeration is poor.
9. Refrigerator cycles but does not freeze ice cubes.
10. Refrigerator cycles; too much ice accumulates on the evaporator.
11. Refrigerator mechanism is very noisy.

Fig. 11-11. Recording thermocouple thermometer is being used to test operating temperatures. It has two sensing bulbs. One is placed in freezer compartment and one in fresh food compartment. (The Dickson Co.)

11-15 LOCATING TROUBLE IN THE HERMETIC REFRIGERATOR-FREEZER

Fig. 11-13 charts common troubles, their causes, symptoms and remedies. This chart should be considered a general guide only. It does not apply to all units.

Experience and judgment are needed when one is trying to

Many refrigerators having a separate door on the freezer compartment also have an electric heater around the door opening to keep ice or moisture from forming there. If this heater is not working, ice build-up may keep the door from closing properly. Then too, moisture may enter and collect on the evaporator. The wiring circuit diagram for the refrigerator will indicate whether or not a "door heater" is used.

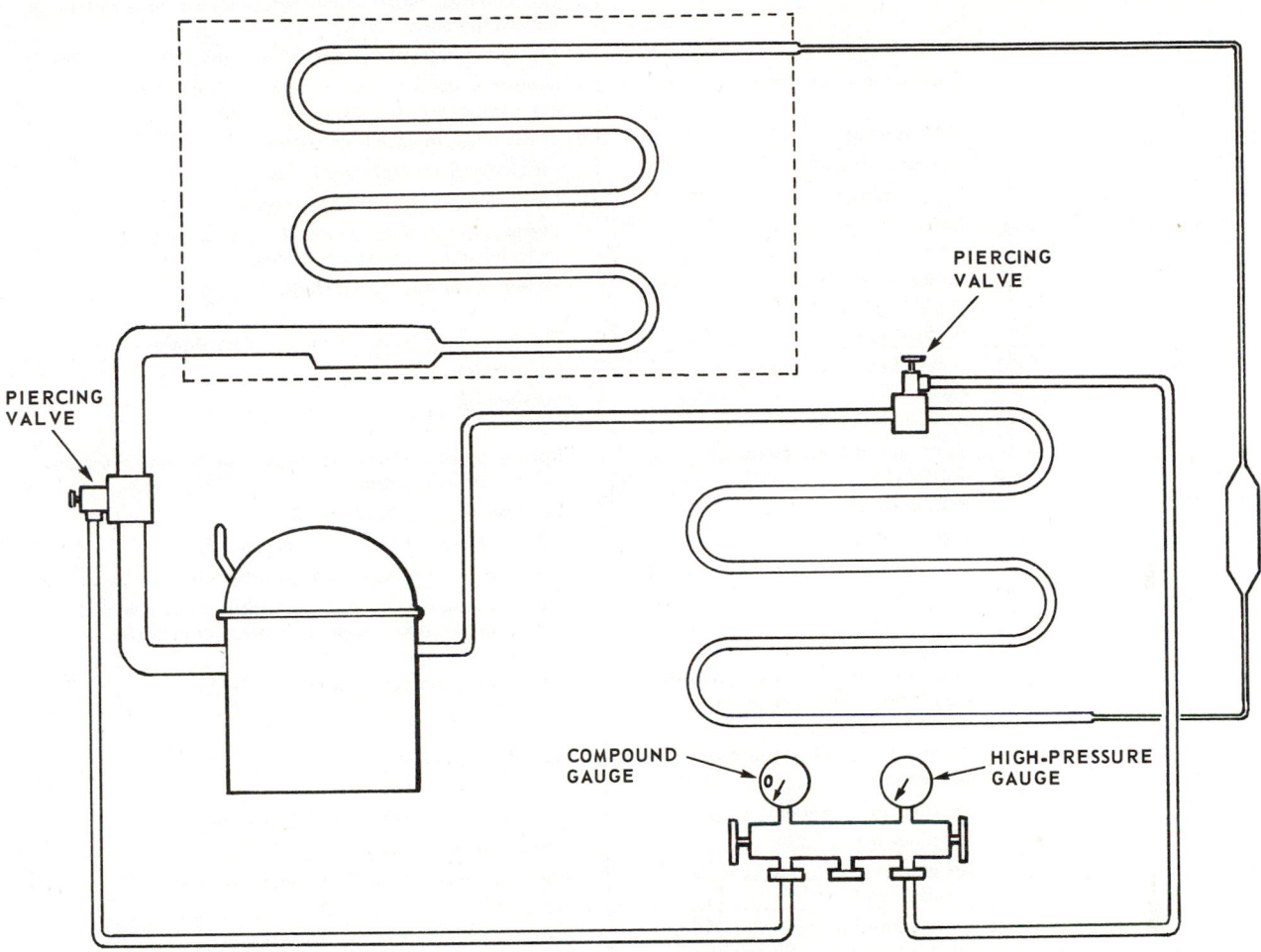

Fig. 11-12. A high-pressure gauge and a compound gauge are connected to a hermetic system with service lines. Piercing valves are used to open the service lines to the system.

find the real cause of poor operation. *Always remember, cooling takes place only when the pressure is low enough and if there is liquid refrigerant present.*

For example, if an evaporator has the correct low pressure but is warm, there is no liquid in the evaporator. Another example, if a drier is frosting, it indicates that there is liquid and also a low pressure in the drier. (Drier is partially clogged.)

11-16 ICE ON THE EVAPORATOR

When large amounts of ice build up in the food storage space, it acts as insulation and may result in poor cooling. The cause usually is a leaky gasket seal, or the defrost system in frost-free or automatic defrost refrigerators is not operating.

It may be that the door seal has lost its flexibility (life) or is broken. In either case, the gasket should be replaced.

A flat slip of paper inserted between the door and the cabinet should be held tightly when the door is closed. If this paper can be pulled out easily, the gasket is not tight enough. The hardware (latch and hinges) must be adjusted or the gasket replaced.

11-17 MOISTURE AND ICE IN THE CABINET INSULATION

Moisture and ice in the cabinet insulation is serious. It means there is an air leak in the outside cabinet seal or shell. The leak allows warm, moist air to enter this space. When the

TROUBLESHOOTING CHART

TROUBLE	COMMON CAUSE	REMEDY
1. Unit will not run.	Blown fuse.	Replace fuse
	Low voltage.	Check outlet with voltmeter, should check 115V plus or minus 10 percent.
		If circuit overloaded, either reduce load or have electrician install separate circuit.
		If unable to remedy any other way, install auto-transformer.
	Broken motor or temperature control.	Jumper across terminals of control. If unit runs and connections are all tight, replace control.
	Broken relay.	Check relay, replace if necessary.
	Broken overload.	Check overload, replace if necessary.
	Broken compressor.	Check compressor, replace if necessary.
	Defective service cord.	Check with test light at unit; if no circuit and current is indicated at outlet, replace or repair.
	Broken lead to compressors, timer or cold control.	Repair or replace broken leads.
	Broken timer.	Check with test light and replace if necessary.
2. Refrigerator section too warm.	Repeated door openings.	Instruct user.
	Overloading of shelves, blocking normal air circulation in cabinet.	Instruct user.
	Warm or hot foods placed in cabinet.	Instruct user to allow foods to cool to room temperature before placing in cabinet.
	Poor door seal.	Level cabinet, adjust door seal.
	Interior light stays on.	Check light switch; if faulty, replace.
	Refrigerator section airflow control.	Turn control knob to colder position. Check airflow heater.
		Check if damper is opening by removing grille. With door open, damper should open. If control inoperative, replace control.
	Cold control knob set at too warm a position, not allowing unit to operate often enough.	Turn knob to colder position.
	Freezer section grille not properly positioned.	Reposition grille.
	Freezer fan not running properly.	Replace fan, fan switch, or defective wiring.
	Defective intake valve.	Replace motor compressor.
	Air duct seal not properly sealed or positioned.	Check and reseal or put in correct position.
3. Refrigerator section too cold.	Refrigerator section airflow control knob turned to coldest position.	Turn control knob to warmer position.
	Airflow control remains open.	Remove obstruction.
	Broken airflow control.	Replace control.
	Broken airflow heater.	Replace heater.
4. Freezer section and refrigerator section too warm.	Fan motor not running.	Check and replace fan motor if necessary.
	Cold control set too warm or broken.	Check and replace if necessary.
	Finned evaporator blocked with ice.	Check defrost heater thermostat or timer. Either one of these could cause this condition.
	Shortage of refrigerant.	Check for leak, repair, evacuate and recharge system.
	Not enough air circulation around cabinet.	Relocate cabinet or provide clearances to allow sufficient circulation.
	Dirty condenser or obstructed condenser ducts.	Clean the condenser and the ducts.
	Poor door seal.	Level cabinet, adjust door seal.
	Too many door openings.	Instruct customer.
5. Freezer section too cold.	Cold control knob improperly set.	Turn knob to warmer position.
	Cold control capillary not properly clamped to evaporator.	Tighten clamp or reposition.
	Broken cold control.	Check control. Replace if necessary.

TROUBLE	COMMON CAUSE	REMEDY
6. Unit runs all the time.	Not enough air circulation around cabinet or air circulation is restricted.	Relocate cabinet or provide proper clearances around cabinet — remove restriction.
	Poor door seal.	Check and make necessary adjustments.
	Freezing large quantities of ice cubes, or heavy loading after shopping.	Explain to customer that heavy loading causes long running time.
	Refrigerant charge.	Undercharge or overcharge — check, evacuate and recharge with proper charge.
	Room temperature too warm.	Ventilate room as much as possible.
	Cold control.	Check control; if it allows unit to operate all the time, replace control.
	Defective light switch.	Check if light goes out. Replace switch if necessary.
	Excessive door openings.	Instruct customer.
7. Noisy operation.	Loose flooring or floor not firm.	Tighten flooring or brace floor.
	Tubing contacting cabinet or other tubing.	Move tubing.
	Cabinet not level.	Level cabinet.
	Drip tray vibrating.	Move tray — place on styrofoam pad if necessary.
	Fan hitting liner or mechanically grounding.	Move fan.
	Compressor mechanically grounded.	Replace compressor mounts.
8. Unit cycles on overload.	Broken relay.	Replace relay.
	Weak overload protector.	Replace overload protector.
	Low voltage.	Check outlet with voltmeter. Underload voltage should be 115V plus or minus 10 percent. Check for several appliances on same circuit or extremely long or under-sized extension cord being used.
	Poor compressor.	Check with test cord and also for ground before replacing.
9. Stuck motor compressor.	Broken valve.	Replace motor compressor.
	Insufficient oil.	Add oil; if unit still will not operate, replace motor compressor.
	Overheated compressor.	If compressor faulty for any reason, replace motor compressor.
10. Frost or ice on finned evaporator.	Broken timer.	Check with test light and replace if necessary.
	Defective defrost heater.	Replace heater.
	Defective thermostat.	Replace thermostat.
11. Ice in drip catcher.	Defective drip catcher heater.	Replace heater.
12. Unit runs all the time, temperature normal.	Ice builds up on the evaporator.	Check door gaskets — replace if necessary.
	Control bulb on thermostat not in contact with evaporator surface.	Place control bulb in contact with the evaporator surface.
13. Freezer runs all the time. Temperature too cold.	Faulty thermostat.	Check thermostat — test and replace if necessary.
14. Freezer runs all the time. Temperature too warm.	Ice buildup in insulation.	Remove breaker strips, stop unit, melt ice and dry insulation, seal outer shell leaks and joints and then assemble.
15. Rapid ice buildup on the evaporator.	Leaky door gasket.	Adjust door hinges. Replace door gasket if cracked, brittle or worn.
16. Door on freezer compartment freezes shut.	Faulty electric gasket heater.	Use alternate gasket heater or install new one.
	Faulty gasket seal.	Inspect and check gasket. If worn, cracked or hardened, replace it.
17. Freezer works then warms up.	Moisture in refrigerator.	Install drier in liquid line.
18. Gradual reduction in freezing capacity.	Wax buildup in capillary tube.	Use capillary tube cleaning tool or replace capillary tube.

Fig. 11-13. Chart lists some common hermetic system troubles, their causes and suggested remedies.

	70 F. Ambient Temperature	90 F. Ambient Temperature	100 F. Ambient
Cabinet Temperature	38°	40°	47°
% Operating Time	38	62	100
Cycles Per Hour	3	2	None
kWhr/24 hrs.	3.8	6.0	9.9
Control Position	4	4	4
Evaporator Air Temperature	1.5°	-1°	0°
Suction Pressure (Psi) (Min-Max)	2"-13	0-13	2"
Watts (Complete System)	390 $\pm$ 20	395 $\pm$ 20	395 $\pm$ 20

Fig. 11-14. Chart showing operation of 18 cu. ft. combination refrigerator-freezer which has 1/3 hp two-pole motor compressor. Note kilowatts per hour changes as ambient temperature changes.

warm air strikes the cold inner liner, it gives up its moisture.

When this occurs in the cabinet of the fresh-food refrigerator, the insulation becomes wet and loses its heat insulating qualities. There will be two indications pointing to this trouble:

1. The condensing unit will run more than normal.
2. The outside surface of the refrigerator will feel colder than normal wherever the insulation is wet.

In the case of a food freezer, the condensed moisture will form ice in the insulation. The symptoms will be much the same as in fresh-food refrigerators. If this condition continues, enough ice will soon be built-up to cause the sides of the cabinet to buckle. The leak in the outside cabinet surface must be located and completely sealed.

Most freezers provide a small opening through the inner lining, connecting the insulated area with the inside of the freezer cabinet. Since in freezers the temperature inside of the cabinet is much lower than the insulation temperature, any moisture will tend to escape through this small opening and condense on the evaporator surface.

11-18 HERMETIC SERVICING GUIDE

To service refrigerating units successfully, one must know how they should perform when in good condition. Fig. 11-14 shows the performance data of a typical domestic unit.

Always check the system data before trying to locate the cause of the trouble. System data for a 1/3 hp two-pole motor unit used in a combination refrigerator-freezer is shown in Fig. 11-15.

	DATA
REFRIGERANT	R-12
CHARGE (IN OUNCES)	10 1/2
COMPRESSOR hp	1/3
COMPRESSOR SPEED rpm	3450
RUNNING AMPERES	5.6
VOLTAGE	120
PHASE	SINGLE

Fig. 11-15. Refrigerant oil and electrical data which is typical of the information usually found on the identification plate mounted on the motor compressor.

Para. 11-15 and Fig. 11-13 provide information which one should have in mind before beginning any service operation. Servicing of hermetic refrigerators may be divided into three major area:

1. External servicing.
2. Internal servicing.
3. Overhaul of hermetic system.

11-19 EXTERNAL SERVICING OPERATIONS

Some of the more common external service operations are:

1. Cabinet hardware.
2. Cleaning.
3. Noise (rattles).
4. Electrical:
   a. Power-in circuit.
   b. Thermostat.
   c. Defrost thermostat.
   d. Interior light and circuit.
   e. Fan motor and circuit.
   f. Damper controls.
   g. Motor compressor:
      Relay and overload protector.
      Capacitor.
      Motor terminals.
   h. Defroster.
   i. Defroster control and circuit.
   j. Defroster heater coil.
   k. Cabinet heaters.
   l. Evaporator fan and circuit.
   m. Condenser fan and circuit.
   n. Light circuit.
   o. Butter conditioner circuit.
5. Ice cube maker.

These external mechanisms, electrical wiring and electrical parts can all be checked for operation quickly by inspection and by using a volt-amp-ohmmeter.

Other external service troubles can be located by checking for ice on evaporators, frost or sweat on suction lines, warm or hot discharge line, ice or sweat on driers and dirty condensers.

It is important to locate the trouble and determine the cause accurately.

11-20 DIAGNOSING EXTERNAL TROUBLES

Some hermetic units have been needlessly replaced because the service technician thought the internal mechanism was faulty when the real trouble was in the external devices.

For example, if the mechanism will not start; if the unit hums but will not start; or if the unit short cycles, the trouble could be in some part of the external electrical circuit:

1. Power-in connections.
2. Thermostat.
3. Wire terminals.
4. Relay.
5. Capacitor (if the unit has one).

Each of these devices should be checked carefully before the unit itself can be considered faulty. These parts can be checked best by removing them from the wiring system. They can then be:

1. Checked independently.
2. Or, temporarily replaced by a test part of the proper size to see if the unit will run. Fig. 11-16 shows a motor

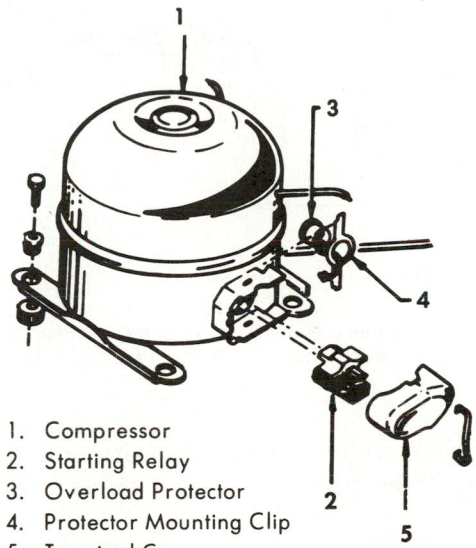

1. Compressor
2. Starting Relay
3. Overload Protector
4. Protector Mounting Clip
5. Terminal Cover

Fig. 11-16. Exploded view of starting relay and overload protector mounted on motor-compressor. (Franklin Mfg. Co.)

compressor with a relay and an overload protector mounted on it. Fig. 11-17 shows various electrical circuits typical of a domestic refrigerator.

Electrical connections must be clean and tight. If loose or dirty, they often overheat. This high temperature will discolor the connection.

The connection may be darkened by oxidation. A blue or greenish tint indicates overheating and corrosion. If the surrounding insulation is charred, overheating has occurred.

Troubles such as open circuits and grounded electrical wires can be checked easily with a test light. A test cord can be used to check four-pole motors, but two-pole motors should only be tested using a proper size relay in the circuit. These motors overheat if the starting circuit is connected for more than two

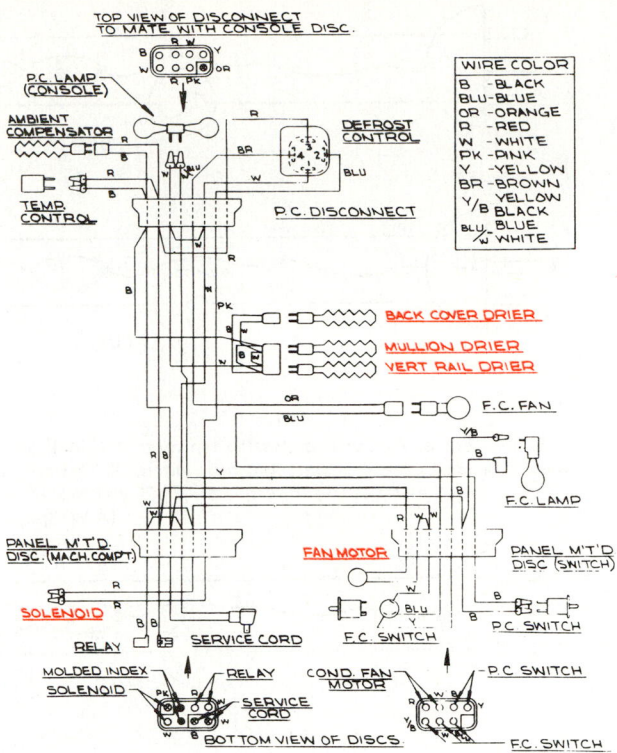

Fig. 11-17. Wiring diagram of domestic system equipped with hot gas defrost control, lights, condenser fan and electric heaters. Color code of wiring assists service technician in tracing circuits. (Kelvinator, Inc.)

or three seconds.

Chapters 6, 7 and 8 should be carefully reviewed before attempting to locate trouble in electrical units.

When checking and servicing hermetic electrical circuits, first make sure that the electrical supply at the outlet is good. Check the appliance voltage specifications.

Using a voltmeter, test the open circuit voltage. Next, plug in the appliance and check the voltage again while it is running. The open circuit voltage is likely to be slightly higher than with the motor running. However, this difference should not be more than 5V.

A difference of 10V or more indicates serious trouble:

1. An overload.
2. Something wrong in the motor windings.
3. Poor wiring to wall outlet.

Most refrigerators and freezers have a wiring diagram attached to the back of the refrigerator-freezer. Locate this diagram and check each circuit independently.

If the compressor fails to start:

1. Find out if electricity is reaching the motor compressor.
2. If it is, check the starting relay and circuit protectors. See Chapter 8.

Next, all wiring should be disconnected from the motor compressor. Then check the motor compressor with a manual start test cord. Such a test cord is shown in Fig. 11-18.

The ground clip labeled 4, must be fastened to the dome. All the clips or connectors should be plastic coated or have a rubber boot over them to protect the technician from shocks.

When the manual switch is pressed, the run circuit closes

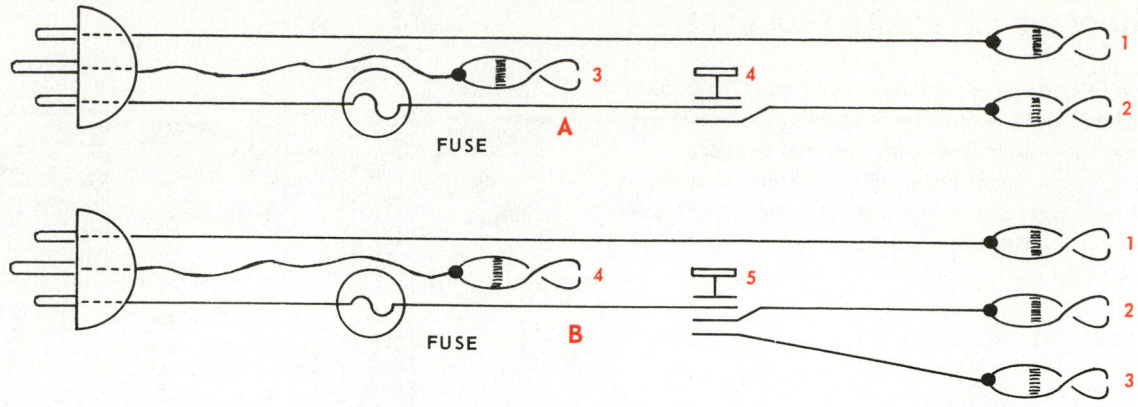

Fig. 11-18. Test cords. A—Cord for testing fan motors. Clip 1 goes to common winding terminal. Clip 2 attaches to running winding terminal. Clip 3 goes to ground. No. 4 is start and run switch. B—Test cord for hermetic motors. Clip 1 goes to hermetic compressor common terminal. Clip 2 attaches to run terminal. Clip 3 goes to start terminal. Clip 4 goes to ground. No. 5 is start and run switch. (A.W. Sperry Instruments, Inc.)

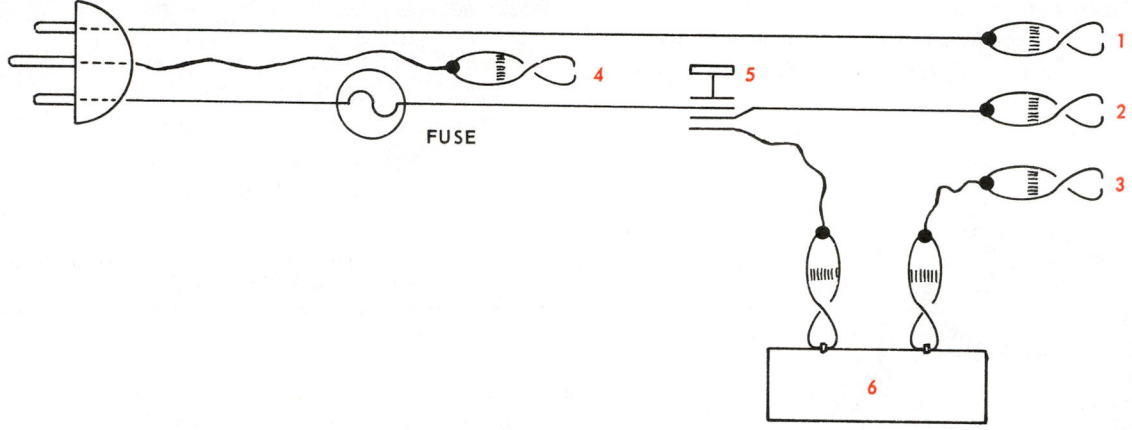

Fig. 11-19. Test cord for capacitor-start, induction-run hermetic motor. 1—Common winding terminal. 2—Running winding terminal. 3—Starting winding terminal. 4—Grounding wire and clip. 5—Start and run switch. 6—Starting capacitor. (A.W. Sperry Instruments, Inc.)

first, then the start circuit. After one to two seconds, lift the switch button just enough to open the start winding. If the motor operates correctly, the problem is in the external circuit. The fuse is located in the black wire of the three-prong plug.

A capacitor-start, induction-run motor is tested as shown in Fig. 11-19. The capacitor should be replaced by a new one of the same voltage and microfarad rating. An extra clip wire or lead is needed. Operate the switch as explained in the previous paragraph. After testing, remember to short the testing capacitor with a 20,000 ohm resistor to avoid a shock.

The test for a capacitor-start, capacitor-run motor is shown in Fig. 11-20. Use new capacitors of the same rating as the ones on the system being tested. Operate the switch as previously explained.

If the motor compressor works, check the electrical system up to the compressor.

If the motor does not operate when tested, further motor checks are needed, as explained in Chapter 7.

The test cords shown in Figs. 11-18, 11-19 and 11-20 can be used for checking continuity and grounds by replacing the fuse with a light bulb.

The trouble may be the evaporator motor or the condenser fan motor. These motors are usually replaced if found to be faulty.

Before removing a fan from a motor shaft, mark the position of the fan hub on the shaft to be sure the fan is located correctly on the new shaft.

An electrical failure in the mullion heater may cause a door gasket to freeze to the cabinet. The heater must be checked with a continuity light or an ohmmeter. Locate the circuit in the wiring diagram. Disconnect both ends of the mullion heater leads and test the heater for continuity.

If it is defective, look for a second heater in the insulation. Most cabinets have one. Test it also. If it is operating properly, use it. If there is no extra heating unit, install one of the same wattage.

If the problem is a faulty wire, use a stiff steel wire to pull new wiring through foamed-in-place insulation. If necessary, drill a hole (up to 1/2 in.) in the back of the refrigerator to help feed wires. Seal the hole after the wire or wires are pulled through.

If the complaint is cabinet temperature not responding properly to the temperature control, the temperature control

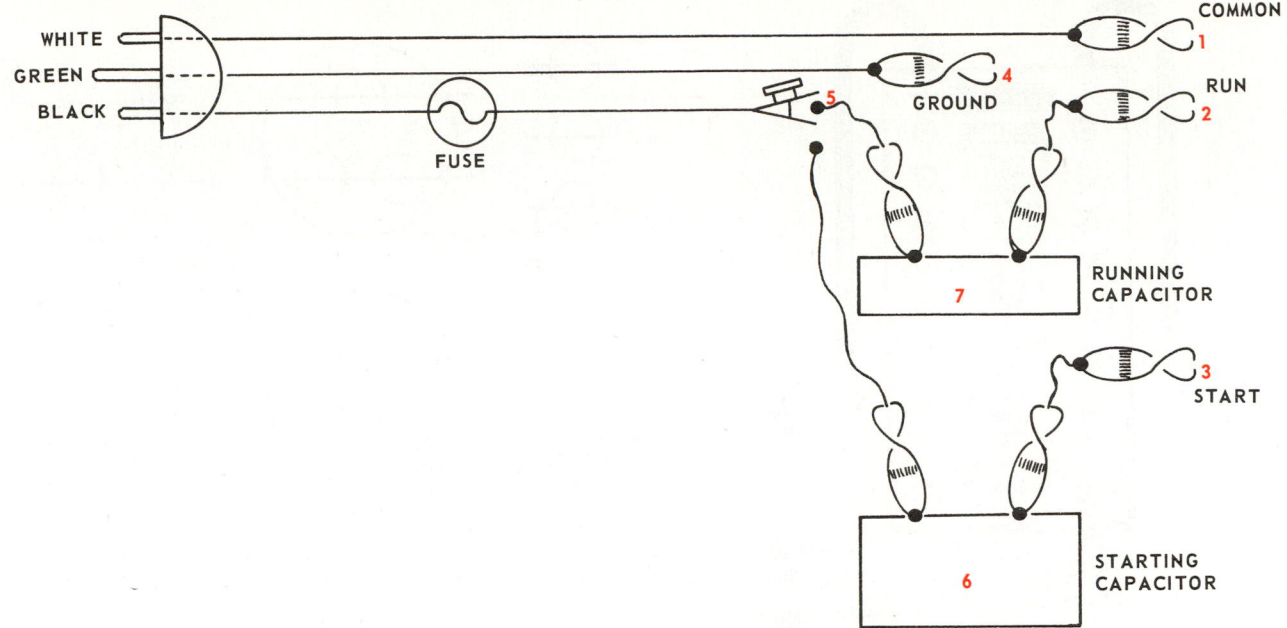

Fig. 11-20. Test cord used to check out a capacitor-start, capacitor-run hermetic motor. 1—Common winding terminal. 2—Running winding terminal. 3—Starting winding terminal. 4—Ground wire and clip. 5—Manual switch. 6—Starting capacitor. 7—Running capacitor. (A.W. Sperry Instruments, Inc.)

mechanism may be faulty. It is advisable to use a recording thermometer. Record temperature and time over a 24-hour period. The temperature-time chart will show if the appliance is operating properly.

11-21 ELECTRICAL TROUBLESHOOTING

The General Electric Co. has developed a rapid method (RED system) to check some of the electrical circuits of their refrigerators. Included are defrost thermostats, heaters, fans and compressors. Simpler circuits, such as lights, being easily tested, are not included in this system.

A multiple circuit connection, mounted at the front bottom of the cabinet, is used. See Fig. 11-21. When this connector is separated, the tester is connected to it. See Fig. 11-22. A diagram of the circuits to be tested is shown in Fig. 11-23. Note number code for test connections.

Always turn the thermostat to the off position before separating the connector. The symbol for the female connectors is >— and for the male connectors, —> . After connecting the tester and turning the power on, the "power on" light should be lit. If not, the power circuit needs repair.

One should refer to the General Electric data and diagrams when performing this test.

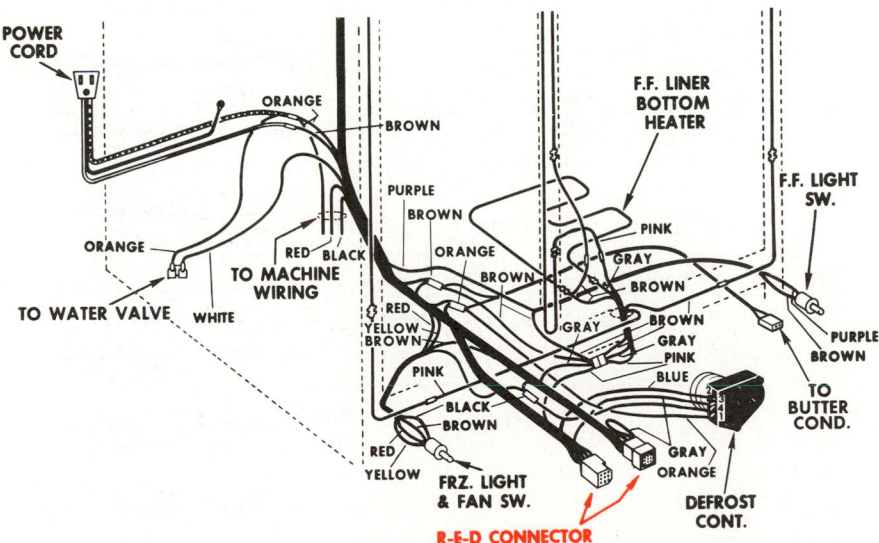

Fig. 11-21. Multi-circuit connector used for RED system simplifies testing of electrical circuits on some General Electric refrigerators.

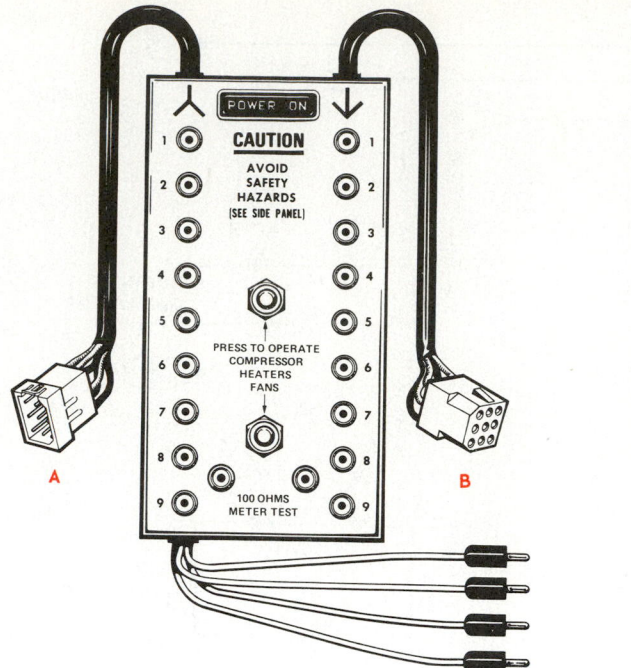

Fig. 11-22. The RED test instrument and adaptor. A—Male connector to electrical system. B—Female connector to electrical system. (General Electric Co.)

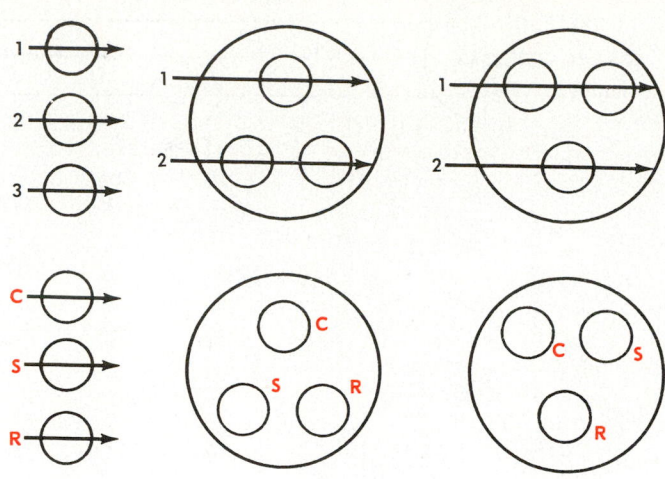

Fig. 11-24. Tecumseh motor compressor terminals are easy to identify. Referring to illustration above, read from left to right and then from top to bottom. Arrows show direction to read. Numbers show order of reading arrows.

All Tecumseh motor compressor terminals are set up to read "Common-Start-Run" as one reads from left to right, going down a line at a time. See Fig. 11-24.

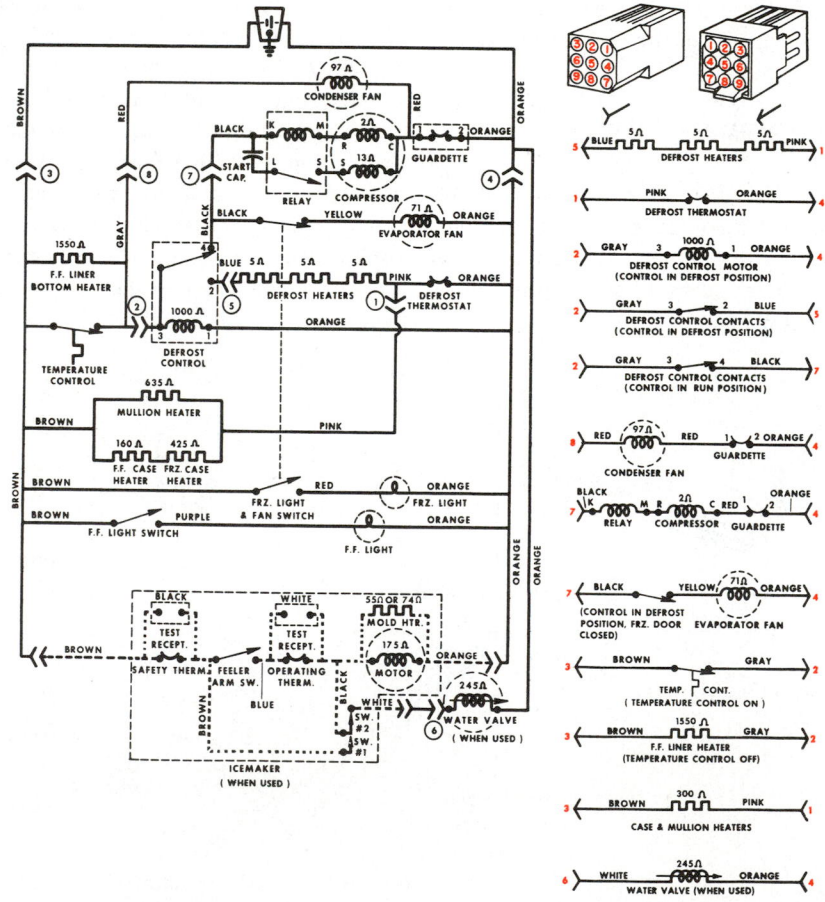

Fig. 11-23. Diagram of General Electric refrigerator circuits with number code where RED instrument and adaptors connect into circuits. Note: Numbers at either ends of the circuits shown at the right correspond to the terminals in the male and female plugs pictured above.

11-22 CLEANING THE EXTERNAL MECHANISM

The hermetic refrigerating system is basically a heat transfer mechanism. Air must circulate around and through the unit and condenser to carry away the heat. Therefore, the condenser and compressor unit must be kept as clean as possible, since dirt and lint act as heat insulation. Completely clean about every three months for economical operation and long life.

The hermetic mechanism can be cleaned where it is, by using a small vacuum cleaner or a special vacuum cleaner nozzle with a brush attachment. The vacuum cleaner eliminates the lint cloud that circulates in the room or settles on the floor. It is also quicker and more thorough than hand brushes or cleaning cloths. If a brush or cleaning cloth is used, place a paper or cloth on the floor under the unit.

If the unit uses a fan on the condensers, be sure to turn off the power or disconnect the unit before proceeding with the cleaning. In some cases it may be necessary to partially remove the unit to do a good cleaning job.

In the shop, high-pressure air, nitrogen or carbon dioxide is often used to blow lint and other dirt from between the fins or coils or where it is difficult to reach. Wear goggles and have good ventilation. See Para. 11-58.

11-23 STARTING A STUCK COMPRESSOR

A unit which does not start when connected to the electric power may have a small piece of dirt in the compressor or the unit has not been run for a long time. To start such a compressor, three methods may be tried:

1. Disconnect the wiring to the motor compressor. Connect a test cord into the electrical circuit of the main winding. Use an extra capacitor, connecting it into the circuit as shown in Fig. 11-25. Turn on the power from one to three seconds. The extra capacitor will try to reverse the compression rotation. CAUTION: this capacitor cannot be

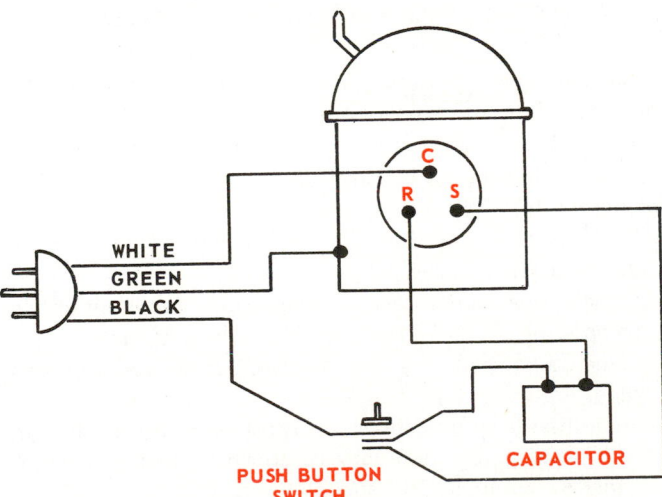

Fig. 11-25. Capacitor is being used in the running winding circuit to reverse rotation of "stuck" motor compressor. C—Common terminal. R—Running winding. S—Starting winding.

left in the circuit for more than a second or two. It will cause the motor to overheat.

If the compressor is successfully reversed, remove the reversing condenser. Try to operate the compressor, using the normal electrical circuit. If the compressor does not start after three or four attempts at reversal as indicated above, it will usually be necessary to rebuild or replace the motor compressor.

2. Another method is to connect the 120V motor compressor into a 240V power circuit using a starter cord. *One must be careful. Press the push button switch for a second at a time or the motor will be damaged.*

The extra voltage may break the stuck compressor loose. If not, the motor compressor will have to be replaced.

3. An extra torque method is to connect a 240V, 100 mfd (microfarad) start capacitor across the terminals of the run capacitor for NO MORE THAN ONE SECOND. (Count "one thousand one.") This may free the compressor.

11-24 SHORT CYCLING

Short cycling means that the unit starts and stops too frequently. Causes may be:
1. Thermostat not mounted securely.
2. Loose connections in the starting relay.

Avoid tapping a relay to check it. The jar may cause points to touch. This short contact may ruin the points and injure the motor. If the relay will not function correctly without being tapped, it should be replaced.

11-25 DIAGNOSING INTERNAL TROUBLES

There are many ways to find the cause of trouble inside a small hermetic system.

A lack of refrigerant is indicated when the evaporator is partially frosted, while another part is heavily frosted.

A sweating or frosted suction line means liquid refrigerant is in the suction line. This may be caused by a broken thermostat or too much refrigerant (if capillary tube is used).

Internal electrical troubles, involving the motor and connections, are very rare (about three cases out of 1000). Most internal problems come from air and moisture in the motor compressor. This causes corrosion and, eventually, a burnout.

In cases where liquid refrigerant reaches the compressor, the liquid may remove the oil as it evaporates in the crankcase and carry the oil with it into the condenser. Valves may be broken as the compressor tries to pump oil or liquid refrigerant.

A restriction on the high side (capillary tube, drier-filter or screen) will be indicated by continuous running, no refrigeration and a condenser cooler than normal.

Para. 11-27 through 11-33 explain how connections from the gauge manifold may be tied into a hermetic system.

11-26 INTERNAL SERVICE OPERATIONS

To remove any part of a hermetic system, or to find out if there is air in the system, a lack of refrigerant, clogged

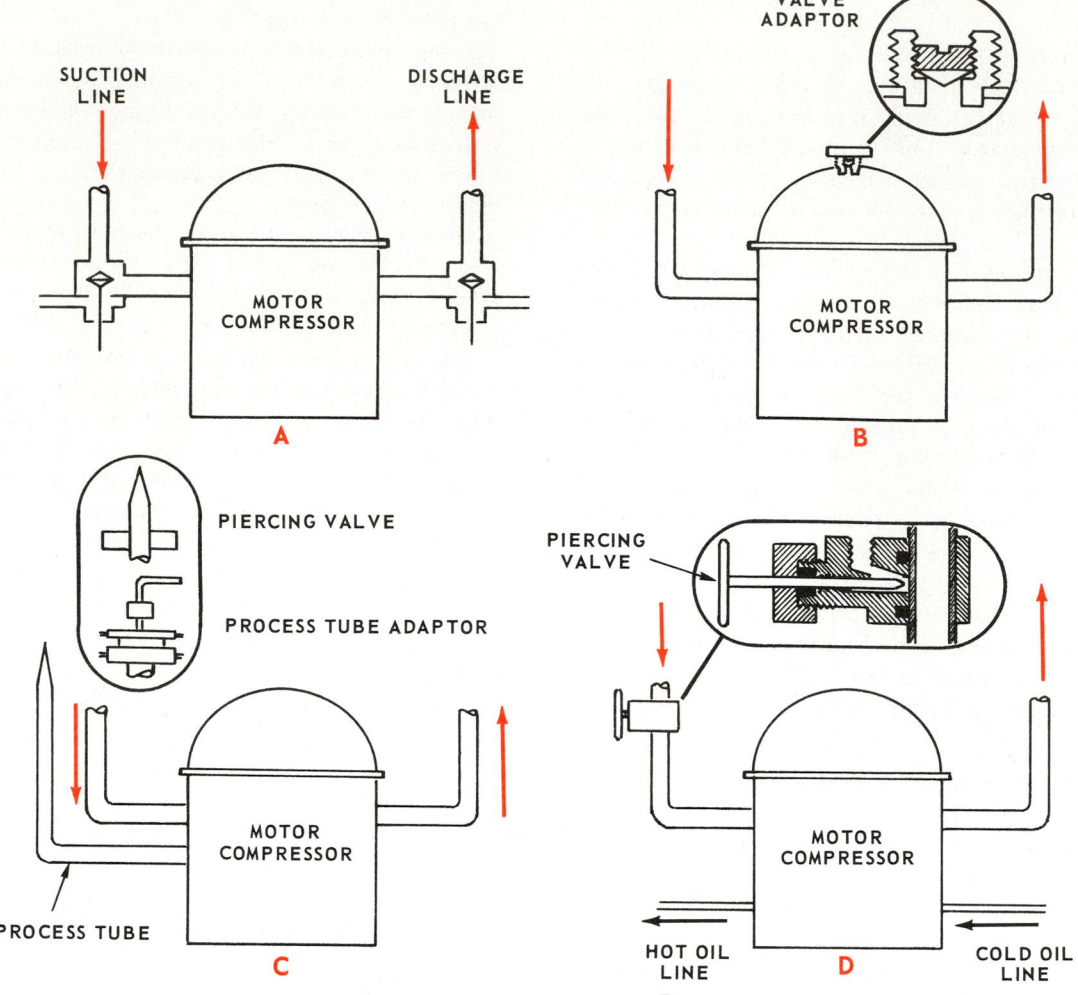

Fig. 11-26. Four different methods for connecting a gauge manifold to a hermetic system.

drier-filter or capillary tube, one must attach gauges and servicing devices. These include vacuum pumps, refrigerant cylinders and so forth.

Before attempting any field service operations which require opening the system:

1. Thoroughly clean all connections and valve fittings.
2. Install a valve adaptor or piercing valve.
3. Install a gauge manifold.

Probably the most frequent service operations will be the following:

1. Locating and repairing refrigerant leaks.
2. Purging, charging and discharging refrigerant.
3. Cleaning or replacing capillary tube.
4. Replacing a compressor.
5. Replacing filter-drier, high side.
6. Installing filter-drier, low side.
7. Evacuating.
8. Adding oil.
9. Replacing a condenser.
10. Use of high vacuum pump.

These same service operations can also be performed in the shop. Shops have better facilities to do the job.

Before performing any service operation, the service technician should study this complete chapter carefully.

11-27 INSTALLING A GAUGE MANIFOLD

To check the pressure in a system, gauges must be connected to the system without allowing air, moisture or dirt to enter. The procedure for connecting gauges to a system depends on the system design. It is different for each system. See Fig. 11-26.

A. Some systems have both a suction service valve and a discharge service valve.

B. Some have a suction service valve adaptor mounted on the compressor.

C. Some do not have any service valves but do have a process tube.

D. Some have a process tube too short or not reachable. In such systems it is necessary to attach a piercing valve to either the liquid line, the suction line or both.

System A (two service valves) is the easiest for attaching gauges. It also permits one to check both the low-side pressure and the high-side pressure. Because this system is most

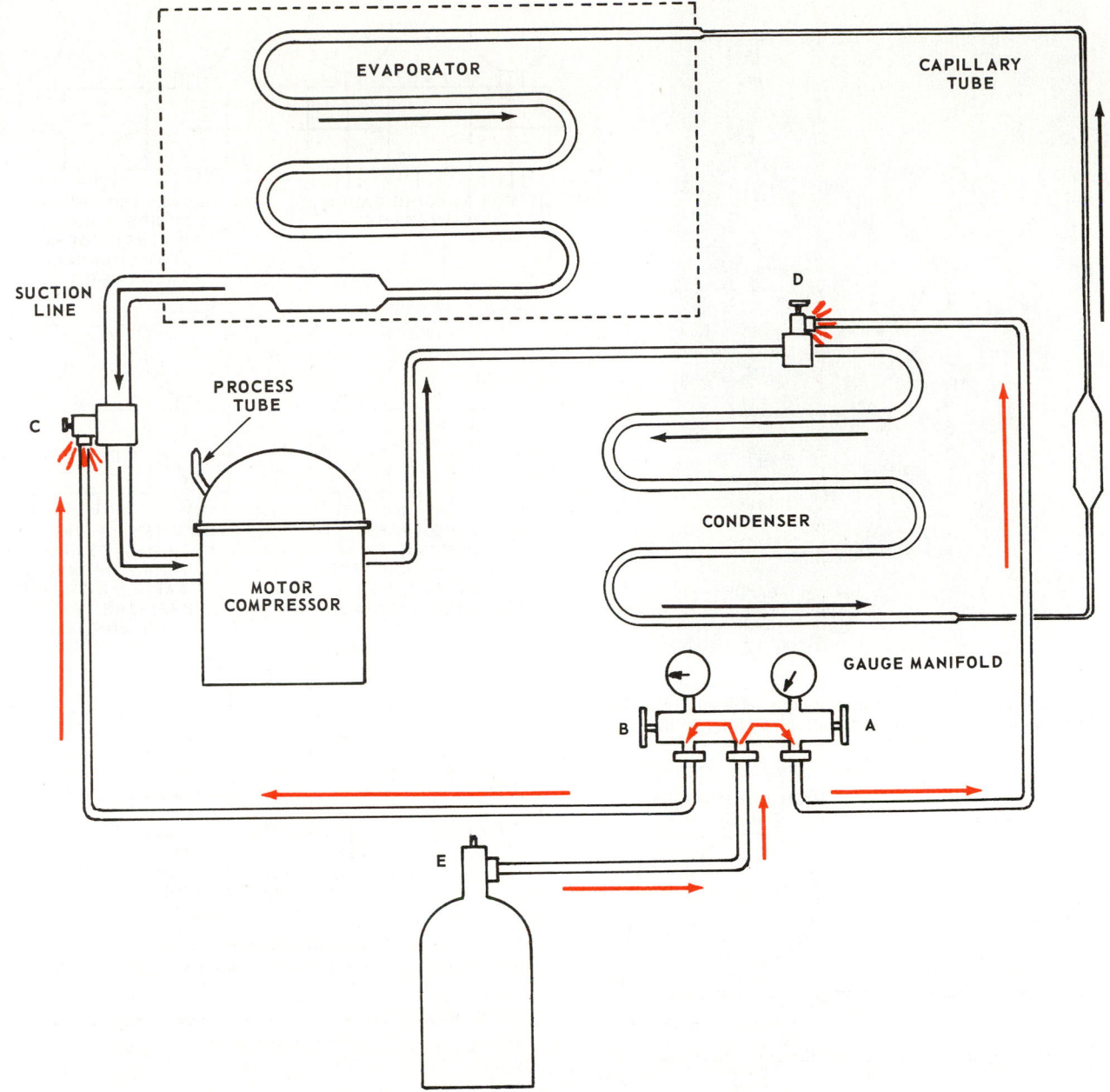

Fig. 11-27. Valves on gauge manifold are opened to purge service lines. Fittings at C and at D are loosened to allow air to leak out.

common on commercial systems, its installation and use is described in Chapter 14.

System B (valve adaptor) is described in Para. 11-30. System C (process tube) is described in Para. 11-31. System D (piercing valve) is described in Para. 11-32.

When connecting refrigerant lines or gauge manifolds to any refrigerating system, one must keep the system clean. The lines, gauges and manifold must be free of dirt, moisture and air. The manifold should be purged with the same refrigerant as is used in the system. The manifold and connecting lines must be purged before the system service valve is opened or before the piercing valve stem makes an opening in the tubing.

Referring to Fig. 11-27, the most popular way to purge the

service lines is to loosen the line fitting on the system service valve at C, open valve B and then open cylinder line valve E just a little. Repeat the same procedure for valve D. The cylinder refrigerant will free or purge all the lines and the manifold of air and moisture.

Usually only one connection is made to the system and this is to the low or suction side, at valve C. The flexible line between B and C is connected to the system valve, C. However, the use of the gauge manifold allows one to check both low-side pressures (valve C is open, valve B is closed) and high-side pressures (valve D is open, valve A is closed).

It also allows one to charge a system. Valves C and B are open; cylinder valve, E, is opened slowly. It can also be used to

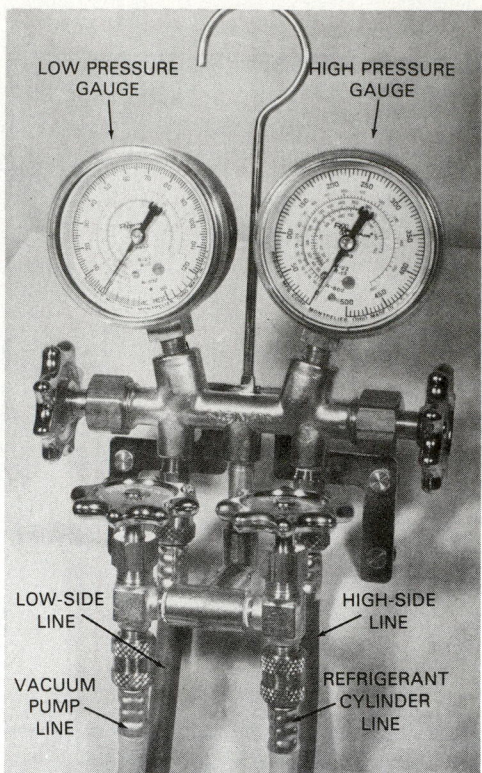

Fig. 11-28. Gauge manifold with vacuum pump valve and refrigerant cylinder valve added to manifold. (Robinair Mfg. Corp.)

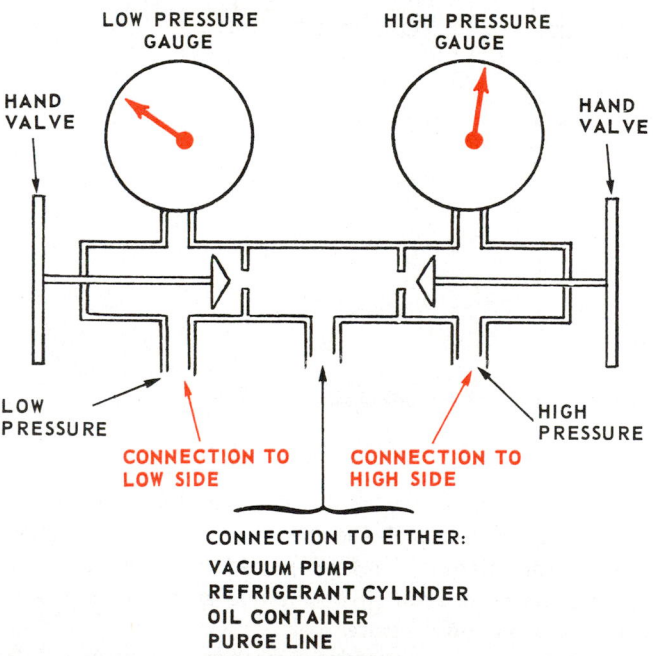

Fig. 11-29. This gauge manifold shows hand valves, gauges and refrigerant openings. Gauges will always show a pressure reading. When low-pressure hand valve is turned all the way in, the low (evaporator) pressure can be checked. When high-pressure hand valve is turned all the way in, high (condensing) pressure can be checked. When both valves are open (turned out by twisting to the left), high-pressure vapor will flow into low side. When only low-pressure valve is open, one can charge the system or evacuate it, put oil in system or clean it.

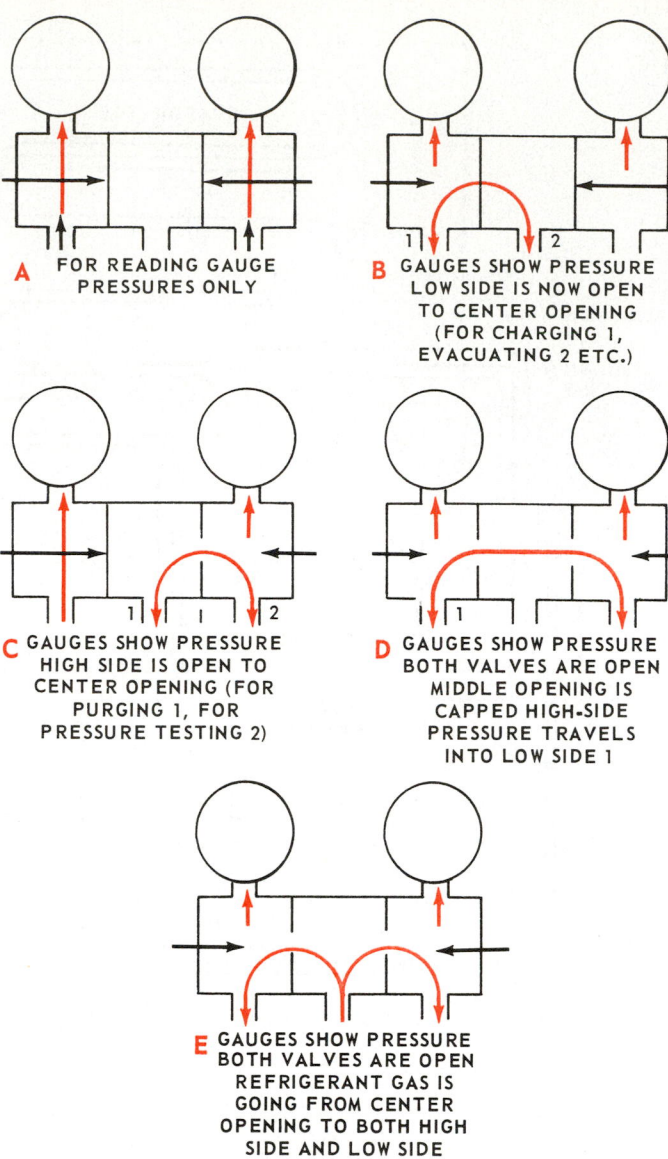

Fig. 11-30. How to use the hand valves of a gauge manifold for various service operations.

evacuate the system. (Vacuum pump line is connected to the middle connection of the gauge manifold; valve C is open and valve B is opened.)

When checking high-side pressure, use a piercing valve if the system is already charged. Braze a process tube into the condenser if system has just been assembled and not charged.

After installing the gauge manifold (if the unit will run), operate the system through at least three operating cycles. Carefully record the suction pressures, condensing pressures, evaporator temperature and the condenser temperature. It will be helpful if a table similar to Fig. 11-15 is recorded for future reference.

11-28 THE GAUGE (SERVICE) MANIFOLD

It is good practice to make all gauge and service connections to a hermetic system through a gauge (service) manifold.

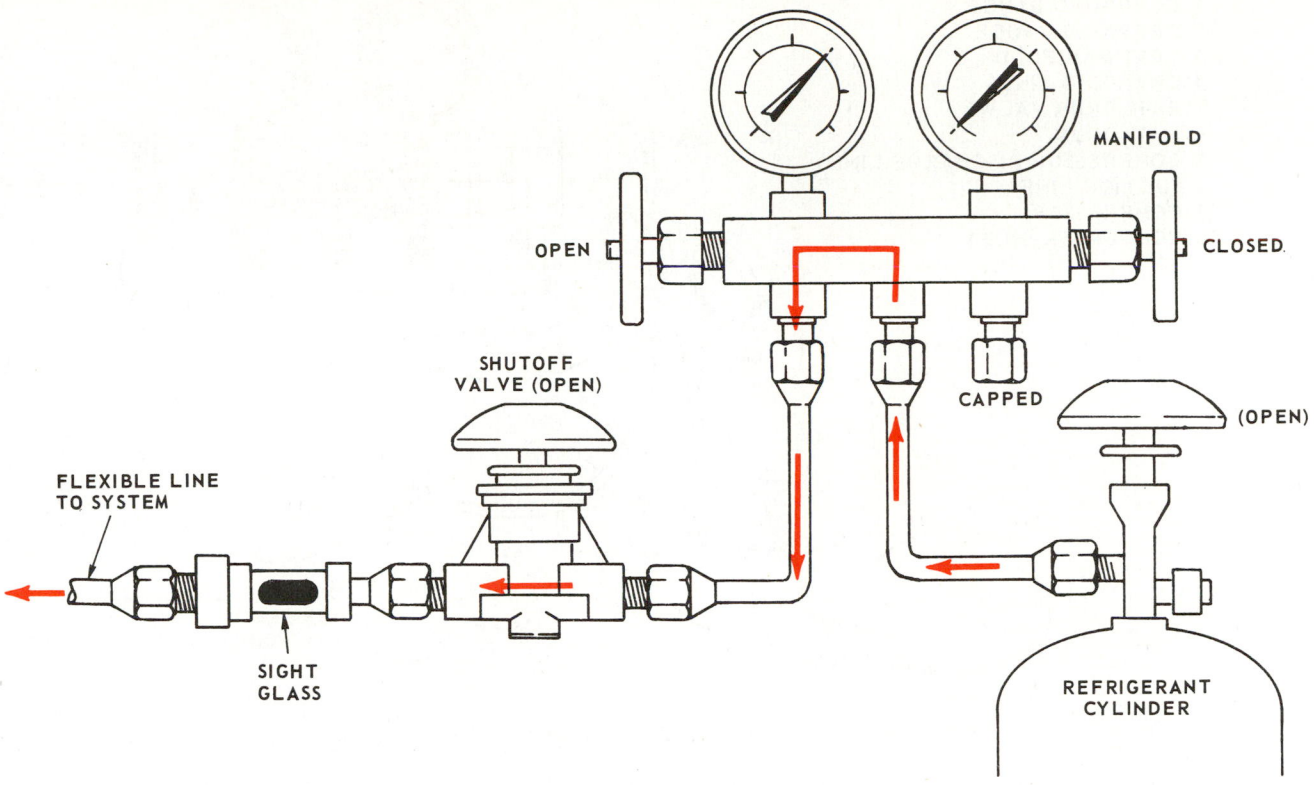

Fig. 11-31. Gauge manifold fitted with flexible refrigerant tubing. This installation is used when charging system.

Most domestic refrigerators and freezers with hermetic systems do not provide for gauge openings. Therefore, special attaching devices must be used to make it possible to use the gauge manifold. See Fig. 11-28. A schematic for this manifold is shown in Fig. 11-29. Fig. 11-30 shows the various uses and valve positions of a gauge manifold.

Separate gauges and hand valves can be used with the special service valves mounted on hermetic systems. However, to purge a system, check pressures or to evacuate and to charge the system would mean removing these devices and using others for each operation. Fig. 11-31 shows the set up for charging a system.

A gauge manifold with two gauges, two hand valves and three separate lengths of flexible refrigerant tubing enables the service technician to perform these operations more easily. The three flexible hoses are equipped with 1/4-in. flare fittings with synthetic rubber gaskets. Thus, connections can be made pressure tight with finger pressure alone. Fig. 11-32 shows the design and construction of flexible service and refrigerant line.

The service technician must learn how to use the gauge and service manifold. The hand valves on the manifold can be used for most operations. Fig. 11-33 shows a test manifold connected to a hermetic system and to a refrigerant cylinder.

11-29 HERMETIC SERVICE VALVES AND ADAPTORS

Most hermetic refrigerators do not have service valves. Some have fittings to which valves may be attached for service operations. The valves are removed when the service work has been completed.

Others have neither service valves nor provision for fitting valves to them. Where no service valves are provided, it is necessary to fit and attach valves to the mechanism. Attachments of various types are available from refrigeration supply wholesalers.

Some hermetic mechanisms have a process tube. One is shown being used in Fig. 11-34. Note the hand valve and process tube fitting. Note also that a charging tube is used.

Once service valves have been mounted on a hermetic system, they can be used for many purposes:

1. To check the internal pressures.

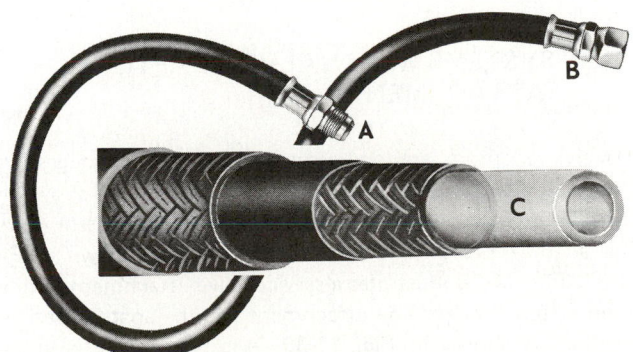

Fig. 11-32. Flexible charging line. A—External flare connection. B—Internal flare connection. C—Cutaway showing wall construction of tubing. (Resistoflex Corp.)

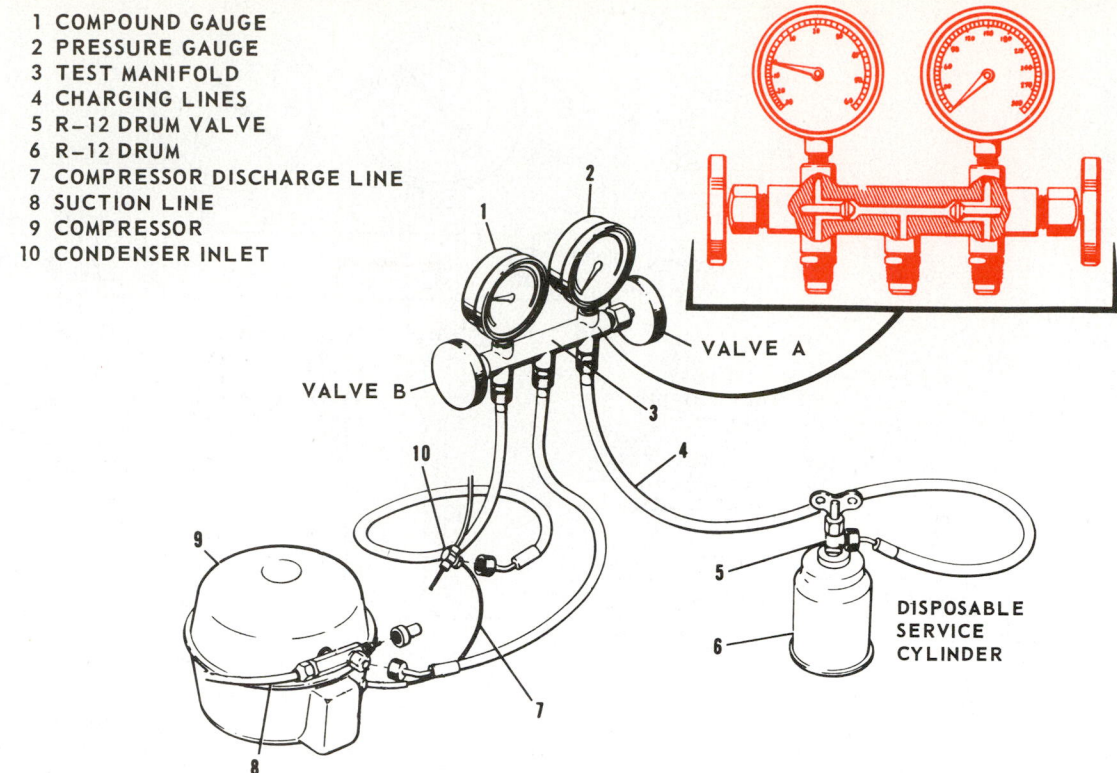

1 COMPOUND GAUGE
2 PRESSURE GAUGE
3 TEST MANIFOLD
4 CHARGING LINES
5 R—12 DRUM VALVE
6 R—12 DRUM
7 COMPRESSOR DISCHARGE LINE
8 SUCTION LINE
9 COMPRESSOR
10 CONDENSER INLET

VALVE A

VALVE B

DISPOSABLE
SERVICE
CYLINDER

Fig. 11-33. Drawing of hermetic system being charged using gauge manifold. Note disposable service cylinder and use of high-side manifold opening for charging.

Charging Cylinder

Hand Valve

Process Tube Adaptor

Fig. 11-34. System being charged with a hand valve and process tube adaptor. Note use of charging cylinder — usually called a charging tube. (Kelvinator, Inc.)

2. To discharge the system or add refrigerant.
3. To add oil.
4. To evacuate the system.
5. To make it easier to replace driers, motor compressors, evaporators and refrigerant controls.
6. To recharge the system.

Usually a flexible charging line is connected to the service valve adaptor and to either a hand valve or a service manifold mounted on the other end of this tubing. This makes service easier. Fig. 11-35 shows this type of service connection set up for charging a hermetic system.

Have the valve loose at F and use the cylinder vapor in G to blow out (purge) the lines. The gauge at F may be located on the compressor dome, on the suction line or on the process tube.

11-30 SYSTEMS WITH VALVE ADAPTORS (ATTACHMENT)

Valve adaptors are one way of connecting gauges and charging cylinders to a hermetic system. Fig. 11-36 shows the part of the adaptor which is fastened to the compressor dome. The adaptor has a removable service valve as shown in Fig. 11-37. Fig. 11-38 illustrates a service valve attachment with an Allen setscrew drive. The attachment and the adaptor connect together as shown in Fig. 11-39. A service valve with two openings is shown in Fig. 11-40.

Note that for these refrigerators, the adaptor provides a means of operating the small needle valve mounted on the motor compressor. It also provides an opening for a service

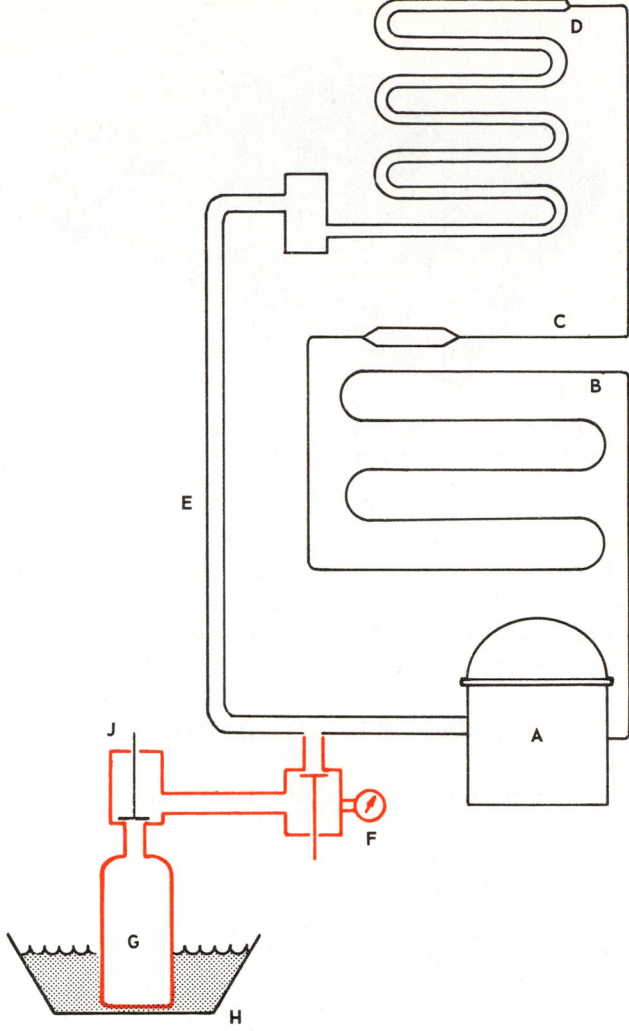

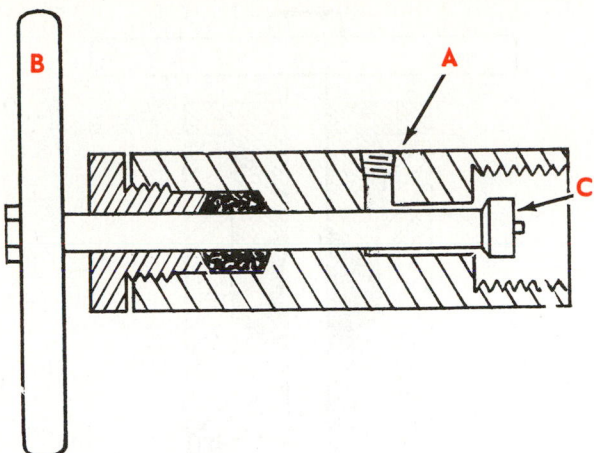

Fig. 11-37. Service valve attachment is installed on the valve adaptor which is fastened to the motor compressor dome. A—Opening for gauge connection and for servicing. B—Hand-wheel. C—Valve adaptor needle valve drive.

Fig. 11-35. Charging a hermetic system. A—Compressor. B—Condenser. C—Capillary tube. D—Evaporator. E—Suction line. F—Valve attachment. G—Refrigerant cylinder. H—Hot water. J—Cylinder valve.

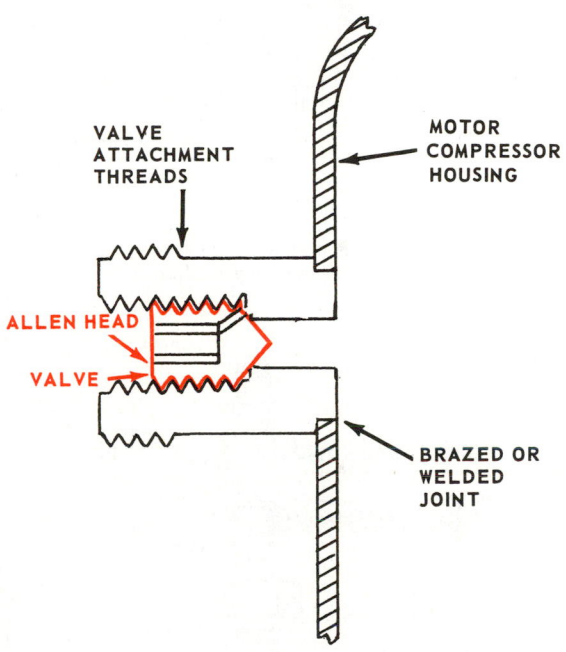

Fig. 11-38. This service valve adaptor makes use of an Allen setscrew valve turning device.

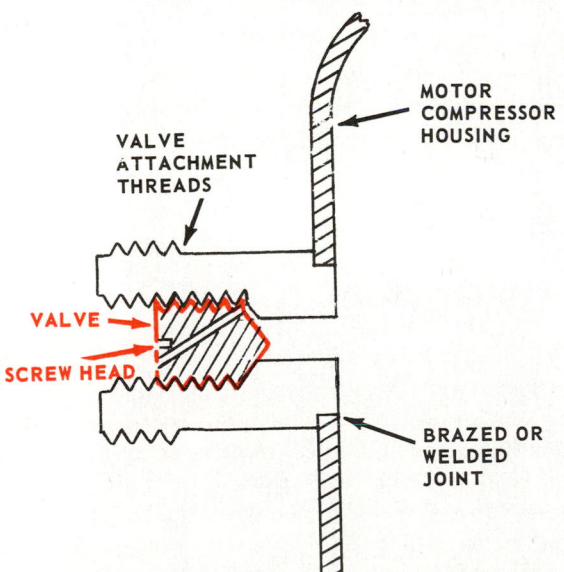

Fig. 11-36. Service valve assembly for hermetic units. Valve attachment must be fastened in place before valve can be opened; otherwise, refrigerant will escape.

gauge or a gauge manifold connection. Synthetic or copper gaskets are used to seal the valve joints. An assortment of valve adaptors is shown in Fig. 11-41.

Certain procedures should be followed when using valve adaptors:

1. Clean the outside.
2. Remove the dust cap from the adaptor mounted on the motor compressor.
3. Choose the correct valve stem drive.
4. Push the service valve stem forward in the body of the valve attachment.
5. Engage the valve stem in the valve adaptor needle.
6. Thread the body unit into the valve adaptor.

Installing and Servicing Small Hermetic Systems / 355

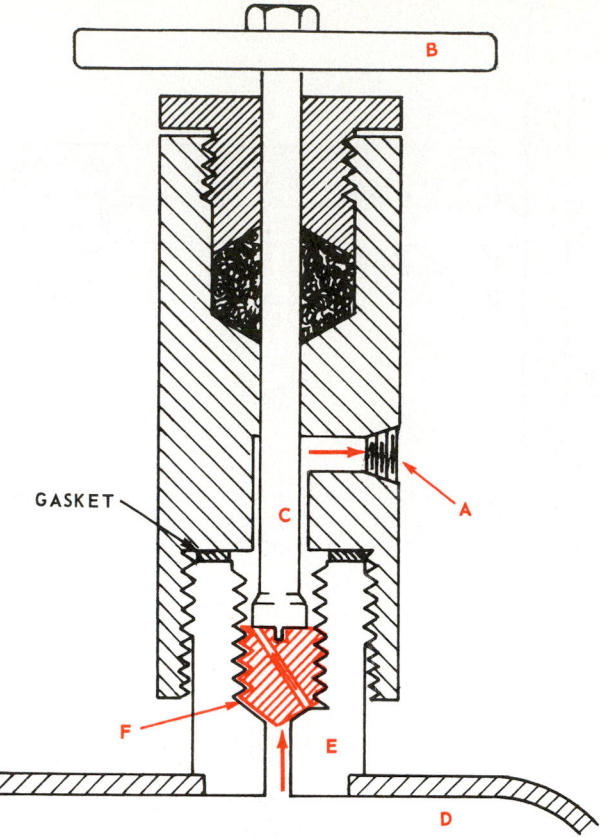

GASKET

A

C

F

E

D

B

Fig. 11-39. The valve attachment as it would appear in cutaway when connected to the valve adaptor.

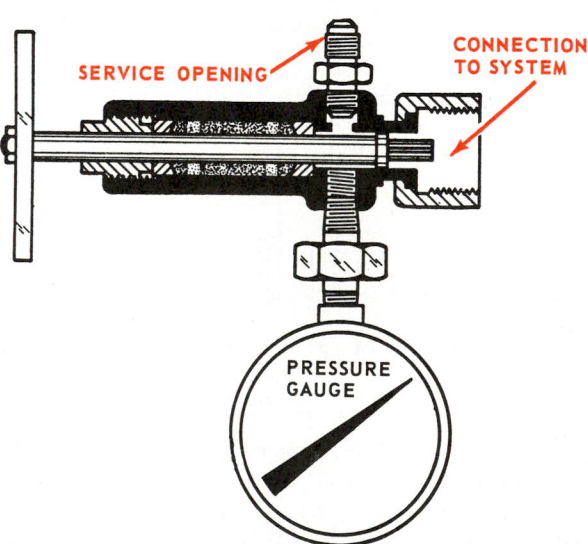

SERVICE OPENING

CONNECTION TO SYSTEM

PRESSURE GAUGE

Fig. 11-40. Service valve attachment. Note there are two openings. One may be used for pressure gauge and the other for performing service operations such as discharging, charging and adding oil. (Fedders Corp., Norge Div.)

7. Use good gaskets.
8. Before opening the valve adaptor needle, tighten the packing unit around the valve stem.
   Blow out the passages (valves, gauge and flexible lines)

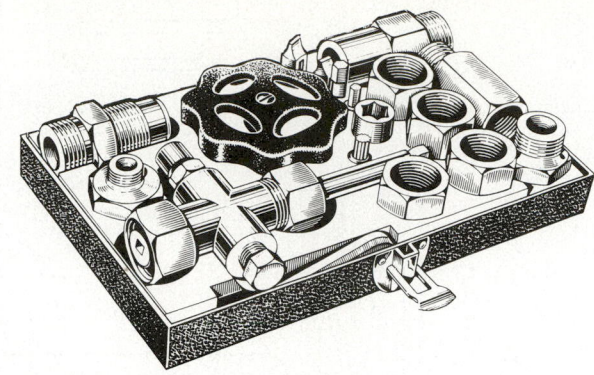

Fig. 11-41. Valve kit and adaptors that can be used on various makes of semihermetic or hermetic refrigeration units. (Mueller Brass Co.)

using the same refrigerant as is in the system. Purge the assembly using gas from a refrigerant cylinder, leaving the flexible line fitting loose at the valve attachment. After purging, tighten the loose connection. Always test the assembly for leaks using a 15 to 20 psi refrigerant pressure.

11-31 PROCESS TUBE AND ADAPTORS

The process tube can be adapted for service of systems in two ways:

1. Install a piercing valve on the process tube. See Para. 11-32.
2. Cut the end off the process tube. (All the refrigerant will escape.) Then mount a process tube adaptor on the clean tube. Caution: Always wear safety goggles when charging or discharging a system.

The process tube left in the system at the factory is used by the manufacturer to evacuate, test and charge the new unit. It can be used by the service technician by brazing an extension

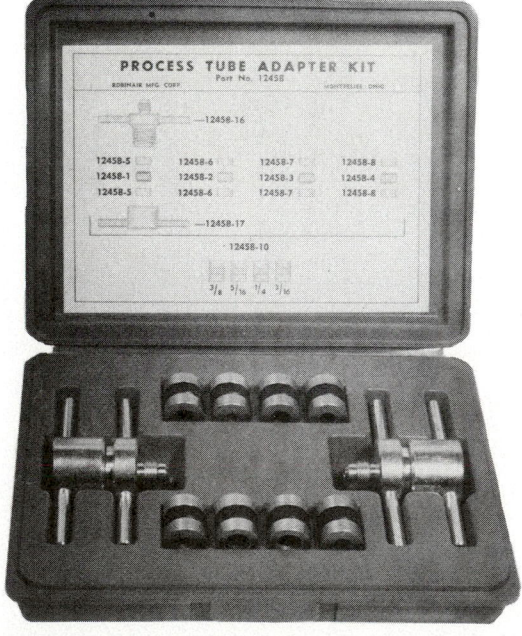

Fig. 11-42. Process tube adaptor and kit. (Robinair Mfg. Corp.)

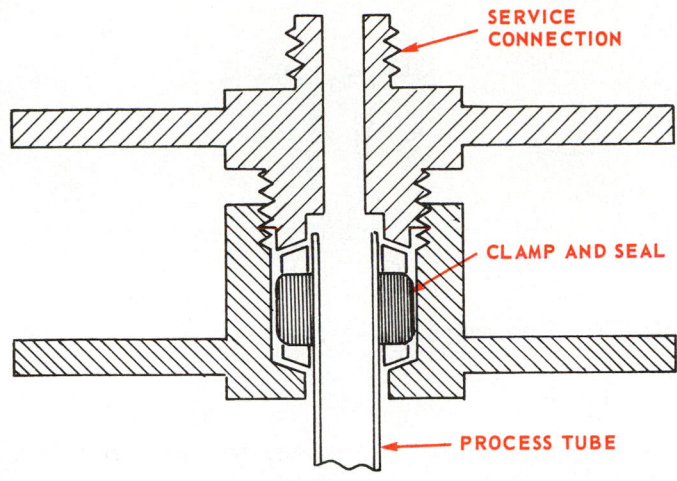

Fig. 11-43. Process tube adaptor.

5. Another method is to cut the process tube with a cutter. **Wear goggles!**

6. Attach the adaptor to the process tube.

After evacuating the system to dry it and to remove air, recharge it. Wait until the system operates correctly, then close off the process tube using a positive pinch-off tool. Remove the adaptor and braze the end of the process tube.

The pinch-off tool is used wherever it is necessary to seal off soft copper tubing up to 3/8 in. OD. (Fig. 11-44 shows one type of pinch-off tool.) It has a screw type action shaft with a ball bearing on the end that presses against the tube. The tool is placed over the copper tubing in the same manner as a tubing cutter. The tubing should be slowly compressed by turning the pinch-off tool handle clockwise.

As the handle is turned, the ball bearing presses into the tubing and compresses it against the die on the bottom of the tool, producing a permanently pinched line. See Fig. 11-45.

to it or by mounting a process tube adaptor on it as a means of attaching a manifold line. The adaptor kit, as shown in Fig. 11-42, makes it possible to use the process tube without soldering an extension to it or without flaring the tubing. It provides a positive seal. See Fig. 11-43.

By using adaptors of various sizes, the tool may be used on 3/16, 1/4, 5/16 or 3/8-in. copper tubing.

To attach the process tube adaptor:

1. Clean the outside of the process tube. Use clean, fine sandpaper to remove paint and dirt.

2. Using cutting pliers, cut off the tip of the process tube to release the refrigerant from the system. Use goggles! A flexible rubber tube should then be placed over the cut end of the process tube to discharge the refrigerant away from the work area.

3. If the process tube is short, it may be advisable to braze an extension to it before attaching the valve adaptor.

4. Maintain some internal pressure in the system to blow out metal chips which may be formed while cutting.

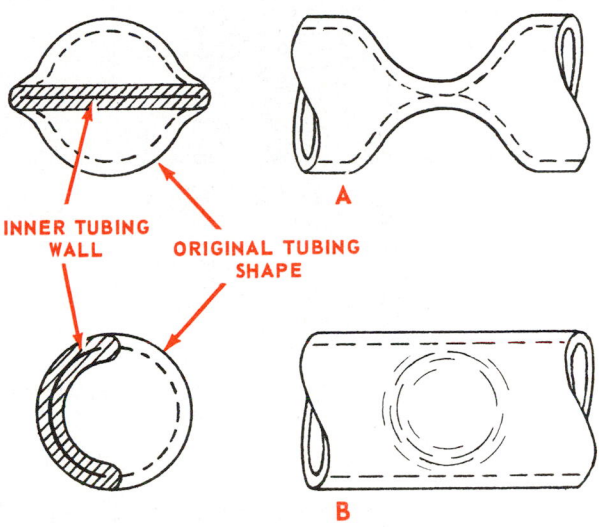

Fig. 11-45. Action of pinch-off tool. A—Pinch-off made with pliers-type tool. B—Pinch off made by tool shown in Fig. 11-44.

Take care that the pinch-off tool is not rotated too far or excessive pressure applied. It is best to leave the tool in place until the adaptor is removed and the tubing end is sealed by brazing.

The pinch-off tool may also be used when an emergency arises, requiring isolation of parts of the refrigeration system (such as a bad leak).

11-32 PIERCING VALVES

A popular way to gain access to a hermetic system is to mount service piercing valves on the suction tubing, on the discharge tubing (tubing to condenser), on both or on the process tube. A piercing valve is shown in Fig. 11-46. Many designs of tubing mounted piercing valves have been developed.

There are two general types:

1. Bolted-on valves.

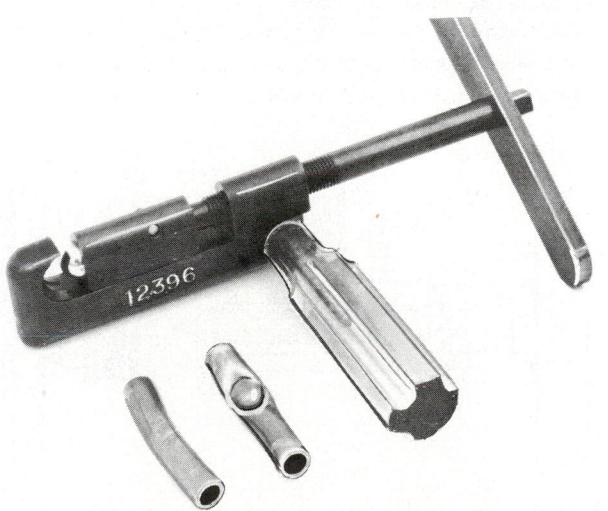

Fig. 11-44. Pinch-off tool is usable on tubing up to 3/8 in. OD. This makes good seal and also keeps tubing strong at pinch-off spot.

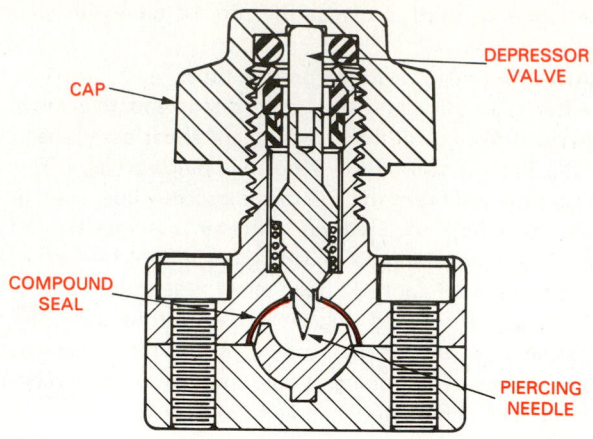

Fig. 11-46. Bolted-on piercing valve. Valve is bolted to line by two socket head screws. Note use of special compound seal. (Watsco, Inc.)

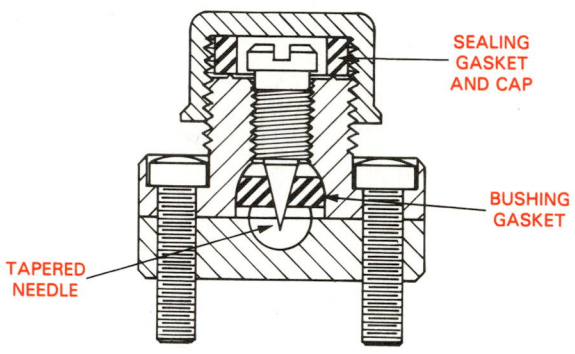

Fig. 11-47. Bolted type tubing-mounted service valve. Gasket seals hole created by piercing needle. (Watsco, Inc.)

2. Brazed-on valves.

Fig. 11-47 shows a bolted-on valve using a bushing gasket. This valve is available in several sizes for various sizes of tubing.

Tubing should be straight and round. It should be carefully

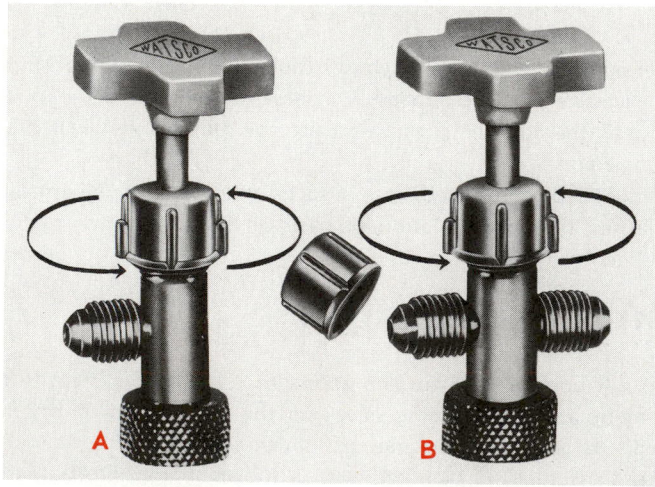

Fig. 11-48. Two types of service valve attachments used with tubing mounted piercing valve adaptors. A—Valve with one 1/4 in. male flare service opening. B—Valve with two 1/4 in. male flare service openings. (Watsco, Inc.)

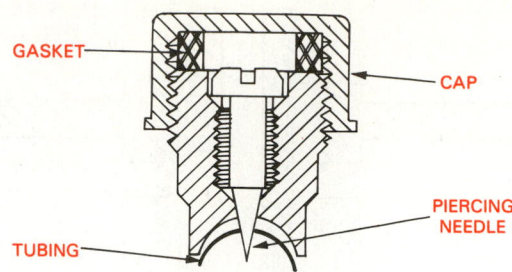

Fig. 11-49. Brazed-mounted tubing piercing valve. Note use of synthetic material sealing gaskets. Preformed silver brazing ring is usually used for brazing. (Watsco, Inc.)

cleaned. (Do not scratch it.) Make sure there are no dents in it. Check also to see there is enough space to operate the attachment valve and that the connecting tubing can be easily mounted on the attachment valve.

Put a little clean refrigerant oil on the tubing. Be sure the sealing synthetic washer is in place and that the needle point piercing valve stem is all the way out. Mount the valve on the tubing. Tighten the unit, clamping screws evenly. These valves are usually left on the system. The attachment valve is similar in design and construction to those shown in Figs. 11-39 and 11-40. Two types of valves are shown in Fig. 11-48.

The brazed-on type (saddle design) is safe to use because both the suction tubing and the condenser tubing have no liquid in them and can be heated to a brazing temperature. See Fig. 11-49. However, make sure there are no flammables or soft-soldered joints close to the brazing.

Be sure the tubing is straight and round at the brazing point. Clean both the saddle and tubing mating surface with clean sandpaper or clean steel wool. Remove the piercing valve stem and the gasket from the saddle. Put clean fresh brazing flux on the saddle (outer edges) or use a phosphorous-copper brazing filler rod.

NOTE: If flux is used, follow manufacturer's specifications. If phosphorous-copper brazing filler rod is used, flux is unnecessary because the phosphorous in the brazing material deoxidizes the copper surface.

Mount the saddle on the tubing. Check to see if there is room (clearance) for mounting the service valve on the tubing mounting valve. Heat both the tubing and the saddle until the

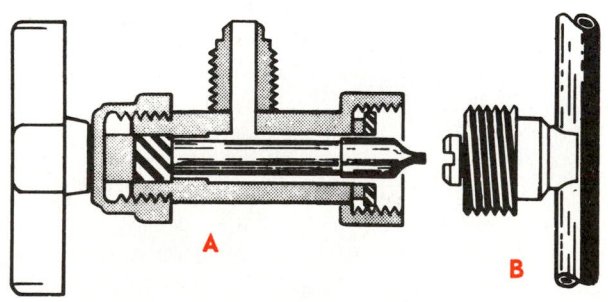

Fig. 11-50. This piercing valve, brazed on the line, may be used on hermetic refrigerator systems. Part A can be removed after servicing to prevent tampering with system. Part B remains on system. (Kelvinator, Inc.)

filler rod material flows around the saddle.

The saddle must not move or shift during the brazing or while the brazed joint is cooling. Some service technicians hold the saddle in place with a small C-clamp during the brazing operation.

Do not overheat the tubing as it may be weakened to the point of failure and may burst. Use goggles during the brazing operation.

Inspect the brazed joint carefully. Use a mirror to check hard-to-see edges.

After the brazed joint has cooled, install the piercing needle and gasket. The unit is then ready for the installation of the service valve attachment as shown in Fig. 11-50. A variety of gasket type piercing valves is shown in Fig. 11-51.

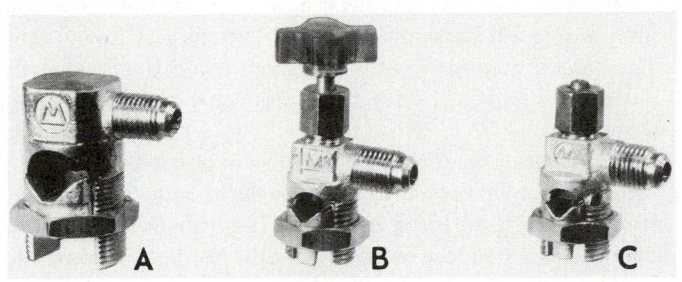

Fig. 11-51. Three types of piercing valves. A—Charge and tap valve. B—Hand valve type. C—Line tap type. (Robinair Mfg. Corp.)

11-33 CORE VALVES

Many systems use a Schrader core valve (Fig. 11-52) to gain access to a hermetic system. This type is similar to the valve cores used in automobile tires.

A clamp-on core valve adaptor is shown in Fig. 11-53. The flexible service tubing fitting or the service valve mounted on this fitting has a pin which depresses the core valve stem as the fitting or service device is mounted.

Some valve adaptors are threaded to the system, some are clamped to the tubing and some are brazed to the tubing.

Some service technicians use a service valve attachment that mounts on the Schrader valve adaptor. This device has a long stem which is used to remove the valve core while evacuating the system. The main reason the core is loosened is to allow more flow of gas as when a vacuum is to be drawn on the hermetic system. Vacuum lines and fittings should be as large as possible. Fig. 11-54 shows the advantage of removing the valve core while evacuating.

11-34 LOCATING INTERNAL TROUBLES

With gauges, thermometers, electrical instruments and by observation, the service technician should be able to locate the cause of almost every problem in a refrigerating system.

The paragraphs that follow discuss most of the reasons why refrigerating systems will not operate correctly; then the repair and testing will be described.

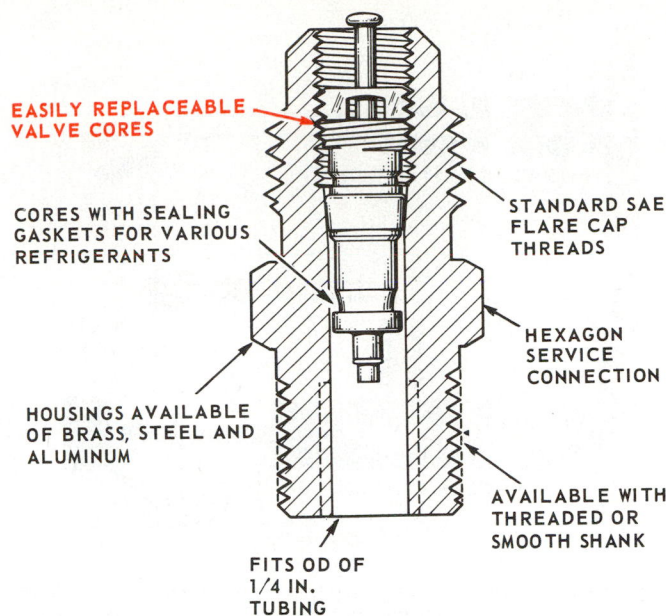

Fig. 11-52. This Schrader valve fitting may be used to mount service lines on hermetic system. When service line fitting is mounted on this fitting, a pin depresses or forces in stem of valve core and opens system for service.

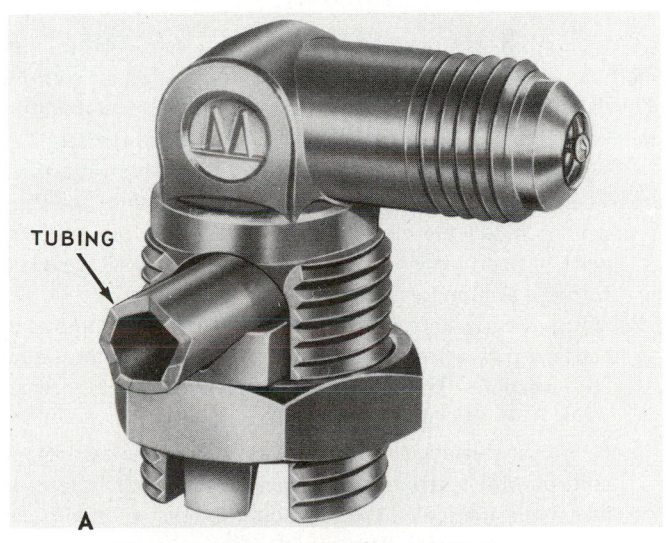

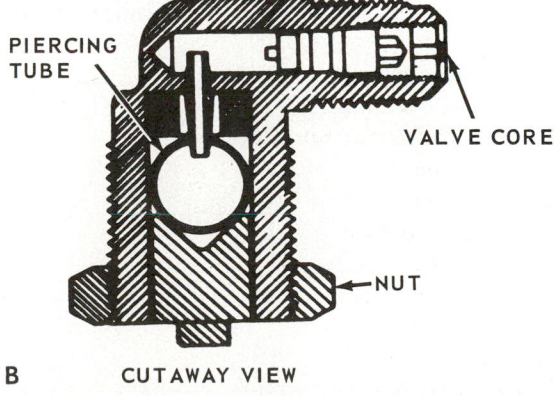

Fig. 11-53. This valve core access valve adaptor clamps on tubing. A—Piercing valve mounted on tubing. B—Cross-section of same adaptor.

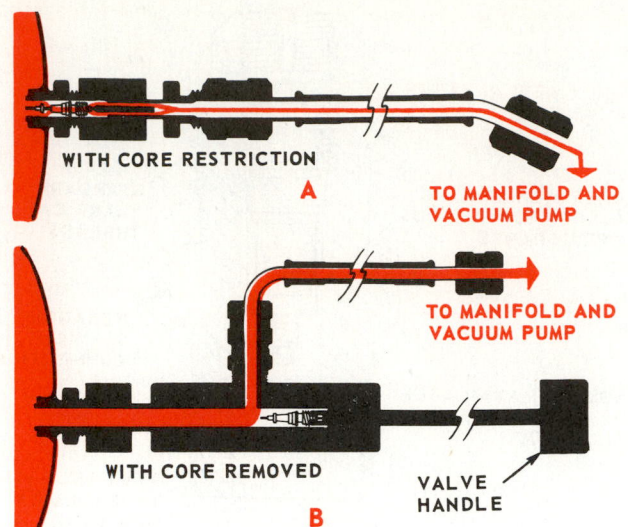

Fig. 11-54. Evacuating passages are larger when valve core is removed from its fitting. A—Small flow with core in place. B—Large flow with core removed.

11-35 MOISTURE IN THE REFRIGERANT CIRCUIT

Moisture in the refrigerant system will cause the unit to malfunction. The moisture forms ice in the refrigerant control. This is at the point where it is expanding into the evaporator. Icing closes the opening, blocking flow into the evaporator.

This condition can be recognized by several observations.

1. The system will completely defrost. Then, since the icing which caused the blockage has disappeared, the unit will work properly again. But only for a while until ice again forms at the refrigerant control.
2. Another symptom is decreasing pressure. The compound gauge shows a steady decrease over several hours — even to a vacuum. Then pressure suddenly becomes normal again. This odd cycle will keep repeating.
3. If, during system shutdown, one warms the refrigerant control with a safe resistance heater (hot pad) or radiant heat bulb, the ice will melt. Should the system then begin to work properly, there is definitely moisture in the refrigerant.

This moisture may be removed by putting a drier in the liquid line.

1. Install gauge manifold.
2. Remove the refrigerant.
3. Dry and clean filter-drier connections.
4. Apply flux.
5. Heat connections.
6. Remove old drier.
7. Install new drier.
8. Braze connections.
9. Test for leaks.
10. Evacuate system.
11. Charge system.
12. Warm the refrigerant control enough to melt the ice. The drier will absorb this moisture as it circulates.

Substances are available which, when placed in the refrigerant circuit, tend to keep moisture from forming ice. A filter-drier is best. It prevents circulation of the moisture through the system and reduces the chance of oil breakdown (sludge and acid).

11-36 WAX

Wax has been removed from refrigeration oil, as described in Para. 9-31. However, small amounts still remain. Some oil circulates with the refrigerant. Sudden expansion at the refrigerant control, accompanied by low temperature and pressure, causes some wax to separate from the oil. The wax collects in the refrigerant control. In time it will restrict flow or clog the control completely.

This condition can be checked with the aid of a piercing valve. A test will show low-side pressure to be very low. At the same time, liquid refrigerant will show up in the condenser. The unit, itself, will have been producing no refrigeration whatever.

Always clean or replace a clogged valve or capillary tube. *It is good service procedure to keep the ice or wax locked in the refrigerant control being removed.* This can be done most easily by packing the control in dry ice before removal. Another method is to very quickly open the joints after the unit is discharged. Use the same steps as in Para. 11-55 but repair and/or replace the refrigerant control. Use only the best low-wax oil when servicing frozen foods equipment.

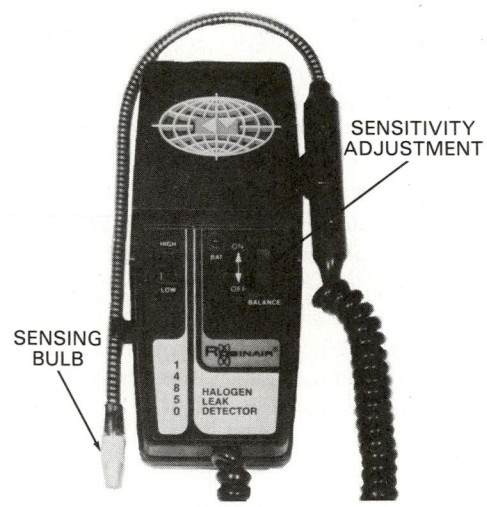

Fig. 11-55. An electronic leak detector which operates on batteries. (Robinair Mfg. Corp.)

11-37 SHORTAGE OF REFRIGERANT

Shortage of refrigerant is a common source of poor refrigeration. Because small systems only have one or two pounds of refrigerant, even the smallest leak will soon cause poor refrigeration. A leak with a loss rate of one ounce a year can be located and must be repaired.

A lack of refrigerant will be shown by:

1. A below normal low-side pressure.

2. Evaporator is warm or the outlet end of the evaporator is warm.
3. A below normal high-side pressure.
4. A piercing valve mounted on the outlet of condenser, when opened, allows only gas to escape (it should be liquid).

If a shortage of refrigerant is found, there is normally a leak. This leak must be found and repaired.

11-38 LOCATING REFRIGERANT LEAKS

Methods of testing for leaks vary with the refrigerant used. However, all methods have one procedure in common: at the start of testing, a positive pressure (greater than atmospheric) of 5 to 30 lb. is necessary throughout the circuit. If no leaks are found, then test again at the normal condensing pressure, for the refrigerant used (i.e., 135 psi for R-12).

Check for leaks before the unit is evacuated. Moisture could enter the system through a leak during evacuation or pump down.

Many companies recommend using the refrigerant in the system to test for leaks. A sensitive leak detector should be employed. An electronic leak detector, shown in Fig. 11-55, is good for this purpose.

If a leak is found, it is very important to recheck the complete unit after repair has been made. This provides a check of the repair and will reveal any additional leaks.

11-39 PRESSURE TESTING FOR LEAKS

With proper care, nitrogen or carbon dioxide may be used safely when pressure testing for leaks. The pressure in the nitrogen cylinder is about 2000 psi and in a carbon dioxide cylinder about 800 psi. *A pressure reducing device which has* *both a pressure regulator and a pressure relief valve must always be used when testing with either of these two gases.* A recommended pressure regulating device is shown in Fig. 11-56.

A refrigerating system would explode if pressure were to build up on the system. Many accidents have been caused by using too much testing pressure.

Before using nitrogen or carbon dioxide to test a system, look at the name plate. In most cases, it will give recommended testing pressures. If these pressures are not known, never go over 170 psi pressure when testing all or part of a hermetic system. See Chapter 14 for pressure testing commercial systems.

Caution: Never use oxygen or acetylene to develop pressure when checking for leaks. Oxygen will cause an explosion in the presence of oil. Acetylene will decompose and explode if it is pressurized over 15 to 30 psi.

11-40 LEAK DETECTING DEVICES

Leaks in a refrigerating system are usually very tiny. Therefore, detecting devices must be very sensitive. Most commonly used are soap bubbles, the halide torch detector and an electronic detector.

11-41 SOAP BUBBLES

The soap bubble method of leak detection uses a water-soap solution. This solution is brushed over an area suspected of leaking. Any gas will cause bubbles.

Patented solutions may be used instead of soap. These provide a stronger, longer-lasting bubble film than the soap solution. Fig. 11-57 illustrates the action of one of these

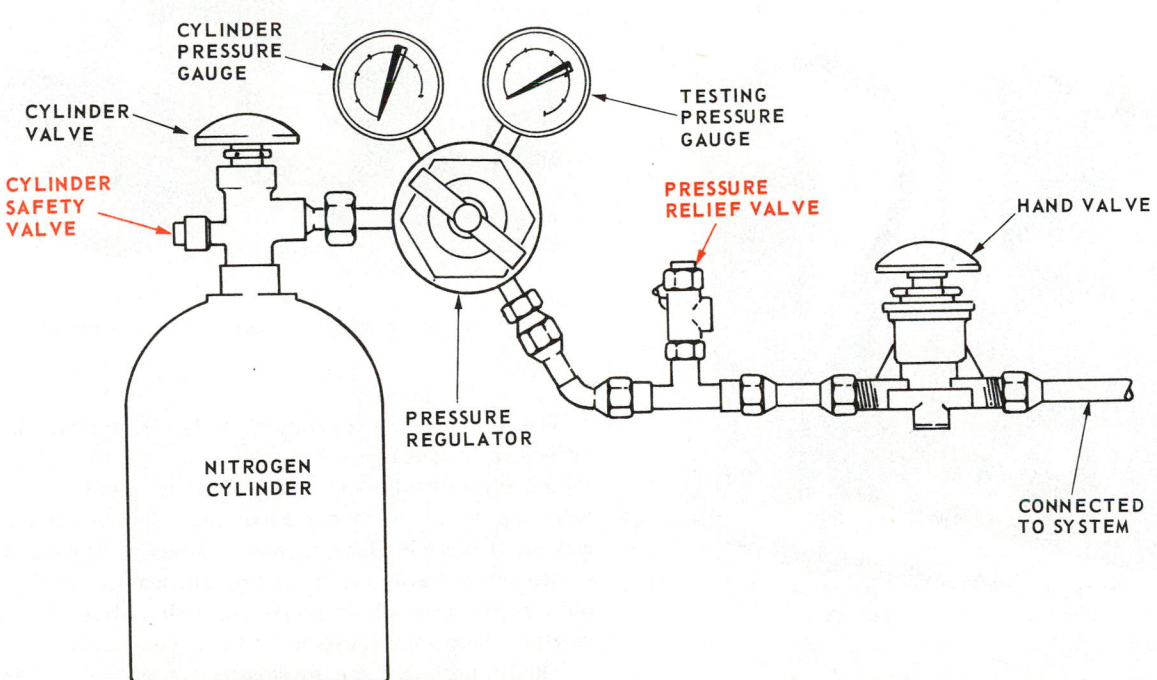

Fig. 11-56. Pressure regulator system. Note that both a regulator and a pressure relief valve are used.

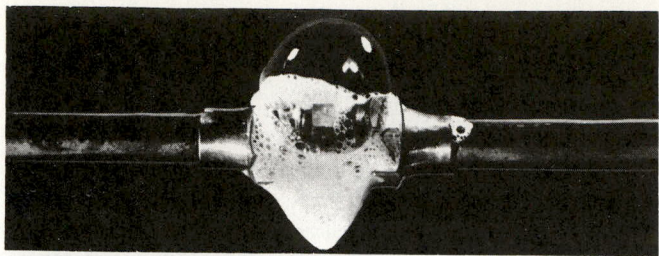

Fig. 11-57. Bubble leak test. Foam is placed on connection. Bubbles indicate refrigerant leakage.

solutions in the presence of a leak.

Both systems — soap solutions or detectors — have strong and weak points. The use of soap bubbles is difficult with large leaks in high-pressure systems. The pressure will break bubbles before they can be seen.

The halide torch and electronic leak detector are difficult to use around urethane insulation. Since urethane uses refrigerant as the expander, such detection devices show a leak trace all the time. In this case, the soap bubble test is best. Be sure to clean the soap solution off the tubing or fitting afterward.

11-42 HALIDE TORCH LEAK DETECTOR

Alcohol, propane, acetylene and most other torches burn with an almost colorless flame. If a strip of copper is placed in this flame, the flame will continue to be almost colorless.

Fig. 11-58. Halide torch used to test for leaks. Green flame showing in burner opening will indicate a leak at sniffer tube opening. (Bernz-O-Matic Corp.)

However, if even the tiniest quantity of a halogen refrigerant (R-12, R-22, R-11, R-500, R-502, etc.) is brought into contact with this heated copper, the flame will immediately take on a light green color. This principle is used in halide torches to detect leaks in refrigeration systems.

See Fig. 11-58. The torch burner is shown at the top. One end of a rubber tube is connected into the base of the burner. The other end is free to be moved about to various parts of the refrigerator system. The rubber tube will draw air from the open end into the burner.

If the open end of the tube is brought near a leaking refrigeration connection, some of the leaking refrigerant vapor will be drawn up the rubber tube into the burner. Immediately, the color of the flame will change to green, indicating a leak.

11-43 ELECTRONIC LEAK DETECTOR

The most sensitive leak detector of all is the electronic type. See Fig. 11-59.

Three types have been used:
1. Ion source detector.
2. Thermistor type (based on change of temperature).
3. Dielectric type (now most in use) measures a balance in surrounding air and then responds only to halogen gas.

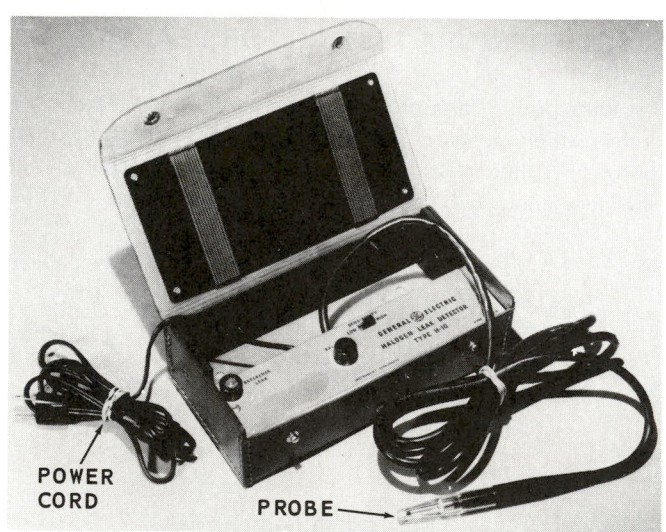

Fig. 11-59. Electronic leak detector.   (General Electric Co.)

The principle of operation is based on the dielectric difference of gases. See Fig. 11-60. In operation, the gun is turned on and adjusted in a normal atmosphere. The leak-detecting probe is then passed over surfaces suspected of leaking. If there is a leak, no matter how tiny, the halogenated refrigerant is drawn into the probe. The leak gun will then give out a piercing sound, or a light will flash, or both, because the new gas changes the resistance in the circuit.

This is probably the most sensitive of any of the leak-detecting devices. It uses transistorized circuitry. Flashlight batteries supply the energy. The plastic tip guard should be

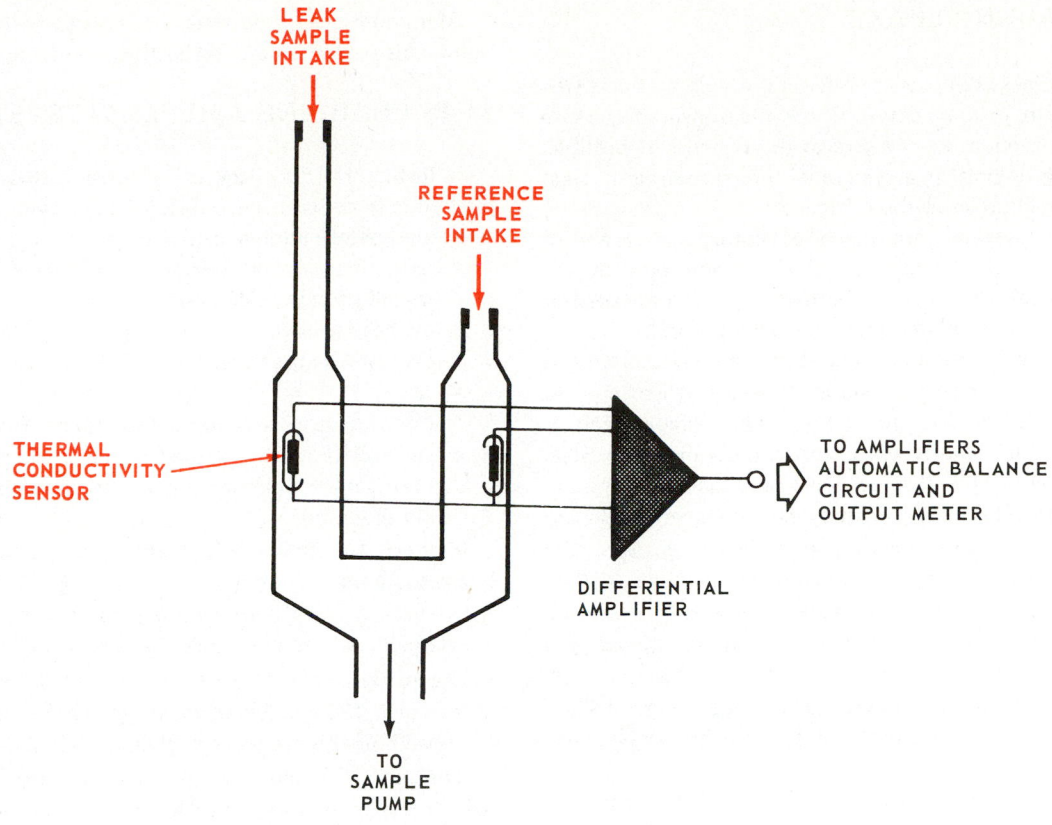

Fig. 11-60. Schematic of electronic leak detector. (USON Corp.)

used only in situations that may be potentially contaminating to the sensing tip. The internal construction of a popular leak detector is shown in Fig. 11-61.

When using an electronic leak detector, minimize drafts by shutting off fans or other devices which cause air movement. Always position the sniffer below the suspected leak. Being heavier than air, refrigerant drifts downward.

Move tip slowly (about one inch per second). One can measure this by moving tip an inch after each "one-thousand" verbal count. Tip adjusted in ambient air will only buzz, instrument will squeal when tip sniffs refrigerant —R-12, R-22, R-500, R-502, R-11 or any of the halogen family refrigerants. Remove plastic tip and clean it before each use. Avoid clogging with dirt and lint.

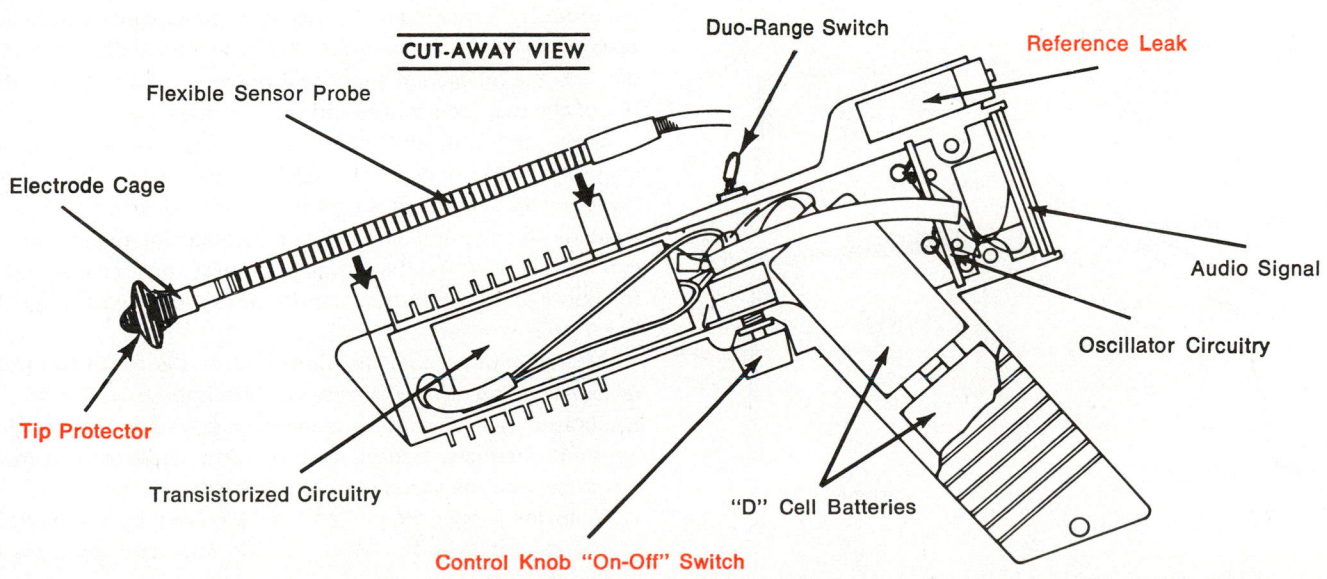

Fig. 11-61. Internal construction of battery-powered electronic leak detector. (Robinair Mfg. Corp.)

11-44 REPAIRING LEAKS

To repair a leak, remove the refrigerant from that part (the complete system, in some cases). Check the pressure to be sure it is 0 psi (no vacuum or pressure in the system). If possible, avoid soldering or brazing a system with refrigerant in it. Heat may cause a breakdown of the refrigerant.

Refrigerator systems are made of copper, steel and/or aluminum materials. Leaks may start in any part of the system. The repair depends on the material that has failed or on the combination of materials at the leaky point.

To find out what metal is used, scrape the surface. Steel is gray-white. It is hard and magnetic. (Use a small magnet to test.) Copper is reddish in color when scraped and is nonmagnetic. Aluminum is white, soft and nonmagnetic. Steel and copper may be silver brazed; aluminum may be aluminum soldered or brazed. Aluminum may be resistance welded to steel or copper or it can be repaired with epoxy cement.

Leaks most often are found at tubing connections. If they occur at a flared connection, the tube flare is not correct, the flare nut has not been tightened securely or threads are stripped. It is best to replace a leaking fitting by making a new flare and using new flared fittings. Leakage at a brazed or silver soldered connection can be repaired by cleaning, coating with flux and reheating.

Steel tubing usually has a lengthwise seam. This seam must be clean for silver brazing. Clean by wire brushing lengthwise or file off enough metal to remove the seam depression.

If the fitting has been taken apart, then reflux, assemble, heat the connection and solder or braze it in place. Avoid overheating other parts of the system. Use metal sheet, as shown in Fig. 11-62, to reflect heat away from the cabinet.

Never heat a drier. Moisture will be driven out into the system. It is better to cut tubing with pliers or tube cutter.

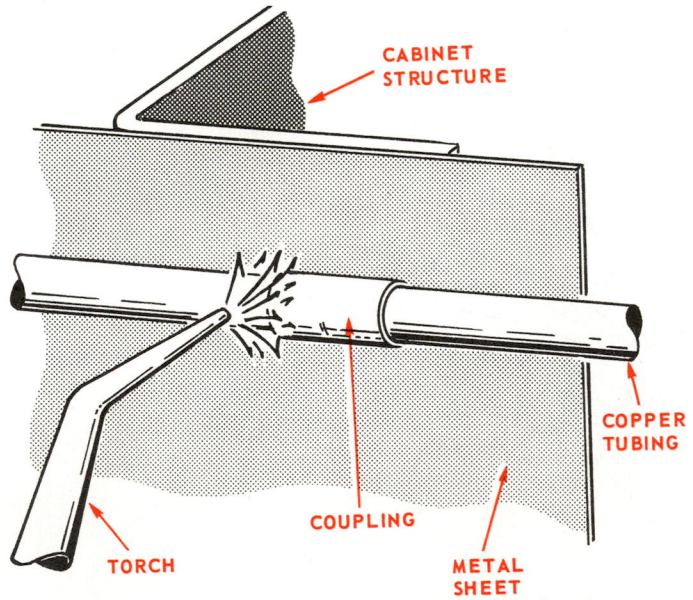

Fig. 11-62. Tubing joint prepared for brazing. Note use of metal sheet to protect other parts of cabinet and system from heat of flame.

Aluminum evaporators can be repaired with epoxy cement. Follow instructions supplied by the manufacturer.

11-45 CHARGING A HERMETIC SYSTEM

If testing indicates lack of refrigerant, this is an indication that there is, or has been, a leak in the system. Be sure the leak is repaired before adding refrigerant.

A hermetic unit needs refrigerant if there is:
1. A partially frosted evaporator.
2. A low head pressure.
3. A low low-side pressure.
4. A leak.
5. The refrigerator is running too much. Remember that a pressure difference is needed to move the refrigerant from the refrigerant cylinder (higher pressure) into the system (lower pressure).

Methods used to add refrigerant are the following:
1. Evacuate the system.
2. Connect a refrigerant cylinder to the charging manifold. Charge only with correct refrigerant vapor.
3. The refrigerant cylinder may be heated with warm water (not over 120 F.). *Never use an open flame for heating.*
4. Install pressure gauges and valves as described in Para. 11-27.

The service connection lines must be clean and free of air (which has moisture). This cleaning can be done by purging refrigerant through the lines before recharging.

When charging a partially charged unit, it is best to add a small quantity of vapor refrigerant. The unit should then be allowed to cycle. A proper charge is indicated best by the frost on the evaporator. When frost starts to come down the suction line, purge out a little of the refrigerant. The system will have the correct charge.

If a system has been evacuated first, it can be charged by replacing the evacuating pump with a refrigerant cylinder or by using valves to close off the pump and then opening the connection to the cylinder.

If the system has not been evacuated, purge both lines and manifold by loosening the service line at the piercing valve and opening the left side manifold valve. When the cylinder valve is opened, the refrigerant vapor will purge air, moisture and dirt out of the two service lines and the manifold.

Start the unit and open the line service valve, gauge manifold valve and refrigerant cylinder valve. Watch the low-pressure side gauge so that not more than 5 to 25 psi is created. This pressure is controlled by adjusting the refrigerant cylinder valve. Allow the refrigerant charge to enter the system for about 3 to 5 minutes. Connection should be as shown in Fig. 11-63.

After the time lapse mentioned, close the gauge manifold valve. Allow the unit to operate and check the frost line on the evaporator. If the frost line is inadequate, repeat the charging for short intervals, rechecking after each. The frost line must not go beyond the accumulator in the suction line.

When the proper amount of frost has been observed, close the refrigerant cylinder valve, adaptor valve and gauge manifold valve. Then:
1. Pinch the process tube between the compressor and the

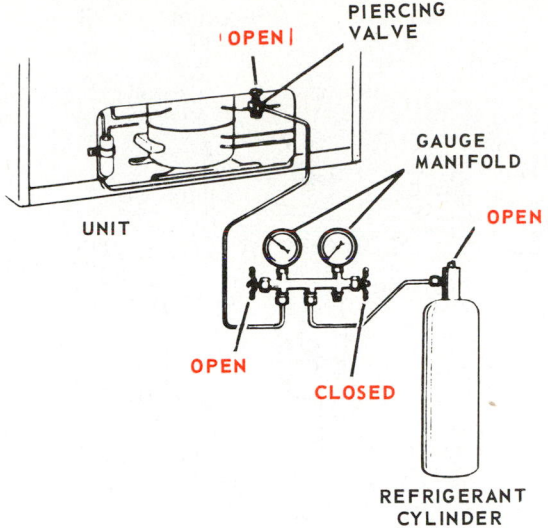

Fig. 11-63. Recharging setup after adjustment of refrigerant cylinder valve. Note open piercing valve, left-side manifold valve and refrigerant cylinder valves.

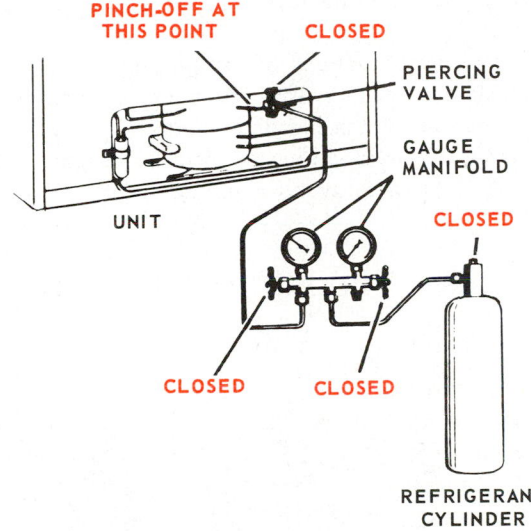

Fig. 11-64. System after piercing valve is closed along with the refrigerant cylinder valve and the gauge manifold valves. Note pinch-off point.

adaptor valve with a pinch-off tool, as shown in Fig. 11-64.
2. Remove the adaptor valve and silver braze the end of the tubing.
3. Check for leaks using a leak detector.
4. Close the adaptor valve, if one was installed on the suction line, and leave it mounted for future service operations.

Refrigerants charged into hermetic refrigeration units must be of top quality. It is very important to transfer refrigerants in chemically clean cylinders and lines. Always charge a unit (except large commercial units) with refrigerant vapor.

Never charge liquid refrigerant into the low side of a domestic or small commercial unit.

Always keep the charging cylinder at room temperature or warm it only with warm water.

There is a charging device which, when mounted between the refrigerant cylinder and the low side, allows liquid refrigerant to flow from the cylinder into this charging device. The refrigerant vaporizes in this device.

When the charging device is first used, the process tube will sweat and may even frost a little. As the system becomes fully charged, this sweating and frost will go away because the suction pressure is higher. The system is now correctly charged.

A fast reading dial thermometer provides a second check on the correct charge. The suction line temperature should be about 20 deg. higher about six to ten inches away from the compressor than at the evaporator outlet.

If the temperature is lower, liquid refrigerant may enter the compressor causing damage. If the temperature is higher, the motor compressor may overheat and "burnout."

A check valve in the charging unit allows the service technician to evacuate the system easily with this unit attached to the line. The charging unit is available in three capacities. The correct size unit must be used.
1. Less than 1 hp.
2. 1 to 4.75 hp.
3. 5 to 10 hp.

Systems which use a capillary tube must have an exact refrigerant charge. If the system is overcharged, the evaporator will be overcharged or flooded.

The use of accumulator spaces at the outlet of these evaporators relieves the problem somewhat, but one must be careful of the amount charged into these systems. A common method is to slowly charge these systems with refrigerant in the vapor state until the suction line starts to sweat and/or frost back. Then they are purged a little at a time until the "frost back" disappears.

Another method is to completely discharge the system. Then, recharge with a cylinder containing exactly the right amount of refrigerant according to the manufacturer's recommendations.

The following points are important and bear repeating:
1. Always charge a system into the low side, if possible.
2. Refrigerant should be put into the system in vapor form. Forcing liquid refrigerant into the system may damage the compressor and injure the service technician.
3. Always remember that if a system is short of refrigerant, there is a leak. Locate and correct the leak before the system is charged.

11-46 CHARGING WITH A PORTABLE CHARGING TUBE

If a portable charging cylinder, Fig. 11-65, is used, the following steps are recommended, after evacuation.
1. Attach a line from the charging cylinder to the center of the gauge manifold and purge with the fitting loose at the center part of the gauge manifold. See Fig. 11-66. Tighten this connection.
2. Open the piercing valve or valve adaptor and gauge manifold valve.
3. Crack the charging cylinder valve and allow the refrigerant

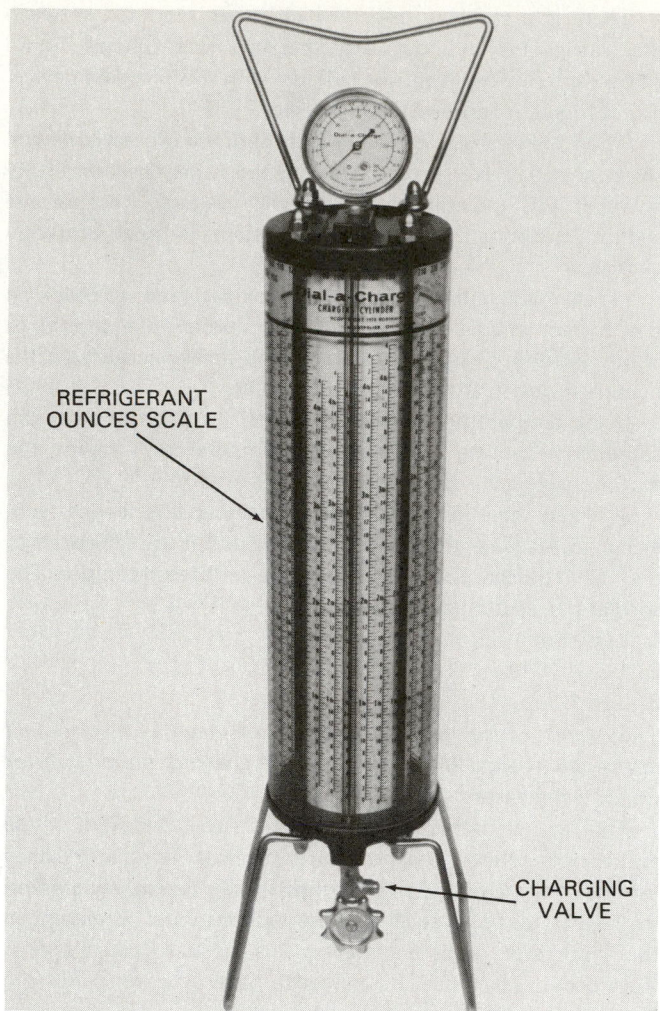

Fig. 11-65. Portable charging cylinder may be used to accurately charge hermetic systems. Cylinder is accurately calibrated in ounces of refrigerant. (Robinair Manufacturing Corp.)

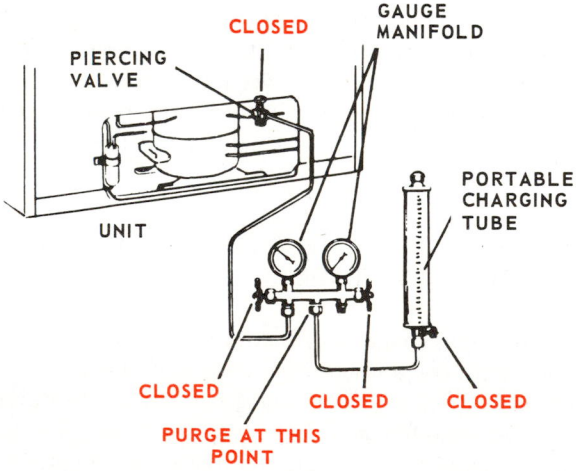

Fig. 11-66. When portable charging tube is used, purge the charge line, (after evacuating a system) by leaving the fitting loose at center part of gauge manifold. If system is not evacuated, then purge all lines up to piercing valve.

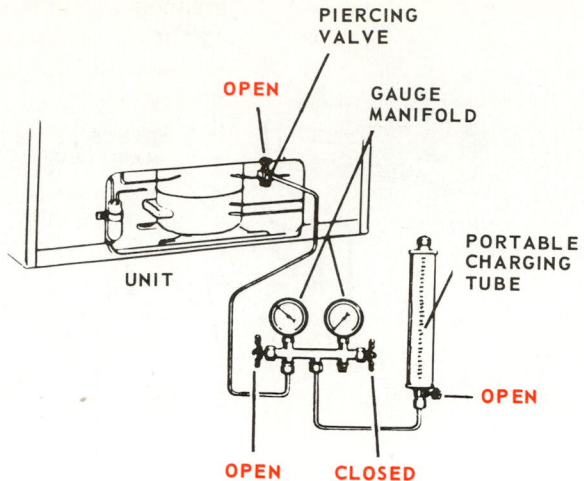

Fig. 11-67. Proper hookup for using the portable charging tube to charge refrigerant into system.

to enter the system, as shown in Fig. 11-67. Know what the new scale reading on the tube must be to put in the correct charge.

4. When the correct amount of refrigerant has entered the system, close the cylinder valve. Amount can be checked by reading the scale on the charging cylinder.

5. Then close the piercing valve or the valve adaptor and the gauge manifold valves, as shown in Fig. 11-68.

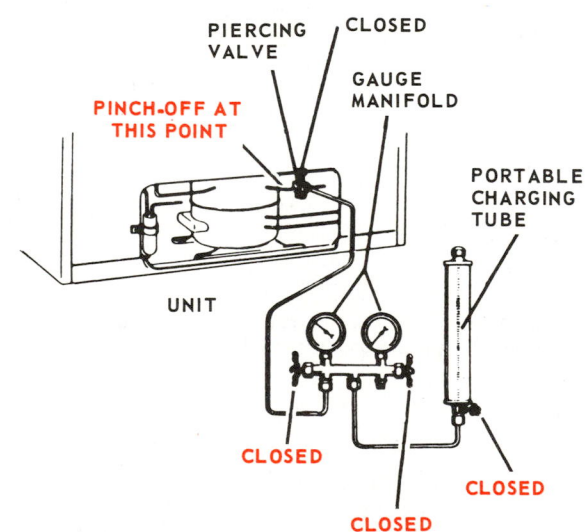

Fig. 11-68. Proper steps and valve settings for removing servicing equipment after system has been charged.

6. Use the pinch-off tool and close off the process tube between the compressor and the valve adaptor.

7. Remove the piercing valve or the valve adaptor.

8. Braze the end of the process tube.

9. Check the system for leaks.

A charging cylinder (measuring tube), with a glass tube

liquid level indicator, allows the technician to transfer refrigerant into a system and measure the amount by reading a scale. Some of these cylinders are electrically heated to speed up the evaporation and maintain pressures in the cylinder.

If electrically heated, both a pressure control and a thermostat should be used to provide safety and control pressure. The system has a pressure gauge and a hand valve on the bottom for filling the charging cylinder (measuring tube) or for charging liquid refrigerant into a system. It also has a valve at the top of the measuring tube. This valve is used for charging refrigerant vapor into the refrigerating system. (This is the best and safest method.)

11-47 ADDING OIL TO THE SYSTEM

Having the correct amount of oil in a system is very important. A lack of oil will shorten the life of the mechanism, increase friction and cause noise. An overcharge will cause the compressor to pump excessive amounts of oil, reducing its refrigerant pumping capacity, and subject the compressor valves to severe strain.

Refrigerant oils are available in several viscosities (ease of flow at different temperatures). Be sure to follow the manufacturer's viscosity recommendations. On a service call, add oil only if there is a sign of oil leakage.

Only rarely is it necessary to add oil to a hermetic system. However, leaking refrigerant always carries some oil with it and this lost oil should be replaced. The conventional method of adding oil to a system may be used if the hermetic unit is completely equipped with service valves. That is, oil can be siphoned in or poured in.

If the system has had a low-side leak, moisture and air may have entered the system. In this case, it is best to replace the

compressor oil. Measure the amount removed and replace it with the same amount of clean, dry oil. See Chapter 14. The unit should be charged in much the same way as when adding refrigerant to the system.

Fig. 11-69 illustrates a practical transparent charging board for accurately measuring the amount of refrigerant or oil charged into a system by volume. This method provides greater accuracy.

Use clean lines. Purge lines of air with clean refrigerant.

There are several ways to connect a transparent charging tube to a system. Fig. 11-69 shows the tube connected to the low side directly, using a compound gauge and a process tube adaptor. The process tube is brazed to the suction line. A method of combining a vacuum pump and a gauge manifold is shown in Fig. 11-70.

Fig. 11-70. A—Vacuum pump and gauge manifold are combined for charging oil into system. They are connected to a charging tube and the low side of the system by a tube adaptor. B—Gauge manifold. C—Charging tube. D—Tube adaptor. (Philco-Ford Corp.)

A pump may be used to put oil into a system. See Fig. 11-71. The charging lines must be purged to remove air, moisture and dirt. This hand pump can build up pressures to 300 psi. Oil can be forced into the system even when the system is under pressure.

To add oil, using the single service line technique, install as shown in Fig. 11-72. The correct amount of oil is put in the service cylinder while a small amount of refrigerant (the same as in the system) is also put in the cylinder to create a pressure. The cylinder is inverted and connected from the cylinder valve to the valve attachment by clean lines. Be sure the cylinder pressure is higher than the system pressure. Then open the cylinder valve and the valve attachment. Oil will be forced into the system.

Fig. 11-69. A charging tube connected to the low side. A—Charging tube. B—Compound gauge. C—Line adaptor. High-pressure gauge is connected to the outlet of the filter-drier by tube adaptor. D—High-pressure gauge. (Philco-Ford Corp.)

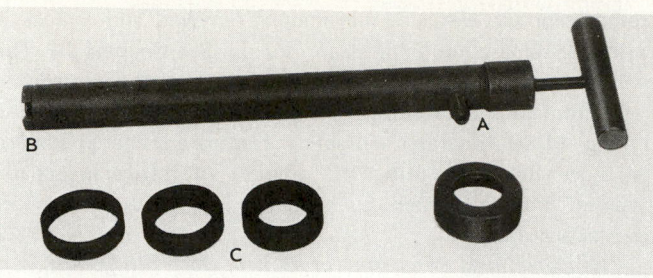

Fig. 11-71. Hand pump is used to force oil into a system. A—Opening to oil charging line. B—End fits into refrigerant oil storage can. C—Rubber adaptors for different oil can sizes. (Robinair Mfg. Corp.)

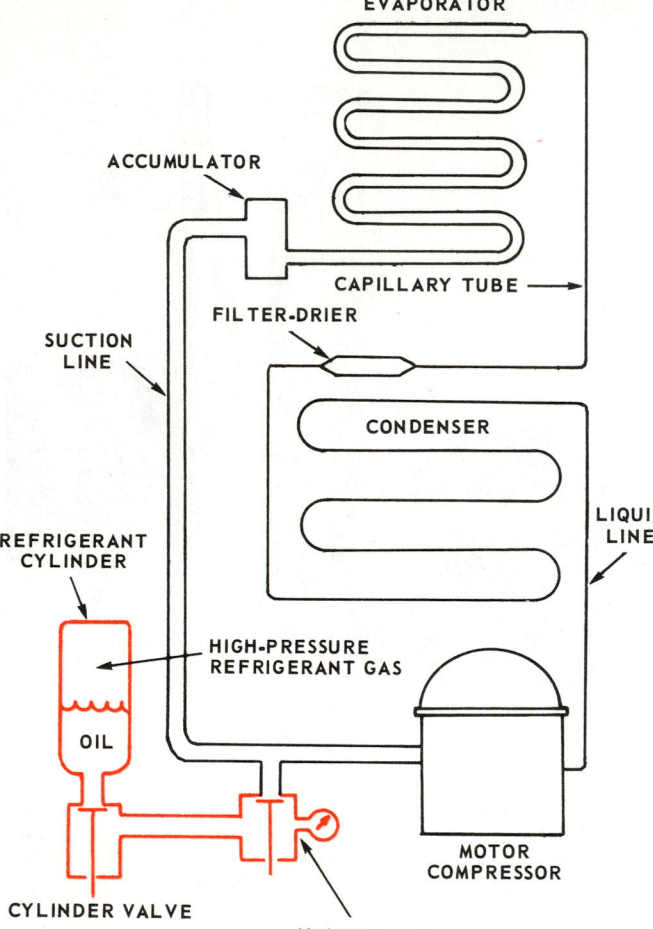

Fig. 11-72. Single line technique can be used for adding oil to a small hermetic system. First place the correct quantity of oil in a service cylinder, along with small amount of refrigerant to create pressure. Then turn the service cylinder upsidedown. Its pressure must be higher than system pressure so that oil will enter the system.

11-48 REMOVING THE SYSTEM

When checking a unit, mechanical or electrical trouble is sometimes found that cannot be easily fixed. In this case, the unit should be removed from the cabinet and sent to a shop for repair.

It is important to protect the refrigerating unit while it is being moved. Cradles or crates should be used to hold mechanisms. Wooden frames and C-clamps will hold the parts down and keep them from being damaged while in transit.

The complete cabinet, if moved in a truck, should be shrouded in a special padded blanket.

The procedure for removing the unit from the cabinet varies because some evaporators are removed from the rear of the cabinet while others are removed from the front (by way of the cabinet door).

The refrigerant lines are sometimes run just under the door frame breaker strips. These breaker strips must be removed in order to remove the lines. Do this carefully. Strips are brittle. If allowed to come up to room temperature, they will be more flexible.

These strips are usually made of a plastic material. Several other designs are shown in Chapter 10. There are several methods used to fasten the breaker strips to the cabinet shell and the liner.

The cabinets which have the evaporator removed from the rear of the cabinet are not difficult to dismantle. Skill and patience are required to remove the unit from the front of the cabinet. The hardest part is removing the breaker strips. When the mechanism is removed, be careful not to kink or buckle the refrigerant lines.

On double-door cabinets the strips between the two compartments must be removed before the unit can be taken out.

The compressor and condenser are sometimes fastened to the rear of the cabinets by four to six mounting screws and bolts. Be careful not to damage the mechanism while removing these devices.

11-49 DIAGNOSING INTERNAL TROUBLES

Before removing any part of the hermetic system, one should be absolutely sure that the part is the one causing the problem. The parts which may cause the trouble are:
1. Motor compressor.
2. Filter-drier.
3. Refrigerant control.
   a. Capillary tube.
   b. AEV.
4. Hot gas defrosting valve.

11-50 LOCATING MOTOR COMPRESSOR FAULTS

Replacing the motor compressor is the most costly service repair. Be sure that it is a bad motor compressor before replacing it.

The fault may be either:
1. Electrical.
2. Mechanical.

Electrical faults are the most common reason for replacing a motor compressor when the unit is actually in good condition.

Other electrical problems often mislead a service technician into thinking the motor compressor is at fault. To check it,

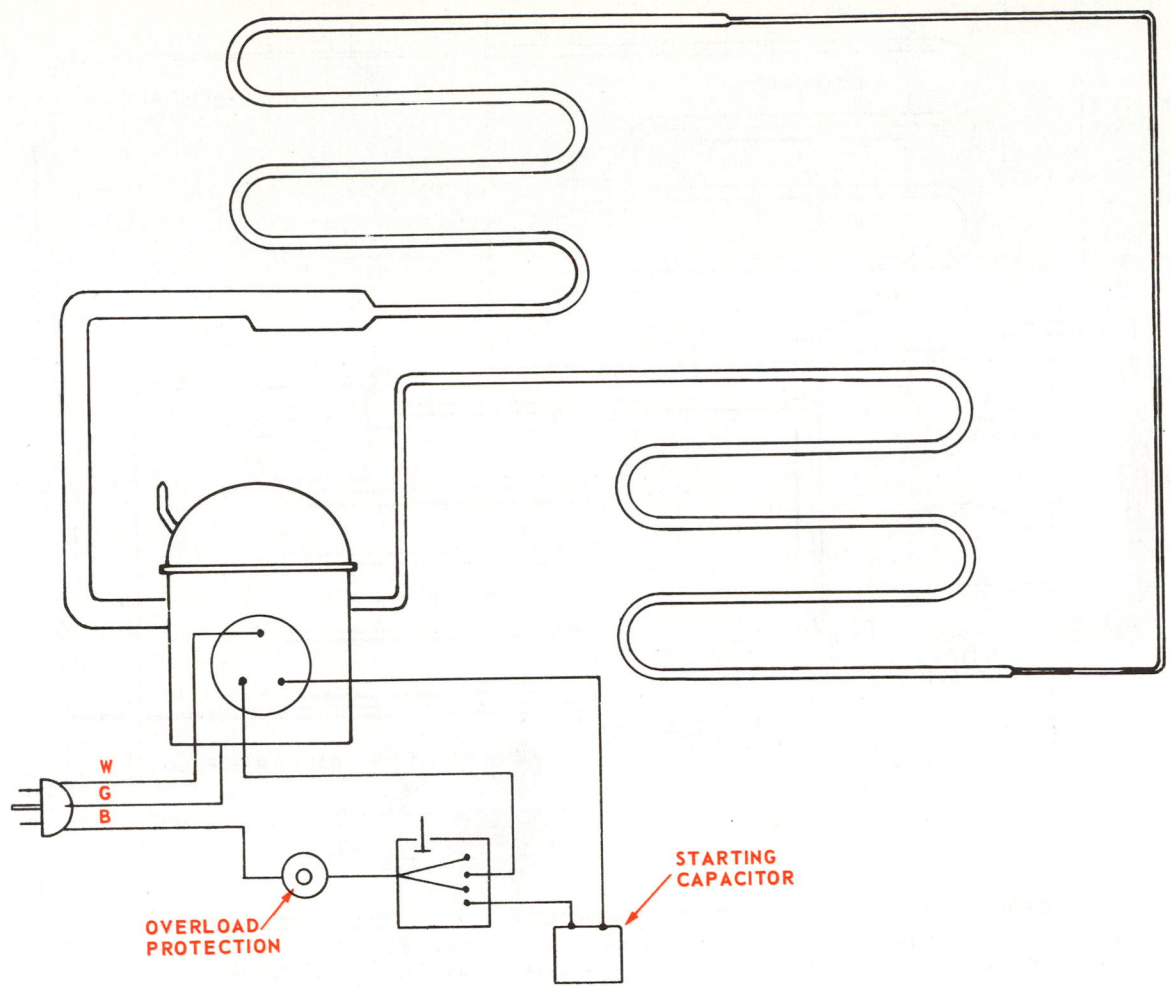

Fig. 11-73. Testing a capacitor start motor compressor after removing all the electrical leads and connecting a test cord.

clean the outside of the compressor dome, remove the electrical connections cover and disconnect the system wiring from the compressor: relay, capacitors, overload and all. Check motor windings for continuity, shorts and grounds. (Use on ohmmeter.) See Chapter 7. If unit checks out correctly, connect a starting circuit to the motor compressor. Use exactly the correct size of capacitors and overload cutout. Connect as shown in Fig. 11-73.

If system starts and operates correctly with these manual start electrical connections, the problem is in the external part of the system (wiring, thermostat, relay or overload). If the internal electrical motor is faulty, the motor compressor must be replaced.

If the electrical system operates correctly, the compressor may not be pumping.

The best check of the motor compressor is its wattage reading at normal low and high pressures. If wattage reading is below motor rating, the pump may be worn out. To check the compressor's pumping ability, install a piercing valve, a gauge manifold and then pinch the suction line as shown in Fig. 11-74. Next, run the unit to determine how high a vacuum it will pull. (It should pull between 25 and 28″ Hg. of vacuum.) Stop the compressor. If the vacuum starts decreasing (moving

toward 20″ Hg., then 10″ Hg.), the exhaust valves of the compressor are leaking. The motor compressor must be replaced or overhauled.

Another method is to install a piercing valve on the process tube, connecting service line and manifold. Purge these lines to clean them. Then pinch the suction line. Start the unit. The motor compressor should pull 16″ Hg. of vacuum in about two minutes. (Do not run any longer with no cool suction gas flowing. The motor will overheat.) Stop the compressor and observe. If low-side pressure increases, the compressor valves are leaking.

11-51 CAPILLARY TUBE SERVICE

Capillary tubes must be of the correct inside diameter (ID) and length for the capacity of the system and for the evaporator temperatures desired. Cut capillary tubing by filing a notch around tubing and then breaking it by small back and forth bending motions. A tube cutter will change the inside diameter too much.

View A of Fig. 11-75 shows a capillary system of correct design while the system in view B has too long a capillary tube or one with an undersized inside diameter (ID). View B also

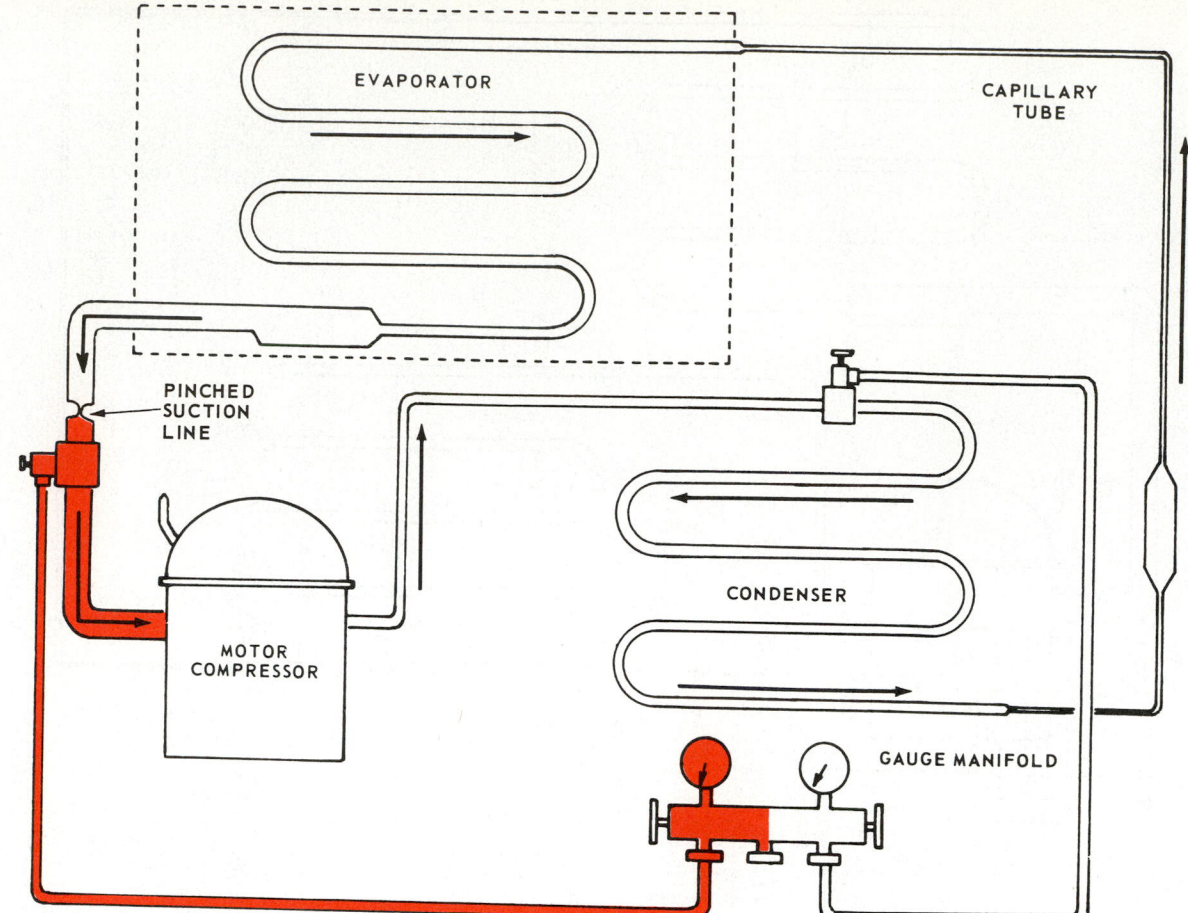

Fig. 11-74. Compressor's pumping capacity can be tested by pinching the suction line. This is a very short run test or the motor will overheat. Compressor should pull 16 to 26 in. of vacuum in a few seconds. Red area indicates a vacuum in the line.

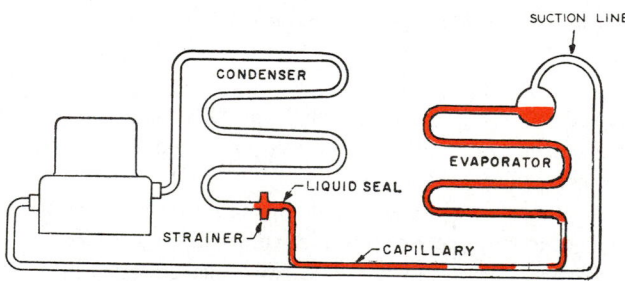

Capillary selected for capacity balance conditions. Liquid seal at capillary inlet but no excess liquid in condenser. Compressor discharge and suction pressures normal. Evaporator properly charged.

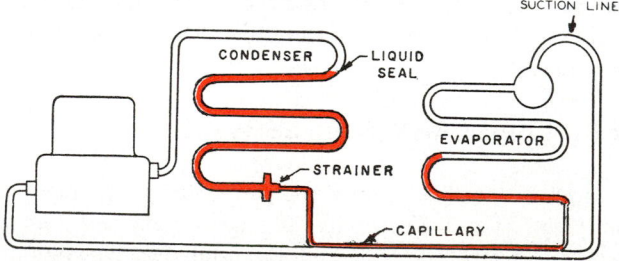

Too much capillary resistance—liquid refrigerant backs up in condenser and causes evaporator to be undercharged. Compressor discharge pressure may be abnormally high. Suction pressure below normal. Bottom of condenser subcooled.

Fig. 11-75. Effect of correct and incorrect capillary tube installation. Above. Correct installation, operation normal. Below. Incorrect. There is too much resistance in the tube and the evaporator is "starved."

illustrates what happens if the system has a starved evaporator, because of a partially clogged filter-drier or capillary tube.

The amount of refrigerant in a capillary tube system is critical. Refer to Fig. 11-76. Notice the change in head pressure as the charge of refrigerant changes. If the system is undercharged as at C, the evaporator will not receive enough refrigerant and the system may run all the time. If the system is overcharged, the liquid refrigerant may flow down the suction line and may cause oil pumping in the motor compressor. Suction line will sweat and even frost up all the way to the motor compressor.

11-52 CHECKING FOR RESTRICTED SYSTEM

To check the capillary tube, run the system for a few minutes. Stop the unit and listen where the capillary tube enters the evaporator. If there is no hissing sound, the capillary tube is clogged.

Heat the evaporator end of the capillary tube with warm water and rag. (Do not use a flame.) If the clogging is from ice, there will be a hissing sound as the ice melts. A clogged strainer or capillary tube will fill the condenser with all the refrigerant. It may stop or overload during a startup.

There is a quick test to determine if the system is either short of refrigerant or if the filter-drier or capillary tube is clogged. First, install a piercing valve on the suction line or on

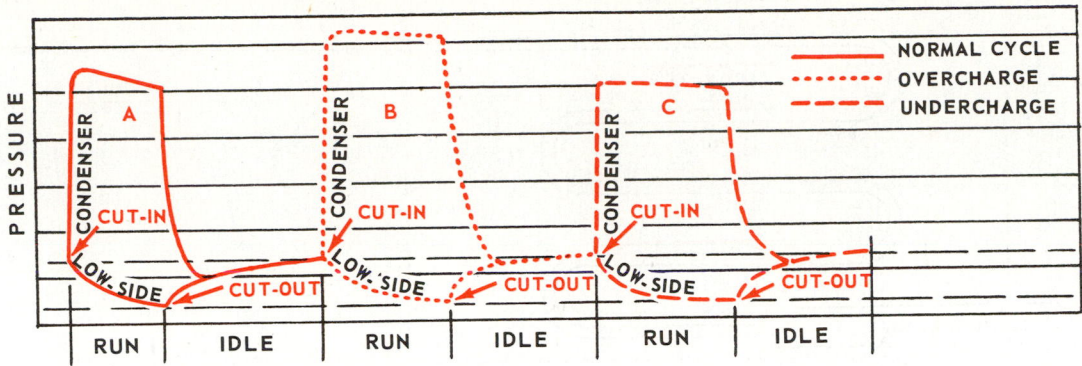

Fig. 11-76. Pressure cycle diagrams for three conditions in a capillary tube system. A—Normal charge. B—Overcharged. C—Undercharged. The overcharge will usually cause a frosted or sweating suction line.   (Allin Mfg. Co., Inc.)

the end of the process tube of the compressor. Purge the service lines using gas from a refrigerant cylinder. Open piercing valve. If the low side is in a vacuum, the evaporator fan is not working or, if the defrost system is not working (ice buildup), the system has a restriction or is low on refrigerant.

Check the evaporator fan. If it is working, check the defrost

system (inspection and electrical test of defrost resistance wire), then check as follows:

1. Install another piercing valve at the condenser outlet. See Fig. 11-77. Open the valve just a little. Allow some refrigerant to escape.

2. If no gas escapes or if only gas escapes, the system is short

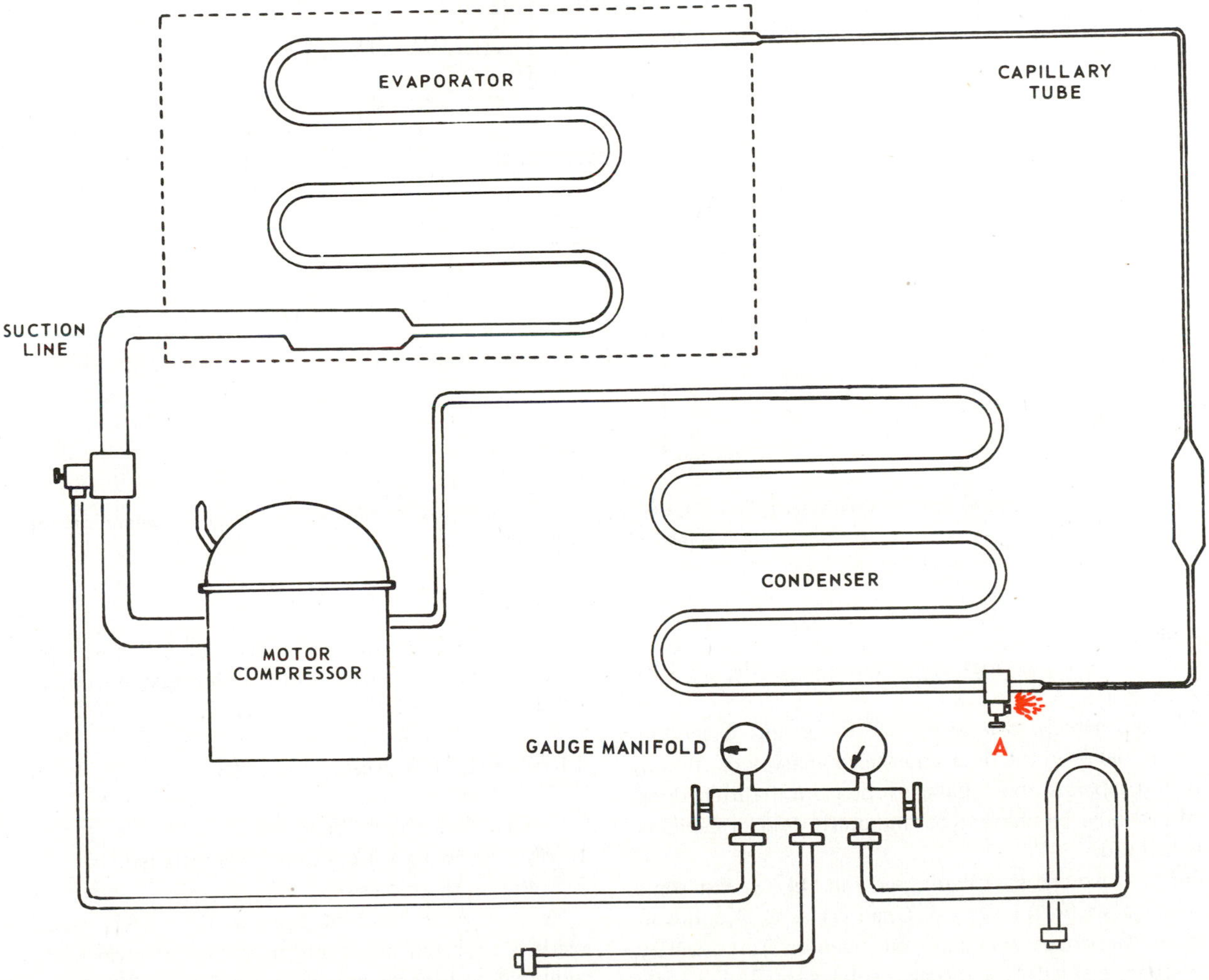

Fig. 11-77. Testing for a clogged drier or capillary or for a lack of refrigerant. With valve A open, escaping gas means there is a lack of refrigerant. If liquid refrigerant escapes, the filter-drier or the capillary tube is clogged.

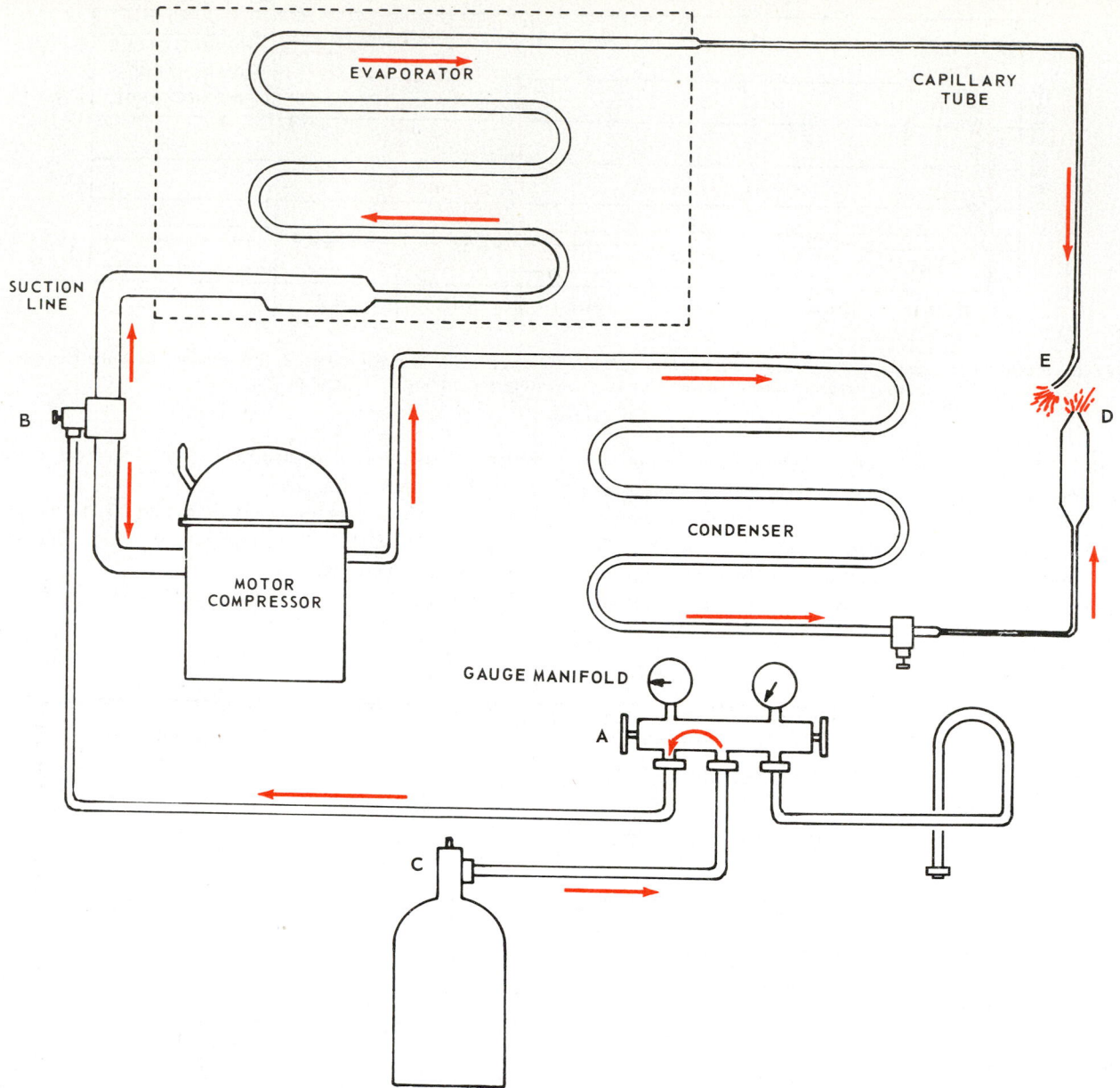

Fig. 11-78. To check if capillary tube or filter-drier is clogged, separate by checking flow at D. Charge some vapor refrigerant into the system by way of valves C, A, and B.

of refrigerant.

3. If liquid refrigerant escapes, the filter-drier or the capillary tube is clogged.

4. To find out which, discharge the refrigerant. Clean the connection between the drier and the capillary tube, flux it, heat it and separate the capillary tube from the filter-drier. System can now be checked to find out if drier or capillary tube is clogged.

5. To find out which is at fault, see Fig. 11-78. Put some vapor refrigerant in the system. Open valves C, A and B. If filter-drier is open, gas will come out opening D. If capillary tube is open, a small flow (because tube is small) will come out its opening at E. If either one is clogged, there will be no flow.

Clogged units must be replaced. Sometimes clogged capillaries can be opened with a high-pressure hydraulic pump. (See Para. 11-63.)

11-53 FILTER AND DRIERS

A drier-filter should be replaced:

1. When a new motor compressor is installed.
2. If the filter is clogged.

A positive method of keeping the system clean and dry inside is to install driers and filters in the refrigerant circuit. A combination filter-drier is shown in Fig. 11-79.

A solid moisture absorbent will usually do a satisfactory job. Silica gel, alumina gel and synthetic silicates are excellent

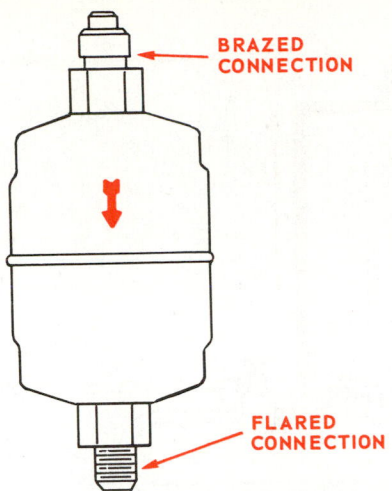

Fig. 11-79. Filter-drier which is used on small hermetic systems. Arrow designates direction of flow of refrigerant.

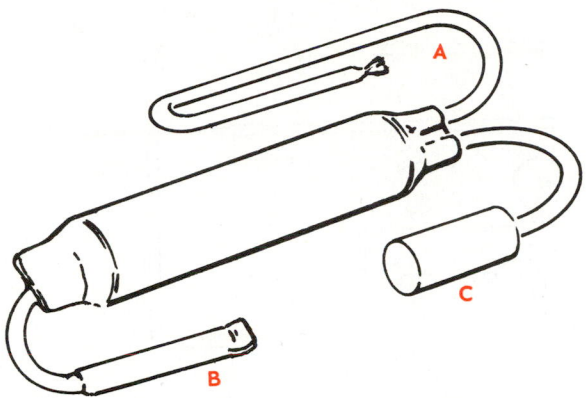

Fig. 11-80. A replacement filter-drier equipped with a valve adaptor fitting. A—Connection to condenser. B—Connection to capillary tube. C—Valve adaptor fitting. (General Electric Co.)

moisture absorbers. A bead silica gel gives good results. A filter-drier with a valve adaptor built into it is shown in Fig. 11-80. The same unit installed is shown in Fig. 11-81.

Never use a liquid drying agent in a unit equipped with a solid desiccant (drying chemical). The liquid dryer chemical

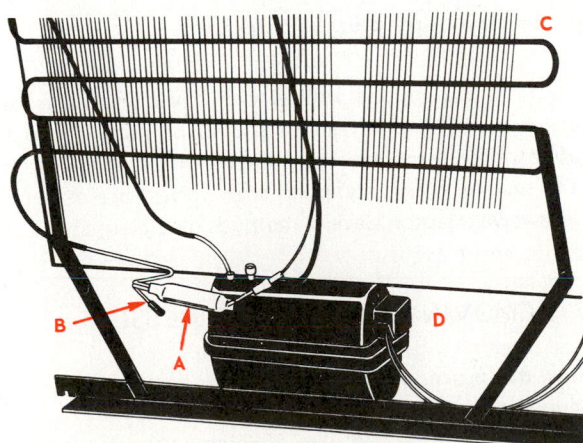

Fig. 11-81. Replacement filter-drier installed. A—Filter-drier. B—Valve adaptor. C—Condenser. D—Compressor.

will release the moisture already trapped in the drier.

For the same reason, a solid drier should not be put in a system that is already using a liquid drier. To avoid this danger, all systems should be labeled as to drying agent in use. The moisture absorbent properties of these desiccants are shown in Fig. 11-82.

DESICCANT (DRYING CHEMICAL)	MESH	ABSORPTION FROM LIQUID, PERCENT OF WEIGHT OF DESICCANT
SILICA GEL	8–20	16
ACTIVATED ALUMINA	8–10	12
SYNTHETIC SILICATES	8–20	16

Fig. 11-82. Moisture absorbing ability of some desiccants in driers. Some driers have mixtures of these chemicals. These driers may be used with R-12, R-22, R-500 and R-502.

DRYER SIZES – DOMESTIC	
Cu. In.	Capacity, HP
2	1/8
3	1/6 to 1/4
6	1/4 to 1/2
9	1/2 to 3/4

Fig. 11-83. Drier capacities in cubic inches recommended for various horsepower systems.

The Refrigeration Electrical Manufacturer's Association has recommended that the capacity of dehydrators be rated. Fig. 11-83 shows the recommended cubic content of drying agents according to horsepower of the unit. All driers are sealed by the manufacturer. Do not remove the sealing caps until just before installation.

Driers absorb water faster at lower temperatures. If at all possible, the drier should be installed just before the refrigerant control. If by an accident the filter-drier becomes heated, the moisture that it has absorbed may be driven out and will recirculate with the refrigerant.

A filter-drier has an arrow stamped or cast on the body to indicate the direction that the refrigerant should flow through it. Be sure that it is installed properly.

It is always recommended that a new filter-drier be installed in the liquid line whenever it is necessary to break into the circuit.

Filter-driers may be installed with either flared or silver-brazed connections. Many service technicians install two filter-driers on a system after repairing it. One is placed on the high side just before the refrigerant control and one on the low side, between the evaporator and compressor. This improves chances of removing all moisture or contaminants which may have entered the system during servicing.

11-54 HOT GAS DEFROST TROUBLES

The hot gas valve is usually operated by a solenoid. If the evaporator is overloaded with ice, the solenoid electric coil

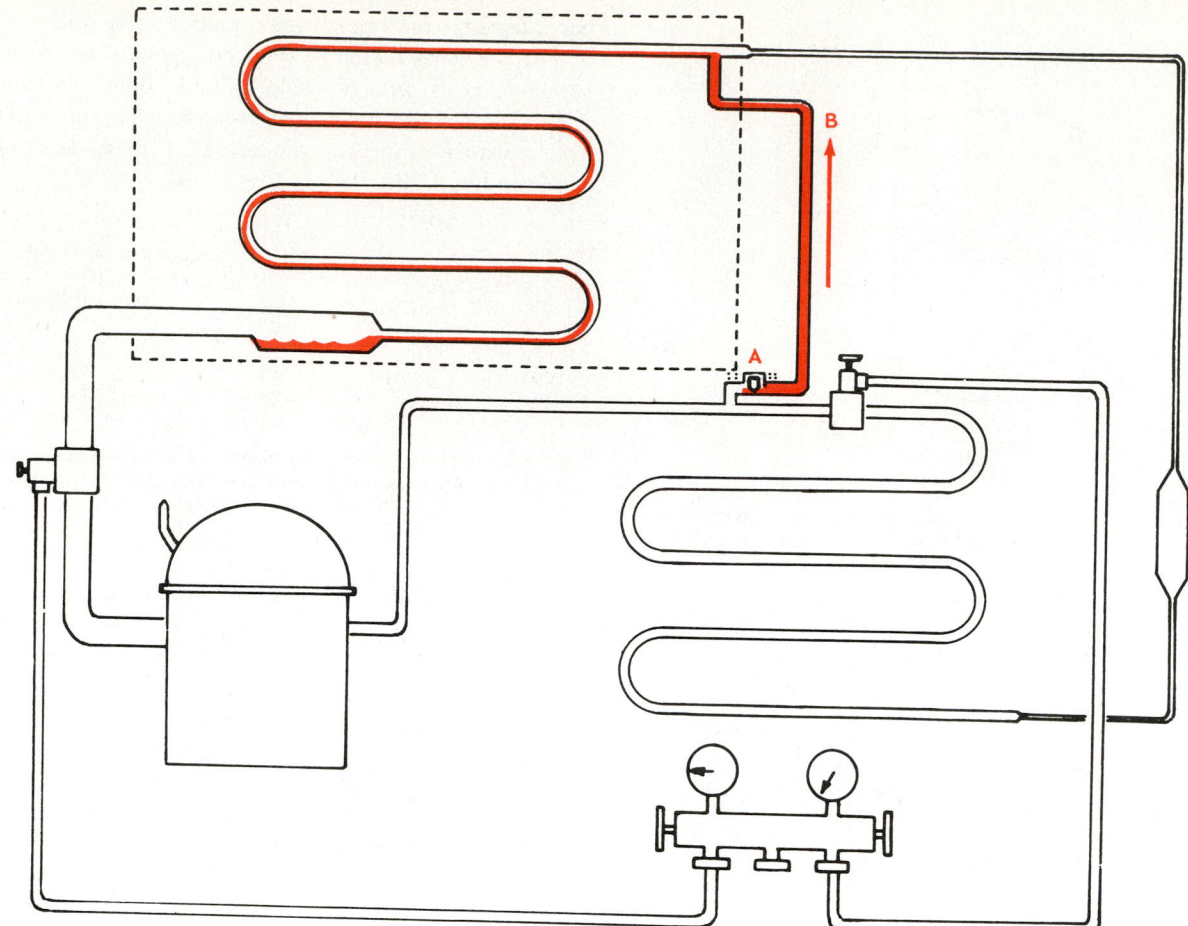

Fig. 11-84. A hot gas system with a stuck open solenoid valve: A—Solenoid valve. B—Hot gas line.

may have failed (open circuit) or the timer may not be operating correctly. Both of these problems can be checked electrically.

If the system is operating correctly, the problem is probably a stuck valve stem in the solenoid. See Fig. 11-84. (Hot gas is shown in color.) Rap the valve body while the defrost timer switch is closed. If the valve breaks loose one can hear the surge of hot gas. The line between the solenoid and the evaporator will also become warm to the touch.

If the evaporator is warm and the line between the hot gas solenoid and evaporator is warm, the valve is stuck open. Again, rap the valve sharply while the timer is on open circuit. If the valve closes, the low-side pressure will start to decrease immediately and the evaporator will start to cool and frost.

If the valve does not operate, it must be removed and replaced with a new one.

11-55 DISMANTLING SYSTEM

Whenever it is necessary to replace parts such as the motor compressor, condenser, capillary tube, evaporator, accumulator, filter-drier and the like, the system should be prepared as follows:

1. Disconnect the electrical circuit.
2. Carefully clean all surfaces. It is well to clean the entire

mechanism, as this will reduce the chance of dirt and other contamination entering the system.

3. Install service valve and gauge manifold.
4. Remove the refrigerant. *It is usually purged to the outdoors.* **Ventilate!** Put an oil trap at the end of purged line to hold oil. If oil is acidic, purge into a caustic solution.
5. Cut the tubing and remove the part to be replaced. **Wear goggles!**

11-56 REMOVING REFRIGERANT

If the unit is equipped with service valves or service valve attachment plugs, the refrigerant should be purged into the open atmosphere through lines equipped with special oil traps.

If the system is not equipped with valves, use a piercing valve or a valve adaptor. Several methods may be used. **Be sure to wear goggles!**

11-57 REMOVING MOTOR COMPRESSOR

Follow this procedure to remove motor compressor:

1. Disconnect the electrical circuit.
2. Install a gauge manifold. Use a piercing valve if necessary.
3. Discharge the refrigerant.
4. Disconnect the lines. There are several methods:

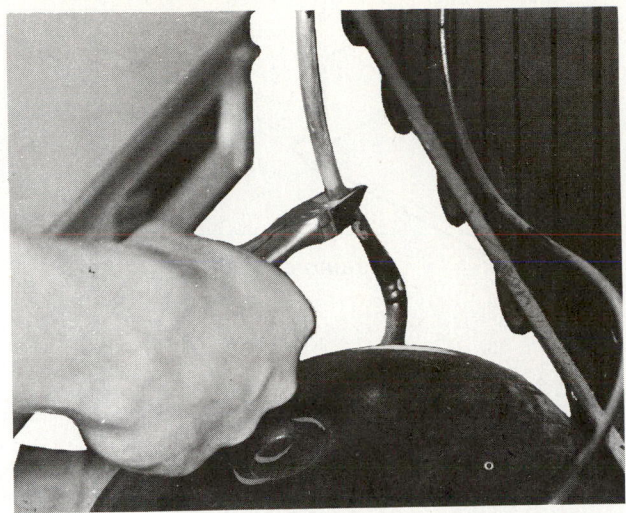

Fig. 11-85. Suction lines are separated by pinching and then breaking. Sawing is not recommended. Chips may enter the system and later cause damage to refrigerant control, compressor or motor. (Philco-Ford Corp.)

a. Use a pinching tool. Pinch the lines as shown in Fig. 11-85. Wrap the tubes with cloth and bend them sharply at the pinch to separate. Fig. 11-86 shows the disconnected compressor. **Wear goggles!**

b. Clean both the suction and discharge tubing on straight sections near the compressor. Use a tube cutter. **Wear goggles!** Plug the lines at once.

c. Clean the tubing or fittings at the compressor. Put brazing flux on the connection. Heat the joint and pull the tubing out of the fittings. **Wear goggles!** Plug openings right away.

5. Remove the motor compressor.

6. If the motor compressor has oil cooler lines, these too must be pinched and broken. The removed compressor tubing openings should be sealed (pinched and brazed).

The unit is now ready to have a replacement motor compressor installed.

11-58 CAUSE OF MOTOR COMPRESSOR BURNOUTS

A certain indication of a burnout is the strong pungent odor of the refrigerant when a piercing valve is opened just a little. Much study has gone into why motor compressors burn out. Moisture, dirt and air in the system are possible causes. Too much current flow from inaccurate safety devices in the electrical circuit, a stiff compressor, low voltage or a lack of refrigerant (poor cooling of motor), may also be the trouble.

High head pressure is one of the most frequent reasons for motor burnout. This pressure creates very high temperatures as the gas passes the discharge valves of the compressor. The high temperature increases chemical action, adding to or creating new corroding elements in the system. Oil breaks down and forms carbon and sludge. If discharge line to condenser reaches 350 F. (177 C.), oil breakdown is taking place.

It is therefore very important that the condenser be large enough, that it be clean, that air flows over it efficiently. (Fans, fan motors, ducts, air-in and air-out must all be in good condition.)

The service technician should check the head pressure of each unit and do all the necessary things to bring this pressure down. Purge the system, completely clean the condenser with high gas pressure (air, carbon dioxide, nitrogen). **Wear goggles!** Brush the condenser. Use long bristle brushes and a vacuum. Check the air-in and air-out passages; they must be in good condition. See Para. 11-39.

11-59 CLEANUP AFTER MOTOR BURNOUT

When a motor burns out, it overheats. This overheating will cause the refrigerant to break down and, if moisture is present, form hydrochloric and hydrofluoric acid. Oil in this condition is said to be "acidic." The acid will cause insulation on motor windings to deteriorate and increase the motor temperature. Eventually, the motor windings will short circuit and burn out.

If a system has a motor compressor burnout, refrigerant controls should be repaired or replaced (AEV, solenoid valves, reversing valves, etc.). Flush the system with R-11 or the same refrigerant used in the system.

Do not touch the oil from a burned-out motor compressor as it will cause a severe acid burn!

If oil cooler lines must be cut, be even more careful. Do not allow oil to run on the floor. Trap it in glass containers. A burned-out motor compressor has a very unpleasant odor.

The burnout can be mild or severe. If severe, the oil will be black and acidic with a very pungent odor. **Caution: Wear goggles and rubber gloves.** If mild, oil will be clear but there will be a pungent odor and a mild acidic condition. If oil is clean and odor free, there is no burnout. The trouble is mechanical.

After replacing a motor compressor, install two filter-driers. See Fig. 4-9. One should be in the suction line between the evaporator and the compressor. The other should be between the condenser and the liquid refrigerant line (probably the capillary tube).

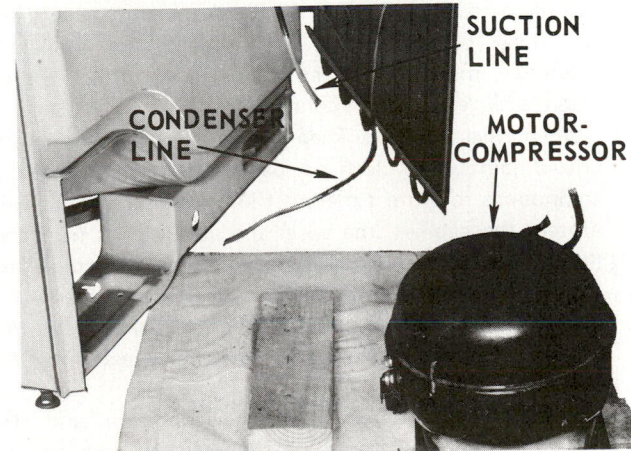

Fig. 11-86. Old motor compressor has been removed and its refrigerant lines sealed by pinching. Unit is now protected against oil spillage and internal corrosion.

One of the many acid test kits available may be used to determine the amount of contamination. See Chapter 14 for further information.

Flushing a system with R-11 refrigerant is still another method of cleaning a burned-out system. The R-11 can be pumped through the system after removing the motor compressor using a rotor or diaphragm pump. Or, a cylinder of R-12 or R-22 may be used to create pressure to push the R-11 through the system.

Repeat this operation until R-11 comes out of the suction line clean. Use a pail to catch it.

R-11 is probably the best substance for cleaning and washing out the inside of the refrigerating mechanism. It has a boiling temperature of 74.8 F. (24 C.), is nonflammable and nontoxic. It leaves no noncondensable residue and does not react with the electrical insulation. R-11 can also be cleaned and reused. It is particularly recommended after a motor burnout.

One can then also purge the system with R-12 gas and even some liquid to be sure that all the R-11 is removed. Wear goggles!

11-60 INSTALLING A MOTOR COMPRESSOR

The replacement motor compressor should be an exact replacement and it must have the same capacity as the one being replaced. The motor compressor must be designed for the low-side pressure desired (low, medium or high).

This is the recommended procedure for installing a replacement motor compressor:

1. Carefully clean the ends of the suction line, discharge line and oil lines (about 2 in.).
2. Carefully clean the suction, discharge and oil line connections on the motor compressor.
3. First, install the piercing valve and valve adaptor and then the gauge manifold — if they have not already been installed. (See Para. 11-32.)
4. Connect the lines. Use adaptor fittings, lengths of tubing or an expander, if necessary, in order to telescope the tubes from the motor compressor into the suction line and discharge line. See Fig. 11-87. A punch-type swaging tool may be used if an expander is not available. Telescope the tubes together, using as little flux as possible. Silver braze the connections.

   When silver brazing, keep the heat away from other brazed joints. Use cloth or special wet heat absorbency compounds to protect the other joints. Metal sheeting will protect the cabinet, the wires and plastic parts from the flame. *CAUTION: If the motor compressor or stub lines on the replacement motor compressor are smaller in diameter than the suction line and/or discharge line of the cabinet unit, the replacement motor compressor may be too small.*

5. Cut the liquid line between the condenser and the refrigerant control. Install a filter-drier of the proper capacity. (See Fig. 11-83.)
6. Put some vapor refrigerant into the system (about 25 psi) and check for leaks.

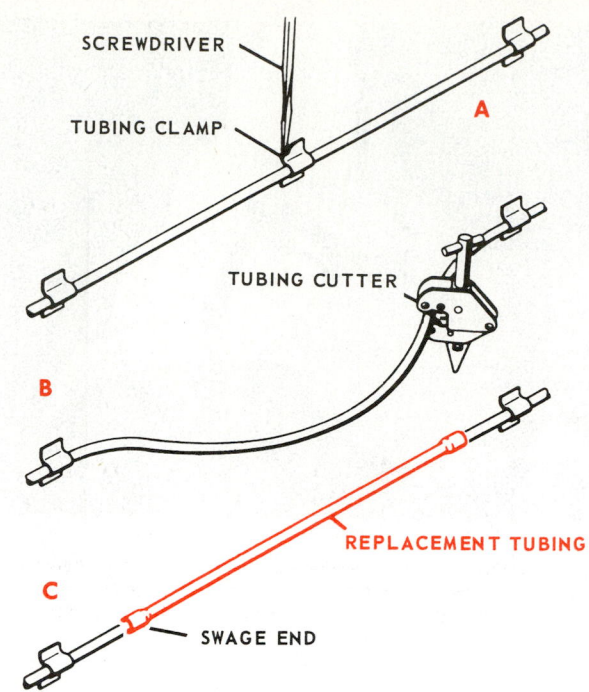

Fig. 11-87. Replacing tubing using swage ends. A—Removing damaged tubing. B—Cutting old tubing with tubing cutter. C—Installing swaged and replacement tubing. Joints are silver brazed. (Franklin Mfg. Co.)

7. Connect a vacuum pump to the system through the gauge manifold. Draw as high a vacuum as possible. Now, with the vacuum pump operating, hold this high vacuum for at least an hour or again put refrigerant in and then evacuate it — the three-step method. (Another way to check for leaks is to close off the connection to the vacuum pump and observe — if the vacuum on the system remains constant. If it does not, it is an indication that there is a leak in the system.)
8. Charge a small amount of R-11 into the system and purge. This will help clean out the refrigerant lines.
9. Charge the lines with a small amount of the refrigerant used in the system. Adjust the pressure to atmospheric or very slightly above.
10. Reconnect the electrical circuit.
11. Charge the system with the correct amount of refrigerant. It is best to overcharge slightly and purge to the desired quantity. The use of the service manifold makes this operation quite easy.
12. Close all manifold valves and plug in the electrical circuit and operate the system through several cycles.
13. Adjust the refrigerant charge as described in Para. 11-46.
14. Remove the gauge manifold and carefully seal all openings into the system, depending upon the gauge manifold connections used.
15. It is good practice to place a recording thermometer in the refrigerator cabinet to check its operation continuously for at least 24-hours.

Fig. 11-88 shows a replacement motor compressor. It has both a low-side and high-side process tube.

In Fig. 11-89, the process tube has had a tubing extension

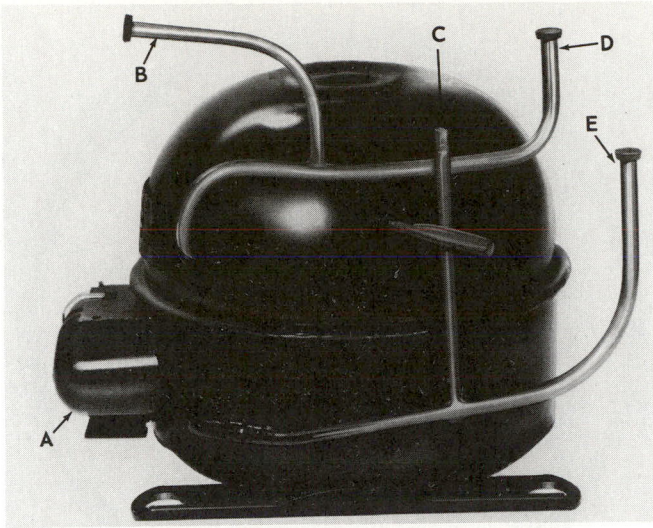

Fig. 11-88. A replacement compressor. A—Relay. B—Suction line process tube. C—Discharge line process tube. D—Suction line. E—Discharge line. (Philco-Ford Corp.)

brazed to it. The right-hand process tube is brazed to the discharge tube.

There are many instances of replacement motor compressors burning out soon after they are installed. Most of these repeated burnouts are due to the system not being clean enough or having moisture in the system. Installation of both a suction line filter and a high-side filter has been found to reduce repeat burnouts.

11-61 CONDENSER TROUBLES

Condenser troubles are usually caused by leaks or the collecting of lint and dirt on the outside.

The service technician may be able to repair leaks without bringing the unit into the shop. He must install piercing valves, purge the refrigerant, repair the leak, test it, evacuate, charge and check out operation. If the leak is not repairable (difficult to get at), the condenser must be replaced.

11-62 REPAIRING EVAPORATORS

Evaporators may be made of either stainless steel or aluminum. Follow the procedure recommended for repairing leaks in each type:

1. For stainless steel, either silver braze or weld (tungsten arc — inert gas system):
   a. Locate the leak.
   b. Discharge the refrigerant.
   c. Clean the metal around the leak.
   d. Purge nitrogen (very low pressure) while brazing or welding.
   e. Polish the weld or clean the brazed joint.
   f. Test for leaks.
2. For aluminum evaporators, repair with aluminum solder, aluminum braze, aluminum weld or a special epoxy.
   a. Locate the leak.

b. Remove the refrigerant.
c. Clean around the leak. The surface oxide is hard and must be removed. Repair right after cleaning as the oxide surface forms quickly. (Sand, file and then clean with epoxy cleaner.) If leak is a large hole, use a clean small screw or metal plug to fill most of the opening.
d. Mix the epoxy and catalyst.
e. Apply with mixing spatula. (Be sure there is no positive pressure in the system. System must be open to atmosphere at some other opening.)
f. Allow at least an hour for hardening.
g. Sand the patch to a smooth finish.
h. Test for leaks. If system still leaks at repaired joint, remove epoxy by filing and/or grinding and install a new patch.

Aluminum foil can be used with the epoxy cement to strengthen the joint and to improve the appearance of the repair. Another method of repair is to heat the tubing and apply a paste mix (or stick) of special epoxies and resins.

3. For repairing with aluminum solder or aluminum brazing, see Chapter 2 — Aluminum.

Avoid brazing aluminum tubing as it overheats too easily. Too much of the tubing will be annealed (softened) weakening the tubing walls. Aluminum tubing is usually 3003 alloy; however, 5005 and 1100 alloy are also used.

Aluminum solder, containing 92 to 100 percent zinc, is used. The melting temperatures are 700 to 800 F. Do not use a flux.

Aluminum may also be welded. Use an inert gas tungsten arc welding system (sometimes called TIG or GTA).

11-63 SERVICING CAPILLARY TUBES

The service technician removes and replaces the capillary tube using the same steps as in Para. 11-56 and Para. 11-62. One step is different — the capillary tube and filter-drier are removed by loosening the brazed joints.

It is sometimes possible to repair a capillary tube by cleaning it. Proceed as follows: Disconnect the capillary tube at both ends if possible.

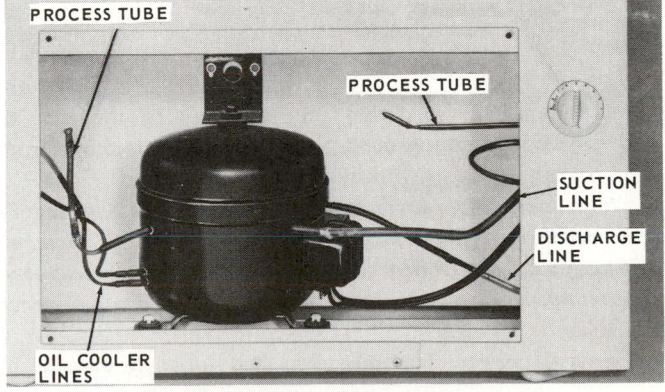

Fig. 11-89. A motor compressor replacement installed showing two process tubes. (Kelvinator, Inc.)

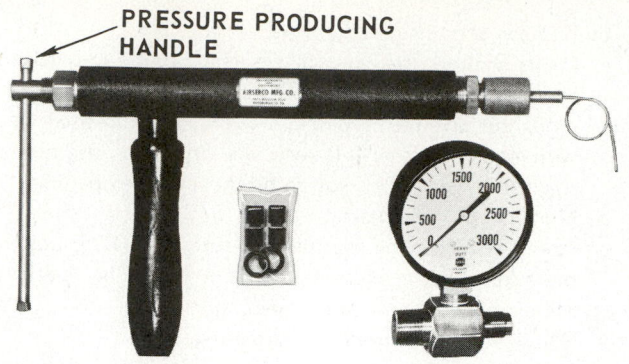

Fig. 11-90. Capillary tube cleaner used to clear obstructions from capillary tubes. It has a capillary tube adaptor fitting. Either oil or R-11 is used as pressure fluid.

1. Attach the capillary tube cleaner to the outlet end of the tube.
2. Build up pressure on the tube to force the wax and dirt out. Fig. 11-90 illustrates a capillary tube cleaner attached to a capillary tube for cleaning. These cleaners are capable of building up pressure to a possible 20,000 psi.
3. After the capillary tube has been cleaned, continue to flush out the tube thoroughly. Use either R-11 or the refrigerant with which the system is charged.
4. Install a new filter-drier and reconnect the capillary tube into the system. If needed, new tubes must have the same inside diameter (ID) and the same length as the one removed. Fig. 11-91 shows a tool for measuring tubing ID.

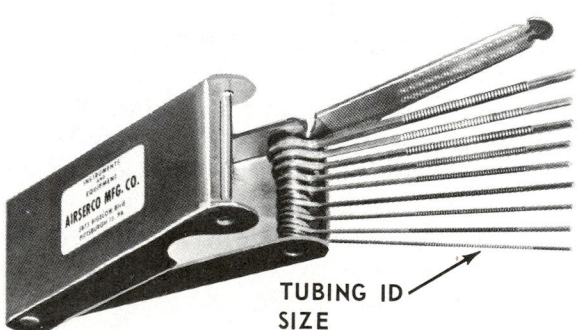

Fig. 11-91. Capillary tube sizing kit. Gauge measures the ID of the capillary tube. (Airserco Mfg. Co.)

If there is an indication that wax caused the tube to become stopped up, probably the oil should be removed from the system. Replace it with fresh, clean, wax-free oil.

Never use a liquid drier (mainly methanol) to stop moisture from freezing at the refrigerant control. These "antifreeze" substances do not remove the moisture, they merely keep it in circulation. In many cases, they may damage the motor insulation.

Some patented replacement capillary tubes on the market may be used as capillary tube replacements when necessary. They are patented tubes arranged to be fitted either with a calibrated wire inside the tube to provide the proper refriger-

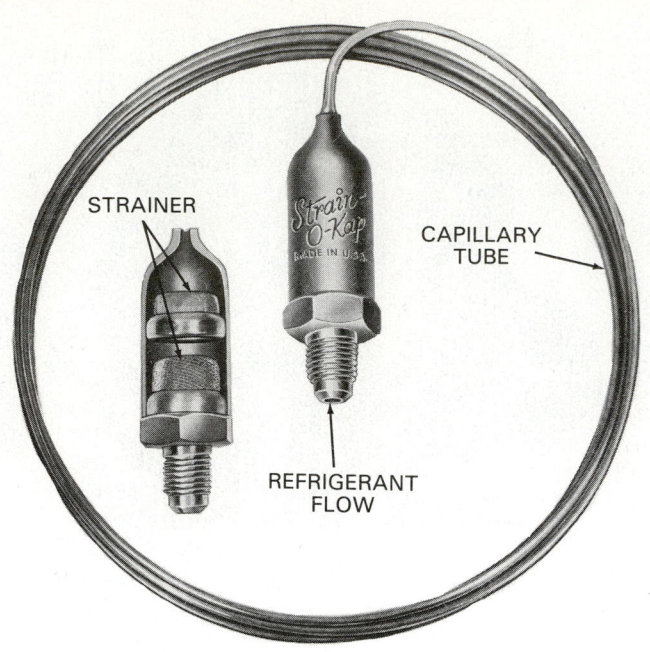

Fig. 11-92. Capillary tube, complete with strainer, meters refrigerant flow. Different strainer size is used for 1/20 to 1/4 hp and 1/3 to 1 hp units. Capillary tube can be installed by either soldering or using standard capillary tube fitting. (Watsco, Inc.)

ant control, or the capillary tube, itself, is adjustable. See Fig. 11-92. Other capillary tube replacements are shown in Figs. 11-93 and 11-94. Fig. 11-95 is a replacement capillary tube with adaptor fittings.

11-64 OVERHAULING A HERMETIC SYSTEM

A complete overhaul of a hermetic system is usually done in a specialty repair shop.

The shop must be well equipped. Some of the equipment needed includes:
1. A compressor opener.
2. A welding system (a tungsten arc welder).
3. A cleaning booth.
4. A paint booth.
5. Benches and vises.
6. Storage cabinets.
7. A lathe.
8. A surface grinder.
9. A compressor test bench.
10. A deep vacuum system.
11. Compressed air (dry and clean).

The shop should be well ventilated with humidity kept at 35 to 45 percent range to prevent moisture getting into the internal assemblies. Para. 11-65 through Para. 11-78 explain recommended repair shop procedures.

11-65 REMOVING OIL

Oil ought to be removed if the compressor dome is cut open either by grinding or by turning on a lathe. If the sealed

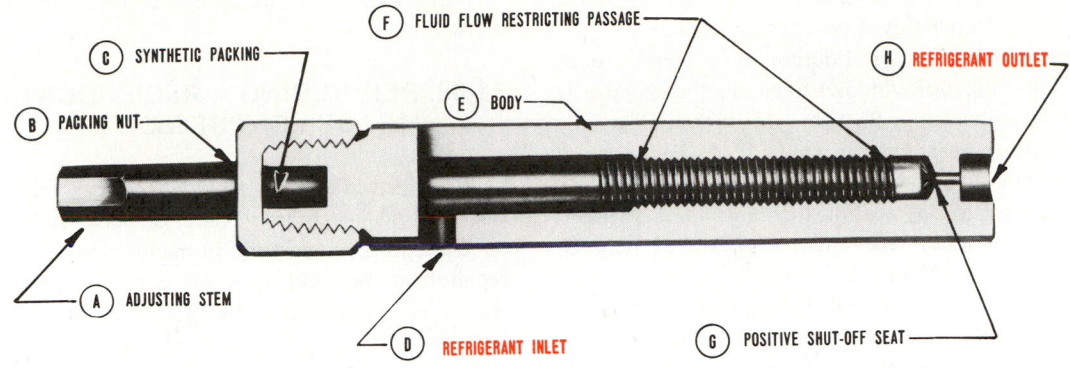

Fig. 11-93. Adjustable capillary valve. Stem can be turned in or out to regulate refrigerant flow at shut-off seat.

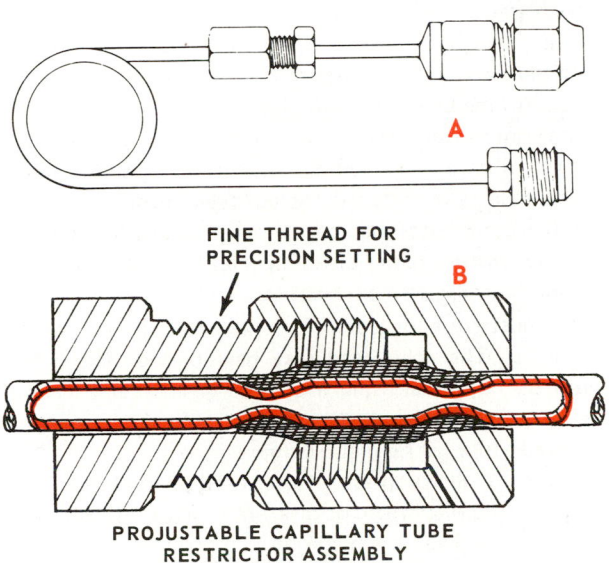

FINE THREAD FOR PRECISION SETTING

PROJUSTABLE CAPILLARY TUBE RESTRICTOR ASSEMBLY

Fig. 11-94. Internal design of adjustment used on capillary tube. A—Complete assembly. B—Details of adjustment.

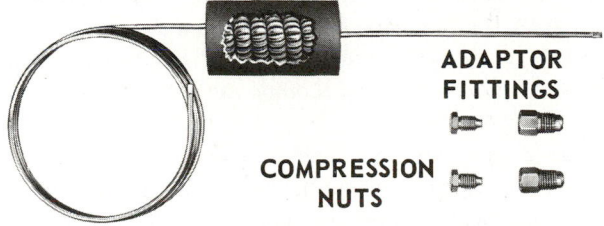

ADAPTOR FITTINGS

COMPRESSION NUTS

Fig. 11-95. Capillary tube refrigerant control replacement kit. (Wabash Corp.)

unit is of bolted construction, the oil may be left in until the unit is dismantled. Old oil should be drained off and carefully measured to assure replacement of the exact quantity. Replace with fresh oil of the proper viscosity. Test the old oil. If acidic, it will usually have an offensive odor and be dark in color. Acidic oil may damage motor windings. Check the windings with a megavolt meter for shorts and grounds. Handle acidic oil with extreme care. Wear rubber gloves and goggles.

11-66 OPENING THE MOTOR COMPRESSOR

Hermetic motor compressors are of two types:
1. Bolted (accessible).
2. Welded.

The bolted or accessible motor compressor overhaul is explained in Chapter 14.

The welded motor compressor is enclosed in a steel dome made of two steel stampings welded together. The process tube, suction line, discharge line and the two oil lines (if used) are silver brazed to this dome.

The welded motor compressor units may be opened easily by:
1. Grinding away the welded seam.
2. Mounting the unit in a lathe and cutting the weld bead away with a cutting tool.

Fig. 11-96 shows a machine used to remove the motor compressor weld by grinding. The motor compressor is clamped to a turntable and the grinding wheel removes the weld as the compressor slowly turns. The wheel automatically

Fig. 11-96. A motor compressor opener. Special dome weld grinding machine will open motor compressor in about 20 minutes. A—Welded bead and grinding wheel. B—Machine controls. (Airserco Mfg. Co.)

Installing and Servicing Small Hermetic Systems / 379

adjusts itself to the contour or shape of the weld seam. The machine will remove top welds as well as side welds.

The opener mechanism is housed in a steel cabinet equipped with shatterproof windows to enable the operator to watch the grinding operation. Both motors of the unit require single-phase, 120V power and use about 11A. The machine automatically grinds the weld.

One motor is used to operate the turntable while a second motor drives the grinding wheel. Grinding wheels may be either 1/8, 1/4 or 3/4-in. thickness.

Remove a minimum of material. Enough base metal must be left for the assembly to be rewelded after overhaul.

After the compressor dome weld is removed, the dome can usually be lifted off the base. The motor and compressor will then be exposed. Clean the inside of the dome thoroughly before continuing.

11-67 CLEANING FACILITIES

Clean the outside of the unit thoroughly. Mineral spirits, obtainable from most oil companies, are used widely as a general cleaner. Carbon tetrachloride should not be used.

A special booth with a grate bottom — having drainage facilities, equipped with a pump and spray nozzle and ventilated — is desirable. Local fire and industrial hygiene regulations must be followed closely. Also follow OSHA (Occupational Safety and Health Administration) standards.

Several degreasing solvents are available which are quite safe. Some of their features are low toxicity (low in poisons), noncorrosive and have a high flash point. (Flashpoint is the temperature at which something will burn.)

Use Class III flammables having a flash point of 140 F. (60 C.) or higher. Use self-closing cleaning pans or tanks. Tanks with a safety fuse link that closes the lid in case of fire are recommended. Suppliers can supply solvents safe to use for this type of cleaning. See Chapter 28 for the classification of cleaning solvents.

11-68 MOTOR REPAIRING

A common trouble with hermetic mechanisms is a burned (overheated) motor. The overheating may be caused by poor cooling, an overload, a short or a ground.

Inspect by looking at it closely. One can tell the condition of the motor. A short (bare wires touching each other) will show up as a dark brown spot. A grounded spot will have the same brown spot. A completely burned-out winding will have all the insulation burned and it will be dark in color.

A system with a burned-out motor should be dismantled and repaired as soon as possible (two or three days). The corroding chemicals in the system will keep increasing the corrosion of the compressor and system parts.

When the windings overheat, insulation is destroyed; stator windings short and ground. If the motor starting winding or running winding is faulty, the stator must be dismantled and rewired (rewound) or replaced. The rewinding should be done by a specialist. Wire must be the same size and have the same insulation as the original.

Replacement or exchange stators are made for most hermetic motors.

11-69 REBUILDING A RECIPROCATING MOTOR COMPRESSOR

Many manufacturers provide exchange service on complete refrigeration machines. These do not include the external controls, fans and cabinet. In many cases, if the unit cannot be repaired in the field, it is advisable to use this type service. But, if a manufacturer has gone out of business, or if it is impossible to get a replacement mechanism, overhaul procedures described in the following paragraphs may be useful.

To repair the motor compressor:
1. Check the work order. It may indicate the fault.
2. Take the compressor apart. Store the parts carefully in clean trays.
3. Clean the parts with a nontoxic, nonflammable solvent. Various spirits and/or liquids may be used. Clean the parts a second time before reassembly.

The compressor repair may include several operations:
1. If valves have been leaking, inspect, check and if necessary, repair the valve plate and replace valve reeds.
2. If the compressor has been noisy, hone the cylinders and replace piston, repair piston pins and connecting rod, repair crank journal and main bearings.

The most common compressor troubles are with valves and valve seats. The valve reed may be replaced. The valve seat should be reground using a surface grinder or a drill press. Next, lap them accurately. Replacements are available for valve reeds on most hermetic compressors.

Practically all compressor valves used on domestic refrigerators are of the diaphragm or disk type. Usually these are called flapper valves. They are made of thin spring steel and are held in place by their own spring tension and a machine screw, or by an auxiliary coil or flat spring.

Compressor bodies and valve seats are usually made of cast iron. After long use, they may be worn. It is a good policy to replace the valve reed when overhauling a compressor. The cost is small. It is also good practice to lap the valve seats.

Be sure to use fine lapping compound and clean all lapped surfaces carefully afterward. Compound left on the surfaces may ruin the compressor.

If raised valve seats are used, a plate glass surface may be used as the lapping tool, or special lapping blocks may be used. Some companies use fine polishing paper clamped to a flat plate as a lapping surface.

If the seat is in bad condition, the valve plate should be mounted in a lathe — either in a chuck or on a face plate — and the whole plate trued. Some repair shops use a surface grinder to true valve seats.

In compressors using a spring steel type valve, the proper surface of the valve must come in contact with the valve seat. The proper surface may be detected by the slightly turned over edge of the opposite surface of the valve. This is caused when the valves are stamped out. The stamping process turns the edge of the valve on one side (burrs it). If this side is placed against the valve seat, it might not seal properly.

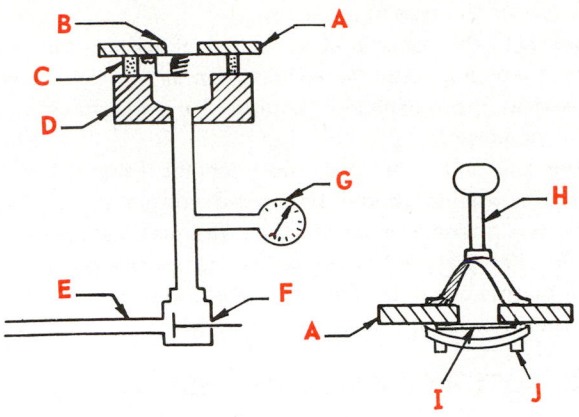

Fig. 11-97. Testing valves. A—Valve plates. B—Refrigerant oil. C—Synthetic rubber gasket. D—Fixture. E—Clean, dry compressed air. F—Air valve. G—Pressure gauge. H—Vacuum cup. I—Valve. J—Valve retainer.

Additional information on valve repairing will be found in Chapter 14. Replacement valve plates are available and their use is recommended.

Valves must be tested before assembling the unit. Use fixtures and synthetic gaskets or a synthetic rubber vacuum cup, as shown in Fig. 11-97. Noisy valves may be caused by too much valve lift (or excessive pressure differences). The valve movement is usually measured in thousandths of an inch. Valve lift must be accurately adjusted. Measure the lift with a dial indicator before dismantling the valve. Too little lift leads to poor compressor operation and overheating.

Poor vacuum indicates a poor intake valve but the compressor will hold the vacuum when it stops. This condition may also be due to worn pistons and rings or the use of a too-thick gasket. Keep in mind, also, that lack of oil will cause poor pumping ability.

A check valve can be examined in much the same way as an intake valve. It is important to keep these units as quiet as possible, so the bearing and bushing clearances must be as small as possible. Check all main bearings, connecting rod bearings, piston pin bushings and piston clearances. Use accurate micrometers and thickness gauges for checking.

The main body or parts of the compressor are usually made of cast iron. Valves are of spring steel and the piston pins of high carbon steel. Some repair shops use drill rod as replacement piston pins. Replacements for moving parts should be the same weight as the old parts. This will keep dynamic vibrations to a minimum.

End play must be kept to a minimum to eliminate end play slap. The pump should be checked for satisfactory operation.

11-70 REBUILDING A ROTARY MOTOR COMPRESSOR

Repairing the rotating vane type of rotary compressor requires some special attention:
1. Positioning of the roller housing to locate accurately the contact spot between the concentric roller and the eccentric housing is important. In Fig. 11-98 the contact point,

X, must be carefully adjusted. Four cap screws hold the housing, A, and an end plate up against the main housing. Point F is a dowel pin. The parts are assembled and the housing is tapped lightly until the rotor, B, binds when the shaft is turned by hand or by using the electric motor. Then the housing is relieved slightly by tapping opposite to X until the binding is released. Dial micrometers may be used to measure the distance.

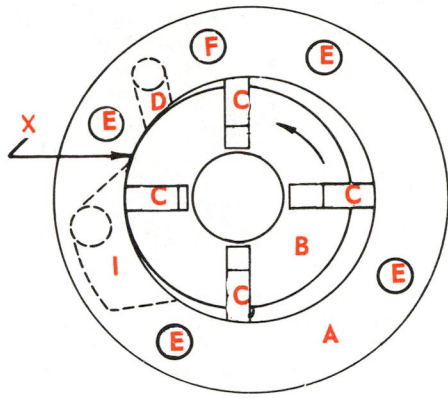

Fig. 11-98. Parts and adjustments on hermetic rotary compressor. A—Housing. B—Rotor. C—Blades. D—Exhaust port. E—Bolts. F—Dowel pin. I—Intake port. X-Contact point.

2. Vanes at C in Fig. 11-98 must be accurately fitted to the roller, B, and housing A. These vanes must accurately match the length of the roller, B, and the housing length, A. This is to prevent blow-by. Vanes must also fit the slots in the roller accurately. If too wide, they will bind as they pass point X. If too narrow, they will permit blow-by (leakage) as they pass point X.

The single or stationary vane type compressor has much the same fit problem as the rotating vane type. Parts must be accurately fitted together. The main bearing or bearings must be in excellent condition. The length of roller and vane must be exactly the same. The housing should be .0001 (one-ten thousandth in.). There must be no evidence of scoring.

Usually the vane is spring loaded to keep it riding on the roller. If this spring is too weak, vapor will bypass the vane and cause a loud clicking noise. If it is too strong, it will place an unnecessary load on the roller and rapid wear will occur. Dowel pins are used to align housing with shaft.

Both compressor types need enough oil to keep the parts in a constant oil bath. During repair, clean all oil passages. Clean or replace the oil metering screws. Use dry compressed air.

In compressors of the stationary blade type compressor, the cylinder is mounted snugly; then the shaft is turned and the cylinder is shifted until there is equal resistance over the complete revolution. Then the cylinder bolts are tightened according to torque specifications.

11-71 ASSEMBLING COMPRESSOR

Insuring cleanliness of internal parts of the mechanism is one of the most important steps in the assembly process. The

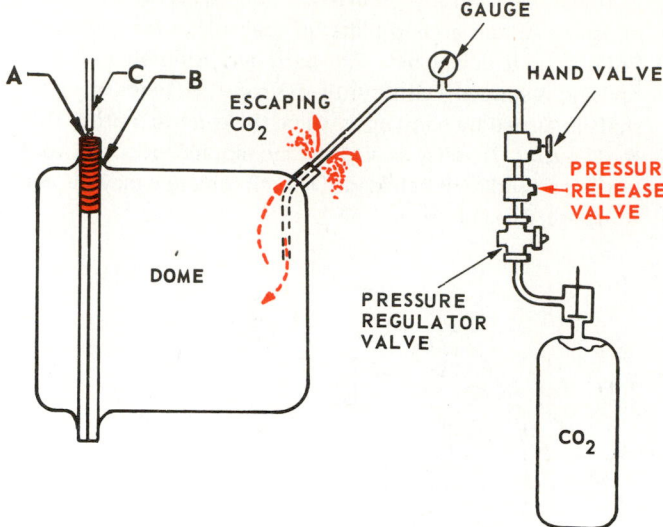

Fig. 11-99. Welding hermetic dome. A—Arc welded seam. B—Dome flange. C—Arc welding electrode. Always clean surfaces before welding. Note use of carbon dioxide or nitrogen. It acts as flushing gas to prevent collection of explosive mixtures and to reduce oxide scale inside dome.

inside of the compressor and the motor windings are usually hardest to clean.

One of the best procedures is to use an air gun and a mineral spirits spray. Immersing (dunking) the parts in a mineral spirits bath is also effective. Carbon tetrachloride should not be used. After this cleaning, iron and steel parts must be oil dipped immediately to prevent rusting.

It is essential to run the compressor after assembly and before welding the dome. It should be run for a few moments to check its pumping ability and its noise level. This can be done without danger of scoring the parts due to lack of oil.

A vacuum gauge equipped with a synthetic rubber tip can be held against the inlet opening of the unit while checking its pumping ability. A similarly equipped high-pressure gauge may also be used on certain models. Running the motor provides a check of the electrical work for opens, shorts, grounds and suitable power.

11-72 WELDING COMPRESSOR DOME

After the compressor has been checked and tested, the dome may be welded in place. Arc welding makes the best seal. Because this requires such a short time, heat does not travel to the interior of the dome, compressor and windings.

It is a good safety practice to bleed the dome with carbon dioxide or nitrogen during welding to prevent an explosive mixture of oil fumes and air from collecting. See Fig. 11-99. Use reverse polarity coated electrodes, tungsten arc or inert gas metal arc during arc welding to keep the motor as cool as possible. Before welding, it is important to clean the metal surfaces for at least one-half inch on each side of the weld area to prevent contamination of the weld with dirt. Dirt may form blow holes that may one day leak.

The welding should be done by an experienced operator. The welding station must be well ventilated. The arc must be

shielded from the eyes of passersby.

Tack weld the dome in at least three equally spaced places before proceeding with the welding. Do all the welding with the weld in the downhand position. Turn the dome as the welding progresses.

After the unit is welded, short lengths (8 to 10 in.) of proper size tubing should be brazed to the dome. Avoid getting flux inside the dome. Cool the metal as quickly as possible after the weld is completed. Use a damp cloth or a stream of cool air. After the welding is completed, the motor should again be tested for running characteristics.

11-73 MOTOR COMPRESSOR TESTING

After the dome has been welded and the tubing brazed to it, test the assembly for leaks. Install tubing adaptors on the suction and discharge lines. Seal the oil cooler lines. Then either charge with refrigerant vapor and test for leaks or use nitrogen gas and immerse the dome in water. Never use more than 150 psi!

After the compressor has been tested for leaks, it should be filled with the correct amount of the proper viscosity oil and then tested again before it is installed. Use the type and quantity of oil recommended by the manufacturer.

The following is one way to put oil in the compressor. Connect a charging line to the process tube, valve adaptor or suction line. Connect this tubing to a gauge manifold and to both a vacuum pump and an oil container. Run the vacuum pump to create a vacuum on the motor compressor with the tubing immersed in fresh refrigerant oil. The oil will flow into the crankcase. When the unit is running, oil will start spraying through the discharge service valve once the compressor has enough oil.

Next, connect a gauge manifold to the compressor. Use a high pressure gauge calibrated from 0 to 160 psi. Use a compound gauge calibrated to 30 in. and 150 psi. If the compressor tests all right, (18″ Hg. to 28″ Hg. of vacuum against a 150 psi head pressure and holds the vacuum), it should be baked for a period of eight hours or more at a temperature of 150 F. to 200 F. with at least a 20″ Hg. vacuum on the system.

It can also be evacuated for several hours at 50 to 500 microns of pressure or lower. This evacuating will remove almost all of the air and moisture from the system.

Test benches are available which may be used to test newly repaired hermetic motor compressors. Fig. 11-100 shows an electrically connected test motor analyzer bench with flexible motor and discharge lines connected to the motor compressor. Inside construction of the unit is shown in Fig. 11-101.

After evacuating, the compressor should be charged with nitrogen or with the correct refrigerant vapor to 15 psi.

The suction line, discharge line, process tube and oil cooler lines should then be pinched and silver brazed if the motor compressor is to be shipped or stored.

11-74 ASSEMBLING THE HERMETIC SYSTEM

After the motor compressor has been repaired and tested, it may be installed in the system. The procedure follows:

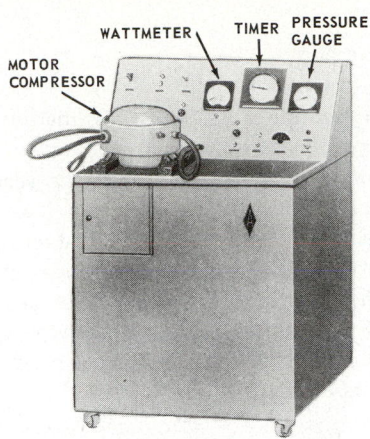

Fig. 11-100. A motor compressor analyzer stand which checks motor wattage, pressures and time to pump certain volume of dry air up to certain pressure (volumetric efficiency).

1. Clean and test the rest of the system. Add the suction line filter-drier and liquid line filter-drier.
2. Clean the capillary tube, evaporator, suction line, oil cooler tubing and hot gas defrost lines. This is done by forcing R-11 through the system until it comes out of the system clear. The filter-drier and capillary tube may have to be removed to permit enough circulation. See Fig. 11-102.

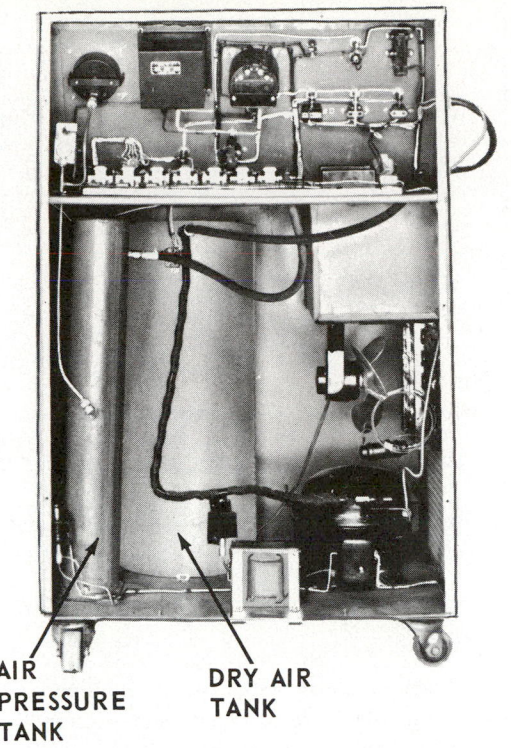

AIR PRESSURE TANK **DRY AIR TANK**

Fig. 11-101. Inside construction of motor compressor analyzer.

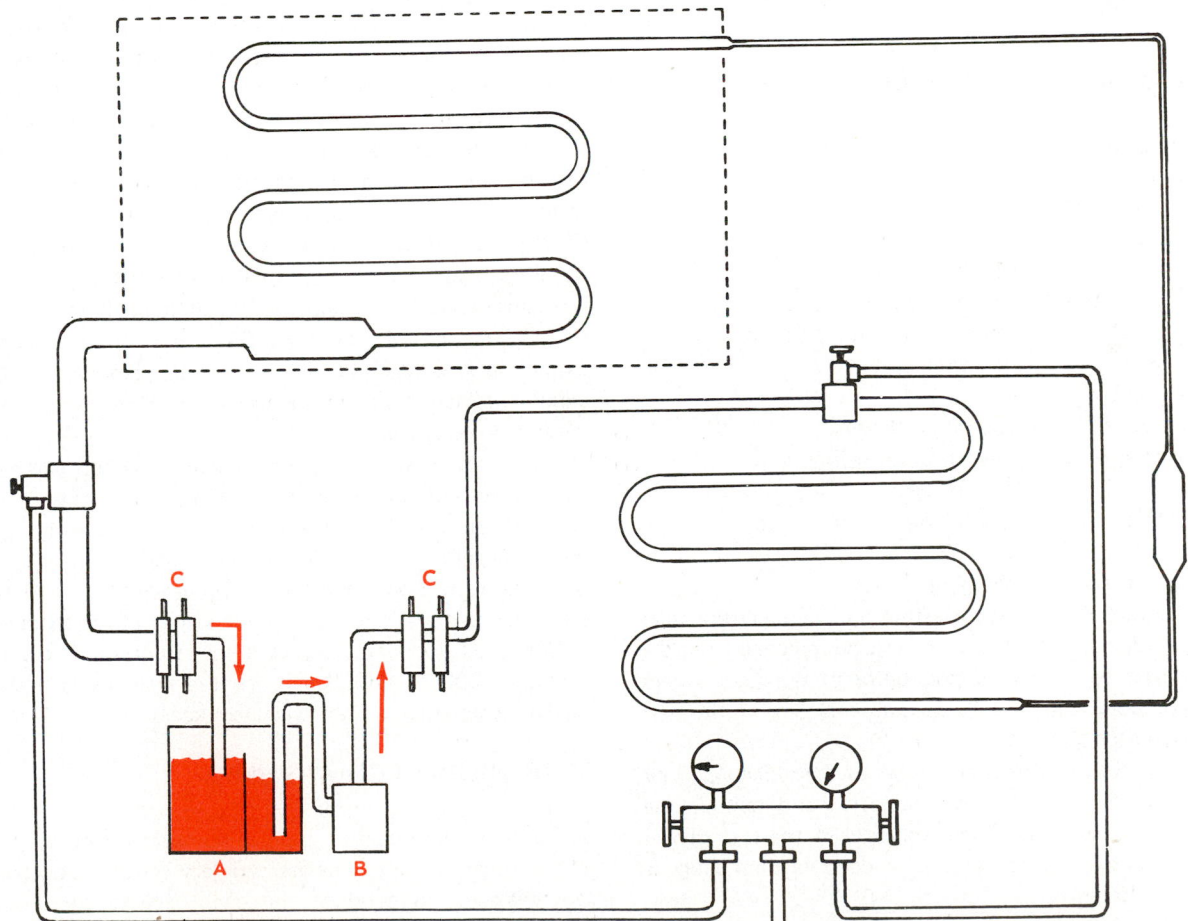

Fig. 11-102. Cleaning out system before installing motor compressor. A—R-11 tank. B—Pump. C—Tubing adaptors.

Installing and Servicing Small Hermetic Systems / 383

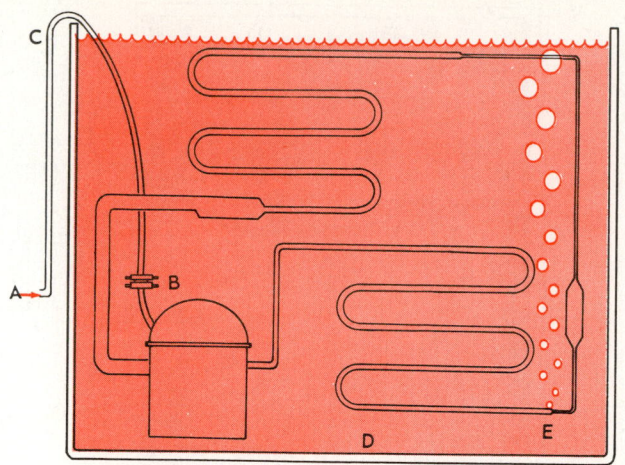

Fig. 11-103. Testing a rebuilt hermetic system for leaks. A—150 to 180 psi nitrogen or air. B—Process tube adaptor. C—Pressure line. D—Water tank. E—Leak indicated by bubbles.

3. When the system is clean, install the suction line filter-drier, the liquid line filter-drier and connect suction line, discharge lines and the oil cooler lines. Silver braze the joints.

4. The assembled system must be given a thorough leak test. The system can be pressurized to about 150 psi with nitrogen. The complete system can then be immersed in a water tank. See Fig. 11-103.

11-75 EVACUATING A SYSTEM

A system opened for any type of repair must be completely evacuated to remove air and moisture.

Two different methods are used.

1. Deep vacuum
2. Triple evacuation method.

In the first method, a vacuum of 500 microns (.5 millimetre) or deeper is pumped on the system until no moisture or other gas remains in the system.

In the second method:

1. A vacuum of 28" Hg. or 50 mm is drawn.
2. System is charged to 0 psi with vapor refrigerant.
3. A vacuum of 28" Hg. is drawn again.
4. System is charged to 0 psi with vapor refrigerant.
5. A vacuum of 28" Hg. is drawn.
6. The system is ready for charging.

R-22 has a higher discharge pressure and temperature than R-12. Dirt, sludge, moisture and air must be removed from the system in order to avoid any possibility of burnouts where R-22 is used. Complete (high) evacuation is one of the best means to clean a system.

Triple evacuation will not remove all of the moisture. Only warmth and a deep vacuum will vaporize and remove the moisture and any solvents. Some solvents in the system will vaporize only under a deep vacuum. A deep vacuum gauge is the best way to determine:

1. If system has water in it.
2. If system has leaks.

To check:

1. Stop vacuum pump.
2. Close valve to vacuum pump.
3. Watch high vacuum gauge. If it rises, there is still moisture in the system.

Large openings must be used for deep vacuum. Piercing valve openings are not large enough.

The system should be pressure tested first. Use either dry nitrogen or dry carbon dioxide. These gases should be used only with a pressure regulator and a large capacity pressure relief valve set to release at 175 psi. Test the system at 150 psi. The system should hold this pressure after gas valve is closed for several hours (no decrease in pressure).

Never heat nitrogen or CO_2 cylinders. The maximum cylinder temperature should be 110 F. (43 C.). A test set up for a triple evacuation is shown in Fig. 11-104.

The evacuating and drying of the unit is a very important part of the assembly work. The system should be as close to 100 percent clear of air, moisture, solvents and other foreign matter as possible.

Remember that the most careful evacuating and purging will not clean a unit that was carelessly put together with dirt in the system.

A vacuum pump should be used to remove as much air as possible from the unit. No pump will remove all air. A pump that produces a 28 in. vacuum (28" Hg.) will take out only about 93 percent of the air. With a 28" Hg. vacuum moisture particles must be heated to 100 F. (38 C.) or above, before they will evaporate and can be pumped out as water vapor.

To remove the moisture, the unit must be heated to a temperature that will not only vaporize the moisture, but will drive the moisture out of all the crevices. For the same reason, the unit should be run for part or all of this operation. This is to be sure all pockets in the compressor and in the bearings are vibrated to release pocketed air. This also warms the motor windings which are an additional source of trapped moisture. It is considered good practice to evacuate for 8 hours at 250 F. (121 C.) or for 24 hours at 150 F. (66 C.). A drying oven, temperature controlled, is of utmost importance. Carbon dioxide attracts moisture and circulating it in the system helps to remove moisture.

To eliminate still more air, charge some refrigerant vapor into the system. Evacuate the system again. This will take out more of the air. Repeat the charging and evacuating. Only about .01 percent of the air will then remain in the system.

However, if a deep vacuum pump draws 50 to 100 microns and holds this pressure, the system is clear of moisture and air.

If a good vacuum pump is used and draws 1 to 2 mm of vacuum (1000 to 2000 microns), the system will need to be partially charged and then evacuated again.

11-76 HIGH-VACUUM PUMPS

Standard reciprocating type air compressors do not create a high enough vacuum; nor are ordinary refrigeration compressors designed to produce the deep vacuum for evacuating refrigeration systems being serviced.

There are two main types of vacuum pumps:

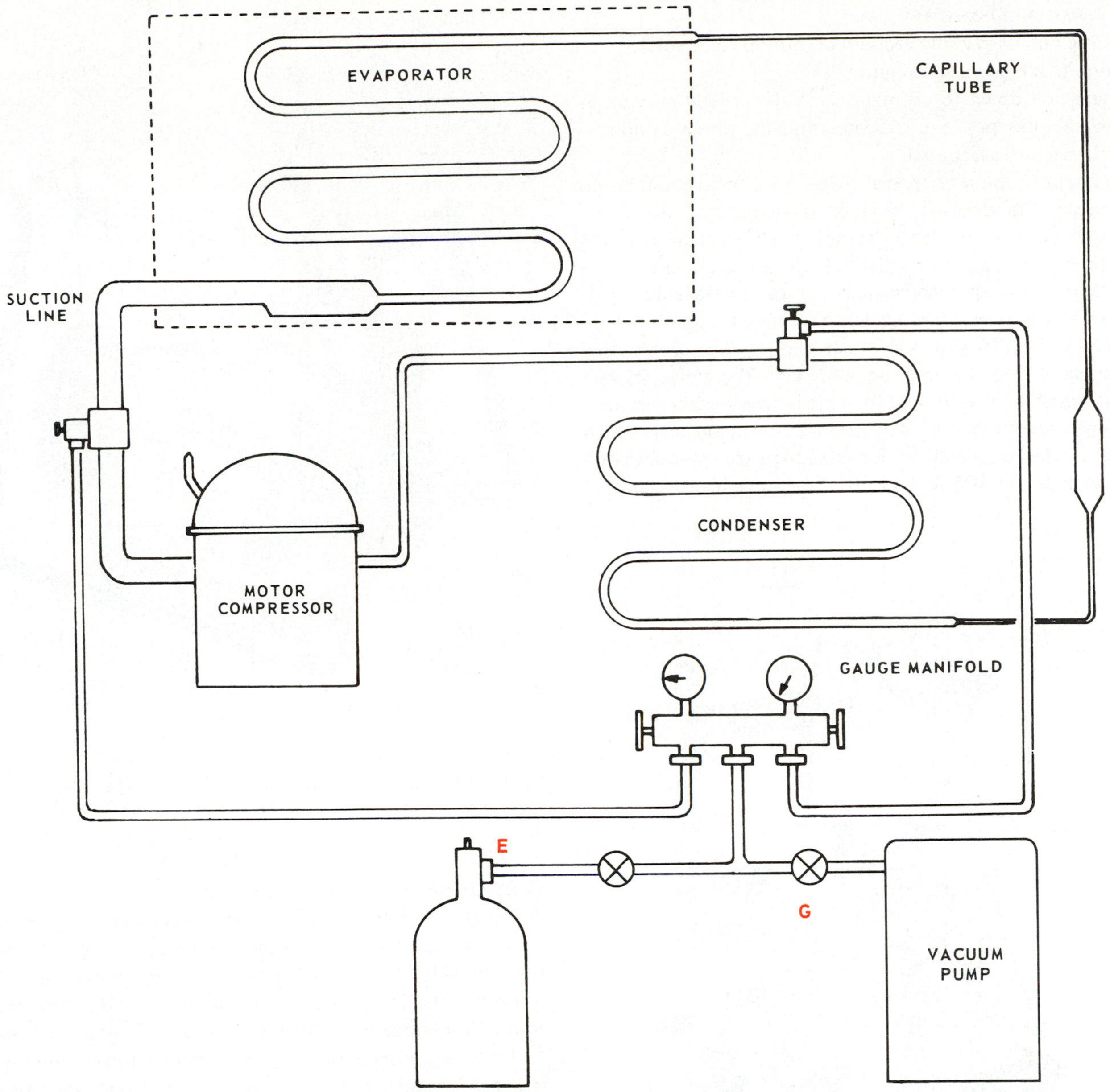

Fig. 11-104. A system with gauge manifold, refrigerant cylinder, and vacuum pump ready for triple evacuation process. 1—Pressure with E. 2—Evacuate with G. Repeat three times.

1. Single-stage.
2. Two-stage.

The single-stage vacuum pumps are used when the triple evacuation method is used. Two-stage vacuum pumps are used when the high-vacuum (deep vacuum) method is used.

There are two designs in high-vacuum pumps:

1. Rotary pump (oil sealed).
2. Vapor pump (diffusion type).

The rotary pump should be able to pull a 50 micron vacuum pressure. (A 50 micron vacuum is equal to .05 torr. A torr equals 1 millimetre.)

The better rotary pumps use two rotors in series (compound pump). See Fig. 11-105. Most refrigeration compressors can pull about 50 to 80 torr (50,000 — 80,000 microns).

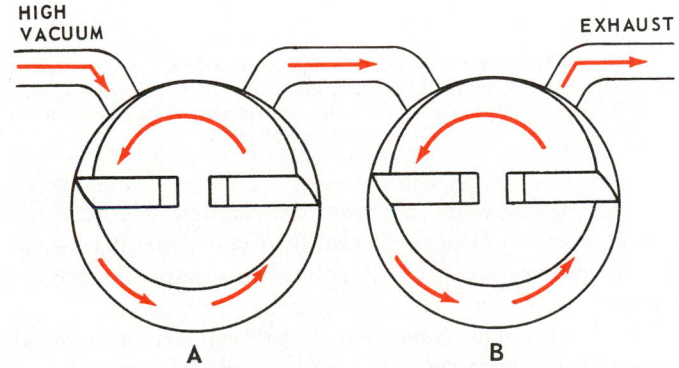

Fig. 11-105. A two-stage rotary high-vacuum pump. A—First stage. B—Second stage.

To evacuate, proceed as follows:

1. Pull vacuum with a low-vacuum pump (50 to 80 torr).
2. Switch to a high-vacuum pump.
3. Repeat pull down to 50 microns. A rise to 300 microns in three minutes or more indicates that the refrigeration system is dry and evacuated.

Use copper tubing or special metal hose for vacuum pump connections. Standard charging or servicing hose (synthetic materials) tend to collapse at this high vacuum and the synthetic material tubing is volatile (will release gases).

Portable two-stage high-vacuum pumps are available which will draw down to less than 1 micron of mercury column.

A micron is 1/1000 or a millimetre (mm), 2.54 centimetres (cm) equal 1 in.; 10 mm equal 1 cm. Therefore, 25,400 microns equal 1 in. One micron is close to a perfect vacuum.

A high-vacuum pump will produce a vacuum lower than 29″ Hg. (inches of mercury). Portable high-vacuum pumps are shown in Figs. 11-106 and 11-107. Vacuums lower than 29″

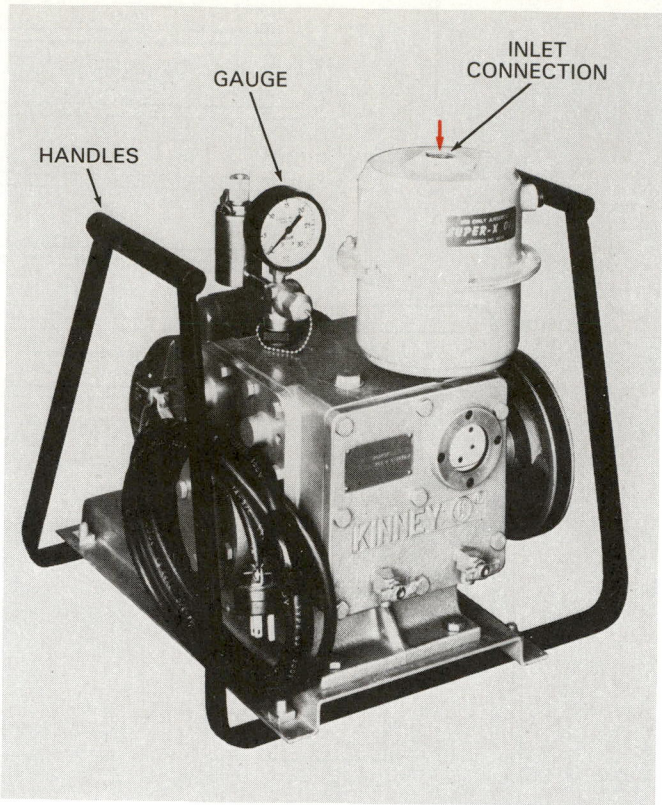

Fig. 11-107. Another portable high-vacuum pump. It may be used to remove air or refrigerant from system. (Airserco Mfg. Co.)

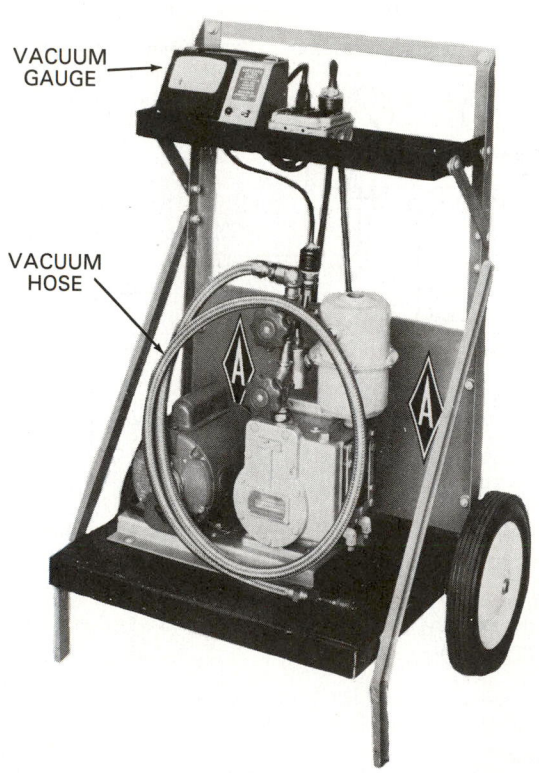

Fig. 11-106. Portable high-vacuum pump mounted on stand for transport. Note flexible metal evacuating line. This is a large capacity line and will not collapse. Two hand valves control evacuating operation. (Airserco Mfg. Co.)

Hg. are necessary for complete dehydration (moisture removal) from the system. The oil sight port permits checking the oil level, as well as the oil color. This special oil should be replaced frequently.

Oil in single-stage pumps rapidly becomes dirty when there is water and solvent vapor in the system. Water in the oil will:
1. Raise the oil level.
2. Turns the oil white and foamy. If this dirty oil is left in the pump, sludge will form in the system.

For good results, change oil before each system pump-down or test pump with valves closed and vacuum gauge connected. If pump will not pull down to high vacuum, change oil.

Compound (two-stage rotary) pumps need oil changes after about 10 pump-downs. They will pull 20 microns or better.

To measure deep vacuum, an electronic high-vacuum gauge is used. A regular compound gauge cannot read accurately to micron levels. A high-vacuum gauge is shown in Fig. 11-108. A

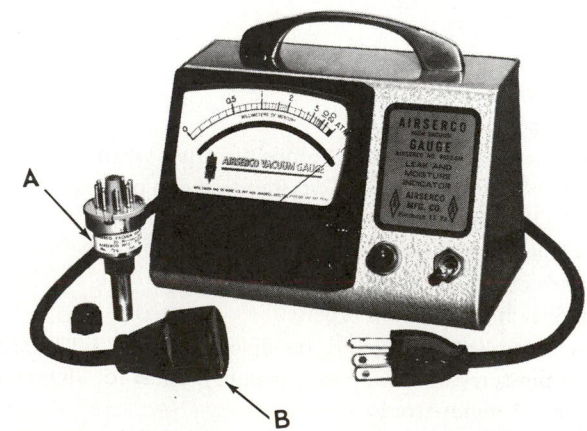

Fig. 11-108. High-vacuum gauge. A—Gauge tube with 1/8-in. pipe threads. B—Gauge tube electrical connector. In operation, gauge tube is connected into system to be checked by using pipe threaded extension.

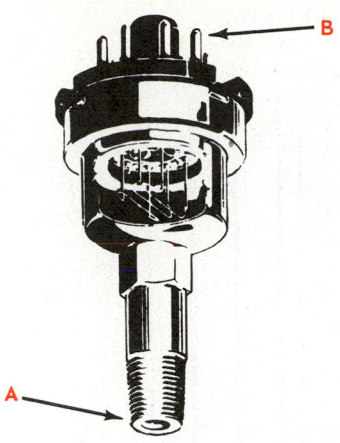

Fig. 11-109. Internal construction of a high-vacuum gauge tube. A—Pipe thread connection to system manifold. B—Electrical connections to vacuum gauge.

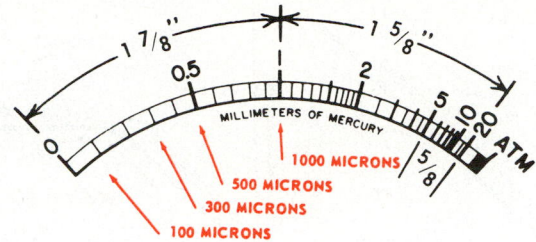

Fig. 11-111. Dial scale of high-vacuum gauge. Note that pressure between 0 and 1000 microns has been greatly expanded for easy reading. (Airserco Mfg. Co.)

vacuum from 29.25″ Hg. (about 18,000 microns) to 29.9″ Hg. (1000 microns) is necessary to allow the moisture inside the system to evaporate at room temperature.

The inside design of a vacuum gauge tube is seen in Fig. 11-109. The sensing element is a thermocouple. It is important to use the tube in an upright position to keep out foreign matter.

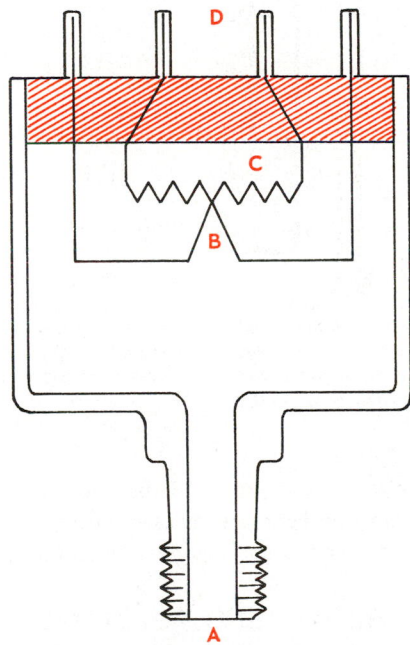

Fig. 11-110. The schematic internal design of the high-vacuum gauge. A—Connection to system. B—Thermocouple. C—Resistance unit. D-Electrical leads.

Gauge's electrical circuitry is shown in Fig. 11-110. The filament gets warmer as air pressure around it decreases. (There are fewer air molecules to remove heat.) The thermocouple gets warmer and the increase in emf is registered on the meter. The meter is calibrated in microns. The vacuum dial scale is shown in Fig. 11-111.

If the vacuum gauge reading levels off at 5000 microns, there is ice or free water in the system. Ice may be located by a cold spot, frost or sweat on outside of system. Stop the pump; allow the ice to melt or use radiant heat.

Never allow the system pressure to enter a high-vacuum gauge.

The thermocouple vacuum gauge has two parts:
1. A meter.
2. A tube which threads into the refrigeration system. The tube should be right side up with threads down.

The tube sometimes collects vapors and/or oil. It can be cleaned by putting a solvent in the opening with an eyedropper. Freon 23 is a good cleaner. Clean as follows:
1. Fill.
2. Rock tube gently.
3. Empty.
4. Repeat the above two or three times.
5. Clean with an alcohol rinse.

Clean the instrument dial covers with soap, water and facial tissues. This crystal is usually plastic and solvents will fog it.

If a system has a leak, the vacuum produced will not be a high vacuum. The dial needle will rise steadily when the valves are closed. If there is moisture in the system, the vacuum produced will also be less than desired. When the valve is closed, the dial needle will rise and level off at a pressure corresponding to the water vapor pressure at that temperature. Fig. 11-112 shows the two conditions of a pressure-time graph. These are made with the valve to the vacuum pump closed.

To make sure the dehydration (drying out) of a wet system is complete, this procedure is recommended for domestic units:
1. Connect a 250W lamp (in place of a fuse) in series with a test cord.
2. Attach the cord to the compressor run and common terminals.
3. Connect to a 120V outlet.

This procedure allows only about 30V to go through the running winding because the lamp becomes a resistor. This warms the compressor windings vaporizing the moisture which may be trapped in the system. *The compressor should not run during this operation.*

If a manifold is used, it must be degassed by pulling a vacuum on it for several days to remove all gases absorbed in the walls and cracks of the manifold. These residual gases may make a service technician believe there is a leak in the system. Evacuate the system from both the high side and low side.

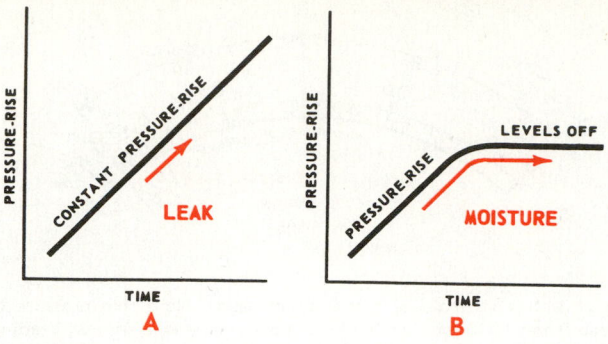

Fig. 11-112. Leak and moisture affect high-vacuum gauge readings. A—Shows effect of a leak. B—Shows effect of moisture in system.

Use as large a diameter in the vacuum line as possible. Keep the line as short as possible. Vacuum pump sizes are given in cubic feet per minute (cfm):

1. 1.5 cfm — good for domestic systems (3-5 tons).
2. 3-5 cfm — medium systems (5-100 tons).
3. 10-15 cfm — large systems (over 100 tons).

Always break the vacuum (allow air in) of a vacuum pump before storing it. Otherwise the cylinder will fill with oil and become oil locked (hard to turn).

A cold trap can be used to keep moisture out of the vacuum pump. This trap is placed in the tubing between the system and the vacuum pump. As moisture vapor contacts the cold surface, it will collect as ice. By closing the valves on each side and unbolting the trap, one can remove the ice. See Fig. 11-113.

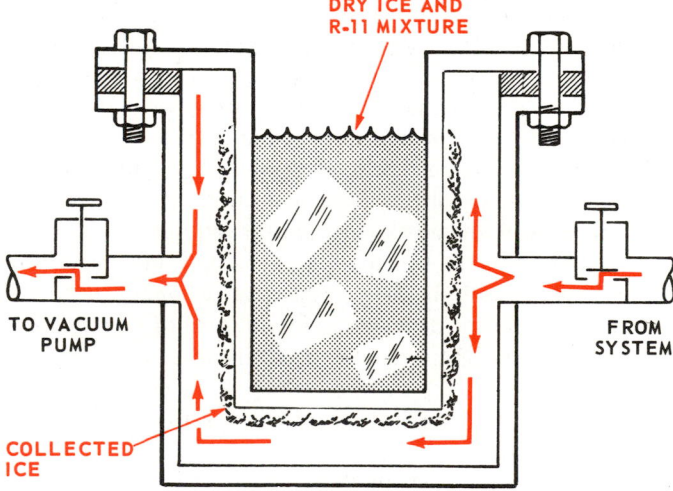

Fig. 11-113. A cold trap used to collect moisture from a system being evacuated. The moisture vapor condenses on the cold surface and then turns into ice.

Fig. 11-114 is a chart of the various pressures that are possible with different type pumps. Note the evaporating temperature of water at the various pressures.

During manufacture, systems are evacuated 50 to 100 microns before oil is added. In the field, when oil is already in the unit, evacuating to 50 microns may cause some of the oil to vaporize.

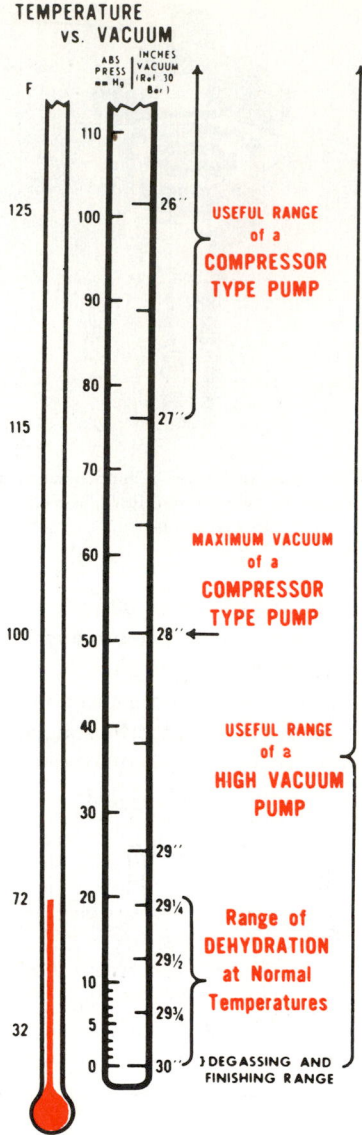

Fig. 11-114. Pressure scales used during high-vacuum evacuation of system. Temperature scale shows evaporating temperature of water at these pressures. For example, water will evaporate (boil) at 72 F. at 20 mm Hg. pressure (20,000 microns). (Airserco Mfg. Co.)

Because the motor compressor depends on gas flow to cool the motor windings and compressor, avoid using it as a vacuum pump. The motor compressor may heat up and be damaged.

11-77 CHARGING REBUILT SYSTEMS

After the system has been evacuated, it is carefully charged with clean, dry oil and refrigerant.

The refrigerant charging tube, the oil cylinder, the connecting lines and the gauge manifold must all be as clean and as free of moisture as the system being charged.

The correct amounts of oil and refrigerant should be checked from the manufacturer's specifications. (The amount of oil was measured when the old system was dismantled.)

A charging cylinder may be used for charging refrigerant or oil into a system. One is shown in Fig. 11-115.

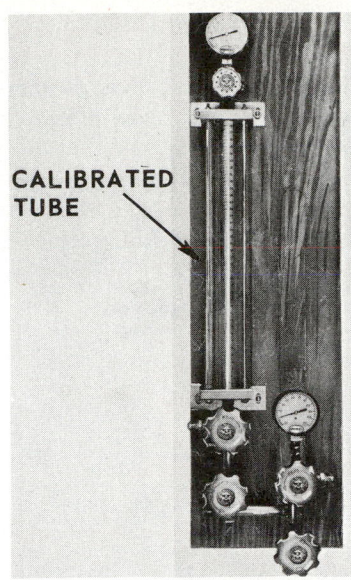

Fig. 11-115. A wall-mounted charging cylinder tube. It is charged by being connected to a large cylinder. It measures refrigerant charged into system using an ounce scale.

11-78 TESTING REBUILT SYSTEMS

After the unit has been assembled, tested, evacuated and charged with oil and refrigerant, it should be run with a thermostatic control for at least 24 hours to determine its behavior. A recording thermometer should be placed in the refrigerator during this test period to enable the technician to know what is happening during the entire test period. This very important part of repair work is often neglected. Fig. 11-116 shows a 24-hour temperature recorder. Recorders with 48-hour and 72-hour chart speeds are also available.

The sensing bulbs are put in the refrigerator and the recorder is connected to a wall plug. The one-day chart shows both the fresh food and the freezer cabinet temperatures at all times. If possible, the cabinet should be placed in a warm room (100 F., 38 C.) for the test. The testing room should be quiet so the unit can be checked for noise.

A 30-hour portable temperature recorder is shown in Fig. 11-117. A self-starting electric clock may be used to record the running time of the unit.

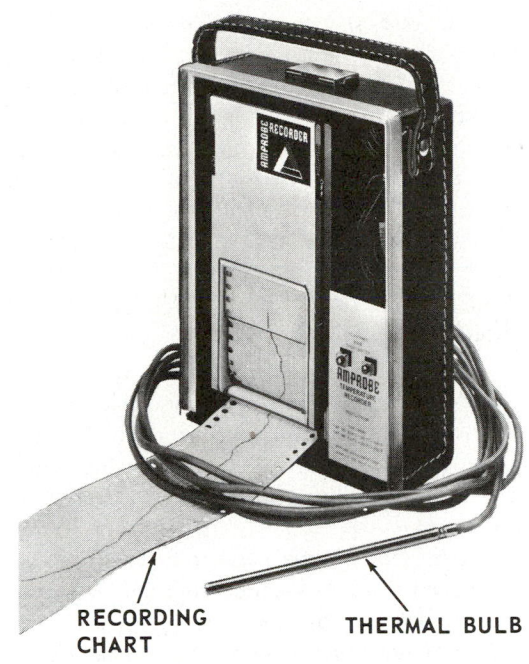

Fig. 11-117. Portable temperature recorder for up to 30 hours recording. The thermal bulb may be placed in the cabinet or clamped to the evaporator. (Amprobe Instrument)

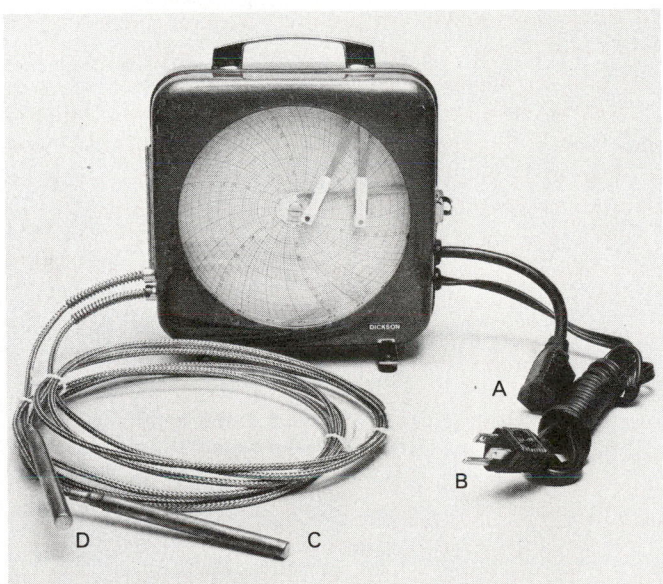

Fig. 11-116. Temperature recorder charts temperature changes of refrigerator-freezer unit over 24-hour period. A—Series connector. B—Power connection for chart motor. C—Sensor for fresh food compartment. D—Sensor for freezer compartment. (The Dickson Co.)

Fig. 11-118. Combination hermetic compressor analyzer and electrical tester. A—Dial has four scales: ohms, volts, microfarads and amperes. (Airserco Mfg. Co.)

The testing should also include checking the temperature of the dome, the lines and the even frosting of the evaporator.

Use an ammeter and voltmeter or wattmeter to check the electrical section of the unit. Many service technicians use a combination hermetic compressor analyzer and electrical tester. There are numerous types available. Fig. 11-118 pictures one type of heavy-duty analyzer. It serves as a hermetic motor analyzer, three-phase tester, motor resistance tester, potential relay tester and insulator. For additional information on the use of electrical instrumentation, see Chapter 8.

One way to test a system for efficiency is to cover the evaporator with insulation after installing a thermometer (thermocouple type) on one of the last coils of the evaporator. Run the system.

At the end of 30 minutes, the evaporator temperature should read as low as −25 F. (−32 C.) if the room air is a normal 70 F. to 75 F. (21 C. to 24 C.) or as low as −18 F. (−28 C.) if the room is 90 F. to 100 F. (32 C. to 38 C.).

11-79 CYCLING TIME FOR REFRIGERATORS AND FREEZERS

Cycling time on home refrigerators and freezers cannot be stated very well in definite limits of time. This will vary depending on the amount of storage space being used, on the outside temperature of the box and on the condition of the compressor.

Placing warm food in the cabinet to be frozen will also affect the cycling time. The condensing unit may run about one-third of the time. In other words, it may run five minutes and be off 10 minutes or it may run an hour and be off two hours. The important point is that any unusual change in cycling time should be investigated immediately. It may indicate that trouble is developing in the system.

11-80 SHUTTING DOWN REFRIGERATORS AND FREEZERS

When shutting down a refrigerator-freezer, special precautions should be taken to prevent rusting and to remove odor. After the electrical plug has been removed or the current shut off by a switch, allow several hours for the unit to completely defrost.

When the defrosting is complete, remove water and wash the inside of the cabinet with a solution of baking soda and water. Thoroughly dry the inside of the box. A portable heater set inside the cabinet will speed up this step. Leave the doors or lids ajar slightly to allow circulation of air during the shut down period.

CAUTION: There is a federal law in effect which states that the cabinet door must be removed from a refrigerator or freezer that is being taken out of service or is being disposed of for any reason. This law was passed because many children have been suffocated when attempting to hide in or play in an unused or carelessly discarded refrigerator or freezer. Therefore, always remove the cabinet door when taking a refrigerator or freezer out of service.

11-81 SERVICING PRECAUTIONS

Pinching lines is a practice to be used in cases of emergency only. Some service technicians follow this practice needlessly and it leads to future trouble.

One must remember that when refrigerant is added to a system, some of the oil in the system will be dissolved in the refrigerant. If the unit becomes noisy soon after the refrigerant has been added, oil should be added.

Electric motors, when installed, should have clean and tight electrical connections.

11-82 REVIEW OF SAFETY

Working safely implies three things:
1. Safety to the service technician.
2. Safe handling of tools, instruments and equipment.
3. Proper preservation of food so that it is kept in a safe condition.

So far as the technician is concerned, there are not many great hazards in the refrigeration service work. The following items are some of the more common hazards which repair or service personnel should always recognize.

Good housekeeping is very important. Keep work area clean. Keep oil and water off the floor.

Since the refrigerator is usually motor driven, electrical supply to the motor and the controls presents some hazard. If not properly insulated and handled, it is possible for the operator to receive a dangerous electrical shock.

Always disconnect the electrical circuit or make sure all electrical devices are safe before starting on a job. An electrical short across a ring or wrist watch can cause a severe burn. It is best to remove rings and wrist watches when working on electrical equipment.

Many electrical shocks occur when the service technician comes in contact with an electrical current and a ground. Avoid working on any electrical circuit if standing on a damp floor or if one hand is touching a water pipe.

In some refrigerating systems, the condensing pressures run quite high. If the condenser is not losing enough radiating heat or if there is a restriction because of dust or lint, head pressures becomes high. These can become dangerous if they reach the bursting point. The service technician should watch the gauges and shut off the system if the pressure is too high.

A hot compressor, exhaust line or condenser can burn the operator.

Freezing of skin is a hazard when handling some refrigerants. If the liquid refrigerant spills on skin, rapid evaporation may lower the skin temperature to considerably below the freezing temperature. Liquid refrigerant on the face or eyes is very critical. *Always wear goggles or a head shield when handling liquid refrigerants.*

Dropping heavy objects on feet or toes is another potential hazard. This can be avoided by using proper trucking and hoisting equipment. The operator should wear safety shoes with metal tips to protect toes.

Back injuries may be caused by attempting to lift heavy objects and by not using arm and leg muscles correctly.

Fire is always a potential hazard. Never use gasoline or any other flammable material when cleaning.

Use face shield and rubber gloves when handling oil which may be acidic. (This may be the case in hermetic systems where motor compressor has burned out.) *If burned with acidic oil, wash with water, apply ice, and see a physician right away.*

There is some danger in using common tools such as screwdrivers and wrenches. If they should slip, knuckles may be skinned. Be careful to handle tools correctly.

Vacuum pumps expose a hazard in that the rotating shaft, fans, belts and the like, may catch on the clothing; or, if a hand is drawn into a pulley, a severe injury may result.

General safety precautions include the following:

1. Wear goggles when working on a charged system. When using a flame, be sure there are no flammables nearby. Be sure the pressure in the system is closed to atmospheric pressure before opening a system.
2. Each time an operation is planned, train yourself to think, "Is there a possible safety hazard in the operation?" Then take the care needed to reduce the hazard.
3. Never, under any circumstances, use carbon tetrachloride for cleaning. Its use is illegal. Remember, its effects are cumulative in the human body and its continued use may be fatal.
4. Avoid testing with excessive pressures. Test systems at 150 psi maximum and wear goggles when testing.

11-83 TEST YOUR KNOWLEDGE

1. Do all hermetic systems have service valves?
2. Which refrigerants are in most common use in hermetic systems being built today?
3. What causes tubing to create noise?
4. How may a motor compressor dome be opened?
5. Describe a good way to clean the outside of the mechanism.
6. How should the hermetic mechanism be prepared for moving?
7. What is the most popular method of sealing tubing joints when servicing hermetic systems?
8. What trouble is indicated by a hot liquid line?
9. Give several reasons why a gauge installation must be tested for leaks.
10. What trouble results if a 250 psi head pressure exists in an air-cooled R-12 system?
11. A frosted suction line with an excessive low-side pressure is the indication of what trouble?
12. To what part of the system is the compound gauge usually connected?
13. To what part of the system is the high-pressure gauge usually connected?
14. Why must the pressures be balanced before a system is opened?
15. What should the average low-side pressure be for R-12, operating at 15 F. (—9 C.) evaporator temperature?
16. What should the average high side pressure be for R-12 operating in a room temperature of 85 F. (29 C.)?
17. What indicates the presence of air in a condenser?
18. Describe one method of adding oil to a system.
19. What is a micron?
20. How many microns are in a millimetre?
21. How does a process tube adaptor fasten to a process tube?
22. At what temperature does water evaporate at a 29" Hg. vacuum?
23. What should be used to trap the clogging material when a capillary tube is unplugged?
24. How does a motor compressor analyzer test for volumetric efficiency?
25. Why is the oil in a burned out motor compressor dangerous for a service technician to handle?
26. What happens to a system if a capillary tube of the same ID, but longer, is installed?
27. What type pump is needed to produce a high vacuum?
28. How can a stuck hermetic compressor be started?
29. Where are piercing valves installed?
30. How can one find out if a defrost heating element is burned out?
31. How are the service lines and manifold cleaned of air and moisture?
32. When should one install a new liquid line filter-drier?
33. If the low-side pressure reads 28" Hg. vacuum, what is wrong?
34. What is the most sensitive leak detector?
35. If the suction line is frosted, what could the trouble be?
36. How can one find out if the capillary tube refrigerant control is clogged?
37. What happens if the hot gas valve is stuck in closed position?
38. Is it necessary to check the service lines and gauge manifold for leaks?
39. Why is the valve core removed when evacuating a system?
40. Is the refrigerant charged into the system in the liquid form or gas form?
41. What are the three steps of the three-step evacuation method?
42. What is done to a silver brazed joint before it is heated and pulled apart?
43. When is a suction line filter-drier installed?
44. What kind of oil is usually found in a motor compressor burnout?
45. What kind of a gauge is used to read a 25 micron pressure?

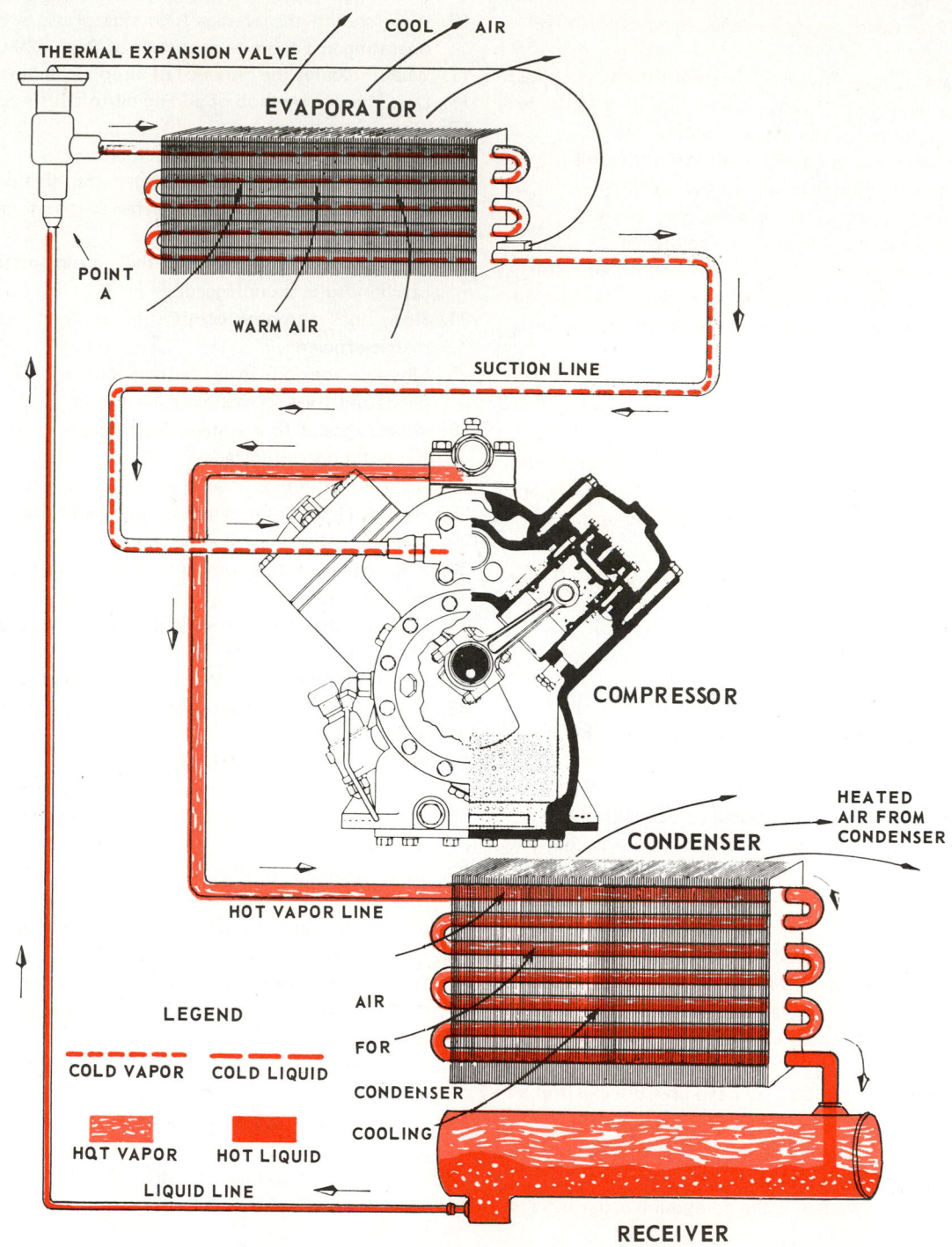

THERMAL EXPANSION VALVE

COOL AIR

EVAPORATOR

POINT A

WARM AIR

SUCTION LINE

COMPRESSOR

HEATED AIR FROM CONDENSER

CONDENSER

HOT VAPOR LINE

AIR FOR CONDENSER COOLING

LEGEND

COLD VAPOR   COLD LIQUID

HOT VAPOR   HOT LIQUID

LIQUID LINE

RECEIVER

Fig. 12-1. A serviceable commercial system with air-cooled condenser, thermostatic expansion valve and V-type compressor. (Carrier Air Conditioning Group, United Technologies Corp.)

Chapter 12

COMMERCIAL SYSTEMS

This chapter will explain the design, construction and operation of commercial systems. There are many different sizes and designs of commercial refrigerators. Some systems as small as 1/8 hp are used for beverage coolers, vending machines and dehumidifiers. Systems of 7 1/2 to 15 hp and more are used for frozen food walk-in cabinets, industrial cooling and the like. Fig. 12-1 illustrates a typical commercial refrigeration system.

12-1 CONSTRUCTION OF REFRIGERATING MECHANISMS

Operating fundamentals of domestic refrigeration systems also apply to commercial systems. But many commercial systems using mechanical cycle mechanisms differ in some way from the domestic mechanism. These differences are chiefly in the following:

1. In the number of evaporators connected to one condensing unit.
2. Compressor design and size.
3. Condenser unit design and size.
4. Motor controls — both temperature and pressure.
5. Refrigerant controls — both liquid and vapor.
6. Piping.
7. Variety of evaporator designs.

8. Defrosting systems.
9. Variety of refrigerants used.

12-2 MECHANICAL CYCLE

Some large commercial compressor systems are of the semihermetic design. These, too, can be completely serviced. Many single unit applications, such as bottle coolers, beverage dispensers and ice cream cabinets use a completely hermetic system. Parts of the system are similar to the designs shown in Chapter 10. The design of the cabinet varies with its application.

In some cases, the cycles are more complicated. This is because they may include unloading and automatic defrost devices, multiple evaporators, additional controls and more complicated piping.

12-3 COMPLETE MECHANICAL MECHANISM

Fig. 12-2 shows a single unit mechanism. It includes:
1. High-pressure side.
   a. Compressor — usually hermetic.
   b. Condenser — usually air-cooled.
   c. Liquid receiver — when a thermostatic expansion valve or automatic expansion valve is used.

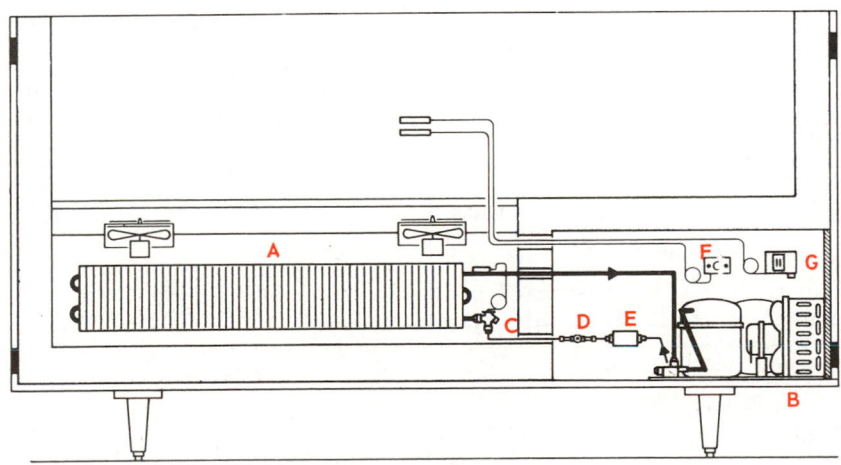

Fig. 12-2. An open top display cabinet with a self-contained refrigeration unit. A—Evaporator. B—Condensing unit. C—Thermostatic expansion valve. D—Sight glass. E—Drier. F—Signal thermostat. G—Cycling thermostat. (Danfoss, Inc.)

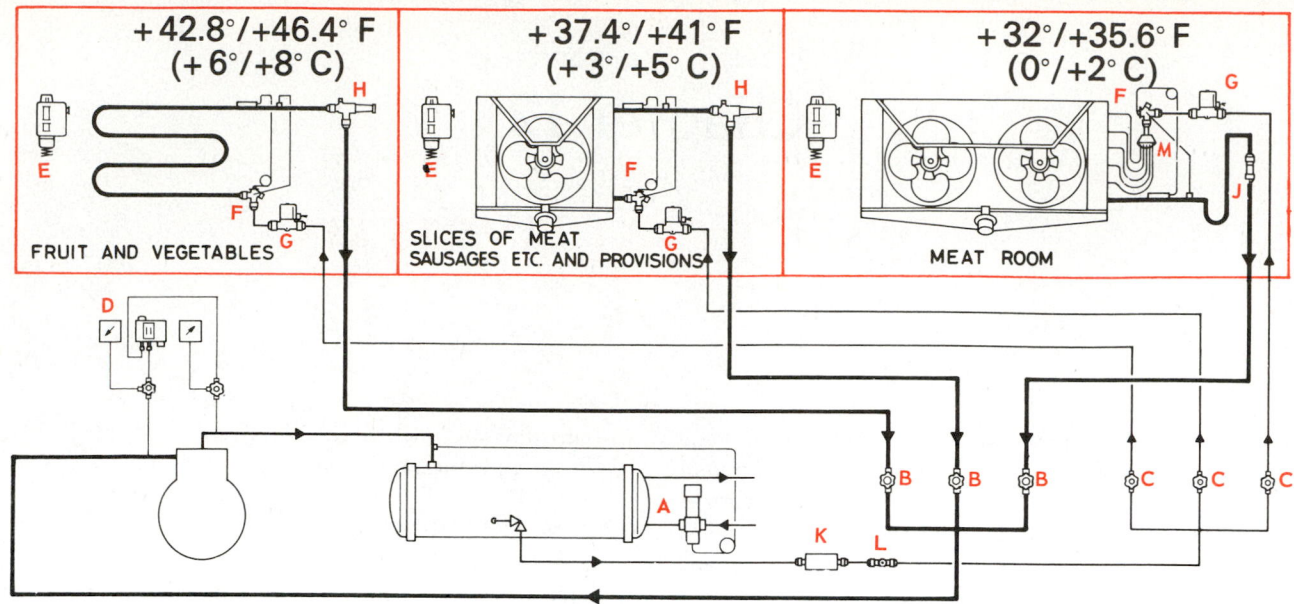

Fig. 12-3. A multiple evaporator system. A—Water valve. B—Suction line shutoff valves. C—Liquid line shutoff valves. D—Low and high-pressure motor control. E—Thermostats. F—Thermostatic expansion valves. G—Liquid line solenoid valves. H—Two-temperature valves. J—Check valve. K—Drier. L—Sight glass. M—Distributor.

d. High-pressure safety motor control.

e. Liquid line — with drier and sight glass.

The refrigerant control is at the division point between the low side and the high side of the system. It will consist of an automatic thermostatic expansion valve or capillary tube.

2. Low-pressure side:

a. Evaporator.

b. Low pressure or temperature motor control.

c. Suction line — some with filter-driers and surge tanks.

The multiple mechanism is shown in Fig. 12-3. It includes:

1. High-pressure side.

a. Compressor — often with an oil separator.

b. Condenser — water or air-cooled.

c. Liquid receiver.

d. High-pressure motor control.

e. Liquid lines — with a drier and a sight glass.

f. Water valve — used with a water-cooled unit.

The refrigerant control is the division point between the high-pressure side and the low-pressure side.

2. Low-pressure side.

a. Refrigerant controls, two or more, usually of the thermostatic expansion valve type.

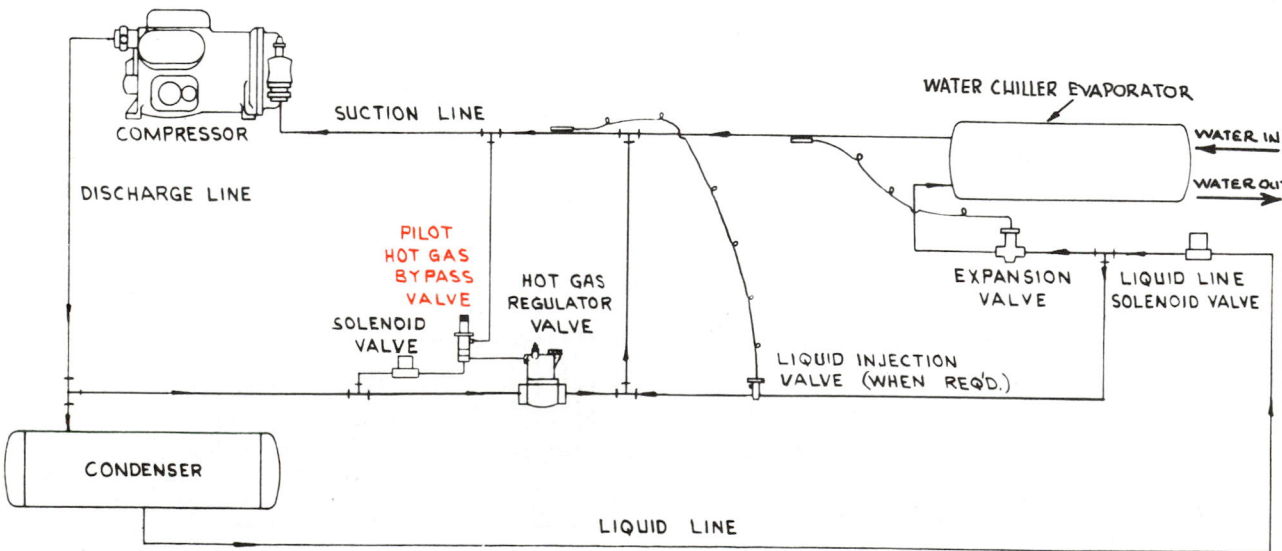

Fig. 12-4. Water chiller. Hot gas bypass keeps low-side pressure high enough to prevent freezing of chilled water. Liquid injection keeps suction vapor cool enough to prevent compressor motor overheating.

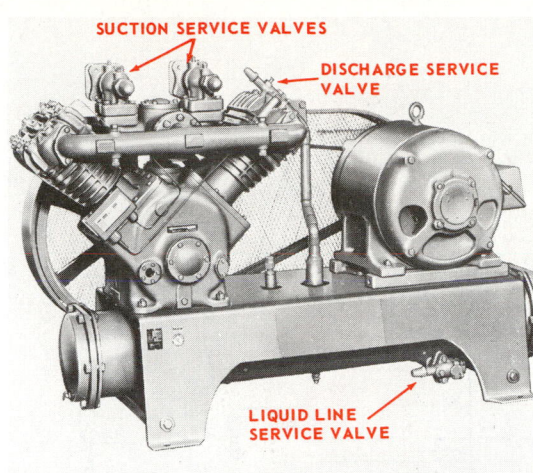

Fig. 12-5. External drive (open system) commercial condensing unit has four cylinder compressor, shell-and-tube water-cooled condenser and flange type service valves.

b. Evaporators — two or more. These may be any of several types: natural convection, forced convection or submerged.

c. Motor control, which is usually of the pressure type.

d. Suction lines with drier and suction pressure regulator.

e. Two-temperature valves for multiple temperature installation.

f. Surge tanks for reducing rapid pressure changes.

g. Check valves for multiple temperature installations.

There are many varieties of commercial systems. A water chiller system is shown in Fig. 12-4.

Condensing units are normally mounted on a steel base. In the external drive unit, the motor is mounted outside the compressor and drives the compressor either directly or with one or more belts as in Fig. 12-5.

In the hermetic unit, the motor is connected directly to the compressor. Fig. 12-6 shows a 25 hp condensing unit which has a 2 1/8-in. sweat type liquid line. It is designed for low-temperature work and has a 40,600 Btu/hr. capacity at —52 F. (—46 C.).

For large commercial installations, compressors are made with three, four, five, six, seven, or more cylinders.

The cylinder arrangement is another method of naming compressors; vertical single, horizontal single, 45 deg. single

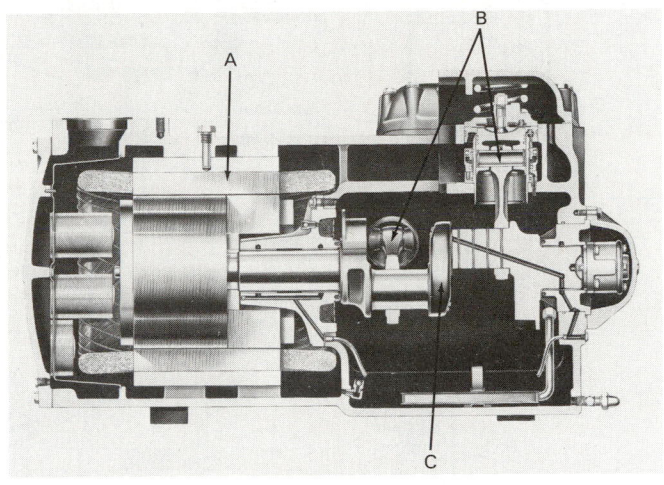

Fig. 12-7. Cutaway view of accessible hermetic motor compressor. Compressor is four cylinder V-type. A—Motor. B—Piston. C—Crankshaft. (The Trane Co.)

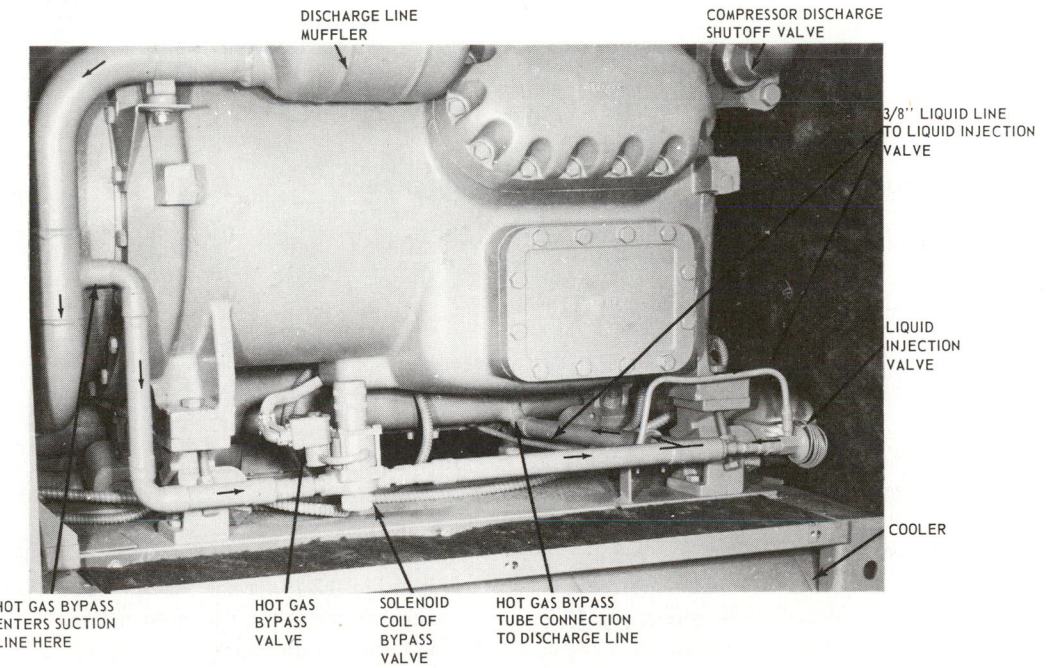

Fig. 12-6. Large commercial chiller. Note hot gas bypass system. (Airtemp Applied Machinery Co.)

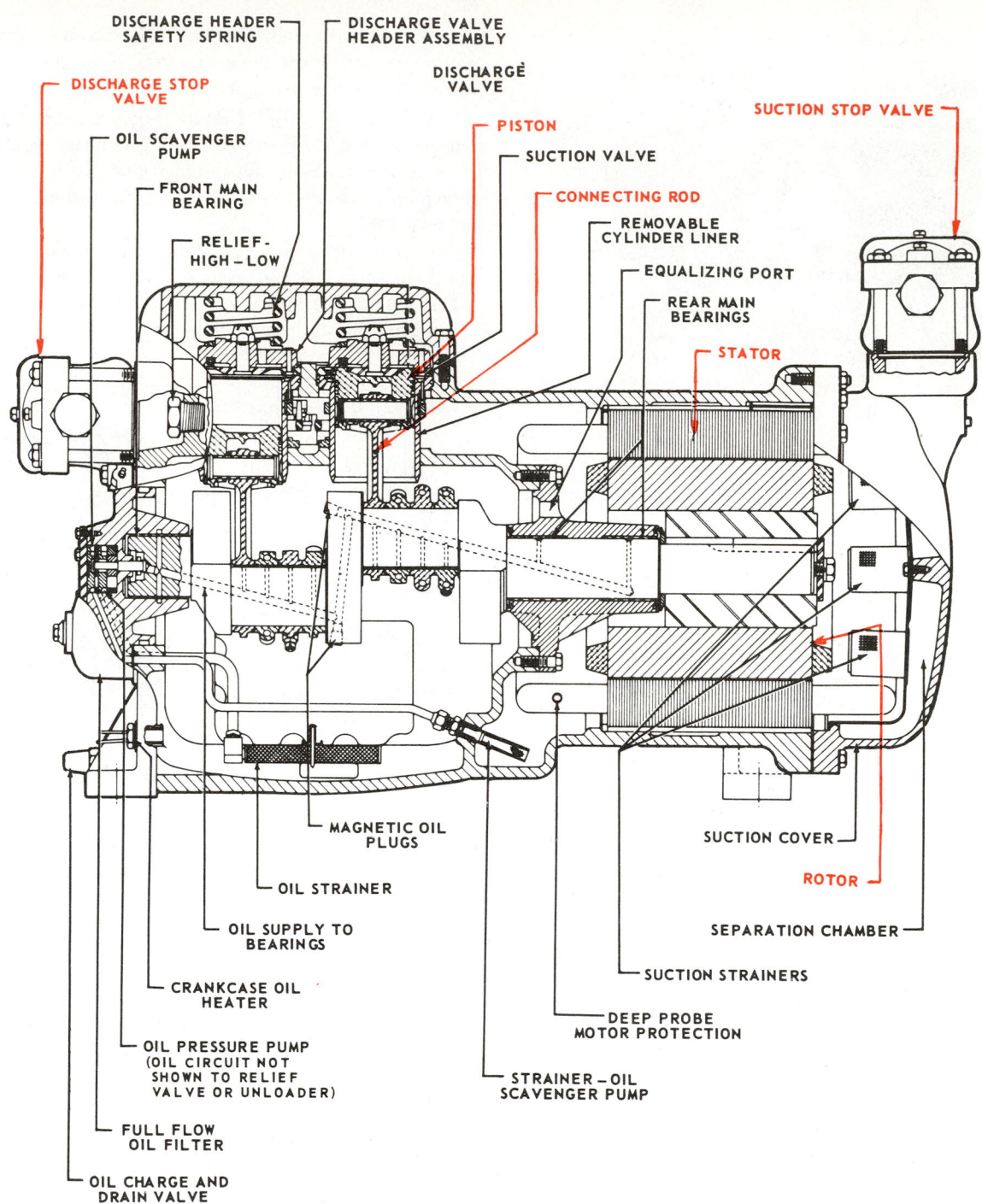

DISCHARGE HEADER
SAFETY SPRING

DISCHARGE VALVE
HEADER ASSEMBLY

DISCHARGE
VALVE

DISCHARGE STOP
VALVE

SUCTION STOP VALVE

OIL SCAVENGER
PUMP

PISTON

SUCTION VALVE

FRONT MAIN
BEARING

CONNECTING ROD

RELIEF-
HIGH – LOW

REMOVABLE
CYLINDER LINER

EQUALIZING PORT

REAR MAIN
BEARINGS

STATOR

MAGNETIC OIL
PLUGS

OIL STRAINER

OIL SUPPLY TO
BEARINGS

CRANKCASE OIL
HEATER

OIL PRESSURE PUMP
(OIL CIRCUIT NOT
SHOWN TO RELIEF
VALVE OR UNLOADER)

FULL FLOW
OIL FILTER

OIL CHARGE AND
DRAIN VALVE

SUCTION COVER

ROTOR

SEPARATION CHAMBER

SUCTION STRAINERS

DEEP PROBE
MOTOR PROTECTION

STRAINER – OIL
SCAVENGER PUMP

Fig. 12-8. Bolted type field serviceable eight cylinder hermetic motor compressor.
(Airtemp Applied Machinery Co.)

(inclined), vertical two cylinder, V-type two cylinder, W-type three cylinder, radial three cylinder, vertical four cylinder and V-type four cylinder, have all been used.

As with cylinder arrangements, there are many crankshaft arrangements. One type of serviceable hermetic motor compressor has an eccentric type crankshaft as shown in Fig. 12-7.

The inside construction of a multiple cylinder serviceable hermetic motor compressor is shown in Fig. 12-8.

To increase the cooling of the compressor and motor of a serviceable motor compressor, one company uses a fan and

motor as in Fig. 12-9. This manufacturer can convert the motor compressor easily into an external drive compressor as illustrated in Fig. 12-10.

A six cylinder compressor with an external drive motor is shown in Fig. 12-11. It has a crank throw type crankshaft.

A steel frame is used to hold the shell and tube condenser on the hermetic unit shown in Fig. 12-12. Note water cooling coil around the electric motor, the oil level sight glass in the lower right of the compressor and the spring mounted hermetic unit.

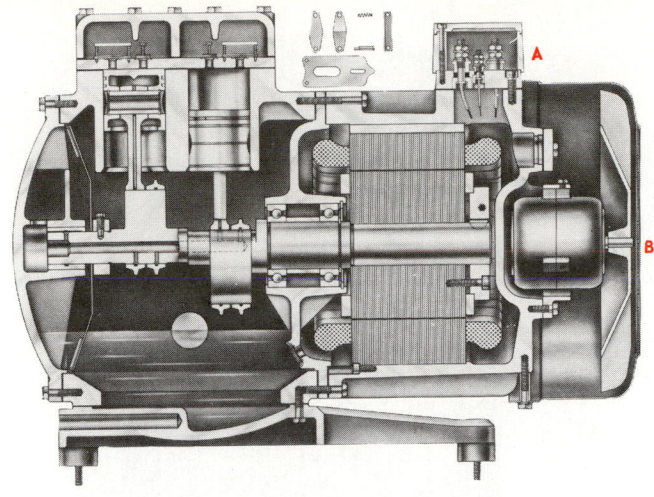

Fig. 12-9. Accessible hermetic motor compressor with four cylinder V-type compressor. Note oil level sight glass in center of crankcase. A—Electrical terminals. B—External fan and motor force cooling air over unit.

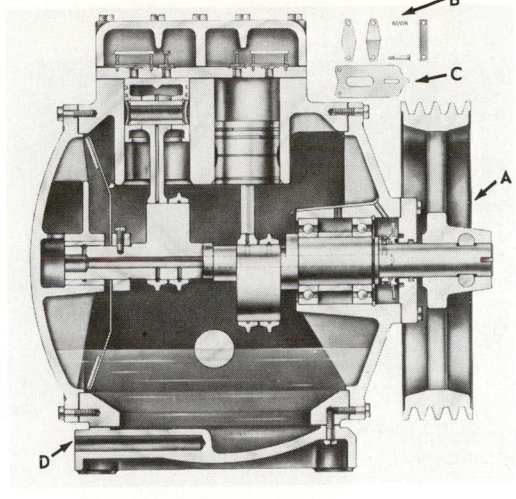

Fig. 12-10. Belt-driven external drive compressor is V-type four cylinder model and uses eccentric type crankshaft. A—Four-belt flywheel. B—Exhaust valve parts. C—Intake valve parts. D—Opening for electric oil heater. (Hans Goldner and Co.)

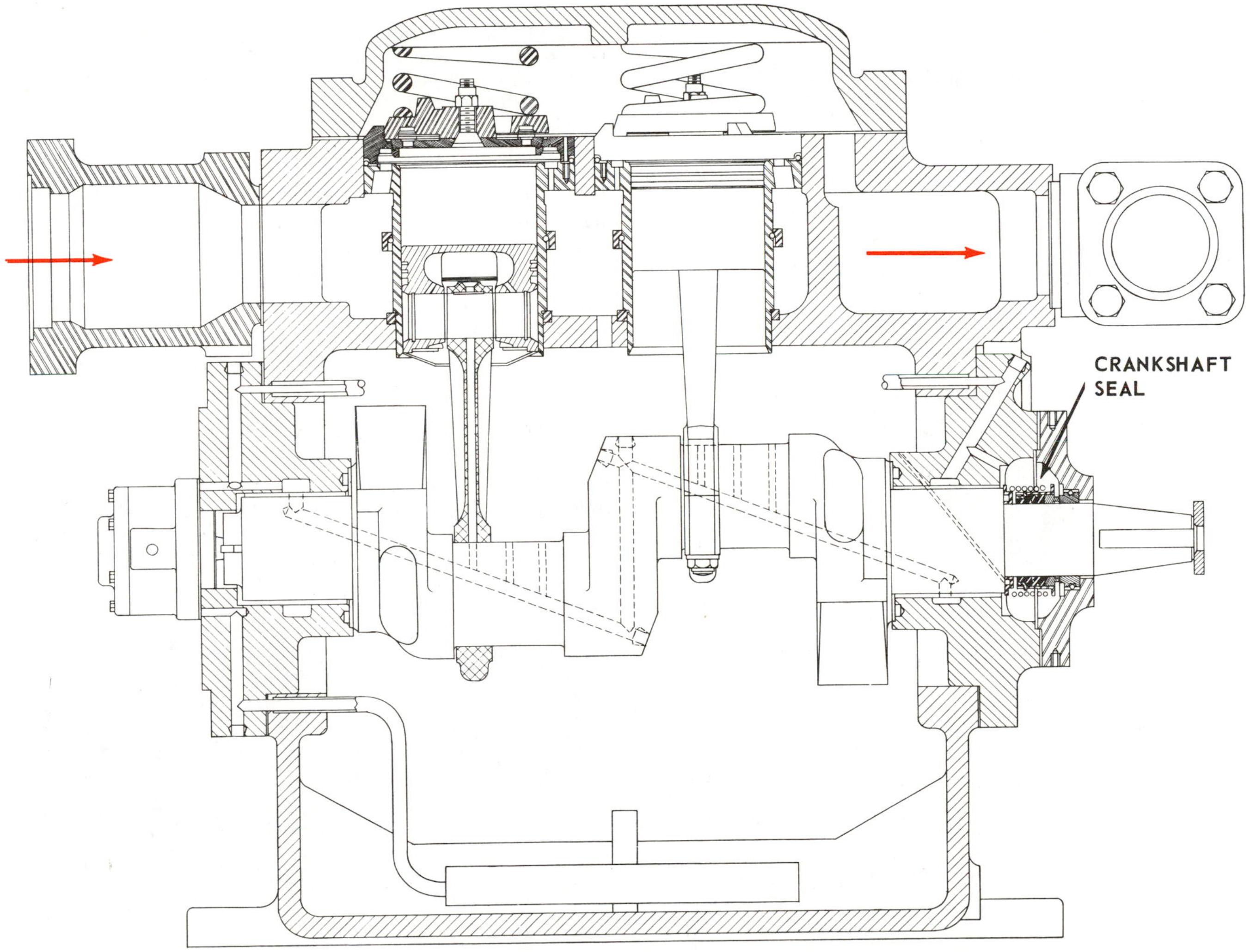

CRANKSHAFT SEAL

Fig. 12-11. Serviceable six cylinder W-type compressor designed for external drive motor (or engine). Note crankshaft seal and drilled crankshaft for pressure lubrication. (Vilter Manufacturing Corp.)

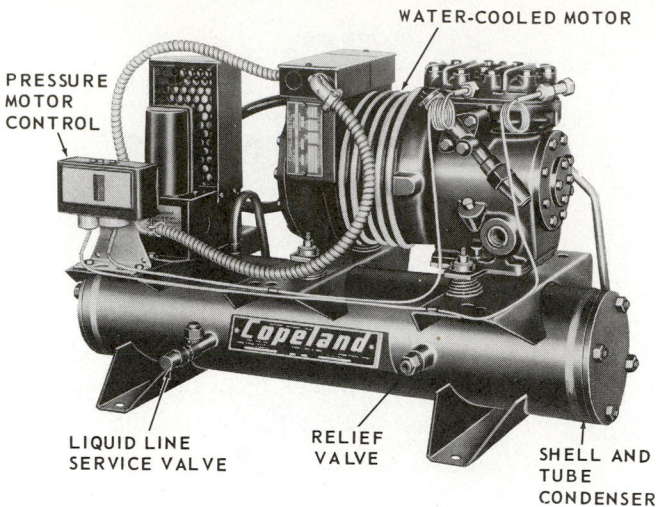

Fig. 12-12. Serviceable commercial hermetic condensing unit. (Copeland Corp.)

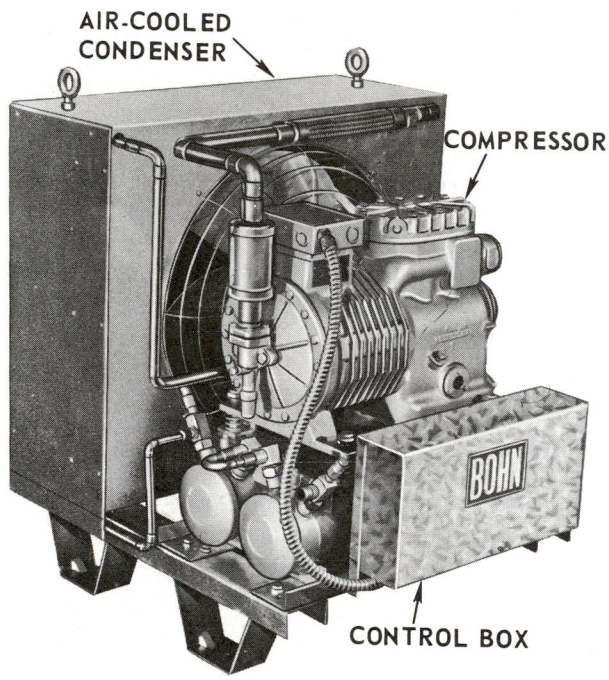

Fig. 12-13. Commercial hermetic condensing unit with air-cooled condenser. Separate motor is used to drive condenser fan. (Bohn Aluminum and Brass Div., Gulf & Western Mfg. Co.)

12-4 COMMERCIAL HERMETIC UNITS

Several companies now produce hermetic units of 20 or more hp. Some of the units are the bolted assembly type. These are often called "field serviceable" or "accessible." Some units are sealed in a welded casing. Both types are equipped with service valves. They may be connected to any type of evaporator and used for many different applications.

An advantage in the use of hermetics in the commercial field is the elimination of the crankshaft seals and belts. Because any trouble in the compressor mechanism involves

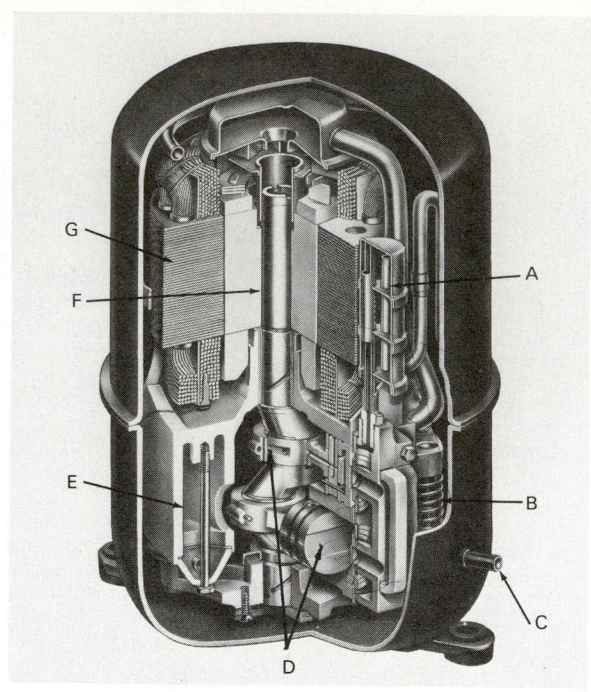

Fig. 12-14. Two cylinder hermetic motor compressor. It is used on air conditioning, heat pump and commercial condensing units. A—Discharge muffler. B—Spring mounting. C—Discharge tube. D—Pistons. E—Suction muffler. F—Crankshaft. G—Motor winding. (Tecumseh Products Co.)

both the compressor and the motor, the service technician must be very careful when working on these hermetic units, to keep moisture and dirt out of the system.

A serviceable motor compressor equipped with a fan condenser, a shroud, and service valves is shown in Fig. 12-13. The inside of a welded hermetic motor compressor is shown in Fig. 12-14. Smaller units have single-phase motors. Units over 1/2 hp generally have three-phase motors, Fig. 12-15.

The condensing units may be installed in many different ways. Some are mounted on the roof, some on the same floor level with the evaporator but in different rooms or outside the building. See Fig. 12-16. Note the suction service valve on the

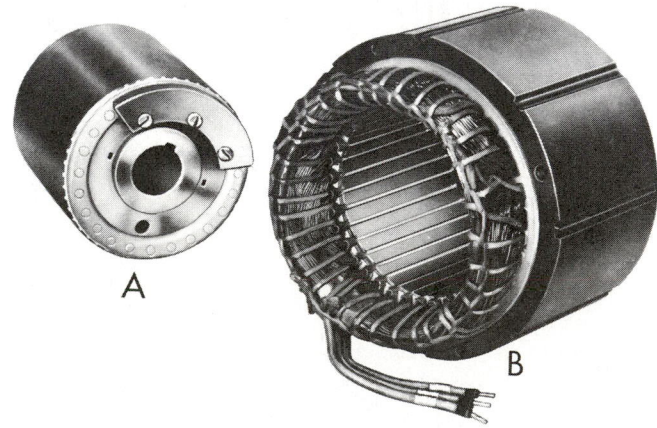

Fig. 12-15. Three-phase electric motor used in larger hermetic refrigeration units. A—Rotor. B—Stator. (Emerson Electric Co.)

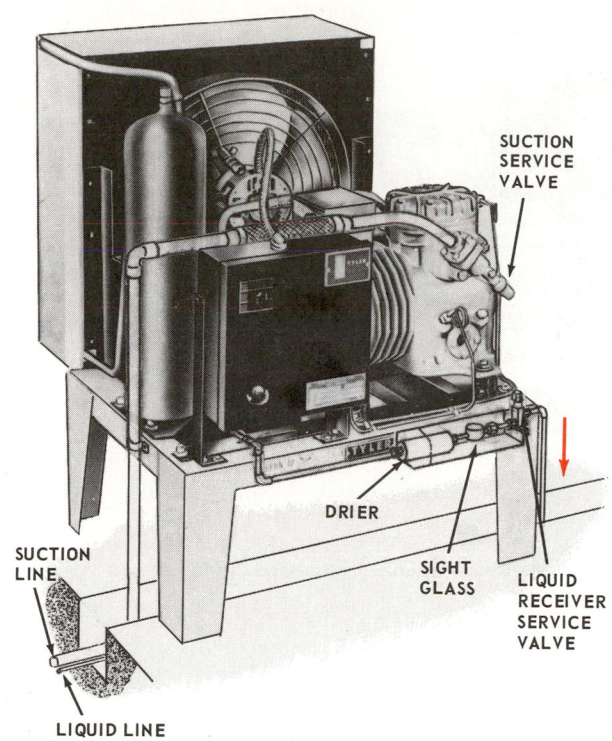

Fig. 12-16. Installed condensing unit. Note trough in floor for suction liquid lines. (Tyler Refrigeration Corp.)

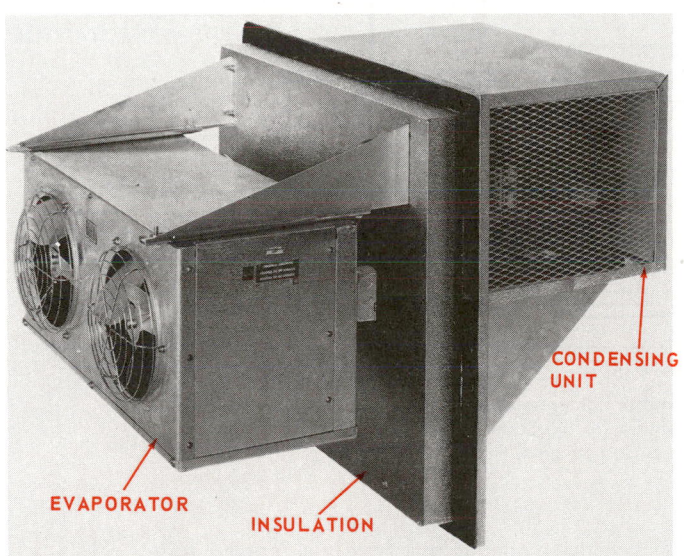

Fig. 12-17. A self-contained commercial refrigeration system. Units are available in 1/2 to 3 hp sizes. Low temperature systems are equipped with defrosting devices. (American Panel Corp.)

compressor, the liquid receiver service valve on the receiver, and the forced convection condenser.

A factory assembled condensing unit and evaporator combination can be installed as shown in Fig. 12-17. These straddle or plug-in units eliminate the need to put in piping between the condensing unit and the evaporator. The installation, therefore, consists of preparing the opening, mounting

the unit, running the electrical lines, and opening the shutoff valves. Fig. 12-18 shows a top mounted unit.

For large installations two-motor compressor designs have been developed to provide greater refrigeration capacity. They are:

1. Tandem assembly motor compressors.
2. Parallel assembly motor compressors.

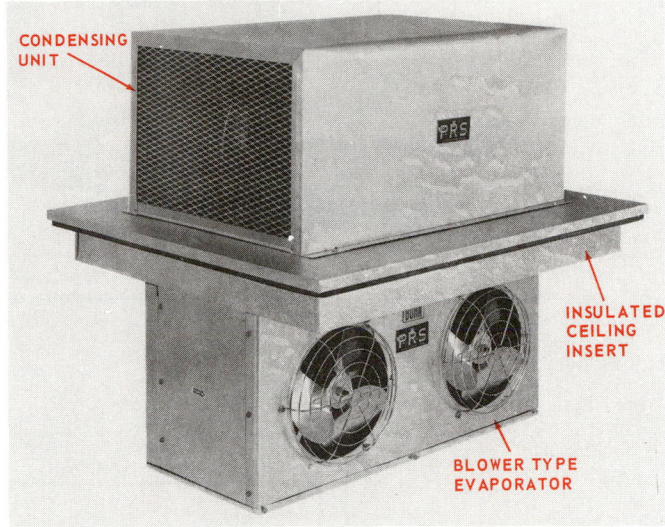

Fig. 12-18. Self-contained through-the-ceiling commercial unit. (American Panel Corp.)

The tandem design connects two motor compressors together at the motor end. See Fig. 12-19. These units can be run separately for low load or together for full load. But, if one motor compressor fails and must be replaced, the complete system must be shut down during the service time.

Parallel design connects two or more units in parallel by piping (Fig. 12-20). The units also require a compressor oil piping system to make certain that all the compressors have the correct amount of oil in each crankcase while in operation.

12-5 OUTDOOR AIR-COOLED CONDENSING UNITS

To save space in commercial buildings and in homes (when air conditioning), there is an increasing use of outdoor air-cooled condensing units, Fig. 12-21. Air-cooled units save the cost of plumbing for water circuits and are also used where chemicals in the water make water cooling impractical.

These units may be mounted outside the building, on the roof, on the outside wall or at ground level. Four major cautions:

1. There must be a head pressure control if the unit is exposed to outdoor weather that may go below the operating cabinet temperature.
2. A method of preventing short cycling must be designed into the system.
3. A means must also be provided to prevent dilution of the

Fig. 12-19. Tandem motor compressor assembly doubles refrigerating capacity of unit. (Copeland Corp.)

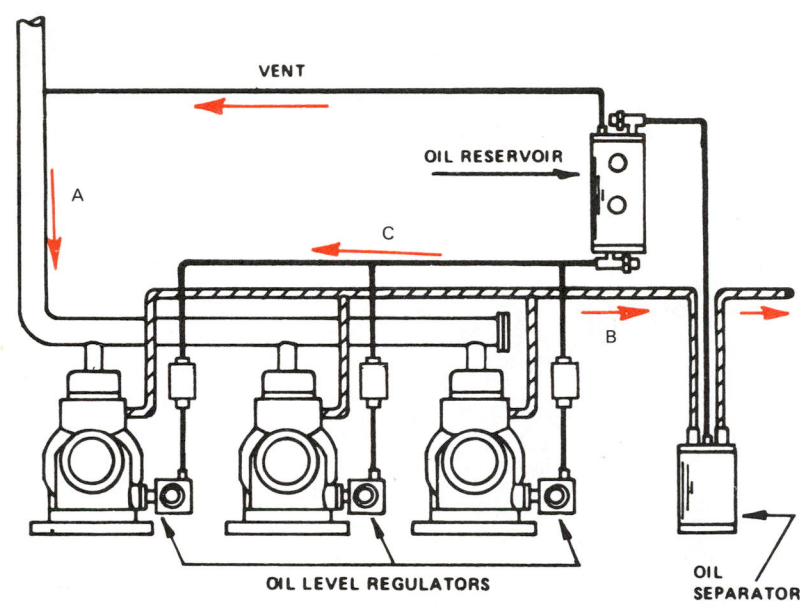

Fig. 12-20. A parallel motor compressor assembly. Three motor compressors are used. Floats in oil reservoirs and in oil level regulators keep proper oil level in each compressor. A—Suction piping. B—Discharge piping. C—Oil level piping. (A C and R Components, Inc.)

compressor oil by liquid refrigerant.

4. The completed condensing unit must be constructed and installed so it is virtually weatherproof.

Low ambient temperatures will cause low head pressures. This pressure may drop so low it may even stop the flow of refrigerant. Four different methods will maintain pressure:

1. Partially fill the condenser with liquid refrigerant.
2. Stop or slow the condenser fans.
3. Partially close or completely close the ambient air louvers.
4. Heat the condenser.

Outdoor units require about 1000 cubic feet per minute (cfm) of condenser air circulation per horsepower. They are less costly to operate than indoor air-cooled units.

One of the main problems with outdoor units is to keep the thermostatic expansion valve operating at full capacity during cold weather. Capacity depends on the pressure difference across the valve. If condensing pressure reduces from 102 psi, 90 F. (32 C.) for R-12 to 56 psi, 30 F. (−2 C.) for R-12, the valve capacity will drop. (Not enough liquid refrigerant will flow.) The refrigerated fixture temperatures may then rise too high. Also, a small pressure difference may cause short cycling of the condensing unit.

The condensing temperature and pressure may be kept at a proper operating level by a design change. The unit is made to

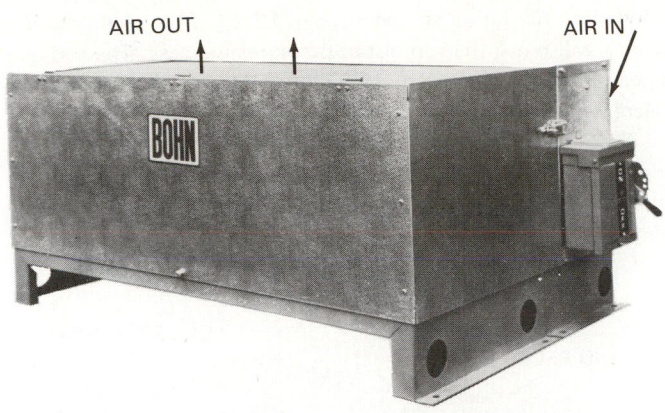

Fig. 12-21. Air-cooled condensing unit for outdoor installation (usually roof-mounted) with housing in place. Air enters through front of condensing unit. (Bohn Aluminum & Brass Div., Gulf & Western Mfg. Co.)

LIMITIZER HEAD PRESSURE CONTROL SYSTEM

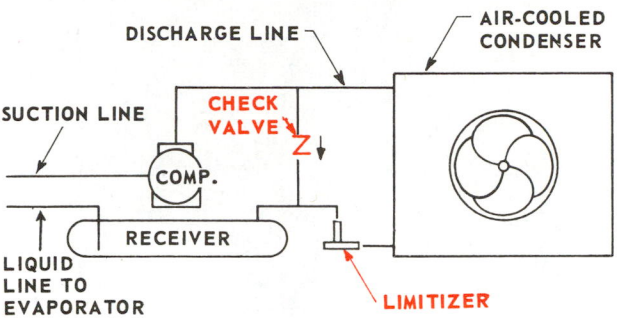

Fig. 12-22. Outdoor condensing system. The check valve and limiter valve insure good condensing pressures during cold weather.

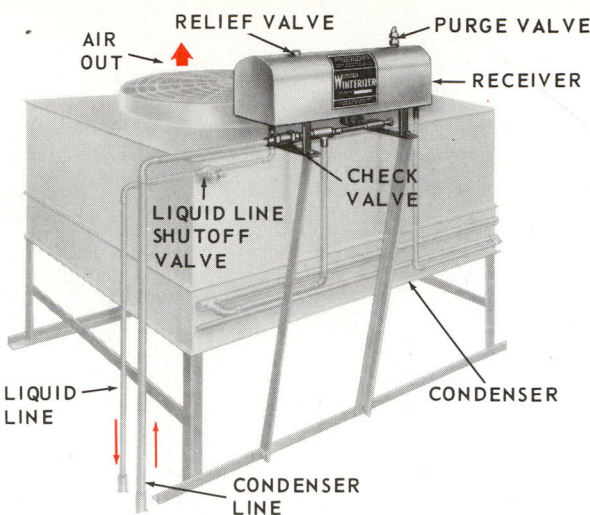

Fig. 12-23. Outside air-cooled condenser designed for cold weather operation. Airflow is vertical. To provide for satisfactory cold weather operation, receiver is insulated and has electric heater. (The Winterizer Co.)

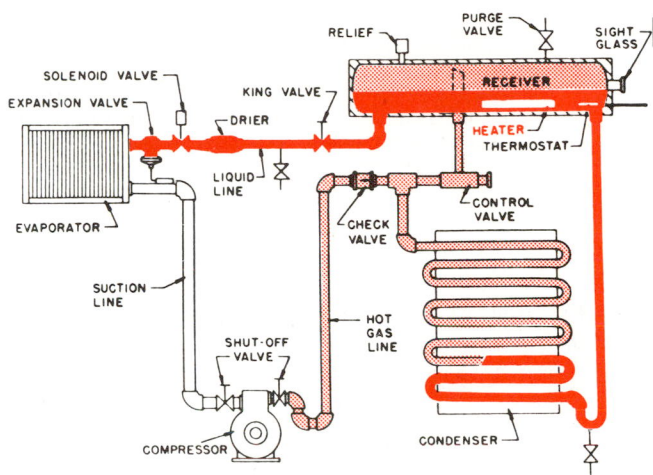

Fig. 12-24. Condensing unit during on cycle. Liquid refrigerant is stored in liquid receiver.

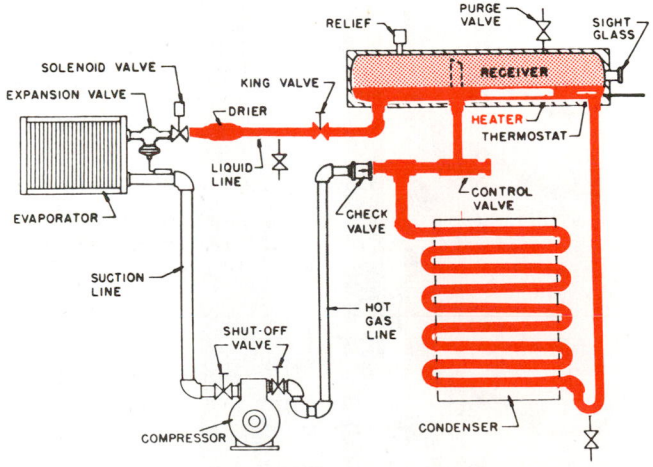

Fig. 12-25. Condensing unit during off cycle. Liquid refrigerant fills condenser tubes.

nearly fill the condenser tubes with liquid. Just enough condensing surface is left to maintain the pressure.

Fig. 12-22 shows how a check valve and limiter valve are used to maintain a head pressure in the condenser even when in low outdoor temperatures. The pressure on the limitizer valve on the outlet of the condenser will not open (and allow liquid to leave the condenser) until the condensing pressure reaches the proper level. The limitizer valve is found at the outlet of the condenser. It opens when pressure rises.

The installation specifications must be carefully checked. The receiver must hold enough liquid refrigerant to flood most of the condenser in the winter and still safely hold the refrigerant during the warm season.

Another system for cold weather operation is shown in Fig. 12-23. Here the receiver is mounted above the condenser. The refrigerant in the receiver is kept at 80 F. (27 C.) minimum by having the receiver insulated and by using an electric heater.

During the on part of the cycle, the unit operates as a normal system, but never at a temperature less than 80 F. (27 C.). This is shown diagrammatically in Fig. 12-24. During the off part of the cycle, the condenser fills with liquid refrigerant (still at the 80 F. temperature-pressure relationship). This condition prevents short cycling and also stops refrigerant from moving out of the low side, through the compressor, into the high side during off cycle. See Fig. 12-25.

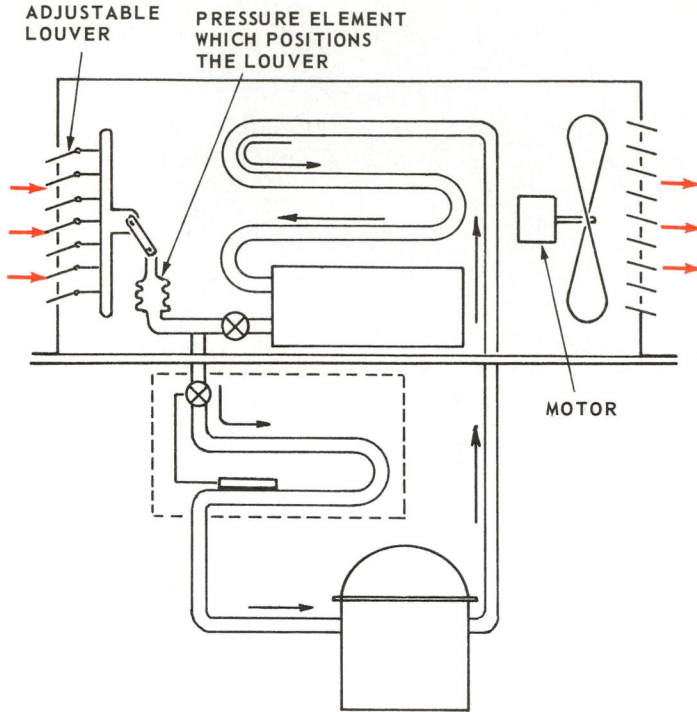

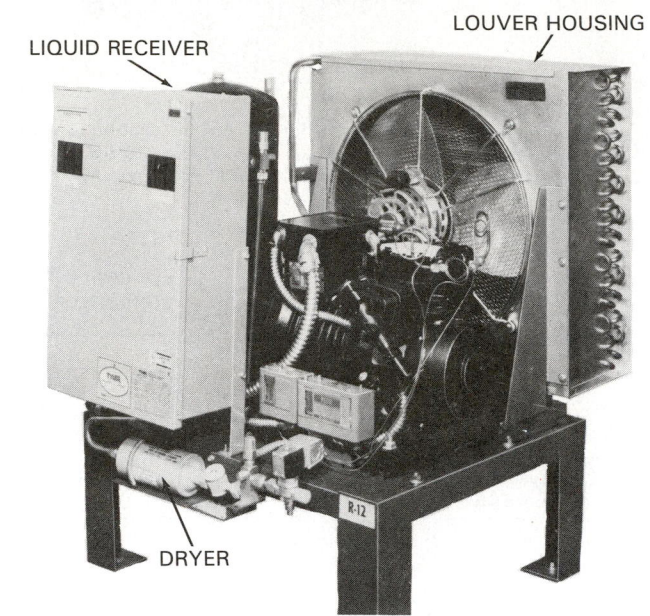

condenser tubing as shown in Fig. 12-26. This head pressure device will move the rod out as pressures increase. The rod will open the louvers, as Fig. 12-27 illustrates. As head pressures decrease, the rod will move back and start to close the louvers.

The condenser fans may either operate when louvers are closed or they may be shut off when the louvers near the closing point. A condensing unit with an ambient-temperature-controlled, adjustable damper is shown in Fig. 12-28.

Other systems shut off the condenser fan when the

Fig. 12-26. An air-cooled condenser with pressure operated louver. As condensing pressure decreases, louver will start to close, reducing condenser airflow.

The condensing pressure may also be controlled by limiting the airflow to the condenser.

Another way to keep the pressures up is to close the condenser housing airflow louvers as the pressure drops. A pressure-sensitive device does this. It is connected into the

Fig. 12-28. A 3 hp air-cooled condensing unit equipped with an adjustable louver to vary airflow as ambient temperature changes. (Tyler Refrigeration Corp.)

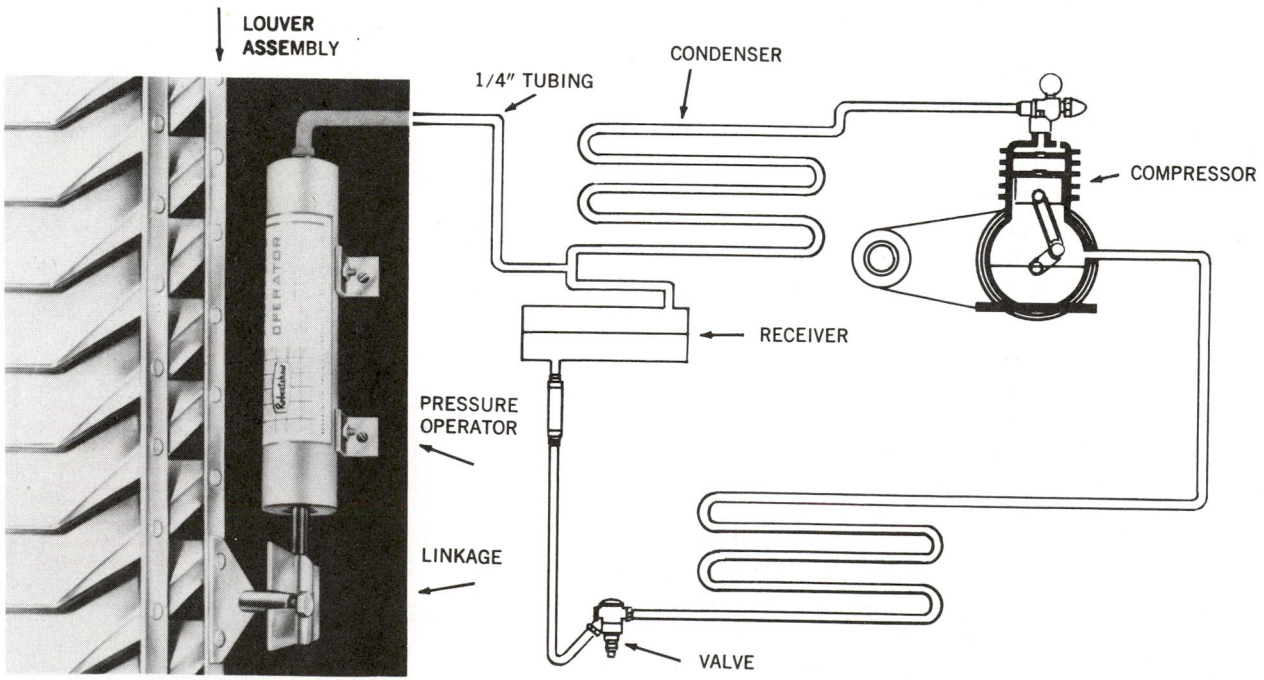

Fig. 12-27. Pressure-operated louver positioning cylinder. (Robertshaw Controls Company)

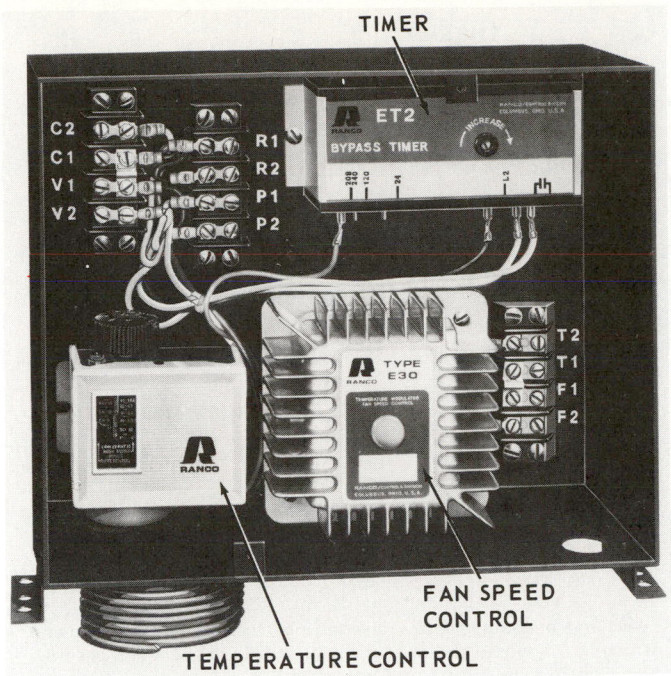

Fig. 12-29. This electronic control system is designed for an outdoor air-cooled condenser system. (Ranco, Inc.)

condenser pressure falls to a minimum level. Some shut off one or more of several fans. Others lower the fan speed when the head pressure drops by using electrically controlled modulated fan speeds. This system operates with a thermistor type sensor on the condenser, a special fan motor, and a timing device. See Fig. 12-29. It permits the system to operate for a few minutes on start-up even though the low-side pressure is low. Fig. 12-30 is the schematic for this system's electrical circuit.

To keep the receiver temperature warmer than the cabinet temperature, electric heating elements are sometimes placed in or around the receiver. If allowed to become too cold, the receiver would act like a condenser.

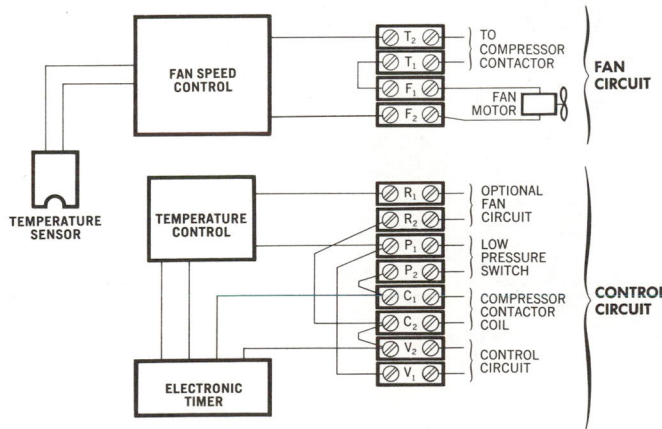

Fig. 12-30. Electrical wiring diagram for fan speed control system. (Ranco, Inc.)

Some systems use a bypass from the compressor to the receiver. This bypass feeds a certain amount of hot refrigerant vapor to the receiver to keep it warm. The bypass has a check valve mounted in it to insure only one-way flow.

The compressor, itself, must be kept warm enough during cold weather to prevent dilution of the oil by the liquid refrigerant. Heat is provided by electric heating elements in or around the motor compressor. They are thermostatically operated to energize the heating element at about 50 F. (10 C.). This heater usually has a 100W to 200W capacity.

The built-in capacity to overcome low ambient temperatures may not be enough if the outdoor unit is exposed to above-normal winds. Windy conditions can prevent damper and fan operation. Or the cooling effect may be more than the electric heating element can overcome. The unit must be installed in a position to avoid the harmful effects of any high-velocity cold winds.

The unit should be weatherproofed as well as possible — particularly the electrical installation. See Fig. 12-31.

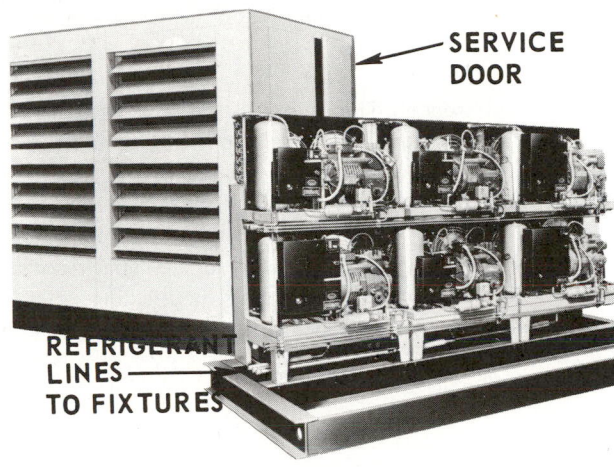

Fig. 12-31. Six commercial air-cooled condensing units mounted on rack. Weather-protecting shroud has been removed and is in background. (Tyler Refrigeration Corp.)

Head pressure control valves are often used. These are usually thermostat operated. A low-pressure switch may not cut in because condenser pressures are below cut-in pressures. All refrigerant may then transfer to the condenser, because it will be the coldest part of the system. A check valve is often used in the condenser outlet to prevent flow of refrigerant to the cold receiver.

As noted before, some systems flood the condenser with liquid refrigerant during low ambient temperatures to maintain high enough condensing pressures. Such systems require valves. When the condenser pressure decreases, one valve closes the outlet of the condenser and the condenser starts to fill with liquid refrigerant. This action reduces the condensing surface and condensing pressure will rise.

Fig. 12-32 is the schematic of a system which uses a valve designed to open as the pressure in the receiver falls. This system allows hot gas to bypass into the receiver (at about 20

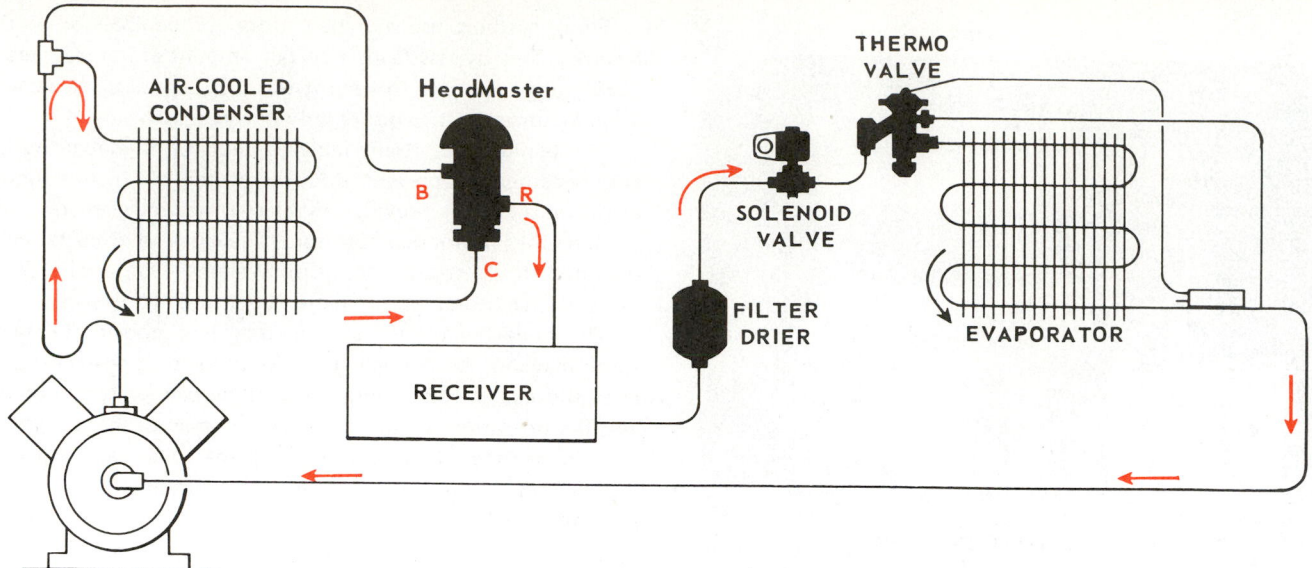

Fig. 12-32. Schematic shows condenser pressure control valve. Compressor discharge pressure is above valve setting. Opening at B is closed and flow is through condenser and openings C and R. During cold weather, opening at C is closed and openings at B and R are open. Hot gas then flows directly from compressor into receiver. (Alco Controls Div., Emerson Electric Co.)

psi pressure difference). This raises the receiver pressure and increases the flow of liquid refrigerant to the evaporators.

The valve has two openings, B and C. As one closes, the other will open. If located in a cold place, the receiver may need an electric heating element to help the system operate efficiently. These valves must be sized to the capacity of the system. Avoid using excessive pressures when testing for leaks. The valve bellows may suffer damage. Keep the pressure at or below 200 psi.

The system usually is charged with twice as much refrigerant as the system would need without the condenser flooding feature. To permit year-around operation, the receiver must have the capacity to store all this extra refrigerant during the summer. Also, for service purposes, the receiver should be twice the normal size in order to hold all the refrigerant.

The compressor may also collect liquid refrigerant during the off cycle. A trap may be needed in the compressor discharge line and an inverted trap at the condenser outlet.

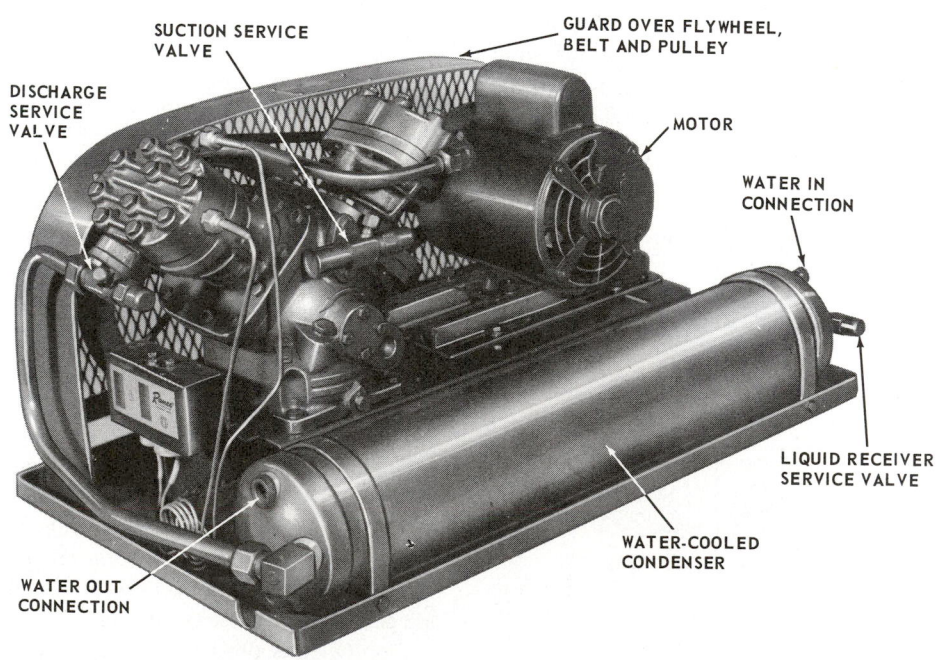

Fig. 12-33. Condensing unit with belt-driven two cylinder air-cooled compressor using water-cooled condenser. (Frick Co.)

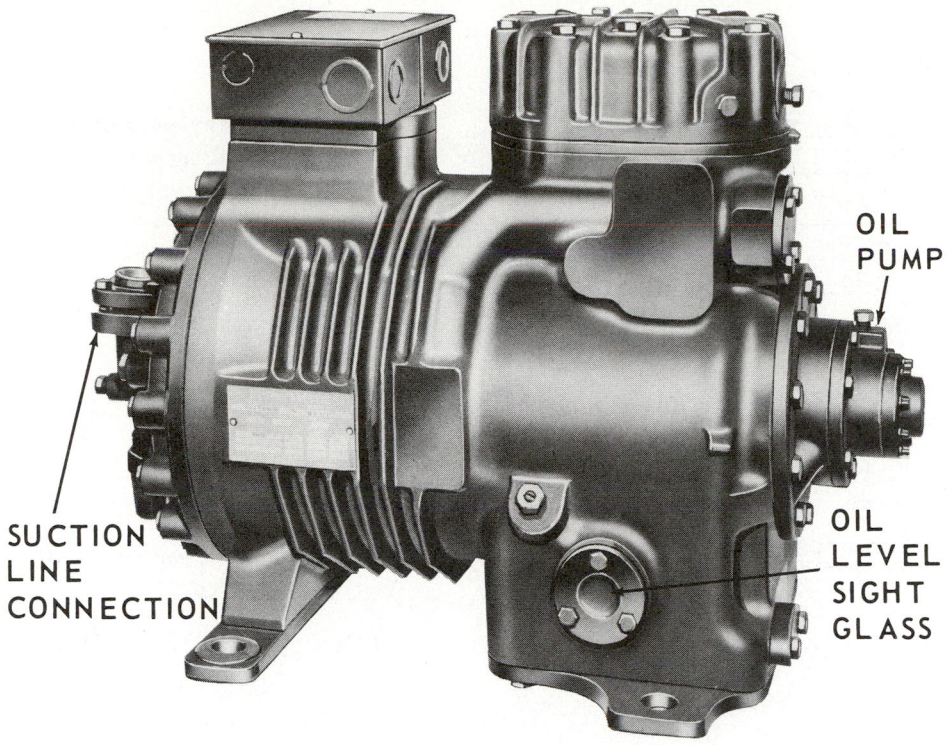

Fig. 12-34. Bolted type serviceable hermetic motor compressor. Note oil pump used for forced lubricating system. (Copeland Corp.)

The labels on the figure read:
- OIL PUMP
- OIL LEVEL SIGHT GLASS
- SUCTION LINE CONNECTION

12-6 THE COMPRESSOR

Commercial compressors are of two general types:
1. External drive.
2. Hermetic.

Commercial compressor designs are explained in Chapter 4. A typical small, belt-driven, external drive compressor is pictured in Fig. 12-33.

There are several types of commercial hermetic compressors. Fig. 12-34 calls attention to some parts of a bolted (serviceable) hermetic two cylinder compressor with a force-feed lubrication system.

Some large units have either hydraulic or electric unloading devices to control the number of cylinders which are pumping. The higher the load, the more cylinders used to pump the vapor. Fig. 12-35 is a schematic of the oil circuit and describes how it is controlled to operate the compressor unloader.

Another type of hermetic compressor unit is the welded motor compressor design, nonfield serviceable. These units are built in sizes from 1/6 hp up to approximately 20 hp. Internal design varies with size and manufacturer. Some are spring mounted internally, while some use outside (external) mounting springs. The smaller units usually have one cylinder, while larger units (1/2 hp and up) have two or more cylinders. Small-unit motors may be either two or four-pole (single-phase). Three-phase motors are generally used in the larger units.

Many combinations, types and sizes of compressors can be combined to provide the pumping for a great variety of evaporator types and sizes. Each compressor is most efficient under definite limits. Each compressor has a minimum and maximum:
1. Revolutions per minute (rpm) for efficiency.
2. Compression ratio (a maximum pressure difference between low side and high side).
3. Discharge temperature.
4. Volume of gas it can pump.

Before using a certain compressor, the manufacturer's operating specifications must be known. See Chapter 15.

Many low-temperature systems use a cascade system. The first stage compressor may be a reciprocating type, but rotary units are also used. The rotary compressor pressure limit is about 45 psi across the compressor. It works very well with a compression ratio of about 4:1 and with a discharge temperature of about 200 F. (93 C.).

The rotary has a high volumetric efficiency. A check valve is usually placed in the discharge to prevent back-up of refrigerant during off cycle. A check valve should be placed in the oil lines for the same reason.

Compressors can have from one to twelve cylinders in many different cylinder arrangements: vertical, V, W, Y, X or radial. See Fig. 12-36.

Internal unloaders are usually operated by oil pressure. A spring holds the intake valve open until the oil pressure builds up causing all intake valves to operate. It is also used to reduce pumping capacity during low-load periods. Solenoid valves are mounted in the oil lines to unloaders. When the solenoid closes, the oil pressure drops in the unloader while the intake

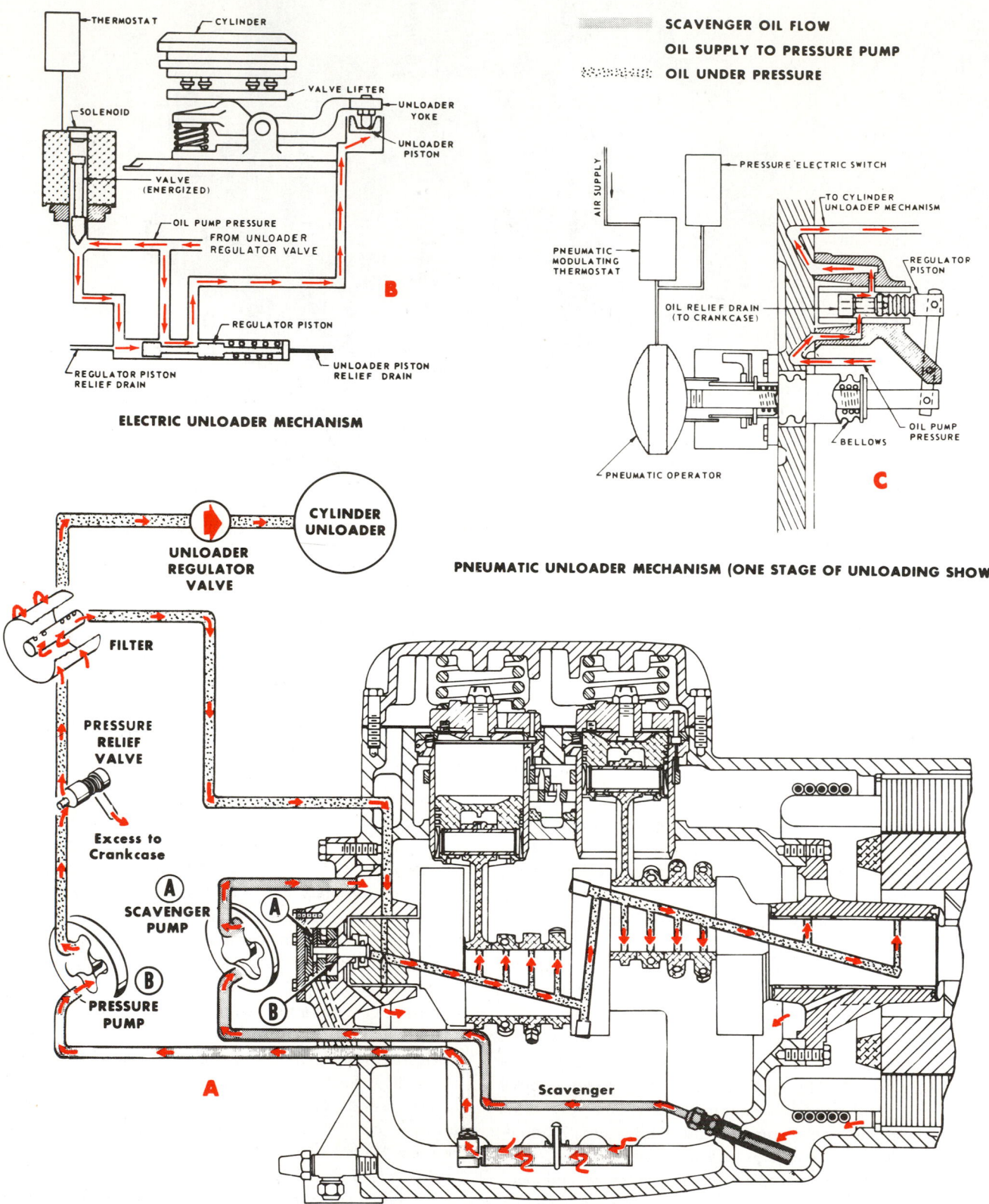

SCAVENGER OIL FLOW
OIL SUPPLY TO PRESSURE PUMP
OIL UNDER PRESSURE

ELECTRIC UNLOADER MECHANISM

THERMOSTAT
CYLINDER
SOLENOID
VALVE LIFTER
UNLOADER YOKE
UNLOADER PISTON
VALVE (ENERGIZED)
OIL PUMP PRESSURE FROM UNLOADER REGULATOR VALVE
B
REGULATOR PISTON
REGULATOR PISTON RELIEF DRAIN
UNLOADER PISTON RELIEF DRAIN

PRESSURE ELECTRIC SWITCH
AIR SUPPLY
TO CYLINDER UNLOADER MECHANISM
PNEUMATIC MODULATING THERMOSTAT
REGULATOR PISTON
OIL RELIEF DRAIN (TO CRANKCASE)
OIL PUMP PRESSURE
BELLOWS
PNEUMATIC OPERATOR
C

PNEUMATIC UNLOADER MECHANISM (ONE STAGE OF UNLOADING SHOWN)

CYLINDER UNLOADER
UNLOADER REGULATOR VALVE
FILTER
PRESSURE RELIEF VALVE
Excess to Crankcase
Ⓐ SCAVENGER PUMP
Ⓑ PRESSURE PUMP
A
Ⓐ
Ⓑ
Scavenger

Fig. 12-35. Oil circuit of an eight cylinder compressor with oil take-off to operate compressor unloader. A—Oil circuit. B—Electrical control. C—Pneumatic control. (Airtemp Applied Machinery Co.)

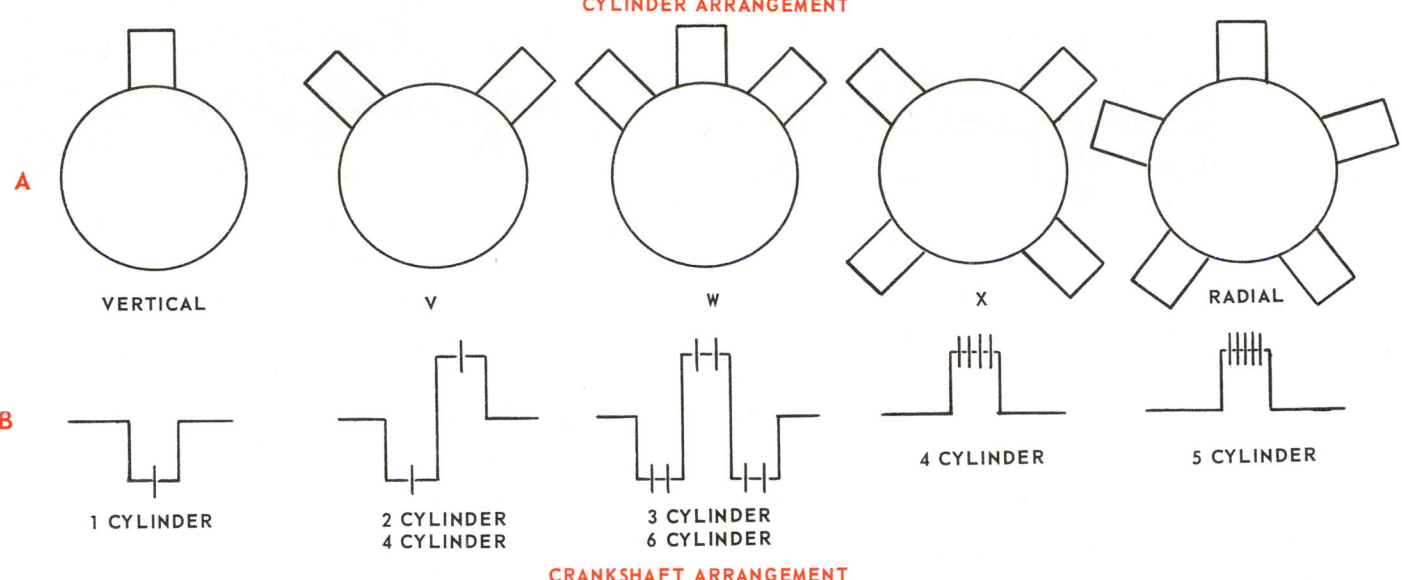

A

VERTICAL   V   W   X   RADIAL

B

1 CYLINDER   2 CYLINDER 4 CYLINDER   3 CYLINDER 6 CYLINDER   4 CYLINDER   5 CYLINDER

CRANKSHAFT ARRANGEMENT

Fig. 12-36. Cylinder arrangements used for reciprocating compressors. A—Cylinder arrangement. B—Crankshaft shape. Markings on crankshaft indicate number of rods which attach to each throw.

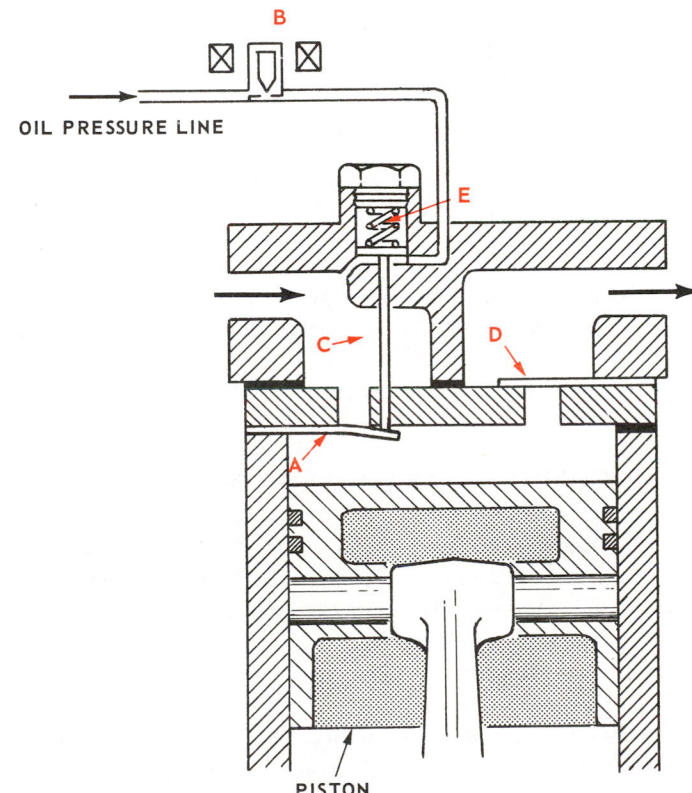

Fig. 12-37. An internal unloader for a compressor. During starting, the capacity needs to be reduced. A—Intake valve. B—Solenoid valve. C—Unloader valve. D—Exhaust valve. E—Spring.

valves are kept open. See Fig. 12-37.

Low-side pressure switches operate the solenoids. A timer bypass pressure switch is used to operate the system at full capacity for about a minute each hour or two. External unloaders use a bypass to the evaporator inlet to make sure suction vapor is cool (de-superheating).

12-7 AIR-COOLED CONDENSER

Air-cooled condensers are quite common in commercial systems. Cooling water may be too expensive or corrosive. Smaller units use static condensers, thermal airflow.

Larger condensers may be cooled by a big fan built onto the motor or into the compressor flywheel on external drive units. Larger hermetic units use separate motors to drive the fans. See Chapter 9 for determining condensing temperatures and pressures when an air-cooled condenser is used.

The efficiency of the fan on an air-cooled condenser may be increased by placing a metal shroud around it. In Fig. 12-38, an air-cooled condenser uses two fans. Air can be drawn, induced (led into) or forced through the condensers. These condensers have fins and frequently use a double or

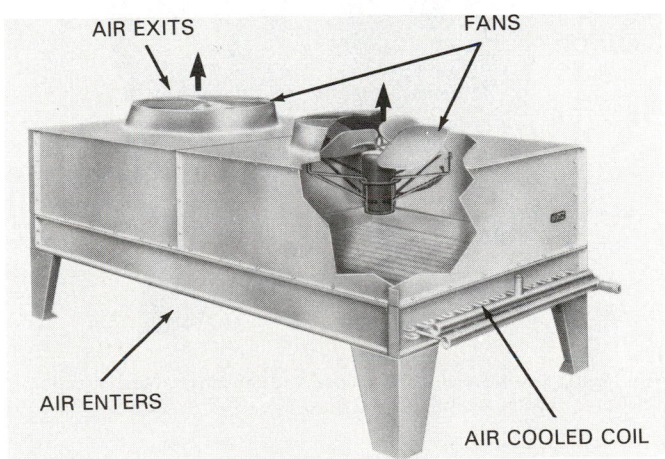

Fig. 12-38. Air-cooled condenser cutaway shows fan, motor, mount, and condenser coil arrangement. Air enters from beneath and leaves at top. (Bohn Aluminum & Brass Div., Gulf & Western Mfg. Co.)

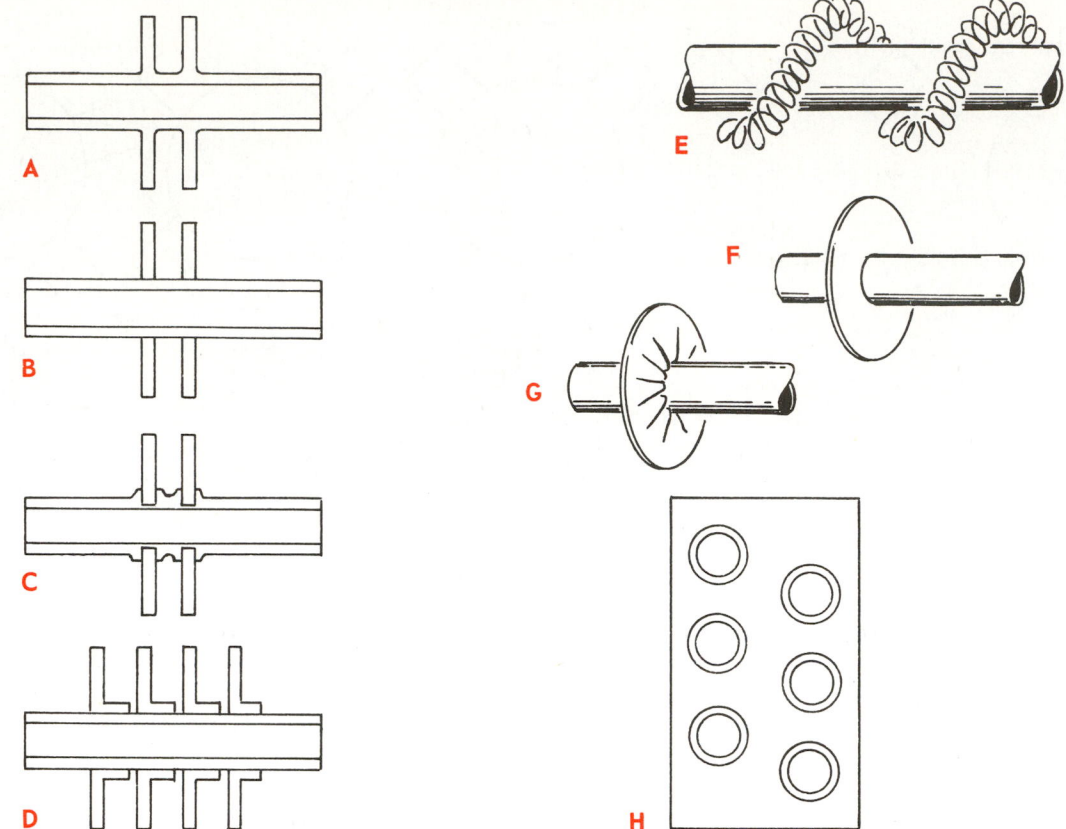

Fig. 12-39. Extended surface fin arrangements used in air-cooled condensers. A—Fin is part of tubing. B—Fin pressed on tubing. C—Fin fastened by crimped tubing. D—Fin flange pressed on tubing. E—Coiled wire used as fin. F—Circular fin. G—Crimped circular fin. H—Large multiple tubing fin.

Fig. 12-40. Commercial system air-cooled condenser. Note direction of airflow. (Bohn Aluminum and Brass Div., Gulf & Western Mfg. Co.)

triple row of tubes. Many fin arrangements and constructions have been used. See Fig. 12-39. In Fig. 12-40 is pictured another type of commercial, air-cooled condenser.

To cool the compressor head and valves, a double air-cooled condenser is sometimes used. Refrigerant leaves the compressor and passes through one condenser. Then it is led back through the motor compressor to help cool it. From there it goes into the second condenser, where it is condensed (cooled) into liquid.

12-8 OUTDOOR AIR-COOLED CONDENSERS

Outdoor air-cooled condensers may be used on systems which have the motor compressor and the liquid receiver inside the building. The condenser only is placed outdoors. The compressor discharge line carries the hot high-pressure vapor to the outdoor air-cooled condenser. Condensed liquid is piped back into the building.

These units use the same devices described in Para. 12-5 to protect the condenser and to maintain good head pressures when air temperature is as low as 50 F. (10 C.) or lower.

12-9 WATER-COOLED CONDENSER

Many large commercial refrigerating units use the water-cooled type condenser. This condenser is built in three styles:
1. Shell and tube.
2. Shell and coil.
3. Tube-within-a-tube.

In the first type, the refrigerant vapor goes directly from the compressor into a tank or shell while the water travels through the tank or shell in straight tubes. The second type also uses a shell but the water travels through the shell in coils of tubing.

The third type uses two pipes or tubes — one inside the other. The refrigerant passes one way through the outer pipe, while the condenser water flows in the opposite direction through the inner tube.

Water velocity should be at least 7 fps (feet per second) but no more than 10 fps or the inside of the water tube may develop corrosion. If flow is too fast, water may remove the oxide coating causing pitting. If the water velocity drops to 3 fps, scaling will take place.

Water-cooled compressors are sometimes used with water-cooled condensers. The water flow, with few exceptions, is through the condenser first, then through the cylinder head and finally into the drain. Water flow may be regulated by means of an automatic water valve. See Para. 12-57.

12-10 SHELL AND TUBE CONDENSER

The shell and tube condenser is a cylinder usually made of steel with copper tubes inside. Water circulates through the tubes and condenses hot vapors in the cylinder into a liquid. The bottom part of the shell serves as the liquid receiver. See Fig. 12-41.

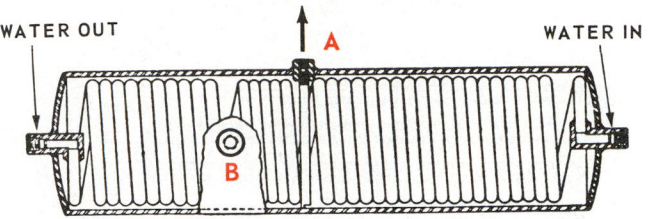

Fig. 12-42. Shell and coil type water-cooled condenser which also serves as liquid receiver. A—Refrigerant out. B—Refrigerant in.

12-11 SHELL AND COIL CONDENSER

The shell and coil condenser is very much like the shell and tube water-cooled condenser. It has a coil of water tubing inside the shell rather than a straight tube. It is often used in smaller commercial units. See Fig. 12-42.

While it is less costly to manufacture it cannot be cleaned mechanically. The water tube is only cleanable with chemicals.

12-12 TUBE-WITHIN-A-TUBE CONDENSER

The tube-within-a-tube water-cooled condenser is popular because it is easy to make. Water passing through the inside

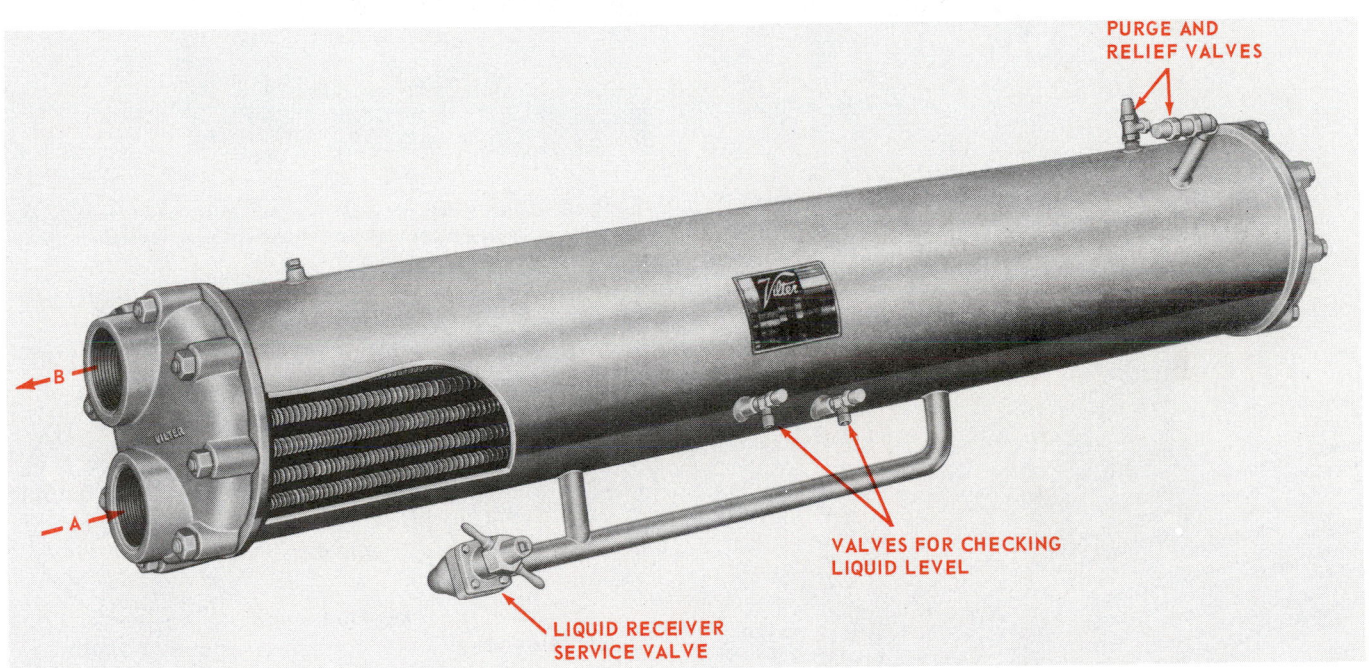

Fig. 12-41. A shell and tube type condenser-liquid receiver. Note the straight water pipes. They have fins for better heat transfer. Ends of receiver are removable to allow access for cleaning the water tube. A—Water in. B—Water out.   (Vilter Mfg. Corp.)

It has some advantages. It is compact, needs no fans and combines the condenser and receiver in one. It uses a number of straight tubes inside the receiver with a water manifold on both ends. When these manifold ends are removed, the water tubes can be cleaned of deposits easily. This is sometimes called a shell and pipe condenser.

tube cools the refrigerant in the outer tube, Fig. 12-43. The outside tubing is also cooled by air in the room. Double cooling improves efficiency.

This type of condenser may be constructed in a cylindrical, spiral or rectangular style. See Fig. 12-44.

Water enters the condenser at the point where the

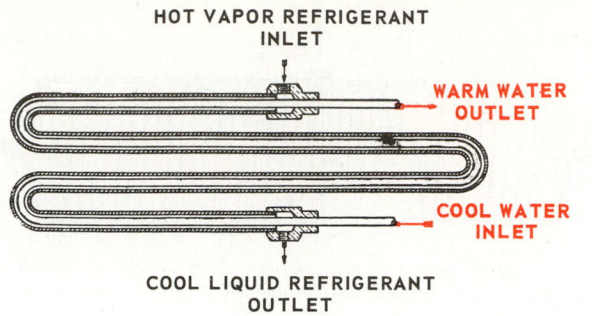

Fig. 12-43. Tube-within-a-tube condenser. Water flows through inner tube. Water flow is opposite vapor flow.

refrigerant leaves the condenser. It leaves the condenser at the point where the hot vapor from the compressor enters the condenser. This is called counterflow design. The warmest water is adjacent to the warmest refrigerant, and the coolest refrigerant to the coolest water. This condenser can also be made with hard copper pipe. See Fig. 12-45.

The rectangular type uses straight, hard copper pipe with manifolds on the ends. When the manifolds are removed, the water pipes can be cleaned mechanically. See Fig. 12-46. To increase heat transfer, the inner tube may be formed with a spiral groove flute (depression). See Fig. 12-47.

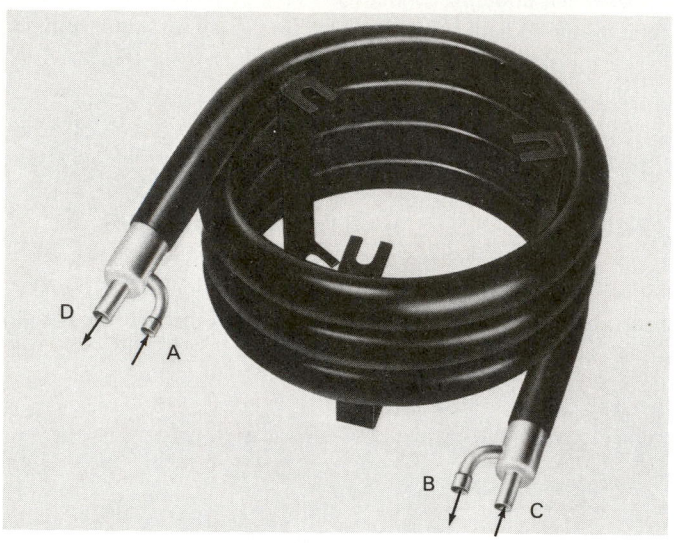

Fig. 12-44. A tube-within-a-tube condenser. A—Refrigerant in. B—Refrigerant out. C—Water in. D—Water out. (Packless Industries)

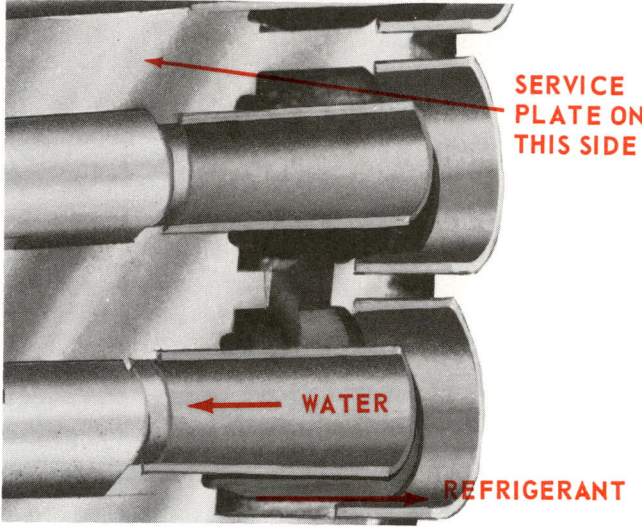

Fig. 12-46. Construction detail of pipe-within-a-pipe condenser. Note how both refrigerant tubes and water tubes are silver brazed into header.

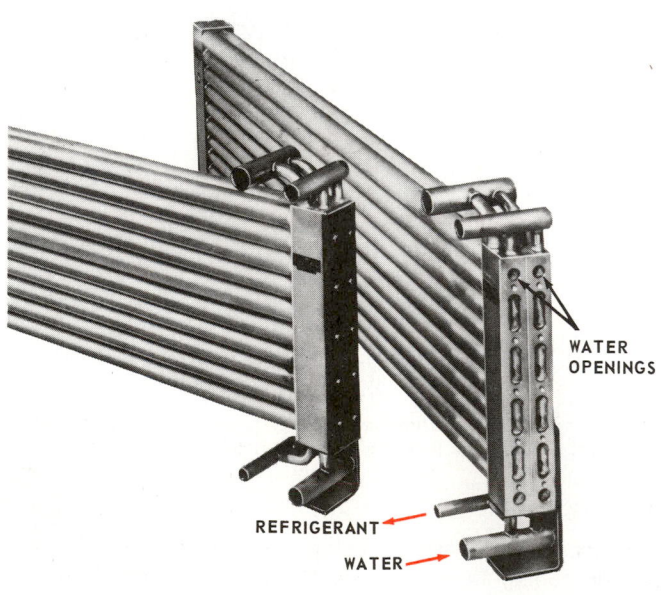

Fig. 12-45. Tube-within-a-tube condenser designed to permit cleaning water tubes (inner tube). Clean-out plate is removed in right-hand view. (Halstead and Mitchell)

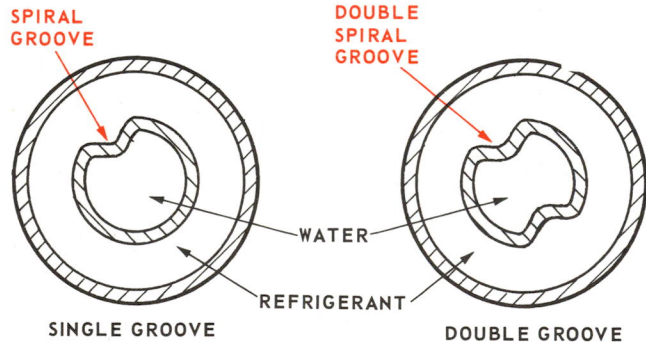

Fig. 12-47. A tube-within-a-tube condenser. Inner tube has a groove design to increase heat transfer.

12-13 COOLING TOWERS

In some areas, water contains chemicals causing it to be unsuitable for use as a coolant. In other localities, water may be very scarce, expensive, or its use may be limited by law.

To permit use of a water-cooled mechanism and to save on water consumption, water-cooling towers are used. These

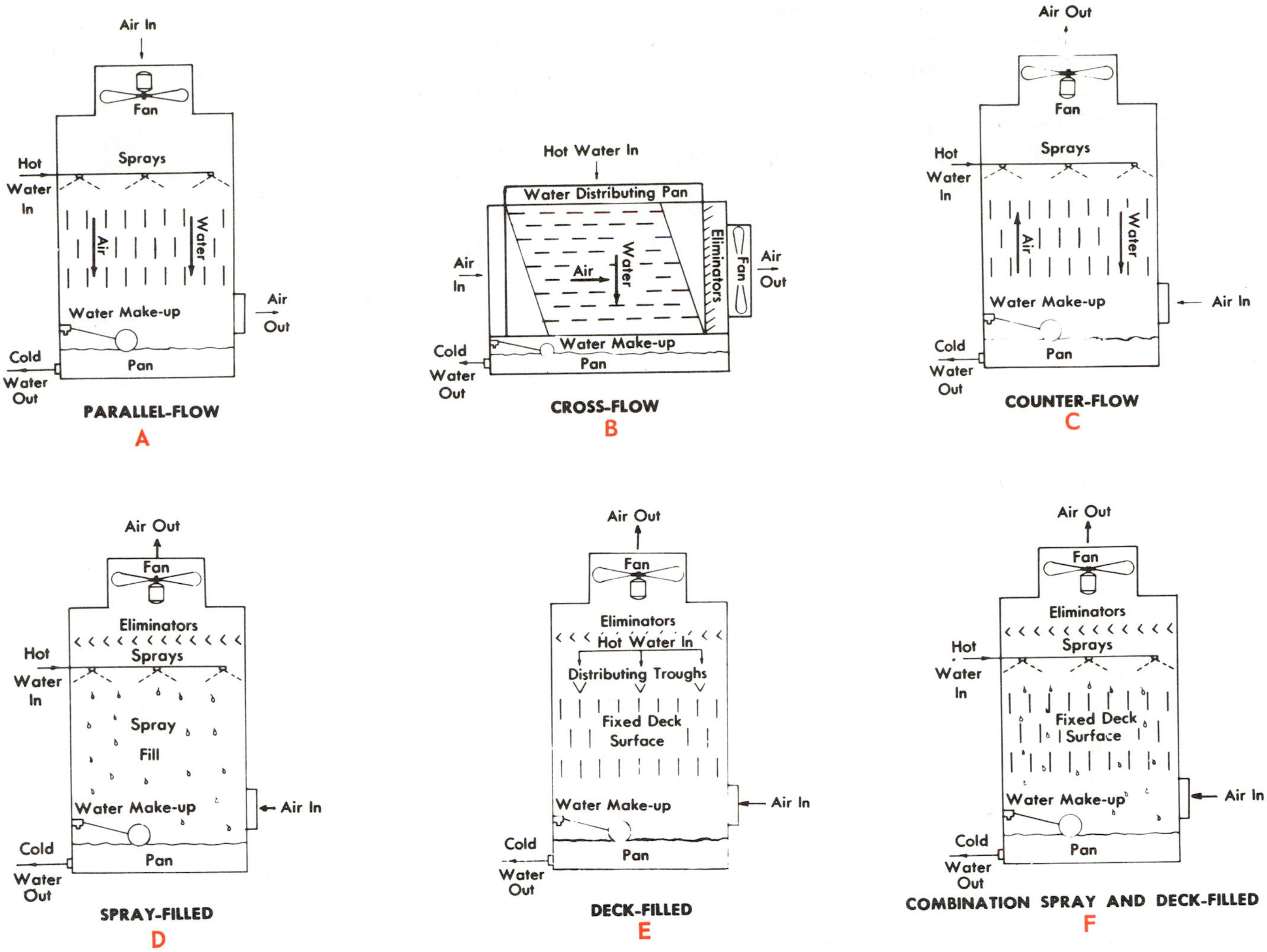

Fig. 12-48 diagrams

A — PARALLEL-FLOW: Air In / Fan / Sprays / Hot Water In / Air / Water / Water Make-up / Air Out / Cold Water Out / Pan

B — CROSS-FLOW: Hot Water In / Water Distributing Pan / Air In / Air / Water / Eliminators / Fan / Air Out / Water Make-up / Cold Water Out / Pan

C — COUNTER-FLOW: Air Out / Fan / Sprays / Hot Water In / Air / Water / Water Make-up / Air In / Cold Water Out / Pan

D — SPRAY-FILLED: Air Out / Fan / Eliminators / Sprays / Hot Water In / Spray Fill / Water Make-up / Air In / Cold Water Out / Pan

E — DECK-FILLED: Air Out / Fan / Eliminators / Hot Water In / Distributing Troughs / Fixed Deck Surface / Water Make-up / Air In / Cold Water Out / Pan

F — COMBINATION SPRAY AND DECK-FILLED: Air Out / Fan / Eliminators / Sprays / Hot Water In / Fixed Deck Surface / Water Make-up / Air In / Cold Water Out / Pan

Fig. 12-48. Six different types of cooling towers. A, B and C show three different airflow patterns. D, E and F show three different ways to vaporize some of the water.   (Baltimore Aircoil Co., Inc.)

towers serve the same purpose as spray towers used in large industrial refrigeration systems.

A variety of cooling tower designs are shown in Fig. 12-48. Cooling towers can be rather noisy. They should, therefore, be located away from noise-sensitive areas such as offices, restaurants and living quarters.

One system connects the water lines of the condenser to a water coil in an enclosure. A pump forces the water through the condenser and then through the coil in the tower. The tower coil is pierced with holes, and the water is sprayed into the enclosure.

Air rushing through the sprayed water evaporates some of it. Evaporation cools the remaining water to the outdoor temperature or even lower (wet bulb temperature).

Motor-driven fans control airflow through a cooling tower. Fig. 12-49 shows a cooling tower fan.

Cooled water collects in the bottom of the enclosure and passes through a screen to remove leaves or other foreign material. Then it is recirculated through the condenser.

A float-controlled valve in the lower water pan adds more water as needed. This float operates like a refrigerant low-side

Fig. 12-49. Base of cooling tower showing motor and fan installation.

DOUBLE V-BELT DRIVE

BELT TENSION ADJUSTMENT

float mechanism. A drain continually bleeds some water out of the water pan to keep water hardness to a minimum. Chemicals may be added to the water to retard rust formation, algae, fungus growth and the like.

Cooling towers are made of corrosion resistant materials. Among these are steel (zinc dipped after assembly), copper, stainless steel, plastic or treated wood.

The more water surface in contact with the air flowing through a cooling tower, the more efficient the cooling action. Most towers have some arrangement so that water flows over materials in thin films. This material, usually called fill, provides a surface for this. These fills are made of many materials: metal fins, wood slats, plastic, asbestos-plastic, and asbestos-cement. The shapes of the surfaces vary from Z-shaped, honeycomb, embossed, flat sheet to corrugated sheet. The cellular (honeycomb) fill is becoming very popular.

The distribution system (nozzles, troughs, V-notches) must be kept clean and must distribute the water evenly to prevent scale buildup.

Cooling towers do not need freezing protection while operating. Electric heat is often used to keep the water from freezing during shutdown periods. Immersion and convection heaters have been used. An electric water heater in the pump circuit may also be used.

Use only coarse screens on pump inlet and be sure all suction lines are below water level in the cooling tower, otherwise air may enter the suction line causing pump volume to drop. Pumps can be damaged in this way.

Put fine screens on pump outlets. The water pump should push water through the system to prevent low water pressures in the condenser tubes or pipes.

The cooling tower has overflow pipes that carry any excess water to the drain system of the building. Some have air control flow to prevent freezing if wet bulb or dry bulb temperature goes below 32 F. (0 C.). The air outlet thermostat operates the fan dampers.

Electric heat, hot water or steam can be used to keep the water reservoir from freezing during the winter. Pipes may need insulation or electric heater tape.

Cooling towers evaporate about two gallons of water every hour for each ton of refrigeration.

One gallon of water weighs about 8.3 lb. About 1000 Btu are needed to evaporate 1 lb. of water. Then, for one gallon of water, it will require 8.3 x 1000 = 8300 Btu. For two gallons, it would require 8300 x 2 = 16,600 Btu.

See manufacturer's catalogs for details of sizes and capacities.

12-14 EVAPORATIVE CONDENSERS

The evaporative condenser system carries the refrigerant into a condenser which is in an enclosure much like a cooling tower. In this system — as the word "evaporative" indicates — water is sprayed or drips over the condenser. This cools it. The water cycle is in the condenser cabinet only. See Fig. 12-50.

Usually, the evaporative condenser is mounted outdoors. However, it may be used indoors if air ducts are provided to the outside.

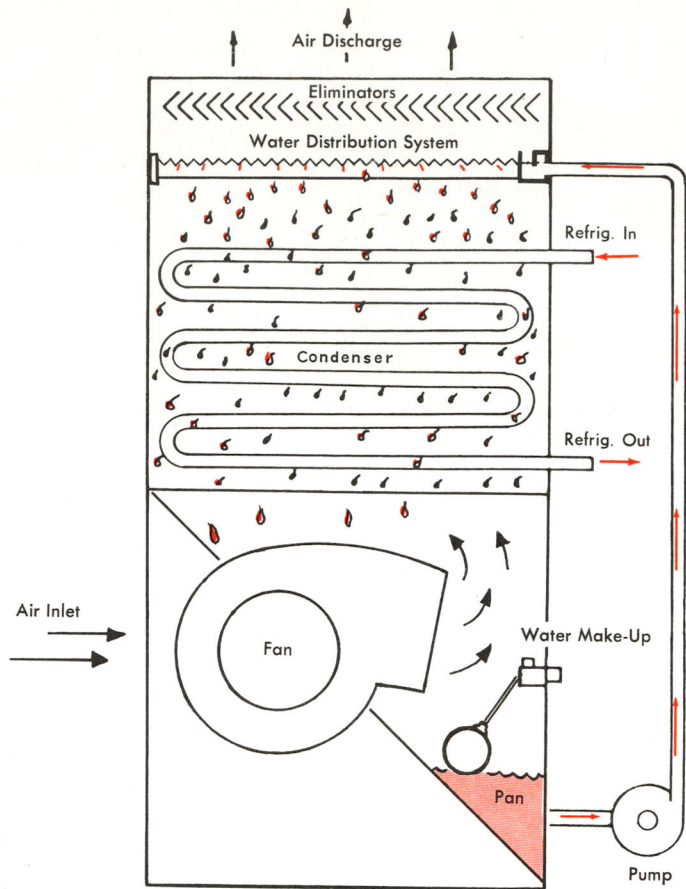

Fig. 12-50. Evaporative condenser. Note air and water flow. Water make-up is controlled by float valve.

Some systems pump water to a trough above the condenser. The water then drips over coils as air is forced through them. A thermostat can be used to control the water flow. A fan blows air over the condenser whenever the condensing unit is operating. The condenser is cooled by air alone until the condenser temperature reaches 80 F. (27 C.) or more. Water cooling is then turned on by a thermostat.

Some units subcool the refrigerant leaving the receiver by piping the receiver outlet (liquid line) into and out of the evaporative condenser. Temperature of the refrigerant can be dropped 10 F. (6 C.) by subcooling. When the temperature reaches 45 F. (7 C.) or lower, the water is shut off. However, the condenser can still carry the load as an air-cooled condenser.

Some have water reservoirs inside the building. For protection from freezing weather, the reservoir must be large enough to hold all the water in the system. Fan dampers are sometimes used to decrease airflow as outside temperature drops.

12-15 LIQUID RECEIVER

The liquid receiver is a steel tank of welded construction. It is usually equipped with two service valves. One is a liquid receiver service valve mounted between the liquid receiver and the condenser. The other is located between the receiver and the liquid line (king valve). These two valves enable the service technician to disconnect the liquid receiver from the system separately.

Receivers should have safety devices. A thermal release plug provides minimal safety. Some receivers have both thermal and pressure releases. See Para. 12-65. A special line should be installed on relief valves to carry released refrigerant outdoors.

Receivers may be mounted either vertically or horizontally. The horizontal style usually hangs underneath the compressor and motor frame. Some are provided with a device — sight glass, magnet floats or valves — for determining the level of the liquid refrigerant.

If there is a water coil inside the liquid receiver, the shell is just like the receivers having no water coil. However, it is usually larger for the same size compressor unit. The receiver should be large enough to hold all the refrigerant in the system.

12-16 COMMERCIAL EVAPORATORS

Because customers demand great variety in commercial refrigeration, special evaporator designs are required for many installations. These evaporators vary from coils of tubing immersed in a sweet water bath to forced-circulation evaporators which have the air blown over them or by a motor-driven fan.

Evaporators may be divided into two main groups:
1. Those used for cooling air; in turn, the air cools the contents of the cabinet.
2. Those submerged in a liquid such as brine or a beverage.

Evaporators for cooling air are of two principal types:
1. Natural convection.
2. Forced convection.

In natural air convection evaporators, air circulation depends on gravitational (warm air rises — cool air descends) or thermal circulation. Natural convection, air cooling evaporators fall into three classes:
1. Frosting.
2. Defrosting.
3. Nonfrosting.

The conditions under which an evaporator must work determine its classification. The governing conditions are the desired temperature range of the cabinet and the temperature difference between the evaporator and the cabinet.

12-17 FROSTING EVAPORATORS

A frosting evaporator builds up frost continuously when in use. It usually operates at 5 F. (—15 C.) refrigerant temperature cut-out and 25 F. (—4 C.) refrigerant temperature cut-in. The machine must be manually or automatically shut down from time-to-time to rid the system of frost.

Frost which forms on the evaporator comes from moisture in the air. This leaves the air in the cabinet dry. Dry air rapidly dries out food.

If the refrigerant temperature is below 28 F., the evaporator needs heat energy of some kind to defrost the evapo-

rator. Or the evaporator must be turned off for a longer time than the normal cycle.

Some evaporators run at extremely low temperatures to keep the fixture cool. This allows frost and ice to build up. As the frost grows thicker, it reduces the cooling efficiency of the evaporator. These evaporators are used in low-temperature and frozen food fixtures.

12-18 DEFROSTING EVAPORATORS

Many evaporators run on what is called a defrosting cycle. While the condensing unit is running, the temperature of the evaporator is low. This causes frost to accumulate on it. But after the compressor shuts off, the coil warms up above 32 F. (0 C.) and the frost melts. During the running of the compressor, the evaporator will remain at temperatures of about 20 to 22 F. (—7 to —6 C.).

This defrosting process is called air defrosting. It clears the evaporator surfaces of frost and provides efficient heat transfer. It also keeps a high humidity — about 90 to 95 percent RH (relative humidity). However, this sacrifices temperature differences between the evaporator refrigerant and the air in the cabinet. Greater evaporator area is needed to make up for this loss.

The defrosting evaporator sometimes presents a problem. In some installations, the top of the evaporator may defrost and moisture flows down the evaporator surface. Before it has time to escape, it may freeze around the lower parts of the evaporator. This ice accumulation on the evaporator fins may eventually block air circulation around the evaporator and interfere with proper refrigeration. Fig. 12-51 is a sketch of a defrosting evaporator.

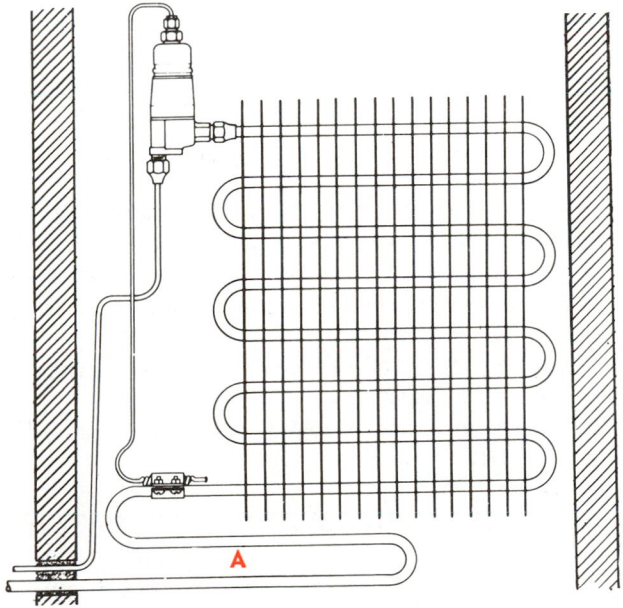

Fig. 12-51. "Drier" coil at A prevents sweating or frosting of suction line outside fixture. This evaporator is usually the defrosting type. Frosted surfaces transfer heat poorly.

12-19 NONFROSTING EVAPORATORS

Nonfrosting evaporators operate at temperatures which do not go much below 32 F. (0 C.). Frost, therefore, does not form on the evaporator.

Occasionally, the evaporator may build up a light coat of frost just before the compressor shuts off. This frost immediately melts on the off-cycle. These evaporators operate at a temperature of 33 F. (.6 C.) to 34 F. (1 C.). However, the temperature of the refrigerant inside the evaporator will be around 20 F. (−7 C.) to 22 F. (−6 C.). Since they do not frost over, they draw only a little moisture from the inside of the cabinet. This is a big advantage, because it becomes possible to maintain a relative humidity (RH) of 75 to 85 percent in these cabinets. This helps to keep the produce fresh and stops shrinking in weight. Nonfrosting evaporators look like the one shown in Fig. 12-52.

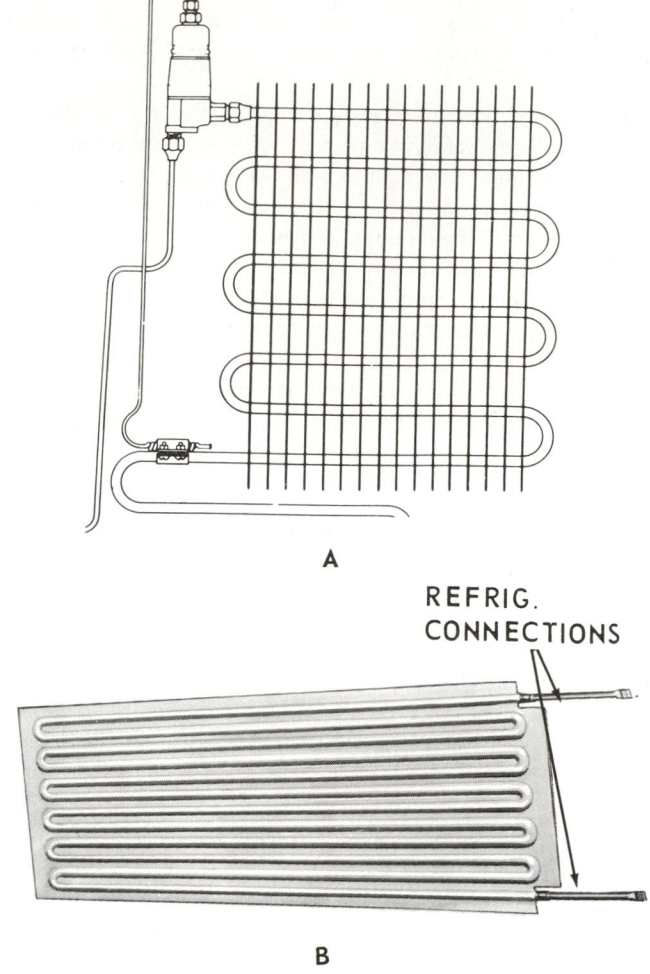

A

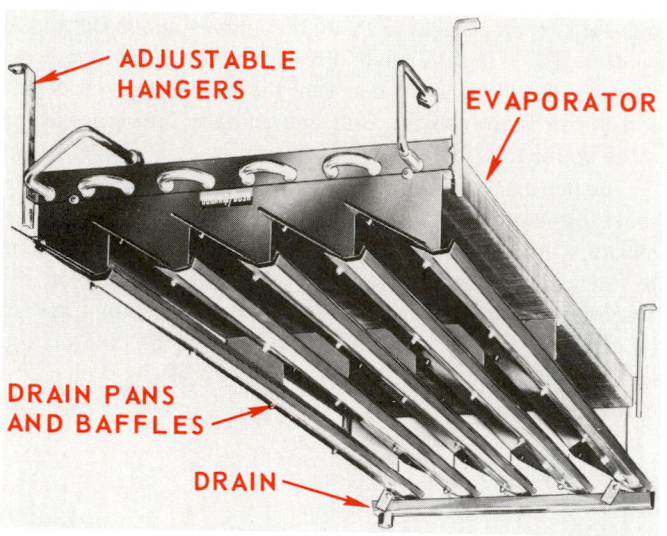

Fig. 12-52. Nonfrosting evaporator. Note baffles used to collect and remove condensate.

B

Fig. 12-53. Two types of evaporators. A—Fin type. (Detroit Controls Corp.) B—Plate type. Refrigerant tubes are formed by two plates welded together. (Kold-Hold Div., Tranter Mfg., Inc.)

Two types of evaporator construction are commonly used. The fin type is shown at A in Fig. 12-53, and the plate type at B in Fig. 12-53. Some of the plate type evaporators use a eutectic solution which helps increase the cycling interval and produces a more constant temperature. (Eutectic means having the lowest melting point possible.) Fig. 12-54 illustrates an eutectic plate type evaporator.

These evaporators may be slightly larger than frosting types of equal capacity.

12-20 FORCED CIRCULATION EVAPORATOR

Forced circulation evaporators are a compact arrangement of refrigerant-cooled tubes and fins. A fan blows air over it. The fan is driven by an electric motor. Both the evaporator and the fan are usually enclosed in a metal housing. The evaporators take up only a small amount of space, and do not need any extra baffling. See Fig. 12-55.

Fig. 12-54. Plate-type evaporator designed to hold eutectic solution. The solution freezes at a certain temperature and latent heat of melting increases cycling time while holding more constant temperature. (Dole Refrigerating Co.)

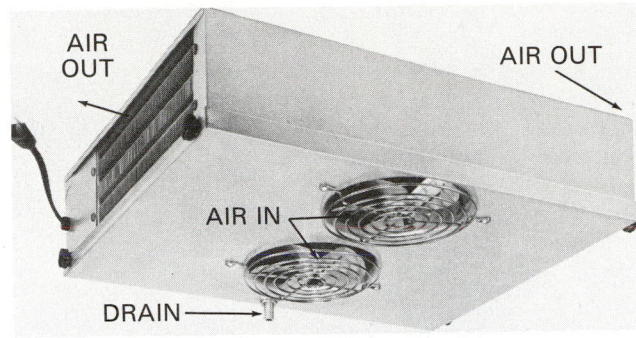

Fig. 12-55. Blower-type evaporator designed for ceiling mounting. Air enters at bottom, is cooled and exits at sides. (Peerless of America, Inc.)

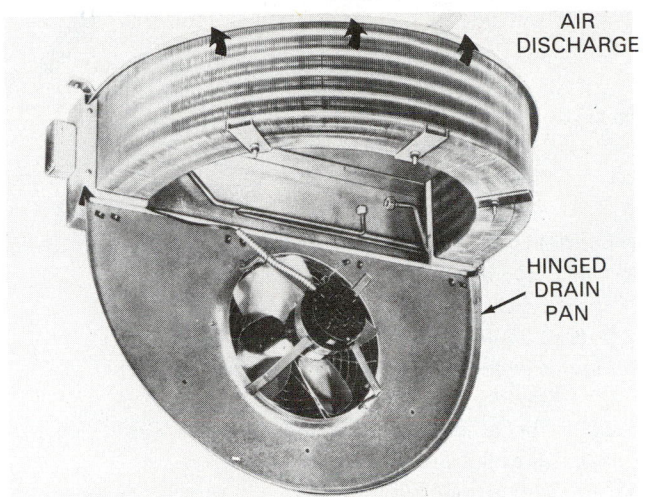

Fig. 12-56. Forced circulation evaporator with hinged drain pan for easy cleaning. Note small amount of space needed for installation. Air enters at fan and discharges 180 deg. at coil. (Bohn Aluminum & Brass Div., Gulf & Western Mfg. Co.)

These evaporators, Fig. 12-56, do have a tendency to cause rapid dehydration (drying) of foods unless special care is taken. If the evaporator is large, operates at a small temperature difference (10 F. to 12 F. or 6 C. to 7 C.), and the air is circulated slowly, the drying of foods will be cut down.

In installations where dehydration is not harmful and air may flow through the evaporator rapidly, small evaporators are practical. They are operated at a greater temperature difference, 20 F. to 30 F. (11 C. to 17 C.). Thermostatic expansion valve refrigerant controls are usually used with forced convection evaporators.

The fan motor may be from 1/50 hp and up and may run continuously. It also may be controlled by the evaporator or fixture temperature. Refrigerant temperature is usually held quite low, but rapid air circulation keeps the evaporator from frosting up. However, considerable sweating does occur and drainage must be provided in the installation.

Blower evaporators operated at low refrigerant temperatures need special defrosting care. Fin spacing is small, the evaporator surface is small and frost or ice may interfere drastically with heat transfer.

Fig. 12-57. Slope front blower evaporator. Compact unit used in low headroom fixtures. (Peerless of America, Inc.)

The forced circulation evaporator will cool a refrigerator cabinet quickly. It is well suited for air conditioning as well as for refrigerator cabinets used to store bottled beverages or foods in sealed containers. Figs. 12-57 and 12-58 illustrate two

Fig. 12-58. Vertical, flat-type blower evaporator designed to be mounted behind either window or door frames of fixture. (Peerless of America, Inc.)

of these evaporators. Figs. 12-59 and 12-60 show evaporators for large fixtures. Each has two fans for air return. A high capacity blower evaporator used for low temperature fixtures is shown in Fig. 12-61.

A condensate pump, Fig. 12-62, may be used to provide a positive method of removing the condensate. The pump is mounted on the drain and is self-priming. It uses only about 10W and operates continuously.

Many evaporators use a hot gas bypass system to maintain above freezing temperatures when the cooling load is de-

Fig. 12-59. Motor drives two propeller-type fans and cooled air exits at ends of this evaporator.

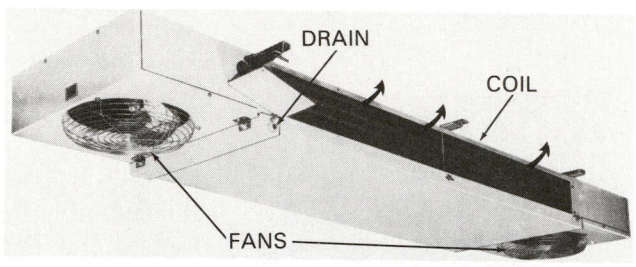

Fig. 12-60. Low velocity blower evaporator draws air at fans and discharges through coil from both sides. (Bohn Aluminum & Brass Div., Gulf & Western Mfg. Co.)

Fig. 12-61. Low temperature blower evaporator. The unit has two direct drive propeller fans drawing air into coil face and discharging through the fan grille. (Bohn Aluminum & Brass Div., Gulf & Western Mfg. Co.)

creased. Fig. 12-63 shows a method of providing the necessary hot gas to keep the evaporator from frosting.

12-21 LIQUID-COOLING EVAPORATOR

Refrigeration makes drinks more enjoyable, preserves liquid, improving a manufacturing process and minimizes

Fig. 12-62. Condensate pump is used on evaporator drains when condensate is to be raised above drain pan level.

evaporation. It is used especially for water, soft drinks, alcoholic beverages and brines.

Three types of liquid cooling evaporators are used:
1. Bottled liquids.
2. Liquids under atmospheric pressure.
3. Liquids under pressure.

When the evaporator is mounted inside the liquid being cooled, it is called an "immersed" evaporator.

12-22 IMMERSED EVAPORATOR (BRINE)

In the immersed type of liquid cooling, the evaporator may be surrounded by a brine, beverage, or water. A small, plain, tube-type evaporator, submerged in a liquid, provides good heat transfer.

Immersion makes the evaporator more efficient because liquids transfer heat to metals faster than air. An efficiency ratio of 50 or 100 to 1 is common. That is, a submerged evaporator can remove 50 to 100 Btu per hour, per degree temperature difference, per square foot of evaporator surface. Air-cooling evaporators can only remove 1 Btu under the same conditions.

These evaporators may use either a low-side float or a thermostatic expansion valve refrigerant control when used in multiple installations. Smaller, self-contained installations normally use a capillary tube refrigerant control.

The cooled liquid may be circulated and used for various purposes.

12-23 IMMERSED EVAPORATOR (SWEET WATER)

Another type immerses the evaporator in ordinary tap water, called a sweet water bath. The water-cooling device used in soda fountains is of this type. A coil of tubing in which

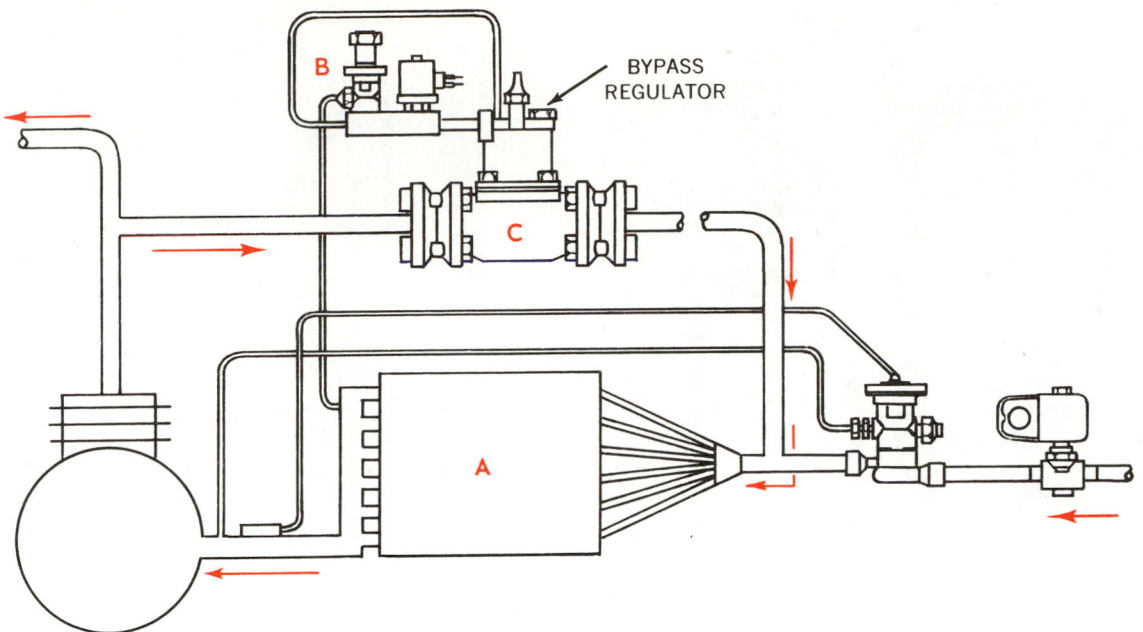

Fig. 12-63. A hot gas bypass system used to prevent a frosting evaporator during low head conditions. When the pressure in the evaporator, A, decreases to freezing conditions, the pilot valve at B opens. This opens the large valve at C, allowing enough hot gas to enter the evaporator to prevent frosting.   (Alco Controls Div., Emerson Electric Co.)

the water is to be cooled and consumed by the customer may also be submerged in the same water.

This design allows the sweet water to freeze around the evaporator during what is called the nonload period. The light ice accumulation acts as a reserve of refrigeration.

These systems usually have a thermostatic expansion valve refrigerant control as shown in Fig. 12-64. The evaporator

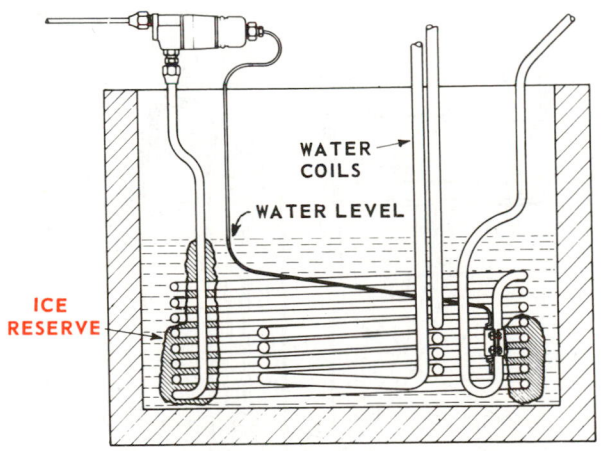

Fig. 12-64. Sweet water (tap water) bath is often used at soda fountains. Ice accumulation is controlled by location of thermostatic expansion valve thermal bulb in unit.

should reach to the bottom of the bath if the temperature of the bath is to be less than 39.1 F. (4 C.). Water, as it cools from 39.1 F. to 32 F. (4 C. to 0 C.) tends to expand and rise. Therefore, the coldest water will be at the top.

12-24 PRESSURE TYPE (BEVERAGE) EVAPORATOR

Included in the submerged class is the evaporator used in pressure type beverage coolers. The liquid refrigerant is carried in a tube submerged in the beverage to be cooled. The beverage itself is under pressure. Construction is similar to that of the liquid receiver type water-cooled condenser. This design is quite common for instantaneous cold water coolers. *Be careful that the temperature of the beverage is never so low that it will freeze to any extent.*

The all-metal beverage evaporator takes advantage of the high heat conductivity of aluminum. Two separate copper tubes are coiled in a helix (spiral) design. They are placed in a permanent mold and liquid aluminum is poured into the mold. After the liquid aluminum cools and becomes solid, the two tubes are completely enclosed in a hollow cylinder of aluminum, as in Fig. 12-65.

Heat transfer is excellent. Refrigerant is evaporated in one coil, and heat will flow from the liquid in the other coil.

Surrounded and encased by the casting, the coil is strong enough to sustain the pressure of water freezing in it.

Stainless steel tubing may also be used. For storing some beverages, it is the only suitable material.

Coolers having three or more separate tubes encased by aluminum are used when more than one liquid must be cooled.

12-25 ICE CUBE MAKER EVAPORATOR

There are many different ice maker mechanisms. The simplest and the one used in domestic refrigeration mechanisms, freezes water in ice cube trays.

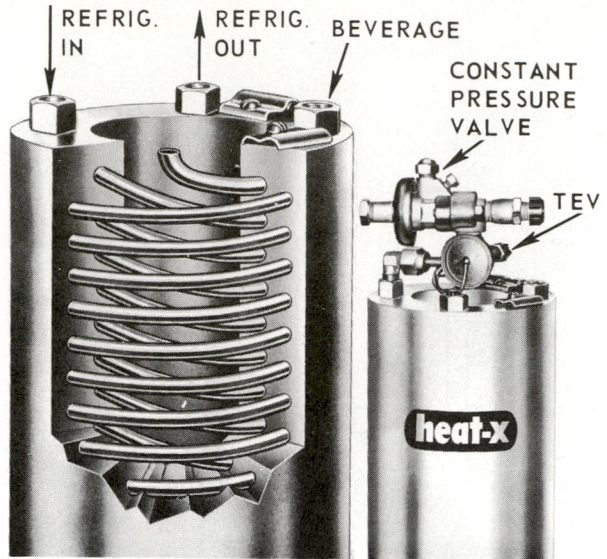

Fig. 12-65. Dry-type beverage cooling evaporator. Aluminum casting, surrounding both refrigerant and beverage coils, permits rapid heat transfer. Refrigerant control is a thermostatic expansion valve. Constant pressure valve prevents freezing of beverage.
(Dunham-Bush, Inc.)

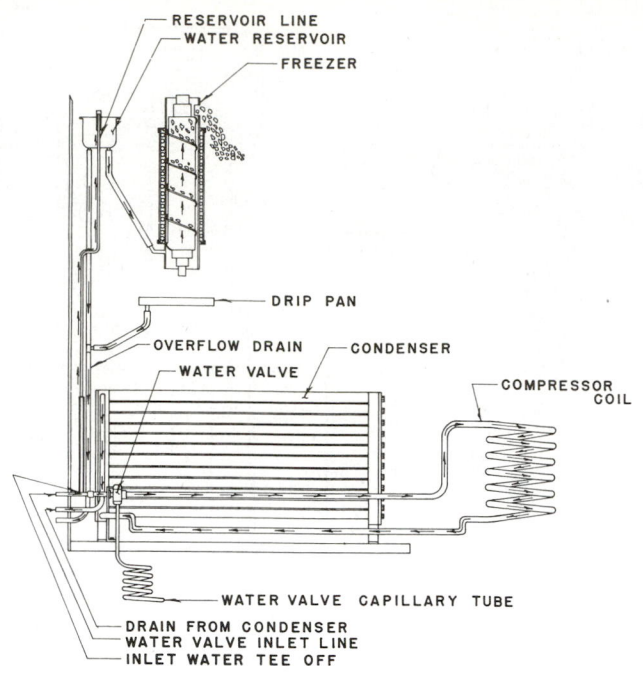

Fig. 12-67. Water circuit of ice flake or ice chip-making unit. Inclined screw scrapes ice from inside of refrigerated cylinder as it turns.
(Queen Products Div., King-Seeley Thermos Co.)

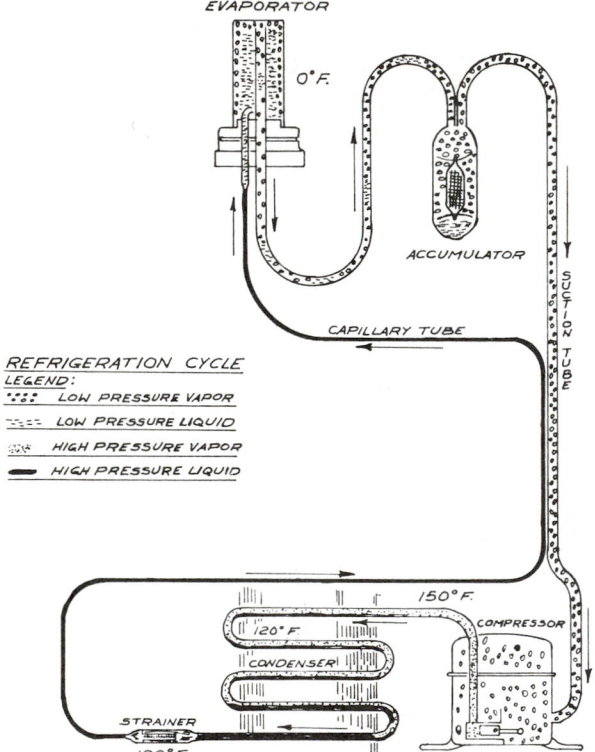

Fig. 12-66. Refrigeration cycle for self-contained ice maker which produces flake ice. (Ross-Temp, Inc.)

Since a large portion of artificial ice is used for beverage cooling, ice in shapes other than cubes and in smaller sizes is frequently desirable. One type produces ice flakes. The cycle diagram for a flake ice maker is shown in Fig. 12-66.

Water is made to flow over an evaporator shaped like a

cylinder. The surface of the evaporator is cold (0 F. or −18 C.) so that the water is rapidly frozen.

Fig. 12-67 shows a similar machine which has a water-cooled condenser. Fig. 12-68 shows how the ice is removed

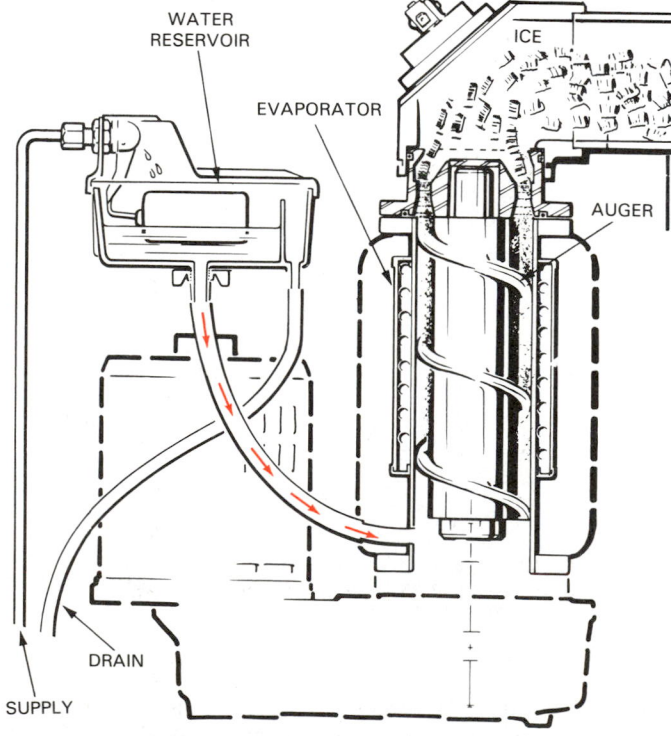

Fig. 12-68. Auger and evaporator of ice making unit. Note water flow to evaporator and ice formation.
(Queen Products Div., King-Seeley Thermos Co.)

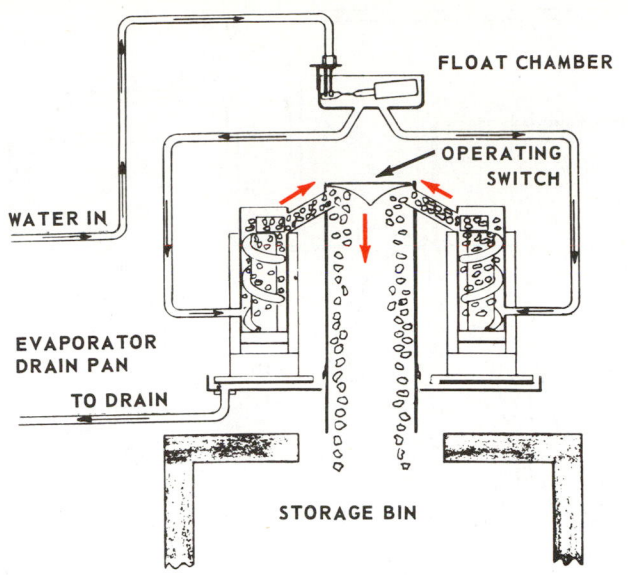

Fig. 12-69. Automatic flake ice system shows water and ice system. Water flow is float controlled. When enough ice flakes have been made, operation switch shuts off refrigerating unit and auger motors. (Ross-Temp, Inc.)

Fig. 12-71. Commercial ice flake maker. (Howe Corp.)

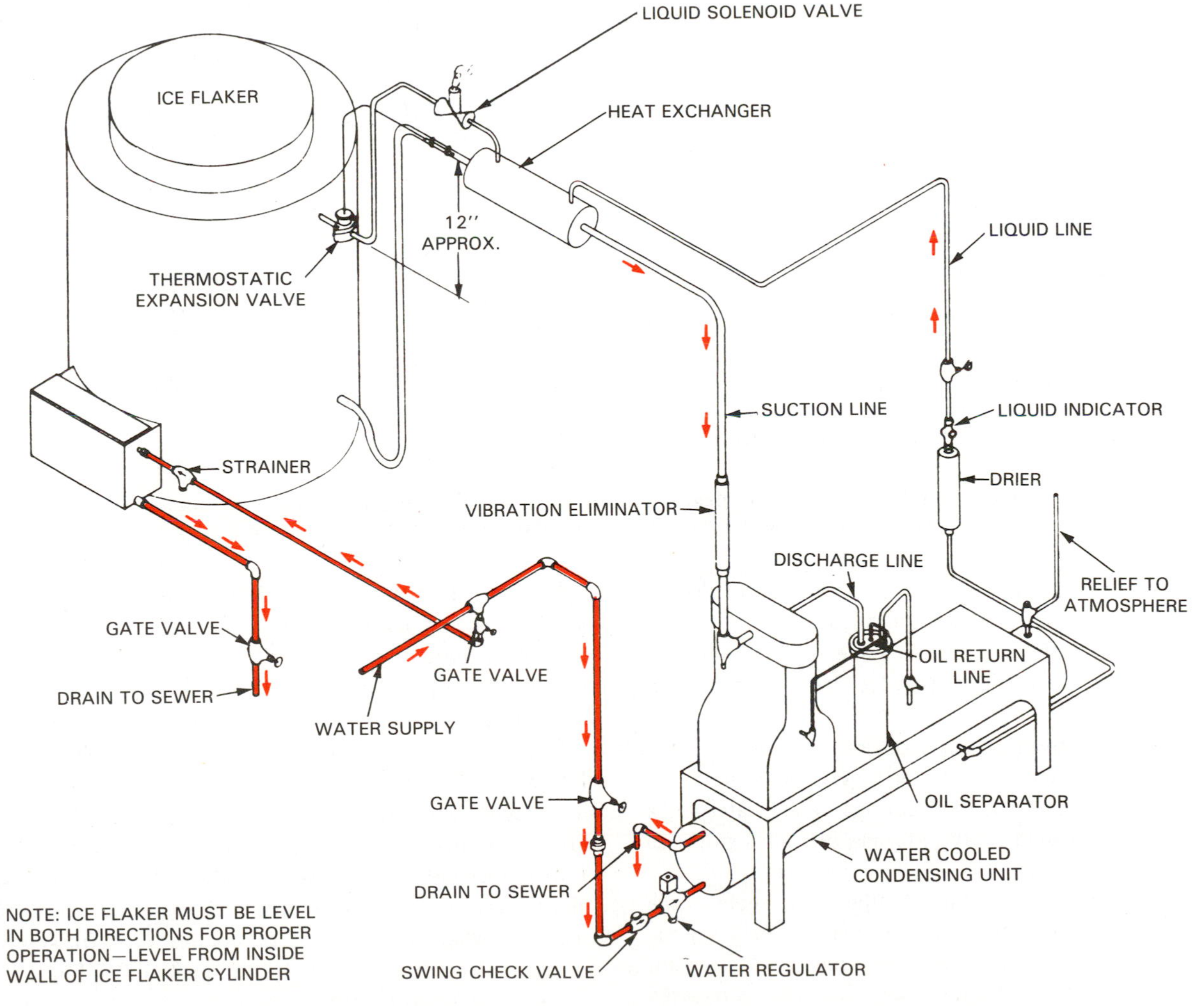

NOTE: ICE FLAKER MUST BE LEVEL IN BOTH DIRECTIONS FOR PROPER OPERATION—LEVEL FROM INSIDE WALL OF ICE FLAKER CYLINDER

Fig. 12-70. Flake ice refrigeration system. Water circuit is in red. (Howe Corp.)

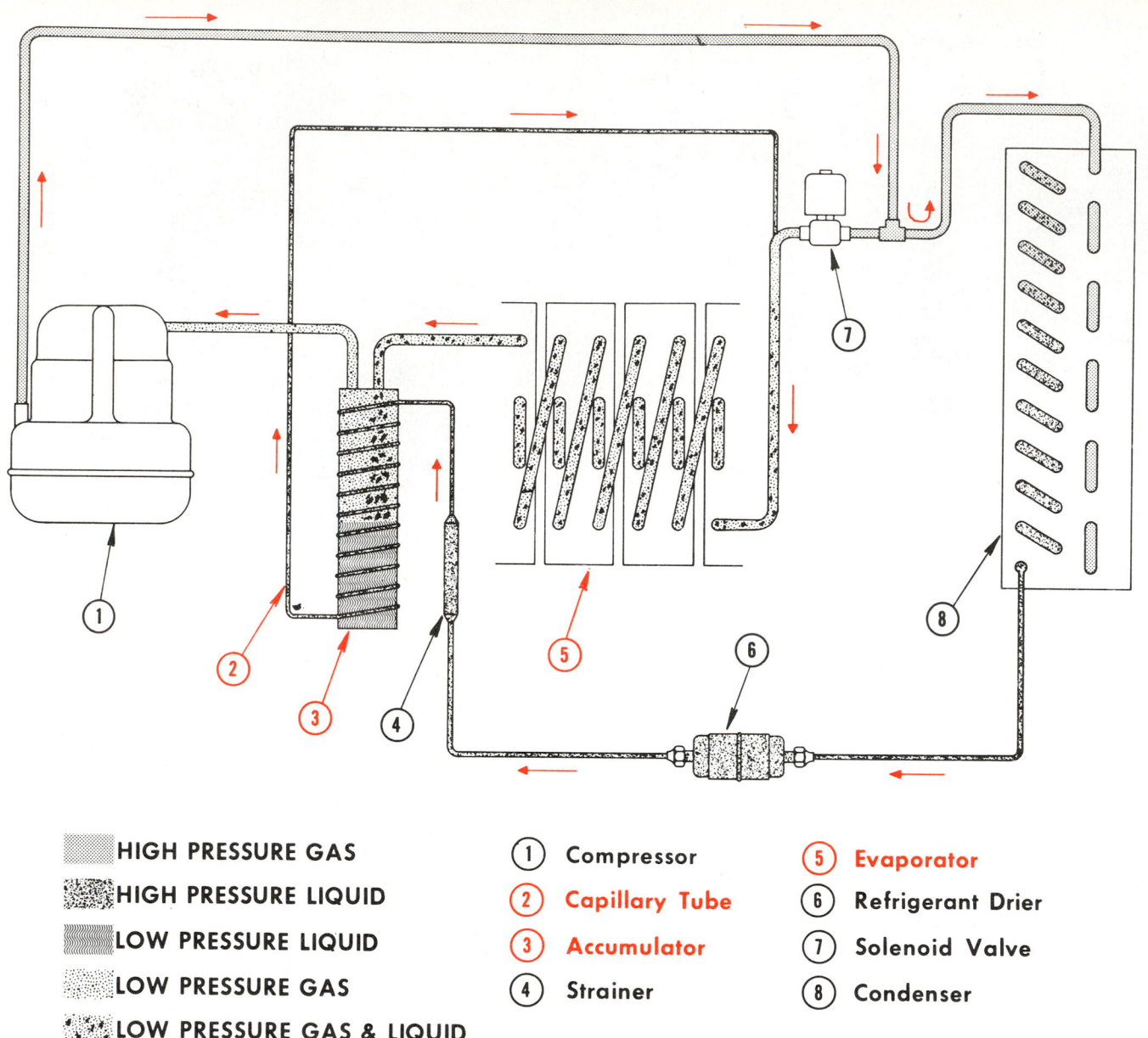

HIGH PRESSURE GAS
HIGH PRESSURE LIQUID
LOW PRESSURE LIQUID
LOW PRESSURE GAS
LOW PRESSURE GAS & LIQUID

① Compressor   ⑤ Evaporator
② Capillary Tube   ⑥ Refrigerant Drier
③ Accumulator   ⑦ Solenoid Valve
④ Strainer   ⑧ Condenser

Fig. 12-72. Ice cube maker refrigeration cycle during freezing process. Note capillary tube and accumulator. Cubes are released from evaporator tubes when solenoid valve at 7 opens and hot gas is admitted to evaporator.

from the evaporator. A heavy steel auger driven by an electric motor cuts and scrapes the ice from the surface.

Fig. 12-69 is a diagram of an ice flake maker using two evaporators and two harvesting augers. Ice flakes are fed into a storage bin.

Fig. 12-70 shows another cycle diagram for an ice flake maker. Fig. 12-71 shows the exterior view. All of these ice flake makers provide for float level control of the water and a shutoff mechanism located in the storage bin. This stops ice making when the bin is full.

The square or cylindrically shaped cube with a hole in the center is formed by running water through a tube-within-a-tube. Ice forms on the inside of the inner tube. The refrigerant flows between the inner and outer tube. The cube is released from the tube during the defrost cycle.

Fig. 12-72 shows the process during the freezing part of the

cycle. Fig. 12-73 shows the defrost cycle (hot gas) while Fig. 12-74 shows the water circuit with the recirculating pump and ice cube-forming tubes.

Another style of ice cube maker is the one in Fig. 12-75. Water flows through vertical stainless steel tubes. When a hollow square length of ice is formed, the refrigeration stops. Hot gas defrosting starts and, as the long square rods of ice slide down the tubes, they are cut into cubes. When all the tubes are empty, the refrigeration cycle starts over again.

In the ice cube maker shown in Fig. 12-76, water is sprayed up into very cold ice cube molds until the molds are filled with clear ice. When hot water is pumped around the metal cube container the cubes fall into the storage bin. See Fig. 12-77.

Electrical circuit controls are needed with automatic ice makers. These include bin level controls, ice cube size controls and water pump controls. The diagram in Fig. 12-78 is typical

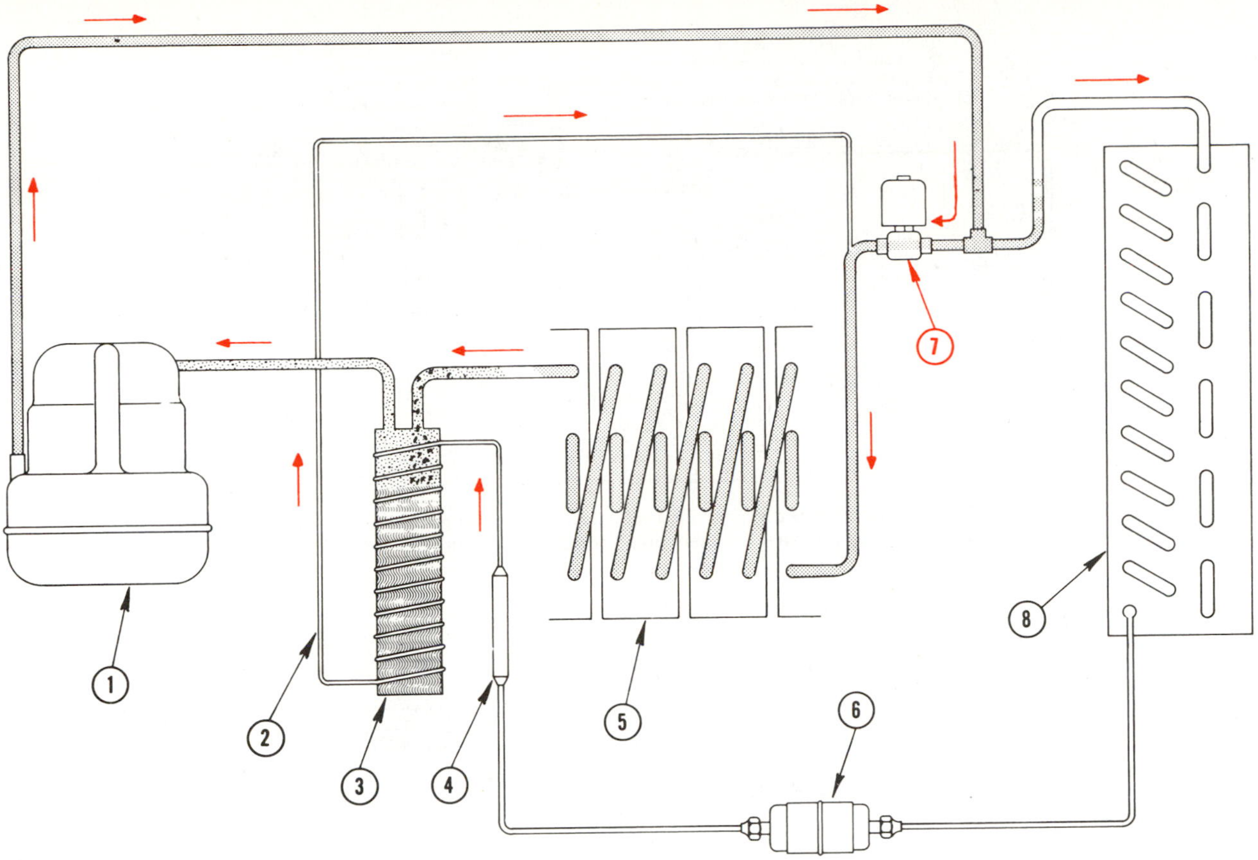

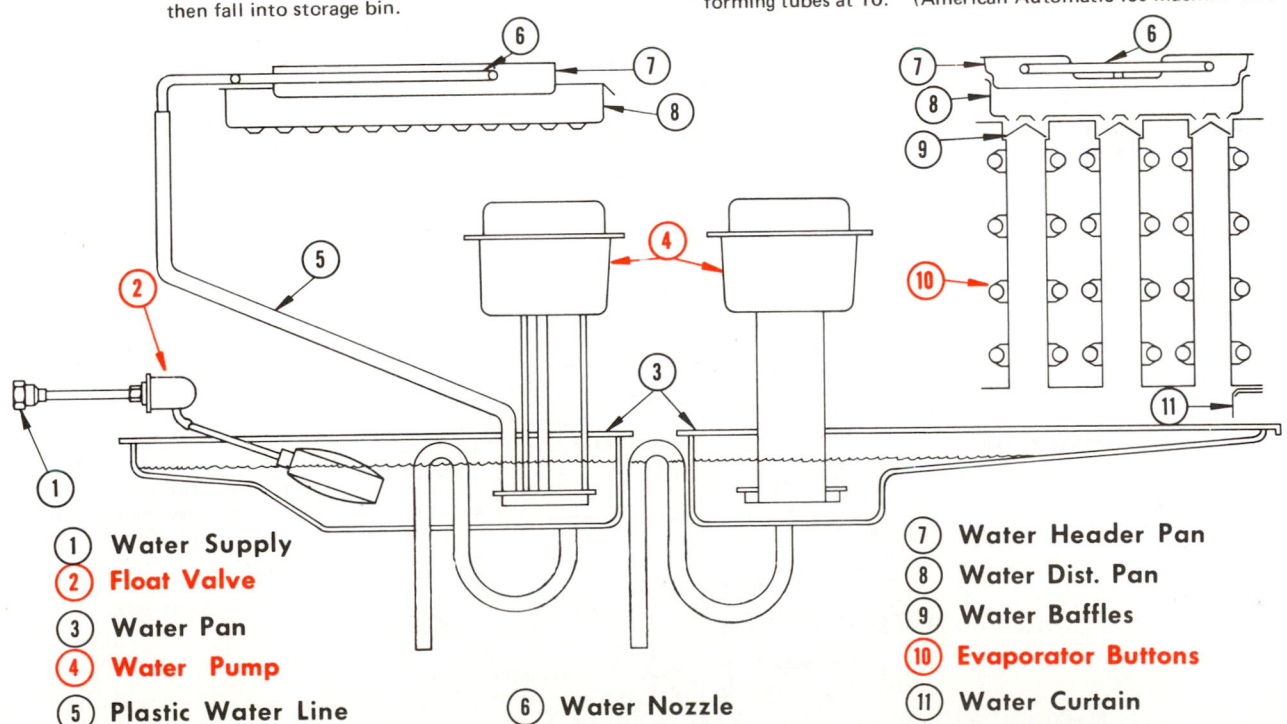

1
2
3
4
5
6
7
8
HIGH PRESSURE GAS
LOW PRESSURE LIQUID
LOW PRESSURE GAS
LOW PRESSURE GAS & LIQUID
1 Compressor
2 Capillary Tube
3 Accumulator
4 Strainer
5 Evaporator
6 Refrigerant Drier
7 Solenoid Valve
8 Condensor

Fig. 12-73. Above. Defrost cycle of automatic ice cube maker. Solenoid valve, at 7, is open. Hot gas heats and releases ice cubes which then fall into storage bin.

Fig. 12-74. Below. Water and ice circuit of automatic ice cube maker. Note water control float at 2, recirculation pumps at 4, and ice cube forming tubes at 10. (American Automatic Ice Machine Co.)

1
2
3
4
5
6
7
8
9
10
11
1 Water Supply
2 Float Valve
3 Water Pan
4 Water Pump
5 Plastic Water Line
6 Water Nozzle
7 Water Header Pan
8 Water Dist. Pan
9 Water Baffles
10 Evaporator Buttons
11 Water Curtain

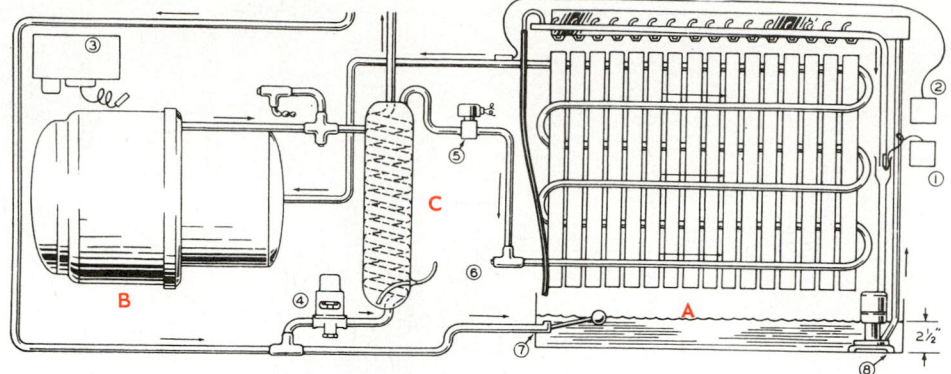

① T-1 OVERFLOW THERMOSTAT
② T-2 SUCTION LINE THERMOSTAT
③ COMBINATION BIN SAFETY THERMOSTAT & H.P. CUT-OUT
④ CONDENSER WATER REGULATING VALVE
⑤ HOT GAS SOLENOID VALVE
⑥ RESTRICTOR TUBE
⑦ MAKE-UP WATER VALVE
⑧ WATER RECIRCULATING PUMP

Fig. 12-75. Automatic ice cube making cycle. System uses hot-gas defrost. A—Evaporator and ice tubes. B—Compressor. C—Water-cooled condenser. (York Div., Borg-Warner Corp.)

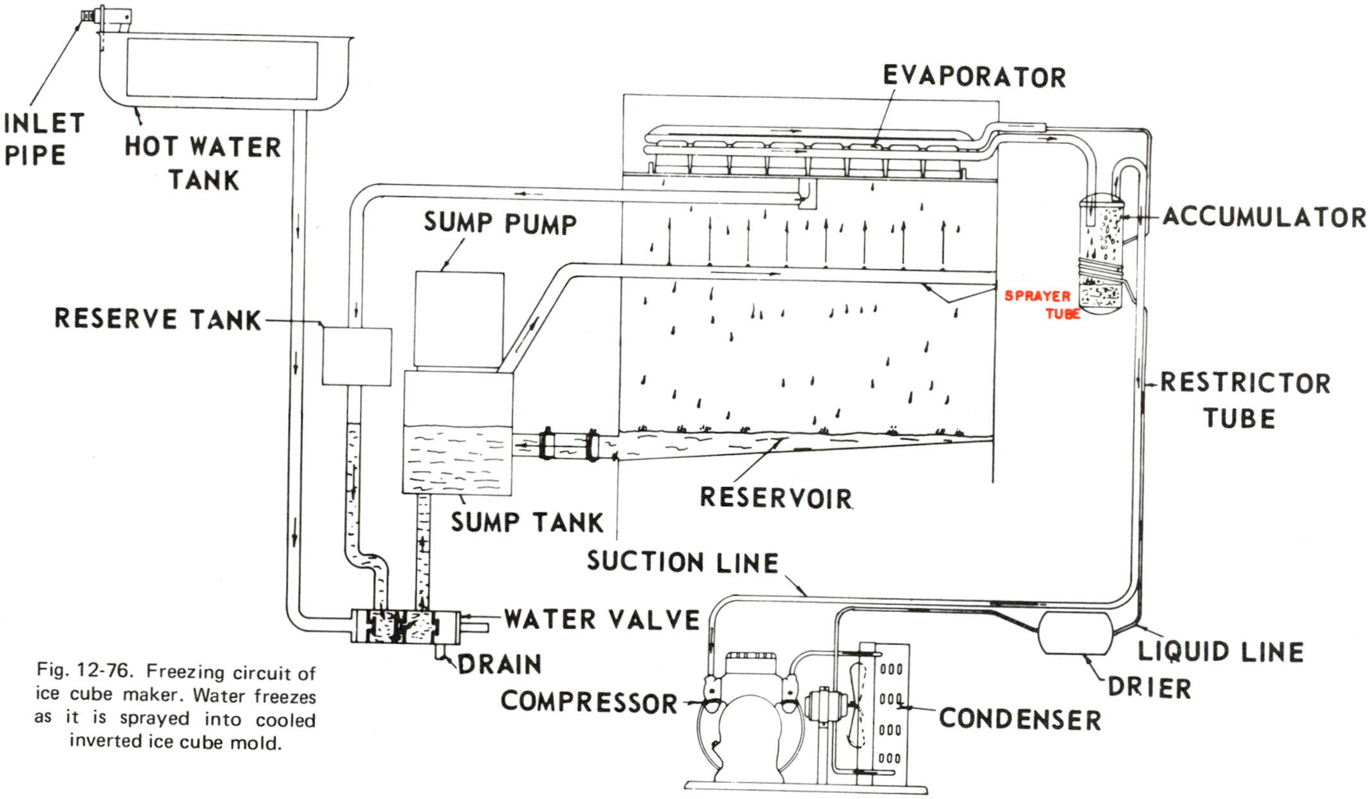

INLET PIPE

HOT WATER TANK

EVAPORATOR

ACCUMULATOR

SUMP PUMP

SPRAYER TUBE

RESERVE TANK

RESTRICTOR TUBE

RESERVOIR

SUMP TANK

SUCTION LINE

WATER VALVE

DRAIN

COMPRESSOR

CONDENSER

LIQUID LINE DRIER

Fig. 12-76. Freezing circuit of ice cube maker. Water freezes as it is sprayed into cooled inverted ice cube mold.

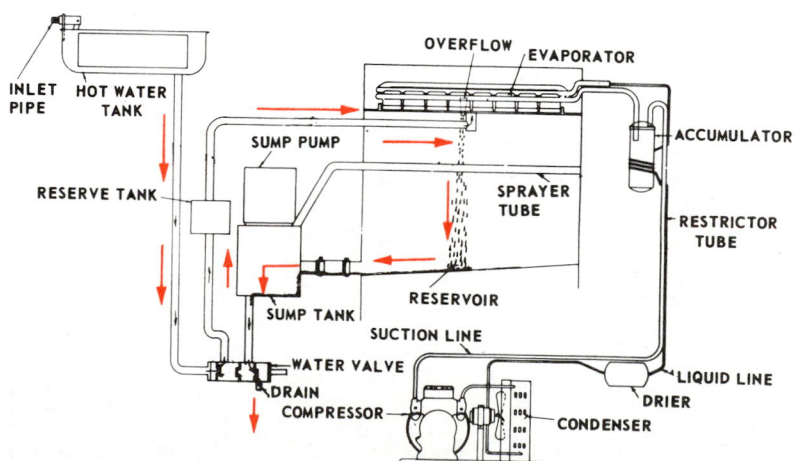

INLET PIPE

HOT WATER TANK

OVERFLOW EVAPORATOR

ACCUMULATOR

SUMP PUMP

SPRAYER TUBE

RESERVE TANK

RESTRICTOR TUBE

RESERVOIR

SUMP TANK

SUCTION LINE

WATER VALVE

DRAIN

COMPRESSOR

CONDENSER

LIQUID LINE DRIER

Fig. 12-77. Hot water defrosting cycle of automatic ice cube maker.

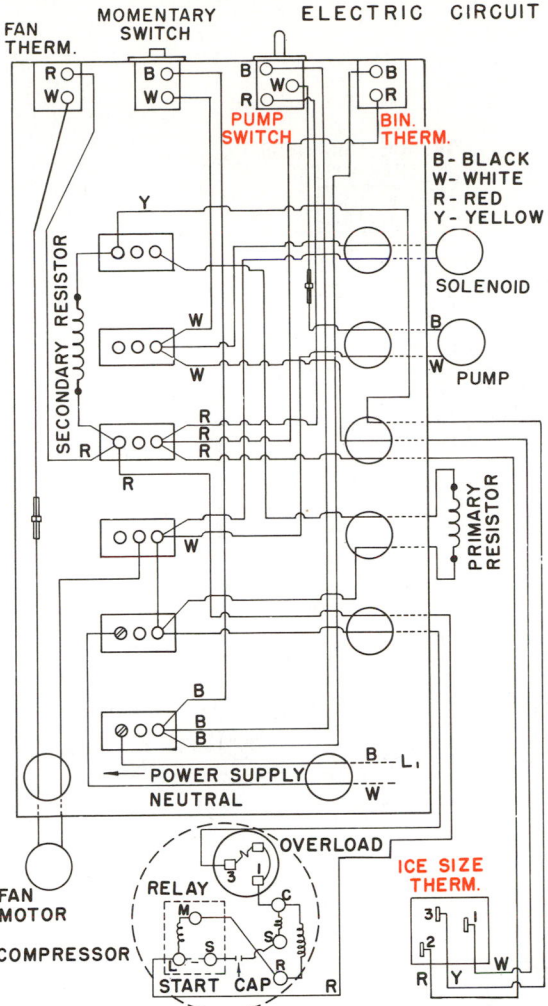

FAN THERM.
MOMENTARY SWITCH
ELECTRIC CIRCUIT

PUMP SWITCH

BIN. THERM.

B - BLACK
W - WHITE
R - RED
Y - YELLOW

SECONDARY RESISTOR

SOLENOID

PUMP

PRIMARY RESISTOR

POWER SUPPLY
NEUTRAL

OVERLOAD

ICE SIZE THERM.

FAN MOTOR

RELAY

COMPRESSOR

START CAP.

Fig. 12-78. Electrical circuits of automatic ice maker. Note the use of a bin thermostat, ice-size thermostat and pump switch.

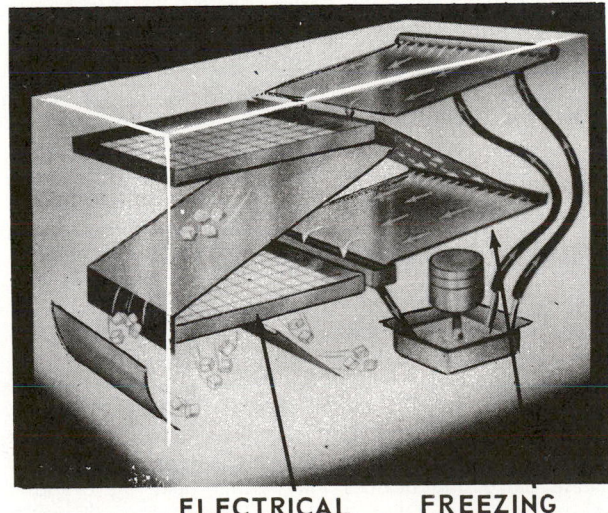

ELECTRICAL GRID   FREEZING PLATE

Fig. 12-79. Automatic ice cube maker makes slab of clear ice and then cuts it into cubes.

of electrical circuitry found in units just described.

The automatic unit shown in Fig. 12-79 makes solid ice cubes. Water flows over an inclined evaporator. A layer of ice is formed about 5/8 in. thick. During the defrost cycle (hot gas), this layer of ice is freed from the plate and, because of gravity, it slides over the electric grid. When the grid wires are heated, the wires cut the plate of ice into cubes.

Design of the electrically heated grid is shown in Fig. 12-80.

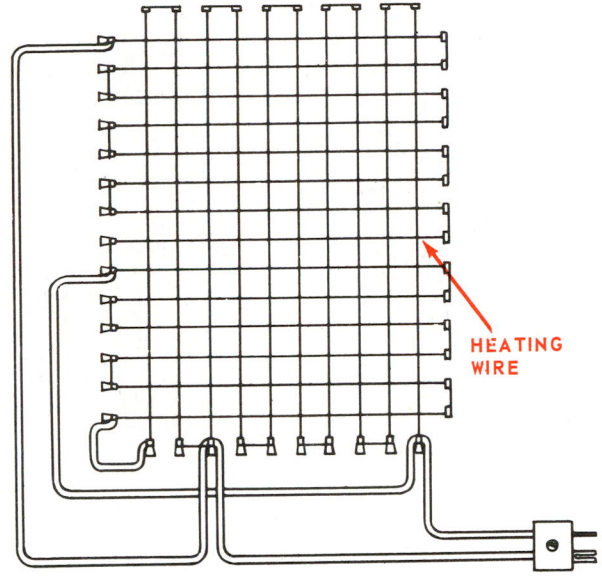

HEATING WIRE

Fig. 12-80. Grid arrangement for changing ice slabs into ice cubes.

Fig. 12-81 is a wiring diagram for this unit. The grid circuit has a transformer to reduce grid voltage to an efficient level.

The most popular refrigerant for these systems is R-12. *It is very important that the drain water be piped to a drain with an air break to prevent reverse syphoning of contaminated water.*

12-26 FREEZING DISPENSER EVAPORATORS

Freezing dispenser evaporators are used to keep the mix cold and then to freeze and churn the mix before it is force fed into cups for the customer. The temperatures are very critical.

The freezing evaporator is a cylinder in which a motor-driven dasher churns the mix. The refrigerant then travels to the premix storage compartment. See Fig. 12-82 for wiring diagram. Note the dasher motor.

For the actual mechanism, see Fig. 12-83. It uses a serviceable hermetic motor compressor, a water-cooled condenser, and a separate motor to drive the mixing blade in the evaporator.

12-27 EVAPORATOR DEFROSTING

Many evaporators operate at temperatures below freezing. The demand for open display cases and frozen food has required these low temperature systems. The evaporators

Commercial Systems / 423

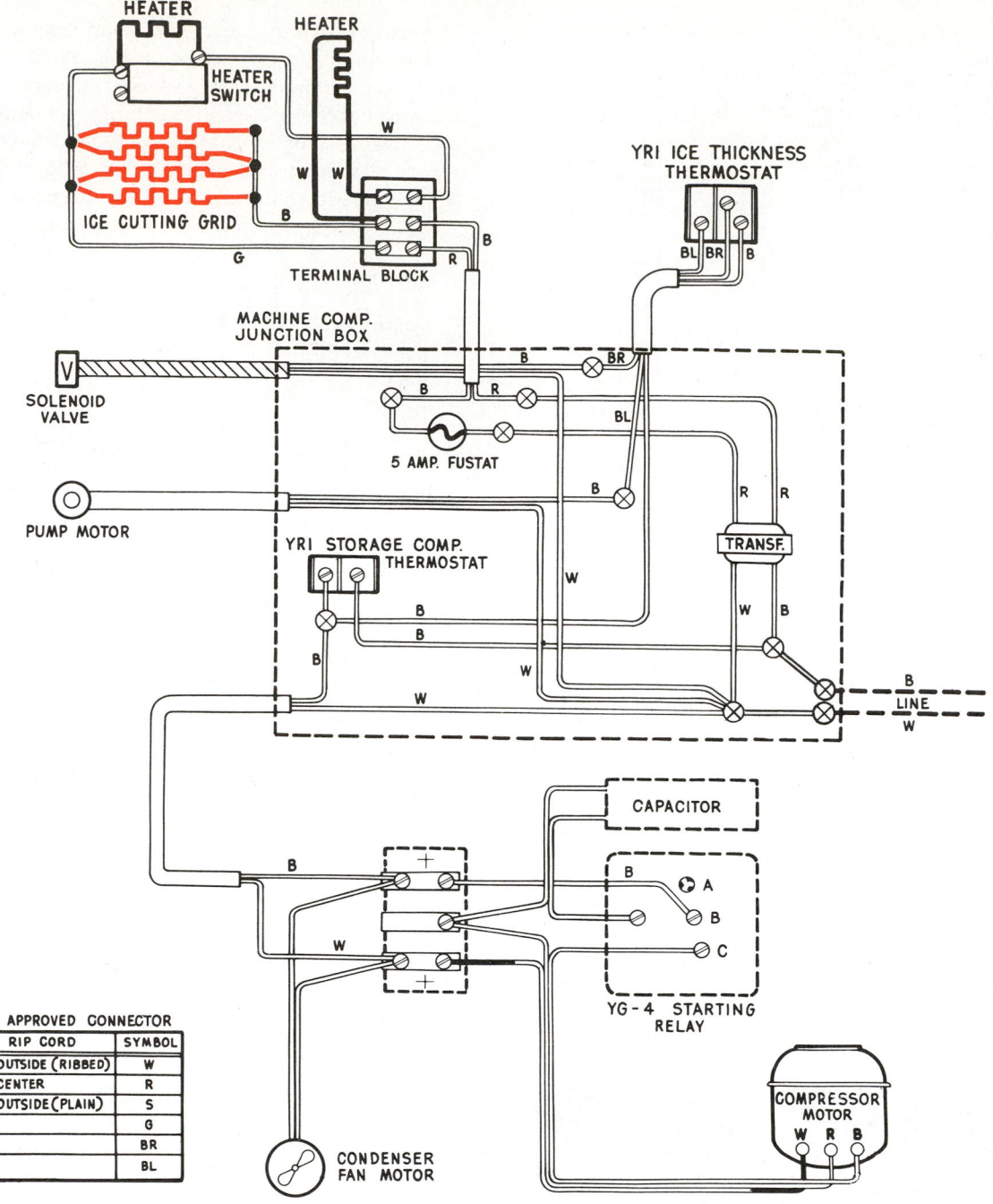

Fig. 12-81. Wiring diagram of automatic ice cube maker which uses electrically heated grid. Ice cutting grid has four parallel circuits. (Frigidaire Co.)

COLOR	RIP CORD	SYMBOL
WHITE	OUTSIDE (RIBBED)	W
RED	CENTER	R
BLACK	OUTSIDE (PLAIN)	S
GREEN		G
BROWN		BR
BLUE		BL

⊗ USE APPROVED CONNECTOR

operate at refrigerant temperatures of 0 F. (−18 C.), −10 F. (−23 C.), and even −20 F. (−29 C.). Blower evaporators are often used.

Low temperatures and small fin spacings make frequent defrosting necessary. Frost accumulation would otherwise soon clog the evaporator. Other types of evaporators also need defrosting even though not so frequently. It is desirable that this be done with very little rise in fixture temperature.

Defrosting is usually automatic. Some defrost during each off part of the cycle. On others, a time clock control either turns on the defrosting mechanism once a day or after a given number of hours of compressor operation.

There are six defrosting methods:
1. Hot refrigerant vapor system.
2. Nonfreezing solution system.
3. Water system.
4. Electric heater system.
5. Reverse cycle defrost system.
6. Warm air system.

These defrosting devices may either heat the evaporator internally (from the inside) or externally (from the outside) to melt the frost.

It is important to clean the evaporator, drain pans and drain lines frequently.

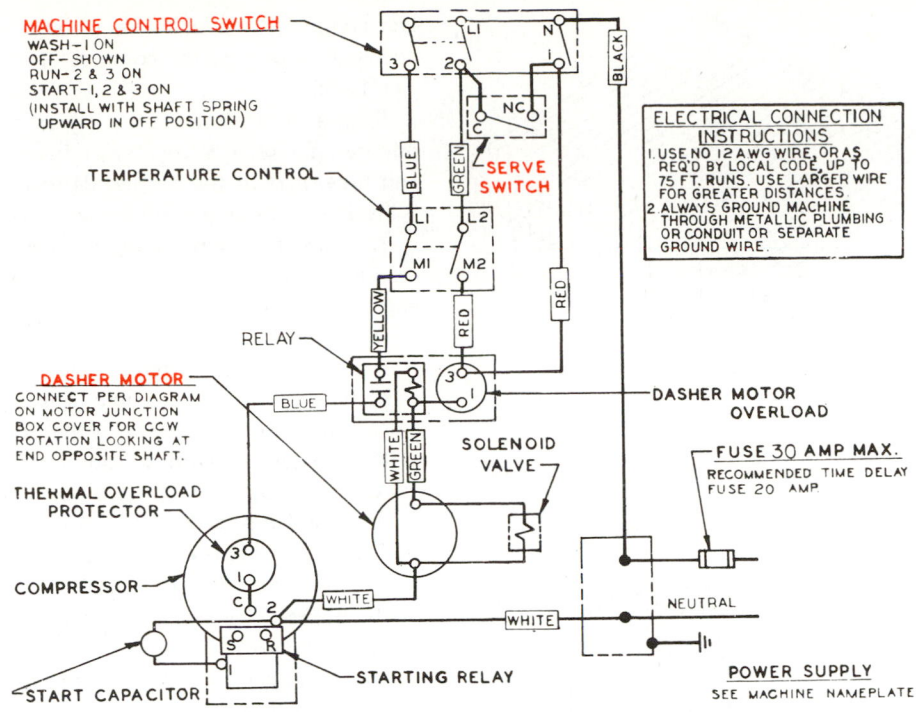

MACHINE CONTROL SWITCH
WASH-1 ON
OFF-SHOWN
RUN-2 & 3 ON
START-1, 2 & 3 ON
(INSTALL WITH SHAFT SPRING UPWARD IN OFF POSITION)

TEMPERATURE CONTROL

SERVE SWITCH

BLUE
GREEN
BLACK

ELECTRICAL CONNECTION INSTRUCTIONS
1. USE NO 12 AWG WIRE, OR AS REQ'D BY LOCAL CODE, UP TO 75 FT. RUNS. USE LARGER WIRE FOR GREATER DISTANCES
2. ALWAYS GROUND MACHINE THROUGH METALLIC PLUMBING OR CONDUIT OR SEPARATE GROUND WIRE.

L1   L2
M1   M2

RELAY

DASHER MOTOR
CONNECT PER DIAGRAM ON MOTOR JUNCTION BOX COVER FOR CCW ROTATION LOOKING AT END OPPOSITE SHAFT.

YELLOW   RED   RED

BLUE

THERMAL OVERLOAD PROTECTOR

WHITE   GREEN

SOLENOID VALVE

DASHER MOTOR OVERLOAD

FUSE 30 AMP MAX.
RECOMMENDED TIME DELAY FUSE 20 AMP

COMPRESSOR

WHITE

NEUTRAL

WHITE

START CAPACITOR

STARTING RELAY

POWER SUPPLY
SEE MACHINE NAMEPLATE

Fig. 12-82. Wiring diagram for a soft ice cream freezing dispenser. Note control switches for serving and for washing. (Sweden Freezer Mfg. Co.)

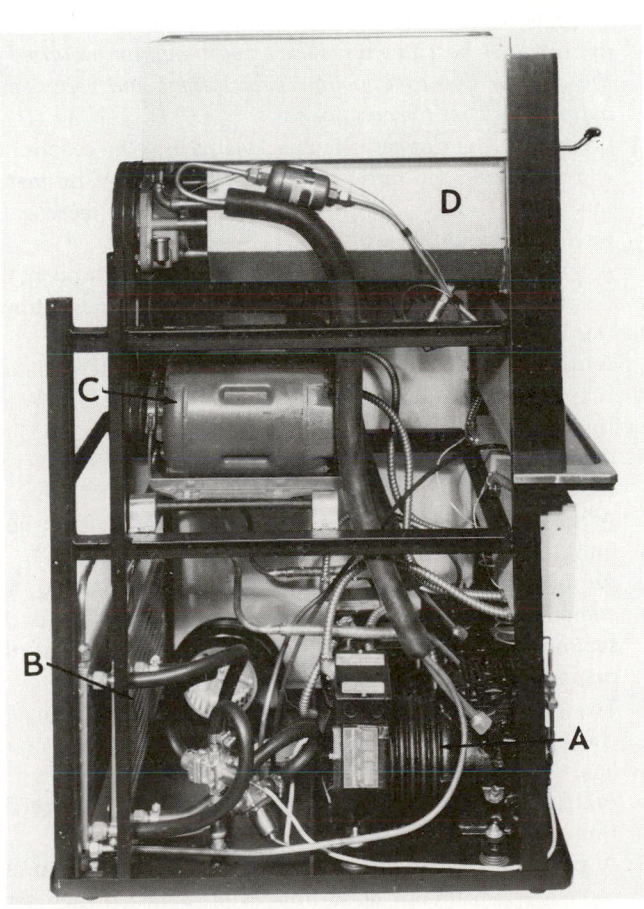

Fig. 12-83. A freezer-dispenser system. A—Motor compressor. B—Water-cooled condenser. C—Freezer mixing blade drive motor. D—Evaporator. (Taylor Freezer)

12-28 EVAPORATOR MOUNTING

The evaporators are mounted in the cabinet by one of three methods. They are:

1. Suspended from the ceiling.
2. Mounted on a pipe fastened to the vertical baffle and the wall of the cabinet.
3. Mounted on stands fastened to the horizontal baffle.

Where the evaporator is vertical the thermostatic expansion valve may be connected to the top, allowing refrigerant and oil to flow by gravity to the suction line down to the compressor. Or, the expansion valve may be connected to a horizontal run of evaporator coil tubing making the refrigerant pass horizontally to come to the suction line as shown in Fig. 12-84.

If the expansion valve is fastened to the upper part of the coil permitting a gravity flow, oil binding will be negligible. However, the fins will be coldest at the top where they are in contact with the warmest air.

12-29 HOT GAS DEFROST SYSTEM

In a "hot gas" system, hot refrigerant vapor is pumped directly through the evaporator tubing. The system has a refrigerant line running directly from the compressor discharge line up to the evaporator. In some cases, this line is connected between the thermostatic expansion valve (TEV) or the capillary tube and the evaporator. The line is opened and closed by a solenoid shutoff valve.

At the predetermined time (usually 12 midnight or 1 A.M.), the time clock closes a circuit which starts the compressor, opens the solenoid valve, and stops the evaporator fan motors.

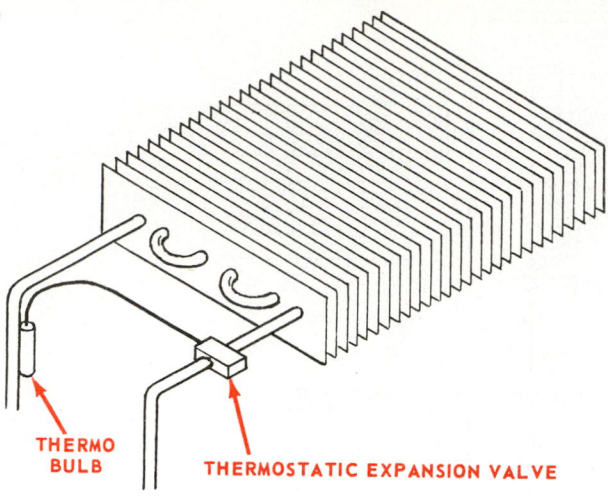

THERMO
BULB

THERMOSTATIC EXPANSION VALVE

THERMOSTATIC
EXPANSION VALVE

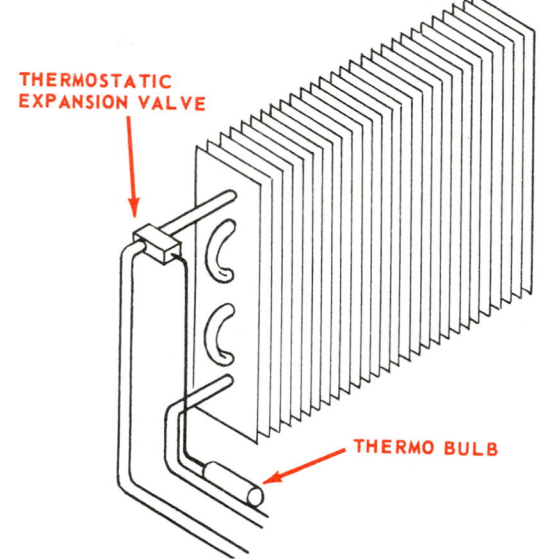

THERMO BULB

Fig. 12-84. Two different evaporator installations and recommended expansion valve mounting for each.

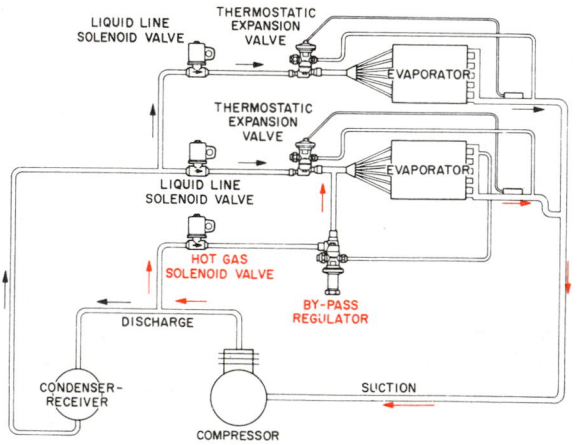

LIQUID LINE
SOLENOID VALVE

THERMOSTATIC
EXPANSION
VALVE

EVAPORATOR

THERMOSTATIC
EXPANSION
VALVE

EVAPORATOR

LIQUID LINE
SOLENOID VALVE

HOT GAS
SOLENOID VALVE

BY-PASS
REGULATOR

DISCHARGE

SUCTION

CONDENSER-
RECEIVER

COMPRESSOR

Fig. 12-85. Schematic of typical "hot gas" bypass from compressor discharge to evaporator inlet. Top evaporator is nonfrost type. (Alco Controls Div., Emerson Electric Co.)

Hot compressed vapor rushes through the evaporator, warms it and then returns to the compressor along the suction line. See Fig. 12-85.

Such a system will usually defrost the evaporator in 5 to 10 minutes. To keep defrost water from freezing in the drain pan and tube, part of the hot gas defrost line or an electric heater is installed under the drain pan and the drain pipe.

It is best to evaporate the refrigerant that condenses in the evaporator during the defrost cycle. The following steps explain how this is done:

1. A defrost bypass puts some hot gas into the suction line to vaporize any liquid refrigerant there. Some hot gas bypass systems also use a liquid injection system. A TEV is mounted on a line between the liquid line and the suction line. Its sensing bulb is mounted on the suction line. If gas returning to the compressor becomes warm, the TEV will open and mix liquid refrigerant with the hot bypass gas. The mix is kept at 45 F. (7 C.) to 65 F. (18 C.) to make sure that the compressor will be cooled. A solenoid valve in the bypass TEV inlet shuts off the TEV when the system is operating normally or on a full load.

2. Heat is applied to vaporize the returning refrigerant. This heat is electric in some cases.

3. A special blower-evaporator may be installed in connection with the suction line. This, plus air forced over the re-evaporator, insures that only vapor can get back to the compressor. The blower works only while the unit is on defrost. *It is best to always use an accumulator mounted in the suction line to trap liquid refrigerant and vaporize it before it reaches the compressor.*

4. Sometimes the hot gas is fed backwards into the evaporator and the condensed refrigerant bypasses the TEV by means of a check valve and forces the liquid into the receiver by way of the liquid line.

A system for forcing the hot gas backwards through the evaporator is shown in Fig. 12-86. The defrost cycle is shown in Fig. 12-87.

A timer starts the defrost action and a termination thermostat returns the system to normal refrigerating. When defrosting starts:

1. A solenoid valve opens a line from the top of the receiver to the suction line.

2. A holdback valve reduces the high-pressure gas as it goes into the compressor.

3. A three-way solenoid closes the suction line to the compressor and opens a valve to allow hot gas up the suction line to the evaporator. The hot gas warms the evaporator and the hot gas condenses.

4. The condensed refrigerant bypasses the TEV through a check valve and travels to the receiver by way of the liquid line.

5. A check valve in the condenser drain tube keeps refrigerant from backing up into the condenser.

6. A pressure regulating valve at the top (entrance to the condenser) maintains good hot gas pressures and temperatures. Fig. 12-88 shows the wiring diagram.

One type of hot gas valve has a connection to the suction line. This valve is adjusted by the low-side pressure of the

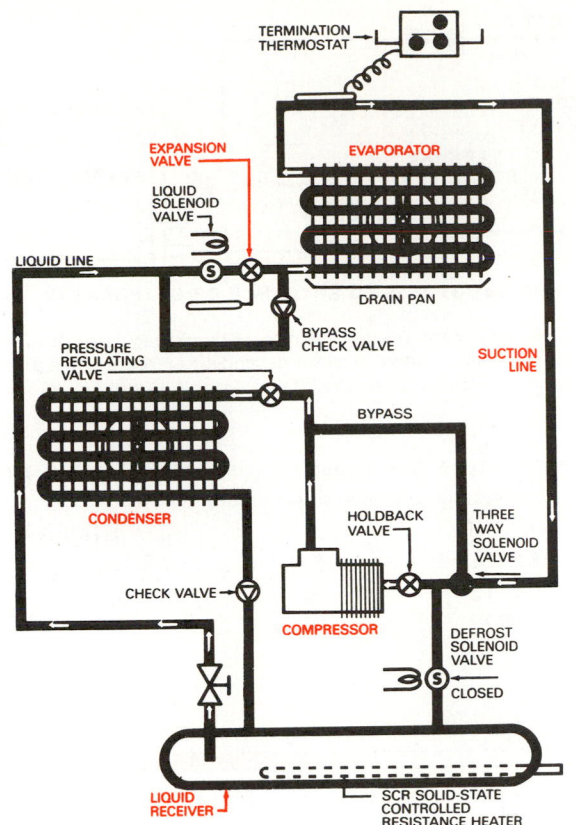

Fig. 12-86. Cooling cycle for normal two-pipe hot gas defrosting system. Note direction of refrigerant flow.

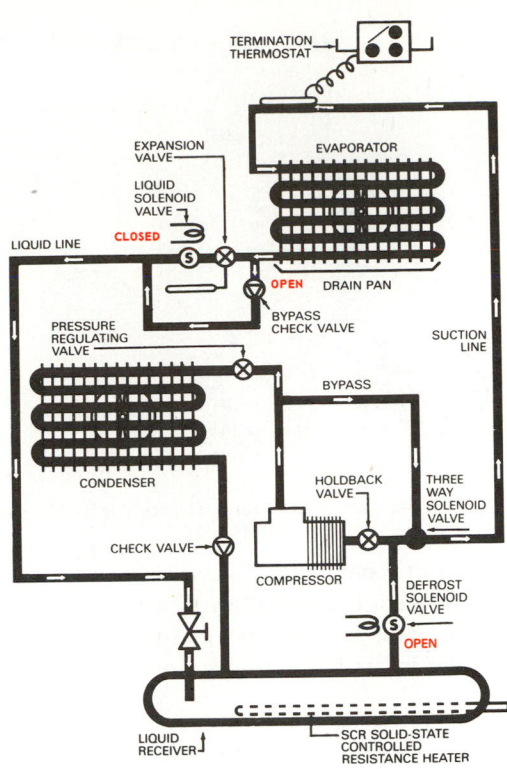

Fig. 12-87. Defrost cycle of the two-pipe hot gas defrost system. Note that hot gas from compressor and receiver is traveling to the evaporator by way of suction line. Condensed liquid refrigerant is traveling around TEV by way of check valve and is returning to receiver by way of liquid line. Heater in receiver provides more hot gas for defrosting.

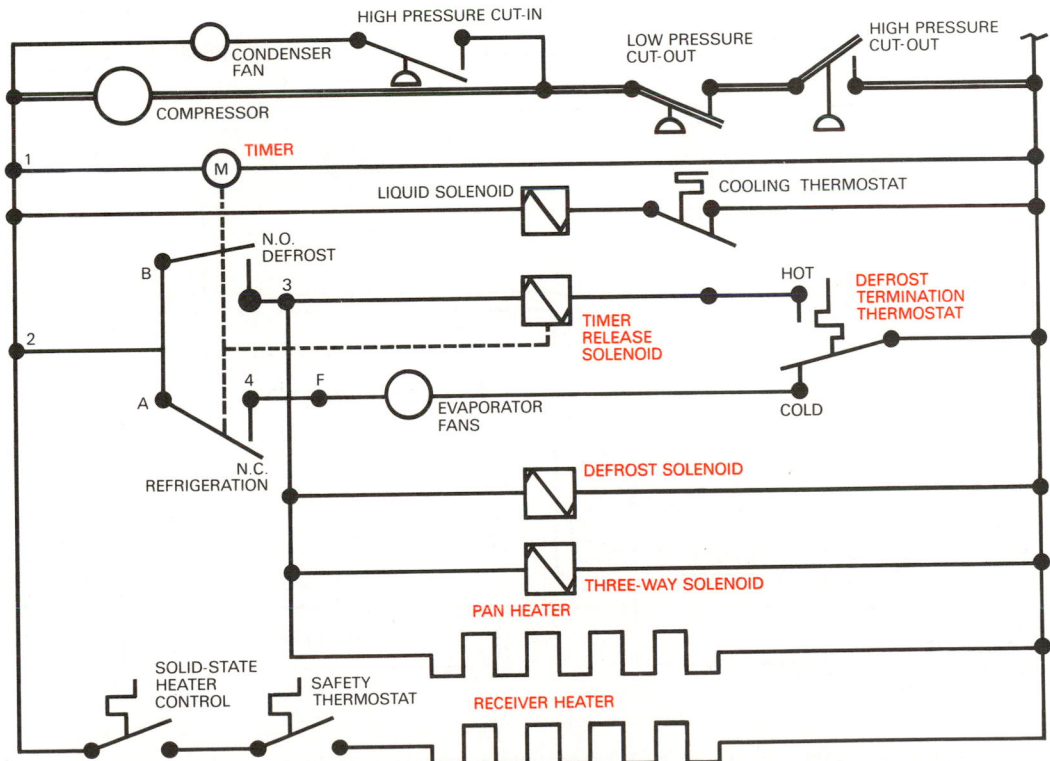

Fig. 12-88. Schematic wiring diagram of a two-pipe hot gas defrosting system. Note the electric pan heater and electric receiver heater. Note, also, that cooling thermostat only controls liquid line solenoid. (Halstead & Mitchell)

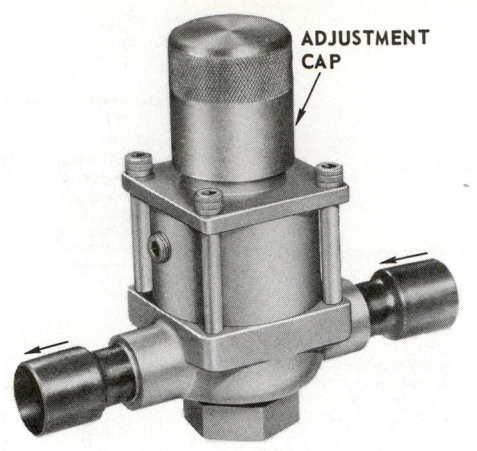

Fig. 12-89. "Hot gas" bypass valve. Capacity is 2 — 11 ton.

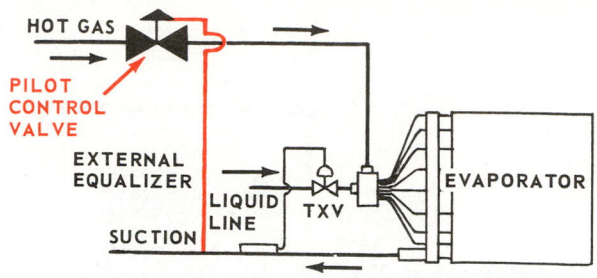

HOT GAS BYPASS TO ENTRANCE OF EVAPORATOR

Fig. 12-90. Schematic diagram of hot gas bypass regulator. Note external equalizer line. It operates valve dependent on suction line pressure. (Jackes-Evans Div., Parker-Hannifin Corp.)

vapor going to the compressor. The valve shown in Fig. 12-89 is pressure operated. It opens wider as suction pressure drops and starts to close as low-side pressure rises. The pilot-controlled valve is installed in a system as shown in Fig. 12-90.

Another type has an adjustable bellows in the sensing element to change the opening pressure:

OPENING PRESSURE	PSI	ADJUSTMENT RANGE
R-12	30	25—35
R-22	58	50—65
R-500	38	32—44

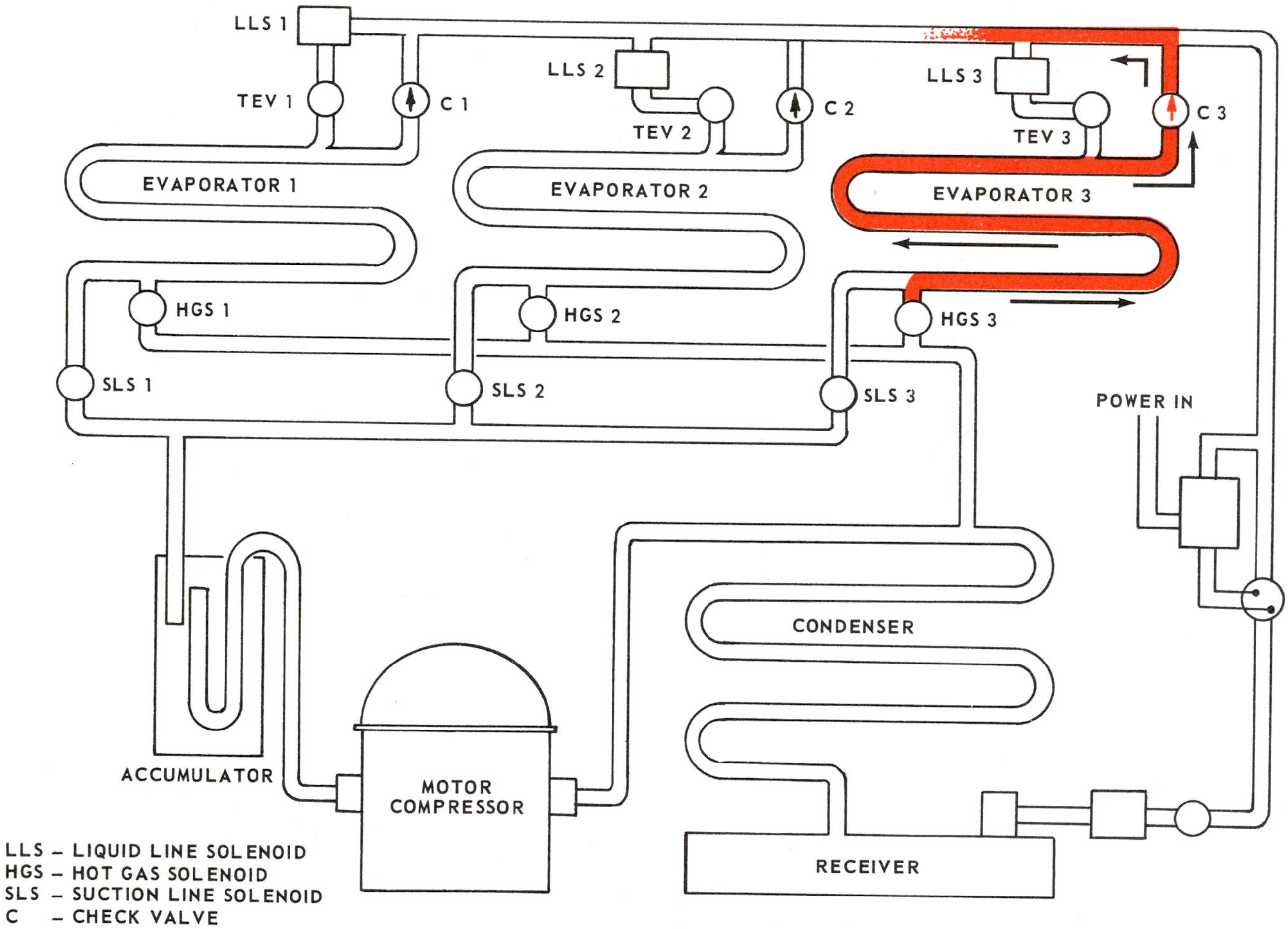

Fig. 12-91. "Hot gas" defrost system for multiple system. Timer controls each evaporator defrost at a different time. Liquid solenoid closes, partially pumps down system. Hot gas solenoid on one evaporator opens suction line, solenoid valve closes and hot gas rushes into evaporator backwards. It enters liquid line through bypass check valve and feeds liquid refrigerant to other two evaporators. If Evaporator 1 is to defrost, Solenoid S1 closes, S2 opens, and S3 closes. Hot gas flows into Evaporator 3, condenses as it melts frost; then condensed liquid goes into liquid line by way of Check Valve C3. This liquid then moves into Evaporators 1 and 2 through TEVs 1 and 2.

The valve sends hot condenser gas into the evaporator if the evaporator refrigerant temperature reaches freeze-up temperatures such as 26 F. (-3 C.). It is used where there are intervals of low-heat load conditions. This system is used on medium tonnage units (approximately five to 30 tons).

Slugs of liquid refrigerant must be stopped from entering the compressor. Re-evaporation should be almost complete before the refrigerant reaches the compressor.

In multiple systems having several evaporators, it is good practice to defrost one evaporator at a time using the other to evaporate the liquid coming from the defrosting evaporator. This method makes sure there will be no liquid refrigerant reaching the low side of the motor compressor. See Fig. 12-91.

A valve may take the place of the two suction line valves on each evaporator. See Fig. 12-92. The evaporator gas at point A

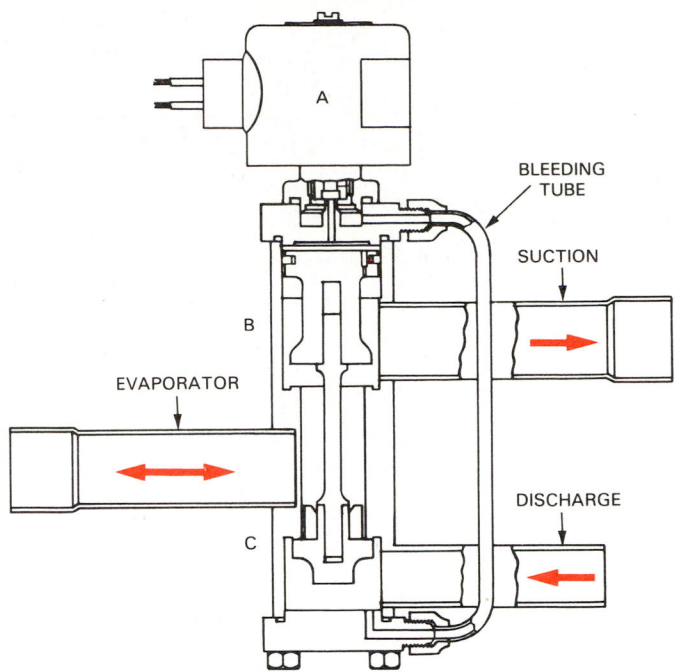

Fig. 12-93. Internal construction of a hot gas defrost solenoid valve. Note small tubing connection from discharge connection to pilot valve bleed. (Jackes-Evans Div., Parker-Hannifin Corp.)

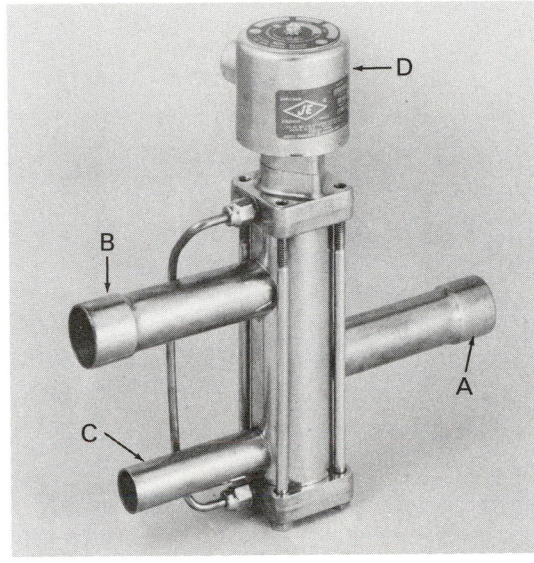

Fig. 12-92. Hot gas defrost solenoid valve. A—Evaporator connection. B—Suction line to compressor. C—Hot gas defrost connection. D—Pilot solenoid valve. (Jackes-Evans Div., Parker-Hannifin Corp.)

travels into the valve during normal operation, and the hot gas travels out of the valve at A during defrosting. The internal construction of the valve can be seen in Fig. 12-93. When the pilot solenoid valve at A is energized, it opens and allows high-pressure gas from the discharge connection to push down on the double valve. This action closes the top valve B. It stops flow from evaporator into suction line, and opens bottom valve, C. High-pressure hot gas flows from the discharge connection up into the evaporator.

12-30 NONFREEZING SOLUTION DEFROST SYSTEM

The "hot fluid" defrost system has been used for years. It uses a container in which a brine (a nonfreezing solution) is stored. The refrigerant vapor from the compressor is pumped through this heat storage container before it goes to the condenser.

The brine in the container may also be electrically heated.

Such heating is provided during the normal running (freezing) part of the refrigerating cycle.

When the refrigerating system shuts off, the defrost timer closes a solenoid valve in a line running from the liquid line to the evaporator. This is the beginning of the defrost cycle. The evaporator fan is usually shut off. The brine solution is pumped through its own piping along the drain line, the drain pan and the evaporator. Then it returns to its container. Fig. 12-94 shows such a defrost cycle.

12-31 WATER DEFROST SYSTEMS

The water defrost system either manually or automatically runs tap water over the evaporator while the system is not running. During this operation, the evaporator louvers are closed. The water is warm enough to melt the ice which drains away into the evaporator drain pan. Water must drain from the water lines before the unit is turned on or this water will freeze.

Either the water is sprayed over the evaporator or it is fed to a pan located over the evaporator. Holes in the pan feed the water evenly over the evaporator.

An electric timer gives this system automatic operation. Fig. 12-95 demonstrates the principle of water defrost and shows the two types of manual water defrost, as well as one automatic defrost system.

Special systems have been designed to defrost by spraying a brine over the evaporator. A pump may be employed to recirculate a lithium chloride brine. Eliminator plates are needed to prevent brine spray from passing into the refrigerated space.

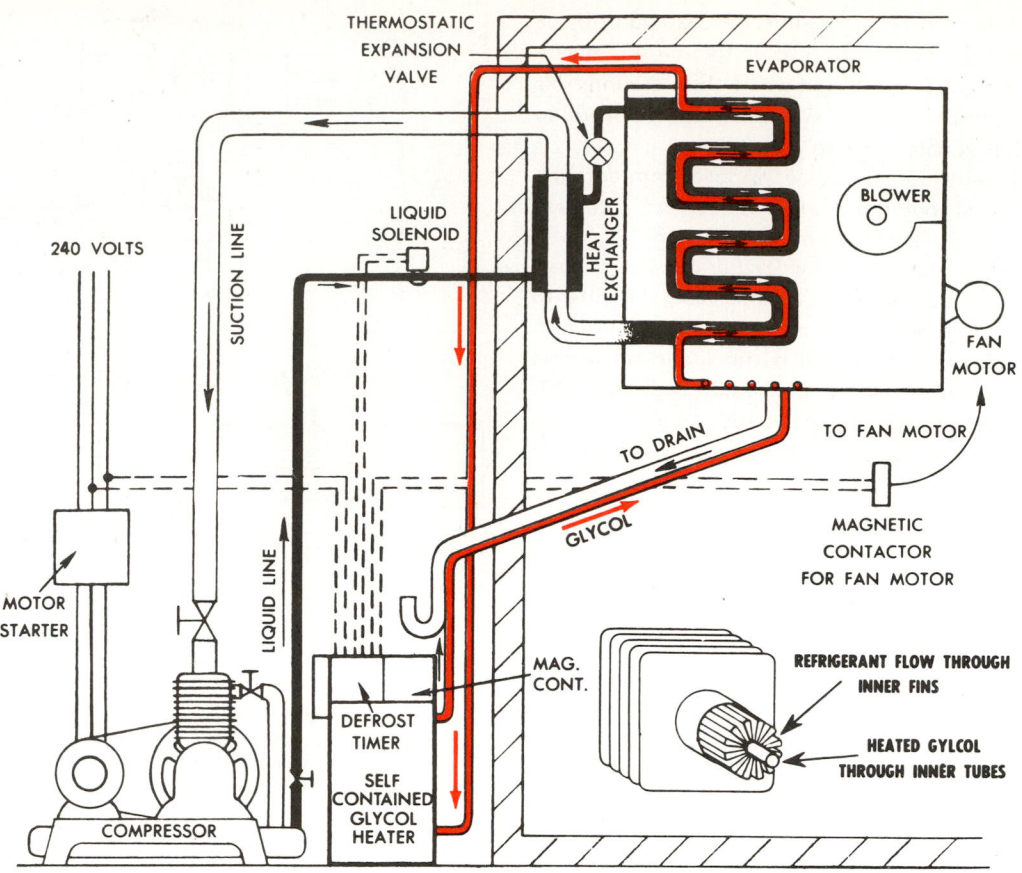

Fig. 12-94. Nonfreeze solution defrosting system. During defrost, glycol solution is pumped through inner tubing of evaporator and along the drain piping.

12-32 ELECTRIC HEATER DEFROST SYSTEM

Electric heat is popular for defrosting low-temperature evaporators. Heating coils are installed in the evaporator, around it or within the refrigerant passages to furnish heat.

One type uses resistance wire heating elements mounted underneath the evaporator, under the drain pan and along the drain pipe. A timer stops the refrigerating unit, closes the liquid line and pumps the refrigerant out of the evaporator. Then the blowers and the electric heaters are turned on.

The heaters quickly melt the frost from the evaporator and the water drains away. When the evaporators are warm enough to insure that all frost is gone, a thermostat on the evaporator returns the system to normal operation.

"Pump down" is a control system in which the thermostat operates a solenoid in the liquid line while a low-pressure switch operates the compressor. Its purpose is to prevent flow of liquid refrigerant from the evaporator to the compressor. It is especially important in cases of electric defrost.

1. When the thermostat is satisfied, it opens and the liquid line solenoid closes.
2. The compressor continues to run and continues to remove the refrigerant vapor from the evaporator and suction line.
3. When the proper low-side pressure is reached, the low-pressure switch opens and the compressor stops.

There should be very little refrigerant in the compressor oil. It may be necessary to use crankcase heaters during off cycle when a pump down system is used for each cycle.

"Pump out" has a similar purpose. However an extra relay is wired into the compressor circuit in parallel to the normal relay. It is connected to the thermostat circuit. The extra relay operates the start button on the normal starting relay. The compressor, therefore, cannot start again until the thermostat points close.

Another electric defrost system uses an immersion type electric heater to heat a separate charge of refrigerant. This warm refrigerant circulates around the evaporator in its own passageways to warm the evaporator and defrost the system. This happens while the refrigerating unit is turned off.

Still another way of using the electric heater defrost system is with a double-tube evaporator. The evaporator refrigerant passes through the passageway between the tubes during normal refrigeration. Electric heating elements are inserted in the center tube. In the defrost operation, the system is stopped and the electric heating elements are turned on. See Fig. 12-96. Thereby, the evaporator tubes cause defrosting from the inside.

12-33 REVERSE CYCLE DEFROST SYSTEM

Another system defrosts evaporators by reversing the flow of refrigerant. This causes the evaporator to become the condenser and the condenser an evaporator. When the evaporator functions as a condenser, it melts the accumulated frost.

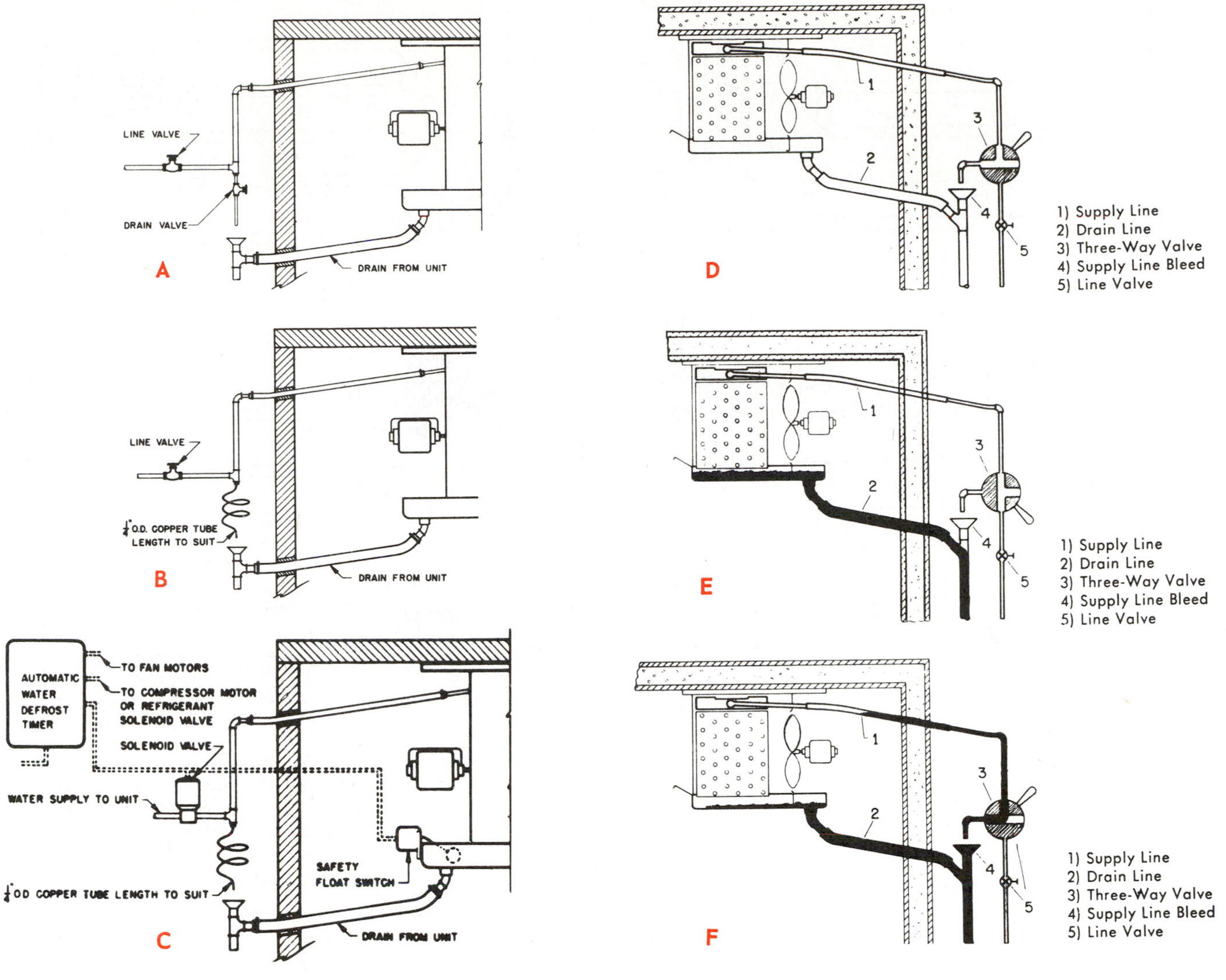

Fig. 12-95. Water spray defrost system principles showing three methods of operation. A—Manual defrost and manual drain. B—Manual defrost and automatic drain. C—Automatic defrost. The three parts of the defrost cycle of view A are shown at D, E and F. D—Unit during refrigeration. E—Defrost operation. F—Water lines and drain being cleared of water at end of defrost operation. (Refrigeration Engineering Corp.)

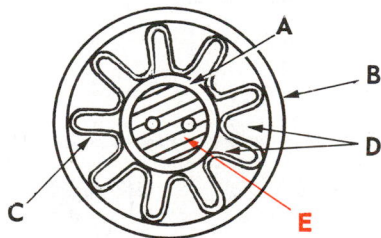

Fig. 12-96. Electric defrost system with electric heating elements installed within evaporator tubing. A—Inner tube. B—Outer tube. C—Inner fin. D—Refrigerant passages. E—Heating element.

This reversing is handled by installing a four-way valve. Chapters 3, 19 and 23, describe the reverse cycle (heat pump) in detail. See Figs. 12-97 and 12-98.

To operate on defrost, the four-way valve is turned either manually or automatically and hot gas from the compressor travels up the suction line. It heats the evaporator when gas condenses in it and bypasses the refrigerant control by means of a check valve. It passes through the receiver. As it leaves the receiver, a check valve bypasses it through another refrigerant control into the condenser. The refrigerant evaporates in the condenser and is returned to the compressor in a vapor state.

The liquid receiver is designed to permit the reverse flow of vapor to travel over the reverse liquid in the receiver. It does not return the vapor to the condenser.

12-34 WARM AIR DEFROSTING

Where there is enough of it, warm air can be used to defrost low-temperature evaporators. Cabinet air at the right temperature can be used for defrosting. The cycles must be frequent enough and long enough to defrost the evaporator completely. Some installations bring in outside air for defrosting, using a controlled duct system with blowers and fan.

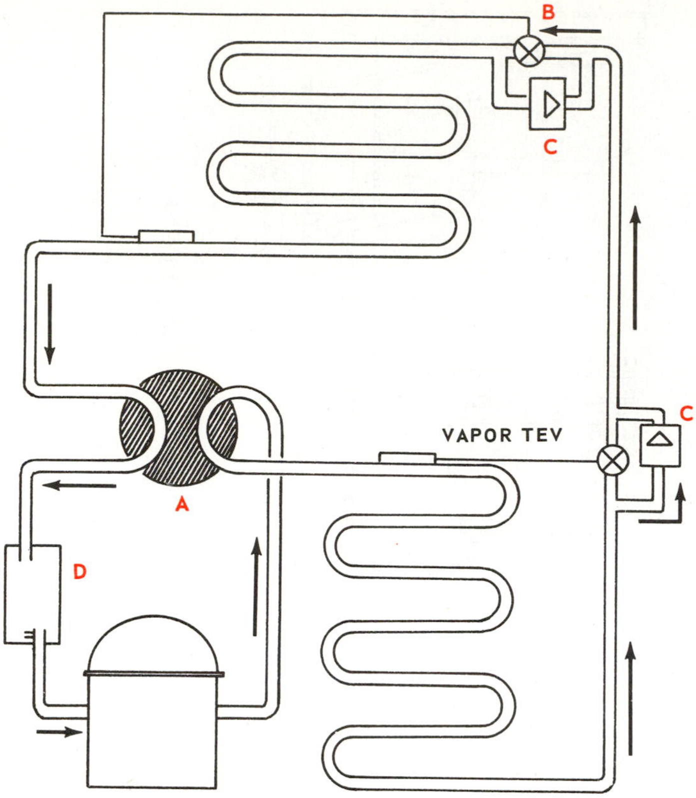

Fig. 12-97. Reverse cycle defrosting system. Four-way valve position shown at A is for normal refrigerating operation. A—Four-way valve. B—TEV (thermostatic expansion valve). C—Check valve. D—Accumulator. Illustration at right shows same system during defrost.

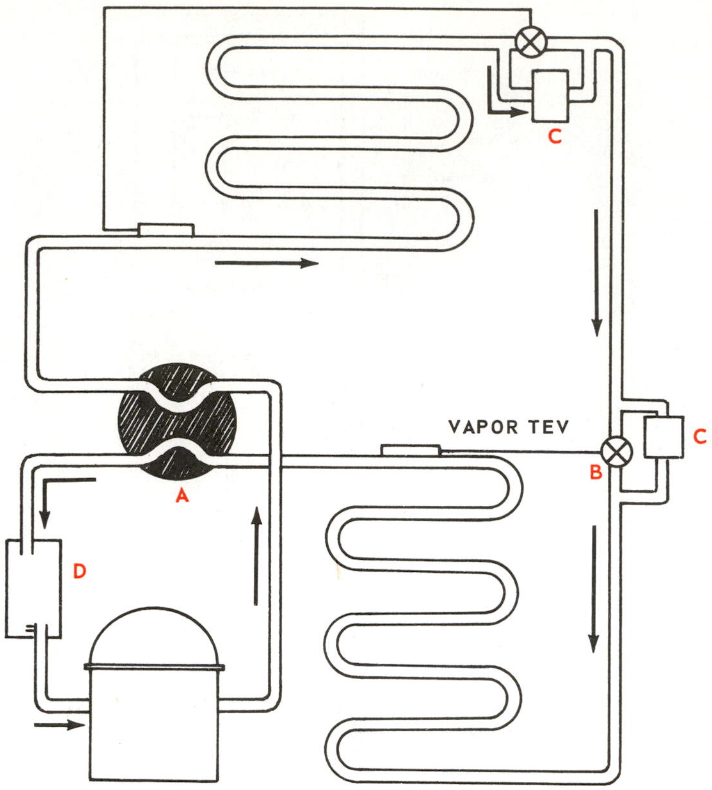

Fig. 12-98. Defrost system, using four-way valve, is shown in defrosting position. Valve is positioned to make evaporator serve as a condenser and condenser as an evaporator. A—Four-way valve. B—Extra TEV. C—Two check valves. D—Accumulator tank.

12-35 HEAT EXCHANGERS

A heat exchanger mounted in the suction and liquid line has three advantages:

1. It subcools the liquid refrigerant and increases operating efficiency.
2. It reduces flash gas in the liquid line.
3. It reduces liquid refrigerant in the suction line.

A heat exchanger like the one in Fig. 12-99 provides for a

prevailing head pressure, it can absorb more latent heat as it changes to a vapor in the evaporator.

The reduction of flash vapor (sometimes called "flash gas") is important. Flash gas (vaporized refrigerant) comes from the sudden change of some of the liquid to a vapor as the refrigerant passes through the refrigerant control. This reduces valve capacity, increases low-side pressure drop and reduces the amount of heat each pound of refrigerant can absorb as it evaporates. The "flash gas" cools the remainder of the liquid to the evaporating temperature.

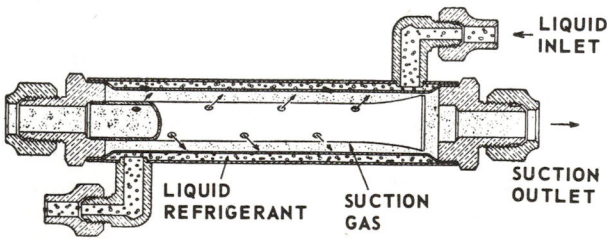

Fig. 12-99. Cross section of heat exchanger used on commercial systems. Note flared connections.   (Mueller Brass Co.)

Fig. 12-100. A heat exchanger with brazed connections leading to liquid and suction lines. A—Liquid in. B—Liquid out. C—Suction vapor in. D—Suction vapor out.

heat transfer from the warmer liquid in the liquid line to the cool vapor coming from the evaporator. Fig. 12-100 shows the outside appearance of a heat exchanger.

If the liquid is cooled 10 to 20 F. (5 to 11 C.) at the

The heat exchanger also helps prevent sweat backs or frost backs on the suction line. If there is low temperature liquid refrigerant present in the returning suction vapor, it will evaporate in the heat exchanger as it absorbs heat from the liquid line.

The subcooled liquid in the liquid line reduces the chance of flash gas forming in the liquid line — especially on warm days or if the liquid line has a long vertical run.

The pressure drop in the suction line portion of the heat exchanger should not be over 2 psi.

12-36 REFRIGERANT CONTROLS

For single installations involving one evaporator and one condensing unit, five types of refrigerant controls can be used: thermostatic expansion valves, automatic expansion valves, high-side floats, low-side floats and capillary tubes. These are explained in Chapter 5.

In multiple installations which cover a greater number of uses, two types of refrigerant controls are usable. They are the low-side float and the thermostatic expansion valve. Thermostatic expansion valves are used extensively but there are also some low-side float systems.

Some of the thermostatic expansion valves have a large capacity. Usually they are constructed with a pilot valve operating a larger valve.

The thermostatic expansion valve is explained in Chapter 5. One should study its design, operation, installation, care and repair before proceeding with this chapter.

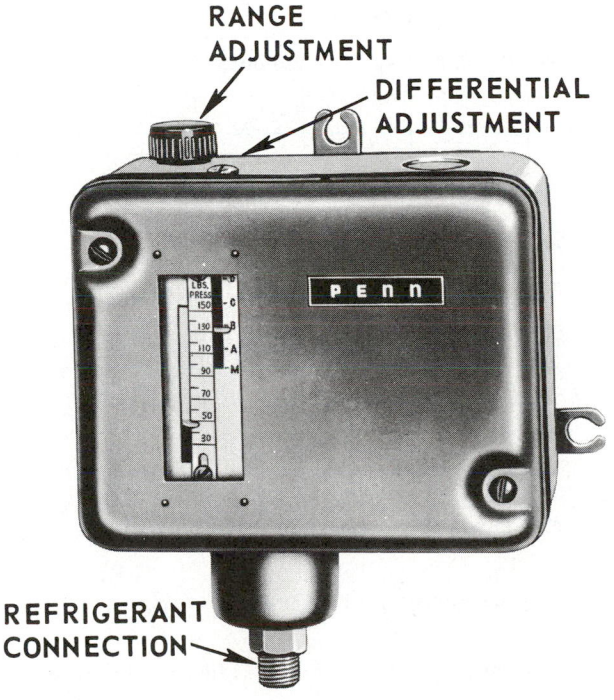

Fig. 12-101. Pressure motor control used on commercial installations.

12-37 MOTOR CONTROLS

Two basic types of motor controls are used in commercial refrigeration.
1. The thermostatic motor control.
2. The pressure motor control.
These are the same ones used in domestic refrigeration.

Large systems use magnetic starters operated by motor controls.

In multiple evaporator commercial work, pressure motor controls are used quite often because:
1. The low-side pressure is an indication of the temperature in the evaporators.
2. One control works well regardless of the number of evaporators connected to it.

The controls provide both range and differential adjustments as shown in Fig. 12-101. Explanation of various range and differential adjustments will be found in Chapter 8. Fig. 12-102 shows the internal construction of such a control.

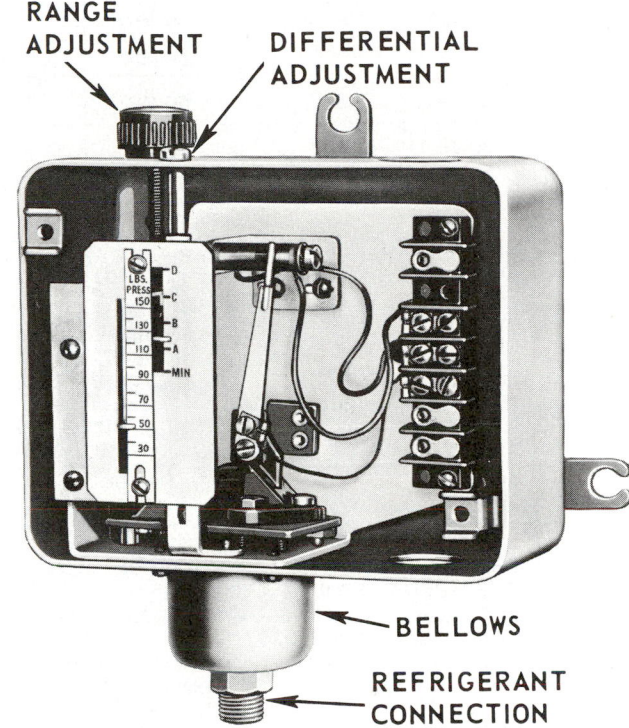

Fig. 12-102. Pressure motor control. Cover is removed to show location of electrical connections and pressure scales. (Penn Controls, Inc.)

The current draw of larger motors on starting is more than control contacts can handle. A motor starter is necessary for single-phase a-c motors over 1 hp.

Three-phase a-c motors also require a starter. The motor control operates the relay in the starter. Electrical work should be done by a licensed electrician and the work should comply with local electrical codes.

Fig. 12-103 lists the average pressure motor control settings for both temperature control and defrosting.

12-38 PRESSURE MOTOR CONTROL

The pressure motor control is usually mounted on the condensing unit. It is operated by low-side pressure. Some companies suggest connecting the control into the low-side suction line about 10 to 15 ft. from the compressor to reduce vibration effect on the control. The range settings vary with

| | DEFROST CONTROL | | EPR VALUE | | PRESSURE CONTROL | | | |
| | FAIL SAFE OR TIMER SETTING | DEFROSTS PER DAY | EPR OR TYPICAL OPERATING PRESSURE | | CUT-OUT | | CUT-IN | |
			R-502	R-12	R-502	R-12	R-502	R-12
FROZEN FOOD CASES	60 MIN.	1	10—15 #	6"V—0 #	5—10 #	4—10"V	20—29 #	3—8 #
ICE CREAM CASES	36 MIN.	1	NOT USED	NOT USED	5—10 #	4—10"V	14—20 #	0—3 #
THREE AND FOUR SHELF CASES	36 MIN.	2 FROZEN FOOD 3 ICE CREAM			0 #		25 #	
FOUR AND FIVE SHELF CASES	36 MIN.							
GLASS DOOR FROZEN FOOD CASES	60 MIN.	1	12—15 #	2"V—0 #	9—12 #	@5"V	30 #	8 #
GLASS DOOR ICE CREAM CASES	60 MIN.	1	8—10 #	4—8"V	6—8 #	@8"V	23 #	4 #
GLASS DOOR REFRIGERATORS, MEDIUM TEMPERATURE	60 MIN.	1	50—55 #	20—22 #	37—50 #	12—20 #	34 #	10 #
OPEN MEAT CASES AND MULTI-SHELF MEAT CASES/FOR BEST RESULTS USE TPR SET AT 23 F. ENTERING CASE AIR STD. DEF. AND ELECT.	STD. 60 MIN. ELECT. 36 MIN.	1 OR 2	40—47 #	14—18 #	31—39 #	9—15 #	50—55 #	20—22 #
MEAT AIR CASES — SET TPR'S FOR 25 F. ENTERING AIR	46 MIN.	3	TPR'S BUILT IN		14 #	0 #	50 #	20 #
MULTI-SHELF DAIRY-DELI CASES { STD. DEF. / ELECT.	60 MIN. 36 MIN.	4 4	50—54 #	19—20 #	41—50 #	15—20 #	60 #	26 #
ROLLING COLD CONVEYOR CASES	60 MIN.	2	45—50 #	17—20 #	37—45 #	12—20 #	60 #	26 #
CLOSED-SERVICE MEAT CASES	AIM FOR 36—40 F. AIR TEMP.		49—60 #	19—25 #	43—54 #	16—19 #	79 #	36 #
ALL OPEN PRODUCE CASES AND REACH-IN REFRIGERATORS	AIM FOR 38—45 F. AIR TEMP.		54—61 #	22—26 #	43—54 #	16—22 #		

DEFROST CHECK LIST:
1. CHECK TO SEE THAT DEFROST CONTACTOR IS WIRED TO CASES ON THE CONDENSING UNIT, THEN CHECK DEFROST TIME AND NUMBER.
2. CHECK HEATER AMPS. DURING AND AFTER DEFROST. IS COIL CLEAR AT TERMINATION? DOES LIMIT SWITCH OPEN TOO SOON?
3. CHECK WASTE OUTLET FOR PROPER HOOK-UP (MAXIMUM OF 12 FT. OF 1 IN. PIPE, 1/4 IN. PER FT. SLOPE) AND SEE THAT IT ISN'T FROZEN BY REFRIG. LINE CONTACT.
4. FLUSH THE WASTE OUTLET WITH A BUCKET OF HOT WATER. ELIMINATE DRAFTS OVER OPEN CASES. (MAXIMUM ALLOWABLE DRAFT — 50 FPM) DO NOT BLOCK AIR DUCTS. OBSERVE LOAD LINES! CASES MUST BE LEVEL TO OPERATE PROPERLY.

STANDS FOR psi.

Fig. 12-103. Recommended motor control pressure settings. These pressures are recommended for various case applications. Pressures are in psi. It may be necessary to change these settings somewhat for a particular installation. (Tyler Refrigeration Corp.)

the application. The cut-out pressure should be set about 10 F. (5 C.) lower than the desired evaporator outside surface temperature. The cut-in pressure should be about the same as the highest allowable evaporator temperature. See Fig. 12-104.

The differential setting will vary depending on the temperature accuracy wanted. A wide pressure difference will allow some variation in cabinet temperature and will lengthen the operating cycle interval of the condensing unit. (This means the compressor would not run as often.) A differential set to close limits will maintain a more uniform cabinet temperature but will shorten the cycling interval of the condensing unit. The unit would run oftener. A common pressure difference between cut-in and cut-out point is about 20 psi for R-12, 22 psi for R-22, 16 psi for R-500 and 25 psi for R-502.

12-39 THERMOSTATIC MOTOR CONTROL

The thermostatic motor control is like the pressure motor control in design except for the sensing bulb and capillary tube. See Fig. 12-105.

This type control is generally used in large single installations. However, satisfactory setups have been made in multiple installations. Hopefully, when the controlled cabinet is at the desired temperature, the others are also. These controls are also used together with a solenoid valve to control each separate cabinet in a multiple installation.

Some are made with a very close differential such as 1 F. (0.5 C.) for use in certain display cases, bulk milk coolers, frost alarms, liquid chillers and refrigerated trucks. Each has a

*INCHES VACUUM	APPLICATION	R-12		R-22		R-502	
		OUT	IN	OUT	IN	OUT	IN
FROZEN FOOD — OPEN TYPE		7*	5	4	17	9	23
FROZEN FOOD — CLOSED TYPE		1	8	11	22	17	29
ICE CUBE MAKER — FLOODED OR DRY TYPE COIL		4	17	16	37	22	47
SWEET WATER BATH — SODA FOUNTAIN		21	29	43	56	52	67
SHOW CASE — FROST CYCLE		10	25	25	50	32	60
SHOW CASE — DEFROST CYCLE		18	34	39	64	49	76
BEER, WATER, MILK COOLER		19	29	40	56	47	67
WALK-IN COOLER — DEFROST CYCLE		12	35	29	66	37	77
ICE CREAM TRUCKS, HARDENING ROOMS		2	15	12	33	17	42
VEGETABLE DISPLAY — DEFROST CYCLE		11	35	27	66	35	77
EUTECTIC BRINE TANK, ICE CREAM TRUCK		1	4	11	16	17	22
REACH-IN COOLER — DEFROST CYCLE		18	36	39	68	47	79
BEER COOLERS — BLOWER DRY TYPE		15	34	33	64	42	76
BEER COOLERS — BARE PIPE DRY TYPE — FROST CYCLE		12	27	29	53	37	64
INSTANTANEOUS BEER COOLERS		12	29	29	56	37	67
RETAIL FLORIST BOX — BLOWER COIL		26	42	51	77	61	88

Fig. 12-104. Typical refrigeration applications and low-side pressure motor control settings for each.

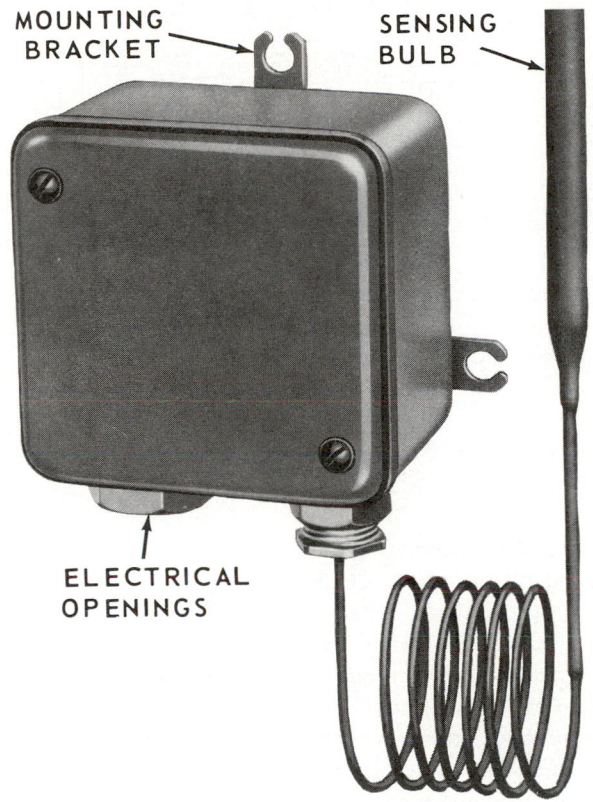

Fig. 12-105. Thermostatic motor control. Note mounting brackets and electrical wiring opening.

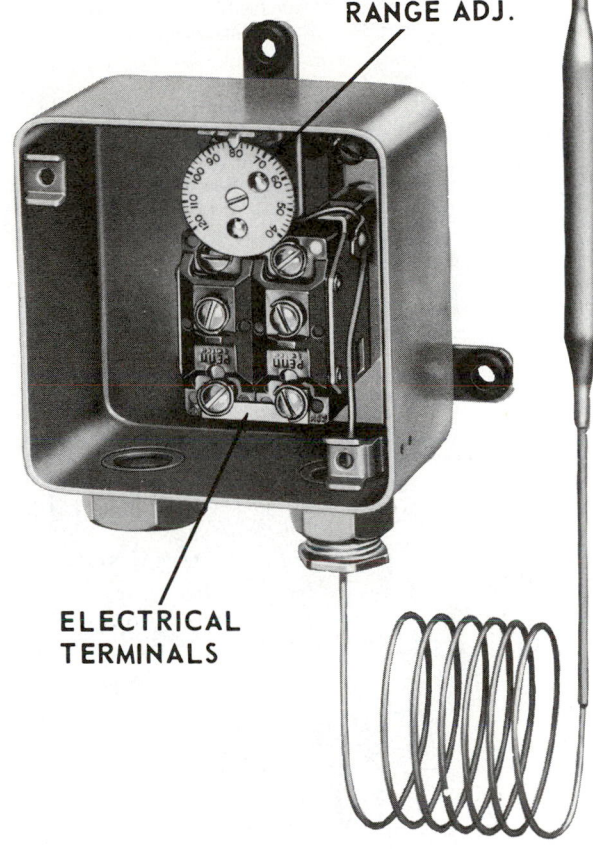

Fig. 12-106. Thermostatic motor control with cover removed. Note temperature range dial with Fahrenheit scale and electrical terminals.

capillary tube and a sensing bulb.

Thermostatic motor controls are popular in brine cooling installations with the sensing bulb being submerged in the brine. Ice cream cabinets are a typical example. When used in single cabinet installations, the sensing bulb is usually mounted in the cabinet 4 ft. up from the floor between the cold and warm air flues and at least 2 in. from the wall. Fig. 12-106 shows the control with the cover removed.

Some are made for wall mounting to be used in walk-in coolers, meat storage rooms, warehouses and florist cabinets. Some have double throw contacts (SPDT) so that the control may also operate other devices (fans and defrost systems).

12-40 SAFETY MOTOR CONTROLS

An important difference between commercial and domestic controls is the fact that many commercial electrical systems also use safety devices known as:

1. A high-pressure safety cut-out.
2. An oil pressure safety cut-out.

The high-pressure safety device is a bellows built into the control. It is connected to the high-pressure side of the system,

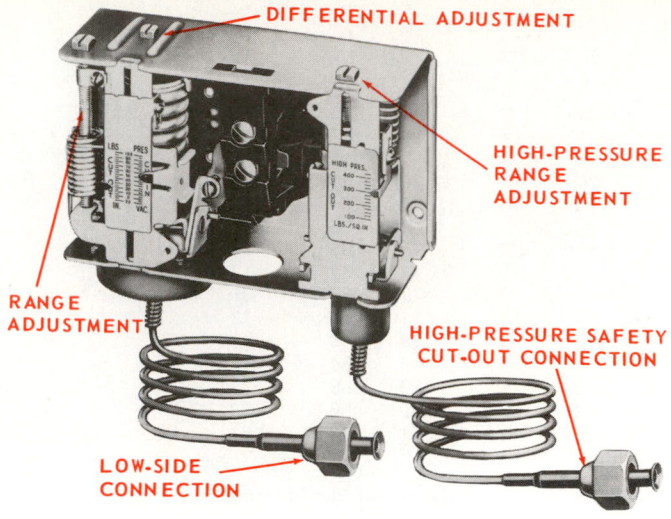

Fig. 12-107. Pressure operated motor control with high-pressure safety cut-out. Note three adjustments.

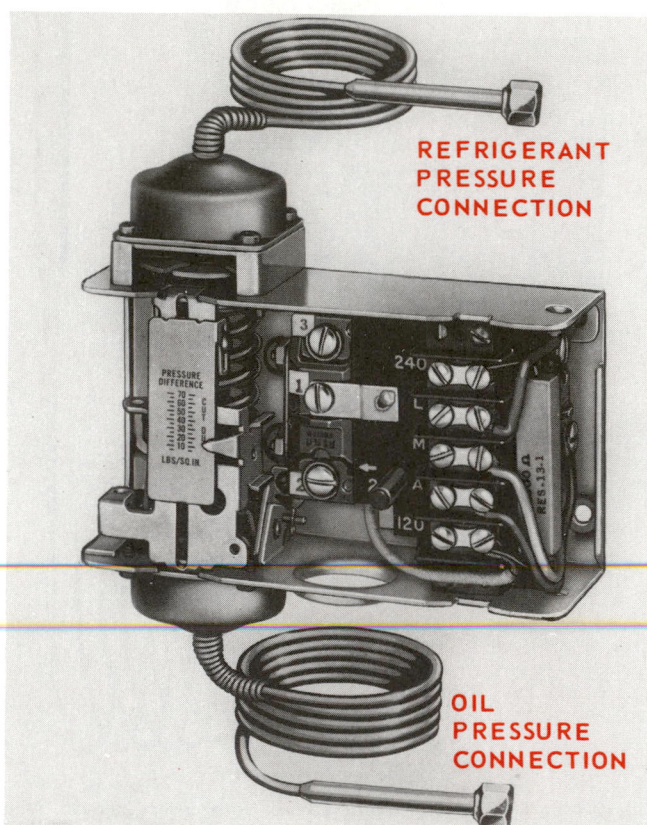

Fig. 12-108. Oil pressure safety cut-out for large commercial systems. It operates on difference between refrigerant and oil pressures.

Fig. 12-107. It is often connected to the cylinder head to permit easy disconnecting of the control from the system.

The bellows is attached to a plunger in such a way that, if the head pressure becomes too high from air in the system, condenser water being shut off or other causes, the bellows will expand, push the plunger against the switch and shut off the motor.

Action of the high-pressure safety device prevents the building up of dangerous pressures within the system. It also prevents ruining the motor through overloading and over-heating.

The control is usually set to cut out at about 20 percent above normal head pressure. In R-12 systems, the control is set at about 150 to 160 psi; R-22, 260 to 270 psi; R-502, 280 to 290 psi; and R-500, 190 to 200 psi.

The oil pressure safety cut-out will shut off the electrical power if the oil pressure fails or drops below normal. It is a differential control, using two bellows. One bellows responds to the low-side pressure and the other responds to the oil pressure. The oil pressure must always be above the low-side pressure for oil to flow. See Fig. 12-108.

The wiring diagram for an oil pressure safety control is shown in Fig. 12-109. The control will open the circuit if the

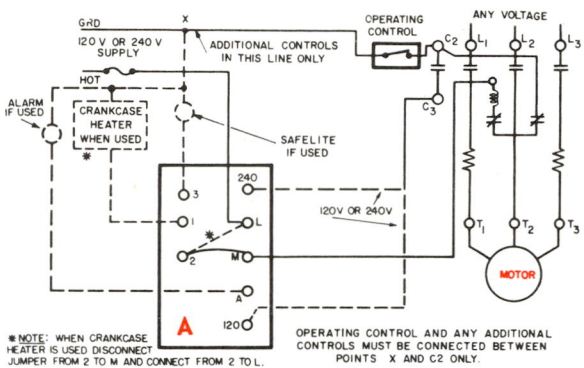

Fig. 12-109. Wiring diagram shows oil pressure safety cut-out at A. Motor is three-phase unit. Note possible use of alarms, safety lights and crankcase heater. (Penn Controls, Inc.)

pressure difference between the two bellows drops below the required oil pressure needed. Large commercial systems use this type. In some systems, the control points are in the compressor motor circuit. In other cases, the points will close and current is sent through a bimetal strip or a resistance heater near the bimetal strip. If this strip heats up before the pressure returns to normal, the power will be disconnected.

A solid state oil pressure safety control is shown installed on a motor compressor in Fig. 12-110. Internal construction and circuit diagram are shown in Fig. 12-111.

Refrigerant level may be kept within safe limits by a float switch. The float may be used for signalling or it may actually control the refrigerant level.

The switch may be used to control the liquid level in:
1. Flooded surge drums.
2. Flooded shell and tube chillers.
3. High and low-pressure receivers.
4. Intercoolers.
5. Transfer vessels.
6. Various kinds of accumulators including liquid recirculating types.

If the refrigerant level is too high, the float switch closes an electrical circuit. The circuit acts to allow a refrigerant flow out of the control device. If the refrigerant level is too low, the float switch will actuate (cause to work) a circuit which will allow refrigerant to flow into the system.

A float control switch is illustrated in Fig. 12-112. This assembly must be mounted vertically.

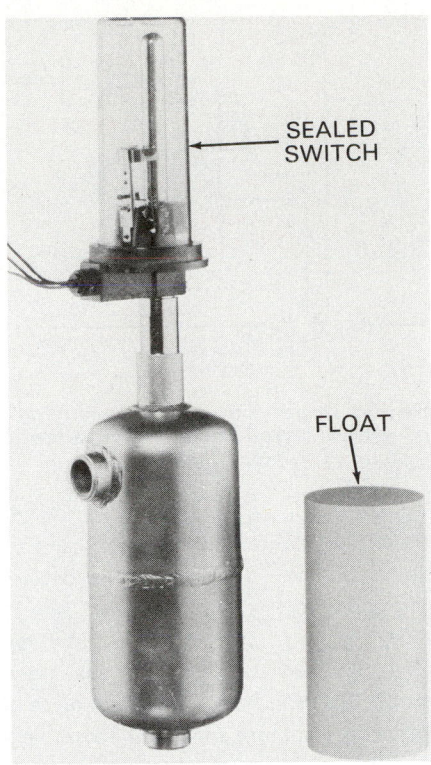

Fig. 12-112. Float-operated switch is designed to control level of liquid refrigerant in system. (Refrigerating Specialties Div., Flo-Con)

12-41 MOTOR STARTERS

The pressure or temperature control contacts — whether open or sealed — are limited in the amount of current they can safely carry. The National Electric Code and local electric codes usually set down the limitations of these controls.

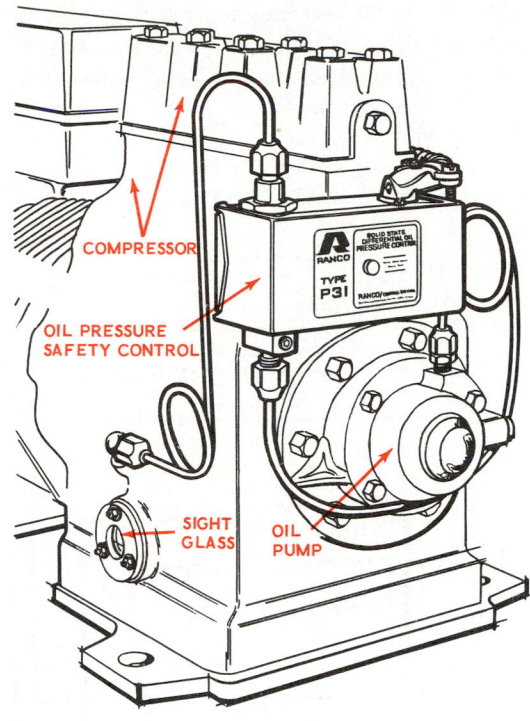

Fig. 12-110. A solid state oil pressure safety control connected to a serviceable hermetic compressor. (Ranco, Inc.)

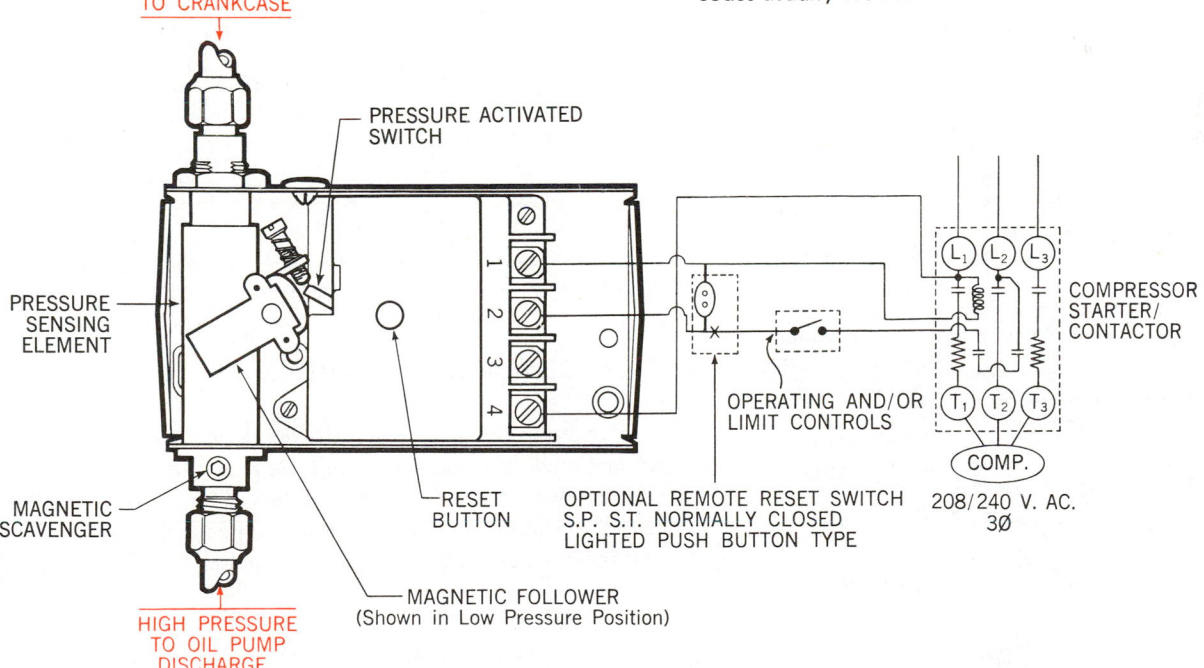

Fig. 12-111. Internal construction and circuit diagram for a solid state oil pressure safety control. (Ranco, Inc.)

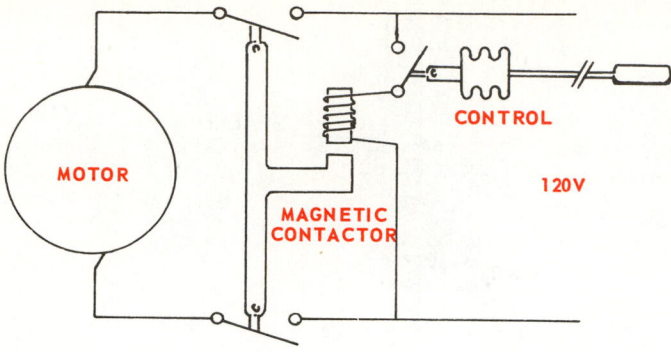

Fig. 12-113. Schematic diagram of automatic control on a magnetic starter. Circuit allows high current flow to motor without overloading control contact points.

However, these same commercial controls can handle larger motors (larger loads) with the help of a device called a magnetic starter (contactor).

The magnetic starter is an electromagnetic device. The magnetism is controlled by the electricity that flows through the motor control. The magnetism attracts a piece of steel (or armature). When this armature moves, it closes large contact points that safely carry the larger current flow needed for the large motors. Fig. 12-113 is a schematic wiring diagram of a magnetic starter.

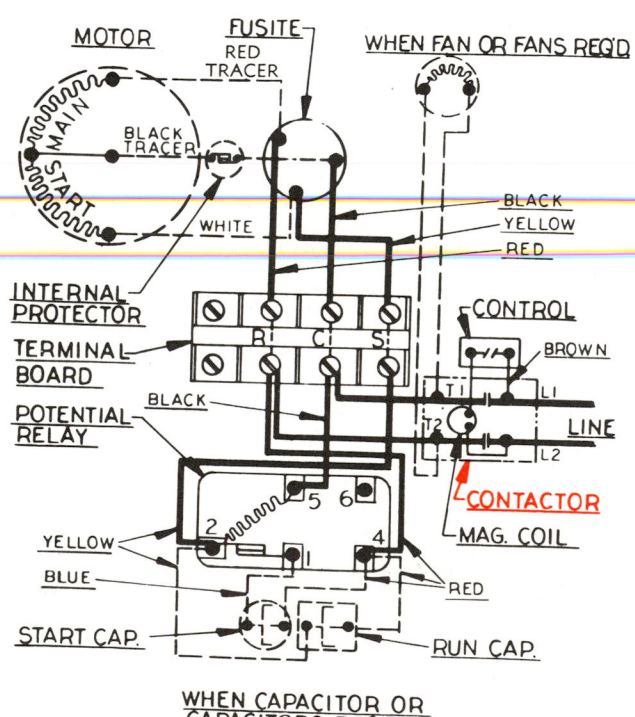

Fig. 12-114. Wiring diagram used for 120—240V single-phase unit. Note contactor (motor starter).

These starters are mounted in an approved metal box with a safety access door. Some units incorporate a manual shutoff switch, fuses and an overload thermal safety breaker switch.

The safety switch is operated by a heating element located in the black lead of the motor circuit inside the contactor box or starter. If the motor demands too much current (shorts, grounds or overloads), this heater will bend a thermal bimetal strip in the control circuit, opening the electromagnet circuit. This action opens the main switch.

Fig. 12-114 shows a wiring diagram of a 120—240V single-phase system using a magnetic starter. A wiring diagram for a three-phase system is shown in Figs. 12-115 and 12-116.

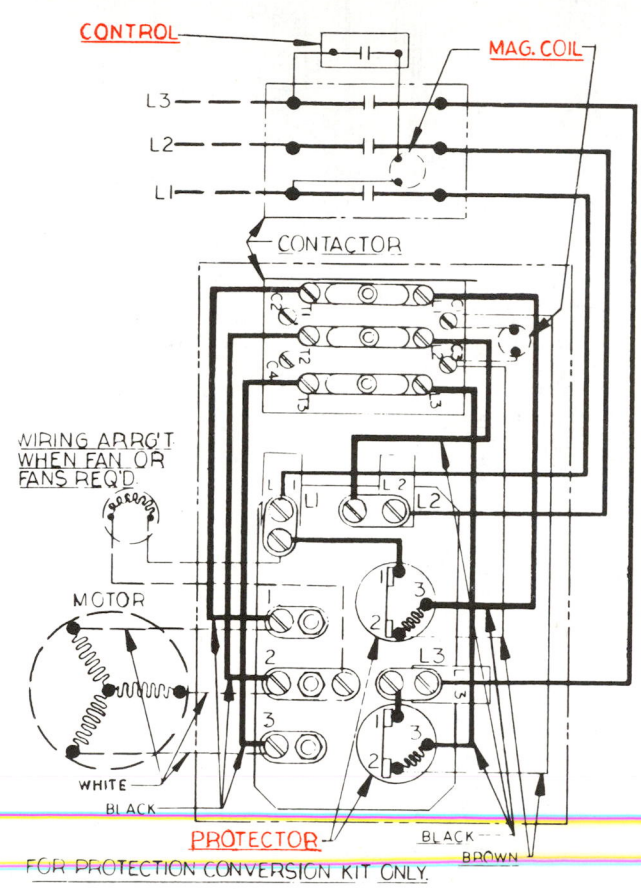

Fig. 12-115. Wiring diagram of three-phase system using magnetic starter. Control is in series with magnetic coil of contactor (starter).

12-42 ICE MAKER CONTROLS

In addition to the usual refrigerant and motor controls, automatic ice cube makers or ice flake makers have controls to stop the system when the storage bin is full. The control is located in the bin. It shuts off the machine until some of the ice is removed or melts. Devices used include:
1. Mechanical levers.
2. Temperature controls.

The mechanical type has a lever or a diaphragm which, when contacted (pressed) by the accumulated ice, opens a switch and stops the unit.

The temperature control shuts off the unit when the

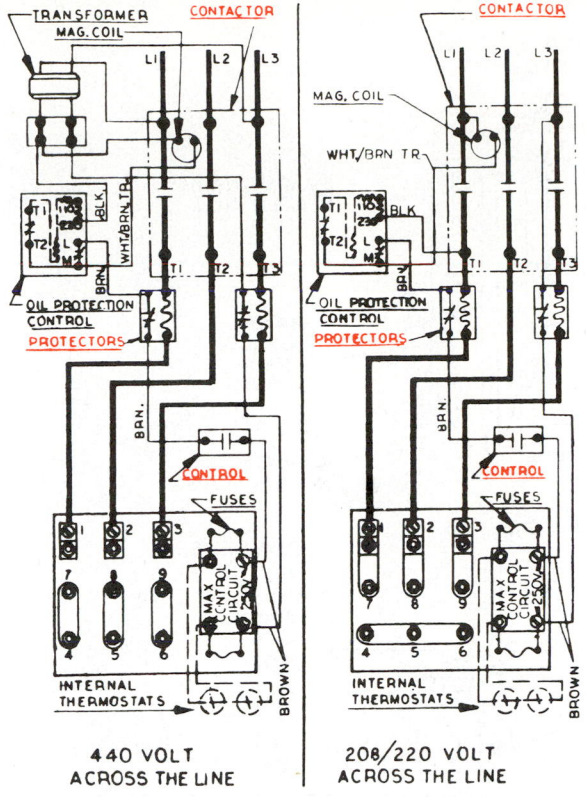

Fig. 12-116. Wiring diagrams of 440V circuit and 208—220V circuit as used with three-phase power. (Copeland Corp.)

control bulb is in direct contact with the ice. Both controls are located at the top of the ice bin.

The wiring diagram, Fig. 12-117, is from a system which freezes cubes and then removes (harvests) the cubes from the freezing grid. The refrigerating system has a hot gas defrosting system, as shown in Fig. 12-118. Fig. 12-119 is an actual wiring diagram. This shows the interlocking of the controls, devices and wiring.

12-43 VENDING MACHINE CONTROLS

Most vending machines which use refrigeration operate automatically. These machines can perform several operations:
1. Heat foods.
2. Cool foods.
3. Select foods.
4. Accept coins to activate the dispenser system.

Some of the units automatically heat, if necessary, and move the items being dispensed (bottles, bulk fluids, packages of ice cream and the like). Thermostats, relays, micro-switches, positioning motors and solenoids are used.

The automatic operation of this unit with its vending motor, magnets, signal lights, relays and so on, makes an elaborate wiring system necessary. Fig. 12-120 shows eight parallel circuits being used in one dispenser. The evaporator fan operates continuously.

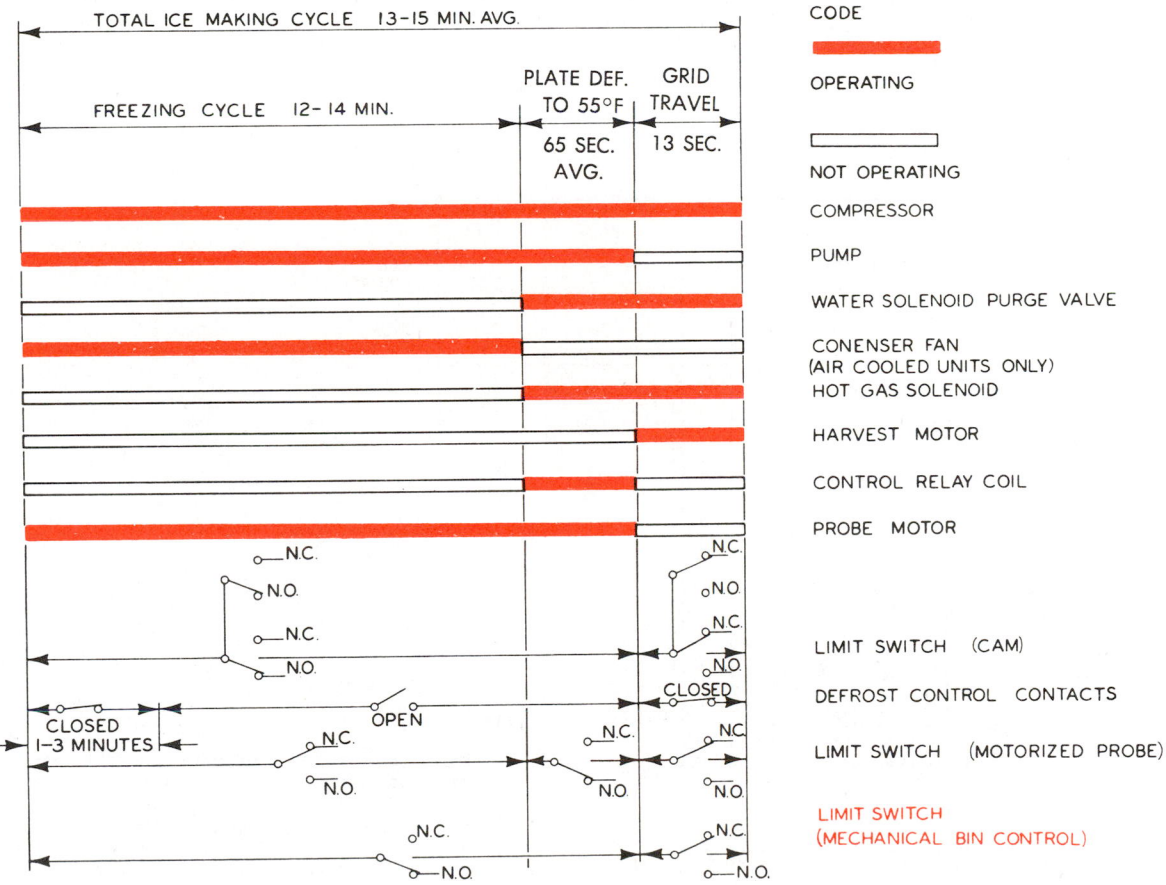

Fig. 12-117. Schematic wiring diagram of ice cube maker which uses harvest motor to remove cubes from the freezing grid. (Mile High Equipment Co.)

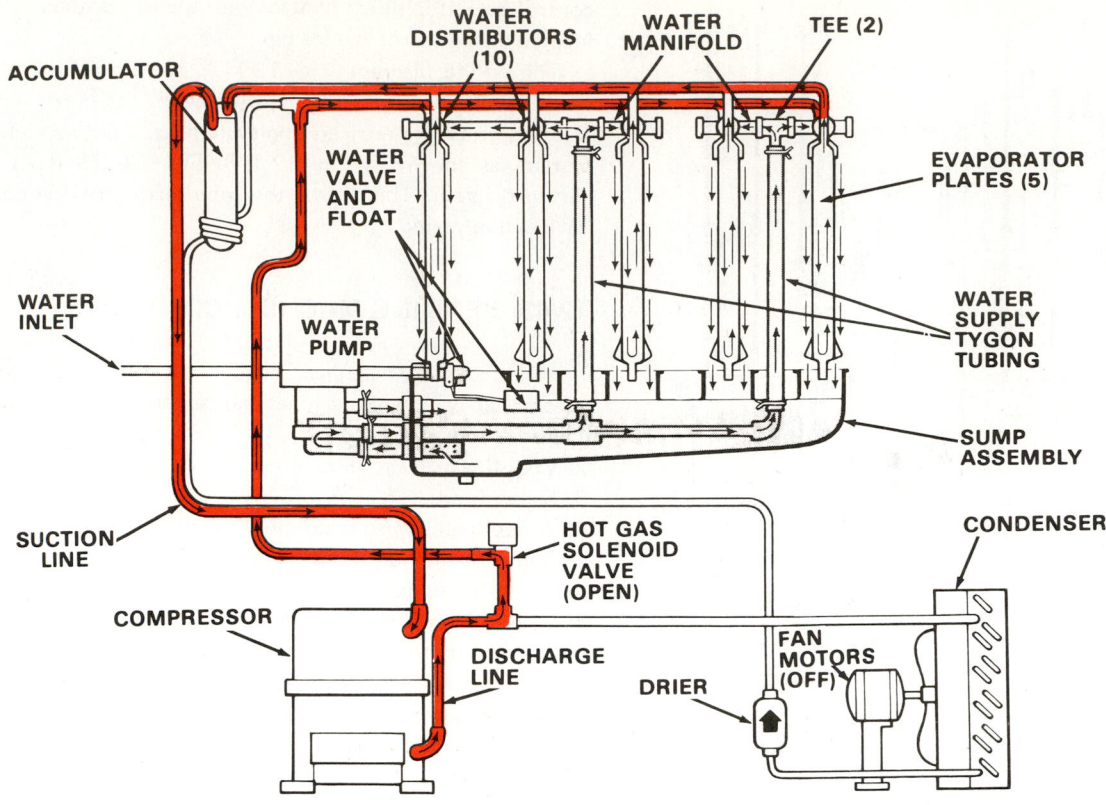

Fig. 12-118. Ice cube maker refrigerating system which uses a hot gas defrosting system. (Queen Products Div., King-Seeley Thermos Co.)

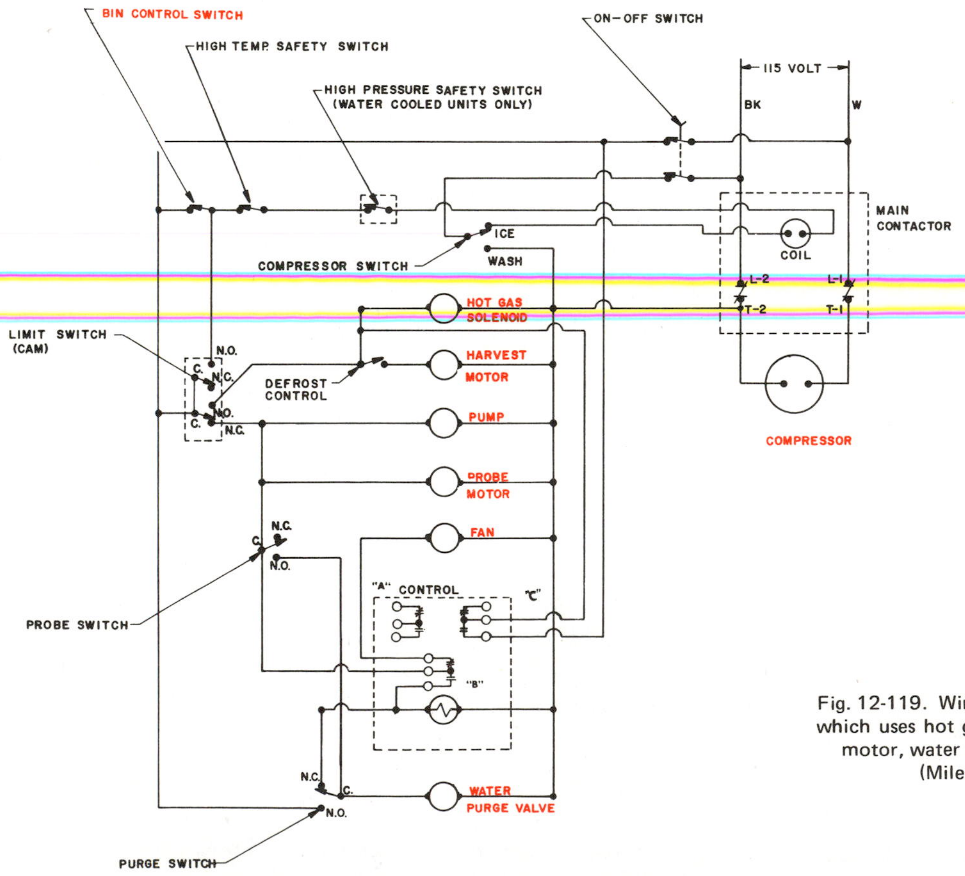

Fig. 12-119. Wiring diagram of ice cube maker which uses hot gas defrost, probe motor, harvest motor, water purge valve and a water pump. (Mile High Equipment Co.)

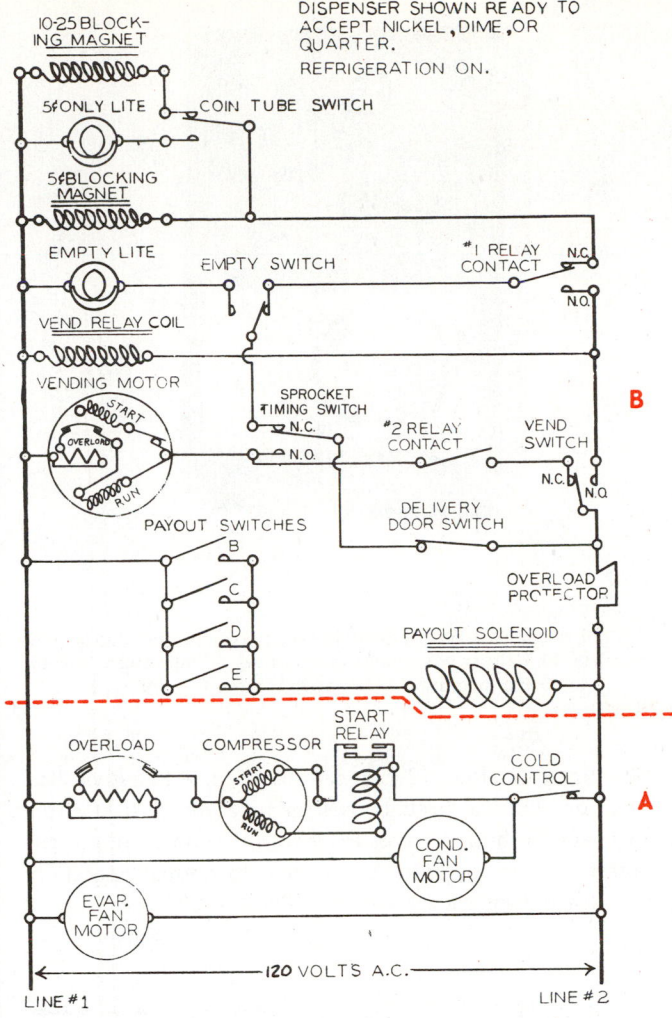

Fig. 12-120. Wiring diagram for refrigerated bottled beverage vending machine. A—Refrigerating unit wiring. B—Vending wiring.

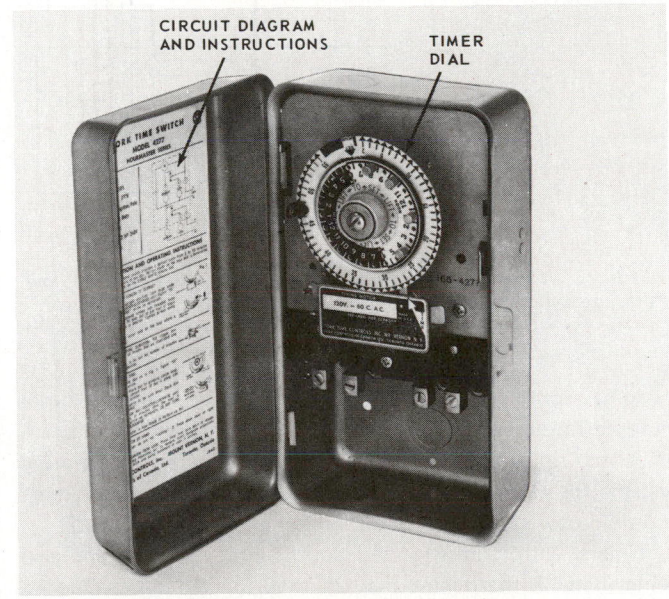

Fig. 12-121. Time switch used for controlling defrost cycles in commercial systems. Note wiring diagram and instructions on inside cover.

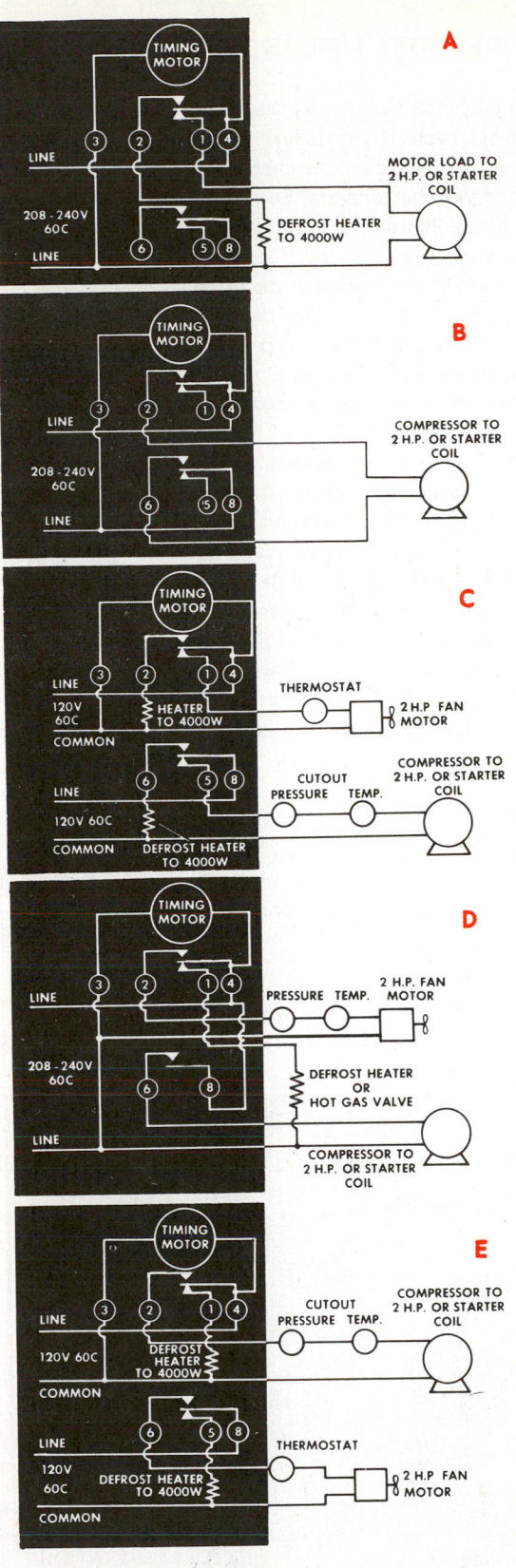

Fig. 12-122. Wiring diagram for several types of defrost control arrangements. A—SPDT (single-pole double-throw) circuit which activates defrost heaters as it shuts off refrigerating unit. B—DPST (double-pole single-throw) which only shuts off refrigerating unit. C—DPDT (double-pole double-throw) which shuts off compressor and fan and turns on two defrost circuits. D—Circuit for delayed fan shutoff during defrost and one defrost circuit. E—DPDT (double-pole double-throw) circuit for delayed fan shutoff and two defrost circuits.

12-44 DEFROST TIMERS

Most automatic defrosters need an automatic device to start the defrost cycle. In installations using a defrost timer, an electric self-starting clock mechanism operates a cam. This cam makes the switches operate. See Fig. 12-121. These timers are usually of the 24-hour or seven-day design.

Some time clocks are connected directly to electric power and will defrost the system at the intervals necessary to keep it working well. Each evaporator design has its own requirements for good operation. Some need to be defrosted during each cycle; some every few hours. Others need defrosting but once a day. The cams can be adjusted to control the length of the defrost cycle.

Some timers are connected to electric power in parallel with the motor. The clock mechanism registers only the running time of the condensing unit. These mechanisms then start the defrost cycle after so many hours of running time.

The timer wiring differs with the type of defrost system. In one "hot gas" system, the timer energizes the solenoid bypass valve (causes it to act). It stops the fan motors, energizes auxiliary electric heater elements and runs the compressor. It also may be used to prevent the normal cycle from starting until the low-side pressure is at normal levels. Some basic electrical circuits using timer controls are shown in Fig. 12-122.

Another type of automatic timer for defrosting is shown in Fig. 12-123. A timer starts the defrost cycle while the temperature bulb returns the unit to normal operation after the evaporator reaches a temperature above 32 F. (0 C.).

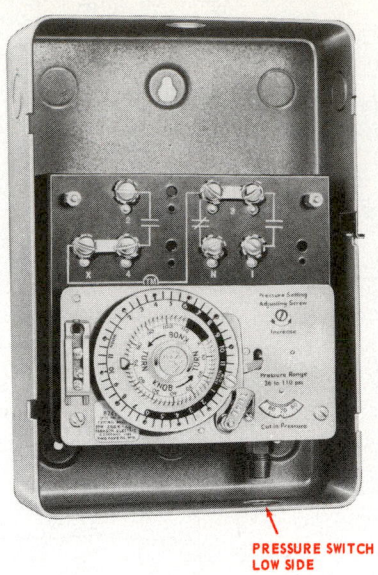

Fig. 12-124. A defrost timer with a low-side pressure operated switch. Timer starts the defrost action and low-pressure switch returns system to normal operation. (Paragon Electric Co., Inc.)

The timer in Fig. 12-124 can be used with either "air defrost" or electrical heat. It uses the timer motor to start the defrost action and a pressure control connected to the low-pressure side to return the system to normal operation. The electrical diagrams are shown in Fig. 12-125.

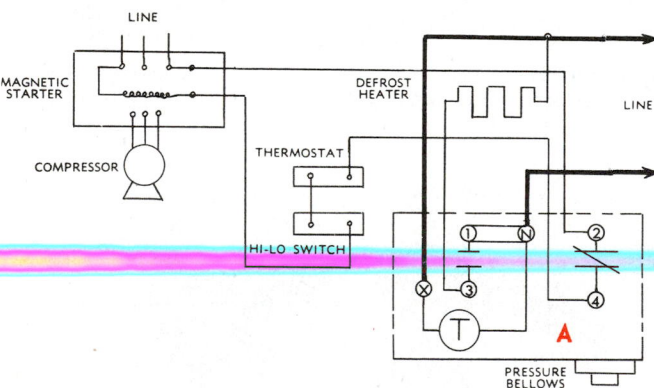

Fig. 12-125. Wiring diagram with timer to start defrost and low-pressure switch to return system to normal operation. A—Control.

Some commercial installations use several timers on a multiple fixture system. Fig. 12-126 shows the wiring diagram of a multiple defrost system with separate refrigerating systems for each fixture. An electrical panel with four timers and six motor starters (contactors) is shown in Fig. 12-127.

Timers with transistorized solid state circuitry are also being used. A thermistor may control the defrost cycle in a display case. A thermistor measures the temperature difference of the air moving through the evaporator. This control replaces a timer. Defrost is controlled only by demand. It measures the

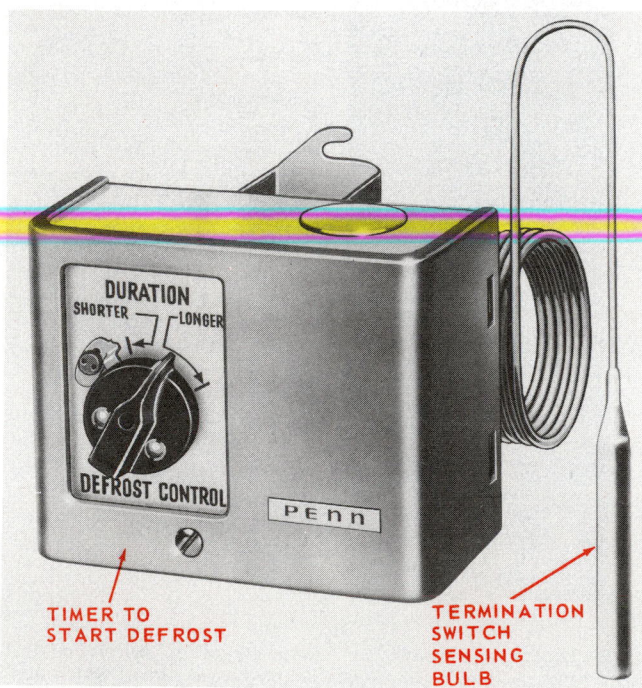

Fig. 12-123. Timer and thermal bulb combination used to control defrost cycle. Thermal bulb is located on evaporator. Timer starts defrost action and thermal sensing bulb returns the system to normal operation after frost has melted. (Penn Controls, Inc.)

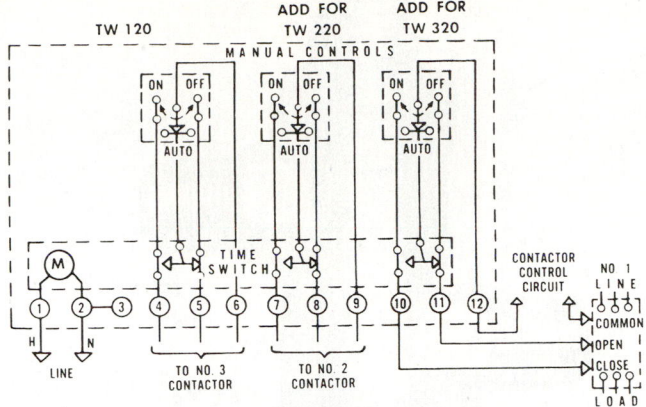

Fig. 12-126. Circuit for automatic defrost timer for several fixtures.

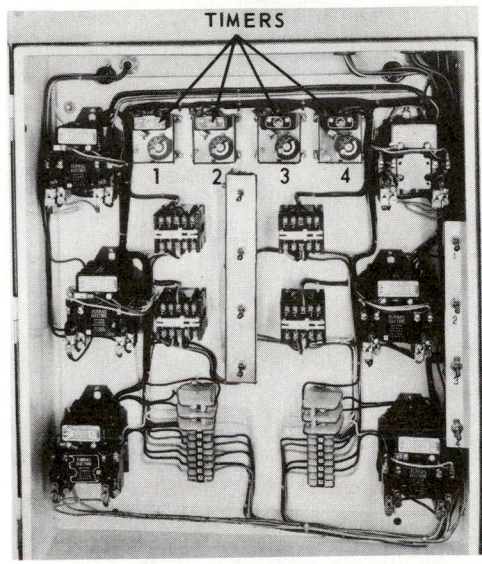

Fig. 12-127. Electric control panel with four timers. Each of the four controls production of 10 tons of ice on each cycle. (Turbo Refrigerating Co.)

temperature differences between the air entering the evaporator and the air leaving the evaporator. If this temperature difference becomes more than 20 F. (11 C.) to 30 F. (16 C.), this unit will start the defrost cycle. A standard thermostat returns the system to a normal cycle when the evaporator temperature measures about 40 F. (4 C.).

12-45 VALVES, PRESSURE REGULATING

Commercial systems use many types of pressure regulating valves. Some of these valves control:
1. Evaporator pressure (two-temperature valves).
2. Crankcase pressure.
3. Discharge bypass pressure with solenoid valve control for pull down (service), to prevent freezing.
   a. Some into suction.
   b. Some into evaporator.
4. Head pressure control valve.

Some of these valves have Schrader service connections for gauge mounting.

12-46 VALVES, TWO-TEMPERATURE

In many multiple installations it is necessary to maintain different temperatures in various evaporators connected in the same system. Thermostatic expansion valves may be used if the temperature differences are not too great — not over 5 F. (3 C.). But in some instances, such as with a storage cabinet and an ice cream cabinet combination, the temperature differences are too great to be taken care of this way. To provide for this, a two-temperature valve is put into the warmest evaporator suction line. This prevents pressure of the warmest evaporator from going below a safe setting.

The controlled evaporator or evaporators should not have more than 40 percent of the total load of a system. If the controlled evaporator is too large, erratic cycling will result. (See surge tanks, Para. 12-53.) If the controlled load amounts to more than 40 percent, separate condensing units should be used.

12-47 TYPES OF TWO-TEMPERATURE VALVES

Two-temperature valves are sometimes called constant pressure valves or pressure reducer valves. They are also used to insure a constant low-side pressure. They are made up of a bellows or diaphragm, a needle and seat arranged in such a way that the bellows are operated by the pressure in the warmest evaporator.

As the compressor pumps the low side down to the desired pressure, the bellows shuts off the valve. This action stops the pressure in the warmest evaporators from going below the pressure desired. As pressure in the evaporator builds up from vaporizing the refrigerant, the bellows again opens the valve passing vapor on to the compressor.

The pressure maintained on the surface of a quantity of liquid refrigerant determines the temperature at which the refrigerant will evaporate. The suction line valve will control the temperature of the evaporator to which the line is attached, even though the suction pressure of the compressor is considerably below the evaporator pressure.

Two general types of two-temperature valves are:
1. Pressure operated.
   a. Metering.
   b. Snap-action.
2. Temperature operated.
   a. Sensing bulb and bellows.
   b. Thermostat and solenoid.

12-48 METERING TYPE TWO-TEMPERATURE VALVE

The metering type, two-temperature valve acts more as a throttling device than as a shutoff valve. See Fig. 12-128. Fig. 12-129 shows a cross-section of this valve.

Some of these valves are equipped with a gauge opening so that the service technician can check and adjust the warmer

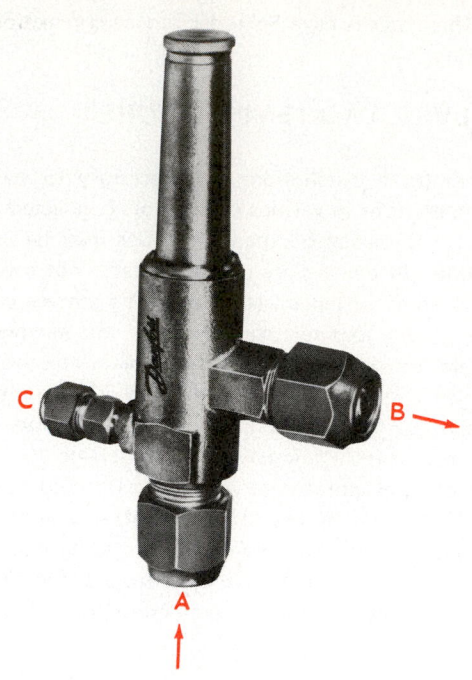

Fig. 12-128. A metering type two-temperature valve. A—Refrigerant vapor in. B—Refrigerant vapor out. C—Gauge connection. (Danfoss, Inc.)

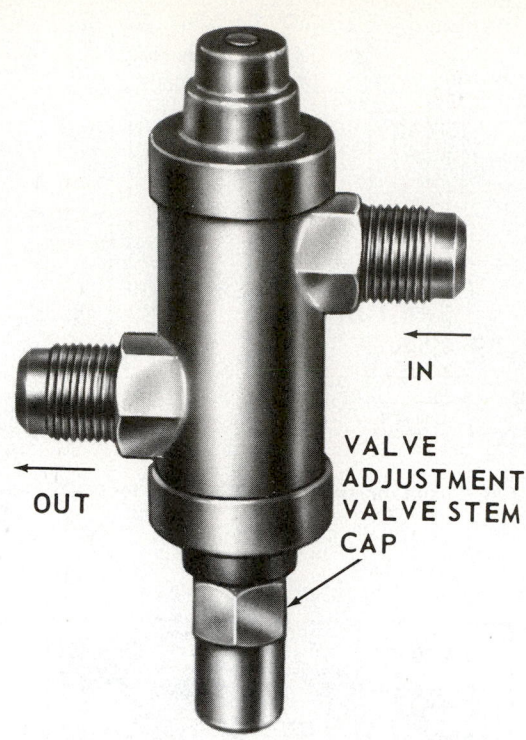

Fig. 12-130. Metering type two-temperature valve with flare fitting connections. This is a pressure operated throttling valve. (A C and R Components, Inc.)

evaporator pressure. It opens and closes when the pressure varies only a fraction of a pound, having no differential. (A differential means one pressure to open it and a different temperature to close it.) Because the bellows pressure area and

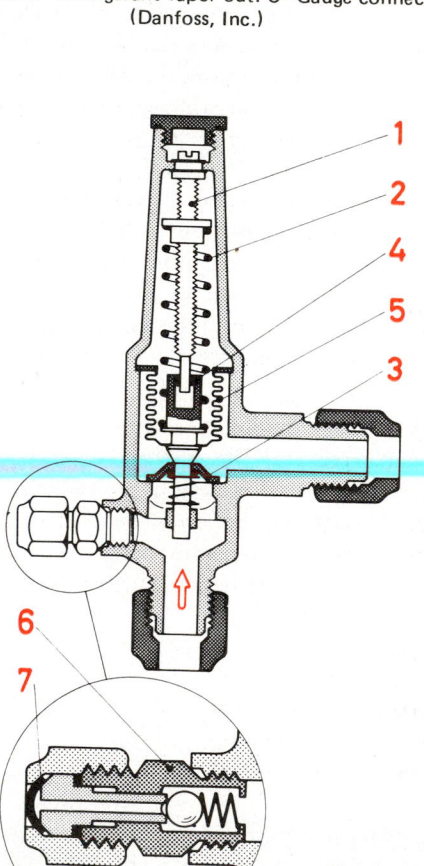

Fig. 12-129. Cross-section through metering type two-temperature valve. 1—Adjusting screw. 2—Adjusting spring. 3—Valve. 4—Valve stem. 5—Bellows. 6—Service valve. 7—Cap.

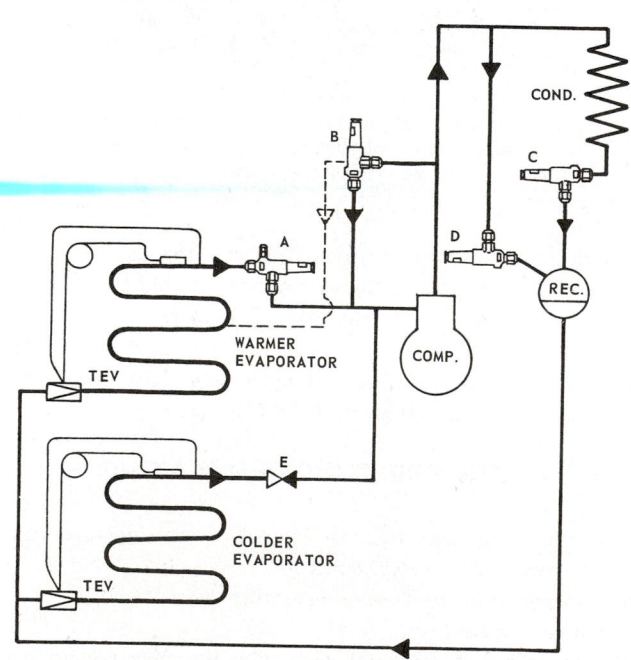

Fig. 12-131. A two-evaporator system equipped with several pressure control valves. A—Two-temperature valve (evaporator pressure regulator). B—Condenser bypass valve. C—Condenser pressure regulating valve. D—Capacity regulating valve. E—Check valve. (Danfoss, Inc.)

the valve area are equal, only the adjustment spring and the warm evaporator pressure changes can operate the valve motion.

The valve openings must be made large enough to offer efficient vapor flow. Many of these metering controls have a small adjustment range, being especially designed to maintain temperature pressures just above crankcase pressures. See Fig. 12-130. A two-evaporator system equipped with several pressure-controlled valves can be seen in Fig. 12-131.

Large systems must rely on forces other than springs to control the pressure for efficient operation. A large capacity two-temperature valve is shown in Fig. 12-132. It has a plastic seat and connections for silver brazing it to the suction line.

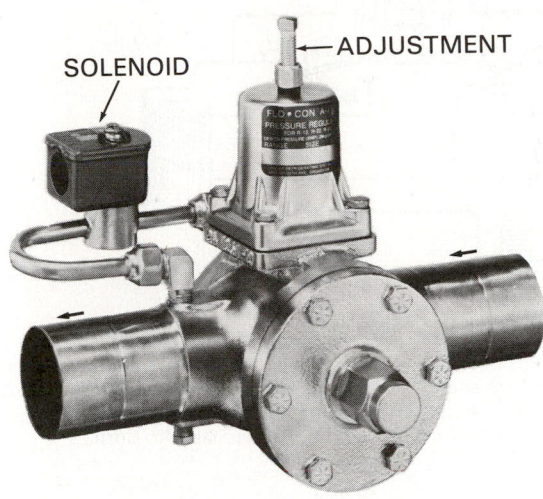

Fig. 12-132. A large capacity evaporator pressure regulator. Note use of metallic fittings for brazed connections to suction line pipe. (Refrigerating Specialties Div., Flo-Con)

12-49 SNAP-ACTION TYPE TWO-TEMPERATURE VALVE

The snap-action type valve is such that, when the valve closes, there is a decided rise in pressure in the warm evaporator before the valve opens again. An important feature of the snap-action pressure-operated valve is that it has a definite cut-in pressure and temperature. It is often used where defrosting is wanted on each cycle.

A snap-action two-temperature valve is normally used with multiple evaporator systems, such as a walk-in cooler and a display case, which do not operate at a wide temperature difference. The valve should be located on the suction line to the display case.

12-50 THERMOSTATIC TYPE TWO-TEMPERATURE VALVE

Another type of two-temperature valve has a temperature control and is built much like a thermostatic expansion valve. It also works much the same because it operates from the temperature of the evaporator or the air entering or leaving. It has a capillary tube and a sensing bulb much like the

thermostatic expansion valve. It also has a bellows to move a rod as different pressures are created in the sensing bulb.

When the evaporator becomes cool enough, the cooling sensing bulb lowers the pressure in the bellows. So the bellows contracts. This pulls on the valve plunger and shuts off the valve. With the valve closed, pressure cannot drop any lower in the evaporator. Then, the valve controls the minimum temperatures of the evaporator.

As the evaporator warms up, the sensing bulb warms up also. The increase in pressure is transmitted to the bellows. It expands, pushes the plunger so that the valve opens. With the valve open, the compressor is able to once more draw vaporized refrigerant from the evaporator. This type of valve is always located in the suction line of the warmest evaporator.

Fig. 12-133 shows a regulator which responds to the air

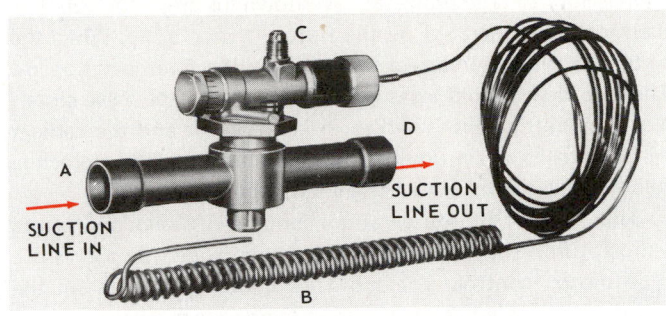

Fig. 12-133. Temperature controlled suction line metering valve. A—Suction line from evaporator. B—Temperature sensing bulb. C—High-side pressure connection. D—Suction line to compressor. (Jackes-Evans Div., Parker-Hannifin Corp.)

temperature as it leaves the evaporator and enters the fixture or case. It may also respond to the temperature of the air entering the evaporator.

It has a sensitive element located in this airstream. The valve body is mounted in the suction line. A liquid line connection is also made to the valve body. This provides the

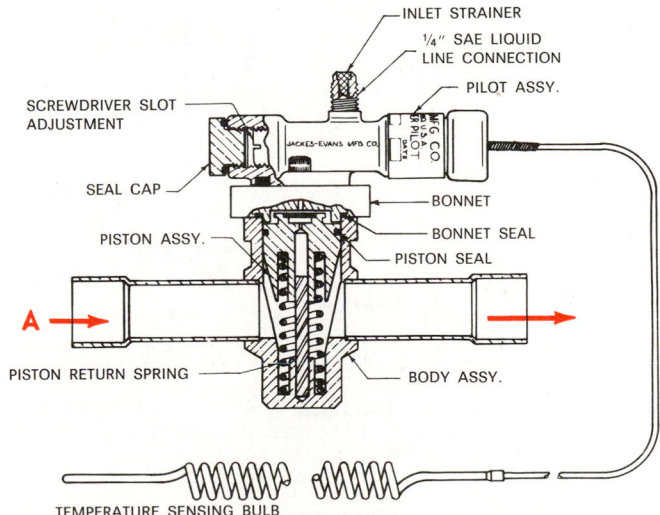

Fig. 12-134. Temperature controlled suction line evaporator pressure control. A—Suction line vapor travel.

pressure needed to open the main valve in the TPR (temperature pressure regulator). See Fig. 12-134.

As the temperature sensing bulb warms, its increase in pressure closes a small valve and the high-side pressure in the valve decreases allowing the main piston to open the suction line more. As the sensing bulb cools, the high-pressure pilot valve is opened and the main piston is forced down closing the suction line passage partially or completely (modulates).

12-51 SOLENOID TWO-TEMPERATURE VALVE

A fourth way to get various fixture temperatures in a multiple installation is to use a thermostat connected in series with a solenoid valve.

The solenoid shutoff valve is usually placed in the liquid line of the evaporator it is to control. It has an electrical connection to a thermostat, as shown in Fig. 12-135. The thermostat is operated by the fixture temperature. When the fixture reaches the correct temperature, the thermostat opens. The electric solenoid loses its magnetism and the valve closes. No more refrigerant is fed to the evaporator and the cabinet will gradually warm up until the thermostat points close, the solenoid valve opens and refrigeration starts again.

This system of refrigeration control is based on fixture temperature. The condensing unit is controlled by a pressure type motor control. The motor will not stop until all the fixtures are cooled to their correct temperature. Some systems have the solenoid valve in the suction line to prevent removing the refrigerant from the evaporator.

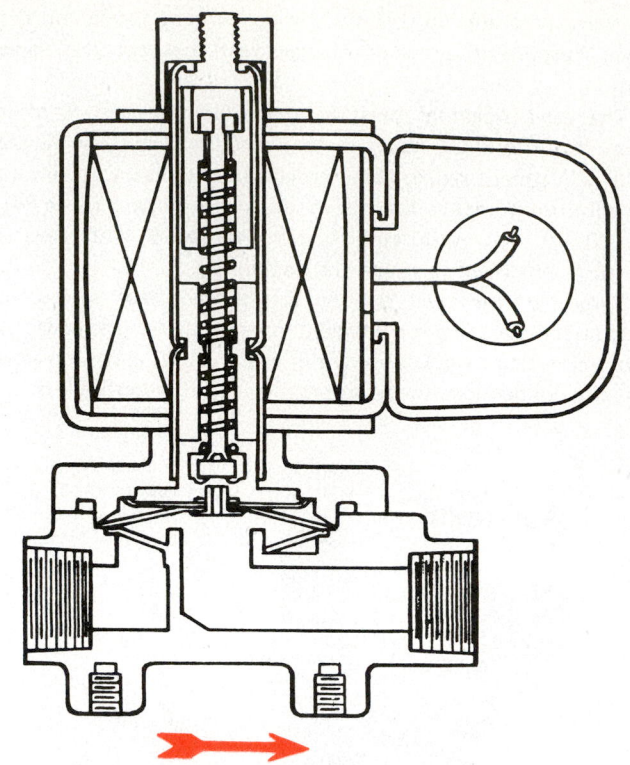

Fig. 12-136. A solenoid valve designed to open the valve when magnet is not energized and close the valve when magnet is energized. Note the threaded pipe connections at either end.
(Jackes-Evans Division, Parker-Hannifin Corp.)

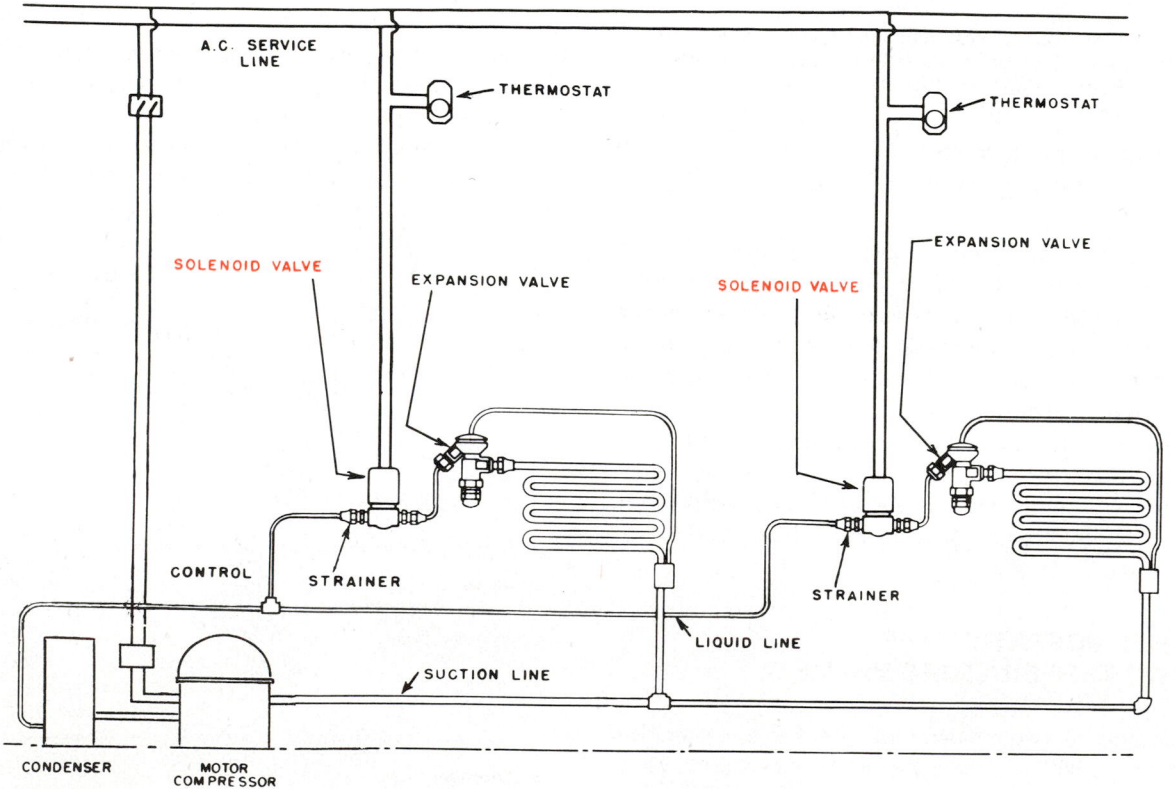

Fig. 12-135. Installation of solenoid valves. Each solenoid controls the cabinet temperature of different cabinets. Valves and thermostats are using line voltages.

This system uses a normally open solenoid and a thermostat that opens the circuit on temperature drop and closes it on temperature rise. When magnetized, the valve closes. See Fig. 12-136.

A solenoid valve is also used to stop flooding the low-pressure side during the off cycle. This solenoid is also located in the liquid line. It is electrically connected in parallel with the pressure motor control, Fig. 12-137.

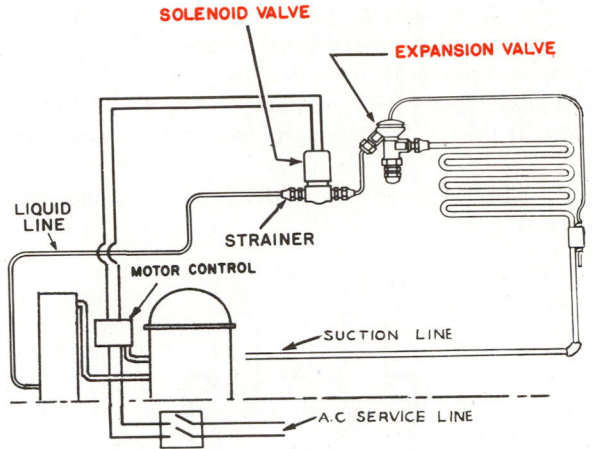

Fig. 12-137. Solenoid valve located in liquid line and controlled by pressure motor control.

12-52 CHECK VALVES

Check valves are used in refrigerating systems to prevent flow of liquid and/or vapor refrigerant in the wrong direction. They are used in two-temperature installations, in defrost systems and to prevent vapor passage during off cycles.

In multiple installations, one condensing unit is connected to several evaporators all of which are at different temperatures. Two-temperature valves are used to obtain the desired temperatures. Check valves, Fig. 12-138, are sometimes put in the suction line of the coldest evaporators to prevent excess warming of the cold evaporator during the off cycle.

After the condensing unit has stopped, one of the two-temperature valves may open before the condensing unit turns on. This may flood the low side with warm refrigerant vapor. This vapor will also travel along the suction line to the coldest evaporator.

If it enters the evaporator it will start condensing, releasing its latent heat. This will make the cold evaporator defrost or at

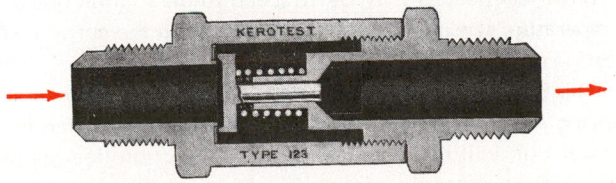

Fig. 12-138. Suction line check valve used on colder evaporators in multiple systems.   (Kerotest Mfg. Co.)

least warm up somewhat. The check valve, when installed in the suction line of the coldest evaporator, will only allow vapor to be drawn from this evaporator.

This check valve must have a tight seat and it must open easily. If the valve is too small or if it opens with difficulty, it will act as a throttling device and cause too much pressure drop. The result will be poor refrigeration in the coldest evaporator.

Fig. 12-139 shows a large-capacity check valve. Reverse cycle systems use check valves. So do some hot-gas defrost systems.

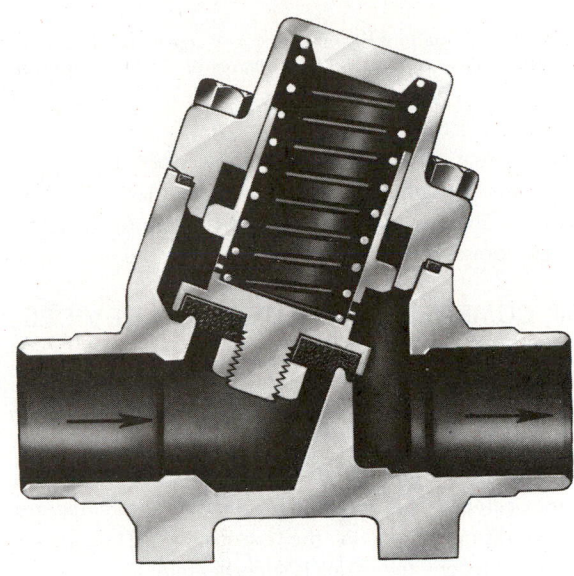

Fig. 12-139. Large-capacity check valve. Note double cylinder design to provide smoother operation.   (Superior Valve Co.)

12-53 SURGE TANKS

Multiple temperature installations may short cycle a pressure-controlled condensing unit. This may be caused by the pressure fluctuations which result from the opening and closing of the two-temperature valves. Pressure is said to fluctuate when it rises and falls repeatedly and in an uncertain pattern.

The following will explain how a short cycle happens:

1. If the two-temperature valve is closed and the condensing unit cools the lowest temperature evaporator enough to open the pressure motor control, the condensing unit will stop.

2. If, just after it stops, a two-temperature valve which controls one of the warmer evaporators opens, the low-side pressure will rise rapidly. The pressure turns on the condensing unit and, thus, causes a short cycle.

To eliminate this trouble, a surge tank or a large cylinder may be installed in the main suction line just ahead of the compressor. The surge tank shown in the Fig. 12-140 schematic is large enough to absorb a pressure buildup. Thus, if the unit is stopped and a two-temperature valve opens, the low-side pressure cannot build up quickly and short cycle the unit. The capacity of the surge tank is great enough to absorb

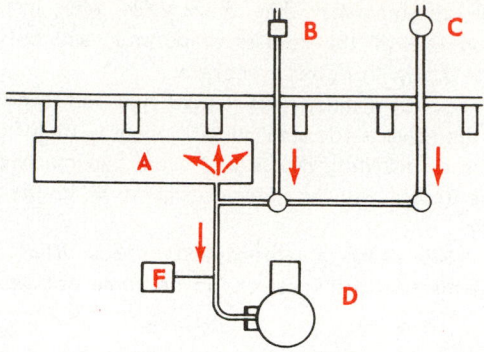

Fig. 12-140. A surge tank installation. A—Surge tank. B—Check valve. C—Two-temperature valve. D—Compressor. F—Motor control.

a large volume of vapor and thereby slow down the rapid pressure changes which would make the motor control turn off and on. The line connected to the bottom of the tank leads to the compressor. It helps return the oil to the compressor.

12-54 COMPRESSOR PROTECTION DEVICES

Many reciprocating compressors are damaged when liquid refrigerant accidentally flows into the compressor from the suction line.

The refrigerant must be in a vaporous state. This means that the vapor temperature must be higher than the temperature of the evaporating liquid in the evaporator. This increase in temperature means the vapor is superheated.

Many devices have been used to prevent or minimize entry of suction line liquid refrigerant into the compressor:

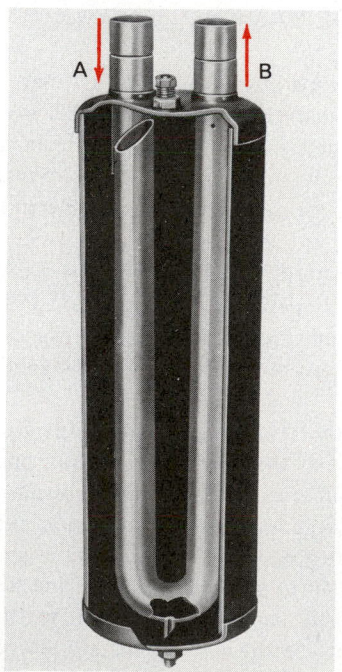

Fig. 12-141. Suction line accumulator. A—Gas inlet. B—Gas outlet. (AC & R Components, Inc.)

1. An accumulator in the suction line.
2. Hot gas bypass valves to move hot gas into the suction line where it can evaporate any liquid.
3. Temperature sensing devices and solenoid valves.
4. Heat exchangers to warm the suction line vapor-liquid.
5. Electrical heaters to warm the suction line vapor-liquid.
6. An evaporator (blower coil) in the suction line.

When a system is using a hot-gas defrost or when a system has a sudden load change, drops and slugs of liquid refrigerant may travel in the suction line. This liquid may damage the compressor. Many systems have an accumulator in the suction line to reduce the danger of flow of liquid refrigerant into the compressor. See Fig. 12-141.

The accumulator will lengthen the cycling interval (gas storage). It usually has aspirating (suction) devices to return oil.

Fig. 12-142 shows an internal design of an accumulator.

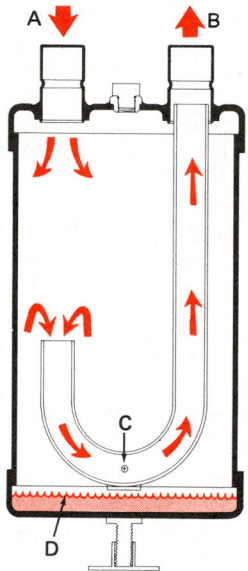

Fig. 12-142. Inside of an accumulator. A—Suction gas in. B—Suction gas out. C—Oil return aspirator hole. D—Liquid refrigerant trapped until it can evaporate. (Virginia Chemicals, Inc.)

The hot gas bypass device depends on the use of a temperature sensor attached to the suction line. This sensor controls a solenoid valve. This valve will open, allowing hot gas to flow into the suction line, when there is danger of liquid refrigerant flowing into the compressor.

Liquid line-suction line heat exchangers are explained in Para. 4-7 and Para. 12-35.

Electrical heaters may be attached to the suction line and a temperature sensing element used to turn on the current when heating of the suction line is needed.

The blower evaporator is usually operated by a temperature sensing element attached to the suction line. The fan in the blower coil will be turned on when the suction line temperature indicates danger of liquid refrigerant flowing into the compressor.

The biggest problem is to have a temperature-operated

device or a pressure device which will detect the presence of liquid quickly enough to turn on mechanisms which will stop the liquid from reaching the compressor.

The best sensor is a thermistor which can react very quickly. It can be connected to an alarm circuit or operating circuit to stop the compressor and take away any danger of damage.

12-55 OIL SEPARATORS

It is very important that the compressor be kept lubricated. It pumps a certain amount of oil along with the refrigerant vapor and there is a possible chance that too much oil may leave the compressor. Therefore, it is important to return the pumped oil to the compressor as quickly as possible.

Refrigeration systems work better when the oil is kept in the compressor. The capacity of each part of the system is increased by 5 to 15 percent. Oil in the condenser and evaporator will reduce efficiency of the unit.

It is important to keep the oil from circulating in low-temperature installations. It thickens in very low temperatures and becomes difficult to move out of the evaporator.

Oil separators remove the oil from the hot compressed vapor as the vapor leaves the compressor. The oil will separate because the vapor flow slows down as it arrives in the separator. See Fig. 12-143. The oil will collect in the separator

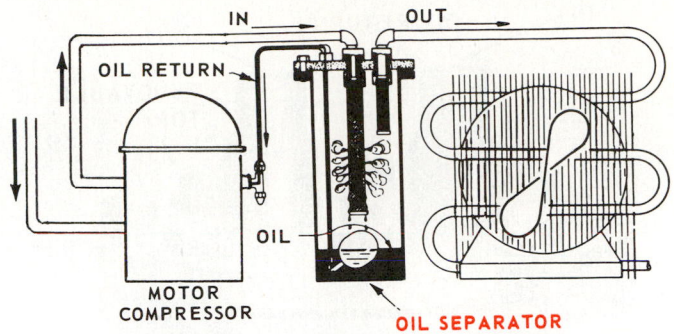

Fig. 12-144. Oil separator installation. Oil is removed from high-temperature, high-pressure, refrigerant and returned to compressor.

until a certain level is reached in the oil separator. Then, a float opens a needle valve. This allows the oil to return to the compressor crankcase.

Oil separators are also placed between the compressor and the condenser, Fig. 12-144. Fig. 12-145 illustrates the inside of a typical oil separator. Note the separator is insulated. This prevents it from acting as a refrigerant condenser.

Many oil separators are serviceable (bolted construction). See Fig. 12-146. On hermetic systems, the oil-return line is usually connected to the suction line near the motor compressor.

Liquid refrigerant may collect in the oil separator during long off-cycles or during long manual shutdown. This liquid

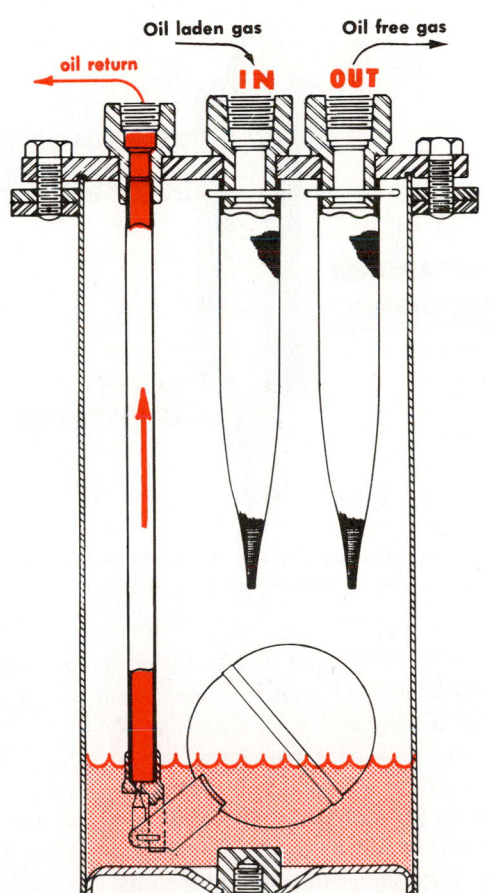

Fig. 12-143. Cutaway of oil separator. Bag-like attachments are screens. (Eaton, Yale & Town, Inc., Dispenser Div., Temprite Products)

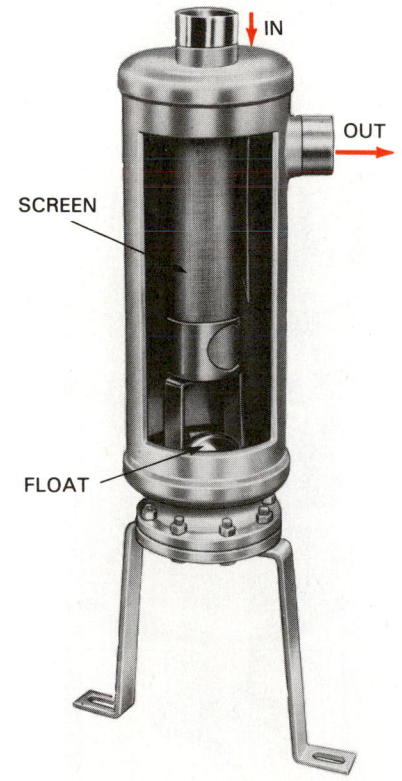

Fig. 12-145. Internal construction of an oil separator. (AC & R Components, Inc.)

Fig. 12-146. Oil separator with bolted assembly top. This type is cleanable and serviceable.

refrigerant, as it returns by way of the oil return line to the compressor, may cause oil pumping. This action may damage the compressor. A check valve in the vapor outlet of the oil separator will reduce this danger. A filter in the oil return line will help keep the oil clean.

A solenoid valve in the oil return line is sometimes used to keep oil or refrigerant returning to the crankcase during the off-cycles. A thermostat controls the solenoid. The thermostat will close only when the oil separator is warm (100 to 130 F.) (38 to 54 C.).

A capillary tube — instead of a 1/4 in. OD tube — is sometimes employed to carry the oil from the oil separator to the compressor crankcase. It reduces oil (or liquid refrigerant) flow to the crankcase and also reduces the chance of oil slugging and damage to the compressor.

Fig. 12-147. Large-capacity oil separator. Refrigerant line connections are 3 1/8 in. OD. (A C and R Components, Inc.)

Large units — up to 150 tons capacity — can use the oil separator shown in Fig. 12-147. It has a 16-in. diameter shell. Note oil return located at the bottom.

12-56 COMPRESSOR LOW-SIDE PRESSURE CONTROL VALVES

Starting a compressor places a heavy load on the motor. It has to overcome inertia of the moving parts, (objects at rest, tend to stay at rest) as well as high crankcase pressure. (In fact, crankcase pressure may be at its highest just then.) It is important, then, to use higher horsepower. Even so, the motors are usually taxed to the limit at the moment of starting — especially against normal or above-normal head pressures.

A compressor low-side pressure control may be used on some installations. This keeps low-side pressures in the crankcase at a reasonable level, even though the rest of the low-side pressure may be high. It may be called a reverse metering two-temperature valve because it never permits the crankcase pressure to exceed a certain safe value.

Crankcase pressure regulating valves are needed where the compressor runs too long before the low side drops to a pressure which does not overload the compressor. If the suction line pressure is higher than the safe pressure, the valve shuts the suction line off from the compressor.

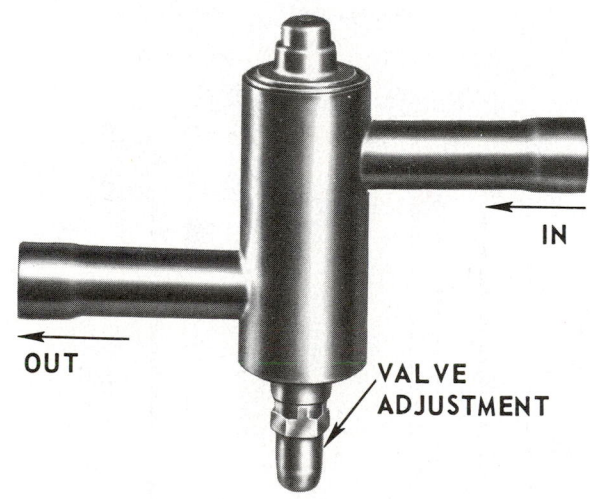

Fig. 12-148. Suction pressure throttling valve. Arrows indicate in which direction gas is flowing.

The body of the valve is usually made of brass, the diaphragm or bellows of phosphor bronze and the needle and seat of wear-resisting steel alloy. See Fig. 12-148. Fig. 12-149 shows the valve installed in the suction line.

For another type of pressure regulating valve, see Fig. 12-150. This valve is nonadjustable. It is preset to open at 11 psi (.77 kg/cm^2) for standard applications. The valve is made to work by a pressure difference between the low side and the sealed-in vapor in the cap.

A refrigerating system suction line with a pilot-operated suction regulator valve is shown in Fig. 12-151.

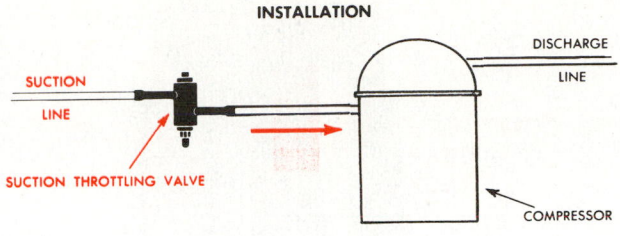

Fig. 12-149. Location of low-pressure side throttling valve.

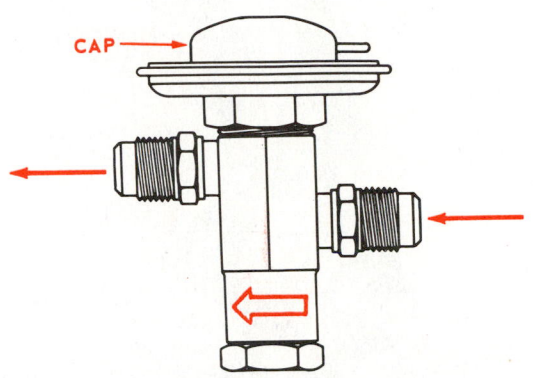

Fig. 12-150. Nonadjustable preset low-side pressure regulating valve. (Sporlan Valve Co.)

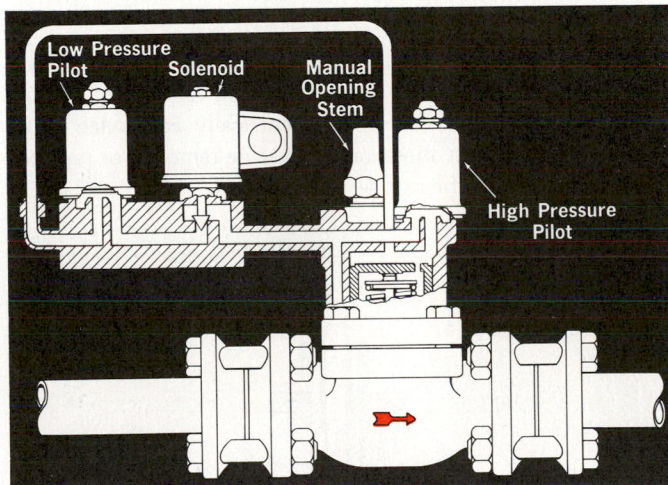

Fig. 12-151. Pilot-operated compressor low-side pressure control. Pilot valve releases pressure above main piston when compressor pressure reaches safe level. This opens main valve allowing evaporator vapor to move to compressor. (Alco Controls Div., Emerson Electric Co.)

12-57 WATER VALVES

Some commercial units of 1/2 hp and up use water-cooled condensers where good, inexpensive water is available. Less power is needed to drive the condensing unit than for the same size air-cooled installation. This is due to better heat transfer as well as the lower condenser temperatures and pressures made possible in a water-cooled condenser. The saving in electrical power makes up somewhat for the cost of the large amounts of water used for cooling.

The water valve turns the water on and off as needed. But it also varies the amount of water as required. Three types of water valves are used:

1. Electric.
2. Pressure.
3. Thermostatic.

It is good practice to install a strainer in the water inlet to the valve. See Fig. 12-152.

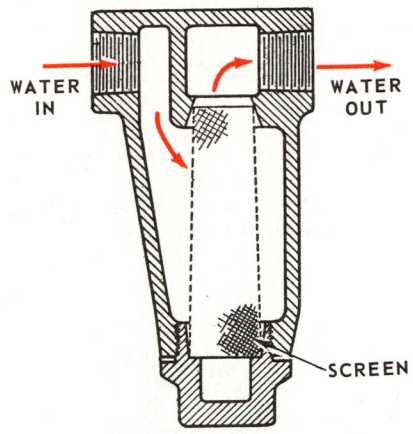

Fig. 12-152. Water line strainer. Screen is removable for cleaning. (Kerotest Mfg. Co.)

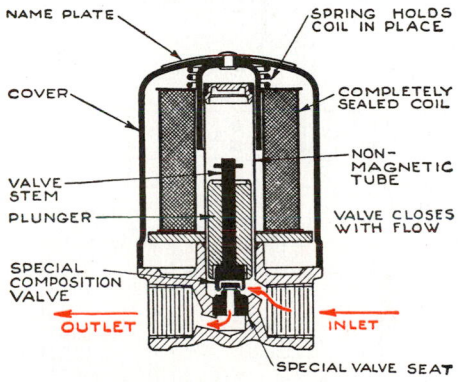

Fig. 12-153. Solenoid-operated water valve. (Controls Co. of America)

12-58 ELECTRIC WATER VALVE

Electrically-operated water valves are of two principal types: solenoid valves and motor-operated valves.

The valve is located between the water supply and the condensing unit. Usually it is mounted on the condensing unit base. The moment the motor starts, this valve opens. When the motor circuit is opened the solenoid is de-energized and the valve closes. See Fig. 12-153.

Electric water valves consume a small amount of current while in operation (6 to 10W). Fig. 12-154 illustrates two typical electric water valve hookups. One uses a low-voltage solenoid valve, and the other a 120V solenoid valve. Most valves require 120V.

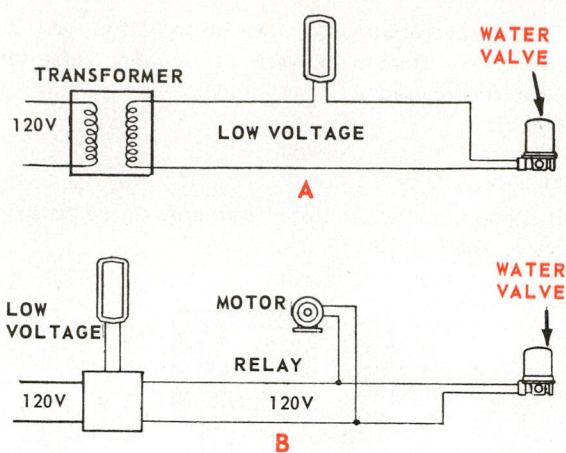

Fig. 12-154. Solenoid-operated water valve wiring diagrams. A—Uses low voltage. B—Uses 120V.

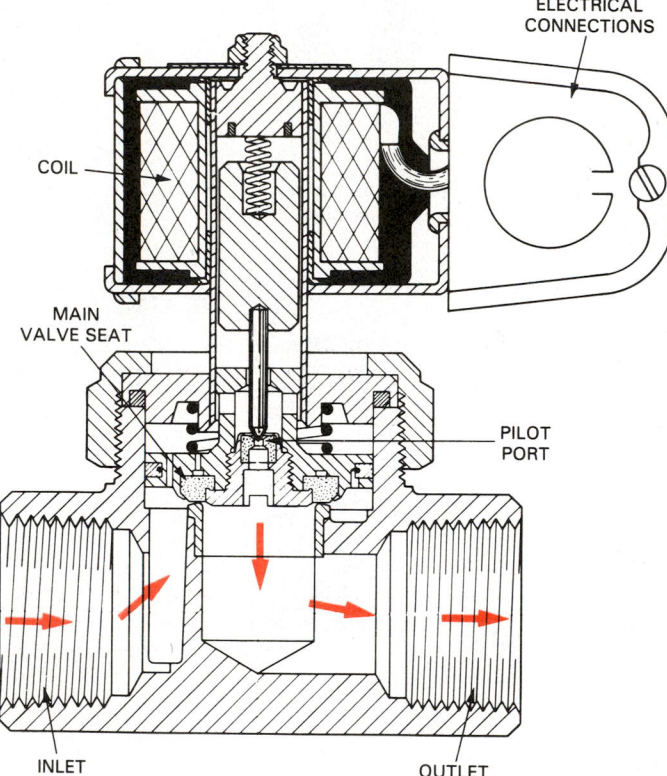

Fig. 12-155. Pilot-operated electric water valve used on large installations. Note that solenoid valve, when open, only decreases water pressure above large piston. (Sporlan Valve Co.)

A solenoid water valve of larger capacity is shown in Fig. 12-155. The body of the valve is brass and is made with either threaded or soldered connections. The plunger is made of noncorrosive steel. The valve seats are usually made of brass or bronze and the valve face is of a special rubber composition.

Water flow is constant in this type of control. The valve stem is loosely connected to the plunger to permit a shock action to open the valve. Gravity and water pressure close the

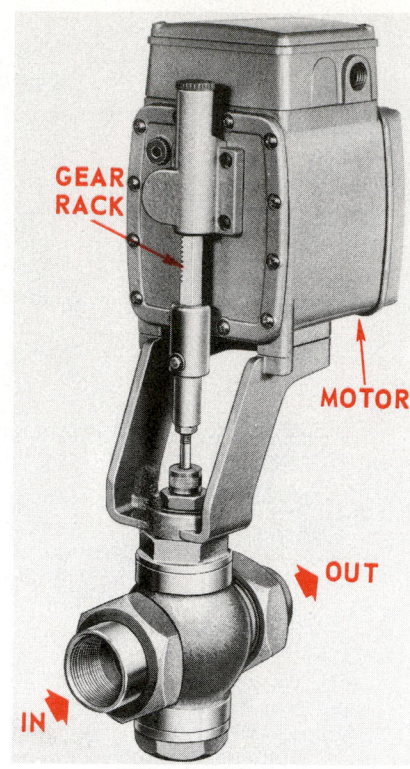

Fig. 12-156. Motorized water valve. Motor raises or lowers valve stem. Note gear rack on stem.

valve when the power is shut off. Large-volume water flow may also be controlled by motor-operated valves, Fig. 12-156. An advantage in the use of the electrically controlled water valve is the fact that these valves may be removed or replaced without disturbing the refrigeration system.

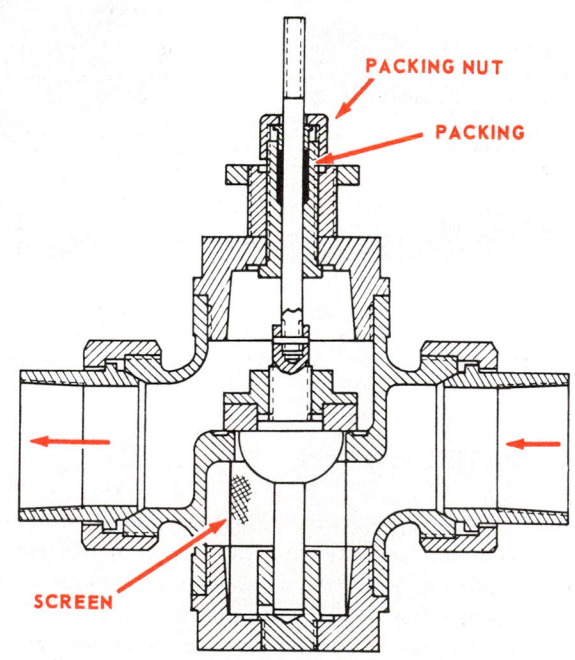

Fig. 12-157. Water valve body of motor-actuated valve. Note direction of flow, screen, valve stem packing and packing nut.

The inside of the motor-actuated water valve is shown in Fig. 12-157. Pipe joints are unions to enable easy removal of the valve. The screen may be serviced by removing the cap on the bottom of the valve. These valves have capacities varying from 1/2 in. to 4 in. pipe size.

12-59 PRESSURE-OPERATED WATER VALVE

The pressure-operated water valve is the most popular. It is a bellows attached to the high-pressure side of the system, preferably to the cylinder head. This bellows operates the water valve, as shown in Fig. 12-158.

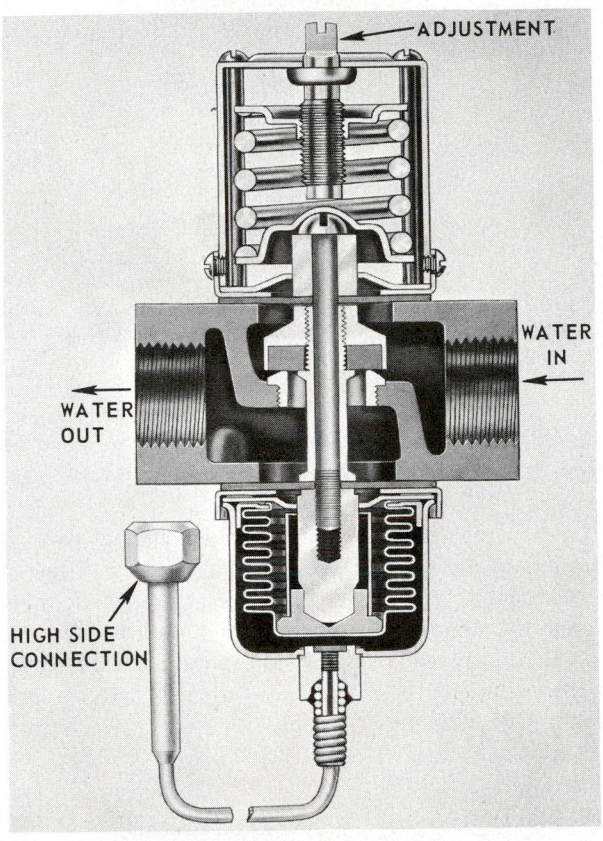

Fig. 12-158. Pressure-controlled water valve is connected to high-pressure side at compressor head. Rate of water flow is adjusted by spring pressure (top) on valve. Water flows through valve from right to left.   (Penn Controls, Inc.)

As condenser pressure rises, the bellows in the water valve contracts. The valve is opened by any of various mechanisms, depending on the specific water valve. Water flows into the condenser to cool the compressed vapor. The valve opens the water circuit only when the water is needed — as the pressure rises. It will keep increasing the water flow just as long as there is a tendency for an increase of pressure in the high side.

These valves may be adjusted by adjusting a heavy spring which presses against the bellows. The valves are set to open at definite head pressures. The pressure depends on the temperature of the water and the refrigerant used. See Chapter 14.

Some pressure-controlled water valve designs require opening the system to remove the valve. The water valve of the

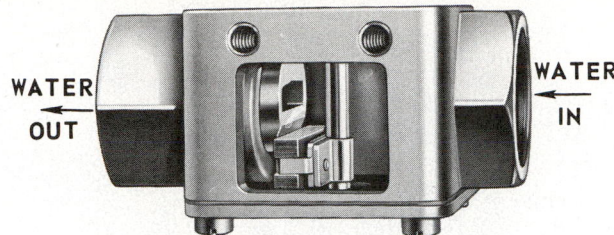

Fig. 12-159. This water valve and seat is self-cleaning. (Controls Co. of America)

control may be removed in some designs without disturbing the refrigeration system. Some water valves are self-cleaning, Fig. 12-159.

A modulated flow of water is obtained with this valve. (Modulated means it can vary the amount of flow.) As the condensing pressures and temperatures increase, the valve opens farther. When the pressures and temperatures drop, the water flow decreases.

The valve faces are a hard rubber composition, Bakelite or fiber. The seat is usually made of copper or brass. The valves are equipped with either a packing gland or with a bellows at the point where the water valve stem goes into the water valve body. The packing must be adjusted occasionally to keep it from leaking. Some valves have a bellows in place of a packing.

These valves usually do not depend on the pipe for support, but are provided with a mounting arrangement or flange. The water-in and the water-out connections are clearly labeled. Valves are usually threaded for standard pipe connections. Most are constructed so that water pressure tends to keep the valve closed. Fig. 12-160 shows a large-capacity valve used on 1-in. lines. This valve has a gear mechanism for adjusting the pressures.

The pressure-operated double water valve in Fig. 12-161 controls flow in two separate circuits.

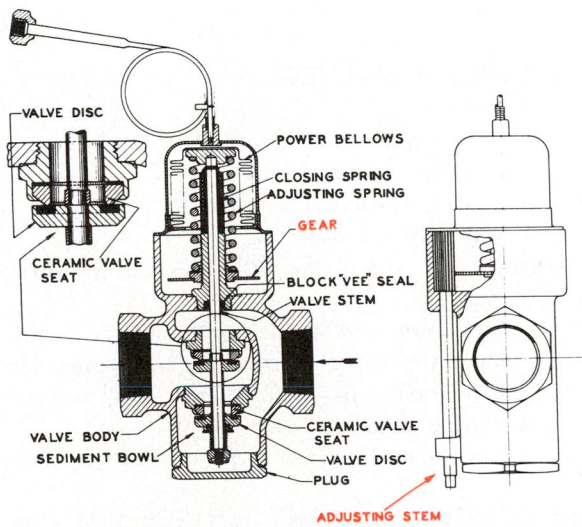

Fig. 12-160. Large capacity pressure-operated water valve. Double valve and seat arrangement balances force from water pressure. One valve is opened and the other is closed by water pressure.

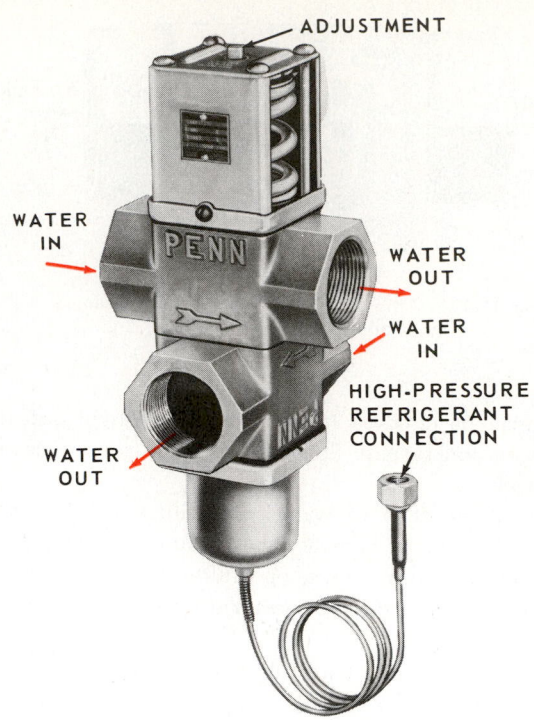

Fig. 12-161. Double water valve, pressure operated. Note water flow direction arrows on two valve bodies.
(Penn Controls, Inc.)

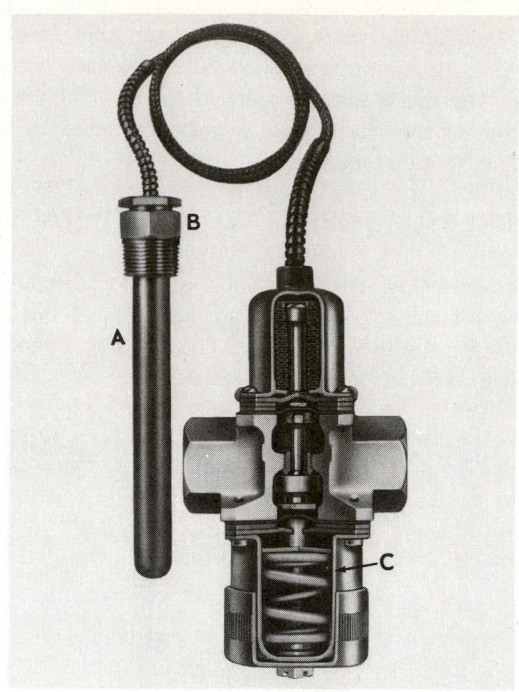

Fig. 12-162. Thermostatic water valve has sensing bulb, A, fitted inside water outlet piping, using fitting at B. Temperature adjustment is at C. (Marsh Instrument Co.)

12-60 THERMOSTATIC WATER VALVE

The thermostatic water valve is controlled by the temperature of the exhaust water. The valve proper is identical to the pressure water valve except for a thermostatic element connected to the bellows operating the valve (Fig. 12-162). The element is charged with a volatile liquid. The power bulb is mounted in the condenser water line. Pressure created by the volatile liquid in the bulb opens the valve when the condenser water becomes warm. It closes the valve as the water cools.

12-61 MANUAL VALVES

Manual servicing valves used on commercial refrigerating systems help the service technician:
1. To determine the operating pressures.
2. To charge or discharge a system.
3. To remove any part of the system without disturbing the other parts.

These hand valves and service valves must be sturdy and designed to withstand frequent opening and closing without leaking. They should be built of the best materials to withstand corrosion and handling.

Valve stems and packing must be handled with care.

12-62 CONDENSING UNIT SERVICE VALVES

Many of the condensing units are equipped with servicing valves of both the two-way and one-way type. See Chapter 2.

Some of these valves are quite large. That is, the valve stems may be 3/8-in. across flats, and larger because liquid lines are as large as 3 1/8-in. OD.

Some of the larger systems may be equipped with additional service valves such as separate valves for installation purposes and/or for servicing. Many systems have a valve between the condenser and the liquid receiver. The gauge connections may be 1/4 or 1/8-in. pipe. Some systems use Schrader valves to connect gauges and to perform service operations. See Chapter 11.

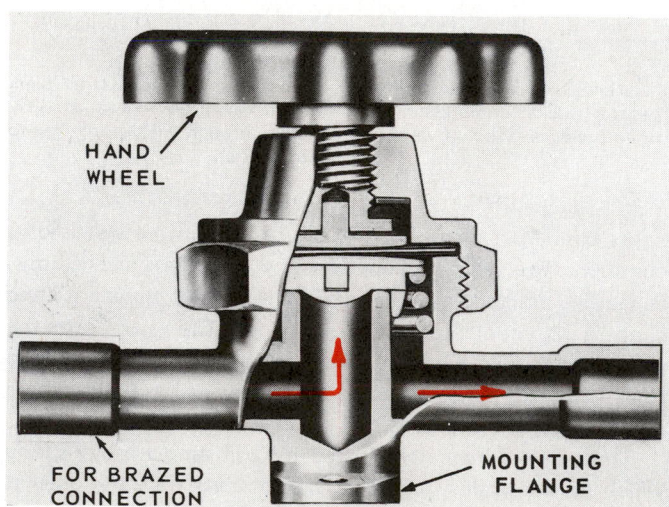

Fig. 12-163. Manual shutoff valve used on multiple installations. Valve uses diaphragm in place of packing. Piping openings are in line.
(Henry Valve Co.)

12-63 MANUAL INSTALLATION VALVES

In addition to the usual service valves, multiple installations are usually equipped with what are called hand shutoff valves, Fig. 12-163. These valves operate by hand and, by law, must be located so that they may be easily turned. They may be classified as riser or manifold valves.

In multiple installations, it is best to run the suction line from the compressor to a manifold. Then have the individual suction lines for each evaporator go from this manifold to the evaporators.

Between each of these suction lines and the manifold, and mounted into the manifold, is a hand-operated shutoff valve. This valve permits any one of the suction lines to be closed without interfering with the operation of the others. A similar manifold device is also provided for the liquid line. Installation procedure usually provides for mounting these valve groupings in a steel box or cabinet or on a special valve board near the condensing unit.

12-64 RISER VALVES

There is another type of valve used as a shutoff valve but also called a riser valve. It is hand-operated with three openings to which refrigerant lines may be connected. Two of these openings are in line with each other on opposite sides of the valve. The third is a little closer to the valve wheel and is at right angles to the other two openings.

By turning the hand valve in, the opening at right angles to the other two is closed. This construction permits mounting of the valve in either a liquid or suction line. One can then connect another evaporator to it. This may be shut off from the remainder of the system by screwing the valve all the way in. Fig. 12-164 shows a multiple installation using two liquid line riser valves and two suction line riser valves.

Service valves are usually made of drop forged brass to hold down seepage through the valve. The valve stem may be of either brass or steel. Packing around the valve stem may be asbestos, lead and graphite, or the valve may be of the packless type. This is the type which uses a bellows or a diaphragm as a sealing device rather than packing. Some valves have self-seating features. This means the valve is easily seated again by tapping the valve stem into the seat. The valve seat is made of a soft lead alloy or Monel metal.

12-65 RELIEF VALVES

A refrigerating system, regardless of size, is a sealed system. It is a pressure container. The pressures vary; but, during shutdowns, fire, extreme temperature conditions, or faulty electrical controls, high pressures could cause some part of the system to explode.

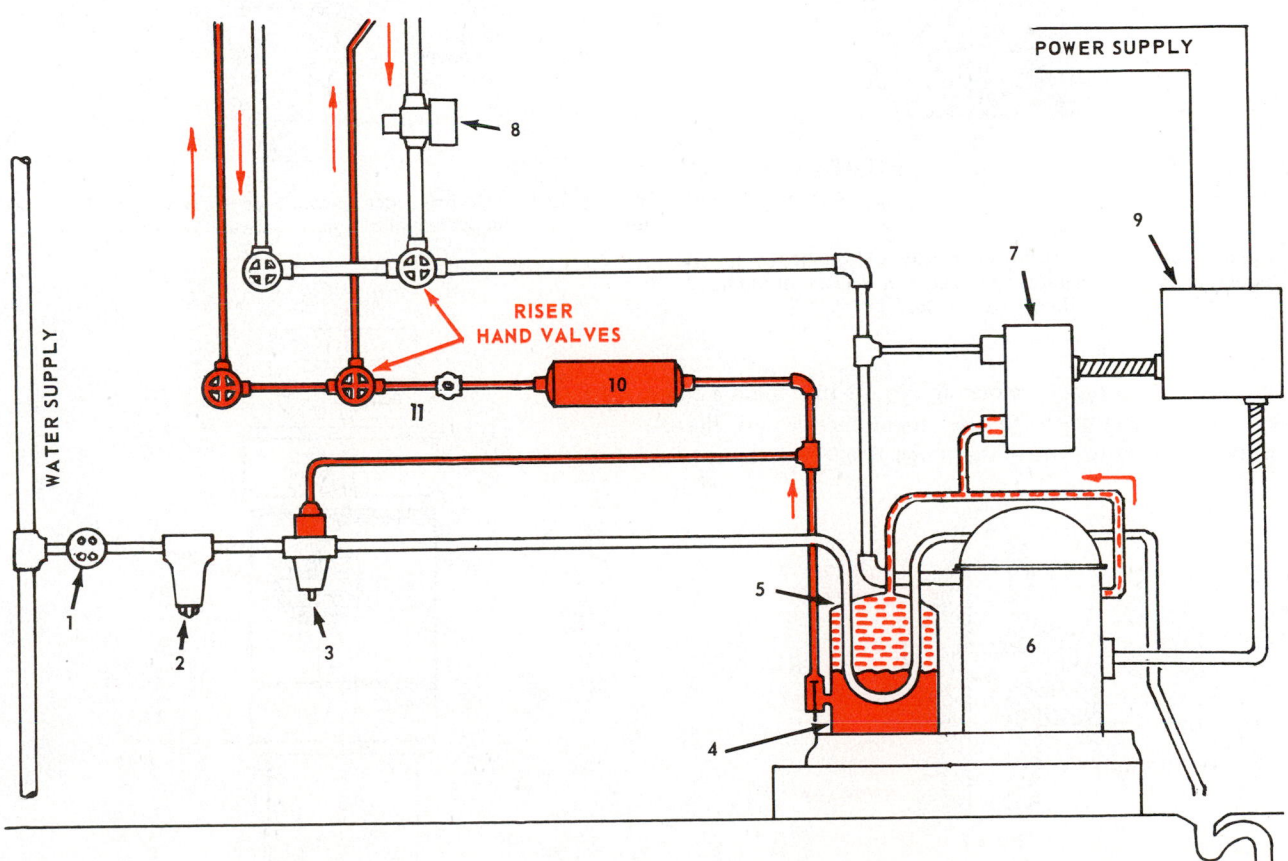

Fig. 12-164. Typical multiple installation showing location of important parts. 1—Water shutoff valve. 2-Strainer. 3-Water valve. 4—Liquid receiver. 5—Condenser. 6—Compressor. 7—Motor control. 8—Two-temperature valve. 9—Magnetic starter. 10—Drier. 11-Sight glass. Note this installation uses four riser valves.

To prevent extreme, dangerous pressures, relief valves are mounted on the units, usually on the liquid receiver. The National Refrigeration Code and most local codes make this a requirement when the unit is of a certain tonnage or more, or if the amount of refrigerant exceeds specified minimums or if the internal volume is large enough.

The relief devices come in three principal types:
1. Fusible plug.
2. Rupture-disk type.
3. Spring-loaded valve.

Hand valves must not be placed between the system and the relief valve.

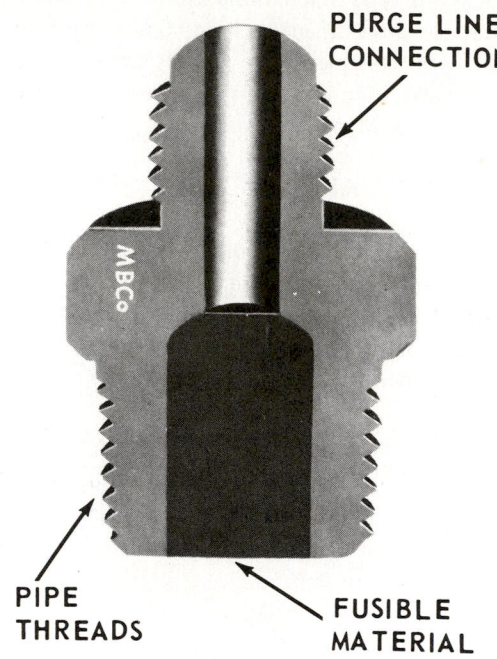

Fig. 12-165. Fusible plug for liquid receivers. Note flared fitting at outlet for connecting purge line to carry refrigerant outdoors. (Mueller Brass Co.)

The fusible plug type is shown in Fig. 12-165. It has a pipe thread for connecting it to the liquid receiver. A flared connection is used to fasten the purge line which is used to

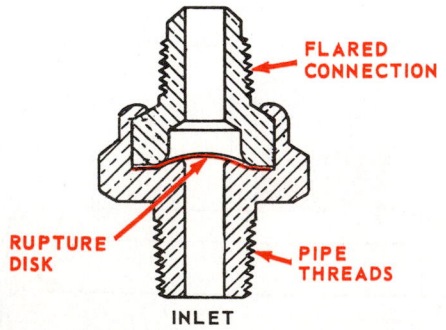

Fig. 12-166. A safety head (rupture disk type) for R-12. Disk is made of silver. Safety head is available with various rupture pressures from 175 to 1000 psi. (Reprinted from ASHRAE Guide and Data Book)

carry the released refrigerant outdoors. The low-temperature alloy in the plug will melt if the receiver temperature rises above a certain temperature. When the alloy melts, all the refrigerant will be released.

The rupture-disk type is shown in Fig. 12-166. This is similar in appearance to the fusible plug, but it has a thin metal disk inside which will burst before the pressure in the system reaches dangerous levels. It too is threaded into the liquid receiver and it is connected to a purge line to carry the released refrigerant outdoors.

The spring-loaded safety valve has the advantage of re-sealing itself or closing when the pressure drops to a safe limit. Fig. 12-167 shows such a valve. Note the spring-loading. The relief pressure is adjustable but, once it is set, the valve is sealed to prevent tampering as shown in Fig. 12-168. It is

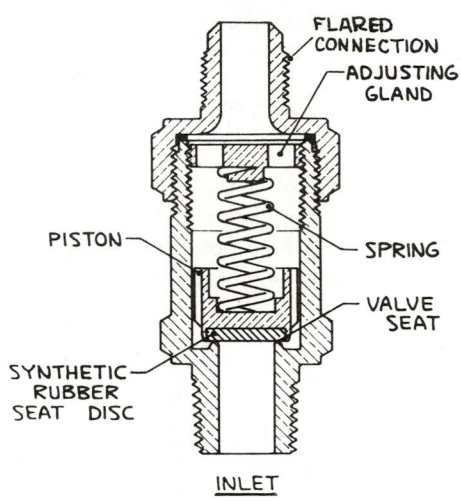

Fig. 12-167. Spring-loaded pressure relief valve. It uses a synthetic rubber seat and is made in several pressure ranges.

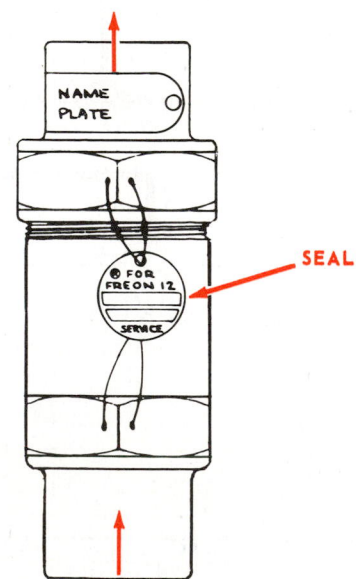

Fig. 12-168. Adjustable pressure relief valve. Note use of wire seal to prevent tampering. (Henry Valve Co.)

important that relief settings not be adjusted in the field. If the seal is broken, the valve should be replaced with a correctly adjusted and sealed valve. Pressure relief valves usually close about 10 to 20 percent below their opening pressure.

12-66 REFRIGERANT LINES

Hard drawn copper pipe is usually used to carry the refrigerant around the system. This pipe is furnished in iron pipe sizes and the fittings are not interchangeable with tubing sizes. See Chapter 2 for copper tubing sizes. Streamline brazed connections are used to connect the fittings to the pipe.

The National Refrigeration Code and local codes require the use of hard copper pipe. Soft copper tubing is permissible at the condensing unit end of the lines and also in the fixtures, but even these short lengths should be eliminated wherever possible.

Because the appearance of an installation is important, the piping should be put in as neatly as possible.

12-67 VIBRATION ABSORBERS

To reduce any condensing unit vibrations that might travel into the lines, vibration absorbers may be installed in the suction and liquid lines near the condensing unit. Fig. 12-169 shows construction of a vibration absorber. Fig. 12-170 shows two designs of vibration absorbers.

For good sound absorption, it is best to put in two vibration absorbers in each line, one vertically and one horizontally.

The flexible absorber should be fastened to the unit or to a

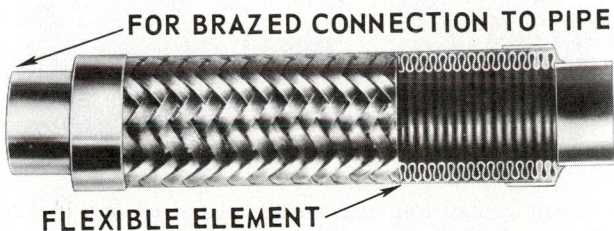

FOR BRAZED CONNECTION TO PIPE

FLEXIBLE ELEMENT

Fig. 12-169. Vibration absorbers, such as above, are installed in liquid line and suction line to prevent condensing unit vibration from traveling along these lines. (Flexonics Div., UOP, Inc.)

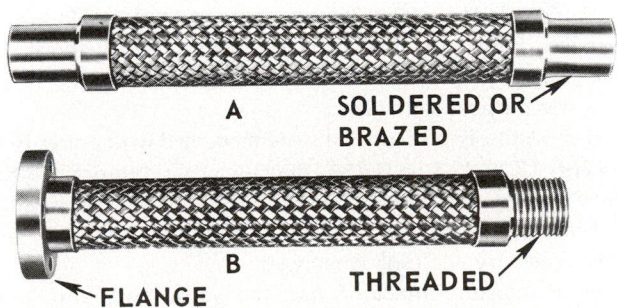

A — **SOLDERED OR BRAZED**

B — **FLANGE** — **THREADED**

Fig. 12-170. Two designs of vibration absorbers. A—Soldered or brazed connection type. B—Flanged and threaded connection type (AC & R Components, Inc.)

wall at the end pointing away from the vibration source. This fastening will prevent the vibration traveling along the pipe. *It is important not to stretch, compress or twist the vibration absorber when installing it.*

12-68 MUFFLERS

Most domestic refrigerating systems have small mufflers built into the refrigerant circuits to break up the pressure pulses which create noise. The mufflers are usually located in the compressor suction and discharge lines within the hermetic unit dome. Compressor action pressure pulses tend to follow the refrigerant lines. The mufflers break up these pulses.

Most commercial refrigerating systems also use mufflers, especially systems used for comfort cooling in air conditioning. These mufflers are installed near the condensing unit. Fig. 12-171 shows one of these mufflers. Mufflers are usually

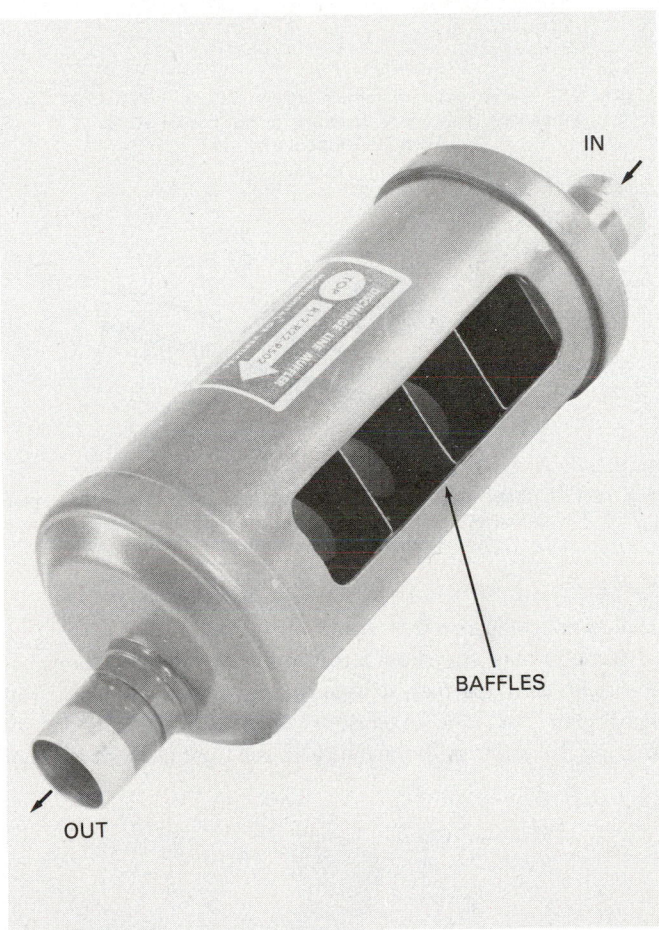

IN

BAFFLES

OUT

Fig. 12-171. Refrigerant gas muffler designed for vertical or horizontal mounting on discharge line. (AC & R Components, Inc.)

installed vertically to provide efficient oil movement. Fig. 12-172 shows a muffler that can be mounted horizontally. It has an aspirator device (dip tube) which uses the velocity of the suction line vapor to create a lower pressure in the dip tube. This reaches the bottom of the muffler, removes oil from the muffler and puts it into the outlet refrigerant line.

12-69 SIGHT GLASSES

Sight glasses are usually installed in liquid lines of commercial installations. The sight glass will show bubbles if the system is low on refrigerant. The one shown in Fig. 12-173 has an internal device which reads "full" when there is enough refrigerant. It shows nothing at all when there is no liquid in the line.

It may show a few bubbles when the system first starts or just as the system stops. These are normal equalizing actions

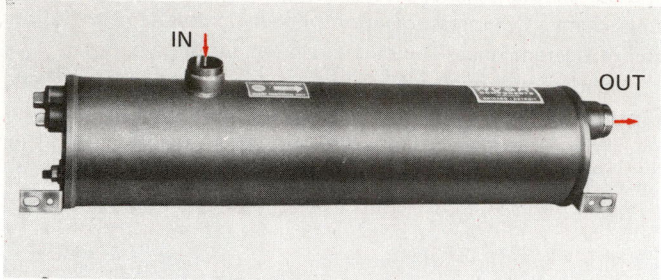

Fig. 12-172. Commercial system mufflers reduce noise from vapor pulsations. This one is designed for horizontal use only. (AC & R Components, Inc.)

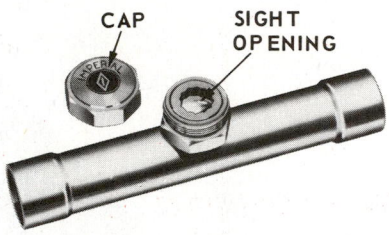

Fig. 12-173. Sight glass designed for soldered or brazed connections. Cap shown at left of sight opening keeps glass clean. (Imperial-Eastman Corp.)

and do not indicate a shortage of refrigerant.

Bubbles may also show a restriction in the circuit ahead of the sight glass (partially clogged drier, screen or filter). This sight glass has long extensions which permit soldering or brazing the joints without injury to the sight glass. Some sight

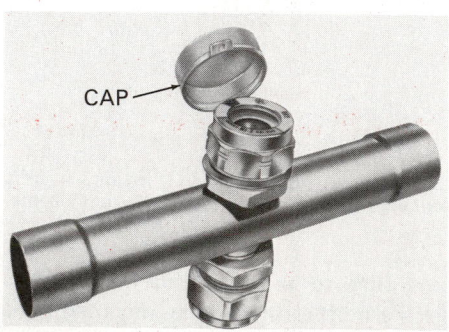

Fig. 12-174. See-through sight glass. Cap is removed from top only in this view. When both caps are removed, one can easily see any bubbles present by looking through liquid refrigerant. (Henry Valve Co.)

glasses are of the see-through type as in Fig. 12-174.

A sight glass may be used on a large liquid line by installing a smaller parallel flow pipe, as shown in Fig. 12-175. If bubbles are in the liquid, some of the bubbles will travel through the sight glass tube and indicate a refrigerant shortage.

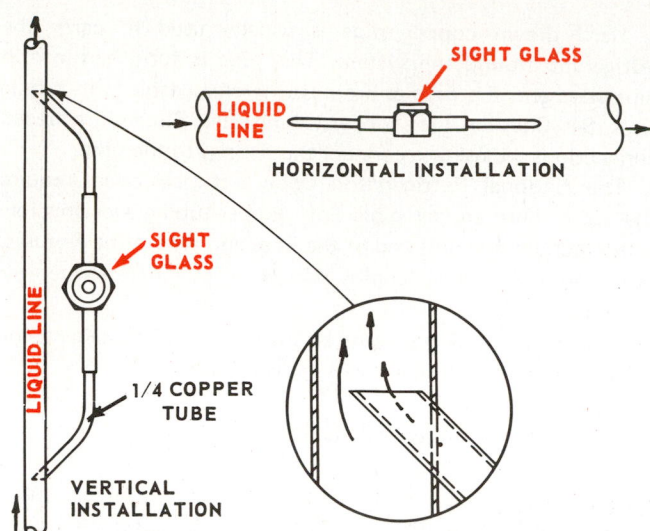

Fig. 12-175. Method of installing parallel sight glass in large liquid line. Joints are usually silver brazed. (Airtemp Applied Machinery Co.)

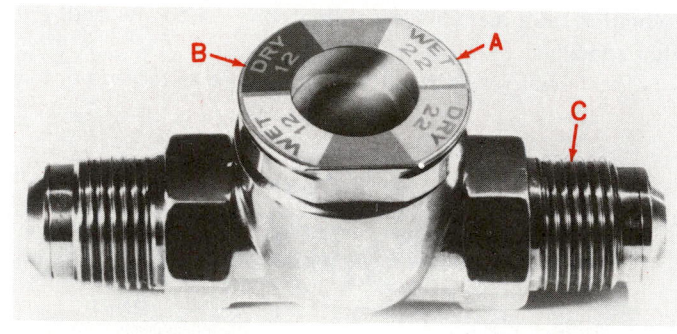

Fig. 12-176. A sight glass may also have moisture indicator. A—For R-22. B—For R-12. C—Flared connections. (Virginia Chemicals, Inc.)

12-70 MOISTURE INDICATORS

Many of the sight glasses have a moisture-indicating chemical built into the sight chamber. The chemical will change color if there is moisture in the system. See Fig. 12-176 for a two purpose sight glass.

If moisture is present in a system charged with either R-11, R-12, R-13, R-113 or R-114, the chemical substance will turn pink. But it turns blue if there is a safe minimum of moisture. With a system charged with R-22, R-500 or R-502, it turns green when dry and pink when wet.

Some moisture indicators have the words "wet" and "dry," which become legible at the same time the chemical changes color. Temperature is important. The higher the liquid temperature, the higher the moisture content needed to

R-502	PPM (PARTS PER MILLION)	
DEGREE F.	TURN BLUE (DRY) BELOW	TURN PINK (WET) ABOVE
75	5	15
100	10	30
125	15	45
R-12	PPM (PARTS PER MILLION)	
DEGREE F.	DRY	WET
75	5	15
100	10	30
125	15	45
R-22 — R-500	PPM (PARTS PER MILLION)	
DEGREE F.	DRY	WET
75	30	120
100	45	180
125	60	240

Fig. 12-177. Effect of temperature on moisture indicators. Note that the amount of water in parts per million can increase as the temperature increases and indicator will still show a "dry" condition. One should install indicator where liquid line will remain cool.

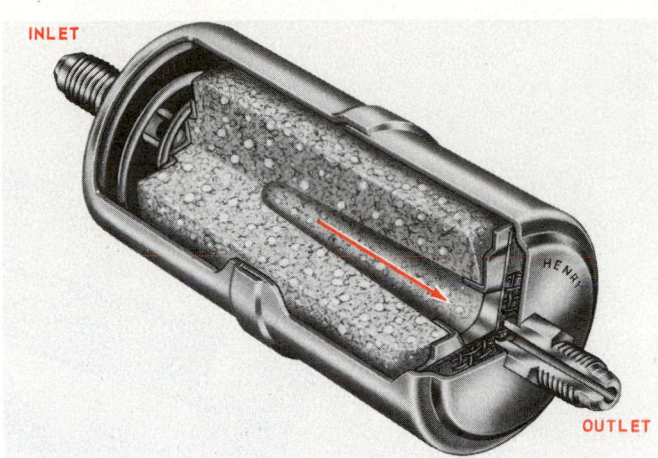

Fig. 12-178. This filter-drier is designed for use in liquid lines. It has a solid core and flared connection. (Henry Valve Co.)

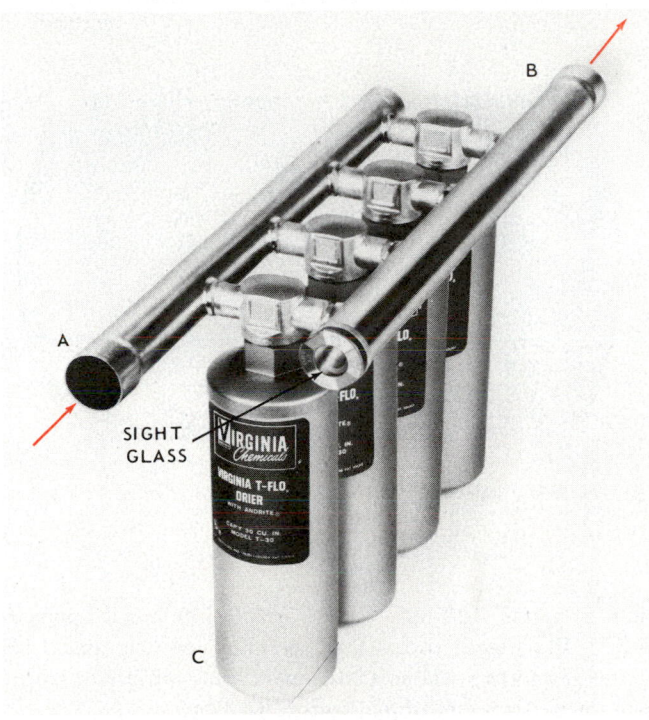

Fig. 12-179. Several filter-driers connected in parallel in liquid line. A—Liquid refrigerant in. B—Liquid refrigerant out. C—Replaceable filter-driers. Note sight glass. (Virginia Chemicals, Inc.)

produce a change of color. An indicator can show "dry" even though the system has too much water if the indicator is hot. See Fig. 12-177. For accurate indicators, the liquid line should be as close to 75 F. (24 C.) as possible.

It takes an hour to get a good reading but about eight hours for the indicator to give an accurate color signal.

Oil may turn moisture indicator chemical a tan color. Flushing indicator with clear refrigerant will remove the color. But, if tan color continues, the system has too much oil.

If alcohol has been placed in the system to absorb the moisture, it will affect the operation of the moisture indicator. Too much water or alcohol in the system will wash the chemicals off the indicator surface. The indicator will have to be replaced after the system is dried and the alcohol removed.

12-71 FILTER-DRIERS (LIQUID LINE)

The efficient operation of a commercial system depends, to a great extent, on the internal cleanliness of the unit. Only clean, dry refrigerant and clean, dry oil should circulate in the system.

Practically all dirt and water must be removed or must be trapped in some part of the system where they can do no harm. Screens, filters and water adsorbents are used. These devices may be in separate units or may be built into a single unit which filters and adsorbs.

A common method of removing moisture is with a liquid line drier (Fig. 12-178). If enough drying material is used for both the high and low moisture ranges and if it is fully activated, it can keep the refrigerant both clean and dry. A multiple filter-drier is shown in Fig. 12-179.

The conventional straight-through drier is a cylinder (brass, copper or steel) filled with a chemical (desiccant) such as activated alumina or silica gel. These materials remove moisture by adsorption. Both ends of the cylinder usually contain filter elements. The end caps are fitted with either flare or soldered connections.

One design of liquid line drier allows the casing to stay in the line. Only the drier cartridge needs to be changed. Fig. 12-180 shows a drier which uses one or more replacement type cartridges. Fig. 12-181 shows the replacement cartridge.

The most common desiccants (chemicals in drier shells) are combinations of activated alumina, silica gel and molecular sieves. These chemicals can adsorb 12 to 16 percent of their weight in water. Driers are usually installed in the liquid line.

The refrigerant should be dried below 15 parts of water per million if R-12 is used; below 25 ppm if R-22 or R-500 is used and 5 ppm for R-502. The beginning of corrosion in R-12 is 15

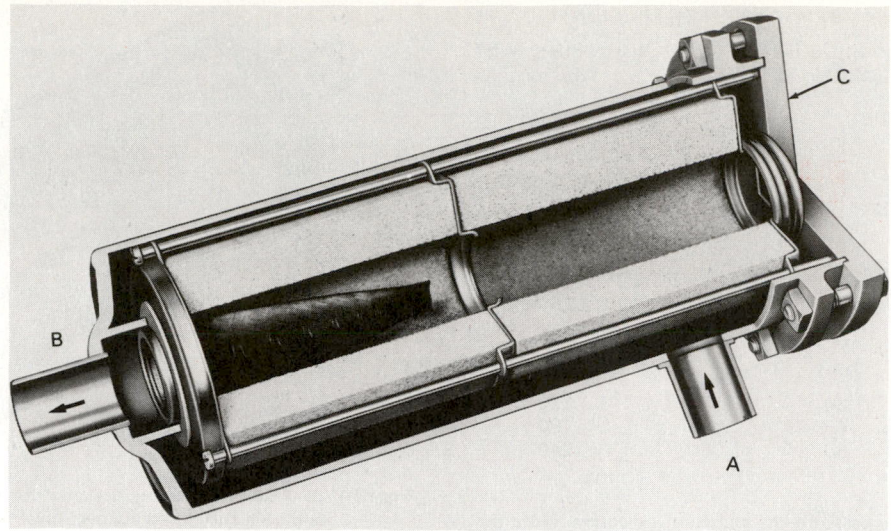

Fig. 12-180. Commercial filter-drier with removable end. A—Refrigerant in. B—Refrigerant out. C—Removable end. Note two drier cartridges. (Sporlan Valve Co.)

Fig. 12-181. Replacement cartridge for large commercial filter-drier. (Henry Valve Co.)

ppm of water, 120 ppm for R-22 or R-500, and 15 ppm for R-502. Experience shows that corrosion, oil breakdown and motor burnouts are almost eliminated if the refrigerant has the safe (or dry) amount of moisture in the system.

When cleaning a refrigeration system, four basic things are to be done:

1. Water removal.
2. Acid removal.
3. Filtering out of circulating solids.
4. Some means of indicating when the drying job is complete. Driers will do the first three. A moisture indicator is required to do the fourth.

Driers should be left in the system permanently since oil loses its moisture slowly. Also, insulation in hermetic compressors and in small crevices may release moisture over a long period of time. A drier is like a sponge; however, it can become saturated and leave the refrigerant still wet if the drier is too small. A moisture indicator is the only sure means of recognizing a wet condition.

Remember that R-22 driers must be three to five times as large as those needed for an equal quantity of R-12. The greater the ability of a refrigerant to hold water, the larger the drier required. R-500 driers need to be as large as R-22 driers and R-502 driers need to be as large as R-12 driers.

Most driers have shaped cores of two or three drier (desiccant) materials. The cores are designed for efficient flow and efficient drying. They are shaped, then fused together at high temperatures into a porous ceramic structure which must be:

1. Capable of absorbing moisture.
2. Noncorrosive.
3. Nonsoluble.
4. Nonreactive with oil.
5. Neutralizer for hydrochloric and hydrofluoric acid.
6. Capable of filtering out particles down to 10 micron in size (.025 in.).

12-72 FILTER-DRIERS (SUCTION LINE)

Filter-driers are often mounted in the suction line to prevent foreign particles of over 5 microns in size as well as acids, sludge and moisture, from entering the compressor. Fig. 12-182 shows the inside of a filter designed for suction line use. Strainers (screens) are usually made of Monel metal.

Only two things should be allowed inside a refrigeration

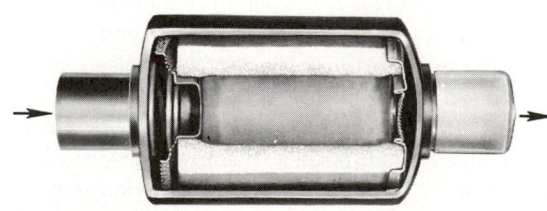

Fig. 12-182. Suction line filter removes particles as small as 5 microns. It is made in sizes up to 1 5/8 OD pipe size.

system: clean, dry refrigerant and good, dry oil. A system which is clean, dry and acid-free will run almost indefinitely without corrosion, freeze-ups, oil breakdown or hermetic motor burnouts. In such a system, there is nothing to filter and plugging is impossible. A clean, dry, acid-free system remains factory bright and trouble free in operation.

A normal system is completely clean. A dirty system is faulty and must be regarded as a mechanical failure, just as much as a faulty valve plate or connecting rod. A large capacity suction line filter-drier is shown in Fig. 12-183.

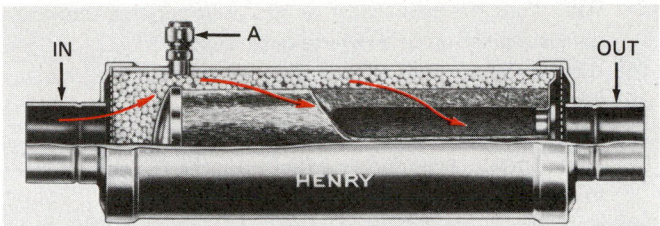

Fig. 12-183. A large capacity suction line filter-drier. A—Schrader valve used to check the pressure drop through the filter-drier. (Henry Valve Co.)

A suction line filter-drier should be replaced if pressure drop is excessive. Replace for R-12 and R-500 refrigerants if it exceeds 2 psi for low-temperature units, up to 8 psi for high-temperature units.

	Low Temp.	Medium Temp.	High Temp.
	psi	psi	psi
R-12 and R-500	2	6	8
R-22 and R-502	3	9	14

Replace for R-22 and R-502 refrigerants if pressure drop exceeds 3 psi for low-temperature units, up to 14 psi for high-temperature units.

The density of the gas increases more with a pressure increase than it does with a temperature rise. The best way to know that a system is dry is to use and depend on a moisture indicator.

12-73 ENGINE-DRIVEN SYSTEMS

Natural gas, gasoline and propane engines may be used to drive refrigerating compressors. The advantages are a variable compressor speed to produce flexible capacity and a comparatively low operating cost. Such units are available in four to 75 ton capacities. Engine-compressor units of one to five tons capacity are available for use on truck units and for air conditioning.

Pressure controls are usually used. The pressure control is connected to the engine's throttle. It is placed in the low-side suction line. The linkage is such that as the suction pressure increases, the engine's throttle is opened to increase the compressor's speed to increase the rate of refrigeration. As the temperature in the evaporator drops the engine will slow down. This should result in a balance between the engine's speed and low-side pressure. It will give the desired temperature in the refrigerated space.

12-74 REVIEW OF SAFETY

Commercial systems vary considerably in size.

The small self-contained units must be handled with all the care and safety as described in the reviews of safety in earlier chapters.

As the units become larger, the safety precautions become increasingly important both because the investment in the machines is greater and the damages are more costly. The larger machines are also more dangerous. The energy output of the larger moving parts and the larger refrigerant containers all are potentially dangerous.

Closing the compressor discharge valve on a 10-ton capacity unit while it is operating would almost instantly ruin the compressor or rupture a gasket. Carelessly opening a receiver valve may cause the loss of hundreds of pounds of refrigerant while possibly injuring the service technician. Trapping liquid refrigerant in any part of the system with no gas space may cause a hydraulic pressure sufficient to burst the container.

One must be positive that the pressures inside are atmospheric and that there is no liquid present before opening any part of the system. Goggles should ALWAYS be worn when working on any unit.

Open the electrical circuits and lock the switches before working on a refrigerating system if one does not want the power on.

Local and national refrigeration and electrical codes must be followed when servicing all systems. Follow OSHA standards.

It is not safe to work on any part of the system unless the pressure and temperature are known and the condition of the refrigerant (liquid or vapor) inside that part and the fundamentals of working on that system. (See Chapters 11 and 14.)

Pressure and temperature relief devices are installed on the units for equipment protection, user's protection and service technician's protection. They should be kept accurate and in good condition. Always keep them in operation.

Never use cylinder oxygen to test any device for leaks. Use either refrigerant, carbon dioxide or nitrogen. CAUTION: See Para. 11-39.

12-75 TEST YOUR KNOWLEDGE

1. What are the advantages of a nonfrosting evaporator?
2. Why must high-pressure motor cut-outs be used with water-cooled condensing units?
3. What part of the compressor usually contains the intake valves?
4. Why is it advisable to connect the high-pressure motor cut-out into the cylinder head of the compressor?
5. Name the three types of water valves.
6. Name the various two-temperature valves.
7. What are the advantages of water-cooled condensers? Air-cooled condensers?
8. What are the advantages of forced circulation evaporators?
9. Why is the pressure type motor control usually used in multiple installations?
10. In multiple installations which use two-temperature valves, where should the check valves be placed?

11. What percent of a refrigeration load may be placed on an evaporator controlled by a two-temperature valve?
12. Why does an evaporative condenser save about 85 percent of the water consumption?
13. Are liquid receivers equipped with safety devices? Why?
14. What is inside the inner tube of a tube-within-a-tube condenser?
15. How is cast aluminum used in a liquid evaporator?
16. What is the counterflow principle in water-cooled condensers?
17. Why is a pressure limiter valve used on some outdoor air-cooled condensers?
18. What principle is used to produce clear ice in an automatic ice cube maker?
19. How is water used to defrost a system?
20. Why must the drain pan and the drain pipe be heated during defrosting cycles?
21. What is the purpose of the check valve in the reverse cycle defrost system?
22. Why is a motor starter necessary?
23. Why is an oil separator insulated?
24. Why do some "hot gas" defrost systems reheat the refrigerant before it returns to the compressor?
25. Which type water valve will not vary the water flow as the refrigeration load changes?
26. Why is a float valve used with a cooling tower?
27. Does a flash cooler for a walk-in meat cabinet need an automatic defrost system? Why?
28. What is the purpose of a surge tank?
29. What is a sweet water bath?
30. Where are vibration dampers installed and why are they needed?
31. How is an auger used to produce flake ice?
32. What shuts off an ice cube maker when the bin is full?
33. How is an electric grid used to produce ice cubes?
34. What does the evaporator temporarily become during the defrosting action of a reverse cycle system?
35. What is the purpose of an accumulator?
36. Why are some systems pumped down before the electric defrost starts?
37. When should a suction line filter-drier be replaced? How is this measured?
38. How does an oil pressure failure shutoff switch operate?
39. What is a tandem compressor assembly?
40. How are adjustable louvers and fans used on outdoor air-cooled condensers?

Chapter 13

COMMERCIAL SYSTEMS APPLICATIONS

That part of commercial refrigeration which uses only automatic systems will be covered in this chapter. This field includes all automatic refrigerating mechanisms other than:

1. The domestic or household type.
2. The comfort cooling field.
3. Industrial refrigeration.

The industrial field is sometimes confused with commercial refrigeration. Industrial refrigeration uses refrigerating machines which need an attendant, usually a licensed refrigeration engineer, on the job constantly. Industrial plants are usually manually-operated refrigeration machines. They are commonly used for ice making, for large storage houses, packing houses, industrial plants, ice cream manufacturing and frozen food processing plants.

13-1 COMMERCIAL CABINET CONSTRUCTION

Commercial cabinets are designed and constructed to suit the service required of them. Surfaces are either metal or plastic and finishes are formulated for easy cleaning. Structural members are steel capable of supporting the evaporator and condensing unit. The insulation is usually polystyrene or urethane and may be applied in slabs or foamed-in-place.

The capacity of the evaporator and condensing unit is such that adequate refrigeration is possible under the most severe service conditions. Heat leakage, in some cases, may cool the cabinet surfaces enough to cause some moisture to condense on them. To avoid this condition, some cabinets have a resistance heating strip around these surfaces to warm them.

Many commercial cabinets are designed to be used with a remote condensing unit. These condensing units may be connected to several cabinets of different temperatures. Most condensing units are air-cooled but some water-cooled condensing units are used.

13-2 GROCERY CABINET (REACH-IN CABINET)

Grocery cabinets have been used for many years to keep perishable products at a satisfactory temperature. They range in size from 20 cu. ft. to 100 cu. ft. inside volume (net capacity). They have from one to three doors with magnetic gaskets. Door widths vary from 30 in. to 85 in. Height range of the cabinet is from 5 1/2 ft. to 6 1/2 ft. Fig. 13-1 shows a typical reach-in unit.

CONDENSING UNIT

SLIDING DOORS

Fig. 13-1. Two-section sliding glass door, reach-in cabinet has self-contained condensing unit mounted at top. (Hobart Corp.)

The inside of a cabinet is shown in Fig. 13-2. A blower evaporator is mounted on the top and cooled air is distributed through a vertical duct. Fig. 13-3 is a cross-section through the same refrigerator.

The space holding the evaporator is usually called the bunker. Blower type evaporators are very popular for grocery cabinets. Hardly any other type is now used. Fig. 13-4 shows a unit with the condensing mechanism located in the base while a blower evaporator is installed inside the cabinet.

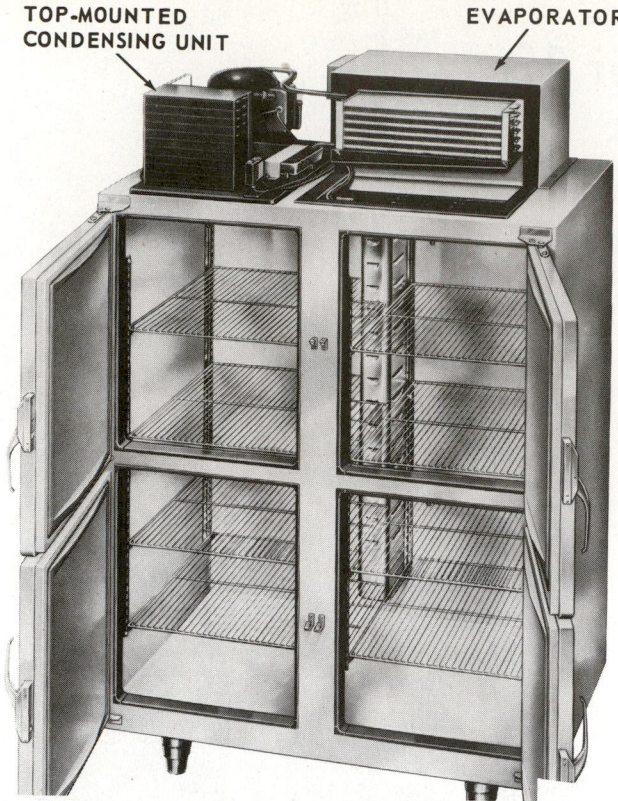

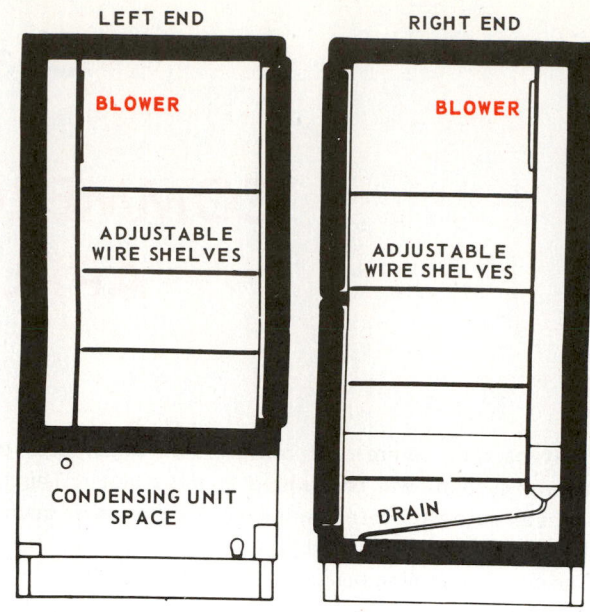

Fig. 13-4. Cross-section of a reach-in cabinet. Note location of blower.

Fig. 13-2. Reach-in refrigerator cabinet with top-mounted condensing unit and evaporator. Cover is removed to show these mechanisms.

The evaporator is usually located in the upper center of the cabinet. The insulation is most commonly foamed in place. Exteriors are aluminum, stainless steel or vinyl.

The temperatures are about the same as in domestic cabinets with a minimum of 35 F. (2 C.) and a maximum of 45 F. (7 C.). A relative humidity of about 80 percent is necessary for salads, desserts and fresh foods.

A reach-in cabinet for small grocery stores and markets is shown in Fig. 13-5. The blower type evaporator is mounted in a vertical position at right angles to the half doors. The evaporator takes little space. The blower circulates the cold air to provide an even temperature through the cabinet.

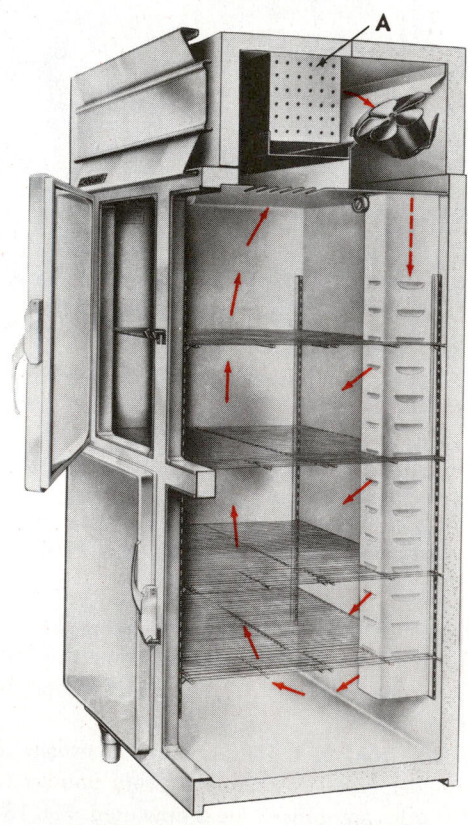

Fig. 13-3. Cross section of reach-in refrigerator. Note wall construction and airflow. A—Top-mounted evaporator.

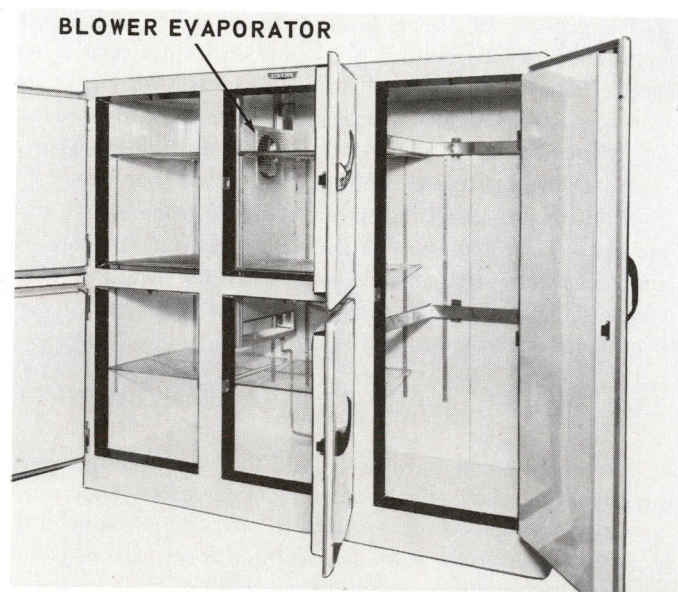

Fig. 13-5. Reach-in cabinet designed to store and refrigerate a variety of foods. (McCall Refrigerator Corp.)

Fig. 13-6. Walk-in cooler. (Tyler Refrigeration Corp.)

This cabinet uses a remote condensing unit. The full height door permits storing of beef quarters.

13-3 WALK-IN CABINET

Establishments such as restaurants and supermarkets, where perishable products are stored, use walk-in cabinets. These cabinets have large doors and windows and are sometimes classified as butcher boxes. Sizes of these cabinets vary but two heights are usually considered standard: 7 ft. 6 in. and 9 ft. 10 in., outside dimensions. See Fig. 13-6. These boxes are the "knockdown" type. This means that they may be taken apart for easier moving. See Fig. 13-7.

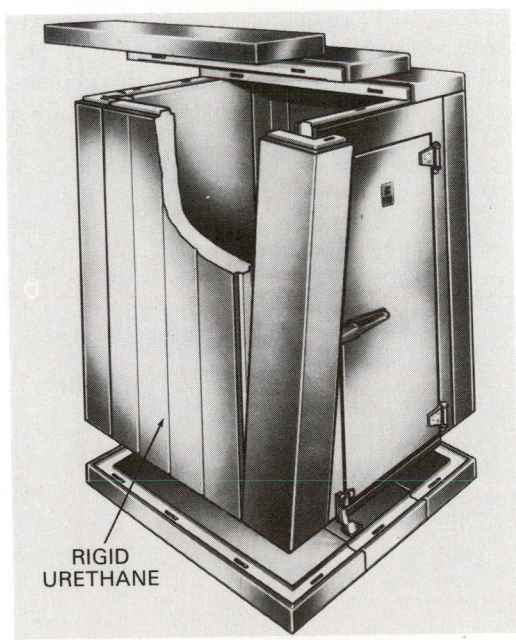

Fig. 13-7. Prefabricated walk-in refrigerator. Sections lock together on site. (Bally Case & Cooler, Inc.)

Typical walk-in cooler sizes are given below:

Length	Width	Height
7 ft.	5 ft.	9 ft. 10 in.
8 ft.	6 ft.	9 ft. 10 in.
8 ft.	8 ft.	9 ft. 10 in.
9 ft.	7 ft.	9 ft. 10 in.
12 ft.	10 ft.	9 ft. 10 in.
6 ft.	5 ft.	7 ft. 6 in.
6 ft.	6 ft.	7 ft. 6 in.
7 ft.	6 ft.	7 ft. 6 in.

Many cabinets are made with metal linings and exteriors. Galvanized steel or aluminum is the usual metal. Vinyl, porcelain and stainless steel are also used extensively.

Cabinet doors are usually of the same construction as the box and are gasketed to make the box airtight. See Fig. 13-8.

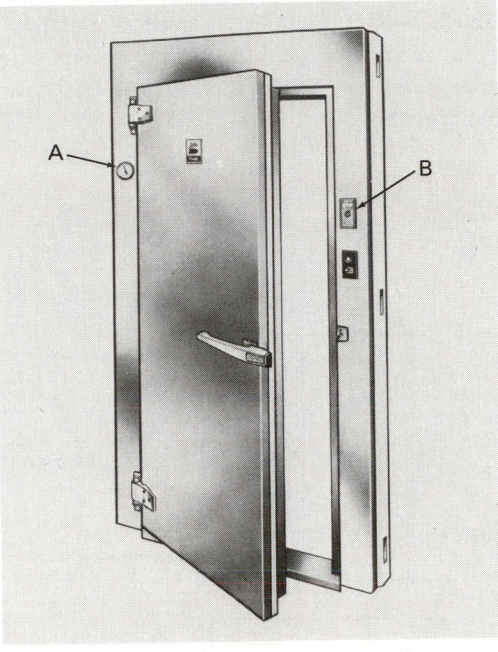

Fig. 13-8. Prefabricated walk-in refrigerator door. Heater wires are built in around door opening to eliminate condensation and freezing. A—Thermometer. B—Condensate control. (Bally Case & Cooler, Inc.)

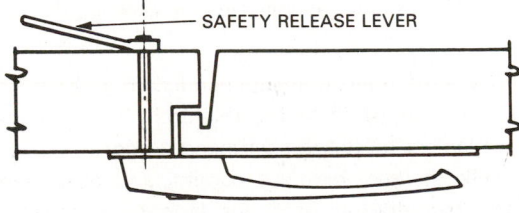

Fig. 13-9. Inside safety release lever is connected to door latch. Inside lever can be used for emergency door opening.

Door latches must be accessible from the inside for safety, Fig. 13-9. The doors may also be provided with heating wires along the edge to eliminate sweating and freezing.

Some walk-in cabinets are dual temperature. These cabinets have both a regular temperature compartment and a frozen foods compartment.

Cabinets may have additional reach-in doors, usually with two, three or four panes of glass. Instead of insulation, these doors have two or three dead air spaces arranged in such a way that they are airtight. Plate glass is usually used. Special chemicals — such as calcium chloride — keep the spaces between the panes free from moisture.

Newer walk-in coolers use rigid polyurethane foamed-in-place insulation. Foamed between the inner and outer walls, such insulation produces a very strong wall and ends the need for metal framing. Insulation is usually 4-in. thick.

Walk-in cabinets usually have a lighting system. Some walk-in cabinets have a wall-mounted evaporator. It is separated from the main part of the cabinet interior by a vertical baffle. Forced convection evaporators are popular.

The temperature in this type of cabinet depends on its use. For meat or fresh produce storage, a temperature of between 35 F. (2 C.) and 40 F. (4 C.) is needed. Relative humidity should be about 80 percent. Air movement is necessary. Ultraviolet lamps may also be used to help keep down bacteria and mold growth.

Because overexposure to ultraviolet rays is dangerous, persons working near these lamps must be protected from the rays. Otherwise, the lamps must be turned off when anyone is in the cabinet.

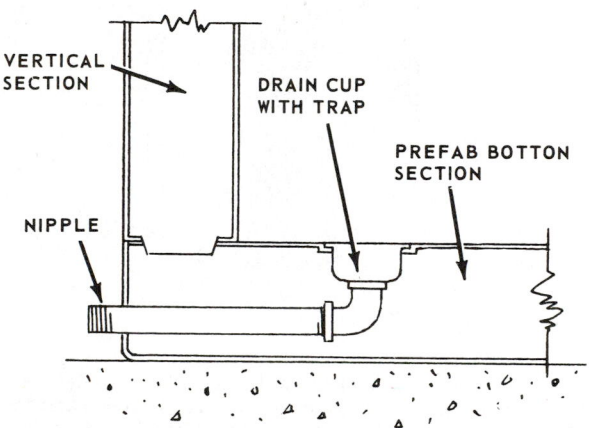

Fig. 13-10. Walk-in cooler drain connection. Note that connection is part of prefabricated bottom section.

Some type of drain is recommended. Fig. 13-10 shows a common method of installing the drain in a prefabricated walk-in which sits on top of a permanent floor.

For milk storage, beverage cooling and other service in which the dehydration of foods is not important, colder temperature may be used as desired and less attention may be paid to relative humidity. Blower evaporators are commonly used in these installations. Walk-in cabinets are also used for storing frozen foods (walk-in freezers).

13-4 FLORIST CABINET

The florist cabinet varies in size and construction and differs from the grocery cabinet and the walk-in cabinet in

Fig. 13-11. Florist's display refrigerator cabinet. (Buchbinder, Chicago, IL)

three principal ways:

1. The temperature within the cabinet may be kept higher than those in the other types of boxes. Temperatures between 55 F. (13 C.) and 58 F. (14 C.) are quite common.
2. The thickness of insulation for a florist cabinet, as a result of the lesser temperature difference, generally is only 1 to 2 in. (2.5 to 5 cm).
3. The cabinet is usually made with an extensive amount of window surface, permitting the display of cut flowers within the cabinet, Fig. 13-11.

Humidity is an important factor in the florist cabinet. It should be kept as high as possible in order to retard evaporation from the surface of the leaves and blooms.

The evaporators have large cooling surfaces to keep the humidity as high as possible. Natural convection evaporators are used in most cases. Also, the motor controls are such that there is but little variation in the cabinet temperature. Many florist cabinets use odor-removing devices to prevent contamination of the flowers. An activated carbon filter, as shown in Fig. 13-12, may be used to reduce mold growth, neutralize ethylene and extract odors given off by flowers.

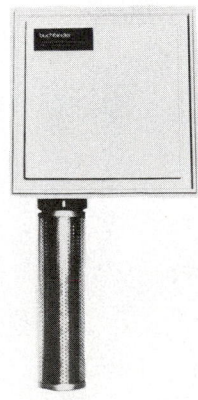

Fig. 13-12. Ethylene purifier automatically filters out odors and ethylene gas that may exist where flowers are stored. (Buchbinder, Chicago, IL)

TYPE FIXTURE	TEMPERATURE—F.		TEMPERATURE—C.	
	MIN.[a]	MAX.[a]	MIN.[a]	MAX.[a]
MEAT, UNWRAPPED				
DISPLAY AREA	35	38	1.7	3.3
STORAGE COMPARTMENT	34	37	1.1	2.8
MEAT, WRAPPED				
DISPLAY AREA	28	36	−2.2	2.2
STORAGE COMPARTMENT	28	35	−2.2	1.7
PRODUCE, DISPLAY AREA	35	45	1.7	7.2
PRODUCE, STORAGE COMPARTMENT	35	45	1.7	7.2
DAIRY	35	42	1.7	5.6
FROZEN FOOD	−b	0	−b	−17.8
ICE CREAM	−b	−12	−b	−24.4

[a] These are air temperatures, with thermometer in refrigerated air stream and not in contact with product.

[b] Minimum temperatures for frozen food and ice cream are not critical; maximum temperature is the important factor for proper preservation of product quality.

Fig. 13-13. Recommended temperatures in display cases. (ASHRAE)

13-5 DISPLAY CASES

To display produce to best advantage, stores frequently use a refrigerated display case. This case is equipped with glass fronts. The purchaser can see the articles handled by the merchant and, at the same time, the food is safely refrigerated. Such refrigeration is necessary to prevent spoiling during the period that the food is stored in the display case.

Temperature in the case is determined by its usage. Fig. 13-13 shows the recommended temperature for some common applications. Electric lights for lighting a display case are usually installed outside the glass case so heat generated by the lights will not increase the refrigerating load.

Display cases vary as to design, length, and height. Three types are:
1. Glass enclosed display case only.
2. Glass enclosed display case and enclosed storage cabinet.
3. Open display case.
  a. Fresh produce.
  b. Frozen foods.
  c. Fresh meats.
  d. Dairy products.

The display case is sometimes classified by the location of the evaporator:
1. Overhead.
2. End.
3. Base.

13-6 SINGLE-DUTY CASE

One popular display case uses an overhead evaporator. In this case the main evaporators are mounted in the upper portion of the display space under the shelf which forms the top. See Fig. 13-14. This provides good refrigerating temperatures all the way through the display space.

Some cases have shelf evaporators. These are called auxiliary evaporators. These are located under the shelves and consist of coils of tinned tubing so placed that each shelf is individually cooled.

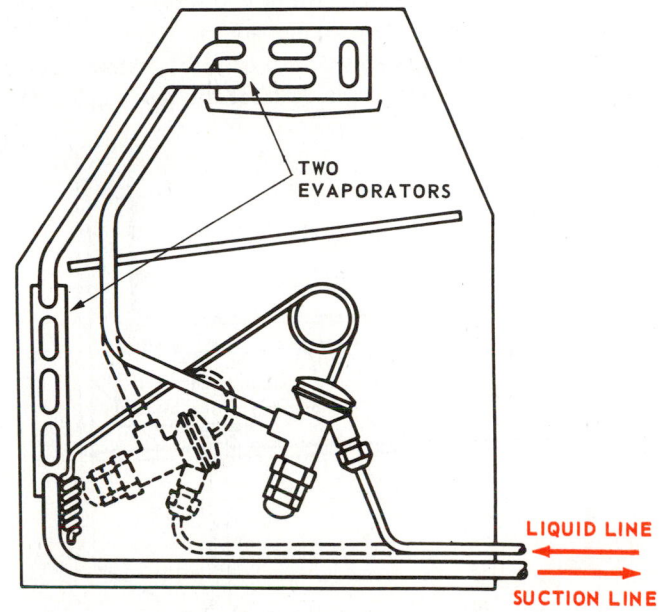

Fig. 13-14. Cross-section of glass-enclosed display case. Two evaporators are connected in series, one is overhead and one is at back of case. Note use of thermostatic expansion valve.

13-7 DOUBLE-DUTY CASE

Some cases have additional storage space beneath the display section of the counter. This also is refrigerated. The evaporators are usually connected in series. Temperatures may be kept at 40 F. (4 C.) to 45 F. (7 C.) in both compartments, because they usually serve as temporary containers for food or produce which is transferred to a walk-in storage cabinet overnight. Evaporators used in these installations must necessarily be narrow and they are made with fins as small as 1 1/4-in. wide. Some of the shelf evaporators are the plain tubing type. Many of these display cases are now using blower evaporators for cooling. These evaporators take little space and, because of the circulating air, provide even refrigeration temperatures throughout the display case.

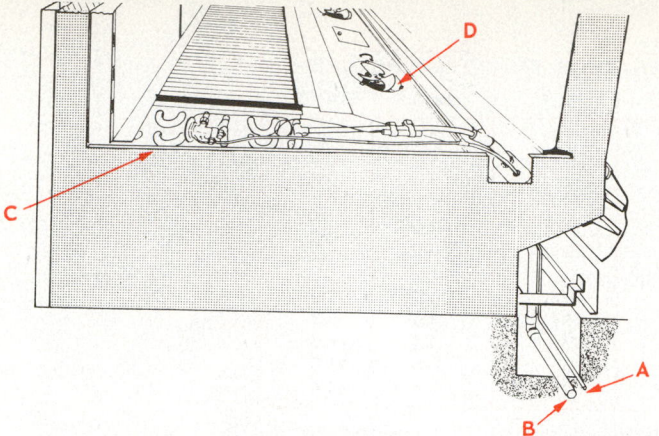

Fig. 13-15. Display case installation. Note trough in floor for refrigeration piping and electrical conduit. A—Liquid line. B—Suction line. C—Evaporator. D—Fan.

13-8 OPEN DISPLAY CASE

To make it easier for the customer to serve himself, open display cases are commonly used in markets.

These cases may be designed with or without storage space in the base of the unit. Storage space at the top is open. The

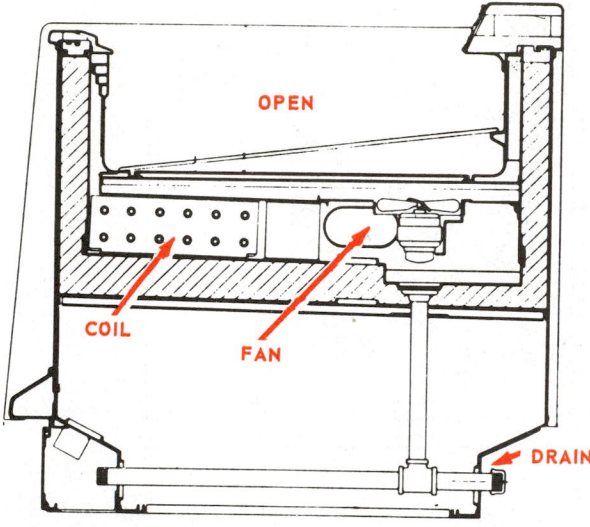

Fig. 13-16. Cross-section of open meat display case. Note evaporator location, fan and drain. (Tyler Refrigeration Corp.)

Fig. 13-17. Open display case with canopy mirror for produce. (Tyler Refrigeration Corp.)

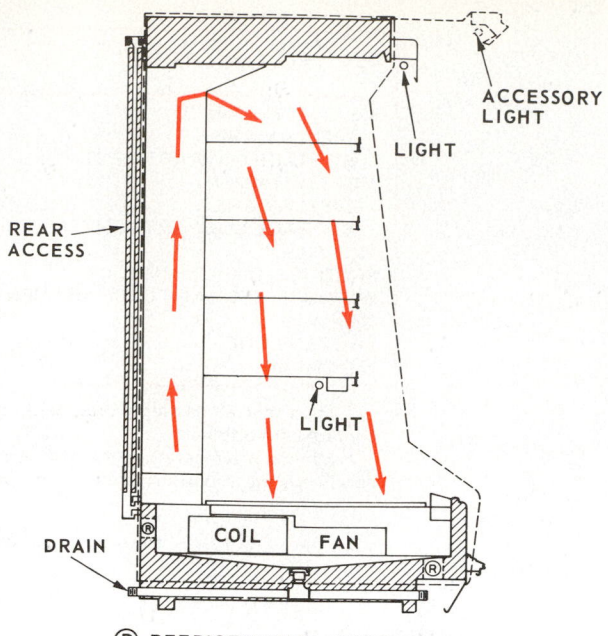

®-REFRIGERATION CONNECTION

Fig. 13-18. Display case for dairy products and delicatessen items. (Warren/Sherer, Div. of Kysor Industrial Corp.)

Fig. 13-19. Front loading dairy case. (Warren/Sherer, Div. of Kysor Industrial Corp.)

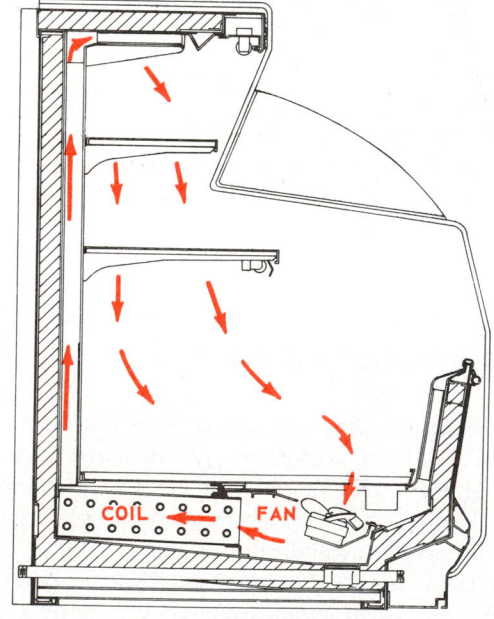

Fig. 13-20. Cross-section of open display case using blower evaporator. Note airflow pattern.

walls, or the upper part of the wall, may be enclosed in three to four layers of glass.

The higher temperature case such as used for fresh meats and dairy products, does not present any special evaporator problems. Blower evaporators are used and ducts carry the cold air through grilles at the rear of the case at the same level as the refrigerated foods. See Fig. 13-15. The warm air returns down the front of the case. A shallow open chest type display case, Fig. 13-16, may be used for frozen foods or for storage of foods at temperatures above freezing.

Many supermarkets have open display cases for produce.

These cases are kept at about 40 F. (4 C.) and at a high humidity. See Fig. 13-17.

A cross-section of an air curtain open display case is shown in Fig. 13-18. Note sliding doors to enable restocking from the rear. An installed unit is shown in Fig. 13-19.

An airflow meter (see Chapter 18) or a chemical smoke may be used to check the airflow patterns in these open cases. The airflow curtain should not touch the shelves or the products. A cross-section through an open display case is shown in Fig. 13-20. Some of the electrical circuits of an open display case are shown in Fig. 13-21.

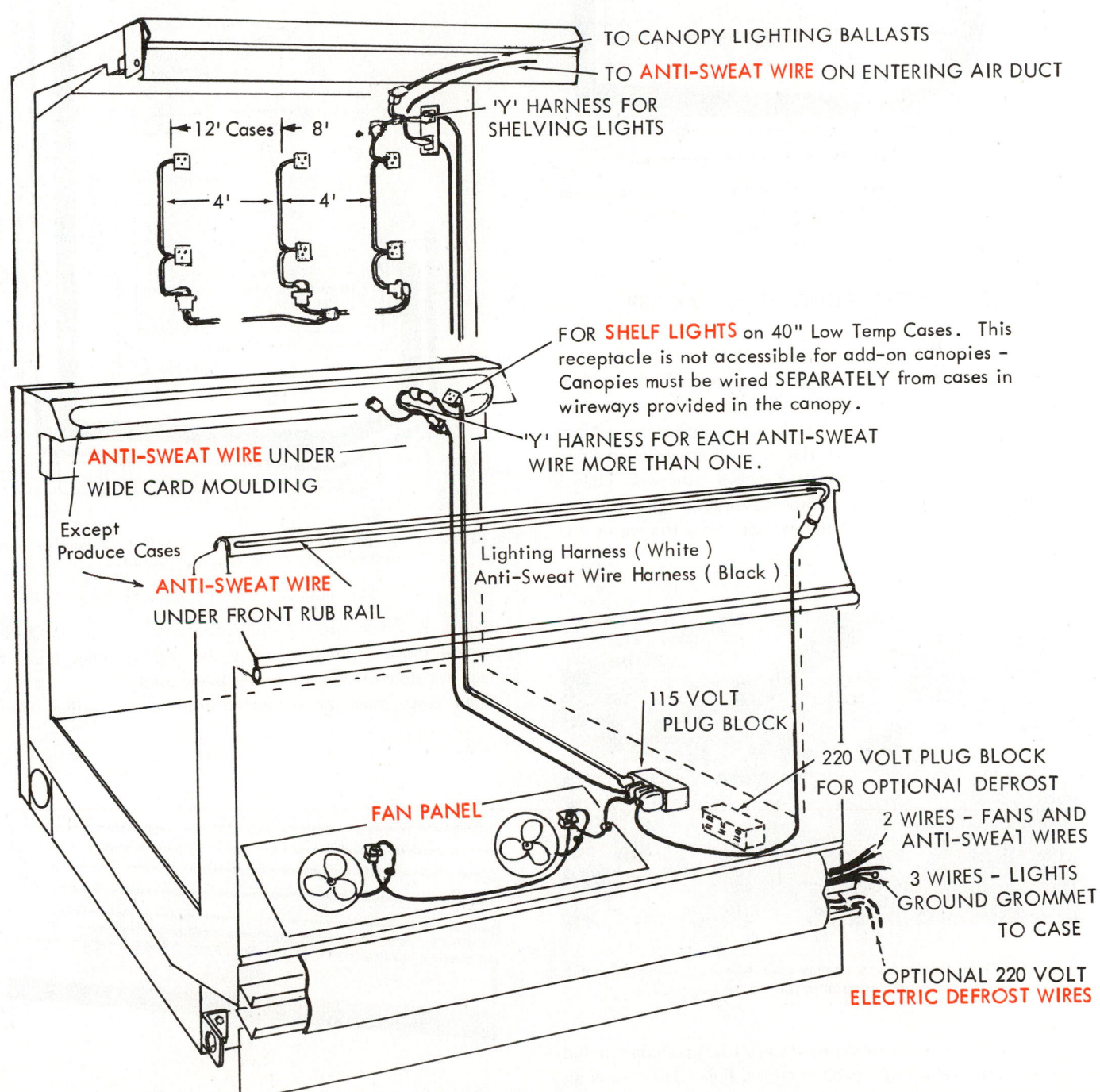

Fig. 13-21. Electrical wiring system used on most open display cases.

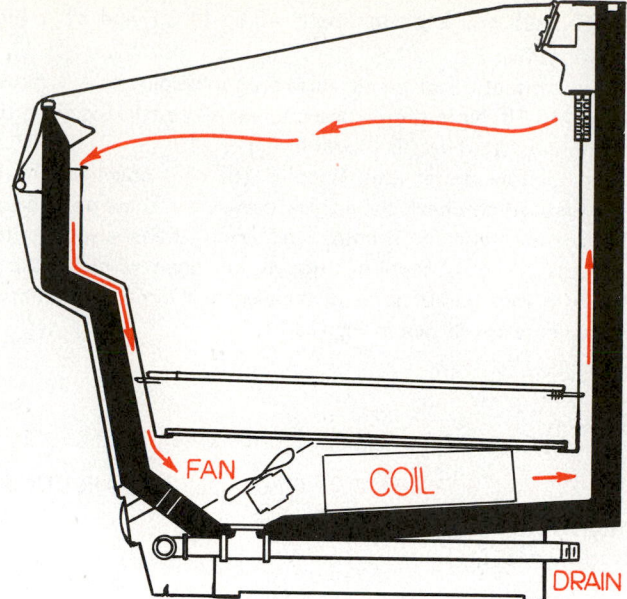

Fig. 13-22. Construction details of open frozen foods case. Notice airflow curtain across top of case (red arrows). (Tyler Refrigeration Corp.)

13-9 OPEN FROZEN FOOD DISPLAY CASE

Storing and displaying of frozen foods in either open or closed cabinets presents some problems. Fig. 13-22 shows such a display case. Temperatures near 0 F. (−18 C.) must be maintained. The evaporators therefore must operate at −10 F. to −15 F. (−23 C. to −26 C.). Heater wires are installed along those parts of display cases that could otherwise collect condensation from the air. Some cases are equipped with alarm systems which warn if the case becomes too warm. See Fig. 13-23.

Fig. 13-23. Open frozen foods display case. Alarm circuit will signal if produce becomes too warm.

Frozen food storage cases and display cases are constructed in both chest and upright cabinet styles. Fig. 13-24 shows an upright case. Since temperatures are very low in these cases, openings are gasketed or sealed. For the same reason,

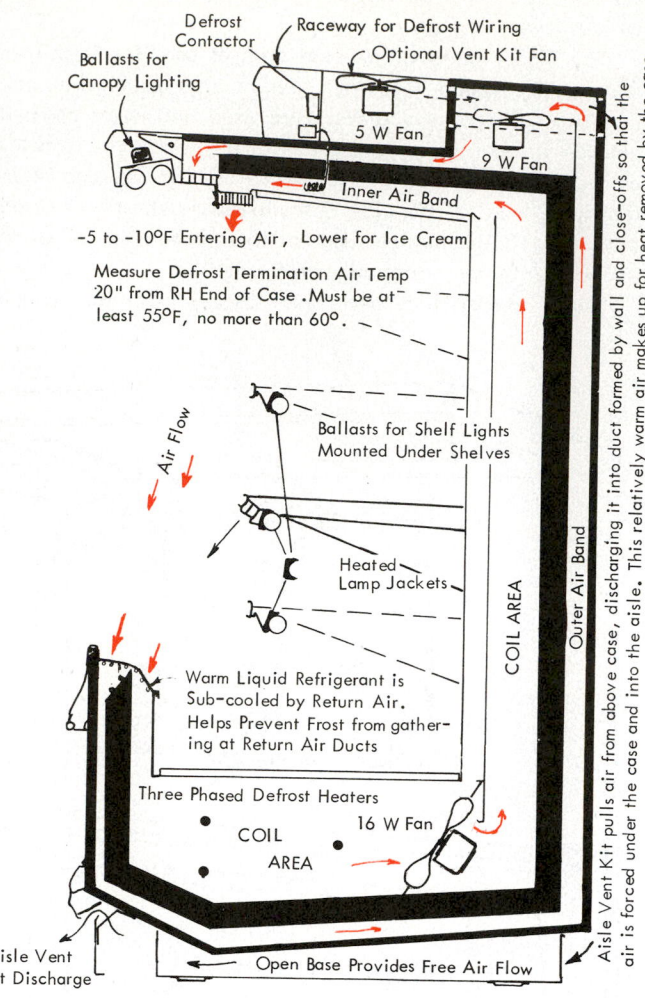

Fig. 13-24. An upright frozen foods display case. Note fans and airflow pattern. (Tyler Refrigeration Corp.)

insulation is thick and carefully hermetically sealed. Chest types are more popular because the top openings prevent spillage of cold air when the case is being used.

Open cases must be protected from drafts, grilles, unit

Fig. 13-25. An open frozen foods display case. Two air curtains flow from top to bottom edge of front section. (Tyler Refrigeration Corp.)

heaters and fans. Drafts will interfere with the air curtain of the case causing higher operation costs and defrosting problems. Several cases are usually connected end-to-end in supermarkets. Total electrical load must be carefully checked to provide enough service.

Need for low temperature presents a difficult evaporator defrosting problem. The evaporator must be defrosted at least once a day. This must be done as quickly as possible to prevent the case from warming up too much. The defrosting must be done automatically. A time clock is usually used. It operates a hot gas defrosting system or an electric heater defroster device. See Chapter 12 for details concerning these systems.

Some cabinets use two or three air curtains. See Fig. 13-25. The principle of operation of this type of case is shown in Fig. 13-26. The construction of this case is shown in Fig. 13-27.

13-10 FROZEN FOOD STORAGE CABINET

The frozen food storage cabinet may be either a chest or upright type with four to six in. (10 to 15 cm) of polyurethane insulation. The doors or access openings are also heavily insulated while double gaskets are usually provided for better sealing. Electrical resistance strip heaters usually surround door frames and other parts which may sweat.

Stainless steel is often used for the inner liner. The outer liner is usually aluminum or stainless steel.

These cases are for storage purposes only and the food is moved to display cases as needed. These cabinets operate at 0 F. (−18 C.) or lower. The refrigerating mechanism is normally installed on top of the cabinet.

13-11 FAST FREEZING CASE

Cases used for freezing foods rapidly are similar to storage cases except that temperatures are maintained at about −20 F. (−29 C.), and food is placed as close to the freezing plates as possible. Some cases use refrigerated shelves to provide more heat transfer surface.

13-12 ICE CREAM CABINET

Ice cream cabinets have a steel framework with a sheet metal exterior. Modern cabinets use polyurethane insulation about 3-in. (8 cm) thick.

The size of the sleeve or tank holder is standard. Therefore, construction of the various makes of bulk ice cream cabinets is similar. The size ranges from one to 12 sleeves.

The bulk ice cream cabinet should be kept at approximately 0 F. (−18 C.). If the temperature is too cold, it is difficult to scoop out the ice cream. Also, a too low temperature tends to crystallize the ice cream.

Dry type evaporators are used with either a capillary tube, a thermostatic expansion valve or an automatic expansion valve refrigerant control. Some of the evaporators are tinned tubing wrapped around and soldered directly to the sleeves. Others are sheet metal which has refrigerant passages formed in it. The sleeve covers (tops which must be raised to get to the ice

cream) are also standardized by manufacturers. Since sleeve openings are at the top, there is no spilling of cold air from the cabinet when opened.

Some ice cream cabinets are self-contained. This means that the refrigerating machine or condensing unit is built into one end of the cabinet. Other cabinets are made with the condensing unit separate (remote type).

In addition to the chest type ice cream cabinet, the upright and open display types are used. Packaged ice cream should be kept at about −10 F. (−23 C.) in order to retain its firmness.

13-13 SODA FOUNTAIN

The soda fountain provides a compact unit for storing and dispensing ice cream, cold water, beverages and syrup. It also makes and stores ice. It is usually attractive and is designed to make serving easy.

A built-in ice cream cabinet usually occupies one part of the fountain. Another part contains a water-cooling mechanism which must be designed to keep water at the right temperature. The evaporator outlet tubing from the ice cream cabinet and the drinking water cooler is passed around the syrups, keeping them relatively cool. Beverages are cooled to the same temperature as the water.

Soda fountains are often difficult to service. Being compact, they leave little space in which to work.

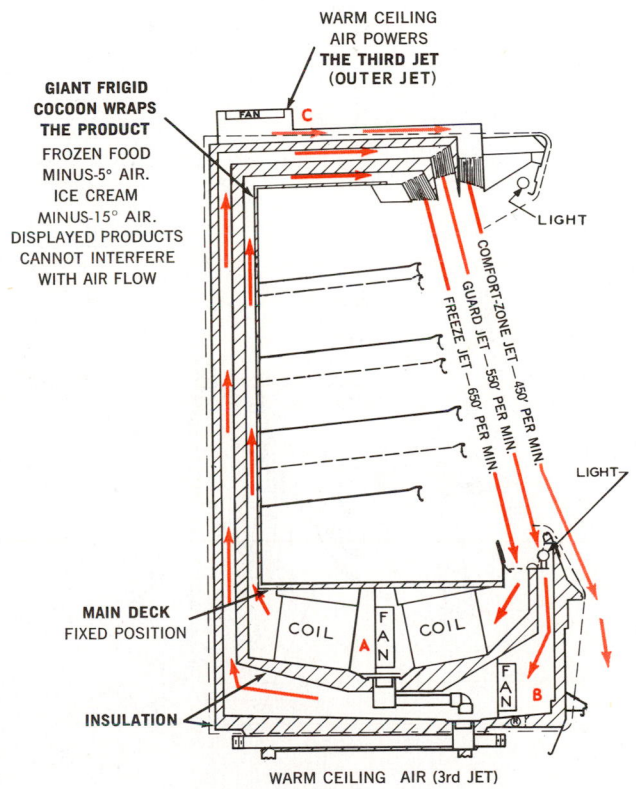

Fig. 13-26. A three air curtain frozen foods case. Note three separate fan systems. Inner curtain is powered by fan at A, middle curtain by fan at B. Fan at C is for outer curtain.
(Warren/Sherer, Div. of Kysor Industrial Corp.)

1. REFRIGERATED DUCT HONEYCOMB.
2. COLD AIR HONEYCOMB ANTI-SWEAT HEATER.
3. NOZZLE HEATER.
4. GUARD DUCT HONEYCOMB.
5. CANOPY HONEYCOMB.
6. FLUORESCENT BULB.
7. CANOPY FAN BLADE.
8. CANOPY FAN MOTOR.
9. BALLAST (2-LAMP).
10. TOP SHELF ASSEMBLY (ACCESSORY SHELF — HIGH OR LOW FRONT).
12. CENTER SHELF ASSEMBLY (HIGH OR LOW FRONT).
14. BOTTOM SHELF ASSEMBLY (HIGH OR LOW FRONT).
16. REFRIGERATED AIR DUCT ANTI-SWEAT HEATER.
17. RETURN AIR GRILLE ANTI-SWEAT HEATER.
18. THERMOPANE CAP ANTI-SWEAT HEATER.
19. RETURN AIR GRILLE.
20. EXPANSION VALVE.
21. RACEWAY BUMPER RAIL.
22. KICK PLATE.

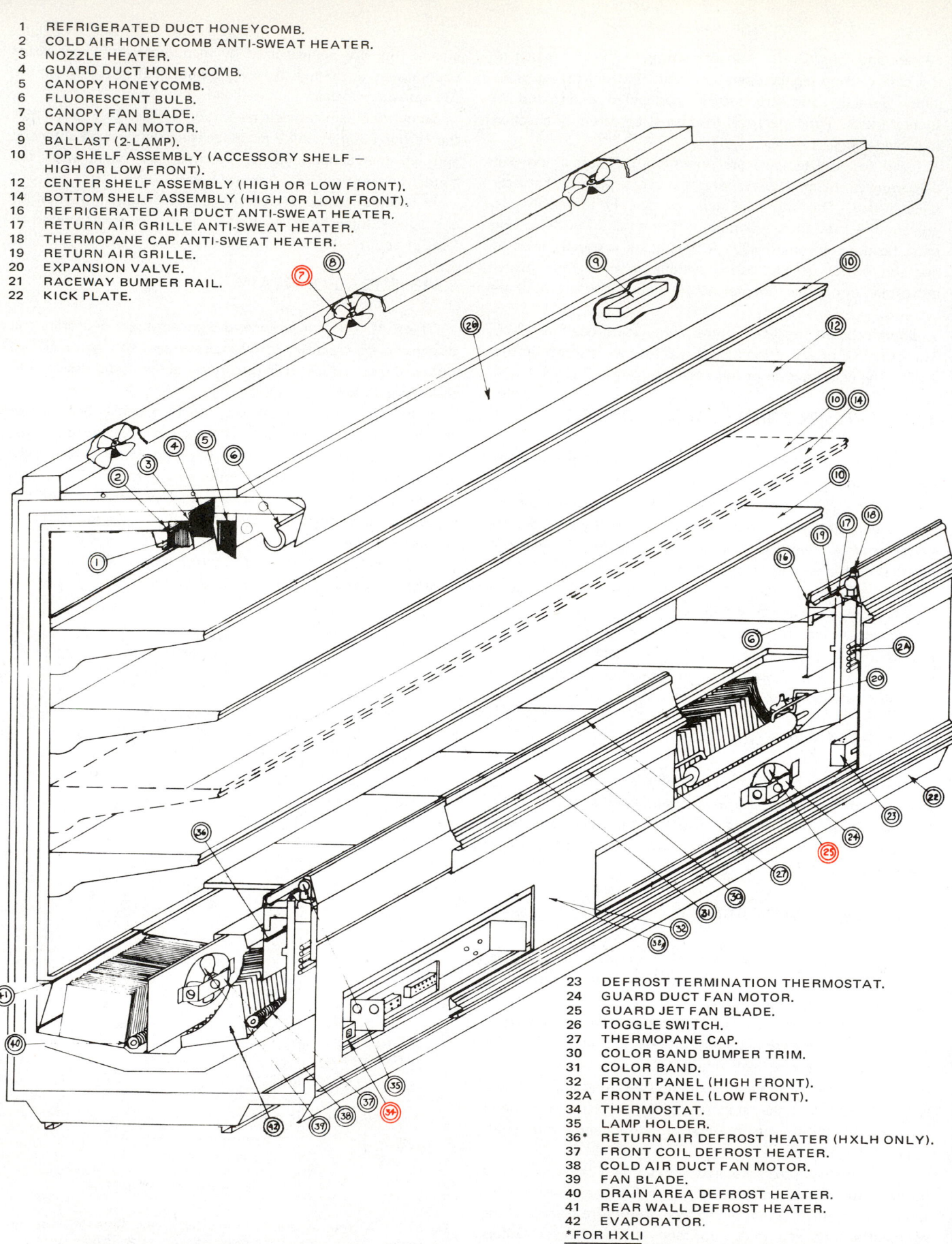

23. DEFROST TERMINATION THERMOSTAT.
24. GUARD DUCT FAN MOTOR.
25. GUARD JET FAN BLADE.
26. TOGGLE SWITCH.
27. THERMOPANE CAP.
30. COLOR BAND BUMPER TRIM.
31. COLOR BAND.
32. FRONT PANEL (HIGH FRONT).
32A FRONT PANEL (LOW FRONT).
34. THERMOSTAT.
35. LAMP HOLDER.
36* RETURN AIR DEFROST HEATER (HXLH ONLY).
37. FRONT COIL DEFROST HEATER.
38. COLD AIR DUCT FAN MOTOR.
39. FAN BLADE.
40. DRAIN AREA DEFROST HEATER.
41. REAR WALL DEFROST HEATER.
42. EVAPORATOR.
*FOR HXLI
36 DUAL VOLTAGE RETURN AIR HEATER.

Fig. 13-27. Construction details of open frozen foods case with three air curtains protecting food.

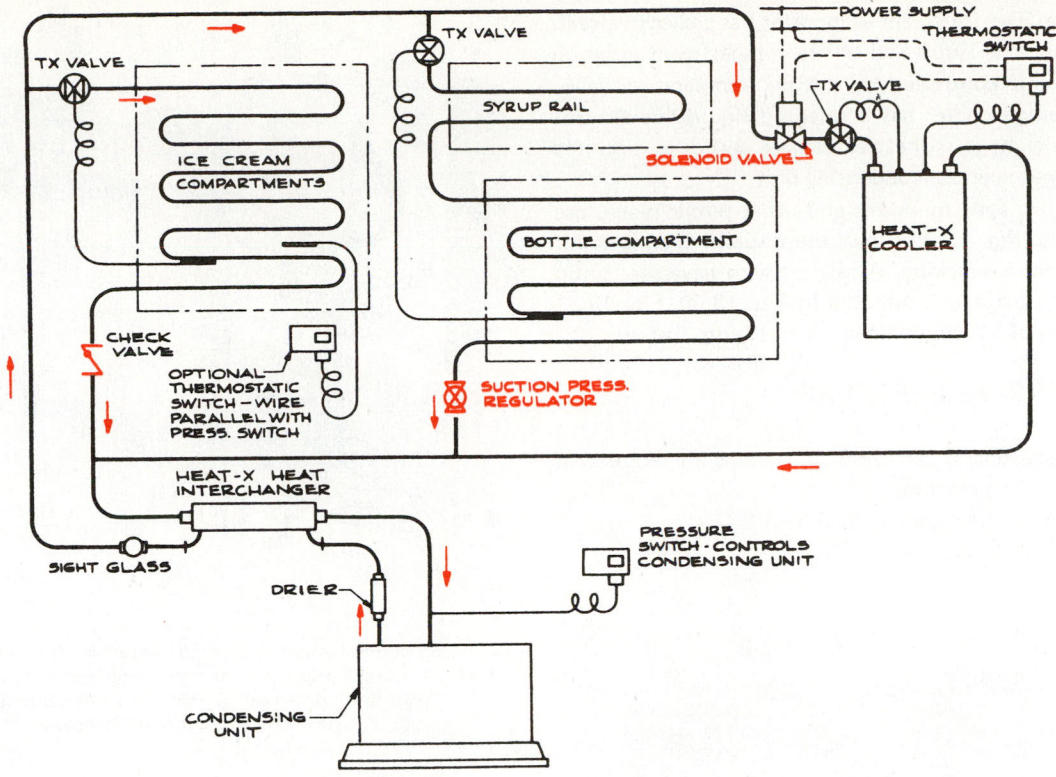

Fig. 13-28. Complete soda fountain cycle diagram. Note solenoid valve, suction pressure regulator and how these valves control temperatures in various parts of installation.   (Dunham-Bush, Inc.)

Syrups should be kept at about 45 F. (7 C.). Water temperature should range between 32 F. and 50 F. (0 to 10 C.). Ice cream, as mentioned, should be kept between 0 F. and 10 F. (−18 C. and −12 C.).

A diagram, Fig. 13-28, shows the complete cycle of a soda fountain refrigerated by one condensing unit. The soda fountain has an ice cream compartment, syrup rail, bottle compartment and beverage coolers. It uses a thermostatic

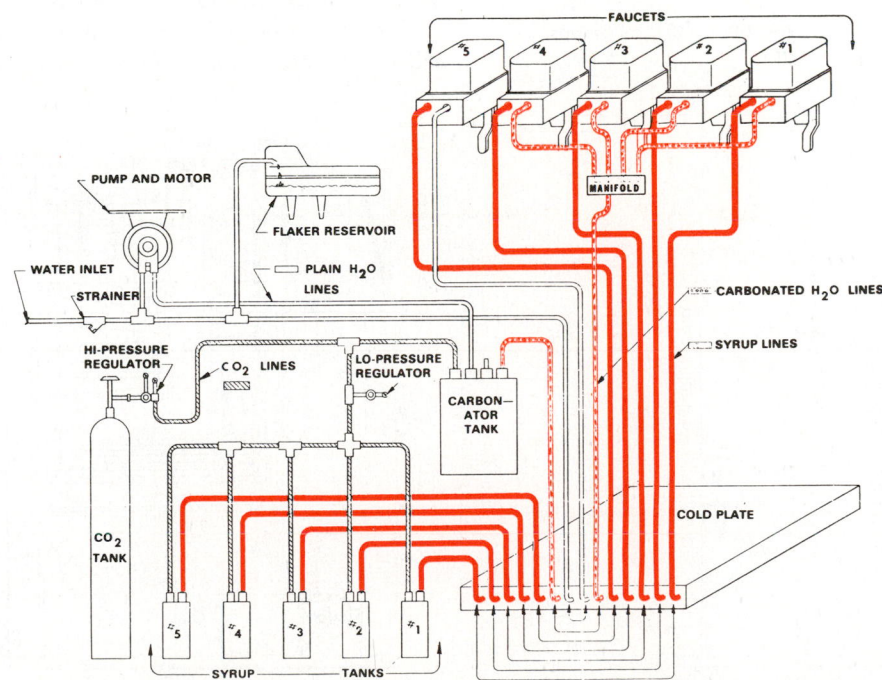

Fig. 13-29. Fluid circuits of drink dispenser. Carbonated water and the syrups are mixed in electric heads. This unit also makes and stores ice flakes.   (Queen Products Div., King-Seeley Thermos Co.)

expansion valve. The ice cream evaporator has a check valve in the suction line. The syrup and bottle compartment evaporator temperature is controlled by the two-temperature valve, while the beverage cooler has a solenoid liquid line shutoff valve. Note the sight glass, heat exchanger and drier mounted in refrigerant lines near the condensing unit.

Many drive-ins, soda fountains and other public places use drink dispensers. Fig. 13-29 shows the dispensing circuits for water, syrups and ice making. A cabinet with beverage, syrup and ice cream dispensing is pictured in Fig. 13-30. Fig. 13-31 shows the design of a carbonated drink and syrup system.

13-14 DISPENSING FREEZERS

Special applications of refrigerating systems are required in soft ice cream making machines:

1. Frozen carbonated beverages, 25 F. (−4 C.).
2. Slushes, 28 F. (−2 C.).

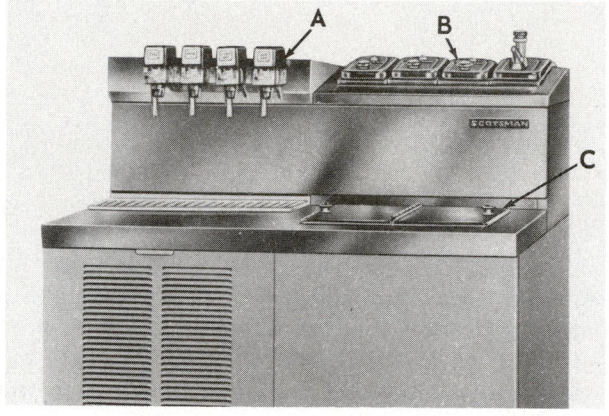

Fig. 13-30. A dispenser cabinet. A—Soft drinks. B—Syrups. C—Ice cream.

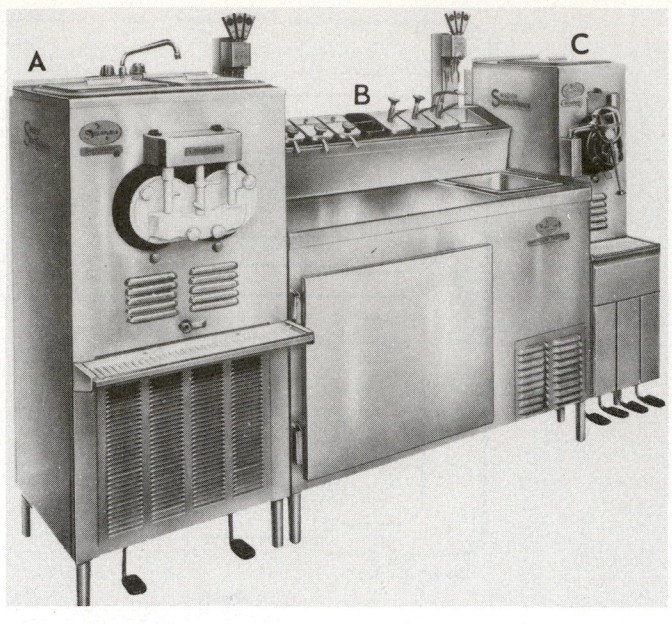

Fig. 13-32. Combination dispensing freezer. A—Two soft ice cream units on left. B—Combination mix storage, syrup and topping rail and three flavor drink dispenser. C—Milk shake dispensing freezer. (Sweden Freezer Mfg. Company)

3. Fruit or water ices, 10 to 20 F. (−12 C. to −7 C.).
4. Soft serve, 21 F. (−6 C.).
5. Milkshakes, 27 F. (−3 C.).
6. Sherbets, 20 F. (−7 C.).

This unit uses a large refrigerating machine to fast freeze or cool the mix before it is fed to the freezing cylinder. The same or a separate motor drives the stirring mechanism (dasher).

A 1/2 hp (746W) refrigerating unit can fast freeze one gallon of custard in about six minutes.

A combination unit is illustrated in Fig. 13-32. Fig. 13-33 shows the construction of a freezing dispenser cabinet.

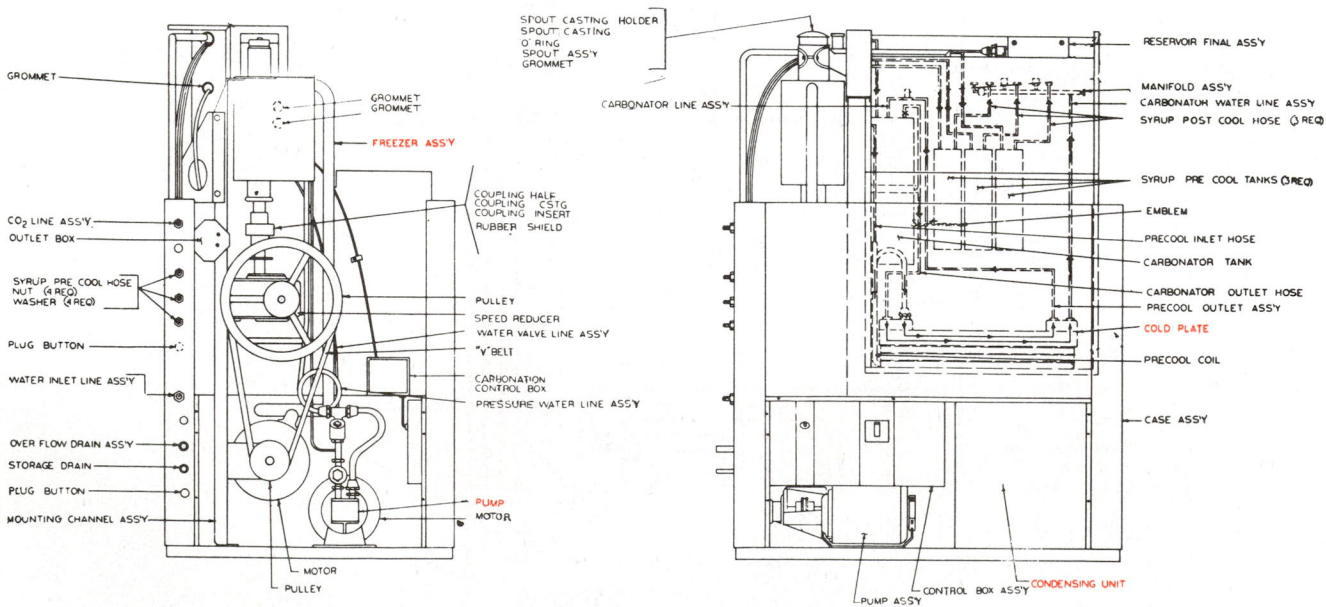

Fig. 13-31. This diagram of a drink dispenser shows construction details and names all parts.

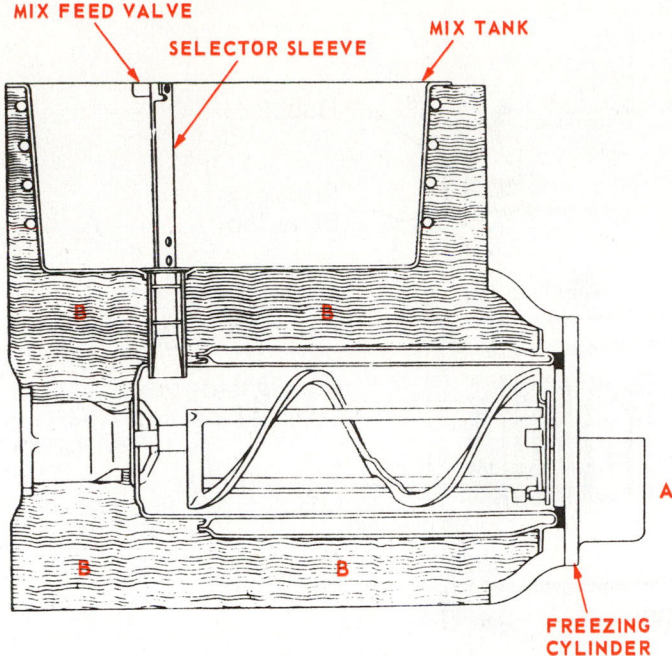

MIX FEED VALVE
SELECTOR SLEEVE
MIX TANK
B
B
A
B
B
FREEZING CYLINDER

Fig. 13-33. A dispensing freezer. A—Dispenser opening. B—Heavy insulation around freezing cylinder. (Sweden Freezer Mfg. Company)

Some units are continuous in operation, because the mix in the freezer often has to be kept within 1 or 2 F. (0.5 to 1 C.) of its correct temperature. Thermostatic expansion valves are usually used as the refrigerant control.

The machine is usually adjusted to deliver the custard or sherbets at 20 F. (−7 C.). A total of 3 hp (2238W) can operate the refrigerating unit and drive the dasher. A unit with a capacity of 12.5 gal. (48 litres) per hour usually has a 2 hp (1492W) dasher motor and a 3 hp (2238W) refrigerating unit. The larger units are water cooled.

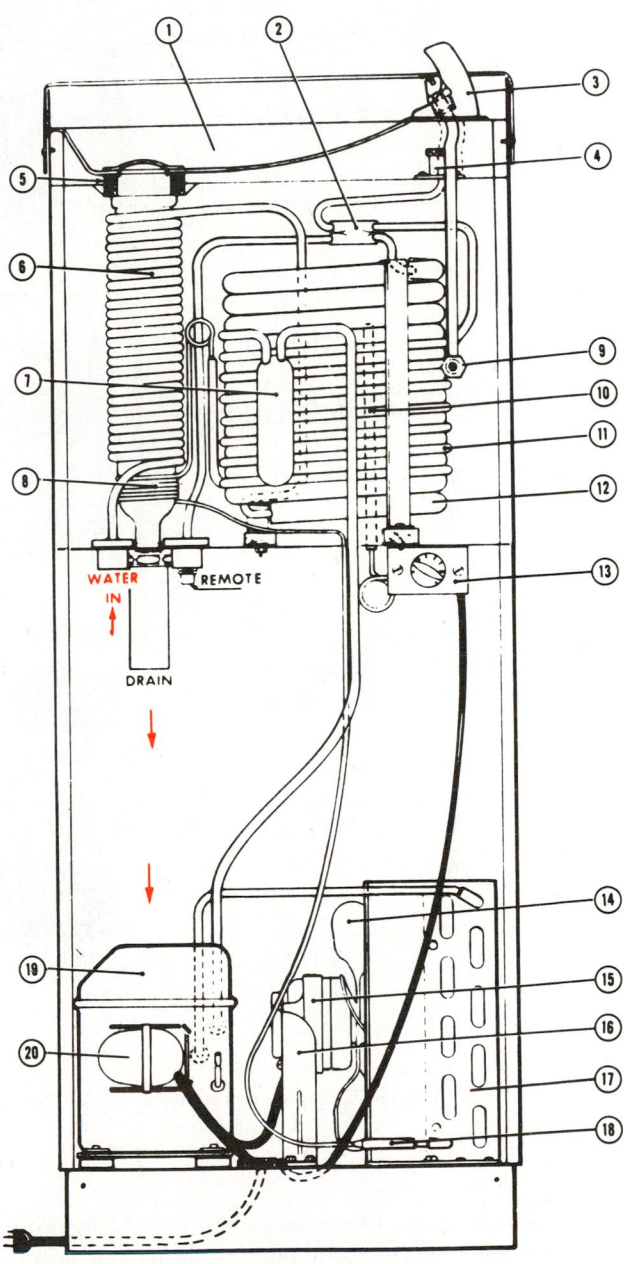

Fig. 13-35. Self-contained water cooler. 1—Top. 2—Cold water distributor. 3—Bubbler guard. 4—Glass filler connection. 5—Drain gasket. 6—Precooler assembly. 7—Accumulator. 8—Capillary tube. 9—Water valve. 10—Thermostat bulb well. 11—Evaporator. 12—Water-cooling coil. 13—Thermostat. 14—Fan blade. 15—Fan motor. 16—Fan bracket. 17—Condenser. 18—Liquid refrigerant strainer. 19—Compressor. 20—Relay, overload protector.
(Temprite Products, Eaton Corp.)

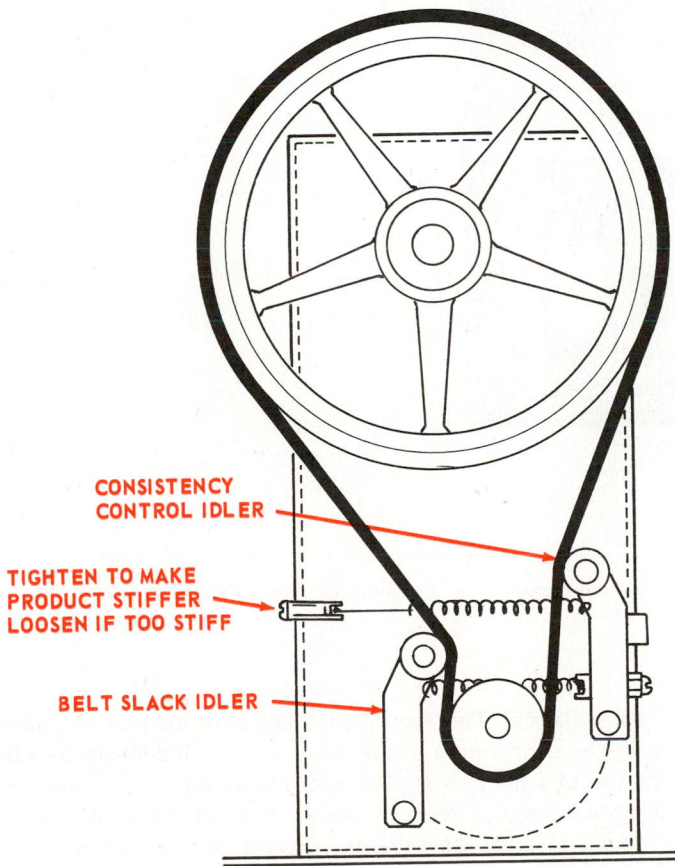

CONSISTENCY CONTROL IDLER

TIGHTEN TO MAKE PRODUCT STIFFER LOOSEN IF TOO STIFF

BELT SLACK IDLER

Fig. 13-34. Shake maker consistency control. As mix hardens, tight side of belt moves consistency control idler and finally opens motor circuit. (Sani-Serv)

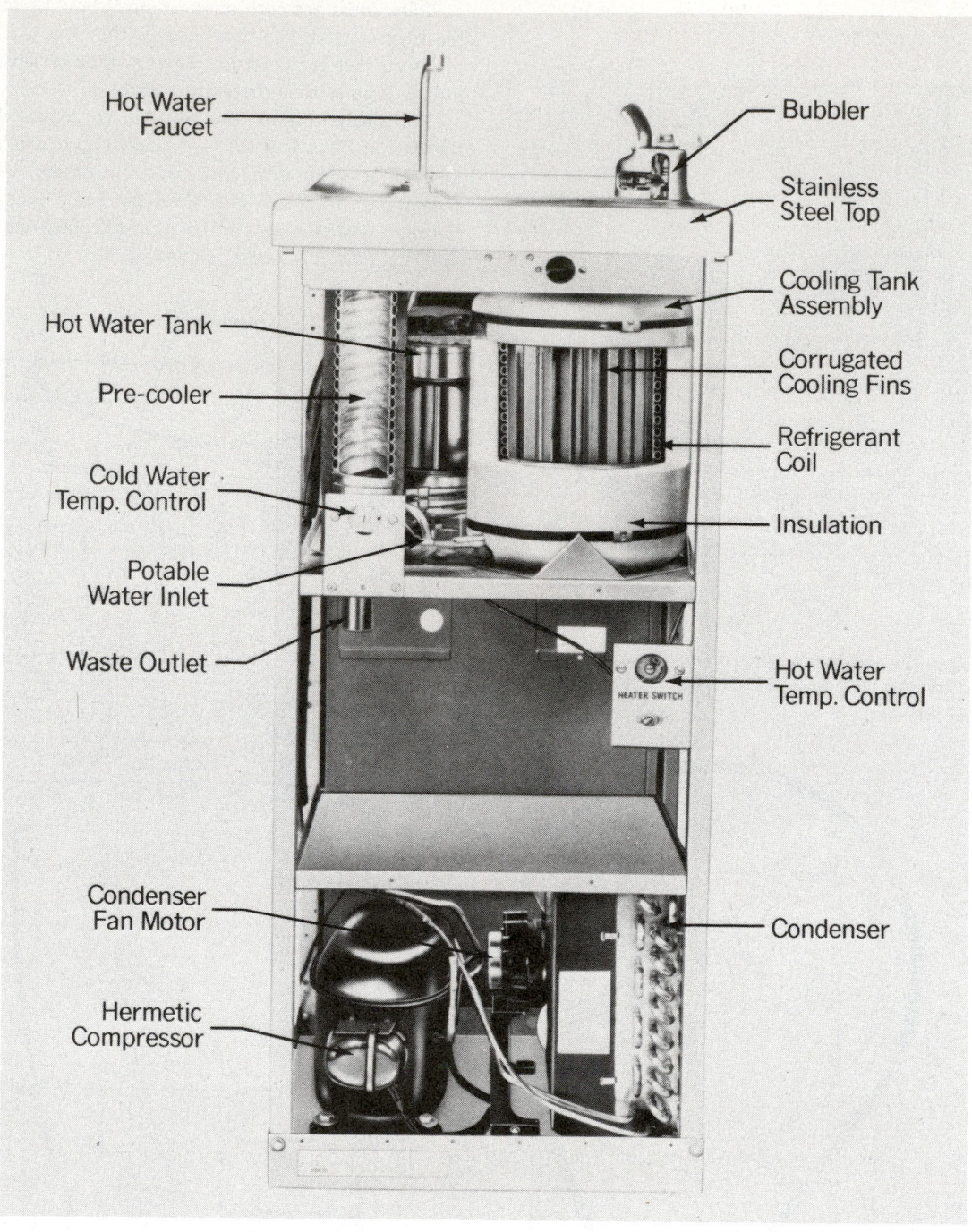

Fig. 13-36. Self-contained water cooler. Note forced convection condenser and location of evaporator. (Ebco Mfg. Co.)

The quality of the mix is very important. Many problems thought to be in the refrigeration have turned out to be poor mixes. The frozen carbonated beverage is about one part syrup and four parts carbonated water. It is served at 22 to 26 F. (−6 to −3 C.). Health codes require keeping the mix containers as well as the freezing cylinder sanitary and clean.

A mix storage or supply cabinet is usually found with three units to store the mix until it is used.

A commercial application of the dispensing freezer is the "shake maker." This machine is filled with the desired shake mix. The temperature of the "shake maker" is controlled with the use of a thermostat. The temperature control is based on the consistency of the mix. As the mix is frozen, the torque required to drive the agitator increases. An idler on the belt drive side is connected to a microswitch. When the mix is frozen to the desired consistency, the belt tightness will move the idler and open the microswitch which shuts off the machine. See diagram in Fig. 13-34.

13-15 WATER COOLER

The water cooler cabinet usually has a sheet metal housing built around a steel framework. Inside this sheet metal housing there is usually a condensing unit, located near the floor. Above is the water-cooling mechanism. The latter is the only part insulated with foamed plastic. The insulation usually is specially formed and between 1 1/2 and 2-in. thick (4 to 5 cm). Some water coolers also have a heater providing hot water. Some also have a refrigerated compartment for storing milk, soft drinks and food.

This cabinet is made so that one or more sides may be easily removed to gain access to the interior. The basin of the water cooler is generally made of porcelain-coated cast iron, porcelain-coated steel or stainless steel.

Heat exchangers are frequently used on water coolers. Some make use of the low temperature of the waste water to precool the fresh water line to the evaporator (Fig. 13-35, items six and eight). The temperature of the cooled water is usually maintained at 50 F. (10 C.). The thermostat controls the temperature of the cooled water.

A water cooler using a plumbing supply and drain connection must be installed according to the National Plumbing Code and local codes. The plumbing should be concealed. A hand shutoff valve should be installed in the fresh water line and a drain pipe, at least 1 1/2 in. in diameter, should be provided. The bubbler opening must be above the drain to eliminate the chance for accidental syphoning of the drain water back into the fresh water system.

The tap water model uses a variety of evaporator designs. Fig. 13-36 shows the evaporator wrapped around the water-cooling tank. Fig. 13-37 is the wiring diagram for a water-cooler having a water heating service.

Temperature of the cooled water may be set by the person regulating the cooling unit. *Refer to Para. 15-26 for recommended temperatures. Clean materials must be used for all water passages.*

In large business establishments, in office buildings or factories, multiple water coolers — instead of individual ones — are popular. These coolers have one large condensing unit supplying refrigeration to many bubblers. These bubblers may be of many different types of construction.

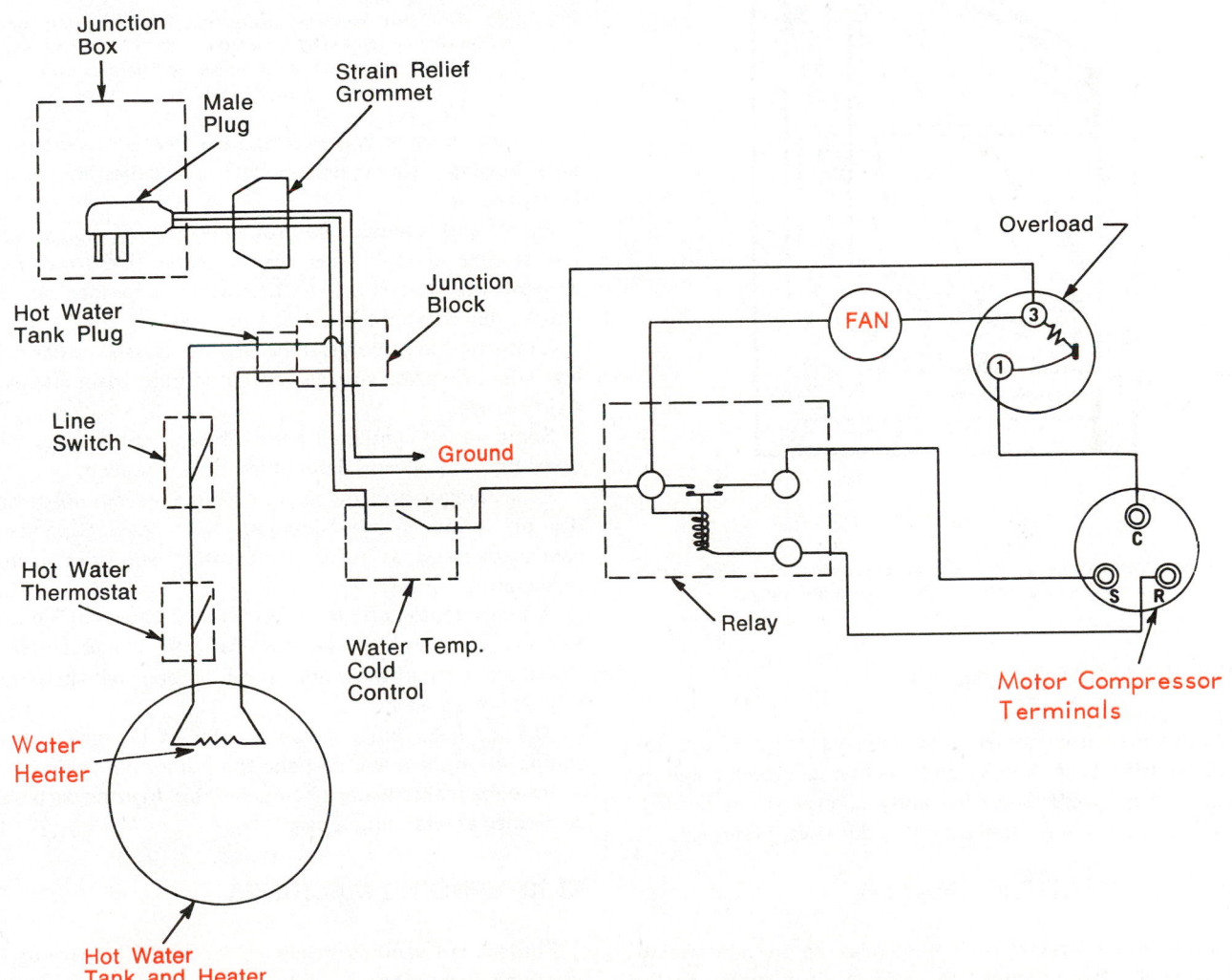

Fig. 13-37. Wiring diagram for water cooler which also provides hot water. Fan is condenser fan.   (Elkay Mfg. Co.)

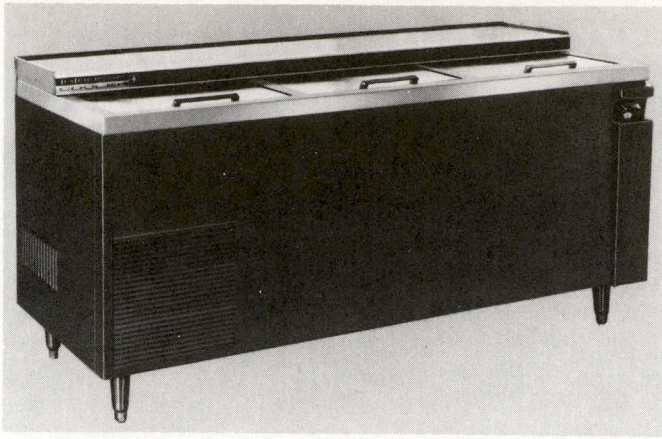

Fig. 13-38. Three-door, dry bottled beverage cabinet. Note how doors slide to rear. It has a condensing unit in the lower, left corner. (LaCrosse Cooler Co.)

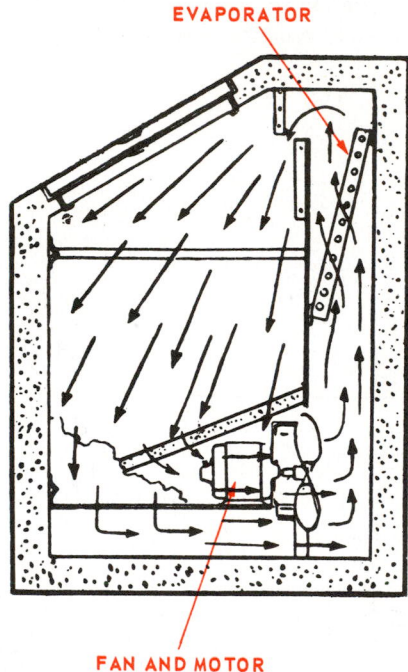

Fig. 13-39. Cross-section of dry bottled beverage cooler, with forced convection evaporator. (National Cooler Corp.)

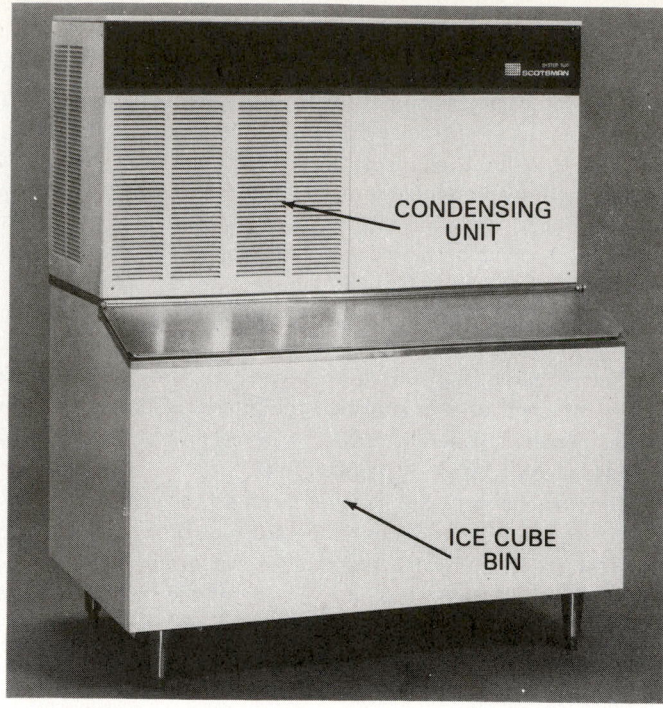

Fig. 13-40. Automatic ice cube maker. The 1 1/2 hp condensing unit is serviceable by removing louvered panels at front and side. (Queen Products Div., King-Seeley Thermos Co.)

13-16 BEVERAGE COOLER

Many stores use special bulk cabinets (Fig. 13-38) for cooling bottled beverages. A cross-section of a bulk cooler is shown in Fig. 13-39. Since humidity control is not needed, high-velocity airflow is often used to reduce evaporator size.

13-17 AUTOMATIC ICE MAKER

Automatic ice makers for commercial use are now widely used. Several different types are available. They automatically control water feed, freeze water into ice, empty the ice into storage facilities and stop when the storage space is full.

The ice cubes or chips formed are clear and sanitary, since only flowing water is used. Cloudy ice cubes are caused by entrapped air.

Floats and solenoids control water flow. Switches operate the storing action when ice is made. Electrical heating elements, hot water, hot gas defrosting or mechanical devices remove the ice from the freezing surfaces.

Urethane foam, polystyrene or fiber glass provides cabinet insulation. Freezing surface and bin storage basin are made of stainless steel.

Some are self-contained while others use remote condensing units. Both supply and drain plumbing are needed.

Capacity can vary among units from a few pounds (grams) a day up to many tons (kilograms) per day. Capacities, of course, decrease as water temperature and/or ambient air temperature increase.

A typical automatic ice cube maker is shown in Fig. 13-40. Another unit, Fig. 13-41, automatically makes chipped or flaked ice. Each of these units produces approximately 900 lb. (408 kg) of ice a day.

The ice flake unit, shown in Fig. 13-42, has ice making equipment in the upper part and the storage bin below.

Ice cube maker water circuits and ice freezing parts should be cleaned at least once a year.

13-18 VENDING MACHINES

Refrigerated vending machines for foods or beverages are becoming increasingly popular. Cold drinks with ice, soft drinks, ice cream, cold food, cold milk and frozen desserts are all automatically dispensed. See Fig. 13-43.

ICE FLAKES BIN

CONDENSING UNIT

Fig. 13-41. Flake or chipped ice unit. It has a 1/2 hp (373W) condensing unit, air cooled. It will store 255 lb. (115 kg) of ice and produces 442 lb. (199 kg) of ice in 24 hours. (Queen Products Div., King-Seeley Thermos Co.)

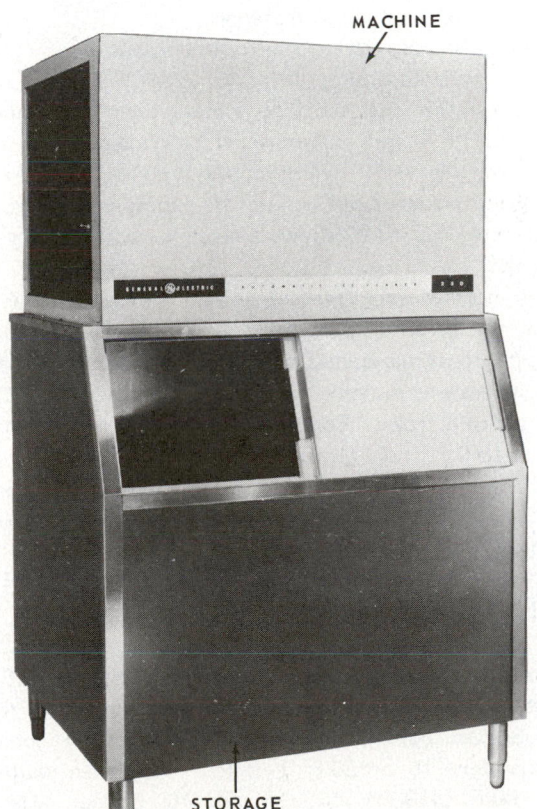

MACHINE

STORAGE

Fig. 13-42. Cabinet for system making ice flakes. Note sliding door access to bin. (General Electric Co.)

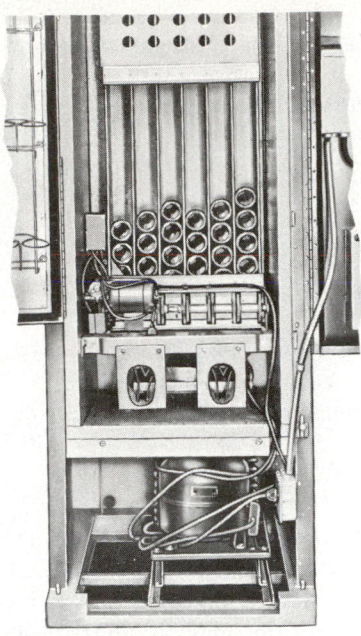

Fig. 13-43. Bottle beverage refrigerated coin operated vending machine. Cabinet doors are open showing storage compartment, controls and refrigeration mechanisms.

Fig. 13-44. Milk vending machine. (K. G. Brown Mfg. Co., Inc.)

Refrigerating systems employed in these units are typical. Most use capillary refrigerant controls, hermetic motor compressors and defrosting devices. Fig. 13-44 illustrates a milk vending machine.

The electrical system generally is part of the overall system that electrically transfers the material and operates the coin and currency devices.

Vending machines have many components necessary for dispensing:
1. Coin and currency devices.
   a. Acceptors.
   b. Rejectors.
   c. Changers.
   d. Steppers and accumulators.
2. Carbon dioxide systems.
3. Cup dispensers.
4. Heating systems.
5. Refrigerating systems that are usually air-cooled using fan evaporators. Some have defrost systems.
6. Transfer systems.

13-19 MILK COOLER

By law, milk must be cooled to 50 F. (10 C.) within an hour after it has been taken from the cow. Furthermore, it must be stored at 40 F. (4 C.) or lower at all times.

Bacterial growth in milk is dramatically affected by temperature. During a 24 hour period, bacteria count will increase to 2400 at 32 F. (0 C.); to 2500 at 39 F. (4 C.); to 3100 at 46 F. (8 C.); to 11,600 at 50 F. (10 C.); to 180,000 at 60 F. (16 C.) and to 1,400,000,000 at 86 F. (30 C.). Bulk type coolers will handle this specific cooling problem.

The stainless steel bulk cooler shown in Fig. 13-45 has polyurethane insulation between the inner and outer tanks. Coolers of this type have a 450 to 6000 gal. (1 703 to 22 712 L) capacity. R-22 is the refrigerant used in this type of system.

Fig. 13-45. Bulk milk cooler. Evaporator is in base of unit, surrounded by circulating water. Water is kept at 32 F. (0 C) and milk is quickly cooled to 38 F. (3 C).   (Dairy Equipment Co.)

Mounted outside the milk room, the condensing unit is usually air-cooled. *It should not be put in the vacuum pump room.* However, air flowing over the condenser can be drawn from the milk room. In winter, where permitted, this warm air can be returned to heat the milk room.

Coolers should be properly grounded. Use a separate safety ground wire the same size as the power wires or larger. Properly installed metal conduits can also serve this purpose. These units also have a crankcase heater.

13-20 MILK DISPENSERS

Many food service places serve milk from milk dispensers. For bulk milk sales, cans or plastic bags holding three to five gallons (11-19 litres) are installed in these dispensers. These units meet all health and sanitation codes. Milk is kept at about 36 F. (2 C.) by a hermetic refrigerating system. Reserve milk containers are kept in a walk-in cooler or in a milk storage refrigerator.

13-21 LABORATORY REFRIGERATED INCUBATORS

An interesting application of refrigeration helps maintain correct temperatures in incubators. Fig. 13-46 shows an incubator which can maintain constant temperatures using a refrigerating system for cooling and electric heating elements for heating. Such a unit can maintain any constant temperature between 36 F. (2 C.) and 158 F. (70 C.).

13-22 BAKERIES

Many raw products used by bakeries must be kept at refrigerated temperatures to preserve or improve quality. The bakeries must:
1. Store the perishable raw materials.
2. Store perishables during interrupted processing.
3. Store finished products.

Frozen ingredients must be stored. Even the water and flour used for bread making must be cooled during certain periods of the year. Health codes require these practices. Creams and custards can be kept for longer periods at a cool temperature. Fig. 13-47 shows a reach-in cabinet designed for baking use.

Both normal and low-temperature refrigeration is used. Normal refrigeration is suitable for butter, eggs, coconut, cream, fat, meat, margarine, nuts, yeast and dough retardation. Low-temperature systems are needed to freeze baked goods which are sold frozen. For example, if bread is fast frozen to −1 F. (−18 C.), it will remain fresh for almost a month.

Air conditioning using refrigerating equipment is also used in bakeries since controlled temperature and humidity is important in many baking processes.

13-23 FUR STORAGE

Furs are stored at a low temperature to prevent moth damage. Moths in the egg, larvae or adult stage, will thus be destroyed. In common practice, furs are first cooled to approximately 15 or 20 F. (−9 to −7 C.) then warmed to above 50 F. (10 C.) for 24 to 48 hours. After this cycle — found to be the most effective — the furs are stored at 35 to 40 F. (2 to 4 C.).

The fur storage cabinet is usually constructed like a cold

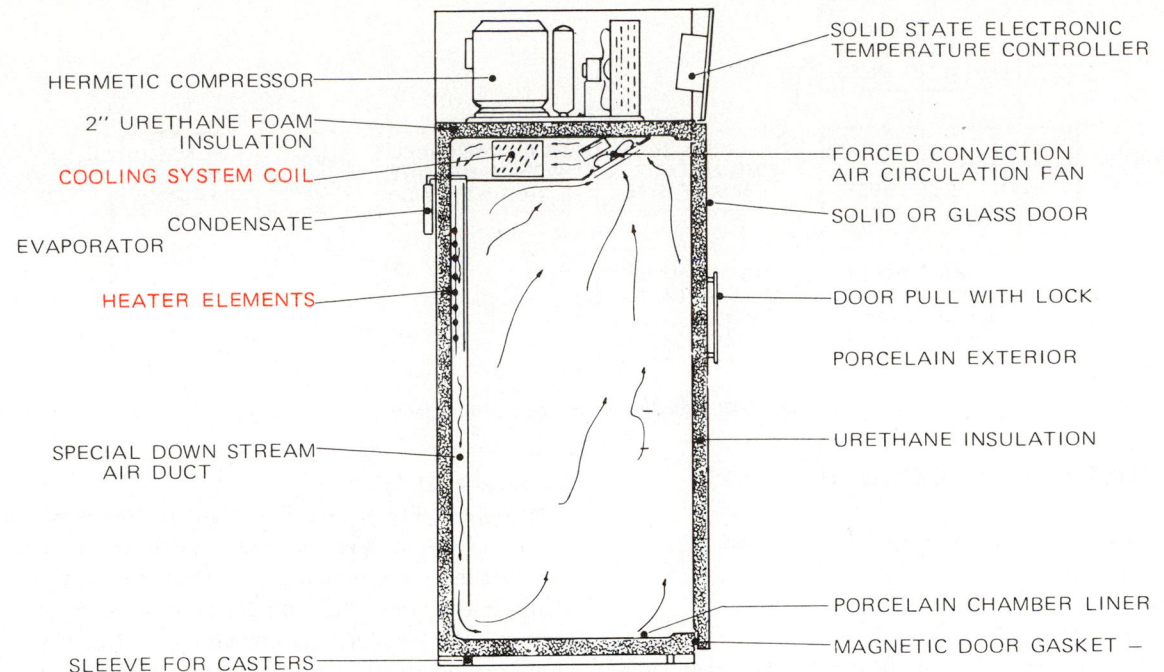

HERMETIC COMPRESSOR

2" URETHANE FOAM INSULATION

COOLING SYSTEM COIL

CONDENSATE EVAPORATOR

HEATER ELEMENTS

SPECIAL DOWN STREAM AIR DUCT

SLEEVE FOR CASTERS

SOLID STATE ELECTRONIC TEMPERATURE CONTROLLER

FORCED CONVECTION AIR CIRCULATION FAN

SOLID OR GLASS DOOR

DOOR PULL WITH LOCK

PORCELAIN EXTERIOR

URETHANE INSULATION

PORCELAIN CHAMBER LINER

MAGNETIC DOOR GASKET —

Fig. 13-46. Incubator has hermetic refrigerating system and electric heat. Temperatures are accurate within 0.5 C. (1 F.). (Puffer-Hubbard Refrigerator Div.)

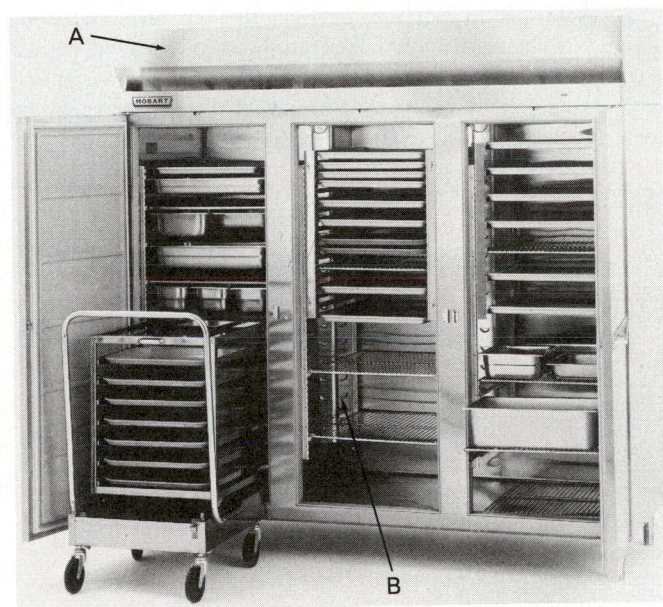

Fig. 13-47. Reach-in cabinet designed for bakery use. A—Condensing unit and blower evaporator. B—Cold air duct. (Hobart Corp.)

storage room. Walls are concrete, thoroughly moisture sealed with asphalt, corkboard insulation or its equivalent. An inside door is built like a refrigerator door; the outer, like a vault door.

Blower evaporators are used to force air circulation around and into the furs. The humidity must be kept at approximately 55 percent to prevent drying of the skins. Vaults should be of fireproof construction.

13-24 INDUSTRIAL APPLICATIONS

Refrigeration has a great variety of applications in manufacturing processes. Smaller units are automatic in operation. Three common uses are:

1. Cooling of water which, in turn, cools electrodes on resistance welders.
2. Cooling of quenching liquids used for cooling metals in heat treating.
3. Cooling compressed air.

This air must be dry since moisture may cause rust and corrosion on air tools or spoil paint spraying. Air, therefore, is cooled to keep its dew point (water condensing temperature) above the coldest spot in the air system. This will prevent moisture in the lines.

Air is cooled, then, with a refrigerating system to bring it below its critical pressure dew point. Then it is reheated. Pressure dew points are about 50 F. (10 C.) above atmospheric dew points at 100 psi (7 kg/cm^2) air pressure. In general practice, the compressed air is cooled to about 35 to 50 F. (1.7 to 10 C.). A 3000 cfm (14,158 m^3/s) air compressor will need about a 20 ton (55.2 kw/hr) capacity refrigerating machine.

If a refrigerating system is near flammable or explosive materials, the system must be explosion proof. Sealed lights and sealed contact points are required.

Refrigerators and freezers of all types are used in research. They are used to maintain constant temperature, constant humidity and low-temperature control. Low-temperature units capable of −140 F. (−96 C.) are available. These systems usually use a 5-in. (13 cm) insulation thickness and a cascade system (two systems connected in a series).

Commercial Systems Applications / 481

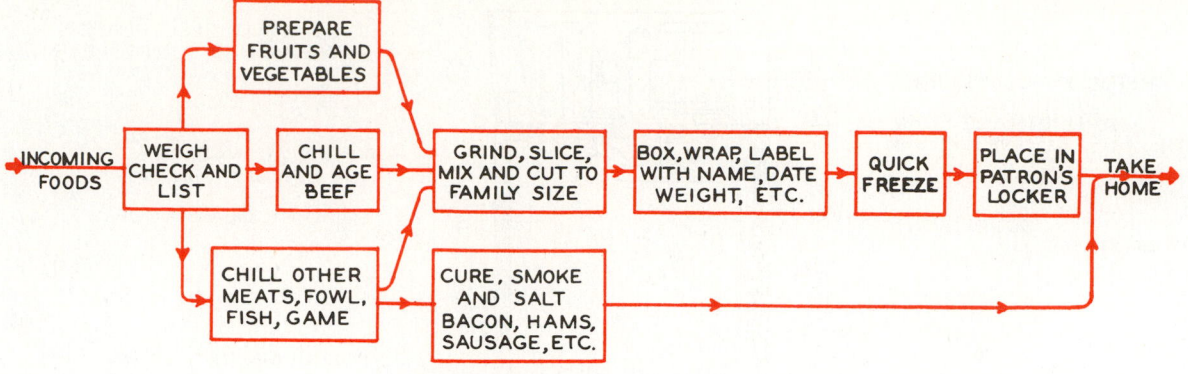

Fig. 13-48. Food flow through freezing plant.

13-25 INDUSTRIAL FREEZING OF FOODS

Industrial freezing of food is carried on in two principal types of establishments:
1. Processing plants.
2. Locker plants.

Processors of frozen foods have freezing centers in many large food-producing areas. For example, fish is packed and frozen along the coast and then shipped to all parts of the country.

A locker plant is a smaller unit designed to prepare, freeze and store a great variety of products. Refrigerating equipment in processing and locker plants vary considerably, but the plan for freezing the food is similar.

Fig. 13-48 shows the flow of produce through a typical plant. The food is weighed and checked for purity and suitability for freezing. Then it moves to the processing room. Here meats are cut, fowl are cleaned and dressed, vegetables blanched and the various items packaged. Next, the packed foods are sent to the freezing section where they are completely frozen and made ready for storage.

Fig. 13-49 shows a typical locker plant floor plan. An ultraviolet ray lamp is sometimes placed in the aging room to kill bacteria.

High humidity is very important in rooms where food is cured and stored. Meat especially tastes better and does not lose weight if relative humidity is kept as close to 100 percent as possible. Temperature should be near 39 F. (4 C.), as this gives the best humidity results with a nonfrosting evaporator.

Processing plants attempt to freeze food as fast as possible. They also attempt to expose as much of the food as possible to the lowest possible temperature. A fast-freezing system usually uses a track to hold the produce which then moves through an ultra low temperature chamber for fast freezing. In other systems, food is pressed between low-temperature plates, exposed to blasts of low-temperature air or sprayed with a low-temperature liquid. Many popular fast-food freezing systems immerse (dip) the food in liquid refrigerant. This may be done through an endless belt system and is very rapid.

13-26 FREEZE DRYING

Freeze drying consists of the following four steps:
1. Food is frozen.
2. Frozen food is put into a vacuum chamber. See Fig. 13-50.

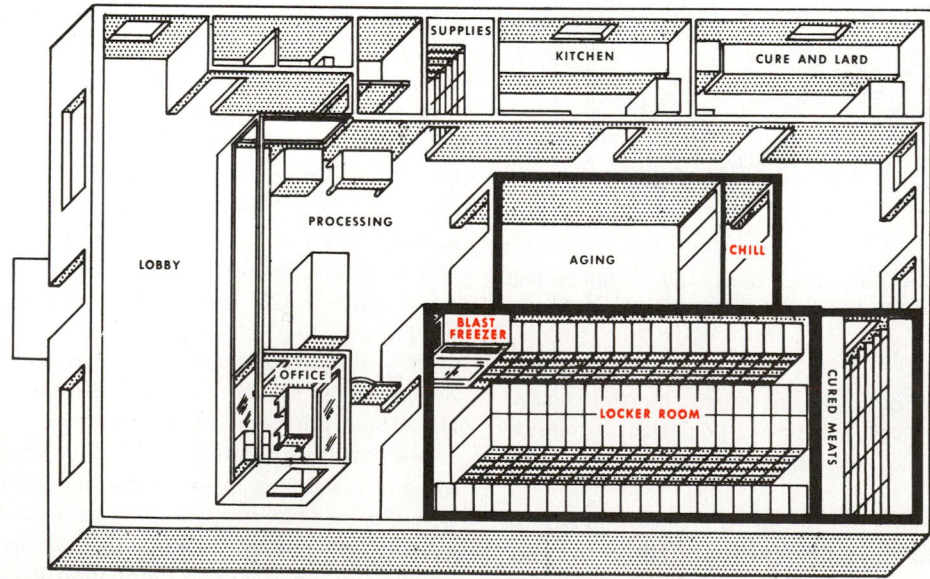

Fig. 13-49. Typical locker plant. Special blast freezer fast freezes food before it is stored in locker room.

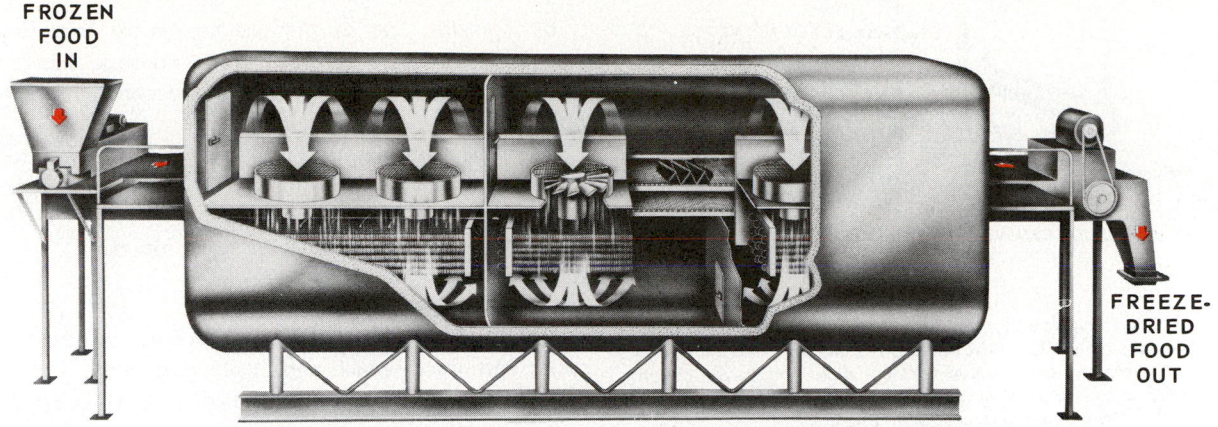

Fig. 13-50. Automatic track fast-freezing system for food. Air in the chamber is about −30 F. (−34 C). Food moves from left to right on conveyor belt.

3. Frozen food is heated to change its frozen water (ice) into vapor. It does not go through the liquid state, since below 4.7 mm Hg. water will change directly from ice to vapor. Freeze drying is usually done between 50 and 500 microns of pressure.

4. Vapor is condensed in the form of frost on a cold evaporator at about −40 F. (−40 C.) in the chamber. This step saves pumping about 100,000 cu. ft. (2832 m³) of water vapor to form 1 lb. (.5 kg) of vapor at 50 microns.

After the product has dried, the evaporator is exposed to heat at atmospheric pressure. The frost melts and is removed. Dry nitrogen is used to raise the chamber pressure from 50 microns to atmospheric pressure.

13-27 INDUSTRIAL STORAGE OF FROZEN FOODS

Storage requirements of most frozen foods are about the same. A temperature of 0 F. to −20 F. (−18 C. to −29 C.) is desirable and a variation of 2 to 3 F. (1 to 2 C.) is normally allowed. A high temperature differential causes harmful "breathing" and volume changes. Humidity in storage rooms should be as high as possible. Storage areas may be cooled by blower evaporators, direct contact plates or brine coils.

In smaller plants, the customer enters the refrigerated area and pulls out a lockable drawer to get frozen food. In other locker plants where special construction makes it possible, frozen foods are delivered from the storage area to the customer without the need to enter the cold area.

13-28 TRUCK REFRIGERATION

Truck refrigeration requires special trailer bodies and refrigerating units. Such bodies are 9 ft. to 18 ft. (3 to 6 m) long and use one refrigerating system of 1 1/2 to 2 hp (1120 to 1490W).

Bodies should be light and well insulated. Construction must be sturdy so that constant vibration and rough handling will not destroy the insulating value of the walls. Fig. 13-51 shows a refrigerated trailer. Note the refrigerating system at A

Fig. 13-51. Refrigerated trailer. A—Refrigeration system may be either electric, gasoline or diesel powered. (Carrier-Transicold Co., Div. of Carrier Corp.)

with its instruments and controls.

Various insulating materials have been used. Most trailers have all-metal bodies with insulation thicknesses depending on the application. Foamed-in-place insulation is most often used. Fig. 13-52 shows such insulation being installed in the sides of the truck body.

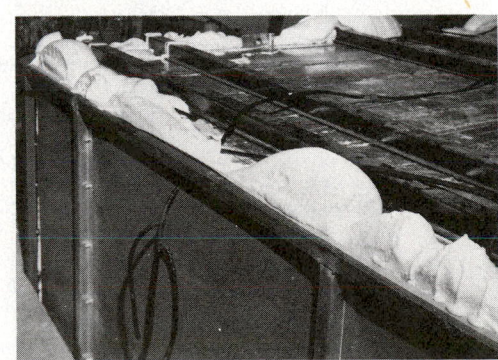

Fig. 13-52. Side of truck body being insulated with plastic foam insulation. Excess foamed-in-place insulation will be trimmed.

STANDBY ELECTRIC MOTOR

A →

POWER TAKE-OFF
FROM TRUCK TRANSMISSION

Fig. 13-53. Truck refrigeration condensing unit. It runs from transmission power take-off while on the road. But a standard electric motor takes over for overnight parking and for cool down. Hydraulic clutch A, operated by a solenoid valve, engages and disengages compressor from power sources automatically.
(Kold-Hold Div., Tranter Mfg. Inc.)

The most interesting thing about trailer refrigeration is the great variety of applications. Each application must be studied before a temperature may be recommended. A truck using dry ice for refrigeration must be insulated for −109 F. (−78 C.). Ice cream trucks must be insulated for −15 F. (−26 C.) while fresh foods require insulation for only 32 to 35 F. (0 to 2 C.) temperatures. Fresh produce, flowers and fruits need accurate control of temperatures, humidity and ventilation due to aspiration of these materials. (Aspiration means the tendency of the materials to lose their fluids to the surrounding air.)

Systems used to produce refrigeration include four main types:
1. Ice.
2. Dry ice.
3. Mechanical.
4. Expendable refrigerant (See Chapter 17).

While ice is still used, its weight and loading requirements make it increasingly less popular.

By contrast, use of dry ice has grown because of this material's lightness, freedom from drainage and servicing problems. However, it is difficult to control temperatures while using it.

Mechanical refrigeration is similar, in most cases, to typical refrigerating units. Major difference is in the nature of the compressor drive. These drives are:
1. Engine-driven electric generator and motor.
2. Driven off transmission shaft.
3. Driven by separate gasoline engines.

Electric generator and motor types use standard voltages, cycles and phases which permit plugging into a wall outlet in the garage if the trailer must be kept cold while off the road.

A transmission power take-off unit, Fig. 13-53, gets its power from the truck motor (through the transmission) when the vehicle is running. When the truck is parked and the engine turned off, a standard electric motor provides refrigeration power.

Units driven by gasoline engines are automatically controlled to start and stop as the system requires.

Some trailers use a dual refrigeration system. Fig. 13-54 shows a refrigerated truck with a refrigerating unit mounted over the cab. This is a standby unit used only when the truck is parked for longer periods. The engine-driven compressor provides refrigeration while on the road.

Fig. 13-55 shows the inside of the standby motor-driven compressor compartment. In addition to this condensing unit, a compressor mounted above (and driven by) the engine may be used. See Fig. 13-56. Most of these systems use a hot gas defrost. In some systems a generator provides power for the evaporator fans.

Typical of a finished installation mounted over the truck cab is the one shown in Fig. 13-57. These units may use plate evaporators as well as blower evaporators. In Fig. 13-58 two

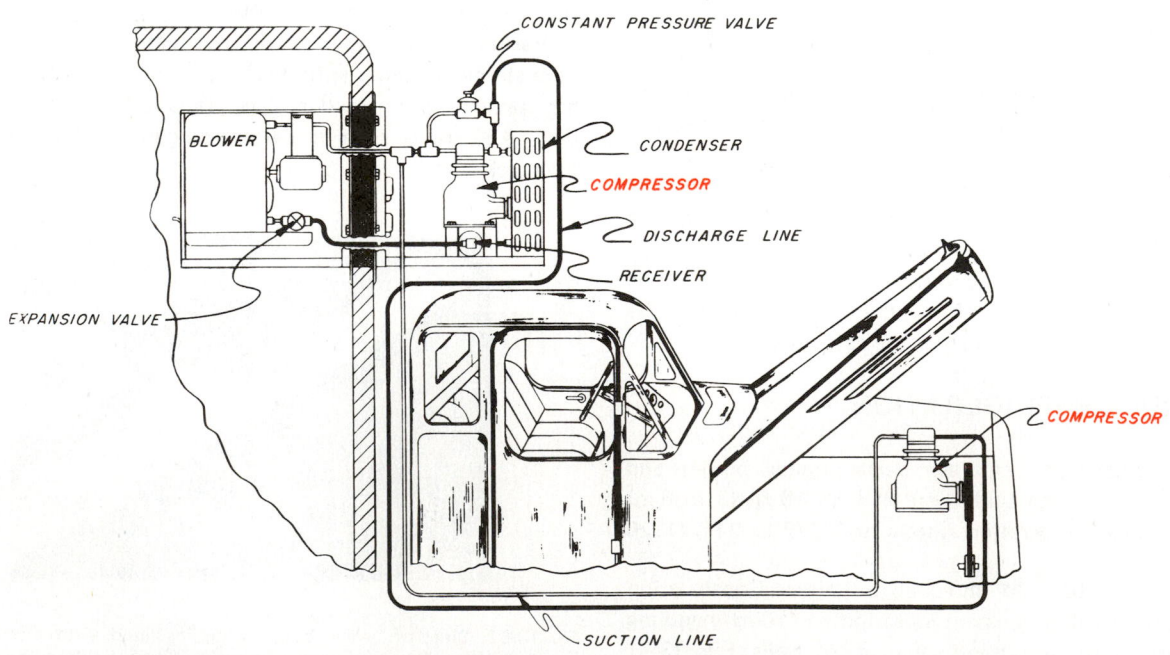

CONSTANT PRESSURE VALVE

BLOWER

CONDENSER

COMPRESSOR

DISCHARGE LINE

RECEIVER

EXPANSION VALVE

COMPRESSOR

SUCTION LINE

Fig. 13-54. Truck refrigerating system with engine-driven compressor and stand-by motor-driven compressor.

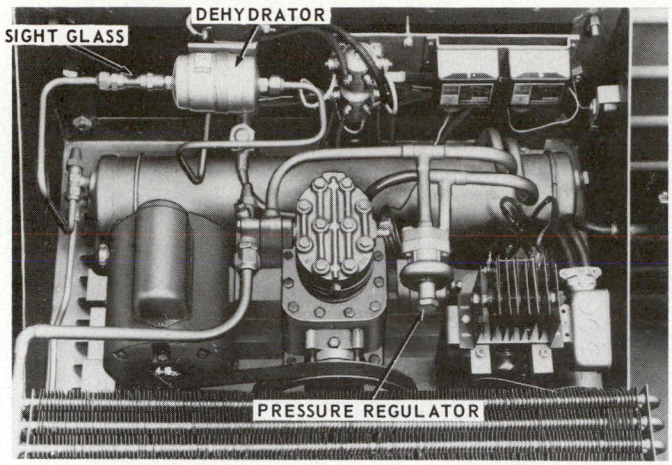

Fig. 13-55. Stand-by truck refrigerating unit.

Fig. 13-56. Truck engine-driven refrigeration compressor.

plate evaporators are used.

Being expendable, liquid nitrogen does not need a condensing unit. It provides excellent low-temperature refrigeration. At atmospheric pressure, it boils at −320 F. or at 140 R. (−196 C. or 77 K.). Temperatures from absolute 0 up to −168 F. (0 absolute to −111 C.) are considered to be the cryogenic range. See Chapter 17.

13-29 RAILWAY CAR REFRIGERATION

Refrigerated railway cars, long used to transport perishable goods, often have mechanical refrigeration. Car construction and insulation is similar to that of the truck trailer.

Two general types of cooling are used on trains — freight refrigeration and air conditioning in passenger cars.

Like truck refrigeration, freight refrigeration involves four methods:
1. Ice.
2. Dry ice.
3. Mechanical (used exclusively on passenger cars).
4. Chemical (expendable refrigerant).

Compressors in mechanical installations are usually driven from the car axle while rail cars are in motion and by an electric motor while cars are in the rail yard.

Some train refrigeration systems use the absorption system. Others have used the steam jet system.

13-30 MARINE REFRIGERATION

Basically, marine refrigeration equipment is the same as for land-type refrigeration. However, refrigerants are restricted to those which are nontoxic and nonflammable. All cabinets are carefully sealed to exclude the possibility of moisture entering the insulation. Refrigerant lines must be installed to permit some vibration and movement without danger of line or joint failure.

Saltwater-cooled units use piping, valves and condenser made of materials designed to minimize the corrosive effects of salt water.

Fig. 13-57. Completed installation of condensing unit on upper front of refrigerated truck body.   (Kold-Hold Div., Tranter Mfg. Inc.)

Fig. 13-58. Two plate evaporators installed in refrigerated truck body. Note hanger devices and refrigerant lines.

Commercial Systems Applications / 485

Fig. 13-59. A snow making machine in operation.　(Larchmont Engineering)

While some automatic refrigerating equipment is used aboard ship, many of the machines are semiautomatic and are interconnected so they can be used either together or separately.

Thus capacity can be adjusted to widely varying loads. R-12 and R-22 are popular as refrigerants aboard ship.

13-31　SNOW MAKING

Many ski slopes add to the natural snow on their runs with artificial snow. Artificial snow is a water spray into which compressed air is added so that the water is broken up into fine mist and freezes very rapidly. See Fig. 13-59.

The low temperature required for snow making is not created by refrigeration mechanism but comes from the surrounding atmosphere. The temperature must be 30 F. (−1 C.) or lower although the maximum temperature which may be used depends on the relative humidity.

If relative humidity is low, snow may be made at temperatures as high as 35 F. (1.7 C.). This is possible because, with low humidity, some of the water evaporates and, in doing so, absorbs heat. This reduces the temperature of the water droplets below the freezing temperature and ice crystals (snow) are formed.

In general, the ground must be frozen before snow making is practical. Usually, manufactured snow is used only in addition to natural snow.

13-32　CRYOGENIC APPLICATIONS

Cryogenics is a branch of physics that relates to producing very cold temperatures. It covers refrigeration in the range of −250 F. to −460 F. or −157 C. to −273 C. (0 K.).

Such cold temperatures are used to liquefy and separate gases such as oxygen, nitrogen and argon. They are also used to liquefy hydrogen and helium.

Three methods are used to create these low temperatures:
1. Expansion process with heat exchangers (Joule-Thompson process).
2. Expansion process with heat exchangers and with gases performing work.
3. Multiple cascade system.

Cryogenics may be used to condense and solidify obnoxious and harmful gases from exhaust products of industrial processes. One example is removing sulphur and ammonia gases from the exhaust gases from burning coke.

Cryogenic techniques are used in the propulsion of liquid fuel rockets and for preserving tissue, blood and whole organs.

Special care must be taken when operating and servicing cryogenic equipment. Liquids are ultra cold and will severely injure anyone coming in contact with them. Certain parts of the system are also ultra cold and could also cause injury.

Caution: The service technician should always wear goggles and gloves when servicing any part of cryogenic equipment. Temperatures being very low, liquid refrigerant will instantly freeze any part of the body it is allowed to touch.

In the presence of such low temperatures, special materials must be used in construction. These include nickel (stainless) steels, copper alloys and aluminum metals. Electrical devices must be of special design and construction.

Lubrication, too, is a special problem because of the extreme cold. Either there is no lubricant at all or a dry type is used (molybdenum sulphide for example).

13-33　REVIEW OF SAFETY

The best way to protect oneself and the public is to check the unit before working on it to be sure the mechanism and cabinet are installed according to local and national fire, electrical, plumbing, building and safety codes.

Move cabinets on rollers. Avoid pinching hands or allowing the cabinets to fall on the feet. Wear safety shoes.

Floor load limits should be checked before heavy commercial equipment is installed. Wiring and plumbing must be adequate, securely mounted and of correct materials. Remember, one must call attention to code violations which might result in serious damage to either the physical structure or to personnel.

Board of Health regulations apply to all equipment which is used around food and beverages. These regulations must be carefully followed. Moving of large commercial cabinets calls for extreme caution. Use handling equipment appropriate for the job and always work carefully.

13-34 TEST YOUR KNOWLEDGE

1. What cabinet temperatures are recommended for the following commercial installations: water-cooling, a walk-in cooler, a florist's cabinet and an ice cream cabinet?
2. Why is it necessary that the doors and windows in commercial cabinets be airtight?
3. What is the purpose of dead air spaces formed between the three and four-pane windows in these cabinets?
4. How is moisture kept from between the glass in multiple pane windows?
5. Why are display case lights sometimes located outside the case?
6. If a thermometer were mounted in a walk-in cabinet, where would the average temperature be found?
7. Why are many commercial refrigeration installations of the multiple type?
8. To what temperature should water be cooled:
   a. If the bubbler were located in the heat treating room of a tool and die factory?
   b. In an office building?
9. What is a wall-mounted refrigerating unit?
10. Why must the humidity be kept high in produce storage cabinets?
11. How is cold air kept from spilling from open display cases?
12. What is a double-duty display case?
13. Where are evaporators located in open display cases?
14. How are frozen foods open display case evaporators defrosted?
15. What are the two basic types of ice makers?
16. What should the malted milk temperature be in a freezing dispenser?
17. Why must milk be cooled immediately after it has been taken from the cow?
18. Are evaporators sometimes mounted outside the refrigerator cabinet?
19. What is the purpose of an activated carbon air filter?
20. How are certain parts of a frozen foods case kept from sweating?
21. How is clear ice manufactured?
22. Why is a high-velocity blower evaporator permissible when cooling bottled beverages?
23. How can electric resistance heat be used in making ice cubes?
24. Why do some trucks have two refrigeration compressors?
25. Why is floor loading important in commercial applications?
26. How many different air streams are common in a frozen foods case?
27. Some water coolers serve two other purposes. What are they?
28. How often should an ice maker be cleaned?
29. What two parts of a freezing dispenser for soft ice cream are refrigerated?
30. Is food heated while being freeze dried?

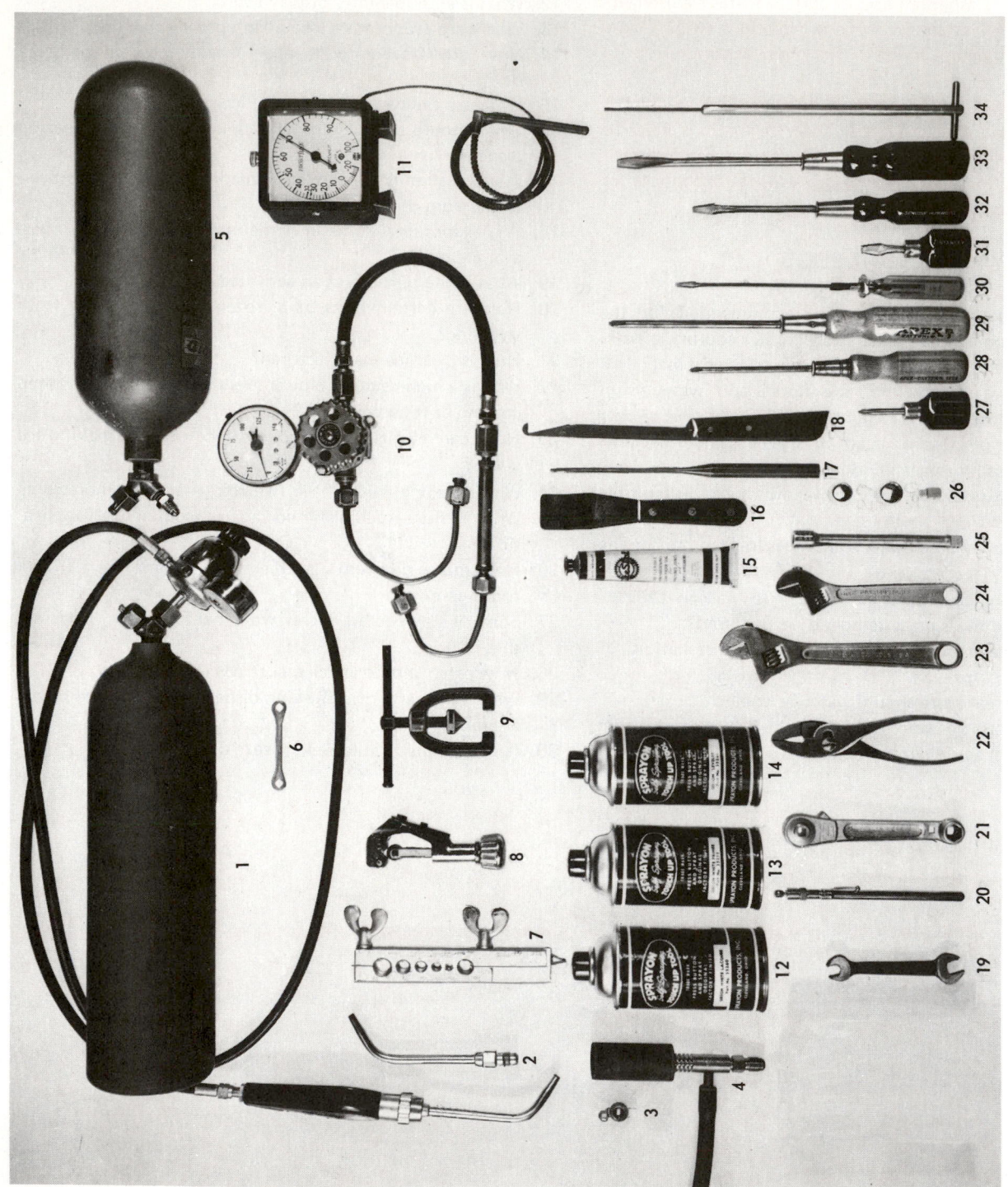

Fig. 14-1. The refrigeration service technician's basic tool kit. 1, 2, 3, 4—Soldering and leak detecting equipment. 5—Refrigerant service cylinder. 6—Cylinder wrench. 7, 8, 9—Tube flaring tools. 10—Compound gauge and connectors. 11—Thermometers. 12, 13, 14—Spray cans of touch-up paint. 15—Sealing material. 16—Scraper. 17—Pin punch. 18—Knife. 19—Open end wrench. 20—Thermometer. 21—Valve stem ratchet wrench. 22—Pliers. 23, 24—Adjustable wrenches. 25, 26—Socket wrench set, 27 through 33—Screwdrivers. 34—Allen setscrew wrench. (Gibson Appliance Co.)

Chapter 14

COMMERCIAL SYSTEMS INSTALLING AND SERVICING

Installing and servicing commercial units is a very important part of the refrigeration industry. If equipment fails, companies that own equipment may suffer severe equipment losses, perishables may spoil and goodwill may be lost.

This chapter covers the installation of both code and noncode systems.

Servicing information in this chapter deals mainly with multiple systems: what goes wrong and how to make repairs. Much of the servicing is similar to that described for domestic refrigerators and the information given in Chapter 11 may be used to a considerable extent. Tools and instruments listed in Chapters 2 and 11 can also be used for commercial servicing.

Servicing involves locating the trouble, removing the part from the system, repairing and reinstalling it.

14-1 TYPES OF COMMERCIAL INSTALLATIONS

Commercial refrigeration installations vary considerably. Following are the classifications of installations commonly used:
1. Self-contained unit installations.
    a. Hermetic.
    b. Conventional.
2. Remote condensing unit installations.
    a. Single cabinet.
    b. Multiple cabinet.

An installation may be as small as a 1/20 hp self-contained unit or as large as a 140 hp unit. Larger units must be assembled on the premises.

Multiple units must be installed to handle the refrigeration load efficiently while eliminating hazardous conditions that might lead to accidents. Two major concerns should be long life and neatness. Many cities have laws and codes covering certain refrigeration systems. Furthermore, most refrigeration manufacturing companies have rules for installing their own equipment.

Where cities and rural communities are not restricted by code, installations are usually made at minimal cost. One must, therefore, consider two types of installations:
1. Noncode installation.
2. Code installation.

Where there is no local code, it is good practice to install units according to the code of the closest city.

14-2 NONCODE INSTALLATIONS

Definite procedures should be followed when assembling a system. This safeguards against mistakes and reduces errors through carelessness. Refrigeration service departments claim that carelessness causes more than 90 percent of the servicing difficulties.

Assume the units are of a size that will efficiently handle the heat load. The installation must then take full advantage of the design. Tubing, safety valves and protective devices should be placed into the system where they will produce operating efficiency, permanency and safety.

All refrigeration installation and service work should be performed with correct tools. Fig. 14-1 shows some that are commonly used. Chapters 2 and 11 describe in more detail such tools and supplies.

14-3 INSTALLING CONDENSING UNITS

First the technician must determine where to place the condensing unit with respect to the cabinets and cases. This location should be as close to the cabinets as possible. A central location is best.

Installation should progress in the following order:
1. Put cabinets in place.
2. Locate place for condensing unit and install it.
3. Install evaporators.
4. Install valves and controls.
5. Install tubing.
6. Check for leaks.
7. Dehydrate installation.
8. Charge system.
9. Start system.
10. Check operation of unit and get 24-hour temperature and pressure records of unit in operation.

Cabinets are bulky and sometimes difficult to handle. A dolly, as shown in Fig. 14-2, is handy for cabinet moving.

It is recommended that the condensing unit be put in the basement or in a room next to the one in which the cabinet is located. *Avoid putting the condensing unit where it will be exposed to the sun or to low or freezing temperatures.* Locating condensing units in the same room with the counters and cabinets is not recommended because of the heat and

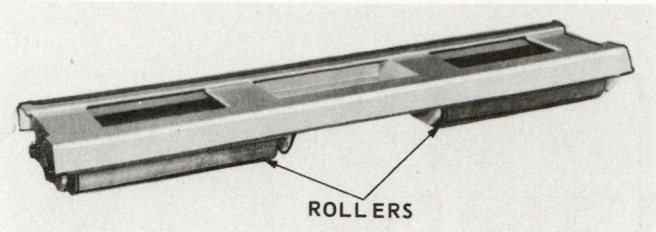

Fig. 14-2. One or more of these dollies under cabinet enables a service technician to easily move and place heaviest and bulkiest cabinets. Dolly shown will carry four ton. (Airserco Mfg. Co.)

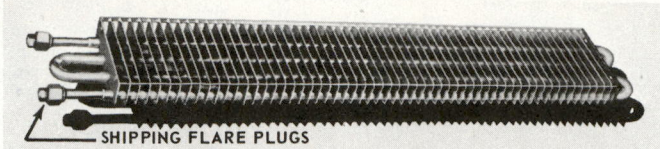

Fig. 14-4. Shelf evaporator. Aluminum fins and copper tubing are mechanically bonded together. Fins are offset. (Peerless of America, Inc.)

noise they produce. There will be some vibration produced by the running condensing unit. Fig. 14-3 illustrates a vibration absorbing type compressor mounting. Protect the condenser by putting a heavy wire cage around it and over it. Install a valve and accessory board on the wall just above the condensing unit to support valves, drier motor controls, two-temperature valves, electrical boxes and service instruction cards.

In some cases, the condenser is installed away from the compressor too. In these instances, traps are sometimes used in the compressor discharge line to keep the oil in the condenser (slant discharge lines down also) and away from the compressor discharge valve. See Fig. 15-80.

14-4 INSTALLING EVAPORATORS

Evaporators for use in commercial refrigeration installations should be carefully mounted, leveled and firmly fastened in place. Type of mounting depends on evaporator type. Fig. 14-4 shows a display case evaporator.

Blower evaporators are usually equipped with hanging brackets, as in Fig. 14-5. Evaporators are usually fastened to the ceiling in florist cabinets and walk-in coolers. A plumb line is used to locate the holder positions. A cardboard template is useful for locating the mounting brackets.

Fig. 14-5. Small blower evaporator. A—Drain connector. B—Drain pan. C—Hanger. Evaporator must be mounted level for efficient operation and good drainage. (Peerless of America, Inc.)

The hanger is attached to the ceiling of the cabinet. Hydraulic or pneumatic adjustable-height platforms lift and hold the evaporators into place until they can be fastened. Regardless of the devices used to mount the evaporator, the unit should be rechecked afterward, to make sure it is level.

All natural convection evaporators should be properly

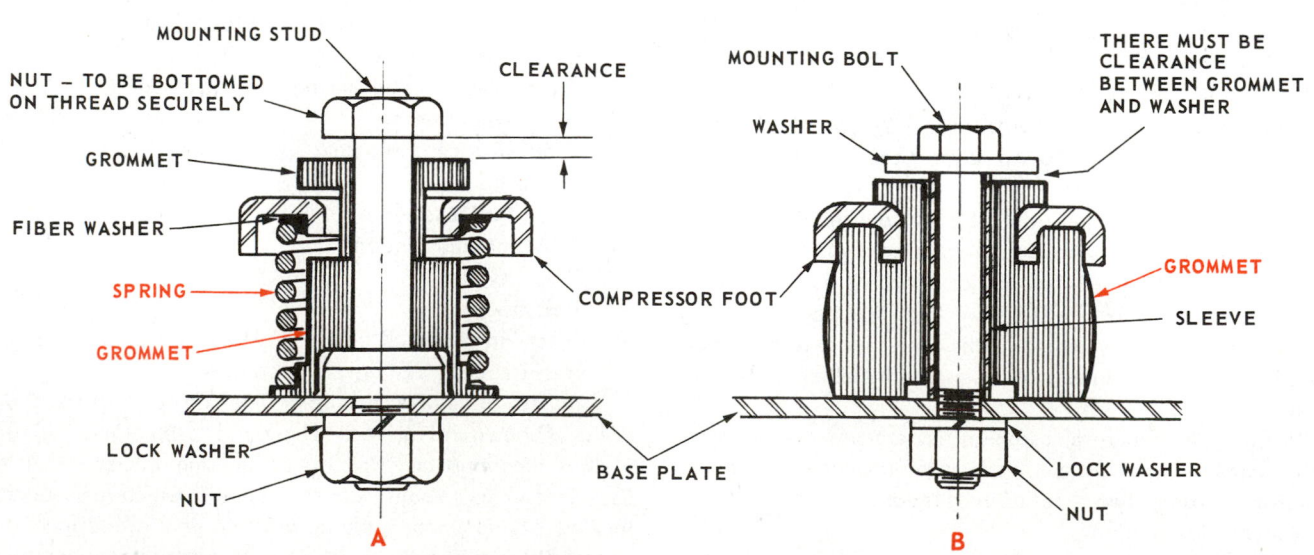

Fig. 14-3. Hermetic motor compressor mountings designed to absorb vibration. A—Synthetic rubber grommet and spring. B—Synthetic rubber grommet only. (Tecumseh Products Co.)

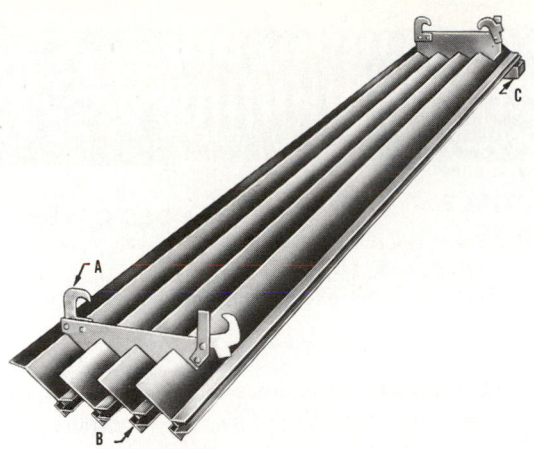

Fig. 14-6. Evaporator baffle and drain pan. It is compact, permits good circulation and condensate drainage. A—Hangers. B—Individual baffle drains. C—Collector drain.

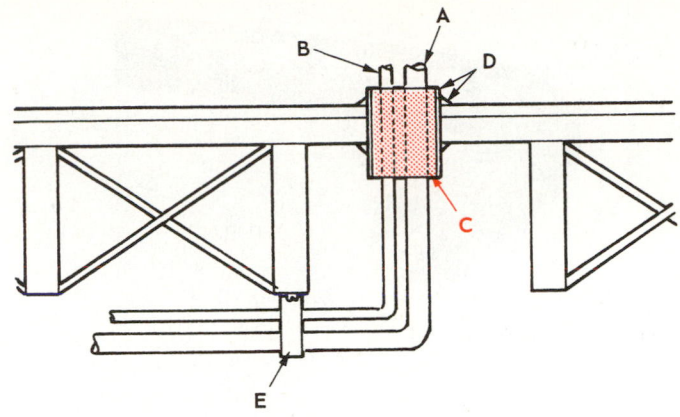

Fig. 14-8. Sleeve protects tubing as it goes through wall or floor. A—Suction line. B—Liquid line. C—Sleeve. D—Sealing compound. E—Clamps.

baffled. Fig. 14-6 shows one type baffle and the hangers that fasten it to the evaporator.

Display counter evaporators are usually supported by stands. These stands or brackets should be provided with leveling adjustments in all directions.

14-5 INSTALLING TUBING

Tubing in noncode installations is usually run along the walls and ceiling with supports at intervals frequent enough to keep tubing straight and firmly in place. Special clamps are available as tubing fasteners but a galvanized 1/2 in. conduit clamp, as shown in Fig. 14-7, is sufficient for most situations. The tubing should be insulated from these clamps or protected from them. Short wrappings of plastic tape will prevent chafing and galvanic action (an erosion or eating away of material caused by two different metals touching in moist air).

Where tubing runs through a floor or wall, it should be protected by short runs of conduit or flexible metal tubing (Greenfield). The ends should be sealed with a sealing compound to avoid chafing and other troubles. See Fig. 14-8. In all cases, tubing should be run horizontally and vertically with neat bends.

The liquid line presents no difficulties as to slope and

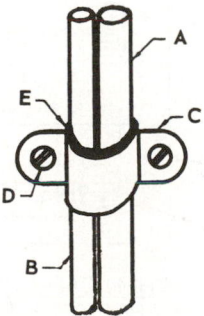

Fig. 14-7. A method of fastening the tubing to the wall. A—Suction line. B—Liquid line. C—Clamp. D—Wood screw. E—Plastic tape.

position, but suction lines must be sloped to drain toward the compressor. Low spots in the suction lines will accumulate return oil which may eventually form a liquid slug in the tubing. Such a slug carried to the compressor will cause disturbance in the crankcase and may cause temporary oil pumping.

If tubing must slant upward from the evaporators to the condensing unit, the run should be made with a steady downward slant of the tubing to a certain point. Then a short U-bend (two street ells) should be made. The U-bend will act as an oil trap, insuring a positive return of oil to the compressor.

Never run tubing near sources of heat, such as hot water lines, steam lines or furnaces. Heat will reduce efficiency.

Copper tubing comes in coils usually of 50-ft. lengths. It is dehydrated and sealed at the ends by the manufacturer. In the average small commercial installation, 1/4 in. tubing is used for the liquid line and 1/2 in. for the suction line. See Chapter 15 for correct line sizes.

During installation, tubing should be kept clean. Never should it be put aside with ends open if it is not to be used for a period of five minutes or more. A practical method of installing tubing is to uncoil about 10 ft. of it at a time, unrolling the coil along the floor. Then run the tubing up through floor openings from underneath, gradually working it into place.

Quarter-inch tubing should not be difficult to install. But 1/2 in. diameter must be carefully handled to prevent buckling when it is bent. A tube bender should be used. See Chapter 2.

In a noncode installation where individual suction lines and liquid lines run to main lines, T-fittings may be used and the valves for shutting off the individual evaporators may be located near the evaporator.

Valves, driers or other heavy objects should not be supported by tubing. These items should be mounted on the wall or some other support. These connections may be of the SAE flare type or of streamline soldered (sweat) fittings.

Newly installed tubing should be sealed immediately after flare or streamline connections are made to keep it clean. The tubing should be attached permanently to the supports along

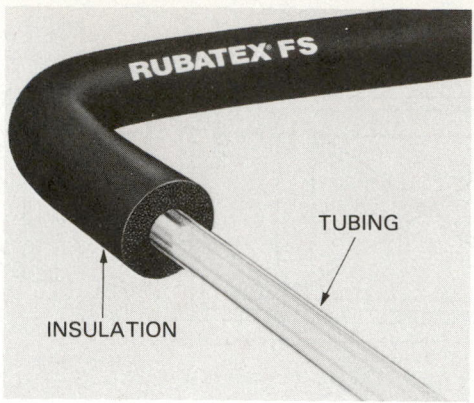

Fig. 14-9. Rubber insulation mounted on suction line. Insulation is usually installed before tubing or pipe connections are made. (Rubatex Corp.)

which it runs.

Always try to run tubing so that the support will protect the tubing from accidents. Many service technicians use a sponge rubber protective covering over the tubing which serves both as a protector and as an insulator, Fig. 14-9. This covering must be placed on the tubing before assembly unless the insulation is split.

At tubing connections or where valves have been installed, it is common practice to cover them with insulating tape, as in Fig. 14-10. Be careful to place the tubing where it will not be

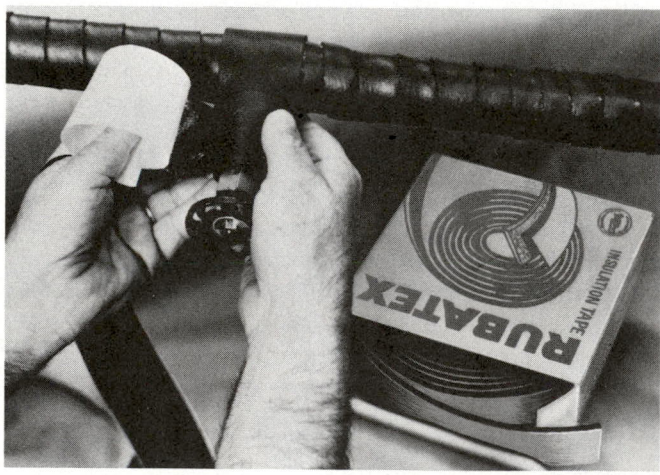

Fig. 14-10. Foamed plastic insulating tape is wrapped around valve to prevent sweating or frosting. (Rubatex Corp.)

damaged by handling of articles in the room. Do not put loops or unsupported bends in the tubing except at the condensing unit.

Install a drier and sight glass in the liquid line at the condensing unit. A vibration damper, as shown in Fig. 14-11, should be included in both lines. One horizontal loop of soft copper tube may be made in the suction and liquid lines if a

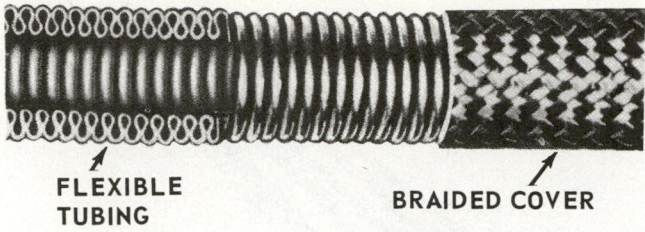

FLEXIBLE TUBING   BRAIDED COVER

Fig. 14-11. Flexible tubing used for vibration dampening. One of these is usually mounted in suction line near condensing unit.

suitable vibration damper is not available. Carefully study the manufacturer's installation and service procedures to make sure that every assembly and adjustment is correct.

14-6 ELECTRICAL CONNECTIONS

Refrigeration units must have enough electrical power of the correct type to operate motor, controls and solenoids. Chapters 6, 7 and 8 explain electrical fundamentals and electrical power variables.

Before installing a condensing unit, be sure the electrical voltage, cycle and phase of the compressor motor are the same as the electrical power source. Smaller units — 1/8 to 1/4 hp — usually use 120V single-phase a-c. Medium size units — 1/4 to 1 hp — may use 208V or 230V single-phase or

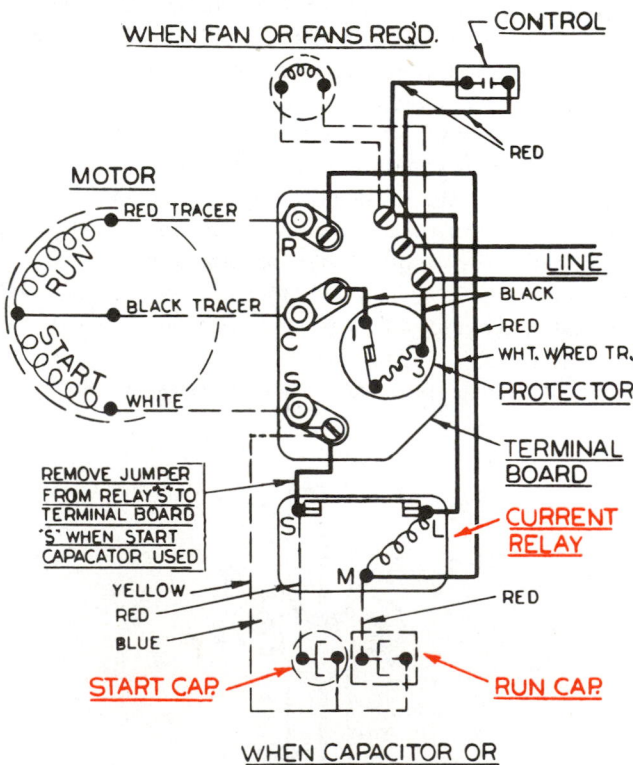

Fig. 14-12. Wiring diagram of single-phase unit using current relay, run capacitor and start capacitor. (Copeland Corp.)

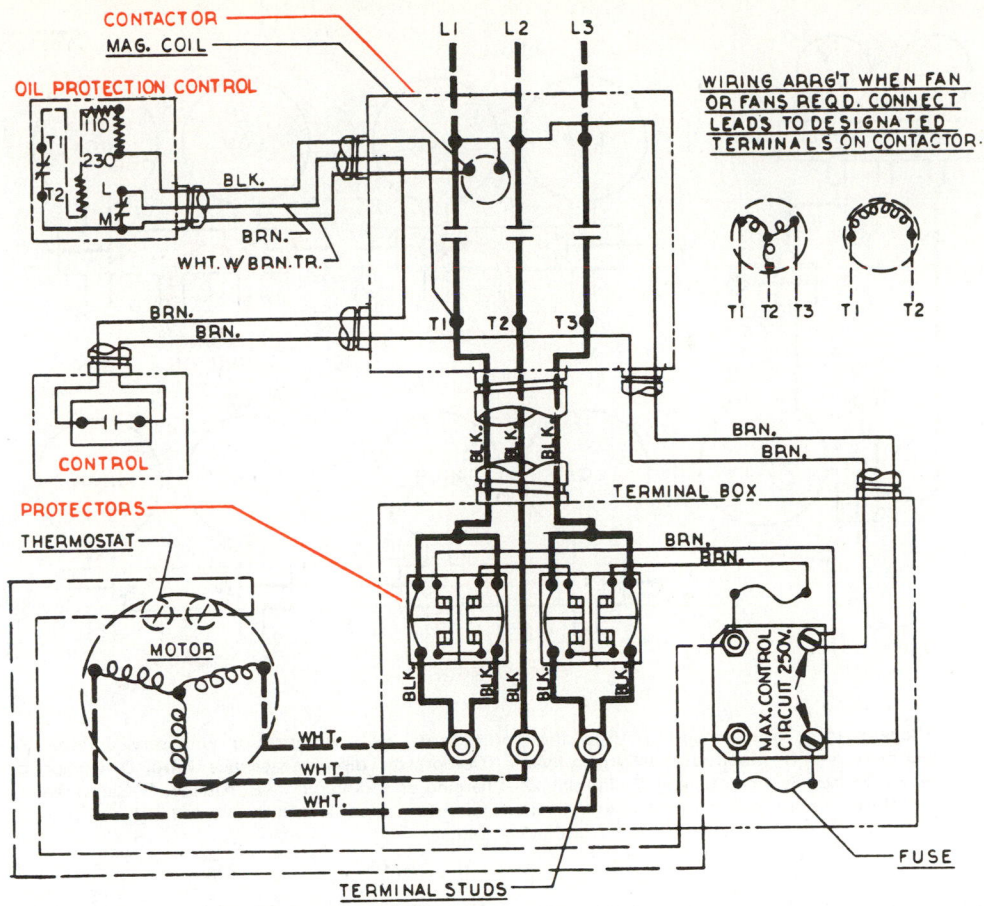

Fig. 14-13. Wiring diagram of three-phase motor refrigeration unit. Note contactor, control, oil protection control and motor circuit protectors.

three-phase a-c. Larger units may use 230V or 440V three-phase a-c.

Fans — both evaporator and condenser — must be checked to be sure the identification plate data matches power available. This is true as well for solenoid valves and controls.

Wire size is important. Wire should have a capacity 50 percent over the load it will carry.

Condensing units have wiring diagrams either fastened to them or supplied in the shipping crate. Fig. 14-12 shows a typical wiring diagram for a single-phase circuit. Fig. 14-13 shows a three-phase refrigeration system wiring diagram. Correct connections are extremely important. Do not turn on the electrical power until the circuits are correct and all connectors are clean and tight.

Overloaded electrical circuits are dangerous. They may cause unit burnouts or electrical wiring fires.

A complete condensing unit installation is shown in Fig. 14-14. Use of an ohmmeter is strongly recommended to check all circuits for continuity before turning on the power.

14-7 GAUGE MANIFOLD

A big help to service technicians is the service gauge and testing manifold. With this piece of equipment, they are able to check low and high-side pressures, charge and discharge a

Fig. 14-14. Typical condensing unit installation. Note vibration damper, drier, sight glass and electrical installation. (Geo. L. Johnston Co.)

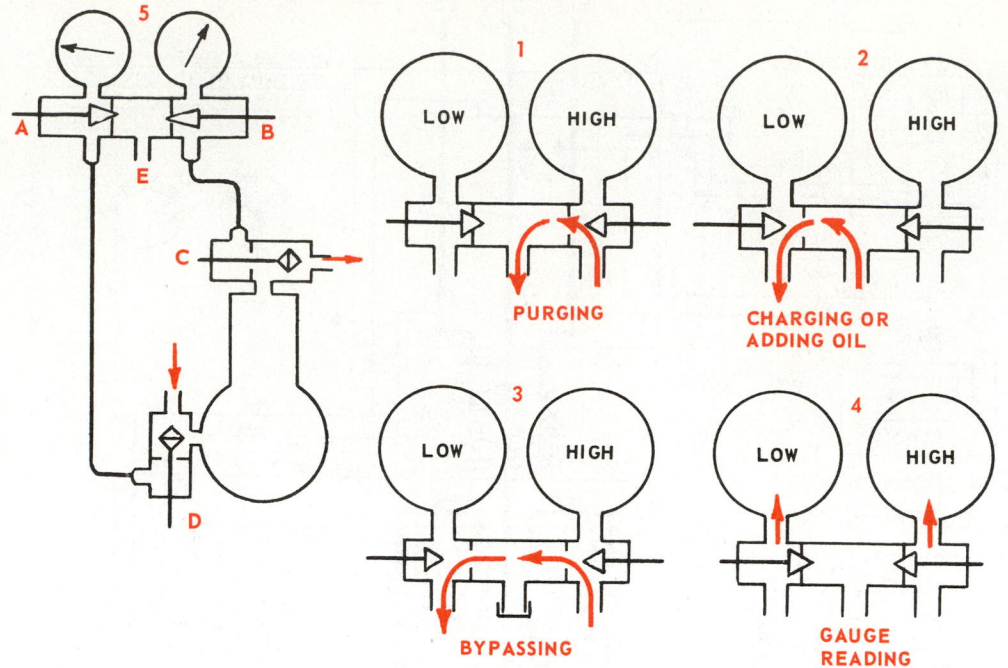

Fig. 14-15. Schematic of gauge manifold installation on external drive compressor with service valves. A—Manifold suction valve. B—Manifold discharge valve. C—Compressor discharge service valve. D—Compressor suction service valve. E—Service opening. 1—Purging. 2—Charging or adding oil. 3—Bypassing. 4—Gauge reading. 5—Both manifold valves are turned all the way in. System is pumping vapor and both low and high-side pressures are being read.

system, check pressures, add oil, bypass the compressor, unload gauge lines of high-pressure liquid and vapor as well as perform many other operations without replacing regular gauges. See Chapter 11 for gauge manifold information.

A typical manifold, as shown in Fig. 14-15, has two gauge openings, three line connections and two shutoff valves that separate the outside connections from the center line connection. Fig. 14-16 shows a compound gauge having a pressure scale and three different refrigerant temperature scales (for R-12, R-22 and R-502).

The manifold has 1/4-in. square drive valve stems or is equipped with handwheels. The three line attachment fittings are usually 1/4-in. MF (male flare).

In Fig. 14-15, a 1/4-in. copper tubing or a flexible line connects the manifold to the SSV (suction service valve) shown at D and the DSV (discharge service valve) at C. Because most service valves on the compressor have 1/8-in. FP (female pipe) gauge openings, two 1/8-in. MP (male pipe) by 1/4-in. MF (male flare) half unions are installed in the service valves. Be sure the compressor service valve stems are turned all the way out and that the outside of the valve is cleaned before removing the pipe line plugs and installing the fittings.

Lines from the manifold are attached to these fittings. The line attached to the SSV at D should be left one to two turns loose while the line to the DSV should be tightened. Then open both the manifold valves at A and B 1/4 to 1/2 turn and cap the middle opening, E.

Now turn the DSV C stem in 1/8 to 1/4 turn for just a moment (crack the valve). A surge of high-pressure refrigerant will then rush through the lines and the manifold and purge to the atmosphere at the loose connection at D, the SSV. This connection may then be tightened. Purging is necessary to remove air and moisture from the manifold and lines.

Carefully test for leaks while the manifold and its lines are under high pressure. **Correct any leak immediately.**

Various service and testing operations may be performed after the testing manifold has been installed:

1. Observe operating pressures by:
   Closing valve A by turning all the way in.
   Closing valve B by turning all the way in.
   Cracking open back seat of valve C.
   Cracking open back seat of valve D.
2. Charge refrigerant into system by:
   Connecting refrigerant cylinder to E (vapor only).
   Opening valve A.
   Closing valve B.
   Closing front seat of valve D slowly.
3. Purge condenser by:
   Closing valve A.
   Opening valve B.
   Cracking open valve C.
4. Charge liquid refrigerant into high side by:
   Connecting refrigerant drum to E.
   Closing valve A.
   Opening valve B.
   Mid-positioning valve C.
5. Build up pressure in low side for control setting or to test for leaks by:
   Sealing E with seal cap.
   Opening valve A.

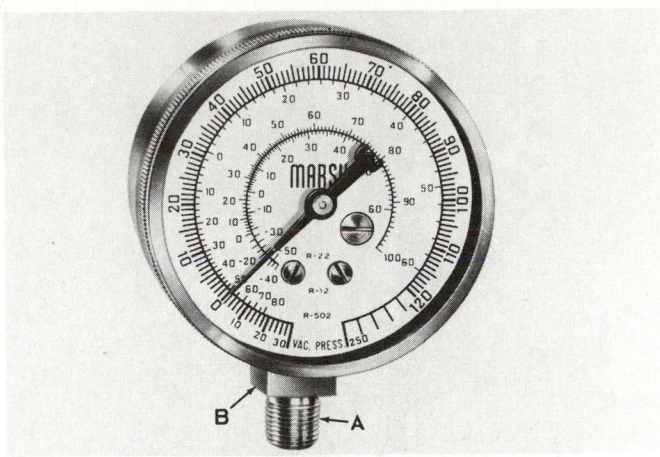

Fig. 14-16. A compound gauge 30 in.—0—120 psi with a retarded scale to 250 psi. Notice three refrigerant temperature scales. These are for R-12, R-22 and R-502. A—Connection is 1/8 in. male pipe thread. B—Wrench flats.   (Marsh Instrument Co.)

Opening valve B.

Back seating then crack open valve C.

Mid-positioning valve D.

6. Charge oil into compressor by:

Connecting oil supply to E.

Opening valve A.

Closing valve B.

Turning valve D all the way in.

After completing service operations, the manifold is removed from the system. This must be done without losing refrigerant or admitting air. Turn the DSV at C all the way out. Then open both manifold valves A and B 1/4 to 1/2 turn. This arrangement will move all the high-pressure refrigerant from the line and the high-pressure gauge and put it into the low side. Now turn the SSV stem at D all the way out and turn both manifold valve stems all the way in. Remove the lines from the service valve. Use soft synthetic fittings for finger tight connections. See Fig. 14-17. Remove the fittings from

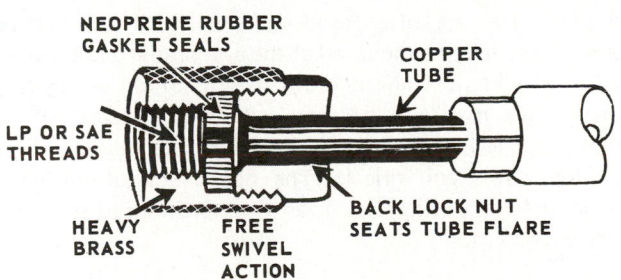

Fig. 14-17. Speed coupling used on charging and purging lines. Synthetic rubber gasket produces leakproof joint when connection is finger tightened.   (Wabash Corp.)

the service valves and install the service valve gauge opening plugs and tighten them. Immediately plug the lines and all other openings on the manifold to keep out dirt, moisture and air. See Fig. 14-18.

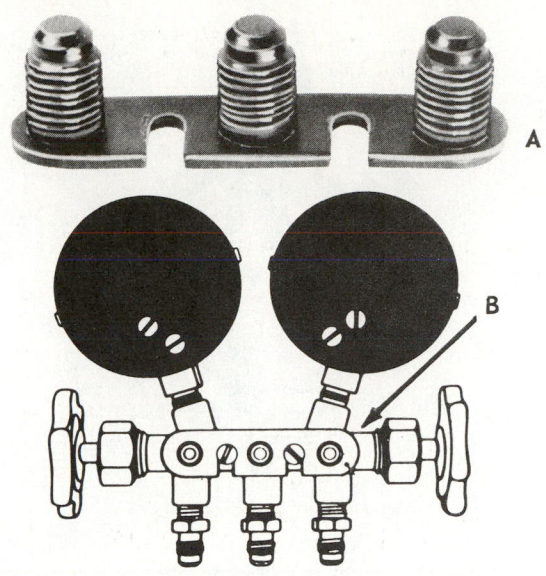

Fig. 14-18. Gauge manifold equipped with three flare plugs. Plugs hold ends of three-gauge manifold flexible lines when manifold is not in use. A—Service line mounting fitting. B—Fitting mounted on manifold. This method is used to keep service lines clean. (Robinair Manufacturing Corp.)

14-8 SERVICE VALVES

In systems equipped with service valves, the valves must be leakproof where the valve stem goes into the valve. The packing is usually made of asbestos, lead and graphite. Packings differ with different valve designs. When replacing them, one must be sure the proper packing is used.

A replacement service valve is shown in Fig. 14-19. It has a slotted flange so the service technician may fit it to a great variety of compressor flange sizes.

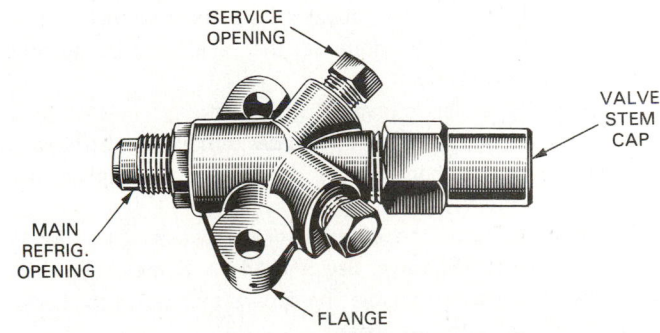

Fig. 14-19. Compressor service valve with slotted flange holes and protective cap over valve stem.   (Mueller Brass Co.)

Practically all service valves have a drop-forged brass body and a steel stem. Stems have a tendency to rust and score the valve gland or packing. Always clean and oil a valve stem before turning it. See Fig. 14-20.

A good way to lessen this corrosion, especially in damp locations, is to fill the valve body with clean and dry

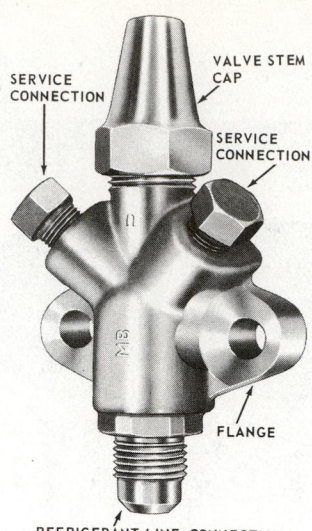

Fig. 14-20. Refrigeration unit service valve with two service openings, two-bolt flange and valve stem cap.

refrigerant oil before replacing the plug each time the service valve is used. This oil should be the specified refrigerant oil for that machine.

Service valves on commercial installations must be kept in good condition. The technician can do three things which will assure good service and valve life:

1. Fit the wrench to the valve stem.
2. Maintain the packing so that the service valve will not leak.
3. Oil the threads of the gauge connections each time gauges are used.

Occasionally, after a period of use, these service valves must be replaced. After flexible line fittings have been mounted in the gauge opening of the valve a number of times, the pipe threads in the valve gauge openings may become worn and leak. If the fittings inserted in these gauge openings are given a thin coat of solder, this trouble can often be eliminated.

When cracking the valve, always use a fixed wrench (not a ratchet wrench). This is done so the valve may be quickly closed again if necessary.

Occasionally a service valve will be found in such bad condition as to be useless. In such cases remove the refrigerant or isolate it in another part of the system and replace the valve.

The third valve of the external drive system is the liquid receiver service (LRS) valve. See Fig. 14-21. Some of the LRS valves are three-way to enable the service technician to charge liquid refrigerant into the system.

When using a system service valve, remove the valve cap — if the valve has one — loosen the service valve packing nut one turn and then clean the valve stem before turning the valve stem. Then turn the valve stem back in about 1/16 of a turn. Tighten the packing nut and replace the valve stem cap.

The purpose of turning the valve stem back in slightly is to prevent the valve from "freezing" against its seat. Such a condition sometimes leads to broken valve stems.

When installing the gauge opening plug, tighten the plug firmly. Never tighten a cold gauge plug into a hot service valve.

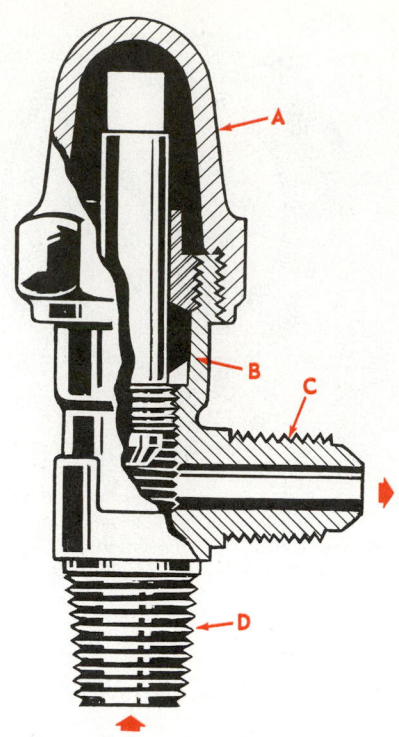

Fig. 14-21. A liquid receiver service valve. A—Cap. B—Packing. C—Connection for liquid line (SAE flare). D—Connection to liquid receiver (pipe threads). (Superior Valve Co.)

This may result in freezing of the plug to its seat. When using a service valve wrench on these valves, apply the turning force gradually. Adjustable end or fixed open end wrenches are not recommended for service valve stems. Only the special socket wrenches — sometimes called keys — are to be used.

If the gauge plug is "frozen" in the service valve, it can be loosened by first heating the outside of the service valve body with flame from a torch. *Be careful not to overheat!* This heating will cause the valve body to expand and will weaken the body thread grip on the plug. The wrench can then be used to loosen the valve stem.

Access valves are often used on commercial refrigerating systems. They may be mounted at the evaporator outlet or the liquid line inlets just ahead (downstream) of the refrigerant control. They may also be mounted on both sides of the automatic valves in the system (solenoid, bypass valves, hot gas defrosting valves and driers). The old policy of having a minimum of connections has given way to the need for more convenient service.

14-9 TESTING FOR LEAKS

After a system is assembled, it must be checked for leaks. Before trying to locate leaks, build up a pressure in all parts of the system. Two methods may be used:

1. Using an inert gas.
2. Using refrigerant under pressure.

In case a low-pressure refrigerant is used, or if the local code specifies a pressure test above the refrigerant's vapor pressure, some other gas may be used for testing. Carbon dioxide, nitrogen or argon are satisfactory. However, the pressure may be dangerous. See Para. 11-39.

Caution: Never use oxygen, air or any flammable gas for this purpose. An explosion may take place.

This testing should include the liquid line, suction line and all other parts installed by the installer. The only exception is the condensing unit itself. It has been pressure tested at the factory. Install a high-pressure gauge only. (A compound gauge may be ruined by the pressure.) After building up a medium (30 to 100 psi) pressure, close the cylinder valve.

If the pressure gauge shows no drop in pressure after an hour or more, raise the test pressure to 170 psi and test again. Do not exceed the pressures prescribed by the code, as a too high pressure may rupture some part of the system. If pressure shows no decrease during a one to 24-hour period, the system is safe to operate.

Purge the test gas from the system, evacuate by the deep vacuum, two or three-step vacuum method and charge the system. The unit should be ready to operate.

Testing for leaks using the system's own refrigerant is the most common noncode practice. It is convenient; there is no need for an inert gas cylinder and the leak testers are a standard part of each service technician's tool kit.

To do this, proceed as follows.

Install a pressure gauge in the system. The liquid line valve (king valve) should be opened just enough to build up a 15 to 30 psi pressure throughout the system.

Test for leaks using one or more of the following:

1. Soapsuds.

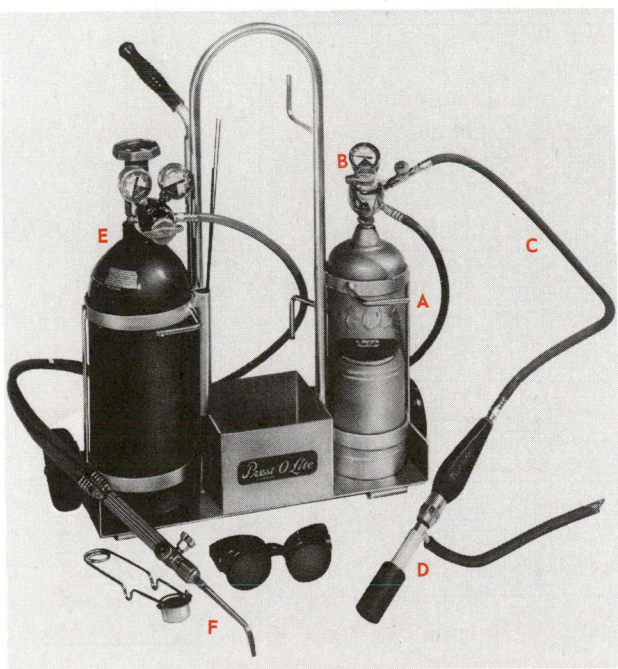

Fig. 14-22. Combination halide leak detector and soldering-brazing unit. A—Acetylene cylinder. B—Regulator. C—Hose. D—Halide leak detector used for checking leaks of most refrigerants. Flame will turn blue-green if there is leak. E—Oxygen cylinder. F—Oxyacetylene welding and brazing torch. (Linde Div., Union Carbide Corp.)

2. Halide torch.
3. Electronic leak detector.
4. Liquid tracer.

Chapter 11 describes these tests. A combination halide torch leak detector, soldering and brazing unit is shown in Fig. 14-22. This apparatus is both an oxygen-acetylene and an air-acetylene unit which can be used for welding, brazing, soldering or leak testing. The acetylene cylinder has a halide leak testing device connected to it.

If no leaks are detected at low pressure, increase system to the full pressure of the refrigerant (vapor only) and again test for leaks.

If a leak is found, purge the system to atmospheric pressure, open the system at the leak point, inspect all parts, replace any defective parts, clean and assemble. If a soldered or brazed joint is leaking: flux, heat and take the joint completely apart. Clean and assemble; then repeat the leak detecting procedure. If no leaks are found, this part of the unit is ready to operate.

When blowing out lines and/or pressure testing with nitrogen or carbon dioxide, use an accurate pressure regulator and a relief valve designed to open at 180 psi. One should not exceed 170 psi for CO_2 or nitrogen, while testing. See Chapter 11 for safe use of high-pressure gases.

Some motor compressor domes are designed to operate under low-side pressure. Tecumseh Products Company warns "Do not exceed 170 psi! Dome may burst!"

14-10 EVACUATING SYSTEM

After it is known that the system is leakproof, remove all air and moisture from the system. Air is pumped out of the lines and the evaporator with a vacuum pump.

Avoid pumping refrigerant vapor into the room where the condensing unit is located. Refrigerant vapors may be harmful to people in the room and will interfere with leak detecting.

Connect a gauge manifold. Open both the discharge service valve and the suction service valve and pump a vacuum on the complete system. Air being removed will be discharged through the vacuum pump.

Fig. 14-23 shows a gauge manifold and a vacuum pump connected to a small commercial hermetic system. See Chapter 11 for evacuating methods.

A 3 cu. ft./min. vacuum pump is large enough for systems up to 10 hp. *Pressure drop is very important! The service lines must be as large and as short as possible.*

It takes eight times longer with a 1/4-in. line as it does with a 1/2-in. line, and it takes twice as long through a 6-ft. line as through a 3-ft. line.

Use heat lamps, electric heaters and blowers to provide heat. Avoid using a torch flame because it may cause local high temperatures which may decompose oil, insulation and refrigerant.

When the pump is shut off (valve closed), pressure will rise a little at once (pressure drop equalizing); take a reading one minute after closing valve and again, 30 minutes later. If there is no pressure rise, it indicates that the system is sealed and is also free of moisture.

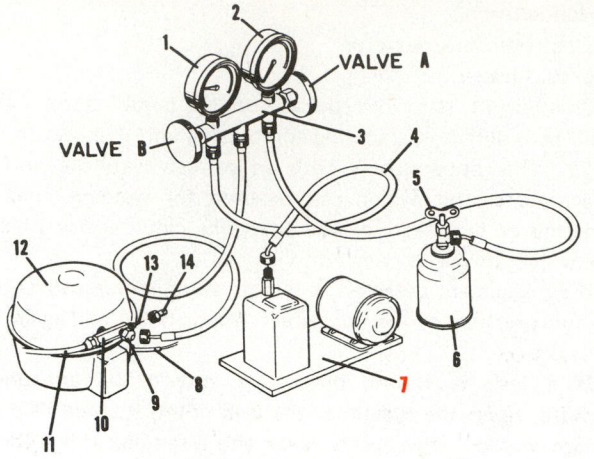

1 COMPOUND GAUGE
2 PRESSURE GAUGE
3 TEST MANIFOLD
4 CHARGING LINES
5 REFRIGERANT CYLINDER VALVE
6 REFRIGERANT CYLINDER
7 VACUUM PUMP
8 COMPRESSOR DISCHARGE LINES
9 SUCTION LINE VALVE SERVICE PORT
10 SUCTION LINE VALVE
11 SUCTION LINE
12 FREEZER COMPRESSOR
13 SUCTION LINE VALVE STEM
14 VALVE SEAL CAP

Fig. 14-23. Gauge manifold, vacuum pump and refrigerant cylinder connected to small commercial hermetic motor compressor.

14-11 CHECKING SYSTEM BEFORE STARTING

If the motor control has not already been adjusted for installation, this should be done before the system is put in operation. The settings of the motor controls will vary with the demands of the cabinets and with various kinds of refrigerant used. See Chapter 8.

Be sure the water is turned on — if it is a water-cooled condensing unit — and check fuses in the electric circuit for proper size.

It is good practice to install recording thermometers, voltmeters and ammeters on a unit for the first 24 to 48 hours of its operation. Records will make adjustments easier.

14-12 CHARGING COMMERCIAL SYSTEMS

There are two basic methods used to charge a system:
1. Low-side method.
2. High-side method.

In the low-side method, charging small quantities of refrigerant into commercial systems is similar to charging domestic machines. It is usually done by charging into the low side (vapor method).

To charge a commercial external drive system equipped with service valves, the storage cylinder should be attached to the gauge manifold, Fig. 14-24. Evacuating and charging apparatus combinations are popular with service technicians. Fig. 14-25 shows a complete unit with charging tube, vacuum pump and vacuum gauge.

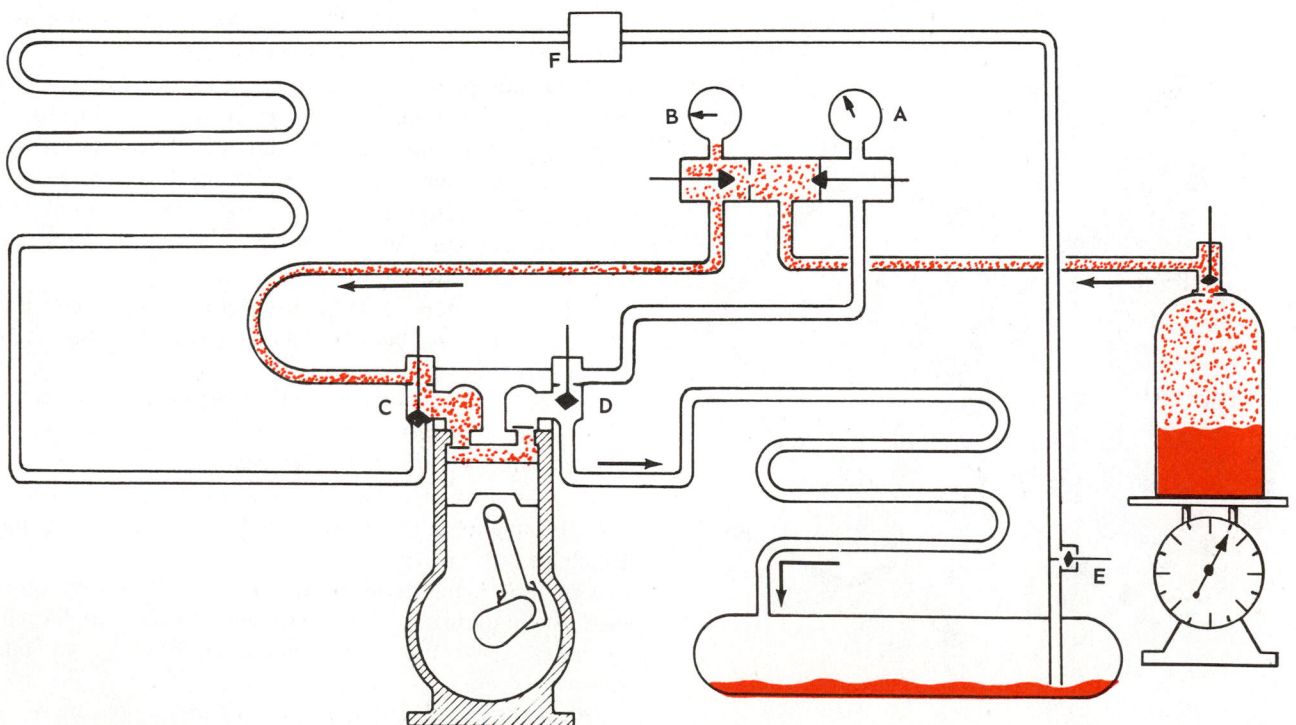

Fig. 14-24. Method of charging small external drive system with refrigerant vapor. Refrigerant cylinder is connected to manifold center opening. After purging charging lines, the SSV is turned almost all the way in. Unit is started and cylinder valve is opened just enough to keep low-side pressure within normal operation safe limits. Scale indicates amount of refrigerant being put into system.

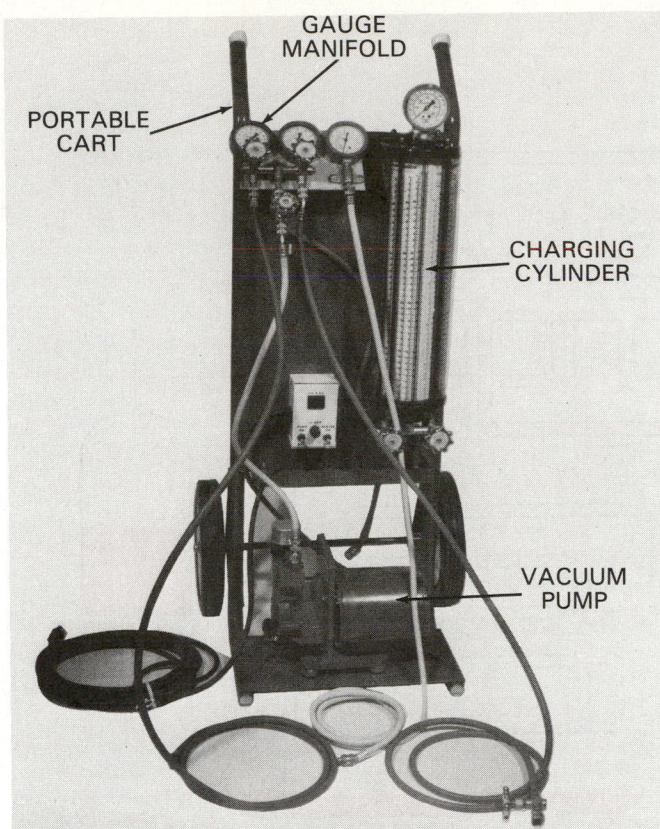

PORTABLE
CART

GAUGE
MANIFOLD

CHARGING
CYLINDER

VACUUM
PUMP

Fig. 14-25. Combination vacuum pump and charging unit.
(Robinair Manufacturing Corp.)

Charging lines must be clean and purged to rid them of air and moisture. Connections must be tested for leaks prior to the actual charging operation. Remember to wear goggles when transferring refrigerants.

In the low-side method, the principle of operation is to use the service cylinder as a temporary evaporator in the system. As the compressor runs, it will remove refrigerant vapor from the cylinder as well as from the evaporator.

Charging may be speeded up by partly closing the suction service valve to reduce flow from the regular evaporators and speed the evaporation from the service cylinder. Hot water may be applied to the service cylinder to help speed the evaporation. Never use a torch to warm a cylinder. The low-side pressure should be kept at normal levels. Too high a pressure may overwork the compressor. Pressures which are too low may cause oil pumping.

The low-side method insures clean refrigerant due to the distilling action during evaporation of the refrigerant. A service technician must be present at all times during the charging. A service cylinder must not be left connected into a system.

It is very important that liquid refrigerant not be allowed to reach the compressor. The liquid is not compressible — and the compressor valves, and even the bearings and rods, may be ruined if the compressor should pump liquid.

Although it is not usually recommended, some service technicians do put liquid refrigerant into the high-pressure side of the system. The compressor should not run while this

charging is being done. Larger systems are equipped with a liquid charging valve on the receiver. This is a dangerous practice because dynamic hydraulic pressures are possible which may rupture the lines causing considerable damage. However, this method can be used to put the initial charge into a system if done very carefully. If one inverts a cylinder and it has a higher pressure than the system, liquid refrigerant will be forced into the system.

One reason this practice is discouraged is that if the compressor exhaust valve is leaking, the liquid may enter the cylinder and the compressor may be damaged when started.

If the unit is water-cooled, the pressure in the liquid receiver with the water flowing will be sufficiently below that of the pressure in the cylinder to permit opening of the two valves after the charging line has been purged. The pressure difference will force refrigerant from the cylinder into the system.

If the unit is air-cooled, the pressure in the refrigerant drum must be increased. This may be done by using the compressor to pump vaporized refrigerant into the cylinder increasing its pressure.

In detail, this method is as follows:

1. Connect the refrigerant cylinder to the gauge manifold with a flexible charging line, as shown in Fig. 14-26. Never use a disposable container here. It may explode!
2. Run the compressor for a few revolutions with the discharge service valve turned all the way in until a pressure of 20 to 30 psi above the condenser pressure is built up in the cylinder.
3. Stop the compressor.
4. Invert the refrigerant cylinder. (Be careful not to injure the line.)
5. Turn the discharge service valve part way out.

High pressure on the surface of the refrigerant in the cylinder will force liquid into the system. While the liquid is flowing into the high side, a gurgling sound may be heard. If this sound stops abruptly, it means that the cylinder has been emptied. Use this method only if all the refrigerant has been removed from the system. Fig. 14-27 is a table of approximate refrigerant amounts which may be added to the system.

General information:

1. The method of checking refrigerant charge is to charge refrigerant vapor into the low side of the system until the evaporator has its normal amount. The liquid line is at room temperature, the bubbles cease at the sight glass and there is no hissing sound at the refrigerant control valve. If one or two extra pounds are put in the unit, this will become a reserve and will be stored in the liquid receiver. The smaller the unit, the less the amount of refrigerant needed as a reserve.
2. Another method of charging liquid refrigerant into a system is to use the liquid receiver service valve — if this valve is a three-way valve. See Fig. 14-28. Connect the cylinder to the service opening of the valve, purge the line, turn the cylinder upside down, open the cylinder valve, and then turn the receiver service valve (king valve) in several turns. The liquid refrigerant will flow into the receiver and into the liquid line. See Fig. 14-29.

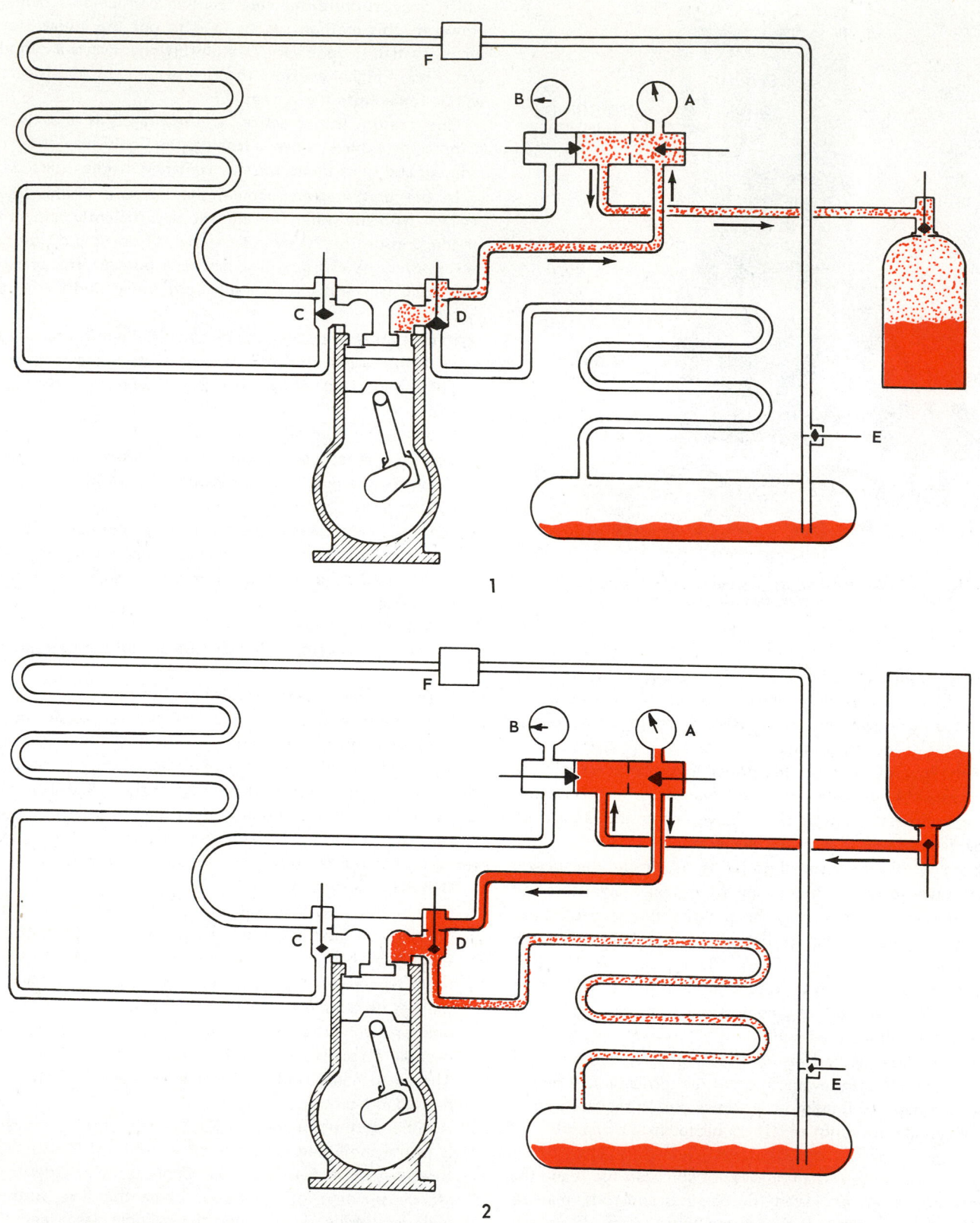

Fig. 14-26. Charging small external drive system through high-pressure side. In part 1, above, compressor is used to build up slightly higher than condensing pressure in cylinder. Then cylinder is inverted and DSV turned out a turn or two as in part 2, below. Small pressure difference will force cylinder refrigerant liquid into condenser receiver.

	R–12		R–22		R–502	
	Flooded	Dry	Flooded	Dry	Flooded	Dry
1/2 hp Unit	3	1 1/2	3	1 1/2	3	1 1/2
1 hp Unit	6	3	6	3	6	3
1 1/2 hp Unit	9	4 1/2	9	4 1/2	9	4 1/2
2 hp Unit	12	6	12	6	12	6

Fig. 14-27. Table of approximate number of pounds of refrigerant which may be safely added to system low on refrigerant.

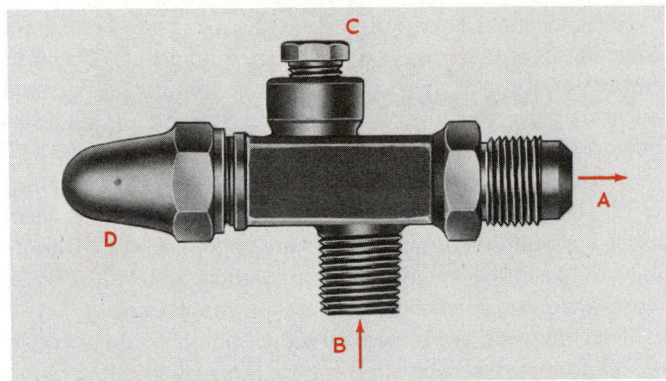

Fig. 14-28. Liquid receiver service valve with service connection. A—Connection to liquid line (flare). B—Connection to receiver (pipe threads). C—Service connection (1/8 in. pipe). D—Valve stem cap. (Superior Valve Co.)

3. A hermetic or serviceable system, not equipped with service valve, is charged as described in Chapter 11. Always remember that a running hermetic motor requires refriger-

ant vapor for cooling the windings or it will overheat. For this reason, it is best to charge these systems with vapor refrigerant into the low side.

14-13 STARTING A SYSTEM

One should follow a planned procedure when starting a new system or starting a system which has been shut down for a period of time. Avoid overloading electrical circuits, compressor and motor.

Make a check first on the electrical characteristics of the power-in circuit.

1. Be sure the phase is correct.
2. Be sure the voltage is correct.
3. Be sure the power leads are large enough. The electric utility company will assist.
4. Connect a voltmeter and an ammeter into the circuit. Recording types are preferred.
5. Install the gauge manifold to check pressures.
6. If the condensing circuit is water-cooled, be sure the water circuit is turned on.

By using hand valves, load the compressor with a normal back pressure during start-up. Remember, it is just as hard on the compressor to have too low a suction pressure as too high (oil pumping).

Once the unit is started, check the electrical meters, the pressure gauges and the water flow as soon as possible. Shut down the unit at the first sign of trouble. If possible, record the electrical characteristics, the pressures and the temperature during the first 24 hours to one week. These charts will serve as a good future reference.

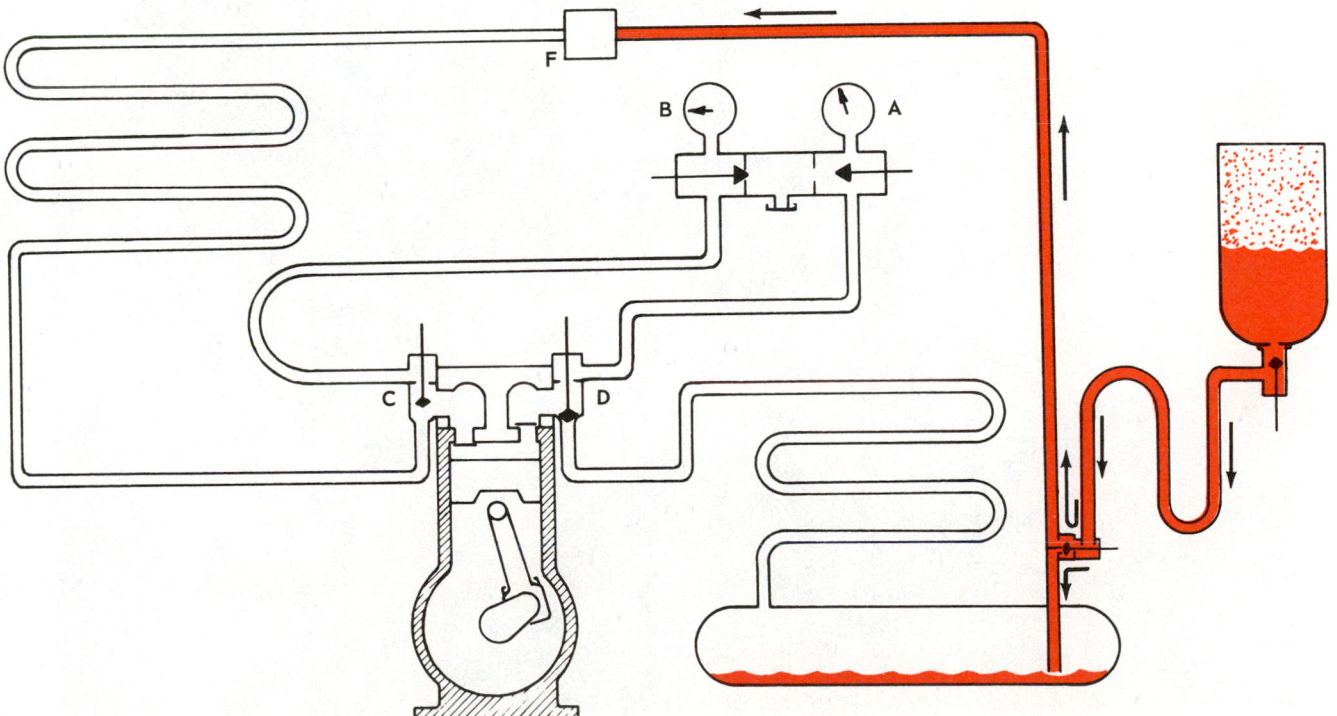

Fig. 14-29. Charging system through two-way liquid receiver service valve.

To start the system with the full load of all the evaporators on it may overload the compressor because the TEVs will open. All evaporator liquid line hand shutoff valves should be closed, therefore, and the condensing unit started. Immediately open just one of the liquid line manifold valves slowly. The low-side pressure should be only a little over the cut-in pressure. It is an important precaution in refrigeration never to overload the compressor even for a short time.

After one evaporator has been opened a few minutes, the evaporator has a chance to cool down somewhat. This tends to make the expansion valve choke off the refrigerant flow. It also gives the compressor a chance to gradually reduce its load. The other evaporators may then be brought into service in the same way — one at a time.

Another way is to almost close the SSV all the way in and then gradually turn it out, keeping the compressor low side at normal pressure.

After starting the unit, the service technician should check high and low-side pressures, amount of water flow in case it is a water-cooled system and operation of each individual expansion valve. He should also determine if the TEV adjustment is correct for each evaporator. That is, frost or sweating on the suction line will indicate whether or not the expansion valve is opened too far or not far enough. Another important procedure or routine to be followed at this time is to determine whether or not the system has enough refrigerant. This can be checked as described in Para. 14-30.

Test for leaks after the unit has operated for 24 hours. Maintain records for use during future maintenance or service operations.

14-14 CODE INSTALLATION

Most localities have definite codes or rules and regulations covering the installation of refrigerating equipment. Domestic systems and some other small capacity self-contained systems are usually not included because these units use only small quantities of refrigerant.

In commercial installations (where the units are assembled on the premises, or where the horsepower or refrigerant needs exceed certain set limits), there are requirements to insure uniform performance and a safe installation. Installation codes also protect the purchaser from careless installations.

The following are some of the high points of these codes:
1. Only licensed refrigeration contractors may install commercial equipment.

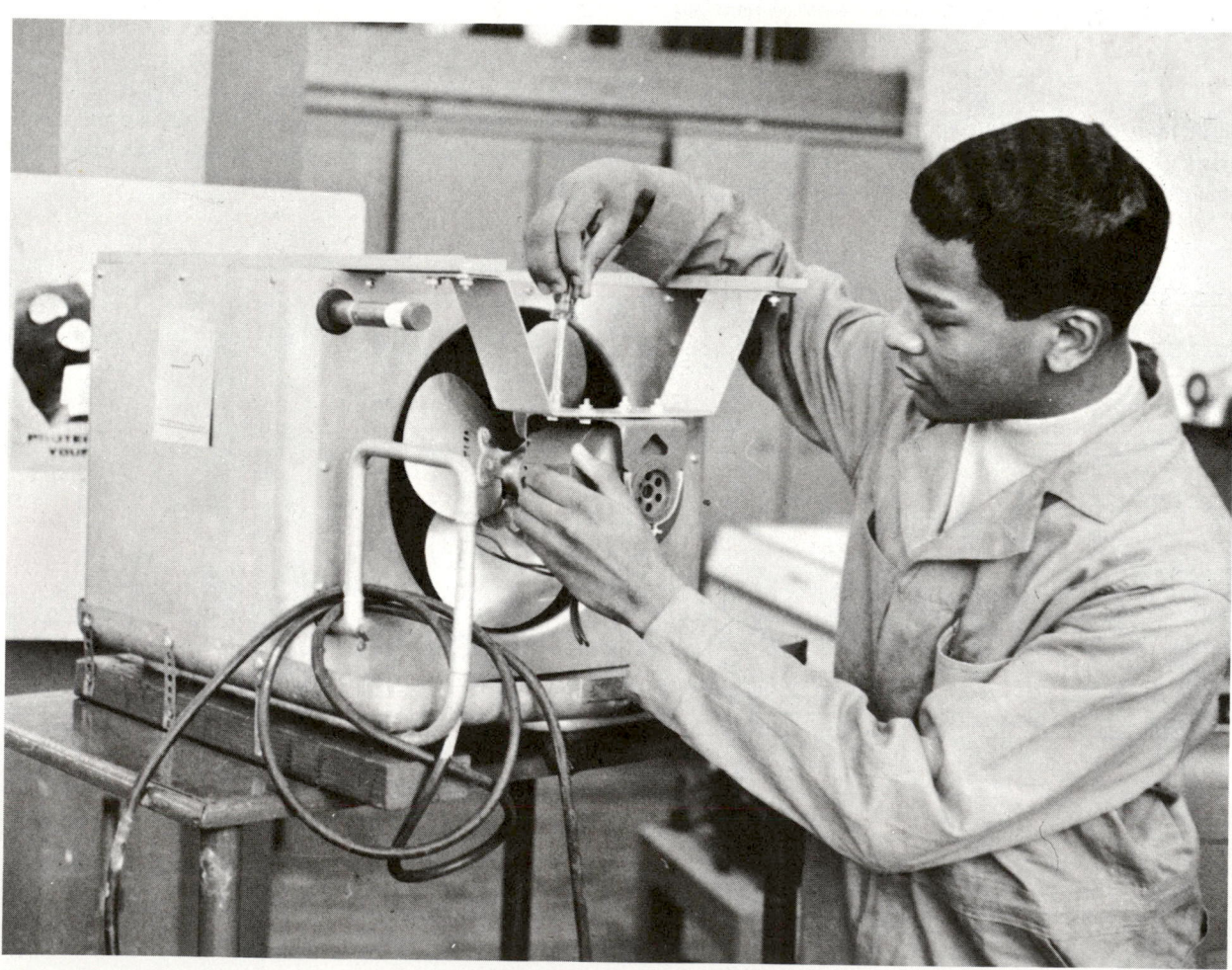

Fig. 14-30. Mounting motor fan on commercial evaporator.
(Detroit Public Schools)

2. A permit must be obtained for each installation.
3. Each installation must be inspected by civic authorities.
4. Lines must be labeled as to refrigerant used.
5. Certain safety devices must be installed in the system.
6. The condensing unit must be installed in a safe place.
7. Electrical and plumbing work must conform to code and must be done by licensed electricians and plumbers.

8. Systems must be tested under certain pressures on both high side and low side, and must be free from leaks.

The national code and local codes are based on experiences with many installations. Codes provide safety for installer, owner, user and public. They should be carefully followed.

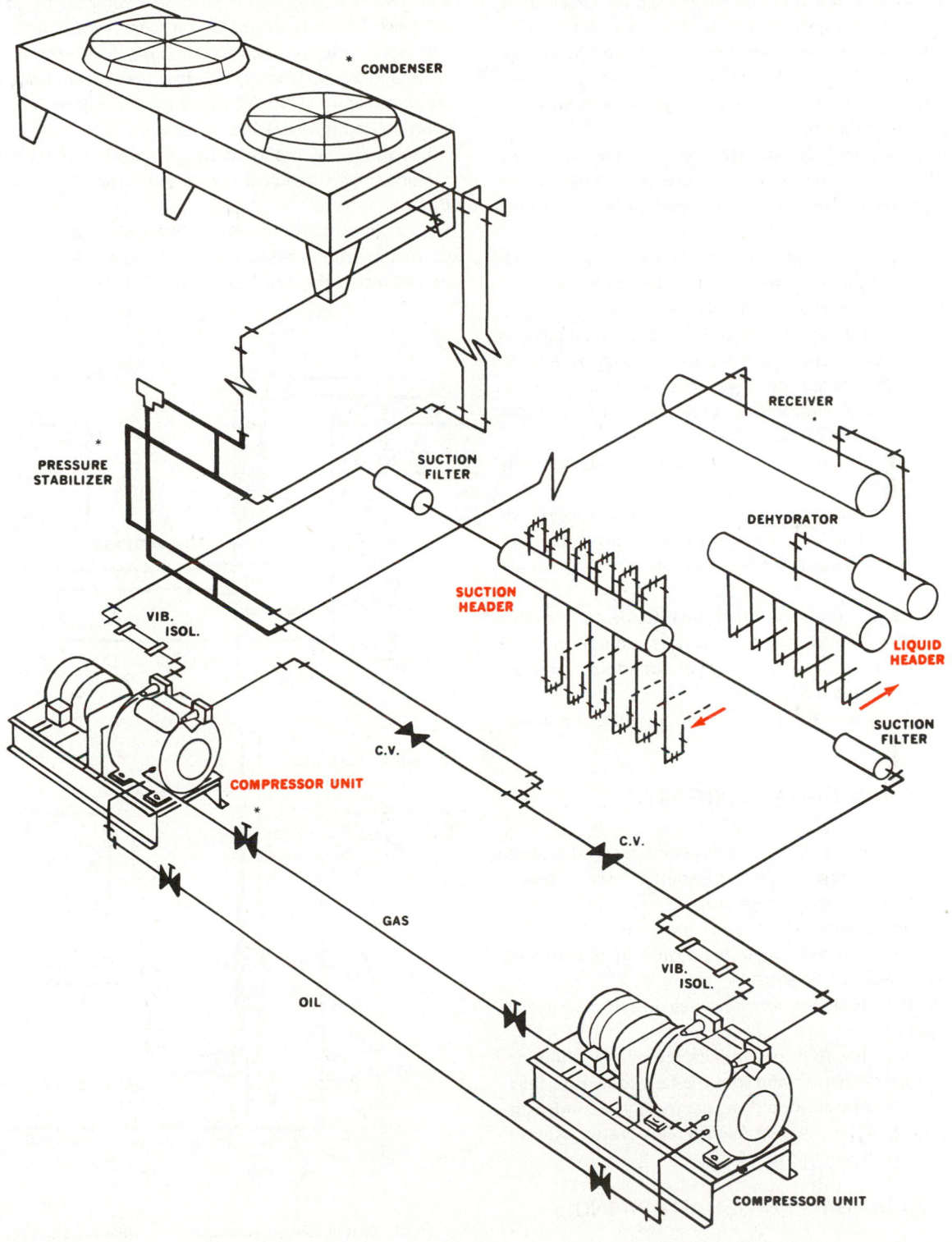

Fig. 14-31. Schematic piping diagram for commercial refrigerating system using roof-mounted air-cooled condenser, two motor compressors and suction and liquid header, each connected to six refrigerant lines. (Dunham-Bush, Inc.)

14-15 INSTALLING A CONDENSING UNIT

Most codes require the condensing unit to be placed where it cannot be damaged. It should have a protective cage around it (small unit) or it should be placed in a separate room. The area should be well ventilated to permit escape of refrigerants should the unit develop a leak. Windows provide this for smaller units while a forced exhaust is needed for larger units. Screen the openings to prevent insects and other objects from entering. Larger units must also be protected from fire damage by fire-resistant self-closing doors. The condensing unit must be electrically grounded. All electrical and plumbing work should be done by a licensed contractor.

If the unit has motors and fans to cool the air-cooled condenser, the shipping blocks used to protect the motor and fan must be removed. Spin the fan by hand to be sure it runs freely.

To prevent violent rupturing or an explosion of the condensing unit due to excessive pressure, the code specifies:
1. High-pressure cut-outs to stop the motor.
2. Pressure relief valves or rupture disks to dissipate discharge slowly. These safety openings are piped outside by way of copper pipe connected by silver brazed joints. Spring-loaded safety valves and/or fuse plugs are used (where the unit may become overheated because of fire).

All refrigerant lines should be permanently labeled with signs identifying refrigerant.

Install the condensing unit where all the parts are accessible for maintenance and service. Keep them away from the sun and away from other heat sources such as steam pipes, hot air grilles and ovens.

Water lines at the unit are either soft copper or flexible plastic pipe. Allow enough piping to permit some movement of the condensing unit. See Para. 14-84 about adding extra oil for long suction and liquid line runs.

Be careful! If a system has too much oil, it will pump liquid oil and be damaged.

14-16 INSTALLING AN EVAPORATOR

The code recommends limits of refrigerant for evaporators that would expose people to the refrigerant in case of leaks. On some installations, large tonnage evaporators installed in air ducts must be cooled with a brine rather than refrigerant.

The evaporator should be mounted firmly in the cabinet and protected to avoid damage to the system.

The evaporator should be leveled when installed. A spirit level may be used.

If the evaporator has motors and fans, remove the shipping blocks which protect them. Hand spin the fans to be sure they run freely. The evaporator should be electrically grounded, if it has a motor and fan. A commercial evaporator using a motor-driven fan is shown in Fig. 14-30.

14-17 INSTALLING REFRIGERANT PIPING

Code specifications require strong piping for refrigerant lines. These should be type K, the strongest, or type L. Piping

should be protected by adequate guards. Some codes recommend that hard copper pipe, where it is exposed, should have at least .065 in. wall thickness. Joints in the refrigerant piping must be placed so they can be easily inspected. The joints must be made with strong fittings and the brazing material used must be of excellent quality.

To avoid pinching or crimping the piping, the code recommends that piping always be supported by the building structure. Pipe should not be run across joists or studs where unsupported sections can be damaged. It is recommended that the piping be at least 7 1/2 ft. above floor level when run across a room. Fig. 14-31 shows a piping system for a roof-mounted refrigerating system.

The suction line should be mounted with a slight drop in horizontal runs toward the compressor. This provides for proper oil return.

It is important for noise control and for piping that experiences rapid temperature changes (defrosting hot gas lines) to install flexible sections in the pipe. See Fig. 14-32.

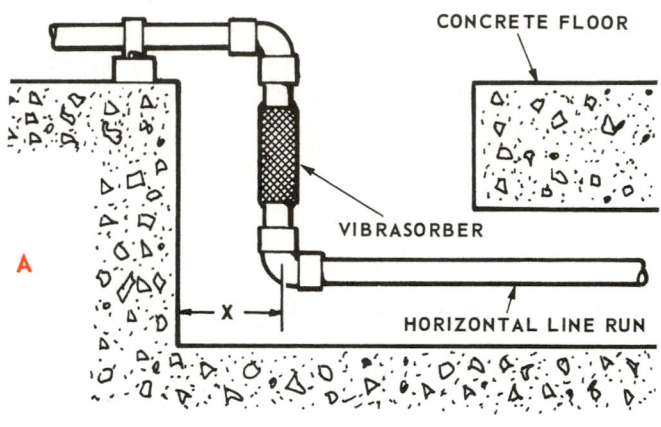

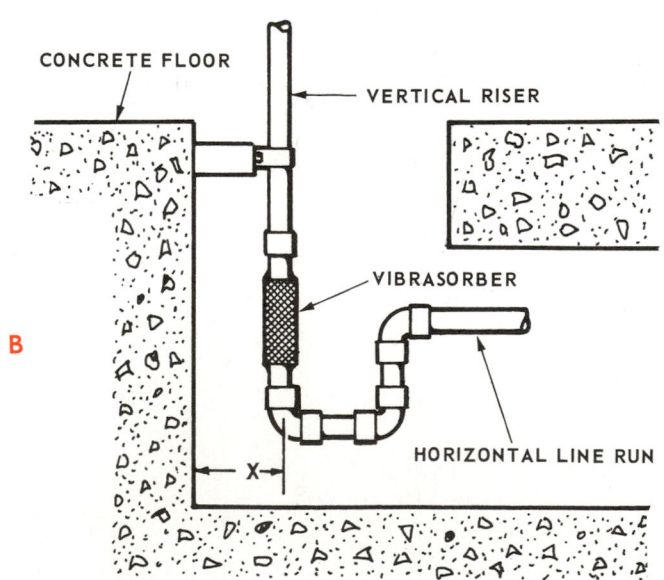

Fig. 14-32. Two instances showing how flexible vibration absorbers are used. Allow a space of 1 1/4 in. at X for each 100 ft. per 100 F. (55 C.) temperature change. A—Horizontal piping. B—Vertical piping. (Tyler Refrigeration Corp.)

14-18 FITTINGS

Commercial system capacity has increased steadily during the past few years. This is especially true with comfort cooling installations in air-conditioning units and in supermarkets.

Sizes of the liquid and suction lines in these installations may be as large as 2 and 6 in. OD. Brazed joints with sweat fittings are used. These fittings, described in Chapter 2, are of drop-forged or extruded copper with a recess large enough to receive the hard copper pipe.

Brazing of hard copper pipe joints must be expertly done or considerable trouble may result (bad joints and leaks). Hard drawn copper pipe which is seamless usually has a greater wall thickness than annealed copper tubing. It comes in 10 or 20 ft. lengths rather than in rolls. Ends are either capped or plugged. When making an installation of this kind, use fluxes and solders recommended by the manufacturer. Joining surfaces should be clean and ends of the tubing should be square to prevent flux or joining metal from running into the tubing.

14-19 SPECIAL TUBING

Copper tubing normally comes with no special finish provided for the inside or the outside surface. Such tubing will corrode if it must be run through liquid, food, beverages or through air saturated with acid fumes or corrosive elements. Where sanitation is of primary importance, tubing with tinned surface may be used.

Where the tubing is used to convey beverages, such as beer, between kegs and dispensers, soft drinks and carbonated beverages, many local codes require the use of special tubing. Stainless steel is usually specified.

14-20 MULTIPLE EVAPORATOR PIPING

There are two common methods of installing piping in a multiple installation.

In one method there is a common liquid line and common suction line. The various evaporator liquid and suction lines tap into it at the most convenient points.

Another method is to use a clustering system. Here, various

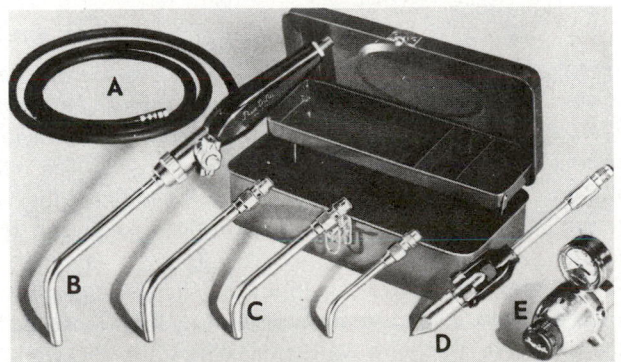

Fig. 14-33. Air-acetylene soldering and brazing kit. A—Hose. B—Large tip and handle. C—Other tips. D—Soldering copper. E—Regulator. (Linde Div., Union Carbide Corp.)

lines are brought to a common point and connected through a hand valve to a manifold. A large suction and liquid line is run from the manifold to the compressor.

This method is not always practical. For example, where one evaporator is at some distance from the box, it would require a duplication of long runs. The manifold of a code installation is located on a wall near the condensing unit.

Keep in mind always that every fitting used, and every bend put into the refrigerant and suction lines, cuts down the efficiency of the installation. Limit their number as much as possible.

Remember that inner parts of valves, driers, filters and sight glasses should be removed while tubing is being brazed to it, or the part wrapped with a wet cloth or some heat absorber. Take care not to allow moisture to enter the valve.

14-21 WELDING EQUIPMENT

Gas welding equipment needed for refrigeration installation consists of an oxygen cylinder, acetylene cylinder, regulators and gauges, hose and a torch, as shown in Fig. 14-22. This equipment may be used for soldering, brazing and welding the various parts of refrigeration systems.

Before doing any welding, the local code on welding should be thoroughly studied and understood. Never operate a welding outfit near flammables.

CAUTION: Never use oxygen, acetylene, or any other welding fuels, for the purpose of developing pressure in refrigeration tubing, piping or equipment.

Carbon dioxide, nitrogen and argon are safe if a pressure regulator and a pressure relief valve are used. They may be used for developing pressures in refrigeration lines.

Never use excessive pressures with any gas. A severe explosion may result.

14-22 BRAZING EQUIPMENT

The refrigeration service technician uses brazing for many jobs. Air-acetylene torches furnish a clean flame at a temperature of 2500 F. With compressed air, the torch flame temperature is about 2500 to 2800 F. Acetylene is supplied in cylinders of 40 cu. ft. capacity or 10 cu. ft. capacity.

Detailed instructions on the construction and use of acetylene-air brazing equipment are in Chapter 2. A kit of this equipment is shown in Fig. 14-33.

It is important to follow these safety precautions:

1. Acetylene pressure should never be over 15 psi. Higher pressure may cause an explosion because acetylene is not stable at higher pressures.
2. Always use the cylinder in a vertical position. It has a porous filler wetted with acetone in which the acetylene is dissolved. If the cylinder is lying down while in use, some acetone may flow out. Acetone causes a dirty flame and may grease up the regulator and valves.

Fuel air torches that use propane or other high-temperature fuel are also used. They are light and work in any position. See Fig. 14-34. Propane fuel is usable down to −10 F. (−23 C). For safety:

1. Keep the flame away from any combustible substance such

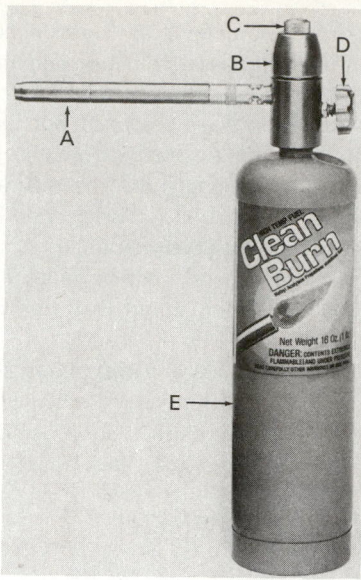

Fig. 14-34. Soldering and brazing torch. A—Tip. B—Regulator. C—Pressure adjustment. D—Valve. E—Disposable fuel cylinder (either propane or high temperature fuel). (Cleanweld Products, Inc.)

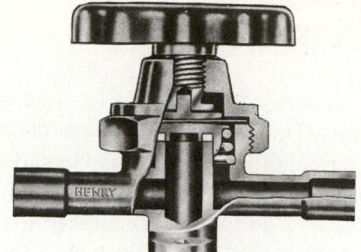

Fig. 14-35. Two-way hand valve designed for brazed connections. (Henry Valve Co.)

MINIMUM DESIGN — TESTING PRESSURE			
	LOW SIDE	HIGH SIDE	
REFRIGERANT		EVAP. OR WATER COOLED	AIR COOLED
R-12	85	127	169
R-22	144	211	278
R-500	102	153	203
R-502	162	232	302
R-717	139	215	293

Fig. 14-36. Recommended minimum design testing pressures are based on Safety Code for Mechanical Refrigeration.

as oil, wood, paper, paint, cleansing fluids, methyl chloride, barrels or cylinders which may have contained flammable material at one time. Use sheet metal or board to protect surfaces that might be discolored or scorched when using the torch, such as assembling piping along a wall.

2. Always light the torch with a flint lighter. Matches or a cigarette lighter may bring your hand too close to the flame.

14-23 REFRIGERANT LINE VALVES

All code line valves are to be hand operated. Codes require that these valves be so constructed that anyone may shut them off without the need of special tools. The valve usually has a handwheel mounted permanently on it. These valves are provided with brackets whereby they may be attached firmly to a panel. Two styles are available:
1. Three-way valves.
2. Two-way valves.

One type of three-way valve shuts off just one of the three connections to the valve. The other two, remaining uncontrolled, permit the passage of refrigerant to the rest of the system.

The other type valve is two-way and stops the flow of refrigerant when turned in clockwise. Fig. 14-35 shows a two-way valve for brazed joints.

14-24 TESTING CODE INSTALLATIONS

Code authorities require that permits be obtained before an installation can be made or a major service operation performed on commercial units. Specifications of the proposed job must be presented. Permits are not issued unless the specifications presented meet code requirements.

On completion of the work, an inspector is called. Approval

must be given by that person before the unit may be run. Some codes require that refrigeration installation personnel be licensed.

The inspector, upon reaching the premises, checks the installation to see if all the work has been done according to specifications and code. Then the system is tested, primarily for leaks and safety. This requires building up the normal pressures in the high and low sides of the system. Nitrogen is usually used for this. These pressures vary with the kind of refrigerant in the system. Fig. 14-36 gives the recommended minimum test pressures to be used for each refrigerant.

Low-side and high-side test pressure are usually specified by the code. One must avoid using higher pressures because the system may rupture.

The same refrigerant that will be used in operation is first charged into the system in a vapor form to create a low pressure (20 to 50 psi). The system is then tested for leaks. Large leaks are easily detected at these low pressures. If there is no leak, nitrogen is then used to build up the pressures to the code pressure.

But the technician or inspector must be very careful! First, the refrigerant cylinder should be disconnected to prevent nitrogen backing up into it. Second, there must be a hand shutoff valve, a pressure regulator, a pressure gauge and a pressure relief valve in the nitrogen charging line. The relief valve should be adjusted to open 1 or 2 psi above the test pressure. A safe method for using nitrogen to pressurize a system is shown in Fig. 11-56.

After pressures are built up in the piping, it is good practice to rap each brazed and mechanical joint with a rubber mallet to make sure the joint will be leakproof under working conditions. (Paint or flux may otherwise temporarily stop a leak.) If no leaks are found, leave the nitrogen pressures in the system for 24 hours.

With nitrogen, the soap bubble test is used to check for leaks. If some R-12 or R-22 is mixed with the nitrogen, the halide torch or electronic leak detector can be used (mix about 1/4 lb./ton of refrigeration).

If a brazed joint leaks, take it apart. Put flux on the joint before heating it to keep the silver brazing material clean. Take the joint completely apart and then assemble and braze again.

If no leaks are indicated at the pressures established, the inspector sometimes checks further by producing a vacuum on all of the system. If this vacuum is maintained over a certain period of time, the installation is approved.

After approval of the system, the installation technician should make a record of the running behavior of the unit for at least 24 hours. Fig. 14-37 shows a recording thermometer. Any variations in temperature reveals need for adjustments.

Fig. 14-37. A 24-hour recording thermometer. Charts should be dated and kept for future reference. (Marshalltown Instruments)

14-25 SERVICING COMMERCIAL UNITS

Modern commercial refrigerating units are available in a great variety of forms. Chapters 12 and 13 describe the design, construction and operation of various mechanisms.

Small units, such as self-contained beverage coolers of hermetic design, are serviced in the same way as the domestic systems. Details are given in Chapter 11.

Some commercial installations use an external drive system with motors and belts. Many use hermetic condensing units.

The servicing of the larger commercial systems is controlled in most communities by the local refrigeration code. Any major repairs or changes to a commercial system can only be done by licensed contractors. When completed, their work done must be checked by the local community refrigeration inspector.

Any plumbing and electrical service work should be subcontracted to licensed plumbers and electrical contractors.

The servicing of commercial installations is much like working on domestic units. However, the use of multiple evaporators on a single compressor is common. Unloading systems and defrosting systems add to service complications.

The troubles encountered come under various headings such as no refrigeration, continuous running, high cost of operation, poor refrigerating temperatures and frosted suction lines.

14-26 SERVICE EQUIPMENT

Two major items of concern are:
1. Obtaining and using high quality tools.
2. Keeping complete records of each job.

Most companies provide a panel truck or pickup truck equipped with major items such as:
1. Vacuum pump.
2. Tubing and piping.
3. Combination soldering, brazing and welding outfit.
4. Supply of replacement parts and materials.
   a. Controls.
   b. Fittings.
   c. Oil.
   d. Refrigerant.
5. Leak detectors — especially electronic tester.
6. Electrical testing instruments.

The service technician is usually expected to furnish his own hand tool kit. Chapter 2 describes many of the tools needed. These should be of good to excellent quality. Three good habits will speed up the work:
1. Keep tools clean. This action will result in better and faster work while stretching tool life.
2. Keep tools together on the job — in a tool kit or in the truck. They should be organized so that the service technician can pick up the desired tool in the correct way without looking.
3. Use good lighting. Keep an extension cord and light that can be safely mounted.

14-27 GENERAL SERVICE INSTRUCTIONS

Servicing, troubleshooting or diagnosing a refrigerating system is mostly common sense plus a thorough knowledge of refrigeration fundamentals. To operate correctly, a system must have the following capabilities:
1. Cooling (low side).
   a. Enough liquid refrigerant must be in the evaporator.
   b. Pressure in the evaporator must be low enough so that the liquid will boil at the correct temperature.
   c. Heat from the items being cooled must transfer to the liquid refrigerant in the evaporator.
2. Condensing (high side).
   a. Vapor must be pumped into condenser at the correct pressure and temperature.
   b. Heat must be removed from condenser (clean condenser, airflow or water flow).
   c. There must be enough vapor space (heat transfer surface) in the condenser.
3. Refrigerant flow in liquid line. Line must be large enough with minimum restrictions (pinched pipe, partially clogged screens, filters or drier). Only liquid refrigerant should be in the liquid line.

4. Vapor and oil flow in the suction line. Only a small pressure drop is allowable. The screens and drier must not be restricted in any way.

The diagnosing starts with the owner's report. Then the service technician should check the low-side and high-side pressures and the evaporator temperature. Check the sight glass for bubbles. Feel the suction line. It should be cool. Feel the liquid line. It should be the temperature of the surrounding air (ambient).

If refrigeration equipment has been exposed to flooding (a flooded basement or a flood), it must be carefully reconditioned before one attempts to start it.

Clean and dry all of the outside of the equipment. Use a detergent and bacteria cleanser. Replace all open motors or have them completely reworked.

Replace all external electrical parts. If one attempts to clean and reuse them, one must use an electric insulation leak inhibitor.

Replace capacitors, relays, overload devices and limit switches. Clean compressor terminals and spray with electrical insulation leak inhibitor. Check the electrical system completely with an ohmmeter. Check especially for grounds.

14-28 SERVICING CONDENSING UNIT

Condensing units come under several divisions:
1. Open (external drive) type compressor.
2. Serviceable hermetic motor compressor (field serviceable compressors).
3. Welded hermetic motor compressor (nonfield serviceable compressors).
   Compressor types may be:
1. Reciprocating.
2. Rotary.
3. Centrifugal.
4. Screw.
   The condensers used may be either:
1. Air-cooled.
2. Water-cooled.

The variety of mechanisms and applications is a great challenge to the service technician. Fortunately, in spite of the great variety, there are certain basic problems all these condensing units have in common:
1. Compressor efficiency.
2. Condenser efficiency (air-cooled or water-cooled).
3. Refrigerant charge.
4. Refrigerant cleanliness.
5. Electric circuit problems.

The compressor may be tested for efficiency as described in Chapter 11.

An air-cooled condenser gives the same symptoms when there is a lack of refrigerant as those explained in Chapter 11. Water-cooled types present a different problem. The water flow should be so adjusted that the temperature rise is no more than 15 deg. as the water goes through the condenser. The water passages must be clean.

If the unit is belt driven, belts should be checked for alignment and tautness.

A decidedly metallic pounding sound occurring regularly in the compressor should be looked into carefully. Check for low oil level or worn parts.

The amount of refrigerant in the system should be carefully checked as explained in Para. 14-30. The motor control should be inspected to determine whether it trips freely and whether the points — if any — are clean. Dirty or pitted contact points should be replaced.

14-29 SERVICING EXTERNAL DRIVE SYSTEMS

As all open (external drive) type compressors are similar, the general instructions which follow will apply to nearly all of them. There are many external drive compressor systems in use. The external drive compressor is either belt driven or a direct drive.

14-30 CHECKING REFRIGERANT CHARGE

The correct refrigerant charge is very important. Several methods may be used to determine whether or not a refrigerator has enough refrigerant.

In undercharged systems the motor operates continuously, the motor compressor is overloaded and there is poor refrigeration. A lack of refrigerant will show up in an increase in liquid line and drier temperature. A heated drier will release some of its moisture and cause a wet system.

Overcharge will cause excessive head pressure in TEV systems. Liquid refrigerant will be forced into the compressor in capillary tube systems.

A dry or expansion valve system is more difficult to check for refrigerant amounts. The appearance of the valve body may be the first sign of low refrigerant. Under normal conditions the body of the valve frosts over evenly, as far back as the liquid line nut. But when the system has too little refrigerant, the expansion valve body next to the liquid line will not frost. This frost method cannot be used for above-freezing evaporator operation conditions.

A common method of determining the amount of refrigerant is to check how much refrigerant is actually in the liquid receiver and condenser. One way to find this out is to determine the high-side head pressure. If the unit is water-cooled, head pressure should correspond to refrigerant temperatures about 10 F. higher than the temperature of the water leaving the condenser. The temperature of the water, in this case, should be checked as it leaves the condenser, not at the end of a long drain pipe. If head pressure indicated on the gauge is below normal as much as 10 psi, lack of refrigerant is indicated.

A very popular way to check for sufficient refrigerant charge is to install a sight glass in the liquid line and note if there are any gas bubbles going up the liquid line. See Chapter 12. Bubbles indicate insufficient refrigerant.

Fig. 14-38 shows a sight glass which also indicates moisture. Figs. 14-39 and 14-40 picture see-through sight glasses for larger liquid lines.

At low head pressure, bubbles may appear regardless of the amount of refrigerant in the system. If no bubbles appear in

Fig. 14-38. Sight glass which also indicates if refrigerant in system is wet or dry. Blue indicates dry system; pink indicates wet system. Bubbles indicate lack of refrigerant. (Henry Valve Co.)

Fig. 14-39. Double port or "see-through" sight glass for larger liquid lines. Unit also indicates dryness of refrigerant. Device has brazed connections. Note dryness code on the seal cap (pink for wet, blue for dry).

the sight gauge, the machine probably has enough refrigerant.

However, if a restriction is in the line ahead of the sight gauge, bubbles may appear even though there is sufficient refrigerant in the system. If possible, the sight glass should be mounted between the drier and the liquid line (upstream from the drier).

Some machines are equipped with refrigerant liquid level indicators such as petcocks mounted in the side of the liquid receiver at definite heights. If the petcock is opened and liquid refrigerant comes out, the level of the refrigerant is at least up to this height. Two petcocks are usually provided. When opened, vapor should come from the top petcock and liquid refrigerant should come from the lower one.

In the liquid receiver type of water-cooled unit, where the water coils are located within the receiver, the amount of refrigerant in the system may be checked by determining the temperature difference on two different points of the receiver

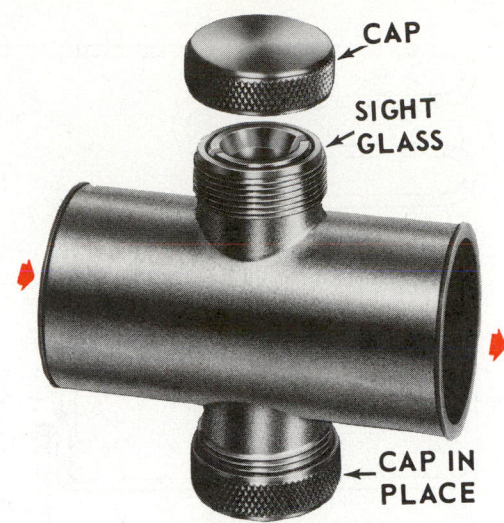

Fig. 14-40. "See-through" or double port sight glass. Both caps are removed, and light source (flashlight) is used to see if there are bubbles in liquid refrigerant flow. (Mueller Brass Co.)

shell. The part of the receiver filled with hot vapor and the part filled with cold liquid refrigerant will be indicated by a temperature difference. This may be easily checked by feeling the receiver with the hand.

One more method for finding the quantity of liquid refrigerant in the liquid receiver involves turning off the cooling water to the condenser and allowing the compressor to operate. If the liquid line warms up quickly, it indicates there is insufficient refrigerant. Another indication is change in head pressure after the water is shut off. It rises quickly. If the compressor is stopped and head pressure drops rapidly, it indicates presence of too little liquid refrigerant in the liquid receiver.

Still another method of checking for the quantity of liquid refrigerant involves shutting the machine down and purging the liquid receiver. Boiling of the refrigerant in the liquid receiver when the pressure is reduced will cause that part of the receiver filled with liquid to get cold, sweat and perhaps frost over. This method should be used only as a last resort because it wastes refrigerant. It may also freeze the water in a water-cooled unit.

Lack of refrigerant is likely due to a leak in the equipment. A careful check should be made of all joints and parts that could possibly leak. Do this before the unit is recharged and put into service. See Chapter 11.

14-31 REMOVING SYSTEM PART

When part of a system needs service, empty the refrigerator cabinet or put contents to one side and cover them. Spread papers or a tarpaulin around and under the mechanism.

Be careful of all surfaces. Porcelain is brittle. Chipping or cracking may necessitate replacing a complete panel. Do not soil enamel finishes with oil or grease.

Tools and materials should be placed in a safe place to prevent injury from tripping. Always arrange for good lighting.

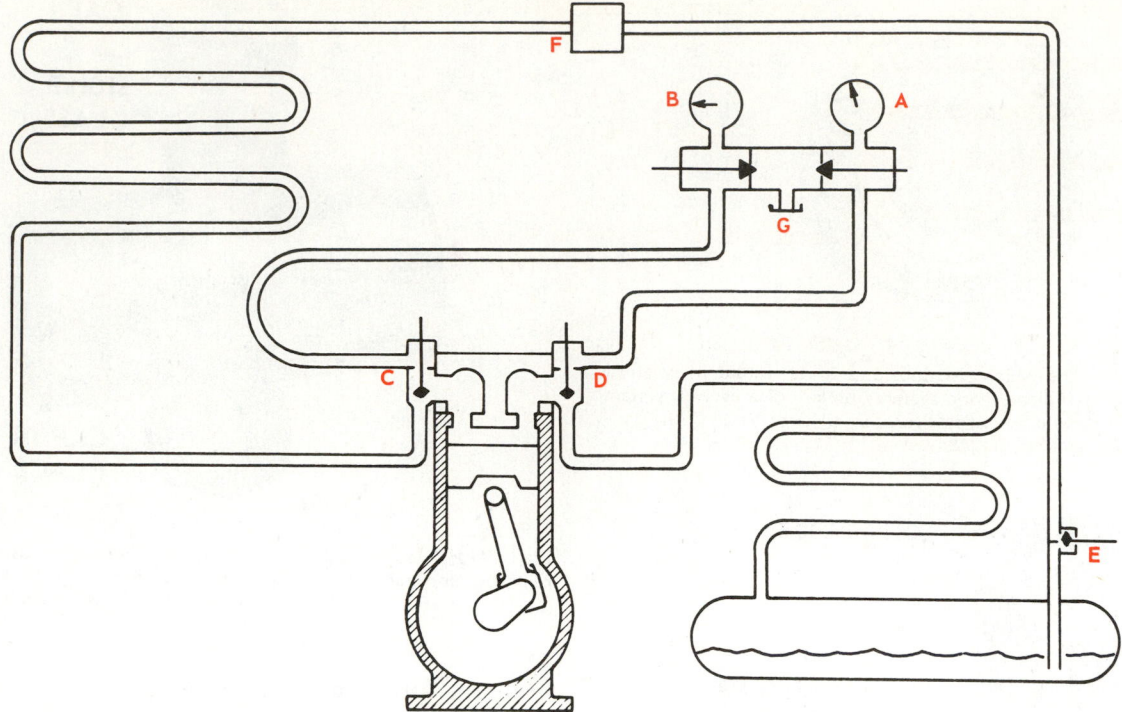

Fig. 14-41. Elementary system shows location of gauges and service valves. A—High-pressure gauge. B—Compound gauge. C—Suction service valve. D—Discharge service valve. E—Liquid receiver service valve. F—Expansion valve. G—Gauge manifold.

When removing any part of any system there are general steps to follow:

1. Remove all refrigerant from part to be opened.
2. Balance pressures in parts just evacuated to 0 psi.
3. Isolate parts to be opened from the rest of the system.
4. Clean and dry joints to be opened.
5. All refrigerant openings should be immediately plugged as soon as they are opened.

Fig. 14-41 shows an elementary unit with the location of the three main service valves marked.

Refrigerant is removed by installing a gauge manifold in the system, by proper adjustment of the service valves and by operation of the compressor.

Removal of any part of the refrigerant is usually a matter of drawing a low pressure (less than atmospheric pressure) on the part to be dismantled in order to evaporate the refrigerant from it. Then pressure is equalized to 0 psi. (This is called balancing with atmospheric pressure.)

The low pressure removes the refrigerant while the equalizing or balancing prevents a rush of air into the mechanism when the system is opened. This last step is very important.

To begin removal of refrigerant, close the inlet service valve to the part to be removed. Run the compressor until the gauge shows a 0 psi or a slight vacuum. Stop the compressor and then, after opening the inlet service valve until the gauge reads zero, close the inlet service valve to that part. Close the outlet service valve to the part. Clean and dry the joints. Remove the part. *Always plug all refrigerant openings immediately after removing the part to keep out dirt and moisture.*

For example, suppose one wished to remove the compres-

sor evaporator or TEV. The refrigerant, then, is stored in the liquid receiver. The liquid receiver service valve is closed. Then the compressor is run until no liquid refrigerant remains in the liquid line, evaporator or suction lines. See Fig. 14-42.

When servicing a refrigerating mechanism, keep in mind the internal part of the machine must be kept as chemically clean as possible. Moisture causes acids, sludge and freezes in low temperature passages. Dirt (solids) will clog screens and cause wearing of control valves, compressor valves and seats.

14-32 REMOVING REFRIGERANT FROM A SYSTEM

If there is no place to store the refrigerant in the system, the refrigerant must be:

1. Stored in an outside cylinder.
2. Thrown away (purged).
   To remove refrigerant from a system:
1. Attach a line from a storage cylinder (if one is to be used) to the middle opening of the manifold.
2. Purge the line leading from the cylinder by sealing the line at the cylinder but leaving it loose at the manifold.
3. Crack the cylinder valve. Escaping gas will force the air out of the line.
4. Seal the line at the manifold, close the discharge line of the compressor by turning the discharge service valve all the way in, test for leaks and start the compressor. The pressure should not exceed the normal condensing pressure for the particular refrigerant. Excessive head pres-

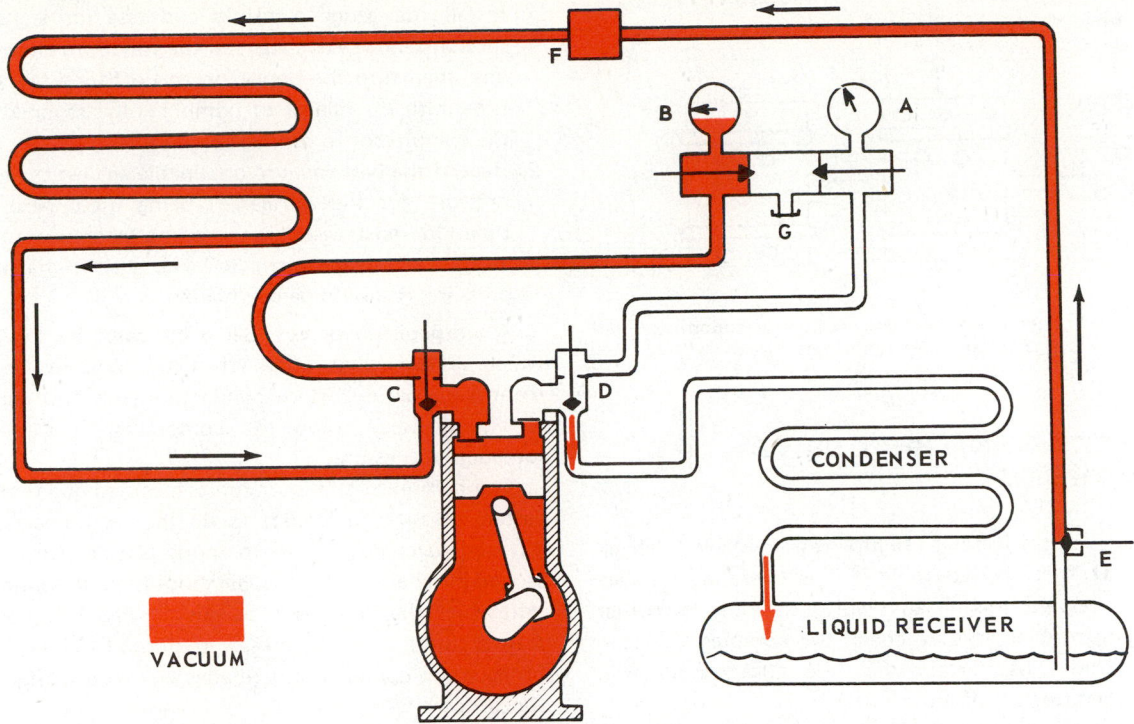

B

A

F

G

C

D

E

CONDENSER

LIQUID RECEIVER

VACUUM

Fig. 14-42. Compressor is evacuating liquid line, evaporator, suction line and compressor crankcase. Refrigerant is being stored in condenser and liquid receiver. Valves A and B are closed, valves C and D are in mid-position and valve E is closed.

sures may be avoided by cooling the refrigerant cylinder with ice or water, or by running the compressor intermittently. (This means intervals of running interrupted by short stop periods.)

5. Allow the compressor to run with all but the discharge service valve open.

6. Shut the compressor off after a constant low pressure has been maintained for several minutes. Never allow the system to pump oil, as the hydraulic pressures may cause serious damage to the compressor and lines.

7. The operation may be speeded up by cautiously applying heat to the liquid receiver and to the evaporator. Use a heat lamp or warm water. Never use a torch, as it may melt the fuse plugs and brazed joints. Never allow any part or spot to become too warm to touch with the hand.

8. After the refrigerant is all pumped from the system and placed in a storage cylinder, stop the compressor.

9. Open the low-side manifold valve until 0 psi is indicated on the compound gauge. This action returns enough vapor refrigerant into the system to balance the pressure in the entire system. Any part of the system may now be removed. Wear goggles! Always leave the gauge manifold connected to the system until the system is opened. One must know the pressure in the system as it is first taken apart. As mentioned before, clean and dry all the connections to be opened.

10. Immediately upon removal of any parts, the refrigerant openings should be carefully plugged.

Saving the refrigerant is not recommended in small systems. The low cost of the refrigerant does not justify the time it takes to store it. The service technician may either exhaust it to the air or into an approved exhaust system. Never dump it into sewers or any other sewage disposal system.

When discarding refrigerant, it is good practice to attach a purging line made of 1/4 in. copper tubing to the manifold center opening. This purging line should have a hand needle valve and a check valve mounted in it at the manifold end. The hand needle valve should be located between the check valve and the manifold. During purging, the manifold high-pressure valve is opened.

Purpose of the hand valve is to control the amount of gas purged. The check valve prevents backing up of air or moisture into the unit after it has been completely purged.

Because all refrigerants being purged have an oil content, purging should be done into an oil trap. Always purge a refrigerant into a well ventilated space.

This method cannot be used for ammonia (R-717) because of the odor. Most communities forbid purging refrigerant into a sewer system.

If the refrigerant is to be put back into the same machine or, if facilities are available for distilling it, it may be stored temporarily in a clean refrigerant cylinder. Remember that the refrigerant will always have an oil content. Some large companies save all refrigerant, redistill it and process it for further use. This is good practice from the standpoint of economy and ecology.

To get rid of refrigerants removed from systems stored in service cylinders, it is possible to purge the refrigerant into perforated containers buried in the ground, as shown in Fig. 14-43. The ground absorbs the oil and the refrigerant.

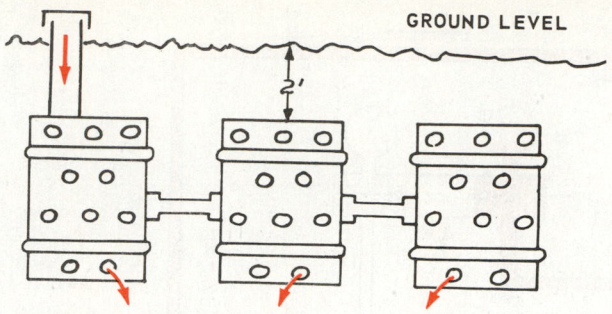

GROUND LEVEL

Fig. 14-43. Buried containers present one method for disposing of used refrigerant and used refrigerant oil.

14-33 CHECKING EXTERNAL DRIVE COMPRESSOR

In most cases of refrigeration failure, it is advisable to check the compressor first to determine if it is operating satisfactorily. Good service technicians will also look for other indications of troubles as they check the compressor. However, they should be certain that the compressor is in satisfactory operating condition.

The amount and the condition of oil in the compressor is important. Some compressors have an oil level sight glass or port. Others have a plug located at the proper oil level.

Two of the most common causes of compressor trouble are faulty valves and seals. Noisy valves may be detected by a sharp clicking noise in the compressor as it operates.

Leaky valves may be detected as follows:

1. Install the gauge manifold and test for leaks. Turn the suction service valve stem all the way in to close the suction line, then turn the power on and off for a few seconds at a time until the danger of pumping oil is stopped; then allow the compressor to run. See Fig. 14-44.
2. Record the best vacuum obtainable against the normal head pressure for the refrigerant being used. Also record the time. In most cases, if the compressor cannot produce a vacuum of greater than 20" Hg. against the normal head pressure, it should be overhauled.

A worn piston or cylinder is indicated by a clicking noise which is somewhat duller than the noisy valve indication mentioned before. Worn connecting rods and main bearings are quite noisy when the compressor is running with a low-suction pressure.

The compressor must pump a specified quantity of gas at a certain pressure difference to do the work necessary. This is difficult to check, so the methods just described are used as secondary checks. Some repair shops use a shop-mounted tank into which the compressor pumps air while being tested. If the time required by a compressor to pump to 150 psi is recorded for each size compressor, a relative volumetric efficiency check is possible.

Compressor testing methods:
1. Vacuum producing ability.
2. Abilty to hold high pressure.
3. Ability to hold both vacuum and head pressure.

If the compressor exhaust valve leaks, there are two ways to check it:
1. The high head pressure will leak back through the exhaust

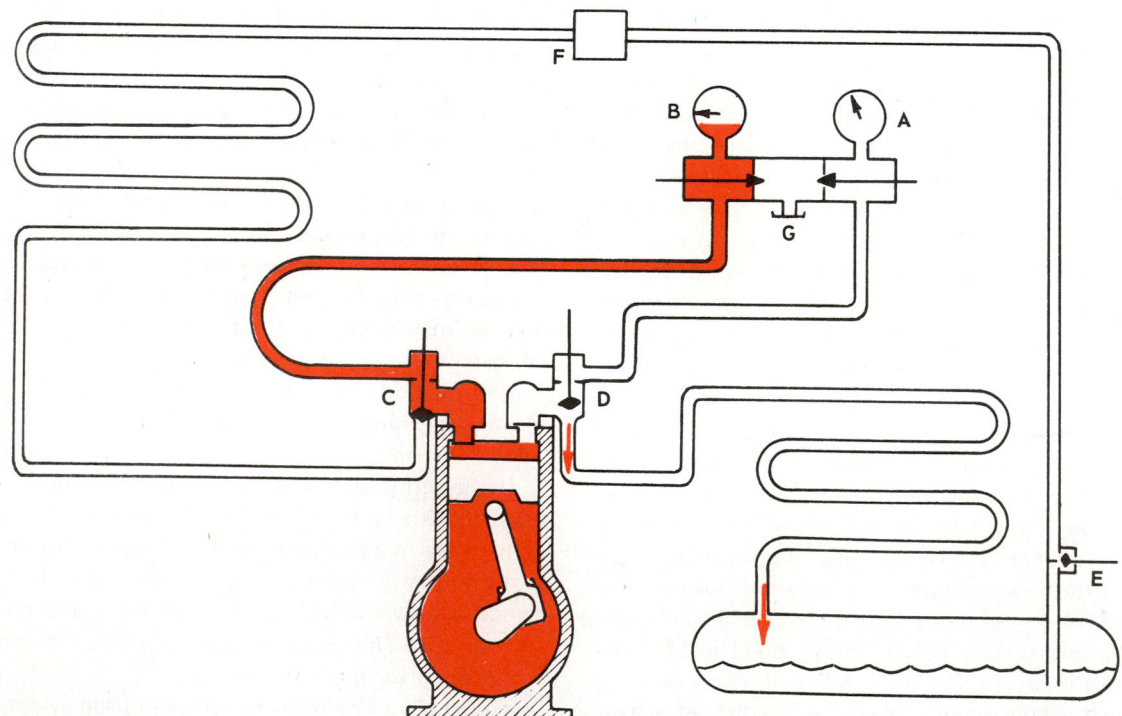

Fig. 14-44. Efficiency test for an external drive compressor. Close suction service valve and run compressor. Pumping efficiency will be indicated by maximum vacuum obtainable and time it takes to develop this vacuum.

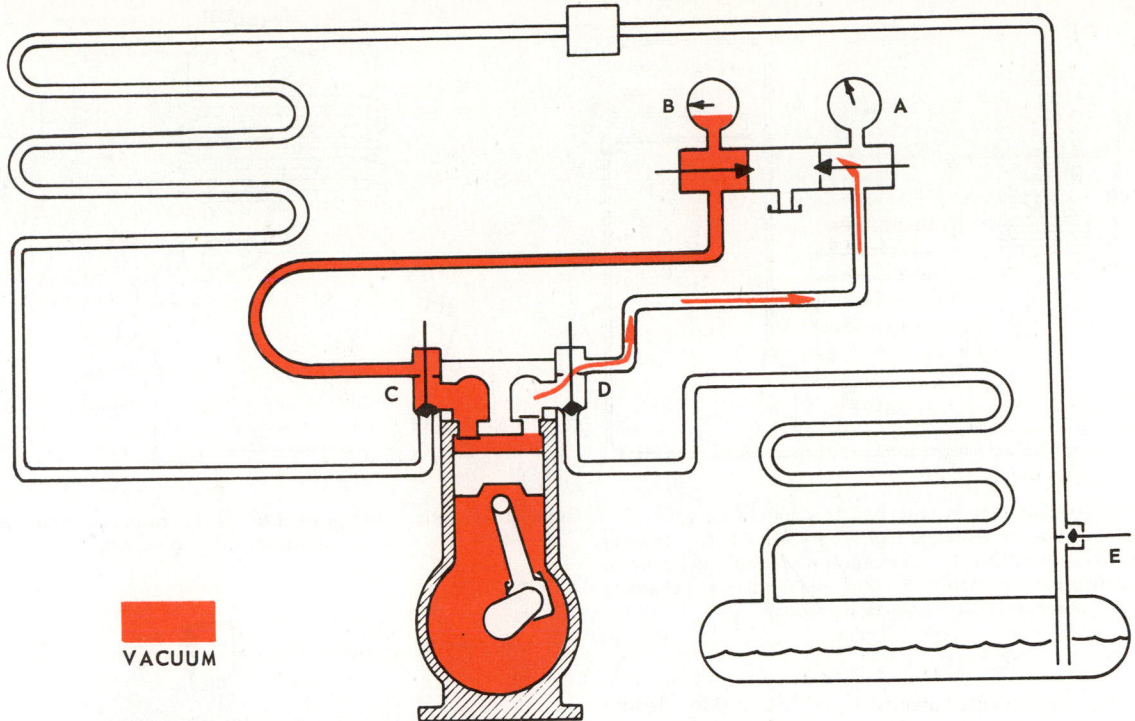

VACUUM

Fig. 14-45. Testing compressor for low-side leaks by determing best vacuum it can create with suction service valve at C turned all the way in, then turning discharge service valve at D all the way in. If air or vapor are entering compressor, high pressure will increase.

valve and produce a pressure above atmospheric on the compound gauge. Therefore, if the compound gauge creeps up above 0 psi when the compressor is idle, it is a sign that the exhaust valve needs repair. If there is a leak on the low side (seal) of the compressor, the pressure will only rise to approximately 0 psi.

2. Turn the discharge service valve stem all the way in. If the discharge valve leaks, the pressure — as indicated on the high-pressure gauge — will decrease. This is especially true if the compressor is turned by hand, because the pressure will drop as the piston goes down. Thus, if the gauge pressure fluctuates considerably, it indicates a leaky exhaust valve. If the pressure merely increases and does not drop back to any extent, the exhaust valve is not leaking.

An intake valve leak is indicated by the inability of the compressor to produce a high vacuum. However, the vacuum produced is maintained after the compressor is shut off, provided the exhaust valve is holding.

This may also be due to too thick a gasket or worn piston and rings. Lack of oil will also result in poor pumping ability.

To determine whether the crankshaft seal is leaking, close the suction service valve and pump as high a vacuum as possible on the crankcase of the compressor. Then turn the discharge service valve all the way in, as shown in Fig. 14-45. Keep the compressor running. If there is a low-side leak, the head pressure (indicated on gauge at A in Fig. 14-45) will gradually increase with the running of the compressor, indicating that gas or air is being drawn in on the low side of the compressor.

A seal leak is usually noticeable by traces of oil at the seal or on the floor underneath. Leak detectors may also be used to check a seal leak. (Crankcase pressure must be over 0 psi.)

14-34 REMOVING SERVICE VALVE

Occasionally a service valve stem will break or the threads will strip. The valve must be replaced.

If it is the suction service valve, one must remove all the refrigerant from the evaporator unit and then balance the pressure — unless the evaporator is furnished with shut off hand valves.

To remove a discharge service valve or a liquid receiver service valve, the refrigerant must be removed from the entire system. Do not pinch the lines to replace valves, as weakened tubing will shortly cause trouble. Service technicians have successfully replaced these valves by super-cooling the refrigerant in the system with dry ice. When dry ice is packed around the refrigerant containing parts of the system and when the gauges show atmospheric pressure, the system can be opened. CAUTION: Wear goggles!

14-35 REMOVING COMPRESSOR

When a compressor must be removed, the procedure is this:
1. Install the gauge manifold, as shown in Fig. 14-46.
2. Carefully test for leaks. Note connection at E is for fastening the vacuum pump to the manifold. Opening has a Schrader valve which closes the opening unless a depressor is connected to the opening and is used for purging or charging.

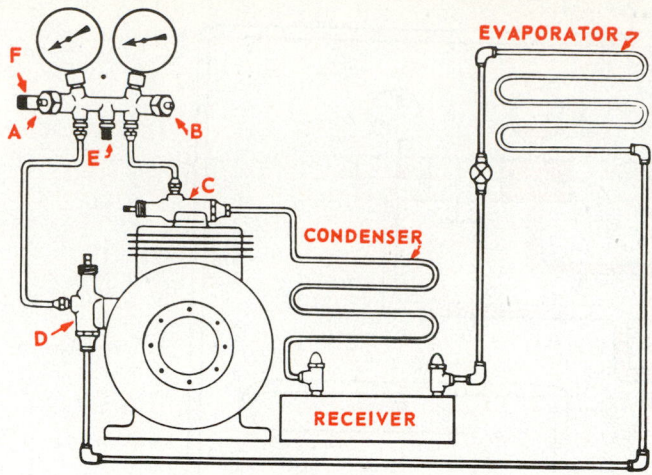

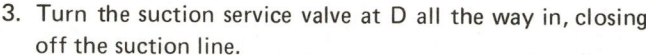

Fig. 14-46. Refrigeration system with gauge manifold installed. A—Manifold low-side valve. B—Manifold high-side valve. C—Compressor discharge service valve (DSV). D—Compressor suction service valve (SSV). E—Vacuum pump connection. F—Manifold purging and charging connection. (Mueller Brass Co.)

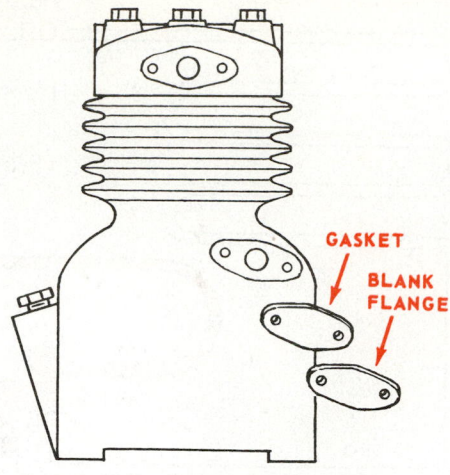

Fig. 14-47. Blank flange used to close compressor valve opening while compressor is being moved.

3. Turn the suction service valve at D all the way in, closing off the suction line.

4. Start the compressor but let it run for only a moment in order to prevent oil pumping. (Oil in the crankcase may bubble vigorously as the refrigerant boils out.) Pumping of oil is indicated by a pounding noise in the compressor and should be avoided.

5. After starting and stopping the unit two or three times, it may finally be run continuously. Keep the unit running for a few minutes after a constant vacuum is reached on the suction gauge.

6. Stop the compressor. Open the two manifold valves at A and B (Fig. 14-46) to allow the high-pressure vapor to build up the crankcase pressure to 0 psi. Then turn the discharge service valve stem at C all the way in.

7. Some service technicians crack the suction service valve at D until the compound gauge reads 0 or 1 psi (equalizing the pressures). Shut off the electric power and lock the switch in the open position. Close manifold valves.

8. Joints should be cleaned with a grease solvent and dried before opening. Unbolt the suction service and discharge service valves from the compressor. *Do not remove the suction and discharge lines from the compressor service valves.* Immediately plug all openings through which refrigerant flows using dry rubber, "cork" stoppers or tape.

9. Disconnect bolts that hold the compressor to the base and remove the belt. The compressor is ready now for removal.

10. The oil should be drained immediately and compressor refrigerant openings plugged. Fig. 14-47 shows a blank flange. These are best for plugging openings. Do not reuse old oil if it is discolored.

To keep the crankshaft seal from being abused, never rest the compressor weight on the flywheel. Always place the compressor on a block so that the flywheel hangs free.

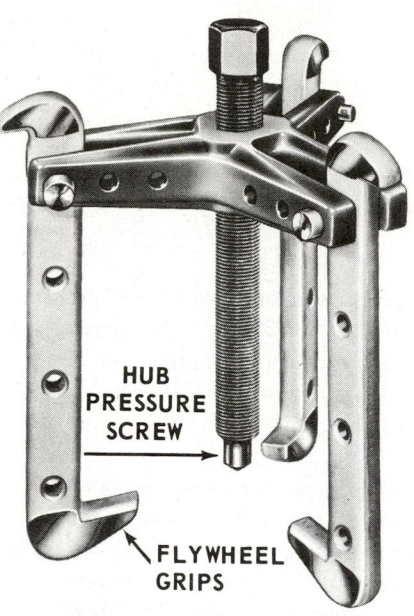

Fig. 14-48. Universal flywheel puller.

If possible, remove the flywheel before removing the compressor. Any undue strain on the flywheel may injure the crankshaft and/or the crankshaft seal.

The flywheel can be removed with a universal flywheel puller. Supplying a little heat to the flywheel hub will help while the wheel puller is drawn up snugly. Fig. 14-48 shows a universal type flywheel puller.

When the compressor from a larger unit must be removed for overhaul, the handling of the compressor — because of its weight — presents a problem. When lifting a compressor, the service technician should avoid strain from assuming an awkward position. Use care not to slip on oil or loose tools. Carts and small hydraulic hoists are available for moving heavy compressors.

Compressors are usually reconditioned by companies specializing in this work.

14-36 OVERHAULING THE COMPRESSOR

If only the valve plate needs repairs or reconditioning, the service technician can do this without removing the compressor. Para. 14-39 explains how to remove a compressor head and the valve plate; how to recondition a valve seat, assemble a valve plate and test it.

Frequently, crankshaft seals can also be replaced and/or repaired on the job without removing the compressor. First the technician removes the flywheel and the old seal. Para. 14-38 explains how to install a crankshaft seal kit.

In cases where a compressor has been removed and is to be reconditioned, this procedure should be followed:

1. Identify (tag) the compressor.
2. Clean the outside of the compressor.
3. Prepare work order form.
4. Take compressor to shop.
5. Take compressor apart.
6. Clean all parts. Place parts in tagged trays. (Fig. 14-49 shows a cleaning station.)

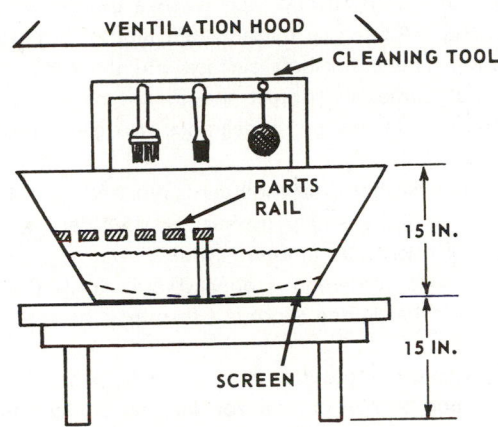

Fig. 14-49. Ventilated cleaning station.

7. Order replacement parts; tag parts needing repairs. (All precision parts must be precision inspected with dial indicators and micrometers. These parts must fit to tolerances of .001 to .003 in.)
8. Recondition parts.
9. Assemble. (Gasket surfaces must be flat, clean and free from burrs.)
10. Test compressor.

Many means have been used to clean parts of a refrigerator. Each method has its advantages. The cleaner should be a good moisture absorber while removing oil and grease quickly. It should be nontoxic, nonflammable and should evaporate quickly.

R-11 is a good cleaning solvent, being nontoxic and nonflammable. It is an excellent cleaning solvent for use in flushing systems which have been contaminated by a hermetic motor burnout. It leaves no noncondensible residue and has no reaction with insulation. R-11 can be recleaned and reused. It has a high boiling point of 74.8 F.

Also available commercially is Virginia No. 10, a degreasing solvent. It has low toxicity, is noncorrosive, and has a high flash point of 165 F.

Hand wire brushes and power wire brushes are excellent tools for removing scale and crusted dirt, but grease and oil should be removed first. **Be sure to wear goggles.**

Fig. 14-50 shows another type of cleaning station. This one uses an acid solution.

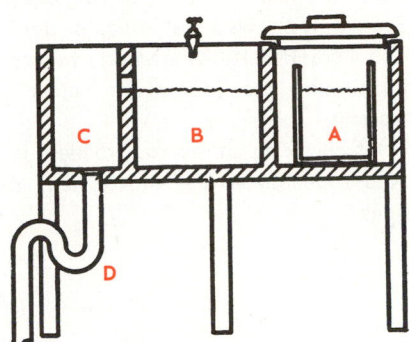

Fig. 14-50. Acid type cleaning station. A—Weak acid. B—Rinsing basin. C—Drain (overflow from B). D—Trap.

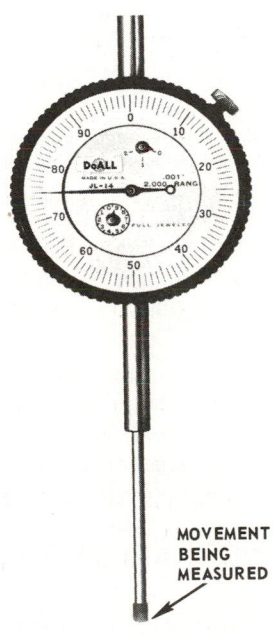

Fig. 14-51. Dial indicator used to check clearance, roundness and accuracy of surfaces. (DoALL Co.)

14-37 REPLACING BEARINGS

Compressor bearings may be of either plain, ball or roller type.

Only rarely must any work be done on the connecting rod or crankshaft bearings of compressors. These are usually plain bearings or bushings. Should work be necessary, it is a replacement and reaming process similar to automobile repair. Old bearing sleeves, usually of brass or bronze, are removed by first splitting them with a cape chisel. The new sleeve is pressed into place and line reamed to fit.

A bushing pressed into a blind hole can be removed by using hydraulic pressure. Fill the cavity with grease and then insert a shaft the same size as the ID of the bushing. Cover the shaft with a cloth to protect against flying grease and then hit the shaft a sharp blow. The hydraulic pressure created will usually push out the bushing.

Always measure a crankshaft at the journals (where it turns in the bearings) for size, taper, out-of-roundness and/or bent crankshaft. Fig. 14-51 shows a dial indicator used to determine the variation in a crankshaft. A variation of more than .001 in. makes reconditioning the journal or replacement of the crankshaft necessary.

A method of checking shafts for trueness is illustrated in Fig. 14-52. Close clearances are usually held in fitting compressor bearings.

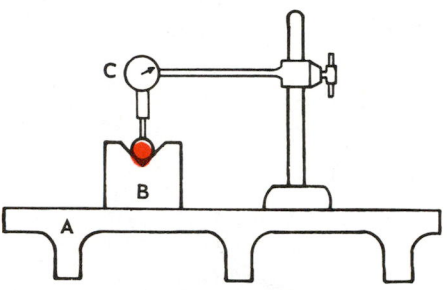

Fig. 14-52. Checking shafts, rollers and cylinders for trueness. A—Surface plate. B—Blocks. C—Dial indicator.

One of the most common sources of noise in a compressor is the piston pin. This pin must be replaced in practically all overhauls. Tolerances of .0005 in. are not too small. The fit must be very snug.

Adjustable reamers, hand operated, may be used to ream the piston and connecting rod in line. If the pin itself is badly worn, a new one must be used. Some automobile piston pins are usable in some of the larger refrigerating compressors. Badly scored eccentric connecting rod bearings must be replaced.

14-38 REPAIRING CRANKSHAFT SEAL

Crankshaft seal design and construction is described in Chapter 4. Repair or replacement required depends on the seal design.

In Sylphon seals, a squeaky noise may be caused by running the compressor with a dry seal surface. A leaky seal may be caused by a scored seal surface.

A noisy seal will soon become a leaky one if not attended to. The trouble may be fixed by the usual process of replacing or lapping the seal surface. It may sometimes be repaired by tapping the seal box lightly with a hammer. Another remedy is to wrap the bellows with oil-soaked wool yarn or string.

A leaking seal may be detected by the usual leak test. Air in the system may be the result of a leaky seal, if the system operates below atmospheric pressure on the low side. This may

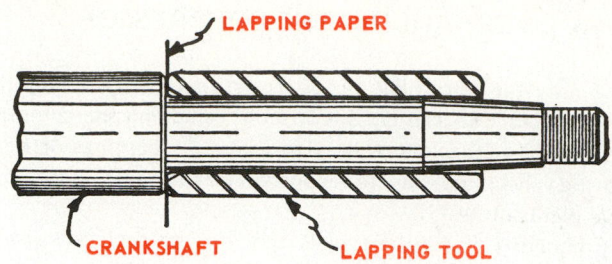

Fig. 14-53. Lapping crankshaft seal shoulder.

be detected by a high head pressure in those systems having a below atmospheric low-side pressure. A seal leak will usually cause a lack of refrigerant in systems using an above-atmospheric low-side pressure. The usual symptoms of this trouble are oil under the compressor, high power bill, constant running and poor refrigeration.

The contact surfaces of the Sylphon ring and the crankshaft shoulder must be perfectly square and polished. To lap in a Sylphon seal properly, the work should be done with a special tool made of case-hardened steel with a ground surface, as shown in Fig. 14-53. Use oil-saturated lapping compound of fine texture. The crankshaft surface and the surface of the Sylphon that comes in contact should not be scratched and should have a burnished appearance to give satisfactory service.

A scored crankshaft seal shoulder is repaired by putting the crankshaft in a lathe and polishing the shoulder face with a high speed grinder. The amount ground away must be small because the case hardening is only .015 in. to .030 in. deep. Once this case hardening is ground through, the seal will not wear long.

After the shaft is ground, it must be lapped in the lathe, using phosphor bronze or cast-iron lapping blocks and a fine lapping compound with a light, even pressure. The seal ring or the seat may be lapped also. Fig. 14-54 shows a kit used for lapping. Seal ring is on pad between pans. The use of a lapping block is shown in Fig. 14-55.

When assembling a new or rebuilt seal, a special tool may be used so that the seal mechanism will be properly aligned with

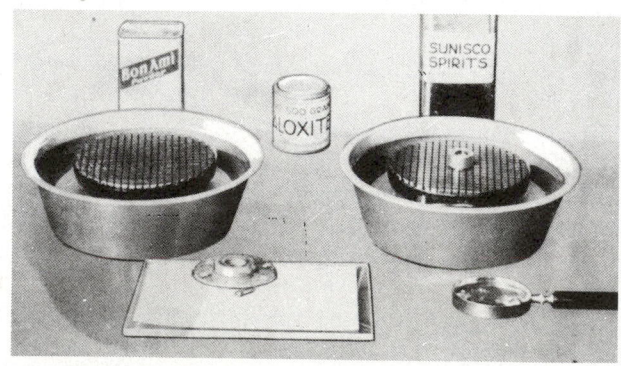

Fig. 14-54. Equipment and supplies needed to lap seals and valve plates. Note magnifying glass used for inspection and checking cleanliness needed for accurate lapping. Lapping block is shown in pan. (Norge Div., Fedders Corp.)

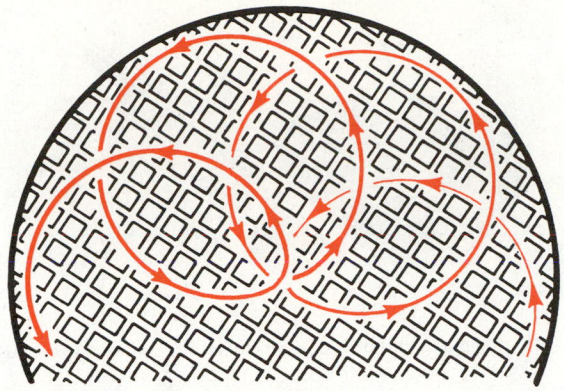

Fig. 14-55. Correct lapping procedure is a circular traveling motion of seal faces or valve plates over the lapping block above.

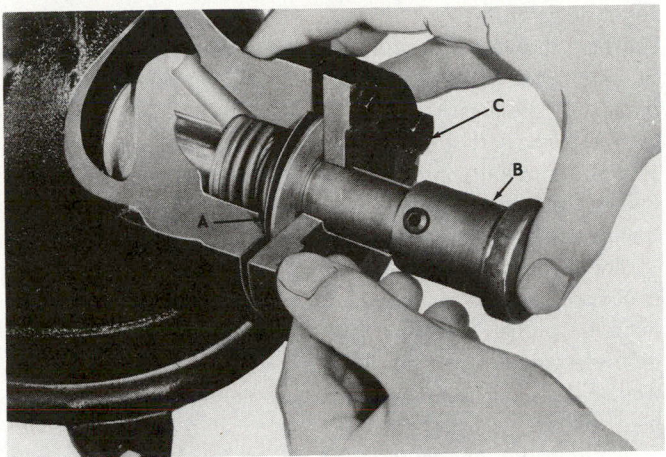

Fig. 14-56. Tool used to align shaft seals. A—Seal. B—Alignment tool. C—Seal ring.

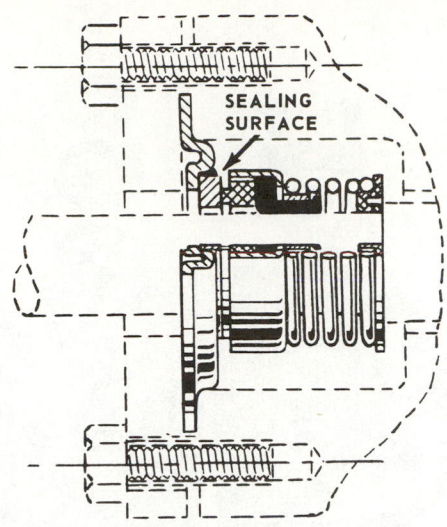

Fig. 14-57. Replacement seal for conventional compressors. Seals of this type are also used on many new compressors.   (Rotary Seal Corp.)

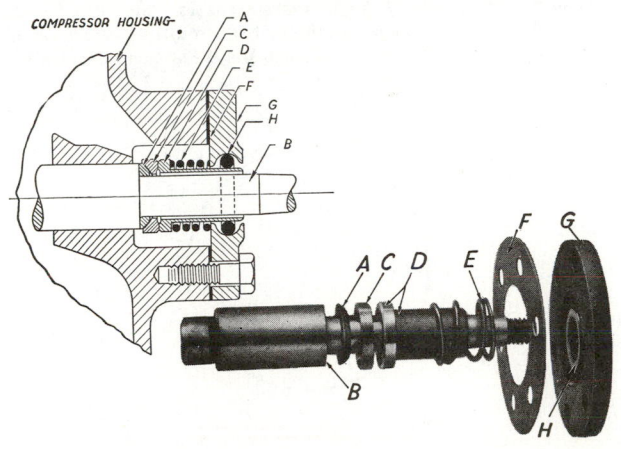

Fig. 14-58. Replacement seal. Such units are available for most external drive compressors and pumps. Letters on photo correspond to letters on cross-section above photo.   (Chicago Valve Plate and Seal Co.)

the crankshaft and seal housing. Fig. 14-56 illustrates a seal alignment tool.

Many service technicians use replacement seal assemblies. These assemblies do not require grinding and polishing. The kit replaces both the crankshaft shoulder and the seal ring.

Fig. 14-57 shows one type of replacement seal. The rotating seal ring is fastened to the crankshaft by a synthetic circular gasket (neoprene or flexible, oil-immune plastic). Another type of seal is shown in Fig. 14-58.

To install a replacement seal:

1. Remove original seal.
2. Clean the shaft with lintless clean cloth.
3. Put clean refrigerant oil on seal surfaces.
4. Install the seal parts in correct order (varies with type of seal).

14-39 REPAIRING COMPRESSOR VALVE PLATE

Carefully clean the valve plate. Resurface the valve seat with great precision. Minor repairs can be made using a lapping block and fine grinding compound. But extensive wear (erosion and pits) are removed best by grinding the complete valve plate surface on a surface grinder. Grinding should

continue until the seat is in good condition. Then finish it by lapping the surface.

The complete valve plate is usually replaced if a new or reconditioned one is available. Fig. 14-59 shows a replacement type valve plate. These units are available for most compressors.

If possible, replace the valve. Chapter 11 describes both the precautions to follow and the testing of these valve plates. Valves are also described in this chapter. There is no practical way to repair the disks or reeds, so they must be replaced if leaking. The drawing in Fig. 14-60 shows two views of a disk valve.

Exhaust valve carbon deposits (coking) indicate that the compressor had a discharge temperature too high for the oil used. The system should be checked for excessive head pressures after the repaired compressor is installed. The trouble may be a low suction pressure, a dirty or undersize condenser.

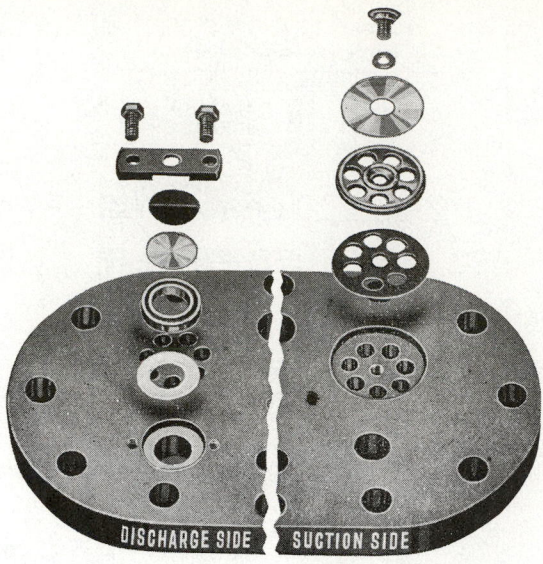

Fig. 14-59. Replacement valve plate. Special valve plates are available for most conventional refrigeration compressors. Valves and valve seats are removable and can be easily repaired. Left picture shows discharge or exhaust valve assembly. Right picture shows plate turned over to reveal intake (suction) valve assembly. Note multiple intake openings and single exhaust port. (Chicago Valve Plate and Seal Co.)

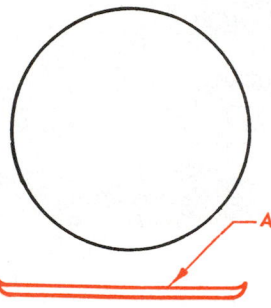

Fig. 14-60. Disk valve used frequently as an exhaust valve in compressor. Surface A should be away from the valve seat; otherwise, small burr may cause valve leak.

14-40 ASSEMBLING COMPRESSOR

Always use new gaskets when assembling a compressor. Old gaskets lose their ability to compress. Lead or special composition gaskets are often used. Thickness of the lead is usually between .010 and .020 in. Composition gaskets, when used, must be of the same thickness as gaskets removed and must be thoroughly dry (dehydrated).

If a gasket has to be made, remember the gasket must be exactly the same thickness as the original. If too thick, it will reduce the compressor efficiency. If too thin, it may cause an annoying knock.

If a compressor has been "frozen" due to a high head pressure or moisture in the refrigerant, it should be carefully cleaned and the piston and cylinder burnished. This precaution should remove all foreign substances.

When a compressor is overhauled, only new refrigerant oil should be put in the crankcase. The compressor must be

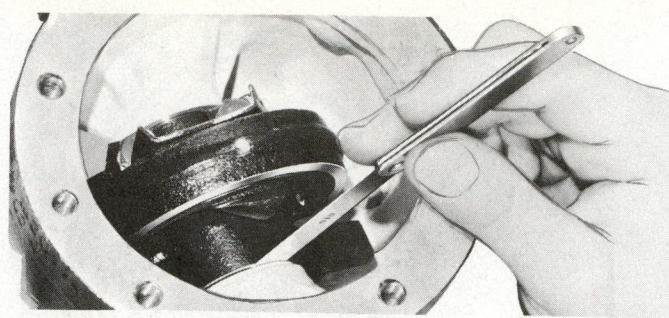

Fig. 14-61. Checking end play of crankshaft. Thickness gauge is .010 in.

thoroughly dehydrated (baked) for eight to 24 hours at 200 F. while subjected to a high vacuum. Only after this has been done can it be used. Manufacturers' specifications should be followed carefully.

There should be no noticeable end play (no more than .010 in.) in the crankshaft. See Fig. 14-61. Some have spring-loaded end thrust bearings. Main bearings must be in line.

The crankshaft seal must be clean when assembled. A drop of refrigerant oil should be put on the two sealing surfaces. Parts must be carefully aligned on the shaft. They also must be free to move to allow the sealing surfaces to press together.

The cylinder head, end bearing housing, seal plate and crankcase must be fastened to the cylinder evenly. Draw up or tighten each cap screw a little and alternate across the center of the assembly until all are tightened evenly. Use a torque wrench for final tightening. Careless tightening may warp or break parts.

All brass and copper parts of a compressor can be cleaned. A weak acid solution is used. Sometimes called a muriatic acid, this is 78 percent water and 22 percent commercial hydrochloric acid, with 1.19 specific gravity or 1/4 oz. inhibitor powder per gallon.

To make the mixture, first put the water in an acid-proof container, then add the inhibitor. Stir until mixed, then slowly add the acid. (The solution will become warm as the acid is added.) A buffing wheel can also be used to polish the parts.

Wear goggles and rubber gloves for acid work and buffing. Use tongs to handle parts being cleaned in the acid bath. Gases are formed as the solution reacts with the deposits. Good ventilation is required to prevent breathing problems and respiratory damage.

It takes from 12 to 24 hours for the solution to thoroughly clean the parts.

14-41 TESTING REPAIRED COMPRESSORS

After a compressor has been repaired, it should be tested, dehydrated, sealed and painted. Test for leaks and pumping efficiency. This may be done more easily on a shop stand, as shown in Fig. 14-62.

To test for leaks on the low-pressure side of a compressor, such as at gaskets, at the suction service valve or at the crankshaft seal, one of these methods may be used:

1. Close the suction service valve and draw as high a vacuum on the compressor as possible. Then turn the discharge

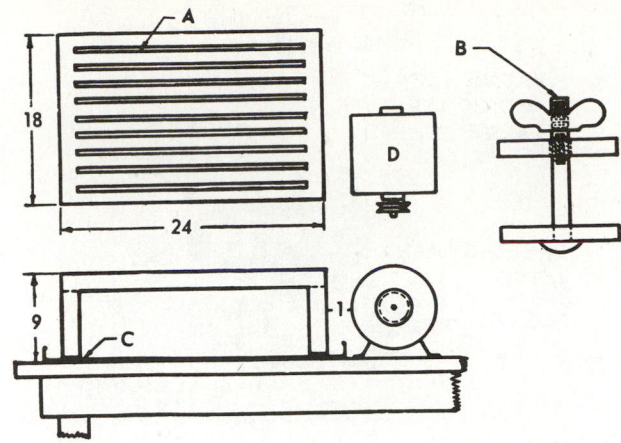

Fig. 14-62. Compressor testing stand. A—Universal clamping plate. B—Clamp down bolts, wing nut and bars. C—Oil pan. D—Motor.

service valve all the way in. Keep the compressor running. If the head pressure rises gradually, air is being drawn into the low side of the system.

2. A better way is to balance pressures in the crankcase and turn the discharge service valve all the way in. Remove the discharge service valve gauge plug and connect a 15 in. length of copper line to this opening or assembly, as shown in view A of Fig. 14-63. Then immerse the end of the copper line into a glass bottle partly filled with oil.

A gauge manifold can also be used during this test. If, when the compressor is running, the tube continuously causes bubbles in the oil, air is being admitted to the low side of the compressor. If there are no leaks, the bubbling will stop immediately after the compressor is started.

To locate the leak, put refrigeration oil around one joint

at a time. If air is leaking in at that point, air bubbles will cease while oil is being drawn in instead of air.

Exhaust valve leaks may be located by one of three ways:

1. By turning the discharge service valve all the way in after mounting the high-pressure gauge or a gauge manifold as shown in view B of Fig. 14-63.

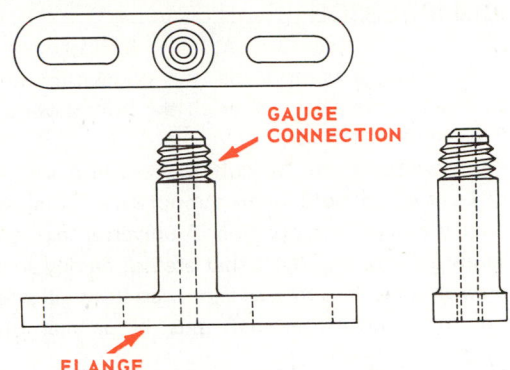

Fig. 14-64. Flange attachment which takes place of service valve on compressor while compressor is being tested.

2. When the compressor is turned over (revolved) a few turns and the head pressure rises rapidly.

3. If the exhaust valve leaks, the pressure will decrease when the compressor is stopped. Any decrease in pressure will indicate a leaking exhaust valve.

Compressor efficiency (intake valve, piston ring fit and valve action) are checked best by running the compressor at a constant low-side pressure while checking the time required to pump a head pressure in a certain size of cylinder.

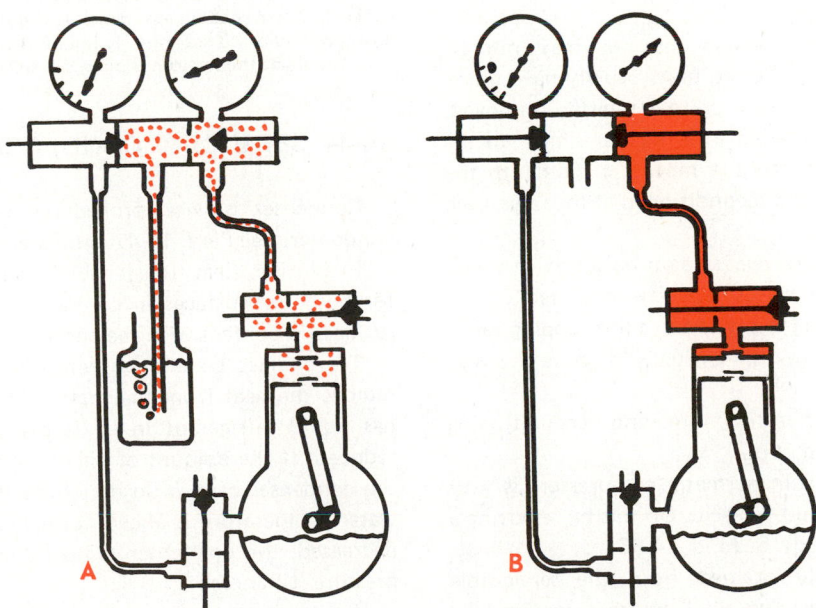

Fig. 14-63. Testing for leaks. A—Testing compressor for low-pressure side leaks. Bubbles indicate air is leaking into compressor at seal, gaskets and/or fittings. B—Testing compressor exhaust valve for leaks. If high-pressure gauge reading drops while compressor is idle, vapor is leaking back into compressor low side.

Another efficiency test is to observe the amount of vacuum a compressor will produce against a standard head pressure. If service valves are not available, a universal flange can be bolted to the compressor in place of the service valve. This is shown in Fig. 14-64.

14-42 INSTALLING EXTERNAL DRIVE COMPRESSOR

Before installing a compressor, the condensing unit base should be cleaned. Be certain that all the hold-down bolts are used and are tightened evenly.

If hold-down bolts are difficult to hold in place, masking tape may be used to hold them temporarily. String may be used to pull the bolts into position. A universal socket wrench is very handy for holding bolts that are not readily accessible.

Compressor and motor must be carefully aligned. The compressor shaft and motor shaft must be parallel. Flywheel and pulley must be in line.

Belt tension is important. The belt should be tight enough to allow only 1 in. of belt deflection with approximately 25 lb. force on it. Multiple belts should all deflect the same amount. New gaskets should always be used when mounting service valves on a compressor.

Mount the compressor on a flat level base, test with a thickness gauge and use shim stock if necessary. An uneven base will put stress on the crankcase and may cause damage or misaligned parts.

14-43 SERVICING HERMETIC COMPRESSORS

Two types of hermetic compressors used in commercial refrigeration are:
1. Welded motor compressors.
2. Bolted motor compressors.

The welded motor compressors may or may not be equipped with service valves. Bolted units usually have them.

The bolted motor compressor units are tested, removed, overhauled, retested and installed in a manner similar to the conventional compressor. Since the motor is built into the housing, it is usually tested and reconditioned at the same time as the compressor.

Welded motor compressors very seldom have service valves in the smaller units. Such valves must be mounted in the system. One method of doing this is to use a line tapping valve, as shown in Fig. 14-65. Larger units usually do have two-way compressor service valves.

Chapter 11 describes the testing, removing, reconditioning and installing of welded hermetics.

The condition of the oil in hermetic compressors is very important. It should be tested for acid. Do this by removing a sample and using an oil test kit. See Fig. 14-66.

An oil sample is easily drained from the serviceable hermetics. Welded hermetics usually require removing the motor compressor and draining some oil out of the suction line opening.

When replacing motor compressors, always use an exact replacement.

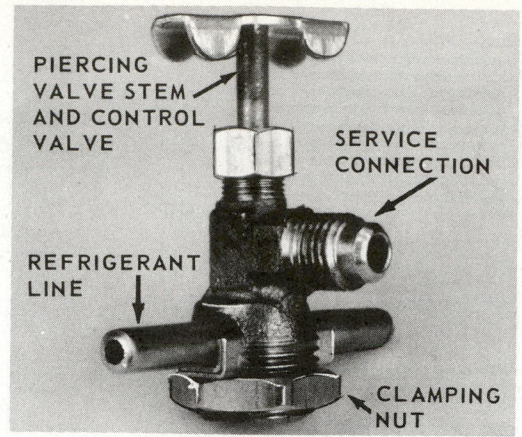

Fig. 14-65. Special valve can be mounted on suction line to pierce refrigerant line for pressure testing, charging and/or discharging purposes. (Mechanical Refrig. Enterprises)

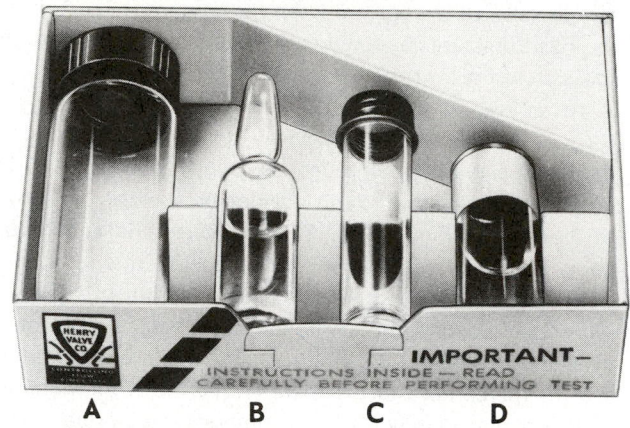

Fig. 14-66. Kit tests oil for acidity. Oil sample is put in container D. Oil and acid test solutions in containers B and C are poured into container A. After shaking, container A is put on bench for a minute or so to allow liquids to separate. If liquid at bottom is clear, oil is acid. If liquid at the bottom is pink, oil is OK. (Henry Valve Co.)

14-44 SERVICING CONDENSERS

Condenser service procedures depend on the type of condenser. See Para. 14-47 and Para. 14-49.

In all cases, heat transfer surfaces must be clean. This is true for both the surfaces in contact with the refrigerant and the surfaces in contact with the cooling medium (air or water).

There must be enough refrigerant vapor space (area) to remove the heat from the vapor. For example, if a condenser has liquid refrigerant in it (is overcharged), cooling will be reduced. If the amount of air or water flowing is not enough, the condenser cannot do its job. If the air temperature or the water temperature is above normal, the condenser capacity is decreased; the condenser temperature will rise and the head pressure will increase.

If there is air in the system, it will collect in the condenser. (It cannot condense and will be held back by the liquid trap in the receiver or at the lower end of the condenser.) Each pound of air pressure will increase the head pressure by 1 lb. This increase in pressure will reduce the pumping efficiency of the compressor and will increase the condensing temperatures.

Excessive head pressures are very hard on a system. High exhaust temperatures will cause formation of sludge, carbon — even acid — and will injure reed valves.

High head pressures may be cause by:
1. Excessive low-side pressure.
2. Poor cooling by air or water.
3. Air in the system.
4. Overcharge.

Check the low-side pressure with a low-side pressure gauge. If the trouble is poor cooling of the condenser, make following checks and adjustments.
1. Air cooled: Condenser must be clean — very clean. Use high pressure jet of air.
   Wear goggles!
   Use mechanical scrubbing or high-pressure water (jet) with detergent.
   Some air-cooled condensers may be cleaned by the use of a vacuum cleaner.
2. Water cooled: Is there enough water flow? Check outlet temperature and inlet temperature. Rise should not be over 15 F. (8 C.). Is there scale in water tube? Inspect inside of water tube. Clean with cleaning solution.

For air in the system:
1. Purge the condenser while compressor is stopped.
2. Take precautions against freezing. This is necessary with water coolers and water chillers.

For an overcharged condition: Purge the system while compressor is stopped.

Remember that the purpose of the condenser is to remove heat. The condenser will fail to do its job if the heat transfer surfaces are inefficient or if the heat removing medium (air or water) is not in the correct volume or temperature.

14-45 SERVICING AIR-COOLED CONDENSERS

The correct pressure in an air-cooled condenser may be determined by first adding 30 F. or 35 F. to the air temperature to get the refrigerant temperature on the inside of the condenser. Then, using this corrected temperature, refer to the refrigerant charts in Chapter 9 for the correct head pressure.

If pressure is above normal and there is enough air movement (both air-in passages and air-out passages must be free), there is a good chance there is air in the system. There is also a chance that the unit is overcharged. If the pressure is below normal, it is possible that the unit is undercharged.

Most commercial air-cooled condensers are of the forced convection type. They have one or more fans for moving air through the condenser. Some larger fans are belt driven and others are direct driven. Fans, motors and belts need regular maintenance and service. Belt service operations are described in Chapters 7 and 22 as well as Para. 14-42.

Multiple fan condensers sometimes have sequenced fans. This means that when more condensing is needed, all fans operate. As the condensing load decreases, first one fan is shut off, then the second, and so on. Sequence controls should be checked if the fans fail to operate.

Other systems use variable speed fans. Some outdoor air-cooled condensers have thermostat controlled louvers which are powered to partly close or completely close as the outdoor temperature decreases. If the head pressure is too low, the refrigerant control capacity is decreased.

14-46 REMOVING AIR-COOLED CONDENSER

If a condenser is leaking or needs replacing for some other reason, it usually must be removed from the system. Before removal, the liquid refrigerant must be removed from the condenser and pressure must be adjusted to atmospheric pressure.

Usually, the procedure is this. (Be sure to wear goggles.) Close the valve between the condenser and the liquid receiver and purge the condenser by removing the gauge plug from the discharge service valve. Then open the valve until the condenser pressure is down to atmospheric.

Be careful. There may be oil in the condenser and it is best to connect a purge line to the gauge opening of the valve and run the purge line outdoors and into a container to trap the oil.

If there is no shutoff valve between the condenser and the liquid receiver, refrigerant can be saved by pumping it into a cylinder, as shown in Fig. 14-67. The refrigerant cylinder usually has to be cooled during this operation or its pressure will quickly rise to dangerous levels. Run the compressor intermittently and put the cylinder in a tub or bucket of running cold water or ice water.

Some service technicians put a spare condenser between the compressor and the service cylinder. This speeds the condensing operation.

The liquid receiver may need to be heated to vaporize the liquid in it. Use warm water. Never use an open flame.

If the refrigerant is not be saved, purge the refrigerant outdoors. Be sure to use an oil trap in the purging line.

To remove a condenser, first clean the condenser as well as possible. Brushes, vacuum cleaner, air or nontoxic refrigerant jets, carbon dioxide and/or nitrogen jets may be used. Wear goggles! CAUTION: See Para. 11-39.

Most air-cooled condensers are housed in a protective shroud which also serves as an air duct. On some of the larger units these sheet metal parts are heavy. Handle with care. Gloves and safety shoes are recommended. Be sure to save sheet metal screws and/or assembly bolts in a container.

Fans, fan brackets, belts and motors may need to be removed on some units. These parts should be labeled and stored for reuse. Be sure the fan blades are not nicked or bent. This may put them out of balance and decrease their efficiency.

If electrical connections are removed, label them. Use masking tape and a marking pencil.

Always clean the connections before disconnecting the condenser from the unit. Immediately plug the refrigerant openings to keep the internal refrigerant passages clean and to avoid oil spills as the condenser is moved.

Avoid damage to condenser fins. Wood or cardboard protectors taped over the corners of the fins will provide protection. Because fins are sharp, always use gloves when lifting or carrying a condenser.

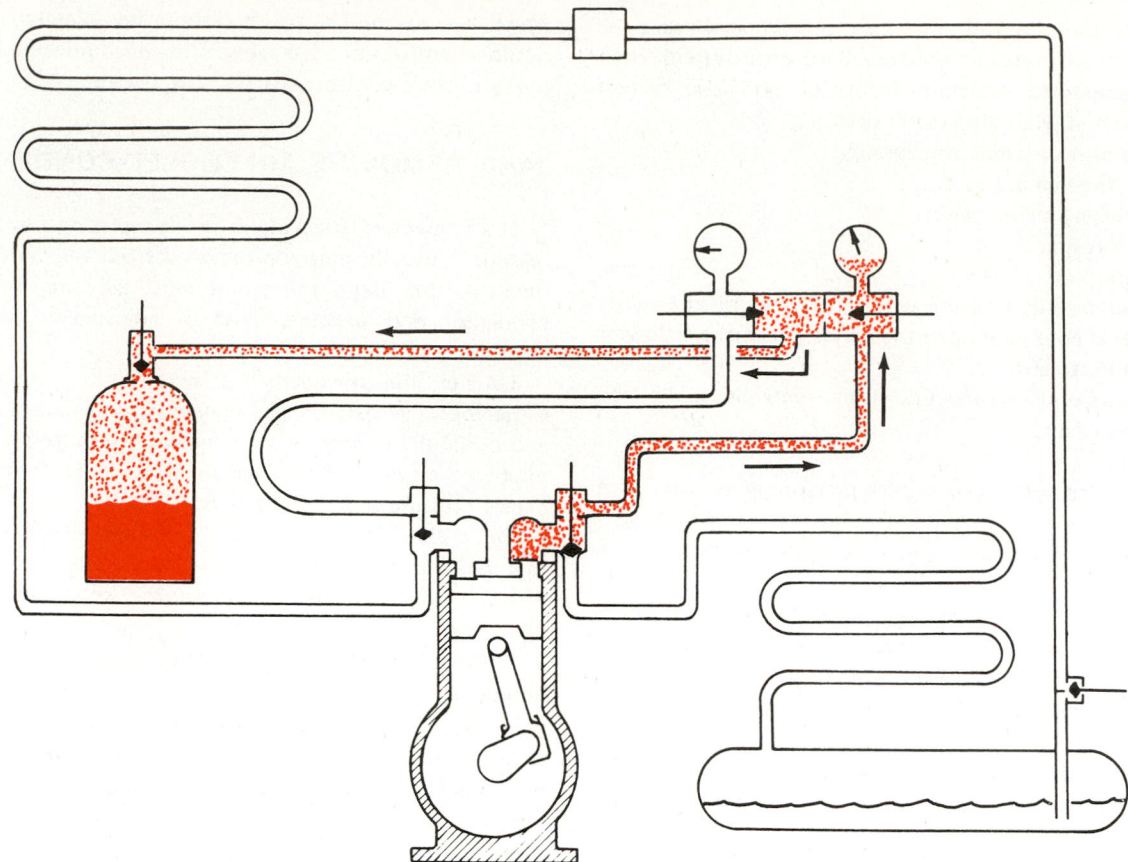

Fig. 14-67. One way to discharge sytem. Refrigerant is being pumped into refrigerant cylinder.

14-47 REPAIRING AIR-COOLED CONDENSER

A leaking condenser can be repaired. First clean the condenser and flush the inside of the refrigerant tubes. If a brazed joint is leaking, clean the outside of the joint, put flux on it, heat it and take the joint apart.

Clean the brazed surfaces, flux the male part of the joint, assemble, support the joint and braze. Remove the flux.

If a tube is cracked, remove the damaged part and replace with a new tube section. Silver braze the new part in place.

Fins may be straightened using a fin comb, as in Fig. 14-68.

To test a condenser, plug one end of the condenser and connect a refrigerant cylinder to the other end. Build up a refrigerant vapor pressure in the condenser and test for leaks. Use one of the following methods:

1. Bubble test.
2. Halide torch.
3. Electronic leak detector.
4. Immerse condenser in water.

Inspect fittings (and flares, if used). These connections must be in good condition.

14-48 INSTALLING AN AIR-COOLED CONDENSER

Protect the condenser fins and the condenser tubing, including return bends, at all times. Mount the condenser

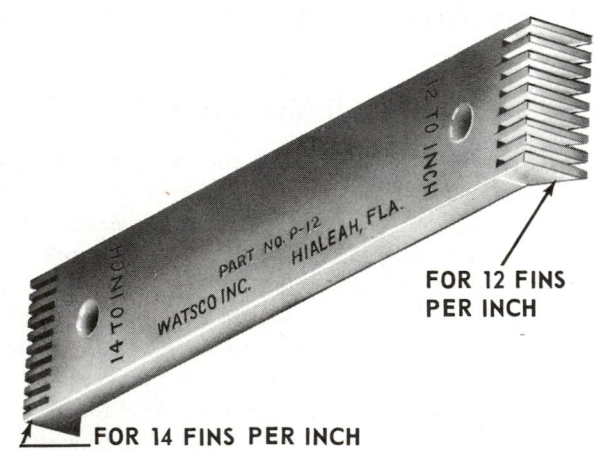

Fig. 14-68. Plastic comb used to straighten condenser and/or evaporator fins. (Watsco, Inc.)

securely in its frame. Install it as level as possible. Connect the condenser to the compressor and liquid receiver, if one is used.

Flare connectors should be carefully aligned. The fittings should be in line so that it is not under tension or forced in any way. If the fittings are out of line or are under strain, threads on the fittings or the flare may be damaged. Brazed connections must also be carefully aligned before brazing.

The surface to be brazed should be cleaned with clean steel

wool, wire brush, or dry sandpaper just before assembling the joint. Flux should be put on the outside of the male part of the fitting only. The joint should be supported during the brazing operation. See Chapter 2 for instructions on brazing. Use asbestos or metal sheets to protect other parts of the unit from the brazing flame.

Use the refrigerant in the system or a service cylinder to purge the air from the condenser. (Purge outdoors or out of the room if possible.) Build up a pressure of about 15 psi in the condenser and test for leaks. If a leak is found, do not attempt to patch the leak, but rather take the joint completely apart and do it over. If no leak is found, increase the pressure to approximately 100 to 170 psi and test for leaks again.

If no leaks are found, install fans, belts and motors as needed to operate the system. All these parts should be cleaned prior to assembling.

Run the system. (Keep clear of the fan, belts and pulleys as they may cause serious injuries.) Check refrigerant charge and operation of the unit.

Test for leaks again. (Shut the unit off to stop airflow.) If the unit is OK, install the shroud or casing. These parts should be securely fastened or rattles and inefficient airflow may result.

14-49 SERVICING WATER-COOLED CONDENSERS

Water removes heat from metal surfaces about 15 times more rapidly than air. Therefore, water-cooled condensers are much smaller than air-cooled condensers. Since the water is usually colder than air, the condenser temperature and pressure can be lower.

Usually the condenser is designed to permit a water temperature rise of 10 F. (6 C.) as it goes through the condenser. If the water-in temperature is high, the condenser temperature will be high and vice versa. Care must be taken to avoid freezing the water circuit of water-cooled units.

If the unit is water-cooled, add 10 F. to 15 F. (6 C. to 8 C.) to the temperature of the water as it is leaving the condenser to determine what the refrigerant temperature should be. The correct head pressure may then be taken from a refrigerant chart. The unit must be running for these conditions to hold true.

If the head pressure exceeds this value by more than 5 lb., stop the compressor. Purge the system through the discharge service valve gauge opening for 10 to 15 seconds. Then run the condensing unit again.

To determine whether the trouble is excess refrigerant, or air in the system, stop the unit and purge as before for 15 to 20 seconds. If the pressure drops somewhat, the trouble was air in the system.

If the pressure does not drop, continue to purge the unit until that part of the condenser and liquid receiver which is full of liquid refrigerant becomes cold. Do not allow the temperature to go lower than 32 F. (0 C.) or the water tubes may freeze and burst.

Some condensing units have small valves that can be used to check liquid levels. A liquid level sight glass is installed in some large units. It is connected to the top and bottom of the receiver. When the pressures are equalized, the liquid level in the sight glass will be the same as that in the receiver. This will quickly reveal the level of the refrigerant in the condenser and liquid receiver, and one may easily judge whether this is the correct amount.

Always leak test the water leaving the condenser; there may be a leak between the water tubing and the refrigerant in the system.

An excess of oil in a system is indicated by erratic refrigeration and a constant slugging or pumping of oil in the compressor, especially just as it starts. (Erratic means that sometimes it works and sometimes it does not work.)

A common water-cooled condenser problem is formation of deposits from the water on the tubing walls. This deposit (carbonate, sulphate and the like) acts as an insulation. See Fig. 14-69. If it cannot be removed the condenser must be replaced.

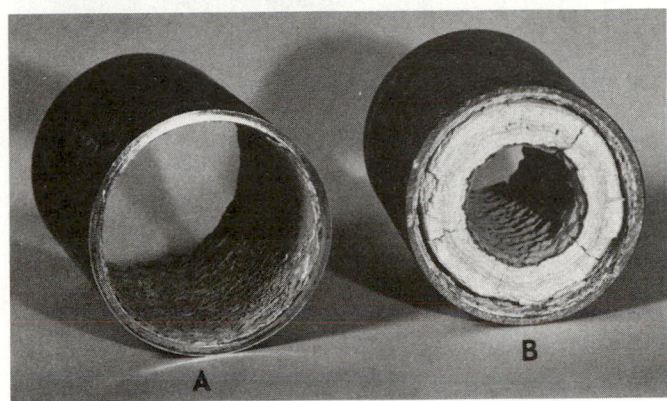

Fig. 14-69. Scale deposits. A—Tubing with treated water. B—Tubing using hard, untreated water.   (Thermal Engineering Co.)

Cleaning condensers has two different operations:
1. Preventing scale.
2. Cleaning the system.

A sulphuric acid-chromate solution is good scale prevention in open systems (cooling towers, evaporative condensers), but chromates are a poison and no significant amounts should be allowed in waterways. (Wear goggles!)

Sulphuric acid in weak solution reacts with steel, while sulphuric acid in strong solution reacts with copper. It is best not to use it at all for scale prevention.

Careless mixes of sulphuric acid and water can ruin a copper piping system and a steel structure very fast. Some do use acids to descale, but it must be done as fast as possible and then cleaned as fast as possible. Wear goggles and rubber gloves.

For cleaning condensers, it is safest to use only prepared chemicals from a good company. Carefully balanced cleaners such as inhibited muriatic acid or sulphuric acid do a good job without damage to the equipment, if properly used.

One can recognize a corroded condenser by checking the liquid line temperatures of the refrigerant. If the amount of refrigerant is correct and if the other troubles just mentioned are not found, a corroded condenser will produce a hot liquid line in a water-cooled condenser. Temperature and pressure in the condenser will be considerably above that to be expected. If all other possible causes of excessive head pressure are eliminated, a badly corroded or dirty condenser is probably the cause.

Soft scale deposits can be removed from some water-cooled condensers by using a power-driven wire brush, as demonstrated in Fig. 14-70. This can usually be done without removing the condenser from the system. However, the water circuit must be closed and the unit shut down. Always use new gaskets and tighten assembly cap screws evenly.

When the condenser water tubes have a hard lime or iron

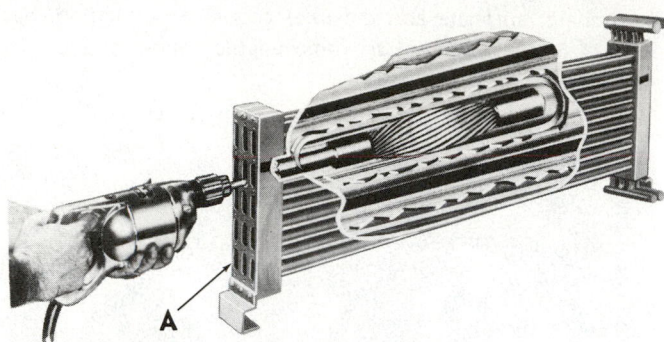

Fig. 14-70. Tube-within-a-tube condenser being cleaned with power-driven wire brush. A—End plates removed. (Halstead and Mitchell)

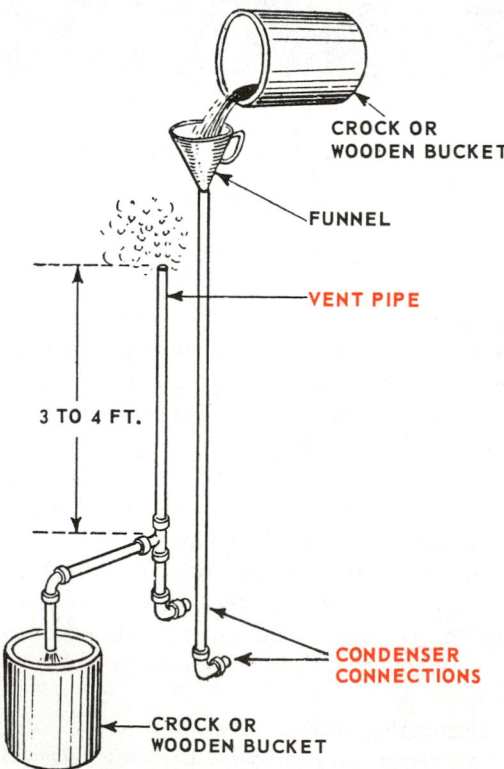

Fig. 14-71. Cleaning water-cooled condenser water tubes. Using dilute hydrochloric acid solution, connect vent pipe to upper condenser connector as shown. (Copyright, Frigidaire Co.)

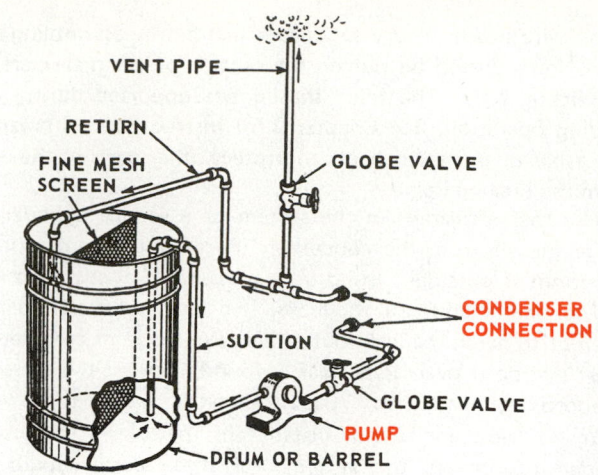

Fig. 14-72. Condenser water tube cleaning using a forced circulation system for acid solution. Note screen in drum to keep scale from entering pump. Drum must be acid proof. Use ceramic, crock, glass or glass-lined tank. (Copyright, Frigidaire Co.)

deposit, they may be cleaned with the acid solution described in Para. 14-40. The procedure is as sketched in Fig. 14-71. CAUTION: Always wear goggles, rubber gloves, and a rubber bib apron when using acid solution.

Fig. 14-72 shows cleaning being done with a forced circulation system. The pump must be designed for an acid solution. Fig. 14-73 pictures a pump which can be put into the solution. Note the use of plastic tubing instead of pipe.

Warm solutions will react faster than cold. But do not heat above a warm temperature. Avoid spilling solution on the skin, clothing or on the floor.

Crocks and barrels used must be made of acid-resisting materials. Wood or ceramic materials are acceptable but do not use galvanized or aluminum containers.

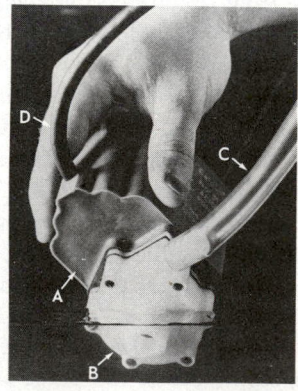

Fig. 14-73. Pump designed to pump cleaning acids through the water tubes of water tube condensers. A—Motor. B—Pump. C—Acid fluid outlet. D—Power cord. (Calgon Corp.)

14-50 REMOVING WATER-COOLED CONDENSER

A leaking condenser should be removed and repaired or replaced. Removing a water-cooled condenser is similar to removing an air-cooled condenser except for the disconnection of the water line. First, close off the water circuits; then disconnect the water lines and drain all the water from the condenser. (It may freeze.) Then remove the refrigerant from the condenser and isolate the condenser. Do not allow freezing temperatures.

If a pressure-operated water valve is used, leave it installed in the system, if possible, because of its refrigerant tubing connection.

Clean the outside of the condenser, wipe or mop the water and clean the condenser connections. Dry the connections thoroughly. If the connections are mechanical, use wrenches of proper size. Wear goggles. Plug the refrigerant openings at once using good plugs. Either synthetic rubber expanding plugs or flared plugs are recommended.

If the condenser is large and heavy, two or more individuals should handle it. Or use a lifting device.

14-51 REPAIRING CONDENSERS AND RECEIVERS

When either a condenser or a receiver does not work properly, it is usually cheaper to replace the unit than to repair it. Welding or brazing is sometimes used to repair leaks, but it requires careful work. If possible, all joints should be welded or silver brazed to insure a lasting leakproof joint.

A welded shell type condenser can be cut open, a new water coil installed, and the shell rewelded. But this is done only in an extreme emergency. The reason is this: pressure vessels should be made under well controlled conditions by a pressure vessel certified welder. Then it must be tested to at least twice the operating pressure.

Liquid receivers in most commercial systems serve as refrigerant storage cylinders. They usually have a welded steel shell.

The shell and coil condenser — liquid receiver that has a water coil built into it — sometimes develops leaks. Because of the corrosive action of the water and refrigerant under certain conditions, the copper tubing used to carry the water is eventually eaten through. The leaking tube lets refrigerant from the system into the cooling water. A leak of this kind may be found by checking for refrigerant at the water drain. Leaks sometimes occur at the joints where the water-cooling coil is attached to the liquid receiver. Such damage may be due to abuse or to corrosive action. In such a case, the condenser should be replaced with a new one.

If new parts cannot be found, a fairly satisfactory repair may be made as follows:
1. The water tubing — if eaten through within the liquid receiver — must be removed.
2. The liquid receiver should be mounted on a lathe and the end of the receiver cut open. This permits removing the old water coil and putting in a new one. The replacement unit must have the same length of tubing as the one removed, or the capacity of the condenser will be changed. The new coil is usually made up by winding it on a drum mounted on a lathe. The tubing is then put in the liquid receiver and the joints are silver brazed. The end of the liquid receiver that

was cut off may now be replaced and welded after the interior has been cleaned thoroughly.

A bolted condenser can be dismantled quite easily to replace the water tubing.

All liquid receivers above a certain size must be equipped with safety release valves. Receiver repair should be attempted only with permission of local inspectors.

14-52 INSTALLING WATER-COOLED CONDENSER

Installing a water-cooled condenser is similar to the installation of an air-cooled unit. The mounting of the condenser, brazing of joints and leak testing, should be done with the same care. On a water-cooled unit, the leak test should include the exhaust water.

Connect water lines according to the local plumbing code. Test the water circuit for leaks also. All parts should be cleaned before assembly. This includes the assembly devices. Be sure to test for leaks after the unit has run for a few hours and the system is in normal operation.

14-53 SERVICING COOLING TOWERS

Evaporative condensers and cooling towers also collect deposits from the cooling water. These deposits must be removed periodically or they will act as insulation. Deposits may be reduced by using water softening chemicals. Such chemicals can be bought from wholesale supply companies.

The treatment of water is necessary. Chemicals in the water are measured by a pH factor. The scale of pH is from 1 to 14, with 1 through 7 indicating an acid solution and 8 through 14 indicating a basic condition. Chemicals may be added to the water to create a 7 or 8 condition in the water.

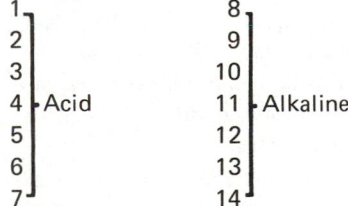

pH
Intensity factor (parts per million in quantity)

1		8	
2		9	
3		10	
4	Acid	11	Alkaline
5		12	
6		13	
7		14	

When testing water for pH factor (acidic or basic), the temperature of the water is important. The warmer the water, the more active the reaction to probes or color testers. It is best to test the water between 70 F. and 80 F. (21 C. to 27 C.). Water near boiling is about 15 percent more active. Very cold water (near freezing) will give readings about 5 percent below the true value.

Other chemicals may be used to lessen algae, mold, and slime growths. If deposits have formed, they can be removed by scraping or by using a weak acid solution. This is followed by a soda solution rinse and wash. A water-cooled condenser must be protected by electric heater or automatic drain controls if it is exposed to freezing temperatures.

If a water-cooled condensing unit is to shut down, and perhaps be subjected to below freezing temperatures, the condenser coils must be completely emptied of water. This may be done by blowing out the coils with air, nitrogen or carbon dioxide. *Do not exceed a 50 to 60 psi pressure or the system may be damaged.*

The water drain valves should be left open to allow drainage of residual water in the piping. Be sure the drain plug of the circulating pump is removed and left loose.

Cooling towers need regular maintenance. Once a year repair any corrosion spots. It is good practice to do the following about once a month:

1. Inspect fan and motor bearings; oil sleeve type; grease (with water inhibitor) ball bearing types.
2. Inspect belt tightness and alignment; adjust, if necessary.
3. Clean strainer.
4. Clean and flush pump.
5. Inspect water level; adjust float, if necessary.
6. Inspect spray nozzles and clean if necessary.
7. Inspect water level bleed; it must be working.
8. Inspect air inlet screens and clean, if necessary.
9. Inspect water for algae or leaves or other particles of dust.

During cool weather, cooling towers and evaporative condensers create a fog exhaust. The high humidity air leaving the unit condenses enough moisture (dew point condition in the air) to create this fog. The fog condition can be reduced by cutting down cooling tower operation in cold weather. It can also be reduced by increasing airflow and reducing water flow.

Cooling tower drain lines should be as carefully planned and installed as the refrigerant and water lines. They must be large enough to be easily cleaned and must have clean-out connections. The joints must be leakproof. Piping must have a down slope all the way (1/4 in. per foot for horizontal runs). If a rise is unavoidable, a pump must be used.

All lines exposed to a freezing temperature should be insulated and heated. Heating tape can be used.

Most outdoor air-cooled condensers, cooling towers and evaporative condensers use motor-driven fans. The motor bearings, belts and fan bearings are exposed to wide temperature changes, moisture and dirt. The plain bearings require a special lubricant that will not wash out under moisture conditions. Silicone oils and greases are very satisfactory. The same lubricants are also good for ice makers, water pumps and hydronic heating pumps.

Good city water systems deliver water with about 120 parts per million (ppm) of dissolved solids. Water of this nature can be cycled through a cooling tower only about six times before scale starts to form. For this reason, most water must be treated. Moreover, there must be a bleed-off device to keep the evaporating water cycle and system clean.

Water which evaporates should be in vapor form. If any small droplets of water are also exhausted from the cooling tower, these droplets will contain solids and add to air pollution.

One of the new ways to treat water is the electrostatic method. The device exposes untreated water to an electrostatic force. The force loosens the bond in the scale forming chemical. The loosened bond prevents scales from forming and prevents the impurities from combining. It even loosens scales already formed on metal surfaces.

The basic theory behind water treatment is to reduce the ion content in the water to a condition where the salts and foreign matter are made electrically neutral. These materials then remain in solution.

The electrostatic treatment must be adjusted to the raw water condition. Electrical instrumentation and chemical analysis is used to determine the adjustment.

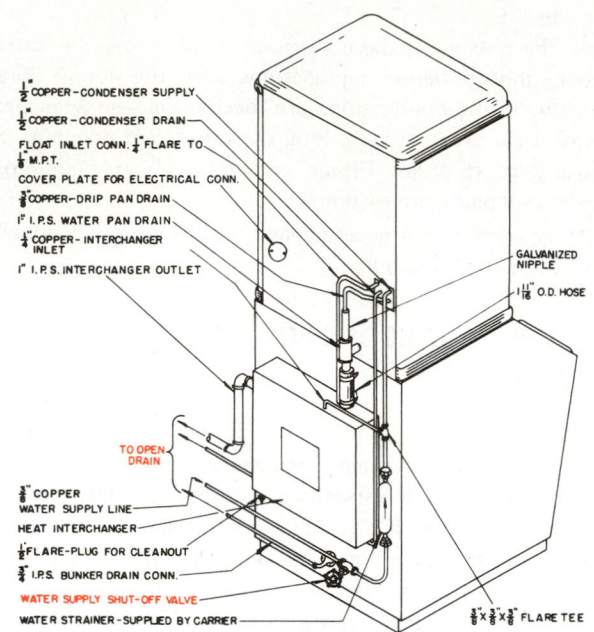

Fig. 14-74. External plumbing installation for automatic ice maker. (Carrier Air Conditioning Group, United Technologies Corp.)

14-54 SERVICING ICE MAKERS

Ice makers have refrigerating systems similar to other cooling applications. However, they have special water circuits, defrosting devices and ice cube or flakes moving devices. Fig. 14-74 shows an ice maker piping installation. Many of these units have a water pump for recirculating the water used for

Fig. 14-75. Water circulating pump and water float control for ice maker. (Ross-Temp Inc.)

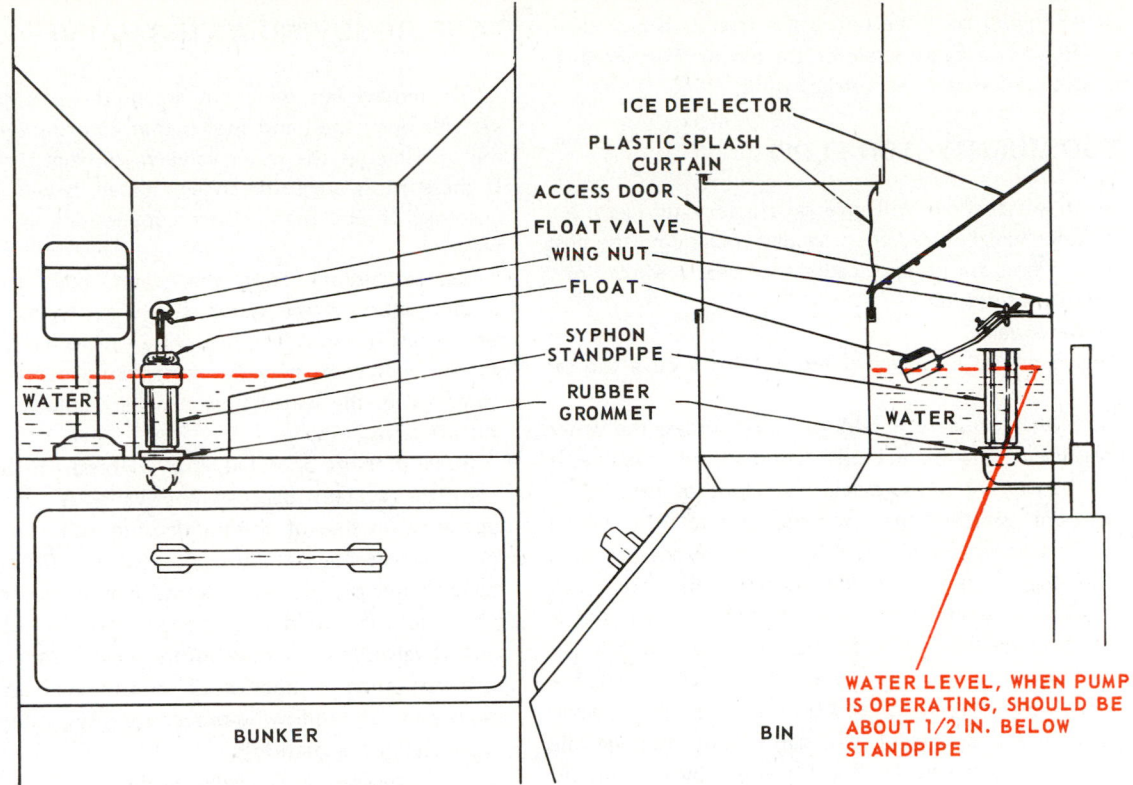

ICE DEFLECTOR
PLASTIC SPLASH
CURTAIN
ACCESS DOOR
FLOAT VALVE
WING NUT
FLOAT
SYPHON
STANDPIPE
RUBBER
GROMMET

WATER

WATER

BUNKER

BIN

WATER LEVEL, WHEN PUMP
IS OPERATING, SHOULD BE
ABOUT 1/2 IN. BELOW
STANDPIPE

Fig. 14-76. Water pump and water level control for typical automatic ice maker.

making ice. Fig. 14-75 shows a water pump and motor installed. Just in front and below the motor, is the float control for water makeup. Motors, pumps and water level float controls should be checked frequently. Fig. 14-76 shows one type of installation.

Those parts of the ice maker which are in contact with water should be cleaned about once each month. Special commercial cleaners for ice makers are available.

14-55 SERVICING WATER VALVES

Water-cooled condensers often require attention because of incorrect water flow. This trouble is sometimes due to the water valve or to the screens in the water circuit. See Chapter 12 for details on how these valves are built.

The main job of the water valve is to provide water when the unit is running and to stop the water flow when the unit is idle. Some of the trouble caused by the water valves are:
1. Too little water flow.
2. Too much water flow.
3. Water flow does not stop when the unit is idle.

A water control valve will only operate correctly if the installation is correctly made. Water must also be clean.

14-56 RESTRICTED WATER FLOW

If water flow is too little, the trouble might be:
1. Leaking valve.
2. Clogged or partially clogged screen.
3. Chattering valve.

4. Valve adjustment turned out too far.
5. Sediment-bound valve.
6. Leaking bellows.

In addition to these, the water-cooling system may present other problems. Some water-cooled systems use a length of rubber or plastic hose between sections of the water pipe.

This hose is run along the wall and the condensing unit water lines to eliminate the transmission of the condensing unit vibration into the plumbing system of the building and the breakage of tubes from this vibration.

The cold water inlet hose connection presents no difficulties and this hose will ordinarily give many years of service. However, the water outlet hose may decompose and clog.

Occasionally, someone may partially or completely shut off the water supply by closing the hand-operated valve installed in the system. A service technician should always put signs near the shutoff valves warning of the effect on the refrigerating unit if these valves are closed.

There are two common complaints which indicate troubles with water circulation. They are a lack of cooling in the condensing unit and too great a consumption of water.

If the water circulation is stopped, the refrigerating system will start to short cycle. The high-pressure switch causes this since these systems are always provided with a high-pressure safety motor control for this purpose. As the head pressure of the machine builds up (due to a lack of cooling), a pressure is soon reached which will cause the control to open a switch. This stops the motor. Once the motor is stopped, the head pressure drops rapidly. This permits the high-pressure control to turn the motor on again.

Short cycling will continue unless the trouble is remedied. Such a condition is a severe strain on the motor. Furthermore, it does not provide satisfactory refrigeration.

14-57 TOO MUCH WATER FLOW

Too much water flow will give satisfactory refrigeration, but more water will be used than needed, increasing the cost of operation. Three things may cause too great a water flow:
1. Water pressure too high.
2. Leaking water valve.
3. Water valve out of adjustment and holding a valve too far open.

Too-high water pressure is seldom found, unless the water supply pressure in uncontrolled. If found in one machine, it will perhaps be true for all the machines in that locality.

To determine whether the condition is due to a leaking valve or to a valve open too far, find out if anyone has been working on the machine. If the machine has not been tampered with; if trouble has just started, the trouble is usually a leaking water valve. A further indication of this is a continuous flow of water on the off cycle. Quite often a leaking water valve may be caused by foreign matter lodging between the valve and its seat. One can usually dislodge this material by flushing the valve. The flushing may be done by using a screwdriver to pry the valve open several times.

14-58 TRACING WATER CIRCUIT TROUBLES

In water circuit troubles, one must determine whether the water valve is at fault, or if it is some other part of the water circulating system. To locate the exact source of the trouble, disconnect the joints where the hose is fastened to the rest of the system.

Disconnecting the water outlet pipe from the machine unit will indicate if the water is flowing as far as this point. If so, the connection should be resealed and the other end of the pipe disconnected from the wall pipe. If the water does not flow up to this point, the trouble is probably in the drain pipe of the system.

Determine whether the water is coming as far as the water valve by disconnecting the pipe used as the inlet connection. If the water flows through the pipe, the sources of trouble are the water valve and the condenser proper. To check these sources reseal the inlet pipe to the system and disconnect the water valve from the condenser.

If the water flows through the water valve, but does not go through the condenser, the trouble is a major one, requiring replacement or cleaning of the condenser water tubes. However, if the water does not flow through the water valve, the water valve must be disconnected from the system and repaired or replaced.

Water hammer is a very noisy condition that is easily recognized. A single, distinct thump or rap is heard in the pipes just as a valve is closed. Generally, this condition can be corrected by placing a short, vertical pipe in the water line just ahead of the valve. It provides a column of air which absorbs the shock of flowing water suddenly being required to stop.

14-59 REMOVING WATER VALVE

To remove an electrical solenoid water valve from the system, open the hand switch that controls the circuit to the motor. Remove the water valve wires from the motor circuit. If these wires are soldered and taped, it will be necessary to unsolder or cut them. Most connections are of the slip-on type.

Before disconnecting the water valve from the water system, shut off the water supply at the hand valve. If the service technician does not have a replacement valve on hand, the water system may be temporarily connected without a water valve, the water flow can be regulated with the hand shutoff valve.

Some pressure-operated water valves are difficult to remove from the system because the valve is connected to the high-pressure side of the condensing unit. The pressure tube for these valves is usually connected into the cylinder head of the compressor. However, some manufacturers connect this tube into the liquid line. Sometimes this tube has a hand shutoff valve. If so, removal of the valve is simple.

If the tube is connected to the cylinder head of the compressor, the following procedure is suggested:
1. Install gauge manifold.
2. Turn suction service valve all the way in.
3. Run compressor until pressure in crankcase reaches 0 psi. *Note: Be very careful of oil pumping which will sometimes occur before 0 psi can be reached.*
4. Heat the water valve line and the water valve bellows carefully with a heat lamp for three or four minutes until both are quite warm to the touch. This operation will move the liquid refrigerant (which has condensed in this tube and valve) back into the condensing unit. Then, only a small quantity of high-pressure vapor will be left in this tube.
5. Turn discharge service valve all the way in.
6. Bypass the high pressure in the manifold and water valve refrigerant line into the low side by opening both manifold valves.
7. Clean joints to be opened.
8. Disconnect pressure tube from water valve. Wear goggles!
9. Plug the refrigerant pressure tubing openings immediately.
10. Gently heat water valve again. Often a quantity of liquid refrigerant becomes oil bound within the bellows chamber and releases with explosive force a few minutes after the valve is opened to the atmosphere. Heating will drive it out.

    Note: Be very careful not to point the refrigerant openings toward anyone because of the danger of being hit by the refrigerant. It is best to wrap the refrigerant openings with several layers of heavy toweling to absorb any refrigerant being thrown from the mechanism.
11. Shut off water supply.
12. Disconnect water valve from water line and replace it with a good one or connect water lines directly. Now the water valve is ready to be dismantled and repaired.

Some water valves are designed to permit removal of the valve body without disturbing the refrigerant connections. To disconnect one of these valves, simply shut off the water and disconnect valve body from water lines and bellows body.

Thermostatic water valves or motorized water valves are easily removed; only electrical connections need to be broken and the water circuit closed.

14-60 REPAIRING WATER VALVES

It is better to replace a worn water valve. Sometimes a water valve only needs cleaning. Wire brushes and a muriatic acid solution work best. Use same precautions as when cleaning the water tubes of a condenser.

If no replacement valve is available, the valve and valve seat must be repaired. The valve seat, brass in most cases, may be lapped in a manner similar to lapping a compressor valve seat.

The valve proper usually is made of fiber, rubber or Bakelite material. It should be replaced. In cases of emergency, however, this valve may be trued up by using a fine grade of sandpaper backed up by a level surface.

Occasionally, where the valve stem passes into the valve body proper, a packing gland is used to seal the joint. This packing is usually asbestos, graphite and lead. If the packing nut has been turned all the way down and this joint still leaks, the packing should be replaced.

Electric water valves may have a faulty electrical coil (either shorted or with an open circuit). Replace it with a coil of the same electrical properties (voltage and wattage). Thermostatic water valves may lose the element charge. If so, replace.

14-61 INSTALLING AND ADJUSTING WATER VALVES

After cleaning and repairing a pressure-operated water valve, it should be tested and adjusted before being placed in service. If the maximum temperature of the water supply is 75 F., the valve should be adjusted to open at the following pressures: 87 psi for R-12; 144 psi for R-22; 152 psi for R-717; 112 psi for R-500 and 173 psi for R-502.

If the water-in temperature is different from this, adjust the valve to its correct opening pressure by referring to Fig. 14-77.

A valve should also be tested for leaks at the same time it is being adjusted. To do this, connect an air pressure line to the inlet water opening of the valve. To adjust and test the pressure operating bellows, which controls the water flow, connect another air line and a pressure gauge to this fitting. No air should flow through the water valve until the correct

control bellows pressure has been reached. Adjustment may be made to obtain this condition.

Solenoid water valves are nonadjustable. Thermostatic water valves may be tested by using an adjustable temperature well to change the bulb temperature and air pressure used to check the valve operation.

After installing the water valve, check it for leaks (both water and refrigerant). Outlet water temperature, water flow and condensing pressure should also be checked.

14-62 SERVICING DIRECT EXPANSION EVAPORATORS

"Dry" evaporators are coils using either an automatic expansion valve (AEV) or a thermostatic expansion valve (TEV) refrigerant control. See Chapter 5.

These evaporators must be clean both inside and out for good heat transfer. They must contain just enough liquid refrigerant at the proper vapor pressure to provide the required cooling. Air or water being cooled must flow in and out of the evaporators efficiently. The evaporator must be leakproof and of the proper size.

A service technician should check these conditions. Ideally, pressure at the inlet of the evaporator (just after the AEV or TEV) should be measured as well as pressure at the outlet.

However, most service technicians check only the low-side pressure at the compressor suction service valve and think that the evaporator pressure is close to this pressure. The pressure drop due to friction in the tubing and bends can be checked by reading the low-side pressure when the unit in running and then reading it again just as the compressor stops. The rise in pressure is the pressure drop. Normally this pressure drop will be 2 to 3 psi.

Evaporator temperature should be checked too. Thermometers can be mounted on tubing using spring-loaded clip-on thermometer holders. Superheat setting of the thermostatic expansion valve can be checked using thermometers. The best setting is when the bulb temperature varies the least while the unit is running. Location of the liquid in the evaporator can also be determined this way.

Frost accumulation acts as an insulation and also tends to reduce the airflow. If found near the TEV, it usually means too great a superheat adjustment along with low-suction pressures. Spotty frost usually means uneven airflow over the evaporator or that some defrosting elements are not working.

Airflow through the evaporator can be checked with an anemometer. This is called an air velocity meter too. See

REFRIGERANT	WATER TEMP. (F.)										
	50	55	60	65	70	75	80	85	90	95	100
	HEAD PRESSURE (PSI)										
R-12	56	62	68	74	80	87	93	101	108	117	125
R-22	95	104	113	123	133	144	155	158	180	194	208
R-717	98	108	119	130	140.5	152	164	177	191	205	220
R-500	71	78	86	94	103	112	121	131	142	153	165
R-502	116	126	137	148	160	173	186	200	214	230	246

Fig. 14-77. Table of head pressures for systems with various refrigerants at various inlet water temperatures.

Chapter 18. If the air outlet or inlet is too small, the evaporator will be starved. Air-in and air-out temperatures can also be checked. The air temperature will usually drop about 15 F. as it passes through the evaporator.

When checking for leaks, the fan and unit should be shut off. The low-side pressure should be at least 15 psi when testing for leaks.

14-63 REMOVING THE EVAPORATOR UNIT (DRY SYSTEM)

If a dry evaporator needs to be removed, first install a gauge manifold and test for leaks. Start the compressor and close the LRSV (liquid receiver service valve). Run the compressor until atmospheric pressure or a constant vacuum has been produced. Continue running the compressor until the evaporator and liquid line are warm. At this point, all the liquid refrigerant is removed. To speed up this operation, heat the evaporator carefully with a heat lamp or hot water. Never allow it to get more than warm to the hand.

A balanced (atmospheric) pressure in the evaporator may be obtained by either warming the unit or by bypassing high pressure back through the gauge manifold. Fig. 14-78 shows an evaporator being pumped down. Turn the suction line service valve, C, all the way in, closing the suction line after balancing.

Check the suction line; if it has a suction pressure regulator in it, or a solenoid valve, be sure it is open.

If a "hot gas" injection unit or a liquid injection unit is connected to the suction line, be sure the solenoid control valves for these units are shut off.

If the system has hand shutoff valves for each evaporator of a multiple installation, use these valves instead of the compressor service valves. Close the liquid line valve first and pump refrigerant out of the evaporator. Be sure there is 0 psi or slightly more in the evaporator. Close the suction line hand valve and the evaporator is ready for removal.

Shut off the electric power to the fan and liquid line solenoid valve (if there is one). Remove the casing or shroud of the evaporator carefully.

If electric defrost elements are mounted in or on the evaporator, disconnect them.

Clean and dry the suction line where it is connected to the evaporator. Also clean the inlet connection. Then unfasten the suction line and liquid line from the evaporator. Plug the openings with appropriate fittings. Wear goggles!

14-64 REPAIRING DIRECT EXPANSION EVAPORATOR

Repairs are usually limited to:
1. Repairing leaks.
2. Repairing or replacing fittings.
3. Straightening fins.
4. Replacing defrosting elements.
5. Repairing or replacing hangers.
6. Repairing or replacing fins and/or motors.

Where leaks occur, completely dismantle that part and clean the surfaces. If it is a silver brazing repair, follow the correct procedure explained in Chapter 2.

Always anneal an old tube before flaring it. Fins can be straightened using a fin comb or wide-jaw pliers.

Electrical defrosting elements should be checked for conti-

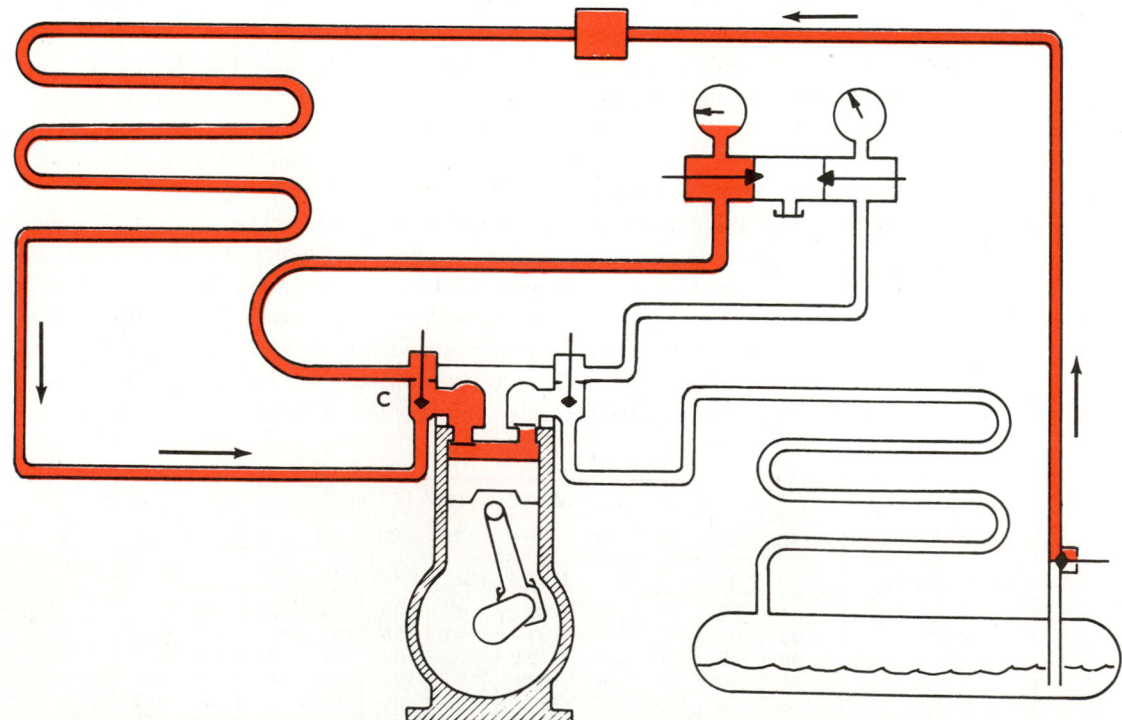

Fig. 14-78. Liquid line, evaporator, and compressor crankcase being pumped down. Refrigerant is being pumped into condenser and receiver.

nuity. Terminals and insulation should be inspected also. Rusty or bent hangers and abused hanger assembly bolts should be replaced.

Check the fan and motor for vibration: check tightness of the fan on the motor shaft, motor end play, motor bearing wear, and condition of lubricant. Small faulty motors should be replaced. Larger motors can be rebuilt. (See Chapter 7.)

All parts should be cleaned before assembly. The evaporator is usually assembled on the job. If leaks have been repaired, the evaporator should be leak tested before it is installed.

14-65 INSTALLING DIRECT EXPANSION EVAPORATOR

In the event the evaporator has been removed, several important things must be remembered during the assembly to insure proper operation. After bolting the evaporator back into the refrigerator and leveling it, remove the plugs on the refrigerant openings and attach the liquid line and suction line to the unit. Be careful during these operations that no moisture enters the lines. It is good practice to dry the surfaces of the lines and evaporator before removing these seals.

The thermostatic expansion valve (TEV) should be installed using a new gasket — if it is of the bolted type. Connect a vacuum pump and evacuate the evaporator, suction line and compressor to the air. Evacuate the evaporator a second time. Test for leaks with 5 to 25 psi (.35 to 1.8 kg/cm^2) pressure; test for leaks again at ambient refrigerant pressure (high-side pressure) before operating the evaporator.

The evaporator may be dehydrated more completely by heating it to a fairly high temperature of 175 to 200 F. (79 to 93 C.) as it is evacuated. This drives out any moisture that may be present. Heat lamps may be used for this purpose.

After the evaporator has been installed and tested for leaks, install the electrical connections for the defrost units. Install the fan and motor. The electrical connections should be tight and moisture-proof. Operate the defrost unit and the fan.

Assemble the casing or shroud. Start the unit and check for normal operation.

14-66 INSTALLING EXPANSION VALVES

Mount the expansion valve and evacuate the liquid line, evaporator and suction line. Carefully test for leaks by first purging, then building up a refrigerant vapor pressure. Install the fan and motor, if used. Open liquid receiver valve or liquid line hand valve, start compressor and observe its operation.

In multiple systems, all dry evaporators using expansion valves should be installed with individual shutoff valves for both the liquid and suction lines to each evaporator.

Attempts to adjust thermostatic expansion valves in an effort to maintain too great a difference in temperatures in various cabinets in the system gives rise to erratic operation. This is true, particularly in the evaporators which are closed off the most. To fix this — in addition to the thermostatic expansion valve — install one or more two-temperature valves in the proper places in the suction line.

Use gauges and thermometers to check for superheat

setting. When the unit is operating correctly, connect the defrost wires and, if used, install casing and shroud.

In multiple commercial installations, in which finned evaporators are used, the expansion valve is sometimes attached to the evaporator with an SAE flared connection. The flare nut in such an installation must be shellacked, or sealed from moisture. This is done after the installation has been made and before the unit starts to operate. Otherwise, ice may form between the nut and the tubing and in a short time will cause the tube to collapse or break. This condition can also occur at the point where the suction line fastens to the evaporator. Fig. 14-79 shows the pinching operation of ice

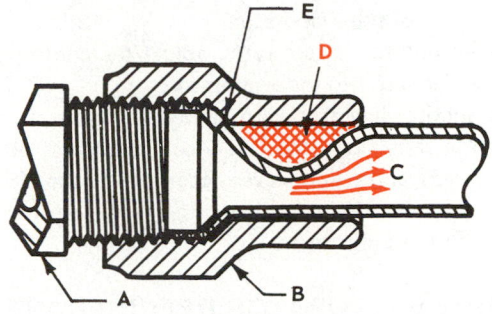

Fig. 14-79. Illustration shows what happens when moisture freezes between nut and tubing. A—Fitting. B—Flare nut and tubing. C—Restricted refrigerant opening. D—Ice formation. E—Flare being pulled out of place.

formation between the flare nut and the tubing.

Other methods have been devised to stop moisture from collecting behind flare nuts. One has a rubber seal at the end of the flare nut. Another method consists of drilling holes through the flare nut. The idea is that moisture will drain out and, if ice does form, it will release its pressure through the holes in the nut rather than against the tubing.

Short shank flare nuts should be used in places where frosting occurs. Fig. 14-80 shows a short flare nut with

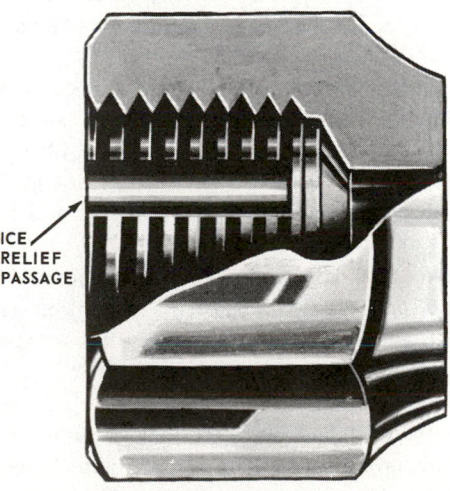

ICE
RELIEF
PASSAGE

Fig. 14-80. Special flare nut used to prevent ice accumulation between flare nut and tubing.   (Superior Valve Co.)

openings across the threads for moisture escape. As mentioned before, the best sort of connections for use in the interior of the cabinets is a silver-brazed flanged connection.

14-67 ADJUSTING THERMOSTATIC EXPANSION VALVES

The sensing element of the thermostatic expansion valve should be clamped to the suction line at the point where it attaches to the evaporator. It is possible to make bench adjustments of this valve.

In service, the thermostatic expansion valve may be adjusted as follows:

1. With the refrigerating condensing unit operating, note the temperature of the evaporator.
2. If the evaporator is too warm, adjust the control to allow more refrigerant into the evaporator.
3. If the suction line frosts up, adjust the control to reduce the refrigerant flow. Most service technicians adjust the thermostatic expansion valves according to the MSS point. See Chapter 5. The amount of superheat is properly taken care of by this adjustment.

14-68 DRY EVAPORATOR REFRIGERANT CONTROLS

The operation of refrigerant controls in commercial systems is described in Chapter 5.

All types of refrigerant controls may be found in commercial systems. If the system has a self-contained hermetic unit, it may be serviced as described in Chapter 11.

The multiple evaporator installations, however, and the larger commercial units, do have several other features that a service technician should know.

14-69 SERVICING THERMOSTATIC EXPANSION VALVES

Dry systems which use the thermostatic expansion valve refrigerant control may be of the single evaporator or multiple evaporator type. The design and operation of this valve is explained in Chapter 5.

The service technician should check the complete system, install the gauge manifold and check for pressure. Check the amount of refrigerant through the sight glass. Check liquid line solenoid valves, two-temperature valves, "hot gas" or liquid injection systems, driers (both suction line and liquid line), temperature or pressure motor controls and electrical supply.

The service technican can check the evaporator by appearance, sound, and temperature of the expansion valve. If the evaporator is frosting back so the suction line is frosted, it may be due to the following troubles:

1. Needle may be leaking.
2. Control may be adjusted for too little superheat.
3. Valve may have the incorrect thermal bulb charge.
4. Valve orifice may be too large.
5. Power element may be attached loosely to suction line.
6. Power element may be located in too warm a position.

7. Dirt in system may be holding valve open.
8. An external equalizer valve may be needed.
9. Screen may be clogged.

It is difficult to determine which of these troubles is responsible for the fault. The best method is to systematically check everything it could be and rule them out one-by-one. The location of the power element may be easily checked and its attachment to the suction line noted. Recommended procedure is to place the thermostatic bulb on top of the suction line rather than beside or below it. In this position the liquid in the bulb will make good thermal contact with the suction line. Fig. 14-81 shows two recommended thermal bulb locations.

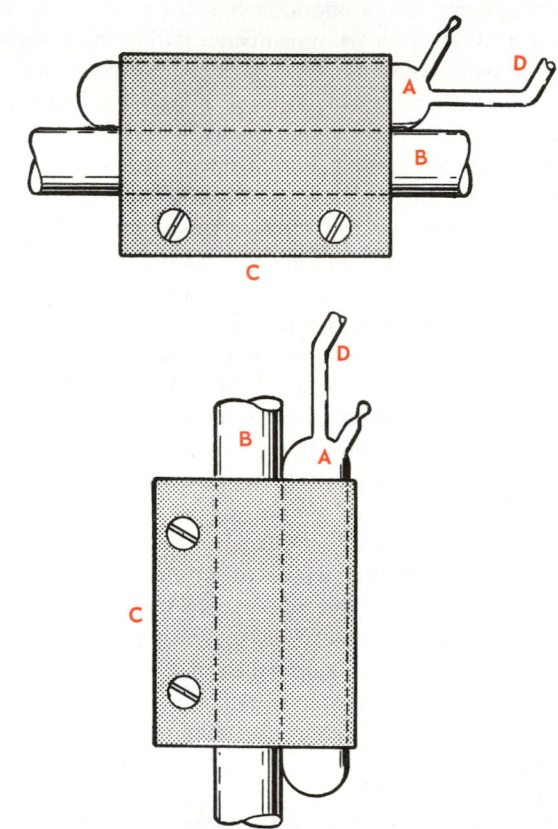

Fig. 14-81. Recommended locations for thermostatic bulbs on suction lines to obtain best operation. Bulb should be on top of horizontal suction line. It should have closed end on bottom when mounted on vertical suction line. A—Thermal bulb. B—Suction line. C—Clamp. D—Thermal bulb capillary tube.

It is rare to find a thermostatic control in which someone has tampered with the adjustment. Usually customers will not try to set them. Therefore, do not attempt to readjust the control at first. Check for other troubles instead. A leaking needle or valve cannot be repaired. It should be replaced.

A starved evaporator (one giving poor cabinet temperatures while frosting unevenly or not sweating properly) may be due to the following:

1. Clogged screen in expansion valve which may give no refrigeration.
2. Loss of refrigerant from power element in expansion valve

which will give erratic (undependable) refrigeration.

3. Moisture in the system which may sometimes give good refrigeration and then none. The moisture will freeze in the orifice to expansion valve and close it, then defrost.
4. Wax in valve. This wax is from the oil and its presence means that the oil used was for a different temperature range or was improperly prepared for refrigeration service.
5. Needle stuck shut. (This is a rare occurrence.)
6. Under-capacity valve orifice.

When an expansion valve gives trouble, it is usually recommended that a new valve be installed.

Too much refrigerant flow will be indicated by a sweating or frosted suction line beyond the thermal bulb position. This condtion may be caused by:

1. Thermal bulb loose from suction line.
2. Thermal bulb in warm airflow.
3. TEV orifice is too large.
4. TEV needle stuck open.
5. Undersize evaporator.
6. Thermal bulb has wrong charge.
7. Pressure drop in evaporator is too great.

If the bulb is loose, is mounted wrong, or is in a warm air stream, remove it; clean it and the tubing; remount it firmly and insulate it if necessary.

An oversize TEV should be replaced with one of the correct size. When a needle is stuck open, the best remedy is to replace the valve.

Only rarely does one find an undersize evaporator (replace it), a TEV with the wrong charge (replace), or too great a pressure drop (replace evaporator).

14-70 REMOVING EXPANSION VALVE

The service technician should remove a faulty expansion valve and replace it with one in good condition. The troublesome valve may then be checked in a shop equipped for this purpose and the trouble accurately determined.

Remove the valve using the procedure described in Para. 14-63.

14-71 REPAIRING CLOGGED TEV SCREEN

A clogged screen may be easily detected by poor refrigeration, sweating or frosting near the TEV only and no refrigerant sound.

Removing an expansion valve which has a clogged screen means removing both the liquid and vaporized refrigerant from the lines to be opened. The liquid line may be carefully heated with a heat lamp, driving the liquid refrigerant back to the nearest shutoff valve. This valve then is closed. The evaporator is already evacuated (indicated by a warm evaporator). If low-side pressure is at atmospheric pressure or higher, the suction line valve may be closed. The screen may be removed after cleaning and drying the connections.

Clean the screen or replace with a new one. This is important. Fine-mesh copper or stainless steel screens may be cleaned fairly successfully with air pressure and a safe solvent. But the best way is by heating it. This must be very carefully performed or the screen will be burned. *Never allow an expansion valve to go into service without a screen being placed in the liquid line entrance.*

Install the screen, assemble the TEV, install it in the system, evacuate the air, test for leaks and return the evaporator to normal operation by opening the suction and liquid line valves.

14-72 TESTING AND ADJUSTING THE THERMOSTATIC EXPANSION VALVE

Thermostatic expansion valves sometimes get out of adjustment; also, certain installations may require a periodic readjustment of this valve. A simple method of testing these valves in the shop will be explained.

A regular service kit having the necessary equipment and setup required is shown in Fig. 14-82. The equipment required is as follows:

1. Service cylinder of R-12 or R-22. In the shop a supply of clean dry air at 75 to 100 psi pressure can be used in place of the service cylinder. The service cylinder or air is needed only to supply pressure.
2. High-pressure and low-pressure gauges. The low-pressure gauge should be accurate and should be in good condition so that the pointer does not have too much lost motion. A high-pressure gauge is recommended in order to show the pressure on the inlet of the valve.
3. A small quantity of finely crushed ice is necessary. One of the most convenient ways of handling this is to keep it in a thermos bottle with a large neck. If such a container is completely filled with crushed ice, it will easily last for 24 hours. Whatever the container is, fill it completely with crushed ice. Do not attempt to make this test with the container full of water and a little crushed ice floating around on the top (32 F. ice floats at the top, while 39 F. water sinks to the bottom).

Proceed as follows to adjust the valve:

1. Connect the valve as shown in Fig. 14-82. The adaptor on the expansion valve outlet provides a small amount of leakage through the No. 80 drill orifice opening. A 10-cu. in. tank is used to reduce pressure fluctuations.
2. Insert the bulb in the crushed ice and allow it to cool.
3. Open the valve on the service cylinder and be sure the cylinder is warm enough to build up a pressure of at least 70 psi on the high-pressure gauge connected in the line to the valve inlet.
4. The expansion valve can now be adjusted. The pressure on the outlet should equal the pressure for the refrigerant at 22 F. The water and ice mixture is 32 F. and the sensitive bulb is also 32 F. If the superheat is to be 10 F., then the temperature of the refrigerant will be 22 F. and the pressure must be adjusted to match this 22 F. temperature. The pressure on the outlet gauge should be different for various refrigerants as follows:

R-12,	22 psi
R-22,	45 psi
R-500	29 psi
R-502	55 psi

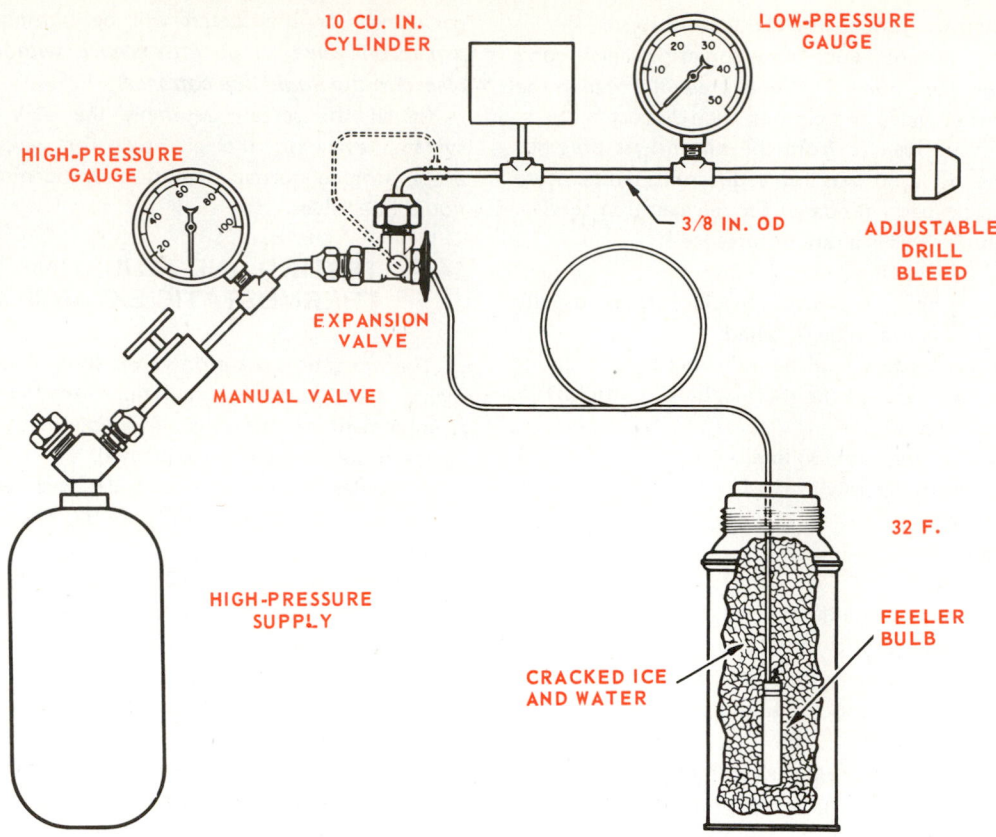

HIGH-PRESSURE
GAUGE

10 CU. IN.
CYLINDER

LOW-PRESSURE
GAUGE

3/8 IN. OD

ADJUSTABLE
DRILL
BLEED

EXPANSION
VALVE

MANUAL VALVE

HIGH-PRESSURE
SUPPLY

32 F.

FEELER
BULB

CRACKED ICE
AND WATER

Fig. 14-82. Recommended setup for testing and adjusting thermostatic expansion valve. Note that liquid refrigerant is supplied from storage cylinder. Control bulb is in thermos bottle filled with cracked ice at temperature of 32 F.

When making the adjustment, be sure there is a small amount of leakage through the low-pressure orifice.

5. Tap the body of the valve lightly in order to determine if the valve is smooth in operation. The needle of the low-pressure gauge should not jump more than one pound.

To test the needle for leaks, close the orifice to stop the leakage and determine if the expansion valve closes off tightly. If the valve is in good condition and not leaking, the pressure will increase a few pounds and then either stop or build up very slowly. With a defective, leaking valve, the pressure will build up rapidly until it equals the inlet pressure.

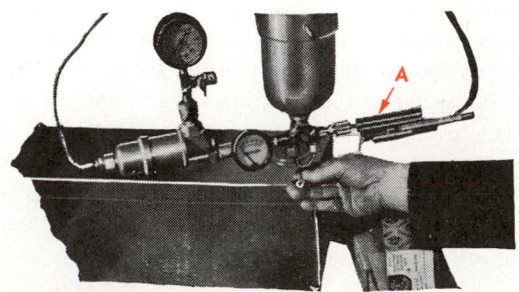

Fig. 14-83. Thermostatic expansion valve being tested. Note refrigerant cylinder, thermometer, low-pressure gauge, high-pressure gauge, bulb temperature control device at A, and expansion valve being tested.

To test the power element, remove the power element bulb from the crushed ice and warm it up with the hand or by putting it in water at about room temperature. The pressure should increase rapidly if the power element is operating.

Another method of checking and adjusting a thermostatic expansion valve (TEV) makes use of a device with a clamp for the TEV bulb, an accurate thermometer and a small cavity for refrigerant. With this instrument mounted on the TEV bulb and the cavity charged with liquid R-12, the small screw adjusts boiling pressure. Therefore, the temperature of the R-12 and the bulb temperature can be quickly and accurately lowered and raised to check the control adjustment and to readjust it if necessary. Fig. 14-83 shows the instrument in use.

Note: With vapor charged expansion valves (charged with a small quantity of refrigerant) the amount of charge in the power element is limited and the pressure will not build up above the specified pressure. This pressure is always marked on the valve body and must be considered when testing vapor-charged valves.

The body bellows is tested with high pressure showing on both gauges as outlined in the preceding paragraph. A leak (escape of vapor) is detected by using a leak detector.

When making this test, it is important that the body of the valve has a fairly high pressure. Gauge connections and other fittings must be tight to eliminate leakage at other points. Leakage can also be detected by use of soap suds.

14-73 REPAIRING EXPANSION VALVES

Three things likely to go wrong with an expansion valve are:
1. Needle and seat become worn.
2. Screen gets dirty; bellows leak. (Described in Para. 14-71.)
3. Thermal element loses charge. See Para. 14-72. If charge is lost, one must replace thermal element or the complete valve. Occasionally, one finds such trouble as a broken spring, but this is rare and the troubles are easy to locate.

As there are many models and types of TEV's on the market, only general instructions are given here. One type is shown in Fig. 14-84. A worn needle and seat is usually due to

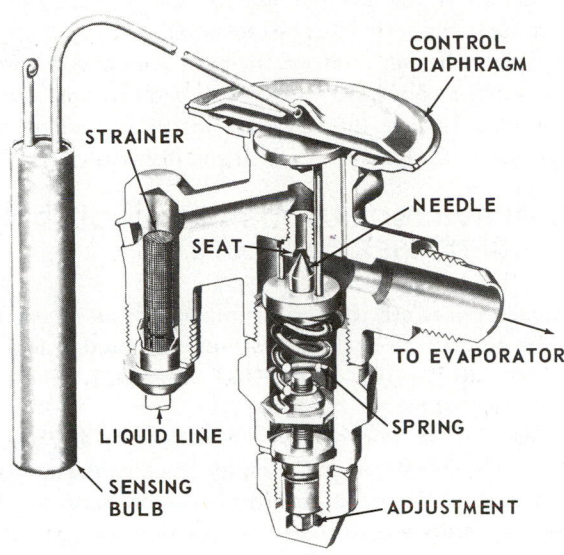

Fig. 14-84. A thermostatic expansion valve with a removable screen, needle and valve seat.   (Alco Controls Div., Emerson Electric Co.)

a lack of refrigerant in the system. Refrigerant does not have a chance to condense before it passes through the expansion valve orifice and this dry, hot gas rapidly cuts or erodes the needle and seat.

The best remedy for a worn needle and seat is replacement. If replacements are not available, the needle must be reground or restored in a lathe or drill press until the shoulder, that has been worn into the needle surface, disappears. It is important that the same taper be kept on the needle point.

The seat, which is nearly always made of softer metal than the needle, may usually be filed, with a dead smooth file, until the older surface against which the needle was seated has disappeared. The seat must be filed square at right angles to the center line of the needle. A file guide is best for this purpose. One may tap the new needle into the new seat during assembly of the valve and get good results.

A leaking bellows may be due either to breaking down of the soldered joint, or to a fracture in the bellows or diaphragm itself. A bellows is a difficult thing to repair. A special fixture is needed to hold all the parts in the proper position. When the technician encounters a bellows in this condition, it is best to replace the expansion valve.

14-74 INSTALLING TWO-TEMPERATURE VALVES

The pressure-operated two-temperature valves should be installed in the suction line of the warmer evaporators. They may be connected at any place on the suction line because their operation is usually not affected by the distance from the evaporator. Frequently valves of this kind are found in the soda fountain. Some are located near the condensing unit. Solenoid valves for two-temperature operations are mounted in the liquid line of the warmer evaporators.

This type of installation may be improved by mounting a check valve in the suction line of the coldest evaporator. This prevents the backing up of the higher pressure gases into it. Also, a surge tank should be mounted near the condensing unit and connected between the compressor and the main suction line to cut down the fluctuations (rapid up and down changes) of the low-side pressure.

In most code installations, the two-temperature valve and the check valve must be mounted near the condensing unit. The surge tank must be mounted on the condensing unit base.

14-75 SERVICING TWO-TEMPERATURE VALVES

As explained in Chapter 12, the two-temperature valves are automatic. They maintain a higher refrigerant pressure in one or more evaporators of a multiple evaporator system. This pressure is higher than pressures in the remainder of the system.

Four types of two-temperature valves are:
1. Metering type.
2. Snap-action type.
3. Thermostatic type.
4. Solenoid valve, which is thermostat controlled.

The four troubles commonly encountered with pressure valves are:
1. Leaky needle.
2. Valve stuck shut.
3. Valve out of adjustment.
4. Frost accumulation on bellows.

If the valve is leaking, the warmer evaporator will be too cold and there will be danger of freezing. This cold temperature may also be due to the two-temperature valve being adjusted too close and at too low a pressure.

To determine which of the two troubles is present, check to see if the valve has been adjusted recently. If the valve has not been tampered with, the trouble is very likely a leaking needle. If the valve has been tampered with, one must readjust it by using a thermometer to get the correct evaporator temperature. The adjusting nut should not be turned more than a half turn at a time. A 15 minute interval should be allowed between each adjustment to permit the evaporator to completely respond to the new pressure.

For accuracy in making adjustments on two-temperature valves, a low-pressure gauge should be installed in the evaporator side of the valve. Sometimes such a gauge opening is available, but in many cases it is not. Service work on such valves will be made easier if the service technician, when installing two-temperature valves, will install a shutoff valve with a gauge opening in the suction line. This will permit the use of a gauge to check the evaporator low-side pressure.

It is rare to have a valve stuck shut. It may be easily recognized by a lack of cooling in the warmer evaporator and an adequate refrigerant supply to the two-temperature valve (but not through it). This refrigerant supply may be checked by cracking the flare nuts on the high-pressure side of the two-temperature valve. Some of these two-temperature valves are provided with screens. A clogged screen will be indicated by a warmer cooling condition of the evaporator with symptoms similar to the stuck needle condition.

Frost will accumulate on the bellows only when the valve is located in or near a freezing compartment. The valve should be removed from the freezing compartment and the bellows covered with light grease.

The same troubles are encountered with the snap-action, two-temperature valve as with the metering type. However, most snap-action valves have a gauge connection in the nature of a one-way service valve mounted on the two-temperature valve body. This gauge connection makes adjustment of the valve simple and therefore it can be determined if the valve is leaking or out of adjustment.

Thermostatic two-temperature valves offer the same troubles, causes and remedies as the other two. In addition, they have troubles resulting from the thermostatic element. These troubles are the same as mentioned for thermostatic expansion valves.

They are:
1. Loss of charge from thermostatic element.
2. Frost on bellows.
3. Poor power element contact with evaporator.
4. Pinched capillary tube.
5. Wrong adjustment.

These troubles are checked in much the same way as one would check the thermostatic element in thermostatic expansion valves.

An electric solenoid valve in the liquid line of the warmer evaporator and electrically connected to a thermostat in the cabinet may have the following troubles:
1. Needle stuck open.
2. Needle stuck shut.
3. Thermostat troubles:
   a. Points stuck together.
   b. Open circuit.
   c. Out of adjustment.
4. Poor wiring.
5. Open solenoid winding.
6. Burned-out solenoid winding.

The solenoid valve should be replaced if the needle is not working properly. When a solenoid coil is replaced, be sure to use the correct coil voltage (24V, 120V or 240V).

14-76 SERVICING HOT GAS BYPASS VALVES

These valves are of two types:
1. Solenoid, timer on — thermostat off.

2. Solenoid pilot operated pressure valve, timer on — thermostat off.

If the evaporator is not defrosting, check the timer. If it is not running, determine the problem and repair or replace. If working properly, turn the timer to the on position and check the hot gas line. If it doesn't become warm, check the electrical circuit with a voltmeter to find out if there is power to the solenoid.

If the timer is faulty, check the points. If they are corroded, replace them. If the solenoid coil is faulty, replace it. To determine condition, disconnect wires after shutting off the power and check for open circuit or grounds.

If the valve is sticking, rap the body of the valve to find out if this action allows the hot gas to flow. If the valve core remains stuck, the valve must be replaced.

To replace the valve, refrigerant must be removed from the system (valve is connected to high side and low side), balance the pressures, remove the valve and install a new one. Then test the system for leaks, evacuate it and recharge.

14-77 SERVICING OUTDOOR AIR-COOLED CONDENSER CONTROLS

Several devices are used to maintain good head pressures when the outdoor air-cooled condenser is exposed to temperature from 50 F. (10 C.) to −10 F. (−23 C.). These are explained in Chapter 12.

Service of these units depends on the type. In all cases, one can check the condenser operation by blocking the condenser air inlet or outlet with cardboard to raise pressures and then remove the cardboard to lower the pressure. Use procedures following, according to type:
1. Modular fans. These units have electrical troubles. Either the fan motor is faulty or the solid-state control. First check the motor for mechanical faults such as loose fan or stuck rotor. Then check for electrical power to the motor with a voltmeter. If it has power, shut off power and check the motor windings or ground. Check the sensing device of the solid-state control for correct position and fastening. If the control is faulty, replace it.
2. Controlled louvers. Check operation by blocking air over condenser to find out if louvers will operate automatically. If the power element is condenser-pressure operated (operated by head pressure) and it is faulty, the refrigerant must be removed from the system unless a shutoff valve was originally installed. If one pinches the line to remove this control, be very careful that the pinched part of the line is well supported to prevent a break at this point.

14-78 SUCTION LINE

Suction lines have parts that may need service:
1. Fittings.
2. Hand valves.
3. Two-temperature valves.
4. Check valves.
5. Surge or accumulation tanks.
6. Hot gas injection unit.

7. Liquid injection unit.
8. Suction line filter-drier.

The fittings are checked as was mentioned before. The hand valves are similar in construction to liquid line hand valves.

It is important to check the suction line size. Then, using the compound gauge, determine the pressure drop in the low side of the system. The pressure drop varies, but it should not exceed 2 psi.

To check the pressure drop, record the low-side pressure when the unit is running and again just as the unit stops. The difference in the readings is the pressure drop unless two-temperature valves are mounted in the suction line.

14-79 ASSEMBLING REFRIGERATION SYSTEMS

When the repairs have been made and the parts, after testing, are found to be all right, the proper assembling of the system is very important.

Four fundamentals must be followed when installing parts of a refrigerating mechanism:
1. Clean and dry each part to be put into the system.
2. Purge and evacuate that part of the system which has been opened.
3. Test for leaks.
4. Start and adjust the unit.

It is almost impossible to assemble a refrigerating mechanism in the field without having some foreign matter (dirt and moisture) get into the system. Filter-driers, installed in both the liquid line and suction line of the system, are needed to remove this matter.

A service call should be made within the next day or two after repair to check on the operating pressures and on the general condition of the refrigerator. If at all possible, connect temperature and pressure recorders into the system for 24 hours to check on unit operation.

14-80 MOISTURE IN SYSTEM

Many troubles in refrigerating systems may be traced to the presence of moisture. As it circulates through the system in the presence of oil and refrigerant at high temperatures (compressor and condenser), moisture causes many complex actions.

Moisture may freeze at the refrigerant control orifice, eventually clogging it. In hermetic systems it may also cause a chemical breakdown between the oil, the refrigerant and the motor winding insulation. Acids are created which ruin the motor windings. If immediate action is taken to remove all moisture or to make the moisture harmless, many troubles may be eliminated.

Fig. 14-85 shows the safe limits of moisture allowable in a system using R-12, R-22, R-500, and R-502.

Many service technicians install large driers in a system on a temporary basis. These units quickly clean the system and remove the moisture. Fig. 14-86 shows one installed in a liquid line.

To remove the unit, close valve No. 1. Wait until the drier has had most of the refrigerant removed, and then close valve No. 3, and open valve No. 2. The system is now back in operation. Then close valves No. 4 and No. 5. Remove the connecting lines and cap the openings. *Be careful; there may be high pressure in the lines and there may be some liquid.*

14-81 USE OF DRIERS (DEHYDRATORS) AND FILTERS

Surveys show that many service calls are either directly caused by or are traceable to moisture. If there is enough moisture, it will form ice in capillary tubes and expansion valves, plugging them.

There are a number of ways to remove moisture. These include:
1. High vacuum.
2. Slow steady flow of hot, dry air.
3. Several alternate (different) moderate vacuums broken with dry refrigerant. (Moderate means not very high nor very low.)
4. Use of filter-driers. (Chapters 11 and 12 describe driers in detail.)

High vacuum, hot, dry air and the several moderate vacuum methods are used just after a system is assembled. Filter-driers are used during the life of the system.

Application	Max Tolerable Moisture, PPM		Moisture in Units, PPM	
	Refrigerant 12 & R-500	Refrigerant 22 & R-502	Refrigerant 12 & R-500	Refrigerant 22 & R-500
Domestic Refrigerators, Freezers	10,10,10,15	– –	5,8,10	– –
Food Display Cases	10,15,15,20	15,20	5,10	5
Frozen Food Cases	10,10,10,10,15	10,15,15	10	– –
Room Air Conditioners	15,25	25,25,40,55,60,10,100	10,10	5,15,15,15,20,60
Residential, Package Air Conditioners	10,15,15,25,50	10,15,25,25,35,40,40,40,60,100,180	7,10,15,40	7,10,15,15,15,20,25,40,50,55,60
Field-Built Systems	15,20,25	15,40,40,100	– –	10,15,60

Fig. 14-85. Maximum moisture allowed in refrigerating system and average amounts found in system. (ASHRAE Guide and Data Book)

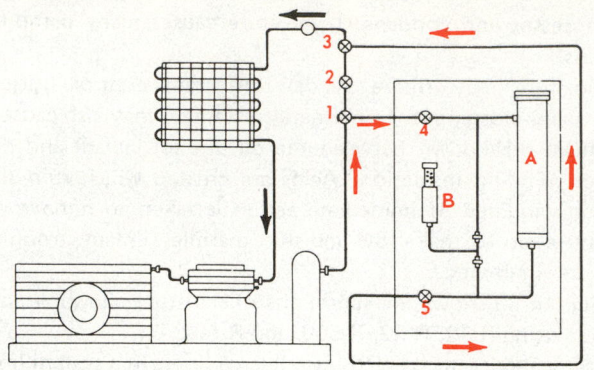

Fig. 14-86. Master drier and filter installed in liquid line. A—Drier. B—Moisture indicator. Valves at 1 and 3 are two-way. Valves at 2, 4 and 5 are one way. To operate, close valve at 2. Open valves at 1, 4, 5 and 3. Red arrows show direction of refrigerant flow. (Superior Valve Co.)

In all commercial systems, liquid line driers are a standard part of the installation. Drier size is usually based on the horsepower of the system, kind of refrigerant and use of system (air conditioning, commercial or low temperature). A 30 cu. in. drier will serve for a 1 hp R-12 or R-500 air conditioning system. A 100 cu. in. (3 in. dia. by 10 in. long) will serve a 5 hp air conditioning system, a 3 hp commercial, or a 3 hp low temperature unit having R-12 or R-500 refrigerant. It will also serve a 10 hp air conditioning system, a 5 hp commercial or a 5 hp low temperature unit using either R-22 or R-502.

Driers will hold more moisture as their temperature lowers. Intall the drier as close to the refrigerant control as possible (inside fixture, if one can).

Remember, if a drier heats up due to shortage of refrigerant or high temperature, it may release some of its absorbed moisture.

Keep the drier sealed until it is installed. If left open, it will absorb moisture from the air.

One can sometimes tell if a drier is absorbing water. The drier will become warmer as the refrigerant flows and as the drier absorbs water from the refrigerant. A sight glass with a moisture indicator is good for determining whether the system is dry.

Install the sight glass in the liquid line between filter-drier and refrigerant control. As the unit starts, some bubbles will usually appear in the sight glass. This is normal. If bubbles continue to appear, the system is low on refrigerant or the filter-drier is partially clogged. If not working, the filter-drier will feel cooler to the touch than normal. (Evaporation is taking place.)

When installing the sight glass moisture indicator, avoid overheating. A temperature of 275 F. (135 C.) or higher will injure the chemical. Put a cold pack on the sight glass indicator when silver brazing connections near it. Avoid getting flux inside the sight glass indicator. Flux chemicals may injure the moisture indicator chemicals.

All driers use screens to trap solids in the refrigerant. These screens or strainers are of several types. The screens are usually made of bronze, brass, stainless steel, or monel wire and should be 100 to 120 mesh. That is, there should be 100

openings along a 1-in. rule length, or, 10,000 holes per sq. in. The popular screens are 100 by 90, 100 by 100, 120 by 108, and 120 by 120. The wire is usually .004 to .005 in. diameter. In this size wire, the openings are about .005 in. square.

Felt filters may be used. These are about 1/8 in. thick and are made of special material. Wool batt is also used as a filter in some driers. One company has a specially processed coarse cotton yarn wound in a diamond pattern over a metal frame. One of the latest filters makes use of powdered metal pressure castings.

Driers and filters to protect the motor compressor from burnout may be installed in the suction line. They should always be used after a hermetic motor burnout. Dirt in a system may cause serious damage. The small solid particles act as an abrasive. They will wear the needles and seats of refrigerant controls, valves and valve seats of compressors, bearings and pistons of compressors. These particles may also wear through the motor insulation and cause a burnout.

Fig. 14-87 shows a filter for a large system. It has a bolted assembly to permit replacement of the filter element. The service connection allows a service technician to check the pressure drop. If this gauge reads more than 2 psi higher than the compressor compound gauge with the unit running, the filter element should be replaced.

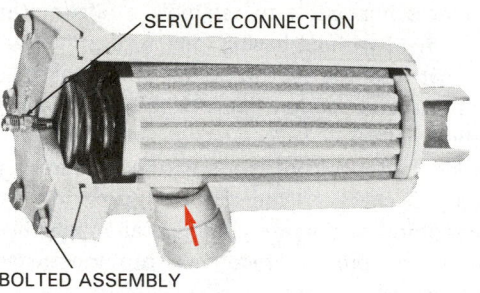

Fig. 14-87. Suction line filter with bolted flange construction permits easy replacement of filter element. Service connection is equipped with Schrader valve. (Superior Valve Co.)

Assume the refrigerant is 2 percent oil.

Each circulation of 10 lb. of refrigerant x .02 = .2 lb. oil. If the unit has 1 qt. of oil or about 1.5 lb.:

It takes $\dfrac{1.5}{\dfrac{2}{10}} = \dfrac{15}{2} = 7.5$

circulations of the refrigerant to clean most of the oil. This means 2 min. x 7.5 = 15 minutes to clean most of the oil.

A suction filter-drier is a must if oil is added to a system. The oil additives in the two oils may react and make a sludge. All of the common refrigerants can be successfully dried with a drier in either the liquid or suction line.

14-82 SERVICING BURNOUTS

A motor burnout is possible with any system using a welded motor compressor or a bolted assembly motor compressor. Moisture, dirt in the system, plus too-high temperature (warmest spot is usually motor windings) may cause an acid condition to develop. In time, the condition could lead to a motor winding short (burnout). See Chapter 11.

To prevent this, the system must be kept free of moisture and dirt. To detect acid formation, the system should be checked regularly.

1. Use a sight glass with moisture indicator.
2. Take an oil sample often. Test it with an oil test kit. The oil sample must be kept sealed until tested. See Para. 14-43, Fig. 14-66.

If a burnout occurs, the motor compressor must be replaced. If it is a small hermetic, discard the refrigerant.

Be careful: the oil coming out with the refrigerant may be acidic. In large systems equipped with shutoff valves, it may be possible to save the refrigerant. Purge the motor compressor. Use goggles, rubber gloves and ventilate the space. The oil may be acidic and cause serious burns — do not get it on the skin. The fumes may also be irritating and toxic.

The system can be reconditioned by:

1. Flushing the complete system with R-11 using regulated CO_2 pressure to circulate the refrigerant or by using a separate pump.
2. Installing a suction filter-drier of excessive capacity (too large) in the suction line.

In the first method, after cleaning, install the new motor compressor; test for leaks; evacuate the system to 50 to 500 microns. Install a drier in the suction line; test for leaks and evacuate again. Charge the system.

Connect the electrical wires. Never solder leads to the compressor terminal. The glass may crack or the terminal may come loose, causing a leak.

Start the unit and check operation. Make frequent oil acid tests. Replace the drier if the oil sample is discolored or shows an acid trace.

In the second method, install the new motor compressor.

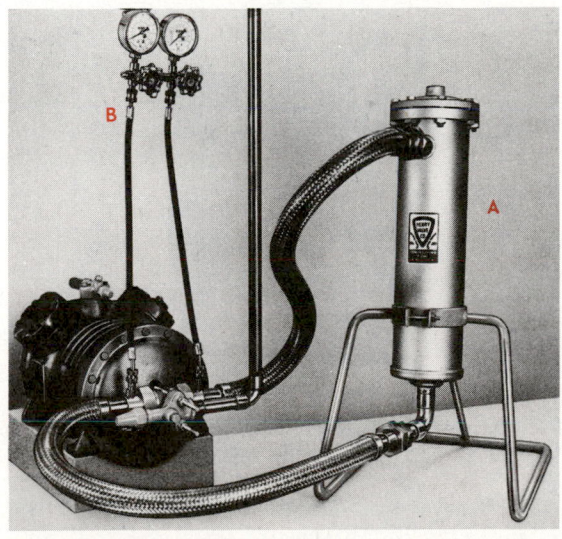

Fig. 14-88. An installation for reconditioning refrigerant after a burn-out. A—Suction line filter-drier is installed on bolted assembly motor compressor. B—Pressure drop across filter-drier is being checked with gauge manifold.

Place an excessive capacity filter-drier in the suction line; test for leaks. Evacuate the system to 50 to 500 microns, charge the system and operate it. The pressure drop across the filter-drier should be checked frequently. Examine oil samples frequently. Fig. 14-88 shows such an installation. The installation of the unit can also be permanent, with quick couplers, as shown in Fig. 14-89.

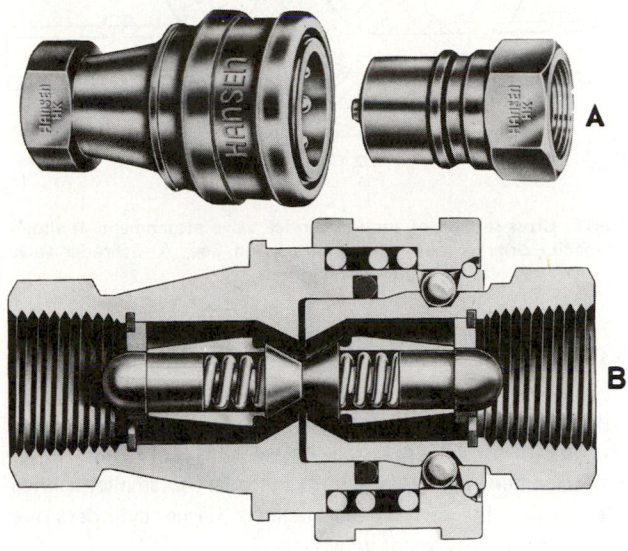

Fig. 14-89. Reusable quick coupler. A—Disconnected. B—Connected. (The Hansen Mfg. Co.)

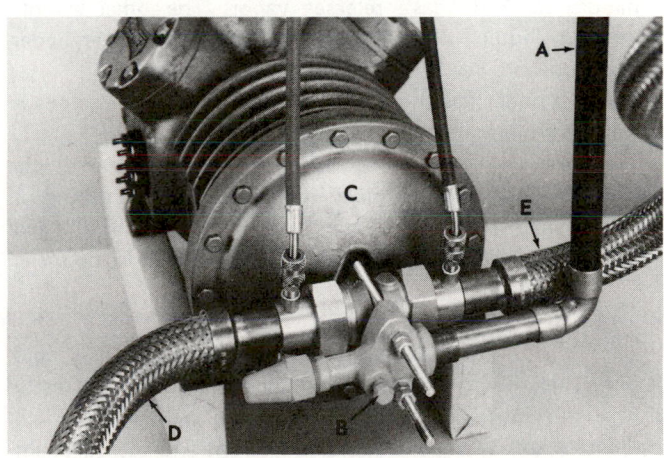

Fig. 14-90. Connecting high capacity filter-drier to suction service valve connector of motor compressor unit. A—Suction line. B—SSV. C—Motor compressor. D—Flexible line bypass to filter-drier. E—Flexible bypass from filter-drier.

Fig. 14-90 shows a bypass or flow diverter block installed between the suction service valve and the compressor. This may be used to divert the flow of vaporized refrigerant through a high-capacity filter-drier.

Fig. 14-91 shows how suction line vapor is bypassed through the drier.

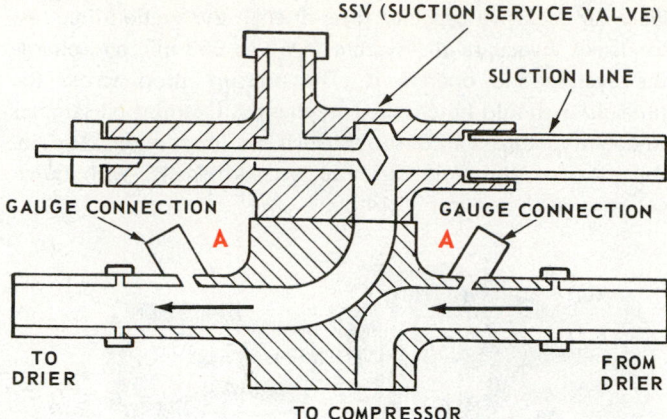

Fig. 14-91. Cross-section of suction service valve attachment. It allows large capacity drier to be installed in suction line. A—Schrader valve gauge connections.

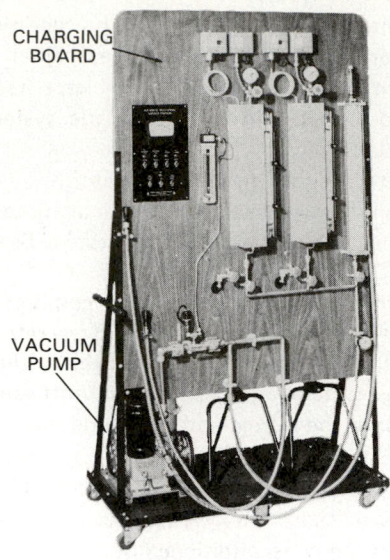

Fig. 14-92. Complete evacuating and charging system. The 8-cfm pump is connected to both the high side and low side of system. Large evacuating lines and vacuum breaker are used. (Airserco Mfg. Co.)

14-83 TRANSFERRING REFRIGERANTS

It is good economy to purchase refrigerants in 50-lb. and 100-lb. quantities. This practice, however, requires that refrigerant be transferred from the large cylinders to smaller service cylinders used by service technicians. Large cylinders are equipped with two types of valves:

1. A regular valve which releases vapor only when the cylinder is upright and the valve open. This cylinder must be turned upside down to remove liquid refrigerant from it.
2. A special valve with two valve stems and handles. The handle marked "gas" releases vapor. The other handle marked "liquid" allows liquid to escape from the cylinder when opened. See Chapter 9.

If the cylinder has the first type valve, liquid refrigerant is removed from it following certain steps.

Turn the cylinder upside down in a special stand and connect it to a small cylinder with a horizontal tubing or flexible charging line at least four feet long. Purge this line and then open both valves after placing the small cylinder on scale.

When transferring refrigerant from a storage cylinder to a service cylinder, never fill a service cylinder completely full of liquid refrigerant. This is particularly dangerous in cold weather since liquid refrigerant expands and contracts greatly with a rise in temperature. If a service cylinder is filled completely full with cold liquid refrigerant and then brought into a warm room, the liquid tends to expand and the resulting increase in hydrostatic pressure may burst the service cylinder.

When the small cylinder is charged with the correct amount, as indicated on the scale, close the valve on the large cylinder. Carefully warm the line with a heating tape or a heat lamp. If neither is available, simply rub the line to warm it. This will force the liquid out of the line. Next, close the valve on the small cylinder. Never allow the line to become more than warm to the hand. Wear goggles.

If the cylinder has a double valve, the same procedure is used except that the cylinder is kept upright.

A charging board provides a more accurate means of transferring refrigerant. This system uses a glass cylinder or

measuring tube to measure the amount of refrigerant. The scale reads in pounds and usually has scales for R-12, R-22, R-500, and R-502. Fig. 14-92 shows a complete evacuating and charging service station. In many cases — especially with small commercial systems — it is better to use disposable containers of refrigerant rather than transferring refrigerant from one cylinder to another. See Fig. 14-93.

Fig. 14-93. Disposable refrigerant container. Containers range from 1 to 30 lb. capacity. (E.I. du Pont de Nemours & Co., Inc.)

14-84 ADDING OIL TO SYSTEM

When a compressor runs too warm or is noisy and it is found that the trouble is a lack of refrigerant oil, there are several ways to add oil. Remember, the oil and all the oil transfer equipment must be clean and dry. The oil must be the proper viscosity for the compressor, the refrigerant and the low-side temperature.

1. The most rapid method is to attach tubing equipped with a hand valve to the middle opening of the gauge manifold. After purging the tubing, immerse it in a clean, dry glass jar nearly filled with refrigerant oil. Then run the compressor. After drawing a vacuum on the low side by turning the suction service valve all the way in, crack the manifold suction valve. Oil will then be drawn into the crankcase. It is important that some of the oil in the glass container be left in the container so the filling tube is always immersed in the oil. Otherwise, air will be drawn into the system. The reason a glass container is used, is to enable the service technician to observe how much oil has been added to the unit. It is a safe policy never to add more than 1/4 pt. of oil at a time to smaller units.

2. Another method of adding oil to the system is to evacuate the crankcase. Then equalize the pressures, remove the oil plug of the crankcase housing and add the oil. Replace the oil plug and then evacuate the compressor.

3. Oil can also be forced into the system by putting the oil in a service cylinder first (draw in by using an evacuated cylinder). Then build up a pressure in the cylinder with refrigerant vapor through the gauge manifold. Invert the cylinder and, by using the low-side manifold valve, the oil can be forced into the compressor.

4. Special pumps are available to hand pump oil into a compressor, even against a high low-side pressure. Because some compressor oil circulates around the system with the refrigerant, one must add oil to the compressor if the refrigerant lines are over 30 ft. long (suction and liquid). Add about 3 fl. oz. of oil for each 10 ft. of tubing installed.

14-85 SERVICING ELECTRICAL CIRCUITS

More and more electrical devices are being used on refrigerating systems. With the addition of electrical defrost systems, solenoid valves, multi-units with electrical interlocks, crankcase heaters, internal motor winding protectors, and various other accessories, a service technician must be knowledgeable about electrical devices and electrical circuits. Chapters 6, 7 and 8 explain the fundamentals of electricity, electric motors and electric controls.

It is important that the service technician have a wiring diagram of the system being serviced. Certain items should always be checked:

1. Are the wires large enough?
2. Is there electrical power up to the machine?
3. Is the voltage correct? (It must not be too low.) See Fig. 14-94.
4. Is the current draw correct? See Fig. 14-95.

With the power on, the current draw and the voltage at the unit can be checked. If the unit will not run, turn off the power and check the circuits for continuity. Use an ohmmeter. See Fig. 14-96.

Locate the break in the circuit continuity by measuring

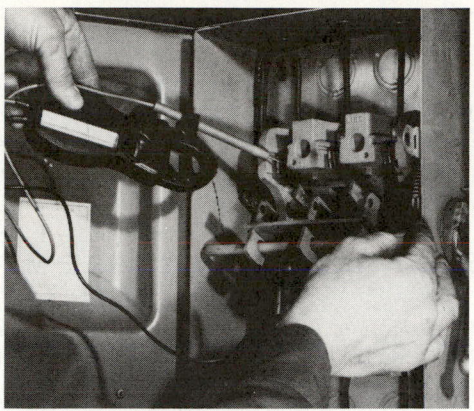
Fig. 14-94. Checking voltage up to main switch. (Amprobe Instrument)

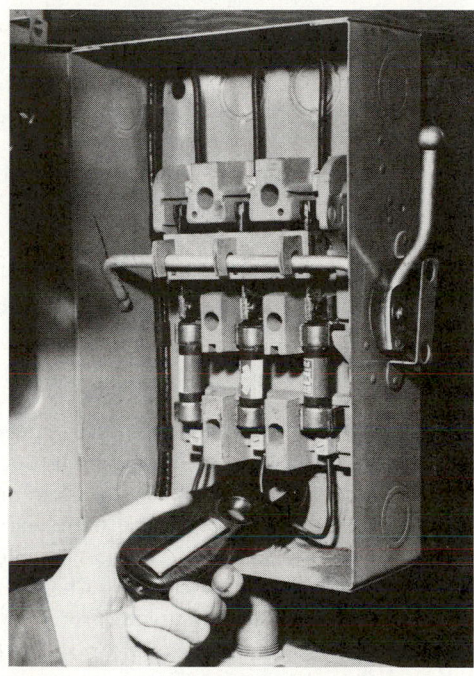
Fig. 14-95. Measuring current flow to motor. Current flow should not be more than motor rating.

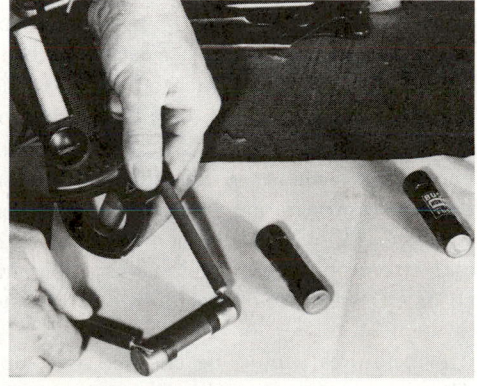
Fig. 14-96. Using ohmmeter to check continuity of fuses.

sections of the circuit with the ohmmeter.
1. Controls.
2. Motor. (If circuit is open and motor is warm, wait for at least an hour to allow the thermal cut-out points to close.)

If the motor hums but will not start, check the starting capacitor with a capacitor tester. Do not start a unit without having overload cut-out in the circuit.

14-86 SERVICING EXTERNAL DRIVE MOTORS

Motors used on external drive commercial systems usually vary in size from 1/12 hp for fans, to 15 hp for compressors. Air conditioning systems require motors of 1/3 hp to 25 hp. These motors are connected to either 120-240V single-phase or 240V three-phase lines.

In addition to the condensing unit motors, commercial systems use motors for fans, water pumps, mixers in ice cream machines and the like.

Many localities require that a licensed electrical contractor remove, repair and install these motors. However, the refrigeration service technician must be able to diagnose motor troubles to locate the fault. But, motor work should be subcontracted.

Motor troubles can be traced to:
1. Mechanical troubles.
2. Electrical troubles.

Mechanical troubles are faults in the bearings, pulleys, out of alignment and excessive end play. Electrical troubles may be further classified as:
1. Internal troubles.
2. External troubles.

The sound of the motor can indicate trouble. Under normal conditions a motor will make a steady low hum. But, in case of worn bearings, rubbing armatures, dry bearings or lack of voltage, erratic beats will be heard in the humming and the rotor may chatter. If the condition of the motor is in doubt, it should be thoroughly checked as described in Chapter 7.

With the motor running, make sure the rotor position is between the two extremes of the rotor end play. If the rotor operates against one extreme of end play, it means that it is trying to assume its magnetic center. Because it cannot, it is running inefficiently. The end play should never exceed 1/16 in. This may be adjusted by using end play washers which are obtainable at electrical supply houses.

Adequate lubrication of the motor bearings is necessary. Normally motors should be oiled twice each year. Use electric motor oil (about SAE 30). Too much oil is just as bad as not enough. Most motors are equipped with overflow openings which eliminate most of the effects of too much oil. Many lubricants are now available in aerosol cans. These spray lubricants around refrigerating systems cause a problem because the pressurizer in the container is usually R-12. If one tests for leaks after using one of these aerosol containers, the halide torch or electronic leak detector will falsely indicate a leak in the system.

Bearing temperatures should be checked after operation. Use a thermometer.

Thoroughly clean the motor occasionally. Dust, dirt and grease accumulations should be removed from inside and outside the motor. Clean commutators and brushes, if used. They must make good contact.

The brush throw-out mechanism should move freely. Brush releasing mechanisms of small motors can be checked by mounting a V-belt on the motor pulley and putting a load on the motor by pulling on the belt.

A torque stand is the best means of determining the real capacity of a motor.

A noisy motor may be caused by a loose pulley, loose fan on the pulley or loose flywheel. These items should be checked when a noise complaint is received.

Fan motors are usually of the shaded-pole type. The most common trouble is worn bearings. Location of these motors usually results in a lack of attention to oiling. Many of these motors are designed to not need lubrication but practice has proven that many do need lubrication periodically.

The pump motors and mixer motors are serviced much like the methods described.

Always be sure the motor is wired correctly and that the voltage at the motor is enough. Always test a motor for grounds and always ground the frame of the motor.

14-87 REMOVING ELECTRIC MOTORS

To remove an electric motor, first disconnect the power line. Then remove the wires from the motor terminals. Label the terminals for easier assembly later. Next, loosen hold-down bolts which attach the motor to the base.

If it is a belt-driven unit, such as a compressor drive or large fan drive, remove the belt from the flywheel first, then from the pulley. The motor can then be lifted out. Do not allow the fan to hit the condenser or catch on the belt. Use a puller to remove the pulley from the motor shaft after loosening the lockscrew.

Fan motors are sometimes difficult to remove. It is best to loosen and remove the fan. Generally, the fan hub is locked in place on the motor shaft by means of an Allen setscrew. Fig. 14-97 shows an Allen setscrew wrench set designed to work in hard-to-reach places.

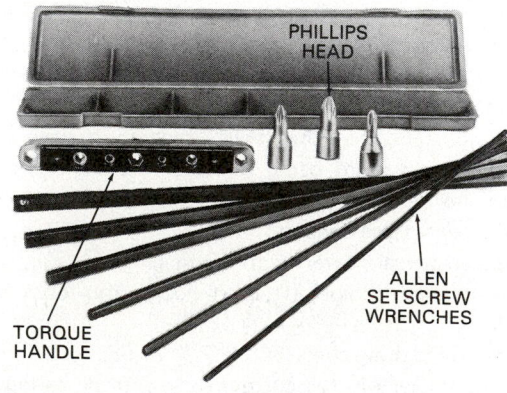

Fig. 14-97. Set of Allen setscrew wrenches and Phillips screwdrivers with turning handle. These tools can reach screw in recess up to 9 in. in depth. (Watsco, Inc.)

14-88 REPAIRING AND TESTING ELECTRIC MOTORS

Conventional motors are described in Chapter 7.

The motor should be cleaned outside, completely dismantled and cleaned inside. Replace the internal switches on capacitor start motors. The bearing walls should be cleaned of oil and the bearings carefully inspected. If they are scored or have more than .001 in. clearance, they should be replaced (ream to size in line). If the contact points are scored or worn, install new points.

When assembling the motor, do not force the parts. Be sure the end bells fit into place. Keep turning the rotor by hand during the assembly to detect any binding before damage is done.

Test a motor before installing.

1. Run under no load.
2. Run under load.

Run the motor, check it for end play (1/16 in. max.), check bearing temperatures and listen for noises other than the normal hum.

Run the motor under load. Fig. 14-98 shows a torque arm

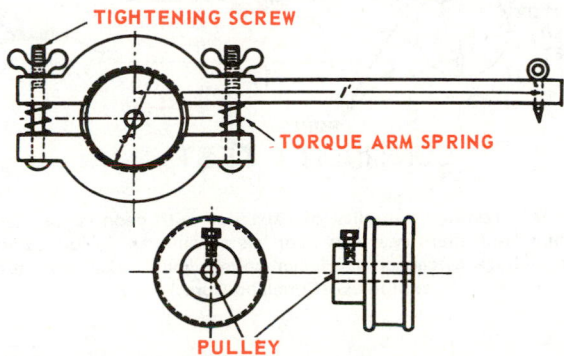

Fig. 14-98. Torque arm and pulley used for testing conventional motors under load.

and pulley for motor testing. Fig. 14-99 shows a motor being tested on it. The electrical load of the motor is checked by using meters connected as shown in Fig. 14-100.

When checking three-phase motors measure the voltage across each compressor terminal when running. The voltage should not vary over 3 percent. If it does, check again at the power-in panel. If satisfactory there, the problem is in the system wiring (loose or dirty connection or wrong wire size). If difference is more than 3 percent at the power-in panel, call the electric utility and have them correct the fault.

14-89 INSTALLING EXTERNAL DRIVE CONVENTIONAL TYPE ELECTRIC MOTOR

External drive type compressor motors may drive a compressor directly or by means of a belt. In either case, the motor must be carefully aligned when installed. Always lock out the power before starting work.

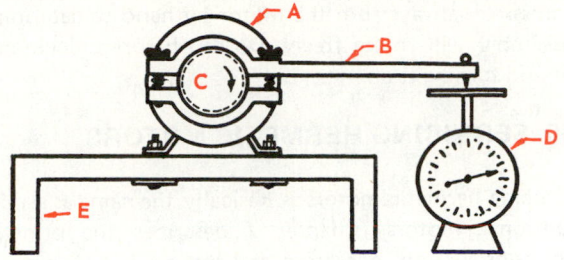

Fig. 14-99. Motor being tested under load. A—Motor. B—Torque arm. C—Motor pulley. D—Torque scale. E—Motor stand.

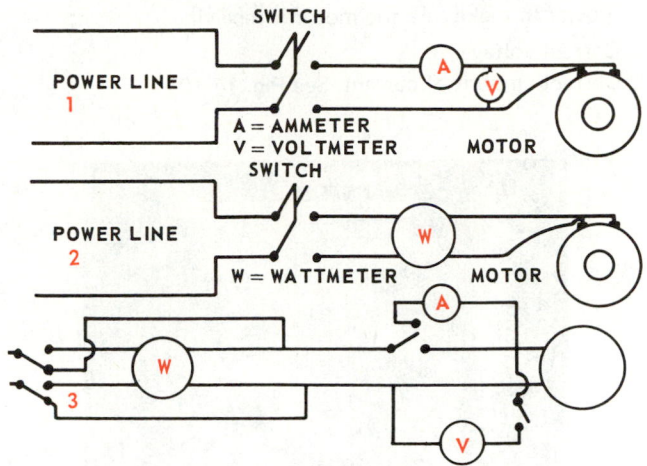

Fig. 14-100. Electric motor testing. 1—Using voltmeter and ammeter. 2—Using wattmeter. 3—Using either voltmeter or combination of voltmeter and ammeter.

Four hold-down bolts are usually used. After installing the bolts loosely, install the belts. Use a lever to move the motor until the belts are tight and the motor is in line. While holding the motor in this position, tighten at least two of the hold-down bolts. Then tighten the others. Motors are usually fastened to their mounting base with four nut, bolt and washer combinations. Motors of 1/2 to several hp are heavy and it is hard to put on the four bolts at one time.

Usually, either the motor base, or the base float it is bolted to, have slots for the bolts. Often it is very difficult to reach under the mounting base to hold the bolt in place or to put a wrench on it.

Some service technicians use caulking compound to hold the bolts in place until the motor is installed and the washers and nuts are started on the bolts. Make clean, tight electrical connections.

In the direct drive units, set the motor on its part of the stand, assemble and install the coupling. Check the alignment carefully because the motor shaft center must be the same height as the compressor shaft center. The two shafts must be in alignment when looking down on them. A dial gauge should be used to check the alignment. Install and tighten the bolts. Water pump motors require the same care.

Fan motor installations sometimes require installing the fan on the shaft before the motor is installed.

If possible, always turn the motor by hand to determine if the assembly will rotate freely. Do this before unlocking the power and turning it on.

14-90 SERVICING HERMETIC MOTORS

Servicing hermetic motors is basically the same as servicing conventional motors. Chapter 7 describes the principles, design, construction, operation and servicing of such motors.

As with all energy devices, first check to determine if the power has the correct characteristics. Voltmeters, ammeters, demand meters and wattmeters are used for this purpose. Check the electrical properties of the motor and then check the power to make sure the motor is receiving:

1. Correct voltage.
2. Correct amount of current. See Fig. 14-101.

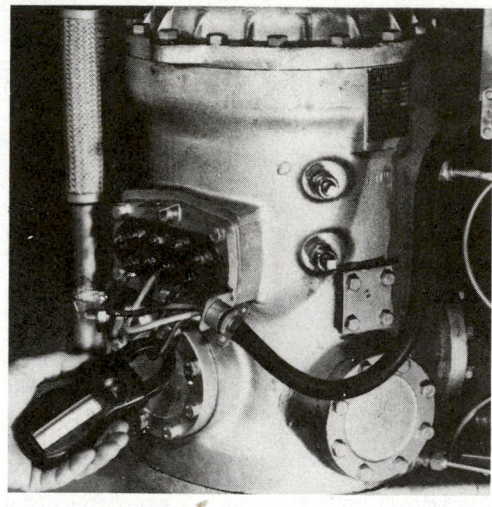

Fig. 14-101. Checking current flow to serviceable hermetic motor using ammeter. (Amprobe Instrument)

Voltage must be within 10 percent of the motor rating. (A 120V motor should have 108V circuit voltage minimum.) This voltage should be read with the motor running. Voltage higher than required is not as critical unless it is 20 percent or more over the rating.

The correct amount of current is vital to good operation. Again, a good instrument is needed. An ammeter of the terminal type (shunt unit) should never be connected across the line (in parallel). It must always be put in series (interrupt one wire only) with the electrical device being checked.

Carefully check external wiring and electrical controls for correct operation before assuming the motor is at fault. An ohmmeter is recommended to check continuity in these circuits. Check each circuit separately. Disconnect if there is a chance of parallel circuits. Be sure the power is locked off.

Capacitors should be checked with a capacitor tester, as shown in Figs. 14-102, 14-103 and 14-104. Study paragraphs 7-37 and 7-38. Do not test capacitors by shorting after charging, as this method is not accurate enough.

The motor should be checked for:

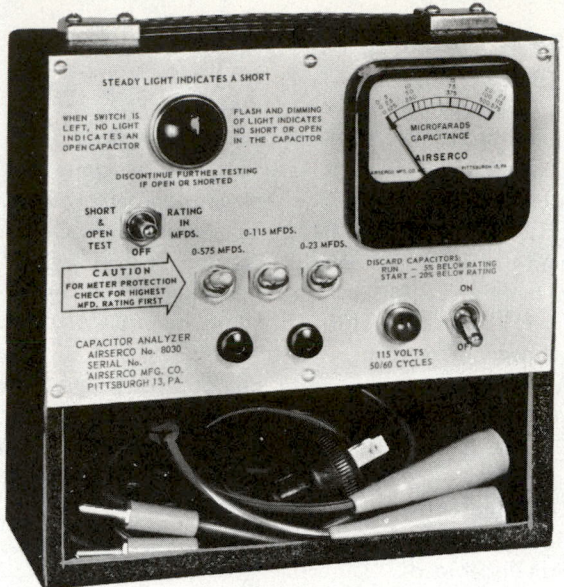

Fig. 14-102. Capacitor tester in range of 0 to 575 mfd. Also tests for open and short circuits. (Airserco Mfg. Co.)

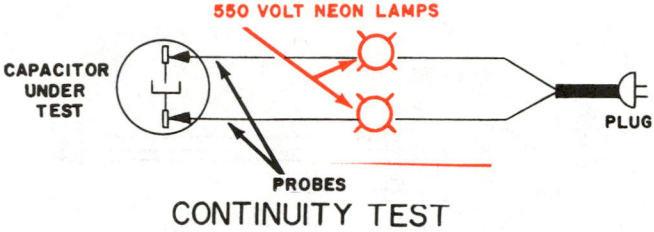

Fig. 14-103. Testing continuity of capacitor with neon lamps. Lamps will light to full intensity if capacitor has continuity. Be sure circuit is of same voltage as capacitor. Higher voltage may cause capacitor to explode. (Copeland Corp.)

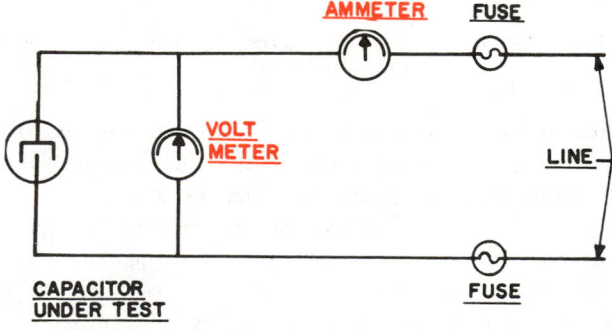

Use this formula for 50 cycle:

$$\mu F = \frac{3180 \times A}{V}$$

Use this formula for 60 cycle:

$$\mu F = \frac{2650 \times A}{V}$$

Fig. 14-104. Wiring diagram and formula used for determining capacity (capacitance) of capacitors. (Copeland Corp.)

1. Open circuits. (Motor should be cool or internal overload in open position will give false readings.)
2. Shorted windings. (Ohmmeter readings should be compared to manufacturer's specifications.)

HERMETIC COMPRESSOR SERVICE CHART

PROBLEMS AND CAUSE	REMEDY
Compressor will not start — no hum.	
1. Open line circuit.	1. Check wiring, fuses, receptacle.
2. Protector open.	2. Wait for reset — check current.
3. Control contacts open.	3. Check control, check pressures.
4. Open circuit in stator.	4. Replace stator or compressor.
Compressor will not start — hums intermittently (cycling on protector).	
1. Improperly wired.	1. Check wiring against diagram.
2. Low line voltage.	2. Check main line voltage, determine location of voltage drop.
3. Open starting capacitor.	3. Replace starting capacitor.
4. Relay contacts not closing.	4. Check by operating manually. Replace relay if defective.
5. Open circuit in starting winding.	5. Check stator leads. If leads are all right, replace compressor.
6. Stator winding grounded (normally will blow fuse).	6. Check stator leads. If leads are all right, replace compressor.
7. High discharge pressure.	7. Eliminate cause of excessive pressure. Make sure discharge shut-off and receiver valves are open.
8. Tight compressor.	8. Check oil level — correct binding condition, if possible. If not, replace compressor.
9. Weak starting capacitor or one weak capacitor of a set.	9. Replace.
Compressor starts, motor will not get off starting winding.	
1. Low line voltage.	1. Bring up voltage.
2. Improperly wired.	2. Check wiring against diagram.
3. Defective relay.	3. Check operation — replace relay if defective.
4. Running capacitor shorted.	4. Check by disconnecting running capacitor.
5. Starting and running windings shorted.	5. Check resistances. Replace compressor if defective.
6. Starting capacitor weak or one of a set open.	6. Check capacitance — replace if defective.
7. High discharge pressure.	7. Check discharge shutoff valves. Check pressure.
8. Tight compressor.	8. Check oil level. Check binding. Replace compressor if necessary.
Compressor starts and runs but cycles on protector.	
1. Low line voltage.	1. Bring up voltage.
2. Additional current passing through protector.	2. Check for added fan motors and pumps connected to wrong side of protector.
3. Suction pressure too high.	3. Check compressor for proper application.
4. Discharge pressure too high.	4. Check ventilation, restrictions and overcharge.
5. Protector weak.	5. Check current — replace protector if defective.
6. Running capacitor defective.	6. Check capacitance — replace if defective.
7. Stator partially shorted or grounded.	7. Check resistances; check for ground — replace if defective.
8. Inadequate motor cooling.	8. Correct cooling system.
9. Compressor tight.	9. Check oil level. Check for binding condition.
10. Unbalanced line (three-phase).	10. Check voltage of each phase. If not equal, correct condition of unbalance.
11. Discharge valve leaking or broken.	11. Replace valve plate.
Starting capacitors burnout.	
1. Short cycling.	1. Reduce number of starts to 20 or less per hour.
2. Prolonged operation on starting winding.	2. Reduce starting load (install crankcase pressure limit valve), increase voltage if low — replace relay if defective.
3. Relay contacts sticking.	3. Clean contacts or replace relay.
4. Improper relay or incorrect relay setting.	4. Replace relay.
5. Improper capacitor.	5. Check parts list for proper capacitor rating — mfd. and voltage.
6. Capacitor voltage rating too low.	6. Install capacitors with recommended voltage rating.
7. Capacitor terminals shorted by water.	7. Install capacitors so terminals will not be wet.
Running capacitors burnout.	
1. Excessive line voltage.	1. Reduce line voltage to not over 10 percent above rating of motor.
2. High line voltage and light load.	2. Reduce voltage if over 10 percent excessive.
3. Capacitor voltage rating too low.	3. Install capacitors with recommended voltage rating.
4. Capacitor terminals shorted by water.	4. Install capacitors so terminals will not be wet.
Relays burnout.	
1. Low line voltage.	1. Increase voltage to not less than 10 percent under compressor motor rating.
2. Excessive line voltage.	2. Reduce voltage to maximum of 10 percent above motor rating.
3. Incorrect running capacitor.	3. Replace running capacitor with correct mfd. capacitance.
4. Short cycling.	4. Reduce number of starts per hour.
5. Relay vibrating.	5. Mount relay rigidly.
6. Incorrect relay.	6. Use relay recommended for specific motor compressor.

3. Grounded windings. First check with ohmmeter and then check with a 500V (megavolt) circuit tester. Insulation breaks can only be checked accurately with this high voltage tester. Handle this instrument carefully to avoid shocks.

Most companies report that a high percentage of motor compressors returned labeled "faulty motor" actually have good motors. This false diagnosis indicates the need for careful checking.

The chart on page 545 is a list of typical motor and circuit troubles, their cause and their remedy.

14-91 REMOVING HERMETIC MOTOR COMPRESSORS

If it is certain that the motor in a hermetic motor compressor unit is faulty, it must be removed. Removal procedure follows:

1. Remove the refrigerant as described in Chapter 11 and as described in this chapter for serviceable hermetics.
2. Open the main circuit switch and lock the switch in open position. Tag the switch to inform others why the switch is locked open.
3. Disconnect the wires from the motor compressor (label them or use color code).
4. Clean outside of motor compressor.
5. Disconnect lines (wear goggles)! The type of disconnect depends on assembly. Unbolt service valves, open brazed joints by heating or cut the lines. Use tube cutter only to avoid chips getting into the system.
6. Remove motor compressor. Do not tilt or oil may be spilled. Avoid lifting over 50 lb. If it is a large heavy unit, use a lifting machine (tripod or fork lift).
7. Plug refrigerant openings.

14-92 REPAIRING HERMETIC MOTORS

Removing a hermetic motor, repairing the bearings, or replacing the windings is a specialty job. Most service technicians replace the complete assembly by getting a replacement unit, either new or rebuilt.

The actual overhaul of a hermetic motor compressor is described in Chapter 11.

14-93 INSTALLING HERMETIC MOTOR COMPRESSORS

A replacement hermetic motor compressor is usually furnished with a starting relay, capacitors, overload protectors and other accessories. Use an exact replacement. These motor compressors are either designed for low, medium or high low-side pressures. Use the same type as the one removed. The refrigerant openings are closed with service valves or the unit is provided with short tubing ends, brazed in place with the ends of the tubing crimped and brazed.

Carefully mount the motor compressor in place. Use all the safety precautions (lifting, safety shoes, protecting floors and equipment). Install the mounting bolts. The springs and grommets must be in the correct position and the hold-down bolt or nut must be positioned correctly.

Install the electrical devices (overload and starting relay) and the electrical wires. All connections must be clean and tight. All wires, including insulation and wire terminals, must be in good condition.

Avoid connecting aluminum wires to copper wires or copper terminals. Rapid corrosion takes place. Special adaptors must be used or the joint will corrode.

Install suction and condenser lines. Units using service valves are installed the same way as described for a conventional compressor.

If brazed connections are used, identify which tubing stubs are the suction connections, the discharge connections, the process tube and the oil cooler connections. Cut the tubing stubs with a tube cutter. Select connector fittings or swage the tubing. Flux the outside of the joints brazed to the compressor dome. Connect the system lines to the compressor lines. Clean the inside of the suction line and condenser line. Silver braze the joints. The area must be well ventilated during brazing. Install a liquid line drier and a suction line drier. Clean the joints with warm water to remove flux.

Install a gauge manifold using charging stub or a valve mounted on the suction line and liquid line.

Evacuate the system, put some vaporized refrigerant in (enough to build a 15 psi pressure), test for leaks (repair any, if found), evacuate to a 50 to 500 micron range for several hours, and then charge the system.

Turn on the power, run motor, check the temperatures, pressures and electrical power. If any operation is not normal, be sure to diagnose and remedy it before leaving the job. Remove the service connections and braze these joints. Clean up the area.

14-94 SERVICING MOTOR CONTROLS

Four main types of motor controls are used in commercial refrigeration: low-side pressure motor control, thermostatic motor control, high-side safety motor control and an oil pressure safety control. Troubles encountered with these include:

1. Corroded points.
2. Broken mercury bulb.
3. Out of adjustment.
4. Corroded or broken operating springs.
5. Out of level.
6. Leaking bellows.

Motor controls normally work year in and year out without giving trouble. However, in cases such as unit overloading, the resultant short cycling will rapidly deteriorate the contact points in the control or will so overload the mercury contact that it will crack and be destroyed. Corroded points should be replaced. They can be temporarily repaired by cleaning with fine sandpaper or a clean fine mill file. Never use emery cloth. A broken mercury bulb must be replaced.

An out-of-adjustment switch is often the result of tampering. A pressure control may be easily checked by installing the gauge manifold. Using the compressor as a vacuum and pressure pump, check the cut-in and cut-out points. Three different methods may be used to build up pressure in the crankcase after the compressor is run to the cut-out point of the control:

1. The suction line may be cracked open again.
2. A bypass may be run from the discharge service valve of the compressor to the suction service valve. (Use the gauge manifold.)
3. A refrigerant service cylinder may be used containing the same kind of refrigerant attached to the gauge manifold. Many service technicians carry a hand vacuum and pressure pump in their tool kit. This tool allows a rapid check of pressure controls. See Fig. 14-105.

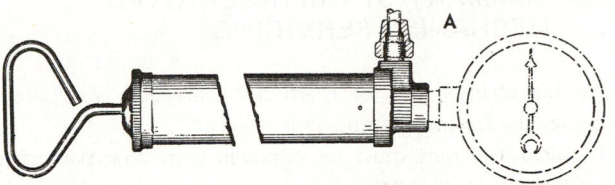

Fig. 14-105. Hand vacuum and pressure pump used for adjusting motor controls. A—Connection to control being tested and adjusted.

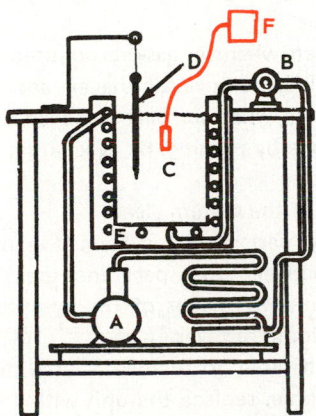

Fig. 14-106. Temperature bath system. Cooling system used to control liquid bath temperature. May be used for checking thermostat cut-out and cut-in temperatures and for checking thermostatic expansion valves. A—Condensing unit. B—Refrigerant control (AEV). C—Temperature bath. D—Thermometer. E—Evaporator. F—Thermostat being tested.

A thermostatic motor control is more difficult to reset. An approved method is to use an ice bath and thermometer or a temperature bath, as shown in Fig. 14-106. The control may then be set to cut in or out at the temperature desired.

The high-pressure motor control presents few difficulties because it is adjusted to work only under extreme conditions. The most common troubles are leaks in the bellows or at the joints of the controls. Occasionally, however, this control may be short cycling the refrigerating mechanism. This short cycling is due to excessive head pressure.

To check a high-pressure motor control element, connect it to the gauge manifold and turn the discharge service valve all the way in. Running the compressor will produce a head

pressure sufficient to cut out the control. The high-pressure gauge will record the pressure at which the cutting out takes place.

Many large compressors use oil pressure lubrication. If oil pressure fails or falls below a certain pressure above the low-side pressure, there is danger of damaging the compressor. An oil pressure safety control protects these systems. It will cut out if the oil pressure drops to a dangerous level.

14-95 SERVICING SOLENOID VALVES

Solenoid valves develop both electrical problems and refrigerant troubles. If the electrical connections are dirty or loose, the coil may not create enough magnetism to raise the valve. Usually the valve should be mounted with the coil on top and with the valve level, or they may stick or chatter. Solenoid valves sometimes will develop a leaky needle and seat. In this case, the valve must be replaced. It is important that the solenoid be of the proper voltage and amperage rating; otherwise, the coil may burn out. A 120V valve cannot be connected to a 240V circuit. The service technician must also be careful to place most solenoid valves right side up and in a vertical position. Otherwise, they will not operate.

To remove a valve from the system, one must follow the same pump-down procedure as for a TEV.

14-96 SERVICING LIQUID LINE

The liquid line contains several important items needing inspection by the service technician:

1. Size of the liquid line.
2. Hand shutoff valves.
3. Sight glass.
4. Moisture indicator.
5. Screen filter.
6. Drier or dehydrator.
7. Vibration absorber.
8. Connections.
9. Solenoid valves.
10. Joints.
11. Pinched or buckled pipe or tubing.

When diagnosing troubles of a system, the service technician's first problem is to determine if each part of the system is the right capacity. It is important that the liquid line be as large as the condensing unit liquid receiver valve connections. Check to be sure that reducer fittings have not been used. See Chapter 15 for recommended liquid line sizes. Many large units use various sizes of liquid lines as the line feeds to various evaporators.

All service technicians must remember to have a least one end of the liquid line valves open when servicing the unit. Otherwise, a temperature rise may create a very high hydrostatic pressure. This pressure may burst the line. Be especially careful if a solenoid valve or a clogged screen or a clogged dehydrator is in the line.

Check the full length of the line for line condition. The line must be protected from abrasion and abuse as objects are moved. The line should be well supported along its full length.

When admitting liquid refrigerant into a liquid line by opening the liquid receiver service valve, always open the valve slowly or the sudden rush of liquid may injure the screen and may pack the desiccant in the drier so firmly that it will soon clog.

Test all joints for leaks. Check the temperature of the liquid line; it should be close to room temperature for its full length.

A partially clogged screen or drier is indicated by a lower-than-normal temperature at the outlet. Sweating and even frosting may be seen. There will also be an excessive pressure drop. This may cause bubbles in the sight glass.

14-97 SERVICING THE SUCTION LINE

Servicing the suction line is much like servicing the liquid line. Parts to be inspected are:
1. Size of the suction line.
2. Hand shutoff valves.
3. Vibration absorber.
4. Check valves.
5. Two-temperature valves.
6. Constant-pressure valves.
7. Filter-drier.
8. Muffler.
9. Accumulator.
10. Joints.
11. Pinched or buckled suction line.

Suction line size is important. If it is too small, it will cause too much pressure drop and high gas velocities will cause noise. The line should be the size of the suction service valve connection or the suction line connection on the hermetic dome.

On multiple installations, the suction line is smaller for each evaporator than the main suction line. For example, the most remote evaporator may have a 1/2 in. OD size, the next size 5/8 in. OD, the next 1 in. OD, the next 1 1/2 in. OD. The line at the compressor may be 2 in. OD. See Chapter 15.

The pressure drop should be checked by installing a gauge at the most remote evaporator and one at the compressor. The pressure drop should be approximately 2 psi. If more, line sizes should be increased.

The pressure change across two-temperature valves and constant pressure valves should be checked. Their pressure drops are separate from the line pressure drops and are not included in the psi pressure drop.

Also determine the pressure drop across the suction line filter-drier. A large pressure drop here indicates a partially clogged filter-drier. In this case, the drier should be replaced.

Test for leaks with 15 psi pressure or more in the lines.

14-98 SERVICE NOTES

Pinching lines is a practice to be used in cases of emergency only. Some service technicians follow this practice needlessly and it only leads to future trouble. Almost all systems are provided with enough valves to service them. However, pinching lines used to service hermetic units, prior to silver brazing the stub, is common practice.

Remember when adding refrigerant to a system that some of the oil will be dissolved in the refrigerant. *If a unit becomes noisy soon after the refrigerant has been added, some refrigerant oil should be added.*

Driers, if heated while in the system (because of a refrigerant shortage, for example), will release water to the system. Install a new drier if the moisture indicator signals moisture.

Crankshaft seals may leak if the compressor has been idle for a long time. Turn the compressor over by hand a few times to allow oil to seep between the rubbing metal surfaces. Also put an ounce of one of the special refrigerant detergent oils into the crankcase to help eliminate this problem.

14-99 SUMMARY OF REFRIGERATOR MECHANISM SERVICING

If this chapter has been studied carefully, one should recognize the following important things:
1. Liquid refrigerant must be removed from that part of the mechanism to be overhauled.
2. It is necessary to equalize pressures in the unit before dismantling. This avoids a rush of air into the unit or of refrigerant out of the unit upon opening the lines.
3. All refrigerant openings should be plugged immediately after dismantling.
4. Put in new gaskets wherever gaskets are used.
5. When reassembling, remove all the air and moisture from the lines and from whatever part has been open to the air. This may be done by purging, by evacuating and purging or by deep evacuating.
6. Keep the inside of the system clean.
7. To remove any part of the system one must close the nearest valve between that part and the liquid receiver. Evacuate the unit, by means of the compressor, into the condenser and liquid receiver.

It is usually advisable to remove a large overhaul to the shop. In the meantime, replace the unit with a temporary one so the owners may have the use of the system.

Where possible, a service technician should review the wiring diagram and the specific service manual before working on a refrigerating system. The complex nature of some of the systems requires that a service manual be used.

14-100 PERIODIC INSPECTIONS

The capital investment in a commercial refrigeration installation amounts to hundreds of dollars. Because of the way most mechanisms are built, one problem in the system will cause others. It is important, therefore, that all commercial machines be completely checked over periodically. The service technician should use a systematic method of doing this. Then no detail is overlooked. All inspections should cover such things as:
1. Electrical connections.
2. Motor and safety devices.
3. Compressor noises.
4. Amount of refrigerant.

5. Dryness of refrigerant.
6. Oil level.
7. Water flow.
8. Gas leaks.
9. Coil conditions.
10. Supports for tubing.
11. Coil supports.
12. Cleanliness.

For a conventional condensing unit, one should check:

1. Belt condition.
2. Belt alignment.
3. Belt tightness.

For a hermetic condensing unit, one should check:

1. Overload cut-out.
2. Relay.
3. Capacitors.

It establishes good will and also builds up a good contact file to prepare a check sheet, one copy of which should be given to the owner. This check sheet is a time-tried system that prevents overlooking important items.

14-101 LOCATING TROUBLES

Methods of testing to locate sources of trouble are based on the operating principles of the mechanism. By checking the pressures, temperatures, running time and current or voltage, the service technician is soon able to pin point which part of the system is causing the trouble.

The service technician must have a thorough knowledge of the fundamentals of refrigeration and of the cycles before he can become reliable and competent at trouble tracing and repair. *Troubles in a refrigerator mechanism must be located before dismantling.* This keeps the cost of servicing at a minimum and assures proper operation of the unit after repair and assembly.

Methods of locating troubles vary with the type of system — whether it has direct expansion or a capillary tube. The call for service should indicate what the trouble is. The owner will probably say that it costs too much to operate, that it is not freezing but is running continuously or that it is freezing but running continuously. From these complaints, the service technician may usually get an idea what the trouble is. Always verify these statements by checking over the refrigerator before attempting any troubleshooting or service work.

In trouble tracing, first classify the type of service call and then determine what caused the trouble described in the service call. The following troubleshooting pointers have been prepared to help the service technician. Naturally, it is impossible to give every detail, but once the technician learns the method of tracing trouble, there should be no difficulty.

One should check all of the following things in a refrigerating mechanism before deciding what is the trouble:

1. Low-side pressure.
2. High-side pressure.
3. Temperature of evaporator.
4. Temperatures of liquid and suction lines.
5. Amount and dryness of refrigerant.
6. Running time of mechanism.
7. Probability of leaks.
8. Noise.

Several basic fundamentals make locating trouble easier. When there is poor refrigeration, or no refrigeration, either or both of two things can be wrong:

1. There is little or no refrigerant.
2. The pump is not moving the refrigerant. Pressures are not correct.

If there is no refrigerant, there will be no liquid refrigerant in the evaporator. This means that the refrigerant has leaked out or is being held in a certain part of the system by clogged needles, clogged screens and pinched lines. Clogging causes a high vacuum reading on the low side.

If there is a lack of refrigerant throughout the system, there will be a hissing sound at the refrigerant control indicating the refrigerant passages are not closed. The sight glass will show bubbles.

A hissing sound at the refrigerant control always indicates a lack of refrigerant because the dry gas going through the restriction will cause the gas noise.

If the pump is not functioning, the low-side pressure will be above normal while the condenser and discharge line from the compressor will be below normal temperature.

To determine what is responsible for a poor condensing condition, proceed as follows: Install the gauge manifold and determine the head pressure. Compare this pressure with what the pressure should be for the refrigerant being used.

14-102 LITTLE OR NO REFRIGERATION AND UNIT RUNS CONTINUOUSLY

DIRECT EXPANSION SYSTEM (TEV)
If the unit has lost all refrigerant, there is, naturally, no refrigeration. To test for this, install gauges and determine evaporating or low-side pressure. If this pressure is normal, the unit probably has little or no refrigerant. If the compound gauge indicates a high vacuum — 20″ Hg. or more — it means:

1. The expansion valve is adjusted so that it draws this vacuum.
2. It is frozen closed.
3. It has a clogged screen.

Clogging can be caused by moisture freezing at the refrigerant control and stopping the flow of refrigerant. The results are the same as a needle stuck closed, or a clogged screen. There is one difference. After the system warms above 32 F. at the valve, frozen moisture will melt and normal refrigeration will return.

The only sure cure for the moisture condition is to remove the moisture by installing a new drier. If it is suspected that moisture in the valve has caused the clogging, heat the valve using a heat lamp. This moisture problem occurs with all refrigerants that do not chemically combine with the water; for example: R-12, R-22, R-500, and R-502.

A high vacuum may also be caused by a clogged or restricted suction line filter. In either case, it is not allowing refrigerant to flow through. There will be a high pressure in the evaporator which gives continuous running with little or no refrigeration.

If the compound pressure gauge shows a high pressure on

the low side — that is, a pressure which will not allow the refrigerant to evaporate at a low temperature — the trouble may be an expansion valve stuck open or out of adjustment. If this is the trouble, there may also be a frosted or sweating suction line. It simply shows that the refrigerant is going into the low side too fast and the liquid will flood both the evaporator and the suction line.

High pressure may also be due to an inefficient compressor. If the expansion valve is stuck open, it may be due to dirt on the needle. To remedy, one may flush the valve by alternately opening and closing the liquid receiver service valve or liquid line hand shutoff valve. Surges of liquid will rush past the expansion valve needle, cleaning it.

CAPILLARY TUBE

If the capillary tube is restricted or completely clogged, if there is moisture frozen in the tube or if the screen is clogged, no refrigerant can pass into the evaporator. This stoppage will cause a high vacuum reading, the evaporator will be warm and the condenser will be cool with a normal or low-head pressure. If refrigerant is low, the capillary tube will have a hissing sound just as the compressor shuts off. Low and high-side pressures will be below normal.

An inefficient pump will be indicated by an above-normal low-side pressure and a normal or below-normal head pressure.

14-103 NO REFRIGERATION: UNIT DOES NOT RUN

Where there is no refrigeration and the unit does not run, the trouble is probably somewhere in the electrical circuit. A test light or a voltmeter will tell if power is being supplied to the motor.

If there is power, the motor may be burned out or the circuit open. An ineffective temperature control — one having a leaking power element, for instance — may prevent the motor from starting. It may be checked out as described in Chapter 8.

Further checking of the circuit may show a manual switch or overload in the off position. Or an internal overload switch may be open.

If the motor compressor is hot, this may well be the trouble. It should be checked with a test light or an ohmmeter. Basically, the problem is in the motor or the electrical power circuit.

14-104 MOTOR RUNNING CONTINUOUSLY WITH NORMAL OR TOO MUCH REFRIGERATION

If the motor compressor runs continuously, and the unit produces normal or too much refrigeration, the probable cause of the trouble is a faulty motor control. It will not cut out at the correct temperature. Remember, there is a relationship between the control cut-out point and the evaporating pressure on the low side. If the control is adjusted to cut out at a temperature or pressure lower than the evaporating pressure, the control cannot stop the electric motor.

DIRECT EXPANSION SYSTEM (TEV)

An undercharged system may provide just enough refrigerant to fill the evaporator partially, but not enough to operate the temperature control cut-out point and shut down the unit. In this case one may get normal refrigeration, but the unit never shuts down. This trouble may be indicated for certain by the way frost or sweat collects on the outlet tubing of the evaporator and seeing if it reaches as far as the TEV sensitive bulb. Another indication of refrigerant shortage is a warm liquid receiver and liquid line. The sight glass should show bubbles.

The trouble may also be caused by an overcharge of refrigerant or air in the condenser. Excessive head pressures lower the efficiency of the compressor so much that continuous operation results.

A leaking expansion valve will sometimes give normal refrigeration but will not allow the pressure to drop to where the motor control will cut out the motor. An improperly adjusted expansion valve may cause the same trouble. This will create a frosted or sweating suction line and an above-normal, fluctuating low-side pressure. (Fluctuating means rapid changing from higher to lower and back.)

An inefficient compressor is another possible cause and can be checked by use of gauges. Expansion valve troubles and the compressor troubles may be checked as mentioned earlier in this chapter.

CAPILLARY TUBE

If there is too much refrigerant in the system, the excess will collect on the low side and may enter the suction line. This liquid may prevent the compressor from producing a low enough pressure to operate the thermostat. Therefore, the unit will run continuously and will produce either a normal refrigeration effect or, more likely, excessive refrigeration.

A slight lack of refrigerant will cause a partially refrigerated evaporator and the refrigerated part may not be close enough to the thermostat to cause it to shut off the motor.

14-105 SHORT CYCLING

Short cycling means that the system runs and then stops every few minutes. It is first necessary to locate the control which is turning the system on and off:
1. Temperature control.
2. Overload controls.
3. High-pressure safety control.
4. Oil pressure safety control.

Short cycling may come about from a rapid pressure rise on the low side of the system where an expansion valve is leaking. This leak will also cause a frosting or sweating of the suction line.

Most units are equipped with an overload device in the electrical system. If the motor becomes too hot or uses too much current, these safety devices will stop the motor and then restart it after it cools. A temperature control which is out of adjustment — that is, one with a small differential — will also cause a short cycle.

If the refrigerator has a pressure motor control, short

cycling may be caused by either a leak in the refrigerant control or poorly seated compressor valves. In either case, the pressure on the low side will rise rapidly during the off part of the cycle causing the motor to start.

Machines equipped with a high-side pressure safety control will sometimes short cycle if the condensing pressure becomes too high because of a high condensing temperature.

14-106 NOISY UNIT

Noise can come from three principal sources: compressor, electric motor and the mounting of the complete condensing unit. A compressor is noisy when it pumps oil or when the valves, the piston pin, the connecting rod and piston have become worn. Sometimes when the compressor gets very warm, it will develop knocks. These are usually hard to remedy. Lack of oil in the compressor may cause it to be noisy.

The metal shaft seal used on most of the open type compressors occasionally becomes noisy and gives out a shrill squeal. This is usually caused by a lack of oil at the seal. If not remedied, it will soon score the seal and cause it to leak.

A conventional electric motor may have noises such as a fan roar, bearing squeak or motor rumble. Occasionally, if the motor is loaded too heavily, the starting winding does not cut out and will cause continuous noisy operation. If allowed to run this way, the motor will burn out. End play in an electric motor is necessary, but too much will cause a dull knock.

In a conventional unit, belt noise may result from a dry belt or pulleys that are out of line. It may be stopped by using a dressing recommended for belts and by realigning the pulleys. *Do not use oil.*

Sometimes the whole machine unit will vibrate excessively, producing a rumbling sound as the unit runs, shaking the cabinet disagreeably. This is probably due to poor mounting or spring suspension. Another cause may be too little movement in the suction and liquid lines. Or, some obstruction may have been put in the compartment which destroys the action of the shock and the noise-absorbing mounting of the condensing unit.

Excessive head pressure will make a unit vibrate more than normal. A badly worn needle or seat will sometimes make a chattering noise while the unit is in operation.

14-107 REFRIGERATION SERVICE CONTRACTING

It is good business to offer contracts for maintenance and service. Many large companies have developed such contracts, and even larger independents are now offering their customers this type of servicing.

The usual contracting plan offers a definite monthly or weekly rate, for which the service company agrees to keep the refrigerating mechanism in good condition. This charge may or may not cover parts.

The success of such a plan depends on large volume in order to offset cost of maintaining extremely bad installations. Contracts may be on a time and materials basis. Two features

of a service contract which appeal to the purchaser are the 24-hours available service clause and an absolute guarantee of work done.

If one has a service contract, a procedure sheet or record sheet should be used to prove service and to insure thorough inspection. This check sheet should include:

1. Test for leaks
2. Check refrigerant charge.
   a. Head pressure.
   b. Low-side pressure.
3. Check oil charge.
4. Check water valve.
5. Check water drain.
6. Check and lubricate motor.
7. Check belt condition and tension.
8. Clean evaporator.
9. Clean condenser.
10. Straighten fins.
11. Voltage reading.
12. Wattage reading.
13. Check circulating fans.

Service records are absolutely essential if one wishes to establish a permanent business. These records should contain details of the ownership, machine, what type work was done and materials used. This record enables "check backs" if the system does not operate correctly. Furthermore, it establishes sales prospects as systems become older.

14-108 SERVICE ESTIMATES

Many organizations operating refrigerating equipment ask for bids when repair, replacement or service is required. A service organization bidding on such work should have a person who specializes in estimating work of this kind. This specialist should be thoroughly acquainted with cost of material, service problems and labor costs. He must be able to judge the time necessary to do the repair. Records kept of service and maintenance work are used as a guide in making estimates.

Needless to say, estimates must include overhead expense such as rent, equipment obsolescence, office and shop services and advertising.

A pleasing personality combined with rapid and accurate estimating ability is essential.

14-109 REVIEW OF SAFETY

A refrigeration service engineer must always be alert to safety. Refrigerating systems have hazards arising from pressure, electricity, power devices, heat, flames, heavy objects and climbing.

The safety program must include:

1. Safety for the mechanism.
2. Safety for the items being refrigerated.
3. Safety for the operator, the installation, the service technician and people who may be near the mechanism.

All parts in a refrigerating system must be absolutely clean before they are installed in the system.

Always know what is inside a pressure vessel and know the pressures. Always wear goggles when working on a pressure vessel (refrigerating unit) and when there is danger of flying particles.

Never breathe fumes of any kind. Do not neglect the use of the gas mask when working in a refrigerant-laden atmosphere. This protective measure also applies to cleansing bath fumes, soldering fumes, brazing fumes and welding fumes. It is true that one's body can and will get rid of certain amounts of strange chemicals and fumes, but some chemicals and fumes accumulate in the body and there may be no ill effects felt for years. Good ventilation is of vital importance.

Avoid exposure to electrical shocks. Keep open electrical terminals covered. Do not work on electrical circuits in damp or wet surroundings.

Avoid spilling liquid refrigerant on any fixture, finished surface or floor. It may ruin the finish. Avoid contact with the liquid refrigerant, especially on the body and eyes. A freeze burn will result.

Put guards on powered moving objects such as flywheels, belts, pulleys and fans. Use the leg muscles, not the back, when lifting.

Before using a flame for leak testing, soldering, brazing or welding, have a fire extinguisher handy. Remove all combustibles from the area and provide good ventilation. A dry chemical fire extinguisher is one of the best for all types of fires. When using a flame to perform tests or make repairs, protect surrounding objects and surfaces with sheet metal or some other flame-resistant shield.

Always test used compressor oil for acid content before allowing any of it to touch the skin. A severe acid burn may result.

Never use air, oxygen or any fuel gases for developing pressure in a system. If gas, other than refrigerant, is desired, use carbon dioxide, nitrogen, helium or argon at controlled pressures and with a pressure relief valve. See Para. 11-39.

Always wear goggles when handling refrigerants, when opening a refrigerating mechanism or at any time when there is danger from flying liquids. Wear rubber gloves when handling substances which may have an acid content. Never use carbon tetrachloride for any cleaning operation. This caution is given because at one time it was common practice to use carbon tetrachloride for cleaning refrigerant parts. Such use is now illegal.

Use care when handling capacitors. A charged capacitor can deliver a severe shock.

The Occupational Safety and Health Act (OSHA) is now in effect with regulations. It is also known as the William Steiger Act of 1970. Most of the regulations became mandatory in March, 1973.

Many of the regulations are based on standards developed by the American National Standards Institute (ANSI); National Fire Protection Association (NFPA); Walsh Healy Act (noise); The Service Contract Act; American Society for Testing and Materials (ASTM); and the American Conference of Governmental and Industrial Hygienics (ACGIH).

The Act covers almost every present safety standard plus about everything that can be sensed. Among areas covered are:

1. Walking and working surfaces.
2. Means of exit (egress).
3. Powered platforms.
4. Occupational health and environment control.
5. Hazardous materials.
6. Personal protective equipment.
7. General environment controls.
8. Medical and first aid.
9. Fire protection.
10. Compressed gases.
11. Material handling.
12. Machinery and machine guarding.

Any employee can ask for an inspection. The inspector can impose fines. The employee cannot be fired for requesting an inspection. The employer is responsible. If safeguards are present but the employee does not use them, the employer will still be fined. If this happens, however, the employer has excellent grounds for releasing the employee who is not using the safety precautions.

Some of the OSHA priorities during inspection are:
1. Asbestos.
2. Carbon monoxide.
3. Cotton dust.
4. Lead.
5. Silica.

Asbestos and carbon monoxide are of special importance to refrigeration and air conditioning.

Some other OSHA concerns are:
1. Beryllium.
2. Heat.
3. Mercury.
4. Ultraviolet radiation.
5. Fibrous glass.
6. Trichloroethylene.
7. Chromic acid.
8. Paratheon.

Heat, ultraviolet radiation, fibrous glass and trichloroethylene are of special interest in refrigeration and air conditioning.

The Threshold Limit Values (TLV) are important. Usually this is the upper safety limit for the hazard over an eight-hour exposure period. Most have been established by ANSI (American National Standards Institute).

Noise is important. The OSHA lists 90 dB on dBA noise scale as maximum for eight hours of exposure, 92 dB for six hours, 95 dB for four hours.

For seated mental performance for eight hours, the room temperature maximum should be 79 F. (26 C.) for men and 76 F. (24 C.), for women. For a four-hour exposure, it can be as high as 87 F. (31 C.). This regulation will have an impact on the ventilation and air conditioning industry.

Above all else report any injury in writing, no matter how slight. Send the report to your employer and keep a copy.

14-110 TEST YOUR KNOWLEDGE

1. Why do code installations require safety release valves on some receivers?

2. Why should hand valves be provided in the refrigerant lines to each individual evaporator?

3. Why is it necessary to run refrigeration tubing parallel with beams rather than across them?

4. Why is it necessary to mount evaporators absolutely level?

5. Where are two-temperature valves usually located?

6. Where may soft tubing be used in a code installation?

7. Explain the method for removing air from a multiple dry evaporator system.

8. Why is it necessary to put a drier on a new system?

9. What would be the purpose of a drier placed in a suction line?

10. Why are hard drawn copper tubing and streamline fittings used?

11. How is a lack of refrigerant in a TEV multiple installation indicated?

12. What may be wrong if a thermal expansion valve evaporator suddenly starts to frost or sweat excessively at the expansion valve, but the evaporator near the suction line connection is dry?

13. How are flare nuts protected so moisture cannot get under them and freeze?

14. Why is it necessary to have an open water drain?

15. What must be done to balance the low-side pressure in a multiple evaporator installation in case the normal low-side system pressure is 5" Hg. (127 mm) vacuum?

16. Are commercial systems normally charged through the high-pressure side or low-pressure side?

17. What safety precautions should a service technician follow when charging a system through the low side?

18. Why are brazed flanged fittings recommended for use in making inside cabinet connections?

19. What is the purpose of the felt and fine mesh screen in a drier?

20. Why must a system be very carefully checked for leaks if a lack of refrigerant is discovered in the system?

21. How is a leaking exhaust valve detected in an external drive compressor?

22. What special precautions must be followed when installing a suction line?

23. What will happen if the lowest temperature evaporator check valve leaks?

24. What cleaning fluid can be used to clean joints before they are opened?

25. What may cause a water-cooled condensing unit to short cycle?

26. Of what material are service valve packings made?

27. List the procedure to be followed when lapping an exhaust valve.

28. Why are suction service valves and discharge service valves called three-way valves?

29. What are four common TEV troubles?

30. What two things can most sight glasses indicate to a service technician?

31. What should be done to the main switch before one works on the electrical parts of a system?

32. Why is a fuse plug used?

33. What is put on a brazed joint before it is taken apart?

34. What may happen if moisture collects on the outside of the bellows of a refrigerant control?

35. What happens if a suction line is undersized?

36. What is done to a water-cooled condenser before the refrigerant is purged from it?

37. How should the contaminated hermetic compressor oil be handled?

38. Can a conventional compressor crankshaft seal be repaired without removing the compressor?

39. When should one wear goggles when working on a refrigerating mechanism?

40. What should one do if a leaky refrigerant joint is located?

41. What are two uses for an ohmmeter?

42. Why do some compressors have a sight glass?

43. How may one put oil in a compressor?

44. What may one do if refrigerant lines are noisy?

45. What treatment should be given to evaporative condenser water?

46. What chemical is sometimes used to clean water lines?

47. What causes a hermetic motor compressor burnout?

48. Why should a suction line have a slight down slope toward the compressor?

49. What is a refrigerant charging tube? How does it indicate the weight of the refrigerant?

50. How often should the water circuit of an ice maker be cleaned?

51. What happens if the TEV orifice is too large?

52. How may a fuse be tested for continuity before installing it?

53. What is done with the refrigerant when a hot gas bypass valve is replaced?

54. Can any 1/2 hp hermetic motor compressor replace a worn out 1/2 hp unit? Why or why not?

55. When a system is purged, what else leaves the system besides refrigerant?

56. What happens when a drier is heated?

57. Can liquid refrigerant be charged into the low side of a system? Why or why not?

58. What happens to the liquid line when a drier is almost clogged?

59. Can aluminum electrical leads be connected to copper leads? Why?

60. How and why is nitrogen used when brazing a connection in a refrigerating system?

Chapter 15

COMMERCIAL SYSTEMS HEAT LOADS AND PIPING

Commercial refrigeration installations must be properly engineered. To accomplish this, a sales engineer usually selects the proper equipment to handle a particular refrigerating load. Information given in this chapter will explain how to determine specific refrigeration loads and what equipment is needed.

To have good refrigeration system performance, four main items must be matched (balanced or made equal to each other):

1. Heat load. Determine the total amount of heat that must be removed for each 24 hours.
2. Condensing unit. Determine what size condensing unit is needed to handle the heat load. To do this, determine whether the unit is to run 16, 18 or 20 hours of each 24 hours.
3. Evaporator. Determine evaporator capacity required to handle the heat load. The evaporator can remove heat only while the condensing unit is running. Therefore, its capacity must be based on the same hours of operation as the hours of running of the condensing unit.
4. Total system. Consider such factors as water supply, temperature control devices, refrigeration line sizes, air circulation and humidity control. Correctly install the components according to code.

When determining the heat load, two main factors must be considered:

1. Heat leakage into the cabinet. Heat leakage is affected by: amount of exposed surface; thickness and kind of insulation; temperature difference between inside and outside of cabinet.
2. Usage or service heat load of the cabinet. This load is determined by: temperature of articles put into the refrigerator; their specific heat; generated heat; latent heat as the requirements demand. Another consideration is the nature of the service required. This includes air changes determined by the number of times per day that doors of the refrigerator are opened, and heat generated inside by fans, lights and other electrical devices.

Total heat load is the sum of:

1. Wall heat transmission load.
2. Air change load.
3. Product load.
4. Miscellaneous loads.

The selection of a condensing unit is usually made from manufacturers' tables of condensing unit capacities.

Evaporators are selected from manufacturers' specifications for capacities to balance the capacity of the condensing unit. Also affecting the selection of the evaporator are the capacity of the refrigerant control, type of temperature control, arrangement for air circulation and specific duty.

The installation of all commercial refrigeration equipment involves a technical understanding of the variables. This is a determining factor in the operation of the system. Technical installation knowledge is also important to service technicians. The more a technician understands about the refrigerant liquid and vapor behavior inside the system, the more accurately a service problem can be diagnosed.

The following paragraphs on commercial refrigeration describe the fundamentals of equipment selection for commercial refrigeration.

15-1 HEAT LOAD

The total heat load consists of the amount of heat to be removed from a cabinet during a certain period. It is dependent on two main factors:

1. Heat leakage load.
2. Heat usage or service load.

The heat leakage load or heat transfer load is the total amount of heat that leaks through the walls, windows, ceiling and floor of the cabinet, per unit of time (usually 24 hours).

The heat usage or service load is the sum of the heat loads of: cooling the contents to cabinet temperature; cooling of air changes; removing respiration heat from fresh or "live" vegetables, and from meat; removing heat released by electric lights, electric motors and the like; removing heat given off by people entering and/or working in the cabinet, per the same unit of time as heat leakage calculations (usually 24 hours).

In this chapter, heat load calculations are in the U. S. Conventional System (pounds, Btu, temperatures F., foot, etc.). Conversion factors used to make these calculations in the SI System (kilograms, kilocalories, temperatures C., centimetres, etc.) are explained in Chapter 28.

15-2 HEAT LEAKAGE VARIABLES

Research organizations, manufacturers and refrigeration associations have determined the amount of heat leakage through walls and other heat loads. Charts and tables based on these calculations are used by engineers and technicians.

SUMMER DESIGN TEMPERATURES

STATE	DESIGN DRY BULB F.	C.	STATE	DESIGN DRY BULB F.	C.
ALABAMA	95	29	MASSACHUSETTS	90	32
ALASKA	74	23	MICHIGAN	88	31
ARIZONA	105	41	MINNESOTA	90	32
ARKANSAS	98	37	MISSISSIPPI	97	36
CALIFORNIA:			MISSOURI	98	37
LOWER	86	30	MONTANA	88	31
MIDDLE	94	34	NEBRASKA	97	36
UPPER	83	28	NEVADA	95	35
COLORADO	92	33	NEW HAMPSHIRE	90	32
CONNECTICUT	88	31	NEW JERSEY	92	33
DELAWARE	93	34	NEW MEXICO	95	35
DIST. OF COL.	94	34	NEW YORK	90	32
FLORIDA:			NORTH CAROLINA	95	35
UPPER	96	36	NORTH DAKOTA	93	34
LOWER	93	34	OHIO	90	32
GEORGIA	95	35	OKLAHOMA	102	39
HAWAII	87	31	OREGON	90	32
IDAHO	94	34	PENNSYLVANIA	92	33
ILLINOIS:			RHODE ISLAND	87	31
UPPER	95	35	SOUTH CAROLINA	95	35
LOWER	97	36	SOUTH DAKOTA	95	35
INDIANA	95	35	TENNESSEE	96	36
IOWA	95	35	TEXAS	101	38
KANSAS:			UTAH	95	35
UPPER	97	36	VERMONT	87	31
LOWER	100	38	VIRGINIA	95	35
KENTUCKY	95	35	WASHINGTON	90	32
LOUISIANA	98	37	WEST VIRGINIA	94	34
MAINE	88	31	WISCONSIN	90	32
MARYLAND	94	34	WYOMING	90	32

Fig. 15-1. Table of summer design temperatures that may be used as ambient temperatures when calculating heat leakage loads.

Five factors (variables) which affect heat leakage are:

1. Time. The longer the period of time, the more heat will leak through a certain wall. The standard time unit is the 24-hour period in refrigeration situations, while the one-hour period is used in air conditioning situations.
2. Temperature difference. The difference in temperature is an important factor in the heat leakage into a container. The greater the temperature difference, the more heat will leak or transfer through the wall. One might compare this idea to pressure: the more pressure, the more water will flow through an opening. The room temperature is usually chosen as the average summer temperature. In the United States, it varies between 90 and 105 F. (32 C. and 40 C.). See Fig. 15-1. This value can be reduced to 75 F. or 80 F. (25 C. or 27 C.) if the room is air conditioned.
3. Thickness of insulation. The thicker the insulation, the less heat will flow through it. Twice as much heat will leak through a wall with 1-in. insulation than through a wall having 2-in. insulation.
4. Kind of insulation. The kind of insulation or the material used is important. Expanded polystyrene, for instance, will insulate approximately six times better than wood. Some insulations, however, are more costly than others.
5. External area of cabinet. The more area through which heat may leak, the greater the heat flow. To compare this with water flow: the size of a pipe determines how much water will flow through it. The bigger the pipe, the more water will flow. The common unit used for determining the heat flow is the total square foot area. This area is always measured on the outside of the cabinet.

15-3 HEAT LEAKAGE – K FACTOR

To bring together the values just discussed, standards have been developed which are used by refrigerating companies. In preparing these standards, the variables have been reduced to unit values. The unit values, in turn, are used to indicate heat leakage of the wall.

The unit or basic values are the thermal conductance (symbol is K) obtained for an area of insulation one square foot in size, one inch thick, with a temperature difference of 1 F., over a period of time of either one hour or 24 hours. Values obtained represent the amount of heat flow through the insulation under these conditions.

Unit values vary with the kind of insulation. This material has no air film, or liquid film, on either side. The symbol is K.

If the insulation is less than or more than 1-in. thick, the heat leakage will be different. In this case, the symbol is K_T. Example:

$$K_T = \frac{K_1}{\text{thickness}}$$

By definition:

K_T = conductance, total

K_1 = conductance for 1 in. thickness

If thickness is 2 in.:

$$K_T = \frac{K_1}{2} \qquad K_T = \frac{1}{2} K_1$$

If thickness is $\frac{1}{2}$ in.:

$$K_T = \frac{K_1}{\frac{1}{2}} \qquad K_T = K_1 \div \frac{1}{2}$$

$$K_T = K_1 \times 2 \qquad K_T = 2K_1$$

A special formula is needed to find heat leakage or thermal

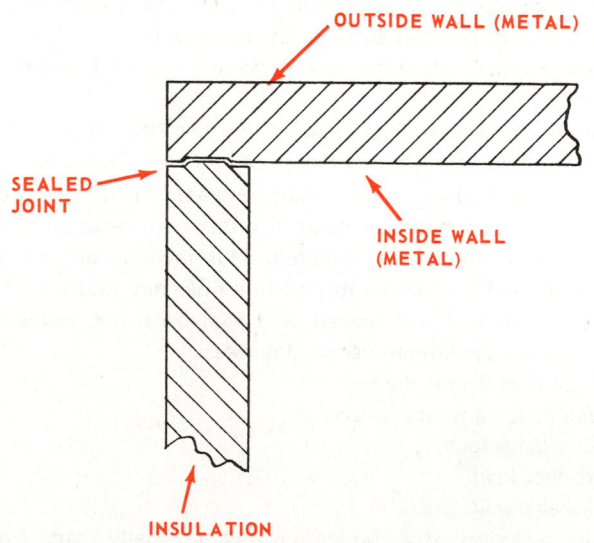

Fig. 15-2. Cross-section of a walk-in cooler wall.

conductance through a composite wall, Fig. 15-2. The formula for computing the K_T or total heat conductance factor is explained as follows:

Resistance to heat flow is known by the symbol R. If the same heat is flowing through two substances, the total resistance is equal to the resistance of each substance.

$R_T = R_1 + R_2$
R_T = resistance total
R_1 = resistance of substance 1
R_2 = resistance of substance 2

R is the opposite, or the reciprocal of K, or $R = \frac{1}{K}$. In the formula, it would be:

$\frac{1}{K_T} = \frac{1}{K_1} + \frac{1}{K_2}$
K_T = conductance, total
K_1 = conductance, substance 1
K_2 = conductance, substance 2

Example:

$K_1 = .6$ and $K_2 = .2$

$$\frac{1}{K_T} = \frac{1}{.6} + \frac{1}{.2} \qquad \frac{1}{K_T} = \frac{1}{\frac{6}{10}} + \frac{1}{\frac{2}{10}}$$

$$\frac{1}{K_T} = \frac{10}{6} + \frac{10}{2} \qquad \frac{1}{K_T} = \frac{5}{3} + \frac{5}{1}$$

$$\frac{1}{K_T} = \frac{5}{3} + \frac{15}{3} \qquad \frac{1}{K_T} = \frac{20}{3}$$

$$K_T = \frac{3}{20} \qquad K_T = .15$$

$$R = \frac{1}{K_T} \qquad R = \frac{1}{.15}$$

$$R = \frac{1 \times 100}{.15 \times 100} = \frac{100}{15} = 6.7$$

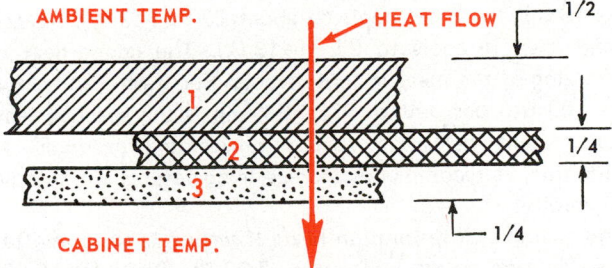

AMBIENT TEMP. → HEAT FLOW —| 1/2

CABINET TEMP.

Fig. 15-3. A composite insulated panel or wall: 1, 2 and 3 are different insulating materials of different thickness.

If the wall is made up of three different materials, Fig. 15-3, overall heat leakage (K) is found as follows:

$$K_T = \frac{1}{\dfrac{\text{thickness of material 1}}{\text{conductivity factor material 1}}} +$$

$$\frac{1}{\dfrac{\text{thickness of material 2}}{\text{conductivity factor material 2}}} +$$

$$\frac{1}{\dfrac{\text{thickness of material 3}}{\text{conductivity factor material 3}}}$$

where Th_1 = thickness of material 1
   Th_2 = thickness of material 2
   Th_3 = thickness of material 3

and K_1 = conductivity factor for material 1
   K_2 = conductivity factor for material 2
   K_3 = conductivity factor for material 3

then the formula becomes:

$$K_T = \frac{1}{\dfrac{Th_1}{K_1} + \dfrac{Th_2}{K_2} + \dfrac{Th_3}{K_3}}$$

To solve for the conductivity for the panel shown in Fig. 15-4, proceed as follows:

$$K_T = \frac{1}{\dfrac{\text{thickness A}}{\text{K for wood}} + \dfrac{\text{thickness B}}{\text{K for Celotex}} + \dfrac{\text{thickness C}}{\text{K for corkboard}}}$$

From Chapter 28, K values are as follows:

K for wood = .80 for 1-in. thickness
K for Celotex = .31 for 1-in. thickness
K for corkboard = .285 for 1-in. thickness

However, in the example, the wood is 1/2-in. thick, the Celotex is 1/4-in. thick and the corkboard is 1/4-in. thick.

Substituting these values in the above formula, we have:

$$K_T = \frac{1}{\dfrac{.5}{.80} + \dfrac{.25}{.31} + \dfrac{.25}{.285}} = \frac{1}{.625 + .807 + .877} = \frac{1}{2.3}$$

K_T = .435, which is the conductivity for the panel in Btu/hr./deg. F./sq. ft.

An air film that clings to the outer and inner surfaces of the cabinet adds to the insulating value of the walls of the cabinet. This added resistance is calculated in the following formula in which the outside air film (F_O) is considered to have a heat transfer value of 6.00 and the inside wall air film (F_I) has a value of 1.65

K_T = unit of conductivity for materials of a composite nature
U = unit of conductivity for materials of a composite nature plus the effect of the air clinging to both the outside (F_O) and the inside (F_I) walls

If the insulating value of the air clinging to the walls is considered, the formula becomes:

$$U = \frac{1}{\dfrac{1}{F_O} + \dfrac{Th_1}{K_1} + \dfrac{Th_2}{K_2} + \dfrac{Th_3}{K_3} + \dfrac{1}{F_I}}$$

If the value of $F_O = 6.0$ and the value of $F_I = 1.65$, then the problem in Fig. 15-4 may be solved as follows:

$$U = \frac{1}{\dfrac{1}{6.0} + \dfrac{.5}{.80} + \dfrac{.25}{.31} + \dfrac{.25}{.285} + \dfrac{1}{1.65}} =$$

$$\frac{1}{.166 + .625 + .807 + .877 + .606} =$$

$$\frac{1}{3.081} = .32 \text{ Btu/hr./deg. F./sq. ft.}$$

The value of K as computed in the previous problem is .435. The value of U as computed in this problem is .32, which shows the additional insulating effect of the air films.

This type of computation is complicated and slow. For this reason, standard tables for computing heat leakage are generally used.

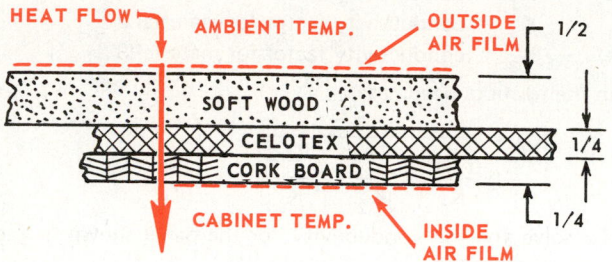

Fig. 15-4. Composite insulating panel of various materials at specified thicknesses, with air film on both sides.

15-4 AIR CHANGE HEAT LOAD

Air that enters a refrigerated space must be cooled. Air has weight, and it also contains moisture. When air enters the refrigerated space, heat must be removed from it.

By Charles's Law, air which enters and is cooled reduces in pressure. If the cabinet is not airtight, air will continue to leak in. Also, each time a service door or a walk-in door is opened, the cold air inside, being heavier, will spill out the bottom of the opening allowing the warmer room air to move into the cabinet. The actions of material moving in or out of the cabinet, a person going into or leaving a cabinet, all result in warm air moving into the space. This action is sometimes called infiltration of air.

Fig. 15-5 shows accepted air change volume values for refrigerated cabinets of various internal volumes. Fig. 15-6 shows the total heat (sensible + latent) to be removed from this air, depending on various outside conditions and refrigerator temperatures.

15-5 PRODUCT HEAT LOAD

Any substance which is warmer than the refrigerator it is placed in will lose heat until it cools to the refrigerator temperature.

Three kinds of heat removal may be involved:
1. Specific heat.
2. Latent heat.
3. Respiration heat.

The total product heat load would be the sum of these three heat loads. An example of all three heat loads would be moist head lettuce at 55 F. (13 C.) being put into a 35 F. (2 C.) refrigerator. The lettuce must be cooled — specific heat — to 35 F. (2 C.). Some of the moisture on the lettuce will evaporate and collect on the evaporators (latent heat). The lettuce, being a live vegetable, would absorb carbon dioxide and release oxygen. This change would release heat energy (respiration heat). Meats go through a slow bacteriological change and, during this action, heat is released (respiration heat).

1. An example of specific heat: If bottled beverages at 50 F. (10 C.) are placed in a 35 F. (2 C.) refrigerator, they will release heat until their temperature reaches 35 F. (2 C.). This action is a specific heat problem. If the bottles were moist or if the bottle cartons were moist, there would be a moisture evaporating problem (latent heat).

2. An example of latent heat: If meat at 50 F. (10 C.) is placed in a refrigerator, and cooled (frozen) to 0 F. (–18 C.), the meat cools to about 27 F. (–3 C.), freezes, and then it cools to 0 F. (–18 C.). The latent heat of freezing of the meat is considerable. For fresh lean beef, it is 100 Btu per pound. Fig. 15-7 shows the specific heats and latent heats of various refrigerated products. In addition, it recommends storage temperatures and relative humidity.

3. An example of respiration heat: If lettuce is store at 40 F. (4 C.), each pound will release 7.99 Btu/24 hr./or 15,980 Btu/ton. Fig. 15-7 also shows the respiration heat of some of the more common vegetables and fruits.

15-6 MISCELLANEOUS HEAT LOAD

All sources of heat not covered by heat leakage, product cooling and respiration load are usually listed as miscellaneous heat loads. Some of the more common miscellaneous heat loads are: lights, electric motors, people, defrosting heat sources and sun (solar) heat.

1. Lights located in the refrigerated space will release heat.

VOLUME CU. FT.	AIR CHANGES PER 24 HR.	VOLUME CU. FT.	AIR CHANGES PER 24 HR.
200	44.0	6,000	6.5
300	34.5	8,000	5.5
400	29.5	10,000	4.9
500	26.0	15,000	3.9
600	23.0	20,000	3.5
800	20.0	25,000	3.0
1,000	17.5	30,000	2.7
1,500	14.0	40,000	2.3
2,000	12.0	50,000	2.0
3,000	9.5	75,000	1.6
4,000	8.2	100,000	1.4
5,000	7.2		

NOTE: For heavy usage multiply the above values by 2. For long storage multiply the above values by 0.6.

Fig. 15-5. Average air changes per 24 hours for storage rooms due to door openings and air infiltration. (ASHRAE Guide and Data Book)

HEAT REMOVED IN COOLING AIR TO STORAGE ROOM CONDITIONS
(BTU PER CU. FT.)

TEMPERATURE OF OUTSIDE AIR, F.

STORAGE ROOM TEMP. F.	85		90		95		100	
RELATIVE HUMIDITY, PERCENT	50	60	50	60	50	60	50	60
65	0.65	0.85	0.93	1.17	1.24	1.54	1.58	1.95
60	0.85	1.03	1.13	1.37	1.44	1.74	1.78	2.15
55	1.12	1.34	1.41	1.66	1.72	2.01	2.06	2.44
50	1.32	1.54	1.62	1.87	1.93	2.22	2.28	2.65
45	1.50	1.73	1.80	2.06	2.12	2.42	2.47	2.85
40	1.69	1.92	2.00	2.26	2.31	2.62	2.67	3.06
35	1.86	2.09	2.17	2.43	2.49	2.79	2.85	3.24
30	2.00	2.24	2.26	2.53	2.64	2.94	2.95	3.35

TEMPERATURE OF OUTSIDE AIR, F.

STORAGE ROOM TEMP. F.	40		50		90		100	
RELATIVE HUMIDITY, PERCENT	70	80	70	80	50	60	50	60
30	0.24	0.29	0.58	0.66	2.26	2.53	2.95	3.35
25	0.41	0.45	0.75	0.83	2.44	2.71	3.14	3.54
20	0.56	0.61	0.91	0.99	2.62	2.90	3.33	3.73
15	0.71	0.75	1.06	1.14	2.80	3.07	3.51	3.92
10	0.85	0.89	1.19	1.27	2.93	3.20	3.64	4.04
5	0.98	1.03	1.34	1.42	3.12	3.40	3.84	4.27
0	1.12	1.17	1.48	1.56	3.28	3.56	4.01	4.43
−5	1.23	1.28	1.59	1.67	3.41	3.69	4.15	4.57
−10	1.35	1.41	1.73	1.81	3.56	3.85	4.31	4.74
−15	1.50	1.53	1.85	1.92	3.67	3.96	4.42	4.86
−20	1.63	1.68	2.01	2.09	3.88	4.18	4.66	5.10
−25	1.77	1.80	2.12	2.21	4.00	4.30	4.78	5.21
−30	1.90	1.95	2.29	2.38	4.21	4.51	4.90	5.44

Fig. 15-6. Chart gives total heat removed in cooling air to storage room in Btu per cu. ft. (ASHRAE Guide and Data Book)

For example, a 100-watt lamp will give off 342 Btu in one hour or

342 x 24 = 8208 Btu/24 hr.

If the work day is eight hours (only time light is on), the heat load would be

342 x 8 = 2736 Btu/24 hr.

2. Electric motors release 2545 Btu/hp/hour on the average. The amount of heat released also depends on motor efficiency; the larger the motor, the more efficient the motor. Fig. 15-8 shows the heat given off by motors and devices they drive. Because forced convection evaporators usually have motors and fans, it should be noted that the total heat release of such a motor is about 4250 Btu/hp/hr. for motors of 1/8 to 1/2 hp. For example, a 1/8-hp motor-fan would release:

4250 Btu/hp/hr.

4250 Btu/hr. x 1/8 x 1 =

4250 ÷ 8 = 531 Btu/hr.

531 x 24 = 12,744 Btu/24 hr. if motor runs continuously

3. When people are inside a refrigerated space, they release heat at varying rates depending on what they are wearing (insulation), the temperature of the cabinet, and on how hard they are working. Fig. 15-9 shows a variation of 720 Btu/hr./person at 50 F. (10 C.) to 1400 Btu/hr./person at

Product	Quick Freeze Temp.	Storage Temp. Long	Storage Temp. Short	Humidity % R.H.	Specific Heat Above Freezing	Specific Heat Below Freezing	Latent Heat	Freezing Point	Respiration BTU/lb. Per Day
Apples	—15	30-32	38-42	85-88	0.92	0.39	91.5	28.4	0.75
Asparagus	—30	32	40	85-90	0.95	0.44	134.0	29.8	....
Bacon, Fresh		0-5	36-40	80	0.55	0.31	30.0	25.0	....
Bananas		56-72	56-72	85-95	0.81	....	108.0	30.2	4.18
Beans, Green		32-34	40-45	85-90	0.92	0.47	128	29.7	3.3
Beans, Dried		36-40	50-60	70	0.30	0.237	18	....	....
Beef, Fresh, Fat	—15	30-32	38-42	84	0.60	0.35	79	....	....
Beef, Fresh Lean	—15	30-32	38-42	85	0.77	0.40	100	....	....
Beets, Topped		32-35	45-50	95-98	0.90	....	....	26.9	2.0
Blackberries	—15	31-32	42-45	80-85	0.89	0.46	125	28.9	....
Broccoli		32-35	40-45	90-95	0.93	....	....	29.2	....
Butter	+15	....	40-45	....	0.64	0.34	15	15.0	....
Cabbage	—30	32	45	90-95	0.93	0.47	130	31.2	....
Carrots, Topped	—30	32	40-45	95-98	0.87	0.45	120	29.6	1.73
Cauliflower		32	40-45	85-90	0.90	....	....	30.1	....
Celery	—30	31-32	45-50	90-95	0.95	0.48	135	29.7	2.27
Cheese	+15	32-38	39-45	....	0.70	....	....	....	....
Cherries		31-32	40	80-85	0.85	....	118	28.0	6.6
Chocolate Coatings		45-50	....	....	0.3	....	....	....	....
Corn, Green		31-32	45	85-90	0.86	....	....	29.0	4.1
Cranberries		36-40	40-45	85-90	0.91	....	....	27.3	....
Cream		34	40-45	....	0.88	0.37	84	....	....
Cucumbers		45-50	45-50	80-85	0.93	....	....	30.5	....
Dates, Cured		28	55-60	50-60	0.83	0.44	104	....	....
Eggs, Fresh	—10	30-31	38-45	....	0.76	0.40	98	31.0	....
Eggplants		45-50	46-50	85-90	0.88	....	....	30.4	....
Flowers		35-40	....	85-90	....	....	....	....	....
Fish, Fresh, Iced	—15	25	25-30	....	0.82	0.41	105	30.0	....
Fish, Dried		30-40	....	60-70	0.56	0.34	65	....	....
Furs		32-34	40-42	40-60	....	....	....	....	....
Furs, To Shock		15	15	....	....	....	....	....	....
Grapefruit		32	32	85-90	0.92	....	111	28.4	0.5
Grapes		30-32	35-40	80-85	0.92	....	111	27.0	0.5
Ham, Fresh		28	36-40	80	0.68	0.38	87	....	....
Honey		31-32	45-50	....	0.35	0.26	26	....	....
Ice Cream	—20	....	0-10	....	.5-.8	0.45	96	....	....
Lard		32-34	40-45	80	0.52	0.31	90	....	....
Lemons		55-58	....	80-85	0.91	0.39	190	28.1	0.4
Lettuce		32	45	90-95	0.90	....	....	31.2	8.0
Liver, Fresh		32-34	36-38	83	0.72	0.42	94	....	....
Lobster, Boiled		25	36-40	....	0.81	0.42	105	....	....
Maple Syrup		31-32	45	....	0.24	0.215	7.0	....	....
Meat, Brined		31-32	40-45	....	0.75	0.36	75.0	....	....
Melons		34-40	40-45	75-85	0.92	0.35	115	28.5	1.0
Milk		34-36	40-45	....	0.92	0.46	124	31.0	....
Mushrooms		32-35	55-60	80-85	0.90	....	....	30.2	....
Mutton		32-34	34-42	82	0.81	0.39	96	29.0	....
Nut Meats		32-50	35-40	65-75	0.30	0.24	14	20.0	....
Oleomargarine		34-36	....	....	0.65	0.34	35	15.0	....
Onions		32	50-60	70-75	0.91	0.46	120	30.1	1.0
Oranges		32-34	50	85-90	0.89	0.40	91.0	27.9	0.7
Oysters		....	32-35	....	0.85	0.45	120.0	....	....
Parsnips	—30	32-34	34-40	90-95	0.82	0.45	120.0	28.9	....
Peaches, Fresh		31-32	50	85-90	0.92	0.42	110	29.4	1.0
Pears, Fresh		29-31	40	85-90	0.90	0.43	106	28.0	6.6
Peas, Green		32	40-45	85-90	0.80	0.42	108	30.0	....
Peas, Dried		35-40	50-60	....	0.28	0.22	14	....	....
Peppers		32	40-45	85-90	0.90	....	....	30.1	2.35
Pineapples, Ripe		40-45	50	85-90	0.90	....	127	29.9	....
Plums		31-32	40-45	80-85	0.83	....	115	28.0	....
Pork, Fresh		30	36-40	85	0.60	0.38	66	28.0	....
Potatoes, White	—30	36-50	45-60	85-90	0.77	0.44	105	28.9	0.85
Poultry, Dressed	—10	28-30	29-32	....	0.80	0.41	99	27	....
Pumpkins		50-55	55-60	70-75	0.90	....	....	30.2	....
Quinces		31-32	40-45	80-85	0.90	....	....	28.1	....
Raspberries		31-32	40-45	80-85	0.89	0.46	125	30.0	3.3
Sardines, Canned		....	35-40	....	0.76	0.410	101	....	....
Sausage, Fresh		31-36	36-40	80	0.89	....	....	....	....
Sauerkraut		33-36	36-38	85	0.91	0.47	128	....	....
Squash		50-55	55-60	70-75	0.90	....	....	29.3	....
Spinach		32	45-50	85	0.92	....	....	30.8	....
Strawberries	—15	31-32	42-45	80-85	0.92	0.48	129	30.0	3.3
Tomatoes, ripe		40-50	55-70	85-90	0.95	....	135	30.4	0.5
Turnips		32	40-45	95-98	0.90	....	....	30.5	1.0
Veal	—15	28-30	36-40	....	0.71	0.39	91	29	....

Fig. 15-7. Temperature, specific heat and latent heat data for some common foods.   (Dunham-Bush, Inc.)

MOTOR HP	CONNECTED LOAD IN REF. SPACE[1]	MOTOR LOSSES OUTSIDE REF. SPACE[2]	CONNECTED LOAD OUTSIDE REF. SPACE[3]
1/8 to 1/2	4,250	2,545	1,700
1/2 to 3	3,700	2,545	1,150
3 to 20	2,950	2,545	400

[1]For use when both useful output and motor losses are dissipated within refrigerated space; motors driving fans for forced circulation unit coolers.
[2]For use when motor losses are dissipated outside refrigerated space and useful work of motor is expended within refrigerated space; pump on a circulating brine or chilled water system, fan motor outside refrigerated space driving fan circulating air within refrigerated space.
[3]For use when motor heat losses are dissipated within refrigerated space and useful work expended outside of refrigerated space; motor in refrigerated space driving pump or fan located outside of space.

Fig. 15-8. Heat released by operating electric motors. (ASHRAE Guide and Data Book)

HEAT RELEASED PER OCCUPANT

COOLER TEMPERATURE F.	HEAT RELEASED/PERSON BTU/HR.
50	720
40	840
30	950
20	1050
10	1200
0	1300
−10	1400

Fig. 15-9. Heat released by a person in the cooled space. (ASHRAE Guide and Data Book)

−10 F. (−23 C.). For example, if one person worked in a 30 F. (−1 C.) refrigerator for eight hours, the heat load would be:

950 Btu/hr. x 8 hr. = 7600 Btu

4. Many refrigerating units have defrosting heat sources, especially if the fixture temperature is 32 F. (0 C.) or lower. Whether the defrost heat source is electric, hot gas or water, the defrosting operation adds heat to the interior of the refrigerator. The amount of heat is difficult to determine because most of the defrosting heat is removed in the defrost drain water. Add approximately 10 percent of the defrosting heat input as part of the heat load.

5. If part of the refrigerator is exposed to the sun, the heat from this source must be considered. Add the following to the room or ambient temperature: If a dark surface, add about 10 F. (6 C.). If it is a medium colored surface, add 5 F. (3 C.). If it is a light surface, add 3 F. (2 C.) to the ambient temperature.

15-7 CABINET AREAS

The area of a cabinet is the outside area. There are six surfaces: four walls, the ceiling and the floor. Usually the floor and ceiling are the same area, and the opposite walls are the same area. To determine the total outside area:

1. Multiply width by length, then multiply by two. These areas are the areas of the floor and ceiling of the cabinet.
2. Multiply width by height, then multiply by two. These areas are the areas of ends of the cabinet.
3. Multiply length by height, then multiply by two. These areas are the areas of sides of cabinet. Add these three values to determine the total external area of the cabinet. By formula:

L = Length W = Width H = Height
$W \times L \times 2$ = area of ceiling and floor
$W \times H \times 2$ = area of ends
$L \times H \times 2$ = area of sides
Total external area = sum of the three areas

Most companies compute total area based on the outside of the cabinet. This is done because the exterior is easier to measure and the results are on the safe side.

After obtaining the external area of the cabinet, subtract the window area to obtain the area of the insulated surface. Window areas are calculated from the measurements of the outside edges of the window frame and must be considered separately.

Total external area minus the window area equals the insulated area.

Use the accompanying table to find the amount of heat that will leak through the insulation per square foot of area per 24 hours for that particular type of wall construction for the temperature difference. Consider a wall made of steel paneling on both sides with a 4-in. slab of cork insulation in between. The tables in Fig. 15-10 will reveal that, at a temperature difference of 60 F. (95 F. −35 F.), 108 Btu will leak through per square foot during the period of 24 hours. Expanded polystyrene insulation would either be 33 percent less or about 2 1/2-in. thick for the same heat loss.

Some synthetic material insulation values are:

	K
Expanded rubber, rigid	0.22
Glass fiber, organic bonded	0.25
Expanded polystyrene (extruded), plain	0.25
Expanded polystyrene (extruded), R-12 expanded, 1-in. thick or greater	0.19
Expanded polystyrene, molded beads	0.28
Expanded polyurethane, R-11 expanded, 1-in. thick or greater	0.16
Silica aerogel, loose fill	0.17

The glass leakage table will similarly give values for the heat leakage through one square foot of glass. If the cabinet has double glass, 660 Btu will leak through at a temperature difference of 60 F. Adding the two heat leakages will give the total heat leakage into the cabinet.

Example: A walk-in cooler is 10 ft. x 9 ft. x 8-ft. high with two double-pane glass windows 1 1/2 ft. x 2 ft. The box is kept at 35 F. (2 C.) in a room with a summer design temperature of 95 F. (35 C.). The wall construction consists of 4-in. cork with metal on each side (or it could be 2 1/2 in. of expanded polystyrene). The windows are of the double-pane construction. The temperature difference is 95 minus 35 = 60 F.

Heat Gain Factors (Walls, Floor and Ceiling)
Btu per (sq ft) (24 hr)

Insulation		Temp difference (ambient temp minus storage temp), F deg																
Cork or equivalent in.	1	40	45	50	55	60	65	70	75	80	85	90	95	100	105	110	115	120
3	2.4	96	108	120	132	144	156	168	180	192	204	216	228	240	252	264	276	288
4	1.8	72	81	90	99	108	117	126	135	144	153	162	171	180	189	198	207	216
5	1.44	58	65	72	79	87	94	101	108	115	122	130	137	144	151	159	166	173
6	1.2	48	54	60	66	72	78	84	90	96	102	108	114	120	126	132	138	144
7	1.03	41	46	52	57	62	67	72	77	82	88	93	98	103	108	113	118	124
8	0.90	36	41	45	50	54	59	63	68	72	77	81	86	90	95	99	104	108
9	0.80	32	36	40	44	48	52	56	60	64	68	72	76	80	84	88	92	96
10	0.72	29	32	36	40	43	47	50	54	58	61	65	68	72	76	79	83	86
11	0.66	26	30	33	36	40	43	46	50	53	56	60	63	66	69	73	76	79
12	0.60	24	27	30	33	36	39	42	45	48	51	54	57	60	63	66	69	72
13	0.55	22	25	28	30	33	36	39	41	44	47	50	52	55	58	61	63	66
14	0.51	20	23	26	28	31	33	36	38	41	43	46	49	51	54	56	59	61
Single glass	27.0	1080	1220	1350	1490	1620	1760	1890	2030	2160	2290	2440	2560	2700	2840	2970	3100	3240
Double glass	11.0	440	500	550	610	660	715	770	825	880	936	990	1050	1100	1160	1210	1270	1320
Triple glass	7.0	280	320	350	390	420	454	490	525	560	595	630	665	700	740	770	810	840

Note: Where wood studs are used multiply the above values by 1.1

Fig. 15-10. Heat gain factors (walls, floor and ceiling).   (ASHRAE Guide and Data Book)

Solution:

Walls: 10 x 9 x 2 = 180 sq. ft. (ceiling and floor)
        9 x 8 x 2 = 144 sq. ft. (ends)
        10 x 8 x 2 = 160 sq. ft. (sides)
        484 sq. ft. of total area

Windows: 1 1/2 x 2 x 2 = 6 sq. ft. of window
          484 − 6 = 478 sq. ft. of insulated wall

From table, Fig. 15-10:

1 sq. ft. of the wall allows 108 Btu per 24 hr.

108 x 478 sq. ft. = 51,624 Btu per 24 hr. through the walls

From table, Fig. 15-10:

1 sq. ft. of window allows transfer of 660 Btu per 24 hr.

660 x 6 sq. ft. = 3960 Btu per 24 hr. through the windows

Or a total heat leakage of:

51,624 + 3960 = 55,584 Btu per 24 hr.

15-8 CABINET VOLUME

Cabinet volume is the volume based on the inside dimensions of the cabinet. The volume is used to help find out the air changes and product load.

In the sample cabinet, which is 10-ft. long x 9-ft. wide x 8-ft. high, the walls are 4-in. thick. Therefore the internal or inside width is 9 ft. minus 4 in. minus 4 in. (there is a wall at each end).

9 ft. − (4 in. + 4 in.) = inside width

or 9 ft. − 8 in. = inside width

or 8 ft. 4 in. = inside width

or 8 1/3 ft. = inside width

The same method is used to calculate other internal dimensions.

The internal dimensions are:

Width 8 1/3 ft., length 9 1/3 ft., height 7 1/3 ft.

The inside volume =

8 1/3 x 9 1/3 x 7 1/3 =

$$\frac{(8 \times 3) + 1}{3} \times \frac{(9 \times 3) + 1}{3} \times \frac{(7 \times 3) + 1}{3} =$$

$$\frac{24 + 1}{3} \times \frac{27 + 1}{3} \times \frac{21 + 1}{3} =$$

$$\frac{25}{3} \times \frac{28}{3} \times \frac{22}{3} =$$

$$\frac{15,400}{27} = 570.4 = 570 \text{ cu. ft.}$$

Usable inside volume is the total inside volume minus shelves, racks and evaporator space. For practical purposes and to be on the safe side, the total internal volume is used when figuring heat loads. Fig. 15-11 uses a net (internal) volume.

15-9 TOTAL HEAT LOAD

Information given in the previous paragraphs can be illustrated by the following problem:

If a metal sheathed walk-in cabinet 10-ft. long x 9-ft. wide x 8-ft. high with 4-in. cork insulation is in an 85 F. (29 C.) 50 percent relative humidity (RH) room, and it cools 2000 lb. of fresh lean beef from 60 F. to 35 F. each day, and the evaporator has two 1/10-hp motors and the cabinet has two 40 watt lamps (8 hr.), and one person works in the cabinet 8 hr. each day, what is the total heat load?

1. Heat leakage load = 55,584 Btu/day (see end of Para. 15-7)
2. Air change load = volume x air changes/24 hr. x Btu/cu. ft.
   (Heat to be removed, cooling air from 85 F. (29 C.) 50 percent RH to 35 F. (2 C.) 60 percent RH)
   Air change load = 570 (see Para. 15-8) x 24 (see Fig. 15-5) x 2.09 (see Fig. 15-6) = 28,591 Btu per day
3. Product load = weight x spec. heat (see Fig. 15-7) x temp. diff.
   Product load = 2000 lb. x .77 spec. heat x 25 F. temp. diff. (14 C.) = 38,500 Btu per day

CABINET		INTERNAL VOLUME 8 FT. HIGH CAPACITY (CU. FT.)								INTERNAL VOLUME 10 FT. HIGH CAPACITY (CU. FT.)						
Lg. & Wd.	Outside Sq. Ft.	2" Cork	2½" Cork	3" Cork	4" Cork	5" Cork	6" Cork	8" Cork	Outside Sq. Ft.	2" Cork	2½" Cork	3" Cork	4" Cork	5" Cork	6" Cork	8" Cork
5x 5	210	137	131	124	112	101	90	71	250	174	167	159	144	131	117	93
5x 6	236	169	161	154	140	127	114	91	280	215	206	197	180	164	148	120
5x 7	262	201	194	184	168	153	138	111	310	256	248	236	216	198	179	146
5x 8	288	233	224	214	196	179	162	131	340	296	286	274	252	232	210	172
6x 6	264	209	201	193	175	160	145	119	312	266	256	247	225	207	188	156
6x 7	292	248	238	228	210	193	176	146	344	316	304	292	270	249	228	192
6x 8	320	286	277	267	245	226	207	173	376	364	353	342	315	292	269	228
6x 9	348	325	315	305	280	259	238	200	408	414	402	390	360	335	309	263
6x10	376	364	353	343	315	292	269	227	440	463	451	439	405	378	350	299
6x12	432	444	432	419	385	358	331	281	504	555	546	536	495	463	430	370
7x 7	322	294	283	272	254	234	214	180	378	374	361	348	326	302	278	237
7x 8	352	341	329	317	294	273	252	214	412	434	420	406	378	353	328	281
7x 9	382	386	374	362	334	312	290	248	446	492	477	463	430	403	377	326
7x10	412	433	420	407	374	346	318	282	480	551	536	521	481	448	413	371
7x12	472	527	512	497	454	414	374	348	548	670	653	635	583	535	486	458
8x 8	384	394	382	369	343	320	296	253	448	501	487	473	441	413	385	333
8x 9	416	448	434	420	392	367	341	294	484	570	553	538	504	474	443	386
8x10	448	503	587	471	441	413	385	335	520	641	748	734	692	653	615	548
8x12	512	610	591	573	539	506	473	417	592	776	755	734	692	653	615	548
8x14	576	718	697	675	637	594	561	499	664	914	889	864	818	768	730	656
9x 9	450	510	489	469	448	420	392	341	522	649	623	600	576	543	510	449
9x10	484	570	554	537	504	473	443	386	560	725	706	686	647	612	575	508
9x12	552	694	674	654	616	581	545	476	636	883	859	836	792	752	708	626
9x14	620	814	793	771	728	687	647	566	712	1035	1011	987	935	888	840	745
10x10	520	638	620	602	567	534	500	440	600	870	790	770	729	680	650	579
10x12	592	776	755	734	693	655	617	547	680	988	962	939	890	847	802	720
10x14	664	912	889	866	818	775	733	654	760	1158	1132	1110	1050	1005	954	860
12x12	672	946	919	893	848	804	760	680	768	1203	1172	1144	1090	1038	988	895
12x14	752	1110	1086	1052	1001	951	900	809	856	1411	1382	1348	1289	1230	1170	1060
14x14	840	1304	1269	1235	1180	1126	1072	968	952	1660	1619	1568	1518	1458	1394	1272

Fig. 15-11. Table of cabinet external areas and internal volumes (capacity).

4. Miscellaneous load =
   a. The two motors' load (continuous operation) = No. of motors x Btu/hp/hr. (see Fig. 15-8) x hp x hr.
      2 x 4250 x 1/10 x 24 = 20,400 Btu
   b. The two lamps' load = No. of lamps x watts x hr. of operation x 3.42 Btu/watt
      2 x 40 x 8 x 3.42 = 2189 Btu
5. Occupancy load = No. of persons x hours of work x heat equivalent per hour (see Fig. 15-9)

$$1 \times 8 \times \frac{950 + 840}{2}$$

$$1 \times 8 \times \frac{1790}{2}$$

$$1 \times 8 \times 895 = 7160$$

Therefore: Total heat load = 55,584 + 28,591 + 38,500 + 20,400 + 2189 + 7160 = 152,424 Btu/24 hr.

Per 16 hr. = $\frac{152,424}{16 \text{ hr.}}$ = 9527 (3/4-ton load) Btu per hr.

based on 16 hours of system operation

The 16 hours of running time will provide a system with 50 percent reserve capacity. If 30 percent reserve capacity is desired, select the equipment which will operate 18 hours per day to handle the load. If the fixture is in an air-conditioned room, and the ambient temperature is 75 F. the year round, one may select equipment to run 20 hours per day.

15-10 DETERMINING HEAT LEAKAGE USING TABLES (SHORT METHOD)

A method used by some manufacturers to determine heat leakage into a cabinet is shown in Fig. 15-10. The table in Fig. 15-11 gives the external area and volume of a cabinet, based on actual experiments and investigations.

To use the tables, proceed as follows (using a walk-in refrigerator box as a sample problem): Visit the establishment and obtain all the data possible about the cabinet and the service. Determine exterior dimensions of the box, dimensions of windows, dimensions of wall, thickness of insulation, kind of insulation, number of panes in windows, how much business user does, temperatures user desires in cabinet, the average summer temperature for the locality and the highest possible water temperature (if a water-cooled installation is to be made).

Specification sheets are available for tabulating data needed for the selection of proper equipment. A sample sheet is shown in Fig. 15-12.

Using cabinet size 9 ft. x 10 ft. x 8 ft., without windows, Fig. 15-11 shows that the area is 484 sq. ft. In Fig. 15-10, note that the Btu leakage per sq. ft./24 hr. (4-in. thickness, 60-deg. temperature difference) is 108 Btu.

Heat leakage = total sq. ft. x leakage per sq. ft.
Heat leakage = 484 x 108 = 52,272 Btu/24 hr.

REFRIGERATION SALES ENGINEERS DATA SHEET

Name _____ Type of Business _____ Date _____

Address _____ City _____ Zone _____ State _____

Person Contacted _____ Title _____ Phone _____

Fixture No. 1-Make _____ Fixture No. 2-Make _____ Fixture No. 3-Make _____

Use _____ Model _____ Use _____ Model _____ Use _____ Model _____

Temperature _____ _____ _____

Humidity _____ _____ _____

Width _____ _____ _____

Length _____ _____ _____

Height _____ _____ _____

Construction _____ _____ _____

Insulation:
  Kind _____ _____ _____

  Thickness _____ _____ _____

Glass:
  Area _____ _____ _____

  No. of Panes _____ _____ _____

Produce _____ _____ _____

Lights _____ _____ _____

Motors _____ _____ _____

Sun Load _____ _____ _____

No. of People in Refrigerator ____ _____ _____

Unusual Temperatures _____ _____ _____

Unusual Service _____ _____ _____

Remarks _____ _____ _____

Use Reverse Side for Sketch of Installation

Salesman _____

Fig. 15-12. Sample date sheet of type used by sales engineers in recording data for refrigeration installation.

Volume Cu. Ft.	Service*	Temperature Difference F. Deg. (Ambient Temp. Minus Storage Room Temp.)										
		1	40	50	55	60	65	70	75	80	90	100
20	Average	4.68	187.	234.	258.	281.	305.	328.	351.	374.	421.	468.
	Heavy	5.51	220.	276.	303.	331.	358.	386.	413.	441.	496.	551.
30	Average	3.30	132.	165.	182.	198.	215.	231.	248.	264.	297.	330.
	Heavy	4.56	182.	228.	251.	274.	297.	319.	342.	365.	410.	456.
50	Average	2.28	91.	114.	126.	137.	148.	160.	171.	182.	205.	228.
	Heavy	3.55	142.	177.	196.	213.	231.	249.	267.	284.	320.	355.
75	Average	1.85	74.	93.	102.	111.	120.	130.	139.	148.	167.	185.
	Heavy	2.88	115.	144.	158.	173.	188.	202.	216.	230.	259.	288.
100	Average	1.61	64.	81.	84.	97.	105.	113.	121.	129.	145.	161.
	Heavy	2.52	101.	126.	139.	151.	164.	176.	189.	202.	227.	252.
200	Average	1.38	55.	69.	76.	83.	90.	97.	103.	110.	124.	138.
	Heavy	2.22	90.	111.	122.	133.	144.	155.	166.	178.	200.	222.
300	Average	1.30	52.0	65.	71.5	78.	84.5	91.	97.5	104.	117.	130.
	Heavy	2.08	83.2	104.	114.	125.	135.	146.	156.	166.	187.	208.
400	Average	1.24	49.6	62.	68.2	74.4	80.6	86.8	93.	99.2	112.	124.
	Heavy	1.96	78.4	98.	108.	118.	128.	137.	147.	157.	176.	196.
500	Average	1.21	48.4	60.5	66.6	72.6	78.7	84.7	90.7	96.8	109.	121.
	Heavy	1.87	74.8	93.5	103.	112.	122.	131.	140.	150.	168.	187.
600	Average	1.17	46.8	58.5	64.	70.	76.	82.	88.	94.	105.	117.
	Heavy	1.85	74.0	92.5	102.	111.	120.	130.	139.	148.	167.	185.
800	Average	1.11	44.4	55.5	61.1	66.6	72.2	77.7	83.3	88.8	100.	111.
	Heavy	1.76	70.4	88.0	96.8	106.	115.	123.	132.	141.	158.	176.
1,000	Average	1.10	44.0	55.0	60.5	66.	71.5	77.	82.5	88.	99.	110.
	Heavy	1.67	66.8	83.5	91.9	100.	108.	117.	125.	134.	150.	167.
1,200	Average	.995	39.8	49.8	54.7	59.7	64.7	69.7	74.7	79.6	89.6	99.5
	Heavy	1.58	63.2	79.0	86.9	94.8	103.	111.	119.	126.	142.	158.
1,500	Average	.920	36.8	46.0	50.6	55.2	59.8	64.4	69.	73.6	82.8	92.
	Heavy	1.50	60.0	75.0	82.5	90.0	97.5	105.	113.	120.	135.	150.
2,000	Average	.835	33.4	41.8	45.9	50.1	54.3	58.5	62.7	66.8	75.2	83.5
	Long storage	.775	31.0	38.8	42.6	46.5	50.4	54.3	58.1	62.	69.8	77.5
3,000	Average	.750	30.0	37.5	41.3	45.0	48.8	52.5	56.2	60.0	67.5	75.0
	Long storage	.576	23.0	28.8	31.7	34.6	37.3	40.3	43.2	46.1	51.8	57.6
5,000	Long storage	.403	16.1	20.2	22.2	24.2	26.2	28.2	30.2	32.2	36.3	40.3
7,500	Long storage	.305	12.2	15.3	16.8	18.3	19.8	21.4	22.9	24.4	27.5	30.5
10,000	Long storage	.240	9.6	12.0	13.2	14.4	15.6	16.8	18.0	19.2	21.6	24.0
20,000	Long storage	.187	7.48	9.35	10.3	11.2	12.2	13.1	14.0	15.0	16.8	18.7
50,000	Long storage	.178	7.12	8.90	9.79	10.7	11.6	12.5	13.4	14.2	16.0	17.8
75,000	Long storage	.176	7.04	8.80	9.68	10.6	11.5	12.3	13.2	14.1	15.8	17.6
100,000	Long storage	.173	6.92	8.65	9.52	10.4	11.2	12.1	13.0	13.8	15.6	17.3

* For average and heavy service, product load is based on product entering at 10 deg. above the refrigerator temperature; for long storage the entering temperature is approximately equal to the refrigerator temperature. Where the product load is unusual, do not use this table.

Fig. 15-13. Usage heat gain table using average storage, heavy storage and long storage variables.

15-11 DETERMINING USAGE LOAD USING TABLES (SHORT METHOD)

The total heat load of the refrigerator cabinet depends upon the heat leaking through the walls and windows. It is also affected by the heat to be removed from articles in the cabinet, the air change and other sources of heat. This heat is called heat usage, or service load. It is caused by changes of air in the cabinet, by produce to be cooled, by lights and motors which may be used inside the box, and by the occupancy of the box.

Refrigeration equipment manufacturers have developed a standard that gives a fairly accurate estimate of the usage heat load. With this method, the cabinet is classified according to the type of service to be performed: florist's cabinets, grocery boxes, normal market coolers, fresh meat cabinets and restaurant short order cabinets. From experience, these companies have found that cabinets used for the same general line of business hold rather closely to the same usage heat load. This load depends on four basic factors:

1. Temperature difference between exterior and interior of cabinet.
2. Volume of cabinet (internal).
3. Type of service.
4. Time.

It is possible to determine the usage heat load of an installation. One must know the amount of food put into the refrigerator, how many times the door is opened and how long the employees are inside the cabinet. This is a difficult process and, unless carefully done, errors are bound to appear in the results.

Data in the tables are based on 1 cu. ft. content at various temperature differences. To determine the usage heat load, proceed as follows:

1. Use temperature difference of the same value used for heat leakage into the cabinet.
2. Figure the volume of the cabinet from inside dimensions.
3. Determine the type of service for which the cabinet is being used, such as average, heavy or long storage. The service load is also dependent on cabinet size. The smaller the cabinet, the more heat load is caused by service. A meat market, for instance, may be one in either a small neighborhood store or it may be in a supermarket.

4. Time, 24 hours.
   a. After the total volume of the box has been found (from Fig. 15-11), the load for each cubic foot is determined by using the table in Fig. 15-13.
   b. If the 9 x 10 x 8-ft. cabinet appears to have average service with a temperature difference of 60 F. (33 C.), and it has a volume of 504 cu. ft., the amount of heat to be removed from each cubic foot will be 78 Btu per 24 hours. Multiply this value by the total volume in cubic feet, and a fairly accurate estimate of the service or usage load may be obtained. The table in Fig. 15-13 gives the heat usage over a period of 24 hours, since this time is the established standard.

   Heat usage = usage Btu per cu. ft. x volume in cu. ft.
   Heat usage = 78 x 504 = 39,312 Btu/24 hr.

Another table based on type of usage of the cabinet is shown in Fig. 15-14. Its values can be substituted for the values in Fig. 15-13 when exact use of cabinet is known.

15-12 TOTAL HEAT LOAD USING TABLES

The total heat load is the sum of the heat leakage load and the heat usage load.

Total heat load = heat leakage + heat usage
From the previous example:

Total heat load = heat leakage + heat usage = 52,272 (Para. 15-10) + 39,312 (Para. 15-11)
Total heat load = 91,584 Btu/24 hr.

The addition of the heat leakage and usage will give the total heat load upon the cabinet for a certain set period of time. This value may be listed either as Btu per 24 hours, or Btu per 1 hour.

Btu/1 hr. = Btu/24 hr. ÷ 24
= 91,584 ÷ 24
= 3816 Btu/hr.

However, in refrigeration applications, the unit to be installed to remove this heat should do so in less than 24 hours. This extra capacity permits reserve refrigeration for heavy loads, wear in the unit and time for defrost operations.

For fixtures above 32 F. (0 C.), it is generally understood the system should operate 16 hours out of 24 (two-thirds of the time). For fixtures below 32 F. (0 C.), unit should operate 18 hours out of 24 (three-fourths of the time).

TEMPERATURE DIFFERENCE IN DEGREES FAHRENHEIT	USE OF REFRIGERATOR			
	FLORIST	GROCERY OR NORMAL MARKET	MARKET WITH HEAVIER SERVICE OR FRESHLY KILLED MEATS	RESTAURANT SHORT ORDER
40°	40.0	65.0	95.0	120.0
50°	50.0	80.0	120.0	150.0
60°	60.0	95.0	145.0	180.0
70°	70.0	114.0	167.0	210.0
80°	80.0	130.0	190.0	240.0
90°	90.0	146.0	214.0	270.0

Fig. 15-14. Usage heat gain in Btu per 24 hours for one cu. ft. interior capacity.   (ASHRAE Guide and Data Book)

Example: if the system were to operate 16 hr./day, the service technician would divide the total Btu heat load for 24 hr. period by 16.

For 16-hr. running:

$$\frac{91,584}{16} = 5724 \text{ Btu/hr. for each hour of running}$$

For 18-hr. running:

$$\frac{91,584}{18} = 5088 \text{ Btu/hr. for each hour of running}$$

15-13 EVAPORATOR AND CONDENSING UNIT CAPACITIES

After calculating heat load, it is necessary to find what size evaporator is needed to furnish adequate refrigeration.

Some important things to remember are:

1. Evaporator removes heat from the cabinet only when the condensing unit is running.
2. Refrigerating unit usually runs from 16 to 20 hours out of each 24 hours. This means that the unit must have a refrigerating capacity in 16 hours of operation equal to the total heat load in 24 hours. If the refrigerator is installed in an air-conditioned room, the design room temperature can be lowered or the running time of the unit can be increased (for example, from 16 hours to 18 or 20 hours).

The evaporator's capacity depends upon three conditions:

1. Cabinet temperature.
2. Refrigerant temperature.
3. Space allowed for the evaporator.

Condensing unit capacity depends on:

1. Low-side pressure.
2. Condensing cooling medium (air or water).
3. Size of compressor and condenser.

It is more important to balance the capacity of the evaporator to the capacity of the condensing unit than to balance either one to the heat load of the cabinet. When balancing the capacity of the condensing unit and the evaporator, all calculations for each must be based on the same low-side pressure.

The same low-side pressure is used because:

1. The capacity of the evaporator increases as the evaporator temperature decreases.
2. The capacity of the condensing unit decreases as the low-side pressure decreases. See Fig. 15-15.

From Fig. 15-16, it can be seen that this particular evaporator matches the condensing unit at a low-side pressure of 31 psi. The evaporator/condenser combination will remove 12,500 Btu/hr. when the evaporator refrigerant temperature is 32 F. (0 C.) and the pressure is 31 psi. The temperature difference for a 42 F. (6 C.) cabinet in this case is 10 F. (6 C.).

The manufacturers of evaporators and condensing units list the capacities of their products in Btu for either one hour, 16 hours or 18 hours of operation. Fig. 15-17 lists typical condensing unit capacities at condensing temperatures of 110 F. (43 C.) and 120 F. (49 C.). Fig. 15-17 also gives four different evaporating temperatures of −30 F., −15 F., 20 F. and 40 F. (−34 C., −26 C., −7 C. and 4 C.).

Note that only at a 40 F. (4 C.) evaporator refrigerant

CONDENSING UNIT		EVAPORATOR	
LOW-SIDE TEMP.	BTU/HR.	TEMP. DIFF.	BTU/HR.
40	6650	1	400
35	6100	10	4000
30	5600	12	4800
25	5100	15	6000
20	4650		
15	4200	300 sq. ft. surface natural convection evaporator.	
10	3800		
5	3400		
0	3000		
− 5	2600		
−10	2250		
−15	1900		
−20	1550		
−25	1250		
−30	950		

90 F. ambient air.
Add 6% for 10 F. drop in air temperature, subtract 6% for each 10 F. rise in ambient temperature.
Liquid line 1/4 in.
Suction line 5/8 in.
Approximately one hp.

Fig. 15-15. Tables of condensing unit and evaporator capacity variations with temperature and pressure. Evaporator temperature difference is the difference between refrigerant temperature and cabinet air temperature.

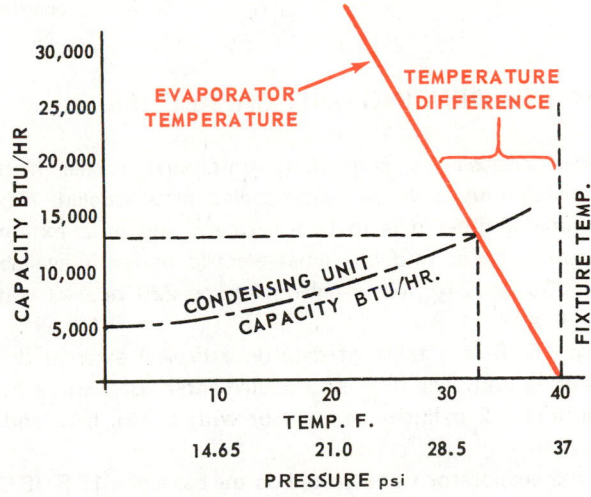

Fig. 15-16. Graph shows relative effect on evaporator and condensing unit capacity at different temperatures inside the evaporator.

temperature does the hp equal the tonnage capacity. Referring to Fig. 15-17, for example, a 3-hp condensing unit, operating at 40 F. (4 C.) evaporator refrigerant temperature and 110 F. (43 C.) condensing temperature, has the capacity of 36,300 Btu/hr. The capacity of a 1-ton machine is 12,000 Btu/hr. Therefore, 36,300 ÷ 12,000 = 3-ton (plus) capacity.

AVERAGE COMPRESSOR CAPACITIES BTU/HR.

	EVAPORATING TEMPERATURES °F.							
	−30°		−15°		+20°		+40°	
	CONDENSING TEMPERATURE °F.							
HP	110°	120°	110°	120°	110°	120°	110°	120°
2	5,200	4,500	9,100	8,200	18,000	16,800	22,800	21,200
3	9,000	8,300	14,300	13,200	22,100	20,800	36,300	34,200
5	14,200	12,500	24,800	22,400	41,700	39,300	62,400	58,500
7 1/2	25,000	20,000	31,000	28,000	53,000	48,000	87,000	81,700
10	31,000	26,000	43,600	44,800	81,000	75,000	120,000	112,000
15	42,600	37,500	74,400	67,200	111,000	102,000	171,600	160,000
20	56,000	44,700	82,000	71,000	154,000	142,000	235,000	218,000
25	70,000	56,000	96,000	85,000	188,000	174,000	283,000	263,000
30	80,000	67,000	116,500	102,500	225,000	210,000	349,000	324,000
40	94,000	75,000	155,000	135,000	325,000	306,000	439,000	406,000
50	122,000	100,000	188,500	159,500	375,000	350,000	585,000	550,000
60	168,000	134,000	240,000	220,000	450,000	420,000	710,000	670,000
70	196,000	156,000	272,000	239,000	571,000	534,000	800,000	742,000
75	210,000	167,000	291,000	256,000	582,000	542,000	855,000	795,000
80	224,000	178,000	310,000	273,000	622,000	578,000	900,000	842,000
90	252,000	201,000	349,000	307,000	750,000	700,000	1,027,000	955,000
100	280,000	223,000	388,000	341,000	777,000	723,000	1,170,000	1,100,000

NOTE: The above figures are only approximate and based on catalog ratings of leading compressor manufacturers. For precise figures, refer to catalog of your compressor manufacturer.

Fig. 15-17. Table of condensing unit capacities in Btu per hour at various evaporator temperatures and at two different condensing temperatures.

15-14 CONDENSING UNIT CAPACITIES

When choosing a condensing unit, first decide if the condensing unit is to be water-cooled or air-cooled. Next, determine whether it is to be a hermetic unit or an external drive unit. Then, find out what electric power is available (120, 208 or 240 volt, single-phase, or 220 or 440 volts, three-phase).

Fig. 15-18 is a table of data on a typical external drive condensing unit. The unit is air-cooled and is used with a 1-hp motor and a 2 cylinder compressor with a 2-in. bore and a 2-in. stroke.

If the evaporator was selected on the basis of a 15 F. (8 C.) temperature difference and a 16 hour running time, this means that if the cabinet is to operate at 38 F. (3 C.) the refrigerant temperature will be 38 minus 15 F., or 23 F. (−5 C.). If it is an R-12 refrigerant unit, the condensing unit capacity must be matched to the evaporator capacity at a low-side pressure of 23.2 psi.

The table, Fig. 15-17, lists the average hourly capacity of various horsepower capacity condensing units. This clearly shows the increase in capacity of a condensing unit as the low-side pressure (and temperature) increases, providing the head pressure remains fairly constant. It also shows the effect of condensing temperature (and pressure) on the capacity of a condensing unit.

The capacity of hermetic condensing units can also be found in tables provided by manufacturers.

15-15 EVAPORATOR INSTALLATIONS

Practically all refrigerator cabinets are built to be used with mechanical refrigeration. These cabinets are specially designed to improve the efficiency of the cooling of the cabinet with evaporators. The refrigeration service engineer should know the theory of air circulation in the cabinet, and what is supposed to happen during the operation of the unit. The following deals with the study of efficient baffling and correct air circulation in cabinets.

Baffles are surfaces, or air ducts, which increase the efficiency of the airflow through the evaporator and throughout the cabinet. They direct the airflow so that it is carried all around the interior of the box, leaving no dead or warm air spots.

A typical evaporator and baffle are shown in Fig. 15-19. The colder air coming from the evaporator is made to flow down the center of the cabinet, while the warmer air is directed up the walls back to the evaporators. The design must

CONDENSING UNIT CAPACITIES

COMPRESSOR RPM	REFRIGERANT (EVAPORATOR) TEMPERATURE	BTU/HR.
	45	11,000
	40	10,200
475	35	9,370
	30	8,530
	25	7,740
	25	8,700
	20	7,960
540	15	7,200
	10	6,430
	5	5,740
	0	5,100
	0	5,780
	−5	5,700
665	−10	4,430
	−15	3,820
	−20	3,280
	−25	2,800

Fig. 15-18. Btu/hr. capacity table for a one horsepower condensing unit, air-cooled, 90 F. ambient temperature, at various low-side temperatures.

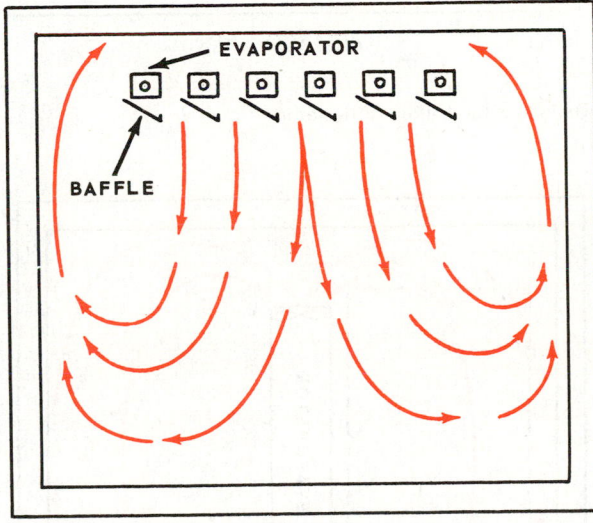

Fig. 15-19. Looking at a refrigerator cabinet from one end. The natural convection airflow inside fixture is shown in red.

be scientifically proportioned to insure that air circulation is unrestricted, with no objects blocking airflow.

Any horizontal baffle or drain pan should be insulated, because the top surface may be in contact with cold air while the under part of the baffle may be in contact with relatively warm air. If it were not insulated, this temperature difference may cause condensation. Eddy currents of air (small circular flows of air) would also disturb airflow in the cabinet.

Multiple baffled evaporators of the natural convection type are often used, and the air flows around the cabinet because of the relative weights of the cold air and the warm air (density). Warm air is lighter per cubic foot of volume and, therefore, rises in the box. This natural circulation must not be hindered or the box temperature will not be constant. Baffling the evaporators promotes this natural circulation of the air and speeds it up. Baffles may also serve as drain pans.

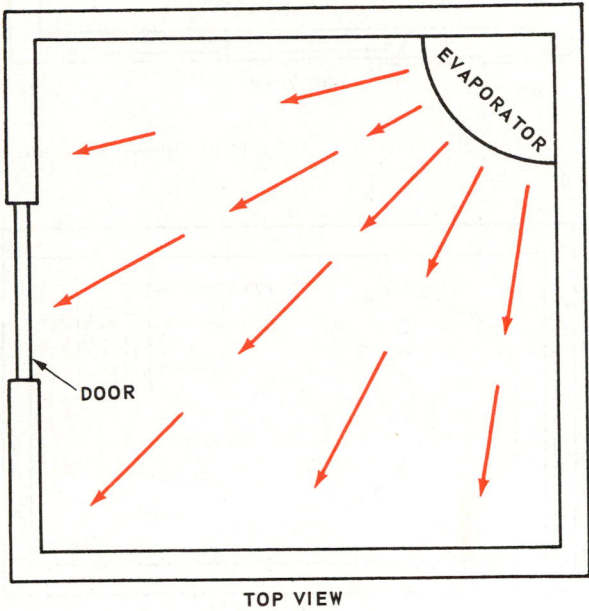

Fig. 15-20. Blower evaporator mounted in upper corner of walk-in cooler. Note air distribution relative to door.

15-16 EVAPORATOR LOCATIONS

Many cabinets do not have room to mount overhead evaporators. A walk-in cooler with an exterior height of approximately 7 1/2 feet is too low for overhead evaporators. This height necessitates the use of a blower evaporator. Two common types of mountings are:

1. Evaporator may be placed in upper corner of the cabinet as far away as possible from the entrance door, Fig. 15-20.
2. Wall evaporators may be mounted against the wall opposite the windows or the reach-in doors of the cabinet, Fig. 15-21.

The evaporators are shrouded and have a motor-fan to circulate the air. They also have built-in drain pans. Also see Figs. 15-22, 15-23 and 15-24.

15-17 EVAPORATORS FOR DISPLAY CASES

Evaporator designs for closed and open type display cases have a difficult air circulation problem to overcome. The cases are narrow and there is less room for the evaporator. Many of the cases are open. Fig. 15-25 shows a dual evaporator installation for a double duty case.

A display case using a blower evaporator is shown in Fig.

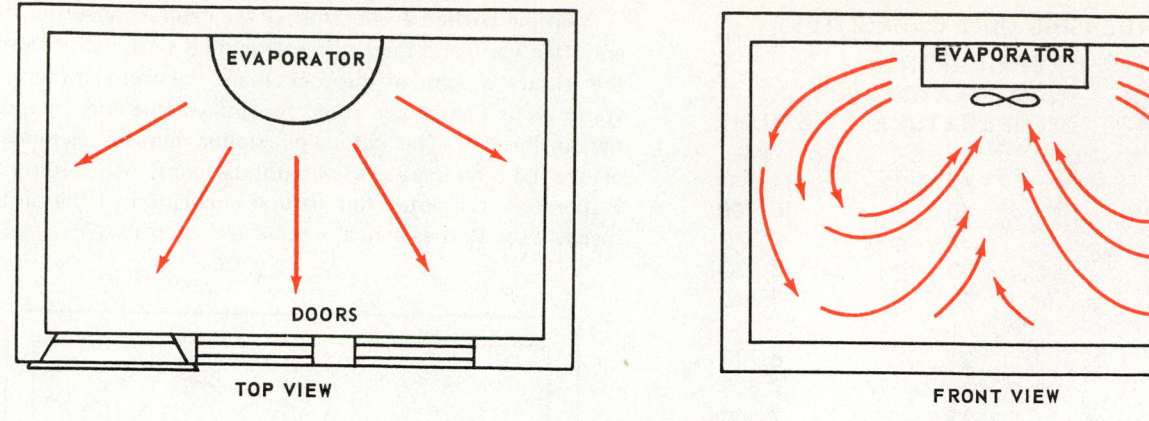

TOP VIEW

FRONT VIEW

Fig. 15-21. Blower evaporator mounted on center back wall of walk-in cooler.

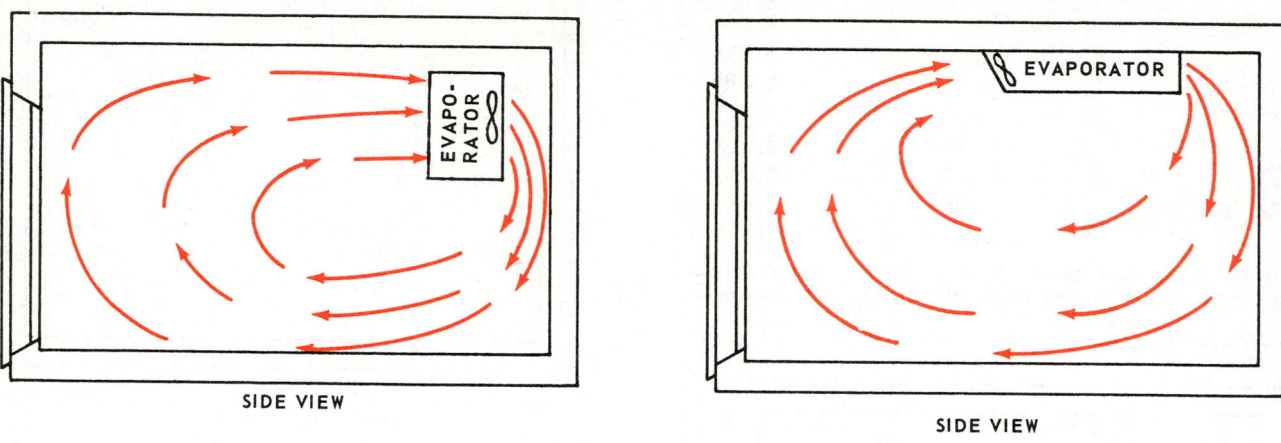

SIDE VIEW

SIDE VIEW

Fig. 15-22. Two different blower evaporator installations for walk-in refrigerator.

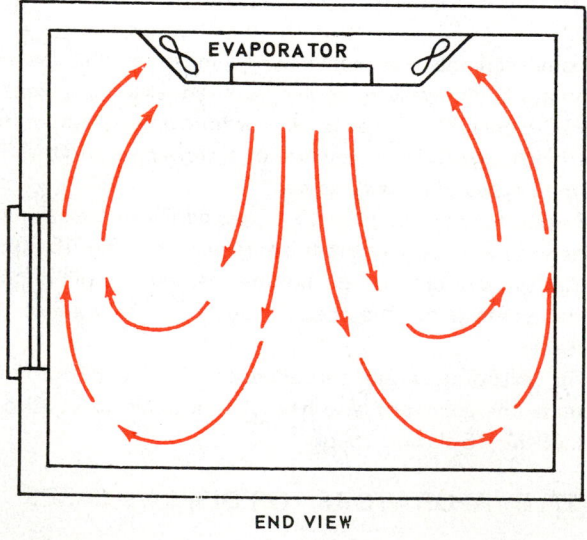

END VIEW

Fig. 15-23. Two motor-driven fans reduce distance air circulates in cabinet.

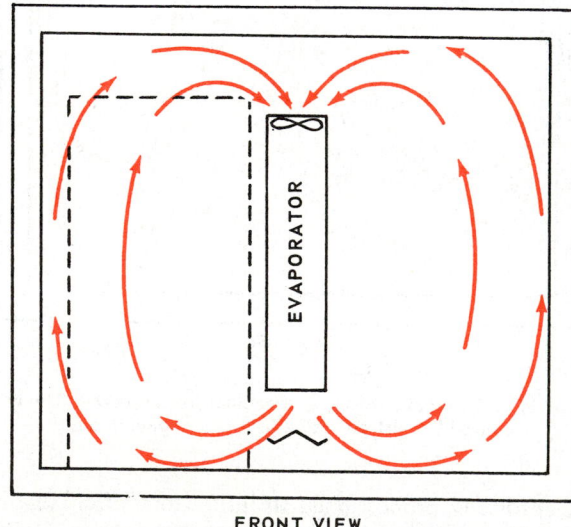

FRONT VIEW

Fig. 15-24. Reach-in cabinet with blower evaporator mounted behind mullion or door frame.

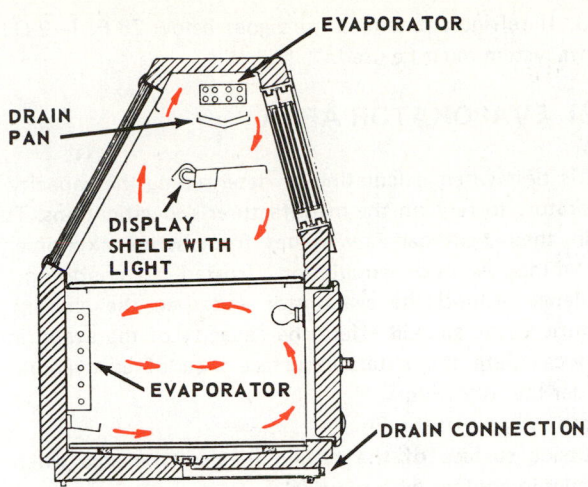

Fig. 15-25. A cross section of a double-duty display case showing two evaporators. Note combination drain pans and baffles. (ASHRAE Guide and Data Book)

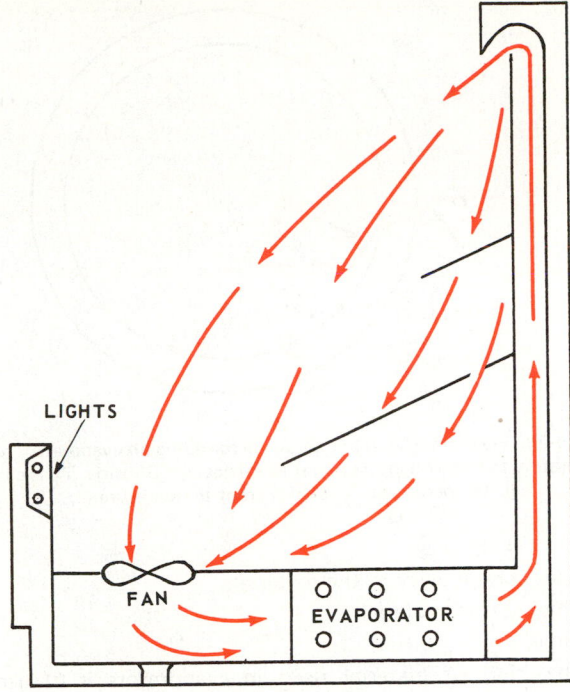

Fig. 15-27. Open display case with a blower evaporator located in base of cabinet.

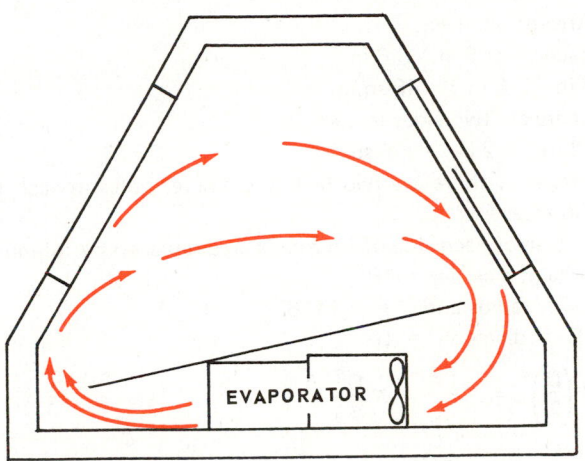

Fig. 15-26. Closed display case with a blower evaporator located below display shelf.

15-26. Note the location of the motor-fan.

A design for an open display case is shown in Fig. 15-27. Notice how ducting is used to control the flow of the refrigerated air.

15-18 EVAPORATOR TYPES

Many kinds of evaporators have been used in mechanical refrigeration. However, there are two basic types in use:
1. The air-cooling evaporator is used to cool the air within the cabinet directly.
2. The liquid-cooling evaporator is used to cool a liquid, which may be consumed or used to cool other substances.

The air-cooling evaporator is classified as the dry type. It may be either frosting, defrosting or nonfrosting type. Some air-cooled units are forced circulation evaporators.

Liquid-cooling evaporators are of two types: the submerged evaporator and the tube-within-a-tube evaporator.

In commercial refrigeration, certain kinds of evaporators are more popular than others. Among these are the dry nonfrosting air-cooling evaporator, forced circulation dry evaporator, and submerged flooded or dry evaporator. See Chapter 12.

15-19 AIR-COOLING EVAPORATOR THEORY

The theory of the air-cooling evaporator involves heat transfer from the air circulating over the evaporator to the refrigerant as follows: As the warmer air comes in contact with the evaporator, air molecules strike the fins and release some of their energy to the fin (transfer some heat to it). This heat, in turn, travels through the fins, then through the tubing and comes in contact with the liquid refrigerant on the inside, transferring the heat energy to it.

Due to the low density of air, heat transfer is slow. After reaching the metal, the heat travels efficiently and rapidly. Upon reaching the interior surface of the tubing, it again has difficulty in reaching the refrigerant in the system. Gas bubbles clinging to the internal surface along with an oil film reduce the heat flow.

15-20 EVAPORATOR CAPACITIES

One of the laws of thermodynamics (heat in action) is that heat always flows from an object at a higher temperature to an object at a lower temperature. As in the case of heat leakage, the amount of heat transfer depends on five variables:
1. Area.
2. Temperature difference.

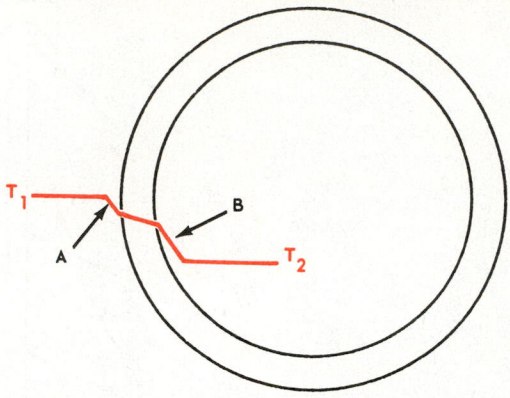

Fig. 15-28. Heat transfer from air A, surrounding an evaporator, to the evaporator, then through the metal to refrigerant B inside. T_1 is cabinet air temperature; T_2 is refrigerant temperature.

3. Heat conductivity of the material.
4. Thickness of material.
5. Time.

The kind of material used in evaporators is of utmost importance because materials used must be good heat conductors.

Heat transmission may be through various materials. For air evaporators, Fig. 15-28, the heat must pass through air film A on the metal surface, then through the metal tubing and oil or liquid refrigerant film on the inside of the evaporator tubing to refrigerant B.

If the air is moved rapidly, heat flow to the metal is greater because more air contacts the metal per unit of time, and the air film is thinner.

If the oil or refrigerant film is moved faster, the film will be thinner due to greater movement.

Generally: the denser the fluid, the greater the heat flow; the faster the fluid motion, the greater the heat flow.

The U factor (heat transfer) for natural convection evaporators is approximately:
    1 Btu/sq. ft./F./hr.
The U factor for blower evaporators is approximately:
    3 Btu/ sq. ft./F./hr.
The U factor for liquid-cooling evaporators is approximately:
    15 Btu/sq. ft./F./hr.

These values are fairly accurate if the temperature difference is taken between the average air or liquid temperature and the refrigerant temperature. See T_1 and T_2 in Fig. 15-28.

For natural convection evaporators, the temperature difference usually selected is 10 F. (5 C.). For example, a 45 F. (7 C.) cabinet would have a refrigerant temperature in the evaporator of 35 F. (2 C.).

The smaller the temperature difference, the higher the relative humidity can be kept. For example, 10 F. to 12 F. (5 C. to 6 C.) temperature difference will keep a 75 to 90 percent RH, while a 20 F. (11 C.) to 30 F. (17 C.) temperature difference will keep a 50 to 70 percent RH.

If the evaporator refrigerant temperature goes below 28 F. (−2 C.), frost will form on the evaporator. The cycle off time must be long enough to permit defrosting, with the air at 35 F.

(2 C.). If refrigerant temperature goes below 28 F. (−2 C.), a defrost system must be used.

15-21 EVAPORATOR AREA

It is best when calculating or determining the capacity of evaporators to rely on the manufacturer's specifications. They obtain their heat capacity values from actual experiments. Such things as poor circulation, frosted fin condition, air turbulence around the evaporator and even the amount of moisture in the air will affect the capacity of the evaporator.

To calculate the external surface area of an evaporator, consider the following:
1. Both surfaces of the fin.
2. Outside surface of the tubing (neglect the area where it comes in contact with the fins).
3. External surface of the tubing bends.

Example: Find the area of a 10-ft. long evaporator with 6 x 8-in. fins .025-in. thick, having 1/2-in. fin spacings and using two 5/8-in. tubes 4 in. apart. See Fig. 15-29 and proceed as follows:
1. Area of one fin:
Each fin is 8 in. x 6 in.
8 in. x 6 in. = 48 sq. in.
There are two sides to each fin
48 in. x 2 in. = 96 sq. in.
2. However, there are two 5/8-in. diameter holes in each fin:
Fin area =
    area of each side of fin minus area replaced by tubing
Area of a circle = πR^2
(R = radius and π = 3.1416)
2R = diameter = D

$$\pi\left(\frac{D}{2}\right)^2 = \pi\frac{D^2}{2^2} = \pi\frac{D^2}{4}$$

$$\text{Area} = \frac{\pi \times D^2}{4} = \frac{\pi \times \dfrac{5}{8}^2}{4}$$

$$= \frac{\pi \times \dfrac{25}{64}}{4} = \frac{\pi \times 25}{4 \times 64} = \frac{3.1416 \times 25}{4 \times 64}$$

$$= \frac{78.54}{256} = .307 \text{ sq. in.}$$

However each hole takes out two fin surfaces this size
.307 x 2 = .614 sq. in.
There are two holes in each fin
.614 sq. in. x 2 = 1.228 sq. in.
Heat removing area for one fin =
96 sq. in. − 1.228 = 94.8 sq. in.
3. The total number of fins:
10 ft. x 12 in./ft. = 120 in. long
2 fins per in. = 120 x 2 = 240 fins
240 fins + 1 extra fin at end = 241 fins
4. Total fin area = 94.8 x 241 = 22,846.8 sq. in.
5. Area of 2 tubes 5/8 in. D; 10 ft. long:
10 ft. x 12 in./ft. = 120 in.
There are two tubes
120 in. x 2 = 240 in.

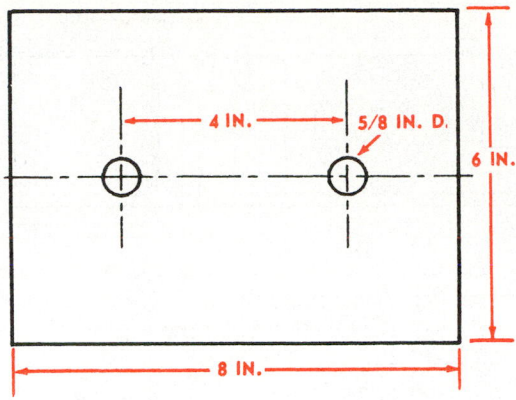

Fig. 15-29. Evaporator fin specifications for sample problem. Fin is .025-in. thick.

The circumference of the tube is
Cir. = π x diameter
    = π x 5/8 in.
    = 3.1416 x 5/8 in.
    = 1.9635 in.
The area = length x circumference
          = 240 in. x 1.9635 in.
          = 471.24 sq. in.

6. The actual tube area is decreased by thickness of fins:
Fin contact area = π 5/8 x .025 x 241 =
1.9635 x .025 x 241 = 11.83 sq. in.
Actual tube area 471.24 − 11.83 = 459.41 sq. in.

7. Tube bend area =
length of bend x circumference length x number of bends (1)
(2 in. radius)
Length = π x D
       = π x 4 in.
       = 3.1416 x 4 in.
       = 12.5664
but it is only 1/2 of a circle so 12.5664 ÷ 2 = 6.2832
Circumference of tube =
5/8 in. x 3.1416 = .625 x 3.1416 = 1.96
Tube bend area = 6.2832 x 1.96 = 12.3 sq. in.

8. Total area: 22,846.8 = total fin area
             459.4 = actual tube area
              12.3 = tube bend area
        23,318.5 sq. in.
or in sq. ft. 23, 318.5 ÷ 144 = 161.9 sq. ft.

The evaporator is unable to remove heat from the cabinet when the compressor is not running. Therefore, the heat removing capacity of the evaporator is calculated on the same running time as that of the compressor.

Heat transfer problem: What is the capacity of a natural convection evaporator having an external area of 15 sq. ft. with a refrigerant temperature of 22 F. (−16 C.), if the average cabinet temperature is 42 F. (6 C.)?

Solution:
First, it is known that 1 sq. ft. will handle 1 Btu per F. per hr. The refrigerant temperature is 22 F. (−6 C.) with the air temperature passing over the evaporator at 42 F. (6 C.) the temperature difference is 20 F. (11 C.). If 1 F. temperature

difference will handle 1 Btu, 20 F. (11 C.) temperature difference will handle 20 Btu. Then multiply this value by the number of sq. ft., to find the total capacity of the evaporator per hr. In this case
Btu = 15 x 20 = 300 Btu per hr.

It is claimed that the maximum effective distance that the fin should extend from a natural convection evaporator is 3 in.

Fig. 15-30 shows an evaporator constructed of rippled fin surfaces and with a flange contact between the prime surface (tubing) and the fin.

Fig. 15-30. Section through tubes and fins of an evaporator. A—Tubing (primary surface). B—Fins (secondary surface).

A typical evaporator capacity table for nonfrosting evaporators is shown in Fig. 15-31.

15-22 EVAPORATOR DESIGN

Many types of evaporators are available, including some which use the following combinations of material: copper tubing and aluminum fins, copper tubing and copper fins, and steel tubing and aluminum fins (for ammonia R-717).

Usually, the evaporator fins are securely bonded to the tubing. However, construction varies. Some manufacturers make the fin to fit the tubing with a drive fit; others use some mechanical devices to attach the fins firmly to the tubing. Some expand the tubing with a mandrel or by hydraulic pressure to force it against the fin. See Fig. 15-32. A method of bonding tubing to off center fins is shown in Fig. 15-33. Another method of mounting fins on tubing spacing them at the same time is shown in Fig. 15-34.

Fin spacing varies between 1/2 and 1 1/2 in. for natural convection evaporators, and 1/16 to 1/4-in. spacing for forced convection. This spacing is a means of varying the capacity of the evaporator, and it is also used to compensate for the depth of the evaporator. The deeper the evaporator, the greater the

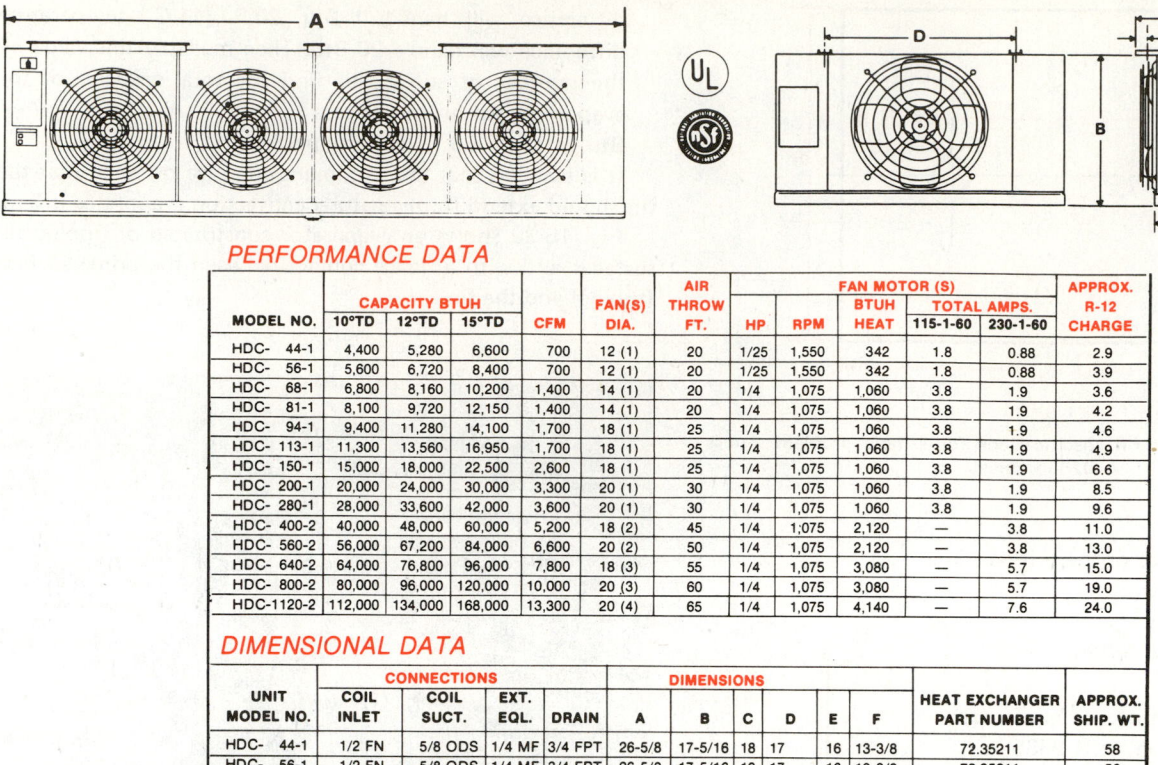

PERFORMANCE DATA

MODEL NO.	CAPACITY BTUH			CFM	FAN(S) DIA.	AIR THROW FT.	FAN MOTOR (S)					APPROX. R-12 CHARGE
	10°TD	12°TD	15°TD				HP	RPM	BTUH HEAT	TOTAL AMPS.		
										115-1-60	230-1-60	
HDC- 44-1	4,400	5,280	6,600	700	12 (1)	20	1/25	1,550	342	1.8	0.88	2.9
HDC- 56-1	5,600	6,720	8,400	700	12 (1)	20	1/25	1,550	342	1.8	0.88	3.9
HDC- 68-1	6,800	8,160	10,200	1,400	14 (1)	20	1/4	1,075	1,060	3.8	1.9	3.6
HDC- 81-1	8,100	9,720	12,150	1,400	14 (1)	20	1/4	1,075	1,060	3.8	1.9	4.2
HDC- 94-1	9,400	11,280	14,100	1,700	18 (1)	25	1/4	1,075	1,060	3.8	1.9	4.6
HDC- 113-1	11,300	13,560	16,950	1,700	18 (1)	25	1/4	1,075	1,060	3.8	1.9	4.9
HDC- 150-1	15,000	18,000	22,500	2,600	18 (1)	25	1/4	1,075	1,060	3.8	1.9	6.6
HDC- 200-1	20,000	24,000	30,000	3,300	20 (1)	30	1/4	1,075	1,060	3.8	1.9	8.5
HDC- 280-1	28,000	33,600	42,000	3,600	20 (1)	30	1/4	1,075	1,060	3.8	1.9	9.6
HDC- 400-2	40,000	48,000	60,000	5,200	18 (2)	45	1/4	1;075	2,120	—	3.8	11.0
HDC- 560-2	56,000	67,200	84,000	6,600	20 (2)	50	1/4	1,075	2,120	—	3.8	13.0
HDC- 640-2	64,000	76,800	96,000	7,800	18 (3)	55	1/4	1,075	3,080	—	5.7	15.0
HDC- 800-2	80,000	96,000	120,000	10,000	20 (3)	60	1/4	1,075	3,080	—	5.7	19.0
HDC-1120-2	112,000	134,000	168,000	13,300	20 (4)	65	1/4	1,075	4,140	—	7.6	24.0

DIMENSIONAL DATA

UNIT MODEL NO.	CONNECTIONS				DIMENSIONS						HEAT EXCHANGER PART NUMBER	APPROX. SHIP. WT.
	COIL INLET	COIL SUCT.	EXT. EQL.	DRAIN	A	B	C	D	E	F		
HDC- 44-1	1/2 FN	5/8 ODS	1/4 MF	3/4 FPT	26-5/8	17-5/16	18	17	16	13-3/8	72.35211	58
HDC- 56-1	1/2 FN	5/8 ODS	1/4 MF	3/4 FPT	26-5/8	17-5/16	18	17	16	13-3/8	72.35211	56
HDC- 68-1	1/2 FN	5/8 ODS	1/4 MF	3/4 FPT	30-5/8	17-5/16	20	21	18	15-3/8	72.35211	80
HDC- 81-1	1/2 FN	5/8 ODS	1/4 MF	3/4 FPT	30-5/8	17-5/16	20	21	18	15-3/8	72.35212	90
HDC- 94-1	1/2 FN	7/8 ODS	1/4 MF	3/4 FPT	34-5/8	25-5/16	20	25	18	15-3/8	72.35212	100
HDC- 113-1	1/2 FN	7/8 ODS	1/4 MF	3/4 FPT	34-5/8	25-5/16	20	25	18	15-3/8	72.35213	104
HDC- 150-1	1/2 FN	7/8 ODS	1/4 MF	3/4 FPT	34-5/8	25-5/16	23	25	21	18-3/8	72.35213	106
HDC- 200-1	1/2 FN	7/8 ODS	1/4 MF	3/4 FPT	42-5/8	25-5/16	23	33	21	18-3/8	72.35214	130
HDC- 280-1	1/2 FN	1-1/8 ODS	1/4 MF	3/4 FPT	42-5/8	28-5/16	23	33	21	18-3/8	72.35214	136
HDC- 400-2	7/8 ODS	1-1/8 ODS	1/4 MF	1 FPT	64-5/8	29-1/2	31	48	—	21-3/16	72.35214	390
HDC- 560-2	1-1/8 ODS	1-3/8 ODS	1/4 MF	1 FPT	72-5/8	29-1/2	31	48	—	21-3/16	72.35215	412
HDC- 640-2	1-1/8 ODS	1-5/8 ODS	1/4 MF	1 FPT	84-5/8	29-1/2	31	48	—	21-3/16	72.35215	524
HDC- 800-2	1-1/8 ODS	2-1/8 ODS	1/4 MF	1 FPT	91-1/8	32-1/2	31	51-3/4	—	21-3/16	72.35216	620
HDC-1120-2	1-5/8 ODS	2-1/8 ODS	1/4 MF	1 FPT	116-5/8	32-1/2	31	*77-1/8	—	21-3/16	72.35217	700

Fig. 15-31. Table of performance and dimensional data for 35 F. (2 C.) evaporators. (The Singer Co.)

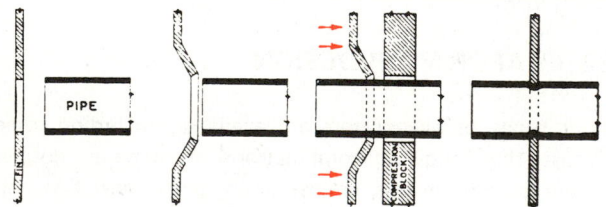

Fig. 15-32. Illustration shows mechanical means of bonding evaporator fins to tubing. Fin is dished to expand tube hole, then tubing is inserted. Mandrel flattens fin, firmly pressing fin into tubing.

fin spacing to minimize air restriction. Evaporators which have 6 to 8-in. depth usually have 1-in. spacing, whereas those with an 18 or 20-in. depth have 1 1/2-in. spacing. Fin spacing of 1 in. or less is said to decrease air turbulence. The tubing used usually is 5/8 in. OD, although 3/4 in. OD tubing is used in large evaporators.

Some companies use one continuous piece of tubing for the complete evaporator; others have all bends made separately. The bends are silver brazed or copper brazed to the straight

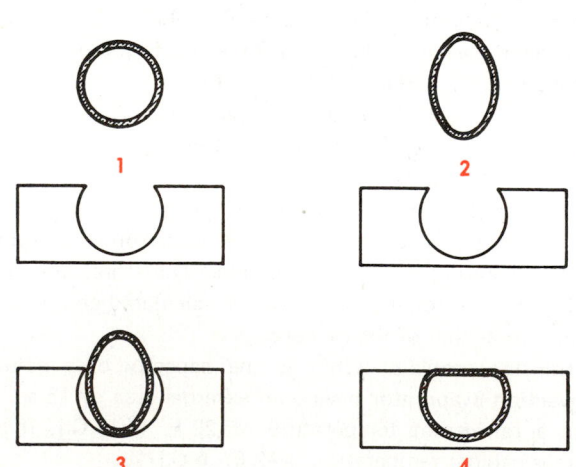

Fig. 15-33. Method of mechanically bonding tubing to off-center fins. 1—Original tubing and fin. 2—Tubing is formed into elliptical shape. 3—Tubing is inserted into fin opening. 4—Fixture holds fins while tube is pressed into fin-opening shape. (Peerless of America, Inc.)

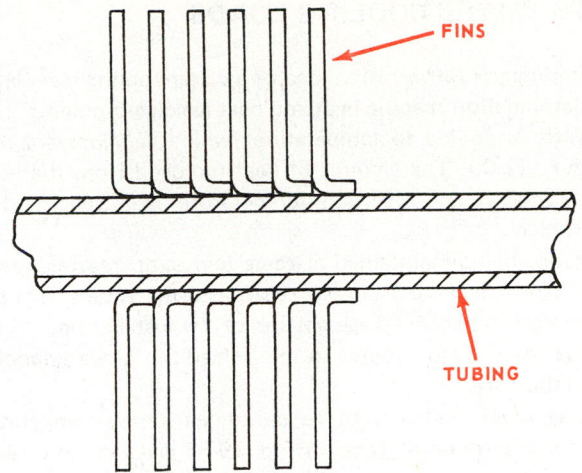

Fig. 15-34. Flanged fins mounted on a tubing. Flanges automatically determine fin spacing.

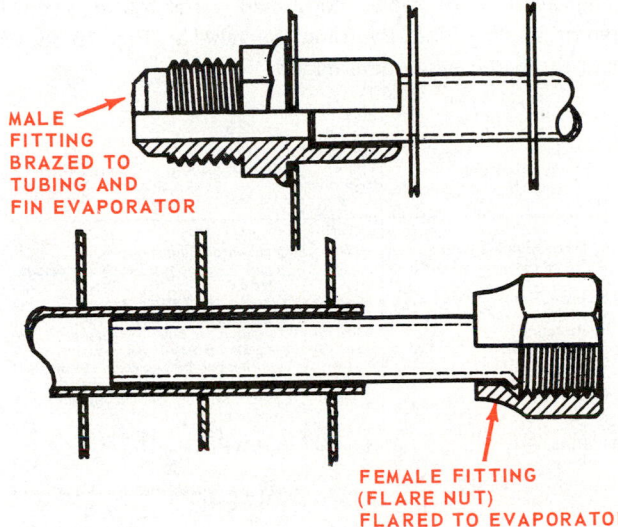

MALE FITTING BRAZED TO TUBING AND FIN EVAPORATOR

FEMALE FITTING (FLARE NUT) FLARED TO EVAPORATOR

Fig. 15-35. Construction of coupling devices (fittings) used in attaching refrigeration lines to evaporator.

lengths. Some manufacturers use devices inside the tubing to swirl the refrigerant to improve heat transfer to the boiling refrigerant.

Fittings to the evaporator are usually 1/2 in. OD tubing brazed to the 5/8-in. tubing, then flared with an external nut mounted on it. Some manufacturers use a 1/2-in. male flare fitting, brazed to the end fin. Fig. 15-35 shows two types.

15-23 FORCED CIRCULATION AIR-COOLING EVAPORATOR CAPACITIES

A forced circulation evaporator is one having an electric fan mounted adjacent to it to increase the airflow. Velocities of 44 feet per minute (fpm) to 2000 fpm are permissible, with 1000 fpm being the average value. Sometimes, variable speed fan motors are used. Draining facilities for condensation removal must be built into the unit.

Because of the large amount of air striking the evaporator per unit of time, the capacity of the evaporator in Btu per square foot per deg. F. hour is increased remarkably. The values vary with the air speed. The table in Fig. 15-36 gives these values.

Evaporators are finned, and the spacing of the fins varies considerably. See Para. 15-22. The fins must be kept straight and equally spaced or airflow will be reduced. See Chapter 14 for service instructions.

15-24 LIQUID EVAPORATOR CAPACITIES

Liquid evaporators, regardless of type, may be calculated for capacity on the basis of a factor of approximately 10 to 120 Btu per square foot per deg. F. per hour. However, the U factor (heat transfer) varies with the fluid velocity, evaporator construction and total temperature difference. Fig. 15-37 shows the average values for a certain evaporator.

Some evaporators are used to cool a brine solution. This is a nonfreeze solution, so it does not provide any frost accumulation problems. Many submerged evaporators use a sweet water bath to provide an ice holdover around the evaporator to maintain good capacity during peak loads.

The capacity of the evaporators varies, depending upon

FORCED CIRCULATION EVAPORATOR CAPACITIES

SURFACE IN SQ. FT.			PARAL-LEL PATHS	COOLING CAPACITIES BTU PER HOUR		MOTOR		FAN	
TUBE	FIN	TOTAL		15 F. TD*	25 F. TD*	HP	SPEED RPM	DIAMETER	CFM
2.94	32.39	35	3	2200	4500	1/80	1800	12"	620
5.88	63.0	68	5	4100	7200	1/80	1800	12"	465
7.94	70.06	78	3	5200	9000	1/10	1140	15 1/2"	1200
13.2	116.8	130	5	8300	12300	1/10	1140	15 1/2"	1000
14.7	83.3	98	3	6500	10500	1/8	1140	17"	2020
17.2	145.8	163	5	10000	14000	1/8	1140	17"	1715

* TEMP. DIFFERENCE

Fig. 15-36. Table of forced circulation evaporator capacities.

EVAPORATOR HEAT TRANSFER DATA					
WATER VELOCITY FT./MIN.	TOTAL TEMPERATURE DIFFERENCE				
	6	8	10	12	15
	BTU TRANSFER/SQ. FT./HR./F.				
150	67	76	83	90	97
200	83	95	103	110	118
250	97	109	115	122	129
300	103	115	123	130	138

Fig. 15-37. Heat transfer in Btu per sq. ft. per hr. for typical flooded liquid evaporator using 5/8 in. OD tubes (Btu/sq. ft./hr./F.). Total temperature difference is between refrigerant and water temperatures.

whether or not they are of the frosting type. This can be understood if the theory of heat transfer is applied. When ice forms around the evaporator, it requires that the heat go through this extra material. Therefore, the cooling capacity of the evaporator is decreased. Furthermore, it makes the heat travel through one extra contact surface, which reduces efficiency. Liquid cooling evaporators are used for beverage cooling, for water chillers for air conditioning, and for cooling brines.

15-25 SPECIAL EVAPORATOR CAPACITIES

A considerable variation of evaporator types are in use, including: cast metal, iron pipe, brine spray, intermediate refrigerant evaporators and cold plates. The problems with these will be the same as the ones discussed before.

The metal used as the refrigerant carrying device does not have much effect upon the heat transfer capacity of the evaporator. Its conductivity is relatively much greater than the contact between the evaporator and the air. The capacity of the cast metal and the iron pipe evaporator may be calculated exactly the same as the method described in Para. 15-21.

Some installations use brine spray. In this system, the brine is forced through a pipe extending into the cooling chamber, and the pipe is perforated with a number of fine holes. The brine sprays out of these holes and mixes with the air flowing over the baffle, removing the heat. The capacity of these systems is calculated upon the temperature difference between the air in the box and the temperature of the brine. It is safe to assume a high efficiency of heat transfer for brine spray installations.

The Baudelot Cooler (milk cooler) runs water, or the liquid to be cooled, over refrigerant cooled pipes or plates. The liquid being cooled is in the open and can be easily controlled. Icing is not critical and, therefore, the liquid can be cooled close to its freezing temperature. See Fig. 15-61.

The cooling capacity of cold plates may also be determined by the area, temperature difference and the materials. The heat transfer for air to metals is about 1 Btu/deg. F./sq. ft./hr. Cold plates usually contain a eutectic solution which freezes at a certain temperature and the plate remains at this constant temperature until all the eutectic solution has melted. The refrigerating system may or may not be operating at the time the eutectic solution is absorbing heat (solution is melting).

15-26 WATER-COOLING LOADS

Finding the refrigeration load of a water-cooling installation is a combination specific heat and heat leakage problem:
1. Water is cooled to temperatures which vary upward from 35 F. (2 C.). The amount of heat removed from the water to cool it to a certain temperature is a specific heat problem.
2. Water, being maintained at these low temperatures, results in a heat leakage from the room into the water. This part involves the heat leakage portion of the installation.

The two major items to be solved in a water-cooling installation are:
1. How much water is to be consumed at the temperature difference desired. Table in Fig. 15-38 gives values of these two variables.
2. Temperature of water for drinking purposes should be regulated according to type of work the consumers are performing. The heavier the work, or the warmer the room temperature, the warmer the drinking water must be.

The amount of water consumed varies quite a bit in different applications. By using the table in Fig. 15-38, the exact heat load is easily determined.

USAGE	FINAL TEMP. REQUIRED °F.	TOTAL AMOUNT OF WATER USED AND WASTED
1. Office Building—Employees	50	1/8 gallon per hour per person
2. Office Building—Transients	50	1/2 gallon per hour for each 250 persons per day
3. Light Manufacturing	50 to 55	1/8 gallon per hour per person
4. Heavy Manufacturing	50 to 55	1/4 gallon per hour per person
5. Restaurant	45 to 50	1/10 gallon per hour per person
6. Cafeteria	45 to 50	1/12 gallon per hour per person
7. Hotels	50	1/2 gallon per day per room (14 hr. day)
8. Theaters	50	1 gallon per hour per 75 seats
9. Stores	50	1 gallon per hour per 100 customers per hour
10. Schools	50 to 55	1/8 gallon per hour per student
11. Hospitals	45 to 50	1/12 gallon per day per bed

NOTE—Total amount of water used and wasted varies with type of installation and kind of service. This table will serve as a basis for determining cooler capacity required.

Fig. 15-38. Drinking water-cooling table gives values of various applications. (Temprite, Eaton Corp.)

Example: Item No. 4 points out that for heavy manufacturing, water to be consumed should be kept within 50 to 55 F. (10 to 13 C.). Also, 1/4 gallon of water per hour per person will be consumed.

A production foundry may be classed as heavy manufacturing. If the foundry employs 50 men for a period of eight hours, the water load per day would be 50 x 8 x 1/4, or 100 gallons of water to be cooled each eight hours.

If the city water is at a temperature of 75 F. (24 C.) in the pipes, it must be cooled 20 F. (11 C.) to reach the temperature of 55 F. (13 C.). Since there are 8.34 lb. of water in 1 gal., the specific heat load can be computed as follows:

Btu = specific heat x weight x temperature difference
Therefore:

Btu = 1 x 100 x 8.34 x 20 = 16,680 Btu per 100 gal. water

The heat leakage for this particular problem is determined by the external area of the insulated parts of the system.

Insulation one to three inches thick is common for water-cooling insulations, with ice water insulation being standard at 1 1/2 in. Heat leakage for water-cooling installations is calculated identically with that of heat leakage for cabinets. The example just described deals only with a unit installation.

Many water-cooling installations involve the circulation of refrigerated water to various fountains. The heat leakage load in a case of this kind is calculated on the basis of gallons of water per hour to be circulated through the system to maintain satisfactory temperatures.

The table in Fig. 15-39 illustrates another method of computing this load.

Gallons per Hour To Be Circulated per 100 Feet of Pipe to Hold Temperature Rise Within 5° F. Add This Amount to Usage Fig. 15-38								
PIPE SIZE	TEMPERATURE DIFFERENCE BETWEEN ROOM AND CIRCULATING WATER							
	20°	25°	30°	35°	40°	45°	50°	55°
¼"	4.8	6.3	7.3	8.2	9.7	10.9	12.3	13.6
⅜"	5.5	6.8	8.2	9.6	11.0	12.3	13.7	15.0
½"	6.3	8.0	9.5	11.2	12.7	14.3	16.0	17.6
¾"	6.7	8.4	10.1	11.8	13.5	15.2	16.9	18.6
1"	7.3	9.1	10.9	12.8	14.6	16.5	18.5	20.6
1¼"	8.6	10.4	12.5	14.6	16.6	18.7	20.4	22.1
1½"	9.0	11.2	13.5	15.7	18.0	20.2	22.5	24.7

Fig. 15-39. Table showing heat gain through insulated cold water pipes. (Temprite, Eaton Corp.)

15-27 ICE CREAM FREEZING AND STORAGE LOAD

Ice cream is manufactured from milk, solids, fat, sugar, gelatin and water. After mixing, the product is cooled to about 27 F. (−3 C.), then frozen. Next, it is cooled rapidly to approximately −20 F. (−29 C.). Brick ice cream is maintained between 0 to 5 F. (−18 to −15 C.). Bulk ice cream is held at 5 to 12 F. (−15 to −11 C.).

The heat values of the various ice creams vary. See Fig. 15-7. Average values are as follows: Specific heat of the mix before freezing is 0.80; latent heat at 27 F. (3 C.) is about 96 Btu per pound. Specific heat of frozen ice cream is 0.45 Btu per pound. Weight of the original mix is about nine pounds per gallon. On freezing it expands and comes to a density of five pounds per gallon if simply flavored, and to six pounds per gallon if it contains fruits or nuts.

Next, determine the size of the refrigerating unit needed to keep ice cream in its cabinet during dispensing. Normally, an ice cream cabinet is designed to hold brick ice cream and flavored ice creams next to the evaporator. However, many cabinets need separate evaporators for the two types. Fig. 15-40 shows an ice cream display cabinet. Metal-finished cabinets with 2 to 3 in. of urethane are used.

The ice cream is delivered at the correct temperature. The heat load, therefore, is composed only of leakage and air entering when the covers are removed. The table in Fig. 15-11 may be used to calculate heat leakage. Use the outside area and add 20 percent to take care of the cover openings.

15-28 THERMODYNAMICS OF THE REFRIGERATION CYCLE

Thermodynamics is the science which deals with the relationships between heat and mechanical action. The refrigeration compression cycle is based on thermodynamics.

One should know how to determine how large a compressor is needed to produce a certain amount of refrigeration, and how large a motor is needed to drive this compressor. To understand how these values are determined, the heat behavior of the refrigerant must be understood.

The refrigerant cycle is simple. The refrigerant is let into the evaporator in the liquid state and at near room temperature. Some of the refrigerant vaporizes under the low pressure in the evaporator and cools the remainder of the refrigerant to the desired refrigerating temperature. Then, as the remainder of the refrigerant evaporates, it removes heat from the evaporator and, therefore, from the cabinet. The total amount of heat absorbed is the total LATENT HEAT of vaporization, while the amount of heat absorbed from the cabinet and evaporator is the EFFECTIVE LATENT HEAT.

The refrigerant vapor formed during evaporation passes down the suction line. As it does so, the vapor decreases a little in pressure (usually 2 psi) and increases in temperature about 10 F. (6 C.). The condition of a vapor warming up or increasing in temperature after it has vaporized is called "superheating of the vapor." The degree of superheat is the temperature difference between the temperature of the vapor at the compressor and its evaporating temperature.

The compressor then takes the slightly superheated vapor and converts (compresses) it to a high-temperature, high-pressure vapor. This condensing temperature sometimes becomes as high as 250 F. (121 C.), depending upon the refrigerant and the conditions.

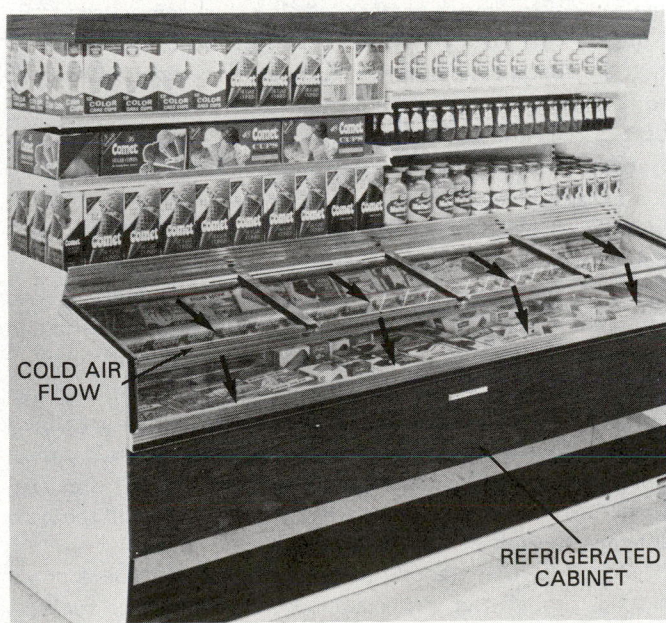

COLD AIR FLOW

REFRIGERATED CABINET

Fig. 15-40. Ice cream display cabinet used to merchandise packaged ice cream. (Hussman Refrigeration, Inc.)

The superheated vapor passes to the condenser and, if its temperature is higher than ambient temperature (water or air), it transfers enough of its heat to the air or water to cool to its vapor pressure-temperature. If the vapor pressure-temperature is above that of the water or air ambient temperature, the refrigerant vapor starts losing some latent heat of evaporation. The quantity of heat it loses determines the amount of the vapor that will condense into a liquid. After it has become liquid, the refrigerant cools down close to the water or air temperature.

The refrigerant then goes to the refrigerant control, where the pressure is reduced. The refrigerant is cooled as it vaporizes (called "flash gas"), and the rest vaporizes to remove heat from the cabinet. Thus the refrigerant cycle is repeated. Fig. 15-41 shows this cycle taking place and the temperatures in various parts of the refrigerating system.

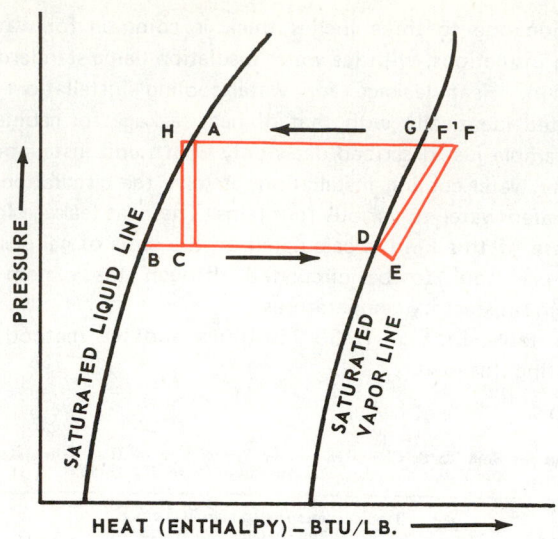

Fig. 15-42. Pressure-heat diagram for a refrigerant. Saturated liquid curve represents heat in liquid refrigerant at various pressures before it will start vaporizing. Saturated vapor curve represents division between superheated gas and point where gas starts condensing into liquid.

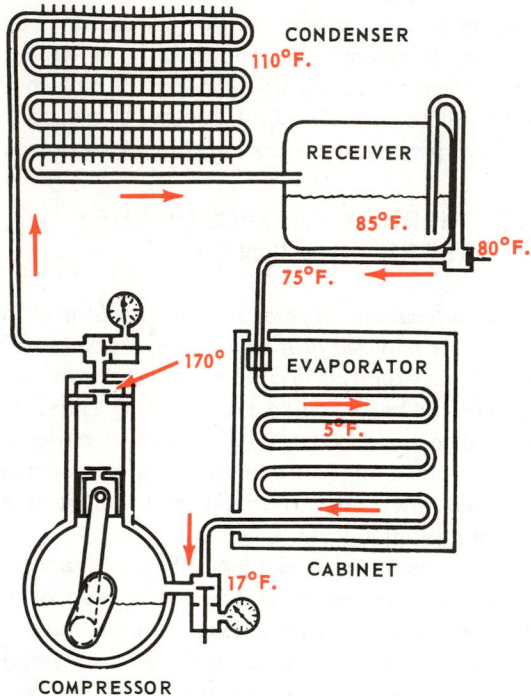

Fig. 15-41. A refrigerating system schematic shows the approximate temperatures of refrigerant in various parts of system.

15-29 PRESSURE — HEAT DIAGRAM

The following discussion of refrigerant behavior is based on one pound of refrigerant regardless of its state (liquid or vapor). The discussion deals only with the pure refrigerant, neglecting the effect of lubricating oils and other influences.

Fig. 15-42 charts the behavior of one pound of refrigerant in a refrigerating machine. The horizontal scale shows the amount of heat present in one pound of refrigerant at all times and under all conditions. The vertical scale shows the pressure imposed upon it.

The graph shown in Fig. 15-42 is commonly called a pressure-heat chart. It is also called a pressure-enthalpy chart.

The heat (Btu) in the pound of refrigerant is usually measured from saturated liquid refrigerant at −40 F. (−40 C.). The pressures will be different for each kind of refrigerant.

Note in the pressure/heat chart that as the refrigerant vaporizes at the lower constant pressure, it passes horizontally from B to D. This line indicates the vaporization of the refrigerant from a liquid into a vapor in the evaporator. The distance D to E represents the heating of this vapor into a superheated condition as it passes down the suction line. Note that only a few Btu of heat have been added and that the pressure has decreased a little.

Point E is the condition of the vapor when it moves into the compressor and is compressed. It is then compressed to F. Note how the pressure increases rapidly and how a few Btu of heat are added to the vapor as the vapor is compressed from E to F. The vapor leaving the compressor is considerably superheated. See Para. 15-34.

Point F represents the condition of the vapor as it leaves the exhaust valve of the compressor. The distance between F and G is the cooling down of this superheated vapor to the point where it starts to condense.

At G the vapor has no superheat and is 100 percent saturated vapor. The line G to A represents the condensation of the refrigerant in the condenser from a vapor into a liquid.

Point A represents the amount of heat in the liquid and the pressure imposed on the liquid as it forms in the condenser. From A to H is the loss of heat from the liquid as it passes along the liquid line to the refrigerant control. This action (sometimes called "subcooling") is because the liquid refrigerant cools to room temperature.

Line H to C represents the throttling of the liquid upon passing through the refrigerant control orifice. The cycle is now ready to be repeated for the pound of refrigerant.

Note that the distance C to D does not represent the total LATENT HEAT of the liquid at the low-side pressure condition. This means that R-12, which has a latent heat of 70 Btu per pound at 5 F. (−15 C.), will not remove all of this heat from the evaporator. Some of it (approximately 19 Btu) is used to cool down the remaining liquid refrigerant to the 5 F. (−15 C.) temperature. The 51 Btu remaining is called the EFFECTIVE LATENT HEAT or the EFFECTIVE REFRIGERATING CAPACITY. This means that 19/51 of the pound (37 percent) of the R-12 flashes into gas at the refrigerant control.

The pressure-heat chart in Fig. 15-42 graphically shows physical property changes of a pound of a refrigerant as it passes around the refrigerating cycle. A thorough knowledge of this chart is helpful to the engineer and service technician.

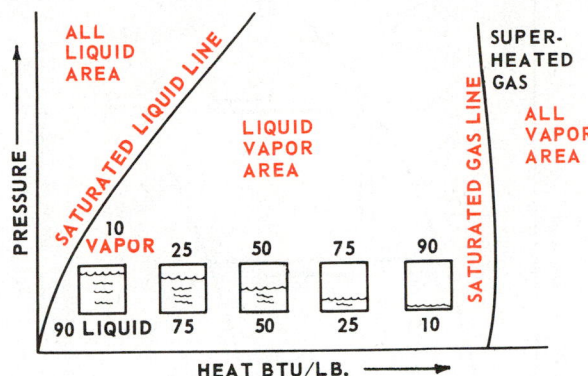

Fig. 15-43. Pressure-heat diagram. Note that as heat is subtracted, refrigerant becomes a liquid. As heat is added, refrigerant becomes a vapor.

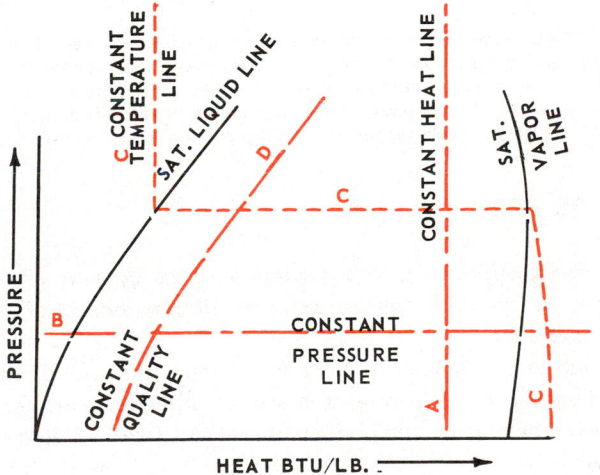

Fig. 15-44. Pressure-heat diagram. A—Constant heat condition with pressure change. B—Constant pressure and condition of refrigerant with changing heat content. C—Constant temperature with changing pressure and heat.

15-30 PRESSURE — HEAT AREAS

The pressure-heat chart in Fig. 15-43 is divided into three main areas. To the left of the saturated liquid line, all of the one pound of refrigerant is liquid. Between the saturated liquid line and the saturated vapor line, the refrigerant is a

mixture of liquid and vapor, as shown in the squares on the drawing.

Close to the saturated liquid line, Fig. 15-43, the refrigerant is almost all liquid; close to the saturated vapor line, the one pound of refrigerant is almost all vapor. Note that the vapor in the area to the right of the saturated gas line is in a superheated condition.

15-31 CONSTANT VALUE LINES OF PRESSURE — HEAT CHART

Many facts can be read from the chart in Fig. 15-44. Along any vertical line such as A, the heat in one pound of refrigerant is the same or constant. Along any horizontal line B, the refrigerant has the same (constant) pressure. The line C, along which the temperature reading is the same (constant), is almost vertical in the liquid area, is horizontal in the liquid and vapor area, and slants down and to the right in the superheated vapor area.

Refrigerant quality means how much of the one pound of refrigerant is liquid and how much is vapor. Ten percent (10%) quality means that the pound of refrigerant is 10 percent vapor and 90 percent liquid. The line showing the same or constant quality is shown at D.

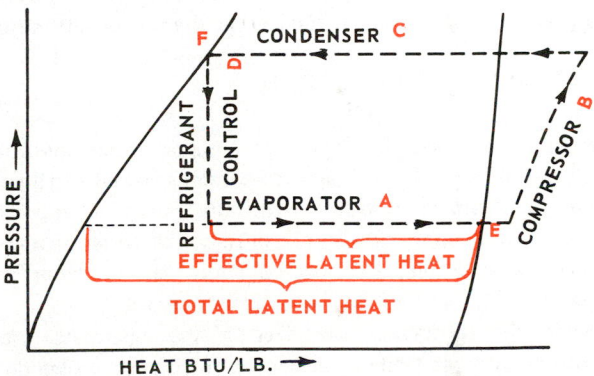

Fig. 15-45. Pressure-heat diagram. A—Liquid is boiling in evaporator, and pressure is constant as heat is being added to refrigerant. B—Compressor raises pressure to condensing pressure, heat of compression is added and temperature rises. C—Condenser cools hot vapor to saturated vapor line, then condenses vapor into a liquid. Pressure is constant and heat is removed. D—Refrigerant control reduces pressure quickly to evaporating or low-side pressure. Temperature of remaining liquid drops as some of liquid refrigerant "flashes" into vapor.

15-32 EFFECT OF PRESSURE ON LATENT HEAT

The value of the latent heat of the refrigerant when vaporizing or when condensing in a refrigerating machine differs at different pressures. At the lower pressure (vaporizing), Fig. 15-45, the TOTAL LATENT HEAT to be added to (be absorbed by) the liquid refrigerant to vaporize it is more than that needed to be subtracted or removed from it to condense it into a liquid at the higher pressure. This is because the liquid, when formed, is at a higher temperature than the liquid at the vaporizing pressure. This difference is represented by the specific heat of the refrigerant multiplied by the temperature difference for the two conditions.

The EFFECTIVE LATENT HEAT of the refrigerant, when vaporizing, is the total latent heat at the low pressure minus the difference in the heat content of the liquid at the high (condensing) pressure, minus the heat content of the liquid at the low (evaporating) pressure. This is because the high-pressure liquid, when throttled in the refrigerant control, must be cooled down to a low-pressure temperature liquid before it can vaporize and remove heat from the surrounding substances. Part of the liquid vaporizes in order to cool the remaining liquid to the lower temperature. The vapor formed during this operation is called "flash gas."

The effective latent heat is the total heat at E in Fig. 15-45 minus the heat of the liquid at F. The effective latent heat is an average value because low-side pressure and high-side pressure vary somewhat during the operation of the system. Other conditions, such as oil in the system, and system efficiencies affect the ideal cycle.

For R-12 the total latent heat of vaporization at 5 F. (−15 C.) is 79 Btu per pound. However, its actual heat-absorbing ability or effective latent heat is only about 51 Btu per pound. These values are based on the standard evaporating temperature of 5 F. (−15 C.) and standard condensing temperature of 86 F. (30 C.). This reduction in heat absorption is due to the fact that some heat is absorbed by the flash gas when refrigerant pressure is reduced at the refrigerant control.

15-33 SATURATED VAPOR

A saturated vapor is a refrigerant vapor under conditions which permit some of the vapor to condense when a little heat is removed from it. Another definition: Saturated vapor is a substance in a vapor form in the presence of some of its own liquid. For example, a saturated vapor is the vapor in a refrigerant cylinder half full of liquid refrigerant.

When the refrigerant vaporizes in the evaporator, it is saturated vapor at first. However, as this vapor passes down the suction line to the compressor, it usually becomes warmer by 5 F. to 15 F. (3 C. to 8 C.). This additional heat and increase in temperature is called superheating the low-side vapor. That is, it is raising the vapor above its saturated condition for this pressure, and the vapor will now obey Charles' or Boyle's Laws. See Chapter 28.

In Fig. 15-43, any vapor in the space marked liquid and vapor area is saturated vapor.

15-34 SUPERHEATED VAPOR

A superheated vapor is a vapor under conditions where the volume and/or pressure of the vapor will decrease, with no condensation, when some heat is removed from it. For example, the vapor that the compressor handles is always superheated — unless a condition arises where the last drop of liquid refrigerant evaporates just as the refrigerant travels past the exhaust valve of the compressor.

After the low-pressure superheated vapor enters the compressor, it is compressed. The energy put into it by the

compressor tremendously increases the temperature and pressure on the vapor, and the amount of superheat is increased.

Superheating of the vapor lowers the efficiency of a machine. The less the superheating, the more efficient the machine will be. The heat added to the vapor is the mechanical energy of compression being converted into heat energy. By knowing how much heat has been added, one may calculate the size of the motor necessary to drive the compressor.

There are two places in the refrigeration cycle where superheating usually takes place: In the suction line (a low-temperature superheat); and in the compressor and top part of condenser (a high-temperature superheat). These two conditions are shown in Fig. 15-46.

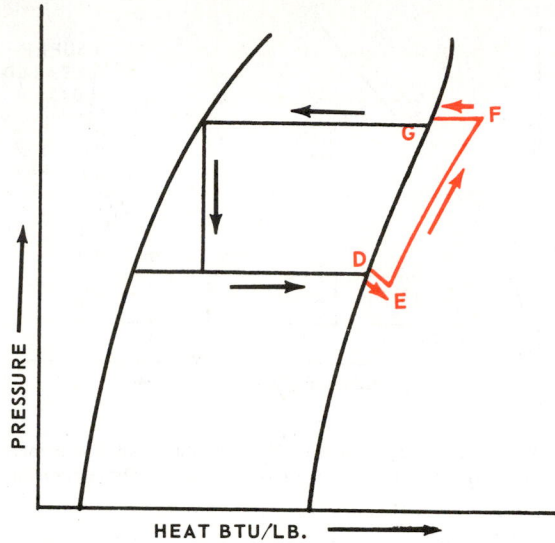

Fig. 15-46. Superheated vapor sections of refrigeration cycle. D to E—Superheat added to one pound of refrigerant vapor as it travels through suction line from evaporator to intake valve of compressor. E to F (condition at exhaust valve)—Superheat added as vapor is compressed. F to G—Superheat removed in top portion of condenser.

15-35 SPECIFIC HEAT

The specific heat of a substance is the amount of heat necessary to raise the temperature of one pound of that substance 1 F.

Substances may exist in three different states (solid, liquid and vapor). Every substance has three different values for its specific heat, depending on whether it is a solid, a liquid or a vapor.

There are two kinds of specific heat of vapor:
1. Specific heat when under a constant pressure.
2. Specific heat when confined to a constant volume.

A vapor under constant pressure has a greater specific heat value than that of the same vapor under constant volume. This is because the vapor heated with a constant pressure upon it will expand and do external work (such as increasing the size of a balloon). This external work naturally requires an additional quantity of heat.

When a compressor compresses one pound of the refrigerant vapor, it does not add heat to it under a constant pressure

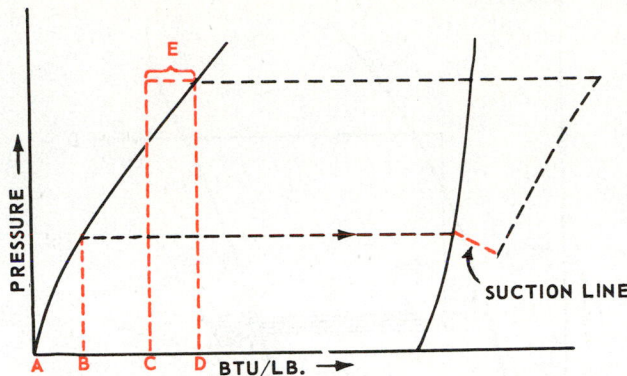

Fig. 15-47. Pressure-heat diagram. By using heat exchanger or installing liquid and suction lines together, heat increase in suction line vapor comes from heat decrease in liquid line. E indicates cooling of liquid in receiver, liquid line and by heat exchangers. Note gain in effective latent heat from C to D.

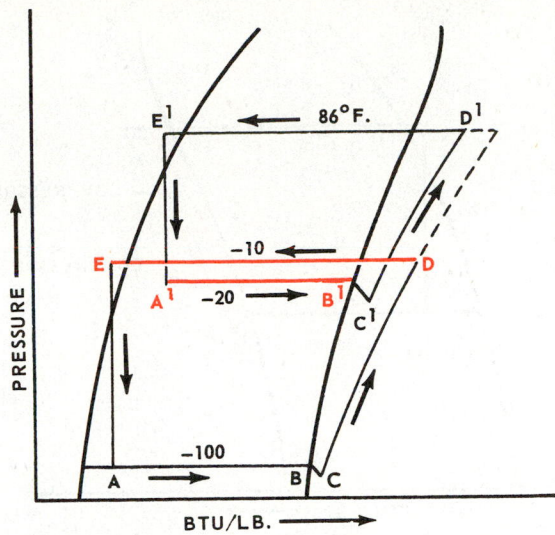

Fig. 15-48. Pressure-heat diagram for cascade system used to obtain ultra-low temperatures. Evaporator A^1 to B^1 removes heat from condenser D to E.

or constant temperature condition. This state of affairs in the compressor is called adiabatic compression, meaning that no heat has been removed from or added to the vapor as it was compressed. Actually, a refrigeration compressor operates almost adiabatically because the compression takes place so rapidly.

The specific heat of liquid refrigerants varies considerably, depending on pressure imposed upon them. As shown in Fig. 15-47 the pressure to which the liquid refrigerant is subjected in the condenser, after it has condensed at D, must be determined before calculating how much heat must be removed from one pound of the liquid at this temperature to further cool it to room temperature C.

Example: One pound of R-12 condensing at 125 F. (52 C.), and then cooled to 75 F. (24 C.) in the line, must lose:
Heat of liquid at 125 F. = 37.28 (Fig. 9-4)
Heat of liquid at 75 F. = 25.20
37.28 − 25.20 = 12.08 Btu/lb. must be removed

After the refrigerant passes through the throttling valve, it is subjected to a lower evaporating pressure. The specific heat of liquid under the evaporating pressure, shown at B, must be determined to find out how much heat must be removed to cool it down to vaporizing temperature. If refrigerant goes through the refrigerant control while at its condensing temperature, it will have A to D liquid specific heat/lb.

15-36 CASCADE SYSTEM

To obtain extremely low temperatures efficiently, two refrigerating systems may be used instead of one. The two systems are connected in series (cascade system). That is, the evaporator of the higher pressure-temperature cycle (first or low stage) removes the heat from the condenser of the lower pressure-temperature cycle (second or high stage). Fig. 15-48 shows the principle of this type system on a pressure-heat diagram.

Many cascade systems use a different refrigerant for the low temperature system than used in the high temperature system.

In actual design, the evaporator of the high temperature system must remove all the condensing heat of the low temperature system. A^1 to B^2 should equal E to D.

15-37 TWO-STAGE COMPRESSOR

Some refrigerating systems, especially ultra-low temperature systems, use two compressors connected in series to pump the very low pressure suction line vapor up to the condensing pressure and temperature condition. The first stage, usually a large cylinder, pumps the vapor up to a midpoint on the compression curve, then the compressed vapor is cooled but kept in the vapor condition. The second cylinder then compresses the cooled intermediate vapor to the final pressure-temperature condition. Fig. 15-49 shows an approximate cycle.

Two-stage compressors are used when the compression ratio is more than 10 : 1. That is, if the low-side pressure is 0 psi and the head pressure is 210 psi, the ratio is:

$$\frac{\text{Head pressure abs}}{\text{Low-side pressure abs}} = \frac{210 + 15}{0 + 15} = \frac{225}{15} = 15 : 1$$

In this case, a two-stage compressor would be used.

$$\text{Stage 1: } \frac{45 + 15}{0 + 15} = \frac{60}{15} = 4 : 1$$

$$\text{Stage 2: } \frac{210 + 15}{45 + 15} = \frac{225}{60} = 3.75 : 1$$

15-38 BYPASS CYCLE

The "hot gas" bypass may be used for any of these purposes:
1. Defrost evaporators.
2. Prevent suction pressure from going too low (when cooling load decreases).
3. Keep liquid refrigerant from entering compressor.

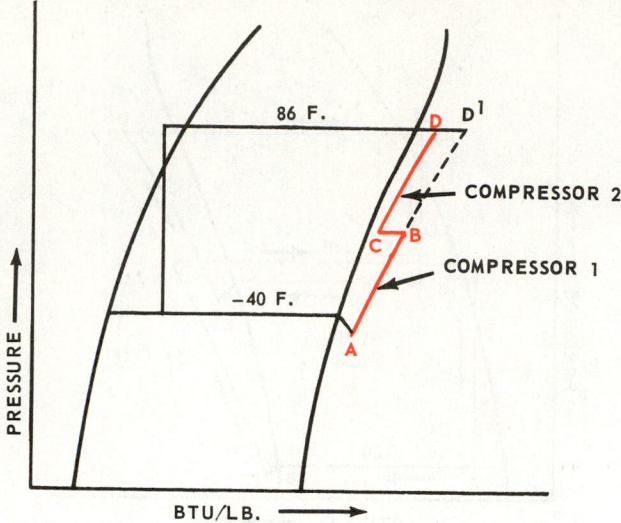

Fig. 15-49. Cycle of a two-stage compressor system. Compressor 1 (first state) compresses vapor from A to B. Vapor is cooled in heat exchanger (air or water) from B to C. Compressor 2 then compresses vapor to condensing pressure from C to D. This action reduces amount of heat of compression at final stage D to D^1 and also reduces superheat temperature at compressor 2 exhaust valve.

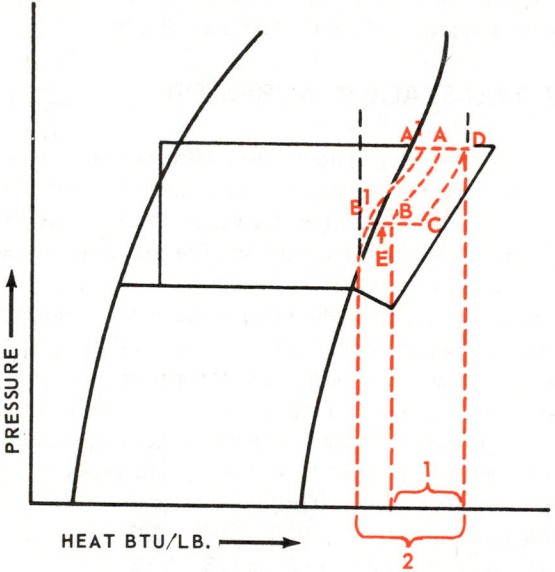

Fig. 15-50. "Hot gas" defrosting cycle. Heat lost D to B as shown at 1 is heat used to defrost evaporator. If defrost hot gas is cooled too much as at B^1, it will become partly liquid. This could cause liquid to enter the compressor, so an accumulator is put at E to insure that only vapor can reach compressor. Defrost cycle is A to B to C to D. Maximum heat for defrosting is shown at 2, unless bypass gas is allowed to condense and is then vaporized (in another evaporator of a multiple system or in a special defrost evaporator).

Many refrigerating systems use an automatic bypass system. Two automatic bypass types are:

1. "Hot gas" bypass.
2. Liquid bypass.

The "hot gas" bypass cycle used to defrost an evaporator is shown in Fig. 15-50. The "hot gas" is traveling through the evaporator from A to B to C.

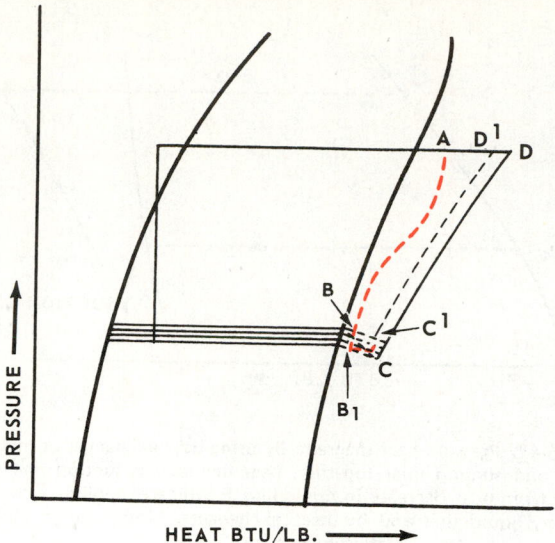

Fig. 15-51. "Hot gas" bypass system allows low side to maintain normal low-side pressure. If pressure tends to drop to point B^1, bypass circuit A to B opens and brings low-side pressure up to normal at C^1. Without bypass, low-side pressure would tend to operate at C.

The cycle for "hot gas" bypass for low-pressure control is shown in Fig. 15-51. The bypass line (controlled by a solenoid valve and a pressure valve) is piped from the "hot gas" part of the condenser into the suction line near the compressor. The bypass circuit is controlled by a pressure control connected to the suction line. The bypass action will return the compression line to approximately C^1 to D^1. The four horizontal evaporator lines represent how the low-pressure side changes from cut-in pressure to cut-out pressure.

A liquid bypass cycle is shown in Fig. 15-52. If the low-side

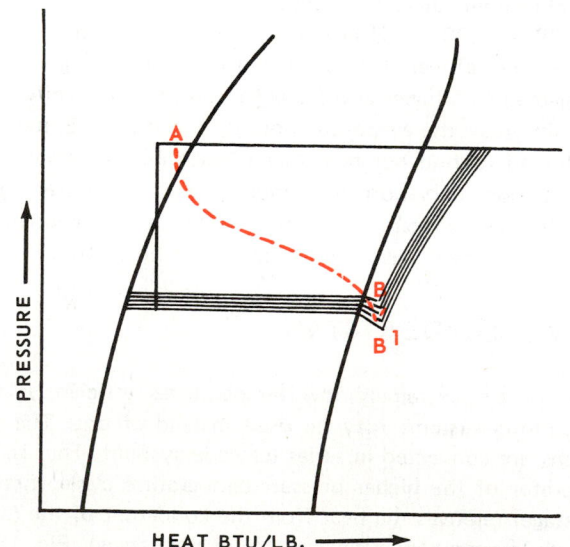

Fig. 15-52. Liquid refrigerant bypass A to B is used to maintain normal low-side pressures. The five parallel evaporator pressure lines and compression lines show changes in the cycle as the low-side pressure changes from cut-in pressure to cut-out pressure.

pressure drops to B¹ because the evaporator is not picking up enough heat, the suction pressure control will open a solenoid valve. This permits liquid refrigerant to bypass the refrigerant control and evaporator, and it is fed directly into the suction line at B. The liquid evaporates quickly and maintains a definite low-side pressure.

15-39 PRACTICAL PRESSURE-HEAT CYCLE

The refrigeration cycles described in previous paragraphs are based on 5 F. (−15 C.) evaporating temperature and 86 F. (30 C.) condensing temperature.

Practical cycles for various other refrigerating systems are somewhat different. For example:

1. With an air-cooled condensing unit for frozen foods, the evaporator refrigerant temperature will be −10 F. (−23 C.).
2. With summer design ambient temperatures of 95 F. (35 C.), the condensing temperature will be 95 F. (35 C.) + 30 F. (−1 C.) or 125 F. (52 C.).

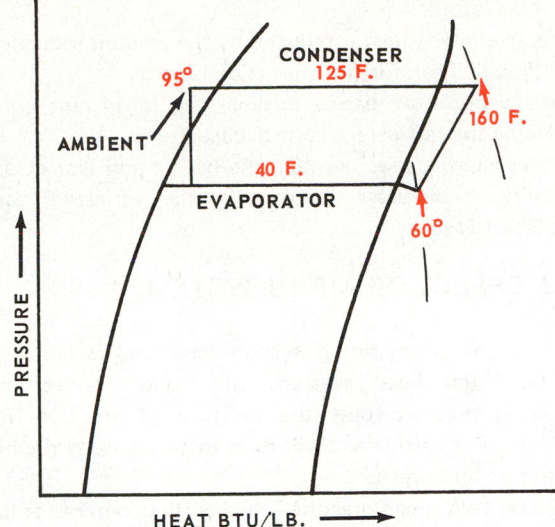

Fig. 15-54. Typical air conditioning, comfort cooling cycle with an evaporator temperature of 40 F. and an ambient temperature of 95 F.

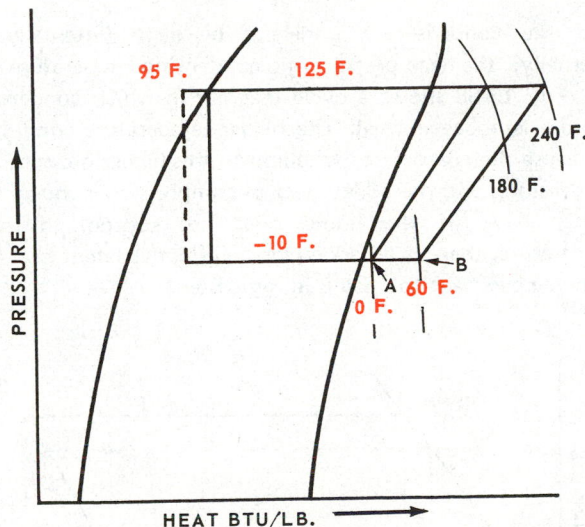

Fig. 15-53. A typical cycle of an air-cooled system for refrigerating frozen foods. Cabinet is kept at 0 F. (Refrig. −10 F.). Air temperature is 95 F. Note effect of the temperature change of suction vapor entering the compressor at Point A (0 F.) and at Point B (60 F.)

In Fig. 15-53, note the difference in compressor performance and refrigerant temperatures between suction line refrigerant temperatures entering the compressor at 0 F. (−18 C.) and 60 F. (16 C.). NOTE: Surface of suction line must be above dew point temperature of air or suction line will sweat, collect moisture and/or collect frost.

An air conditioning (comfort cooling) cycle is shown in Fig. 15-54. A water-cooled unit with 70 F. (21 C.) and, therefore, 80−90 F. (27−32 C.) refrigerant condensing temperature is similar to the standard cycle.

15-40 REFRIGERATION TROUBLESHOOTING

To locate trouble, the service technician must be able to accurately determine what is going on inside a refrigerating

system. The system is sealed. So the technician uses gauges to check the pressure and thermometers to measure evaporator, line and condenser temperatures. He also uses the system sight glass to check the amount of refrigerant and its dryness.

Much of the investigation has to be by logic. The technician needs to know what is supposed to be going on inside the system. He must be able to visualize the behavior of the refrigerant and what each part of the system is supposed to do. The pressure-heat diagram provides considerable aid in this area.

The following paragraphs show the effect of some of the more common troubles on the pressure-heat cycle.

15-41 EFFECT OF LACK OF REFRIGERANT

If the system is undercharged, each pound of refrigerant does not completely liquefy before it passes through the refrigerant control, as shown at part A in Fig. 15-55. The

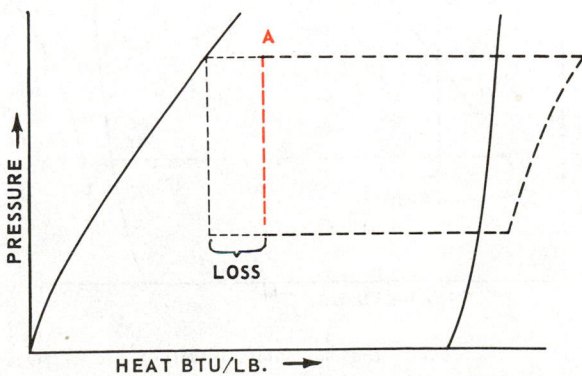

Fig. 15-55. Pressure-heat diagram shows effect of insufficient refrigerant in system. Note loss of effective latent heat. This loss means unit will have to run longer to remove same amount of heat.

result is threefold:

1. Effective latent heat is reduced by the amount indicated by the "loss." Therefore, refrigeration is poor.
2. Some vapor now passes through the refrigerant control, reducing the refrigerant control capacity.
3. As this vapor passes between the needle and seat at a high velocity, it increases the wear on the refrigerant control needle and seat.

15-42 EFFECT OF AIR IN SYSTEM

Air in the refrigerating system increases the total head pressure. Total head pressure will equal the refrigerant condensing pressure plus the pressure of the air in the condenser. The refrigerant will have to condense at the higher temperature and pressure.

Because total head pressure is higher, the compressor has to pump the vapor to a higher temperature and pressure. The extra work performed by the compressor is illustrated in Fig. 15-56. The heat added to do this is a "loss." The cylinder head

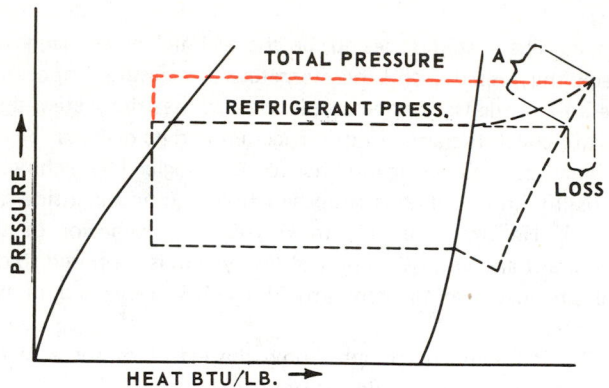

Fig. 15-56. Pressure-heat diagram shows effect of air in system. Point A indicates increase in cylinder head and exhaust valve temperature. The "loss" is heat energy put into the vapor by compressor and is electric motor energy waste.

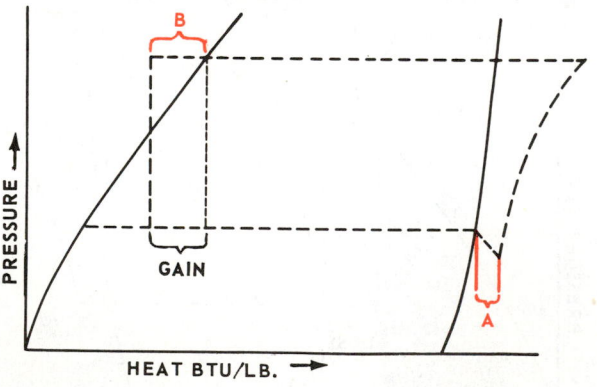

Fig. 15-57. Pressure-heat diagram shows effect of use of heat exchanger. A—Slight amount of decrease in intake pressure and an increase in temperature. B—Amount of effective heat gain. In addition, there is a reduction of flash gas, which improves operation of refrigerant control. Heat exchanger also minimizes chance of liquid refrigerant in suction line reaching compressor.

(especially the exhaust valve) and the top tube of the condenser will be at above normal condensing temperatures, which may also harm the oil.

15-43 EFFECT OF HEAT EXCHANGER

After the suction line vapor leaves the evaporator, it travels down the suction line and into the compressor. During this part of the cycle, the vapor usually warms up somewhat, as shown at part A in Fig. 15-57.

Since the low-pressure vapor picks up heat in most cycles, system efficiency can be improved if excess heat is removed during some part of the cycle. The heat exchange is done by putting the suction line in contact with the liquid refrigerant line just before the liquid goes into the refrigerant control.

Fig. 15-57 illustrates the removal of heat at B. The result is a gain in effective latent heat and a reduction in "flash gas," which will serve to increase the life of the refrigerant control.

15-44 EXCESSIVE CONDENSING PRESSURE

If the condenser is undersize or dirty (internally or externally), the head pressure and condensing temperature will rise. Fig. 15-58 shows a cycle diagram in which condensing pressure is above normal. The higher temperature conditions will cause the compressor to pump to this higher pressure and temperature, and the added heat of compression is shown as a "loss" at A. If the liquid does not subcool to room temperature, there is an added loss in effective latent heat and an increase in flash gas. See B in Fig. 15-58.

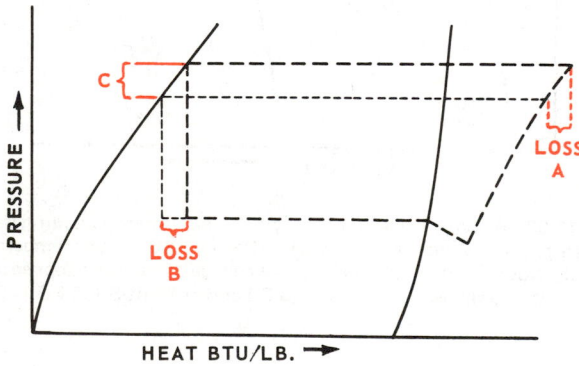

Fig. 15-58. Pressure-heat diagram shows effect of a dirty or undersize condenser, or an above-average room temperature. A—Loss due to unnecessary added heat of compression. B—Loss in effective latent heat of liquid. C—Loss due to work done to compress vapor at higher pressure.

15-45 EXCESSIVE SUCTION LINE PRESSURE DROP

If the pressure of the vapor going into the compressor decreases, the compressor will pump less weight of vapor per stroke and, therefore, per minute. The less vapor pumped, the lower the capacity of the system.

Fig. 15-59 shows the effect of excessive suction line pressure drop on the cycle. Note that as the volume of the

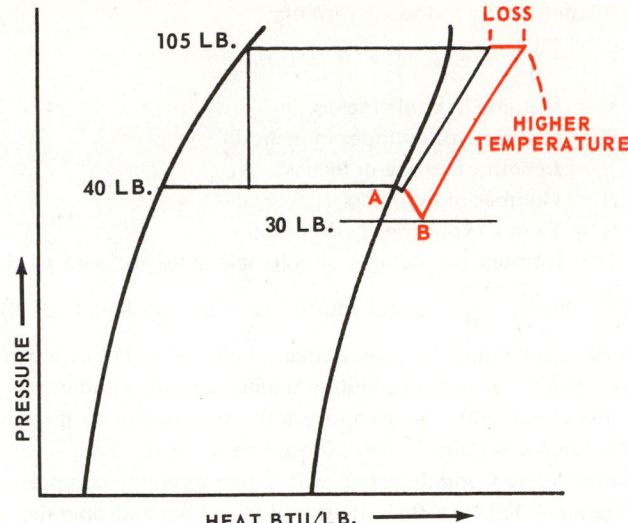

Fig. 15-59. Typical refrigeration cycle is shown, except there is an excessive pressure drop and temperature rise in suction line A to B. This causes an excessive temperature at the exhaust valve and cylinder head.

vapor increases: there is more volume per pound of vapor; the vapor picks up heat, increasing its temperature. Therefore, the exhaust valve temperature increases, and the condenser must remove more heat from each pound of vapor. When the exhaust valve temperature becomes too high, there is danger that the oil will deteriorate.

The effect of a partially clogged filter-drier in the suction line is shown in Fig. 15-60.

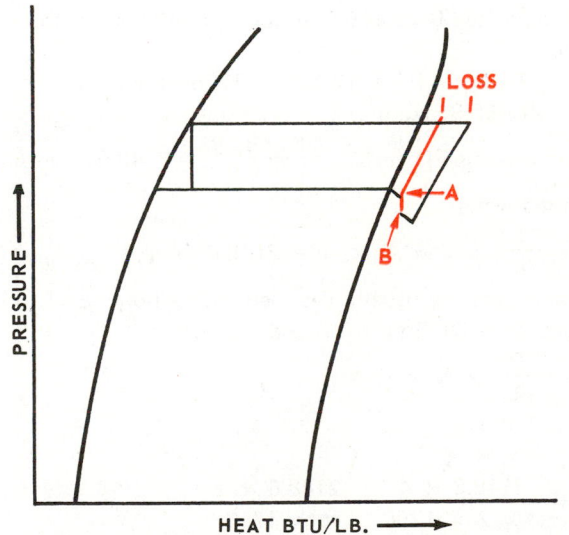

Fig. 15-60. Effect of partially clogged suction filter and/or drier. Drastic pressure drop occurs at A to B.

15-46 SYSTEM CAPACITY

There are several ways to find out the capacity of a system:
1. Measure the refrigerant flow.
   a. Amount — weight.
   b. Flow meters.
2. Measure the heat gain of the condenser cooling water.
3. Measure the heat gain of the air flowing over the condenser.

Next, record the condensing temperature and pressure, and the evaporating temperature and pressure of the refrigerant in the cycle. Use the refrigerant's pressure-heat diagram to find the Btu pickup in the evaporator.

1. If one knows the pounds/hour of refrigerant flow:
   a. Draw the system cycle on the pressure-heat diagram.
   b. Multiply the Btu/pound heat pickup by the pounds circulated to find the Btu removal capacity of the system.

   NOTE: Take the temperature of the refrigerant in the liquid line just before it reaches the refrigerant control. Use a good thermometer in a thermometer well on the line or a thermocouple thermometer fastened to the line.

2. The heat gain in water:

   lb./hr. x temp. change x spec. heat = Btu lb./hr. = gal./hr. x 8.34 lb./gal. (1 gal. water = 8.34 lb.)

   Temp. change = temp. out minus temp. in specific heat = 1

   Example: If a system uses 100 gal. per hour of 70 F. (21 C.) water that warms to 80 F. (27 C.) at the condenser outlet, how much heat is being removed?

   (300 gals. x 8.34) x (80 F. − 70 F.) x 1 = Btu

   2502.00 x 10 x 1 = Btu

   25,020 = Btu

   A one-ton machine would be about 12,000 Btu/hr. Then, 25,020 divided by 12,000 equals approximately two. So this will be approximately a two-ton machine.

3. Measure air temperatures in to and out of the condenser. Measure the average air velocity in ft./hr. in or out (smoothest airflow). Take as many as sixteen readings at different places across the airflow, then average them. (See Chapter 18.)

   Measure the length and width of the condenser in inches.

   Air temp. rise = air temp. out − air temp. in

   Condenser area = in. in width x in. in length =

   $\dfrac{\text{Area in sq. in.}}{144}$ = area sq. ft.

   Volume of air = area in sq. ft. x average velocity (ft./hr.)

   About 14 cu. ft. of air = 1 lb.

   Spec. heat of air = .24 Btu/lb.

   Btu = weight of air per hr. x spec. heat x temp. change

   = $\dfrac{\text{volume of air/hr.}}{14 \text{ cu. ft./lb.}}$ x .24 Btu/F. x (temp. out minus temp. in)

   Example: An air-cooled condenser has an air velocity of 300 ft./min., an air-in temperature of 80 F. and an air-out temperature of 90 F. The condenser measurements are 25 in. x 40 in. Calculate the Btu/hr. released by the condenser.

   Btu = weight of air/hr. x spec. heat x temp. change

   Weight of air:
1. Condenser area = 25 x 40 = 1000 sq. in.

   $\dfrac{1000}{144}$ = 6.94 sq. ft.

2. Volume = 6.94 sq. ft. x (300 ft./in. x 60 min./hr.)

   = 6.94 x 18,000

Volume = 124,920 cu. ft./hr.
Weight = 124,920 ÷ 14 cu. ft./lb.
    = 8922.86 lb. of air/hr.
Btu = 8922.86 x .24 x (90 − 80)
  = 8922.86 x .24 x 10
  = 8922.86 x 2.4
  = 21,414.86 Btu (about a 1 3/4-ton system)

15-47 COMPRESSOR CAPACITIES

The compressor is the heart of the refrigerating machine. It is the means whereby mechanical energy, produced by an electric motor, is used to pump the refrigerant through the cycle which picks up heat at one place and releases it at another place.

The most efficient construction possible is to have a compressor built just large enough to handle the necessary amount of refrigeration. If the compressor is too large, energy is lost in excess friction, starting, etc. If the compressor is too small, it will not produce the amount of refrigeration required.

Basically, the compressor must remove the vapor fast enough from the evaporator to enable the refrigerant to vaporize at the correct low pressure. To do this, it must remove the refrigerant vapor as fast as heat goes into the evaporator and vaporizes the refrigerant.

The method of determining compressor size may be simply stated as follows: An evaporator is usually designed to remove the 24-hour heat load in a 16 or 18-hour running period. The amount of time depends on the factors described in previous paragraphs.

Assume that the effective heat removing ability of the refrigerant is 60 Btu per pound. This means that as each pound of refrigerant vaporizes in the evaporator, it picks up 60 Btu of heat from the evaporator. To remove this much heat, the compressor must handle all the vapor formed. Refrigerant tables give these values called "specific volumes." The specific volume values mean that at certain pressures one pound of refrigerant, as it is vaporizing, will form a certain number of cubic feet of vapor.

Assume, for example, that one pound of R-12 vaporizing at 9.17 psi and 0 F. forms 1.637 cubic feet of vapor in 10 minutes. The compressor, then, must remove the same amount of heat from the evaporator during the period that the evaporator removes heat from the cabinet. That is, 1.637 cubic feet of vapor in 10 minutes.

The size of the compressor needed to do this depends on the volume of vapor pumped per revolution of the compressor. This is dependent upon the bore, stroke, number of cylinders, speed of the compressor (rpm) and its volumetric efficiency.

As the crankshaft of the compressor completes one revolution, the piston reaches lower dead center of its travel, and the low-pressure vapor fills up the space between the top of the piston and the head of the cylinder. As the crankshaft completes its revolution, the piston compresses this vapor on the upstroke and pushes it through the exhaust valve into the high-pressure side of the system. The volume handled in each case is the volume displaced by the piston as it moves from upper dead center to lower dead center. This volume may be calculated by the following formula:

$$V = \frac{\pi D^2}{4} \; x \; S \; x \; N \; x \; R$$

V = Volume in cubic inches
D = Diameter of cylinder in inches
S = Length of stroke in inches
N = Number of cylinders
R = Rpm (revolutions per minute)

This formula for volume simply calculates the area of the piston head $\frac{\pi D^2}{4}$, and multiplies it by the length of the stroke producing the displacement volume in cubic inches. Next, this figure is multiplied by the number of cylinders, then by the revolutions per minute of the compressor to give the total volume in cubic inches pumped per minute.

Example: How much vapor will a two cylinder compressor pump if it has a 2-in. bore, a 2-in. stroke and operates at 400 rpm? Using the following formula:

$$V = \frac{\pi D^2}{4} \; x \; 2 \; x \; 2 \; x \; R$$

$$= \frac{3.1416 \; x \; 4}{4} \; x \; 4 \; x \; 400 = 3.1416 \; x \; 4 \; x \; 400$$

= 3.1416 x 1600 = 5026.56 cu. in./min.
1728 cu. in. = 1 cu. ft.
The volume in cu. ft. will be

$$V = \frac{5026.56}{1728} = 2.9 \; (\text{plus}) \; \text{cu. ft./min.}$$

Compressor problem: Calculate the bore and stroke of a two cylinder compressor operating at 1750 rpm, which will compress in 10 minutes the refrigerant vapor formed by vaporizing 10 pounds of R-22 at 5 F. Note from the table that vaporized R-22 at 5 F. occupies 1.2434 cu. ft./lb.

Solution:

V = 12.434 cu. ft. = 12.434 x 1728 cu. in./cu. ft. = 21,485.95 cu. in.

Volume pumped per min. = $\frac{21,485.95}{10}$ = 2148.6 cu. in./min.

Using the formula:

$$V = \frac{\pi D^2}{4} \; x \; S \; x \; N \; x \; R = 2148.6 \; \text{cu. in.}$$

Compressors are usually designed with a bore equal to the stroke (S = D). So the formula becomes:

$$V = \frac{\pi D^3}{4} \; x \; N \; x \; R =$$

$$D^3 = \frac{V \; x \; 4}{\pi \; x \; N \; x \; R}$$

$$D^3 = \frac{2148.6 \; x \; 4}{\pi \; x \; 2 \; x \; 1750} = \frac{2148.6 \; x \; 2}{\pi \; x \; 1750}$$

$$D^3 = \frac{2148.6}{\pi \; x \; 875}$$

D^3 = .8 cu. in.
D = .93 in. (Approx.)

Therefore, this compressor has a bore of .93 in. and a stroke of .93 in. However, this value is the theoretical amount of vapor pumped by the compressor. The actual amount will be less and will depend on the volumetric efficiency of the compressor. See Para. 15-48.

15-48 VOLUMETRIC EFFICIENCY

Volumetric efficiency is the ratio between the volume actually pumped per revolution, divided by the volume calculated from the bore and stroke.

If the refrigerating unit is maintaining a 150 psi head pressure and a 0 psi low-side pressure, the following things happen:

1. When the piston is on its upward stroke, it compresses this 0 psi vapor until the pressure of the vapor in the cylinder reaches 150 psi. When this pressure is reached, the vapor should start passing through the exhaust valve into the condenser. However, in addition to reaching high-side pressure, it must overcome exhaust valve spring force or weight. This means a slight additional increase in pressure is required.

2. After the piston reaches upper dead center, there is still a little volume of the vapor at a high pressure between the piston head and the exhaust valve. The space it occupies is necessary for clearance between the piston and cylinder head. Otherwise, the piston would pound against the cylinder head when it reaches upper dead center.

The little volume of residue vapor is under a high pressure of 150 psi or more. When the piston goes down to receive a new charge of vapor, this high-pressure vapor expands and partially fills the cylinder chamber. The residue vapor decreases the amount of vapor that may move into the chamber from the low-pressure side of the system. The necessary space, or "clearance volume," varies between 4 and 9 percent of the piston displacement.

3. At speeds of 300 to 3400 rpm or more, the piston is traveling so fast that the inertia or weight of the vapor prevents it from filling the cylinder chamber completely, so there are losses. The pressure in the cylinder may be 2" Hg., instead of 0 psi. The resistance to vapor flow through the valve openings is called "wire drawing." The pressure in the cylinder never gets as high as the pressure in the suction line during the suction stroke. The higher the speed of the compressor, the less vapor will be pumped per stroke.

4. Just as the exhaust valve offers a restriction to the vapor flow, so does the intake valve with its force and the weight of the valve parts.

5. The compressor runs at a warm temperature. Some of this heat warms the vapor as it enters the cylinder, causing an expansion of the vapor that also keeps a complete load of vapor from entering the cylinder.

6. Other losses, such as the leaking of the vapor past the piston and rings into the crankcase, etc., also explain why compressors cannot pump the amount of vapor calculated by the bore and stroke formula.

For small compressors used in domestic refrigeration (bore and stroke of about 1 1/2 in.), the volumetric efficiency varies between 40 and 75 percent, with 60 percent being an average value. The larger commercial compressors, depending on their size and speed, have volumetric efficiencies between 50 and 80 percent, with 70 percent being an average value. A value of 60 to 65 percent volumetric efficiency should be used if the unit is air-cooled.

Example: If the compressor explained in Para. 15-47 has a volumetric efficiency of 60 percent, the size of the compressor would be increased as follows:

D^3 = .8 (As calculated in Para. 15-47.)

D^3 = corrected = $\frac{.8}{.60}$ = 1.33 cu. in.

D = cube root of 1.33
= 1.1 x 1.1 x 1.1
= 1.33 in.

D corrected = 1.1 in.

Therefore, a bore and stroke of 1.1 in. x 1.1 in. would be required. Note that the correction for volumetric efficiency was made on the displacement volume of the cylinder, not on the calculated bore and stroke.

The volumetric efficiency of a compressor depends on the difference between low-side pressure and high-side pressure of the machine. For instance, the compressor in an R-12 system used for domestic purposes will be more efficient than if it were converted over into an ice cream system. This is because the decrease in low-side pressure from 10 psi down to 10" Hg. with the same head pressure reduces the actual pumping capacity of the compressor. The low-pressure vapor expands when the cylinder is filled at low-side pressure. Therefore, only a small amount, by weight, is pumped. All of the various items affecting efficiency — like increasing head pressure, increasing speed, using thicker gaskets and overheating the compressor — will reduce the pumping efficiency of the compressor.

15-49 COEFFICIENT OF PERFORMANCE

Coefficient of performance (cop) is the ratio of output divided by input. In refrigeration work: the output is the amount of heat absorbed by the system; the input is the amount of energy required to produce this output.

The cooling effect in Btu values in a refrigeration cycle compared to the Btu equivalent of the energy put into the system is called the coefficient of performance.

For example, if one pound of refrigerant has an effective latent heat of 50 Btu, and the compressor pumping energy is equivalent to 10 Btu per pound, the coefficient of performance is 50 to 10 or 5 : 1.

The heat input by the compressor is less than the electrical energy put into the motor. This is because the motor is not 100 percent efficient, and there are also compressor friction losses. Usually, the overall coefficient of performance will be approximately 60 percent of the theoretical. The actual coefficient, then, is near 3 : 1 rather than 5 : 1.

This means, however, that three times the amount of heat would be obtained by using the compressor than by using electricity to produce the heat. This explains the advantage of a heat pump. It also explains why hot gas defrost is used in some large systems. The cost of the extra piping and valves is soon paid for in the savings in cost of defrosting.

15-50 MOTOR SIZES

The size of an electric motor necessary to drive a compressor in a refrigerating machine may be calculated by different methods:

1. Mean effective pressure method. The horsepower (hp) of the motor may be calculated by determining the hp put into the compressor. This hp is based on the speed of the compressor and the mean effective pressure (mep) of the vapor in the compressor.
2. Heat input method. Motor size may be determined by using the amount of heat added to the vapor in the compressor as being the energy taken out of the motor.

15-51 MOTOR SIZE — MEAN EFFECTIVE PRESSURE METHOD

The mean effective pressure (mep) of the vapor is the median (average) pressure bearing down on the piston head. It is the pressure to be overcome by the electric motor when driving the compressor.

The mep of the vapor is determined by a formula which uses:

1. Low-side pressure.
2. High-side pressure.
3. Ratio of specific heat of constant pressure to specific heat of constant volume $\frac{C_p}{C_v}$ for the refrigerant used.

The formula for mep and the solution will be found in Chapter 28, Technical Characteristics.

15-52 MOTOR SIZE — HEAT INPUT METHOD

From the pressure heat charts, one can determine the amount of heat (Btu) added to a vapor when compressed by the compressor. Referring to the R-12 pressure heat chart in Chapter 9, note that approximately 10 Btu are added to the one pound of gas if the low-side pressure is 10.81 psi and the high-side pressure is 93.2 psi.

It is known that 2545.7 Btu per hour is equal to 1 hp and is also equal to 746 watts.

Example: Calculate the hp required to drive the above compressor. Add 10.0 Btu per pound. Suppose 10 pounds of refrigerant are compressed in two minutes.

Solution:

The Btu rate per hour = $\frac{100}{2}$ x 60 = 3000 Btu per hour

The hp required (Btu method) = $\frac{3000}{2546}$ = 1.18 or, in round numbers, 1.2 hp

hp = 1.2 hp

This is the mechanical equivalent of the heat energy put into the vapor. If the compressor friction were 0, this would be the size of the motor necessary to drive the compressor. However, about 50 percent must be added to this value to allow for compressor friction and motor losses. Therefore, this system would require about: 1.2 x .50 = .6

1.2 + .6 = 1.8 or 2 hp

15-53 MOTOR EFFICIENCY

Theoretically, an electric motor should produce 1 hp of mechanical energy for every 746 watts of electrical energy put

into it. That is, a 1-hp electric motor on a 120V circuit should only consume 6.8A. This situation, however, is not encountered because of bearing friction, magnetic eddies, magnetic air gaps and the power factor of the motor.

The efficiency of the motor is the mechanical energy delivered at the motor shaft divided by the power input to the motor. For small domestic motors of approximately 1/6 hp, the efficiency of the motor is only 40 to 60 percent. For 1 hp to 2 hp motors, efficiency increases to 75 to 80 percent. As the size of the motor increases, the efficiency increases. This is because friction losses and air gap losses are constant, even though the size of the motor increases. Large motors have an efficiency of 90 to 95 percent.

Example: The motor example used in Para. 15-52.

Solution: With 1.8 hp needed, the electrical input to the motor would need to be more. If 1.8 hp is 75 percent of motor input, the motor input would be:

Total input x .75 = 1.8 hp

Total input = $\frac{1.8}{.75}$ = 2.4 hp

2.4 x 746 watts/hp = 1790 watts input

15-54 CONDENSER CAPACITIES

The calculation of the heat transfer capacity of a condenser is similar in many ways to the problem of figuring the capacity of an evaporator. The condenser must remove the heat from the vapor fast enough so that just as much vapor condenses in the condenser as is being pumped into it by the compressor in a given unit of time. When this condition is reached, the head pressure will have built up until the temperature rises to the point where the heat removed will equal the heat put into the condenser.

The problem of figuring the capacity of the condenser varies according to the type of condenser being used. Condensers may be divided under the following headings:

1. Air-cooled.
   a. Plain tubing.
   b. Finned tubing.
      Natural convection.
      Forced convection.
2. Water-cooled.
   a. Tube and shell type.
   b. Pipe and shell type.
   c. Tube-within-a-tube type.

The method of calculating condenser capacities is explained in Para. 15-55 and Para. 15-56.

15-55   AIR-COOLED CONDENSER CAPACITIES

The capacity of an air-cooled condenser may be calculated by one of two basic methods:

1. Using the total external area of the condenser to compute its heat dissipating ability.
2. Computations based upon the frontal area of the condenser.

Using the total external area of the condenser for dissipating heat depends upon the following variables:

1. External area.
2. Temperature difference.
3. Time.
4. Air velocity.

Using these values, the capacity of an air-cooled condenser varies between 1 and 4 Btu per square foot per deg. F. per hour. The effect of air velocity is to increase the condenser's capacity as the air speed is increased. The fans drive air through the condenser at speeds between 400 to 1000 feet per minute. When an air speed of 400 feet per minute is used, a 2.5 Btu per square foot per deg. F. per hour value will be found satisfactory. This value will increase up to approximately 4 Btu with a 1000 feet per minute air velocity. The calculation of the area of the condenser is the same as that for a finned evaporator. (See Para. 15-21.)

Example: If a condenser has 60 square feet of surface with a heat removal of 2.5 Btu/sq. ft./hr./deg. F., what refrigerant temperature is necessary to dissipate 5000 Btu per hour if the room temperature is 75 F.?

Formula:

Area x Btu/sq. ft./hr./F. x temp. diff. = Btu/hr.

60 x 2.5 x temp. diff. = 5000

150 x temp. diff. = 5000

$$\text{temp. diff.} = \frac{5000}{150}$$

temp. diff. = 33.3 F.

Assuming an ambient temperature of 75 F., the refrigerant temperature = 33.3 + 75 = 108.3 F.

The same problem using a 75 sq. ft. condenser:

75 x 2.5 x temp. diff. = 5000

$$\text{temp. diff.} = \frac{5000}{187.5}$$

temp. diff. = 26.7 F.

Therefore:

The refrigerant temperature = 26.7 + 75 = 101.7 F.

The heat to be removed by the condenser for each pound of vapor is the heat content of the vapor as it leaves the compressor, minus the heat of the liquid at the condensing pressure.

If a condenser is under capacity, the compressor head pressure will rise in proportion in order to dissipate the required amount of heat. Therefore, condensers of various sizes can be used with the same compressor. If too small a condenser is used, it will result in a decrease in compressor efficiency, an increase in motor load and a decrease in the life of the unit. The examples given illustrate this principle.

When the capacity of the condenser is based upon frontal area, it is claimed that the air being blown through the condenser is removing heat only from the surface which it strikes directly. Also, the turbulent flow against the rear surfaces reportedly makes the heat removal from these surfaces negligible.

A single row tube condenser has a total area 20 times its frontal area. The capacity per square foot of frontal area naturally is greater than the value stated. It is between 6 to 10 Btu per square foot per deg. F. per hour, depending on the air speed. For air-cooling, the dry bulb temperature of the room should be used in the calculation.

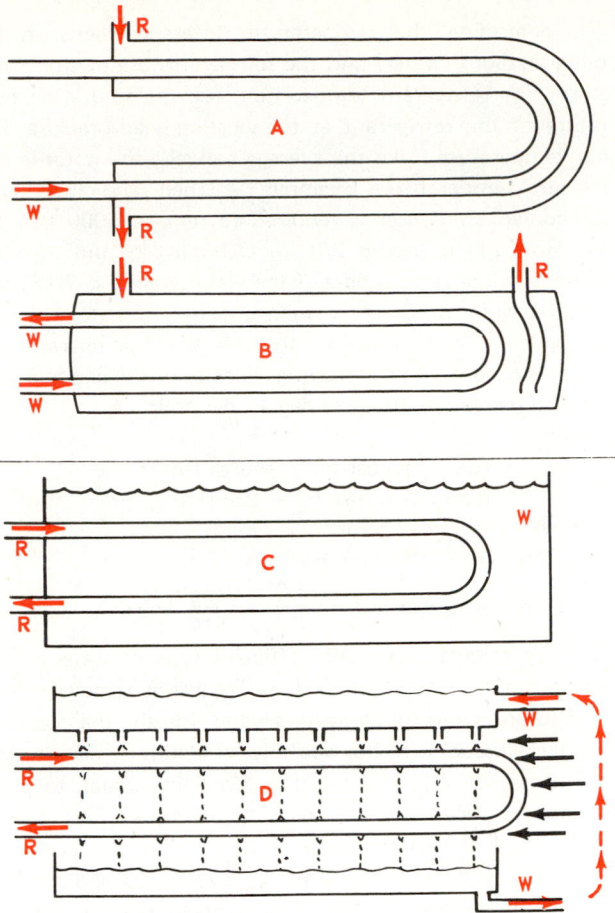

Fig. 15-61. Water-cooled heat exchangers. A—Tube-within-a-tube. B—Shell and tube. C—Tank. D—Baudelot. R—Refrigerant. W—Water.

15-56 WATER-COOLED CONDENSER CAPACITIES

The capacity of a water-cooled condenser is high because of the good thermal contact between the cooling medium and the refrigerant. Different types of water-cooled condensers are in common use. See Fig. 15-61. Capacity will vary with type of water-cooled condenser used.

The heat transfer varies directly with the amount of water passed through the condenser. If the water flow is fast, more heat will be removed; if the flow is slow, heat removal will be less. At 50 feet per minute, water will remove about 185 Btu/sq. ft./hr./deg. F. At 200 feet per minute, the water will remove about 330 Btu/sq. ft./hr./deg. F. The heat removing capacity of these units varies between 30 and 50 Btu per square foot per deg. F. per hour in the smaller machines. For machines of one ton capacity or more, this value may be increased up to 90 Btu per square foot per deg. F. per hour or 330 + 90 = 420 Btu/sq. ft/deg. F./hr.

In addition to heat removal by water, the air-cooling surface of the condenser must also be calculated to reach the correct capacity. Include such things as external area of the shell, or the external area of the refrigerant tubing in a tube-within-a-tube type.

To determine the temperature difference between the cooling medium (water) and the refrigerant, use the following factors: For refrigerant temperature, use the saturation temperature of the refrigerant at the existing head pressure. For water temperature, take the average between the water-in and water-out temperature. Example: A shell and tube type water-cooled condenser is required to remove 5000 Btu per hour. How much tubing 3/8 in. OD must be put into the receiver to remove this heat if the water supply is 70 F. and the outlet water is 80 F.? How many gallons of water per hour must be circulated? Consider the refrigerant temperature at 100 F. Assume the heat removing capacity of the condenser to be 40 Btu per square foot per deg. F. per hour.

Formula:

Tube area = condenser capacity = area (sq. ft.) x temp. diff. deg. F. x Btu rate x time

Area = circumference x length

Area = π D x length

Area in square feet = $\dfrac{\pi D \times length (in.)}{144}$

Cooling towers are a very efficient type of water-cooled condenser. One pound of evaporating water removes about 1050 Btu and cools the remaining water. Ideally, the evaporating water will cool it to the wet bulb temperature. See Chapter 18. But, practically, it cools the remaining water to some temperature above the wet bulb temperature. This can be measured with a thermometer.

About 3 percent of the water is evaporated and must be made up by using a float valve water feed as a control. The float valve also makes up for run-off water about 3 percent. (Used to keep water chemicals to a minimum.) The water pump used to circulate this water should be about 1 percent of the condensing unit horsepower, or about 2 percent if the water pipes are long (high total head).

15-57 LIQUID RECEIVER SIZES

Liquid receivers for a commercial system should be 15 percent larger than all the liquid volume in the system. This practice is recommended for service operations and as a safety measure should the refrigerant circuit become restricted (clogged filter or screen).

Fig. 15-62 shows recommended minimum sizes of receivers based on horsepower capacity of the system. The receivers may have to be larger than this, depending on the refrigerant, piping lengths and other factors.

Liquid receivers are a service addition to a system. The systems would operate efficiently without them, but practical problems of refrigerant reserve and a convenient service storage makes them a common part of most systems.

15-58 REFRIGERANT LINES AND PIPING

The liquid lines, suction lines, compressor discharge lines and "hot gas" lines on refrigerating machines must be of sufficient size to handle the amount of the liquid or vapor required. To calculate the capacities of these lines, first determine the maximum velocity allowed in the line. Then,

RECOMMENDED LIQUID RECEIVER VOLUMES					
HP	VOLUME CU. IN.	WEIGHT — LB. REFRIGERANT			
		R-12	R-22	R-500	R-502
1/2	150	6.8	6.2	5.9	6.3
3/4	225	10.3	9.3	8.9	9.4
1	300	13.7	12.4	11.9	12.9
1 1/2	450	20.5	18.6	17.9	19.3
2	600	27.4	24.8	23.8	25.8
3	750	35	32	31.8	33.0
5	900	41	37	35.5	38.5
7 1/2	1500	70	64	61.6	66.0

Fig. 15-62. Minimum net recommended liquid receiver volume for four common refrigerants: R-12, R-22, R-500 and R-502. (ASHRAE Guide and Data Book)

knowing the amount of liquid or vapor to be handled, the internal cross-section of the line may be calculated.

Generally speaking, for R-12, R-22, R-500 and R-502, liquid velocities are about 100 fpm, suction lines have about 1500 fpm, discharge lines about 3000 fpm and hot gas lines about 3000 fpm.

Fig. 15-63 shows a graph of refrigerant line capacities for R-12. For a 6-ton capacity unit: suction line size at −40 F. (−40 C.) evaporating temperature at 2000 fpm is 3 1/8-in. dia.; liquid line size is 3/4-in. dia.; discharge line size is 7/8-in. dia.

Fig. 15-64 is for R-22, while Fig. 15-65 is for R-500.

15-59 REFRIGERANT (LIQUID) LINE CAPACITIES

Velocities in the liquid line of a refrigerating unit vary with the density of the liquid and with its viscosity. These velocities may vary between 50 to 200 feet per minute (fpm), depending on the refrigerant used. (R-12 should have velocities no greater than 100 fpm.)

Example: If 75 cu. in. of liquid were used a minute, the internal cross-section area of the liquid line to keep line velocity at 100 ft. per minute or below would be:

Cross-section area = $\dfrac{volume\ in\ cu.\ in.}{velocity\ in\ in./min.}$ = $\dfrac{75}{100 \times 12}$

Cross-section area = .063 sq. in.

Area = $\dfrac{\pi D^2}{4}$ $D^2 = \dfrac{area \times 4}{\pi}$ $D = \sqrt{\dfrac{area \times 4}{\pi}}$

The inside diameter = $\sqrt{\dfrac{.063 \times 4}{\pi}}$ = .28 in.

Use 3/8-in. OD tubing (ID = .307).

It is important that refrigerant carrying lines are of sufficient capacity. The cost of increasing the tubing size is so small in comparison to the total cost of the machine that there is no real necessity for calculating tubing size to close limits. At least a 10 to 20 percent oversize liquid line is recommended.

If the liquid line is too small, or has too many restrictions, pressure drop may reduce refrigerant flow at the capacity of the refrigerant control below the capacity of the evaporator.

"FREON" 12 REFRIGERANT
VELOCITY IN LINES (65°F EVAP. OUTLET)

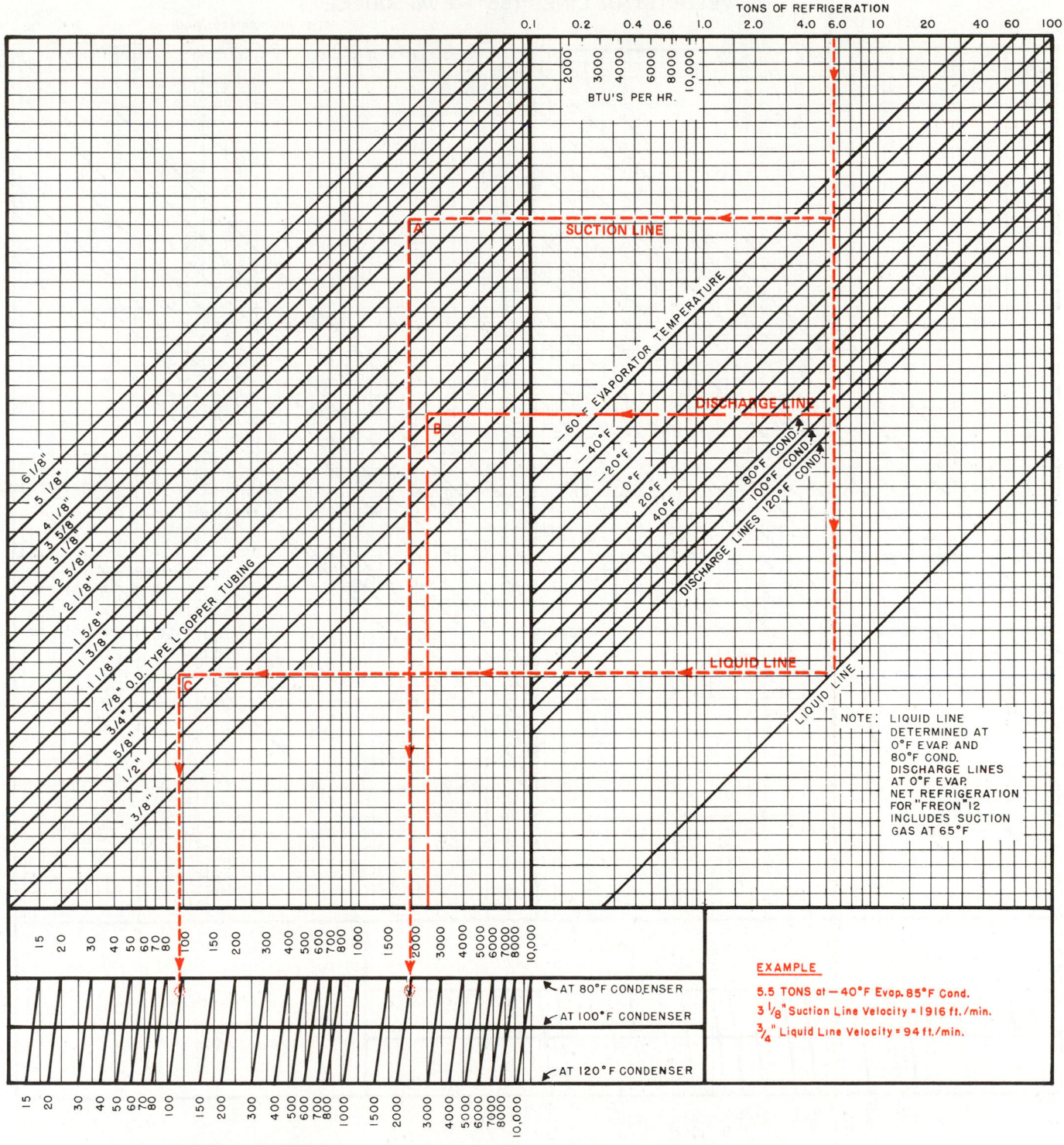

Fig. 15-63. Chart shows liquid and suction line sizes for R-12 systems from 0.1 ton to 100-ton capacity.
(E. I. du Pont de Nemours & Co., Inc., Freon Products Div.)

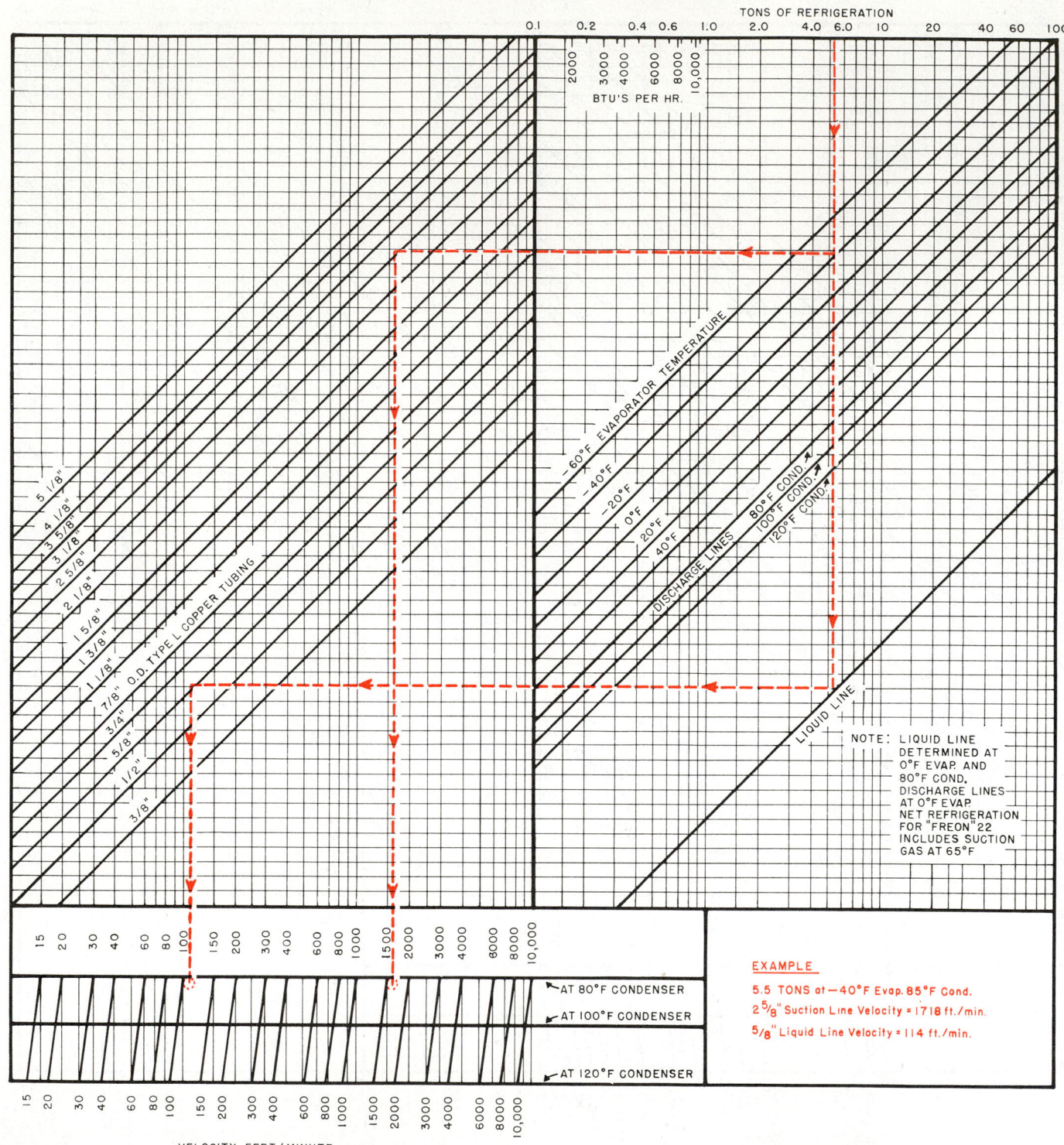

"FREON" 22 REFRIGERANT
VELOCITY IN LINES (65°F EVAP. OUTLET)

TONS OF REFRIGERATION

BTU'S PER HR.

O.D. TYPE L COPPER TUBING

EVAPORATOR TEMPERATURE

DISCHARGE LINES

LIQUID LINE

AT 80°F CONDENSER
AT 100°F CONDENSER
AT 120°F CONDENSER

VELOCITY, FEET/MINUTE

NOTE: LIQUID LINE DETERMINED AT 0°F EVAP. AND 80°F COND. DISCHARGE LINES AT 0°F EVAP NET REFRIGERATION FOR "FREON" 22 INCLUDES SUCTION GAS AT 65°F

EXAMPLE
5.5 TONS at –40°F Evap. 85°F Cond.
2 5/8" Suction Line Velocity = 1718 ft./min.
5/8" Liquid Line Velocity = 114 ft./min.

Fig. 15-64. Chart used to find liquid and suction line sizes for systems using R-22.

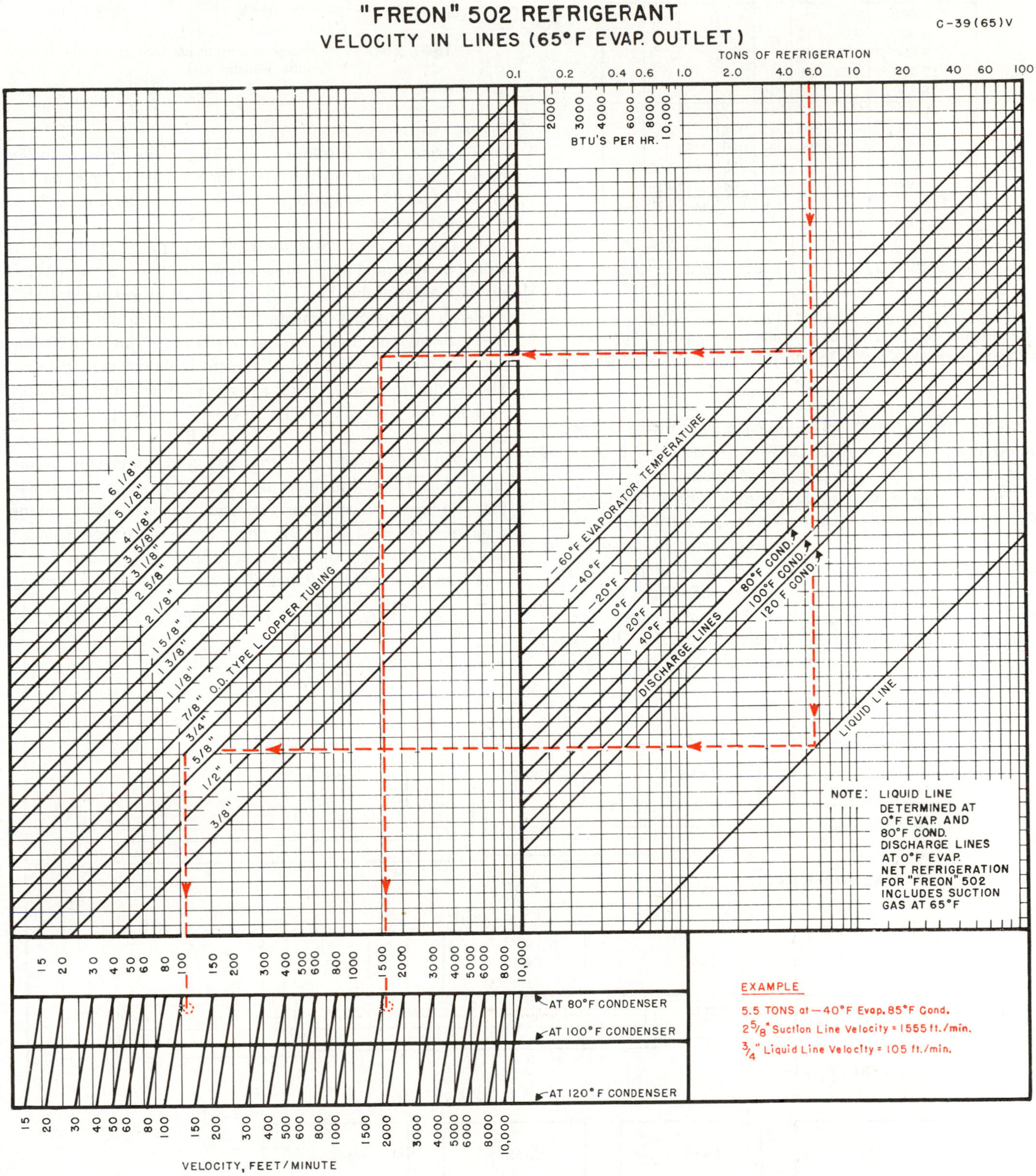

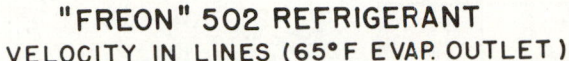

"FREON" 502 REFRIGERANT
VELOCITY IN LINES (65°F EVAP. OUTLET)
C-39(65)V
TONS OF REFRIGERATION
0.1 0.2 0.4 0.6 1.0 2.0 4.0 6.0 10 20 40 60 100
2000 3000 4000 6000 8000 10,000
BTU'S PER HR.
6 1/8" 5 1/8" 4 1/8" 3 5/8" 3 1/8" 2 5/8" 2 1/8" 1 5/8" 1 3/8" 1 1/8" 7/8" 3/4" 5/8" 1/2" 3/8"
O.D. TYPE L COPPER TUBING
-60°F EVAPORATOR TEMPERATURE
-40°F -20°F 0°F 20°F 40°F
DISCHARGE LINES
80°F COND. 100°F COND. 120°F COND.
LIQUID LINE
NOTE: LIQUID LINE DETERMINED AT 0°F EVAP. AND 80°F COND. DISCHARGE LINES AT 0°F EVAP. NET REFRIGERATION FOR "FREON" 502 INCLUDES SUCTION GAS AT 65°F
15 20 30 40 50 60 80 100 150 200 300 400 500 600 800 1000 1500 2000 3000 4000 5000 6000 8000 10,000
AT 80°F CONDENSER
AT 100°F CONDENSER
AT 120°F CONDENSER
EXAMPLE
5.5 TONS at -40°F Evap. 85°F Cond.
2 5/8" Suction Line Velocity = 1555 ft./min.
3/4" Liquid Line Velocity = 105 ft./min.
VELOCITY, FEET/MINUTE

Fig. 15-65. Chart for finding liquid and suction line sizes for systems using R-502.

| Load Btu/hr. | TUBE SIZE OD |||||||
	¼"	⅜"	½"	⅝"	¾"	⅞"	1⅛"
3,000	.035						
6,000	.120	.011					
9,000	.250	.021					
12,000	.420	.036					
18,000		.075	.010				
24,000		.127	.016				
36,000		.260	.033	.012			
48,000		.450	.054	.020	.010		
60,000			.080	.030	.014	.009	
84,000			.150	.054	.025	.015	
120,000			.280	.100	.049	.028	.009
240,000				.350	.160	.095	.029
360,000					.340	.200	.058
480,000						.340	.100

Fig. 15-66. Pressure drop in an R-12 liquid line in psi per foot length of tube. Table is based on size of tube and load in Btu/hr. Total pressure drop will be indicated pressure drop per foot multiplied by length in feet.

¼"	⅜"	½"	⅝"	¾"	⅞"	1⅛"
.015	.043	.086	.134	.202	.269	.458

Fig. 15-67. Refrigerant charge in pounds per foot of liquid line. (Dunham-Bush, Inc.)

for each fitting or valve.

If normal head pressure were 125 psi, the pressure in the liquid line near the thermostatic expansion valve would be 125 minus 42 or 83 psi. At 83 psi, the boiling temperature is 79 F. and sweating might occur in a humid 90 F. room.

In large systems, bear in mind how much refrigerant is stored in the liquid line. This amount also affects pressure based on the weight of the liquid (static head), as shown in Fig. 15-67.

Bends and fittings increase the resistance to the fluid flow. It is claimed that there is as much friction to flow in a 90 deg. elbow as there is in five feet of straight tubing of the same size. The friction in bends, fittings and normal friction of fluid flow through the tubing must be calculated when figuring fluid velocities.

In multiple installations that use series-connected evaporators, Fig. 15-68, the liquid line usually varies in diameter as the number of evaporators it feeds changes. Note in the diagram that two 1/2 in. OD liquid lines do not feed from a 1 in. OD line. Instead, the cross-sectional areas are added. Line 1 area + line 2 area = line 3 area. The wall thickness is neglected.

$$\frac{\pi D_1^2}{4} + \frac{\pi D_2^2}{4} = \frac{\pi D_3^2}{4}$$

$$D_1^2 + D_2^2 = D_3^2$$

Extremes of this condition are revealed by sweating or frosting liquid lines when excessive pressure drops occur because of partially clogged driers, strainers or pinched lines.

The pressure drop in a liquid line carrying R-12 is shown in Fig. 15-66. Note that if one tried to use a 1/4 in. OD liquid line for a 12,000 Btu/hr. or one-ton load, the pressure drop would be .42 psi per ft. A 100-ft. equivalent length liquid line would then have a total pressure drop of 42 psi. Equivalent length is the actual length of the piping, plus the pressure drop in the bends and fittings, as expressed in feet. Looking ahead, Fig. 15-73 lists values that should be added to the tube length

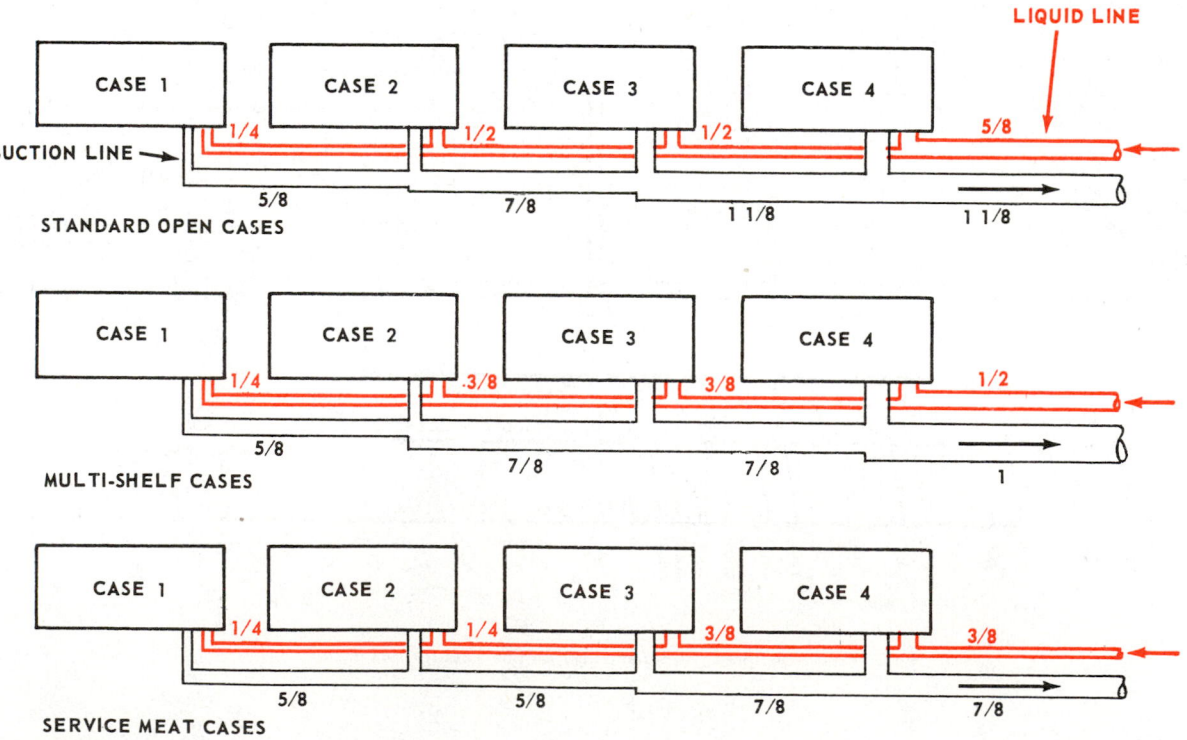

Fig. 15-68. Liquid line sizes. The size of liquid line must increase as the number of evaporators increases. Line sizes shown are for normal temperature refrigeration installations. Note change in diameter of liquid lines.

$$\sqrt{D_1{}^2 + D_2{}^2} = D_3$$

$$\sqrt{\frac{1^2}{2} + \frac{1^2}{2}} = D_3$$

$$\sqrt{\frac{1}{4} + \frac{1}{4}} = D_3$$

$$\sqrt{\frac{2}{4}} = D_3$$

$$\sqrt{\frac{1}{2}} = D_3$$

$$\sqrt{.5} = D_3$$

$$.70 = D_3$$

The liquid line presents no problem except size, unless it has a considerable static head (vertical run). In case of a large static head, pressure at the refrigerant control end of the liquid line must be high enough to maintain pressures above the flash point of the refrigerant at the temperature at the refrigerant control. Due to the weight of the liquid refrigerant, the pressure in a vertical liquid line will drop. The amount of the pressure drop per foot of vertical rise is shown in Fig. 15-69.

PRESSURE DROP IN VERTICAL LINES	
REFRIGERANT	PRESSURE DROP PER FOOT OF VERTICAL COLUMN
R-12	.56
R-22	.50
R-500	.50
R-502	.50

Fig. 15-69. Pressure drop in vertical liquid refrigerant line. This table is for refrigerant at 100 F. (37.8 C.). At lower temperatures, pressure drop will be slightly more.

Example: If a 50-ft. rise in a 3/4 in. R-12 liquid line is needed, the pressure drop will be .56 x 50 = 28 lb. If pressure at the condensing unit is 100 lb., pressure at the 50-ft. level will be 100 minus 28 = 72 lb. If the liquid line temperature goes over 87 F., some flash gas may form in the liquid line. (See R-12 tables in Chapter 9.) It is best to arrange the liquid line piping in multiple evaporator installations to make any flash gas formed go to the nearest evaporator. See

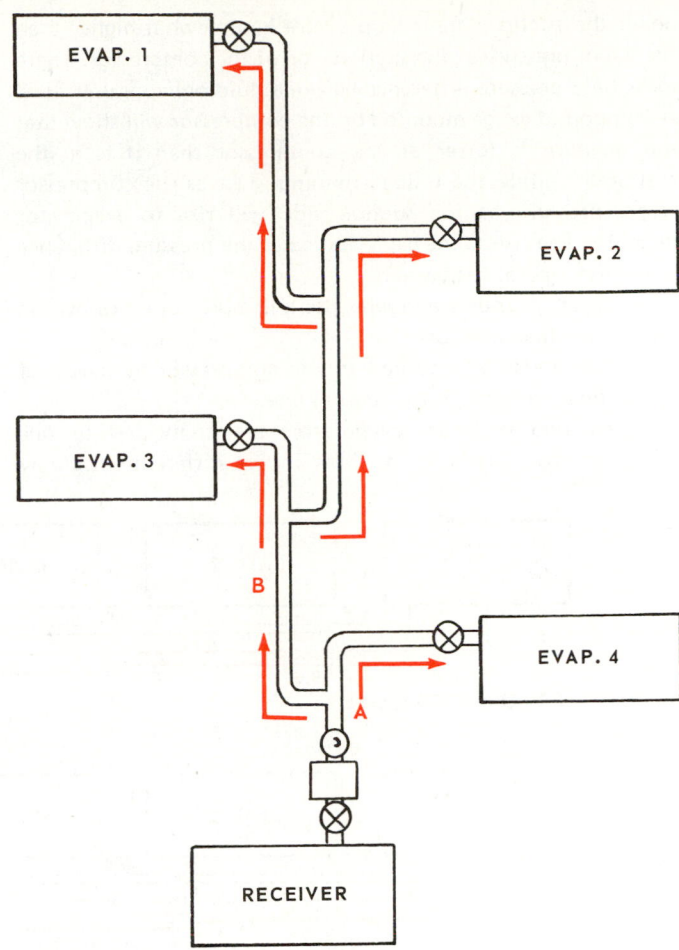

Fig. 15-70. Multiple evaporator liquid line installation. This design distributes flash gas to the nearest evaporator rather than to the most remote evaporator. Example: Flash gas at A will go to Evaporator No. 4. Flash gas at B will go to Evaporator No. 3, etc.

Fig. 15-70. This design prevents one evaporator from receiving all the flash gas.

15-60 REFRIGERANT (SUCTION) LINE CAPACITIES

Suction line velocities usually are between 1500 and 2000 feet per minute (fpm), as shown in Fig. 15-71. As the refrigerant evaporates in the evaporator, the vapor will go

	ALLOWABLE VELOCITIES IN FEET PER MINUTE		
REFRIGERANT	LIQUID LINE VEL. (LIQUID)	SUCTION LINE (VAPOR)	CONDENSER VEL. (VAPOR)
Ammonia	100–250	4000–5000	5000–6000
R-12	80–100	1500–1800	1800–2250
R-22	100–125	1500–2000	1800–2200
R-500	100–125	1500–2000	1800–2200
R-502	100–125	1500–2000	1800–2200
Water	100–250	30–50 (Liquid)	. . . .

Fig. 15-71. Table of allowable refrigerant velocities in feet per minute. Note that there is very little difference in allowable velocities of the different refrigerants.

down the suction line if its pressure is somewhat higher than the vapor pressure at the suction side of the compressor. There must be a pressure difference before a fluid or vapor will flow. A compound gauge mounted on the compressor will show that the pressure is lower at the compressor than it is in the evaporator while the unit is running. Just as the compressor stops, pressure at the suction side will rise to evaporator pressure. The rise at the gauge indicates the pressure difference in the low side of the system.

In order to understand why there is a pressure drop within the suction line, consider that:

1. The oil must be returned to the compressor by means of the flowing vapor in the suction line.
2. When two or more evaporators are connected to one compressor, resistance to flow in the different lines may cause the pressure on the surface of the refrigerant in the different evaporators to vary as much as 2 to 3 psi.

Fig. 15-72 shows suction and liquid line systems for multiple refrigerating systems with four cases. The friction in a 1/2 in. OD suction line is approximately .25 in. mercury ("Hg.) per 10 ft. of length with a gas velocity of 1000 ft. per minute. The varying pressure drops pose the problem of balancing evaporator capacities.

Total pressure drop becomes greater:

1. As the length of the suction line increases.
2. As the cross-sectional area of the tubing decreases.
3. As the number of bends and fittings in the suction line increases.

Fig. 15-73 shows the values to be added to the actual length of the pipe in order to obtain the equivalent length.

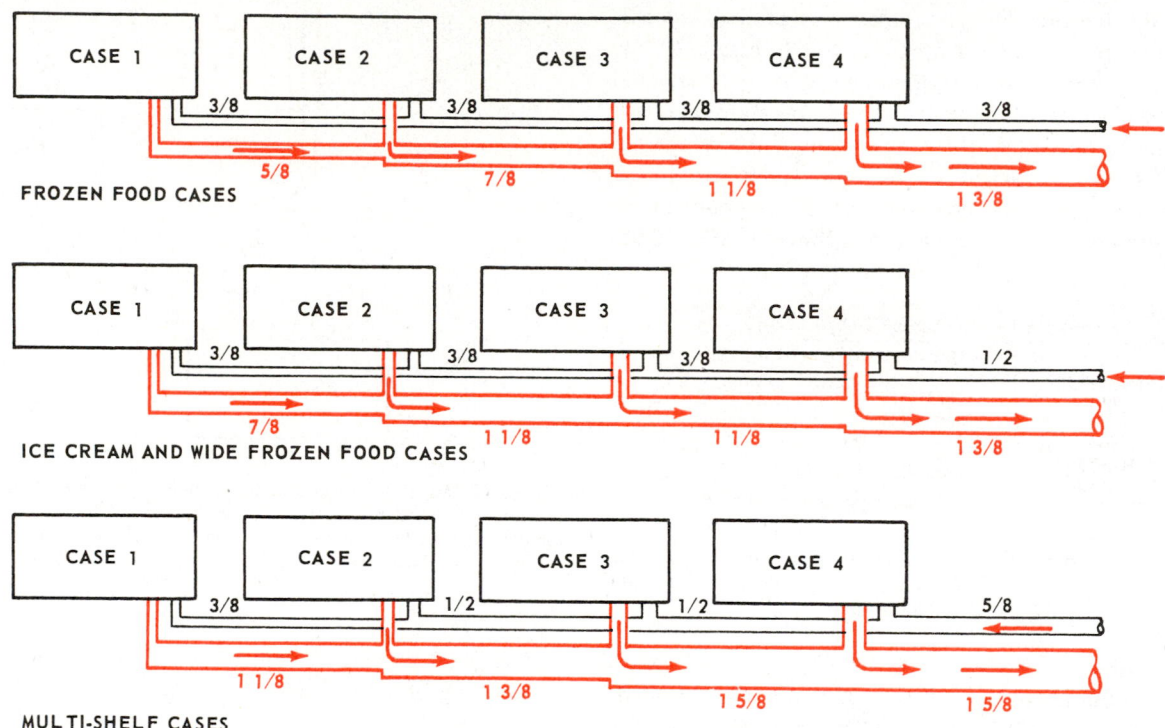

Fig. 15-72. Three typical supermarket piping diagrams for low-temperature fixtures. Note how size of suction lines increases as vapor accumulates from each case.

FITTING	TUBE AND PIPE SIZE, OD													
	¼	⅜	½	⅝	¾	⅞	1⅛	1⅜	1⅝	2⅛	2⅝	3⅛	3⅝	4⅛
	FEET TO BE ADDED FOR EACH FITTING													
Valve	1.5	1.5	2	2	2.5	3.0	4.0	5.0	6.0	7.5	9.0	11.0	13.0	15.0
Elbow (90°)	.75	.75	1	1	1.5	1.5	2	2.5	3	4.0	5.0	5.5	6.5	7.5
Tee	1.5	1.5	2	2	2.5	3.0	4.0	5.0	6.0	7.5	9.0	11.0	13.0	15.0

Fig. 15-73. Table for amount of resistance of valves, elbows and tees over straight length pipe. Values for each fitting shown should be added to length of pipe in order to obtain equivalent length. Note that valve has as much resistance as 2 ft. of 1/2-in. pipe. (Dunham-Bush, Inc.)

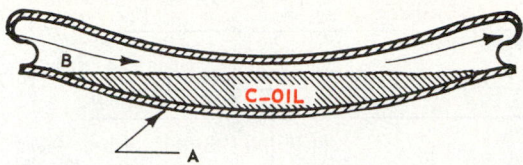

Fig. 15-74. Illustrating how vapor flow is restricted due to oil collecting in low spot in suction line. A—Suction line. B—Refrigerant vapor. C—Oil.

The correct size of suction lines is important in refrigerating units. An excessive pressure drop in a suction line is similar to operating the compressor at a lower pressure, thereby reducing its capacity.

Example: A pressure drop in an R-12 system of 2 psi at —15 F. (2.45 psi) is the same as trying to operate at —20 F. (1 psi), and the condensing unit capacity is lowered approximately 8 percent.

(2.45 + 15) — (1 + 15) =

17.45 — 16 = 1.45

1.45 ÷ 17.45 = .08 or 8 percent

The oil return should be taken care of by slanting the suction line consistently downward from the evaporator to the compressor to permit the oil to drain naturally into the compressor. If a low spot is constructed into the suction line, the oil will accumulate there. Fig. 15-74 shows how this oil buildup will decrease the cross-sectional area of the tubing, causing an orifice action that decreases the efficiency of the gas flow. Furthermore, when this low spot eventually becomes filled with oil, the pressure difference builds up and oil is slugged into the compressor. Slugging of the oil into the crankcase of the compressor accelerates oil pumping momentarily and may damage the compressor.

Fig. 15-75 shows the capacity of various size suction lines using R-12 refrigerant. Note that a 1-in. nominal or 1 1/8 in. OD suction line can carry from 1/2 to 3 tons of capacity, depending on the pressure drop. However, the best choice would probably be between .02 and .03 psi pressure drop per foot, and this pipe should be used for 2 to 2 1/2-ton units. If the suction temperatures are lower or higher than 20 F., the pressure drops must be corrected. See Fig. 15-76. Denser vapor should have lower velocities, and vice versa.

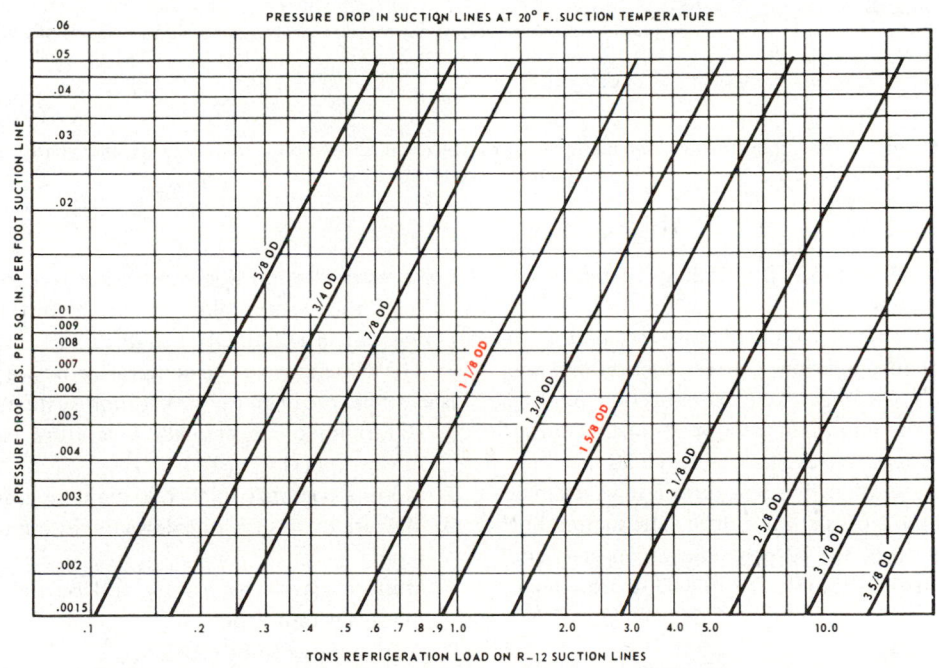

Fig. 15-75. Graph of suction line capacities for R-12 refrigerant. Choose suction line size by using a pressure drop of .02 to .03 psi per ft. Chart is based on 20 F. If lower temperatures are used, larger suction lines are needed, and vice versa.

CORRECTION FACTORS FOR OTHER SUCTION TEMPERATURES

Suction Temp	—40	—30	—20	—10	0	10	20	30	40	50
Correction Factor	2.70	2.28	1.90	1.62	1.38	1.18	1.00	0.88	0.75	0.64

Fig. 15-76. Table of correction values for pressure drops in a suction line. If suction temperatures exceed 20 F., pressure drop is decreased. If temperatures are 0 F., pressure drop increases by 1.38. Equivalent length is to be multiplied by correction factor.

RECOMMENDED REFRIGERANT LINE SIZES

COMPRESSOR CAPACITY BTU/HR.	LENGTH OF RUN				
	15 FT. TUBE DIA.	25 FT. TUBE DIA.	35 FT. TUBE DIA.	50 FT.* TUBE DIA.	100 FT.* TUBE DIA.
SUCTION LINE					
18,500—20,000	5/8	5/8	5/8		
20,000—22,000	5/8	5/8	5/8		
22,000—24,000	5/8	5/8	5/8	3/4	3/4
24,000—34,000	5/8	5/8	3/4	3/4	3/4
38,000—40,000	3/4	3/4	3/4	7/8	7/8
40,000—44,000	3/4	7/8	7/8	7/8	7/8
44,000—51,000	7/8	7/8	7/8	7/8	7/8
53,000—66,000	7/8	7/8	7/8	1 1/8	1 1/8
LIQUID LINE					
18,500—20,000	5/16	5/16	5/16		
20,000—22,000	5/16	5/16	5/16		
22,000—24,000	5/16	3/8	3/8	3/8	3/8
24,000—34,000	5/16	3/8	3/8	3/8	3/8
38,000—40,000	5/16	3/8	3/8	3/8	3/8
40,000—44,000	3/8	3/8	3/8	3/8	3/8
44,000—51,000	3/8	3/8	3/8	3/8	3/8
53,000—66,000	1/2	1/2	1/2	1/2	1/2
DISCHARGE LINE					
18,500—20,000	5/16	3/8	3/8		
20,000—22,000	3/8	3/8	3/8		
22,000—24,000	3/8	3/8	3/8	1/2	1/2
24,000—34,000	3/8	3/8	1/2	1/2	1/2
38,000—40,000	3/8	1/2	1/2	1/2	1/2
40,000—44,000	3/8	1/2	1/2	1/2	1/2
44,000—51,000	3/8	1/2	1/2	1/2	5/8
53,000—66,000	1/2	1/2	5/8	5/8	3/4

(1) These recommendations are based on the use of standard refrigeration tubing with .028 or .032 wall thickness.
(2) Line sizes listed are outside tube dimensions.
(3) These suggestions do not include consideration for additional pressure drop due to elbows, valves or reduced joint sizes.
* = Add 3 fluid ounces for each 10 ft. of pipe over 35 ft.

Fig. 15-77. Suction line, discharge line and liquid line sizes. Selection is based on system capacity and length of pipe. (Tecumseh Products Co.)

Fig. 15-77 shows a method of finding suction line, discharge line and liquid line sizes.

One usually knows the capacity of the installation in Btu/hour or in tons of refrigeration. Correct suction line size can be estimated by first getting an approximate size from Fig. 15-75. Next, calculate the equivalent length of pipe from Fig. 15-73. Then correct for temperature using Fig. 15-76.

Example: To determine suction line size for a 5-ton system at 0 F., allow for a total 2 psi pressure drop. The suction line has 30 ft. of straight run, six 90 deg. elbows, one tee, and one valve, at 0 F. Assume that 1 1/8 in. OD suction line is to be used.

Suction line	30 ft.
6 elbows x 2	12 ft.
1 tee x 4	4 ft.
1 valve x 4	4 ft.
The equivalent length	50 ft.

If a total of 2 psi pressure drop is desired, then 2 ÷ 50 = .04 psi per foot.

However, this line operates at 0 F. instead of 20 F. Therefore, .04 ÷ 1.38 = .029 psi per foot of length.

Now, referring back to Fig. 15-75, a 5-ton load with a .029 psi per foot pressure drop needs a 1 5/8 in. OD pipe for the suction line.

It is important to always use piping as large as the fittings that are on the evaporator and on the compressor. If a compressor suction line connection is 1 in. OD, it is advisable to use this size piping. If the liquid receiver liquid line connection is 1/2 in. OD, use this size.

The oil return to the compressor by way of the suction line is a critical problem in refrigeration systems.

1. Returning oil is actually a mixture (solution) of oil and refrigerant.
2. Mixture becomes thicker as the temperature is lowered.
3. Mixture travels mainly along the inside wall of the tubing or piping.
4. Mixture travels by gravity and by the action (velocity) of the refrigerant vapor.

Also, observe these cautions:

1. Keep the mixture as fluid as possible.
2. Slant the horizontal suction line downward, toward the compressor.
3. It is important to leave enough refrigerant vapor velocity to push the mixture along the pipe.

Many experiments have shown that the velocity in a horizontal line should be at least 500 to 750 fpm. If the refrigerant vapor must flow up a suction line, the velocity in the vertical tubing or pipe must be at least 1000 to 1500 fpm. (The vapor velocity must overcome both gravity and viscosity.)

Be sure to install small U traps at the base of the vertical up-flow suction lines. (Small traps prevent a large slug of oil returning to the compressor during start-up of system.)

The viscosity of refrigerant oil determines how easily it flows. See Chapter 28.

Oil with dissolved refrigerant has a lower viscosity (flows easier). As the oil travels in the suction line, it becomes warmer (suction superheat), but it also loses some of its dissolved refrigerant. Tests show that the viscosity of the oil actually increases as it travels through the suction line toward the compressor. This means that the longer the suction line, the more careful one must be to provide proper suction line velocities and oil traps.

In low-temperature systems, the refrigerant dissolved in the oil is the one main factor that keeps the viscosity low enough to allow the return of oil.

Example: 150 USU oil at −20 F. has 100,000 USU viscosity. However, with R-12 dissolved in it, 150 USU oil has a viscosity of about 50 USU and it flows with relative ease. As temperatures in the suction line rise, viscosity of the oil first increases (as refrigerant content decreases), then it finally starts to decrease (as the oil becomes warmer).

Refrigerant vapor velocity will vary in the suction line as the heat load changes. At maximum heat load, the amount of vapor produced will be maximum, vapor velocities will be high and the oil return will be good. But as the vapor volume decreases and the compressor unloads (one or more compressors stop if it is a compound system or one or more cylinders of a modulated compressor stop pumping), then vapor velocity will slow down and oil return will be more difficult.

One solution to suction line problems under varying heat loads is to use a double suction line, one with an oil trap and one without. See Fig. 15-78. When the system is at full capacity, both suction lines A and B will carry the refrigerant vapor at about 1500 ft./min. As load decreases and compressor pumping is reduced, the vapor velocity will slow and the oil trap C will fill with oil, closing line B. Now the vapor velocity in suction line A will be high enough to carry the oil back to the compressor. When the system returns to full capacity, the oil at C will be moved back to the compressor by way of line B. D in Fig. 15-78 is the evaporator.

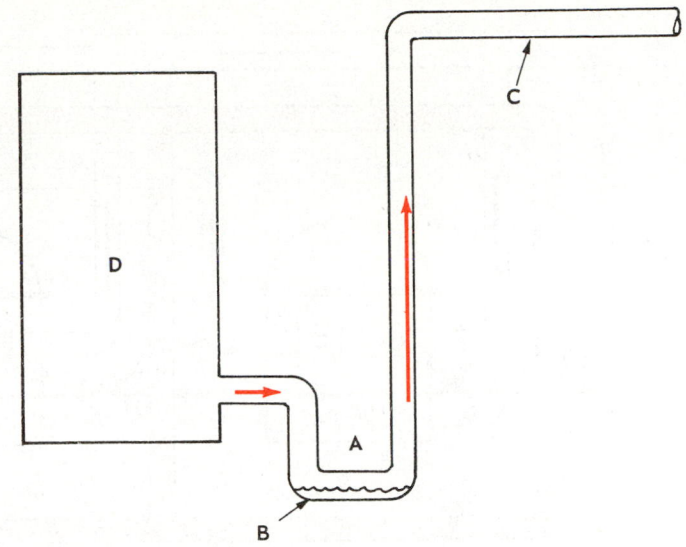

Fig. 15-79. An oil trap installed in a vertical rise suction line. A—Oil trap. B—Oil. C—Horizontal pipe slanted down toward compressor (about 1/4 in. per each 10 ft. of pipe). D—Evaporator.

Remember that all horizontal suction lines must slant down to the compressor (about 1/4 in. per 10-ft. length of pipe). If the suction line rises, a trap must be installed at the bottom of the rise. See Fig. 15-79.

Discharge lines from a compressor to a remote condenser which have an 8 ft. or more vertical rise must also have an oil trap. This is to keep oil from returning by gravity flow to fill the space above the exhaust valves of the compressor with oil during the off cycle. This oil may damage the valves when the

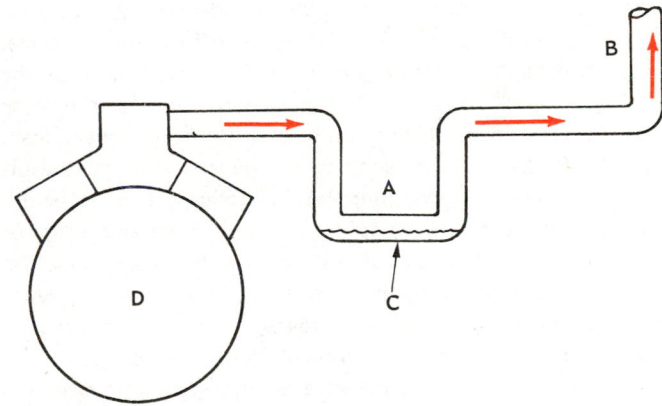

Fig. 15-80. An oil trap installed in the discharge line of a compressor which has a remote condenser keeps oil from draining back to exhaust valves of the compressor. A—Oil trap. B—Discharge line. C—Oil collection during off cycle. D—Compressor.

compressor first starts up. See Fig. 15-80. A horizontal discharge line must slant down toward the condenser (about 1/4 in. per 10-ft. length of pipe).

Those systems using compound compressors have an oil balance to maintain. Otherwise, one compressor may collect

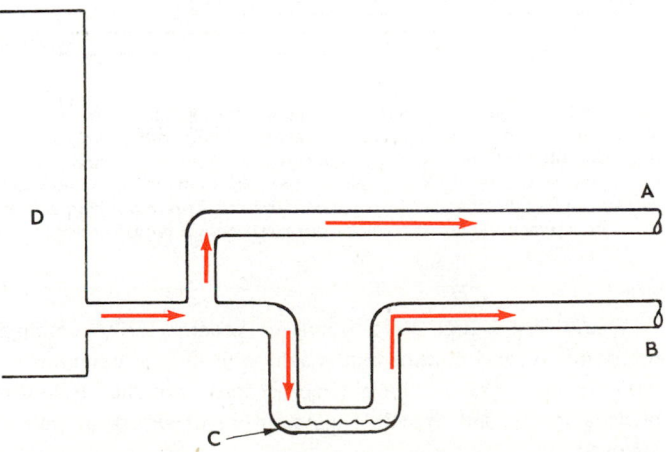

Fig. 15-78. Double suction line. A—Suction line direct to compressor. B—Suction line with oil trap. C—Oil trap. D—Evaporator.

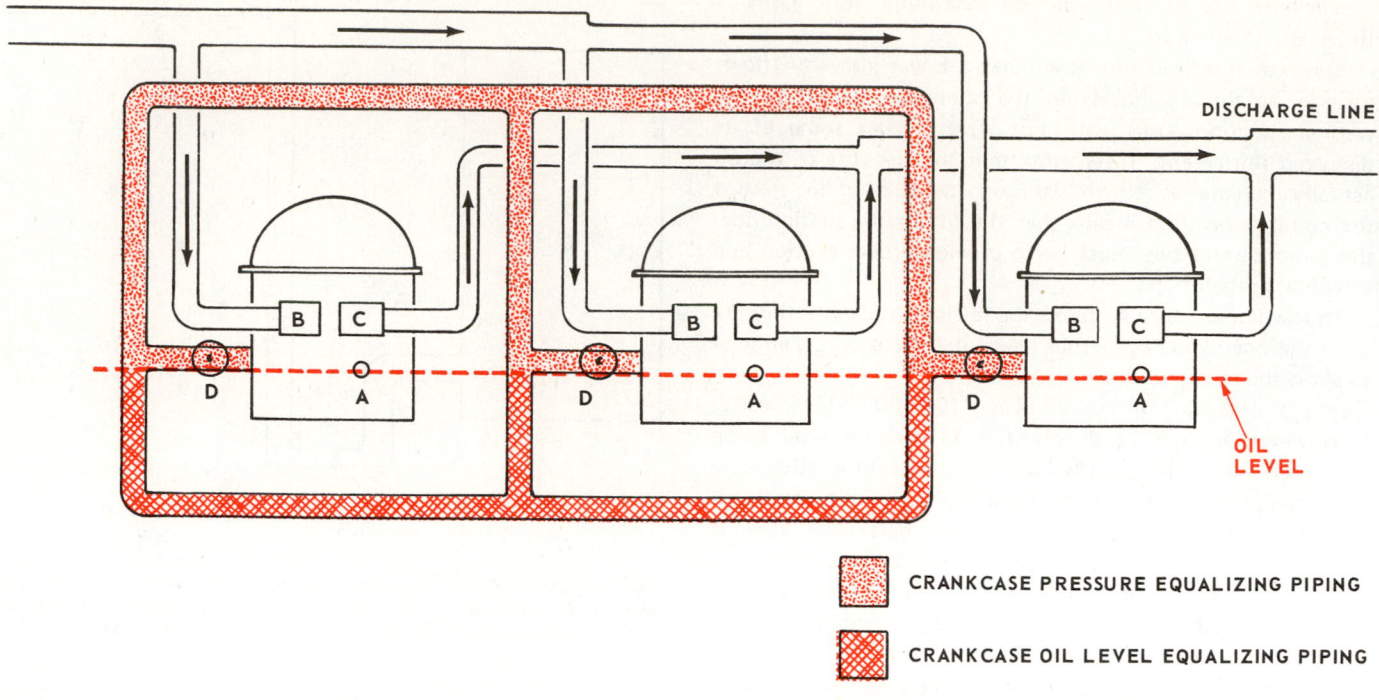

SUCTION LINE

DISCHARGE LINE

OIL LEVEL

▒ CRANKCASE PRESSURE EQUALIZING PIPING

▦ CRANKCASE OIL LEVEL EQUALIZING PIPING

Fig. 15-81. The piping system used to keep an equal amount of oil in each motor compressor crankcase. A—Oil level sight glass. B—Suction service valve. C—Discharge service valve. D—Shutoff valve (closed only when removing motor compressor).

too much oil and pump oil. Another compressor may be low on oil and be damaged. Fig. 15-81 shows oil piping which provides equal distribution of oil.

15-61 DISCHARGE LINE PIPING

When compressor discharge vapor is piped to a remote condenser, it is possible that the condenser may become warmer than the compressor during the off-part of the cycle. When this happens, refrigerant vapor may move back from the condenser and condense in the head of the reciprocating compressor. Then, if the compressor discharge valves leak, liquid refrigerant will collect in the cylinders. This could result in the compressor pumping liquid refrigerant on start-up, which would reduce lubrication of the pistons and valves or even break them. If the compressor valves do not leak, the collection of liquid refrigerant in the cylinder head may cause damage when the compressor starts up because of dynamic hydraulic pressure on the compressor head and piping.

A check valve installed in the discharge line near the condenser will eliminate this potential danger.

15-62 REFRIGERANT CONTROL CAPACITY

Two popular refrigerant controls are the thermal expansion valve (TEV) and the capillary tube. The TEV feeds refrigerant to the evaporator. The size of its orifice (opening) is controlled by a needle and must be carefully calculated. See Fig. 15-82. TEV orifice size depends on the shape of the opening, viscosity of liquid passing through it, and upon the pressure difference as the fluid passes through the orifice.

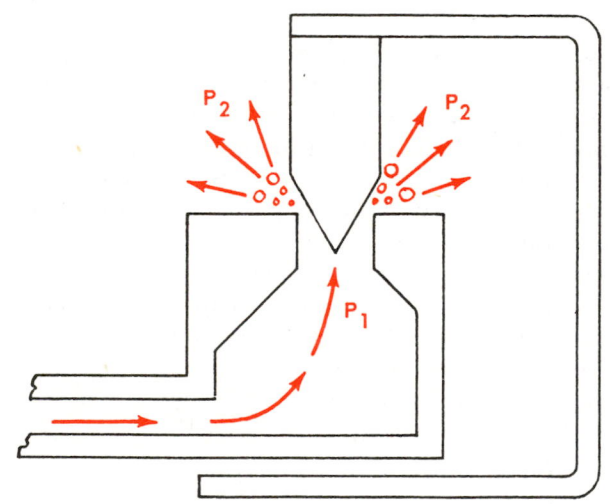

Fig. 15-82. Action of refrigerant as it passes through orifice of an automatic or thermostatic expansion valve. Liquid refrigerant at P_1 (high-side pressure) is forced through orifice and almost at once reaches P_2 (low-side pressure). As pressure changes, some liquid (about 30 percent) instantly changes into vapor (flashes). This vaporizing action cools remaining liquid to evaporator refrigerant temperatures.

These TEV orifice sizes, however, become fairly standard for domestic and commercial machinery. Orifice openings of 0.93 in. and .156 in. have become most popular. If larger orifices are needed, multiple installations of expansion valves are used.

If the orifice is undersize (too small), not enough refrigerant can pass through the valve and the evaporator will be

RECOMMENDED CAPILLARY TUBE LENGTH AND DIAMETER

COMPRESSOR CAPACITY BTU/HR.	CONDENSER TYPE	NORMAL EVAPORATING TEMPERATURE			
		−10 TO +5	+5 TO +20	+20 TO +35	+35 TO +50
R-12 LOW TEMPERATURE					
200—300	STATIC (FAN)	16 FT. — .026 IN.	10 FT. — .026 IN.		
300—400	STATIC (FAN)	12 FT. — .026 IN.	12 FT. — .031 IN.		
400—700	STATIC	12 FT. — .031 IN.	12 FT. — .036 IN.		
	FAN	10 FT. — .031 IN.	10 FT. — .036 IN.		
700—1,100	STATIC	12 FT. — .036 IN.			
	FAN	10 FT. — .036 IN.			
1,100—1,300	STATIC	10 FT. — .036 IN.			
	FAN	8 FT. — .036 IN.			
1,300—1,700	STATIC	12 FT. — .042 IN.			
	FAN	10 FT. — .042 IN.			
1,700—2,000	STATIC	12 FT. — .049 IN.			
	FAN	10 FT. — .042 IN.			
2,000—3,000	FAN	10 FT. — .054 IN.	15 FT. — .059 IN.		
3,000—4,000	FAN	10 FT. — .059 IN.	12 FT. — .064 IN.		
4,000—4,500	FAN	12 FT. — .064 IN.	12 FT. — .070 IN.		
4,500—5,000	FAN	10 FT. — .070 IN.	12 FT. — .080 IN.		
5,000—7,000	FAN	10 FT. — .059 IN. (2 PCS.)	12 FT. — .064 IN. (2 PCS.)		
7,000—9,000	FAN	10 FT. — .064 IN. (2 PCS.)	10 FT. — .070 IN. (2 PCS.)		
9,000—12,000	FAN	10 FT. — .070 IN. (2 PCS.)	12 FT. — .080 IN. (2 PCS.)		
12,000—15,000	FAN	10 FT. — .070 IN. (3 PCS.)	12 FT. — .080 IN. (3 PCS.)		
R-22 LOW TEMPERATURE					
1,000—2,000	FAN	10 FT. — .036 IN.	12 FT. — .042 IN.		
2,000—3,000	FAN	12 FT. — .042 IN.	15 FT. — .049 IN.		
3,000—4,000	FAN	10 FT. — .054 IN.	15 FT. — .059 IN.		
4,000—5,000	FAN	10 FT. — .064 IN.	15 FT. — .070 IN.		
R-12 MEDIUM AND HIGH TEMPERATURE					
1,400—1,600	FAN		12 FT. — .036 IN.	8 FT. — .036 IN.	8 FT. — .042 IN.
1,600—1,800	FAN		10 FT. — .036 IN.	12 FT. — .042 IN.	
1,800—2,500	FAN		12 FT. — .042 IN.	12 FT. — .049 IN.	8 FT. — .049 IN.
2,500—3,500	FAN		10 FT. — .042 IN.	10 FT. — .049 IN.	
3,500—4,000	FAN		12 FT. — .049 IN.	10 FT. — .054 IN.	
4,000—5,000	FAN		10 FT. — .054 IN.	10 FT. — .059 IN.	
5,000—6,000	FAN		12 FT. — .059 IN.	12 FT. — .064 IN.	
6,000—7,000	FAN		10 FT. — .059 IN.	10 FT. — .064 IN.	12 FT. — .070 IN.
7,000—10,000	FAN		12 FT. — .070 IN.	12 FT. — .080 IN.	
			12 FT. — .054 IN. (2 PCS.)	10 FT. — .059 IN. (2 PCS.)	
10,000—13,000	FAN		12 FT. — .059 IN. (2 PCS.)	10 FT. — .064 IN. (2 PCS.)	
13,000—16,000	FAN		12 FT. — .070 IN. (2 PCS.)	10 FT. — .080 IN. (2 PCS.)	
16,000—25,000	FAN		12 FT. — .080 IN. (2 PCS.)	10 FT. — .085 IN. (2 PCS.)	
25,000—40,000	FAN		10 FT. — .070 IN. (4 PCS.)	12 FT. — .080 IN. (4 PCS.)	
40,000—60,000	FAN		10 FT. — .070 IN. (5 PCS.)	12 FT. — .080 IN. (5 PCS.)	

Fig. 15-83. Capillary tube sizing. Length and diameter of the capillary tube is based on the kind of refrigerant, system capacity, type of condenser and evaporator temperature. These capacities are based on the assumption that not less than 3 feet of the capillary tube is attached to the suction line (heat exchanger). There is no subcooling of the liquid below ambient temperature. (Tecumseh Products Co.)

starved. Also, the evaporator pressure will drop too fast.

If the orifice size is too large, the valve will feed too much refrigerant too fast, causing a "sweat back" or "frost back" down the suction line. The result will be a "searching" or "hunting" action, causing alternate flooding and then starving of the evaporator.

The pressure difference is important. As the difference increases, TEV capacity increases. Therefore, if head pressure is high, the valve may feed refrigerant too fast and cause a sweat back or frost back. If the pressure is too low, the valve will feed too little refrigerant and the evaporator will starve.

Low pressure can be caused by:
1. Low head pressure.
2. Liquid line is too long or has too many bends or fittings.
3. Liquid line is too small.

4. TEV is too high above the liquid receiver.

Capillary tube capacity is determined by the pressure difference, length of the tube and inside diameter of the tube. Fig. 15-83 lists suggested capillary tube sizes for two different temperature applications. It covers low-temperature and medium-to-high temperature evaporators, based on capacity of the system in Btu/hr.

15-63 ENERGY EFFICIENCY RATIO

By law, refrigeration and air conditioning units sold for household use must be rated for efficiency by the U. S. Department of Commerce. This rating, called an "Energy Efficiency Ratio" (EER), must be indicated on the machine. See Fig. 15-84.

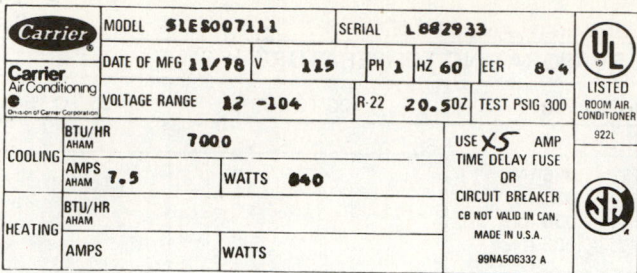

Fig. 15-84. Label indicating cooling capacity and efficiency must be carried by all air conditioning and refrigeration units.

The energy efficiency ratio is the rated cooling capacity in Btu per hour divided by the electrical power input in watts. For the label shown:

EER = 8000 Btu/hr ÷ 860 watts = 9.3.

Window air conditioners have EER's varying from 5.4 to 9.9. The higher the EER the more efficient the machine. An EER of 9.3 indicates a very efficient machine, but ratings of 9.9 are available in some air conditioners of this size.

The Coefficient of Performance (COP) of a machine can be found by multplying the EER by a factor of 0.293:

COP = EER x 0.293.

The COP for the machine indicated is:

COP = 9.3 x 0.293 = 2.72.

Example: Find the COP and EER for a refrigerator which has a cooling capacity of 10 000 Btu/hr and requires 800 watts of electrical energy:

Solution:

EER = 10 000 ÷ 800 = 12.5

COP = EER x 0.293 = (12.5) x (0.293) = 3.66

Example: What is the EER of a 2-ton air conditioner which required 1.5 kW of electrical power?

Solution: A ton of refrigeration is 12 000 Btu/hr. The cooling rate is then 2 x 12 000 = 24 000 Btu/hr. The EER is then:

EER = 24 000 ÷ 1 500 = 16.

15-64 REVIEW OF SAFETY

Because excessive temperatures and excessive pressures are dangerous, refrigeration service technicians must be aware of the danger of guessing at piping sizes, condenser sizes, evaporator sizes and motor sizes to be used on a refrigeration system. Use manufacturer's specification sheets and recommendations in all cases.

Carefully compute the sizes of each item according to the methods described in this chapter. Improper sizing of the unit or any part of the unit may create a damaging or dangerous condition.

Always carefully review any calculations having to do with pressures, velocities and capacities. An error might cause too high a pressure or otherwise cause a system to fail.

15-65 TEST YOUR KNOWLEDGE

1. What is a Btu?
2. What is the Btu equivalent of a hp?
3. What is sensible heat?
4. What is latent heat?
5. What is specific heat?
6. Why is it important when calculating the heat load of a cabinet to know what the cabinet is used for?
7. Why is the heat leakage into a cabinet based upon the external area?
8. What is the purpose of a baffle?
9. What is usage heat load?
10. What materials are used for the interior walls of newer cabinets?
11. What is the purpose of the ice holdover in the sweet water bath?
12. Why must a pressure type water cooler never be allowed to accumulate too much ice?
13. What is the specific heat of ice cream?
14. At what temperature should brick ice cream be maintained?
15. What is meant by the superheating of the vapor in the suction line?
16. What is the "effective latent heat" of a refrigerant? Why is it different from latent heat?
17. What superheats vapor passing through the compressor?
18. Why may one use the Btu added to the vapor as it goes through the compressor to calculate the size of the motor to drive the unit?
19. What is meant by volumetric efficiency, and what variables influence it?
20. What happens to the compressor capacity as the low-side pressure increases?
21. What is meant by motor efficiency, and what are the usual efficiencies?
22. Explain how heat travels from one carrying medium to the next, starting with the refrigerated cabinet and continuing to the suction line vapor, describing the ease of transfer from one medium to the other.
23. What is the product heat load?
24. How many heat leakage surfaces does a cabinet have?
25. How does evaporator capacity vary as refrigerant temperature decreases?
26. If evaporator capacity is "rated" at a temperature difference of 10 F. (5.5 C.) what two temperatures are involved in establishing this rating?
27. What is a nonfrost evaporator?
28. What is the coefficient of performance?
29. Why are two suction lines used in some single evaporator systems?
30. What is the best way to determine liquid line and suction line sizes?

Chapter 16

ABSORPTION SYSTEMS PRINCIPLES AND APPLICATIONS

The absorption system is different from the compression system. It uses heat energy instead of mechanical to make a change in the conditions necessary to complete a refrigeration cycle. The absorption system may use natural gas, LP gas, kerosene, steam or an electric heating element as a source of heat.

The system has few moving parts. Smaller units have moving parts only in the heat source valves and controls which are used. Some larger units also use circulating pumps and fans.

Absorption systems operate quietly. They are used in commercial installations and in domestic installations as well. In recent years, absorption units have been widely used in recreation vehicles, campers, trailers — and boats. The fact that an absorption unit can be operated from a small gas cylinder has increased its use in portable refrigerators.

16-1 THE ABSORPTION SYSTEM

The condenser, receiver and evaporator (cooling coil) are quite similar to those used in the compression system. The compressor has been replaced by a heater and generator. The systems shown here have been simplified by leaving out various controls. These will be covered later.

Fig. 16-1 illustrates a basic absorption system of the liquid absorbent type. This unit uses a water-cooled condenser.

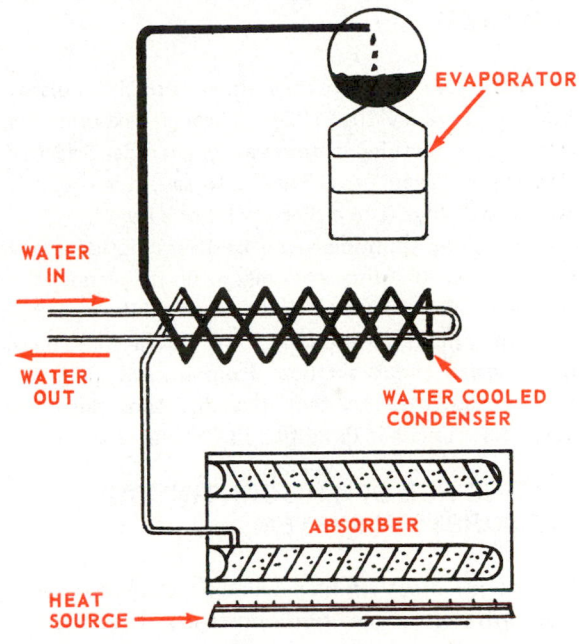

Fig. 16-2. Elementary solid absorbent refrigeration unit. Note water-cooled condenser.

Fig. 16-2 illustrates the fundamentals of a basic absorption system. This diagram shows the solid absorbent type.

16-2 TYPES OF ABSORPTION SYSTEMS

Absorption systems are based on several combinations of substances which have an unusual property. One substance will absorb the other without any chemical action taking place. It will absorb the other substance when cool and release it when heated. *If the substance is a solid, the process is sometimes called adsorbing; if the substance is a liquid, the process is called absorbing.*

There are two types of absorption refrigerators. One uses a solid absorbent material; the other uses a liquid absorbent.

Absorption systems are further classified as:

1. Intermittent systems.
2. Continuous systems.

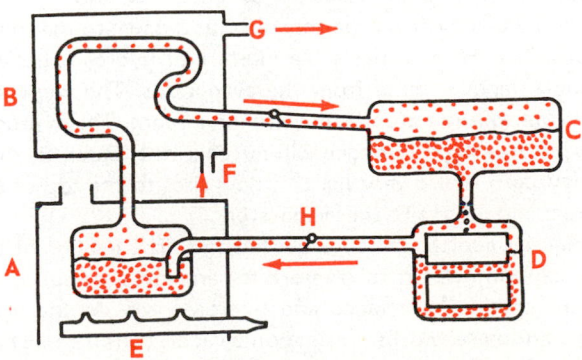

Fig. 16-1. Simple liquid absorbent refrigerating unit. A—Generator. B—Condenser. C—Receiver. D—Evaporator. E—Burner. F—Water in. G—Water out. H—Check valve.

Absorption systems have had several applications:
1. Domestic.
2. Recreation vehicle.
3. Industrial.
4. Air conditioning.
   Absorption systems are also identified by heat source:
1. Kerosene.
2. Natural gas.
3. Steam.
4. Electrical heat.
5. Solar energy.

Some absorption units used in family trailers and mobile homes may be heated either electrically or by LP fuel. These systems are explained in this chapter.

16-3 PRINCIPLE OF THE SOLID ADSORPTION SYSTEM

Solid adsorption systems operate on principles discovered by Michael Faraday in 1824. Through experiments, he succeeded in liquefying ammonia, which scientists had believed to be a "fixed" gas. That is to say, it was considered impossible to change it to either a solid or a liquid.

He exposed the ammonia vapor to silver chloride, a powder. When the silver chloride had taken all the vapor it could adsorb, he applied heat and got a liquid. But when the heat was removed, he discovered that the liquid soon began to "boil," vaporize and draw heat from its surroundings. The present-day adsorption system uses this same phenomenon. Faraday's experiment is described in Chapter 3 and Fig. 3-21.

16-4 PRINCIPLE OF THE CONTINUOUS ABSORPTION SYSTEM

The absorption system uses ammonia, water and hydrogen. When it provides refrigeration constantly it is called a continuous absorption system. A continuous refrigerating cycle operates automatically through the use of automatic controls.

Although many companies have variations of the basic system, the principle of operation remains the same. When the burner is lighted and its heat applied to the generator at 1 in Fig. 16-3, ammonia vapor is released from the solution. This hot vapor passes upward through the percolator tube at 2. As the hot ammonia vapor rises through this tube, it carries the solution to the upper level of the separator at 3.

Most of the liquid solution settles in the bottom of the separator and flows into the absorber. The hot ammonia vapor, being light, rises to the top of tube marked 4 into the condenser. The hot ammonia vapor then condenses into a liquid. The ammonia is now in a pure state and it flows by gravity into the evaporator.

Because a liquid will always seek its own level, the liquid ammonia flows through the liquid ammonia tube and spills into the evaporator. There it forms in large shallow pools on a series of horizontal baffle plates.

The hydrogen gas that is being fed to the evaporator in large quantities permits the liquid ammonia to evaporate at a low pressure and at a low temperature (Dalton's principle).

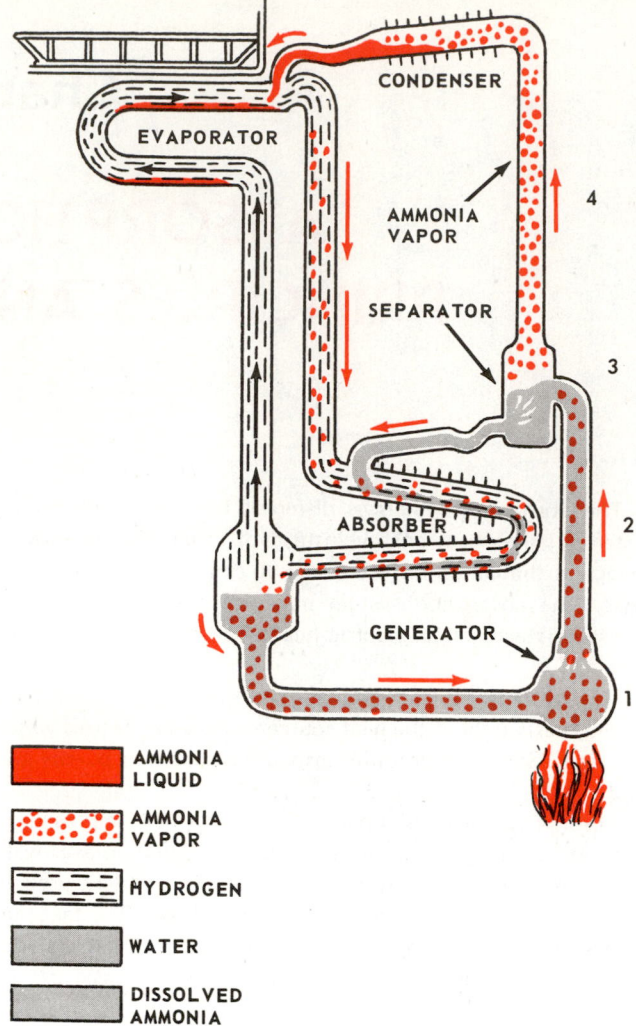

	AMMONIA LIQUID
	AMMONIA VAPOR
	HYDROGEN
	WATER
	DISSOLVED AMMONIA

Fig. 16-3. Air-cooled continuous refrigeration cycle. Note water circuit, ammonia flow and hydrogen circuit.

During this process of evaporation, the ammonia absorbs heat from the food compartment of the refrigerator and causes the water in the ice cube containers to freeze. The more hydrogen and less ammonia, the lower the temperature. The evaporated ammonia mixes with the hydrogen gas.

Meanwhile, a weak solution of ammonia and water is flowing by gravity from the separator at 3 down to the top of the absorber. Here it meets the mixture of hydrogen gas and ammonia vapor coming from the evaporator. The weak and fairly cool solution absorbs the ammonia vapor. The hydrogen gas is left free since hydrogen will not mix with water. Because the hydrogen is also very light, it now rises to the top of the absorber and returns to the evaporator.

Being air cooled, the absorber has fins. The cooling of the weak solution helps it to reabsorb the ammonia gas out of the mixture of ammonia vapor and hydrogen gas. As the weak water solution reabsorbs the ammonia vapor, considerable heat is liberated. The air-cooled fins remove this heat to permit refrigeration to continue.

The ammonia liquid and water mixture flows back to the

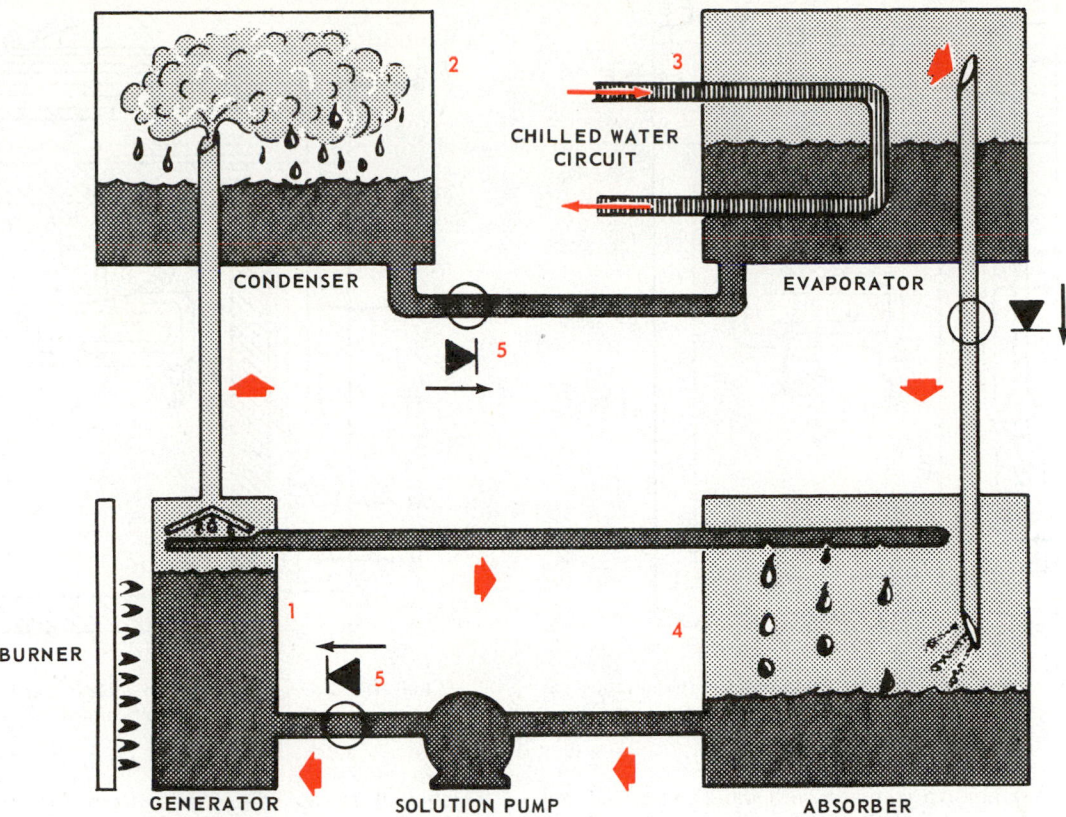

Fig. 16-4. Continuous absorption system uses pump to maintain pressure difference between low-pressure side and high-pressure side of system. Same pump transfers strong-in-ammonia, weak-in-water solution. 1—Generator. 2—Condenser. 3—Evaporator. 4—Absorber. 5—Check valves. (Arkla Air Conditioning Co.)

generator where it again starts its cycle.

The apparatus is a welded assembly. There are no moving parts to wear out or get out of adjustment. Since the total pressure throughout the cycle is about 400 psi at a room (ambient) temperature of 100 F. (38 C.), construction must be rugged to insure a long life.

To produce a temperature of 0 F. (−18 C.) in the evaporator, the ammonia must boil at 15.7 psi, which means that the hydrogen must make up the remainder of the pressure (384.3 psi), if the total pressure is 400 psi. This refrigerator is considered to be unique among domestic refrigerators.

16-5 CONTINUOUS ABSORPTION SYSTEM WITH PUMP

The continuous absorption refrigerating system with a pump, Fig. 16-4, uses ammonia as the refrigerant and an aqueous ammonia solution as the absorbent. Any means of heating can be used, but natural gas, steam or LP gas are the most popular.

The system operates under two pressures. The high-side pressure is from 200 to 300 psi and the low-side, 40 to 60 psi. The high and low sides are separated by check valves, liquid traps, a pump or other controlling devices.

The operational system can be divided into four sections including generator, condenser, evaporator and absorber.

The generator at 1 in Fig. 16-4 is heated by a vertical burner. Heat causes the liquid to boil and the ammonia in the liquid turns into vapor. The vapor will rise up through the tube to the air-cooled condenser. In the condenser, heat from the vapor is removed by the cooler air passing across it. The vapor will condense into a liquid which then acts as the refrigerant.

The liquid refrigerant now passes, at a high pressure, to the evaporator at 3. In the evaporator, water carrying heat from the cooled area passes through tubes. The heat from the water tubes is transferred to the refrigerant liquid. The water in the tubes returns to the area that needs to be cooled. Being at a low temperature, the water can absorb heat from the area that is to be cooled.

The heat that the refrigerant has absorbed from the chilled water circuit causes it to boil and turn into a vapor. This vapor refrigerant is drawn back to the solution-cooled absorber at 4, where the heat is transferred to the outside air.

The liquid refrigerant is then pumped back — preheated — by the solution pump to the generator, where the procedure is repeated.

16-6 PRINCIPLE OF INTERMITTENT ABSORPTION SYSTEM

A convenient refrigerator cycle for localities having neither gas nor electricity is the Superfex and Trukold cycle. The Superfex cycle is about the same as the Faraday principle but it has some different features.

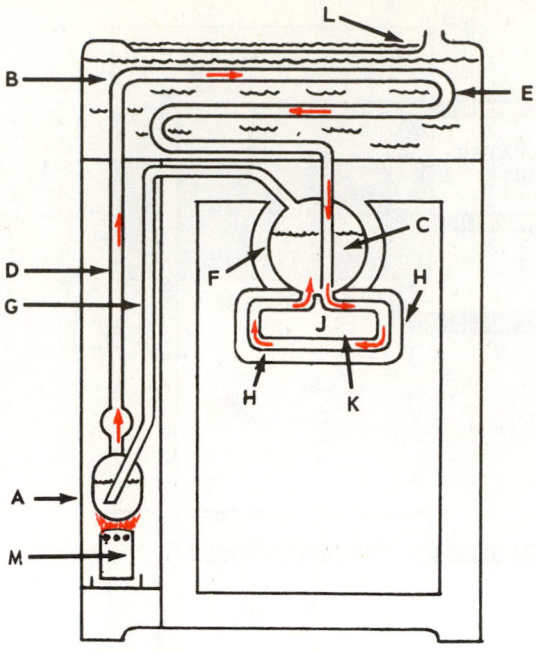

Fig. 16-5. Generation or heating interval in typical intermittent type absorption refrigerator.

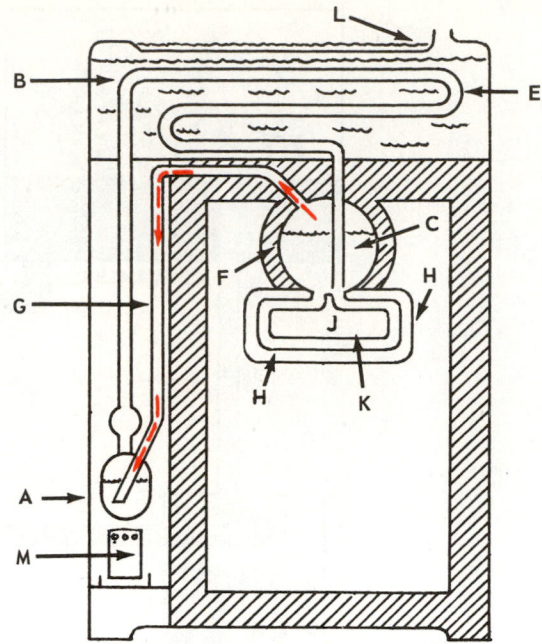

Fig. 16-6. Refrigerating part of cycle in intermittent type absorption refrigerator.

In Fig. 16-5, ammonia is mixed with water in a sealed tank or generator shown at A. Underneath, a kerosene burner, M, heats it. Heat from the burner drives the ammonia, in vapor form, out of the mixture. This vapor is forced up the pipe marked D and through a condenser, E, which is immersed in water from a tank at B on top of the refrigerator.

Cooling effect of the water causes the ammonia vapor to change back into a liquid (condenses) at the high generating pressure. This liquid ammonia drops through a pipe into the liquid receiver, C, and from here it passes to the evaporator, K, which is surrounded by a brine at H. The liquid receiver is insulated at F to prevent this container from overcooling the food compartment by acting as the evaporator.

The process continues for a short time until the kerosene is used up and the burner goes out.

As the absorber cools to room temperature, the ammonia will evaporate at a low temperature in the evaporator because, as the generator cools, it tends to reabsorb the ammonia vapor. This reduces the pressure and permits the liquid ammonia in the evaporator to boil at a low temperature. This evaporation causes the cooling effect or refrigeration required to preserve the contents of the food compartment. See Fig. 16-6.

In other words, heat from the oil burner drives the ammonia from the generator, A, into the evaporator, K, in a short time. The ammonia in the evaporator vaporizes and passes back to the generator slowly over a period of 24 to 36 hours. The vaporization of the ammonia in the evaporator produces a refrigerating effect.

For additional efficiency in unusually hot climates or for handling extra large loads, a depression, L, in the top of the condenser tank may be filled with water which will evaporate rapidly and aid the cooling of the condenser.

Absorption mechanisms are provided with a fuse plug which will release the charge from the mechanism if the temperature of the unit becomes excessive (175 F. to 200 F. or 79 C. to 93 C.), as might be experienced in a fire. This prevents the mechanism from exploding.

16-7 DOMESTIC ABSORPTION TYPE SYSTEMS

There has been an increase in the production of absorption units for use in recreational vehicles. Continuous absorption types of refrigerators have four main sections: the generator, condenser, evaporator and absorber. The four sections are made of steel connected by steel tubes. The entire system is welded together.

A typical unit is shown in Fig. 16-7. It is charged with ammonia, water and hydrogen. The amount of this combined solution is at a pressure which will allow the ammonia to condense at room temperature.

The necessary heat for the generator is obtained by using either a gas burner underneath the tube at A, or an electric heating element inserted in tube B.

When the unit begins operation, some of the ammonia is in a strong solution with water in the generator. When heat is applied to the generator, ammonia vapor separates from the water and rises. It carries with it some weak ammonia solution through the system pump or percolator, C.

The weak ammonia solution enters the smaller inside tube, D, and the ammonia vapor enters the larger outer tube, E. At point F, it passes through the strong liquid ammonia solution. As it passes through the strong ammonia solution, it absorbs more ammonia and rises up through tube, G, as a vapor and passes into the water separator tube. In the water separator, water vapor condenses and drains back into the generator, leaving dry ammonia vapor to continue into the condenser.

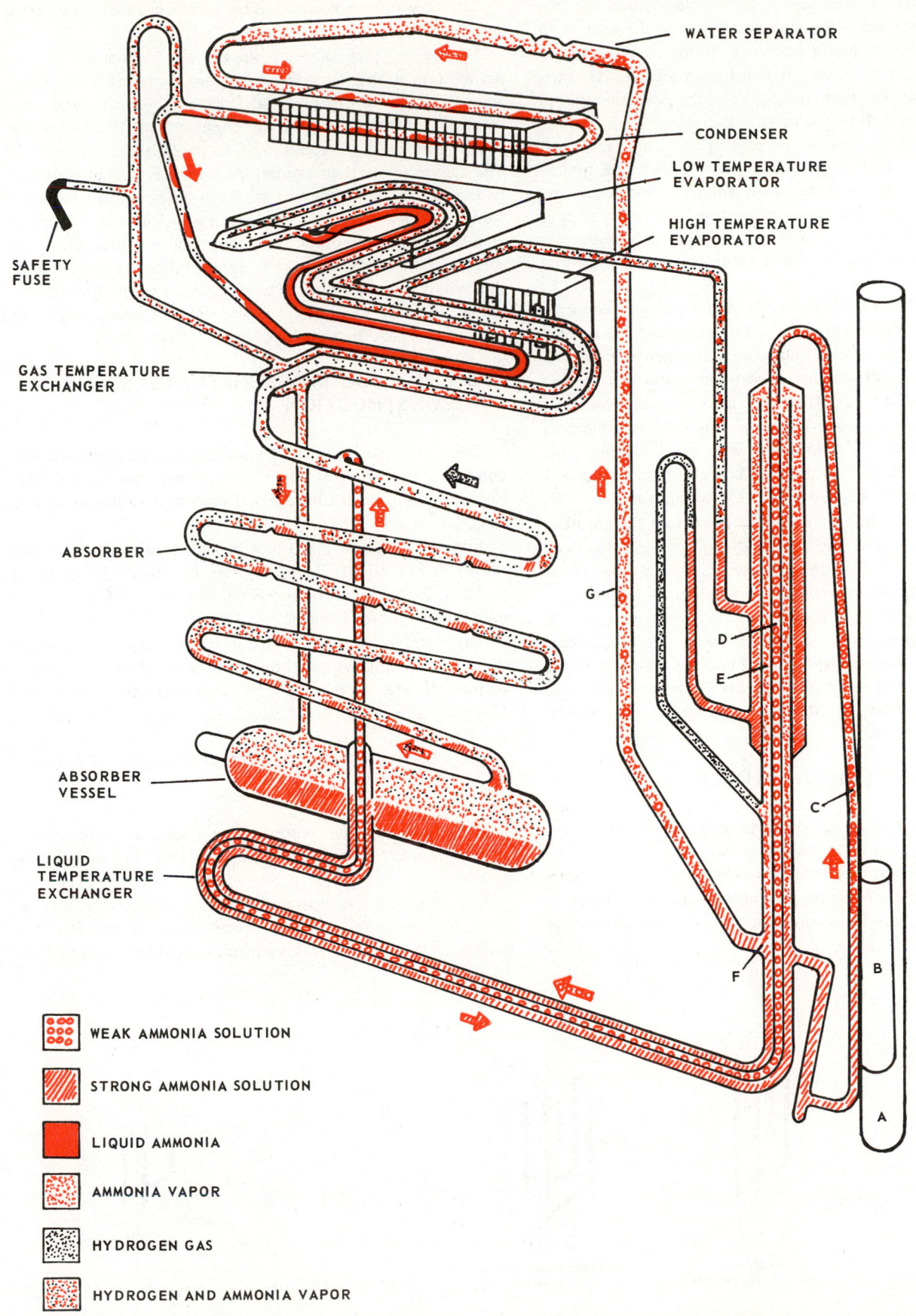

SAFETY
FUSE

WATER SEPARATOR

CONDENSER

LOW TEMPERATURE
EVAPORATOR

HIGH TEMPERATURE
EVAPORATOR

GAS TEMPERATURE
EXCHANGER

ABSORBER

ABSORBER
VESSEL

LIQUID
TEMPERATURE
EXCHANGER

G

D

E

C

F

B

A

WEAK AMMONIA SOLUTION

STRONG AMMONIA SOLUTION

LIQUID AMMONIA

AMMONIA VAPOR

HYDROGEN GAS

HYDROGEN AND AMMONIA VAPOR

Fig. 16-7. Continuous absorption cycle used in domestic and mobile refrigerators. (A.B. Electrolux, S-105-45, Stockholm, Sweden)

At the condenser, the outside air circulating over the fins removes the heat from the ammonia vapor and causes it to turn (condense) into liquid ammonia. It then flows into the low-temperature evaporator in the frozen storage compartment. The ammonia then passes from the low-temperature evaporator to the high-temperature evaporator.

As the ammonia vapor is passing into the low and high-temperature evaporator, hydrogen is added. The hydrogen passes across the surface of the ammonia and allows the ammonia vapor partial pressure to become low enough so that the liquid ammonia begins to evaporate at a low temperature. As this evaporation begins, heat is taken from the food inside the cabinet.

Since the mixture of ammonia and hydrogen vapor is heavier than the hydrogen alone, it settles through the vertical pipe to the absorber vessel. Entering the inclined absorber is a continuous and small amount of weak ammonia solution fed by gravity from tube D. The weak ammonia solution, as it passes through the inclined absorber, absorbs all ammonia vapor returning to the evaporators with the hydrogen. This separates the hydrogen so that it can return (being lighter) through the inclined absorber coil to the evaporator.

Note that the hydrogen only circulates from the absorber to the evaporator and back. The strong ammonia solution, produced in the absorber, flows down to the absorber vessel and back to the boiler (generator), completing the cycle.

It is important to understand that the entire cycle is carried out entirely by gravity flow of the refrigerant. *It is important that the unit remain in a level, upright position. It is important that the heat generated in the absorber be removed and that the heat removed by the condenser be carried away by the surrounding atmosphere.*

16-8 AUTOMATIC DEFROSTING

Automatic defrosting on domestic continuous absorption units is usually done by the hot gas method.

Hot gas is brought from the generator directly to the fresh food evaporator rather than to the freezer evaporator. This hot gas melts the ice on the evaporator fins. Defrost water runs into a drip tray.

The operation is controlled by a siphon tube in the boiler (generator). See Fig. 16-8.

The bypass pipe outlet from the siphon chamber during normal operation is closed by the strong ammonia-water liquid solution pictured in part A. During a normal cycle, the solution collects slowly in the chamber until it reaches the siphon tube outlet — about each 15 to 25 hours. See view B. The siphon action then empties the siphon system at view C of its liquid and allows the hot gas from the boiler to go directly from the generator to the evaporator, in view D. This circulation will continue for about 30 minutes until the solution fills the siphon system again and covers the outlet to the bypass pipe in view E. The circulation of the hot gas will repeat when the liquid level in the siphon chamber rises high enough to repeat the siphon action.

16-9 CONTINUOUS ABSORPTION SYSTEMS CONSTRUCTION

There are several types of continuous absorption systems. Some are heated with fossil fuels. Some are heated with electricity. An electrically heated generator is shown in Fig. 16-9.

Either fuel gas or electricity can be used to heat the unit shown in Fig. 16-10. Electricity may be either 12V (battery) or 120V (house current). A switch, as shown in Fig. 16-11, is used to change over from 12V to 120V.

The combination gas/electrical systems use two thermostats, one when gas is used, the other when electricity is used. Wiring diagrams for the 120V system and for the combination 12V and 120V system are shown in Fig. 16-12.

16-10 INSTALLING A DOMESTIC ABSORPTION SYSTEM

Installation of the absorption refrigerator depends on whether it is to be placed in a home, a recreational or a mobile vehicle.

The absorption refrigerator, like all refrigerator mechanisms, is a device to move heat from inside the cabinet to the outside of the cabinet. This warmed air must be removed from

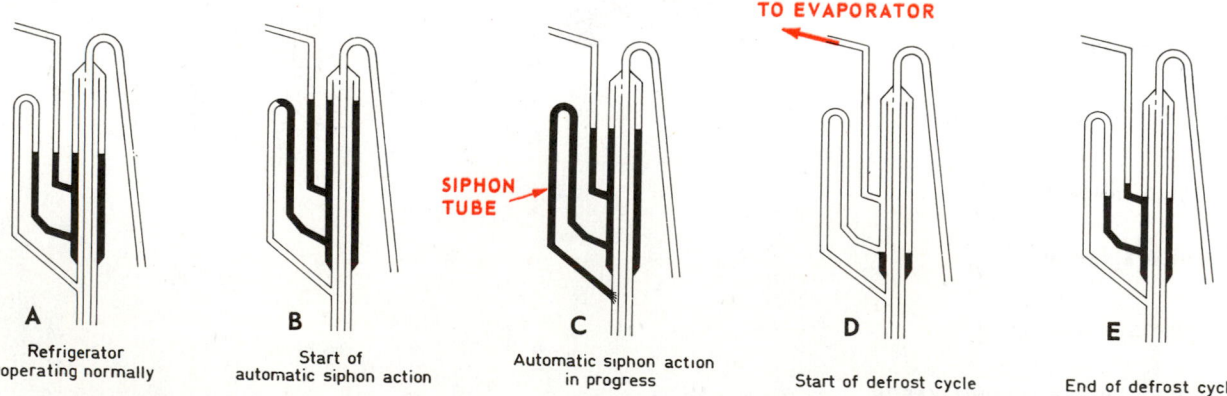

Fig. 16-8. One type of automatic defrost system for domestic continuous cycle absorption system. Liquid gradually builds up in right side of siphon tube and every 15 to 24 hours spills over and allows hot vapor to move into higher temperature evaporator. Defrosting stops when enough liquid collects in outer tube to close tube to evaporator. (A.B. Electrolux)

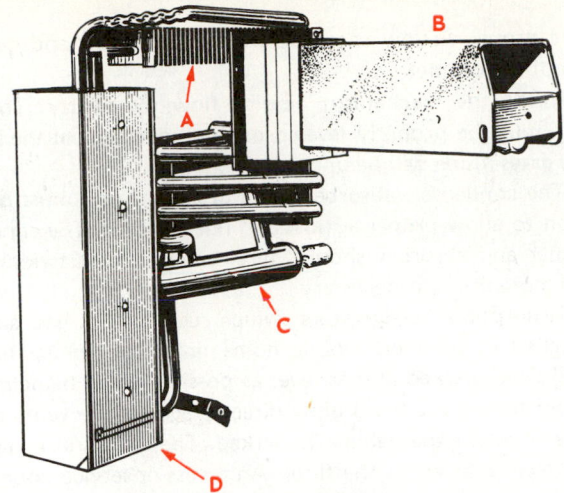

Fig. 16-9. Continuous absorption refrigerating mechanism using electricity as heat source. A—Condenser. B—Evaporator. C—Absorber. D—Electrically heated generator.

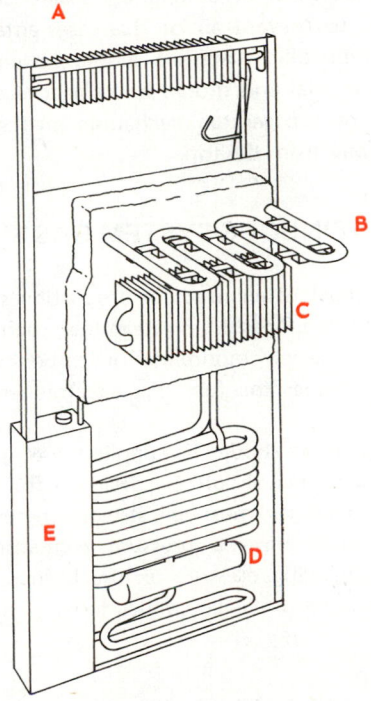

Fig. 16-10. Continuous operation absorption unit can be heated by fuel gas or electricity. A—Condenser. B—Low-temperature evaporator. C—High-temperature evaporator. D—Absorber. E—Generator.

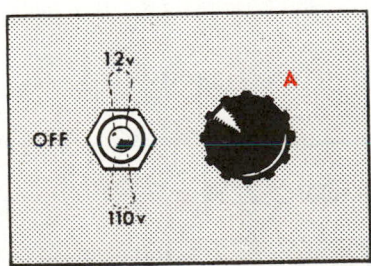

Fig. 16-11. Electrical switch and fuse used to shut off electrical power or change from 120V to 12V. If switch is on 12V when system is connected to wall outlet, fuse at A will open circuit (blow) to protect heater from damage. (A.B. Electrolux)

WIRING DIAGRAM 120V

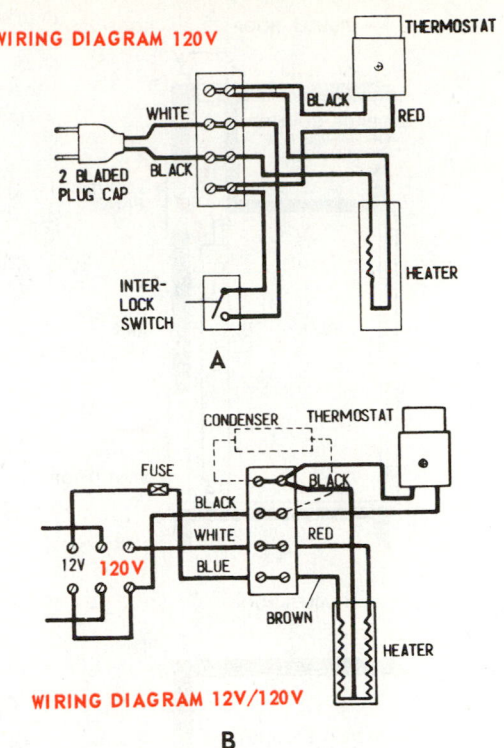

WIRING DIAGRAM 12V/120V

Fig. 16-12. Wiring diagrams for an electrically heated continuous operation absorption system. A—120V circuit. B—12V or 120V circuit.

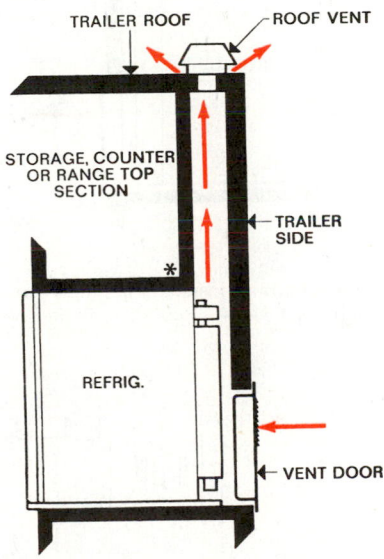

Fig. 16-13. Recommended method for installing fuel gas-heated absorption refrigerator.

near the cabinet to allow cooler air to continue to receive heat from the condensers, as shown in Fig. 16-13.

Kerosene, natural gas or LP gas, when burned, forms carbon dioxide gas (harmless) and steam vapor (harmless). If burner is not burning all the fuel, carbon monoxide may be formed and is dangerous! To allow good airflow past the burner, these products of combustion must be moved away from the cabinet. Because warm gases rise, the absorption cabinets must

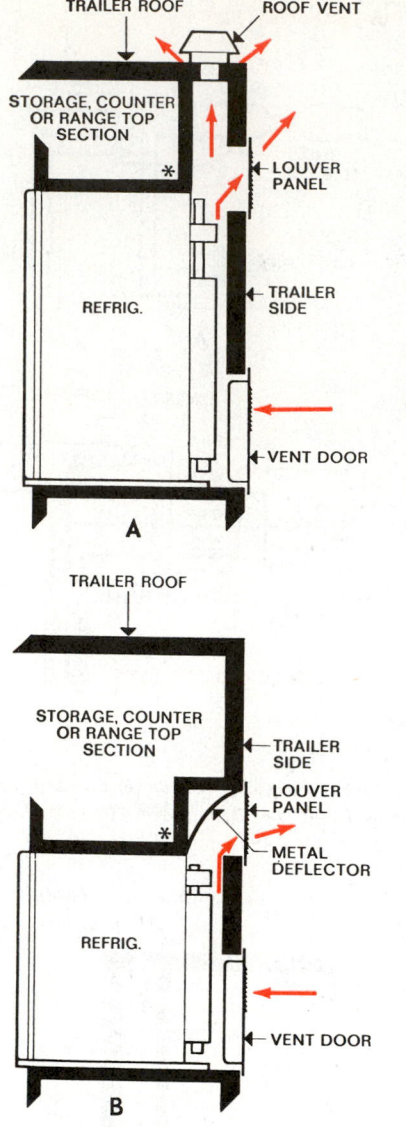

have proper airflow space beneath, in back of, and over the top of the cabinet.

Inside the mechanism, liquids flow by gravity. The unit then must be properly leveled or the movement of the liquids and gases inside will be uncertain.

The condenser, absorber and flues of the unit must be kept clean to allow proper airflow and flue gas flow. The condenser burner and absorber should be cleaned at least twice a year and more often, if necessary.

Absorption refrigerators which use a fuel gas and are installed in a trailer, mobile home or motor home, must be level when parked and as level as possible when being moved. Avoid having the wind blow directly against the vents on the outside when the vehicle is parked. The refrigerator must be securely fastened to the floor. An access or service door in the outside of the vehicle must be provided. This door usually serves also as the inlet vent. There should also be an exhaust vent. See Fig. 16-14.

The cabinet must be completely sealed on the top, sides and bottom, to prevent air or flue gases entering the mobile vehicle. Combustible materials should be covered with fire protection material and should be kept at least 1 in. (25 mm) away from the refrigerator mechanism on the sides and 7 in. (180 mm) away from the top.

16-11 CONTINUOUS SYSTEM GAS SUPPLY

The fuel most often used for absorption system refrigerators is natural or LP gas. Using clean fuel, such as gas, prevents formation of carbon monoxide or carbon deposits in the burner. The cleaner the fuel, the less frequently the burners will need cleaning.

Gas refrigerators should be supplied with gas under steady pressure. The burner should be designed for the type of gas being used. The gas must be strained before entering the burners. It must have a pressure regulator to provide a constant, unchanging pressure on the burner. Fig. 16-15 is a diagram showing suitable gas connections.

Fig. 16-14. Two methods of installing gas-fired absorption system in mobile vehicle. A—Roof vent. B—Side vent.

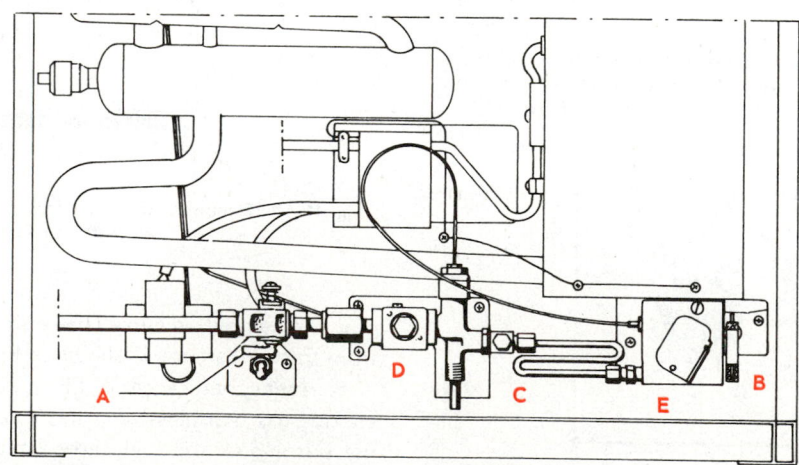

Fig. 16-15. Gas connections as used on absorption type refrigerator. A—Shutoff valve. B—Lighter. C—Safety shutoff. D—Thermostat valve. E—Burner cover. (A.B. Electrolux)

It is important to know the local laws and codes about installation of these refrigerators before trying to install them.

In case of burner difficulty, a check should be made to make sure the burner is the correct one for the gas used.

Three different gases are used as fuel for gas refrigerator units. These fuels are discussed in more detail in Chapters 20 and 28. Heating quality of each is given as follows:
1. Artificial gas (500 to 600 Btu/cu. ft.).
2. Natural gas (1000 to 1100 Btu/cu. ft.).
3. Liquid petroleum LP gas (2500 to 3200 Btu/cu. ft.).

16-12 CONTINUOUS SYSTEM CONTROLS

Aside from cabinet care and cleaning, service on the absorption refrigerator is generally limited to the heating controls and air circulation equipment. Since they determine the efficiency of the unit, adjustments must be carefully made.

Heating gas valves are used in the absorption system to automatically control the amount of heating gas burned. The systems use a continuous gas flow and the flame size varies depending on the demand.

Continuous operation systems require a gas volume control and a safety control. A bulb pressure-temperature control, located at the evaporator, regulates the amount of gas burned. It senses the needs of the refrigerator. Fig. 16-16 shows the

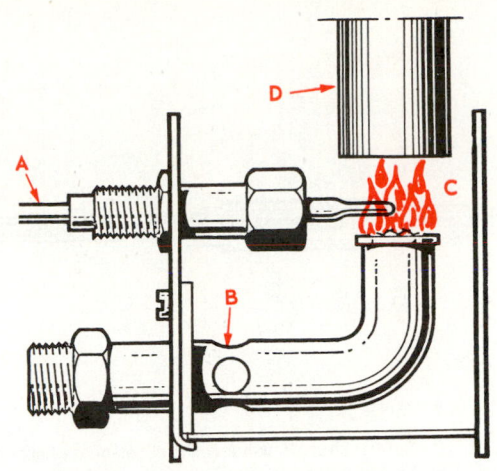

Fig. 16-17. Burner, flue and thermocouple safety device. A—Safety thermocouple. B—Burner. C—Flame. D—Flue.

heat is continuous. However, to take care of changes in demands on the refrigerator itself, the size of the gas flame must be automatically controlled. This is possible with a control valve operated by a power element located at the evaporator. As the refrigerator warms up, gases in the power element expand and press on a diaphragm in the control valve

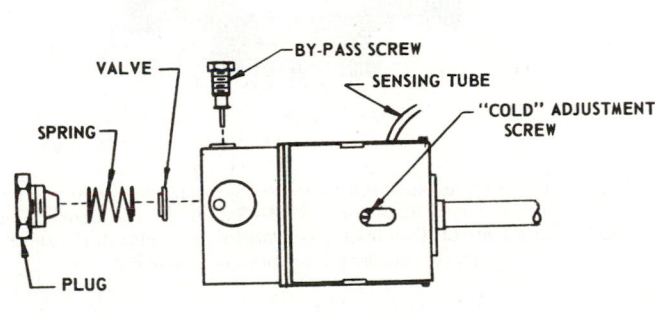

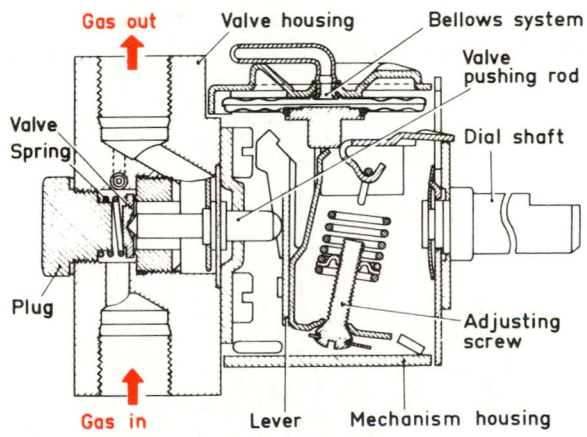

Fig. 16-16. Thermostat used in gas-fired small continuous operation absorption systems. Valve has small bypass screw opening which allows small flame to burn even when main valve is closed (thermostat on 0 when sensitive bulb is 40 F. or colder).

control in cross-section. The temperature of the evaporator affects the size of the flame.

Successful operation of domestic refrigerators depends largely on the way the automatic control valves work. The service technician must be thoroughly familiar with their operation and servicing.

Heat energy, for the continuous operation absorption unit, usually is supplied by a gas burner. See Fig. 16-17. Several different methods have been used for regulation and control of this gas. A manual shutoff valve, a strainer and a pressure-regulating valve are placed between the gas main and the operating controls of this type of refrigerator.

As noted, the unit operates on the continuous cycle and

to open the gas control. This allows more gas to flow to the burner.

The larger flame speeds up the refrigerating cycle and continues to speed it up until the evaporator has cooled.

As the evaporator cools, the power (control) element on the evaporator will cool, reducing the pressure on the gas valve. This reduction closes the heating gas opening and reduces the size of the flame. Turning the adjustment clockwise, or in, increases the gas supply and produces more refrigeration.

A safety valve will shut off the gas in case the flame goes out. This valve is a thermoelectric type. See Fig. 16-18. A thermocouple is placed close to the flame. It remains hot as

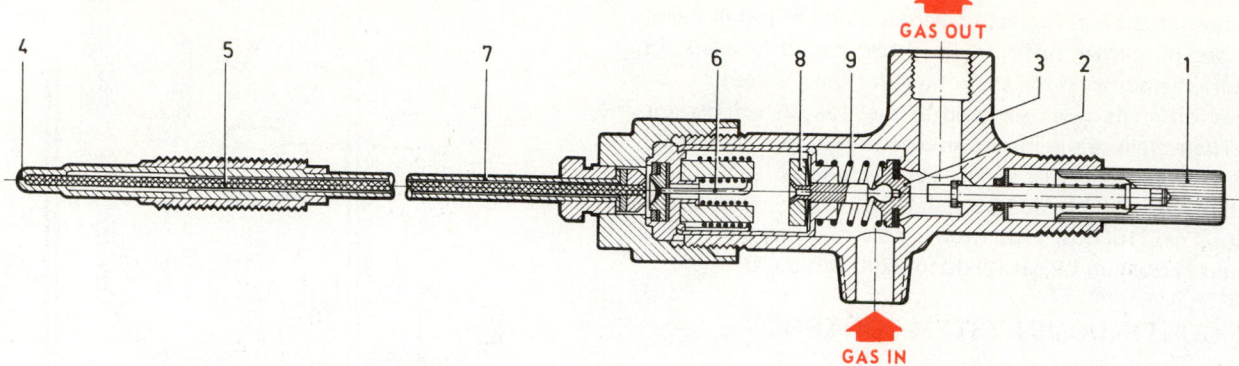

Fig. 16-18. Safety shutoff valve for fuel gas-heated absorption system. 1—Manual valve opener. 2—Valve. 3—Valve body. 4—Thermocouple. 5—Thermocouple mounting fitting. 6—Thermocouple electromagnet. 7—Thermocouple tube. 8—Thermocouple induced magnet (fastened to valve). 9—Valve spring.

long as the gas is ignited. If the flame goes out, the thermocouple at 4 will cool. On cooling, it stops creating electricity. The magnetic coil at 6 will lose its strength, allowing the spring, 9, to close the valve at 2. This action completely shuts off the supply of the gas. If this happens, the gas is re-ignited by pushing in the button at 1 which will hold the valve at 2 open. Then operate the spark lighter mounted on the burner housing, Fig. 16-19, to ignite the fuel gas.

Each burner unit has an automatic pressure control for maintaining a constant gas pressure. Through this valve, changes in the pressures from the gas supply are kept as small as possible.

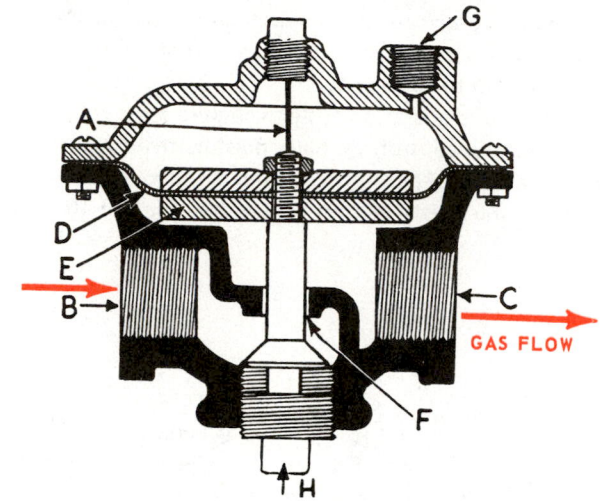

Fig. 16-20. Pressure regulating valve. A—Shipping pin (this is removed when valve is placed in operation). B—Gas in. C—Gas connection to thermostatic control. D—Flexible diaphragm. E—Weights. F—Valve and seat. G—Bleeder connection. H—Valve cap.

provides a constant gas pressure. See Fig. 16-20.

Liquid petroleum (LP) gases do not need a pressure regulator at the refrigerator if a pressure regulator is mounted on the LP cylinder. It performs the same duty.

The pressure regulator operates much the same as expansion valve. Pressure at the outlet presses against a diaphragm (synthetic rubber or fabric reinforced). If the pressure begins to drop, the diaphragm will move, opening the gas valve to allow more gas to flow. The increased gas flow will press the diaphragm up closing the valve. The pressure regulator should be accurate to about 1/100 in. (.25 mm) of water pressure.

Pressures which the regulator must maintain vary from 1.6 to 3.9 in. (40 mm to 99 mm) of water. Pressure needs vary with gas flow in cubic feet per hour. This gas flow is controlled by the orifice size in the burner.

Pressure also varies with the density or specific gravity of the gas. The greater the gas flow the greater the pressure needed. Pressures must be adjusted to within .1 in. (2.5 mm) water pressure for good results.

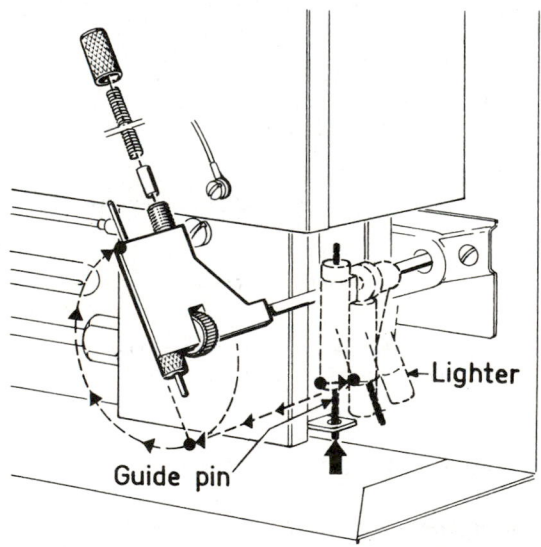

Fig. 16-19. Flint gas igniter in recreational vehicle refrigerating unit.

16-13 PRESSURE REGULATING VALVES

The pressure regulating valve supplies a steady flow of gas to the burner. Without a regulating valve, changing gas pressure would change the flame and it might go out.

The pressure regulator both reduces the pressure and

Fig. 16-21. Portable absorption refrigerator which can use propane gas cylinder or electric heating element as its energy source. A—Cabinet. B—Hinged cover. C—Unit.

16-14 PORTABLE ABSORPTION REFRIGERATORS

A portable refrigerator shown in Fig. 16-21 holds 1.1 cu. ft. of food and weighs only 42 lb. The cabinet is plastic and uses a foam insulation.

The refrigeration system is the self-contained continuous absorption type, using bottled gas, 120V a-c or 12V d-c as a source of heat to the generator. Construction details are shown in Fig. 16-22.

The pull-down time for this refrigerator is from two to six hours, depending on the ambient temperature. The small propane cylinder will provide approximately 70 hours of continuous operation.

Another portable absorption refrigerator is shown in Fig. 16-23. The gas cylinder, electrical connections, and controls for this refrigerator are shown in Fig. 16-24.

When operating from a propane cylinder:

1. Attach gas connection to gas inlet fitting.
2. Open valve on gas bottle.
3. Press the red button on the safety valve and hold it for 10 to 15 seconds. (This is to clear air from the gas line.)

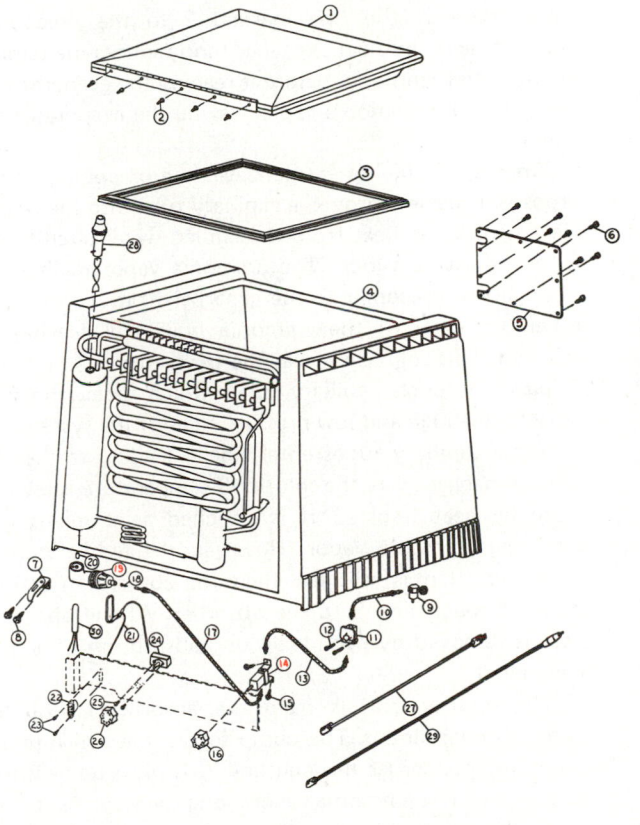

NO.	PART NAME
1	Cover Assembly
2	Pan Head, Tapping Screw
3	Closure Gasket
4	Refrigerator Case
5	Plastic Case Closure
6	No. 7-16 "Tapit" Screw, 1/2 in. Long
7	Burner Bracket Assembly
8	No. 6-18 "Tapit" Screw, 3/8 in. Long
9	Valve Assembly
10	Hose Assembly
11	Gas Regulator
12	Pan Head, Self Tapping Screw
13	Hose Assembly
14	Gas Thermostat
15	No. 6-18 "Tapit" Screw, 3/8 in. Long
16	Knob—Gas Thermostat
17	Hose Assembly
18	Orifice (.008)
19	Burner Assembly
20	Burner Tip
21	Heating Element
22	Service Cord Receptacle
23	Pan Head, Tapping Screw
24	Electric Thermostat
25	Round Head Machine Screw
26	Knob—Electric Thermostat
27	Service Cord, 115V
28	Windshield, Spiral Assembly
29	Service Cord, 12V
30	Combination Cartridge Heater

Fig. 16-22. Schematic of portable absorption refrigerator, such as might be used on a recreational vehicle. 19—Bottle gas burner. 21—Heating element for 120V. 30—Heating element for 12V. Separate thermostats are used depending on heat source. One is on gas supply and the other on electrical supply.

Absorption Systems, Principles and Applications / 613

4. Light match. Press red button again and apply flame to burner. Keep red button depressed for 20 seconds after burner is lit.

The safety valve will automatically shut off gas supply, should the flame go out for any reason. A pressure regulator will maintain an 11-in. (279 mm) water column pressure on the burner. When using the gas burner, be sure no combustible material or vapors are near the refrigerator.

When operating with electricity, separate leads are supplied for the 120V a-c and the 12V d-c connection.

Fig. 16-23. Portable absorption refrigerator. Heating source used in this refrigerator may be either LP gas, 120V a-c or 12V d-c current. A—Cabinet. B—Hinged cover. C—Unit. (Paulin Products Co.)

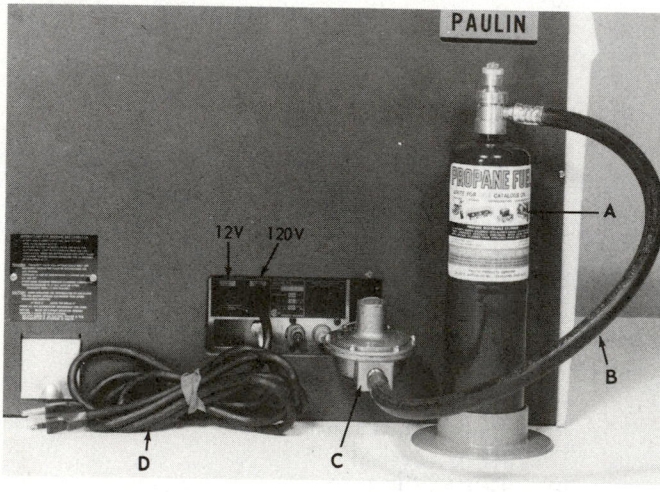

Fig. 16-24. Portable absorption refrigerator. A—Propane fuel cylinder. B—Flexible gas hose. C—Pressure regulator. D—Electrical power cord. Refrigerator can be operated with either gas flame or electric heating unit of 12V or 120V. (Paulin Products Co.).

16-15 ABSORPTION REFRIGERATORS FOR MOBILE HOMES

Mobile homes and travel trailers often use an absorption type refrigerator. The units are usually designed with both an electric heating element and a gas burner to heat the generator of the continuous unit. Gas heat is used when electricity is not available. It is important that the refrigerator be mounted level. If this is not done, the gravity-controlled flow of the

fluids will not operate well.

The installations must be carefully designed with air ventilation to cool the condenser and to provide air for the flame and outside exhaust for combustion gases.

16-16 RESIDENTIAL ABSORPTION AIR CONDITIONERS

The number of absorption type air conditioning systems used in residential and commercial buildings has increased. Their basic operation is similar to that described in Para. 16-5. One of the major changes for such use is the addition of a pump to transfer the weak solution from the absorber to the high side of the cycle. Either a hydraulic diaphragm pulse pump or an electric motor driven magnet pump is sealed in the system.

A typical chilled water residential air conditioning system is shown in Fig. 16-25. Rated at five ton, it has two generators connected in parallel. Each generator has a gas burner which heats a mixture of ammonia and water. The boiling point of ammonia is lower than that of water; therefore, it becomes a vapor and flows through the line marked 1 to the condenser as a high-temperature, high-pressure gas.

As outside air passes over the condenser, it removes heat from the ammonia. The ammonia condenses to a liquid and passes through the line marked 2 to the precooler. The precooler acts as a heat exchanger and reduces the temperature of the liquid ammonia before it reaches the evaporator. It also heats the cold ammonia vapor leaving the evaporator through line 3.

When the liquid ammonia leaves the precooler, its pressure drops as it passes through a capillary tube into the evaporator. Here it picks up heat from the chilled water circuit and boils to an ammonia vapor. The ammonia vapor in line 3 passes through the precooler to the absorber heat exchanger. In the generator, most of the ammonia boils out, leaving a weak solution. This solution leaves the generator at a high pressure. It passes through capillary tubes which meter the flow and separate the high and low-pressure sides of the system.

These capillary tubes enter line 4 leading to the absorber heat exchanger. Here the solution's temperature is lowered still more by heat transfer. It is enriched as it mixes with the returning ammonia vapor. However, the temperature of the mixed solution is still too high for absorption action so it passes through line 5 to the absorber. At the absorber end, heat is removed by outside air and absorption of ammonia is completed.

The solution travels from the absorber to the solution pump through line 6. The pump forces it at a high pressure to the reflux condenser through line 7. It picks up heat from the surrounding hot ammonia vapor along the way. As it continues on to the absorber heat exchanger through line 8, more heat is picked up. This preheated solution returns through line 9 to the generator. The cycle begins all over again. The same action occurs in smaller units except that only one burner is used.

Another system cycle of slightly different design is shown in Fig. 16-26. The weak solution (strong in ammonia content) is returned to the generator by the solution pump. The burner

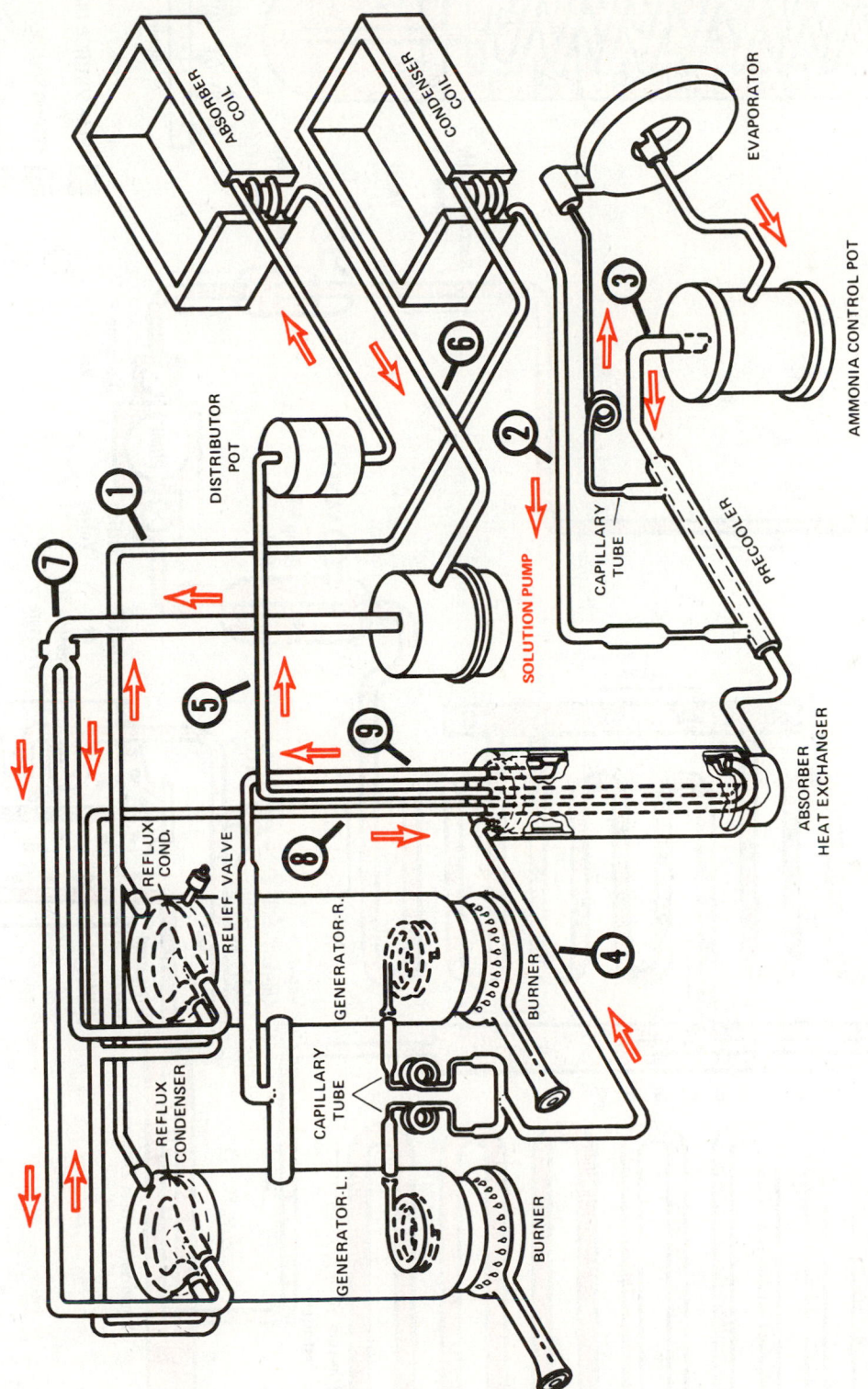

Fig. 16-25. Absorption refrigeration cycle. This refrigeration cycle uses pump to circulate ammonia liquid through cycle. (Whirlpool Corp.)

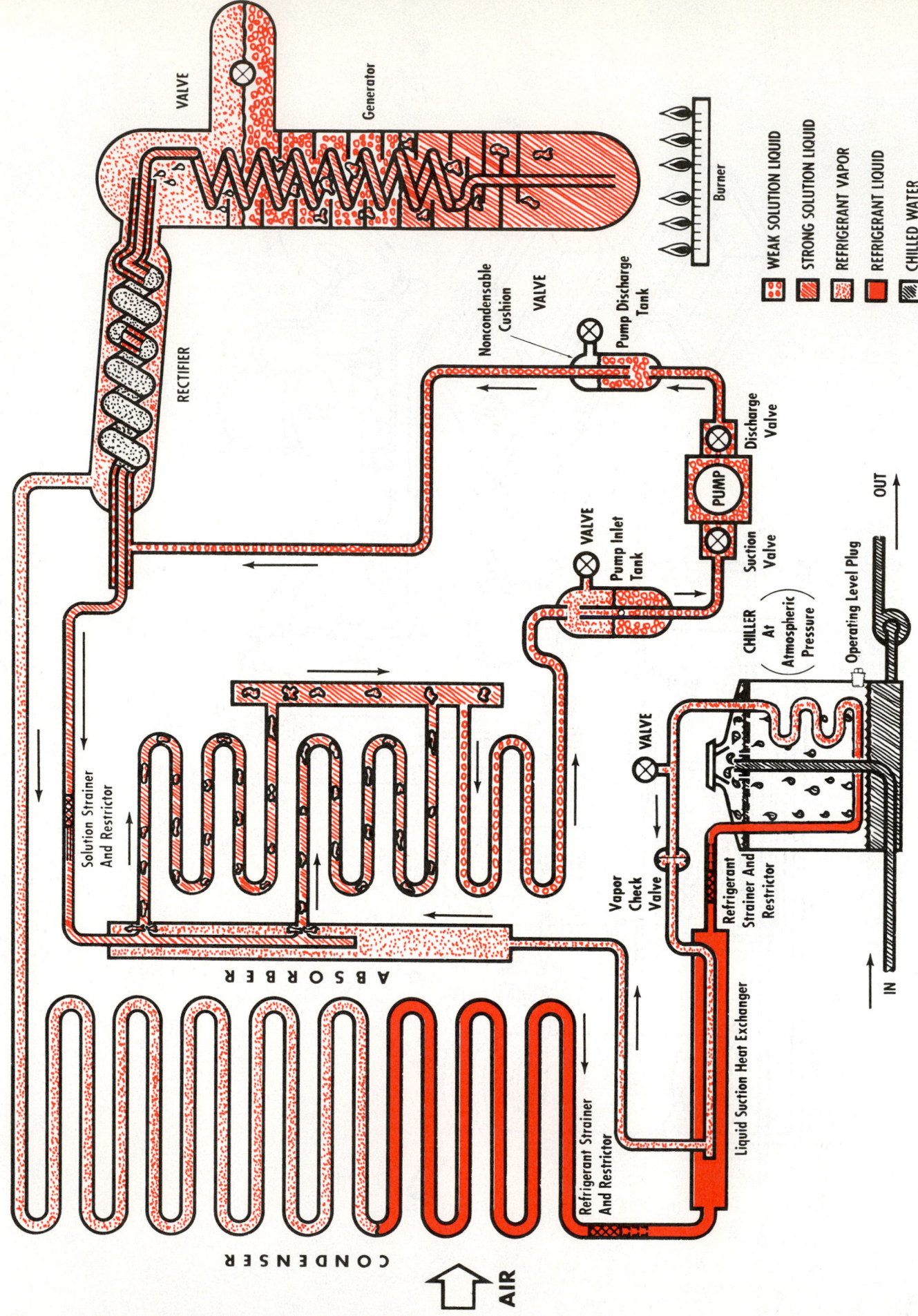

Fig. 16-26. Absorption cycle used for small air conditioning systems. Notice service valves on system.

heats it and drives off the vapor. The vapor, a mixture of ammonia and water, flows through the generator assembly and comes into direct contact with the weak solution coming into the generator. During the process, the vapor is partially purified. Passing through the rectifier assembly, the vapor is further purified as it comes into contact with the cooler weak solution flowing in the opposite direction through a coil.

The rectifier contains rings and, as the ammonia vapor passes through, it comes into direct contact with the rings. The rings, having a large surface contact area, help remove any water vapor left in the ammonia vapor.

The purified ammonia vapor then flows into the condenser tube cooled by air moving across the condenser. As the hot ammonia vapor in the condenser gives up heat to the flow of air, it is liquefied.

Liquid refrigerant leaves the condenser and passes through the first restrictor where pressure and temperature drop somewhat. It then flows through the outer tube of the liquid suction heat exchanger, giving up heat to cooler ammonia vapor flowing through the inner tube in the opposite direction.

From the liquid suction heat exchanger, the liquid ammonia enters the second restrictor, where pressure and temperature are further reduced. The low-pressure, low-temperature liquid refrigerant then flows through the chiller coil (evaporator). Here a glycol and water solution cascades down across the evaporator, giving up heat and vaporizing liquid refrigerant.

The refrigerant vapor leaves the chiller and passes through the inner tube of the liquid suction heat exchanger. In passing, it picks up heat from the liquid refrigerant flowing through the outer tube in the opposite direction.

Next, the vapor enters the absorber header. This completes the refrigeration circuit.

Hot, strong solution, left behind as the ammonia vapor was driven out of the weak solution in the generator, passes up through a coil in the generator. The strong solution leaves the generator and enters the inner coil of the rectifier. Here the strong solution comes into thermal contact with the weak solution flowing in the opposite direction through the outer coil. As the strong solution leaves the rectifier, it passes through the strong-solution restrictor. Here the restrictor reduces the solution from high-side to low-side pressure.

As the strong solution leaves the restrictor, it enters the absorber header. This is a tube-within-a-tube. The inner tube contains the strong solution; the outer tube contains ammonia vapor which is being returned from the chiller (evaporator). The strong-solution tube contained within the absorber header has two small holes that will allow strong solution to flow out and come into direct contact with the ammonia vapor surrounding it. At this point the strong solution and ammonia vapor begin to form a weak solution.

Strong solution and ammonia vapor then leave the absorber header by way of the two tubes and begin to flow into the absorber. Throughout the absorber, the strong solution completely absorbs the vapor, forming a weak solution.

As the weak solution leaves the absorber, it is picked up by the solution pump, which will move the solution back to the high-pressure side of the unit. As the weak solution leaves the solution pump, it passes through the rectifier picking up heat from strong solution vapor flowing in the opposite direction.

By preheating the weak solution as it flows through the rectifier, the need for heat input is reduced and the overall efficiency of the cycle is increased. The weak solution leaves the rectifier and drips back into the generator analyzer assembly to start another cycle.

16-17 RESIDENTIAL ABSORPTION AIR CONDITIONER CONSTRUCTION

In this self-contained unit, the insulated evaporator cools a glycol and water solution. The solution then circulates through a heat exchange coil in the furnace bonnet. It may also be used in a separate air circulation system within the building. This absorption system is shown in Fig. 16-27. Controls are shown in Fig. 16-28.

Two motors are used in the system. One, shown in Fig. 16-29, drives the fan which cools both the absorber and the

Fig. 16-27. Absorption system air conditioner with housing removed. A—Burner. B—Condenser. C—Solution pump assembly. (Arkla Industries, Inc.)

Fig. 16-28. Controls of absorption air conditioner are easy to reach for service when unit's housing is removed. A—Gas burner. B—Gas controls. C—Electrical controls. (Arkla Industries, Inc.)

Fig. 16-29. View of air conditioner absorption system showing absorber-condenser and other parts. A—Chiller tank. B—Generator. C—Fan and motor. D—Hydraulic pump.

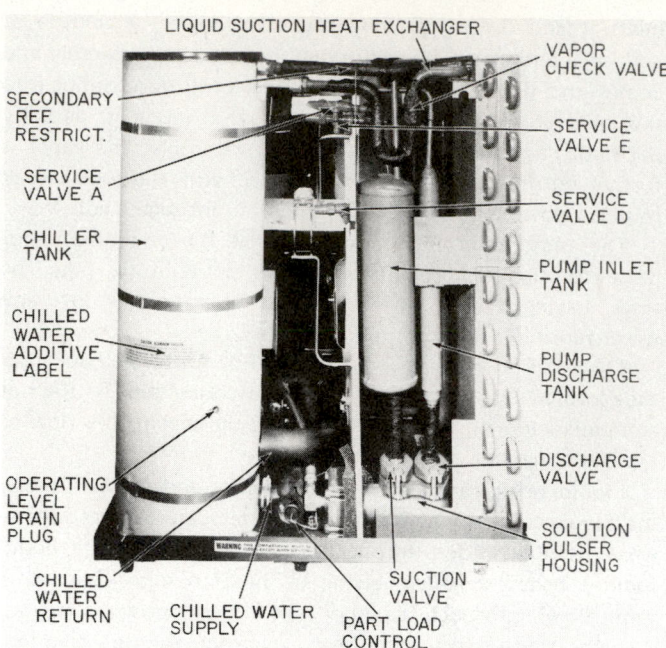

Fig. 16-30. Absorption system for air conditioning. Service valve A is mounted on solution-cooled absorber. Service valve D is mounted on pump inlet tank. Service valve E is mounted on pump discharge tank. Service valve C is not shown.

condenser. Another operates the chilled water pump and the hydraulic pump. The hydraulic pump, in turn, operates the solution diaphragm pump. Magnetic pumps are also used. The solution pump has a strainer at its inlet. This pump keeps working for about three minutes after the unit stops to help cool the generator. Generator is on the left in Fig. 16-29.

These systems are made of steel and aluminum. Use of copper or copper alloys is very dangerous. An explosion may result.

A three-ton capacity uses about 8 to 9 lb. of ammonia. The chilled water circuit uses an antifreeze solution. There is a strainer in the return line. The electric power circuit uses 120—240V while the control circuit is 24V. The gas flame flue gases are exhausted with the condenser's cooling air.

16-18 RESIDENTIAL ABSORPTION AIR CONDITIONER SERVICE

Most of the residential absorption systems are serviceable. They are equipped with service valves. However, one should be trained by the manufacturer before attempting to service these systems.

Ammonia is toxic and flammable when mixed at certain ratios with air. Wear a face shield or safety goggles.

Ammonia reacts with some metals. Use only steel or aluminim tubing, gauges, fittings and manifolds.

Fig. 16-30 shows a system equipped with four service valves. Valves A and D are on the low-pressure side. Valves C and E are on the high-pressure side of the system. Valve C is not shown. Valves D and E are gauge mounts for checking pressures on the system. Fig. 16-26 also shows the location of the four service valves.

Anhydrous ammonia is available in 25-lb. cylinders. The cylinder has both a vapor and a liquid valve.

The system is charged with a solution of ammonia, distilled water (ph 6.0+) and a corrosion inhibitor. A solution cylinder is used to charge the system with the solution. This cylinder

usually has about 45-lb. capacity. The solution charge is about 35 lb. of distilled water, about seven packages of inhibitor and 15 lb. of anhydrous ammonia.

The solution cylinder is filled as follows:

1. The distilled water and inhibitor solution (yellow in color) is put in the solution cylinder first through the fill plug. If there is a white powder in solution, discard it and make a new solution.

2. Fig. 16-31 shows a solution cylinder being charged with anhydrous ammonia. The pail holds some water and is

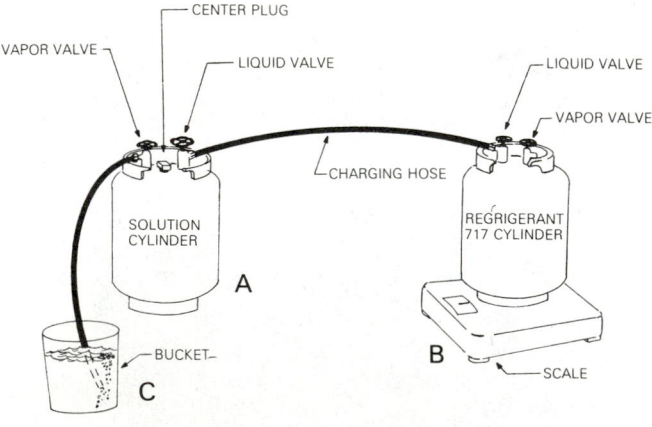

Fig. 16-31. Cylinders set up for recharging an absorption system. A—Solution cylinder has liquid valve, vapor valve and center plug. It is being charged with liquid ammonia from cylinder at right (1 lb. ammonia for each 2 lb. distilled water). B—Ammonia cylinder. C—Pail holds water and a purge line connected to gas valve of cylinder A. Purging decreases pressure in A to allow flow from B. (Arkla Industries, Inc.)

Fig. 16-32. All-steel gauge manifold is connected to valve C (high-pressure side) and valve A (low-pressure side). Valve E is for removal and checking of solution.

connected to the vapor valve of the solution cylinder. Vapor is purged from the solution cylinder if necessary, to lower the pressure so that anhydrous ammonia will flow into the solution cylinder. The water in the pail will absorb the small amount of ammonia purged. Note that the charging line is connected to the liquid valve of the anhydrous ammonia cylinder.

The system must have the correct pressures, the correct amount of anhydrous ammonia and the correct amount of distilled water and inhibitor.

To check the pressures, install a 100 percent steel constructed gauge manifold, steel lines and steel fittings. Do not use copper or brass. Fig. 16-32 shows a gauge manifold installed. Note the steel manifold. It is constructed of standard steel fittings and steel valves. The manifold operates exactly like the manifolds described in Chapters 11 and 14.

The four service valves on the system are used for a number of service operations.

Valve A:

1. Checks absorber pressure (low-side pressure).
2. Purges ammonia vapor.
3. Adds ammonia liquid or vapor.
4. Adds solution.
5. Reduces system pressure to atmospheric pressure.

Valve C:

1. Checks high-side pressure.

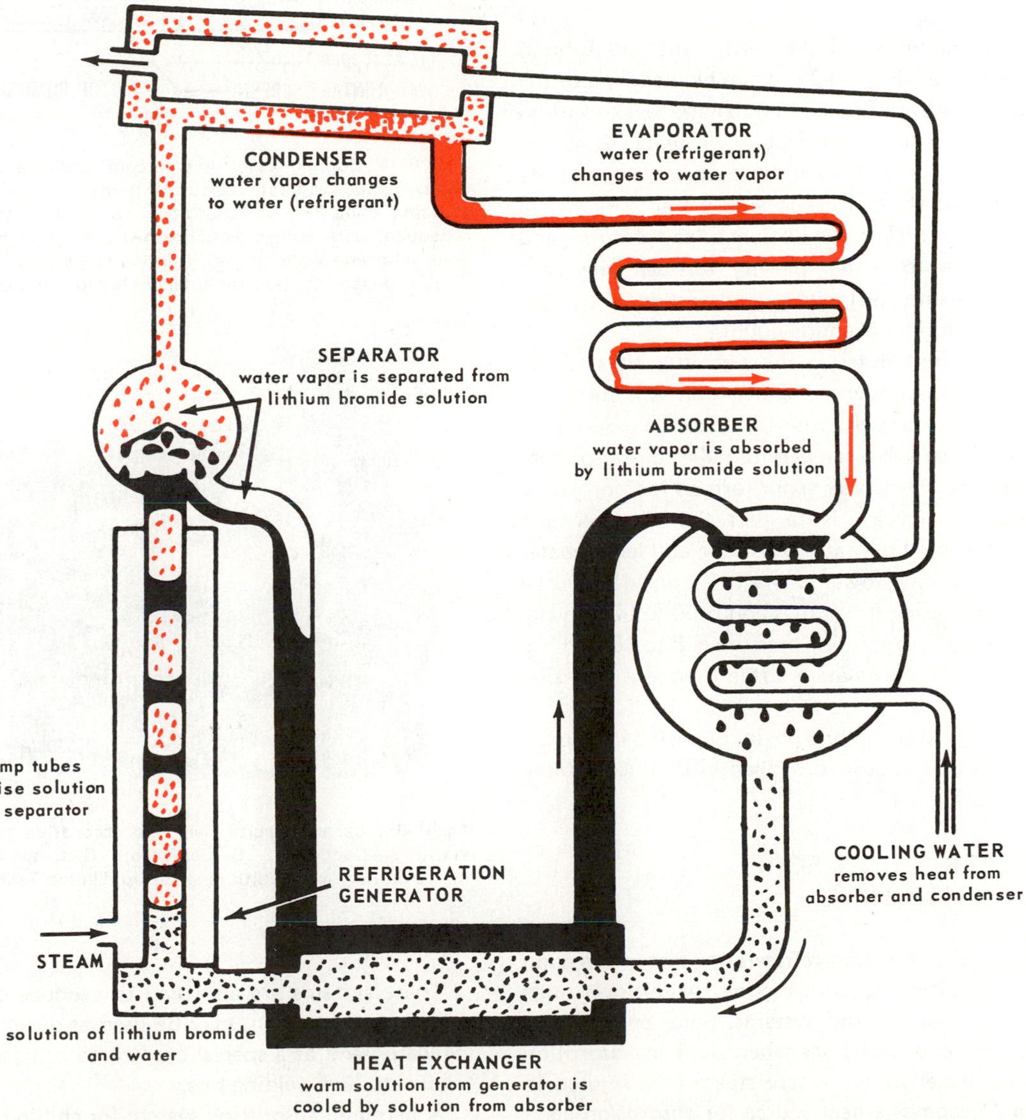

CONDENSER
water vapor changes
to water (refrigerant)

EVAPORATOR
water (refrigerant)
changes to water vapor

SEPARATOR
water vapor is separated from
lithium bromide solution

ABSORBER
water vapor is absorbed
by lithium bromide solution

pump tubes
raise solution
to separator

REFRIGERATION
GENERATOR

COOLING WATER
removes heat from
absorber and condenser

STEAM

solution of lithium bromide
and water

HEAT EXCHANGER
warm solution from generator is
cooled by solution from absorber

Fig. 16-33. Absorption refrigeration cycle which uses water as refrigerant and lithium bromide as absorbent.

2. Checks solution level.
3. Removes excess solution.
4. Adds solution after repairs.
5. Reduces system pressure to atmospheric.

Valve D:

1. Purges air.
2. Adds air.
3. Adds solution.
4. Removes solution.

Valve E:

1. Removes large amounts of solution.
2. Determines if discharge chamber has proper amount of noncondensables.

16-19 COMMERCIAL ABSORPTION SYSTEM

Absorption systems are used successfully for air conditioning comfort cooling installations. But such systems may also be used for heating.

Some units use the ammonia-water-hydrogen continuous cycle. Others use water as the refrigerant, and various chemicals as the absorber.

In a system using water as the refrigerant and lithium bromide as the absorber, Fig. 16-33, steam heat applied to the generator percolates water vapor (red dots) and weak solution up to the separator. The liquid lithium bromide, shown in black, then flows by gravity through the heat exchanger to the absorber where it absorbs the evaporated water. The strong solution (black dots) settles to the bottom of the absorber and returns to the generator after passing through the heat exchanger. The pressure difference is maintained by the pressure head of the lithium bromide liquid.

The water vapor (red dots) in the separator rises to the condenser where it is condensed and becomes water. The water flows by gravity through an orifice into the evaporator. The water evaporates at a low temperature due to near-perfect vacuum in the system. The water vapor formed is absorbed by the lithium bromide (black). Note that the absorber and condenser are both cooled by water coils. The condenser water is then taken to a cooling tower where it is cooled and used over again. The condensing pressure is about 50 to 60 mm Hg. (about 3 psia), and the evaporating pressure is 8 to 10 mm Hg. (about .5 psia). Lithium chromate is often used as a corrosion inhibitor.

A typical cooling tower is shown in Fig. 16-34. A basic water tower and its operation is described in Chapter 3. More detail is shown in Chapter 12.

16-20 ABSORPTION UNIT FOR AIR CONDITIONING

The application of absorption refrigerating systems in comfort cooling air conditioning is increasing. Absorption units have advantages in solar energy systems. Solar energy, as a source of heat, can cool buildings when used in absorption systems. In other installations, where steam heat is used in winter, the steam becomes a heat source for absorption cooling in the summer.

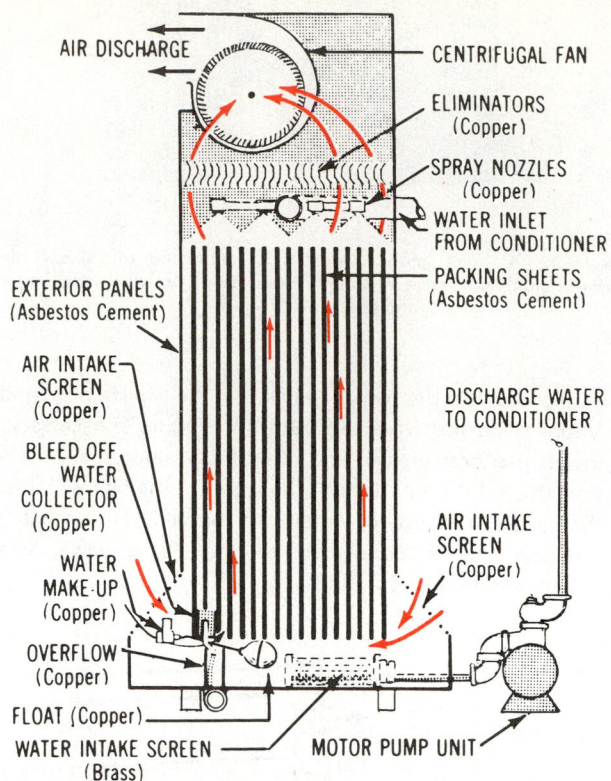

Fig. 16-34. Cooling tower used to cool condenser and absorber cooling water. Tower evaporates about 15 percent of condenser water and, in so doing, cools rest of water down to wet bulb temperature of air. It consists of water sprays, asbestos sheets, overflow tubes, make-up water float valve and water pump. Eliminator plates keep water from being drawn into fan. Air enters at bottom and leaves at top.

Fig. 16-35. Large capacity hermetic absorption system used to cool water. A—Condenser. B—Generator. C—Evaporator. D—Absorber. (Carrier Air Conditioning Group, United Technologies Corp.)

These systems are also used to produce chilled water. The chilled water, in turn, may be used as quenching baths, drinking water, and as a special coolant to bring down the working temperature of welding tips.

A hermetic absorption system for chilling water is shown in Fig. 16-35. The cooling cycle is shown in Fig. 16-36. The condenser water cools the absorber before passing through the condenser.

16-21 EFFICIENCY OF ABSORPTION SYSTEMS

In absorption cooling systems performance is evaluated in two ways:
1. The energy efficiency. This is the cooling effect produced divided by the heat energy supplied to the absorber.
2. The effectiveness. This is the cooling effect produced divided by the work equivalent of the heat supplied to the absorber.

The second evaluation helps in comparing absorption systems with vapor compression systems where the input energy is work not heat.

16-22 INSTALLING ABSORPTION SYSTEM REFRIGERATORS

Absorption refrigerators must be carefully installed. The absorption unit must be carefully leveled to operate correctly. It will be necessary to install a gas supply line from the house gas piping to the refrigerator. This gas line must be tested for leaks. **Use only soap suds!** Gas pressure must be carefully adjusted. The minimum flame and maximum flame adjustments must be made. A water column manometer is usually used. See Chapter 20 for more information on fuel gas servicing.

Some of the refrigerators have electric circuits. The electrical service must be carefully checked. Some city codes specify that the fuse plug opening in the refrigerant system must be vented to the outside to prevent any chance of discharging the refrigerant into the house.

Always provide enough air inlet and exhaust for proper combustion, condenser and absorber cooling.

16-23 SERVICING ABSORPTION REFRIGERATORS

When servicing gas-fired absorption refrigerators, check to be sure that the gas supply is at the correct pressure. Check the gas pressure using a water-filled manometer, as shown in Fig. 16-37. (The amount of gas fed to the refrigerator may be checked by the size of the flame.) The safety valve body has a manometer connection. The flue must be kept clean to allow good transfer of heat. Brushes should be used to clean the flue.

Fins on the ammonia condenser must be cleaned periodically to make sure there is good heat removal from these surfaces.

If a service call indicates that the refrigerator is too cold, it should be checked as follows:
1. First, check the temperature control dial. It may be set too cold.
2. The temperature of the evaporator unit may be lower than that indicated by the temperature control dial setting. A time-temperature graph of the evaporator temperature should be taken.

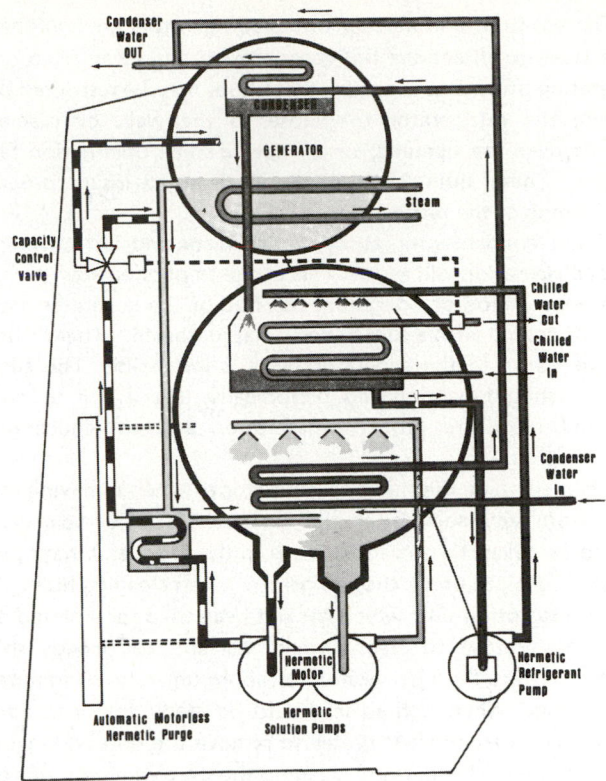

Fig. 16-36. Schematic of hermetic absorption system used to cool water. Chilled water is then used for air conditioning or other cooling.

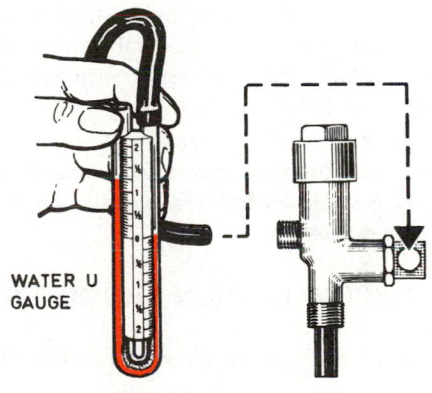

Fig. 16-37. Water-filled manometer measures gas pressure to burner. (General Electric Co.)

Perhaps the most common service call will be "little or no refrigeration." Following are some possible causes of this trouble:
1. Overloaded cabinet.
2. Improper condensing temperatures.
3. Little or no heating of the generating unit.
4. The gas supply has been turned off or restricted. If the line has become clogged, resulting in a small consumption of gas, there is, of course, little or no refrigeration. This trouble may be traced by checking the pressures at the burner.
5. Restricted or dirty gas flue.

The gas-fired and kerosene-fired refrigerators are equipped with flues to direct the hot gases around and away from the generating units. Occasionally, these flues may be restricted by placing the refrigerator too close to the wall, by placing objects over the opening, or by having some obstruction fall into it. These flues must be kept clean to insure proper functioning of the refrigerator.

If the condenser or absorber are dirty and lint covered, poor refrigeration will result. This is due to poor airflow.

After a period of operation, the flue of the generator may become coated with a soot deposit. Rapid transfer of heat from the gas flame to the generator body is impossible. This soot deposit should be removed periodically (every one to two months) to insure proper refrigeration and to reduce gas consumption.

When scraping the flue of a generator or when removing the soot from any surface of the generator, considerable care should be taken to prevent damage to the surface. Always put papers or a cloth under the refrigerator when cleaning flues.

An absorption unit which has not been used for a period of time may refuse to freeze when started. To remedy this trouble, invert the refrigerator for approximately 30 minutes to an hour. When righted it should be ready for operation. Some service technicians prefer to remove the unit and invert the unit only. This permits cleaning the unit while it is out of the cabinet.

If system is overheated, the pipe going to the condenser will overheat and the percolation pump will stop working. If the paint on the pipe to the condenser is blistered, overheating has taken place. To remedy, shut off the heat, allow the system to cool, turn the unit upside down several times to put the fluids in their proper place, and then restart with a lower heat input to the boiler (generator).

If there is a leak, a yellow deposit will collect at the leak. If the leak occurs at the evaporators, an ammonia odor will be noticeable. A burning sulphur candle will produce a white smoke at an ammonia leak.

16-24 SERVICING LITHIUM BROMIDE SYSTEMS

Evacuation of the system is necessary after the system is opened:
1. To be able to reach 40 F. (4 C.).
2. To remove noncondensables.
Evacuation is needed if the pressure in the system is 1" Hg. (25,400 microns) or more. The pressure is determined by a manometer which is connected by means of a service valve.

Lithium bromide is a nontoxic, nonflammable, non-explosive and chemically stable substance which is used as a liquid. It can be handled in open containers but becomes corrosive when exposed to air. It may irritate skin, eyes and mucous membranes. Octyl alcohol is sometimes added to reduce surface tension of lithium bromide because it acts as a wetting agent.

Important note: "Strong" (concentrated) solution means strong in ability to absorb. "Weak" (dilute) means weak in ability to absorb.

Sixty five percent lithium bromide by weight will start to crystallize at 110 F. (43 C.). Solution must not be allowed to reach high concentrations or low temperatures which allow crystallization.

The typical charge is a:
1. Lithium bromide solution   120 gal.
2. Inhibitor charge   1 pt.
3. Refrigerant (water)   35 gal.
4. Octyl alcohol (two ethyl hexanol)   1 gal.

The solution becomes thicker as the amount of lithium bromide increases. This will cause a greater temperature difference between refrigerator temperature and chilled water temperature. If solution concentration gets too high, the refrigerant will turn solid and must be dissolved. If the absorber becomes too cold (below 85 F.), solidification can also occur.

The system is charged with R-13 vapor to test for leaks (not soluble in water) and an electronic leak detector is used. Then the system is evacuated completely. Helium is also used but it requires a special leak detector or soap bubbles.

16-25 REVIEW OF SAFETY

The refrigerant most commonly used in the small absorption refrigerating units is ammonia. Its odor is pungent (sharp or irritating) and tends to restrict breathing. It is toxic and injurious to the skin and eyes. Avoid puncturing the system or creating too high a pressure in the system or a leak may result.

Caution: Never cut or drill into an absorption refrigerating mechanism. The high-pressure ammonia solutions are dangerous and may cause blindness if the fluid gets into the eyes.

Many of the absorption units are heated with LP gas or natural gas. The gas piping system must be leakproof. Always use soapsuds to check for leaks. Never use an open flame such as a match. An explosion may occur. The burner flues should be cleaned periodically or poor flame action may occur.

The flame safety device should be checked. To do this, smother the flame and check to determine if the safety valve closes.

The condenser duct system and the condenser should be cleaned at least each six months or excessive condenser pressures may result.

Some absorption systems use electrical current as well as fuel gas. The usual precautions should be used in handling these circuits.

It is a good idea to ground these refrigerators to eliminate any danger of receiving a shock should a circuit become grounded to the frame of a cabinet or part of the mechanism.

16-26 TEST YOUR KNOWLEDGE

1. Who discovered the absorption principle?
2. What localities would probably need kerosene-fired intermittent absorption refrigerators?
3. Name three substances which have been used in absorption refrigerators to absorb refrigerant gas.

4. In absorption refrigerators, does the liquefication of the refrigerant depend upon compression?

5. Why is the ammonia and water combination popular?

6. What purpose does hydrogen serve in the continuous absorption system?

7. Why is the storage cylinder or receiver in the intermittent absorption refrigerator insulated?

8. Why are the mechanisms provided with a fuse plug?

9. How can burning more gas in the continuous absorption cycle produce more cold?

10. What three fluids are used inside a continuous absorption cycle mechanism?

11. How is the cabinet temperature adjusted in the continuous absorption cycle refrigerator?

12. Why must the absorption unit be level?

13. Is kerosene sometimes used as the fuel for continuous systems?

14. Why is a pressure regulator not required in the base of the cabinet for LP systems?

15. How is the temperature regulated in a piped gas continuous system?

16. Why doesn't the flame go out when the cabinet is cold enough in a continuous LP gas system?

17. What are two basic causes for too little refrigeration in a continuous system?

18. How are some continuous cycle systems automatically defrosted?

19. What is the purpose of lithium bromide in an absorption system?

20. Do any absorption systems operate under a vacuum?

21. What does the solution pump do in an ammonia water five-ton air conditioner?

22. What causes the liquid refrigerant to flow in a domestic absorption system?

23. Does the generator percolate a weak-in-ammonia and strong-in-water solution?

24. Can a weak-in-ammonia solution be used to defrost the freezer evaporator?

25. Is hot ammonia vapor sometimes used to defrost the freezer evaporator?

Chapter 17

SPECIAL REFRIGERATION SYSTEMS AND APPLICATIONS

Previous chapters have covered the common compression and absorption systems of refrigeration. These systems include most of the mechanisms in present use and the majority of the refrigerants. There are a few unusual devices which have special applications. Some of these are covered in this chapter.

17-1 EXPENDABLE REFRIGERANT SYSTEMS

Use of liquid nitrogen or liquid carbon dioxide for cooling transportation vehicles (truck bodies) is rapidly increasing. Expendable refrigerant systems are also used in the cooling of railroad cars and shipping containers used to transport perishable items. Basic system uses a liquid nontoxic low-temperature refrigerant. It is the same as any vapor system but has no condensing unit. The low cost liquid can be used as a refrigerant, then is released to the atmosphere. This is called chemical or open cycle refrigeration.

Latent heat of vaporization of the refrigerant determines its heat absorbing ability. The three most common refrigerants and their latent heat of vaporization are shown in Fig. 17-1.

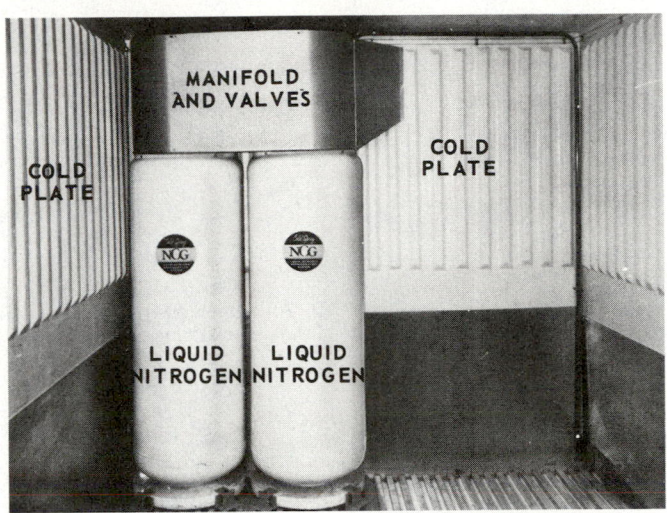

Fig. 17-2. Expendable refrigerant system. Two nitrogen cylinders located inside truck body are connected by a manifold to regulators and to temperature control solenoid valves. Vaporizing liquid nitrogen flows into vaporizers or cold plates to refrigerate truck box.

REFRIGERANT	BOILING TEMPERATURE		LATENT HEAT OF VAPORIZATION Btu/lb.
	F.	C.	AT 32 F. (0 C.) AND 0 psi
AMMONIA	−28.1	−33	590.4
CARBON DIOXIDE	−109.3	−78.5	275.
NITROGEN	−320.5	−196	173.

Fig. 17-1. Heat absorbing ability of expendable refrigerants.

All are fairly high in latent heat of vaporization. If properly processed, they create little air pollution when released into the atmosphere.

Refrigerant is supplied in large cylinders. These are replaced as the refrigerant is used up.

Two basic cooling mechanisms are in common use. One is cold plate cooling; the other is spray cooling.

17-2 EXPENDABLE REFRIGERANT EVAPORATOR SYSTEM

Liquid refrigerant is kept in large metal insulated cylinders. These are really large thermos bottles. Sometimes they are

located in the front of the cargo vehicle, as shown in Fig. 17-2. Each unit has a temperature control providing a temperature range of −20 F. (−29 C.) to 60 F. (16 C.).

The temperature control is connected to a temperature sensor much the same as in a standard thermostatic motor control. As temperature rises, the switch operating the control valve is opened and liquid refrigerant flows into the evaporators. The evaporators may be blower coils, plates or eutectic plates. (See Chapter 12.) As liquid refrigerant passes through the plates, it vaporizes. Vapor is forced through the plates by the pressure difference. When the desired temperature is reached, the refrigerant valve closes. The used vapor leaves the evaporator at approximately the same temperature as the air in the cargo space. With this method, no refrigerant mixes with the air in the interior of the vehicle.

17-3 EXPENDABLE REFRIGERANT SPRAY SYSTEM

Transport vehicles are effectively cooled by spraying liquid nitrogen or carbon dioxide directly into the refrigerated space. The nitrogen turns into vapor inside the cargo area. Fig. 17-3

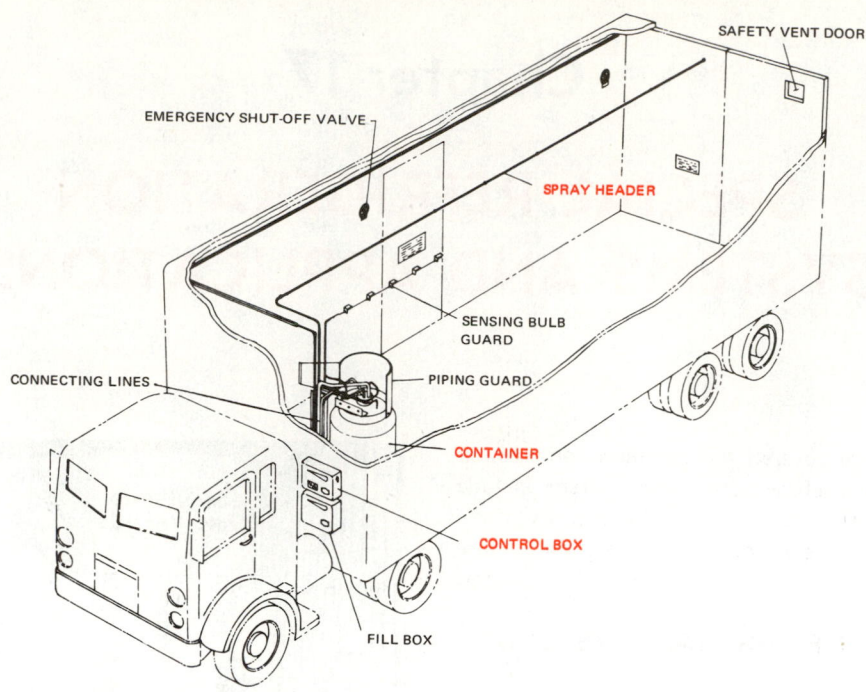

EMERGENCY SHUT-OFF VALVE

SAFETY VENT DOOR

SPRAY HEADER

SENSING BULB GUARD

CONNECTING LINES

PIPING GUARD

CONTAINER

CONTROL BOX

FILL BOX

Fig. 17-3. An expendable refrigerant system for a refrigerated truck. Liquid nitrogen is in the insulated container. The container is installed vertically inside the truck body.

shows a complete spray system with the cylinder inside the truck body. Details of the system are shown in Fig. 17-4.

The liquid spray method has many of the same parts as the cold plate method — liquid containers, control box and fill box, for examples. It also requires additional devices not necessary in the plate method, such as spray headers, emergency switches and safety vents. Another type of spray cooling system is shown in Fig. 17-5. Details of the horizontal cylinder system are shown in Fig. 17-6.

This is how the system operates:

1. Liquid nitrogen is pumped into the storage cylinder by way of the fill box. (Fig. 17-7 shows the details of the container piping connections.)

2. When the containers are filled and the cargo space is loaded, the temperature is selected at the main control. A temperature sensing device anticipates temperature changes in the cargo space. See Fig. 17-8.

3. When the cargo temperature rises above the setting, the temperature controller opens the liquid line solenoid valve. This allows liquid refrigerant to enter into the spray header where it becomes a vapor and maintains the desired temperature.

Some units have two or more containers — a primary container and a secondary container — which are filled in series. As the first or primary container is filled, liquid refrigerant will overflow into the second container and so forth. Fig. 17-9 is a wiring diagram for a typical installation.

The spray header system is a perforated pipe usually mounted along the roof at the center of the cargo compartment. The nitrogen tanks are equipped with safety valves

which will release nitrogen to the outside of the vehicle if the pressure in the containers rises above 22 psi.

All units have a safety vent, Fig. 17-10, which allows gas to exhaust to the atmosphere when inside cargo space pressure increases above atmospheric. This safety vent is a spring loaded device and will close when excess pressure has been released. In addition, each door has a safety switch connected to it which will automatically shut down the unit when a door is opened and before anyone can enter. A system which maintains four compartments in the truck body at different temperatures is shown in Fig. 17-11.

Some truck bodies have electric heaters. These heaters are used when the truck is exposed to temperatures below the temperature wanted inside the truck. See Fig. 17-12.

Always read the warning signs on refrigerated vehicles before entering them. Caution: the temperature of liquid nitrogen, as it comes from the spray nozzles — depending on the thermostat setting — is much below 0 F. (—18 C.). Were it to hit any part of a human body, the flesh would be frozen instantly. Be sure that no living animal or human is in the refrigerated space when the doors are closed.

The spray cooling system which uses nitrogen or carbon dioxide has other advantages beyond the ease in providing necessary refrigerating temperatures. Since these gases take the place of oxygen in the storage space, fruits, vegetables, meats and fish — either in transit or in storage — are preserved by the inert atmosphere.

The inside of the expendable refrigerant systems must be kept free of dirt and moisture. To pressure test the piping, use only nitrogen or helium. CAUTION: See Para. 11-39.

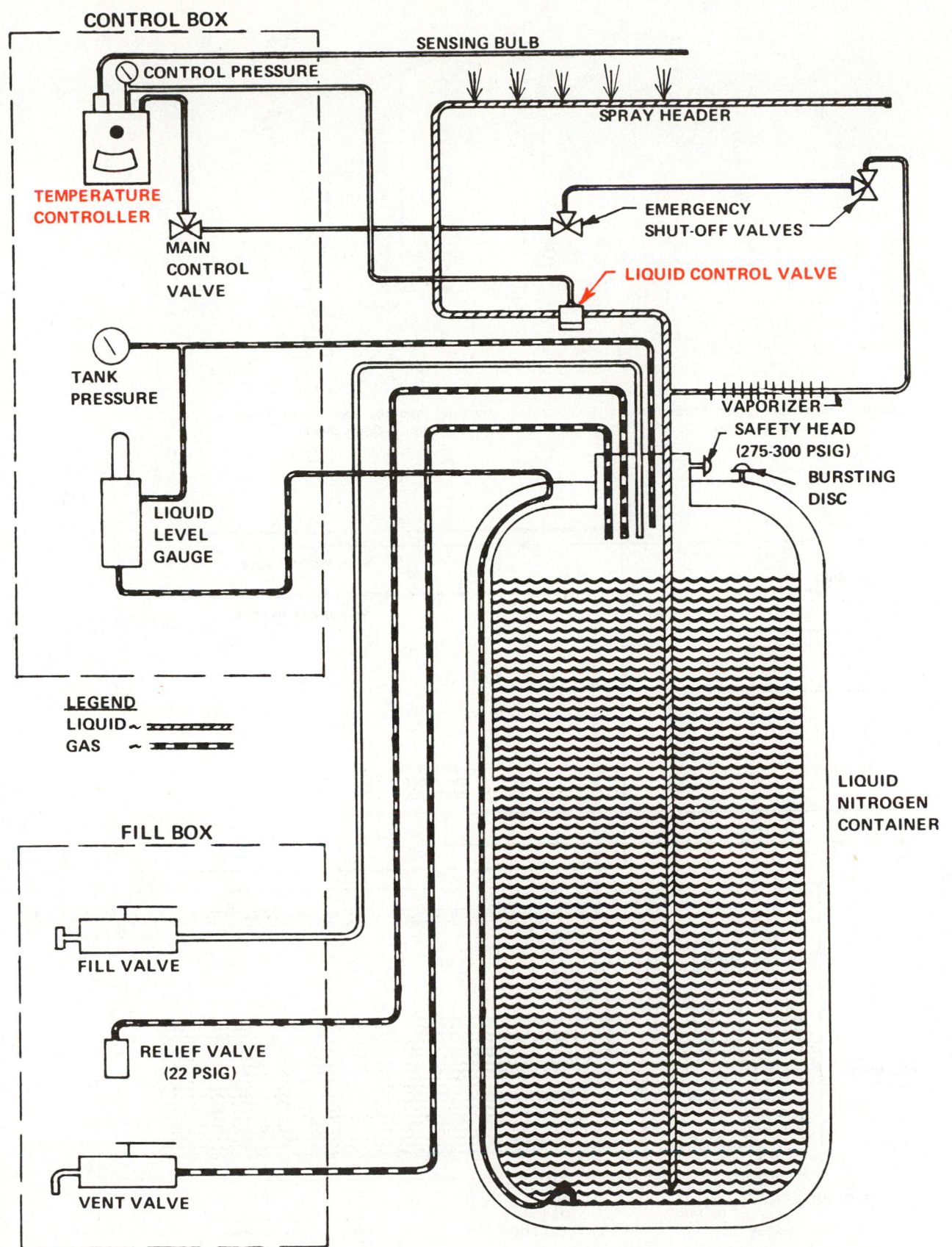

CONTROL BOX

SENSING BULB

CONTROL PRESSURE

SPRAY HEADER

TEMPERATURE CONTROLLER

MAIN CONTROL VALVE

EMERGENCY SHUT-OFF VALVES

LIQUID CONTROL VALVE

TANK PRESSURE

VAPORIZER

SAFETY HEAD (275-300 PSIG)

BURSTING DISC

LIQUID LEVEL GAUGE

LEGEND
LIQUID ~
GAS ~

LIQUID NITROGEN CONTAINER

FILL BOX

FILL VALVE

RELIEF VALVE (22 PSIG)

VENT VALVE

Fig. 17-4. A vertical container expendable refrigerant system. Note controls, safety devices and filling piping. (Union Carbide Corp., Linde Div.)

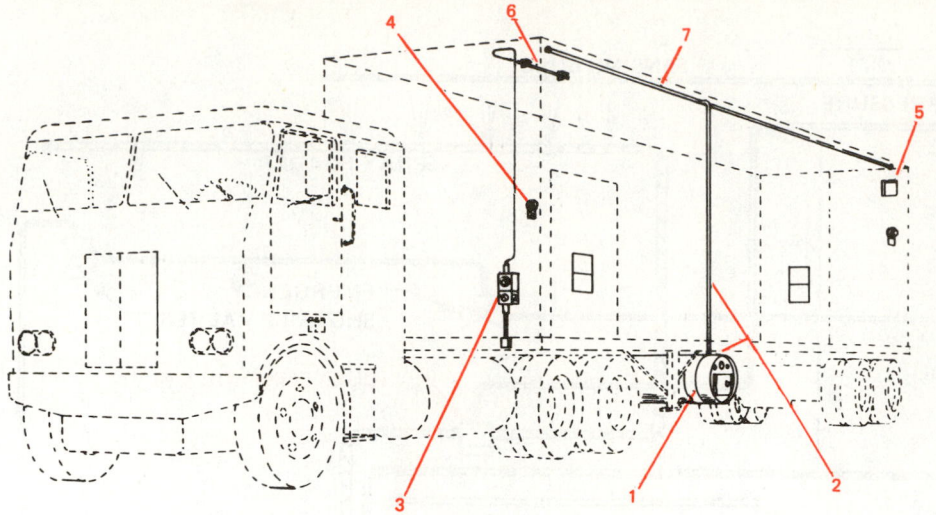

Fig. 17-5. An expendable refrigerant system using nitrogen. 1—Horizontal container mounted under truck body. 2—Liquid line. 3—Controls. 4—Shutoff valve. 5—Safety vent door. 6—Sensing bulb. 7—Spray header.

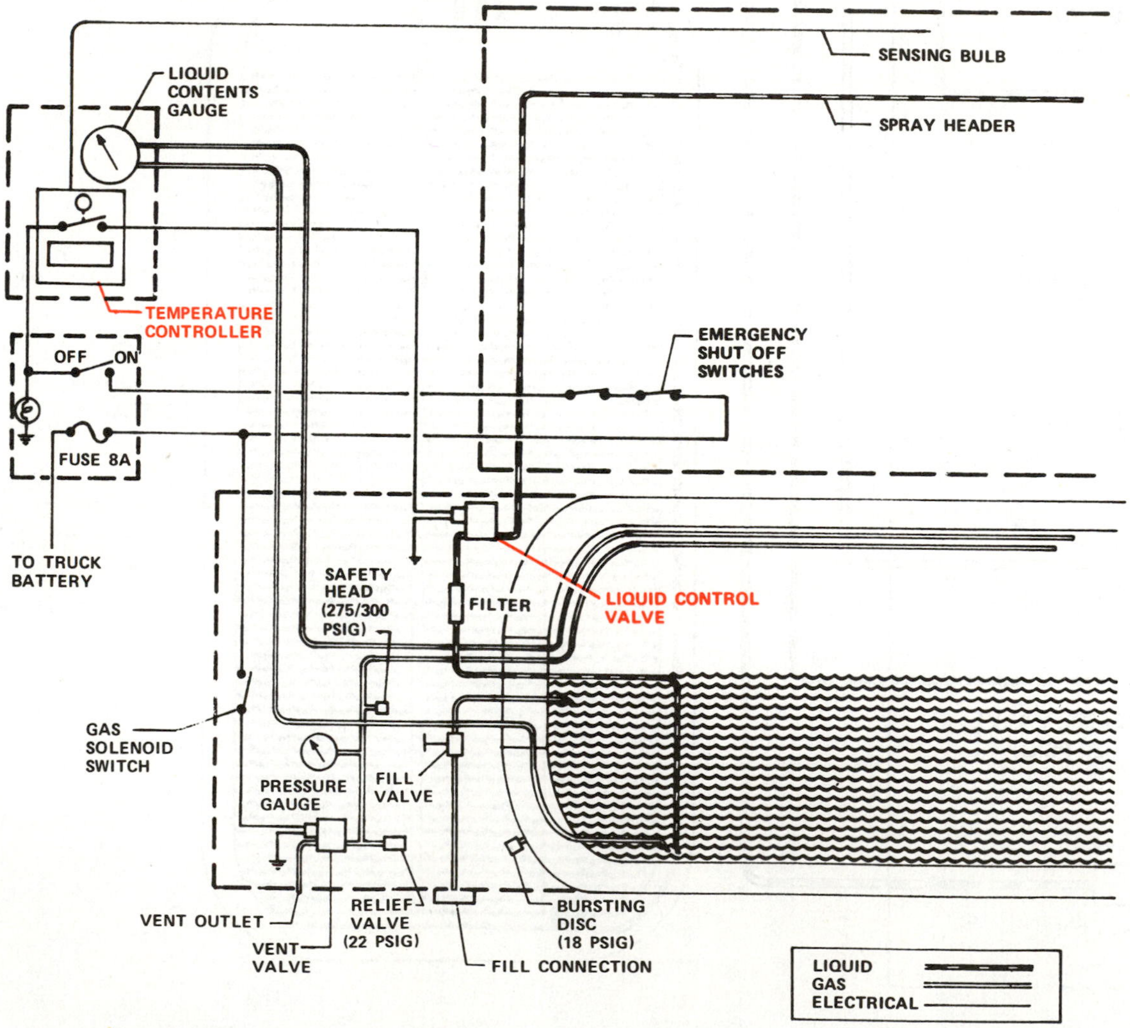

Fig. 17-6. Expendable refrigerant system showing piping, safety devices and controls.

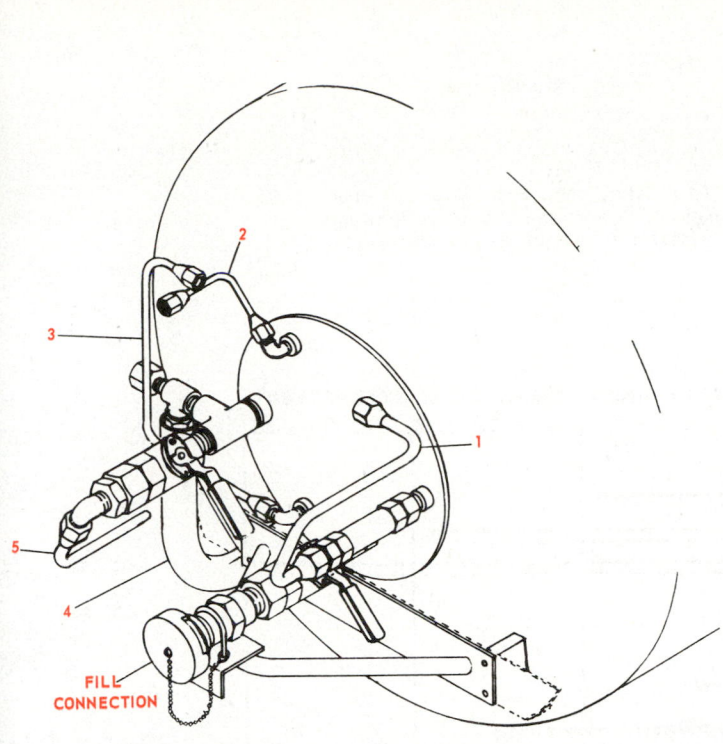

Fig. 17-7. Piping connections to container of expendable refrigerants. 1—Liquid line connection. 2—Liquid level gauge connection, vapor phase. 3—Liquid level gauge connection, liquid phase. 4—Polyethylene tubing. 5—Line assembly. (Union Carbide Corp., Linde Div.)

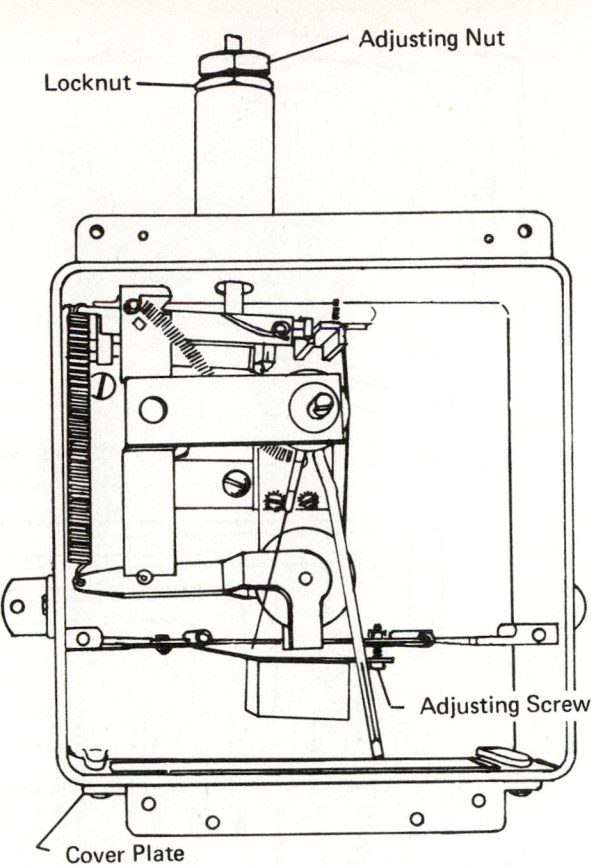

Fig. 17-8. Temperature controller used with expendable refrigerant systems. Unit has both temperature (range) and differential adjustments.

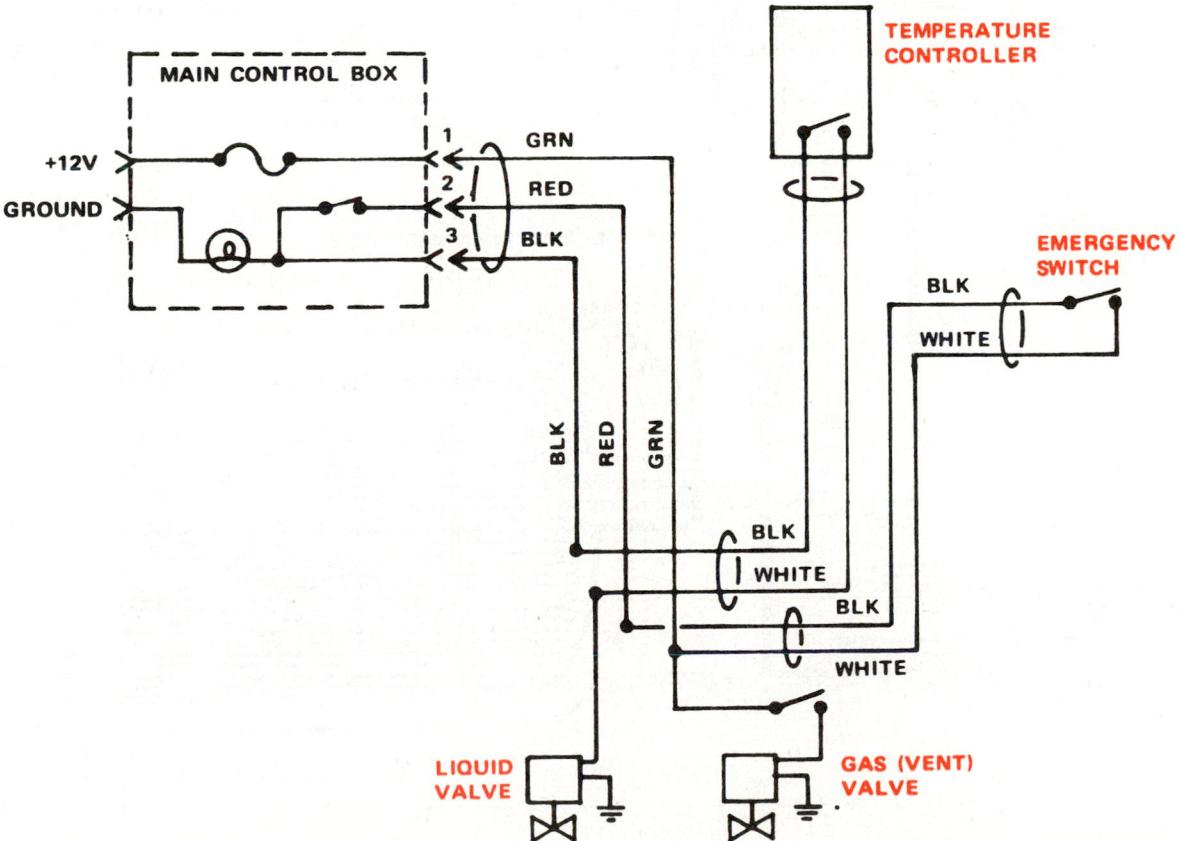

Fig. 17-9. Diagram of circuit for automatic expendable refrigerant system.

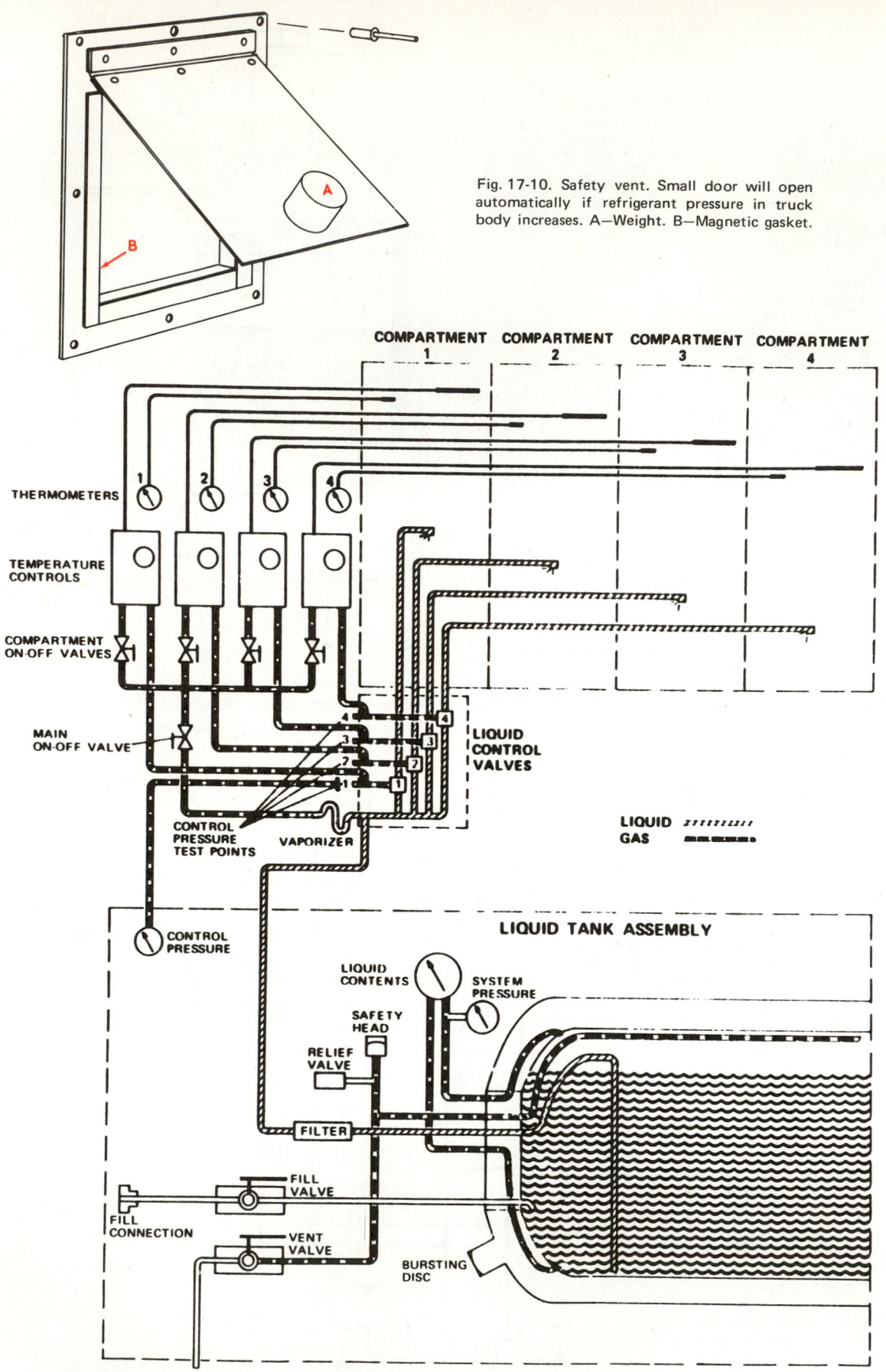

Fig. 17-10. Safety vent. Small door will open automatically if refrigerant pressure in truck body increases. A—Weight. B—Magnetic gasket.

COMPARTMENT 1   COMPARTMENT 2   COMPARTMENT 3   COMPARTMENT 4

THERMOMETERS

TEMPERATURE CONTROLS

COMPARTMENT ON-OFF VALVES

MAIN ON-OFF VALVE

LIQUID CONTROL VALVES

CONTROL PRESSURE TEST POINTS

VAPORIZER

LIQUID
GAS

CONTROL PRESSURE

LIQUID TANK ASSEMBLY

LIQUID CONTENTS
SYSTEM PRESSURE
SAFETY HEAD
RELIEF VALVE
FILTER

FILL VALVE
FILL CONNECTION
VENT VALVE
BURSTING DISC

Fig. 17-11. Expendable system with four separately controlled sprayers and controls. Four compartments in truck body can be kept at four different temperatures.

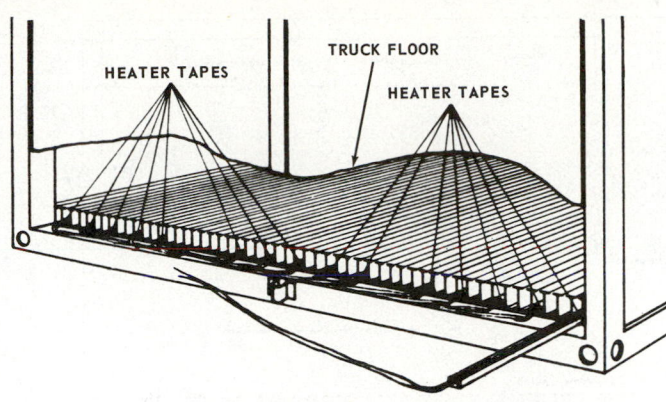

Fig. 17-12. Electric heating system for truck bodies. (Union Carbide Corp., Linde Div.)

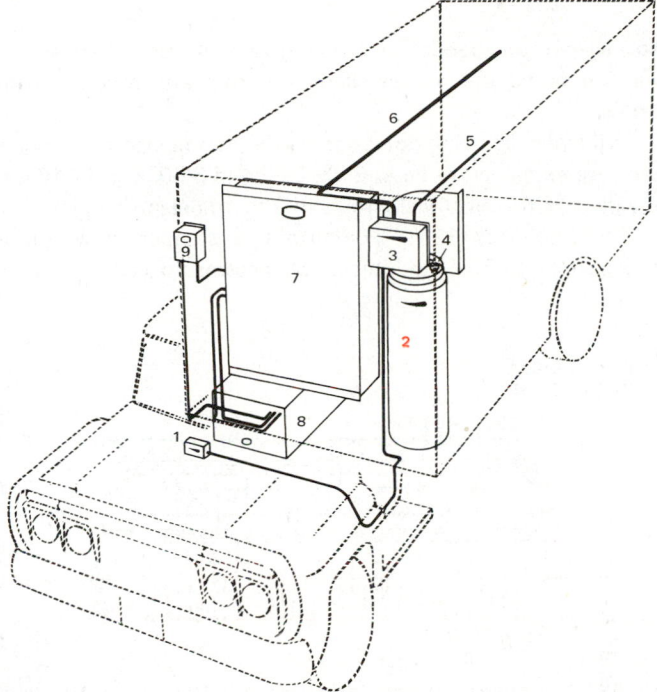

Fig. 17-13. Truck body with both expendable refrigerating system and compression refrigerating system. 1—Cab main switch. 2—Nitrogen container. 3—Fill and control box. 4—Nitrogen control valve. 5—Thermostat bulb. 6—Spray header. 7—Blower evaporator. 8—Condensing unit. 9—Control box. (Dole Refrigerating Co.)

17-4 COMBINATION SPRAY AND COMPRESSOR SYSTEM

A nitrogen spray header system may be combined with a compression system. This system is used in trucks. Liquid nitrogen sprayers refrigerate the product and absorb the heat leakage during normal delivery. There is also a plug-in compression refrigeration system and a blower evaporator for use during the night and weekend hours when the vehicle is not on the highway. This combination is shown in Fig. 17-13, as it would be installed in an ice cream delivery truck. The

parts of the system are:
1. Main switch box mounted on dash board.
2. Liquid nitrogen container.
3. Tank fill control.
4. Liquid nitrogen control valve.
5. Temperature sensing bulb.
6. Nitrogen spray header.
7. Pancake blower coil.
8. Condensing unit.
9. Front-mounted control box for plugging in electrical power.

17-5 OPEN CYCLE AMMONIA

There are two different methods of using ammonia in open cycle refrigeration.
1. Burn the exhaust ammonia vapor in the atmosphere.
2. Burn the ammonia in an internal combustion engine.

Ammonia has a very high latent heat of vaporization. This makes it a good heat-absorbing substance. Large quantities of commercial grade ammonia are used in agriculture. This makes possible a rather inexpensive product.

Ammonia vapors are very irritating and offensive. They cannot be tolerated by humans or animals. In an open cycle refrigeration system, using ammonia, the exhaust must be processed in some way so that the irritating quality is removed.

The simplest way to do this is to burn the vapor in an open flame. The chemical formula is NH_3; this means that ammonia is made up of nitrogen and hydrogen. When ammonia burns in an open flame, the hydrogen combines with oxygen from the air and forms water. The odorless nitrogen is released to the atmosphere.

Another system of processing expendable ammonia vapor is to burn the vapor in an internal combustion engine. In this way the vapor produces some fuel and, as in the open flame, the products of combustion will be released as nitrogen and water vapor.

17-6 DRY ICE — CO_2 PELLETS

Dry ice pellets are used in many types of food dispensing devices: ice cream dispensing carts, truck transports, railroad dining cars and camping equipment. Another use is in fresh food freezing.

Dry ice pellets are small pieces of solid carbon dioxide. These pieces sublime at a temperature of approximately −109 F. (−78 C.)

Sublime means that the solid dry ice does not pass through the liquid stage. It goes directly from a solid to a carbon dioxide vapor form. In going from the solid to the vapor state, one pound of solid carbon dioxide will absorb 248 Btu of heat.

A popular type of carbon dioxide refrigerator consists of a bin or container with a compartment to hold the carbon dioxide pellets. See Para. 17-13. A refrigerating cycle serves as a heat sink. Liquid refrigerant flows by gravity from the condenser into the evaporators where it absorbs heat from the compartment being cooled. This arrangement provides a wide range of refrigerating temperatures ranging from −10 F. (−23 C.) to 40 F. (4 C.).

This type of container may be charged with CO_2 pellets and the storage space filled with frozen cargo. Then, loaded aboard a truck or airplane, it will maintain the desired temperature until it reaches its destination. A simplified drawing of such a device is shown in Fig. 17-14. Its refrigerating cycle is shown in Fig. 17-15.

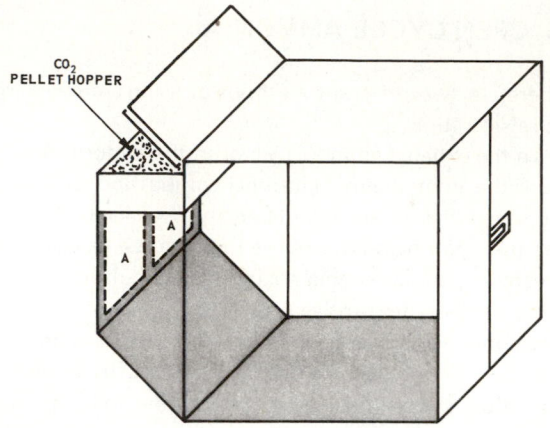

Fig. 17-14. A carbon dioxide pellet refrigerated shipping container. It uses closed-circuit refrigerant circuit shown at A for cooling container. This unit holds 200 lb. of dry ice and will maintain 34 F. (1 C.) for four days or 10 F. (−12 C.) for two days. (Airco Industrial Gases)

17-7 THERMOELECTRIC REFRIGERATION

The thermoelectric process is a means of removing heat from one area and putting it in another area using electrical energy rather than refrigerant as a "carrier". It has been used mainly in portable refrigerators and luxury type stationary domestic refrigerators, water coolers and for cooling the scientific apparatus used in space explorations and in aircraft.

Thermoelectric cooling requires none of the conventional equipment necessary in a vapor system. There is no compres-

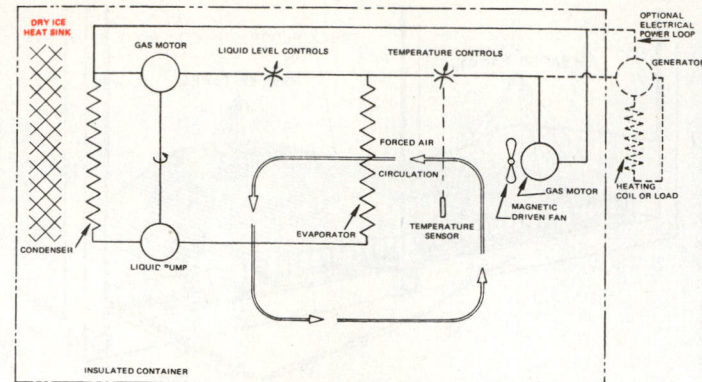

Fig. 17-15. Carbon dioxide pellet refrigerating system. In the secondary refrigerating system, refrigerant vapor is used to operate fan and liquid pump. (Statham Instruments, Inc.)

sor, evaporator, condenser or refrigerant. In fact, there are no moving parts; the unit is silent, compact and requires little service.

Principles covering operation of the thermoelectric refrigerator are explained in Para. 6-25, 6-52 and 6-60. Fig. 17-16 is a diagram of the electrical circuit in a thermoelectric operation.

The input of 120V a-c current is stepped down in a transformer to 20V a-c. This current passes through a rectifier

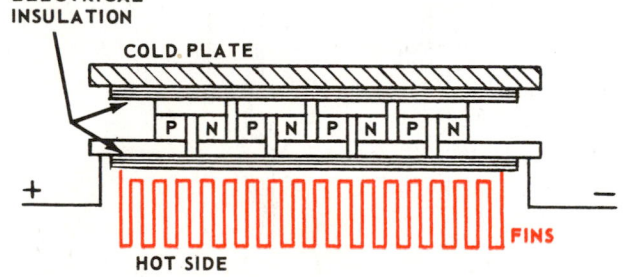

Fig. 17-17. Assembled thermoelectric module. Note use of fins (hot side) to speed up removal of heat which has been absorbed from surface of the cold plate. (Koolatron Industries.)

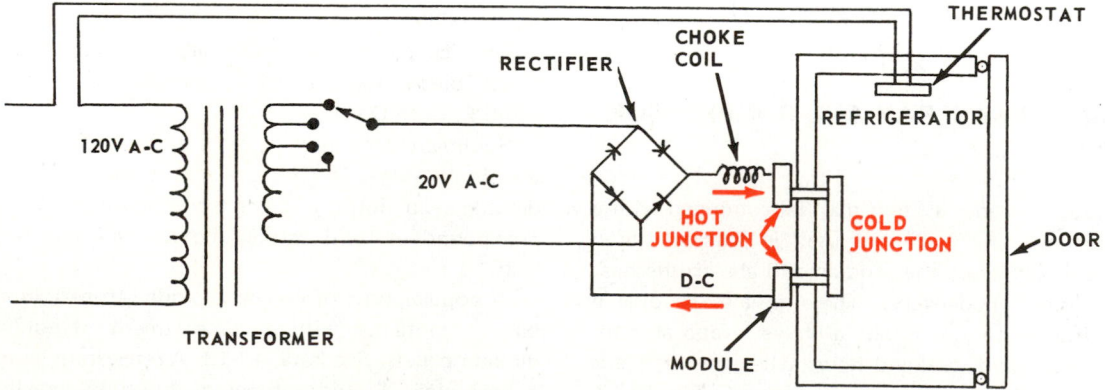

Fig. 17-16. Electrical circuit for thermoelectric refrigerator.

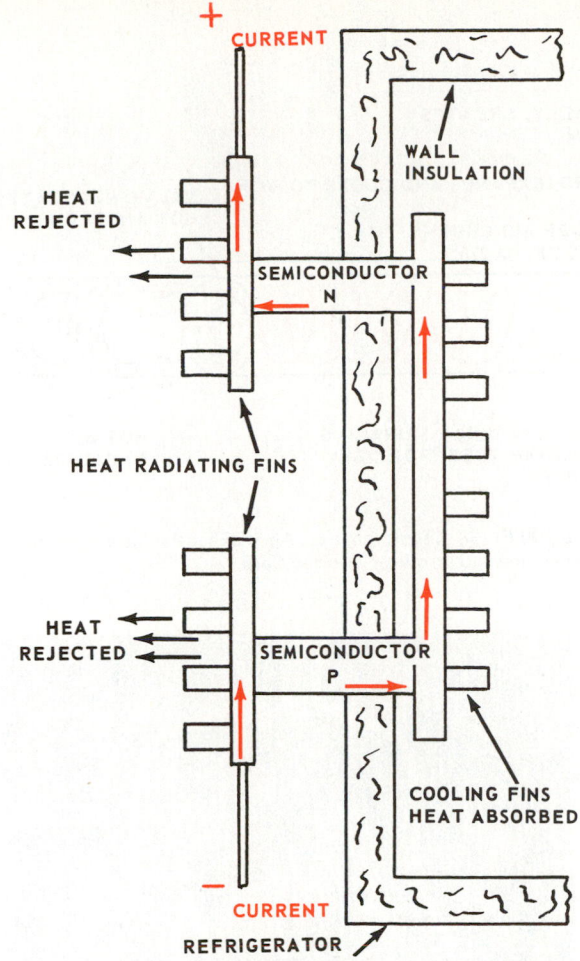

Fig. 17-18. Diagram of simplified thermoelectric system used in cooling small areas. Note flow of current to produce cooling within box.

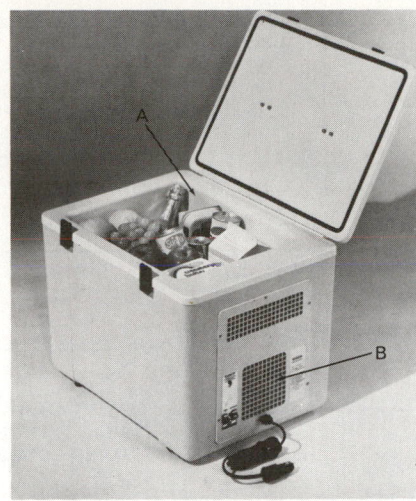

Fig. 17-19. Small portable refrigerator operates from normal 120V household current or from 12V battery. A—Cooling surface. B—Airflow openings for cooling hot junctions. (Koolatron Industries)

and is changed to 20V d-c. This direct current is then passed through the thermoelectric module so that the junction inside the refrigerator becomes cold and the junction outside the refrigerator is warmed.

Thermoelectric cooling units, when used in refrigeration, are called modules. A module consists of several cold and hot junctions in series. The diagram of such a module is shown in Fig. 17-17. *The letters P and N, used in thermoelectric application, do not refer to current polarity positive (+) and negative (−).* P and N, in thermoelectric units, refer to the properties of the semiconductor materials. The materials are also designated as positive or negative depending on how the semiconductor electrons behave under the influence of current flow. Construction of a module attached to a refrigerator is shown schematically in Fig. 17-18. The direction of the direct current flow into the module determines whether the junction will be warmed or cooled.

The reversing switch is usually made a part of the electrical current. This makes it possible to cause the cabinet to either cool the food or warm it.

The use of thermoelectric refrigerators is increasing, even though they have a low coefficient of performance (COP). The unit has versatility. It can operate on 12V d-c power or from a 110V a-c power adaptor. This has made the unit popular for picnics and camping use. One of these refrigerators is illustrated in Fig. 17-19.

17-8 VORTEX TUBE

The vortex tube is an interesting device capable of providing both cooling and heating at the same time. See Chapter 19. Its source of energy is compressed air. It converts the compressed airflow into two streams of air, one hot and one cold.

The amount of hot or cold air released from the outlets can be varied. For example, a unit with a 100 psi air supply at 70 F. (21 C.) can be adjusted to cool half the air to −29 F. (−34 C.) while heating the other half to 91 F. (33 C.).

Fig. 17-20 is a schematic drawing of a vortex tube. Compressed air enters at A. It then goes into a number of

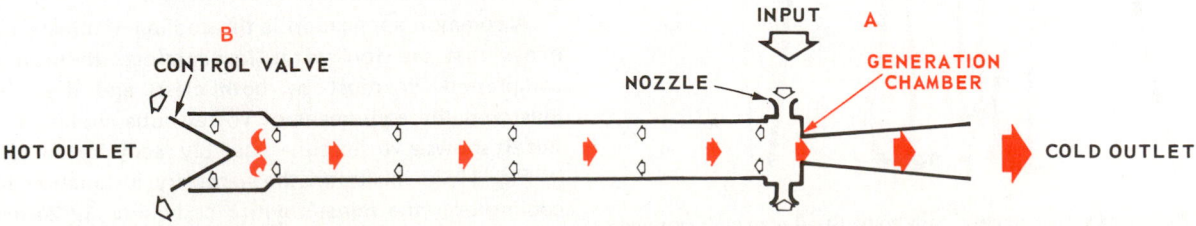

Fig. 17-20. Schematic drawing of vortex tube. A—Compressed air inlet. B—Control valve. (Vortec Corp.)

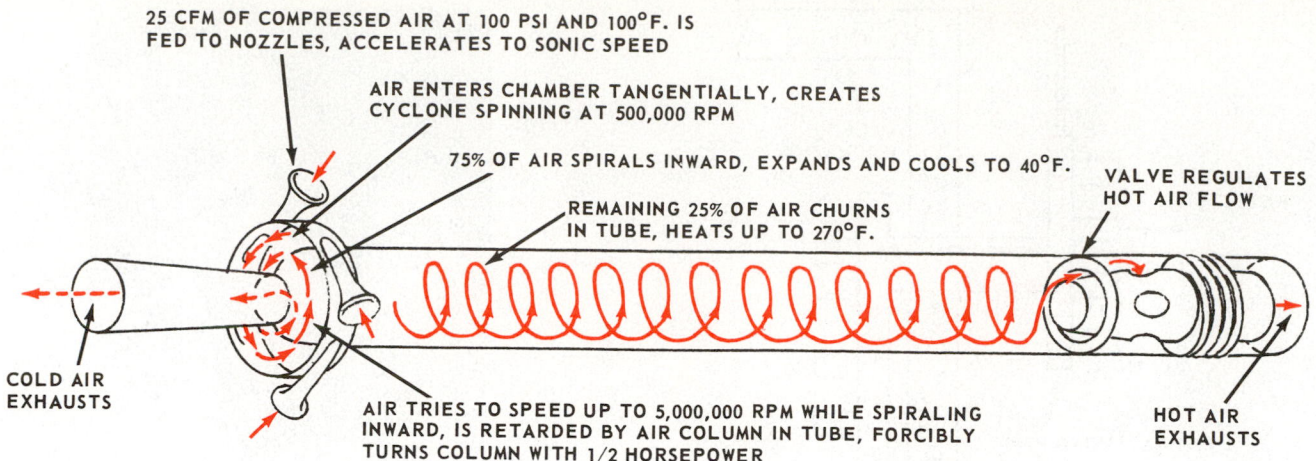

25 CFM OF COMPRESSED AIR AT 100 PSI AND 100°F. IS FED TO NOZZLES, ACCELERATES TO SONIC SPEED

AIR ENTERS CHAMBER TANGENTIALLY, CREATES CYCLONE SPINNING AT 500,000 RPM

75% OF AIR SPIRALS INWARD, EXPANDS AND COOLS TO 40°F.

REMAINING 25% OF AIR CHURNS IN TUBE, HEATS UP TO 270°F.

VALVE REGULATES HOT AIR FLOW

COLD AIR EXHAUSTS

AIR TRIES TO SPEED UP TO 5,000,000 RPM WHILE SPIRALING INWARD, IS RETARDED BY AIR COLUMN IN TUBE, FORCIBLY TURNS COLUMN WITH 1/2 HORSEPOWER

HOT AIR EXHAUSTS

Fig. 17-21. Diagrammatic view of vortex tube. Note flow of air at 100 psi and 100 F. (38 C.) into nozzles. Air is exhausted at temperature of 40 F. (4 C.) at cold end. Cold air is shown by the broken line and arrows. (Vortec Corp.)

small nozzles where it will lose some of its original high pressure. As it expands, it moves at near sonic speed (velocity). This is shown in Fig. 17-21.

Nozzles are arranged so that the air is injected tangentially (around the inner surface) to the circumference of the generation chamber. This makes the air swirl or spin like a cyclone. The control valve at the end of the hot air tube controls the flow of both the heated air and the cooled air. The position of the control valve determines how much air will leave the hot end and how much air will be forced out at the cold end.

There are two streams of spinning air going through the generation chamber. The outside layer moves toward the hot outlet and the inside layer moves towards the cold outlet. In this process, the air traveling through the center becomes very cold and leaves the vortex tube at the cold opening providing the required cooling.

Vortex tube cooling has many different applications. Since

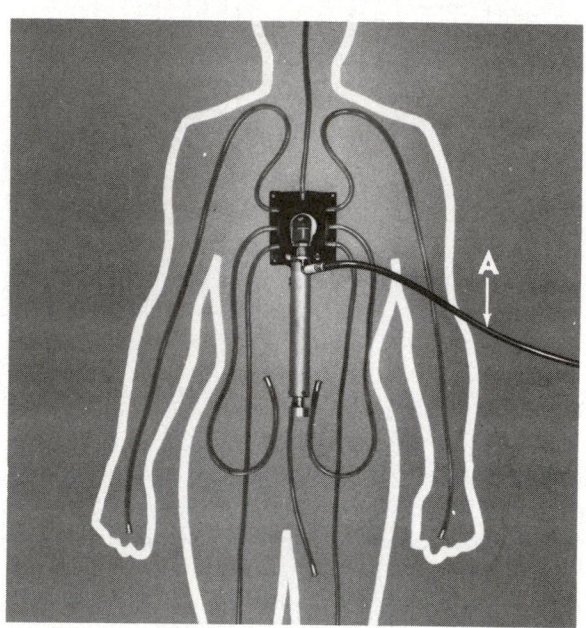

Fig. 17-23. Vortex tube cooling device fitted inside miner's clothing. Note distribution tubes which carry cooled air to various parts of wearer's body. A—Compressed air supply.
(Mine Safety Appliances Co.)

it must depend upon the supply of compressed air at fairly high pressure, it is often used where the air exhaust — either hot or cold — for the vortex tube can be used for other purposes also. It is particularly desirable in locations where both ventilation and cooling are needed.

A common application is the cooling of miners' clothing in mines that are too warm for comfort. In most uses, the compressed air must be both clean and dry. Fig. 17-22 illustrates the equipment of vortex tube cooling in a miner's suit. It shows a vortex tube assembly ready for the suit.

Fig. 17-23 illustrates the necessary installation to provide cooling over the miner's entire body. Fig. 17-24 illustrates a garment and hood fitted with a vortex tube heating and cooling device.

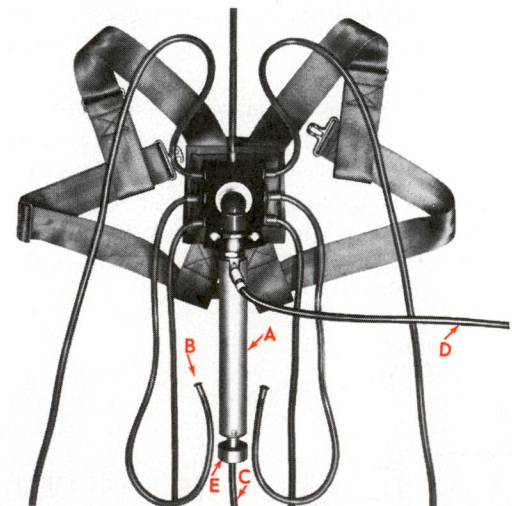

Fig. 17-22. Vortex tube assembly ready to be fitted to miner's clothing. A—Vortex tube device. B—Cooled air outlet (seven). C—Exhaust warm air. D—Compressed air lines. E—Temperature adjustment valve.

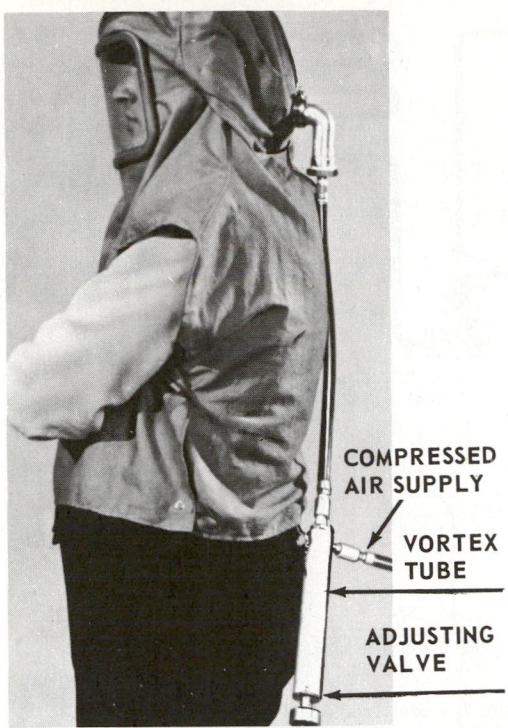

Fig. 17-24. Garment and hood fitted with vortex tube cooling and heating device.

Many industries use the vortex tube for cooling tool bits and in similar applications where the use of liquid coolants would be undesirable. It is particulary useful in installations where the stream of cold air may be used for both cooling and removing chips or in installations where the exhaust air will provide good air for the operator to breathe. Typical vortex tubes are illustrated in Fig. 17-25.

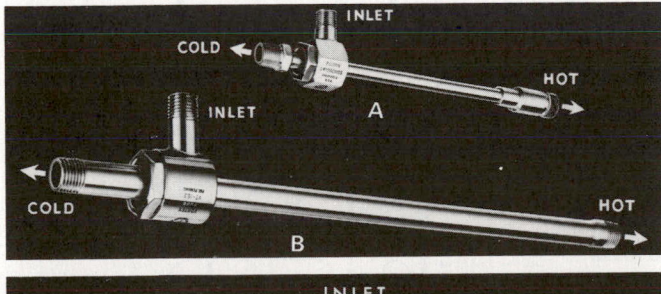

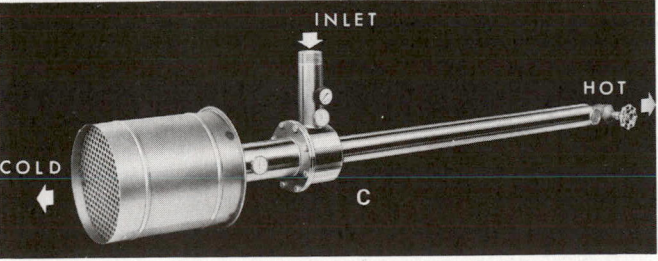

Fig. 17-25. Three vortex tubes. A—Tube 6 1/2 in. long weighing 2 1/2 oz. B—Tube 10 in. long and weighs 6 oz. C—Large capacity vortex tube. Cold end is 10 in. in diameter. Inlet tube has dial thermometer and pressure gauge. Cold and hot ends have dial thermometers. (Vortec Corp.)

17-9 STEAM JET COOLING SYSTEM

The steam jet system consists mainly of a venturi in the steam jet ejector, a condenser, evaporator or flash chamber, chilled water circulating pump and condensate pump.

The principle of operation is based on the fact that water under a high vacuum boils at a relatively low temperature. This causes evaporation to occur and reduces the temperature. Chapter 19 shows a table of water boiling temperatures under varying vacuum pressures. However, since water is the refrigerant used in steam jet applications, only temperatures down to about 40 F. (4 C.) are possible.

The steam jet system is used primarily as a means of cooling water which will in turn be used for either comfort cooling or process temperature cooling. Thus, it is used in air conditioning, cold water gas absorption in chemical plants, beverage cooling in distilleries and many other applications where a steam power plant is needed for other uses. Its operation is normally limited to installations where there is an abundance of steam and condensing water at a low cost and where the desired temperatures are in the 40 F. to 50 F. (4 C. to 10 C.) range.

Since a relatively low steam pressure is required, steam jet cooling is often used in plants which use high-pressure steam for operating machines. The exhaust steam is right for steam jet refrigeration.

17-10 MULTISTAGE SYSTEMS

Multistage systems are used where ultralow temperatures are desired but cannot be obtained economically through the use of a single-stage system. This is because the compression ratios would be too high to get the necessary evaporating and condensing vapor temperatures.

The name, multistage, applies to any refrigeration system which contains more than one stage of compression. There are two general types: cascade and compound.

In the cascade system, as shown in Fig. 17-26, two separate refrigerant systems are interconnected so that the evaporator from one unit is used to cool the condenser of the other unit. This arrangement allows one of the units to operate at a lower temperature and pressure than would be possible with the same size single-stage system.

Cascade refrigeration systems may be used to produce temperatures as low as −250 F. (−157 C.). See Cryogenics in Para. 1-56. The cascade system is actually two independent units. It allows, if desired, the use of two different refrigerants.

Compound systems obtain low temperatures by using several compressors connected in series to the same refrigeration system. Fig. 17-27 is a diagram of such a unit. It can increase the performance and efficiency of low-temperature refrigeration systems. Vapor in the evaporator of a low-temperature system has a high specific volume at a low temperature. In a single-stage system this would require a longer-than-normal compressor piston stroke operating at high speeds. Because of the temperature involved, it would also reduce the volumetric efficiency to such a point that it would cost too much. In the compound system, the first stage

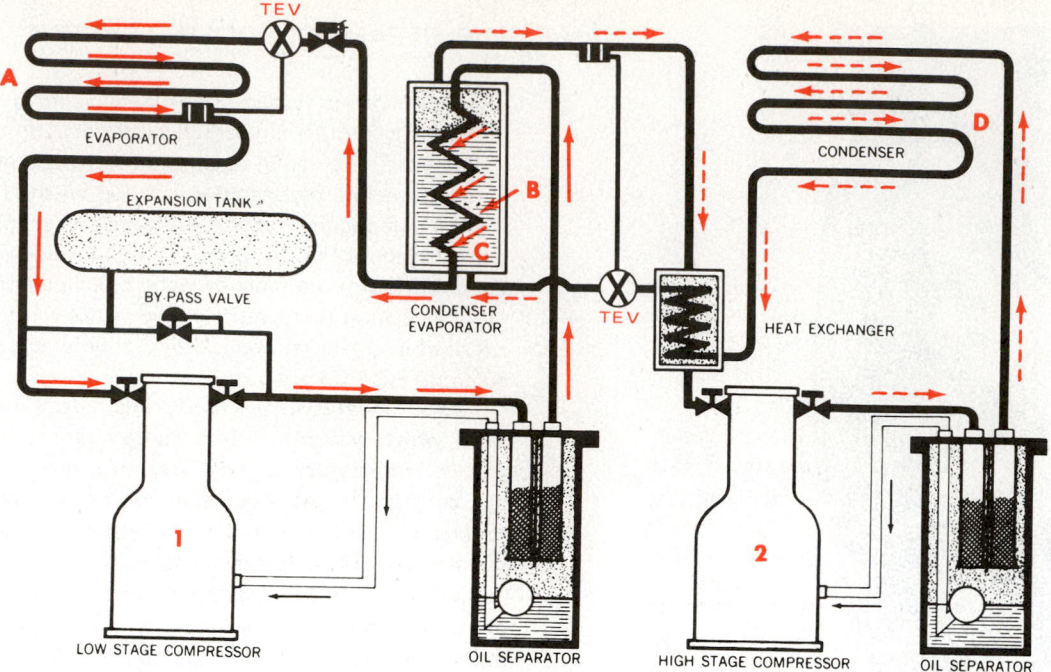

Fig. 17-26. Cascade refrigerating system. Condenser B of system No. 1 is being cooled by evaporator C of system No. 2. This arrangement enables ultralow temperatures in evaporator A of system No. 1. D—Condenser of system No. 2. TEV—Refrigerant controls. Note use of oil separators to minimize circulation of oil.

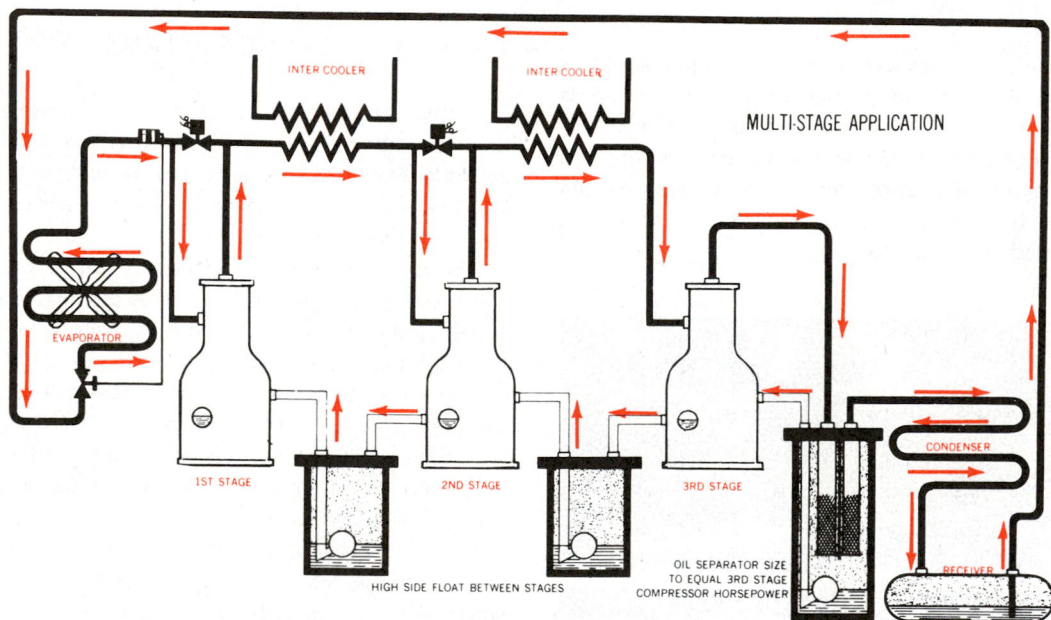

Fig. 17-27. Multistage refrigerating system using three compressors (stages). Compressor No. 1 pumps vapor into intercooler and then into intake of compressor No. 2. This operation is repeated between second and third stages. In third stage, refrigerant vapor is further cooled and travels to evaporator for specific cooling use.

compressor is larger than the secondary stage compressor and each stage finds the compressor getting smaller. This is because each higher stage handles denser vapor.

Compound refrigeration systems using two-stage compression equipment can be used to produce temperatures from −20 F. (−29 C.) to about −80 F. (−62 C.). If three-stage equipment is used (three compressors in series), temperatures down to −135 F. (−93 C.) can be maintained.

17-11 HEAT PIPE

R. S. Gaugler developed the basic principle of the heat pipe in 1942. More recently, it has been used in aerospace work as well as in industrial and domestic applications. Its purpose is to transfer heat from one location to another.

The heat pipe is an evacuated, hermetically sealed chamber containing volatile working fluid. Its inside walls are lined with

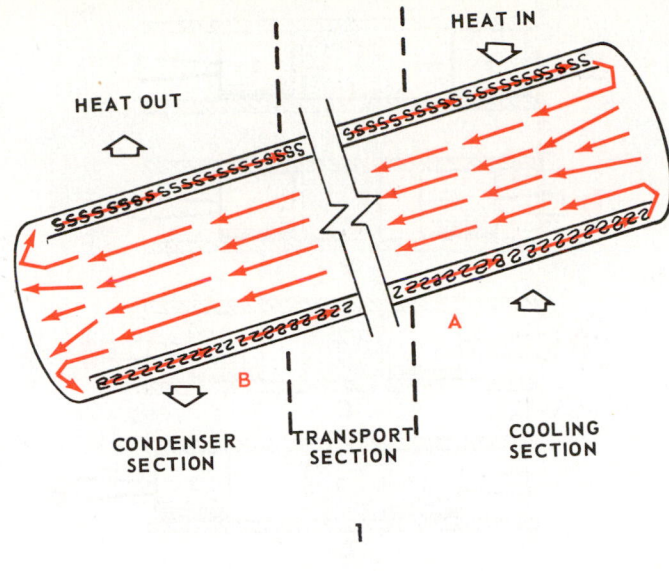

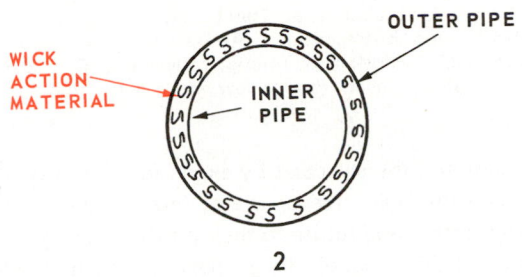

Fig. 17-28. 1—Basic design of heat tube. Fluid evaporates at A as heat travels into tube. Vapor moves length of inner tube to condenser B where fluid releases enough heat to become liquid. Liquid then travels through wick (by capillary action) back to evaporator. Transport section does not gain or lose heat. 2—Cross section shows end view of tube.

a porous substance called a wick. Fig. 17-28 is a cross-section through such a pipe.

Heat is applied to one end (heat in). The liquid vaporizes and vapor flows to the opposite end. When the heat is removed, the vapor condenses into a liquid again. The liquid returns to the evaporator or hot end of the pipe through the wick. This completes the cycle.

Return of fluid as a liquid to the evaporator section is normally accomplished by a gravity return. The pumping action of the wick is the major breakthrough. This occurs as a result of the liquid vaporizing in the evaporator, thereby leaving voids in the wick's porous structure. The attractive force between the wick material and liquid, combined with the surface tension of the liquid, fills these voids with liquid next to them, thereby allowing liquid to continuously flow through the wick from the condenser and to the evaporator section.

The heat pipe has been used in aerospace projects where it cools instruments or devices. Another successful application has been found in cooking. A heat pipe (pin) may be inserted into a piece of meat and, when placed in a conventional oven, cooks the meat from the inside.

Still another application of the heat pipe is for heat recovery from the flue of a furnace or boiler. With the heat

pipe, the high stack temperature in a flue may provide additional heat for basements, garages and the like. Fig. 17-29, view A, shows how a heat pipe can be used to recover heat from exhaust air during the heating season. In view B it is used to cool incoming air during the cooling season.

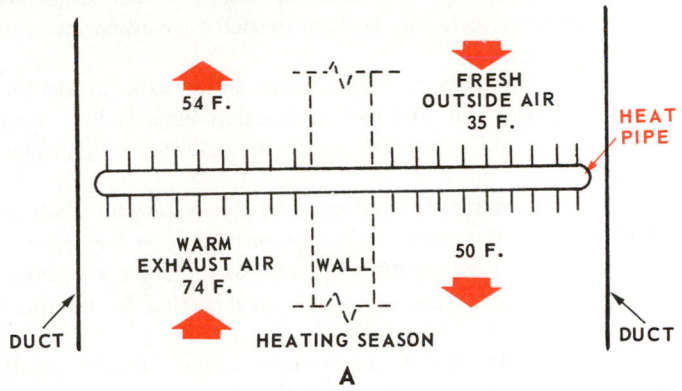

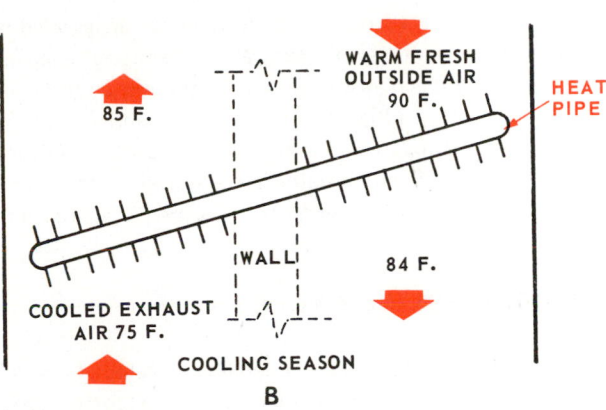

Fig. 17-29. Heat pipe application. Pipe is used to transfer heat between building exhaust air and fresh incoming air.

17-12 IMMERSION (FAST FREEZE)

Immersion freezing consists of dipping articles to be frozen into liquid refrigerant.

For fast freezing of food, the usual refrigerant is liquid R-12, specially prepared for this purpose. For some very low-temperature applications, liquid carbon dioxide or liquid nitrogen may be used.

The temperature is so low that none of the refrigerant should ever be allowed to touch the worker. It would result in immediate freezing of the skin.

In this freezing system, some liquid refrigerant boils and is vaporized in absorbing heat from the food. Some larger installations recover the vaporized refrigerant. R-12 refrigerant is recovered by using a refrigerated, finned coil placed over the liquid refrigerant. This will condense the vapor so it may be collected and returned to the refrigerant storage tank. With liquid carbon dioxide or nitrogen, however, the refrigerant is lost and becomes expendable.

Special Refrigeration Systems and Applications / 637

17-13 REFRIGERATED CONTAINERS

Refrigerated containers are being used aboard ships, trucks, railroad cars and airplanes. Perishable commodities are often gathered in the fields or orchards and stored immediately in refrigerated containers. Some containers may be as large as 8 x 8 x 20 ft. For long distance shipment the containers may have a refrigeration mechanism such as an evaporator and a condensing unit.

Condensing units may be driven either with an electric motor or with an internal combustion engine. For short distances, a cold plate with a eutectic solution is used. (See Chapter 28.)

The eutectic solution is frozen by the evaporator which is part of the cold plate. It is not necessary to operate the condensing unit during the short trip, as the eutectic solution in the cold plate provides considerable cooling for the short time the shipment is in transit.

Containers shipped by air often use carbon dioxide pellets as the refrigeration. See Para. 17-6. In such installations both the structure and the refrigerating equipment can be kept quite lightweight.

Containers used on shipboard may be refrigerated by being connected to the ship's central refrigerating system. If the refrigerating mechanism uses an electric motor and condensing unit, it may be plugged into an outlet and driven by power aboard the ship. In such cases, the condensing unit is usually water cooled. When containers are stored on the dock or in a warehouse, the condensing unit may be plugged into the local power supply.

17-14 STERLING CYCLE

A refrigerating cycle originally developed in 1816 by Robert Sterling is now being used in some refrigeration installations which operate at −110 F. (−79 C.) down to −300 F. (−184 C.). This cycle, when used as a three-stage system, can produce temperatures down to −450 F. (−268 C.). These units are very compact. Some of the gases used are helium and hydrogen.

The ideal system will pick up heat only at the lowest temperature and discard it only at the highest temperature. There can be no heat gain or heat loss between these two temperatures.

The Sterling cycle is almost as good because it conserves the energy and uses it in another part of the cycle. The Sterling cycle was adapted for refrigeration by John Herschel in 1834. The first practical machine was built in 1845.

The system uses one cylinder and two pistons with a stationary regenerator in between the pistons. See Fig. 17-30. In view A, piston No. 2 is stationary (standing still) and up to the regenerator, while piston No. 1 is at the beginning of compression. Then in view B, piston No. 1 compresses the gas and this gas is cooled (no temperature rise).

In view C, piston No. 1 now completes its stroke. At the same time, piston No. 2 moves to the right. The volume of trapped gas thus remains the same. The regenerator collects heat during this operation. In view D, piston No. 2 moves to

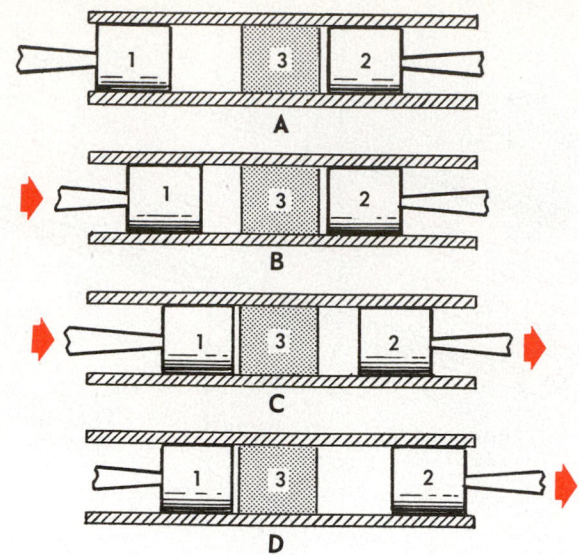

Fig. 17-30. Basic actions of Sterling Cycle system. 1—Left piston. 2—Right piston. 3—Regenerator. A—Start position. B—Compressing gas and cooling. C—Moving gas through regenerator. D—Expanding gas (it absorbs heat). Pistons now return to positions in A.

the right and the gas cools by expansion. This gas is heated by warming the right side of the cylinder. Finally both pistons move together and return to their position in view A.

During this time, all the gas passes through the regenerator. Heat is added to the gas from the regenerator during this action.

In practice, the left end of the cylinder is water-cooled and the right end of the cylinder is the cooling unit. The regenerator must allow flow of the gas and it must have a good heat absorbing ability.

17-15 REVIEW OF SAFETY

When working with electrical equipment it is always recommended that the equipment being tested be fully grounded through either the use of a polarity plug or a ground wire. See Chapter 8.

Local and national refrigeration and electrical codes should be followed when servicing and installing all units.

Before checking a thermoelectric system, get a wiring diagram of the system to be certain that the polarity is not reversed.

When checking expendable refrigerant systems, always make certain that the safety doors are opened and the truck body vented before entering the conditioned space. Before entering a unit that is being cooled, make certain that all refrigerant flow control valves have been closed.

As has been repeated throughout the text, a service technician should always wear goggles when checking a unit which uses a refrigerant.

It is not safe to work on any part of an expendable refrigerant system unless one knows the required pressures and the nature of the safety valves and controls used in the system.

As indicated in Para. 17-10, temperatures in the cryogenic

range are −250 F. (−157 C.) and below. These temperatures are dangerous. The rate at which heat will be removed from the surface of the body at these temperatures is so great that the flesh may be severely frozen before one feels the cold.

Remember, most expendable refrigeration systems use carbon dioxide or nitrogen. Humans and animals cannot live in atmospheres of either of these substances.

In handling any type of refrigerants, the operator should wear gloves, face shield or goggles, and make sure that no liquid refrigerant is ever allowed to touch the skin. It is particularly dangerous to handle such refrigerants as liquid nitrogen, liquid air and liquid carbon dioxide.

17-16 TEST YOUR KNOWLEDGE

1. What are the advantages of a thermoelectric system as compared to a compression system?
2. Does a thermoelectric system use alternating current throughout the system?
3. How can a thermoelectric module used for cooling be converted into a heating unit?
4. What refrigerants are used in an expendable refrigerant cycle?
5. Name two basic types of expendable refrigerant systems available today.
6. What is the purpose of the safety vent in an expendable refrigerant system cabinet?
7. Name one application of a cascade system.
8. In most portable cargo containers, what is the primary cooler?
9. What is the main advantage of the containerized system used for transporting goods?
10. What are the least number of compressors a multistage system will use?
11. How are the refrigerating systems in a cascade system placed in reference to each other?
12. What are the major advantages of liquid nitrogen in a fast-freeze system?
13. How is the fluid in a heat pipe returned to the evaporator or heat source?
14. What determines the amount of hot air released from a vortex tube system?
15. What is used as a coolant in a vortex tube system?
16. What is the temperature of carbon dioxide pellets?
17. What supplies the energy to a venturi tube?
18. Can a Sterling cycle refrigeration system produce very low temperatures?
19. What is the purpose of the regenerator in a Sterling cycle?
20. Why does an expendable refrigerant spray system require an electrical system?

Chapter 18

FUNDAMENTALS OF AIR CONDITIONING

Fundamental principles of temperature measurement and refrigeration covered in Chapter 1 should be carefully reviewed as a foundation for the study of air conditioning.

This chapter on the fundamentals of air conditioning includes additional principles and definitions such as: atmosphere, humidity measurement and control, and the use of psychrometric charts and tables. This chapter also explains the principles of operation of air conditioning instruments and how they are used.

Basically, the first part of the chapter covers the physical principles of air movement and humidity. The second part deals with the various factors having to do with comfort, health and the methods of controlling these factors.

18-1 DEFINITION OF AIR CONDITIONING

The American Society of Heating, Refrigerating and Air Conditioning Engineers (ASHRAE) defines air conditioning as: "The process of treating air so as to control simultaneously its temperature, humidity, cleanliness and distribution to meet the requirements of the conditioned space."

As indicated in the definition, the important actions involved in the operation of an air conditioning system are:
1. Temperature contol.
2. Humidity control.
3. Air filtering, cleaning and purification.
4. Air movement and circulation.

Complete air conditioning provides automatic control of these conditions for both summer and winter.

Temperature control for winter heating conditions requires automatic control of the heating source as a means of maintaining desired room temperatures.

Temperature control for summer cooling conditions requires automatic control of the refrigerating system to maintain the desired room temperatures.

Humidity control for winter conditions usually requires automatic control of the addition of moisture to the heating system (by use of a humidifier).

Humidity control for summer conditions requires the automatic control of dehumidifiers. Usually, this is done at the time the air to be cooled is passed over the cold evaporator surfaces.

In general, air filtering is the same for both summer and winter air conditioning. Air filtering equipment usually consists of very fine porous substances air is drawn through to remove contaminating particles. Filters using activated carbon and electrostatic precipitators may be added to the usual filtering mechanisms to improve air cleaning. The air pollutants, and methods used to remove them from the air, will be covered in later paragraphs.

In addition to the comfort provided, many industries air condition their plants for more complete control of manufacturing processes and material, which improves the quality of the finished product.

18-2 AIR — ATMOSPHERE

Air is an invisible, odorless and tasteless mixture of gases which surround the earth. Air surrounding the earth is called the atmosphere. It extends above the earth about 400 miles and is divided into several layers. The layer closest to the earth is called the lower atmosphere. It extends from sea level up to about 30 000 ft. The next layer, called the troposphere, is from 30 000 to 50 000 ft. The layer extending from 50 000 ft. up to 200 miles is called the stratosphere. The layer from 200 miles upward is called the ionosphere.

Atmospheric air is a mixture of oxygen, nitrogen, carbon dioxide, hydrogen, sulphur dioxide, water vapor (moisture), and a very small percentage of rare gases. Fig. 18-1 gives the percentages of these gases, both by volume and by weight. Each of these gases behaves as though it occupied the space alone (Dalton's Law).

1. Oxygen. The atmosphere is approximately 23 percent oxygen by weight. Oxygen readily combines with many substances. When fuels such as wood, coal, or oil are burned, the oxygen of the atmosphere combines with the carbon and hydrogen in the fuel to form carbon dioxide and water. The oxygen in the atmosphere is replenished by growing plants. The roots of the plants absorb moisture from the soil; the leaves absorb carbon dioxide from the air. Part of the moisture (H_2O) absorbed by the plant combines with the carbon in carbon dioxide. Cellulose, the plant structure (stem, leaves, flowers, etc.) is produced in this way. Oxygen from the water intake of the plant is released by the leaves. Note in Fig. 18-1 that, because oxygen (O_2) is a heavier gas, it has a higher percentage by weight than by volume.

2. Nitrogen. About three-fourths of the earth's atmosphere by weight consists of nitrogen, a gaseous element that does

Name	Chemical Symbol	DRY AIR Amount by Weight %	DRY AIR Amount by Volume %
Nitrogen	N_2	75.47	78.03
Oxygen	O_2	23.19	20.99
Carbon Dioxide	CO_2	.04	.03
Hydrogen	H_2	.00	.01
Water	H_2O	.00	.00
Dust		.00	.00
Rare Gases		1.30	.94

Fig. 18-1. Gases and substances that make up the air in the atmosphere.

not readily combine with other substances. If combined with other elements, nitrogen usually is unstable and tends to separate from the other elements. Compounds of nitrogen make up most explosives. Nitrogen is combined commercially with hydrogen to form ammonia. Ammonia produced in this way is the basis of most fertilizers. Ammonia is also an important refrigerant (NH_3) (R-717). Liquid nitrogen obtained by cooling of air is also a special purpose expendable refrigerant.

3. Carbon dioxide. The atmosphere contains approximately 0.03 to 0.04 percent carbon dioxide. Carbon dioxide is a combination of carbon and oxygen. Absorbed by growing plants, it becomes one of the "building blocks" in the development of plant cells.

4. Hydrogen. Hydrogen (H_2), very light gas, does not show in weight percentage. However, it is shown as volume in Fig. 18-1. Hydrogen makes up a very small part of the atmosphere. It is present in most fuels. When burned, it combines with oxygen to form water (H_2O) in steam and vapor form.

5. Sulphur dioxide. The most common gaseous contaminant, sulphur dioxide is formed by combustion of fuels which contain sulphur. Many large power plants now have facilities for removing sulphur from these fuel sources, and also for removing sulphur dioxide from the stack gases.

6. Water vapor (moisture). The amount of water vapor in the atmosphere varies with the temperature. It is not indicated in percentage, but rather by the term "relative humidity."

7. Rare gases. Rare gases make up from 0.9 to 1.3 percent of the atmosphere by weight. These gases include neon, argon, helium, krypton, and xenon. Some are used to light bulbs and tubes and in certain industrial processes.

In addition to these substances, air contains a variety of contaminants. These are so variable that they cannot be given any definite value in a table. However, contaminants in the air are of great importance in air conditioning.

18-3 PHYSICAL PROPERTIES OF AIR

Air has weight, density, temperature, specific heat and heat conductivity. In motion, it has momentum and inertia. It holds substances in suspension and in solution.

Air pressure at the surface of the earth is due to the weight of the air above the earth. Air pressure decreases with altitude due to the reduction of the weight of the air above. Air presses

against the earth at sea level with a force of 14.7 psi (1.0 kg/cm²).

Because air has weight, energy is required to move it. Once in motion, air has energy of its own (kinetic energy). The weight of moving air turns windmills. The mills convert the kinetic energy to mechanical energy.

The kinetic energy of air in motion is equal to half the mass of the air multiplied by the square of the velocity (speed). Velocity is measured in feet per second or metres per second. According to Bernoulli's Equation, increasing the velocity decreases the pressure. In a tornado, the velocity is very high, reducing the pressure. The low pressure in a tornado causes much of the damage to buildings. Outside pressure is lowered rapidly. Pressures inside the building push outward in an "explosion" of air.

Tiny particles of dust may be picked up and held in suspension in moving air for long periods of time. It is possible to measure the amount of particles so suspended.

The density of air varies with the atmospheric pressure and humidity. One pound of air at sea level occupies a space of approximately 14 cu. ft. (0.40 m³). It has a density of 0.0725 lb./cu. ft. (0.00098 kg/m³). The density of gases is indicated in pounds per cubic foot or in kilograms per cubic metre.

Air temperatures may be measured with either the Fahrenheit scale or the Celsius scale. Under ordinary conditions, the familiar glass-stemmed thermometers are satisfactory. Expanding metals (solids) such as bimetal strips or rods are also used for ordinary circumstances. When making measurement of very low temperatures, thermocouple thermometers or resistance temperature detectors are used. Thermocouple thermometers may be used for measuring high air temperatures. Thermistor thermometers and pyrometers are also popular.

The specific heat of air is the amount of heat required to raise the temperature of one pound of air one degree Fahrenheit or one kilogram of air one degree Celsius. The specific heat of air at sea level is 0.24 Btu per pound.

Air is a poor conductor of heat. For this reason, air spaces are often used for insulating purposes.

For computation purposes, certain pressure, temperature and density are required. The requirements are defined under Standard Air in Chapter 28.

18-4 HUMIDITY

Humidity is a term used to describe the presence of moisture or water vapor in the air. The amount of moisture that the air will hold depends upon the temperature of the air. Warm air will hold more moisture than cold air.

The amount of humidity in the air affects the rate of evaporation of perspiration from the body. Dry air causes rapid evaporation, which makes the surface feel cool. Moist (humid) air prevents rapid evaporation of perspiration, making it feel warmer than the temperature indicated by a thermometer. Remember that this moisture (humidity) is in vapor form, and it is invisible.

How much moisture the air will hold is shown in Fig. 18-2 and explained in Para. 18-5.

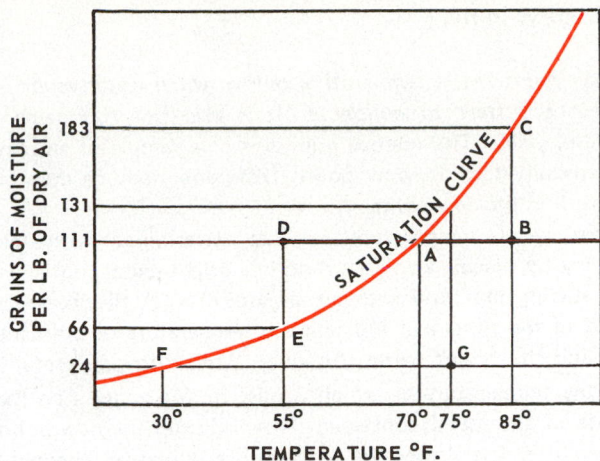

Fig. 18-2. Typical water vapor saturation curve for air. As temperature increases, amount of moisture that air will hold also increases.

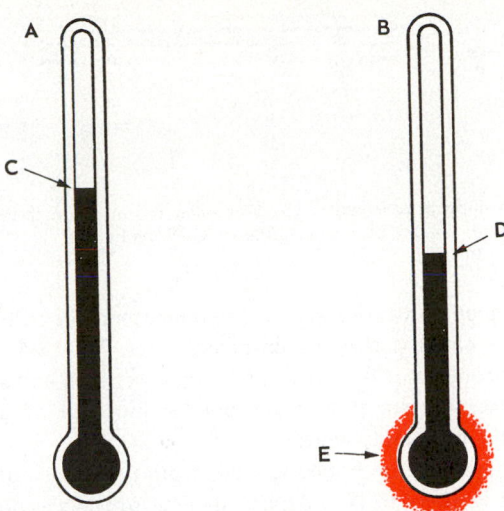

Fig. 18-3. Dry bulb and wet bulb thermometers. A—Dry bulb thermometer. B—Wet bulb thermometer. C—Dry bulb temperature. D—Wet bulb temperature. E—Wick surrounding wet bulb. Note that the temperature shown on the wet bulb thermometer is considerably lower than the dry bulb thermometer.

18-5 RELATIVE HUMIDITY

Relative humidity (rh) is a term used to express the amount of moisture in a given sample of air in comparison with the amount of moisture the air would hold if totally saturated at the temperature of the sample. Relative humidity is stated in percentage, such as 30 percent, 75 percent, 85 percent, etc.

Referring to the water vapor saturation graph in Fig. 18-2, Point B contains 111 grains of moisture per pound of dry air at 85 F. The saturated condition at C for the same temperature is 183 grains of moisture per pound of air. Therefore, the relative humidity at Point B is:

$$\frac{111}{183} \times 100 = rh$$
$$.6 \times 100 = rh$$
$$60 \text{ percent} = rh$$

Line A to B in Fig. 18-2 represents what happens when saturated air is warmed. Point D represents what happens when saturated air is cooled. The distance D to E represents the moisture condensed out of the air, since saturated air at the same temperature will hold only 66 grains of moisture. The amount condensed is 111 minus 66 = 45 grains.

A typical outdoor condition in winter is represented at Point F in Fig. 18-2. Air is taken indoors at 30 F. (−1 C.) and 100 percent relative humidity. It holds 24 grains of moisture. If this air is heated to 75 F. (24 C.) and no moisture is added, its new condition will be as shown at G. The saturated condition at Point G would be 131 grains. Since the original air had only 24 grains of moisture, the relative humidity is 24 ÷ 131 x 100 = 18.3 percent.

18-6 DRY BULB THERMOMETER

Human comfort and health depend a great deal on the air temperature. In air conditioning, the air temperature indicated usually is dry bulb temperature (db) taken with the sensitive element of the thermometer in a dry condition. It is the temperature measured by thermometers in the home.

18-7 WET BULB THERMOMETER

If a moist wick is placed over a thermometer bulb, the evaporation of moisture from the wick will lower the thermometer reading (temperature). This temperature is known as "wet bulb" temperature. If the air surrounding a wet bulb thermometer is dry, evaporation from the moist wick will be more rapid than if the air is quite moist. Fig. 18-3 compares dry bulb temperature and wet bulb temperature taken at the same place and at the same time.

When the air is saturated with moisture, no water will evaporate from the cloth wick, and the temperature on the wet bulb thermometer will be the same as the reading on a dry bulb thermometer near it.

However, if the air is not saturated, water will evaporate from the wick. In doing so, it will lower the wick temperature. Then, heat will flow from the mercury to the wet wick and the reading will be lower.

The accuracy of the wet bulb reading depends on how fast the air passes over the bulb. Speeds up to 5000 ft./min. or 60 mi./hr. are best but dangerous if the thermometer is moved at this speed. Also, the wet bulb should be protected from heat radiation surfaces (radiator, sun, electric heater, etc.). Errors as high as 15 percent may be made if the air movement is too slow, or if too much radiant heat is present.

A hygrometer is an instrument used to measure the amount of moisture in the air. The hygrometer uses both a dry bulb thermometer and a wet bulb thermometer. By using a psychrometric chart, Fig. 18-8, the relative humidity can be found. Also see Para. 18-15 and Para. 18-16.

18-8 PSYCHROMETER

To insure that the recorded wet bulb temperature is accurate, airflow over the wet bulb should be quite rapid. A

Fig. 18-4. A sling psychrometer. A—Wet wick mounted on thermometer. B—Dry bulb thermometer. C—Sling handle.

device designed to whirl a pair of thermometers, dry bulb and wet bulb, is called a sling psychrometer. See Fig. 18-4. This instrument consists of two thermometers, a wet bulb and a dry bulb. The wick on a sling psychrometer must be of clean cotton fabric, preferably white.

Because evaporation is taking place from the surface of the wick, there is likely to be a deposit of lime substances on the wick. Therefore, to get accurate measurements, a clean wick should be used. Also, use distilled water on the wick. Sling psychrometers come in a variety of sizes.

There are certain places in which it is difficult to spin the psychrometer (narrow passages, etc.). To obtain accurate results in these places, an aspirating psychrometer is used. With this instrument, Fig. 18-5, the air sample is blown over the wet and dry bulb thermometer by suction created by an air pump.

A battery-operated aspirating psychrometer is shown in Fig. 18-6. It has illuminated thermometer scales and a fan which draws air over the thermometer sensitive bulbs.

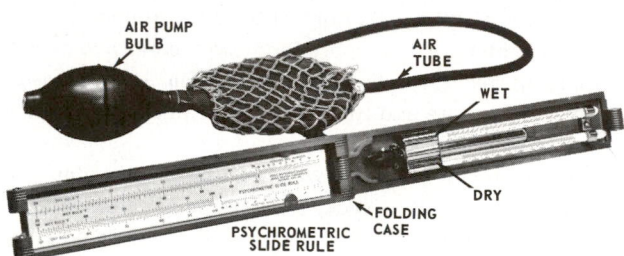

Fig. 18-5. Aspirating psychrometer. Air samples are drawn over thermometer bulbs by air pump. Note handy calculating slide rule incorporated in case. (Bendix Corp.)

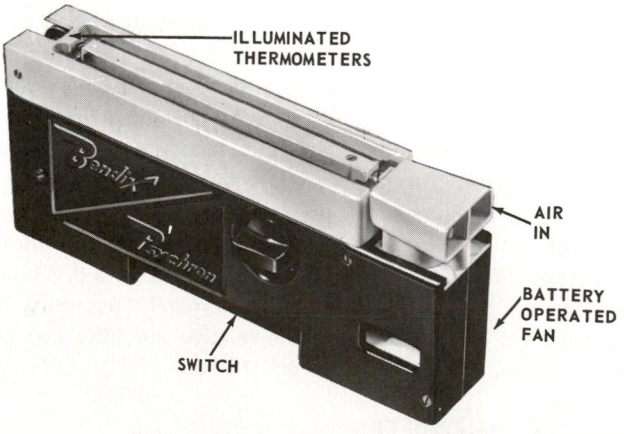

Fig. 18-6. Aspirating psychrometer; a battery powered unit with illuminated thermometer scales. Motorized fan draws air over wet bulb and dry bulb thermometers.

18-9 DEW POINT

Dew point is the temperature below which water vapor in the air will start to condense. It is also the 100 percent humidity point. The relative humidity of a sample of air may be determined by its dew point. Different methods may be used to find the dew point.

Dew point temperature can be determined with fair accuracy by placing a volatile fluid in a bright metal container, then stirring the fluid with an air aspirator. A thermometer placed in the fluid will indicate the temperature of both the fluid and the bright metal container. While stirring, carefully note the temperature at which a mist or fog appears on the outside of the metal container. This indicates the dew point temperature. The lower the temperature is lowered, the more accurate the reading. Flammable or toxic volatile fluids must not be used for this experiment.

A commercial instrument for determining dew point is illustrated in Fig. 18-7. This unit can measure dew point

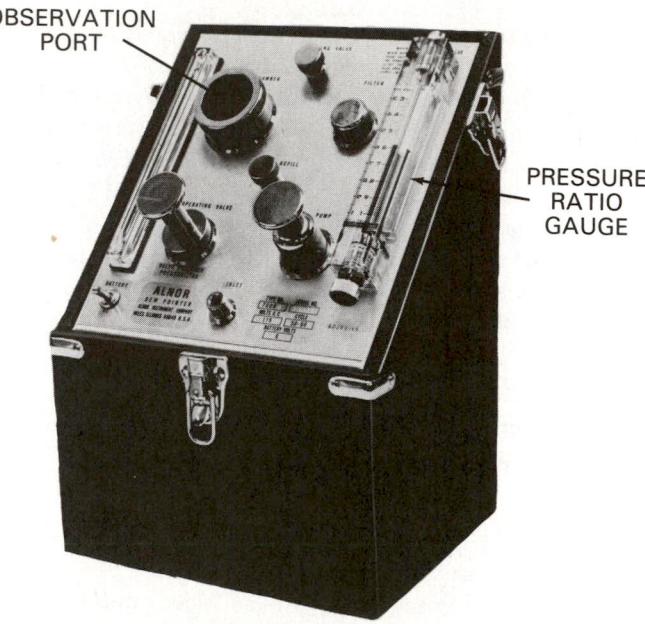

Fig. 18-7. An instrument for determining dew point temperature. Note observation port. (Alnor Instrument Co.)

temperatures from room temperature to as low as −80 F. (−62 C.). The principle of operation is to pump a sample of air into the observation chamber of the instrument. The pressure is above atmospheric.

The pressure ratio gauge on the right in Fig. 18-7 adjusts for this pressure, then the valve is manipulated to exhaust the air. The lighted observation window will indicate a fog when the sample is cooled to its dew point. This window is lighted and a "sunbeam" effect is noted if any fog exists. The pressure ratio determines the dew point temperature.

Use the psychrometric chart in Fig. 18-8 to determine the dew point.

Example: Compute the dew point for a sample of air in which

Fig. 18-8. Psychrometric chart. Vertical lines are the dry bulb temperature; oblique lines are the wet bulb temperature. Curved lines show relative humidity. (Carrier Air Conditioning Group, United Technologies Corp.)

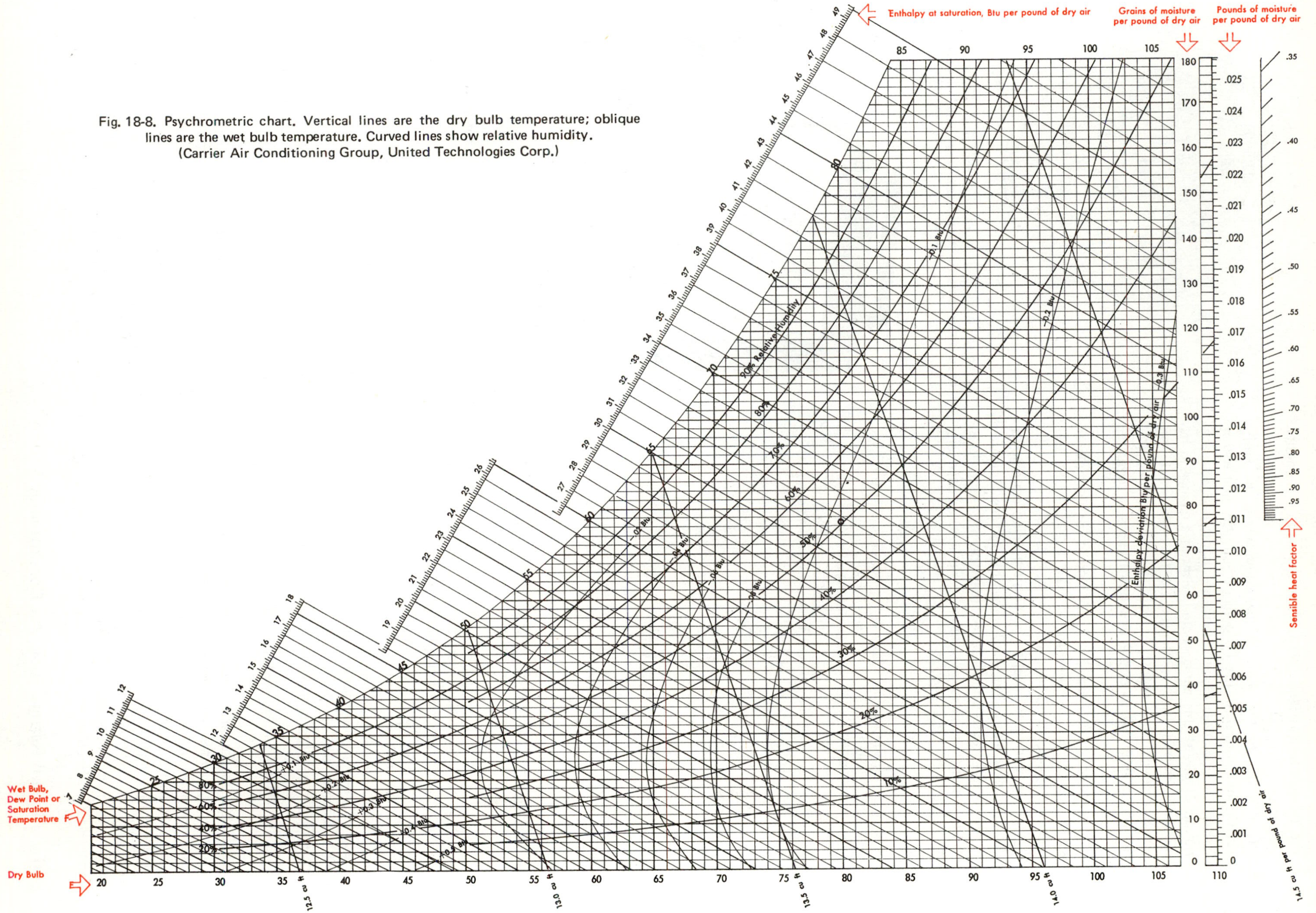

Below 32 F, properties and enthalpy deviation lines are for ice.

the temperature is 80 F. (27 C.) and the relative humidity 60 percent.

Solution: Refer to the chart at the point where the 80 F. (27 C.) line intersects the 60 percent humidity line which represents the quantity of moisture contained in each pound of air. If this air is cooled without a change in its moisture content (as represented by the horizontal line going through this point), it will be found that the horizontal line intersects the dew point line at a temperature of approximately 64 F. (18 C.).

Therefore, 64 F. (18 C.) is the dew point for a sample of air in which the temperature is 80 F. (27 C.) and the relative humidity is 60 percent.

A window during the winter heating season offers a good example of dew point. Fig. 18-9 shows the surface temperature which will cause condensation (dew point) for various humidity conditions. The two room temperatures used are 70 F. and 80 F. (21 C. and 27 C.).

RELATIVE HUMIDITY OF AIR (PERCENT)	DRY BULB TEMPERATURE OF SURFACE WHEN CONDENSATION STARTS	
	70 F. ROOM AIR TEMP.	80 F. ROOM AIR TEMP.
100	70	80
90	67	77
80	64	73
70	60	69
60	56	65
50	51	60
40	45	54
30	37	46
20	28	35

Fig. 18-9. Table gives the temperature to which a surface must be cooled to have condensation start. The table is based on the ambient temperature of the air at either 70 F. or 80 F. (21 C. or 27 C.).

18-10 INDICATORS OF LOW HUMIDITY

Low atmospheric humidity will be indicated by an increase in the amount of noticeable electrostatic energy. As one moves about and touches grounded metal objects, a spark jumps from the hand or fingers to the object. Also, human hair tends to become unmanageable. Furniture joints shrink and become loose. Woodwork, such as doors and floors, crack open. The surface of the skin becomes dry, and membranes in the nose tend to become dry. To feel more comfortable, it is usually necessary to raise the ambient temperature (db) above normal.

18-11 HUMIDITY MEASUREMENT

Humidity measurement using a dry bulb and wet bulb thermometer is explained in Para. 18-6 and Para. 18-7. The use of the dry bulb and wet bulb thermometer means one must use psychrometric charts.

Instruments have been developed which give a direct reading of relative humidity. See Fig. 18-10. The operation of these instruments depends upon the property of some substances to absorb moisture and then change their shape or size,

Fig. 18-10. Wall type hygrometer shown is calibrated in percent of relative humidity. A—Air movement openings.   (Abbeon Cal, Inc.)

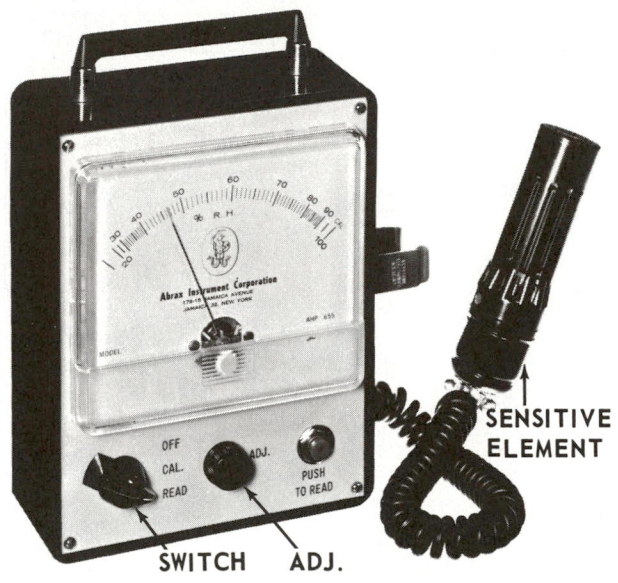

Fig. 18-11. Electric hygrometer. It is possible to check the calibration of this instrument. If necessary, an adjustment can be made against a dry bulb-wet bulb instrument reading.   (Abrax Instrument Corp.)

depending upon the relative humidity of the atmosphere. Human hair, wood and other substances may be used.

It is also possible to measure relative humidity electronically. This is done by using a substance in which the electrical conductivity changes with the moisture content. Such an instrument is shown in Fig. 18-11. In operation, the sensing element is placed in the space in which the relative humidity is to be measured.

Fig. 18-12 illustrates an easy-to-use electronic relative humidity measuring instrument. To use the meter, expose sensing element A to the air sample. Turn the switch on. Rotate the calibrated thumb wheel in the direction of whichever of the two indicator lamps on the barrel is lighted. When both lamps are out, the percent humidity may be read directly from the thumb wheel.

As in measuring pressure and temperature, it is sometimes

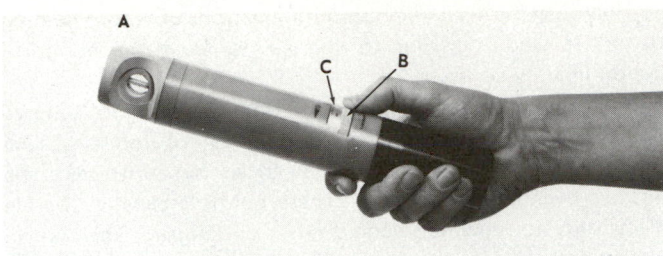

Fig. 18-12. An electronic relative humidity meter. A—Sensing element. B—Thumb wheel. C—Relative humidity reading. (Beckman Instruments, Inc.)

Fig. 18-13. A recording type hygrometer with sensing element at right. The pen records the relative humidity over a 24-hour period.

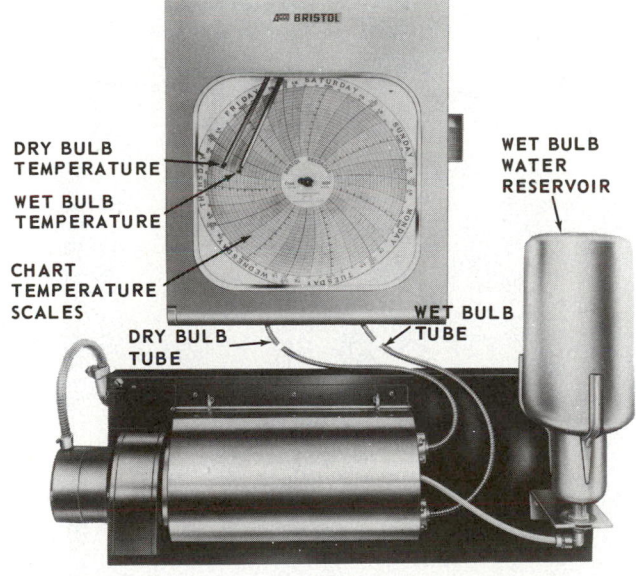

Fig. 18-14. Seven-day recording dry bulb-wet bulb thermometer. Small motor forces air over dry and wet bulbs. Note bottle on right, used to wet bulb wick. (Bristol Babcock Inc.)

helpful to have a 24-hour or 7-day record of the humidity in a controlled space. A 24-hour recording type hygrometer is shown in Fig. 18-13. A 7-day recording type dry bulb and wet bulb thermometer is shown in Fig. 18-14. When using this instrument, it is necessary to refer to a psychrometric chart to find the relative humidity.

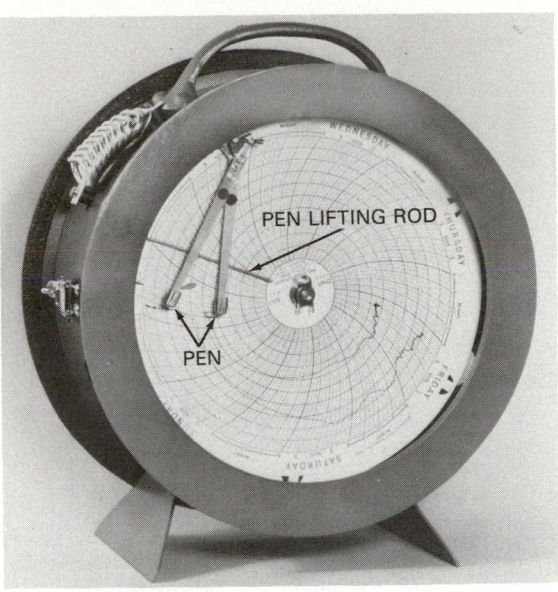

Fig. 18-15. Seven-day recorder for both temperature and humidity. These records are important as a means of checking efficiency of air conditioning systems. (Bristol Babcock Inc.)

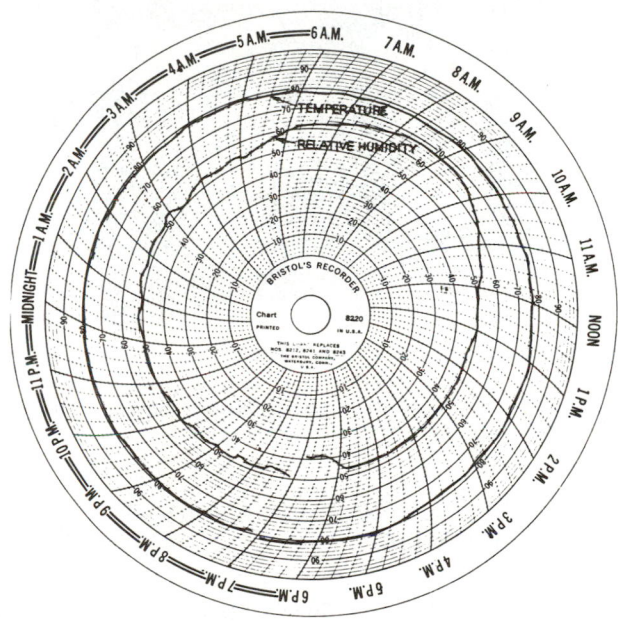

Fig. 18-16. Chart from 24-hour temperature and humidity recorder. Note how the relative humidity changes as much as 5 percent to 10 percent during the 24-hour period.

Fig. 18-15 pictures a 7-day instrument that records both temperature and humidity. Fig. 18-16 shows how temperature and humidity is charted by a 24-hour recorder.

A different type of temperature and humidity recorder is shown in Fig. 18-17. The charts, printed on stiff paper, move down as the recording proceeds. A common humidity sensitive element is usually made of multiple strands of human hair.

The unit in Fig. 18-18 is a portable temperature and humidity recorder. It can be set to record both dry bulb temperature and relative humidity for one day or for one week.

Fundamentals of Air Conditioning / 647

Fig. 18-17. Temperature-humidity recorder. Various time clock records are available, such as 10-hour or 30-hour charts. A—Roll-type chart. B—Air openings. (Bendix Corp.)

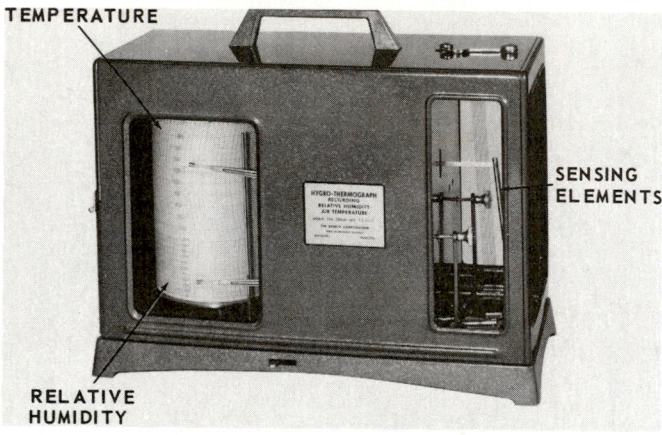

Fig. 18-18. Portable temperature and relative humidity recorder.

18-12 HYGROSCOPIC SUBSTANCES — DESICCANTS

Substances that have the ability to absorb moisture from the air are called desiccants. Some common desiccants are: activated alumina, silica gel, calcium sulfate and zeolites. Many desiccants can be reactivated (dried out) by heating.

Many instruments are packaged in containers with a package of desiccant. The desiccant tends to absorb the moisture in the container and keeps the instrument dry to reduce corrosion.

18-13 HUMIDITY CONTROLS

Health studies indicate that humidity control is an important factor in air conditioning.

Humidity controls are used to keep the relative humidity of air conditioned rooms at a satisfactory level. These controls determine the hygrometric state of the air.

Humidity controls operate during periods of winter heating season to add moisture to the air to keep the humidity approximately constant.

Humidity controls operate in the summer to remove moisture from the air. For the removal of moisture, the humidity control usually operates an air bypass to vary the airflow over the evaporators. These controls usually operate electrically to regulate solenoid valves or dampers. The control element may be a man-made fiber or human hair, which are sensitive to the amount of moisture in the air. Fig. 18-19 shows construction principles of a humidity control device.

In computer rooms and other installations which require close humidity control, thermo-humidigraphs (temperature

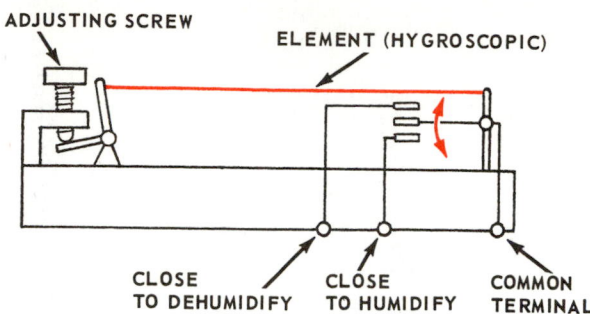

Fig. 18-19. A schematic diagram of a humidity control, showing the operating mechanism.

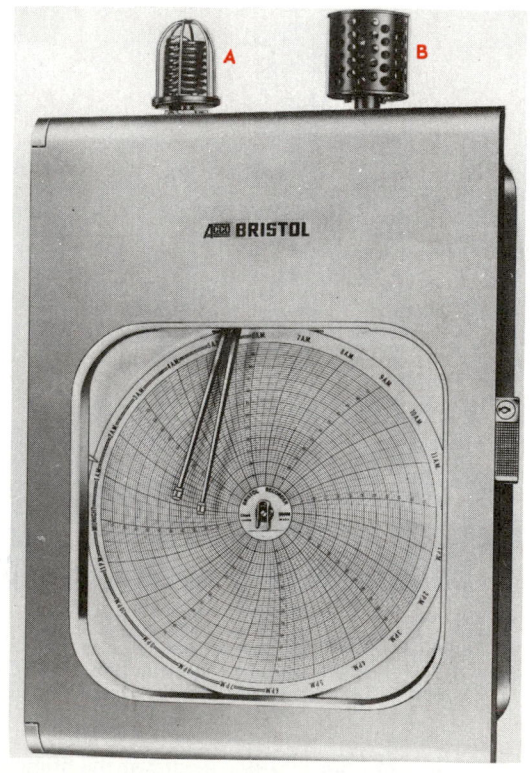

Fig. 18-20. A temperature-relative humidity recorder fitted with contact points. The points may be connected to an electric alarm system to provide a signal if the temperature or humidity is not kept within required limits. A—Temperature sensor. B—Humidity sensor. (Bristol Babcock, Inc.)

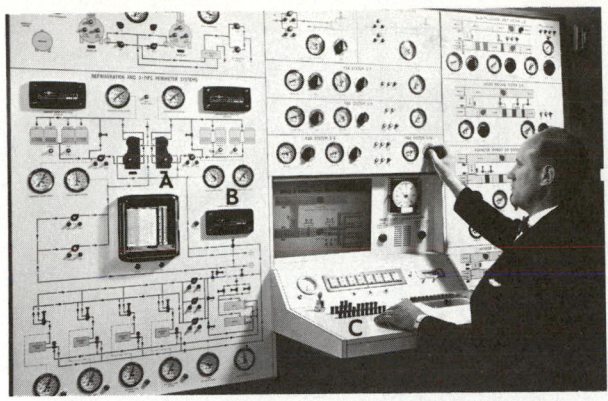

Fig. 18-21. Modern control panel being used to determine and control environmental conditions throughout building. A—Recorder. B—Direct reading instruments. C—Master control panel. (Johnson Service Co.)

and humidity recorders) are fitted with alarms which will alert attendants in the event the temperature or humidity fails to remain at the proper level. Fig. 18-20 shows a thermo-humidigraph fitted with alarms.

18-14 PSYCHROMETRIC PROPERTIES OF AIR

Psychrometry is the science and practice of dealing with air mixtures and their control. The science deals mainly with dry air and water vapor mixtures. Fig. 18-21 shows the control panel used to determine and control the condition of the air in a large building complex.

Psychrometry deals with the specific heat of dry air and its volume. It also deals with the heat of water, heat of vaporization or condensation, and the specific heat of steam in reference to moisture mixed with dry air.

Tables and graphs have been developed to show the pressure, temperature, heat content (enthalpy), and volume of air and its steam content. The tables and charts are based on one pound of dry air, plus the water vapor to produce the air conditions being studied.

A standard pressure of 29.92" Hg. (76 cm Hg.) of mercury is used as the standard atmospheric pressure.

18-15 PSYCHROMETRIC CHART

The psychrometric chart is a graph of the temperature-pressure relationship of steam (water vapor). The horizontal scale (abscissa) is the temperature, while the vertical side (ordinate) is the water vapor pressure scale. A basic psychrometric chart is shown in Fig. 18-22. The dry bulb temperature for Point A is 80 F. (27 C.). Wet bulb temperature for Point A is 60 F. (16 C.). This indicates that the relative humidity at Point A is 30 percent.

Several types of psychrometric charts are available. Fig. 18-8 illustrates a type of chart used frequently. Fig. 18-23 is a

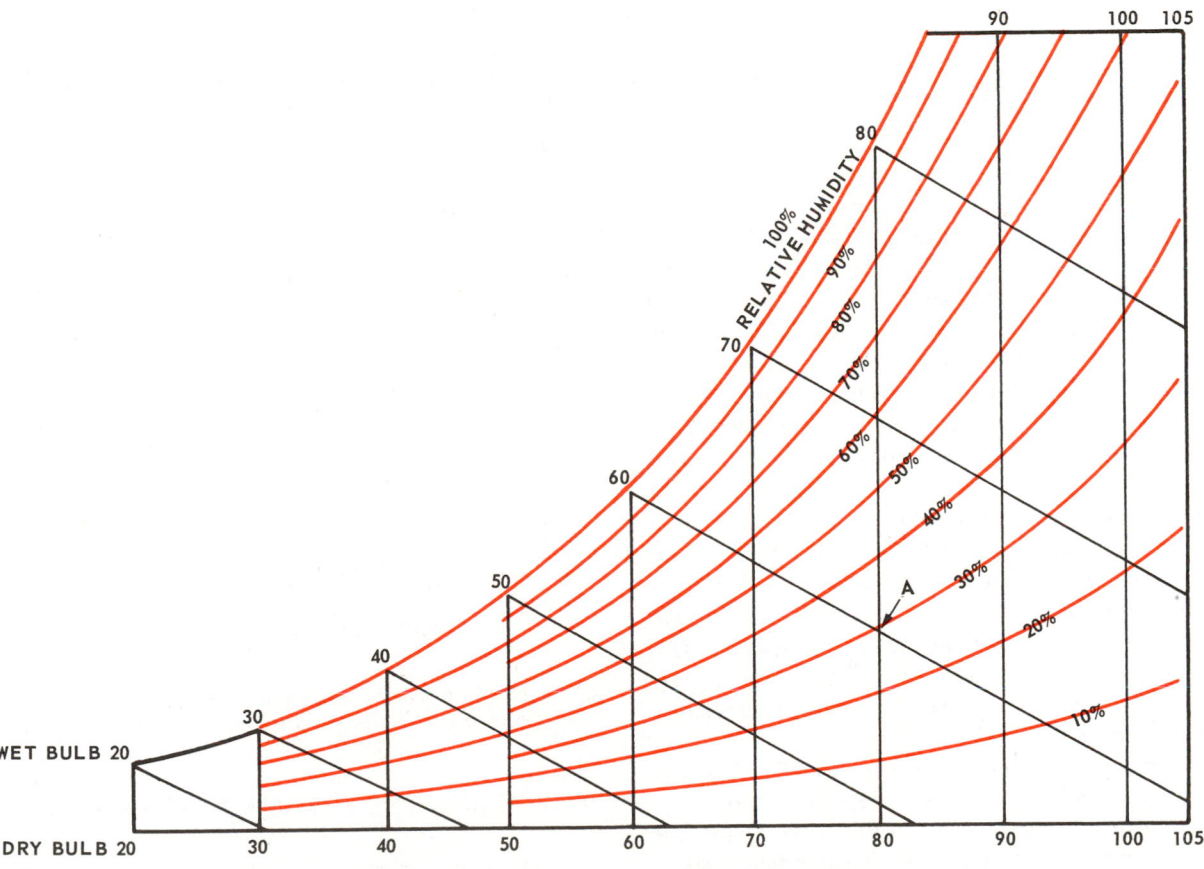

Fig. 18-22. A basic psychrometric chart. Red lines indicate relative humidity in percent. Dry bulb temperature is shown at the bottom. Point A indicates the relative humidity of 30 percent.

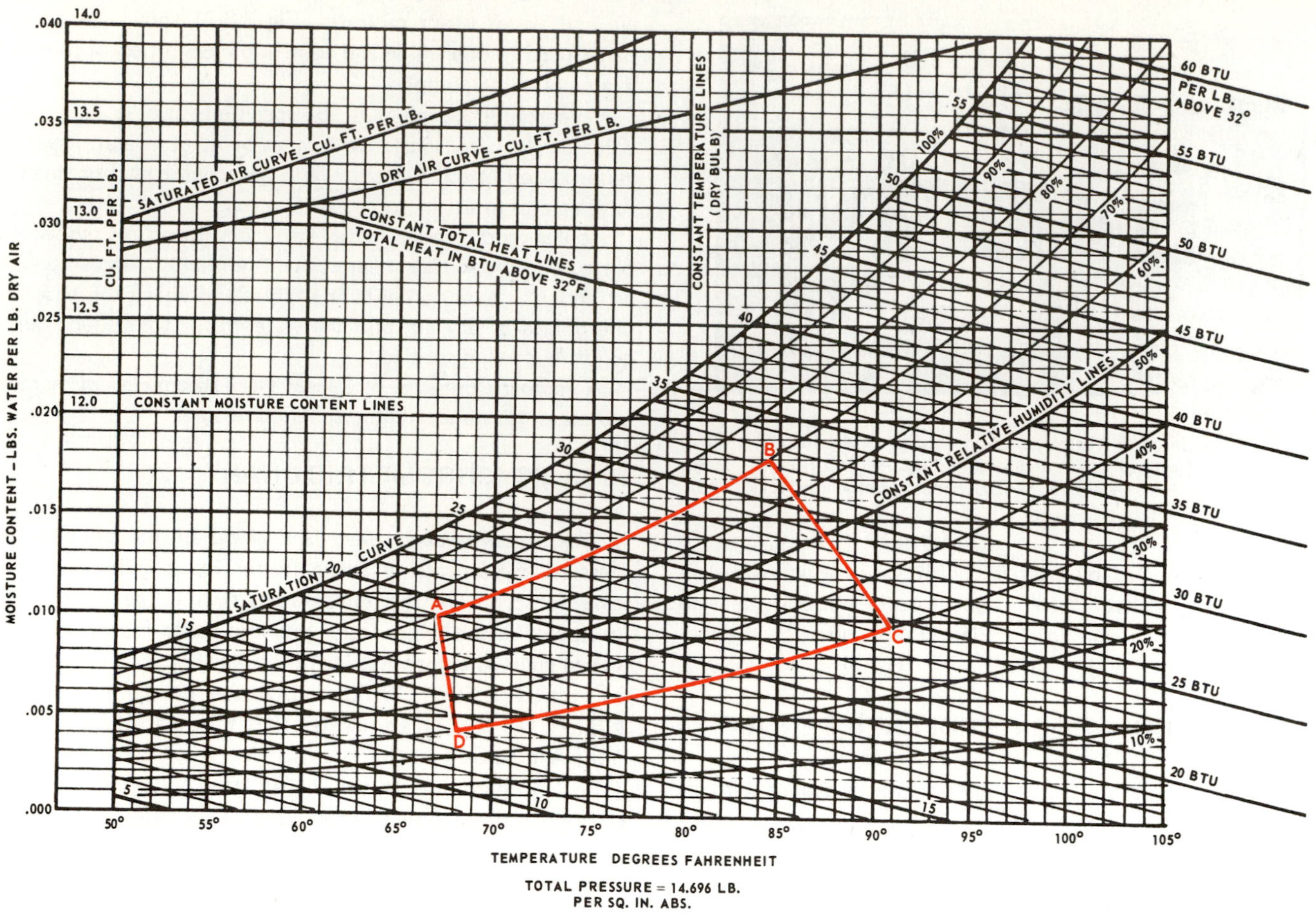

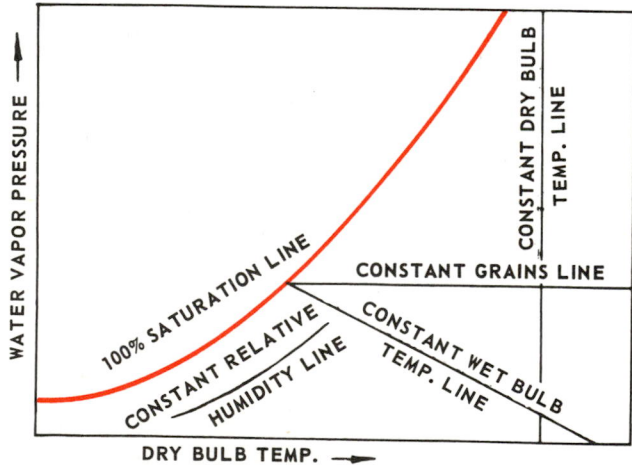

Fig. 18-23. Psychrometric chart shows different variables based on one pound of dry air. Space marked A, B, C and D is most comfortable for majority of people. (Kelvinator, Inc.)

special chart that shows vapor pressure values on the left side. The area bounded by A, B, C and D is the area in which most people feel comfortable either in winter or summer.

The chart should be studied carefully. It is a means for showing air at various conditions and can be used to determine the results of mixing air having various properties. Constant value lines are shown in Fig. 18-24.

Many of the air conditioning problems in this text will involve the use of a psychrometric chart. Some examples of the use of the graph, Fig. 18-8, are included in the following paragraphs.

The numbers along the horizontal scale are dry bulb F. temperatures. The numbers along the vertical scale represent grains of moisture per pound of dry air. The 100 percent humidity curved line (line of saturation) is indicated. The wet bulb temperatures are also indicated along this same line. This psychrometric chart may be used in connection with the wet and dry bulb thermometers to determine the relative humidity under various conditions.

Example: Given the dry bulb temperatures of 75 F.: If the wet bulb temperature is 60 F., what is the relative humidity?

Refer to the psychrometric chart in Fig. 18-8. Follow the vertical line corresponding to the 75 F. dry bulb temperature. Then follow the 60 F. wet bulb temperature line. This crosses the 75 F. vertical line at approximately 41 percent humidity.

Fig. 18-24. Psychrometric chart line nomenclature. The 100 percent saturation line is the pressure-temperature curve for water.

18-16 USING THE PSYCHROMETRIC CHART

Psychrometric charts give a considerable range of temperature and humidity conditions. A typical psychrometric chart is shown in Fig. 18-22; a special chart is presented in Fig. 18-23.

As explained earlier, the human body will be comfortable under a variety of combinations of temperature and humidity. This is shown in Fig. 18-23. In using the psychrometric chart, remember that the warmer the air, the more moisture it will hold. Also, as pressure is reduced, air absorbs more moisture.

To determine the amount of moisture in a pound of dry air at 67 F., locate the temperature of 67 F. at the bottom of the chart in Fig. 18-23, then move upward to the saturation curve. This will indicate that, at this temperature, a pound of air will hold approximately .014 pounds of water, or 98 grains.

Most people are comfortable in an atmosphere where the relative humidity is between 30 and 70 percent, and the temperature is between 70 and 85 F. If one locates these points on the psychrometric chart, it will be found to fall within the area outlined in red in Fig. 18-23.

18-17 VAPOR BARRIERS

Water vapor flows easily through all porous substances. Water in vapor form will remain a vapor as long as its temperature is above the dew point. However, as soon as its temperature drops to the dew point, the water vapor will condense into water droplets.

In modern housing, water vapor is kept from passing through walls and toward surfaces where it might condense by the use of moisture-proof materials. Aluminum foil, plastic sheeting, etc., have been used. These materials form an effective vapor barrier which keeps water vapor from passing from warm surfaces toward cold surfaces. Vapor barriers should always be installed on the warm side of an air conditioned space.

The lack of proper vapor barriers is often indicated by the peeling of the paint on a house near kitchen and bathroom areas. The moisture from these areas travels through the wall structure and, on contacting the cold under-surface of the paint, forms droplets of water. This condition, with freezing and thawing, may cause the paint to peel.

18-18 AIR MOVEMENT

Air movement is an important condition affecting comfort. If cool, dry air is circulated past a warm body, heat flow from the body will speed up and evaporation will increase. This tends to cool the body.

During cold days in winter, early spring or late fall, a person exposed to outside atmospheric conditions often feels much colder than the thermometer indicates. This chilling effect is due to wind velocity and relative humidity. The term "wind chill" applies to this uncomfortable feeling. See Para. 18-30.

Air movement in a conditioned space is also very important. Air movement is very necessary to supply fresh air to a controlled space. If the air moves too fast, persons feel uncomfortable (a draft). If the air movement is too slow, the air becomes stale (contaminated) and lacks oxygen.

18-19 AIR VELOCITY MEASUREMENT

Outside air velocity (wind) is measured in miles per hour (mph) or in knots.

Air velocity (distance traveled per unit of time) is usually expressed in feet per minute (fpm). If the air velocity is multiplied by the cross-section area of a duct, it is possible to calculate the volume of air flowing through the duct in cubic feet per minute.

Different methods may be used to measure air velocity:
1. Anemometer (rotating).
2. Velocimeter (swinging vane).
3. Velocity pressure (pitot tube).
4. Anemometer (hot wire).

The rotating anemometer, the direct-reading velocimeter and the pitot tube are not accurate at very low air velocities.

Fig. 18-25. Anemometer used for measuring airflow. Large dial reads airflow up to 100 ft./sec.; lower left dial reads in 100-ft. graduations (spaces); lower right dial reads in 1000-ft. graduations. (Taylor Instrument, Consumer/Industrial Div., Sybron Corp.)

18-20 ANEMOMETER — ROTATING AND HOT WIRE

If a small propeller is placed in an airstream, it will revolve as the air flows past the blades. If the propeller is connected to a dial calibrated in feet, it will indicate the feet of flow. See Fig. 18-25. Devices of this type are called "anemometers."

Anemometers generally have a start lever and a return-to-zero lever. To use the instrument, carefully place it in the airstream at right angles to the airflow. Allow it to reach a constant speed (takes about one minute). Then trip the registering mechanism and, at the same time, start a stopwatch. Record the reading and the time. From this data, compute the velocity of the air in feet per minute. For example, if the reading is 236 for 1/2 min., the velocity will be 472 ft./min.

It is advisable to take several readings, then compute the average to insure greater accuracy.

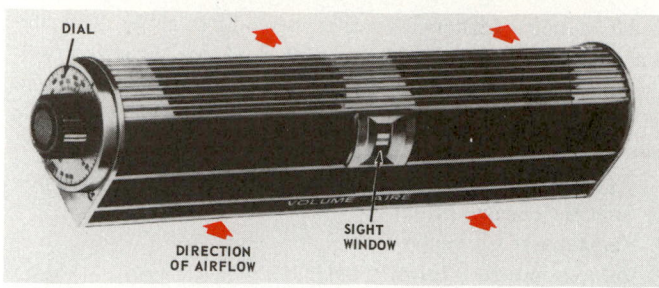

Fig. 18-26. This anemometer directly reads air velocity in fpm. (Thermal Industries of Florida, Inc.)

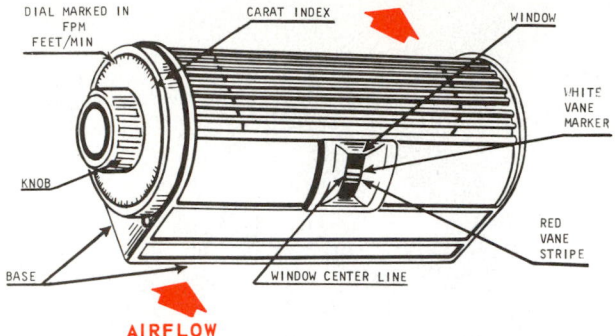

Fig. 18-27. Names of parts of the anemometer shown in Fig. 18-26.

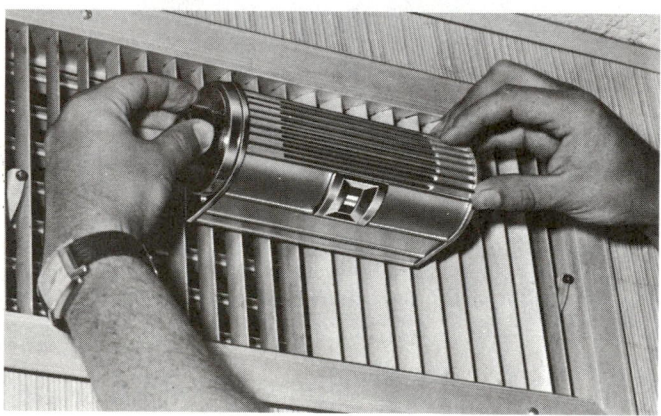

Fig. 18-28. Anemometer in use, checking discharge air velocity from a rectangular grille.

Another type of anemometer, shown in Figs. 18-26 and 18-27, is used mainly for measuring discharge air velocities at grilles. To use the unit: base is placed against grille; knob is turned until white vane marker is in center of window; then knob is turned until red vane is just under center line. The dial will indicate air velocity in fpm. For rectangular grilles, the unit is used as shown in Fig. 18-28. For circular grilles, the anemometer should be positioned on the radius of the grille, as shown in Fig. 18-29.

The operation of the hot wire anemometer depends upon the cooling effect of air flowing over an electrically heated wire. A hot wire instrument is shown in Fig. 18-30.

Fig. 18-29. Anemometer being used to measure the discharge air velocity from a circular grille. (Thermal Industries of Florida, Inc.)

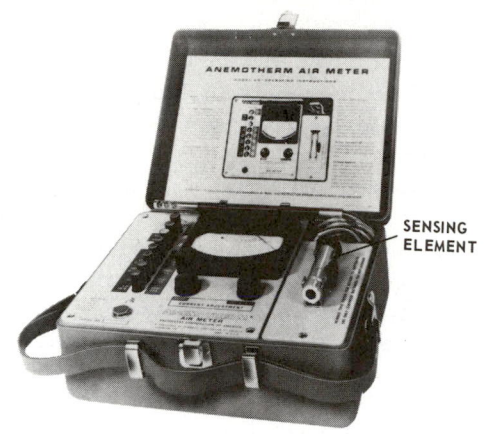

Fig. 18-30. A direct-reading, air velocity-indicating instrument of hot wire type. (Anemostat Products Div.)

18-21 VELOCIMETERS (SWINGING VANE)

Some velocity measuring instruments read directly in feet per minute (fpm). See Fig. 18-31. Service technicians often use these instruments to avoid the arithmetic which is necessary when using the pitot tube instrument.

In using a swinging vane velocimeter, incoming air pushes on a small vane that tilts at different angles as the air velocity or speed increases. The instrument is put directly in the airstream with the left side facing the airflow. The velocimeter shown in Fig. 18-31 has two velocity scales, 0-200 and 0-800 feet per minute.

For velocity readings where it is difficult to place the instrument in the airstream, special jets are used to adapt the instrument to these conditions.

The direct-reading instrument can be used to measure grille air velocities, as shown in Fig. 18-32. Note that a special jet is attached to the air inlet of the instrument by means of a flexible tube.

The velocimeter is also used to measure air velocities in main ducts and branch ducts. See Fig. 18-33. An instrument of this type is necessary to balance air distribution systems.

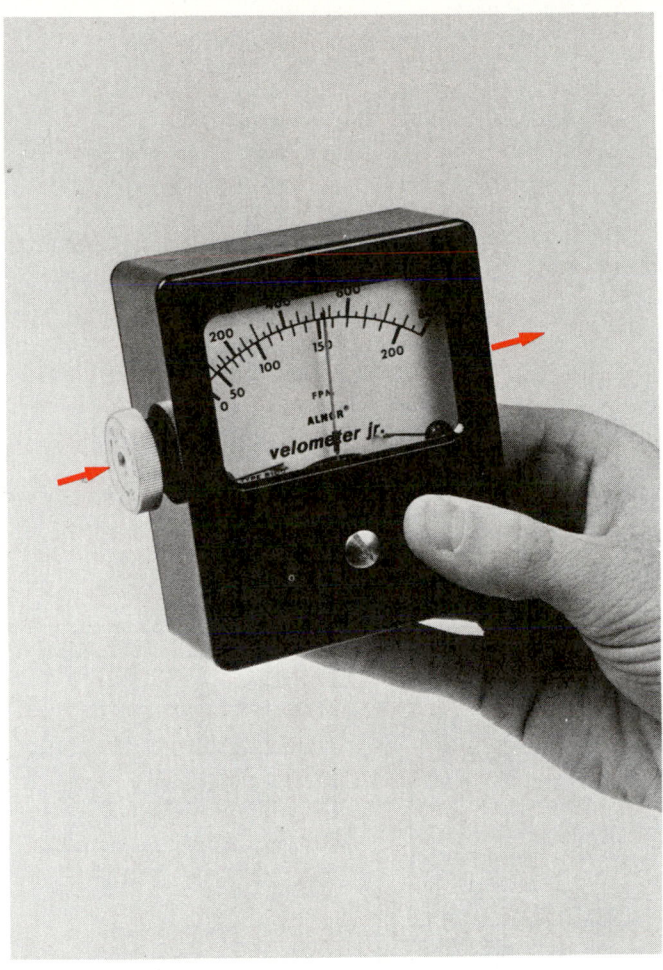

Fig. 18-31. Direct-reading air velocity meter. Note airflow from left to right. (Alnor Instrument Co.)

The instrument is calibrated for use at a temperature of 68 F. Corrections may be made if the duct temperature is not at 68 F. Formula for correction:

$$fpm = \frac{460 + T}{460 + 68} \times \text{instrument reading}$$

(T = temperature Fahrenheit of air in duct.)

18-22 VELOCITY PRESSURE (PITOT TUBE)

The velocity pressure method of measuring air velocity makes use of an instrument called a pitot tube, Fig. 18-34. Air contacting the nose of the pitot tube creates a total pressure. The outer tube, with the holes on the side, measures the static pressure. When these two pressures are connected to the end of a manometer, the difference in the pressures is the velocity pressure. An incline manometer is used with the pitot tube, as shown in Fig. 18-35.

This pressure difference is measured in inches of water.
Formula:

Velocity = 4050 x square root of velocity pressure in inches of water.

Example: If the velocity pressure is 1 in. of water, what is the velocity?

Velocity = $4050 \sqrt{1 \text{ in.}}$

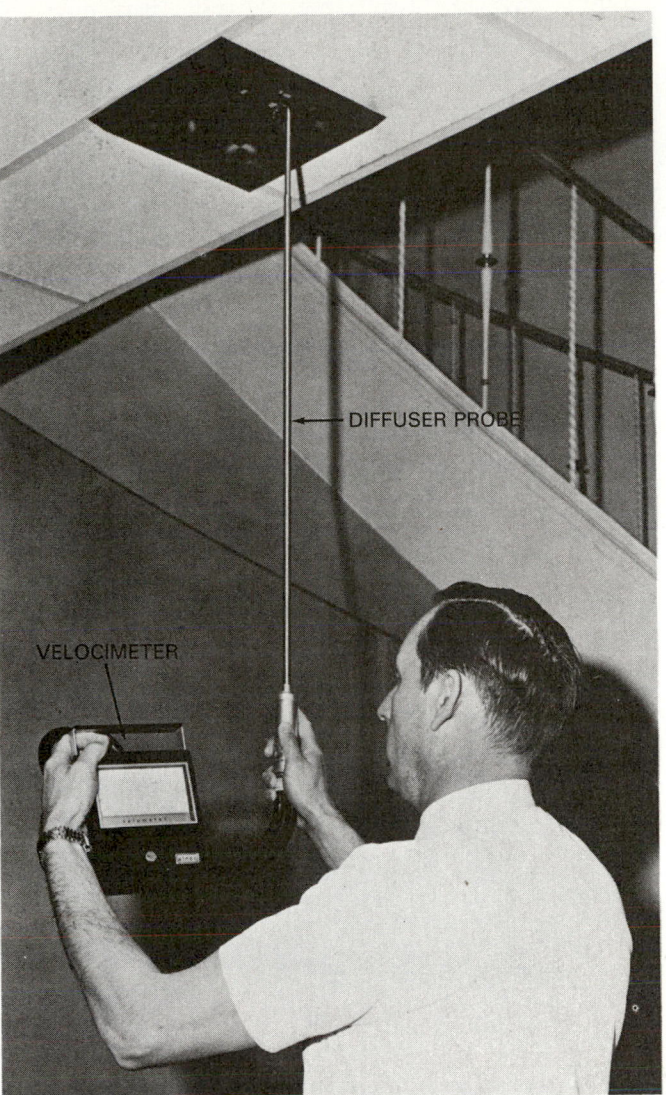

Fig. 18-32. Measuring air velocity at an air conditioning diffuser. Note use of diffuser probe. (Alnor Instrument Co.)

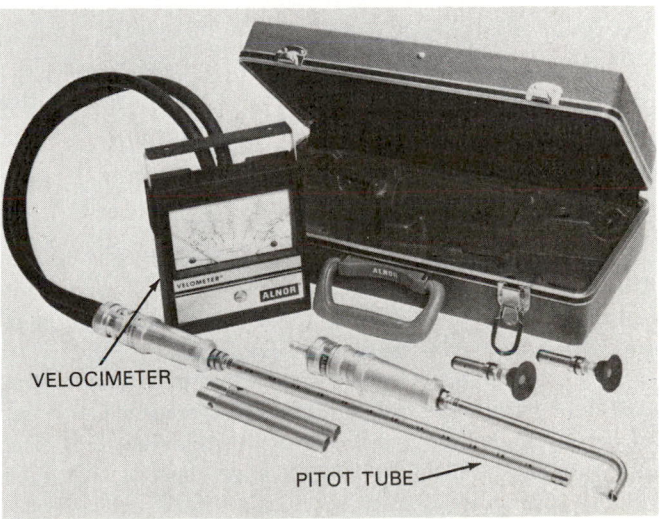

Fig. 18-33. Direct-reading airflow meter which may be used to determine air velocity inside a duct. (Alnor Instrument Co.)

Fundamentals of Air Conditioning / 653

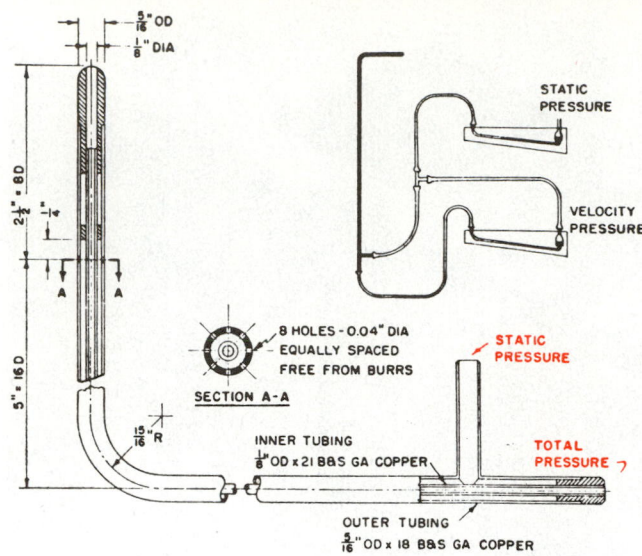

Fig. 18-34. A pitot tube connected to two inclined manometers measures air velocity. (ASHRAE Guide and Data Book)

Volume Cu.ft./lb.*	Velocity Constant
11.5	3720
12.1	3818
13.2	3980
**13.4	4010
14.1	4118
15.1	4260
16.2	4410
17.1	4530

*Values for any conditions may be read from the psychrometric chart.
**Standard.

Fig. 18-36. The velocity correction factor changes with a change in air density (effect of temperature and altitude).

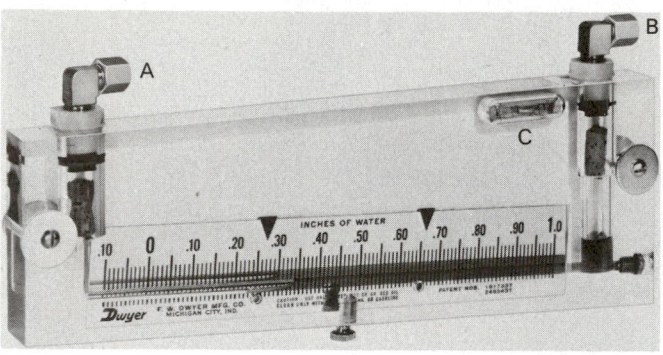

Fig. 18-35. Inclined gauge for use with pitot tube. A—Total pressure connection. B—Static pressure connection. This gauge may also be used for measuring filter pressure drops. C—Built-in spirit level. Liquid level must be adjusted to zero reading to level unit. (Dwyer Instruments, Inc.)

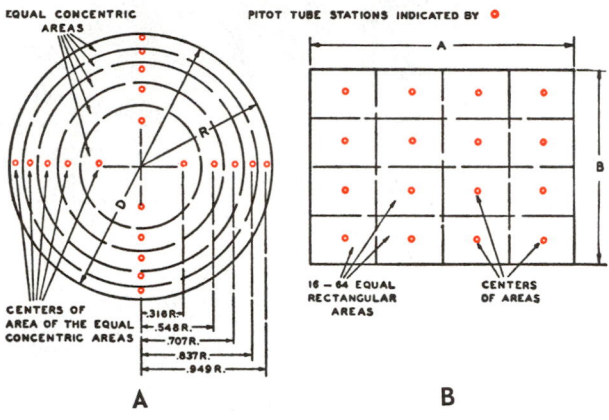

Fig. 18-37. Locations for velocity readings in duct. Average of readings will produce average duct velocity. A—Round duct (20 readings). B—Rectangular duct (16 readings). (ASHRAE Guide and Data Book)

location of each of the structure points is as recommended by the American Society of Heating, Refrigerating and Air Conditioning Engineers (ASHRAE).

18-23 DRAFTS AND DRAFT MEASUREMENT

If the heating and cooling arrangement in a structure is set up in such a manner that air flows through the occupied space at a rate of over 15 to 20 feet per minute, occupants of the space will feel uncomfortable. They feel a draft.

It is difficult to develop accurate instruments to measure drafts. The usual method is to use a smoke generator and a stopwatch to time the flow of the smoke through the space.

18-24 VENTILATION

Ventilation is a term applied to changing the air in a work place or living place. In any space occupied by people, the breathing of the air reduces the oxygen content. Activities in the space may add some pollutants to the environment. The most economical way to maintain the health and comfort

Velocity = 4050 x 1

Velocity = 4050 ft./min.

If the velocity pressure is .25 in. of water, what is the velocity?

Velocity = $4050 \sqrt{.25}$

Velocity = 4050 x .5

Velocity = 2025 ft./min.

The constant 4050 is for approximately 80 F. at a 500 ft. altitude. Other values are shown in Fig. 18-36. The constant changes are based on density of the air. The manometer must be mounted level to obtain accurate readings.

To obtain correct velocity readings in a duct, take several readings in various parts of the duct and average the readings. The recommended method to use is shown in Fig. 18-37. The rectangular duct is divided into equal areas, then the pitot tube is put in the center of each small area. The 16 velocities are averaged to obtain the overall average velocity.

The round duct is more difficult to measure, because it must be divided into equal circular areas. See Fig. 18-37. The

conditions of the space is by replacing the air. This is done by bringing in outside air to ventilate the space.

Sometimes it is desirable to quickly replace all the air in the confined space. This is done by opening windows and doors to flush the space completely with 100 percent outside air.

In most heating systems, provisions are made to continually replace a small percent of the air in the conditioned space. This is done by slowly exhausting some of the air and bringing in makeup air. In homes, the gradual change may not be noticeable, since there is always a small amount of air entering and leaving the home. This movement of air is by way of the cracks around windows and doors and through doorways each time doors are opened.

A considerable amount of air also filters in and out of a building because some building materials are porous (many tiny holes). The amount of infiltration depends chiefly upon the wind velocity and the temperature difference inside and outside the building. Air tends to enter the building on the upwind side and leave the building on the downwind side.

Since warm air is lighter than cold air, it tends to rise in a room. If the building has more than one story, the warm air tends to rise from the lower floors to the upper floors. The rising air will create a slight pressure that causes some warm air to escape through the upper surfaces of the building. This lost air is replaced by cold air entering the lower levels.

With summer air conditioning, the opposite situation occurs. Cold air tends to flow downward and may leave the building at the lower levels. The cold air, then, is replaced by warmer air entering at the upper levels.

Whenever air is exhausted from a space, such as a room in a home or factory building, air must be brought into the room to replace the air exhausted. If the outdoor temperature is either too high or too low for comfort, the air brought into the area to replace exhausted air must be conditioned; either heated or cooled, and cleaned, to provide a comfortable room environment. This conditioned air brought into the room is called "makeup air."

A structure which maintains an inside air pressure of slightly above atmospheric pressure is said to have a positive

pressure. A structure which maintains an air pressure slightly below atmospheric pressure is said to have a negative pressure.

Fuel-burning furnaces, stoves and fireplaces operating in the winter tend to cause a negative pressure in a building. This means that there will be a considerable amount of air leakage through the walls and through cracks. Positive pressure can be maintained only if a fan or blower of some kind is used to bring in fresh air.

With heated tall structures, there may be a tendency for upper rooms to be slightly above atmospheric pressure, and lower rooms to be slightly below atmospheric pressure, because warm air rises.

18-25 CLIMATE, OUTDOOR — INDOOR

Climate usually is defined as the weather conditions of a region. These conditions include temperature, humidity, sunshine, pressure and air movement.

It is evident that outdoor climate cannot be affected much by any type of air conditioning (heating, cooling, humidifying). In an enclosed space, however, the above factors may be controlled and an "indoor" climate provided to meet any desired conditions.

Indoors, it is possible to completely control the factors which determine comfort in an enclosed space. There is a definite relationship between comfort and the conditions of temperature, humidity and air movement. Fig. 18-38 illustrates the constant comfort condition with varying temperatures and humidities. Many homes and work places are completely air conditioned.

Increasing the air movement tends to give a cooling effect on the human body. If the heating system provides too much air movement (over 15 to 20 fpm), it may be necessary to increase the temperature somewhat in order to maintain a comfortable indoor climate.

18-26 WEATHER

Weather usually is defined as the conditions in the atmosphere such as temperature, wind velocity and direction,

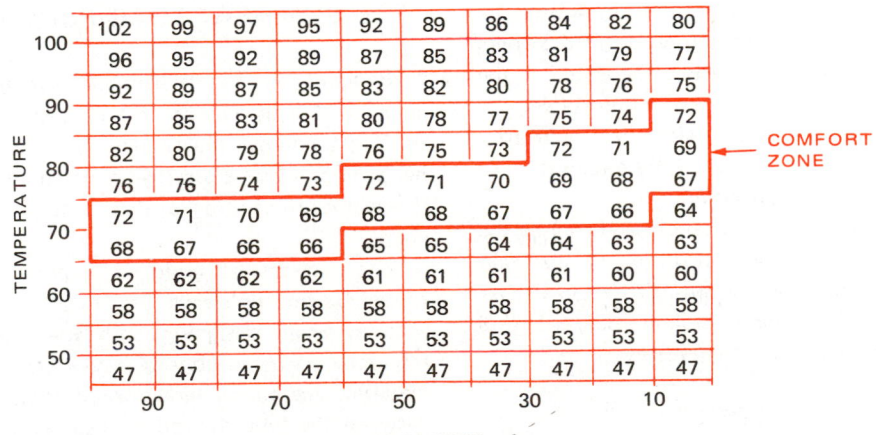

Fig. 18-38. Comfort zone. Area inside red lines indicates the usual temperature and humidity range under which most people are comfortable. Note that with a high humidity, one is comfortable at a lower temperature than in the case under low humidity conditions.   (Westclox)

		AMBIENT TEMPERATURE													
	°F.	40	35	30	25	20	15	10	5	0	−5	−10	−15	−20	−25	−30
	°C.	4	2	−1	−4	−7	−9	−12	−15	−18	−21	−23	−26	−29	−32	−34
WIND VELOCITY IN MPH		EQUIVALENT TEMPERATURE IN STILL AIR													
CALM (0)	°F.	40	35	30	25	20	15	10	5	0	−5	−10	−15	−20	−25	−30
	°C.	4	2	−1	−4	−7	−9	−12	−15	−18	−21	−23	−26	−29	−32	−34
5	°F.	37	33	27	21	16	12	7	1	−6	−11	−15	−20	−26	−31	−35
	°C.	3	1	−3	−6	−9	−18	−14	−17	−21	−24	−26	−29	−32	−35	−37
10	°F.	28	21	16	9	2	−2	−9	−15	−22	−27	−31	−38	−45	−52	−58
	°C.	−2	−6	−9	−13	−17	−19	−23	−26	−30	−33	−35	−39	−43	−47	−50
15	°F.	22	16	11	1	−6	−11	−18	−25	−33	−40	−45	−51	−60	−65	−70
	°C.	−6	−9	−11	−17	−22	−24	−28	−32	−36	−40	−43	−46	−51	−54	−57
20	°F.	18	12	3	−4	−9	−17	−24	−32	−40	−46	−52	−60	−68	−76	−81
	°C.	−8	−11	−16	−20	−23	−27	−31	−36	−40	−43	−47	−51	−56	−60	−63
25	°F.	16	7	0	−7	−15	−22	−29	−37	−45	−52	−58	−67	−75	−83	−89
	°C.	−9	−14	−18	−22	−26	−30	−34	−38	−43	−47	−50	−55	−59	−64	−67
30	°F.	13	5	−2	−11	−18	−26	−33	−41	−49	−56	−63	−70	−78	−87	−94
	°C.	−16	−15	−19	−24	−22	−32	−36	−41	−45	−49	−53	−57	−61	−66	−70
35	°F.	11	3	−4	−13	−20	−27	−35	−43	−52	−60	−67	−72	−83	−90	−98
	°C.	−11	−16	−20	−25	−29	−33	−37	−42	−47	−51	−55	−58	−64	−68	−72
40	°F.	10	1	−6	−15	−22	−29	−36	−45	−54	−62	−69	−76	−87	−94	−101
	°C.	−12	−17	−21	−26	−30	−34	−38	−43	−48	−52	−56	−60	−66	−70	−74

(Overlaid diagonal zone labels: COOL, COLD, VERY COLD, BITTER COLD, EXTREME COLD)

Fig. 18-39. Wind chill index reveals that increasing the wind velocity greatly increases the chill effect. With a thermometer reading of 0 F. and a wind velocity of 20 mph, the effect on the human body is the same as it would be if the person were in a temperature of −40 F. (−40 C.). (Michigan Farmer)

clouds, moisture and atmospheric pressure. Weather affects the need and requirements for air conditioning.

18-27 AIR TEMPERATURE

Air temperatures in the United States vary from a low of about −55 F. (−48 C.) to a high of around 120 F. (49 C.). The normal, desirable temperature is said to be 72 F. (22 C.).

Normally, the human body temperature is 98.6 F. (37 C.). Skin temperature is lower, about 91 F. (33 C.). In temperate zones, the average atmospheric temperatures in winter are below the body temperature, so clothing is required to help conserve the body heat. Also, heat needs to be added to the occupied space so that the occupants may be comfortable.

In the summer, the human body can lose heat only after the air temperature exceeds 98.6 F. (37 C.), by evaporation of perspiration and respiration from the body.

Basically, heating the air in some instances and cooling the air in other instances is necessary in order to maintain temperatures that are comfortable. The specific heat of dry air is .24 Btu per lb. Either for heating or cooling, then, energy is required to bring about the desired temperatures.

18-28 TEMPERATURE CONTROLS

When heating an air conditioned space, the amount of heat supplied to the conditioned space regulates the temperature. The temperature regulating devices control the flow of heat-carrying media (substances). Usually, the heat-carrying medium is either warm air, warm water, steam or water vapor. Typical heat sources are gas flame, oil flame, electric resistance or coal fire.

When cooling an air conditioned space, the amount of cooling will depend upon the temperature of the cooling surface (evaporator), the rate of flow of air over the cooling surface, and the initial temperature of the air in the conditioned space. All of these variables may be controlled.

18-29 SUN HEAT LOAD FUNDAMENTALS

Radiant heat (light) from the sun furnishes a tremendous amount of heat energy. If a glassed-in surface is exposed to this light energy, the energy will enter a space and become heat. Since glass is a rather poor conductor of heat, the heat that comes in as a ray of light is trapped in the room as heat energy.

The surfaces of buildings exposed to sunlight are also heated by the sun's rays. This heat source must be taken into account when designing both the heating and the cooling requirements of air conditioning systems. Since many building materials are poor conductors of heat, there is a lag or delay between the time the radiant heat energy strikes the building surface and the time the heat enters the air conditioned space.

Color has a considerable effect on the amount of heat absorbed from the sun's rays. Black and red absorb much more heat than white and yellow. Likewise, surfaces that radiate heat are much more efficient if painted in dark colors than if painted in light colors. Light reflecting surfaces, such as polished metal, chrome and nickel plate do not absorb heat easily; neither do they efficiently radiate heat from their surfaces.

18-30 WIND CHILL INDEX (CHILL FACTOR)

During the winter months, the chill index temperature, or "chill factor," combines temperature and wind speed. The chill factor is calculated and released by United States Weather Bureau through the usual weather forcasting channels.

The chill factor is based on both temperature and wind speed in mph. For example, at a temperature of 0 F. and a wind speed of 10 mph, the chill index temperature is -22 F. Human flesh exposed to the atmosphere freezes at about -25 F., making the chill factor an important consideration.

Fig. 18-39 shows the chill factor in degrees corresponding to the wind speed. Note that, at a temperature of 0 F., with a wind speed of 40 mph, the chill factor mounts to -54 F. This temperature will quickly freeze exposed flesh.

18-31 BEAUFORT SCALE

The Beaufort Scale is frequently used by the Weather Bureau in indicating wind velocity. Fig. 18-40 gives wind velocity values and effects according to the Beaufort Scale.

Increasing the wind velocity increases the heat loss of a heated structure. The calculated heat load for a structure should include provisions for the maximum wind velocity expected for the area.

18-32 COMFORT CONDITIONS

Comfortable conditions result from a desirable combination of temperature, humidity, air movement and air cleanliness. However, one may have comfort under varying values of these factors. For instance, high relative humidity which tends to be uncomfortable may be counteracted by a relatively low temperature and rapid air movement. In many homes in wintertime, a low relative humidity is compensated for by an increase in room temperature and slight air movement. Fig. 18-41 illustrates what is commonly accepted as the comfort

BEAUFORT NUMBER	WIND VELOCITY		OBSERVED WIND EFFECTS	TERMS USED IN USWB FORECASTS
	MPH	KNOTS		
0	LESS THAN 1	LESS THAN 1	CALM; SMOKE RISES VERTICALLY	
1	1–3	1–3	DIRECTION OF WIND SHOWN BY SMOKE DRIFT; BUT NOT BY WIND VANES	LIGHT
2	4–7	4–6	WIND FELT ON FACE; LEAVES RUSTLE; ORDINARY VANE MOVED BY WIND	
3	8–12	7–10	LEAVES AND SMALL TWIGS IN CONSTANT MOTION; WIND EXTENDS LIGHT FLAG	GENTLE
4	13–18	11–16	RAISES DUST, LOOSE PAPER; SMALL BRANCHES ARE MOVED	MODERATE
5	19–24	17–21	SMALL TREES IN LEAF BEGIN TO SWAY; CRESTED WAVELETS FORM ON INLAND WATERS	FRESH
6	25–31	22–27	LARGE BRANCHES IN MOTION; WHISTLING HEARD IN TELEPHONE WIRES; UMBRELLAS USED WITH DIFFICULTY	STRONG
7	32–38	28–33	WHOLE TREES IN MOTION; INCONVENIENCE FELT WALKING AGAINST WIND	
8	39–46	34–40	BREAKS TWIGS OFF TREES; GENERALLY IMPEDES PROGRESS	GALE
9	47–54	41–47	SLIGHT STRUCTURAL DAMAGE OCCURS; LEAVES, BRANCHES BLOWN FROM TREES	
10	55–63	48–55	SELDOM EXPERIENCED INLAND; TREES UPROOTED; CONSIDERABLE STRUCTURAL DAMAGE OCCURS	WHOLE GALE
11	64–72	56–63	VERY RARELY EXPERIENCED; ACCOMPANIED BY WIDESPREAD DAMAGE	
12 OR HIGHER	73 OR HIGHER	64 OR HIGHER	VERY RARELY EXPERIENCED; ACCOMPANIED BY WIDESPREAD DAMAGE	HURRICANE – TYPHOON

Fig. 18-40. Beaufort Scale of wind velocity.

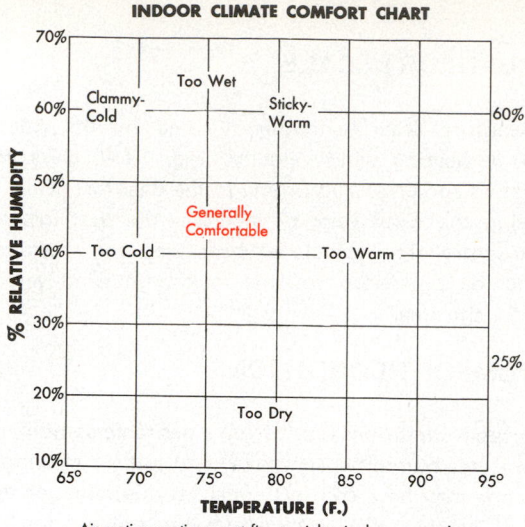

Fig. 18-41. Indoor comfort chart. Most people will feel comfortable at the temperature and humidity indicated in red.
(Lennox Industries, Inc.)

zone for the various conditions.

A more technical graph showing the comfort zones for both winter and summer is illustrated in Fig. 18-42. This comfort zone represents a considerable area. However, experiments indicate that any point in this area gives approximately equal comfort under equal conditions of clothing and work. These areas are sometimes defined as effective temperature (ET). Effective temperature is the combined effect of dry bulb temperature, wet bulb temperature and air movement, which provides an equal sensation of warmth or cold.

The human body is able to accustom itself only to a certain amount of change in a given length of time. Therefore, it becomes necessary to regulate air conditioning equipment so that it will produce only a certain output. While this will be less comfortable, it will not subject the person to too great a shock on entering this conditioned space, or when going out into the normal outdoor atmosphere.

In the summertime, air conditioned buildings are usually maintained at a temperature which is not over 10 degrees

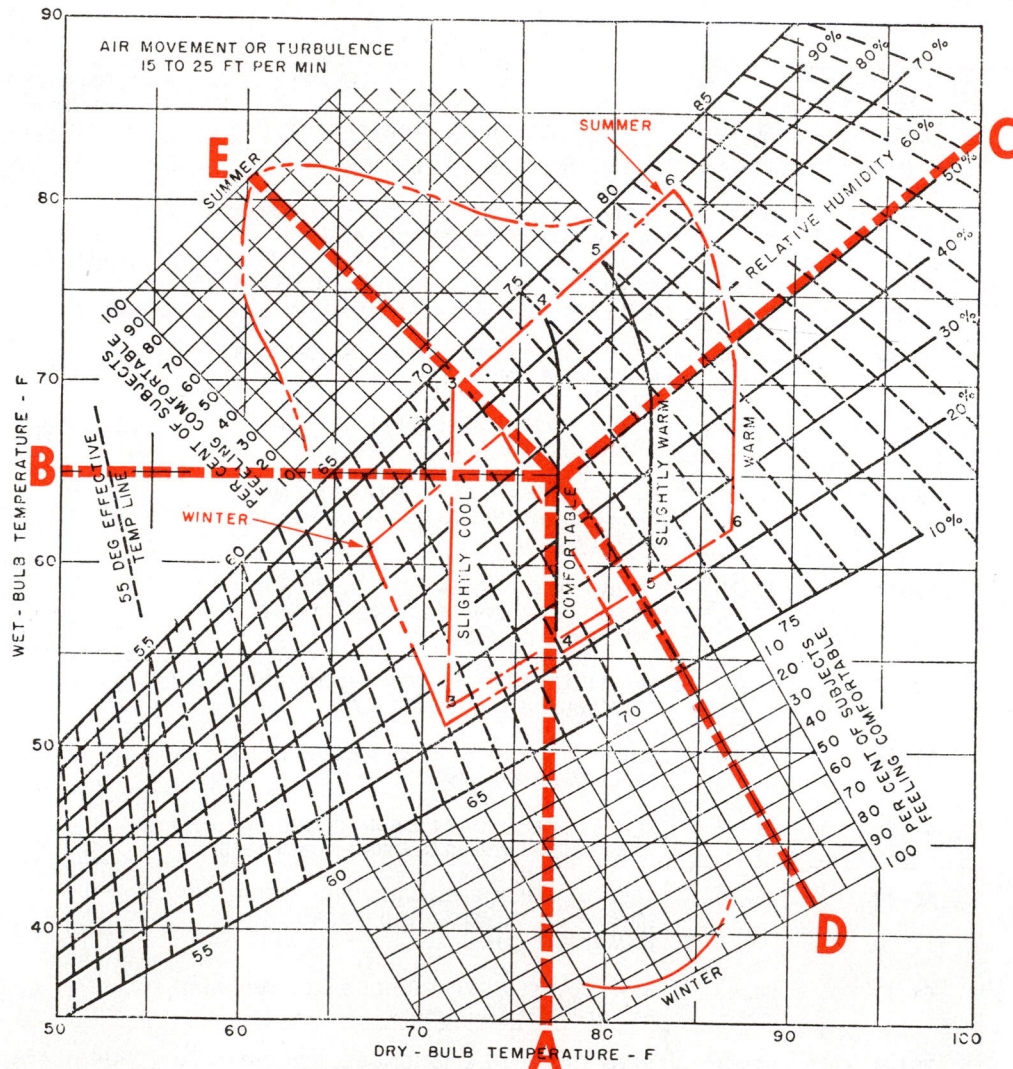

Fig. 18-42. A graph of the comfort zone. A—Constant dry bulb temperature line. B—Constant wet bulb temperature line. C—Constant relative humidity line. D—Constant winter effective temperature line. E—Constant summer effective temperature line.
(ASHRAE Guide and Data Book)

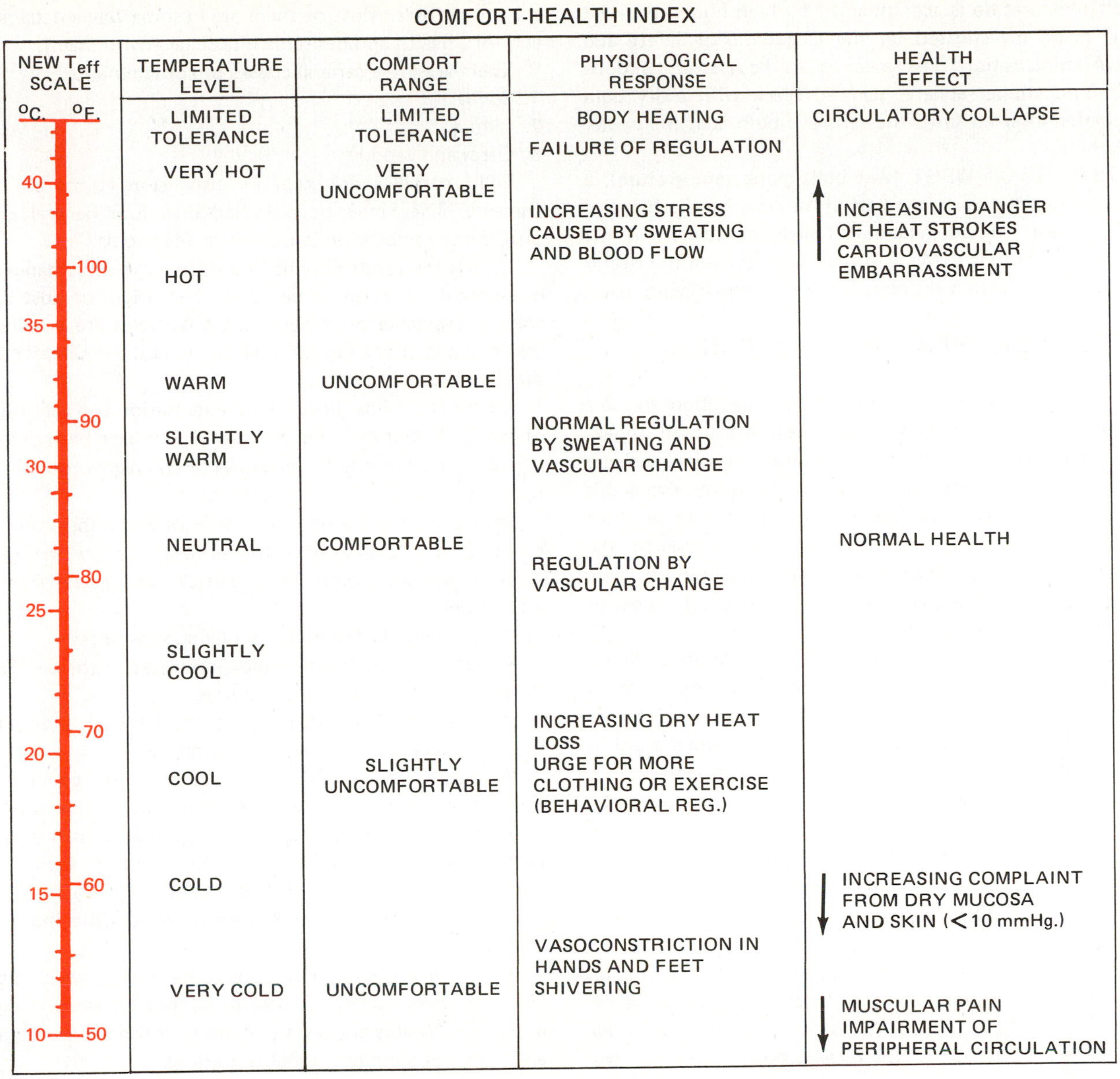

NEW T_{eff} SCALE °C. °F.	TEMPERATURE LEVEL	COMFORT RANGE	PHYSIOLOGICAL RESPONSE	HEALTH EFFECT
	LIMITED TOLERANCE	LIMITED TOLERANCE	BODY HEATING FAILURE OF REGULATION	CIRCULATORY COLLAPSE
40 — 100 —	VERY HOT	VERY UNCOMFORTABLE	INCREASING STRESS CAUSED BY SWEATING AND BLOOD FLOW	↑ INCREASING DANGER OF HEAT STROKES CARDIOVASCULAR EMBARRASSMENT
	HOT			
35 —	WARM	UNCOMFORTABLE		
— 90	SLIGHTLY WARM		NORMAL REGULATION BY SWEATING AND VASCULAR CHANGE	
30 —				
— 80	NEUTRAL	COMFORTABLE	REGULATION BY VASCULAR CHANGE	NORMAL HEALTH
25 —	SLIGHTLY COOL			
— 70	COOL	SLIGHTLY UNCOMFORTABLE	INCREASING DRY HEAT LOSS URGE FOR MORE CLOTHING OR EXERCISE (BEHAVIORAL REG.)	
20 —				
15 — — 60	COLD			↓ INCREASING COMPLAINT FROM DRY MUCOSA AND SKIN ($<$10 mmHg.)
	VERY COLD	UNCOMFORTABLE	VASOCONSTRICTION IN HANDS AND FEET SHIVERING	
10 — — 50				↓ MUSCULAR PAIN IMPAIRMENT OF PERIPHERAL CIRCULATION

Fig. 18-43. Comfort-Health Index indicates related sensory, physiological and health responses by people to prolonged exposures.   (ASHRAE Handbook of Fundamentals)

below the outside temperature. Some people are quite sensitive to thermal shock when entering or leaving an air conditioned space. The danger of this thermal shock is lessened if the difference between inside and outside temperatures is reduced, or if a person will put on a sweater or coat when entering a cooled air conditioned space, and remove it when returning to the warm outdoors.

Fig. 18-42 indicates that in the summer most people are comfortable between 72 F. db (dry bulb temperature) and 100 percent RH (relative humidity) up to 85 F. db and 10 percent RH. During the winter, most people feel equally comfortable between 66 F. db and 100 percent RH and 74 F. db and 10 percent RH.

Experiments show that the average person is most comfortable under normal temperature and humidity conditions if the skin surface temperature is approximately 91 F. (33 C.). This

skin temperature is usually maintained in cold weather by wearing clothing. In hot weather, the temperature is maintained by the evaporation of moisture (sweat) from the skin surface and from radiation from the skin surface.

The skin temperature may drop below this figure in hot, humid weather because of rapid evaporation of moisture from the skin surface. Even though this skin temperature may be considerably below what is considered to be a comfortable temperature, the person is not uncomfortable because the heat generated in his body is being released by moisture evaporation from the skin surface.

Human ills due to thermal environment are sometimes called "thermal disorders." In cold climates, it is possible for the body temperature to drop a few degrees below normal. This is chiefly due to the reduction in the rate of metabolism.

High temperatures may cause human illness, particularly if

Fundamentals of Air Conditioning / 659

the high temperature is accompanied by high humidity. Heat stress is being investigated by the Occupational Safety and Health Administration. The measure of heat stress is done using a 6-in. copper sphere, painted black with a dry bulb thermometer inserted until the sensitive bulb is at the center of the sphere.

At 79 F. (26 C.) WBGT (wet bulb globe temperature), a person should only work one half of the time for the first five days. A person should take salt tablets moderately if the WBGT is 79 F. (26 C.). Many studies have proven the health benefit obtained from a properly air conditioned living space.

18-33 COMFORT-HEALTH INDEX (CHI)

The American Society of Heating, Refrigeration and Air Conditioning Engineers recognizes a Comfort-Health Index. Fig. 18-43 indicates the temperature, the sensation and the effect on the physiology and health of the body. From this chart, it may be seen that, under long exposure to very hot conditions, the human body attempts to adjust to the conditions by increasing sweating and the flow of blood. These physiological conditions may result in an increased danger of heat strokes and cardiovascular difficulty.

The chart shows that, at comfortable temperatures, there is no sensation of either warmth or cold. Also, there are no apparent physiological effects, and the body is in the condition of normal health. Moving down in temperature to very cold conditions, the body is uncomfortable. Also, physiologically, the body attempts to correct this condition by shivering. From the standpoint of health, this may cause an increase in mortality, particularly in older people.

18-34 DEGREE DAYS

Degree days is a term used to help indicate the heating or cooling needed for any certain day. Calculations are based on a temperature of 65 F. (18 C.). The degree day is computed by taking the mean (average) of the highest temperature and the lowest temperature for a day and subtracting it from 65 F. (18 C.).

Example: The lowest recorded temperature for a certain day was 28 F. (−2 C.).

The highest recorded temperature for the same day was 36 F. (2 C.).

The mean temperature for the day was

$$\frac{28 + 36}{2} = 32 \text{ F.} \qquad 65 - 32 = 33 \text{ F. degree days.}$$

In Celsius degrees;

$$\frac{-2 + 2}{2} = 0 \text{ C. or } 18 \text{ C.} - 0 \text{ C.} = 18 \text{ C. degree days.}$$

Degree days may be added by weeks, months or for a season to give a comparison of the heating needs for different years.

18-35 AIR CONTAMINANTS

Air contains substances other than those described in Para. 18-2. Called contaminants, these substances are not normal in the atmosphere. Most of them are in some way a detriment to comfort, health and desirable industrial environment.

There are three general classes of contaminants:
1. Solid.
2. Liquid.
3. Gases and vapors.

Solid particles are kept in suspension in the air by air currents. They may be classified into four general groups: dust; fumes; smoke; pollen, bacteria and molds.

Dust is the result of wind, a sudden earth disturbance or by mechanical work on some solid. The origin of dust can be animal, vegetable or mineral. Dust particles are usually over 600 microns in size (about .004 in. in dia.). See Chapter 28 for the definition of micron.

Fumes are solids formed by condensation and solidification of materials that are ordinarily solids but have been put into a gaseous state (usually an industrial or chemical process). These particles are about 1 micron in size.

Smoke is the result of incomplete combustion. Solid particles are carried into the atmosphere by the gaseous products of combustion. These particles vary in size from .1 to 13 microns.

Pollen, bacteria and molds are living substances.

In addition, there are liquid impurities in the air. Two of the most common are mists and fogs.

Mists are small liquid particles, mechanically ejected into the air by splashing, mixing, atomizing, etc.

Fogs are small liquid particles formed by condensation. Fogs mean that the atmosphere has reached the saturation state for that particular chemical. A fog may consist of minute particles of water, sometimes contaminated with sulphur dioxide, fumes, smoke and dust particles. It may form a very obnoxious atmosphere to be breathed by either humans or animals.

There are also gases and vapor contaminants which may act like true gases. There is little difference between these two impurities. Vapors are gases that have condensing temperatures and pressures close to normal conditions.

Not all contaminants are objectionable or harmful. Perfumes and deodorizers have been used for years to make air more pleasant to smell or to conceal objectionable odors.

Pollen grains come from vegetation growth such as weeds, grasses and trees. Their presence in the air is usually responsible for hay fever, rose fever and other respiratory conditions. Removal of pollen from the air has been an important contribution of air conditioning. These particles vary in size from 10 to 50 microns.

Bacteria are microorganisms that are responsible for the transfer of many diseases. Many manufacturing processes require the removal of these bacteria. Hospital rooms and some refrigerators use bacteria-removing devices.

Mold is a growth of minute fungi which forms on vegetable and animal matter. Many typical air conditioning applications may provide a favorable environment for the growth and development of many molds (particularly if some moisture is present). There have been instances where the spores from these molds have caused illness to the occupants of the air conditioned space.

Mold-killing sprays are available which may be used to spray air conditioning evaporators to make them free of any possible mold development. See Chapter 28 for information on solid particle sizes.

18-36 POLLUTANTS

The Federal Clean Air Act of 1963, and as amended annually since then, gives the United States Department of Health, Education and Welfare power to establish and enforce standards for clean air. The pollutants, in addition to those previously named, include particulates, carbon monoxide, photochemical oxidants and nitrogen oxides. The photochemical oxidants result from the effect of sunlight on hydrocarbons and nitrogen oxides, which causes them to react to produce smog. The word smog combines the terms "smoke" and "fog." Terpene, a hydrocarbon released from growing trees, is sometimes considered a pollutant. Methane comes from the decomposition of vegetable matter.

Particulates include fogs, mists, molds, pollen, dust, fly ash, asbestos and larger bacteria. In addition to the vapors and particulates, the environment often includes bacteria, viruses and fungi. Remember that outside air under the most favorable conditions always includes some of the above items. In some cases, the construction and condition of operation of some air conditioning equipment may increase rather than decrease these pollutants. For instance, humidifiers kept at a lukewarm temperature may provide an excellent breeding place for bacteria. Also, surfaces of dehumidifiers may become breeding places for bacteria and fungi. This means that these surfaces must be sterilized from time to time to maintain a safe operating condition.

Sulphur dioxide, which frequently results from the burning of coal, gas and oil, is a common gaseous pollutant. Hydrogen sulphide results from some industrial processes, particularly papermaking. In addition, chlorine, odors from paints, insecticides and other volatile solvents release polluting vapors into the atmosphere. Many vapor-caused illnesses may be difficult to identify. However, it has been established that when patients are removed from these environments their illnesses disappear.

Carbon monoxide is the result of incomplete combustion of fuel. A common source of carbon monoxide is automobile engine exhaust. Fuel-burning furnaces also produce carbon monoxide, and it is present in the combustion chamber, heat exchanger, the flue and the stack. It is very dangerous. It is an odorless, tasteless and colorless gas.

The effect of carbon monoxide on people is that it produces headaches, nausea and vomiting. However, if a person is exposed to a large amount of carbon monoxide, it is possible that none of these symptoms will have time to develop. Instead, the person suddenly becomes unconscious, leading to death. Carbon monoxide replaces the oxygen in the red corpuscles.

Nitrogen oxide is formed at high temperatures. The nitrogen and oxygen atoms combine to form the compound nitrogen oxide. This combination can take place in automobile engine combustion. This substance, nitrogen oxide, is one of

Fig. 18-44. An odor measuring device. A—Air in. B—Nasal openings. (Barnebey-Cheney)

the substances common in photochemical oxidants which produce smog. Nitrogen oxide is unstable, meaning that it changes back to nitrogen and oxygen easily at lower temperatures.

Organic vapors are a major source of air pollution. These vapors can be absorbed by activated charcoal. This material also can be used to measure the level of these vapors in the air. For example, the instrument pictured in Fig. 18-44 has two chambers. The air to be measured is passed through one chamber. An air sample with a known level of odor is passed into the other. Then two nasal tubes are used to detect the difference or sameness of the two air samples.

18-37 OZONE

Ozone is a form of oxygen. The chemical formula for oxygen is O_2. Ozone is O_3. However, latest research indicates that ozone may consist of both oxygen and nitrogen.

Ozone is produced in nature photostatically. In the upper atmosphere, ozone is made by ultraviolet light reacting with oxygen. Or, it may be produced by an electric discharge in air (lightning). Ozone is a disinfectant. It is sometimes used to purify water, maintain a sterile atmosphere, and to remove odors in such places as cold storage rooms and hospitals.

No universal agreement exists concerning the benefits or possible hazards in the use of ozone in a conditioned space. It is generally considered that a very small amount of ozone in the air is beneficial. However, in larger concentrations, it may be detrimental to health. Ozone may be one of the reasons for irritation caused by smog. There is no evidence that ozone is accumulatively harmful.

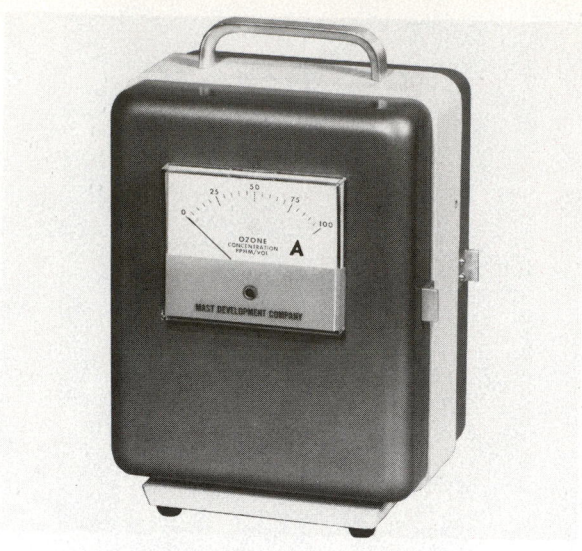

Fig. 18-45. Ozone monitor indicates the ozone content of a space in parts per hundred million. A—Gauge, 0 to 100 pphm side. (Mast Development Co.)

temperature. The effect doubles for each 15 F. (7 C.) increase in temperature. It is said that the use of of some electronic air cleaners may slightly increase the ozone content of the conditioned space.

Reliable instruments have been developed for measuring the concentration of ozone in a conditioned space. This measurement is made in parts per million (ppm) or in parts per hundred million (pphm). A typical ozone monitor is illustrated in Fig. 18-45.

18-38 COMMON AIR CONTAMINANTS — MICRON SIZES

In filtering out contaminants, the size of the particle determines to a great extent the nature of the filter required. Fig. 18-46 illustrates some common contaminants in relation to their micron size. Refer to Para. 1-12 for information concerning the size of a micron.

Many different types of filters may be used for removing solid particles from the air. See Chapter 22.

18-39 MEASURING FILTER EFFICIENCIES

Three filter efficiencies are recognized by the National Bureau of Standards (NBS) using the NBS Discoloration Test. This test is also called the Atmospheric Dust Spot Method. The 55 percent efficiency filter removes particles from 10

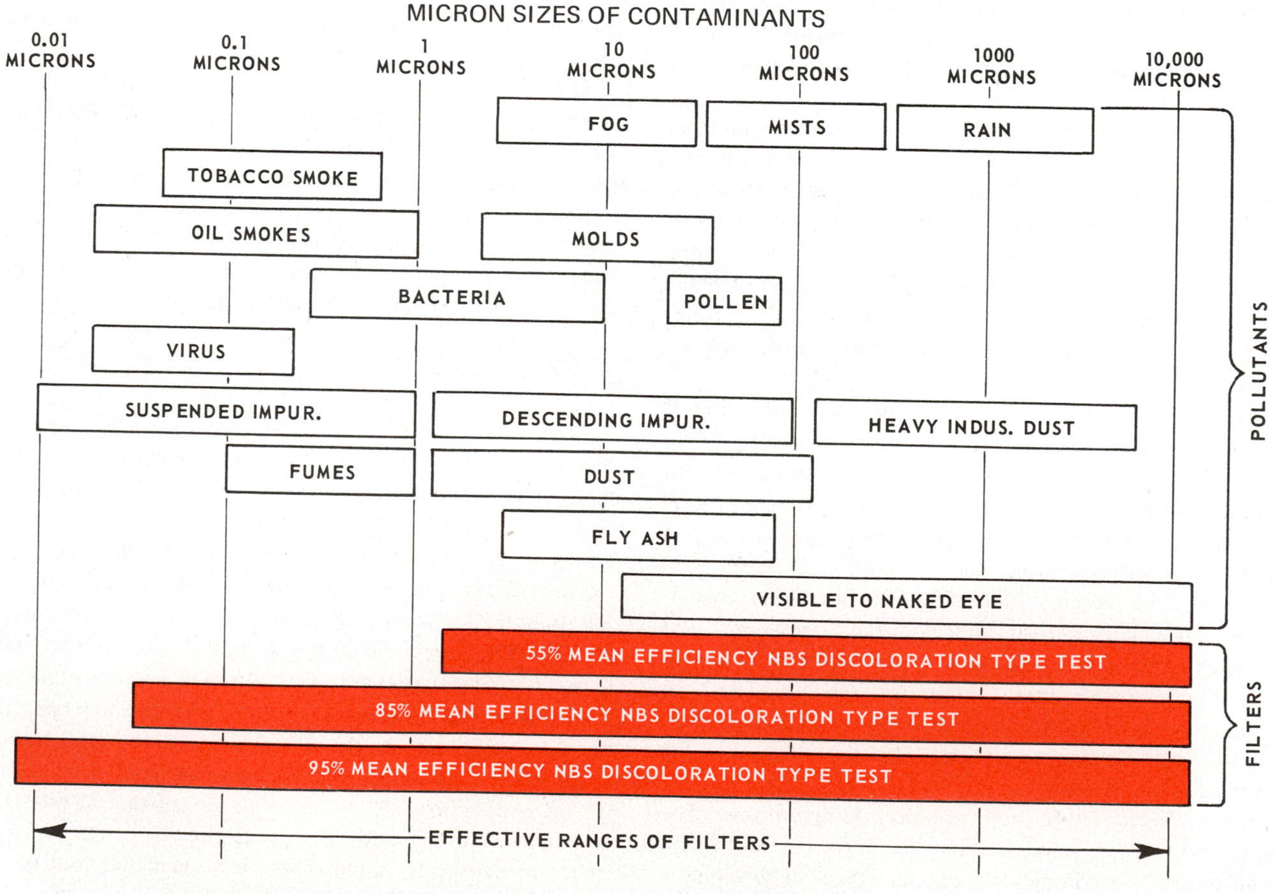

It is generally considered that concentrations of ozone of 0.1 parts per million (ppm) is the maximum permissible for eight-hour exposure. For continuous occupancy, ozone should not exceed 0.01 ppm. The effect of ozone increases with

MICRON SIZES OF CONTAMINANTS

0.01 MICRONS	0.1 MICRONS	1 MICRONS	10 MICRONS	100 MICRONS	1000 MICRONS	10,000 MICRONS

FOG
MISTS
RAIN

TOBACCO SMOKE

OIL SMOKES
MOLDS

BACTERIA
POLLEN

VIRUS

SUSPENDED IMPUR.
DESCENDING IMPUR.
HEAVY INDUS. DUST

FUMES
DUST

FLY ASH

VISIBLE TO NAKED EYE

POLLUTANTS

55% MEAN EFFICIENCY NBS DISCOLORATION TYPE TEST

85% MEAN EFFICIENCY NBS DISCOLORATION TYPE TEST

95% MEAN EFFICIENCY NBS DISCOLORATION TYPE TEST

FILTERS

← EFFECTIVE RANGES OF FILTERS →

Fig. 18-46. Micron sizes of common air contaminants. (Arco Mfg. Corp.)

microns to 10,000 microns in size. The 85 percent efficiency filter removes particles 0.1 micron to 10,000 microns in size. The 95 percent efficiency filter removes particles down to as small as .01 of a micron.

Another filter efficiency test is called the DOP Smoke Penetration Method. The name of the test comes from the name of the testing particles DiOctylPhthalate. This test is used mainly with high efficiency filters. Particles of 0.3 microns are sprayed into the inlet duct of the filter being tested. Small sample white filters collect some of this dust from the airstream before the filter being tested (inlet). Other sample filters collect dust from the air leaving the filter being tested. The difference in the sampling filters by color or weight determines the filter efficiency. For example, the outlet sample filter may collect only 1 percent of the particles as the sample filter in the inlet. This filter would then have 99 percent efficiency.

18-40 POLLUTANT STANDARDS INDEX (P.S.I.) (OLD MURC INDEX)

Most county health departments regularly measure the contaminants and gases in the atmosphere. Results are reported daily in newspapers and on television.

As of January, 1978, the term MURC Index was dropped, and the new terminology of air contamination is Pollutant Standards Index (P.S.I.).

The P.S.I. measures not only dust and dirt, but four gaseous pollutants — sulphur dioxide, carbon dioxide, nitrogen dioxide, and ozone. P.S.I. is a way of making the citizens aware of pollutants by use of numbers.

Under the P.S.I., good clean air is 0-100, which equals 0.146 ppm. Any measurement over 100 is considered bad. The range is from 0 to 500.

Pollutants are measured by pulling a measured volume of

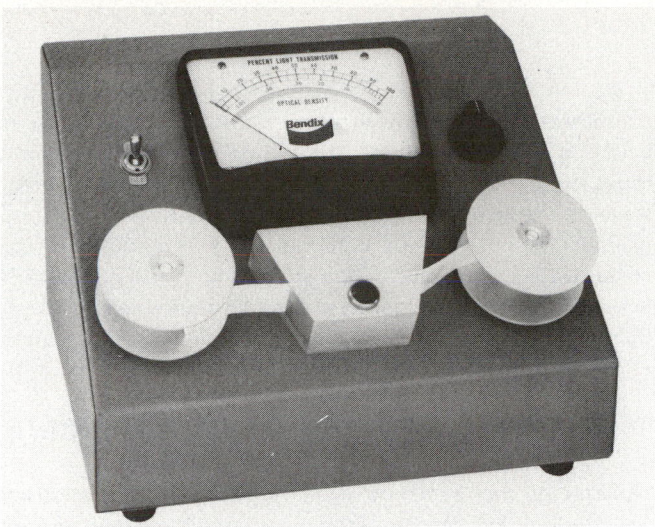

Fig. 18-48. A paper tape densitometer measures the percent of light transmission through the spots on the tape. The greater the density of contaminants on the tape, the less light will be transmitted through it. (Bendix Corp., Environmental Science Div.)

air through a filter paper tape (amount drawn through per hour approximates amount of air a human being would breathe in one hour). In drawing air through the filter paper, any solid particles suspended in the atmosphere are deposited on the paper. This darkens the filter paper in the area through which the air is drawn. The darker the area, the greater the contamination.

Contamination is measured by how much of the intensity of a beam of light is lost as it passes through the darkened area of the filter paper. This measurement is then translated into a numerical reading through the use of a mathematical formula. The contaminants measured by this method are suspended particles. These cause haziness and soiling and, most important, are the type people breathe. The mechanism for drawing air through the filter tape is shown in Fig. 18-47. The instrument used to translate the amount of contamination is shown in Fig. 18-48. Fig. 18-49 shows the degree of pollution.

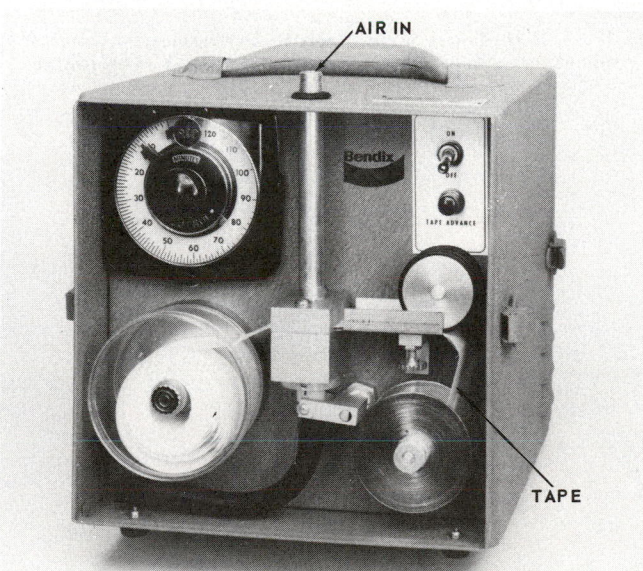

Fig. 18-47. Paper tape air sample automatically draws air through the tape. Contaminants are deposited on the tape, making dark spots. (Bendix Corp., Environmental Science Div.)

POLLUTANT STANDARD INDEX (P. S. I.)	DEGREE OF POLLUTION
0-49	minimal
50-99	light
100-199	medium
200-299	heavy
Over 300	extremely heavy

Fig. 18-49. P.S.I. numbers indicate the degree of pollution.

18-41 POLLEN COUNT

During certain times of the year, some plants create a concentration of pollen in the atmosphere which may be irritating to many people. The plants that most commonly cause the trouble are ragweed, timothy and, to a lesser extent, some flowering plants such as goldenrod and roses.

Pollen count may be determined by exposing a surface coated with an adhesive to the atmosphere for a period of 24 hours. The number of pollen grains in a square centimetre are then counted, and this becomes the pollen count for the past 24 hours.

18-42 THERMOMETERS — AIR CONDITIONING

Common thermometers regularly used by refrigeration and air conditioning service technicians are illustrated and explained in Para. 2-51. The air conditioning technician frequently needs special thermometers to accurately determine the operating temperatures. A quick-reading instrument is shown in Fig. 18-50. It has a scale from 0 to 600 F. (—18 to 316 C.)

Fig. 18-51. An electric digital thermometer. Scale is from —140 F. to 1700 F. (—95 C. to 930 C.). (Alnor Instrument Co.)

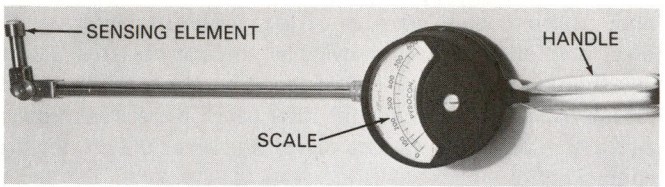

Fig. 18-50. Temperature measuring instrument used to obtain air and surface temperatures. (Alnor Instrument Co.)

and operates on the thermocouple principle. Other temperature scales are available. To use the thermometer, place the end of the adjustable probe against the surface where the temperature is to be measured. The temperature will be shown on the chart.

A thermistor type electric thermometer is shown in Fig. 18-51. These units may be either battery or 120V a-c powered. The probe reacts quickly and accurately; the scale is calibrated in both Fahrenheit and Celsius degrees.

A recording type thermometer helps locate malfunctions by making 24-hour or 7-day temperature records. Fig. 18-52 illustrates a recording type thermometer.

The wet globe thermometer shown in Fig. 18-53 has been developed to measure overall comfort conditions in hot work places. The instrument consists of a 2 3/8-inch hollow copper sphere that is painted black and covered with a double layer of black cloth. A five-inch aluminum tube connects to the copper sphere. This tube is filled with water and is capped at the other end. It keeps the globe wet. The steam of a dial thermometer passes through the center line of the water reservoir tube and into the globe. When placed in a hot area, the globe is warmed by the surrounding air and by heat radiating from hot surfaces. It is cooled by evaporation from the globe surface, which depends on air velocity and relative humidity.

The wet globe reaches an equilibrium temperature after a

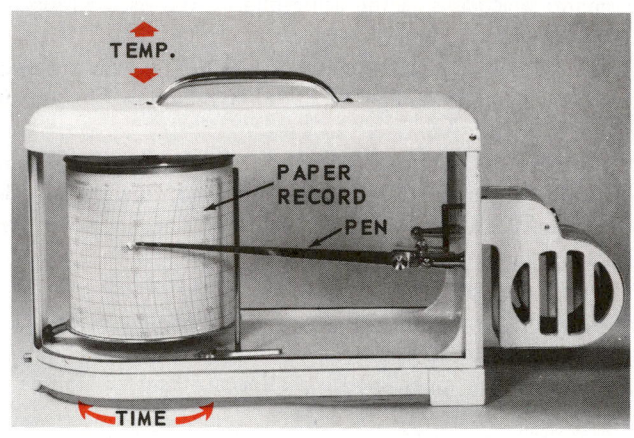

Fig. 18-52. Recording type thermometer. Sensing element is at right of cabinet. Pen at left records the temperature over a 24-hr. period.

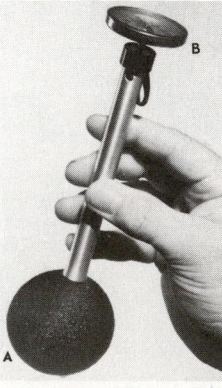

Fig. 18-53. A wet globe thermometer. Black cloth covered sphere will reach a temperature which is the balance of dry bulb, wet bulb and radiation. A—Cloth covered sphere. B—Dial thermometer. (Howard Engineering Co.)

few minutes, at which time the heating and cooling effects are in balance. The dial thermometer reading will indicate the wet globe temperature and provide a direct physical measurement of the thermal environment. The wet globe thermometer provides an excellent index of human responses to heat.

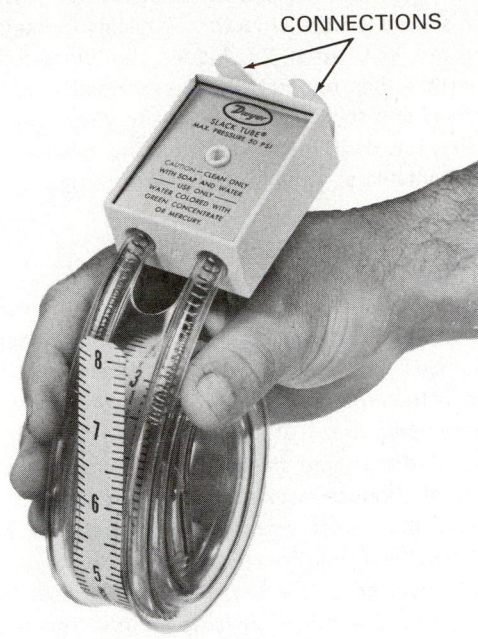

Fig. 18-54. Manometer often used for measuring air pressure in ducts. Flexible tube permits easy storage. (Dwyer Instruments, Inc.)

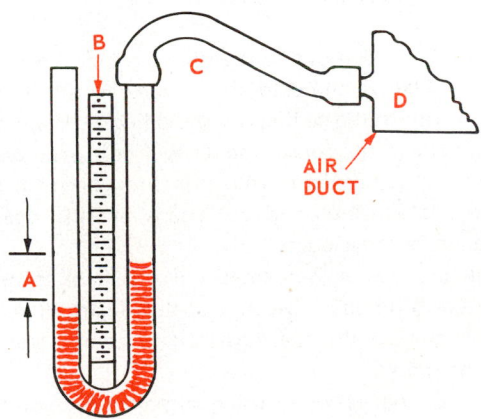

Fig. 18-55. Simple manometer in operation. A—Pressure is indicated by difference in level of liquid in two sides of manometer. Usually it is measured in inches. B—Scale in inches. C—Rubber connecting tube. D—Pressure being measured.

18-43 MANOMETERS

The principle of operation of the manometer is explained in Para. 1-17. A manometer used in air conditioning work is shown in Fig. 18-54. Made with flexible tubing, it can be rolled or folded into a small space for carrying.

Fig. 18-55 illustrates a method of connecting a manometer

to an air duct to determine its pressure. Measuring duct pressures usually calls for a water manometer. Scale B usually is movable, making it easier to adjust for the neutral point. Sudden pressure changes must be avoided or the liquid may be forced out of the manometer.

Some manometers measure the pressure difference between two different places in a duct. An example of this is a manometer used to measure the pressure drop across a filter in an airflow system.

Manometer scales are based on the following data:

14.7 psi = 29.9" Hg. = 34 ft. water
1" Hg. = .492 psi
1 psi = 2.034" Hg.
1 psi = 2.31 ft. water
1 ft. water = .432 psi
1 in. water = .036 psi

A dial type manometer is shown in Fig. 18-56. These

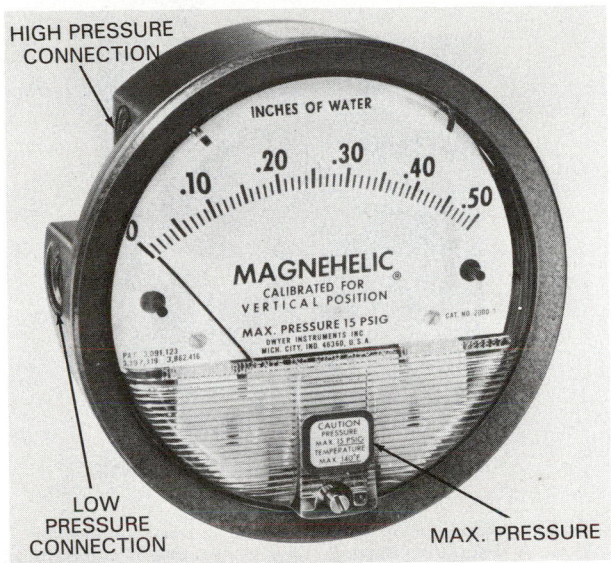

Fig. 18-56. Dial type manometer. Note that the calibration is in inches of water, maximum pressure is 15 PSIG. (Dwyer Instruments, Inc.)

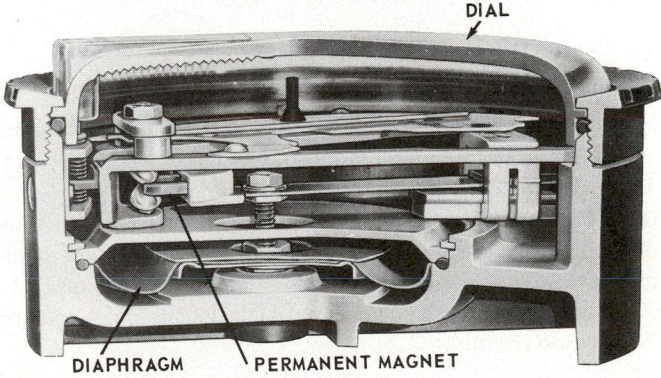

Fig. 18-57. Cross-section of a dial type manometer. Moving magnet moves permanent magnet which rotates helix attached to needle. Diaphragm separates two pressure chambers. (Dwyer Instruments, Inc.)

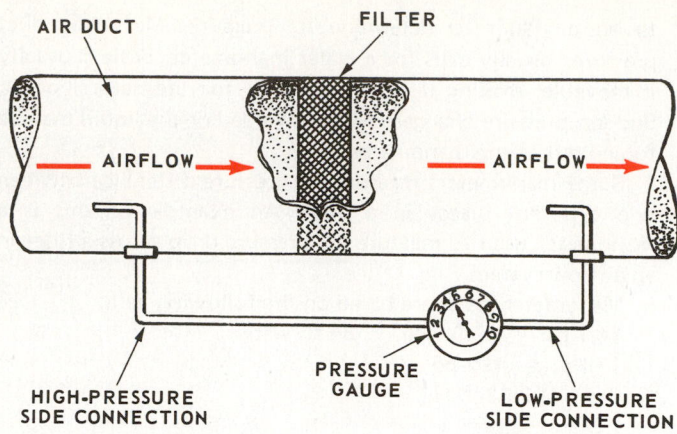

Fig. 18-58. One of many uses of the dial type manometer. The greater the pressure difference in this case, the more resistance (clogging) is indicated at filter.

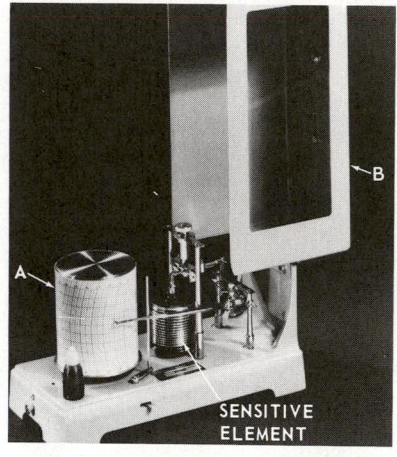

Fig. 18-59. A 24-hour recording barometer (barograph) with cover lifted. A—Recording chart. B—Cover. (Weather Measure Corp.)

manometers may have different scales indicating pressure from 0 to 5 in. of water or 0 to 5 psi.

The inside of a manometer is shown in Fig. 18-57. Fig. 18-58 shows a dial type manometer in use.

18-44 BAROMETERS

Barometers are used to measure atmospheric pressure. The simple mercury barometer is illustrated and explained in Para. 1-17. Barometers used in air conditioning usually are of a different type. The barometric pressure is measured by the deflection of a bellows or diaphragm. A recording type baromenter commonly used in air conditioning work is shown in Fig. 18-59. This instrument is available in either 24-hour or 7-day drum rotation.

18-45 HEAT INSULATION

In order to economically maintain desired air conditioned temperatures in climates of either extreme hot or cold, it is desirable to use materials which do not transfer heat readily. One method is to reduce heat conductivity. Usually, spaces in the structure can be filled with insulating material, which helps to stop or prevent the flow of heat through the structure. Heat flow is measured in Btu's per square foot of area per degree F. per hour temperature difference between the inside and outside. This is called the U factor. Problems in heat leakage are worked out in Chapter 25. Modern buildings are usually insulated with either mineral wool, expanded mica, balsam wool, urethane and sometimes cork in either sheet or granular form. Usually, it is desirable to use insulating materials that are either nonflammable or at least very slow burning.

18-46 HEAT SINK

A warm body is always giving off heat rays. If these heat rays strike another body or surface of the same temperature, the rays bounce back or reflect back. Since there is no increase or decrease in the amount of heat, the radiating body remains at the same temperature.

However, if the radiant heat strikes a colder surface, heat rays do not all bounce back. Some of the radiant heat is absorbed into the colder surface, and this surface becomes what is called a "heat sink."

On a cold day, heat in a heated room will flow from the room through cold window surfaces. If one sits close to a window under such conditions, one will feel cold due to the loss of heat into the heat sink. This will also be true if the walls of the room are not well insulated. If the walls are cold, they become a heat sink.

18-47 STRATIFICATION

If there is no air movement within a room, the air may tend to stratify. That is, the cold air will sink to the floor and the warmer air rise to the ceiling. By providing a certain amount of air movement in the room, the air will be stirred up so that a more uniform temperature will exist throughout the room. Air movement is accomplished by means of fans located in air conditioners or in air ducts.

If the thermostat is located quite high in the room, the temperature difference (because of stratification) will be more noticeable than if the thermostat is located nearer to sitting level in the room.

This also applies when using summer air conditioning to cool a room.

18-48 HEAT EXCHANGE

Methods of heat transfer are covered in some detail in Chapter 1. However, some additional information on heat exchange may be necessary in the study of air conditioning.

When radiant heat exchange takes place, heat is removed or it travels from one body that is surrounded by environment at a higher temperature to one at a lower temperature. Radiant heat exchange means that the body is radiating heat. If the heat being radiated strikes a body or substance at a lower temperature, this heat is lost to the lower temperature substance. If, on the other hand, a body is sourrounded by surfaces at a higher temperature, its temperature will tend to increase. Therefore, it may be necessary to mechanically cool this body to maintain its temperature. This is because the radiant heat it receives is greater than the amount it gives out.

In addition to heat gain or loss by radiation, a body may either gain or lose heat by convection. A person sitting at a desk loses heat by convection. The warm body tends to warm the surrounding air, and the heat rises and floats away. This is one of the values of clothing; it slows up the convective flow of heat from the human body.

In addition to radiant and convective heat transfer, evaporative heat exchange also takes place from the human body. Moisture is fed to the skin from the sweat glands, and evaporation of this moisture tends to lower the skin temperature. Respiration also exhausts moisture from the body. The evaporative moisture constitutes a considerable heat exchange from the human body.

The principle of heat exchange is used to provide more comfortable living environment. Rather than supplying a small heat source at a high temperature, the tendency now is to heat or warm walls, floors and ceilings to a medium temperature. These large warmed surfaces do not absorb the heat of the body, but bounce back the heat, which gives a person the feeling of a comfortable temperature. This also applied to comfort cooling or summer air conditioning.

18-49 NOISE

The air conditioning system must be designed for handling both load and the fresh air required. It must do these functions in a manner that will not be annoying to the occupants. Two causes of annoyance are objectionable noise and drafts.

Noise is unwanted sound. Complaints of unpleasant noise connected with air conditioning are often directed to equipment vendors and service technicians.

The noise problem can be divided into three types:
1. Noise source.
2. Noise carriers.
3. Noise amplifiers or reflectors.

The noise source is an audible vibration. This vibration may start in the heating unit, cooling unit, fan mechanism, air turbulence, duct panels or hangers, or grilles.

Sound or noise is produced by movement of an object. This movement may be caused by vibration of the object or by the movement of air against the object, as in air conditioning ducts.

A vibrating duct panel will create alternate waves of low-pressure and high-pressure air and produce a sound similar to the hum of a mosquito.

Another common complaint is noise caused by high speed air traveling through the ducts, causing air turbulence. Often, this is the result of an undersize unit or duct, and the blower has been speeded up in an attempt to make up for the unit's small size.

Noise or vibration carriers are rigid structures that carry vibrations to places where they may be annoying. Floors,

ceilings, ducts, doors and pipes may carry these vibrations.

Noise amplifiers or reflectors are usually hard, smooth surfaces in conditioned space. Walls, ceilings, floors and furnishings may pick up a small vibration and reflect it at such a frequency and in such a direction that all or certain parts of the space may be made uncomfortable. Problems involving acoustics and maintenance of a low decibel noise level are constantly being studied and improvements are being made.

Soft fabrics such as drapes and curtains and fabric covered furniture are noise absorbers. Felt lined or soft insulation lined ducts also absorb noise.

Some communities have codes regulating how noisy a mechanism may be. For example, one city is requiring that the decibel level of a window unit or outdoor condensing unit must not exceed 50 decibels at a 10-foot distance. More sound-deadening devices may be needed to meet this standard.

The air velocity (feet/minute) permissible is somewhat dependent on the type of building being air conditioned — hospital, church, hall, residential, etc. Where noise is a factor, the velocity should be kept to a minimum. If the velocity cannot be decreased, noise may be reduced by using acoustical discharge chambers or by lining or wrapping the ducts with sound absorbing material such as felt or other soft material.

18-50 NOISE MEASUREMENT

Sound waves are considered to be rapid changes of air pressure. The amplitude or strength of the sound pressure waves can be measured. Sound strength and sound pressure level (SPL) are rated in decibels (dB). In engineering practice, two words are used in connection with the definition of sound:
1. Sound strength, in decibels, is the total amount of sound coming from a unit.
2. Sound pressure is the strength, in decibels, of sound after it travels a specified distance from a source.

For instance, a truck engine gives off a measurable sound strength at the engine. Sound, after it has traveled 50 feet to a storefront, is measured in sound pressure.

The number of vibrations per second of sound waves is also measurable. The usual measuring instrument is an amplifying microphone. The international unit for sound frequencies is the Hertz (Hz), which means cycles per second (cps).

Sound travels on sound waves through the air. Sound travels from its source in all directions and its strength diminishes with the distance from the source. The first measurement of the strength of sound is usually taken about three feet from its source. The amplitude of the sound waves will be reduced by the cube of the distance that the receiving or recording instrument is away from the sound source.

Sound power is shown in Fig. 18-60. Increasing the sound frequency tends to increase the apparent loudness, although small changes in frequency cannot be detected by the human ear. This effect is illustrated in Fig. 18-61. Sounds in the frequency range of 5000 to 10,000 Hz are the easiest to hear.

In measuring the loudness of sound, the meters generally read in dB(A) or dBA. The dBA scale loudness means that a standardized "A" filter has been placed in the microphone

SOURCE OF SOUND	POWER OUTPUT WATTS
JET AIRLINER AT TAKEOFF	100,000
CHIPPING HAMMER	1
AUTOMOBILE	0.1
LOUD SHOUT	0.001
WHISPER	0.000000001

Fig. 18-60. Sound power measured in watts. (ASHRAE Guide and Data Book)

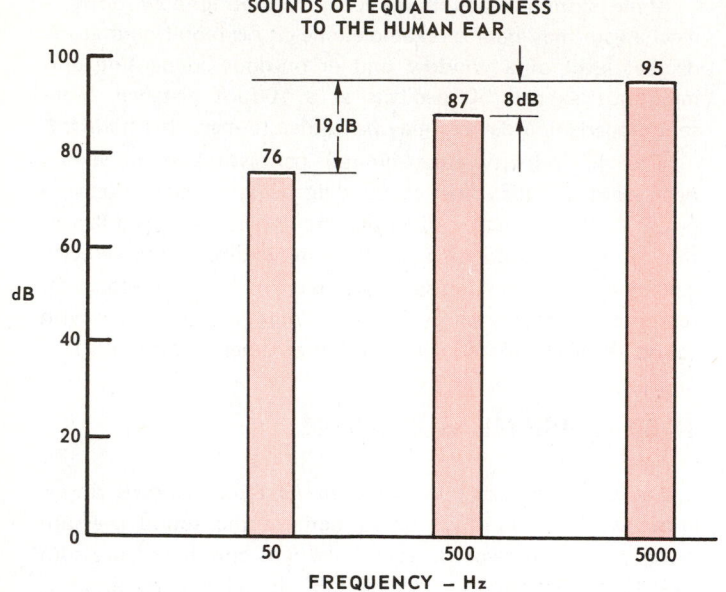

Fig. 18-61. Effect of frequency on the ear's ability to hear sounds. These three sounds all appear of the same strength to the human ear, yet the decibel rating for each of the frequencies is different. (Vickers Div. of Sperry-Rand)

LOUDNESS SCALES

	FREQUENCY	dB	dB(A)
A	5000 Hz	76	76
	500	87	84
	50	95	65

RECOMMENDED MAXIMUM SOUND

		DECIBELS	
	FREQUENCIES (Hz)	NIGHT	DAY
B	63	66	76
	125	59	69
	250	52	62
	500	46	56
	1000	42	52
	2000	40	50
	4000	38	48
	8000	37	47

Fig. 18-63. A—Effect of using an "A" filter on sound loudness for the three frequencies shown in Fig. 18-61. B—A city code maximum noise level standard (Milwaukee, Wisconsin).

TOLERANCE TO SOUND

NOISE LEVEL dBA	MAXIMUM EXPOSURE HRS.
90	UNLIMITED
90 TO 92	6
92 TO 95	4
95 TO 97	3
97 TO 100	2
100 TO 102	1.5
102 TO 105	1
105 TO 110	0.5
110 TO 115	0.25
ABOVE 115	NONE

Fig. 18-64. Limits of human tolerance to sound. Note that there is no limit, providing the dBA is 90 or below. Also, the unprotected human ear should not be exposed to sound over 115 dBA. (Walsh-Healy Act of 1969.)

circuit to reduce the intensity of the low frequencies. Fig. 18-62 illustrates the effect of using an "A" filter in the microphone circuit.

A comparison of loudness measurements using the dB and the dBA scales shows the effect of this filter in Part A in Fig. 18-63. Note that equal loudness requires a higher dB rating than the dBA rating in the lower frequencies. Part B shows one city's code standards.

Laws regulating permissible sound or noise levels usually are

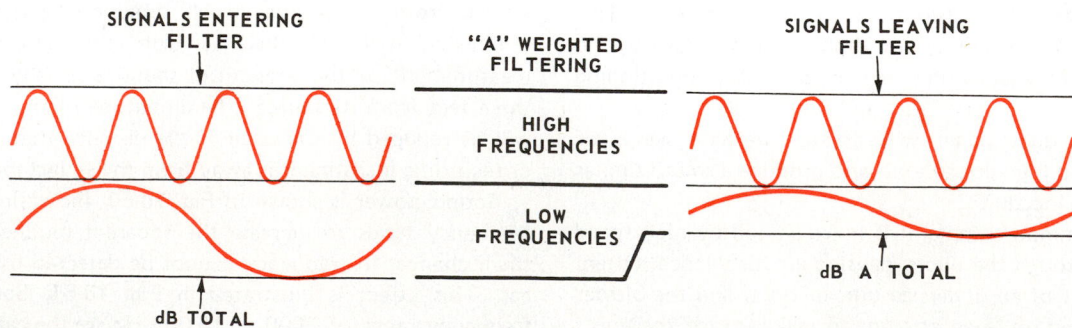

Fig. 18-62. Effect of using an "A" filter in a microphone circuit. Left. Unfiltered sound waves. Right. Sound waves after passing through an "A" filter. Note that only low frequency waves have been reduced in amplitude. (Vickers Div. of Sperry-Rand)

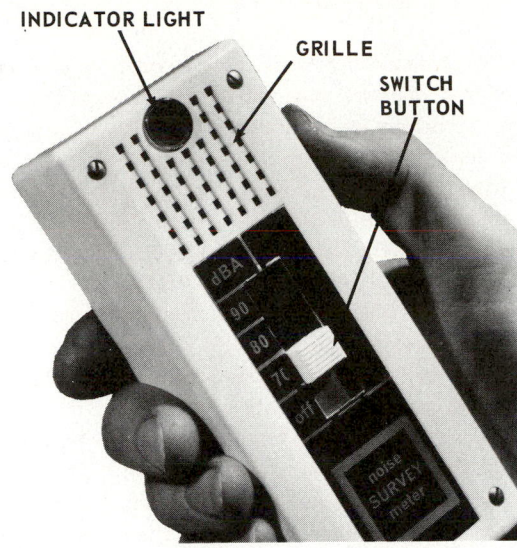

Fig. 18-65. Noise survey meter. If the red light comes on at the setting of the switch button, sound level is at or above switch button setting. (R-Deck, Inc.)

pointed toward the sound source. Start by setting the switch button at 70 dBA. If the red light does not come on, it is an indication that the sound level is below 70 dBA. If the light does come on and stays on, move the switch button to 80 dBA. If the light goes off, the sound level is between 70 and 80 dBA. This instrument has a range of 70 to 90 dBA. Sound levels below 70 are considered not too objectionable. Sound levels between 80 and 90 are objectionable, and sound levels at 90 or above should be avoided.

The meter shown in Fig. 18-66 makes sound level measurements over a range of 40 to 140 dB. It also provides an output

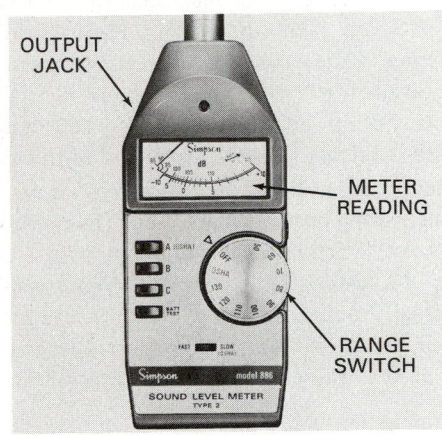

Fig. 18-66. Sound level meter measures sound pressure level over a range of 40 to 140 dB. (Simpson Electric Co.)

written around the "A" scale. The Walsh-Healy Act of 1969 establishes limits on exposure time under which workmen may work at various sound levels. A table of these limits is shown in Fig. 18-64.

An instrument for measuring sound level (dBA) is illustrated in Fig. 18-65. To use the instrument, hold it with the front

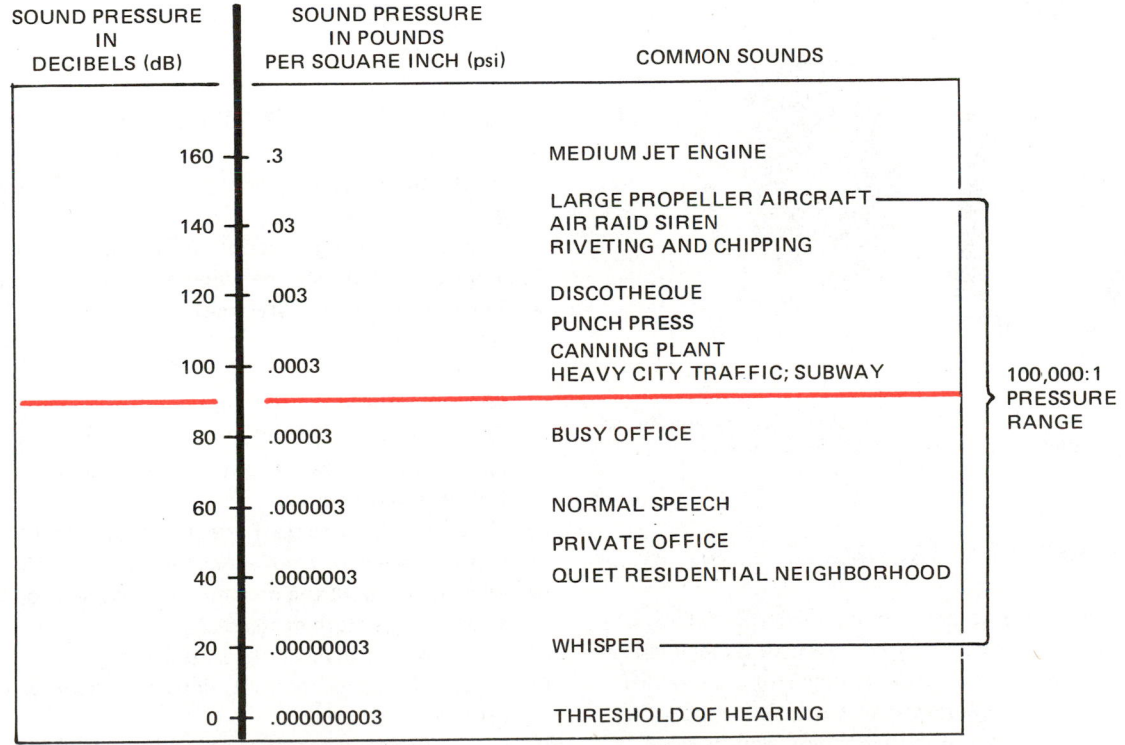

SOUND PRESSURE IN DECIBELS (dB)	SOUND PRESSURE IN POUNDS PER SQUARE INCH (psi)	COMMON SOUNDS	
160	.3	MEDIUM JET ENGINE	
140	.03	LARGE PROPELLER AIRCRAFT AIR RAID SIREN RIVETING AND CHIPPING	
120	.003	DISCOTHEQUE PUNCH PRESS	
100	.0003	CANNING PLANT HEAVY CITY TRAFFIC; SUBWAY	100,000:1 PRESSURE RANGE
80	.00003	BUSY OFFICE	
60	.000003	NORMAL SPEECH	
40	.0000003	PRIVATE OFFICE QUIET RESIDENTIAL NEIGHBORHOOD	
20	.00000003	WHISPER	
0	.000000003	THRESHOLD OF HEARING	

Fig. 18-67. Decibel (dB) rating of some common sounds. Line in color indicates usual upper limit to which the human ear may be continuously exposed. (Vickers Div. of Sperry-Rand)

jack which permits the use of a meter with a recorder.

Usually, a noise is made up of a number of different vibrations per second (not a pure tone). It is possible, with the use of filters, to separate noise into octave bands. The sound level in each octave band is recorded in decibels. By adding the decibels for all the bands, the sones value is obtained:
Octave: A series of eight tones extending from a given tone to a tone on the eighth degree from the given tone.
Sone: A calculated sound loudness rating.

The sone rating is particularly useful in comparing machine noise levels.

Noise sources include fans, compressors, high velocity air, the a-c hum of a motor and high velocity refrigerant flows (especially at sharp pipe turns). Fig. 18-67 illustrates some typical ratings and their corresponding sound pressure ratings.

Sound pressures in the 0 to 90 dB range generally are not objectionable. Sound levels in the 100 to 160 range are very objectionable, particularly in the upper end of this range.

To protect the hearing of people in noisy situations, earmuff type hearing protectors are used. These protectors will reduce the noise (attenuation) from about 20 percent for low frequencies, 40 percent for medium frequencies, and 30 percent for high frequencies. The Federal Safety Law (OSHA) requires their use in certain places. Always wear them when working on or near noisy machinery. See Fig. 18-68.

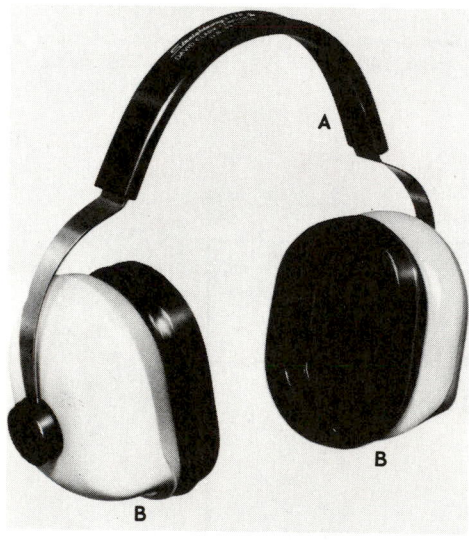

Fig. 18-68. A hearing protector should be used when working on or near painful noise. A—Head band. B—Ear covers.
(David Clark Co., Inc.)

18-51 ECOLOGY — ENVIRONMENT

"Ecology" has come to mean more than the initial definition, which states: "Ecology is the branch of biology which treats the relationships between organisms and their environment." In this text, ecology means air temperature, humidity, pollution, air movement, oxygen content and percent of noxious vapors. Webster defines environment as "the aggregate of surrounding things, conditions or influences."

Air to breathe should be as clean as possible and have the correct oxygen content. It is absolutely essential that none of the fumes (products of combustion) become mixed with the air being sent to the rooms. Air conditioners must provide enough fresh air to the rooms being conditioned to keep the oxygen content of the air within allowable limits.

Kitchen ventilating fans tend to produce a slightly lowered pressure in a house. Leakage into the house must make up for this exhausted air. Under certain conditions of prevailing winds and building construction, it is possible that products of combustion may be drawn back into the house if the air pressure is too low.

Work or experiments performed in connection with this chapter require the use of instruments. Conditions of safety must be met. If the instruments are connected to an electrical or compressed air supply line, carefully check the installations so there is no danger to the operator handling these supplies.

Instruments used in connection with work or experiments in this chapter are very delicate. They must be handled carefully, never dropped, and many must be kept in an upright position. Some instruments, such as hygrometers and psychrometers, are made by using some glass tubing. Use care in handling these instruments. Do not break the tubing and perhaps cause the operator to be cut by broken glass.

Many of these instruments are very expensive. If connected incorrectly or not handled carefully, the instrument may not read accurately or it may be severely damaged. The person handling the instrument must use great care.

In this chapter, the operation and use of instruments are explained in some detail. This information and knowledge will be used in performing experiments and work in Chapters 20, 21, 22, 23 and 24.

18-53 TEST YOUR KNOWLEDGE

1. What four important actions are involved in a complete air conditioning system?
2. Is air conditioning used only for human comfort?
3. What gases make up the living portion of the atmosphere?
4. In what form does water exist in the air?
5. What is fog?
6. What is a grain?
7. What is a micron?
8. What do the dry bulb temperature and the wet bulb temperature indicate about any particular air sample?
9. Define psychrometry.
10. How does air movement affect one's comfort?
11. List two ways to remove moisture from the air.
12. What happens to the moisture absorption properties of air as the temperature decreases?
13. How should a psychrometer be used?
14. What values are constant along a horizontal line of the psychrometric chart?
15. Does air contain moisture below 32 F. (0 C.)?
16. When air is heated, what happens to relative humidity?
17. Under what conditions are most people comfortable in

the summer?

18. How large are dust particles?
19. What are fumes?
20. What is the percentage of oxygen in the atmosphere?
21. What three pressures are found when using a pitot tube?
22. Why is a vapor barrier needed in building walls?
23. What is the moisture condition in air when the dew point is reached?
24. Does vegetation provide oxygen in the atmosphere?
25. What is a hygroscopic material?
26. What is a common sensor material in a hygrometer?
27. The chemical formula for oxygen is O_2. What is the chemical formula for ozone?
28. At what velocity does a draft become uncomfortable?
29. How many dew points can one sample of air have?
30. What is an aspirating psychrometer?

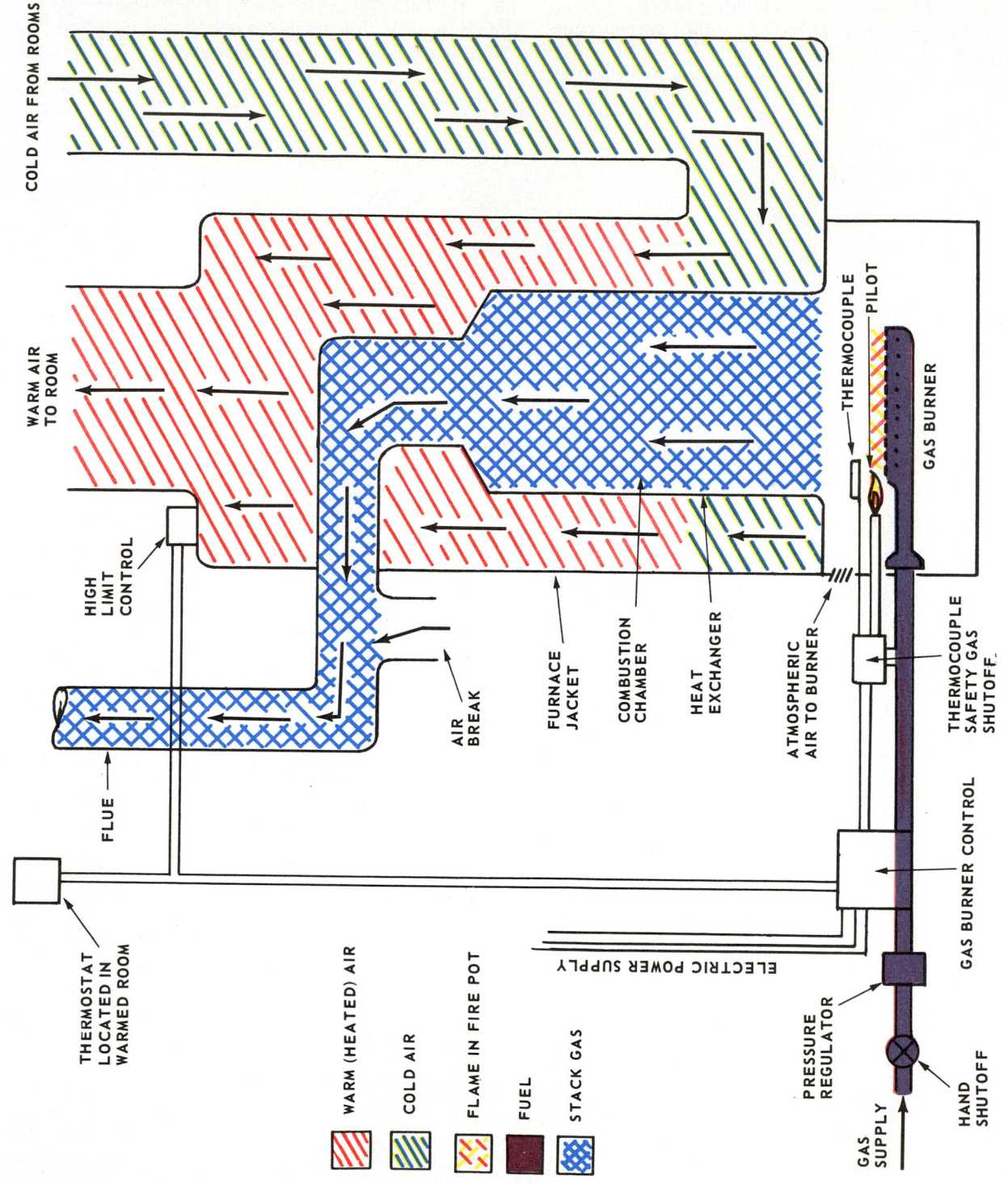

COLD AIR FROM ROOMS

WARM AIR TO ROOM

HIGH LIMIT CONTROL

AIR BREAK

FLUE

FURNACE JACKET

COMBUSTION CHAMBER

HEAT EXCHANGER

ATMOSPHERIC AIR TO BURNER

THERMOCOUPLE

PILOT

GAS BURNER

THERMOCOUPLE SAFETY GAS SHUTOFF

GAS BURNER CONTROL

PRESSURE REGULATOR

HAND SHUTOFF

GAS SUPPLY

ELECTRIC POWER SUPPLY

THERMOSTAT LOCATED IN WARMED ROOM

WARM (HEATED) AIR

COLD AIR

FLAME IN FIRE POT

FUEL

STACK GAS

Fig. 19-1. Gravity warm air furnace. Air heated at furnace rises. Colder air from rooms sinks to take its place. This natural convection circulates air through rooms.

Chapter 19

BASIC
AIR CONDITIONING SYSTEMS

Basic design of most air conditioning systems will be described and illustrated in color in this chapter. Each basic air conditioning system will be explained as follows:
1. The name of the system.
2. A four-color schematic diagram of the system, showing the parts of the system and the controls.
3. A short description of how the system works.

Detailed application, installation, maintenance and servicing data of each system will be explained in later chapters.

19-1 GRAVITY WARM AIR FURNACE — FUEL GAS ATMOSPHERIC BURNER

Fuel gas, controlled by a pressure regulator, is fed to the burner in this heating system. The fuel is under constant low pressure. An atmospheric type burner is used. Fuel may be natural gas, propane, LP gas or artificial gas.

The room thermostat controls the operation of the burner through a solenoid-controlled gas valve. A pilot light, which burns continuously, ignites the gas in the burner whenever the solenoid valve opens the gas line. See Fig. 19-1. Some systems use electric ignition. A thermocouple is connected in series with a safety gas control solenoid. It will shut off the gas if the pilot light goes out and the thermocouple cools.

Products of burning, carbon dioxide and water vapor, flow through the stack into the chimney. An air break helps keep a constant pressure in the combustion (burning) chamber.

The heat made in the combustion chamber (heat exchanger) is conducted (carried) through the combustion chamber wall and is carried or radiated into the air surrounding the combustion chamber.

Air around the combustion chamber heats up and naturally rises. It flows through the warm air ducts into the rooms through the warm air registers. As air cools in the rooms, it becomes heavier and flows down through the cold air duct and back into the bottom of the furnace.

A high-limit control (safety stat) is located in the bonnet of the furnace and will automatically shut off the gas if the bonnet temperature goes higher than the high-limit control temperature setting.

The thermostat operation keeps the temperature of the room within about 2 F. (1.1 C.) of the desired temperature.

19-2 FORCED WARM AIR — FUEL GAS POWER BURNER

In this heating system, fuel gas is fed to the burner under constant low pressure controlled by a pressure regulator. It is burned in a power-type burner.

The room thermostat controls the operation of the burner through a solenoid-controlled gas valve. The thermostat also turns on a combustion air blower which forces air into the combustion chamber. A pilot light, or an electric spark, ignites the gas in the burner at the instant the solenoid valve opens the gas line and the power burner starts. See Fig. 19-2.

The flame heats a thermocouple which, in turn, controls a safety shutoff valve. A thermocouple solenoid, located above the pilot light, will shut off the gas control if the pilot light goes out.

Products of combustion flow through the stack into the chimney. An air break helps keep constant pressure in the combustion chamber.

Heat generated in the combustion chamber (heat exchanger) is conducted through the combustion chamber wall and is radiated into the air surrounding the combustion chamber. As the air around the combustion chamber heats, it rises and warms the bonnet fan control.

As soon as the bonnet temperature is hot enough, the fan in the cold air duct return starts and moves air through the heating system. This air is drawn from the cold air register in the floor above, through the air filter and then through the furnace. Warm air is distributed through the ducts and through the warm air registers or diffusers into the space to be heated.

A high-limit control (sometimes a part of the bonnet fan control) will automatically shut off the burner if the bonnet temperature exceeds the high-limit control temperature setting.

19-3 FUEL GAS ATMOSPHERIC BURNER — HYDRONIC SYSTEMS

This heating system burns fuel gas under low and constant pressure. An automatic pressure regulator maintains constant gas pressure on the atmospheric type burner. See Fig. 19-3. The room thermostat controls the operation of the water

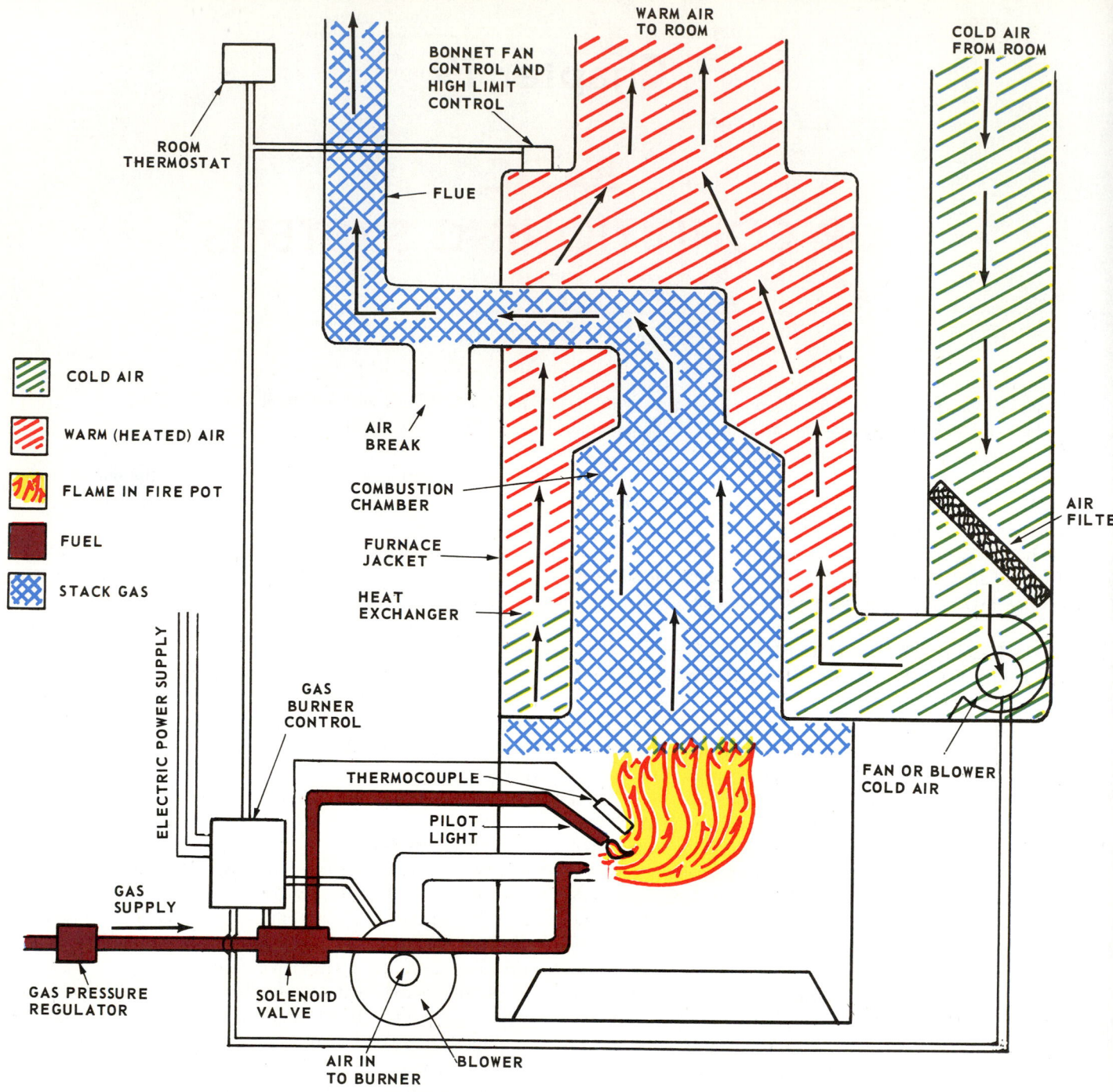

COLD AIR

WARM (HEATED) AIR

FLAME IN FIRE POT

FUEL

STACK GAS

ROOM THERMOSTAT

BONNET FAN CONTROL AND HIGH LIMIT CONTROL

WARM AIR TO ROOM

COLD AIR FROM ROOM

FLUE

AIR BREAK

COMBUSTION CHAMBER

FURNACE JACKET

HEAT EXCHANGER

AIR FILTE

ELECTRIC POWER SUPPLY

GAS BURNER CONTROL

THERMOCOUPLE

PILOT LIGHT

FAN OR BLOWER COLD AIR

GAS SUPPLY

GAS PRESSURE REGULATOR

SOLENOID VALVE

AIR IN TO BURNER

BLOWER

Fig. 19-2. Forced air circulation heating system using a gas fuel power burner.

pump. The pump circulates the warm water through the room radiators and returns it to the boiler.

Water temperature in the boiler is controlled by a temperature and pressure-sensing element in the top of the boiler. This sensing element is connected into the electrical system in such a way that the burner turns on when the temperature drops below the required level. It also turns off when the temperature reaches this level. A pilot light ignites the burner. A thermocouple, located at the pilot light, is connected into the

electrical controls and automatically shuts off the gas if the pilot light goes out.

The heat generated (made) in the combustion chamber is carried through the boiler wall and into the water. The products of burning (smoke) flow through the stack into the chimney. An air break helps maintain a constant pressure in the combustion chamber.

A high-limit control (safety stat) is attached to the warm water outlet of the boiler and automatically shuts off the gas,

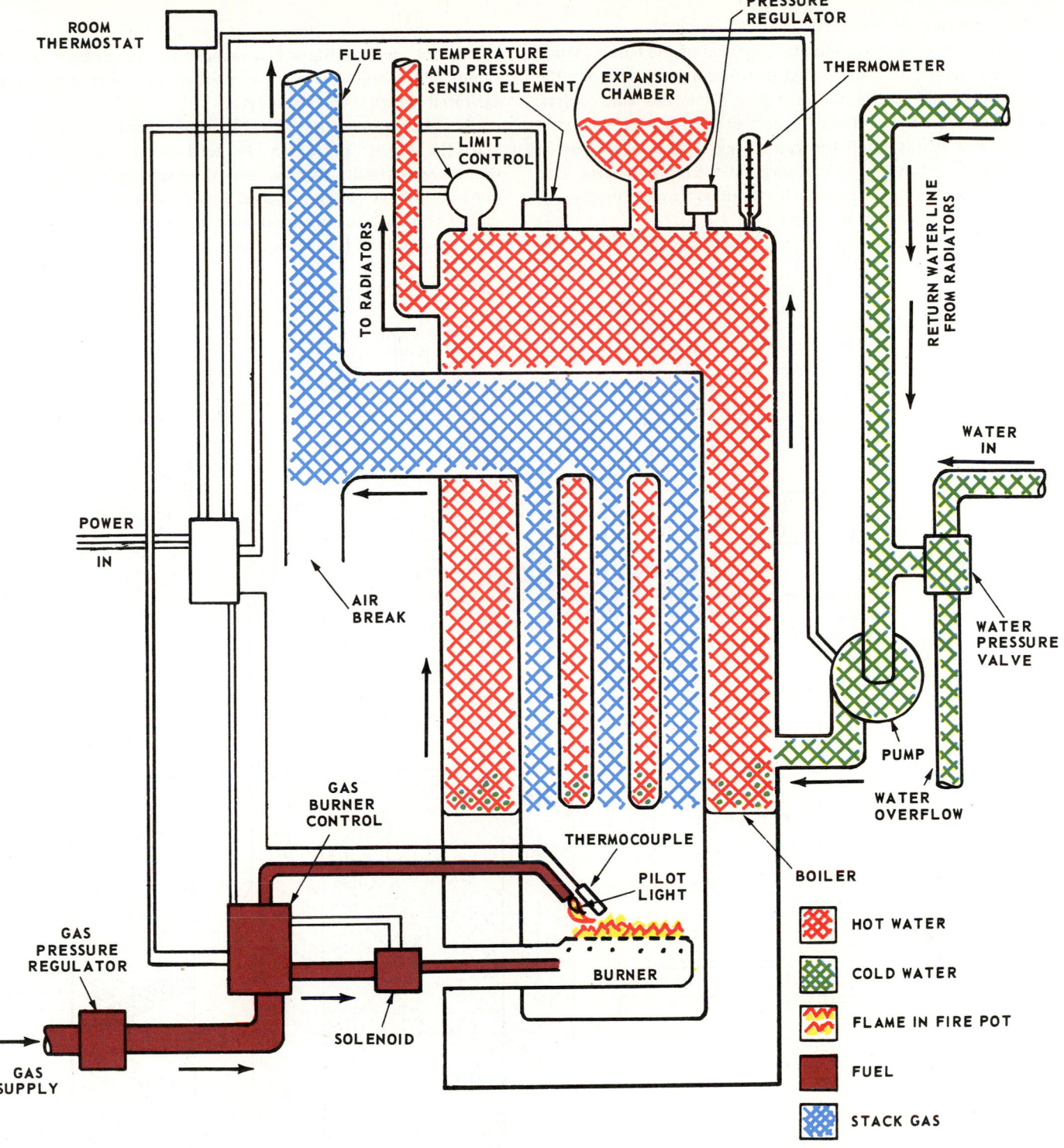

Fig. 19-3. Hydronic heating system using atmospheric fuel gas burner as source of heat.

should the water temperature or pressure get too high.

The system also has a pressure relief valve which prevents build up of dangerous pressures in the boiler. An expansion tank permits the water to expand and contract in volume as it heats and cools. The expansion tank should be located at the highest place in the heating system. It is possible to use a pressure regulator in place of an expansion tank.

19-4 OIL BURNER — FORCED WARM AIR

Where fuel oil is the heat source, a gun-type oil burner throws a flame into a firepot lined with refractory (fire resistant) material.

Fuel oil is stored in a tank, either inside or outside the building. Fuel is drawn through a pump into the burner nozzle under a pressure of about 100 psi (7.03 kg/cm^2).

A room thermostat controls operation of the burner. When heat is required, the thermostat operates a relay. It closes the electrical circuit to the burner motor. When the burner starts, a high-tension transformer is connected into the electrical circuit and sparks jump the gap of two electrodes located at the edge of the atomized fuel spray of the burner nozzle. This spark ignites the fuel, causing a continuous flame in the firepot as long as the burner is operating.

A stack stat senses the temperature of the gases leaving the furnace. Its control is such that if, for any reason, the atomizer fuel is not ignited after a few seconds of pump operation, the pump will stop. Normally, a manual reset will have to be operated before it will again cycle.

A temperature-sensing device in the furnace bonnet will start the blower or fan in the cold-air duct as soon as the bonnet temperature reaches its desired setting. A temperature-controlled limit switch is placed in the bonnet. It will open the circuit and stop the burner if the bonnet temperature goes too high. This type of furnace is shown in Fig. 19-4.

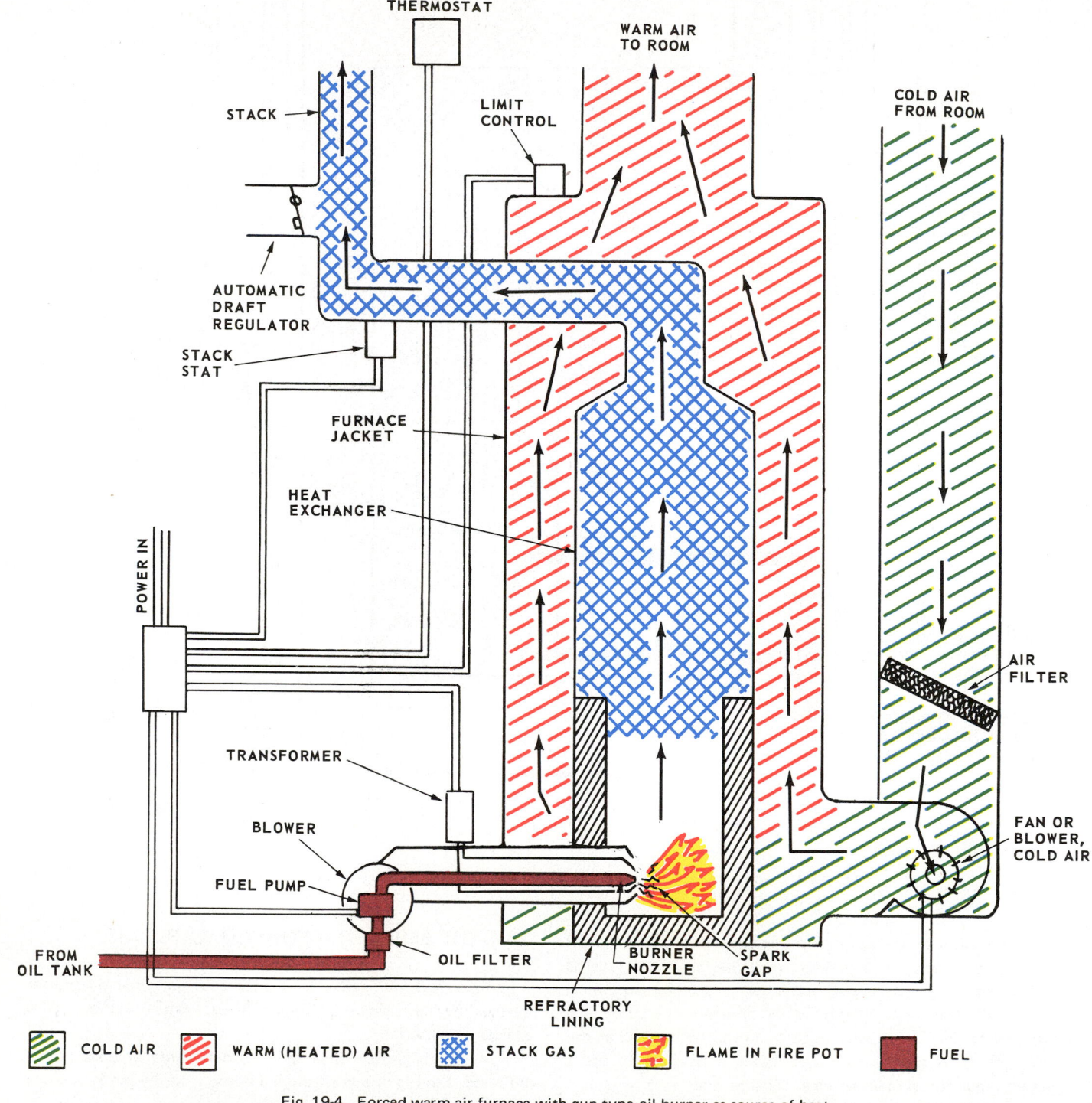

Fig. 19-4. Forced warm air furnace with gun type oil burner as source of heat.

19-5 OIL BURNER — HYDRONICS

In oil-fired hydronic heating systems, fuel oil is burned in a gun type burner. See Fig. 19-5. The firepot is lined with refractory material. Fuel oil is stored in a tank which may be located outside the building. Fuel oil is drawn through a filter and pumped into the burner nozzle under a pressure of about 100 psi. (7.03 kg/cm²).

A room thermostat controls the water pump which circulates the warm water through the room radiators and returns it to the boiler. The temperature of the boiler water is controlled by a temperature and pressure-sensing element in the boiler top. This sensing element is connected into the

electrical system. It turns the burner on when the temperature drops below the required level, and shuts it off when the temperature reaches the desired level.

As the burner starts, a high tension transformer is connected into the electrical circuit. Sparks jump across the spark gap just at the edge of the atomized (broken into small drops) fuel spray of the burner nozzle. This spark ignites the atomized fuel, causing a flame in the firepot as long as the burner is operating.

A stack stat senses the temperature of the gases leaving the furnace. If the atomized fuel is not lighted after a few seconds of pump operation, the pump will stop. Probably a manual reset will have to be operated before it will again cycle.

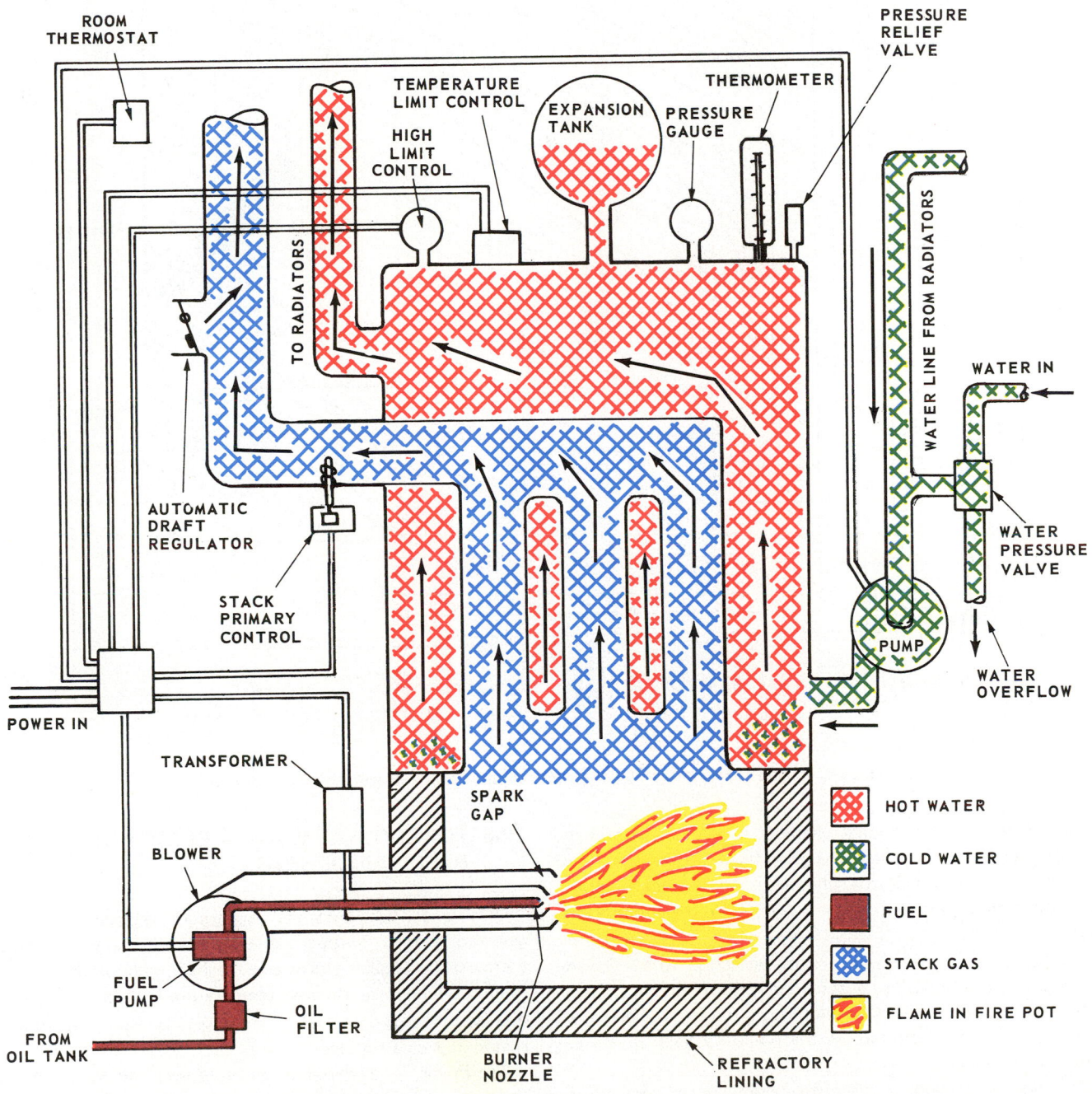

Fig. 19-5. Hydronic heating system with gun type oil burner as source of heat.

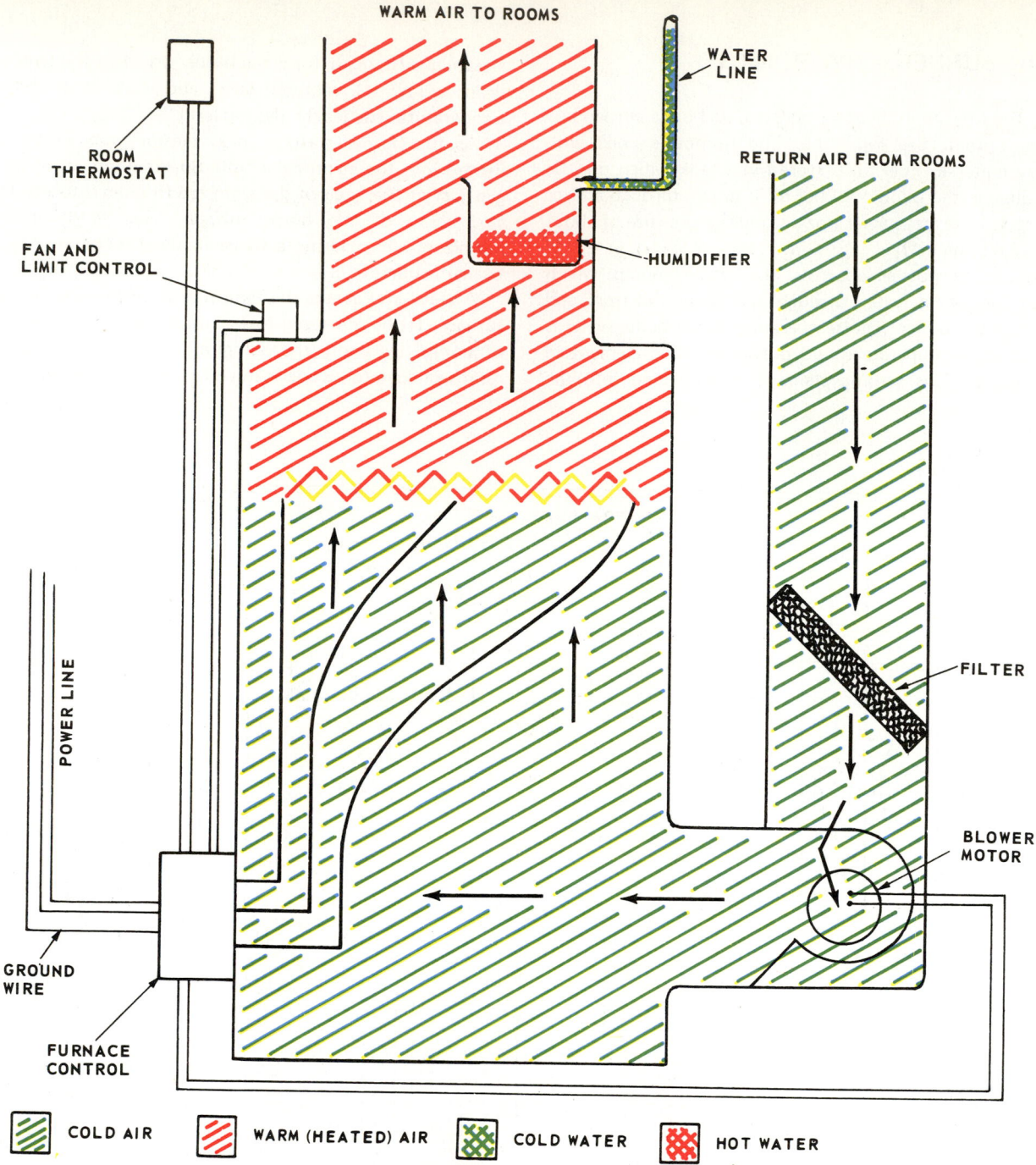

WARM AIR TO ROOMS

WATER LINE

ROOM THERMOSTAT

RETURN AIR FROM ROOMS

FAN AND LIMIT CONTROL

HUMIDIFIER

POWER LINE

FILTER

BLOWER MOTOR

GROUND WIRE

FURNACE CONTROL

COLD AIR   WARM (HEATED) AIR   COLD WATER   HOT WATER

Fig. 19-6. Forced warm air heating system with electric heating elements as source of heat.

Heat from the combustion chamber is conducted through the boiler wall into the water. The gases from burning fuel flow through the stack into the chimney. An automatic draft regulator helps maintain a constant pressure in the combustion chamber (firepot).

A high-limit control (safety stat), attached to the warm water outlet of the boiler, automatically shuts off the fuel if water temperature or pressure should get too high. A pressure relief valve is also mounted on the boiler to keep pressures down to a safe level.

An expansion tank is used to take care of expanding (warm) or contracting (cool) water.

19-6 FORCED WARM AIR — ELECTRIC RESISTANCE HEAT

Resistance heating units sometimes replace the oil or gas-fired flame. Through a system of relays, the room thermostat energizes (turns on) the heating unit when heat is needed. When desired room temperature is reached, the same thermostat turns off power to the resistance unit. Fig. 19-6 is a sketch of such a furnace.

Warm air is distributed by a blower. It forces the air through the resistance unit where it picks up heat and then distributes it to the registers.

A combination bonnet, high-limit and blower control turns on the blower as soon as bonnet temperature reaches a predetermined setting. The blower will continue to run as long as the bonnet temperature remains above a minimum temperature setting. In the event of poor air circulation through the bonnet causing the bonnet temperature to exceed a predetermined setting, the high-limit control will shut off the resistance unit. This protects furnace and ducts from overheating.

A filter is placed between the cold-air duct and blower. A humidifying device is usually placed in the warm-air duct leading out of the furnace. The humidifying device operates whenever the blower is operating.

Individual circuit breakers should be put in the power line electrical resistance heating units. No other appliance or lights *should be connected into these circuits. This applies to all heating and cooling system circuits.*

This furnace does not require a stack or chimney.

19-7 HYDRONICS — ELECTRIC RESISTANCE HEAT

In hydronic systems, the electrical resistance heating units are inside the boiler. The boiler is very small. See Fig. 19-7.

A high-limit and safety control is attached to the warm-water outlet of the boiler. It automatically turns on one or all three stages of the resistance heating units when the temperature of the water in the top of the boiler drops below a minimum setting. It also turns off the electrical heating units when the temperature of the water in the top of the boiler

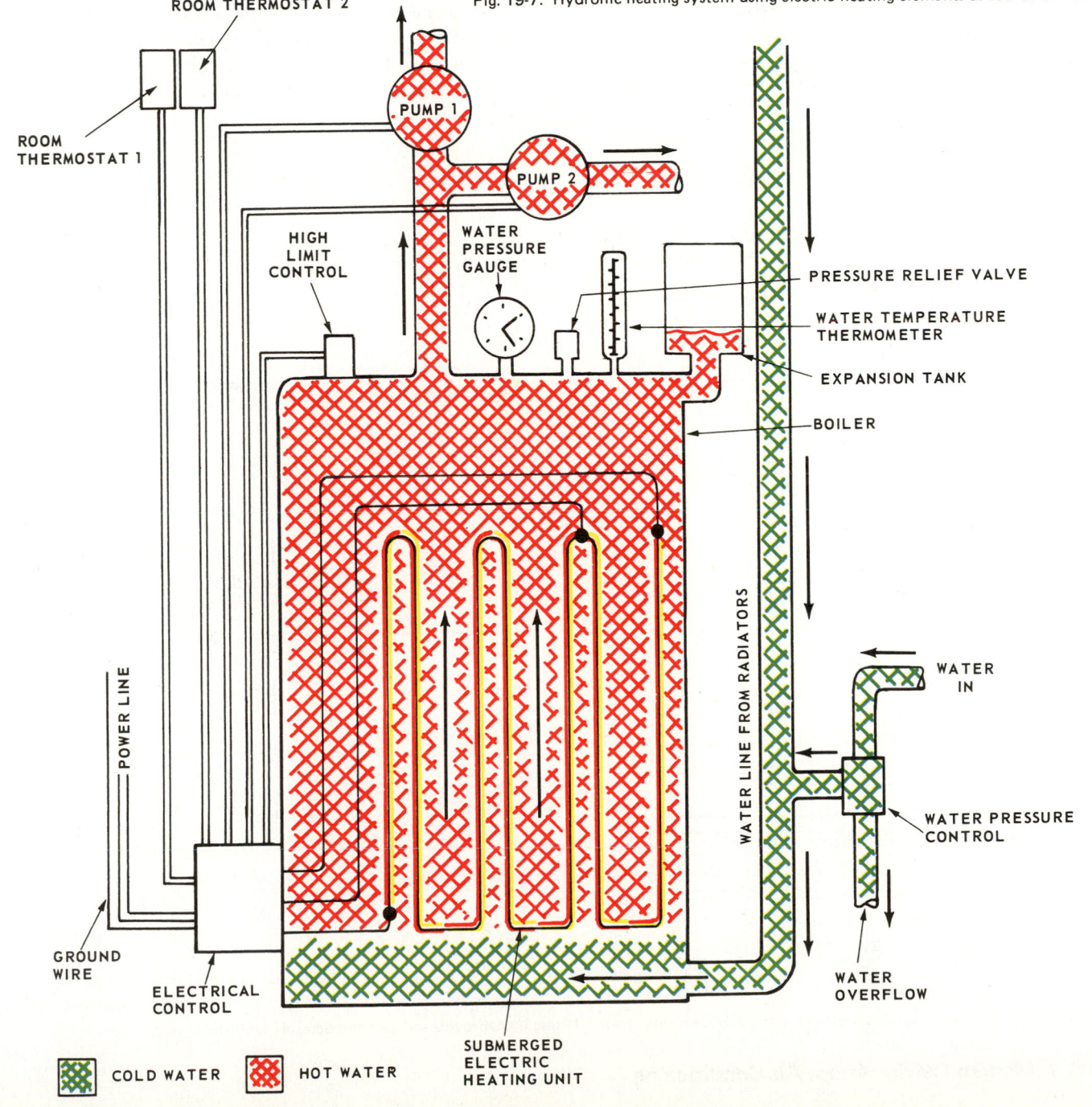

Fig. 19-7. Hydronic heating system using electric heating elements as source of heat.

reaches the upper setting. The same control becomes a safety device shutting off heating elements in the event no water is circulating through the radiators.

In some cases there are pressure controls as well. They will shut off the heating unit should the water pressure in the boiler exceed a preset limit. See Fig. 19-7.

The room thermostat controls the operation of the pump or pumps which force the warm water through the room radiators. With this system, it is possible to use more than one pump. In this case, a separate thermostat controls the temperature of the space served by each pump.

Individual circuit breakers should be installed in the power line electrical resistance heating units. No other appliance should be connected into these circuits. As with the system described in Para. 19-6, no stack or chimney is required.

19-8 ROOM HEATING UNITS — ELECTRICAL RESISTANCE

Where this heating system is installed, electrical resistance units are located in each room. Electrical power is brought to

the units from an electrical heating power panel. The power supply is usually 240V. One advantage of this type of heating is that the temperature of each room is regulated by its own thermostat. There are four different ways that the control may be accomplished. These different ways are illustrated in Fig. 19-8 at A, B, C and D.

At A, electrical power is taken directly from the power panel to the baseboard heating unit. The baseboard heating unit has an individual thermostat attached to it. Electrical power is supplied to the thermostat and, if heating is required, the thermostat connects the power supply to the resistance heating unit.

At B, a room thermostat controls the power supply at the power panel. When heat is required, a relay in the power panel connects the baseboard resistance unit to the electrical power supply.

At C, a room thermostat is mounted on the wall. The power supply from the power panel is connected through the thermostat to the baseboard heating unit.

In the case of A and C, all of the current used by the heater flows through the thermostat points. In the case of B, the

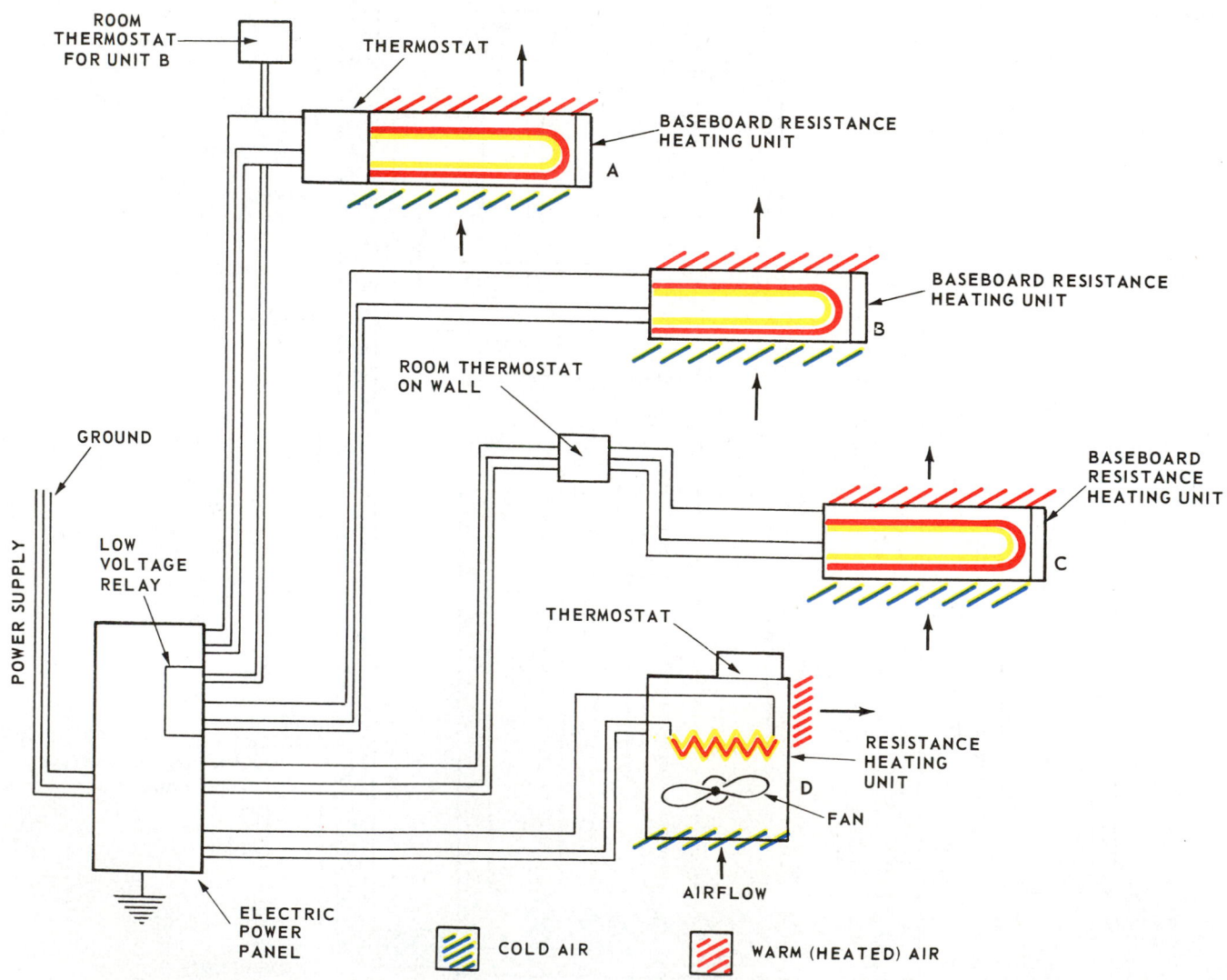

Fig. 19-8. Four different types of electric heating systems. These units provide separate temperature control for each room.

room thermostat is handling only a small amount of low-voltage current. The relay, connected to the thermostat and located in the power panel, switches the current to the room resistance heater.

The electric heating units are always grounded. The green wire indicates the cabinet ground.

At D, the unit is a resistance heater and a fan. The thermostat controlling this unit is usually mounted on top of the heater. It controls the current to both the heating unit and the fan. The fan either runs any time that the heating unit is on or it can be run separately.

Circuit breakers should be installed in the power line to each electrical resistance heating unit. No other appliances should be connected into these circuits.

19-9 AIR CONDITIONER, COOLING — WINDOW OR THROUGH-THE-WALL

Window or through-the-wall air conditioners consist of three basic parts:
1. A hermetic compressor.
2. Condenser.
3. Evaporator using a capillary tube refrigerant control.

In the schematic diagram, Fig. 19-9, red indicates high-pressure liquid refrigerant; green, low-pressure liquid refrigerant; yellow, low-pressure vapor and blue, high-pressure vapor.

Liquid refrigerant collects in the lower coils of the condenser and flows through the capillary tube refrigerant control into the evaporator. When the unit is in operation, this is under low pressure. The liquid refrigerant rapidly boils and picks up heat from the evaporator surface. A motor-driven fan draws air from inside the room, through a filter and forces it over the evaporator. Here it is cooled and goes back into the room. Arrows in Fig. 19-9 show the airflow pattern.

Low-pressure vapor is drawn from the evaporator through the suction line back to the compressor. Compressed to the high-side pressure, it is forced into the condenser to be cooled and condensed to a liquid. The cycle then repeats.

An adjustable thermostat, mounted on the control panel, provides the necessary control. The thermostat has an on-and-off switch.

Compressor and condenser are mounted in such a way that the fan in the compressor-condenser compartment draws outdoor air in, circulates it over the condenser and discharges it outside. Air flowing through the evaporator is cooled and, to some extent, dehumidified.

Moisture which collects on the evaporator drains to a drip pan under the evaporator. In some machines it flows into a

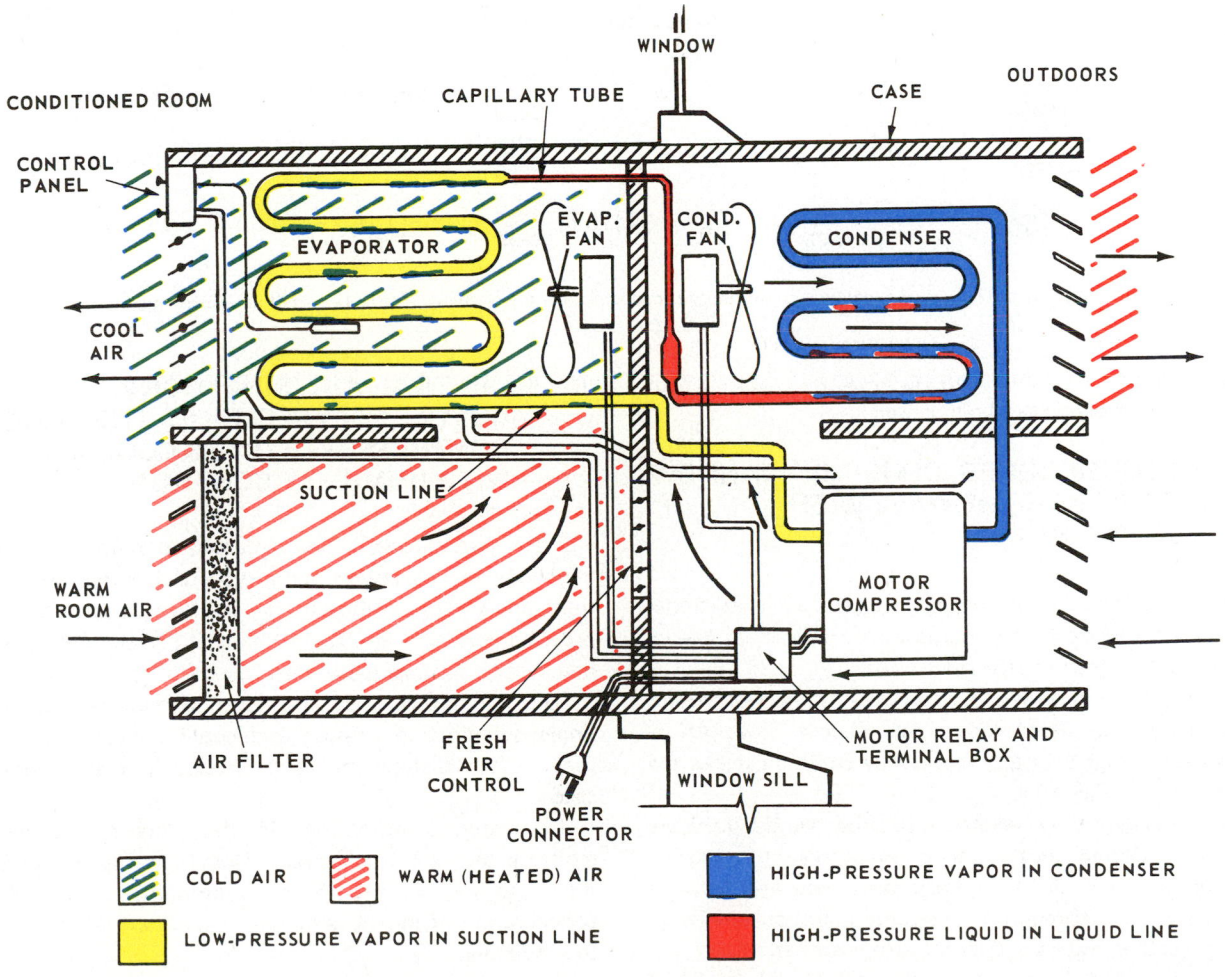

Fig. 19-9. Window type air conditioning comfort cooling system.

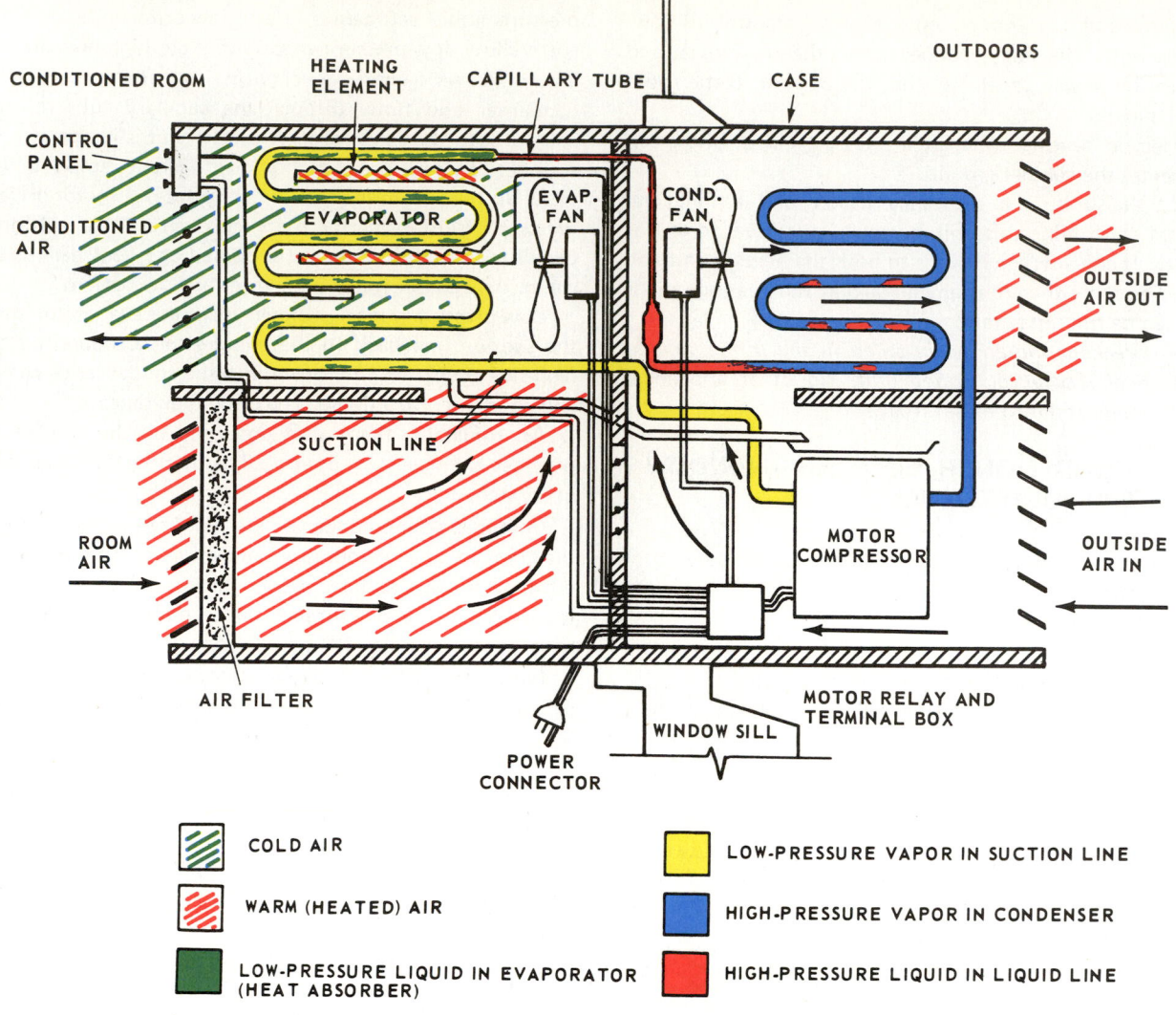

CONTROL PANEL

CONDITIONED AIR

EVAPORATOR

EVAP. FAN   COND. FAN

OUTSIDE AIR OUT

SUCTION LINE

ROOM AIR

MOTOR COMPRESSOR

OUTSIDE AIR IN

AIR FILTER

MOTOR RELAY AND TERMINAL BOX

WINDOW SILL

POWER CONNECTOR

COLD AIR

WARM (HEATED) AIR

LOW-PRESSURE LIQUID IN EVAPORATOR (HEAT ABSORBER)

LOW-PRESSURE VAPOR IN SUCTION LINE

HIGH-PRESSURE VAPOR IN CONDENSER

HIGH-PRESSURE LIQUID IN LIQUID LINE

Fig. 19-10. Window or through-the-wall air conditioner with electric heating elements. These provide heat during cold weather.

pan in the compressor compartment. Here, in evaporating, it helps cool the compressor and condenser.

19-10 AIR CONDITIONER, COOLING — WINDOW OR THROUGH-THE-WALL WITH ELECTRIC HEAT

Generally, a window or through-the-wall air conditioner consists of a hermetic motor, compressor, condenser, evaporator and capillary tube refrigerant control.

In Fig. 19-10, red shows high-pressure liquid refrigerant; green, low-pressure liquid refrigerant; yellow, low-pressure vapor and blue, high-pressure vapor. The cooling cycle is the same as the one in Fig. 19-9.

In this air conditioner, electric resistance heating units are included and, during cold weather, the refrigerating mechanism is turned off while the electric resistance heating units and room air fan is turned on. The same fan circulates warm air in cold weather and cooled air in warm weather.

These air conditioners are usually connected to 240V circuits. A control provides a choice of temperatures.

19-11 CENTRAL AIR CONDITIONER, COMPLETE SYSTEM — GAS HEATING, COMPRESSION SYSTEM COOLING, WITH HUMIDITY CONTROL

Fuel gas, burned in an atmospheric burner, is used for heating in this air conditioning system. A compression system using an A-frame evaporator in the furnace plenum chamber provides cooling. Figs. 19-11A and 19-11B show the system in heating and cooling operations.

The condensing unit is located outside the building. A single combination heating and cooling thermostat is often used. A humidistat controls the humidity in the conditioned space.

In winter, a humidifier in the plenum chamber adds moisture to the heated spaces. Details of the humidifier are shown in Fig. 19-11C. Summer humidity is controlled by condensation of moisture on the evaporator. A drain removes this moisture.

Warm air from the furnace is forced into the rooms by a blower. This is located beneath the filter in the cold air return.

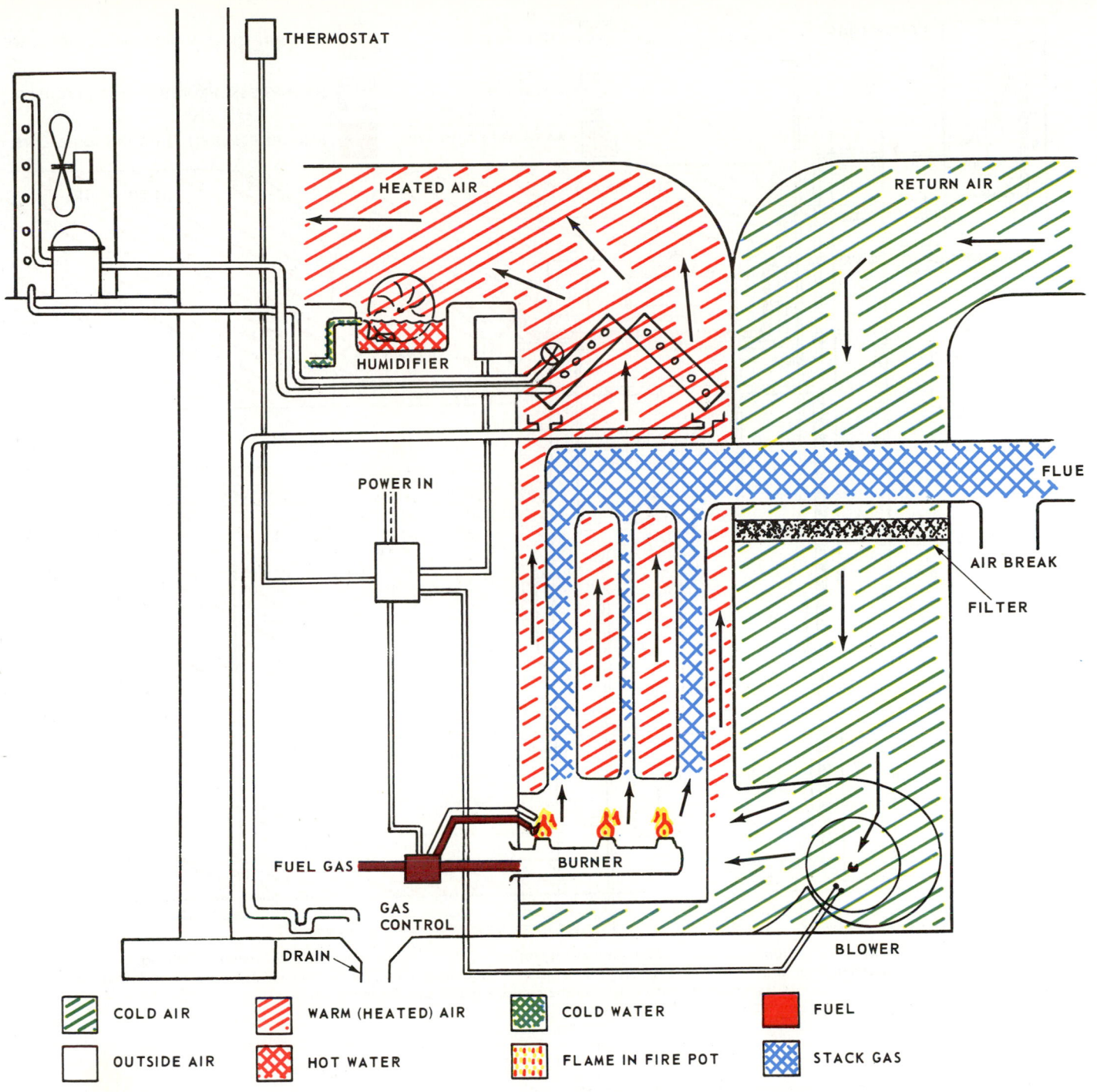

THERMOSTAT

HEATED AIR

RETURN AIR

HUMIDIFIER

POWER IN

FLUE

AIR BREAK

FILTER

FUEL GAS

BURNER

GAS CONTROL

DRAIN

BLOWER

	COLD AIR		WARM (HEATED) AIR		COLD WATER		FUEL
	OUTSIDE AIR		HOT WATER		FLAME IN FIRE POT		STACK GAS

Fig. 19-11A. Complete air conditioning unit providing both heating and cooling. Winter heating is supplied by gas burner. Humidity is supplied by humidifier in plenum chamber. Same blower and filter are used for both summer and winter operation.

A control in the top of the furnace turns on the blower when the desired bonnet temperature is reached. This control will also turn off the furnace in the event the temperature in the bonnet goes higher than a predetermined setting. This is a safety device to keep the furnace from overheating. Electrical power to the furnace is turned on and off by a control panel located on the outside wall of the furnace.

An arrangement is sometimes provided to bring in outside fresh air as needed. This may be either thermostatically or manually controlled.

The pilot light is controlled by a thermocouple connected in series with a solenoid valve in the gas supply line. In the event the pilot light goes out, the gas supply to the burner will be shut off.

When the centrally located thermostat calls for cooling the same airflow, cleaning and distribution occur. However, instead of the forced air passing through a heated chamber, it passes across the cooled evaporator. This lowers the temperature of the air and, at the same time, it removes some moisture to reduce the humidity.

Basic Air Conditioning Systems / 683

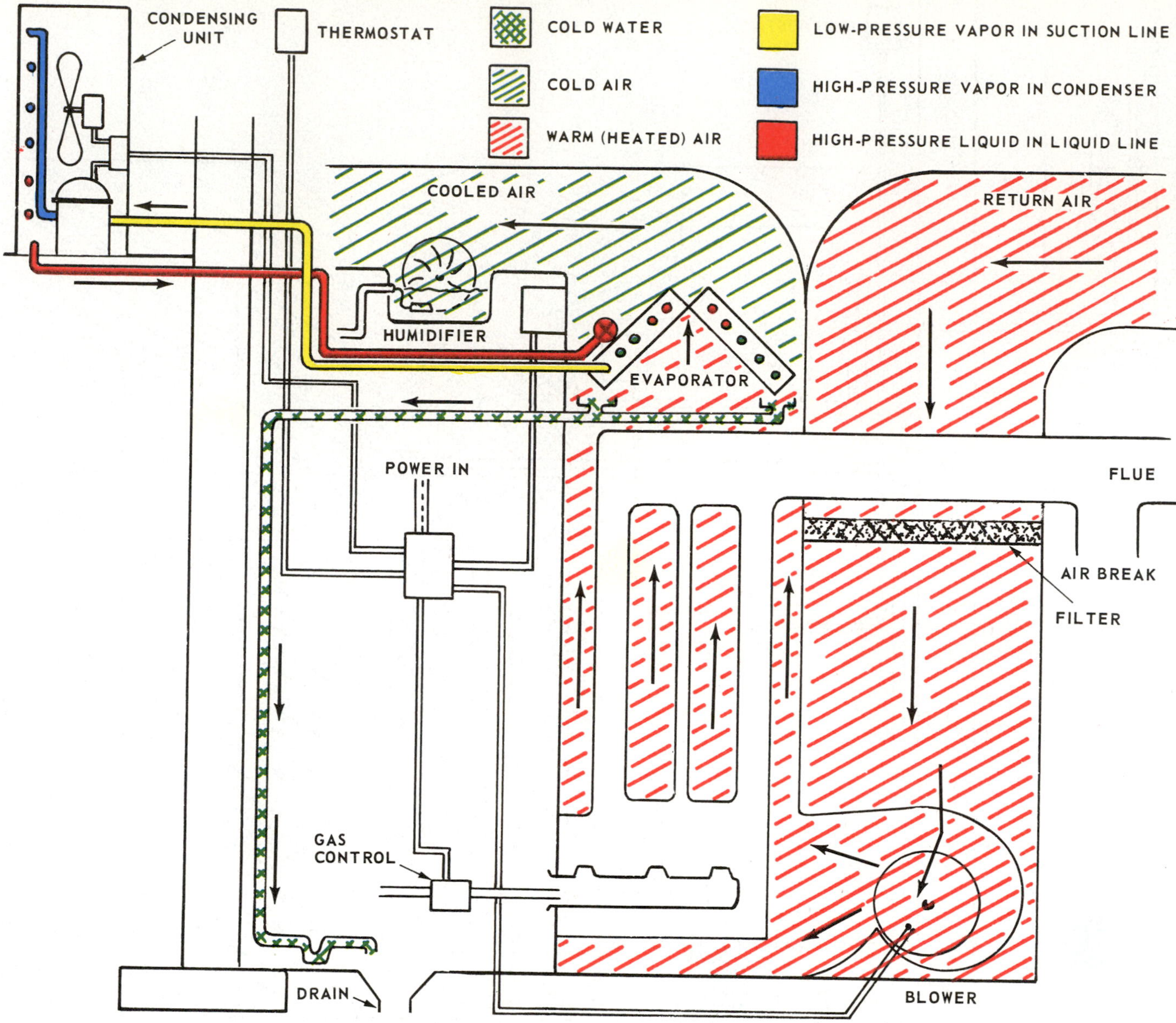

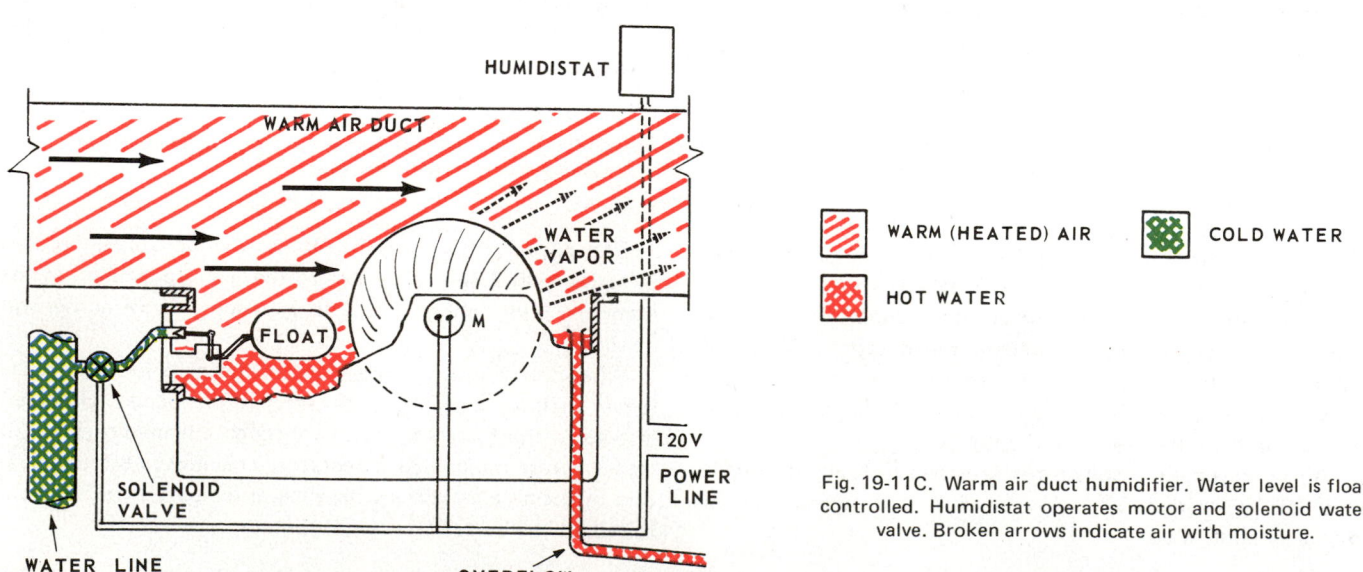

CONDENSING UNIT   **THERMOSTAT**

⊠ **COLD WATER**   ▨ **LOW-PRESSURE VAPOR IN SUCTION LINE**

▨ **COLD AIR**   ▨ **HIGH-PRESSURE VAPOR IN CONDENSER**

▨ **WARM (HEATED) AIR**   ■ **HIGH-PRESSURE LIQUID IN LIQUID LINE**

COOLED AIR   **RETURN AIR**

HUMIDIFIER   **FLUE**

EVAPORATOR   **AIR BREAK**

POWER IN   **FILTER**

GAS CONTROL

DRAIN   **BLOWER**

Fig. 19-11B. Complete air conditioning system during summer operation. A-frame evaporator in plenum cools air forced through it by blower. Outside condensing unit disposes of heat absorbed in evaporator. Summer humidity is removed by condensing of moisture on evaporator surface. Drain tube carries away condensed moisture.

HUMIDISTAT

WARM AIR DUCT

WATER VAPOR   ▨ **WARM (HEATED) AIR**   ⊠ **COLD WATER**

FLOAT   **M**   ▨ **HOT WATER**

120 V

SOLENOID VALVE   **POWER LINE**

Fig. 19-11C. Warm air duct humidifier. Water level is float controlled. Humidistat operates motor and solenoid water valve. Broken arrows indicate air with moisture.

WATER LINE   **OVERFLOW**

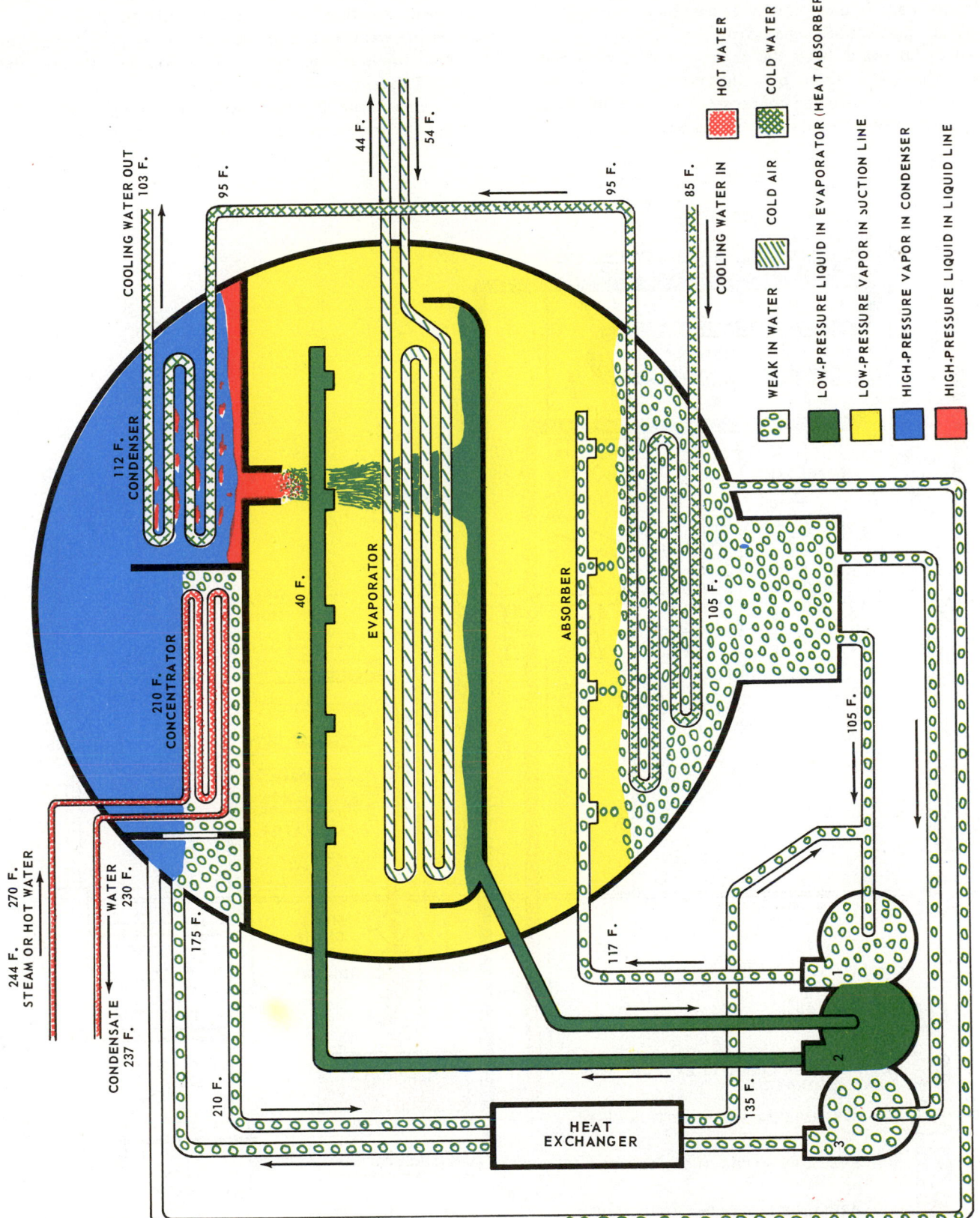

COOLING WATER OUT 103 F.

95 F.

44 F.

54 F.

95 F.

85 F.

COOLING WATER IN

112 F. CONDENSER

210 F. CONCENTRATOR

40 F.

EVAPORATOR

ABSORBER

105 F.

105 F.

117 F.

1

2

3

135 F.

HEAT EXCHANGER

210 F.

175 F.

WATER 230 F.

CONDENSATE 237 F.

270 F. STEAM OR HOT WATER

244 F.

HOT WATER

COLD WATER

COLD AIR

WEAK IN WATER

LOW-PRESSURE LIQUID IN EVAPORATOR (HEAT ABSORBER)

LOW-PRESSURE VAPOR IN SUCTION LINE

HIGH-PRESSURE VAPOR IN CONDENSER

HIGH-PRESSURE LIQUID IN LIQUID LINE

Fig. 19-12. Absorption type air conditioning system. It uses water as refrigerant and liquid lithium bromide as absorber. (The Trane Co.)

19-12 ABSORPTION CYCLE

Most large absorption air conditioning systems use water as the refrigerant and lithium bromide as the absorber. Fig. 19-12 is a schematic diagram of such a system.

Steam or hot water heats the water and lithium bromide solution. The water turns to water vapor which is then condensed by a water-cooled condenser. The water then flows into the evaporator where it evaporates and is absorbed by the lithium bromide at the absorber.

Three pumps maintain the pressure difference. Pump No. 1 moves the weak solution back to the absorber and removes more weak solution from the concentrator. Pump No. 2 recycles the water not evaporated in the evaporator back to the spray heads in the evaporator. Pump No. 3 moves the strong absorber liquid up to the concentrator.

The temperature changes in the system are marked on the drawing. Cooling water leaves the evaporator at 44 F. (7 C.)

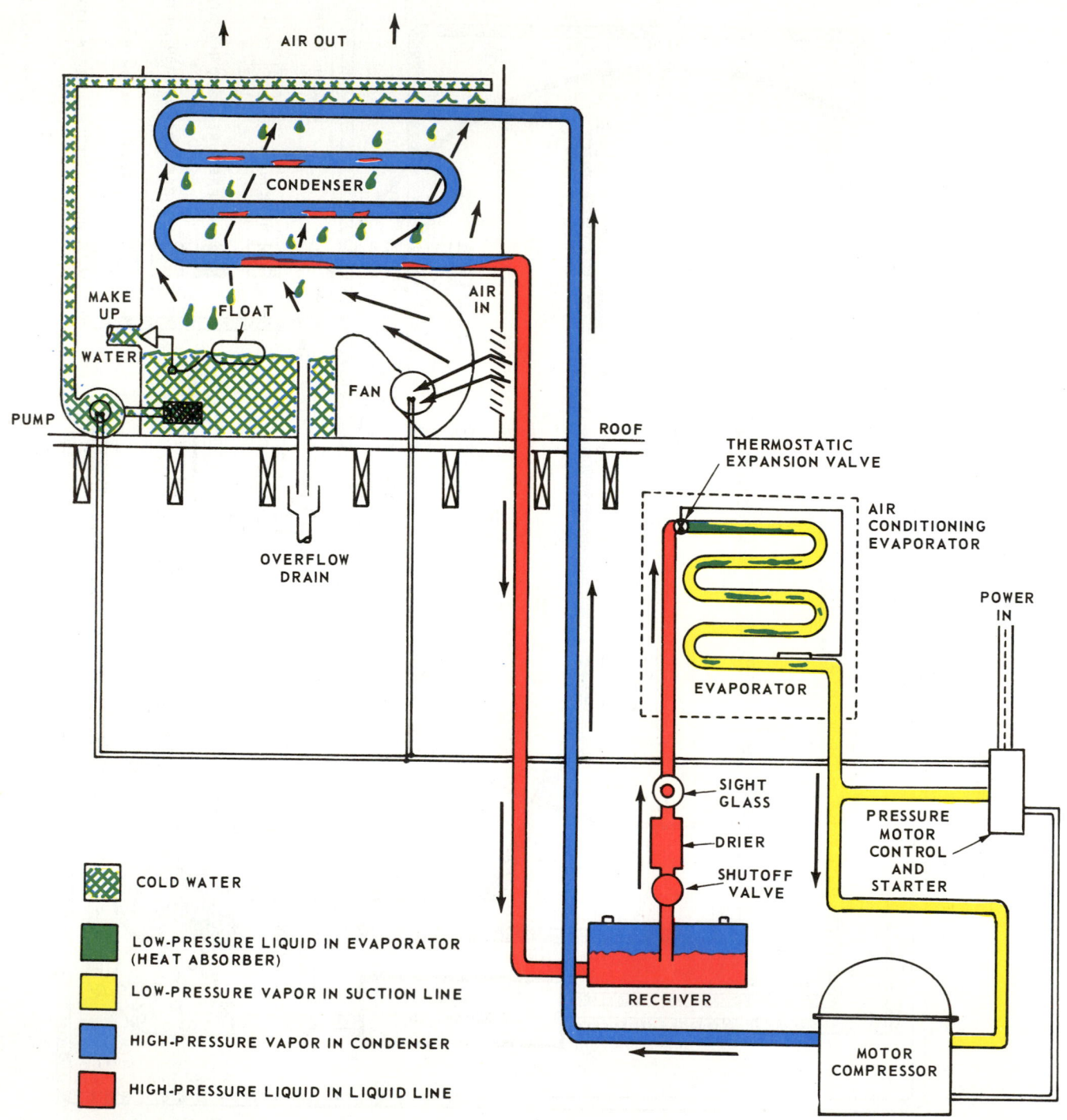

Fig. 19-13. An air cooling unit which uses an evaporative condenser. Note condenser is located outside of conditioned space.

and travels through the cooling coils located in the rooms to be air conditioned (comfort cooled). The water then returns to the evaporator at 54 F. (12 C.). The lithium bromide always stays in liquid form. The condenser cooling water also cools the absorber.

19-13 EVAPORATIVE CONDENSER

Many air conditioning systems use water-cooled condensers. An evaporative condenser, as shown in Fig. 19-13, may be used to cool the condenser.

In this system, a conventional motor compressor, condenser, liquid receiver, drier, thermostatic expansion valve and evaporator are used. The hot compressed refrigerant vapor is piped to the evaporative condenser. This part of the system is usually located on the roof or outside the building, as shown.

In this mechanism, the water supply is piped to a holding tank and a float mechanism maintains a constant level of water in the tank. A water pump circulates and sprays water over the refrigeration condenser.

A fan draws in air through the side of the evaporative condenser housing and forces it upward through the top. The

water droplets are cooled by evaporation and then flow over the condenser. Some water is used up by the evaporative process. This is automatically replaced using a holding tank and a float mechanism. A pressure motor control is used in this instance.

19-14 COOLING TOWER

Many refrigeration and air conditioning systems have water-cooled condensers. These are very efficient and do not take very much space. Often water-cooled condensers have tap water circulated through them. This water is then discharged into the sewer. Such an arrangement uses large amounts of water and may be expensive. Moreover, many places do not allow the use of tap water for cooling air conditioner condensers.

In such cases, cooling towers can be employed to cool the water. In this way, the water is cooled and recirculated through the condenser and sometimes through the outer shell of the compressor. Some makeup water will be required to replace the water lost by evaporation. A schematic of a water cooling tower is shown in Fig. 19-14A.

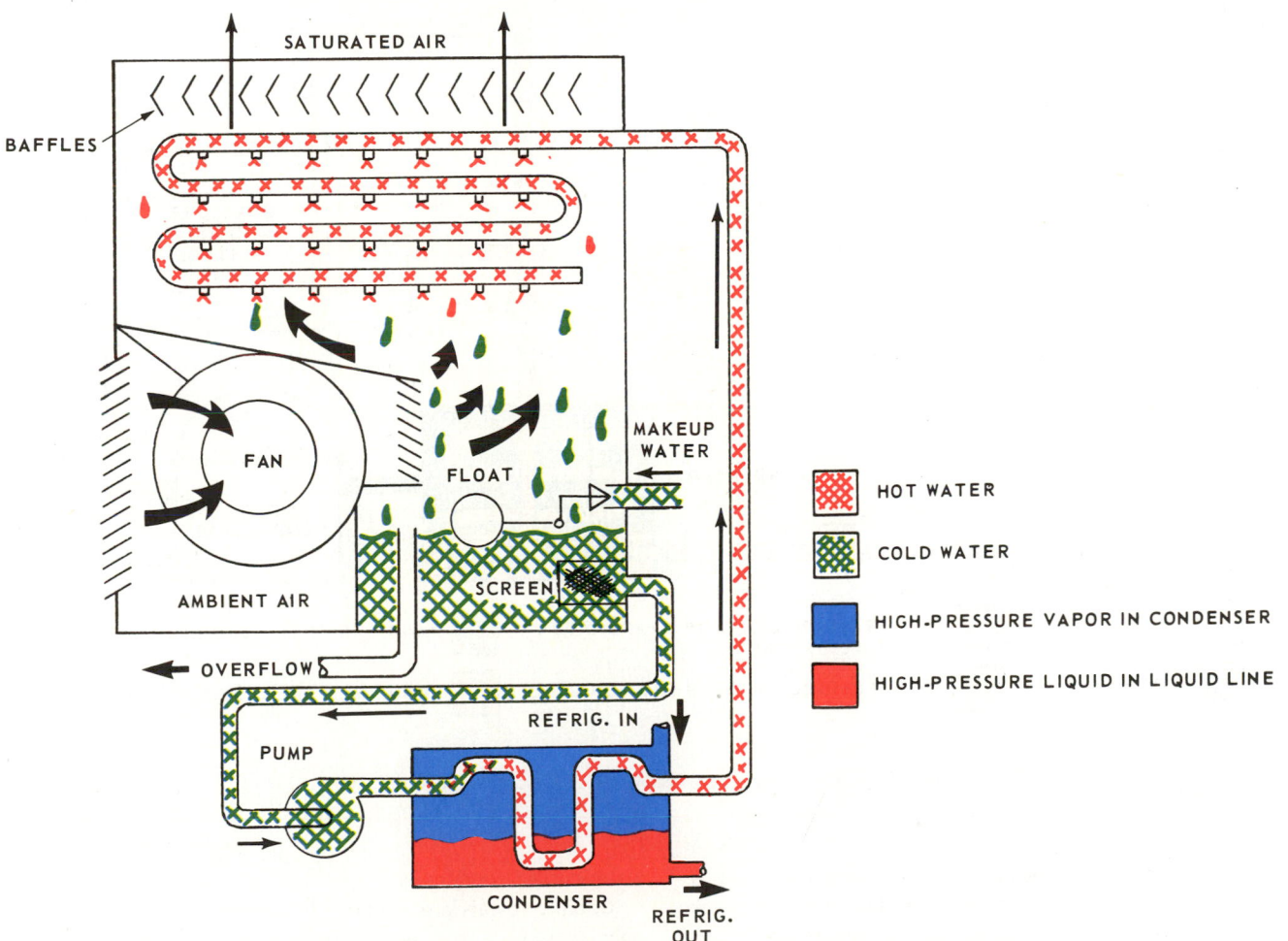

Fig. 19-14A. Cooling tower in which recirculated water is cooled several degrees in tower and then is circulated through water-cooled condenser.

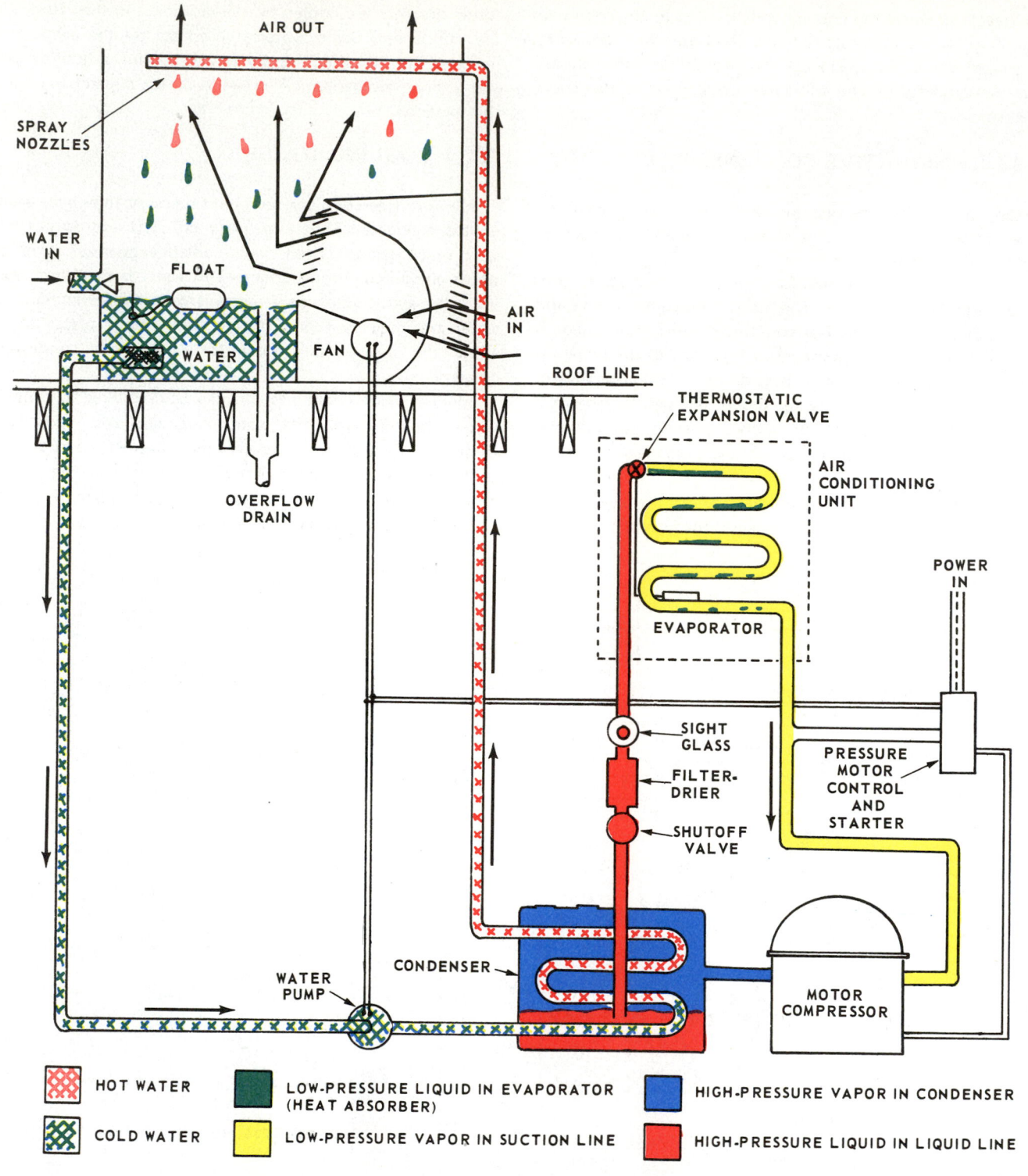

AIR OUT

SPRAY NOZZLES

WATER IN

FLOAT

WATER

FAN

AIR IN

ROOF LINE

THERMOSTATIC EXPANSION VALVE

AIR CONDITIONING UNIT

EVAPORATOR

POWER IN

OVERFLOW DRAIN

SIGHT GLASS

FILTER-DRIER

SHUTOFF VALVE

PRESSURE MOTOR CONTROL AND STARTER

WATER PUMP

CONDENSER

MOTOR COMPRESSOR

HOT WATER

COLD WATER

LOW-PRESSURE LIQUID IN EVAPORATOR (HEAT ABSORBER)

LOW-PRESSURE VAPOR IN SUCTION LINE

HIGH-PRESSURE VAPOR IN CONDENSER

HIGH-PRESSURE LIQUID IN LIQUID LINE

Fig. 19-14B. Air conditioning comfort cooling installation using cooling tower.

The cooling tower is a housing or shed into which air is drawn. It has a water spray arrangement and baffles. The water sprayed over the baffles is exposed to the stream of air and becomes cool. A float mechanism connected to the water spray maintains a constant water level in the water reserve tank. The pump circulates the cooled water through the water-cooled condenser.

Cooling towers are available in a great range of sizes. Small ones may cool the water-cooled condensers for home air conditioners. Very large ones are required for cooling the condensers used in large steam power plants. Fig. 19-14B shows a complete system.

19-15 ROOM HUMIDIFIER

Room humidifiers may be used to maintain room humidity. These humidifiers are housed in a cabinet which is located in the room or space in which humidity is to be increased. The cabinet is supplied with air-in and air-out louvers. A fan circulates air through the cabinet. A water pan or trough is used. A rotating screen or filter (wetted surfaces) dips into a water pan or trough and then exposes the wetted surfaces to the air stream.

An electric heating element is sometimes used to warm the water for greater evaporation. The controls consist of an on-and-off switch, a humidity control and an indicator light to signal the need for more water. The water in the trough may be renewed by pouring more water into it. It may be connected to the building water supply by means of an automatic float mechanism. A typical humidifier is shown in Fig. 19-15.

19-16 ROOM DEHUMIDIFIER

Typically, a dehumidifier consists of a hermetic compressor, condenser and evaporator using a capillary tube refrigerant control. See Fig. 19-16. In the schematic diagram, red indicates high-pressure liquid; green, low-pressure liquid; yellow, low-pressure vapor and blue, high-pressure vapor.

Liquid refrigerant collects in the lower coils of the condenser and flows through the filter into the capillary tube. Then it moves into the evaporator which is under low pressure. In the evaporator, the liquid refrigerant boils rapidly and picks up heat from the evaporator surface. A motor-driven fan forces large amounts of air through the evaporator.

Because of the low temperature of the evaporator, the moisture carried in the air condenses on the evaporator surfaces. The moisture drips to the bottom of the evaporator and into the condensate trough. Air flowing through the evaporator is both cooled and dehumidified. Cooled air is then

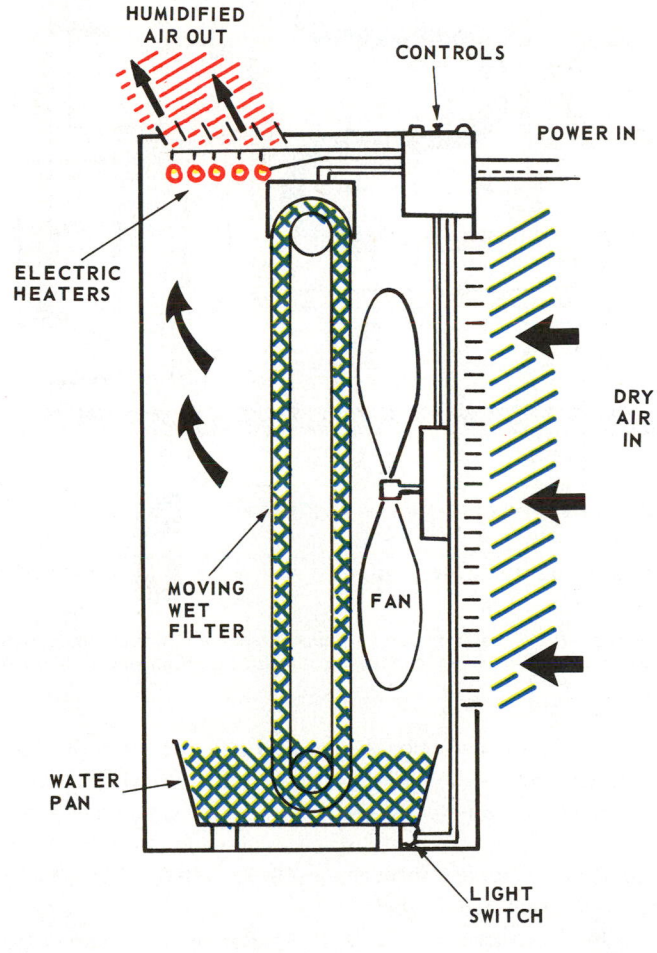

 COLD WATER  WARM (HEATED) AIR  COLD AIR

Fig. 19-15. Room humidifier. Porous belt slowly moves through water in a water pan. Fan forces air through moist porous belt and humidity is increased. Some units have electric heaters to reheat humidified air. Humidistat controls operation of unit. Signal light goes on when water pan is empty.

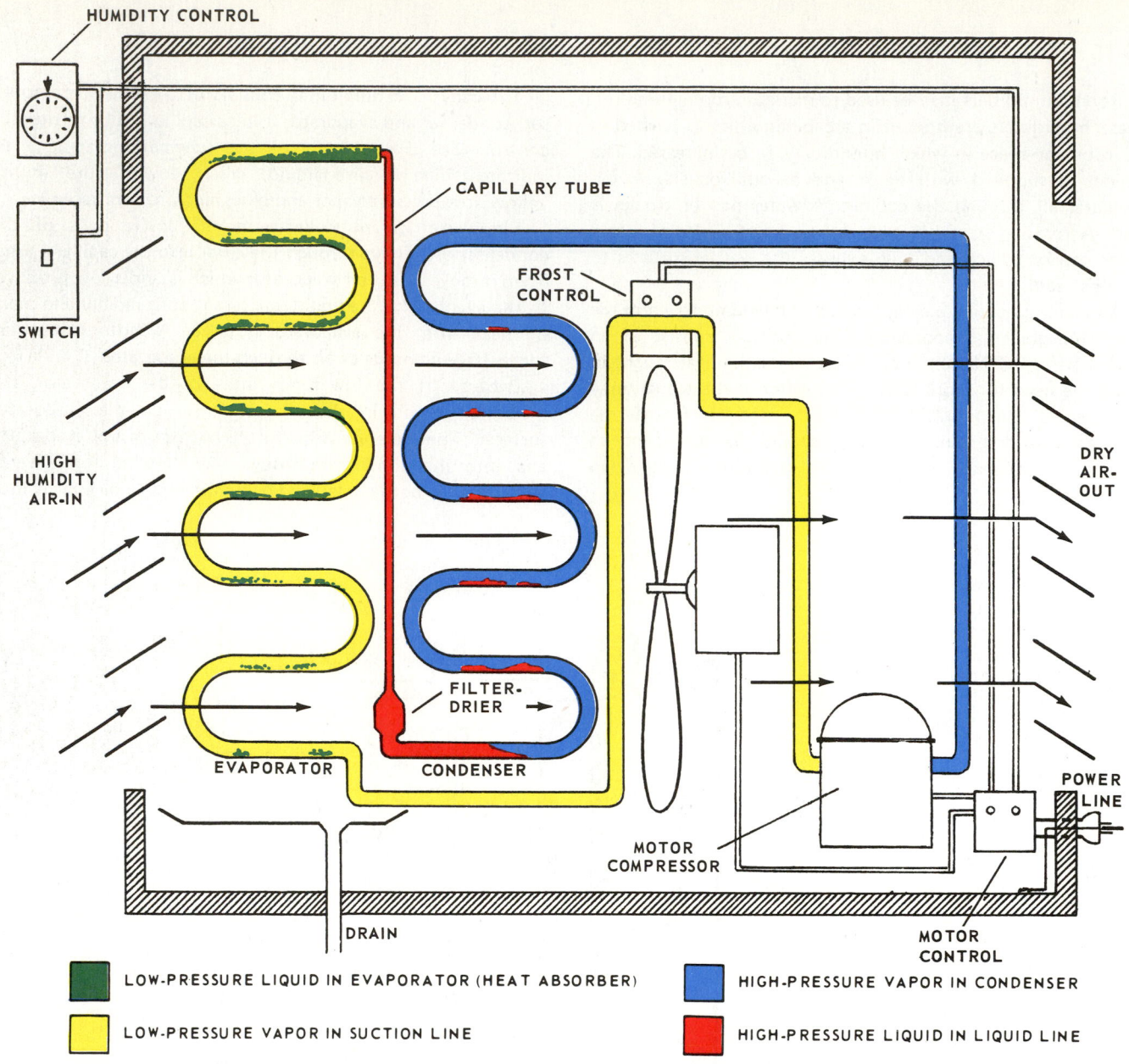

HUMIDITY CONTROL

SWITCH

HIGH
HUMIDITY
AIR-IN

CAPILLARY TUBE

FROST
CONTROL

EVAPORATOR

FILTER-
DRIER

CONDENSER

MOTOR
COMPRESSOR

DRY
AIR-
OUT

POWER
LINE

MOTOR
CONTROL

DRAIN

◼ LOW-PRESSURE LIQUID IN EVAPORATOR (HEAT ABSORBER)   ◼ HIGH-PRESSURE VAPOR IN CONDENSER

◼ LOW-PRESSURE VAPOR IN SUCTION LINE   ◼ HIGH-PRESSURE LIQUID IN LIQUID LINE

Fig. 19-16. Room dehumidifier. Room air is cooled as it flows through evaporator. Considerable water vapor is condensed on evaporator surface and drains away. Air is reheated as it flows through and cools condenser.

forced through the condenser, where it cools the condenser and again picks up heat, so the air leaving the dehumidifier is about the same temperature as it was when it entered but it has a lower humidity.

Low-pressure vapor is drawn from the evaporator through the suction line to the compressor. It is again compressed to high-side pressure and is forced into the condenser. Here it is cooled, becomes a liquid and the cycle is repeated.

In addition to an on-and-off switch, dehumidifiers usually have two other controls. One is for humidity. It permits the dehumidifier to operate until the desired relative humidity is reached; then the control shuts the machine off. The other is a frost control element placed in the suction line between the evaporator and the compressor. It stops the motor compressor at a high enough temperature so the evaporator will not freeze

over and stop the flow of air through it.

In the drawing, arrows in black show the direction of airflow through the dehumidifier.

19-17 HEAT PUMP — AIR-TO-AIR

Used in homes and in some industries, the "heat pump" is a heat-moving mechanism. Heat is absorbed in an evaporator in one location and released through a condenser in another location. The system can reverse its operation so that the evaporator becomes the condenser and the condenser becomes the evaporator. Heat flow is reversed. Thus, using a special reversing valve, the mechanism either heats or cools the conditioned space. *The flow through the compressor is always in the same direction.*

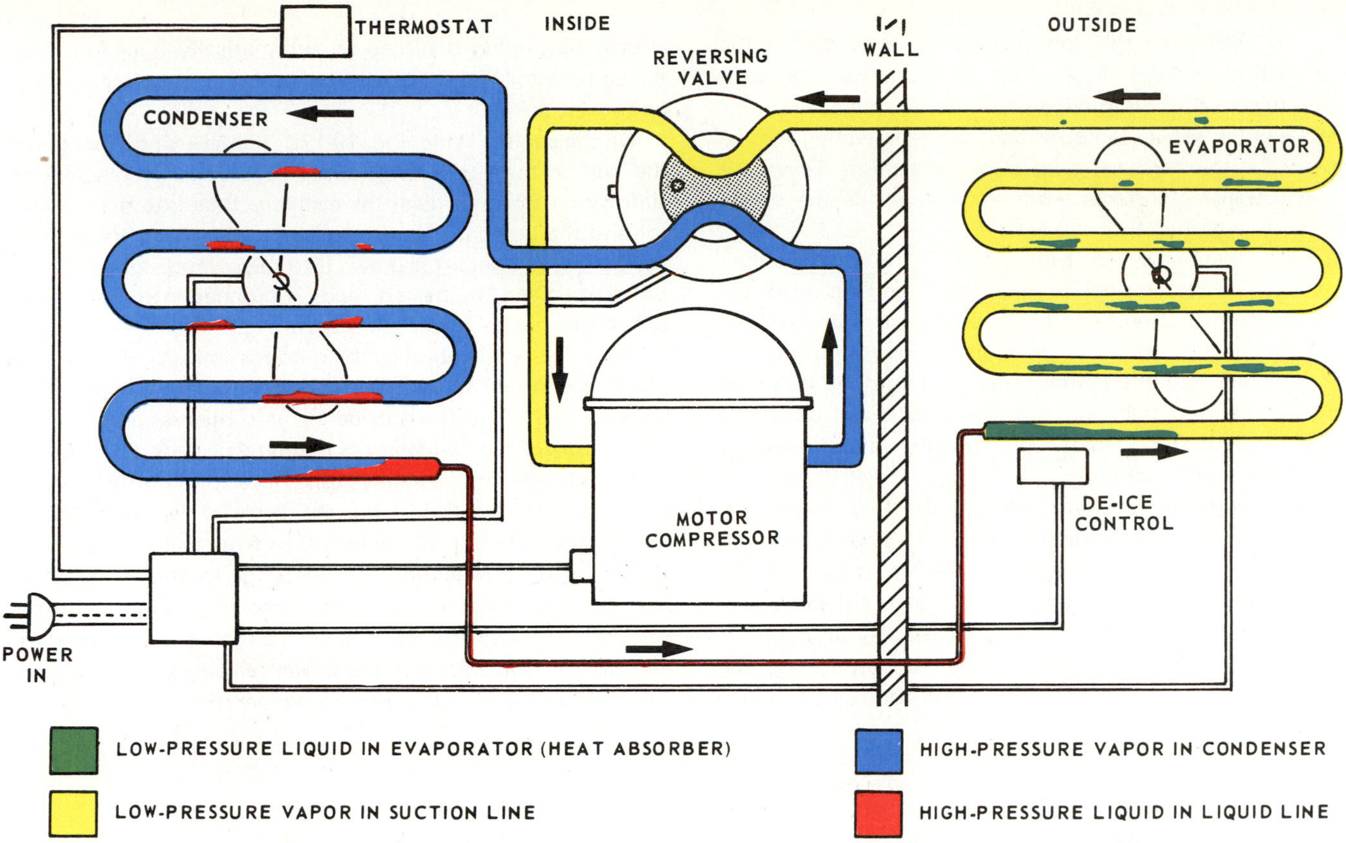

THERMOSTAT INSIDE REVERSING VALVE WALL OUTSIDE

CONDENSER

EVAPORATOR

MOTOR COMPRESSOR

DE-ICE CONTROL

POWER IN

■ (green) LOW-PRESSURE LIQUID IN EVAPORATOR (HEAT ABSORBER)

■ (yellow) LOW-PRESSURE VAPOR IN SUCTION LINE

■ (blue) HIGH-PRESSURE VAPOR IN CONDENSER

■ (red) HIGH-PRESSURE LIQUID IN LIQUID LINE

Fig. 19-17A. Air-to-air heat pump illustrating heating cycle. Reversing valve is set so that coil on outside is acting as an evaporator. Heat absorbed in evaporator is released by condenser inside house.

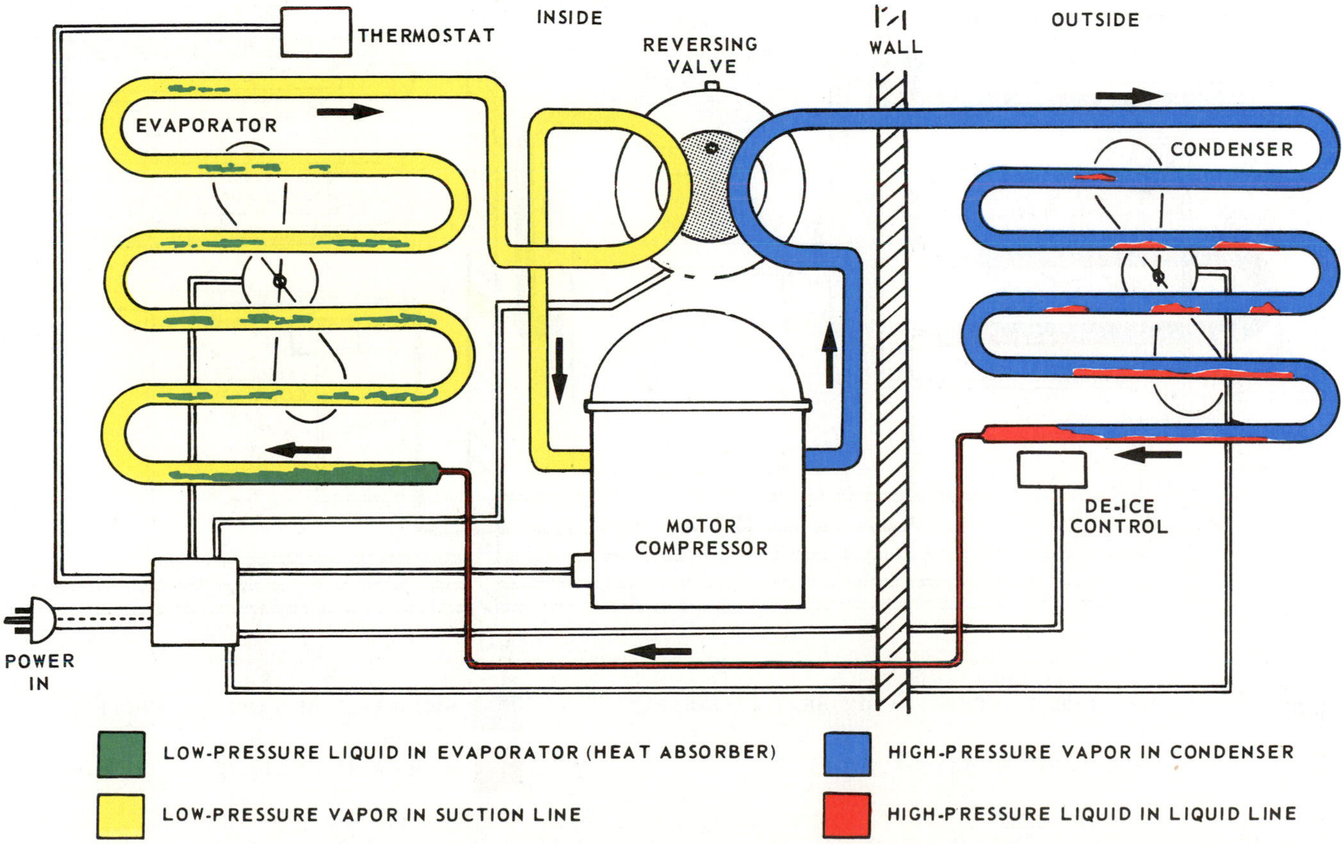

THERMOSTAT INSIDE REVERSING VALVE WALL OUTSIDE

EVAPORATOR

CONDENSER

MOTOR COMPRESSOR

DE-ICE CONTROL

POWER IN

■ (green) LOW-PRESSURE LIQUID IN EVAPORATOR (HEAT ABSORBER)

■ (yellow) LOW-PRESSURE VAPOR IN SUCTION LINE

■ (blue) HIGH-PRESSURE VAPOR IN CONDENSER

■ (red) HIGH-PRESSURE LIQUID IN LIQUID LINE

Fig. 19-17B. An air-to-air heat pump during cooling cycle. Valve is set so that coil on inside is acting as an evaporator. Heat absorbed in evaporator is released by condenser outside house.

Fig. 19-17A shows the flow through the valve causing the conditioned space to be heated. Fig. 19-17B shows the valve in position to cool the conditioned space.

Heat pumps use a compression type refrigerating mechanism, similar to a regular refrigerating mechanism. They have two heat transfer surfaces — one located inside the conditioned space and the other out-of-doors.

On the heating cycle, Fig. 19-17A, the outdoor coil becomes an evaporator while the indoor coil becomes the condenser. In operation, liquid refrigerant enters the outdoor coil, picks up heat from out-of-doors and is vaporized. The vapor is drawn into the compressor, is compressed to a high temperature and is pumped into the indoor coil. Since its temperature is higher than the indoor temperature, heat will be released into the room.

Compressed refrigerant vapors will condense upon giving up their heat of vaporization and will return to a liquid state. The liquid then flows back through the capillary tube into the evaporator, and the cycle is repeated. Since the outdoor coil is colder than the outdoor surrounding air, ice may form on it if the outdoor temperature is rather low. Therefore, outdoor coils are fitted with de-ice controls. These controls are either electric heating units turned on automatically if ice forms; or, the compressor stops, allowing the evaporator surface to warm up to melt the ice.

On the cooling cycle, Fig. 19-17B, the reversing valve causes the coil in the conditioned space to become an evaporator. Refrigerant flows through the capillary tube into the evaporator and the liquid refrigerant boils absorbing heat. Vapor from the boiling refrigerant is drawn into the compressor where it is compressed and the heated vapor is pumped into the outdoor coil, which has become a condenser.

Since the air surrounding the outdoor coil is cooler than the compressed vapor in the coil, the compressed refrigerant vapor gives up its heat to the outside air. It condenses and flows to the bottom of the condenser as liquid refrigerant. From here it flows through the capillary tube into the bottom of the evaporator. From this point the cycle is repeated. Motor-driven fans on both coils aid heat flow from coil surfaces.

Heat pump installations are ideal for locations where the heat load in winter is almost the same as the cooling load in summer. *Air-to-air installations are most satisfactory when the ambient air temperature in the winter remains above — or only occasionally below — the freezing temperature.*

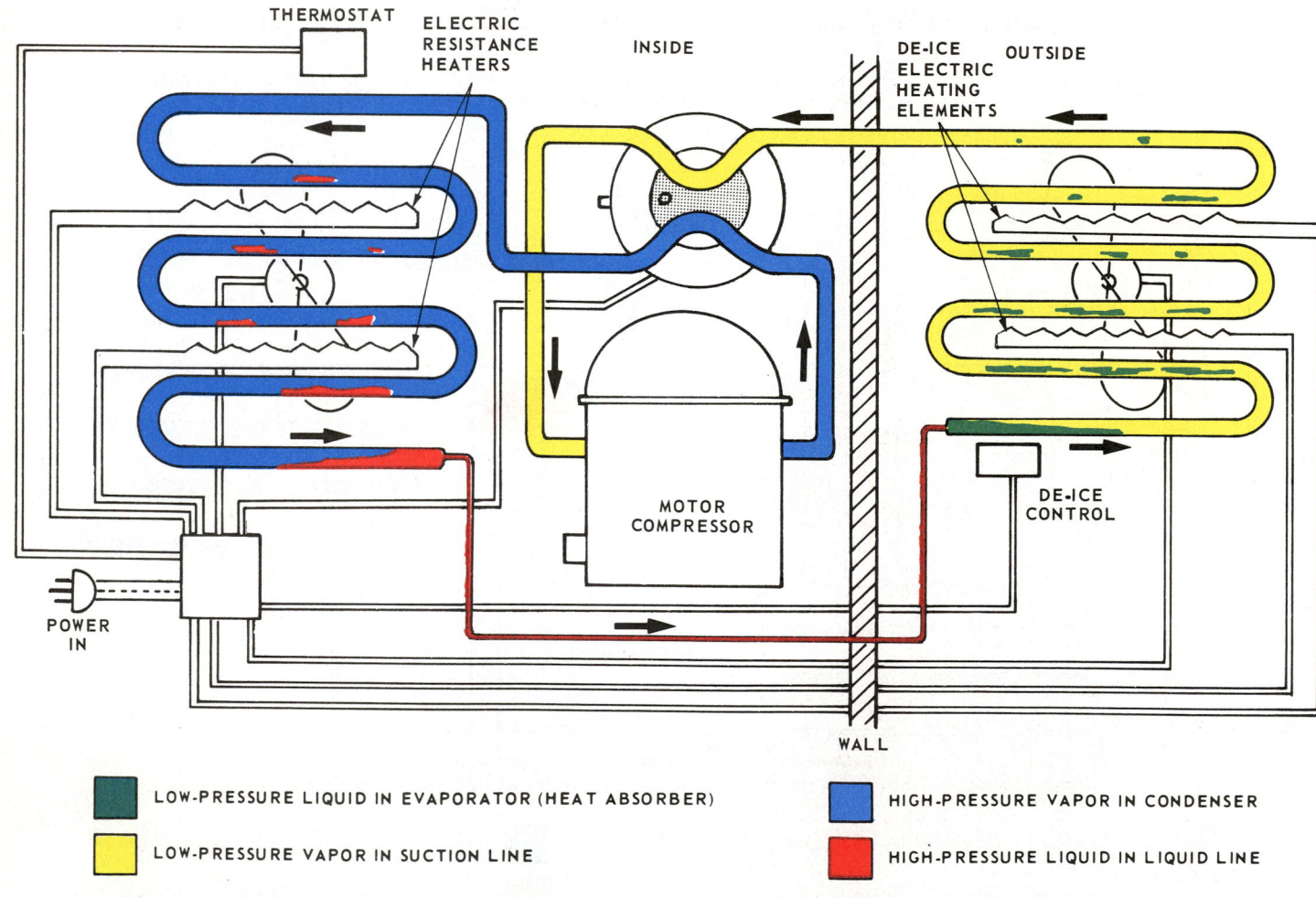

Fig. 19-18. An air-to-air heat pump with electric resistance heating elements. The heating cycle is on in this diagram. Electric heaters provide additional heat if needed.

19-18 HEAT PUMP — WITH ELECTRIC HEATERS

Air-to-air heat pump installations operate efficiently when outside air temperature is above freezing. However, when the outside temperature drops down to or below freezing, efficiency drops off rapidly.

To make up for this loss of efficiency, the indoor section is often fitted with electric resistance heating units. When the thermostat calls for more heat than the heat pump is able to deliver, the electric heating elements will turn on.

Heat pump operation of the heating and cooling cycle is the same as explained in Para. 19-17. Fig. 19-18 is a heat pump cycle diagram. Note the resistance heating units.

19-19 HEAT PUMP — GROUND OR WATER COIL

As explained in Para. 19-18, the efficiency of the air-to-air heat pump depends greatly on outdoor temperature. To improve this efficiency, some installations use a coil buried in the ground beneath the frost level, rather than a coil in the atmosphere. If the coil is buried at some depth and a long

enough coil is used, the efficiency of the heat pump may be very good.

On the heating cycle, liquid refrigerant flows through a refrigerant control and into the ground coil. Since the refrigerant in the ground coil is under low pressure, it boils, absorbing heat from the ground surrounding the coil.

The vaporized refrigerant is then drawn into the compressor. It is compressed and discharged into the condenser, which, in this case, is the heating coil for the heating system. The condenser changes the vaporized refrigerant to a liquid. It returns to the refrigerant control to repeat the cycle. A schematic of the cycle is shown in Fig. 19-19A.

The same mechanism may be used to cool the building in summer. The cycle is reversed to move heat from the building to the outdoors. In this case, the inside coil serves as the evaporator and the ground coil becomes the condenser. The ground absorbs the heat from the vaporized refrigerant. Fig. 19-19B illustrates the heat pump with the valves set for cooling the conditioned space.

The four-way valve is electrically controlled by the thermostat. If heat is called for, the valve will allow fluid flow as

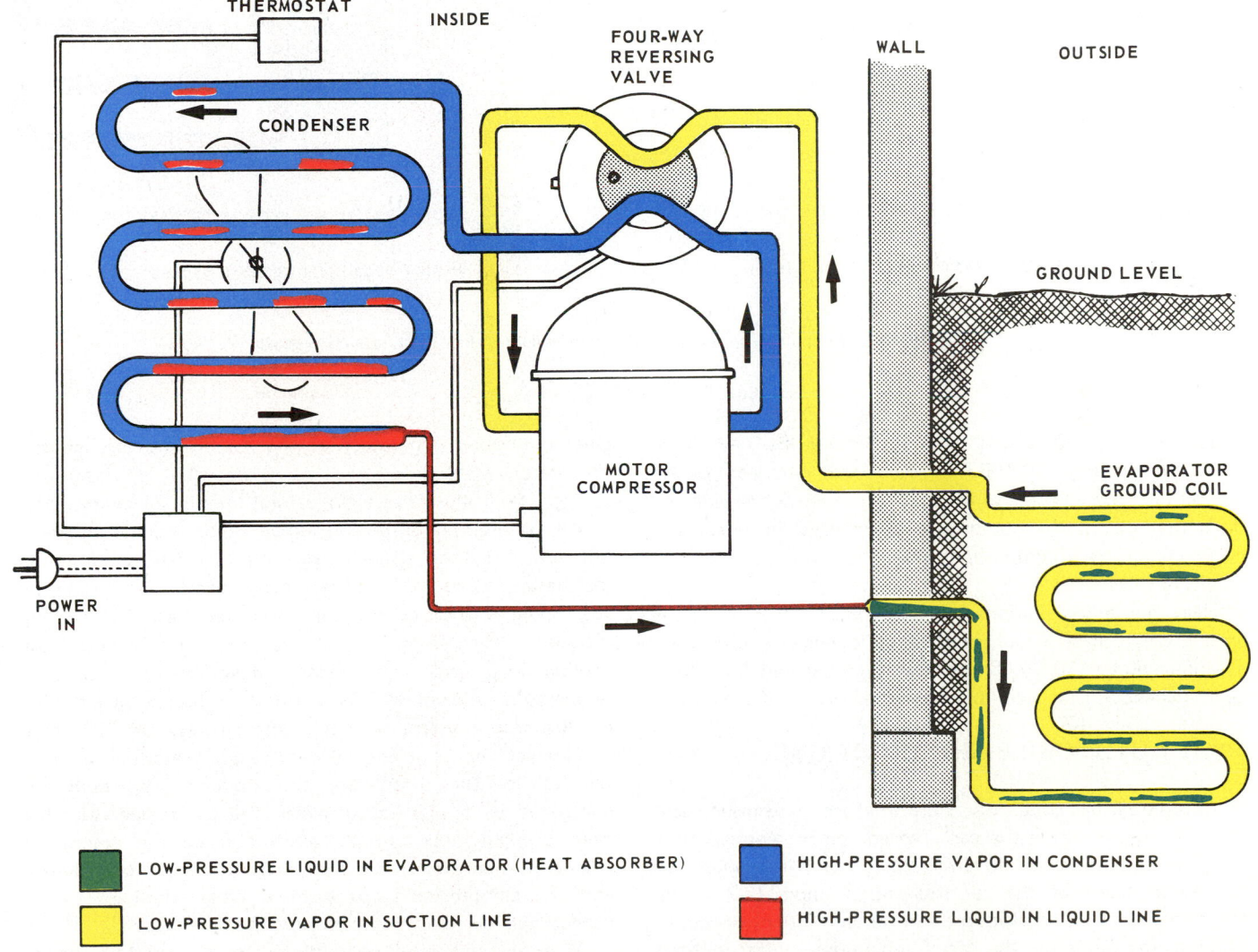

🟩	LOW-PRESSURE LIQUID IN EVAPORATOR (HEAT ABSORBER)	🟦	HIGH-PRESSURE VAPOR IN CONDENSER
🟨	LOW-PRESSURE VAPOR IN SUCTION LINE	🟥	HIGH-PRESSURE LIQUID IN LIQUID LINE

Fig. 19-19A. Heat pump using a ground coil (or coil in a well or lake). Heating cycle is shown.

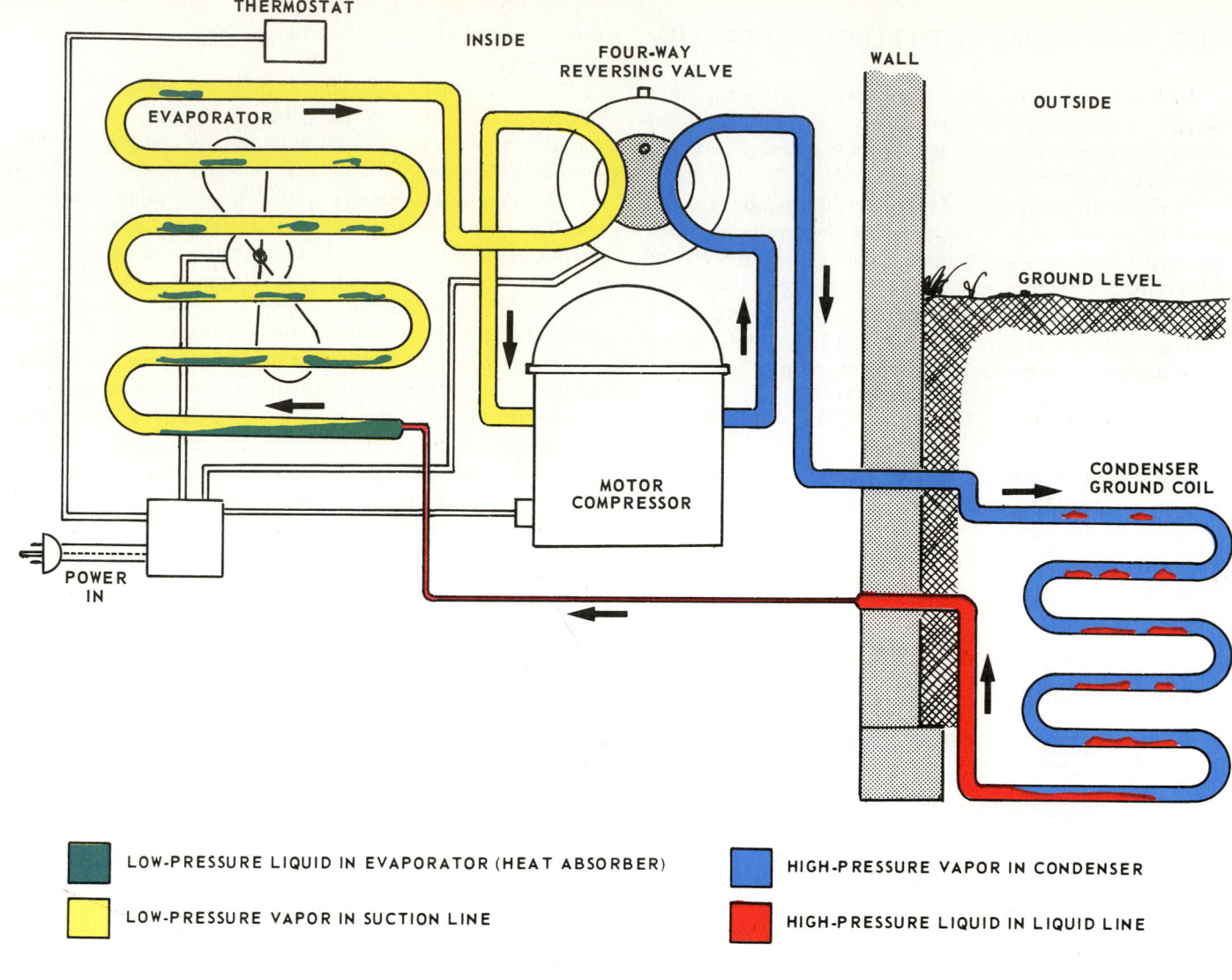

Fig. 19-19B. Heat pump using ground coil (or coil in a well or lake) Cooling cycle is shown.

indicated in Fig. 19-19A. If cooling is needed, the flow will be as shown in Fig. 19-19B. In each case, the compressor refrigerant flow is in the same direction. The suction side and discharge side of the compressor are always the same. The cycle change is accomplished entirely by operation of the four-way valve.

Heat pump installations are even more efficient if the ground coil is placed in a spring or in a flowing well with water at about 50 F. (10 C.). Some heat pump installations have been successful using a coil placed in the bottom of a lake.

19-20 AUTOMOBILE AIR CONDITIONING

Automobile air conditioners use a refrigerating mechanism to cool the air inside the car. The equipment consists of a refrigerating compressor driven by the car engine, a condenser located in front of the car radiator, a liquid line to the refrigerant control, an evaporator, a fan and a duct system to circulate the air inside the car. Temperature control is based

chiefly on the temperature of the air flowing through the evaporator.

Fig. 19-20 illustrates a typical automobile air conditioner.

Air conditioning a moving vehicle presents some problems not found in the usual refrigeration or air conditioning installation. Since the compressor is driven by the engine, its speed will change as the car speed changes. The cooling capacity of the system is usually sufficient to take care of the cooling load under the most unfavorable conditions of temperature and speed. As a result, under normal driving conditions, the system has much more capacity than is needed

This problem is solved by providing a magnetic clutch in the hub of the compressor drive pulley. This clutch is controlled by a thermostat. When the car temperature has been brought down to the desired level, the thermostat releases the magnetic clutch on the compressor drive pulley, and the compressor stops turning. When cooling is again needed, the magnetic clutch engages the compressor.

There is still another problem in air conditioning an

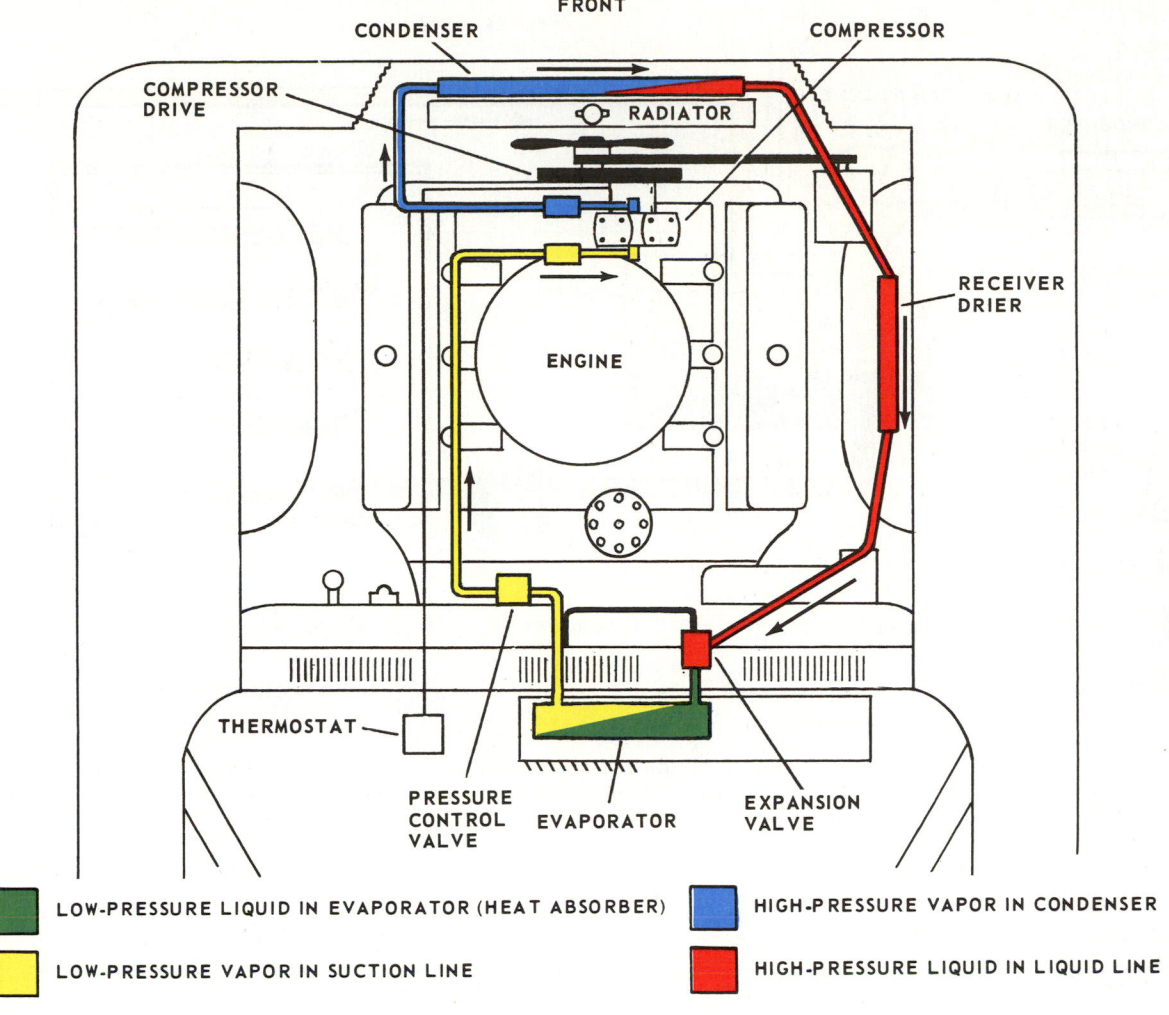

CONDENSER   COMPRESSOR

COMPRESSOR DRIVE

RADIATOR

RECEIVER DRIER

ENGINE

THERMOSTAT

PRESSURE CONTROL VALVE   EVAPORATOR   EXPANSION VALVE

LOW-PRESSURE LIQUID IN EVAPORATOR (HEAT ABSORBER)

LOW-PRESSURE VAPOR IN SUCTION LINE

HIGH-PRESSURE VAPOR IN CONDENSER

HIGH-PRESSURE LIQUID IN LIQUID LINE

Fig. 19-20. Automobile air conditioner. Red indicates high-pressure liquid refrigerant; green, low-pressure liquid refrigerant; yellow, low-pressure vapor; blue, high-pressure vapor. Compressor is driven by engine. Condenser coil is located ahead of car radiator. Evaporator is located in duct system in passenger compartment.

automobile. The evaporator condenses moisture from the air. If the evaporator temperature is maintained at or below freezing temperature, this moisture will freeze and adhere to the evaporator. Soon the evaporator will be completely frozen over. No air can circulate through it. This problem is solved by placing a suction throttling valve in the suction line. This valve maintains the pressure in the evaporator slightly above the pressure at which the temperature of the boiling refrigerant in the evaporator will cause the moisture to freeze to the evaporator surface.

See Chapter 26 for full details concerning automobile air conditioning.

19-21 STEAM JET COOLING

Steam jet cooling uses water as the refrigerant. Pressure on the surface of water is reduced to lower its boiling temperature. This is shown in Fig. 19-21A. At 0.2 psia, the boiling temperature of water is 53 F. (12 C.).

WATER BOILING TEMPERATURES

PSIA	BOILING TEMPERATURE		PSIA	BOILING TEMPERATURE	
	F.	C.		F.	C.
.1	35	2	5.	162	72
.2	53	12	6.	170	77
.3	64	18	7.	177	81
.4	73	23	8.	183	84
.5	80	27	9.	188	87
.6	85	29	10.	193	89
.7	90	32	11.	198	92
.8	94	34	12.	202	94
.9	98	37	13.	206	97
1.	102	39	14.	209	98
2.	126	52	14.7	212	100
3.	141	61	15.	213	101
4.	153	67	20.	228	109

Fig. 19-21A. Table shows boiling temperature of water at various pressures. Note that pressures are in pounds per square inch absolute (psia). Atmospheric pressure is 14.7 psia.

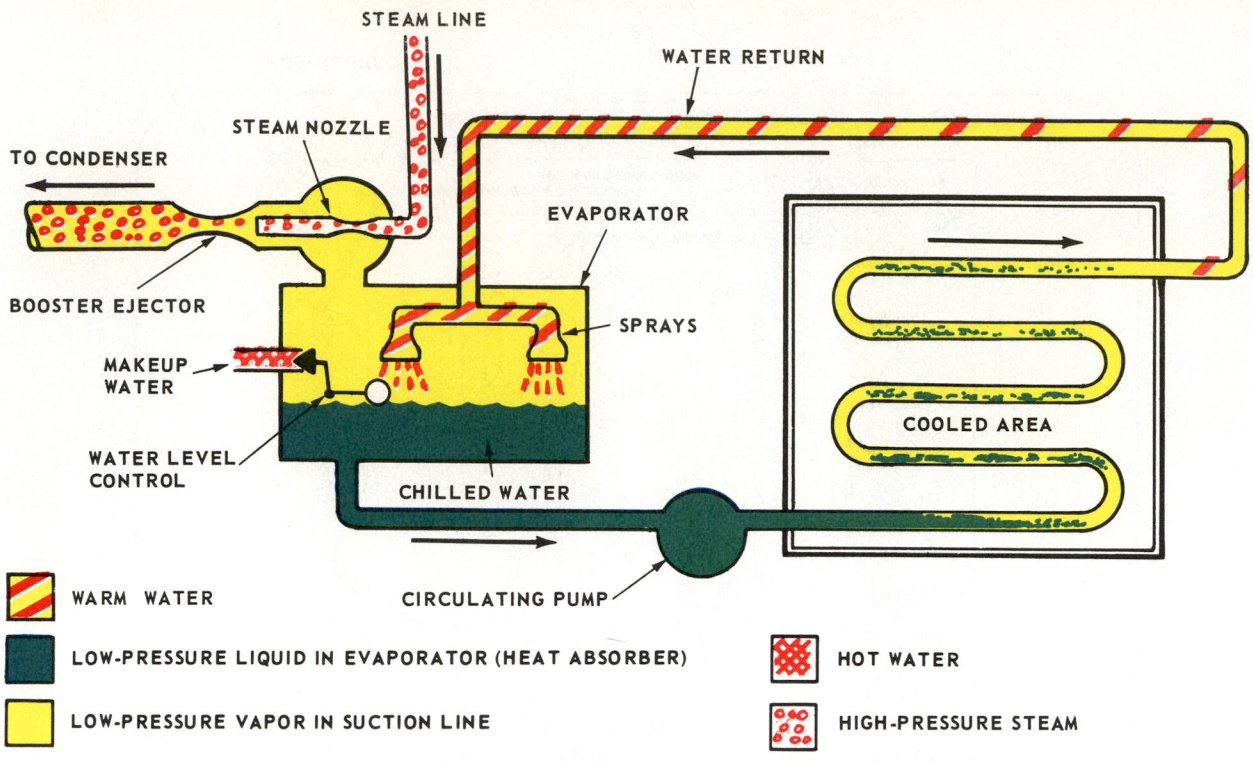

Fig. 19-21B. Steam jet refrigeration. Steam escaping through nozzle in ejector causes low-pressure condition over surface of water in evaporator. This low pressure on surface of water causes it to evaporate rapidly, absorbing heat and reducing temperature of water in evaporator. Chilled water may be circulated where needed for cooling purposes.

WARM WATER

LOW-PRESSURE LIQUID IN EVAPORATOR (HEAT ABSORBER)

LOW-PRESSURE VAPOR IN SUCTION LINE

HOT WATER

HIGH-PRESSURE STEAM

A steam jet is shown in Fig. 19-21B. An ejector sucks or draws water vapor from the surface of the water in the evaporator, causing the pressure in the evaporator to drop. The ejector reduces the pressure in the evaporator to a point at which the water will vaporize at the desired temperature. While vaporizing, it absorbs heat and cools the rest of the water in the evaporator.

Steam pressure at the ejector nozzle should be about 150 psia. The pressure in the condenser, not shown in the illustration, will be about 3 psia.

Evaporation of some of the water in the evaporator reduces the temperature of the remaining water. Pumps circulate this cold water at 40 to 70 F. (4 to 21 C.), to the area to be cooled.

Because of the need for a large supply of steam under a fairly high pressure and for a large supply of water for cooling the condenser, these systems usually have a large capacity — 100 tons and over.

Steam jet systems are often used in air conditioning and for cooling water used in certain chemical plants for gas absorption. The cooling temperatures provided by the steam jet mechanism are usually between 40 and 70 F. (4 and 21 C.). Temperatures below 40 F. (4 C.) are impractical due to the danger of freezing.

19-22 VORTEX TUBE COOLING

An interesting device called a vortex tube uses compressed air to produce low temperature. Pressurized air is directed smoothly along the inner surface of a tube or cylinder. One end of the tube is completely open. The other end is closed off except for a small-diameter tube. During operation, warm air leaves through the unrestricted (wide open) end while cold air leaves through the small tube at the other end.

Fig. 19-22 shows the operation of the vortex tube. There are three openings. One is an inlet while the other two, as explained above, are outlets. The inlet opening at C is a jet nozzle. It is connected to the compressed air source B (red). This jet is arranged to inject the air into the tube at an angle.

Due to the design of the jet and the high pressure of the air, the air swirls rapidly in a corkscrew pattern inside the large tube. Both openings, D and E, are at or near atmospheric pressure. The temperature of the air, as it leaves through tube E, will be greatly reduced (green). Temperatures below zero are easily obtained with this device.

This system is particularly useful where both fresh air and cooling are desired at the same time. A large quantity of compressed air must be used.

A typical application of the vortex tube principle is in the cooling of suits for industrial workers who must do their jobs in bad atmospheres or in very hot places. Suits cooled by vortex tubes are used in mining and foundry work. In most cases, applications of this kind are called for only when the miner or foundry worker is performing a particular job.

Vortex tube devices are designed to operate on a continuous basis. No thermostatic control of any type is used. However, there is a manual control that allows the vortex tube operator to adjust outlet air temperature to suit the need.

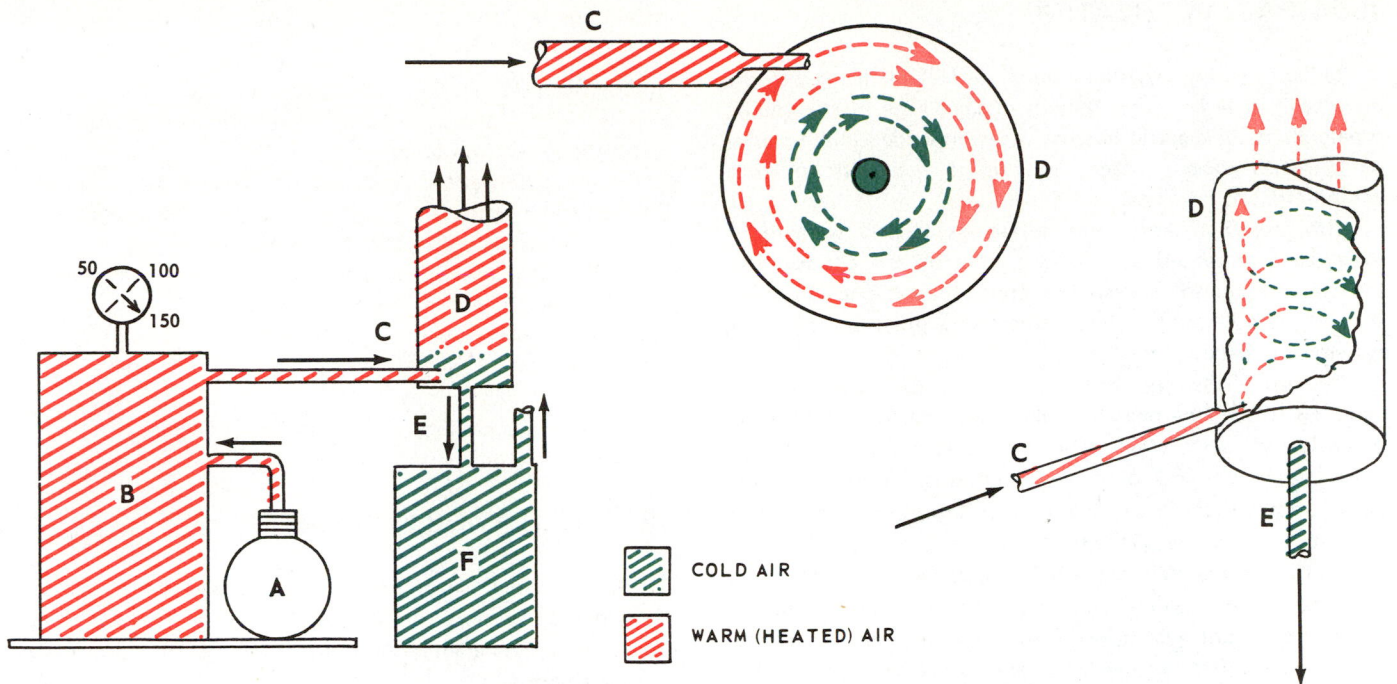

Fig. 19-22. Simplified illustration of vortex tube refrigeration. A—Air compressor. B—Compressed air storage tank. C—Compressed air nozzle. Cold air is produced at E, and flows into space F, which is to be cooled. Warm air is expelled through large tube D

COLD AIR

WARM (HEATED) AIR

19-23 EVAPORATIVE COOLING WITH SWAMP COOLER OR PONDED ROOF

Evaporative cooling is often used in the summertime in bright sunshine, when humidity is low, to maintain a safe temperature for plants growing in greenhouses. See Fig. 19-23.

One end of the greenhouse will have a lattice of fibers such as excelsior. Water pipes with small holes are arranged along the top of the lattice, and the water from the small holes flows downward, wetting the lattice material. A large fan at the other end of the greenhouse draws air through the lattice and discharges it outdoors.

Due to the evaporation of the water on the lattice surfaces, the air entering the greenhouse will be cooler than the outside air. Since plants do best under conditions of high humidity, this type of cooling is ideal. It provides a high humidity as well as a lower temperature inside the greenhouse.

Foundries sometimes use this system where a few degrees of cooling are desirable. In such installations, the evaporation usually takes place in a structure on the roof of the building, and the cooled air is brought down into the work area.

It is also possible to provide a shallow pool of water, 2 or 3 in. deep, on a flat roof for cooling purposes. Where there is considerable bright sunshine and a relatively low humidity, the ponded roof will help greatly in maintaining a comfortable temperature inside the building.

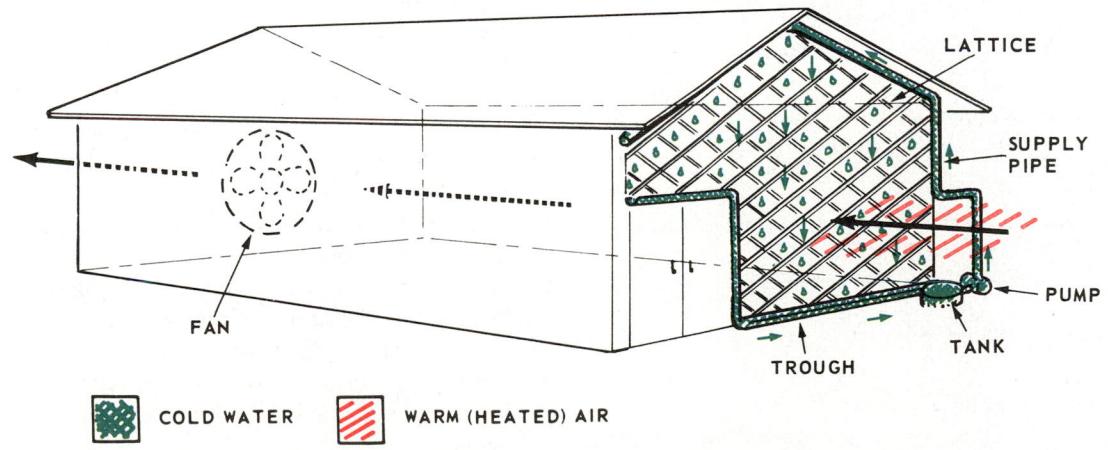

LATTICE

SUPPLY PIPE

PUMP

TANK

TROUGH

FAN

COLD WATER

WARM (HEATED) AIR

Fig. 19-23. An evaporative cooler used as a swamp cooler for a greenhouse.

19-24 RADIANT HEATING

Radiant heating provides a very comfortable living environment with little or no equipment in sight. The most common type consists of electric heating wires imbedded in the plaster of ceilings, walls, in the floor of a room or in some combination of the three.

With this installation, the surface is slightly warmed. A thermostat, mounted on the wall, controls the current flow through the heating wires, thus controlling the temperature. Actually, the amount of heat radiated by this type of heating system can heat a room.

This type of heating works the opposite of a heat sink. (See Para. 18-46.) In the heat sink, heat radiated from the human body is lost. With radiant heating, heat radiated from the human body bounces back and is not lost. Accordingly, the body is comfortable in ambient temperatures somewhat lower than the temperature of the human body.

In order to maintain these building surface temperatures, it is necessary to heavily insulate the walls, floors or ceilings which use radiant heat. Fig. 19-24 illustrates such a structure. Often radiant heat installations are supplemented by other heat sources to provide a complete heating and air conditioning system with enough capacity.

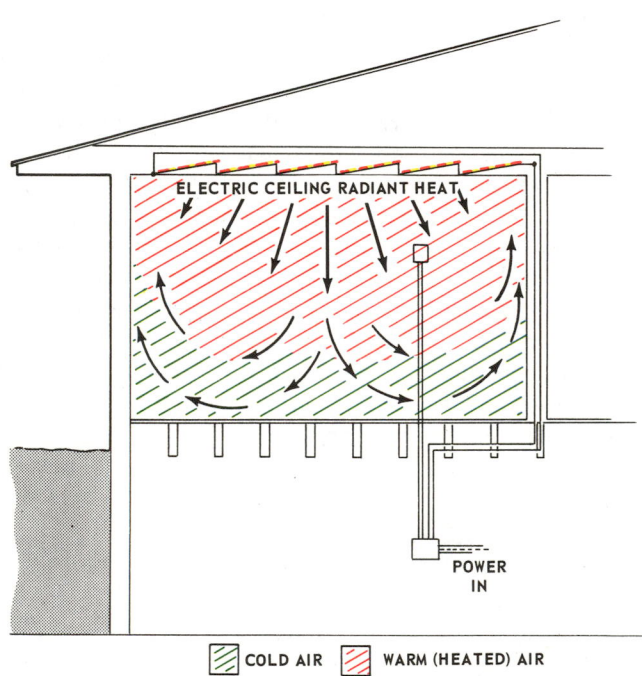

ELECTRIC CEILING RADIANT HEAT

POWER IN

▨ COLD AIR   ▨ WARM (HEATED) AIR

Fig. 19-24. Radiant heat is supplied by electrical resistance heating unit imbedded in the plaster on the ceiling.

19-25 REVIEW OF SAFETY

It is very important to be experienced (have some on-the-job training as an assistant to a service technician) before installing, maintaining or servicing air conditioning systems.

It is vital to know the regional and local legal codes and regulations in force before doing any type of work on air conditioning systems.

Whenever one works with equipment which has electrical circuits, fuels, vapor or liquids under pressure, safety is the first thing to be considered.

One should always use instruments to check the equipment. Only by knowing what the voltages are, what the pressures are, can one be reasonably certain of one's safety.

Avoid using any ignition source where there is any chance of a fuel being present. A friction spark, an electric spark (opening or closing switches) or a flame can ignite a mixture of fuel vapors and air which could be fatal and also damaging to property.

Chapters 20, 21, 22, 23, 24 and 25 all have specific safety precautions relating to air conditioning systems.

19-26 TEST YOUR KNOWLEDGE

1. Where is steam jet refrigeration most used?
2. What is the most common application of the heat pump?
3. In automobile air conditioning, how is the compressor usually driven?
4. What is the basic principle of operation of a room air conditioner installed in a window opening?
5. In automobile air conditioning, why is it necessary to use a magnetic clutch on the compressor drive pulley?
6. What changes from a liquid to a vapor in a humidifier?
7. Will a dehumidifier, operating in a room, change the temperature within the room?
8. In a heat pump installation, is the direction of flow of refrigerant vapor reversed through the compressor when the cycle is reversed from heating to cooling?
9. What is a common application of the vortex tube cooling system?
10. Why are some furnaces known as gravity type?
11. Can an air filter be used on a forced warm air heating plant?
12. What is a hydronic heating system?
13. What fluid action takes place to provide cooling in a cooling tower?
14. Why is an automatic draft control needed on an oil burner installation?
15. Name three common ways of heating the water in a hydronic heating system.
16. In a window or through-the-wall air conditioner, what becomes of the heat absorbed by the evaporator?
17. In a complete central air conditioning system, where is the condensing unit usually located?
18. In an air-to-air pump installation, where is the heat for warming the room obtained?
19. Why is it necessary to provide supplementary electric heating on some heat pump installations?
20. Is steam jet refrigeration recommended for use where temperatures below freezing are desired?
21. What is one of the chief advantages when using a cooling tower?
22. Where are the heating elements usually placed in radiant heating installations?

Chapter 20

AIR CONDITIONING SYSTEMS HEATING AND HUMIDIFYING

Air conditioning systems for heating and humidifying provide heat sources, heat distribution, heat control, filtering and cleaning equipment and humidifying devices.

The fundamentals of design and operation of the heat and humidifying air conditioning systems and their control are explained in this chapter. Filtering devices are explained in Chapter 22; controls and instruments in Chapter 24.

20-1 TYPES OF SYSTEMS

Air conditioning systems that are able to perform all air conditioning functions are not in common use. Most systems do only part of the job. The two most difficult jobs to be

20-2 TYPES OF HEATING AND HUMIDIFYING SYSTEMS

There are many types of heating equipment in use. Regardless of type, the heating source must be economical and safe. Systems which require a minimum of attention by the user are most desirable.

Sources of heat may be classified by fuels: oil, gas and electric including resistance, light (radiant) or heat pump. Wood and coal-fueled furnaces are becoming obsolete and are not covered in this text.

The three most commonly used heat sources as applied to furnaces are shown in Fig. 20-1.

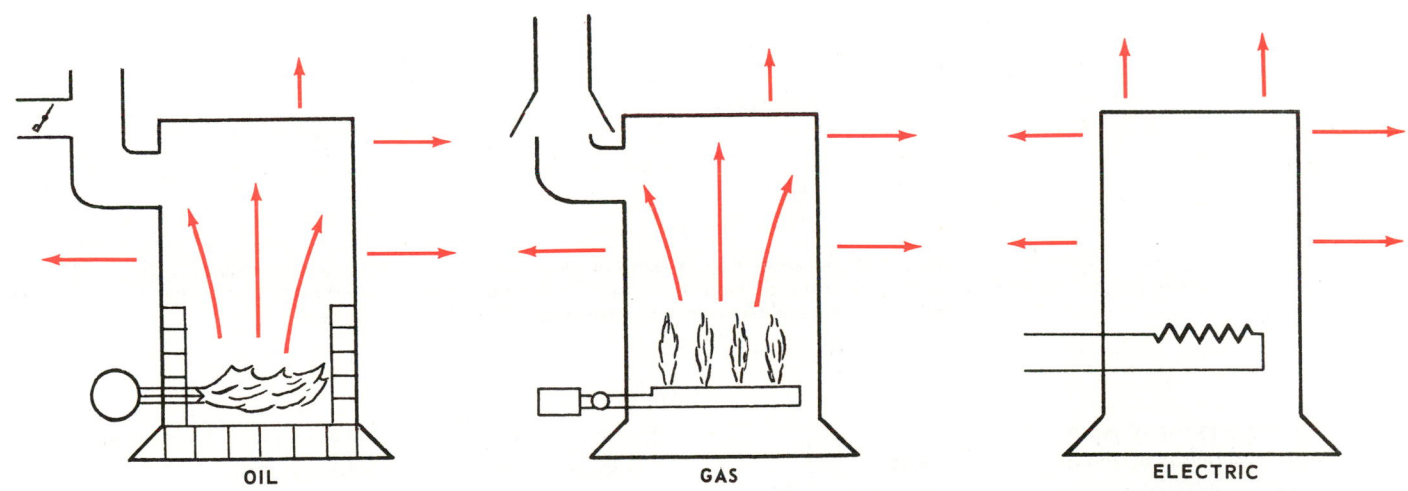

Fig. 20-1. Three popular types of heat sources used in furnaces.

performed are satisfactory air cleaning and humidity control.

Most so-called air conditioning systems are only partial air conditioning systems. That is, the system is designed for heating, humidifying, cleaning and distribution; or, it is designed for cooling, dehumidifying, cleaning and distributing. The operation of these systems usually is automatic.

Humidifying devices provide a means of turning water into water vapor and mixing this vapor with air in the occupied space. Fig. 20-2 shows six methods:
1. Exposing a large surface of water to air being humidified.
2. Spraying atomized water into air being humidified.
3. Water tray on radiator.

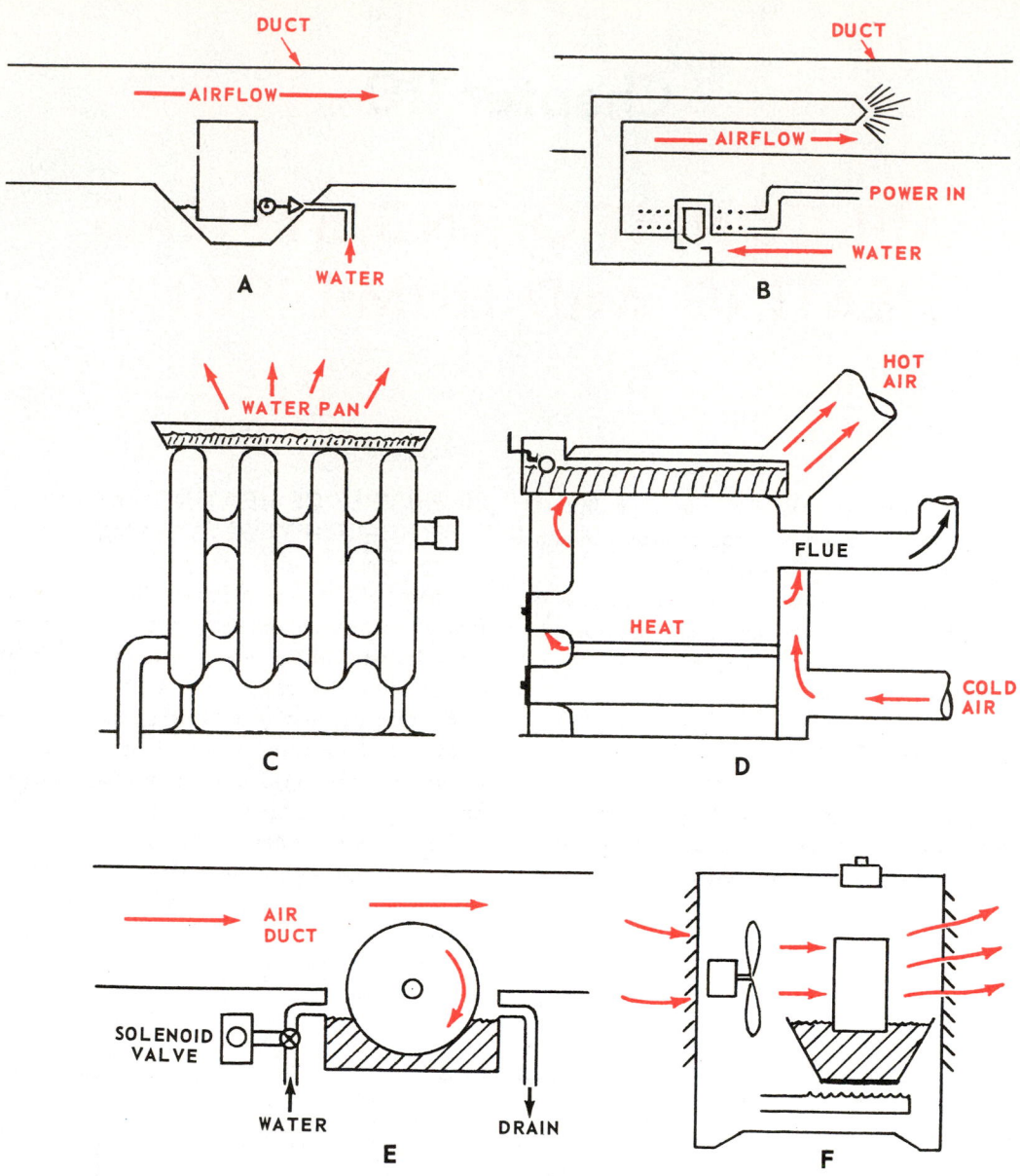

Fig. 20-2. Types of humidifiers. A—Open water tray in warm air duct. B—Spray nozzle in warm air duct. C—Water pan on radiator. D—Water pan in top of warm air furnace. E—Wetted revolving screen in warm air duct. F—Space humidifier (not part of heating system). Water is vaporized by electric heat.

4. Water tray in top of warm air furnace.
5. Wetted revolving screen in a warm air duct.
6. Space humidifier, not part of a heating system. Water is vaporized by electric heat.
   Heat energy may be distributed by:
1. Circulated air through ducts.
2. Warm water thermal circulation.
3. Hydronic systems; warm water circulated by means of a pump.
4. Steam lines and radiation.
5. Electric heat radiation from electric grids or infrared lights.
   Basics of these heating systems are shown in Fig. 20-3. Transferring heat to occupied space using steam or warm water

heating is shown in Fig. 20-4. All heating furnaces should have Underwriter's Laboratory (UL) approval.

20-3 HEATING — HUMIDIFYING EQUIPMENT

Complete air conditioning equipment may be divided into six main parts:
1. Heating equipment.
2. Humidifying equipment.
3. Filtering and cleaning equipment.
4. Circulating equipment.
5. Dehumidifying equipment.
6. Cooling equipment.

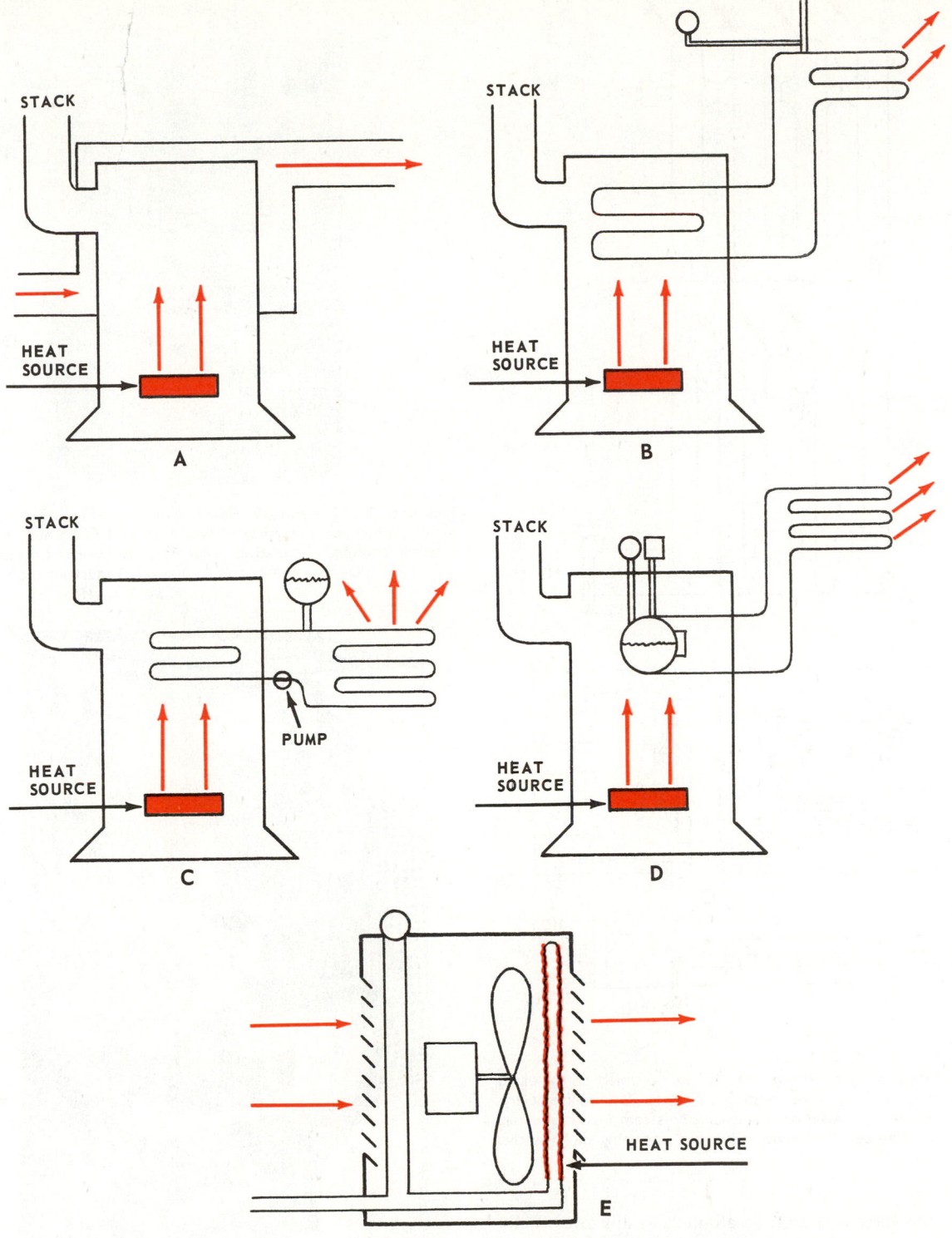

Fig. 20-3. Five basic types of heating systems. A—Natural convection warm air. B—Natural convection warm water. C—Forced convection warm water. D—Low-pressure steam. E—Electric resistance space heater.

During the winter or heating season, heating and humidifying equipment usually is used. Other equipment, such as filtering, cleaning and circulating equipment, is used the year round.

Complete air conditioning equipment may be of three different types:

1. Entire air conditioning plant may be located in a remote area and conditioned air circulated throughout the building through the use of ducts. See Fig. 20-5. A warm air furnace using an oil burner is shown in Fig. 20-6.

2. Heating may be located in a remote place, but pipelines are used throughout the building to carry the warm water or

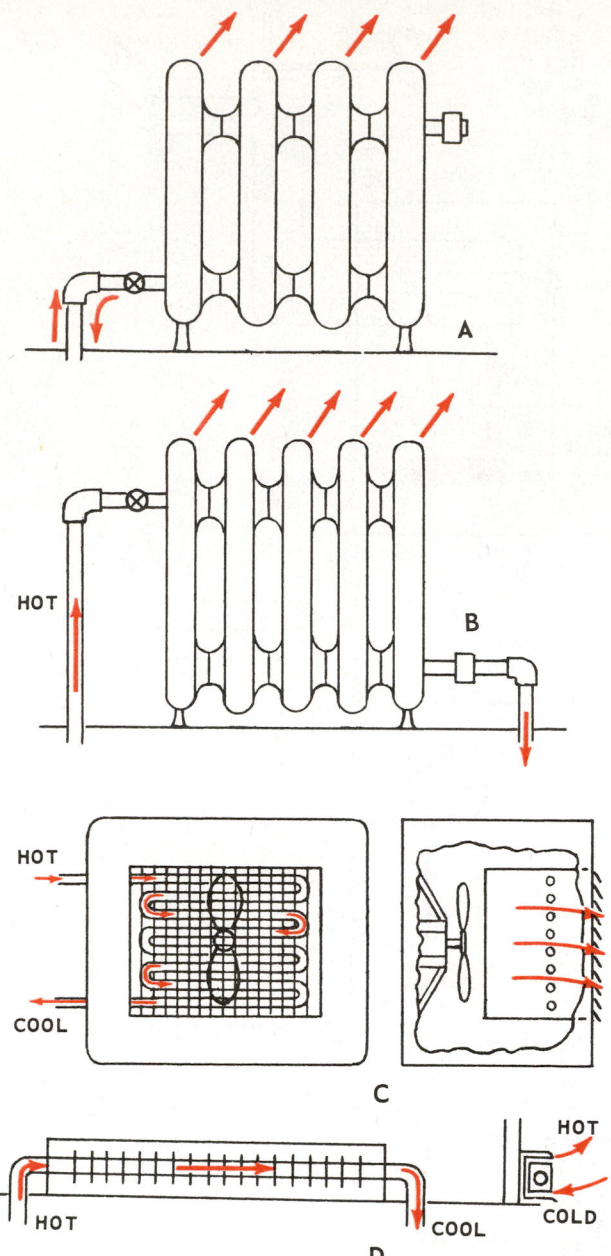

Fig. 20-4. Methods of transferring heat to an occupied space, using steam or warm water as heating medium. A—One-pipe radiator using steam or vapor heat. B—Two-pipe radiator using steam or warm water. C—Forced air heating coil. D—Baseboard convector using warm water.

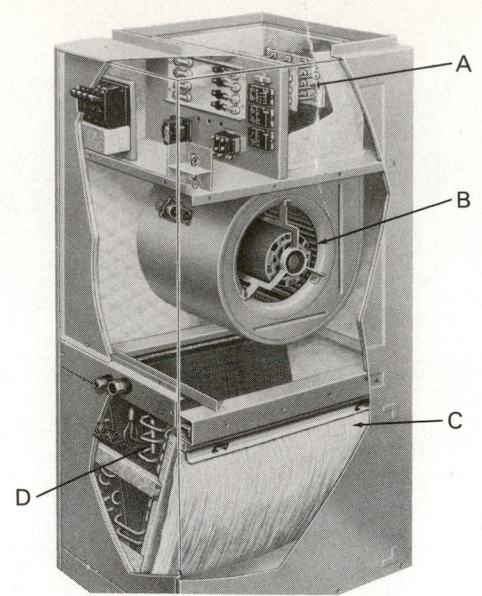

Fig. 20-5. Total electrical heating and cooling unit. Electricity in heating chamber provides winter heating. Evaporator coil provides summer cooling. A—Heating element. B—Blower and motor. C—Filter. D—Evaporator coil. (Lennox Industries Inc.)

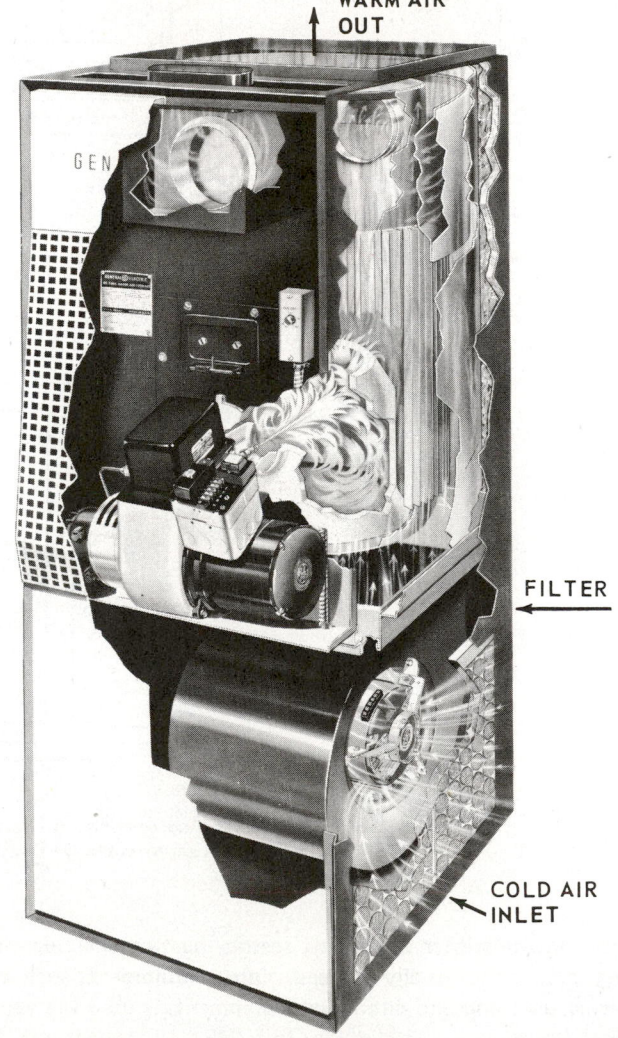

Fig. 20-6. Sectional view of warm air furnace using an oil burner. Note blower at cold air entrance at base of furnace. (General Electric Co.)

steam to the various rooms. Small ducts in this case usually provide ample ventilation and air circulation. Fig. 20-7 shows a hydronic furnace using a gun type oil burner.

3. Unit installations have a complete heating unit located in each room (unitary equipment). Such plants usually contain humidifying equipment, as well as heating and air circulating equipment.

20-4 HEATING EQUIPMENT

Most heating equipment is automatic in operation. Usually, the source of the heat is oil, gas, electricity or heat pump. The

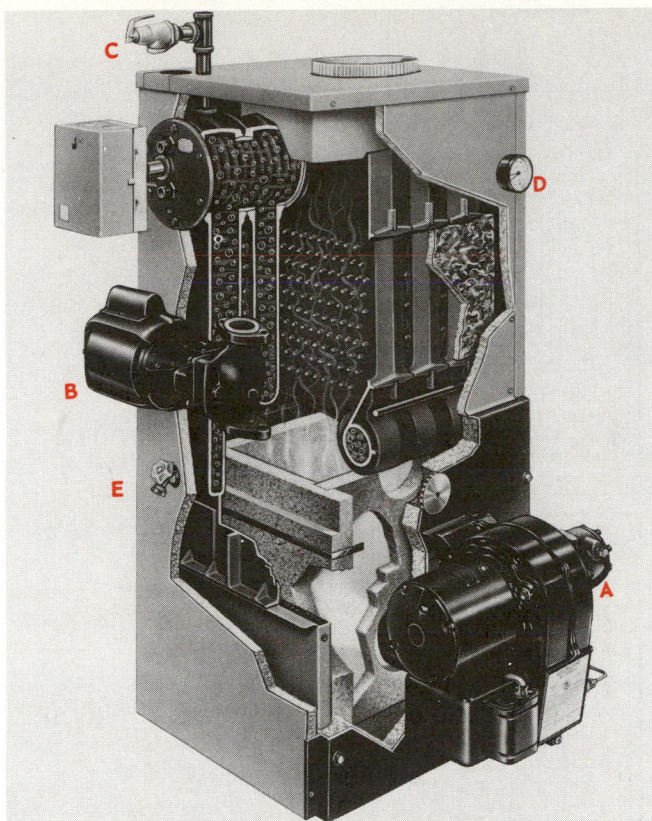

Fig. 20-7. Gun type oil burner used in a furnace which is part of a hydronic heating system. A—Oil burner. B—Pump. C—Pressure relief valve. D—Pressure gauge. E—Drain valve.

choice will be governed largely by the location and equipment available.

In systems having the central plant located in a remote place such as a basement, heat is circulated throughout the structure. With steam or water heating plants, radiators or heating coils, the structure may have room air conditioning units. In systems so equipped, each room unit does the necessary humidifying, air circulating and filtering.

The furnace in which the fuel is burned must be of safe design. Products of combustion must be vented outside the building to prevent health hazards. The assembly must be equipped with devices to close the unit down under the following circumstances:

1. If combustion is delayed or ceases.
2. If the furnace overheats.
3. If air, steam or water stops circulating.
4. When the heated space becomes warm enough.

The efficiency of a furnace is reduced if, during the "off" part of the cycle, the air which usually flows through the firepot to support combustion is permitted to travel into the furnace, through the combustion chamber, through the heat exchanger, through the flue and up the chimney. This air will tend to cool these parts to room temperature; then the furnace must reheat them during the next "on" part of the cycle. Forced draft or induced draft combustion air devices reduce this loss. Examples are gun type oil burners and forced convection (power) gas burners.

20-5 FURNACE DESIGN AND CONSTRUCTION

Furnace design is based on the fuel used and on the heat transfer medium (air, water or steam).

Furnace construction includes a combustion chamber (except electric furnaces), a fuel feeding and burning device, a flue and a heat exchanger. The combustion chamber must be leakproof. It must provide efficient heat transfer to the air or water. It must be able to change temperatures from room temperature to almost 2000 F. (1093 C.) with expansion or contraction stresses which may cause cracks.

The heat transfer surface must be large enough to remove enough heat from the combustion gases and from radiation to reduce the stack or flue temperature to an efficient level. The heat transfer surface usually consists of the combustion chamber surface and the heat exchanger. See Fig. 20-8. Some systems design the flue with the heat exchange surface. Some even have part of the flue installed in the cold air or cold water return to improve efficiency.

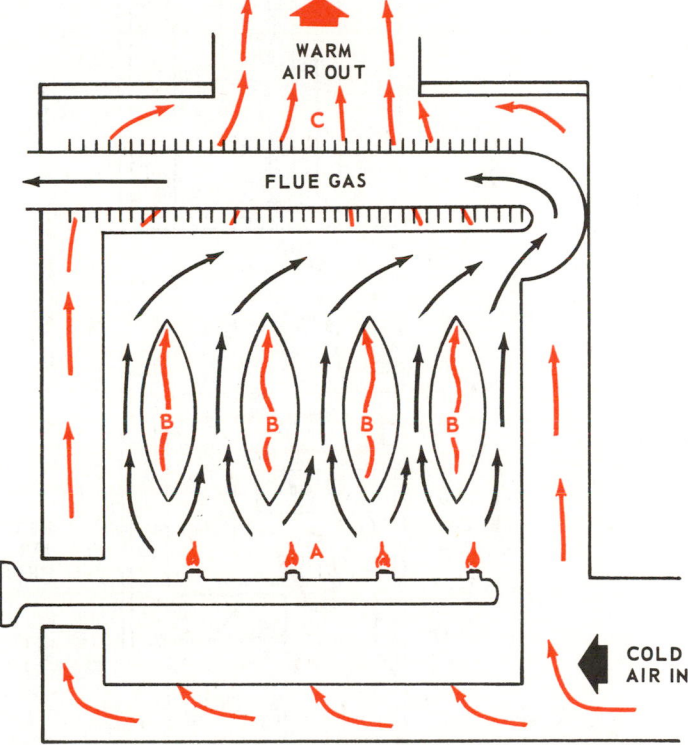

Fig. 20-8. A warm air furnace design. A—Combustion chamber. B—Clam shell-shaped heat exchanger. C—Flue heat economizer.

The design of the combustion chamber, heat exchanger flue and chimney is determined by the type of fuel used.

Heat exchangers have been made of cast iron. Today, most are made of 12 to 14 gauge steel with a ceramic coating on those surfaces contacted by the combustion gases. Fig. 20-9 shows some basic designs.

Some heat exchangers are made of stainless steel. In all cases, the heat exchanger must be designed to eliminate noise

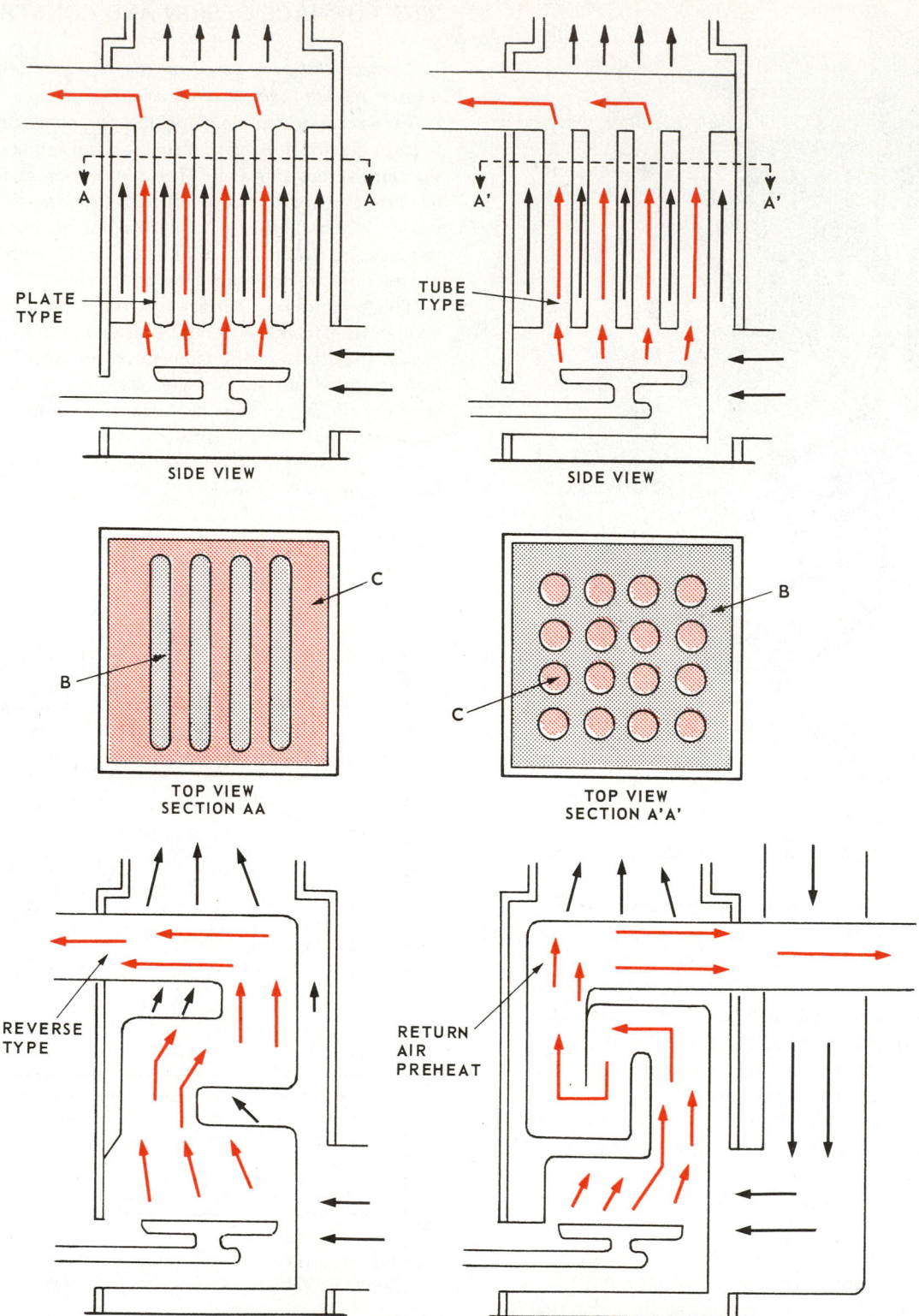

PLATE TYPE

SIDE VIEW

TUBE TYPE

SIDE VIEW

TOP VIEW
SECTION AA

TOP VIEW
SECTION A'A'

REVERSE TYPE

RETURN AIR PREHEAT

Fig. 20-9. Four types of heat exchangers for warm air furnaces. AA and A^1A^1—Section lines. B—Air. C—Flue gas.

as it expands and contracts. The unit must be corrosion resistant and rust resistant. Moisture from pilot light condensation and from a comfort cooling evaporator creates rust problems. A poorly placed or maladjusted humidifier may also expose the heat exchanger to moisture. If the burner is incorrectly placed, a hot spot may form on the heat exchanger and shorten its life. Most manufacturers ripple the metal and place the welds where expansion and contraction will not cause the heat exchanger to crack and leak (very dangerous because of fumes).

Poor airflow will cause a heat exchanger to overheat. A dirty filter, loose fan belts and/or undersize ducts will reduce airflow. Corrosive vapors in the circulated air also will increase the corrosion of furnace heat exchangers.

Furnaces are going through a great change due to technological advances and pollution reduction requirements. Most modern warm air furnaces have two-speed motors driving the blower that circulates air through the rooms: high speed for summer comfort cooling; slow speed for winter heating. Some motors are three and four-speed units to enable the service technician or owner to select the speed wanted for the air volume (cfm) desired.

The flow of combustion gases in a flue and chimney has considerable effect on the efficiency of a heating system. Fuel losses of 4 to 15 percent are possible, plus air pollution.

The flow of combustion gases affects the amount of air entering the furnace for combustion purposes. This flow is affected by the difference in pressure between the combustion air entering a furnace and combustion air leaving the flue or chimney. Both the pressure in the building and atmospheric pressure affect the flow of combustion gases. The temperature of the combustion gases also has an effect (too cold — slow flow, too hot — fast flow).

A draft control device is used to keep the flue pressure constant. With this device, it is possible for a heating system to maintain proper combustion air, even with changes in atmospheric pressure. The pressure may change in the building if the flue temperature changes or if the wind changes the flue pressure dynamically. The draft control improves the flow of flue gases by varying dilution air into the flue as the flue pressure tries to change. The basic design of a draft control is shown in Fig. 20-10. A draft control for an oil burner is shown in Fig. 20-11.

Combustion gas vent pipes (flues) should always be the same size as the furnace vent opening. Horizontal length

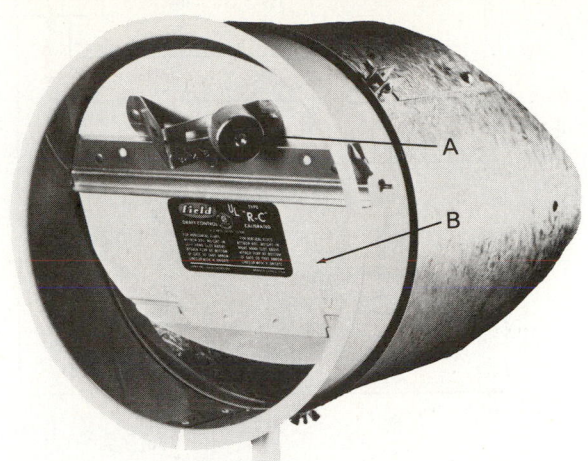

Fig. 20-11. Draft control for oil or coal-fired furnaces and boilers. A—Adjustable weight for low, medium or high draft. B—Gate. (Field Control Div. of Heico, Inc.)

should not exceed 20 ft. (6 m), and it is best to have it as short as possible. Any horizontal run should slant up (pitch) toward the chimney 1/2 in. (13 mm) for each foot of pipe to improve flue gas flow. Minimize the use of elbows as they add resistance to flow.

The chimney or vent pipe should extend about 2 ft. (.6 m) above the highest roof part.

A typical chimney with a 450 F. (232 C.) temperature average, and 40 F. (4 C.) outside temperature will produce .14 in. of water draft pressure if the chimney is 30 ft. (9 m) high (measured from floor of furnace room).

The draft control must be located so that the flue gases will not flow against it. See Fig. 20-12. All controls must be located between the furnace and the draft control.

Gas furnaces use an air break to control flue gas flow and to prevent back pressure from a gust of wind from reaching the furnace flame or pilot light. Fig. 20-13 shows an air break installed on a gas furnace. The construction of gas furnace control used on some furnace installations is shown in Fig. 20-14.

Make very sure that there is no leak from the furnace flue or chimney into the building. Carbon monoxide is deadly, and carbon dioxide reduces oxygen in the house. Check the flue draft control to be sure the air is entering the flue. One safe way to do this is by hanging thin strips of light paper by the flue opening. If the paper bends into the flue, the draft control is good. If the paper bends away from the flue, the gases are entering the house. Air out the building at once. Determine the cause of the back flow (partially blocked chimney, etc.,). Repair at once.

If the flue gas flow is too slow, a flue or smoke-pipe booster fan can be installed near the chimney. These fans run only when the furnace is operating.

20-6 FUEL OILS

Fuel oils vary considerably. Generally, they contain about 85 percent carbon (C), 12 percent hydrogen (H) and various

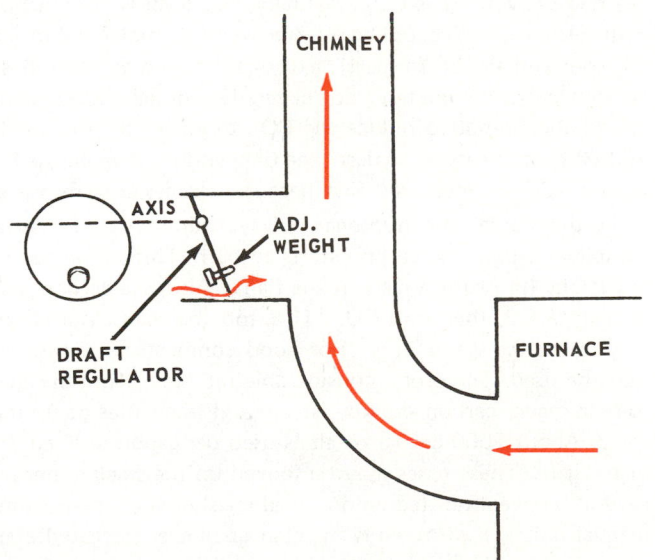

Fig. 20-10. Schematic of draft regulator as used on coal, coke and/or oil furnaces. Valve will open if draft tends to increase and will close as flue stack draft decreases, thereby keeping a constant draft in the furnace.

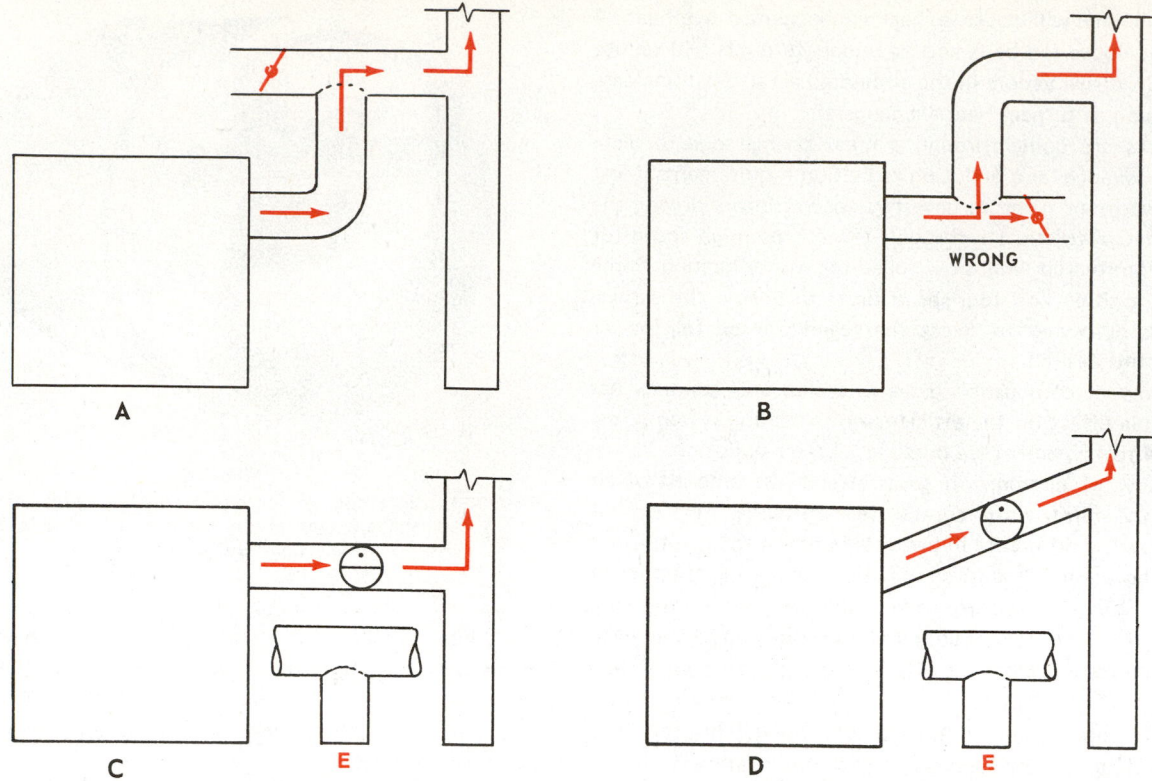

Fig. 20-12. Locating draft control in a flue. A—Good location. B—Wrong location (hot gases flow against draft control).
C and D—Designs for oil furnaces. E—Top views of C and D.

other elements in the remaining 3 percent. During combustion, carbon and hydrogen combine with the oxygen (O) in the air to produce carbon dioxide (CO_2) and water (H_2O). Fuel oil grades which are established by the U.S. Department of Commerce conform to ASTM specifications. A very low sulphur(S) content is very important as the sulphur turns into corrosive gases and liquids.

Fuel oil grades 1 and 2 are used in domestic and small commercial furnaces. Grade 1 is used in pot type oil burners; Grade 2 is the most popular domestic fuel oil (about 140,000 Btu/gal.). It has a flash point of 100 F. (38 C.), a Sayboldt viscosity of 40 (compared to Grade 4 with a viscosity of 125. Grades 4, 5 and 6 are used in industrial applications. The heavier oils 3, 4, 5 and 6 provide slightly more heat per gallon.

Products of combustion should be carbon dioxide (CO_2) and water in vapor form. Actually, there is also carbon monoxide (CO) (very toxic — dangerous), plus sulphur dioxide (SO_2) and sulphur dioxide vapors. About 106 lb. of air are required for each gallon of Grade 2 fuel oil consumed. Multiplying 106 x 14 (cu. ft. per lb.) equals about 1500 cu. ft., which is the quantity of air that must be fed to a furnace for each gallon of fuel oil consumed. This means that 1500 cu. ft. of air must enter the building for each gallon burned (about each two hours of oil burner running time for the average home).

It takes about 104 cu. ft. of air (7.43 lb.) to burn 7 lb. of oil (1 gal.). The 7 lb. of oil in one gallon are made up of 6 lb. of carbon and 1 lb. of hydrogen. The 106 lb. of air equal about 1500 cu. ft. of air. Of the 106 lb. of air, about 84 lb. are

nitrogen. Nitrogen does not produce burning, but acts as a gas to lower temperature and to waste heat as it is warmed and goes up the chimney.

Of the 106 lb. of air, about 22 lb. are oxygen (O) that combine with the oil to form about 20 lb. of CO_2 and 9 lb. of water (steam).

When perfect combustion takes place, about 15 percent of the flue gas volume is CO_2. Actually, this level is not reached with oil burners. Because of the heavy carbon molecules in the oil (soot and smoke formers), excess air is used to burn these carbons more completely. So, generally, enough excess air is fed to the firepot to reduce the CO_2 to about 10 percent. If 100 percent excess air is used, the CO_2 content reduces to 7.5 percent. CO_2 is measured first because it is easier to measure.

In the flame, the hydrogen always burns first. It burns completely, then the carbon starts to burn. This action causes pulsations (pressure waves) in the flame. Also, the carbon first turns into CO, then into CO_2. This, too, may cause pulsation.

Combustion gases vary. For good combustion, excess air must be used. Therefore, considerable nitrogen (from the air), some oxygen, carbon dioxide, steam and impurities go up the stack. About 2000 cu. ft. of air is used per gallon (400 cu. ft. of oxygen). These gases may be moved up the stack either by natural convection (common in domestic and small commercial units), by forcing with a fan or blower (forced draft), or by drawing the gases up the chimney (induced draft).

It is important to keep flue gases warm. Otherwise, condensation will take place in the stack and flue, causing severe corrosion. One corrosion agent will be sulphurous acid

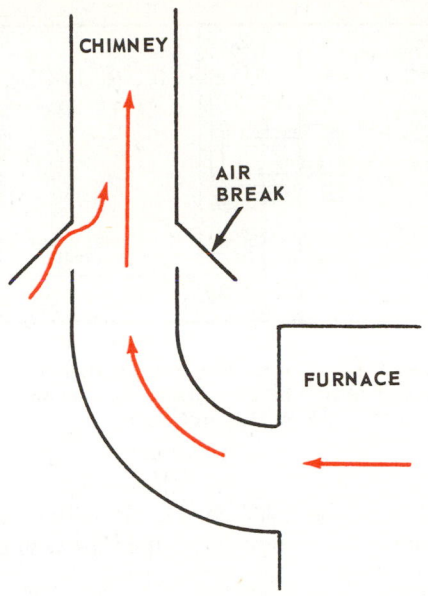

Fig. 20-13. Air break system used on gas furnace flues (stacks) to maintain a constant draft in furnace.

Fig. 20-14. Gas furnace draft control mounted on the furnace flue. A—Amount of flue draft is adjusted by washer weights on the chain. (Field Control Div. of Heico, Inc.)

(H_2SO_3), which corrodes steel rapidly and discolors brick and stone. Most good fuel oil additives will: reduce sulphur dioxide about 50 percent; keep the flue and chimney cleaner (65 percent cleaner); cause less opaque (visible) smoke; reduce soot blowing of tubes. If there is incomplete combustion, the flame is white rather than the normal orange color (more complete combustion).

An oil furnace in good condition should not release visible smoke from the flue, chimney or stack. However, there may be soot deposits and fly ash, which should be removed annually. The soot may be removed by using air pressure, mechanical cleaning, vacuum cleaning or chemicals.

Oil sludge which clogs filters and nozzles may be caused by bacteria which multiply if water is present in the oil. An additive can be used to kill the bacteria.

The combustion chamber must be kept in good condition. Deposits on the refractory lining must be kept to a minimum.

Fuel oil additives reduce deposits in the combustion chamber, heat exchangers and flue.

Sulphur trioxide is more odorous than sulphur dioxide and is minimized by using additives and a higher temperature.

Number 2 fuel oil viscosity changes from between 50 and 100 at 0 F. (—18 C.) to 35 and 45 at 70 F. (21 C.). This means that gun type oil burners may have pumping and combustion problems when the oil is cold.

Efforts to reduce air pollution have increased the use of Number 2 fuel oil distillate in commercial buildings and in industry. This grade burns more completely and cleaner than Number 3, 4, 5 and 6 fuel oils. Some fuel oils are called distillate, because they are products of a distillation or cracking process at the oil refinery. The Number 1 and Number 2 oils are called distillates because they are oils which were vaporized, then condensed in the refining system.

20-7 OIL FURNACES

The most common type of oil burner is the gun type. Rotary type and pot type oil burners are rarely used.

The gun type burner forces oil under pressure through orifices of definite size. The oil is broken into finely divided particles (atomized), mixed with air and forced into the combustion chamber by a blower. Fig. 20-15 shows a gun type

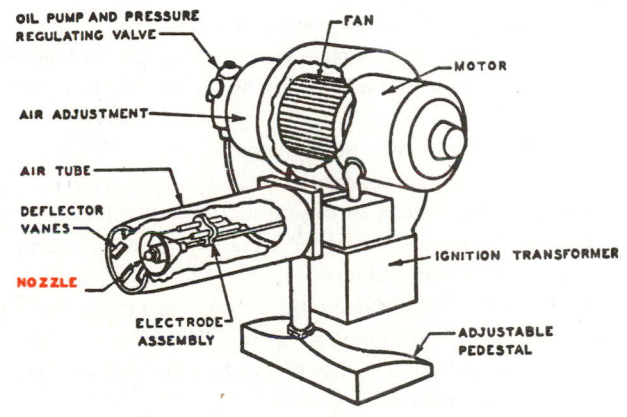

Fig. 20-15. Gun type oil burner. Oil is atomized as it leaves nozzle of burner. (ASHRAE Guide and Data Book)

oil burner. A cutaway of a pedestal-mounted gun type oil burner is shown in Fig. 20-16.

Remember that oil will not burn while it is in the liquid form. It must be vaporized and turned into a gas. To vaporize oil, heat must be added to the oil (latent heat of vaporization). The oil will turn into a gas quicker and easier if it is finely divided (sprayed). This spraying process is called atomizing. Gun type oil burners accomplish atomizing by forcing oil into a twisting, spiraling and turbulent air stream. A small amount of heat (electric spark) will turn a few of the finely divided particles into gas and the burning will start.

Fig. 20-16. Cutaway of a pedestal-mounted gun type oil burner. A—Motor. B—Oil pump. C—Nozzle. D—Pedestal adjustment. E—Combustion air blower. (The Carlin Co.)

Some large industrial furnaces use combination oil and gas burners.

Pulsation in an oil furnace is usually caused by positive pressure in the combustion chamber (not enough draft). Draft should be .02 in. water pressure. Too much positive pressure may be caused by:

1. Too much air (air shutter open too much).
2. Chimney is too small, partially blocked or not high enough (2 ft. above everything on building within 10 ft. subject to downdraft).
3. A faulty nozzle (distorted flame pattern).

Oil on the floor of the furnace room is dangerous. It may be caused by an air leak in the oil suction line (air causes drip at nozzle). It may be caused by loose compression fittings or unions, or by a pump seal leak.

Check for air in the system by connecting a pressure gauge. If the gauge needle fluctuates, it is sign of air. For small leaks: put oil outlet tube in a bottle of oil; bubbles will indicate an air leak in the suction line.

If there is soot in the boiler flue passages:
1. Clean.
2. Clean blower blades with a brush; clean blower tube.

An oil furnace blowback is usually caused by delayed ignition. The most common reasons for blowback are:
1. Electrodes improperly spaced.
2. Distorted pattern away from electrodes.

If the oil nozzle is in poor condition, replace it with an exact replacement unit (same orifice, spray angle and solid or hollow cone as originally used).

Large oil burner furnaces may have a metal combustion chamber. Smaller furnaces use refractory cement liners for the combustion chamber. This cement consists of 80 percent dry asbestos, 20 percent Portland cement and enough water to make mixture workable. Avoid putting this cement on edge of

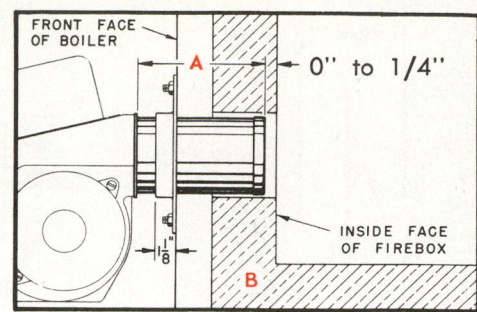

Fig. 20-17. Typical gun type oil burner installation in a boiler. A—Length of air tube. B—Refractory insulation. (R. W. Beckett Corp.)

or inside of air cone (will change air pattern and cause inefficient burning). It is best to fill the tube with a rag while applying the cement.

Many refractory liners are preformed, then installed. Fig. 20-17 shows a burner installed in a boiler.

20-8 GUN TYPE OIL BURNER

Gun type oil burners are available in two types:
1. High-pressure type.
2. Low-pressure type.

In the high-pressure type, oil is fed under 80 to 100 psi, to a nozzle. Air is forced into the furnace through a tube that surrounds this nozzle. Usually, the air is twisted in one direction; the oil spray is given a twist in the opposite direction. Fig. 20-18 shows a gun type oil burner. The nozzle should be carefully centered in the housing. The ignition transformer furnishes a high-tension spark between two electrodes located near the front of the nozzle. Fig. 20-19

Fig. 20-18. Flange mounted gun type oil burner. A—Oil pump. B—Air adjustment. C—Motor. D—Transformer. (R.W. Beckett Corp.)

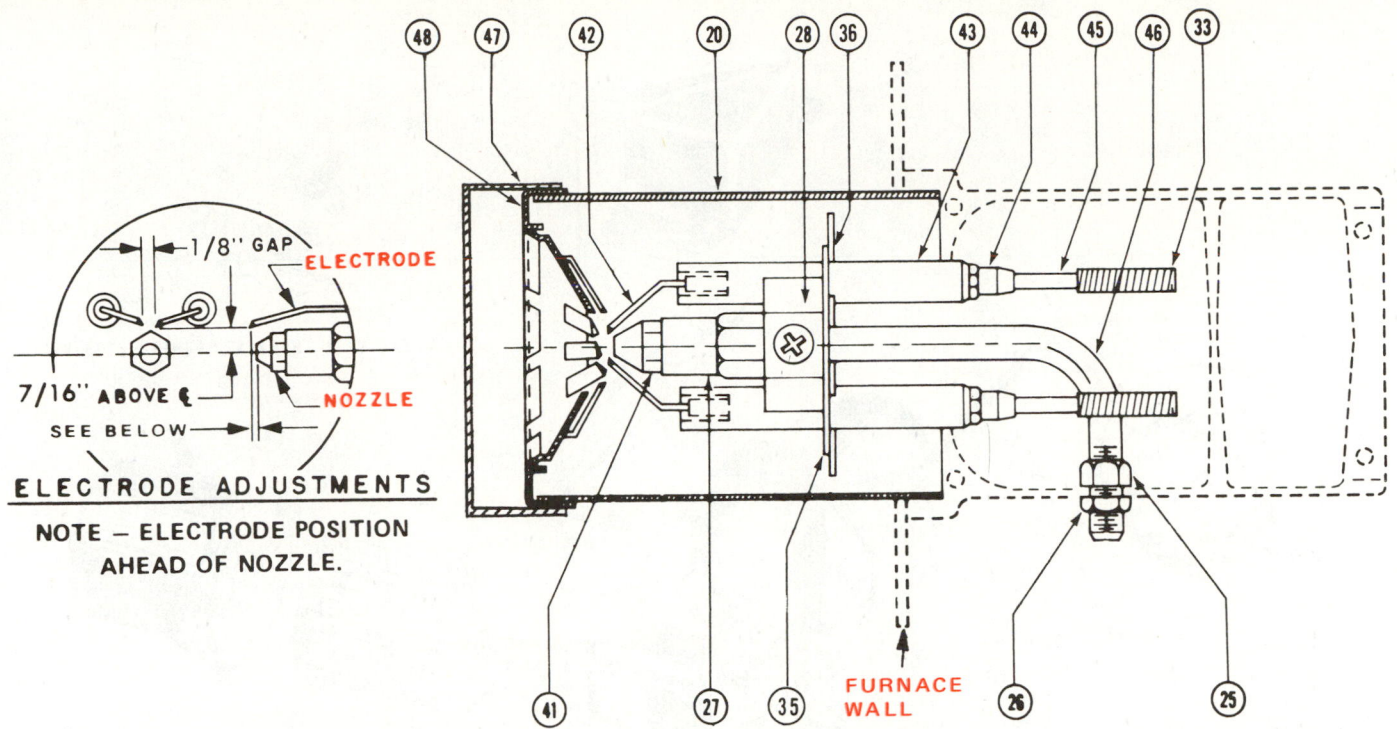

Fig. 20-19. Design of the air tube, oil nozzle and electrode assembly of a gun type oil burner. 20—Air tube. 25—Nozzle line fitting (pump end). 26—Locknut, nozzle line fitting. 27—Nozzle adaptor—single. 28—Electrode clamp. 33—Contact springs, static plate and nozzle line, support assembly. 35—Centering spider. 36—Static plate, static plate holding screws. 41—Nozzle. 42—Electrode rod and tip. 43—Porcelain. 44—Electrode rod extension adaptor, as required. 45—Electrode rod extension, as required. 46—Nozzle line and vent plug. 47—Burner head. 48—Choke assembly.   (R.W. Beckett Corp.)

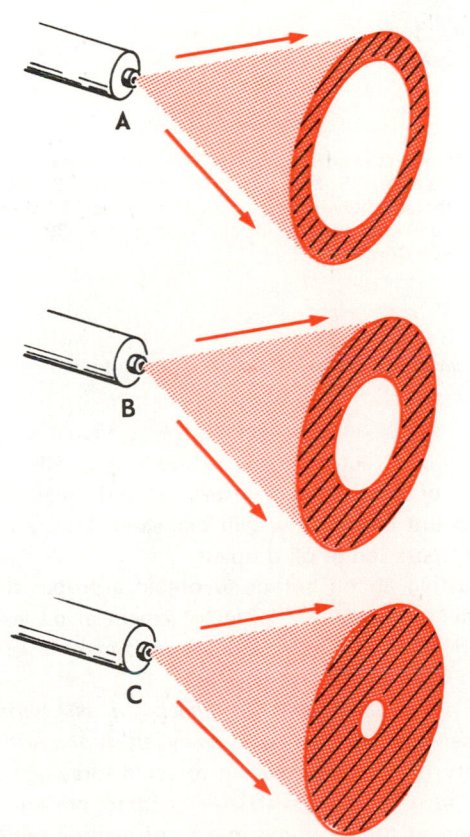

Fig. 20-20. Three different oil nozzle spray patterns. A—Hollow cone (most popular). B—Semihollow cone. C—Solid cone.

shows the nozzle and electrode assembly. Nozzle spray patterns are shown in Fig. 20-20.

The low-pressure type burner uses oil at 1 to 4 psi. Oil is mixed with air before it reaches the nozzle.

The main parts of a gun type burner are: motor, oil pump, fans, nozzle, choke, air tube and ignition system. Fig. 20-21 shows the various parts of a gun type oil burner.

The choke is a tapered down or smaller opening at the end of the air tube. It is located just past the oil nozzle. The choke has swirl strips (vanes) to increase the twist and turbulence of the air to obtain better mixing of the oil spray and air for more efficient burning. The flame shape can be changed by moving the choke closer to or away from the nozzle.

The air tube has a disc mounted inside it. This disc disturbs the airflow and creates air turbulence for better mixing. This disc is usually called the static pressure disc.

The oil moving through the nozzle is also given a twisting movement as it travels through very small holes drilled at an angle to the nozzle.

The oil burner motor is usually a split-phase 1/6 hp unit that provides power for both the fan and the fuel pump. The motor is electrically connected to the master oil burner stack control and uses 120V 60 cycle electricity. Fig. 20-22 shows a flange-mounted gun type oil burner.

Motor speeds may be 1750 rpm (60 Hz) fan speed and pump speed, or 3450 rpm (60 Hz) fan speed and pump speed, depending on whether the motor is a 4-pole or 2-pole type. On a 50 Hz, a 4-pole motor will run at 2850 rpm.

The fan is usually the radial flow type with adjustable air

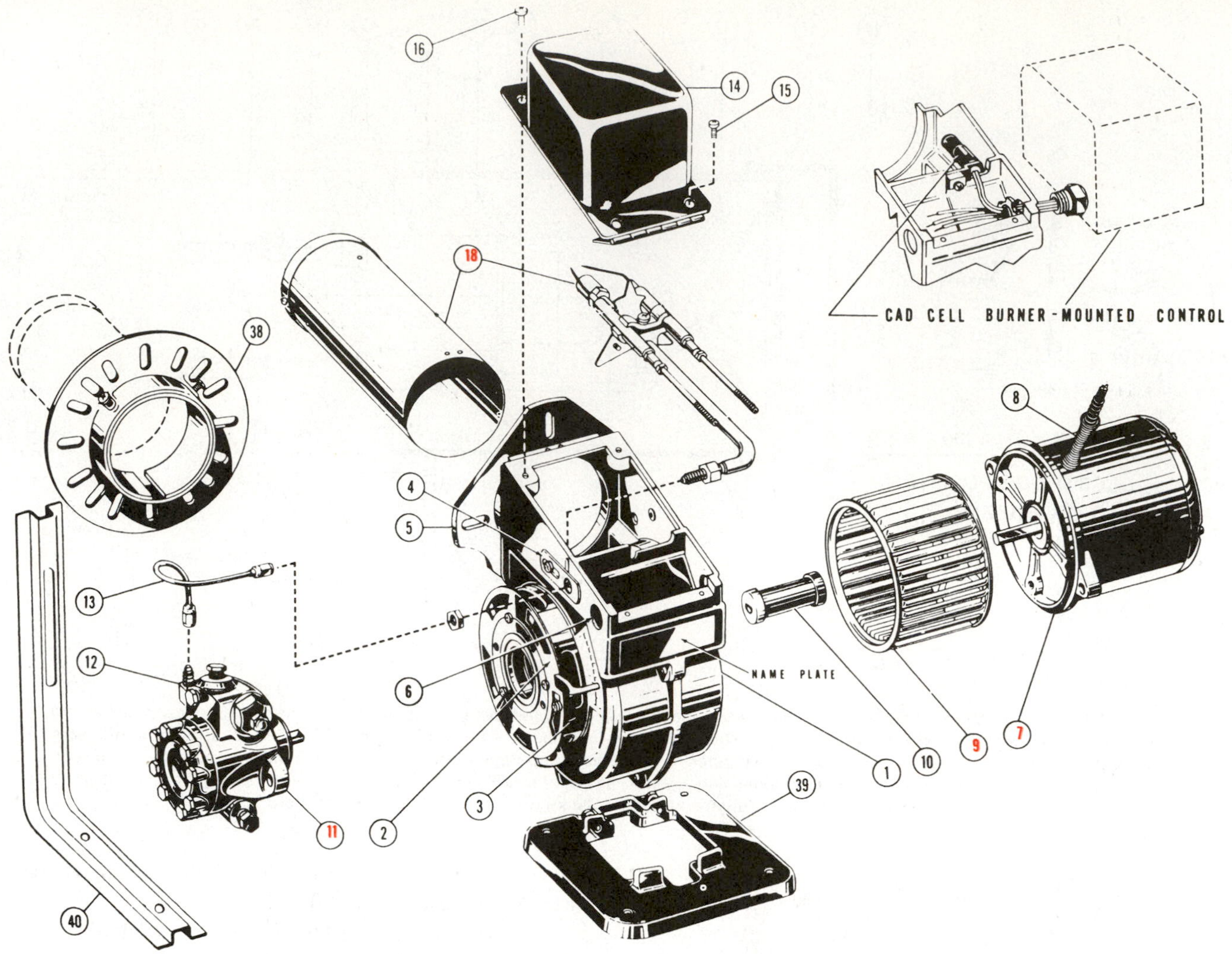

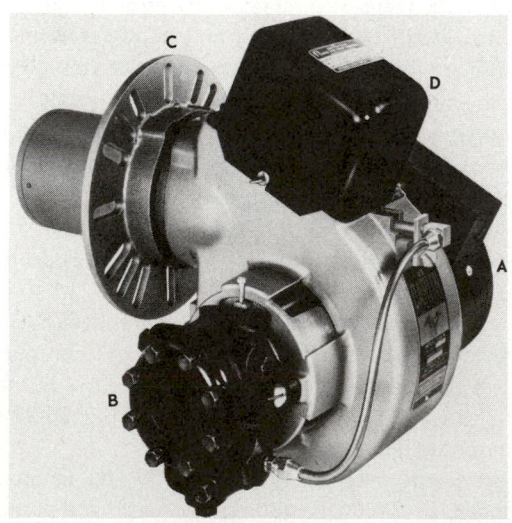

Fig. 20-21. Exploded view of a gun type oil burner. 1—Burner housing with inlet bell. 2—End air shutter. 3—Bulk air band. 4—Nozzle line escutcheon plate. 5—Unit flange or square plate. 6—Hole plug – wiring box. 7—Drive motor. 8—Motor lead and wire guard. 9—Blower wheel. 10—Flexible coupling. 11—Fuel pump. 12—Pump outlet fitting. 13—Connector tube assembly. 14—Ignition transformer. 15—Hinge screws. 16—Holding screws. 17—Contact spring terminals (not shown). 18—Air tube and electrodes. 38—Air tube flange. 39—Burner base. 40—Mounting bracket. (R. W. Beckett Corp.)

Fig. 20-22. Flange-mounted gun type oil burner. A—Motor. B—Oil pump. C—Flange mounting. D—Transformer. (The Carlin Co.)

inlet openings. The openings are adjusted until the flame burns a yellow color.

Some excess air is needed to insure enough oxygen for the oil (the flame action is fast!) and also to allow for airflow decrease between furnace inspection and cleaning (fan blades pick up lint and airflow will decrease). Excess air will slow down evaporation of oil droplets.

Adjusting an oil burner to obtain a proper flame can be misleading. A dirty nozzle, impingement or oil leakage during the "off" part of the cycle will tend to make a flame look like it needs more air when it does not.

The only good way to check an oil burner is with instruments (O analyzer, CO_2 analyzer, smoke test, etc.).

Safety devices are installed to avoid spraying unburned oil into a furnace **(DANGEROUS)** and to prevent continuous operation of the oil pump in case of ignition failure or if the oil flame goes out. The stack control is one method. See Chapter 24 for oil burner controls and wiring circuits.

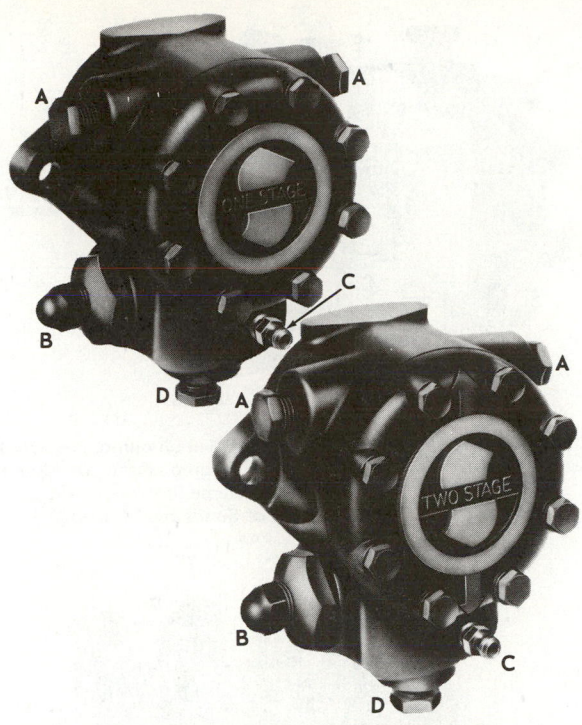

Fig. 20-23. Gun type oil burner fuel pumps. Both single-stage and two-stage pumps are shown. A—Oil inlet connections. B—Pressure regulator. C—Air bleed and gauge connection. D—Oil outlet connection.

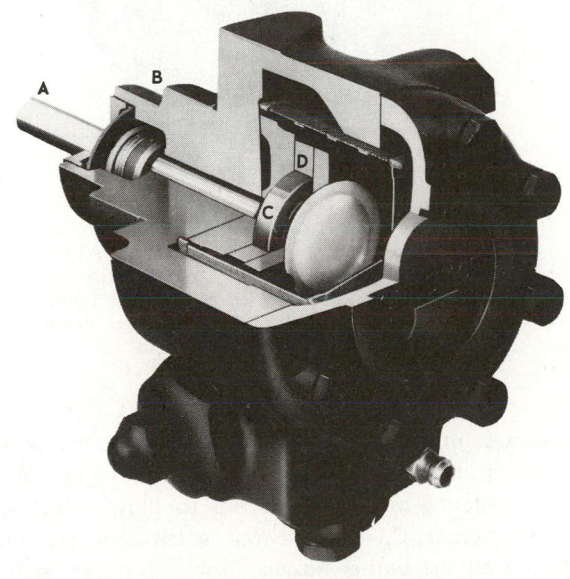

Fig. 20-24. Single-stage rotary fuel pump for gun type oil burners. A—Shaft. B—Shaft seal. C—Pump rotor. D—Pump housing. (Sundstrand Hydraulics, Div. of Sundstrand Corp.)

20-9 GUN TYPE OIL BURNER PUMPS

Several types of oil pumps are used in gun oil burners, including the gear type and the rotary type. These pumps come in either single-stage or two-stage models. See Fig. 20-23. The internal construction of a single-stage fuel oil pump is shown in Fig. 20-24.

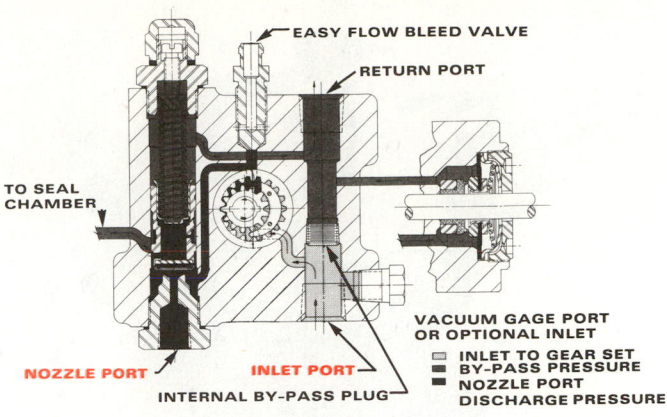

Fig. 20-25. Schematic diagram of oil flow through single-stage oil pump. Excess oil delivered by pump escapes past pressure-controlling piston and returns to pump through intake strainer.

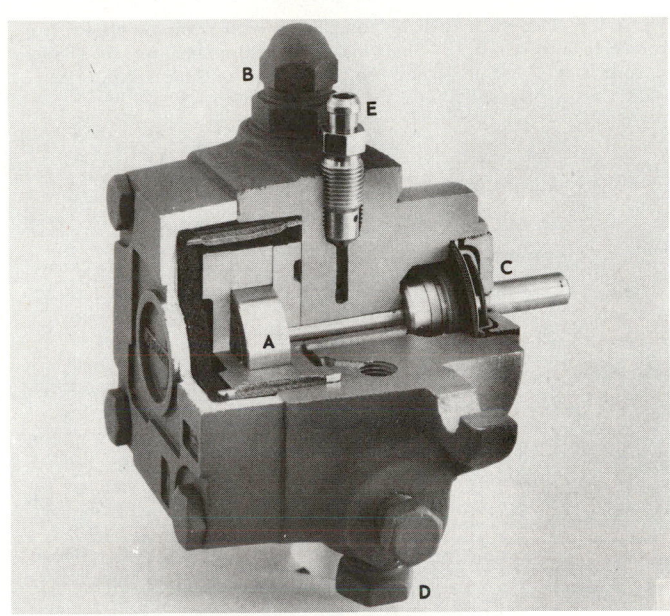

Fig. 20-26. Single-stage rotary fuel oil pump for oil burners. A—Rotor. B—Regulator. C—Shaft seal. D—Oil inlet connection. E—Air bleed and pressure gauge connection.

The oil supply system carries fuel oil from the tank through a filter in the line, through the inlet screen in the oil burner, to the pump and into the pressure regulator and relief valve. The pump rotates counterclockwise and oil flow is from left to right. The oil leaves the upper center of the pressure regulator and passes to the gun nozzle, as shown in Fig. 20-25.

Another design for a single-stage rotary fuel oil pump is shown in Fig. 20-26.

Many systems are equipped with the two-stage fuel oil pump when the two-pipe system is used and part of the oil is returned to the fuel tank. This type of pump is necessary where the pump has to lift oil, even a few inches, above the bottom of the fuel tank. Fig. 20-27 shows a two-stage oil pump. Its principle of operation is shown in Fig. 20-28. The intake from the tank is at the top. Oil passes through the first

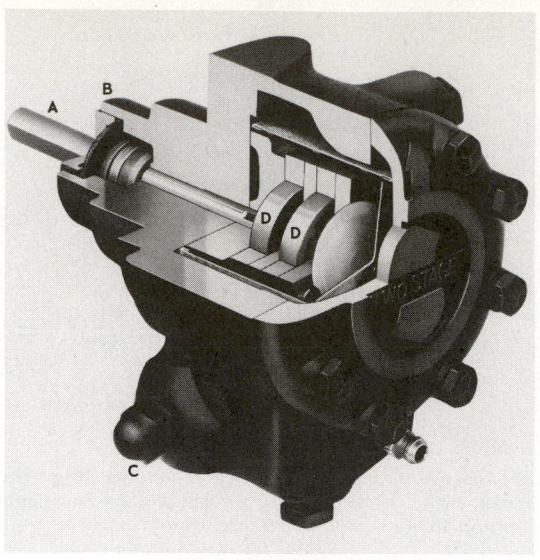

Fig. 20-27. Two-stage fuel oil pump used with two-pipe system from storage tank. A—Shaft. B—Shaft seal. C—Pressure regulator. D—The two rotor stages. (Sundstrand Hydraulics, Div. of Sundstand Corp.)

Fig. 20-29. Cutaway view of relief valve for fuel oil pump. A—Outlet to nozzle. B—Pressure regulating screw. C—Pump shaft. D—Mounting flange. E—Oil pressure release line. (Sundstrand Hydraulics, Div. of Sundstrand Corp.)

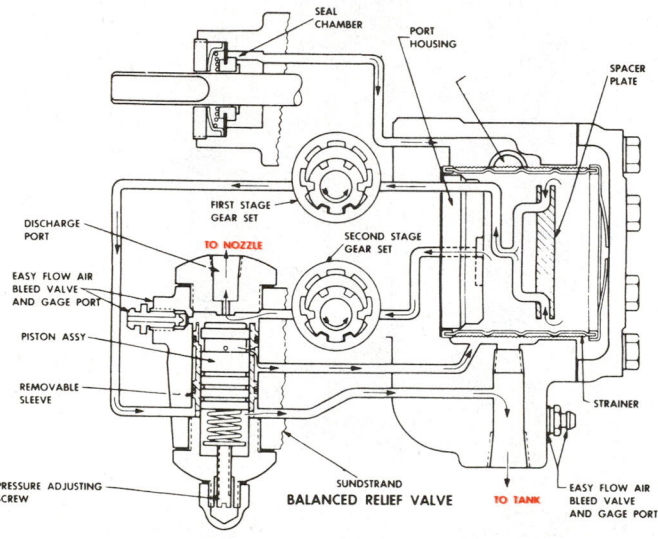

Fig. 20-28. Two-stage fuel oil pump for oil burners, used with two-pipe system from storage tank.

Fig. 20-30. Single-stage gear pump for gun type oil burners. A—Inlet. B—Nozzle connection. C—Pressure gauge connection. D—Pressure adjustment.

stage of the pump, into the regulator and back to the tank. The second stage removes only oil from the strainer chamber and pumps it into the nozzle. Excess oil is returned to the strainer chamber.

Details of the relief valve are shown in Fig. 20-29. Oil pressure creates a force against the piston. When this force equals the compression spring force, the piston moves down and permits oil to flow back into the pump inlet.

The gear type oil pump is available in both single-stage and two-stage models. Fig. 20-30 shows the external appearance of a single-stage gear pump. Single-stage pump operation is shown in Fig. 20-31. Fuel oil enters through inlet into strainer D, flows upward through silencer orifice O into vacuum chamber

A. It passes into the gear pump, then enters the pressure equalizing chamber from which oil flows to the nozzle. The pressure regulating piston valve opens to allow excess oil to flow into bypass C. In this way a constant pressure is maintained on the burner nozzle. Cushions, shown at B, are used to insure more even oil flow.

The nozzles generally are the 80 deg. hollow cone type of .75 to 1.75 gph (gallons per hour) capacity, or the 60 deg. hollow cone type of 1.75 to 12.00 gph (1 U.S. gallon = 3.79 litres). To help prevent oil dripping from the nozzle during off cycles, some burners have a solenoid valve shutoff in the oil line to the nozzle. See Fig. 20-32.

The gun type oil burner is an efficient heating unit. However, it must be properly maintained to give peak performance. An experienced service technician should check, clean and adjust the system each year. Some of the important items to check are shown in Fig. 20-33.

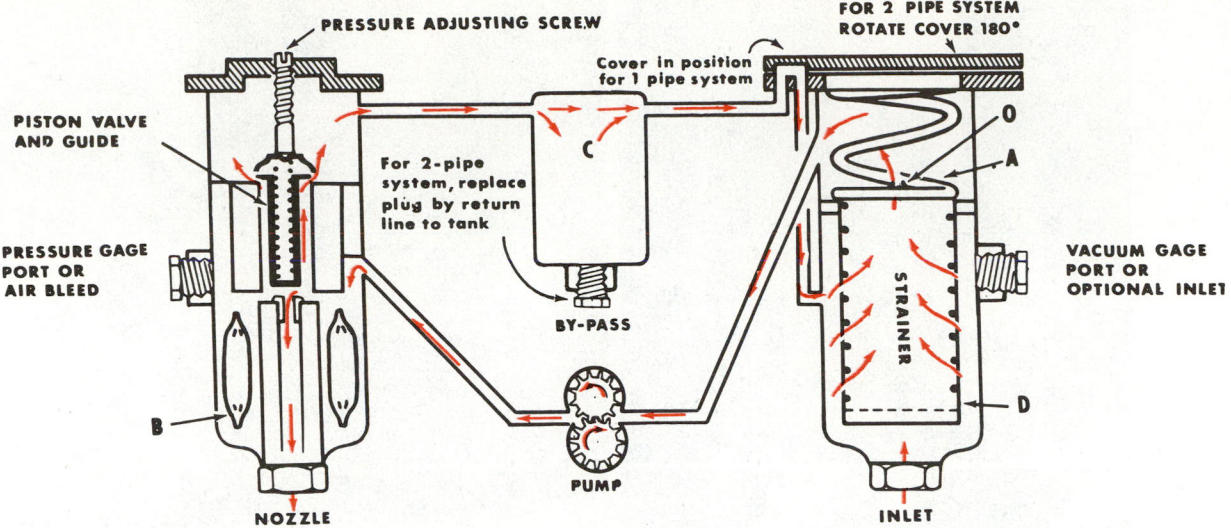

Fig. 20-31. Diagram shows oil flow through single-stage, gear type fuel oil pump. A—Vacuum chamber. B—Oil pressure cushions. C—Bypass system. D—Strainer. O—Silencer orifice. Note how the unit can be changed to a two-pipe system. (Webster Electric Co., Inc.)

Fig. 20-32. Gun type oil burner with a solenoid valve in oil line to nozzle to reduce oil dripping at nozzle. A—Solenoid valve. B—Pedestal mounting adjustment. (The Carlin Co.)

20-10 ELECTRICAL IGNITION

Gun type oil burners generally use electrical ignition. Ignition controls are described in Chapter 24. The system includes a transformer and two electrodes. The transformer is mounted on the oil burner. It transforms 120V a-c to about 10,000V a-c. The ignition system must raise the oil temperature to 700 F. (371 C.) for burning to take place. The electrodes, made of stainless steel, are mounted in ceramic insulators. No part of the electrodes should be less than 1/4 in. (6 mm) away from metal parts.

The electrode ends are positioned in front and above the nozzle. As the atomized oil swirls out of the nozzle and mixes with the turbulent air, a spark jumps between the electrode ends and ignites the mixture. The ignition may be continuous while the oil burner is in operation, or it may operate only until the fuel ignites.

The electrode gap should be between 1/8 in. (3 mm) and 3/16 in. (4 mm). The electrode ends should be approximately 1/2 to 5/8 in. (13 to 16 mm) above the nozzle and 5/16 to 1/2 in. (8 to 13 mm) in front of the nozzle. For over 45 deg. nozzles, this last dimension should be approximately 1/2 in. (13 mm) for 45 deg., 5/16 in. (8 mm) for 30 deg. Refer to the manufacturer's service manual for exact setting specifications. The porcelain insulators must be kept clean or the high voltage will "short." The ignition should be powerful enough to jump a 1-in. (25 mm) gap without the blower being turned on.

A flame mirror, Fig. 20-34, can be used to observe ignition action and spray action, to check if operation is normal.

Weak ignition, wrong position of the electrodes or poor insulation may cause delayed ignition and a puffback. Fig. 20-35 shows the interior of an ignition transformer. A transformer and line voltage testing instrument are shown in Fig. 20-36. A puffback is the ignition of a large amount of vaporized oil in the firepot. Sometimes, it will blow soot into the furnace room and into the living quarters, making a major cleaning job necessary. In no case should the electrode ends be touching the oil spray. If they do, the electrodes will become carbon coated.

Moisture in oil will retard combustion and may even cause a "flame out." A flame out is when combustion stops, but fuel continues to feed into the combustion chamber. If moisture is causing poor combustion, continuous ignition is usually recommended.

20-11 OIL TANK INSTALLATION

Oil burners must be installed with great care. A complete installation includes a 200 to 1000 gal. tank, hand shutoff valve, filter and trap combination, and copper tubing oil line.

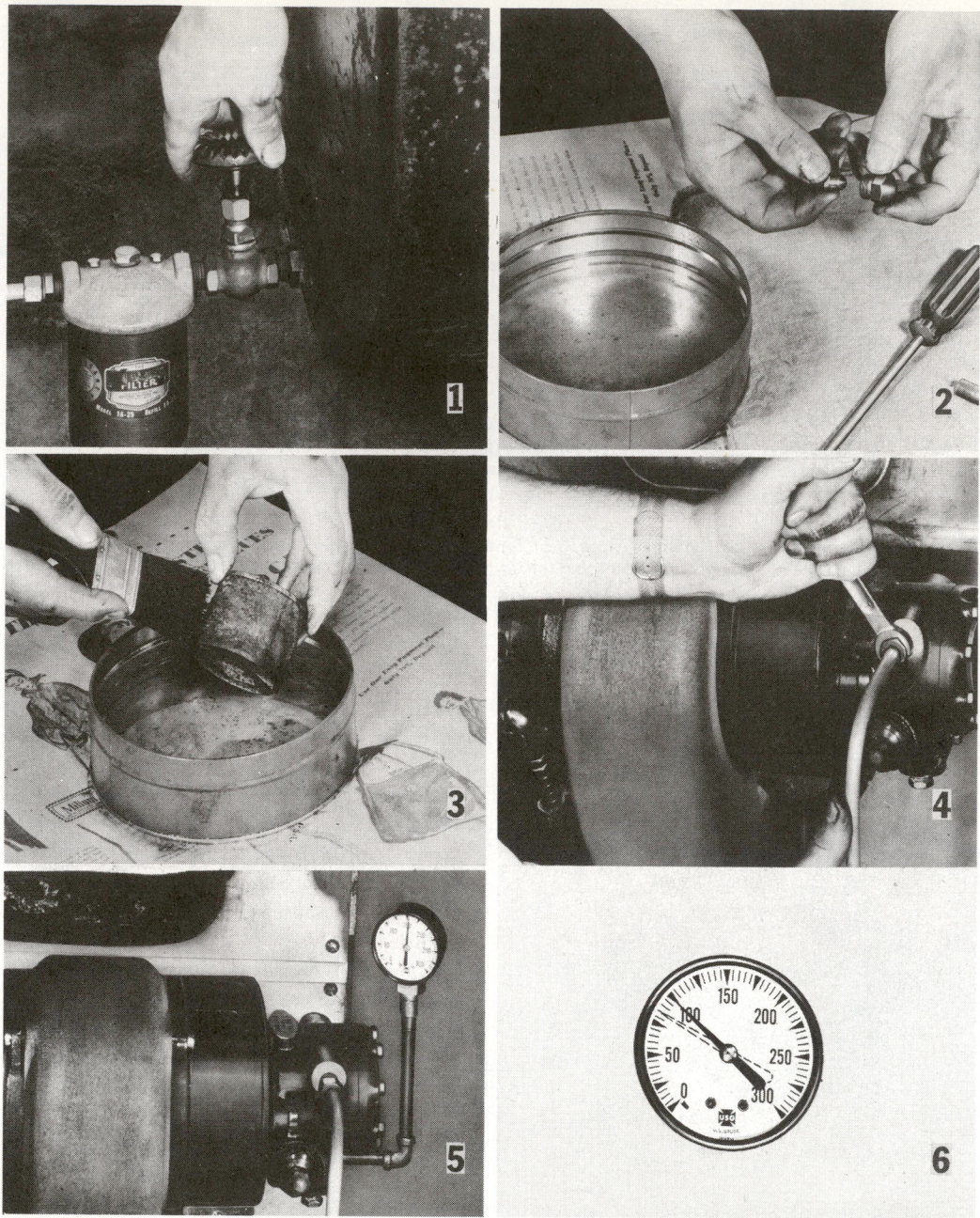

Fig. 20-33. Six main items to be checked at start of each heating season. 1—Shutoff valve and line filter. Replace filter cartridge. 2—Check and clean nozzle assembly. Follow manufacturer's recommendations. 3—Clean strainer using clean fuel oil or kerosene. 4—Check connections for tightness. 5—Insert pressure gauge into pressure port. Start burner and adjust pressure setting to manufacturer's specifications, usually about 100 psi. 6—Pressure gauge reading for correct pressure setting. (Sundstrand Hydraulics, Div. of Sundstrand Corp.)

Fig. 20-37 shows an installation for a one-pipe system with the storage tank located in the room with the furnace. **Remember that the storage tank should be at least 7 ft. (2 m) away from the furnace.** In this installation, oil feeds by gravity to the oil burner. The storage tank should be elevated less than 25 ft. (7 m) above the burner to keep the feed line pressure below 10 psi.

In some installations, the fuel tank is placed outside the building; sometimes underground. The two most common installations are:

1. With the tank above the oil burner, as in a residence with a basement, Fig. 20-38.
2. With the tank below the level of the oil burner, as in a home without a basement, Fig. 20-39.

These installations should have the tank located within a reasonable distance of the oil burner. On runs of 50 to 100 ft. (15 to 30 m), 3/8-in. (10 mm) tubing should be used. For runs of 200 to 300 ft. (60 to 90 m), 1/2-in. (13 mm) tubing should be used. The manufacturer's specifications should be checked if the oil must be raised above the tank.

Fig. 20-34. Inspection mirror used to check nozzle condition, electrode condition and flame. A—Telescoping handle. B—Hinge. C—Metal mirror.

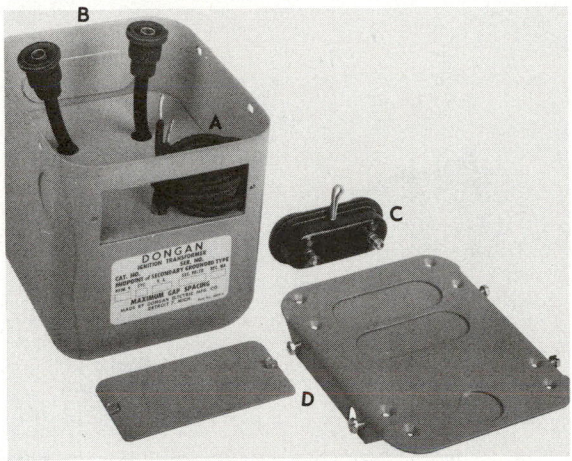

Fig. 20-35. Ignition transformer used on gun type oil burners. A—Primary leads. B—Secondary leads. C—Secondary bushing. D—Cover plates.

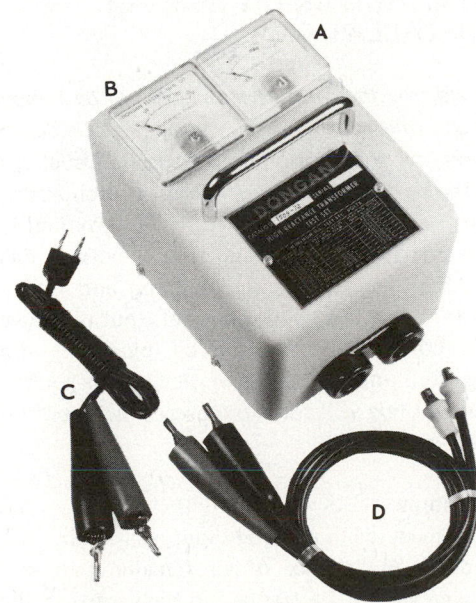

Fig. 20-36. An ignition transformer testing instrument. A—High voltage meter. B—Line voltage meter. C—Line testing leads. D—Transformer high voltage testing leads. (Dongan Electric Mfg. Co.)

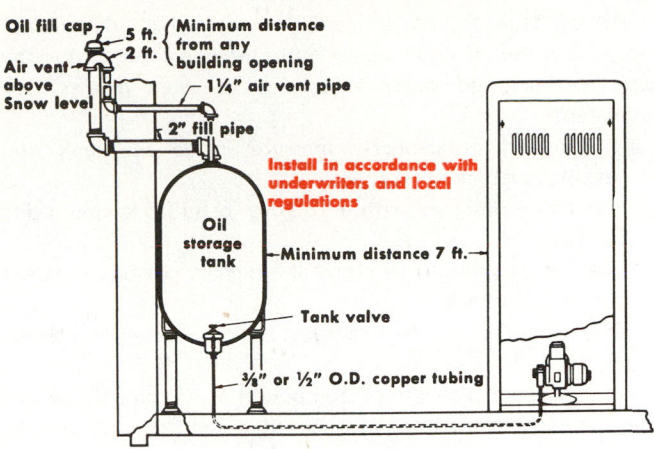

Fig. 20-37. Typical gun type oil burner installation. Note fill pipe, vent pipe and oil line installation. (Webster Electric Co., Inc.)

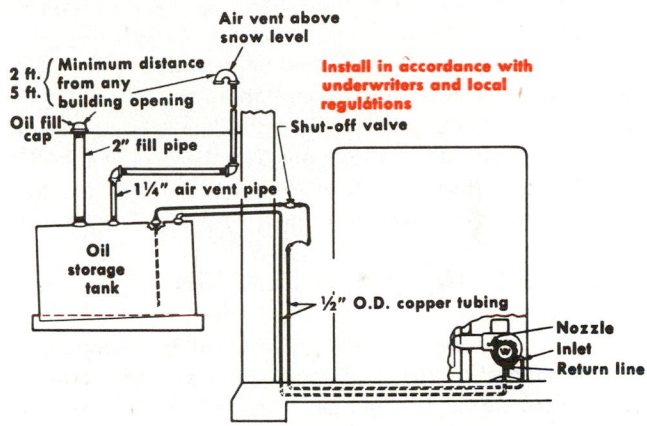

Fig. 20-38. Gun type oil burner installation with storage tank installed underground but above oil burner. Note two oil lines.

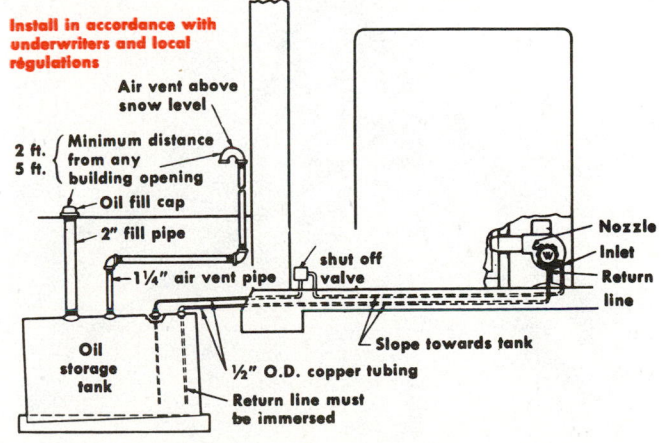

Fig. 20-39. Gun type oil burner installation having storage tank underground and below level of oil burner. Note special precautions to be observed with oil lines. (Webster Electric Co., Inc.)

An oil tank should be installed with a slight down slant away from the oil line connection to provide a low spot in the tank for dirt and water to collect. The vent pipe is very important:

1. It provides atmospheric pressure inside the tank and permits volatiles to escape.
2. It must be designed with a 180-deg. bend (to keep out dirt and rain).
3. The opening should be above the highest possible snowfall or other blockage.

The oil fill cap should always be in place except when filling the tank.

Always use a pipe thread compound on the pipe threads of the fittings. This compound should be of the oil-resistant, nonhardening type. During storage and while installing tubing, keep tubing ends sealed with tape to keep out dirt and moisture. Remove tape at the time the tubing is connected. The 3/8 or 1/2-in. (10 or 13 mm) OD copper tubing is attached to the fittings with standard SAE 45 flares. Flaring techniques are described in Chapter 2. Tight, leakproof connections are essential.

The oil lines in the tanks should be mounted so the tubing opening is 3 to 4 in. (7.5 to 10 cm) above the bottom of the tank. If system has a return oil line, this line does not need to go near bottom of tank for light oils (Number 1 or Number 2). It should go within 4 to 5 in. (10 to 15 cm) of the tank bottom for the heavy fuel oil to help keep the oil more fluid.

20-12 OIL BURNER INSTALLATION

The oil burner must be carefully installed. Installation must be made in accordance with local codes and the manufacturer's instructions. The burner must be the correct height above the bottom of the combustion chamber. Burners are mounted either on adjustable legs or on a flange bolted to the furnace. Burner air tube and nozzle must be inserted into the combustion chamber the exact distance the manufacturer recommends. The opening into the furnace must be carefully sealed (some have flange adaptors) to prevent air leaks.

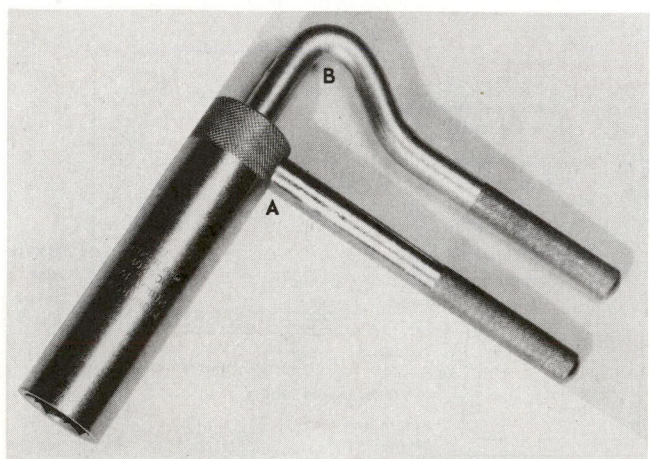

Fig. 20-40. A nozzle removing and installing tool. A—Nozzle tube socket wrench and handle. B—Nozzle socket wrench and handle. (Monarch Mfg. Works, Inc.)

The nozzle must be the correct size, and it must be in good condition. The size of the hole in the nozzle and the amount of oil pressure determines the rate at which fuel oil is burned and, therefore, the rate at which heat is produced. The size of the nozzle selected must match the heating requirements of the heated space. If the nozzle is too small, the burner may not heat the space adequately. If the nozzle is too large, there will be a tendency for the burner to come on and off quite frequently.

Nozzles are usually supplied with a fine filter at their fuel oil inlet. The filter is designed to eliminate the possibility of dirt entering and plugging the nozzle. This filter should be cleaned and replaced whenever the oil burner is serviced.

Be careful not to twist the tube or move the nozzle tube out of line. A tool for safely removing the nozzle is shown in Fig. 20-40. A tool used to remove and replace nozzle filters is shown in Fig. 20-41.

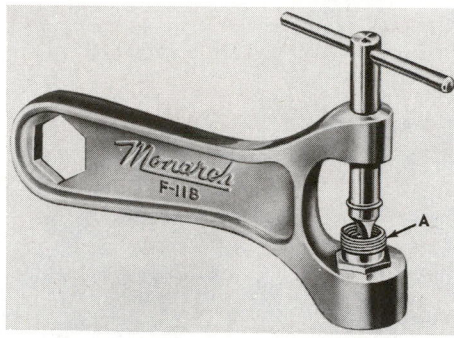

Fig. 20-41. A fuel oil nozzle filter being removed from the nozzle. A—Nozzle.

20-13 STARTING AN OIL BURNER INSTALLATION

Before starting the oil burner, air should be removed from the lines and the pump. A vent plug (air bleeder fitting) is mounted on the top of the pump housing. Usually, this vent plug seals the port used for pressure gauge installations.

If enough oil and air collects in the firebox and is ignited, anything may happen — anything from a puff of flame to an explosion that may wreck the building and maim or kill. Inspect the firepot. If any oil is present, shut off oil valves and vent the firebox. Then remove the oil (by means of a suction pump, rags, etc.) until all danger of oil fumes is gone.

Air in an oil line will form bubbles, which could result in:
1. Oil not being pumped.
2. Blow backs.
3. Flame failures.

The line must be completely purged of air. A two-pipe system reduces the chance of air remaining in the system. However, air can still be trapped in high spots in the line. A leak in the oil line will almost always cause air-in-line troubles.

Always check the fuel oil nozzle to be sure it is the correct size, and that it is in the center of the gun air duct. The

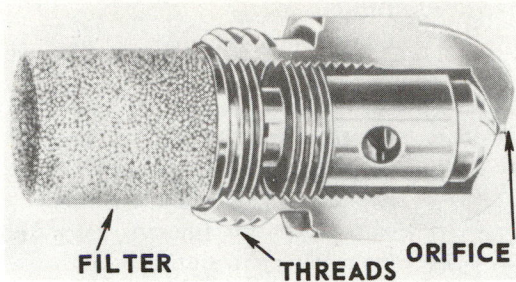

Fig. 20-42. Stainless steel nozzle used with gun type oil burners. Note fine filter at entrance to nozzle. (Monarch Mfg. Works, Inc.)

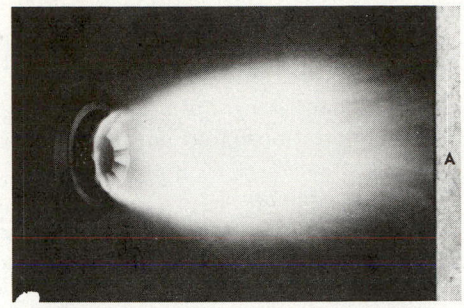

Fig. 20-43. Gun type oil burner flame. A—Refractory insulation. (The Carlin Co.)

electrodes must be kept clean and in correct relation to the nozzle. Fig. 20-42 shows a typical oil burner nozzle. These nozzles come in various capacities, all based on gallons per hour (gph) at 100 psi (from .40 to 28 gph). Some nozzles are large enough to feed 100 gph. Remember that a 1 gph nozzle delivers 140,000 Btu/hr. If the overall efficiency is 60 percent, the useful heat would be 84,000 Btu/hr. Poor oil delivery may be the result of the main filter, the pump filter or the nozzle filter being partially clogged. Check all three filtering devices when servicing the unit.

Flame failure may be caused by one or more of the following:

1. Oil tank out of oil.
2. Oil tank not vented.
3. Clogged filter in oil line.
4. Ice in fuel line.
5. Loose oil line connection (air in line).
6. Dirt in supply line.
7. Water in supply line.
8. Loose wiring or connections.
9. Motor not running (check reset button).
10. Defective pump.
11. Pump losing prime.
12. Changing pressure or low pressure at pump (slipping coupling).
13. Clogged nozzle.
14. Damaged nozzle.
15. No spark at electrodes:
    a. Loose wiring.
    b. Bad transformer.
    c. Low voltage.
    d. Crack in electrode porcelain.
    e. Electrodes carboned.
    f. Electrodes spacing too far or too close.
    g. High voltage wiring loose.

Proper flame appearance is luminous (mainly yellow). If there is insufficient air, the flame turns dull orange or red, and there may be smoky tips to the flame. Fig. 20-43 illustrates a properly adjusted flame.

The draft in the firepot is measured by the air pressure drop (in the firepot). It should be about 0.02 to 0.05 in. (.5 to 1.3 mm) of water. (Use an inclined water tube manometer.) See Fig. 1-16. This check will also help determine if the automatic draft is working satisfactorily.

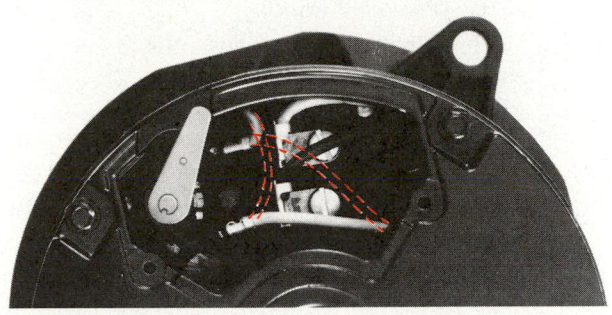

Fig. 20-44. Method of reversing direction of rotation of oil burner motor. This is accomplished by reconnecting two wires as shown by broken lines.

Some oil burner motors are reversible. Fig. 20-44 shows the method of reversing one type of oil burner motor. Controls for oil burners and testing instruments are described in Chapter 24.

Inspect the electrode wires. If cracked or brittle, replace the wires. Inspect the electrode tubular ceramic insulators. If the ceramic tubes are cracked, replace them.

Most soot deposits in a firebox collect when the unit first starts. The more often the unit starts (cycles frequently), the greater the soot deposit. A correctly sized oil burner which operates less frequently will leave less soot in the furnace and stack.

If an oil line is dirty or clogged, blow it out using nitrogen gas (always use a pressure regulator and relief valve) or R-12 or R-22 refrigerant gas. Never use compressed air or oxygen, because a violent explosion may result.

20-14 FUEL OIL BURNER SERVICING

Oil burner problems, symptoms and possible causes are given in the following troubleshooting outline.

I. Burner motor does not start, starts and locks out, or cycles.
   A. Does not start.
      1. Relay does not close (will not close or contacts dirty).
      2. Safety lock out stays open.
      3. Bad relay coil.

4. Low voltage.
5. Open high limit control.
6. Broken wires or loose connections.
7. Relay transformer open.
8. Thermostat open (dirt on contacts, loose or dirty connections).
9. Stack switch open.
10. Heat sensing contacts out of place or open.
11. Motor overload open (burned out, dirty contacts).

B. Starts, but locks out.
1. No fuel oil out of nozzle.
   a. Clogged.
   b. Pressure too low.
   c. Pump not working.
   d. Loose motor coupling.
   e. Air leaks in fuel line.
   f. Fuel oil line hand valve closed.
   g. Strains or screens clogged (filter, pump screen or nozzle strainer).
   h. Pressure regulator stuck open.
   i. Vent on fuel oil tank closed.
   j. Empty fuel oil tank.
2. Fuel oil coming out of nozzle but no ignition.
   a. Electrodes not positioned correctly.
   b. Insulators cracked.
   c. Ignition wires worn, loose or with dirty connections.
   d. Transformer not operating.
   e. Primary wires worn, loose or with dirty connections.
   f. Low line voltage.
3. Fuel oil to nozzle, ignition OK, but no flame.
   a. Clogged nozzle.
   b. Clogged nozzle strainer.
   c. Nozzle loose.
   d. Pressure too low.
   e. Fuel oil too heavy (wrong oil or too cold).
   f. Excessive air or too much draft.
   g. Electrodes in wrong position.
4. Flame only burns a few seconds.
   a. Flame sensor not in correct position.
   b. Stack switch not operating correctly.
   c. Excessive air or air too cold.
   d. Flame is too lean.

C. Cycles, but not on lockout.
1. Thermostat differential too close.
2. Anticipator set too close.
3. Limit switch set too low.
4. Overfired.

II. Burner does not operate correctly.
A. Smoke, soot, odors and/or pulsating sound.
1. Wrong oil pressure.
2. Flame touches combustion chamber.
3. Not enough draft.
   a. Dirty chimney.
   b. Draft control out of adjustment or stuck open.

c. Dirty flue.
d. Combustion chamber or heat exchanger leaks.
4. Poor mixing of air and oil.
   a. Nozzle is worn, loose, dirty, or drips.
   b. Oil pressure too low or high.
   c. Poor air velocity and turbulence.
   d. Not enough air (shutter closed too much, fan binding or tight bearings).

B. Puffs back.
1. Water in oil.
2. Delayed ignition.
   a. Electrodes not positioned correctly or loose.
   b. Insulators carbonized.
   c. Nozzle worn, loose, dirty, or drips.
   d. Voltage drop when burner starts.
   e. Oil pressure too low or too high.
   f. Transformer leads loose or dirty.
   g. Transformer not operating correctly.
   h. Excessive air or high draft.

C. Noise.
1. Loose fan.
2. Loose shutter.
3. Worn pump.
4. Dirty strainer.
5. Air in oil line.
6. Transformer hum.
7. Draft control vibrates.
8. Motor coupling worn.
9. Motor and pump not lined up correctly.
10. Relay contacts not seating tightly.
11. Oil suction line restricted.
12. Motor mounting loose.
13. Tight motor bearings.
14. Tank hum.

D. Fuel oil consumption is too high.
1. Nozzle is worn, loose.
2. Combustion chamber is dirty.
3. Too much combustion air.
4. Poor mixing of air and oil.
5. Not enough draft over fire.
6. Air leaks into combustion chamber.
7. Oil pressure too high or too low.
8. Stack temperature too high.

There are two ways to check combustion efficiency: carbon dioxide (CO_2) analysis of flue gas; temperature of flue gases.
1. Use a carbon dioxide analyzer to check a sample of flue gas. It should be 10 to 12 percent CO_2 without visible smoke. If the reading is too low, it means too much air:
   With 6 percent CO_2, 155 percent excess air is used.
   With 8 percent CO_2, 85 percent excess air is used.
   With 10 percent CO_2, 50 percent excess air is used.
   With 12 percent CO_2, 26 percent excess air is used.
2. If the temperature of the flue gas is too high, much heat is being wasted. If the temperature of the flue gas is too low, water vapor will condense in the flue or chimney and the small amount of sulphur will form sulphurous acid (H_2SO_3) which is very corrosive.

With a 10—12 percent CO_2 stack analysis, combustion efficiency is as follows:

TEMP. F.	TEMP. C.	PERCENTAGE OF EFFICIENCY
1000	538	65 to 69
800	427	70 to 73
600	316	76 to 79
400	204	82 to 84

These are flue gas temperatures minus furnace room air temperature.

20-15 GAS FURNACES

Gas is being used at an ever-increasing rate as a heating fuel. The gas is piped into the building. Usually, it is fed to the furnace at 4 to 6-in. (100 to 150 mm) water column pressure. A pressure regulator reduces this pressure to approximately 2 in. Note that about 27-in. water column pressure equals 1 psi.

A solenoid valve generally is used to control the main gas flow to the furnace burners. This solenoid valve may be operated by 120V current. Some operate through a transformer at a reduced voltage (24V). Some solenoid valves are operated by current generated by a thermocouple located near the pilot light. These do not require any connection to the house current. See Chapter 24.

The burner may be made of steel pipe or a casting. It may be either of the multiple jet or orifice type, or it may be the one-opening type with a deflector plate. An elementary gas burner is shown in Fig. 20-45. These units usually operate on

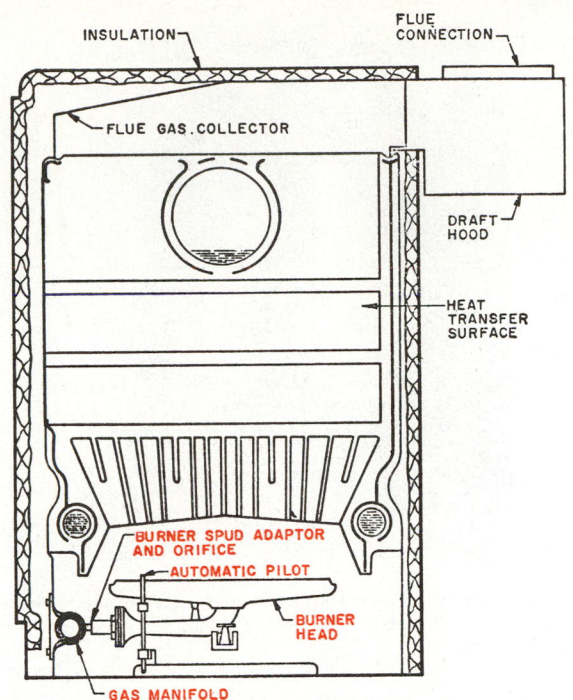

Fig. 20-46. Gas-fired domestic heating boiler. Automatic pilot is electrically or pressure connected to safety controls that will shut off gas flow to burner if the pilot is extinguished.
(ASHRAE Guide and Data Book)

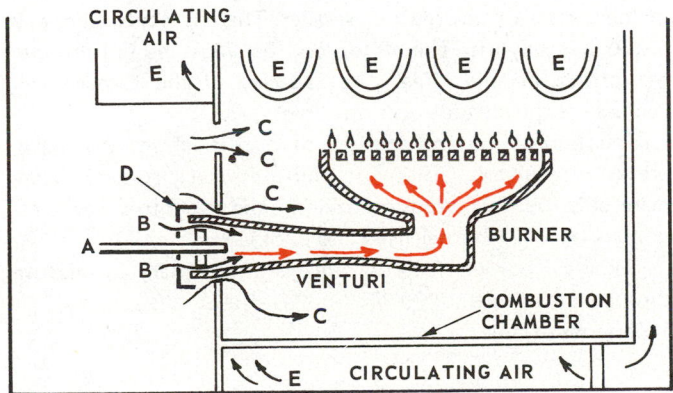

Fig. 20-45. Schematic of a gas burner installed in a warm air furnace. A—Fuel gas. B—Primary air. C—Secondary air. D—Primary air adjustment. E—Circulating air for heating.

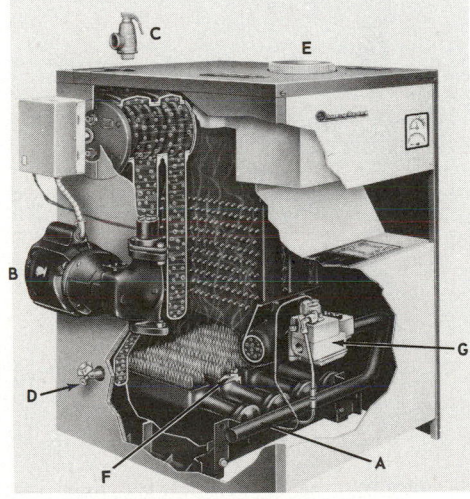

Fig. 20-47. Complete gas-fired boiler. A—Burners. B—Water pump. C—Relief valve. D—Drain valve. E—Flue. F—Pilot light. G—Control.

the Bunsen burner principle. Fig. 20-46 shows a gas-fired boiler. Note the gas manifold, burner head, automatic pilot and the burner spud (nozzle) adaptor and orifice combination.

A complete gas-fired furnace is shown in Fig. 20-47. Note the four burners, control and pilot light gas line. A diverter arrests back pressure in the chimney. Back pressure may be caused by wind gusts that might blow out the gas flame in the furnace. Fig. 20-48 shows a diverter for a round stack.

New furnace designs include ceramic heat exchangers, forced draft pulse combustion (burning a mixture of gas and

air at timed intervals), and combustion air ducted from outside to the furnace and then ducted to the outside.

Fig. 20-49 shows a two-burner gas warm air furnace.

20-16 GAS BURNERS

Gas burners usually have a simple design. Gas is fed through an orifice and mixed with a certain amount of air (primary air). This mixture passes to the burner head, where com-

Fig. 20-48. Diverter designed for use on gas furnace stack. (The Excelsior Steel Furnace Co.)

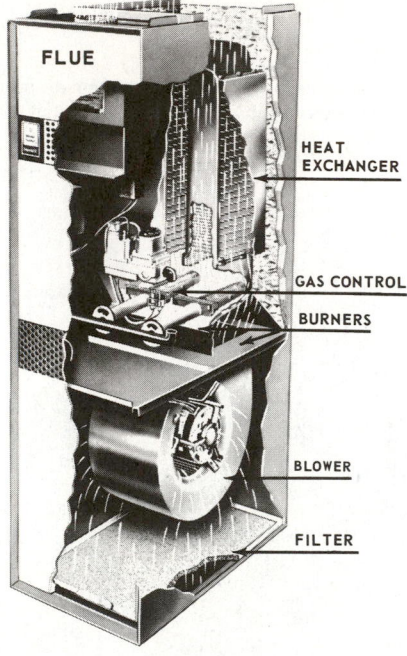

Fig. 20-49. Sectioned warm air furnace. Heat exchanger is designed to increase heating efficiency. (General Electric Co.)

bustion takes place and where the gas mixes with the secondary air. As much as 35 percent excess air is fed to the burner to insure thorough combustion.

There are several types of gas burners:

1. Atmospheric injection.
2. Luminous flame.
3. Power burner.

These burners are used in:

1. Unit heaters.
2. Furnaces.

Furnace type gas burners may be used with warm air, hot water or steam. The warm air furnace may be one of the

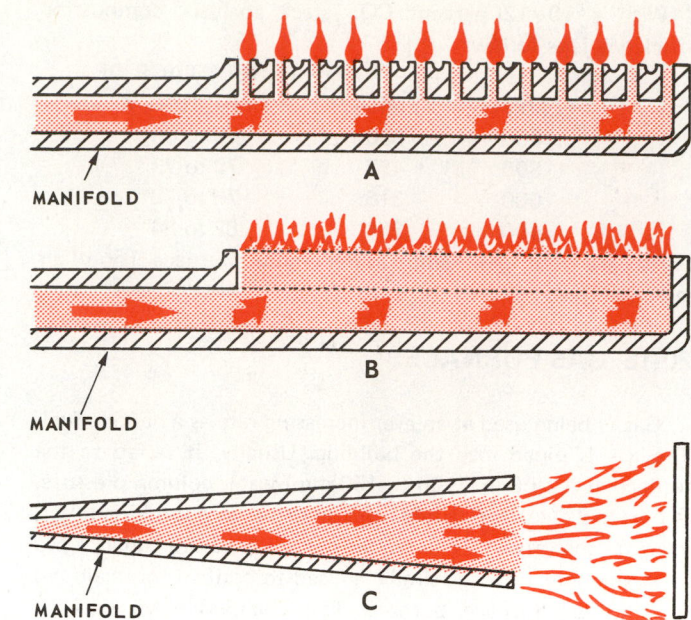

Fig. 20-50. Three gas burners of the atmospheric type. A—Burner has holes. B—Burner has slots. C—Burner flame hits a target plate.

following three types:

1. Airflow up through heat exchanger.
2. Airflow down through heat exchanger.
3. Airflow across the heat exchanger.

There must be a burner orifice correction made when the altitude is over 2000 ft. (610 m). As altitude increases, the size of the burner orifice must be smaller. There is less air and, as a result, less oxygen. Therefore, less fuel must be fed through the orifice at one time. The capacity of the furnace will decrease as the altitude goes up.

The burner system consists of a manual shutoff valve, pressure regulator, automatic shutoff valve, control valve, manifold, burner spuds (nozzles) and adaptors, orifices, primary air inlet, burner head and pilot valve.

Some gas burners feed the fuel gas and primary air mixture through a manifold to:

1. A series of holes.

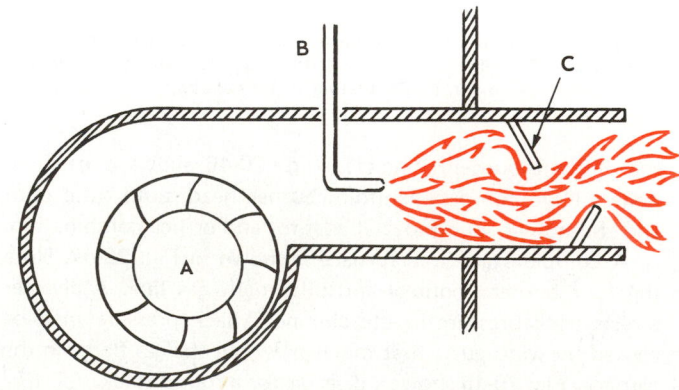

Fig. 20-51. A power burner. A—Blower air inlet. B—Fuel gas inlet. C—Deflector plates.

2. A series of narrow slots.
3. A large flame hitting a target (inshot burner). See Fig. 20-50.

The power burner uses a blower to force both primary air and secondary air into the burner. See Fig. 20-51. The burner tube usually has angular vanes to spin or twirl the flame for more efficient burning.

Fig. 20-52 shows a gas burner designed to replace either a stoker or an oil burner. It is an atmospheric pressure type burner. A power conversion gas burner is shown in Fig. 20-53. A different design power gas burner is shown in Fig. 20-54.

Many commercial buildings now have combination gas furnaces and cooling systems installed on the roof. Many use fans to force or draw air through the combustion chamber. Keep these fans clean. No stack is needed. Fans reduce the chance of downdrafts. Filters are used to clean the combustion air, and they require service. Clean the heat exchange surfaces each year.

To check gas pressure to manifold (main burner must be on), install a water manometer by attaching it to the manifold gauge opening located on the outlet of the pressure regulator (1/8-in. pipe).

It is possible to use LP-Gas as a standby fuel source if the natural gas is mixed with air before the LP-Gas is burned in the heating device. The mixing of the proper amount of air with 1

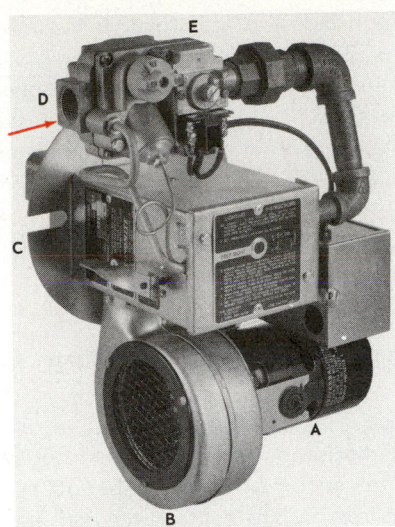

Fig. 20-54. Power gas burner. A—Motor. B—Blower. C—Mounting flange. D—Gas inlet connection. E—Controls. (Midco International Inc.)

cu. ft. of LP-Gas (in vapor form) allows use of the same primary air opening, the same secondary air opening, and the same burner holes or slots. In other words, the furnace needs no adjustment if fed the air and LP-Gas mixture. The final air and LP-Gas mixture should have about 1000 Btu/cu. ft. to be usable in a natural gas furnace.

20-17 FUEL GASES

There are three types of fuel gases:
1. Natural.
2. Manufactured.
3. Liquefied petroleum (LP-Gas).

Natural gas is obtained from gas deposits in the ground. Manufactured gas is made by distilling or cracking coal or oil, and by other processes. Natural gas consists of about 84 percent CH_4 (methane) and 16 percent C_2H_6 with a heating value of 1000 to 1100 Btu/cu. ft. To burn 1 cu. ft. of natural gas, 8 cu. ft. of air is required. However, some excess air is needed, so about 11 cu. ft. of air is used for each cu. ft. of natural gas (30 percent excess air).

As the gas burns, it yields about 1 cu. ft. carbon dioxide, 1 cu. ft. nitrogen, 2 cu. ft. of water vapor and about 25 to 50 percent excess air. If there is insufficient primary air, it will produce a yellow flame (incomplete combustion). If there is too much primary air, the flame will be noisy, and it will jump around above the burner. If the stack has CO (carbon monoxide), more primary air or secondary air is needed (secondary air is usually fixed).

Manufactured (coal) gas varies in content. It contains about 50 percent H_2 (hydrogen), 8 percent CH_4 (methane) and other gases. It has a heating value of approximately 500 to 600 Btu/cu. ft.

Liquefied petroleum gas (LP-Gas) usually is propane with a little butane added. It can be liquefied, stored and transported in cylinders or tanks. LP-Gas vaporizes easily and is changed

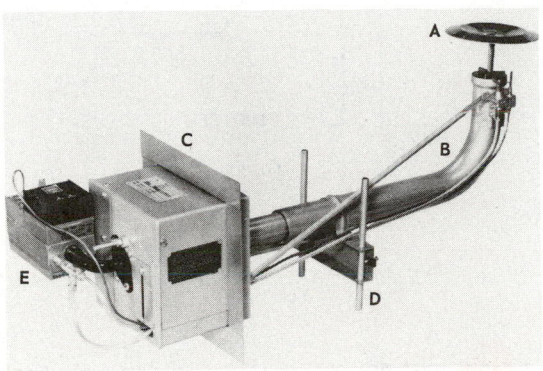

Fig. 20-52. Replacement type natural draft (atmospheric air) natural gas or propane gas burner for air or hydronic furnaces. A—Flame spreader. B—Burner tube. C—Mounting flange. D—Adjustable legs. E—Controls.

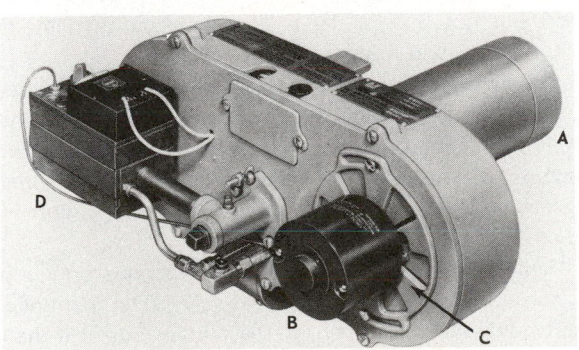

Fig. 20-53. Forced draft gas burner for furnaces. A—Burner tube. B—Blower motor. C—Adjustable air inlets. D—Controls. (White-Rodgers Div., Emerson Electric Co.)

into its gas form before it is burned. It has a heating value varying between 2500 and 3200 Btu/cu. ft.

Propane boils at −40 F. (−40 C.) at atmospheric pressure. Butane boils at 32 F. (0 C.) at atmospheric pressure. It is fed to the burner at about 11-in. water column pressure.

Propane and butane are dangerous if carelessly used. Both are heavier than air and will collect in the firebox and in the basement. All these gaseous fuels have an odor added. If an odor is detected, the gas is present. If gas odor is found, again, shut off the main fuel valve and thoroughly vent the area.

20-18 PILOT LIGHTS AND ELECTRIC IGNITION

A gas furnace is designed to heat a building to a comfortable temperature regardless of the outdoor temperature. Usually, the gas furnace must be shut off at certain times to prevent overheating. The controls required are explained in Chapter 24.

When the main gas flame is shut off, some device must be used to ignite the fuel gas and air mixture when heat is needed. Two devices are used:

1. Pilot light.
2. Electric ignition.
   a. Spark.
   b. Glow coil.

Pilot lights keep burning. Then, when the thermostat turns on the heating system, the pilot light ignites the gas. See Fig. 20-55. The pilot light obtains its fuel gas from a tube attached to the combination pressure regulator and gas control valve.

The pilot light is also equipped with a safety device, either a thermal fluid bulb or a thermocouple. The thermal bulb allows the main valve to open only if the pressure is high enough (heated by the pilot flame). This safety device generally is connected to the main gas regulator.

The thermocouple unit is operated by its generation of electricity. When a thermocouple is heated by the pilot flame,

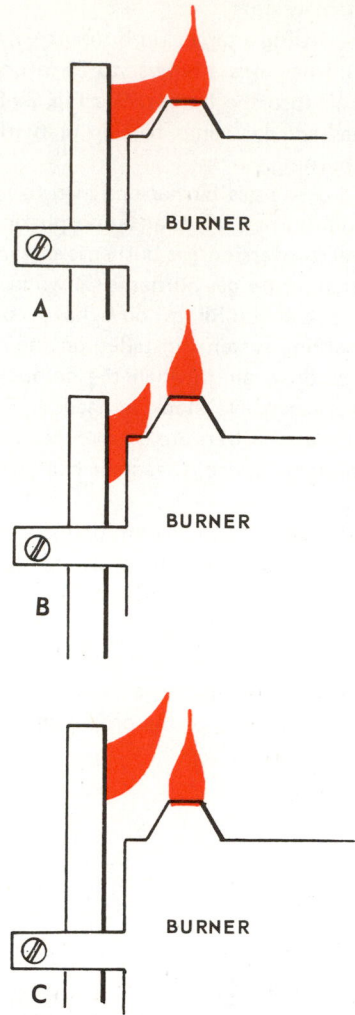

Fig. 20-56. Pilot light flame positions. A—Correct. B—Too low. C—Too high.

about 25 millivolts are generated to operate a pilot solenoid valve. Only when the thermocouple is heated will the solenoid valve stay open and allow the main gas valve to open.

One system has a 24V a-c through flame which rectifies the current to d-c. It uses the same principle as a vacuum tube rectifier, where electrons will flow from a point to a flat surface, but not from a flat surface to a point. If the flame goes out, no d-c will flow.

Some thermocouple assemblies consist of several thermocouples electrically connected in series. These units generate from 125 to 750 millivolts of electricity, which is powerful enough to operate the main solenoid valve. Therefore, the furnace is independent of the regular electrical circuit in the building.

In large systems, the thermocouple operates only a small pilot solenoid valve in the main gas valve. The thermocouple current holds the valve open. If the pilot is out or if the pilot flame is low, the valve will close.

A pilot flame should be blue. If it is yellow, the primary air inlet to the pilot is probably dirty and should be cleaned. The pilot should be cleaned at least once a year.

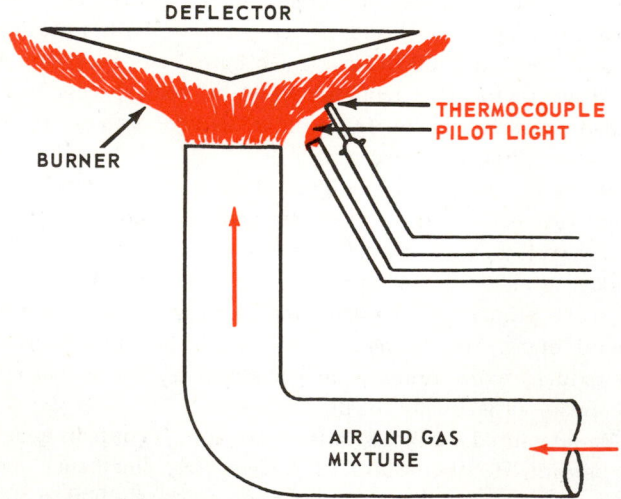

Fig. 20-55. Schematic drawing of safety thermocouple and pilot light for gas furnace. Thermocouple generates small electric current which actuates a control that will shut off gas supply to furnace if pilot light is extinguished.

To test a system, shut off pilot flame while burner is on. The main burner should shut off in about two minutes. The pilot light flame should be on 3/8 to 1/2 in. (10 to 12 mm) from the thermocouple tip. The tip should be very hot and dull red. See Fig. 20-56. The pilot light should be located at the edge of the main burner flame. The location varies with various manufacturers.

An electronic spark igniter (about 19,000V) may be used to keep the pilot light burning. Ignition spark starts either in less than a second or after five minutes (to allow gas to empty from furnace) if the flame goes out. It stops if the pilot light burns (usually in 20 seconds). If flame does not go on, ignition shuts off in 15 seconds.

Electric ignition depends on an electric spark to light the gas. The system includes electrodes, transformer, flame sensor, safety switch and purge timer. It is used where pilot lights would be difficult to service or where air drafts may make it difficult to service or where air drafts may make it difficult to keep a pilot flame burning.

Electric ignition systems can operate in temperatures of −40 F. (−40 C.). The gas valve and pressure regulator must also be designed for low-temperature operation. From 4000 to 12,000V are used with a .040 in. (1 mm) to .190-in. (5 mm) gap. The electrodes are extended so that the spark gap is within 1/4-in. (6 mm) of the gas stream just outside the

burner. One electrode is insulated, the other is grounded. These electrodes can operate satisfactorily at high temperatures. The ceramic insulators are designed for about 1200 F. (649 C.). If the main flame goes out, the flame sensor cools enough in four seconds to operate the system shutoff controls. This unit is rod-and-tube design of about 4 1/2 in. (114 mm) length.

Still another method is used to ignite a gas burner in an automatic system. It makes use of a glow coil. When the thermostat points close, a transformer feeds 2.5V a-c to the glow coil. As soon as the glow coil turns red, it ignites the pilot. The flame detector then opens the glow coil switch circuit and turns on the main gas valve. The burner fuel gas is ignited.

When working on this system, always open the main electrical switch. See Fig. 20-57.

20-19 THERMOCOUPLE CIRCUITS

A thermocouple may be used to prevent the main gas valve from opening if the pilot light goes out. The principle of operation of the thermocouple is described in Para. 6-62.

The thermocouple may be used to shut off the gas supply to the burner in the event that the pilot light goes out.

The hot junction of the thermocouple is located in the pilot

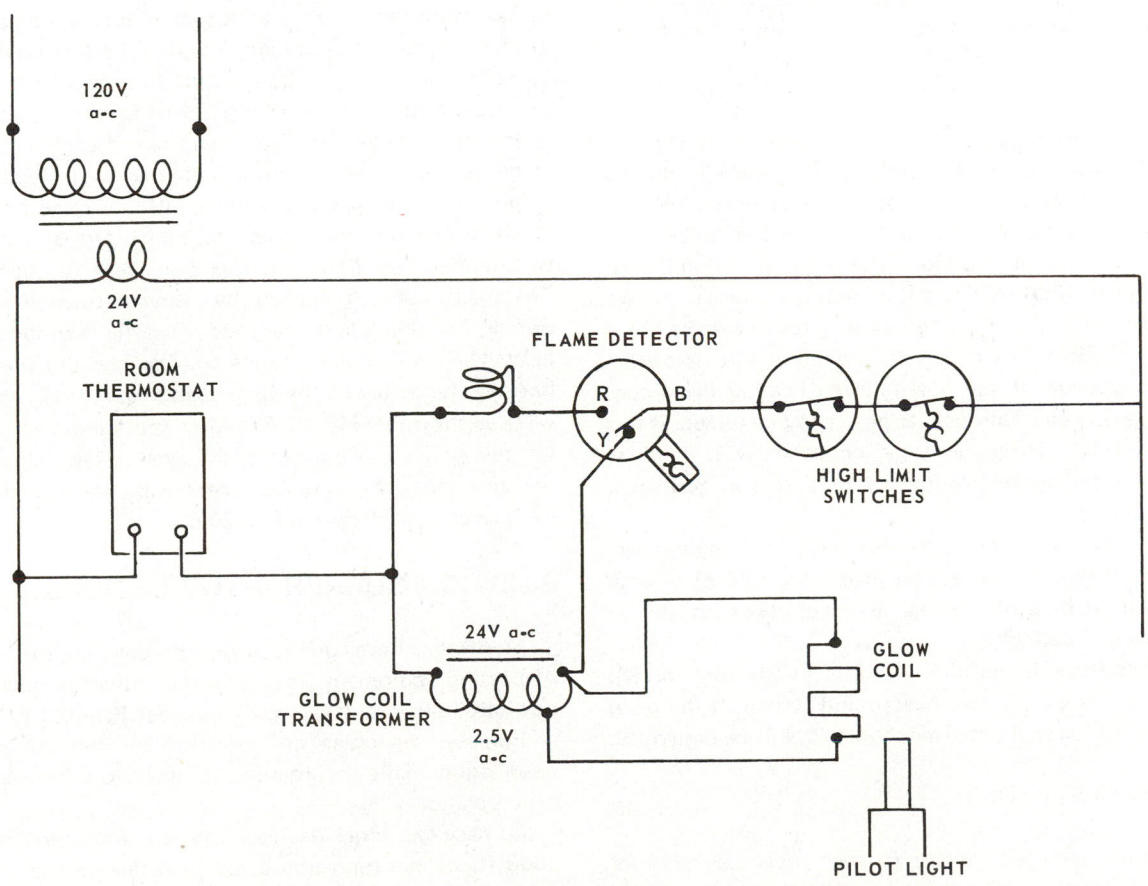

Fig. 20-57. A glow coil circuit. When points close, glow coil heats to a red temperature and ignites pilot. Flame detector heats up and moves switch from Y to R, opening main gas valve and burner starts.

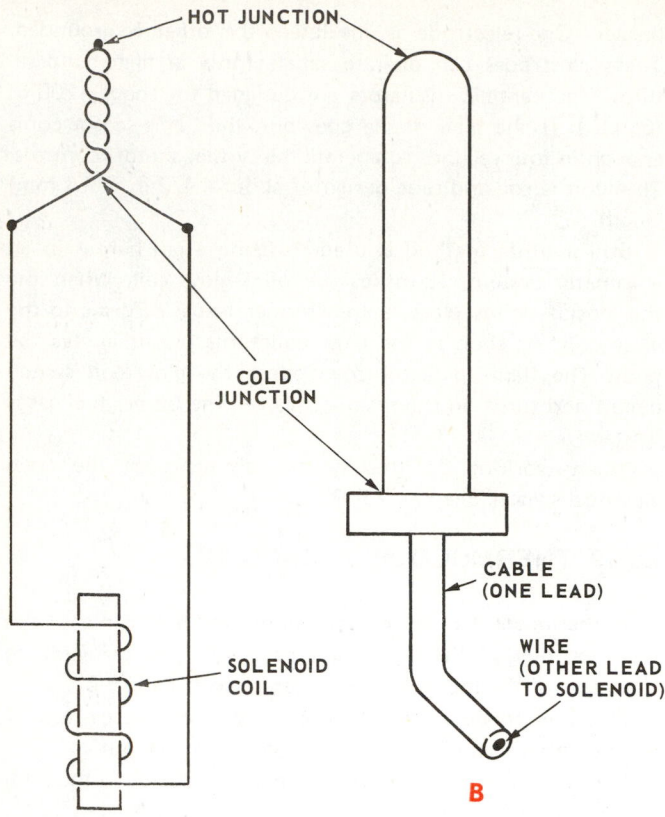

Fig. 20-58. A thermocouple. A—Schematic of an electrical circuit of a thermocouple pilot light safety control. B—Construction of the thermocouple tip.

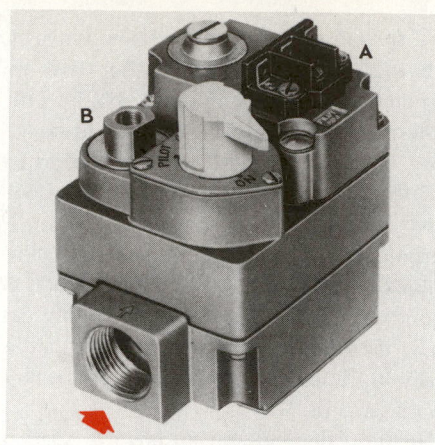

Fig. 20-59. A complete gas control. It has a shutoff valve, pressure regulator, gas flow control and pilot light control. A—Terminals. B—Thermocouple connection.
(White-Rodgers Div., Emerson Electric Co.)

flame. A small electrical current is generated in the thermocouple. This current opens the pilot solenoid valve in the gas supply line when the thermostat contacts are closed. The pilot solenoid valve operates the main gas valve. See Fig. 20-58.

On open circuit, the thermocouple generates 30 millivolts (one volt equals 1000 millivolts). A millivolt reading of two millivolts or more above or below this reading indicates a faulty thermocouple. The pilot solenoid requires at least seven millivolts to operate. If the hot junction heating flame goes out, the magnetic coil will lose its power in one minute if it is operating correctly (although some codes allow as much as three minutes for system shutdown in domestic and small commercial units).

Aluminum tubing is preferred for pilot light gas connections (copper has a flaking action inside the tubing). Aluminum tubing must be protected against electrolysis (usually by a plastic exterior coating).

If the pilot flame is too high (there is an adjusting screw), the thermocouple may be overheated and ruined. If the main burner flame contacts the thermocouple, it will be destroyed.

20-20 GAS CONTROLS

The air-fuel gas ratio to the burner must be carefully maintained. Since the fuel gas must be supplied at a constant pressure, a pressure regulator generally is mounted in the inlet to the gas manifold. The regulator operates much like an expansion valve, opening and closing a valve in response to the outlet pressure. This pressure is measured in inches of water column. Most pressure regulators are adjustable. However, they should be adjusted only when a water column manometer is connected to the outlet.

The pressure regulator has a tapped hole in the bonnet for two reasons:

1. To insure that there is atmospheric pressure on the outside of the regulator diaphragm for good pressure control.
2. To permit tubing to be installed from this bonnet opening to the outdoors or to the pilot light. This is a safety device in case the regulator diaphragm leaks. Escaping gas will then go outdoors or be burned at the pilot light. The best practice is to pipe it outdoors with a weatherproof outlet.

Most pressure regulators are built into a complete gas control. See Fig. 20-59. Internal design is shown in Fig. 20-60. The main valve is opened by using a solenoid controlled orifice. The solenoid is energized when the thermostat calls for heat. The cycling valve opens and gas pressure travels to the bottom, (underneath the large diaphragm), opening the main valve as shown in Fig. 20-61. After the thermostat opens and the relay coil is de-energized, the cycling valve closes and gas pressure under the diaphragm decreases. This causes the main valve to close as shown in Fig. 20-62.

20-21 GAS BURNER INSTALLATIONS

Most cities have code requirements covering the installation of heating equipment. It is important to know this code and carefully follow it. **Gas is dangerous.** See Para. 20-17.

Furnace capacities are certified by the American Gas Association. This information is given on furnace identification plates.

Furnace capacities decrease at about four percent for each 1000 ft. of elevation above sea level due to the decrease in atmospheric pressure (air becomes less dense). This rating change is effective above 2000 ft. elevation.

When installing gas line pressure regulator valves and

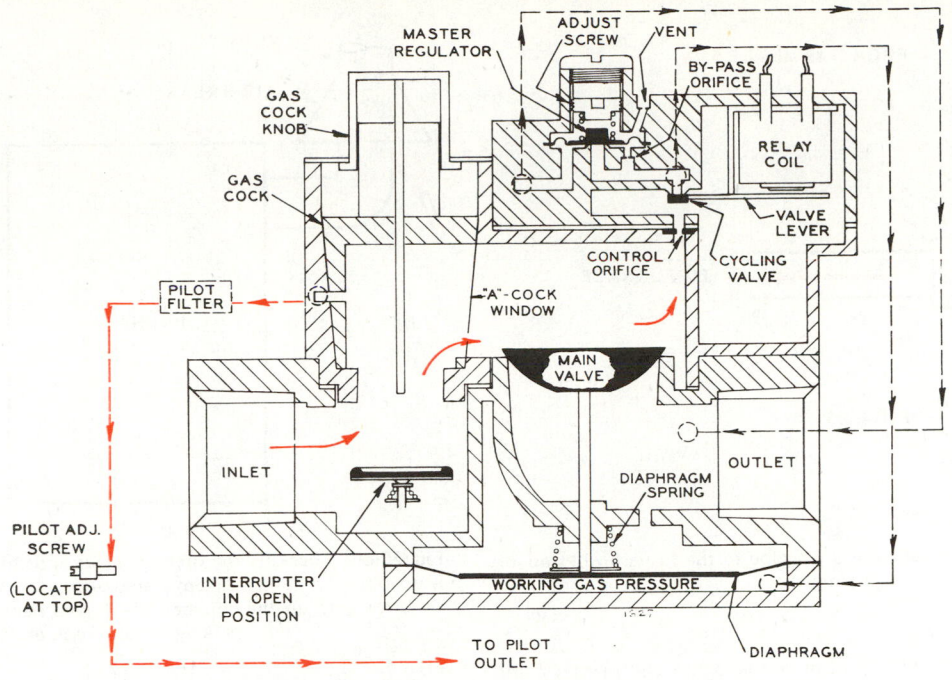

Fig. 20-60. Schematic of gas control in closed (no flow) condition.

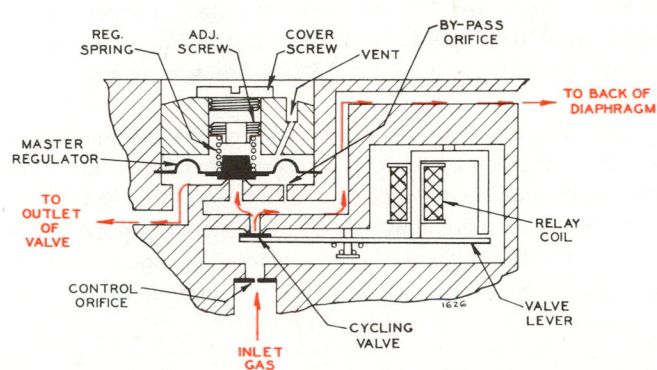

Fig. 20-61. Bypass control system for opening a gas control main valve. Relay coil is energized. (White-Rodgers Div., Emerson Electric Co.)

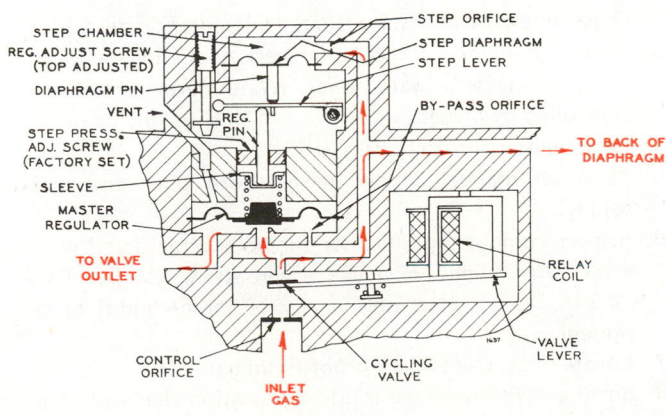

Fig. 20-62. Bypass control system used to close main valve. Relay coil is not energized.

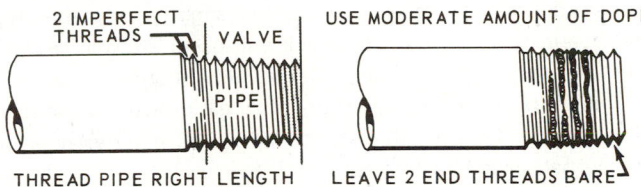

Fig. 20-63. Diagram shows proper way to put pipe compound on threads. (Honeywell, Inc.)

control valves, cut the pipe accurately and apply the thread sealing compound correctly. See Fig. 20-63. Tighten pipe joints with a moderate torque. The regulator and valve housings are usually made of a zinc die casting, and excessive force will crack them.

It is good practice to install a pipe drip leg in the gas pipe to the furnace to keep dirt and moisture from entering the pressure regulator and gas controls. See Fig. 20-64.

When threading the pipe into the pressure regulators and controls, place the wrench holding the regulator or control near the opening where the pipe connects to the control. Otherwise, the die cast bodies may be bent out of shape, and may even crack.

When starting a gas furnace unit, first purge the air from the installed gas piping up to the furnace gas valve. Allow the purge air (it will have some gas mixed in with it) to clear the room. Then turn on the furnace hand valve, light the pilot light and start the system.

Only use soap suds when checking for gas line leaks. Use of a match or any open flame may cause an explosion. Another safe practice when looking for gas leaks is to turn off the electric power and use a spark-proof flashlight.

Air Conditioning Systems, Heating and Humidifying / 725

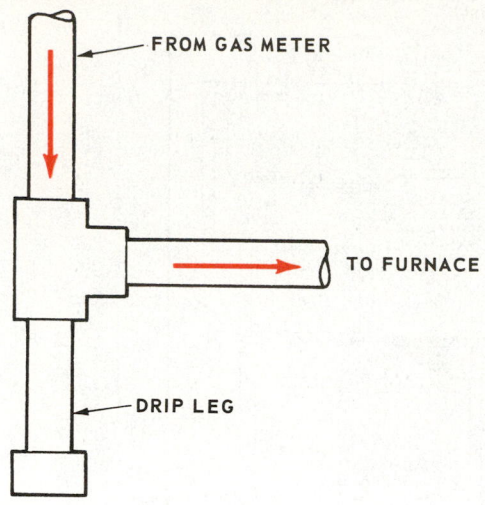

Fig. 20-64. Drip leg installed in a gas pipe to the furnace will trap dirt and moisture.

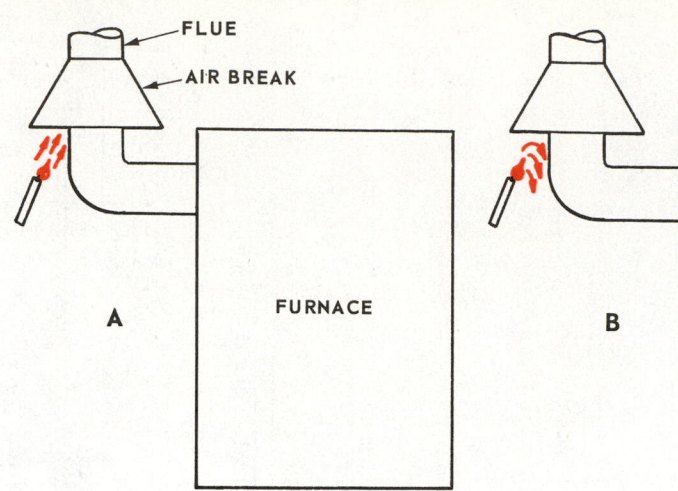

Fig. 20-66. Checking the draft in a flue and chimney when the furnace is operating. A—Smoke from a smoke candle should be drawn into the air break and up the chimney. B—Smoke being blown away from the flue indicates a back draft.

Gas burners burn with a blue flame when the primary and secondary air adjustments are correct. A yellowish flame indicates a lack of primary air, and perhaps secondary air. A collection of soot usually indicates a lack of secondary air.

Flame propagation (speed of burning) is important. The primary air and gas mixture must flow slightly faster than the flame burns. Otherwise, flashback will occur and the gas will start burning at the primary inlet (spud). Flashback occurs if too much primary air is used or if fuel gas pressure is too low. If primary airflow is too fast, the flame will blow away from the burner (called "lifting").

Annually check the primary air inlet, fuel gas pressure, flame color, pilot lights and burners. Fig. 20-65 shows a water column manometer which is used to measure gas pressures.

Incomplete combustion may be the result of:

1. Poor mixing of fuel gas and air.
2. Partial lack of air.
3. Temperature too low to produce ignition, or to keep combustion going.

Flue gas temperatures must be maintained above the condensation temperature (dew point) of the flue gas. To check the flow of gases in the flue and in the stack, use a smoke candle or a lighted match. See Fig. 20-66.

For natural gas, the recommended stack temperature is 300 F. to 900 F. (149 C. to 482 C.), depending on the length of the chimney. Values for LP-Gas are approximately the same.

Excessive temperatures in any part of the heating system will quickly erode the unit. It is recommended that no part of the system should exceed 830 to 1230 F. (443 to 666 C.).

In all cases there must be a sufficient air supply to the furnace for combustion. Approximately one square inch of opening is needed for each 1000 Btu per hour capacity of the furnace.

20-22 STARTING A GAS FURNACE

The service technician should take the following steps in starting a gas furnace:

1. Adjust thermostat to its lowest setting.
2. Determine type of system (forced air, water or steam).
3. Check installation for completeness, including flue, controls, wiring and water level. (Water level of boiler is controlled by owner.)
4. Read lighting instructions.
5. Close all manual supply valves, both pilot and main supply.
6. Inspect combustion chamber for holes, cracks and water leaks. Use a light, or make a salt spray test (see Para. 20-24). Do not start unit if any of these conditions are present.
7. Locate pilot. Use a spark-proof flashlight.
8. Smell and listen for gas inside combustion chamber. If gas is present, do not proceed until this condition is corrected.
9. If pilot gas line is connected before main gas shutoff, light

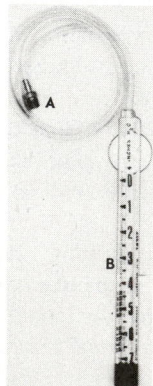

Fig. 20-65. A manometer used to measure gas pressures up to 7 1/2-in. water column. Large scale enables easy reading. A—Connection to gas line. B—Scale.   (Halbro Products)

the pilot.

10. If pilot gas line is after main shutoff, close pilot valve, open main shutoff, and listen for gas entering the furnace. If leakage is noticed, turn main valve off and determine cause of leakage and repair. If no leakage is noticed, open pilot valve and light pilot.
11. Check pilot flame for size, color, stability and position to insure quick, safe lighting of main burner. If pilot flame is yellow, clean pilot light.
12. Check main burner, including position of any baffles.
13. Determine type of safety system. If manual, reset engage mechanism.
14. Turn on main valve to check thermostat, then turn it off.
15. Turn thermostat above room temperature.
16. Turn on main valve.
17. Check "off time" after pilot has heated up.
18. Inspect all controls and operate them.
19. Check gas pressure with a manometer.
20. Adjust thermostat to temperature customer desires.

20-23 GAS FURNACE SERVICING

To locate the source of a problem in a gas furnace, use the following steps:

I. Gas valve doesn't operate.
  A. Pilot light not burning like it should.
    1. Low gas pressure.
    2. Excessive draft.
    3. Partially clogged pilot orifice or filter.
  B. Pilot lights, but goes out when reset button is released.
    1. Loose or dirty thermocouple.
    2. Defective thermocouple.
    3. Defective power unit.
    4. Loose or dirty electrical connection.
  C. Pilot lights, but will not burn normally.
    1. Yellow flame, because pilot primary air is partially clogged.
    2. Blue wavering flame, caused by too much draft.
    3. Flame too small, because orifice is partially clogged or pressure is too low.
    4. Flame blows off pilot, because pressure is too high.
    5. Flame too sharp in shape, because orifice is too small.
II. Gas valve operates, but poor heating.
  A. Delayed ignition.
    1. Weak pilot.
    2. Pilot not mounted correctly.
    3. Pilot flame not located near burner.
  B. Flashback to spud.
    1. Not enough primary air.
    2. Pressure too low in manifold.
    3. Burner inserts damaged.
  C. Not enough heat.
    1. Thermostat set too low or out of calibration.
    2. Some heat source placed near thermostat.
    3. Limit control cut-out switch set too low.
    4. Furnace dirty; soot, lint or dirty filter.
    5. Ducts dirty.
    6. Low manifold pressure.
    7. Pilot goes out while unit operates.
  D. Too much heat.
    1. Thermostat set too high or out of calibration.
    2. Burner keeps operating when thermostat opens.
  E. Soot and/or fumes.
    1. Puffs back because of delayed ignition or overfiring.
    2. Not enough primary air.
    3. Firepot or heat exchanger leaks, resulting from not enough duct air or too much heat.
  F. Cycles too often.
    1. Thermostat anticipator loose or burned out.
    2. Limit control.
  G. Does not cycle often enough.
    1. Faulty thermostat.
    2. Dirty contacts.
    3. Loose or dirty electrical connections.

20-24 FURNACE INSPECTION (WARM AIR)

The condition of the furnace should be checked at least once each year. Usually, this checkup is made before the heating season starts:

1. Clean and light the pilot.
2. Test pilot safety.
3. Check and oil fan motor.
4. Check blower and fan belt; oil blower bearings.
5. Test fan control operation.
6. Test limit control operation.
7. Test gas valve operation.
8. Check burner adjustment.
9. Check thermostat operation.
10. Check filters (usually replace).
11. Check heat exchanger (usually clean it).
12. Check combustion chamber for cracks.
  a. Spray a salt water solution (6 tsp. salt per pint of water) into the combustion chamber with the flame lighted.
  b. Direct a small propane torch flame into warmed air inside furnace jacket. If combustion chamber is cracked, color of torch flame will turn from blue to yellow.
  c. With warm air systems, combustion chamber must be airtight. Any leakage through combustion chamber may allow products of combustion to enter occupied space through warm air registers.
13. Operate system and make a stack gas analysis.
14. Make adjustments if necessary.

20-25 WARM AIR HEATING SYSTEM

The forced warm air system or mechanical warm air system uses a motor blower to increase the flow of heated air to the needed areas. This system usually includes an air filter. Fig. 20-67 shows a forced warm air furnace with a gas fuel heating unit.

With this heating arrangement, a system of ducts delivers warm air to the various spaces to be heated. The ducts are carefully sized to provide the correct amount of heat to each

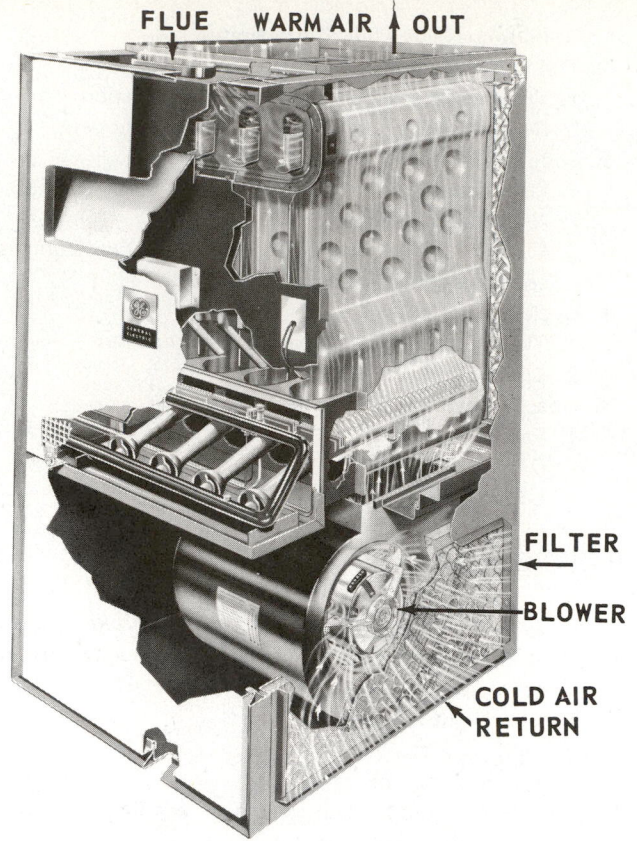

FLUE   WARM AIR ↑ OUT

FILTER ←

BLOWER →

COLD AIR RETURN →

Fig. 20-67. Gas-fired, forced warm air furnace has a capacity of up to 180,000 Btu per hr. (General Electric Co.)

Fig. 20-69. Hot water boiler with a combination gas and oil burner. A—Boiler section. B—Insulated jacket. C—Flue.

20-26 HYDRONIC HEATING SYSTEM

Hot water systems for carrying heat to occupied spaces have been in use for many years. In some systems, the hot water circulates by thermal convection. The circulating water is under atmospheric pressure, and changes in volume are provided for by an expansion tank.

Most hot water systems use a circulating pump. The pump increases water flow and carries more heat per unit of time to the room heat transfer units. Having a pump in the system permits the use of a smaller furnace. Fig. 20-68 shows a cross-section of a gas-fired furnace for a hydronic installation. The gas burner is constructed to provide a wide surface of

room. See Chapter 22 for duct sizing. Another system of ducts returns the cool air to the furnace for reheating. To provide the correct humidity conditions, a humidifier usually is installed in the airflow, either in or near the furnace.

All ducts collect dust and dirt during use. They should be vacuum cleaned about each five years, using special equipment designed for this purpose.

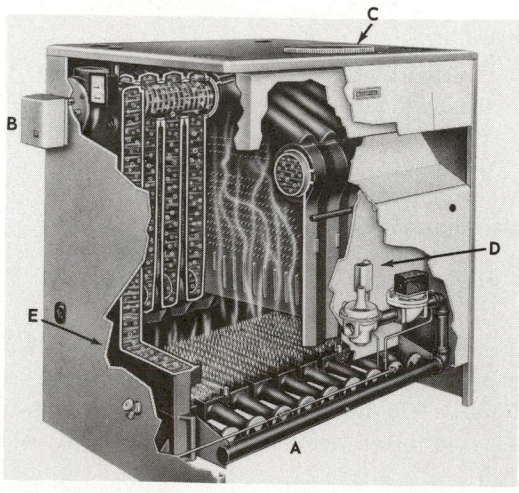

Fig. 20-68. Cross section view of a nine-burner, gas fired furnace used with a hydronic home heating system. A—Burners. B—Controls. C—Flue. D—Gas controls. E—Insulated jacket.

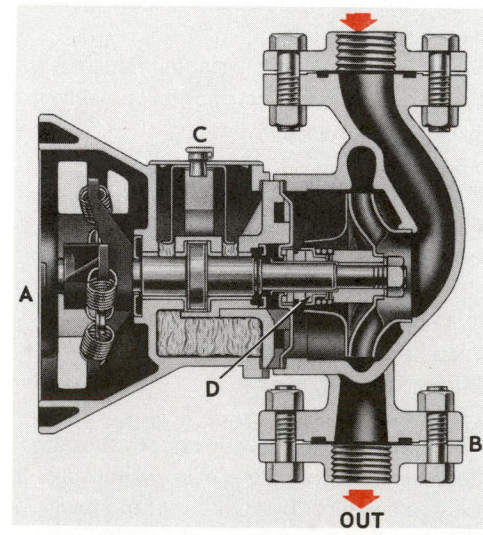

OUT

Fig. 20-70. Cross section of a hydronic system pump. A—Motor mount. B—Flange for threaded pipe connection. C—Oil cup. D—Seal. (Bell and Gossett, ITT)

Fig. 20-71. Exterior view of hydronic water pump. A—Motor. B—Capacitor. C—Inlet shutoff valve. D—Outlet shutoff valve. A cutaway view is shown in Fig. 20-72.

flame to heat the water quickly and evenly.

A combination fuel gas-air and oil-air hot water boiler is shown in Fig. 20-69. Switches on the burner control permit the choice of fuel being burned, either gas or light oil.

Fig. 20-70 shows a circulating pump of the centrifugal type. A shaft seal is located where the pump shaft leaves the casing.

A hermetic type of hydronic pump is shown in Fig. 20-71. It will pump 10 gpm at a 6-ft. water head. The motor is a permanent split capacitor type. Its rotor turns in the water, and the shaft seal has been eliminated. See Fig. 20-72.

To remove the pump, turn the shutoff valves all the way in. See Fig. 20-73. A piping arrangement when one pump is used is shown in Fig. 20-74. A system with three zones is illustrated in Fig. 20-75. It has a pressure relief valve, which is required in all pressurized heating systems.

Fig. 20-72. Internal view of a hermetic hydronic pump. A—Turn acorn nut all the way out to check for proper rotation. Then turn the nut all the way back in.

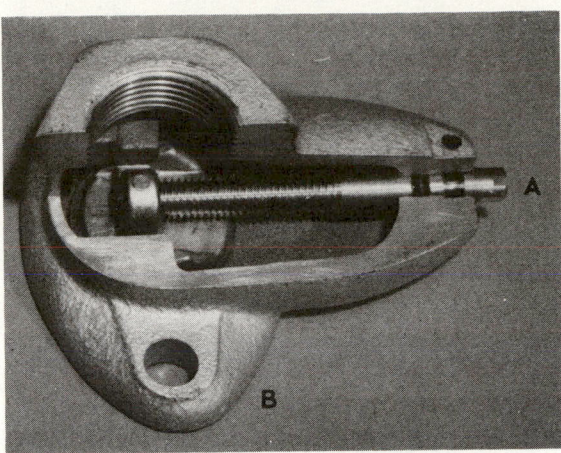

Fig. 20-73. Inside construction of shutoff valve for hydronic pump. A—Valve stem. B—Bolt flange. (Sundstrand Hydraulics, Div. of Sundstrand Corp.)

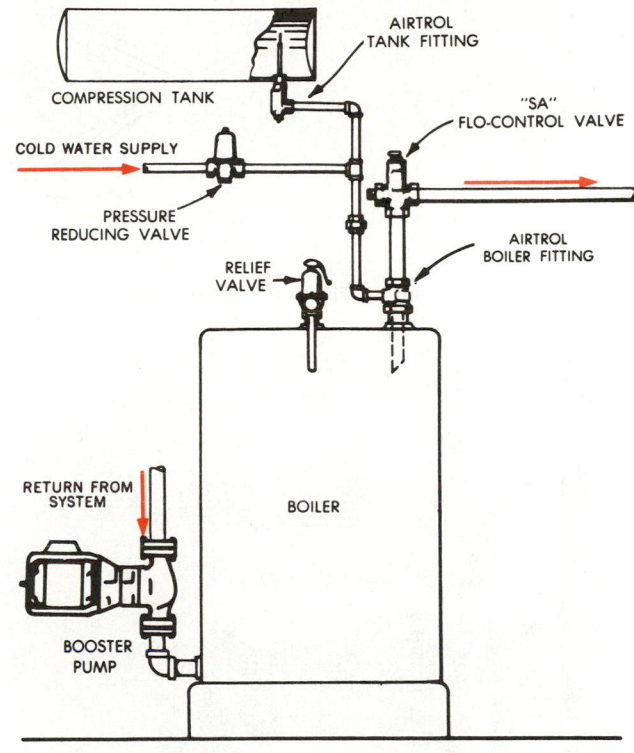

Fig. 20-74. Diagram of hydronic system shows piping and location of pump, compression tank and relief valve. (ITT Bell and Gossett)

Hot water heating systems may use one of several different temperature control devices:

1. Single control, which starts and stops the pump.
2. Zone control, using two or more controls. Each control operates one pump.
3. Individual radiator controls for individual room control.

Fig. 20-76 illustrates a hydraulic type radiator control. Temperature control is adjustable. Note that the control operates the flow valve, and the sensitive bulb is located in the cold air inlet to the radiator. It is a modulating control

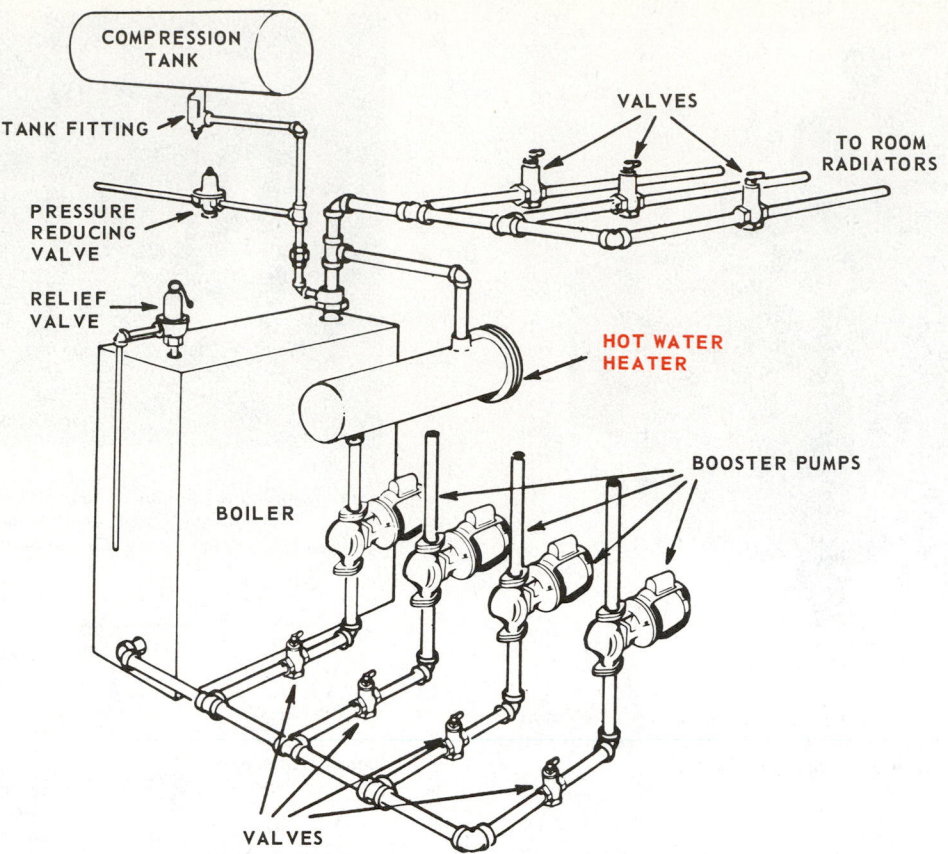

Fig. 20-75. A three-zone hydronic system with four booster pumps. Fourth pump circulates water through heater to provide hot water.

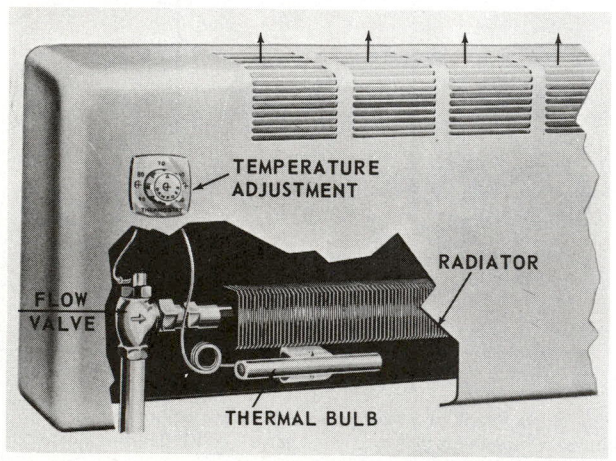

Fig. 20-76. A thermal control valve mounted on convector heating system. This control provides individual room temperature control with either circulating hot water (hydronic) or steam as heat source. (Danfoss, Inc.)

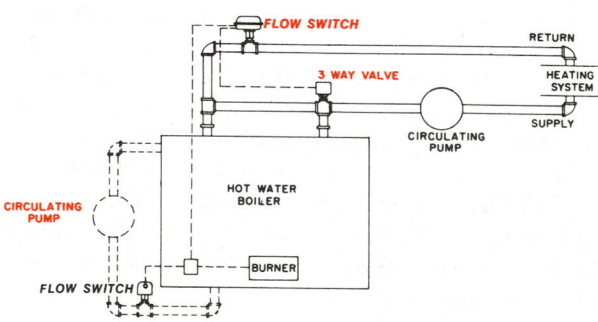

Fig. 20-77. Schematic diagram of hot water system using a circulating pump, three-way valve and flow switch. (ITT McDonnell & Miller)

(variable volume flow) that can be used for either steam systems or hot water systems (hydronic).

A hot water system using a flow switch control system is shown in Fig. 20-77. The main circuit is shown in solid lines. If the water ceases to flow, the flow switch will open the electrical circuit and shut off the burner. Some systems use a recirculating system within the boiler, as shown in the broken line piping diagram. This recirculation of water maintains a more constant water temperature within the boiler. If water flow stops in this water circuit, the flow switch shown will shut down the system.

Fig. 20-78 shows a hot water system used to heat the incoming outside air of a ventilating system.

A system which mixes outside air with return air is shown in Fig. 20-79. The air exhaust, recirculated air and fresh air intake are controlled together to insure proper air conditions

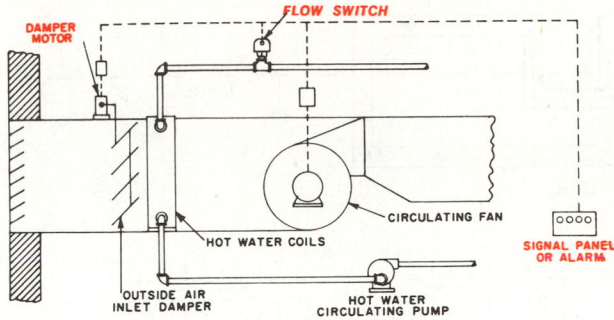

Fig. 20-78. Hot water heating system used to heat incoming air of ventilating system. Note motorized damper control, flow switch and signal panel. (ITT McDonnell & Miller)

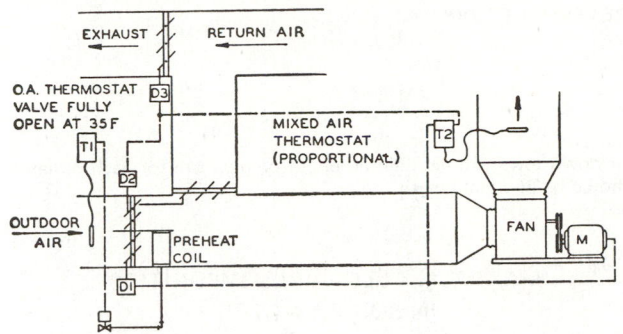

Fig. 20-79. Schematic diagram of controlled air system. Exhaust air, fresh air intake and recirculated air are proportionally mixed to obtain correct atmosphere for occupied space. (ASHRAE Guide and Data Book)

in the occupied space.

In many establishments, it is desirable to maintain different temperatures in different rooms or parts of the establishment. This is also desirable in some homes. For instance, the bedrooms may be kept at a different temperature than the living room; the laundry at a different temperature than the kitchen. This is accomplished by thermostatically controlling the heating or cooling media for these areas.

Hydronic systems are controlled in four basic ways:

1. The heat is turned on at the same time the pump is turned up.
2. The pump is cycled to provide heat.
3. The pump is operated continuously and the zone valves are cycled.
4. The zone valves are cycled and turn on the pump or heater.

In forced circulation closed systems, the high-temperature water is above atmospheric pressure and smaller pipes can be used. The heat load is based on 20 degrees differential between water-in and water-out temperature. The top of the tubing should be even. Use eccentric reducer fittings. See Fig. 20-80.

Two-pipe systems are the most common:

1. Direct return (less piping). Each circuit is a different design.
2. Reverse return. Easier to balance and each pipe circuit is the same length. See Fig. 20-81.

When installing hot water heating systems, one must allow for pipe expansion. Expansion for steel is 3/4 in./100 Ft./

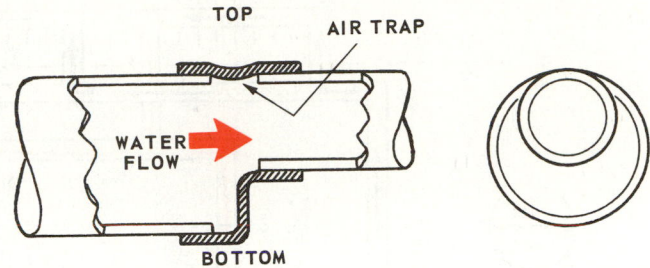

Fig. 20-80. An eccentric fitting should be used in water circulation systems to reduce danger of air pockets when pipe size is reduced.

100 F.; for copper, 1 1/16 in./100 ft./100 F. Where riser pipes connect to a horizontal run, allow for a flexible joint. Also install an expansion joint at the boiler.

When a hydronic system delivers hot water to two or more heating units (radiators), the water flow must be balanced to insure that each radiator receives its design quantity of water per unit of time. This balancing is usually done by installing gate valves in the piping. Opening and/or closing these valves obtains balance. Flow meters are used to accurately measure quantity of flow.

Water used in hydronic systems usually has substances added which lower the freezing temperature and raise the boiling temperature. These substances may also keep the water from forming deposits in the pipe. Tap water has impurities that may cause:

1. Scale.
2. Corrosion.
3. Embrittlement.

Scale is formed from salts in the water. The salts settle on metal surfaces as the water passes through temperature changes. These salts should be removed before the water enters the heater; or, chemicals should be added to form a sludge with these salts. The salts can then be purged from the system.

The salts are usually calcium carbonate, calcium sulphate, calcium chloride; magnesium carbonate, magnesium sulphate and magnesium chloride; sodium carbonate, sodium hydroxide and silica oxide. Iron and manganese may also form deposits in the boiler.

Corrosion takes place when the water is acidic or gases are dissolved in the water. Corrosion is reduced by neutralizing the acid condition with an alkali and by removing the gases by de-aeration (release of gases dissolved in a liquid). Chemical scavengers and corrosion inhibitors are also used.

The usual dissolved gases are hydrogen sulfide, carbon dioxide and oxygen.

Organic matter and oil cause foaming of the water.

Some companies specialize in the treatment of boiler water.

Embrittlement causes metal failure along drum seams, under rivets and at tube ends. Water flashing to steam through any small leaks in these highly stressed areas allows any sodium hydroxide in water to concentrate. See Fig. 20-82. Embrittlement can be checked by maintaining low hydroxide alkalinity, avoiding leaks at stressed metal and using special inhibiting agents.

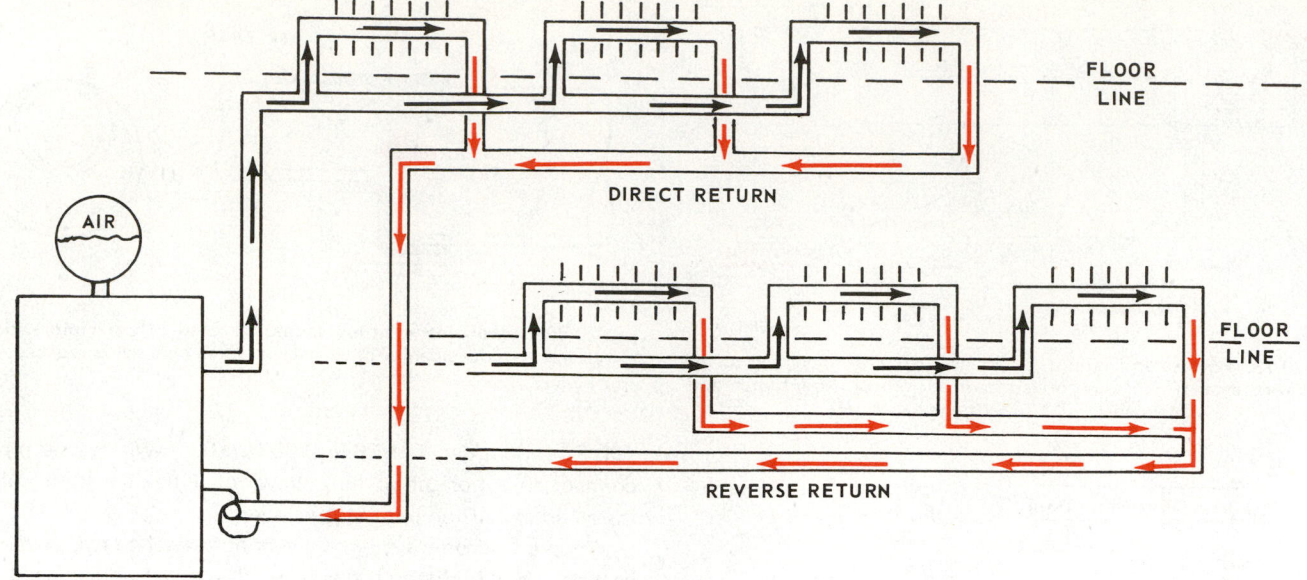

Fig. 20-81. Horizontal pipes in a hydronic steam system should always slant down toward boiler. Total length of pipe run for each radiator (both hot water and return water) should be of equal length.

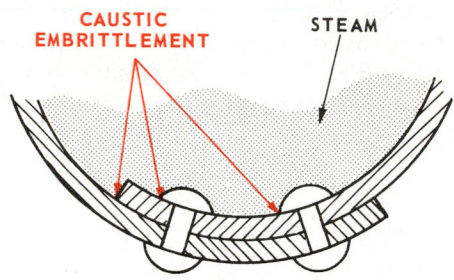

Fig. 20-82. A steam container (pipe or boiler) showing where caustic embrittlement can take place. Activity is greatest where the metal is stressed and where caustic material (sodium hydroxide) can collect.

MAXIMUM ALLOWABLE IMPURITIES IN BOILER WATER		
CHEMICAL NAME	CHEMICAL SYMBOL	PPM
SODIUM SULPHIDE	$NaSO_3$	1.0
SODIUM CHLORIDE	NaCl	10.0
PHOSPHOROUS OXIDE	PO_4	5.0
SODIUM SULPHATE	$NaSO_4$	25.0
SILICA OXIDE	SiO_2	0.20
TOTAL DISSOLVED SOLIDS		50.0

Fig. 20-83. Maximum allowable amount of certain impurities in good quality boiler water listed in parts per million (ppm).

Safe levels of impurities in boiler water for various chemicals are shown in Fig. 20-83. The pH level should be about 10. See Para. 14-53.

20-27 INSTALLING HYDRONIC SYSTEMS

The boilers for hydronic systems must be mounted level. All local code requirements must be checked and must be followed.

A new boiler should be boiled out before it is put in service. The water to be used should be analyzed and a preventative maintenance treatment procedure should be recommended.

After a system is installed, and before putting it into service, always flush the system. The fluxes, pipe joint compounds and cutting oils sometimes form gases in a system. Dirt, sand, steel thread chips, sawing chips, filing chips and solder bits may erode the system and clog screens; also ruin valves and pump seals.

Fill the system with water, add about one pound of tri-sodium phosphate for each 50 gallons of water. Circulate

for about four hours, then drain. Clean the screens and fill the system with water. It is ready to operate.

Unless the pump seals leak and the vents are used quite often, avoid treating the water with chemicals. The chemicals may injure seals and valves. Avoid using phosphates and polyphosphates, and do not use over 300 ppm of chromates and over 500 ppm of nitrites.

Organic growth can be controlled by using sodium penta-chlorophenate. It is best to consult a water treatment expert before attempting boiler water treatment.

20-28 SERVICING HYDRONIC SYSTEMS

Air in a hydronic system is a common cause of trouble. Air will cause noise in the system, reduce the system water supply and interfere with water circulation. The air acts as a brake on the circulation of water. Series systems can be purged of air by putting an outlet hose in a bucket. When bubbles stop appearing, the air has been removed.

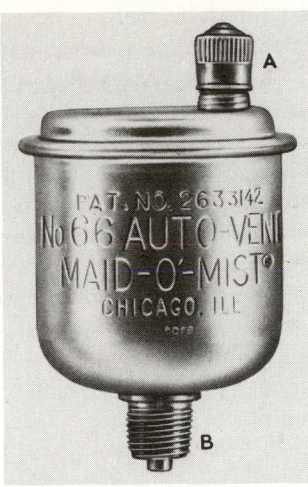

Fig. 20-84. An automatic air vent for hot water heating systems (also on chilled water systems) is installed at high point of convector, baseboard units or radiator. A—Vent. B—Connection to system.

Other systems must have a manual or automatic air vent at the high points in the system. See Fig. 20-84. A standpipe (drain) should be provided for each air vent.

Thermostats are another common source of heating problems for these reasons:
1. Vibration.
2. Poor contacts.
3. Broken wires.
4. Improper temperature settings.
   Other common problems include:
1. Air in hydronic systems using natural gas. Also pump motor failure and water leaks around the pump packing.
2. Water pump motor and pump bearings (noise) made of carbon or teflon (water lubricated).
3. Cavitation.
4. Turbulence.
5. Poor supports (noise).
   Hydronic heating problems include:
1. Uneven heating. (Room-by-room heat loss calculation needed. Balancing valves needed.)
2. Velocity.
3. Vibration of parts of system.
4. Pipe expansion noises (do not clamp tightly; pipes should not touch edges of openings through floors). Expansion joints are of considerable help.
   Some controls needed are:
1. Air control devices (very much needed).
2. Air vents (piping pitched up to the vent). Automatic air vents are a problem source.
3. Balancing valves (to control flow rates).
4. Check valves that prevent reverse flow during low heat load periods.
5. Gate valves.

Venting a system requires more water to be added to a system. This water contains air in solution, plus some corrosive chemicals. It is best to trap this new air in the compression tank. Avoid venting a compression tank if at all possible. The air compressed in this tank permits the water volume in the system to expand and contract, and it permits higher water temperatures.

When the system is working properly, one should be able to hold a hand on the compression tank. Higher temperatures indicate problems, and the relief valve may open. Because the compression tank water is cooler, it absorbs air and flows into the boiler. When in the boiler, the water is heated and the air is released. Some air may go into the pipes and end up in the vents. The tank may gradually lose its air, then the relief valve will open. If the relief valve spills water on each heating cycle, the compression tank lacks air. Most air is released from the water when the water velocity is lowest, and where the temperature is highest (in the boiler).

Another problem is when the heat is off. The water will cool and the pressure will drop. The reducing valve will allow more fresh water (containing more air) into the system. If this continues, the inside of the system will become corroded from the chemicals that were in the "makeup" water.

Cut the exhaust end of the pipe from the relief valve at an angle to prevent someone plugging it or capping it.

20-29 PREPARING BOILER SYSTEM FOR HEATING SEASON

Use the following checking procedure on a boiler system before the heating season begins:
1. Clean burner (gas or oil).
2. Clean nozzle of oil burner. Use a cloth and solvent. Do not use wire brushes (bristles may scratch orifice).
3. Clean and adjust electrodes. Inspect insulation. Replace if cracked.
4. Clean flame detector lens. Operate by closing fuel valve; detector controls should lock out.
5. Clean pilot light if a gas burner is used.
6. Tighten all connections.
7. Oil motor.
8. Check motor temperatures. If warmer than normal, cleaning or new bearings may be necessary.
9. Inspect tubes for soot or fly ash. Clean.
10. Inspect breeching. Clean.
11. Cycle controls. Shut off water feed to check for low water cut off.
12. Operate all valves and cocks to check operating condition.
13. Operate safety and relief valves.

20-30 STEAM HEATING SYSTEMS

Steam heating is a means of distributing heat to occupied areas. Steam is generated in a boiler. The steam (vapor), being lighter, travels to the upper parts of the piping circuit. This steam is at 212 F. (100 C.) or higher, except for vacuum systems (rare). As it releases its heat to the occupied area, the steam condenses. The condensed water, being heavier, returns to the furnace boiler.

The steam releases about 1000 Btu for each pound that condenses. The heat exchange devices located in the room are called "radiators."

Two basic systems are in use. The single-pipe system uses

Fig. 20-85. Steam boiler. A—Hot water heater. B—Cast iron sections. (Weil-McLain Co., Inc.)

the same pipe to carry steam to the radiators and return the condensate. The two-pipe system uses one pipe to carry the steam to the radiator and another pipe to return the condensate.

For domestic purposes, these systems operate at low pressures or at a partial vacuum. The units are tested at 50 psi for safety purposes. Commercial and industrial systems use progressively higher pressures that approach 1000 psi.

A steam boiler must have pressure safety valves, water level gauge, pressure gauge and temperature gauge. Fig. 20-85 shows a steam boiler.

A hot water heating unit that may be inserted in a boiler is shown in Fig. 20-86.

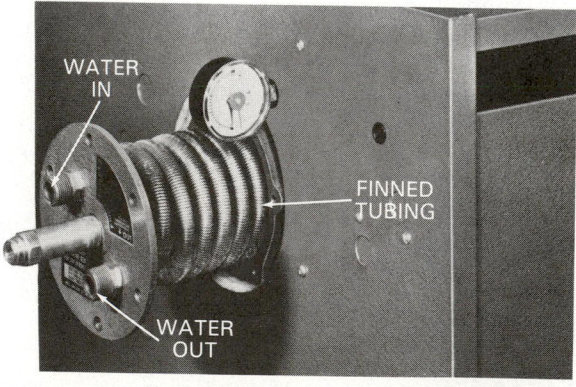

Fig. 20-86. Hot water heater being mounted in a steam furnace. (Weil-McLain Co., Inc.)

20-31 STEAM HEATING INSTALLATION

Steam heating systems are excellent heating systems. When room radiators are used, the system does not produce humidity or humidify or clean the air. Air circulation is by thermal movement (hot air rises). Separate humidity systems

and cleaning systems must be used.

Many installations have the radiator installed in a forced convection duct. This duct system may also have a filtering system and humidifying system.

All steam heating systems must be installed according to code regulations. As with hot water boilers, the steam boiler must be mounted level. The pipes must be mounted with a slope down to the boiler. The piping must be designed to provide for expansion of the pipe. Air vents must be located at the high points of the system. Each radiator should have an air vent and a steam trap.

The steam trap keeps the steam in the radiators and only allows condensate to return to the boiler. There are three types of steam traps:
1. Mechanical.
2. Thermostatic.
3. Impulse.

After installation, the system must be flushed to remove all dirt. The system must be leak tested as per code regulations before it may be operated. After system startup, perform the safety inspection detailed in Para. 20-33.

20-32 STEAM HEATING SYSTEM SERVICE

Servicing a steam heating system should be done with great care. Escaping steam or condensate can cause severe body burns. A boiler will explode if the steam pressure is permitted to exceed the boiler safe pressure.

Check the water level gauge and the pressure gauge. Shut down the system at once if the water level does not show in the level gauge, or if the pressure gauge is above normal.

Servicing the gas burner or the oil burner has been explained in previous paragraphs.

If one radiator of the system is cool while the others are hot, this radiator is not receiving steam. This problem may be caused by:
1. Thermal valve to radiator is closed (thermostat for valve or valve not working).
2. Radiator may be air bound (air vent not working).
3. Radiator may be filled with condensate (steam trap not working).

Lightly rapping the control with a rubber mallet may loosen the valve momentarily and allow the system to work. If this happens:
1. Shut down system.
2. Reduce pressures to atmospheric by purging. (BE CAREFUL! Stand to one side to prevent burns from steam or condensate water.)
3. Replace thermal valve, air vent or steam trap.
4. Start up system.

If the system is low on water, and the boiler is hot, cool the boiler before opening or repairing the boiler feed valve. Cold water may crack a hot boiler section.

Some boilers are made with tubing heat exchangers. These boilers may use either a water tube or a fire tube design.

Short cycling times of the furnace cause rapid temperature changes in the boiler. After several thousand cycles, the boiler walls or tubes may crack.

20-33 STEAM HEATING SAFETY INSPECTION

A steam heating system should be checked once each month during the heating season.

Check the water level. The water level gauge must show that the boiler water level is one-third to one-half full. If the water level is higher, drain the system to the correct lower level. If the water level is low, but still shows in the level gauge, add water until the level is correct. Close the fill valve completely. If no water shows in the water level gauge, shut the system off at once. Add water only after the boiler has cooled.

The water level sight glass may show water if its openings are clogged. Trust the sight glass reading only if the system is clean, and if the sight level glass and its connections have been cleaned recently. Water level petcocks (when used) are a good way to check the sight level glass tube reading. **Be sure the petcocks do not spill steam or hot water on anyone. Wear goggles!**

Operate the safety to be sure it is not stuck closed. Again, be careful not to allow the escaping steam to hit anyone. Wear goggles! If no steam or water comes out, shut off the system at once and have the relief valve serviced or replaced.

Check the low level shutoff by opening the drain valve; the shutoff should operate right away. If it does not operate, shut down the unit and service the low water level shutoff.

20-34 ELECTRIC HEAT

The use of electricity to heat homes, stores, commercial buildings and factories is steadily growing in popularity.

Some advantages of electric heat are:

1. Low first cost.
2. Electric heating devices need no oxygen and, therefore, need no air supply.
3. Highest temperatures needed are below the ignition temperature of most materials. Therefore, the system is considered safer than other heating systems.
4. Because of the absence of combustion and combustion gases, there is less danger of toxic conditions arising. No chimney is needed.
5. Equipment normally requires less space.
6. Individual room temperature control is easily obtained.

The electric heating process has some disadvantages:

1. Cost per unit heat is higher than for some other fuels.
2. Humidity control problems may occur.
3. Added electrical circuits are required.

20-35 PRINCIPLES OF ELECTRIC HEATING

Electricity is a form of energy. Since energies can be changed to other forms of energy, electrical energy can be changed to heat energy.

Heating with electricity can be done either directly or indirectly:

1. Direct heat.
   a. Resistance heating, accomplished by passing a fluid over an electrically heated element. (Air is the fluid, although some units use water.)
   b. Radiant heating, accomplished by heating an element to a temperature high enough to give off heat.
   c. Thermoelectric heating. (See Chapter 17.)
2. Indirect heat, by using a heat pump. (See Chapter 23.)

20-36 APPLICATIONS OF ELECTRIC HEATING

Electric heating has a wide range of applications. It may be used for heating operations in industrial processes, such as fast drying of paints and melting low temperature metals or metal alloys. It has been used extensively for domestic and commercial cooking and baking. It is widely used for providing hot water.

For residential use, electric heating has a growing application for use as:

1. The only source of heat.
2. Supplementary heat, even though some other system provides much of the heat in the residence, or where the heat load is low. Heat pump systems may use electric resistance heating in climates where the heating energy required during cold weather is more than the heat pump can supply.
3. Resistance heating is used to provide heat in parts of a building that are unsatisfactorily heated by the standard system. It also may be used to heat additions to buildings where the present system does not have enough capacity to heat the addition, or where the extension of the present system would be too costly.

20-37 PRINCIPLES OF ELECTRIC RESISTANCE HEATING

In electric resistance heating, metals are generally used as heating elements. The metals are designed to permit a certain current to flow at either 120V or 240V to provide the heat required.

Some units are designed to operate at incandescent temperatures; some are mounted in protected cabinets. Types of systems used are baseboard units or wires installed in floors, walls and/or ceilings. These systems are described as non-incandescent temperature units.

The higher the temperature of the heating element, the smaller the space it needs to occupy. Some units are designed with high temperature heating elements to release both air heating energy and radiant energy.

Electrical energy is changed into heat energy in the following ratios:

$$1 \text{ watt} = 3.415 \text{ Btu}$$
$$100 \text{ watts} = 341.5 \text{ Btu}$$
$$1000 \text{ watts} = 3415 \text{ Btu}$$

The voltage multiplied by the amperage flow in a circuit equals the watts.

At 20 amperes maximum input, a 120V circuit will provide:

2400 watts (20 x 120 = 2400) or
8196 Btu (2400 x 3.415 = 8196)

Example: A home with a need for 50,000 Btu/hr. for heating, therefore, needs the following amount of electricity:

Btu = Btu/watt x watts

Btu = 3.415 x volts x amperes, or about 60 amperes service at 240 volts or 14,400 watts or 14.4 kilowatts

Btu = 3.415 x 240 x amperes

50,000 = 3.415 x 240 x amperes

Amperes = $\frac{50,000}{3.415 \times 240} = \frac{50,000}{891.6}$ = 61 amperes

Watts = 240 x 61

Watts = 14,640 = 14.64 kilowatts

For an hour = 14.64 kilowatts

= 14.64 kW/hr.

For most computations, 1 watt equals 3.4 Btu

20-38 BUILDING DESIGN FOR ELECTRIC RESISTANCE HEATING

In some cases, converting a coal, oil or gas-fired heating system in a building into an electric resistance heating system is possible and practical if the building is modified to reduce heat transfer and air infiltration. The walls and ceiling should be insulated as thoroughly as possible. Basement walls or the slab of buildings without a basement must be insulated. Windows should be double glazed. Wood or plastic window and door frames are preferred, rather than metal. Walls should have 4-in. insulation; ceilings, 7-in. insulation. The floor slab should be insulated 4-in. thick and 42-in. deep. The basement wall should have 2 to 4-in. insulation. Follow the latest specifications available from manufacturers.

Where electric resistance heating is desired, the building should be designed to reduce heat losses to cut operating cost to a minimum.

When possible, advantage should be taken of solar heat as it contacts the east, south and west exposures.

Humidity control may require dehumidification rather than humidification.

20-39 ELECTRIC HEATING ELEMENTS

There are three types of electric resistance heating wires:
1. Open wire.
2. Open ribbon.
3. Tubular cased wire.

The open wire type usually consists of nickel chromium resistance wire mounted on ceramic or mica insulation. The open wires must be carefully protected to keep them from being contacted by metal objects and/or by humans or animals to avoid the danger of burns or electrical shock. Some heating wire designs are shown in Fig. 20-87.

Ribbon type resistance heating wires are made of the same material as open wires, and they are mounted in the same general way. It, too, must be carefully covered by a grid to prevent burns or shock. The ribbon design provides more surface exposure for air contact.

The tubular heating element is similar to the heating elements used in electric stoves. Usually, nickel chromium resistance wire is surrounded by a magnesium oxide powder. The wire and powder are enclosed in a heat and corrosion-resistant steel tube.

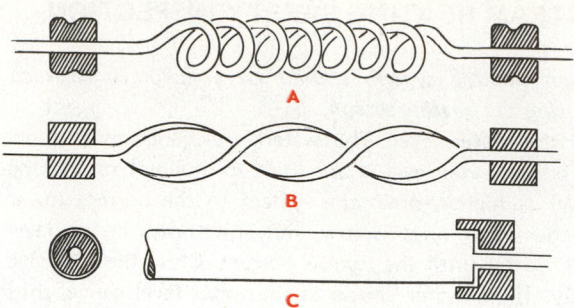

Fig. 20-87. Three types of electric heating elements. A—Open wire. B—Open ribbon. C—Tube encased wire.

The tubular cased wire design protects against electrical shock. However, the element may reach rather high temperatures. To increase the heating surface, and to reduce the danger of high-temperature wiring, tubular covered elements are sometimes placed in fin type aluminum castings.

Tubular heating elements are usually made into:

1. A helical wound coil of resistance wire that is supported by ceramic insulators. These open coils are made of nickel chromium (nichrome) wire and are wound on insulation (ceramic and mica).

2. Metal foil is expanded to form mesh-like resistance heating strips. These operate on lower temperatures [black heat range about 1450 F. (788 C.)] than the resistance wire.

3. A small diameter coil of resistance wire is inserted in a metal tube and powder forms the insulation. The metal tube is pressed and the powder (magnesium oxide) becomes

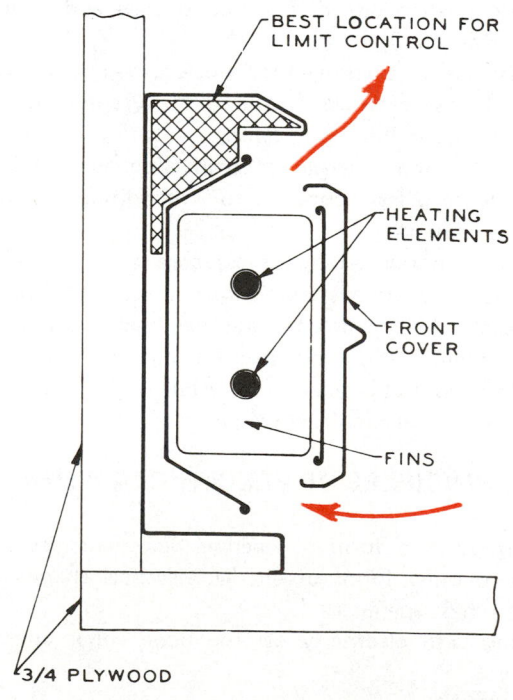

Fig. 20-88. Common construction of a baseboard natural convection electric resistance heating unit. (White-Rodgers Div., Emerson Electric Co.)

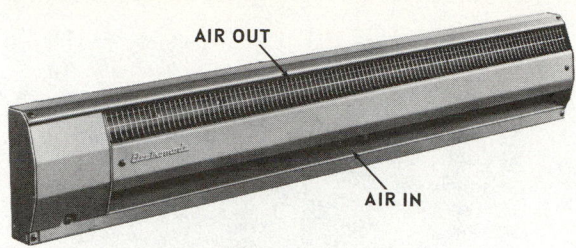

Fig. 20-89. Section of a baseboard electric resistance heating unit. Covered left end may be used to house junction box and thermostat. (The Singer Co., Climate Control Div.)

rigid. Heated to 1550 F. (843 C.), these tubes become dull red. These tubes are then formed into many shapes.

20-40 BASEBOARD ELECTRIC RESISTANCE HEATING

Baseboard heating is a popular form of natural convection heating. It has the electrical resistance heating unit mounted in a casing that is designed to efficiently move air over the heating element by natural convection. See Fig. 20-88. Warmed air is lighter (air expands when heated) and rises. The colder air, being heavier, settles to the lower opening and enters the unit to replace the rising heated air. The units used are shaped much like a regular baseboard. They are mounted on the wall close to the floor, usually under windows. Fig. 20-89 shows a section of a baseboard heater. The construction of a baseboard heater is shown in Fig. 20-90.

In most cases a baseboard heating unit should be mounted on the wall. If built into the wall, dust in the heated air coming from the unit may streak the wall and necessitate frequent cleaning. It is important to keep the air passages clear to prevent poor airflow. The temperature of the heating element may become too high if the air passages are blocked.

Each baseboard heating unit may be thermostatically operated, which permits individual room temperature control. The units are easy to install and take up a minimum of space. See Fig. 20-91. Since there are no moving parts, they are noiseless. Installation of a safety (limit) switch in each unit is

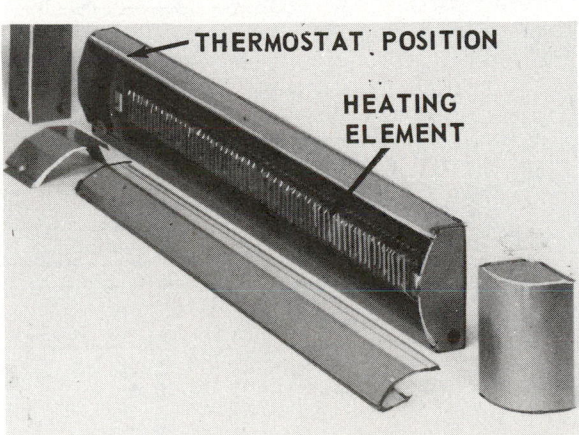

Fig. 20-90. Baseboard electric resistance heating unit with front cover removed. Note thermostat location, corner block and end piece.

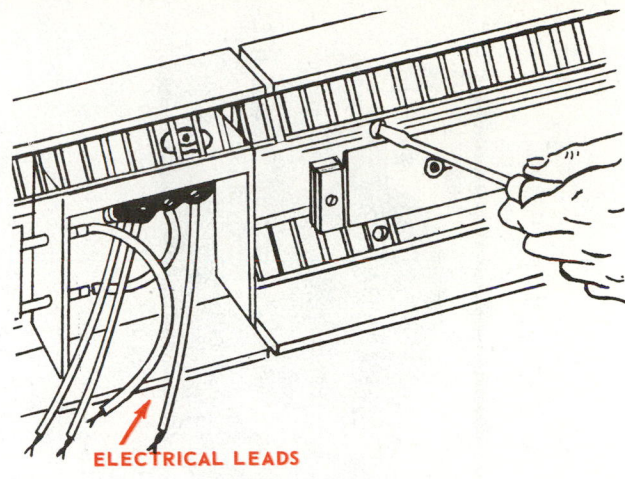

Fig. 20-91. Baseboard electric resistance heating unit with decorative panel removed, showing mounting screw.

strongly recommended. This switch will open the electrical circuit if any part of the heater reaches an above normal temperature. See Chapter 24.

Room temperature variation will be greater using baseboard electric heat than with ceiling cable heat. Baseboard heat temperature variation between floor and ceiling can be 4 to 15 F. (2 to 8 C.) with about two heating periods per hour. Cycles of five to six per hour are recommended. Ceiling cable temperature variation between floor and ceiling can be 4 to 5 F. (2 to 3 C.), with about one heating period per hour (as recommended). With ceiling cable heat, the floor is warm.

Baseboard units are available in lengths from approximately 36 to 100 in. Some units have only one heating element. Others have two or more elements connected in parallel.

On most installations, it is good practice to keep the current load to 20A or less per circuit. The use of 240V circuits, where practical, is desirable. Smaller wires can be used in 240V circuits than in 120V circuits.

20-41 ELECTRIC FURNACES

Electricity may also be used to provide heat for either warm air or hydronic central heating systems. These furnaces have capacities of 34,000 Btu/hr. (10 kW) up to 120,000 Btu/hr. (30 kW).

A forced warm air furnace that uses tubular heating elements is shown in Fig. 20-92. This furnace can be installed for upflow, downflow or horizontal airflow. The heating elements are used in stages of 5 kW or 10kW. Their use is sequenced (turned on one at a time). The electrical circuits are shown in Fig. 20-93. Note that the power in is 240V and controls are 24V.

A hydronic furnace is shown in Fig. 20-94. This furnace has a capacity of 34,130 Btu/hr. (10 kW) to 81,912 Btu/hr. (24 kW). It is a very efficient unit measuring about 10 1/2-in. wide by 21 3/4-in. high and 23 3/4-in. long.

Electric heat furnaces designed for warm air usually have several heating elements. These elements vary from 5 to 10 kW

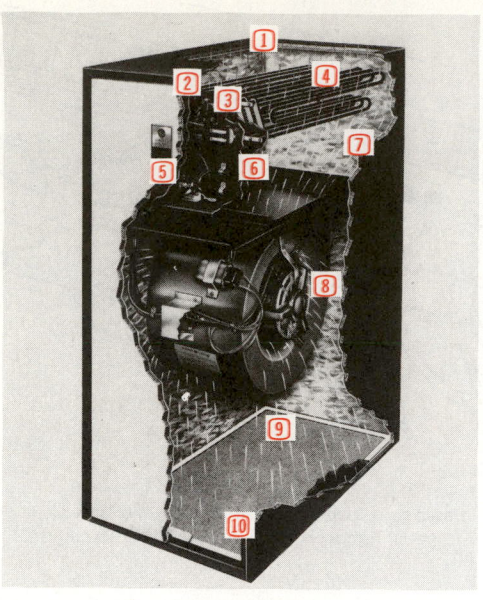

Fig. 20-92. A forced warm air furnace having tubular electric heating elements. 1—Warm air outlets. 2—Insulated jacket. 3—Fan and limit controls. 4—Electric heating elements. 5—Operating controls. 6—Fan outlet. 7—Heating chamber. 8—Fan and motor. 9—Filter. 10—Furnace base.

Fig. 20-94. An electrically heated hydronic boiler. A—Heating element. B—Expansion tank. C—Relief valve. D—Drain valve. E—Water return. F—Water outlet.

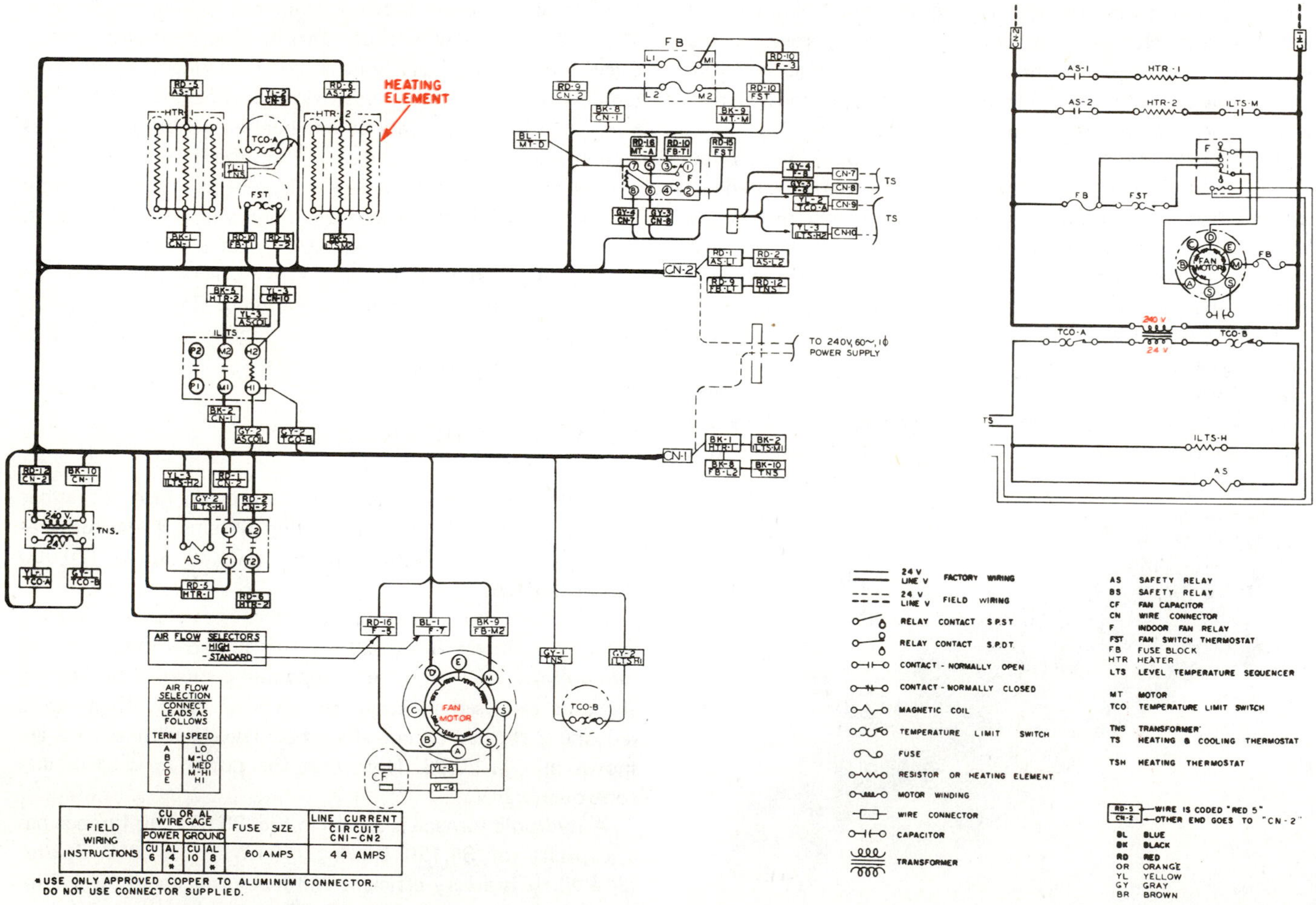

Fig. 20-93. A wiring diagram of an electric furnace, with schematic diagram.   (General Electric Co.)

each (17,000 to 34,000 Btu). These elements are usually sequenced by a relay panel which energizes them about 30 seconds apart, one after the other. The thermostat is usually the low-voltage type with two heat stages. These heat stages are put into operation from .5 to 1.5 F. (.3 to 1 C.) apart.

Humidifiers are rarely needed with electric heat.

The total furnace output should be about 20 percent over the design heat load. The extra 20 percent can be wired and controlled to turn on only during those very rare times when the heat load exceeds the design load.

20-42 DUCT HEATERS

Spaces can be heated by installing electric heating elements in existing duct systems or in ducts of comfort cooling installations.

Fig. 20-95 shows a unit, complete with controls, designed for a duct installation. A duct heat unit being installed in a duct system is shown in Fig. 20-96.

Fig. 20-97. A forced convection electric resistance heating unit. (The Singer Co., Climate Control Div.)

Fig. 20-95. An electric heating unit designed for duct installation. A—Heating elements. B—Control panel. (Brasch Mfg. Co., Inc.)

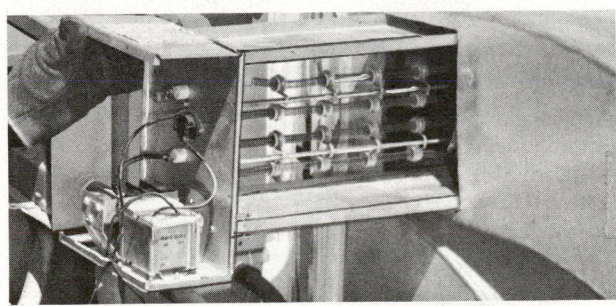

Fig. 20-96. Electric heating unit being installed in roof duct. (Tutco Inc.)

20-43 SUPPLEMENTARY ELECTRIC HEATING

Electricity is often used to heat building additions which have a heating plant without enough capacity to carry the extra load. Also, it is used where extending the present system would be difficult.

Fig. 20-98. Wall-mounted electric heater which may be used for either primary heating or supplementary heating. It has a thermostat, off-and-on switch and fan switch. (The Singer Co., Climate Control Div.)

Radiant heat panels built into the wall, baseboard heaters and resistance heating wire imbedded in the ceiling or wall plaster may be used for this purpose. Added bedrooms, family rooms and utility rooms may be heated in this manner.

Fig. 20-97 shows a unit which may be used either as a primary heating source or as a supplementary heating source. Air enters at the top, and a fan forces the air down over the electrically heated aluminum element. Air leaves the unit through the lower part of the grille. Figs. 20-98 and 20-99 show another design, while Fig. 20-100 shows a schematic wiring diagram for the unit.

Fig. 20-99. Parts of electric resistance space heater. A—Frame. B—Front cover. C—Heating element and fan. D—Back case.

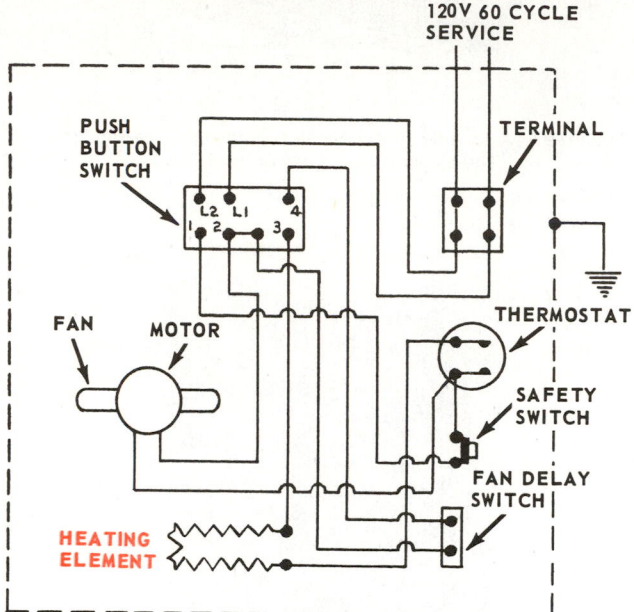

Fig. 20-100. Wiring diagram of a wall-mounted heater. (The Singer Co., Climate Control Div.)

20-44 RADIANT HEAT

Radiant heat usually is the impact of the infrared band of light energy waves against an object. These rays are at a frequency of 915 MHz to 2450 MHz (M = million) and wave length of 4.0 microns or less. The object absorbs the rays and becomes warmer. When a furnace door is opened, even if one stands back from the door, the heat impact which is felt is usually an infrared ray impact. This principle can be used for comfort heating. The energy source may be any fuel, although gas-fired and electrically heated elements are the chief sources.

Radiant energy sources may be either:

1. A lamp source or high-temperature electric element or ceramic source.
2. A low-temperature electric cable or hot water source.

If radiant heat rays are focused on an individual having several square feet of surface to absorb the rays, this person can be kept quite comfortable even though the ambient temperature is below the comfort range. A large warehouse with a few small areas where employees work is a typical example of where radiant heat is often used to good advantage.

Radiant heat decreases as the square of the distance. That is, an object twice as far away from the radiant heat source will receive only one fourth as much heat.

20-45 RADIANT CEILING

Radiant ceiling installations are most popular for homes. In these installations, electric heating cables are enclosed in the ceiling plaster. The ceiling cables are about 500 to 5000 watts, and they are 1/8 to 1/4-in. diameter.

Ceiling cables usually release 2.75 watts per foot and are spaced at 1 1/2-in. centers. Their temperature is about 150 F. (66 C.). This heats the ceiling surface to about 120 F. (49 C.) on 1 1/2-in. centers, but wider spacing will reduce the ceiling surface to about 100 F. (38 C.). Dry wall uses lower watts/foot ratings (about 2.2 watts/foot). About 60 percent of the heat released is by radiation.

Floor cables are heavier and are rated at 2.75 watts/foot, with a 1 1/2 to 3-in. thickness of concrete. About 45 percent is radiant heat.

20-46 RADIANT LAMPS AND GLASS PANEL HEATERS

Some electrically heated fixtures use quartz lamps that are either Vycor or metal sheathed. Some use open resistance wire or ribbon. See Fig. 20-101. Quartz lamps usually consume 800 to 2500 watts of power. The lamps and wires reach temperatures of approximately 1200 F. (649 C.).

Radiant heat lamps may have either 90 deg. and 60 deg. or 45 deg. reflection. They radiate about 50 watts per square foot from at least two directions. Install one watt for each degree desired above lowest expected temperature (minimum 12W per sq. ft.). Add five percent for each foot distance if the source is farther than 10 feet from a person.

When these lamps are used outdoors, use wind shields and about two watts/degree above coldest temperature expected.

Quartz tubes usually use a nickel chromium coil. They are

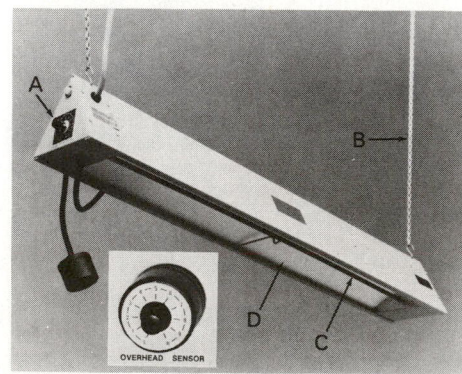

Fig. 20-101. Radiant heat lamp. Quartz tube is mounted in fixture. A—Overhead sensor. B—Hanger. C—Quartz tube. D—Reflector. (Kalglo Electronics Co., Inc.)

about 60 percent efficient. The air-filled open tube operates at 1500 to 1700 F. (816 to 927 C.), and it gives off light infrared rays. A quartz lamp filled with inert gas operates at 4000 F. (2204 C.). It uses a tungsten coil and is about 85 percent efficient.

Glass panel heaters are also available. They give off about 60 percent radiant, 40 percent convection heat. The glass is borosilicate. Electrical conductors are installed on the back of the glass and covered with a reflector surface. The element operates at about 500 F. (260 C.). The glass surface is about 350 F. (177 C.). Glass panel heaters are used on the wall under windows.

A useful application for radiant heat in the temperate zones is for snow melting. Generally, 100 to 200 watts are needed for each square foot of surface to be serviced.

Gas may also be used for radiant heating. Gas-fired radiant heat usually is provided by heating ceramic elements to incandescence and using a reflector to focus this heat. In gas-fired units, about 50 percent of the heat energy is converted into radiant heat. These units operate at about 700 F. to 1600 F. (371 C. to 871 C.).

20-47 HEATING COIL INSTALLATION

Electric heating elements must be installed in accordance with both the electrical codes and the manufacturer's recommendations.

The electric heating installation must be checked to be sure it is safe with reference to fires, safe with reference to humans and animals, and it must be efficient.

Baseboard heaters must have unhampered air circulation. The unit heaters must not be installed dangerously close to flammables. Furnace units must be shielded electrically and heat protected. All metal parts of the units must be grounded.

The electrical service must use wire sizes according to the voltage and amperage of the circuit. The circuit must be

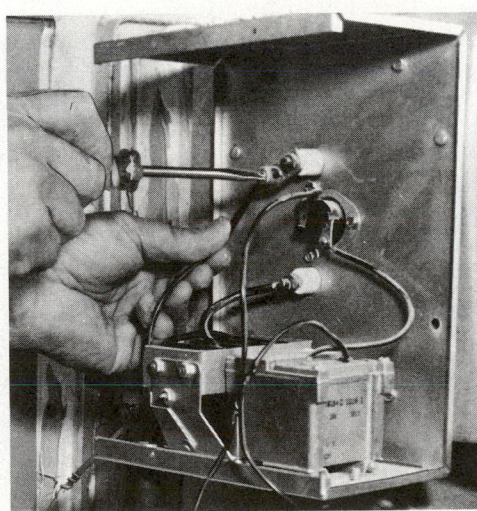

Fig. 20-102. Service technician connecting line to terminals of duct-mounted electric resistance heating unit.
(Tutco Inc.)

properly fused and provided with adequate limit and safety controls. All controls must be for the correct voltage and current. Fig. 20-102 shows electrical connections being made to an open wire, supplementary heating coil that has been installed in a duct.

20-48 HEATING COIL SERVICE

Electric resistance heating requires a minimum of service. Air passages over the heating units must be kept clean. Grilles, ducts, heating coils and fins should be cleaned at least once each year. Brushing and vacuuming are recommended. It is extremely dangerous to use flammable fluids for cleaning electric resistance heating units.

Check the terminals for tightness and cleanliness. A voltmeter or ohmmeter may be used to determine if connections are loose or corroded. The ohmmeter is preferred, because it may be used for checking with the power off.

The thermostat, limit control and relay are possible service problems. If the circuit does not function, the following typical electrical circuit diagnosis is recommended:

1. Is there power to the fuse box or circuit breaker box?
2. Is the fuse in good condition and are the connections electrically good?
3. Is the thermostat operating? (Check opening and closing temperatures.)
4. Is the limit switch operating? (Check opening and closing temperatures.)
5. Are the relay coils in good condition and operating?
6. Are the relay contact points clean and operating?
7. Does the electrical heating coil circuit have continuity?

Some troubles and possible causes are as follows:

A. Blower turns on and off. If heater strips heat and cool as blower runs and stops, the thermostat is short cycling. Perhaps too high rating of the anticipator.
   1. Limit control may be opening and closing.
   2. Incorrect low voltage.
   3. Motor overload may be opening and closing.
B. Blower runs, but there is not enough heat.
   1. Dirty filters.
   2. Voltage too low.
   3. Only some heater elements are energized.
     a. Open fuse or circuit breaker.
     b. Element burned out.
   4. Sequence switch not operating.
   5. Second stage of two-stage thermostat not operating.
C. No power or low voltage.
   1. Sequence switch defective (open).
   2. Thermostat open.
   3. Low line voltage.
   4. Low transformer output voltage.
   5. Motor not operating.
     a. Defective capacitor.
     b. Defective internal overload.
D. Proper voltage at motor.
   1. Defective motor (open circuit).
   2. Defective overload (open circuit).
   3. Defective motor capacitor (shorted open).

20-49 UNIT HEATERS

Many stores, commercial buildings and factories use unit heaters to heat certain rooms or spaces. These heaters can be gas-fired, oil-fired or use hot water coils or steam coils. They are suspended from the ceiling and use a motor-fan to force the heated air in a controlled direction. Many units have adjustable louvers to help direct airflow. The different sizes handle from 300 cfm to about 6000 cfm. Their capacities range from 20,000 Btu to 360,000 Btu/hr. Gas-fired and oil-fired unit heaters use a flue to carry the products of combustion outdoors.

Special unit heaters are available for vertical down warm airflow and also for high velocity airflow (about 2500 fpm) against large door openings to keep out cold outside air. Usually, they are mounted about three feet above and four feet away from the door opening. They are often used on car and truck doors or shipping and receiving doors. The unit is operated by either a door switch or a thermostat connected in parallel. Doors less than 8-ft. high and 10-ft. wide usually are not protected by these devices. Some users operate the fan only during the summer months to keep out dust and insects.

20-50 HUMIDIFIERS

When air is heated, it can absorb more water vapor. Human comfort requires a relative humidity (RH) of about 35 percent. When outside air at 30 F. (−1 C.), 90 percent RH, is heated to 72 F. (22 C.), its RH drops to about 18 percent. See Fig. 20-103.

A dry atmosphere causes dry skin, breathing dryness and loss of moisture from hygroscopic materials such as wood natural fibers (wood furniture and woodwork) and most foods. A dry condition also creates static electricity conditions. Moisture must be added to the air. Moisture sources are from plumbing devices, cooking and perspiration. In addition, moisture in the air can be increased and controlled by using humidifiers.

Humidifiers add water vapor (low temperature steam) to

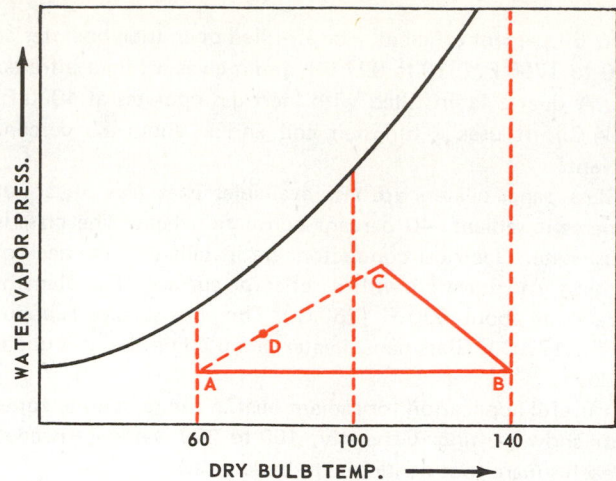

Fig. 20-104. Psychrometric chart indicates warm air recirculating heating cycle. A—Cold air return. A to B—Heating in furnace. B to C—Humidifying air. C to A—Mixing of air with room air. D—Final conditions after mixing.

the air. If the return to a warm air furnace is about 60 F. (16 C.) and 25 percent RH, and the furnace heats the air to 140 F. (60 C.), a humidifier may be used to add moisture to the warmed air. This heated air is then mixed with the air in the room. In Fig. 20-104, A to B indicates the air being heated. From B to C, this warm air is passing over the humidifier (total heat is constant). Between C and A, the heated and the humidified air are mixed. D indicates the final condition of the air as it is delivered to the conditioned space. Remember, it requires about 1000 Btu to vaporize each pound of water (7000 grains).

Most humidifiers in warm air systems are part of the furnace or the duct work. Fig. 20-2 shows some basic designs.

Some humidifiers are water sprays with air pressure used to increase the "atomizing" of the water. See Fig. 20-105. Air pressure is used to syphon water out of a container for

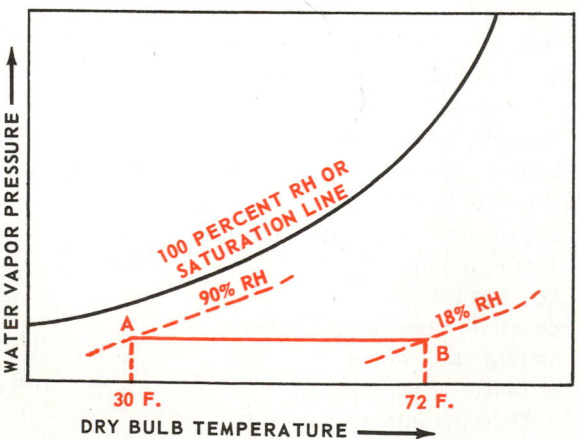

Fig. 20-103. Graph showing decrease in relative humidity. A—Air sample is heated from 30 F. (−1 C.) at 90 percent RH, to 72 F. (22 C.) at 18 percent RH. B—Without any water vapor being added to air sample.

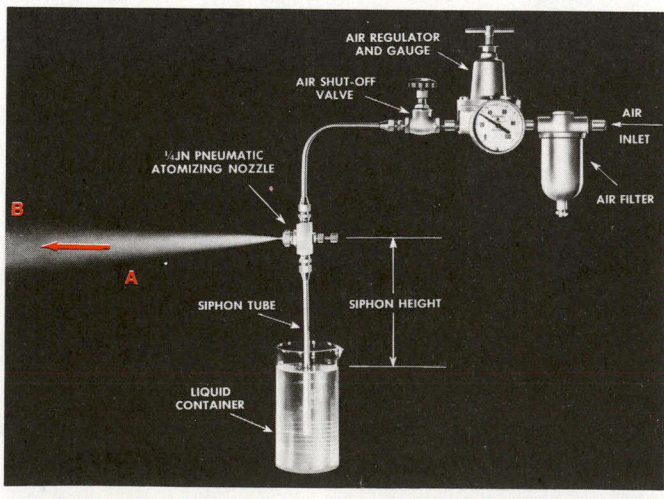

Fig. 20-105. Water spray humidifier uses an air pressure jet to lift (syphon) water out of the container. A—Water is atomized. B—Water evaporates.

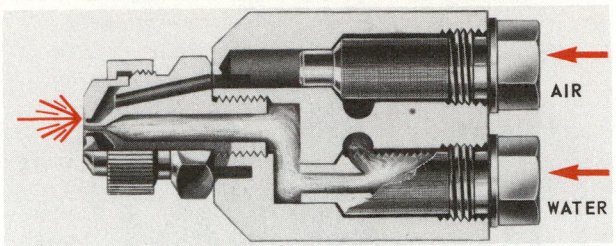

Fig. 20-106. Sectional view shows operation of an atomizing water spray nozzle of a humidifier. Air and water mix to form a spray of very small water particles which quickly evaporate.

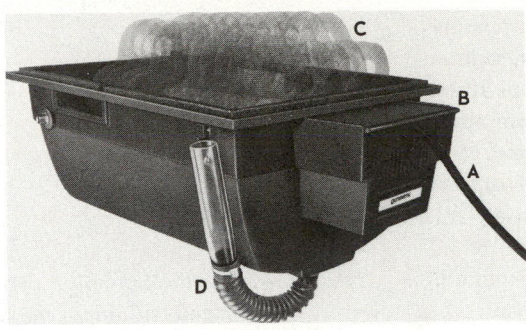

RELATIVE HUMIDITY (PERCENT)			
OUTSIDE TEMPERATURE	OUTSIDE RH	INSIDE RH	SAFE INSIDE RH
−10	70	2	20
0	70	5	25
10	70	7	30
20	70	11	35
30	70	17	35

Fig. 20-107. Chart shows humidity change in air as it is brought into a building from outside; also recommended inside humidity based on outside temperature. Example: 0 F. (−18 C.) outside air at 70 percent humidity when brought into a building and heated to 72 F. (22 C.) will have a relative humidity of 5 percent. It should have a relative humidity of 25 percent.

atomizing purposes. Construction and operation of the nozzle are shown in Fig. 20-106.

Humidifiers easily can be added to warm air heating systems. Hydronic heating systems, steam heating systems and most electric heating systems, however, require a separate cabinet type humidifier if the needed relative humidity is to be maintained.

Humidifiers of various types are used:

1. Plate type humidifier (low capacity).
2. Rotating drum type. (For restricted spaces.)
3. Rotating disc type.
4. Fixed filter type.
5. Fan type.
6. Plenum/warm air duct humidifier. (Slings the water.)
7. Plenum/duct electric type.

Humidistats are used to control the level of humidity. Too much humidity may cause swelling of hygroscopic materials (wood products) and may cause condensation on cold surfaces such as windows, window frames and doors. It may also condense inside the outside walls.

Excess humidity should be avoided because mold and rot can occur at 70 percent RH.

If outside air at 80 percent RH filters into a building and is heated to 70 F. (21 C.), this heated air will have the humidity shown in Fig. 20-107. All water services in the home, plus perspiration and respiration of the humans and animals in the home, will increase the relative humidity. Even if these moisture sources double the relative humidity, it would still be too low in all cases except at 30 F. (−1 C.) outside air.

Humidifiers are used to add the needed moisture (water vapor) to the room air. A ranch home 25 ft. x 60 ft., with 8 ft. ceilings, has a volume of 12,000 cu. ft. and requires about four (tight building) to 16 (loose building) gallons per day, depending on the number of changes of air in the building:

1. A tight building has .5 air changes/hour.
2. An average building has 1.0 air changes/hour.
3. An average loose building has 1.5 air changes/hour.
4. A loose building has 2.0 air changes/hour.

Regular water (city mains or well water) contains different foreign matter. If used, this foreign matter must be treated or removed. Fig. 20-108 shows a humidifier with a water level indicator, which can be used to drain the water out of the humidifier.

The quality of the water varies with its source.

1. Soft water.

Fig. 20-108. A humidifier designed to fasten to the bottom of a duct. A—Motor to drive moisture pickup wire mesh discs. B—Flange for mounting unit to duct. C—Wire mesh moisture discs. D—Water level tube and water drain.

a. Natural, untreated water with low mineral content has about 5 grains of hardness per gallon and no chlorides. The natural source of soft water is rain water.

b. Softened water is treated by the ion exchange process (water softener) to remove hardness and minerals, then replace with water soluble sodium salts.

2. Demineralized water has been treated to remove minerals.
3. Medium hard water is untreated water with 5 to 15 grain/gallon of mineral content (well water).
4. Very hard water (well water) is untreated water with over 15 grains/gallon of mineral content.

The perfect water for humidifiers would be distilled water, which does not put any foreign vapors into the air. In addition, distilled water does not leave any deposits in the humidifier or in the duct system.

20-51 INSTALLING AND SERVICING HUMIDIFIERS

Humidifiers should be carefully installed. Mount the water reservoir level in all directions (use a spirit level). The humidifier should have the capacity to add enough moisture to the building at design conditions. See Chapter 25.

The water feed line should be tapped into a cold water line in the same manner as an ice cube maker line. See Chapters 11 and 14. The overflow drain tube (bleeder tube) should slant down all the way to the outlet. The outlet should be at least one inch above the open drain.

If the unit is power driven by a fan or rotary unit (or both), the electric system can be low-voltage controlled by a humidistat and relay, or it can be controlled by a 120V humidistat. Electrical installation should be done according to electrical codes in effect.

After installation, the unit should be turned on (electrical and water) and checked for proper operation.

20-52 ELECTRIC HUMIDIFIERS

When a house or business facility is electrically heated, there is a tendency to have too much humidity. Water vapor sources include:

1. Cooking.
2. Wash basins.
3. Lavatories.
4. Respiration.
5. Perspiration.
6. Laundry.
7. Showers and/or bathing.
8. House plants.
9. Pets.

A dehumidifier may be needed if the humidity is too high. See Chapter 21. However, most heated buildings need humidifying. A water vapor producing device is usually needed during the winter reason in the temperate zones. An electric humidifier may be used for this purpose.

Buildings which have steam heated, hydronic heated or warm air heated equipment are examples of installations that can use an electric humidifier. Warm air heating units sometimes are equipped with a humidifier that uses the warm air as a source of heat. However, if the humidifier water is separately heated, the amount of humidification can be more accurately controlled.

The electric humidifier has the advantage of ease of installation, flexibility of location and accuracy of humidity

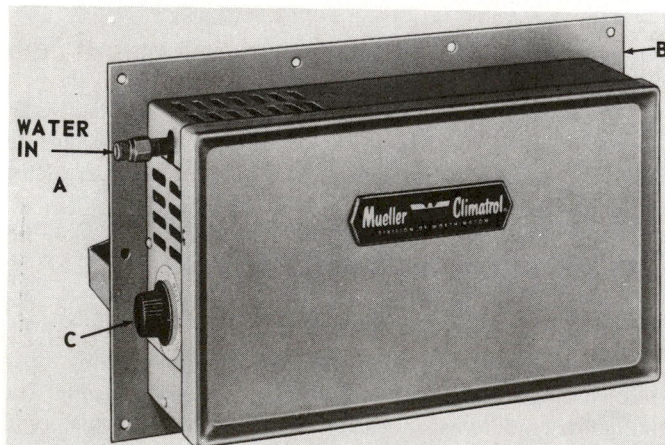

Fig. 20-109. Electric humidifier designed for installation in warm air heating duct. A—Water connection. B—Flange for mounting on duct. C—Electric heater control.   (Climatrol Industries, Inc.)

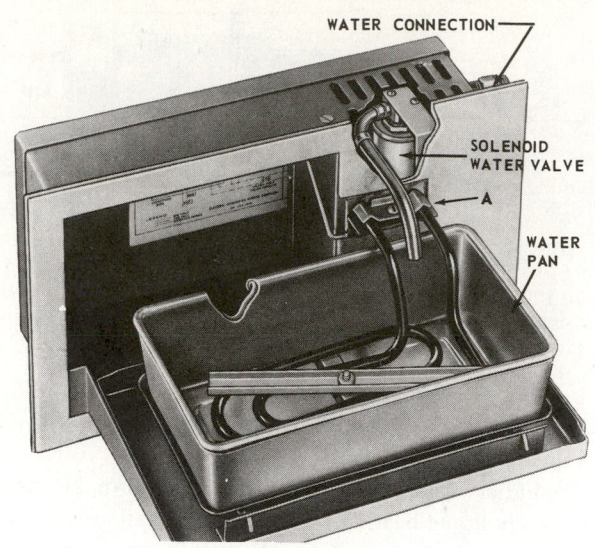

Fig. 20-110. Details of construction of electric humidifier. A—Electric heating element.

control. Fig. 20-109 shows an electric humidifier designed for use in a duct or plenum chamber installation. This unit has an 800 watt electric heating coil which is controlled by an adjustable humidistat. Water level is controlled by a pan type float which operates a switch in the solenoid water valve circuit. Fig. 20-110 shows the construction of the electric humidifier.

20-53 CABINET TYPE HUMIDIFIERS

The use of humidifiers is required in many situations that are separate from and independent of the heating system. They are popular when a hydronic or steam heating system is used. Many are used with warm air heating systems.

Independent humidifying units are housed in an attractive cabinet and usually consist of a large wheel or drum which has a rim of porous material. A humidistat controls a fan motor, which drives the large wheel that is wetted as it dips into a water reservoir. As the air flows through the wet, porous rim, most of the moisture evaporates and adds moisture (water vapor) to the air. See Fig. 20-111. Some of the units have an electric heater to warm the outlet air. The water reservoir is manually filled with water. A float operated switch will turn on a signal light when the water level is low. The unit is easily maintained. The water tank and the porous wheel rim can be removed and cleaned.

20-54 MOBILE HOME AIR CONDITIONING

Most mobile home heating is with LP-Gas, although some mobile home heating equipment may use oil. These furnaces usually heat the mobile home through an air distribution system. Ducts for this type of system are shown in Fig. 20-112. An evaporator may be placed in the plenum chamber, and a condensing unit is located outside the mobile home to provide cooling through the heating duct installation.

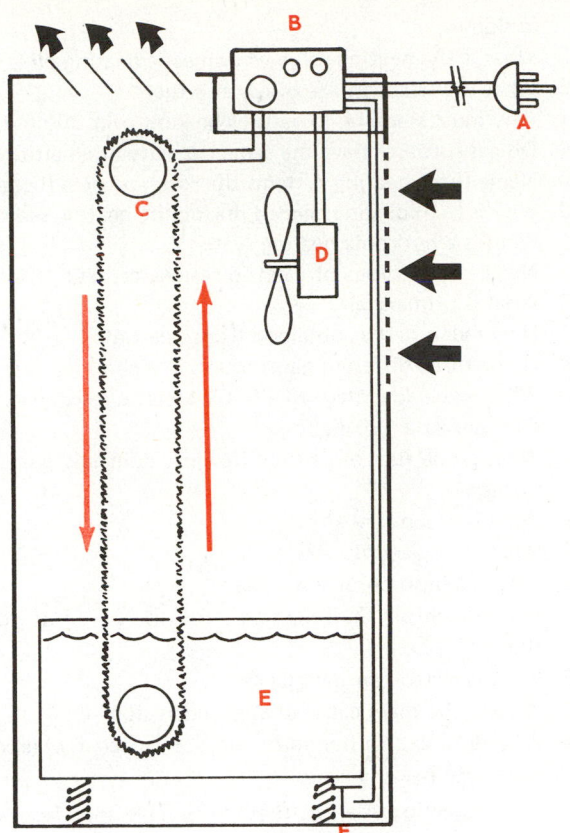

Fig. 20-111. A cabinet type humidifier in which a porous filter revolves and picks up moisture. Air blown through porous material vaporizes some of the water, raising the humidity in the room. A—Power cord. B—Humidistat, switch, indicator lights. C—Moisture belt and filter. D—Fan and motor. E—Water tank. F—Switch light which, when lit, indicates that water tank is empty.

20-55 REVIEW OF SAFETY

Safety must be strongly emphasized when installing and operating heating systems. Oil and fuel gas furnaces are fires in a confined area.

1. Fuel must be stored safely.
2. Fuel must be fed in a safe manner to firepot of furnace.
3. Means must be taken to shut down the unit:
   a. If fuel flow ceases.
   b. If any part of system overheats.
   c. If products of combustion are blocked from leaving building.

Avoid tampering with safety controls; they must be in excellent condition and properly adjusted. Tampering may cause delayed ignition, and an explosion may damage equipment and injure persons. Avoid turning on the electric ignition or lighting a pilot light until the firepot has been examined for gas or oil.

Remember, if fuel is in the presence of air and an ignition source exists, fire will result.

Know exactly what is being done. Follow manufacturers' specifications carefully. Always follow the building and safety codes in effect in the locality.

Electric heating is considered to be a safe heating method.

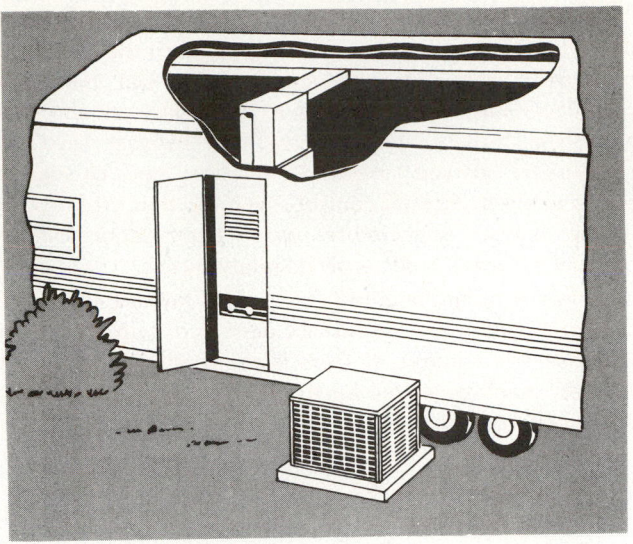

Fig. 20-112. A mobile home year-round air conditioner. The condensing unit for summer cooling is located on the ground outside the home.

The elimination of flames, pilot flames, electric spark ignition and sparks reduces many sources of danger. There is, however, possible danger from electrical shock and the chance of burns or of combustion from the heating elements.

All electrical devices must be installed according to local, state and national electrical codes. The equipment should have Underwriters Laboratory approval (UL). The installation should be made by a licensed electrician.

All nonelectrical metal parts of the unit must be safely grounded. Combustibles must not come in contact with heating elements. The heating elements must be mounted to eliminate the chance of persons or animals contacting the heating grids or elements.

Special gas heaters for tents and cabins should be installed with a stack and a safety pilot light which will shut off the fuel gas flow in case the pilot flame is extinguished.

Testing for leaks should be done with soapsuds. If a leak is suspected, do not use open flames. Turn off the electric power. Use an explosion-proof flashlight only.

The propellant in a hair spray aerosol can and the chemical methylene chloride found in varnish removers have an intensely corrosive effect on metal surfaces. This action is accelerated with an increase in temperature, so it may shorten the life of the heat exchanger in warm air furnaces.

An improperly vented furnace, regardless of the kind of fuel used, presents two dangers. If the combustion is incomplete, a considerable amount of carbon monoxide may be generated. This is a very poisonous gas, and inhaling it may result in illness and death. In low concentrations, the symptoms may be headaches, fatigue, dizziness or loss of muscular control.

If the combustion is complete, a considerable amount of carbon dioxide (CO_2) may be formed. If CO_2 is not vented, it replaces the oxygen in the air. Animals require oxygen in the air they breath, so if carbon dioxide has replaced the oxygen, the body will suffer from the lack of oxygen.

All heating furnaces should be equipped with safety devices. If, for any reason, the fuel fed into the combustion chamber is not ignited, the safety device will immediately shut down the furnace. This is to avoid an explosion in the event the furnace and flues were to become filled with combustible gas and ignited. All furnaces must be equipped with a temperature limit control which will shut off the burner if the bonnet temperature, water temperature or steam pressure exceeds a predetermined safety setting.

Before lighting the pilot light in a gas furnace, shut off the gas supply and allow the furnace door to remain open for not less than five minutes. If there is any combustible gas in the furnace, it will be vented by this procedure, making it safe to light the pilot.

Any gas odor or any sign of oil in the combustion chamber is a strong signal to be super-cautious.

A hot water or steam system is a pressure vessel. The safety devices must work or the plant may explode. Always operate the safety devices as a part of system inspection. Low water failures are a common cause of trouble; always inspect the low-water controls and safety drains.

Many materials are both flammable, explosive and/or toxic (harmful) to living things. But all these materials are safe until a threshold limit value (TLV) is reached. For example: The TLV of a gas may be 200 ppm while the lowest explosive limit (LEL) is 6.7 percent (67,000 ppm).

The National Fire Protection Agency explains explosive limits as the minimum amount of the gas in air or oxygen that can be ignited by an ignition source and keep on burning (self-propagating). Remember that slow burning is a fire, while very rapid burning is an explosion.

Many devices have been used to detect the presence of these gases, including birds, animals, plants and instruments.

Many electrical instruments are now available for not only quickly detecting the presence of these gases, but also measuring the amount of these gases in the space being tested. These instruments have greatly increased the safety conditions. However, there are many situations in service work where these instruments are not available or the conditions change quickly. *One of the best health safeguards and fire preventions is to completely and thoroughly ventilate the space in which one is working.*

20-56 TEST YOUR KNOWLEDGE

1. Name three types of heat sources.
2. How does a tight building affect combustion air?
3. How does a bypass air system operate?
4. List three most common sources of energy for heating buildings.
5. What is the heating value of domestic heating oil?
6. What is one advantage of gas as a fuel?
7. Why must stack temperatures be kept over 300 F.?
8. Do gas furnaces have the same capacity at all altitudes?
9. Name three heating systems that do not use a flame.
10. Why is humidifying needed during the heating season?
11. What is a hydronic heating system?
12. Name three types of electric resistance heating elements.
13. What is primary air?
14. How much heat is obtained from one watt?
15. Name three different gases found in a chimney flue.
16. What is the advantage of a 240V electric resistance heating circuit over a 120V circuit?
17. What grade fuel oil is used in most domestic gun type oil burners?
18. What is secondary air?
19. What is induced draft?
20. Why is a limit control needed?
21. How much air is needed to burn one gallon of domestic fuel oil?
22. What is a stack air diverter?
23. What is the main cause of stack corrosion?
24. Why does excess humidity sometimes occur when electric resistance heating is used?
25. What chemicals can be used to remove soot from a flue?
26. What is meant by supplementary electric resistance heating?
27. How does a gun type oil burner obtain combustion air?
28. What will happen if the pump of a hydronic system will not pump?
29. What is the color of the combustion of a well-adjusted gun type oil burner?
30. How many oil filters or screens does a gun type oil burner system use?
31. What safety device is part of a pilot light?
32. Where is a thermocouple used in a fuel gas system?
33. What is the color of a correctly adjusted fuel gas flame?
34. What happens if there is air in a hydronic system?
35. What is the most common service problem with hydronic system pumps?
36. Some furnace blowers are directly driven. How are others driven?
37. Where should the pilot light flame be when it is measured from the thermocouple?
38. Are all gas burner main automatic control valves "solenoid valves?"
39. What is meant by zone control heating systems?
40. What flue gas is most dangerous to humans and animals?

Chapter 21

AIR CONDITIONING SYSTEMS COOLING AND DEHUMIDIFYING

Refrigeration is the heart of the comfort cooling part of air conditioning. All comfort cooling systems use one of the standard refrigerating cycles, standard refrigerants, standard types of compressors, condensers, piping, refrigerant controls, motor controls and evaporators. The refrigerant evaporating temperature of most comfort cooling systems is about 40 to 50 F. (4 to 10 C.). This chapter describes most designs which use mechanical refrigeration for cooling and dehumidifying.

Absorption systems are becoming more popular for comfort cooling purposes. These units are described in Chapter 16.

Some air conditioning cooling installations involve engineering problems. These calculations must be made accurately. If the right capacities are not provided, the system will not operate properly. See Chapter 25.

21-1 PRINCIPLES OF ATMOSPHERE COOLING

As explained in Chapter 18, one's comfort as well as the success of certain industrial operations (dealing with hygroscopic materials and processes) depend upon temperature and humidity.

The first mechanical atmosphere cooling and humidity control used cooled water, both to reduce temperature of the air and dehumidify it. Air was passed over water-cooled coils or through cooled water sprays.

Fig. 21-1 shows the basic operation of such cooling units. Return air is mixed with some fresh air. Then this air mixture is filtered and cooled. Moisture is removed before it is redistributed into the building.

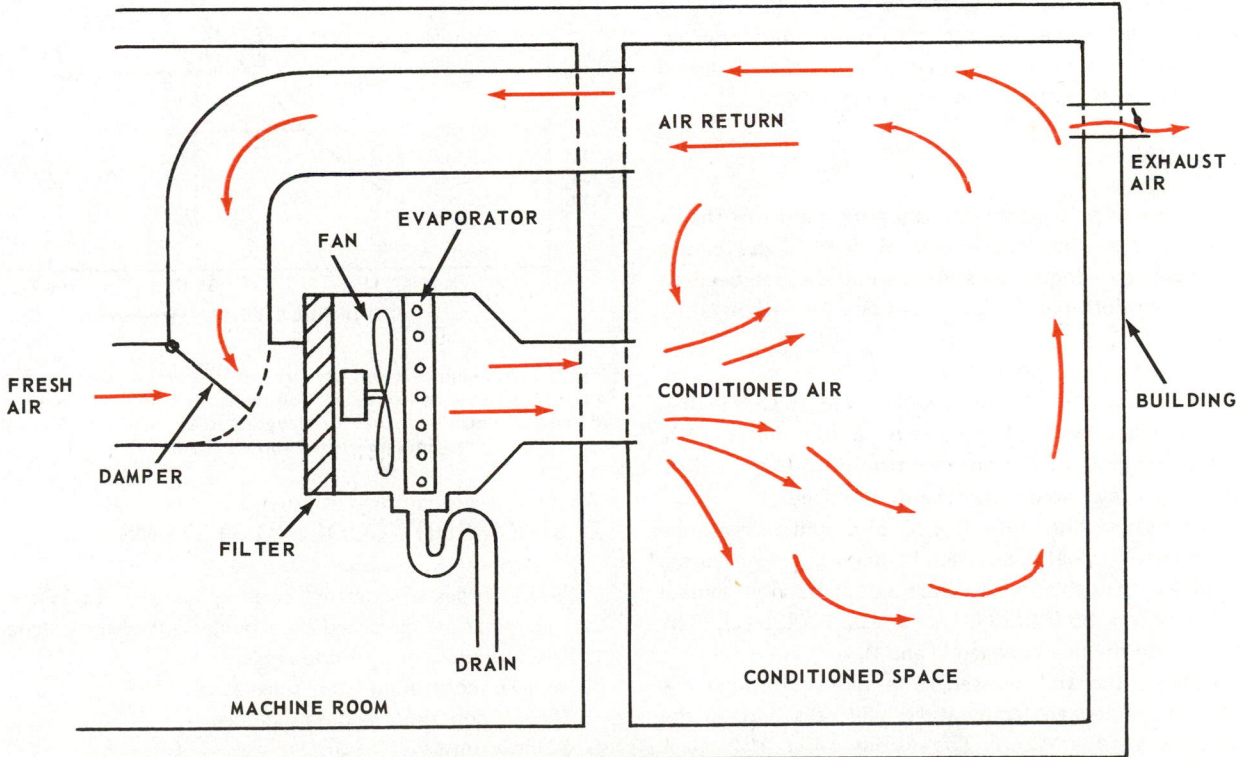

Fig. 21-1. Basic operation of a conditioner which cools air, and removes moisture from fresh and recirculated air.

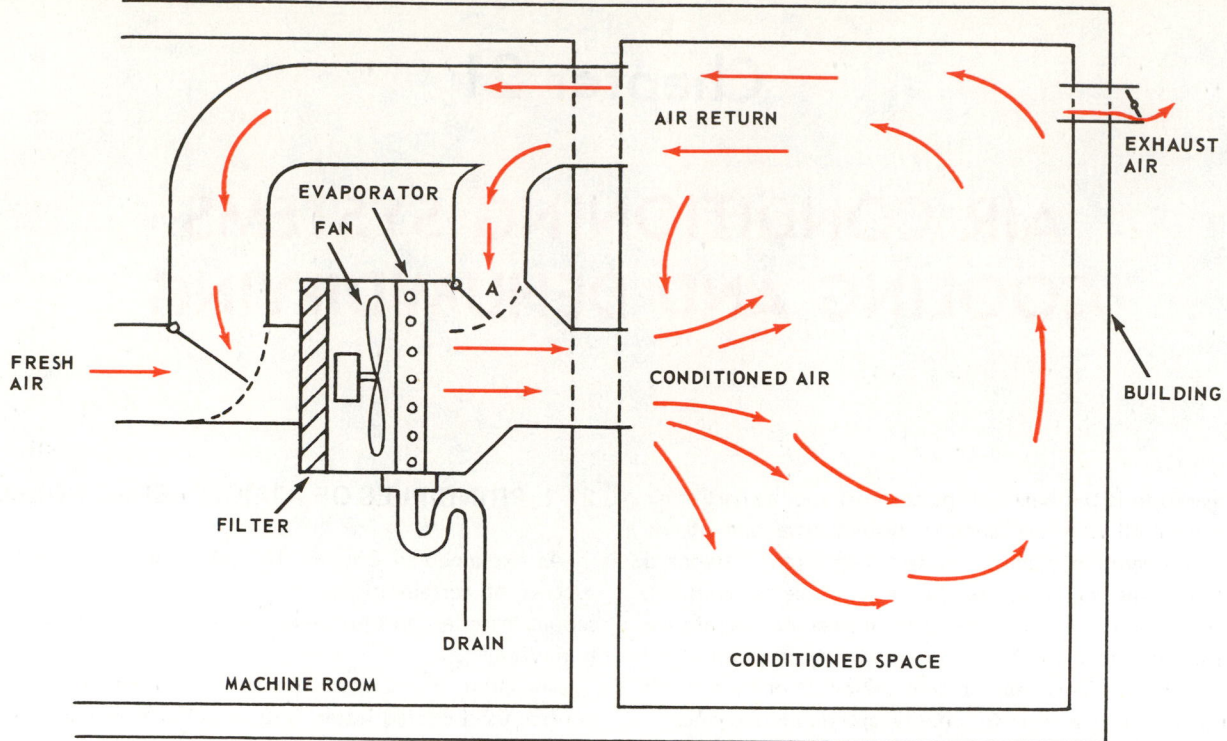

Fig. 21-2. Comfort cooling installation with return air bypass, A, for relative humidity control.

Cool air leaving the evaporator is at 100 percent relative humidity. This saturated air, as it mixes with air in the conditioned space, warms up somewhat. Thus, relative humidity is brought down to a comfortable level.

There is a more positive way to control relative humidity. It involves bypassing some of the return air into the air conditioner outlet to warm up the cooled air before it leaves the duct system. This method is shown in Fig. 21-2.

21-2 COOLING

In a cooling cycle, the dry bulb (db) temperature of the air is lowered. When this happens (Fig. 21-3, A to B) the relative humidity increases. Some moisture should be removed to make this air comfortable. Moisture can be removed by either of two methods:

1. Dehydrate the air with chemicals.
2. Cool the air down to the saturation curve at C and then remove moisture by condensing it on a cool surface. See line C to D, Fig. 21-3. The distance from C to D is the drop in vapor pressure or grains of moisture removed.

Reheating along a horizontal line, D to E, will decrease the humidity. However, what most often happens is the air leaving at D is mixed with the room air which is at some intermediate condition between 85 F. (29 C.) and 100 F. (38 C.). The mixture meets on the line between D and A.

If a third of the air, by weight, is passed through the evaporator, the mixed air temperatures will be a third of the way from D to A — that is to F. The system in Fig. 21-2 mixes the air inside the air conditioner and also introduces fresh air into the air conditioner.

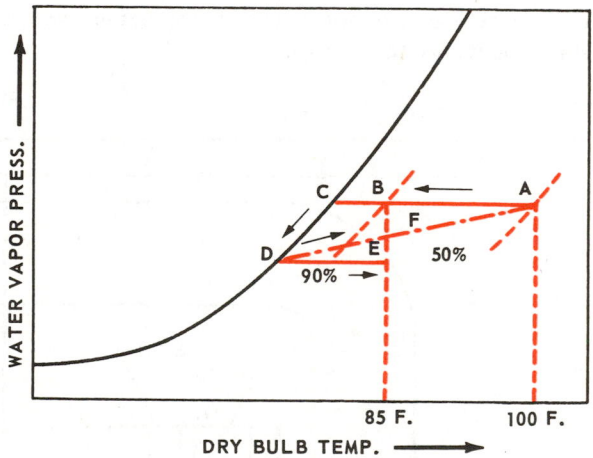

Fig. 21-3. Cooling cycle on psychrometric chart. A—Condition of outside air. B—Partly cooled air. C—Air cooled to saturation. D—Air cooled to remove some moisture. E—Dehydrated air reheated. F—Result of mixing treated and untreated air.

21-3 COMFORT COOLING SYSTEMS

Several types of comfort cooling systems are in common use. They can be classified by arrangement of the mechanism:
1. Self-contained or unit coolers.
2. Remote (controlled from a distance).
   The self-contained system includes:
1. Window units.
2. Through-the-wall units.
3. Cabinet units.

Remote units are of two types:

1. The condensing unit is remote. The evaporator is installed in the room to be conditioned or in the main duct.
2. The central air conditioning plant. The condensing unit and the evaporator are installed away from the place being conditioned. A cooled brine or water is circulated to heat exchangers in the various spaces to be conditioned.

21-4 COOLING EQUIPMENT FOR AIR CONDITIONING PURPOSES

Cooling equipment usually consists of evaporators. These are kept at refrigerant temperatures of 40 to 50 F. (4 C. to 10 C.) and air to be conditioned is moved through them.

There are several ways to cool the air:

1. By mechanical refrigeration.
2. By absorption refrigeration. (Absorption system air conditioning is described in Chapter 16.)
3. With ice.

Ice may be the cooling medium, particularly if cooling is only needed for a few days of the year. Cold water from streams or wells may also be used. The water should be 50 F. (10 C.) or cooler to produce satisfactory dehumidification (removal of moisture from the air).

But most air conditioning installations use automatic mechanical or absorption refrigeration for cooling. The mechanical system consists principally of:

1. Condensing unit.
2. Evaporator.
3. Motor-driven fans.
4. Filters.
5. Ducts and airflow controls.
6. Motor controls.

7. Temperature and humidity controls.
8. Piping.
9. Refrigerant.

21-5 UNIT COMFORT COOLERS

Unit or self-contained air conditioners are of two types:

1. Comfort cooling only for summer.
2. Both summer cooling and winter heating. (See Chapter 23.)

Both types have a complete refrigeration plant. This includes condensing unit, refrigerant valves and evaporators. Filtering equipment is also a part of the system. Individual room thermostats provide control. These units may also be classified as window or in-the-wall units, or console units.

The window units and in-the-wall units are air-cooled, easily mounted and operate from 120V or 240V single-phase circuits. Capacity varies from 4000 Btu/hr. to 40,000 Btu/hr.

Console units may be either air-cooled or water-cooled. They are installed in the room to be conditioned or in an adjacent room with short ducts to deliver air and provide for return air.

21-6 WINDOW UNITS

The window or wall-mounted comfort cooler is very popular. The window unit mounts on a windowsill and installation is relatively easy. The condenser is located in the section of the cabinet that is outside the building. Outside air is forced over the condenser by a fan. Inside the room another fan draws air in through a filter and forces it over the evaporator.

The two airflow fans may be driven by the same motor or each may have its own motor. Fig. 21-4 shows the airflow in a

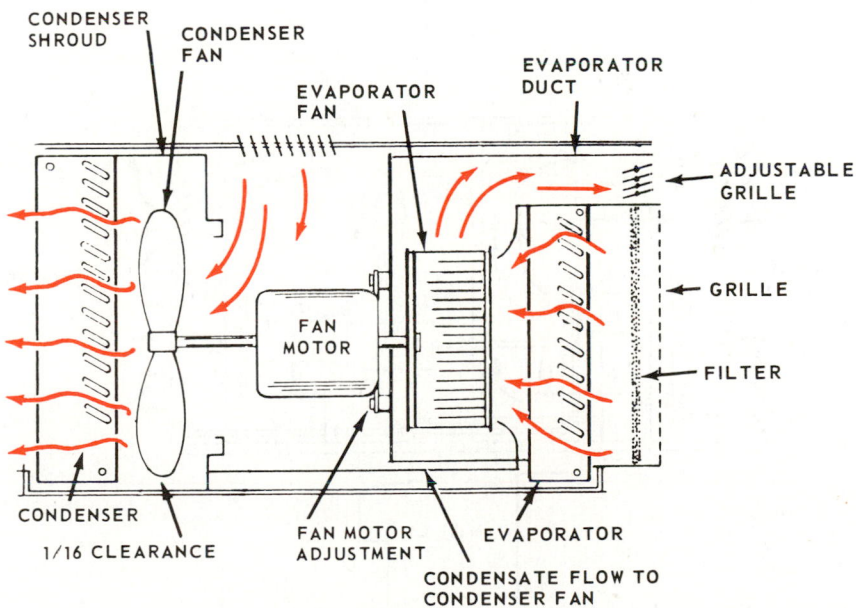

Fig. 21-4. Schematic drawing of air circuits in window air conditioner. Axial type fan is used at condenser. Evaporator air circuit uses radial flow fan. Same motor drives both fans. (Hupp, Inc.)

Fig. 21-5. Window air conditioner. A—Air inlet and filter. B—Cooled air outlet with adjustable grilles. C—Expandable partitions to fit various width window openings. D—Thermostat. E—Fresh air damper control. F—Switch for fan only (two-speed) or for high or low cooling. (Kelvinator, Inc.)

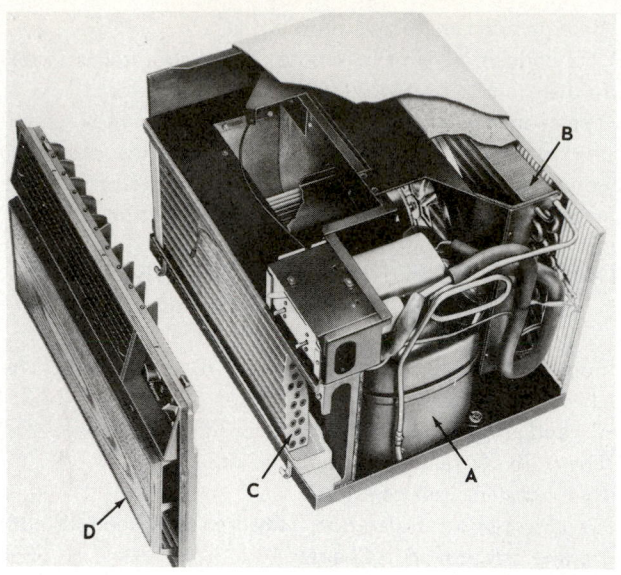

Fig. 21-6. Internal construction of window-mounted air conditioner. A—Hermetic motor compressor. B—Condenser. C—Evaporator. D—Removable front panel. Return grille, filter and outlet grille are built into it.

window air conditioner. Fig. 21-5 illustrates a modern window air conditioner. The internal construction of a unit is shown in Fig. 21-6.

Elements of a window comfort cooler are shown in Fig. 21-7. A more detailed schematic is provided in Fig. 21-8.

Window units are available in several types. One type cools and filters the air and has a fresh air intake. Another type has these same devices but, in addition, has an electrical resistance heating unit to furnish heat. A third type uses a reverse cycle system (heat pump) to permit the use of the refrigerating units both for comfort cooling and heating. See Chapter 23.

Window units may be obtained to fit double hung windows, casement windows or they can be installed in special wall openings.

The condensate from the evaporator is often drained to the base of the motor compressor and the condenser where it helps to cool these parts. A capillary tube or a bypass type AEV refrigerant control is usually used.

Some units change the cooled airflow from side to side as the unit runs.

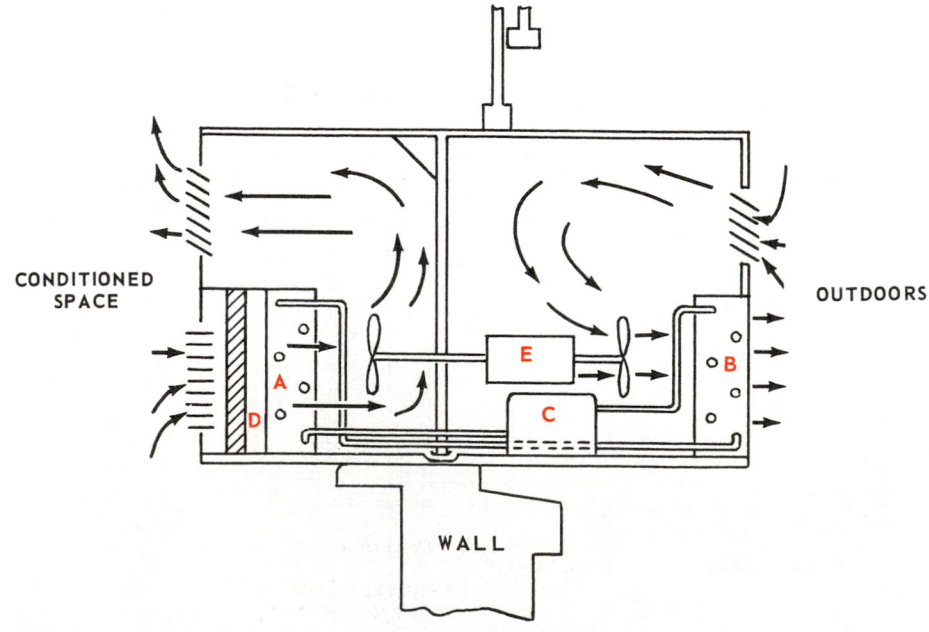

Fig. 21-7. Schematic drawing of a window comfort cooler installation. A—Evaporator coil. B—Condenser. C—Motor compressor. D—Filter. E—Fan motor.

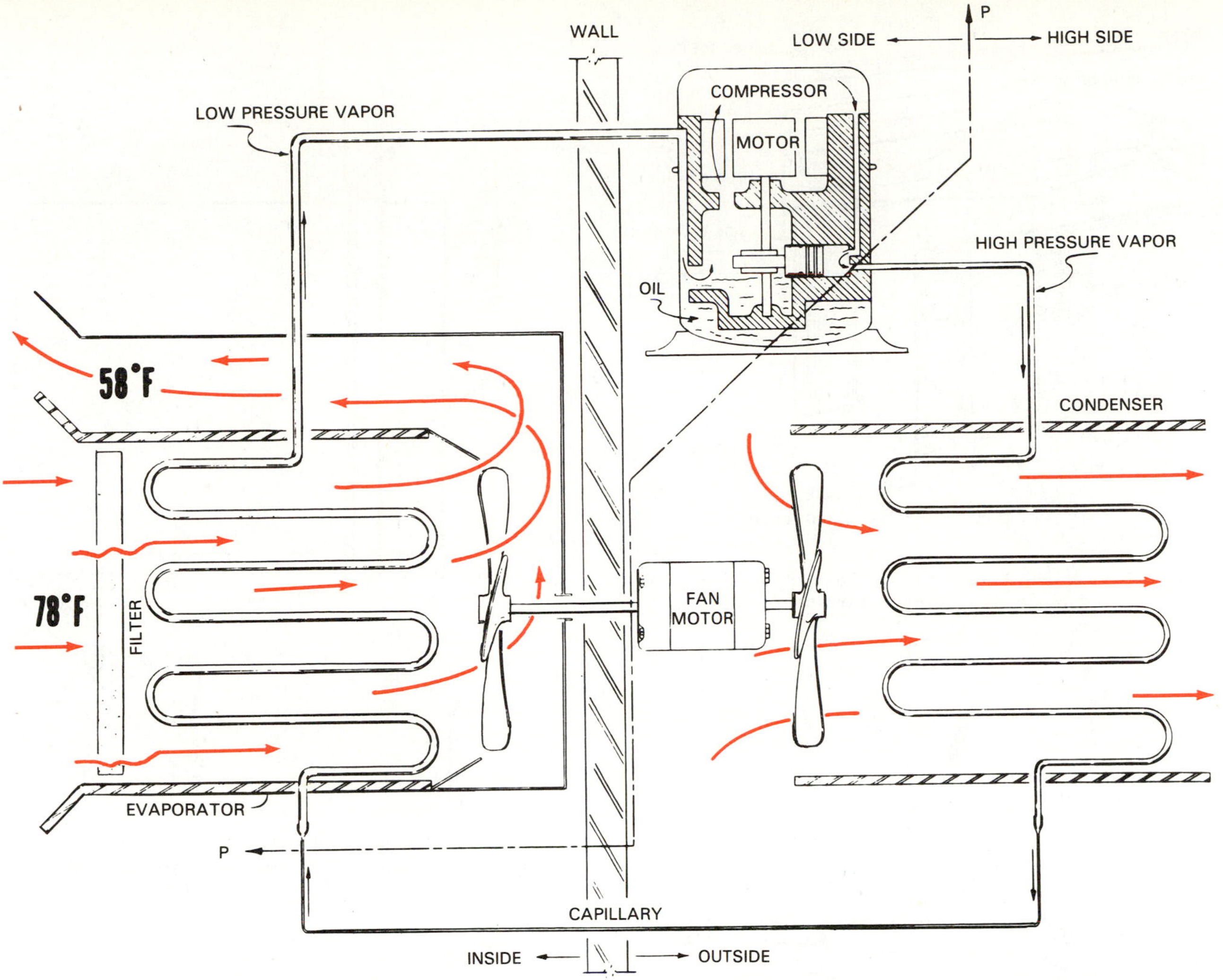

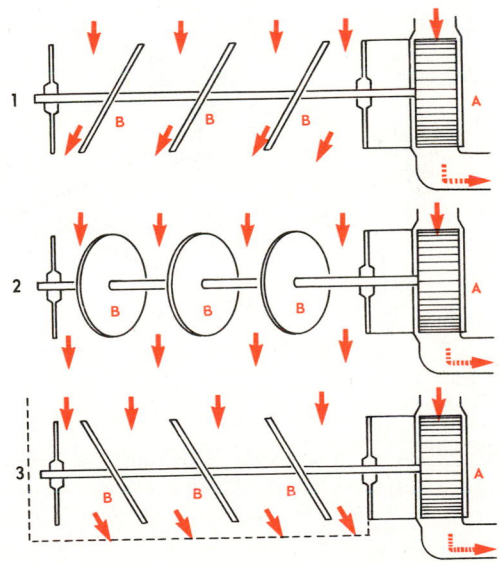

Fig. 21-8. Schematic of window or in-the-wall comfort cooling unit. Note conditioned airflow and separate condenser airflow.

Fig. 21-9. An oscillating grille deflector. The conditioned air outlet for an air conditioner which continually sweeps the air from side to side as long as the fan runs. An air turbine, A, revolves. As it does, it turns deflector plate at B. 1—Air is deflected to the left. 2—Air flows directly ahead. 3—Air flows to the right.

This is done by rotating angle deflector plates. The plates are mounted on a shaft which is turned by an air-operated rotor in the exhaust air. This is shown in Fig. 21-9.

The systems are controlled by thermostats. The sensitive bulb is usually mounted at the inlet of the evaporator. A differential of about 5 F. (3 C.) is normal.

If the part of the bulb farthest from the evaporator is insulated, the bulb will respond better to the evaporator temperatures. It will cool sooner and stop the unit before it overcools. It will also stop the unit if the evaporator ices up and will prevent it from starting again until the ice melts.

21-7 INSTALLING WINDOW UNITS

Window units must be installed with the outside tilted down for condensate drain. They must be securely fastened in place to prevent the unit from falling out of the window. All edges should be sealed to minimize air infiltration and the window must be secured in the proper position.

When units are installed in windows, metal plates, rubber gaskets, and sealing compounds are used to seal the unit into the opening. The unit and the parts needed to install it are shown in Fig. 21-10. Fig. 21-11 shows a windowsill with leveling bracket and security bracket mounted. Another

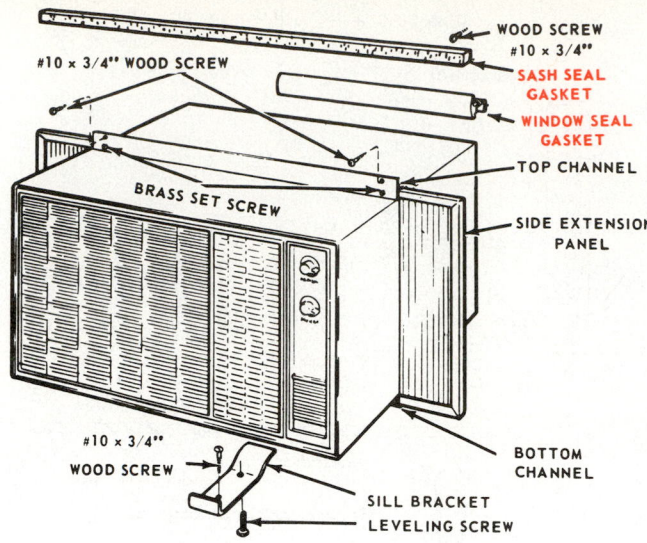

Fig. 21-10. Window unit showing necessary parts to safely mount unit in window and seal openings. (Kelvinator, Inc.)

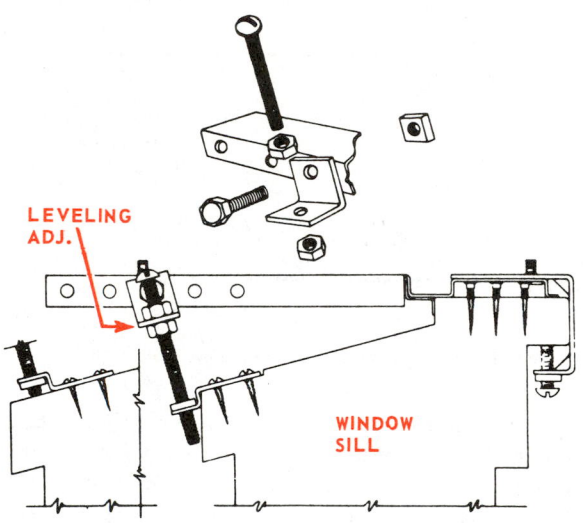

Fig. 21-11. Bracing used to hold window air conditioner on a sill. (The Singer Co.)

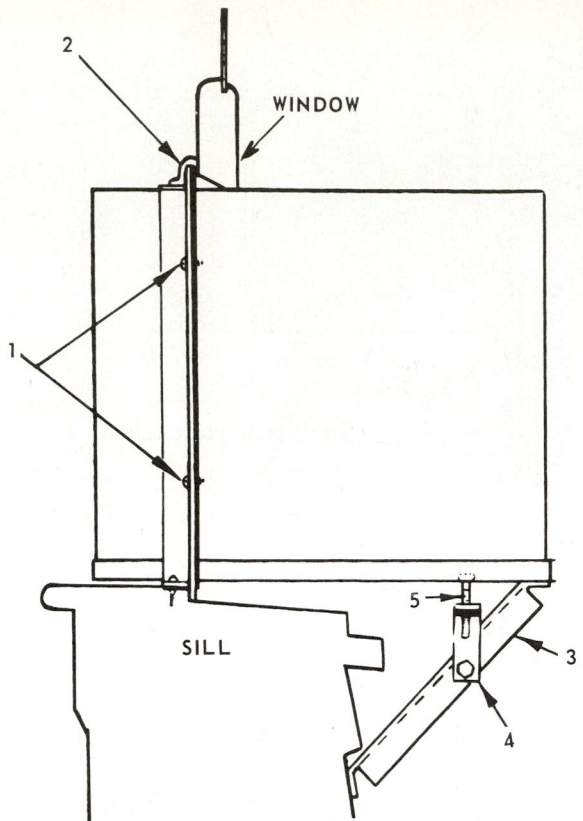

Fig. 21-12. Window unit with indoor flush mounting. 1—Sheet metal screws to hold side closure panels. 2—Gasket. 3—Bracket. 4—Clamp pivot. 5—Clamp adjustment.

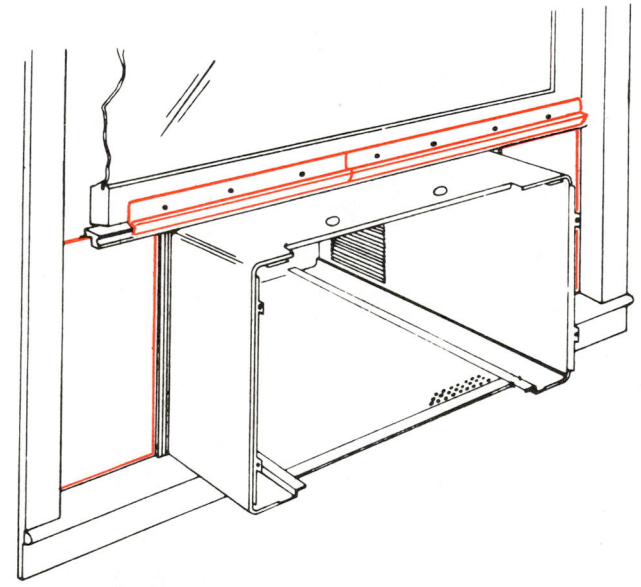

Fig. 21-13. Window air conditioning casing installed, showing rubber seal strips and filler boards.

method of bracing and leveling a comfort cooling unit is shown in Fig. 21-12.

The unit housing should be adjusted to tilt downward about 1/4 in. on the outside. This is enough to provide condensate drainage. A sponge rubber or plastic strip is usually placed between the housing and the windowsill to help make a leakproof joint. After the sill brackets are installed and the unit housing installed, the rubber seal strips and the filler boards are put in place, as shown in Fig. 21-13. Manufacturers each have slightly different methods for mounting window units.

Where the lower sash is raised to make room for the air conditioner, an air gap will exist between the two sashes. This opening may be sealed with a sponge rubber or styrofoam, as shown in Fig. 21-14.

The housing must be securely fastened in place before the unit is put in place. The filler boards (between the unit housing and side of the window) are usually sealed with sponge rubber strips or styrofoam and held in place with sheet metal screws and with spring clips. Fig. 21-15 shows one method of installing them. A typical window unit installation

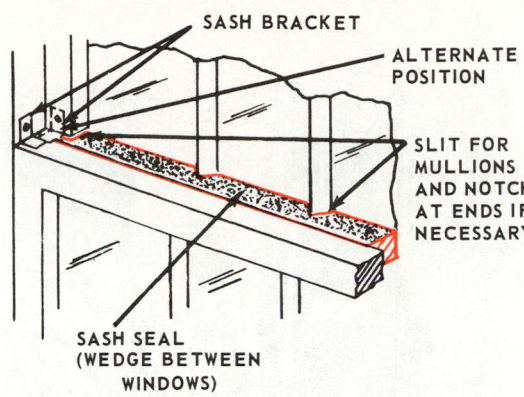

Fig. 21-14. Sponge rubber seal placed between upper edge of lower sash and upper sash of double hung window. Sash bracket keeps lower sash locked. (Philco-Ford Corp.)

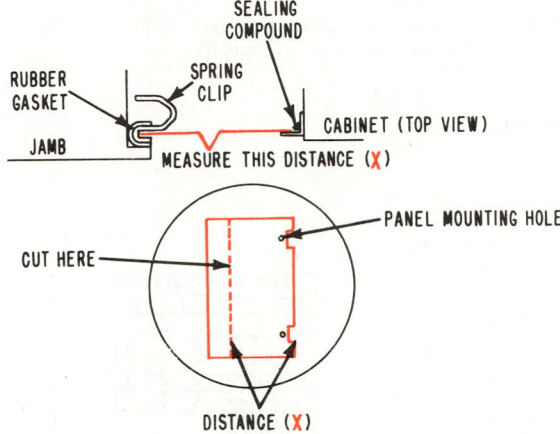

Fig. 21-15. Filler panel between unit housing and window casing.

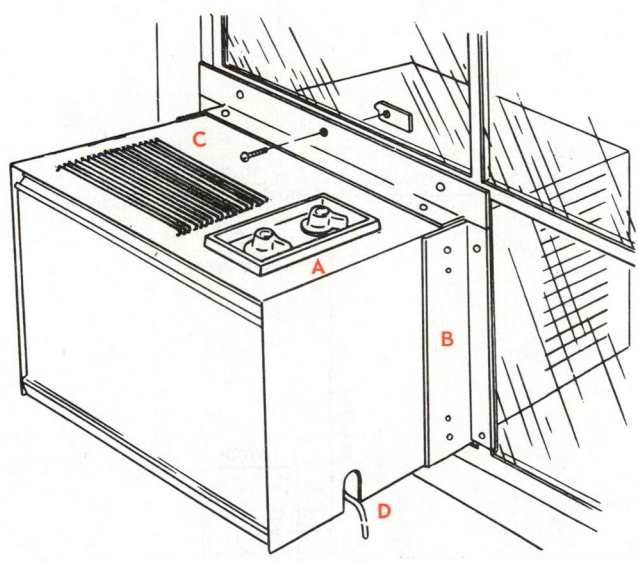

Fig. 21-16. Method of mounting window unit in casement window. A—Controls. B—Angle plate, usually enameled steel, fastened to both casing and window frame. C—Machine screw to hold angle plate to window frame. D—Electrical. (Kelvinator, Inc.)

in a casement window is shown in Fig. 21-16.

The inside mechanism is heavy. It should be moved using a dolly or special carrier. Avoid moving or lifting the unit by using the tubing or coils as hand grips. Carry the unit by holding onto the bottom pan.

Avoid forcing the unit into the casing. Check to be sure the refrigerant lines and the wiring are free and clear as the unit moves into the casing. The front grille, filter and control knobs are easily installed.

As a final step, check all joints for tightness. Caulk seams which show light leaks or which one suspects may not be airtight.

When making the electrical hookup, use a separate circuit. A polarized plug (one with a ground wire) is required.

Thermostats are used with most window units. They are adjustable to cut out between 56 F. (13 C.) and 60 F. (16 C.) and cut in between 77 F. (25 C.) and 80 F. (27 C.). Their differentials vary between 3 F. (2 C.) to 8 F. (4 C.). If a thermostat fails, the unit will not start. To test the operation of a thermostat, cover the air-out and air-in with a cloth. The air will now recirculate into the unit and the temperature will quickly drop to the cut-out temperature. Use a thermometer.

Units which mount through the wall are popular in new apartment units. There is no interference with windows and comfort cooling can be provided as desired. Fig. 21-17 shows a typical installation.

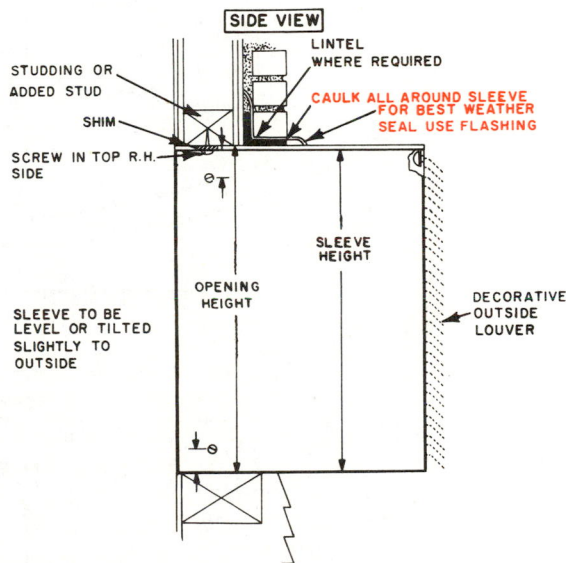

Fig. 21-17. Typical through-the-wall unit air conditioning installation.

21-8 SERVICING WINDOW UNITS

Servicing window units is similar to the servicing of hermetic refrigerating units. Chapters 11 and 14 describe most of the servicing operations.

Some of the external service operations are as follows:
1. Semiannual cleaning or replacement of the filter (usually done by the owner). Fig. 21-18 shows a filter design.

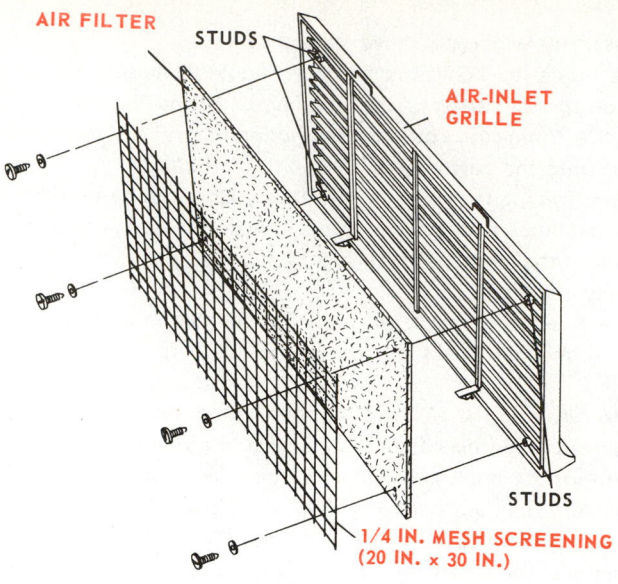

Fig. 21-18. Typical filter installation for window air conditioner.

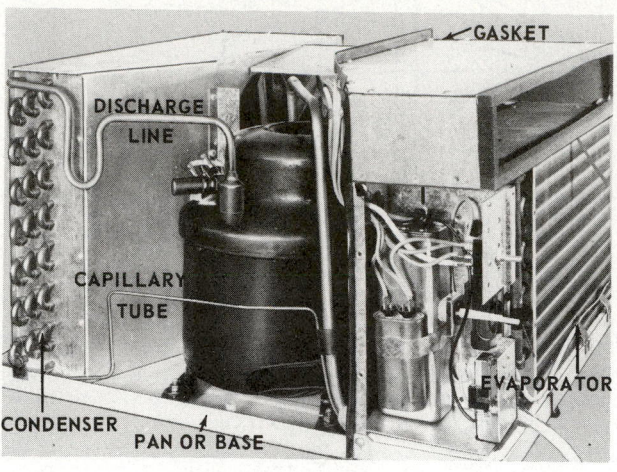

Fig. 21-19. Window air conditioner unit removed from its cabinet and ready for cleaning. (Frigidaire Co.)

Fig. 21-20. Wiring diagram of 120V window air conditioner with a 7000 Btu/hr. capacity. Note two-speed fan motor. (Fedders Corp., Norge Div.)

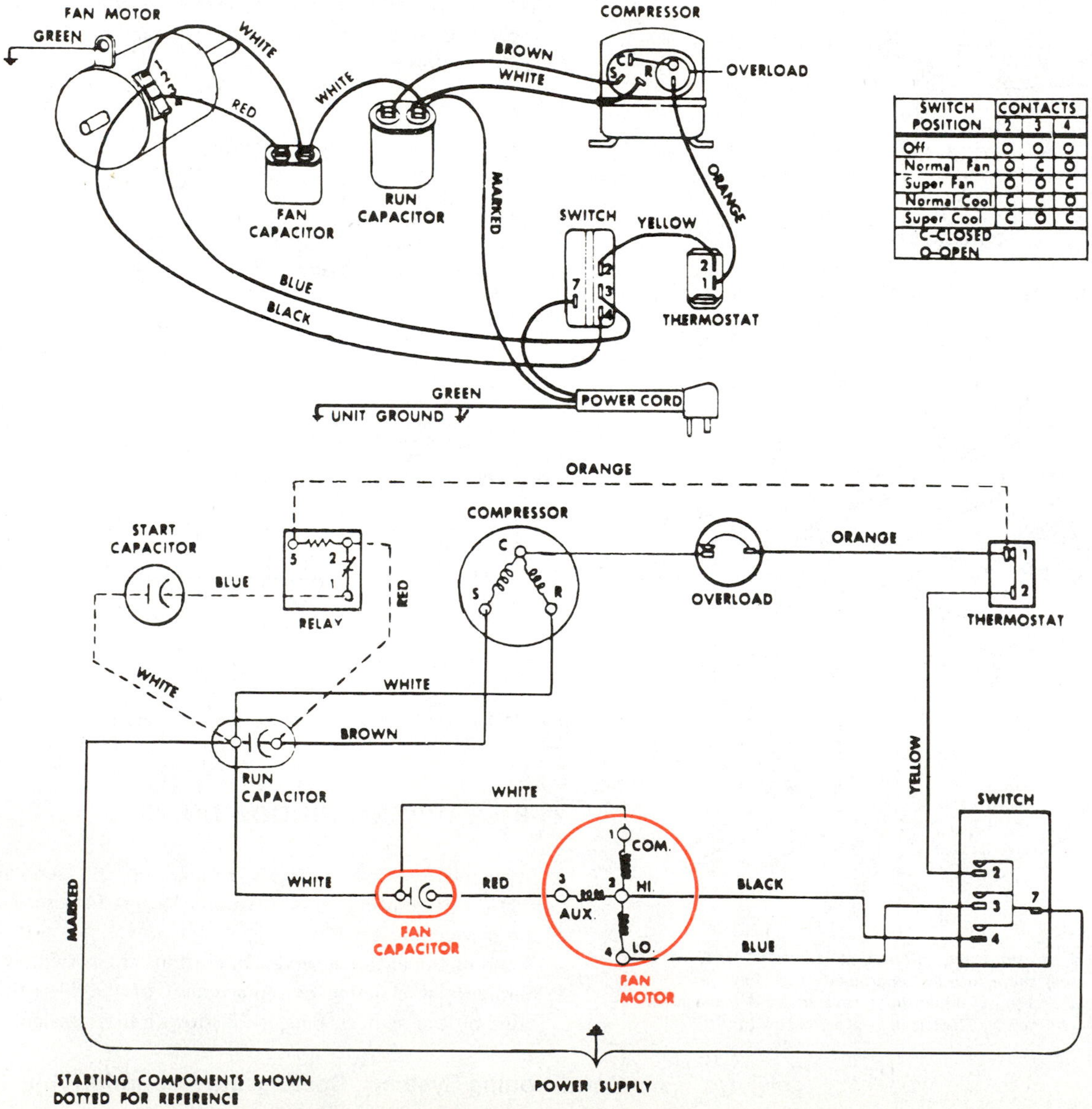

SWITCH POSITION	CONTACTS		
	2	3	4
Off	O	O	O
Normal Fan	O	C	O
Super Fan	O	O	C
Normal Cool	C	C	O
Super Cool	C	O	C
C—CLOSED O—OPEN			

STARTING COMPONENTS SHOWN DOTTED FOR REFERENCE

2. Annual cleaning of the evaporator, condenser, fan blades, fan motor, motor compressor, and casing. The unit is removed from its casing for these operations, as shown in Fig. 21-19.
3. Inspect fan motor or motors and lubricate them unless they have sealed-for-life bearings. Always wipe away excess oil. Oil mist on the fan blades collects lint and reduces air movement efficiency.

Place a tarpaulin or newspapers on the floor and remove or tie back curtains or drapes before cleaning the unit. Use a commercial model vacuum cleaner with a brush-equipped nozzle to clean the inside of the cabinet.

Finned evaporators and condensers are difficult to clean. The fin spacing prevents the vacuum brush from reaching the lint and dirt. In such cases, plastic blades along with a powerful vacuum will remove most of the dirt. Such dirt must be removed if the unit is to continue working efficiently. *Never use metal blades for cleaning; they may cause leaks.*

If the unit can be taken outside (and this is always more desirable), a powerful water and detergent spray does a good job. Coils must be cleaned thoroughly. Fins, if bent, should be straightened.

When servicing fan motors, make certain that fans are tight on the shaft. They should be carefully positioned in the shroud for efficient air movement. Avoid bending the fan blades or twisting them. An off-balance fan will soon wear out the motor bearings and will be noisy because of vibration. Replace an abused fan.

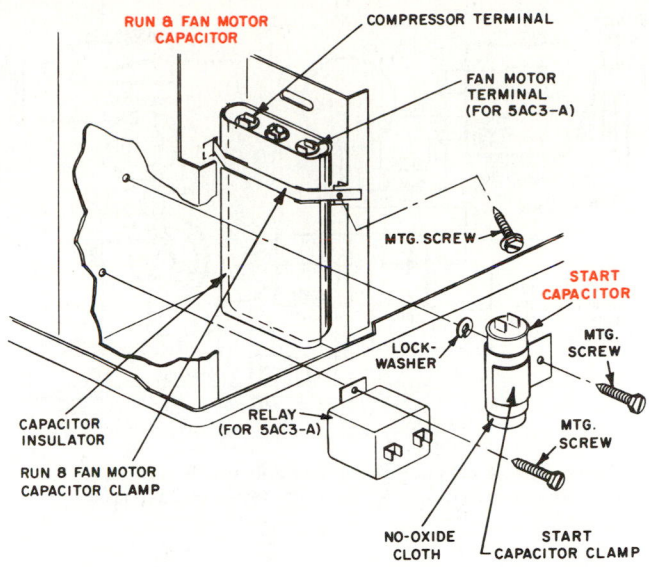

Fig. 21-22. An arrangement of start, run and fan motor capacitors. Note that fan motor run capacitor and compressor motor run capacitor are in same container.

Inspect the drain. It must be clean. Remove lint from the drain hole and tube using a soft wire. Check all bolts, nuts and screws for tightness.

Before replacing the unit in the cabinet, run it to check for noise. Find its source and stop the noise.

Always put a cloth over the outlet of the air conditioner when it is first started after cleaning. Loosened dirt that the vacuum cleaner failed to pick up will be blown out the adjustable grille.

The wiring of a window unit is very similar to other refrigerating units. See Fig. 21-20. External electrical servicing procedures are usually the same as for domestic and commercial units except that:
1. Fan motors usually have two or three speeds.
2. Some systems have three capacitors: starting capacitor, running capacitor and fan motor capacitor.

Shown in Fig. 21-21 is a unit with a starting capacitor and a running capacitor. An arrangement with three capacitors is shown in Fig. 21-22. The position of the fan control switch, thermostat, capacitors and wiring are seen in Fig. 21-23.

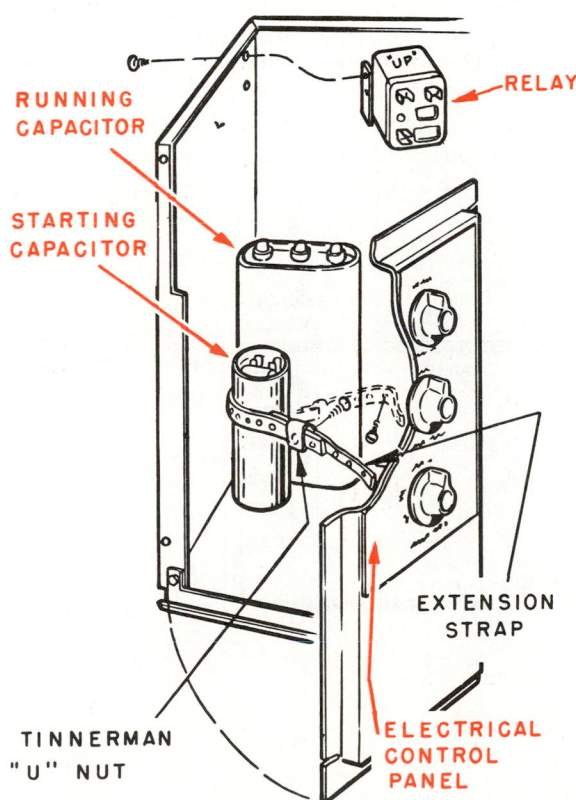

Fig. 21-21. Window unit showing location of parts such as capacitors, control panel and relay.

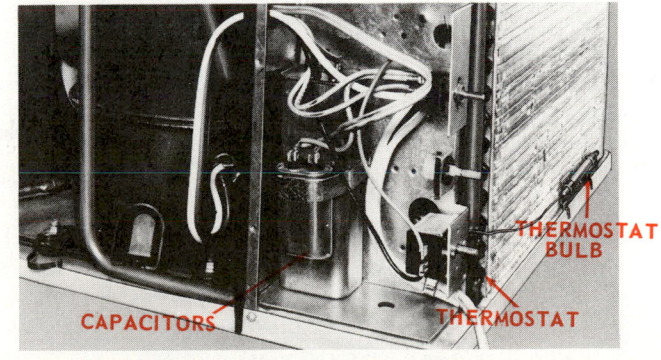

Fig. 21-23. Window unit showing location of capacitors and thermostat.

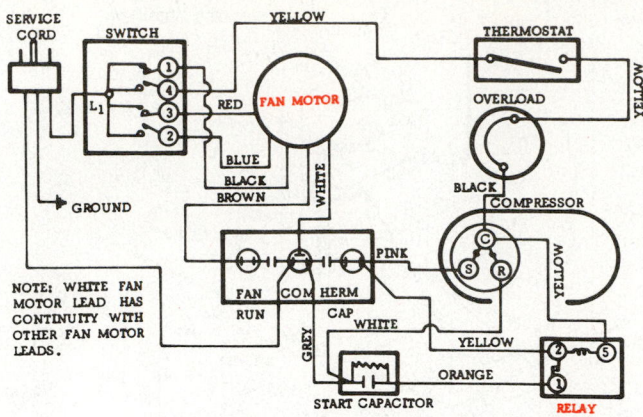

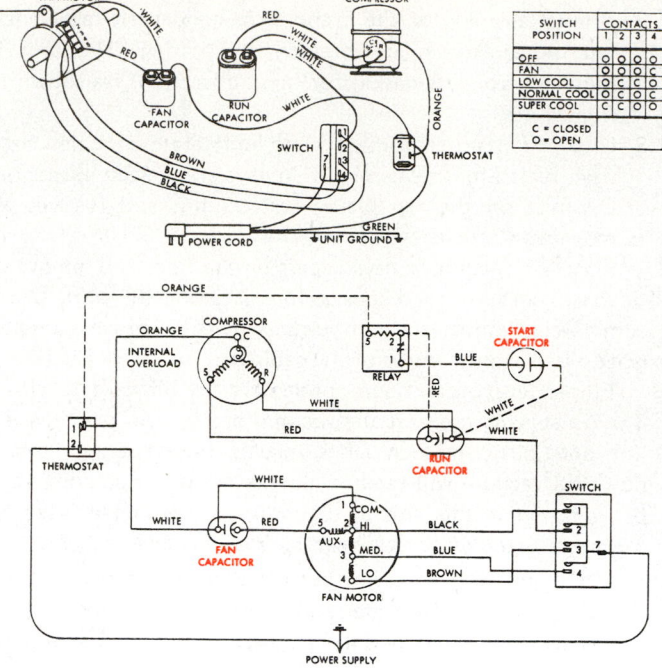

Fig. 21-24. Wiring diagram of window unit which uses R-22. It draws approximately 11.7A at a power factor of 90 percent.

The wiring diagram in Fig. 21-24 is for a window unit with a variable speed fan motor and a thermostat. A wiring diagram for a three-speed fan motor system of 21,000 Btu/hr. (6.53 watts) capacity using a 240V circuit is shown in Fig. 21-25.

Testing of outside electrical parts is described in Chapters 6, 7 and 8. Fan and compressor motor testing is described in Chapter 7. All but the motor compressor can be repaired on-the-job. Before doing internal service work on the unit, be sure the malfunction is not in the external circuit. Test for power in. Check the thermostat, the relay, the capacitors and the overload protectors (both electrical and temperature).

Troubles in the unit may include:

1. Lack of refrigerant.
2. Stuck compressor.
3. Inefficient compressor.
4. Clogged refrigerant circuit.
5. Shorted, open circuit or grounded motor windings.

Fig. 21-25. Wiring diagram of three-speed fan system. (Fedders Corp.)

Condition of the motor can be checked with a continuity light or with an ohmmeter. To check for lack of refrigerant or clogged refrigerant lines, install a gauge manifold as shown in Fig. 21-26.

Determining (diagnosing) trouble is explained in Chapters 11 and 14. The unit should be moved to the shop if the motor compressor needs repairs. Motor compressor can be replaced on the owner's premises. If the unit lacks refrigerant, locate the leak and repair it before recharging.

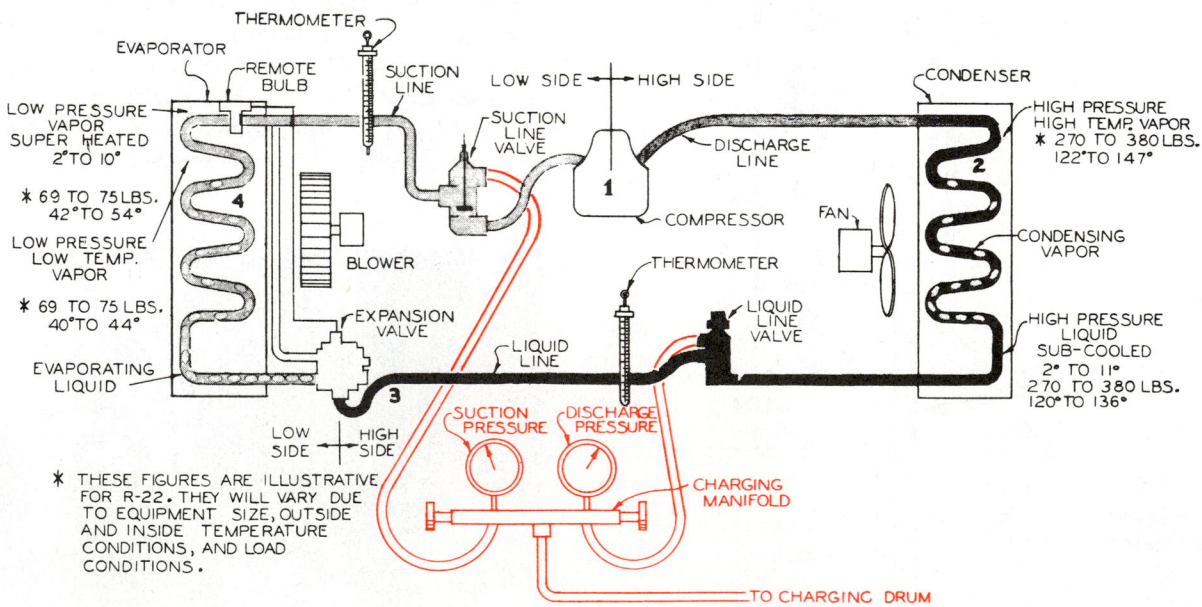

Fig. 21-26. Air conditioning unit cycle showing gauge manifold installed. (Tappan Air Conditioning Div.)

Many window air conditioners use PSC (permanent split capacitor) compressor motors. These motors do not use a relay for starting. If supplied voltage is low (10 percent or more) they will start with great difficulty. Many service technicians install a starting capacitor and a relay to overcome this problem. The capacitor and the relay must be exactly the right size for the motor.

The best way to determine the correct size is to follow the manufacturer's recommendation. If this is not available, the table in Fig. 21-27 will help in selecting the correct electrical starting system.

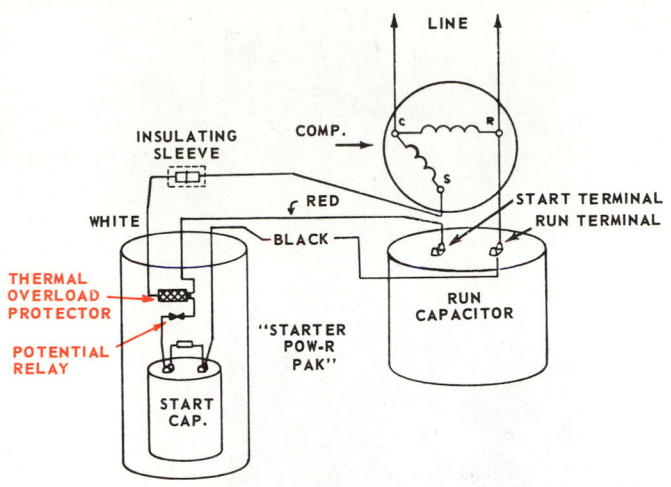

Fig. 21-28. Motor compressor starting kit which may be used on PSC motor compressors. (Sealed Units Parts Co., Inc.)

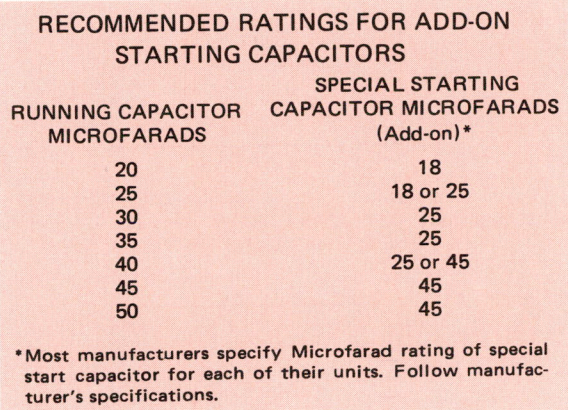

RECOMMENDED RATINGS FOR ADD-ON STARTING CAPACITORS

RUNNING CAPACITOR MICROFARADS	SPECIAL STARTING CAPACITOR MICROFARADS (Add-on)*
20	18
25	18 or 25
30	25
35	25
40	25 or 45
45	45
50	45

*Most manufacturers specify Microfarad rating of special start capacitor for each of their units. Follow manufacturer's specifications.

Fig. 21-27. Table of special starting capacitor sizes to be used on PSC motor compressor when starting difficulties are found.

When the motor compressor reaches a satisfactory speed, the starting capacitor needs a potential type relay to open the starting capacitor circuit. Special starting kits are available for PSC motor compressors. One is illustrated in Fig. 21-28.

Window units are often removed during the winter season. But, if this is inconvenient, the unit can be winterized by blocking the air-in and the air-out grilles with cardboard or flexible plastic sheeting. Storm sash can be custom built to fit around the air conditioner or one may use plywood held in place with caulking or rubber grommets.

Replacement capillary tubes for window units must be very accurately picked for size. Fig. 21-29 gives correct sizes when R-22 refrigerant is used.

If the window unit drips water into the room, it is not correctly installed. Check the slope of the unit from inside to outside with a spirit level. It must slope to the outside (condenser edge) about 1/4 in. Condensate water will then run to the depression in the unit base under the condenser fan and condenser. Make sure the drain hole is open, then level the unit along its other dimension. Finally, recheck the unit installation for airtight sealing in the window opening.

COMPRESSOR CAPACITY Btu/Hr.	NO. OF CAPILLARIES	CAPILLARY SIZE		COIL CIRCUITS	
		SHORT	LONG	3/8 IN. TUBE	1/2 IN. TUBE
4500	1	36 IN. x .042	80 IN. x .049	1	
5000	1	25 IN. x .042	64 IN. x .049	1	
5500	1	20 IN. x .042	52 IN. x .049	1	
6000	1	40 IN. x .049	75 IN. x .054	1	
6500	1	35 IN. x .049	65 IN. x .054	1	
7000	1	28 IN. x .049	52 IN. x .054	1	
8000	1	36 IN. x .054	65 IN. x .059		1
9000	1	28 IN. x .054	48 IN. x .059	2	1
10,000	1	36 IN. x .059	64 IN. x .064	2	1
11,000	1	28 IN. x .059	50 IN. x .064	2	1
12,000	1	40 IN. x .064	68 IN. x .070	2	1
13,000	1	32 IN. x .064	56 IN. x .070	2	1
14,000	1	44 IN. x .070	70 IN. x .075	2	1
15,000	1	36 IN. x .070	56 IN. x .075	3	2
16,000	1	30 IN. x .070	48 IN. x .075	3	2
17,000	1	38 IN. x .075	65 IN. x .080	3	2
18,000	1	35 IN. x .075	55 IN. x .080	3	2
19,000	1	28 IN. x .075	48 IN. x .080	3	2
20,000	1	40 IN. x .080	58 IN. x .085	3	2

Fig. 21-29. Capillary tube sizes for window air conditioners which use R-22 refrigerant. Coil circuits are the size of the tubing in the evaporator. (Tecumseh Products Co.)

Fig. 21-30. This self-contained console type comfort cooling air conditioner has a water-cooled condenser. Cooled air is delivered at upper grille.

21-9 CONSOLE AIR CONDITIONERS

In console air conditioners whole systems are mounted in a cabinet. They vary in capacity from 2 hp to 10 hp. Such units are often used in small commercial establishments such as restaurants, stores and banks.

Console models may have either water-cooled or air-cooled condensing units. Air-cooled models, needed in some localities because of water restrictions, must have air ducts to the outdoors for condenser cooling.

Fig. 21-30 shows a water-cooled console unit. Return air enters the lower grille. Cooled air is discharged at the upper grilles. Ducts can be connected to portions or all of the upper section. These are needed when partitions interfere with cooled air distribution. The condensing unit is mounted in the bottom of the console. Air blowers are in the middle, while the evaporator is in the top of the cabinet.

Console units also have adjustable fresh air intakes and evaporator bypass controls as shown in Fig. 21-31. All must have drains to remove the condensate flowing from the evaporator.

Most of the console models have a complete refrigerating system, filtering system and evaporator. Fig. 21-32 is a schematic of a console unit. A large self-contained comfort cooler with an air-cooled condenser is shown in Fig. 21-33.

Water-cooled units require plumbing connections to both fresh water and a drain. The drain also receives the moisture condensed out of the air by the evaporator in the summer.

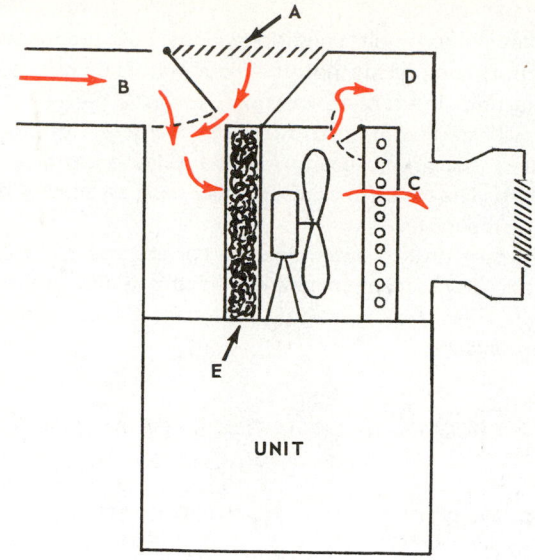

Fig. 21-31. Air circulation in console comfort cooling air conditioner. A—Recirculated air. B—Fresh air. C—Cooled air. D—Recirculated untreated air. E—Filter.

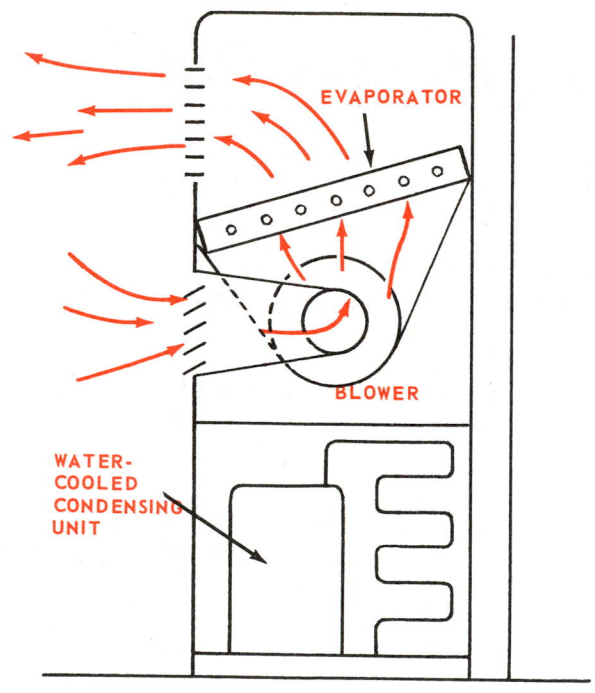

Fig. 21-32. Schematic of console type air conditioner.

One company has developed three openings in one rubber hose for this purpose. This permits moving the unit quickly from room to room. Such units usually do not provide for winter conditioning facilities.

21-10 INSTALLING CONSOLE AIR CONDITIONERS

Console units are factory assembled. After moving the unit into place and leveling it, plumbing and electrical connections are required. Such connections must conform to local codes.

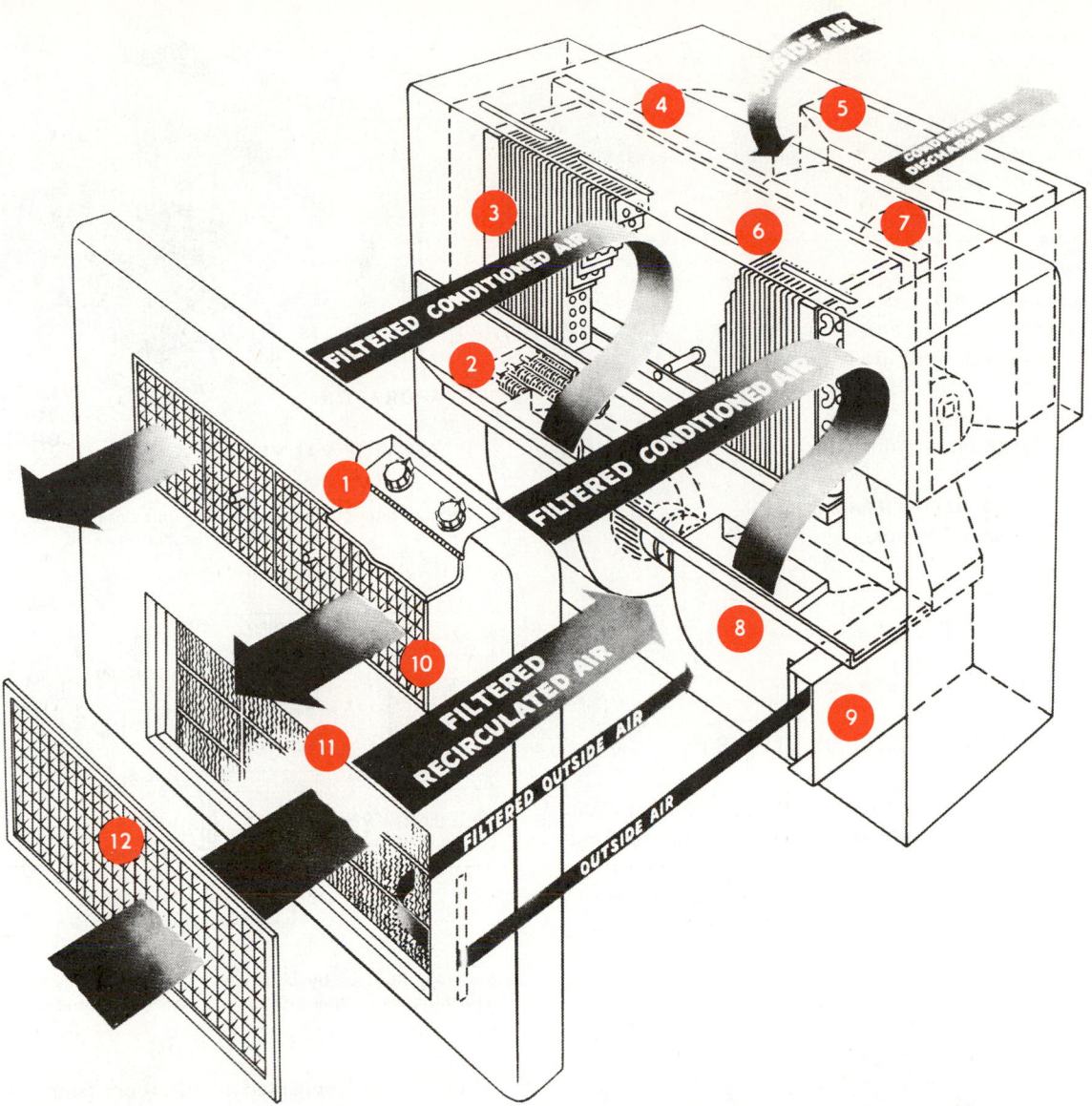

Fig. 21-33. Console unit with air-cooled condenser. 1—Controls. 2—Heating coil. 3—Evaporator. 4—Compressor. 5—Air cooled condenser. 6—Draft diverters. 7—Condenser fan. 8—Evaporator fan. 9—Ventilating air duct. 10—Discharge grilles. 11—Air filter. 12—Return air grille. (The Singer Co.)

Usually the motor compressor is hermetic and the refrigerant control is generally a thermostatic expansion valve. The unit should be thoroughly checked. The air temperature both in and out, the electrical load and the operating pressures should be checked and recorded for future reference.

21-11 SERVICING CONSOLE AIR CONDITIONERS

Panels must be removed to work on internal parts of the unit. Periodic maintenance duties include replacing the filter or cleaning it, cleaning the evaporator and fins, cleaning the fan motor and oiling it (unless it has sealed bearings), cleaning the drain pan and drain tube. The inner lining of the cabinet sometimes gathers lint. This should be removed by vacuuming.

Servicing of the refrigerating unit and the condenser water circuit is explained in Chapter 14. *It is important to check the refrigerant charge, the operation of the thermostatic expansion valve and the water flow.*

A regular maintenance schedule is necessary if the owner is to receive long and satisfactory service from the air-conditioning system. Different parts of the system must be checked more frequently than others. A record should be kept of all checks including both the data and the date.

Check the following weekly:
1. V-belts.
2. Fan speeds.
3. Pump speeds.
4. All standby units.
5. Water leaks.
6. Controls (pressure, temperature and airflow).
7. Lubrication.
8. Canvas connectors on ducts.

9. Cooling tower.
10. Water treatment.
11. Bleed off.

Monthly checks should be made on these:
1. Refrigerating system (charge, purge, test for leaks; check strainers and dryers).
2. Filters.
3. Humidifier.
4. Safety valves.
5. Cooling tower pump.
6. Duct dampers, registers and diffusers.
7. Piping (insulation, vibration and wear).

Every six months:
1. Clean fans and casings.
2. Clean duct registers and diffusers.

Every year:
1. Efficiency check of compressors.
2. Efficiency check of pumps.
3. Damper operation.
4. Clean water circuits.
5. Operate all hand valves.

Every two years:
1. Inspect condenser wet surfaces.

21-12 REMOTE COMFORT SYSTEMS

In remote air conditioning systems the refrigerating equipment is located away from the space to be conditioned. These units vary in capacity from two tons to thousands of tons.

Some units condition the air which is then distributed by ducts to the space or spaces, as shown in Fig. 21-34.

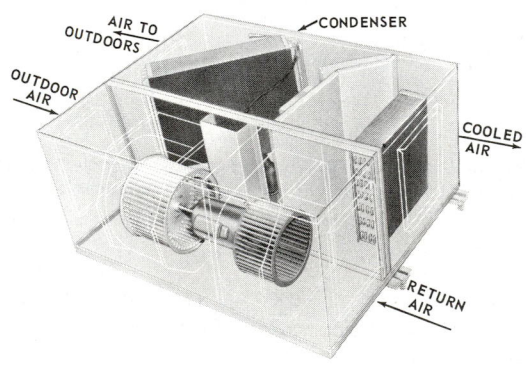

Fig. 21-34. Air-cooled refrigerating unit for duct distribution of comfort-cooled air. This sytem is designed for duct system cooled air distribution. (Fedders Corp.)

A water-cooled system which has service valves, a tube-within-a-tube condenser and a water valve is shown in Fig. 21-35. A chilled water system used in combination with a duct distribution system is shown in Fig. 21-36.

Refrigeration lines are run to air conditioning units in each room to be air conditioned. Individual thermostats are connected to each room unit. A solenoid valve usually controls the flow to each unit.

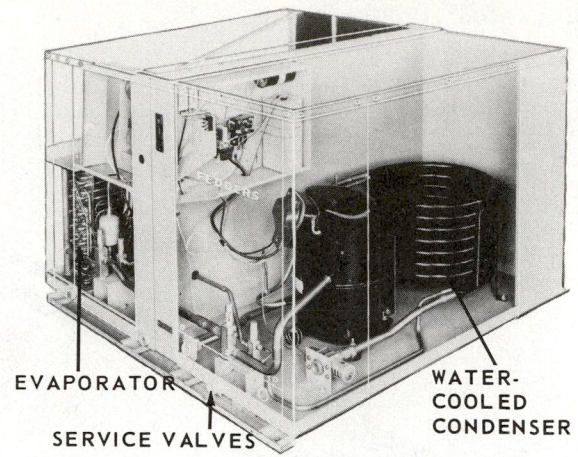

Fig. 21-35. Water-cooled condensing unit designed for central comfort cooling installation.

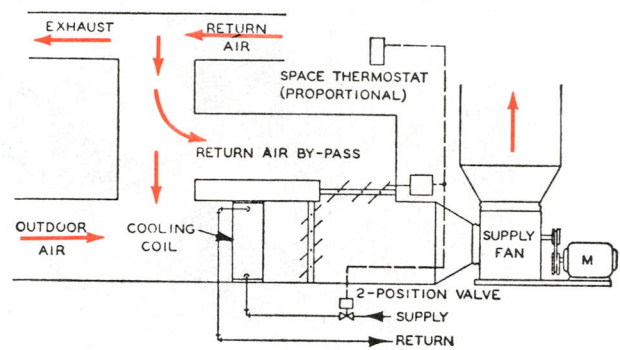

Fig. 21-36. Control by chilled water coil of space temperature using return bypass damper and valve. (ASHRAE Guide and Data Book)

This valve is connected to the room temperature control. This controls both a fan and the solenoid refrigerant valve in the room unit. Chilled water starts to flow to the room unit when the fan starts in that unit. It is shut off when the fan stops. Air filters are also sometimes installed in these units.

21-13 MOBILE HOME AIR CONDITIONING

Air conditioners similar to the home window units are being used more and more on house trailers, mobile homes and larger pleasure boats. Central systems are often found in mobile homes. The evaporator is mounted in the furnace plenum as in domestic units. Conditioned air travels through the heating ducts. See Chapter 23.

Self-contained units are usually mounted on the roof of the mobile home. See Fig. 21-37. The three main sections of a roof-mounted unit are shown in Fig. 21-38.

Since a 120V electrical supply is needed, most can be used only when the vehicles are parked or the boats moored at the dock. But, where the vehicle or boat has an a-c generator, the air conditioning system can be used on the move. Because of corrosion problems, marine air conditioners use a great deal of copper and plastic.

Fig. 21-37. A roof-mounted travel trailer air conditioner. A—Roof line. B—Condensing unit mounted on top of trailer. C—Inside grille. (International Metal Products Div., McGraw-Edison Co.)

21-14 EVAPORATIVE COOLING

In dry climates, evaporative comfort cooling is very desirable and practical. If air at 105 F. (41 C.) and 20 percent relative humidity is moved rapidly over water at the same temperature, some of the water will evaporate and the remaining water can cool to as low as 55 F. (13 C.). See the psychrometric chart in Chapter 18. Practically speaking, the water will cool down to a range of about 65 F. (18 C.) to 70 F. (21 C.). Now, if other air is forced around the water container and then distributed into the space to be cooled, it will cool down the air leaving the unit to around 75 F. (24 C.) or 80 F. (27 C.) at about 40 percent relative humidity. The conditioned room would be quite comfortable. Such a system is illustrated in Fig. 21-39.

These units use two fans, one for the outside air and one for the conditioned air. Small units have been built to mount in the windows of automobiles.

21-15 STORING COLD FOR COMFORT COOLING

There are many situations where comfort cooling is only needed part of the day. Stores, funeral homes, churches, tourist homes and theaters need cooling for only an eight-hour period.

In such cases, one can use a smaller capacity system. Store cold during 16 hours of operation. This allows the smaller unit to have twice as much air cooling capacity during the eight hours it is needed.

These storage systems may be tanks of water, cooled to 32 F. (0 C.) with some ice forming. They may also be tanks using a eutectic solution — see Chapter 28 — designed to freeze or partly freeze at a low temperature.

The tanks or plates store up the refrigeration during the time cooling is not needed. When the extra cooling is needed,

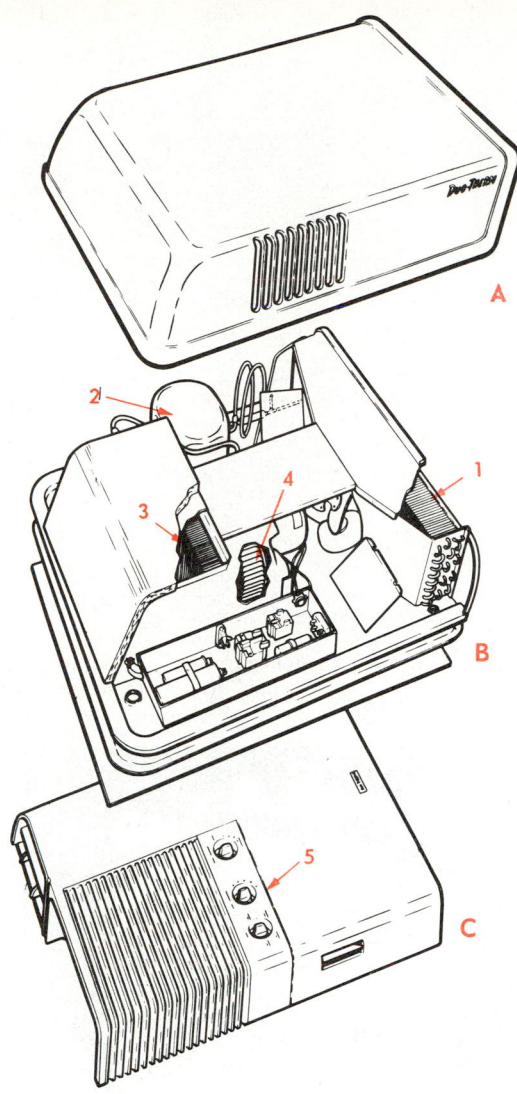

Fig. 21-38. Mobile air conditioning system is of compact design. A—Outside cover. B—Cooling system. C—Interior unit. 1—Condenser. 2—Motor compressor. 3—Evaporator. 4—Fan. 5—Controls. (Duo-Therm, Div. of Motor Wheel Corp.)

the tanks or plates can release the stored cooling, adding it to the normal cooling of operation.

21-16 DEHUMIDIFYING EQUIPMENT

A mechanism which dehumidifies (removes moisture) is known as a dehumidifier. Such equipment depends upon a cold coil over which air is blown. Moisture is condensed out by coming in contact with the cold surface.

When coil surfaces are at a temperature below the dew point of the air, moisture will condense out of the air. The coil surface temperature must be kept above freezing. Frost or ice formation would block airflow.

A dehumidifier, as shown in Fig. 21-40, is usually a small hermetic refrigerating system that has both condenser and evaporator in a cabinet. Air is drawn over the evaporator. As the air touches the cold surface of the evaporator it cools below its dew point. Water condenses out of the air and

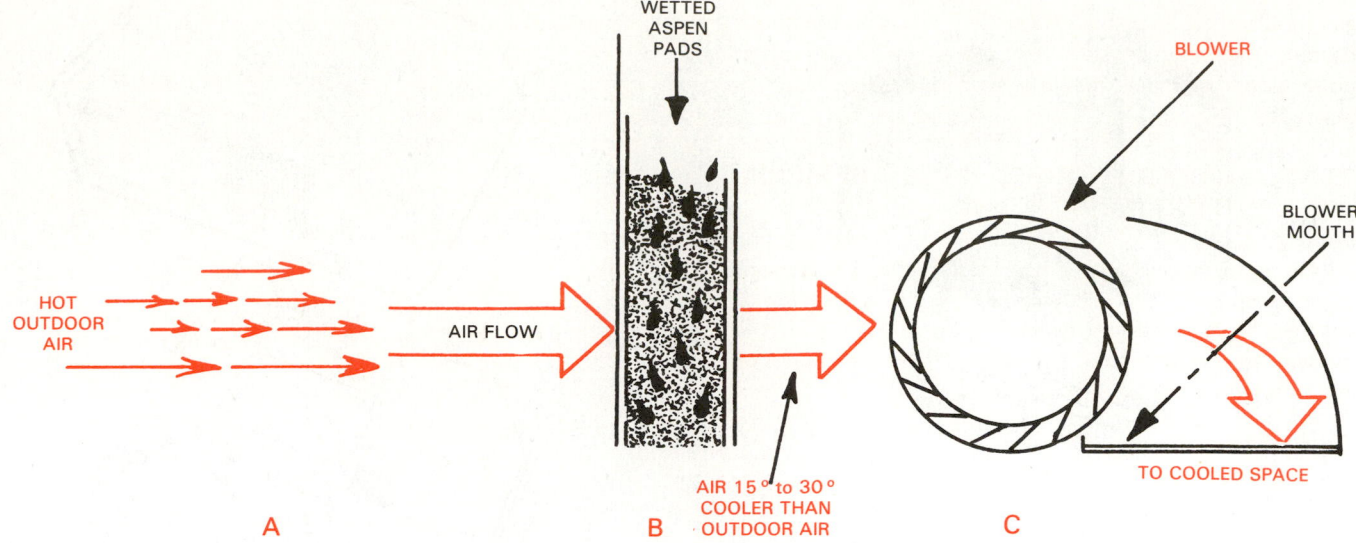

Fig. 21-39. Evaporative cooling process. A—Outside air. B—Filtering through saturated evaporative media and cooled by evaporation. C—Circulated by blower. (International Metal Products Div., McGraw-Edison Co.)

Fig. 21-40. Dehumidifier installed in basement. Air enters at front of cabinet and is exhausted at back (note arrows).

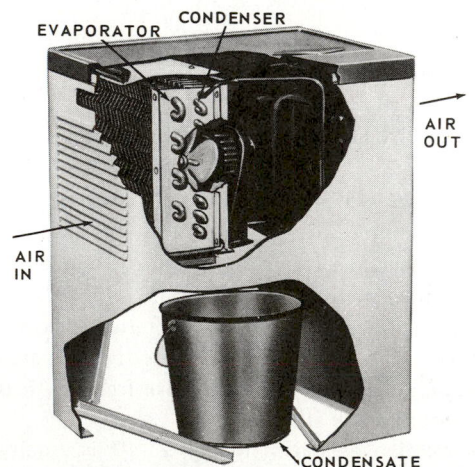

Fig. 21-41. Section view of dehumidifier. First pass of finned coil near fan is the cooling and dehumidifying coil while next pass is the condenser. Either a pail or a direct drain can be used to remove condensate. (International Harvester Co.)

Fig. 21-42. Dehumidifier equipped with humidistat. Humidistat automatically stops dehumidifier when desired humidity has been reached. (Fedders Corp.)

collects on the evaporator. The cooled air is then moved over the condenser to reheat it to a reasonable relative humidity.

The device is used to "dry" the air. It is useful in basements and other damp places. Fig. 21-41 shows a sectioned view of such a unit.

These units usually have a humidistat. This is a device that senses moisture in air. It is shown in Fig. 21-42. A container is used to collect the condensate or a drain tube carries it off.

In some installations, certain chemicals are used which can absorb moisture from the air. Where used, the chemicals are usually cycled in such a way that first the moisture from the air is absorbed into the chemical. Then the chemicals are heated. Moisture driven from the chemical is exhausted out-of-doors, and the chemicals are ready to once more absorb moisture.

21-17 AIR CONDITIONER ENERGY EFFICIENCY RATING (EER)

Air conditioner energy efficiency rating (EER) is measured by the amount of cooling (heat absorbed) in proportion to the amount of electrical energy used. (See Chapter 15.) It is calculated by dividing the Btu/hr. rating by the watts per hour. It is expressed by the term, "Btu per watt hour."

The formula is: $\dfrac{Btu/hour}{watts} = EER$

In an attempt to conserve energy, the United States Department of Commerce, National Bureau of Standards, has recommended that an energy guide label be attached to all air conditioners. This label gives the information needed to calculate the EER.

Suppose that the following data is given for an air conditioner: 8000 Btu/hr. capacity; 860 watts.

$$EER = \frac{Btu/hr.}{watts} = \frac{8000}{860} = 9.3 \text{ Btu/watt hr.}$$

The efficiency of an air conditioning unit goes up as the Btu/watt hr. value increases. Generally, an air conditioner with a high efficiency rating will be less expensive to operate than a lower rated air conditioner that produces the same amount of cooling action.

Therefore, when planning to purchase an air conditioner, the EER of the unit should be an important consideration.

21-18 REVIEW OF SAFETY

All of the safety practices described in Chapters 11 and 14 also apply to comfort cooling units.

Window units should be handled with care. Heavy units should be moved and lifted with hand trucks and lifts. Many window units are installed in upper floors. With double-hung windows, the upper sash is lowered to the air conditioner and aids in keeping the air conditioner in place.

Sometimes the upper sash is opened accidentally. Then, since most of the weight of the air conditioner is outside the building, the unit may fall to the ground. It should be securely attached to the windowsill and braced! When removing mechanisms be careful not to drop them. They may be slippery. Safety shoes are recommended. Carefully follow installation instructions supplied by the manufacturer.

Remote systems should be sturdily mounted. Suction lines and liquid lines should be protected from abuse.

Always review instructions on larger units before performing any service work or installation work on them.

Window air conditioners are usually available for use on either 120V or 240V circuits. It is advisable — particularly with the higher Btu rating — to use a 240V circuit. This is because the voltage drop between the power panel and the air conditioner will be less when using a 240V motor. A separate air conditioner circuit should be provided. Make certain all systems are properly grounded.

Whenever it becomes necessary to add a refrigerant to a system, always be sure that the refrigerant added is the same as the refrigerant already in the unit.

Always wear goggles when testing for leaks and adding refrigerant.

Much of the sheet metal used in air conditioning has very sharp edges. Be careful not to cut your hands or fingers when making repairs.

21-19 TEST YOUR KNOWLEDGE

1. What refrigerant controls are usually used on air conditioning evaporators?
2. Why are solenoid refrigerant control valves used on some air conditioning evaporators?
3. Why is air leaving an evaporator considered to be damp air?
4. Can air be dehumidified by any other method than by cooling?
5. How is condensate handled in a unit comfort cooler?
6. What air is used to cool the condenser of an air-cooled comfort cooler?
7. What is the relative humidity of the air just as it leaves the evaporator?
8. How many fans does a window type comfort cooler have?
9. How many coils does air pass through in a dehumidifier?
10. What device varies the capacity of the centrifugal compressor?
11. Do window comfort coolers have different types of electrical extension plugs? Why?
12. How are the installation joints sealed?
13. Why is a window comfort cooler slanted toward the outside of the home?
14. What is the normal cut-out setting of a window unit thermostat?
15. Why is the dew point reading important when using a dehumidifier?
16. Can some of the return air be used to reduce the humidity of the cooled air? If so, how?
17. Is it possible to have a hydronic comfort cooling system?
18. Describe the two ways window air conditioners may be installed.
19. How is the cooled air moved from side to side as the window unit operates?
20. What may happen if a window unit is shut off and then immediately turned on again?
21. If the window comfort cooling unit will not start, what should one check first?
22. Why should one avoid bending or twisting fan blades?
23. How many capacitors does a window comfort cooler have?
24. How can a console air conditioner be installed for condenser air cooling?
25. What is happening if a window unit drips water inside the room?
26. What electrical power is most used with mobile air

conditioning systems?

27. What chamber is usually located in the upper part of a console air conditioner?

28. Why is drain needed on air-cooled comfort cooling system?

29. Can one use a duct system with a chilled water system? Explain.

30. Is it permissible to plug a window air conditioner into the utility outlet?

Chapter 22

AIR CONDITIONING SYSTEMS DISTRIBUTING AND CLEANING

Air conditioning mechanisms are designed to condition the air and distribute it to the proper place, in the proper amounts with the most comfort to the occupants.

When a radiator system or a room convector system — such as a steam heating plant or a hot water plant — is used, air distribution is simple. The heat exchange units are located along the outside walls. During the heating season the heated air rising from the radiator along the wall mixes with the cold air adjacent to the cold wall. Then natural air currents (convection) distribute the air mixture throughout the room. Many heating systems use motor-driven fans to help circulate the air.

During the cooling season, chilled water may be circulated in these same convectors while a blower moves air over the coils. Many installations are combinations of a hot water (hydronic) system and a warm air duct system.

An important operation of a good air conditioning system is to deliver clean air to the space being conditioned. Air quality control is becoming one of the most important parts of air conditioning. Chapter 18 describes the impurities in air.

22-1 CONDITIONED AIR

In air conditioning practice, air is passed through the mechanism where it is heated or cooled, humidified or dehumidified, cleaned and distributed where it is needed. An air conditioning system, properly designed and installed, will provide proper amounts of conditioned air at the proper temperature and humidity. The distributed air must:

1. Be clean.
2. Provide the proper amount of ventilation.
3. Carry enough heat to keep the conditioned spaces warm or be able to absorb enough heat to cool the conditioned spaces.

22-2 WEIGHT OF AIR

Air has definite weight. Although it is invisible, its gases have a definite mass and it takes energy to move it. Fig. 22-1 gives the weight of air under various temperature and humidity conditions. One lb. (0.454 kg) of dry air at 70 F. (21 C.) at standard atmospheric pressure will occupy a space of 13.35 cu. ft. (0.378 m^3). If there is 50 percent relative humidity, 13.51 cu. ft. (0.382 m^3) of air and moisture mixture weighs 1 lb. (.454 kg). Because air is a gas, it responds closely to Boyle's and Charles's Laws. Therefore, as the temperature rises, it takes more cubic feet to weigh one pound. As the pressure drops, it takes more cubic feet to weigh one pound. As humidity increases, it takes more cubic feet to weigh one pound.

22-3 HEAT IN AIR

Because air is a physical substance, it can carry heat. It will remove heat from or take it to a space.

The psychrometric properties already studied in Chapter 18

Air Temp. F.	Cu.ft./lb. Dry Air	Lb./cu.ft. Dry Air	Cu.ft./lb. 50% Saturated	Lb./cu.ft. 50% Saturated	Cu.ft./lb. 100% Saturated	Lb./cu.ft. 100% Saturated
0	11.58	.08635	11.585	.08632	11.59	.08628
50	12.84	.0778	12.915	.0774	12.99	.0769
70	13.35	.075	13.51	.074	13.68	.0731
100	14.10	.0709	14.585	.06856	15.07	.06635
120	14.60	.0685	15.55	.0643	16.50	.0606
150	15.3	.0652	17.7	.0565		
200	16.7	.0600				

Fig. 22-1. Weight of air at various temperatures and humidities.

show how heat content can change as the temperature and humidity change. The specific heat of dry air is .24 Btu/lb. Converted to metric, this is equal to 0.56 joules per kilogram (0.56 J/kg). The additional heat due to the moisture in the air varies considerably, depending on the amount of saturation. For example, from 0 F. (−18 C.) to 100 F. (38 C.), 24 Btu are added to 1 lb. of dry air. There may be .04293 lb. of moisture added to this 1 lb. of dry air to saturate it, but the heat in this moisture is 47.4 Btu (latent heat and sensible heat). The total heat in 1.04293 lb. is 71.4 Btu. However, as far as distributing the air is concerned, only sensible heat need be taken into account. This is because vaporizing or condensing of water should not take place in the ducts, or in the room being conditioned.

There are 7000 grains to a pound; thus, because it requires about 1000 Btu/lb. or 2326 joules per kilogram (J/kg) to change water to water vapor, each grain of water changed to vapor requires a latent heat of 1000/7000 or .143 Btu per grain. One grain equals about 0.065 grams. If condensation takes place in a cooled air duct or on the outside surface, some heat will be released. The temperature of the air delivered to the conditioned space may be changed. It may cause a failure in the operation of the equipment. Only sensible heat changes should take place outside of the heating or cooling system.

22-4 BASIC VENTILATION REQUIREMENTS

As noted before, air is a mixture of gases. Normally air contains about 21 percent oxygen. A human system requires that a certain oxygen content be contained in the air:
1. To maintain life.
2. To be comfortable.

If a room is tightly sealed, any human in that room would slowly consume the oxygen and increase the amounts of carbon dioxide, water vapor and various impurities. This could cause drowsiness or even death.

One must remember that space for human living must have air with a good oxygen content and that this air must be kept at a reasonable temperature. It is of utmost importance that fresh air be admitted to provide the oxygen.

In the past, this fresh air entered the space by infiltration (leakage) from the outside at door and window openings and through cracks in the structure. However, modern construction is reducing this air leakage. Air conditioning apparatus, then, must furnish fresh air. Modern units have a controlled fresh-air intake. This fresh air is conditioned and mixed with the recirculated air before it reaches the room.

Some conditioned air leaves a building through doors, windows and other construction joints. Some also leaves by exfiltration. (This means leaking out or being blown out by mechanical means.) Any kind of exhaust fan removes conditioned air. Some of this air is replaced by infiltration on those sides of the building exposed to wind pressure.

It is best to bring in replacement fresh air through a makeup air system. When this is done:
1. The makeup air can be cleaned.
2. The makeup air can be cooled or heated.
3. A positive pressure can be maintained in the building to

keep out airborne dirt, dust and pollen. (A negative pressure reduces the efficiency of exhaust fans and fuel-fired furnace.)
4. A definite amount of fresh air is brought into the building for health purposes (oxygen content).

Certain areas of a building should have a slightly less positive pressure (5 to 10 percent) than the rest of the building to reduce the spread of odors. Such areas would include the kitchen, lavatories and where certain industrial operations produce fumes.

The presence of positive or negative pressures can be measured using a manometer. See Chapter 24.

The amount of fresh air required depends on the use of the space and the amount of fresh air admitted by infiltration. One basic rule is to provide at least 4 cfm of fresh air per person to provide enough oxygen and to remove carbon dioxide. If six people occupy this space, there is $10,000 \div 6$, or 1667 cu. ft. per hour for each person, or $1667 \div 60 = 27.7$ cu. ft. per minute (.78 m^3/min.). This meets or exceeds ventilating code requirements.

One must remember that the air can be handled either to produce positive pressure (higher than atmospheric pressure) in a building or negative pressure (below atmospheric pressure). A positive pressure will eliminate infiltration of air from outside or from other spaces. It is done by using special air intakes to the blowers. A positive pressure assures that all air entering a building can be filtered and cleaned before reaching the occupied space. Negative pressure increases the infiltration at windows and doors. This air is untreated and may be dirty.

Residential homes which use fuel-burning furnaces need air for combustion. Combustion air, leaving by way of the chimney, might leave the interior of the house under a slightly negative pressure. View B of Fig. 22-2 shows a basic diagram of negative pressure conditions in a home.

If the amount of impurities in the air — such as odor, smoke and bacteria — is great enough to require air cleaning, the remedy may be either ventilation, using fresh air, or improved air cleaning.

Ventilation is usually based on air changes per hour for the conditioned space. If the space is 1000 cu. ft., for example, three changes per hour would mean 3000 cu. ft. per hour or 50 cu. ft./min. Three changes every hour is the minimum for a residence during the heating season. As high as 12 changes an hour (in the above case, 200 cu. ft./min.) are recommended for cooling. Fig. 22-3 shows typical air changes for both the heating season and the cooling season.

It is good practice to keep the air blowers running all the time to provide good ventilation to all parts of the building. Variable speed blowers are sometimes used. They provide more air movement when the heating or cooling system is running; less movement when the systems are off.

An adequate air supply is the best way to control comfort. Body comfort is controlled by evaporation, convection, radiation and respiration. One must, therefore, control the temperature of the walls, floor or ceilings to make sure they are not too warm or cold (radiation). One must also supply enough air to promote good respiration, evaporation and convection. If the specific conditions are not known, it is best

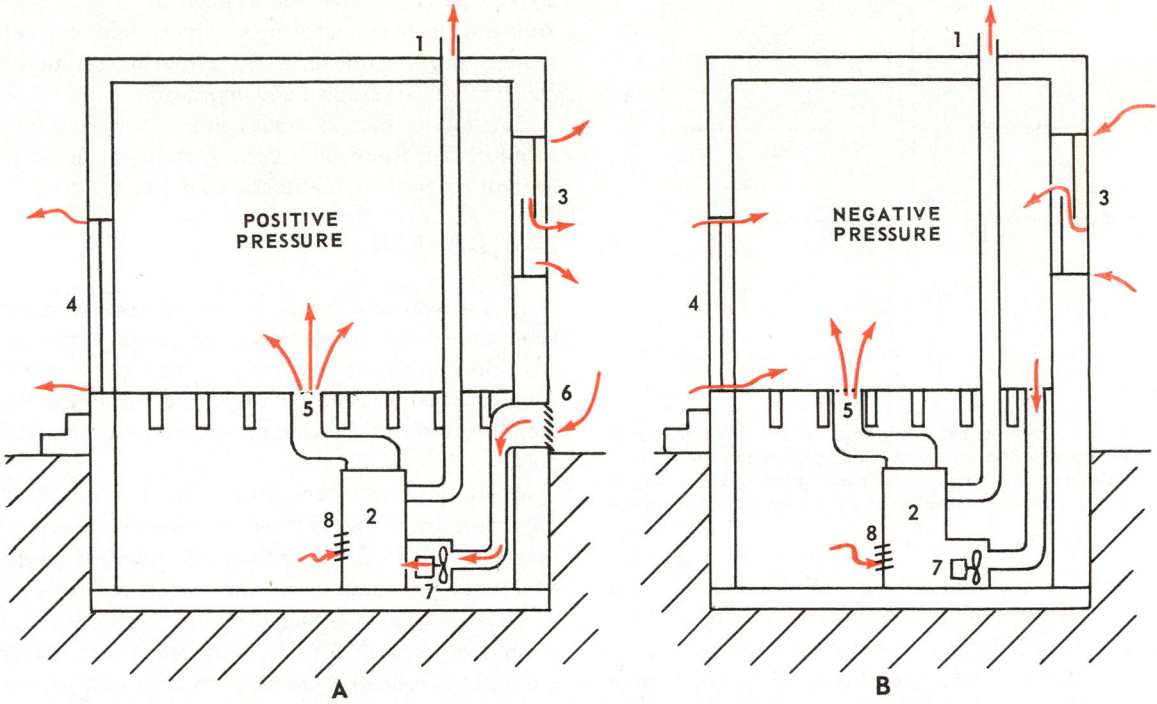

Fig. 22-2. Simplified diagram of airflow into and out of a building during heating season. A—Positive air pressure. B—Negative air pressure. 1—Chimney. 2—Furnace. 3—Window. 4—Door. 5—Warm air grille. 6—Fresh air intake. 7—Fresh air fan. 8—Furnace draft control. The red arrows indicate the airflow — both within the building and into and out of the building.

Use	Air Changes/Hour	
	Heating	Cooling
Homes	3–6	6–9
Offices Stores	5–8	6–12
Public Assembly	5–10	6–12

Fig. 22-3. Recommended air changes for various types of occupancies.

to design for 2 cfm/sq. ft. and/or 12 changes/hr. It is also very important to remember that people occupying a closed space give off considerable heat. A sleeping person gives off about 200 Btu/hr., while a person doing heavy work gives off up to 2400 Btu/hr. One Btu = 252 calories.

Another way to determine ventilation requirements is to design for 4 cfm to 6 cfm of fresh air per person and for about 25 cfm to 40 cfm of recirculated air per person. This means the system should handle a total of 29 cfm to 46 cfm per person. One cfm = 0.0283 cu. m/min.

22-5 NOISE

Air distribution systems must be designed to circulate a certain required amount of clean air. Yet, the movement of air must not annoy occupants or cause them discomfort. Two common problems that must be overcome are objectionable noise and drafts. Noise is produced by movement or vibration of an object.

Three factors affect noise. These are:
1. Noise source.
2. Noise carriers.
3. Noise amplifiers or reflectors.

The noise source is a vibration that is loud enough to be heard. This vibration may originate in the heating system, cooling system or fan mechanism. A vibrating duct panel will create alternate waves of low-pressure and high-pressure air and produce sound.

In measuring the loudness of sound, the unit used is the decibel. Decibel meters register noise level. See Chapter 18. *Decibels (dB) refer to the frequency of the pressure fluctuations in the air and the amplitude or size of these vibrations. Airborne sound is usually expressed in cycles per second (cps).*

Both mechanisms and the motion of matter cause noise in systems. Sources include:
1. Fans.
2. High-velocity air traveling through ducts and causing turbulence.
3. The a-c hum of a motor.
4. Sucking and throbbing noises produced by compressors.
5. High-velocity refrigerant flow, especially at sharp bends in piping.

Noise caused by high-speed air is often the result of an undersize unit or duct. Such noise problems develop where the blower has been speeded up to compensate (make up for) the duct's or the unit's lack of capacity.

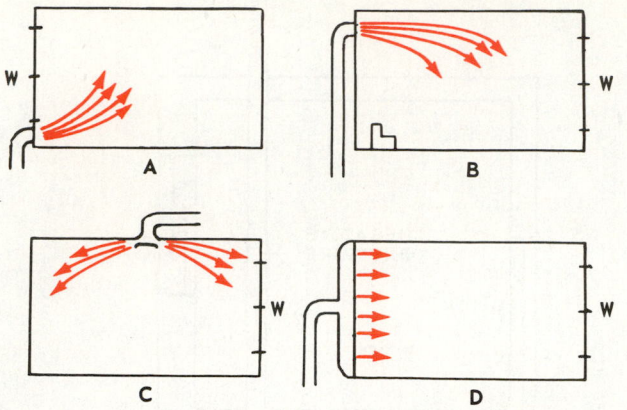

Fig. 22-4. Location of heating grilles minimizes drafts in living areas of room. A—People are exposed to drafts. B—High velocity air is above living level. C—Center location permits lower grille velocity; higher velocity is above living level. D—Ideal large grille opening. W—Windows.

Noise or vibration carriers are rigid structures that carry vibrations to places where they may be annoying. Floors, ceilings, ducts, doors and pipes may carry vibrations.

Hard, smooth surfaces in the space being conditioned often reflect or amplify sound (make it louder). Walls, ceilings, floors and furnishings may pick up a small vibration and reflect it at such a frequency and in such a direction that all or

certain parts of the space may be made uncomfortable. Problems involving acoustics (absorption and reflection of sound) and maintenance of a low-decibel noise level are constantly being studied and improved.

Soft fabrics such as drapes and curtains and fabric-covered furniture are noise absorbers. Felt lined or soft insulation lined or covered ducts also absorb noise.

22-6 DRAFTS

It is a relatively simple matter to provide ducts and fans large enough to give a room the correct amount of air for conditioning. However, the air must enter the room, and circulate to all parts of it without interfering with flow of air to the air return. Neither should there be objectionable drafts or noise.

When air moves past people faster than 25 fpm (about 1/3 mph), most people feel an annoying draft. This means that, if the air flows faster than 1/3 mph through the length of a 25-foot living room, an uncomfortable feeling results.

To have a grille outlet designed to throw the air into the room a distance of 8 to 13 ft., a velocity of 500 fpm (about 5.5 mph) is needed. Therefore, to keep that part of the space occupied by humans at a 25 fpm velocity, the location of grilles or outlets must be carefully selected.

To move air across a long space at a reasonable velocity, the

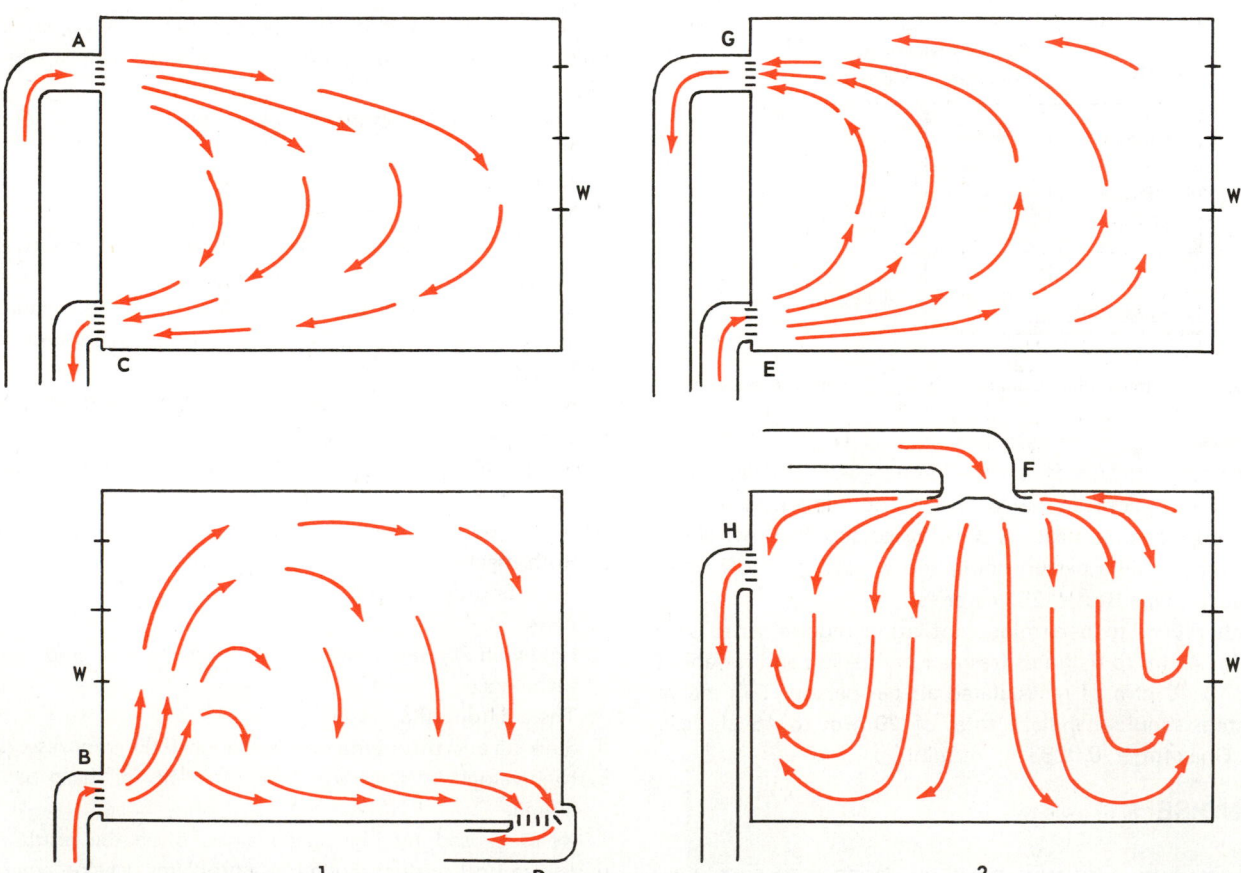

Fig. 22-5. Location of return air grilles in residential installation. 1—For heating: A and B warm air in, C and D cold air return. 2—For cooling: E and F cold air in, G and H warm air return. W—Window.

location of the air returns is important. The air returns should be located on the opposite side of the space from the air entrance.

Air returns should be located high on the wall if warm air return is desired (cooling season) and low on the wall (or in the floor) if cold air return is desired (heating season). Fig. 22-4 shows some typical airflow patterns. The return air openings often used are shown in Fig. 22-5.

22-7 STRATIFICATION OF AIR

Warm air tends to rise. Cold air tends to settle. In a room where the air is not deliberately moved, the air will assume levels according to its temperature, as shown in Fig. 22-6. This is called stratification. *Air in an occupied space must be kept moving in order to eliminate stratification.*

It is important to place thermostats and humidistats at the proper level because of this stratification. Another thing,

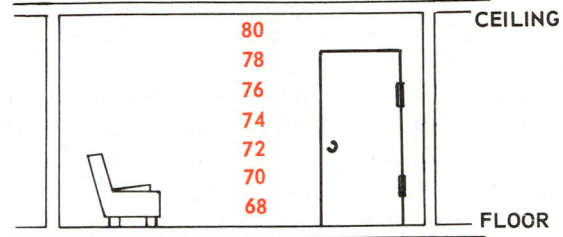

Fig. 22-6. Various temperature levels (stratification) found in room with little or no air circulation.

stratification tends to make smoke haze hover in layers.

Unfortunately, some grilles are poorly located. Then air moves only in certain parts of the room and becomes stagnant (not moving) in other parts. Then too, furnishings obstruct air movement in the room. For this reason, and to enable higher grille velocities, some grilles are located 6 ft. high in the room or in the ceilings. In these locations the grilles must be attractive in appearance or concealed entirely. In Fig. 22-7, a

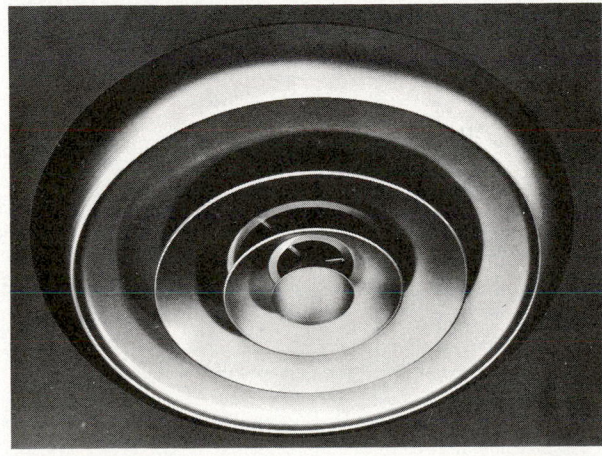

Fig. 22-7. Ceiling diffusion grille distributes air in all directions in occupied space.

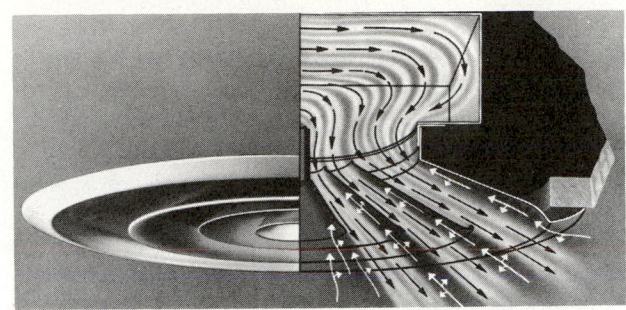

Fig. 22-8. Airflow and air mix of ceiling grille. Black arrows indicate airflow from duct. White arrows indicate room air moving into grille to mix with duct air.   (Anemostat Products Div.)

diffusion grille promotes mixing of some of the room air with the entering air in the grille. The mixing principle is shown in Fig. 22-8.

22-8 AIR DUCTS

To deliver air to the conditioned space, air carriers are needed. These carriers are called ducts. They are made of sheet metal or some structural material that will not burn (non-combustible).

Ducts work on the principle of air pressure difference. If a pressure difference exists, air will move from the higher pressure area to the lower pressure places. The greater this pressure difference, the faster the air will flow.

Ducts are made of many materials. Pressure in the ducts is small, so materials with a great deal of strength are unnecessary. Originally, hot air ducts were thin, tinned sheet steel. Later, galvanized sheet steel, aluminum sheet and, finally, insulated ducts made from materials such as asbestos and fiberboard, were developed. Passageways formed by studs or joists are sometimes used for return air where a fire hazard

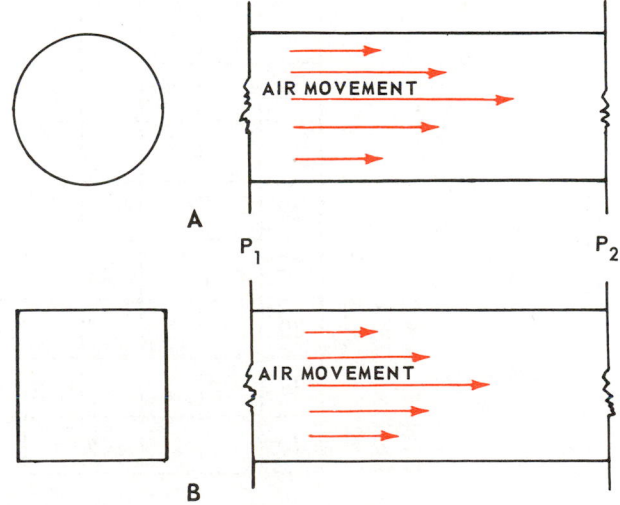

Fig. 22-9. Shape of ducts vary. A—Round. B—Rectangular. Arrows indicate that the closer air is to inside surfaces of duct the slower the flow. P_1 and P_2 indicate pressure along a duct. In order to create movement, P_1 must be greater than P_2.

does not exist.

There are three common classifications of ducts:
1. Conditioned-air ducts.
2. Recirculating-air ducts.
3. Fresh-air ducts.

Ducts commonly used for carrying air are round, square or rectangular. See Fig. 22-9. Round ducts are more efficient based on volume of air handled per perimeter distance (distance around). That is, less material is needed for the same capacity as a square or rectangular duct. Resistance to airflow is also less.

The square or rectangular duct conforms better to building construction. It fits into walls and ceilings better than round ducts. It is easier to install rectangular ducts between joists and studs.

Tables have been developed to compare carrying capacities of rectangular and round ducts. See Figs. 22-10 and 22-11. There are several round duct equivalent sizes to choose from. The one selected depends on the one side dimension desired. For example, ducts 11 in. high may be wanted to improve appearance, or to fit in between joists (a 14 in. distance). Fig. 22-11 shows the equivalents.

An 18 in. round duct has a perimeter (distance around) of:

diameter x π = circumference (perimeter).

18 x 3.1416 = 56.55 in.

A rectangular duct of equal capacity is 17 x 16 in. It has a perimeter twice the width (W) plus twice the depth (D), or 2W + 2D = perimeter.

17 + 17 + 16 + 16 = perimeter

34 + 32 = 66 in.

66 in. − 56.55 = 9.45 in. more of duct material
per unit of duct length.

In other words, for every foot of length, the rectangular duct requires 12 in. x 9.45 in. or 113.4 sq. in. more of material than the round duct.

22-9 TYPES OF DUCT SYSTEMS

Air ducts deliver air to a room or rooms and then return the air from this room or the rooms to the heating (furnace) or cooling (evaporator) system. Fig. 22-12 shows a typical residential air duct system.

There are several types of supply duct systems:
1. Individual round pipe system.
2. Extended plenum system.
3. Reducing trunk system.

Some of the installations, like that shown in Fig. 22-12, are combinations of two of the systems or of all of the systems.

Return air systems are usually of two types:
1. Single return system.
2. Multiple return system.

The return systems can also be combinations of the two systems. Fig. 22-13 shows the basic designs. Some types of duct connections are shown in Fig. 22-14.

Duct systems may be installed in basements, in crawl spaces, in attics and in concrete floors (slabs) of homes without basements. Fig. 22-15 shows an overhead duct system.

In basements, the conditioned-air main duct is run across and just under the floor joists. The branch ducts, round or rectangular, are then run between the joists to the grille or diffuser openings. The return ducts usually use the joists and the floorboards for three sides of the branch ducts (panned joist space) and then run the main return duct alongside the conditioned air duct.

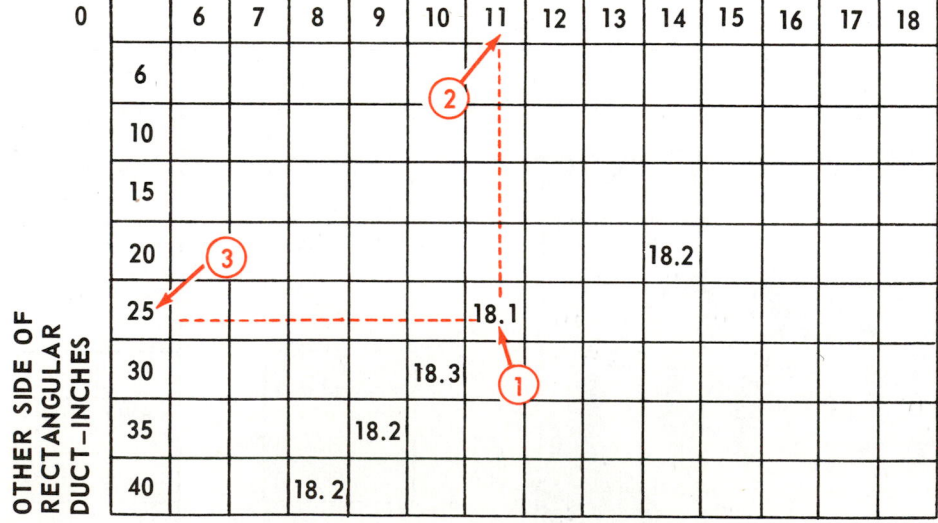

Fig. 22-10. Chart for converting round ducts to square or rectangular duct measurements. Numbers represent inches. 1—Locate round duct size. 2—Go up vertically and read the number at the top in the corresponding square. This value is for one side of rectangular duct. 3—Moving horizontally from the round duct size, read number at left side. This value is the measurement of the other side of rectangular duct.

Side Rectangular Duct	4.0	4.5	5.0	5.5	6.0	6.5	7.0	7.5	8.0	9.0	10.0	11.0	12.0	13.0	14.0	15.0	16.0
3.0	3.8	4.0	4.2	4.4	4.6	4.8	4.9	5.1	5.2	5.5	5.7	6.0	6.2	6.4	6.6	6.8	7.0
3.5	4.1	4.3	4.6	4.8	5.0	5.2	5.3	5.5	5.7	6.0	6.3	6.5	6.8	7.0	7.2	7.4	7.6
4.0	4.4	4.6	4.9	5.1	5.3	5.5	5.7	5.9	6.1	6.4	6.8	7.1	7.3	7.6	7.8	8.1	8.3
4.5	4.6	4.9	5.2	5.4	5.6	5.9	6.1	6.3	6.5	6.9	7.2	7.5	7.8	8.1	8.4	8.6	8.9
5.0	4.9	5.2	5.5	5.7	6.0	6.2	6.4	6.7	6.9	7.3	7.6	8.0	8.3	8.6	8.9	9.1	9.4
5.5	5.1	5.4	5.7	6.0	6.3	6.5	6.8	7.0	7.2	7.6	8.0	8.4	8.7	9.0	9.4	9.6	9.8

Side Rectangular Duct	6	7	8	9	10	11	12	13	14	15	16	17	18	19	20	22	24	26	28	30	Side Rectangular Duct
6	6.6																				6
7	7.1	7.7																			7
8	7.5	8.2	8.8																		8
9	8.0	8.6	9.3	9.9																	9
10	8.4	9.1	9.8	10.4	10.9																10
11	8.8	9.5	10.2	10.8	11.4	12.0															11
12	9.1	9.9	10.7	11.3	11.9	12.5	13.1														12
13	9.5	10.3	11.1	11.8	12.4	13.0	13.6	14.2													13
14	9.8	10.7	11.5	12.2	12.9	13.5	14.2	14.7	15.3												14
15	10.1	11.0	11.8	12.6	13.3	14.0	14.6	15.3	15.8	16.4											15
16	10.4	11.4	12.2	13.0	13.7	14.4	15.1	15.7	16.3	16.9	17.5										16
17	10.7	11.7	12.5	13.4	14.1	14.9	15.5	16.1	16.8	17.4	18.0	18.6									17
18	11.0	11.9	12.9	13.7	14.5	15.3	16.0	16.6	17.3	17.9	18.5	19.1	19.7								18
19	11.2	12.2	13.2	14.1	14.9	15.6	16.4	17.1	17.8	18.4	19.0	19.6	20.2	20.8							19
20	11.5	12.5	13.5	14.4	15.2	15.9	16.8	17.5	18.2	18.8	19.5	20.1	20.7	21.3	21.9						20
22	12.0	13.1	14.1	15.0	15.9	16.7	17.6	18.3	19.1	19.7	20.4	21.0	21.7	22.3	22.9	24.1					22
24	12.4	13.6	14.6	15.6	16.6	17.5	18.3	19.1	19.8	20.6	21.3	21.9	22.6	23.2	23.9	25.1	26.2				24
26	12.8	14.1	15.2	16.2	17.2	18.1	19.0	19.8	20.6	21.4	22.1	22.8	23.5	24.1	24.8	26.1	27.2	28.4			26
28	13.2	14.5	15.6	16.7	17.7	18.7	19.6	20.5	21.3	22.1	22.9	23.6	24.4	25.0	25.7	27.1	28.2	29.5	30.6		28
30	13.6	14.9	16.1	17.2	18.3	19.3	20.2	21.1	22.0	22.9	23.7	24.4	25.2	25.9	26.7	28.0	29.3	30.5	31.6	32.8	30
32	14.0	15.3	16.5	17.7	18.8	19.8	20.8	21.8	22.7	23.6	24.4	25.2	26.0	26.7	27.5	28.9	30.1	31.4	32.6	33.8	32
34	14.4	15.7	17.0	18.2	19.3	20.4	21.4	22.4	23.3	24.2	25.1	25.9	26.7	27.5	28.3	29.7	31.0	32.3	33.6	34.8	34
36	14.7	16.1	17.4	18.6	19.8	20.9	21.9	23.0	23.9	24.8	25.8	26.6	27.4	28.3	29.0	30.5	32.0	33.0	34.6	35.8	36
38	15.0	16.4	17.8	19.0	20.3	21.4	22.5	23.5	24.5	25.4	26.4	27.3	28.1	29.0	29.8	31.4	32.8	34.2	35.5	36.7	38
40	15.3	16.8	18.2	19.4	20.7	21.9	23.0	24.0	25.1	26.0	27.0	27.9	28.8	29.7	30.5	32.1	33.6	35.1	36.4	37.6	40
42	15.6	17.1	18.5	19.8	21.1	22.3	23.4	24.5	25.6	26.6	27.6	28.5	29.4	30.4	31.2	32.8	34.4	35.9	37.3	38.6	42
44	15.9	17.5	18.9	20.2	21.5	22.7	23.9	25.0	26.1	27.2	28.2	29.1	30.0	31.0	31.9	33.5	35.2	36.7	38.1	39.5	44
46	16.2	17.8	19.2	20.6	21.9	23.2	24.3	25.5	26.7	27.7	28.7	29.7	30.6	31.6	32.5	34.2	35.9	37.4	38.9	40.3	46
48	16.5	18.1	19.6	20.9	22.3	23.6	24.8	26.0	27.2	28.2	29.2	30.2	31.2	32.2	33.1	34.9	36.6	38.2	39.7	41.2	48
50	16.8	18.4	19.9	21.3	22.7	24.0	25.2	26.4	27.6	28.7	29.8	30.8	31.8	32.8	33.7	35.5	37.3	38.9	40.4	42.0	50
52	17.0	18.7	20.2	21.6	23.1	24.4	25.6	26.8	28.1	29.2	30.3	31.4	32.4	33.4	34.3	36.2	38.0	39.6	41.2	42.8	52
54	17.3	19.0	20.5	22.0	23.4	24.8	26.1	27.3	28.5	29.7	30.8	31.9	32.9	33.9	34.9	36.8	38.7	40.3	42.0	43.6	54
56	17.6	19.3	20.9	22.4	23.8	25.2	26.5	27.7	28.9	30.1	31.2	32.4	33.4	34.5	35.5	37.4	39.3	41.0	42.7	44.3	56
58	17.8	19.5	21.1	22.7	24.2	25.5	26.9	28.2	29.3	30.5	31.7	32.9	33.9	35.0	36.0	38.0	39.8	41.7	43.4	45.0	58
60	18.1	19.8	21.4	23.0	24.5	25.8	27.3	28.7	29.8	31.0	32.2	33.4	34.5	35.5	36.5	38.6	40.4	42.3	44.0	45.8	60
62	18.3	20.1	21.7	23.3	24.8	26.2	27.6	29.0	30.2	31.4	32.6	33.8	35.0	36.0	37.1	39.2	41.0	42.9	44.7	46.5	62
64	18.6	20.3	22.0	23.6	25.2	26.5	27.9	29.3	30.6	31.8	33.1	34.2	35.5	36.5	37.6	39.7	41.6	43.5	45.4	47.2	64
66	18.8	20.6	22.3	23.9	25.5	26.9	28.3	29.7	31.0	32.2	33.5	34.7	35.9	37.0	38.1	40.2	42.2	44.1	46.0	47.8	66
68	19.0	20.8	22.5	24.2	25.8	27.3	28.7	30.1	31.4	32.6	33.9	35.1	36.3	37.5	38.6	40.7	42.8	44.7	46.6	48.4	68
70	19.2	21.	22.8	24.5	26.1	27.6	29.1	30.4	31.8	33.1	34.3	35.6	36.8	37.9	39.1	41.3	43.3	45.3	47.2	49.0	70
72															39.6	41.8	43.8	45.9	47.8	49.7	72
74															40.0	42.3	44.4	46.4	48.4	50.3	74
76															40.5	42.8	44.9	47.0	49.0	50.8	76
78															40.9	43.3	45.5	47.5	49.5	51.5	78
80															41.3	43.8	46.0	48.0	50.1	52.0	80
82															41.8	44.2	46.4	48.6	50.6	52.6	82
84															42.2	44.6	46.9	49.2	51.1	53.2	84
86															42.6	45.0	47.4	49.6	51.6	53.7	86
88															43.0	45.4	47.9	50.1	52.2	54.3	88
90															43.4	45.9	48.3	50.6	52.8	54.8	90
92															43.8	46.3	48.7	51.1	53.4	55.4	92
96															44.6	47.2	49.5	52.0	54.4	56.3	96

Equation for Circular Equivalent of a Rectangular Duct:[5]

$$d_c = 1.30 \frac{(ab)^{0.625}}{(a+b)^{0.250}} = 1.30 \sqrt[8]{\frac{(ab)^5}{(a+b)^2}}$$

where

a = length of one side of rectangular duct, inches.

b = length of adjacent side of rectangular duct, inches.

d_c = circular equivalent of a rectangular duct for equal friction and capacity, inches.

Fig. 22-11. Chart determines sizes of rectangular ducts necessary to equal carrying capacity of round ducts. To use, find the diameter of the round pipe in the chart. Then find one side of rectangular duct by reading up. Find the other side by reading left to the first row of numbers representing the other side of the rectangular pipe.   (ASHRAE Guide and Data Book)

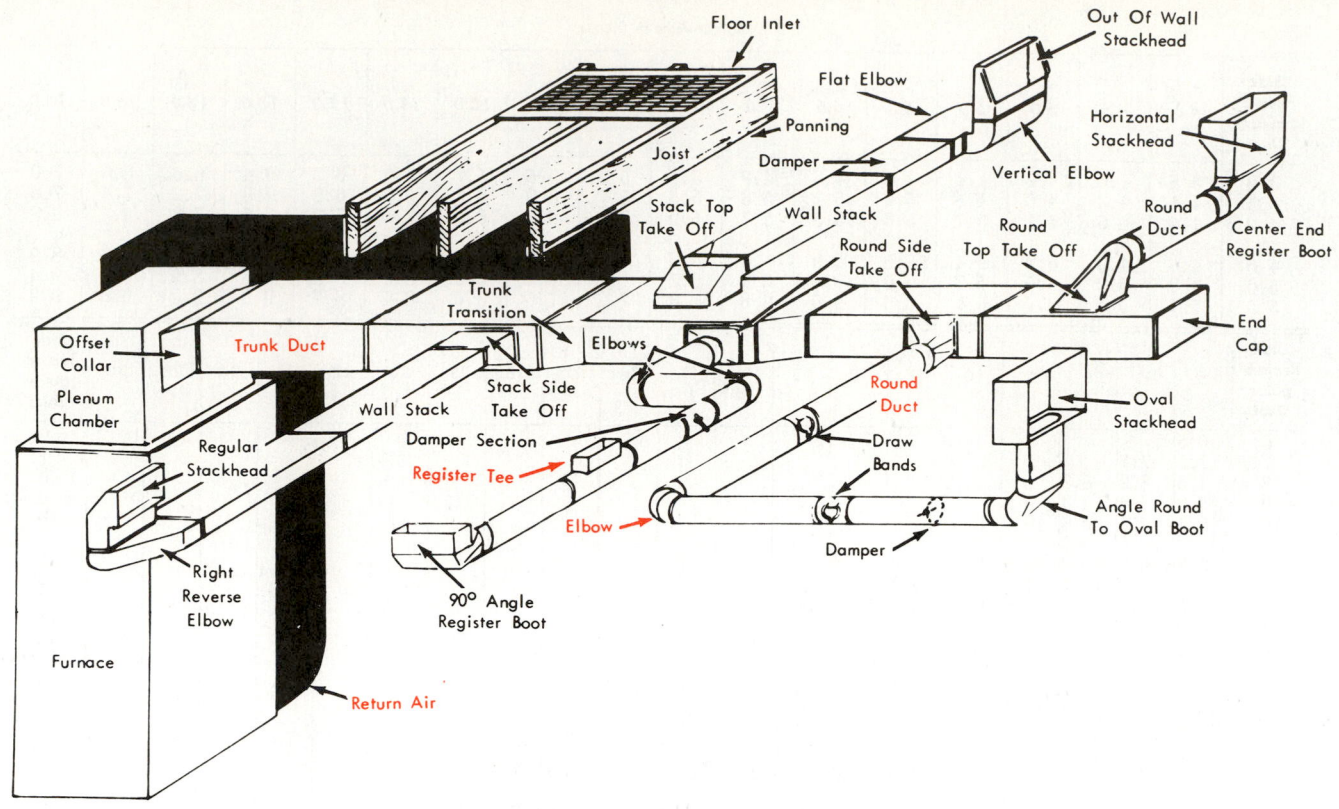

Fig. 22-12. Residential layout showing most sheet metal parts used in conditioned air duct system.
(National Environmental Systems Contractors Assoc.)

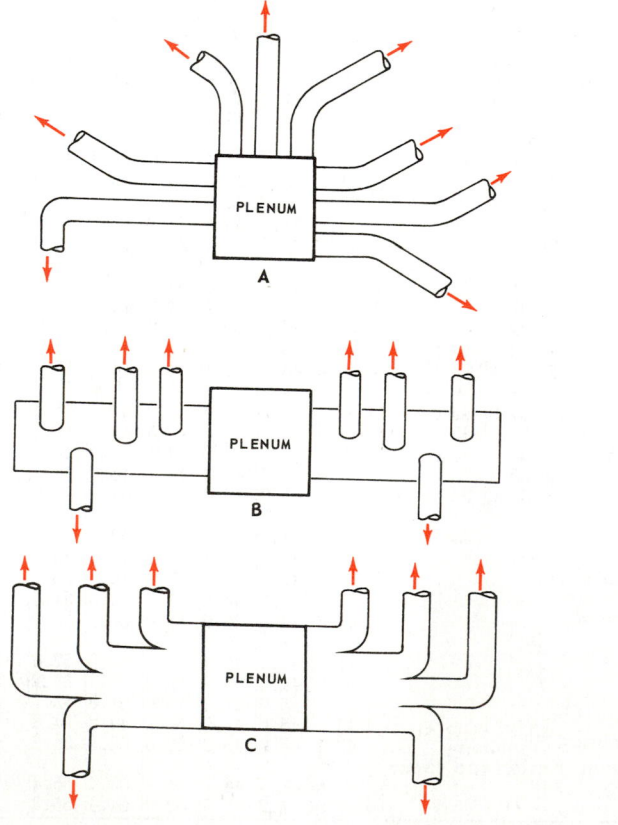

Fig. 22-13. Top view of three types of duct systems commonly used:
A—Individual round pipe system. B—Extended plenum system. C—
Reducing trunk system.

Fig. 22-14. Typical branch duct designs: A—Elbow. B—T-fitting.
C—Reducing T. D—Cross. E—Lateral.

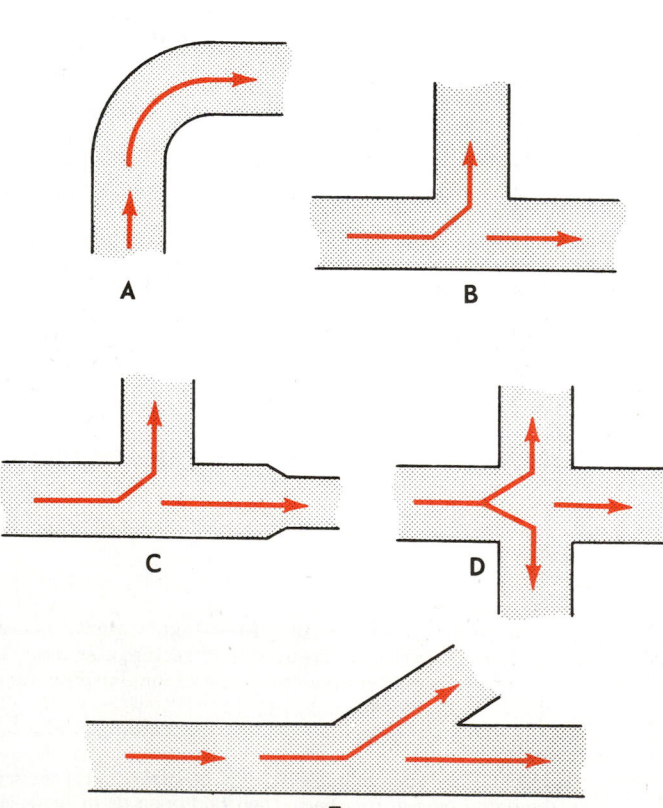

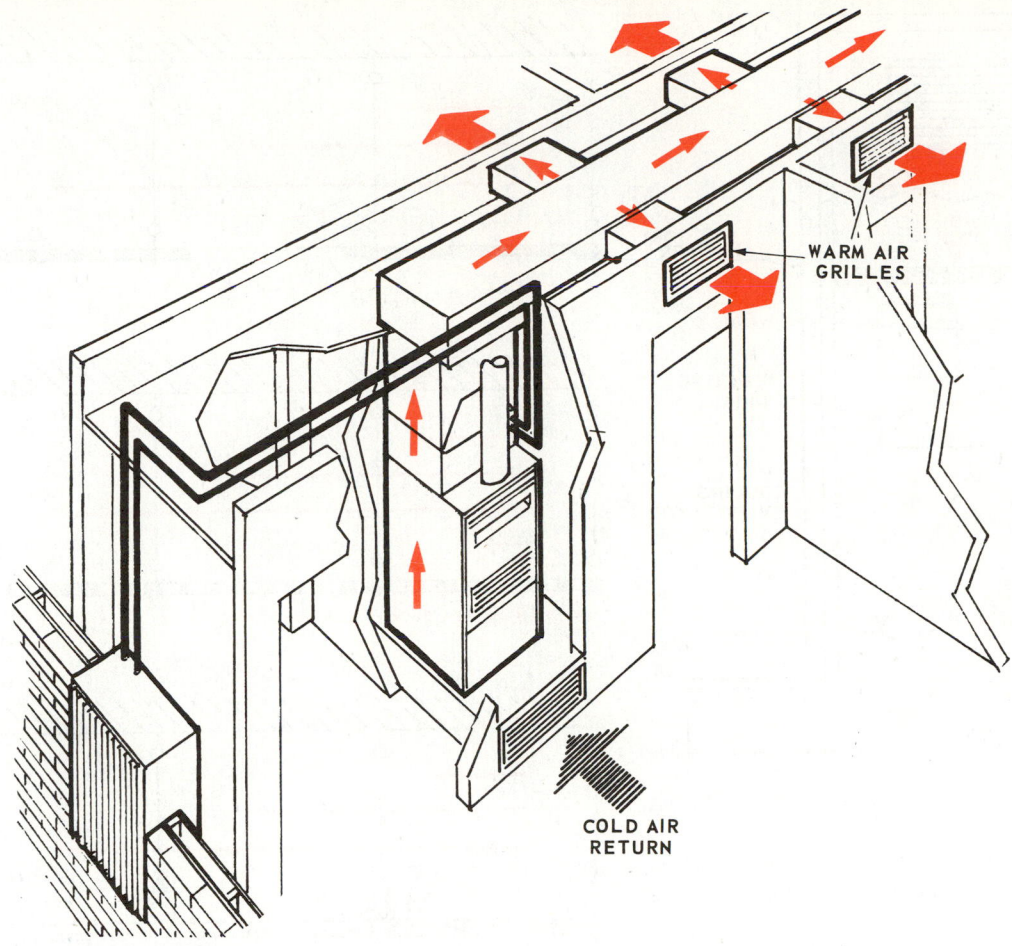

WARM AIR GRILLES

COLD AIR RETURN

Fig. 22-15. Overhead extended plenum duct system. (Lennox Industries, Inc.)

Fig. 22-16. Top view of perimeter loop system for conditioned air ducts. Ducts are cast into the concrete floor slab of the building.

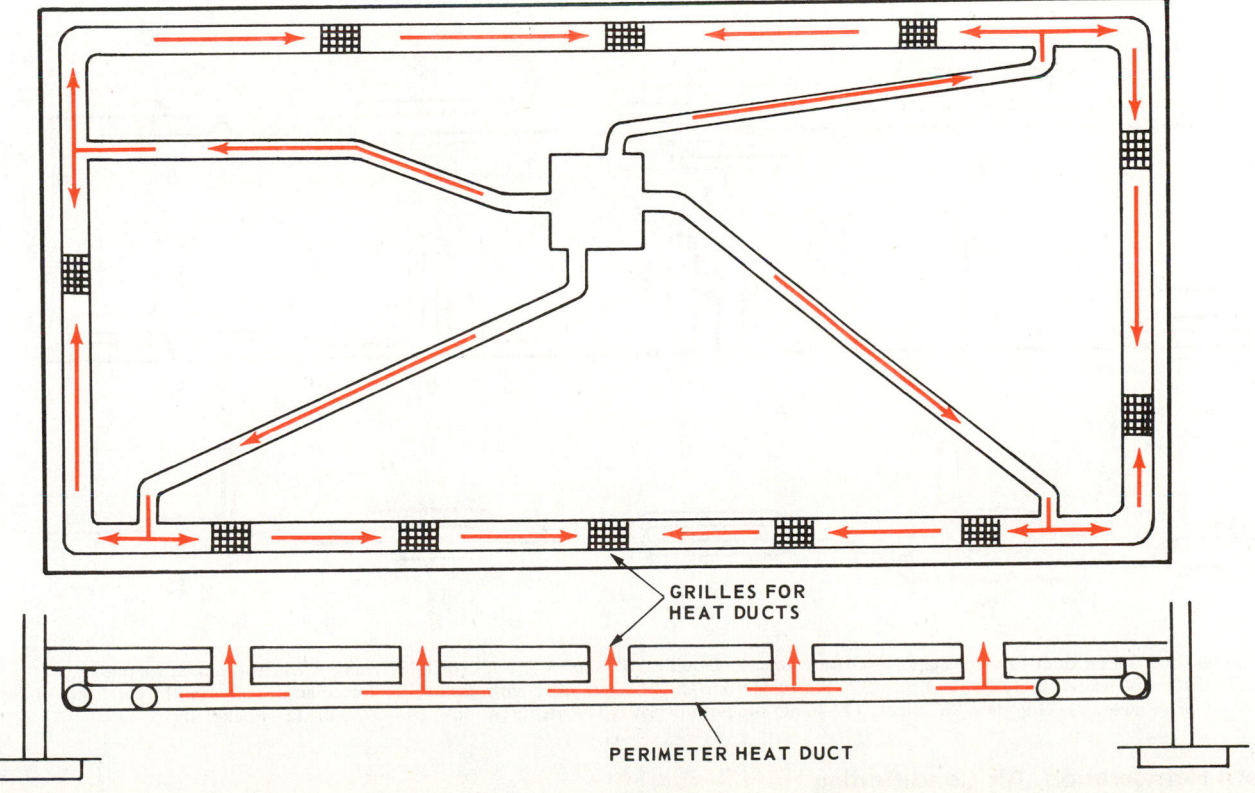

GRILLES FOR HEAT DUCTS

PERIMETER HEAT DUCT

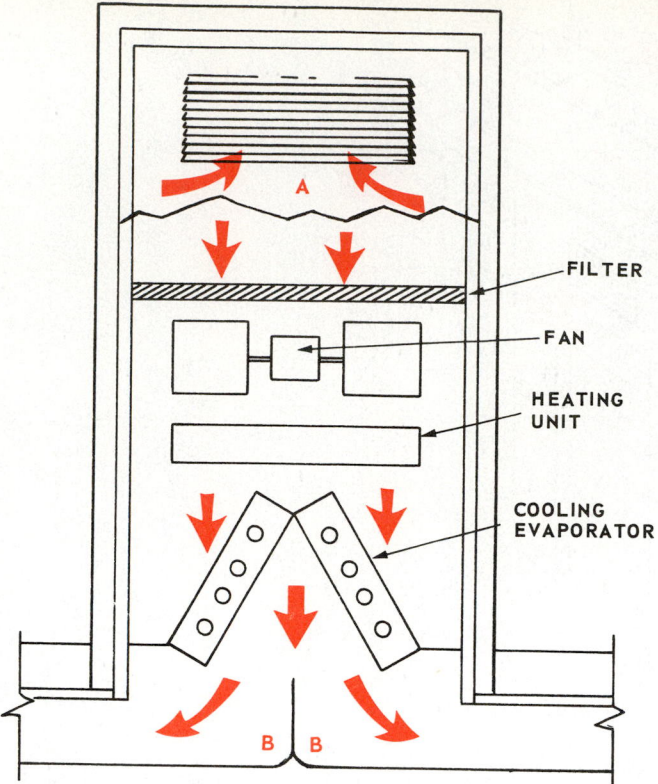

Fig. 22-17. A downflow furnace. A—Air return. B—Air outlet.

FILTER

FAN

HEATING UNIT

COOLING EVAPORATOR

When the duct system is located in a crawl space or attic, the ducts are usually insulated. Otherwise heat loss would be too great.

When installed in the concrete slab, ducts are usually made of metal, plastic or ceramic. The branch ducts usually connect to a perimeter duct. Grilles or diffusers are connected to the perimeter duct at intervals along the floor. See Fig. 22-16. A downflow type furnace is used as shown in Fig. 22-17.

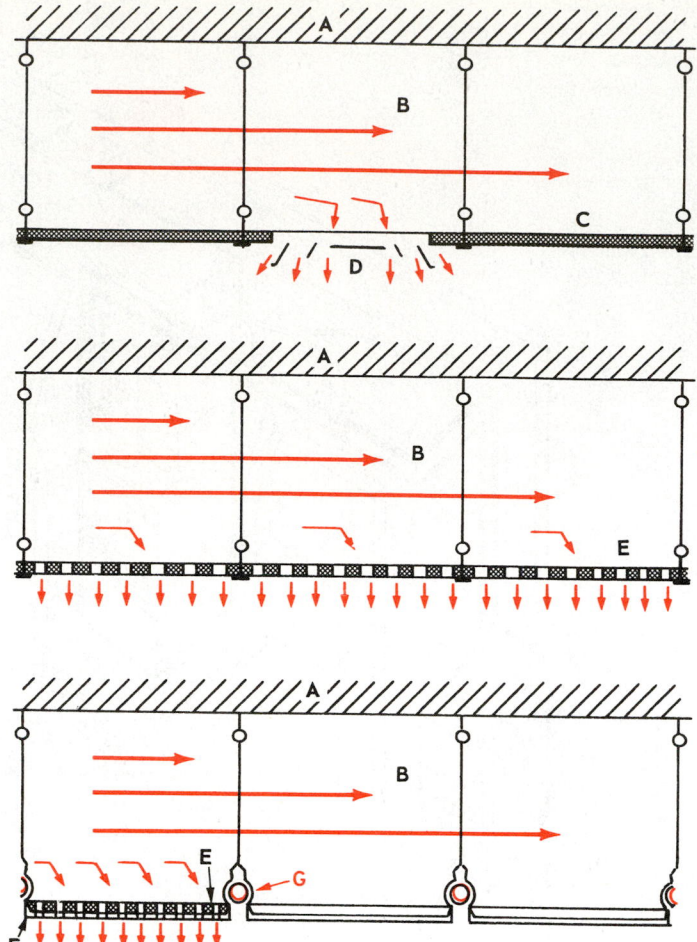

Fig. 22-18. False ceiling can be used to form a plenum chamber for either cool or warm air. A—Ceiling. B—Plenum chamber. C—False ceiling. D—Diffuser. E—False ceiling with fresh air holes. F—False ceiling with metal panels heated or cooled by water at G.

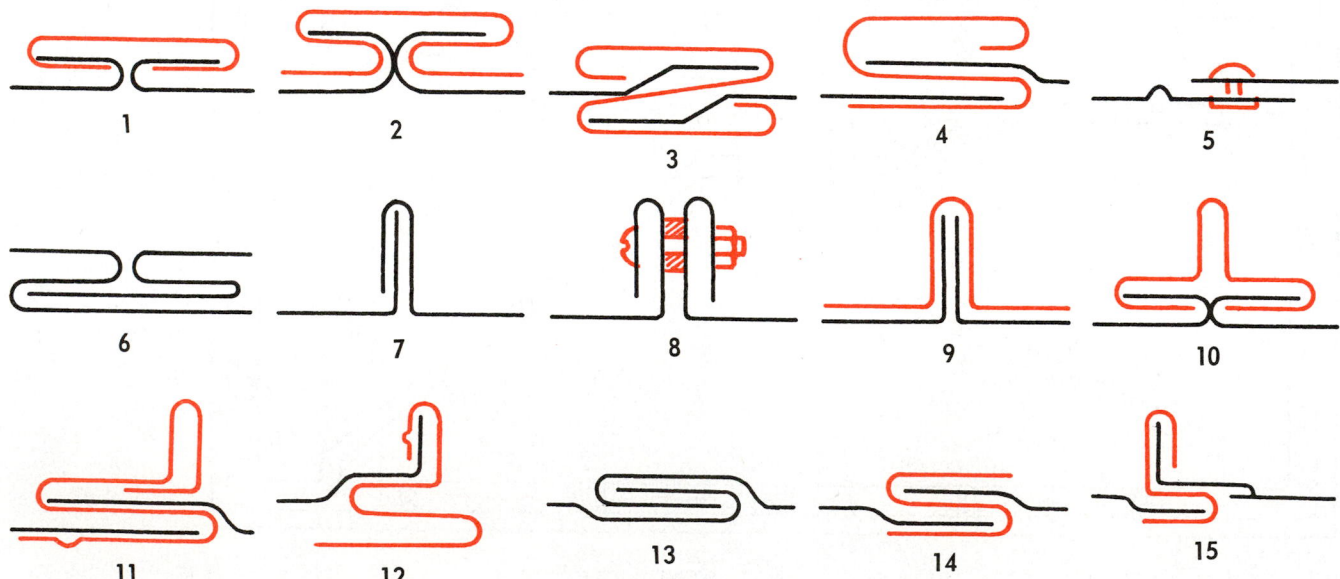

Fig. 22-19. Some sheet metal duct joint designs. 1—Drive cleats. 2—Double "S" slip. 3—Reinforced "S" cleat. 4—Button punch snap lock. 5—Button punch, rivet or screw. 6—Inside slip joint. 7—Standing seam. 8—Flanged joint with gasket. 9—T-connector. 10—Reinforced "on center" drive cleat. 11—Standing "S" cleat. 12—Angle slip with offset. 13—Pittsburgh lock. 14—"S" slip. 15—Pocket slip.

Some buildings are designed with unit ventilators that automatically draw air from the conditioned space and, at the same time, bring in makeup air from the outside.

False ceilings are often used to conceal piping, wiring, heat exchangers and ducts. The false ceiling may have holes to allow movement of conditioned air into the room below or diffusers can be used. Space between the false ceiling and the real ceiling may be used as a fresh air plenum chamber. Some systems use heated panels to provide either radiant heat or radiant cooling. See Fig. 22-18.

22-10 DUCT CONSTRUCTION

Ducts may be made of metal, wood, ceramic or plastic materials. Metal ducts are used for warm air distribution and for exhaust air ducts. The metal is usually sheet steel coated with zinc (galvanized steel). Some ducts are made of aluminum to reduce weight. Sheet lead is used when the duct must carry corrosive gases.

Ducts made of aluminum, glass fiber or plastic are not approved in some codes for duct installations where fire spread is possible. Many flexible ducts made of wire and fabric are not permitted where fire could spread to other areas.

Sheet metal brakes and formers are used for making ducts. Elbows and other connections such as branches are designed using geometric principles. Sheet metal work is a specialty trade and is done by skilled sheet metal workers.

Sheet metal ducts expand and contract as they heat and cool. Fabric joints are often used to absorb this movement. To prevent most of the fan and furnace noise from traveling along the duct metal, fabric joints should also be used where ducts fasten to a furnace or air conditioner. But, in fact, most duct joints are made of sheet metal.

Several types of sheet metal joints have been developed. See Figs. 22-19 and 22-20. The joint should be airtight and strong. Many joints are riveted for added strength and tightness. Fig. 22-21 shows a tool used to rivet from one side only.

To use this tool:
1. Drill a hole.
2. Place the rivet blank in the hole.
3. Attach the tool.
4. Work the handle which withdraws the expander.
5. When the rivet head is completely formed, break off the expander flush with the inside surface of the rivet. Cost of the tool is quite inexpensive considering the great saving in time when making riveted connections.

Many of these joints are also sealed with special duct tape to make them leakproof. Sealants are put in the duct seam for the same purpose.

Duct joints in forced warm air system should be as leakproof as possible. Leaking joints cut down on air volume delivered at outlets at the end of long runs. Connect all duct sections with good sheet metal joints.

Leaks may also make cold air ducts inefficient. Joints and

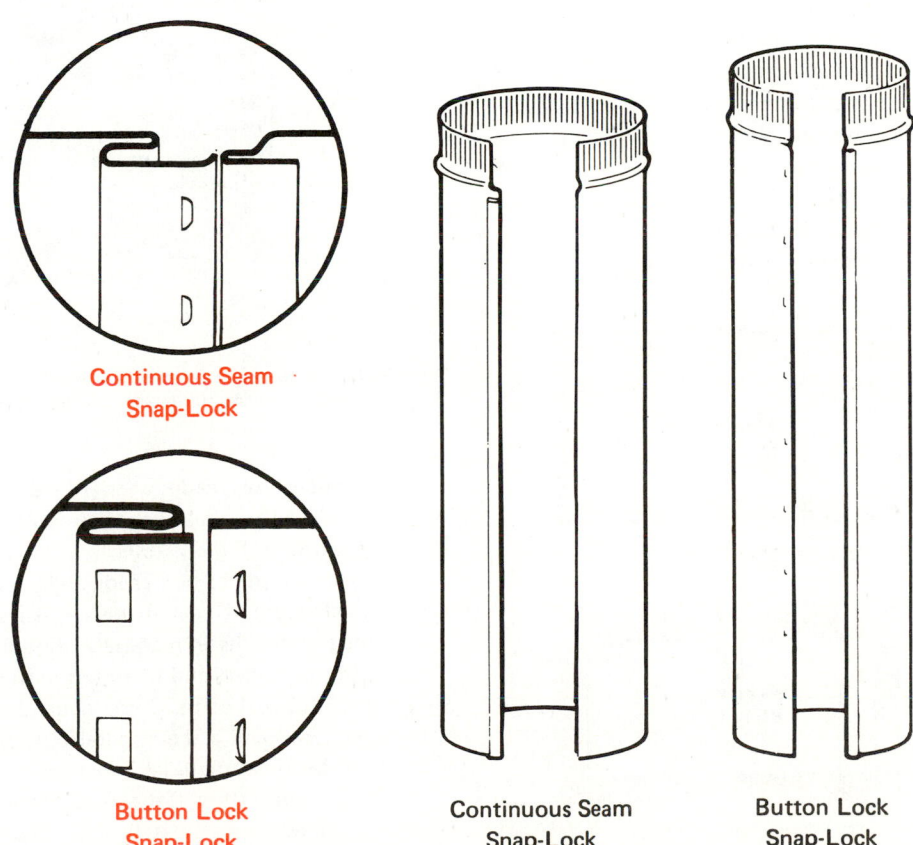

Continuous Seam Snap-Lock

Button Lock Snap-Lock

Continuous Seam Snap-Lock

Button Lock Snap-Lock

Fig. 22-20. Two seamlock methods. 1—Snap lock. 2—Button lock. (National Environmental Systems Contractors Assoc.)

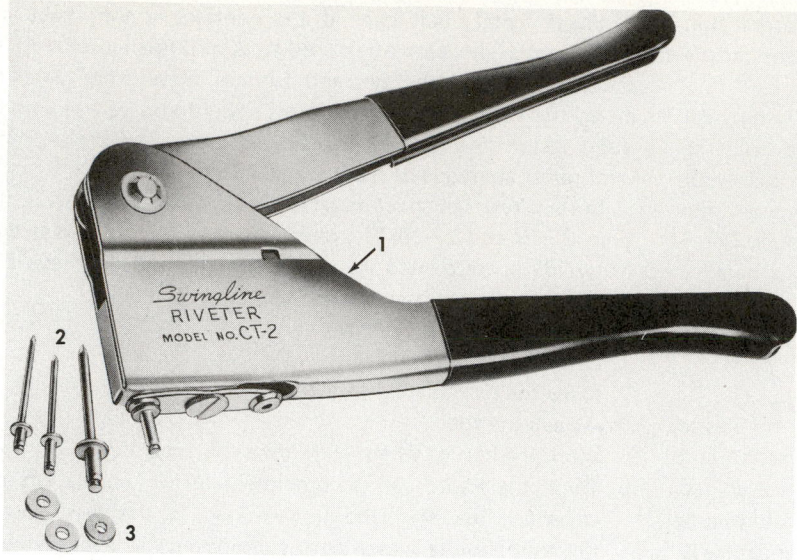
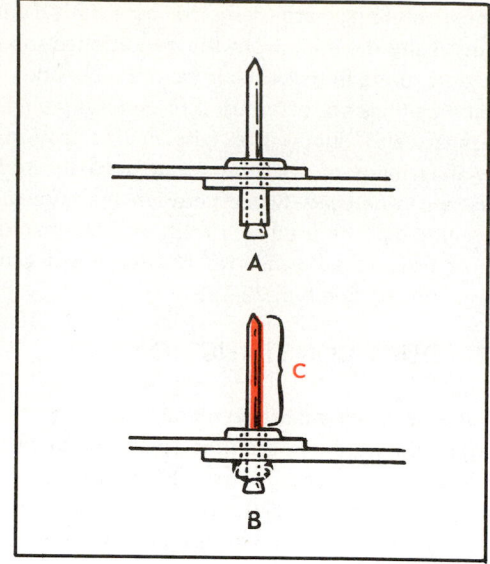

Fig. 22-21. Useful tool for riveting sheet metal and duct work in places where it would be difficult to install a solid rivet. 1—Riveting pliers. 2—Rivets. 3—Washers. Inserts illustrate riveting procedure. A—Hole drilled and rivet inserted. B—Riveting completed. C—This part is cut off. (Swingline, Inc.)

seams should also be sealed the same as warm air ducting.

If duct systems are noisy, air velocity may be too high. There may be turbulence and metal edges may produce wind noise. Some ducts are lined with an acoustic material to produce more silent duct system. Duct sections equipped with silencers are also available. Such units act as mufflers.

Many ducts are insulated — either on the inside or outside — to reduce noise as well as heat transfer. See Fig. 22-22. The insulation is fastened to the duct with adhesives. In some cases metal clips hold the insulation in place. Fig. 22-23 shows a sheet metal connector insulated on the inside.

Round ducts made of plastic and spring steel wire are often used. They need no elbows and are easily installed around other piping, beams and joints. Plastic used will not burn.

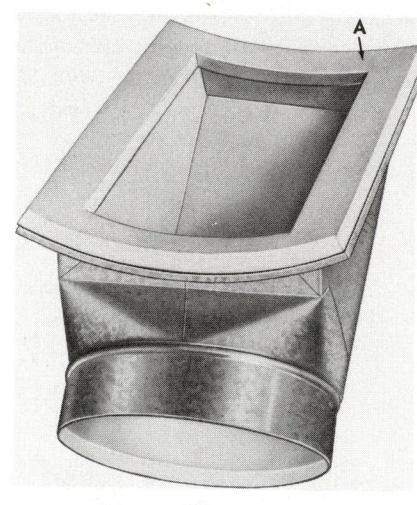

Fig. 22-23. Round duct branch elbow made of sheet metal with inside insulation. A—Insulation. (The Williamson Co.)

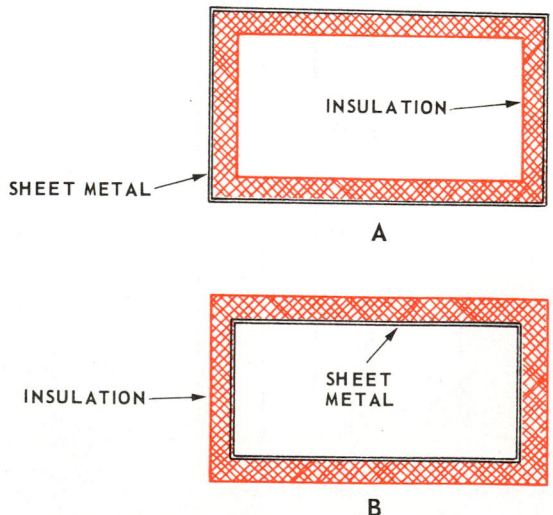

Fig. 22-22. Insulated sheet metal duct. A—Insulation inside the duct. B—Insulation outside the duct.

Some ducts are made of nonmetals such as fiberglass, urethane, plastic or sheetrock. See Fig. 22-24. A duct with a branch is shown in Fig. 22-25.

Duct material must not erode (release) any loose material into the airstream. Ducts made of asbestos or fiberglass or ducts lined with these materials should be plastic coated because asbestos fibers and fiberglass particles released into the air are harmful to health. Some ducts are made with a sheet metal inner wall and a sheet metal outer wall with insulation and sound absorbent material in between.

Any reduction in the size of a duct should be tapered. The taper should be approximately 1 in. for every 4 to 7 in. in length. When changing from small to large, use a change rate of 1 in. in every 7 in. of length.

Rectangular ducts must be carefully chosen. If the width is

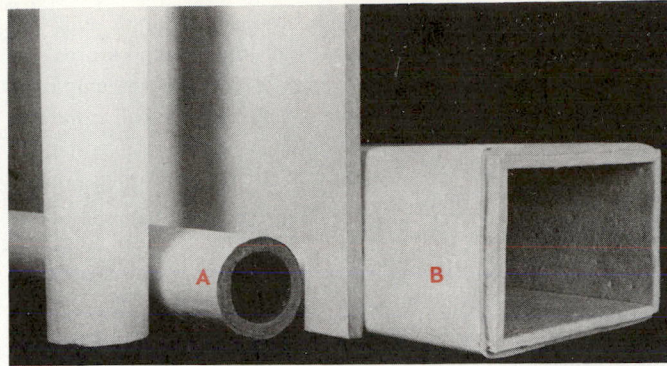

Fig. 22-24. Fiberglass ducts. A—Round duct. B—Rectangular duct. (Owens-Corning Fiberglas Corp.)

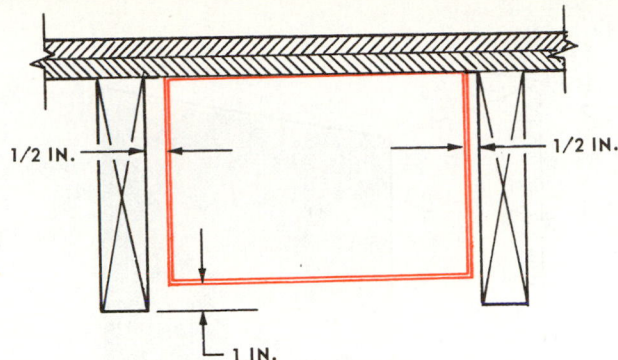

Fig. 22-27. Duct installation between floor joists. Note clearance space allowed between duct and joists.

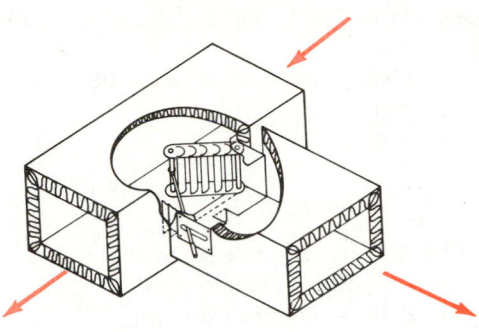

Fig. 22-25. Fiberglass duct connection with adjustable vane air diverter.

much more than the height, the amount of metal needed for the cross-sectional area becomes excessive. The ratio of the wide side to the narrow side is called the aspect ratio. For example, if a duct is 18 in. wide and 6 in. deep, the aspect ratio is $\frac{18}{6}$ or 3 to 1. Costs of making the duct also increase.

To show how much more metal is used as the aspect ratio increases: Assume a duct is 10 by 10 in. (aspect ratio 1:1). It has a cross-sectional area of 100 sq. in. while the distance around (perimeter) is 10 + 10 + 10 + 10 = 40 inches. Another duct of 100 sq. in. is made 20 in. wide and 5 in. deep. It has a perimeter of 20 + 5 + 20 + 5 = 50 in. or a 25 percent increase in metal for each unit of length. Sheet metal contractors classify ducts by their aspect ratio and estimate costs by class. Fig. 22-26 shows some common aspect ratios.

One should not try to put a duct in a stud or joist space when its size is equal to the inside dimensions of the stud or

DUCT CLASS (ASPECT RATIO)	WIDE SIDE INCHES	1/2 OF PERIMETER
1	6–18	10–23
2	12–24	24–46
3	26–40	32–46
4	24–88	48–94
5	48–90	96–176
6	90–144	96–238

Fig. 22-26. Duct aspect ratio. This ratio is determined by dividing width of duct by its height.

joist space. There must be room for fittings and allowance for out-of-alignment. It is better to make the ducts 1/2 in. smaller both ways. Fig. 22-27 illustrates a satisfactory duct installation between floor joists.

Fig. 22-28 gives the recommended gauge thicknesses for

ROUND DUCT DIAMETER	COMMERCIAL SHEET STEEL GALVANIZED GAUGE	RESIDENTIAL SHEET STEEL GALVANIZED GAUGE
Up to 12	26	30
13 to 18	24	28
19 to 28	22	
27 to 36	20	
35 to 52	18	

Fig. 22-28. Recommended gauge thicknesses for round metal ducts.

RECTANGULAR DUCTS WIDE SIDE INCHES	COMMERCIAL SHEET STEEL GALVANIZED	ALUMINUM	RESIDENTIAL SHEET METAL GALVANIZED
Up to 12	26	.020	.28
13–23	24	.025	.26
24–30	24	.025	.24
31–42	22	.032	
43–54	22	.032	
55–60	20	.040	
61–84	20	.040	
85–96	18	.050	
Over 96	18	.050	

Fig. 22-29. Prescribed gauge thicknesses for rectangular metal ducts.

various sizes of round ducts. Fig. 22-29 lists recommended gauge thicknesses for various size rectangular ducts. Ducts going through floors should be protected on the corners by angle iron. Large sections of sheet metal or ducts must be cross broken. This reduces panel vibration. See Fig. 22-30.

The use of controlled fresh air in a building is increasing. It has been found that natural infiltration is unreliable and too variable. It is very necessary that this fresh air be filtered. The outdoor intake end of the fresh air duct should be fitted with a strong screen and bar combination to eliminate debris, birds, animals, insects and illegal entry.

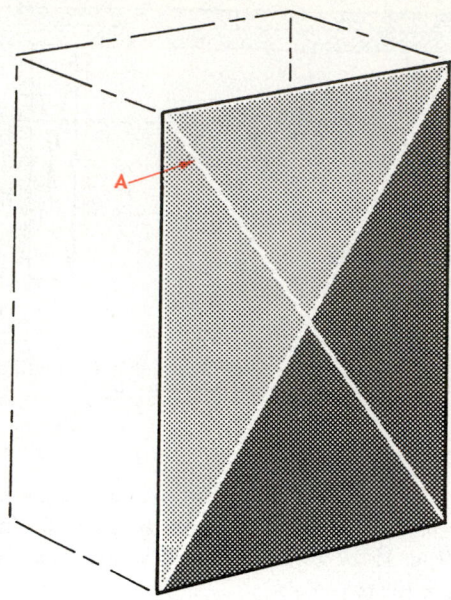

Fig. 22-30. Break lines (ridges) are formed on large sheet metal panels or large ducts to make them more rigid. A—A typical break line.

22-11 DUCT SIZES

Before determining the size duct needed to carry air to a room, one must know the volume of air that is to be delivered. This volume depends on the amount of heat the air must deliver to the room during the heating season or the amount of heat to be removed during the cooling season.

Normally, these calculations are not difficult but situations such as split heating systems complicate the problem. In any case, the amount of air delivered to a room must always equal or exceed the minimum fresh air ventilation requirements.

To reduce duct size and to save space in a building, smaller size ducts are now being used. Such ducts operate with about twice the normal air velocity. This increase in pressure and velocity requires more powerful fans. In turn, there is more noise.

The size of the ducts is based on the heating needs in the colder climates and on cooling needs in the warmer climates. The same duct system serves both heating and cooling but for the cooling season fan capacity is raised by increasing its speed.

22-12 AIR VOLUMES FOR HEATING

If the furnace is the only source of heat for a room, only three things need be known to be able to calculate the air volume:
1. Heat load.
2. Room temperature.
3. Duct temperature.

The heat load can be determined by methods described in Chapter 25. The room temperature is decided by the designer. Normally it is 72 F. (22 C.) dry bulb temperature (dbt).

The duct air temperature is more difficult to decide. If a low duct temperature is used, large air volumes will be necessary to carry enough heat. If high duct temperatures are used, the furnace will have to operate with higher chimney (stack) temperatures and the ducts may have to be insulated.

Engineers recommend that the grille temperatures be at least 125 F. (52 C.) and that duct air temperature be near 140 F. (60 C.). The lowest temperature at which these results are possible depends on the lengths of the ducts. Knowing that the specific heat of air is .24 Btu/lb., the weight of air needed is easily found by using the specific heat equation:

Room heat load = .24 x wt. of air x the temperature difference.

For example:

What is the weight of air needed for one hour if a room has a heat load of 20,000 Btu/hr.?

With a room temperature of 72 F. (22 C.) and a duct temperature of 140 F. (60 C.), the temperature difference is 68 F. (38 C.).

$$20,000 = 24 \times \text{wt. of air} \times 68$$

$$\frac{20,000}{.24 \times 68} = \text{wt. of air}$$

$$\frac{20,000}{16.32} = \text{wt. of air}$$

$$1225.5 \text{ lb.} = \text{wt. of air per hour}$$

Divide by 60 to obtain lb./min.

$$20.425 \text{ lb.} = \text{wt. of air per min.}$$

To find the volume, we must first find the volume of one pound of air at the duct temperatures. This value is obtained from the psychrometric chart, Fig. 18-8.

One lb. of air has a volume of 17.1 cu. ft. (0.48 cu.m). If the chart does not read as high as 140 F. (60 C.), we may calculate this volume by using Charles's Law (Chapter 1) and by knowing the volume at 72 F. (22 C.) dbt and 50 percent relative humidity (13.55 cu. ft.).

$$\frac{V_o}{V_m} = \frac{Fa_o}{Fa_m}$$

$$\frac{13.55}{\text{Vol.}} = \frac{460 + 72}{460 + 140}$$

$$\frac{13.55}{\text{Vol}_m} = \frac{532}{600}$$

$$13.55 \times 600 = \text{Vol.} \times 532$$

$$\frac{13.55 \times 600}{532} = \text{Vol.}$$

$$\frac{8130}{532} = \text{Vol.}$$

$$15.28 = \text{Vol.}$$

The volume of air/min. is 20.425 lb./min. x 15.28 cu. ft.
Vol. = 312.24 cu. ft./min. (8.8 m³/min.)

Now, to determine the duct size, two separate items must be considered. If the space is limited, the area of the duct is already fixed. For example, if the duct is to run between studs in a partition, the space available is no more than 14 x 3 1/4 in. (This fits between 2 x 4 studding on 16 in. centers.) Such a duct has an area of:

14 x 3 1/4 = 45.5 sq. in. or 293.6 cm²

Convert this to square feet, dividing by 144, thus:

$$\frac{45.5}{144} = .316 \text{ sq. ft. or } 0.029 \text{ m}^2$$

Using the volume of 312.24 cu. ft./min.:

The velocity in ft./min. multiplied by .316 sq. ft., must provide a volume of 312.24 cu. ft./min.

Since velocity is not known, one must find it by dividing volume by area, thus:

$$\text{Velocity} = \frac{312.24}{.316} = 988 \text{ ft./min. or } 301 \text{ m/min.}$$

Since this velocity would produce air turbulence noise, one must use two ducts measuring 14 x 3 1/4 in.

Resulting velocity will now be 494 ft./min. (151 m/min.). This should be satisfactory.

22-13 AIR VOLUME FOR COOLING

To get air volume needed, it is better to use large ducts, less air pressure (velocity) and, therefore, less power. Duct cost is a one-time cost, but a larger fan and motor mean continuous higher power costs. The higher velocity needed with smaller ducts also creates more noise.

A short method for determining air volume is to use 1 cfm for each square foot of floor space, excluding basement. If a home has 1500 sq. ft. (139 m^2), based on outside dimensions, the fan capacity should be 1500 cu. ft. (42 m^3) per minute.

To determine the cooling load, consider that the typical uninsulated home may need 12,000 Btu of cooling per hour (1 ton) for each 400 sq. ft. (37 m^2) of floor space. The 1500 sq. ft. (139 m^2) should then use $\frac{1500}{400}$ x 12,000 Btu or about 45,000 Btu/hr. (4 ton).

Number of air changes are important; there should be 6 to 10 per hour. Volume of the 1500 sq. ft. room with an 8-ft. ceiling is 1500 x 8 = 12,000 cu. ft. (340 cu.m). The 1500 cu. ft./min. fan will move 1500 x 60 min./hr. = 90,000 cu. ft./hr.

$$\frac{90,000}{12,000} = 7.5 \text{ air changes per hour.}$$

Return air ducts must be as efficient as the delivery air ducts. The return air is warmer, is at a lower pressure and, therefore, occupies more volume. Return duct work should be about 20 percent larger in cross-section area than the delivery duct because it is induced (low pressure) flow.

If the delivery duct to a room has a cross-section of 30 sq. in. (194 cm^2), the return air duct should be 30 + (30 x .20) = 30 + 6 = 36 sq. in. (232 cm^2) or 30 x 120 percent = 20 x 1.2 = 36 sq. in.

Air volume for cooling is calculated in much the same way as for heating. Use the same specific heat question. (See Chapter 1.)

Btu = sp. ht. of air x wt. of air x (temp. diff.).

Knowing heat load, specific heat of air and temperature difference, the weight of air needed can be determined. If the weight is known, the air volume can be determined and the duct sizes selected.

For example, using a heat load of one ton of cooling (12,000 Btu/hr.) and a specific heat value of .24 Btu/lb., the air temperature in the duct and the air temperature in the room must be determined. If the designer wants 75 F. (24 C.) in the room and the comfort cooling unit is designed for 65 F. (18 C.) duct air temperature:

12,000 Btu/hr. = .24 x wt. of air x
(diff. between 75 F. and 65 F.)

12,000 Btu/hr. = .24 x wt. of air x 10

$$\frac{12,000}{.24 \times 10} = \text{wt. of air}$$

$$\frac{12,000}{2.4} = \text{wt. of air}$$

5000 lb./hr. = wt. of air

$$\frac{5000}{60} = 83.33 \text{ lb./min. (37.8 kg/min.)}$$

At 60 F. (16 C.) db and 60 percent relative humidity (from the psychrometric chart), 13.4 cu. ft. (.38 m^3) of air weighs 1 lb.

To change 86.7 lb./min. to cfm:

Cu. ft. per min. = 86.7 x 13.4 = 1162 cu. ft./min. or
(33 m^3/min.)

To determine duct size for this volume of air moving into the room at 500 ft./min. refer to Figs. 22-31 and 22-32.

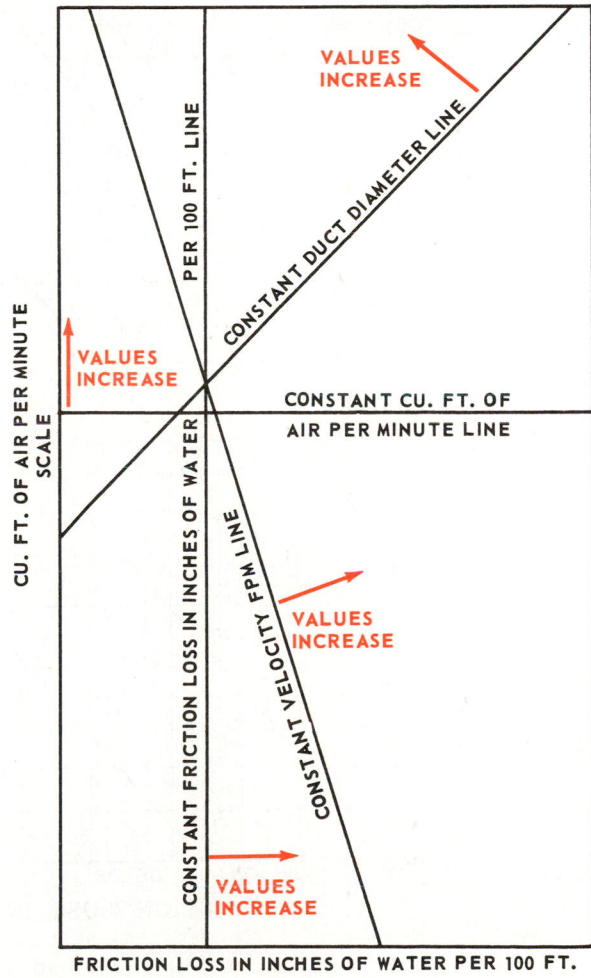

Fig. 22-31. Diagram shows how to read value lines in Fig. 22-32. Volume of air is read off scale on the left-hand side. By locating air volume per minute line one can find duct diameter, velocity and friction by following along the appropriate lines to the scales on the edges.

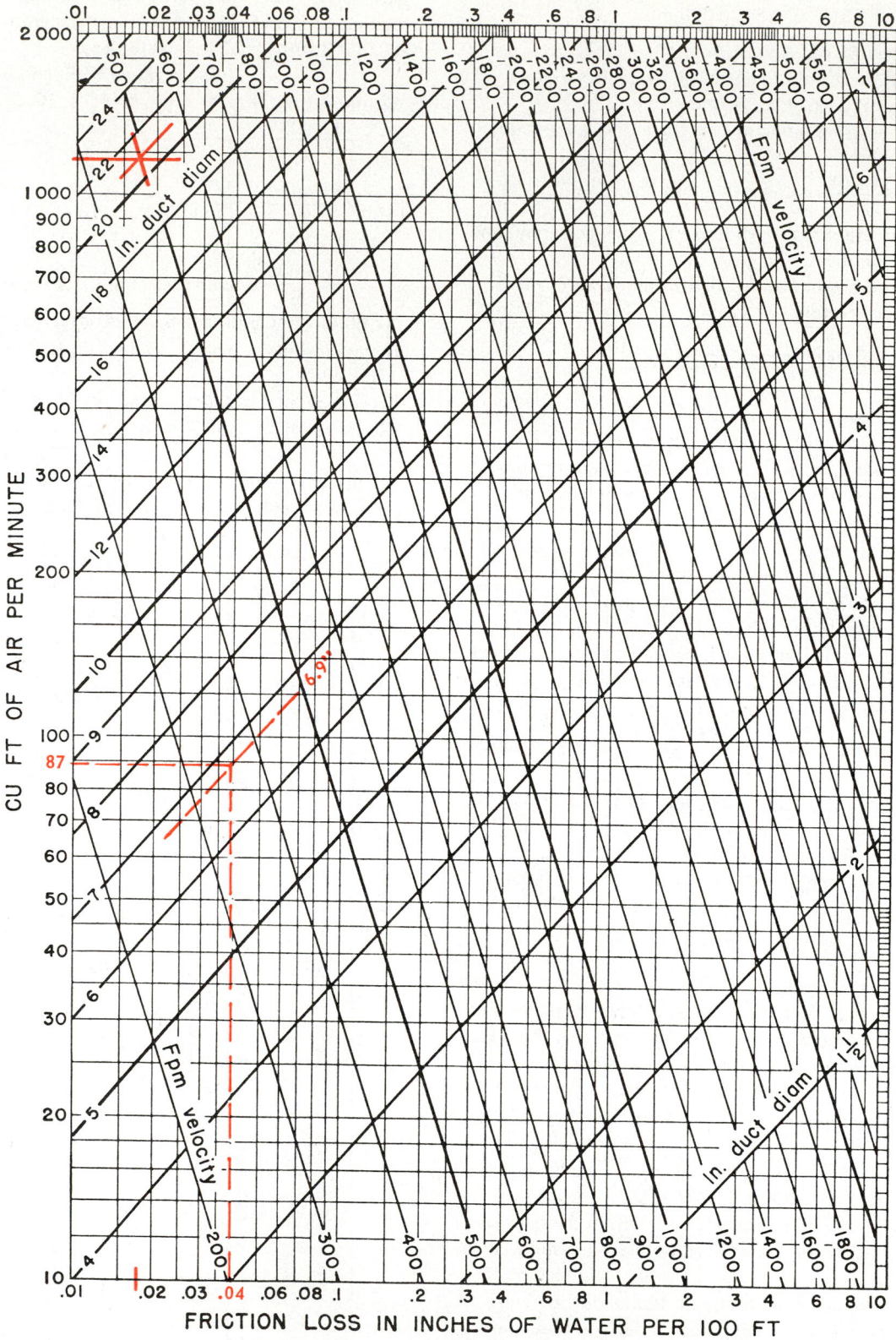

Fig. 22-32. Friction chart for low volumes of 10 to 2000 cfm airflow in ducts. (ASHRAE Guide and Data Book)

Locate the point where the 1162 cu. ft./min. line crosses the 500 ft./min. velocity line. This point will show the several alternative sizes:

1. A 21 in. (53 cm) round duct (at .018 in. of friction loss).
2. A 22 x 17 in. (56 x 43 cm) rectangular duct (using conversion table in Fig. 22-11).
3. A 30 x 13 in. (76 x 33 cm) duct.

22-14 AIR CIRCULATION

In warm air heating three basic systems are used to circulate the air:

1. Gravity.
2. Intermittent forced air.
3. Continuous forced air.

The gravity system is no longer popular. A Standard Code for Installation of Gravity Warm Air Heating Systems is published by the National Warm Air Heating and Air Conditioning Association.

This code recommends register or grille temperatures of 175 F. (79 C.). It has tables that indicate the Btu-carrying capacity of five different duct combinations for either first or second floor registers. The same association has developed a code for mechanical warm air systems.

Most installations use the intermittent forced air system. A thermostat in the furnace plenum chamber controls the fan.

Becoming more popular is the continuous blower system. It provides a more constant temperature in rooms.

Systems designed to provide cooling as well as heating need additional capacity to move air. This is because a cubic foot of cooled air will not change the room temperature as much as the same amount of warmed air.

The reason for this is the temperature difference. The warmed air comes out of the duct many degrees warmer than the room air it is replacing; the cooled air is not all that much cooler than the room air it is replacing. Therefore, greater quantities of it need to be moved into the room to get the desired effect.

This need is somewhat offset by the lower cooling load. In average conditions, a 30 to 50 percent airflow increase is required. Either of two methods will increase the airflow when needed:

1. Use a two-speed blower motor (for directly-driven blowers).
2. Install a two-speed pulley on the motor if a belt-driven unit is used.

22-15 ROOM AIR MOVEMENT

Air entering a conditioned space through ducts must circulate throughout the room without causing annoying drafts. This depends on the number and size of the air inlet grilles as well as on the velocity (speed) of the air moving through them.

Air coming into the room pushes against and mixes with air already in the room. *Combined air that moves at 150 ft. per minute is called primary air. The distance the air from the grille travels before it slows down to 50 ft. per minute (terminal velocity) is called the throw.* The outlet velocity is the speed of the duct air as it leaves the grille.

Overall size of the grille is not important. Total area of the air openings in the grille determines the grille capacity. The spread of the air that leaves the grille is very important. The return air grille or grilles should be located where the slowest movement of air takes place in the room (stratified air).

22-16 DIFFUSERS, GRILLES AND REGISTERS

Room openings to ducts have several different devices:

1. To control the airflow.
2. To keep large objects out of the duct.

These opening devices are called:

1. Diffusers.
2. Grilles.
3. Registers.

Diffusers deliver widespread fan-shaped flows of air into a room. Some diffusers cause the duct air to mix with some room air in the diffuser. Grilles are usually used as covers for the return air duct and have no adjustments.

Registers are used to deliver concentrated air streams into a room. Many have one-way or two-way adjustable air stream deflectors.

Grilles control the distance, height and spread of airthrow, as well as the amount of air. Grilles offer some resistance to airflow. Grille cross-section pieces block about 30 percent of the air. For this reason — and also because they must use slow air movement to reduce noise — the duct cross-section is usually enlarged at the grilles. Fig. 22-33 shows a typical warm air grille.

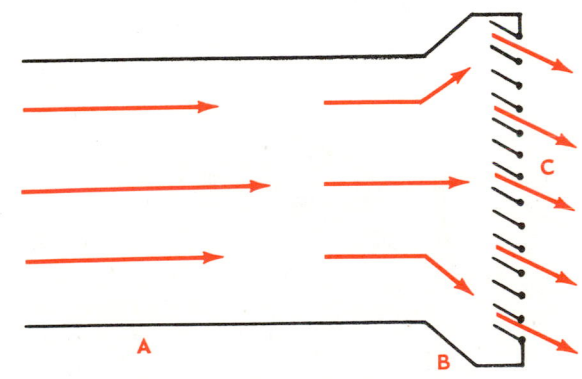

Fig. 22-33. Typical warm air grille. Many are supplied with controls for adjusting size of opening, and to redirect airflow. A—Duct. B—Enlarged duct. C—Grille with adjustable vanes.

Grilles have many different designs. Some are fixed and can direct the air only in one direction. Others are adjustable and can be set to send air in different directions.

A rectangular grille with adjustable direction airflow vanes is shown in Fig. 22-34. Some grille designs have air vanes to direct the airflow three different directions at one time. See Fig. 22-35. Others are made with adjustable dampers, horizontal flow vanes and vertical flow vanes. A ceiling-mounted rectangular air diffuser, Fig. 22-36, directs air in four directions along the ceiling.

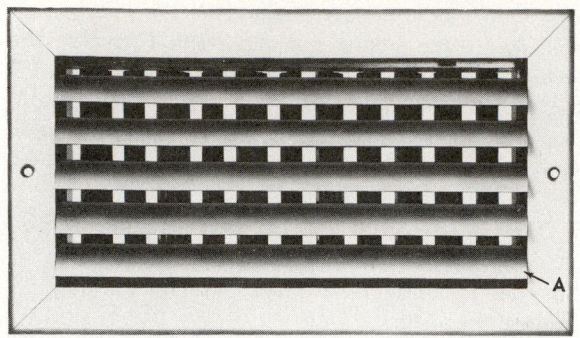

Fig. 22-34. Rectangular grille. A—Adjustable airflow direction vanes. (Hart & Cooley Mfg. Co.)

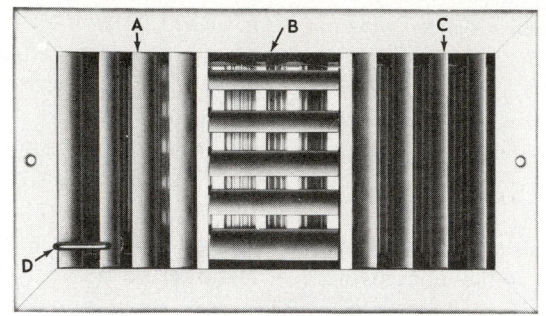

Fig. 22-35. Rectangular grille with adjustable airflow vanes. A—Horizontal flow. B—Vertical flow. C—Horizontal flow. D—Adjustment.

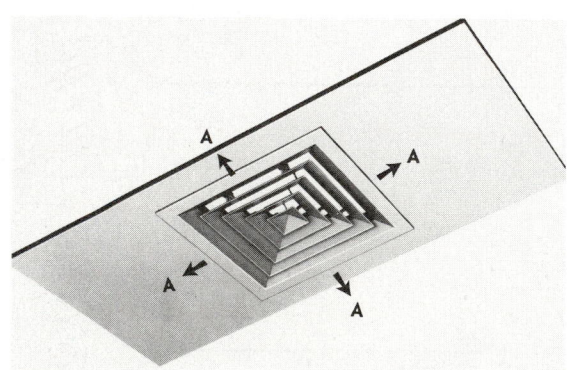

Fig. 22-36. Ceiling-mounted air diffuser. A—Air flows four ways.

22-17 DAMPERS

Without some way of controlling airflow in forced air systems, some spaces would receive too much air while others would not get enough. One method of getting even air distribution is through the use of duct dampers.

These dampers balance airflows or they can shut off or open certain ducts for zone control. Such a damper is shown in Fig. 22-37. Some are located in the diffuser or grille, some are in the duct itself. They are of three types:

1. Butterfly.
2. Multiple blade.

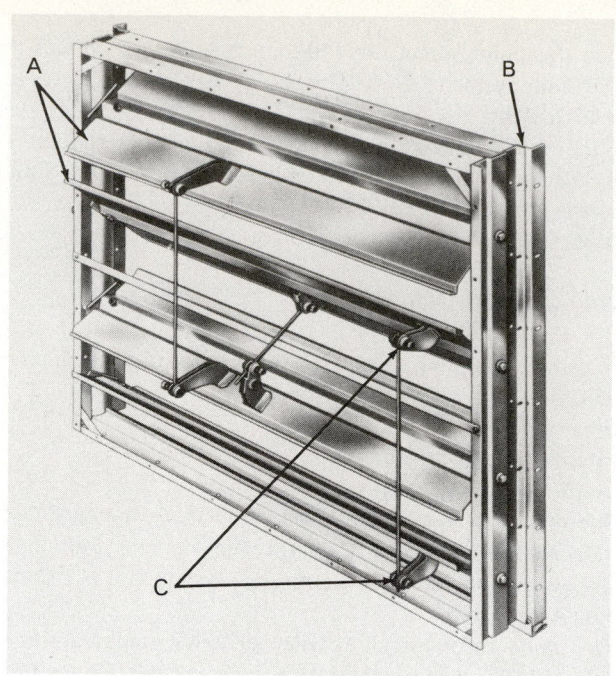

Fig. 22-37. Four-vane adjustable damper used to control airflow in duct. These dampers may be operated either manually or automatically. A—Damper vanes. B—Duct frame. C—Linkage. (Honeywell Inc.)

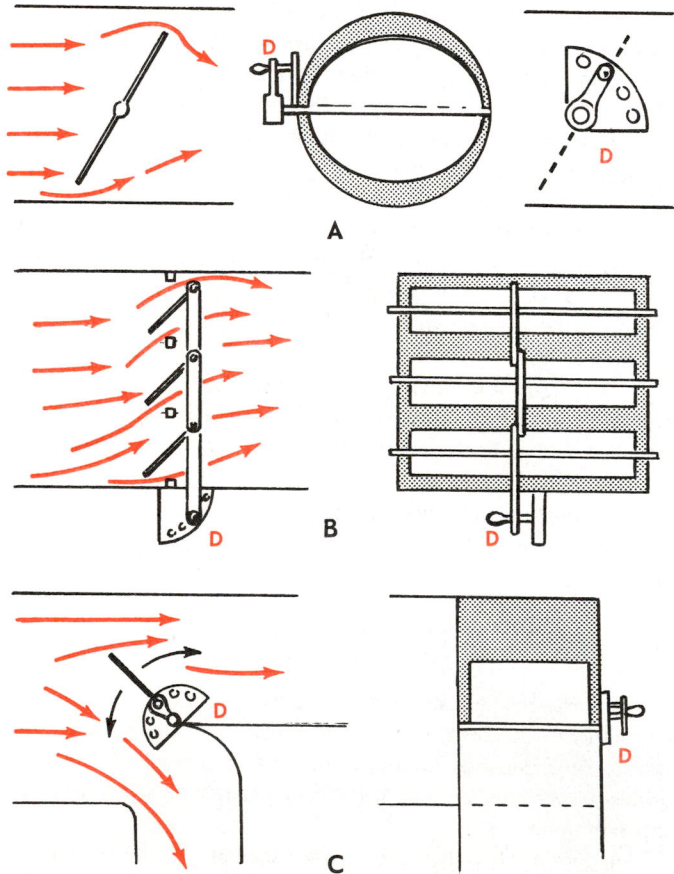

Fig. 22-38. Three types of duct airflow controls. These dampers are used to adjust airflow volumes to help service technician balance system. A—Butterfly damper. B—Multiple vane damper. C—Splitter damper. D—Adjustment handle.

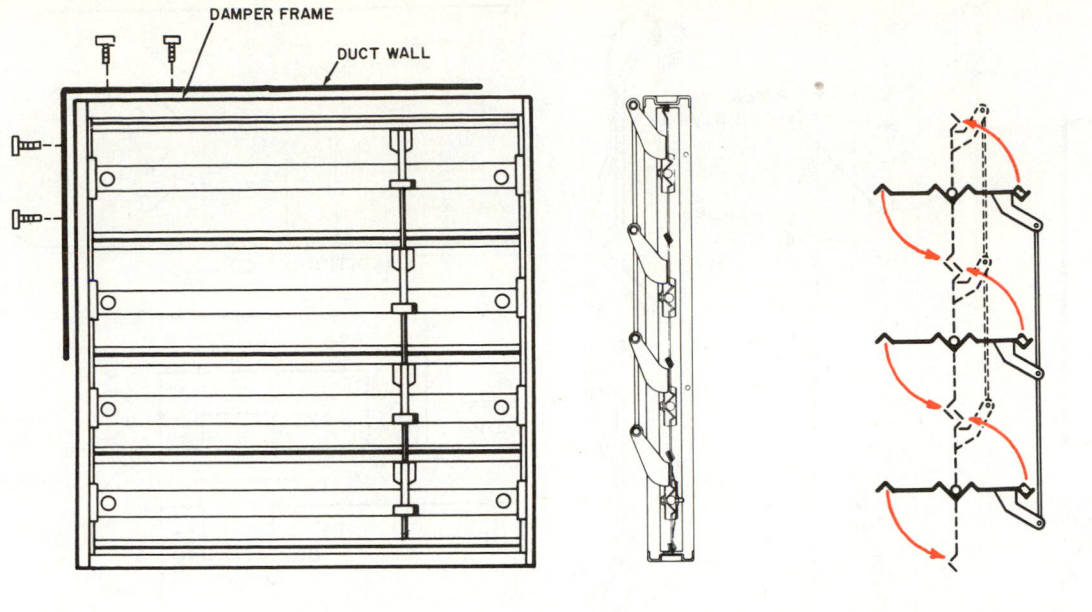

PARALLEL BLADE ACTION

Fig. 22-39. Construction detail of parallel blade damper.
(Honeywell, Inc.)

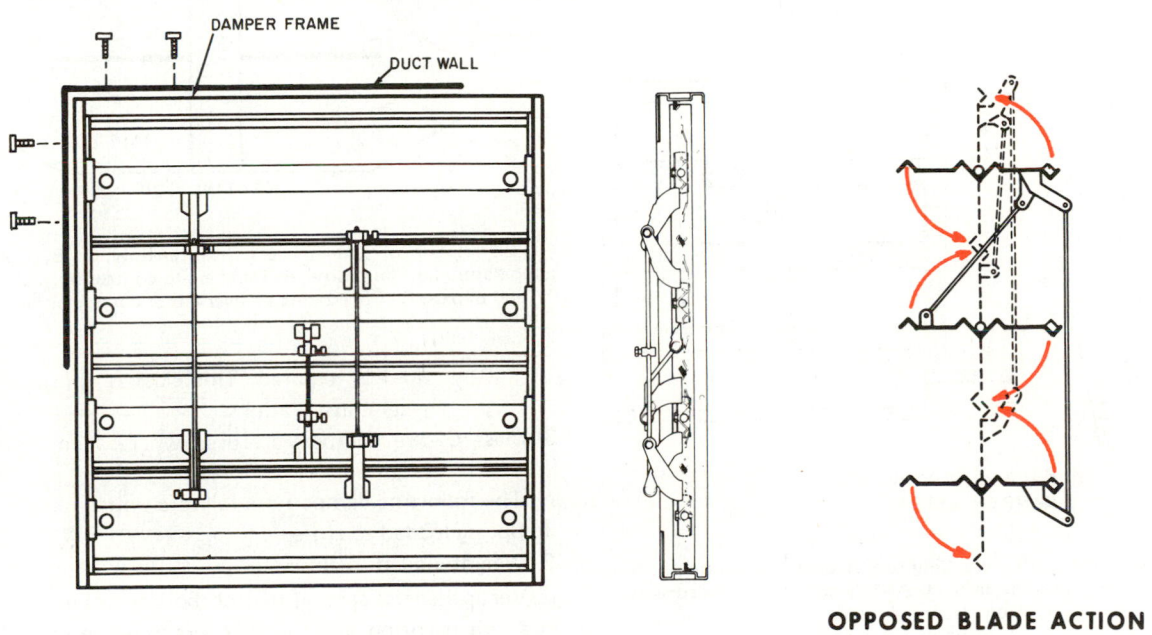

OPPOSED BLADE ACTION

Fig. 22-40. Opposed blade damper with blades open.
(Honeywell, Inc.)

3. Split damper.

All three types are shown in Fig. 22-38. There are two types of multiple blade dampers: the parallel blade type, as in Fig. 22-39, and the opposed blade type, Fig. 22-40. Detail of the blade design is shown in Fig. 22-41. When installing a damper, always draw a line on the end of the damper shaft that extends out of the duct to show the position of the damper. See Fig. 22-42.

For accurate air control these dampers should be tight fitting with minimum leakage. Many are automatically controlled for either zone heating or cooling. Automatic controls are also used to mix two airflows for either fresh air and recirculated air mixes, for humidity control or for temperature control. Fig. 22-43 is a diagram of a typical damper installation. These multiple dampers are interlocked by controls to provide different mixes and to maintain correct total airflow.

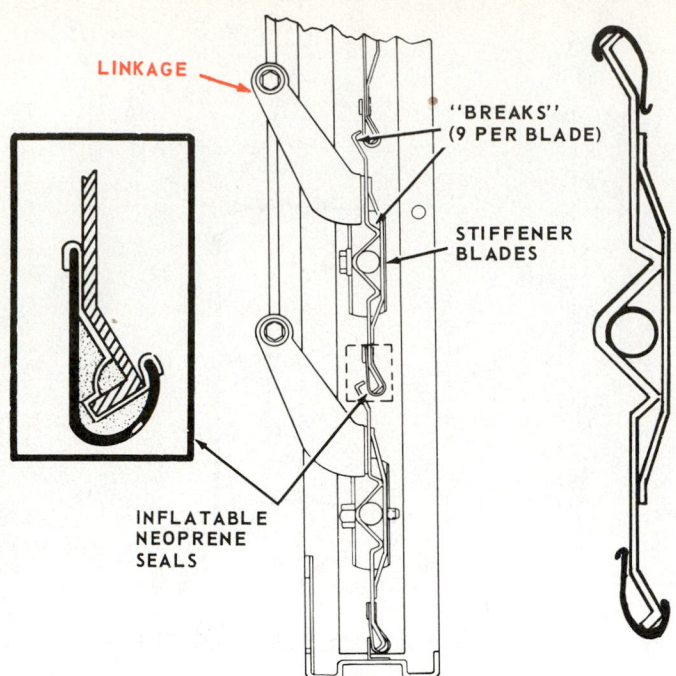

Fig. 22-41. Damper blades are of parallel design. (Honeywell, Inc.)

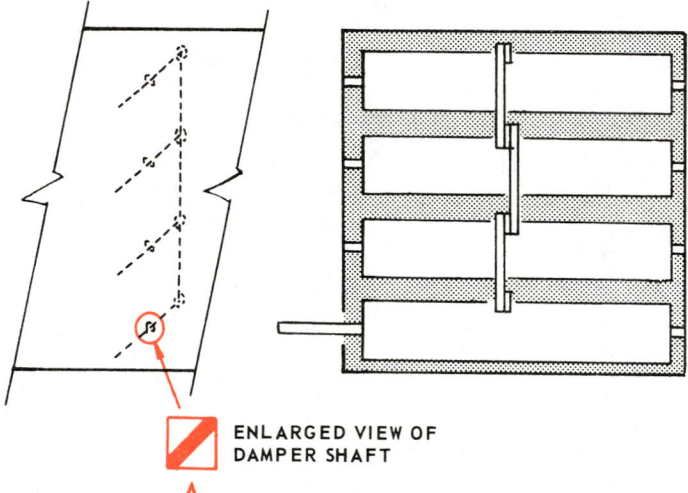

Fig. 22-42. Mark the shaft extending out of duct to indicate position of hidden damper blades inside duct. A—File cut or hacksaw cut shows position of damper blades.

22-18 FIRE DAMPERS

Automatic fire dampers should be installed in all vertical ducts in commercial and industrial buildings. Ducts, especially vertical ones, will carry fumes and flames from fires. These dampers should be inspected and tested at least once a year to be sure that they are in good operating condition.

Ducts going into or through a fire wall must have fire dampers. These openings are rated as follows:

1. Class A—opening will hold back a fire indefinitely.
2. Class B—fire dampers may be used when a two to four-hour

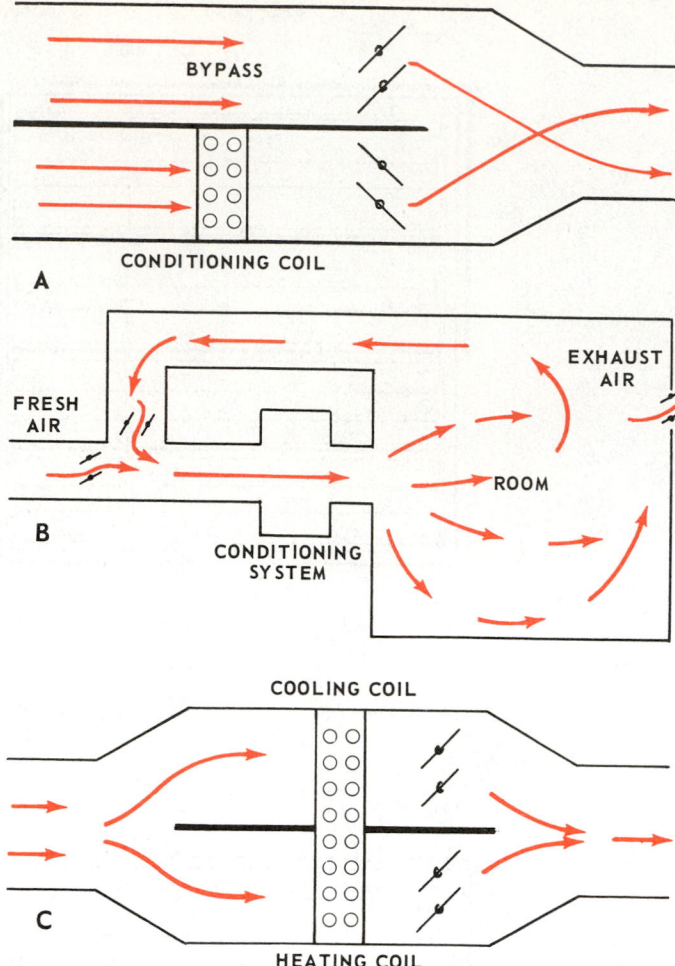

Fig. 22-43. Use of dampers to control airflow. A—Bypass to control temperature and humidity. B—Dampers to control fresh air, exhaust air and bypass. C—Controlling air over either cooling or heating coil.

hold is required. This class is approved for most general installations.
3. Class C—fire dampers are used where a one-hour hold is required.

The following dampers are fail-safe units:
1. Spring loaded to close.
2. Weight loaded to close.

Vertical shafts serving two or more floors must be enclosed in a fire partition and fire dampers must be used. Fig. 22-44 shows a fire damper. Ducts of less than 20 sq. in. (129 sq. cm) area do not require a fire damper.

Fire dampers are usually held open by a fusible link. Heat will melt the link and damper will close either by gravity, weights or springs. See Fig. 22-45. Some fire dampers are made to close by electronic sensors. In these, the damper blade latches operate either pneumatic (compressed air) or electric power devices which close the damper.

Smoke dampers use a photoelectric device to detect smoke. An electronic device will trip a holding device and the damper will close.

Before one makes a duct installation, local smoke and fire damper regulations must be checked.

Fig. 22-44. Fire damper in closed position. A—Blades. B—Duct frame.
(Air Balance Inc.)

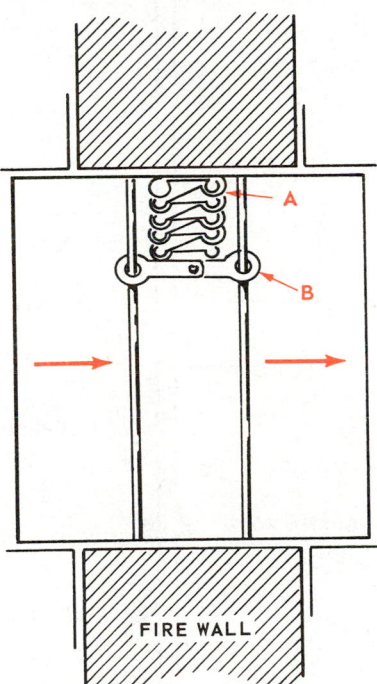

FIRE WALL

Fig. 22-45. Fire damper in open position. A—Damper vanes. B—Fusible
link. If link is heated, it will melt and vanes will fall, closing duct.
(Air Balance Inc.)

22-19 DUCT CALCULATIONS

In a heating or cooling system where a duct serves more
than one room, the duct must be designed so that each room
being served receives the correct amount of air. If the
distribution is not balanced, one room will be too warm while
another will be too cold.

Two methods used to calculate the proper size plenum
chambers, main ducts, branch ducts and grilles are:
1. Unit pressure drop system.
2. Total pressure drop system.

22-20 UNIT PRESSURE DROP SYSTEM

Air forced through a duct follows the path of least
resistance. Many air conditioning duct systems have several
openings (grilles) for the air to escape from the duct. A duct
with low resistance will allow most of the air to flow through
it while other ducts with higher resistance will not carry the
correct amount of air.

In the past, many duct installations were made that fed too
much air to some rooms and did not heat or cool other rooms
sufficiently.

The unit pressure drop calculating system uses the same
pressure drop for each length of duct throughout the system.

For example, suppose that the total heat load during the
heating season is 80,000 Btu/hr. and that there are six rooms
with heat loads as follows:
1. Living Room   = 25,000 Btu/hr.
2. Dining Room   = 15,000 Btu/hr.
3. Kitchen   = 5000 Btu/hr.
4. Bathroom   = 8000 Btu/hr.
5. Bedroom No. 1   = 15,000 Btu/hr.
6. Bedroom No. 2   = 12,000 Btu/hr.

Referring back to the problem in Para. 22-12, one can work
out the air volume needed to heat these six rooms. Recall the
specific heat of air is .24 and that the volume of one pound of
air is 15.28 cu. ft. Therefore:

Btu/cu. ft./hr. for a 68 F. (38 C.) change =

$$\frac{15.28}{.24 \times 68} = .936 \text{ Btu/cu. ft./hr.}$$

Divide by 60 (to get the answer into minutes) one gets
.0156 Btu/cu. ft./min.

Knowing the amount of heat that must be carried to each
room per minute, one can work out the air volumes required
per minute for each room:
1. Living Room   25,000 x .0156 = 390 cu. ft./min.
2. Dining Room   15,000 x .0156 = 234 cu. ft./min.
3. Kitchen   5000 x .0156 = 78 cu. ft./min.
4. Bathroom   8000 x .0156 = 124.8 cu. ft./min.
5. Bedroom No. 1   15,000 x .0156 = 234 cu. ft./min.
6. Bedroom No. 2   12,000 x .0156 = 187.2 cu. ft./min.
The total air volume is 1248 cu. ft./min.

To determine duct sizes that will handle the air volumes
specified in this problem, data must be obtained about airflow.
Figs. 22-32 and 22-46 are friction air charts for straight ducts.
Values were obtained by research. These charts have four
variables:
1. Friction loss in inches of water on the horizontal scale
   (equal value lines are vertical).
2. Cubic feet of air/min. on the vertical scale (equal value lines
   are horizontal).
3. Velocity on scale lines that slant down to right.
4. Round duct diameter on scale lines that slant down to left.
   To continue with the problem, the main duct must handle

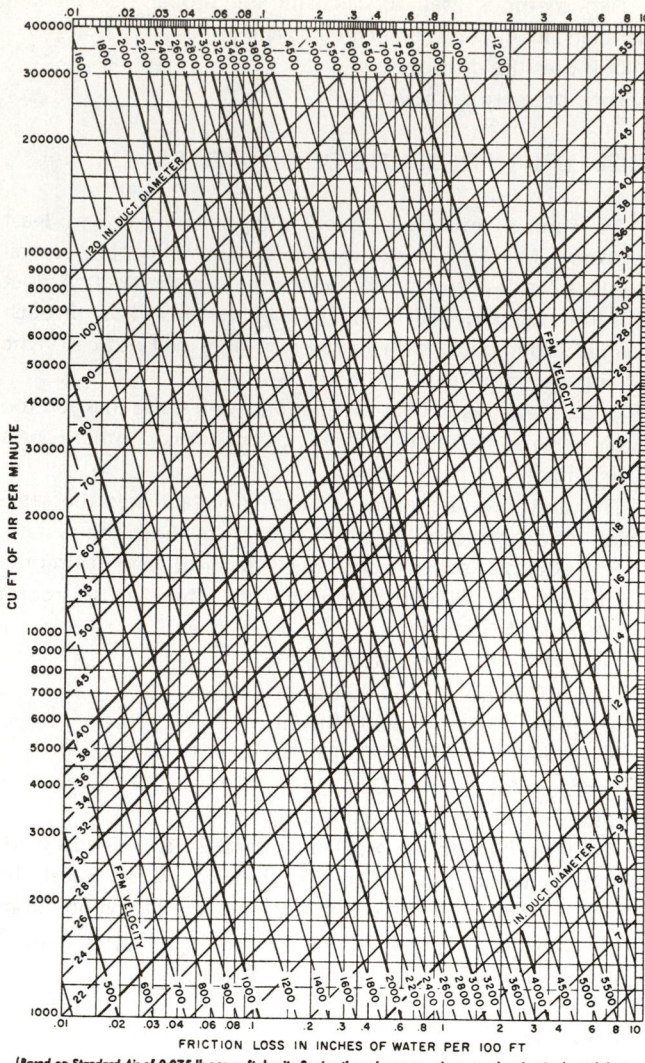

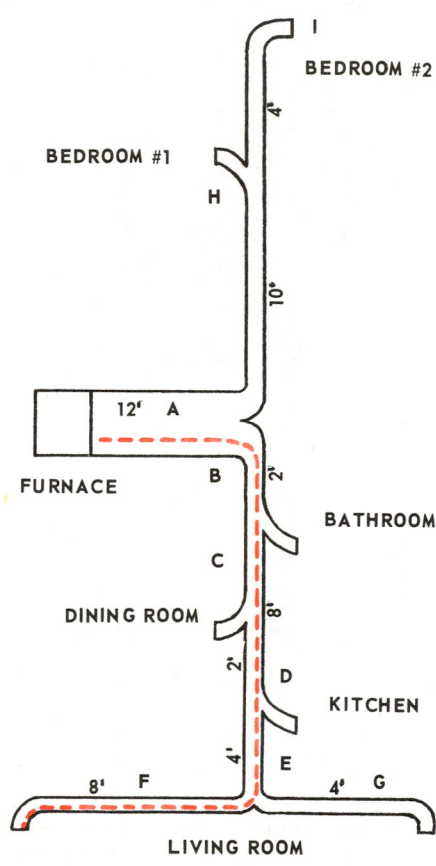

Fig. 22-46. Friction chart for high-volume airflow in ducts.
(ASHRAE Guide and Data Book)

(Based on Standard Air of 0.075 lb per cu ft density flowing through average, clean, round, galvanized metal ducts having approximately 40 joints per 100 ft.)

work. However, the total pressure drop system is a more accurate method. See Para. 22-21.

Round duct diameters may be changed to rectangular duct sizes using the table in Fig. 22-11. *When changing round duct sizes to rectangular duct sizes, remember that partition ducts cannot exceed 3 1/4 in. deep and 14 in. wide.* All ducts in the basement should have the same depth for better appearance. This also makes concealment of the ducts easier where the basement is used as recreation space or living quarters. In most installations, basement ducts should not exceed 8 in. in depth.

22-21 TOTAL PRESSURE DROP SYSTEM

A more accurate method of calculating proper sizes of ducts is based on having the same total pressure drop from the fan to each outlet. Fig. 22-47 shows the duct system used with the rooms as calculated in Para. 22-20. To keep the various ducts identified, it is good practice to letter each different size of duct.

Fig. 22-47. A typical duct installation. Longest air path is shown in red.

1248 cu. ft./min. To keep the velocity to a low noise level, a friction loss of .04 in water column per 100 ft. (30.5 m) should be used.

On the chart, these two values meet and show that the velocity will be 700 ft./min. (214 m/min.), and the round duct will be 18 in. (46 cm) in diameter.

Using the same friction loss for the branch ducts, one gets the following round duct sizes:

1. Living Room = 550 ft./min. and 12 in. dia. (30.5 cm)
2. Dining Room = 480 ft./min. and 10 in. dia. (25.4 cm)
3. Kitchen = 370 ft./min. and 6.9 in. dia. (17.5 cm)
4. Bathroom = 400 ft./min. and 8 in. dia. (20.3 cm)
5. Bedroom No. 1 = 480 ft./min. and 10 in. dia. (25.4 cm)
6. Bedroom No. 2 = 450 ft./min. and 9.3 in. dia. (23.6 cm)

(In converting these diameters to metric remember that 1 in. = 2.54 cm.)

These velocities are reasonably low and the system would

In Fig. 22-47, suppose that the following air volumes must be carried:

Duct	Air Volumes in cu. ft./min.
A	1404
B	930

C	790
D	526
E	439
F	220
G	220
H	474
I	211

To be sure that the correct air volume leaves each outlet, each must have correct, equal amount of total air pressure drop.

The method followed is to determine the longest and most complicated duct. This combination includes ducts A, B, C, D, E and F.

Assume a total pressure drop of .04 in. (0.1 cm) of water. This total means that the pressure drop to each room outlet must be .04 in. For example, the opening to the bathroom is the shortest overall distance. It must have the same total pressure drop as the longest run through F.

An important part of this duct design is that the bends and elbows must be considered when determining pressure drop. *Generally speaking, the pressure drop of one elbow is equal to 10 diameters of the duct.* Assuming there is one large bend above the furnace and that the grilles are located at the 7-ft. level in the room, the total length of duct A, B, C, D, E and F is approximately:

Elbow (18 x 10)	15 ft.	(4.6 m)
A	12 ft.	(3.8 m)
Elbow (16 x 10)	13 ft.	(4.0 m)
B	2 ft.	(0.6 m)
C	8 ft.	(2.4 m)
D	2 ft.	(0.6 m)
E	4 ft.	(1.2 m)
Elbow (9 1/2 x 10)	8 ft.	(2.4 m)
F	8 ft.	(2.4 m)
Elbow	8 ft.	(2.4 m)
Vertical rise	7 ft.	(2.1 m)
Elbow	8 ft.	(2.4 m)
Total	95 ft.	(28.9 m)

The total length (equivalent) is 95 ft. Because the .04 in. pressure drop was for 100 ft. (30.5 m), the new pressure drop for the equivalent feet of duct is:

$$\frac{100}{95} \times .04 = \frac{20}{19} \times .04 = \frac{.8}{19} = .042 \text{ in. (.105 cm) water}$$

However, more important than this factor is the pressure drop in each section of the longest duct:

Pressure drop in each part equals pressure drop/100 ft. multiplied by the ratio of length of part to longest equivalent length.

$$\text{Pressure drop for each part} = .04 \times \frac{\text{Length of part}}{\text{Total equivalent length}}$$

Elbow (18 x 10)	$= .04 \times \frac{15}{95} =$	.0063
A	$= .04 \times \frac{12}{95} =$	.0050
Elbow (16 in. x 10)	$= .04 \times \frac{13}{95} =$	.0055
B	$= .04 \times \frac{2}{95} =$	.0008
C	$= .04 \times \frac{8}{95} =$	.0034
D	$= .04 \times \frac{2}{95} =$	.0008
E	$= .04 \times \frac{4}{95} =$	.0016
Elbow (9 1/2 x 10)	$= .04 \times \frac{8}{95} =$	.0034
F	$= .04 \times \frac{8}{95} =$	.0034
Elbow	$= .04 \times \frac{8}{95} =$	.0034
Vertical rise	$= .04 \times \frac{7}{95} =$	.0029
Elbow	$= .04 \times \frac{8}{95} =$	.0034
Total		= .0399 =

.04 in. (0.1 cm) of water column pressure drop.

Knowing the pressure drop in each part of the longest duct, now determine the pressure loss up to each branch duct. Then, from this value and the length of the branch duct, determine the pressure loss per 100 ft. (30.5 m) for the branch duct.

For example, consider the kitchen duct:

The pressure loss up to the kitchen branch duct is the sum of all pressure losses along the way. Thus, .0063 + .0050 + .0055 + .0008 + .0034 + .0008 = .0218, which is the total pressure drop.

If the total pressure drop to the outlet at the kitchen must equal .04 in. (0.1 cm) water, then, .0400 − .0218 = .0182 as the pressure drop in the kitchen branch. Assuming 87 cu. ft./min. (2.5 cu. m/min.) volume and .04 in. (0.1 cm) pressure drop, the kitchen branch has length equal to the following:

Elbow (6.9 in. x 10)	= 6 ft.	(1.8 m)
Riser	= 7 ft.	(2.1 m)
Elbow (6.9 in. x 10)	= 6 ft.	(1.8 m)
Total	19 ft.	(5.7 m)

If the pressure drop in 19 ft. is .0182, the pressure drop per 100 ft. = $.0182 \times \frac{100}{19} = \frac{1.82}{19} = .096$ in. water/100 ft. From the graph, using a volume of 87.73 cu. ft./min. and the resistance of .096, the following data is obtained: Size = 5.8 in. (14.7 cm) dia. Velocity = 530 cu. ft./min. (15 cu. m/min.)

Compare this method with the unit pressure drop values. Notice how these values differ. One calculates comfort cooling air in about the same way. If the cold air duct is exposed to warm, moist air, condensation on the outside surface of the duct may cause corrosion. Moisture may also drip on structural parts and cause damage. In such cases, ducts should be insulated.

22-22 RETURN AIR DUCTS

Return air ducts, as already mentioned, are important. Flow of air through these ducts is almost always the result of the "pulling" action of a fan or blower. If the return airflow is not matched with the airflow into a room, the flow of air in cubic feet per minute will not be properly balanced.

If there is more return air than air going into a room, there may be a negative pressure in that room. Thus, more air-in will be used by this room. In turn, other rooms may starve for air. During the heating season rooms starved for air will be too cold.

Duct return grilles should be placed in the stratified (stagnant) air zone of a room. During the heating season this area is along the floor and during the cooling season this place is near the ceiling. Ideally there should be two places for return air grilles. In all cases, the place is the maximum distance from the air-in grille.

22-23 ELBOWS

Air has inertia. That is to say, air has weight and it obeys Newton's laws of motion. In other words, once set in motion, air tends to continue on in the direction it is moving. In addition, air is compressible and, because of these laws, air in motion has the following characteristics.

It takes energy to make airflow change its direction. The air wants to flow in a straight line. On turns, therefore, it crowds against the outside, as shown in Fig. 22-48. View A shows a typical elbow. It has a short bend radius. Pressure drop as the air goes through the elbow is about 10 times greater than in an equal length of duct. View C shows a better airflow design but cost of the duct and the room needed to install it make it impractical for many installations. View B is a turbulent air duct design while view D is the same type of duct with a much better airflow design. The duct elbow at E has vanes placed at the bend. These vanes, located at point F in view E, help reduce the pressure drop.

22-24 BALANCING THE SYSTEM

Balancing means sizing the ducts and adjusting the dampers to insure that each room receives the correct amount of air. Conditioned air must be fed in the right amounts to each different room of a multiple room system (home or office). Also the correct amount of air must be returned. If the system is not balanced, rooms will maintain different temperatures; some ducts will be noisy; some will have incorrect humidity; and some will have stale air.

To total air balance (TAB) a system, one must measure the air velocity leaving each grille and determine how much "free" area the grille or diffuser has. The "free" area is the actual size of the air openings.

To balance a system do the following:
1. Inspect the complete system; locate all ducts, openings and dampers.
2. Open all dampers in the ducts and at the grilles.
3. Check the velocities at each outlet.
4. Measure the "free" grille area.
5. Calculate the volume at each outlet.
    Velocity x area = volume

    $$fpm \times \frac{area\ in\ sq.\ in.}{144} = cu.\ ft./min.$$
    (volume per minute)
6. Total the cu. ft./min.
7. Determine the floor areas of each room. Add to determine

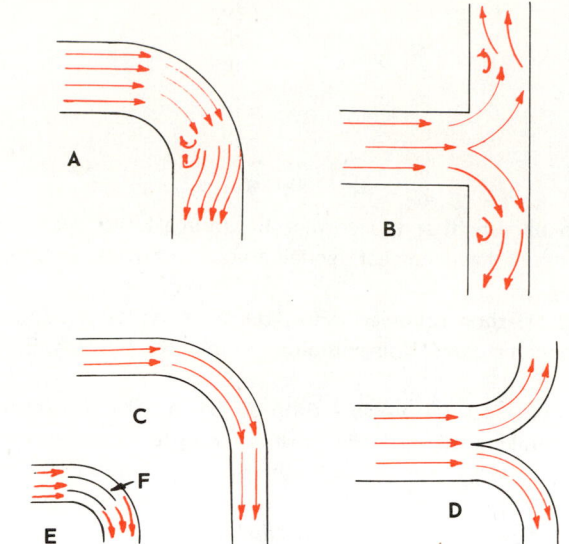

Fig. 22-48. Airflow in duct bends and elbows. A—Turbulence in air. B—Turbulence in air. C—Smooth airflow. D—Smooth airflow. E—Smooth airflow using vanes at F.

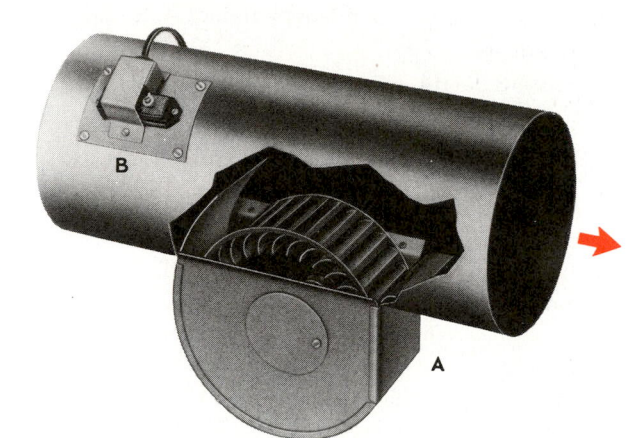

Fig. 22-49. Duct fan increases airflow in conditioned air ducts. A—Fan. B—Control.   (Tjernlund Mfg. Co.)

total area.
8. Find out the proportion each room should have.
    $$\frac{Area\ of\ room}{Total\ floor\ area} \times total\ cfm = cfm\ for\ room$$
9. Adjust duct dampers and grille dampers to obtain these values.
10. Recheck all outlet grilles.

In some cases, it may be necessary to overcome excess duct resistance by installing an air duct booster. These are fans used to increase airflow when a duct is too small, too long, or has too many elbows. A booster fan is shown in Fig. 22-49.

An effective but simpler technique can be used to balance airflow to the different rooms:

Mount accurate thermometers in folded cardboard as shown in Fig. 22-50. Place one of these thermometers in each room. Locate them on a table away from sunlight, lamps or

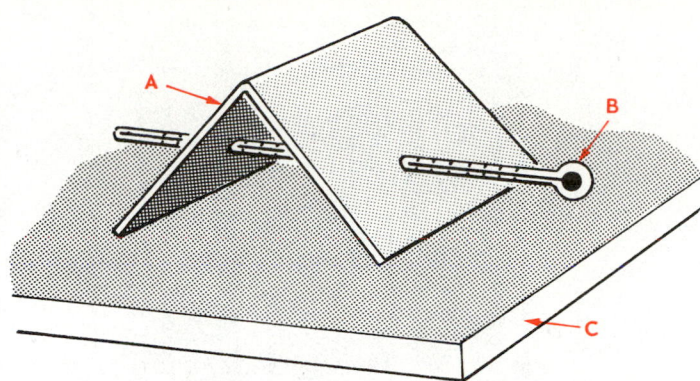

Fig. 22-50. How to use glass stem thermometer to measure room temperature. One is placed in each room. A—Folded heavy paper. B—Thermometer. C—Table top.

any extra heat source. Adjust dampers until each room has the temperature desired. It is best to allow several hours for the system to adjust to any damper change.

Remember, if one room is too warm, one or more other rooms should be too cool. Close damper to the warm room a little and open dampers a little to the room or rooms that are too cool.

22-25 DUCT NOISE

Duct noise can be very disturbing. Noises may be:
1. A high pitch sound usually caused by too-high air velocity or by air hitting sharp metal edges.
2. A low pitch rumble usually caused by fan and motor sounds traveling along the duct system.
3. A popping sound when the unit starts or stops. This is caused by expansion (becoming larger) or contraction (shrinking) of the duct as it warms up or cools.

To locate the source of the high-pitched sound, remove the grille or diffuser. If the noise stops, it is caused by sharp edges in the grille. If it continues, the air velocity is too high (use an anemometer to measure) or there is a sharp edge in the duct system. Locate and correct.

22-26 FANS

Air movement is usually produced by some type of fan. Usually fans are located at the inlet of the air conditioner. Air can be moved by either creating an above-atmosphere pressure (positive pressure) or a below-atmosphere pressure (negative pressure). All fans produce both conditions; the air inlet to a fan is below atmospheric pressure, while the exhaust of the fan is above atmospheric pressure. See Fig. 22-51. *The air feed into a fan is called induced draft and the air exhaust from a fan is called forced draft.*

There are several types of fans, but the two most popular are:
1. Axial flow (propeller).
2. Radial flow (squirrel cage).

Basic construction of the fan shows which type it is. If air

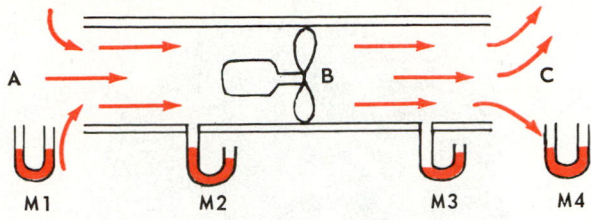

Fig. 22-51. Pressure conditions in simple duct and fan installation. A—Intake. B—Fan and motor. C—Exhaust. M1—Atmospheric pressure. M2—Negative pressure. M3—Positive pressure. M4—Atmospheric pressure.

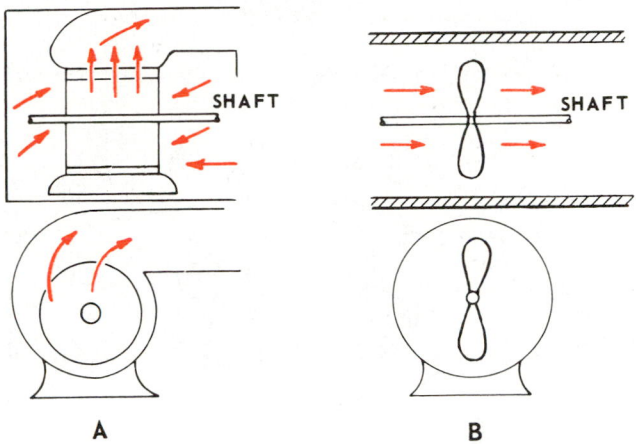

Fig. 22-52. Principal types of fans. A—Radial flow. B—Axial flow.

Fig. 22-53. Blade for axial flow fan.   (Torin Corp., Torrington, CT)

flows along the direction the axle is pointing, it is called axial flow. If the flow is at right angles to the axle (radius), it is called radial flow. This is shown in Fig. 22-52.

A four-blade axial flow fan is shown in Fig. 22-53. Rotation is clockwise. The axial flow fan is usually direct driven by mounting the fan blades on the motor shaft. These fan blades should be handled carefully. If they are bent or twisted, the fan should be replaced.

Part of a radial flow fan is shown in Fig. 22-54. A belt-driven radial flow fan assembly is shown in Fig. 22-55.

Fig. 22-54. Rotor for radial flow fan. (Torin Corp., Torrington, CT)

Fig. 22-56. Radial flow belt-driven fan shown in complete assembly including housing and filters. A—Motor. B—Fan belt. C—Fan. (International Metal Products Div., McGraw-Edison Co.)

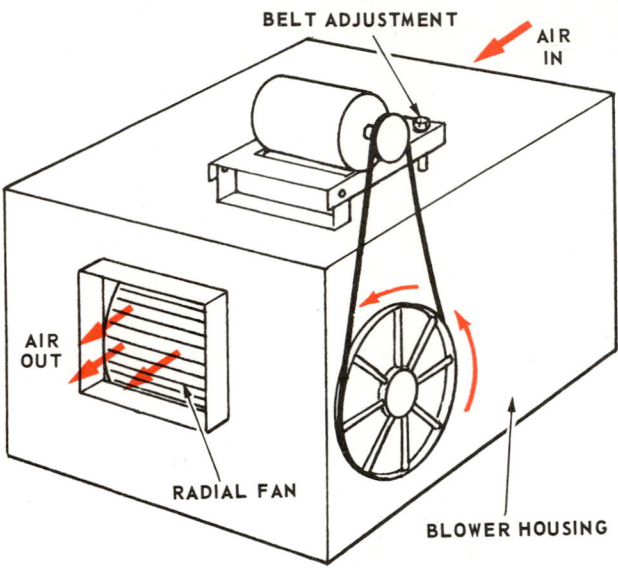

Fig. 22-55. Typical blower housing using belt-driven radial flow fan. Note direction of rotation of fan pulley and belt adjustment.

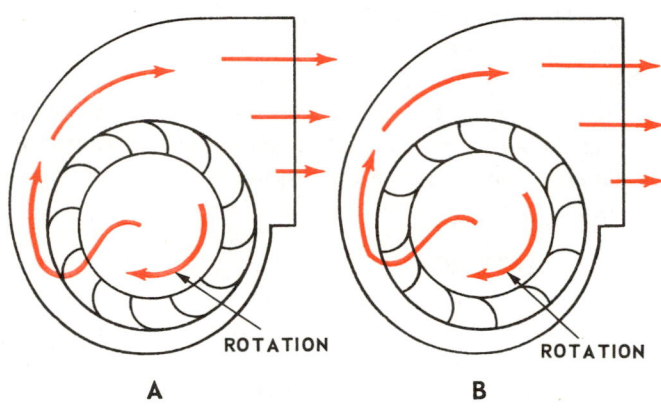

Fig. 22-57. Basic radial flow fan designs. A—Backward-inclined blades. B—Forward-inclined blades.

The radial flow fan is most often used on large installations. It is either directly driven or belt driven. A belt-driven unit is shown in Fig. 22-56.

These units are made in the four classes of air systems listed by the Air Moving and Conditioning Association (AMCA):

Low pressure, Class 1	Up to 3 3/4 in. (9.5 cm) water column total pressure.
Medium pressure, Class II	Up to 6 3/4 in. (17 cm) water column total pressure.
High pressure, Class III	Up to 12 1/4 in. (31 cm) water column total pressure.
High pressure, Class IV	Over 12 1/4 in. (31 cm) water column total pressure.

Centrifugal flow or radial flow fans are made in different designs:

1. Backward-inclined blades (small units).
2. Forward-inclinded blades (large units). See Fig. 22-57.

Static pressure increases as the square of the ratio of change of the cfm.

1000 to 2000 cfm = two times.

$2 \times 2 (2^2)$ = four times more static pressure.

This means that the static pressure is the square of the ratio change of rpm. Different ways to measure the various outputs of a fan (total pressure, static pressure and velocity pressure) are shown in Fig. 22-58.

To determine the fan capacity for a furnance, use the following formula:

$$cfm = \frac{Btu/hr. \text{ output of furnace}}{1.1 \times \text{air temperature rise in F.}}$$

For cooling, use 400 cfm/ton of capacity. The total pressure drop in the ducts should be about .2 in. of water column. (This equals pressure rise across the furnace.) It should be .4 to .5 in. of water column for furnaces with a cooling unit installed. (This leaves .2 in. for all but cooling unit.)

If possible, belt tension should be on the lower belt section to provide more efficient belt drive, as shown in Fig. 22-59. Belt tension is about right when the belt can be pushed out of line a distance equal to its width. It is often desirable to change fan speeds to obtain more or less airflow in cubic feet per second. One way to vary speed is to use adjustable (variable pitch) pulleys. The one shown in Fig. 22-60 is for a two-belt drive.

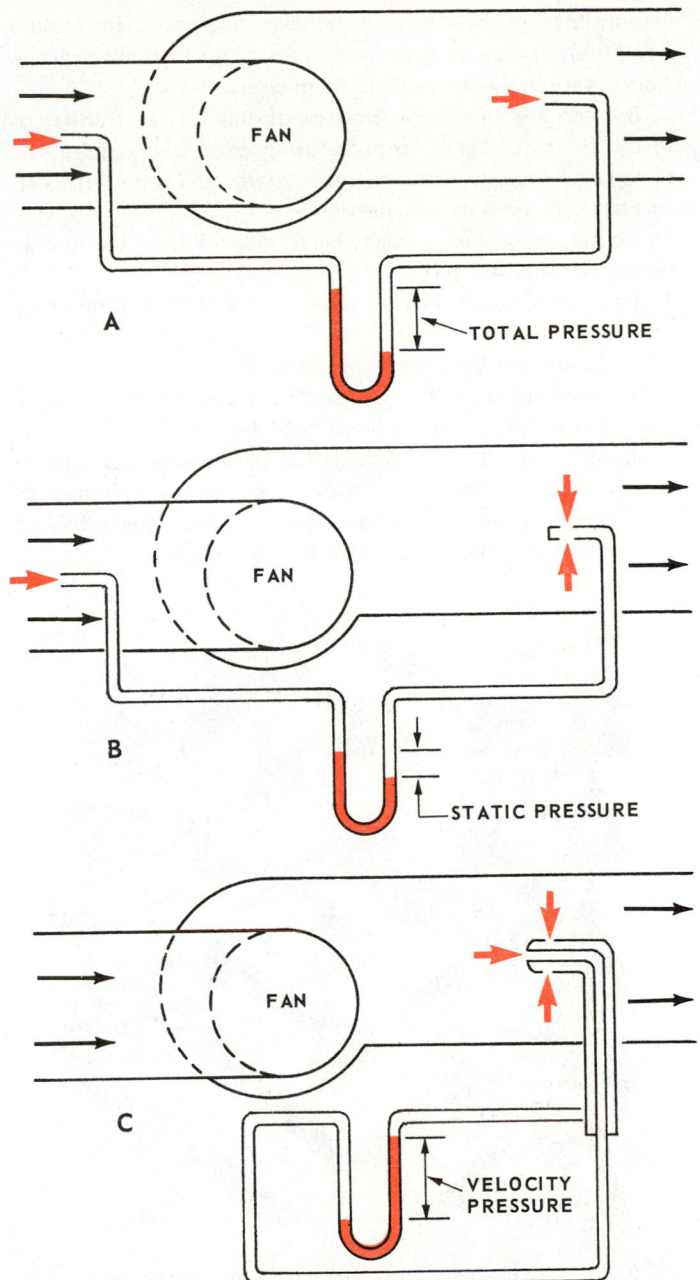

Fig. 22-58. Three ways to measure performance of fan. A—Total pressure. B—Static pressure. C—Velocity pressure.

Any fan which moves air collects lint and dirt reducing the efficiency of the fan. Dirt should be removed every six months. Remove the fan and scrape, rub or vacuum the dirt off the blades. Blades will collect dirt even more quickly. If fan bearings are oiled too much, this extra oil coats the fan blades. Each bearing requires but one or two drops of oil each year. See Chapter 7 for more fan motor information.

22-27 ATTIC FAN

Many buildings use exhaust fans to remove extremely hot air that collects in attic spaces above insulated ceilings during

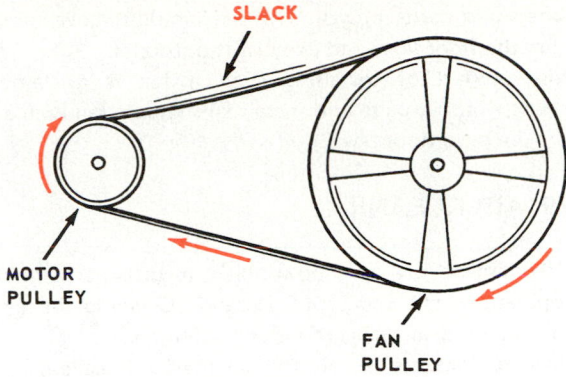

Fig. 22-59. Motor pulley and fan pulley. Note tension on lower part of belt and slack at upper. This is how properly tensioned belt should look in operation.

the summer. Some attic fans may also be used to bring cooler evening air into a building.

These fans have large capacities ranging from 1000 cfm to 4000 cfm. It is usually desirable to have a fan large enough to make a complete change of air in the building every 8 to 10 minutes.

Example: A building measures 30 ft. x 60 ft. with an 8-ft. ceiling.

$$\text{The volume} = \text{Length x width x height}$$
$$= 60 \times 30 \times 8$$
$$= 1800 \times 8 = 14,400 \text{ cu. ft.}$$

An exhaust fan of 1440 cfm capacity will change air every 10 minutes.

$$\frac{\text{cu. ft. of space}}{\text{cu. ft./min.}} = \frac{14,400}{1440} = 10 \text{ minutes/change.}$$

22-28 BASEMENT FAN

Since basements tend to be cool and damp in the summer, mold and odors are a problem. An exhaust fan will reduce the

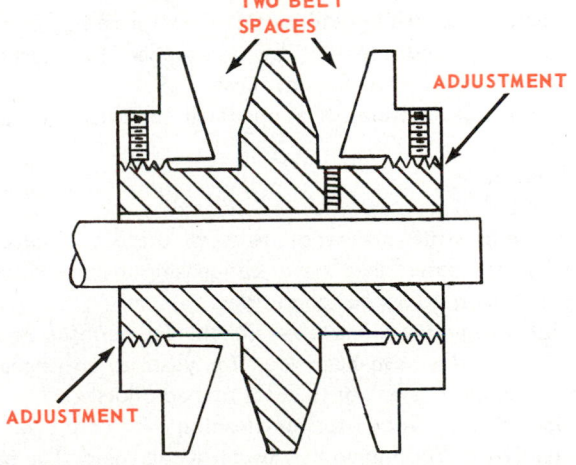

Fig. 22-60. Variable pitch pulley. Note location of setscrews. Pulley is used on dual belt system and can provide fan variation of about 100 rpm through adjustment.

dampness and mold growth. This fan should remove basement air from the floor level and exhaust it outdoors.

One method of installing such a fan is to remove a basement window pane and install an exhaust fan with a duct leading down to floor level.

22-29 AIR CLEANING

Air pollution is a growing problem as urban areas increase in population and industries expand. Cleaning the air has become an important part of air conditioning.

Air contaminants, as all foreign matter is called, include solids, liquids, gases and vapors. Efficient air conditioning systems will remove 75 to 95 percent of these contaminants.

Solid particles, kept in suspension in the air by air currents, fall into three general groups:

1. Dust. Usually, this results from wind, a sudden earth disturbance or from mechanical work on some solid. Dust can have its origin in animal, vegetable or mineral matter.

   Dust particles are usually over 600 microns in size (about .004 in. in diameter. Coal dust particles are usually 1 to 100 microns while atmospheric dust is .001 to 30 microns in size.

2. Fumes. These are formed from materials that are ordinarily solids but have been put into a gaseous state usually by an industrial or chemical process. Such particles are about 1 micron in size.

3. Smoke. Caused by incomplete combustion, smoke consists of solid particles carried into the atmosphere by the gaseous products of combustion. These particles vary in size from .01 to 13 microns. Oil and tobacco smoke particles range from .01 to 1 microns in size.

Gases are on the increase as air contaminants. They include carbon monoxide, sulphur oxides, nitrogen oxides and hydrocarbons. These gases form smog when combined with water vapor.

There are also liquid impurities in the air. Two of the most common are:

1. Mists. These small liquid particles are mechanically ejected into the air by splashing, mixing or atomizing.

2. Fogs. These small liquid particles are formed by condensation. Fogs indicate that the atmosphere has reached the saturation state for that chemical.

A third general classification of air impurities act as true gases:

1. Vapors.

2. Gases.

There is little difference between these two impurities. Vapors are gases that have condensing temperatures and pressures close to normal conditions.

Not all impurities are objectionable or harmful. Perfumes and deodorizers have been used for years to either make air more pleasant to smell or to cover up bad odors.

Special applications for air cleaning may be provided for: pollen (10 to 1000 microns), bacteria and mold. The best air conditioning practice is to clean the air of these materials.

Pollen grains come from vegetation growth such as weeds, grasses and trees. Their presence in the air is usually responsible for hay fever, rose fever and other respiratory conditions. These particles vary in size from 10 to 50 microns. Spores vary in size from 10 to 30 microns.

Bacteria are microorganisms responsible for the transfer of many diseases. Many manufacturing processes require the removal of these bacteria. Hospital rooms and some refrigerators use bacteria-removing devices.

Air may be cleaned in many ways, depending on the foreign matter contaminating it.

1. To remove solids such as dust, soot and smoke, one may resort to:

   a. Centrifugal force (for large particles).

   b. Washing the air (for particles that are wettable).

   c. Screens (to block the larger particles).

   d. Adhesives. The air impinges on (strikes against) a tacky or sticky surface and the dirt particles in the air stick to the adhesive. Fig. 22-61 shows a filter which has an adhesive material on a honeycomb surface.

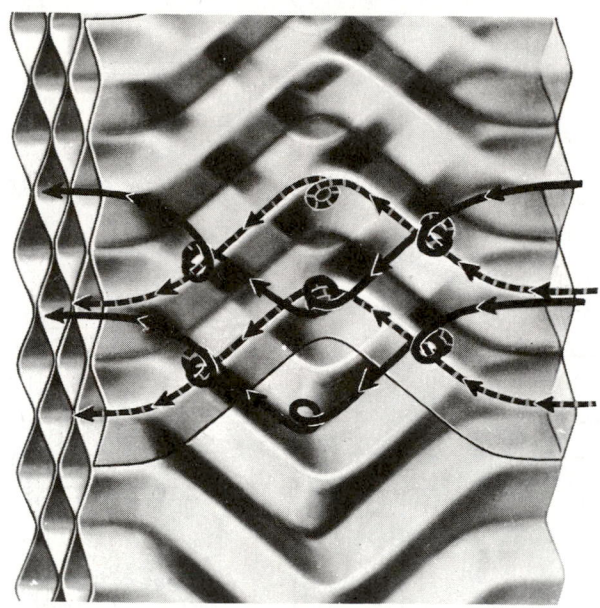

Fig. 22-61. Washable aluminum corrugated filter used to remove dust from forced air system. Aluminum is coated with adhesive material so dust particles will adhere to surface. Note airflow caused by honeycomb type filters. Air turbulence insures trapping of dust particles. (Continental Air Filters, Inc.)

   e. Electrostatic (electrically charging the particles and adhering these particles to an opposite charge surface). Fig. 22-62 shows an electronic air cleaner. Most of these cleaners have a screen to trap large particles, an electronic unit to remove particles as small as .001 micron, and a mat to trap the electron-treated particles. They are usually equipped with a pressure drop indicator and controls.

2. To remove liquids:

   a. Liquid absorbents (chemicals to absorb or react with the liquid).

   b. Deflector plates.

Fig. 22-62. Electronic air cleaner. Unit is installed in return air duct. A—Control panel.　(Electro-Air Div., Emerson Electric Co.)

c. Settlement chambers.
3. To remove gases and vapors (these are molecular size impurities):
   a. Condensation (cool the contaminant gas to its dew point and remove as a liquid).
   b. Chemical reaction (to react with the gas).
   c. Dilution.

It is possible to remove almost 100 percent of the contaminants in the air, but to do so is expensive. Removal of 90 to 95 percent is much more common and practical.

Filter efficiency is measured by:
1. Total weight of dirt it collects.
2. Size of the smallest particle it will remove.
3. Checking for discoloration on the exhaust side of the filter being tested.

Many rooms gradually collect a brown-yellow color on walls, windows and light-colored drapes. This deposit does not come from the furnace or ducts. It is carried in the vapors from smoking and open cooking such as for meats. These gases and small particles collect on the cooler surfaces of a room, especially when air movement is slow.

22-30 ADHESIVE FILTERS

Adhesive filters are made of various fibers — glass, cotton, synthetic material and aluminum. There are two classes of these filters.
1. Class 1:　fire resistance when clean.
2. Class 2:　nonfire resistant.
Most homes use Class 2 filters.

Fibers of adhesive filters are coated with adhesive liquid or oil. Air is forced to change direction and lose speed as it passes through the filter. This results in trapping of the particles of lint and dust as they contact the adhesive surfaces. The filter material is also packed tighter at the outlet side of the filter to improve its dirt-holding capacity.

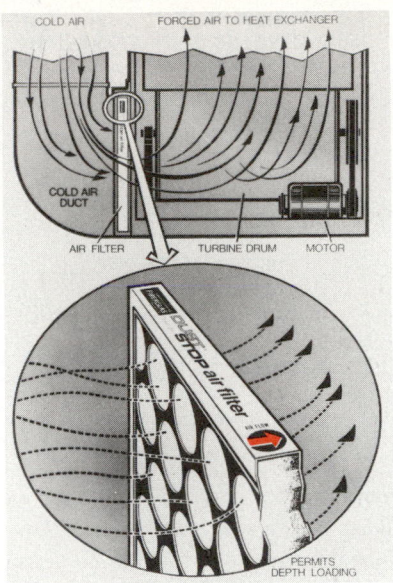

Fig. 22-63.　Throwaway paper frame glass fiber filter. (Owens-Corning Fiberglas Corp.)

These filters will remove as much as 90 percent of the dirt if they do not become "loaded," or if air velocity is not too high. The more common filters are of the throwaway or disposable type. See Fig. 22-63. *These filters should be renewed twice each year or more frequently if the dust conditions are high.* The frames are usually made of rustproof steel or cardboard with wire reinforcement. Fig. 22-64 shows how a fiber filter is installed in a furnace.

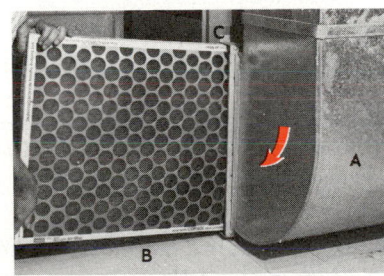

Fig. 22-64. Throwaway type filter being installed in a furnace. A—Air return. B—Filter. C—Furnace.　(Owens-Corning Fiberglas Corp.)

Another method used to determine if a filter needs replacement is to use a water manometer. The two manometer openings are connected to measure the airflow on the two opposite sides of the filter. The filter should be replaced if the pressure drop exceeds .5 in. of water across the filter.

The system is usually designed to allow the filter pressure drop (resistance) to be about a fourth of the total pressure drop (pressure rise across the fan).

For example, if the total pressure rise across the fan is 4.0 in. (10 cm) of water column, the allowed pressure drop across the filter is 1.0 in. of water column. These pressures are measured with a water manometer. See Chapter 18.

NEW NBS EFFICIENCY PERCENTAGE	NEW INITIAL PRESS DROP IN IN. H$_2$O	DIRTY FINAL PRESSURE DROP IN IN. H$_2$O
40	.15	.7
60	.25	.8
85	.35	.9
95	.40	1.0

Fig. 22-65. Table of filter efficiencies.

The National Bureau of Standards (NBS) rates filter efficiency as follows:

1. 40 percent.
2. 60 percent.
3. 85 percent.
4. 95 percent.

It is important to remember that filters are more efficient dirt removers when they are dirty, but the airflow decreases. A 40 percent efficient filter is one that has 20 percent efficiency when clean and 60 percent efficiency when dirty. The NBS standard is the mean or average efficiency. Fig. 22-65 is a table of filter efficiencies.

Filters are tested in laboratores (hot DOP) and they can also be tested on the job (cold DOP). See Para. 18-39. One test is to find out how many 0.3 micron size particles the filter can remove. The test is made by measuring the interference with diffused (scattered) light.

To increase filtering surface or area in filters, many different designs are used. A popular method to increase the area is to use pockets to trap the air, as shown in Fig. 22-66.

Most electronic filter systems also have cleanable filters. Remove filter. Be sure power is off first! Clean with a mild detergent-warm water mixture, rinse and replace. Turn on the power. These filters last indefinitely.

22-31 THROWAWAY FILTERS — SERVICING

Filters should be replaced when they lose their efficiency or when they are so clogged that they produce too much pressure drop across the filter.

Visual inspection is one way to decide that filters need replacement. If they have turned black, if the frame is bent or warped or if the filtering medium is punctured, replace the filter. If the housing shows signs of corrosion, clean it by sand blasting and repaint.

Checking the pressure drop across the filter, as described in Para. 22-30, is another way to decide whether the filter should be replaced. *When the pressure drop across the filter is more than 25 percent of the pressure drop across the fan, the filter should be changed.*

Always replace filters with the arrows (printed on the frame) pointing in the direction of airflow. The side towards the blower has more adhesive and must be on the air-out side of the filter. If this is not done, the filter will quickly load with dirt and clog.

When replacing filters make these two checks:

1. Inspect filter for tears or holes. Place a strong light on one side of filter and look through filter from other side.
2. Use manometer to check pressure drop.

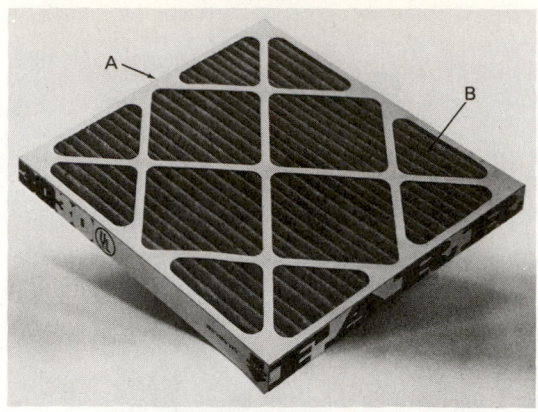

Fig. 22-66. Pocket type disposable air filter. Air pockets increase filtering surface considerably. A—Frame. B—Folded filtering material. (Farr Co.)

22-32 ELECTROSTATIC (ELECTRONIC) THEORY OF CLEANING

Static electricity has an important effect on dirt and dust clinging to walls, drapes and ceilings of rooms. Static electricity is created by two surfaces coming into rubbing contact and then separating.

Rubbing a cloth on a nonconductor will build up surplus electrons on one of the nonconductors (negative charge). The other material will lack electrons (positive charge). This excess of electrons is called static electricity. When this static electrical charge jumps an air gap, the spark can cause a fire or explosion.

Static electricity also attracts dirt and dust to vertical and overhead surfaces.

Ionizing the air will neutralize this static electrical charge. (Ionizing means to break it down into positive and negative particles or charges.) When this is done, dirt and dust will settle to the floor and the danger of a static electricity spark is over.

Instruments are used to measure static electricity. They can "read" both a lack of electrons (positive charged) and a surplus of electrons (negative charged).

Materials vary in their ability to generate static electricity. Fig. 22-67 lists a variety of materials and their abilities to generate static electricity. The farther apart the substances are on the list, the greater their ability to generate static electricity. The material at the top assumes a positive (+) charge, while the material at the bottom assumes a negative (−) charge.

There are three ways to ionize air:

1. Power.
2. Nonpower.
3. Nuclear.

The power unit is the electronic filter. A nonpower unit uses metal to remove the static electrical charge. The nuclear power unit gives off a double positive or alpha particle.

Basically, the electrostatic filter puts a static electrical charge on all particles that pass through it. These charged

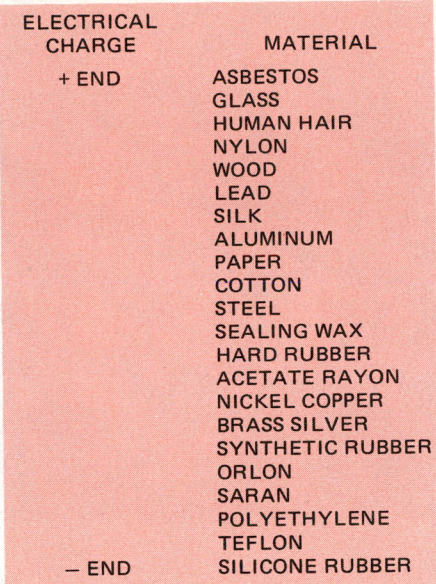

ELECTRICAL CHARGE	MATERIAL
+ END	ASBESTOS
	GLASS
	HUMAN HAIR
	NYLON
	WOOD
	LEAD
	SILK
	ALUMINUM
	PAPER
	COTTON
	STEEL
	SEALING WAX
	HARD RUBBER
	ACETATE RAYON
	NICKEL COPPER
	BRASS SILVER
	SYNTHETIC RUBBER
	ORLON
	SARAN
	POLYETHYLENE
	TEFLON
− END	SILICONE RUBBER

Fig. 22-67. Ability of various materials to generate electricity. Materials high on list, when brought into contact with materials lower on list, provide efficient generation of static electricity.

particles are then attached to collector plates with an opposite electrical charge.

Usually the air is first passed through a throw away filter to remove most of the larger particles of dirt. Then it is fed through the electrostatic filter. In this filter, the air first passes through a highly ionized field. A wire with a high positive voltage is suspended between ground wires.

Electrons passing through the air space put a positive electrical charge on any particle that attempts to pass through the ionized field. This particle is then drawn to the grounded plates (negative potential).

Potentials of as high as 12,000V are used. Fig. 22-68 shows a diagrammatic view of an electrostatic filter. A typical wiring diagram is shown in Fig. 22-69.

Because of the high voltages used, the electrostatic filter may be dangerous. The units should be designed to shut off

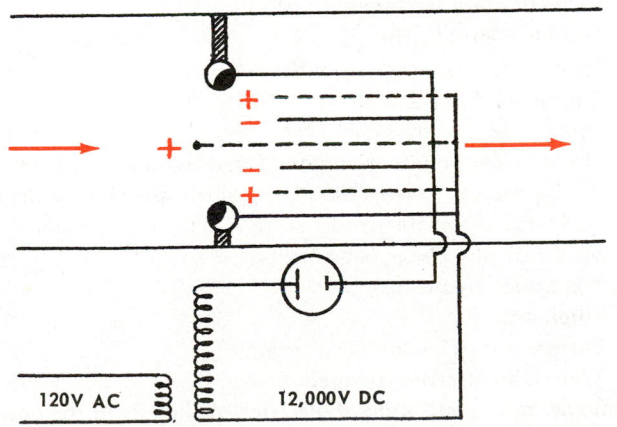

Fig. 22-68. Electrical circuit and airflow in simple electrostatic type air filter for a forced-air furnace.

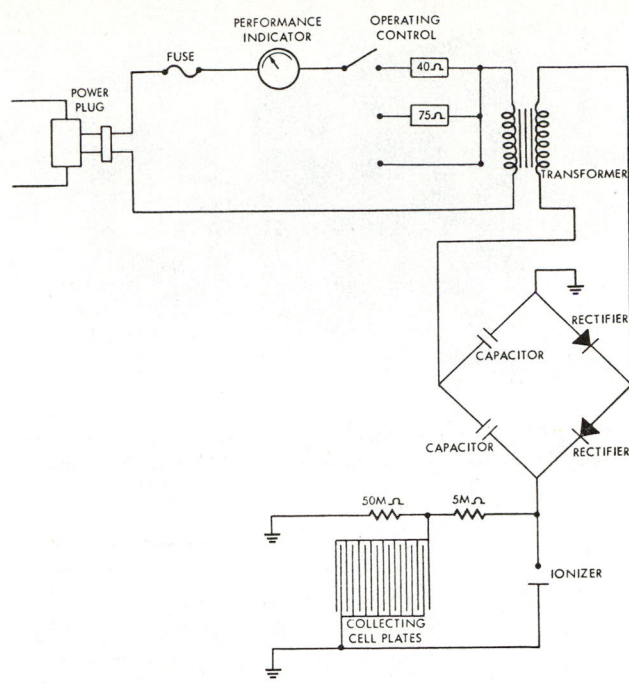

Fig. 22-69. Wiring diagram of an electronic air filter. (White-Rodgers Div., Emerson Electric Co.)

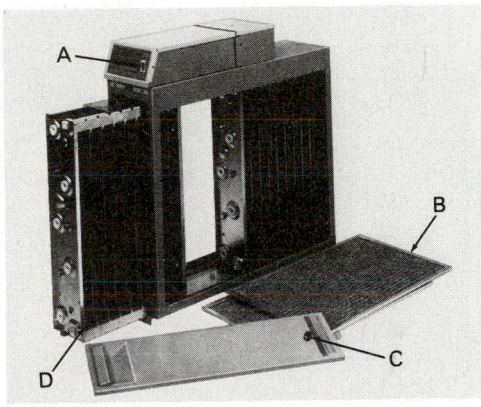

Fig. 22-70. Electronic filter designed for installation in air duct. A—Switch. B—Protective screens. C—Cleaning indicator light. D—Electronic cell unit. (Honeywell Inc.)

automatically when the service doors are opened to gain access to the units. Fig. 22-70 shows an electronic cleaner designed for duct installation.

Filters which carry small static electricity charges are also available. They usually are of the same size as throw away filters. These filters remove dirt particles by attracting them to surfaces charged with static electricity. Such filters are cleaned by washing in water. Be sure to open the power circuit before washing. See Fig. 22-71.

Electronic filters may be installed in duct systems and are also used as separate cabinet units. The filters can be placed in the return airflow ducts in several ways. See Fig. 22-72.

The electronic cells and protective screens must be cleaned

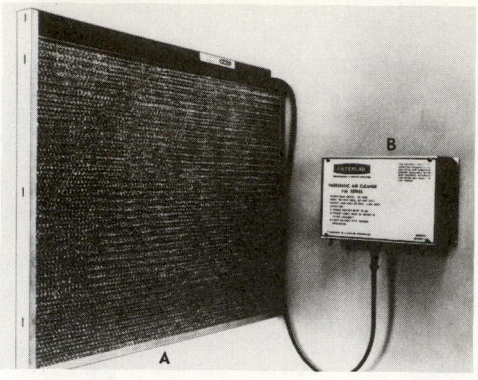

Fig. 22-71. An electronic filter which can be used in place of throw-away filters. A—Filter. B—Power unit. (Union Carbide Corp.)

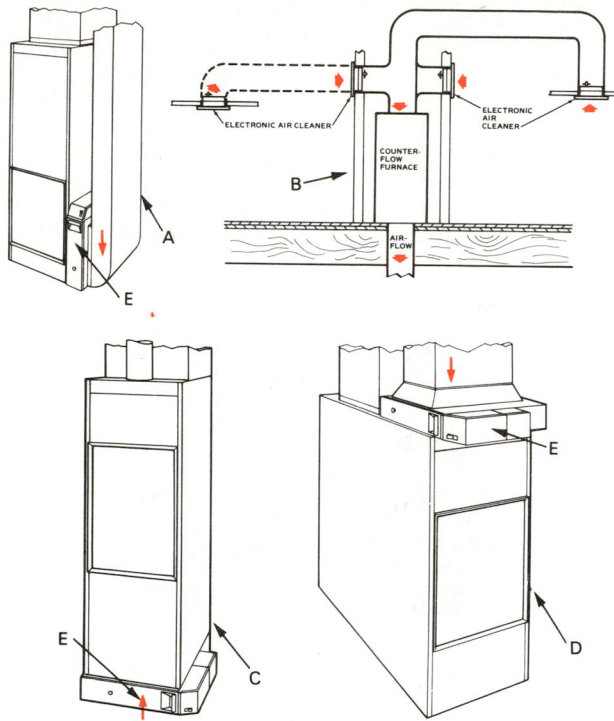

Fig. 22-72. Various installation positions for electronic filter. A—Cold air return from the side. B—Independent room units. C—Upflow. D—Downflow. E—Electronic filter. (Honeywell Inc.)

every two to three months. Material collected is black in color. Continue cleaning with water, until cleaning water is clear. Use recommended detergents and water-detergent solutions.

Fig. 27-73 shows a unit being washed with a built-in washer. After cleaning, rinse away all detergent solutions. The drying action takes place after assembly and operation. Some electronic filters have a built-in wash design. Some are manually operated with a hand valve. Some are semiautomatic. A push-button switch shuts off the system, opens the water solenoid valve and starts a timer which will put the system back in operation after the wash cycle. The full automatic system uses a timer to turn wash cycle on and off. All built-in washing systems require a water supply and a drain. It must be installed according to code.

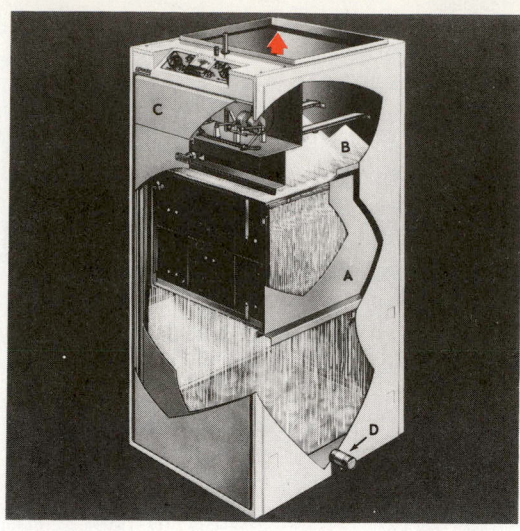

Fig. 22-73. An electronic filter being cleaned by its water and detergent spray solution. A—Collector cell. B—Spray. C—Controls. D—Drain. (Lennox Industries, Inc.)

Electronic filters have four main parts:
1. Frame.
2. Power supply.
3. Prefilter and airflow distributor.
4. Electronic cell.

The unit must be installed level and plumb for proper drainage and more efficient airflow. The hot water line to the washer should have a strainer.

Airflow should be evenly distributed across the face of the air cleaner for maximum efficiency. If the unit is near an airflow elbow, use movable air vanes or baffles.

Excessive lint interferes with electronic air cleaner (EAC) operation. A fine-mesh screen or filter should be installed ahead (upstream) of the filter.

22-33 SERVICING ELECTRONIC FILTERS

Electronic air cleaners (EAC) need service when:
1. Unit does not arc.
2. Meter (if used) reads low.
3. Trouble lights remain on.
4. Strong ozone odor is detected.
5. Rooms are dusty and dirty.
6. The unit arcs all the time.

Check owner on last cleaning of filter. Be sure power switch is on. Be sure the power doors or panels are closed. Check fuses. Check the meter readings. (Refer to service manual or owner's manual.) These meters will show if:
1. Conditions are normal.
2. Filter is dirty.
3. Filter is wet. (Operate "dry" switch.)
4. There is an electrical failure.

Some meters will show if electrical trouble is in the power source circuit or in the high-voltage circuit.

If the trouble is in the high-voltage circuit, inspect and electrically test the "power pack" capacitors and collecting

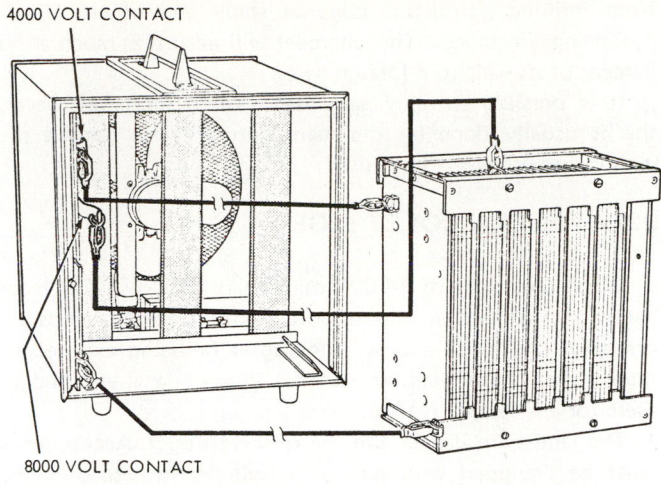

4000 VOLT CONTACT

8000 VOLT CONTACT

Fig. 22-74. Portable electronic filter showing collector cell removed. Electrical leads are being used to test high voltage circuit. (White Rodgers Div., Emerson Electric Co.)

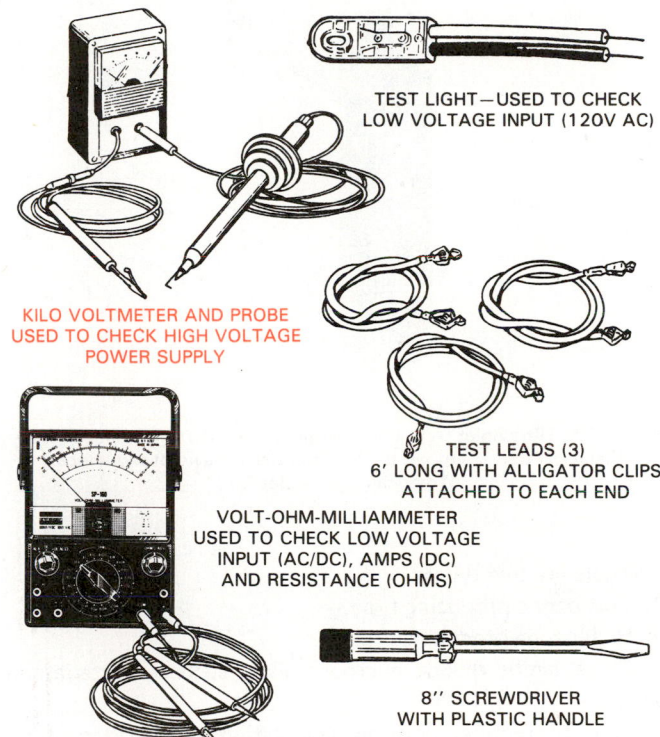

TEST LIGHT—USED TO CHECK LOW VOLTAGE INPUT (120V AC)

KILO VOLTMETER AND PROBE USED TO CHECK HIGH VOLTAGE POWER SUPPLY

TEST LEADS (3) 6' LONG WITH ALLIGATOR CLIPS ATTACHED TO EACH END

VOLT-OHM-MILLIAMMETER USED TO CHECK LOW VOLTAGE INPUT (AC/DC), AMPS (DC) AND RESISTANCE (OHMS)

8'' SCREWDRIVER WITH PLASTIC HANDLE

Fig. 22-75. Instruments, leads and tools used to check out electrical circuits of an electronic air filter. (A.W. Sperry Instruments, Inc.)

cell ionizing wires. A portable electronic filter with the collector cell removed and connected for testing purposes is shown in Fig. 22-74. This is a dangerous operation.

In the power pack, inspect the low side first with either a test light or voltmeter. Test each part starting with the wall outlet or power source.

Then check the transformer, rectifiers (a-c to d-c), the

capacitor resistors (should discharge capacitors in about 10 seconds). The capacitor can be checked by replacement. Tools and instruments used during servicing are shown in Fig. 22-75.

In the collector section, inspect the bent plates, plates out of position, dirt bridging the gap between ionizing wires and the plates, broken insulators and broken wires. Plates must be straight. Remove and replace broken insulators and ionizing wires.

One should inspect the building and complete air-handling system. New carpeting, for example, may temporarily cause an overload on the filter. Leaking duct systems and untreated concrete floors are all unusual high load conditions. Dusty construction work in the vicinity may also overload the unit.

A properly operating unit will be indicated by black water when the cell is cleaned. A properly operating unit allows only fine white dust to leave the ducts. A cheesecloth over a grille — if it becomes discolored — shows that the EAC is not working properly.

22-34 DIRT ON WALLS AND DRAPES

Dirt collecting on walls, ceilings, curtains and drapes of a conditioned space is always a problem. In most cases, this dirt does not come from the ducts but is already in the room. Room air movement, also called convection air current, is responsible for carrying it to the room surfaces.

The collection of dirt around warm-air grilles is called thermal precipitation. Warm air coming out of the grille picks up dirt from the room air. As this air hits cooler surfaces, the dirt settles (precipitates) on the surfaces. This precipitation takes place on windows also. The cooler the surface, the more dirt it collects; therefore, insulation and storm sashes reduce the amount of dirt settling out of the air.

"Clean" rooms are now in use for surgery, for research, and for manufacturing, repairing, and servicing critical items such as instruments.

A clean room has three devices to maintain the "clean" requirements:
1. Extremely high efficiency filters.
2. Laminar airflow.
3. Anticontamination devices.

Ideally, the future of residential living will be air cleaning similar to the "clean room" standards.

Filters are ineffective against gases. Activated charcoal will adsorb gases and so will suitable liquids. Water removes gas effectively.

22-35 WATER SPRAYS

Large air conditioners use water sprays to remove wettable solid contaminants, liquid contaminants and water soluble gas contaminants from the air. Some of these gases are sulphur dioxide, nitrogen oxides and carbon monoxide. Water does not remove soot.

Usually the water is sprayed in a pattern which produces 100 percent duct cross-section coverage. A drain pan catches the water while eliminator plates in the duct collect any water droplets which travel down the duct. The water drain pan is usually equipped with a float–controlled makeup water connection.

A continual overflow drainoff is used to remove dust and dirt as it collects on the surface of the water. Water in the drain pan is recirculated by centrifugal pump. A screen is located at the pump inlet to prevent dirt particles from clogging the spray nozzles. These air washers are popular during the heating season.

Care must be taken to avoid freezing temperatures in the spray chamber. A preheat coil is usually used to keep the temperatures above freezing. The water spray, in addition to cleaning the air, also serves as a humidifier.

Comfort cooling systems which condense moisture out of the air make use of the wet evaporator surfaces as a filtering device.

22-36 ODOR

Vapors and odors frequently form a large part of atmospheric air contaminants. Most odors are gases, and filters, even electrostatic ones, will not remove them.

Some odors can be removed by cooling the gases to their condensation or freezing temperature. Some can be removed by oxidation (and by ultraviolet ray treatment).

Others can be removed by chemically combining them with other chemicals, by diluting them with air or by absorbing them into a liquid or adsorbing them into a solid.

Both vapors and odors can be removed with activated charcoal. Activated alumina with potassium permanganate has also been used. There are other chemicals which have the ability to absorb and/or destroy odors.

22-37 CARBON FILTERS

A filter made of activated carbon will remove solid particles as well as odor-causing gases and bacteria. This type of filter is being used in air conditioners and in refrigerators with considerable success. Fig. 22-76 shows an activated carbon filter assembly. The carbon in activated charcoal form is made from various substances, including such materials as carbon

Fig. 22-76. Activated carbon filter and air purifier. A—Frame. B—Activated carbon elements.   (Connor Engineering Corp.)

from refining petroleum, coconut shells and other carbon-producing substances. This charcoal will adsorb as much as 50 percent of its weight in foreign gases.

It is possible to rejuvenate used carbon filters. However, this is usually done by the manufacturers who remove the carbon and process it for reuse.

22-38 ULTRAVIOLET LIGHT

Ultraviolet light of 14,000 microwatt CM^2 will kill most bacteria in a fraction of a second. Such ultraviolet lamps are available. One should avoid looking at or being exposed to these rays. The effects are harmful when one is exposed to them for any length of time.

The lamps are installed in the return air duct. Access doors must be equipped with safety-off switches in case someone opens these doors. The rays must cover the full cross-section of the duct to be effective. Fig. 22-77 shows a typical ultraviolet ray lamp.

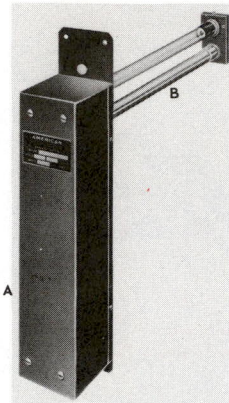

Fig. 22-77. Ultraviolet ray unit designed for duct installation. A—Electrical power supply. B—Two tubular ultraviolet ray lamps. (American Ultraviolet Co.)

There are two types:
1. Low ozone-producing type.
2. High ozone-producing type.

These lamps reduce microorganisms such as bacteria and viruses.

Pure air requires both air sanitization and deodorization. Ultraviolet light is available in two sizes, domestic and commercial. Each type can use either one, two or three lamps.

The odor control types produce ozone. About 20 percent of ultraviolet ray lamps should be high ozone lamps. However, some authorities claim one should not exceed .05 ppm of ozone for continuous occupancy. Install the lamp tubes across the line of airflow. See Fig. 22-78.

Put the unit in the longest main return duct of the system where the air speed is the slowest. It is best to aluminize the inside of the duct to reflect rays back into the air stream.

Lamps should remove about 90 percent of bacteria in domestic and commercial systems and 98 percent of the bacteria in hospital systems. In operating rooms and recovery

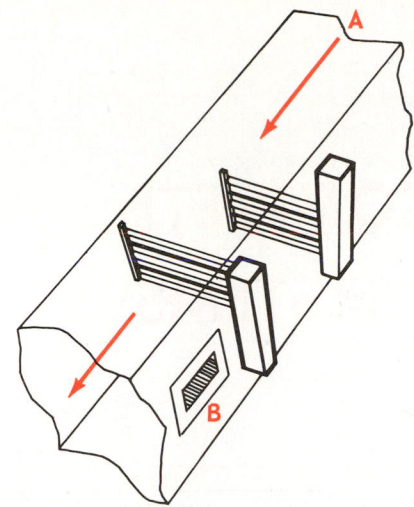

STERILE CONDITIONING UNITS INSTALLED
ACROSS AIR FLOW OF TYPICAL
AIR CONDITIONING SYSTEM.

Fig. 22-78. Two banks of ultraviolet lamps installed in cold air return duct. Rays are directed the length of duct. A—Airflow. B—Protected inspection opening.

rooms, 100 percent removal is needed. An inspection glass window may be used because glass does filter out the ultraviolet rays. One should clean the tubes once a month. The lamps operate about 7500 hours before replacement is needed. The air to be treated should be 70 F. (21 C.) minimum or special transformers are needed.

22-39 AIR CURTAINS

Air curtains are often used at garage doors and the like. They may be used on doors opened frequently for either winter use in cold climates or for summer use in hot climates.

A rather powerful blower is connected to a source of warm or cold air. As a door is opened, this air is directed through openings which provide a narrow stream of warm or cooled air across the entire door area. This air flow stops any natural flow of air from the building to the outside air or from the outside air into the building.

22-40 REVIEW OF SAFETY

It is especially important to remember that the conditioned air must contain enough oxygen to support life. The carbon dioxide content must be kept to a minimum.

Always be careful when working with or handling metal duct material. Use gloves with metal inserts when handling the material. Use stepladders with nonskid bases.

Be sure that the pressure in a duct is low before opening a duct door. If the door bursts open, it may injure someone.

Fans, motors and belts are potential safety hazards. When these units are operating, protective shields or guards should be provided for protection. When adjusting, be sure main power switch is off and locked in off position before handling.

Be careful that objects do not fall into a revolving fan. The object (nut, bolt, tool) may become a dangerous projectile. Always spin a fan by hand to check if it is free to move before turning on the power.

Electrostatic air filters require high voltage to charge the dust particles. Be sure that the current is turned off before servicing them.

All air conditioning equipment is provided with safety controls which cut out the burner if the bonnet (plenum chamber) temperatures get too high. If filters become clogged, the airflow may be reduced, which may cause the temperature limit control to cut off the heat source.

Fan belts and fan motors sometimes fail. A failure in the fan drive will result in an overheated furnace. Check to make sure that the temperature limit controls are in satisfactory condition. Avoid exposure to ultraviolet lights.

22-41 TEST YOUR KNOWLEDGE

1. Is it possible to have a hydronic system and duct system combination?
2. What is meant by negative pressure?
3. What type of air contaminant is molecular in size?
4. What is the volume of 70 F. (21 C.) air at atmospheric pressure if it has 50 percent relative humidity?
5. Does air have specific heat?
6. How is latent heat removed during the cooling season?
7. Is increasing the relative humidity at a constant db temperature an evaporation process?
8. When a comfort cooling air duct has condensation on its outer surface, does the duct air become warmer?
9. How much fresh air should each person receive?
10. What three methods cool a person?
11. What is voltage of ionizing wires of an electronic filter?
12. How efficient are 1-in. throwaway filters?
13. Why is an air velocity of 25 ft. per minute important?
14. Does air have inertia?
15. List two ways smaller ducts may be used to distribute heated air and maintain good heating season temperatures.
16. What is meant by the unit pressure drop system?
17. What is meant by .05 in. (1.3 mm) water column?
18. What is meant by a total pressure drop system?
19. What is a manometer?
20. What is induced draft?
21. What is forced draft?
22. What is the principal value of a variable pitch pulley?
23. What is the purpose of a water spray in an air conditioning system?
24. What is the purpose of the fire damper?
25. List three uses for air ducts.
26. What is the purpose of a grille?
27. Why must one be careful when using ultraviolet light?
28. What is the principal impurity removed by an activated carbon filter?
29. How is electricity used to remove dust from the air?
30. How are most electronic filters cleaned?

Chapter 23

HEAT PUMPS AND COMPLETE AIR CONDITIONING SYSTEMS

All refrigerating systems are heat pumps. They move heat from one place to another. The refrigerating unit picks up heat at a low temperature and releases it at high temperature.

"Climate-control" is the term often used to describe a space in which an ideal climate is maintained. This is accomplished by controlling the movement, temperature, humidity, and cleanliness of the air.

This controlled climate can be maintained for the benefit of people, structures, furnishings, animals, vegetation, or food.

23-1 HEAT PUMP

Heat pump theory rests on the principle that heat will move from a higher temperature to a lower temperature. Thus, if a heat transfer coil can be kept at a lower temperature than its surroundings, it will pick up heat.

If the evaporator of a refrigerating system is mounted outdoors and is operated at a refrigerant temperature of −18°C (0 F), it will remove heat from the air even when the outside temperature is −12° to −9°C (10 to 15 F). If, after it has evaporated, the refrigerant is compressed to a temperature of 49° to 60°C (120 to 140 F), the hot refrigerant will release heat to surrounding space (inside a building).

Then, if, by using a system of valves, the evaporator is changed into the condenser and the condenser becomes the evaporator, heat can be removed from the living zone during hot weather and discharged outdoors. Fig. 23-1 shows the pressure-heat diagram for a heat pump.

From the foregoing, one can see that evaporators and condensors are heat transfer devices which can be used for cooling (picking up heat) or heating (releasing heat).

The principle of using a refrigeration unit as a heating mechanism too was first proposed by Lord Kelvin over 100 years ago.

The heat pump is sometimes called a reverse-cycle mechanism. However, the cycle is not actually reversed, only the evaporator and condenser are interchanged. Therefore, the name "reverse-cycle" is not technically correct.

The heat pump may be used for many purposes. Heating water and heat recovery from industrial processes are but two. Even the defrosting of evaporators, using the hot gas defrosting method, is a form of heat pump. Both the compression system and the absorption system can be adapted as heat pumps.

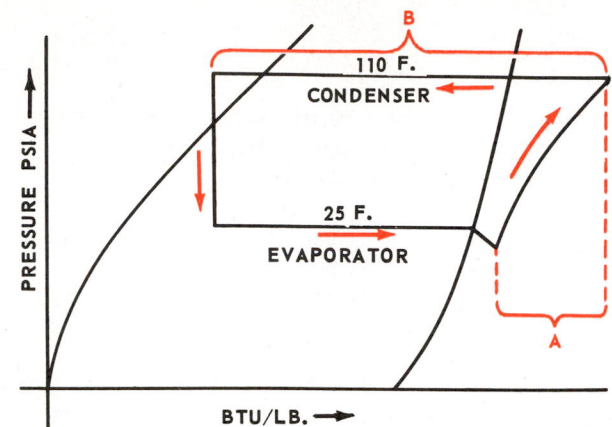

Fig. 23-1. Typical heat pump cycle used for heating. Refrigerant evaporates outside at −4°C (25 F.) in 2°C (35 F.) air. Same refrigerant condenses at 43°C (110 F.) in condenser located in air duct. A—Heat of compression. B—Heat released to house.

Fig. 23-2 illustrates a typical heat pump installation with the heat pump operating on a heating cycle. Fig. 23-3 illustrates the same heat pump operating on a cooling cycle. The basic principles are described in Chapter 19.

Large systems of 100 to 1000-ton capacity are in use as are window units of 1/2 to 2 tons. Self-contained systems of 2 to 25 tons are common.

23-2 HEAT PUMP OPERATION

Operation of the heat pump is like any other compression cycle. The principal parts of the systems are:
1. Compressor.
2. Condenser.
3. Liquid line.
4. Two refrigerant controls.
5. Evaporator.
6. Suction line.
7. Motor control.
8. Reversing valve.
9. Two check valves.

Notice that two check valves, a reversing valve and two refrigerant controls are needed in the simpler of the heat pumps to change from summer cooling to winter heating. Fig. 23-4

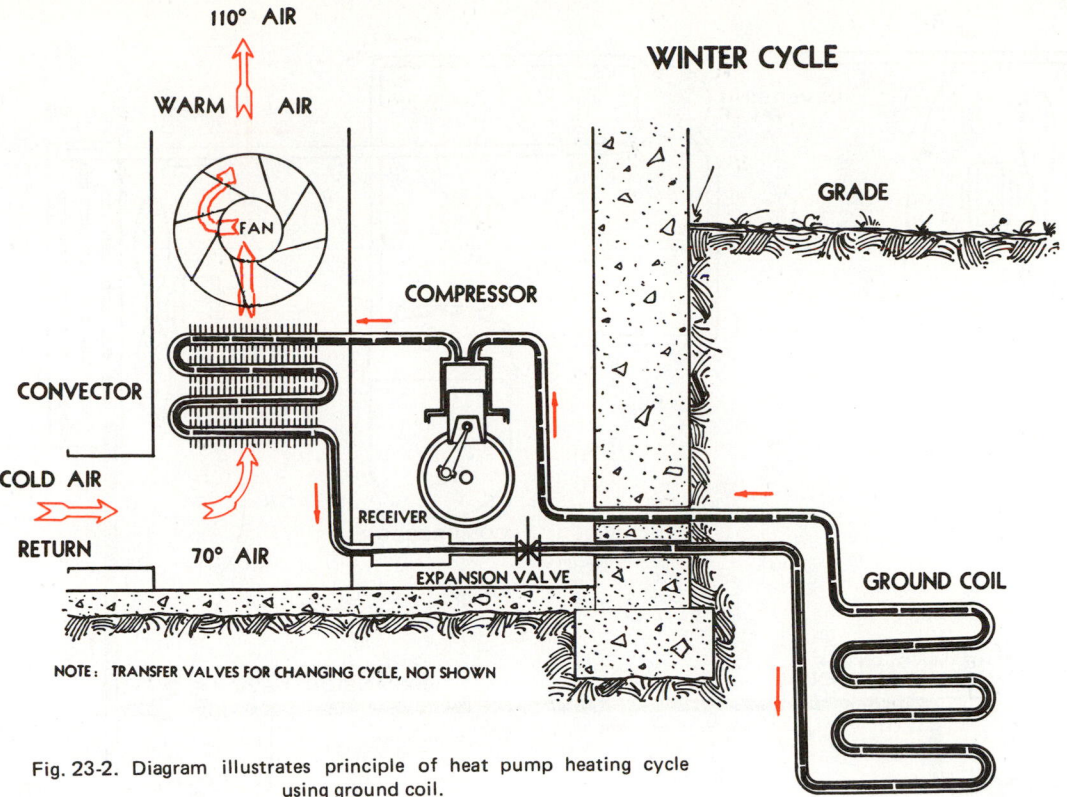

110° AIR

WARM AIR

FAN

COMPRESSOR

CONVECTOR

COLD AIR

RETURN

70° AIR

RECEIVER

EXPANSION VALVE

GRADE

GROUND COIL

NOTE: TRANSFER VALVES FOR CHANGING CYCLE, NOT SHOWN

Fig. 23-2. Diagram illustrates principle of heat pump heating cycle using ground coil.

shows a heat pump operating as a comfort cooler. The liquid refrigerant bypasses the TEV at the lower left. (It goes through the check valve bypass.) The TEV at the lower right is the refrigerant control in use.

By turning the reversing valve 90 deg., refrigerant flow is reversed in all lines except the two lines leading into the compressor. These valves are often called four-way valves because of the number of openings.

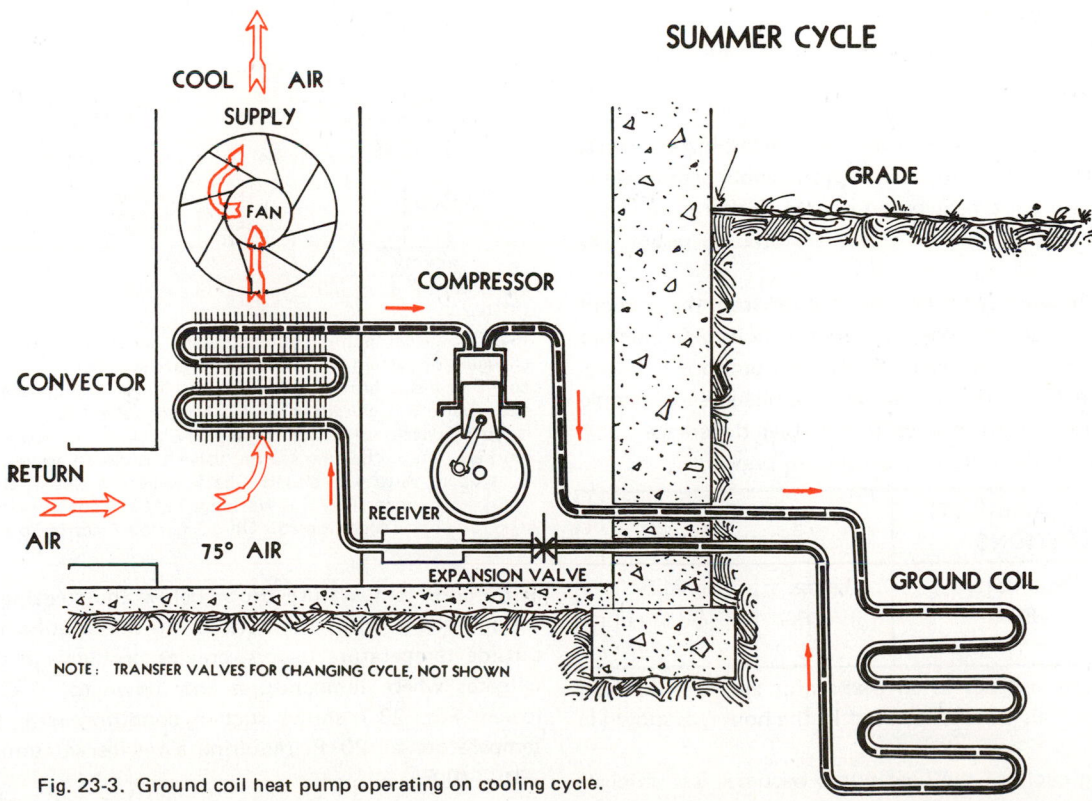

SUMMER CYCLE

COOL AIR

SUPPLY

FAN

COMPRESSOR

CONVECTOR

RETURN

AIR

75° AIR

RECEIVER

EXPANSION VALVE

GRADE

GROUND COIL

NOTE: TRANSFER VALVES FOR CHANGING CYCLE, NOT SHOWN

Fig. 23-3. Ground coil heat pump operating on cooling cycle.

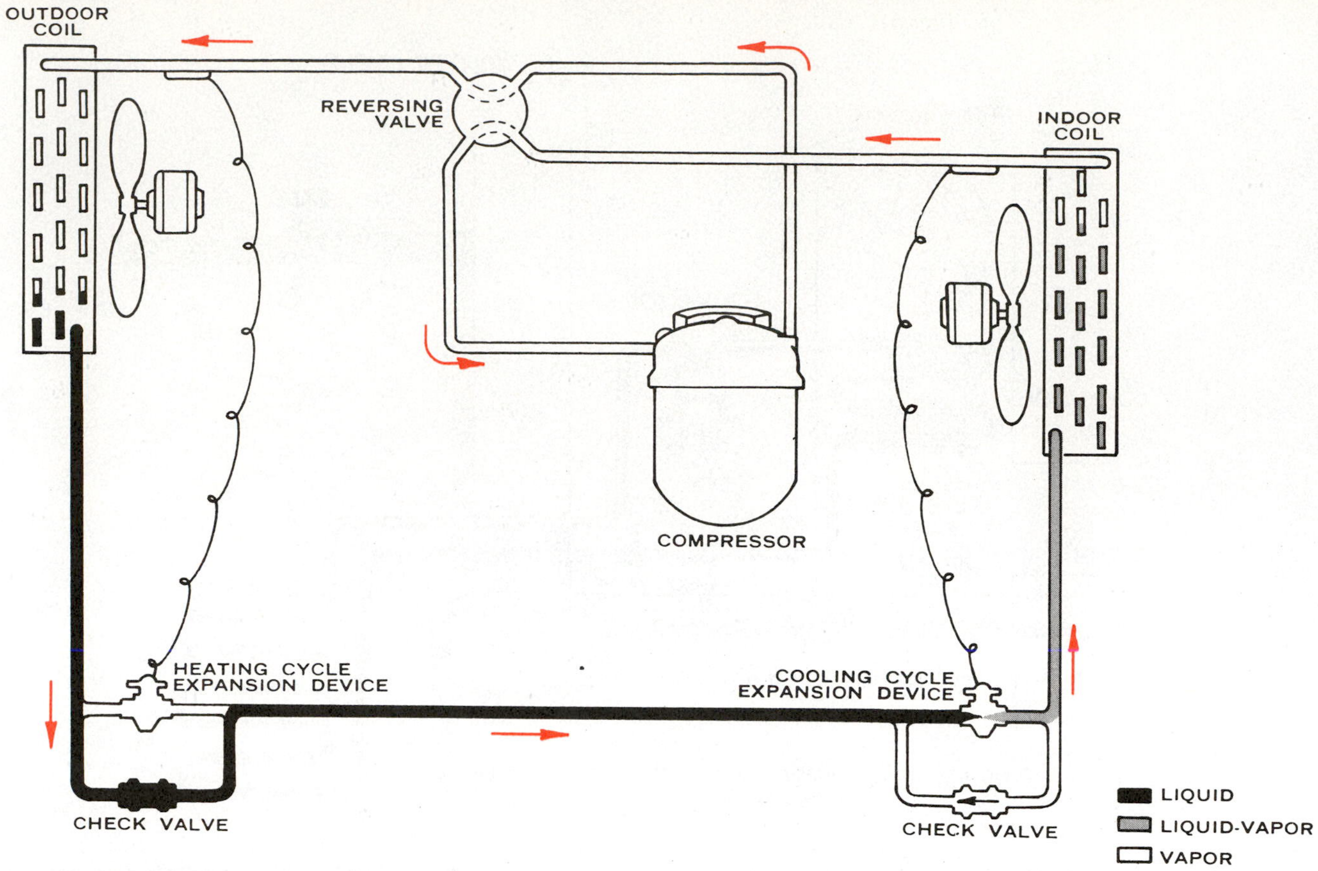

Fig. 23-4. Heat pump. Both heat transfer coils are blower coils. TEV refrigerant controls are used. Note four-way reversing valve. System is operating as comfort cooling unit with outdoor coil as condenser and indoor coil as evaporator. (Westinghouse Electric Corp.)

23-3 HEAT PUMP CYCLES

Actually, the heat pump operates in two different cycles:
1. Heating cycle.
2. Cooling cycle.

The same mechanism is used for both cycles, but the travel of refrigerant is reversed to change from cooling to heating. Fig. 23-5 shows a basic heat pump system. Either hand valves or thermostatically controlled valves are used to reverse the cycle.

During the heating cycle, heat is removed from the ambient (surrounding) air and released inside the house. Remember that heating is not usually needed until the outdoor temperature is less than 18°C (65 F.). Heat from appliances and people living in the house usually makes up this small difference.

For example, if the following conditions prevail,

OUTSIDE (AMBIENT) CONDITIONS		INSIDE CONDITIONS	
Temp. 50 F. 10°C	Humidity 80%	Temp. 72 F. 22°C	Humidity 50%

the outdoor coil will act as an evaporator and pick up heat from outdoors. This heat is released in the house, as shown in Fig 23-6.

The heating cycle of the heat pump becomes less efficient

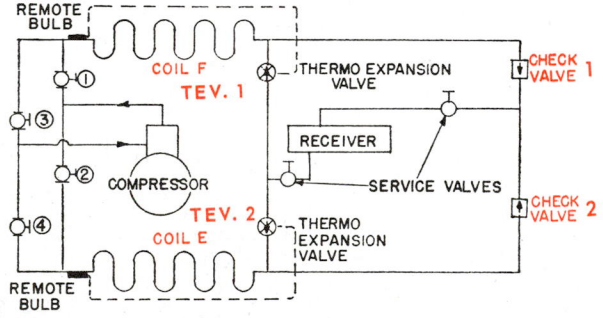

Fig. 23-5. Heat pump equipped with hand valves to permit manual changing of system from heating to cooling. Coil F is outside coil and coil E is inside heat transfer surface. During cooling season, valve 1 is open; valve 3 is closed; valve 2 is closed; valve 4 is open. Check valve 1 is open, check valve 2 is closed. TEV 1 is not working; TEV 2 is working. During heating season, valve 1 is closed, valve 3 is open, valve 2 is open, valve 4 is closed. Check valve 1 is closed; check valve 2 is open; TEV 1 is working; TEV 2 is not working. (Alco Controls Div., Emerson Electric Co.)

as outdoor temperature drops below the freezing level. This action, along with an increase in the heating load as the outside temperature lowers, creates problems in those colder climates where temperatures cool down to −7°C (20 F.) or lower. Fig. 23-7 shows such a condition with the outside temperature at 20 F. requiring a refrigerant temperature of −18°C (0 F.).

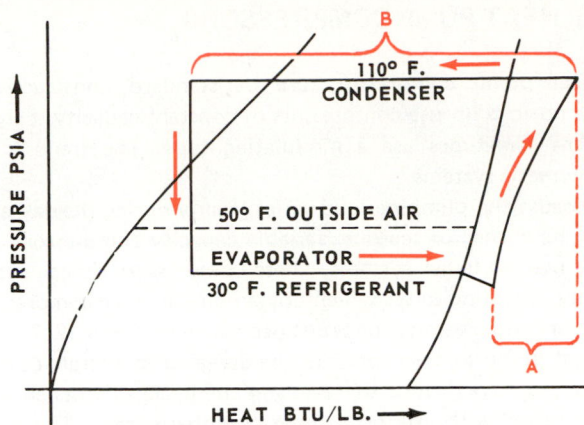

Fig. 23-6. Pressure-heat chart for a heat pump serving as heating system with outside temperature at 1°C (50 F.). Refrigerant is evaporating at −1°C (30 F.). Refrigerant is condensing at 43°C (110 F.). The Btu ratio of A to B is coefficient of performance (COP).

Note that A has increased with very little increase in B, which means that the coefficient of performance is less. Also, with the refrigerant boiling at −18°C (0 F.), the evaporator will frost rapidly. Some way of frequent defrosting would have to be found.

If the ambient conditions are within 5°C (10 F.) and 10 percent relative humidity (RH) of these conditions, very little treatment is needed due to the heat lag and time lag in controlling the variables. But if the ambient temperatures were to increase over this amount, for example:

OUTSIDE (AMBIENT) CONDITIONS		INSIDE CONDITIONS	
Temp.	Humidity	Temp.	Humidity
85 F. 29 °C	75%	72 F. 22°C	50%

a cooling cycle would be required. Heat and humidity would be removed from the inside of the house and released outdoors as shown in Fig. 23-8.

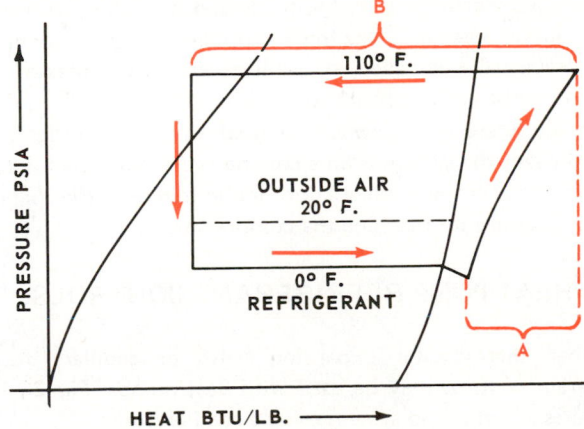

Fig. 23-7. Pressure-heat chart of heat pump heating cycle in operation when outside (ambient) temperature is −7°C (20 F.). This requires a −18°C (0 F.) refrigerant temperature. Note increase in heat of compression.

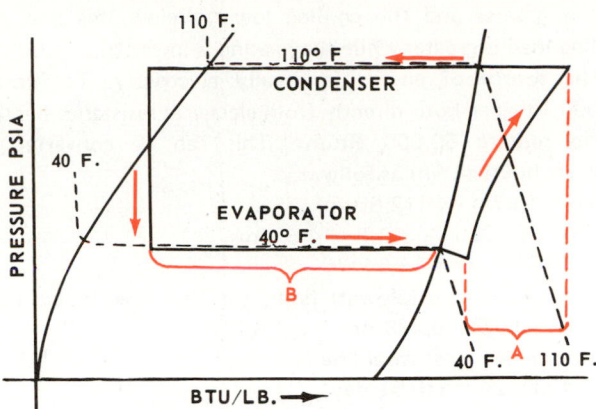

Fig. 23-8. A pressure-heat chart showing cooling cycle for heat pump. A—Heat energy of compression. B—Heat energy removed from air (cooling and dehumidifying). Note that A is about one-third of B, meaning that three times more heat is moved than is used to pump. This is a Coefficient of Performance (COP) of 3:1.

On the psychrometric chart, the apparatus would affect the air conditions as shown in Fig. 23-9. Point A is the condition of the ambient (surrounding) air and represents air at 29°C (85 F.) and 75 percent RH. The line from A to B shows how the air is cooled to 100 percent humidity. Line B to C shows how the air is further cooled and how moisture is removed from the air.

If 100 percent fresh air is being conditioned, point C represents the air as it leaves the evaporator. Then, as the air mixes with the air in the house, or as it mixes with air being brought into the duct system (recirculated air), point D is reached.

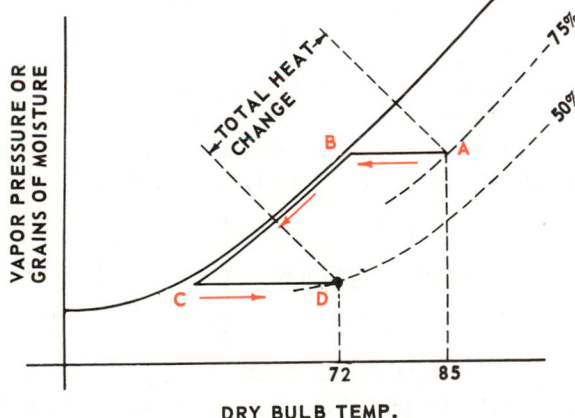

Fig. 23-9. A psychrometric chart showing effect on air in house when using cooling cycle of heat pump.

23-4 HEAT PUMP PERFORMANCE

Heat pump units have been designed for single rooms, for complete houses and for industrial uses. The unit takes the place of separate comfort cooling and heating apparatus.

Cooling load is about 60 percent of heating load for most of the United States. Above the 37 deg. parallel, the heating load is greater and the cooling lower. Below this line, the cooling load is greater while the heating is lower.

The source of energy is usually electricity. To furnish 50,000 Btu per hour directly from electrical resistance heating would require 50,000 Btu/hr. This can be converted to kilowatt hours (kWh) as follows:

Since 1 kWh = 3412 Btu/hr,

kWh = 50,000 ÷ 3412 = 14.6, thus,

50,000 Btu/hr = 14.6 kWh

At three cents a kilowatt hour, the heat load would cost

14.6 × 0.03 = 0.438/hr.

For one day, cost would be

0.438 × 24 = $10.51/day.

For one 30-day month, the cost would be

$10.51 × 30 = $315.30/mo.

This cost appears high. However, the actual cost is much less than this, as the heat load for a house averages much less than this value. The average outdoor temperature is much higher than the design temperature.

The 50,000 Btu/hr is based on 21°C (70 F.) indoor and −18°C (0 F.) outdoor. If the average outdoor temperature is 2°C or 35 F. (quite common), the heat load would be 25,000 Btu/hr ($\frac{70}{35}$). The cost would, therefore, be cut in half, or $157.65/mo. The heat pump reduces this cost considerably.

Electricity drives the heat pump compressor. The refrigeration cycle — if the proper temperatures are used — permits the condenser to release three to four times as much heat as it takes in electrical energy to drive the compressor. This coefficient of performance (COP) means that one kilowatt hour of electrical energy driving the compressor can, by using the heat pump, release not 3412 Btu/hr, but 3412 × 3, or 10,236.0 Btu/hr.

This coefficient of performance can be further increased by using a heat source that is warmer than the outside air. For example, well water, lake water, and the ground itself may be used to provide the heat for the heat pump.

If a well furnishing water at 16°C (60 F.) can be found, the coefficient of performance can be raised to as much as 4 or 5. This factor will lower the cost of heating with a heat pump.

23-5 HEAT PUMP SYSTEMS

Use of heat pumps for residential heating and cooling is increasing. Hermetic units have been developed in many designs and in a wide range of sizes.

The hermetic system is ideal for heat pump installations as it is so simple to install. Some are equipped with service valves, suction and discharge mufflers and other special features which make for quiet, reliable operation and long life.

The coil mounted inside the house is usually a standard finned type with a blower. The outside coil comes in a variety of designs. Choice depends on the medium (substance) in which the coil is placed — air, water or earth.

The "outside" coil is not always located outside the structure. It may be located inside but ducted to the outside if it is an air coil. See Para. 23-10 and Fig. 23-24.

23-6 HEAT PUMP COMPRESSORS

Heat pump compressors are of standard construction. Units up to 5 hp use compressors of constant capacity. Larger systems sometimes use a modulating type. The trend is to use hermetic systems.

Because the pumping load varies greatly during the day and with the change of seasons, variable capacity compressors are being used in larger systems. These compressors change capacity by operating valves which unload one or more compressor cylinders into clearance pockets. See Chapters 4 and 12.

Heat pump compressors must be designed to operate during the unusual conditions of reversing the cycle. Motors should be protected with internal temperature thermostats. The compressor must be designed to handle some liquid slugging without injury. Crankcase heaters are often used to protect the compressor from liquid refrigerant buildup during low temperature operating periods. Suction line accumulators are often used to protect the motor compressor.

23-7 HEAT PUMP MOTORS

Heat pumps of 1/3 to 1-ton capacity use standard single-phase motors. They usually have a starting capacitor. Single-phase motors should be operated at 240V, if possible, so that smaller wires can be used. Three-phase motors are preferred in units over one ton capacity, mainly for electrical economy.

Motors in hermetic systems are usually well insulated. They can tolerate some voltage change and have temperature sensor protectors built into the motor windings.

23-8 HEAT PUMP MOTOR CONTROLS

A double set of automatic controls is usually required for the heat pump. One thermostat must be designed for heating during cold weather conditions. Another is needed for warm-weather cooling. In most cases, these two controls are mounted in one casing or housing. Humidistats are not used on all models. However, they are needed for complete automatic control.

Lines are normally fitted with brazed or welded connections. Silver brazing is the most popular. Flexible connections are usually installed in the lines at the compressor to eliminate noise and to absorb some vibration.

Usually there is a receiver to hold the extra refrigerant needed when the system is on a cooling cycle. An accumulator is often used on the main suction line to minimize the chance of liquid refrigerant reaching the compressor.

23-9 HEAT PUMP REFRIGERANT CONTROLS

Either thermostatic expansion valve or capillary tube refrigerant controls may be used with heat pumps. Fig. 23-10 illustrates a heat pump system using a capillary tube.

Two thermostatic expansion valves or capillary tubes may be used on a heat pump system. The thermostatic expansion valve is usually the pressure-limiting type. Only one control will be operational at a time. The other will be in the liquid line.

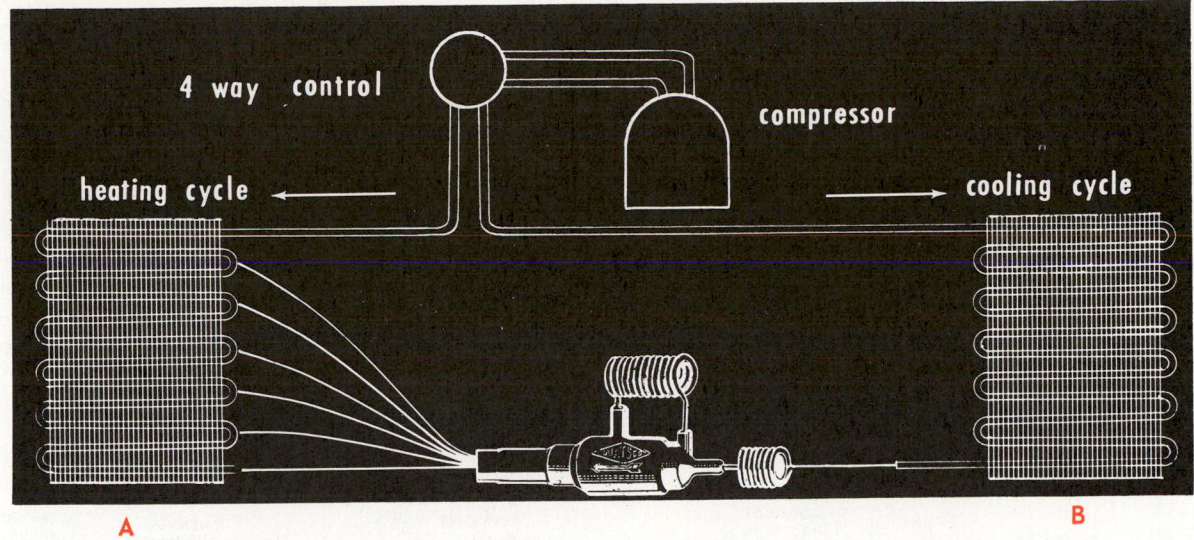

Fig. 23-10. Heat pump system using capillary tube refrigerant controls. A—Indoor coil. B—Outdoor coil. Arrows show refrigerant flow to the left when system is used for heating and to the right when system is used for cooling.   (Watsco, Inc.)

If an air coil is used, the valve should be cross charged to provide more efficient operation at low air temperatures. The changeover valves are much simpler in units that use the capillary tube control, because the flow of refrigerant can be reversed through the tube for pressure-reducing purposes. In this type of installation, a strainer must be installed at both ends of the tube. Both the thermostatic expansion valve (TEV) and the capillary tube control are described in Chapter 5.

Fig. 23-11 is a diagram of a heat pump using a capillary tube and a solenoid reversing valve. As shown, it is operating on the cooling cycle. The wiring diagram in Fig 23-12 shows the control contacts in position for this cooling cycle. It also shows the interlocking of the four-way solenoid-operated reversing valve and the compressor.

Fig 23-13 is a simple diagram of a heat pump using a capillary tube and a solenoid-operated reversing valve. It is shown operating on a heating cycle. The wiring diagram in

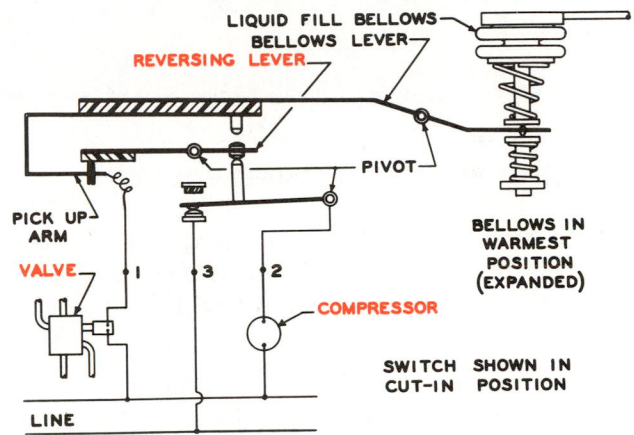

Fig. 23-12. Wiring diagram of heat pump showing electrical connections for cooling cycle.

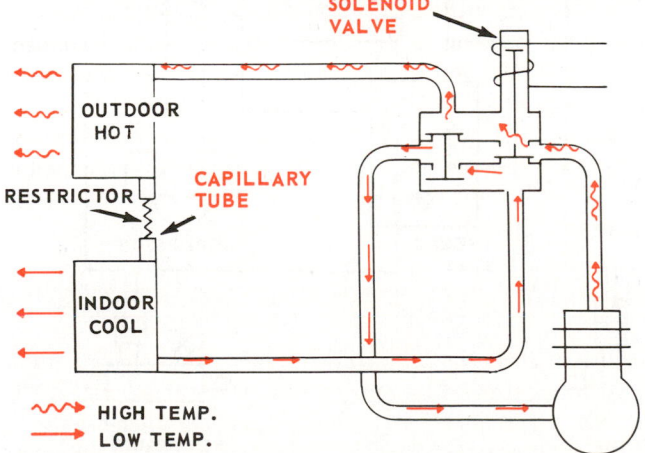

Fig. 23-11. Heat pump with capillary tube refrigerant control operating on cooling cycle (indoor coil). Bent arrows mean high temperature. Straight arrow lines mean low temperature.   (Ranco, Inc.)

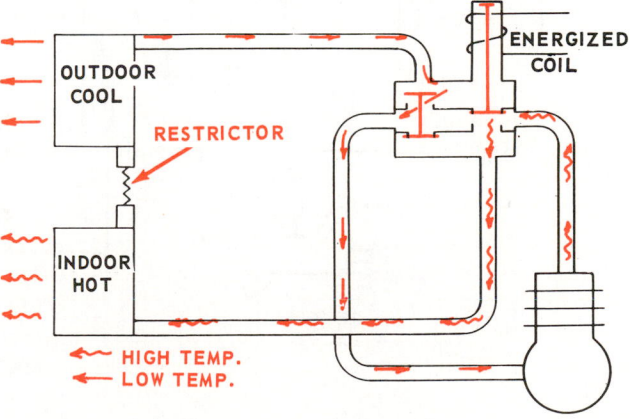

Fig. 23-13. Heat pump operating on heating cycle. Crooked arrows indicate high temperature. Straight arrows indicate low temperatures. Note use of four-way valve.

Heat Pumps and Complete Air Conditioning Systems / 805

Fig. 23-14 shows the control contacts for this heating cycle. Note the change in contact position of the reversing lever.

The thermostat in Fig. 23-15 operates the four-way valve and also controls temperature. Note the three terminals. The wiring of heat pumps depends mainly on whether the motor is connected directly to the line or whether some kind of starter (magnetic or transformer) is used. Note parts A, B and C in Fig. 23-15.

may be operated automatically, manually or electrically (through solenoids).

Some units use three-way valves either manually or electrically operated. These valves have one opening to the compressor, one to the condenser, and one to the evaporator. Two of these valves are needed to operate the unit.

Other units use a four-way valve to reverse the flow of refrigerant. The valve is operated by the movement of one valve stem. This stem closes and opens several ports in one valve body. Its operation may be manual or electrical. The complete system is easily reversed with one of these valves. They are popular in small tonnage conditioners such as win-

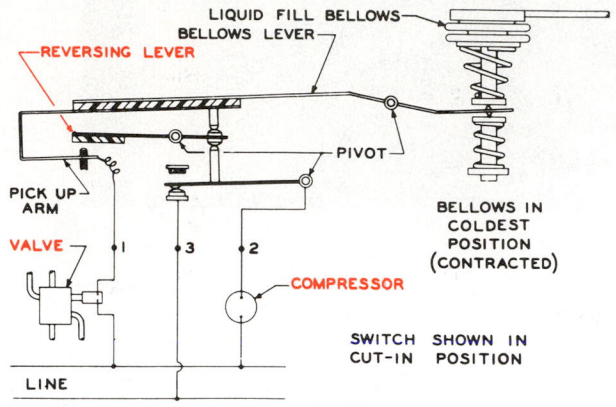

Fig. 23-14. Electrical connections during heating cycle of heat pump.

23-10 HEAT PUMP REVERSING VALVES

Several different types of special reversing valves are used in heat pumps. If refrigerant flow is manually reversed, at least six one-way valves are needed. The special reversing valves

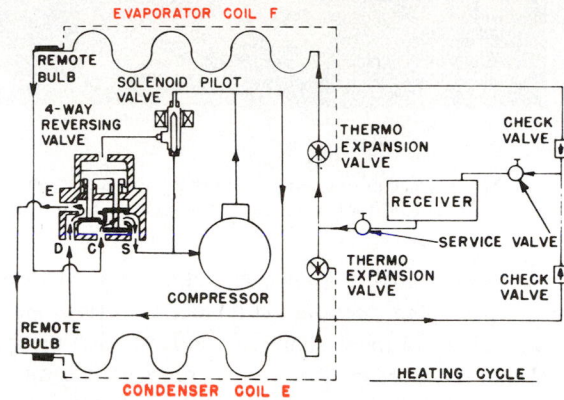

REVERSE CYCLE SYSTEM USING A 4-WAY REVERSING VALVE

Fig. 23-16. Heat pump schematic diagram of system which uses one four-way reversing valve. A solenoid pilot valve is used to operate valve. It is operating on heating cycle. E—Indoor coil. F—Outdoor coil. (Alco Controls Div., Emerson Electric Co.)

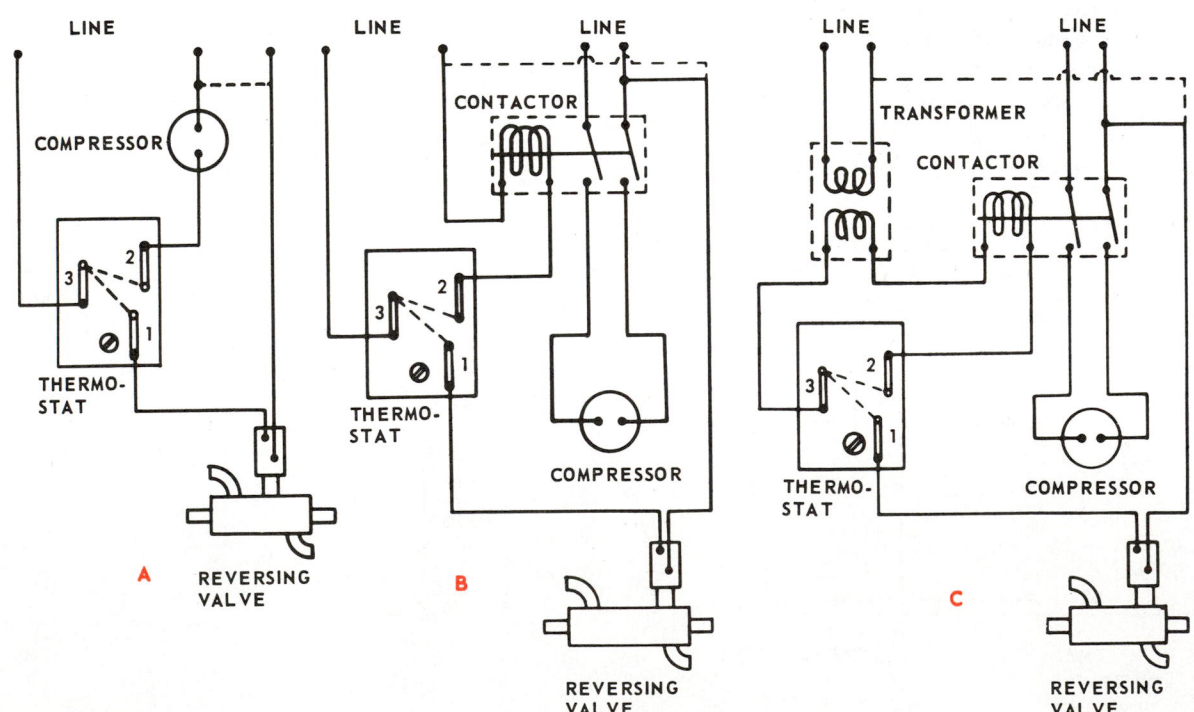

Fig. 23-15. Three wiring diagrams for heat pumps: A—Thermostat controls motor circuit. B—Circuit uses line voltage coil magnetic contactor. C—Circuit diagram using low-voltage coil magnetic contactor.

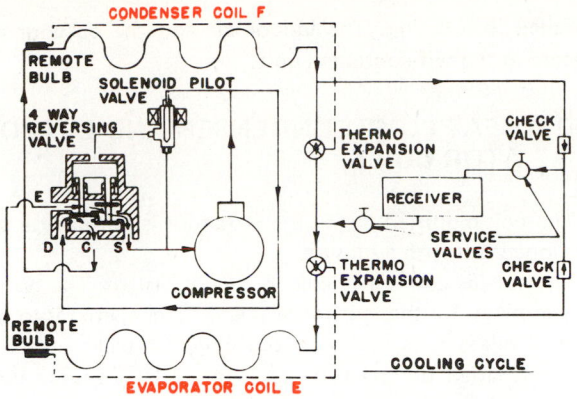

REVERSE CYCLE SYSTEM USING A 4-WAY REVERSING VALVE

Fig. 23-17. Heat pump schematic diagram shows special four-way reversing valve in use. It is operating on cooling cycle. E—Indoor coil. F—Outdoor coil. (Alco Controls Div., Emerson Electric Co.)

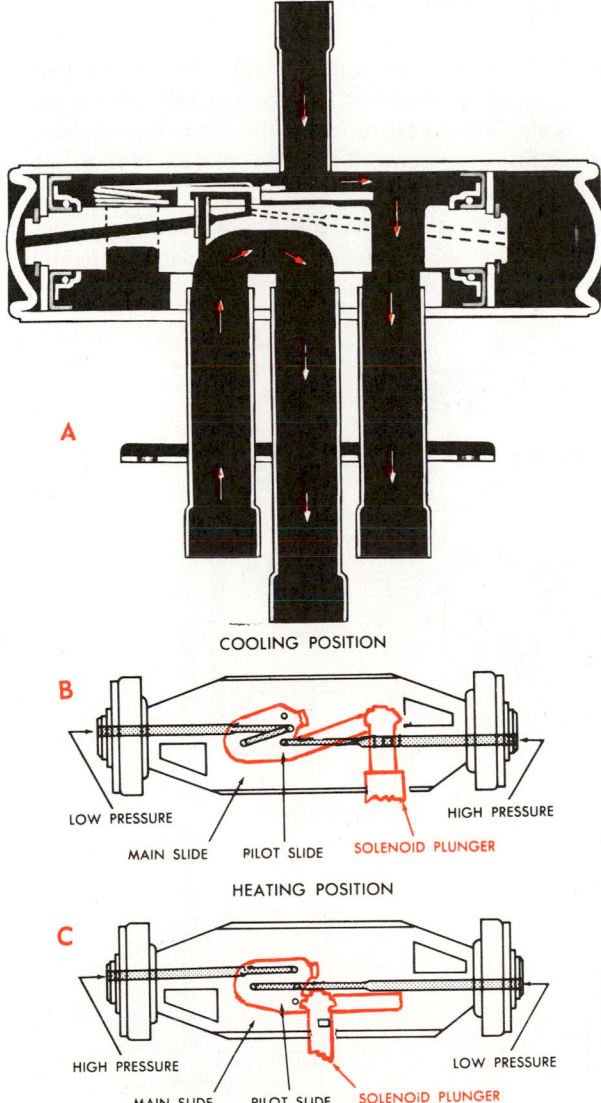

Fig. 23-18. Cutaway of pilot valve mechanism used to operate four-way reversing valve. Note position of solenoid plunger in heating position when solenoid is energized. A—Side view of four-way valve. B—Pilot mechanism and piston in cooling position. C—Pilot mechanism and piston shown in heating position. (Robertshaw Controls Co., Milford Div.)

dow units and other air-to-air systems.

One kind of four-way valve operated by pressure is shown in Fig. 23-16. A solenoid valve controls the pressure at the top portion of the reversing valve. When the solenoid pilot valve is energized (caused to move), the compressor low-side pressure pushes on the four-way valve. All three internal valves are lifted, producing the heat cycle. Coil E is the indoor coil, while the coil at F is outdoor.

When the solenoid pilot valve circuit is opened, either manually or by a thermostat, the solenoid valve closes and the three valves inside the four-way valve drop. Refrigerant flow to the coils is reversed.

The heat pump has a thermostatic expansion valve and a check valve on each coil. When the system is reversed the refrigerant flow bypasses the thermostatic expansion valve of the coil serving as the condenser.

During the cooling cycle, the upper thermostatic expansion valve is used. The refrigerant cannot travel through the companion check valve.

Fig. 23-17 illustrates this same system operating as a cooling unit. Coil F becomes the condenser and is releasing heat to the outdoors. Coil E becomes the evaporator.

Another popular reversing control is the solenoid pilot-operated sliding port valve. It is controlled by a solenoid valve. The solenoid coil is energized on the heating cycle. The pilot slide pivots and changes the flow of pressure. Pressure moves the main slide as seen in Fig 23-18. In the cooling cycle the solenoid is not energized. Fig. 23-19 shows the reversing valve installed in a heat pump system. The system is on a cooling cycle. When valved for a heating cycle, the system is set up as in Fig 23-20. Figs. 23-21 and 23-22 show this sliding valve in more detail.

Fig. 23-23 is a diagrammatic sketch of a self-contained system. Two separate circulating fans are used. In each illustration, the white coil is the warm one (condenser) and the black

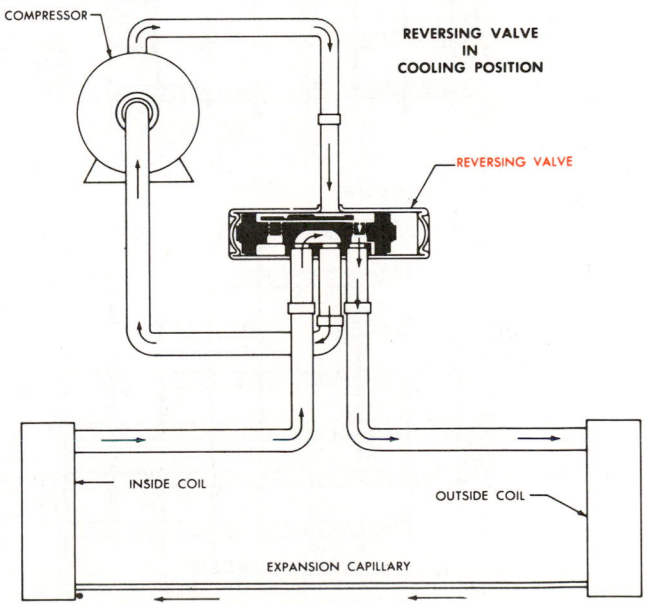

Fig. 23-19. Basic heat pump cycle using reversing valve. Note flow of refrigerant when valve is set for cooling cycle.

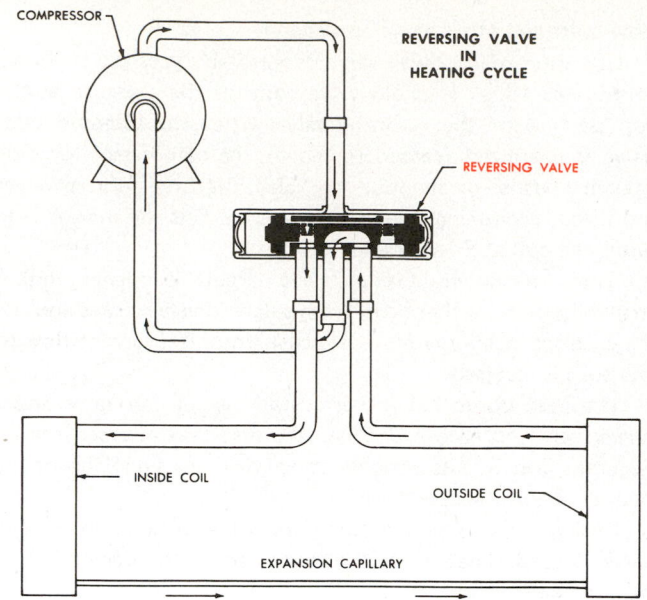

Fig. 23-20. Heat pump circuit with reversing valve set for heating cycle.

coil is the evaporator.

These units are installed either in the basement or on the first floor of a home. Fig 23-24 shows a typical first-floor utility room installation. Both coils and the motor compressor are installed in a casing. The indoor air and the outdoor air are ducted to the self-contained unit.

23-11 HEAT PUMP CONDENSER AND EVAPORATOR COILS

The coil mounted inside the structure is normally a standard finned coil with a blower.

The outside coil is available in several designs. The type used is determined by the substance (water, stone, etc.) into which the coil releases its heat or from which it picks up its heat. Coils, classified by the type of external heat source used to obtain the heat of evaporation, are:
1. Air coil.
2. Water coil.
3. Ground coil.

23-12 HEAT PUMP AIR COIL

Of all the heat pump outdoor coils, the least expensive is one that picks up heat from (or releases heat to) the outside air. It is also the easiest to install. In climates where outdoor temperatures range from −7°C (20 F.) to 43°C (110 F.), the air coil has a number of advantages.

A standard heat transfer coil, the air coil has tubing for primary surface with extended fins bonded to the tubing.

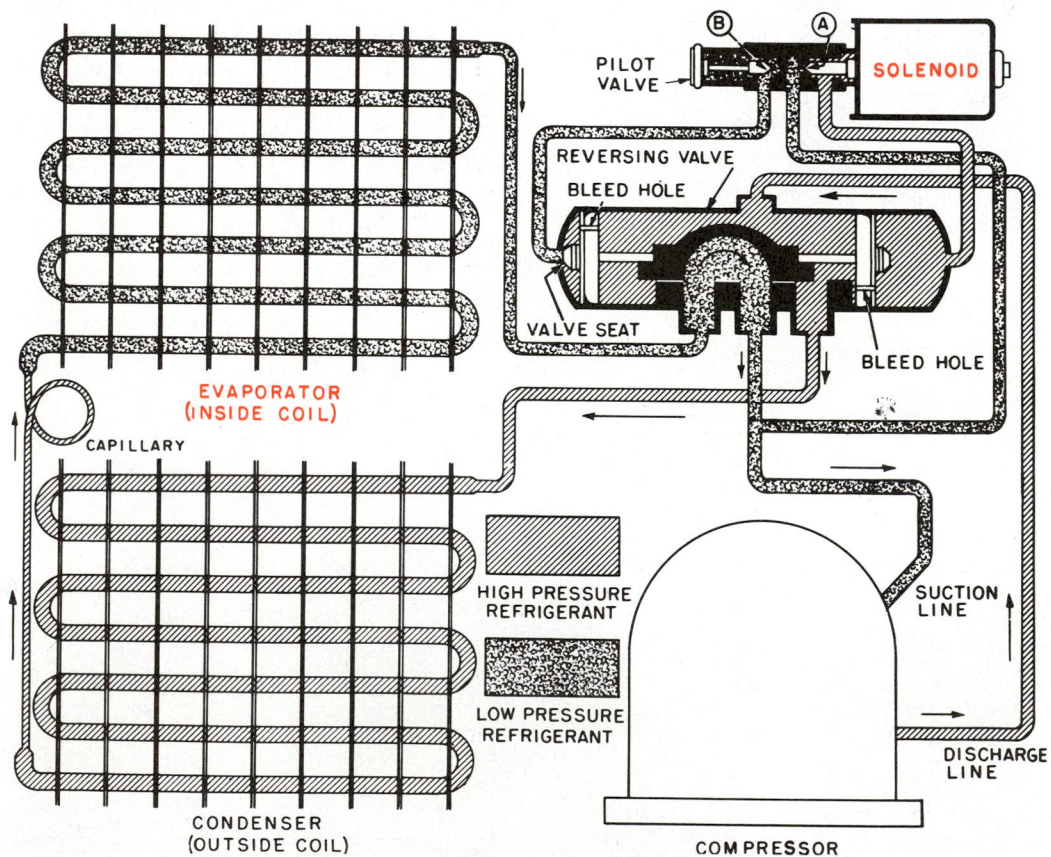

Fig. 23-21. Capillary tube refrigerant control heat pump using solenoid valve controlled four-way valve. Indoor coil is serving as an evaporator.

PILOT
VALVE
SOLENOID
B
A
REVERSING VALVE
BLEED HOLE
BLEED HOLE
VALVE SEAT
CONDENSER
(INSIDE COIL)
CAPILLARY
HIGH PRESSURE
REFRIGERANT
LOW PRESSURE
REFRIGERANT
SUCTION
LINE
DISCHARGE
LINE
EVAPORATOR
(OUTSIDE COIL)
COMPRESSOR

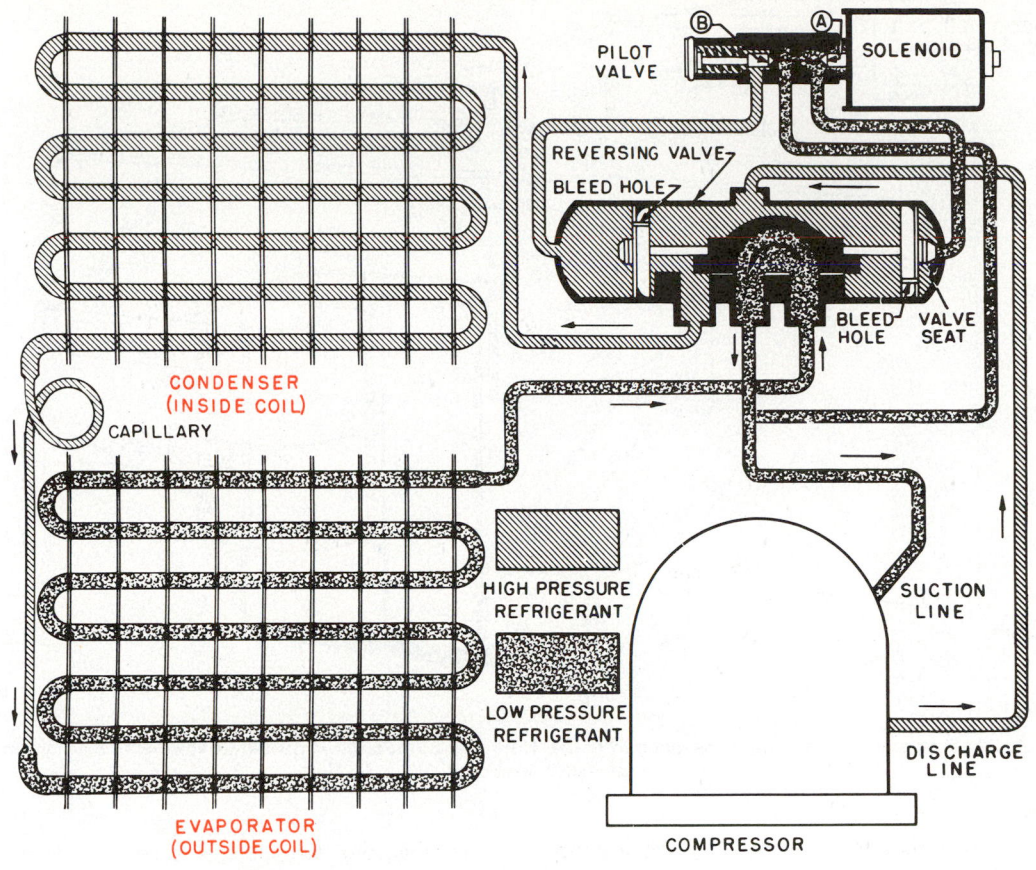

Fig. 23-22. Capillary tube type heat pump using indoor coil as condenser.

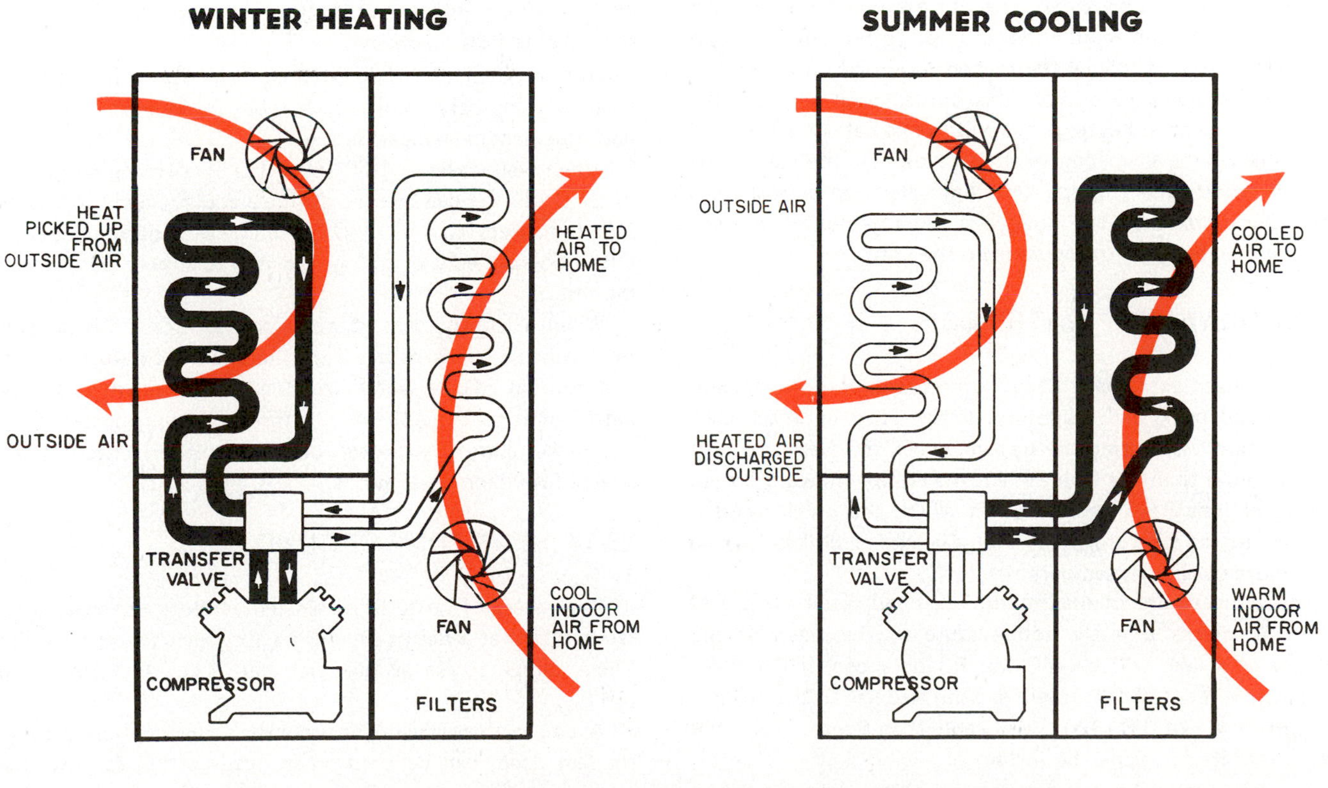

WINTER HEATING
FAN
HEAT
PICKED UP
FROM
OUTSIDE AIR
HEATED
AIR TO
HOME
OUTSIDE AIR
TRANSFER
VALVE
COOL
INDOOR
AIR FROM
HOME
FAN
COMPRESSOR
FILTERS
SUMMER COOLING
FAN
OUTSIDE AIR
COOLED
AIR TO
HOME
HEATED AIR
DISCHARGED
OUTSIDE
TRANSFER
VALVE
WARM
INDOOR
AIR FROM
HOME
FAN
COMPRESSOR
FILTERS

Fig. 23-23. Air and refrigerant circuits in heat pump during heating and cooling operation. (Westinghouse Electric Corp.)

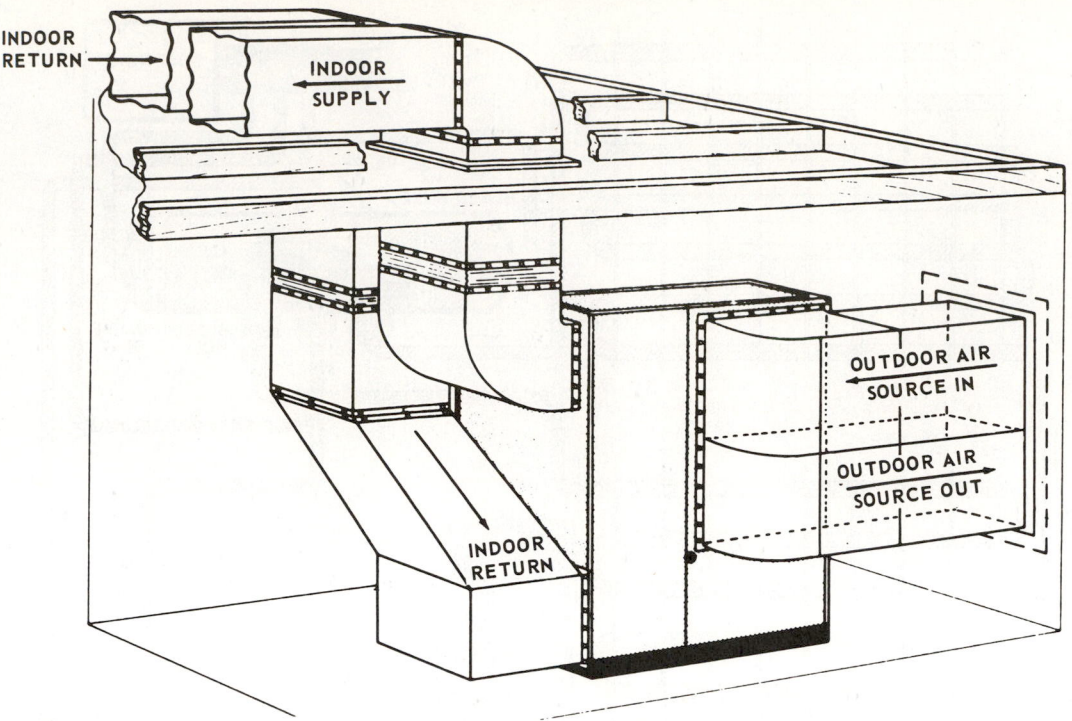

Fig. 23-24. Diagram of heat pump installation in house. Note that outdoor air is ducted to heat pump mechanism located inside house.

Along with a blower, it is mounted in a housing that protects it from the weather. The housing may be mounted on the outside wall of the building, on the roof or in a shelter near the building.

During the heating cycle, the outside coil tends to frost and ice up in certain weather conditions. To prevent this, most heat pumps have a special thermostat mounted on the coil. If frost or ice starts to gather, this control will shut off the system and electric heating coils will start a defrost action. Or, the cycle will reverse long enough to defrost the coil. Some systems use a timer. It operates the system on defrost for a time. During defrost, the drain also must be heated to make sure drain water can flow away from the unit.

23-13 HEAT PUMP WATER COIL

As has been explained, a heat pump becomes more efficient as the condensing and evaporator temperature approach each other. Therefore, during the heating cycle, if a warmer source can be found than the ambient air, the pump will work better.

Installations have been tried in which the outside coil is lowered to the bottom of a nearby lake. The coil so located has a more consistent temperature.

The temperature of maximum density of water is 4°C (39 F.). This means that the temperature of the water at the bottom of a lake will be 4°C (39 F.) or above. If the water cools below 4°C (39 F.), it will expand and rise to the surface. A temperature of 0°C (32 F.) will cause it to freeze. This is the reason that ice forms on the surface of a lake.

The lake would have to freeze solid before the temperature of the water at the bottom could be lower than 4°C (39 F.).

From this, one can see that a lake — even one covered with ice — may be a reservoir of heat. So too with well water; it may provide an efficient heat pickup unit for heating and a good heat dissipator during the cooling cycle. The cost of the well is a disadvantage, but this will probably soon be offset by the lower cost of operation.

In one popular method, well water is drawn from the well, used, and quickly returned. A tube-within-a-tube or a shell-and-tube heat exchanger is used.

When well water is 16°C (60 F.), a condensing temperature of 27°C (80 F.) can be used during a cooling cycle. An evaporator temperature of 4°C (40 F.) can be used during heat cycle. Often flowing wells or springs are used to supply water for heat pumps.

When a lake or ground water coil installation is used, one of the major concerns has been the chemical nature and mineral content of the water. It often was unknown and proved detrimental to the coil. A coil made of cupronickel (copper and nickel alloy) has proven to be corrosion resistant and desirable for water installations.

23-14 HEAT PUMP GROUND COIL

Regardless of latitude or air temperature changes, ground temperature at a depth of four to six feet changes very little. These temperatures average between 4°C (40 F.) and 16°C (60 F.).

A coil with sufficient heat transfer surface, buried four to six feet deep, can be used an an outdoor coil for both heating and cooling cycles. Fig. 23-2 illustrates a ground coil system as it operates during the winter. Actually, ground coils are installed flat. Also, the air return is usually a split-air system with a fresh air makeup duct.

Fig. 23-3 illustrates the same basic system used as a comfort cooling mechanism. In practice, valves would be used to flow the condensed refrigerant through the liquid receiver. From there the refrigerant would go through the expansion valve from left to right before the reduced-pressure refrigerant enters the evaporator in the duct.

Installations have been made using a combination of air coils and either a lake coil, well coil or ground coil. The air coil is used alone when the outdoor temperature permits efficient operation. But, when the outside air becomes too cold or too warm, the auxiliary coil may be connected into the heat pump air coil system.

23-15 HEAT PUMP DEFROSTING

The defrosting cycles on many heat pumps are accomplished at the beginning of the defrost cycle by supplementary electric heat. Also in common use today are defrost controls based upon temperature differential, air pressure differential and time and temperature differentials. Another recent method is a unit that depends upon the electrical loading of the heat pump's condenser fan motor. It electrically measures the mechanical load of an induction motor. The cutout signal is obtained from a remote thermistor clipped to the liquid line or from a pressure or temperature control. This method is used in order to maintain a comfortable indoor temperature during the defrosting process, which is the reversal of refrigerant flow.

The reversing of the heat pump cycle, for defrosting when the compressor is running, causes a strain on the compressor and motor. A heat pump on low ambient temperatures cools the compressor motor with liquid refrigerant. If the defrosting system fails to return the liquid rapidly from the outdoor coil at the start of the defrost cycle, it fails to provide adequate and rapid defrosting. Also, when the unit shifts to defrost, a large surge of liquid is returned to the compressor, often causing liquid floodback or slugging.

To compensate for the liquid line floodback, suction line accumulators are used by many manufacturers (see Fig. 23-25). The accumulators act as a receiver for the excess liquid and return it to the compressor in a vapor form.

23-16 SUPPLEMENTAL RESISTANCE HEATERS

It is standard practice to size heat pumps to handle the cooling load. This size may or may not be adequate to handle heating needs.

In southern climates, a unit big enough for cooling can probably supply adequate heat with little supplementary aid. However, in very cold weather, additional heat must be supplied from another source. The supplementary electric resistance heating unit is one such source. It requires little space. Blowers distribute the heated air evenly throughout a room. Such units may be installed in the blower-coil section of the heat pump or directly in the supply air duct of an indoor system.

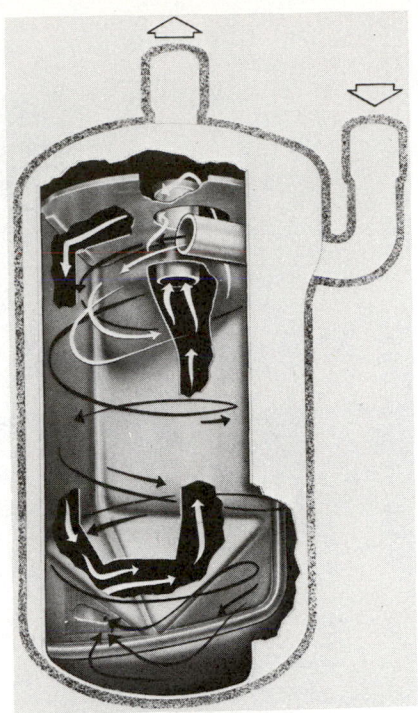

Fig. 23-25. Suction line accumulator. Note that inlet is near top of accumulator and suction line vapor is made to swirl as it leaves accumulator at top. This prevents liquid refrigerant from entering the compressor. (Tecumseh Products Co.)

23-17 HEAT PUMP INSTALLATION

Basically, a heat pump is the same as a refrigerating system. Most of the installation instructions in Chapters 11 and 14 apply to heat pumps as well. The units must be carefully leveled. Electrical service must be of the correct voltage and phase. Wiring must be large enough to carry the load without a critical voltage drop.

In cold climates the heat pump should be installed above the anticipated total snow height for that area or approximately 8 to 18 in. above ground. It is also desirable in extremely low-temperature areas to install a defrost timer instead of standard outdoor thermostats. Always advise the owner to shovel the snow away from the condensing unit.

Commonly misunderstood is the register output temperature of heat pumps. The heat pump output at a register is 41°C (105 F.), whereas oil and gas produce approximately 52°C (125 F.) register temperature. Therefore, the running time will be longer. This causes many first-call customer complaints.

Window and wall units require the same casing mounting techniques described in Chapter 21. It is important to closely follow the manufacturer's instructions.

Fig. 23-26 pictures a heat pump unit which can be installed in an exterior wall of an office or apartment. This type extracts "natural heat" from the outside in the winter for heating the room or apartment and cools it in the summer by absorbing heat from the room and discharging it to the outside atmosphere.

Through-the-wall heat pumps are easy to maintain. The heat pump chassis may be removed for servicing. The indi-

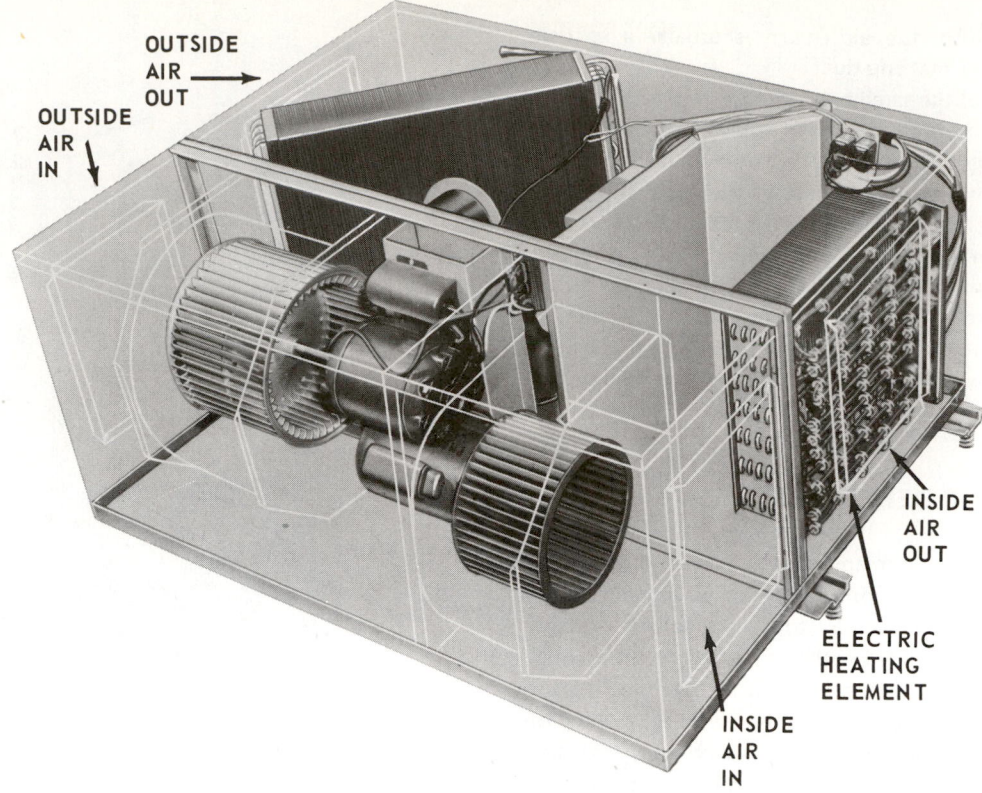

Fig. 23-26. A 3 hp heat pump which can be installed through wall. Note supplementary electric resistance heating element used to increase amount of heating during unusually cold weather. (Fedders Corp.)

vidual heat pump unit is well adapted to apartment units. The owner of a multiple unit does not have to operate a central air conditioning system. Each tenant can set the thermostat to give the heating or cooling desired without affecting others living in the building.

23-18 HEAT PUMP SERVICE

Because the heat pump system is a refrigerating unit, the service techniques of troubleshooting and repair are much the same as the service described in Chapters 11 and 14.

Routine maintenance requires that voltage, current and refrigerant pressures be checked. The heat pump should be cleaned. Blow out the coils and fins, clean the duct passages, oil the fan bearings and fan motor bearings. Clean the fan blades. The unit should be run on both cycles to check the operation of the reversing valve.

Service calls often start with a customer complaint, such as:

1. Unit will not operate on either cooling or heating.
2. Unit runs too much.
3. System is noisy.

The technician must find the cause of the trouble and repair it. For example, a lack of refrigerant indicates a leak. The leak must be located and repaired. A four-way valve that will not operate must be replaced.

Certain checks should be routinely made:

1. Check defrost control.
2. Check and lubricate condenser fan motors.

3. Check refrigerant charge.
4. Check all electrical connections for tightness.

23-19 TWO-SPEED HEAT PUMPS

Two-speed heat pumps, Fig 23-27, are produced by some manufacturers to get a higher energy efficiency ratio (EER).

Fig. 23-27. A two-speed heat pump. It operates at low speed for normal loads and at high speed for heavy loads. (Lennox Industries, Inc.)

For normal heating or cooling loads the compressor operates at low speed. Under exceptional conditions when the temperature is very high, the compressor will operate at high speed for maximum cooling. Likewise, when the temperature is very low, the compressor will operate at high speed for maximum heating.

One of the major differences between a single-speed motor and a two-speed motor is its adaptability. At high speeds it operates as a two-pole motor; at low speeds it is a four-pole motor.

23-20 HI/RE/LI® SYSTEM

An innovation in heat pump cycles is the Hi/Re/Li® system developed by Westinghouse. It differs from the conventional heat pump system in many ways. Fig. 23-28 illustrates some of the main operational differences. They are as follows:

1. The location of the refrigerant control is not on the evaporator but on the condensing unit with the thermal bulb mounted on the liquid line. Thereby, liquid is controlled as it leaves the condenser rather than when it enters the evaporator.
2. The heat exchanger-accumulator in the suction line allows more subcooling of the refrigerant. It separates liquid-vapor and allows dry vapor and oil to enter the compressor because the liquid is boiled off by the heat exchanger.
3. A suction line-heat exchanger subcools the condensed liquid refrigerant before it enters the expansion valve.

Through subcooling, the amount of "flash gas" and pressure drop is lowered.

4. The evaporator is completely flooded and utilized for cooling since it is not required to superheat "suction gas" as it is in the normal system.
5. The condenser system can be reduced because it is not necessary to store refrigerant during periods of low load.

The Hi/Re/Li® system uses a reversing valve to change the direction of flow of the refrigerant to fit the needs of either heating or cooling. In addition, it uses a check valve to direct the flow of refrigerant in the proper direction dependent upon whether a heating or cooling cycle is desired. A diagram of this system in its cooling cycle is shown in Fig 23-29.

When charging the system, it is best to first remove all of the refrigerant. Then charge it with the correct amount as indicated in the manufacturer's manual. Install service valves on the suction lines and discharge lines fairly close to the compressor to get the correct operating pressures.

23-21 SOLAR—HEAT PUMP COMBINATION

Solar heat systems (Chapter 27) and heat pumps are being linked for residential use. Several designs are being manufactured.

One system is based on air solar collectors. These collect heat which can be:

1. Routed directly into the residential area.
2. Stored in a bin of rocks or in a thermal storage tank.

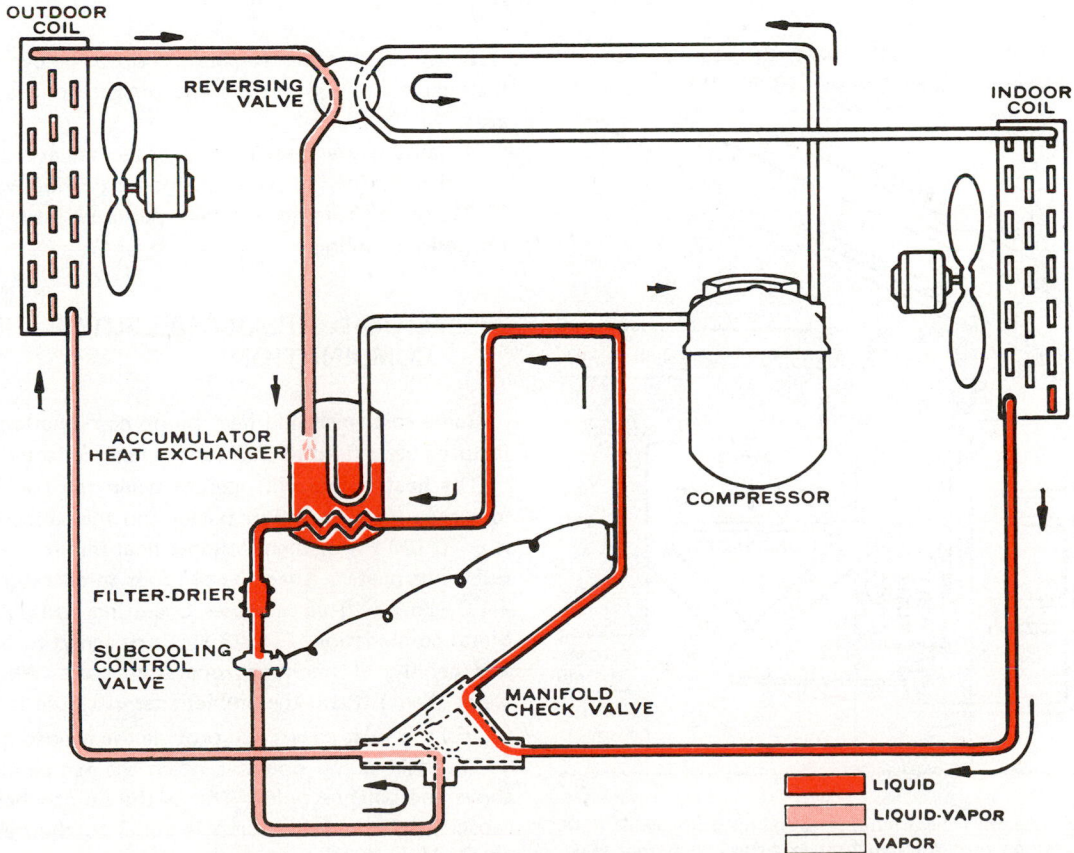

Fig. 23-28. Hi/Re/Li® heat pump system operating as a heating system.

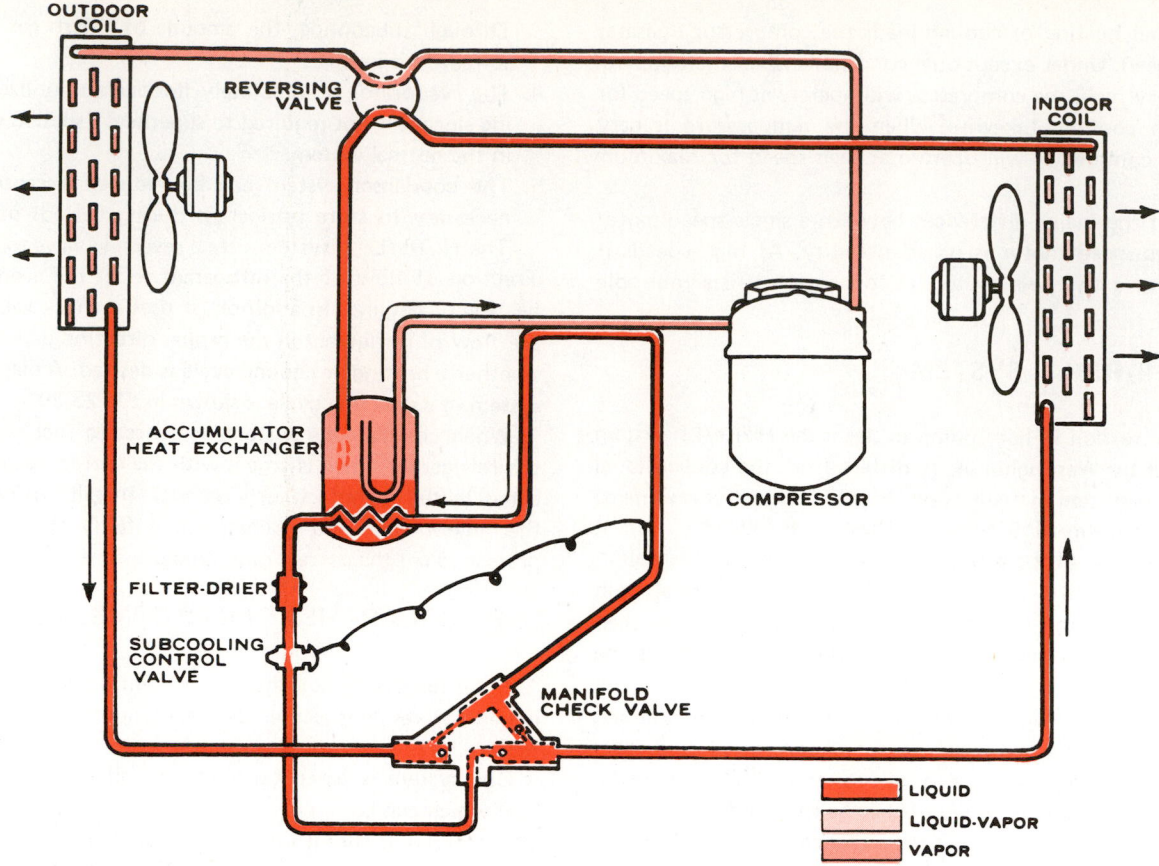

Fig. 23-29. Hi/Re/Li® heat pump system operating as a cooling system. Note how manifold check valve changes or reverses refrigerant flow.

When there is no direct solar heat, the storage area will supply heat for the living area. See Fig. 23-30.

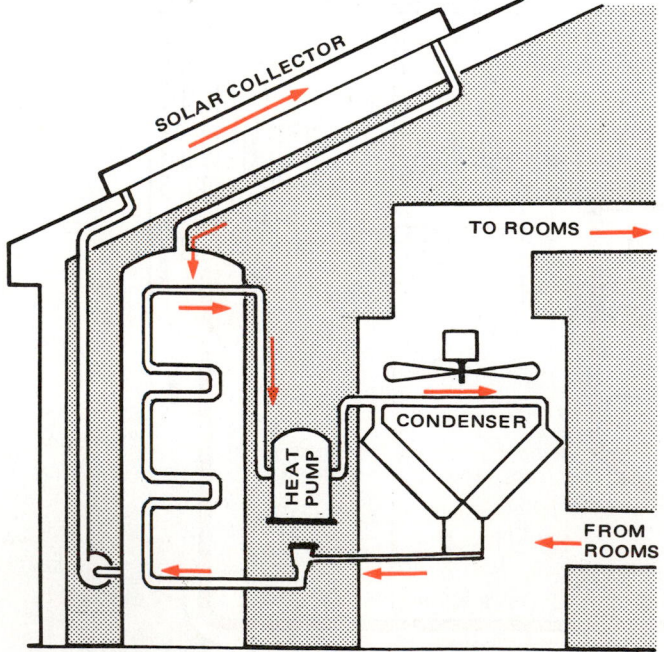

Fig. 23-30. Solar collector — heat pump. Heat pump is circulating fluid from solar collector on roof through heat exchanger in storage tank. Heat pump evaporator is in the storage tank. Condenser is in plenum chamber of furnace.

When the storage temperature drops below 38°C (100 F.), the heat pump will operate and produce the required temperature.

An air-to-water heat exchanger located at the output of the air collectors can provide the domestic hot water heat, Fig. 23-31. In the summer the heat pump will operate to provide the desired cooling.

23-22 LIQUID SOLAR COLLECTOR—HEAT COMBINATION

Some solar heat and heat pump combinations being manufactured use a fluid to transfer the heat collected.

The heat pump will operate when the coefficient of performance (COP) is 1.0 or better and the ambient temperature is −7°C (20 F.) or higher. Some heat pumps use supplemental resistance heaters when the outdoor temperature drops below −4°C (25 F.). This increases operating costs. A solar — heat pump combination, Fig. 23-32, is designed to provide heating and cooling at minimum operating costs. When the COP is lower than 1.0 and the ambient temperature is less than −7°C (20 F.) the solar panels will provide the needed heat.

The heat pump operates when the temperatures are at or above the balance point. This is the point where the heating capacity of the heat pump is equal to the heat loss of the home.

The solar panels are sized to meet only the supplemental

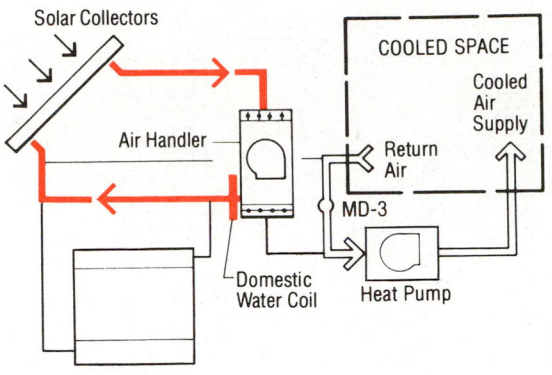

Fig. 23-31. This air collector solar heat pump system has provisions for domestic hot water supply. (Solaron Corp.)

heating needs of the heat pump. The heat from the solar panels is stored in a water storage tank. The stored heat is used when needed. An electric immersion heater provides supplemental heat if needed.

The heat pump will draw heat from the solar panels until the outdoor temperature drops to the point where supplementary heat is required. Then the supplementary heat will be provided by the solar energy that has been collected and stored.

'The system is a basic standard heat pump unit installed

in a residential home. The solar panel collector is installed outside, and a hydronic coil is placed in the supply air duct. The hydronic coil receives its heat from the thermal energy storage tank which holds the heated water.

The solar supplemented heat pump system combines five basic systems:

1. A standard heat pump system that will provide cooling when required.
2. A standard heat pump system that will produce heat and operate whenever the COP is above 1.0.
3. A solar circuit. This unit consists of solar panels containing a solution of ethylene glycol and water that passes through the shell side of a tube-and-shell heat exchanger. The heat that has been collected by the solar panels is not stored in the heat exchanger.
4. A storage circuit. A thermal energy storage tank uses a storage pump to circulate water through the tube side of the heat exchanger. Here the water is heated and returned to the storage tank.
5. A hydronic circuit. A hydronic pump circulates the heated water from the storage tank to the hydronic coil in the supply duct.

23-23 COMPLETE AIR CONDITIONING SYSTEMS

The complete air conditioning system controls the temperature, humidity, air movement, and air cleanliness. Details of the construction and location of the components of the complete air conditioning system vary with the design of the system. Some of these are described in the following paragraphs.

23-24 THROUGH-THE-WALL SYSTEMS

The complete through-the-wall system (heating, cooling, filtering, and moving air) is mounted in the wall. Also called a unitary system, it is popular in moderate climates. Some models are mounted flush to the outside of the building while some extend outside a few inches. They must be mounted firmly to the studding of the building and leveled in all directions. Joints must be weather stripped and leakproof. Use caulking or gaskets.

Through-the-wall units which also have electric heat usually require 240V power. Gas-fired or oil-fired systems use 120V power. Sometimes a separate 240V power circuit is used for the refrigerating system.

All of these units trap the comfort-cooling condensate and evaporate it to help cool the air-cooled condenser. The plenum chambers of some are designed to allow direct airflow into the room. Others use a duct for air distribution.

Through-the-wall units must be installed away from combustible material. They provide fresh air intake, combustion air intake, and a chimney flue. They are factory assembled.

23-25 OUTSIDE COMPLETE SYSTEMS

One type of all-season air conditioning for homes has all the systems, heating, cooling fan and filtering, factory assem-

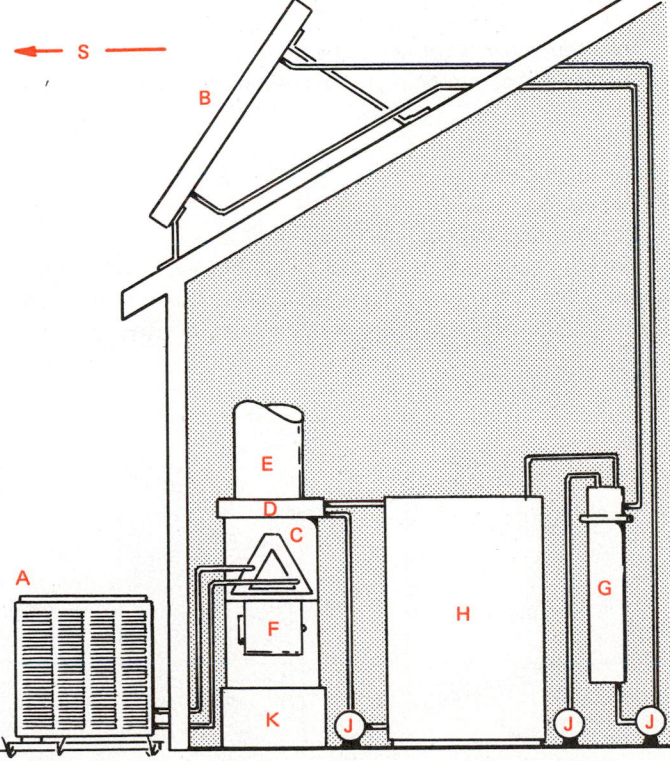

Fig. 23-32. A solar—heat pump combination that provides both summer cooling and winter heating. A—Outdoor section. B—Solar heat collector panel. C—Indoor heat exchanger. D—Hydronic coil. E—Supply duct. F—Blower. G—Heat exchanger. H—Heat storage tank. J—Pumps. K—Return duct. (General Electric Co.)

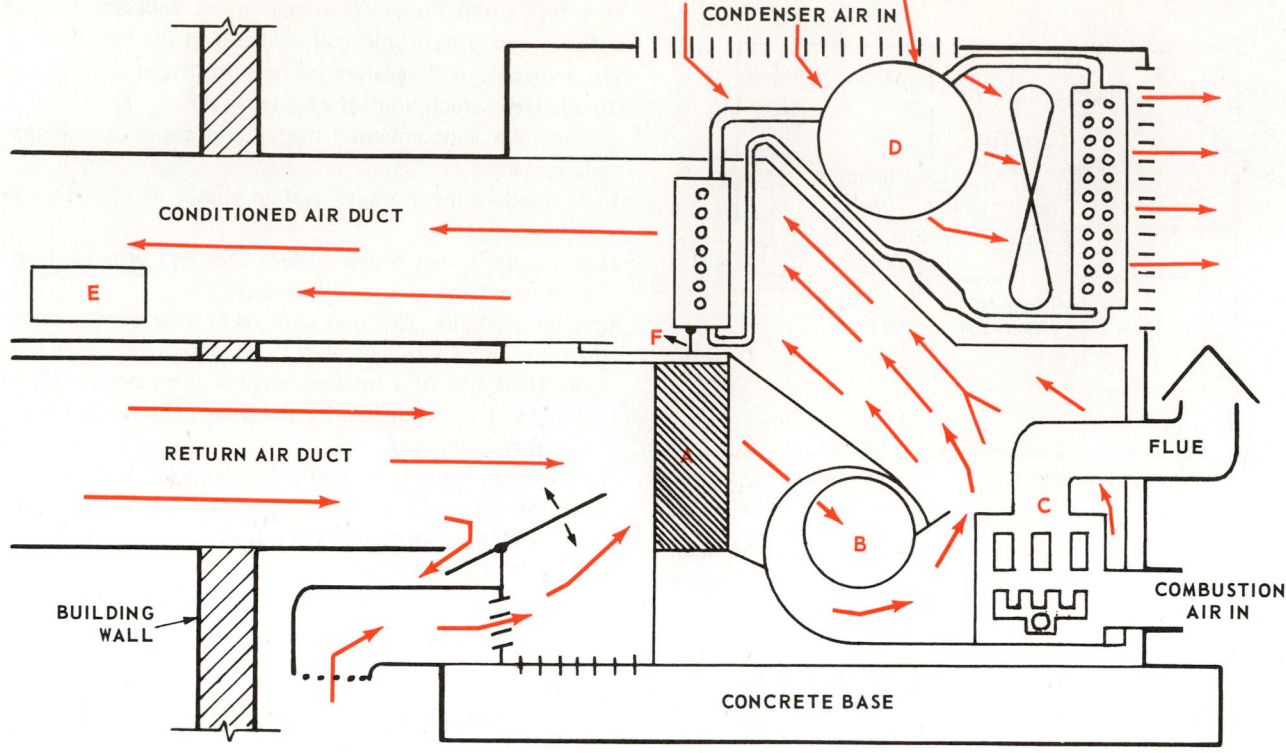

Fig. 23-33. An all-season air conditioning system installed outside house or office. A—Filter. B—Fan. C—Furnace. D—Cooling system. E—Humidifier. F—Air bypass for humidity control during cooling.

bled into one unit. The unit is installed outdoors. Then it is connected to the duct system of the house or building. A conditioned air duct connection and a return duct connection are all that is needed. These systems are an adaptation of rooftop units commonly manufactured for commercial and industrial use. See Fig. 23-33.

The advantages are:

1. The heating system, cooling system, and fan system are factory assembled and tested.
2. The installation consists of mounting the unit, connecting the electric power, the gas line, and the two duct connections. Fig. 23-34 shows two views of this system.

23-26 ROOFTOP UNITS

Rooftop systems for heating and cooling were first developed for flatroof commercial structures such as supermarkets and office buildings.

There are three types:

1. Heating only.
2. Cooling only.
3. Both heating and cooling.

Being factory assembled and tested, the systems are economical. They are easily installed once hoisted to the site. However, the roof must be strong enough to carry the extra

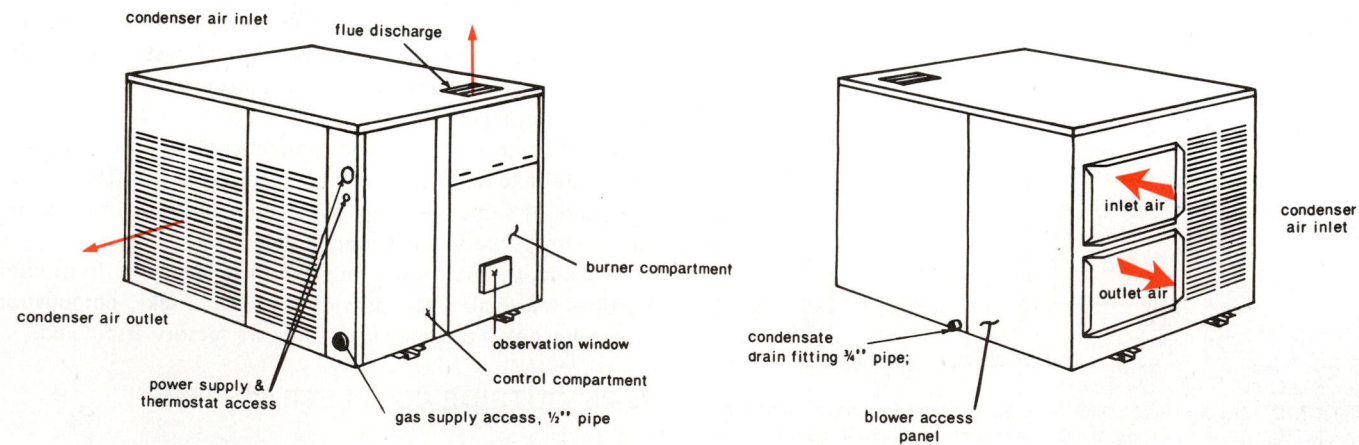

Fig. 23-34. Complete air conditioning system designed for connection to building air conditioning system. (Fedders Corp.)

weight. Openings must be leakproof. Units must be designed for ease of service even in very cold or inclement weather. They are also exposed to the wind which may affect their

Fig. 23-35. Rooftop unit for both cooling and heating. This unit has gas-fired furnace. A—Condenser air-out. B—Condenser air-in. C—Furnace flue. D—Rooftop mounting curb.
(BDP Co., Div. of Carrier Corp.)

operation. The service technician must carry tools and supplies to the roof. Equipment must be adequate to service the unit in all kinds of weather. Fig. 23-35 shows a typical rooftop unit for both heating and cooling.

Another assembly is shown in Fig. 23-36. These units are completely weatherproofed since they are exposed to all kinds of weather. Access doors for inspection are designed to operate safely under high wind conditions. Fig. 23-37 is a pictorial drawing of the refrigerating mechanism of a rooftop unit. A complete heating and cooling system is shown in Fig. 23-38.

Gas furnaces usually use an electric ignition system because gas pilot flames tend to blow out during high winds or gusts. The gas furnace design in Fig. 23-39 is for a rooftop. One or more of these furnaces can be used in a single unit.

Rooftop units sometimes use oil furnaces. Fuel is stored in an underground tank. A lift pump draws oil to a second smaller tank. This tank may be at the same level as or slightly lower than the oil burner. But, under no circumstance should the tank be more than 10 ft. below the burner.

Electric heat may also be used in rooftop units. In Fig. 23-40 the electric heating elements are partially removed from the housing. These heating elements are available in about 20, 40 and 60 kW capacities. Heating elements are carefully pro-

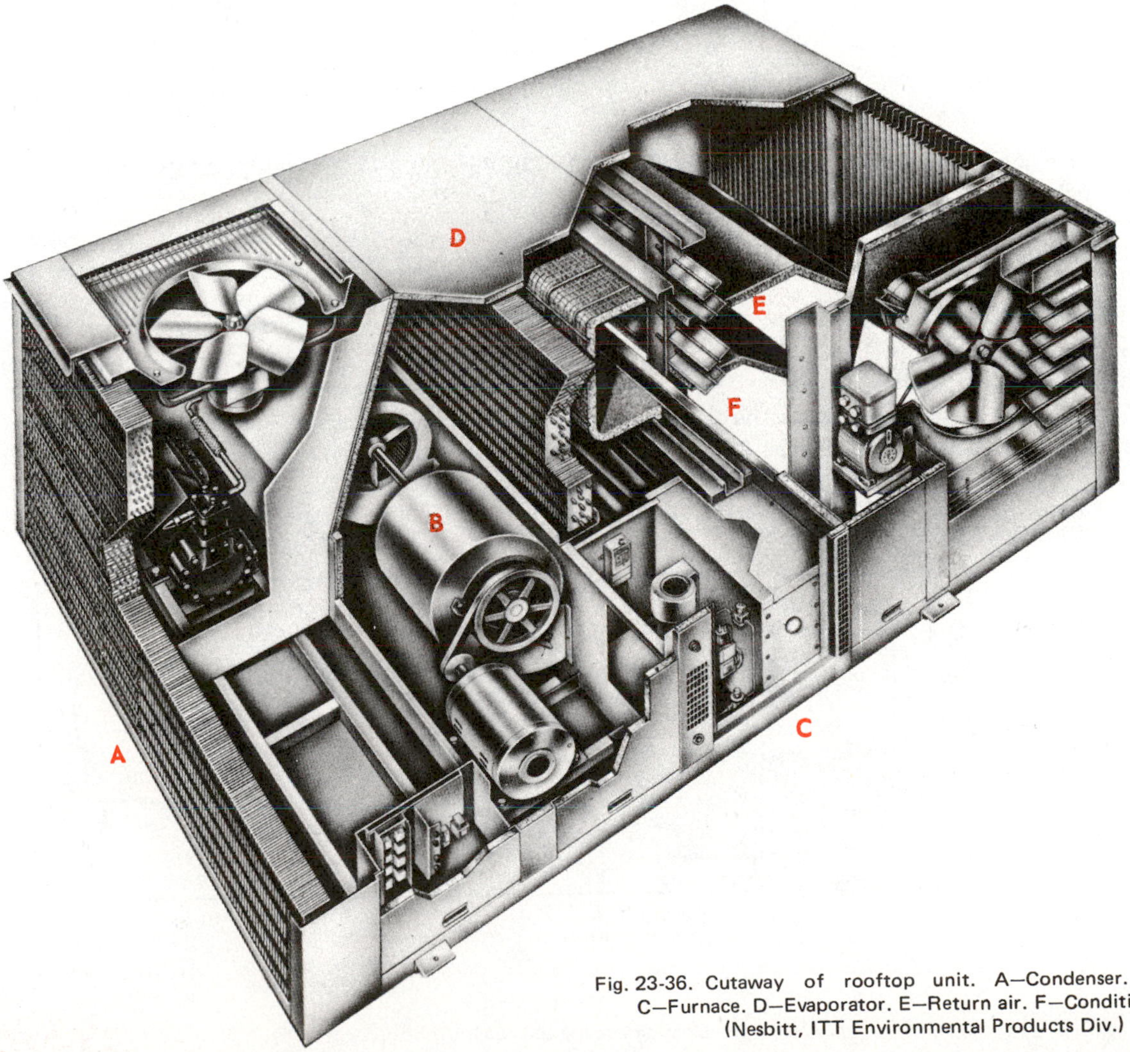

Fig. 23-36. Cutaway of rooftop unit. A—Condenser. B—Blowers. C—Furnace. D—Evaporator. E—Return air. F—Conditioned air.
(Nesbitt, ITT Environmental Products Div.)

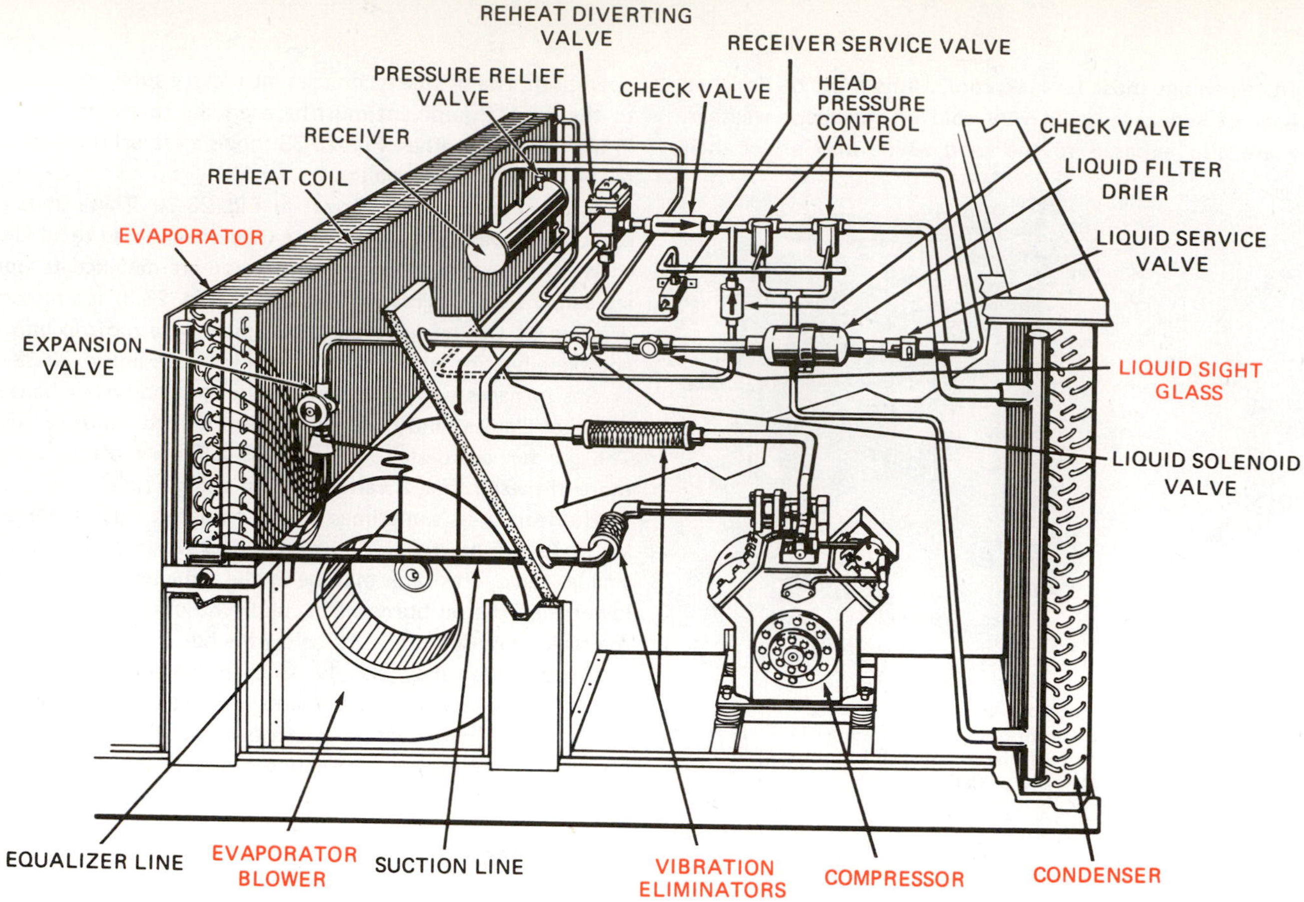

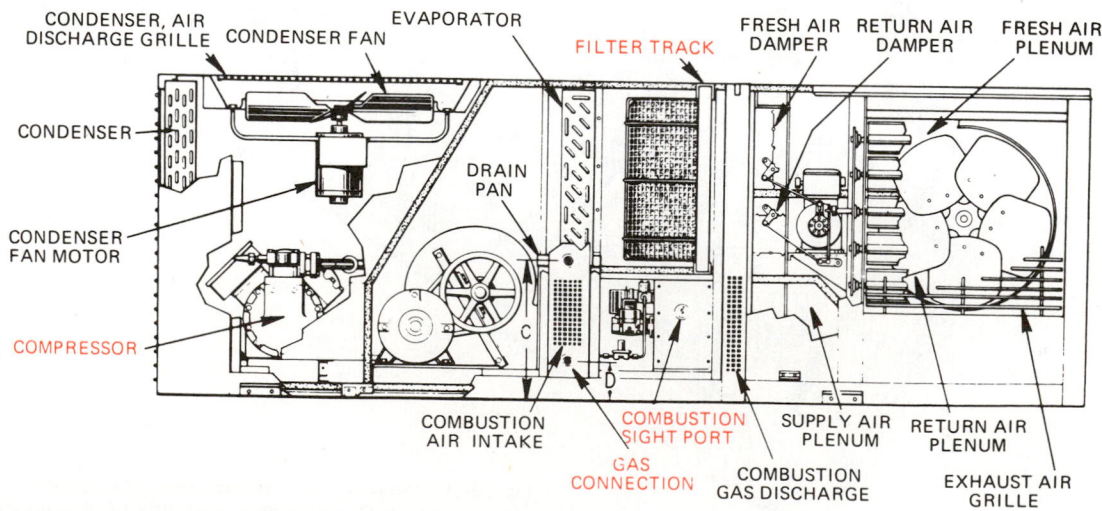

Fig. 23-37. Refrigerating circuit of rooftop system. Notice service valves, sight glass and vibration eliminators.

REHEAT DIVERTING VALVE
PRESSURE RELIEF VALVE
RECEIVER SERVICE VALVE
CHECK VALVE
HEAD PRESSURE CONTROL VALVE
RECEIVER
REHEAT COIL
CHECK VALVE
LIQUID FILTER DRIER
EVAPORATOR
LIQUID SERVICE VALVE
EXPANSION VALVE
LIQUID SIGHT GLASS
LIQUID SOLENOID VALVE
EQUALIZER LINE
EVAPORATOR BLOWER
SUCTION LINE
VIBRATION ELIMINATORS
COMPRESSOR
CONDENSER

tected by ceramic insulators. The circuit has fuses, fusible links, and thermal cutouts.

Air is controlled by dampers. Sensing elements in the mixed-air duct control a motor which operates:
1. The return damper.
2. The conditioned air damper.
3. The exhaust air damper.
4. The fresh air damper.

Fig. 23-41 shows the air control system. The air is filtered as in other air circulation systems. See Chapter 22.

23-27 ROOFTOP UNITS—INSTALLATION

Roof top units must be installed on a leakproof roof connection. Fig. 23-42 shows how the rooftop system is connected to a duct distribution system. Actual installation is usually as shown in Fig. 23-43.

One method used to make the roof leakproof is shown in Fig. 23-44. Notice the use of acoustical material for deadening sound. The resilient (flexible) gasket installed between the rooftop base rail and the roof edge is very important. There

CONDENSER, AIR DISCHARGE GRILLE
CONDENSER FAN
EVAPORATOR
FILTER TRACK
FRESH AIR DAMPER
RETURN AIR DAMPER
FRESH AIR PLENUM
CONDENSER
CONDENSER FAN MOTOR
DRAIN PAN
COMPRESSOR
COMBUSTION AIR INTAKE
COMBUSTION SIGHT PORT
GAS CONNECTION
COMBUSTION GAS DISCHARGE
SUPPLY AIR PLENUM
RETURN AIR PLENUM
EXHAUST AIR GRILLE

Fig. 23-38. Combination heating and cooling rooftop system.

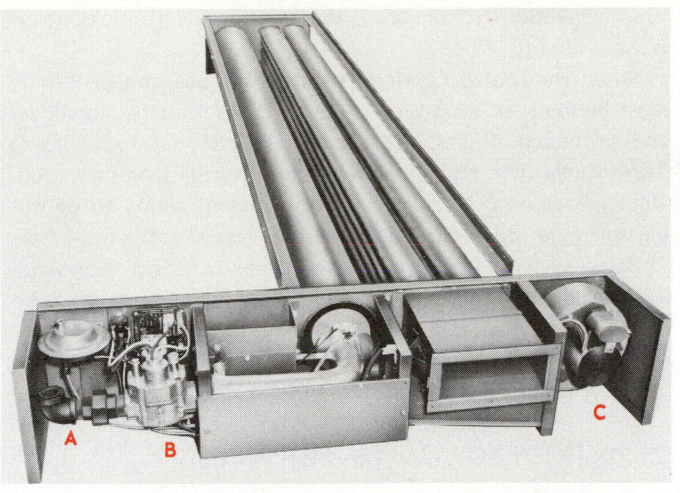

Fig. 23-39. Gas furnace designed for use in a rooftop heating-cooling system. A—Gas connection. B—Controls. C—Forced draft fan. (Nesbitt, ITT Environmental Products Div.)

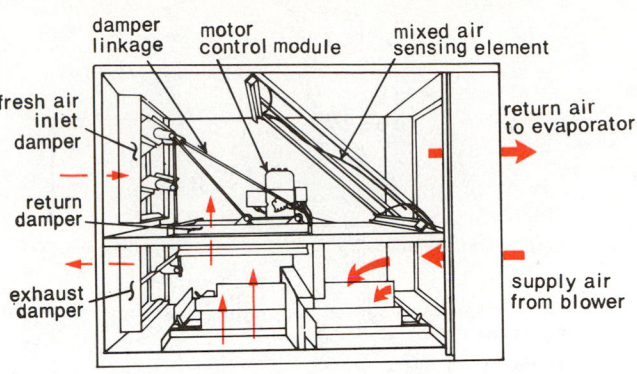

Fig. 23-41. An air control system for rooftop unit. Dampers are automatically controlled to always provide desired air mixture.

Fig. 23-40. Rooftop system equipped with electric heat. A—Control panel. B—Electric heating elements. (Fedders Corp.)

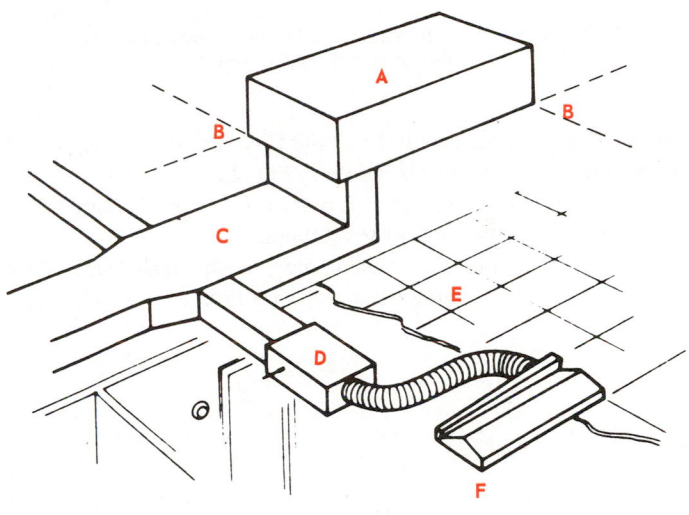

Fig. 23-42. Rooftop unit connected to duct distribution system. A—Rooftop unit. B—Roof line. C—Main duct. D—Branch duct. E—False ceiling. F—Diffuser. (Nesbitt, ITT Environmental Products Div.)

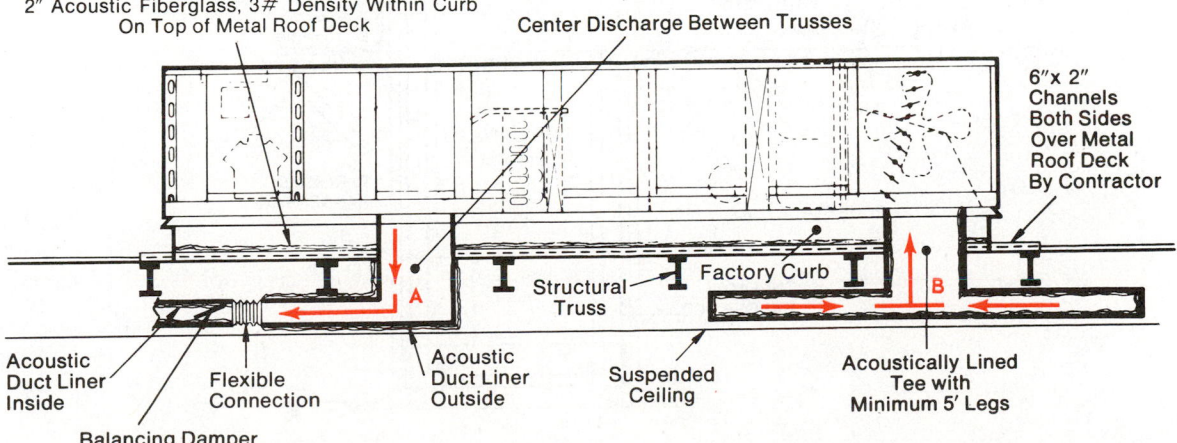

Fig. 23-43. Installation details of a rooftop complete air conditioning system. A—Conditioned-air duct. B—Return-air duct.

Heat Pumps and Complete Air Conditioning Systems / 819

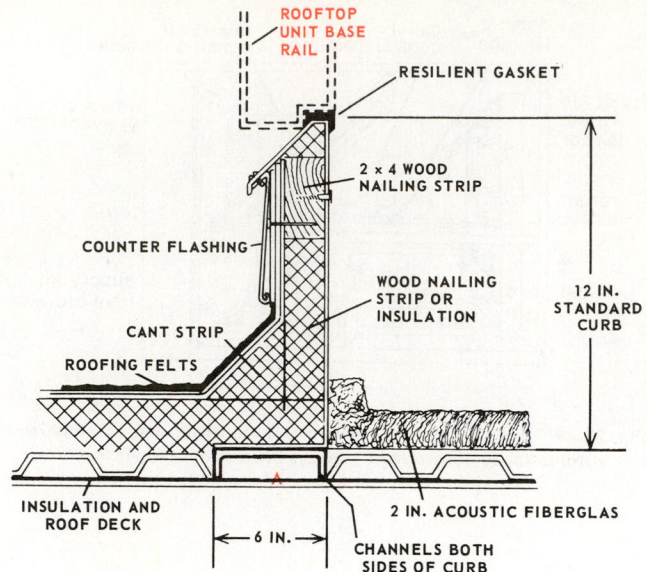

Fig. 23-44. Method used to make roof leakproof when rooftop unit is installed. A—Roof reinforcement.

may be leaks if this gasket is not tight. A duct installation connected to a rooftop unit is shown in Fig. 23-45.

Rooftop units are heavy. Riggers are usually required for installing, removing and replacing them.

The electrical service must comply with local codes. The control system is usually the responsibility of the refrigera-

tion installation technician. A system using electronic controls is shown in Fig. 23-46.

When the rooftop system has a gas furnace, the gas piping must be installed according to code. Pipes must be supported and protected. Fig. 23-47 shows a typical installation for a 1 1/4 in. gas line. Note that two hand shutoff valves are used. One is located outside the rooftop system casing to permit emergency shutdown of the furnace in case of accident or fire.

Some systems use piped-in hot water or steam to provide the heating required. The piping of these systems must be carefully installed to avoid air traps and freezing. Hot water systems usually use a nonfreezing solution of water and glycol. Fig. 23-48 shows the piping for a steam system.

23-28 ROOFTOP UNITS—SERVICING

Certain chapters have been devoted to servicing different systems. For more information on:
1. Refrigerating systems, refer to Chapter 14.
2. Heating systems, refer to Chapter 20.
3. Blowers and filters, refer to Chapter 22.

However, there are some very important service operations special to rooftop systems. It is very important that the service technician use all possible safety precautions when climbing to the roof.

If a portable ladder is used, it must be securely and firmly based on the ground. It must be inclined (leaned against the building) at an angle that will keep it from falling away from the building. It must extend two or three rungs above roof edge.

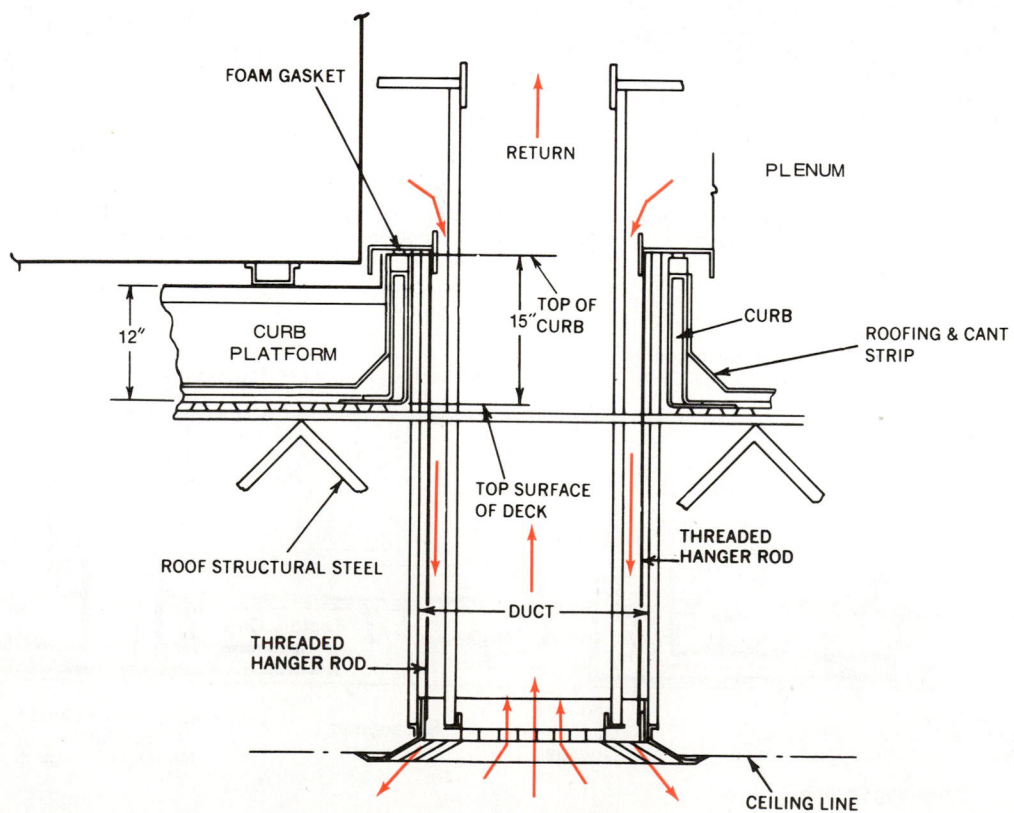

Fig. 23-45. Method of connecting ducts to rooftop unit. (Fedders Corp.)

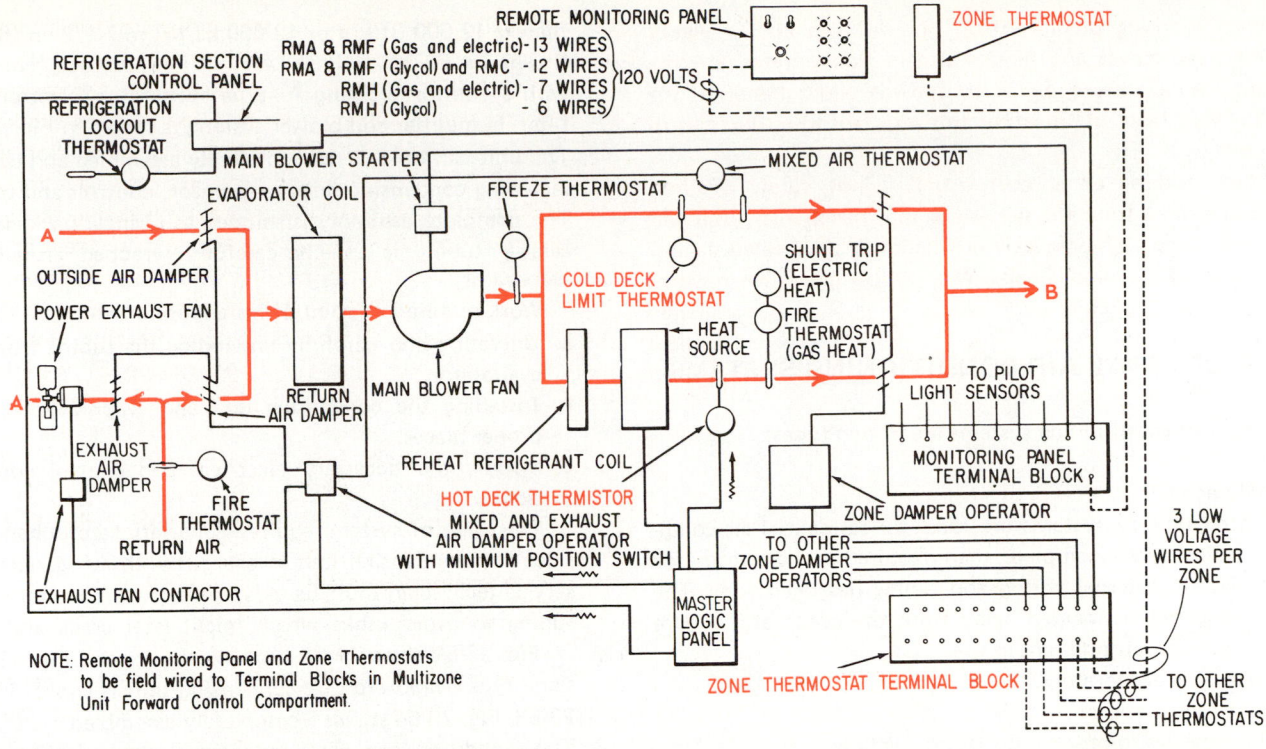

Fig. 23-46. An electronic control system for rooftop unit. A—Return air. B—Conditioned air. Red arrows indicate the direction of airflow.

The service technician must use both hands on the ladder when climbing or coming down the ladder. The portable ladder should not be used in a high or gusty wind.

Be especially careful it if is raining or snowing. Lift tool boxes, refrigerant cylinders, and other objects with a rope. The rope should be guided from below also.

When the roof is wet or snow covered, be extremely cautious while working on or near electrical circuits. Hinged panels must be secured in the open position or the wind may swing them violently and injure someone. Loose panels must be held down securely to prevent a wind blowing them off the roof. If the system has a gas or oil burner, avoid breathing fumes coming out of the flue.

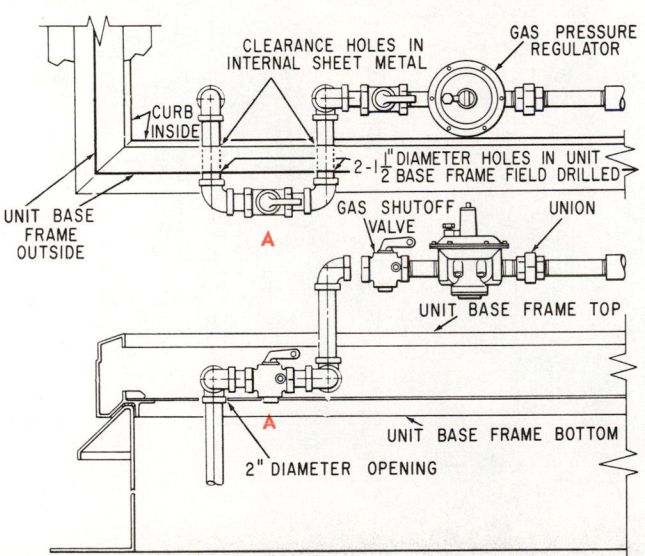

Fig. 23-47. Fuel gas line installation used on rooftop system. A—Hand shutoff valve is mounted outside this casing for easy access. (Nesbitt, ITT Environmental Products Div.)

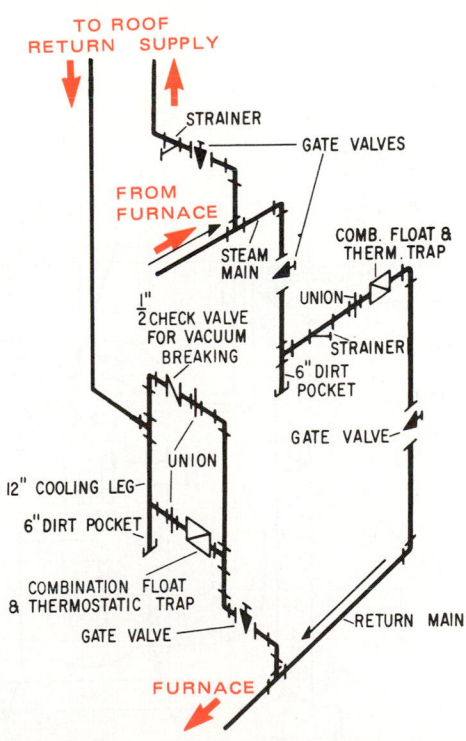

Fig. 23-48. Piping installation for steam heating system used on rooftop complete air conditioning system.

23-29 CENTRAL AIR CONDITIONING SYSTEMS

Cental air conditioning systems are of two types:
1. Unitary.
2. Field-erected.

Central unitary systems are ideal for residential air conditioning. They are a complete, manufactured package ready for assembly. All internal wiring and piping has been done. The condensing unit is located away from the evaporator. There are three evaporator designs in use:
1. An A-type evaporator.
2. Slant type evaporator.
3. Flat type evaporator for horizontal airflow.

Three methods are used to install central air conditioning:
1. One can purchase the condensing unit, evaporator, controls and tubing, assembling the conditioner on the customer's premises. Most of these systems are installed in forced warm air heating systems. Fig. 23-49 is a drawing of such an installation.

These units vary in capacity from 1 1/2 hp (approxi-

mately 12,000 Btu/hr or 12 660 kJ/hr) to 7 1/2 hp (60,000 Btu/hr or 63 300 kJ/hr). An oil-fired furnace, complete with a comfort cooling A—type evaporator, electronic air filter, humidifier and blower system, is shown in Fig. 23-50.

2. The unit can be ordered completely assembled and charged including condensing unit, evaporator, controls and tubing. The complete assembly is shipped as a single package. Necessary tubing is usually carefully wrapped around the evaporator.

Work required at point of installation includes:
a. Uncrating and carefully unwinding the tubing from the evaporator.
b. Installing the condensing unit and evaporator in their proper places.
c. Making the necessary electrical and control connecttions.

This type of system requires some very careful handling, since the condensing unit is charged with refrigerant. The service technician must be very careful while uncrating the tubing to avoid kinks which might later crack and leak.

Fig. 23-51 shows an A-type evaporator and Fig. 23-52 a slant type evaporator. A condensing unit is shown in Fig. 23-53. Fig. 23-54 shows a completely assembled system.

3. The technician can get a completely charged evaporator, condensing unit and lines. However, the condensing unit, evaporator and connecting tubing are shipped as separate items. See Fig. 23-55. The parts are connected with "quick couplers." This system is easy to assemble since the quick

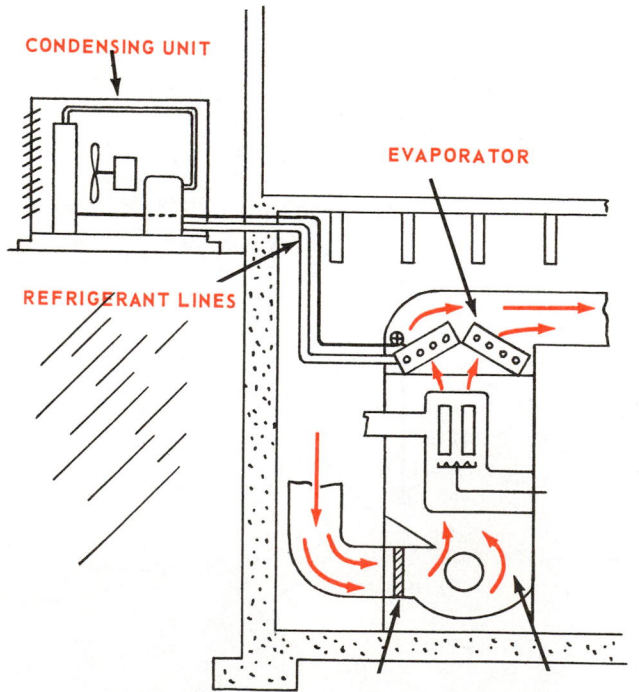

Fig. 23-49. Schematic of residential central comfort-cooling system installed in forced-air furnace. Condensing unit is mounted outdoors.

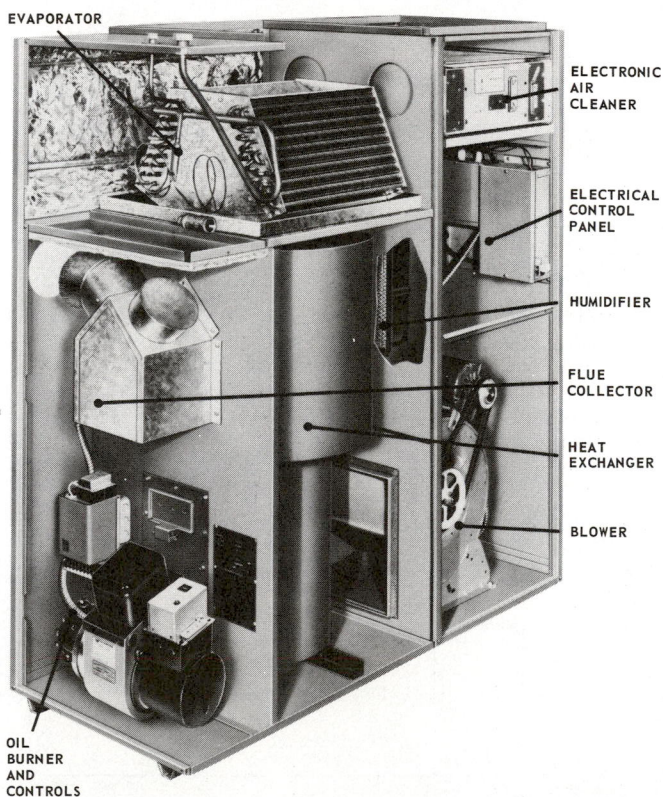

Fig. 23-50. All-season air conditioner with oil burner, evaporator, electronic air cleaner and humidifier. (The Williamson Co.)

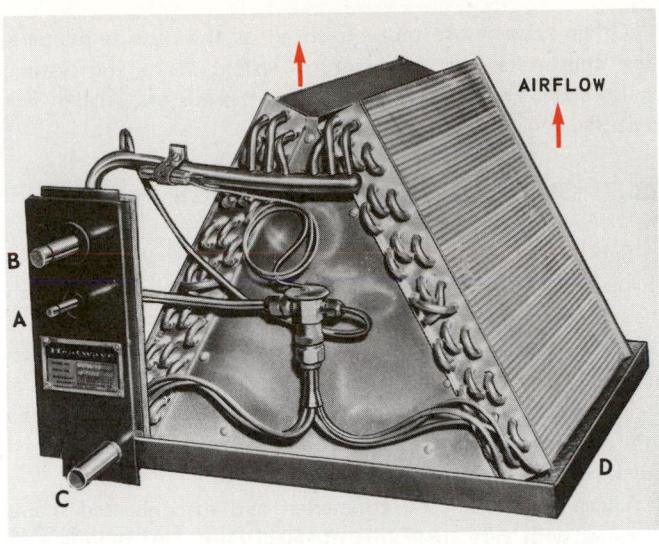

Fig. 23-51. An A-shaped evaporator designed for installation in warm-air furnace plenum chamber. A—Liquid line connection. B—Suction line connection. C—Drain connection. D—Drain pan. (Southwest Mfg. Co.)

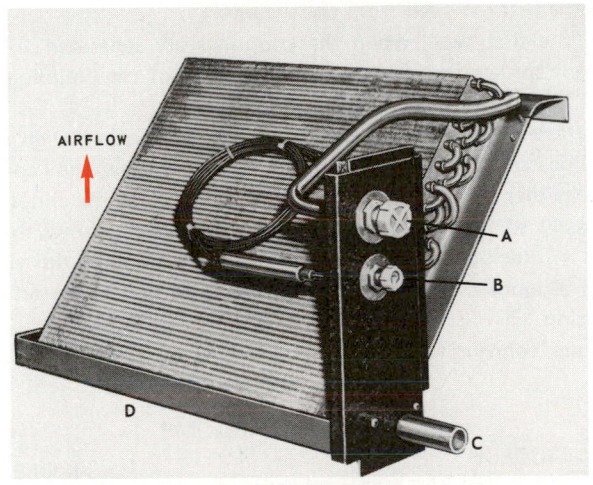

Fig. 23-52. Slant-type evaporator designed for installation in plenum chamber of warm-air furnace. A—Suction line connection. B—Liquid line connection. C—Drain connection. D—Drain pan.

couplers may be connected without losing refrigerant or getting air into the system. The complete system is shown in Fig. 23-56.

Some systems use the liquid line as the capillary tube. This larger bore (ID) tubing reduces the chance of clogging from dirt or moisture. *It is very important not to shorten or lengthen this liquid line capillary tube combination when installing this unit.*

Many systems use evaporators with aluminum fins mechanically bonded to copper tubing. Plastic grilles are often used on the condensing unit to avoid corrosion problems. Fig. 23-57 shows the internal details of the condensing unit.

Field-erected air conditioning systems are systems in which all the components — motor, compressor, receiver, evaporator, piping, and controls — are assembled and erected at the spot where the system is to be used. Chapter 14 covers most of the

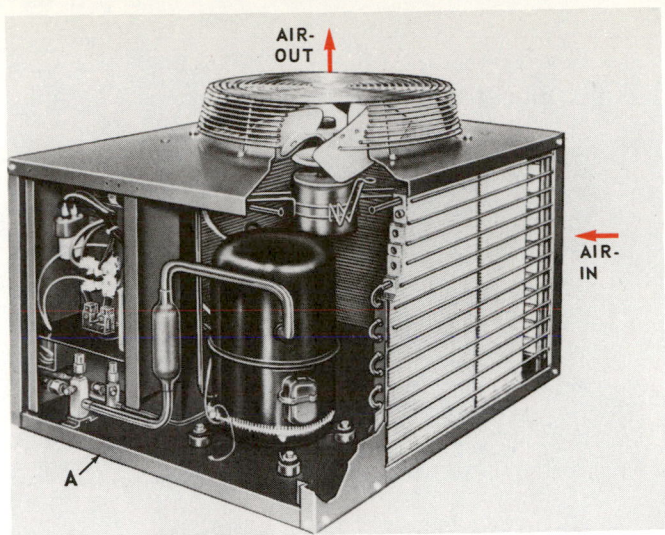

Fig. 23-53. Condensing unit is for cooling system which uses an evaporator installed in plenum chamber of warm-air furnace. A—Service valves. (Southwest Mfg. Co.)

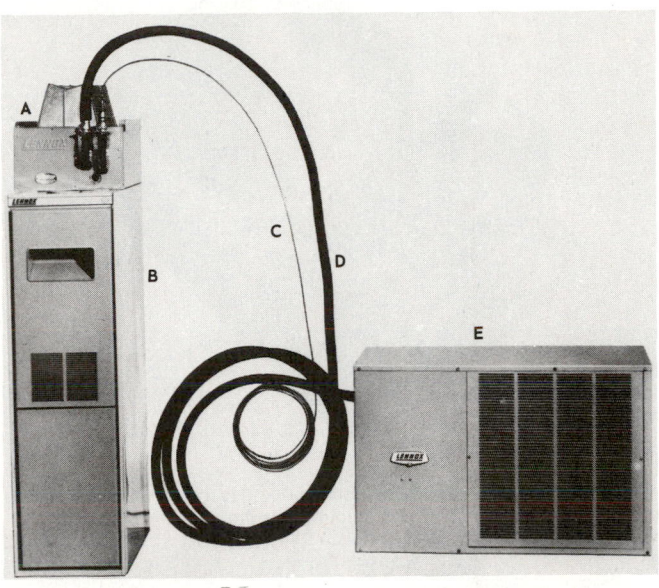

Fig. 23-54. Central comfort-cooling system for a home. A—Evaporator. B—Plenum chamber of warm-air furnace. C—Liquid line. D—Insulated suction line. E—Outdoor air-cooled condensing unit. (Lennox Industries, Inc.)

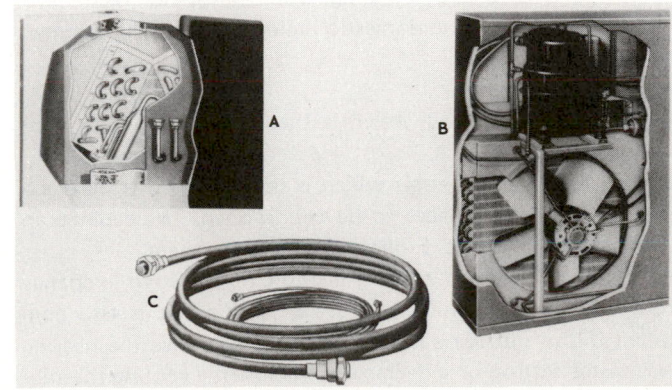

Fig. 23-55. Domestic central comfort cooling system uses quick-connect refrigerant lines. A—Evaporator. B—Air-cooled condensing unit. C—Suction and liquid lines.

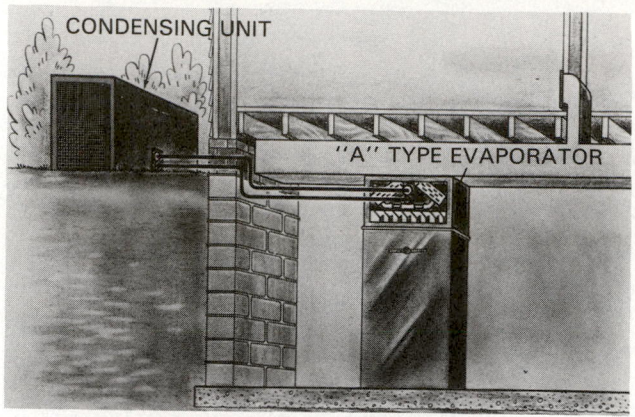

Fig. 23-56. This is a typical installation of a complete air conditioning system. (Aeroquip Corp.)

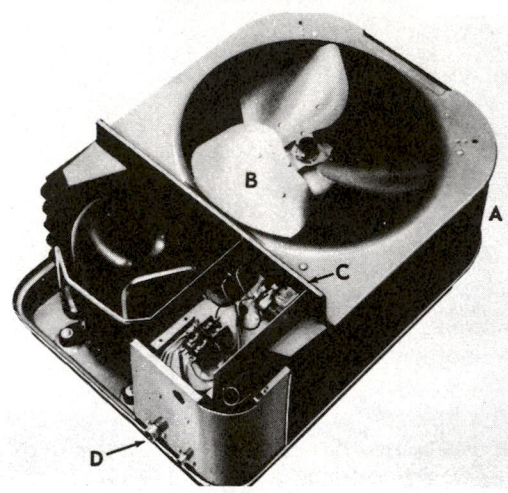

Fig. 23-57. Vertical airflow outdoor air-cooled condensing unit made for domestic central comfort cooling system. A—Condenser. B—Fan with grille removed. C—Controls. D—Refrigerant line connections. (Addison Products Co.)

installation instructions.

There are a variety of central field-erected systems. Some are large systems which heat and cool a number of buildings or various parts of a large building. Others are field-erected systems which service one domestic building or a small commercial building.

These systems may:
1. Cool and/or heat air which is then distributed by a dual system.
2. Cool and/or heat water which is then pumped to heat exchangers in the spaces to be conditioned. The cabinets in the spaces have fins, filters and controls in them.

Automatic controls make it possible for a system to change from heating to cooling. The use of outside air is also controlled if the outside air is a degree or so above the heating thermostat setting or a degree or so below the cooling thermostat setting. The system then becomes an air-distribution and air-cleaning system only. This permits greatest economy of operation.

Other systems use more fresh air as the outside temperature approaches the temperature desired inside the system. Solid state controls, operated by thermistors, make this possible.

23-30 QUICK CONNECT COUPLINGS

Self-sealing couplings enable manufacturers to produce precharged refrigeration and air conditioning units — along with the necessary tubing — in separate packages. These separate units may be assembled at the installation site and are ready to operate without evacuating, charging or cleaning.

The self-sealing coupling fittings are brazed directly to the tubing; flared joints are not needed. There are two types of quick connect fittings:
1. Those which can be connected and disconnected many times with very little loss of refrigerant. This type is very seldom used.
2. Those which can only be quick-connected once. This type uses diaphragms and, when the fittings are attached, the diaphragms are punctured to allow refrigerant to pass. These couplings cannot be disconnected unless the refrigerant is first removed from the system.

In the first type, when the couplings are separated, independent springs force valves in both halves of the coupling to close. This prevents the escape of refrigerant.

To assemble either type of quick-connect fitting, align the couplings and tighten the coupling nut. This draws the coupling together and pierces the sealing diaphragm of the coupling internally so the refrigerant can flow. Fig. 23-58 shows three views of quick-connect coupling. Fig. 23-59 is an exterior view of an assembled quick-connect as it would appear on an installation.

Quick-connect fittings are used mostly on precharged resi-

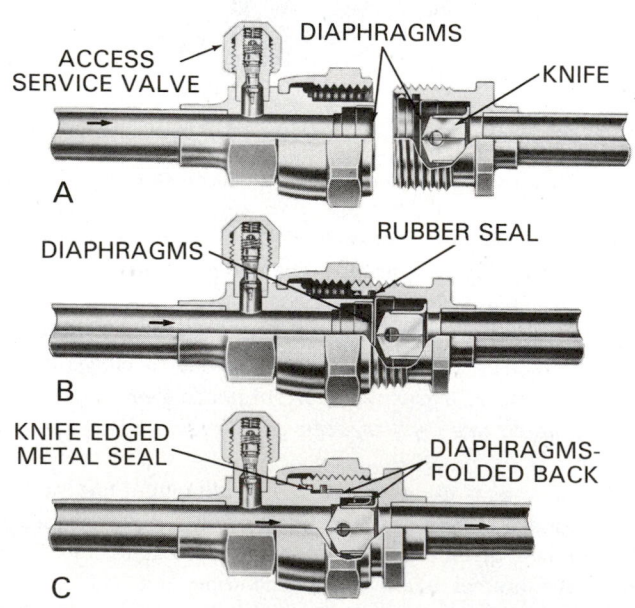

Fig. 23-58. Quick-connect-and-disconnect coupling, with access service valve. Three views of assembly are shown. A—Disconnected. B—Partially assembled. C—Connected and refrigerant passage opened. (Aeroquip Corp.)

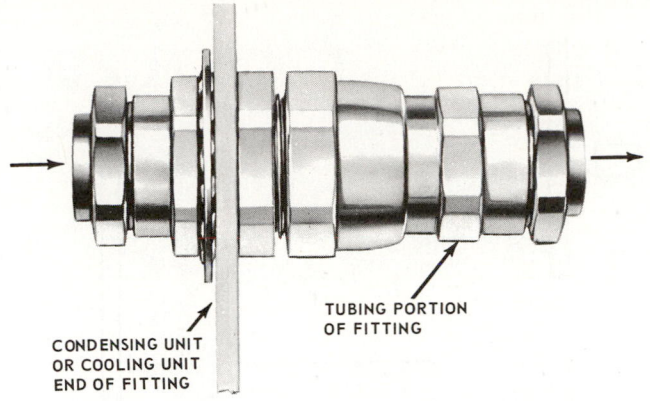

Fig 23-59. An assembled quick-connect-and-disconnect fitting.

dential air conditioning systems and precharged transportation units. Units are usually charged at the factory with the condensing unit, refrigerant lines and evaporator being charged separately. Fig. 23-60 shows a liquid line equipped with an access (service) port.

It is recommended that the gasket which joins the quick-connect-disconnect fittings (often made of neoprene and asbestos) be covered with clean, dry refrigerant oil just before assembly. Avoid excessive wrench pressure because a distorted fitting may leak. Fig. 23-61 shows two wrenches being used to tighten the fittings. Most quick-connects-and-disconnects will reseal themselves several times.

23-31 INSTALLING RESIDENTIAL CENTRAL AIR CONDITIONING SYSTEMS

Cooling systems should be installed only in furnaces less than 15 years old. Older furnaces will need to be replaced.

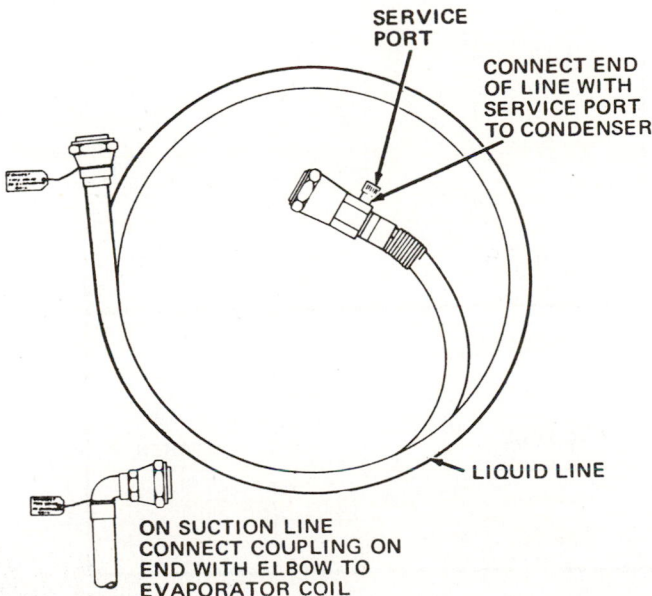

Fig. 23-60. Quick-coupler liquid line. Service port is used for making service manifold high-side pressure guage connection. (The Coleman Co., Inc.)

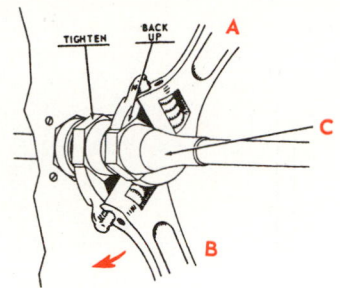

Fig. 23-61. Correct way to tighten a quick coupler. Wrench A is held firmly while wrench B is turned. Use of two wrenches prevents twisting of tubing, C.

Units are assembled on-site in four steps:
1. Install condensing unit.
2. Install evaporator.
3. Install suction and liquid lines.
4. Install electrical wiring.

The condensing unit uses outdoor air to cool the condenser. Some homes have water-cooled condensers.

Many installation methods have been used. Some units are mounted inside the building with ducts bringing outdoor air to the condenser and discharging warm air outside. Some units have been mounted on an outside wall. A more popular practice is to mount the unit on a concrete slab or prefabricated slab 12 to 24 in. (30 to 60 cm) from the building. A concrete slab at least 4 in. (10 cm) thick and reinforced with steel mesh is recommended. Fig. 23-62 shows various installation methods. It is desirable for the outlet air from the condensing unit to move in the same direction as the prevailing summer winds.

Location of the condensing unit is very important. The suction line, liquid line and power lines should be as short as possible. The condensing unit should be carefully located:
1. Away from neighbors (noise).
2. Away from bedrooms (noise).
3. At least 24 in. away from wall (air circulation).
4. Away from inside corners (air circulation).
5. Not under eaves (airflow).
6. Beyond overhang (air circulation).
7. Away from patio (noise).

Fig. 23-63 shows suggested condensing unit location.

The condensing unit should be mounted level. A typical installation is shown in Fig. 23-64.

The evaporator is mounted level and firm in the bonnet or plenum chamber of the furnace. Design of the evaporator and its condensate drain depend on the type of furnace (upflow, downflow, or horizontal flow). Removable panels are needed for periodic cleaning and servicing as required. A slant evaporator installation is shown in Fig. 23-65.

The plenum chamber is blanked off to make sure all the air goes through the evaporator. The condensate drain should be piped to an open drain with an air break at the drain. If the drain pipe is plastic, keep it away from the warmer parts of the furnace. Some technicians install a 4 in. U-trap in the drain line to stop airflow through the line. The drain pan is built into the evaporator as is the drain connection. Check the

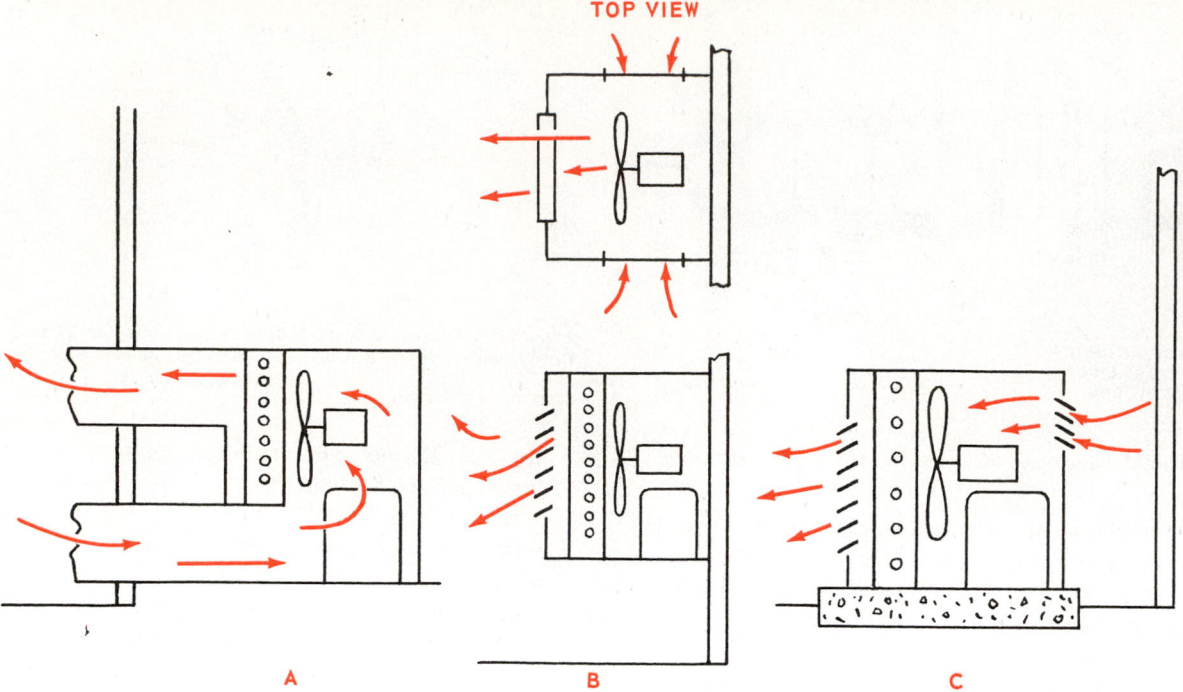

Fig. 23-62. Three types of air-cooled condensing units for residential air conditioning installations. A—Condensing unit inside building. B—On outer wall. C—On concrete slab outside building

local building code for proper installation of all units. In some installations, it may be necessary to install a drain pump to remove the condensate to the outdoors. Fig. 23-66 shows an A-type evaporator being installed in a furnace plenum chamber.

Suction and liquid line connections may be:

1. Flared connections.

2. Brazed connections.

3. Quick-connect-and-disconnect couplings.

The condensing unit in Fig. 23-67 has flared or brazed tubing connections equipped with service valves. Refrigerant control is a thermostatic expansion valve. This unit would be installed as described in Chapter 14.

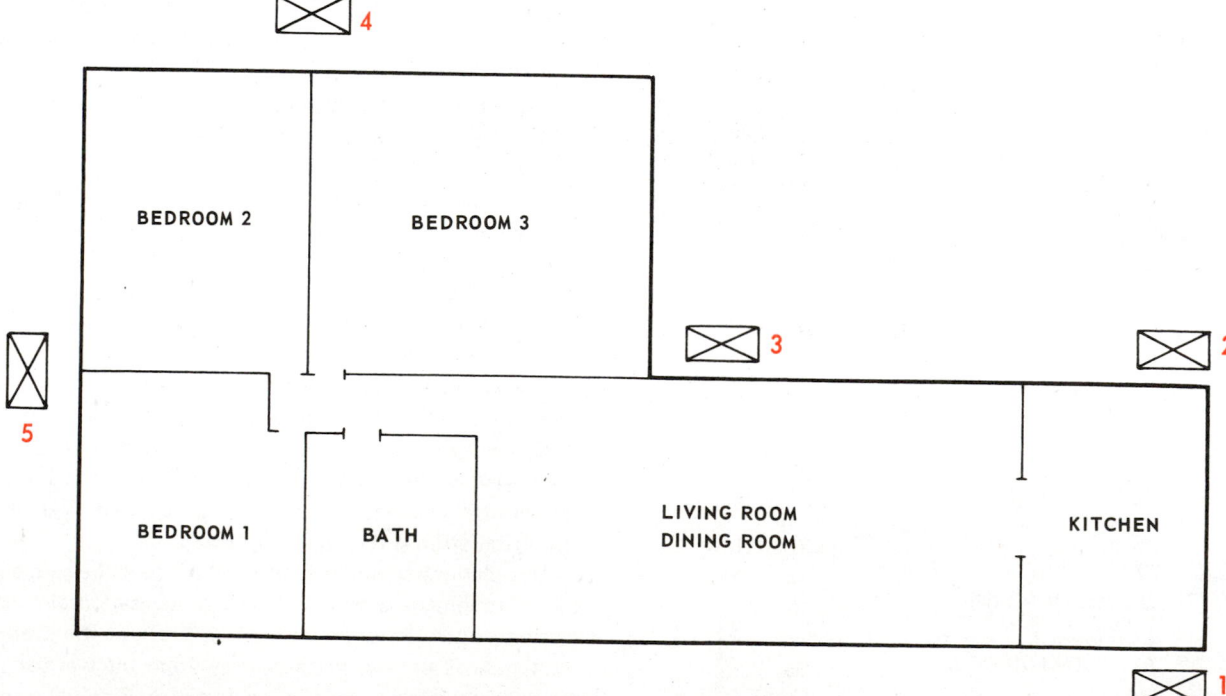

Fig. 23-63. Location of outdoor air-cooled condensing unit. 1 and 2 are in good positions; 3 is not recommended as it is in an air pocket and near the bedroom; 4 and 5 are also too near bedrooms.

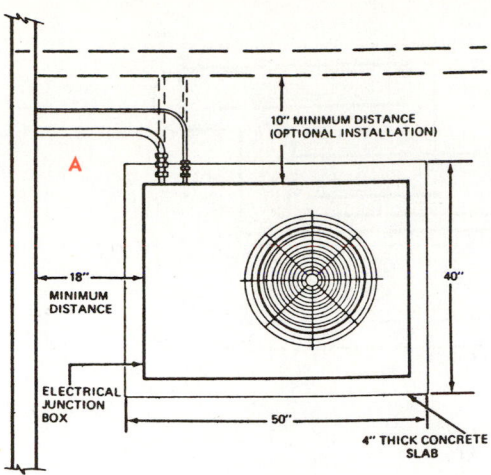

Fig. 23-64. Top view of typical installation of outdoor condensing unit. A—Refrigerant lines. (The Coleman Co., Inc.)

Many condensing units are installed above the evaporator; therefore, a U-bend should be put in the suction line (to assist oil return). The suction line should slope downward slightly toward the condensing unit.

A filter-drier and a sight glass should also be put in the liquid line. Many service technicians also place a filter-drier in the suction line to prevent motor compressor burnouts.

That part of the suction line installed inside the building should be insulated. Use 1/2 in. line for hot, humid conditions and 1/4 to 3/8 in. for normal conditions. Without insulation, moisture from the air will condense on the suction line and drip. Lines should be supported and free of kinks. Openings in the furnace duct and the wall should be sealed with weatherproof, nonhardening sealing compound and tape.

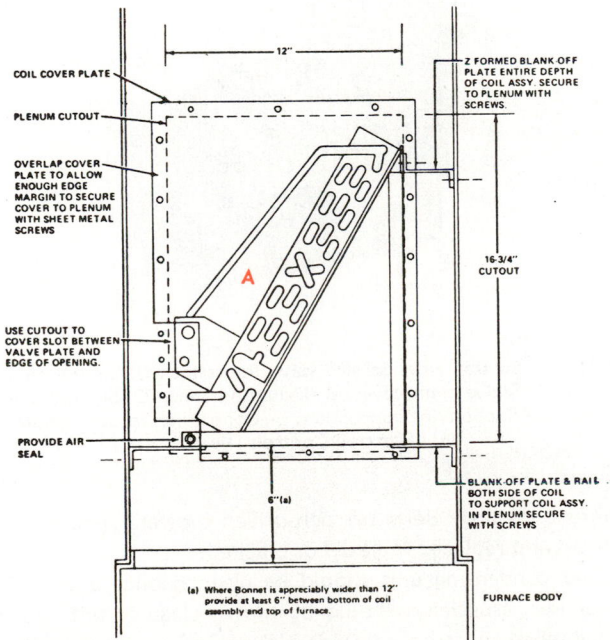

Fig. 23-65. Slant evaporator installed in furnace plenum. Note blank-off plate to make all air go through evaporator. A—Evaporator.

Fig. 23-66. An A-type evaporator being installed in the plenum chamber of an upflow warm furnace. (Fedders Corp.)

When the quick-connect system is used, the lines are first installed. The correct length of prefabricated line should be used. When the quick-connects are made, the unit is ready to operate.

In all cases, the system should be thoroughly tested for leaks while the pressures inside the system are near ambient temperature-pressure conditions.

The electrical installation must follow the wiring diagram furnished with the unit. Fig. 23-68 shows a wiring diagram for a single-phase 230V a-c system.

The electrical circuit must follow the National Electrical and local codes. Consult with the local electrical utility concerning the primary service capacity. With 240V circuits, smaller wires can be used.

Fig. 23-67. Condensing unit designed for residental central comfort cooling system. A—Condenser. B—Fan. C—Controls. D—Refrigerant line connections. E—Process tubes. (The Williamson Co.)

23-32 INSPECTING RESIDENTIAL CENTRAL AIR CONDITIONING SYSTEMS

The complete system should be inspected and serviced each season before the system is put in use.

Services which the owner can perform include:

1. Energize crankcase heater 24 hours before starting system.
2. Clean condenser and fans.

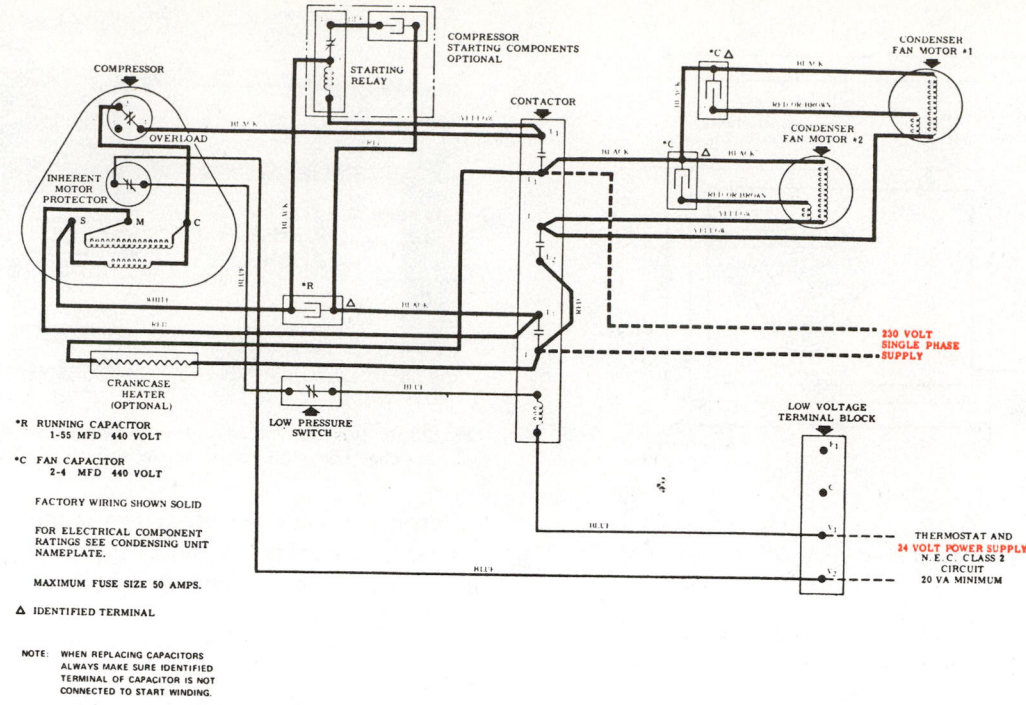

Fig. 23-68. Wiring diagram for central system condensing unit. (The Coleman Co., Inc.)

3. Check dampers in ducts.
4. Replace filters.
5. Lubricate motor and fan bearings.
6. Check fan belt for cracks and glaze (replace if necessary). Adjust belt tension.
7. Check and clean drain pans and drain pipe.

Service handled by service technician:

8. Clean thermostat points.
9. Check system pressures.
10. Check voltage and full load amperage.
11. Check refrigerant charge.
12. Check suction line sweating or frosting.

23-33 SERVICING RESIDENTIAL CENTRAL AIR CONDITIONING SYSTEMS

Servicing procedures and troubleshooting diagnosis for residential central systems are similar to those described in Chapters 11 and 14.

Check for leaks, proper refrigerant charge, malfunctioning refrigerant controls and motor controls, and moisture in the system. Some systems use service valves. Fig. 23-69 shows one type of valve. Fig. 23-70 is a cross-section showing internal construction. An access type of service valve with a quick-connect tubing connection is shown in Fig. 23-71.

Often the system uses the same blower, motor, filter, and duct system used during the heating season. It is essential that the blower be cleaned once a year; that the motor be lubricated (a few drops of 30 SAE oil) once or twice a year, that the filter be replaced or cleaned at least twice a year (beginning of heating season and beginning of cooling season).

The motor-blower speed should be increased for summer

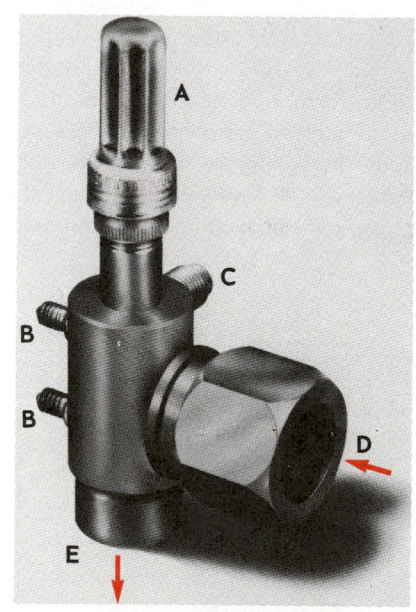

Fig. 23-69. Suction lines service valve for residential central air conditioner. A—Valve stem cover. B—Mounting studs. C—Service connection. D—Suction line connection. E—Connection to compressor. (Chatleff Controls Div.)

comfort cooling. Belts on belt-driven blowers should be inspected and replaced if glazed or cracked.

The condensing unit should be cleaned once a year. The condenser, especially, should be blown clean of lint and fins straightened. A carbon dioxide blower and/or vacuum cleaner may be used for cleaning the unit. The A-style evaporator should be cleaned of any lint and its fins straightened, if bent.

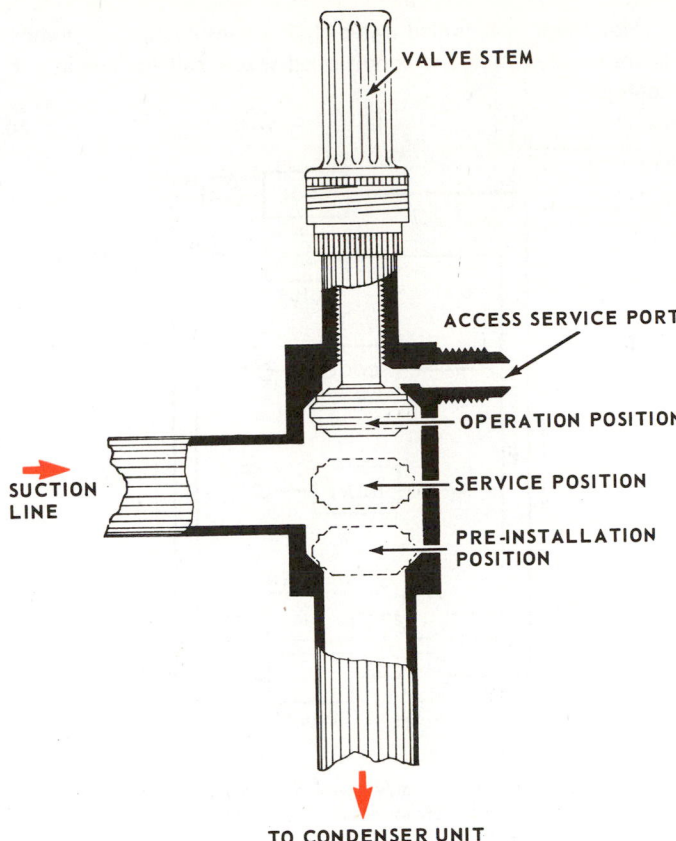

Fig. 23-70. Internal construction of central system air conditioner suction service valve.

Check condensate drainage. Any condensate which escapes the drain pan may drip on the furnace heat exchanger and corrode (rust) it. The best way to clean these coils is to remove the coil, plug all connections at once, and then either steam clean or use hot water and detergent to clean the fins and the tubes. Steam is the best cleaner, although high-pressure warm water and detergent will do a fair job.

Do not adjust the thermostat too low for summer cooling. The evaporator may freeze the condensate. This will stop air-flow through the evaporator. The evaporator may continue to collect a lot of ice.

A special charging system for use with R-22 refrigerant has been developed. See Fig. 23-72. Note that the charging

cylinder is upside down and liquid refrigerant goes to the charging control. An automatic control feeds the refrigerant into the system's suction line until the system is correctly charged. The suction line then cools and the charging action is automatically stopped.

There may be several reasons for service calls:

1. No heat or insufficient heating. See Chapter 20.
2. No cooling or insufficient cooling. See Chapters 11, 14, 21.
3. Humidity too high. See Chapter 20.
4. Air in house is stuffy (stale). See Chapter 22.
5. Excessive noise. See Chapters 20 and 21:
   a. Indoors.
   b. Outdoors.
6. High cost of operation. See Chapters 20 and 21.
7. Unit will not start. See Chapters 20 and 21.

23-34 AIR CIRCULATION SYSTEMS AND HUMIDITY CONTROL

Good, complete air conditioning systems must provide heat, remove heat, clean and circulate the air. Most systems accomplish each of these. However, many systems do not completely control the humidity. To control the humidity one must have in the system two devices ready to be used at any time, winter or summer:

1. A device to add water vapor to the air. See Chapter 20.
2. A device to remove water vapor from the air. See Chapter 21.

Another method is to always have a supply of cool air with normal humidity (50 percent) and a supply of warm air with normal humidity (50 percent). By mixing these air volumes one can produce the temperature and humidity conditions needed.

Duct systems are used when the heating and cooling system are remote (far away) from the space to be conditioned.

Some systems confine the air distribution to the space being conditioned. A cabinet with fans, filters, grilles and registers is located in the room.

23-35 TWO-DUCT SYSTEMS

The two-duct system uses two supply ducts, one with cool dehumidified air and the other with warm humid air. These

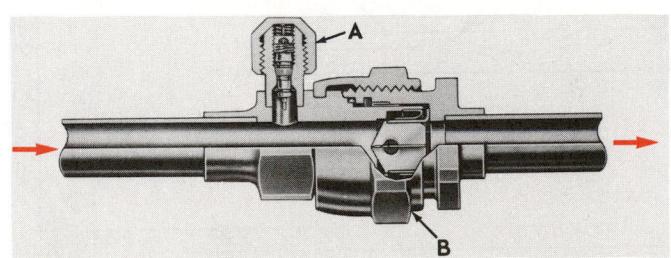

Fig. 23-71. Access type service valve and connection. A—Access service valve. B—Quick-connect fitting for precharged tubing. (Aeroquip Corp.)

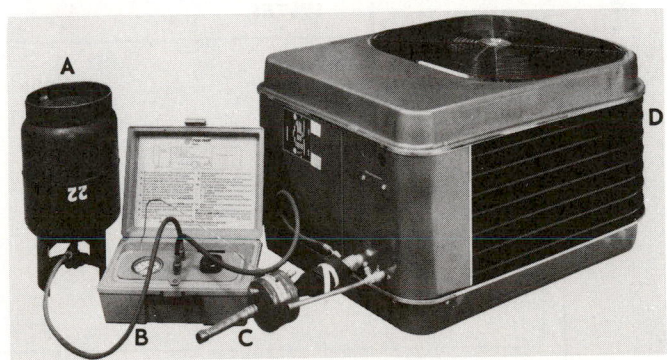

Fig. 23-72. Special charging system for R-22 refrigerant only. A—Charging cylinder. B—Charging control unit. C—Suction line. D—Condensing unit. (Addison Products Co.)

separate airstreams are mixed just before they reach the space to be conditioned. Through dampers that control and balance air, each different space in the building can be conditioned as needed. A single-duct air return is used. See Fig. 23-73. The mixture of the two airflows takes place at B.

The air control is excellent in these sytems. However, the ducts are large in cross-section and take up considerable space. Some space savings may be obtained by using high-velocity ducts, but a noise problem then develops.

23-36 FOUR-PIPE SYSTEMS

Four-pipe complete air conditioning systems have a hot water supply pipe, a chilled water supply pipe, and two return pipes. Four-pipe systems are of two types. One has separate heating and cooling coils in the space to be conditioned. The other uses the same coil for both warm water and the chilled water.

A system using separate heating and cooling coils is shown in Fig. 23-74. The heating circuit is completely separate from the cooling circuit. In this case, the heating fluid (heat carrier) may be either water or steam. One two-way valve is used for each coil.

Some systems use the same space heat transfer coil for both heating and cooling as shown in Fig. 23-75. Two separate three-way control valves are needed for each heat transfer coil in the conditioned space.

Hot water and chilled water pipes are insulated. The pumps operate only when the conditioned spaces call for heating or cooling.

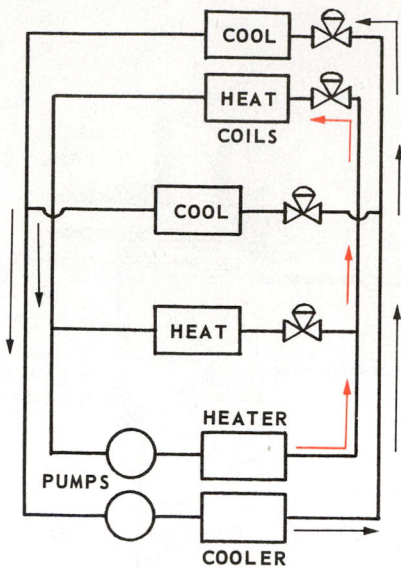

Fig. 23-74. Complete four-pipe system using separate heating and cooling coils. Only one thermostatically controlled valve is used for each heat transfer coil. Heating circuit and cooling circuit are separate.

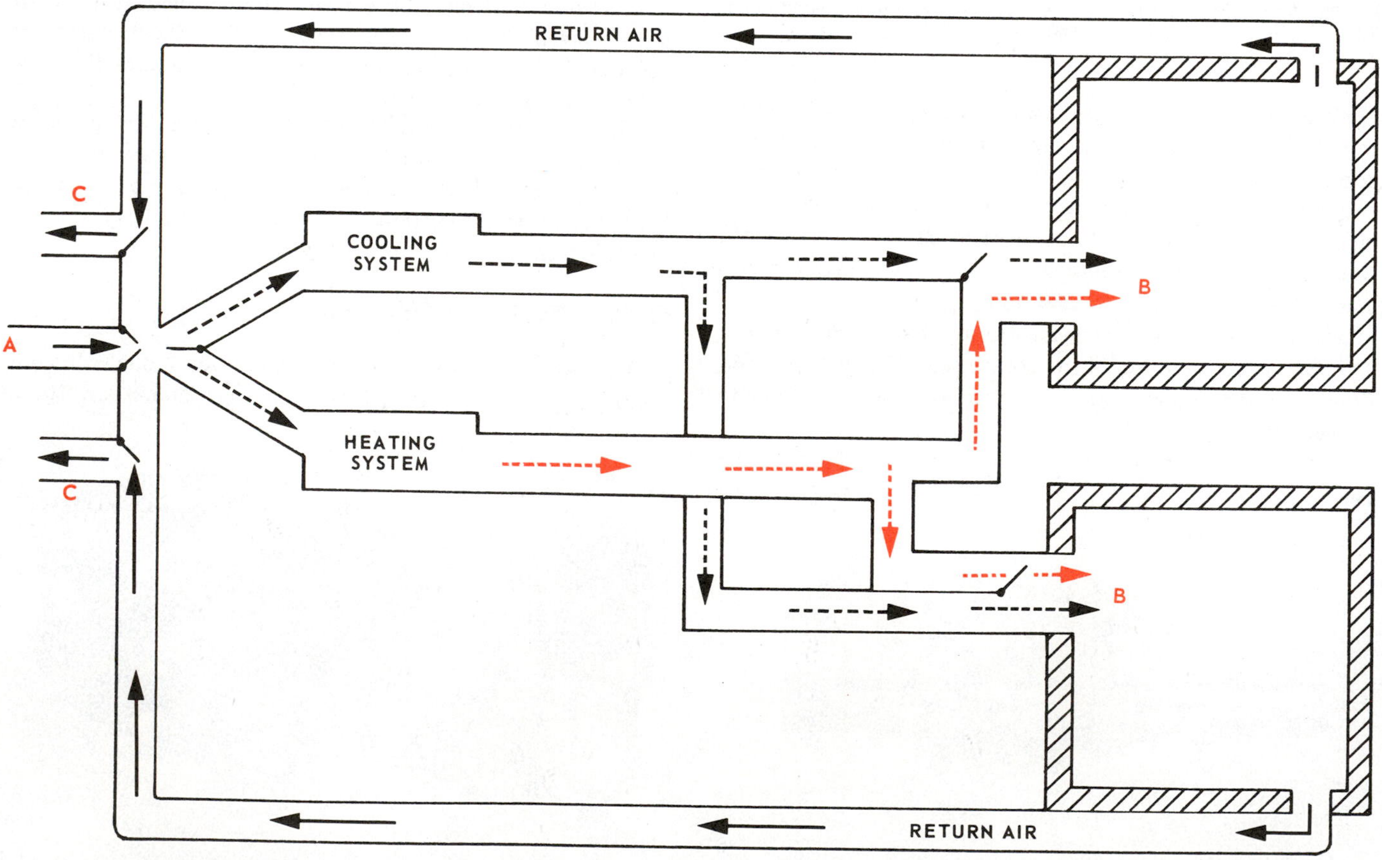

Fig. 23-73. This system has two supply ducts for complete air conditioning. A—Fresh air intake. B—Mixing plenums and diffusers. C—Exhaust air.

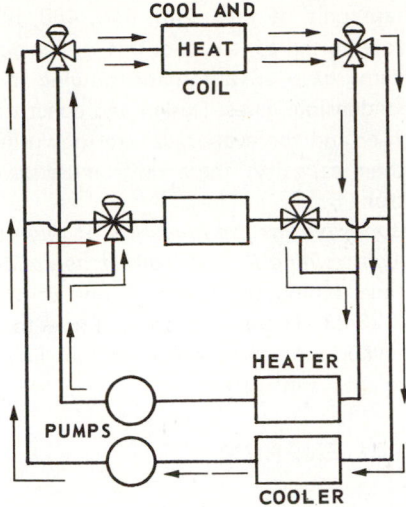

Fig. 23-75. Complete four-pipe system using water or a water and glycol mixture to move heat. Same coil is used for heating and cooling. Note four valves (two at each heat transfer unit). These valves, controlled by room thermostats, may be off-on or modulating type.

23-37 LARGE COMFORT SYSTEMS

Central station comfort cooling, part of a complete system, is available in many styles. One unit uses shell-and-tube construction in both the water-cooled condenser and water-chiller evaporator. This unit has a variable capacity system which unloads cylinders as the load decreases. These hydraulic controls also unload the compressor to minimize starting load.

Another air-conditioning system circulates chilled water to

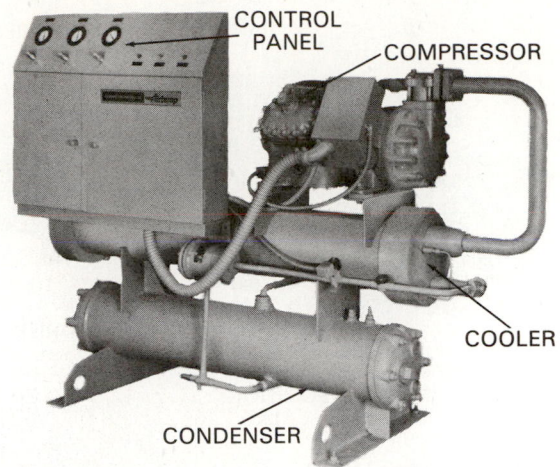

Fig. 23-76. Complete water chiller unit is typical of those used for industrial processing. (Airtemp Applied Machinery Co.)

the various cooling coils in the multiple installation. Chilled water systems are used for many industrial processing installations. One of these units is shown in Fig. 23-76. The three gauges shown are for suction pressure, oil pressure and high pressure. A compressor of 10 to 15-ton capacity is shown in Fig. 23-77.

A water chiller using a serviceable hermetic compressor is shown in Fig. 23-78. Its design permits the unit to be moved through regular doors. The chilled water evaporator has built-in freeze protection. The motor compressor is a six-cylinder unit with two cylinders in each bank. Fig. 23-79 shows internal construction.

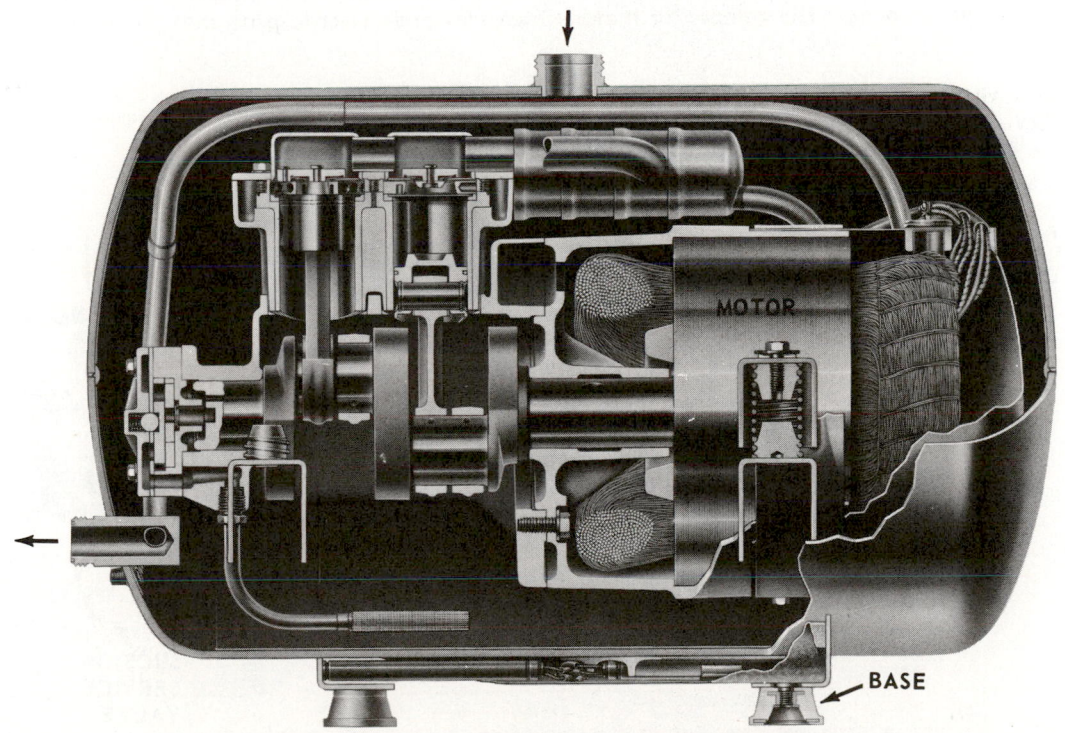

Fig. 23-77. Cross-section of "Hermeticom" hermetic compressor. It is for use in units ranging from 10 ton to 15 ton capacity.

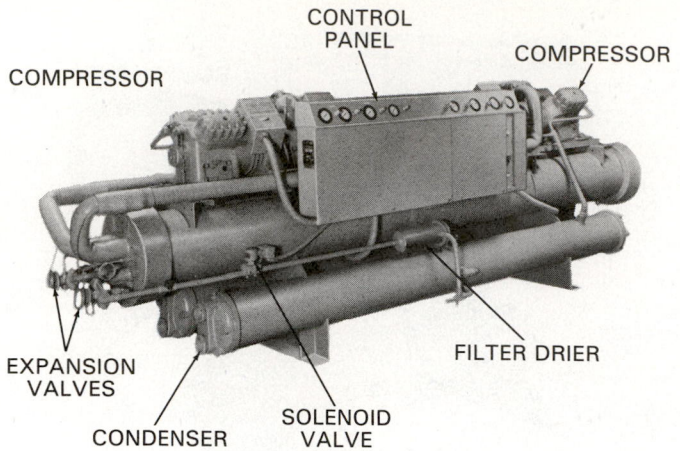

Fig. 23-78. This water chiller uses serviceable hermetic compressors. (Airtemp Applied Machinery Co.)

Many large comfort cooling installations use centrifugal type compressors. Large centrifugal units are frequently designed with capacities of 100 to 2000 tons. Their basic design is shown in Fig. 23-80. These systems use low-pressure refrigerants, and the evaporator operates at below-atmospheric pressures. Both condenser and evaporator are the shell-and-tube type. Water lines vary in size from 6 ips to 14 ips (internal pipe size). The compressor is two-stage centrifugal, driven by a hermetically sealed motor. Capacity is controlled by inlet vanes to the two-stage centrifugal compressor. The vanes may be either electronically or pneumatically controlled and hydraulically operated. During starting, vanes are closed to reduce the starting load. Fig. 23-81 shows compressor construction. This compressor has a forced lubrication system. A separate motor drives the oil pump. The compressor motor

is a three-phase unit of 208, 240, 440, 480, 550, 2300 or 4160 volts. Note bolted construction for service purposes.

These systems have an automatic purging device for removing noncondensible gases. Design and construction details of the condenser and the evaporator are shown in Fig. 23-82. Because of their capacity, these units must have thorough, accurate control.

Persons responsible for the operation of these units should receive thorough training in their correct operation. The complete wiring and piping schematic of one of these units is shown in Fig. 23-83. These units are also used to cool process liquids. The evaporator, compressor suction line and chilled liquid lines are always insulated.

23-38 TOTAL ENERGY

In recent years, many large buildings have been constructed using "total energy" or "single energy" systems. All the energy-use devices are designed to use up all the energy of combustion before discarding the byproducts. Thus, all the exhaust air, exhaust gases and water leaving the building are at ambient temperature conditions. The system tries to use each Btu or kilojoule and each watt of energy input.

In total energy systems, all the energy consumed is purchased in one form. This may be liquid, gas or solid fuel. The system generates whatever electricity is needed in the building.

The advantage is twofold. In the first place, a lower fuel rate can be obtained because of the large volume of fuel supplied. Secondly, more complete use is made of the energy released by burning this fuel.

In a total energy system, electricity is generated with reciprocating gas engines, gas turbines or steam turbines. The engines drive electric generators in the building. Hot water

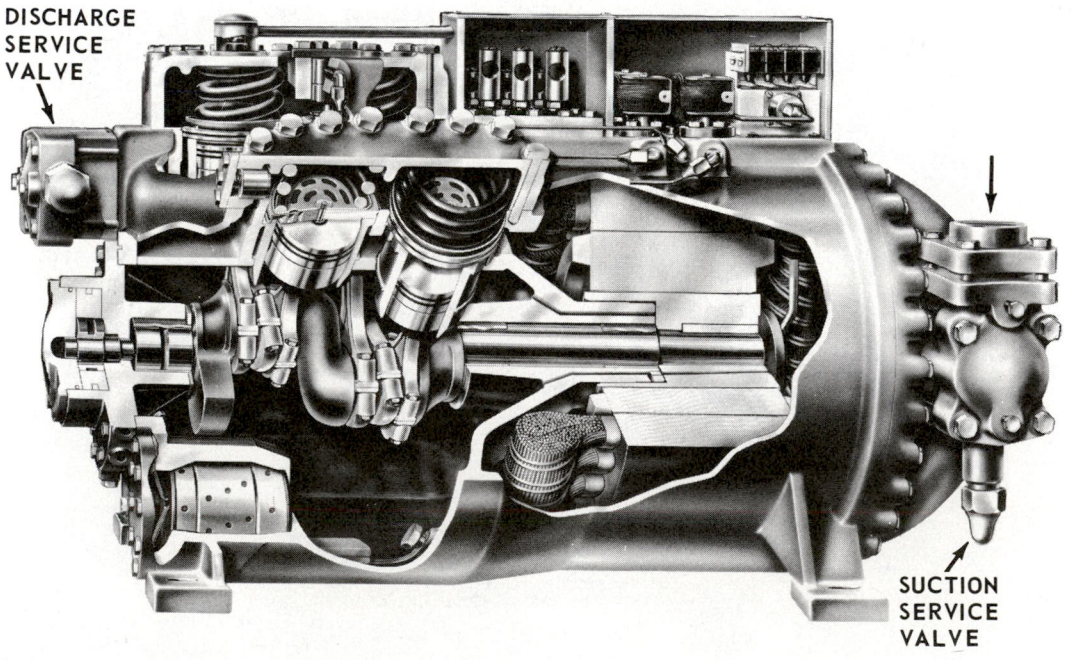

Fig. 23-79. Cutaway of a six cylinder serviceable hermetic compressor motor. Note arrangement of three piston rods on each crank throw. (Westinghouse Electric Corp., Air Conditioning Div.)

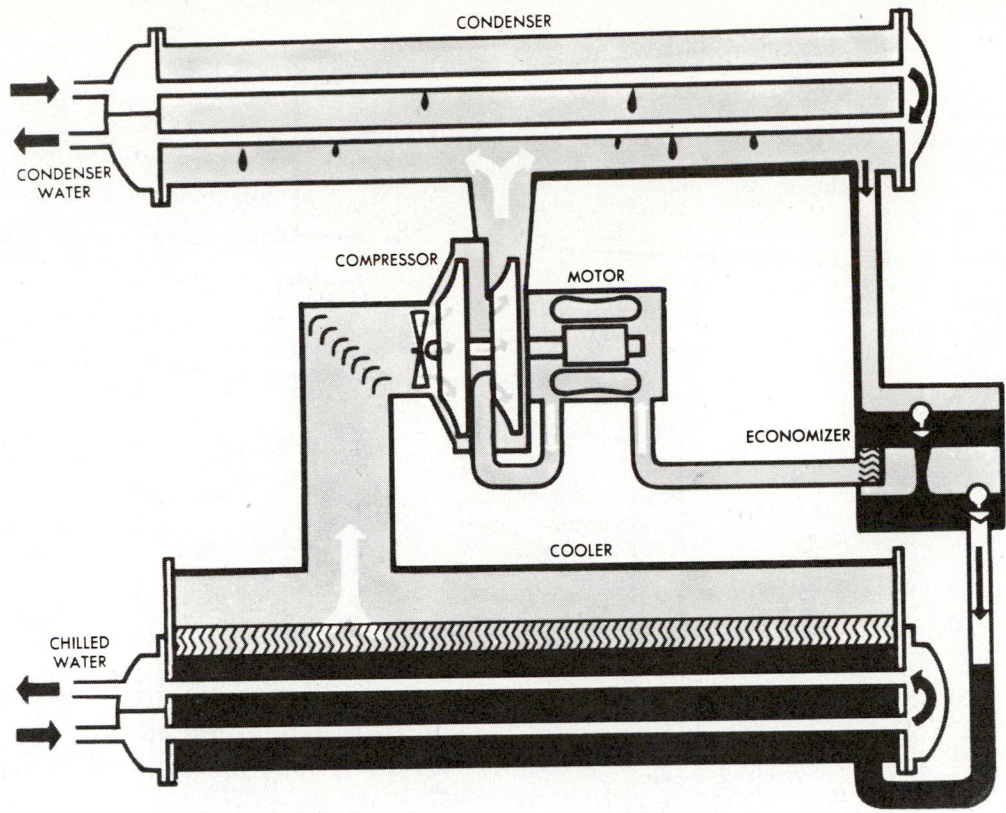

Fig. 23-80. Centrifugal compressor for hermetic chilled water system
uses rotors instead of pistons.
(Carrier Air Conditioning Group, United Technologies Corp.)

from the engine's cooling jacket is used as a prime source of heat. The exhaust gas is another major heat source. A gas engine or turbine uses about 33 percent of its fuel energy to generate electricity. Another 30 percent is used to heat the water-cooling jacket and about 30 percent is released in the exhaust gases. About 7 percent is used to heat the lubricating oil or is lost by radiation.

Heat in the jacket hot water, in the exhaust gases and in the hot lubricating oil can be converted to useful purposes such as:

1. Air conditioning (comfort cooling). Absorption systems often use the exhaust heat. At present, because exhaust gases cannot be cooled below 121°C (250 F.), only about 50 or 60 percent of the exhaust gas heat can be used.

2. Exhaust heat from turbines may also be used to raise the

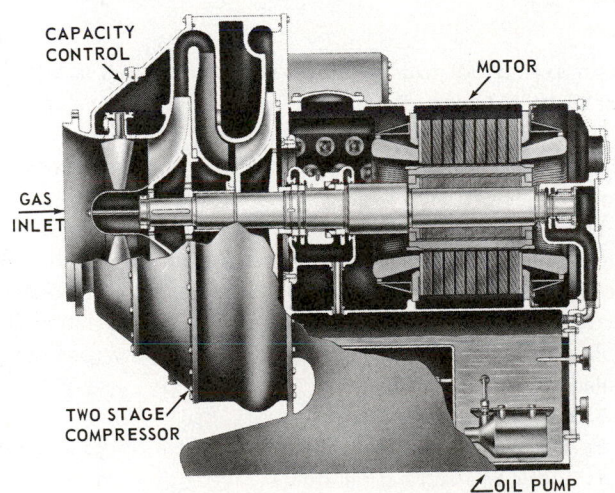

Fig. 23-81. Two-stage centrifugal compressor. Oil pump is driven by
separate power source.

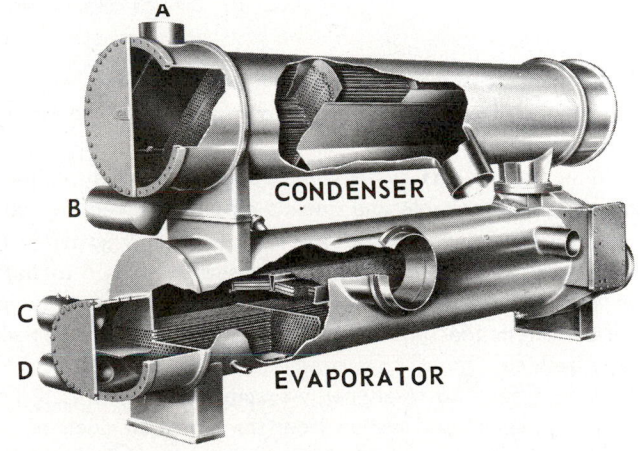

Fig. 23-82. Condenser and evaporator of large chilled water system which uses centrifugal compressor. A—Condenser water in. B—Condenser water out. C—Chilled water out. D—Chilled water return.
(Carrier Air Conditioning Group, United Technologies Corp.)

Heat Pumps and Complete Air Conditioning Systems / 833

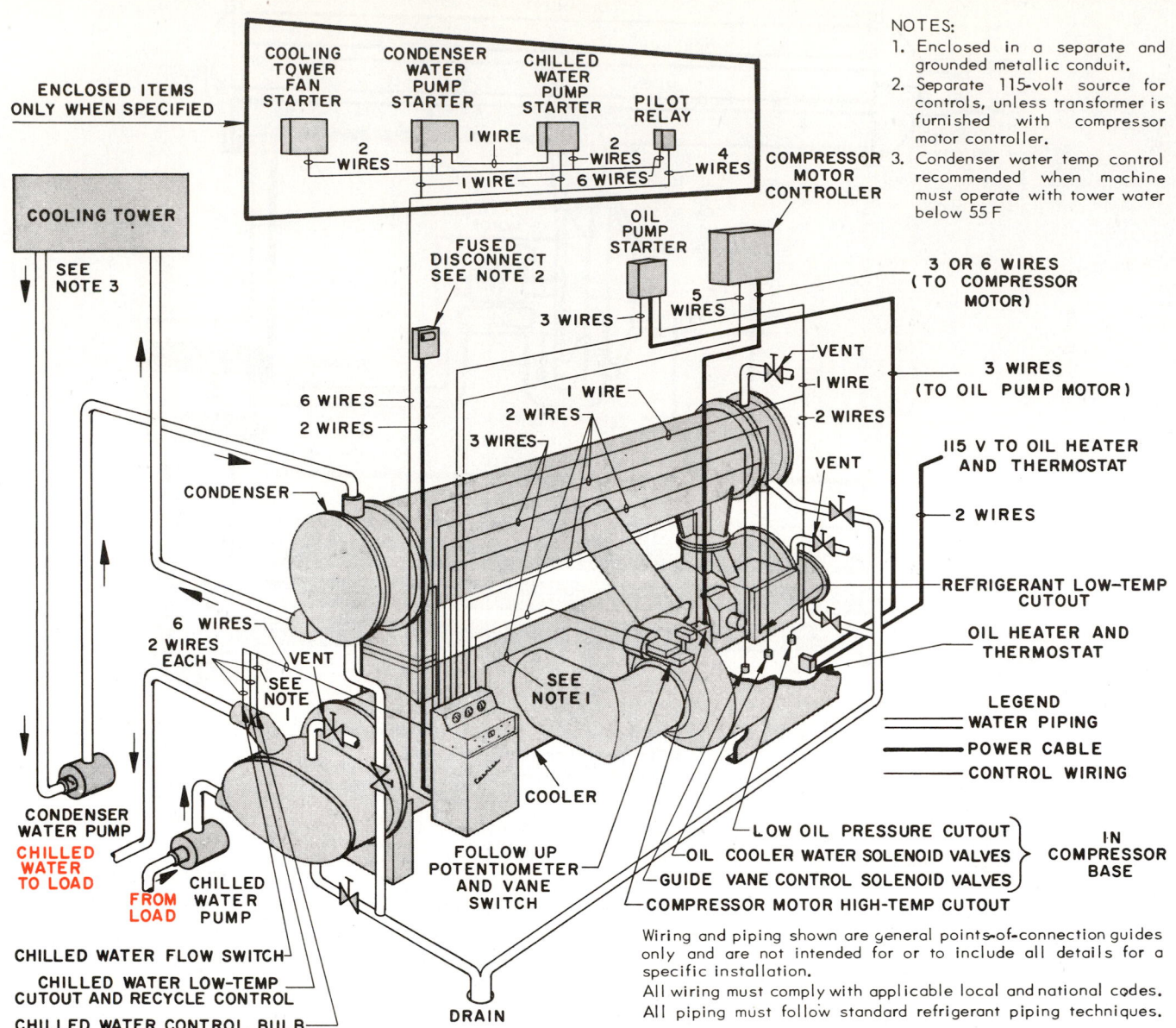

NOTES:
1. Enclosed in a separate and grounded metallic conduit.
2. Separate 115-volt source for controls, unless transformer is furnished with compressor motor controller.
3. Condenser water temp control recommended when machine must operate with tower water below 55 F

ENCLOSED ITEMS ONLY WHEN SPECIFIED

COOLING TOWER FAN STARTER
CONDENSER WATER PUMP STARTER
CHILLED WATER PUMP STARTER
PILOT RELAY

1 WIRE
2 WIRES
2 WIRES
4 WIRES

1 WIRE
6 WIRES

COMPRESSOR MOTOR CONTROLLER

COOLING TOWER

SEE NOTE 3

FUSED DISCONNECT SEE NOTE 2

OIL PUMP STARTER

3 OR 6 WIRES (TO COMPRESSOR MOTOR)

3 WIRES
5 WIRES

CONDENSER

VENT
1 WIRE
2 WIRES
VENT

3 WIRES (TO OIL PUMP MOTOR)

115 V TO OIL HEATER AND THERMOSTAT

2 WIRES

6 WIRES
2 WIRES
3 WIRES

1 WIRE
2 WIRES

REFRIGERANT LOW-TEMP CUTOUT

OIL HEATER AND THERMOSTAT

6 WIRES
2 WIRES EACH
VENT
SEE NOTE 1

SEE NOTE 1

COOLER

CONDENSER WATER PUMP

CHILLED WATER TO LOAD

FROM LOAD

CHILLED WATER PUMP

FOLLOW UP POTENTIOMETER AND VANE SWITCH

LEGEND
WATER PIPING
POWER CABLE
CONTROL WIRING

LOW OIL PRESSURE CUTOUT
OIL COOLER WATER SOLENOID VALVES
GUIDE VANE CONTROL SOLENOID VALVES
COMPRESSOR MOTOR HIGH-TEMP CUTOUT

IN COMPRESSOR BASE

CHILLED WATER FLOW SWITCH
CHILLED WATER LOW-TEMP CUTOUT AND RECYCLE CONTROL
CHILLED WATER CONTROL BULB

DRAIN

Wiring and piping shown are general points-of-connection guides only and are not intended for or to include all details for a specific installation.
All wiring must comply with applicable local and national codes.
All piping must follow standard refrigerant piping techniques.

Fig. 23-83. Schematic shows wiring and piping of large-capacity water chiller.

temperature of water during the heating season.

3. The heat from the water jackets can be used to furnish hot water.

In a total energy system, the heat produced by lighting, solar radiation, people and outside air are all taken into consideration. Each is a heat source for total energy system. The warm exhaust air, during the heating season, is used to warm the cold replacement fresh air entering the building. Cooled exhaust air in the summer cools the incoming warm replacement fresh air.

Many aspects of total energy systems have been used. For example, heat in one system from the return air ducts of the heating system is released through channels in lighting fixtures. The cool air passes across the lighted fixtures where it is heated. The needed heat for that room is allowed to enter through a duct. The surplus heat is passed on back to the primary duct system for use where needed in the building.

An example of total energy system is one used in a multi-apartment complex. A separate hot and chilled water plant is used for each independent apartment complex. A single electric power generator is used for the entire apartment complex.

The hot and chilled water plant provides the heating, air conditioning, and domestic hot water. A four-pipe heating and cooling system is used in each complex. The domestic water is circulated to each use area and then back for reheating. The cooling may be activated through high-efficiency gas-fired air cooled absorption chillers.

Three types of systems are in use:
1. Fuel is burned to create steam; the steam drives a turbine to power an electric generator.
2. Fuel is burned to drive an electric generator:
   a. Gas engine.
   b. Turbine engine.

3. Nuclear fuel is used to create steam.

In most cases, the exhaust gases are used to heat one or more of the following:
1. Hot water for heating or for consumption, or both.
2. Absorption cooling systems.
3. To heat or cool the fresh air intake.

In most cases, exhaust air (either cooled or heated) is used to cool or heat incoming fresh air. Likewise, waste cool or warm water is most often used to cool or heat incoming fresh water. Heat from lighting is often used to heat water or to heat air.

23-39 REVIEW OF SAFETY

Working on a complete air conditioning system involves most of the dangers which exist in refrigerating systems and heating systems.

Always wear goggles when working on these systems. Be sure the electrical service is off (and locked off) before working on the electrical circuits or on electrically powered parts such as the controls and fans.

Always install pressure gauges when checking the system, charging it or adding oil.

When working on rooftop units, use only a safe ladder and use it safely. Never carry equipment, tools or supplies up or down a ladder. Lift and lower these items using a rope with a safety tie-down handled by someone on the roof and another person on the ground. Avoid live electrical circuits when the roof is wet or snow covered.

Avoid having loose panels or loose light materials on the roof during a high wind or when it is gusty. Fasten the hinged panels of a rooftop unit to prevent uncontrolled swinging of these panels during a wind.

23-40 TEST YOUR KNOWLEDGE

1. How can heat be removed from 0 F. air?
2. What are the two operating cycles of a heat pump?
3. Under what conditions is a heat pump most efficient?
4. What is coefficient of performance?
5. Is it less expensive to heat with electrical resistors or with a heat pump? Why?
6. How many Btu/hr are generated by one kW·h through direct electrical resistance?
7. What are the various types of heat pump outdoor coils?
8. List one disadvantage of each type of outside heat pump coil.
9. What type of heat pump coil is used indoors?
10. What are the three ways to install the condensing unit of a residential central system?
11. What is the high speed used for in a two-stage heat pump?
12. What does the TEV control in the Hi/Re/Li system?
13. What does the check valve do in the Hi/Re/Li system?
14. Describe the five systems which are combined in a solar-supplemented heat pump system.
15. Why is the capillary tube especially advantageous to use in heat pumps?
16. In a heat pump, how many thermostatic expansion valves are used?
17. How many refrigerant connections does a four-way reversing valve have?
18. Explain the operation of a four-way valve used to reverse the refrigerant flow in a heat pump.
19. Describe the two ways a solenoid is used to operate a reversing valve.
20. How many check valves are used in a heat pump system equipped with thermostatic expansion valves?
21. What is done to remove frost or ice from an air outdoor coil?
22. How does well water release heat to or pick up heat from the heat pump?
23. When is the operation of the motor compressor most critical in a heat pump system?
24. What two electrical devices does a heat pump thermostat control?
25. What is an A-coil? Where is it used?
26. What parts of a warm-air furnace may be used with a central comfort cooling system?
27. What is carried in the ducts of a two-duct system?
28. What is carried in each pipe of a four-pipe system?
29. Name the main parts of a packaged-terminal air conditioner.
30. How many wrenches should be used when tightening a quick coupler connection?
31. Why do rooftop gas burners usually use electric ignition?

Chapter 24

AIR CONDITIONING
CONTROLS, CIRCUITS AND INSTRUMENTS

Modern homes, offices and work places are kept comfortable and healthful with automatically controlled temperature, humidity, air movement, air cleaning and air sterilization. Three important developments have improved heating and cooling systems:

1. Automatic controls developed to operate systems.
2. The electrical circuits which control and operate the automatic systems.
3. New instruments for the installation and service technician to aid in checking operation of systems.

It is important to understand the principles, design, construction, operation and correct use of these controls, electrical circuits and instruments. Unless the technician thoroughly understand them, many wasted hours may be spent trying to diagnose the problem.

24-1 CONTROLS

Controls are used on heating systems, cooling systems, humidifying systems, dehumidifying systems, combustion and flue systems. They control airflow and filter operation. Main types include:

1. Those that respond to temperature change.
2. Those that respond to pressure change.
3. Those that respond to liquid flow and/or gas flow.
4. Those that respond to liquid level.
5. Those that respond to time.

These controls have made it possible to develop safe, automatic systems.

Several types of control systems are in use:

1. Electric.
2. Pneumatic.
3. Electronic.
4. Fluidic.
5. Combinations.

These control mechanisms automatically turn the system on and off, modulate (adjust) certain operations and signal conditions. Such mechanisms are interlocked with safety devices. These systems control the correct balance of airflow and/or water flow.

The basic principles of many of these controls have been described in earlier chapters.

24-2 AIR CONDITIONING CONTROLS

The chief value of complete air conditioning lies in its accurate automatic operation. This operation requires many controls such as:

A. Temperature controls.
   1. Heating controls for:
      a. Coal heat.
      b. Oil heat.
      c. Gas heat.
      d. Electric heat.
      e. Steam.
      f. Water.
   2. Cooling and icing controls.
B. Humidifying controls.
C. Dehumidifying controls.
D. Airflow controls.
E. Filter controls.
F. Defrost controls.
G. Safety controls.

There are three basic groups of controls:

1. Operating controls.
2. Primary controls.
3. Limit Controls.

Operating controls are usually the thermostats which signal the start or the stop of a heating or cooling system. Primary controls insure a safe start and safe operation of the system. Limit controls are for safety. They will not permit a system to run unless all the safety conditions are in good order.

Operating controls start and stop a system through the primary controls when the limit controls permit these actions.

24-3 THERMOSTATS

Because heating and cooling systems affect temperatures, they need an operating control which starts and stops the system when correct temperature conditions are reached. Some thermostats modulate (increase or decrease the effect) instead of starting and stopping the system.

There are two basic types:

1. Heating thermostat.
2. Cooling thermostat.

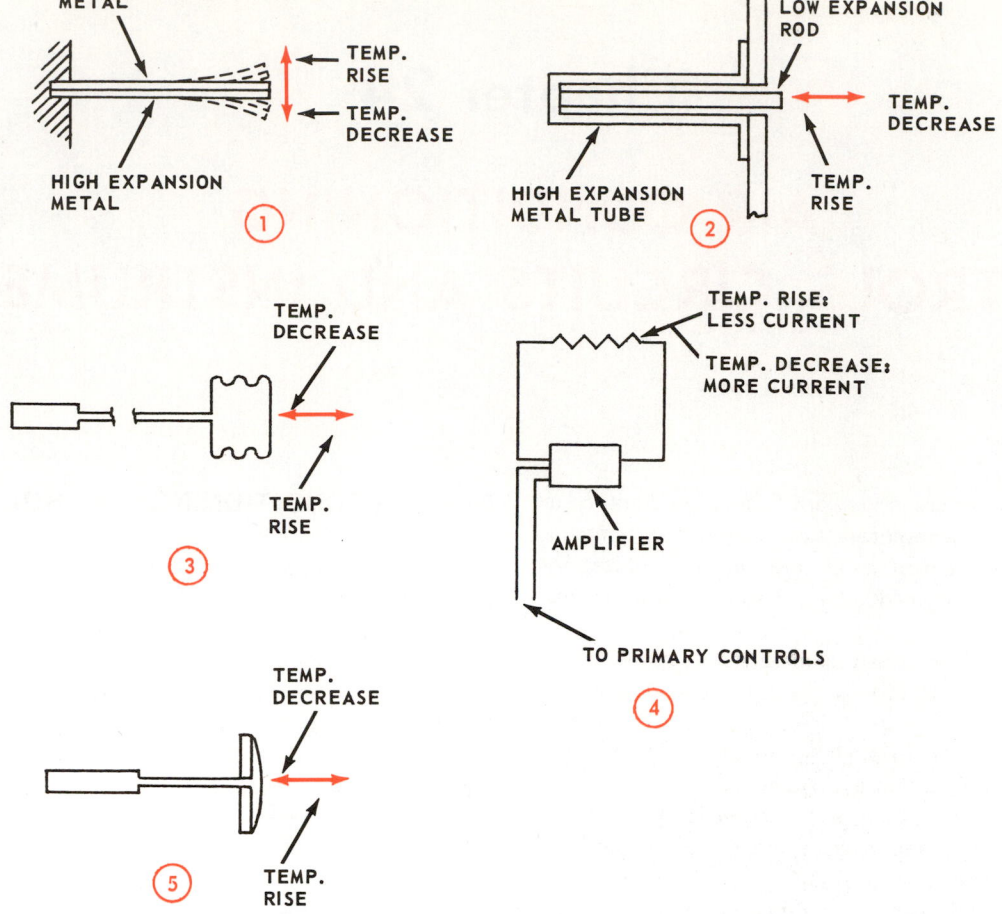

Fig. 24-1. Five basic types of thermostat operating elements. 1—Operates on principle of different expansion properties of two different metals. 2—Operates on principle of expansion of metal with heat. 3—Operates on principle of expansion of gas with heat. 4—Operates on principle of change of resistance in conductor or semiconductor with change in temperature. 5—Hydraulically operated diaphragm (100 percent liquid).

These may be combined into one unit which is called a heating-cooling thermostat. Some combination thermostats change over manually, while others do it automatically.

Time-operated thermostats are coupled with a clock mechanism. They will change the off-on settings for night and reset it again for day. Some units will also change the on-off settings for different days of the week (for example, off during Saturdays and Sundays).

Automatic furnaces and other heating devices are controlled by safety devices which turn the system on and off or stop it if abnormal conditions arise. These devices are mainly electrical, activated (tripped) by temperature, pressure or time. Relays are often used either to control full voltages by low voltage signals or to interlock signal devices for safe operation.

All systems use room thermostats. Warm air systems use a bonnet safety thermostat to shut off the system if the plenum chamber overheats. If automatic humidity is desired, humidistats are used.

Oil burners sometimes use a stack thermostat which will shut off the burner if the stack temperature does not rise within a few seconds after the oil burner is turned on.

Pressurestats are used in steam systems to shut them off if pressure becomes dangerously high. Each type of heating system has its own special automatic devices.

The device in the thermostat which reacts to temperature change may be one of several types:

1. Bimetal strip.
2. Rod and tube.
3. Bellows or diaphragm.
4. Electrical resistance.
5. Hydraulic.

Fig. 24-1 shows the elements of these five types. Another type of element for heating systems (boilers and hot water systems) is shown in Fig. 24-2.

Thermostats are further classified as:

1. Line voltage thermostats.
2. Low voltage thermostats.

Usually, line voltage thermostats are used directly in the electrical operating circuit as shown in Fig. 24-3. They are designed to be mounted on the baseboard or on the wall four to five ft. above the floor. Most have a manual "off" switch.

Low voltage thermostats are found in a low voltage circuit which is connected to a solenoid coil of relay switch. *These thermostats will be damaged if connected into a 120V line circuit.* They get their electrical power from a step-down transformer.

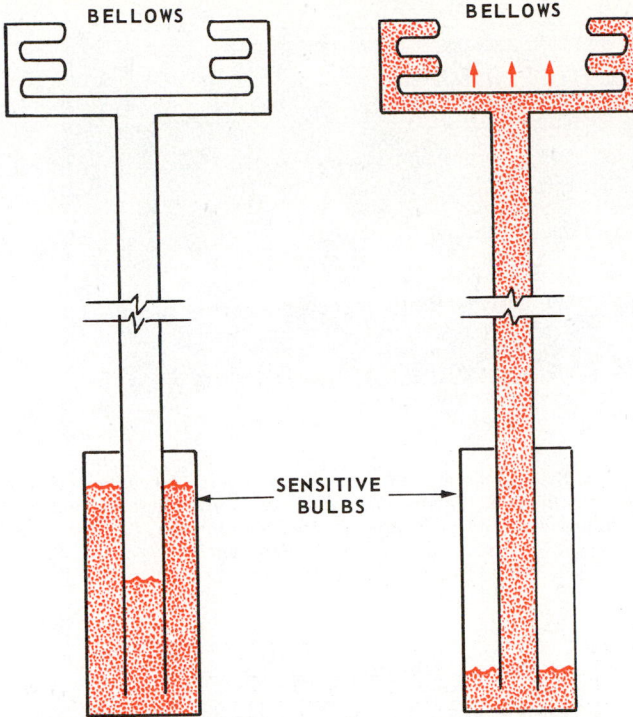

Fig. 24-2. Temperature sensing element for heating units such as boilers. Bellows is at the control and is the coolest. Sensitive bulb is placed where temperature is to be controlled. As bulb warms, it forces liquid into bellows area and moves bellows.

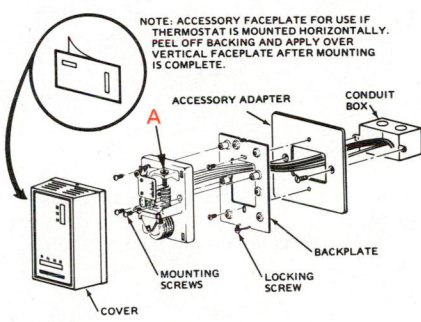

NOTE: ACCESSORY FACEPLATE FOR USE IF THERMOSTAT IS MOUNTED HORIZONTALLY. PEEL OFF BACKING AND APPLY OVER VERTICAL FACEPLATE AFTER MOUNTING IS COMPLETE.

ACCESSORY ADAPTER

CONDUIT BOX

A

BACKPLATE

MOUNTING SCREWS

LOCKING SCREW

COVER

Fig. 24-3. Line voltage heavy-duty thermostat, showing installation details. It controls electric heating circuit. A—Temperature adjustment. (Honeywell Inc.)

There are three types of heating thermostats:
1. Control of electrical circuits.
2. Control of air circuits (pneumatic).
3. Control of combination electric and pneumatic circuits.

Some of the newer thermostats are making use of solid state electronic devices. These include triads, transistors and amplifiers which control the functions of systems such as:
1. Power circuits.
2. Airflow.
3. Water flow.
4. Steam flow.
5. Damper operation.

This solid state control will provide the best operation for highest efficiency of the system. These same systems will

indicate, by instruments, lights and recorders, the conditions of all the variables in the system. Such complete controls are expensive. Therefore, until costs become more reasonable, they will be found only on large systems.

The unit which replaces the thermostat usually consists of a thermistor temperature sensor, a potentiometer temperature adjustment and an SCR (silicon controlled rectifier).

There are many types and models of thermostats designed for heating and cooling systems:
1. Voltage:
   a. Low voltage types — 24V.
   b. Line voltage types — 120V, 120/240V or 240V (also used for 208V).
2. Points:
   a. SPST (single-pole single-throw) two wire.
   b. SPDT (single-pole double-throw) three wire.
3. Selector switches:
   a. None.
   b. Winter — summer.
   c. Heat — fan.
   d. Heat on — heat off.
4. Thermometer:
   a. Yes.
   b. No.
5. Solid state thermostats.

Thermostats are supplied with or without heat anticipators. Their use varies. Some applications are:
1. Air — outside, inside, combination.
2. Cooling coil.
3. Heating coil.
4. Fan (exhaust or fresh air).
5. Fan — coil.

If thermostats have a large range of operation, they should have ambient temperature correction. For example, a thermostat designed to operate from 50 F. to 250 F. (10 C. to 121 C.), will not be accurate throughout this range unless it has a built-in correction. The correction is needed because all parts of the thermostat expand as they become warmer. The correction can be made by using a temperature-compensating bimetal strip which balances the expansion by moving slightly in the opposite direction.

Some three-wire thermostats have two sets of contacts. One set is for "pull-in" and the other is for "hold-in." The design has more stable contact connections.

The bimetal strip is usually wound in a spiral to get more length. This is needed to give more contact movement with each degree of temperature change.

These controls are designed to reduce contact points bounce (causes arcing) and contact "walking" (sliding over each other). The latter action also causes arcing.

Mercury contacts are used. Sealed tubes prevent dust, lint and oxygen from wearing the contacts. Inside the sealed glass tube is a ball or globule of liquid mercury and two or more electrical probes fused into the glass. Fig. 24-4 shows an SPST mercury switch, while Fig. 24-5 shows an SPDT switch.

The mercury tube thermostat wiring diagram shown in Fig. 24-6 is for a zone valve. When connections No. 5 and No. 4 are closed, the motorized zone valve opens. When the bimetal of

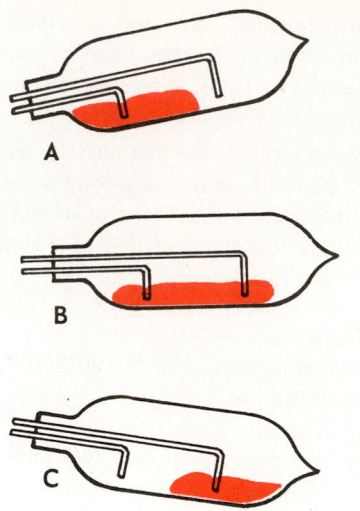

Fig. 24-4. How mercury switch works. A—Off for heating. B—On. C—Off for cooling system.

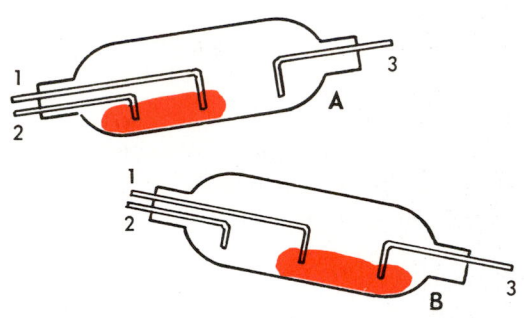

Fig. 24-5. A single-pole double-throw (SPDT) mercury switch. Mercury tube is usually tilted or moved by bimetal coil. A—Leads 1 and 2 are connected by mercury globule. B—Leads 1 and 3 are connected by mercury globule.

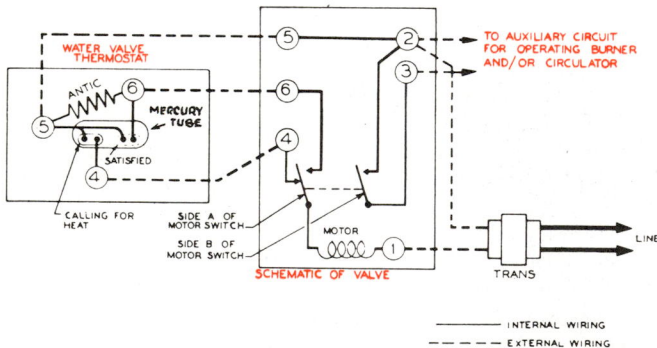

Fig. 24-6. Single-pole double-throw thermostat mercury switch. This thermostat opens and closes motorized zone valve in hydronic or steam heating system. This system can also control zone valve in chilled-water cooling system.

the thermostat tilts the mercury tube and connections No. 5 and No. 6 are closed, the motorized valve closes. A bimetal-operated thermostat with four mercury switches is shown in Figs. 24-7 and 24-8. Two are for stage No. 1 and stage No. 2

Fig. 24-7. Staging thermostat controls two stages of heating and two stages of cooling. A—Thermostat temperature adjustment, marked "C" for cooling and "H" for heating. B—Dial thermometer. C—Heat, cool, automatic switch. D—Fan switch.
(White-Rodgers, Div. Emerson Electric Co.)

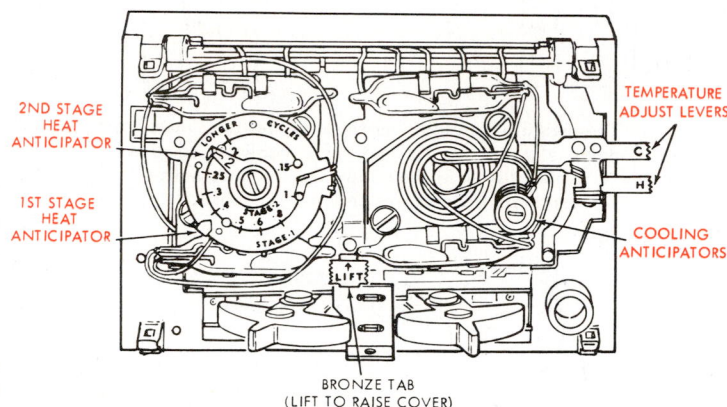

Fig. 24-8. Internal construction of staging thermostat. Two heating control mercury tubes are on the left and two cooling control mercury tubes are on the right.

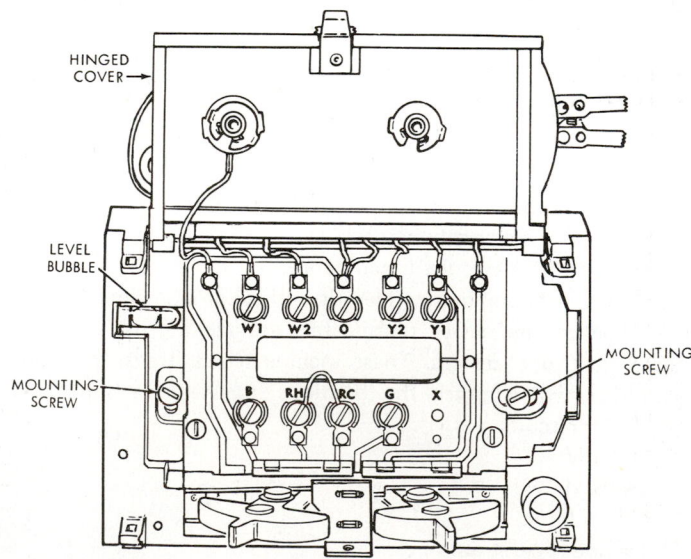

Fig. 24-9. Wiring terminals of staging thermostat.
(White-Rodgers, Div. Emerson Electric Co.)

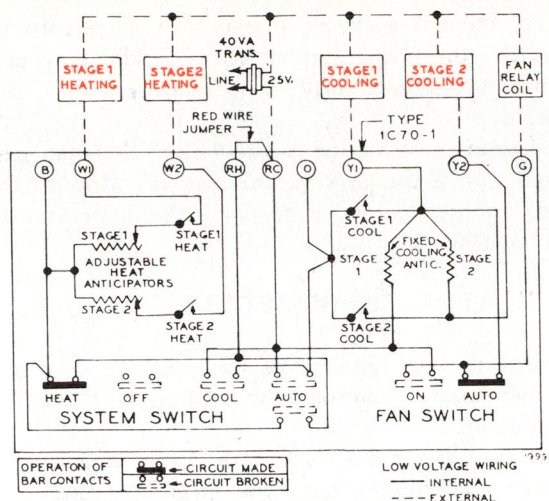

Fig. 24-10. Schematic wiring diagram of staging thermostat heating and cooling system.

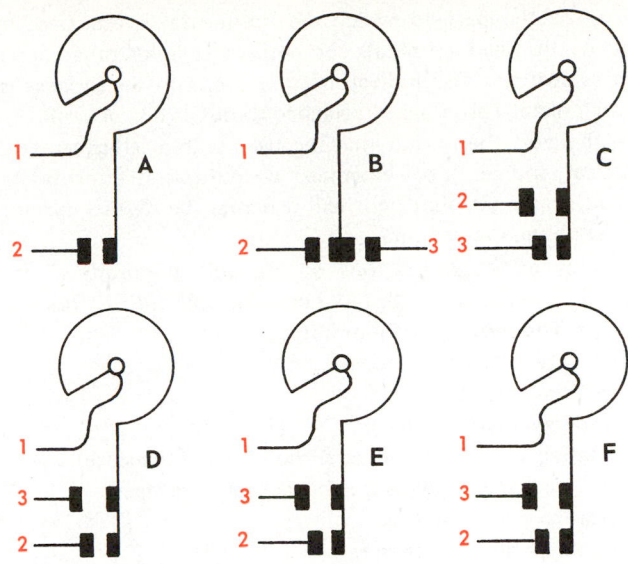

Fig. 24-11. Schematic of thermostat classification: A—Single line break thermostat. B—Heating-cooling thermostat. C—Double line break disconnect. D—Double line break cycling. E—Two-stage thermostat. F—Two-circuit thermostat. 1—Common lead. 2—First controlled lead. 3—Second controlled lead.

cooling. Two are stage No. 1 heating and stage No. 2 heating. The wiring connections for this control are shown in Fig. 24-9, while the schematic wiring diagram is as shown in Fig. 24-10.

Some heavy-duty thermostats are designed to carry high ampere flow. They can carry as much as 16A at 120V and will not be damaged by as much as 96A locked rotor amperes for a short time.

24-4 THERMOSTATS — LINE — ELECTRIC

Line thermostats are either 120 or 240V. They are designed to control as much as 22A. However, many electric heating experts recommend low voltage thermostats for most electric heating systems.

Wiring to line thermostats must be installed according to the local code. Thermostats may have a fan switch mounted on the side. The fan has an off-on automatic setting. Off position means blower is off. On position means the fan is on continuous operation. "Auto" (automatic) means the fan will be controlled by the fan thermostat in the furnace plenum chamber.

The thermostat operates the electrical load directly or through a line voltage operating control. They are designed for 31 to 300V and 24A maximum for noninductive circuits (resistances only — no coils such as solenoids or transformers or motor windings are in the circuit).

Some are designed for 31 to 300V with six different amperage capacities for inductive circuits (solenoids, transformers and motors). These capacities are: 3, 6, 8, 10, 14 and 16A at 120V, full load.

The capacities are: 18, 36, 48, 60, 84 and 96A at 120V — locked rotor (starting load) on motor.

At 240V, one will use half these ampere ratings. Thermostat Classifications:

A. Single line break (single-pole) thermostat:
1. Heating only.
2. Cooling only.
B. Heating and cooling.

C. Double line break disconnect thermostat.
D. Double line break cycling thermostat.
E. Two-stage thermostat.
F. Two-circuit thermostat.
    Fig. 24-11 is a schematic of thermostat classifications.

24-5 THERMOSTATS — LOW VOLTAGE

Low voltage thermostats carry only a small amount of electrical power. This is an advantage because line voltage thermostats have problems with the electrical power heating the thermostat and opening the contacts too soon. This extra heating temperature rise is called thermostat droop.

Example: Thermostat is set for 72 F. (22 C.) and room is at 68 F. (20 C.). As room heats up, the thermostat becomes warmer than room temperature because of electric resistance heat of electrical parts in the thermostat. The thermostat then reaches 72 F. (22 C.) and cuts off too soon (about the time the room is 69 F. or 70 F.).

Low voltage thermostats should use No. 18 American Wire Gauge (AWG) for distances up to 50 ft. Use No. 16 AWG wire for distances over 50 ft.

Many thermostats have three wires or four wires. Each of these wires is a different color. Red, white, green, yellow and black are popular. *Be sure to wire as recommended in the wiring diagrams provided by the manufacturer. This is required.*

These controls are rated as follows:
0 to 15V — Maximum of 5A
15 to 30V — Maximum of 3.2A
They control the applied energy to a low voltage control of a Class 2 (power) circuit.

The thermostat mounting usually requires other parts. The

subbase is the part to which the thermostat is attached. It contains the lead terminals. Sometimes it holds other parts such as switches and indicator lamps. Beneath the subbase is the wall plate. This is a part attached to outlet box or wall.

A millivolt thermostat may be used with a self-generating electrical source. For example, a multiple thermocouple located near a gas pilot light will generate the needed current to operate the thermostat.

Groups and classifications of millivolt thermostats are:
Group 1. Thermostats with fixed or adjustable anticipators.
Group 2. Thermostats without anticipators.

There are 12 classifications:
1. Heating only (one-stage).
2. Cooling only (one-stage).
3. Heating and cooling, manual changeover (one-stage).
4. Heating and cooling, auto changeover (one-stage).
5. Heating only (two-stage).
6. Cooling only (two-stage).
7. Heating (one-stage); cooling (two-stage); manual changeover.
8. Heating (one-stage); cooling (two-stage); auto changeover.
9. Heating (two-stage); cooling (one-stage); manual changeover.
10. Heating (two-stage); cooling (one-stage); auto changeover.
11. Heating (two-stage); cooling (two-stage); manual changeover.
12. Heating (two-stage); cooling (two-stage); auto changeover.

24-6 THERMOSTAT — THERMOMETER TYPE

Thermometer type thermostats consist of a glass tube and a mercury-filled thermometer. Two leads are inserted in the glass column and contact the inner tubing in which the mercury column moves. When the mercury column rises as the temperature increases it will close the connection between the two contacts. The closed circuit can either energize (turn on) an alarm, or equipment such as a cooling system, or disconnect a heating system by energizing a cut-off relay, Fig. 8-70.

24-7 PORTABLE THERMOSTATS

Portable thermostats are available which may control a furnace or air conditioner. The thermostat electronically activates a responder on the furnace or air conditioner up to a distance of about 120 ft.

The thermostat may be moved from room to room and responds to the temperature of that room. Since it is an electronic battery-powered unit, it has no moving parts. It has an accurate adjustable differential from 0 to about 5 F. (0 to about 3 C.) with a range of 45 to 90 F. (7 to 32 C.).

Once every minute, the thermostat senses the temperature. If heating (or cooling) is necessary, the thermostat sends a radio signal to the responder.

24-8 THERMOSTAT GUARD

One of the biggest problems with thermostats is their ease of adjustment. Whenever there is more than one person in the building, there is a chance that two or more persons will change the thermostat setting. These actions may produce uncomfortable results, may be irritating and may even be dangerous.

One solution is to use a thermostat that can only be adjusted with a special key. This key is controlled by one person. Another method is to cover the thermostat with a clear plastic lockable cover.

24-9 HEATING THERMOSTATS

Smaller heating systems usually have the open contact point thermostat. It controls either a 24 or 120V circuit to the primary controls.

Fig. 24-12 shows the inside of a typical room thermostat designed to mount on a wall. A coiled bimetal strip reacts to temperature change. It has a range of 35 or 50 F. low setting to 90 or 95 F. high setting.

Fig. 24-12. Inside view of thermostat shows coiled bimetal element. (Penn Controls, Inc.)

These thermostats usually operate on a .5 F. to 2 F. differential or smaller. This allows close temperature control. Fig. 24-13 shows more details of a thermostat operated by a bimetal strip.

Generally, the sensitive element is a bimetal strip. A small magnet is sometimes used to produce snap action in the points. A range adjustment is usually a direct force on the bimetal, while the differential usually consists of moving the small magnet either closer to or farther away from the bimetal strip.

Some thermostats are hydraulically operated. Fig. 24-14 shows a diaphragm mechanism. The diaphragm moves as a liquid expands and contracts. The dial adjusts the setting and the small magnet produces snap action in the thermostat.

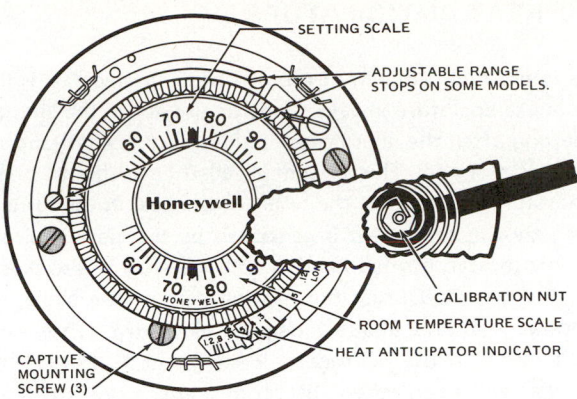

Fig. 24-13. Heating thermostat which uses bimetal strip. (Honeywell, Inc.)

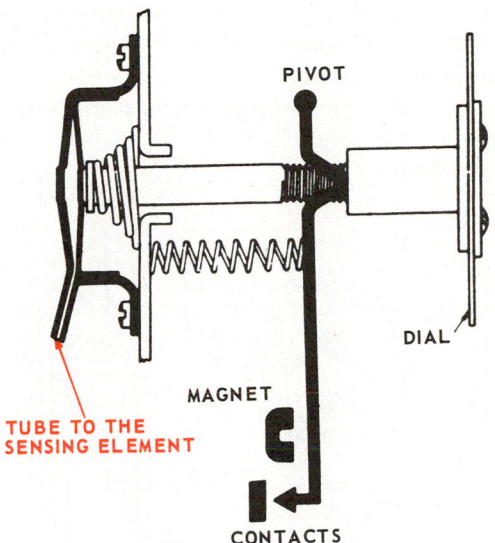

Fig. 24-14. Schematic diagram of hydraulically operated thermostat. (White-Rodgers Div., Emerson Electric Co.)

Electronic measurement of an electric resistance element may be used for temperature control. It is possible to obtain differentials of .01 deg. using this method.

Thermostats are rated by the voltage they carry and also by the controls to which they are electrically connected. These controls open the circuit when the temperature rises and complete the circuit when the temperature falls.

One 24V thermostat has three wires connected to it. The wires make a contact on temperature drop and close a holding circuit. They will open the circuit on a rise of temperature. This thermostat must be used with a relay. The relay opens and closes the 120V circuit for the oil burner, gas burner or electric heater.

Another 24V thermostat has separate controls. One opens the contacts on rise in temperature and one closes contacts on a rise of temperature. It is used on gas burners, oil burners or electric heaters.

A third type is a two-wire 24V thermostat that controls 24V furnace devices directly.

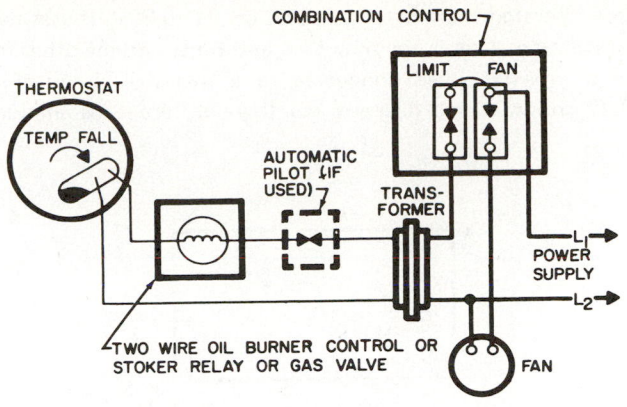

Fig. 24-15. Wiring circuit of a heating thermostat. Note that automatic pilot and fuel control are in the low-voltage (24V) circuit.

A wiring circuit is shown in Fig. 24-15 for a control which uses a mercury bulb contact device. The circuit is designed for forced warm-air heating.

The location of thermostats is important. They must not be exposed to drafts, to hot or cold walls or to sunlight. They must be placed in an average-temperature location. An inner wall is a popular place. Mounting should be about five feet above the floor. If closer to the floor, the temperature will be more uniform; however, the thermostat is more likely to be disturbed by children or furniture.

If lint should lodge between the points, an open circuit will result. To clean contact points use a piece of clean paper. If a mercury tube is used instead of contact points, the thermostat must be carefully leveled. Use a plumb line or spirit level.

Another style of thermostat has been developed for air conditioning systems. These units have easily operated settings. They are designed with all the various types of electrical bases to enable the thermostat to control either heating, cooling or both heating and cooling.

Some heating-cooling systems use a clock-operated thermostat. It changes the temperature range during the night and returns it to normal during the day. Fig. 24-16 shows a

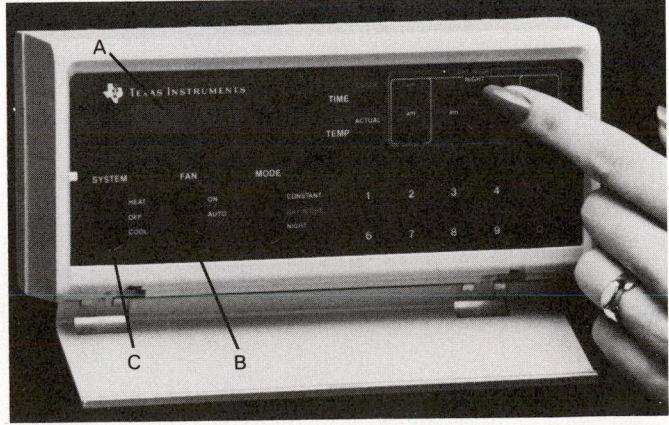

Fig. 24-16. Electronic digital thermostat. This may be used for either heating or cooling. A—Digital time-temperature display. B—Auto-on fan control. C—Cool-off-heat switch.   (Texas Instruments Inc.)

clock-operated thermostat powered on 24 to 30V. It uses two transformers. One is for the clock and timer and the other for the thermostat when connected to a two-wire circuit. Fig. 24-17 shows wiring diagrams for three different installations

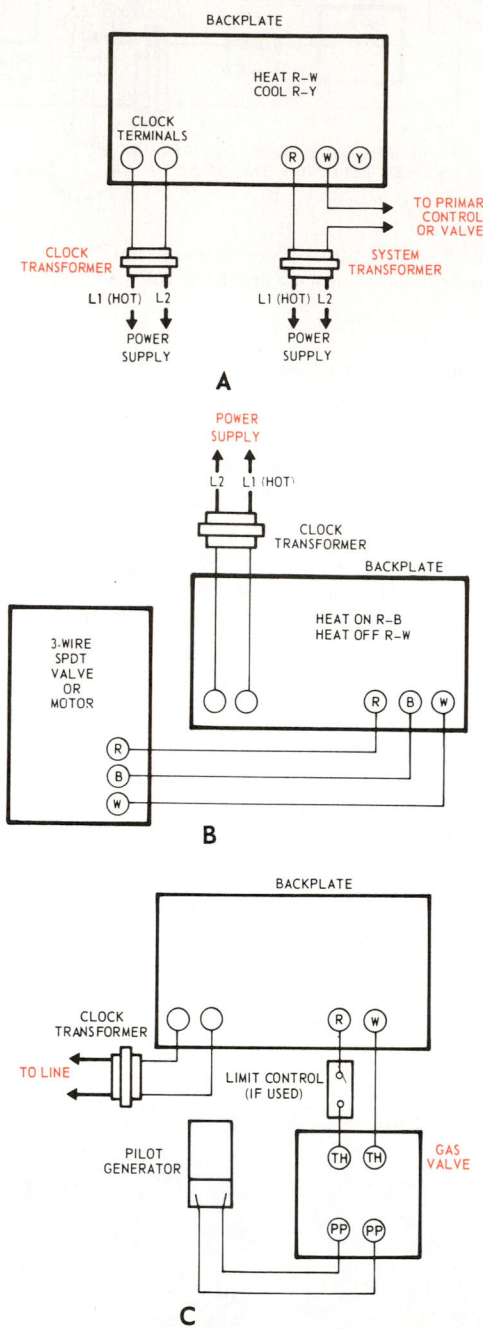

Fig. 24-17. Three different clock thermostat wiring diagrams. A—Two-wire low-voltage system. B—Three-wire system. C—Multivolt pilot generator system. (Honeywell, Inc.)

using two and three-wire clock thermostats.

Frequently a unit will short cycle because the thermostat is exposed to vibration (on a shaky wall, or stairs). *Mount the thermostat in a firm position.*

24-10 HEAT ANTICIPATORS

An interesting problem in room heating thermostats is that the room temperature always tends to rise above the thermostat setting after the thermostat points have opened and the burner has stopped. This action is called "overshoot" of the thermostat. This action is the result of the heat in the furnace.

Even though some heat is generated by the electric circuit in the thermostat, a small heating coil (resistor) is also placed in the thermostat. During the heating part of the cycle, the thermostat is always about 1 deg. F. (0.56 deg. C.) warmer than the room. If the thermostat is set for 74 F. (23 C.), the thermostat will open when the room temperature is actually 73 F. (22.8 C.). Then, after the thermostat has opened, the room temperature will still rise to 74 F. (23 C.), because of the heat in the already heated furnace.

These small resistance coils are usually called heat anticipators. Fig. 24-18 shows an electric heating thermostat equipped with an anticipator. The heat anticipator is connected in series with the operating contact points.

Both heating and cooling anticipators for thermostats are of two types:

1. Fixed anticipators.

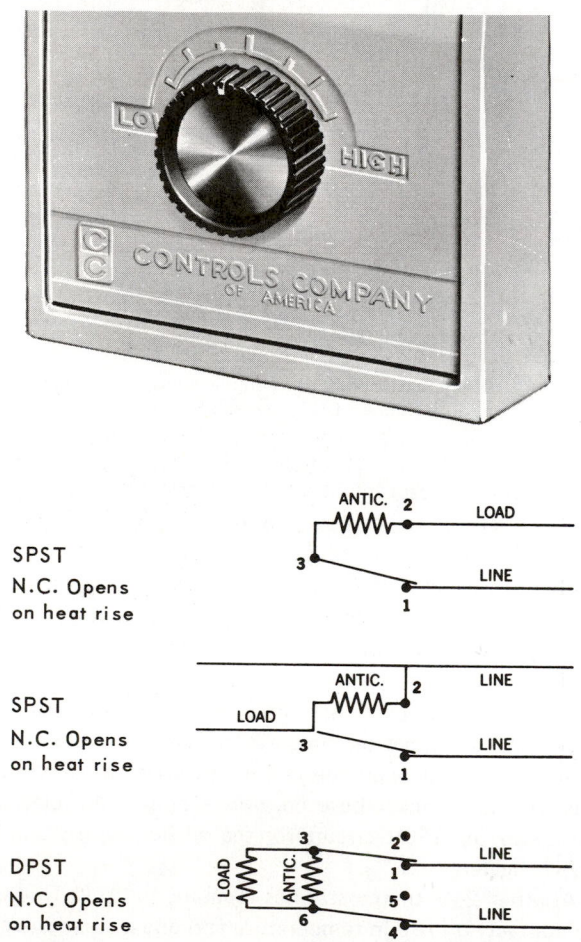

Fig. 24-18. Thermostat used in electrical heating systems. Above. External view. Below. Wiring diagrams for thermostat. Note that it is equipped with an anticipator. The SPST indicates that control has single-pole single-throw switch. (Controls Co. of America)

2. Adjustable anticipators.

They are most often used on low voltage thermostats and on heating-cooling combination thermostats. The size of the anticipator (current capacity) is related to the current (amperes) that flows through the thermostat contacts. This current varies between .15 to 1.0 amperes depending on the controls connected to the thermostat.

Fixed anticipators' current flow should equal the control ampere flow.

The adjustable anticipator should be regulated to the same ampere flow as the control circuit. The current flow of the control circuit is shown on the label of the control.

If the anticipator has a higher rating than the control circuit, it will warm the thermostat more slowly since there will be less anticipator heat. The reverse is also true. Anticipators are used in 24 V systems and also in 750 millivolt systems.

Cooling thermostats may have a heat anticipator connected in parallel with the thermostat points. The anticipator becomes warm during the off cycle and will close the thermostat points about 1 F. (.56 C.) warmer than the room temperature. If control is set to cut in at 78 F. (26 C.), the control will actually start the system at 77 F. (25 C.). This will allow the system to start cooling the room before it becomes too warm and humid.

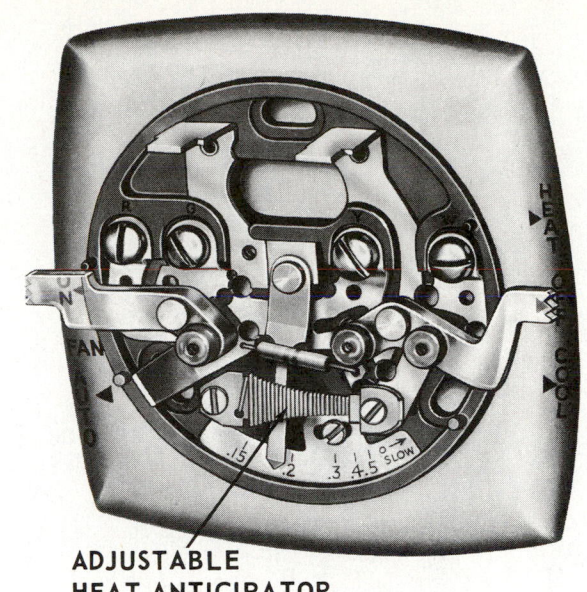

ADJUSTABLE HEAT ANTICIPATOR

Fig. 24-20. Inside of thermostat. Variable heat anticipator is shown. (Penn Controls, Inc.)

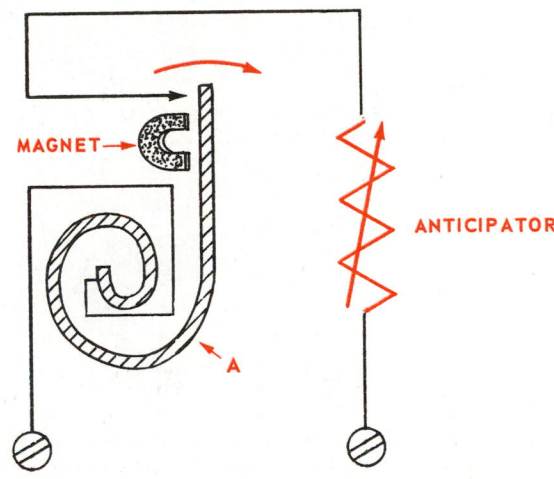

Fig. 24-21. Schematic wiring diagram of heating thermostat equipped with anticipator. A—Bimetal strip.

Fig. 24-19. Combination heating and cooling thermostat which has heat anticipator. A—Switch setting for heating. B—Off position for heating-cooling switch. C—Switch position during cooling season. D—Control switch for fan operation. Top scale is temperature setting or adjusting scale 50—90. Bottom scale is room temperature (72 F.).

Different anticipator heaters are used depending on the system. Fig. 24-19 shows the outside of a combination thermostat with a heat anticipator. The inside of this thermostat is shown in Fig. 24-20. The more residual heat in the heating system, the larger the capacity of the anticipator. Fig. 24-21 shows a heating thermostat wiring schematic diagram.

At the end of the off part of the heating cycle, when the thermostat points close, it takes time for the furnace to heat and to move this heat to the room. The room temperature

may fall about 1 F. (.56 C.), during this start-up lag. This action is called "system lag." The cut-in temperature is usually set about 1 F. above the lowest temperature desired.

24-11 ELECTRIC HEATING THERMOSTATS

Electric heating units use various controls:
1. Thermostats.
2. Relays.
3. Sequence relays.
4. Limit switches.

Thermostats are generally used for temperature control. Some are line thermostats and some are low voltage thermostats. Line thermostats are usually rated at 5000W at 120V a-c. See Fig. 24-22. The inside of such a control, operated by sensitive bulb, is shown in Fig. 24-23.

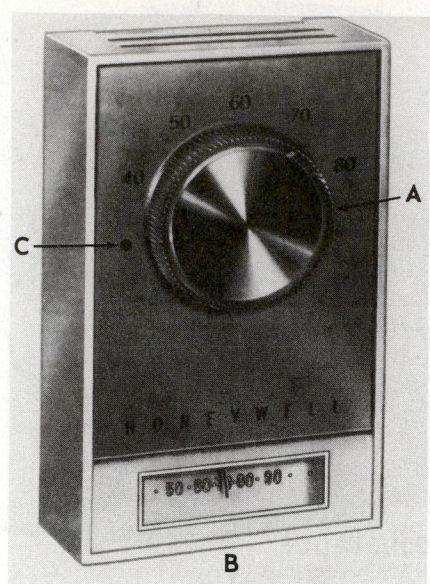

Fig. 24-22. Two-switch wall thermostat used for controlling electric heating circuit. It is line voltage control. One switch is operated by thermostat and one opens manually when setting is turned to "off." A—Temperature setting. B—Thermometer. C—Off position.

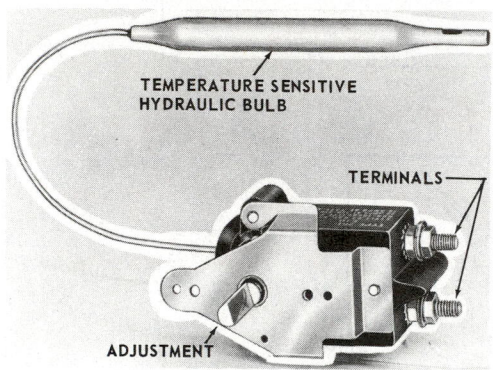

Fig. 24-23. Hydraulic type thermostat used on electric resistance heating units. Rated at 240 to 277V it can control up to 5000W. Its 1 1/2 deg. differential provides uniform temperature control.
(White-Rodgers Div., Emerson Electric Co.)

Units are equipped with such safety controls as temperature limit switches. Large-capacity units use a relay along with low voltage thermostats to control high-wattage loads. Some prefer to use low voltage thermostats with a relay on all electric heating systems. It reduces "heat lag." Also sometimes used is the sequence control. It is made up of a block of relays. Each relay controls a separate heating circuit. The relays are activated one at a time. This sequence timing prevents placing the full electrical load on the main building circuit at one time.

Some codes require that both leads to the heating element be opened by thermostatic control. Fig. 24-24 shows such a thermostat.

Thermostats mounted on baseboard heaters are usually encased in attractive covers and have a calibrated dial mounted on the adjustment knob. Fig. 24-25 shows such an installation.

Some line voltage thermostats produce a wide swing of temperatures causing room temperatures to vary too much.

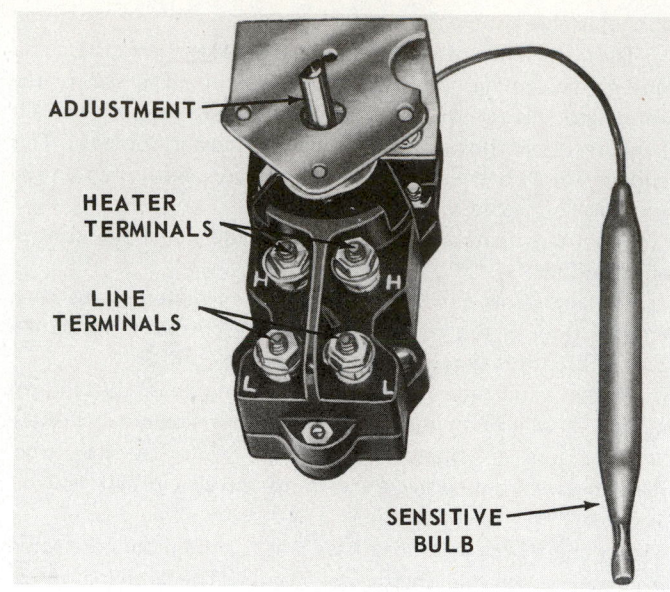

Fig. 24-24. Hydraulic type thermostat used on electric resistance heating. It controls both electrical leads to heating unit.

Others may have a temperature droop and the temperature of the room goes too low before heating effect is felt.

For example, if the thermostat is set for 75 F. with a 1 F. differential but room cools to 70 F. and then warms to 80 F., the 10 F. difference is too much swing.

If the thermostat is set for 75 F. with a 1 F. differential and room only heats to 65 F. when the thermostat shuts off, the 10 F. below normal is a droop. This condition is usually caused by too much heat release in the thermostat from wiring or too much heat from the anticipator.

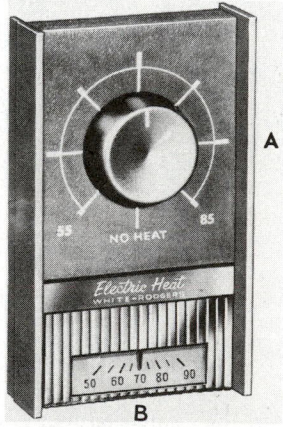

Fig. 24-25. Thermostat designed to be mounted on electric resistance baseboard heater casing. A—Temperature setting. B—Thermometer. Note "no heat" or off position.

When adjusting a thermostat, be careful to avoid heating the thermostat with the hands or by breathing on the element. If the thermostat is warmed above the room temperature, wrong settings of the thermostat will result.

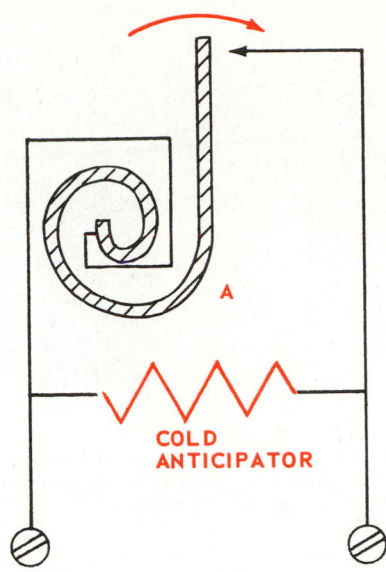

Fig. 24-26. Schematic wiring diagram of cooling thermostat equipped with cold anticipator. A—Bimetal strip.

24-12 COOLING THERMOSTATS

Comfort cooling thermostats are similar in design to heating thermostats except that the contacts open as the room cools and close as the room warms up.

The popular bimetal thermostats usually have a 1 F. (.6 C.) differential. Both the 24V and 120V controls are available.

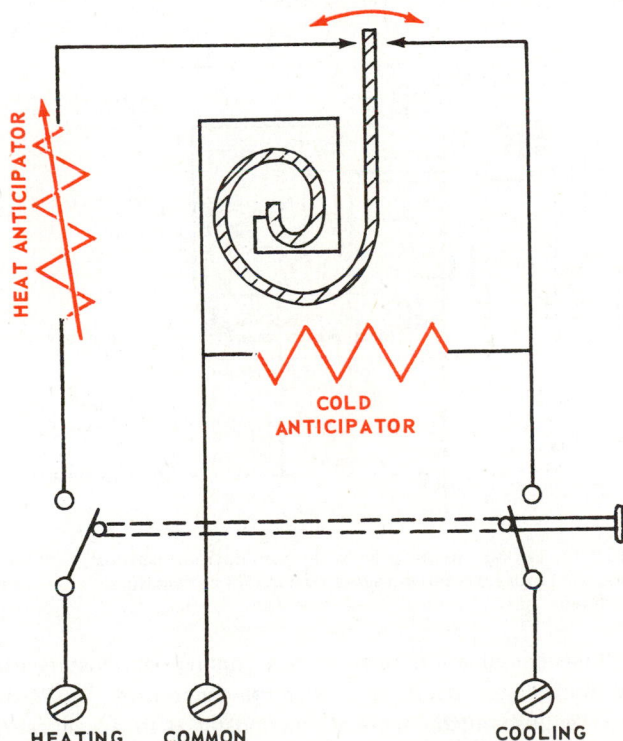

Fig. 24-27. Schematic of bimetal heating-cooling thermostat. Unit is equipped with heat anticipator and cold anticipator.

Some units have cold anticipators. These small electric resistance elements are in parallel with the bimetal strip. They tend to close the bimetal-controlled points just a little before the unit reaches the room cut-in temperature.

Fig. 24-26 shows a simplified wiring diagram of a cooling thermostat. The cold anticipator warms the bimetal strip during the off-cycle but is bypassed on the on-cycle. This heat during the off-cycle tends to turn the system on just a little ahead of normal.

24-13 COMBINATION THERMOSTATS

Some thermostats have control mechanisms for both heating and cooling. These thermostats are used with heat pumps or with other installations having both a heating and a cooling system. Fig. 24-27 is a simplified wiring diagram for a combination thermostat using a bimetal strip, heat anticipator and cold anticipator.

The appearance of a combination heating-cooling thermostat is shown in Fig. 24-28. This control is for 24 to 30V electrical circuits. Inside view is shown in Fig. 24-29.

Fig. 24-28. A 24V thermostat that may be used for both heating and cooling. Fan control switch is at A and a heating or cooling switch is located at B. Range adjustment is on upper temperature scale. Bottom scale indicates room temperature.   (Honeywell, Inc.)

Fig. 24-29. Construction of combination heating-cooling thermostat, shown in Fig. 24-28. Filter warning light is shown at A. Bulb lights when filter becomes dirty.

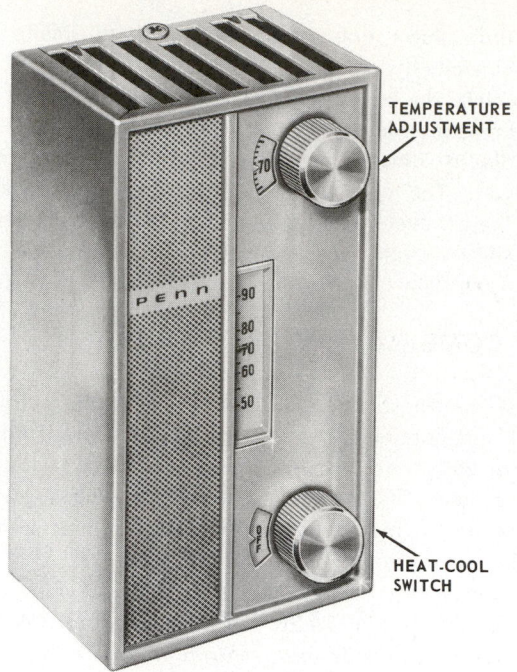

Fig. 24-30. Combination thermostat used for controlling both heating and cooling units. (Penn Controls, Inc.)

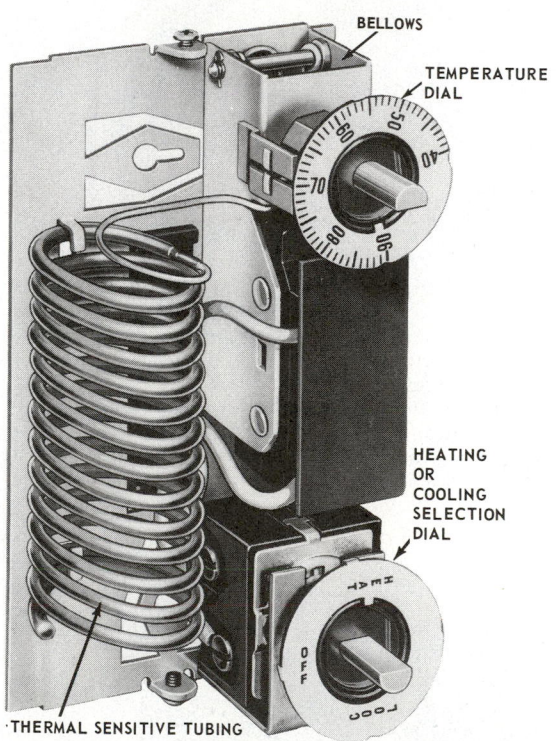

Fig. 24-31. Inside of cooling-heating thermostat.

A thermostat using a bellows operating mechanism is shown in Fig. 24-30. Its construction is shown in Fig. 24-31.

Controls are made that go beyond control of cooling and heating. Some regulate humidity, indicate the condition of the filter and control an odor system. These controls usually operate on a 24V circuit.

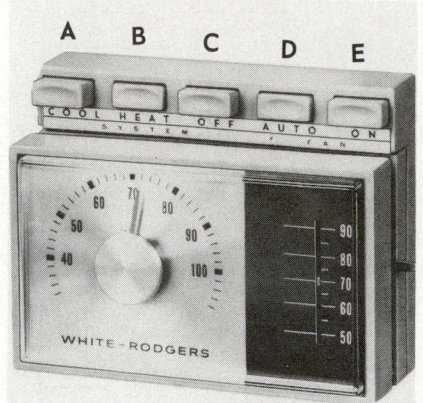

Fig. 24-32. Push-button-operated combination heating-cooling thermostat. A—Cooling. B—Heating. C—Fan off. D—Fan automatic. E—Fan on. (White-Rodgers Div., Emerson Electric Co.)

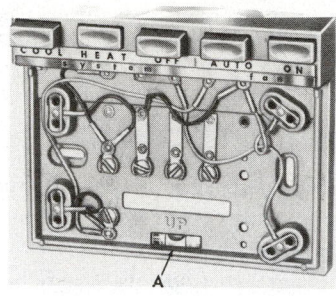

Fig. 24-33. Base of push-button combination thermostat. Notice four terminals. A—Spirit level used to help check the level of base while it is being installed.

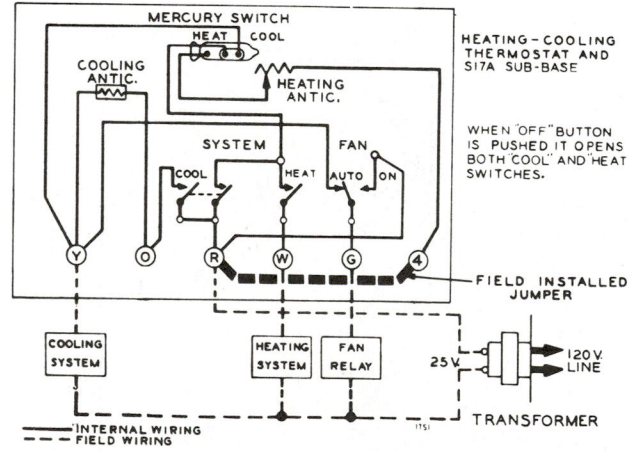

Fig. 24-34. Wiring circuit used with push-button thermostat. Notice four push-button switches and two anticipators.

Other combination heating and cooling thermostats use push buttons to operate the systems under control. Fig. 24-32 shows a push-button-operated thermostat with a calibrated temperature-setting dial and a thermometer. The thermostat base, showing the mounting holes, terminals and push buttons, is shown in Fig. 24-33. A wiring diagram for this push-button

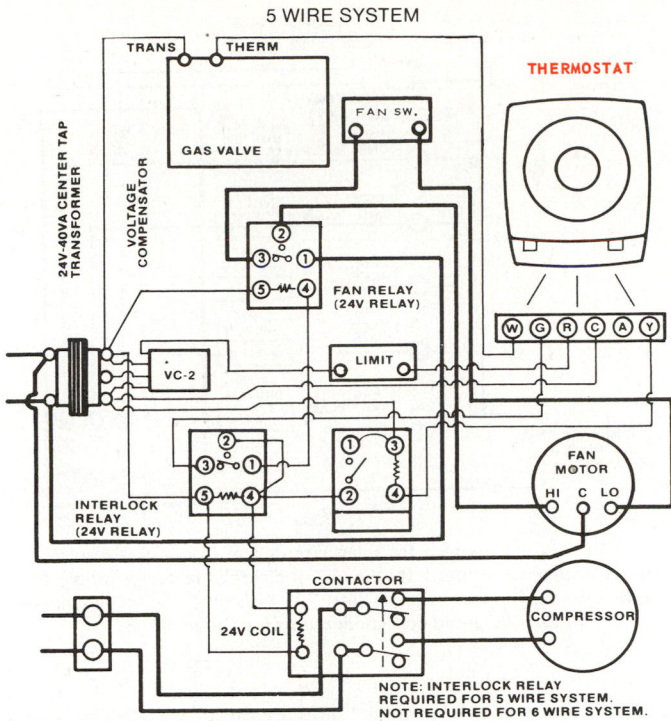

Fig. 24-35. Electrical wiring diagram for an electronic thermostat. (Robertshaw Controls Co.)

installed in offices need not run at normal temperatures on Saturdays and Sundays or during other nonworking hours. Many users of air conditioners want the units to start functioning at a certain time before the premises are occupied.

Heating systems, too, can provide reduced temperatures during those hours or days the building is not occupied.

Fig. 24-37 illustrates an automatic timer that operates on a seven-day schedule. The "trippers" on the edge of the time disk can be set to turn on the air conditioner and shut it off again at any set time each day.

Fig. 24-37. Timer used to control operation of heating or cooling system on seven-day schedule. It can be used for different circuit control every three hours up to a total of 28 actions a week. A—Trippers mounted on disk. (Paragon Electric Co., Inc.)

thermostat is seen in Fig. 24-34. The control system is low voltage. Relays are used to operate the electrical circuits of the fan, the heating system and the cooling system.

An electronic circuit using a combination thermostat is shown in Fig. 24-35. Five leads go to the thermostat. Used with a modulating gas control, it keeps the heating temperature range within 1 deg. F. Cooling is held within a 3 deg. F. range. The thermostat is shown in Fig. 24-36. An electronic system is mounted on the thermostat base.

24-14 TIMER-THERMOSTATS

There are many air conditioning installations that can be automatically clock controlled. For example, cooling systems

24-15 MULTISTAGE THERMOSTATS

Multistage thermostats may be designed for low voltage or line voltage. They will operate two or three separate circuits in sequence and are made for either heating (close on temperature fall) or for cooling (close on temperature rise).

Fig. 24-38 is a schematic of a multistage thermostat

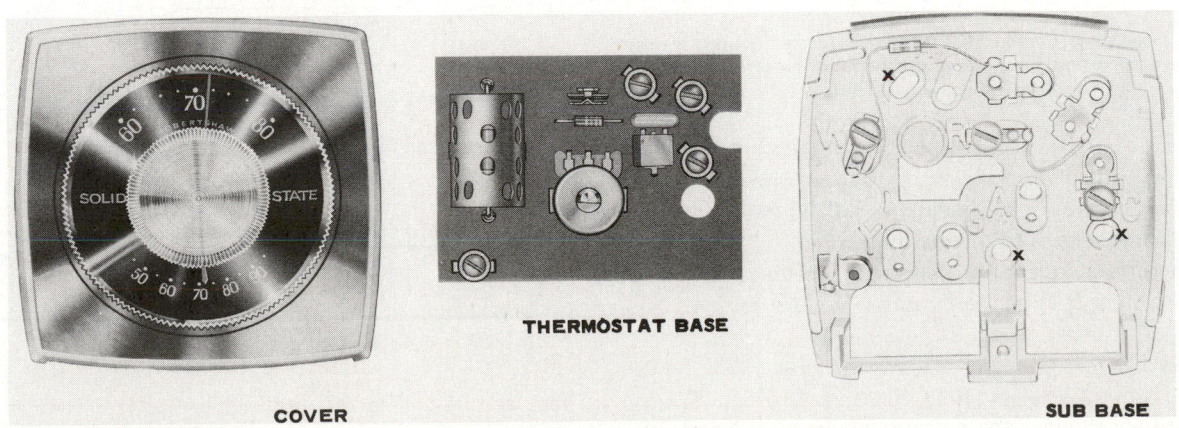

Fig. 24-36. Electronic thermostat used on combination heating and cooling system.

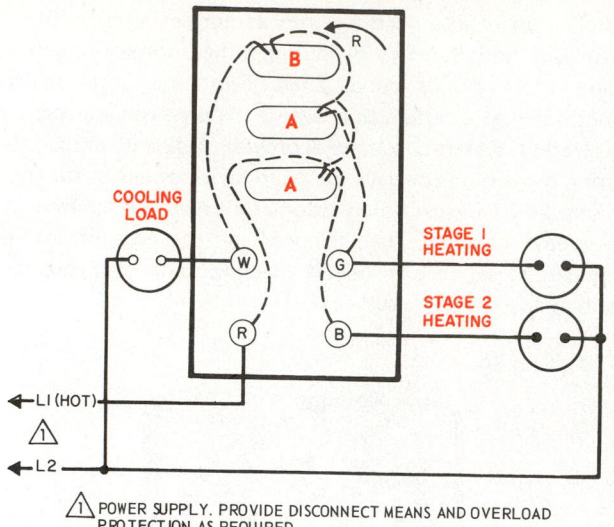

Fig. 24-38. This multistage thermostat controls two heating stages and one cooling stage. A—Heating contacts. B—Cooling contacts. (Honeywell, Inc.)

⚠ POWER SUPPLY. PROVIDE DISCONNECT MEANS AND OVERLOAD PROTECTION AS REQUIRED.

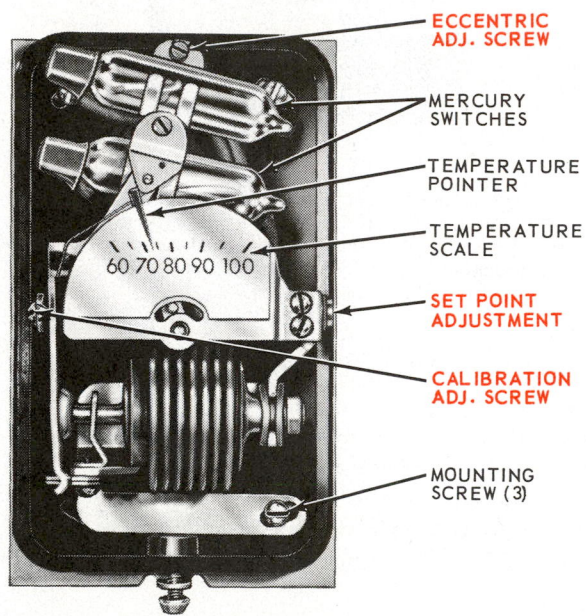

Fig. 24-39. Two-stage thermostat operated by sealed bellows element and using mercury switches. The eccentric adjusting screw adjusts the differential (usually 1 F.).

designed to control two stages of heating and one stage of cooling. Three mercury tube switches (SPST) are used. Notice the four electrical leads to the thermostat.

The thermostat operating element is a bellows. Fig. 24-39 shows a two-stage heating thermostat.

24-16 RELAYS

When using a relay, the current required for operating a system does not flow through the thermostat. Thus the main

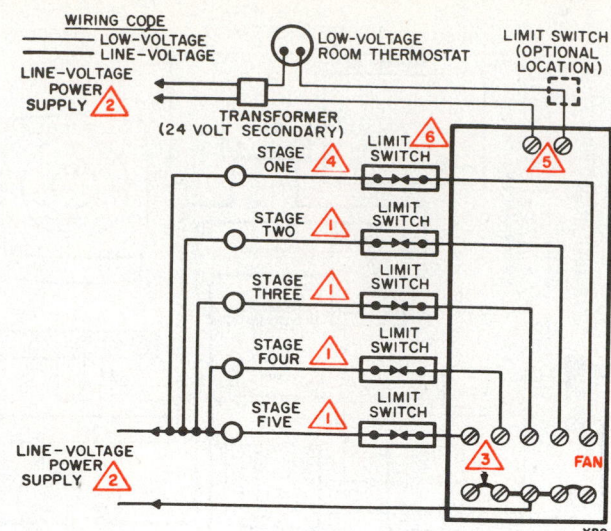

Fig. 24-40. Wiring diagram for relay circuit controlling five electric resistance heaters. In upper left, the "2" indicates the power supply to a step-down transformer. In lower left, "2" indicates line voltage power supply. Solenoid coil connections are shown at "5."

circuit can be made as short and as direct as possible. See Chapter 8. Relays are sometimes called contactors.

The solenoid which creates magnetism to close the points in the heating circuit or circuits may be energized (turned on) by either 24, 120 or 240 volts. Voltage of the thermostat and the relay solenoid coil must be the same. Fig. 24-40 shows a schematic wiring diagram of a relay which controls five separate electric heating circuits and one fan motor circuit.

When the solenoid is energized, it moves (activates) a lever which operates all of the snap switches. The snap switches are calibrated (set) to make and break (close and open) in sequence. The snap switch controlling the fan motor makes (closes) first and breaks (opens) last.

24-17 LIMIT CONTROLS

Heating devices must have built-in safety features to limit the highest temperature the device may reach. If for any reason the electric heating baseboard chamber, the hydronic

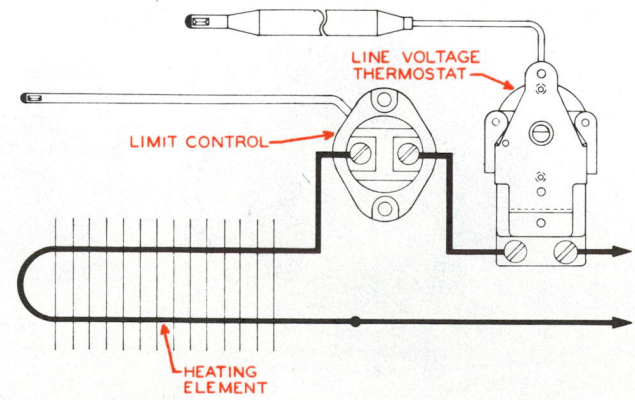

Fig. 24-41. Diagram of electric resistance heating circuit. Note that thermostat and limit control are connected in series with power line.

system or furnace plenum chamber reaches a dangerous temperature, pressures created in the limit switch element will open the switch points and protect the equipment and the building. Fig. 24-41 shows a wiring diagram of an electric heating element, thermostat and limit control circuit.

One type of limit control mounts the temperature sensitive element the full length of the baseboard chamber. Fig. 24-42 shows such a safety switch. The limit control is usually set to open the circuit at about 20 to 40 F. above the highest thermostat setting.

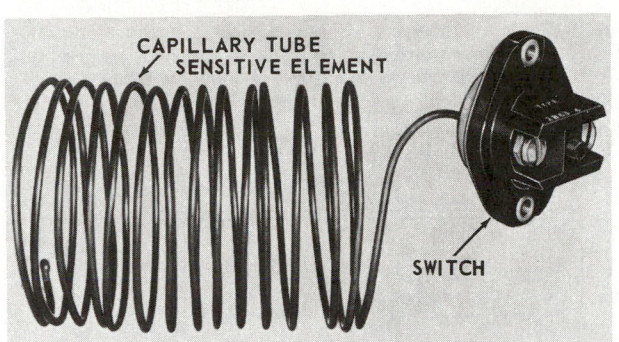

Fig. 24-42. Safety limit switch may be connected in series in electric heater circuit. If temperature exceeds safe limit, switch will open circuit.

24-18 SEQUENCE CONTROLS

Sequence controls are used to vary heat input during the heating season or to vary cooling capacity during the cooling season. Large electrical resistance heating units usually have several heating coils. For example, one must avoid putting a sudden high kilowatt load on electrical service lines which might cause a voltage drop and flickering lights, radio interference or TV flicker. Using sequence controls when heat is called for, a control system closes the circuit to one heating element or grid at a time until all of the heating circuits are functioning. This is called "sequencing."

Methods used to sequence separate multiple heating elements are:
1. Thermal element switches which stay open until one element heats up and then close the circuit for the next element. This action goes on until all the circuits are operating.
2. Relays which, as each one closes, also activate the relay for the next heating element.
3. Timers which rotate to close a series of contacts closing the circuit for one heating element at a time.

The electrical controls of an electrically heated boiler are shown in Fig. 24-43.

24-19 ELECTRICAL CIRCUITS

Heating and cooling systems use four types of electrical circuits:
1. The 120V circuit. (In a few cases 240V up through several voltages and phases are used for large commercial and industrial systems.)

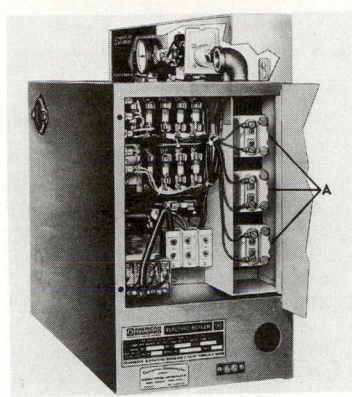

Fig. 24-43. Electrical control system of electrically heated boiler. A—Heating element.

2. A 24V circuit. (Such circuits operate relays which, in turn, control the main electrical circuits.)
3. Thermocouple circuits (a few millivolts to about 700 millivolts) used in some pilot light safety devices and in some gas heating systems.
4. Electronic circuits (which use thermistors, diodes and transistors).

These circuits must have overload safety devices. These may be fuses and/or circuit breakers.

Smaller motors usually have built-in overload devices. Larger motors use external overloads usually located in the magnetic starter. In all cases, the wiring and the devices must conform to the local electrical code.

24-20 PRIMARY CONTROLS

Primary controls are devices in an automatic system which safely turn on and operate the system on command from the operating controls (thermostats in most cases). These primary controls differ from one another depending on the type of heating and/or cooling system used.

The kind of controls used depends on the type of heat energy used: oil, gas or electricity. Another factor is the kind of heat distribution system used: steam, water or air. These controls will be explained.

Fig. 24-44 is the wiring diagram for a gun type oil burner

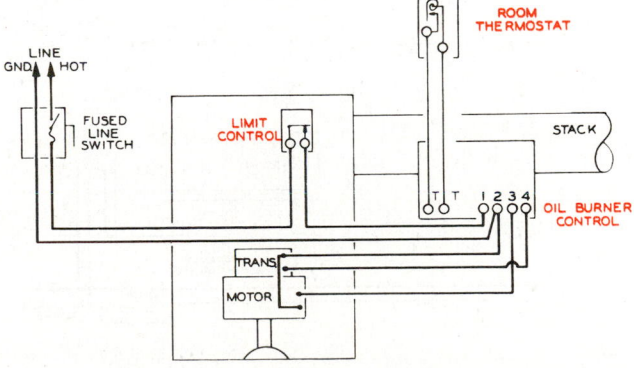

Fig. 24-44. Wiring diagram for gun type oil burner. (White-Rodgers Div., Emerson Electric Co.)

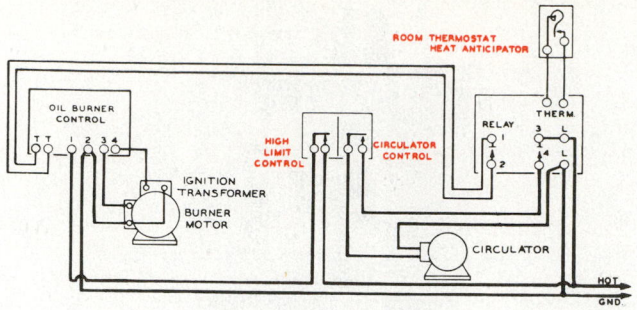

Fig. 24-45. Wiring diagram for gun type oil burner, hydronic installation. Separate thermostat controls water circulating pump.

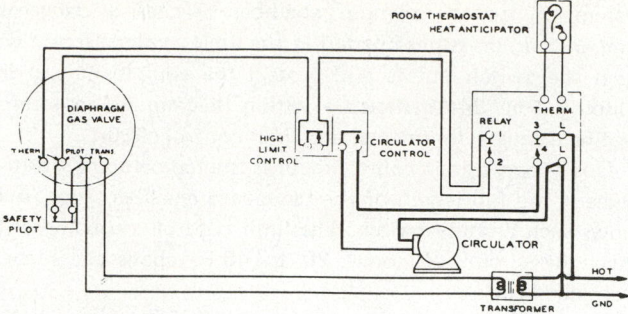

Fig. 24-48. Wiring diagram for diaphragm type control on gas-fired furnace. Note room thermostat controls operation of water-circulating pump. This system uses a low-limit control.

system. Note thermostat, oil burner control and limit control. An oil burner circuit used with a circulation pump (hydronic system) is shown in Fig. 24-45. The heavy wires are high-voltage (120V) lines. The light wires are 24V lines.

Fig. 24-46 shows a wiring diagram for a gun type oil burner installation that is somewhat different. The circulator motor is turned on by the room thermostat when it calls for heat.

Gas furnaces use many varieties of electrical devices. Some operate on 120V, some on 24V and some on current generated by a thermocouple (25 to 700 millivolts).

The wiring diagram in Fig. 24-47 is designed for a solenoid-operated gas line valve. It is installed on a hydronic heating system.

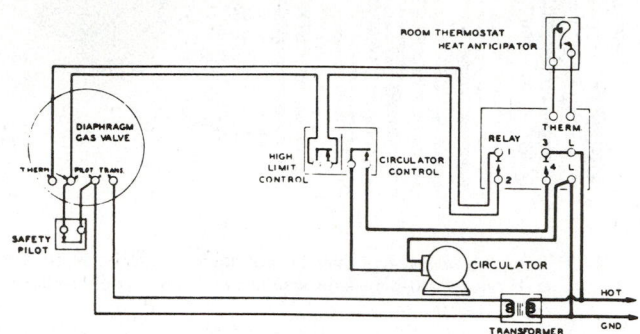

Fig. 24-49. Wiring diagram of diaphragm type gas valve heating control. Heavy lines represent line voltage; light lines represent low voltage from step-down transformer.

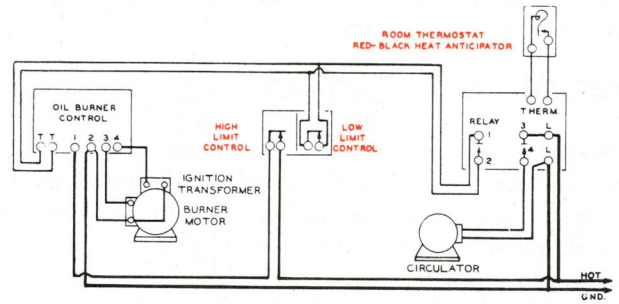

Fig. 24-46. Wiring diagram for gun type oil burner, hydronic installation using motor-driven circulator pump. Room temperature controls hydronic pump operation.

Wiring diagram as shown in Fig. 24-48 is used on a system with a diaphragm gas valve. Circulator motor operates all the time the burner is on. Fig. 24-49 shows another circulator motor wiring arrangement.

24-21 GAS FURNACE PRIMARY CONTROLS

Gas furnaces have primary controls to insure safe starting and safe operation of gas burners. Controls used to operate a gas fuel heating system include:

1. Room thermostat or thermostat with outdoor air-sensing adjustment.
2. Pilot light temperature sensor, such as a thermocouple.
3. Limit controls.

Fig. 24-50 shows three wiring diagrams for a gas furnace. Note location of the limit controls in the line.

These controls allow gas to flow only if the pilot light is burning. A thermal element is located near the pilot light. It must either produce thermocouple electrical energy or develop a sensitive bulb pressure to open valves that allow main gas flow. See Fig. 24-51. This thermal element will also shut off the system if the pilot light stops working during the operating cycle of the unit.

If a blower or water pump motor is used in the distributing system, it is turned on by a relay. Such a relay may have instant action or there may be a short delay.

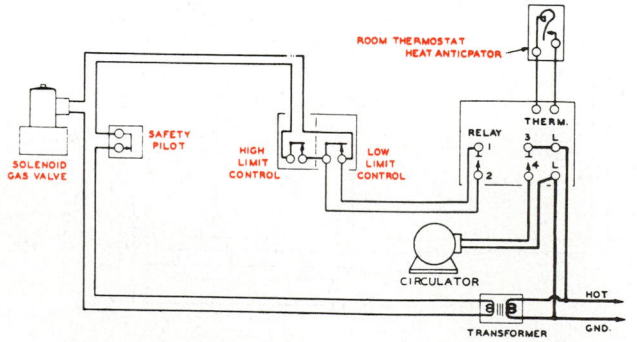

Fig. 24-47. Solenoid-operated gas line valve. This is a low voltage type valve. Note step-down transformer at lower right-hand corner.

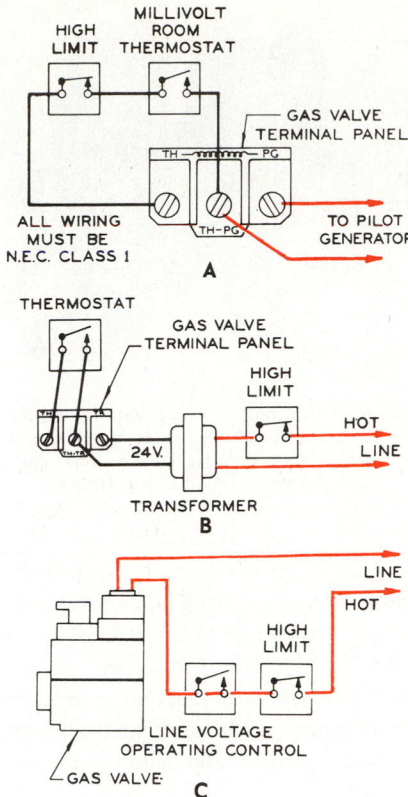

Fig. 24-50. Electrical circuits used with gas controls. A—Millivolt circuit. B—24V circuit. C—120V circuit. A high limit control is used on each. (White-Rodgers Div., Emerson Electric Co.)

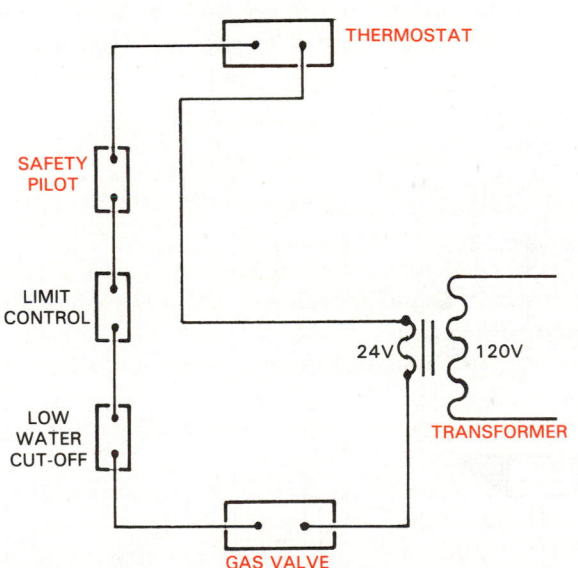

Fig. 24-51. Schematic for low-voltage electrical circuit used on gas furnace without blower.

Primary controls operate on a series of electric interlocks. All conditions must be safe before the interlocks are in the correct position to allow the system to operate. Mechanical or thermal sensors are usually considered adequate for the smaller capacity domestic burners.

Commercial and industrial systems use more elaborate controls due to the larger flow of fuel. Flame sensors are generally used. These shut off the system by reacting very fast (in parts of a second) should a flame fail. The sensors are electronic. They use a flame rod and a photocell. The photocell is either sensitive to the radiant energy or ultraviolet rays of the flame. Some installations use a lead sulphite cell which responds to the infrared rays from the gas flame.

Electric ignition for gas systems is used when the gas furnace is located outside the building or is in a hard-to-reach position. In this system, an electrical spark ignites the gas at the main burner. It also automatically shuts off the gas supply to the main burner if the main burner does not ignite. Fig. 24-52 shows an electrical ignition system for a gas burner. The wiring diagram shown in Fig. 24-53 is for such a system.

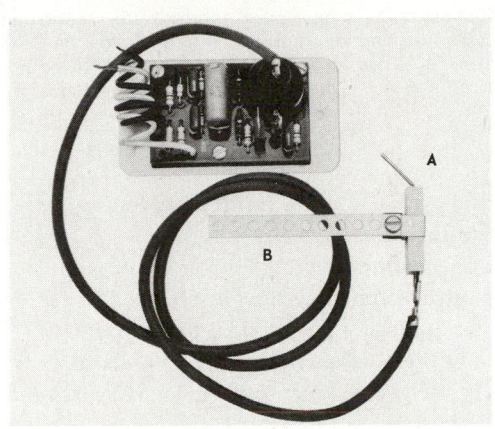

Fig. 24-52. Solid state electric ignition system for gas furnace burner. A—Spark gap electrodes. B—Mounting bracket. (General Controls, ITT)

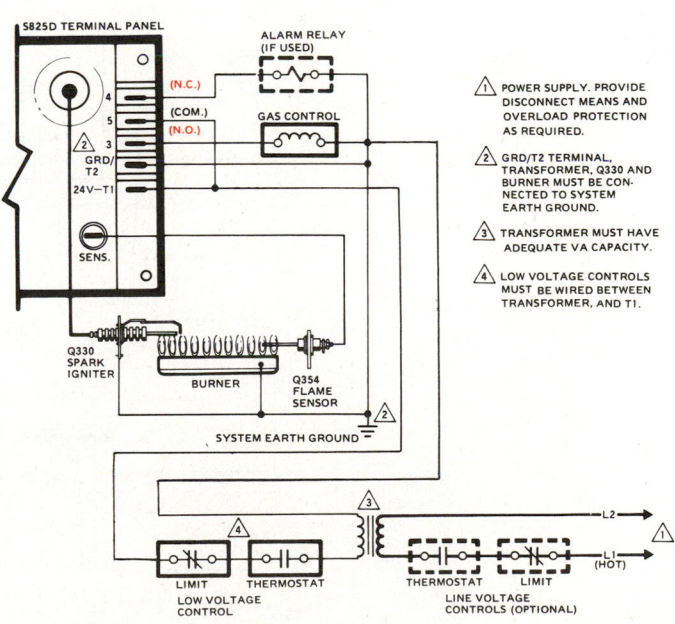

Fig. 24-53. Wiring circuit for gas furnace electric ignition system. "N.C." means normally closed. "N.O." means normally open. (Honeywell Inc.)

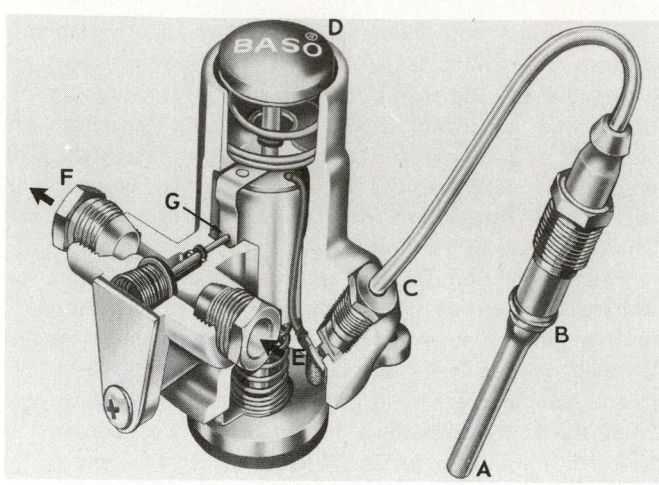

Fig. 24-54. Automatic safety pilot thermocouple construction. A—Thermocouple hot junction, mounted in pilot flame. B—Cold junction and mounting. C—Thermocouple connection to valve body. D—Reset button for pilot gas. E—Pilot flame gas in. F—Pilot flame gas out. G—Pilot flame gas valve operating lever.

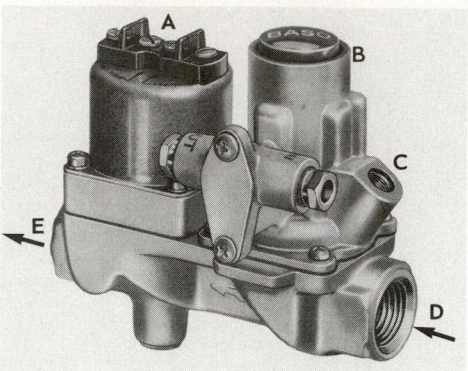

Fig. 24-55. Combination gas valve and automatic pilot for 240,000 to 550,000 Btu furnaces. A—24V solenoid-operated gas valve. B—Automatic pilot unit. C—Thermocouple connections. D—Gas in. E—Gas to burner. (Penn Controls, Inc.)

The main gas line to a gas furnace has four valves:
1. Hand shutoff valve.
2. Pressure regulator.
3. Safety valve operated by the pilot flame. See Fig. 24-54.
4. Automatic gas valve operated by thermostat.

Originally these valves were separate units. About 1957 the pilot safety valve and the automatic gas valve were made into one unit or body.

About 1959, the pressure regulator, the pilot safety valve and the automatic gas valve were made into one unit or body,

mainly for ease of installation, fewer joints to leak, space saving purposes and economy. Fig. 24-55 shows a combination valve. This combination gas control is often called a "CGC." Fig. 24-56 is a cross-section of the combination gas control.

The automatic gas valve portion is operated by either a solenoid, an electrical resistance-heated bimetal blade or by a sensing bulb, capillary tube and bellow combination (hydraulic thermal element).

24-22 ELECTRONIC CONTROL SYSTEM FOR GAS FUEL

Electronic circuits, sometimes used to control heating systems, can vary (modulate) the size of the gas flame. The

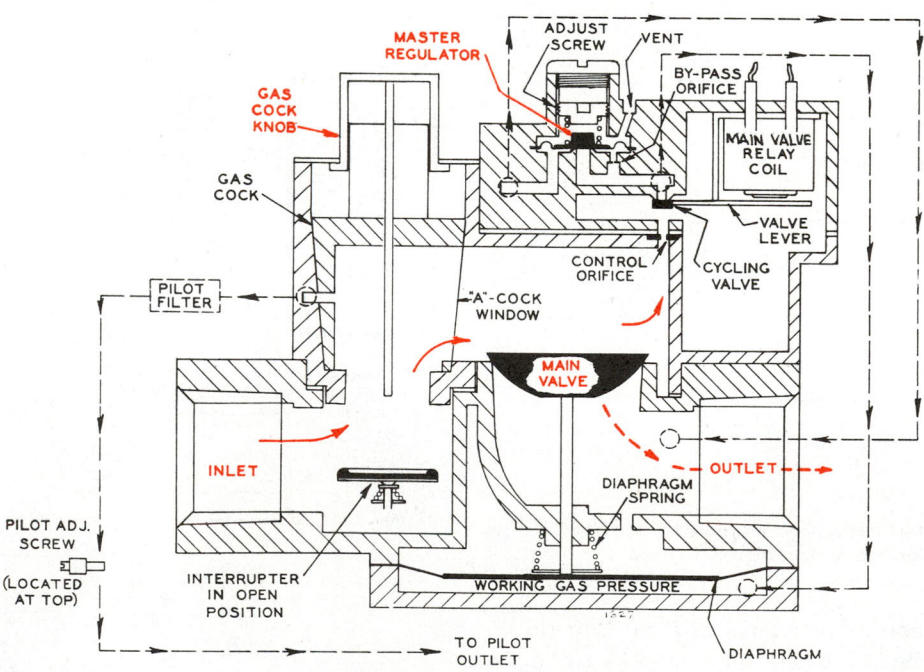

Fig. 24-56. Combination gas control. It has hand shutoff valve, pilot light control, bypass-operated main valve and pressure regulator. (White-Rodgers Div., Emerson Electric Co.)

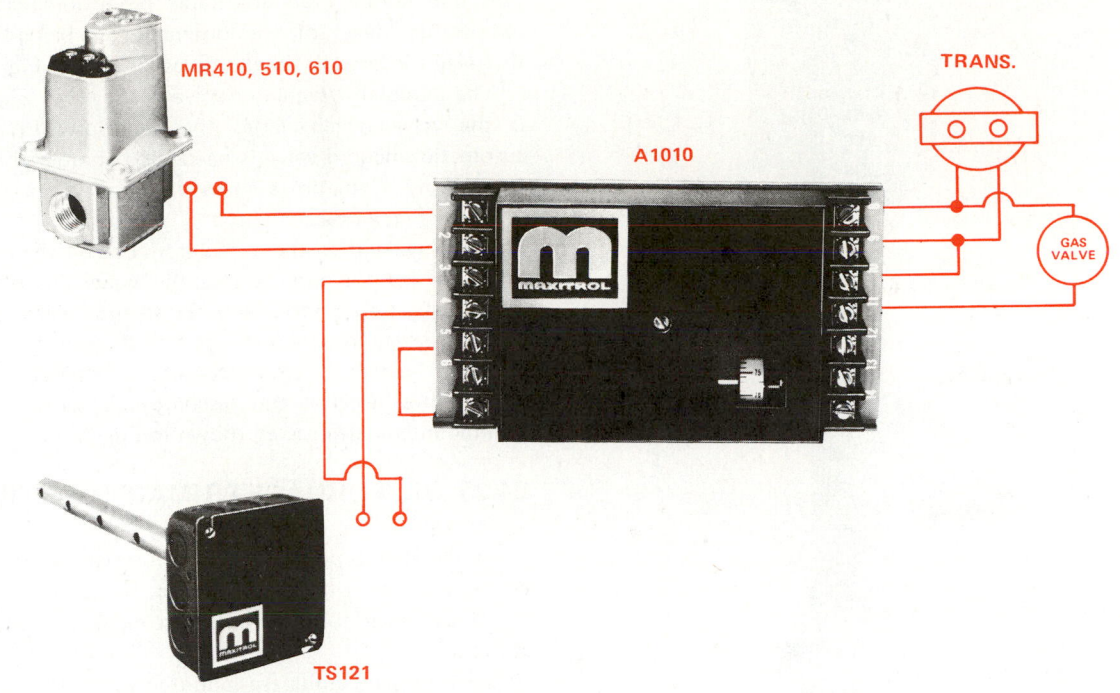

Fig. 24-57. Wiring diagram showing indirect fired makeup air application. (Maxitrol Co.)

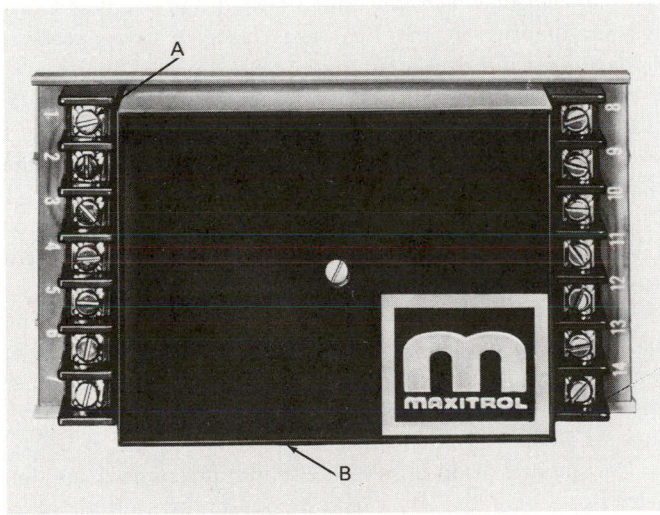

Fig. 24-58. Solid state electronic amplifier may be placed at any convenient location. A—Electrical terminals. B—Cover plate. (Maxitrol Co.)

Fig. 24-57 illustrates a wiring circuit used with a makeup air system. The 120V a-c power is reduced to 24V a-c, using a transformer. A rectifier in the amplifier changes this current to d-c. Three control devices are used:

1. The remote temperature selector.
2. The discharge air sensor.
3. The duct stat (for safety).

These controls signal the amplifier, which in turn operates the solenoid and modulating regulator on the main gas burner line.

Fig. 24-58 shows the amplifier unit. It holds the rectifiers and solid state components for the control circuits. It also contains an adjustable potentiometer for calibration.

The remote temperature selector is shown in Fig. 24-59.

modulating system uses a solid state thermostat. It uses a thermistor and several transistors.

The size of the gas flame depends on temperature difference between the thermostat setting and room temperature. The flame is larger with a greater temperature difference and gets smaller as room temperature approaches the thermostat setting. The unit starts up the flame at about 20 to 50 percent of capacity; and then it adjusts the flame to temperature differences. The system's three main parts are the thermostat, amplifier and a modulating gas valve.

Fig. 24-59. Remote temperature selector is not temperature sensitive and may be placed in any convenient location.

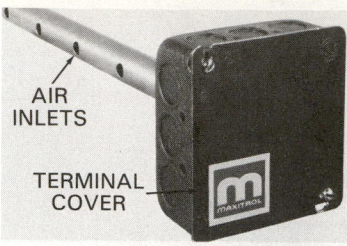

Fig. 24-60. Discharge air sensor which signals the amplifier of any temperature change from set point. (Maxitrol Co.)

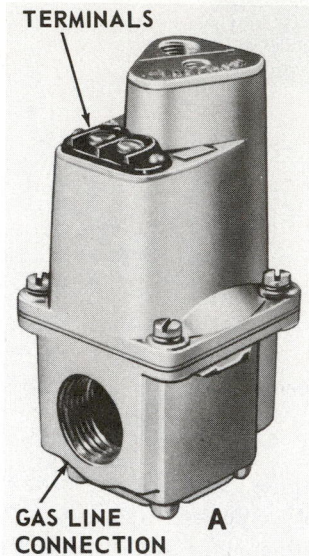

Fig. 24-61. Modulator/regulator valve performs both regulation and modulation to vary burner flame size (or burning rate). A—Outside view. Note gas line connection and electrical terminals. B—Cutaway of same modulator/regulator valve. Note adjusting screws used to vary amount of pressure required to operate valve. (Maxitrol Co.)

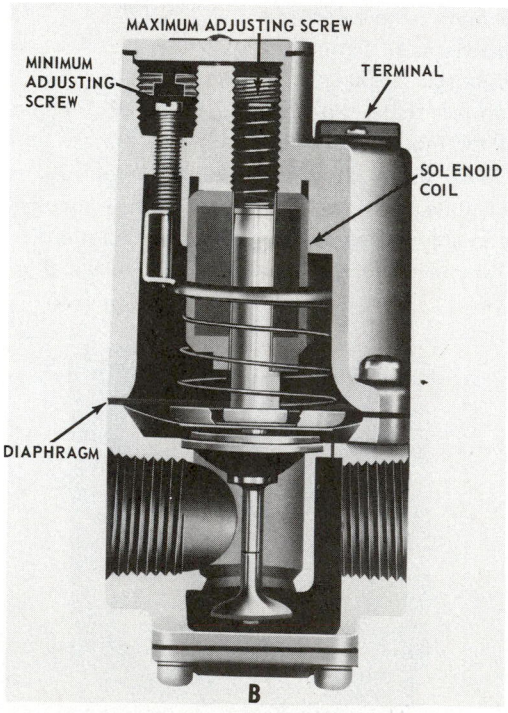

This unit contains an adjustable potentiometer. It sets the temperature level of the discharge air being sensed by a thermistor located in the discharge air sensor, Fig. 24-60.

The modulator/regulator valve is shown in Fig. 24-61. This is the valve which varies the gas flow. In addition, an automatic solenoid valve is needed to completely shut off the fuel supply. Sometimes these valves are in a single unit as shown in Fig. 24-62.

Direct current to the modulator controls the amount of gas flow. The less the current flow, the higher the flame. The duct stat is connected in series with the solenoid valve and is used as a safety device in case too high a duct temperature is reached.

Space heating is accomplished by combining the remote temperature selector and discharge air sensor into a single wall-mounted Selectrastat, shown in Fig. 24-63.

24-23 OIL FURNACE PRIMARY CONTROLS

In the gun type oil burner a primary control is usually employed to start the burner motor. If the flame goes out or if the flue temperature becomes too high, the control will also stop the unit.

Some primary units are mounted in the flue. Upon a signal from the thermostat for heat (closing of points), the primary control will start the gun type oil burner motor and turn on the ignition.

This control is shown in Fig. 24-64. It has a temperature sensing element which will shut down the unit unless a fast temperature rise in the flue takes place in a few seconds (indicating that the oil is burning).

This same sensor will constantly check for flame temperatures and will shut off the system if the flame goes out. Or it will shut off the system if the thermostat, or any one of the limit controls, opens the circuit.

Fig. 24-65 shows an electrical circuit for a gun type oil burner with continuous ignition. Fig. 24-66 is an intermittent ignition system.

Many oil burners also use a flame out safety control. The type shown in Fig. 24-67 is an ultraviolet safety control. The scanner is mounted in the wall of the firebox and is aimed at the oil fire. The circuit for this type of safety device is shown in Fig. 24-68.

New models of oil burners use solid state controls. Ignition (electrical) may be either continuous or intermittent. Fig. 24-69 shows a gun type oil burner primary control. One model operates on low-voltage input while another model operates on line voltage. Both controls use a low voltage photosensitive flame detector. When the thermostat calls for heat, the gun oil burner motor turns on, the oil pump starts and the ignition is turned on. If the fuel does not ignite in about 40 seconds, the flame detector stops the burner motor and closes the solenoid valve if one is used. Fig. 24-70 shows the circuits.

A solid state primary control, as shown in Fig. 24-71, can be used as a replacement for older model primary controls. One must be absolutely certain that the power is off before installing the unit. The unit is wired as shown in Fig. 24-72. Mounting of the cadmium cell must be very carefully done. The cell must "see" the flame as well as be heated over 140 F.

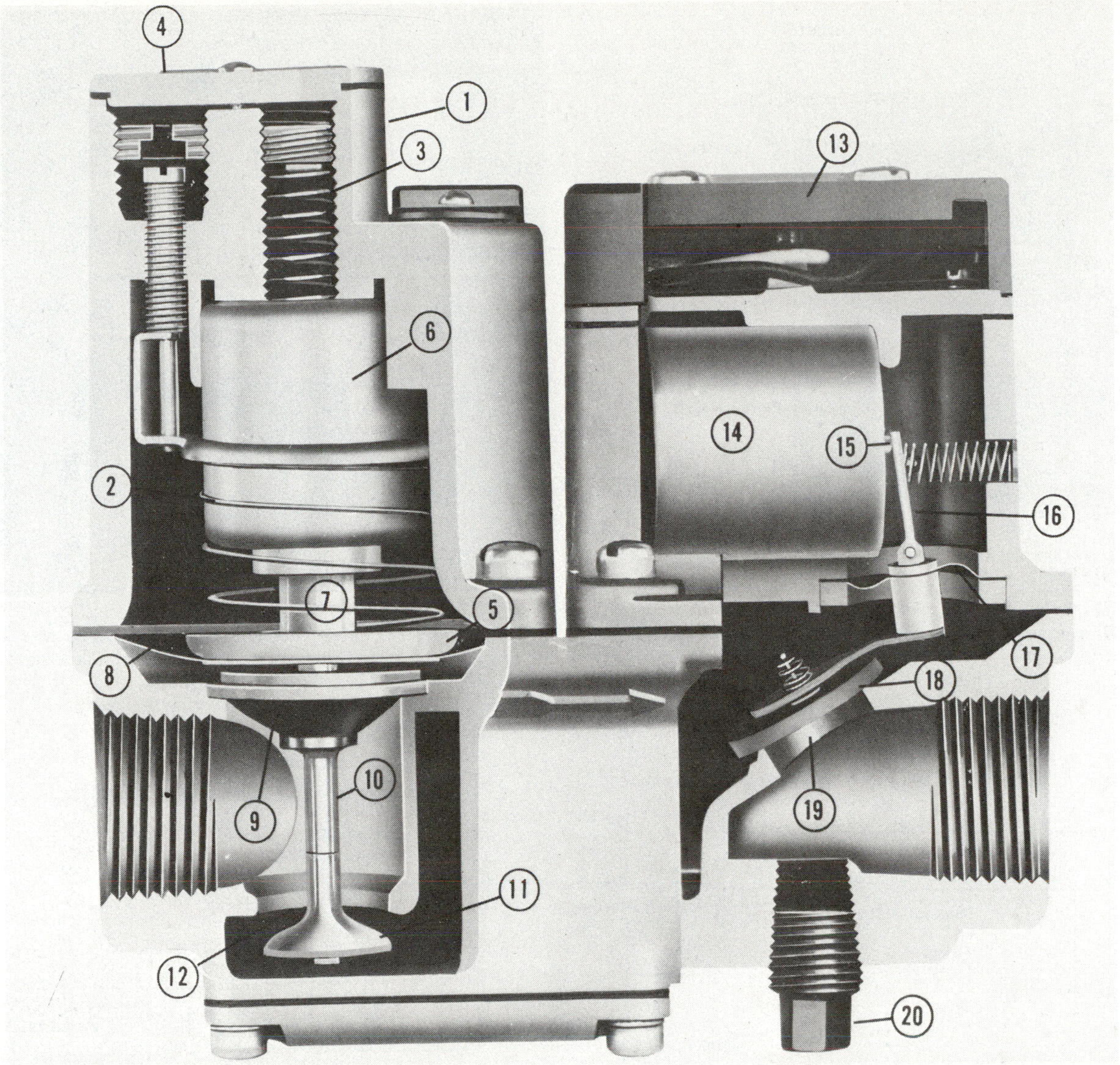

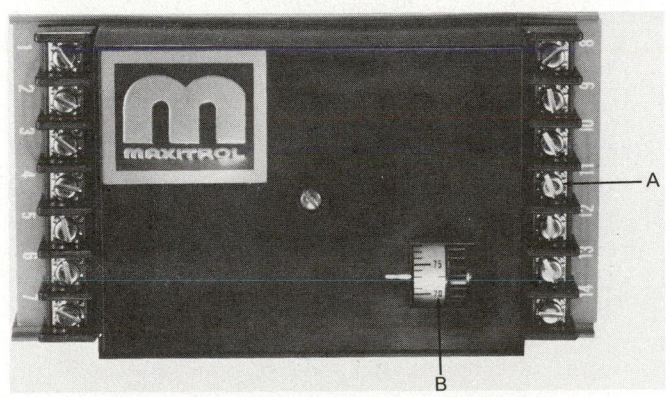

Fig. 24-62. Unitized manifold control valve of an electronic modulator control system for gas furnaces. Flame is modulated from full flame to 20 percent of full flame. Control is a combination of modulator/regulator valve and a solenoid shutoff valve. 1—Body. 2—Body. 3—Maximum adjustment. 4—Cover. 5—Diaphragm attachment. 6—Solenoid coil. 7—Solenoid core. 8—Diaphragm. 9—Valve guide. 10—Modulating valve. 11—Gas passageway. 12—Valve port. 13—Cover. 14—Solenoid. 15—Solenoid plunger. 16—Lever. 17—Diaphragm. 18—Valve. 19—Valve port. 20—Pipe plug (service connection).

Fig. 24-63. Combination temperature selector and amplifier. This control maintains precise room temperature conditions. A—Terminals. B—Temperature selector. (Maxitrol Co.)

(60 C.). The correct mounting is shown in Fig. 24-73. Another photocell flame detector mounting is shown in Fig. 24-74. The flame detector must be lined up with an opening in the static pressure disk to be able to "see" the flame.

The transformers have a primary winding of 120V, 240V or 208V. The secondary winding usually provides 10,000V. Some systems use 12,000V.

It is recommended that when 10,000V transformers need replacement, a 12,000V be used. This is advisable especially in cold air or cold oil situations or when line voltage drops are known or suspected.

24-24 ELECTRIC HEAT PRIMARY CONTROLS

Baseboard units usually have individual thermostats. Primary controls are most often relays and limit controls. Units

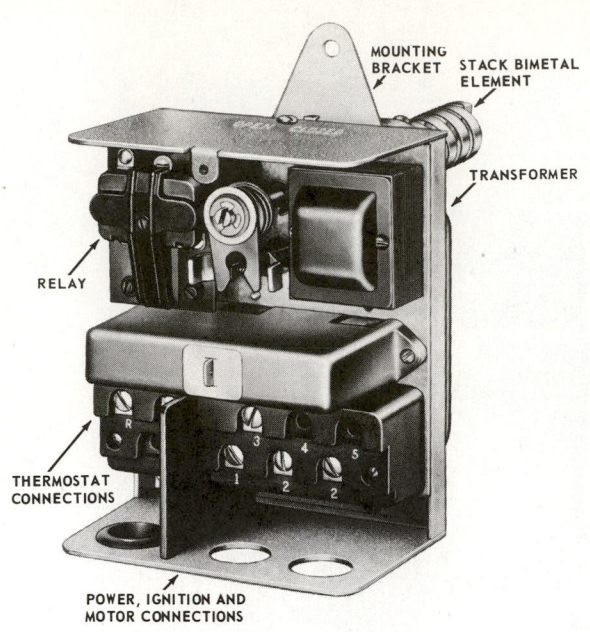

Fig. 24-64. Oil burner primary control. This control will cycle burner, operate electric ignition system, shut off unit if ignition fails and scavenge unit after each cycle. (Penn Controls, Inc.)

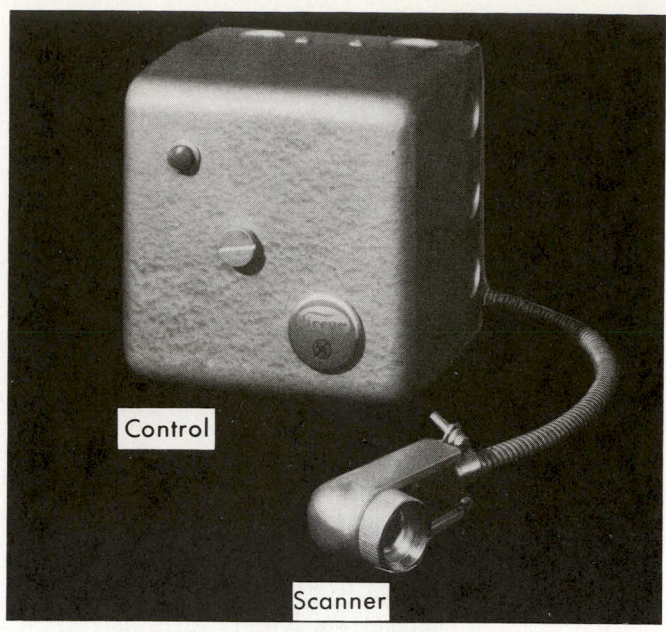

Fig. 24-67. Safety control used with large oil or gas burners is operated by ultraviolet rays. If combustion is delayed or flame goes out, unit safely shuts down heating system. (Electronics Corp. of America)

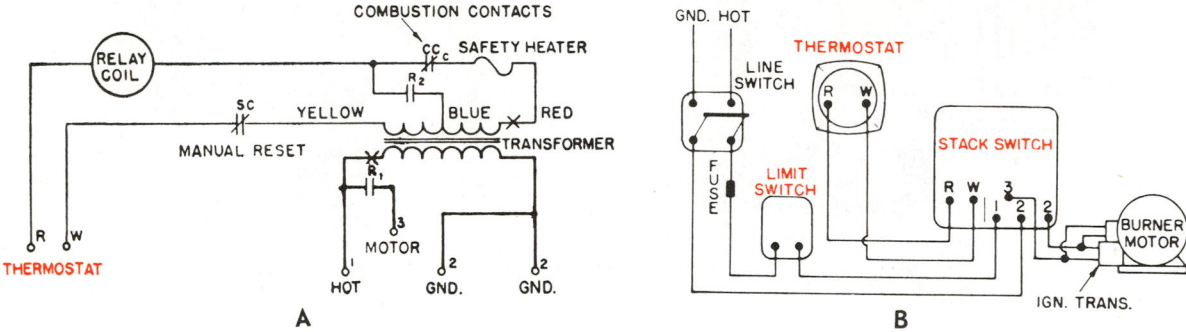

Fig. 24-65. Wiring diagrams of continuous ignition gun type oil burner stack control. A—Internal wiring diagram. B—External wiring diagram.

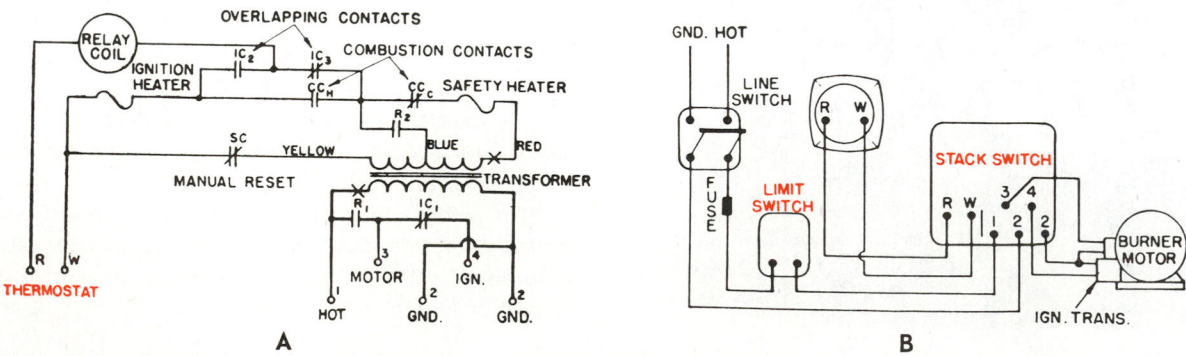

Fig. 24-66. Internal and external wiring diagrams of an intermittent ignition system used on a gun type oil burner stack control. A—Schematic wiring diagram for intermittent ignition stack control. B—External wiring diagram for intermittent ignition stack control. (Penn Controls, Inc.)

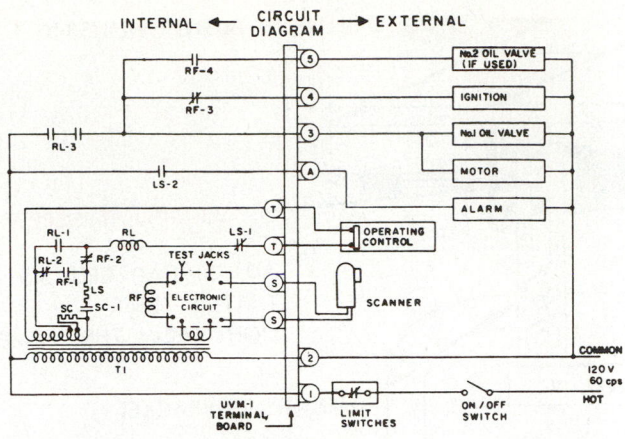

Fig. 24-68. Wiring diagram of ultraviolet sensitive safety device for large oil burners.

Fig. 24-69. Low-voltage solid state gun type oil burner primary control. It has a continuous ignition system and a photosensitive unit for flame detection. If ignition fails, the flame detector will shut off the system in two seconds. A—Photocell. B—Leads to furnace. (Penn Controls, Inc.)

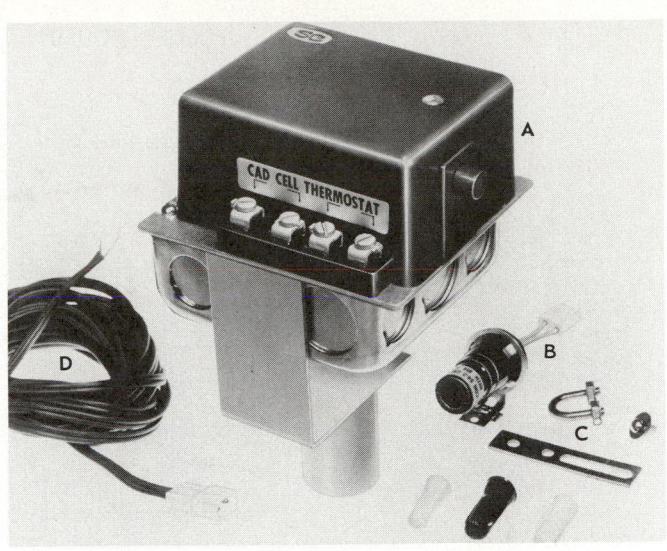

Fig. 24-71. Solid state primary control for gun type oil burner. A—Primary control. B—Cadmium cell flame detector. C—Cadmium cell mounting bracket and fittings. D—Connecting lead. (Simicon Co.)

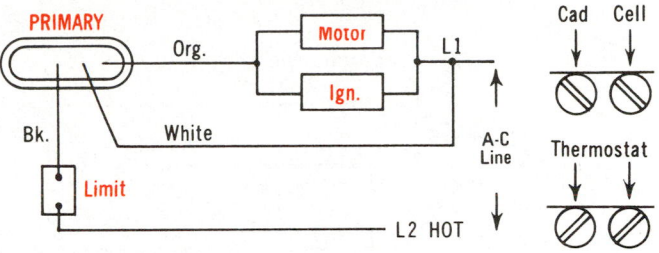

Fig. 24-72. Schematic wiring diagram for solid state primary control for gun type oil burner.

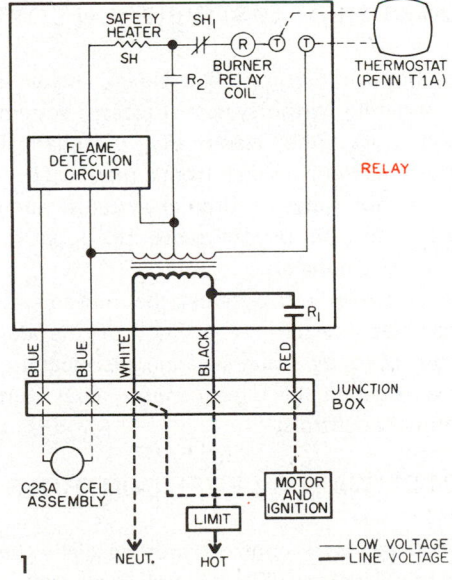

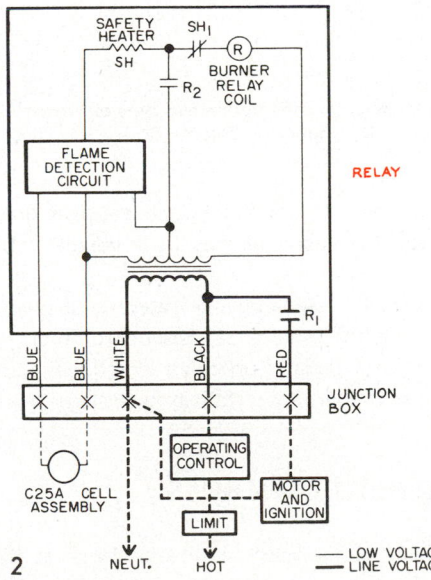

Fig. 24-70. Wiring circuits for solid state control equipped with photosensitive flame detector. 1—Low-voltage system. 2—Line voltage system.

Air Conditioning, Controls, Circuits and Instruments / 859

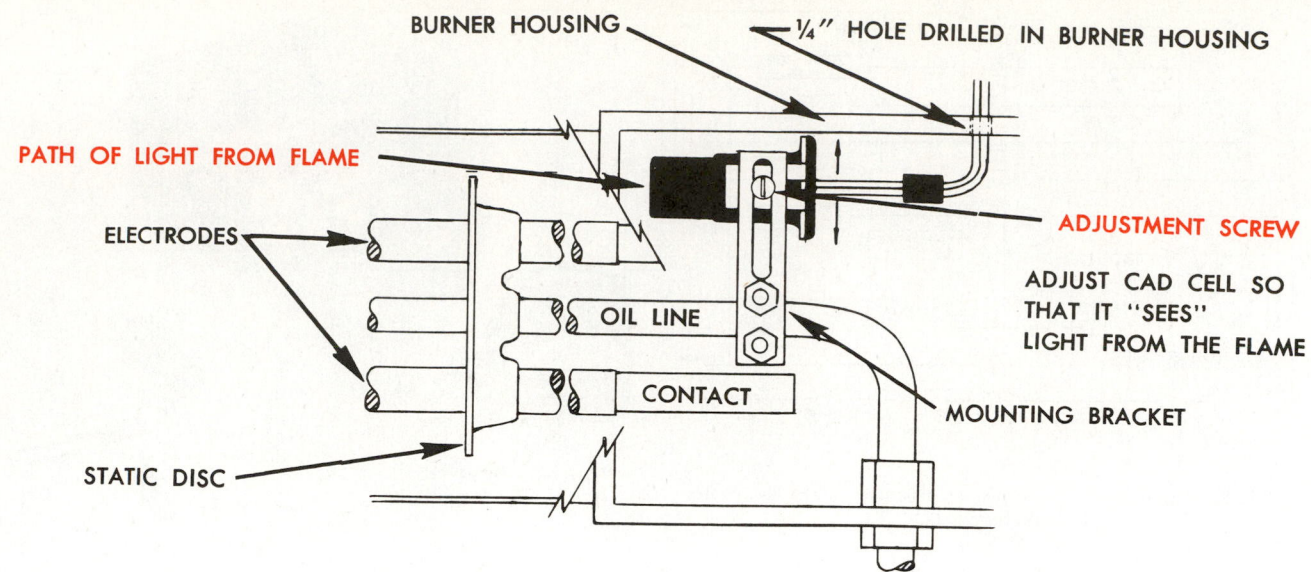

Fig. 24-73. Method of mounting photoelectric flame sensor in gun type oil burner. (Simicon Co.)

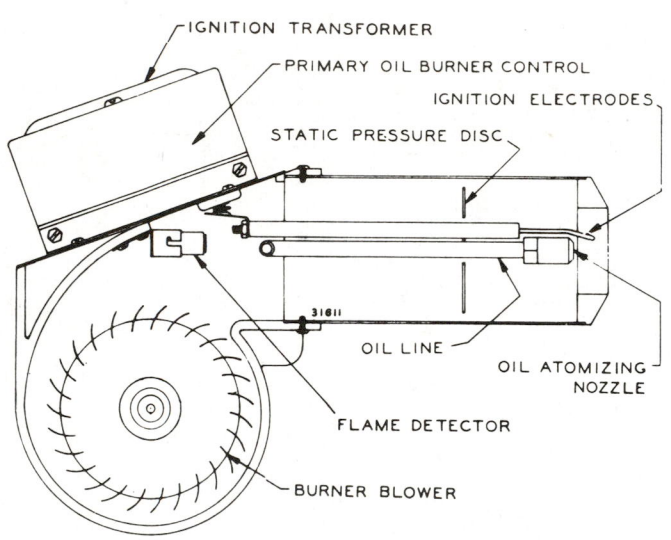

Fig. 24-74. Flame sensor detector mounted in gun type oil burner. (White-Rodgers Div., Emerson Electric Co.)

using blowers or fans have fan controls. Such controls may operate from a separate thermostat or may be in parallel with the heating element.

Central systems normally use sequence relays as primary controls. Blowers operate the same as on baseboard units. In addition, a safety control is usually provided which will shut off the heating elements if the blower fails to operate or if air fails to circulate.

24-25 INFRARED HEAT CONTROLS

Electric infrared heat lamps may require as much as 32 kilowatts (kW). The heating unit may have as many as 16, 2000W lamps (32,000W). The electric circuit, in such cases, is controlled by a large-capacity relay (contactor). At 240V a-c,

the line capacity will need to carry 175A. The 175A main circuit is usually divided into four or more separate circuits, each with a relay (contactor).

Either thermostat or solid state sensors energize the operating coil of the contactors. The contactors are usually sequenced so that only one closes at a time. (In other words, sequencing means that the contactors are set to close one after the other — not all at once. This is usually done by a time-delay on each contactor switch which operates the coil of the next contactor.)

Solid state temperature controls modulate the current flow with thermistors and triacs. They reduce part of each sine wave of the a-c flow to maintain a constant temperature. See Chapter 6.

24-26 HEAT DISTRIBUTION CONTROLS

Controls insure that steam, water or air is properly circulating in the system. In steam systems, the zone control valves are either electrically, pneumatically or hydraulically operated upon a signal from a thermostat.

In hot water (hydronic) systems, pumps and valves must work in the proper sequence upon command from the operating controls.

In warm air systems, the movement of the air in the complete system or in part of the duct system may be controlled by using separate thermostats. These thermostats may operate the blower motor, duct controls or the furnace primary controls.

24-27 WATER LEVEL CONTROLS

Water level controls are especially important on steam heating systems. The control is generally a switch turned on and off by a float. If the water level drops near the dangerous point, the float drops and cuts out the electrical circuit to the

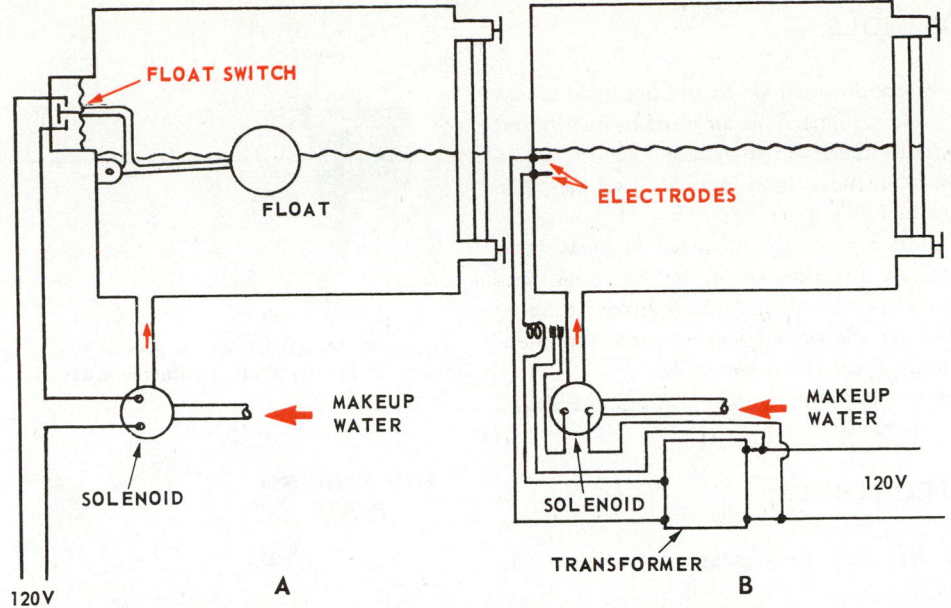

Fig. 24-75. Two types of water level controls. A—Float control. Float valve will open makeup water valve when water level drops. Some are connected to electric switch which will shut off unit if water level is too low. B—Probe type water level control. Solenoid opens when current stops traveling across electrodes.

operating controls stopping the furnace. The switch generally operates in two stages. Lowering of the float will trip a switch to turn on the feed water pump or feed water solenoid valve. If the float drops still more, the system is shut off.

Some water level controls are of the probe type. This type immerses two electrodes in the water. As long as water covers both electrodes, a small current flowing in the water between the two will energize a holding relay and allow the system to run. If the water level falls below the upper probe, the current flow will cease and the operating controls will shut down.

The operating principle of these two water level controls is shown in Fig. 24-75.

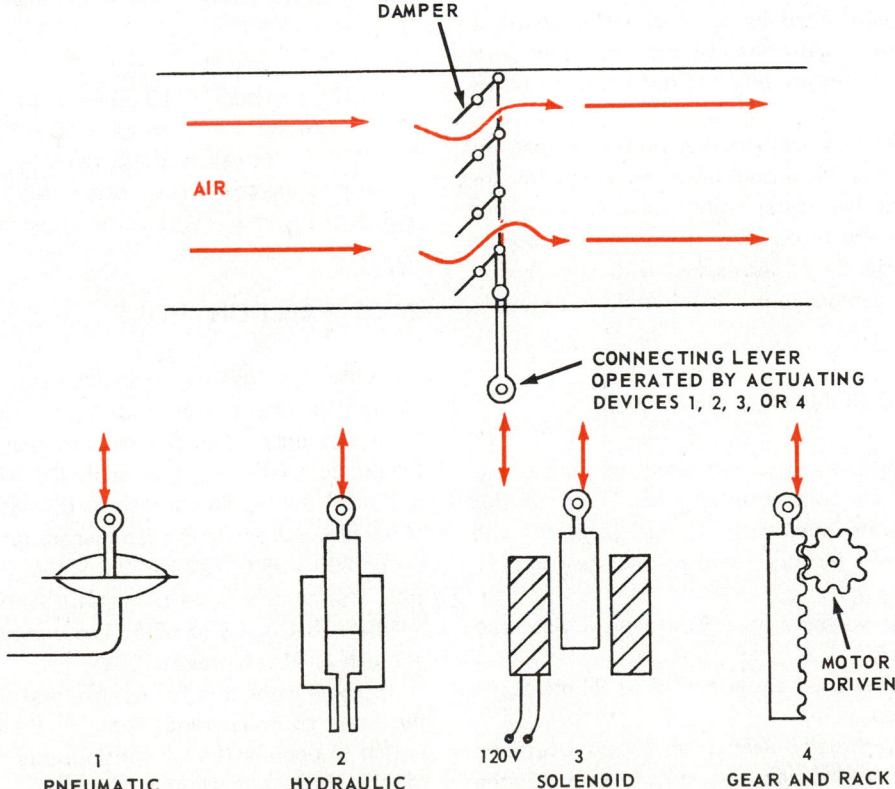

Fig. 24-76. Duct damper for controlling airflow and power devices used to operate it. 1—Pneumatic motor. 2—Hydraulic motor. 3—Solenoid. 4—Motor driven gear and rack

24-28 DUCT CONTROLS

Safe distribution of conditioned air to the occupied spaces is important in warm air systems. This air must be distributed in the proper amounts to condition the space. It must not be too cold or too hot. It must have enough fresh air (for oxygen). The air must be flowing.

Many warm air units are zone controlled. These systems have dampers to regulate the flow of air to the zones. The dampers open and close upon command of a thermostat. Some dampers are powered by motorized valves, some by pneumatics, some by hydraulics, others by solenoids.

Fig. 24-76 is a sketch of a typical duct damper and shows four ways to control damper action.

24-29 AIRFLOW CONTROLS

Airflow controls are used to regulate amount of air, temperature and flow.

The amount of air is controlled by power-operated dampers. If outside air is too cold, thermostats will close the outside air damper. Thermostats will also react if the mix of recirculated air and outside air is out of balance. They will open one damper and close the other enough to produce the correct condition. Damper motors are usually used, although a heated vapor element may also be used, as may air pressure or vacuum (pneumatic).

If the air temperature is too high or too low — in the airflow to the room, in the recirculated air, in the fresh air or in the exhaust air — the thermostat will react to adjust the dampers to a correct condition.

Pneumatic motors controlled by an outside thermostat or electric modulating motors controlled by the outside temperature are used to vary the supply of air as outside temperature changes.

One motor for operating dampers uses power to open the damper. It works against a spring pressure and holds the dampers open until the thermostat points open. A spring then closes the damper. The motor operates the damper through a gear reduction train. Fig. 24-77 shows such a damper-control motor and shaft. This damper motor is mounted in a duct as shown in Fig. 24-78.

24-30 PNEUMATIC CONTROL SYSTEMS

Pneumatic (air) systems are often used to control air conditioning. Thermostats control an air line. The air in this line can, in turn, operate pneumatic motors (a piston and cylinder or a diaphragm). Motors, in turn, operate dampers, valves and switches.

The system's two main parts include sensing devices and penumatic controllers.

The operating pressures used are either 12 or 24 psi. Some systems operate at a vacuum.

The 12 psi system actually uses 3 to 15 psi operating pressure. The 24 psi system actually uses 3 to 27 psi operating pressure.

The air pressure supplied to these systems is as follows:

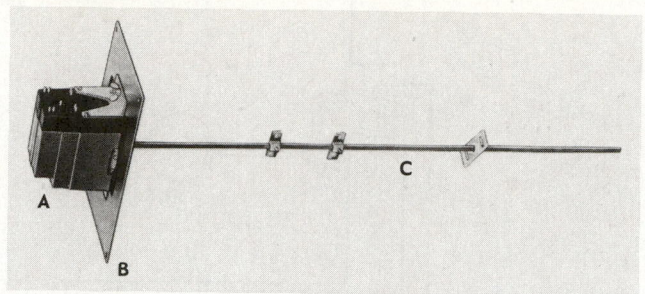

Fig. 24-77. An electric motor damper control. A—Motor. B—Mounting plate. C—Damper shaft.　(White-Rodgers Div., Emerson Electric Co.)

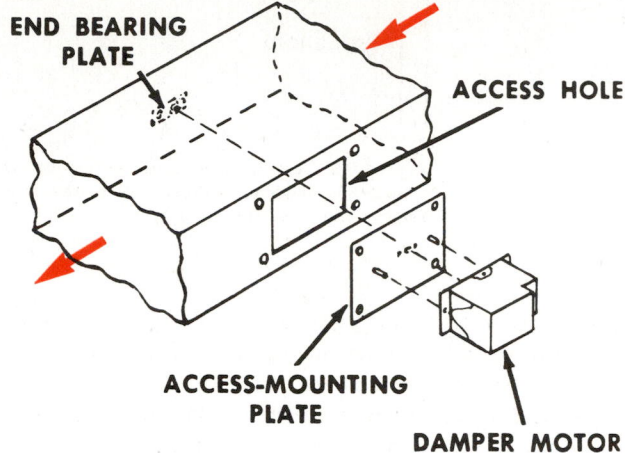

Fig. 24-78. Motor-driven damper installation. Access mounting plate allows large enough opening to install damper without removing duct.

12 psi system = 18 psi to 20 psi (1.3-1.4 kg/cm^2)
24 psi system = 30 psi to 35 psi (2.1-2.4 kg/cm^2)
(1 psi = .0703 kg/cm^2)

Air pressure controls are often used in large commercial and industrial systems. These control systems should be thoroughly checked each month.

24-31 FAN CONTROLS

Forced air heating systems have a thermostat on the plenum chamber to start and stop the fan motor (blower). The fan starts only when the plenum chamber reaches a certain temperature (about 140 F. or 60 C.). The blower continues to run when the heating unit shuts off until the plenum chamber temperature drops to the fan thermostat cut-out temperature— about 80 to 90 F. (27 to 32 C.). Fig. 24-79 shows a fan control. This control has a temperature range of 70 to 150 F. and a differential of 15 to 55 F. These controls are built to carry as much as 14 amperes at 120V.

In some cases, the fan control and the limit control are in the same control casing. Some of these fan controls have a switch to operate the fan continuously during the hot summer months. Recent developments in this area include combination motors and fan controls that operate the motors at two or more speeds or at modulating speeds.

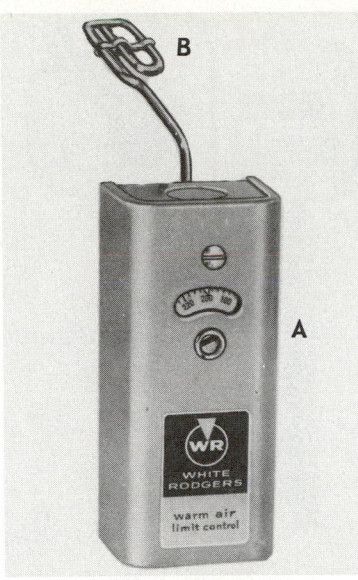

Fig. 24-79. Fan control thermostat for forced warm air furnace. A—Body of thermostat mounted on plenum chamber. B—Thermal element located inside plenum chamber.

24-32 LIMIT CONTROLS

Limit controls guard against damage from abnormal conditions in a heating system. They will stop the operation if:
1. The temperature becomes excessive.
2. The water level is too low in a steam heating system.
3. The bonnet or plenum chamber temperature is too high in a warm air system.
4. The flue temperature in any system becomes too high.
These controls will shut off the system at some point beyond the operating controls' settings.

In warm air furnaces, the limit control is generally a coiled bimetal strip. It is connected in series with the fuel controls. Located in the plenum chamber, it is usually adjusted to open the circuit if the furnace bonnet or plenum chamber reaches about 190 F. (88 C.). It cuts back in again about 160 F. (71 C.).

It will have a fixed differential in most cases. The cut-out setting is usually about 40 F. (22 C.) above the cut-in temperature of the fan control. The controls have either a 25 F. (14 C.) or a 35 F. (19 C.) nonadjustable differential.

Hot water (hydronic) limit controls are usually mounted in the boiler. The thermal element is in the boiler water. Some sensing devices are bimetal and some have a hydraulic sensing element. These controls usually have a fixed differential of about 5 F. (3 C.) and a range of 100 F. to 225 F. (38 C. to 107 C.). In most instances they are adjusted to about 150 F. (66 C.).

Water temperature limit controls are connected in series with the thermostat. They open the circuit when the temperature rises to the cut-out setting.

Steam systems usually use a pressure limit control. The control is connected by a pipe fitting to the top of the furnace. When the steam pressure in low pressure systems rises to between 0 psi and 10 psi, the operating circuit is opened and the furnace cannot operate until the pressure drops. Cast iron steam boilers usually operate between 2 to 5 psi. One psi = .0703 kg/cm^2.

24-33 MODULATING CONTROLS

In most air conditioning systems conditioned space is heated by turning on the burner or heat source to warm the space. When a preset upper temperature is reached, the heat is turned off. As the heated space cools to a preset low temperature, the heat source again turns on. This action alternately heats the space and allows it to cool. Such a system does not provide a constant temperature.

Modulating controls are designed to provide a heat input more or less equal to the heat lost. A properly modulated control will, therefore, maintain a constant temperature in the conditioned space.

Solid state electronics make possible this accurate modulating of temperature. These controls may be applied to either heating or cooling.

The modulator is usually controlled by a thermostat, a remote bulb and bellows or diaphragm, pressurestats or humidistats. These controllers, in most cases, operate a variable potentiometer. The variable resistance of the potentiometer is operated by the controller.

The modulator motor usually combines a reversible capacitor motor with a balancing relay, a feedback potentiometer and a gear train. See the schematic wiring diagram of the motor control in Fig. 24-80. The current fed to the

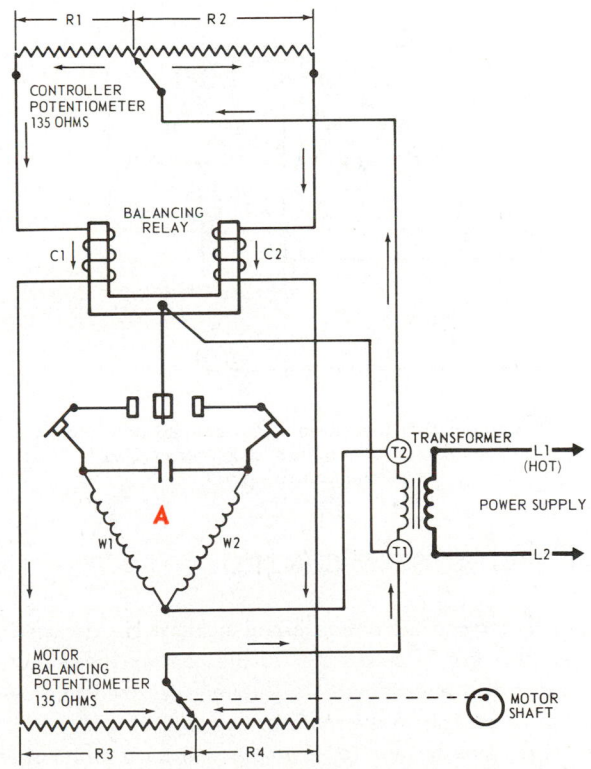

Fig. 24-80. Wiring diagram for modulating electric motor used for controlling valves and dampers. A—Motor windings. (Honeywell, Inc.)

balancing relay is controlled by a three-wire thermostat.

Motors are also designed especially to move only to two separate positions. These two-position motors use cams to open and close switches when the driven mechanism is operated. For such use, motors are usually equipped with electric brakes. They are operated by the motor position switches and lock the rotor in position. Fig. 24-81 shows a solid state controlled-positioning motor.

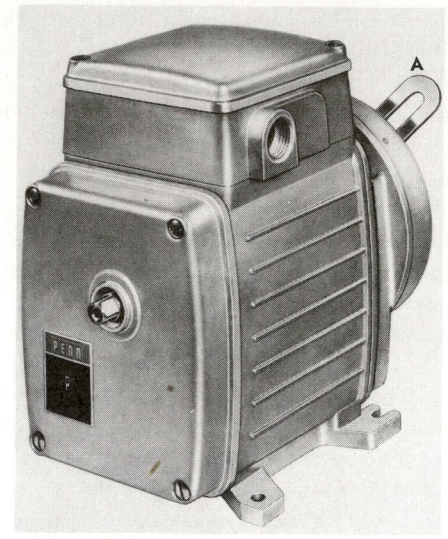

Fig. 24-81. A solid state controlled-positioning motor.
A—Damper lever.

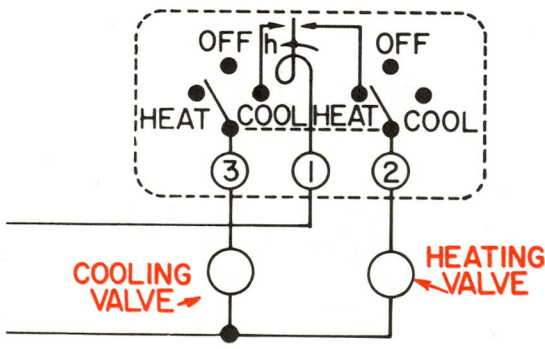

Fig. 24-82. Combination thermostat connected to two valves of split-system, one for cooling and one for heating.
(Penn Controls, Inc.)

24-34 SPLIT SYSTEM CONTROLS

More and more air conditioning systems are designed to provide either hot or cold water to the heat exchange unit in the space to be conditioned. Or, they are designed to provide either hot or cold air to the space to be conditioned.

The three-pipe system, for example, will carry both hot and cold water to a heat exchanger. The third pipe is a return line for either. Fig. 24-82 is a diagrammatic sketch of a thermostat connected to the two valves of a split system.

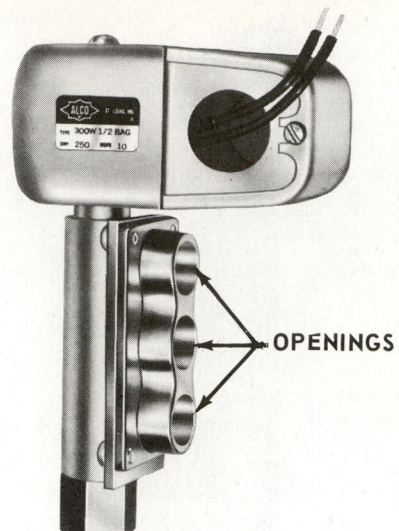

Fig. 24-83. Three-way valve which bypasses cold or hot water when desired. (Alco Controls Div., Emerson Electric Co.)

Many water heating or cooling systems bypass the heat exchange coil when no heating or cooling is needed. This action is obtained through a special solenoid valve. Fig. 24-83 shows such a valve. The inside is shown in Fig. 24-84.

24-35 ZONE CONTROLS

Many heating and cooling systems have zone controls to maintain each zone at the desired temperature. For heating, the zone controls may control the conditioned air, the flow of hot water, the flow of steam, or the electrical heating circuit.

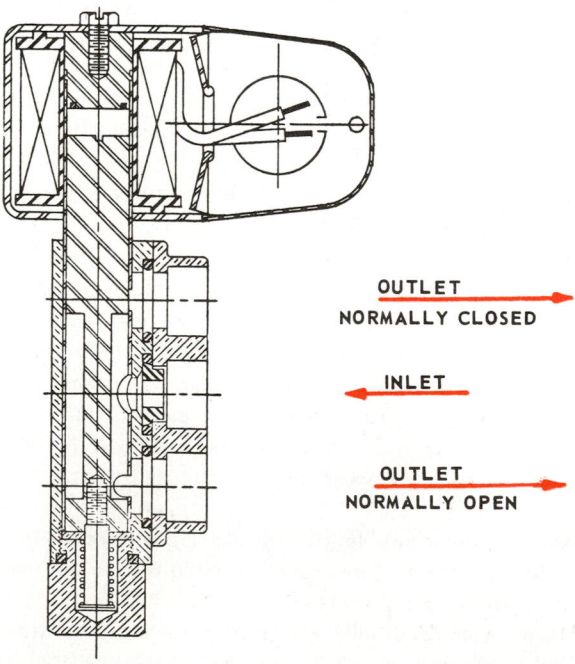

Fig. 24-84. Internal construction of solenoid-operated three-way water valve.

Zone cooling is arranged by using damper controls in ducts or automatic valves in the chilled-water circuit.

Some of these zone controls are:

1. Thermostat or sensors.
2. Proportional thermostats or sensors.
3. Damper motors.
4. Proportional damper positioner.

24-36 COMFORT COOLING CONTROLS

Controls for comfort cooling are of the same basic types as those in heating. There are operating controls, primary controls and limit controls.

The operating controls are thermostats, pressurestats and humidistats. Primary controls include motor starters and starting relays. Limit controls are overload circuit breakers, thermal overloads, internal motor overloads, refrigerant pressure limit controls and oil pressure limit controls. Most of these controls are described in Chapters 8 and 12.

The two most popular refrigerant controls are the thermostatic expansion valve (TEV) and the capillary tube. These controls are described in Chapters 5 and 12.

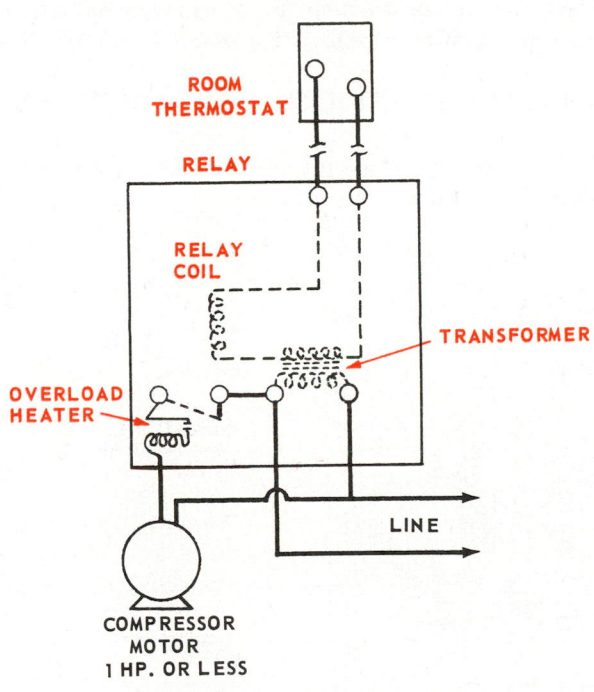

Fig. 24-85. Wiring diagram of thermostat relay combination for small comfort cooling units.

The schematic diagram in Fig. 24-85 is for a circuit used in a comfort cooling unit. This system uses a low voltage, two-wire thermostat. The thermostat controls a relay which will close the motor circuit. If the motor cannot be connected directly to the line, and if pressure safety devices are to be put in the system, the wiring will be somewhat like that shown in Fig. 24-86. The high pressure safety cut-out is wired in series

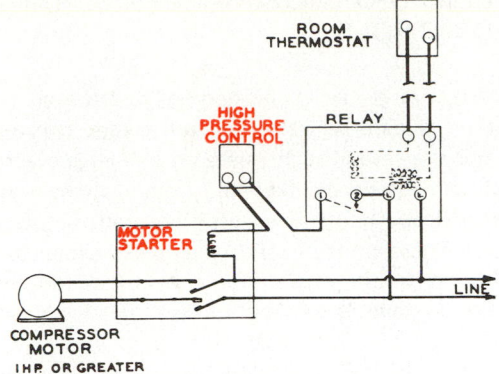

DIAGRAM USING A LARGE REFRIGERATION UNIT.

IF THE INSTALLATION HAS A BLOWER, IT IS USUALLY WIRED TO RUN CONTINUOUSLY DURING THE COOLING SEASON.

Fig. 24-86. Wiring diagram of comfort cooling unit which uses high-pressure safety cut-out and motor starter. (White-Rodgers Div., Emerson Electric Co.)

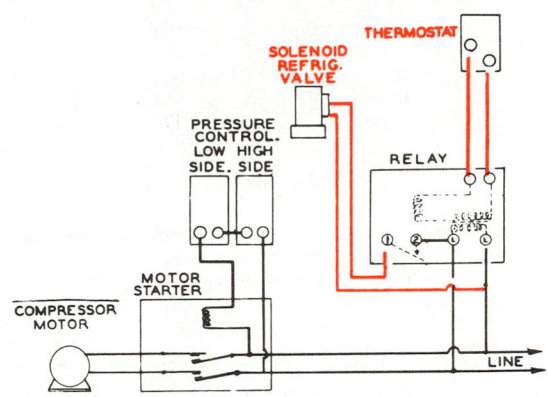

Fig. 24-87. Comfort cooling system wiring diagram. Note that thermostat operates solenoid valve. System cycles as low-side pressures vary.

with the starter coil. It will open the circuit if pressures become too high.

Some systems cycle on command from a low-side pressure control. The thermostat operates a solenoid valve mounted in the liquid or suction line. When the thermostat temperature is satisfied, the solenoid valve will close. When the low-side pressure drops enough the motor circuit will be opened by means of the pressure control connected to a magnetic starter, as shown in Fig. 24-87. The unit will then stop. A high-side switch is also provided in this control. This control will stop the compressor in the event the high-side pressure exceeds a preset limit.

The controls found in comfort cooling systems are:

1. Thermostat (24V or 120V) or thermistor sensor.
   a. One-stage.
   b. Two-stage.
2. Evaporator icing control (freeze-up control); two-wire 24V or 120V to control dampers, valves, or compressors when evaporator approaches 32 F.
3. Multiple compressor sequence starting controls.
4. Multi-cylinder compressor unloading sequencing controls.

24-37 COMFORT COOLING REFRIGERANT CONTROLS

Generally, large air conditioning evaporators use the thermostatic expansion valve for refrigerant control. Some installations use several such valves on one large evaporator to get maximum efficiency. Self-contained systems — especially those hermetically built — may use the capillary tube refrigerant control. These controls are described in Chapter 5.

In addition to the refrigerant control used on automatic refrigeration systems, a solenoid refrigerant valve is sometimes placed in the liquid line. This automatically stops the flow of refrigerant to the evaporator:

1. The instant the condensing unit stops.
2. When the low-side pressure control opens.
   This is done in order that the evaporator:
1. Will not become flooded with refrigerant while the condensing unit is idle.
2. To pump down the evaporator.

Fig. 24-88 illustrates a solenoid refrigerant control valve.

This valve uses 5W at 120V a-c. It has a .100 in. diameter orifice. It has a refrigerating capacity, at a 5 lb. pressure drop across the orifice, of 1.3 ton for R-12, 2.1 ton for R-22, and 1.62 ton for R-500.

24-38 COMFORT COOLING MOTOR CONTROLS

In addition to the operating controls (thermostats), comfort cooling systems have primary controls and limit controls.

Larger refrigerating units are usually equipped with pressure controls. The low and high-pressure controls are usually designed to lock the circuit open if any unusual pressures occur. The operator must then manually turn the system on. This allows a careful check for faults.

The internal construction of a combination low and high-pressure control is shown in Fig. 24-89.

24-39 COMFORT COOLING LIMIT CONTROLS

Comfort cooling systems use several types of limit controls:
1. Motor limit controls.

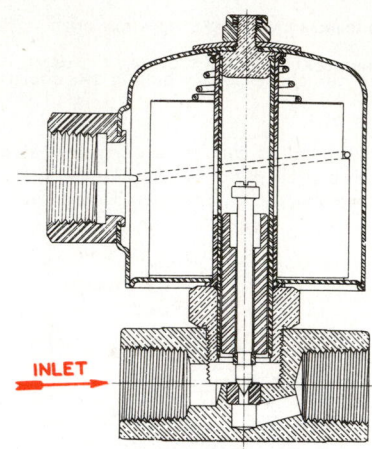

Fig. 24-88. Cross-section of solenoid refrigerant control valve used in liquid lines on air conditioning systems. (Controls Co. of America)

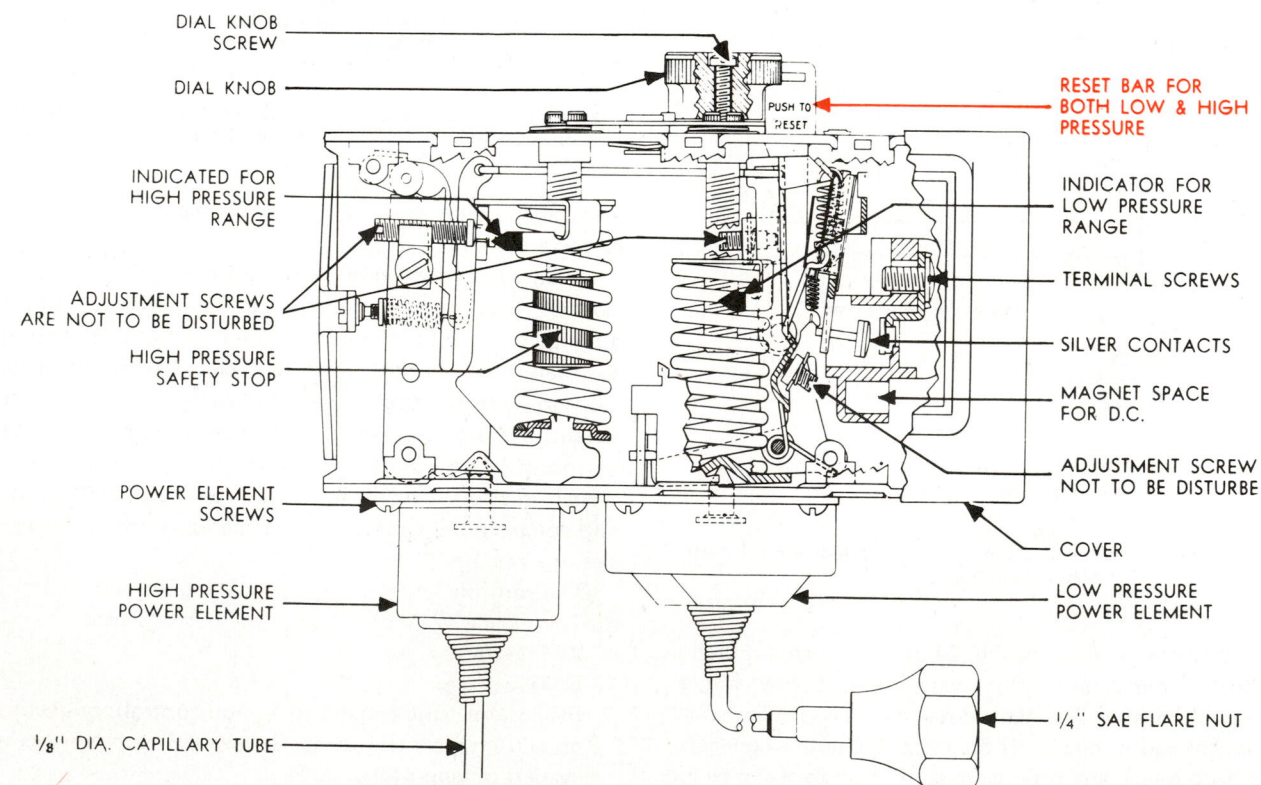

Fig. 24-89. Inner mechanism of air conditioning condensing unit pressure motor control. (Ranco, Inc.)

2. Pressure limit controls.
3. Temperature limit controls.
4. Fluid flow limit controls.

Motor limit controls are described in Chapter 8. Such controls will stop the unit if current draw becomes too high or if motor temperature rises to dangerous levels.

Another type of temperature limit control (other than the motor thermistor or bimetal protector) is an anti-icing control on the evaporator. Should ice accumulate there, this control will open and stop the system.

Fluid flow controls stop system in case chilled water flow ceases or if conditioned airflow stops or slows to inefficient

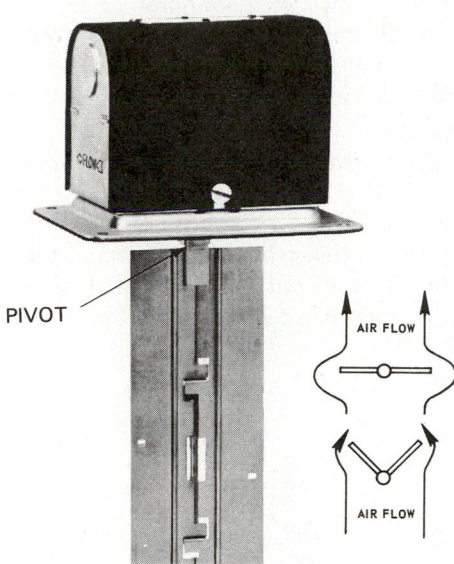

Fig. 24-90. Airflow has hinged paddle to prevent damage when used in high velocity ducts. When airflow strikes the paddle, it moves back through an arc on the pivot, tripping the electrical switch.
(ITT McDonnell & Miller)

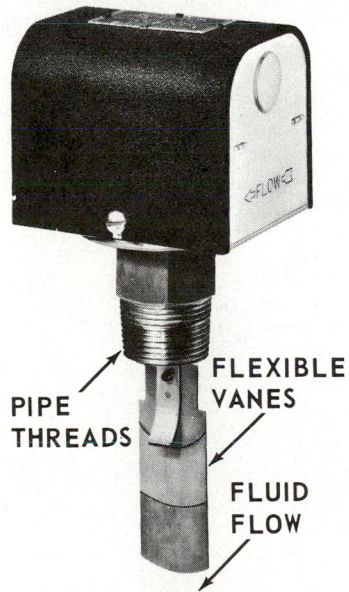

Fig. 24-91. Liquid flow switch. If chilled water or condenser water flow is not sufficient, this unit will close signal circuit, shut off unit, or both.

amounts. Fig. 24-90 shows an airflow signal switch or shutoff switch or both. If airflow is greater than standard, paddle mounted on pivoted arm moves back through small arc tripping an electrical switch. Fig. 24-91 shows similar switch for liquid flow such as chilled water or condenser water.

24-40 HUMIDITY CONTROL SYSTEMS

Humidity control systems are almost always electric, although pneumatic systems are usually used for large systems.

The electrical system uses a humidistat control and an electrical power source. Either a solenoid or a motor is used to operate a valve or damper.

The pneumatic system uses a humidistat control, piping and a vacuum or pressure source. The controlled vacuum or pressure then acts upon a diaphragm-operated valve or a diaphragm-operated damper.

Humidity is most often made higher by adding water vapor to the air. It is most often lowered by cooling air below its dew point temperature. This condenses moisture out of the air.

24-41 HUMIDISTATS

A humidistat is the control device in a humidity control system. It responds to humidity changes and, in doing so, opens or closes a control system.

The sensing element of the humidistat is called a hygroscopic element. The most commonly used hygroscopic elements are: human hair, wood, nylon ribbon, membranes, and electronic solid state sensors. Wood, human hair, membranes, and nylon ribbon stretches in length as the moisture content of the air increases.

In electronic solid state sensors, the sensor changes its resistance as the moisture content of the air changes.

Some sensors used are:
1. Hygroscopic salt (for example lithium chloride).
2. Carbon particles imbedded in a hygroscopic material. In these substances, the resistance decreases as the humidity increases.

The change in size or shape or electrical resistance of the sensing element is used to operate a switch or a pneumatic system. Fig. 24-92 shows a humidistat which uses a multiple hair element. The control should be kept dust free and the cover must permit free air circulation over the element.

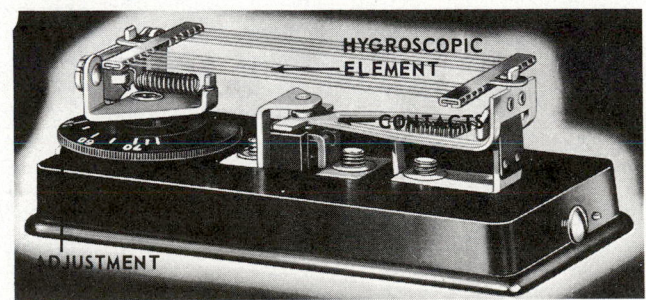

Fig. 24-92. A humidity control showing operating mechanism.

See Para. 18-7 and Para. 18-11 for more information on humidity measurement.

24-42 ELECTRONIC CLEANER CONTROLS

Electronic cleaners have an electrical circuit which converts 120V a-c to about 9000V d-c. This circuit also has safety devices such as door interlocks and service devices such as automatic cleaning and circuits.

Fig. 24-93 shows an electronic cleaner electrical circuit. Note the two step-up transformers, the door interlock, the full wave rectifiers, and the neon light circuit. The actual wiring diagram is shown in Fig. 24-94.

The electronic cleaner is connected into the electrical circuit of a heating-cooling system which uses a single speed fan. There is 120V service to the cleaner and 240V service to the fan motor. This is shown in Fig. 24-95.

Because of the high voltages used in these electronic cleaners, they should be serviced only by one who has had special training on the model being serviced. The automatic water cleaning system has a washing cycle with detergent, a rinsing cycle, and a drying cycle. See Fig. 24-96.

Service to electronic cleaners includes checking rectifiers and high d-c voltage. Fig. 24-97 shows how the rectifiers are

removed and how the high voltage is checked. Note the heavy insulators on the meter leads to protect one from this very dangerous voltage.

24-43 TROUBLESHOOTING CONTROLS

Locating the real fault in a system quickly saves time and prevents mistakes. Some actual service calls and their faults are listed.

A. System fails to start:

1. Check the fuses or circuit breakers and the power-in with a suitable test light or voltmeter.
2. Check the main switch and thermostat switch to make sure they are in the closed position.
3. Check the thermostat setting. It must be on a setting which will close thermostat switch.
4. Check the contactor. It must be pulled in. If not, the contactor circuit is open.
   a. Contactor open but not buzzing. Contactor coil not powered. Check control circuit using voltmeter and ohmmeter.
   b. Contactor open and buzzing. Contactor coil is operating, but armature not pulled in. Either the armature is stuck or control coil voltage is too low.

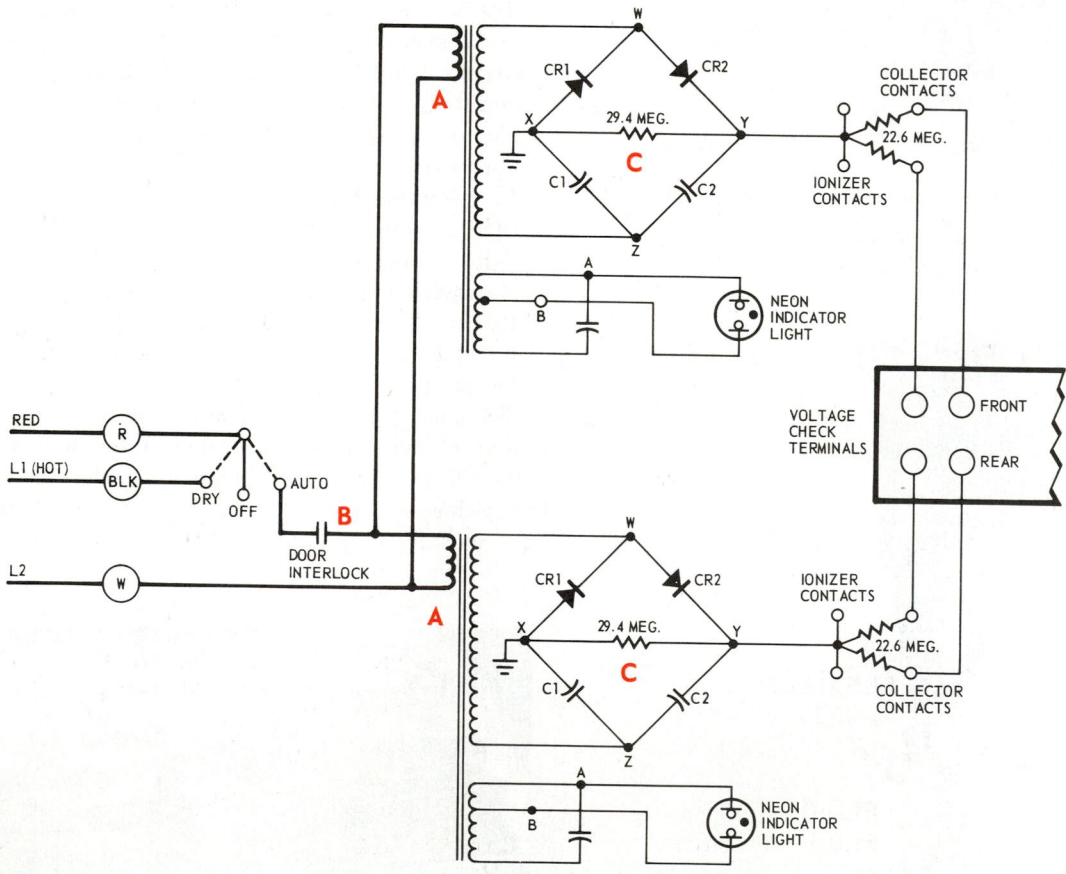

Fig. 24-93. Wiring circuit for an electronic air cleaner. A—Transformers. B—Door interlock (safety switch). C—Rectifiers. (Honeywell, Inc.)

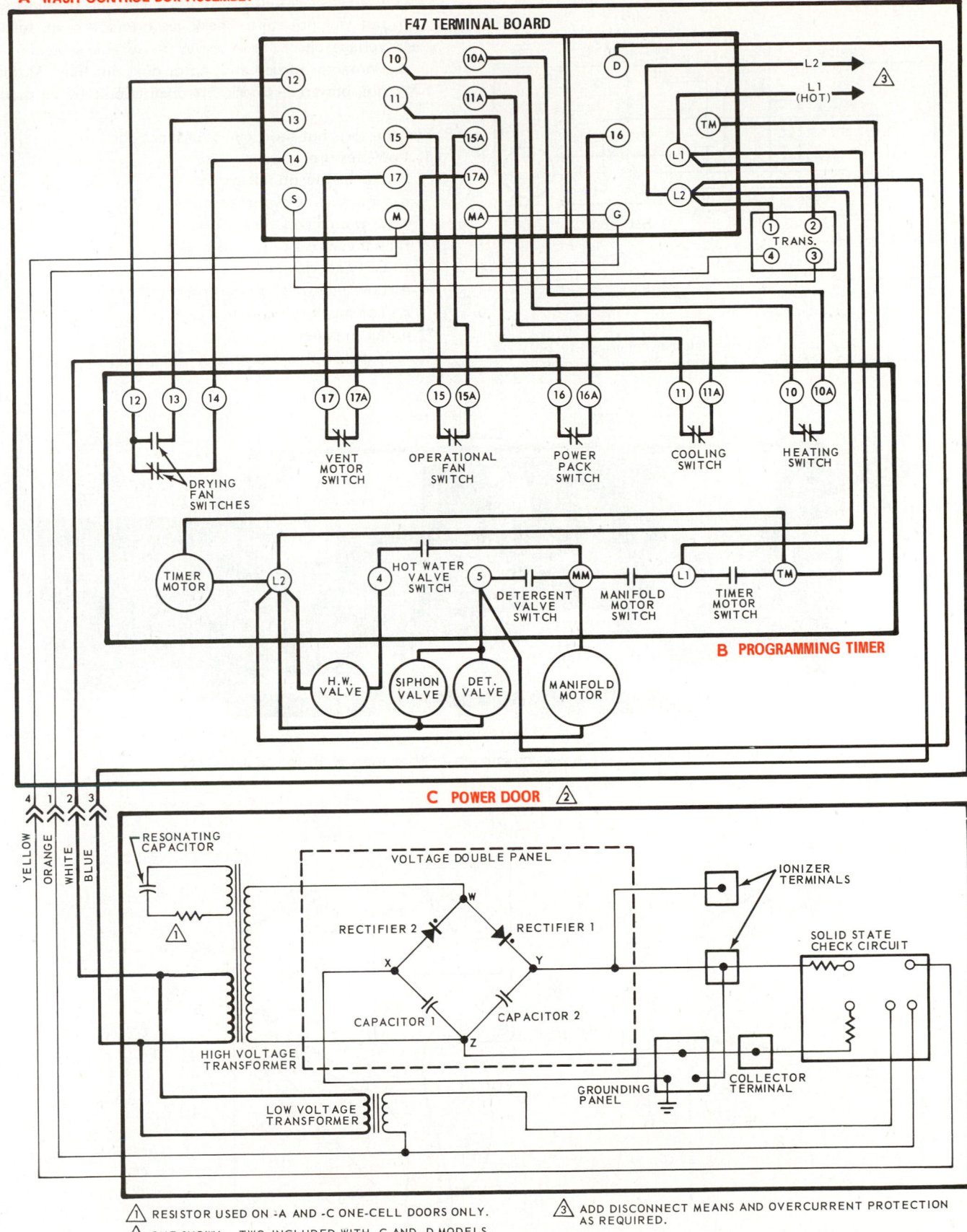

A WASH CONTROL BOX ASSEMBLY

F47 TERMINAL BOARD

L2
L1 (HOT)
③

TRANS.

DRYING FAN SWITCHES

VENT MOTOR SWITCH
OPERATIONAL FAN SWITCH
POWER PACK SWITCH
COOLING SWITCH
HEATING SWITCH

TIMER MOTOR
HOT WATER VALVE SWITCH
DETERGENT VALVE SWITCH
MANIFOLD MOTOR SWITCH
TIMER MOTOR SWITCH

H.W. VALVE
SIPHON VALVE
DET. VALVE
MANIFOLD MOTOR

B PROGRAMMING TIMER

C POWER DOOR ②

YELLOW ORANGE WHITE BLUE

RESONATING CAPACITOR

VOLTAGE DOUBLE PANEL

IONIZER TERMINALS

RECTIFIER 2   RECTIFIER 1

SOLID STATE CHECK CIRCUIT

CAPACITOR 1   CAPACITOR 2

HIGH VOLTAGE TRANSFORMER

COLLECTOR TERMINAL

GROUNDING PANEL

LOW VOLTAGE TRANSFORMER

① RESISTOR USED ON -A AND -C ONE-CELL DOORS ONLY.
② ONE SHOWN — TWO INCLUDED WITH -C AND -D MODELS.
③ ADD DISCONNECT MEANS AND OVERCURRENT PROTECTION AS REQUIRED.

Fig. 24-94. Complete wiring diagram of electronic air cleaner. Three main circuits are: A—Wash control box assembly. B—Programming timer. C—Power door.

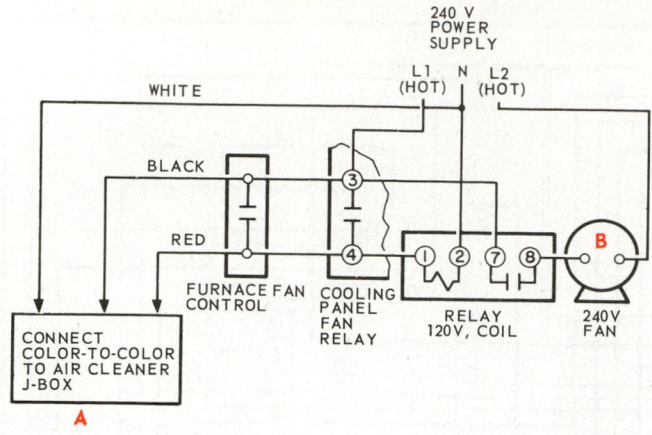

Fig. 24-95. A 120V service electronic cleaner connected to 240V system with 240V fan motor. A—Electronic cleaner. B—Fan motor.

c. Contactor closed and motor hums. Motor is powered but will not start. Check capacitors, if used, for low voltage; check for excessive pressures in system.

d. Contactor closed and motor does not hum. Motor is not powered. Check for open circuit by continuity test.

B. System starts but short cycles or locks out:
1. Low pressure.
   a. Can be low on refrigerant.
   b. Low airflow over evaporator.
   c. Low outside temperature.
2. High pressure.
   a. Dirty condenser.
   b. Low airflow over condenser.
   c. Too much refrigerant.
   d. Air in system.
   e. High low-side pressure.

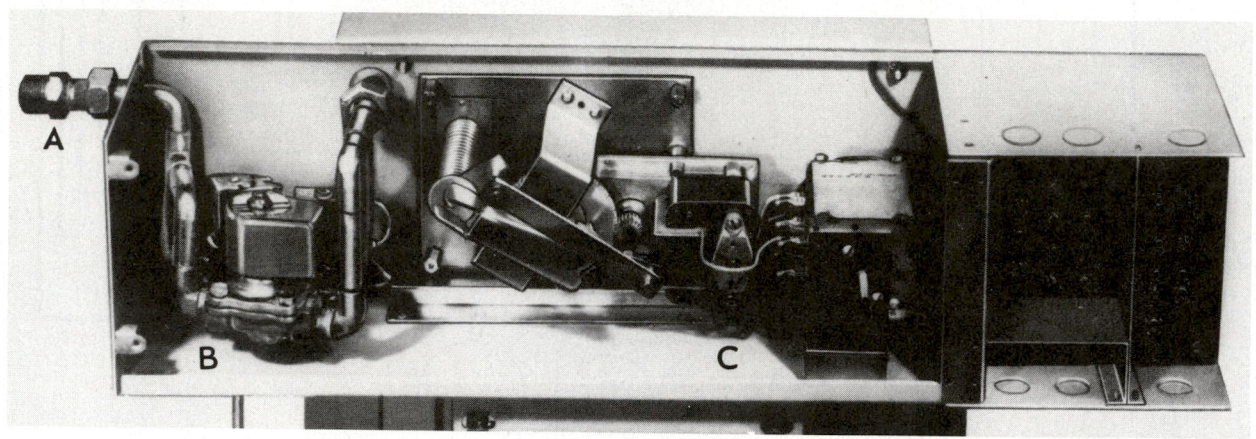

Fig. 24-96. Electronic cleaner washing controls. A—Warm water in. B—Solenoid. C—Timer.

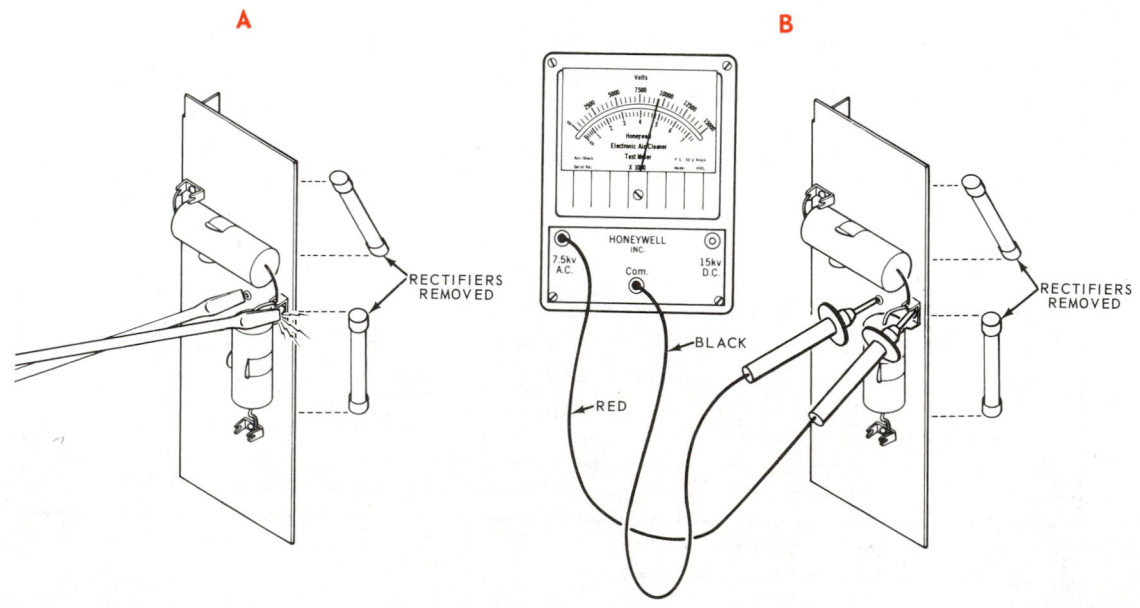

Fig. 24-97. Servicing an electronic air cleaner. A—Removing the rectifiers. B—Testing the high voltage.   (Honeywell, Inc.)

3. Overload cycling.
   a. Improperly wired.
   b. Compressor tight or burned out.
   c. Defective capacitor.
   d. Defective start relay.
   e. High current flow.
   f. High head pressure.
4. Faulty thermostat.
   a. Dirty contacts in thermostat.
   b. Dust clogged.
   c. Cooling anticipator not operating correctly.
   d. Wiring faulty.
   e. Transformer voltage too low.
   f. Thermostat improperly located.

C. System works but not correctly:
   1. Low on refrigerant.
   2. Inefficient compressor.
   3. Faulty airflow.
      a. Filter dirty.
      b. Evaporator iced or dirty.
      c. Fan and/or motor belt not working correctly.
      d. Controls out of adjustment.

24-44 INSTRUMENTS

Instruments are all important to refrigeration, heating and air conditioning installation and service personnel. It is almost impossible to check on the electrical characteristics of a unit

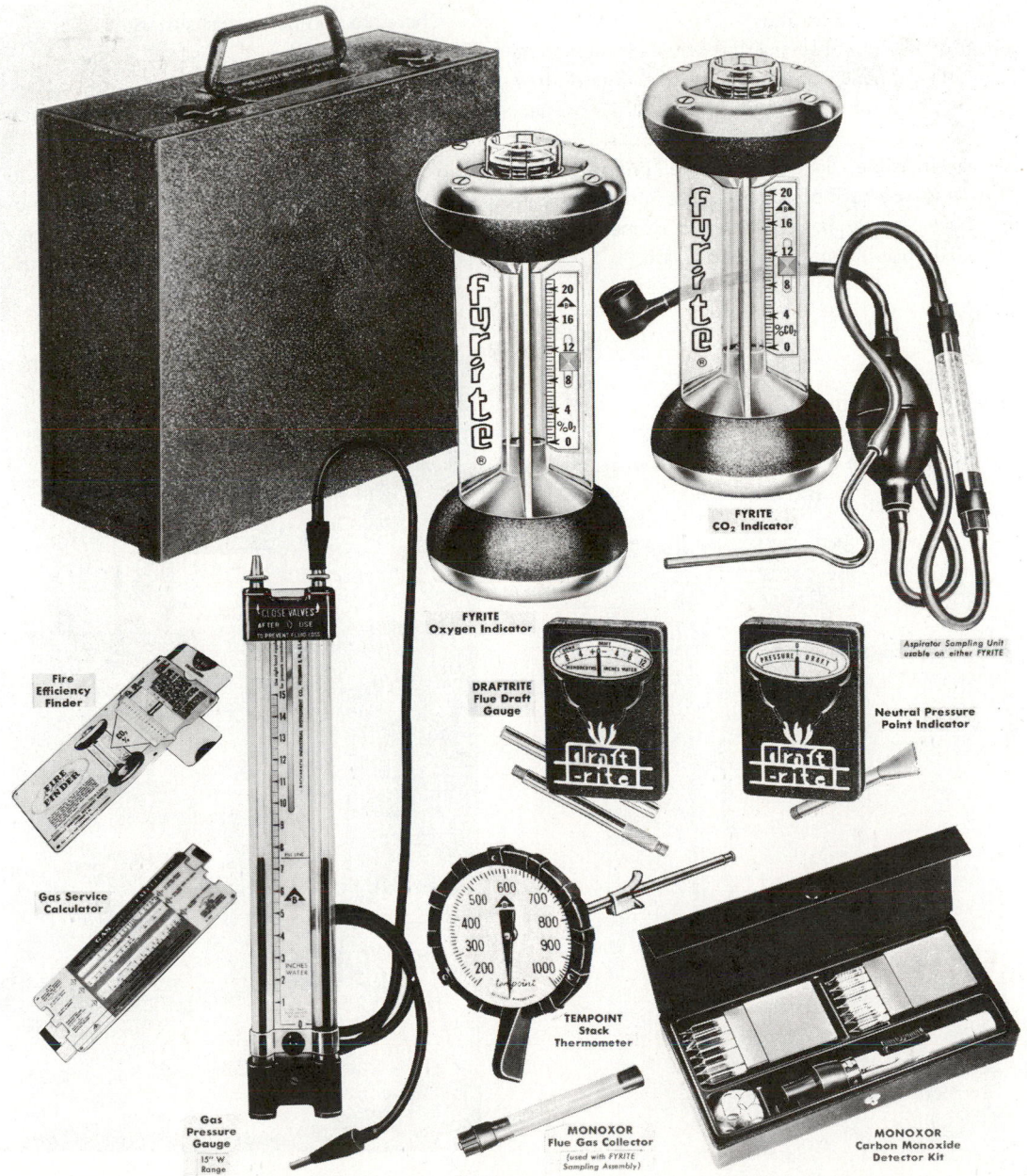

Fig. 24-98. Service technician's kit used to check combustion characteristics of burner and combustion chamber. Note calculator charts which may be used in interpreting instrument readings. (Bacharach Inst. Co., Div. of AMBAC Industries, Inc.)

without using ammeters, voltmeters, test lights and ohmmeters. Instruments are explained in Chapters 6, 7 and 8.

Likewise, what takes place inside a refrigerating system is almost impossible to check without pressure gauges and thermometers. These instruments are discussed in Chapters 2, 11 and 14.

Instruments should be used well within their scale range. They will maintain accuracy longer if use never exceeds more than a half to two-thirds of their scale range.

In heating systems, the combustion process and the flue gas properties can only be accurately determined by the use of instruments. A set of these instruments is shown and identified in Fig. 24-98.

The flow of liquids should be checked as carefully as possible. The velocity and amount of water flowing in hot water systems, chilled water systems, condensers and cooling towers should be accurately measured.

The amount and velocity of refrigerant liquids is important to know during maintenance and repair work. These flow meters can be installed in the piping temporarily or permanently.

Because air is invisible, instruments are important to measure its flow and pressure conditions. Thermometers and pressure gauges have already been explained. Some different instruments especially useful for airflow study are:
1. Manometer.
2. Barometer.
3. Pitot tube.
4. Anemometer.
5. Smoke as a velocity indicator.
6. Hot wire anemometer.

A velocity meter is shown in Fig. 24-99. This instrument is

used with a pitot tube. The inclined scale is for low pressures and velocities. The vertical scale is used for high pressures and high velocities.

Many of these air-measuring instruments are explained in Chapter 18.

24-45 RECORDING INSTRUMENTS

Service technicians can easily take "spot" checks of volts, amperes and watts in a circuit or parts of a circuit. However, most causes of electrical faults may occur when the service technician is not on hand with his instruments. It is possible to use recording instruments. These instruments may provide a 24-hour record or even a seven-day record of electrical values as well as temperature, pressure, and humidity. These latter instruments are illustrated in Chapter 18.

There are two basic charts in use:
1. Circular chart.
2. Strip chart (used in roll form).

There are two types of electrical recording instruments:
1. Galvanometric.
2. Potentiometer.

The galvanometric type uses a meter movement (usually d'Arsonval type) and is most used by service technicians. The potentiometer type uses a potentiometer with a variable (what is being measured) against a standard value.

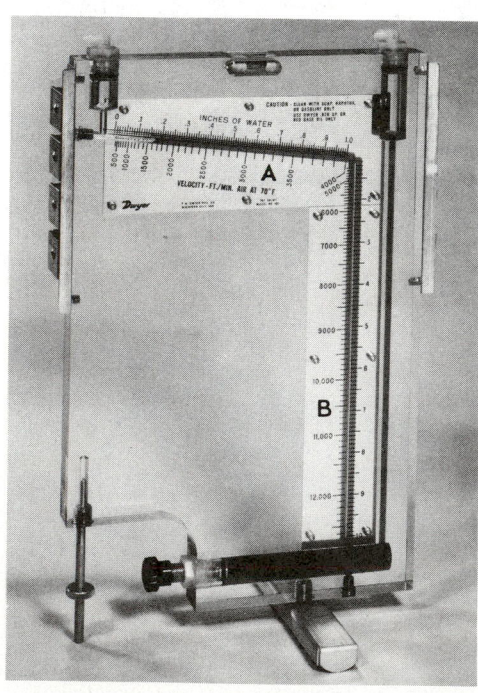

Fig. 24-99. Manometer used with a pitot tube. A—Scale for velocities. B—Scale for pressures. (Dwyer Instruments, Inc.)

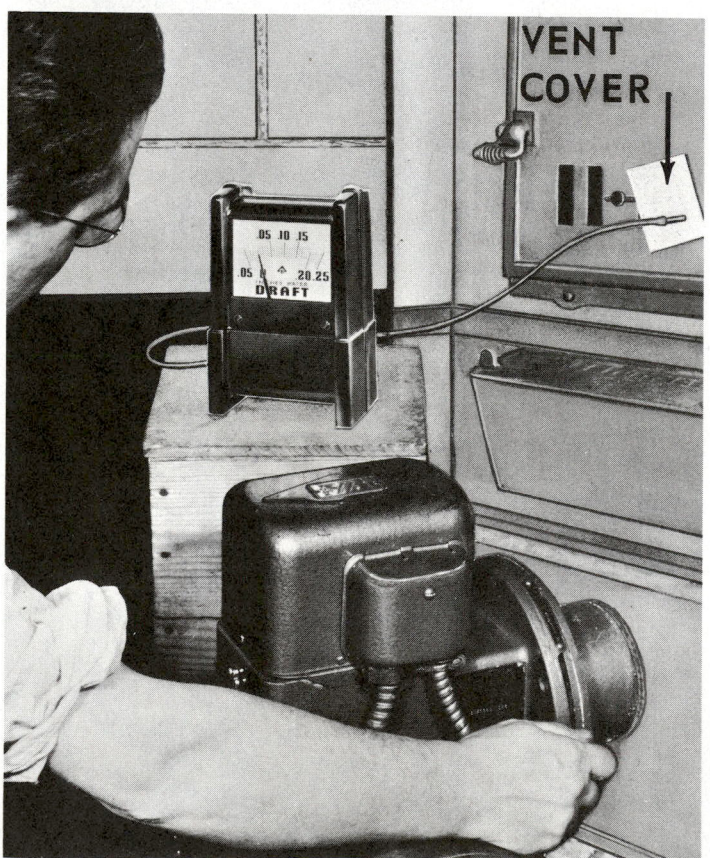

Fig. 24-100. A draft indicator shows combustion gas flow. Note protector over vent in furnace door to prevent inaccurate reading that would occur if room air entered combustion chamber.

The galvanometric type has a meter needle or pointer which makes an impression on a moving piece of paper (circular or strip).

The trace on the paper is produced in several ways:
1. Ink trace.
2. Pressure trace (paper sensitive to pointer pressure).
3. Electrical trace (paper conducts electric current and paper color changes).
4. Thermal trace (pointer is hot and paper is sensitive to heat).

The paper is moved by a spring-loaded clock mechanism or a geared timer motor.

24-46 DRAFT CONTROL INSTRUMENTS

Efficient and safe combustion in a furnace needs accurate draft control (control of flue gas movement). Flue gas flow depends on two things:
1. The density of flue gas versus the density of the air.
2. The pressure difference between the inside of the building and the outside of the building.

Flue gas conditions vary according to the type of fuel used (oil or gas).

A draft gauge is generally used to determine the efficiency of flue gas (combustion gas) flow. Fig. 24-100 shows a draft gauge being used. The stack temperature is also an indicator of draft efficiency. Stack temperatures vary from 300 F. to 900 F. (149 C. to 482 C.). Fig. 24-101 shows a stack thermometer being used to find the temperature of the flue gas.

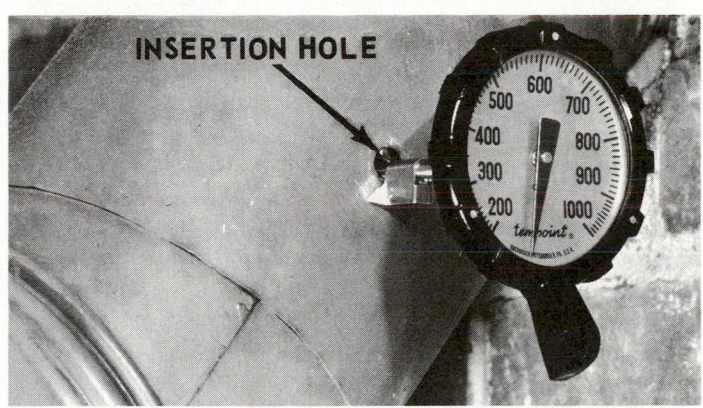

Fig. 24-101. Stack thermometer used to determine flue gas temperature. Note small hole made for putting instrument in flue.

24-47 COMBUSTION EFFICIENCY INSTRUMENTS

Combustion efficiency can be determined by measuring the amount of carbon dioxide (CO_2) in the flue gas. A sample of the combustion gas is exposed to a chemical which absorbs carbon dioxide only. Fig. 24-102 shows a CO_2 indicator being used.

If 10 cc (cubic centimetres) of the flue gas reduces to 9 cc of gas after exposure to the chemical, the flue gas contained 1 cc or 10 percent of carbon dioxide.

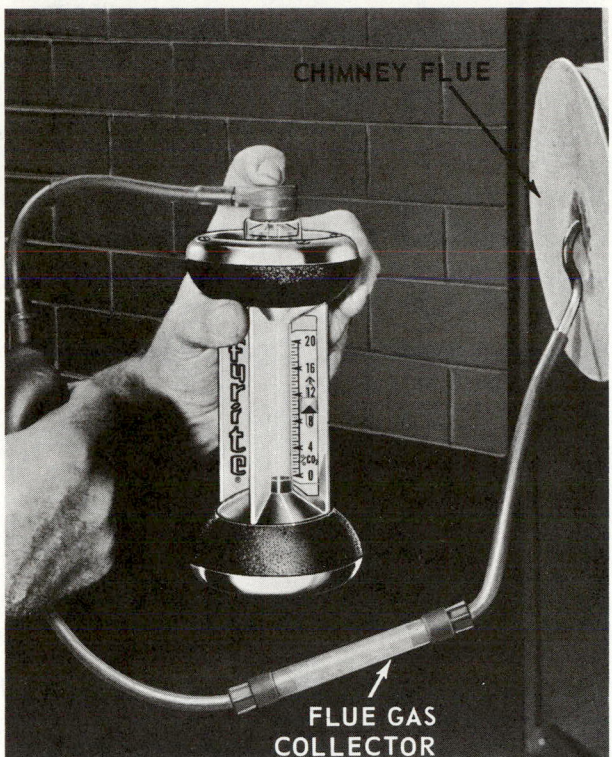

Fig. 24-102. An analyzer is used to check carbon dioxide content of flue gas. Note flue gas collector and connection into chimney.

If the gas cools during this operation, Charles's Law will affect the answer. When the gas cools, its volume reduces, and the service technician will get a too-high reading.

The flue gas temperature before and after must be known and a correction must be made for accurate results. Tables are provided by equipment manufacturers.

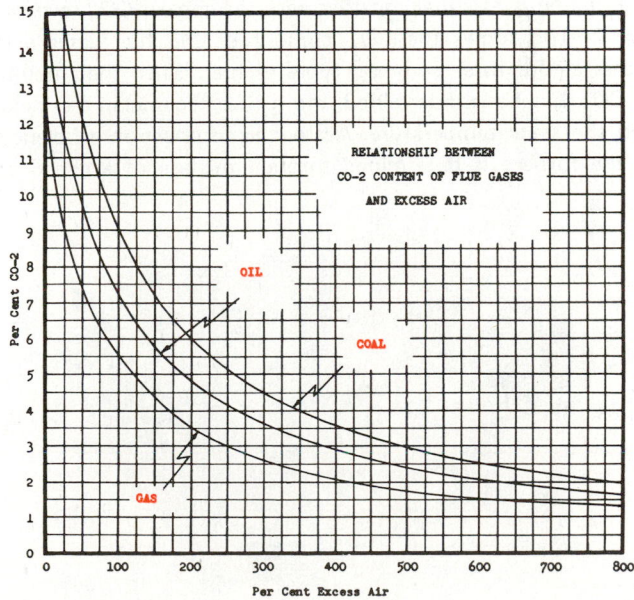

Fig. 24-103. Graph showing change in stack gas carbon dioxide content for various fuels as amount of excess air changes from 0 to 800 percent.

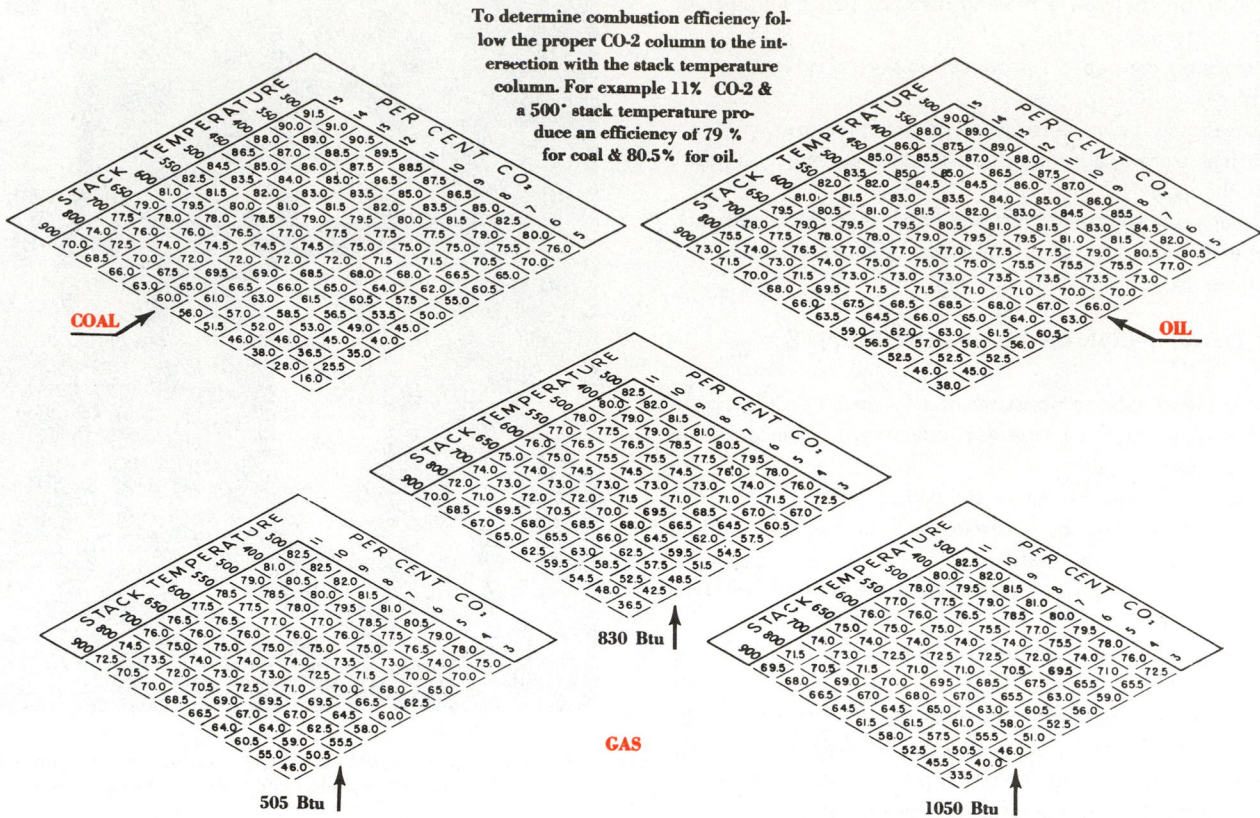

To determine combustion efficiency follow the proper CO-2 column to the intersection with the stack temperature column. For example 11% CO-2 & a 500° stack temperature produce an efficiency of 79 % for coal & 80.5% for oil.

COAL

OIL

GAS

505 Btu

830 Btu

1050 Btu

Fig. 24-104. Charts show combustion efficiencies at various stack temperatures and carbon dioxide percentages. Note that chart gives efficiencies for coal, oil and three different Btu gas qualities. (Dwyer Instruments, Inc.)

Systems will vary in CO_2 content. Some are operating correctly with as low as 8 percent CO_2; some as high as 12 percent CO_2. The manufacturer's service manual will give the correct amount for the system being tested. *Remember, a clean flame is essential together with the correct CO_2 reading.* Fig. 24-103 shows the CO_2 content of the flue gases and excess air for three common types of fuel. When the amount of CO_2 has been determined, the service technician can next find the stack temperature. Again, the combustion efficiency of the furnace is determined through the use of a chart, as shown in Fig. 24-104.

Some analyzing instruments use electronic sensors, circuits and indicators. They are used to test for air pollution, boiler efficiency and stack gas analysis. These instruments use thermal conductivity and a combustion chamber with catalysts to measure the gas sample. See Fig. 24-105. Some use air to carry the sample; some use helium, argon or nitrogen. These instruments can accurately analyze for hydrogen, oxygen, carbon monoxide, methane, ethane and carbon dioxide.

Separate silica gel and molecular sieves are used to partition and adsorb certain gases. A pump is used to move the gases and a built-in recorder records the results. In about three minutes, this unit will give accurate flue gas analysis for oxygen, carbon monoxide and carbon dioxide.

24-48 SMOKE TEST

A smoke test is an excellent way to check combustion efficiency such as air-fuel ratio, primary air, secondary air and draft.

The test is an empirical test (one which depends on experience and observation) rather than a scientific comparison. A white filter paper is inserted in the flue. The flue gas sample is drawn through the filter and the smoke deposit on the filter paper is compared with a sample chart.

Another method inserts a tube in the flue and an aspirator

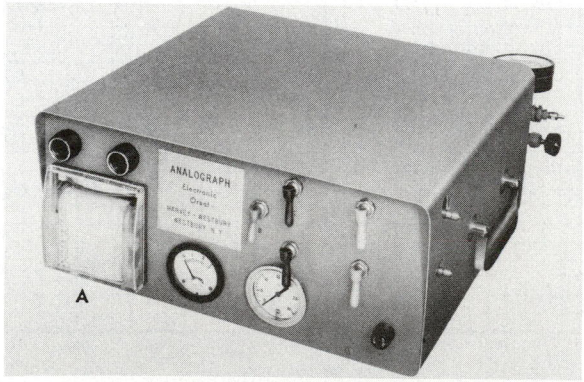

Fig. 24-105. An electronic orsat apparatus used for measuring air pollution and flue gases. A—Recorder. (Harvey-Westbury Corp.)

bulb pulls flue gas samples through a filter mounted in a fixture (about 30 aspirator bulb squeezes are made). The smoke deposit on the filter is then compared to a master comparison scale. This scale (the Ringelmann Scale) rates smoke samples by numbers from 1 to 4.

Photoelectric cells may also be used to check smoke density in large systems or in laboratories. A smoke monitoring system is shown in Fig. 24-106. The reflector can be mounted in

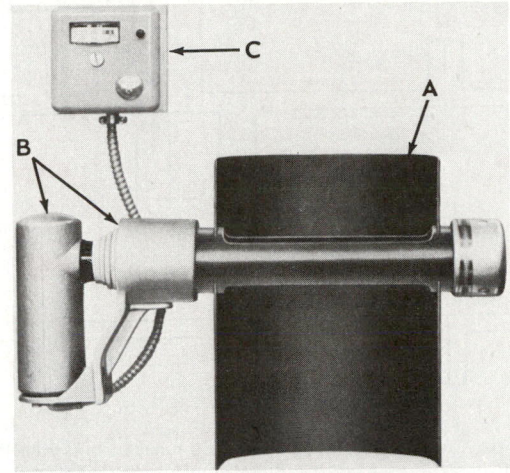

Fig. 24-106. Photoelectric system for measuring smoke density in a chimney. A—Chimney. B—Photoelectric cell and beam tube. C—Indicating instrument calibrated in the Ringelmann Scale. (Electronics Corp. of America)

chimneys from 1.5 ft. diameter to 10 ft. diameter. The photoelectric signal is amplified and the Ringelmann Scale reading is shown on the upper left instrument.

The readings can also be recorded. An alarm signal is given when the smoke density reaches a preset maximum. This unit can also shut down the system or start up flue blowers or afterburners. Fig. 24-107 is the wiring diagram for this system. In no case should there be smoke appearing at the chimney exit.

24-49 AIR VOLUMES

Because air is invisible, instruments are usually used to measure its flow. Chapter 18 explains most of these instruments.

A direct reading instrument for cubic feet per minute (cfm) has been developed. One first measures the grille area in square inches. This area is measured to the outside edge of the grille openings and does not subtract the grille mesh or cross bars. The method of measuring airflow in cubic feet per minute is shown in Fig. 24-108.

24-50 VISIBLE AIRFLOW INDICATORS

Drafts of 15 to 25 ft./min. are allowable and are comfortable for most installations. If air movement is less than this, air stagnation results. If it is more, persons exposed to the draft are uncomfortable.

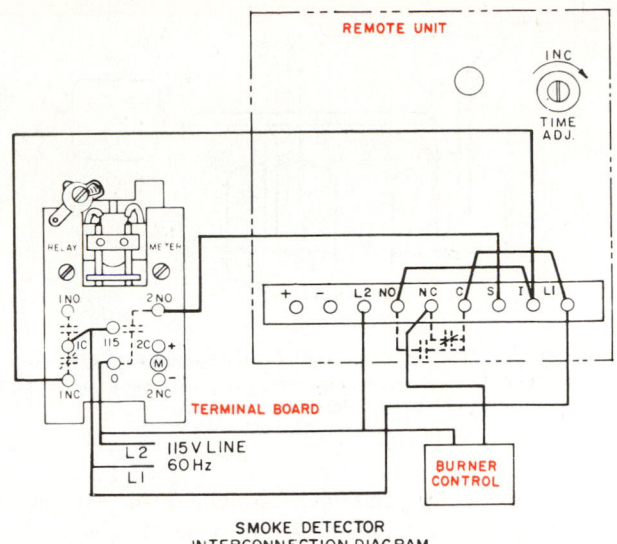

SMOKE DETECTOR
INTERCONNECTION DIAGRAM

Fig. 24-107. A smoke detector wiring diagram.

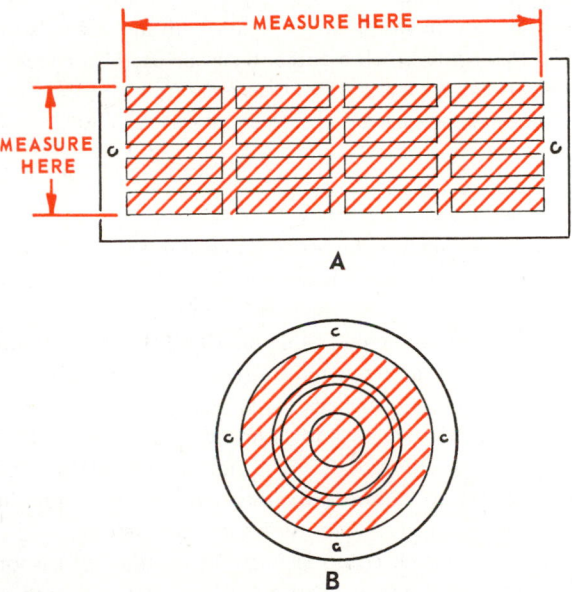

Fig. 24-108. The area to be measured when figuring the volume of air leaving grille. Notice that area does not include rim of grille but does include strips of metal inside. A—Rectangular grille. B—Round grille.

To determine the amount of the draft and the direction, the most successful method has been to use smoke (visible vapor). Smoke generators release small puffs of smoke into the space being tested. The distance the puffs move, in 30 to 60 seconds, is observed. Several readings must be taken and averaged to obtain a degree of accuracy.

One type of smoke generator is shown in Fig. 24-109. Each of the two bottles contains a liquid. The aspirator forces the vapors from the two bottles to mix at the nozzle. The mixing of the two gases forms a white smoke that has a density not much greater than the density of air. The liquids used are diluted hydrochloric acid and aqua ammonia. The smoke

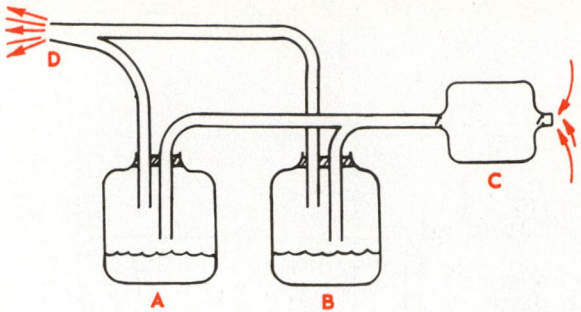

Fig. 24-109. A smoke generator used in determining airflow. A—Hydrochloric acid container. B—Aqueous ammonia. C—Rubber aspirator bulb. Smoke is released at nozzle D.

formed is ammonium chloride. Caution: Both hydrochloric acid and aqua ammonia are dangerous to use. Wear goggles and rubber gloves when handling this material.

A mixture of titanium tetrachloride and air moisture forms a dense, lasting white smoke and can be used for checking air movement and air leaks.

A zinc stearate powder, when mixed with air, may also be used to form a white, small cloud and can be used for checking air movement and air leaks.

Other devices can be used to add smoke to the air:
1. Smoke sticks.
2. Smoke guns.
3. Smoke candles.

Sticks and candles are ignited and placed in the intake of an air distribution system with filter removed. Distribution of air and air balance can then be observed in the system.

A three-minute candle will generate 40,000 cu. ft. of visible smoke. Half-minute and five-minute sizes are also made. The smoke can also be used to check for window and door leaks or to check refrigerator door and window seals. The smoke is nontoxic but long exposure to it is not recommended. It is likewise harmless to clothing or the contents of a building normally exposed to air. Avoid using too much smoke.

The smoke is a zinc chloride mist with a trace of carbon. While they usually produce a white smoke, candles to produce yellow or orange smoke are also available.

24-51 WATER ANALYSIS INSTRUMENTS

Whenever a test of water is involved, its condition is expressed in "pH." This factor indicates the activity of the hydrogen ion whenever there is moisture present. The term "pH" followed by a number is used to indicate whether water tends to be acidic or alkaline. Distilled water at 77 F. (25 C.), neither acidic nor alkaline, has a pH of 7. It is said to be neutral. Numbers above 7 and up to 14 express increasing alkalinity. Numbers decreasing from 7 (6 to 0) indicate increasing acidity.

Since electrical properties of water change as pH changes, acidity or alkalinity of water can be measured with electronic sensors, circuits and meters. An increase in pH causes a decrease in millivolts. A water analysis instrument has an amplifier and two electrodes that are immersed in the water.

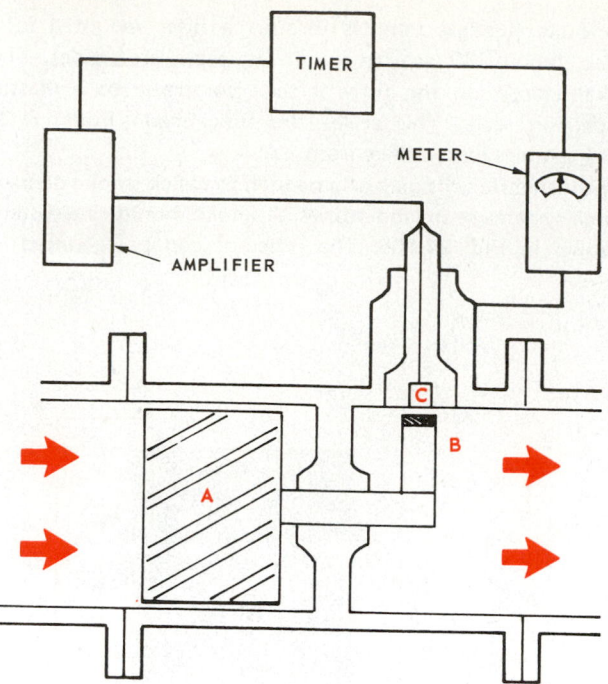

Fig. 24-110. Turbine flow meter. As liquid turns turbine wheels at A, a magnet, B, induces an electrical pulse in pickup at C. This pulse is then amplified, timed and operates a meter which will register flow.

24-52 FLOW METERS

Flow meters have many different designs and are made for many purposes. The gas meter and the water meter are the most common. Both are used to determine consumption.

All heating and cooling systems use moving air, moving liquids or both (condenser water, hydronic water and chilled water). It is important that flows be known if one is to install the equipment correctly and service it accurately.

Airflow is measured with pitot tubes and electronic direct reading meters. See Chapter 18.

Flow of liquids in pipes can also be measured. A variety of liquid flow meters are available. These meters are usually accurate to 1 percent.

One is known as the turbine flow meter. It has a turbine (wheel) turned by a moving liquid. A magnetic pickup gives an electrical pulse each time the turbine wheel makes one turn. These electrical pulses are connected to an electronic circuit with a timing device and a motor. This meter, Fig. 24-110, can be calibrated in cu. ft./min., gal./min., or cc/sec.

Fig. 24-111 shows a meter for measuring water flow. This type measures flowage from 1/4 gal./min. up to 5 gal./min. Fig. 24-112 shows the typical flow in gallons per minute for each of the scale markings.

24-53 REVIEW OF SAFETY

Use a master instrument to check any test instrument that has been dropped. If necessary, have it repaired.

Safety controls for a heating or a cooling system should never be removed and bypassed to keep a system operating.

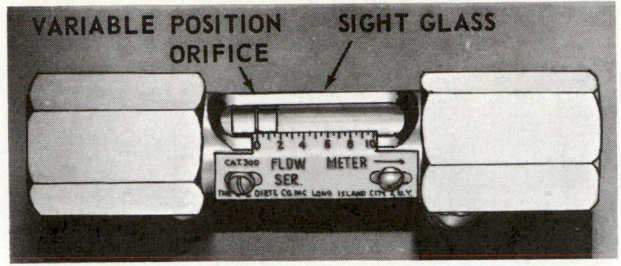

Fig. 24-111. Flow meter for indicating water flow and for measuring volume of water flow. (The Henry G. Dietz Co., Inc.)

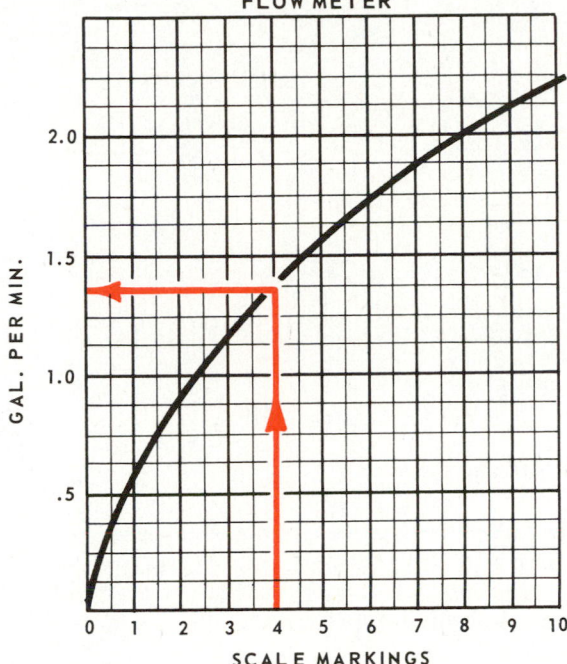

Fig. 24-112. A typical flow meter graph. Flow in gallons per minute are shown for each scale marking on the meter. For example, when orifice moves to the 4 on marking scale, flow is 1.4 gallons per minute.

Dangerous conditions or damage to the system may result.

The handling and use of instruments usually does not present any great hazard to the service technician. However, if he does not read or interpret the instrument readings correctly, the installation may be faulty. In turn, equipment can be seriously damaged and persons within the structure seriously injured.

Always determine the pressure in that part of a system on which work is to be done. Always measure the temperature of the parts of the system which will be repaired, adjusted or touched. Do not guess pressure or temperature.

Use instruments to check electrical circuits. Never assume that the power is off.

Carbon monoxide (CO) is dangerous! It is always present in the combustion gases especially if combustion is not complete.

A substance called palladium chloride may be used to measure the presence of CO. To use it, place a small amount, the size of a dime, on 2 in. by 2 in. plastic tabs. Substance will

darken when exposed to CO. Amounts of CO can be determined by color:

30 to 70 ppm will cause slight darkening.

80 to 120 ppm cause a grey color.

Over 130 ppm result in a black color.

Remember that only a few milliamperes can injure and even kill a person. Always use rubber gloves and rubber boots or shoes when working on electrical devices or circuits. The tools, such as wrenches, pliers and screwdrivers should have insulated handles.

Instruments must be used sensibly. They can be permanently ruined if used beyond their scale readings. Electrical instruments must have clean, tight connections.

24-54 TEST YOUR KNOWLEDGE

1. What refrigerant controls are generally used on air conditioning evaporators?
2. Why are solenoid refrigerant control valves installed on some air conditioning evaporators?
3. What is a bimetal thermostat?
4. How fast must air be moved to produce a noticeable draft?
5. What is the approximate voltage of a low-voltage thermostat?
6. What type thermostat must be carefully leveled?
7. Do millivolt systems have thermostats?
8. What is a pneumatic control?
9. Why is a timer a part of some thermostats?
10. Why is a flame detector used in an oil burner system?
11. What is a limit control?
12. How does a rod and tube thermostat operate?
13. In what controls may a bimetal strip be used?
14. What is the usual bonnet or plenum chamber limit control cut-out temperature setting?
15. Describe a cold anticipator.
16. Describe a heat anticipator.
17. Is there such a control as a combination heating and cooling thermostat?
18. If a flow switch has two sets of contact points, describe the use of each set of points.
19. Is there such a device as an electronic thermostat?
20. Where does the CO_2 come from in flue gas?
21. What is the average CO_2 content of an oil furnace flue gas?
22. How does a smoke tester operate?
23. What is a temperature-compensating bimetal strip?
24. Is a mercury tube switch with two leads called an SPST switch?
25. How does a hydraulic thermostat operate?
26. What does a 24V gun type oil burner thermostat operate when the points close?
27. How are thermometers used as thermostats?
28. What is "system lag" in a heating system?
29. How may a thermostat also indicate the condition of the system's air filter?
30. How many controls must be mounted in the gas line of a gas-fired furnace?

Chapter 25

AIR CONDITIONING SYSTEMS HEAT LOADS

Heating systems require an installation that will give enough heat to keep the occupied space at a comfortable temperature and humidity. This requirement should hold true even at record low temperatures for the given locality. In much of the United States, typical conditions are: a 70 F. (21 C.) inside temperature when it is 0 F. (−18 C.) outside and a 15 mph wind is blowing.

Comfort cooling installations have the same challenge as heating installations. The cooling unit must have sufficient capacity to cool the occupied space to the desired temperature and humidity on the warmest day of the season.

25-1 HEAT LOADS

An air conditioning system must put enough heat into a space to make up for the heat losses during heating. It must remove as much heat as the space accumulates during cooling.

Whenever a temperature difference exists, heat energy will flow from the higher temperature level to the lower temperature level. In the case of heating, it is necessary to retard this heat flow as much as possible since the amount of heat lost must be replaced. In the case of cooling, the system must remove the amount of heat gained.

The most common method of computation is to determine maximum heat load (lost or gained) for a period of one hour.

25-2 TYPES OF HEAT LOADS

There are several major heat loads:
1. Heat that is transmitted through walls, ceilings, and floors (conduction).
   a. From inside to outdoors (heating).
   b. From outdoors to inside (cooling).
2. Heat necessary to control moisture content in the air.
   a. Adding moisture (humidifying requires additional heat).
   b. Removing moisture (dehumidifying requires removal of heat).
3. Conditioning the air that enters the building by leakage and for ventilation.
4. The sun also produces heat in buildings, directly through the windows, and by heating the surfaces it strikes (a cooling load).
5. Energy devices such as light fixtures, electric motors,

electric stoves or gas stoves, all produce heat. People, too, release a considerable amount of heat.

In all cases, the heat load can be described as either sensible heat load (temperature change) or latent heat load (moisture), evaporating or condensing.

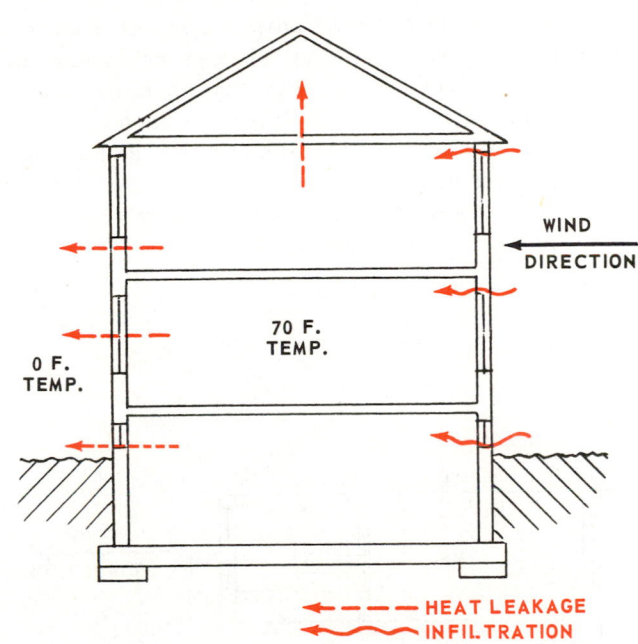

Fig. 25-1. Large heat losses occur during the heating season because of cold air filtering into a building and warm air filtering out. There is also considerable heat loss through walls, floors, windows and ceilings.

25-3 HEAT LOADS FOR HEATING

Heat loads for heating consist of all means by which heat will be lost from a building or to the warming of cooler substances brought into the building. This heat transfer generally is called heat loss. See Fig. 25-1.

The two major heat losses are:
1. Conduction through walls, ceilings and floors of the structure.
2. Air leaking out of the building (exfiltration), and that

which leaks into the building (infiltration).

Combustion air going out the flue from gas or oil furnaces, or from fireplaces, is also a heat loss. Normally, all other heat losses are ignored since they are relatively too small to affect the size of the unit to be installed.

Humidification requires heat, utilizing about 1000 Btu per lb. to vaporize water.

Heat gain from lights, motors, appliances and people are often taken into account in commercial and industrial structures, but usually not in domestic heating calculations.

25-4 HEAT LOADS FOR COOLING

There are definite sources of heat gain in warm weather:
1. Heat leakage into the building.
2. Air leakage into the building or ventilation air.
3. Sun load.
4. Heat from appliances, including lights.
5. Heat gain from occupants.

Heat gain is the term applied to heat gained by a space that is being cooled. This heat must be removed to keep the temperature and humidity at the values desired.

Heat gain in warm weather is produced by heat conduction through the walls, ceilings, floors, windows and doors of the enclosure. Also, heat moves into the room by way of infiltrated air. People or animals in the room also give off heat.

Miscellaneous sources of heat are electrical devices (lights and motors), gas burning devices and steam tables. Another source of heat that may be considerable in some cases is heat from the sun or sun effect. See Fig. 25-2.

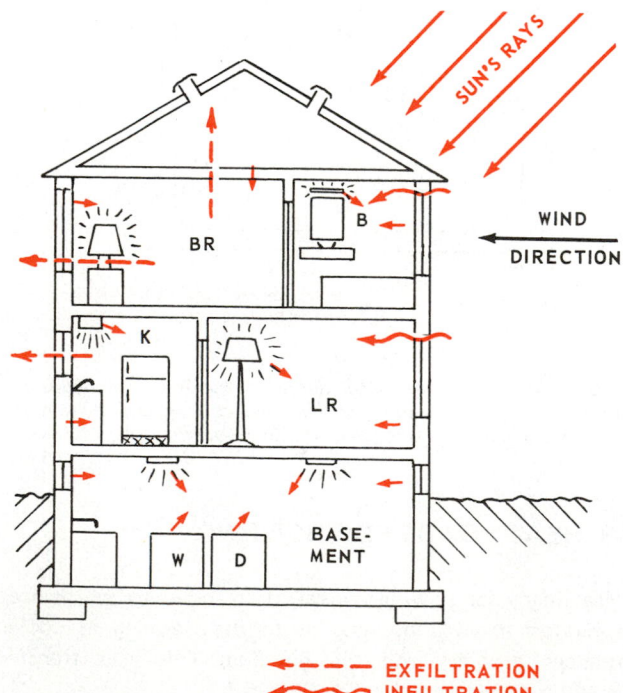

Fig. 25-2. Heat gains in a building during cooling season. Note heat leakage, air infiltration, sun load, lights, appliances and moisture sources. B—Bathroom. BR—Bedroom. K—Kitchen. LR—Living Room. W—Washer. D—Drier.

25-5 HEAT LEAKAGE

Heat leakage is heat that is conducted through the walls, ceilings and floors of a building. To compute heat leakage, first determine how much heat will pass from the air, through a wall and into the air for each square foot, for each degree F. temperature difference, and for each hour (commonly called the U value). Next, find the area of each type of surface through which the heat is leaking. Then, by simple multiplication, find the total heat leakage. U is the symbol for heat leakage, air-to-air, through a complex structure.

Another way to figure heat leakage is to first determine the thermal resistance of the structure, then use this value to compute the amount of leakage. *Thermal resistance is known as the R value. It is the reciprocal of conductance (C) or the overall heat transfer (U).*

All building materials and structures have been carefully measured and tested in laboratories. From this information, one can obtain the amount of heat that will transfer through almost any enclosure surface being used today. Fig. 25-3 shows how the temperature changes during the heating season for a typical wood siding residence wall, both noninsulated and insulated. This heat transfer is called conductivity.

Since there are three general conductivity conditions, the following terms are used:

The letter K represents the Btu that will be transmitted through one square foot of the wall (or surface) in one hour if there is a temperature difference of one degree F., if the material is one-inch thick.
1. The units of K are Btu/sq. ft./deg. F./hr./1-in. thickness.
2. The letter C is used to indicate heat transfer through a wall made of different substances.

$$\frac{1}{C} = \frac{X_1}{K_1} + \frac{X_2}{K_2} + \frac{X_3}{K_3}$$

X is thickness of material in inches

$$C = \frac{1}{\frac{X_1}{K_1} + \frac{X_2}{K_2} + \frac{X_3}{K_3}}$$

3. The letter U is used to represent heat leakage from the air on one side of the wall to the air on the other side of the wall. The meaning and values for U are explained in the next paragraph.

25-6 U FACTOR FOR HEAT LEAKAGE

In computing heat transfer, the letter U has almost the same value as the letter C. The value of U, however, represents the additional insulating effect of the air film that always exists on each side of the surface. See Fig. 25-4.

$$U = \frac{1}{\frac{1}{F_i} + \frac{X_1}{K_1} + \frac{X_2}{K_2} + \frac{X_3}{K_3} + \frac{1}{F_o}}$$

Where F_i is the heat transfer through the inside dead air film and F_o is the heat transfer through the outside air film.

The U value is a common term used to indicate the amount of heat transferred through a structure (wall):

U = Btu/sq. ft./deg. F./hour.

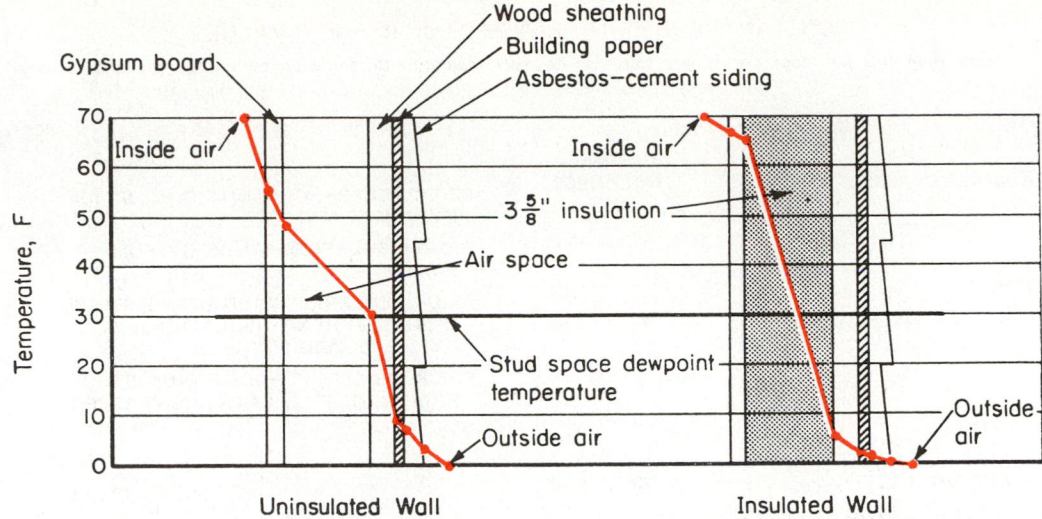

Fig. 25-3. Temperature change through an uninsulated wall and an insulated wall.

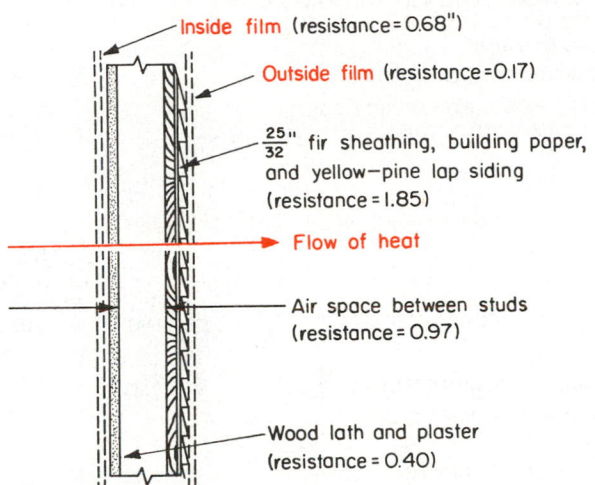

Fig. 25-4. Schematic of an outside wall section, showing inside air film and outside air film. C values are the reciprocal of the R value of the material, or C = 1 ÷ R. (Edison Electric Institute)

This U value is based on a 15 mph wind on the outside and a 15 fpm (1/6 mph) draft on the inside wall surface.

The U value for almost every type of construction can be obtained from reference data books published by the American Society of Heating, Refrigerating and Air Conditioning Engineers (ASHRAE). Fig. 25-5 is a condensed table for some of the more common constructions. Also see Para. 25-7.

Given the U factor, the design temperature conditions of 70 F. (21 C.) indoors and 0 F. (−18 C.) outdoors, and the area, calculate the heat load as follows:

Heat load = area x temp. diff. x U factor

Total heat transfer (Q) = U x total surface x temp. diff.

Example: Structure has 400 sq. ft. of surface. The temperature difference is 70 F. Structure has a brick veneer wall, no insulation and has a U factor of .25.

Solution: This U value means that .25 Btu will transfer through each square foot of wall for each one degree F.

temperature difference in one hour.

Total heat transfer (Q) = 400 x (70 F. − 0 F.) x .25
= 400 x 70 x .25
= 28,000 x .25
(Q) = 7000 Btu/hr.

Example: If the total surface area is 1200 square feet, heat transfer can be determined as follows:

Q = 1200 x (70 F. − 0 F.) x .25
= 1200 x 70 x .25
= 300 x 70
Q = 21,000 Btu/hr.

The outdoor or ambient temperature is different for each locality. Fig. 25-6 shows design conditions for calculating heat loads for heating and cooling for various regions.

In the metric system, the same method of computing heat leakage is used but with metric units.

Example: Celsius (Centigrade) temperatures are used. Square metres are used for area and U values are based on calories/square metre/C./hour or watts. However, the trend is to calculate the heat load in watts instead of calories/hour. Fig. 25-7 shows typical heat leakages listed in watts per hour per square foot. To obtain the watt load per degree F., divide the "35 column" by 35.

Some common metric system heat transmission units are: joules/second; kilocalories/hour; watts. These values are usable on a square metre or square centimetre areas.

Heat transmission using watts per square metre is probably the most popular method. The watt unit is both an English and a metric unit. It may be easily applied to heating and cooling unit capacities.

The watt per degree Celsius unit is written W/deg. C. For a square metre, it is written W/m^2/deg. C.

For either U or C values:

To change English units to metric units, multiply Btu/hr./sq. ft./deg. F. by 3.678 to obtain W/m^2 deg. C.

For R values:

To change English units to metric units, multiply Btu/hr./sq. ft./deg. F. by 0.1761 to obtain W/m^2/deg. C.

HEAT LOADS
CONSTANTS FOR HEAT TRANSMISSION (U AND R VALUES)
Expressed in Btu per hour per square foot per degree temperature difference, based on 15 mph wind velocity.

MASONRY CONSTRUCTION

GENERAL WALL CLASSIFICATION	MASONRY THICKNESS 8" U	R
BRICK — PLAIN		
1/2" WALLBOARD	.29	3.51
1" POLYSTYRENE, 1/2" WALLBOARD	.13	7.54

FRAME CONSTRUCTION

	U	R
WOOD SIDING ON 1" WOOD SHEATHING, STUDS, GYPSUM BOARD 1/2"	.23	4.40
WOOD SIDING, SHEATHING, STUDS, GYPSUM BOARD 1/2", 1" POLYSTYRENE	.07	14.43
WOOD SIDING, SHEATHING, STUDS, 1/2" FLEXIBLE INSULATION, GYPSUM BOARD 1/2"	.15	6.7
WOOD SIDING, SHEATHING, STUDS, ROCK WOOL FILL, LATH AND PLASTER	.072	13.9
NOTE: FRAME WALLS WITH SHINGLE EXTERIOR FINISH SAME AS WALLS WITH WOOD SIDING		
STUCCO, WOOD SIDING, STUDS, GYPSUM BOARD	.30	3.32
STUCCO ON 25/32" RIGID INSULATION, STUDS, 1/2" RIGID INSULATION AND PLASTER	.20	5.0
STUCCO ON 1/2" RIGID INSULATION, STUDS, ROCK WOOL FILL, LATH AND PLASTER	.074	13.5

CONCRETE FLOORS AND CEILINGS

	U	R
4" THICK CONCRETE, NO FINISH	.65	1.54
4" CONCRETE, SUSPENDED PLASTER CEILING	.37	2.7
4" CONCRETE, METAL LATH AND PLASTER CEILING, HARDWOOD FLOOR ON PINE SUB-FLOORING	.23	4.35
4" CONCRETE, HARDWOOD AND PINE FLOOR, NO CEILING	.31	3.23

PITCHED ROOFS

	U	R
ASBESTOS SHINGLES ON WOOD SHEATHING	.48	2.07
ASBESTOS SHINGLES, 1" FLEXIBLE INSULATION	.13	7.7
ASBESTOS SHINGLES, POLYSTYRENE 1"	.14	7.33

WINDOWS AND SKYLIGHTS

	U	R
SINGLE GLASS	1.13	.89
SINGLE GLASS OR STORM WINDOW	.56	1.79
DOUBLE GLASS, INTERMEDIATE AIR SPACE 1/2"	.58	1.73
HOLLOW GLASS TILE WALL. 6" x 6" x 4" BLOCKS	.60	1.67

BRICK VENEER ON FRAME CONSTRUCTION

	U	R
BRICK VENEER, 1" WOOD SIDING, STUDS, GYPSUM BOARD 1/2"	.27	3.71
RIGID INSULATION, STUDS, GYPSUM BOARD 1/2"	.25	4.00
BRICK VENEER, 1" WOOD SIDING, STUDS, 1" POLYSTYRENE INSULATION, GYPSUM BOARD 1/2"	.07	14.43
BRICK VENEER, 1" WOOD SIDING, STUDS, ROCK WOOL FILL, LATH AND PLASTER	.074	13.5

INTERIOR WALLS

	U	R
NOTE: IN GENERAL FOR COOLING COMPUTATIONS, BASE THE CALCULATIONS FOR HEAT GAIN FROM ADJOINING NON-CONDITIONED ROOMS ON A DIFFERENTIAL EQUAL TO 1/2 THE DIFFERENTIAL TO OUTSIDE.		
GYPSUM BOARD 1/2" ON BOTH SIDES	.31	3.23
GYPSUM BOARD 1/2" BOTH SIDES, FOAM INSULATION	.08	13.26

FLAT ROOFS WITH BUILT-UP ROOFING

DECK MATERIAL	NO CEILING U	R	METAL LATH AND PLASTER CEILING U	R
PRECAST CEMENT TILE	.81	1.23	.43	2.33
PRECAST CEMENT TILE, 1" INSULATION	.24	4.7	.19	5.26
4" THICK CONCRETE	.72	1.39	.40	2.50
4" THICK CONCRETE, 1" INSULATION	.23	4.35	.18	5.55
2" WOOD	.32	3.13	.24	4.17
FLAT METAL ROOFS	.95	1.05	.46	2.18
FLAT METAL ROOFS, 1" INSULATION	.25	4.00	.19	5.26

FRAME FLOORS AND CEILINGS

	U	R
HARDWOOD AND PINE FLOORING ON JOISTS, METAL LATH AND PLASTER CEILING	.23	4.37
ROUGH PINE FLOOR, WOOD LATH AND PLASTER CEILING	.28	3.57
NO FLOOR, LATH AND PLASTER CEILING	.62	1.61
NO FLOOR, METAL LATH AND PLASTER CEILING, 3 5/8" ROCK WOOL FILL	.079	12.65
NO FLOOR, LATH AND PLASTER CEILING, 1" FLEXIBLE INSULATION	.17	5.9

Fig. 25-5. U and R values are given for walls, ceilings, floors and partitions for various types of construction and for various thicknesses. (ASHRAE)

DESIGN TEMPERATURES

State	City	Extreme Temperatures		Mean Temperatures		Design Conditions		
						Winter	Summer	
		Low	High	January	July	Dry Bulb	Dry Bulb	Wet Bulb
Ala.	Mobile	— 1°	103°	52°	81°	15°	95°	79°
Ariz.	Phoenix	16	119	51	90	25	110	75
Ark.	Little Rock	—12	108	41	81	5	96	78
Calif.	San Francisco	27	101	50	58	30	90	65
Colo.	Denver	—29	105	30	72	—10	95	72
Conn.	New Haven	—14	101	28	72	0	95	75
D. C.	Washington	—15	106	33	77	0	96	78
Fla.	Jacksonville	10	104	55	82	30	95	79
Ga.	Atlanta	— 8	103	43	78	5	95	78
Idaho	Boise	—28	121	30	73	—10	100	70
Ill.	Chicago	—23	103	25	74	—10	95	75
Ind.	Indianapolis	—25	106	28	76	— 5	96	76
Iowa	Dubuque	—32	106	19	74	—15	96	75
Kan.	Wichita	—22	107	31	79	0	100	75
Ky.	Louisville	—20	107	34	79	0	98	77
La.	New Orleans	7	102	54	82	25	96	80
Maine	Portland	—21	103	22	68	—10	92	75
Md.	Baltimore	— 7	105	34	77	0	96	78
Mass.	Boston	—18	104	28	72	— 5	95	75
Mich.	Detroit	—24	104	24	72	—10	95	75
Minn.	St. Paul	—41	104	12	72	—20	95	75
Miss.	Vicksburg	— 1	104	48	81	15	96	80
Mo.	St. Louis	—22	108	31	79	0	98	79
Mont.	Helena	—42	103	20	66	—20	90	68
Neb.	Omaha	—32	111	22	77	—15	100	75
Nev.	Winnemucca	—28	104	29	71	—10	95	70
N. C.	Charlotte	— 5	103	41	78	10	96	79
N. D.	Bismarck	—45	108	8	70	—25	98	70
N. H.	Concord	—35	102	22	68	—15	95	75
N. J.	Atlantic City	— 7	104	32	72	0	95	75
N. M.	Santa Fe	—13	97	29	69	0	92	70
N. Y.	New York City	—14	102	31	74	0	95	77
Ohio	Cincinnati	—17	105	30	75	0	98	77
Okla.	Oklahoma City	—17	108	36	81	0	100	76
Ore.	Portland	— 2	104	39	67	10	95	70
Penna.	Philadelphia	— 6	106	33	76	0	95	78
R. I.	Providence	—12	101	29	72	0	95	75
S. C.	Charleston	7	104	50	81	15	96	80
S. D.	Pierre	—40	112	16	75	—20	100	72
Tenn.	Nashville	—13	106	39	79	5	98	79
Texas	Galveston	8	101	54	83	25	95	78
Utah	Salt Lake City	—20	105	29	76	— 5	95	70
Vt.	Burlington	—28	100	19	70	—15	92	73
Va.	Norfolk	2	105	41	79	10	98	78
Wash.	Seattle	3	98	40	63	10	90	67
W. Va.	Parkersburg	—27	106	32	75	— 5	96	77
Wis.	Milwaukee	—25	102	21	70	—15	94	75
Wyo.	Cheyenne	—38	100	26	67	—15	92	70

Fig. 25-6. Design temperature conditions used for calculating heat loads for heating or cooling for various regions of the U.S. (ASHRAE Handbook of Fundamentals, 1972)

HEAT LOSS — WATTS/HR./SQ. FT.

CONSTRUCTION	DTD 35	45	55	65	75	85	95	105
EXTERIOR WALLS								
Masonry (8″ concrete block)	3.60	4.63	5.66	6.68	7.71	8.74	9.77	10.80
Masonry (8″ concrete block; ½″ gypsum board) with R-7 (2¼″) Fiberglas	1.05	1.35	1.65	1.95	2.25	2.55	2.85	3.15
Brick Veneer (½″ gypsum sheathing, ½″ gypsum board) with R-11 (3½″) Fiberglas	.75	.97	1.18	1.39	1.61	1.82	2.04	2.25
with R-13 (3⅝″) Fiberglas	.65	.83	1.02	1.20	1.39	1.57	1.76	1.94
Frame (Woodrock siding ½″ gypsum sheathing, ½″ gypsum board) with R-11 (3½″) Fiberglas	.85	1.09	1.33	1.58	1.82	2.06	2.30	2.54
with R-13 (3⅝″) Fiberglas	.75	.97	1.18	1.39	1.61	1.82	2.04	2.25
Frame (wood clapboard or shingles, plywood or wood fiber sheathing, ½″ gypsum board) with R-11 (3½″) Fiberglas	.70	.90	1.10	1.30	1.50	1.70	1.90	2.10
with R-13 (3⅝″) Fiberglas	.60	.77	.94	1.11	1.28	1.45	1.62	1.79
WINDOWS								
Single Light	12.65	16.19	19.73	23.28	27.07	30.61	34.41	37.95
Double Glazed	6.80	8.70	10.61	12.51	14.55	16.46	18.50	20.40
Single with Storms	5.80	7.42	9.05	10.67	12.41	14.04	15.78	17.40
DOORS								
Door only	5.10	6.53	7.96	9.38	10.91	12.34	13.87	15.30
with Storm Door	3.40	4.35	5.30	6.26	7.28	8.23	9.24	10.20
FLOORS								
Concrete Slab No insulation	8.30	10.67	13.04	15.41	17.78	20.15	22.52	24.89
with R-7 (1″) Zer-O-Cel Urethane Perimeter (watts/lin. ft. exposed edge)	2.40	3.08	3.77	4.45	5.14	5.82	6.51	7.19
Wood (⅝″ plywood) over vented space with R-13 (3⅝″) Fiberglas	.65	.83	1.02	1.20	1.39	1.57	1.76	1.94
with R-19 (6″) Fiberglas	.50	.64	.78	.93	1.07	1.21	1.35	1.49
over unheated basement with R-7 (2¼″) Fiberglas	.95	1.22	1.49	1.76	2.03	2.30	2.57	2.84
with R-11 (3½″) Fiberglas	.70	.90	1.10	1.30	1.50	1.70	1.90	2.10
CEILINGS								
with R-19 (6″) Fiberglas	.49	.63	.77	.91	1.05	1.19	1.33	1.47
with R-22 (6½″) Fiberglas	.45	.58	.70	.83	.96	1.09	1.22	1.34
FIREPLACES (Watts/Hr.)								
tight damper	205	264	322	381	440	498	557	615
average damper	510	657	803	949	1095	1241	1387	1533
BASEMENT WALL, above grade* (U = .10, 70°F basement)	1.02	1.31	1.60	1.89	2.18	2.47	2.76	3.05

NOTE: The below grade portion of a heated basement must be calculated separately since heat loss relates to ground or ground water temperature rather than air temperature. The U value of .10 represents typical concrete construction with insulation and a furred finish wall. Add above-grade figure to below-grade figure for total basement heat loss.

*BELOW GRADE
HEAT LOSS, WATTS/HR./SQ. FT. (U = .10, 70°F basement)

Ground Water† Temperature, °F	Floor	Wall
40	.879	1.758
50	.586	1.172
60	.293	.586

†Ground water temperature is available from the local weather bureau.

Fig. 25-7. Table of heat leakage lists watts per hour per square foot. The DTD (Design Temperature Difference) is listed from 35 F. to 105 F. in the vertical column. (National Gypsum Co.)

25-7 R FACTOR FOR HEAT LEAKAGE

Another way to calculate heat leakage is by using the thermal resistance or R factor. See Fig. 25-4. R is known as the unit of thermal resistance. It is also called "ru," which means resistance unit.

If U is the heat transfer, then R is the reciprocal of heat transfer. The symbol R is the reciprocal of C or $R = \dfrac{1}{C}$ and, in case of overall heat transfer, $R = \dfrac{1}{U}$. For a composite wall (a typical building), the total R equals the sum of the individual reciprocals of the C values.

$$R_T = \frac{1}{C_1} + \frac{1}{C_2} + \frac{1}{C_3} + \frac{1}{C_4} + \frac{1}{C_5} \text{ or}$$

$$R_T = R_1 + R_2 + R_3 + R_4 + R_5$$

This method is becoming popular because any composite wall can be easily calculated. By knowing the individual R values, they can be totaled. Then, U will equal the reciprocal of the total R. The R values can be found by taking reciprocals of the heat conductance, (C values from tables) or by using tables showing the R values. See Fig. 25-8.

CONSTRUCTION	(R) RESISTANCE VALUE
1. SURFACE (STILL AIR)	.68
2. AIR SPACE	.97
3. GYPSUM WALLBOARD 3/8 IN.	.32
4. OUTSIDE SURFACE (15 mph WIND)	.17
5. FACE BRICK	.39
6. CONCRETE BLOCK 4 IN.	1.11
7. URETHANE INSULATION	9.1
8. SIDING (WOOD) 1/2 IN. x 8 IN.	.85
9. BUILDING PAPER	.06
10. WOOD SHEATHING	.98
11. WOOD FLOOR 1 IN.	.98
12. LINOLEUM OR TILE	.05
13. ASPHALT SHINGLES OR PLYWOOD	.95

Fig. 25-8. Table of typical thermal resistance (R) values for various parts of a building.   (ASHRAE Guide and Data Book)

Example: R values for a typical brick veneer wall are as follows:

	R
Outside air film	.17
Face brick veneer	.39
Wood siding and building paper	.86
Air space	.97
1/2-in. plaster (.09) on gypsum lath (.32)	.41
Inside air film	.68
Total R =	3.48

$$U = \frac{1}{R} = \frac{1}{3.48} = .287$$

Since conservation of energy becomes more and more important, it is recommended that homes and apartments have thermal insulation. See Fig. 25-9.

A comparison of English and metric system values are as follows:

	ENGLISH	METRIC
Specific heat at constant pressure	Btu/lb./deg. F.	kj/kgk
Internal film coefficient	Btu/hr. sq. ft. deg. F.	W/m²k
Total heat flow	Btu/hr.	watts or kcal/hr.
R—Total resistance to heat flow	hr. sq. ft. deg. F./Btu	m²k/W
U—Overall heat transfer coefficient	Btu/hr. sq. ft. deg. F.	W/m²k
Velocity	ft./hr.	m/s
Density of fluid in tube	lb./ft.³	Pa/s

THERMAL INSULATION VALUES		
AREA	U	R
CEILING WITH UNHEATED SPACE ABOVE	.08	12
EXTERIOR WALL	.07	14
WALL WITH UNHEATED SPACE ON ONE SIDE	.10	10
FLOOR OVER UNHEATED SPACE	.07	14

Fig. 25-9. Some recommended thermal insulation values are charted for homes and apartments.

25-8 WALL HEAT LEAKAGE AREAS

In addition to finding the several U values or R values for the building structure, the area of the walls with each different U or R value will need to be calculated:

Wall heat leakage = U x wall area x temp. diff.

Areas to be measured are the outside dimensions of the building. These measurements will result in slightly higher heat leakage loads than if inside dimensions are used. However, U values based on outside dimensions are on the safer side.

To estimate the heat load, measure the complete building: walls, windows, ceilings and floors.

To measure the walls, take the outside length and width of

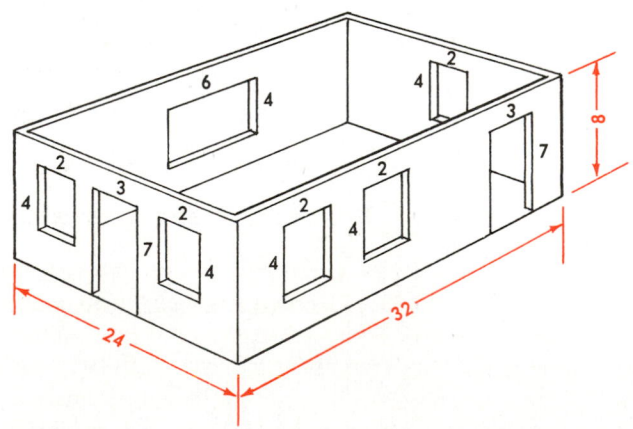

Fig. 25-10. Diagram of simplified one story home, showing wall areas, windows and door areas. In the metric system, the building is 9.8 m long by 7.3 m wide. Room height is 2.45 m. Five windows are 1.22 m high by .61 m wide. Two doors are 2.15 m high by .91 m wide. Large window is 1.22 m high by 1.65 m wide.

the house and the inside ceiling height. To determine the total surface area, first measure the distance around the house. This will be the length plus width, plus length, plus width.

L + W + L + W = perimeter

When these values are added together and multiplied by the wall height, the total wall area is obtained. For example, a house 24 ft. x 32 ft. (outside), as shown in Fig. 25-10, with an 8 ft. ceiling, has a total area of:

Perimeter = L + W + L + W
          = 32 + 24 + 32 + 24
          = 112 ft.
Area = perimeter x height
Area = 112 x 8
Area = 896 sq. ft.

This is total wall area; window and door areas must be subtracted.

In the metric system, total wall area is found by using the same formula:

Area = perimeter x height
Area = (9.8 m + 7.3 m + 9.8 m + 7.3 m) x 2.45 m
Area = 34.2 m x 2.45 m
Area = 84 m^2

25-9 WINDOWS AND DOORS

The area of each window is determined by measuring the opening in the wall. In a brick veneer wall, this would be the distance to the brick edges as shown in Fig. 25-11.

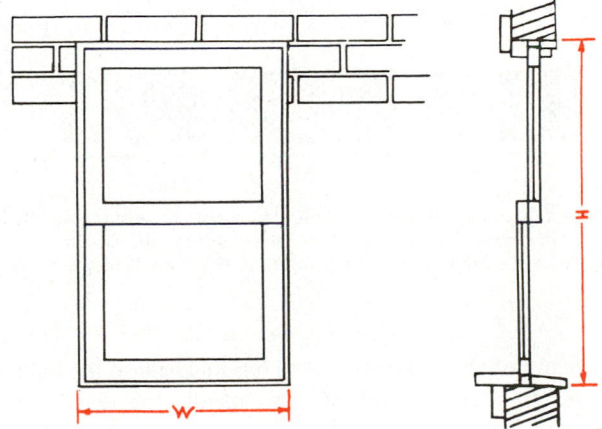

Fig. 25-11. Typical double hung window showing width and height of window opening.

The windows may be either single pane or double pane. The single pane window is referred to as the "primary" window.

A storm window that is not sealed to the primary but is movable is called the "storm" window or "secondary" pane. Many installations have an assembly whereby the primary and secondary panes are inserted in a single frame with an air space between them (often referred to as "thermopane" windows). This air space usually contains nitrogen or some other dry gas to prevent sweating (condensation). See Fig. 25-12. The R and U values of various window pane assemblies are shown in Fig.

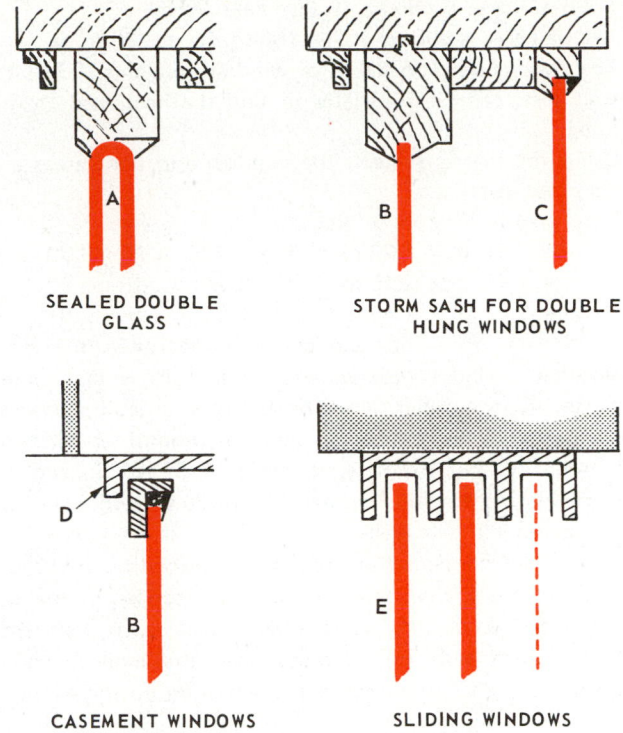

Fig. 25-12. Window construction. A—Dry air. B—Primary glass. C—Storm glass. D—Metal sash. E—Sliding window and storm with screen.

R AND U VALUES		
TYPE WINDOW	U	R
Single Glass	1.13	.9
Double Glass 1/2 in. Space	.65	1.54
Triple Glass 1/2 in. Space	.36	2.79
Storm Windows	.56	1.79

Fig. 25-13. Thermal resistance (R) and overall heat leakage (U) of various window pane assemblies are specified. (ASHRAE Handbook of Fundamentals)

25-13. Remember that R stands for total resistance to heat flow; U represents overall heat leakage.

When computing wall heat leakage area (see Para. 25-8), add the area of the doors in the outside walls to the area of the windows. Then, subtract this amount from the total wall area. Example: In Fig. 25-10, there are five windows measuring 2 ft. x 4 ft., two doors measuring 3 ft. x 7 ft. and one window measuring 4 ft. x 6 ft.

Total area = 2 x 4 x 5 = 8 x 5 = 40 sq. ft.
             3 x 7 x 2 = 21 x 2 = 42 sq. ft.
             4 x 6 x 1 = 24   = 24 sq. ft.
Total opening area = 106 sq. ft.

Total wall area is 896 sq. ft. (See Para. 25-8.)

Net wall area is 896 sq. ft. − 106 sq. ft. = 790 sq. ft.

The two values, 106 sq. ft. of window area and 790 sq. ft. of wall area, can be used later to find the heat load on the building.

Using the metric system, the window and door areas are found as follows:

Total area = W x H x No.

= .61 m x 1.22 m x 5 = .745 x 5 = 3.73

= .91 m x 2.15 m x 2 = 1.96 x 2 = 3.92

= 1.22 m x 1.65 m = 2.01 = 2.01

Total window and door area = 9.66 m^2

Another problem with windows and walls is that water vapor in warm air will condense on cool or cold surfaces. Condensation in walls occurs as a combination of temperature and humidity. For example, if air at 70 F. (21 C.) and 40 percent relative humidity contacts a window at 45 F. (7 C.), condensation will take place.

To prevent condensation, reduce the humidity. However, this action is not always comfortable or practical. Condensation can be avoided by using a vapor barrier on the warm surface. Storm sash on windows keep the inner window temperature higher to eliminate condensation on the window surface.

25-10 CEILINGS

Ceilings generally are made by fastening plasterboard to the joists. Variations in construction will not change the heat transfer to any great extent. Fig. 25-14 shows several typical ceiling constructions. If the joists do not have a floor over them, or if there is no insulation between the joists, heat leakage will be considerable.

See Fig. 25-5 for U values for ceilings.

In Fig. 25-10, the ceiling area is calculated as follows:

Ceiling area = W x L

= 24 ft. x 32 ft.

= 768 sq. ft.

In the metric system, the ceiling area is measured in square metres as follows:

Ceiling area = W x L

= 7.3 m x 9.8 m

= 71.5 m^2

25-11 DESIGN TEMPERATURES

Design temperatures shown in Fig. 25-6 are the result of considerable testing and the accumulation of much data. A study of the table reveals a country-wide range of design temperatures. Contact the local weather bureau or local chapter of the Society of Heating, Refrigerating and Air Conditioning Engineers (ASHRAE) for local data.

Always choose design ambient temperatures on the low side. Heating plants that are overworked cause excessive stack and chimney temperatures and may cause fires. Oversized units will be less efficient and waste energy.

The design temperature is never as low as the lowest temperature recorded for the area, since these extreme lows

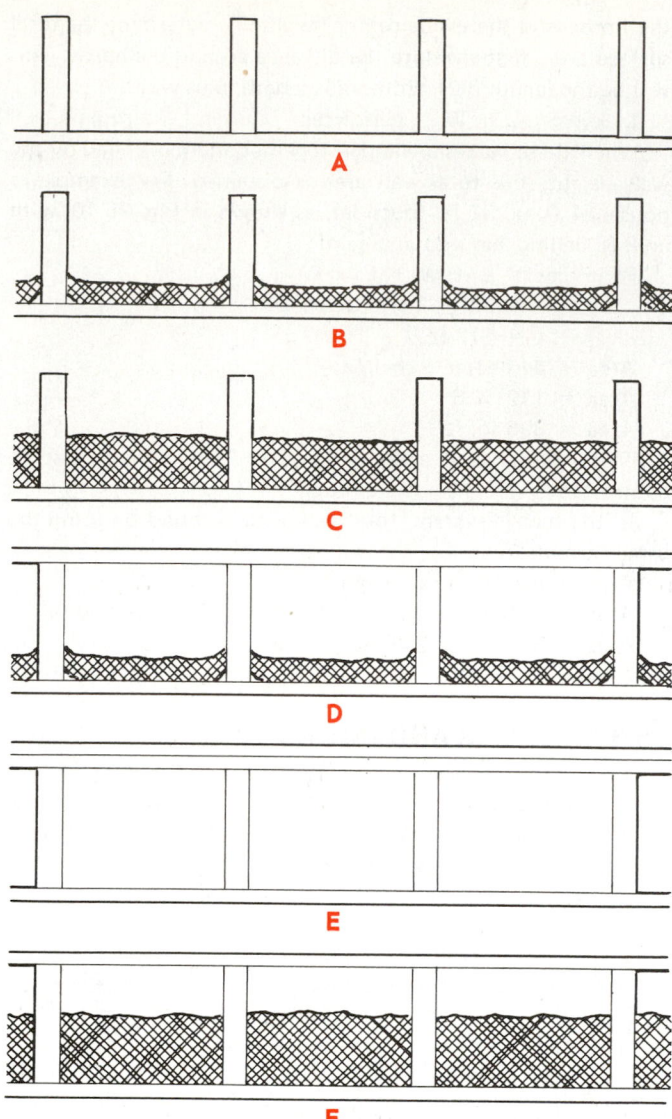

Fig. 25-14. Ceiling construction. A—No floor or insulation. B—No floor, 2-in. insulation. C—No floor, 4-in. insulation. D—Floor, 2-in. insulation. E—Double floor, no insulation. F—Floor, 6-in. insulation.

are usually of short duration. The residual heat in the building enables the design temperature to handle the load in most cases.

ASHRAE has a new method of listing outside design temperatures (ODT). In this method, ODT varies with latitude, and new listings have been selected for most urban areas. When working with design temperatures, use these latest values. It may mean a difference of as much as 33 percent in equipment capacity.

ASHRAE charts give three different ODT values for each locality. The lowest temperature is for small (domestic and office) uninsulated buildings. The 99 percent temperature means that the outdoor temperature is at or above this temperature 99 percent of the time. It can be used for well constructed and well insulated buildings having a standard number of windows.

The 97 1/2 percent means that the outside temperature is

above the value listed 97 1/2 percent of the time. This value is used for large buildings with considerable thermal capacity and small total window area.

Example: Detroit once used −10 F. as ODT for all buildings.

Now it is recommended that 0 F. be used for small, uninsulated buildings; + 4 F. (99 percent factor) for well constructed, insulated buildings with a standard window area; + 8 F. (97 1/2 percent factor) for large buildings with a standard number of windows.

If IDT (Inside Design Temperature) is 72 F., the 0 F. means a 72 F. temperature difference (TD). The 99 percent (4+) means a 68 F. TD. The 97 1/2 percent means a 64 F. TD.

$$\frac{72 - 64}{72} \times 100 = \frac{8}{72} \times 100 = 11 \text{ percent.}$$

This means an 11 percent savings in equipment size.

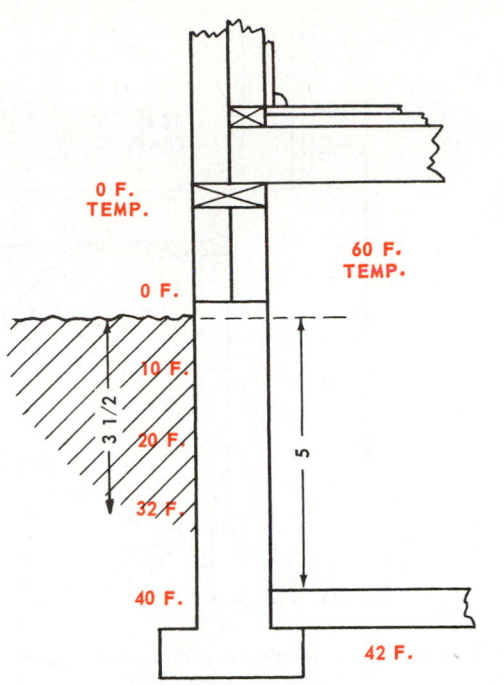

Fig. 25-16. Temperature conditions and construction of a building with basement.

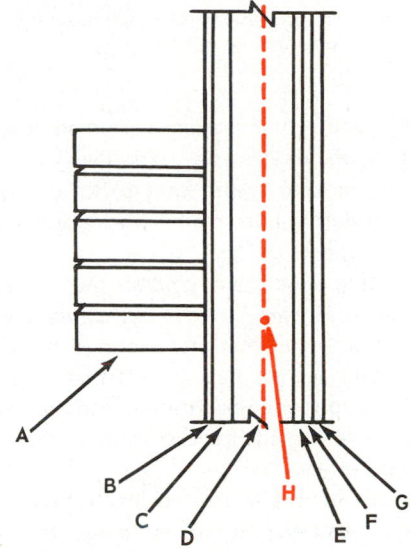

Fig. 25-15. Brick veneer wall construction. A—Brick. B—Outside vapor barrier. C—Sheathing. D—Stud and insulation. E—Inside vapor barrier. F—Rock lath. G—Plaster. H—Approximate dew point location.

25-12 WALL CONSTRUCTION

Building wall construction has been altered the past few years to reduce heat leakage and moisture passage through the wall structure. Fig. 25-15 shows a typical brick veneer outside wall construction. During the heating season, the inside vapor barrier is necessary; in summer, the outside vapor barrier is required. If dew point temperature occurs at any time of the year, it must be kept between these two vapor barriers, or condensation will take place.

The vapor barriers should be as tightly sealed as possible, even to the extent of tarring the breaks in the seal. The barriers may be made of tarred paper, aluminum foil or plastic film. Aluminum foil has a reflection value as well as being a vapor-tight seal. Ideally, use a hermetically sealed (airtight) insulation in those places where dew point temperatures may occur during either the heating or cooling season.

25-13 BASEMENT HEAT LOSS

Heat losses or gains for basements vary widely. The heat loss for a basement built approximately five feet into the ground is shown in Fig. 25-16. The deeper the basement goes into the ground, the less the heat loss. It is usually assumed that a basement is at 60 F. (16 C.). This means a floor loss and a basement wall loss. Leakage through the basement floor is usually not calculated. The usual heat leakage load is calculated through the first floor of the building and the basement ceiling.

Buildings built on a concrete slab have heat losses of a different nature than those built with a basement. See Fig. 25-17. With the building on a slab, the heat loss above ground is calculated in the typical manner. Another solution is to use a rigid urethane slab at least 2-in. thick installed 2 to 4 ft. in the ground. Fig. 25-18 shows a typical installation.

Heat losses for the slab design are usually calculated by determining the perimeter of the building and multiplying the total length by 18 Btu/hr. for each foot of length (0 F. design temperature). Refer to the recommendations of the American Society of Heating, Refrigeration and Air Conditioning Engineers for additional data for other conditions.

25-14 UNHEATED SPACES

In many buildings, certain areas such as closets, hallways and attics are unheated. These particular spaces receive their heat from heat leakage through partitions, ceilings and floors. Generally, the unheated areas are assumed to be at a temperature half the distance between the indoor design temperature and the ambient design temperature.

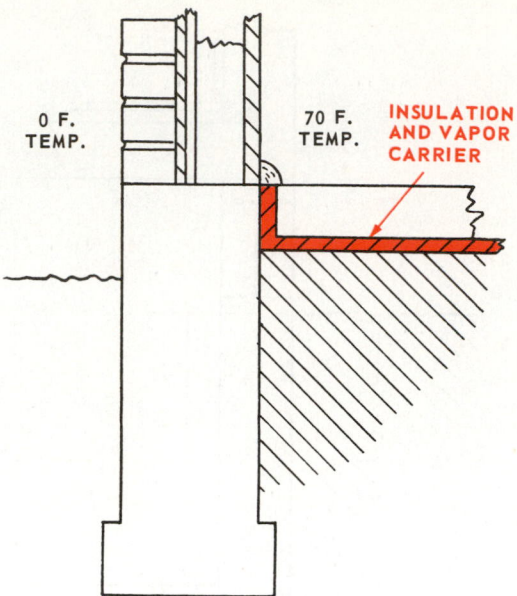

Fig. 25-17. Temperature conditions and construction of a building built on a concrete slab.

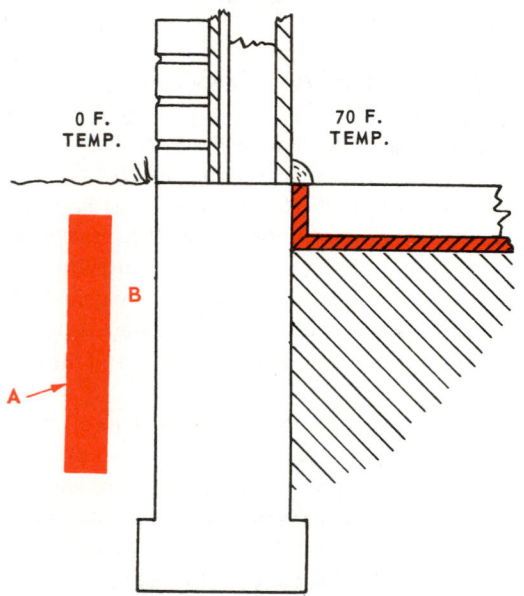

Fig. 25-18. A method is shown for preventing ice and frost from forming around the perimeter of a building built on a slab. A—Rigid urethane insulation. B—Install as close to building as convenient.

25-15 INFILTRATION

Since buildings are not airtight, air will leak in if there is any difference between inside air pressure and outside air pressure. Air also leaks out of the building under the same conditions.

The air pressure difference is usually caused by wind. Parts of the building that the wind is pressing against are those areas through which the air leaks in. The remaining areas are those areas through which the air leaks out. See Fig. 25-19.

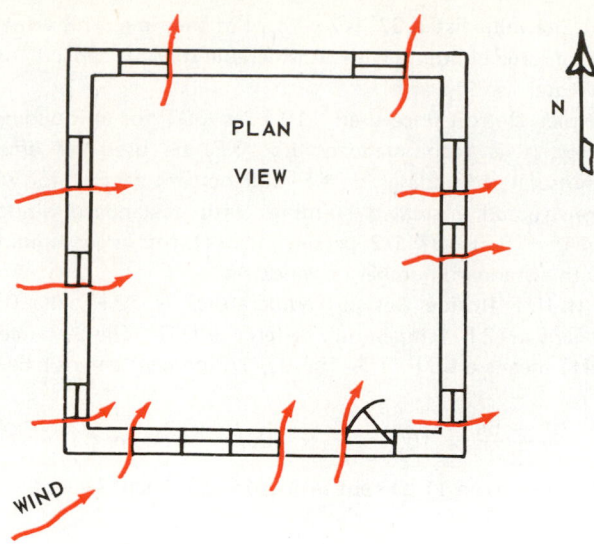

Fig. 25-19. Diagram illustrating how wind direction affects air leakage into and out of house.

Infiltration has one very important part in air conditioning. It provides the fresh air necessary for living comfort and health. It is important to insure that sufficient fresh air is entering a building during both the heating season and cooling season for health purposes.

During the heating season, any cold air that filters in must be heated, and air that leaks out represents lost heat. During the cooling season, warm air that filters in must be cooled, and cooled air that filters out is a heat loss. If the building can be sealed, this infiltration and exfiltration can be minimized. However, take care to provide enough fresh air for ventilation and combustion purposes.

Another way to prevent unwanted infiltration is to maintain a positive air pressure within the building. The pressurized air will filter out at the cracks and openings in the building. This practice necessitates a special fresh air intake, so incoming air can be conditioned before it is admitted to the rooms in the building.

Infiltration calculations can be based on the total volume of the building, or by measuring the length and size of all the cracks in the building. Fig. 25-20 lists the air changes in buildings. If a building has a volume of 10,000 cu. ft., it will have at least 10,000 cu. ft. of fresh air infiltration per hour. If six people occupy this space, there is 10,000 ÷ 6 or 1667 cu.

TYPES OF SPACE	NO. OF AIR CHANGES/HR.
1 Side Exposed	1
2 Sides Exposed	1 1/2
3 Sides Exposed	2
4 Sides Exposed	2
Entrances	2–3

Fig. 25-20. Approximate number of air changes desirable per hour for various room exposures.

SURFACE	AREA	R	U	TEMP. DIFF.	HEAT LEAKAGE Btu/hr.
WALL, GROSS	996				
WINDOW	116	.89	1.13	70	9176
WALL, NET	880	4.0	.25	70	15400
CEILING	768	1.64	.62	35	16666
FLOOR	768	2.94	.34	25	6528
				TOTAL	47770

Fig. 25-21. Typical heat load calculation for 24 × 32-ft. home having an 8-ft. ceiling height.

ft. per hour for each person, or 1667 ÷ 60 = 27.7 cu. ft./min. (.78 m^3/min.), which is a good ventilating value. If this building is constructed with vapor barriers, and all doors and windows are fitted with weather stripping, the air change will be reduced considerably. It may even be reduced to the point of unsafe ventilation (too little oxygen in the air).

25-16 TOTAL HEAT LOAD FOR HEATING

It is best to set up total heat load calculations in table form. Fig. 25-21 shows a typical heat load calculation for a 24 by

RESIDENTIAL FORCED AIR SYSTEM DESIGN GUIDE (For Estimating Purposes Only)

ROOM VO. CU. FT.	4 A/C CFM	WINTER HEAT LOSSES HEATING BTU @ SUPPLY °F. 120°	130°	140°	150°	160°	170°	SUMMER HEAT GAINS COOLING BTU @ SUPPLY °F. 60°	55°	50°
200	14	750	910	1060	1210	1360	1510	300	375	450
300	20	1080	1300	1510	1730	1945	2160	430	540	650
400	27	1460	1750	2040	2330	2625	2915	580	730	875
500	34	1830	2200	2570	2940	3300	3670	735	920	1100
600	40	2160	2590	3025	3455	3890	4320	865	1080	1290
700	47	2540	3025	3550	4060	4565	5080	1015	1270	1530
800	55	2970	3565	4155	4750	5345	5940	1190	1485	1780
900	60	3240	3880	4535	5185	5830	6480	1295	1620	1940
1000	65	3500	4210	4915	5615	6320	7000	1400	1755	2100
1100	75	4050	4860	5670	6480	7290	8100	1620	2000	2430
1200	80	4320	5200	6040	6910	7780	8640	1730	2160	2600
1300	87	4700	5635	6570	7520	8455	9400	1880	2350	2820
1400	95	5130	6155	7180	8200	9235	10260	2050	2560	3080
1500	100	5400	6480	7560	8640	9720	10800	2160	2700	3240
1600	107	5775	6930	8090	9245	10400	11550	2310	2890	3460
1700	113	6100	7320	8540	9760	10950	12200	2440	3050	3660
1800	120	6480	7775	9070	10360	11665	12960	2590	3240	3880
1900	125	6750	8100	9450	10800	12150	13500	2700	3370	4050
2000	135	7300	8750	10200	11660	13120	14600	2920	3645	4370
3000	200	10800	12960	15120	17280	19440	21600	4320	5400	6480
4000	265	14300	17170	20035	22890	25750	28600	5720	7150	8580
5000	335	18100	21700	25325	28940	32560	36200	7240	9040	10850
6000	400	21600	25920	30240	34560	38880	43200	8640	10800	12960
7000	465	25100	30130	35150	40170	45200	50200	10040	12550	15060
8000	535	26900	34670	40450	46220	52000	53800	10760	14440	17330
10000	670	36180	43410	50650	57890	65125	72360	14470	18090	21700
12000	800	43200	51840	60480	69120	77760	86400	17280	21600	25920
14000	935	50500	60590	70680	80780	90880	101000	20200	25240	30290
16000	1065	57500	69000	80510	92010	103520	115000	23000	28750	34500
18000	1200	64800	77760	90720	103680	116640	129600	25920	32400	38780
20000	1335	72100	86500	100920	115340	129760	144200	28840	36040	43250
25000	1670	90200	108215	126250	144290	162320	180360	36070	45090	54100

MAX. AIR ON ONE OUTLET

BASED ON 70 F. RETURN AIR ← → BASED ON 80 F. RETURN AIR

Fig. 25-22. An approximate heat load chart for winter heating and summer cooling based on volume of conditioned space. (Detroit Edison Co.)

Air Conditioning Systems, Heat Loads / 889

32-ft. (7.3 by 9.8 m) house. Note that the temperature difference for the ceiling is only 35 F. (19 C.). Also consider that the roof serves as added insulation and keeps the attic temperature from equaling the outdoor temperature. The attic temperature can be accurately calculated by making the heat leaking into the attic equal the heat leaking out.

Ceiling area x (70 F. attic temp.) x U_c =
Roof area x (attic temp. −0 F.) x U_r.

Most homes have an 8-ft. ceiling, although many homes are now being constructed with 7 1/2-ft. (2.3 m) ceilings.

Each heat leakage value is obtained by means of the following formula:

Heat leakage = area x U x temp. diff.

It is sometimes desirable to calculate the total heat loss for one degree, then multiply this value by the design temperature difference to obtain the total heat loss.

A very fast method used to estimate total heat loads is shown in Fig. 25-22. Note that the heat load is based on room volume. The table also includes cooling. A method for estimating duct sizes is shown in Fig. 25-23. This table is used as a companion to the heat load table.

25-17 TOTAL HEAT GAIN FOR COOLING

Heat gain calculations for a building to determine the total cooling load are similar to the calculations for heat loss.

The temperature difference is based on the locality being considered. The indoor temperature usually is designed to be

RESIDENTIAL FORCED AIR SYSTEM DESIGN GUIDE

ROOM VOLUME CU. FT.	SUPPLY DUCT (IN.)		OUTLET (IN.)			RETURN (IN.)	RETURN DUCT (IN.)	
	R'ND.	EQUIV.	FLOOR	WALL	CEILING	GRILLE	R'ND.	EQUIV.
	(DIA.)				(DIA.)		(DIA.)	
200	4	4 1/2 x 3			4	6 x 10	6	8 x 4
300	4	4 1/2 x 3			4	6 x 10	6	
400	4	4 1/2 x 3			4	6 x 10	6	
500	4	4 1/2 x 3			4	6 x 10	6	
600	5	10 x 2 1/4	2 1/4 x 10		4	6 x 10	6	
700	5	8 x 3 1/4	2 1/4 x 10	4 x 10	6	6 x 10	6	
800	5	5 x 4	2 1/4 x 10		6	6 x 10	6	
900	6	14 x 2 1/4	2 1/4 x 10		6	6 x 10	6	
1000	6	10 x 3 1/4	2 1/4 x 10		6	6 x 10	7	8 x 6
1100	6	8 x 4	2 1/4 x 12		6	6 x 10	7	
1200	6	6 x 5	2 1/4 x 12	10 x 6	6	6 x 10	7	
1300	6	6 x 5	2 1/4 x 12		6	6 x 10	7	
1400	7	14 x 3 1/4	2 1/4 x 14		6	6 x 14	8	8 x 7
1500	7	11 x 4	2 1/4 x 14	12 x 6	8	6 x 14	8	
1600	7	8 x 5	4 x 10	14 x 6	8	6 x 14	8	
1700	7	7 x 6	4 x 10	14 x 6	8	6 x 14	8	
1800	7		4 x 12	14 x 6	8	6 x 14	8	
1900	7		4 x 12	14 x 6	8	6 x 14	8	
2000	7		4 x 12	14 x 6	8	6 x 14	8	
3000	7 1/2	13 x 4				8 x 14	10	8 x 11
4000	9	8 x 8				6 x 24	11	8 x 13
5000	10	8 x 11				6 x 30	12	8 x 16
6000	11	8 x 13				8 x 30	13	8 x 18
7000	11 1/2	8 x 14				8 x 30	14	8 x 22
8000	12	8 x 16				18 x 18	15	8 x 24
10000	13	8 x 18				18 x 18	16	8 x 28
12000	14	8 x 22				18 x 24	18	8 x 36
14000	14 1/2	8 x 24				24 x 24	20	8 x 46
16000	15	8 x 26				24 x 24	20	
18000	16	8 x 30				24 x 30	20	
20000	17	8 x 34				24 x 30	22	8 x 60
25000	18	8 x 39				24 x 30	22	

Fig. 25-23. Duct sizes based on estimated values found in Fig. 25-22. (Detroit Edison Co.)

75 F. (24 C.) at 50 percent relative humidity (RH). Therefore, if the summer design temperature is 100 F. (38 C.), the temperature difference is 25 F. (14 C.). This temperature difference is for load calculations only. In practice, a 10 to 15 F. (6 to 8 C.) difference is recommended.

Miscellaneous heat sources must be considered. Sun load, electrical load, and occupants are large enough sources of heat to be included in the heat load calculations. The following paragraphs describe some of the specifics of these heat sources.

25-18 WINDOW HEAT LOAD FOR COOLING

Heat flow through ordinary window glass is approximately four times as great as heat flow through ordinary residential roofs and ceilings. The U factors (see Para. 25-6) are: ordinary glass, 1:13; residential roofs, .31. Therefore, air conditioning of areas containing a large amount of ordinary glass can become a problem. To reduce the heat conductivity through glass, storm sash is used. To reduce the solar heat through glass, it is advisable to use special types of glass which have high heat reflecting qualities.

Special heat absorbing glass can reduce the solar heat load by as much as 30 percent. Another method is to use glass tinted a bluish gray to reduce the solar glare and cooling load.

Roof extensions over a window will reduce the area exposed to the sun. Double glazed windows also exposed to sun rays reduce solar heat absorption approximately 15 percent.

Awnings to shade glass windows exposed to the sun are also recommended.

25-19 SUN HEAT LOAD

The heat energy that comes from the sun adds considerable heat load to the total load during the summer. The sun's rays in the northern hemisphere shine on the east wall, south wall, west wall and on those roof sections that are exposed to its rays. Therefore, when computing total heat load, the heat from the sun must be considered on the east wall in the morning, on the south wall all day long, and on the west wall in the afternoon, as shown in Fig. 25-24.

The sun releases different amounts of heat to surfaces, depending upon the part of the world in which the building is located. The approximate maximum heat pickup or heat gain from the sun is 330 Btu per hr. per sq. ft. (97 watts/sq. ft.). This condition exists for a black surface at right angles to the sun's rays near the equator (tropic). Any other color or any surface at an angle to the sun's rays will receive less than this amount of heat.

At the 42nd parallel (a line going through New York City, Cleveland and Salt Lake City), the maximum heat from the sun's rays is about 315 Btu per hr. per sq. ft. (92 watts/sq. ft.). Much of the heat from the sun is reflected back into the atmosphere. Heat gain through windows (that must be removed with air conditioning cooling) is listed in Fig. 25-25.

EFFECT OF SUN ON WINDOWS	
EXPOSURE	HEAT ABSORPTION BTU/HR./SQ.FT.
Southwest	110
West	100
South	75
East	55
Single Skylights	110
Double Skylights	60

Fig. 25-25. Heat absorption from sun when sun is shining on windows.

If the windows are not protected with awnings, use a temperature of 15 F. (8 C.) higher than outside ambient temperature to get correct results. The effect of the sun shining on walls may also be taken care of by adding 15 F. (8 C.) to the ambient temperature.

The approximate values obtained by using the 15 F. (8 C.) temperature correction generally are usable. However, there are many special cases that require careful study. Of considerable interest is the changing position of the sun relative to the surfaces of the building and the time lag required for this heat to reach the interior of the building.

25-20 HEAT LAG

When a substance is heated on one side, it takes time for the heat to travel through the substance. This traveling time is called "heat lag." When the sun heats the outside wall of a building, several hours elapse before this heat reaches the inner surfaces of that wall. In normal buildings, this time varies between three and four hours. If the wall is insulated well enough, or is thick enough, the sun will be gone by the time the heat penetrates or "soaks" through.

In the Southwest, adobe walls are made thick enough so that the sun heat moves into the wall while the sun shines. The wall is thick enough to prevent the heat from reaching the interior. The heat flow reverses itself and travels out again during the night when outdoor temperatures fall below inside temperatures.

Except for windows, the effect of heat from the sun on an east wall is as follows:

1. Sun from 8 to 9 a.m., heat goes into room 11-12 a.m.

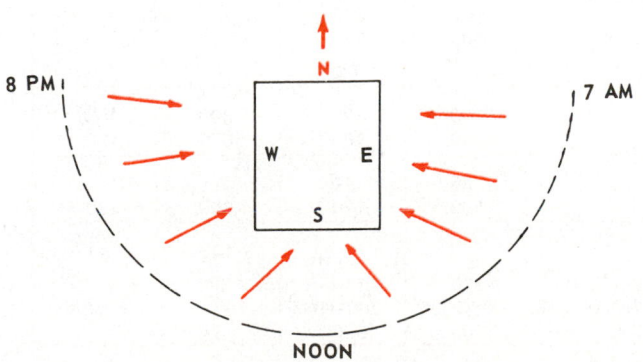

Fig. 25-24. Sun rays and their impact on walls of a building during a 12-hr. period.

2. Sun from 9 to 10 a.m., heat goes into rooms 12-1 p.m.

3. Sun from 10 to 11 a.m., heat goes into rooms 1-2 p.m.

The south wall is affected from 8 a.m. to 7 p.m., but not as strongly as the east and west walls because the rays come from overhead.

Likewise, the west wall receives the sun's rays from 4 to 7 p.m. and acts as follows:

1. Sun from 4 to 5 p.m., heat goes into rooms 7-8 p.m.

2. Sun from 5 to 6 p.m., heat goes into rooms 8-9 p.m.

3. Sun from 6 to 7 p.m., heat goes into rooms 9-10 p.m.

This heat lag causes the rooms to be heated even after the sun goes below the horizon, and when the outdoor temperature drops. See Fig. 25-26. Many people complain of uncomfortable heat in their bedrooms up to as late as 2 a.m.

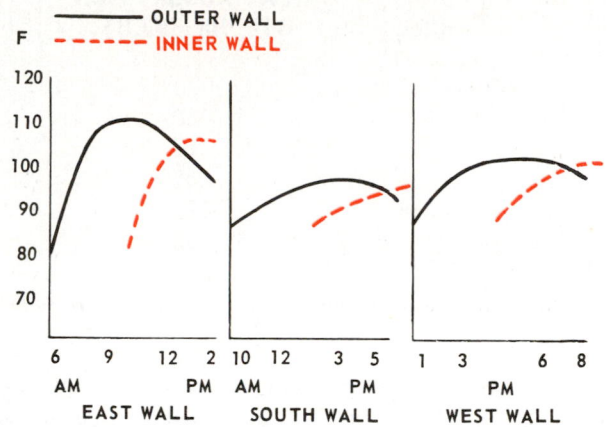

Fig. 25-26. Lag in interior wall temperature following exposure to sun.

25-21 HEAT SOURCES IN BUILDINGS

There are several sources of heat besides infiltration and sunload that must be considered when figuring the comfort cooling load.

During the heating season, on the other hand, the heating plant is aided by many other sources of heat. Practically all energy expended in the building finally becomes heat. Yet, these heat sources usually are ignored when figuring the heat load of small buildings in the winter. They are small amounts compared to the total heat load in temperate zones and are an additional safety factor.

However, when figuring the summer heat load (cooling heat load), all sources of heat energy must be carefully considered. Items such as the heat released by human beings, stoves, lights and electric motors must be computed in the final heat load. Fig. 25-27 shows some of these heat sources. Notice that the two sources of heat are itemized, the sensible heat gain and the latent heat gain.

During the cooling season, the heat released by persons must be taken into account. The heat released by one person weighing about 150 lb. (68 kg) is 74 watts (250 Btu/hr.) when at rest, and about 440 watts (1500 Btu/hr.) when working. About 25 to 45 percent of this heat is by evaporation of moisture from the respiratory system and from the skin.

25-22 INSULATION AND VAPOR BARRIERS

A large number of different insulating materials have been developed for buildings. It is essential that the insulation is designed to reduce heat loss by conduction, convection and

HEAT LOADS

ENERGY SOURCE	DEVICE	HP	HEAT			
			SENSIBLE		LATENT	
			BTU/HR.	WATTS	BTU/HR.	WATTS
ELECTRIC	LIGHTS/KWHR.		3415	1020		
	MOTORS, ELECTRIC/HP IN ROOM	UP TO 1/2 MAX.	4200	1230		
		UP TO 3 MAX.	3700	1080		
		UP TO 20 MAX.	2950	880		
	MOTORS, ELECTRIC/HP OUT OF ROOM	1/2	1700	500		
		3	1150	340		
		20	400	120		
	STOVES, ELECTRIC/KWHR.		3415	1020		
GAS	NATURAL GAS/CU. FT.		1100	320	300	88
	ARTIFICIAL GAS/CU. FT.		550	160	675	198
GENERAL	HEAT FROM MEALS/MEAL		36	10		
	STEAM TABLES/SQ. FT.		400	120	800	230
HUMANS					140	41
		SITTING	370	110		
		WORKING	700—1500	200—440		
		DANCING	2000	590		

Fig. 25-27. Heat released by various energy sources within building in Btu per hour.

radiation. In addition, vapor barriers should be included in the insulation or in the walls to reduce moisture travel through the wall.

The insulation selected for installation should have sufficient strength to support itself and not shrink or settle. It must not deteriorate in the presence of moisture, and it must not have any unpleasant odor. The insulation should be vermin proof and fire resistant.

The type of insulation to use in a building depends, to some extent, on method of application. For example, a bulk type, easy flowing insulation can be blown into the space between the studs of a building already constructed. For new buildings, rigid insulation such as plaster base can be used as a part of the building wall or as a substitute for the sheeting.

Flexible insulations are easy to install, and they conform to any irregularities in the construction. Batts of rock wool and blankets of pulverized wood are examples of this practice. Fig. 25-28 shows a batt type insulation being installed between studs.

Fig. 25-28. Batt type flexible insulation being installed between studs of residence. (Owens-Corning Fiberglas Corp.)

It is exceedingly important that all hygroscopic (moisture absorbent) insulations are hermetically sealed. Aluminum sheet or tarred paper are two popular sealing materials. Even those insulations not affected by moisture should be vapor sealed, since the insulation will lose much of its insulating value if it fills up with moisture. This is particularly important in summer cooling applications.

The National Mineral Wool Association recommends that electrically heated homes and/or air conditioned homes have the following insulation installed:

Ceilings	R-19 or R-22
Walls	R-11 or R-13
Floors over unheated spaces	R-11

R-22 and R-13 are recommended for special situations where extra insulation is necessary. R = Resistance to heat flow. See Para. 25-7.

25-23 PONDED ROOF

An ordinary roof may be heated by the sun to a temperature of 100 F. to 150 F. (38 C. to 66 C.). Ceilings under such roofs will also become quite warm and will radiate this heat through the space below.

Many buildings with flat roofs are now provided with some summer comfort cooling by having a 2 to 3-in. (5 to 6.8 cm) pond of water covering the roof surface. This type of cooling is particularly well suited to one-story factory and market buildings. To be effective, the roof area should be almost as large as the floor area.

The cooling effect comes from the evaporation of water from the roof. Naturally, ponded systems are most effective in areas having high summer time temperature, low humidity and bright sunshine. By ponding, the roof temperature may be kept at, or below, the temperature of the surrounding atmosphere.

Ponded roofs must be provided with a means of maintaining a constant level of water on the roof. Drains are needed to take away excess water due to rain. If the roof is large, wave breakers are needed to prevent waves forming under high winds which might result in a large quantity of water being blown off the edge of the roof.

Buildings having ponded roofs may also be fitted with complete air conditioning equipment. The use of a ponded roof may reduce the required air conditioning capacity by as much as 30 percent.

25-24 HUMIDIFIER HEAT LOAD

During the heating season, water vapor needs to be added to the air to provide comfortable conditions. The heat to produce the water vapor comes from the heated air or from the furnace heat or electric heat.

The amount of heat needed is figured as follows:

Volume changes/hour must be known. Generally, for homes, one change per hour is satisfactory.

The number of grains to be added per pound of air to obtain the required relative humidity must be known. See Chapter 18.

Pounds of air per 24 hours x increase in grains = grains/day.

It is known that:

$$\frac{gr./day}{7000 \ gr.} = lb. \ of \ water/day$$

$$\frac{lb. \ of \ water/day}{8.34 \ lb.} = gal./day$$

To calculate:

$$\frac{24 \times volume \ of \ house \times changes \ per \ hour}{13.55 \ cu. \ ft. \ of \ air/lb.} \times$$

$$\frac{\text{gr./lb. indoors} - \text{gr./lb. outdoors}}{7000\ \text{gr./lb.} \times 8.34\ \text{lb./gal.}} = \text{gal./day}$$

$$\frac{\text{Volume} \times \text{changes/hr.} \times (\text{gr.}_{I} - \text{gr.}_{O})}{33,000} = \text{gal./day}$$

Example: A home has 12,000 cu. ft., and the grains to be added per pound of air to change the air from 35 F. and 90 percent RH to 72 F. and 40 percent are 21. Total gallons of water to be evaporated per day is calculated as follows:

$$\frac{12,000 \times 1 \times 21}{33,000} = \frac{252,000}{33,000} = 7.6\ \text{gal./day}$$

25-25 UNIT AIR CONDITIONER HEAT LOAD CALCULATIONS

A relatively easy way to calculate the summer cooling heat load per hour on a room is to use the following form:

Name _____

Address _____

Zone _____ Phone _____

Space used for _____

Interior room dimensions:

Length _____ Width _____ Height _____

Windows:

No. _____ Facing _____ Size _____ x _____

No. _____ Facing _____ Size _____ x _____

No. _____ Facing _____ Size _____ x _____

WINDOW LOAD

1. Sun exposed (interior shades only)
   west side _____ sq. ft. x 60 = _____ Btu/hr. _____ watts.
2. Sun exposed (interior shades)
   south side _____ sq. ft. x 40 = _____ Btu/hr. _____ watts.
3. Sun exposed (awnings)
   _____ sq. ft. x 35 = _____ Btu/hr. _____ watts.
4. East exposure, north exposure or shaded _____ sq. ft. x 15 = _____ Btu/hr. _____ watts.

WALL LOAD

1. Sun exposed, south and west walls
   _____ sq. ft. x 8 = _____ Btu/hr. _____ watts.
2. East or north exposure
   _____ sq. ft. x 5 = _____ Btu/hr. _____ watts.
3. All exposures, thin walls
   _____ sq. ft. x 10 = _____ Btu/hr. _____ watts.
4. Interior walls
   _____ sq. ft. x 4 = _____ Btu/hr. _____ watts.
5. Interior glass partition
   _____ sq. ft. x 10 = _____ Btu/hr. _____ watts.

FLOOR LOAD

_____ sq. ft. x 3 = _____ Btu/hr. _____ watts.

CEILING LOAD

1. Occupied above
   _____ sq. ft. x 3 = _____ Btu/hr. _____ watts.
2. Insulated roof
   _____ sq. ft. x 8 = _____ Btu/hr. _____ watts.
3. Uninsulated roof
   _____ sq. ft. x 20 = _____ Btu/hr. _____ watts.

VENTILATION LOAD

_____ cu. ft. x 4 = _____ Btu/hr. _____ watts.

OCCUPANCY LOAD

No. of people _____ x 400 = _____ Btu/hr. _____ watts.

MISCELLANEOUS LOAD

Electrical watts _____ x 3.4 = _____ Btu/hr. _____ watts.

Other _____ x _____ = _____ Btu/hr. _____ watts.

Total Btu per hour = _____ Btu/hr. _____ watts.

The multipliers in the tabulation are obtained by multiplying a typical U factor by the temperature difference. For example: The windows (no sun) have a U factor of 1.25. If the temperature difference is about 12 F. (7 C.), the multiplier becomes 15.

12 x 1.25 = 15

A rough estimate or way to check one's calculations is to remember that, on the average, a medium size room needs 5000 to 6000 Btu/hr. cooling.

25-26 BUILDING INSULATION AND VENTILATION FOR ELECTRIC HEATING

Electric heating of residences and commercial establishments is increasing rapidly. Problems in electric heating usually involve the need for insulation or ventilation. Excess humidity also can pose a problem.

Cost of heating is another factor to be considered. Usually, it costs more to produce a unit of heat electrically than with most other types of fuels. If the heat loss through windows, walls, floors and ceilings can be reduced, the cost of electric heating likewise will be reduced. Therefore, the installation of efficient insulating materials and double glazed windows helps cut the heat load.

Electric space heating, as indicated, requires a well insulated and tight structure. Sometimes, however, this adds to the ventilation problem. If the building is a space occupied by a great number of people, provisions must be made for frequent air changes for ventilation purposes.

Spaces which are comfort cooled as well as electrically heated usually have the electric heating elements in the plenum chamber or in the air duct system. Some systems also have electric elements located in each room to enable individual control of the spaces being heated.

Humidity control in electrically heated buildings is usually different than is experienced with fuel-burning heating equipment. With fuel-burning equipment, a fairly large quantity of air passes through the furnace and out the stack. Usually, "make-up" air enters the building by leakage around the doors, windows and cracks in the floor. With most electrically heated structures, building construction permits little infiltration of this nature. Consequently, there is a possibility of too high humidity, since vapor formed in the space cannot readily escape from the occupied space. This may require the use of dehumidifying equipment under certain weather and occupancy conditions.

There are other factors too, which help make the cost of heating with electricity favorable. Since there is no dirt or smoke generated in connection with the heating, the drapes, upholstery and woodwork remain clean longer, and the cost of cleaning and redecorating may be reduced.

See Chapter 28 for more information concerning building construction for electric heating.

25-27 DEGREE — DAY METHOD

The method commonly used to determine fuel consumption and/or the cost of heating during a season is called the "degree-day method." It determines, by previous weather records, the average temperature of each day during the season. It also keeps a record of daily temperatures during the season under study.

The degree-day method uses a 65 F. (18 C.) indoor temperature as a standard.

If the average temperature outside for a particular day was 15 F. (−9.4 C.), the temperature difference was 50 F. (28 C.). Therefore, that day had 50 degree days. Each degree day requires a certain Btu load to keep the building at 65 F.

If one knows the number of degree days since a fuel oil tank was filled, one can accurately calculate how much fuel is left in the tank. If one knows the degree days for a certain season, one can calculate the cost of heating for that season.

25-28 REVIEW OF SAFETY

It is essential to safety that a heating system have enough capacity to heat a structure without taxing or overheating the heating system. The heating system should be slightly oversize; never undersize.

All designs of heating systems and cooling systems should be checked carefully to insure that enough fresh air enters the structure to provide adequate ventilation for the maximum number of occupants (and for combustion air during the heating season).

In some well-insulated modern homes, it may be necessary to open a basement window or provide some other air inlet so that the furnace and fireplace flues may draw properly and not release smoke or carbon monoxide into the building.

25-29 TEST YOUR KNOWLEDGE

1. What are the two main portions of the heating load?
2. What main sources of heat are a part of cooling load?
3. What are the variables for calculating heat leakage?
4. How are window size dimensions determined?
5. Is the heating design outdoor temperature the coldest temperature recorded?
6. How are storage closets handled when figuring the heating load?
7. How are unheated spaces accounted for when figuring heating loads?
8. Describe infiltration and exfiltration.
9. What is meant by having a positive air pressure in the building?
10. What building walls are affected by the sun (42nd parallel)?
11. How is the sun effect usually included in cooling load calculations?
12. Does sun load on the walls and windows affect the heat load at the same time?
13. What is heat lag in a building?
14. If the outside temperature is 20 F. (−7 C.) for 24 hours, how many degree days accumulate?
15. What is an air film?
16. What is the unit of thermal resistance?
17. If one knows the U factor, how is the R (ru) factor determined?
18. Do people release heat when sitting idle?
19. How much heat does it take to vaporize one pound of water in a humidifier?
20. What is the usual design wind velocity?
21. What is conductance?
22. Are outside or inside dimensions used when determining wall areas?
23. Which has the greatest heat flow resistance, inside air film or outside air film?
24. Describe perimeter heat loss for grade level slabs.
25. What is a vapor barrier?

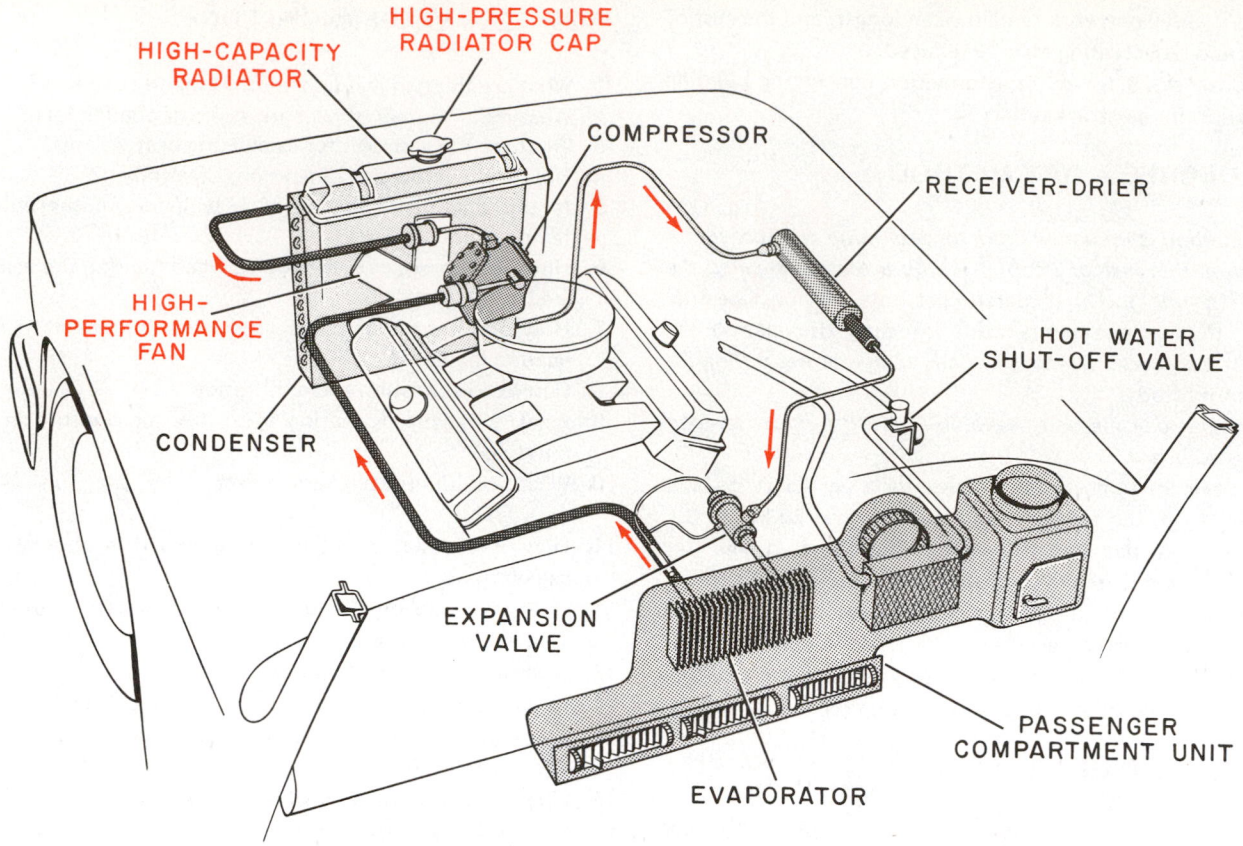

Fig. 26-1. An automotive air conditioning system. Arrows indicate direction of refrigerant flow. Note use of high-pressure radiator cap, large fan and high-capacity radiator. Also, note heating system at right. (Dodge Div., Chrysler Corp.)

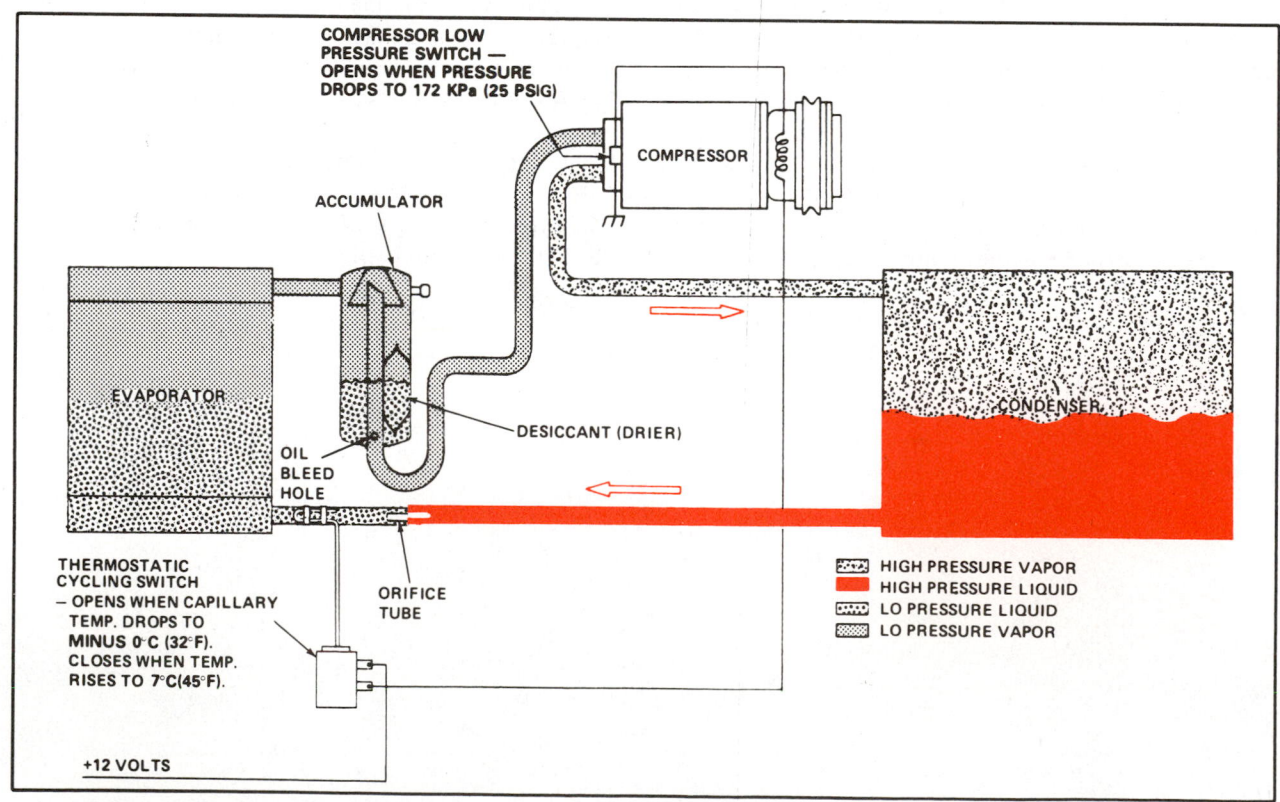

Fig. 26-2. Schematic drawing of modern automotive air conditioning system. (Buick Motor Div., General Motors Corp.)

Chapter 26

AUTOMOBILE
AIR CONDITIONING

Few new principles are involved in automobile air conditioning, but there are many unique applications. Air conditioning a moving automobile presents problems not found in most refrigeration or air conditioning system installations.

Automobile air conditioning involves heating, cooling and dehumidification. The heat required to warm the passenger compartment usually is provided by circulating warm coolant from the engine through a heater core. When a cooling effect is desired, a refrigerating system is brought into operation, causing an evaporator in the plenum chamber of the system to cool the air that is to be circulated through the passenger compartment.

Automatic controls used in automobile air conditioning are of the temperature, pressure and vacuum type. They control the flow of air, warm water and refrigerant. Truck cabs, taxis, tractors, and earth movers use this same type of system and controls.

In preparation for working on automobile air conditioning system, study Chapters 1, 2, 4, 5, 6, 9, 14, 18 and 19 of this text (unless you have had previous refrigeration and air conditioning experience).

26-1 PURPOSE OF AUTOMOTIVE AIR CONDITIONING

There has been a rapid increase in the use of air conditioning in automobiles. An automobile is relatively small. Yet, when it travels at high speed on a hot day, in summer, it will require a considerable amount of refrigerating capacity to keep the interior at a comfortable temperature level. Likewise, the same car traveling on a cold day in winter will require considerable heating capacity to keep it warm.

The automobile air conditioner uses a refrigerating system driven by the car's engine to furnish the cooling action desired. In most cases, warm water from the engine cooling system is used for heating purposes. See Fig. 26-1.

The mechanisms and controls of a factory installed air conditioning system are arranged to ease the task of selecting and controlling car temperature. Fig. 26-2 shows the makeup of a typical automotive refrigerating system used for air conditioning.

With the air conditioner in operation in summer, the humidity of the air inside the car is reduced. In addition,

moisture (condensate) formed on the evaporator surfaces collects much of the dust and pollen. These entrapped particles are carried away by the condensate as it drains from the evaporator underneath the car. In this way, the summer air conditioner serves to clean the air as well as control its temperature. The winter air conditioner does not use an air filtering system.

26-2 AUTOMOBILE AIR CONDITIONER OPERATION

A comfort cooling unit for automobiles is shown in Fig. 26-3. In this installation, the compressor is mounted on the engine and is driven by a belt. The condenser is mounted ahead of the car radiator.

In operation, liquid refrigerant flows from the condenser to the liquid receiver, which dries and filters it. The liquid refrigerant travels through a refrigerant control to the evaporator where it is vaporized and heat is absorbed. The vaporized refrigerant flows back through the suction line to the compressor.

Meanwhile, a blower forces air from the inside of the car through the evaporator. The cool air produced is circulated to the interior of the car by means of ducts and grilles at each

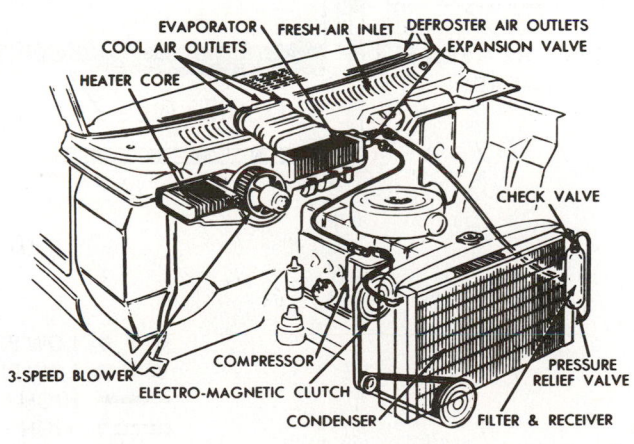

Fig. 26-3. This all season air conditioning system combines cooling, heating, ventilating and defrosting in one unit. (American Motors Corp.)

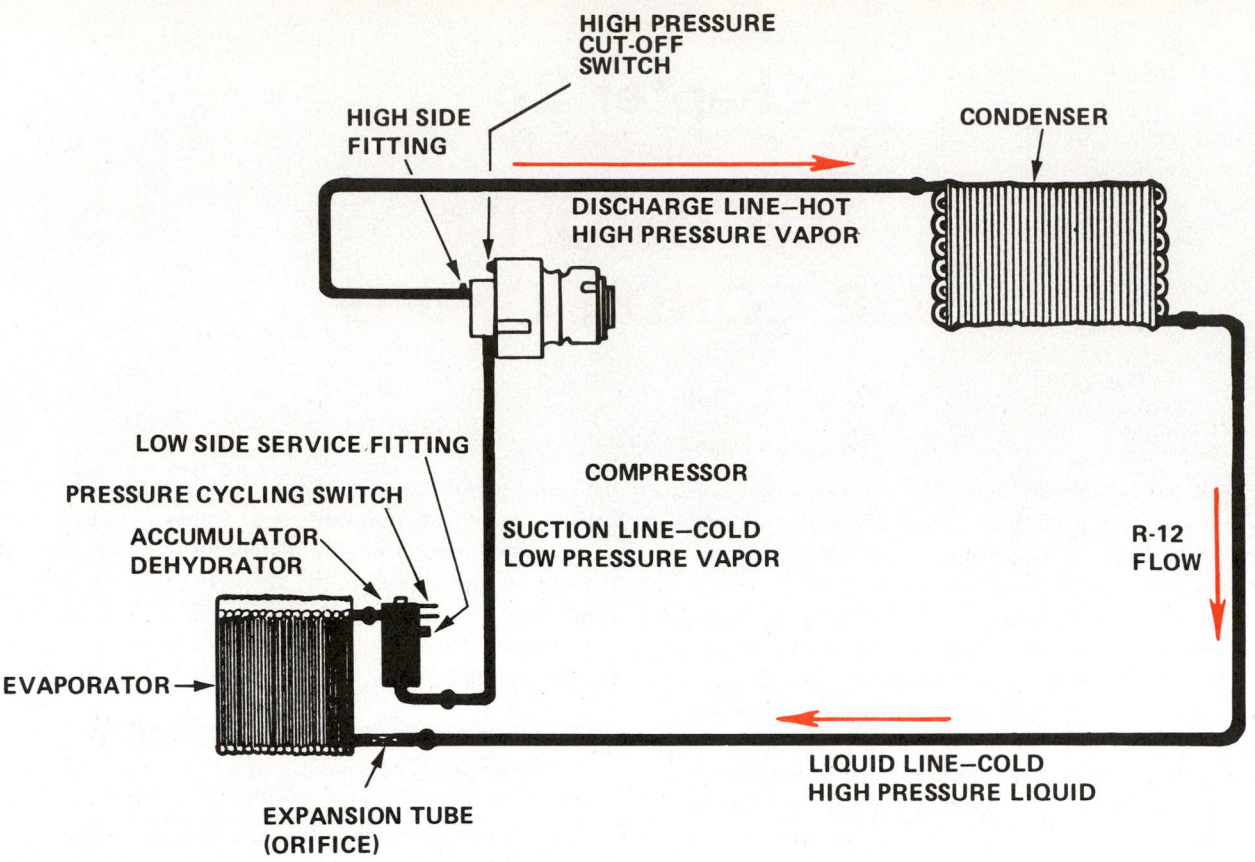

Fig. 26-4. Basic refrigeration cycle of a General Motors air conditioning system.
(Buick Motor Div., General Motors Corp.)

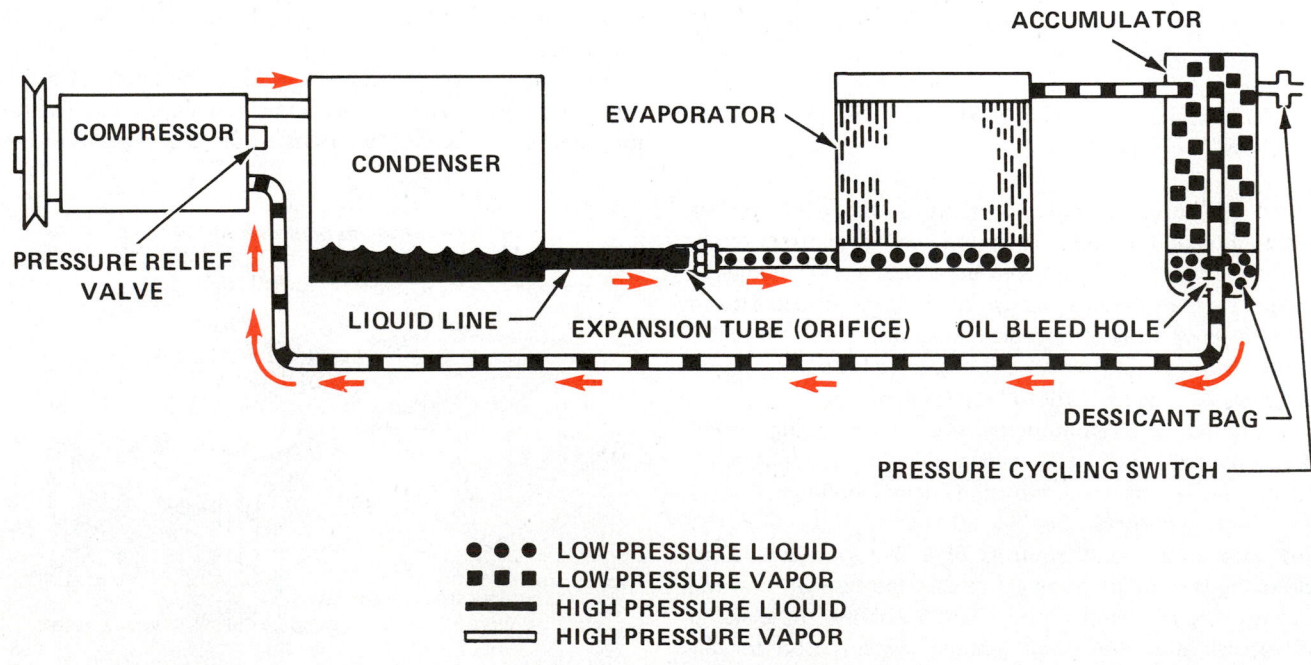

● ● ● LOW PRESSURE LIQUID
■ ■ ■ LOW PRESSURE VAPOR
▬▬▬ HIGH PRESSURE LIQUID
▭▭▭ HIGH PRESSURE VAPOR

Fig. 26-5. Typical automobile air conditioning system showing condition or state of refrigerant in each part of the cycle.
(Delco Air Conditioning Div., General Motors Corp.)

end of the instrument panel. A basic refrigeration system for an auto air conditioner is shown in Fig. 26-4.

Low-pressure refrigerant vapor enters the compressor through the suction service valve (low side). The vapor is drawn into the cylinder where it is compressed by the piston, then discharged through the discharge service valve into the condenser (high side). The heat of compression and the latent heat of vaporization absorbed by the refrigerant are given up to the air flowing past the condenser. The refrigerant is again liquefied, and it moves to the liquid receiver. Fig. 26-5 reveals the condition of the refrigerant in each part of the refrigeration cycle.

If the air conditioning system ran continuously, the temperature in the car would drop to an uncomfortable level, and the evaporator would frost over. To prevent this condition, most systems use a magnetic clutch mechanism that causes "free wheeling." The clutch is operated by a thermostat that opens the electrical circuit to an electromagnetic clutch on the compressor. This allows the compressor pulley to rotate while the compressor crankshaft remains stationary.

When the air conditioning system is not turned on (in winter when the heater is used) the refrigerating mechanism is not in operation. The electromagnetic clutch on the compressor is not energized. Instead, hot water is circulated through the heater core, and the same blower and most of the same duct system are utilized for heating as are used for cooling. See Fig. 26-6. A thermostat in the engine cooling system is used to obtain a quick warmup of the engine cooling liquid. On some cars with air-cooled engines, it is necessary to supply an electrically driven fan to force air through the condenser. See Fig. 26-7.

26-3 OPERATING CONDITIONS

The automobile air conditioner must provide comfort and control conditions in the car during cold, mild, damp and hot weather. It must provide heating, defogging and deicing. It must remove dust, smoke and odor.

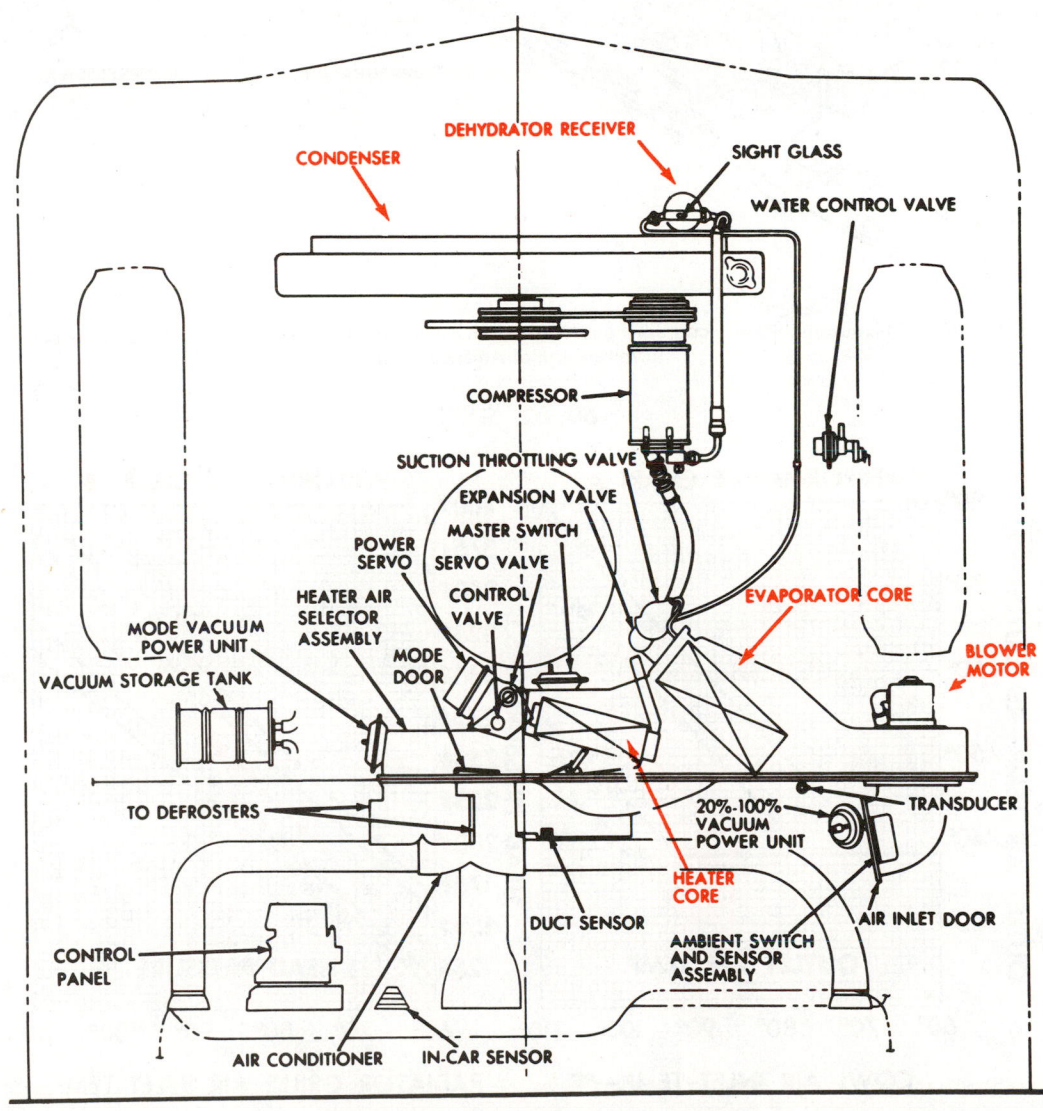

Fig. 26-6. Top view of year-round, all season air conditioning system showing main components.
(Cadillac Motor Car Div., General Motors Corp.)

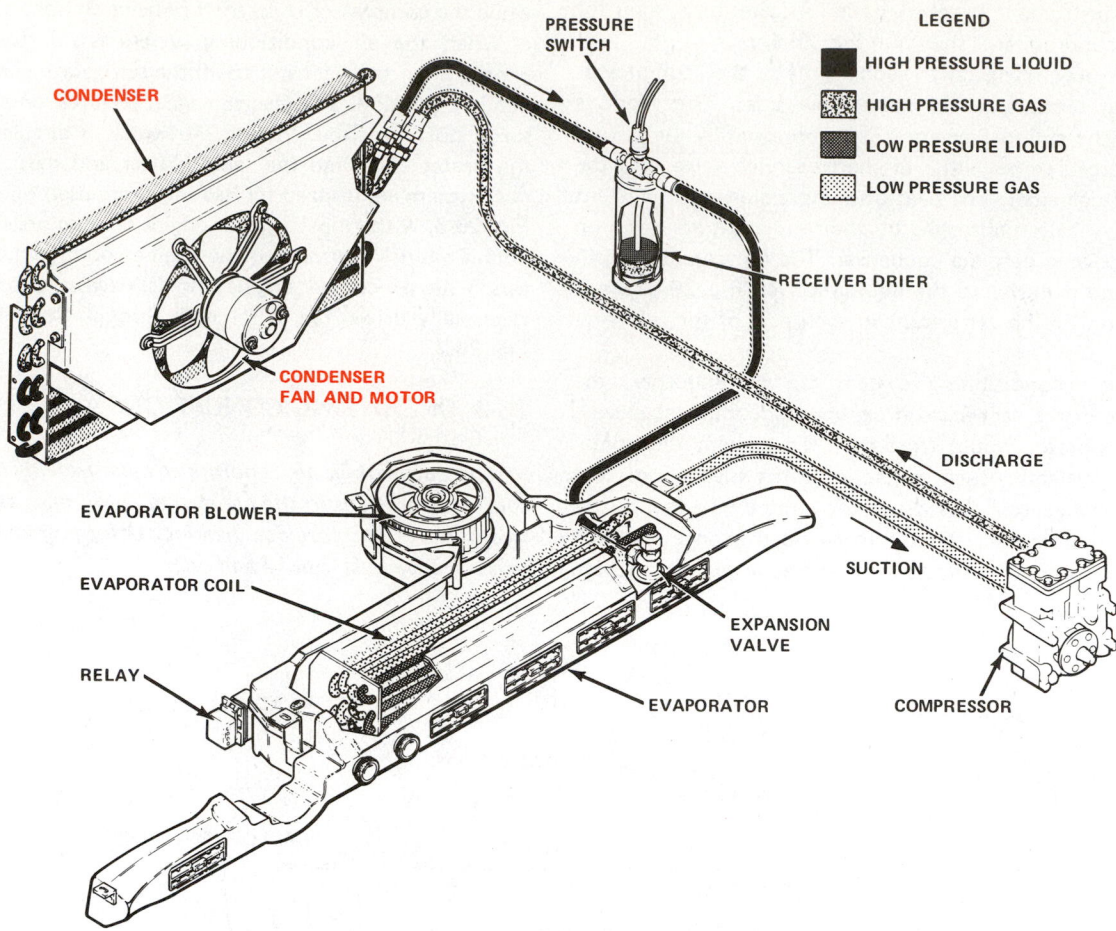

CONDENSER

PRESSURE SWITCH

CONDENSER FAN AND MOTOR

EVAPORATOR BLOWER

EVAPORATOR COIL

RELAY

RECEIVER DRIER

DISCHARGE

SUCTION

EXPANSION VALVE

EVAPORATOR

COMPRESSOR

LEGEND

HIGH PRESSURE LIQUID
HIGH PRESSURE GAS
LOW PRESSURE LIQUID
LOW PRESSURE GAS

Fig. 26-7. This automobile air conditioning system uses an electric motor driven fan to cool the condenser. (Volkswagen of America, Inc.)

60, 62, SERIES

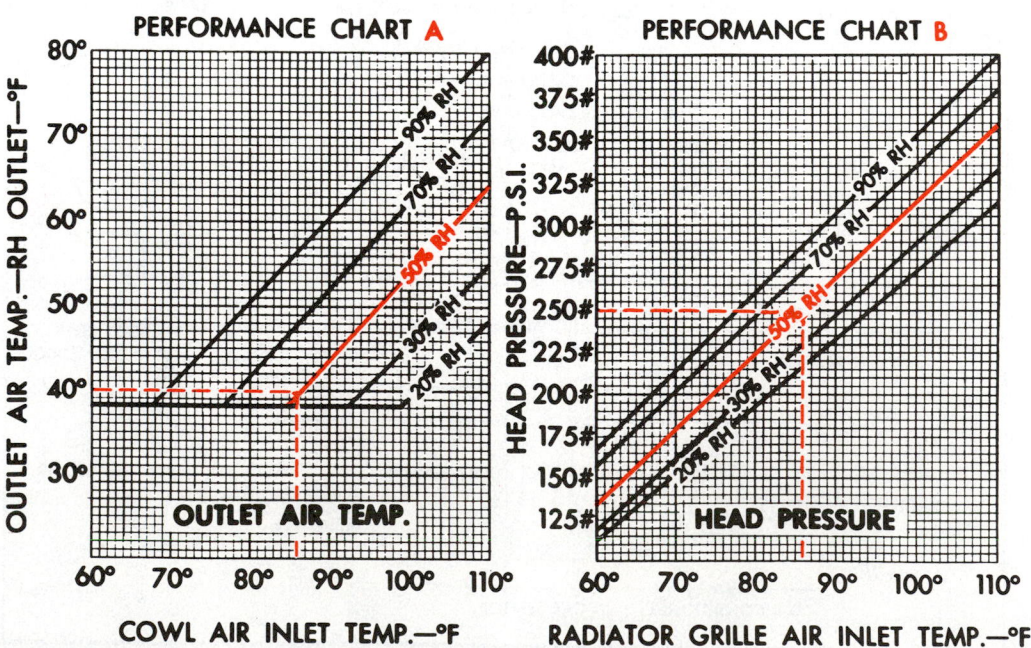

PERFORMANCE CHART A

OUTLET AIR TEMP.—RH OUTLET—°F

90% RH
70% RH
50% RH
30% RH
20% RH

OUTLET AIR TEMP.

COWL AIR INLET TEMP.—°F

PERFORMANCE CHART B

HEAD PRESSURE—P.S.I.

90% RH
70% RH
50% RH
30% RH
20% RH

HEAD PRESSURE

RADIATOR GRILLE AIR INLET TEMP.—°F

Fig. 26-8. Typical automobile air conditioning performance charts. A—Air temperature as it leaves cooling system. Red line is for an 86 F. (30 C.) air inlet. B—Head pressure of system. Red line indicates a 250 psi if ambient air temperature is 86 F. (30 C.).

Because the compressor is belt driven from the engine, compressor speed will vary with engine speed. The system must have enough capacity to provide sufficient cooling at idling speed on the hottest day, in the sun, and under side-wind conditions. This setup will provide considerable excess capacity for normal speed driving, particularly under cool weather conditions. Fig. 26-8 shows typical performance curves for an air conditioning system.

Varying weather conditions can cause problems, both with the control of the temperature and the refrigerant flow (both liquid and vapor) within the system. If the compressor is operating, and little or no refrigeration is needed, the low-side pressure may drop too low.

Decreasing the low-side pressure lowers the evaporator temperature. The evaporator surface temperature should not be allowed to drop below 33 F. (.5 C.). If the evaporator should operate at a temperature of 32 F. (0 C.) or lower for any length of time, its surface will frost over and may become covered with ice. This will stop the air circulation through it.

Also, operating the system with low-side pressure too low may cause oil pumping. This condition may damage the compressor valves and, if continued, may burn out the compressor.

Various cycle systems and mechanical systems have been devised to overcome these problems. Remember that fresh air ducts must be closed during high-heat loads in the cooling season to obtain maximum cooling.

A typical automobile air conditioning system will cool an automobile from 110 F. (43 C.) down to 85 F. (29 C.) in about ten minutes. The inside of the car may reach 150 F. (66 C.) when parked in the sun with the windows closed. The greatest heat load or heat gain is the sun load and heat conducted through the car windows.

Automobile air conditioning systems use anywhere from no fresh air (all recirculated) to 100 percent fresh air. The fans use approximately 200 watts and deliver from 250 to 275 cfm. Air scoops or rams may be used to increase the airflow.

The use of an air conditioning system in a car may reduce the gas mileage by as much as 10 percent when the air conditioning unit is in operation.

26-4 COOLING CAPACITY

Automobile air conditioning units range in size from one to four-ton cooling capacity. A capacity of 12,000 Btu/hr. is minimum. This is equivalent to a one-ton machine. See Para. 1-38. Capacities up to 48,000 Btu/hr. are available.

The capacity of the air conditioning unit should match car size. Undercapacity will reduce cooling. Overcapacity is uneconomical and causes too frequent cycling. The units are usually designed to keep the inside of the automobile 15 to 20 F. (8 to 11 C.) below the outside (ambient) temperature when the car is traveling about 30 mph. Fig. 26-9 shows how horsepower required varies as car speed changes.

As the automobile speeds up, the capacity of the compressor will increase. As it slows down, capacity will decrease. This variation in output is somewhat parallel to the changing heat load, except when the car is parked or is in slow moving

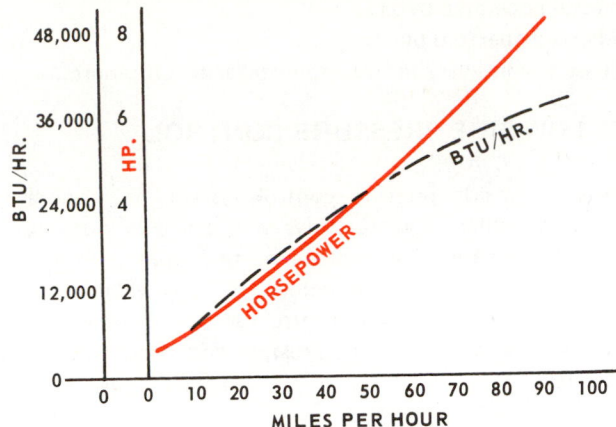

Fig. 26-9. Curves show relationship between car speed, heat load and horsepower required to drive automobile cooling mechanism.

traffic. At these critical times, compressor capacity may be below normal. A partial solution is to idle the engine at a higher speed; another is to travel in traffic in intermediate gear to obtain higher engine speeds.

Larger air conditioning systems can consume as much as 8 hp from the engine at high speeds. Capacity at this speed will be approximately 48,000 Btu/hr. for a four-ton unit. This means that 2 hp is used for each ton of refrigeration. Compare this to the use of 1 hp for each ton of refrigeration in a motor-driven, constant-speed compressor comparably built and with the evaporator and condenser more ideally located.

The typical hot water core using engine heat installed in the air duct supplies heating when required. The same fans may be used during both the cooling and heating cycles.

26-5 TYPICAL INSTALLATION

Most automobile manufacturers provide, as an option, a factory installed air conditioner. There are certain accessory manufacturers who make and market an "add on" or "hang on" type of cooling mechanism for installation in any standard make of automobile. Instructions given in this chapter deal with both types of air conditioners.

Factory air conditioning installations include a larger radiator and a stronger front suspension. Factory systems also use a small plenum or mixing chamber under the dash which holds both the evaporator and heater core. The conditioned air travels from this chamber to the passenger compartment.

Some automatic temperature control systems operate both the evaporator and the heater core. Air travels through the evaporator first and is cooled. Then part or all of this air travels through the heater core and is reheated to the desired temperature. The reheating also reduces relative humidity.

26-6 TYPES OF SYSTEMS

Three different basic cycle and mechanical air conditioning systems have been used on automobiles:
1. Pressure operated low-side pressure regulators.

2. Pressure operated bypass.
3. Solenoid operated bypass.
These systems vary in size and installation procedures.

26-7 LOW-SIDE PRESSURE CONTROL

In the low-side pressure control system, an evaporator pressure controlled regulator valve is installed in the suction line of the system. The purpose of this valve is to hold a constant pressure in the evaporator. The valve will close if the evaporator tends to go below a certain setting, thereby holding the evaporator at a constant pressure and temperature. See Fig. 26-10.

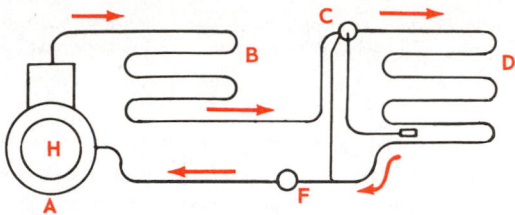

Fig. 26-10. System with evaporator pressure-controlled suction line valve. A—Compressor. B—Condenser. C—Thermostatic expansion valve. D—Evaporator. F—Evaporator pressure control. H—Magnetic clutch.

The low-side pressure control sometimes causes the compressor to produce a high vacuum at high speeds. This might cause the compressor to lose its oil. Fig. 26-11 shows a system with a POA (pressure operated altitude) valve, which maintains a constant evaporator refrigerant pressure.

To overcome this vacuum problem, some manufacturers use an automatic expansion valve bypass in the system. This valve

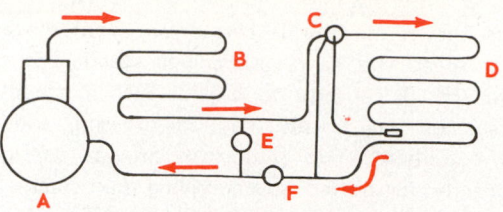

Fig. 26-12. Cycle diagram of pressure operated bypass and low-side pressure control refrigerator system. A—Compressor. B—Condenser. C—Thermostatic expansion valve. D—Evaporator. E—Low-side pressure operated bypass. F—Evaporator pressure control.

has a small bleeder hole in the orifice to allow a small amount of refrigerant to enter the suction line and to prevent too high a vacuum from forming. See Fig. 26-12. In this system, the compressor may operate continuously as long as air conditioning is needed.

In a refinement of the bypass system, an evaporator pressure controlled valve is installed in the body of the compressor or next to it. To avoid vacuum buildup, the valve is designed not to close completely. It allows enough refrigerant to enter the compressor to maintain a positive pressure.

Other variations have been used.

Other low-pressure control systems use an instrument panel Bowden cable attached to a low-side pressure control, as shown in Fig. 26-13. Pulling the control knob out decreases pressure in the evaporator, resulting in a colder evaporator temperature.

26-8 PRESSURE OPERATED HOT GAS BYPASS VALVE

A cycle diagram of a pressure operated bypass is shown in Fig. 26-14. In this system, the pressure operated bypass valve

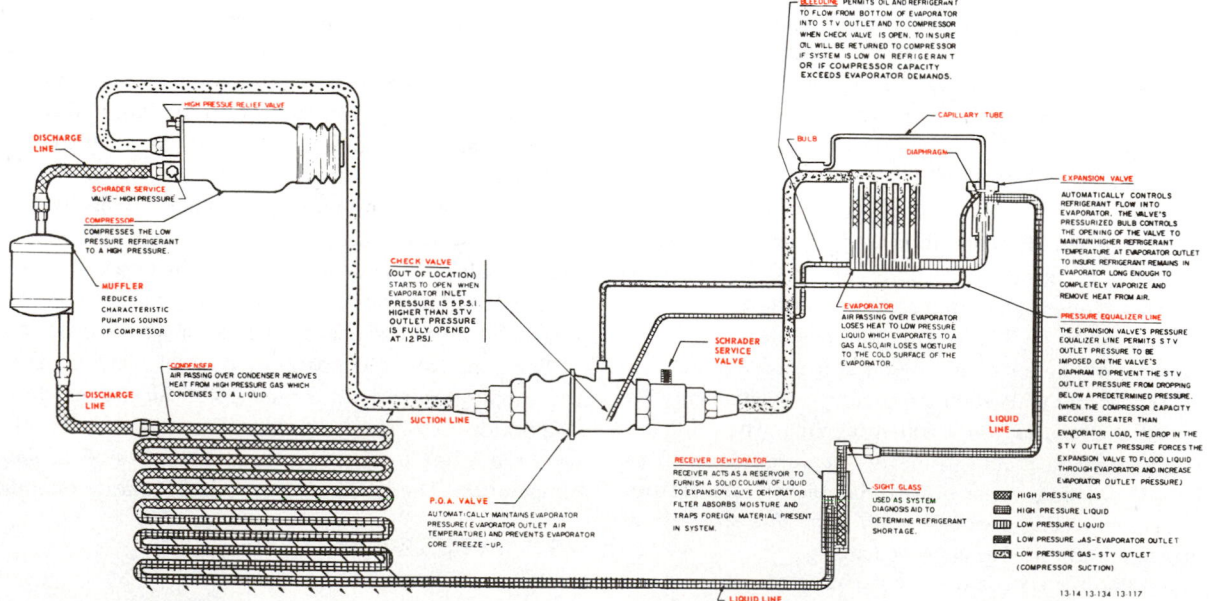

Fig. 26-11. Operation of refrigerating system, showing high and low pressures, POA valve and state of refrigerant. (Buick Motor Div., General Motors Corp.)

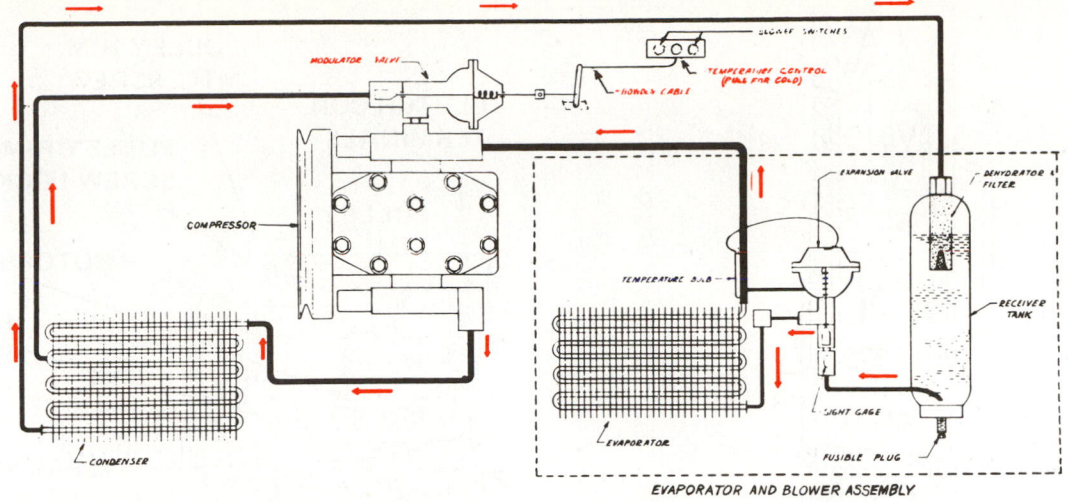

Fig. 26-13. Cycle diagram for manually adjusted evaporator pressure operated bypass. Modulator valve is controlled by the driver. Note solid core dehydrator and filter at entrance to liquid receiver.

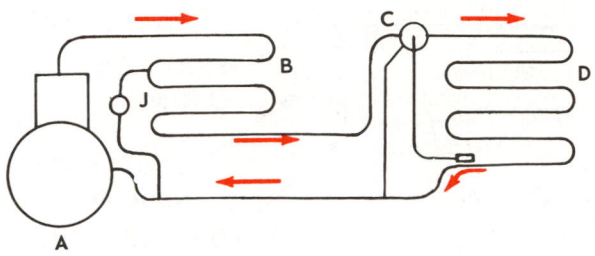

Fig. 26-14. Cycle diagram of a pressure operated bypass system. A—Compressor. B—Condenser. C—Thermostatic expansion valve. D—Evaporator. J—Pressure operated bypass valve.

is connected between the compressor discharge (high side) and the compressor suction line (low side). It is set to open and bypass hot vapor from the high side to the low side when the pressure difference reaches the setting of the valve. This valve opens when there is a lowering of suction line pressure. It closes when there is an increase in suction pressure. Hot gas (vapor) is fed into the low-pressure side to maintain a specific pressure in the evaporator.

26-9 SOLENOID OPERATED HOT GAS BYPASS

In a solenoid operated bypass system, a thermostat mounted on the evaporator opens a solenoid valve to bypass hot gas from the high side to the low side when the temperature of the evaporator drops to 33 F. (.5 C.).

The thermostat is mounted with the sensing bulb located in the evaporator outlet airflow. The air temperature is lowered to 32 F. (0 C.), satisfying the thermostat. It opens the solenoid and allows hot gas from the condenser to bypass back into the suction line.

Since the solenoid valve is either closed or wide open, it does not give the throttling effect of the pressure operated valve described in Para. 26-8.

The solenoid is in a closed position when the thermostat

points are open (air is warm). The valve opens when the circuit is closed. The thermostat points, therefore, close on temperature drop.

26-10 MAGNETIC CLUTCH

Automotive air conditioning compressors have a mechanism that permits the engine to run with the compressor disengaged. A clutch is used to engage the compressor belt drive pulley to the compressor crankshaft or to disengage it. The clutch is operated by forcing a clutch disk against the pulley through the use of electromagnetism. This device is called a magnetic clutch. Its principle of operation is shown in Fig. 26-15. General construction, including the magnetic field circuit, is shown in Fig. 26-16.

Two basic designs of magnetic clutches have been used:
1. Revolving magnetic coil revolves when the compressor

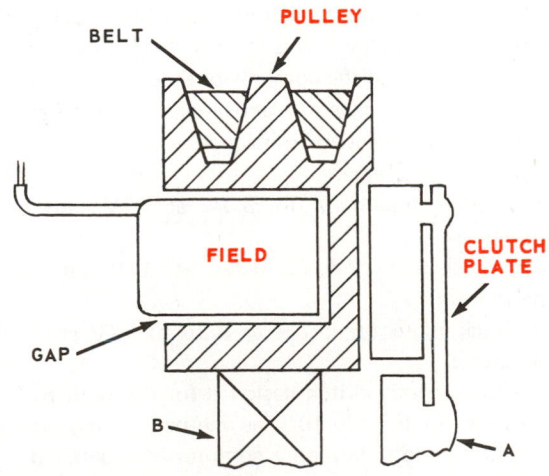

Fig. 26-15. Schematic of an electromagnetic clutch. Note stationary magnetic field. A—Clutch plate. B—Pulley rides on bearing.

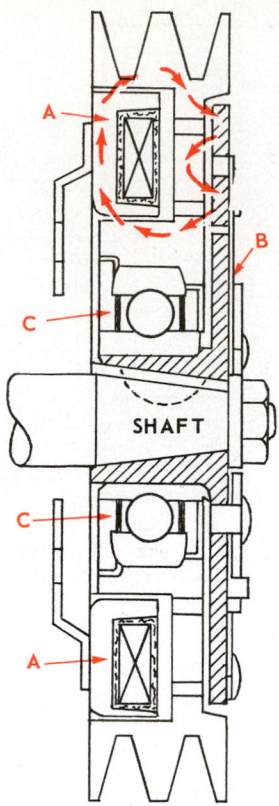

Fig. 26-16. Compressor pulley with stationary electromagnet, showing magnetic field. A—Electromagnetic coil. B—Clutch disk. C—Bearings.

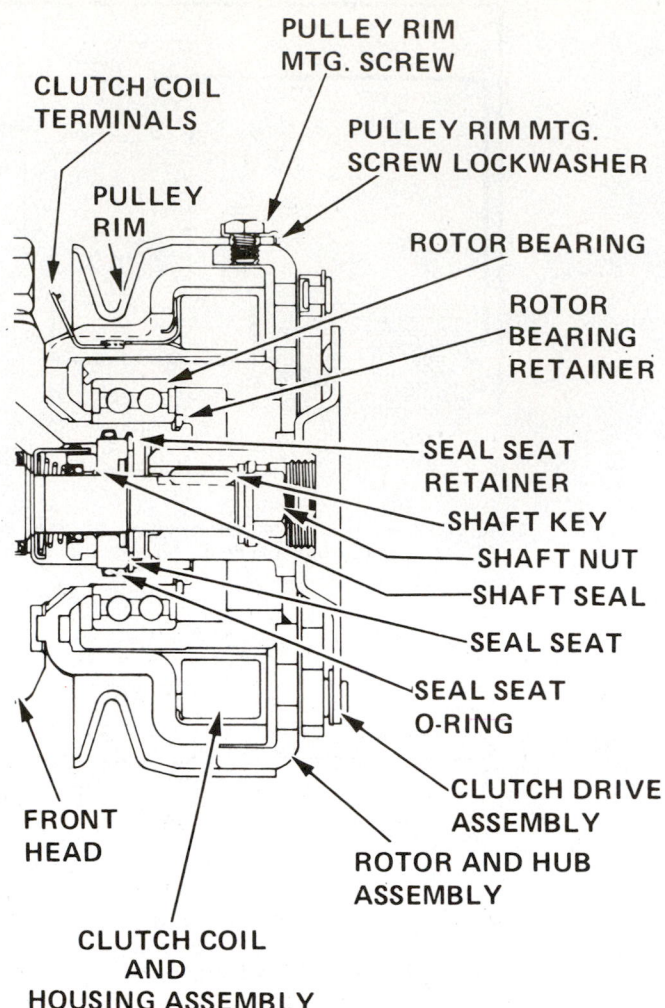

Fig. 26-17. Cutaway view of automotive air conditioning compressor magnetic clutch. When stationary coil is energized, armature revolves with pulley and turns compressor shaft.
(Pontiac Motor Div., General Motors Corp.)

revolves. It has two carbon brushes that are in contact with two copper rings mounted on the coil.

2. Stationary magnetic coil is mounted on the compressor body. It has two electrical leads, one from the control and one to ground. The electromagnetic type is operated by a thermostat. Fig. 26-17 shows the design of one type of stationary coil electromagnetic clutch.

When the temperature of the return air to the evaporator is brought down to a predetermined setting, the thermostat opens the electric circuit to the magnetic clutch on the compressor drive pulley. This causes the pulley to "free wheel" on its shaft, and the compressor stops.

There are several makes of electromagnetic clutches on the market. The current needed to magnetize the various clutches is approximately the same.

1. Warner plate type. 2.7 to 3.3A at 12V and normal temperatures.
2. Warner Heli-Grip. 2.7 to 3.1A at 12V and normal temperatures.
3. Electro-lock plate type. 2.9 to 3.3A at 12V and normal temperatures.

An electromagnetic clutch designed for use with two drive belts is shown in Fig. 26-18. The method of mounting the magnetic clutch and pulley on a compressor is detailed in Fig. 26-19.

Some systems use a thermostat or sensor to open the compressor clutch circuit when the ambient (outside) tempera-

ture is 32 F. (0 C.) or lower. Some owners disconnect the clutch wire during winter to save on fuel consumption. When this is done, the system will not operate at the best comfort level.

26-11 COMPRESSOR

Compressors used at present are belt driven from the engine. They operate at slightly above engine rpm. Therefore, with the engine idling (500 to 900 rpm), the compressor will revolve at 600 to 1000 rpm, depending upon engine application. At maximum engine speeds, the compressor will revolve at near 5000 rpm.

There are two types of compressors in general use in automobile air conditioning:

1. Conventional reciprocating type with crankshaft, connecting rod, piston and cylinder.
2. Swash plate type, which uses a different reciprocating piston and cylinder arrangement.

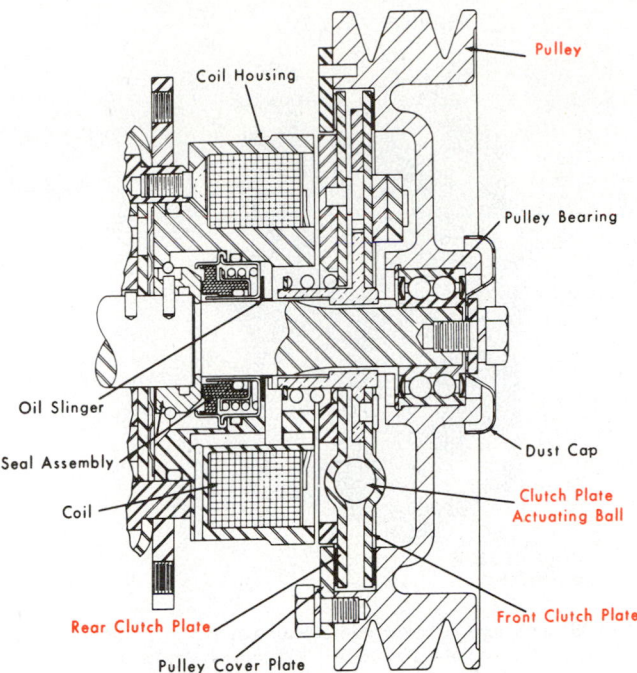

Fig. 26-18. Electromagnetic clutch has two-belt pulley that "free wheels" on compressor shaft until magnetism clamps shaft-mounted clutch plate between front and rear clutch plates.

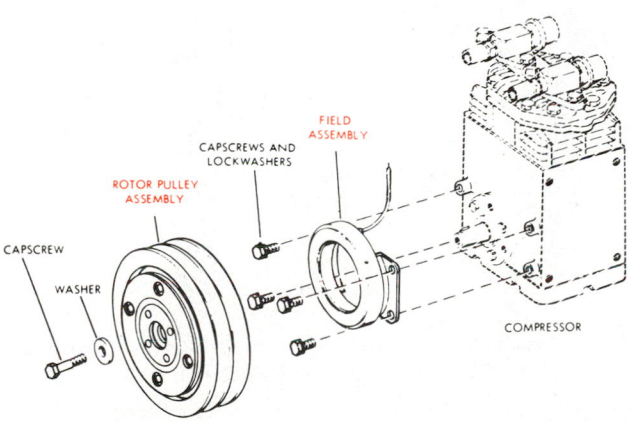

Fig. 26-19. Exploded view of electromagnetic pulley and clutch assembly. Note method of mounting field assembly (magnetic coil) and compressor pulley. (Climatic Air Sales, Inc.)

The swash plate (or "wobble" plate) compressor has a straight shaft and a "swash plate" mounted at an angle to the shaft. Double acting pistons are fitted over the swash plate. As the shaft and swash plate revolve, the pistons are caused to reciprocate in the cylinders (which are parallel to the shaft). This design is also called a "barrel compressor." Housing is die cast aluminum. Parts are mostly steel. Conventional piston type compressors are made of cast iron or aluminum alloy.

One method of mounting the compressor on the engine is shown in Fig. 26-20.

26-12 TWO CYLINDER COMPRESSOR

In general, compressors used on automobile air conditioners (except the swash plate compressor) are similar in operation to

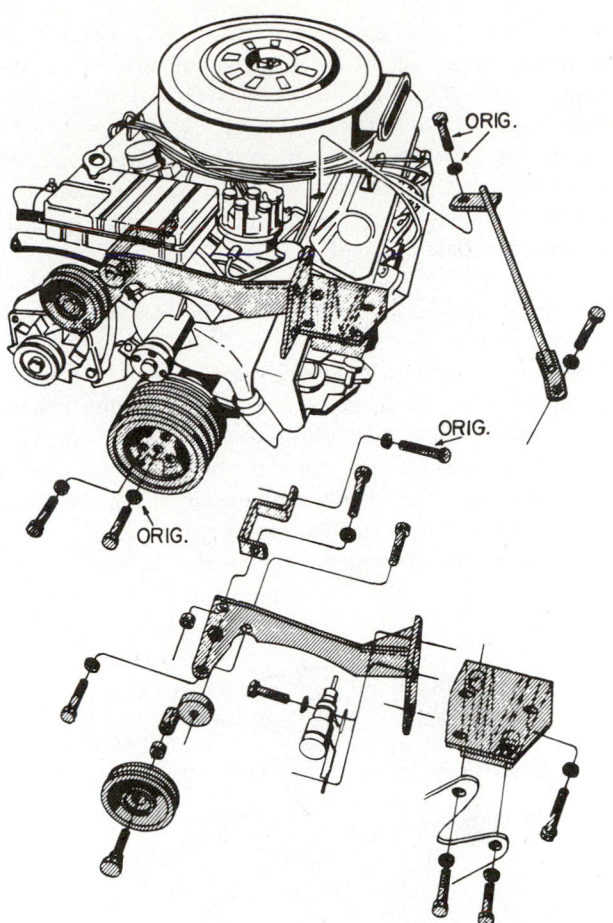

Fig. 26-20. Mounting brackets are needed to install the compressor for an "add on" air conditioning system.

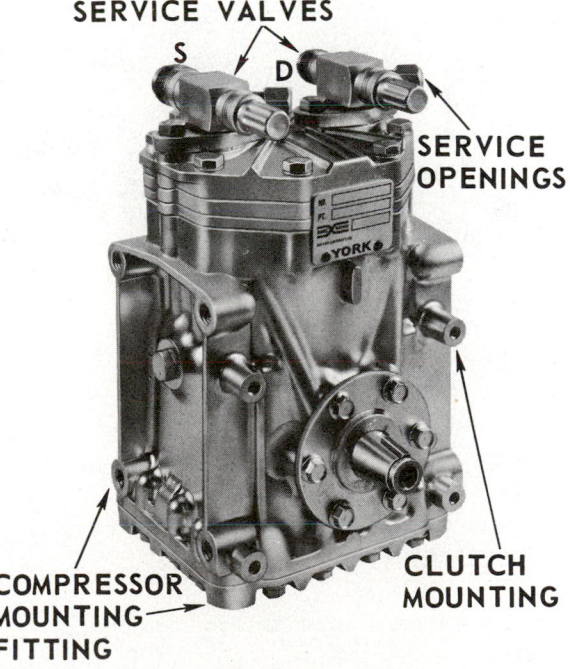

Fig. 26-21. Two cylinder reciprocating compressor. Compressor body has variety of mounting holes to permit vertical, horizontal or inclined mounting. S—Suction service valve. D—Discharge service valve. Service valves are flange mounted, using two cap screws. (York Div. of Borg-Warner Corp.)

those explained in Chapter 4. These compressors use 500 viscosity special refrigerant oil. The amount varies from three to seven oz., depending on the model. A lack of oil may cause bearing, seal and valve trouble.

Crankshaft seals must be of the heavy-duty type. The moving parts must be balanced for all the varying speeds. The compressor volumetric efficiency must remain at a certain minimum, regardless of speed.

Service valves or access valves are provided for mounting gauges and/or servicing the low-pressure and high-pressure sides of the system.

Two cylinder reciprocating type units are used by Ford and American Motors and by some companies that manufacture "add on" units. Fig. 26-21 shows a two cylinder unit, while Fig. 26-22 reveals internal construction of a two cylinder compressor. The two cylinder compressor in Fig. 26-23 has service valves attached to the compressor with a swivel fastener. Some compressors are of the two cylinder V-type, Fig. 26-24.

Fig. 26-23. This two cylinder compressor has swivel mounted service valves for ease of installation. (Tecumseh Products Co.)

Fig. 26-22. Cutaway view of a two cylinder compressor.

26-13 SWASH PLATE COMPRESSOR

There are two types of swash plate compressors:
1. Five cylinder model.
2. Six cylinder model.

The five cylinder model is pictured in Fig. 26-25. Its internal construction is shown in Fig. 26-26. The pistons use a connecting rod, which is fastened to the swash plate by ball

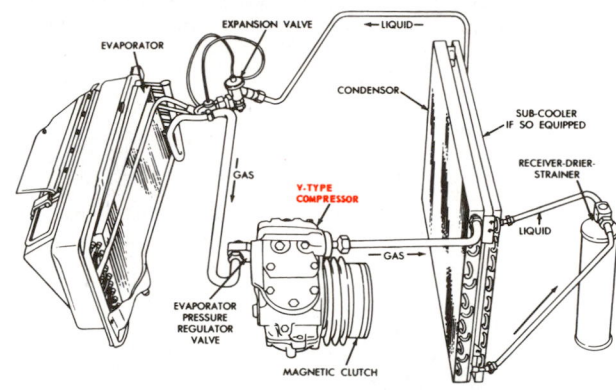

Fig. 26-24. Typical V-type, two cylinder compressor mounted in air conditioning system. (Chrysler Corp.)

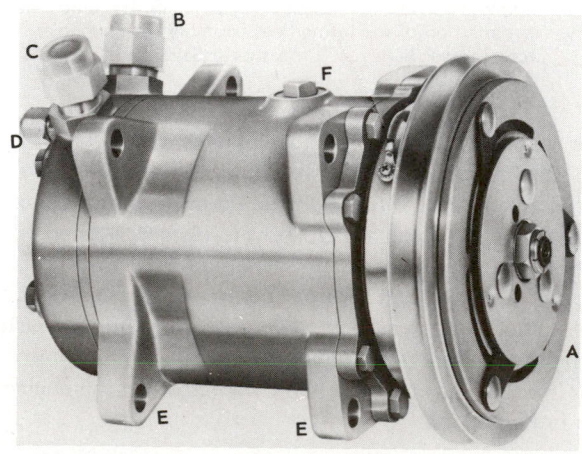

Fig. 26-25. Five cylinder swash plate compressor. A—Clutch. B—Suction line connection. C—Discharge line connection. D—Discharge access service port. E—Mounting flanges. F—Oil filler plug. (Abacus International)

Fig. 26-26. Internal construction of typical five cylinder swash plate compressor. A—Piston. B—Swash plate. C—Pulley bearing. (Abacus International)

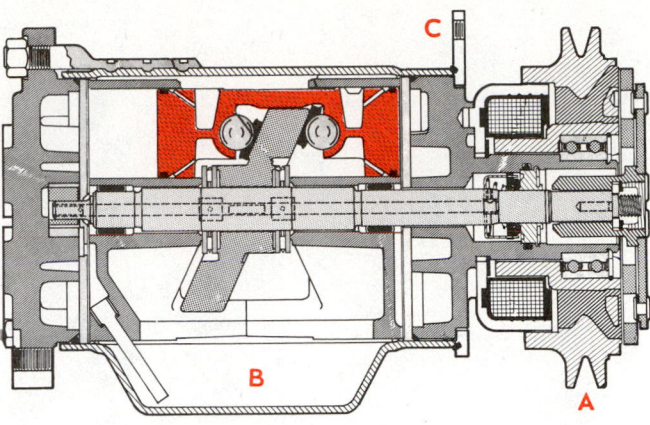

Fig. 26-28. Diagram showing components of a six cylinder axial type (swash plate) compressor. Red color highlights one double acting piston. A—Pulley. B—Crankcase. C—Mounting flange.

joints. These compressors use either the SAE flare connections or the roto-lock connections. This compressor may be rotated either way. It uses seven ounces of 500 viscosity oil; four ounces remain in the compressor and three ounces circulate with the refrigerant.

The compressor body is made of die cast aluminum. The shaft seal is a carbon ring rotating against a cast iron seat. The sealing surfaces are made to a very true tolerance. Do not touch the seal surface with the fingers (to avoid corroding the seal faces). This five cylinder compressor uses metric size cap screws and nuts.

The six cylinder General Motors compressor is shown in Fig. 26-27. It consists of three sets of opposing cylinders and

three double end pistons. Fig. 26-28 shows the compressor with one double-acting piston labeled in red. The parts of a six cylinder swash plate type compressor are shown in Fig. 26-29.

26-14 COMPRESSOR SEAL

All automotive air conditioning compressors have a crankshaft seal. In some cases, sealing is done by means of a carbon ring with a smooth surface rubbing against a flat, smooth cast iron surface. This ring is bolted to and sealed on the body of the compressor. Some compressors use a rotating carbon ring rubbing against a stationary carbon ring. A synthetic rubber O-ring seals the joint between the carbon ring and the shaft.

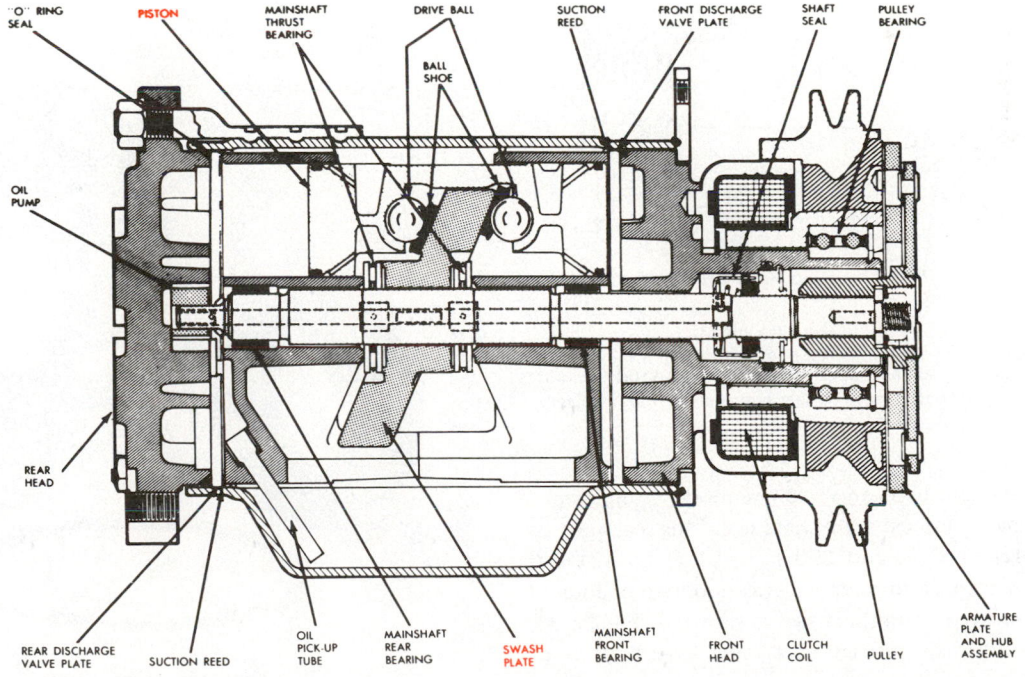

Fig. 26-27. Six cylinder axial type automotive air conditioning compressor. Revolving swash plate set at an angle drives three double acting pistons.

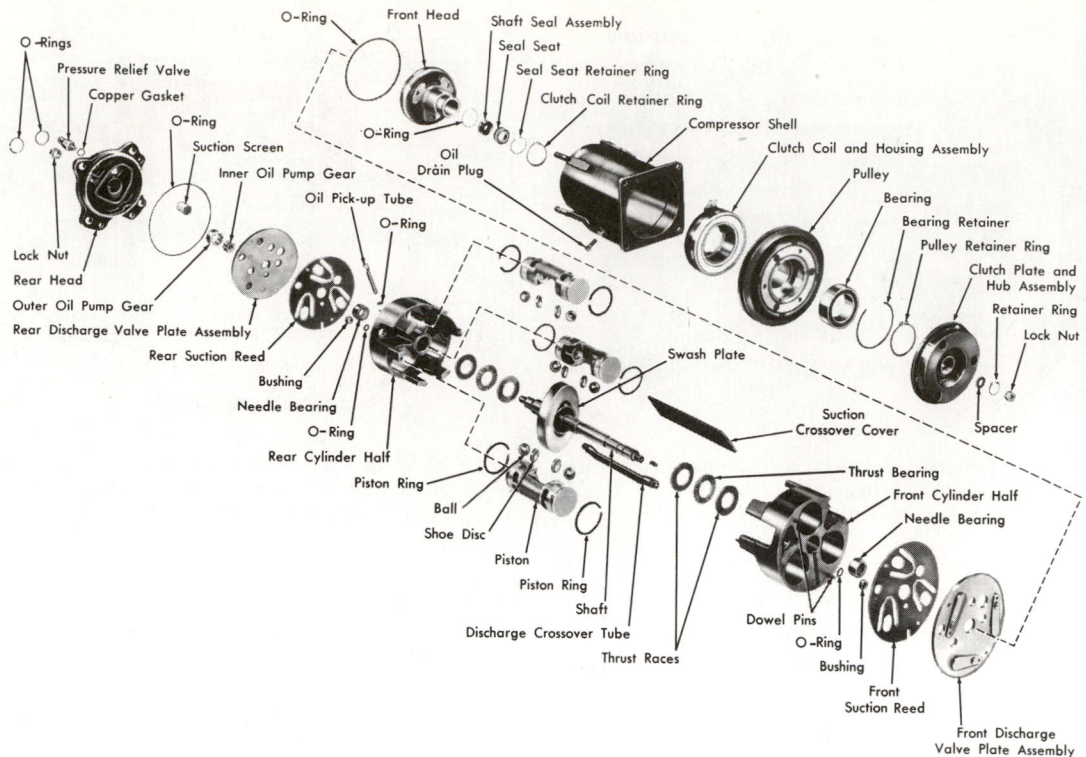

Fig. 26-29. Exploded view of a swash plate axial type six cylinder automotive air conditioning compressor. (Cadillac Motor Car Div., General Motors Corp.)

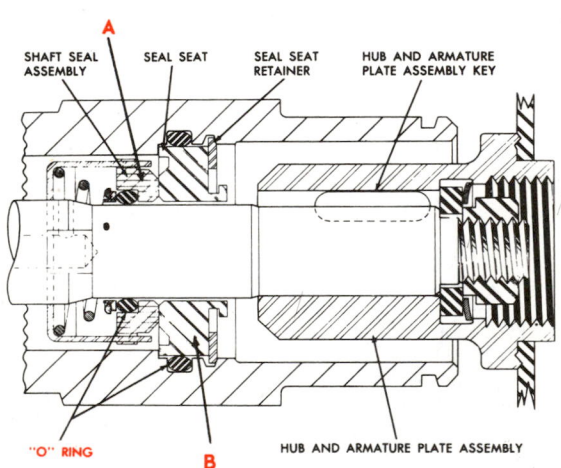

Fig. 26-30. Compressor crankshaft seal. A—Rotating carbon ring. B—Cast iron seat. Rubbing surfaces should be smooth and flat (accurate to .000001 in.). Note use of O-rings to seal two sealing rings in place.

week throughout the year. If the compressor is out of operation for a long period of time, the oil may drain from the compressor seal surfaces. If these surfaces become dry, refrigerant vapor may leak out through the seal.

Fig. 26-31. Replacement crankshaft seal for automotive compressors. (EG&G Sealol Inc.)

Teflon seal surfaces are used on some late model compressors.

Fig. 26-30 shows a typical crankshaft seal. This seal must be leakproof between −60 F. and 250 F. (−51 C. to 121 C.), and from one micron vacuum to several hundred pounds of pressure. Another type crankshaft seal is shown in Fig. 26-31. Seals of this design may be used as a replacement for worn seals on automotive compressors.

Many manufacturers recommend that drivers operate the air conditioning compressor for a few minutes at least once a

In the case of the loss of refrigerant, it is possible to purchase a sealed container which contains the correct amount of refrigerant for any particular make of car or model. This container may also provide the correct amount of refrigerant oil for the system. To make use of these containers, it is necessary to first drain all the refrigerant and oil from the compressor. Then add the refrigerant and oil from the container.

26-15 BELTS

One or two belts may be used to drive the air conditioning compressor. The belt is mounted on the pulley of the electromagnetic clutch and is driven by a pulley on the engine crankshaft. An adjustable idler pulley may be used to maintain correct belt tension.

The compressor belt used must be designed and constructed for automotive type service. It must run true on the pulleys, and the pulleys must be in line. The belt must have the correct tension for efficient use and long life.

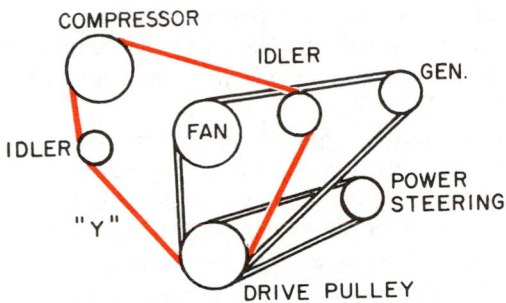

Fig. 26-32. Diagrammatic view of belts used in typical air conditioning installation. This application has belt drive for power steering pump.

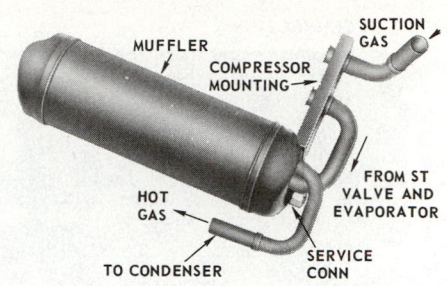

Fig. 26-34. Muffler used on air conditioning system. Note compressor mounting, service connection and how muffler is installed in system.

For a new belt, the tension should be from 140 to 145 lb. Use a belt tension gauge to measure the tension. Always recheck a new belt for tension a day or two after installation. A used belt (when overhauling a system that has been in use) should have a 55 to 75 lb. tension.

One way to roughly check belt tension is to apply firm hand pressure in the middle of the longest belt span when the engine is stopped. If the belt is correctly tensioned, it should depress about 1/2 in. out of line. Also, a correctly tensioned belt will twist 1/4 to 1/2 in. by using a firm grip and some firm twisting.

The belts may also power the water pump and the fan, as shown in Fig. 26-32. Some fans have a clutch which connects the fan to engine only after the engine compartment has reached a correct temperature (for quicker engine warmup). See details of one installation in Fig. 26-33.

26-16 CONDENSER

The condenser usually is mounted in front of the car radiator. The discharge line from the compressor to the

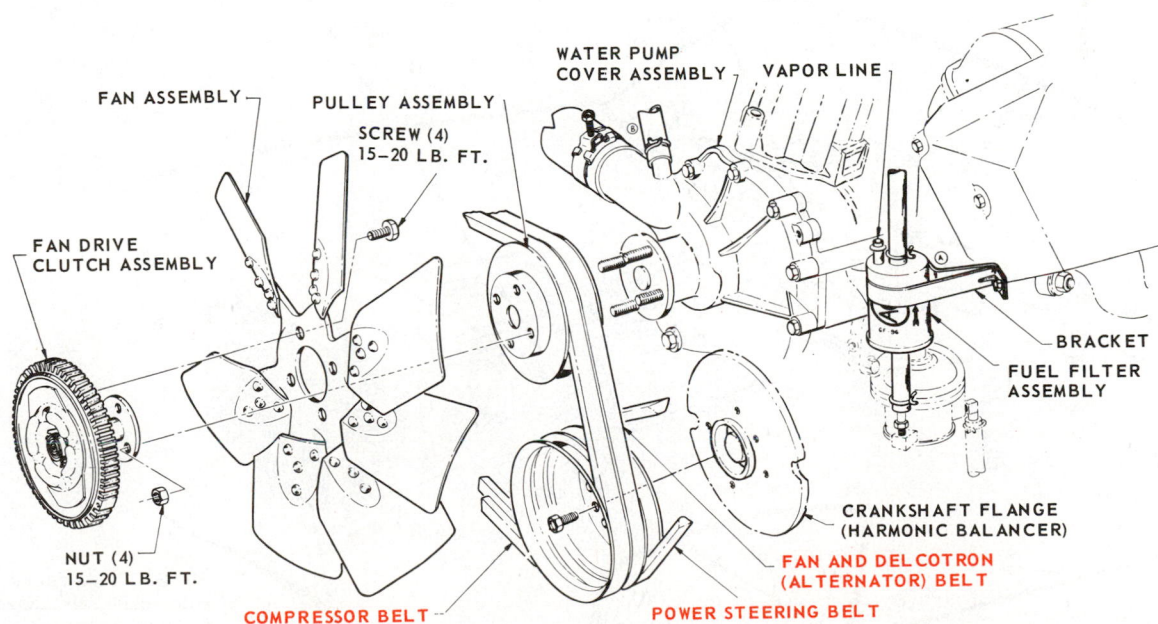

Fig. 26-33. Belts driving water pump, fan, power steering, alternator, and air conditioning compressor. (Buick Motor Div., General Motors Corp.)

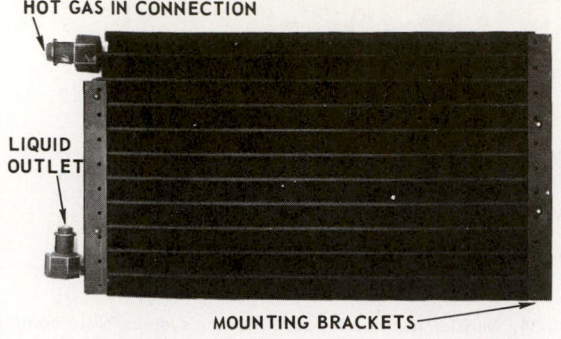

Fig. 26-35. Automotive air conditioning condenser. Usually, condenser is placed in front of cooling system radiator.

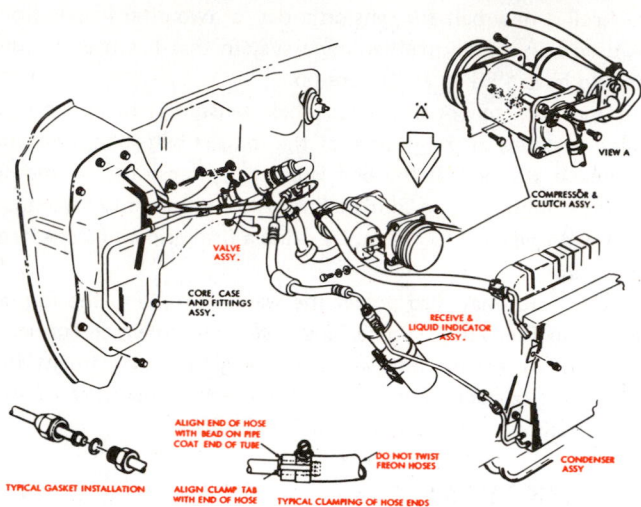

Fig. 26-36. Typical condenser mounting is shown, along with routing of refrigerant lines and line connections.

condenser may have a vibration absorber mounted in it, or the line may be flexible. Many systems also mount a muffler in this line. See Fig. 26-34.

The condenser may be a one, two or three pass finned tube type, made of copper or aluminum. Fig. 26-35 shows a typical automotive condenser. It is firmly fastened to the radiator shell, using rubber grommets, washers and screws. Air going through the radiator and into the engine compartment usually goes through the condenser first. Fig. 26-36 shows a typical condenser installation.

Fig. 26-37 shows one method of mounting the condenser on the front of the radiator. Since the condenser is mounted ahead of the radiator, it may collect a considerable amount of leaves, bugs and lint. Keep the outside of the condenser clean. A partially clogged airflow will affect the efficiency of the condenser and car radiator.

26-17 RECEIVER-DRIER

Most automotive air conditioning systems use a receiver located between the condenser and evaporator. Its purpose is to store liquid refrigerant during service operations. The receiver also allows for some changes in refrigerant charge and in liquid volume (caused by expansion and contraction of the refrigerant as temperatures change). A receiver-drier is shown in Fig. 26-38.

The receiver usually has a drier chemical placed inside. This drier chemical (desiccant) will remove moisture from the liquid and hold the moisture, unless the chemical is heated to a high temperature or unless the water is replaced by a more active chemical, such as alcohol. Most driers have a strainer to remove dirt particles from the refrigerant and oil. Liquid receivers usually have a safety fusible plug that will open and release the refrigerant if heated to about 350 F. (177 C.).

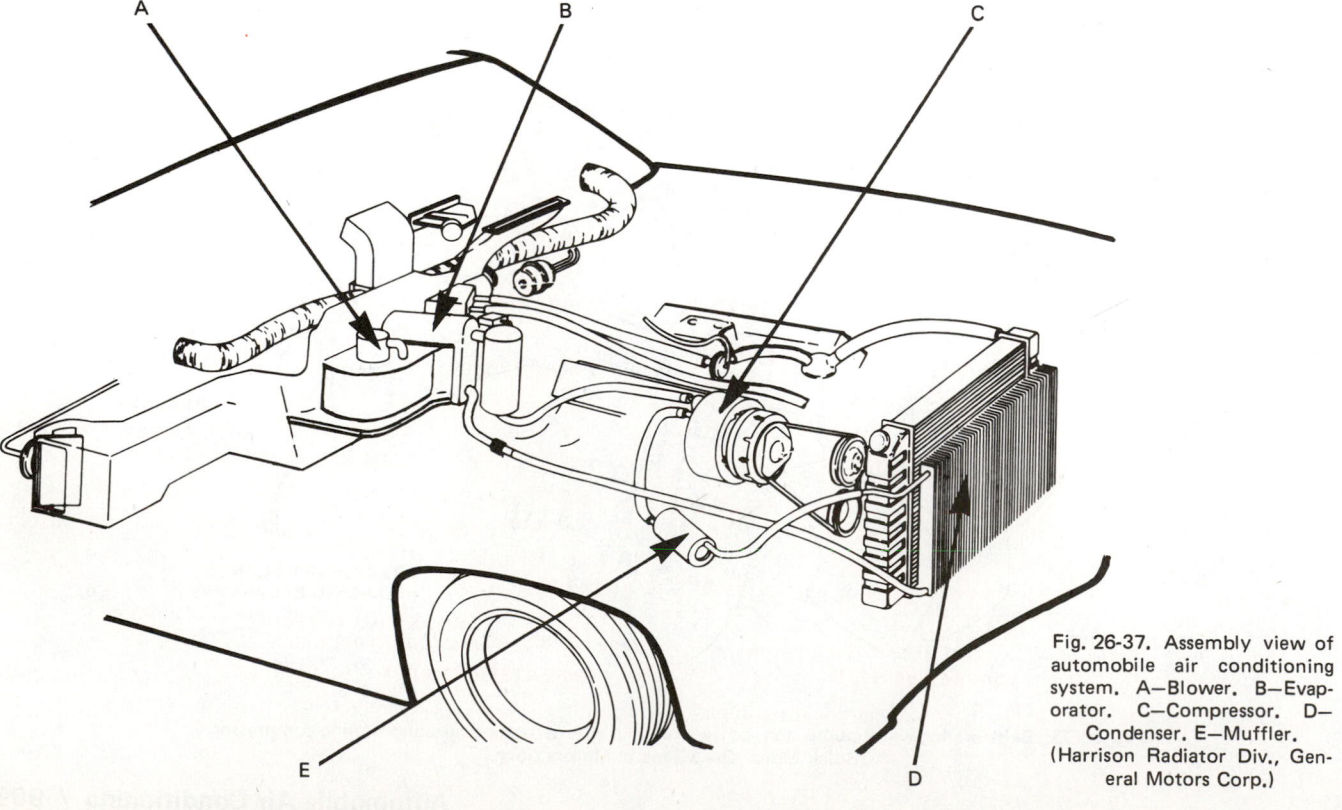

Fig. 26-37. Assembly view of automobile air conditioning system. A—Blower. B—Evaporator. C—Compressor. D—Condenser. E—Muffler. (Harrison Radiator Div., General Motors Corp.)

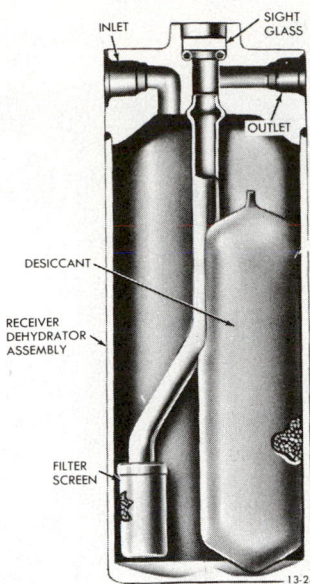

Fig. 26-38. Cutaway view of liquid receiver-drier that holds about 10 cu. in. of desiccant. Sight glass is located in liquid refrigerant outlet. (Buick Motor Div., General Motors Corp.)

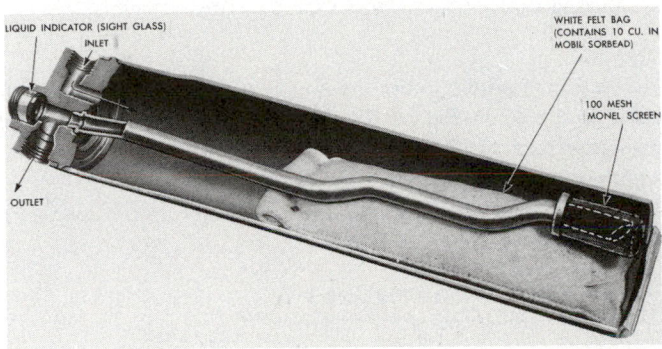

Fig. 26-39. Cross section of combination liquid receiver, filter, drier and sight glass.

Some of these receivers also have a sight glass as part of the liquid line outlet. See Fig. 26-39. After the system has been operating for a few minutes, bubbles may appear in the sight glass. These vapor bubbles signal that the system is short of refrigerant (no reserve liquid in receiver). Usually, the bubbles indicate that the system has lost over half of its charge.

26-18 REFRIGERANT LINES

Special flexible refrigerant lines are used in automobile air conditioning applications:

1. To carry liquid refrigerant from receiver-drier to evaporator expansion valve (liquid line).
2. To carry vapor refrigerant from evaporator to compressor (suction line).
3. To carry hot compressed vapor from compressor to condenser.
4. To carry liquid refrigerant from condenser to liquid receiver-drier (on some units).

Flexible refrigerant lines are also called hoses. They are commonly covered with a braid to protect them against injury. These hoses are designed and constructed to be flexible and vibration proof.

Refrigerant lines are made of steel or copper, or the lines may be flexible. Double flare fittings are sometimes used where units must be disconnected for servicing. Flexible lines also use other types of fittings. Refrigerant lines should be carefully routed to prevent them from rubbing against any part of the car. Wear and corrosion would quickly cause leaks at the points of contact.

Refrigerant lines are fastened to the system parts in various ways:

1. Flared fitting.
2. O-ring fitting.
3. Hose clamp fitting.

These fittings are shown in Fig. 26-40.

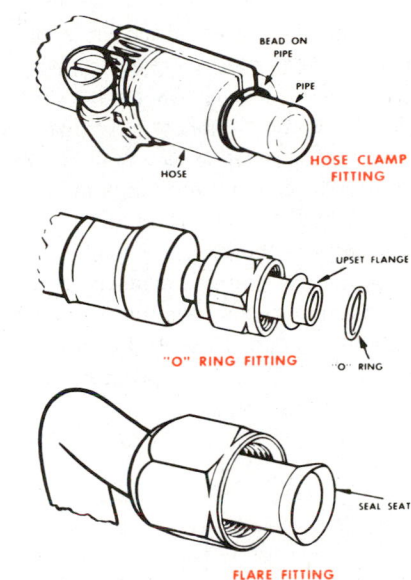

Fig. 26-40. Typical refrigerant line connection fittings are hose clamp, O-ring and flare. (Harrison Radiator Div., General Motors Corp.)

The flexible lines or hoses vary in size from 3/8 to 5/8 in., depending on the capacity of the unit and on the state of the refrigerant. A fitting often used with plastic flexible tubing is shown in Fig. 26-41. Vapor-carrying lines are larger. The size of the lines must match the fittings supplied by the manufacturer, so that system capacity will not be reduced.

Refrigerant lines should have large bends, and they should be supported, grommeted and clamped to prevent wear by chafing and to prevent them from touching hot engine parts. Leave enough slack in the hose at the compressor end of the lines to allow for movement of the engine on its vibration absorber mounts.

The lines come equipped with caps and plugs to keep the inside of the lines clean and dry. Remove these plugs and caps just before installing the lines. All assembly threads and fittings should have clean, fresh refrigerant oil put on them, just before assembly.

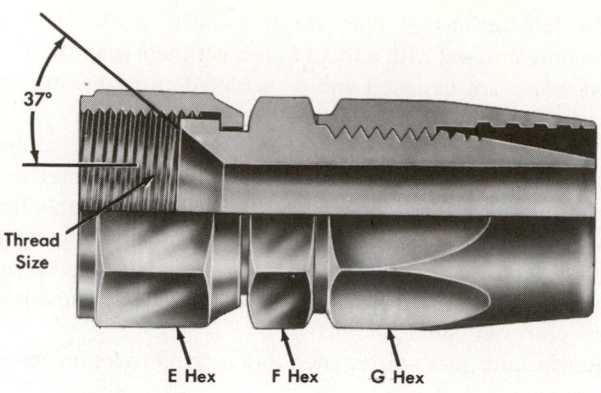

Fig. 26-41. Fitting used in connecting plastic flexible tubing. (Polymer Corp.)

26-19 EVAPORATOR

Evaporators usually are mounted in a plenum chamber attached to the engine compartment, fire wall or dashboard. The evaporator of an automotive air conditioner is of the finned, forced convection type. It is enclosed in a metal or plastic housing that also serves as a duct for the conditioned air. A moisture drain pan and drainpipe must be incorporated in the unit. See Fig. 26-42.

On some systems in the past, the evaporator was mounted in the trunk of the car; others locate it in the front fender well. Most installations today have the evaporator mounted on the engine side of the dash (fire wall) or on the passenger side of the dash (fire wall). An evaporator, blower and duct

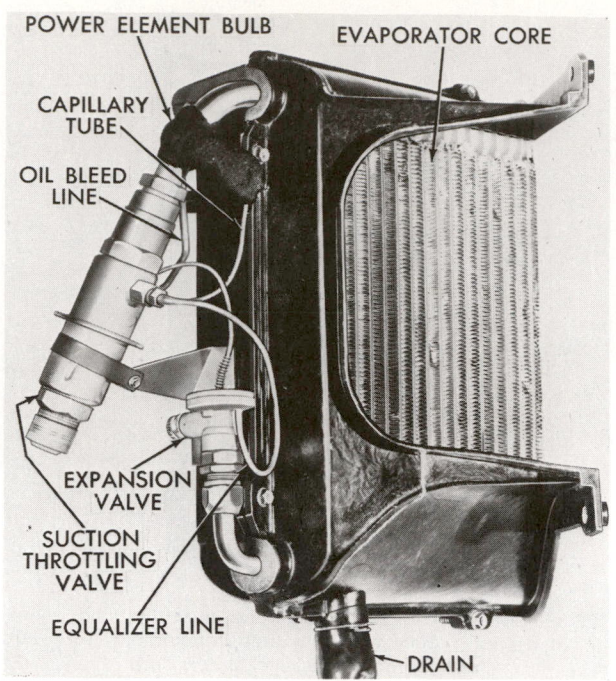

Fig. 26-42. Evaporator complete with housing, drain TEV and suction throttling valve. (Cadillac Motor Car Div., General Motors Corp.)

installation mounted under the hood and under the instrument panel is shown in Fig. 26-43. The heating coil is mounted in the same duct system, which fastens to the fire wall or dash. Another design is shown in Fig. 26-44.

The design and installation of an "add on" unit evaporator,

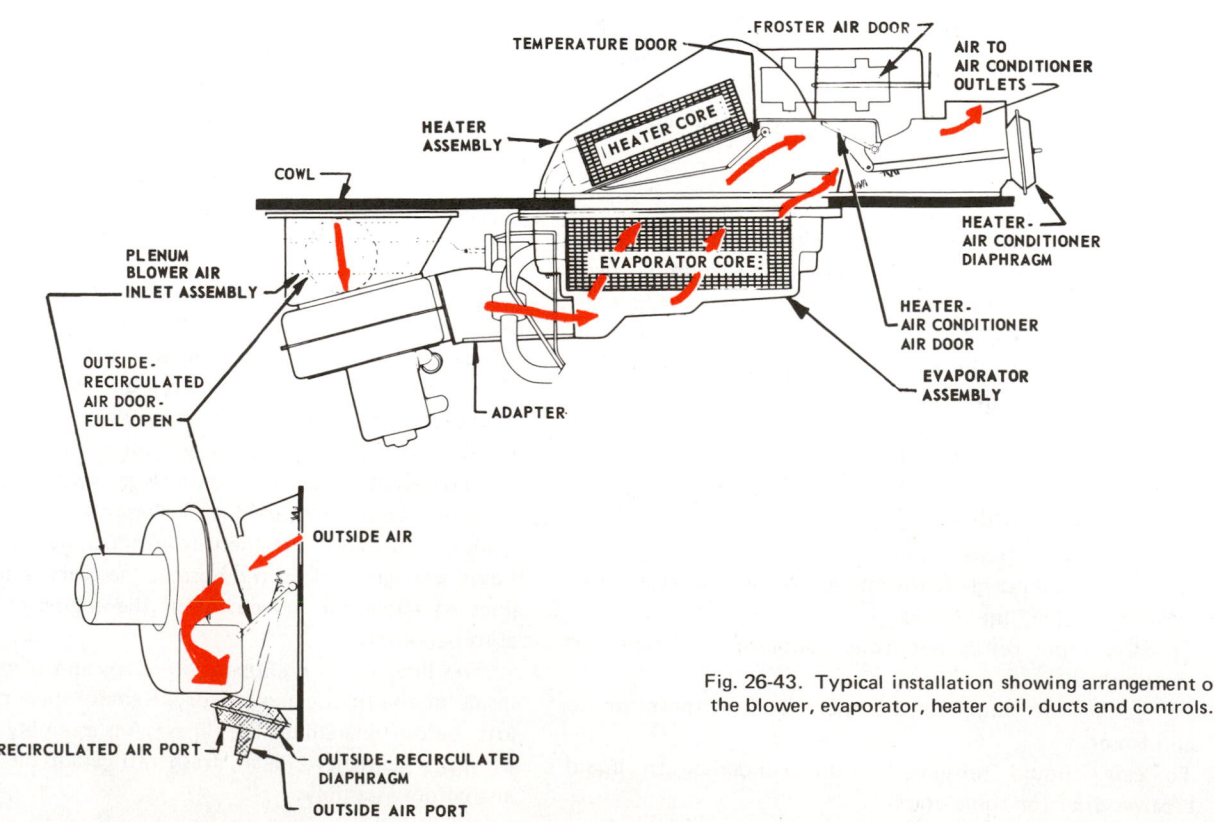

Fig. 26-43. Typical installation showing arrangement of the blower, evaporator, heater coil, ducts and controls.

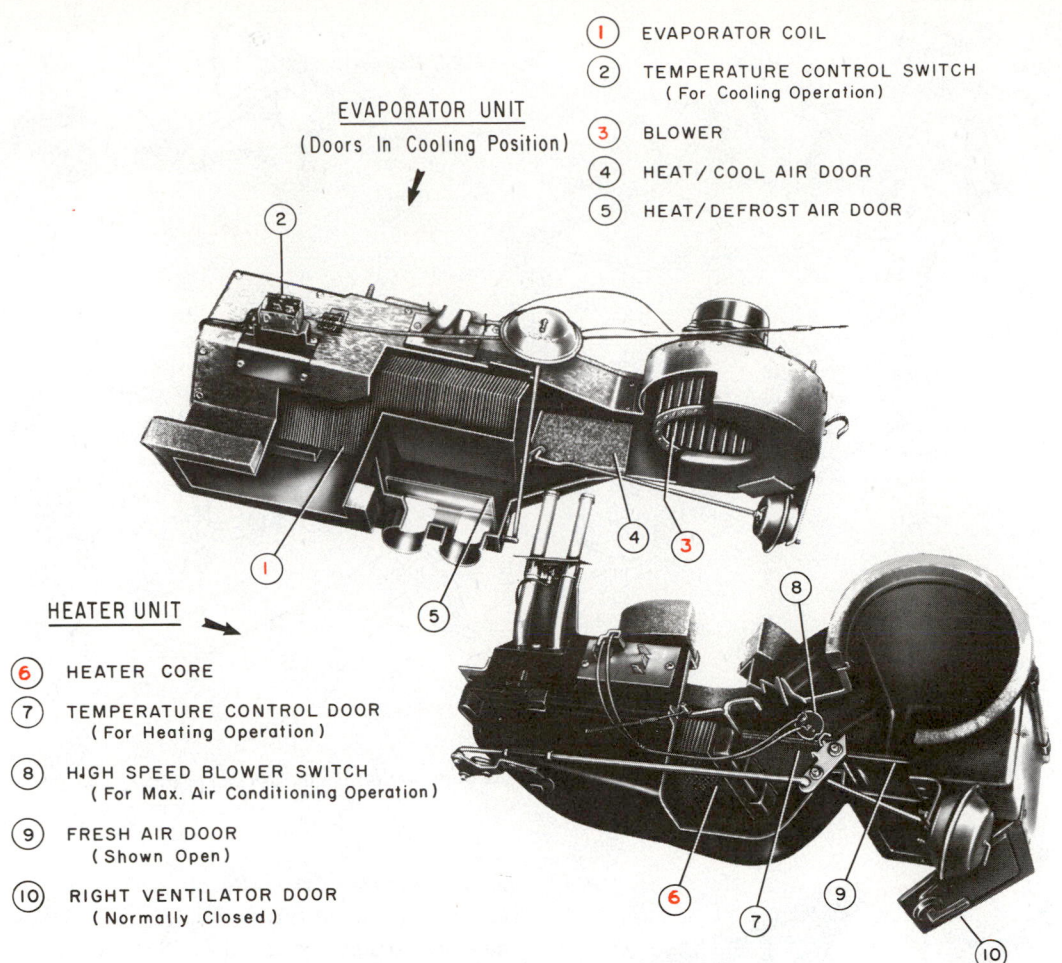

EVAPORATOR UNIT
(Doors In Cooling Position)

1. EVAPORATOR COIL
2. TEMPERATURE CONTROL SWITCH
   (For Cooling Operation)
3. BLOWER
4. HEAT/COOL AIR DOOR
5. HEAT/DEFROST AIR DOOR

HEATER UNIT

6. HEATER CORE
7. TEMPERATURE CONTROL DOOR
   (For Heating Operation)
8. HIGH SPEED BLOWER SWITCH
   (For Max. Air Conditioning Operation)
9. FRESH AIR DOOR
   (Shown Open)
10. RIGHT VENTILATOR DOOR
    (Normally Closed)

Fig. 26-44. Two views of combination heating, cooling and duct system. 1—Evaporator. 3—Blower. 6—Heater core. (Dodge Div., Chrysler Corp.)

evaporator housing and blower unit is shown in Fig. 26-45. The evaporator position can be reversed inside its housing. The blower assembly can also be inverted, as shown, to permit greater flexibility of installation.

Some independent units are mounted in the trunk of the vehicle, because of lack of space under the hood or because of the style of the car (limousine).

A hose schematic and wiring diagram of a typical trunk unit

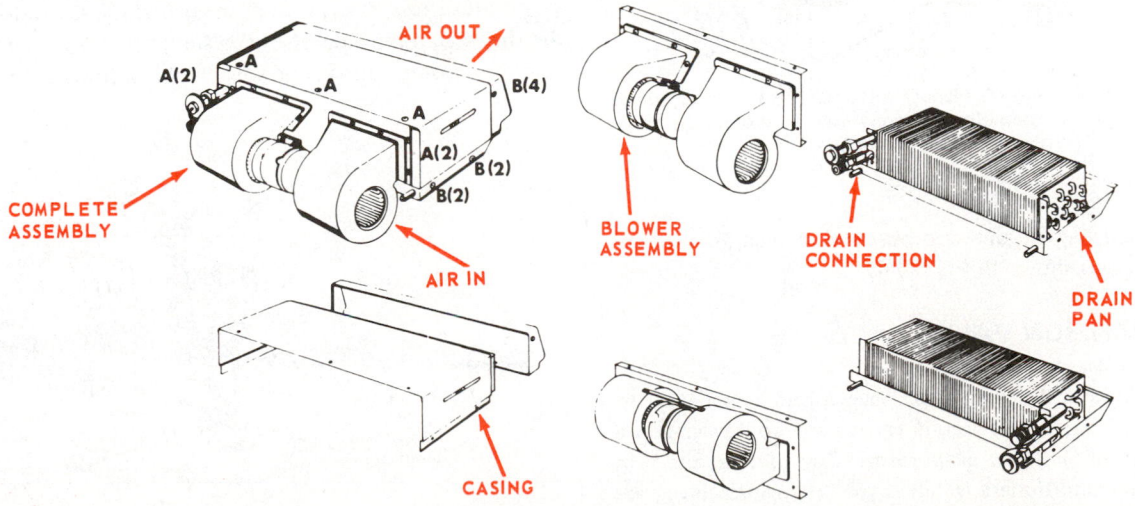

Fig. 26-45. An "add on" unit evaporator and blower assembly. Both evaporator and blower assembly may be inverted if necessary. (Climatic Air Sales, Inc.)

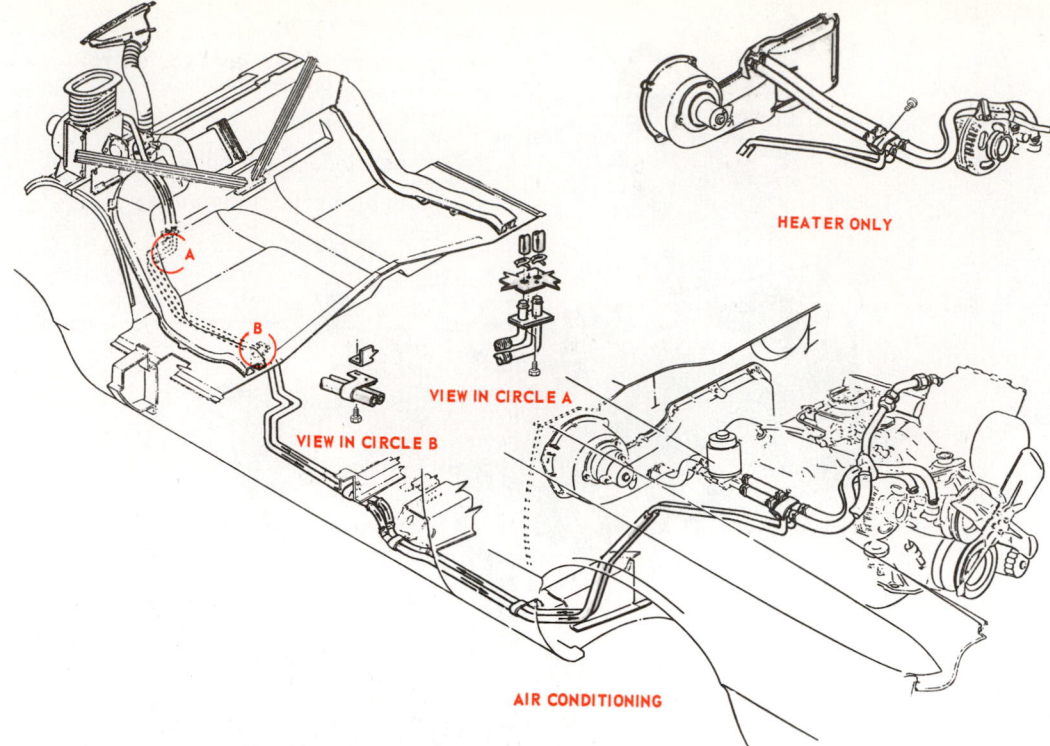

Fig. 26-46. Typical trunk unit installation. Note use of blower system in dash and in trunk.

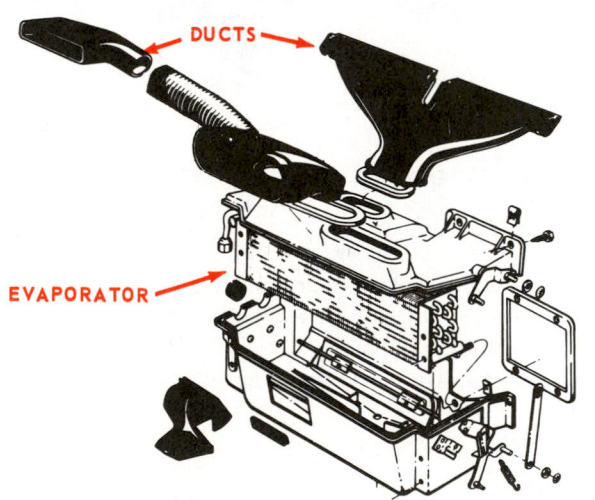

Fig. 26-47. Exploded view of evaporator, evaporator housing, ducts and air dampers. Note assembly cap screw used to mount the upper evaporator housing to firewall.

are pictured in Fig. 26-46.

An exploded view of an evaporator and its plenum chamber and duct system is shown in Fig. 26-47.

26-20 EXPANSION VALVE

Air conditioning systems must have a reducing device to throttle the high-pressure liquid refrigerant to low-pressure liquid refrigerant in the evaporator. The device used in automobile air conditioners is called the "thermostatic expansion valve." This control valve responds to the temperature of the evaporator outlet as well as low-side (suction) pressure.

Chapter 5 explains the principle, design, construction, installation, and service of this type valve. Some TEVs are adjustable; some are preset for a 15 F. (8 C.) superheat and are not adjustable.

The capacity of the expansion valve (orifice size) must match the capacity of the air conditioning unit. A valve that is too small will reduce the capacity, while a valve that is too large will "hunt" or alternately flood and starve the evaporator of liquid refrigerant.

Most automotive TEVs have a pressure equalizer connection from the valve body to the outlet of the evaporator. The equalizer tube is needed because of the pressure drop through the evaporator. This makes it necessary to use the outlet evaporator pressure only to push against the low-side facing of the diaphragm for accurate, consistent operation. A nonadjustable thermostatic expansion valve is shown in Fig. 26-48.

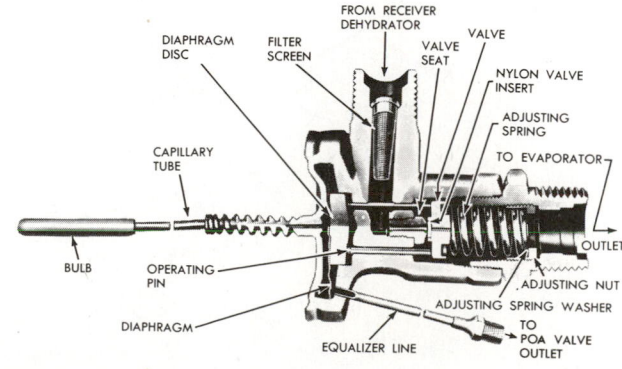

Fig. 26-48. Nonadjustable thermostatic expansion valve. Note inlet connections and outlet connections.

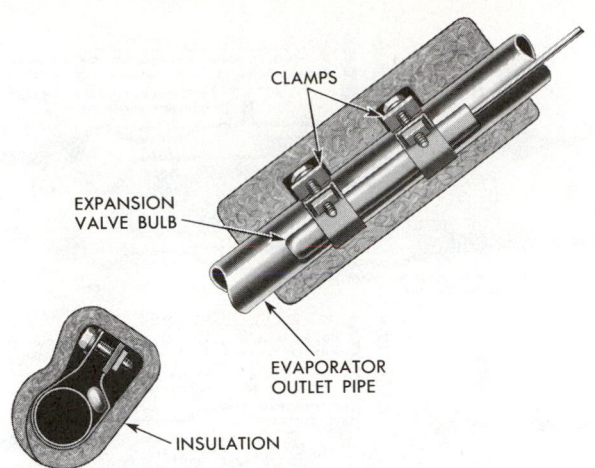

Fig. 26-49. Details of mounting thermostatic expansion valve thermal bulb on outlet of evaporator.

The expansion valve is located on the inlet to the evaporator. The sensitive bulb must be clean. The outlet tube to which it is clamped must be clean, and the clamps must be tight. After mounting the bulb, the assembly should be covered with insulation. In this way, only the evaporator outlet temperature will affect valve operation. Fig. 26-49 shows a recommended bulb mounting, with the capillary tube connection at the top.

Some thermostatic expansion valves are equipped with a built-in sight glass on the inlet side to show the refrigerant flow. When the sight glass is part of the TEV, there is no need for a sight glass in the liquid receiver. Most TEVs with a built-in sight glass are of the nonadjustable type.

26-21 SUCTION PRESSURE CONTROL VALVE

Some systems use suction pressure valves to maintain a certain pressure in the evaporator. This pressure is independent of the compressor low-side pressure and independent of the

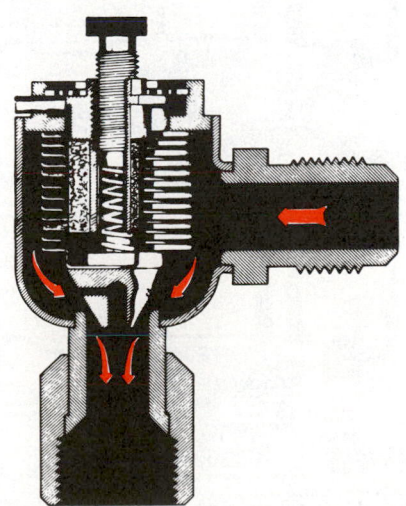

Fig. 26-50. Suction pressure regulator valve with a manual pressure adjustment to control evaporator temperature. (Frigiking, Inc.)

cooling demand. In most cases, a diaphragm or bellows in the valve responds only to the pressure in the evaporator. When the evaporator pressure is above 29 to 31 psi (R-12), the valve opens. The valve closes if the pressure tends to go below these settings. Fig. 26-50 shows a cross-section of a suction pressure valve with a manual pressure adjustment.

Several types of suction pressure valves have been used:
1. Suction throttling valves (STV).
2. Evaporator pressure regulators (EPR).
3. Pressure operated altitude valves (POA) on General Motors' systems.

Most of these valves are adjustable. Some have valve core access ports, which are used to connect service lines and gauge manifolds.

The main purpose of these suction pressure valves is to keep the evaporator above a freezing temperature. This prevents the moisture that condenses on the evaporator from freezing as the air flows through it.

Some suction pressure valves maintain a 28 psi (2 kg/cm^2) pressure in the evaporator until the air temperature becomes too cold. Then, a manual adjustment (by the operator) or a vacuum adjustment raises the pressure to about 30 psi (2.2 kg/cm^2). Fig. 26-51 shows a vacuum control unit. At zero

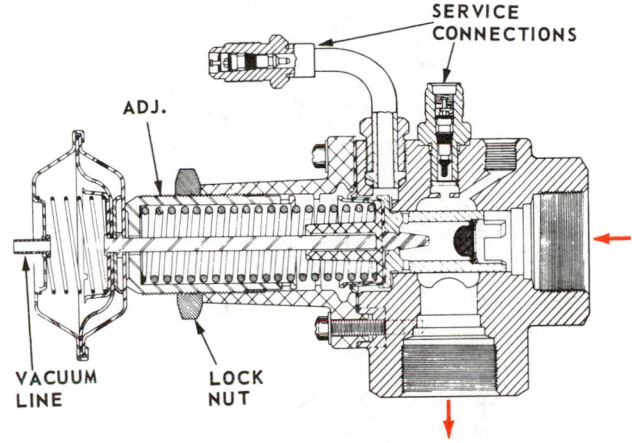

Fig. 26-51. Vacuum operated suction pressure regulator valve.

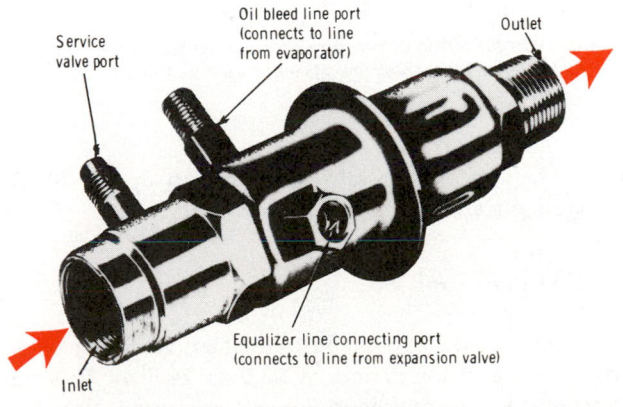

Fig. 26-52. Pressure operated altitude suction throttling valve. (Buick Motor Div., General Motors Corp.)

vacuum, the valve will operate at 3 psi (.2 kg/cm^2) higher evaporator pressure. These valves usually have a service attachment to permit a pressure check of the evaporator. Notice that the vacuum attachment body can be adjusted by threading it in to or out of the main body of the valve. The adjustment and lock nut are shown in Fig. 26-51.

The POA developed by General Motors Corp. uses a sealed pressure element that maintains a constant pressure independent of the altitude of the car. See Fig. 26-52. This valve is equipped with an access valve service connection, an oil bleed connection (to remove oil from evaporator), and a thermostatic expansion valve equalizer connection.

PILOT NEEDLE
VALVE SEAT
PILOT NEEDLE
VALVE
PILOT NEEDLE
VALVE SPRING
BELLOWS
PISTON RING
PISTON
BLEED HOLE
BELLOWS
CHAMBER
SPRING
VAPOR PASSAGE
SCREEN
LIQUID BLEED
CYLINDER
PISTON
IMPACT BAFFLE
EVAPORATOR
GAGE
INLET

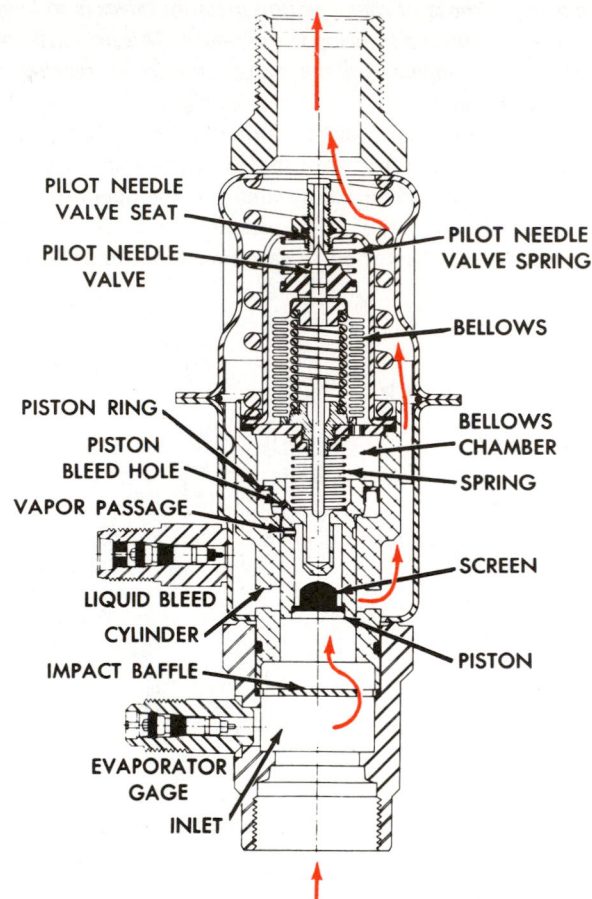

Fig. 26-53. Cross section of pressure operated altitude (POA) valve. (Cadillac Motor Car Div., General Motors Corp.)

A cross section of this valve is illustrated in Fig. 26-53; operating steps are shown in Fig. 26-54.

26-22 BYPASS VALVE

Some systems use a bypass control that automatically keeps the low pressure in the system at 23.5 to 24.5 psi (1.66 to 1.77 kg/cm^2). It accomplishes this by releasing vapor from the high-pressure side into the low side, if the pressure tries to go below preset pressure. Fig. 26-55 shows a bypass control.

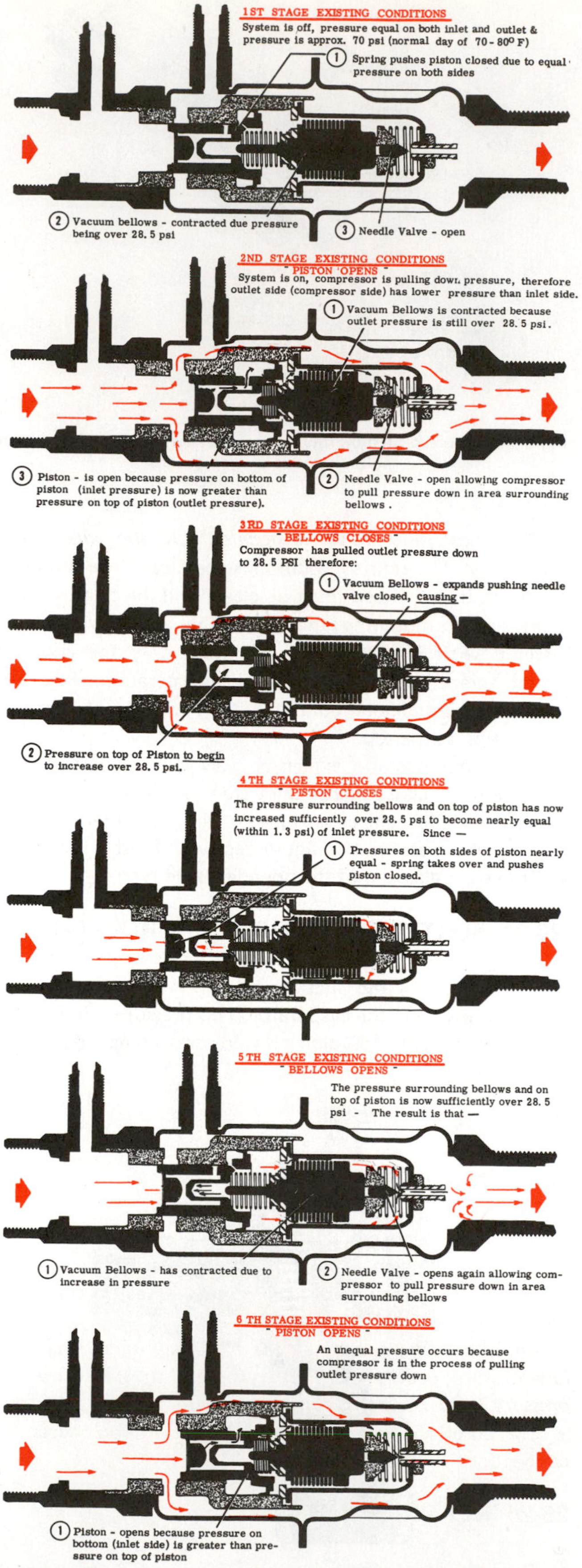

1ST STAGE EXISTING CONDITIONS
System is off, pressure equal on both inlet and outlet & pressure is approx. 70 psi (normal day of 70 - 80° F)
1 Spring pushes piston closed due to equal pressure on both sides
2 Vacuum bellows - contracted due pressure being over 28. 5 psi
3 Needle Valve - open
2ND STAGE EXISTING CONDITIONS
- PISTON OPENS -
System is on, compressor is pulling down pressure, therefore outlet side (compressor side) has lower pressure than inlet side.
1 Vacuum Bellows is contracted because outlet pressure is still over 28. 5 psi.
3 Piston - is open because pressure on bottom of piston (inlet pressure) is now greater than pressure on top of piston (outlet pressure).
2 Needle Valve - open allowing compressor to pull pressure down in area surrounding bellows .
3RD STAGE EXISTING CONDITIONS
- BELLOWS CLOSES -
Compressor has pulled outlet pressure down to 28. 5 PSI therefore:
1 Vacuum Bellows - expands pushing needle valve closed, causing —
2 Pressure on top of Piston to begin to increase over 28. 5 psi.
4 TH STAGE EXISTING CONDITIONS
- PISTON CLOSES -
The pressure surrounding bellows and on top of piston has now increased sufficiently over 28. 5 psi to become nearly equal (within 1. 3 psi) of inlet pressure. Since —
1 Pressures on both sides of piston nearly equal - spring takes over and pushes piston closed.
5 TH STAGE EXISTING CONDITIONS
- BELLOWS OPENS -
The pressure surrounding bellows and on top of piston is now sufficiently over 28. 5 psi - The result is that —
1 Vacuum Bellows - has contracted due to increase in pressure
2 Needle Valve - opens again allowing compressor to pull pressure down in area surrounding bellows
6 TH STAGE EXISTING CONDITIONS
- PISTON OPENS -
An unequal pressure occurs because compressor is in the process of pulling outlet pressure down
1 Piston - opens because pressure on bottom (inlet side) is greater than pressure on top of piston

Fig. 26-54. Six operational stages of POA suction throttling valve.

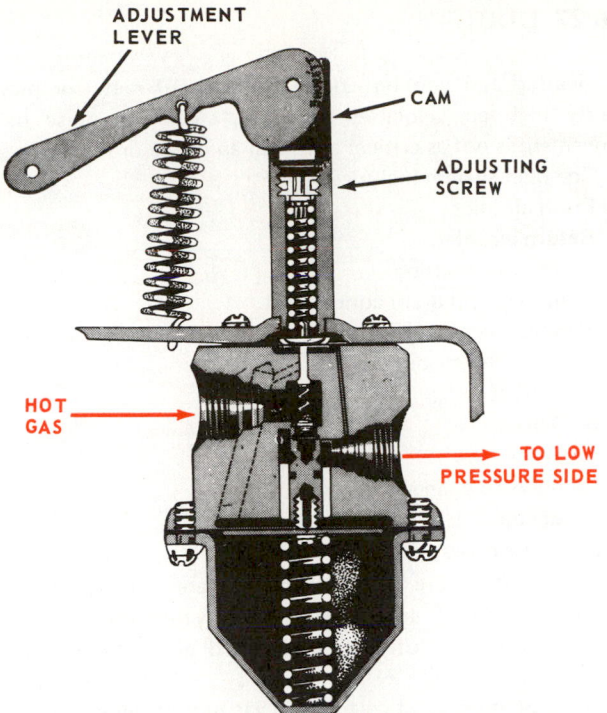

Fig. 26-55. Bypass valve used to keep low-side pressure at 25 psi (1.8 kg/cm²) or higher, independent of running conditions. (John E. Mitchell Co.)

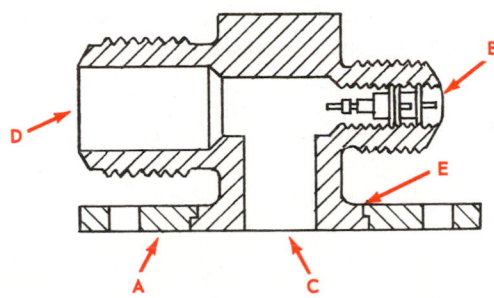

Fig. 26-56. A compressor mounting equipped with Schrader valve attachment. A—Flange. B—Refrigerant service connection with Schrader valve. C—Opening to compressor. D—Refrigerant line opening. E—Swivel joint.

26-23 SERVICE VALVE

A variety of service valves have been used on automotive air conditioning systems. Some are the standard front and back seat (one-way and two-way) service valves (see Chapters 2 and 14). Some are access valves (see Chapter 11).

All systems have some means by which high-pressure and low-pressure gauges can be connected to the system for pressure testing to permit charging or adding oil to the system. Fig. 26-56 shows a compressor fitting with an access valve connection.

26-24 REFRIGERANT

The refrigerant used in automobile air conditioning systems is R-12. It is one of the safest refrigerants used. Its properties

are described in Chapter 9.

The refrigerant operates at 28 F. to 30 F. (−2 C. to −1 C.) in the evaporator, and condenses at about 120 F. (49 C.) in the condenser.

The low-side evaporating pressure is about 28 psi (2 kg/cm²). The condensing pressure varies from 130 psi (9.13 kg/cm²) on cool days up to 200 psi (14.06 kg/cm²) on hot days.

26-25 OIL

Only a specially prepared oil should be used in the refrigerating system. This oil circulates throughout the system with the refrigerant. Most of it stays in the compressor. It must lubricate whether very cold or very hot. It must be dry. Even a very small amount of moisture will freeze at the TEV and may also form a sludge. The oil must be wax free. Chapters 2 and 9 both describe refrigerant oil properties. The automotive systems usually use 500 viscosity oil.

26-26 AIR DISTRIBUTION

The distribution of air within an automobile body offers several unusual problems:
1. Space is confined.
2. Seats present a restriction to airflow.
3. Low roof increases possibility of drafts.

One method of air distribution is to provide two air duct openings in back of the back seat. From these ducts, conditioned air is blown forward along the roof of the passenger compartment where it mixes (by turbulence) with the air in the car and settles over the occupants. This system may produce noticeable drafts under certain conditions.

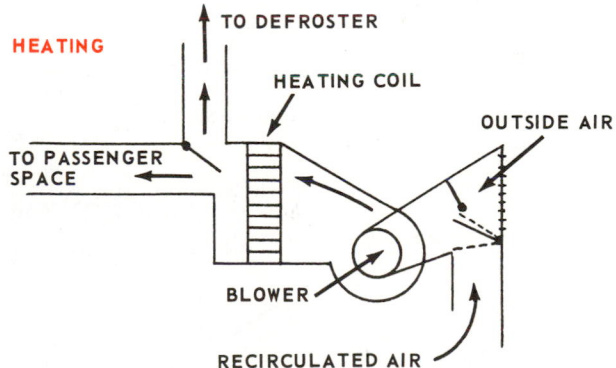

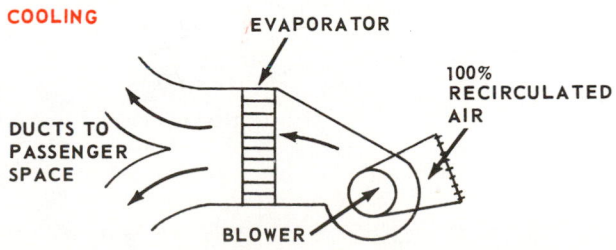

Fig. 26-57. Automobile air conditioning system with separate heating and cooling systems.

A popular method of air distribution has the air conditioning system under the instrument panel on the passenger side of the fire wall (some are on engine side of fire wall). This type of system has an arrangement of ducts to control the airflow.

The main parts of the duct system are dependent on type of system:
1. Separate cooling and heating system.
2. Combination cooling and heating system.

Separate cooling and heating is shown diagrammatically in Fig. 26-57. This popular system has separate blower motors, a separate duct system and separate damper controls. Note in Fig. 26-57 that only the heating system has a fresh air intake and a windshield defrost duct system.

The combination system has the heating coil and the evaporator in the same duct system. It uses only one blower system. Fig. 26-58 shows the principle of operation of a combination system.

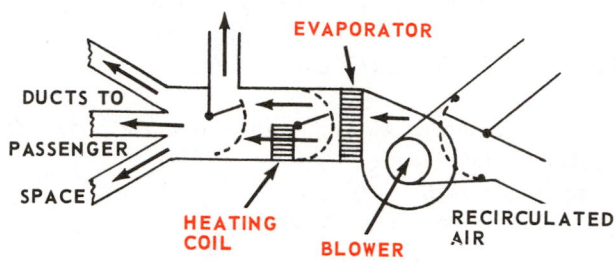

Fig. 26-58. Combination cooling and heating system.

26-27 DUCTS

Heating and cooling ducts are made of metal or plastic. Fairly high air velocities are used, since the noise of air movement is not as critical as it is in an office or residence.

The duct system includes:
1. Fresh air inlet.
2. Return air inlet.
3. Evaporator housing.
4. Drain pan and drain connection.
5. Plenum chamber.
6. Conditioned air outlets.
   a. Defrost.
   b. Deice.
   c. Grilles.
7. Dampers to change airflow.
   a. Manual.
   b. Power operated.

The heater core and evaporator are usually in series (regarding airflow), and the same duct system is used for both systems. The path of the conditioned air is shown in Fig. 26-59.

It is common practice to use a vacuum powered diaphragm to control air dampers (doors) in the ducts. The diaphragm moves the damper in one direction; a spring moves the damper in the other direction. These units are also called servo motors. Fig. 26-60 shows the application of vacuum powered dampers. A vacuum operated power servo motor is shown in Fig. 26-61. The damper diverter valve controls door movement.

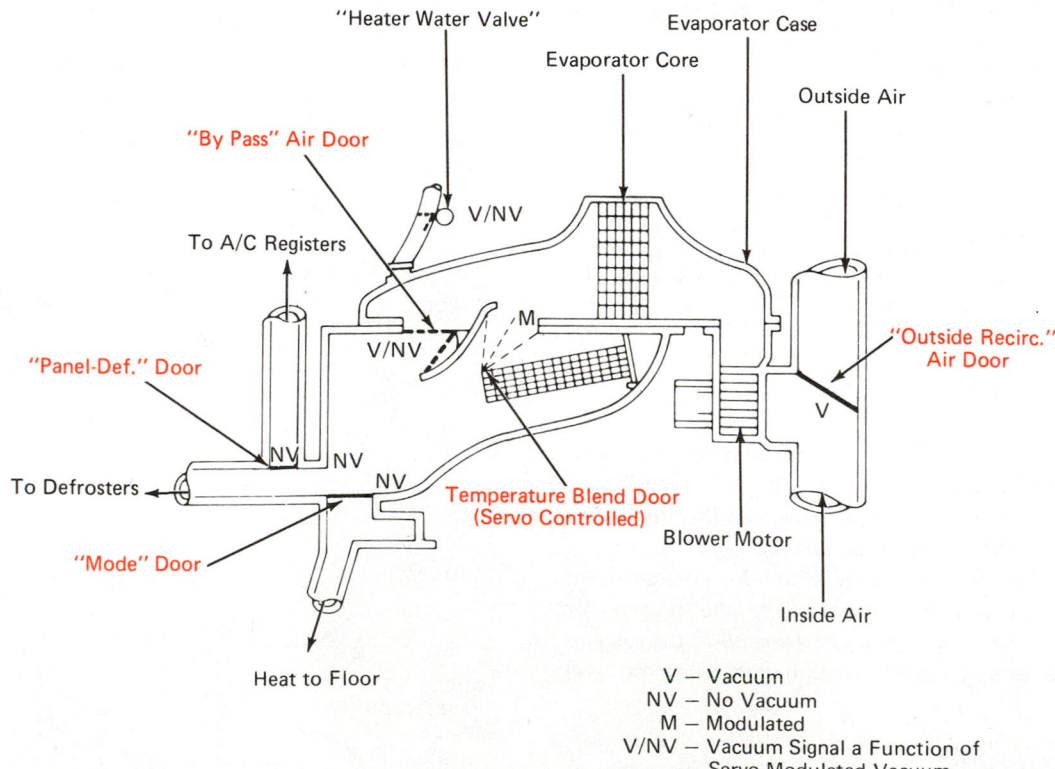

Fig. 26-59. A duct system and plenum chamber. Note use of several doors to control direction of airflow in ducts. (Ford Customer Service Div.)

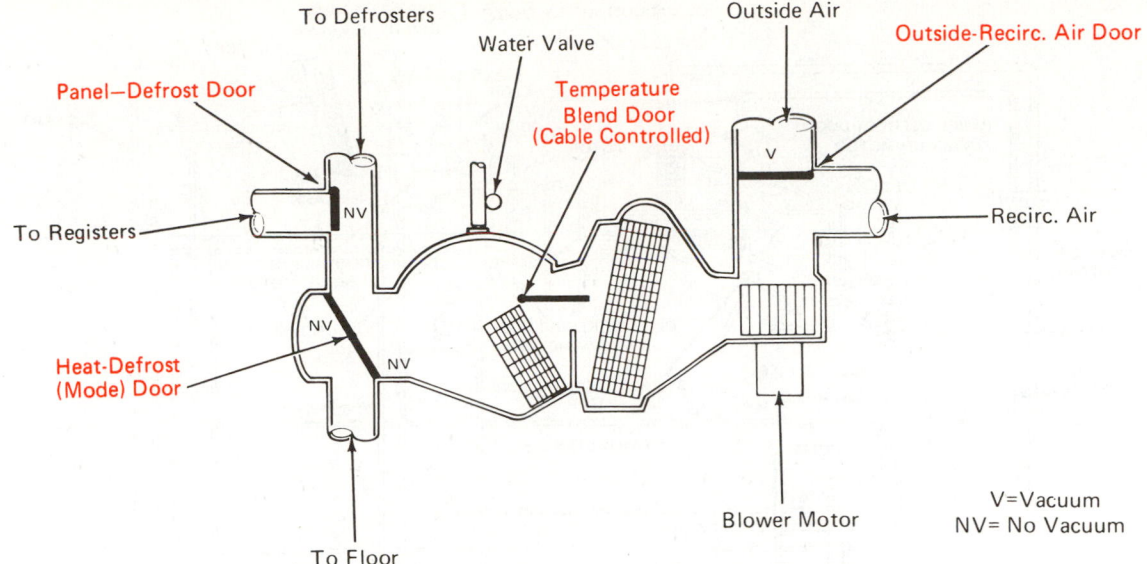

Fig. 26-60. A plenum chamber and duct system showing vacuum operated doors (dampers).

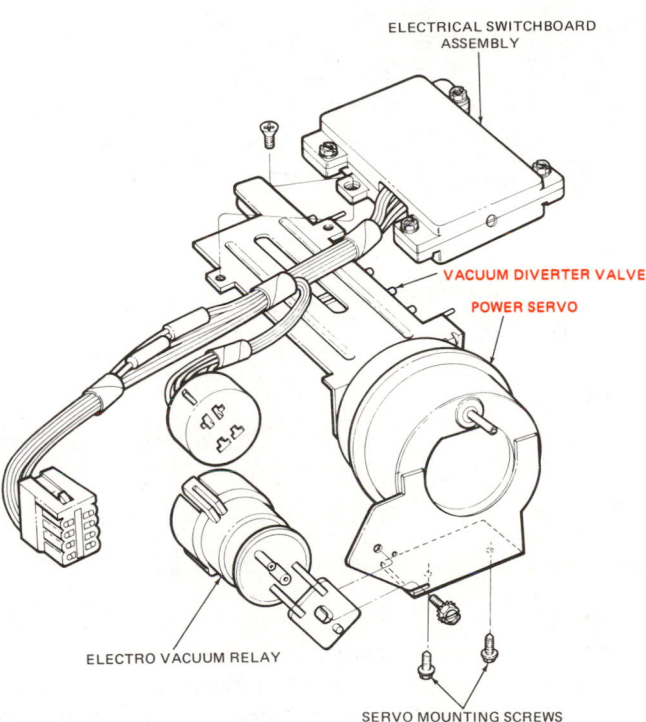

Fig. 26-61. A power servo unit that controls the vacuum diverter valve.

A complete automatic system including ducts and vacuum powered controls is shown in Fig. 26-62. A view of heating airflow, defrost airflow and the damper positions for these actions is detailed in the duct system pictured in Fig. 26-63. A diagrammatic view of a dual air conditioning system for a station wagon style body is shown in Fig. 26-64. Note that this application features a roof-mounted air conditioning unit. The dual system includes two evaporators, two thermostatic expansion valves and an auxiliary blower.

26-28 FANS

Fans used to circulate air in automobile air conditioning applications are radial flow type (squirrel cage or centrifugal). The fans are driven by d-c motors, usually 12V. Generally, the motors are flexibly mounted to reduce noise. They are of several models, including single speed, two speed and three speed.

The fan motors are either single shaft or double shaft, with sealed-in bearings that usually do not require oiling. If a motor develops what appears to be a bearing noise, the trouble may be worn bearings or too much end play. Worn bearings necessitate motor replacement. Too much end play often can be remedied by installing end play washers.

The motors' electrical load can be checked by inserting an ammeter in the line to the blower motors. If the ammeter reading is higher than the specifications in the manufacturer's manual (and if the motor is overheated), the windings may be shorted. Remove the motor, check it again and replace it if faulty. If the reading is lower than the specifications, there may be a loose or dirty connection. Use a voltmeter to check for voltage drops. Correct the cause, when excessive voltage drop is located.

26-29 INSULATION

Most car bodies are insulated. Fiber glass, glass wool and various low K value, nonsettling, flexible insulations are used.

The large amount of window area in an automobile allows considerable heat leakage and also a high sun load. However, the use of tinted glass reduces the radiant heat load considerably. Light colored cars will absorb considerably less radiant heat than dark colored or black cars.

To avoid air conditioning the trunk space, insulation usually is placed over the back of the rear seat of the car. This also reduces the noise level of the blower unit when a trunk

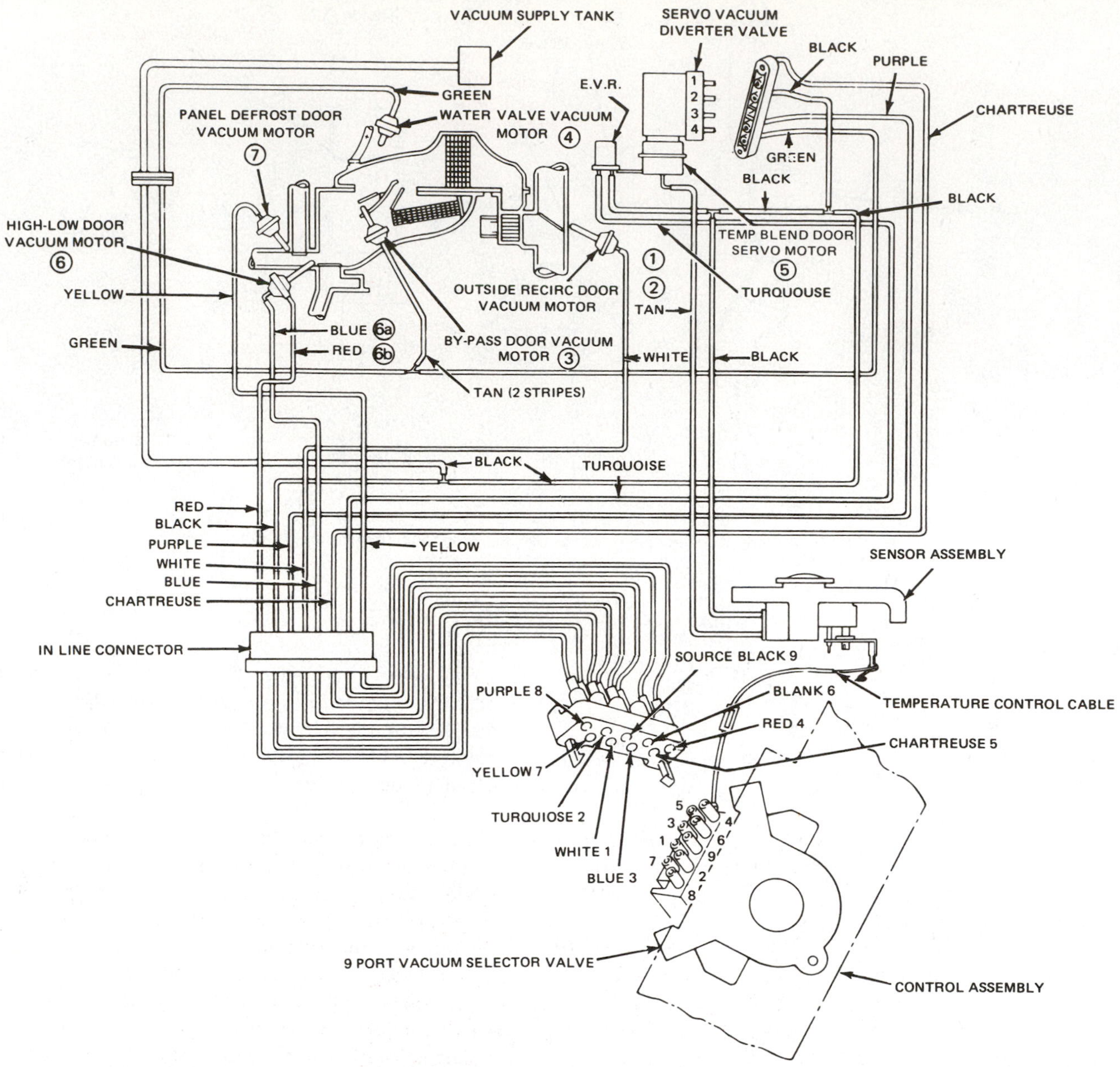

VACUUM SUPPLY TANK

SERVO VACUUM DIVERTER VALVE

BLACK   PURPLE

CHARTREUSE

GREEN

E.V.R.

WATER VALVE VACUUM MOTOR ④

PANEL DEFROST DOOR VACUUM MOTOR ⑦

GREEN

BLACK   BLACK

TEMP BLEND DOOR SERVO MOTOR ⑤

HIGH-LOW DOOR VACUUM MOTOR ⑥

YELLOW

GREEN

BLUE ⑥a

RED ⑥b

OUTSIDE RECIRC DOOR VACUUM MOTOR

① ②   TAN

TURQUOUSE

BY-PASS DOOR VACUUM MOTOR ③

TAN (2 STRIPES)

WHITE   BLACK

BLACK   TURQUOISE

RED
BLACK
PURPLE
WHITE
BLUE
CHARTREUSE

YELLOW

IN LINE CONNECTOR

SENSOR ASSEMBLY

SOURCE BLACK 9

PURPLE 8   BLANK 6

RED 4

TEMPERATURE CONTROL CABLE

CHARTREUSE 5

YELLOW 7

TURQUIOSE 2

WHITE 1

BLUE 3

9 PORT VACUUM SELECTOR VALVE

CONTROL ASSEMBLY

Fig. 26-62. An automatic air conditioning system showing vacuum controls. (Ford Customer Service Div.)

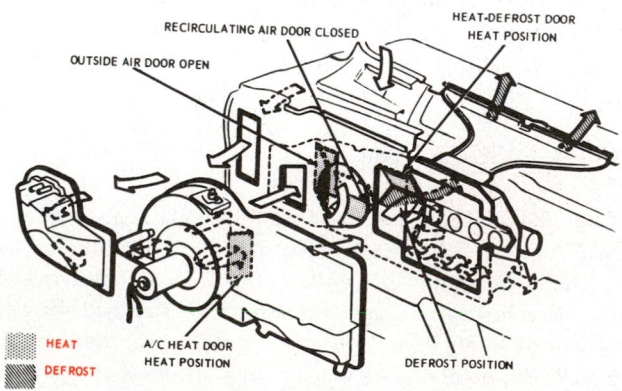

RECIRCULATING AIR DOOR CLOSED

HEAT-DEFROST DOOR HEAT POSITION

OUTSIDE AIR DOOR OPEN

HEAT

DEFROST

A/C HEAT DOOR HEAT POSITION

DEFROST POSITION

Fig. 26-63. Cutaway view of duct system which has blower and coils on engine side of fire wall. Positions of various dampers are shown. (Ford Motor Co.)

mounted evaporator is used.

The body of the car must be tightly sealed at all joints. Door gaskets must be in good condition. In addition to the usual water tightness test, the car body also should be tested for air tightness.

26-30 ELECTRICAL CIRCUITS

The electrical system of the automobile air conditioner is unique in that a single wire system is used. The frame of the car serves as the return wire or ground.

The electrical system varies with type of air conditioning system:

1. Separate system.
2. Combination system.
3. Full automatic control (fixed dry bulb setting).

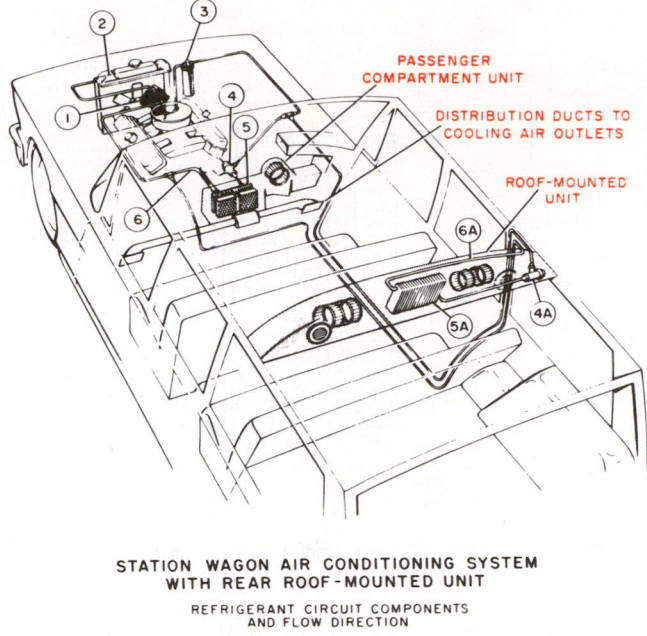

**STATION WAGON AIR CONDITIONING SYSTEM
WITH REAR ROOF-MOUNTED UNIT**

REFRIGERANT CIRCUIT COMPONENTS
AND FLOW DIRECTION

COMPRESSOR ① TO CONDENSER ②
CONDENSER TO RECEIVER DRIER ③
RECEIVER DRIER TO EXPANSION VALVES ④ AND ④A
EXPANSION VALVES TO EVAPORATORS ⑤ AND ⑤A
RETURN LINES ⑥ AND ⑥A FROM EVAPORATORS
TO COMPRESSOR

Fig. 26-64. Complete air conditioning system, using a roof-mounted duct and grille system.

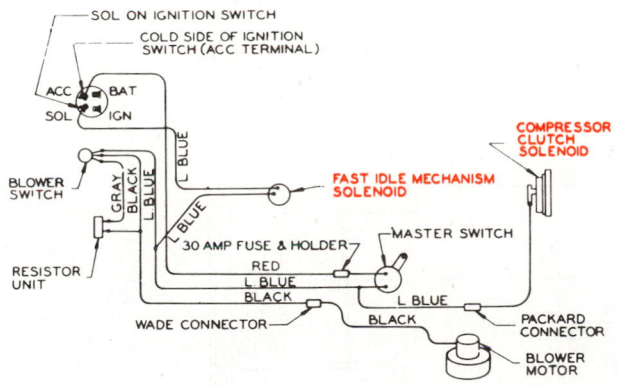

Fig. 26-65. Wiring diagram of automobile air conditioning unit that uses a fast idle solenoid and a compressor pulley solenoid clutch.

The main parts of the electrical system are:
1. Blower or blowers.
2. Blower switch.
3. Thermostat.
4. Electromagnetic clutch.

A typical wiring diagram is illustrated in Fig. 26-65. A two-blower system is shown in Fig. 26-66. The electrical connections must be kept in good condition, because the voltages are low (6 to 8V or 12 to 14V).

The system shown in Fig. 26-65 uses a fast idle solenoid to permit sufficient comfort cooling when the car is parked or waiting for traffic conditions to clear. The solenoid valve usually opens a vacuum control that automatically adjusts

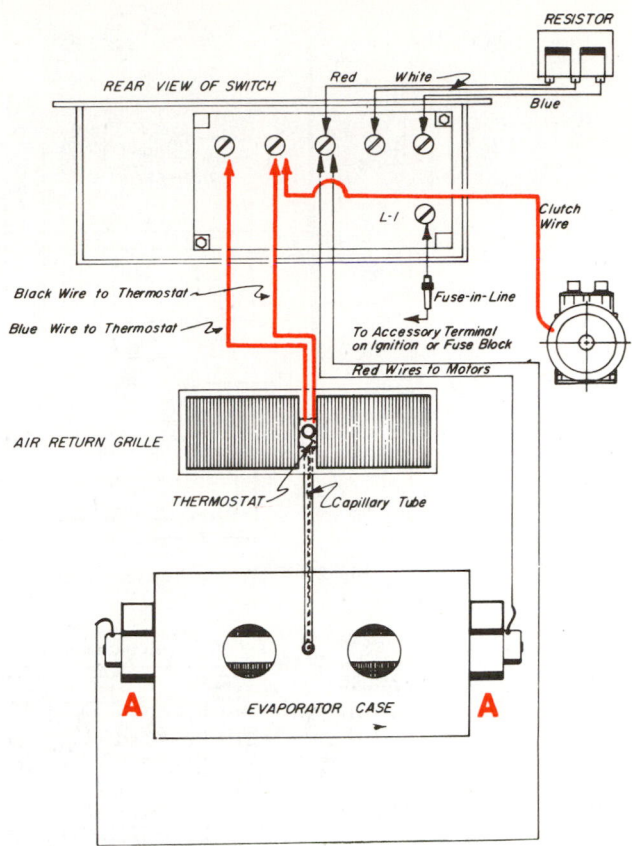

Fig. 26-66. Wiring diagram of "add on" unit with the evaporator mounted in the car trunk. Note thermostat and clutch circuit in red. A—Evaporator blower motors.   (Climatic Air Sales, Inc.)

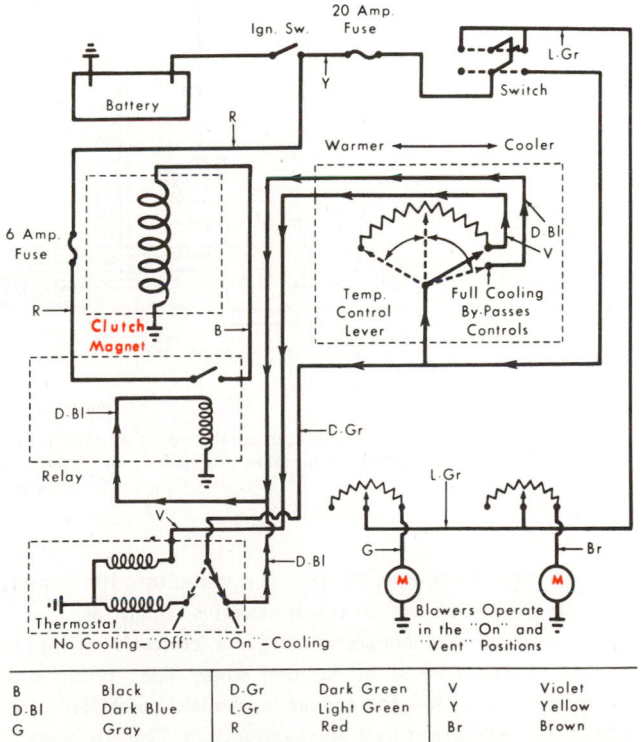

B	Black	D-Gr	Dark Green	V	Violet
D-Bl	Dark Blue	L-Gr	Light Green	Y	Yellow
G	Gray	R	Red	Br	Brown

Fig. 26-67. Wiring diagram of automobile air conditioning system. This unit has a magnetic clutch, two blower motors with variable speed rheostats, a thermostat and a manually operated rheostat in the thermostat circuit.

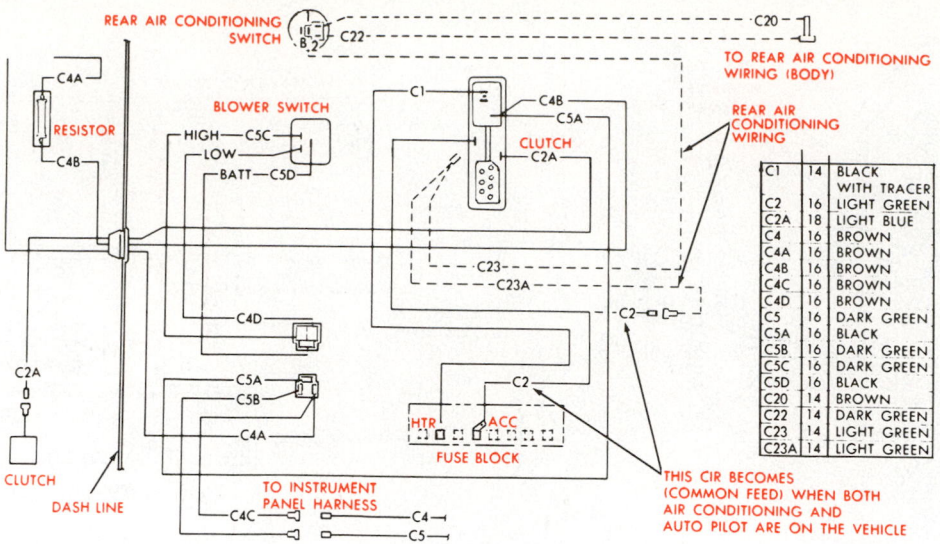

Fig. 26-68. Wiring diagram of station wagon air conditioning system. (Chrysler Corp.)

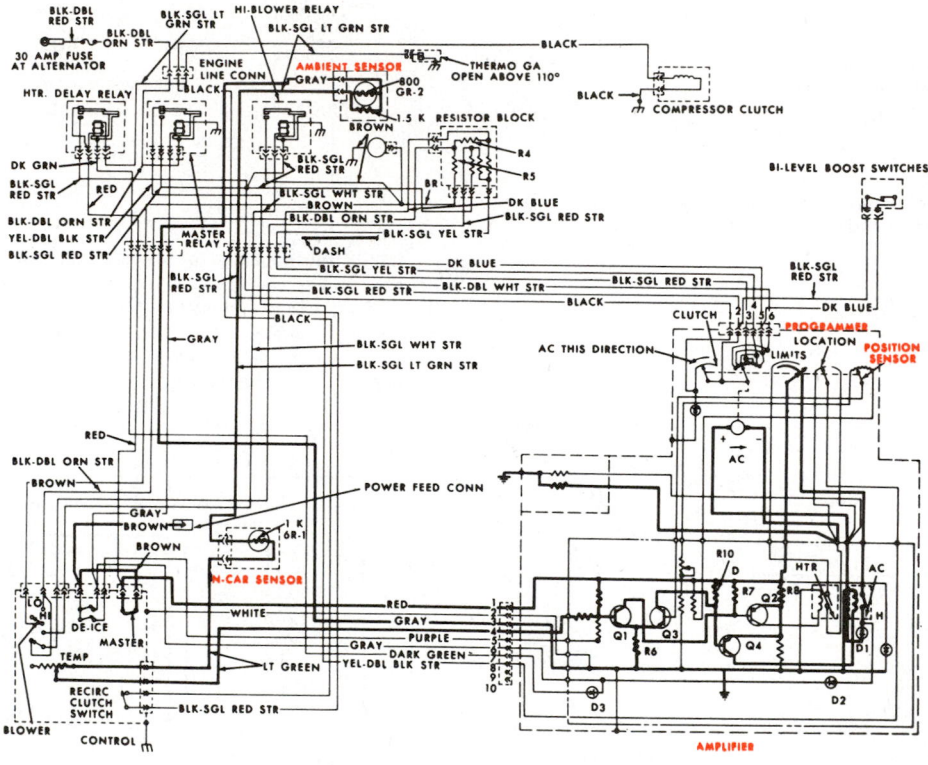

Fig. 26-69. Electrical-electronic system for fully automatic, year-round air conditioner. This system uses temperature sensors, amplifier and programmer to help obtain complete sequence of automatic operation.

engine idle speed to about 750 rpm. It is wired into the system to function only when the car transmission is in neutral.

The refrigerating mechanism and its controls maintain evaporator temperature within certain limits. The amount and degree of cooling within the car are controlled by the amount or rate of air circulation over the evaporators. The air speed is controlled by the speed of the blowers. Some blowers have two or more speed steps; others are controlled by a variable rheostat and a great range of speeds may be provided.

Fig. 26-67 shows a wiring diagram for an automobile air conditioning system that uses a magnetic clutch and two blower motors with variable speed rheostat. The thermostat circuit includes a manually operated rheostat.

Combination systems have more elaborate electrical wiring systems and components. See Fig. 26-68.

A wiring system using sensors, an amplifier and a programmer is detailed in Fig. 26-69. A typical fully automatic system has an electrical system shown in Fig. 26-70.

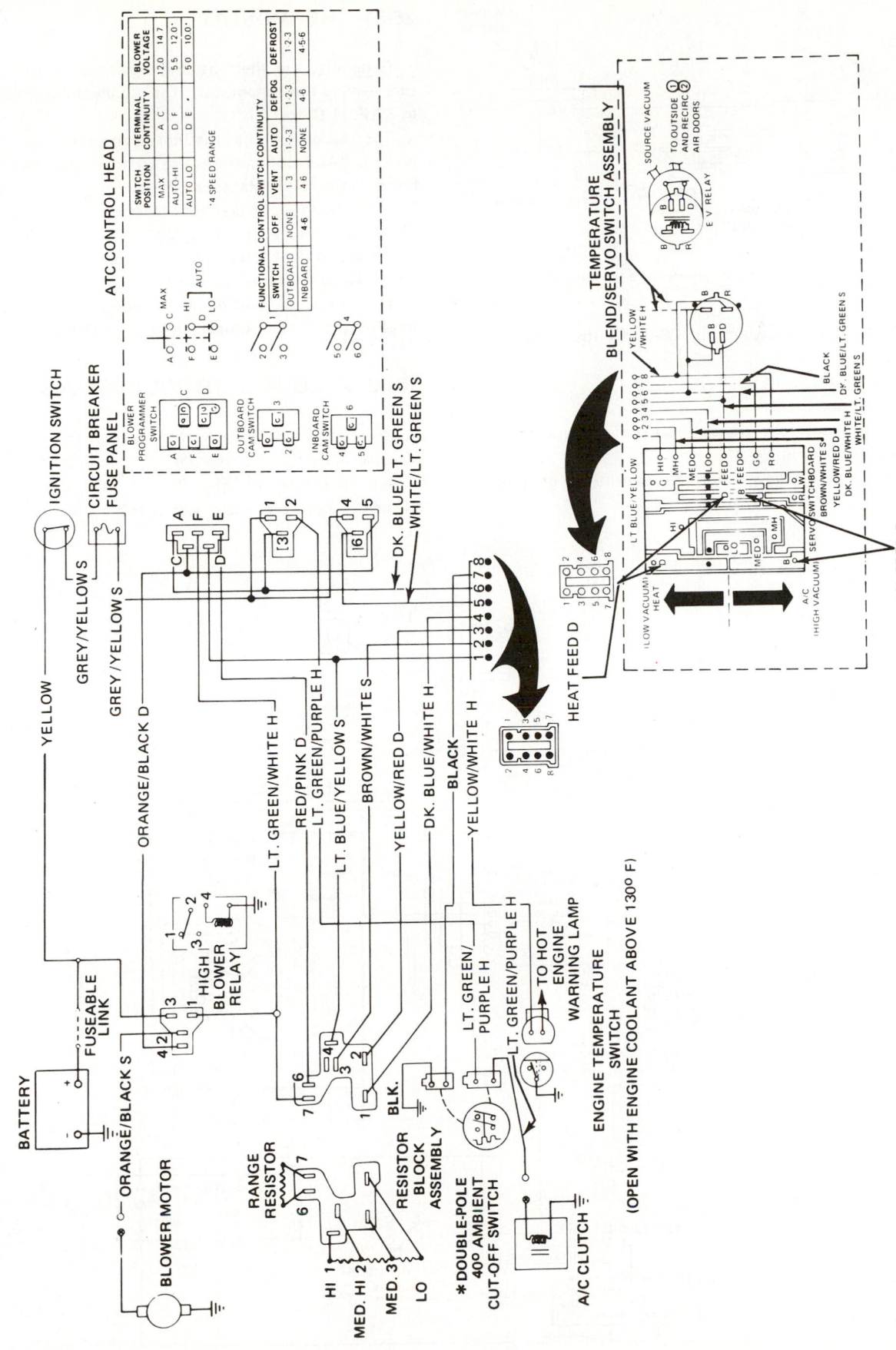

Fig. 26-70. Wiring diagram of an automatic automotive air conditioning system. It has some operation checking information. (Ford Customer Service Div.)

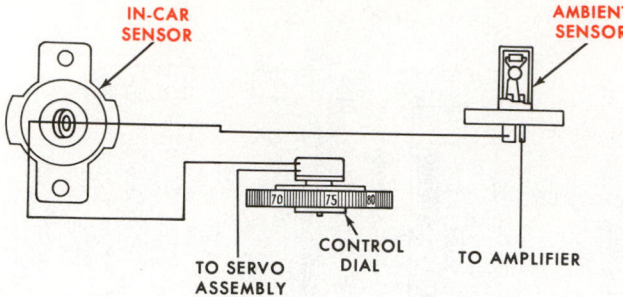

Fig. 26-71. Construction of temperature-sensitive sensors.

The three sensors shown in Fig. 26-69 are thermistors:
1. Outside air.
2. Car temperature.
3. Position sensor.

The amplifier uses four transistors. The appearance of the sensors is shown in Fig. 26-71.

Another type of year-round air conditioning system is shown in Fig. 26-72.

26-31 THERMOSTAT

Generally, the thermostat is operated by the air temperature leaving the evaporator. It is usually adjustable from 34 F. to 57 F. (1 C. to 14 C.).

The design of the thermostat depends on the type of system. "Add on" systems use an adjustable thermostat with the sensitive bulb placed at the air outlet of the evaporator. The thermostat controls the magnetic coil in the clutch pulley.

In the combined systems, the same thermostat may also be connected to the heater water flow control. Fig. 26-73 shows a solid state temperature control.

The completely automatic year-round systems have sensors, amplifiers, and programmers in place of the thermostat.

26-32 VACUUM CONTROL SYSTEMS

Many air conditioning systems use a vacuum power system (vacuum actuators) to operate dampers and valves. A vacuum system is shown in Fig. 26-74. The manual controls or automatic controls select which tubing will be subjected to

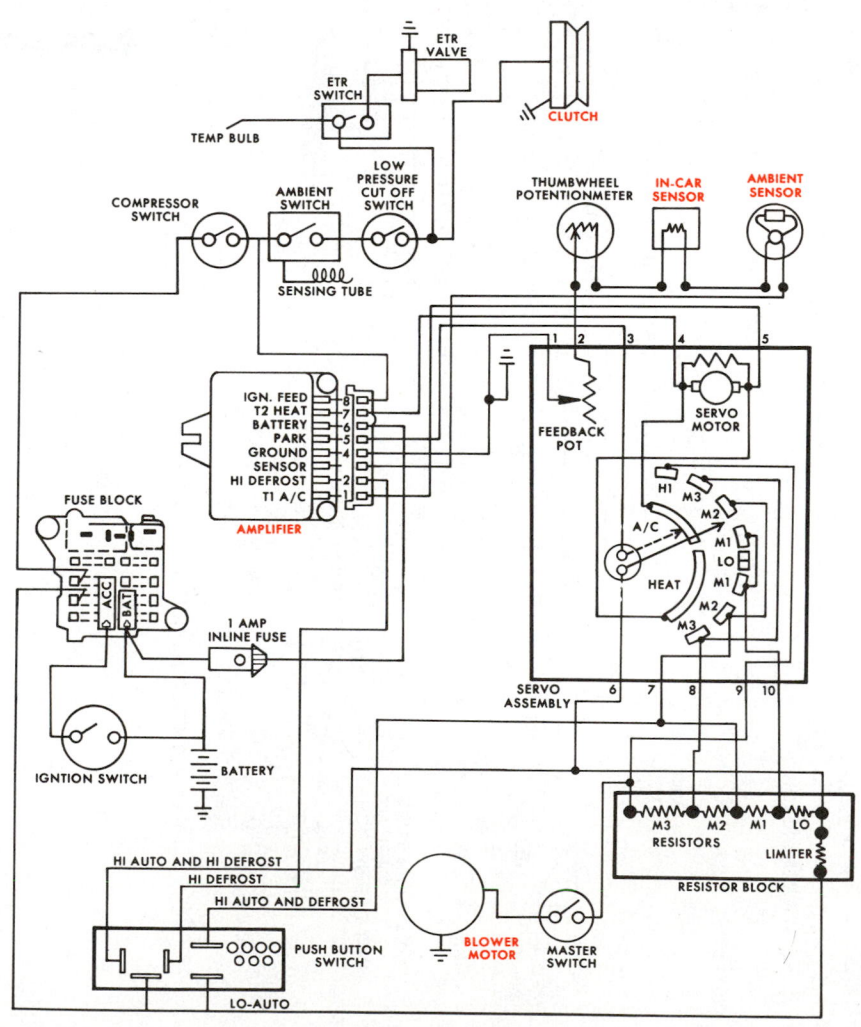

Fig. 26-72. A year-round air conditioning system wiring diagram. (Chrysler Corp.)

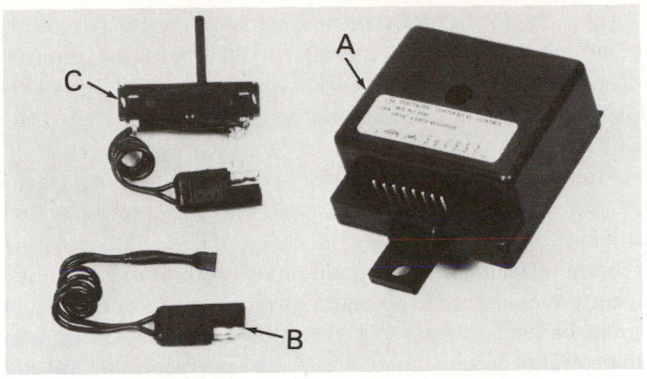

Fig. 26-73. Solid state temperature control. A—Control module. B—Thermistor probe. C—Potentiometer. (Eaton Corp.)

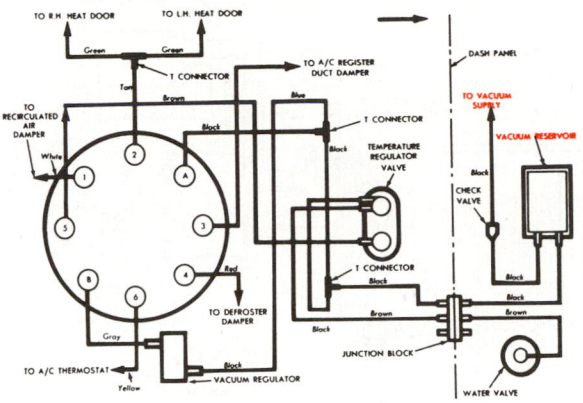

Fig. 26-74. Vacuum piping and vacuum actuators used in automotive air conditioning system. (Ford Motor Co.)

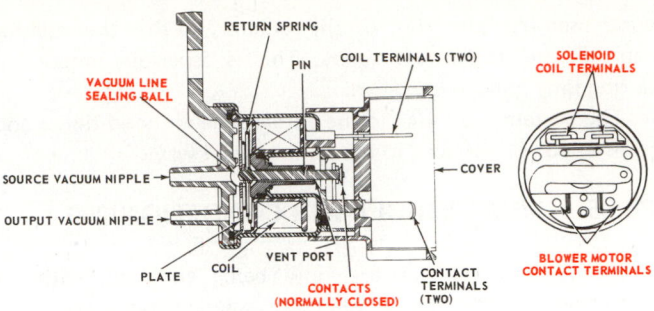

Fig. 26-75. An electro-vacuum relay. When solenoid coil is energized, it opens circuit to blower motors. It also closes vacuum line to fresh air door, causing door to close. (Ford Customer Service Div.)

vacuum. The diaphragms react to the vacuum and move against spring pressure to move dampers or to open or close valves. The springs return the mechanism to its former position when vacuum is released.

The vacuum source for the vacuum power system is the engine intake manifold, which operates at from 5 to 20″ Hg. (127 mm to 508 mm). Generally, this system is designed to operate at 5″ Hg. (127 mm).

An electro-vacuum relay is shown in Fig. 26-75. Its solenoid coil operates a vacuum line valve and also controls blower operation. During cold engine operation, it is used to stop blower operation by opening the contacts. The solenoid also closes the fresh air door by using the sealing ball to close the vacuum line. During maximum heat load, it closes the fresh air door to provide all recirculating air.

To check for leaks in the vacuum control system, connect a vacuum gauge into the system. Run the engine until the system is at 16″ Hg. (406 mm). Stop the engine. If the vacuum gauge starts to creep back toward zero (atmospheric), there is a leak in the system, or the check valve in the line to the intake manifold is leaking.

To check actuator operation, intall the vacuum gauge and run the engine until the vacuum is 16″ Hg. (406 mm). Stop the engine and watch the vacuum gauge as each actuator is operated. The vacuum should creep a little toward zero as each actuator is valved into operation. If the vacuum does not decrease:

1. That actuator has a pinched tube.
2. The line is plugged.
3. The damper is binding.

The dampers usually have an adjustment for full closing or opening.

Vacuum control systems have a vacuum reservoir tank so that the dampers will keep working during engine shutoff intervals.

26-33 BUS AIR CONDITIONING

The air conditioning of buses has progressed rapidly. Due to the large size of the unit, most bus air conditioning systems use a separate gasoline engine with an automatic starting device to drive the compressor. The system is standard in construction, except for the condensing unit. It is made as compact as possible and generally is installed in the bus so that it can be easily reached for servicing.

Condensing units are often mounted on rails with flexible suction and liquid lines to permit sliding the condensing unit out of the bus body to aid in servicing.

Air-cooled condensers are used. Thermostatic expansion valve refrigerant controls are standard. Finned blower evaporators are used.

The duct system usually runs between a false ceiling and the roof of the bus. The ducts, usually one on each side of the bus, have grilles at the passenger seats. Grille opening and closing may be contolled by the passengers.

26-34 TRUCK AIR CONDITIONING

The cabs of many truck tractors and long distance hauling trucks, farm tractor cabs and earth mover cabs are air conditioned. Most of this equipment is of the "hang on" type and is installed after the cab has been made.

Some truck air conditioning units have two evaporators, one for the cab and one for the relief driver's quarters back of the driver. Some systems use a remote condenser mounted on the roof of the cab. This type of installation removes the condenser from in front of the radiator, so that the radiator can operate at full efficiency. This is especially important during long pulls in low gear.

The system is similar to the automobile air conditioner and is installed and serviced in the same general way.

26-35 FARM VEHICLE AIR CONDITIONING

Many farm vehicles are now being equipped with air conditioned cabs. Such vehicles include combines, corn pickers, cotton pickers, hay balers, tractors and windrowers.

Air conditioners used are of many designs. One special feature is a better fresh air inlet filtering system to keep out dust. About 12 percent of the farm vehicles are now air conditioned, and the number is increasing rapidly.

26-36 INSTALLING A "HANG-ON" SYSTEM

Before installing a "hang-on" unit, check to determine:
1. If the unit is the correct capacity.
2. If the unit is designed to fit that model car.

The installation manual should give these specifics.

When installing the evaporator, wear goggles while drilling. Be sure the drill does not contact electrical wires, vacuum lines or fluid lines. Careless drilling may damage car parts and even cause fires.

The refrigerant lines furnished are clean and capped. Remove the caps and/or plugs just before installing them to keep the lines as clean as possible. Put clean and dry (no moisture content) refrigerant oil on the fittings before installing the lines.

Refrigerant line sizes usually are:
1. From compressor to condenser — 1/2 in.
2. From condenser to receiver — 3/8 in.
3. From receiver to evaporator — 3/8 in.
4. From evaporator to compressor — 5/8 in.

Remember, too, that obtaining good tools and using them properly are important factors in efficient installation.

When mounting the lines and tightening the fittings, use two wrenches to prevent twisting the lines. The flexible lines must be loose enough to absorb vibration and engine movement. In addition, they must be mounted to prevent scraping or touching hot parts of the engine (exhaust manifold). Always use two (and sometimes three) wrenches when tightening screw-on type refrigerant lines.

The electrical wiring must be carefully installed to insure against scraping the wire (this may cause a fire) and to insure clean and tight electrical connections.

Air in the system causes above normal condensing pressures as the air collects in the condenser (air is noncondensable). Air also contains oxygen, which will oxidize the oil in the system and form other oxides.

Water must be kept out of the system, or it will form a sludge with the oil. Also, it may collect and freeze at the expansion valve orifice, possibly clogging it.

Oil in the compressor must be up to the correct level. A normal charge is 3 to 7 oz. (85 to 200 cm^3). Be sure to check the manufacturer's specifications for the correct oil charge.

The compressor must be installed with care. It must be firmly mounted. The compressor pulley must be carefully aligned with the idler pulley and engine pulleys. Install the drive belt and adjust the belt tightener to 100 to 140 lb. (45.4 to 63.5 kg) tension.

After installing the evaporator, evaporator housing, electrical lines and refrigerant lines, seal the body openings with caulking compound.

After installing an air conditioning system on a car, use a separate vacuum pump to pump down the system. The system should be held under a high vacuum 1 to 24 hours at room temperature before charging. A gauge manifold, vacuum pump, charging cylinder and connecting refrigerant lines are used. Fig. 26-76 shows a vacuum pump installation.

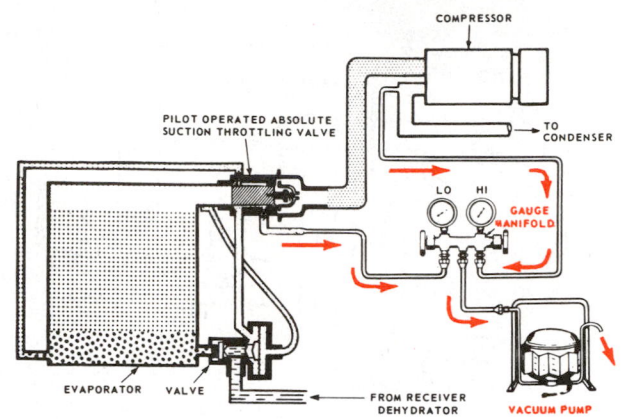

Fig. 26-76. Vacuum pump installation. Both hand valves on gauge manifold are open to insure evacuating both high-pressure and low-pressure sides of system.

Refer to the manufacturer's service manual for the correct amount of refrigerant to be used in charging a particular system. The correct charge varies from 1.25 to approximately 5 lb. (.6 to 2.3 kg). Use gauges to check operating pressures. In most cases, low head pressures indicate lack of refrigerant. Fig. 26-77 shows a system being charged. The failure of the compressor shaft seal is probably the greatest cause of loss of refrigerant. A blown seal usually is indicated by a leakage of oil around the shaft. Check for leaks each time the unit is serviced.

The outside of the system should be clean, especially condenser and evaporator surfaces. Use brushes with detergent and water. A vacuum cleaner will remove some of the lint and dirt. *Pressurized air can be used, but wear goggles. Use great care when working with nitrogen.* See Para. 11-39.

The fins must be clean and straight. Use a fin comb to straighten them.

26-37 SERVICING AUTOMOBILE AIR CONDITIONERS

Servicing the automobile air conditioner is about the same as servicing standard air conditioning systems and commercial systems. Study Chapters 11 and 14 carefully before trying any

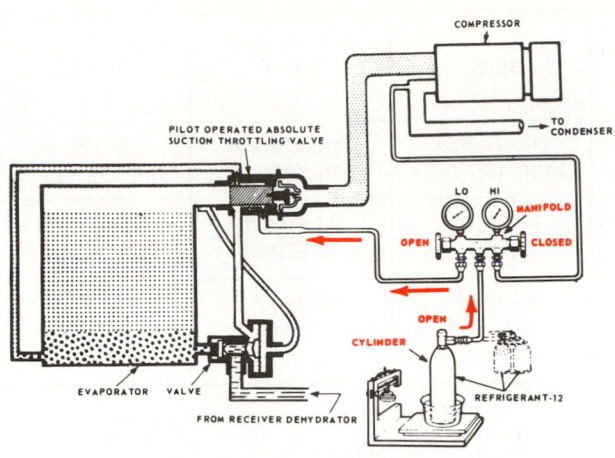

Fig. 26-77. Diagram shows how refrigerant cylinder is connected to air conditioning unit, using gauge and service manifold. Dotted lines indicate use of disposable refrigerant containers. (Pontiac Motor Div., General Motors Corp.)

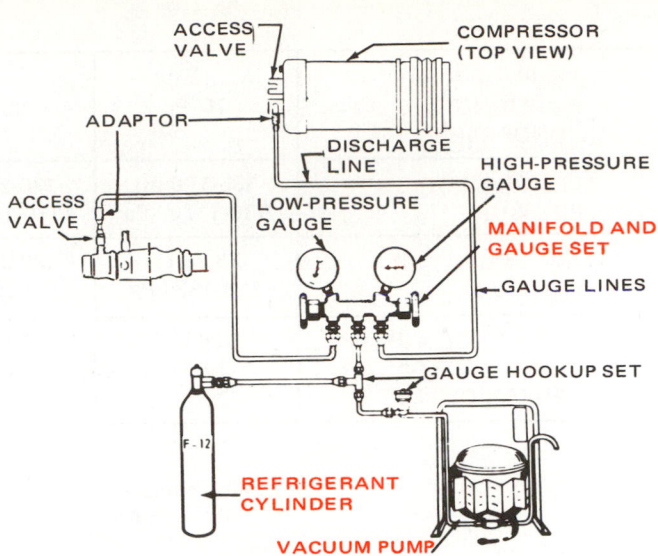

Fig. 26-79. Gauge and service manifold installed in General Motors' system. Both charging cylinder and vacuum pump are connected to middle opening of manifold.

work on an automobile refrigerating system.

Servicing usually begins with a customer's complaint or during an annual check of the system. Owners' complaints received most are:

1. No cooling.
2. Noise.
3. Cooling intermittently.
4. Vibration.

There may be several causes for each complaint. Check the system thoroughly to find the correct cause.

The method of installing the gauge manifold is similar to the procedure described in Chapters 11 and 14. Shut off the engine when installing gauges to avoid injuries. Always clean the connections before removing any caps or plugs.

Fig. 26-78 pictures a set of automotive air conditioning

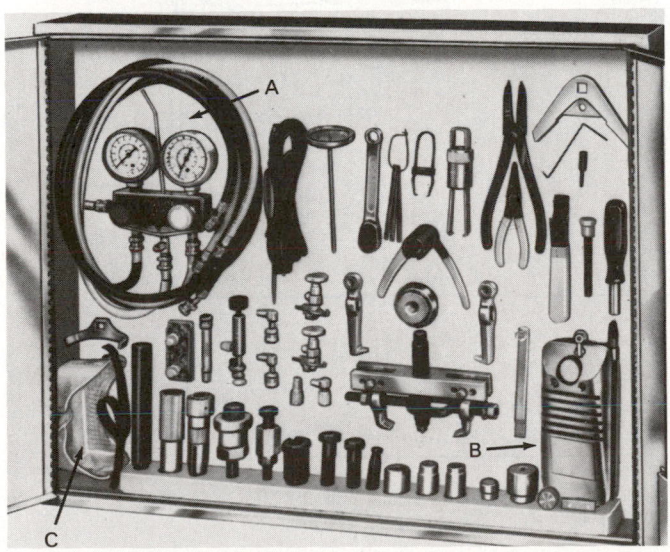

Fig. 26-78. An automotive air conditioning system tool cabinet and tools. A—Service gauges and lines. B—Leak detector. C—Safety goggles. (Snap-on Tools Corp.)

tools. Fig. 26-79 shows a gauge manifold installed.

For personal protection when removing access type valves or lines, use a pad made from several layers of cloth. Liquid refrigerants will freeze the skin or eyes. The gauge manifold and lines must be clean and dry, both inside and outside.

For personal safety, always wear goggles when working on the refrigeration unit or handling service cylinders or disposable refrigerant containers. Avoid breathing any escaped refrigerant. Although the refrigerant is virtually harmless, the fact that it excludes oxygen makes it dangerous.

Always attach a gauge manifold to the automobile air conditioning system before trying to service it. Never use a manifold set that has been left open to the air until after the manifold and lines have been cleaned and dried. Some compressors are fitted with gauge openings at both the suction service valve and the discharge service valve. The service valve stems on some compressors must be back seated to seal the gauge openings. Some compressors use the valve core method, with Schrader or Dill cores.

Before servicing an automobile air conditioning system, know what performance to expect from the unit. Fig. 26-80 is a table of operating conditions of a unit at various temperatures. Abnormal low-side and/or high-side pressures or noise will signal the need for service.

Altitude affects the operation of an air conditioning system. Changing pressure affects vacuum actuators and some suction pressure controls, particularly if any part of the bellows or diaphragm movement is exposed to the pressure of the atmosphere. Fig. 26-81 shows the steps necessary to check a system, to charge refrigerant, to purge the receiver and to evacuate the system.

26-38 ADJUSTING AND REPLACING BELTS

An engine having several belt-driven accessories may be equipped with three or four belts. The various belt tensions are adjusted by moving the generator, power steering pump and

TEMP. OF AIR ENTERING THE CONDENSER	70°F.	80°F.	90°F.	100°F.	110°F.
*COMPRESSOR OUT PRESSURE	135-170 psi 930-1170 kPa	170-205 psi 1170-1410 kPa	205-245 psi 1410-1690 kPa	245-285 psi 1690-1960 kPa	285-320 psi 1960-2100 kPa
*ACCUMULATOR PRESSURE	23-28 psi 158-193 kPa	23-29 psi 158-200 kPa	26-35 psi 185-240 kPa	31-40 psi 215-275 kPa	35-44 psi 185-305 kPa
DISCHARGE AIR TEMP. — LEFT CENTER OUTLET	36-43° F. 1-6° C.	36-43° F. 1-6° C.	36-43° F. 1-6° C.	42-48° F. 5-9° C.	48-53° F. 9-11° C.

Fig. 26-80. Performance chart for a unit operating at various temperatures.
(Buick Motor Div., General Motors Corp.)

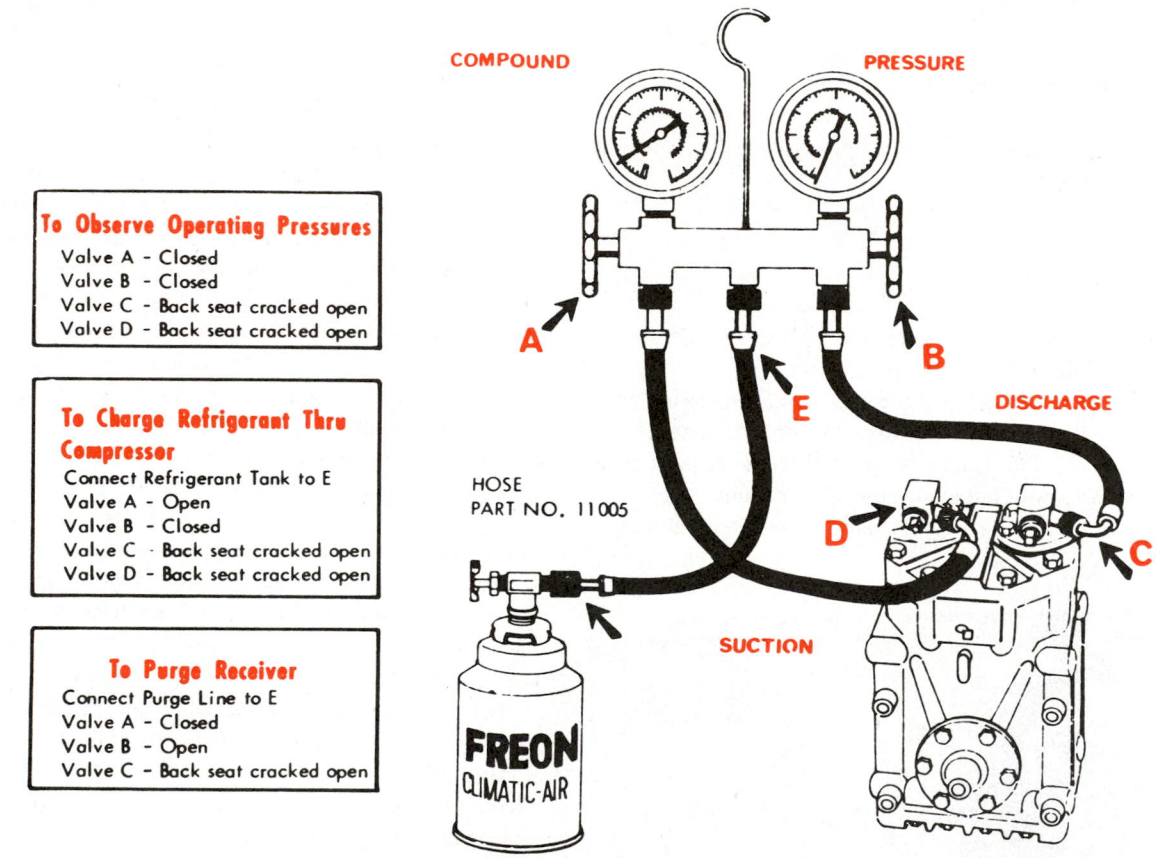

To Observe Operating Pressures
Valve A – Closed
Valve B – Closed
Valve C – Back seat cracked open
Valve D – Back seat cracked open

To Charge Refrigerant Thru Compressor
Connect Refrigerant Tank to E
Valve A – Open
Valve B – Closed
Valve C – Back seat cracked open
Valve D – Back seat cracked open

To Purge Receiver
Connect Purge Line to E
Valve A – Closed
Valve B – Open
Valve C – Back seat cracked open

HOSE PART NO. 11005

FREON CLIMATIC-AIR

TO EVACUATE (REMOVE AIR) SYSTEM
Valve A – Open ..... Valve B – Closed ..... Valve C – Open, back-seated ..... Valve D – Mid position
Connect Hoses: (1) from Valve A to Valve D, (2) from Valve B to Valve C, (3) from E to Vacuum Pump.
Pump vacuum, Close Valve A. Remove hose from vacuum pump and attach to refrigerant tank.
Open Valve A. Open refrigerant tank to break vacuum, and use Step 2 for charging system.

NOTE: CHECK EQUIPMENT MANUFACTURER'S CATALOG OR INSTRUCTION SHEET FOR SPECIFIC RECOMMENDATIONS ON REFRIGERANT CHARGE, OIL CHARGE AND SERVICE PROCEDURES FOR ANY PARTICULAR PIECE OF EQUIPMENT.

Fig. 26-81. Procedures to follow for observing operating pressures, method of charging a system, steps necessary to purge receiver and procedure for evacuating air. (Climatic Air Sales, Inc.)

the idler pulley. Belts stretch in use. They should be periodically checked for tightness and adjusted to approximately 75 lb. (34 kg) tension.

A loose belt will soon fail. Pulleys fail, too, due to wear caused by belt slippage. The belt should be dry. Remove excessive oil from the belt and the pulleys. A shrill squeal when engine speed is increased will indicate loose belts or glazed belt surfaces. A few drops of penetrating oil on a glazed belt may stop the squeal temporarily. However, it is best to replace the belt.

Compressor belts with any sign of cracks or frayed edges should be replaced.

Always remember that a "short life" belt or a broken belt may be the result of an unusual overload (excessive pressures), pulley out of line, wrong type belt or incorrect tension. Determine the cause of the failure and remedy it.

Always loosen a belt before removing it. Forcing a belt over a pulley may injure and weaken it.

If the air conditioning system has double belts, always replace both belts if one fails (lengths of used and new belts are different).

26-39 TESTING FOR LEAKS

Check for refrigerant leaks by using a trace chemical, halide torch, electronic leak detector, foam leak detector (soap bubbles) or pressure rise method. These leak testing techniques are described in Chapters 11 and 14.

Some technicians put reddish dye in the refrigerant in the system. Then, if the system has a refrigerant leak, red dye on the metal surfaces will reveal the source.

Most frequently, leaks are pinpointed by using the halide torch leak detector. This detector will locate a leak so small that it would result in the loss of only 1 lb. (.45 kg) of refrigerant in about 14 years. In using a halide torch, an exploring tube end sniffer is placed near the joint being checked. If there is a leak, some escaping refrigerant is drawn up the tube where it passes over a propane or acetylene heated copper element. If there is refrigerant gas in the air sample, the flame will turn green in color.

The electronic leak detector is a sensitive device designed to locate very small refrigerant leaks. It will detect a leak that would cause the system to lose a pound of refrigerant in about 40 years.

A service technician can check a system for leaks at the time vacuum is being drawn on the system. With the vacuum pump running, shut off the vacuum valve on the manifold. If the vacuum gauge needle starts to creep back toward 0 (atmospheric pressure), there is a leak in the system. This leak must be located and corrected before completing the vacuum operation for drying out the system. Leak detection is done by pressuring the system with R-12, then using the torch type leak detector, electronic leak detector or foam leak detector (soap bubbles).

26-40 TESTING COMPRESSOR

The compressor must pump efficiently. If the capacity of the compressor decreases (worn valves or worn rings), maxi-

mum cooling effect will not be obtained.

A compressor should pump vacuum to 15" Hg. (381 mm) in a short time against normal head pressure. If this cannot be done, the pistons, rings or intake valves are leaking (worn).

A good compressor should quickly pump a 200 psi (14.2 kg/cm^2) head pressure with the discharge service valve closed (turned in), and it should hold this pressure after the compressor is stopped. Be careful. Do this pumping only with the cranking motor. Do not run the engine or the pressure will rise to dangerous levels too quickly. Another safer method is to allow the system to run. Stop the engine, then turn the discharge service valve all the way in. If the head pressure drops, the exhaust valve in the compressor is leaking.

R-12 is used for testing for leaks; gauges are used to test the pumping capacity and valve condition of the compressor.

Note: Never run a compressor unless it has the correct amount of clean refrigerant oil in it.

Some late model York and Tecumseh compressors have a cylindrical screen installed in the compressor body under the suction line service valve mounting. This screen removes foreign particles such as dirt, sand and metal chips to prevent damage to the compressor. It should be removed and cleaned each time the refrigerant is removed from the system. If the screen is blocked (clogged), or almost completely blocked, the system will not refrigerate. In addition, the compressor crankcase would be under a vacuum, low-side pressure would be above normal and high-side pressure would be below normal (very little or no refrigerant would circulate).

26-41 ADDING OIL TO SYSTEM

The compressor must have the correct amount of oil of the right specificiation. Usually 500 viscosity oil with additive is used. It must be clean and dry (no moisture content). Too much oil will cause continuous oil pumping, reducing the efficiency of the system and possibly causing damage to the compressor valves. Too little oil will cause rapid wear of the compressor bearings, pistons, rings and valves. It will also cause scoring of the shaft seal.

Therefore, it is important to check the oil level each time a unit is serviced.

Proceed as follows: Install the gauge manifold, purge the lines, turn the suction service valve all the way in and run the compressor until the compound (low-side) gauge reads 0 psi. Then, turn the discharge service valve all the way in, remove the oil level plug and check the oil level. If the oil is too high, it will drain out. If too low, add oil by syphoning it into the compressor with the compressor crankcase at a partial vacuum. Use either a vacuum pump or the compressor to create this vacuum.

Some compressors must be removed from the system to check the oil level.

26-42 CHARGING THE SYSTEM

When the system is operating normally, there will be no bubbles in the sight glass after the system has run for a few minutes. If the system is short of refrigerant, bubbles will

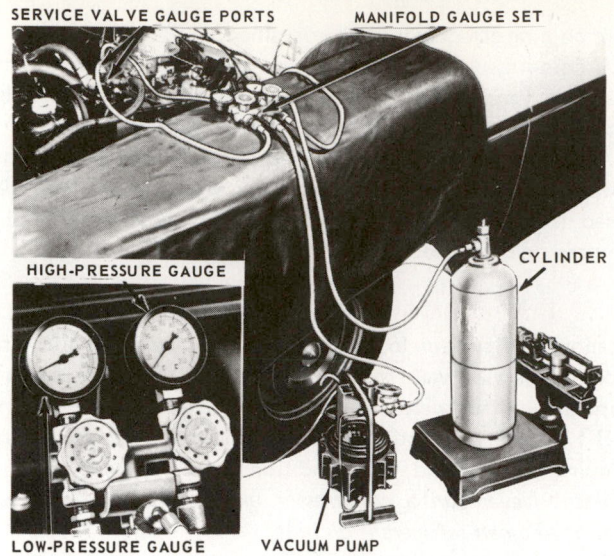

Fig. 26-82. Gauge and service manifold is connected to the system, to the vacuum pump and to the charging cylinder. Note that fender is protected by a cover. (Ford Motor Co.)

appear regularly in the sight glass. A system without any refrigerant or very little refrigerant may not have enough liquid to form bubble formations.

An overcharged unit may be detected by excessive head pressure, but this condition will not show in the sight glass. If high head pressure is shown on the high-side gauge, determine the correct system pressure and purge the excessive refrigerant.

The system is charged by means of the service manifold. Connect the manifold to the system, then connect the charging cylinder to the manifold and purge the lines. See Fig. 26-82.

If only a small amount of refrigerant is to be added to the system, charge through the low side of the system with the cylinder upright. If the complete charge of the cylinder is to

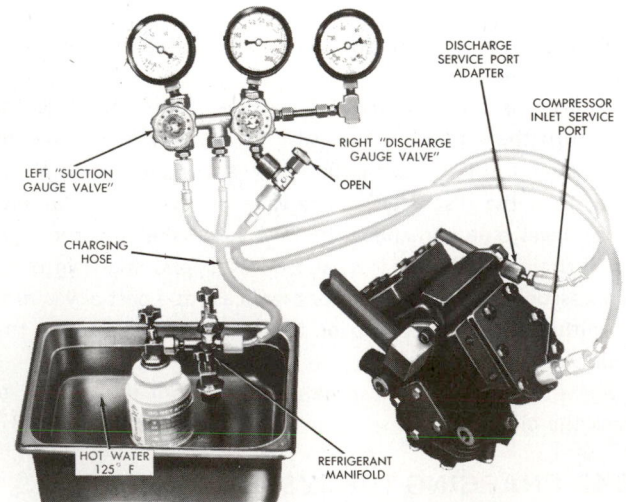

Fig. 26-83. Gauge and service manifold assembly is connected to a Chrysler compressor. Note third gauge on manifold which is used for evacuating purposes. (Chrysler Corp.)

be put in the system, and if the system is under a vacuum, charge the refrigerant into the high side of the system in liquid form. Do this by inverting the cylinder and opening the service valves. Chapter 11 gives full instructions on the operation of the gauge manifold.

Charging a system by using a disposable refrigerant container is shown in Fig. 26-83. In this example, the container is being warmed, using hot water. The gauge manifold has three gauges. The two gauges back of the manifold handwheels are the suction gauge and the high-pressure gauge. The gauge on the right is used during vacuum operations. The compressor is equipped with access valve service connections. A portable unit having a gauge manifold, vacuum pump, charging tube and leak detector is shown in Fig. 26-84.

Automobile air conditioning systems have a receiver for holding some refrigerant. Therefore, the charge can vary somewhat and system operation will still be efficient. If the charge is 4 lb. (1.8 kg), the system may operate on as low as 2 1/2 lb. (1.1 kg) of refrigerant.

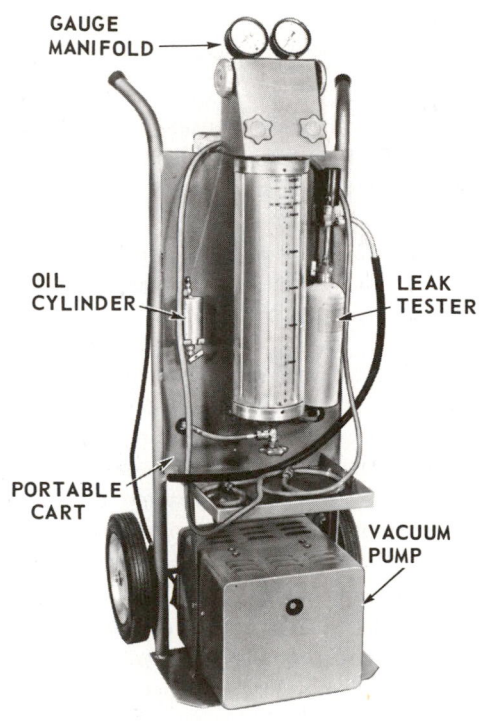

Fig. 26-84. Cart is equipped with devices needed to test air conditioning unit for leaks, to evacuate the system, and to charge it.

26-43 PERIODIC MAINTENANCE

The car owner should run the air conditioning unit for a few minutes once each month in fall, winter and spring to keep compressor parts (especially the shaft seal) lubricated.

The owner should check the unit each spring and fall as follows:
1. Condenser (clean fins and tubes of leaves, lint and insects).
2. Refrigerant lines (should show no signs of chafing or wear).
3. Belt tension.

The service technician should check the unit each spring

and fall or each 10,000 miles as follows:

1. Clean all parts externally, including condenser and evaporator.
2. Straighten fins on condenser and evaporator.
3. Check refrigerant charge.
   a. Sight glass.
   b. Pressures in system.
4. Check oil level in compressor.
5. Check for leaks, using a leak detector or color trace.

When a vehicle equipped with an air conditioning system is stored for a long period, the air conditioner should be started very carefully. Sometimes the compressor binds during storage. It is best to raise the hood and watch the compressor and belt while turning on the air conditioner. If the belt starts slipping, stop the engine at once. This indicates that the compressor is turning with difficulty or is "frozen" (stuck). Try to "free" the compressor by slow and careful turning. If it will not turn, remove it at once for reconditioning.

26-44 REVIEW OF SAFETY

There are several very important safety precautions to be observed when working on automotive air conditioning units.

The engine must be running to provide power for the air conditioner and air for the condenser. Therefore, connect the car's tail pipe to an air exhaust system. Do not touch the exhaust manifold (or serious burns may result). Be cautious to avoid putting tools, hands or clothing in contact with the revolving fan. Moving belts are dangerous, too. Loose clothing presents a hazard around any moving parts.

It is best not to wear rings, a bracelet or a wristwatch. These metal parts may short an electric circuit and burn the wearer, or may catch in a moving part and cause injury. One solution is to put a temporary shield over the fan. Usually, they are made of plastic and fasten to the radiator of the car.

Be careful about putting hands or tools near the spark plugs. The electrical shock is not harmful in itself, but the shock may cause one to jump against something, fall down or jerk into moving parts of the engine. If it is at all possible, stop the engine before performing any work on air conditioner parts that are under the hood. Always block the wheels of the car when running the engine.

Be sure to wear goggles when charging the system or when working on parts that contain refrigerant. R-12 boils at −21.7 F. (−29 C.). When it is spilled on the skin or gets in the eye, it will cause freeze damage.

Keep a protective cap on the refrigerant cylinder when it is not in use. When using the cylinder, fasten it to a part of the vehicle or to some sturdy stand. Otherwise, it may fall and break the connections or the valve.

If it is necessary to heat a refrigerant cylinder, use warm water only. Never use a torch, electric heat, steam heat, stove or radiator because the cylinder may explode.

Avoid welding, brazing, or steam cleaning near the air conditioning system unless the refrigerant has been removed. Otherwise, the excessive pressures may damage the unit and injure people near the unit.

Breathing quantities of any refrigerant is harmful to a human or an animal. Ventilate the area to keep the vapor concentration to a minimum.

When discharging a system near an open flame, the refrigerant tends to break down and form toxic gases. These gases also tarnish metal and plated surfaces.

Most engine cooling systems are pressurized. If a pressure cap is removed when the engine is hot, hot water will erupt out of the radiator and cause severe burn.

Blower fans can cause painful injuries to the hands. The sharp fins on the coils of the condenser and evaporator can cause deep cuts.

Manifold service lines must be kept clear of pulleys, belts and fans.

26-45 TEST YOUR KNOWLEDGE

1. Does the heat load increase or decrease as the speed of the automobile increases?
2. How are air conditioning compressors driven?
3. Where is the condenser mounted?
4. What type of suction line connections are used in automobile air conditioning units?
5. Are automotive systems provided with service connections?
6. What kind of an evaporator fan motor is usually used?
7. Why is air distribution an unusual problem in the automobile?
8. Do automobile comfort cooling units have a fresh air intake?
9. How do most automobile air conditioners clean the air?
10. What is the proper compressor belt tension?
11. What is the most popular refrigerant for automotive applications?
12. Which refrigerant control is used on automobile comfort cooling evaporators?
13. Where is the evaporator usually located?
14. Where is the plenum chamber located?
15. At what speeds do engine-mounted compressors operate compared to engine speed?
16. What is a popular type of electromagnetic clutch?
17. What materials are used to make compressor seal surfaces?
18. How is the evaporator kept from collecting frost or ice?
19. What is the purpose of the drier?
20. Name three compressor designs used in automobiles.
21. What materials are usually used in condensers?
22. Why do some automotive air conditioners use mufflers?
23. Why are flexible refrigerant lines used in automobiles?
24. Where is the sensitive bulb of the thermostat usually located?
25. How are duct dampers usually operated?
26. What type of temperature sensing device is used in conjunction with transistors and programmers?
27. What is the source of vacuum used for vacuum-actuated dampers?
28. Why do thermostatic expansion valves have pressure equalizer connections?
29. Where are O-rings used in compressor seals?
30. What happens at the evaporator when evaporator pressure is 20 psi (1.4 kg/cm^2)?

Energy Flow Diagram

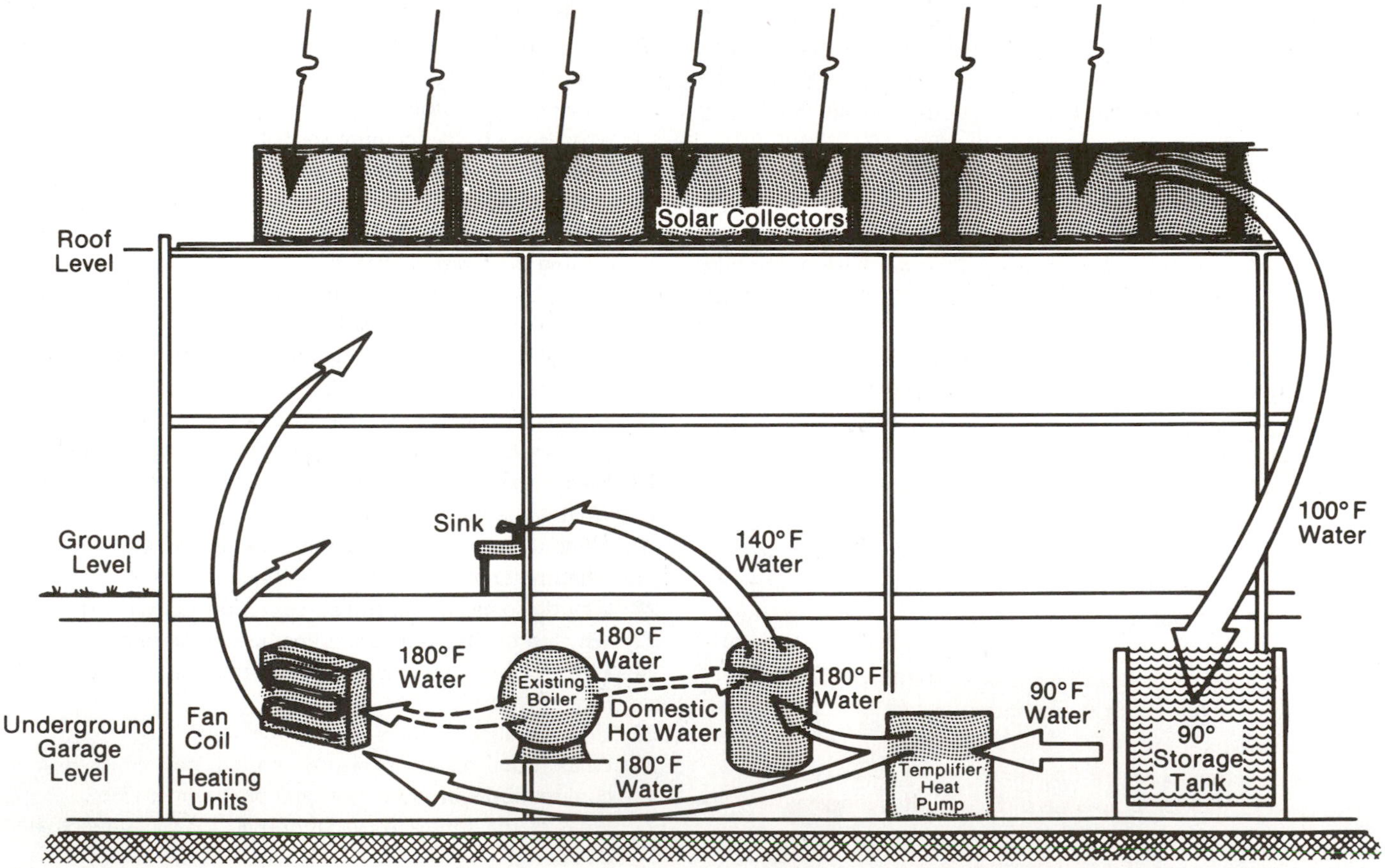

Solar Collectors

Roof
Level

100° F
Water

Ground
Level

Sink

140° F
Water

180° F
Water

180° F
Water

180° F
Water

180° F
Water

90° F
Water

Underground
Garage
Level

Fan
Coil

Heating
Units

Existing
Boiler

Domestic
Hot Water

180° F
Water

Templifier
Heat
Pump

90°
Storage
Tank

This heat-pump-assisted solar energy installation is designed for cold climates. It provides space heat and hot water for townhouses in Ottawa, Canada, as part of demonstration project funded by the National Research Council. Top left. Flat plate collectors — 2300 sq. ft. of them — heat water to 21 °C (70 F.) or 38 °C (100 F.). Heated water is stored in a 7000 gal. storage tank. Top right. High temperature heat pump draws heat from stored water to heat space and raise hot water temperature for domestic use. Bottom. Energy flow diagram and parts of the system. (Westinghouse Electric Corp.)

Chapter 27

SOLAR ENERGY

This chapter deals with the nature of solar energy, ways of collecting it, and present methods of using it. It provides a fundamental understanding of the nature and efficient handling of solar energy.

27-1 WHAT SOLAR ENERGY IS

Solar energy is energy from the sun. The amount of such energy can be measured. Usually it is recorded in watts per square metre radiated to the absorbing surface.

One of the simplest illustrations of the measurement of solar energy is the fact that a dark surface of about one square metre exposed to the sun on a bright day will absorb about one horsepower (746 watts) of energy. This is illustrated in Fig. 27-1.

The effects or benefits of solar energy reach humankind in many ways. Solar energy causes the evaporation of water from the ocean and from lakes and streams. This moisture is absorbed by the atmosphere and later returns as rain or snow. It is this moisture that provides the head of water (kinetic energy) to turn the electric generators at Niagara Falls, Grand Coulee Dam, Hoover Dam, and other water power installations.

The wood we burn in our stoves and fireplaces is giving back, in the form of heat, the energy that was absorbed by the tree as it grew. Another source of energy is the coal mine. Again, the coal is a product of vegetation, and the vegetation has been supported in its plant-growing period by energy from the sun. By some unknown process the oil and gas stored in the earth came from plant or animal life which was, again, supplied with energy from the sun.

27-2 THE NATURE OF SOLAR ENERGY

The light one "sees" on a sunny day is only a small fraction of the radiant energy from the sun. In fact, our eyes are sensitive to only about 25 percent of the wavelengths of energy from the sun.

Radiation is electromagnetic energy. It behaves both as waves and particles. When describing radiation as particles, we call the particles photons. From most heating applications, the wave behavior is the easiest to understand.

An electromagnetic wave, as shown in Fig. 27-2, has both wavelength and amplitude. The wavelength is the distance between two peaks. The amplitude is the height of the wave shown. For any given amplitude of radiant energy, the shorter the wavelength, the greater the energy of radiaton. The wavelength is usually expressed in microns. The symbol for micron

Fig. 27-1. The projected surface of a small horse is about one square metre. In bright sunshine, the black horse will absorb about one horsepower in radiant energy.

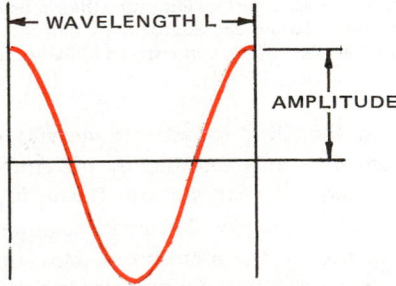

Fig. 27-2. Electromagnetic radiation is described as a wave with wavelength, L, and amplitude as shown.

is (μ). It is pronounced "mu." A millimetre is 1000 microns.

The energy of a single light wave is expressed in the formula:

$$E = \frac{hc}{L}$$

h = Planck's constant which is 6.626×10^{-34} watt seconds squared ($W \cdot s^2$)

c = Speed of light which is 3×10^8 metres per second (m/s)

L = wavelength in metres

The energy (answer) is given in watt seconds ($W \cdot s$) or joules (J) per wave.

The total radiant energy from the sun is known as the solar energy flux. Line A in Fig. 27-3 shows the radiant energy flux from the sun plotted against the wavelength of the radiation. The area under the curve represents the total energy flux. The energy flux is approximately 1.35 kW/m². This is known as the solar constant. This figure also shows the visible part of the sun's spectrum from the blue (about 0.4 micron) to red (about 0.73 micron). Radiation with shorter wavelength is called ultraviolet radiation. Radiation with longer wavelength is called infrared radiation. It may be noted from this diagram that there is more energy in the visible light part of the spectrum than either the ultraviolet or infrared part.

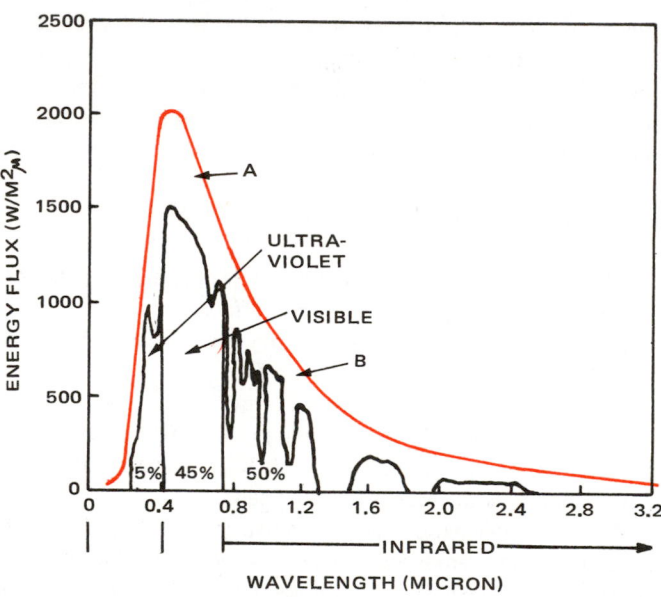

Fig. 27-3. Solar radiation at different wavelengths of light. Curve A illustrates the solar radiation outside the atmosphere which is approximately a 5900 K source temperature. Curve B illustrates an approximate radiation flux at the earth surface indicating the atmospheric absorption of light. Also indicated are the approximate percentages of energy in the ultraviolet, visible and infrared radiation regions.

Curve B in Fig. 27-3 shows that most of the ultraviolet radiation from the sun is absorbed by the atmosphere so that it does not reach the earth surface. Ozone (O_3) is the main absorber of this radiation. Infrared radiation as shown is partially absorbed by the atmosphere. Most of this energy is absorbed by water vapor (H_2O) and carbon dioxide (CO_2).

With this absorption of radiation by the atmosphere, the highest energy flux observed on a clear day at the earth sur-

face is about 0.9 kW/m².

The radiation given off by a hot stove is infrared radiation and not visible. Fig. 27-4 shows how the energy radiated from a black stove at 400 K would be distributed over different wavelengths. The energy given off in the visible range is so small it is not observed. If an object is heated so that its temperature is increased, it will emit more energy. In addition to emitting more energy, the wavelength at which the maximum energy is emitted gets shorter as shown. As an object gets hotter, it becomes visible as a red glow. Increasing the temperature further changes the color to yellow, then to blue. If it reached the temperature of the surface of the sun, the maximum energy emitted would be at about 0.5 microns wavelength, with a temperature of about 5900 K.

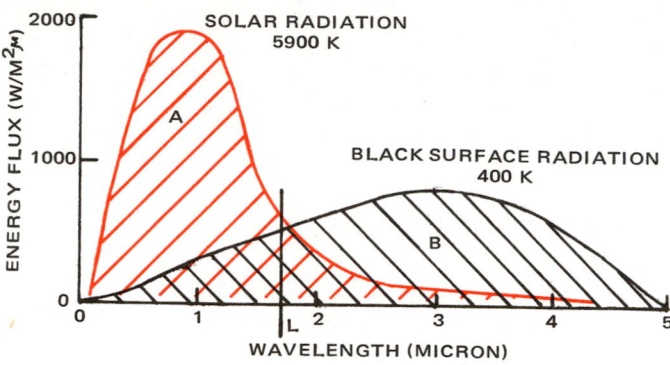

Fig. 27-4. Illustration of the comparison of the energy of radiant rays from the sun with the heat energy of a heated object. An ideal collector would absorb the solar energy with wavelength less than L but not give off radiation for wavelength greater than L.

27-3 PASSIVE AND ACTIVE SOLAR ENERGY SYSTEMS

There are two types of solar energy systems. Passive solar energy systems depend on the solar radiation striking directly on the surface or area to be heated. The best example of passive solar heating is the conventional greenhouse where the energy flows through the glass into the area where the plants are growing. Passive solar energy systems usually require no auxiliary pumps or blowers to distribute the heat collected.

In active solar energy systems, the solar energy is absorbed into a collector. The energy is then transferred from the collector, stored and distributed by an auxiliary circulation system.

27-4 REQUIRED COMPONENTS OF A SOLAR ENERGY HEATING SYSTEM

A solar energy heating system must have parts which will provide the following functions:
1. Collect solar radiation.
2. Circulate heat from the collector and move it to the space being heated.
3. Store heat for later circulation when production of solar

energy is insufficient to heat the space.

Passive solar systems have parts which collect the radiation. However, they depend on natural radiation to circulate the heat through the space being heated.

Active solar heating systems have mechanisms which can store the collected heat. Then, during the night or on cloudy days, a circulating system draws heat from storage and moves it to the space.

Existing houses usually require an active solar heating system. But, in the design of new homes, passive systems can be more easily incorporated. They require less energy input because pumps and blowers are not needed to circulate heat.

27-5 SOLAR COLLECTORS

The rays from the solar energy flux are converted to heat upon striking a dark surface. This applies to visible rays, infrared rays, and ultraviolet rays.

Heat from this solar radiation can be trapped. The simplest trap is an insulated black surface. A surface looks black because it absorbs visible light. Black surfaces will also absorb the infrared and ultraviolet radiation as well. Coatings of lampblack or fine carbon powder produce surfaces which are very close to all black. In the sun they will get much hotter than white or shiny surfaces, because they absorb more radiation.

If an object is insulated on the back and placed in the sun, it will absorb radiant energy and get hot. The temperature will increase until the radiation emitted by the object is equal to the radiation received from the sun. If the object is black, the emitted radiation will be distributed over the spectrum as shown in Fig. 27-4. In this figure, the energy from the sun striking the surface is shown in red. The energy emitted by the surface is shown in black. When the surface reaches its maximum temperature, the area under the red curve will equal the area under the black curve.

The maximum surface temperature a black object can reach in full sunlight is about 123°C (253 F.) or 396 K. There are ways to increase this temperature which involve trapping the radiation emitted by the surface. One way is to place a glass

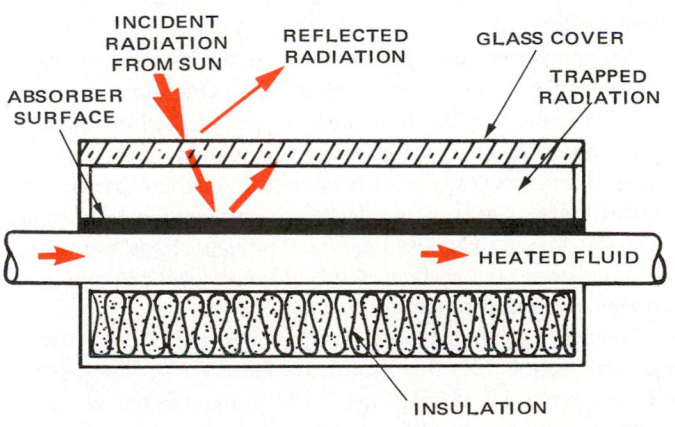

Fig. 27-5. Solar collector elements. Trapping of radiation by glass cover and insulation on the back of the collector is illustrated. Solar radiation passes through the glass, but the radiation from the absorber is trapped.

cover over the black surface. A sketch of a solar collector using this trapping principle is shown in Fig. 27-5. The glass allows radiation with wavelength less than about 2 microns to pass through and be absorbed by the black surface. The black surface at temperatures below 396 K (123°C) emits most of its energy at wavelengths greater than 2 microns. This radiation is not transmitted out by the glass and is trapped inside the collector. Temperatures of 600 K (327°C) are obtainable in collectors of this type. Fig. 27-6 is a drawing of a typical flat plate solar collector showing construction details.

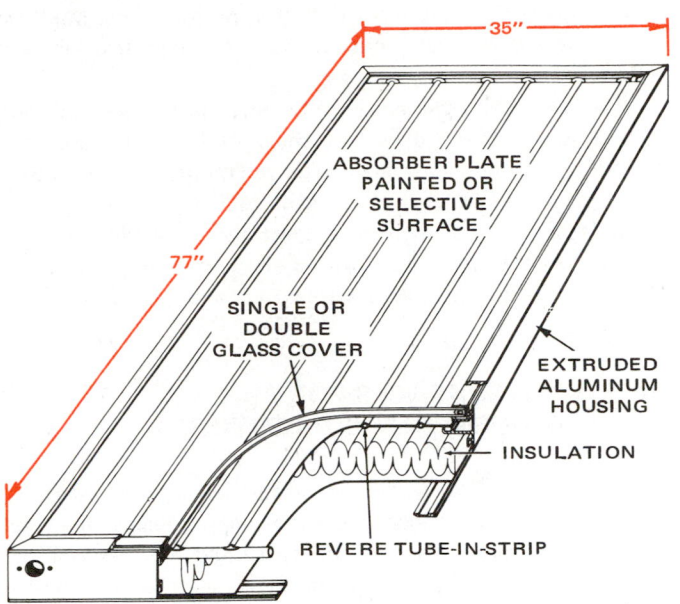

Fig. 27-6. Drawing shows dimensions and design of typical solar collector. (Revere Solar and Architectural Products, Inc.)

27-6 SELECTIVE ABSORBER SURFACES

A single absorber surface is sometimes used to increase the temperature of a collector. Such a design is called a selective absorber surface. Acting much like the combined glass plate and absorber surface, this design absorbs most of the radiation from the sun which has wavelengths less than some "cutoff wavelength" (around 2 microns). It does not, however, absorb longer wavelength radiation.

It can be shown that a surface which absorbs poorly at a given wavelength also emits (gives off) poorly at that wavelength. The surface then emits infrared radiation very poorly and the temperature increases.

Many special paints and surfaces are being developed with this characteristic. They selectively absorb most of the sunlight but not infrared rays. Their performance is specified (rated) by an absorption to emission ratio. This ratio is the absorptivity to sunlight or radiation with wavelength shorter than the cutoff wavelength, divided by the absorptivity for longer wavelength radiation. (Absorptivity means ability to absorb.) Commercial surface materials are produced which have an absorption to emission ratio of about 20 to 1.

The way these surfaces work is similar to the absorption of

sound by a surface with small holes in it. Sound waves with wavelengths smaller than the holes are absorbed by passing into the holes. Sound waves with wavelengths greater than the hole size are reflected.

Selective absorber surfaces do not absorb as much energy as black surfaces at wavelengths shorter than the cutoff wavelength and may appear gray. This is more than offset by the fact that they do not allow energy at long wavelengths to escape. Most newer solar collectors use selective absorber surfaces rather than black surfaces.

The combination of a glass cover and a selective absorber surface gives the best solar collector performance. Multiple glass covers with special antireflecting coatings also increase the radiation trapping efficiency.

The closer the radiation is to being perpendicular (at right angles) to the glass surface of the absorber the more it is transmitted through the glass. The radiation not transmitted is reflected and lost for heating purposes.

Collectors with a single glass cover and a selective absorber surface have been shown to be as efficient in trapping radiation as collectors with two glass covers and a black absorber surface.

27-7 GOOD SURFACES FOR SOLAR ENERGY COLLECTORS

The solar collector surface should be black or very dark. It should be slightly roughened—somewhat rougher than an eggshell condition. The roughness should not exceed 25 millimetres (one-eighth in.) from the highest to the lowest point on the surface. A smooth and shiny surface would reflect the radiant energy away and not absorb it.

The black surface is usually best applied as an electro-deposition process. (In electrodeposition, metallic particles are applied to another metal surface through use of an electric current.) This process bonds the black material to the metal conduction plate. Paints are generally not satisfactory as they will peel and crack at the high temperature of the collector.

27-8 SOLAR COLLECTOR COVERS

A transparent cover is used on most solar collectors. This cover has three basic functions:
1. To protect the absorber surface from the weather.
2. To transmit sunlight to the absorber surface.
3. To prevent the escape of heat collected by the absorber surface.

Glass is the most widely used cover material. It provides good light transmission and remains clear indefinitely. Typical glass covers are 1/8 to 1/4 in. thick. Special glass materials transmit more sunlight and allow less collector heat to escape. Iron free glass is generally used for this purpose.

Glass poses problems in sealing the cover to the other parts of the collector. Glass and metal do not expand at the same rate with temperature change. A flexible sealing material or rubber gasket is necessary at the glass-to-metal joint to avoid breaking of the glass as the materials expand and contract.

Plastic cover materials usually soften at high temperatures.

Plastic also becomes brittle and opaque from absorption of ultraviolet light.

Some covers are formed in a bubble shape rather than flat to provide more structural support and to collect more radiation when the sun is near the horizon. Some special collectors are formed as a tube and are sealed with a vacuum inside. These collectors are very efficient and do not lose much of the heat collected. Fig. 27-7 shows a typical vacuum tube collector.

Periodic washing of the glass will improve the light transmission. Most installations however require washing only once or twice a year.

Fig. 27-7. Vacuum tube solar collectors are used for high-temperature application. A—Vacuum tubes. B—Reflectors. C—Hydronic connectors. (General Electric Co.)

27-9 SOLAR ENERGY STORAGE SYSTEMS

Large bodies of water are a natural solar energy storage system. The surface of a lake may be frozen over in the winter; but in the spring, as the sun shines on it, the ice melts and the water warms. This heat then is given to the atmosphere in the fall or winter as the water on the surface again freezes.

We sometimes use this system by storing heated water in an insulated container for later use. One very common example is the heating of water in a tank located on the roof of a house. The tank is connected to the water system in the house. Many homes, particularly in the southern part of the United States, use this system entirely for heating bath water.

Rocks may also be used for heat storage. If the sun shines on a big stone, the surface of it will be warmed and may stay somewhat warm throughout the night.

The specific heat of water is about 1 Btu/lb. F. and that of the rock about 0.25 Btu/lb. F. (See Para. 1-31). The specific heat of water in SI metric is 4.187 kilojoules per kilogram times degrees kelvin. The equation is written kJ/kg·K. This means that 1 lb. of water will store as much heat as 4 lb. of rocks if they are both heated to the same temperature. The same heat stored in crushed rock requires about four times the volume of a water tank.

27-10 HEAT ENERGY STORAGE IN A CLOSED WATER SYSTEM

To use solar energy more efficiently in heating, water solar collectors may be constructed for heating water. The system gathers, circulates, and stores absorbed energy in water. The amount of solar energy absorbed depends upon the area and color of the absorbing surface exposed to the sun. In a complex system, collector lenses are used to heat the water to a very high temperature.

As a practical application, a heat-absorbing solar energy system usually consists of:

1. A large solar panel facing the sun.
2. Water circulated through the panel to storage.
3. A large insulated holding tank.

Heat from the water may be used by either of two methods:

1. The water may give up its heat directly to the space.
2. The water becomes a source of heat for a heat pump. The pump will then be used to transfer the heat to the space as desired.

Ethylene glycol is added as an antifreeze to protect the system from freezing in colder climates. In mild climates a drain-down system is used which drains the collector on cold nights.

Heated water may be stored in a large insulated tank. Typical solar heating systems have storage tank capacity of from 50 to 100 litres for every square metre of collector surface.

27-11 HEAT ENERGY STORAGE IN A WARM AIR SYSTEM

Solar energy may also be absorbed at the collector and transferred to warm air as heat. Air is circulated through the solar panel, and the heat is stored in crushed stone. The crushed stone is held in place by a porous insulated container so that the air heated by the sun may be circulated through the container. At night, or at other times when heat is needed, air in the house is circulated through the heated, crushed stone. The stone then gives up its heat to the home.

The crushed stones should be from 1 to 3 cm in diameter. Too much energy is required to circulate the air around smaller stones. On the other hand, stones that are too large will not be heated completely. Only part of the stone will be effective in storing heat.

Fig. 27-8 illustrates a rock storage container. The size of this container is about 0.1 to 0.4 m^3 in volume for a square metre of collector area. The height of the container is usually limited to less than 2 metres to allow the air to be circulated from the bottom to the top with a small blower.

27-12 ANGLE OF THE COLLECTOR

It would be ideal if a mechanism could be arranged to always present a collector surface exactly perpendicular to the rays of the sun. A device to provide this aiming — except for extreme conditions — is too expensive. The position of the collector surface, therefore, usually remains fixed. The angle

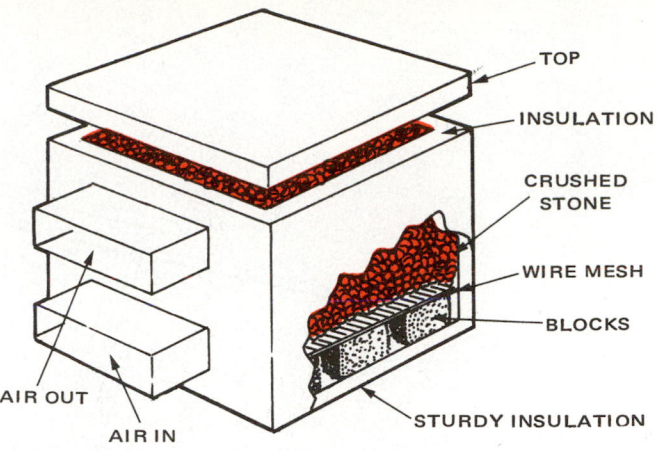

Fig. 27-8. Typical crushed rock heat storage container. Room air, blown in bottom, comes out top heated.

of the sun's rays to the horizon varies during the year. It is, therefore, necessary to establish an angle for the collector which will give the best average collecting effect.

If the collected solar energy is used chiefly for home heating, the angle needs to be set near a position which will absorb the greatest amount of energy possible during the heating season. This setting is approximately 45 to 55 deg. from the horizontal for use across the mid-section of the United States. The collector should face south but an angle from 30 deg. east or west of south only reduces the energy collected by about 10 percent.

Naturally, it is important that there be no obstructions, such as buildings or trees, which might create a shadow across the collector surface.

27-13 HEAT INSULATION OF THE COLLECTOR SURFACE

A collector surface will both absorb solar energy and re-lease heat energy unless it is properly insulated. The solar energy coming into the collector is changed to heat. The surface will then radiate heat the same as any other warmed surface. This means that the collector surface needs to be insulated to cut down the radiation of heat to the atmosphere as much as possible. This can best be done by one or more layers of glass over the front surface to hold in the energy as it is converted from solar radiation to sensible heat. The back surface must also be insulated to hold the heat from escaping through the back. About 4 in. of conventional insulating material will usually provide sufficient insulation.

27-14 SOME TYPICAL SOLAR SPACE HEATING INSTALLATIONS

There are three types of effective solar heating installations in common use:

1. Natural convection closed water system. Heat is absorbed in the collector and, since warm water tends to rise, the storage tank must be located above the collector. The stor-

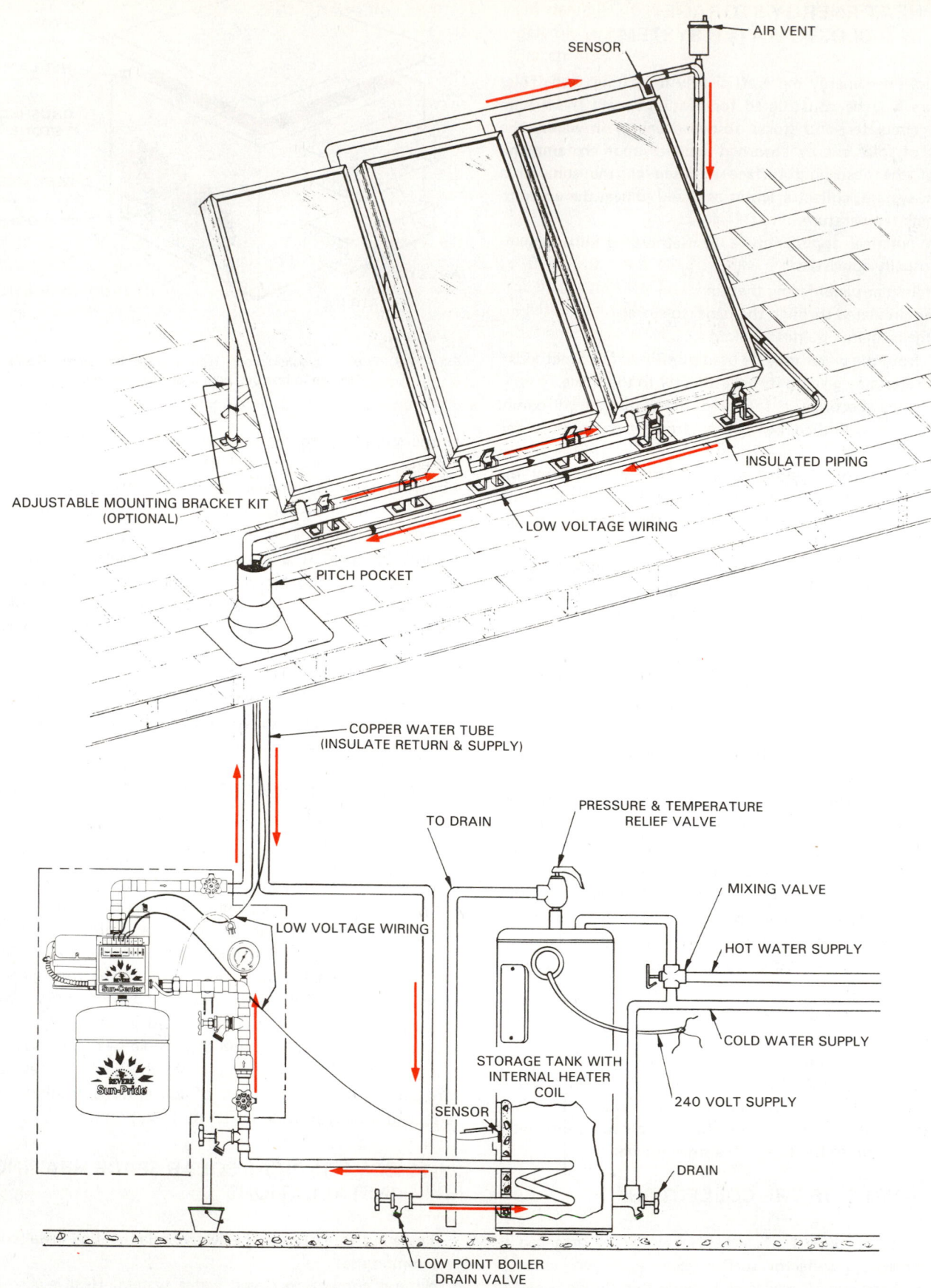

AIR VENT

SENSOR

INSULATED PIPING

ADJUSTABLE MOUNTING BRACKET KIT
(OPTIONAL)

LOW VOLTAGE WIRING

PITCH POCKET

COPPER WATER TUBE
(INSULATE RETURN & SUPPLY)

TO DRAIN

PRESSURE & TEMPERATURE
RELIEF VALVE

MIXING VALVE

HOT WATER SUPPLY

LOW VOLTAGE WIRING

COLD WATER SUPPLY

STORAGE TANK WITH
INTERNAL HEATER
COIL

240 VOLT SUPPLY

SENSOR

DRAIN

LOW POINT BOILER
DRAIN VALVE

Fig. 27-9. Components in forced convection solar heated closed water system. One tank stores heat and supplies hot water.
(Revere Solar and Architectural Products, Inc.)

age tank should be quite heavily insulated to eliminate heat loss. The radiator through which the warm water circulates to heat the desired space must be located with its inlet about even with the warm water supply. As the heat from the warm water is radiated into the room, the water will be cooled — and it will flow down and be carried back to the bottom of the heat storage tank. There is no temperature control with this system. With bright sunshine the temperature will continue to rise. The heated space could become uncomfortable.

2. Forced convection in a closed water system. In this system a pump circulates the water. The warm water storage tank may be located where it is convenient without regard for location of the collector or the radiator. Water is heated in the collector and is circulated to the storage tank and radiator. The circulation through the radiator may be controlled with a thermostat in order to maintain a certain room temperature. Fig. 27-9 illustrates the components of a forced convection solar heated closed water system. The differential thermostat turns off the pump when the temperature in the solar collector is lower than the temperature in the storage tank.

3. Warm air heating system. A warm air heating system using forced air circulation is illustrated in Fig. 27-10. The solar energy is converted to heat in the solar collector. It must be insulated so that only a small portion of the heat is lost. The heated air is circulated through the rock bed. This bed consists of crushed stones approximately 1 to 3 cm in size. Air from the collector is circulated through the rock bed. Heat is stored in the rocks. When heat is needed in the

room, air is circulated through the warmed rock bed, which gives up its heat to the room air. Rocks for this purpose are very carefully washed and cleaned, so that no dust or sediment circulates through the room. A thermostat controls the airflow. Supplementary heat, as needed, may be provided as indicated in the supplementary heat paragraph, Para. 27-17.

27-15 SOLAR DOMESTIC WATER HEATING

Perhaps the most-used solar heating system is one which supplies heated hot water for domestic supply. The heated water for dishwashing, showers and washing can be supplied easily using a solar heating system to preheat the water. If the water supply is from a well, the supply temperature is fairly constant throughout the year, ranging from 4.5 to 10°C (40 to 50 F.). If the supply is through a large municipal water supply system, the water is either drawn from a large reservoir or storage tanks. In some locations the temperature of the water may range from 2°C (35 F.) in the winter to 27°C (80 F.) in the summer. The heat required to supply domestic hot water at 49 to 60°C (120 to 140 F.) can vary considerably throughout the year, dependent upon its geographic location.

Solar domestic water heating systems are more cost effective than space heating. There is a demand for hot water in a home the year around, whereas space heating is only used in the winter season.

Two types of solar water heating are in common use:
1. One tank system.
2. Two tank system.

The one tank system consists of a single hot water storage tank. The water, preheated in the solar collector, is circulated through a heat exchanger in the bottom of the hot water heater tank. An auxiliary heater coil in the top of the tank provides the final heating to bring the water to the desired temperature. This system requires replacement of the present hot water tank in retrofit (see Chapter 30) situations.

The two tank system is shown in Fig. 27-11. It uses a separate storage tank in which the water is preheated by the solar heated liquid through a heat exchanger. Water is circulated to the conventional hot water heater from this storage tank. This system with a separate storage tank is easily added to the present hot water system in a house.

The two tank system is usually from 30 to 50 percent more efficient than a single tank system. The cost is usually higher than a single tank system if a new separate water heater must be purchased.

Domestic hot water systems are most economically designed to supply from 40 to 75 percent of the required hot water heat. Auxiliary heating is then required to supply the final heat to reach to 49 to 60°C (120 to 140 F.) temperature required.

Domestic hot water use is approximately 25 gallons a day for each person in a family. Solar collector sizes range from 1/2 to 2 sq. ft. per gallon of hot water required per day, depending on location. In a two tank system the low temperature storage tank is usually designed so that it is only

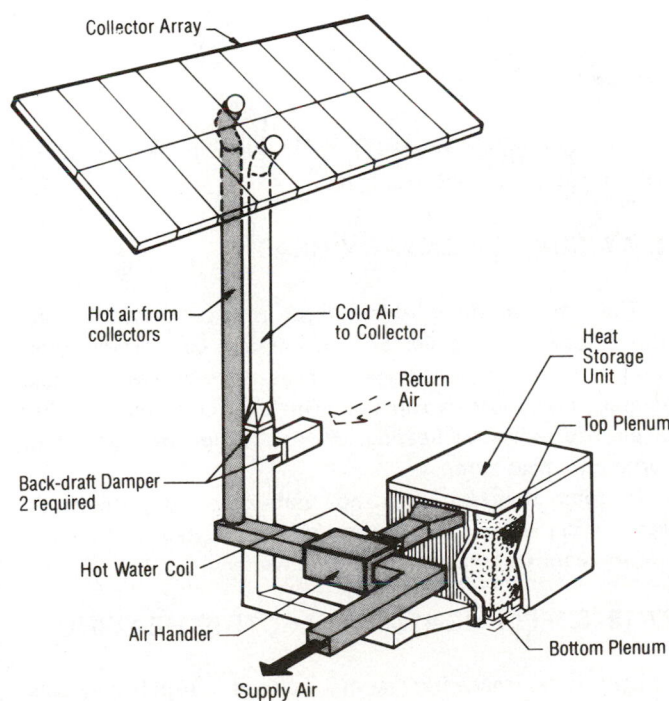

Fig. 27-10. Warm air solar heating system with forced air circulation. Heat is stored in crushed stone. (Solaron Corporation)

Collector Array

Hot air from collectors

Cold Air to Collector

Return Air

Heat Storage Unit

Top Plenum

Back-draft Damper 2 required

Hot Water Coil

Air Handler

Supply Air

Bottom Plenum

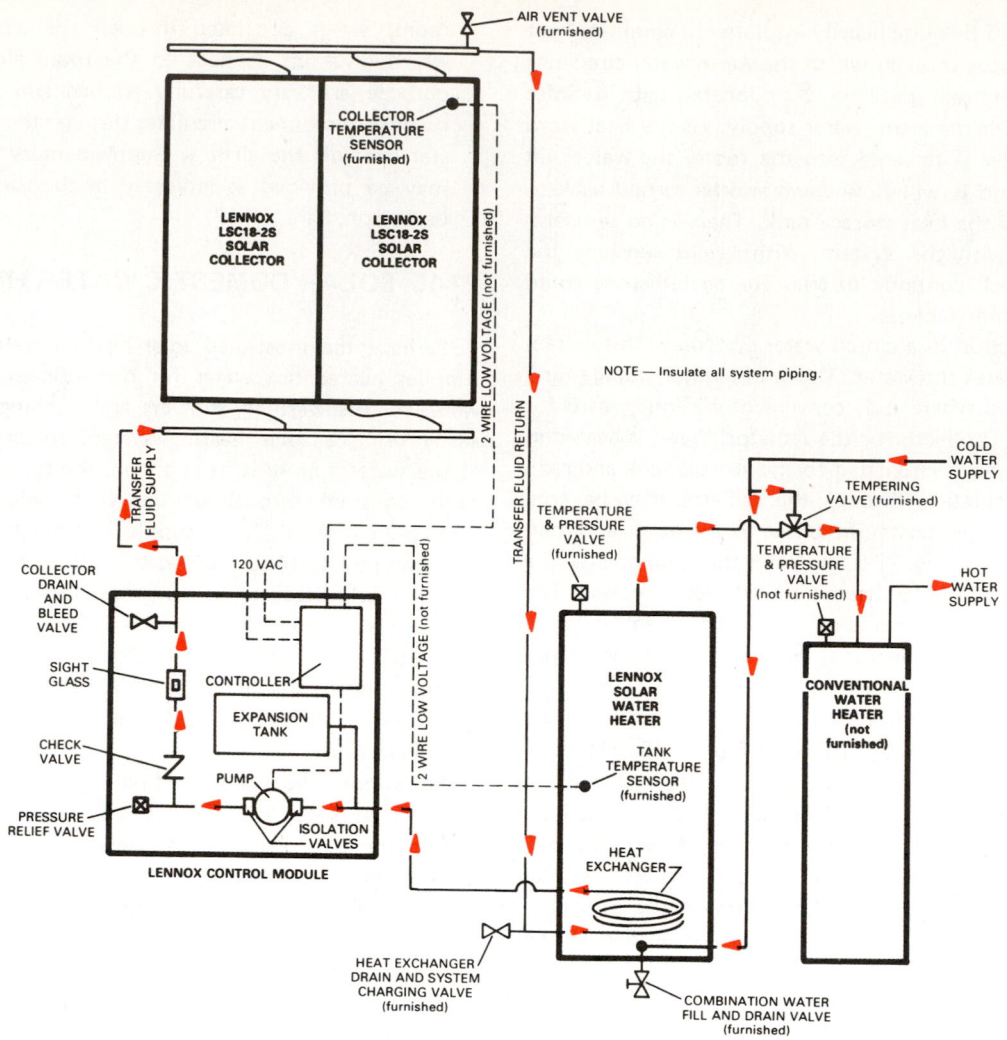

AIR VENT VALVE
(furnished)

COLLECTOR
TEMPERATURE
SENSOR
(furnished)

LENNOX
LSC18-2S
SOLAR
COLLECTOR

LENNOX
LSC18-2S
SOLAR
COLLECTOR

NOTE — Insulate all system piping.

2 WIRE LOW VOLTAGE (not furnished)

TRANSFER FLUID RETURN

TRANSFER FLUID SUPPLY

COLLECTOR
DRAIN
AND
BLEED
VALVE

120 VAC

TEMPERATURE
& PRESSURE
VALVE
(furnished)

COLD
WATER
SUPPLY

TEMPERING
VALVE (furnished)

TEMPERATURE
& PRESSURE
VALVE
(not furnished)

HOT
WATER
SUPPLY

SIGHT
GLASS

CONTROLLER

EXPANSION
TANK

CHECK
VALVE

PUMP

2 WIRE LOW VOLTAGE (not furnished)

LENNOX
SOLAR
WATER
HEATER

TANK
TEMPERATURE
SENSOR
(furnished)

CONVENTIONAL
WATER
HEATER
(not
furnished)

PRESSURE
RELIEF VALVE

ISOLATION
VALVES

HEAT
EXCHANGER

LENNOX CONTROL MODULE

HEAT EXCHANGER
DRAIN AND SYSTEM
CHARGING VALVE
(furnished)

COMBINATION WATER
FILL AND DRAIN VALVE
(furnished)

Fig. 27-11. Two tank domestic hot water heating system. Conventional water heating tank supplies auxiliary heat. (Lennox Industries, Inc.)

about half the size of the main storage tank.

27-16 SOLAR HEATING OF WATER FOR SWIMMING POOLS

Heating of water for swimming pools can be used to extend the swimming season in many areas. The solar heating system for a swimming pool can be very simple and inexpensive. Swimming pool solar heaters can be just a set of black or transparent plastic bags. The reason for the simplicity is that:
1. No storage is required.
2. The water pressure that the collector must withstand can be much lower than in space or domestic water heating systems.
3. No freezing protection is necessary.

These swimming pool heaters require only a low pressure pump, plastic solar collectors and piping.

The size of a solar collector for a swimming pool is usually approximately half the surface area of the swimming pool to provide good spring and fall heating. The efficiency of swimming pool solar heaters can be 70 to 80 percent because

of the low temperature (27 °C or 80 F.) water that the collectors are required to supply.

27-17 SUPPLEMENTARY HEAT

There may be times when the sun is not supplying enough heat energy to the collector to maintain a comfortable room condition even with storage of heat. In such cases, supplementary heat must be supplied. This may be through the use of electric resistance heating, the application of a gas or oil burner or a heat pump.

In some domestic solar hot water heating systems, the water is preheated by solar energy then heated to the temperature required by a supplementary heater.

27-18 SUPPLEMENTARY ELECTRIC HEATING

Electric resistance heating may be used to supplement solar heat. The nature of this installation will be governed by the nature of the heating system to which it is applied. With a warm air heating system, a direct electric resistance radia-

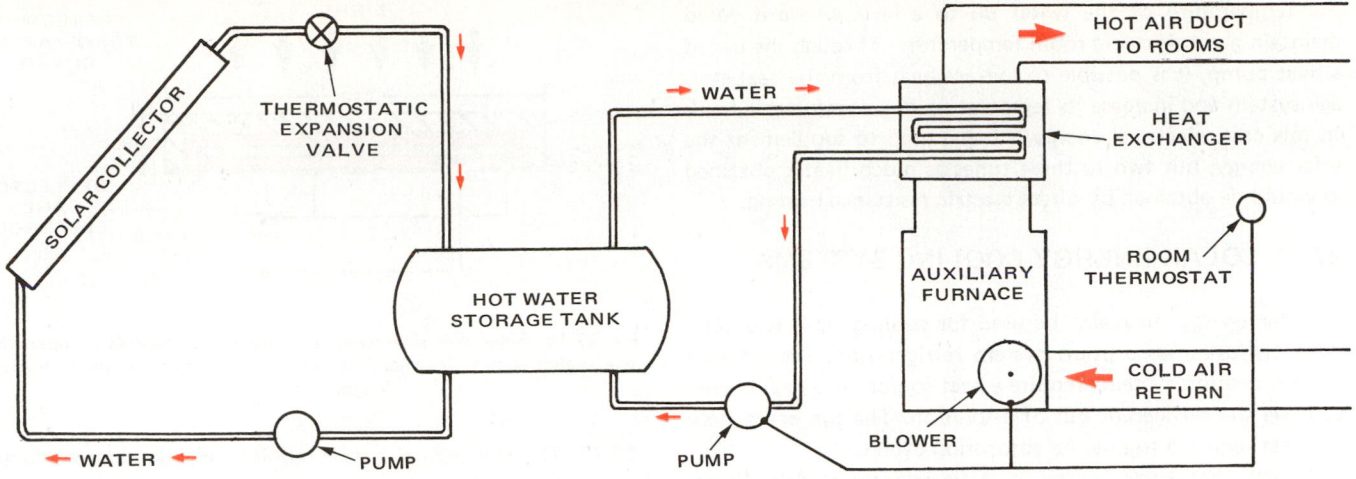

Fig. 27-12. A schematic of a liquid solar collector system used in forced air heating application. The heat exchanger is shown located in the hot air duct of a conventional furnace which supplies auxiliary heat.

tor may be installed in the air duct. A thermostat may be used to control the amount of electric heat needed.

In a hot water space heating system, an electrical heating element may be installed in an auxiliary tank to provide supplementary heating when the heat stored in the main tank is exhausted. This is indicated in Fig. 27-11.

In some heating situations, individual room resistance heating radiators may be installed.

27-19 SUPPLEMENTARY OIL AND GAS HEATING

In the event that oil or gas is used for supplementary heat, a separate furnace is installed as shown in Fig. 27-12. The burner will be controlled by a thermostat which will turn on the burner when the amount of heat from the solar source is below the amount of heat required. If the system uses warm water radiators, the burners may be used to heat the water in an auxiliary tank.

27-20 HEAT PUMPS

The use of a heat pump is probably the most efficient

method of supplying additional heat in a solar heat system. The heat pump may also be used in connection with an air conditioning system to take away heat. The heat pump uses a refrigerant fluid the same as an air conditioner. (See Chapter 23.) It can be arranged to either add heat to a room or absorb heat from a room. Fig. 27-13 illustrates a heat pump installation.

The use of a heat pump in connection with a solar energy collector is the most efficient means of maintaining a desired temperature with the fluctuating heat supplied by the solar collector. Theoretically, a heat pump can be efficient on a 4-to-1 basis — that is, the amount of heat delivered can be four times the heating value of the electrical current required to drive the mechanism. In application, this theoretical value is never reached. A ratio of 3-to-1 is considered to be very good; that is, three times as much heat is obtained from the heat pump as the heating value of the electricity consumed represents if converted directly to heat. In many practical applications, this ratio may not be over 1.5-to-1.

The heat pump is a natural device to be used in connection with solar heat. It is possible to use solar energy to warm water in a forced convection system; the solar energy may not bring

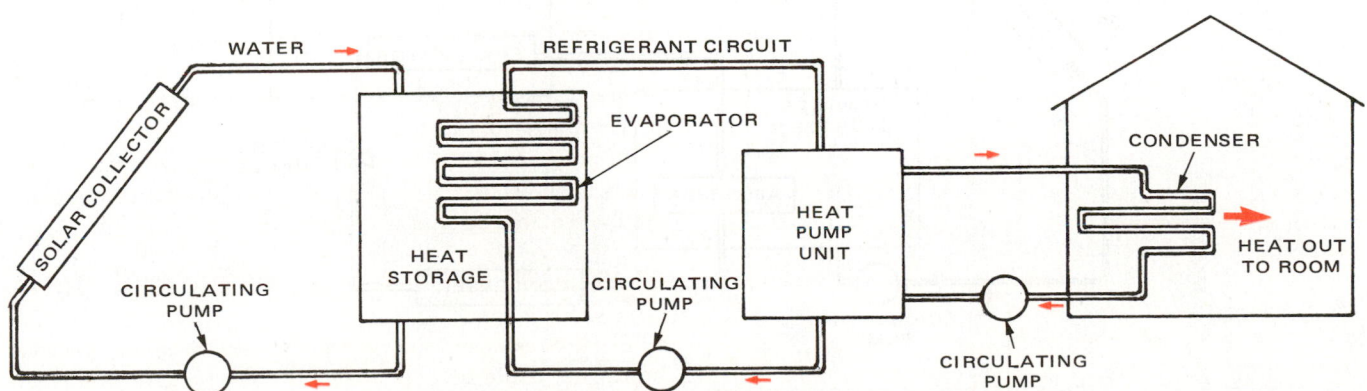

Fig. 27-13. Solar heating with supplementary heat supplied by a heat pump. Heating cycle is shown.

the temperature of the water up to a level where it could maintain a comfortable room temperature. Through the use of a heat pump, it is possible to extract heat from the heat storage system and increase its temperature to a comfortable level. In this case, electrical energy is being used to supplement the solar energy, but two to three times as much heat is obtained as would be obtained by direct electric resistance heating.

27-21 SOLAR ENERGY COOLING SYSTEMS

Solar energy may also be used for cooling. This is usually done by using absorption system refrigeration. See Chapter 16. Absorption systems require a heat source. The heat is used to drive the refrigerant out of a solution. The sun can supply the heat required to operate absorption cycles.

Usually the solar energy must be concentrated by lenses or mirrors to increase the solar radiation temperature to whatever is required by the absorption system.

At present, solar energy is chiefly used for heating. The required mechanisms for cooling are quite expensive.

Experimental systems using solar energy as the heat source have been built. The system elements are shown in Fig. 27-14.

27-22 CONVERTING SOLAR ENERGY TO ELECTRICITY

Solar energy may be converted directly to electricity. Radiation from the sun is electromagnetic energy. There is a close relationship between magnetism and electricity. Solar cells make use of this relationship to convert solar energy directly into electricity.

27-23 SOLAR CELL CONSTRUCTION

The components of a typical solar cell are shown in Fig.

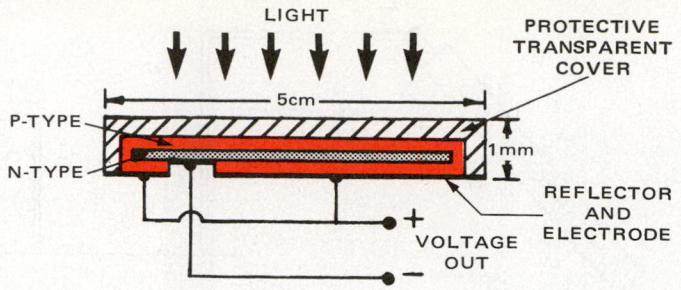

Fig. 27-15. Solar cell schematic indicating the physical components. The sketch is not to scale. The cells are very thin compared to their length and width.

27-15. The principal elements required are two semiconductor materials called an N-type semiconductor and a P-type semiconductor. An N-type semiconductor carries current by means of electrons or Negative charges. A P-type semiconductor carries charges by Positive charges, sometimes referred to as "holes." "Holes" refer to the particle remaining after an electron has been removed from an atom.

Photons in the light striking a solar cell give up their energy to electrons at the junction between the P-type and N-type materials. The electrons with excess energy are then stored in the N material, giving it a net negative charge and leaving a net positive charge on the P-type material. Electrodes are attached to each material and conduct the charge to an external load circuit.

The cell is usually made by starting with a thin glass plate and, by electrochemical means, depositing very thin layers of P-type, N-type, and finally electrode materials on the back.

P-type and N-type materials in most solar cells are mainly silicon. The difference between the two materials is that the silicon of the N-type material contains a very small quantity of a material which, compared to silicon, has an excess of electrons. Nitrogen, phosphorus, or arsenic are materials

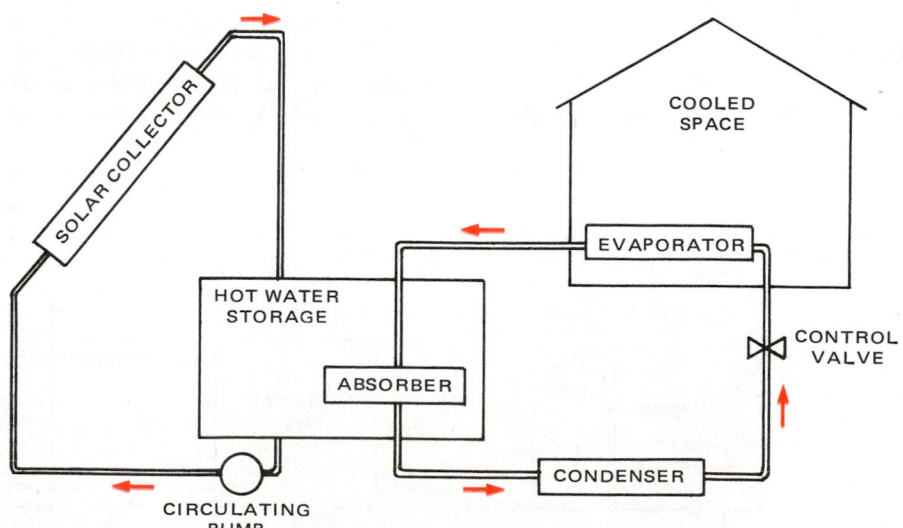

Fig. 27-14. Solar energy air conditioning system utilizing an absorption refrigeration unit. Solar energy supplies heat for the absorption unit. A minimum external power supply is necessary in this cooling unit. Cooling cycle is shown.

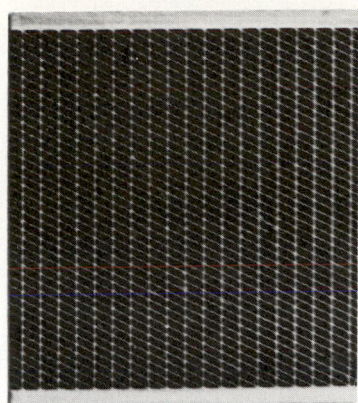

Fig. 27-16. Photograph of a typical solar cell. It converts solar energy into electrical energy. (Solarex Corp.)

having an excess of electrons.

The silicon of the P-type material has a small quantity of material which, compared to silicon, has a deficit of electrons. Aluminum, boron, gallium or indium are such materials.

The protective cover is either transparent plastic or glass. It must:

1. Allow the light to enter the cell.
2. Be cleanable.
3. Resist the effect of weather.

The electrodes on the back are also reflectors so that any light that goes through the cell is passed back through again so that more light can be absorbed.

The P-type material must also be transparent to allow the light to reach the junction between the P and N-type materials. The figure shows the P-type material wrapped around the N-type material. This allows the light to pass through a second junction, giving higher efficiency. Not all cells have the P-type material on the back of the N-type material. If the N-type material is not very transparent, the light will not make it through, and the second P layer is ineffective. Fig. 27-16 is a photograph of a typical solar cell.

27-24 PHOTOVOLTAIC SOLAR CELL APPLICATIONS

Electrically, solar cells behave very much like batteries. A single cell produces about one volt of electricity. This is nearly constant under normal operating conditions. It does not change very much with light intensity. However, under operating conditions, the current produced increases as sunlight intensity increases. At very high light intensity, the current reaches a saturation point and, if the cell is not cooled, the current will decrease.

Solar cells are presently used in specialized applications. Some interesting applications include:

1. Providing electrical energy for remote national monuments.
2. Space satellite power for radio and television communications.
3. Electrical control of motors which must deliver a different amount of work depending upon the sun intensity. This application includes control of air conditioning equipment

and solar heating systems.

Photovoltaic semiconductor devices (solar cells) have no moving parts and remain active for years without maintenance.

If a higher voltage is required, solar cells may be connected in series like a battery. The total voltage will then be the sum of the voltages produced by each cell connected. Fig. 27-17 shows an electrical circuit for delivery of power to an electrical load. If the load is an electrical motor, it would deliver more work when the sunlight intensity is high at noon than in the late afternoon. (At night the motor would not run at all.)

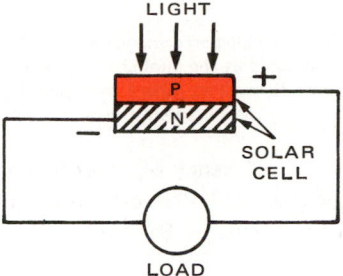

Fig. 27-17. Simple solar cell circuit. Power produced will depend on sunlight intensity.

Solar cells usually need to be connected to an electrical storage device like a storage battery. A motor can draw power from the solar cell directly in bright sunlight and from the storage battery when there is little or no sunlight. The control system would be designed to store electrical energy in the battery in the brightest sunlight. This energy would then be drawn out of the battery when light intensity is low as on cloudy days or at night.

An electrical circuit to provide controlled power would include a storage battery. It would be connected as shown in Fig. 27-18.

A solar cell produces d-c power like a battery. If a-c power is required to operate an electrical appliance or a-c motor, an inverter must be added to the circuit. Modern inverters use solid state semiconductors to produce a-c power from d-c

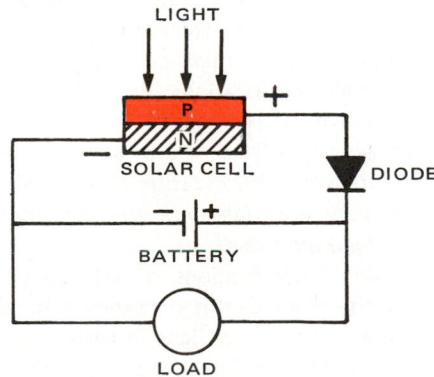

Fig. 27-18. Solar cell circuit provides for electrical energy storage in a battery. But it still provides a constant flow of energy to a load.

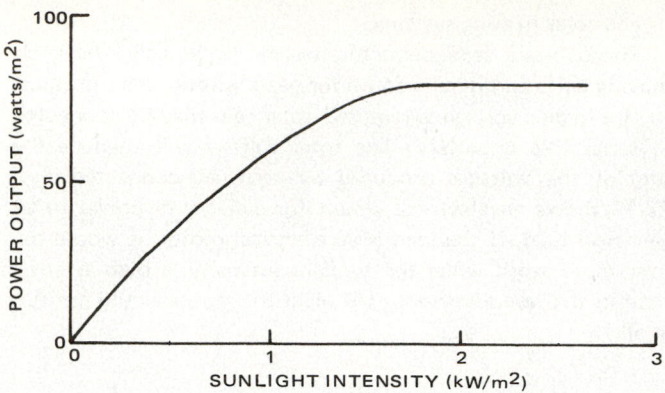

Fig. 27-19. Typical solar cell performance curve. This illustrates how the power output increases as the sunlight intensity increases. At high light intensity, a limit is reached where increased light intensity does not increase output.

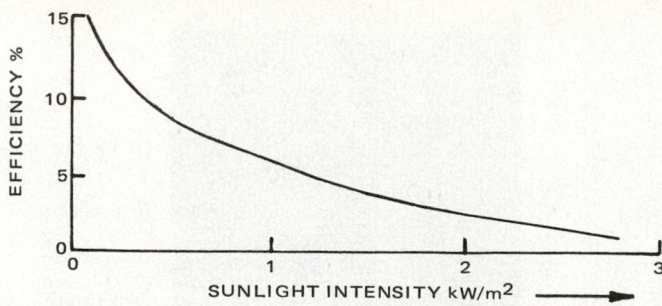

Fig. 27-20. Efficiency of a typical solar cell varies with light intensity.

power. They have an efficiency of about 90 percent. Inverters are expensive, however, and often d-c motors are used directly to avoid the cost of an inverter. See Para. 6-56.

27-25 SOLAR CELL PERFORMANCE

Solar cell performance depends heavily upon the intensity of sunlight and the temperature of the solar cell. Fig. 27-19 shows a typical performance curve for a working cell. The electrical power delivered by the cell increases with the sunlight intensity. If intensity continues to increase, the output power eventually levels off as shown. The leveling off is the result of excessive heat buildup in the cell. This heat can be carried away by circulating a cooling fluid in back of the cell or cells.

Current output can be increased by using a focusing collector to increase the light intensity on the cell. Presently, most practical applications are designed without a focusing system. Experimental focusing systems with intensity up to 10 times full sunlight have operated successfully when the cells are cooled. But most research is directed at producing cells which do not require focusing. Since a focusing system must follow or "track" the sun across the sky, it is expensive.

While the performance curve indicates that power output increases with solar intensity, the accompanying increase in power becomes smaller and smaller the greater the intensity. Often, manufacturers of solar cells will indicate solar cell efficiency in their specifications. Fig. 27-20 shows how this efficiency varies with solar intensity.

The efficiency is defined as the output power divided by the sunlight intensity. This indicates that the efficiency can be very high at very low light intensity but that it decreases rapidly as the light increases.

Manufacturers' specifications of efficiency should show the solar intensity at which this efficiency is achieved.

Typical solar cells at low light intensity can have an efficiency of 15 percent. At full sunlight, the efficiency may be only 5 percent.

A solar energy system having a concentrating collector with cooled cells can be used to provide both electrical energy and

heating. These systems are now in the experimental stage.

27-26 REVIEW OF SAFETY

There may be some hazards connected with the handling of solar energy equipment. Some plumbing connections may be required. If the pipes are not well made, and connectors are not tight, some leakage may result. Also, some plastic materials may not withstand the temperatures and pressures to which they are subjected over a long period of time.

Some attention needs to be given to the use of dissimilar (unlike) metals in a system. Electrolysis may corrode the two materials at the point where they are in contact.

Some insulating materials used with solar systems are flammable. Use care when soldering, brazing, or welding near these materials.

Caution must be used in domestic solar water heating systems that use ethylene glycol or other additives to make sure that there are no leaks into the water in the hot water tank. A pressure regulation system which maintains the hot water at higher pressure than the fluid in the solar collector circulation helps provide some protection.

When installing solar collectors, protect the eyes from bright sunlight reflecting off surfaces. Some vacuum tube solar collectors can also implode (burst inward) if broken. Wear eye protection when installing collectors.

27-27 TEST YOUR KNOWLEDGE

1. Is solar energy a new source of energy?
2. Are all solar rays visible?
3. What sun rays are absorbed by the atmosphere?
4. Wavelengths of the sun's rays are measured in microns. What is a micron?
5. What percentage of the sun rays are visible?
6. What are the names of the sun rays that are not visible?
7. Do the invisible rays produce any heat?
8. What is the approximate value in watts of the sun rays on a clear day?
9. What is a passive solar energy system?
10. Can the sun's rays be converted directly into electricity?
11. What is a solar collector?
12. Is the wave length of solar energy the same as the wavelength of heat rays?
13. Why is it necessary to provide heat insulation to solar

collectors?

14. What color absorbs solar energy best?
15. How may solar energy be trapped?
16. What is a selective absorption surface?
17. What is the best angle for the sun's rays to strike the absorber?
18. What should be the nature of the surface of a solar collector?
19. Why is glass usually used to cover a solar collector?
20. Why are some solar collectors sealed in a vacuum tube?
21. What are common ways of storing solar energy?
22. Is solar energy stored as solar energy or as heat?
23. What is the advantage of using crushed rock as a heat storage system?
24. How may water be used as a heat storage system?
25. What are the advantages of using water as a heat storage system?
26. What solution may be used in cold climates to eliminate freezing in the liquid filled collector?
27. If water is used for a heat storage system, how much water volume should be supplied for each square metre of collector surface?
28. At what angle to the horizontal should the collector surface be set, for best effect in the midsection of the United States?

29. Why is it necessary to insulate the backs of most solar collectors?
30. When should a differential thermostat be used?
31. What is the most common heat storage system used with a warm air heating system?
32. Can solar energy be used to heat domestic water?
33. Does the recommended domestic warm water system use one or two tanks?
34. What are some of the advantages of using solar energy for heating swimming pools?
35. What is meant by supplementary heat?
36. What is the most common supplementary heat?
37. What is a heat exchanger?
38. What is a heat pump?
39. Can solar energy be used for cooling?
40. Why is solar energy not being employed more often to create electricity?
41. What is the main substance used in making a solar cell?
42. What is the voltage provided in a solar cell?
43. How should solar cells be connected to produce a higher voltage?
44. Should solar cells be kept cold or heated?
45. Is the solar cell efficiency greatest with high or low sunlight intensity?

Chapter 28

TECHNICAL CHARACTERISTICS

28-1 KATA THERMOMETER

A Kata thermometer is used to measure air currents in open spaces. It is an alcohol thermometer with a Fahrenheit scale etched in the glass. There are two scales available. One reads from 95 to 100 F.; the other reads from 125 to 130 F.

To use the thermometer, heat it to the higher value in a hot water bath. Thoroughly dry the thermometer and suspend it in the air current. Then record the time it takes, in seconds, to cool to the lower reading on the F. scale. Using this information, determine the air movement in feet per minute from a Kata thermometer table.

Fig. 28-1. An infrared thermometer. It has two scales, 0 to 300 F (−18 to 149 C) and 200 to 600 F (93 to 316 C). A—Optical sight. B—Meter. C—Trigger. D—Hi-Low side switch and battery scale. E—Zero adjustment. F—Adjustment for type of surface. Batteries are in handle. (William Wahl Corp.)

28-2 INFRARED THERMOMETER

Temperatures may be measured with an infrared thermometer, Fig. 28-1. This optical-electronic instrument gives almost instantaneous readings. It is aimed at the surface for which the temperature is to be determined. The temperature is read directly from the temperature scale.

Different materials and types of surfaces give off heat at different rates. It is therefore necessary to set the instrument according to the nature of the surface. The adjustment for this is called the emissivity adjustment. Some emissivity adjustments are given in Fig. 28-2.

INFRARED THERMOMETER ADJUSTMENT VALUES

MATERIAL		EMISSIVITY
ALUMINUM	BRIGHT	0.09
	ANODIZED	0.55
	OXIDIZED	0.2 to 0.3
BRASS	BRIGHT	0.03
	OXIDIZED	0.61
CHROMIUM	POLISHED	0.08
COPPER	BRIGHT	0.05
	OXIDIZED	0.78
IRON AND STEEL	POLISHED	0.55
	OXIDIZED	0.85
NICKEL	POLISHED	0.05
	OXIDIZED	0.95
ZINC	BRIGHT	0.23
	OXIDIZED	0.23
BRICK	BUILDING	0.45
PAINTS	WHITE	0.9
	BLACK	0.86
	OIL PAINTS (ALL COLORS)	0.92
ROOFING PAPER		0.91
RUBBER		0.94
SILICA		0.42 to 0.62
WATER		0.92

Fig. 28-2. Emissivity of various surfaces.

28-3 WEIGHTS AND SPECIFIC HEATS OF SUBSTANCES

MATERIAL	WEIGHT LBS./CU.FT.	SPECIFIC HEAT Btu/LB.
GASES		
Air (normal temp.)	.075	.24
METALS		
Aluminum	166.5	.214
Copper	552	.094
Iron	480·	.118
Lead	710	.030
Mercury	847	.033
Steel	492	.117
Zinc	446	.096
LIQUIDS		
Alcohol	49.6	.60
Glycerine	83.6	.576
Oil	57.5	.400
Water	62.4	1.000
OTHERS		
Concrete	147	.19
Cork	15	.48
Glass	164	.199
Ice	57.5	.504
Masonry	112	.200
Paper	58	.324
Rubber	59	.48
Sand	100	.195
Stone	138–200	.20
Tar	75	.35
Wood, Oak	48	.57
Wood, Pine	38	.47

28-4 ENERGY

There are two kinds of energy — potential and kinetic.

Potential energy is like a body of water controlled by a dam. The water has a great potential for doing work; but, until the water flows through a water wheel, no energy is produced.

Kinetic energy is the energy of a moving body. The formula is $KE = 1/2\,MV^2$ = Foot Pounds. M = Weight, divided by 32 (the acceleration due to gravity). V = Velocity in feet per second (fps).

28-5 ENERGY EQUIVALENTS — ENGLISH

1 Btu	= 778 ft. lb.
	= 252 gram-calories
	= 1054.8 joules
1 Horsepower	= 33,000 ft. lb./min.
	= 550 ft. lb./sec.
	= 746 watts
	= 2545.6 Btu/hr.
	= 42.42 Btu/min.
	= 1.014 hp (metric)
1 Horsepower hour	= 1 horsepower for 1 hr.
	= 1,980,000 ft. lb.
	= 746 watt hours
	= .746 kilowatt hours
	= 2545.6 Btu
1 Kilowatt	= 1000 watts
	= 1.34 horsepower
1 Kilowatt hour	= 1 kilowatt for 1 hr.
	= 1000 watt hours

ICE MELTING EFFECT (IME)

1 ton of Refrigeration	= 288,000 Btu/day
	= 12,000 Btu/hr.
	= 200 Btu/min.
	= 83.3 lb. of ice/hr.

28-6 ENERGY EQUIVALENTS — METRIC

ENERGY

1 dyne cm = 1 erg = 0.001 g·cm = 7.38 x 10.8 ft. lb.
1 g·cm = 980.6 ergs = 7.233 x 10.5 ft. lb.
1 ft. lb. = 13,557,300 ergs = 13,825.5 g·cm
1 therm = 100,000 Btu

RATE OF ENERGY

1 erg/sec. = 1 dyne cm = 7.38 x 10.8 ft. lb./sec.
1 g·cm = 980.6 ergs/sec. = 7.24 x 10.5 ft. lb./sec.
1 ft. lb./sec. = 13,557,300 ergs/sec. = 13,800 g·cm/sec.

SI metric unit for energy is the joule. The erg and the dyne are part of the CGS metric system. CGS stands for centimetre-gram-second, a metric system that is widely used in all branches of science throughout the world before adoption of SI units.

28-7 LINEAR MEASURE EQUIVALENTS — ENGLISH — METRIC

1 inch		= 2.54 centimetres
		= 25.4 millimetres
		= 25 400 microns
1 foot	= 12 inches	= .304 metres
		= 30.48 centimetres
1 yard	= 3 feet	= .914 metres
		= 91.44 centimetres
1 micron		= .000 394 inches
1 millimetre	= 1000 microns	= .0394 inches
1 centimetre	= 10 millimetres	= .3937 inches
1 metre	= 100 centimetres	= 39.37 inches
1 kilometre	= 1000 metres	= .62137 miles

28-8 FRACTIONS, DECIMALS AND MILLIMETRES OF THE PARTS OF AN INCH

INCH	DECIMAL INCH	MILLIMETER
1/64	0.0156	0.3967
1/32	0.0312	0.7937
3/64	0.0468	1.1906
1/16	0.0625	1.5875
5/64	0.0781	1.9843
3/32	0.0937	2.3812
7/64	0.1093	2.7781
1/8	0.125	3.175
9/64	0.1406	3.5718
5/32	0.1562	3.9687
11/64	0.1718	4.3656
3/16	0.1875	4.7625
13/64	0.2031	5.1593
7/32	0.2187	5.5562
15/64	0.2343	5.9531
1/4	0.25	6.5
17/64	0.2656	6.7468
9/32	0.2812	7.1437

Continued

INCH	DECIMAL INCH	MILLIMETER
19/64	0.2968	7.5406
5/16	0.3125	7.9375
21/64	0.3281	8.3343
11/32	0.3437	8.7312
23/64	0.3593	9.1281
3/8	0.375	9.525
25/64	0.3906	9.9218
13/32	0.4062	10.3187
27/64	0.4218	10.7156
7/16	0.4375	11.1125
29/64	0.4531	11.5093
15/32	0.4687	11.9062
31/64	0.4843	12.3031
1/2	0.50	12.7
33/64	0.5162	13.0968
17/32	0.5312	13.4937
35/64	0.5468	13.8906
9/16	0.5625	14.2875
37/64	0.5781	14.6843
19/32	0.5937	15.0812
39/64	0.6093	15.4781
5/8	0.625	15.875
41/64	0.6406	16.2718
21/32	0.6562	16.6687
43/64	0.6718	17.0656
11/16	0.6875	17.4625
45/64	0.7031	17.8593
23/32	0.7187	18.2562
47/64	0.7343	18.6531
3/4	0.75	19.05
49/64	0.7656	19.4468
25/32	0.7812	19.8437
51/64	0.7968	20.2406
13/16	0.8125	20.6375
53/64	0.8281	21.0343
27/32	0.8437	21.4312
55/64	0.8593	21.8281
7/8	0.875	22.225
57/64	0.8906	22.6218
29/32	0.9062	23.0187
59/64	0.9218	23.4156
15/16	0.9375	23.8125
61/64	0.9531	24.2093
31/32	0.9687	24.6062
63/64	0.9843	25.0031
1	1.0000	25.4

28-9 AREA EQUIVALENTS

1 sq. in.   = .0065 sq. metres (m^2)
1 sq. ft. = 144 sq. in. = .093 sq. metres (m^2)
1 sq. yd. = 9 sq. ft.   = .836 sq. metres (m^2)
1 sq. yd. = 1296 sq. in.

28-10 VOLUME EQUIVALENTS

1 cu. in.   = .016 litres
                        = 16.39 cm^3

1 cu. ft. = 1728 cu. in. = 28.317 litres = .0283 m^3
      = 7.481 gal.   = 28 317.00 cm^3

1 cu. yd. = 27 cu. ft.
       = 46,656 cu. in.

1 gal.   = .1337 cu. ft. = 3.79 litres
         = 231 cu. in.   = 3785 cm^3

1 cm^3 = .155 cu. in.

1 litre   = 61.03 cu. in. = 1 000 cm^3
       = .2642 gal.

28-11 PRESSURE EQUIVALENTS

1 psi   = 0.068 atmosphere = .0703 kg/cm^2
          = 144 lb./sq. ft.   = .703 metres water
          = 2.036 in. of mercury = 70.3 cm water
          = 2.307 ft. of water   = 51.7 mm Hg.
          = 27.7 in. of water   = 6.9 kPa

1 oz./sq. in.   = .128 in. of mercury
          = 1.73 in. of water

1 in. of mercury = .0334 atmosphere = .0345 kg/cm^2
          = .491 psi   = 25.4 mm Hg.
          = 1.13 ft. of water   = .3453 m water
          = 13.6 in. of water
          = 70.73 psf

1 ft. of water   = .0295 atmosphere = .03 kg/cm^2
          = .434 psi   = 22.42 mm Hg.
          = 62.43 lb./sq. ft. = .305 m water
          = .03 atmosphere
          = .883 in. of mercury (Hg.)

1 atmosphere   = 29.92 in. of mercury = 1.03 kg/cm^2
          = 33.94 ft. of water = 760 mm Hg.
          = 14.696 psi   = 10.33 m water
          = 2116.35 psf

1 psf   = .007 psi   = 4.88 x 10^4 g/cm
          = 4.725 x 10.4   = .359 mm Hg.
            atmosphere
          = .01414 in. Hg.   = .0049 m water
          = .016 ft. water

1 kilogram/sq. cm = 14.22 psi   = 10 metres of water
          = 2048.17 psf
          = .967 atmosphere
          = 28.96 in. Hg.
          = 32.8083 in. water

1 metre of water = 1.42 psi   = 73.55 mm Hg.
          = 204.8 psf   = 10 kg/cm^2
          = .097 atmosphere
          = 2.896 in. Hg.
          = 3.28 ft. water

1 mm Hg.   = .019 psi   = .001 36 kg/cm^2
          = 2.78 psf   = .0136 m water
          = .001316 atmosphere
          = .039 in. Hg.
          = .0446 ft. water

28-12 VELOCITY EQUIVALENTS

1 mi./hr.   = 1.47 ft./sec. = 1.61 km/hr.
         = .87 knots   = .45 metres/sec.

1 ft./sec.   = .68 mi./hr.   = 1.1 km/hr.
         = 60 ft./min. = .305 metres/sec.
         = .59 knots

1 metre/sec. = 3.28 ft./sec. = 3.6 km/hr.
         = 2.24 mi./hr.
         = 1.94 knots

1 km/hr.   = .91 ft./sec.   = .28 metres/sec.
         = .62 mi./hr.
         = .54 knots

28-13 LIQUID MEASURE EQUIVALENTS

Liquid Measure	U.S.	Metric
1 pint	= 16 ounces	= .473 litres
1 quart	= 2 pints	= .946 litres
1 quart	= 32 ounces	
1 gallon	= 4 quarts	= 3.785 litres
1 gallon	= 8 pints	
1 gallon	= 231 cubic inches	
1 cubic foot	= 7.48 gallons	
1 gallon	= 8.34 pounds of water	
1.136 quart		= 1 litre

28-14 WEIGHT EQUIVALENTS

AVOIRDUPOIS

1 ounce	= 473 grains	= 28.35 grams
		= .028 kilogram
1 pound (lb.)	= 7000 grains	= .4536 kilograms
		= 453.6 grams
1 pound	= 16 ounces	= 453.6 grams
1 grain	= .00043 pounds	= .064 80 grams
1 ton	= 2000 pounds	= 909.09 kilograms
1 gram	= 15.43 grains	= .001 kilogram
	= .03527 ounces	
	= .002205 pounds	
1 kilogram	= 2.2 pounds	

SPECIFIC WEIGHTS (DENSITY)

1 lb./cu. in.	= 1728 lb./cu. ft.	= 27.68 grams per cubic metre (g/m^3)
1 lb./cu. ft.	= 5.787 x 10.4 lb./cu. in.	= .016 g/cm^3
1 gm/cm^3	= 62.43 lb./cu. ft.	
1 kg/m^3	= .06243 lb./cu. ft.	

28-15 FLOW EQUIVALENTS

1 cu. ft. per min.	= 7.481 gal./min.	= 28 317 cm^3/min.
	= 449 gal./hr.	= 28.32 litres/min. (l/min.)
		= 1 700.00 l/hr.
1 cu. ft. per hour	= .0167 cu. ft./min.	= .472 l/min.
	= .1247 gal./min.	= 28.317 l/hr.
	= 7.481 gal./hr.	= 472 cm^3/min.
1 gal. per min.	= .1337 cu. ft./min.	= 3.79 l/min.
	= 8.022 cu. ft./hr.	= 3785 cm^3/min.
1 litre per min.	= .0353 cu. ft./min.	= 1 000 cm^3/min.
	= 2.118 cu. ft./hr.	
	= .2642 gal./min.	
	= 15.852 gal./hr.	

28-16 TEMPERATURE CONVERSION TABLE

To use the following table, find the temperature to be converted in the center column. If converting to Celsius, read to the left. If converting to Fahrenheit, read to the right.

°C		F.
−273.15	−459.67	
−268	−450	
−262	−440	
−257	−430	
−251	−420	

°C		F.
−246	−410	
−240	−400	
−234	−390	
−229	−380	
−223	−370	

°C		F.
−218	−360	
−212	−350	
−207	−340	
−201	−330	
−196	−320	
−190	−310	
−184	−300	
−179	−290	
−173	−280	
−169	−273	−459.4
−168	−270	−454
−162	−260	−436
−157	−250	−418
−151	−240	−400
−146	−230	−382
−140	−220	−364
−134	−210	−346
−129	−200	−328
−123	−190	−310
−118	−180	−292
−112	−170	−274
−107	−160	−256
−101	−150	−238
−95.6	−140	−220
−90.0	−130	−202
−84.4	−120	−184
−78.9	−110	−166
−73.3	−100	−148
−67.8	−90	−130
−62.2	−80	−112
−56.7	−70	−94
−51.1	−60	−76
−45.6	−50	−58
−40.0	−40	−40
−34.4	−30	−22
−28.9	−20	−4
−23.3	−10	14
−17.8	0	32
−17.2	1	33.8
−16.7	2	35.6
−16.1	3	37.4
−15.6	4	39.2
−15.0	5	41.0
−14.4	6	42.8
−13.9	7	44.6
−13.3	8	46.4
−12.8	9	48.2
−12.2	10	50.0
−11.7	11	51.8
−11.1	12	53.6
−10.6	13	55.4
−10.0	14	57.2
−9.4	15	59.0
−8.9	16	60.8
−8.3	17	62.6
−7.8	18	64.4
−7.2	19	66.2
−6.7	20	68.0
−6.1	21	69.8
−5.6	22	71.6
−5.0	23	73.4
−4.4	24	75.2
−3.9	25	77.0
−3.3	26	78.8
−2.8	27	80.6
−2.2	28	82.4
−1.7	29	84.2
−1.1	30	86.0

°C		F.
−0.6	31	87.8
0	32	89.6
0.6	33	91.4
1.1	34	93.2
1.7	35	95.0
2.2	36	96.8
2.8	37	98.6
3.3	38	100.4
3.9	39	102.2
4.4	40	104.0
5.0	41	105.8
5.6	42	107.6
6.1	43	109.4
6.7	44	111.2
7.2	45	113.0
7.8	46	114.8
8.3	47	116.6
8.9	48	118.4
9.4	49	120.2
10.0	50	122.0
10.6	51	123.8
11.1	52	125.6
11.7	53	127.4
12.2	54	129.2
12.8	55	131.0
13.3	56	132.8
13.9	57	134.6
14.4	58	136.4
15.0	59	138.2
15.6	60	140.0
16.1	61	141.8
16.7	62	143.6
17.2	63	145.4
17.8	64	147.2
18.3	65	149.0
18.9	66	150.8
19.4	67	152.6
20.0	68	154.4
20.6	69	156.2
21.1	70	158.0
21.7	71	159.8
22.2	72	161.6
22.8	73	163.4
23.3	74	165.2
23.9	75	167.0
24.4	76	168.8
25.0	77	170.6
25.6	78	172.4
26.1	79	174.2
26.7	80	176.0
27.2	81	177.8
27.8	82	179.6
28.3	83	181.4
28.9	84	183.2
29.4	85	185.0
30.0	86	186.8
30.6	87	188.6
31.1	88	190.4
31.7	89	192.2
32.2	90	194.0
32.8	91	195.8
33.3	92	197.6
33.9	93	199.4
34.4	94	201.2
35.0	95	203.0
35.6	96	204.8
36.1	97	206.6
36.7	98	208.4
37.2	99	210.2
37.8	100	212.0

°C		F.
43	110	230
49	120	248
54	130	266
60	140	284
66	150	302
71	160	320
77	170	338
82	180	356
88	190	374
93	200	392
99	210	410
100	212	413
104	220	428
110	230	446
116	240	464
121	250	482
127	260	500
132	270	518
138	280	536
143	290	554
149	300	572
154	310	590
160	320	608
166	330	626
171	340	644
177	350	662
182	360	680
188	370	698
193	380	716
199	390	734
204	400	752
210	410	770
216	420	788
221	430	806
227	440	824
232	450	842
238	460	860
243	470	878
249	480	896
254	490	914
260	500	932
266	510	950
271	520	968
277	530	986
282	540	1004
288	550	1022
293	560	1040
299	570	1058
304	580	1076
310	590	1094
316	600	1112
321	610	1130
327	620	1148
332	630	1166
338	640	1184
343	650	1202
349	660	1220
354	670	1238
360	680	1256
366	690	1274
371	700	1292
377	710	1310
382	720	1328
388	730	1346
393	740	1364
399	750	1382

°C		F.
404	760	1400
410	770	1418
416	780	1436
421	790	1454
427	800	1472
432	810	1490
438	820	1508
443	830	1526
449	840	1544
454	850	1562
460	860	1580
466	870	1598
471	880	1616
477	890	1634
482	900	1652
488	910	1670
493	920	1688
499	930	1703
504	940	1724
510	950	1742
516	960	1760
521	970	1778
527	980	1796
532	990	1814
538	1000	1832
543	1010	1850
549	1020	1868
554	1030	1886
560	1040	1904
566	1050	1922
571	1060	1940
577	1070	1958
582	1080	1976
588	1090	1994
593	1100	2012
599	1110	2030
604	1120	2048
610	1130	2066
616	1140	2084
621	1150	2102
627	1160	2120
632	1170	2138
638	1180	2156
643	1190	2174
649	1200	2192
704	1300	2372
760	1400	2552
816	1500	2732
871	1600	2912
927	1700	3092
982	1800	3272
1038	1900	3452
1093	2000	3632
1149	2100	3812
1204	2200	3992
1260	2300	4172
1316	2400	4352
1371	2500	4532
1427	2600	4712
1482	2700	4892
1538	2800	5072
1593	2900	5252
1649	3000	5432

°C		F.		°C		F.
0.56	1	1.8		3.33	6	10.8
1.11	2	3.6		3.89	7	12.6
1.67	3	5.4		4.44	8	14.4
2.22	4	7.2		5.00	9	16.2
2.78	5	9.0		5.56	10	18.0

Courtesy of Thermatron Corporation

28-17 HEAT EQUIVALENTS

1 Btu	=	252 calories
1 kilocalorie	=	1000 calories
1 Btu/lb.	=	.55 kcal/kg
1 Btu/lb.	=	2.326 kJ/kg
1 kcal/kg	=	1.8 Btu/lb.
1 Btu/hr.	=	0.2931 watts

Unit of heat — pound Celsius or Centigrade heat unit is the heat required to raise one pound of water 1 C. degree and equals 1.8 Btu.

1 Btu/hr./sq. ft./F. $= 4.8$ kgcal/hr./m^2/C.

1 kgcal/hr./m^2/C. $= .205$ Btu/hr./sq. ft./F.

28-18 COMPRESSION RATIO

Compression ratio is the ratio of the volume of the clearance space in a compressor to the total volume of the cylinder.

For Example: A compressor has a 2-in. bore and a 2-in. stroke and a clearance space of .010 in. The volume in the valve ports equals .05 cu. in. The piston displacement is $\pi r^2 \times S$.

$\pi = 3.1416$

$r = 1$

$S = 2$

The volume equals $3.1416 \times 1 \times 2 = 6.2832$ cu. in.

The volume of the clearance space equals $\pi r^2 \times r \times S$

$\pi = 3.1416$

$r = 1$

$S = .010$

The volume equals $3.1416 \times 1 \times .010 = .031416 = .03$ cu. in.

Total clearance space: $.05 + .03 = .08$ cu. in.

The total volume of the cylinder is:

$6.2832 + .08 = 6.3632 = 6.36$

The compression ratio is $6.36 \div .08 = 79.5{:}1$

28-19 PUMPING RATIO

The pumping ratio is the ratio of the suction pressure to the condensing or heat pressure expressed in absolute pressures. Absolute pressure equals gauge pressure plus 15.

For example, if the low-side pressure of a system is 15 psi:

$15 + 15 = 30$ psi;

and the high-side pressure is 150 psi:

$150 + 15 = 165$ psi,

the pumping ratio equals $\frac{30}{165} = \frac{1}{5.5} = 5.5{:}1$.

The pumping ratio should not exceed a certain value for each particular refrigerant. If the pumping ratio is too high, the temperature of the high-pressure vapor going through the exhaust valve would overheat the mechanism and cause some of the oil in the exhaust pocket to become carbonized.

28-20 LAWS OF THERMODYNAMICS

FIRST LAW OF THERMODYNAMICS is a formula for the conversion of heat into work or work into heat. The formula is: 778 foot pounds of work is equivalent to the heat energy of 1 Btu.

SECOND LAW OF THERMODYNAMICS states that heat will only flow from a body at a certain temperature to another body which is at a lower temperature.

28-21 ENTROPY

Entropy is the heat available measured in Btu per pound degree change for a substance.

Entropy calculations are made from generally accepted temperature bases. For heating and steam power using water as the medium, the accepted base is 32 F. (0 C.). For domestic and most commercial refrigeration calculations, the base is —40 F. (—40 C.). For research and very low temperature work, a base of a lower temperature may be selected.

Entropy is generally used only in engineering calculations. Entropy tables have been worked out and are contained in most engineering handbooks.

28-22 THEORY OF MATTER

To understand electricity and electonics, one must know the properties of matter. All materials are made up of molecules. A molecule is the smallest part of a substance that has all the properties of that substance. All molecules are made up of atoms. The molecules may contain as few as one atom or up to hundreds of atoms.

Atoms are made mainly of electrons, neutrons and protons. All the material in the universe is made from only about 110 different atoms. Each of these is different, depending on the number of electrons, protons, and neutrons in it.

All substances have mass, which is determined by the amount of matter contained in a substance. Mass of a body, then, is determined by dividing the weight of the body by the acceleration due to gravity, 32.2

The total of mass and energy is constant. Mass can be turned into energy. Energy can be turned into mass. The equation (A. Einstein's) is: $E = MC^2$. E = energy. M = mass. C = speed of light.

28-23 ELECTRON THEORY

Various principles of construction and operation of electrical equipment are based on theories which have come to be accepted principles of operation.

Scientific studies indicate that all substances are made up of molecules. Molecules are the smallest part of a substance that has all the properties of the substance. Molecules are made up of atoms. Atoms are made up of electrons, neutrons and protons. Electrons are negative charges of electricity. Neutrons

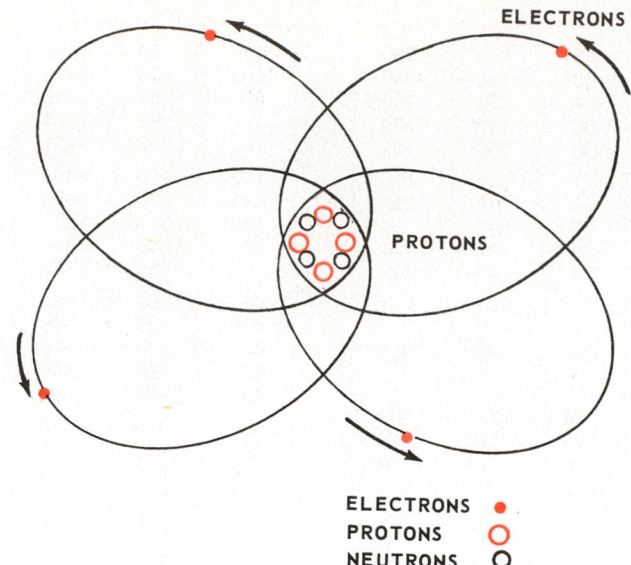

Fig. 28-3. Makeup of an atom, showing electrons, protons, neutrons and their relative motion.

are neither negatively nor positively charged. Protons are positive charges of electricity.

The electrons revolve around the nucleus (center). The nucleus is made of protons and neutrons. This theory is illustrated in Fig. 28-3. The revolving electrons are of two types:

1. Free electrons.
2. Bound electrons.

If an atom has free electrons, the atom can transfer or conduct electrical energy. If an atom collects an electron charge for an instant, it is negatively charged. If the atom loses an electron, it becomes positively charged. The electrons travel from atoms having extra electrons to atoms having a lack of electrons.

28-24 THE ATOM

An elementary drawing of an atom, Fig. 28-3, looks like the solar system, with a sun and planets. The electrons revolve around the nucleus (center). The nucleus is made of protons and neutrons. Atoms also seem to have many other particles, but these rare particles are, at present, only of interest and use to research and nuclear scientists.

The electrons seem to be little clouds of energy rather than a small sphere. Two electrons opposite each other seem to be shaped like a dumbbell. Some of the smaller particles may be what scientists see in a cloud chamber as they do research.

The revolving electrons are attracted to the center (nucleus), yet their speed of travel keeps them a small distance from the nucleus by centrifugal force. The force trying to pull the electron into the center is called the negative electrical charge (negative sign is —). The nucleus attraction is called the positive electrical charge (positive sign is +). Based on this attraction, electricity is movement of the electrons (—) toward the nucleus (+).

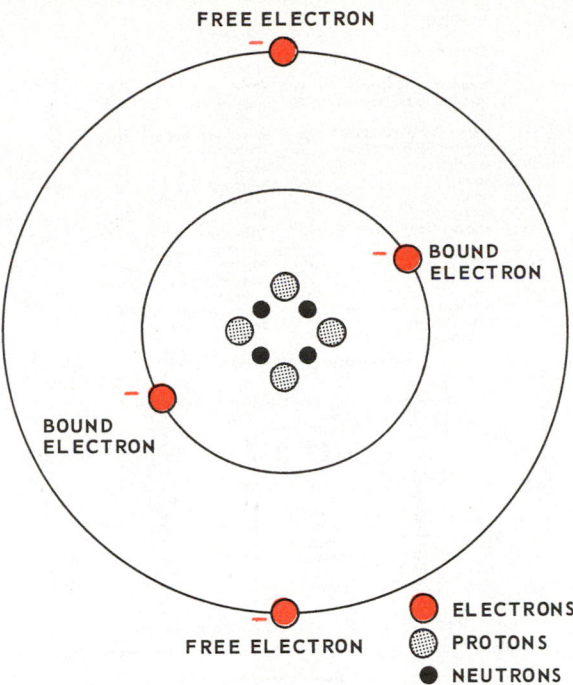

FREE ELECTRON

BOUND
ELECTRON

BOUND
ELECTRON

FREE ELECTRON

● ELECTRONS
◦ PROTONS
● NEUTRONS

Fig. 28-4. An atom with two free electrons, two bound electrons, four protons and four neutrons.

The revolving electrons are of two types.

1. Free electrons.
2. Bound electrons.

If an atom has free electrons, the atoms can transfer or conduct electricity (electrons can move from one atom to another). See Fig. 28-4.

When an atom enters a chemical change, it does so in the form of a charged particle. These particles are called ions. Ions are of two types.

1. Negative.
2. Positive.

A negative ion is an atom that is negatively charged (has an extra electron). A positive ion is an atom that is positively charged (lacks an electron). Like charged ions repel. Unlike charged ions attract.

In the lighter elements, the atom contains one proton for each neutron. In the heavier elements, the atom contains more neutrons than protons. For example:

Helium   =   2 protons,   2 neutrons
Boron   =   5 protons,   5 neutrons
Mercury = 80 protons, 120 neutrons

A proton and a neutron are equal in mass. Either one is about 1845 times as great in mass as the electron mass. For example, if one can imagine that the outer electrons of a typical atom structure are in an orbit 200 yards in diameter, then the nucleus in the center would be 1/2 in. in diameter.

In most atoms, the number of electrons equals the number of protons in the atom. See Fig. 28-5. The negative charges equal the proton charges, and the resulting (net) charge of the atom is zero (0) if no chemical action is taking place or if no electrical flow is taking place.

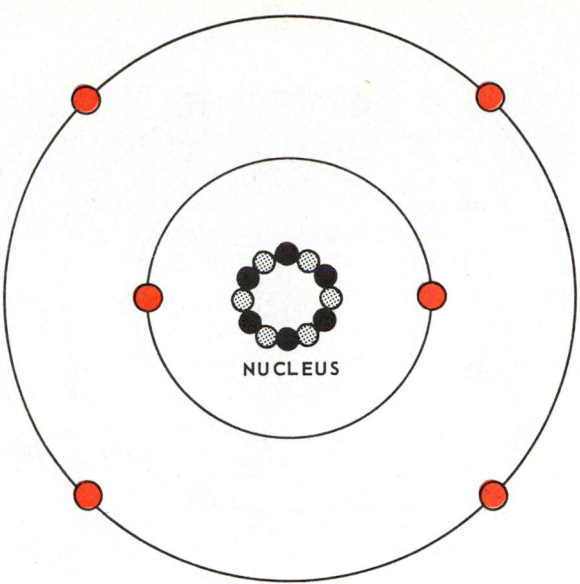

NUCLEUS

Fig. 28-5. A carbon atom with four free electrons, two bound electrons, six protons and six neutrons.

If an external force is applied to an atom, one or more of the outermost electrons may be removed. These removable electrons are commonly called "free" electrons.

When the outer electrons of an atom interact with electrons of another atom, they combine to form a compound.

In some substances, each atom will retain all its electrons. In other substances, one or more outer atoms will be gained or lost as a result of the bonding.

The nucleus does not enter into chemical or electrical processes. To disrupt the nucleus, a vast amount of energy is needed, such as in an explosion of an atom bomb.

Two forces act on revolving electrons:

1. Centrifugal force.
2. Centripetal force.

Centrifugal force tends to make electrons move away from protons.

Centripetal force tends to make electrons move toward protons. (Electron has a negative charge; proton has a positive charge.)

The electron orbit is a balance of these two forces.

The electron has two energies:

1. Kinetic (motion).
2. Potential (position).

These two energies together determine the radius of the electron orbit. To stay in a certain orbit, the electron must neither gain nor lose energy.

28-25 THE ELECTRON

Electrons are negative charges of electricity. Fig. 28-6 shows an atom including its nucleus. Protons are positive charges of electricity. If an atom collects an extra free electron for a moment, it is negatively charged. If an atom loses a free electron, it is positively charged. The electrons travel from atoms with extra electrons to atoms with a lack of electrons.

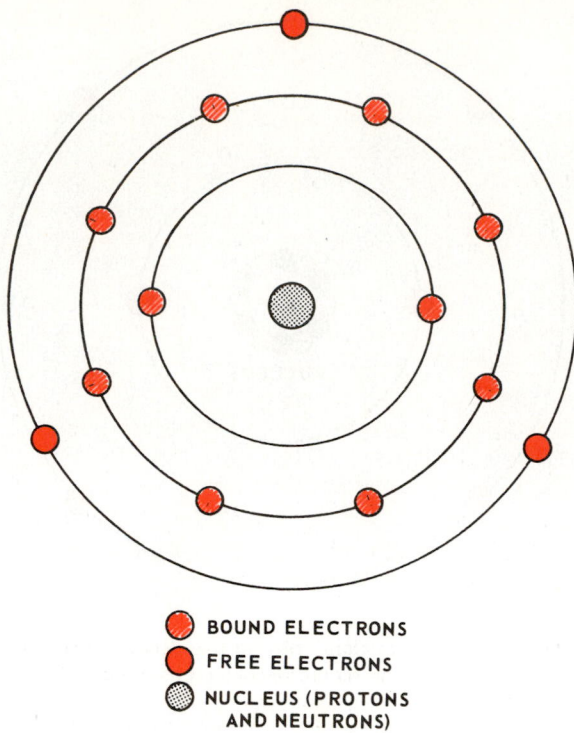

BOUND ELECTRONS
FREE ELECTRONS
NUCLEUS (PROTONS
AND NEUTRONS)

Fig. 28-6. An atom with three free electrons, ten bound electrons and a nucleus of 13 protons and 13 neutrons.

Electricity is the flow of these electrons (or negative charges). The number of electrons that flow is extremely high. The term coulombs is used to represent the flow of 6.24 x 10^{18} or 6,240,000,000,000,000,000 electrons. One coulomb flowing for one second equals one ampere. The ampere is known as the quantity of electricity.

If there are too many protons, and electrons move to make up the difference, this movement or pressure to move is called electromotive force. Electric current consists of electrons traveling in a conductor, which may be made of a solid, a liquid or a gas.

28-26 ELECTRICAL UNITS AND SYMBOLS

ELECTRICAL ENGINEERING UNITS AND CONSTANTS
As adopted by NBS

Symbols and Units

Quantity	Symbol	Unit	Symbol
charge	Q	coulomb	C
current	I	ampere	A
voltage, potential difference	V	volt	V
electromotive force	$\mathscr{E}$	volt	V
resistance	R	ohm	Ω
conductance	G	mho (siemens)	A/V, or mho (S)
reactance	X	ohm	Ω
susceptance	B	mho	A/V, or mho
impedance	Z	ohm	Ω
admittance	Y	mho	A/V, or mho
capacitance	C	farad	F
inductance	L	henry	H
energy, work	W	joule	J
power	P	watt	W
resistivity	ρ	ohm-meter	Ωm
conductivity	σ	mho per meter	mho/m
electric displacement	D	coulomb per sq. meter	C/m²
electric field strength	E	volt per meter	V/m
permittivity (absolute)	ϵ	farad per meter	F/m
relative permittivity	ϵ_r	(numeric)	
magnetic flux	Φ	weber	Wb
magnetomotive force	$\mathscr{F}$	ampere (ampere-turn)	A
reluctance	$\mathscr{R}$	ampere per weber	A/Wb
permeance	$\mathscr{P}$	weber per ampere	Wb/A
magnetic flux density	B	tesla	T
magnetic field strength	H	ampere per meter	A/m
permeability (absolute)	μ	henry per meter	H/m
relative permeability	μ_r	(numeric)	
length	l	meter	m
mass	m	kilogram	kg
time	t	second	s
frequency	f	hertz	Hz
angular frequency	ω	radian per second	rad/s
force	F	newton	N
pressure	p	newton per sq. meter	N/m²
temperature (absolute)	T	degree Kelvin	K
temperature (International)	t	degree Celsius	C

RECOMMENDED UNIT PREFIXES

Multiples and submultiples	Prefixes	Symbols	Pronunciation
10^{12}	tera	T	těr' á
10^{9}	giga	G	jĭ'gá
10^{6}	mega	M	mĕg' á
10^{3}	kilo	k	kĭl' ŏ
10^{2}	hecto	h	hěk' tŏ
10	deka'	da	děk' á
10^{-1}	deci	d	děs' ĭ
10^{-2}	centi	c	sĕn' tĭ
10^{-3}	milli	m	mĭl' ĭ
10^{-6}	micro	μ	mī' krŏ
10^{-9}	nano	n	năn' ŏ
10^{-12}	pico	p	pē' kŏ
10^{-15}	femto	f	fĕm' tŏ
10^{-18}	atto	a	ăt' tŏ

DEFINED VALUES AND CONVERSION FACTORS

Meter..........................	1 650 763.73 wavelengths of the transition $2p_{10} - 5d_5$ in ^{86}Kr
Kilogram....................	mass of the international kilogram
Second.......................	1/31 556 925.974 7 of the tropical year 1900
Degree Kelvin..............	In the thermodynamic scale, 273.16 °K = triple point of water (fp, 273.15 °K=0 °C)
Unified atomic mass unit, u	1/12 the mass of an atom of the ^{12}C nuclide
Standard acceleration of free fall	9.806 65 m s⁻², 980.665 cm s⁻²
Normal atmosphere........	101 325 N m⁻², 1 013 250 dyn cm⁻²
Thermochemical calorie .	4.1840 J, 4.1840×10⁷ erg
Int. Steam Table calorie .	4.1868 J, 4.1868×10⁷ erg
Liter..........................	0.001 000 028 m³, 1 000.028 cm³ (recommended by CIPM, 1950)
Inch..........................	0.0254 m, 2.54 cm
Pound (avdp.).............	0.453 592 37 kg, 453.592 37 g

28-27 MOTOR SIZE CALCULATIONS— MEAN EFFECTIVE PRESSURE METHOD

Horsepower of a motor required to drive a refrigeration compressor can be calculated. The usual method uses the mean effective pressure (mep) in the cylinder. This is the pressure which is bearing down on the piston head and which must be overcome by the electric motor when driving the compressor. The mep is determined by a formula. Basic variables of the formula are the low-side pressure, the high-side pressure and the ratio of the specific heat of constant pressure to the specific heat of constant volume $\frac{C_p}{C_v}$ = K (for the kind of refrigerant). The formula follows:

$$mep = P_1 \times \frac{K}{K-1}\left[\left(\frac{P_2}{P_1}\right)^{\frac{K-1}{K}} -1\right]$$

P_1 = suction pressure, psia
P_2 = condenser pressure, psia
$K = \frac{C_p}{C_v}$

This value, when multiplied by the area of the piston, by the length of the stroke, and by the rpm, will give the foot-pounds per minute needed to drive the compressor.

Ft. lb. per min. = mep x $\frac{\pi D^2}{4}$ x S x N x R

D = piston diameter

S = stroke in inches

N = number of cylinders

R = rpm

Convert the foot-pound per min. into hp by dividing by 33,000.

$$hp = \frac{ft. \; lb./min.}{33,000}$$

The mep may be determined from the above or may determined by obtaining the indicator card of the compressor being studied. Engineers' handbooks set forth the methods of using and obtaining indicator cards.

For example, if the indicator card shows a mep of 30 lb. per sq. in., the indicated hp necessary to drive a one cylinder compressor with a 2-in. bore and a 2-in. stroke, running at 200 rpm, would be as follows:

$$hp = \frac{30 \; psi \; x \; area \; x \; stroke \; x \; rpm}{33,000}$$

The dimensions have to be in foot-pounds = pressure x area. The resistance is the number of compression strokes x rpm =

$$\frac{30 \; psi \; x \; \frac{\pi D^2}{4} \; x \; \frac{2}{12} \; x \; 200}{33,000} =$$

$$\frac{30 \; x \; \pi D^2 \; x \; 2 \; x \; 200}{4 \; x \; 33,000 \; x \; 12} = \frac{30 \; x \; \pi 4 \; x \; 2 \; x \; 200}{4 \; x \; 33,000 \; x \; 12} =$$

$$\frac{\pi \; x \; 200}{6600} = \frac{\pi}{33} = .0952 \; or$$

approximately .10 hp.

This is the theoretical hp and neglects friction, oil pumping, starting load and drive losses. Up to a 1-ton machine, one should double this hp. For example, 1/4 hp, calculated, will need a 1/2 hp motor. Up to 5 tons, this ratio gradually tapers off until adding 30 percent at 5 tons capacity will take into consideration the above losses. The reason for the decrease is that some of the losses remain constant, while others do not increase as rapidly in proportion to the increase in the size of the unit.

28-28 LOWERING VOLTAGE SAVES POWER

During times of power shortage, some utilities reduce the voltage supplied to the customer. This may result in some power saving. The largest saving will come from noninductive loads, such as lighting, resistance heating and electric cooking. In these appliances, a reduction in voltage applied will cause some drop in the rate of current flow. Electrical power is equal to V x C — E x I equals watts — therefore, if both the voltage and current is reduced, the E x I will result in the reduction in the amount of power used.

In inductive loads there will not be a great saving. This means there will be very little saving in the operation of such appliances as refrigerators, air conditioners and ventilating equipment. If the voltage is dropped, the current required will automatically increase. Therefore, power consumed, E x I equals W, will be affected very little. If "E" goes down and "I" goes up, power required will be affected very little. If motor speed is reduced, there will be some saving.

28-29 RESISTANCES OF CONDUCTORS, SEMI-CONDUCTORS AND NONCONDUCTORS

The value, in ohms, of the resistance of conductors, semiconductors and nonconductors (insulators) varies from very large numbers to very small numbers. Instead of working with long numbers and long decimal numbers, the usual practice is to use the significant number multiplied by a power of 10:

10^3 means 10 x 10 x 10 = 1000

10^2 means 10 x 10 = 100

10^1 means 10

10^0 means 1

For decimals, a negative power of 10 is used (−10):

10^{-1} means .1

10^{-2} means .1 x .1 = .01

10^{-3} means .1 x .1 x .1 = .001

The following table lists the approximate resistance of conductors, semiconductors and nonconductors in ohms per centimetre of length:

For conductors, the resistance in ohms usually varies from:

$.000\;001 = 10^{-6}$

$.000\;01 = 10^{-5}$

$.000\;1 = 10^{-4}$

$.001 = 10^{-3}$

For semiconductors, the resistance in ohms usually varies from:

$1 = 10^0$

$10 = 10^1$

$100 = 10^2$

$1000 = 10^3$

$10,000 = 10^4$

$100,000 = 10^5$

$1,000,000 = 10^6$

For nonconductors, the resistance in ohms usually varies from:

10^9 to 10^{18}

28-30 RESISTANCE IN SERIES

The total resistance of resistances in series is the sum of the separate resistances.

28-31 RESISTANCES IN PARALLEL

The total resistance of resistances in parallel is equal to the reciprocal of the sum of the reciprocals of the separate resistances. This is illustrated in Fig. 28-7.

28-32 ELECTRICAL CODE

The United States Electrical Code and the Canadian Electrical Code each deal with the specifications having to do with refrigerating and air conditioning electrical circuits. In addition, most local communities have codes dealing with these circuits.

The refrigeration and air conditioning service technician should either have or have access to both the national codes

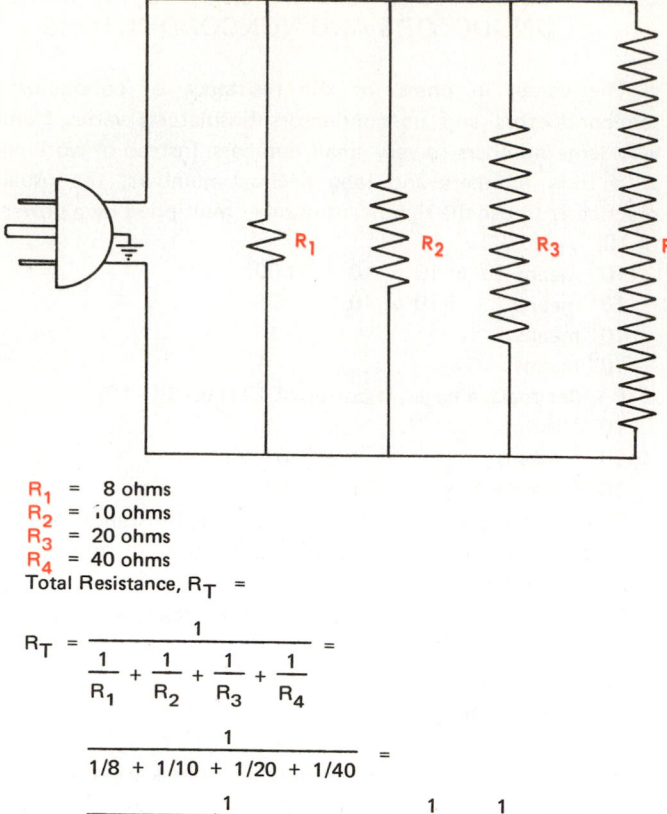

R₁ = 8 ohms → R_1 = 8 ohms
R_1 = 8 ohms
R_2 = 10 ohms
R_3 = 20 ohms
R_4 = 40 ohms
Total Resistance, R_T =

$$R_T = \cfrac{1}{\cfrac{1}{R_1} + \cfrac{1}{R_2} + \cfrac{1}{R_3} + \cfrac{1}{R_4}} =$$

$$\cfrac{1}{1/8 + 1/10 + 1/20 + 1/40} =$$

$$\cfrac{1}{5/40 + 4/40 + 2/40 + 1/40} = \frac{1}{12/40} = \frac{1}{.3} = 3.33 \text{ ohms}$$

R_T = 3.33 ohms

Fig. 28-7. Resistances in parallel.

and the local codes. In most cases, the refrigeration service technicians do not install electrical circuits. They should, however, be familiar with the codes to the extent that they will know whether the supply circuit is properly installed and safe to use.

28-33 AIR MOISTURE HOLDING PROPERTIES

Moisture holding capacity of air depends on its temperature. The warmer the temperature, the greater amount of moisture it will hold. Fig. 28-8 shows a table of moisture holding properties of air. Fig. 28-9 shows the saturation pressure, the heat in the liquid and the total heat after vaporization for mixtures of air and water vapor.

28-34 MOISTURE EVAPORATION SOURCES

OPERATION	POUNDS OF MOISTURE PER DAY
BATHING	.1 to 5
CLOTHES:	
Drying — average family — unvented	26.0
Washing — average family	4.0
COOKING:	
Breakfast	.9
Lunch	1.2
Dinner	2.7
DISHWASHING:	
Breakfast and Lunch	.2
Dinner	.7
HUMANS:	
Average	.4
MOPPING per 100 sq. ft.	3.0

28-35 DESICCANTS

Desiccants are used in driers installed in refrigerator liquid and/or suction lines to collect and remove moisture (water) from the system. Some common desiccants are:

Activated alumina.
Calcium sulfate.
Silica gel.
Alumina gel.
Molecular sieve.

Most driers will remove moisture, sediment and acids from the circulating refrigerant.

Driers are usually placed in the liquid line close to the refrigerant control valve. They should always be placed in a cool place in the line because heat may cause moisture to leave the desiccant and continue to circulate.

Liquid line driers are usually installed permanently and are only replaced when they lose their effectiveness.

Following a burnout, a large-capacity drier is usually placed in the suction line. Suction line driers should be removed as soon as the cleanup is completed.

DEW POINT TEMPERATURE F.	POUNDS OF MOISTURE PER POUND OF DRY AIR	GRAINS OF MOISTURE PER POUND OF DRY AIR	PPM (WT.)
120	.08	570	80,000
110	.06	400	60,000
100	.044	300	42,000
90	.03	210	30,000
80	.022	150	22,000
70	.015	110	15,000
60	.011	76	11,000
50	.0075	53	7,500
40	.005	36	5,000
30	.0033	24	3,300
20	.002	15	2,000
10	.0014	9	1,400
0	.0008	5.5	800

Fig. 28-8. Moisture holding properties of air. (ASHRAE Handbook of Fundamentals)

Temp. F.	Sat. Press. in Hg.	ICE, WATER, WATER VAPOR Heat in Liquid Btu/lb.	Total Heat after Vaporization Btu/lb.	DRY AIR Vol. of Water Vapor cu.ft./lb.	Volume cu.ft./lb.	Specific Heat Btu/lb.	Amount of Water Vapor Saturate Grains
−40	3.790×10^{-3}	−177.1	1043.4	1.343×10^5	10.567	−9.61	.5508
−30	7.503×10^{-3}	−172.7	1047.8	7.441×10^4	10.820	−7.21	1.018
−20	1.259×10^{-2}	−168.2	1052.3	4.237×10^4	11.073	−4.81	1.830
−10	2.203×10^{-2}	−163.6	1056.7	2.475×10^4	11.326	−2.40	3.206
0	3.764×10^{-2}	−159.0	1061.1	1.481×10^4	11.579	0.00	5.480
10	6.286×10^{-2}	−154.2	1065.5	9060	11.832	2.40	9.161
20	.1027	−149.4	1069.9	5662	12.085	4.81	14.99
30	.1645	−144.4	1074.3	3608	12.338	7.21	24.07
32	.1803	−143.4	1075.2	3305	12.389	7.69	26.40
35	.2034	0.0	1076.5	2948	12.464	8.41	29.80
40	.2477	8.0	1078.7	2445	12.591	9.61	36.34
45	.3002	13.1	1080.9	2037	12.717	10.82	44.14
50	.3624	18.1	1083.1	1704	12.844	12.02	53.40
55	.4356	23.1	1085.2	1431	12.970	13.22	64.36
60	.5216	28.1	1087.4	1207	13.096	14.42	77.29
65	.6221	33.1	1089.6	1022	13.223	15.62	92.51
70	.7392	38.1	1091.8	868.0	13.349	16.83	110.4
72	.7911	40.1	1092.6	814.0	13.399	17.31	118.4
75	.8751	43.1	1093.9	740.0	13.475	18.03	131.3
80	1.0323	48.1	1096.1	633.0	13.602	19.23	155.8
85	1.2136	53.1	1098.3	543.3	13.738	20.43	184.4
90	1.4219	58.0	1100.4	467.9	13.854	21.64	217.6
95	1.6607	63.0	1102.6	404.2	13.981	22.84	256.4
100	1.9334	68.0	1104.7	350.2	14.107	24.04	301.5
120	3.4477	88.0	1113.3	203.2	14.612	28.85	569.0
140	5.8842	108.0	1121.7	123.0	15.117	33.67	1071.
160	9.6556	128.0	1129.9	77.27	15.622	38.48	2090.
180	15.295	148.0	1137.9	50.22	16.128	43.30	4598.
200	23.468	168.1	1145.8	33.64	16.632	48.12	16052.
212	29.921	180.1	1150.4	26.80	16.900	50.00	− − −

Fig. 28-9. Air-water vapor values. Properties of a mixture of air and water vapor for various temperatures from −40 F. to 212 F. (−40 C. to 100 C.) are shown. Note large volume occupied by 1 lb. of water vapor at the lower temperature.

Driers which have been used in a refrigerator system should not be reactivated by heating.

28-36 GAS AND VAPOR

A true gas exists as a gas at standard temperatures and pressures.

Vapor is the gaseous state of a substance which is a liquid at standard atmospheric pressures and temperatures (such as water). However, vapor is also the term applied to refrigerants in the gaseous state inside the refrigerating system.

28-37 CHARACTERISTICS OF LITTLE-USED REFRIGERANTS

Characteristics of the most used refrigerants are described in Chapter 9.

Some refrigerants which were in common use at one time are listed in this paragraph. Their physical properties and use as refrigerants are explained in the ASHRAE Handbook of Fundamentals. These refrigerants are:

R-13	Monochlorotrifluoromethane	$CClF_3$
R-13 Bl	Monobromotrifluoromethane	$CBrF_3$
R-21	Dichloromonofluoromethane	$CHCl_2F$
R-30	Methylene Chloride	CH_2Cl_2
R-40	Methyl Chloride	CH_3Cl
R-113	Trichlorotrifluoroethane	CCl_2FCClF_2
R-114	Dichlorotetrafluoroethane	$CClF_2CClF_2$
R-160	Ethyl Chloride	C_2H_5Cl
R-290	Propane	C_3H_8
R-600	Butane	C_4H_{10}
R-717	Ammonia	NH_3
R-744	Carbon Dioxide	CO_2
R-764	Sulphur Dioxide	SO_2

28-38 METRIC PRESSURE-HEAT DIAGRAMS

With increased use of the SI (metric) system of units, metric pressure-heat diagrams for refrigerants are also being used more extensively.

The metric system pressure-heat diagram for R-12 is shown in Fig. 28-10. Fig. 28-11 shows the metric system pressure-heat diagram for R-22.

The standard refrigerating cycle of −15°C evaporating temperature and 30°C condensing temperature is shown on each diagram.

The English unit pressure-heat diagrams for these two refrigerants are shown in Chapter 9.

28-39 MOISTURE IN LIQUID REFRIGERANTS

Moisture is soluble in most liquid refrigerants, and more of it may be safely present in some refrigerants than in others.

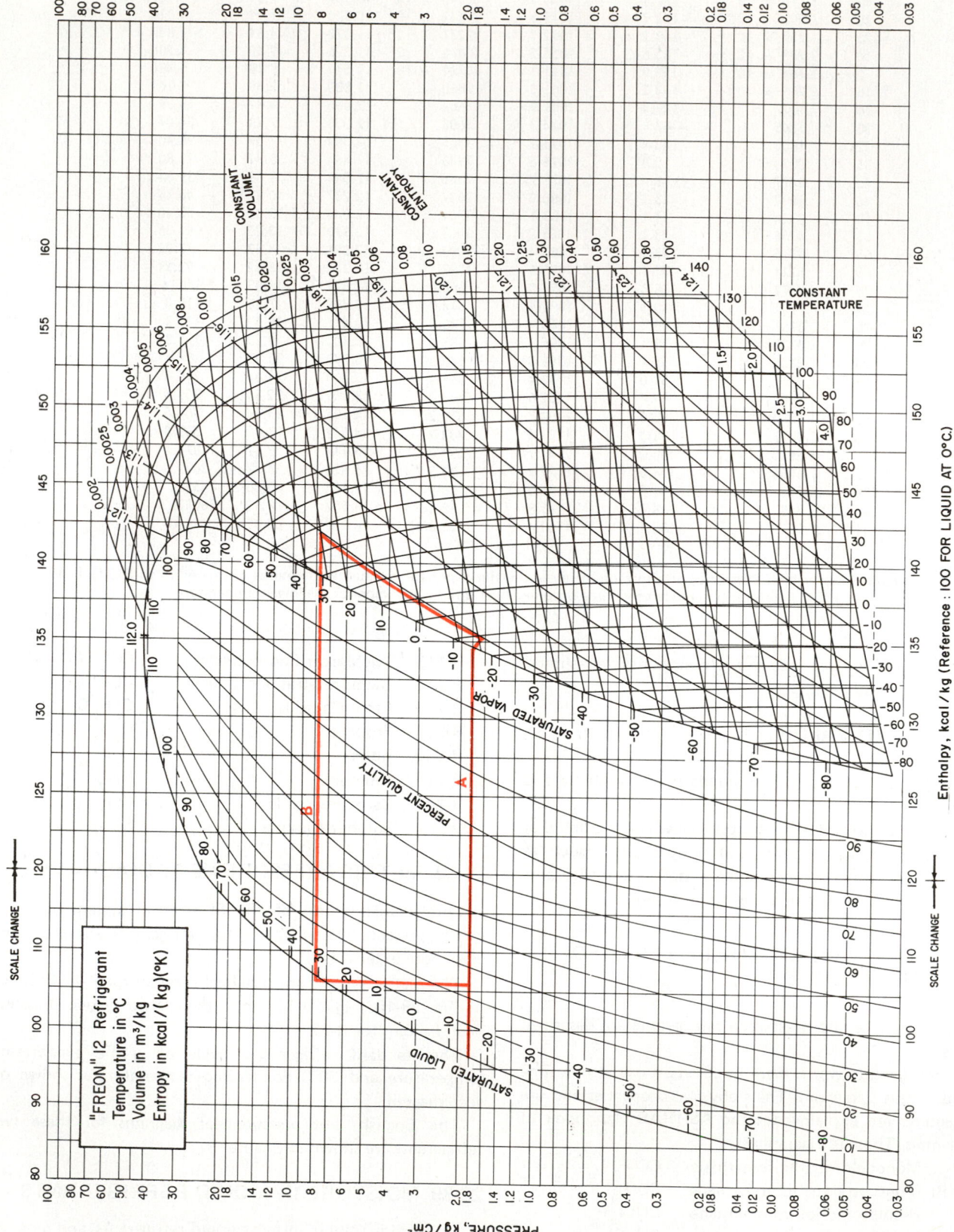

CONSTANT VOLUME

CONSTANT ENTROPY

CONSTANT TEMPERATURE

SATURATED VAPOR

PERCENT QUALITY

SATURATED LIQUID

A

B

"FREON" 12 Refrigerant
Temperature in °C
Volume in m³/kg
Entropy in kcal/(kg)(°K)

SCALE CHANGE

SCALE CHANGE

PRESSURE, kg/cm²

Enthalpy, kcal/kg (Reference : I00 FOR LIQUID AT 0°C.)

Fig. 28-10. Pressure-heat diagram for R-12 expressed in metric units. The standard refrigerating cycle of evaporating temperature is shown at A and condensing temperature at B. (E. I. du Pont de Nemours & Co., Inc.)

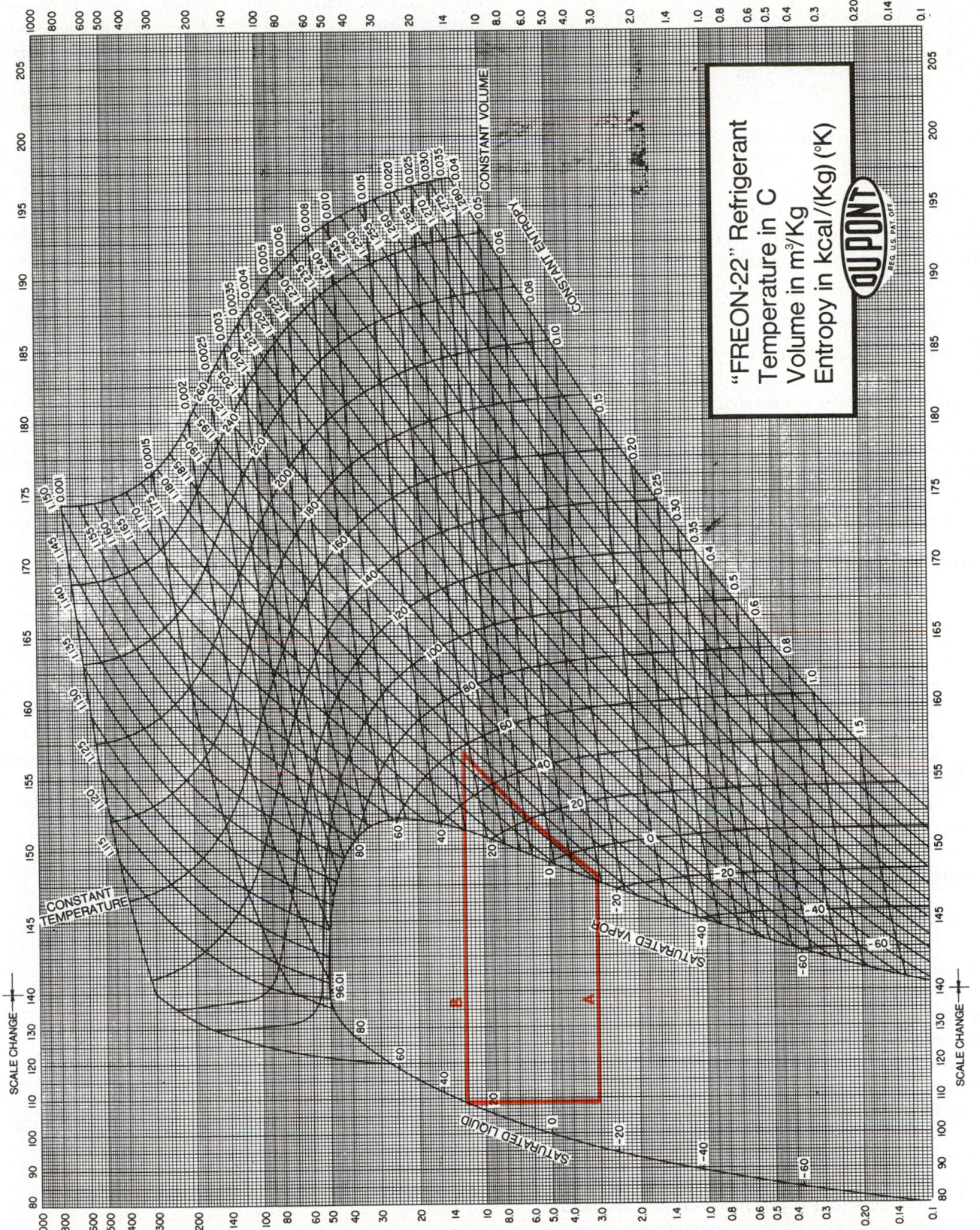

Fig. 28-11. Pressure-heat diagram for R-22 expressed in metric units. The standard refrigerating cycle of evaporating temperature is shown at A and condensing temperature is shown at B.

SAFE MOISTURE SOLUBILITY LIMIT (ppm)					
TEMP.	R–12	R–22	R–500	R–502	R–13B1
−40	1.7	120	48	40	2
−20	3.8	195	81	69	5
0	8.3	308	129	115	10
20	17	472	200	180	21
40	32	690	293	278	40

Fig. 28-12. Safe moisture content in five popular refrigerants. Content is indicated in parts per million (ppm).

The amount of the allowable moisture varies with the temperature. Fig. 28-12 shows the maximum allowable moisture that may be present in five of the common refrigerants. In low temperature applications, a very small amount of moisture may cause trouble.

28-40 DRYNESS OF REFRIGERANTS

Any gas holding water in the vapor state will, when cooled, reach a temperature at which free moisture (liquid water) will appear. The greater the concentration of water vapor in a gas, the higher the temperature at which water will condense out (free water).

This freeing of moisture is readily shown by frost on cold lines, beads of water on glasses of cold liquids and, in the refrigeration system, by ice in the orifice of the expansion valve. Free water must be present in a refrigerating system before ice can be formed.

The highest temperature at which free water is liberated, on cooling, is called the dew point. There is only one dew point temperature for a given moisture concentration in a gas such as air or a refrigerant vapor. Knowing the dew point of a refrigerant determines the lowest temperature at which the refrigerant control will operate satisfactorily.

To prevent the formation of sludges and plating, the dew point should be 10 F. (6 C.) lower than the low-side refrigerant temperatures. To prevent refrigerant control freeze-up, the refrigerant should have a dew point of 5 F. (3 C.) lower than the low-side refrigerant temperature.

28-41 REFRIGERANT OILS

Requirements for a satisfactory refrigerant oil are rather severe. It must provide good lubricating qualities. It must be of the correct viscosity for the refrigerant and the machine in which it is being used. It must be free of moisture.

The viscosity of a refrigerant oil is measured using the Saybolt Universal Viscosity Test. A Saybolt Viscosimeter should be used. In this instrument, oil at 100 F. (38 C.) is allowed to flow through the standard Saybolt orifice. The time required in seconds for 60 cubic centimetres (cm³) of the oil to flow through this orifice is recorded.

The amount of moisture in refrigerant oil may be measured by its resistance to the flow of a current of electricity through it. This is known as its dielectric property. Good refrigerant oil should have a dielectric value of 25,000V minimum.

Another test which may be given to refrigerant oil is known as the floc test. This test applies to oils which are used with completely miscible (mixable) refrigerants, such as R-11, R-12 and R-22. The test is conducted by mixing 10 percent refrigerant with 90 percent oil and sealing it in a glass tube. Then it is cooled slowly until a flocculent (cloudy) precipitate of wax appears. The maximum temperature at which this occurs is recorded as the floc point.

Since the viscosity of oil changes with the temperature, oils at very low temperatures may not pour and may become a plastic solid. The temperature at which oil will just flow is called the pour point. This temperature is recorded either in Fahrenheit or Celsius as the pour point.

The flash point of an oil may be determined by slowly heating a quantity of the oil in an open container. Use a thermometer to determine the temperature of the oil. Bring a lighted candle or other flame to the surface of the oil as it is being heated. The temperature at which the vapors from the surface of the oil burn or flash is the flash point of the oil.

Refrigerant oils are sometimes given a very small amount of antifoam inhibitor to reduce foaming. Compressor parts are sometimes given a phosphating treatment to improve lubrication. Tricresyl phosphate has also been added to refrigerant oils to improve lubrication.

28-42 EUTECTIC

Proprietary compounds (made and sold by only one firm) are available that provide an active thermal energy source to maintain a specified temperature range. This is called thermal energy storage (TES). It is created by the heat of fusion or heat of crystallization.

These eutectic compounds can absorb or release heat with little or no change in temperature while they are changing from one physical state to another. The melting of ice is an example. In the process of cooling water, 1 Btu of heat is removed to lower the temperature of 1 pound of water 1 degree. This is true down to the freezing temperature. At the

CHARACTERISTICS OF THERMAL ENERGY STORAGE COMPOUNDS			
TES MATERIAL	MELTING TEMPERATURE F.	C.	HEAT OF FUSION Btu/lb.
*TRANS TEMP 12	12	−11	115
N-DODECANE	10	−12	47
ETHYLENE GLYCOL	9	−13	63
*TRANS TEMP 27	27	−3	142
H₂O	32	0	144
*TRANS TEMP 40	40	4	75
N-TETRADECANE	42	6	98
*TRANS TEMP 65	65	18	80
N-HEXADECANE	62	17	102
ACETIC ACID	62	17	80
*TRANS TEMP 130	130	54	72
PARAFFIN WAX	130	54	63
TRISTEARIN	133	56	82
DRY ICE	−109	−78	241
*PROPRIETARY MATERIAL — ROYAL INDUSTRIES			

Fig. 28-13. Characteristics of some eutectic thermal energy storage (TES) materials. (Royal Industries)

freezing temperature, 144 Btu per pound must be removed to freeze 1 pound of water. The temperature remains constant until the water is frozen.

A separate compound or mixture is needed for different temperature ranges. Fig 28-13 is a table of some substances which may be used to maintain various temperatures.

Heat is absorbed or expelled by the eutectic as it changes its state. For low temperatures, the heat of fusion is used; for high temperatures, the heat of crystallization.

Eutectic compounds used in eutectic plates, as described in Chapter 12, contain other substances to prevent expansion on freezing. Freezing might otherwise rupture the containers. They also contain substances to reduce oxidization of the container materials. Useful life of the materials is thus extended.

28-43 CRYOGENIC TEMPERATURES

The cryogenic temperature range extends from −250.0 F. to Absolute 0 (−459.69 F.). The cryogenic ranges may also be expressed as follows:

−156.6 C. to −273.16 C.
116.5 K. to 0 K.
209.69 R. to 0 R.
F. = Fahrenheit.
C. = Celsius.
K. = Kelvin.
R. = Rankine.

Do not use low carbon steel for piping or pressure parts. The metal becomes very brittle when cold. Some nickel and/or some chromium improves the properties of steel at very low temperature.

28-44 BRINE FREEZING TEMPERATURES

Brines are water mixed with a substance which will go into solution with the water. The mixture will provide a fluid which can readily flow at temperatures below 32 F. (0 C.). See Fig. 28-14.

Brines are of several types: alcohol, salt and glycol.
1. Alcohol brines are usually made of ethyl alcohol.
2. Salt brines are usually made of sodium chloride and/or calcium chloride. Eutectic point for sodium chloride (common salt) solution is −6 F. (−21 C.) and for calcium chloride, −60 F. (−51 C.).
3. Glycol brines with noncorrosive properties are usually made of glycerine, ethylene glycol and/or propylene glycol.

A hydrometer can measure the density of brine solutions. The freezing temperature may be worked out from the hydrometer reading.

This is the same as measuring the freezing temperature of the cooling solution in an automobile radiator. If the hydrometer reading indicates that the freezing temperature of the solution is too high, it may be lowered by adding more of the antifreeze compound to the solution. Note that in the case of alcohol brine — since alcohol is lighter than water — the density decreases with the lowering of the freezing temperature. A hydrometer with an alcohol testing scale must be used.

28-45 TOTAL ENERGY — SINGLE ENERGY

In some buildings, only a single source of energy — such as natural gas — is supplied. All other needed energies are generated in the building from this gas. As an example, gas-burning engines may be used to drive electric generators. The waste heat from the cooling of the engines and the exhaust may be used either for heating water, space heating, or as heat for industrial processes. Also, the heat from lighting may be used in heating water or air.

28-46 SOLVENTS AND CLEANING

From use or repair operations, parts may become coated with lubricating oils, greases, oxides, dirt, metallic particles or abrasives. Many methods may be used successfully to clean these parts. Lubricants or greases made from animal or vegetable oils or fats — such as tallow, lard oil, palm oil, olive oil — can usually be removed by saponification (making a soap of the oil or fat). Parts are soaked in an alkaline solution where the oils react with the alkali to form water soluble soap compounds.

Mineral oils which cannot be made into a soap solution — such as kerosene, machine oil, cylinder oil and general lubricating oils — are usually cleaned by an emulsification

	FREEZING TEMPERATURE F. AND C.						
BRINE SPECIFIC GRAVITY	20(−7 C.)	10(−12 C.)	0(−18 C.)	−10(−23 C.)	−20(−29 C.)	−30(−34 C.)	−40(−40 C.)
ALCOHOL (FORMULA NO. 1) SPECIFIC GRAVITY AT 60 F.		.9691	.9592	.9486	.9345		
CALCIUM CHLORIDE SPECIFIC GRAVITY AT 60 F.	1.090	1.140	1.175	1.201	1.227	1.254	1.265
PERCENT OF CHEMICAL	10	17	20.5	23	25	27	28
SODIUM CHLORIDE USABLE ONLY DOWN TO 0 F. SPECIFIC GRAVITY AT 60 F.	1.072	1.118	1.158				
PERCENT OF CHEMICAL	10	16	21				
ETHYLENE GLYCOL SPECIFIC GRAVITY AT 60 F.	1.05	1.07	1.075	1.08	1.09	1.096	1.105
PERCENT OF CHEMICAL	32	40	43	45	50	53	57

Fig. 28-14. Table of properties of various brines.

process using soaps, wetting agents and dispersing agents. (Emulsify means to make a liquid of solids by mixing them with a liquid. The solids do not dissolve but are suspended in the liquid and make it thick.)

Dirt, abrasives, metal dust and inert materials are generally removed by one or both of these processes.

SOLVENT CLEANING is used for removing most of the oils from coated pieces. They are dipped (immersed) in a solvent such as mineral spirits. Solvent tanks should have safety lids and should be hooded and vented.

VAPOR CLEANING is also used to remove oils. Parts are held in a container where solvent vapors can condense on the parts. The condensed solvent washes away the oily coating, leaving surfaces dry and nearly clean.

Production degreasing machines have two or three compartments. The work is immersed in the first compartment containing a boiling solution of the solvent. It is then dipped into the second section, which contains clean cold solvent. Finally, it is hung in the third section where only clean vapors condense on and wash over the work. The degreasing unit is selfpurifying. Oils and waste gather at the bottom of the third section.

Job shop cleaning uses the third section only. Solvents should be those approved by the Occupational Safety and Health Act (OSHA). Venting is extremely important for safety to the operator.

ALKALINE CLEANING baths are used primarily for the removal of oils, greases, solid particles of dirt and metal particles, by immersing pieces in hot alkaline solutions. The chemicals saponify (make soap of) vegetable and animal oils and fats, emulsify mineral oils and greases and suspend the solid material.

The combination of heat, active chemicals and agitations (shaking) are important. Soap is either added or is formed by the saponification of vegetable or animal fats. Caustic soda, soda ash and causticized soda are the cheapest and most direct methods of producing alkalinity in the bath. However, such materials, as a general rule, have less surface activity than more complex materials. Sodium metasilicate trisodium phosphate, and similar salts are often used to obtain alkalinity in a solution.

A number of proprietary preparations are used for EMULSION CLEANING. They comprise an emulsification agent which disperses organic solvents in water solutions. Emulsifiable cleaners are miscible (mix) with oils and can be washed off with water; however a film of oil may remain on the work and make subsequent alkali cleaning necessary. Dragout costs are high.

Alkaline materials are used in ELECTROLYTIC CLEANING. The bath is maintained at as near boiling as possible without heavy tarnishing. The gas evolved tends to lift off the soil, making a clean surface for subsequent operations. The work to be cleaned is usually made the cathode. There are many formulas available for this work, but the one used depends upon the nature of the materials to be used and the degree of tarnish permissible. In many cases — particularly with carbon steel or cast iron — unusual results can be obtained by switching polarity several times during cleaning.

FUEL	HEAT RELEASED — Btu Per Lb.
COAL:	
BITUMINOUS	12,000 TO 15,000
ANTHRACITE	13,000 TO 14,000

Fig. 28-15. Heating value of coal. As noted, the heating value of coal varies considerably depending upon the moisture, sulphur content and other impurities.

28-47 HEATING VALUE OF FUELS

Fuels commonly used for heating purposes are coal, oil, gas and wood. Burning these fuels in atmospheric air produces the heat needed. The heating value of coal is given in Fig. 28-15.

The American Petroleum Institute establishes the commercial grades of heating oil. Grade numbers are 1 through 6, although No. 3 and No. 4 are very little used. The grade of heating oil may be determined with a special heating oil hydrometer. Grade No. 1 has a gravity of between 45 and 38, as shown in Fig. 28-16. The heating value of oil is also shown in this figure.

COMMERCIAL GRADE NO.	Btu PER GALLON	GRAVITY RANGE DEG. API	AVERAGE WEIGHT PER GALLONS IRS
1	137,000	45—38	6.8
2	140,000	40—30	7.1
3	140,000*		
4	141,000*	32—12	7.7
5	148,000	20—8	8.1
6	152,000	18—6	8.2

*NOT IN COMMON USE

Fig. 28-16. Table of heating values of fuel oils. Numbers 5 and 6 are high viscosity oils and need preheating before use. API means American Petroleum Institute.

FUEL	HEAT RELEASED — Btu PER CUBIC FOOT
GAS:	
NATURAL	1000 TO 1100*
MANUFACTURED	500 TO 600
LP (LIQUID PETROLEUM)	2500 TO 3200

*CHECK WITH THE LOCAL GAS COMPANY

Fig. 28-17. Heating value of fuel gases.

The heating value of gas is indicated in Fig. 28-17. (The heating value of natural gas varies greatly, depending upon its composition.)

The heating value of wood is approximately 6200 Btu/lb.

28-48 TWIST DRILL SIZES

Sizes of twist drills may be given either in fractions, decimals, numbers or letters. Metric drill sizes are indicated in millimetres and decimals of a millimetre. As metric usage increases, more metric size drills and taps will be used.

The service technician will use mostly fractional, numbered

DECIMAL SIZES OF NUMBERED DRILLS

No.	Size of Drill In Inches	No.	Size of Drill In Inches	No.	Size of Drill In Inches	No.	Size of Drill In Inches
1	.2280	21	.1590	41	.0960	61	.0390
2	.2210	22	.1570	42	.0935	62	.0380
3	.2130	23	.1540	43	.0890	63	.0370
4	.2090	24	.1520	44	.0860	64	.0360
5	.2055	25	.1495	45	.0820	65	.0350
6	.2040	26	.1470	46	.0810	66	.0330
7	.2010	27	.1440	47	.0785	67	.0320
8	.1990	28	.1405	48	.0760	68	.0310
9	.1960	29	.1360	49	.0730	69	.0292
10	.1935	30	.1285	50	.0700	70	.0280
11	.1910	31	.1200	51	.0670	71	.0260
12	.1890	32	.1160	52	.0635	72	.0250
13	.1850	33	.1130	53	.0595	73	.0240
14	.1820	34	.1110	54	.0550	74	.0225
15	.1800	35	.1100	55	.0520	75	.0210
16	.1770	36	.1065	56	.0465	76	.0200
17	.1730	37	.1040	57	.0430	77	.0180
18	.1695	38	.1015	58	.0420	78	.0160
19	.1660	39	.0995	59	.0410	79	.0145
20	.1610	40	.0980	60	.0400	80	.0135

Fig. 28-18. Numbered twist drill sizes.

Letter	Dia. In.	Letter	Dia. In.	Letter	Dia. In.
A	0.234	J	0.277	S	0.348
B	0.238	K	0.281	T	0.358
C	0.242	L	0.290	U	0.368
D	0.246	M	0.295	V	0.377
E	0.250	N	0.302	W	0.386
F	0.257	O	0.316	X	0.397
G	0.261	P	0.323	Y	0.404
H	0.266	Q	0.332	Z	0.413
I	0.272	R	0.339		

Fig. 28-19. Lettered twist drill sizes.

and lettered drills. If it is necessary to do very much tapping and threading, drills will be used from each of these categories.

Fractional size drills start at 1/16 in. and increase by 1/64 in. through the required range. Decimal size drills are mostly used in manufacturing; hardly ever by the service technician. Number size drills are marked from 1 through 80. The range extends from .0135 in. up to less than 1/4 in. A table of numbered drills is shown in Fig. 28-18.

Letter size drills extend from A through Z, roughly from 1/4 in. through .413 in. A table of the letter size twist drills is shown in Fig. 28-19. Tap drill sizes are shown in Fig. 2-61.

28-49 COLOR CODE FOR PIPING

USE	COLOR
Fire protection equipment	Red
Safe material	Green (or, if needed, white, gray or aluminum)
Protective material	Bright blue
Extra valuable material	Deep purple
Dangerous material	Yellow or orange

28-50 FOOD PRESERVATION BY RADIATION TREATMENT

Experiments have been conducted using atomic energy radiations to preserve food. Fresh foods have been put in sealed containers. These containers were exposed to a form of atomic energy radiation. It was found, upon inspection, that the food became sterile. It remained in its fresh state with no change in appearance, flavor or food value, as a result of being treated in this manner.

28-51 CLEAN ROOMS

In many laboratory situations, it is necessary to provide absolute control of the temperature, humidity, air cleanliness and air chemistry. Such rooms are called clean rooms. Space industry is the primary user of clean room technology. Also some assembly operations must be conducted in clean rooms.

Dust particles too small to be seen may interfere with the operation of some very complicated mechanisms if allowed to enter the mechanisms during assembly. These devices are such that particles in the range of 20 to 50 microns will interfere with the operation. The use of clean rooms is increasing both in hospitals and industry.

28-52 SOLDERS AND BRAZING METALS

The following table lists some common solder alloys used in refrigeration work:

SOLDER	MELTING PT. F.	FLOW PT. F.	SHEAR STRENGTH psi
50-50 Tin-Lead	358	414	83.4
95-5 Tin-Antimony	450	465	327.0
Silver Solder 45 Ag., 15 Cu, 24 Cd, 16 Zn	1120	1145	8340
Phosphorous-Copper	1310	1650	8340

28-53 HEAT CONDUCTIVITY

HEAT CONDUCTIVITY OF MISCELLANEOUS SUBSTANCES

MATERIAL	k*	R**
Air	.175	5.714
Concrete wall	8.00	.125
Glass	5.0	.2
Lead	243.0	.004
Vacuum, High	.004	250.

HEAT CONDUCTIVITY OF MISCELLANEOUS INSULATING MATERIALS

MATERIAL	Density Lb./cu. ft.	k* Conductivity	R**
Asbestos, loose	29.3	.94	1.064
Cork, granulated	8.1	.34	2.941
Cork, granulated, impregnated with pitch	17.79	0.428	2.336
Balsa	7.05	0.32	3.125
Felt	16.9	.25	4.0
Glass wool (curled pyrex)	4.0	0.29	3.448
Kapok	.87	.24	4.167
Mineral (slag) wool, loose packed	12.0	0.26	3.846
Rock wool (fibrous rock, also felted)	6.0	0.26	3.846
Rubber, cellular	5.0	.37	2.703
Sawdust, pine	18.76	0.57	1.754
Straw fibers, pressed	8.67	0.32	3.125
Wood fibers (kingia australis)	8.4	0.33	3.03
Wool, pure	4.99	0.26	3.846

HEAT CONDUCTIVITY OF PROPRIETARY MATERIALS

TRADE NAME	Density Lb./cu. ft.	k* Conductivity	R**
Armstrong's corkboard	7.3	0.285	3.509
Celotex	13.2	0.31	3.226
Dry-Zero	1.0	0.24	4.167
Nu Wood	15.0	0.32	3.125
United's 100 percent pure corkboard	9.0	0.27	3.704
U. S. mineral wool	12.0	0.26	3.846
Ferro Therm metal sheet (4 sheets)	4 oz./sq. ft./ sheet	0.226	4.425

Note: k* = Btu/hr./sq. ft./in. thickness/degree F.
R** = Reciprocal of K.
(American Society of Heating Refrigerating and Air Conditioning Engineers)

Fig. 28-20. Galvanic action sequence.

28-54 GALVANIC ACTION SEQUENCE

Certain materials produce electricity by chemical action. This is known as galvanic action. Some materials are much more active than others. Magnesium is one of the most active and platinum is one of the least active.

Galvanic action causes rapid deterioration at the place where galvanic action is occurring. To minimize galvanic action, joints where two unlike galvanic materials touch should be electrically insulated from one another.

Fig. 28-20 lists the galvanic action sequence for a series of common materials. The items at the top of the list are the most active and most likely to corrode. In galvanic action, these materials become an anode-positive. The items at the end of the list are least likely to corrode and in galvanic action these become a cathode-negative.

The corrosion of materials (usually metals) is often due to galvanic action.

28-55 ELECTROLYSIS

Electrolysis is the movement of d-c electricity through a substance which causes chemical change in the substance or its container. Electrolysis can be used for producing useful results

such as storage battery operation and electroplating.

However, it can also cause severe damage. Whenever there are two electrical conductors and moisture, d-c electricity will flow if the conductors are of different activity levels. As the d-c electricity flow continues (electrons from — to +), one of the conductors will become coated with a new chemical and the other will have material removed (pits and holes). The pitted and "eaten away" areas are called corrosion areas. To avoid electrolytic corrosion, one must remove the moisture, seal off (insulate) the conductors or use conductors of the same activity level.

28-56 STANDARD TEMPERATURE

The weight of a certain volume of air depends on its temperature. The standard temperature used for determining the weight of air is 32 F. (0 C.). This is the temperature of melting ice.

28-57 STANDARD PRESSURE

The weight of a certain volume of air depends on its pressure. The standard pressure for determining the weight of air is 29.92" Hg. or 760 mm Hg. This is the pressure of one atmosphere.

28-58 STANDARD TEMPERATURE AND PRESSURE

A sample of air at standard temperature and under standard pressure is known as a sample at STP conditions.

28-59 STANDARD AIR

Standard air and standard conditions are the same. Standard conditions include pressure, temperature and air density of 0.0725 lb./cu. ft. These values are:

ENGLISH SYSTEM	METRIC SYSTEM
Pressure = 29.92" Hg. = 14.7 psia = 0 psi	760 mm Hg. 1.03 kg/cm^2
Temperature = 69.8 F.	21. C.
Specific Volume = 13.33 cu. ft./lb.	0.833 m^3/kg

28-60 REFRIGERANT PROPERTIES

Several refrigerant tables and pressure-heat enthalpy diagrams are given in this section. Most of the physical properties of refrigerants may be found in these tables and charts.

Chapter 9 describes the characteristics and uses of the more popular refrigerants in more detail.

Fig. 28-21 is a table showing the comparable properties of many of the refrigerants.

28-61 FLARE NUT WRENCH SIZES

A table of flare nut wrench sizes is shown in Fig. 28-22. Old and new wrench sizes across the flats of the flare nut are given. Tube size is based on outside diameter.

Refrigerant Number	Name	Chemical Symbol	Trade Name	Molecular Weight	Odor	Toxicity	Flammability	Pressure psia at 5 F.	Pressure psia at 86 F.	Latent Heat at 5 F.	Sp. Heat of Liquid at 5 F.	Critical Temperature of	Critical Pressure psia	Sp. Volume of Gas at 5 F.	Density of Liquid in a 5 F./#/cu. ft.	CP/CV Ratio	Sp. Heat of Vapor at 86 F.
R-764	Sulphur Dioxide	SO_2		64.06	Pungent	High	Non	11.81	66.45	172.3	.34	314.8	1141.5	6.421	92	1.256 *	.34
R-40	Methyl Chloride	CH_3Cl		50.489	Sweet	Med.	Slight	20.89	95.53	180.6	.45	289.6	969.2	4.530	61	1.20	.4
R-717	Ammonia	NH_3		17.031	Pungent	High	Slight	34.27	169.2	565	1.10	271.4	1651	8.150	41.11	1.247	1.10
R-160	Ethyl Chloride	C_2H_5Cl	Alcozol	64.51	Etheral	Med.	Yes	4.65	27.10	177	.47	369	764	17.55	59.00	1.13	.42
R-12	R-12	CCl_2F_2		120.9	Sweet	Low	Non	26.61	107.9	68.2	.215	232.6	582.0	1.485	90	. . . .	.243
R-13	R-13	$CClF_3$		104.46	Sweet	Low	Non	. . . .	. . . .	63.85 (−115 F.)	.247 (−22 F.)	84	561	.431	77 (0 F.)	1.172	. . . .
R-744	Carbon Dioxide	CO_2		44.005	Non	Low	Non	334.4	1039.0	116	.5	87.8	1066.2	.2673	61.22	1.30 **	1.95
R-611	Methyl Formate	$C_2H_3O_2$		60.04	Slight	. . .	Slight	1.96	13.69	236	.515	417	870	46.7		. . . .	.515
R-30	Methylene Chloride	CH_2Cl_2	Carrene No. 1	84.9	Sweet	. . .	Yes	1.17	10.6	162.1	.34	421	670	50.58		. . . .	.34
R-21		$CHCl_2F$	Thermon	102.92	Sweet	Low	Non	5.5	30.5	105.5	.26	353.3	750.0	8.83	90.1	. . . .	.26
R-22		$CHClF_2$		86.48	Sweet	Low	Non	43.02	174.5	93.43	11.97	204.8	716	1.246	83.34	. . . .	.34
R-11		CCl_3F	Carrene No. 2	137.38	Sweet	Low	Non	29.31	18.28	84.0	.197	388.4	635.0	12.3	97.8	. . . .	.20
R-114		$CClF_2CClF_2$		170.93	Sweet	Low	Non	.98	7.86	58.9	.238	294	474	.488	73.1	1.088	.160
R-113		$C_2Cl_3F_3$		187.4	Sweet	Low	Non		128.14	70.62	.199	417.4	495	27.04	103	4.61	.26
R-500		CCl_2F_2/CH_3CHF_2		99.29	Sweet	Low	Non	31.07	175.1	85.03	11.83	221.1	631	1.5227	80.10	4.37	
R-502		$CHClF_2 + CClF_2CF_3$		111.64	Sweet	Low	Non	50.68		68.86	.027	194.1	618.7	.825	87.24		.07
R-290	Propane	C_3H_8		44.06	Sweet	Low	Yes	41.9	155	170.2	.56	302	661.5	2.48	34.33		.55
R-171	Ethane	C_2H_6		30.04	Sweet	Low	Yes	236.0	675.0	150.5	.66	90.1	730	.533	26.96		.83
R-600	Butane	C_4H_{10}		58.12	Sweet	Low	Yes	8.2	41.6	170.7	.51	308	529	9.98	38.41		.51
R-13B1	Kulene 131	CF_3Br			Etheral	Low	Non	77.93	261.8	44.88	.182	1535	587	.3854	112		.10
R-115	Monochloropenta-fluoroethane	$CClF_2CF_3$		154.48	Sweet	Low	Non	38	148.9			175.9		.82		3.76	. . . .

* at 70 F.   ** at 32 F.   # at 70 F.

Fig. 28-21. Table of properties of refrigerants.

TUBE SIZE O.D.	WRENCH SIZE ACROSS FLATS	
	OLD	NEW
1/4 IN.	3/4 IN.	5/8 IN.
3/8 IN.	7/8 IN.	13/16 IN.
1/2 IN.	1 IN.	15/16 IN.

Fig. 28-22. A table of flare nut wrench sizes.

28-62 BOYLE'S LAW

Boyle's Law expresses a very interesting relationship between the pressure and volume of a gas. It is stated as follows:

"The volume of a gas varies inversely as the pressure, provided the temperature remains constant."

This means that if a quantity of gas has its pressure doubled, the volume becomes half of what it originally was. Or, if the volume is doubled, the gas has its pressure reduced by half. If a perfect gas is considered, Boyle's Law may be expressed as a formula:

Pressure x Volume = A constant number.

This being true, one can say that when either the pressure or the volume is changed, the corresponding pressure or volume is changed in the opposite direction. Therefore:

Old Pressure x Old Volume = the New Pressure x the New Volume

Expressed in letter form:

$P_o \times V_o = P_n \times V_n$

P_o = Old Absolute Pressure
P_n = New Absolute Pressure
V_o = Old Volume
V_n = New Volume

Note: This formula will be true only if the pressures are expressed as absolute pressures (psia).

Example: What will be the new volume of 5 cubic feet (cu. ft.) of gas at 20 psi if it is compressed to 60 psi, providing the temperature remains constant?

For calculation purposes, atmospheric pressure = 15 psi

$P_o \times V_o = P_n \times V_n$

P_o = 20 psi (atmospheric pressure)
   = (20 + 15) = 35 psia (absolute pressure)
P_n = 60 psi + atmospheric pressure =
   (60 + 15) = 75 psia

$P_o \times V_o = P_n \times V_n$, substituting the preceding values in this formula

$35 \times 5 = 75 \times V_n$ or

$\dfrac{35 \times 5}{75} = V_n$

$V_n = \dfrac{35}{15}$

V_n = 2.33 cu. ft. at 60 psi

In the metric system, volume is metres cubed (m^3).

The pressures can be either kilograms per centimetre square = kg/cm^2 or newtons per metre squared = N/m^2 = 1 pascal = 1 Pa.

Atmospheric pressure = 101 kPa
One bar = atmospheric pressure
1 bar = 100,000 Pa = 10^5 Pa

Be careful to use the same volume units and pressure units in the same problem.

28-63 CHARLES'S LAW

Gases behave consistently with temperature changes. This is stated in Charles's Law:

"At a constant pressure the volume of a confined gas varies directly as the absolute temperature; and at a constant volume, the pressure varies directly as the absolute temperature."

Absolute pressures and absolute temperatures must always be used in the equations. In the equation form:

AT CONSTANT VOLUME

The Old Absolute Pressure (P_o) x the New Absolute Temperature (T_n) = the New Absolute Pressure (P_n) x the Old Absolute Temperature (T_o).

$P_o \times T_n = P_n \times T_o$

Example: What is the pressure of a quantity of confined gas when raised to 60 F. (16 C.) if its original pressure was 35 pounds per square inch gauge (psig) and temperature 40 F. (4 C.), at a constant volume?

Solution: (Note that absolute temperatures and pressures must be used.)

$P_o \times T_n = P_n \times T_o$
P_o = (35 + 15) = 50
T_n = (60 + 460) = 520
T_o = (40 + 460) = 500

$P_n = \dfrac{P_o \times T_n}{T_o}$

$P_n = \dfrac{50 \times 520}{500} = 52$ psia

$P_n = 52 - 15 = 37$ psi

AT CONSTANT PRESSURE

The Old Volume (V_o) x the New Absolute Temperature (T_n) = the New Volume (V_n) x the Old Absolute Temperature (T_o).

$V_o \times T_n = V_n \times T_o$

Example: 5 cubic feet of gas at 37 F. is raised to 90 F. at constant pressure. What is the new volume?

Solution:

$V_o \times T_n = V_n \times T_o$
$5 \times (90 + 460) = V_n \times (37 + 460)$
$5 \times 550 = V_n \times 497$

$\dfrac{5 \times 550}{497} = \dfrac{2750}{497} = V_n$

V_n = 5.53 cubic feet

In the Metric system:

1. The volumes will be in metres cubed (m^3).

2. The pressures will be in kilograms per square centimetre

(kg/cm^2) or newtons per square metre (N/m^2) which is called pascal (Pa).

1 N/m^2 = 1 Pa

28-64 GAS LAW

Boyle's Law and Charles's Law may be combined to solve true gas problems. The formula is:

$$\frac{Po \times Vo}{To} = \frac{Pn \times Vn}{Tn}$$

To = Old Absolute temperature
Tn = New Absolute temperature
Po, Vo and To represent original conditions
Pn, Vn and Tn represent new conditions

Absolute values for temperature and pressure must be used in this equation.

Example: If 2 cu. ft. of gas at 15 psi and 140 F. is stored in a 4 cu. ft. container at 30 psi, what is the new temperature?

Formula:

$$\frac{Po \times Vo}{To} = \frac{Pn \times Vn}{Tn}$$

Po = (15 + 15) = 30
Pn = (30 + 15) = 45
Vo = 2
Vn = 4
To = (140 + 460) = 600

$$Tn = \frac{Pn \times Vn \times To}{Po \times Vo}$$

$$Tn = \frac{45 \times 4 \times 600}{30 \times 2} = \frac{108,000}{60} = 1800 \text{ R. (F}_A\text{.)}$$

$$= 1800 - 460 = 1340 \text{ F.}$$

In the metric system the same formula is used but the units are different:

1. The volume is in cubic metres (m^3).
2. The pressure is in kilograms per square centimetre — (kg/cm^2); or newtons per square metre — (N/m^2) or pascal = (Pa).

28-65 ADIABATIC EXPANSION AND CONTRACTION OF A GAS

When the term adiabatic is used it refers to a natural process in a gas. This process or property allows the gas to be expanded or contracted (compressed) without absorbing heat from outside or without transferring heat out of the gas.

Adiabatic expansion and compression of gases would occur if the gas were placed in a perfectly insulated cylinder with a frictionless piston so that heat could not enter the gas during expansion or escape during compression.

During adiabatic compression and expansion of ideal gases, the work performed (compression and expansion) is obtained FROM THE GAS. When work is done on a gas as it is adiabatically compressed, the heat generated is not lost, but increases the temperature and therefore the pressure of the confined gas. During expansion, the pressure and temperature of the ideal gas decreases.

28-66 ISOTHERMAL EXPANSION AND CONTRACTION OF A GAS

An isothermal condition exists when the expansion or contraction of a gas occurs without a change in temperature (Boyle's Law, Para. 28-62). This condition can occur either during the expansion or the compression of a gas.

During expansion, the gas is cooled and the heat needed to keep the gas at a constant temperature must come from an outside source. The heat so obtained must be exactly equal to that given up by the gas during its expansion. This is necessary to keep the temperature constant.

A similar condition must exist during the isothermal compression of a gas. Heat must be removed in an amount equal to the heat energy of compression in order to maintain a constant temperature. Regardless of whether the compressor is air-cooled or water-cooled, the heat removed must be equal to the heat input from the work of compressing the gas.

Gas compression and expansion operations cannot be performed without a change in the temperature of the confined gas. If the action results in raising the temperature of the confined gas, it is called EXOTHERMAL. If the action results in lowering the temperature of the confined gas, it is called ENDOTHERMAL.

28-67 ROOT MEAN SQUARED (rms)

The root mean squared (rms) values of voltage and current are used to calculate the power in a circuit when the current and voltage are not in phase. In this calculation a mean voltage and mean current are used.

To find the root means squared value of a quantity which varies over a cycle, the following steps must be done:
1. Find the instantaneous values of the current and voltage and SQUARE them.
2. Find the MEAN over one cycle of the SQUARE of the instantaneous values.
3. Find the square ROOT of this value.

This gives the ROOT MEAN SQUARED value.

If the current in a circuit varies as a sine curve, as shown in Fig. 28-23, then the rms current (I$_{rms}$) is:

$$I_{rms} = I_{max} \times \left(\sqrt{\frac{2}{2}}\right) = I_{max} \times 0.707.$$

If the voltage varies as a sine curve, then the rms voltage is:

$$V_{rms} = V_{max} \times 0.707.$$

The average power is then:

$$P = I_{rms} \times V_{rms}$$

$$= (I_{max} \times 0.707) \times (V_{max} \times 0.707)$$

$$P = 0.5 \times I_{max} V_{max}.$$

28-68 DANGER IN THE USE OF FLUORO REFRIGERANTS

Scientists are concerned about the possible release of chlorinated refrigerants into the atmosphere. There is some evidence that the balance of nature which controls the layer of ozone above the earth may be affected by the release of these refrigerants.

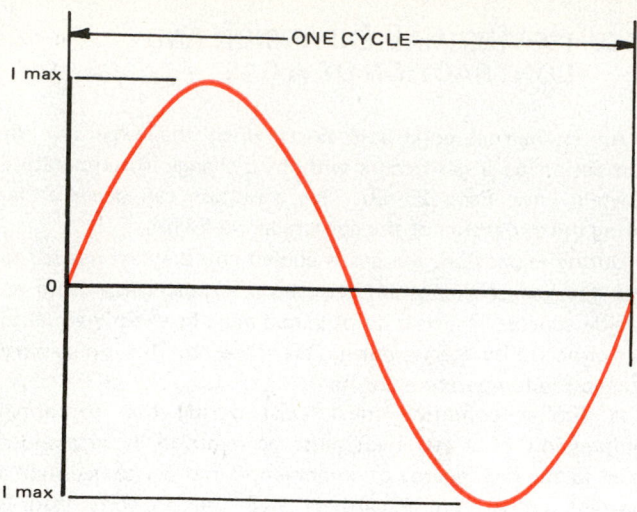

Fig. 28-23. Sine curve represents the current in a circuit.

28-69 SAFETY

A safety code for mechanical refrigeration is sponsored by the American Society of Heating, Refrigeration and Air Conditioning Engineers (ASHRAE). This code has been developed according to the rules and regulations of the American Standards Association. Copies of this safety code are available from the American Society of Heating, Refrigeration and Air Conditioning Engineers.

The Occupational Safety and Health Act provides many regulations and controls for the installation, operation and service of refrigerating and air conditioning mechanisms.

The American Conference of Governmental Industrial Hygienists have published Threshold Limit Values for substances that may be toxic under certain conditions and length of exposure. The threshold values for some of these substances for an eight-hour exposure time are:

SUBSTANCE	ppm
Acetone	1000
Ammonia	50
Carbon dioxide	5000
Carbon monoxide	100
Carbon tetrachloride-Skin	10
Chlorine	1
Dichlorodifluoromethane	1000
Dichloromonofluoromethane	1000
Ethyl ether	400
Fluorine	0.1
Gasoline	500
Methyl acetylene	1000
Methyl alcohol (Methanol)	200
Methyl chloride	100
Ozone	0.1
Sulfur dioxide	5
Xylene (xylol)	200

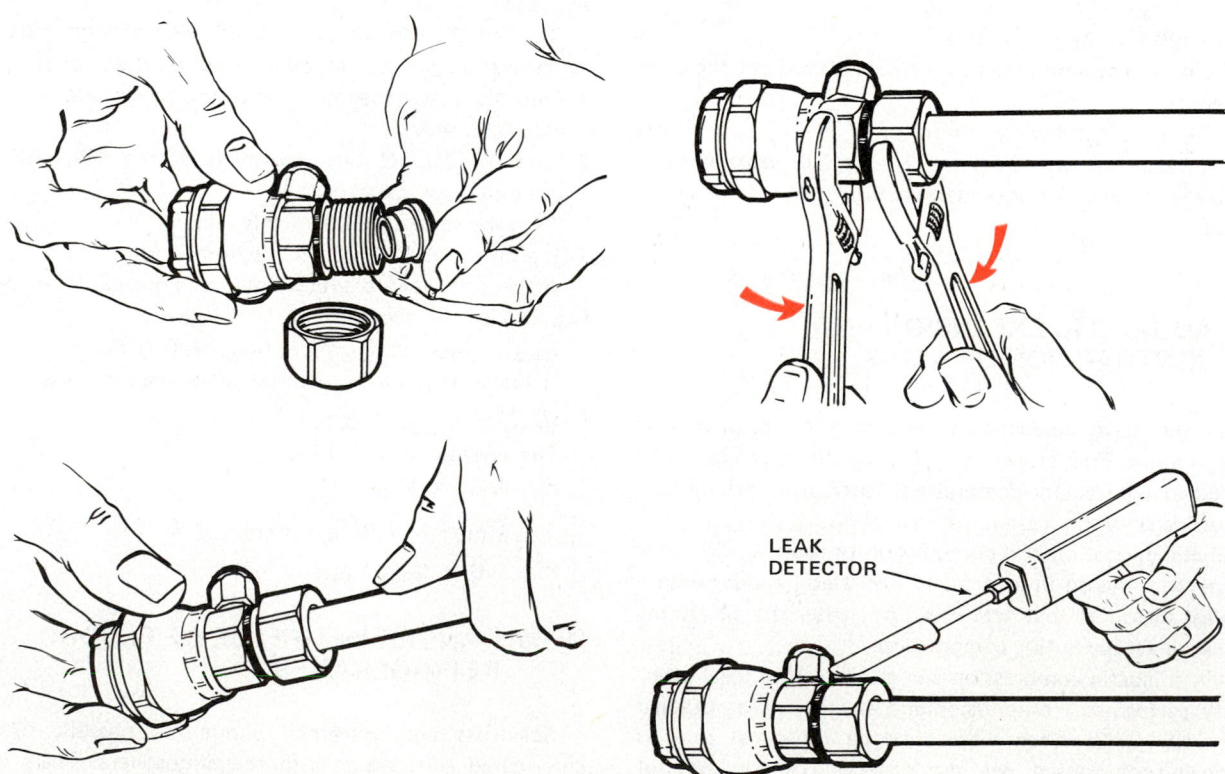

Successful installation and service persons work a great deal with piping. They must be skilled in the use of tools and testing devices. (Aeroquip Corp.)

Chapter 29

CAREER OPPORTUNITIES IN REFRIGERATION AND AIR CONDITIONING

Hardly any phase of modern living is untouched by modern refrigeration and air conditioning. Business, commercial operations, manufacturing processes, storage and shipping are nearly all now carried on under controlled temperature conditions.

IMPORTANCE OF REFRIGERATION AND AIR CONDITIONING

The refrigeration and air conditioning industry helps make possible this system of living. Many fruits and vegetables are refrigerated immediately upon being harvested. The quality of such products is much better for this fact. Air conditioning has improved business and industrial efficiency while adding to human comfort. More and more factories and heavy industries are being air conditioned. The present scale of farming is made possible, to a great extent, by the use of air conditioned tractor cabs and refrigerated harvesting equipment.

Cooling and freezing of meat and meat products makes possible their handling in a much more sanitary way than would be possible without mechanical refrigeration. Beverages, desserts, as well as staple foods, are all at least partially processed by refrigeration equipment.

THE CHALLENGE

Designing, manufacturing, selling, installing and maintaining this equipment provides for many, many jobs that did not exist less than a generation ago. Since refrigeration is used in so many enterprises, it follows that everyone who has to work in these industries has to be familiar with the basic air conditioning and refrigeration processes.

Opportunities for employment in writing specifications for refrigeration and air conditioning equipment and selling this equipment have naturally grown with the industry.

All the careers are available to anyone interested, regardless of race, creed, color or sex.

THE AIR CONDITIONING AND REFRIGERATION INDUSTRY

The air conditioning and refrigeration industry is usually divided into three industries:
1. Domestic.
2. Commercial.
3. Industrial.

The domestic field covers home refrigerators, freezers and window air conditioners.

The commercial field includes all small automatic systems such as for stores, supermarkets, domestic central air conditioning, water coolers, beverage coolers and truck refrigeration systems.

The industrial field includes the large processing systems and air conditioning systems, packing plants, cold storage and ice rinks. These systems require the attention of a refrigeration operating engineer.

CAREERS

Some of the employment opportunities include:
1. Jobs at various levels:
   a. Senior skilled.
   b. Skilled.
   c. Technicians.
   d. Supervisors.
   e. Professional personnel.
2. Various specialties (partial list):
   a. Engineers.
   b. Technicians.
   c. Test technicians.
   d. Sales engineers.
   e. Application engineers.
   f. Installers.
   g. Testers.
   h. Maintenance technicians.
   i. Service persons.
   j. Repair specialists.
      Helpers.
   k. Wholesalers.
      Sales engineers.
      Sales persons.
      Counter persons.
      Parts persons.
      Shipping and receiving persons.
   l. Operating engineers.
      Refrigeration.
      Industrial.
   m. Sheet metal experts.
      Helpers.

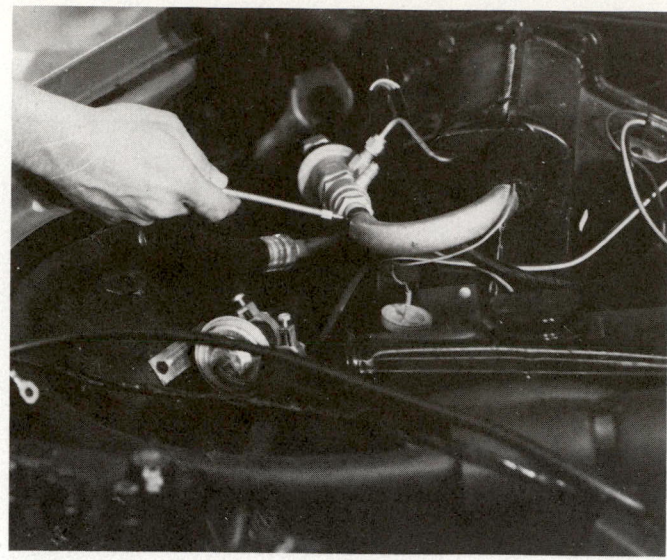

Installation and service of vehicle air conditioning systems is one of the careers open to persons with knowledge and hands-on training in refrigerating and air conditioning. This service technician must also have a good level of skill in automotive mechanics. (Robinair Mfg. Corp.)

Modern supermarket "machine" room. A skilled sales engineer was required to select and specify the equipment. A skilled installer set it up and put it in operation. A skilled service person is required to maintain and service the equipment.

THE WORK THEY DO

As one might expect, the responsibilities of the person working in the air conditioning and heating industry vary greatly. So does the kind of work that person will do.

Consider the air conditioning, heating and refrigeration technicians, for example. Working under the supervision of engineers, they help design, manufacture, sell and service equipment. But, often they specialize in one area, like research and development.

Those working in manufacturing of equipment may design and test or supervise production. They may also work as manufacturer's representatives or field sales persons. In such cases, responsibilities would include supplying contractors and engineering firms with data on installation, maintenance, operating costs and performance specifications of equipment.

Some technicians are employed by contractors to help design and prepare installation instructions. Others work in customer relations and may be responsible for supervision of installation and maintenance of equipment.

Another group employed by the industry works on installation and service. They travel about in service trucks and service units in homes, offices, schools and other buildings. This group includes:

1. Air conditioning and refrigeration mechanics who install and repair units ranging in size from small window air conditioners up to large central systems. They must follow blueprints and specifications to install motors, compressors, evaporators and other components. They will connect ducts, refrigerant lines and piping as well as make power hookups. In event of breakdown, they find the cause and make repairs.

2. Furnace installers or heating equipment installers. They read blueprints and specifications and install oil, gas and electric heating. Installation work includes placement of fuel supply lines, air ducts, pumps and other parts of a heating system. After connecting the electrical wiring and controls, they check units for proper operation.

3. Oil burner mechanics who keep oil-fueled heating systems in good working order. Their work varies with the season. In fall and winter they service and adjust burners. During the summer they service the heating unit, replace oil and air filters, vacuum vents, ducts and other parts of the system.

4. Gas burner mechanics or gas appliance service persons. Duties are similar to those of the oil burner mechanic. They determine why a burner will not work and adjust or repair.

Cooling and heating systems are sometimes installed or repaired by other types of mechanics or craftspeople. For example, ductwork on a large heating or air conditioning job may be done by sheet metalworkers; electrical work by an electrician and piping by pipe fitters. This is often the case on large installations where unions might be involved.

EDUCATION REQUIREMENTS

To qualify for employment, one should have good communication skills, a knowledge of practical mathematics and some physics and chemistry.

One should also take at least a one-year training program in the theory and hands-on laboratory work on refrigerating, heating, and air conditioning systems.

Additional employment information is available from the nearest branch of the United States Employment Service and the local State Employment Service. Local schools or public libraries also offer reference materials such as the DICTIONARY OF OCCUPATIONAL TITLES and the OCCUPATIONAL OUTLOOK HANDBOOK.

Chapter 30

DICTIONARY OF TECHNICAL TERMS

ABSOLUTE HUMIDITY: Amount of moisture in the air, indicated in grains per cu. ft.

ABSOLUTE PRESSURE: Gauge pressure plus atmospheric pressure (14.7 lb. per sq. in.), equals absolute pressure.

ABSOLUTE TEMPERATURE: Temperature measured from absolute zero.

ABSOLUTE ZERO TEMPERATURE: Temperature at which all molecular motion ceases. (—460 F. and —273 C.).

ABSORBENT: Substance with the ability to take up or absorb another substance.

ABSORPTION REFRIGERATOR: Refrigerator which creates low temperatures by using the cooling effect formed when a refrigerant is absorbed by chemical substance.

ACCESSIBLE HERMETIC: Assembly of motor and compressor inside a single bolted housing unit.

ACCUMULATOR: Storage tank which receives liquid refrigerant from evaporator and prevents it from flowing into suction line before vaporizing.

ACID CONDITION IN SYSTEM: Condition in which refrigerant or oil in system is mixed with fluids that are acid in nature.

ACR TUBING: Tubing used in air conditioning and refrigeration. Ends are sealed to keep tubing clean and dry.

ACTIVATED ALUMINA: Chemical which is a form of aluminum oxide. It is used as a drier or desiccant.

ACTIVATED CARBON: Specially processed carbon used as a filter-drier; commonly used to clean air.

ACTUATOR: That portion of a regulating valve which converts mechanical fluid, thermal energy or electrical energy into mechanical motion to open or close the valve seats.

ADIABATIC COMPRESSION: Compressing refrigerant gas without removing or adding heat.

ADSORBENT: Substance with the property to hold molecules of fluids without causing a chemical or physical change.

AERATION: Act of combining substance with air.

AGITATOR: Device used to cause motion in confined fluid.

AIR CLEANER: Device used for removal of airborne impurities.

AIR COIL: Coil on some types of heat pumps used either as an evaporator or a condenser.

AIR CONDITIONER: Device used to control temperature, humidity, cleanliness and movement of air in conditioned space.

AIR CONDITIONING: Control of the temperature, humidity, air movement and cleaning of air in a confined space.

AIR-COOLED CONDENSER: Heat of compression is transferred from condensing coils to surrounding air. This may be done either by convection or by a fan or blower.

AIR COOLER: Mechanism designed to lower temperature of air passing through it.

AIR CORE SOLENOID: Solenoid which has a hollow core instead of a solid core.

AIR DIFFUSER: Air distribution outlet or grille designed to direct airflow into desired patterns.

AIR GAP: The space between magnetic poles or between rotating and stationary assemblies in a motor or generator.

AIR HANDLER: Fan-blower, heat transfer coil, filter and housing parts of a system.

AIR INFILTRATION: Leakage of air into rooms through cracks, windows, doors and other openings.

AIR-SENSING THERMOSTAT: Thermostat unit in which sensing element is located in refrigerated space.

AIR, STANDARD: Air having a temperature of 68 F. (20 C.), a relative humidity of 36 percent and under pressure of 14.70 psia. The gas industry usually considers 60 F. (16 C.) as the temperature of standard air.

AIR VENT: Valve, either manual or automatic, to remove air from the highest point of a coil or piping assembly.

AIR WASHER: Device used to clean air while increasing or lowering its humidity.

ALCOHOL BRINE: Water and alcohol solution which remains a liquid below 32 F. (0 C.).

ALGAE: Low form of plant life, found floating free in water.

ALLEN-TYPE SCREW: Screw with recessed hex-shaped head.

ALTERNATING CURRENT (a-c): Electric current in which direction of flow alternates or reverses. In 60-cycle (Hertz) current, direction of flow reverses every 1/120th of a second.

AMBIENT TEMPERATURE: Temperature of fluid (usually air) which surrounds object on all sides.

AMERICAN STANDARD PIPE THREAD: Type of screw thread commonly used on pipe and fittings to assure a tight seal.

AMMETER: Electric meter, calibrated in amperes, used to measure current.

AMMONIA: Chemical combination of nitrogen and hydrogen (NH_3). Ammonia refrigerant is identified as R—117.

AMPERAGE: Electron or current flow of one coulomb per second past given point in circuit.

AMPERE: Unit of electric current equivalent to flow of one coulomb per second.

AMPERE TURNS: Term used to measure magnetic force. Represents product of amperes times number of turns in coil of electromagnet.

AMPLIFIER: Electrical device which increases electron flow in a circuit.

ANEMOMETER: Instrument for measuring the rate of airflow or motion.

ANGLE VALVE: Type of globe valve design, having pipe openings at right angles to each other. Usually, one opening is on the horizontal plane and one is on the vertical plane.

ANHYDROUS CALCIUM SULPHATE: Dry chemical made of calcium, sulphur and oxygen ($CaSO_4$).

ANNEALING: Process of heat treating metal to get desired properties of softness and ductility (easy to form into new shape).

ANODE: Positive terminal of electrolytic cell.

ARMATURE: Part of an electric motor, generator or other device moved by magnetism.

A.S.A.: Formerly, abbreviation for American Standards Association. Now known as American National Standards Institute.

A.S.M.E. BOILER CODE: Standard specifications issued by the American Society of Mechanical Engineers for the construction of boilers.

ASPECT RATIO: Ratio of length to width of rectangular air grille or duct.

ASPIRATING PSYCHROMETER: Device which draws sample of air through it to measure humidity.

ASPIRATION: Movement produced in a fluid by suction.

A.S.T.M. STANDARDS: Standards issued by the American Society of Testing Materials.

ATMOSPHERIC PRESSURE: Pressure that gases in air exert upon the earth; measured in pounds per square inch, (Grams/per square centimetre).

ATOM: Smallest particle of element that can exist alone or in combination.

ATOMIZE: Process of changing a liquid to minute particles or a fine spray.

AUTOMATIC CONTROL: Valve action reached through self-operated or self-actuated means, not requiring manual adjustment.

AUTOMATIC DEFROST: System of removing ice and frost from evaporators automatically.

AUTOMATIC EXPANSION VALVE (AEV): Pressure-controlled valve which reduces high-pressure liquid refrigerant to low-pressure liquid refrigerant.

AUTOMATIC ICE CUBE MAKER: Refrigerating mechanism designed to automatically produce ice cubes in quantity.

AUTOTRANSFORMER: Transformer in which both primary and secondary coils have turns in common. Step-up or step-down of voltage is accomplished by taps on common winding.

AZEOTROPE: Having constant maximum and minimum boiling points.

AZEOTROPIC MIXTURE: Example of azeotropic mixture — refrigerant R−502 is mixture consisting of 48.8 percent refrigerant R−22, and 51.2 percent R−115. The refrigerants do not combine chemically, yet azeotropic mixture provides refrigerant characteristics desired.

BACK PRESSURE: Pressure in low side of refrigerating system; also called suction pressure or low-side pressure.

BACK SEATING: Fluid opening/closing such as a gauge opening or to seal the joint where the valve stem goes through the valve body.

BAFFLE: Plate or vane used to direct or control movement of fluid or air within confined area.

BALL CHECK VALVE: Valve assembly (ball) which permits flow of fluid in one direction only.

BAR: Unit of pressure. One bar equals .9869 atmospheres.

BAROMETER: Instrument for measuring atmospheric pressure. It may be calibrated in pounds per square inch, in inches of mercury in a column, or millimetres.

BATH: Liquid solution used for cleaning, plating or maintaining a specified temperature.

BATTERY: Electricity-producing cells which use interaction of metals and chemicals to create electrical current flow.

BAUDELOT COOLER: Heat exchanger in which water flows by gravity over the outside of the tubes or plates.

BEARING: Low friction device for supporting and aligning a moving part.

BELLOWS: Corrugated cylindrical container which moves as pressures change, or provides a seal during movement of parts.

BELLOWS SEAL: Method of sealing the valve stem. The ends of the sealing material are fastened to the bonnet and to the stem. Seal expands and contracts with the stem level.

BENDING SPRING: Coil spring which is placed on inside or outside of tubing to keep it from collapsing while bending it.

BERNOULLI'S THEOREM: In stream of liquid, the sum of elevation head, pressure head and velocity remains constant along any line of flow provided no work is done by or upon liquid in course of its flow, and decreases in proportion to energy lost in flow.

BIMETAL STRIP: Temperature regulating or indicating device which works on principle that two dissimilar metals with unequal expansion rates, welded together, will bend as temperatures change.

BLAST FREEZER: Low-temperature evaporator which uses a fan to force air rapidly over the evaporator surface.

BLEEDING: Slowly reducing the pressure of liquid or gas from a system or cylinder by slightly opening a valve.

BLEED-VALVE: Valve with small opening inside which permits a minimum fluid flow when valve is closed.

BOILER: Closed container in which a liquid may be heated and vaporized.

BOILER, HIGH-PRESSURE: Boiler furnishing steam at pressures of 15 pounds per square inch gauge or higher (1.05 kg/cm^2).

BOILER HORSEPOWER: Term now seldom used, meaning equivalent to a heating capacity of 33,475 Btu/hr. (983 watts).

BOILER, HOT-WATER AND LOW-PRESSURE STEAM: A boiler furnishing hot water at pressures not more than 30 pounds per square inch gauge (2.12 kg/cm^2) or steam at pressures not more than 15 pounds per square inch gauge (1.06 kg/cm^2).

BOILING POINT: Boiling temperature of a liquid under a pressure of 14.7 psia (760 mm).

BOILING TEMPERATURE: Temperature at which a fluid changes from a liquid to a gas.

BOOSTER: Common term applied to the use of a compressor when used as the first stage in the cascade refrigerating system.

BORE: Inside diameter of a cylinder.

BOURDON TUBE: Thin-walled tube of elastic metal flattened and bent into circular shape, which tends to straighten as pressure inside is increased. Used in pressure gauges.

BOYLE'S LAW: Law of physics — volume of a gas varies as pressure varies, if temperature remains the same. Example: If absolute pressure is doubled on quantity of gas, volume is reduced one half. If volume becomes doubled, gas has its pressure reduced by half.

BRAZING: Method of joining metals with nonferrous filler (without iron) using heat between 800 F. (427 C.) and melting point of base metals.

BREAKER STRIP: Strip of wood or plastic used to cover joint between outside case and inside liner of refrigerator.

BREECHING: Space in hot water or steam boilers between the end of the tubing and the jacket.

BRINE: Water saturated with a chemical such as salt.

BRITISH THERMAL UNIT (Btu): Quantity of heat required to raise temperature of one pound of water one degree Fahrenheit.

BULB, SENSITIVE: Part of sealed fluid device which reacts to temperature. Used to measure temperature or to control a mechanism.

BUNKER: Space where ice or cooling element is placed in commercial installations.

BURNER: Device in which burning of fuel takes place.

BUTANE: Liquid hydrocarbon (C_4H_{10}) commonly used as fuel for heating purposes.

BYPASS: Passage at one side of, or around, a regular passage.

CALCIUM SULFATE: Chemical compound ($CaSO_4$) which is used as a drying agent or desiccant in liquid line driers.

CALIBRATE: Position indicators to determine accurate measurements.

CALORIE: Two different calorie units are used by scientists. The calorie used by medical science is a small heat unit. It equals the heat required to raise the temperature of one gram of water one degree C. The calorie used by engineering science is a large heat unit. It is equal to the amount of heat required to raise the temperature of one kilogram of water one degree C. In the SI system it is recommended that the Joule unit of energy be used in place of the calorie.

CALORIMETER: Device used to measure quantities of heat or determine specific heats.

CAPACITANCE (C): Property of a nonconductor (condenser or capacitor) that permits storage of electrical energy in an electrostatic field.

CAPACITIVE REACTANCE: The opposition or resistance to an alternating current as a result of capacitance; expressed in ohms.

CAPACITOR: Type of electrical storage device used in starting and/or running circuits on many electric motors.

CAPACITOR-START MOTOR: Motor which has a capacitor in the starting circuit.

CAPACITY: Refrigeration rating system. Usually measured in Btu per hour or watts (metric).

CARBON DIOXIDE: Compound of carbon and oxygen (CO_2) which is sometimes used as a refrigerant. Refrigerant number is R−744.

CARBON FILTER: Air filter using activated carbon as air cleansing agent.

CARBON MONOXIDE: Colorless, odorless and poisonous gas (CO) produced when carbon or carbonaceous fuels are burned with too little air.

CARBON TETRACHLORIDE: Colorless nonflammable and very toxic liquid used as a solvent. It should never be allowed to touch skin and fumes must not be inhaled.

CARRENE: Refrigerant in Group One (R−11). Chemical combination of carbon, chlorine and fluorine.

CASCADE SYSTEMS: Arrangement in which two or more refrigerating systems are used in series; uses evaporator of one machine to cool condenser of other machine. Produces ultra-low temperatures.

CATHODE: Negative terminal of an electrical device. Electrons leave at this terminal.

CAVITATION: Localized gaseous condition that is found within a liquid stream.

CELSIUS TEMPERATURE SCALE: Temperature scale used in metric system. Freezing point of water is 0; boiling point is 100.

CENTIGRADE TEMPERATURE SCALE: See CELSIUS TEMPERATURE SCALE.

CENTIMETRE: Metric unit of linear measurement which equals .3937 in.

CENTRAL STATION: Central location of condensing unit with either wet or air-cooled condenser. Evaporator located as needed and connected to the central condensing unit.

CENTRIFUGAL COMPRESSOR: Pump which compresses gaseous refrigerants by centrifugal force.

CHANGE OF STATE: Condition in which a substance changes from a solid to a liquid or a liquid to a gas caused by the addition of heat. Or the reverse, in which a substance changes from a gas to a liquid, or a liquid to a solid, caused by the removal of heat.

CHARGE: Amount of refrigerant placed in a refrigerating unit.

CHARGING BOARD: Specially designed panel or cabinet fitted with gauges, valves and refrigerant cylinders used for charging refrigerant and oil into refrigerating mechanisms.

CHARLES'S LAW: Volume of a given mass of gas at a constant pressure varies according to its temperature.

CHECK VALVE: Device which permits fluid flow in one direction.

CHEMICAL REFRIGERATION: System of cooling using a disposable refrigerant. Also called an expendable refrigerant system.

CHILL FACTOR: Calculated number based on temperature and wind velocity.

CHIMNEY: Vertical shaft enclosing one or more flues for carrying flue gases to the outside atmosphere.

CHIMNEY CONNECTOR: Conduit (pipe) connecting the heating appliance (furnace) with the vertical flue.

CHIMNEY EFFECT: Tendency of air or gas to rise when heated.

CHIMNEY FLUE: Flue gas passageway in a chimney.

CHOKE TUBE: Throttling device used to maintain correct pressure difference between high-side and low-side in refrigerating mechanism. Capillary tubes are sometimes called choke tubes.

CIRCUIT: Tubing, piping or electrical wire installation which permits flow from the energy source back to energy source.

CIRCUIT BREAKER: Safety device which automatically opens an electrical circuit if overloaded.

CIRCUIT, PARALLEL: Arrangement of electrical devices in which the current divides and travels through two or more paths and then returns through a common path.

CIRCUIT, PILOT: Secondary circuit used to control a main circuit or a device in the main circuit.

CIRCUIT, SERIES: Electrical wiring; electrical path (circuit) in which electricity to operate second lamp or device must pass through first; current flow travels, in turn, through all devices connected together.

CLEARANCE POCKET COMPRESSOR: Small space in a cylinder from which compressed gas is not completely expelled. This space is called the compressor clearance space or pocket. For effective operation, compressors are designed to have as small a clearance space as possible.

CLOSED CIRCUIT: Electrical circuit in which electrons are flowing.

CLOSED CONTAINER: Container sealed by means of a lid or other device so that neither liquid nor vapor will escape from it at ordinary temperatures.

CLUTCH, MAGNETIC: Clutch built into automobile compressor flywheel, operated magnetically, which allows pulley to revolve without driving compressor when refrigerating effect is not required.

CODE INSTALLATION: Refrigeration or air conditioning installation which conforms to the local code and/or the national code for safe and efficient installations.

COEFFICIENT OF CONDUCTIVITY: Measure of the relative rate at which different materials conduct heat. Copper is a good conductor of heat and, therefore, has a high coefficient of conductivity.

COEFFICIENT OF EXPANSION: Increase in unit length, area or volume for one degree rise in temperature.

COEFFICIENT OF PERFORMANCE (cop): Ratio of work performed or accomplished as compared to the energy used.

CO_2 INDICATOR: Instrument used to indicate the percentage of carbon dioxide in stack gases.

COLD: The absence of heat; a temperature considerably below normal.

COLD JUNCTION: That part of a thermoelectric system which absorbs heat as the system operates.

COLD WALL: Refrigerator construction which has the inner lining of refrigerator serving as the cooling surface.

COLLOIDS: Miniature cells peculiar to meats, fish and poulty which, if disrupted, cause food to become rancid. Low temperatures minimize this action.

COMBUSTIBLE LIQUIDS: Liquid having a flash point at or above 140 F. (60 C.), and shall be known as Class 3 liquids.

COMFORT CHART: Chart used in air conditioning to show the dry bulb temperature, humidity and air movement for human comfort conditions.

COMFORT COOLER: System used to reduce the temperature in the living space in homes. These systems are not complete air conditioners as they do not provide complete control of heating, humidifying, dehumidification, and air circulation.

COMFORT ZONE: Area on psychrometric chart which shows conditions of temperature, humidity and sometimes air movement in which most people are comfortable.

COMMUTATOR: Part of rotor in electric motor which conveys electric current to rotor windings.

COMPOUND GAUGE: Instrument for measuring pressures both above and below atmospheric pressure.

COMPOUND REFRIGERATING SYSTEMS: System which has several compressors or compressor cylinders in series. The system is used to pump low-pressure vapors to condensing pressures.

COMPRESSION: Term used to denote increase of pressure on a fluid by using mechanical energy.

COMPRESSION GAUGE: Instrument used to measure positive pressures (pressures above atmospheric pressures) only. These gauges are usually calibrated from 0 to 300 lb. per sq. in. gauge, (psig) $(0-21.1 \text{ kg/cm}^2)$

COMPRESSION RATIO: Ratio of the volume of the clearance space to the total volume of the cylinder. In refrigeration it is also used as the ratio of the absolute low-side pressure to the absolute high-side pressure.

COMPRESSOR: Pump of a refrigerating mechanism which draws a low pressure on cooling side of refrigerant cycle and squeezes or compresses the gas into the high-pressure or condensing side of the cycle.

COMPRESSOR DISPLACEMENT: Volume, in cubic inches, represented by the area of the compressor piston head or heads multiplied by the length of the stroke.

COMPRESSOR, EXTERNAL DRIVE: See COMPRESSOR, OPEN TYPE.

COMPRESSOR, HERMETIC: Compressor in which the driving motor is sealed in the same dome or housing as the compressor.

COMPRESSOR, MULTIPLE STAGE: Compressor having two or more compressive steps. Discharge from each step is the intake pressure of the next in series.

COMPRESSOR, OPEN TYPE: Compressor in which the crankshaft extends through the crankcase and is driven by an outside motor. Commonly called external drive compressor.

COMPRESSOR, RECIPROCATING: Compressor which uses a piston and cylinder mechanism to provide pumping action.

COMPRESSOR, ROTARY: Compressor which uses vanes, eccentric mechanisms or other rotating devices to provide pumping action.

COMPRESSOR, SINGLE-STAGE: Compressor having only one compressive step between low-side pressure and high-side pressure.

COMPRESSOR SEAL: Leakproof seal between crankshaft and compressor body in open type compressors.

CONDENSATE: A fluid formed when a gas is cooled to its liquid state.

CONDENSATE PUMP: Device to remove water condensate that collects beneath an evaporator.

CONDENSATION: Liquid or droplets which form when a gas or vapor is cooled below its dew point.

CONDENSE: Action of changing a gas or vapor to a liquid.

CONDENSER: The part of refrigeration mechanism which receives hot, high-pressure refrigerant gas from compressor and cools gaseous refrigerant until it returns to its liquid state.

CONDENSER, AIR-COOLED: Heat exchanger which transfers heat to surrounding air.

CONDENSER COMB: Comb-like device, metal or plastic, used to straighten the metal fins on condensers or evaporators.

CONDENSER FAN: Forced air device used to move air through air-cooled condenser.

CONDENSER, WATER-COOLED: Heat exchanger designed to transfer heat from hot gaseous refrigerant to water.

CONDENSING PRESSURE: Pressure inside a condenser at which refrigerant vapor gives up its latent heat of vaporization and becomes a liquid. This varies with the temperature.

CONDENSING TEMPERATURE: Temperature inside a condenser at which refrigerant vapor gives up its latent heat of vaporization and becomes a liquid. This varies with the pressure.

CONDENSING UNIT: Part of a refrigerating mechanism which pumps vaporized refrigerant from the evaporator, compresses it, liquefies it in the condenser and returns it to the refrigerant control.

CONDENSING UNIT SERVICE VALVES: Shutoff valves mounted on condensing unit to enable service technicians to install and/or service unit.

CONDUCTIVITY: Ability of a substance to conduct or transmit heat and/or electricity.

CONDUCTOR: Substance or body capable of transmitting electricity or heat.

CONSTRICTOR: Tube or orifice used to restrict flow of a gas or a liquid.

CONTAMINANT: Substance such as dirt, moisture, or other matter foreign to refrigerant or refrigerant oil in system.

CONTINUOUS CYCLE ABSORPTION SYSTEM: System which has a

continuous flow of energy input.

CONTROL: Automatic or manual device used to stop, start and/or regulate flow of gas, liquid and/or electricity.

CONTROL, COMPRESSOR: See MOTOR CONTROL.

CONTROL, DEFROSTING: Device to automatically defrost evaporator. It may operate by means of a clock, door cycling mechanism or during "off" portion of refrigerating cycle.

CONTROL, LOW PRESSURE: Cycling device connected to low-pressure side of system.

CONTROL, MOTOR: Temperature or pressure-operated device used to control running of motor.

CONTROL, PRESSURE MOTOR: High or low-pressure control connected into the electrical circuit and used to start and stop motor. It is activated by demand for refrigeration or for safety.

CONTROL, REFRIGERANT: Device used to regulate flow of liquid refrigerant into evaporator. Can be a capillary tube, expansion valves, or high-side and low-side float valves.

CONTROL SYSTEM: All of the components required for the automatic control of a process variable.

CONTROL, TEMPERATURE: Temperature-operated thermostatic device which automatically opens or closes a circuit.

CONTROL VALVE: Valve which regulates the flow or pressure of a medium which affects a controlled process. Control valves are operated by remote signals from independent devices using any of a number of control media such as pneumatic, electric or electrohydraulic.

CONVECTION: Transfer of heat by means of movement or flow of a fluid or gas.

CONVECTION, FORCED: Transfer of heat resulting from forced movement of liquid or gas by means of a fan or pump.

CONVECTION, NATURAL: Circulation of a gas or liquid due to difference in density resulting from temperature differences.

CONVERSION FACTORS: Force and power may be expressed in more than one way. A horsepower is equivalent to 33,000 ft. lb. of work per minute, 746 watts, or 2546 Btu per hour. These values can be used for changing horsepower into foot pounds, British thermal units or watts.

COOLER: Heat exchanger which removes heat from a substance.

COOLING TOWER: Device which cools by water evaporation in air. Water is cooled to wet bulb temperature of air.

COPPER PLATING: Abnormal condition developing in some units in which copper is electrolytically deposited on some compressor surfaces.

CORE, AIR: Coil of wire not having a metal core.

CORE, MAGNETIC: Magnetic center of a magnetic field.

CORROSION: Deterioration of materials from chemical action.

COULOMB: The quantity of electricity transferred by an electric current of one ampere in one second.

COUNTER EMF: Tendency for reverse electrical flow as magnetic field changes in an induction coil.

COUNTERFLOW: Flow in opposite direction.

COUPLINGS: Mechanical device joining refrigerant lines.

CRACKAGE: Joint in a structure which permits movement of a gas or vapor through it, even under a small pressure difference.

"CRACKING" A VALVE: Opening a valve a small amount.

CRANK THROW: Distance between center line of main bearing journal and center line of the crankpin or eccentric.

CRANKSHAFT SEAL: Leakproof joint between crankshaft and compressor body.

CRISPER: Drawer or compartment in refrigerator designed to provide high humidity along with low temperature to keep vegetables — especially leafy vegetables — cold and crisp.

CRITICAL PRESSURE: Compressed condition of refrigerant which gives liquid and gas the same properties.

CRITICAL TEMPERATURE: Temperature at which vapor and liquid have same properties.

CRITICAL VIBRATION: Vibration which is noticeable and harmful to structure.

CROSS CHARGED: Sealed container of two fluids which together create a desired pressure-temperature curve.

CRYOGENIC FLUID: Substance which exists as a liquid or gas at ultra-low temperatures (−250 F. or lower).

CRYOGENICS: Refrigeration which deals with producing temperatures of 250 F. below zero and lower.

CURRENT: Transfer of electrical energy in a conductor by means of electrons changing position.

CURRENT RELAY: Device which opens or closes a circuit. It is made to act by a change of current flow in that circuit.

CUT-IN: The temperature value or the pressure value at which the

control circuit closes.

CUT-OUT: Temperature value or pressure value at which the control circuit opens.

CYCLE: Series of events or operations which have tendency to repeat in the same order.

CYLINDER: 1—Device which converts fluid power into linear mechanical force and motion. This usually consists of movable elements such as a piston and piston rod, plunger or ram, operating within a cylindrical bore. 2—Closed container for fluids.

CYLINDER HEAD: Plate or cap which encloses compression end of compressor cylinder.

CYLINDER, REFRIGERANT: Cylinder in which refrigerant is stored and dispensed. Color code painted on cylinder indicates kind of refrigerant.

CYLINDRICAL COMMUTATOR: Commutator with contact surfaces parallel to the rotor shaft.

DALTON'S LAW: Vapor pressure created in a container by a mixture of gases is equal to sum of individual vapor pressures of the gases contained in mixture.

DAMPER: Device for controlling airflow.

DEAERATION: Act of separating air from substances.

DECIBEL (dB): Unit used for measuring relative loudness of sounds. One decibel is equal to approximate difference of loudness ordinarily detectable by human ear, the range of which is about 130 decibels on scale beginning with one for faintest audible sound.

DECK (COIL DECK): Insulated horizontal partition between refrigerated space and evaporator space.

DEFROST CYCLE: Refrigerating cycle in which evaporator frost and ice accumulation is melted.

DEFROST TIMER: Device connected into electrical circuit which shuts unit off long enough to permit ice and frost accumulation on evaporator to melt.

DEFROSTING: Process of removing frost accumulation from evaporators.

DEFROSTING TYPE EVAPORATOR: Evaporator operating at such temperatures that ice and frost on surface melts during off part of operating cycle.

DEGREASING: Solution or solvent used to remove oil or grease from refrigerator parts.

DEGREE-DAY: Unit that represents one degree of difference from inside temperature and the average outdoor temperature for one day and is often used in estimating fuel requirements for a building.

DEHUMIDIFIER: Device used to remove moisture from air.

DEHYDRATED OIL: Lubricant which has had most of its water content removed (dry oil).

DEHYDRATOR: See DRIER.

DEHYDRATOR-RECEIVER: Small tank which serves as liquid refrigerant reservoir and which also contains a desiccant to remove moisture. Used on most automobile air conditioning installations.

DEICE CONTROL: Device for operating a refrigerating system in such a way as to provide melting of the accumulated ice and frost.

DELTA TRANSFORMER: Three-phase electrical transformer which has ends of each of three windings electrically connected to form a triangle.

DEMAND METER: Instrument which measures the kilowatt-hour usage of a circuit or group of circuits.

DENSITY: Closeness of texture or consistency of particles within a given substance. The weight per unit volume.

DEODORIZER: Device which absorbs or adsorbs various odors, usually by principle of absorption. Activated charcoal is commonly used.

DESICCANT: Substance used to collect and hold moisture in refrigerating system. A drying agent. Common desiccants are activated alumina and silica gel.

DESIGN PRESSURE: Highest or most severe pressure expected during operation. Sometimes used as the calculated operating pressure plus an allowance for safety.

DETECTOR, LEAK: Device used to detect and locate refrigerant leaks.

DEW: Condensed atmospheric moisture deposited in small drops on cool surfaces.

DEW POINT: Temperature at which vapor (at 100 percent humidity) begins to condense and deposit as liquid.

DIAPHRAGM: Flexible material usually made of thin metal, rubber or plastic.

DICHLORODIFLUOROMETHANE: Refrigerant commonly known as R—12.

DIE CASTING: Process of molding low-melting-temperature metals in accurately shaped metal molds.

DIELECTRIC FLUID: Fluid with high electrical resistance.

DIFFERENTIAL: The temperature or pressure difference between cut-in and cut-out temperature or pressure of a control.

DIODE: Two-element electron tube which will allow more electron flow in one direction in a circuit than in the other direction; tube which serves as a rectifier.

DIRECT CURRENT (d-c): Electron flow which moves continuously in one direction in circuit.

DIRECT EXPANSION EVAPORATOR: Evaporator using either an automatic expansion valve (AEV) or a thermostatic expansion valve (TEV) refrigerant control.

DISPLACEMENT, PISTON: Volume obtained by multiplying area of cylinder bore by length of piston stroke.

DISTILLING APPARATUS: Fluid-reclaiming device used to reclaim used refrigerants. Reclaiming is usually done by vaporizing and then recondensing refrigerant.

DOME-HAT: Sealed metal container for the motor compressor of a refrigerating unit.

DOUBLE DUTY CASE: Commercial refrigerator in which a part of space is for refrigerated storage and part is equipped with glass windows for display purposes.

DOUBLE THICKNESS FLARE: Copper, aluminum or steel tubing end which has been formed into two-wall thickness, 37 to 45 deg. bell mouth or flare.

DOWEL PIN: Accurately dimensioned pin pressed into one assembly part and slipped into another assembly part to insure accurate alignment.

DRAFT GAUGE: Instrument used to measure air movement by measuring air pressure differences.

DRAFT INDICATOR: Instrument used to indicate or measure chimney draft or combustion gas movement. Draft is measured in units of .1 in. of water column.

DRAFT REGULATOR: Device which maintains a desired draft in a combustion-heated appliance by automatically controlling the chimney draft to the desired value.

DRIER: Substance or device used to remove moisture from a refrigeration system.

DRIP PAN: Pan-shaped panel or trough used to collect condensate from evaporator.

DRY BULB: An instrument with a sensitive element to measure ambient air temperature.

DRY BULB TEMPERATURE: Air temperature as indicated by an ordinary thermometer.

DRY CAPACITOR CONDENSER: Electrical device made of dry metal and dry insulation; used to store electrons.

DRY CELL BATTERY: Electrical device used to provide d-c electricity, having no liquid in the cells.

DRY ICE: Refrigerating substance made of solid carbon dioxide which changes directly from a solid to a gas (sublimates). Its subliming temperature is —109 F. (—78 C.).

DRY SYSTEM: Refrigeration system which has the evaporator liquid refrigerant mainly in the atomized or droplet condition.

DUCT: Tube or channel through which air is conveyed or moved.

DYNAMOMETER: Device for measuring power output or power input of a mechanism.

ECCENTRIC: Circle or disk mounted off center on a shaft.

ECOLOGY: Science of life balance on earth.

EDDY CURRENTS: Induced currents flowing in a core.

EFFECTIVE AREA: Actual flow area of an air inlet or outlet. Gross area minus area of vanes or grille bars.

EFFECTIVE TEMPERATURE: Overall effect on a human of air temperature, humidity and air movement.

EJECTOR: Device which uses high fluid velocity, such as a venturi, to create low pressure or vacuum at its throat to draw in fluid from another source.

ELECTRIC DEFROSTING: Use of electric resistance heating coils to melt ice and frost off evaporators during defrosting.

ELECTRIC HEATING: System in which heat from electrical resistance units is used to heat the building.

ELECTRIC WATER VALVE: Solenoid type (electrically operated) valve used to turn water flow on and off.

ELECTROLYSIS: Movement of electricity through a substance which causes a chemical change in the substance or its container.

ELECTROLYTIC CONDENSER-CAPACITOR: Plate or surface capable of storing small electrical charges.

ELECTROMAGNET: Coil of wire wound around a soft iron core. When electric current flows through wire, the assembly becomes a magnet.

ELECTROMOTIVE FORCE (EMF) VOLTAGE: Electrical force which causes current (free electrons) to flow or move in an electrical circuit. Unit of measurement is the volt.

ELECTRON: Elementary particle or portion of an atom which carries a negative charge.

ELECTRONIC LEAK DETECTOR: Electronic instrument which measures electronic flow across gas gap. Electronic flow changes indicate presence of refrigerant gas molecules.

ELECTRONICS: Field of science dealing with electron devices and their uses.

ELECTROSTATIC FILTER: For cleaning air, a type of filter which gives dust particles an electric charge. This causes particles to be attracted to a plate so they can be removed from air.

END BELL: End structure or plate of electric motor which usually holds motor bearings.

END PLAY: Slight movement of shaft along its center line.

ENDOTHERMAL: Chemical reaction in which heat is absorbed.

ENERGY: Actual or potential ability to do work.

ENTHALPY: Total amount of heat in one pound of a substance calculated from accepted temperature base. Temperature of 32 F. (0 C.) is accepted base for water vapor calculation. For refrigerator calculations, accepted base is —40 F. (—40 C.).

ENTROPY: Mathematical factor used in engineering calculations. Energy in a system.

ENVIRONMENT: The surrounding conditions.

ENZYME: Complex organic substance, originating from living cells, that speeds up chemical changes in foods. Enzyme action is slowed by cooling.

EPOXY (RESINS): A synthetic plastic adhesive.

ETHANE (R—170): Refrigerant sometimes added to other refrigerants to improve oil circulation.

EUTECTIC: That certain mixture of two substances providing lowest melting temperature of all the various mixes of the two substances.

EUTECTIC POINT: Freezing temperature for eutectic solutions.

EVACUATION: Removal of air (gas) and moisture from a refrigeration or air conditioning system.

EVAPORATION: Term applied to the changing of a liquid to a gas. Heat is absorbed in this process.

EVAPORATIVE CONDENSER: Device which uses open spray or spill water to cool a condenser. Evaporation of some of the water cools the condenser water and reduces water consumption.

EVAPORATOR: Part of a refrigerating mechanism in which the refrigerant vaporizes and absorbs heat.

EVAPORATOR, DRY TYPE: Evaporator in which the refrigerant is in the liquid droplet form.

EVAPORATOR, FLOODED: Evaporator containing liquid refrigerant at all times.

EVAPORATOR FAN: Fan which increases airflow over the heat exchange surface of evaporators.

EXFILTRATION: Slow flow of air from the building to the outdoors.

EXHAUST PORT: That opening which carries the fluid to the downstream pressure of a fluid system.

EXHAUST VALVE: A movable port which provides an outlet for the cylinder gases in a compressor or engine.

EXOTHERMAL: Chemical reaction in which heat is released.

EXPANSION JOINT: Device in piping designed to allow movement of the pipe caused by the pipe's expansion and contraction.

EXPANSION VALVE: Device in refrigerating system which reduces the pressure from the high side to the low side and is operated by pressure.

EXPENDABLE REFRIGERANT SYSTEM: System which discards the refrigerant after it has evaporated.

EXTERNAL DRIVE: Term used to indicate a compressor driven directly from the shaft or by a belt using an external motor. Compressor and motor are serviceable separately.

EXTERNAL EQUALIZER: Tube connected to low-pressure side of a thermostatic expansion valve diaphragm and to exit end of evaporator.

FAHRENHEIT SCALE: On a Fahrenheit thermometer, under standard atmospheric pressure, boiling point of water is 212 deg. and freezing point is 32 deg. above zero.

FAIL-SAFE CONTROL: Device which opens a circuit when sensing element loses its pressure.

FAN: Radial or axial flow device used for moving or producing flow of gases.

FARAD: Unit of electrical capacity; capacity of a condenser which, when charged with one coulomb of electricity, gives difference of potential of one volt.

FARADAY EXPERIMENT: Silver chloride absorbs ammonia when

cool and releases it when heated. This is basis on which some absorption refrigerators operate.

FEMALE THREAD: The internal thread on fittings, valves, machine bodies and the like.

FIELD POLE: Part of stator of motor which concentrates magnetic field of field winding.

FILTER: Device for removing small foreign particles from a fluid.

FLAME TEST FOR LEAKS: Tool which is principally a torch. When a halogen mixture is fed to the flame, this flame will change color in the presence of heated copper.

FLAMMABLE LIQUIDS: Liquids having a flash point below 140 F. (60 C.) and a vapor pressure not exceeding 40 psi (absolute) (2.81 kg/cm^2) at 100 F. (38 C.).

FLAPPER VALVE: Thin metal valve used in refrigeration compressors which allows gaseous refrigerants to flow in only one direction.

FLARE: An enlargement at the end of a piece of flexible tubing by which the tubing is connected to a fitting or another piece of tubing. This enlargement is made at about a 45 deg. angle. Fittings grip it firmly to make the joint leakproof and strong.

FLARE NUT: Fitting used to clamp tubing flare against another fitting.

FLASH GAS: Instantaneous evaporation of some liquid refrigerant in evaporator which cools remaining liquid refrigerant to desired evaporation temperature.

FLASH POINT: Temperature at which flammable liquid will give off sufficient vapor to support a flash flame but will not support continuous combustion.

FLASH WELD: Resistance type weld in which mating parts are brought together under considerable pressure while a heavy electrical current is passed through the joint to be welded.

FLOAT VALVE: Type of valve which is operated by sphere or pan which floats on liquid surface and controls level of liquid.

FLOODED SYSTEM: Type of refrigerating system in which liquid refrigerant fills most of the evaporator.

FLOODED SYSTEM, HIGH-SIDE FLOAT: Refrigeration system which has a float operated by the level of the high-side liquid refrigerant.

FLOODED SYSTEM, LOW-SIDE FLOAT: Refrigerating system which has a low-side float refrigerant control.

FLOODING: Act of allowing a liquid to flow into a part of a system.

FLOW METER: Instrument used to measure velocity or volume of fluid movement.

FLUE: Gas or air passage which usually depends on natural convection to cause the combustion gases to flow through it. Forced convection may sometimes be used.

FLUID: Substance in either a liquid or gaseous state; substance containing particles which move and change position without separation of the mass.

FLUSH: Operation to remove any material or fluids from refrigeration system parts by purging them to the atmosphere using refrigerant or other fluids.

FLUX (BRAZING, SOLDERING): Substance applied to surfaces to be joined by brazing or soldering to keep oxides from forming and to produce joints.

FLUX, MAGNETIC: Lines of force of a magnet.

FOAM LEAK DETECTOR: System of soap bubbles or special foaming liquids brushed over joints and connections to locate leaks.

FOAMING: Formation of a foam in an oil-refrigerant mixture due to rapid evaporation of refrigerant dissolved in the oil. This is most likely to occur when the compressor starts and the pressure is suddenly reduced.

FOOT POUND: Unit of work. A foot pound is the amount of work done in lifting one pound one foot.

FORCE: Force is accumulated pressure and is expressed in pounds. If the pressure is 10 psi on a plate 10 in. sq., the force is 100 lb. If pressure is 10 kg/cm^2 on a plate 10 cm^2 in area, the force is 100 kg.

FORCED CONVECTION: Movement of fluid by mechanical force such as fans or pumps.

FORCE-FEED OILING: Lubrication system which uses a pump to force oil to surfaces of moving parts.

FREEZER ALARM: A bell or buzzer used in many freezers which sounds an alarm when freezer temperature rises above safe limit.

FREEZER BURN: Condition applied to food which has not been properly wrapped and that has become hard, dry and discolored.

FREEZE-UP: 1—Formation of ice in the refrigerant control device which may stop the flow of refrigerant into the evaporator. 2—Frost formation on an evaporator which may stop the airflow through the evaporator.

FREEZING: Change of state from liquid to solid.

FREEZING POINT: Temperature at which a liquid will solidify upon removal of heat. The freezing temperature for water is 32 F. (0 C.) at atmospheric pressure.

FREON: Trade name for a family of synthetic chemical refrigerants manufactured by E. I. du Pont de Nemours & Co., Inc.

FROST BACK: Condition in which liquid refrigerant flows from evaporator into suction line; usually indicated by sweating or frosting of the suction line.

FROST CONTROL, AUTOMATIC: Control which automatically cycles refrigerating system to remove frost formation on evaporator.

FROST CONTROL, MANUAL: Manual control used to change operation of refrigerating system to produce defrosting conditions.

FROST CONTROL, SEMIAUTOMATIC: Control which starts defrost part of a cycle manually and then returns system to normal operation automatically.

FROST FREE REFRIGERATOR: Refrigerated cabinet which operates with an automatic defrost during each cycle.

FROSTING TYPE EVAPORATOR: Refrigerating system which maintains the evaporator at frosting temperatures during all phases of cycle.

FROZEN: 1—Water in its solid state. 2—Seized (as in machine parts) due to lack of lubrication. The term "freeze-up" is often applied to this situation.

FUEL OIL: Kerosene or any hydrocarbon oil as specified by U.S. Department of Commerce Commercial Standard CS12 or ASTM D296, or the Canadian Government Specification Board, 3-GP-28, and having a flash point not less than 100 F. (38 C.).

FULL FLOATING: Mechanism construction in which a shaft is free to turn in all the parts in which it is inserted.

FURNACE, CENTRAL WARM AIR: Self-contained appliance designed to supply heated air through ducts to spaces remote from or adjacent to the appliance location.

FUSE: Electrical safety device consisting of strip of fusible metal in circuit which melts when circuit is overloaded.

FUSIBLE PLUG: Plug or fitting made with a metal of a known low-melting temperature. Used as safety device to release pressures in case of fire.

GALVANIC ACTION: Wasting away of two unlike metals due to electrical current passing between them. The action is increased in the presence of moisture.

GAS: Vapor phase or state of a substance.

GAS, NONCONDENSABLE: Gas which will not form into a liquid under the operating pressure-temperature conditions.

GAS VALVE: Device in a pipeline for starting, stopping or regulating flow of gas.

GASKET: Resilient (spongy) or flexible material used between mating surfaces of refrigerating unit parts, or on refrigerator doors, to give a leakproof seal.

GASKET, FOAM: Joint sealing material made of rubber or plastic foam strips.

GAUGE, COMPOUND: Instrument for measuring pressures both above and below atmospheric pressure.

GAUGE, HIGH-PRESSURE: Instrument for measuring pressures in range of 0 psi to 500 psi (0 kg/cm^2 to 35.2 kg/cm^2).

GAUGE, LOW-PRESSURE: Instrument for measuring pressures in range of 0 psi to 50 psi (0 kg/cm^2 to 3.52 kg/cm^2).

GAUGE MANIFOLD: Chamber device constructed to hold both compound and high-pressure gauges. Valves control flow of fluids through it.

GAUGE PORT: Opening or connection provided for a service technician to install a gauge.

GAUGE VACUUM: Instrument used to measure pressures below atmospheric pressure.

GRAIN: Unit of weight and equal to 1/7000 lb. It is used to indicate the amount of moisture in the air.

GRILLE: Ornamental or louvered opening placed in a room at the end of an air passageway.

GROMMET: Plastic, metal or rubber doughnut-shaped protectors which line holes where wires or tubing pass through panels.

GROUND COIL: Heat exchanger buried in the ground. May be used either as an evaporator or as a condenser.

GROUND, SHORT CIRCUIT: Fault in an electrical circuit allowing electricity to flow into the metal parts of a mechanism.

GROUND WIRE: Electrical wire which will safely conduct electricity from a structure into the ground.

HALIDE REFRIGERANTS: Family of refrigerants containing halogen chemicals.

HALIDE TORCH: Type of torch used to safely detect halogen refrigerant leaks in system.

HALOGENS: Substance containing fluorine, chlorine, bromine and iodine.

HANGER: Device attached to walls or other structure for support of pipe lines.

HEAD: Pressure, usually expressed in feet of water, inches of mercury, or millimetres of mercury.

HEAD, STATIC: Pressure of fluid expressed in terms of height of column of the fluid, such as water or mercury.

HEAD, TOTAL STATIC: Static head from the surface of the supply source to the free discharge surface.

HEAD FRICTION: Head required to overcome friction of the interior surface of a conductor and between fluid particles in motion.

HEAD PRESSURE: Pressure which exists in condensing side of refrigerating system.

HEAD PRESSURE CONTROL: Pressure-operated control which opens electrical circuit if high-side pressure becomes too high.

HEAD VELOCITY: Height of fluid equivalent to its velocity pressure in flowing fluid.

HEADER: Length of pipe or vessel to which two or more pipe lines are joined carries fluid from a common source to various points of use.

HEAT: Form of energy which acts on substances to raise their temperature; energy associated with random motion of molecules.

HEAT EXCHANGER: Device used to transfer heat from a warm or hot surface to a cold or cooler surface. (Evaporators and condensers are heat exchangers.)

HEAT INTENSITY: Heat concentration in a substance as indicated by the temperature of the substance through use of a thermometer.

HEAT LAG: The time it takes for heat to travel through a substance heated on one side.

HEAT LEAKAGE: Flow of heat through a substance.

HEAT LOAD: Amount of heat, measured in Btu or watts, which is removed during a period of 24 hours.

HEAT OF COMPRESSION: Mechanical energy of pressure changed into energy of heat.

HEAT OF FUSION: Heat released from a substance to change it from a liquid state to a solid state. The heat of fusion of ice is 144 Btu per pound (151.9 joules).

HEAT OF RESPIRATION: Process by which oxygen and carbohydrates are assimilated by a substance; also when carbon dioxide and water are given off by a substance.

HEAT PUMP: Compression cycle system used to supply heat to a temperature-controlled space. Same system can also remove heat from the same space.

HEAT SINK: Relatively cold surface capable of absorbing heat.

HEAT TRANSFER: Movement of heat from one body or substance to another. Heat may be transferred by radiation, conduction, convection or a combination of these three methods.

HEATING COIL: Heat transfer device consisting of a coil of piping, which releases heat.

HEATING CONTROL: Device which controls temperature of a heat transfer unit which releases heat.

HEATING VALUE: Amount of heat which may be obtained by burning a fuel. It is usually expressed in Btu per lb., Btu per gal., or calories per gram.

HERMETIC COMPRESSOR: Compressor which has the driving motor sealed inside the compressor housing. The motor operates in an atmosphere of the refrigerant.

HERMETIC MOTOR: Compressor drive motor sealed within same casing which contains compressor.

HERMETIC SYSTEM: Refrigeration system which has a compressor driven by a motor contained in compressor dome or housing.

HERTZ (Hz): Correct terminology for cycles per second.

Hg. (MERCURY): Heavy silver-white metallic element; only metal that is liquid at ordinary room temperature.

HIGH-PRESSURE CUT-OUT: Electrical control switch operated by the high-side pressure which automatically opens electrical circuit if too high pressure is reached.

HIGH SIDE: Parts of a refrigerating system which are under condensing or high-side pressure.

HIGH-SIDE FLOAT: Refrigerant control mechanism which controls the level of the liquid refrigerant in the high-pressure side of mechanism.

HIGH-VACUUM PUMP: Mechanism which can create a vacuum in the 1000 to 1 micron range.

HOLLOW-TUBE GASKET: Sealing device made of rubber or plastic with tubular cross-section.

HORSEPOWER: Unit of power equal to 33,000 ft. lb. of work per minute. One electrical horsepower equals 746W.

HOT GAS BYPASS: Piping system in refrigerating unit which moves hot refrigerant gas from condenser into low-pressure side.

HOT GAS DEFROST: Defrosting system in which hot refrigerant gas from the high side is directed through evaporator for short period of time and at predetermined intervals in order to remove frost from evaporator.

HOT JUNCTION: That part of thermoelectric circuit which releases heat.

HOT WATER HEATING SYSTEM: System in which water is circulated through heating coils.

HOT WIRE: 1—Resistance wire in an electrical relay which expands when heated and contracts when cooled. 2—Electrical lead which has a voltage difference between it and the grounds.

HUMIDIFIERS: Device used to add to and control humidity.

HUMIDISTAT: Electrical control which is operated by changing humidity.

HUMIDITY: Moisture. Dampness of air.

HYDRAULICS: Branch of physics having to do with the mechanical properties of water and other liquids in motion.

HYDROCARBONS: Organic compounds containing only hydrogen and carbon atoms in various combinations.

HYDROMETER: Floating instrument used to measure specific gravity of a liquid.

HYDRONIC: Heating system which circulates a heated fluid, usually water, through baseboard coils by means of a circulating pump which is controlled by a thermostat.

HYGROMETER: Instrument used to measure degree of moisture in the atmosphere.

HYGROSCOPIC: Ability of a substance to absorb and release moisture and change physical dimensions as its moisture content changes.

ICC — INTERSTATE COMMERCE COMMISSION: Government body which controls the design and construction of pressure containers.

ICE CREAM CABINET: Commercial refrigerator which operates at approximately 0 F. (—18 C.); used for storage of ice cream.

ICE MELTING EQUIVALENT (IME — ICE MELTING EFFECT): Amount of heat absorbed by melting ice at 32 F. (0 C.) is 144 Btu per pound of ice or 288,000 Btu per ton.

IDLER: Pulley used on some belt drives to provide proper belt tension and to eliminate belt vibration.

IGNITION TRANSFORMER: Transformer designed to provide a high-voltage current. Used in many heating systems to ignite fuel, it provides a spark gap.

IMPEDANCE: Opposition in an electrical circuit to the flow of an alternating current that is similar to the electrical resistance to a direct current.

IMPELLER: Rotating part of a pump.

INDUCED MAGNETISM: Ability of a magnetic field to produce magnetism in a metal.

INDUCTION MOTOR: An a-c motor which operates on principle of rotating magnetic field. Rotor has no electrical connection, but receives electrical energy by transformer action from field windings.

INDUCTIVE REACTANCE: Electromagnetic induction in a circuit creates a counter or reverse (counter) emf (voltage) as the original current changes. It opposes the flow of alternating current.

INFRARED LAMP: Electrical device which emits infrared rays; invisible rays just beyond red in the visible spectrum.

INHIBITOR: Substance which prevents chemical reaction such as corrosion or oxidation.

INSTRUMENT: Used broadly to denote a device that has measuring, recording, indicating and/or controlling abilities.

INSULATION, ELECTRIC: Substance which has almost no free electrons.

INSULATION, THERMAL: Material which is a poor conductor of heat; used to retard or slow down flow of heat through wall or partition.

INTERMITTENT CYCLE: Cycle which repeats itself at varying time intervals.

ION: Group of atoms or an atom electrically charged.

IR DROP: Electrical term indicating the loss in a circuit expressed in amperes times resistance (I x R) or voltage drop.

ISOTHERMAL: Changes of volume or pressure under conditions of constant temperature.

ISOTHERMAL EXPANSION AND CONTRACTION: Action which takes place without a temperature change.

JOINT: Connecting point as between two pipes.

JOULE-THOMSON EFFECT: The change in the temperature of a gas on its expansion through a porous plug from a higher pressure to a lower pressure.

JOURNAL, CRANKSHAFT: Part of shaft which contacts the bearing on the large end of the piston rod.

JUNCTION BOX: Box or container housing group of electrical terminals.

KATA THERMOMETER: Large-bulb alcohol thermometer used to measure air speed or atmospheric conditions by means of cooling effect.

KELVIN SCALE (K): Thermometer scale on which unit of measurement equals the Celsius degree and according to which absolute zero is 0 degree, the equivalent of −273.16 C. Water freezes at 273.16 K. and boils at 373.16 K.

KILOMETRE: Metric unit of linear measurement = 1000 metres.

KILOWATT: Unit of electrical power, equal to 1000 watts.

KING VALVE: Liquid receiver service valve.

LACQUER: Protective coating or finish which dries to form a film by evaporation of a volatile (easily goes from liquid to gas) constituent.

LAG: Delay in response.

LAMPS, STERI: Lamp which has a high-intensity ultraviolet ray used to kill bacteria. It is also used in food storage cabinets and in air ducts.

LAPPING: Smoothing a metal surface to high degree of refinement or accuracy using a fine abrasive.

LATENT HEAT: Heat energy absorbed in process of changing form of substance (melting, vaporization, fusion) without change in temperature or pressure.

LATENT HEAT OF CONDENSATION: Amount of heat released (lost) by a pound of a substance to change its state from a vapor (gas) to a liquid.

LATENT HEAT OF VAPORIZATION: Amount of heat, required per pound of a substance to change its state from a liquid to a vapor (gas).

LEAK DETECTOR: Device or instrument such as a halide torch, an electronic sniffer; or soap solution used to detect leaks.

LIMIT CONTROL: Control used to open or close electrical circuits as temperature or pressure limits are reached.

LIQUID: Substance whose molecules move freely among themselves, but do not tend to separate like those of gases.

LIQUID ABSORBENT: Chemical in liquid form which has the property to "take on" or absorb other fluids.

LIQUID INDICATOR: Device located in liquid line which provides a glass window through which liquid flow may be watched.

LIQUID LINE: Tube which carries liquid refrigerant from the condenser or liquid receiver to the refrigerant control mechanism.

LIQUID NITROGEN: Nitrogen in liquid form which is used as a low-temperature refrigerant in expendable or chemical refrigerating systems.

LIQUID RECEIVER: Cylinder (container) connected to condenser outlet for storage of liquid refrigerant in a system.

LIQUID RECEIVER SERVICE VALVE: Two or three-way manual valve located at the outlet of the receiver and used for installation and service purposes. It is sometimes called the king valve.

LIQUID-VAPOR VALVE REFRIGERANT CYLINDER: Dual hand valve on refrigerant cylinders which is used to release either gas or liquid refrigerant from the cylinder.

LIQUOR: Solution used in absorption refrigeration.

LITRE: Metric unit of volume which equals 61.03 cu. in.

LOW SIDE: That portion of a refrigerating system which is below evaporating pressure.

LOW-SIDE FLOAT VALVE: Refrigerant control valve operated by level of liquid refrigerant in low-pressure side of system.

LOW-SIDE PRESSURE: Pressure in cooling side of refrigerating cycle.

LOW-SIDE PRESSURE CONTROL: Device used to keep low-side evaporating pressure from dropping below certain pressure.

LP FUEL: Liquefied petroleum used as a fuel gas.

MACHINE ROOM: Area where commercial and industrial refrigeration machinery — except the evaporators — is located.

MAGNETIC CLUTCH: Device operated by magnetism to connect or disconnect a power drive.

MAGNETIC FIELD: Space in which magnetic lines of force exist.

MAGNETIC GASKET: Door-sealing material which keeps door tightly closed with small magnets inserted in gasket.

MAGNETISM: A field of force which causes a magnet to attract materials made of iron, nickel-cobalt or other ferrous material.

MALE THREAD: External thread on pipe, fittings and valves for making connections that screw together.

MANIFOLD, SERVICE: Chamber equipped with gauges and manual valves, used by service technicians to service refrigerating systems.

MANOMETER: Instrument for measuring pressure of gases and vapors.

Gas pressure is balanced against column of liquid, such as mercury, in U-shaped tube.

MASS: Quantity of matter held together so as to form one body.

MBH: Thousands of British Thermal Units (82 MBH = 82,000 Btu).

MEAN EFFECTIVE PRESSURE (mep): Average pressure on a surface when a changing pressure condition exists.

MECHANICAL CYCLE: Cycle which is a repetitive series of mechanical events.

MEGOHM: A unit of measure for electrical resistance. One megohm is equal to a million ohms.

MEGOHMMETER: Instrument for measuring extremely high resistances (in the millions of ohms ranges).

MELTING POINT: Temperature at atmospheric pressure at which a substance will melt.

MERCOID BULB: Electrical circuit switch which uses a small quantity of mercury in a sealed glass tube to make or break electrical contact with terminals within the tube.

MET: Term applied to the heat release from a human at rest. It equals 18.4 Btu/sq. ft./hr. or 50 kcal/m^2/hr.

METRIC SYSTEM: Decimal system of measuring.

MICRO: One millionth part of unit specified.

MICROFARAD: Unit of condenser electrical capacity equal to 1/1,000,000 farad.

MICROMETER: Precision measuring instrument used for making measurements accurate to .001 to .0001 in. or .001 cm.

MICRON: Unit of length in metric system; a thousandth part (1/1 000) of one millimetre.

MICRON GAUGE: Instrument for measuring vacuums very close to a perfect vacuum.

MILLI: Combining form denoting one thousandth (1/1000); for example, millivolt means one thousandth of a volt.

MINIMUM STABLE SIGNAL (MSS): Correct setting for an expansion valve where it is utilizing the evaporator efficiently but remains free from "hunting."

MISCIBILITY: Substances that are capable of being mixed.

MODULATING: Type of device or control which tends to adjust by increments (minute changes) rather than by either "full on" or "full off" operation.

MODULATING REFRIGERATION CYCLE: Refrigerating system of variable capacity.

MOISTURE INDICATOR: Instrument used to measure moisture content of a refrigerant.

MOLECULE: Smallest portion of an element or compound that retains chemical identity with the substance in mass.

MOLLIERS DIAGRAM: Graph of refrigerant pressure, heat and temperature properties.

MONOCHLORODIFLUOROMETHANE: Refrigerant better known as Freon 22 or R—22. Chemical formula is $CHClF_2$. Cylinder color code is green.

MOTOR: Rotating machine that transforms fluid or electric energy into a mechanical motion.

MOTOR, CAPACITOR: Single-phase induction motor with an auxiliary starting winding connected in series with a condenser (capacitor) for better starting characteristics.

MOTOR, FOUR-POLE: 1800 rpm, 60 Hz electric motor (synchronous speed).

MOTOR, TWO-POLE: 3600 rpm, 60 Hz electric motor (synchronous speed).

MOTOR BURNOUT: Condition in which the insulation of an electric motor has deteriorated (become poor in quality) by overheating.

MOTOR CONTROL: Device to start and/or stop a motor or hermetic motor compressor at certain temperature or pressure conditions.

MOTOR STARTER: High-capacity electric switches usually operated by electromagnets.

MSS POINT (MINIMUM STABLE SIGNAL): It is the best superheat setting which will provide constant or little temperature change at the thermostatic expansion valve temperature sensing element while the system is running.

MUFFLER, COMPRESSOR: Sound absorber chamber in refrigeration system. Used to reduce sound of gas pulsations.

MULLION: Stationary frame member between two doors.

MULLION HEATER: Electrical heating element mounted in the mullion. Used to keep mullion from sweating or frosting.

MULTIPLE SYSTEM: Refrigerating mechanism in which several evaporators are connected to one condensing unit.

NATURAL CONVECTION: Movement of a fluid caused only by temperature differences (density changes).

NEEDLE POINT VALVE: Type of valve having a needle point plug and a small seat orifice for low-flow metering.

NEOPRENE: Synthetic rubber which is resistant to hydrocarbon oil and gas.

NEUTRALIZER: Substance used to counteract acids in refrigeration system.

NEUTRON: That part of an atom core which has no electrical potential; electrically neutral.

NITROGEN DIOXIDE: Mildly poisonous gas (NO_2) often found in smog or automobile exhaust fumes.

NO-FROST FREEZER: Low-temperature refrigerator cabinet in which no frost or ice collects on freezer surfaces or materials stored in cabinet.

NOMINAL SIZE TUBING: Tubing measurement which has an inside diameter the same as iron pipe of the same stated size.

NON-CODE INSTALLATION: Functional refrigerating system installed where there are no local, state, or national refrigeration codes in force.

NONCONDENSABLE GAS: Gas which does not change into a liquid at operating temperatures and pressures.

NONFERROUS: Group of metals and metal alloys which contain no iron.

NONFROSTING EVAPORATOR: Evaporator which never collects frost or ice on its surface.

NORMAL CHARGE: Thermal element charge which is part liquid and part gas under all operating conditions.

NORTH POLE, MAGNETIC: End of magnet out of which magnetic lines of force flow.

OCTAVE: Frequency difference between harmonic vibrations. It is the doubling of the frequency of sound.

OCTYL ALCOHOL — ETHYL HEXANOL: Additive in absorption machines to reduce surface tension in the absorber.

ODOR: That property of air contaminants that affect the sense of smell.

OFF CYCLE: That part of a refrigeration cycle when the system is not operating.

OHM (R): Unit of measurement of electrical resistance. One ohm exists when one volt causes a flow of one ampere.

OHMMETER: Instrument for measuring resistance in ohms.

OHM'S LAW: Mathematical relationship between voltage, current and resistance in an electric circuit; discovered by George Simon Ohm. It is stated as follows: voltage (E) equals amperes (I) times ohms (R); or $E = I \times R$.

OIL BINDING: Condition in which an oil layer on top of refrigerant liquid may prevent if from evaporating at its normal pressure-temperature.

OIL, REFRIGERATION: Specially prepared oil used in refrigerator mechanism which circulates, to some extent, with refrigerant.

OIL SEPARATOR: Device used to remove oil from gaseous refrigerant.

OPEN CIRCUIT: Interrupted electrical circuit which stops flow of electricity.

OPEN COMPRESSOR: Term used to indicate an external drive compressor. Not hermetic.

OPEN DISPLAY CASE: Commercial refrigerator designed to maintain its contents at refrigerating temperatures even though the contents are in an open case.

OPEN TYPE SYSTEM: Refrigerating system which uses a belt-driven or a coupling-driven compressor.

OPERATING PRESSURE: Actual pressure at which the system works under normal conditions. This pressure may be positive or negative (vacuum).

ORGANIC: Pertaining to or derived from living organisms.

ORIFICE: Accurate size opening for controlling fluid flow.

OSCILLOSCOPE: Fluorescent-coated tube which visually shows an electrical wave.

OVERLOAD: Load greater than that for which system or mechanism was intended.

OVERLOAD PROTECTOR: Device, either temperature, pressure or current operated, which will stop operation of unit if dangerous conditions arise.

OZONE: Bluish gaseous form of oxygen usually obtained by silent discharge of electricity in oxygen or air (O_3).

PACKAGE UNITS: Complete refrigerating system including compressor, condenser and evaporator located in the refrigerated space.

PACKING: Sealing device consisting of soft material or one or more mating soft elements. Reshaped by manually adjustable compression to obtain or maintain a leak-proof seal.

PARTIAL PRESSURES: Condition where two or more gases occupy a space and each one creates part of the total pressure.

PASCAL'S LAW: Pressure imposed upon a fluid is transmitted equally in all directions.

PELTIER EFFECT: When direct current is passed through two adjacent metals one junction will become cooler and the other will become warmer. This principle is the basis of thermoelectric refrigeration.

PERMANENT MAGNET: Material which has its molecules aligned and has its own magnetic field; bar of metal which has been permanently magnetized.

PH: Measurement of the free hydrogen ion concentration in an aqueous solution.

PHASE: Distinct functional operation during a cycle.

PHIAL: Term sometimes used to denote the sensing element on a thermostatic expansion valve.

PHOTOELECTRICITY: Physical action wherein an electrical flow is generated by light waves.

PISTON: Close-fitting part or plug which moves up and down in a cylinder.

PISTON DISPLACEMENT: Volume displaced by piston as it travels the full length of its stroke.

PITOT TUBE: Tube used to measure air velocities.

PLENUM CHAMBER: Chamber or container for moving air or other gas under a slight positive pressure.

POLYPHASE MOTOR: Electrical motor designed to be used with a three or four-phase electrical circuit.

POLYSTYRENE: Plastic used as an insulation in some refrigerated structures.

PONDED ROOF: Flat roof designed to hold a quantity of water which acts as a cooling device.

PORCELAIN: Ceramic china-like coating applied to steel surfaces.

POTENTIAL, ELECTRICAL: Electrical force which moves, or attempts to move, electrons along a conductor or resistance.

POTENTIAL RELAY: Electrical switch which opens on high voltage and closes on low voltage.

POTENTIOMETER: Instrument for measuring or controlling by sensing small changes in electrical resistance.

POUR POINT: Lowest temperature at which a liquid will pour or flow.

POWER: 1—Time rate at which work is done or energy emitted. 2—Source or means of supplying energy.

POWER ELEMENT: Sensitive element of a temperature-operated control.

POWER FACTOR: Correction coefficient for the changing current and voltage values of a-c power.

PPM (PARTS PER MILLION): Unit of concentration (for example, of solutions).

PRESSURE: Energy impact on a unit area; force or thrust on a surface.

PRESSURE, ABSOLUTE: See ABSOLUTE PRESSURE.

PRESSURE, ATMOSPHERIC: See ATMOSPHERIC PRESSURE.

PRESSURE, BACK: See BACK PRESSURE.

PRESSURE, GAUGE: Reading in pounds per square inch (psi) above atmospheric pressure.

PRESSURE, HEAD: Force caused by the weight of a column or body of fluids. Expressed in feet, inches or psi.

PRESSURE, OPERATING: Pressure at which a system is operating.

PRESSURE, SUCTION: Pressure in low-pressure side of a refrigerating system.

PRESSURE DROP: Pressure difference at two ends of a circuit, or part of a circuit, the two sides of a filter.

PRESSURE GAUGE: Instrument for measuring the pressure exerted by the contents on its container.

PRESSURE-HEAT DIAGRAM: Graph of refrigerant pressure, heat and temperature properties. (Mollier's diagram.)

PRESSURE, LIMITER: Device which remains closed until a certain pressure is reached and then opens and releases fluid to another part of system or breaks an electric circuit.

PRESSURE MOTOR CONTROL: Device which opens and closes an electrical circuit as pressures change.

PRESSURE-OPERATED ALTITUDE (POA) VALVE: Device which maintains a constant low-side pressure independent of altitude of operation.

PRESSURE REGULATOR, EVAPORATOR: Automatic pressure regulating valve mounted in suction line between evaporator outlet and compressor inlet. Its purpose is to maintain a predetermined pressure and temperature in the evaporator.

PRESSURE SWITCH: Switch operated by a rise or drop in pressure.

PRESSURE WATER VALVE: Device used to control water flow. It is responsive to head pressure of refrigerating system.

PRIMARY CONTROL: Device which directly controls operation of

PROCESS TUBE: Length of tubing fastened to hermetic unit dome, used for servicing unit.

PROPANE: Volatile hydrocarbon used as a fuel or as a refrigerant.

PROTECTOR, CIRCUIT: Electrical device which will open an electrical circuit if excessive electrical conditions occur.

PROTON: Particle of an atom with a positive charge.

PSI: Symbol or initials used to indicate pressure measured in pounds per square inch.

PSIA: Symbol or initials used to indicate pressure measured in pounds per square inch absolute. Absolute pressure equals gauge pressure plus atmospheric pressure.

PSIG: Symbol or initials used to indicate pressure in pounds per square inch gauge. The "g" indicates that is is gauge pressure and not absolute pressure.

PSYCHROMETER OR WET BULB HYGROMETER: Instrument for measuring the relative humidity of atmospheric air.

PSYCHROMETRIC CHART: Chart that shows relationship between the temperature, pressure and moisture content of the air.

PSYCHROMETRIC MEASUREMENT: Measurement of temperature pressure and humidity using a psychrometric chart.

PUMP: Any one of various machines which force a gas or liquid into — or draw it out of — something as by suction or pressure.

PUMP DOWN: The act of using a compressor or a pump to reduce the pressure in a container or a system.

PUMP, CENTRIFUGAL: Pump which produces fluid velocity and converts it to pressure head.

PUMP, FIXED DISPLACEMENT: A pump in which the displacement per cycle cannot be varied.

PUMP, RECIPROCATING SINGLE PISTON: A pump having a single reciprocating (moving up and down or back and forth) piston.

PUMP, SCREW: Pump having two interlocking screws rotating in a housing.

PURGING: Releasing compressed gas to atmosphere through some part or parts for the purpose of removing contaminants from that part or parts.

PYROMETER: Instrument for measuring high temperatures.

QUENCHING: Submerging hot solid object in cooling fluid.

QUICK-CONNECT COUPLING: A device which permits easy and fast connecting of two fluid lines.

R—11, TRICHLOROMONOFLUOROMETHANE: Low pressure, synthetic chemical refrigerant which is also used as a cleaning fluid.

R—12, DICHLORODIFLUOROMETHANE: Popular refrigerant known as Freon 12.

R—22, MONOCHLORODIFLUOROMETHANE: Popular low temperature refrigerant with a boiling point of —41 F. at atmospheric pressure.

R—113, TRICHLOROTRIFLUOROETHANE: Synthetic chemical refrigerant which is nontoxic and nonflammable.

R—160, ETHYL CHLORIDE: Toxic refrigerant now seldom used.

R—170, ETHANE: Low temperature application refrigerant.

R—290, PROPANE: Low temperature application refrigerant.

R—500: Refrigerant which is an azeotropic mixture of R—12 and R—152a.

R—502: Refrigerant which is azeotropic mixture of R—22 and R—115.

R—503: Refrigerant which is azeotropic mixture of R—23 and R—13.

R—504: Refrigerant which is azeotropic mixture of R—32 and R—115.

R—600, BUTANE: Low-temperature application refrigerant; also used as a fuel.

R—611, METHYL FORMATE: Low pressure refrigerant.

R—717, AMMONIA: Popular refrigerant for industrial refrigerating systems; also a popular absorption system refrigerant.

RADIAL COMMUTATOR: Electrical contact surface on a rotor which is perpendicular or at right angles to the shaft center line.

RADIANT HEATING: Heating system in which warm or hot surfaces are used to radiate heat into the space to be conditioned.

RADIATION: Transfer of heat by heat rays.

RAM AIR: Air forced through the condenser due to the rapid movement of the vehicle along the highway.

RANGE: Pressure or temperature settings of a control; change within limits.

RANKINE SCALE: Name given the absolute (Fahrenheit) scale. Zero (0 R.) on this scale is —460 F.

REACTANCE: That part of the impedance of an alternating current circuit due to capacitance or inductance or both.

RECEIVER-DRIER: Cylinder (container) in a refrigerating system for storing liquid refrigerant and which also holds a quantity of desiccant.

RECEIVER HEATING ELEMENT: Electrical resistance mounted in or around liquid receiver. It is used to maintain head pressures when ambient temperature is low.

RECIPROCATING: Back and forth motion in a straight line.

RECORDING AMMETER: Electrical instrument which uses a pen to record amount of current flow on a moving paper chart.

RECORDING THERMOMETER: Temperature measuring instrument which has a pen marking a moving chart.

RECTIFIER, ELECTRIC: Electrical device for converting a-c to d-c.

REED VALVE: Thin, flat, tempered steel plate fastened at one end.

REFRIGERANT: Substance used in refrigerating mechanism. It absorbs heat in evaporator by change of state from a liquid to a gas, and releases its heat in a condenser as the substance returns from the gaseous state back to a liquid state.

REFRIGERANT CHARGE: Quantity of refrigerant in a system.

REFRIGERANT CONTROL: Device which meters flow of refrigerant between two areas of a refrigerating system. It also maintains pressure difference between high-pressure and low-pressure side of the mechanical refrigerating system while unit is running.

REGISTER: Combination grille and damper assembly covering an air opening or the end of an air duct.

RELATIVE HUMIDITY: Ratio of (difference between) amount of water vapor present in air to greatest amount possible at same temperature.

RELAY: An electromagnetic mechanism moved by a small electrical current in a control circuit. It operates a valve or switch in an operating circuit.

RELIEF VALVE: Safety device on a sealed system. It opens to release fluids before dangerous pressure is reached.

RELUCTANCE: A force working against the passage of magnetic lines of force (flux) through a magnetic substance.

REMOTE POWER ELEMENT CONTROL: Device with sensing element located apart from operating mechanism it controls.

REMOTE SYSTEM: Refrigerating system in which condensing unit is away from space to be cooled.

REPULSION-START INDUCTION MOTOR: An electric motor type which has an electrical winding on the rotor for starting purposes.

RESISTANCE: An opposition to flow or movement; a coefficient of friction.

RESISTANCE, (R) ELECTRICAL: The difficulty electrons have moving through a conductor or substance.

RESTRICTOR: A device for producing a deliberate pressure drop or resistance in a line by reducing the cross-sectional flow area.

RETROFIT: Term used in describing reworking an older installation to bring it up to date with modern equipment or to meet new code requirements.

REVERSE CYCLE DEFROST: Method of heating evaporator for defrosting. Valves move hot gas from compressor into evaporator.

REVERSING VALVE: Device used to reverse direction of the refrigerant flow depending upon whether heating or cooling is desired.

REYNOLDS NUMBERS: A numerical ratio of the dynamic forces of mass flow to the shear stress due to viscosity.

RINGELMANN SCALE: Device for measuring smoke density.

RISER VALVE: Device used to manually control flow of refrigerant in vertical piping.

ROTARY BLADE COMPRESSOR: Mechanism for pumping fluid by revolving blades inside cylindrical housing.

ROTARY COMPRESSOR: Mechanism which pumps fluid by using rotating motion.

ROTOR: Rotating or turning part of a mechanism.

RUNNING TIME: Amount of time a condensing unit is run per hour or per 24 hours.

RUNNING WINDING: Electrical winding of motor which has current flowing through it during normal operation of motor.

SADDLE VALVE (TAP-A-LINE): Valve body shaped so it may be silver brazed or clamped onto a refrigerant tubing surface.

SAFETY CAN: Approved container of not more than 5-gal. capacity. It has a spring-closing lid and spout cover. It is designed to relieve internal pressure safely when exposed to fire.

SAFETY CONTROL: Device to stop refrigerating unit if unsafe pressure and/or temperatures and/or dangerous conditions are reached.

SAFETY MOTOR CONTROL: Electrical device used to open circuit to motor if temperature, pressure, and/or current flow exceed safe conditions.

SAFETY PLUG: Device which will release the contents of a container before rupture pressures are reached.

SAFETY VALVE: Self-operated quick opening valve used for fast relief of excessive pressures.

SATURATION: Condition existing when substance contains all of another substance it can hold for that temperature and pressure.

SCAVENGER PUMP: Mechanism used to remove fluid from sump or container.

SCHRADER VALVE: Spring-loaded device which permits fluid flow in one direction when a center pin is depressed; in other direction when a pressure difference exists.

SCOTCH YOKE: Mechanism used to change reciprocating motion into rotary motion or vice-versa. Used to connect crankshaft to piston in refrigeration compressor.

SCREW PUMP: Compressor constructed of two mated revolving screws.

SEALED UNIT: See HERMETIC SYSTEM. Motor compressor assembly in which motor and compressor operate inside sealed housing.

SEAL LEAK: Escape of oil and/or refrigerant at the junction where a shaft enters a housing.

SEAL, SHAFT: Device used to prevent leakage between shaft and housing.

SEAT: That portion of a valve mechanism against which the valve presses to effect shutoff.

SECOND LAW OF THERMODYNAMICS: Heat will flow only from material at higher temperature to material at lower temperature.

SECONDARY REFRIGERATING SYSTEM: Refrigerating system in which condenser is cooled by evaporator of another or primary refrigerating system.

SEEBECK EFFECT: When two different adjacent metals are heated, an electric current is generated between the metals.

SELF-INDUCTANCE: Magnetic field induced in conductor carrying the current.

SEMICONDUCTOR: A class of solids whose ability to conduct electricity lies between that of a conductor and that of an insulator.

SEMIHERMETIC COMPRESSOR: Hermetic compressor with service valves.

SENSIBLE HEAT: Heat which causes a change in temperature of a substance.

SENSOR: Material or device which goes through a physical change or an electronic characteristic change as surrounding conditions change.

SEPARATOR: Device to separate one substance from another.

SEPARATOR, OIL: Device to separate refrigerant oil from refrigerant gas and return the oil to compressor crankcase.

SEQUENCE CONTROLS: Group of devices which act in series (one after another) or in time order.

SERVICE VALVE: Manually operated valve mounted on refrigerating systems used for service operation.

SERVICEABLE HERMETIC: Hermetic unit housing containing motor and compressor assembly by use of bolts or cap screws.

SERVO: A servo motor supplies power to a servo-mechanism. A servomechanism is a low-power device (electrical, hydraulic or pneumatic) used to put in operation and control a more complex or a more powerful mechanism.

SHADED-POLE MOTOR: Small a-c motor designed to start under light loads.

SHELL-AND-TUBE FLOODED EVAPORATOR: Device which flows water through tubes built into cylindrical evaporator or vice versa.

SHELL TYPE CONDENSER: Cylinder or receiver which contains condensing water coils or tubes.

SHORT CIRCUIT: Electrical condition where part of circuit touches another part of circuit and causes all or part of current to take wrong path.

SHORT CYCLING: Refrigerating system that starts and stops more frequently than it should.

SHROUD: Housing over condenser, evaporator or fan.

SIGHT GLASS: Glass tube or glass window in refrigerating mechanism. It shows amount of refrigerant or oil in system and indicates presence of gas bubbles in liquid line.

SILICA GEL: Absorbent chemical compound used as a drier. When heated, moisture is released and compound may be reused.

SILICON-CONTROLLED RECTIFIER (SCR): Electronic semiconductor which contains silicon.

SILVER BRAZING: Brazing process in which brazing alloy contains some silver as part of joining alloy.

SINE WAVE, A-C CURRENT: Wave form of single frequency alternating current; wave whose displacement is sine of angle proportional to time or distance.

SINGLE-PHASE MOTOR: Electric motor which operates on single-phase alternating current.

SINGLE-POLE, DOUBLE-THROW SWITCH, (SPDT): Electric switch with one blade and two contact points.

SINGLE-POLE, SINGLE-THROW SWITCH, (SPST): Electric switch with one blade and one contact point.

SINGLE-STAGE COMPRESSOR: Compressor having only one compressive step between inlet and outlet.

SKIN CONDENSER: Condenser using the outer surface of the cabinet as the heat radiating medium.

SLING PSYCHROMETER: Measuring device with wet and dry bulb thermometers. Moved rapidly through air it measures humidity.

SLUG: 1—Unit of mass equal to the weight (English units) of object divided by 32.2 (acceleration due to the force of gravity). 2—Detached mass of liquid or oil which causes an impact or hammer in a circulating system.

SLUGGING: Condition in which mass of liquid enters compressor causing hammering.

SMOKE TEST: Test made to determine completeness of combustion.

SOLAR HEAT: Heat created by visible and invisible energy waves from the sun.

SOLDERING: Joining two metals by adhesion of a metal with a melting temperature of less than 800 F. (427 C.).

SOLENOID VALVE: Electromagnet with a moving core. It serves as a valve or operates a valve.

SONE: Calculated sound loudness rating.

SOUTH POLE, MAGNETIC: That part of magnet into which magnetic flux lines flow.

SPECIFIC GRAVITY: Weight of a liquid compared to water which is assigned value of 1.0.

SPECIFIC HEAT: Ratio of quantity of heat required to raise temperature of a body 1-deg. to that required to raise temperature of equal mass of water 1-deg.

SPECIFIC VOLUME: Volume per unit mass of a substance.

SPLASH SYSTEM, OILING: Method of lubricating moving parts by agitating or splashing oil in the crankcase.

SPLIT-PHASE MOTOR: Motor with two stator windings. Both windings are in use while starting. One is disconnected by centrifugal switch after motor attains speed. Motor then operates on other winding only.

SPLIT SYSTEM: Refrigeration or air conditioning installation which places condensing unit outside or away from evaporator. Also applicable to heat pump installations.

SPRAY COOLING: Method of refrigerating by spraying expendable refrigerant or by spraying refrigerated water.

SQUIRREL CAGE: Fan which has blades parallel to fan axis and moves air at right angles or perpendicular to fan axis.

STANDARD ATMOSPHERE: Condition when air is at 14.7 psia pressure, at 68 F. (20 C.) temperature and a relative humidity of 36 percent.

STANDARD CONDITIONS: Used as a basis for air conditioning calculations: temperature of 68 F. (20 C.), pressure of 29.92 inches of mercury (Hg.) and relative humidity of 30 percent.

STARTING RELAY: Electrical device which connects and/or disconnects starting winding of electric motor.

STARTING WINDING: Winding in electric motor used only briefly while motor is starting.

STATIONARY BLADE COMPRESSOR: Rotary pump which uses a nonrotating blade inside pump to separate intake chamber from exhaust chamber.

STATOR, MOTOR: Stationary part of electric motor.

STEAM: Water in vapor state.

STEAM HEATING: Heating system in which steam from a boiler is piped to radiators in space to be heated.

STEAM JET REFRIGERATION: Refrigerating system which uses a steam venturi to create high vacuum (low pressure) on a water container causing water to evaporate at low temperature.

STEAM TRAP: Automatic valve which traps air but allows condensate to pass while preventing passage of steam.

STETHOSCOPE: Instrument used to detect sounds and locate their origin.

STRAINER: Device such as a screen or filter used to retain solid particles while liquid passes through.

STRATIFICATION OF AIR: Condition in which there is little or no air movement in room; air lies in temperature layers.

STRIKE: Metal plate fastened to frame and into which the bolt of a latch or lock slides.

SUBCOOLING: Cooling of liquid refrigerant below its condensing temperature.

SUBLIMATION: Condition where a substance changes from a solid to a gas without becoming a liquid.

SUBSTANCE: Any form of matter or material.

SUCTION LINE: Tube or pipe used to carry refrigerant gas from evaporator to compressor.

SUCTION PRESSURE CONTROL VALVE: Device located in the suction line which maintains constant pressure in evaporator during running portion of cycle.

SUCTION SERVICE VALVE: Two-way manually operated valve located at the inlet to compressor. It controls suction gas flow and is used to service unit.

SUCTION SIDE: Low-pressure side of the system extending from the refrigerant control through the evaporator to the inlet valve of the compressor.

SUPERHEAT: 1—Temperature of vapor above its boiling temperature as a liquid at that pressure. 2—The difference between the temperature at the evaporator outlet and the lower temperature of the refrigerant evaporating in the evaporator.

SUPERHEATER: Heat exchanger arranged to take heat from liquid going to evaporator and using it to superheat vapor leaving evaporator.

SURFACE PLATE: Tool with a very accurate flat surface. It is used for measuring purposes and for lapping flat surfaces.

SURGE: Regulating action of temperature or pressure before it reaches its final value or setting.

SURGE TANK: Container connected to the low-pressure side of a refrigerating system which increases gas volume and reduces rate of pressure change.

SWAGING: Enlarging one tube end so end of other tube of same size will fit within.

SWAMP COOLER: Evaporative type cooler in which air is drawn through porous mats soaked with water.

SWASH PLATE-WOBBLE PLATE: Device used to change rotary motion to reciprocating motion. Used in some refrigeration compressors.

SWEATING: 1—Condensation of moisture from air on cold surface. 2—Method of soldering in which the parts to be joined are first coated with a thin layer of solder.

SWEET WATER: Term sometimes used to describe tap water.

SYLPHON SEAL: Corrugated metal tubing used to hold seal ring and provide leakproof connection between seal ring and compressor body or shaft.

SYNTHETIC RUBBER, NEOPRENE: Soft resilient material made of a synthetic chemical compound.

TAIL PIPE: Outlet pipe from the evaporator.

TANK, SUPPLY: Separate tank connected directly or by a pump to the oil-burning appliance.

TAP (SCREW THREAD): Tool used to cut internal threads.

TEMPERATURE: 1—Degree of hotness or coldness as measured by a thermometer. 2—Measurement of speed of motion of molecules.

TEMPERATURE-HUMIDITY INDEX: Actual temperature and humidity of air sample compared to air at standard conditions.

TEST LIGHT: Light provided with test leads. Used to test or probe electrical circuits to determine if they have electricity.

THERM: Quantity of heat equal to 100,000 Btu.

THERMAL RELAY (HOT WIRE RELAY): Heat-operated electrical control used to open or close a refrigeration system electrical circuit. This system uses a resistance wire to convert electrical energy into heat energy.

THERMISTOR: Basically a semiconductor which has electrical resistance that varies with temperature.

THERMOCOUPLE: Device which generates electricity, using principle that if two unlike metals are welded together and junction is heated, a voltage will develop across the open ends.

THERMOCOUPLE THERMOMETER: Electrical instrument using thermocouple as source of electrical flow, connected to milliammeter calibrated in temperature degrees.

THERMODISK DEFROST CONTROL: Electrical switch with bimetal disk controlled by temperature changes.

THERMODYNAMICS: Part of science which deals with the relationships between heat and mechanical action.

THERMOELECTRIC REFRIGERATION: Refrigerator mechanism that depends on Peltier effect. Direct current flowing through electrical junction between unlike metals provides heating or cooling effect depending on direction of flow of current.

THERMOMETER: Device for measuring temperatures.

THERMOMODULE: Number of thermocouples used in parallel to achieve low temperatures.

THERMOPILE: Number of thermocouples used in series to create a higher voltage.

THERMOSTAT: Device which senses ambient temperature conditions and, in turn, acts to control a circuit.

THERMOSTATIC CONTROL: Device which operates system or part of system based on temperature change.

THERMOSTATIC EXPANSION VALVE: Control valve operated by temperature and pressure within evaporator. It controls flow of refrigerant. Control bulb is attached to outlet of evaporator.

THERMOSTATIC MOTOR CONTROL: Device used to control cycling of unit through use of control bulb. Bulb reacts to temperature changes.

THERMOSTATIC VALVE: Valve controlled by temperature change response elements.

THERMOSTATIC WATER VALVE: Valve used to control flow of water through system, actuated (made to work) by temperature difference. Used in units such as water-cooled compressor and/or condenser.

THREE-PHASE: Operating by means of combination of three alternating current circuits which differ in phase by one-third of a cycle.

THREE-WAY VALVE: Multi-orifice (opening) flow control valve with three fluid flow openings.

THROTTLING: Expansion of gas through orifice or controlled opening without gas performing any work as it expands.

TIMERS: Clock-operated mechanism used to control opening and closing of an electrical circuit.

TIMER-THERMOSTAT: Thermostat control which includes a clock mechanism. Unit automatically controls room temperature and changes temperature range depending on time of day.

TON OF REFRIGERATION: Refrigerating effect equal to the melting of 1 ton of ice in 24-hours. This may be expressed as follows: 288,000 Btu/24 hr., 12,000 Btu/1 hr., 200 Btu/min.

TON REFRIGERATION UNIT: Unit which removes same amount of heat in 24-hours as melting of 1 ton of ice.

TORQUE: Turning or twisting force.

TORQUE, FULL LOAD: Maximum torque delivered without overheating.

TORQUE, STALL: Torque developed when starting.

TORQUE, STARTING: Amount of torque available, when at 0 speed, to start and accelerate the load.

TORQUE WRENCHES: Wrench which may be used to measure torque or pressure applied to a nut or bolt.

TOTAL HEAT: Sum of both the sensible and latent heat.

TRANSDUCER: Device turned on by change of power from one source for purpose of supplying power in another form to second system.

TRANSFORMER: Electromagnetic device which transfers electrical energy from primary circuit into variations of voltage in secondary circuit.

TRANSFORMER-RECTIFIER: Combination transformer and rectifier in which input a-c current may be varied and then rectified into d-c current.

TRANSISTOR: Electronic device commonly used for amplification. Similar in use to electron tube. Depends on conducting properties of semiconductors in which electrons moving in one direction are considered as leaving holes that serve as carriers of positive electricity in opposite direction.

TRICHLOROTRIFLUOROETHANE: Complete name of refrigerant R—113. Group 1 refrigerant in rather common use. Chemical compounds which make up this refrigerant are chlorine, fluorine and ethane.

TRIPLE POINT: Pressure-temperature condition in which a substance is in equilibrium (balance) in solid, liquid and vapor states.

TROPOSHERE: Part of the atmosphere immediately above the earth's surface in which occur most weather disturbances.

TRUCK, REFRIGERATED: Commercial vehicle equipped to maintain below-ambient temperatures.

TUBE, CONSTRICTED: Tubing reduced in diameter.

TUBE-WITHIN-A-TUBE: Water-cooled condensing unit in which a small tube is placed inside large unit. Refrigerant passes through outer tube, water through the inner tube.

TUBING: Fluid-carrying pipe which has a thin wall.

TWO-TEMPERATURE VALVE: Pressure-opened valve used in suction line on multiple refrigerator installations which maintains evaporators in system at different temperatures.

TWO-WAY VALVE: Valve with one inlet port and one outlet port.

ULTRAVIOLET: Invisible radiation waves with frequencies shorter than wave lengths of visible light and longer than X-ray.

UNITARY SYSTEM: A heating/cooling system factory assembled in one package and usually designed for conditioning one space or room.

UNIVERSAL MOTOR: Electric motor which will operate on either a-c or d-c.

URETHANE FOAM: Type of insulation which is foamed in between inner and outer walls of a container.

VACUUM: Pressure lower than atmospheric pressure.

VACUUM CONTROL SYSTEM: Intake manifold vacuum is used to operate dampers and controls in some automobile systems.

VACUUM PUMP: Special high efficiency device used for creating high vacuums for testing or drying purposes.

VALVE: Device used for controlling fluid flow.

VALVE, EXPANSION: Type of refrigerant control which maintains constant pressure in the low side of refrigerating mechanism. Valve is caused to operate by pressure in low or suction side. Often referred to as an automatic expansion valve or AEV.

VALVE, SERVICE: Device used to check pressures, service and charge refrigerating systems.

VALVE, SOLENOID: Valve made to work by magnetic action through an electrically energized coil.

VALVE, SUCTION: Valve in refrigeration compressor which allows vaporized refrigerant to enter cylinder from suction line and prevents its return.

VALVE, WATER: In most water cooling units, a valve that provides a flow of water to cool the system while it is running.

VALVE PLATE: Part of compressor located between top of compressor body and head. It contains compressor valves and ports.

VAPOR: 1—Vaporized refrigerant is preferred to the word gas. 2—A gas which is often found in its liquid state while in use.

VAPOR, SATURATED: Vapor condition which will result in condensation into droplets of liquid if vapor temperature is reduced.

VAPOR BARRIER: Thin plastic or metal foil sheet used in air-conditioned structures to prevent water vapor from penetrating insulating material.

VAPOR LOCK: Condition where liquid is trapped in line because of bend or improper installation. Such vapor prevents liquid flow.

VAPOR PRESSURE: Pressure imposed by either a vapor or gas.

VAPOR PRESSURE CURVE: Graphic presentation of various pressures produced by refrigerant under various temperatures.

VAPORIZATION: Change of liquid into a gaseous state.

VARIABLE PITCH PULLEY: Pulley which can be adjusted to provide different pulley drive ratios.

V-BELT: Type of belt commonly used in refrigeration work. It has a contact surface with the pulley which is in the shape of letter V.

VELOCIMETER: Instrument used to measure air speeds using a direct-reading air speed indicating scale.

VELOCITY: Quickness or rapidity of motion, swiftness, speed.

VIBRATION ARRESTORS: Soft or flexible substance or device which will reduce the transmission of a vibration.

VOLTAGE: 1—Term used to indicate the electrical potential or electromotive force in an electrical circuit. 2—Voltage or electrical pressure which causes current to flow. 3—Electromotive force.

VOLTAGE CONTROL: Device used to provide some electrical circuits with uniform or constant voltage.

VOLTMETER: Instrument for measuring voltage in electrical circuit.

VOLUMETRIC EFFICIENCY: Term used to express the relationship between the actual performance of a compressor or of a vacuum pump and calculated performance of the pump based on its displacement.

VORTEX TUBE: Mechanism for cooling or refrigerating which accomplishes cooling effect by releasing compressed air through a specially designed tube.

VORTEX TUBE REFRIGERATION: Refrigerating or cooling device using principle of vortex tube, as in mining suits.

WALK-IN COOLER: Larger commercially refrigerated space kept below room temperature. Often found in supermarkets or wholesale meat distribution centers.

WATER-COOLED CONDENSER: Condensing unit which is cooled through use of water flow.

WATER DEFROSTING: Use of water to melt ice and frost from evaporator during off-cycle.

WATT: Unit of electrical power.

WAX: Ingredient in many lubricating oils which may separate from the oil if cooled enough.

WET BULB: Device used in measurement of relative humidity. Evaporation of moisture lowers temperature of wet bulb compared to dry bulb temperature of same air sample.

WET BULB TEMPERATURE: Measure of the degree of moisture. It is the temperature of evaporation for an air sample.

WET CELL BATTERY: Cell or connected group of cells that converts chemical energy into electrical energy by reversible chemical reactions.

WET HEAT: Heating system using hot water (hydronic) heat or steam heat.

WINDOW UNIT: Air conditioner which is placed in a window.

WOBBLE PLATE-SWASH PLATE: Type of compressor designed to compress gas, with piston motion parallel to crankshaft.

ZERO ICE: Trade name for dry ice. See DRY ICE.

INDEX

A

Abbreviations, 35
Abrasives, 66
Absolute,
 pressures, 17, 18, 19, 20
 temperature scales, 9, 10
 temperatures, 9, 10
 zero, 10
Absorbent, 93
Absorber, 93
Absorber, selective, 935, 936
Absorbers, vibration, 457
Absorption chemicals, 93, 604
Absorption cycle, 89, 91, 93, 604, 686
 air conditioning, 620
 continuous, 93, 604
 intermittent, 89, 605
 servicing, 608, 621
Absorption generator, 89, 91
Absorption mechanism, 89, 91, 93
 air conditioning, 614
 automatic defrost, 608
 burners, 603, 604
 electrical circuits, 608
 gas controls, 611, 612
 gas supply, 610
 installing, 621
 servicing, 618, 621
 thermostats, 611
Absorption refrigerators, 89, 90, 91, 92, 93
 installing, 620, 621
 servicing, 618, 621
 thermostats, 611
Absorption system, 89, 91, 603
 commercial, 620
 construction, 604, 608
 controls, 611, 612
 domestic, 606
 efficiency, 621
 Electrolux, 92, 607, 608, 609, 610
 Faraday, 93, 604
 installing, 608
 refrigerants, 605
 Superfex, 605
 troubles, 621
 TruKold, 605

Absorption to emissions ratio, 935, 936
Access port, 825
Accumulator, 100, 102, 448, 804
Acetylene air soldering, 46, 47
Acetylene torch, 46, 47
Acid,
 cleaning metal, 374
 conditons in system, 375, 378
Acoustics, 767
Activated alumina, 956
Activated carbon, 798
Adaptors, 353, 354, 356
 service valve wrench, 56
Adding oil to system, 367, 540, 929
Additives, 707
Add-on system, 926
Adhesion, 45
Adhesive filters, 793
Adiabatic compression, 967
Adiabatic expansion, 967
Adjustable wrench, 55
Adjusting,
 burner, 708
 controls, 250
 expansion valves, 78
 motor control, 250
 screw, 249
 thermostatic expansion valves, 532, 533
 water valves, 529
Adjustments,
 differential, 247
 emissivity, 947
 flame, 716
 motor controls, 245, 246
 range, 246
 refrigerant controls, 533
Adsorber, 152
Adsorption, 152
 system, 604
AEV, 78, 141
Agitator, 480
Air,
 air-to-air, 311, 690
 air-to-water, 814
 balance systems, 788
 break, 703
 changes, 749

characteristics, 788
chemical reaction, 791
circulation, 781, 785
circulation systems, 829
cleaner, 792
cleaning, 792
coil, 808, 810
colloids, 303
compressed, 633
condensation, 792
conditioned, 765
conditioner, energy efficiency ratio, 763
conditioners, 681, 897
contaminants, 660
cooler, 747
cooling, 748
curtains, 799
defrosting, 413
diffuser, 769
dilution, 792
distribution, 769, 779, 917
ducts, 769
exfiltration, 766
film, 880
filtering, 792
flow, 788
forced air, 781
fumes, 792
gravity, 781
heat in, 765
in systems, 364, 382, 497, 521, 584
infiltration, 766
inlet grilles, 781
leakage, 888
mists, 792
moisture determination, 644
moisture-holding properties, 956
movement, 105, 568, 651, 781
noise, 667
outlet velocity, 781
pollutants, 661
pressure, 766
properties of, 642, 765, 956
psychrometric chart, 649
psychrometric properties of, 649
purification, 660
removing from, 364, 382, 497, 521

commercial unit, 497
condenser, 384
domestic unit, 384
flooded, 384
line, 548
return air ducts, 787
sensing thermostat, 729
smoke, 792
spill-over, 304
standard conditions of, 642, 964
stratification of, 769
temperatures, 656
terminal velocity, 781
throw, 781
velocity measurement, 651
volumes, 875
    for cooling, 779, 880
    for heating, 778, 879
washer, 797
weight, 765
Air conditioning,
    absorption unit, 614, 620
    air cycle, 633
    air filtering, 641
    air movement, 641
    automobile, 694, 897
    bus, 925
    careers, 969
    central plant, 822-824
    circuits, 837
    classification, 841, 842
    comfort coolers, 749
    complete systems, 800, 815
    console, 758
    controls, 252, 837, 850
    cooling units, 681, 682, 749
    cycles, 84, 748, 802
    definition of, 641
    dehumidifiers, 689
    education requirements, 970
    electric controls, 867
    equipment, 700, 748, 749, 761
    fans, 789
    farm vehicles, 926
    fundamentals, 641
    heat loads, 879, 880
    heating equipment, 702
    history, 641
    humidifiers, 699, 742, 743, 744
    humidity control, 641
    industry, 969
    installation, 825
    instruments, 837, 871, 872, 873, 874
    insulation, 892
    maintenance, 930
    mobile home, 744, 760
    plant, central, 822

plant, remote, 760
polarized plug, 753
principles, 747
purpose of, 699
refrigerant controls, 866
safety devices, 837
servicing, 759
systems, 673, 699, 700, 747, 760
    cleaning, 765
    distributing, 765
temperature, 657
temperature control, 837, 866
testing operation, 753
thermometers, 664
thermostats, 842, 844, 845, 847, 849
through-the-wall, 681, 682, 815
timers, 849
truck, 925
unit, 699, 700, 747, 749
wall plate, 842
window, 681, 749
year-around, 747, 765, 815
Air-cooled condensers, 392, 399, 407, 408,
    521, 522, 536
    capacities, 588
    commercial, 107, 407
    console, 758
    domestic, 105
Airflow, 788
    controls, 862
    indicators, 875
Alarm system, temperature, 273
Alcohol, 67
Alcohol brine, 961
Alignment, belts, 237
Allen setscrews, 68, 69
Alternating current, 171
    characteristics, 170
    cycles, 183
Altitude adjustment, 250, 251
Alumina, activated, 373, 956
Alumina gel, 956
Ambient compensator, 308
Ambient temperature, 28
American Petroleum Institute, 962
Ammeter, 176, 233
Ammonia,
    analyzer tube, 291
    open cycle refrigeration, 631
    properties of, 291
    testing for leaks, 291
Amperage, 176
    relay, 260
Ampere turns, 185, 186
Amperes, 176
Amplifier, 193, 197
Amount of refrigerant required, 300

Analogy, water, 171, 172
Anemometer, 529, 651
Angle deflector plates, 751
Angular measurements, 17
Annealed, 37
Annealing, 44
    tubing, 44
Anticipators, 842, 844
    current, 844
Antifoam inhibitor, 960
Appliance truck, 336
Application of latent heat, 25, 579
Arc, 17
Arcing, 839
Area,
    cabinet, 561
    door, 885
    equivalents, 949
    evaporator, 572
    measurement, 17
    of a circle, 17
    room, 563, 889
    wall, 884
    window, 885
Argon, 293
Arithmetic, basic, 11
Armature (see Rotor), 157, 190, 191
    electric motor, 190, 191, 211
Artificial ice, 7
ASA, 50
Aspect ratio, 777
Aspirating devices, 448
Aspirating psychrometer, 644
Aspiration, 643
Assembling,
    compressor, 381, 518
    hermetic unit, 382
    refrigeration systems, 537
    swaged tubing, 45
Assembly devices, 68
Atmosphere, 18, 641
Atmospheric pressure, 17, 18, 19, 20
Atom, 9, 952, 953
Attachment plug configurations, 206, 207
Attachment, service valve, 353, 537
Attic fan, 791
A-type evaporator, 822, 823
Automatic controls,
    air conditioning, 252, 837, 845
    defrosting, 265, 267, 268
    electric, 243
    electric refrigerators, 311
    expansion valve, 78, 110, 141
    motor, 243, 245, 246, 257, 258, 259
    refrigerant, 141, 304, 307, 311, 313, 842
Automatic defrost, 243, 245, 309, 313
Automatic defrosting, 608

Automatic expansion valve, 78, 110
  adjusting, 141
  bellows type, 142
  bypass type, 143
  diaphragm type, 143
Automatic ice cube maker, 85, 478
Automobile air conditioning, 694, 897
  add-on, 926
  air distribution, 917
  ammeter, 919
  automatic, 924
  belts, 909, 927
  bulb mounting, 915
  bus air conditioning, 925
  bypass valves, 916
  comfort cooling, 897, 901
  compressor seal, 904
  compressors, 904
  condenser, 909
  controls, 924
  coolant tubes, 912
  cooling, 917
  cooling capacity, 901
  crankshaft, 903, 904, 906
  dual system, 918
  ducts, 918
  electrical systems, 920
  evaporators, 912
  expansion valves, 914
  fans, 919
  farm vehicle, 926
  fastening, 911
  filters, 910
  fin combs, 926
  finned rotor, 912
  fire wall, 912
  flared, 911
  hang-on, 901, 925
  heating coils, 897
  heating controls, 915
  heating systems, 901
  hoses, 911
  installing, 926
  insulation, 919
  leaks, 925, 929
  lines, 911
  loss of refrigerant, 907
  magnetic clutch, 903
  magnetic coils, 903, 924
  mounting, 904, 912
  O-ring, 911
  oil, 917
  operating conditions, 899
  pressure, low-side, 899, 902
  pressure-operated hot gas bypass valve,
      902, 903
  pulley, 909
  receiver-drier, 910
  refrigerant lines, 911
  rheostat, 922
  SAE flared connection, 907
  service valves, 911
  servicing, 926, 927, 929, 930
  sight glass, 902, 915
  solenoid-operated hot gas bypass valve,
      903
  storing, 930
  suction pressure control valves, 915
  swash plate, 904
  systems, types of, 901, 902
  teflon seal, 907, 908
  TEV, 914
  thermostats, 924
  truck air conditioning, 925
  typical installation, 901
  vacuum actuators, 924
  vacuum control systems, 924
  V-type compressor, 905, 906
Autotransformer, 203, 204
Auxiliary evaporators, 467
Avoirdupois, 950
Awnings, 891
Axial flow, 789
Azeotropic mixture, 285

B

Bacteria, 7, 792
Baffle, 568
Bakeries, refrigeration for, 480
Balance point, 814
Balancing the system, 788
Bar, 18
Barometer, 19, 666
Base, paraffin, 960
Baseboard heating, 737
Basement,
  fan, 791
  heat loss, 887
  specifications, 887
Basic arithmetic, 11
Basic cycles, electricity, 169
Basic refrigeration systems, 73
Bath, ice and water, 533
Battery, 169
Batts, 892, 893
Baudelot coolers, 589
Bearings, 235, 515
  compressor, 118, 515
  electric motors, 235
  lubrication, 234
  replacing, 515
  servicing, 233, 235
Beaufort scale, 657
Bellows operated low-pressure control,
258, 535
Bellows type AEV, 142
Belts,
  alignment, 237
  automobile air conditioning, 909, 927
  care of, 237
  double, 927
  motors, electric, 115, 236
  sizes, 237
  tension, 237
  V, 237
Bending,
  springs, 40
  tools, 40
  tubing, 40
Bernoulli's Theorem, 642
Beverage coolers, 416, 478, 576
  brine, 416
  sweet water, 416
Beverage cooling, 7
Bimetal motor controls, 257
Bimetal strip, 61
Blade,
  compressor, 130
  construction, vane, 130
  rotary, 127
  stationary, 127
Blankflange, 514
Blocks, lapping, 516
Blower control, 678
Blower plates, 625
Blower type evaporator coil, 463
Board, charging, 540
Bodies, 483
Boiler (see Generator)
Boiler, 729
Boiling pressure, 26
Boiling temperature, 10, 26
Bolts, 68
Bonnet, 673
Boost-and-buck transformer, 203,204
Booster compressor, 128
Bore, 586
Borosilicate, 741
Bottle (see Cylinder)
Bottle coolers, 477
Bound electrons, 952
Bourdon tube, 62
Bowden cable, 902
Box wrench, 54
Boxes, refrigerator (see Cabinets)
Boyle's Law, 966
Brazed tube fittings, 44
Brazing, 505, 963
  joint, cleaning, 45, 47, 364, 505
  silver, 38, 44, 46, 47, 364, 505
Breaker strip, 362
Breaker trim, 304

Breathing, 483
Brine, 32, 961
  alcohol, 961
  calcium chloride, 32, 961
  composition, 961
  freezing temperatures, 961
  glycol, 961
  immersed evaporator, 416
  sodium chloride, 32, 961
  specific gravity, 21
British thermal units, 7, 9, 555
Brushes, 191
  cleaning, 66
Btu (see British thermal unit)
Building, heat sources, 892
Bulb, sensitive, 110, 144, 245
Bunker, 463, 568
Burn, freezer, 304
Burner, 604
  adjusting, 708
  Bunsen, 719
  electric ignition, 722
  gas, 673, 719
  gun type, 708
  installation 713, 716
  kerosene, 604
  oil, 675, 677, 708
  pump, 708
  system parts, 719
Burnouts, 375, 538
  clean up, 375
Bus air conditioning, 925
Butane, 292, 610
Butt welded, 38
Butter conditioner, 331
Bypass, 95, 97, 143, 426
  automatic expansion valve, 143
  cycle, 581
  systems, 94, 96, 425

C

Cabinet, 303, 555
  humidifiers, 744
Cabinets,
  accessories, 331
  areas, 561
  care of, 332
  chest type, 323
  commercial, 463, 555
  construction, 303, 463, 555
  dairy, 468, 480
  display, 468
  domestic refrigerator, 303, 307, 311,
    313, 317
  door construction, 463
  finishes, 304, 332, 463

florist, 466
freezer, 330
fresh food, 304
frozen food, 323, 470, 471
gaskets, 332
grocery, 463
hardware, 331
ice accumulation, 330
ice cream, 471, 577
insulation, 304, 341
leveling, 338
reach-in, 463
sizes, 465
soda fountain, 471
temperature, 296, 334
thermometers, 334
upright frozen food, 326
volume, 562
walk-in, 465
Calcium chloride, 32, 961
Calcium sulphate, 956
Calculations,
  air conditioning, 763, 879, 880
  commercial, 555
  duct, 785
  heat loads, 893, 894
  of infiltration, 888
  wattage, 177
Calibrate, 60, 63
  gauges, 60, 62, 63, 64
Calibration,
  cut-in, 248
  cutout, 248
Calorie, 23
Cap screw, 68
Capacitance, 188, 189
Capacitor, 188, 189
  design, 189
  effect, 209, 218
  motor, 211
  running, 211
  selector unit, 232
  servicing, 232, 233
  start, capacitor-run motor, 209, 218
  start, motor, 218
Capacity,
  balancing, 567
  compressor, 586
  condenser, 567, 568, 588, 589
  condensing unit, 568
  evaporator, 567, 571, 575, 576
  liquid lines, 590
  refrigerant control, 600
  refrigerant line, 595
  specific heat, 24
  suction lines, 590, 595
  system, 585

valves, thermostatic, 156, 160
varying system, 84
Capillary tube, 79, 100, 109, 163, 311
  adjustable, 165
  capacities, 164
  clogged, 296
  controls, 109, 163, 369
  designs, 79, 109, 163
  fittings, 165
  operation, 368
  principles of, 163
  refrigerant control, 79, 109, 163, 369
  replacing with automatic expansion valve,
    143
  servicing, 369, 377
  systems, 79, 109, 163, 369
Car, railway, refrigerated, 485
Carbon dioxide, 279, 292, 631, 637
  properties of, 279, 292, 637
  solidified, 292, 637
Carbon filter, 466, 798
Cards, indicator, 955
Care of,
  belts, 75, 237
  cabinets, 330
  gauges, 64, 350, 493
  refrigerant cylinders, 70, 293, 364, 498
  refrigerators, 330
  service valves, 56, 70, 350, 454, 493, 495
Career opportunities, 969
Carrene (see R-11)
Cascade systems, 82, 83, 304, 581, 635
Case-hardened, 516
Cases (see Cabinets)
  dairy, 468, 480
  display, 467, 468, 470, 569
  double-duty, 467
  fast freezing, 471
  freezer, 330
  single-duty, 467
Casings, 133
Catalytic agents, 303
Ceilings, 886
  radiant, 740
Cell, solar, 943, 944
Celsius, 9, 10, 11, 13
Celsius, Anders, 10
Center, magnetic, 238
Centigrade temperature scale, 10
Centimetre, 13, 14, 948
Central air conditioning, 682, 822
  inspecting, 827
  installing, 825
  servicing, 828
Centrifugal, 508
  compressor, 131
  force, 131, 133

units, 832
Changing refrigerants, 300
Charging a system,
  apparatus, 498
  automobile, 929
  checking, 508
  commercial, 498
  domestic, 384
  hermetic, 364
  high side, 498
  low side, 498
  method, 300
  portable tube, 365
  rebuilt, 388
Charging board, 540
Charles's Law, 966
Charts, comfort, 657, 658, 659, 660
Charts, psychrometric, 649
Check valves, 121, 127, 137, 166, 447
  commercial, 447
  domestic, 121, 127
Checking,
  airflow, 875
  compressor, 512
  door gaskets, 332
  leaks, 361, 362, 496
  oil charge, 540
  refrigerant charge, 508
  relays, 263
Chemical,
  dehumidifier, 761
  refrigeration, 86, 485, 625
  scavengers, 731
  stability, 298
Chest-type freezer, 323, 326
Chill factor, 656, 657
Chilled water, 831
  pump, 209
Chilling units (see Evaporator)
Chisels, cold, 67
Chlorotrifluoromethane (see R-13)
Choke, 709
Chromate, 523
Circle,
  arc of, 17
  area of, 14, 15
Circuit,
  breaker, 223
  closed, 172
  Delta, 202
  fundamentals, 171
  open, 171
  protection, 206
  servicing, 541
  short, 172
  solar cell, 943
  star (Wye), 202, 203

symbols, 173
testing instruments, 207
thermocouple, 723
troubles, water, 527, 528, 529
Circuits, electrical, 172, 182, 201, 243, 269, 851, 920
Circular charts, 872
Circular mils, 206
Circulation of air, 781
  systems, 829
Classification of refrigerants, 278
Clean room, 797, 963
Cleaning,
  air, 660, 765, 792
  bath, 66
  brazed joint, 49
  cabinets, 330
  carbon tetrachloride, 67
  caustic solution, 66
  compressor, 378, 380
  condenser parts, 349
  electrostatic, 794
  external mechanism, 349
  facilities, 380
  filters, 793
  gasoline, 67
  hermetic unit, 344
  lacquers, 332
  metal, 332
  methods, 66
  motors, 235
  oleum or mineral spirits, 66
  parts, 66
  porcelain, 332
  refrigerators, 66
  solvents, 66, 961, 962
  steam, 66
  vacuum, 727
  vapor degreasing, 67
Clearance, compressor, 136
Clearance space, 118
Climate, outdoor and indoor, 655
Climatic control, 800, 815
  system installation, 825
Clocks, defrosting, 271
Clogged capillary tube, 370
Clogged screens, servicing, 143
Closed circuit, 172
Closed Delta, 202, 203
Closed water system, solar, 937, 938, 939
  forced convection, 939
  natural convection, 937
Clutch,
  armature, 903, 904
  field, 903, 904
  magnetic, 903, 904
  plate, 903, 904

Coal furnaces, 699
Code, electrical, 205, 955
Code for piping, 963
Code installations, 502
  requirements, 724
  testing, 506
Code, size No., 44
Coefficient of performance (COP), 587, 602, 804
Coil,
  blower, 625, 631
  finned, 637
  heat pump, 804
    air, 808
    ground, 810
    inside/outside, 808
    water, 693, 810
  heating, installation, 741
  heating, service, 741
  helical wound, 736
  magnetic, 903, 924
  plate type, 904
Cold, 9, 761
  ban, 304
  bath for thermostatic controls, 272, 547
  chisels, 67
  junction, 198, 625
  plate, 638
  plate cooling, 625
  storage temperatures, 558
Collector, solar, 934, 935
  angle, 937
  covers, 936
  flat plate, 935
  heat insulation, 937
  lens, 942
  liquid, 814
  panel, 814
  surface, 936
  swimming pool, 940
  vacuum tube, 936
Colloids, 303
Color code,
  pilot light, 722
  piping, 963
  refrigerant cylinders, 296
Column,
  mercury, 19
  water, 19, 653, 724, 872
Combination furnace and cooling, 721
Combination thermostats, 847
Combustion efficiency, 873
Comfort charts, 658
Comfort conditions, 657
Comfort cooler, 700, 831, 865, 866
  console type, 758
  window unit, 749

Comfort cooling, 748, 865, 866
  large, 760, 831
  remote, 760
Comfort-Health Index (CHI), 660
Comfort zone, 657
Commercial,
  absorption system, 620
  applications of refrigeration, 478
  cabinets, 463
  calculations, 555, 562, 566
  compressor, 398, 405
  compressor control, 393, 398, 399,
    405, 408
  condenser, 407, 408, 409
  condensing units, servicing, 489, 508
  construction of refrigeration mechanisms,
    393
  controls, 433—439, 450
  defrost systems, 413, 423, 425, 429, 430
  driers, 459, 460
  equipment, installation of, 393
  evaporators, 413
  heat loads, 555, 562, 566
  hermetic systems, 398
  insulation, 491
  liquid receivers, 412
  low-pressure control, 433
  motor controls, 433, 434, 435
  motors, 110, 209, 210, 215
  refrigerant controls, 433
  systems, 94, 393
    application, 463
    charging, 498
    discharging, 510
    installing, 489
    moisture in, 537
    servicing, 489, 507
Commutator, 191, 211
Complete systems, 682, 800, 815, 816
Composition gaskets, 518
Compound,
  gauge, 19, 63
  refrigerating systems, 81, 82
  refrigeration cycle, 81, 82
  wound motors, 219
Compression cycle,
  operation of, 100
  parts of, 100
  pressure, 100, 104
  rings, 118
  system, 99
  temperature, 100
  typical system, 100
Compression gauge, 62, 63, 64
Compression, heat of, 28
Compression ratio, 137, 817, 951

Compressor, 26, 28, 29, 31, 74, 99, 105,
  405
  assembling, 381, 515, 518
  automobile, 901, 904, 905, 906
  bearings, 124, 234
  booster, 128
  burnouts, 375
  capacities, 586
  centrifugal, 115, 131
  checking, 368, 381, 512, 515
  checking leaks, 518
  cleaning, 515
  commerical, 393
  commercial hermetic, 398
  connecting rods, 115, 120, 402, 515
  construction, 115, 405
  cooling, 136
  crankshafts, 115, 122, 405
  cylinders, 117, 405
  dehydrating of, 384, 518, 520
  design for, 115, 405, 586
  dismantling, 374, 515
  dome, 115, 382
  double acting, 126
  drives, 122
  efficiency, 587
  evacuating, 384, 510, 520
  external drive, 75, 115, 122, 399, 512
  flywheel, 122
  gaskets, 120, 138, 518
  heat pump, 690, 804
  hermetic, 115, 405
    reciprocating, 115, 380, 405
    repairing, 374, 378, 520
    rotary, 127, 381
  housing, crankcase, 126
  installing, 376
  installing open type, 489, 520
  lubrication, 136
  modulating, 804
  motor faults, 368
  muffler, 135, 457
  multi-cylinder reciprocating, 117, 120
  noisy, 512
  oil (see Refrigerant oil)
  open type, 115, 117, 122
  opening, 379
  overhauling, 378, 515
  pistons, 118
  protection devices, 448
  purpose of, 30, 31, 99
  reciprocating, 117
  removing, 374, 513
  rotary, 115, 127
  rotating blade, 130
  screw type, 115, 130

  seal, 122, 516, 907
  service valve, 70, 350, 495, 513
  servicing, 344, 378, 520
  stationary blade type, 129
  swash plate, 126, 906
  testing commercial, 512, 520
  testing domestic, 382, 389
  testing of compressor, 518, 929
  three compressor, 635, 636
  two-speed motor, 211
  two-stage, 83, 581
  types of, 115
  valves, 121, 450
    repair, 380, 517
  volumetric efficiency, 136, 584
  wobble plate, 126, 905
Condensate pump, 209, 415
Condensation, 100
Condensed, 100
Condenser, 8, 100, 105
  air-cooled, 399, 407, 408, 588
  automobile, 909
  capacity of air-cooled, 588
  capacity of water-cooled, 589
  cleaning, 349
  coils, heat pump, 808
  comb, 522
  commercial, air-cooled, 407
  commercial, water-cooled, 408
  construction, 407, 408, 409
  design of, 304, 305
  domestic, 105
  evaporative, 412, 687
    servicing, 525
  fan, 209
  frozen food unit, 323, 326
  installing, air-cooled, 522
  installing, water-cooled, 525
  removing, air-cooled, 521
  removing, water-cooled, 524
  repairing, 525
  repairing, air-cooled, 522
  servicing, air-cooled, 521, 536
  servicing, commercial, 507, 520
  servicing, water-cooled, 523
  shell type, 409
  standard temperatures, 277
  static, 106
  troubles, 377
  tube-within-a-tube, 409
  type of, 407, 408, 409
  unloader, 393
  water pump, 209
  water-cooled, 408
Condensing,
  pressure, 296, 584

temperature, 298, 299
unit,
  air-cooled, 399
  capacities, 568
  frozen foods, 317
  installing, 489, 504
  location, 336
  service valves, 454
  servicing, 525
Conditioners, unit air, 749
Conduction, 31
Conductivity,
  electricity, 180, 181
  heat, 32, 963
  of insulating materials, 556
Conductor, 180, 955
Conical spiral, 50
Connecting rods, 120
Connecting tubing, 41
Connections,
  flared, 41
  silver brazing, 47
  soldering, 45
Console type comfort cooler, 758
  installation, 758
  servicing, 759
Constant pressure valve, 142
Constants, electrical, 953
Constrictor, tubing, 50,
Construction, building, 887, 961
Consumption, water, 576, 589
Containerized refrigeration, 638
Contaminants, 372
  air, 642, 660
  micron size, 662
Continuity, 237
Continuous cycle, 91, 604
Continuous cycle absorption system,
  91, 604, 605
  controls, 611
  gas supply, 610
Contracting, service, 551
Contracts, 551
Control adjustments, 246, 247, 249,
  250
Control, humidity, 829
Controls,
  absorption, 611
  air conditioning, 253, 837
  airflow, 862
  anti-icing, 866
  automatic, 110, 114, 141
  bellows operated, low-pressure, 258
  blower, 678
  capillary tube, 163
  circuits, 223, 243

comfort cooling, 252, 865, 866
commercial, 141, 243, 433, 435, 438
compressor (see Motor control)
condenser, 536
damper motor, 862
defrosting, 265, 267, 268, 304, 307,
  311, 313
deicing, 270
draft, 873
dual duct, 862
dual electric heating, 845
duct, 862
electric, 243, 867
electronic, 854
  cleaner, 868
expansion valve, 141
fan, 274, 862
float, high-pressure, 76
float, low-pressure, 74
fluid flow, 867
frost control, 690
frozen foods, 251
furnace, 837, 850, 851
gas, 604, 719, 724, 837
heat distribution, 860
heat pump, 804
hermetic, 243, 245, 259
high limit, 673, 677, 678
high pressure, 243, 433, 435
humidity, 271, 648, 867
hydraulic, 831
ice bank, 269
ice maker, 254, 438
icing, 865
infrared heat, 860
limit, 850, 863, 866
liquid nitrogen, 631
low-side pressure, 450, 902
modulating, 854, 863
motor, 79, 114, 433
  freezer, 307, 309, 311, 313, 317,
    319, 323, 326
  servicing, 546
  temperature, 79, 334
oil furnace, 856
operating, 865
pneumatic, 862
pressure, 863
pressure, motor, 74, 258, 433
primary, 851, 852, 856, 857
refrigerant, 30, 141, 165, 304, 305, 320,
  323, 326, 433, 532, 600, 804, 866
relays, 850
safety, 845, 846
safety, motor, 258, 435
sequence, 851

solid state, 856
split system, 864
temperature, 656, 837
tester, 532, 533
thermostatic, 434
  expansion valve, 78, 110, 144
  motor, 243, 245, 257, 434
troubleshooting, 868
two-temperature valve, 443, 445, 535
vacuum control system, 924
valve, 634
vapor pressure, 246
vending machine, 439
water cooler, 254
water level, 860
water valves, 451, 453, 527, 528
zone, 864
Convection,
  forced, 32
  natural, 32
  of heat, 32
Conventional, 489
Conversion-temperature, 950
Coolers,
  beverage, 416, 478, 576
  bottle, 478
  comfort, 748
  milk, 480
  swamp, 697
  unit, 748
  walk-in, 465
  water, 416, 477
Cooling,
  air volume, 760
  coil (see Evaporator), 99
  cold plate, 625
  comfort, 748
  compression system, 99
  cycle, 30, 603, 748
  equipment, 749
  evaporative, 697, 761
  heat pump, 802
  ice cream, 577
  loads, 576, 577, 803, 880, 890, 891
  solar energy, 942
  spray, 625
  steam jet, 635, 695
  systems, 747, 748
  thermostat, 847
  tower, 410, 525, 687
  vortex tube, 696
  water, 85, 576
Cooling equipment for air conditioners, 748
Cooling unit (see Evaporator coils)
Copper pipe,
  hard drawn, 38

nominal size, 39
soft, 37
tubing, 37
swaging, 49
Core, magnetic, 185, 186
Core valves, 359
Corrosion, 38, 727
CO_2 indicator, 873
Coulomb, 176
Counter emf, 192
Counterflow, 409, 410
Coupling, quick connect, 539, 824
Couplings, 51
pressure, 165
reducing, 51
self-sealing, 824
Cracking a valve, 55, 71, 494
Crank throw, 122, 126
Crankcase, 117
compressor housing, 126
heater, 138
Crankshaft, 122
crank throw type, 122
eccentric type, 120
Scotch yoke type, 126
Crankshaft seal, 122, 516
repair, 516
Crispers, 304
Critical pressure, 32
Critical temperature, 32
Critical vibration, 122, 124
Cross-charged, 144, 151
Cryogenic fluids, 293
temperature range, 289
Cryogenics, 10, 33, 292, 293, 486
boiling temperature, 292
temperature, 961
Cubic measurement, 14, 15, 16
Cupronickel coil, 810
Current, 114, 170, 176
alternating, 171
alternating cycles, 183
consumption, 232
controls, 243
direct, 170
electric, 169, 170
relay, 217, 260
starting, 213
types, 170
Cut-in, 245
Cutout, 245
Cutter, tubing, 40
Cycles, 99
absorption, 89, 91, 603. 686
air conditioning, 681, 747, 802
automobile air conditioning, 897
bypass, 581

commercial mechanical, 393
compression, 99
continuous, 91, 603, 604
cooling, 73, 99, 603
cycling time, 390
heat pump, 802
heating, 699
hermetic, 304, 307, 311, 313
intermittent, 89, 603, 605
modulating, 84
pressure heat, 583, 584
refrigerating for automobile, 897
reverse, 802
secondary refrigeration, 304, 307,
311, 313, 390
short, 399, 550
single-phase, 201
sterling, 638
three-phase, 201
Cylinder,
charging, 293, 294, 295
compressor, 104, 115, 117
dimensions, 293, 294
displacement, 586
disposable, 293, 294
head, 120, 518
horizontal, 626, 628
returnable, 293, 294
rotary, 129
service, 294
storage, 293
volume of, 17
Cylinders, refrigerant, 293, 294, 296
Cylindrical commutator, 211

D

Dairy cases, 468
Dalton's Law, 34
Damper, 782, 784
fire, 784
motor control, 862
dBA Scale, 667, 668, 669
DC (see Direct current)
Deaeration, 727
Decimals, 948
Decimetre, 14
Defrost cycle, 306, 307, 309, 311, 313,
315, 316
Defrost systems, 311, 313, 316
clocks, 271
commercial, 413, 423, 430, 431
cycle, 429
domestic, 304, 306, 307, 430
electric heating, 97, 306, 430
hermetic, 306, 307
hot brine, 429

hot fluid, 429
hot gas, 96, 271, 306, 373, 425, 800
nonfreezing solution, 429
reverse cycle, 430
timers, 442
warm air, 431
water, 429
Defrosting,
automatic, 245, 608
clocks, 271, 306, 313, 430
controls, 265, 267, 268
evaporators, 413, 423
frozen food units, 307, 311, 313
heat pump, 811
heat sources, 558
solid state, 442
type evaporator, 413
warm air, 431
Degreasing, 515
Degree days, 660, 895
Degrees (see Temperature)
absolute, 10
Celsius, 10
Centigrade (see Celsius)
Fahrenheit, 10
Kelvin, 10
Rankine, 10
Degrees of a circle, 17
Dehumidifiers, 271, 689
Dehumidifying equipment, 270, 761
Dehumidifying systems, 747
Dehydrated oil, 118, 121, 136
Dehydration, 465
Dehydrators (see Drier)
Dehydrator-receiver
(see Drier-receiver)
Deicing controls, 270
Delta, 11
transformer, 202
Demand meter, 203
Density, 21
Deodorizers, 798
Depression, wet bulb, 643
Desert bag, 73, 74
Desiccants, 372, 373, 648, 956
Design,
baffle, 568
condensers, 105
evaporator, 573
furnace, 703
temperatures, 886
Detector, leak, 361
Determining,
dryness of refrigerants, 960
heat leakage, 563
usage load, 566
Development of refrigeration, 7

Devices, fastening, 68, 361
Devices, gas control, 724
Devices, humidifying, 678, 679
Dew point, 644
Diagnosing troubles (see Troubleshooting)
Diaphragm, 143, 535
  expansion valve, 143
Diameter, 17
Dichlorodifluoromethane, R-12, 279
  characteristics, 279, 965
Dichloroethylene, R-1130, 289
Dichloromonofluoromethane, R-21, 965
Dichlorotetrafluoromethane, R-114, 965
Dictionary of Technical Terms, 971
Dielectric of oil, 960
Dielectric value, 960
Dies, 60
  threading, 60
Diestock, 60
Differential, 247
  adjustment, 247, 249
  adjustment, mechanisms, 249
  cut-in type, 247, 249
  cutout type, 247, 248
  double type, 248
  thermostatic expansion valve, 532
  types of, 248
Diffusers, 781
Digital controls signals, 184
Digital voltmeter, 174
Digits, 11
Dimensions, 13
  cylinder, 293, 294
Diodes, 181, 193, 194
Dip tube, 457
Direct current, 170
  motors, 222
Direct expansion system, 529, 530, 531
Discharge line piping, 600
Discharge valves, 120, 121, 122
Discharging refrigerant,
  commercial system, 497
  domestic system, 374, 377
Disk, valve, 120, 121, 122
Dismantling,
  compressor, 380, 513
  general instructions, 374, 509, 510
  hermetic units, 374
  refrigerating system, 374, 509, 510
  system, 374
Dispenser, milk, 480
Dispenser, oil, 374
Dispensing freezer, 474
Displacement, compressor, 136, 586
Displacement, piston, 136, 586
Display case, 467, 569
  dairy products, 467, 480

frozen foods, 470
  open, 468, 470
Disposable cylinders, 294
Distribution, air, 765
Diverter, 717
Dividing block, 129
Dome, hat, 115
Domestic cabinets, 303
Domestic freezers, 303
Domestic hot water use, 939
Domestic refrigeration, 303
  refrigerator humidity, 303
  refrigerator temperature, 296, 303
Domestic system,
  absorption, 603, 606, 608
  charging, 367, 388
  compression, 99
  compressors, 115
  condensers, 105
  discharging, 364, 378
  evaporators, 101, 304
  installing, 337
  motor controls, 114, 243, 347
  refrigerant controls, 79, 109
Domestic water heating, solar,
  one tank system, 939
  two tank system, 939
Door construction, gaskets, 332
Doors, heat leakage, 884
Double-acting compressor, 126
Double-duty case, 467
Double-thickness flare, 43
Dowel pin, 381
Draft, 654, 768, 873
  control, 703, 705
  forced, 703, 789
  gauge, 654
  indicator, 654
  induced, 705
  measurement, 654
Drain heater, 810
Drier, 95, 370, 372, 459, 537, 910
  use of, 103, 108, 382, 459
Drill bits, 58
Drills, 58
  fractional-size, 58
  letter-size, 58
  number-size, 58
  tap, 58, 59
  twist, 58, 962
Drink dispenser, 474, 478, 479, 480
Drinking water cooler, 85
Drip pan, 305
Drive pulley, 694
Drives, compressor motor, 122
Drop, pressure, 159, 360, 529, 785, 786
Drums, refrigerant (see Cylinders)

Dry bulb temperature, 643
Dry bulb thermometers, 643
Dry capacitor, 211, 212, 218, 232
Dry cell battery, 169
Dry evaporators, 529
  refrigerant controls, 532
Dry ice, 32, 88, 631
  refrigeration, 88, 631
Dry system (see Domestic refrigeration), 101
  removing, 530
Dryers (see Driers)
Drying, hermetic system, 360, 382
  oven, 382
Dryness, determining, 384, 458, 643
Dryness of refrigerants, 960
Dual air compressor systems, 918
Duct, 673, 769, 918
  air, 769
  automobile air conditioning, 897, 918
  calculations, 785
  constructions, 775
  controls, 862
  elbows, 788
  heaters, 739
  installations, 737
  insulation, 776
  joint riverter, 775
  loose, 338
  noise, 789
  perimeter, 769, 770
  rectangular, 770
  resistance, 785
  return air, 787
    system, 770
  round, 776
  sizes, 778
  stat, 855
  systems, 694, 770, 829
  tapering, 776
  temperature, 778
  ventilators, 770
  volumes, 778
Dust, 642, 660
Dynamometer, 238

E

Eccentric, 115
Ecology, 670
Education requirements, 970
Effect, ice melting, 948
Effect of pressure .26, 27
Effective latent heat, 579
Efficiency,
  compressor, 587
  motor, 235, 543, 588
  volumetric, 136, 587
Ejector, 696

nozzle, 696
Elbows, 51, 788
Electric,
  circuit, 243, 632
  controls, 243
  current, 169, 170
  cycle, 171
  defrosting, 97, 311, 430
  furnaces, 736
  heater defrost system, 430
  heaters, 138, 678, 693, 735
  heating, 678, 735
    baseboard, 737
    building insulation, 894
    elements, open wire, ribbon type,
      tubular cased, 736
    primary controls, 857
    supplementary, 739, 940
  humidifiers, 744
  principles, 660
    application, 735
  resistance heat,
    forced warm air, 678
    hydronic, 679
    room units, 680
  sensors, 852
  supply, 198
  thermostats, 841, 845
  water valve, 451, 526
Electric motors, 209
  applications, 209
  armatures (see Rotor)
  bearings, 234
  belts, 236
  capacitor, 211
  circuits, 202
  cleaning, 235
  commercial, 398
  commutator, 211
  compressors, 311
  condenser, 311
  connections, 214
  construction, 213
  conventional, 134
  design, 209, 214
  direct current, 222
  domestic, 217
  elementary, 191
  external troubles, 238, 345
  fan, 227
  four-pole, 212
  fuses, 223
  grounding wire, 224
  hermetic, 134, 216
  horsepower, 223, 588
  induction, 210
  lubrication, 234

open type, 134, 210, 227
operation of, 191, 210
overhauling, 234, 238, 380, 542, 544
PSC (permanent split capacitor), 219
polyphase, 219
principles, 191, 210
protection, 224
radio interference, 232
relay, 223, 259
removing, 234, 238, 378, 542
repulsion start-induction, 211
rotation, 213, 222, 240
rotor, 130
servicing, 232, 238, 347, 375, 541
single-phase, 201
split-phase, 211
standard data, 227
starter, 259, 437
starting and running windings, 213
stator, 133
temperature, 226
test stand, 237, 543
three-phase, 201, 219, 221
transistorized, 230
troubles, external, 234, 239, 345, 542
two-pole, 212
types, 134, 209
wattage, 213, 223
windings, 213, 219
Electric refrigerators (see Domestic system)
Electrical,
  burnout, 134
  circuits, 171, 172, 182, 201, 243, 306
    308, 313, 316, 317, 318, 322, 323, 329,
    398, 541, 851, 920
  codes, 205, 955
  connections, 206, 492
  constant, 953
  controls, 114, 243
  energy, 22, 169, 948
  generator, 188
  grounding, 200
  heaters, 138, 678, 735
  ignition, gun type, 713
  instruments, 174, 176, 178, 232, 271,
    382, 541
  magnetic fundamentals, 169
  power, 177, 198
  supply, 337
  switches, 207
  symbols, 172, 954
  system, automobiles, 920
  troubles, 207, 232, 238, 263
  troubleshooting, 347
  units, 954
  wiring, 181, 206, 243, 304, 347
Electricity,

basic, 169
current, 170
from solar energy, 942
generating, 169
static, 169
types, 169
Electrolux system, 7, 91, 607
Electrolysis, 722, 964
Electrolytic capacitor, 188
Electromagnetic,
  amplitude, 933
  induction, 187
  wavelength, 933
Electromagnetism, 185, 187, 903
Electrometer, 169
Electromotive force, 174, 198, 953
Electron, 169, 953
  flow, 169, 953
  theory, 180, 185, 952
Electronic,
  air cleaner (EAC), 794
  cleaner control, 868
  control system, 854
  leak detector, 362
  variable speed motors, 230
Electronics, 169, 193
Electroscope, 169
Electrostatic electricity, 169
Electrostatic filter, 794
Element, power, 144
Elementary,
  absorption, 89, 603
  electric motor, 191
  refrigerator, 29
Ell, street, 51
Emissivity adjustment, 947
Enamel, 332
End bell, 209
End play, 234
Endothermal, 967
Energy,
  conversion factors, 22, 29, 948
  electrical, 22, 169, 948
  heat, 22, 99, 948
  kinetic, 22, 948
  mechanical, 22, 99, 948
  potential, 948
  solar, 933
  stored, 22
  total, 832, 961
  units, 29
Energy Efficiency Ratio (EER), 601, 763
Engine-driven systems, 461
English units (see U.S. conventional)
English micrometer, 64
Enthalpy, 33, 578
Entropy, 952

Environment, 670
Enzymes, 303
Epoxy resins, 51
Equalizer tube, 159
Equipment, soldering, 44, 505
  welding, 47, 505
Equivalents,
  area, 14, 949
  energy conversion, 22, 29, 948
  flow, 950
  heat, 9, 23, 24, 25, 26, 951
  latent heat, 24, 25, 32
  linear measure, 13, 948
  liquid measure, 950
  mass, 17
  power, 22
  pressure, 17, 949
  refrigeration effect, 26, 27, 28
  specific heat, 24
  temperature, 9, 10, 11, 13
  velocity, 949
  volume, 14, 949
  weight, 17, 950
Estimates, service, 551
Ethyl chloride (R-160), 289, 965
Ethyl hexanol, 622
Ethylene glycol, 937
Eutectic, 960
  plates, 414, 625
  solution, 576, 638
Evacuating, 71
  compressor, 384, 497, 513
  equipment, 384, 497
  evaporator, 384, 530
  hermetic system, 384
  liquid line, 497, 530, 547
  suction line, 536, 548
  system, 384, 497, 546
Evaporating,
  cooling, 761
  pressures, 26, 297
  temperatures, 26, 297
Evaporation, 26
  source, 956
Evaporative,
  condenser, 412, 687
  cooling, 697, 761
  refrigeration, 73, 74
Evaporator, 8, 26, 34
  air conditioning, 748
  air-cooling, 571
  area, 567, 572
  A-type, 822
  automobile air conditioning, 912
  beverage, 416, 417
  blower-type, 414
  capacities and calculations, 567, 571,

575, 576
  coils, heat pump, 808, 809, 810
  commercial, 413, 569
  construction, 101, 413, 571
  defrosting, 309, 413, 423
  design, 101, 413, 571, 573
  domestic, 101, 304, 309
  dry-type, 101, 413
  expansion, 529, 530, 531
  expendable refrigerant system, 625
  fan, 102, 310, 311, 414, 415
  fins, 414, 571
  fin-type, 414
  flat-type, 822
  flooded, 813
  forced circulation, 414
  forced convection, 414
  freezer, 304, 324, 413
  freezing dispensers, 423
  frosting, 341, 413, 803
  frozen food, 326, 413
  ice cube maker, 417
  immersed, 416
  installing, 376, 490, 504, 568
  liquid cooling, 416
  location of, 569
  mounting, 425, 490, 504
  multiple, 80
  noncode installation, 491
  nonfrosting, 414
  plate-type, 414, 625
  pressure-type, 417
  removing, 374, 509, 530
  repairing, 51, 377, 530
  servicing, 377, 530
  slant-type, 822, 823
  standard temperatures, 277
  starved, 142
  submerged, 416
  system, 625
  types, 101, 413, 571
Evaporator coils (see Evaporator)
Exchanger, heat, 79, 432, 814
Exfiltration, 750, 884, 888
Exhaust muffler, 135
Exhaust ports, 127
Exothermal, 967
Expansion valve, 78, 110, 141
  adjustments, 142
  automatic, 110, 141
  automatic bypass, 143
  automotive, 914
  bellows type, 142
  construction, 110, 142, 143
  diaphragm type, 147
  installation, 144, 531
  operation of, 110

  pressure controlled, 110
  pressure limited, 154
  removing, 530, 533
  repairing, 533, 535
  servicing, 532
  systems, 78, 141, 144
  temperature controlled, 110, 144
  tester, 533
  thermostatic, 110, 144, 807
Expendable refrigerant, 30, 292, 625
  refrigeration, 86, 485, 625
Exponent, 11
Extension cord, insulated, 243
External compressors, 115
  cleaning, 349
  diagnosing troubles, 345
  installing, 520, 543
  servicing, 508, 512
External drive, 122, 756
  system, 75
External equalizer, 160, 161
External motors, 122, 210, 234
  installing, 543
  servicing, 542
External servicing, 344
  guide, 344

F

Fahrenheit,
  absolute scale, 10
  conversion formulas, 10
  temperature scale, 10
Fans,
  air conditioning systems, 789
  attic, 791
  automobile air conditioning, 919
  axial flow, 789
  basement, 791
  capacity, 790
  cleaning, 792
  controls, 274
  motors, 227
  pressure, 790
  radial fow, 789
  servicing, 238
  speeds, 790
Farad, 188
Faraday experiment, 93
Faraday, Michael, 7, 93
Farm vehicle air conditioning, 926
Fast freeze, 9, 637
Fast freezers, 471
Fast freezing case, 471
Fastening devices, 68
Featheredging, 332
Field erected, 822
Field pole, 186, 193

File card, 67
Files, 67
  flat, hand, mill, 67
Filter-drier, 103, 108, 459, 460
Filters, 79, 372, 459, 537, 793, 798
  adhesive, 793
  air, 792
  automobile, 910
  carbon, 466, 798
  chambers, 711
  cleaning, 792, 793, 794
  efficiency, 662
  electrostatic, 794, 796
  hermetic, 377
  loaded, 793
  measuring efficiencies, 662
  oil saturated, 793
  servicing, 794
  throw-away, 794
  water, 451, 527
Filters and strainers, 712, 717
Fin tube, 102, 588
Finishes, cabinet, 332
Fins, 413, 530, 573
Fire dampers, 784
First Law of Thermodynamics, 22, 952
Fittings,
  capillary tube, 165
  commercial, 505
  compression, 44
  flared, 44
  flexible hose, 39, 44
  metric size, 44
  pipe, 44, 50
  reusable, 39
  self-sealing coupling, 824
  soldered, 45
  tubing, 44
Fixed gas, 604
Fixture temperature, 296
Flake ice, 478
Flame,
  adjustment, 716, 717
  color, 722
  mirror, 713
  out, 713, 716
  propagation, 726
  test, leaks, 362
Flange,
  blank, 514
  mounting, 708
  Flapper valve, 380
Flare, 41, 43
  double thickness, 43
  making, 41
  nut, 42
  nut wrench, 54, 964

plug, 531
single thickness, 41
testing, 47
Flared connection, 41
Flared tubing fittings, 44
Flaring tool, 41
  adaptors, 43
  tubing, 37
Flash,
  chamber, 635
  gas, 148, 432
  vapor, 432
Flashback, 726
Flashpoint of oil, 960
Flat plate solar collector, 935
Flat-type evaporator, 822
Flexible insulation, 893
Flexible tubing, fittings, 39
Flexible tubing, hose, 38
Float, high-side, 113
  low-side, 113
Float control, high-pressure side, 76
  low-pressure side, 162, 417
Float switch, 435
Float valves, 162, 163
  high pressure, 163
  low pressure, 162
  operation of, 162, 163
  systems, 74, 162, 163
Floats, 163
  pan type, 162
Floc test, 960
Floodback, 811
Flooded evaporator, 813
Flooded system, 74, 76, 162, 163
  coil, 417
  control, 74, 162, 163, 417
Flooding, 74, 499
Florist cabinets, 466
  construction, 466
Flow,
  control, refrigerant, 108
  equivalents, 950
  gases, 703
  meter, 876
  of heat, 9, 32
  switch, 727
Flue, 610, 719
  absorption system, 608
Fluid (see Liquid)
Fluid bulb, 722
Fluid solar system, 814
Fluoro refrigerants (danger), 967
Flush, 378
Flux,
  magnetic, 184
  silver brazing, 546

soldering, 44
Flywheel, 75
  puller, 122
Foam leak detector, 361
Foamed-in-place insulation, 466
Fogs, 792
Food, cold preserves, 303
Foot, 14
Foot pounds, 7, 22
Force, 21
  centrifugal, 953
  centripetal, 953
  electromotive, 174
Forced,
  circulation evaporator, 414, 747
  convection, 106, 399, 414, 558, 575, 739
  draft, 705
  feed oiling, 136
  warm air furnace, 673, 675, 678
Fountains, soda, 471
Four-pipe system, 830
Four-way valve, 801
Fractions, 948
Frame, motor, 209
Free electrons, 952
Freezants, 292
Freeze drying, 482
Freezer,
  alarms, 326
  burns, 304
  care of, 330
  chest type, 323
  cycling time, 390
  dispensing, 474
  fast, 303, 482
  installing, 336
  insulation, 304
  push-on type mount, 325
  shutting down, 390
  uncrating, 336
  upright, 326
Freezing, 7, 9, 303
  dispenser evaporator, 423
  food, industrial, 482
  ice cream, 577
  plant, 482
  point, 10
  preservation radiation, 963
  storage data, 296
  temperature, 10, 26, 27
  temperature, brine, 32, 961
Fresh food cabinet, 304
Frost back, 360
Frost control,
  automatic, 265, 442
  element, 690
  semiautomatic, 267

Frost removal, 97
Frost-free refrigerator, 245, 317, 319
Frosting, suction line, 364
Frosting type evaporator, 413
Frozen food compartments, 303
Frozen foods,
    cabinets, 313, 468, 470, 471
    condensing units, 323, 326, 393
    controls, 303, 304, 307, 311, 326
    display case, 470
    evaporator design, 323, 326
    fish, 560
    freezer burn, 304
    fruits, 560
    ice accumulation, 330
    industrial, 482
    insulation, 304
    locker plants, 482
    meat, 560
    motor controls, 79
    motors, 209, 210, 211
    poultry, 560
    preparation, 303
    refrigerant controls, 313
    refrigerants, 298
    servicing, 330
    storage cabinet, 471
    storage time, 304
    temperature, 560
    vegetables, 560
    wax in system, 343, 360
Fuel,
    gas, 721, 962
    gas burner, 673
    oil, 705, 962
    oil additives, 707
Full floating, 118
Fundamentals of refrigeration, 7, 8
Fur storage, 480
Furnace,
    construction, 703
    controls, 741, 837, 857, 860, 867
    designs, 703
    inspection, 727
    puffback, 713
Furnaces,
    electric, 737
    forced warm air, 673
    gas, 719
    gravity, warm air, 673
    hydronic, 700
    oil, 707
    tubular, 736
Fuse, electric motor, 182, 223
Fuses and circuit breakers, 223
Fusible links, 818
Fusible plugs, 293, 456

G

Gallon, 14
Galvanic action, 964
Galvanometric, 872
Gas, 21, 33, 34, 957
    adiabatic expansion and contraction, 967
    burner, 719
    charged sensing element, 146
    charged thermostatic expansion valve, 151
      collapsible, 154
    constant, 33, 35
    control, 611, 612, 724
    cross charge, 144
    definition of, 21
    flash, 148
    flow, 703
    fuel, 721
    furnaces, 719
      electronic controls, 854
      fired radiant heat, 740
      installation, 724
      primary controls, 852
      servicing, 727
      starting, 726
    heating, supplementary, 941
    hot gas defrosting, 96, 271, 313, 393
    hydrogen, 638
    ideal, 33
    isothermal expansion and contraction, 967
    law, 967
    noncondensible, 832
    pressure, 719
    specific gravity, 21
    supply, 610, 719, 721
    valve, 610
Gas and vapor, 957
    saturated, 34
Gaskets, 70, 332
    compressor, 138
    door, 332
    paper, lead, aluminum, and plastic, 70
Gauge pressures, 18, 19
Gauges, 18, 19, 60, 62, 63, 64
    calibration, 64
    care of, 60, 64
    compound, 19, 62, 63, 64
    construction, 64
    high pressure, 63
    low pressure, 63
    manifold for testing and servicing,
      350, 352, 493
    McLeod, 63
    pressure, 62, 63, 64
    principles of, 64
    removing, 495
    use of, 493

    vacuum, 63
Gauss, 186
Gear compressor, 130
    drive, 122
Generator, absorption, 91
Generator, electrical, 190
Genetrohs (see Refrigerants)
Gland, packing, 70
Glass panel heaters, 740
Glass, sight, 457, 493, 508
Glycol, 617
    brine, 961
Goggles, safety, 58, 927
Grain, 17, 642, 742
Gravity, specific, 21
Gravity warm air furnace, 673
Grille, 781
    controls, 781
    plastic, 823
Grinder, 379
Grocery cabinets, 463
Ground coil, 693, 810
Ground fault protector, 200
    short circuit, 172
    wire, 200
Grounding, 200, 224, 274, 337
Gun, oil burner, 702
Gun type, 708
    wiring diagram, 851

H

Hacksaws, 67
Halide,
    leak detector, 278, 362
    refrigerants, 278
    testing for leaks, 362
    torch, 285, 362
Halogen gas, 362
Hammers, 56
Hand tools, 53
Hand valves, 454, 455
Hang-ons, 924, 926
Hard hat, 822
Hardware, 331
Hastelloy, 147
Head pressure, 296, 399
    control, 399
    for common refrigerants, 529
Heat,
    absorbing capacity of ice, 25, 26, 27
    anticipators, 844
    area, 579
    chart, 578, 579
    comfort range, 650
    conduction, 31

conductivity, 963
convection, 32
definition of, 9
diagram, 281, 578
distribution controls, 860
effective latent, 579
electric, 735
energy, 22
enthalpy chart, 578
equivalents, 951
exchange, 666
exchanger, 432, 584, 704
flow, 9, 32
gain, 880
in air, 765
insulation, 666
lag, 891
latent, 24, 25, 579
leakage, 555, 556, 563, 880, 884
  air conditioning, 880
  calculations, 555
  water cooling, 576
load, 555, 558, 562, 566, 879
  889, 891, 893
  calculations, 562, 563, 566, 894
  commercial, 555
  cooling, 880
  heating, 879
  sun, 656, 891
loss, 879
  basement, 887
  unheated spaces, 887
measurement, 9
mechanical equivalent of, 29
pipe, 636
radiant, 740
radiation, 31
reducing, 891
removing, 8
residual, 845
resistance, 735
sensible, 24
sink, 666, 698
specific, 24, 580, 948
storage, solar energy, 934, 936
  closed water system, 937
  rocks, 936
  warm air system, 937
sublimation, 88
superheat, 144, 580
supplementary, 735, 940, 941
unit, 23
Heat of,
  compression, 28
  fusion, 27
Heat pump, 690, 693, 800, 937, 941
  air coil, 808

coils, 804
compressors, 804
condensers and evaporators, 808
controls, 804
cooling cycle, 692, 802
cycles, 692, 802
defrosting, 811
exchanger, 584, 814
ground coil, 810
heating cycle, 690, 691, 692, 802
Hi/Re/Li system, 813
installation, 800, 811
lake coil, 810
motors, 804
mounting, 811
operation, 690, 802
performance, 803
purpose, 690, 800
reversing valves, 806
servicing, 812
solar, 814, 937, 941
systems, 802, 804
theory, 800
through-the-wall, 811
two-speed, 812
valves, 806
well water coil, 810
wiring diagrams, 805
Heat sources, buildings, 892
Heat sources, defrosting, 558
Heat transfer, 31
  coefficients, 951
  conduction, 31
  convection, 32
  radiation, 31
  regenerator, 638
  respiration, 558
  wire, 308
Heaters,
  accessory, 138
  automatic electric defrost, 311
  baseboard electric, 737
  crankcase, 138
  drain, 810
  duct, 739
  electrical 308, 737, 811
  forced convection electric, 739
  glass panel, 740
  mullion, 312, 313
  radiant electric, 740
  unit. 742
Heating,
  coil, 741
  cycle, 742, 802
  electric, 693, 735
  equipment, 700, 702
  equipment for air conditioning, 700

gun type oil, 702, 703
installation, 741
radiant, 698
service, 741
solar energy, 934
supplementary, 693, 739, 811
thermostat, 842, 845
types, 699
value, 706, 962
Helic design, 417
Helical rotors, 130
Heli-grip, 904
Helium, 293, 638, 874
Hermetic refrigeration,
  adaptors, 353
  adding oil, 367, 388
  assembling, 382
  charging, 364, 365
  circuit wiring, 306, 313, 316
  cleaning, 349
  commercial, 398
  compressors, 115, 405
  condensers, 305
  controls, 224, 242 through 255
  core valves, 359
  cycles, 304, 305
  defrosting systems, 309, 423
  diagnosing electrical troubles, 347
  dismantling, 374
  driers and filters, 372
  electrical circuit, 306
  evacuating, 384
  installations, 489, 502
  leaks, repairing, 364
  leaks, testing for, 361, 362
  mechanism, 398
  motor controls, 114
  motors, 134, 215 through 220
    capacitor start-capacitor run, 209, 210
     217, 218
    capacitor start-induction run, 210, 211
     217, 218
    electrical characteristics, 216
    hermetic split-phase, 217
    installing, 546
    polyphase, 219
    polyphase three-phase, 221, 433
    removing, 546
    repairing, 238
    servicing, 238, 544, 545, 546
    starting a stuck motor, 240
    terminals, 221
    types, 216
  oil, 367, 962
  overhauling, 378
    equipment, 378
  reciprocating compressor, 122

refrigerants, 276
removing the unit, 368
repairing compressors, 380, 381, 520
repairing motors, 380
repairing units, 520
rotary compressor, 127, 381
service valve, 353, 354
servicing, 520
  capillary tubes, 360
  check valves, 447
  compressor testing after overhaul, 382, 518
  diagnosing electrical troubles, 347
  diagnosing mechanical troubles, 349
  in-the-field service operations, 341, 342, 343
  mounting tubing service valves, 354, 356, 357
  overhauling sealed units, 378
  process tubes, 356
  rewelding compressor dome, 382
  sealed, 636
  service valves and adaptors, 353
  tools and supplies, 335, 336, 361
  troubleshooting, 342
  use of gauge and service manifold, 350, 352
  use of high vacuum pump, 384, 385, 386, 387, 388
  using Schrader valves, 359
systems, 304, 307, 311, 313
testing, 389
tubing, 364
units, 398
welding, 382
wiring, 306, 313, 316
Hertz, a-c cycles, 183
Hg (mercury), 18, 19
High head pressure, 520
High limit control, 673, 678
High vacuum pump, 384
High-pressure cutout, 547
High-pressure gauges, 63
High-pressure side, 8, 100
  float, 76, 99, 113, 141
    principles, 163
    system, 76
  flooded system, 113
Hinges, 331
Hi/Re/Li system, 813
Horsepower, 7, 22, 176
  heat equivalent of, 29, 948
  hour, 948
Hose, flexible tubing, 38
Hot brine defrost, 429
Hot gas,
  bypass valve, 531, 536, 902, 903

defrost, 96, 268, 271, 313, 415, 425, 800
driers, 313
mechanisms, 313
troubles, 373
Hot junction, 197
Hot water heating, 728, 939
Hot wire, 243
Hot wire anemometer, 651
How a mechanical refrigerator works, 8
Humidification requirements, 742
Humidifier, 271, 689, 742, 744
  cabinet type, 744
  heat load, 893
  installing, 742
  servicing, 743
Humidifying equipment, 700
  operation, 699
  systems, 699
Humidistat, 254, 867
Humidity, 7, 35, 642
  control, 815
  controls, 271, 648, 867
  definition of, 35
  domestic refrigerator, 303
  excessive, 742
  increasing, 742
  low, 646
  measurement, 646
  recorder, 649
  relative, 35, 643, 650, 666
  water source, 742
Hunting, 161
Hydraulic,
  cleaning, 727
  controls, 831, 832
  pressure, 573
Hydrogen, 293, 638
Hydronic, 728
  furnace, 673, 677, 679, 700
  installation, 732
  servicing, 732
Hygrometer, 646
Hygroscopic substances, 648

I

ICC (Interstate Commerce Commission), 293
Ice,
  accumulation, 8, 330, 341
  as a refrigerant, 73
  bank controls, 269
  controls, 254
  crystals, 9
  cube bin, 85
  cube defrost cycle, 420
  cube freezing cycle, 417
  cube maker, 85, 417, 478, 526

controls, 438
  installing, 338
  dry, 32, 88
  in cabinet insulation, 341, 344
  maker, 85
    control, 438
  manufactured, 478
  melting capacity, 27,28
    effect, 948
    equivalent, 27,28
    IME, 948
    temperature, 9
  refrigeration, 73
  zero, 32
Ice and salt mixture, 28
Ice cream,
  cabinets, 471
  freezing, 577
  makers, 474, 478
Ignition,
  electrical, 713, 722
  method, 722
  transformer, 713
Immersed evaporator coil, 416
  brine, 32, 416
  sweet water, 32, 416
Immersion, 637
Impedance, 178
Impeller, 133
Improper refrigeration, 549, 550
Incandescent units, 278, 735
Inch, 13, 948
Inch pound, 22
Incubators, laboratory, refrigerated, 480
Index, discomfort (CHI) comfort, health, 660
Indicator cards, 955
Indicators, airflow, 875
Indicators, moisture, 458, 459
Indoor climate, 655
  airflow, 875
Induced magnetism, 185
Inductance, 192
  mutual, 193
  self, 193
Induction, electromagnetic, 187
  capacitor, 188, 211
  motors, 210
  repulsion-start, 209, 211
  split-phase, 211
Inductive loads, 177
Inductors, 193
Industrial,
  applications, 481
  freezing of food, 482
  storing of foods, 483
Inefficient unit, 549

Inertia, 788
Infiltration, 888
Infrared, 10
  heat controls, 860
  lamp, 740
  radiation, 933
  thermometer, 947
Inside coil, 808
Inside diameter, (ID), 164
Inspecting residential central air
    conditioning, 827
Inspection, periodic, 548
  steam heating, 735
  warm air furnace, 727
Installation, 337, 489
  absorption refrigerator, 621
  air conditioning, 825
  code, 502, 506
  condensing unit, 489, 504
  domestic refrigerator, 337
  ducts, 775
  evaporator coil, 504, 568, 569
  gas burner, 724
  heat pump, 800, 811
  heating coil, 741
  hydronic system, 732
  noncode, 489
  oil burner, 716
  refrigerant lines, 491
  rooftop units, 818
  solar space heating, 937
  steam heating, 734
  testing,
    code, 506
    noncode, 489
  tubing, 491
  window unit, 749, 751
Installing,
  absorption refrigerators, 621
  air-cooled condenser, 522
  commercial systems, 489
  console air conditioners, 758
  cooling unit (see Evaporator)
  direct expansion evaporators, 531
  domestic units, 335, 336, 337, 608, 621
  evaporators, 490
  expansion valve, 144, 531
  external drive compressor, 520
    motor, 543
  gauge, manifold, 350
  gauges, 62, 63, 64
  "hang-on" system, 926
  hermetic motor compressor, 546
  humidifiers, 743
  ice cube maker, 338
  motor compressor, 376

oil tank, 713
open type compressor (see external
    drive compressor)
refrigerant lines, 504
residential central air conditioning, 825
thermostats, 271
tubing, 37, 491
two-temperature valves, 535
water valves, 451, 529
water-cooled condenser, 408
Instructions to the owner, 330
Instruments, 60
  air conditioning, 871
  air distribution, 871
  airflow measuring, 871, 872, 875
  ammeter, 176, 232, 233
  anemometers, 529, 651, 871, 872
  barometer, 19, 666, 871
  circuit testing, 207
  combustion efficiency, 873
  connecting and handling, 179
  draft control, 873
  electrical, 174, 176, 178, 179, 239, 389
  gauges, 62, 63, 64
  hygrometer, 646
  manometer, 20, 871, 872
  pitot tube, 653
  recording, 653, 872
  thermometer, 9, 10, 11, 60
  velocimeter, 652
  voltmeter, 174, 232, 233
  water analysis, 876
  wattmeter, 178, 232
Insulation,
  air conditioning, 880
  automobile, 919
  batts, 892, 893
  building, 880
  commercial, 555, 963
  domestic, 304
  fiber glass, 304
  frozen food, 304
  motor, 216
  safety, 642
  sheeting, 892
  thermal, 892
  tubing, 491
  urethane, 304
Insulators, 181
Intake,
  muffler, 135
  ports, 127
  valves, 99, 120, 121, 122
Interference, radio, tv, 232
Intermittent cycle, 89, 605
Internal,

combustion engine, 638
operation, 349
trouble, 349, 359, 368
International system of units (see
    Le Systeme International d'unites)
Inverter, 194
Ionizing air, 794
Ions, 953
IR drop, 183, 198
Isothermal expansion and contraction,
    967

J

Jet nozzle, 696
Joule, 9, 22
Journal, crankshaft, 120, 126
Junction,
  cold, 632
  hot, 632

K

"K" factor, 556, 880
Kata thermometer, 947
Kelvin, 10, 11, 13, 24
  temperature scale, 10
Kelvinator, 7
Kerosene burners, 89, 90
Kerosene-fired refrigerators, 89, 605, 606
Kilocalorie, 23
Kilojoule, 23
Kilometre, 13, 948
Kilopascal (kPa), 17
Kilovolt, 174
Kilovolt amperes (KVA), 203, 204
Kilowatt, 176, 198
  hour, 948
Kinetic energy, 22, 948
Klixon valve, 263

L

Laboratory, refrigerated incubators, 480
Lag, 845, 891
Lamps,
  quartz, 740
  radiant heat, 740
  ultraviolet, 466, 798
Lapping,
  blocks, 517
  compound, 380, 516
Latches, 331
  magnetic, 331, 333
Latent heat, 24, 579
  application of, 25, 579

effective, 577
of condensation, 25
of fusion, 24
of sublimation, 32
of vaporization, 24
refrigerants, 25, 578, 965
values, 276, 578, 964
Law, Ohm's, 181
Laws of refrigeration, 99
Boyle's, 966
Charles's, 966
Dalton's, 34
gas, 967
general, 952
Lenz's, 190
Peltier, 87
Laws of thermodynamics, 9, 22, 952
Le Systeme International d'unites, 7
Leak detector, 361, 362, 496
electronic, 285, 361, 362, 507
flame, 278, 285, 361, 362, 507
procedure, 496
Leakage, heat, 555, 556, 563, 880
Leaks,
locating, 361
repairing, 364
testing for, 285, 361, 496, 929
Lens, solar collector, 942
Lenz's Law, 190
Letter drill, 962, 963
sizes, 962, 963
Leveling screw, 336, 338
Lift pump oil circuits, 817
Light,
diffused, 794
pilot, 722
speed of, 952
Lights, testing, 207
Limit controls, 850, 863
Limiters, pressure, 154, 273, 401, 433
Limitizer valve, 401
Line,
liquid capacities, 590
refrigerant, 457, 590, 911
suction capacities, 595
thermostats, 837
voltage transformer, 203
Linear measure,
equivalents, 948
symbols, 14
Liquid,
absorbent, 604, 792
charged power, 245, 246, 837
collector, 814
cooling, 416
cross charge, 144, 151

equivalents, 950
evaporator, 416
floodback, 811
helium, 292
impurities, 792
indicator, 80, 82, 898
line, 100, 108, 547
servicing, 547
measure equivalents, 950
moisture, 957
nitrogen, 30, 484, 625, 626
nitrogen control valve, 631
receiver, 108, 412
sizes, 590
refrigerant, 8
sensing element, 150
slugging, 804, 811
Liquids, 20
volatile, 636
Liquid-vapor valve, refrigerant cylinder, 293
Lithium bromide system, 619, 620, 622
Litre, 14, 950
Loads,
cooling, 555, 576, 577, 803, 879, 880, 891
heat, 555, 558, 562, 566, 879,
889, 891, 893
inductive, 177
noninductive, 177
Locating trouble, 341, 359, 368, 507, 549
Location, condensing unit, 393, 489
Location of evaporator, 569
Location of refrigerator, 336, 489
Locker plants, 482
Loose baffles and ducts, 338
Louvers, 536
Low humidity, 646
Low pressure, gun type, 708
Low voltage thermostat, 841
Low wax content, 298
Low-pressure side, 74, 101, 113, 141
filter-drier, 103
float control, 113, 162
design, 162
principles, 162
pressure, 100, 296
controls, 450, 902
limiter, 273
systems, 74, 162
valve, 70, 162
LP fuel, 603, 721
LRSV, 530
Lubrication, compressor, 136, 298, 367,
540, 960
ball bearing, 234
centrifugal, 131, 132, 133
motor, 234, 542

reciprocating, 117
ring, 234
screw-type, 136

M

Machine screws, 68
Magnet,
coils, 722
permanent, 185, 246
pump, 614
Magnetic,
center, 238
clutch, 903
field, 184, 186
flux, 184
gasket, 331
latches, 331
line, 246
relay,
amperage, 260
clutch, 903
current, 185, 217, 260
electronic, 263
hot wire, 263
potential, 261
solid state, 263
starter, 438
thermal, 263
voltage, 261
Magnetism, 184, 185
Magnets, electro, 157, 185
Mallets, 57
Manifold,
gauge, 350, 352, 493
lines, 356, 357
service, 350, 352, 493
Manometer, 20, 665
Manual defrost, 304, 306
Manual valve, 454, 455
Marine refrigeration, 485
Market, coolers (see Walk-in coolers),
465
Mass, 17
Materials, refrigeration, 37
Matter, theory of, 952
Mean effective pressure, 588, 954
Measurement,
angular, 17
area, 14
draft, 654
filter efficiencies, 662
heat, 9, 10
humidity, 642 through 648
linear, 13, 948
noise, 670
pressure, 17

temperature, 10
volume, 14
Measuring,
rules, 64
tapes, 64
tube, 366, 367
Meat, frozen, 303, 560
Mechanical,
cycle, 393
domestic refrigeration, 8
energy, conversion factors, 29
refrigeration, advantages of, 7, 99
refrigerator, operation, 8, 30
Mechanism,
absorption, 603
air conditioning, 699, 747, 765, 897
automotive, 897
commercial, 393
domestic, 304, 307, 311, 313,
317, 335, 336
hermetic refrigerator, 398
hot gas refrigerator, 313
refrigerator, 304
Megavolt, 546
Megawatt, 198
Melting,
capacity, ice, 27
point, 27
temperature of ice, 9
MEP, 588, 954
Mercoid bulb, 546
Mercury column (Hg), 19
Metal cleaning, 515
Metering type two-temperature valve, 443
Methods of adding refrigerants, 364
Methyl chloride, 289
pressure-temperature curve, 291
properties of, 957
Methyl formate, 964
Methylene chloride, 957
Metre, 13, 948
Metric,
micrometers, 66
pressures, 296
system (see Equivalents)
tube fittings, 44
tube sizes, 38
units, metric, 7
Microampere, 179
Microfarad, 188
Micrometer, 64
English or metric, 64, 66
Micrometre, 14
Micron, 384, 933
Microorganism, 303
Microvolt, 174

Mil, circular, 206
Mile, 14
Milk cooler, 480
Milk dispensers, 480
Millimetre, 948
Millivolt, 174, 722, 841, 842
Mixtures, ice and salt, 28, 416, 961
Mobile home air conditioning, 744, 760
Mobile home refrigerators, 614
Modulating, 84
controls, 863
cycle, 84
Module, 87, 633
Moisture,
in air, 642, 956
in cabinet insulation, 341
in compressor, 349
in installation, 330
in systems, 360, 459, 460, 537
indicator, 458, 459, 460, 960
liquid refrigerants, 957, 960
refrigerant oils, 960
removal, 537
sources, 956
table of safe moisture contents, 299
Mold, 792
Molecular,
motion, 100
sieve, 459, 956
theory, 100
Molecules, 9, 20, 100
Monitor top, 7
Monochlorodifluoromethane (see R-12),
279, 281, 339
Motor,
burnout, 375, 460, 461
capacitor, 218
characteristics, 223
compressor fault, 368
connections, 214
controls, 243, 245, 246, 433, 546
pressure, 433
safety, 435
temperature, 226
thermostatic, 434
speed, 212
Motorized water valves, 528
Motors, electric (see Electric motors),
capacitor, 211
circuits, 202
cleaning, 235
compressors, 134, 311
controls, 114
differential adjustment, 247
range adjustment, 247
data, 227

design, 191, 209, 215, 216
direct current, 222
direction of rotation, 214
domestic, 134, 191, 210, 215
drive, 210, 234, 789
efficiency, 588
fan motors, 227
fans, 789
fuses, 223
heat pump, 804
hermetic, 115, 215
horsepower, 22, 223
induction, 210, 217
installing, 546
insulation, 215
lubrication of, 234
magnetic center, 238
MEP, 588, 954
oiling, 234
open type, 209, 543
overloads, 223, 225
permanent split capacitor, 219
polyphase, 219
pressure control, 74, 76, 114, 144,
243, 245, 433, 435
principles, 209
protection, 224, 225
puller bearings, 235, 515
removing, 542, 546
repair, 234, 238, 380
replacement terminals, 221
repulsion-start, induction, 211
rotation, direction of, 213
rotor, 193, 209, 398
safety control, 224, 225, 435
servicing, 232, 234, 238, 543, 544,
545, 546
shaded pole, 229
size,
heat input method, 588
mean effective pressure method, 954
sizes, 587, 588
split-phase, 211, 217
starters, 433, 437
starting stuck motor, 240
stator, 209, 398
stuck, 240
temperatures, 226
terminal, 221
testing, 543
testing stand, 237
theory, 191
thermostatic motor controls, 243,
306, 433
troubles, 232, 234, 238, 543
types, 209

universal, 222
variable speed, 230
winding, insulation, 213
Mount, flange, 708
Moving instructions, 489
Muffler, compressor, 135, 457
Mufflers, 457
Mullion heater, 320
Multiple systems, 80
belts, 122
code installations, 502
installations, 489, 505
servicing dry coils, 529
Multistage systems, 635
Multistage thermostats, 849
Murc, 663
Mutual inductance, 193

N

National fine thread (SAE), 44
National pipe thread (NP), 44
National safety code, 968
Natural convection, 737
evaporators, 413
Natural gas, 610, 721
Natural or manufactured ice, 7
Neon, 293
Newton, 17, 21
Nipple, 51
Nitrogen, 37, 293, 625, 637, 874
dry, 483
liquid, 484, 625, 626
use of, 361
No-frost freezer, 319
Noise, 551, 667, 767
acoustics, 767
compressor, 135
duct, 767, 789
elimination, 338
expansion valve, 532
hermetic, 545
levels, 767
locating, 338
measurement, 667
reflectors, 767
seal, 516
vibration, 767
Nominal size tubing, 37, 39
Noncandescent units, 735
Noncode installation, 489
Noncondensible gas, 521, 832
Nonconductors, 180, 181, 955
Nonfreezing solution defrost system, 429
Nonfrosting evaporator, 414
Nonfrosting system, 414

Noninductive loads, 177
North pole, magnetic, 184
Nozzle,
ejector, 696
jet, 696
Nuclear power, 794
Nuts, flare, 44
Nylon yarn hose, 38

O

O-rings, 39, 138
Octave, 670
Octyl alcohol, 622
Odor removing, 466, 798
Off cycle, 79
Ohm, 178
Ohmmeter, 178
connections, 179
Ohm's Law, 181
Oil,
acidity, 298
adding to the system, 367, 540, 929
additives, 136
antifoam inhibitor, 960
binding, 113
burner, 707, 708
electric ignition, 713
flame failure, 716
forced warm air, 675
gun type, 708
high pressure, 708
hydronic, 677
installation, 713, 716
low pressure, 708
nozzle, 716
primary controls, 856
pump, 711
servicing, 717
starting, 716
vent plug, 716
charging apparatus, 350, 352
compressor, 136, 298, 367, 540, 960
dielectric property, 960
dispenser, 374
filter, 711
flash point, 960
floc test, 960
fuel, 705
furnaces, 707
grades, 706
heating supplementary, 941
hermetic, 298, 367, 960
level, sight glass, 512
moisture in, 299
motor, 234

operation, 104
pour point, 70, 298
pressure safety control, 435
pump, 136
gear type, 711
scavenger, 136, 137
refrigerant, 70, 298, 917, 960
removing, 378
rings, 118
sampler, 520
separator, 84, 104, 449
tank installation, 713
traps, 598, 599, (see Oil separators)
viscosity, 70, 298, 526, 960
Oilless bushings, 235
On-and-off switch, 689
Opaque, 31
Open,
circuit, 172
cycle ammonia, 631
cycle refrigeration, 631
Delta, 202
display case, 468, 470
end wrench, 53, 55
external drive, 508
frozen display case, 470
motor,
cleaning, 234
compressor, 115, 117
lubrication, 234
type system, 75, 115, 508
Opening a motor compressor, 379
Organism, micro-, 303
Orifice, 600
sizes, 600
Ounce, 17
Outdoor,
air-cooled condenser, 408, 536
climate, 655
condensing unit, 399, 408
Outside coil, 808
Outside complete systems, 815
Overhauling,
compressor, 515
hermetic system, 378
Overload, 225, 232
motor, 225
protector, 225, 259
relief valve, 455
Overshoot, 844
Oxygen, 293, 641
Ozone, 661

P

Packing gland, 70

Pan type float control, 162
Panel, solar, 937
Parallel circuits, 182
Parallel motor compressors, 399
Partial pressure, 34
Partial vacuum, 18, 19
Parts, cleaning, 66, 378, 515
Parts per million (ppm), 299
Pascal (Pa), 17
Pascal's Law, 17
Peen, 56
Peltier effect, 87
Percolation pump, 622
Perfect gas equation, 33
Performance, compressor, 586
  coefficient, 587, 804
Perimeter drier, 308
Permanent magnet, 185
Permanent split capacitor motor (PSC), 219
Permeability-reluctance, 188
Petcocks, 509, 735
pH, 876
  factor, 525
Phase, single, 201
Phase, three, 201
Photocell, 181, 198, 853, 857
Photoconductors, 198
Photoelectricity, 193, 194
Photoemissive devices, 198
Photons, 933
Photosensitive detector, 856
Photovoltaic devices, 198
Photovoltaic solar cell, 943
Physics of refrigeration, 8
Piercing valves, 357
Pilot controlled, 146
Pilot light, 722
  flame color, 721
Pinch-off tool, 357
Pipe,
  copper, 37
  fittings, 50
  heat, 636
  threads, 50
  wrench, 55
Piping,
  color code, 963
  diagrams, 596
  heat loads, 555
  refrigerant, 590
Piston, 117, 118
  compressor, 115, 116
  cylinder, 117
  pins, 118
  rings, 118
Pitot tube, 653, 872

Plain tubing, 588
Plastic grilles, 823
Plastic tubing, 38
Plate, condenser, 106
Plates, eutectic, 414, 625
Plates, evaporator, 414
Plenum chamber, 897
Pliers, 57
  cutting, 57
  duckbill, 57
  gas, 57
  nut, 57
  round nose, 57
  slim nose, 57
Plug configurations, 206
Plugs, fusible, 456
Pneumatic control, 405, 862
Polarity of electromagnets, 187
Pollen count, 664
Pollutants Standards Index (PSI), 663
Polyethylene tubing, 38
Polyphase motor, 209, 219
Polyurethane, 304, 466,
Ponded roof, 697, 893
Porcelain, 332
Porcelain finishes,
  cleaning, 332
  repairing, 332
Port valve, 806
Portable,
  absorption refrigerators, 613
  charging cylinder, 365
  thermometer, 10
  thermostat, 842
Potential electrical, 174, 190
Potential energy, 22, 948
Potential relay, 261
Potentiometer, 230, 872
Pound, 17
Pounds per square inch (see Psi)
Pour point, 70, 298
Power, 22
  circuits, 201
  electrical, 198
  element, 150, 151
  element mounting, 155
  factor, 177, 199
  loss, 183
  nuclear, 794
ppm, 299
Preferred storage practice, 296, 560
Prefixes, unit, 11, 954
Preparation of frozen food, 303
Pressure, 7, 17
  absolute, 18
  atmospheric, 17, 18, 166
  barometric, 19

compression cycle, 100
compressor valve, 120, 121
condensing, 296
control, 258, 259
critical, 32
definition of, 17
disk, 709
drop, 584, 590, 785, 786
effect on boiling point, 25, 26, 297
effect on freezing temperature, 26
effect on latent heat, 25, 579
equivalents, 949
evaporating, 296
expansion valve, 110, 141
gas, 21, 719
gauge, 62, 63, 64
head (see Head pressure), 296
heat,
  areas, 579
  cycle, 583
  diagrams, 583
  enthalpy chart, 280 through 290, 578
  graphs, 578, 580, 581, 583, 584
    957, 958, 959
  hydraulic, 573
  latent heat, 579
  limiters, 154, 273
  low head, 926
  low-side, 296
  lubrication, 136
  mean effective, 588
  measurement, 17, 18, 19, 20
  metric, 296
  motor control, 245, 258, 393, 399, 433
    commercial, 433
    domestic, 243
  negative, 766
  operated altitude valves, 250, 251
  positive, 766
  reducer valve, 443
  refrigerant, 296
  regulator, 724
    evaporator, 443, 445, 901
  scales, 18, 19
  standard, 964
  suction, 75
  suction line, drop, 584
  testing for leaks, 361
  type evaporator, 417
  units, 17, 18, 19
  velocity, 653
  water valve, 453
Pressure of condensation, 296
Pressurestats, 837, 863, 865
Pressure-temperature,
  charts, 297
  conditions, 339

curves, 277, 278
tables, 297
Primary controls, 851
air, 781
container, 625
electric heat, 857
gas, 852
oil, 856
power, 202
Process tube, 356
adaptor, 356
Processing ammonia vapor, 631
Processing plant, 482
Propane, 292
Protector, circuit, 206, 219
Protector, overload, 225, 232, 259
PSC (Permanent Split Capacitor), 757
PSI (see Pollutant Standards Index)
Psi, 17, 18
Psia, 19
Psig, 19
Psychrometer, 643, 644
aspirating, 644
recording, sling, 643, 644
Psychrometric,
charts, comfort, 649, 650
measurement, 647, 649, 650
properties of air, 649
Puffback, 713, 727
Pull down, 151
Puller, wheel, 514
Pulley, 236
drive, 694
Pulsation, 708
Pulse wave, 184
Pump,
circulating, 728
condensate, 414, 415
down, 384, 430, 497
cooling cycle, 901
heating cycle, 901
oil burner, 711
gun type, 708
out, 430
rotary, 385
vacuum, 384, 507
Pumping ratio, 951
Punches, 67
center, 67
drift, 67
pin, 67
prick, 67
Purging, 71, 494

Q

Quality, refrigerant, 579
Quantity, refrigerant, 300

Quart, 14
Quartz lamps, 740
Quick-connect coupling, 539

R

R factor, 884
R-11, 276, 283, 298
R-12, 276, 279, 280, 298, 299
R-13, 957
R-13B1, 957
R-21, 957
R-22, 276, 281, 282, 298, 299
R-23, 279
R-30, 957
R-32, 279
R-40, 289, 957
R-113, 298, 957
R-114, 957
R-115, 279
R-160, 289, 957
R-170, 292
R-290, 292, 957
R-500, 276, 284, 285, 298
R-502, 276, 285, 286, 298, 299
R-503, 276, 287, 288, 298
R-504, 276, 289, 290
R-600, 292, 957
R-611, 289
R-702, 292, 293
R-704, 293
R-717, 276, 289, 291, 299, 957
R-720, 293
R-728, 292, 293
R-729, 293
R-732, 293
R-740, 293
R-744, 292, 957
R-764, 289, 299, 957
R-1130, 289
Radial commutator, 211
Radiant heating, 698, 740
gas fired, 741
installation, 739
lamps, 740
source, 740
Radiation,
infrared, 933, 934
of heat, 31, 933
preservation of food, 963
ultraviolet, 933, 934
Radio interference, 232
Radiometer, 10
Radius, 14
Railway car construction, 485
Railway car refrigeration, 485
Range adjustment, 246

controls, 250
mechanisms, 247
screw, 249, 250
Rankine,
conversion formulas, 13
temperature scale, 10
Rapid pressure balancing, 148
Ratchet wrench, 54, 55
reversible, 56
Reach-in cabinet, 463
Reactance, 189
Reamer, 235
bearing, 235
Rebuilt systems, charging, 388
Rebuilt systems, testing, 389
Receiver, 525
drier, 910
liquid, 108, 412
sizes, 590
repairing, 525
Reciprocating compressor, 117, 118, 120,
136, 138, 508
rebuilding, 380
Recommended temperatures for
commercial refrigerators, 296, 558, 560
Recording,
ammeter, 203
instruments, 872
pressure gauge, 60, 388, 507
thermometer, 389, 507
Rectifier, electrical, 181, 193, 194
RED system (G.E. data), 347
Reed valve, 120, 122
Refractory cement liner, 708
Refractory material, 675
Refrigerant control, 30, 108, 141,
165, 433
automatic expansion valve, 141
capacity, 600
capillary, 163
characteristics, 141
dry evaporator, 532
freezer, 326
heat pump, 804
high-side float, 163
low-side float, 162
thermostatic expansion valve, 144
Refrigerants (see "R-" numbers), 8, 29,
70, 276, 917
ammonia (see R-717)
amount required in system, 300
applications, 298
azeotropic mixtures, 285
butane (see R-600)
carbon dioxide (see R-744)
Carrene No. 2 (see R-11)
changing, 300

characteristics of, 957

charge, 364, 508

checking, 300

classification, 278

container, 638

controls, 141, 304, 311, 313, 866

cryogenic fluids, 293

curves, use of, 277

cylinders, 293, 294, 540
  color code, 296

dichlorodifluoromethane (see R-12)

discharging, 510

dryness of, 960

ethane, (see R-170)

ethyl chloride, (see R-160)

expendable, 30, 86, 292, 625

fluoro (danger), 967

gas, 276, 957

group No. 1, 279

group No. 2, 289

group No. 3, 292

halide, 278

head pressure, 529

identification, 276, 277, 299

lack of, 583

latent heat of, 24, 25

leaks, 361

line capacities, liquid, 595
  suction, 595

lines, 108, 457, 590, 596, 911
  installing, 504
  servicing, 547, 548

little used, 957

loss, 907

methyl chloride (see R-40)

methyl formate (see R-611)

moisture in, 299, 360, 957

monochlorodifluoromethane (see R-22)

numbers, 276

oil, 70, 298, 960

piping, 504, 590

pressure temperature curves, 277, 278

pressure temperature tables, 297

pressures, 298

propane (see R-290)

properties, 276, 577, 964

quantity, 300

removing, 368, 374, 510

requirements, 276

safety, 300

saturated vapor, 580

shortage of, 360

substituting, 300

sulphur dioxide (see R-764)

superheated, 580

tables, use of, 276, 580

transferring, 540

trichloromonofluoromethane (see R-11)

vapor, 276, 957

water, 292

Refrigerating effect of ice, 26, 27, 28

Refrigerating mechanism, water chiller,
  395

Refrigeration,
  basic systems, 73
  basis of, mechanical, 8
  career opportunities, 969
  compression, 99, 100
  cycle, absorption, 603
  dry ice, 88
  education requirements, 969
  electric heaters, 311
  elementary, 29
  evaporative, 73, 74
  expendable, 625
  fittings, 44
  history of, 7
  ice, 73
  industry, 970
  installation, 489
  laws of, 99
  marine, 485
  materials, 37
  mechanical, 99
  oil, 70, 298
  open cycle, 625
  operation, 30
  railway car, 485
  scope of, 7
  solid absorbent, 93
  supplies, 66
  systems,
    cascade, 83
    commercial, 393, 537
    compound, 82
  thermodynamics, 577
  thermoelectric, 87, 632
  ton, 28, 948
  tools, 37
  troubleshooting, 583
  truck, 483
  unit, assembling, 382, 489
    supplies, 66

Refrigerator,
  automatic electric, 311
  cabinet,
    construction, 303
    finishes, 332
    insulation, 304
    location, 336, 489
    temperatures, 296
  capillary tubes, 311

care of, 330

charging, 364, 493

cleaning, 330

compressor, 311

condenser, 311

cycling time, 390

defrosting, 330

doors, 331

elementary, 29

fresh food compartment, 304

frozen food cabinets, 303, 467, 468,
  470, 471

frozen food compartment, 306

function of, 9

how a mechanical refrigerator works, 8

installing, 336, 337, 338

leveling, 338

manual defrost, 304, 306

mechanisms, 304

shutting down, 390

starting, 338

uncrating, 336

wiring, 304, 308, 312, 313, 316, 317

Registers, 781

Regulating valves, pressure, 612

Regulators, gas pressure, 724

Relative density of gases, 21

Relative humidity, 35, 643

Relay, 205, 259, 260, 850
  contactors, 850
  current, magnetic, 260
  electronic, 263
  potential, magnetic, 261
  servicing, 263
  starting motor, 259, 308
  testing, 263
  thermal, hot wire, 263
  types, 259, 260

Relief valve, 455

Reluctance, 188

Remote,
  air conditioner, 760
  condensing unit, 465
  system, 760
  temperature sensing element, 257

Removing,
  air from system, 384, 497, 523
  air-cooled condenser, 521
  compressor, 513
  electric motor, 542
  evaporator unit, 530
  expansion valve, 533
  gauges, 495
  hermetic motor compressors, 546
  motor compressor, 374
  oil, 378

refrigerant, 374, 510
service valve, 513
system, 368, 509
water valve, 528
water-cooled condenser, 524
Repairing,
air-cooled condenser, 521
clogged TEV screen, 533
condensers and receivers, 525
crankshaft seal, 516
electric motors, 543
electronic filters (see Servicing)
evaporators, 377
expansion valves, 535
finishes, 332
hardware, 331
hermetic motors, 546
leak, 362, 364
motor, 380
refrigerating systems, 335, 489
threads, 51
valve plate, 517
water valves, 529
Repulsion start-induction motors,
209, 211
Resin, epoxy, 51
Resistance,
electrical, 178, 955
heaters, 308, 811
heating, 678, 679, 680, 735
in parallel, 955
in series, 955
valve, elbow, tee, 596
Resistor-heating, 678, 735
Resistor-impedance, 178
Restricted system, 370
Restricted water flow, 527
Restrictor, 617
Return air ducts, 787
Reverse cycle, 800
defrost, 424, 430
Reversing valve, 805, 806, 807
Rheostat, 922
Riggers, 820
Ringelmann scale, 875
Riser valve, 455
Rocks, heat storage, 936, 937
Rooftop units, 816, 818, 820
Room, clean, 963
Room dehumidifiers, 689
Room heating, electrical resistance,
680
Room humidifiers, 689
Root mean square (rms), 199, 967
Rotary compressor, 127, 128, 518
rebuilding, 381

Rotating anemometer, 651
Rotation, motor, 214, 215
Rotor, 191
construction, 129, 133
Rottenstone, 332
Rounding to whole numbers, 11
RPB, 148
Rules, measuring, 64
Running winding, 211, 213
Rupture disk, 456

S

Saddle, tap line, 358
SAE, flared connection, 41 through 44
SAE, 1008 mild steel, 38
Safety,
code for mechanical refrigeration, 968
control, 845, 846
goggles, 58, 927
motor control, 258, 435
OSHA, 461
plug, fusible, 455, 456
release, 465
valve, 612
vents, 626
Safety-stat, 673
Salt and ice mixture, 28, 416, 961
Salt spray test, 726
Saturated vapor, 34, 580
Saturation, 643
Saw cutter, 40
Saybolt viscosimeter, 960
Saybolt viscosity test, 960
Scales,
Beaufort, 657
dBA, 669
pressure, 18, 297
temperature, 10, 80
Scavenger pump, 136
Schrader valve, 359, 494, 538
Scotch yoke, 126
SCR, 181, 194, 839
Screens, 538
Screw type compressors, 130
Screwdrivers, 57
offset, 57
Phillips, 57
stubby, 57
Screws, machine, 68
Seal,
compressor, 122, 907
leak, 512
repairing crankshaft, 516
shaft, 122
Sealants, duct, 775

Sealed unit, 304, 307, 311, 313
Sealing compound, 51
Second Law of Thermodynamics, 9, 952
Secondary, container, 626
Secondary refrigeration system, 631, 638
Seebeck effect, 197, 198
Selective absorber, 935
Seletrastat, 856
Self inductance, 193
Semiautomatic defrost controls, 267
Semiconductors, 180, 181, 193, 194,
267, 955
Sensible heat, 24
Sensing bulb, 78, 110, 141, 155, 631
Sensing devices, 863
Sensing element, 150, 151, 257
Sensors, 193, 196
Separators, oil, 82, 104, 449
Sequence controls, 845, 851
Series circuit, 182
Service,
contracting, 551
cylinders, 294
engineer's equipment, 507
estimates, 551
instructions, 507
notes, 548
Service valve, 70, 104, 134, 353, 454,
495, 513, 917
attachment, 354, 359
wrench, 55
adaptors, 56
Serviceable hermetic, 115
Servicing,
absorption systems, 608, 618, 621
air conditioning systems, 699, 747, 765,
828, 897
air-cooled condenser, 521
automobile, 897, 926
burnout, 538
capillary tubes, 369, 377
commercial systems, 489, 507, 548
condensing units, 508, 520, 525
console air conditioners, 759
cooling towers, 525
direct expansion evaporators, 529, 530
domestic systems, 335, 336
electric motors, 232
electrical circuits, 541
electronic filters, 796
evaporative condensers, 525
evaporator units, 494, 529, 530
external drive motors, 542
gas furnace, 727
heat pump systems, 812
heating coil, 741

hermetic motors, 544
hermetic systems, 344, 520
hot gas bypass valves, 536
humidifiers, 743
hydronics system, 732
ice makers, 526
liquid line, 547
lithium bromide systems, 622
motor controls, 546
oil burner, 717
open systems, 508
outdoor air-cooled condenser controls, 536
rooftop units, 820
solenoid valves, 547
steam heating system, 734
suction line, 548
thermoelectric units, 632
thermostatic expansion valves, 532, 535
two-temperature valve, 535
water valves, 527, 528, 529
water-cooled condenser, 523
window units, 753
Setscrews, Allen, 68
Shaded-pole motor, 210, 229
Shell and coil condenser, 409
Shell and tube condenser, 409, 410
Shell and tube flooded evaporator, 83, 409
Shell-type condenser, 409
Shelving, 304
Short circuit, 172
Short cycling, 349, 399, 550, 734
Shortage of refrigerant, 360
Shroud, 106
Shunt, 179
Shutting down refrigerators and freezers, 390
SI (see Le Systeme International d'unites)
Side-by-side arrangement, 117, 319
Sight glass, 458, 508
Silica gel, 372, 956
Silicon rectifier, 181, 194
Silver alloys, 47
Silver brazing, 41, 47
 connections, 47
Sine wave, 183, 184
Single energy, 961
Single flare, 41
Single-phase motor, 201, 209
Single-pole, double-throw switch, 253, 839
Single-pole, single-throw switch, 253, 839
Single-stage, two-stage pumps, 385
Sink, heat, 698
Siphon action, 608
Slant type evaporators, 822, 823
Sliding port valve, 807
Slugging, 804, 811

Smoke, 792
 candle, 726, 876
 test, 872, 874
Snap-action switch, 114
Snap-action toggle, 246
Snap-action two-temperature valve, 445
Sniffer, 929
Snow making, 74, 486
Soap bubbles, 361
Socket wrench, 53, 56
Soda fountain, 471
Sodium chloride (NaCl), 32, 961
Soft ice cream maker, 474
Solar,
 cell, 942, 943, 944
  circuit, 943, 944
  construction, 942, 943
  efficiency, 944
  performance, 944
  photovoltaic, 943, 944
 collector, 934, 935
  angle, 937
  cover, 936
  flat plate, 935
  heat insulating, 937
  lens, 942
  panel, 814, 937
  surface, 936
  swimming pool, 940
 electric energy inverter, 194
 energy, 933
  converting to electricity, 942
  definition, 933
  domestic water heating, 940
   one tank system, 939
   two tank system, 939
  electromagnetic, 933
  flux, 933, 934
  Nature of, 933
  space heating installations, 937
  storage, 934, 936
  systems,
   active, 934
   cooling, 942
   heating, 934
   passive, 934
  trapping principles, 935
 heat, 891
 heat pump combination, 813, 815, 937, 941
 panel, 815, 937
Soldered tube fittings, 44
Soldering, 45
 equipment, 505
 soft, 44
Solders, 963
Solenoid, 187

Solenoid plungers, 158
Solenoid valves, 70, 157
 reversing, 805
 servicing, 547
 two-temperature valves, 446
 types, 158
Solid absorbent, 93
Solid adsorbent, 604
Solid state, 153
 defrost controls, 442
 electronic relays, 263
Solids, 20
Solution, eutectic, 576, 638
Solvents, cleaning, 66, 961
Sones, 670
Source detector-ion, 362
South pole, magnetic, 184, 185
Space heating, solar, installations, 937
Space thermostats, 806
Special fittings, 44
Special refrigeration systems and applications, 625
Special tubing, 505
Specific enthalpy, 33
Specific gravity, 21
 brine, 21
 gases, 21
Specific heat, 24, 580
 common substances, 948
Specific volume, 21
Spectrum, sun's, 933
Speeds, motor, 212
Spirit level, 757
Splash oiling, 136
Splash system, oil, 136
Split system, 864
Split-phase motor, 193, 210, 211, 217
Spray cooling, 638
Spray headers, 626, 631
Sprays, water, 797
Spring bender, 40
Spring loaded valve, 456
Square,
 centimetre, 13, 14
 decimetre, 13, 14
 feet, 13, 14
 inches, 13, 14
 metre, 13, 14
 yard, 14
Squirrel cage, 211
Stack thermometer, 873
Stainless steel tubing, 38
Stamps, 57
Standard,
 atmosphere, 964
 conditions of air, 964
 pressure, 964

temperature, 277, 964
Star transformer, 202
Starters, motor, 437
Starting,
dry system, 501
evaporator, 150
gas furnace, 726
oil burners, 716
refrigerator-freezer, 338
stuck compressors, 240, 349
system, 498, 501
Starting current, 213
Starting relays, 259
Starting winding, 210, 213
Stat, stack, 674
Static electricity, 169
Static head, 594
Static pressure disk, 709
Stationary blade, 127
Stator, compressor, 133
Stator, motor, 191, 192, 211
Steam, 9, 733
heating, 733
inspection, 735
installation, 734
service, 734
testing, 733
troubles, 733
Steam-jet cooling, 635, 695
Steel tubing, 38
Stellite, 147
Sterling cycle, 638
Stethoscope, 338
Storage,
battery, 169
cylinders, 293
fresh foods, 303
frozen foods, 303, 471, 483
fur, 480
solar energy, 934, 936
tank, thermal energy, 813, 814, 815
Storing, cold, 761
Straddle-plugged-in units, 399
Strainer (see Filter)
Strainer chamber, 712
Stratification of air, 666, 769
Strike, 332
Strip charts, 872
Strong solution, 622
Sub-base, 842
Sublimation, 32
latent heat of, 32
Substituting refrigerants, 300
Suction line, 100, 103, 536
accumulators, 804
filter, 459, 460
pressure drop, 584

servicing, 548
Suction pressure valve, 166, 450, 915
Suction service valve, 104, 398
Sulphur dioxide, 289, 299
Summer cooling, 800
Sun (see Solar)
Sun effect, 880
Sun heat load, 656, 891
Sun's spectrum, 933
Superfex, 605
Superheat, 144, 150, 580
Superheated vapor, 580
Superheating, 577
Supplemental resistance heaters, 811
Supplementary heat, 940
electric, 739, 940
oil and gas, 941
Supplies, 37, 66, 393, 398
Surge tanks, 447
Swage, 546
Swaging, 49
Swamp cooler, 697
Swash plate, 126, 905, 906
Sweat joint, 45
Sweating, 465
Sweet water, 32, 416
bath, 416
Swimming pool heating, 940
Swing temperature, 846
Swinging vane velocimeter, 652
Switches, 207
limit, 845
power saver, 313, 323
reversing, 633
SPDT, 253, 839
SPST, 253, 839
thermal element, 851
Symbols, 35, 172, 954
Synthetic rubber, 907
Synthetic silicates, 372
Systems,
absorption, 89, 91, 603, 606, 620
adding oil, 540
adsorption, 604
air circulation, 829
air conditioning, 673, 699, 747, 760
alarm, 326
automobile, 897, 901
basic refrigeration, 73
capacity, 585
capillary tube, 79, 109, 147, 360, 369,
377, 378, 379
cascade, 83, 581
central, 822
charging, 388, 498
chilled water, 831, 832
climatic control, 800, 815

comfort cooling, 747, 748, 831
commercial, 94, 393
complete, 682, 800, 815
compound, 82
compression, 99
continuous absorption, 603 through 611
conventional servicing, 489
defrost, 425, 429, 430
discharging, 121, 374, 378
duct, 727, 770, 829
engine-driven, 461
evacuating, 384, 497
evaporator, 625
expendable refrigerant, 86, 625
external drive, 75
Faraday, 93, 604
flooded, 109, 113
forced feed, 136
four-pipe system, 830
heat pump, 803, 804
hermetic, 304, 307, 311, 313
high-side float, 100, 109
Hi/Re/Li, 813
hydronic, 728, 850, 851, 852
intermittent, 603, 605
lithium bromide, 622
low-side float, 100, 109
mechanical refrigerating, 30
multiple, 80, 394
multi-stage, 635
odor, 848
outside, 815
restricted, 370
solar energy, 934, 935, 942
solar heat pump, 813, 814, 937, 941
spray, 631
starting, 498, 501
steam heating, 733
steam-jet, 635
testing, 389
thermoelectric, 632
through-the-wall, 815
total energy, 832, 961
two-duct system, 829
unitary, 815
vacuum control, 924
warm air, 727
Systeme International d'unites (see Le
Systeme International d'unites)

T

TAB, 788
Tables, refrigerants, 276
Tandem motor, compressor, 399
Tanks,
holding, 687

oil, installation, 713, 714
refrigerant (see Cylinder)
surge, 447
Tap drills, 58
  sizes, 59
Tap water type cooler, 477
Tap wrench, 50
Tapes, measuring, 64
Taps, 58
  bottoming, 59
  plug, 59
  taper, 59
Technical characteristics, 947
Technical terms, 971
Teflon, 122
  seal, 908
Temperature, 7, 9, 10, 11
  absolute, 10
  air, 642, 656
  air conditioning, 641
  alarm system, 273
  ambient, 28
  brine freezing, 961
  cabinet, 296, 334
  Celsius, 10
  Centigrade, 10
  change, 837
  compression cycle, 100
  condensing, 297
  control, 243, 245, 656
  control alarm system, 273
  conversion of, 13, 950
  critical, 32
  cryogenic, 961
  curves, 277
  design temperatures, 886
  differences, 13, 32
  droop, 846
  dry bulb, 642
  effective, 658
  equation, 13
  evaporating, 297
  Fahrenheit, 10
  fresh food, 9
  frozen food, 560
  measurement, 9
  motor, 226
  of maximum density of water, 810
  of mixes, 474
  pressure charts, 297
  pressure conditions, 339
  pressure tables, 297
  ranges, 141
  refrigerator, 296
  room and duct, 778
  scales, 10

sensing bulb, 78, 110, 141, 155, 245, 631
standard, 277, 964
storage case, 296
swing, 846
symbols, 9, 10
system lag, 845
tables, low-side pressure, 296
wet bulb, 643
Terminals, motor, 221
Terminals, plug, 206
TES, 960
Test lights, 207
Test stand, 237
Tester, expansion valve, 533
Tester, leak, 496
Testing,
  code installation, 506
  electric motors, 543
  for leaks, 285, 496, 522, 533, 929
  rebuilt systems, 389
  stand, compressor, 237, 518
  thermostatic expansion valve, 533
TEV, 78, 144, 549, 804, 914
Thermal,
  cutouts, 245
  electric, 144, 153
  energy storage tank, 815
  insulation, 963
  laws of refrigeration, 99
  relay, 263
  resistance, 880
  stability, 298
  unit, British thermal, 9, 23
  unit, kilojoule, 23
Thermistor, 10, 61, 193, 196
  sensor, 838
Thermocouple, 10, 197, 723
Thermodynamics, 9, 22, 577, 952
Thermoelectric,
  compared to compressor system, 632
  expansion valve, 153
  operation, 87, 632
  power supply, 632
  refrigeration, 87, 632
  thermocouple, 61, 197
Thermo-humidigraphs, 648
Thermopile, 197
Thermometer, 10
  scales, 10
  type thermostat, 842
Thermometers, 9, 60, 664
  air conditioning, 664
  alcohol, 9
  cabinet, 334
  comparison, 10
  dial stem, 60

dry bulb, 643
glass stem, 10, 60
infrared, 947
kata, 947
mercury, 9
radiometer, 10
recording, 507
stack, 873
wet bulb, 643
Thermomodule, 632, 633
Thermopane, 885
Thermostat, 31, 837
  air coil, 810
  anticipators, 842
  automobile air conditioning, 924
  combination, 847
  cooling, 847
  droop, 846
  electric heating, 842, 843, 845
  guard, 842
  heating, 842, 844, 845
  installing, 271
  line voltage, 841
  locations and mountings, 843
  low voltage, 841
  millivolt, 842
  multistage, 849
  portable, 842
  sensing bulb, 271
  sensitive element, 842
  servicing, 272
  special, 271
  testing, 272
  thermometer type, 842
  timer, 849
  troubles, 273
  types, 842
  unit, 680
  vapor pressure type, 245, 246
  wall, 680
Thermostatic control, 114, 243, 434
Thermostatic expansion valve, 78, 110, 144
  adjusting, 532, 533
  capacities, 156, 160
  construction, 146
  control, 146
  cross charged power element, 151
  design, 146
  gas charged power element, 151
  liquid charged power element, 150
  normal charged power element, 150
  power element mounting, 155
  pressure limiters, 154
  principles, 144
  servicing, 532, 533, 535
  special, 160

temperature, 78
testing and adjusting, 533
Thermostatic motor controls, 243, 245, 246, 434
Thermostatic two-temperature valves, 445
Thermostatic water valve, 454
Thread cutting compound, 51
Threads, repairing, 51
Three physical states, 20
Three-phase motor, 201, 204
Three-wire thermostat, 839
Through-the-wall air conditioners, 681, 682, 811, 815
Throwaway filters, 794
Timers,
  defrost, 442
  thermostat, 849
Toggles, 114, 246
Ton, 17
"Ton" of refrigeration, 28, 948
Tools, 37
  hand, 52, 53
  installation, 716
  servicing, 335, 336
Torque, 56, 223
  wrench, 56
Torr, 18
Total energy, 834, 961
Total heat load, 562, 566
Towers, cooling, 410, 525, 687
Tracing, water circuit troubles, 528
Transducer, 193, 197
Transfer of heat, 31
Transferring refrigerants, 540
Transformer,
  circuits, 202, 203
  line voltage, 203, 204
  principles, 202
  rectifier, 87
  three-phase, four-wire, 204, 205
Transistor, 181, 193, 195, 196
Translucent, 31
Transmission power take-off, 484
Trapping principle, solar energy, 935
Triacs, 860
Trichloromonofluoromethane, 276, 283
Trichlorotrifluoroethane, 279
Triple evacuation, 384
Tripper, 849
Trouble signals, 340, 341'
Troubleshooting, 335, 336, 347, 377, 545, 549, 583, 868
  chart, 342, 343
Truck air conditioning, 925
Truck refrigeration, 483
TruKold, 605

Tube,
  bleeder, 743
  capillary, 79, 109, 163, 164, 165
  constrictor, 50
  pitot, 653, 872
  process, 356
  venturi, 635
  vortex, 633, 696
  within-a-tube condenser, 409
Tubing, 37, 38, 39
  ACR, 37
  annealed, 37
  annealing, 44
  bending, 40, 491
  brazing, 47
  connecting, 41
  coolant, 912
  copper, hard, 37
  copper, soft, 37
  cutting, 40
  finned, 588
  fittings, 44
  flaring, 41, 43
  flexible, hose, 38
  hard drawn, 38
  installing, 491
  insulation, 492
  length, 589
  normal size, 39
  OD size, 37
  plain, 588
  plastic, 38
  polyethylene, 38
  sizes, 38
  soldering, 44
  stainless steel, 38
  steel, 38
  swaging, 49
  winding, 163
Turbine flow meters, 876
Twist drill sizes, 962
Two evaporating temperatures, 80
Two-duct system, 727, 829
Two-speed motor compressor, 212, 812
Two-stage compressor, 581
Two-stage heating, 850
Two-stage oil pump, 711
Two-temperature valve, 443, 445, 446, 535
  servicing, 535
Types of compressors, 115

U

U factor, 557, 880
U.S. conventional units, 7, 13, 14, 15, 16
Ultralow temperature, 635

Untraviolet, 243, 798, 933, 934
Uncrating cabinets, 336
Unheated spaces, 887
Unit,
  air conditioning, 747
  area, 14
  centrifugal, 832
  condensing, 631
  energy, 29
  heat, 23
  heaters, 742
  prefixes, electrical, 954
  pressure, 17
  rooftop, 816, 817, 818, 819
  temperature, 9
  thermostat, 680
  through-the-wall, 815
Unitary systems, 815, 822
Units and symbols, electrical, 954
Universal motors, 222
Unloader, 138
Upright freezer, 326
Upright frozen food case, 470
Urethane insulation, 304, 471
Usage load, 566

V

Vacuum, 19
  control systems, 925
  gauges, 63
  pump, 384, 507
  tube collector, 936
Values, interpolation, 951
Valve,
  absorption system, 612
  adaptor, 354
  automatic expansion, 78, 110, 141
  ball check, 294, 295
  bellows-type expansion, 142
  bolted-on, 357
  brazed-on, 358
  burr, 380
  bypass, 143, 916
  capacity, 142
  check, 137, 166, 447
  compressor, 121, 122
    low-side pressure control, 450, 915
  core, 359
  cracking, 71
  diaphragm type, 143
  Dill core, 927
  discharge, 120, 121
  disk, 122
  electric water, 451
  exhaust, 121

expansion, 141, 142, 143, 144, 531, 532, 914
  servicing, 535
flapper, 380
float, 162, 163
four-way, 158, 801, 805, 806
gas, 673
hand, 454, 455
heat pump reversing, 806
hot gas bypass, 902, 903
  servicing, 536
intake, 121, 122
leaking, 380, 512
liquid vapor, 294
manual, 70, 454, 455
one-way, 70, 293, 806
piercing, 357
plate, 121, 517
port type, 807
pressure controlled expansion, 78, 141
pressure regulating, 443, 612
reed, 122
refrigerant line, 506
relief, 455
repairing compressor, 517
reversing, 690, 805, 806
riser, 455
safety release, 294, 295
Schrader core, 359, 927
service, 70, 104, 134, 137, 454, 495, 513, 917
snap-action type two-temperature, 445
solenoid, 157, 158, 805
  servicing, 547
solid state, 153
suction pressure, 78, 141, 166, 915
suction throttling, 695, 915
thermal-electric, 153
thermostatic expansion, 110, 144, 146
three-way, 158, 506, 806
two-temperature, 443, 445, 446, 535
two-way, 70, 158, 506
type, 70
water, 451, 453, 454
weight, 76
Vanes, 127, 381
Vapor, 9, 34, 957
  barrier, 651, 887, 892
  bubbles, 300
  charged, 150, 151
  lock, 163
  pressure, 30
    curve, 298
  refrigerant, 8
  saturated, 34, 580
  superheated, 150, 580

vaporized refrigerant, 75
Variable pitch pulley, 236
Varistors, 193
V-belt, 236, 909
  drive, 122
Vegetables, fast freezing, 560
Velocimeter, 652
Velocity,
  equivalents, 949
  gas, 595
  measurement, 651
  pressure, 653
Vending machine, 478
  controls, 439
Vent plug, 716
Ventilation, 654
  requirements, 766
Venting, job shop, 962
Vents, 962
  safety, 626
Venturi tube, 635
Vertical baffle, 466
Vibration, 338, 527
  absorbers, 457
  dampers, 94
Viscosity, 70, 960
  Saybolt test, 960
Vises, 57
Volt, 174
Voltage, 174, 175
  drop, 183
  lowering voltage, 261, 955
  relay, 217, 259
Voltmeter, 174, 175, 179
Volume, 14
  air, 875
  cabinet, 562
  constant, 966
  equivalents, 949
  measurement, 13
  of a cylinder, 14
  specific, 21
Volumetric efficiency, 136, 587
  check, 512
Vortex tube, 633
  cooling, 696
  refrigeration, 633
V-type compressor, 405

W

Walk-in cabinets, 465
Walk-in coolers, 465
Wall,
  construction, 887
  heat leakage, 884

insulation, 892, 963
plate, 342
thermostat, 680
type evaporator, 102
Warm air defrost, 431, 442
Warm air furnace, 673, 675, 678
Warm air heating, 727
  inspection, 727
Warm air system, solar, 937, 939
Warmer plate type coil, 904
Water,
  analysis instruments, 876
  boiler water treatment, 731
  circuit troubles, tracing, 527, 528
  coil, 693, 810
  column, 19, 20
  cooled condenser, 408, 523, 524, 525, 589
  cooled tubing, 588
  cooled units, 136
  cooler, 85, 477
  cooling loads, 576
  defrost system, 429
  defrosting, 429
  evaporative refrigeration, 73, 74
  flow, restricted, 527
  flow, too much, 528
  heating, solar, 937, 939, 940
  ice, 7, 9, 18
  latent heat, 24, 25, 26, 27
  lever controls, 860
  manometer, 20, 621, 719, 721
  refrigerant, 292
  sprays, 797
  strainer, 451
  temperature of maximum density, 810
  valve, 451, 527, 528, 529
    electric, 451
    pressure, 453
    pressure setting, 453
    thermostatic, 454
  vapor, 18
Watt, 176, 198, 199
  reading, 232
Wattage, calculating, 177
Wattmeter, 178
Wave, pulse, 184
Wave, sine, 183, 184
Wax, 360
Weak solution, 622
Weather, 655
Weight and mass, 17
Weight equivalents, 950
Weight of air, 765
Weight of substances, 948
Welding compressor dome, 382

Welding equipment, 505
Wet bulb, 643, 646
  temperature, 646
Wet color, 299
Wick, 636
Wind chill, 656, 657
Windings, 186
  main, 212
  running, 210, 212, 213
  starting, 211, 212, 213
  squirrel cage, 211
Windmills, 642
Window unit, 681, 682, 749
  installing, 751
  servicing, 753
Windows, 885, 963
  heat load, 891
  primary, 885
  secondary, 885
  storm, 885
  thermopane, 885

Winterizing air conditioning unit, 757
Wire,
  arcing, 839
  drawing, 587
  sizes, 206
  "walking," 839
Wiring, 306, 308, 313, 316, 317, 320, 323
Wobble plate, 126
Wool insulation, 963
Work (W), 22
Work hardening, 37
Wrenches, 53
  adaptors, 56
  adjustable, 55
  box, 54
  combination, 55
  flare nut, 54, 964
  metric, 54
  offset, 54
  open end, 55
  pipe, 53, 55

  ratchet handle, 53
  service valve, 55
  socket, 53
  swivel handle, 53
  T-handle, 53
  torque, 56
  used by service technician, 53
Wye transformer, 202, 203

Y

Yard, 14
Year-around air conditioner, 641
Yoke, Scotch, 119, 126

Z

Zero,
  absolute, 10
  ice, 32
  pressure, 18
Zone, comfort, 655
  control, 864